Physical Biology of the Cell

Physical Biology of the Cell

Second Edition

Rob Phillips
Jane Kondev
Julie Theriot
Hernan G. Garcia

Illustrated by
Nigel Orme

LONDON AND NEW YORK

Garland Science
Vice President: Denise Schanck
Editor: Summers Scholl
Senior Editorial Assistant: Kelly O'Connor
Cover design and illustrations: Nigel Orme
Production Editor: Natasha Wolfe
Copyeditor: Mac Clarke
Proofreader: Sally Huish
Typesetting: TechSet Composition India (P) Ltd.

Rob Phillips is the Fred and Nancy Morris Professor of Biophysics and Biology at the California Institute of Technology. He received a PhD in Physics from Washington University.

Jane Kondev is a Professor in the Department of Physics and within the Graduate Program in Quantitative Biology at Brandeis University. He attended the Mathematical High School in Belgrade, Serbia, received his Physics BS degree from the University of Belgrade, and his PhD from Cornell University.

Julie Theriot is a Professor in the Department of Biochemistry and the Department of Microbiology and Immunology at the Stanford University School of Medicine. She received concurrent BS degrees in Physics and Biology from the Massachusetts Institute of Technology, and a PhD in Cell Biology from the University of California at San Francisco.

Hernan G. Garcia is a Dicke Fellow in the Department of Physics at Princeton University. He received a BS in Physics from the University of Buenos Aires and a PhD in Physics from the California Institute of Technology.

ISBN 978-0-8153-4450-6

Library of Congress Cataloging-in-Publication Data

Phillips, Rob, 1960-
Physical biology of the cell. – Second edition / Rob Phillips, Jane Kondev, Julie Theriot, Hernan G. Garcia.
pages cm
ISBN 978-0-8153-4450-6 (pbk.)
1. Biophysics. 2. Cytology. I. Title.
QH505.P455 2013
571.6–dc23

2012030733

Published by Garland Science, Taylor & Francis Group, LLC, an informa business,
711 Third Avenue, New York, NY, 10017, USA, and 3 Park Square, Milton Park, Abingdon, OX14 4RN, UK.

Printed in Great Britain by Ashford Colour Press Ltd, Gosport, Hants
15 14 13 12 11 10 9 8 7 6 5 4

Visit our web site at http://www.garlandscience.com

Dedicated to our friend Jon Widom

Preface

"The map is not the territory."
Alfred Korzybski

The last 50 years in biology have seen an explosion of both data and understanding that rivals the fertile period between Tycho Brahe's definitive naked-eye investigations of the heavens and Newton's introduction of the "System of the World." One of the consequences of these stunning advances is the danger of becoming overwhelmed by the vast quantities of data coming at us from quarters ranging from next-generation sequencing to quantitative microscopy. For example, at the time of this writing, there are in excess of two million ribosomal RNA sequences deposited on publically accessible databases. But what does it all mean? A central role of scientific textbooks is to attempt to come to terms with broad areas of progress and to organize and distill the vast amounts of available information in a conceptually useful manner. In our view, an effective textbook can act as a map to help curious people discover unfamiliar territories. As with real maps, different purposes are served by different kinds of abstraction. Some maps show roads, some show topography, with each being useful in its own context.

A number of textbook writers have undertaken the formidable task of writing excellent, comprehensive surveys of cell and molecular biology, although each one of these books serves as a slightly different kind of map for the same overlapping territory. Although we cover some of the same material as a typical molecular and cell biology book, our goal in this book is fundamentally different. There is no single, correct way to construct a conceptually simplified map for a huge and complex field such as cell and molecular biology. Most modern biology textbooks organize ideas, facts, and experimental data based on their conceptual proximity for some particular biological function. In contrast, this book examines the same set of information from the distinct perspective of *physical biology*. We have therefore adopted an organization in which the proximity of topics is based on the physical concepts that unite a given set of biological phenomena, instead of the cell biology perspective. By analogy to a map of the United States, a cell biology textbook might describe the plains of Eastern Colorado in the same chapter as the mountains of Western Colorado, whereas our physical biology book would group Eastern Colorado with the rolling fields of Iowa, and Western Colorado with mountainous West Virginia.

This book does not assume extensive prior knowledge on the part of the reader, though a grounding in both calculus and elementary physics is essential. The material covered here is appropriate for a first course in physical biology or biophysics for either undergraduates or graduate students. It is also intended for any scientist interested in learning the basic principles and applications of physical modeling for research in biology, and aims to provide a novel perspective even to scientists who are already familiar with some of the material. Throughout the book, our organization of ideas and data based on proximity in physical biology space juxtaposes topics that are not obviously related in cell biology space. For example, DNA wrapping around nucleosomes in the eukaryotic nucleus, DNA looping induced by the binding of transcriptional repressors in the context of bacterial gene regulation, and DNA packing into the narrow confines of bacteriophage capsids all appear in the same chapter because they are related by the mechanical rules governing the bending of DNA. Next, the physical and mathematical treatment we derive for DNA bending is directly applied to other kinds of long, thin, biological structures, including the filaments of the cytoskeleton. This organizational principle brings into focus the central thesis of this book, namely, that the appropriate application of a relatively small number of fundamental physical models can serve as the foundation of whole bodies of quantitative biological intuition, broadly useful across a wide range of apparently unrelated biological problems.

During the 12-year journey that led to this book, we benefited immeasurably from the generosity and enthusiasm of hundreds of scientific colleagues who graciously shared their data, ideas, and perspectives. Indeed, in many respects, we view our book as an exercise in *quantitative journalism*, based upon extensive "interviews" with these various scientists in a wide range of disciplines. We offer this book as a report from the front, to share some of the most interesting things that we have learned from our colleagues with any and all inquiring individuals who wish to think both deeply and broadly about the connections between biology and the physical sciences. Our imagined audience spans the range from 18-year-old mechanical engineering undergraduates curious about the application of their discipline to medicine, to 40-year-old string theorists wishing to apply their mathematical and physical talents to living matter, to 70-year-old renowned biologists wondering whether their insights into living systems might be improved by a mathematical treatment.

Although the claim that a handful of simple physical models can shed more than superficial light on complex biological processes might seem naive, the biological research literature is teeming with examples where important quantitative insight into questions of pressing interest has been gained by the application of such models. In every chapter, we have chosen specific examples from classic and current research papers where quantitative measurements on biological systems can be largely understood by recourse to simple, fundamental, physical ideas. In cases where the simplest possible physical models fail to fit the data, the specific quantitative nature of the disparities can often lead to testable new biological hypotheses. For example, a simple calculation estimating the amount of time it would take for a newly synthesized protein to diffuse from the cell body of a motor neuron in the spinal cord to the synapse formed by the same neuron in the foot proves that diffusion is far too slow to get the job done, and an active transport process must occur. Inevitably, researchers performing experiments on biological systems must have physical models explicitly or implicitly in mind, whether imagining how changes in the rate of transcription initiation for a particular gene will lead to changes in the overall amount of the gene product in the cell, or picturing the ways that signaling molecules move through cellular space to encounter their targets, or envisioning how cell movements during embryogenesis lead to the final three-dimensional structures of organs and limbs. In this book, we aim to provide a physical and mathematical toolkit so that people used to thinking deeply about biological problems can make this kind of quantitative intuition explicit; we also hope to provide a perspective on biology that may inspire people from a background more heavily based in physics or mathematics to seek out new biological problems that are particularly appropriate for this kind of quantitative analysis.

Our general approach follows four steps. First, we introduce a biological phenomenon; second, we perform simple order-of-magnitude estimates to develop a "feeling for the numbers" involved in that process; third, we demonstrate the application of an extremely simple first-pass model; and finally, where possible, we present a refinement of the oversimplified model to better approximate biological reality. Our goal is to share the pleasure in seeing the extent to which simple models can be tailored to reveal the *complexity* of observed phenomena. For our examples, we have chosen particular biological cases that we believe to be worthy illustrations of the concepts at hand and that have captured our imaginations, often because of particularly elegant or clever experiments that were designed to generate intriguing sets of quantitative data. While we have been conscientious in our exploration of these facts and in our construction of simple models, it is inevitable that we will have made errors due to our ignorance and also due to the fact that, in many cases, new discoveries may change the particulars of our case studies. (A list of errors and their corrections will be posted on the book's website as well as the website of one of the authors (R.P.).) Nevertheless, because our goal is to demonstrate the power of applying simple models to complex systems, even when some details are wrong or missing, we hope that any particular lapses will not obscure the overall message. Furthermore, in many cases, we have described phenomena that are still awaiting a satisfying physical model. We hope that many of our readers will seize upon the holes and errors in our exploration of physical biology and take these as challenges and opportunities for launching exciting original work.

Our second edition builds upon the foundations laid in the previous edition, with the addition of two new chapters that focus on central themes of modern biology, namely, light and life and the emergence of patterns in living organisms. The new Chapter 18 focuses on several key ways in which light is central in biology. We begin with an analysis of photosynthesis that illustrates the quantum mechanical underpinnings of both the absorption of light and the transfer of energy and electrons through the photosynthetic apparatus. The second part of our story in that chapter considers the rich and beautiful subject of vision. The new Chapter 20 uses insights garnered throughout the book to ask how it is that organisms ranging from flies to plants can build up such exquisite patterns. Here we explore Turing's famed model of several interacting chemical species undergoing chemical reactions and diffusion and other more recent advances in thinking about problems ranging from somitogenesis to phyllotaxis.

The book is made up of four major parts. Part 1, *The Facts of Life*, largely focuses on introducing biological phenomena. For biology readers already familiar with this material, the hope is that the quantitative spin will be enlightening. For physics readers, the goal is to get a sense of the biological systems themselves. Part 2, *Life at Rest*, explores those problems in biology that can be attacked using quantitative models without any explicit reference to time. Part 3, *Life in Motion*, tackles head-on the enhanced complexity of time-dependent systems exhibiting dynamic behavior. Finally, Part 4, *The Meaning of Life*, addresses various kinds of information processing by biological systems.

Because our hope is that you, our readers, represent a broad diversity of backgrounds and interests, throughout the book we try as much as possible to introduce the origin of the facts and principles that we exploit. We are reluctant to ever simply assert biological "facts" or physical "results," and would not expect you to blindly accept our assertions if we did. Therefore, we often describe classical observations by biologists over the past centuries as well as the most recent exciting results, and illustrate how current thinking about complex biological problems has been shaped by a progression of observations and insights. Extended discussions of this kind are separated from the main text in sections labeled *Experiments Behind the Facts*. In a complementary way, whenever we find it necessary to derive mathematical equations, we proceed step by step through the derivation and explain how each line leads to the next, so that readers lacking a strong background in mathematics can nevertheless follow every step of the logic and not be forced to take our word for any result. Specific sections labeled *The Math Behind the Models* and *The Tricks Behind the Math* provide summaries for the mathematical techniques that are used repeatedly throughout the book; many readers trained in physics will already be familiar with this material, but biologists may benefit from a brief refresher or introduction. In addition, we include sections labeled *Estimate* that help to develop a "feeling for the numbers" for particularly interesting cases.

Another critical new element in our second edition is a feature called *Computational Exploration*. The idea of these excursions is to show how simple computer analyses can help us attack problems that are otherwise inaccessible. In the first edition, we underemphasized "computation" because we wanted to combat the spurious idea that theory in biology is synonymous with computation. While we made this exaggeration to make a point, we did so at a price, because computation is not only useful, but downright indispensable in some problems. Further, one of the beauties of turning a model into a specific numerical computation is that to get a computer to produce a meaningful number, nothing can be left unspecified. The Computational Explorations are offered as a way for the reader to develop a particular habit of mind, and none of them should be viewed as illustrating the state of the art for making such calculations. Matlab and Mathematica code related to most of these explorations is provided on the book's website.

Although we review the basic information necessary to follow the exposition of each topic, you may also find it useful to have recourse to a textbook or reference book covering the details of scientific areas among biology, physics, chemistry, and mathematics, with which you consider yourself less familiar. Some references that are among our favorites in these fields are suggested at the end of each chapter. More generally, our references to the literature are treated in two distinct ways. Our suggestions for *Further Reading* reflect our own tastes. Often, the choices that appear at a chapter's end are chosen because of uniqueness of viewpoint or presentation. We make no attempt at completeness. The second class of *References* reflect work that has explicitly touched the content of each chapter, either through introducing us to a model, providing a figure, or constructing an argument.

At the end of each chapter, we include a series of problems that expand the material in the chapter or give the opportunity to attempt model-building for other case studies. In the second edition, we have considerably expanded the scope of the end-of-chapter problems. These problems can be used within formal courses or by individual readers. A complete *Solutions Manual*, covering all problems in the book, is available for instructors. There are several different types of problems. Some, whose goal is to develop a "feeling for the numbers," are arithmetically simple, and primarily intended to develop a sense of order-of-magnitude biology. Others request difficult mathematical derivations that we could not include in the text. Still others, perhaps our favorites, invite the readers to apply quantitative model-building to provocative experimental data from the primary research literature. In each chapter, we have loosely identified the different problems with the aforementioned categories in order to assist the reader in choosing which one to attack depending on particular need. The book's website also includes *Hints for the Reader* for some of the more difficult problems.

Our book relies heavily on original data, both in the figures that appear throughout the book and in the various end-of-chapter problems. To make these data easily accessible to interested readers, the book's website includes the original experimental data used to make all the figures in the book that are based upon published measurements. Similarly, the data associated with the end-of-chapter problems are also provided on the book's website. It is our hope that you will use these data in order to perform your own calculations for fitting the many models introduced throughout the book to the relevant primary data, and perhaps refining the models in your own original work.

Student and Instructor Resources

Figures and PowerPoint® Presentations

The figures from the book are available in two convenient formats: PowerPoint and JPEG. There is one PowerPoint presentation for each chapter, and the JPEGs have been optimized for display on a computer.

Data Sets

The original data used to create both the figures and homework problems are available in Excel® spreadsheets. With this data, the reader can extend the theoretical tools developed in the book to fit experimental data for a wide range of problems. The data files contain explicit statement of all relevant units, and include references to the original sources.

Hints for Problems

This PDF provides both hints and strategies for attacking some of the more difficult end-of-chapter problems. In some cases, the hints provide intuition about how to set up the problem; in other cases, the hints provide explicit mathematical instructions on how to carry through more tricky manipulations.

Matlab® and Mathematica® Code

These files contain code for the Computational Explorations sidebars located throughout the book.

Movies

The movies complement the figures and discussion from the book by illustrating the rich dynamics exhibited by living organisms and the molecules that make them tick.

Solutions Manual

This PDF contains solutions to all problems in the book. It is available only to qualified instructors.

With the exception of the *Solutions Manual*, these resources are available on the *Physical Biology of the Cell*, 2nd Edition, media website:
 http://microsite.garlandscience.com/pboc2

Access to the Solutions Manual is available to qualified instructors by emailing science@garland.com.

Acknowledgments

This book would not have been possible without a wide range of support from both people and institutions. We are grateful for the support of the Aspen Center for Physics, the Kavli Institute for Theoretical Physics at the University of California, Santa Barbara and the ESPCI in Paris, where some of the writing was done. Our funding during the course of this project was provided by the National Science Foundation, the National Institutes of Health, The Research Corporation, the Howard Hughes Medical Institute, and the MacArthur Foundation. We also particularly acknowledge the support of the NIH Director's Pioneer Award and La Fondation Pierre Gilles de Gennes granted to R.P and The Donna and Benjamin M. Rosen Center for Bioengineering at Caltech, all of which provided broad financial support for many facets of this project.

Our book would never have achieved its present incarnation without the close and expert collaboration of our gifted illustrator, Nigel Orme, who is responsible for the clarity and visual appeal of the hundreds of figures found in these pages, as well as the overall design of the book. We also had the pleasure of working with David Goodsell who produced many illustrations throughout the book showing detailed molecular structures. Genya Frenkel also provided assistance on the problems and their solutions. Amy Phillips assisted with editing, responding to reader comments, and obtaining permission for use of many of the previously published images in the figures. Maureen Storey (first edition) and Mac Clarke (second edition) improved our clarity and respectability with their expert copy editing. Our editors Mike Morales (first edition) and Summers Scholl (second edition) have offered great support through the entirety of the project. Simon Hill (first edition) and Natasha Wolfe's (second edition) expert assistance in the production process has been an impressive pleasure.

One of the most pleasurable parts of our experience of writing this book has been our interaction with generous friends and colleagues who have shared their insights, stories, prejudices, and likes and dislikes about biology, physics, chemistry, and their overlap. We are deeply grateful to all our colleagues who have contributed ideas directly or indirectly through these many enjoyable conversations over the past twelve years. Elio Schaechter told us the secret to maintaining a happy collaboration. Lubert Stryer inspired the overall section organization and section titles, and gave us much-needed practical advice on how to actually finish the book. Numerous others have helped us directly or indirectly through inspiration, extended lab visits, teaching us about whole fields, or just by influential interactions along the way. It is very important to note that in some cases these people explicitly disagreed with some of our particular conclusions, and deserve no blame for our mistakes and misjudgments. We specifically wish to thank: Gary Ackers, Bruce Alberts, Olaf Andersen, David Baltimore, Robert Bao, David Bensimon, Seymour Benzer, Howard Berg, Paul Berg, Maja Bialecka, Bill Bialek, Lacra Bintu, Pamela Bjorkman, Steve Block, Seth Blumberg, David Boal, James Boedicker, Rob Brewster, Robijn Bruinsma, Zev Bryant, Steve Burden, Carlos Bustamante, Anders Carlsson, Sherwood Casjens, Yi-Ju Chen, Kristina Dakos, Eric Davidson, Scott Delp, Micah Dembo, Michael Dickinson, Ken Dill, Marileen Dogterom, David Dunlap, Michael Elowitz, Evan Evans, Stan Falkow, Julio Fernandez, Jim Ferrell, Laura Finzi, Daniel Fisher, Dan Fletcher, Henrik Flyvbjerg, Seth Fraden, Scott Fraser, Ben Freund, Andrew J. Galambos, Ethan Garner, Bill Gelbart, Jeff Gelles, Kings Ghosh, Dan Gillespie, Yale Goldman, Bruce Goode, Paul Grayson, Thomas Gregor, Jim Haber, Mike Hagan, Randy Hampton, Lin Han, Pehr Harbury, Dan Herschlag, John Heuser, Joe Howard, KC Huang, Terry Hwa, Grant Jensen, Jack Johnson, Daniel Jones, Jason Kahn, Dale Kaiser, Suzanne Amador Kane, Sarah Keller, Doro Kern, Karla Kirkegaard, Marc Kirschner, Bill Klug, Chuck Knobler, Tolya Kolomeisky, Corinne Ladous, Jared Leadbetter, Heun Jin Lee, Henry Lester, Julian Lewis, Jennifer Lippincott-Schwartz, Sanjoy Mahajan, Jim Maher, Carmen Mannella, William Martin, Bob Meyer, Elliot Meyerowitz, Chris Miller, Ken Miller, Tim Mitchison, Alex Mogilner, Cathy Morris, Dyche Mullins, Richard Murray, Kees Murre, David Nelson, James Nelson, Phil Nelson, Keir Neuman, Dianne Newman, Lene Oddershede, Garry Odell, George Oster, Adrian Parsegian, Iva Perovic, Eduardo Perozo, Eric Peterson, Suzanne

Pfeffer, Tom Pollard, Dan Portnoy, Tom Powers, Ashok Prasad, Mark Ptashne, Prashant Purohit, Steve Quake, Sharad Ramanathan, Samuel Rauhala, Michael Reddy, Doug Rees, Dan Reeves, Joy Rimchala, Ellen Rothenberg, Michael Roukes, Dave Rutledge, Peter Sarnow, Klaus Schulten, Bob Schleif, Darren Segall, Udo Seifert, Paul Selvin, Lucy Shapiro, Boris Shraiman, Steve Small, Doug Smith, Steve Smith, Andy Spakowitz, Jim Spudich, Alasdair Steven, Sergei Sukharev, Christian Sulloway, Joel Swanson, Boo Shan Tseng, Tristan Ursell, Ron Vale, David Van Valen, Elizabeth Villa, Zhen-Gang Wang, Clare Waterman, Annemarie Weber, Jon Widom, Eric Wieschaus, Paul Wiggins, Ned Wingreen, Zeba Wunderlich, Ahmed Zewail, and Kai Zinn.

Finally, we are deeply grateful to the individuals who have given us critical feedback on the manuscript in its various stages, including the many students in our courses offered at Caltech, Brandeis, and Stanford over the last twelve years. They have all done their best to save us from error and any remaining mistakes are entirely our responsibility. We are indebted to all of them for their generosity with their time and expertise. A few hardy individuals read the entire first edition: Laila Ashegian, Andre Brown, Genya Frenkel, Steve Privitera, Alvaro Sanchez, and Sylvain Zorman. We thank them for their many insightful comments and remarkable stamina. For the second edition, we had the great fortune to have Howard Berg read every word of our book always providing pointed and thoughtful commentary. Similarly, Ron Milo has been a constant source of critical commentary, and encouragement throughout the process. Velocity Hughes and Madhav Mani were a tremendous help in reading the entire book in its near final form and providing critical comments at every turn. Justin Bois has also been a source of numerous critical insights. Niles Pierce has also provided his unflagging support throughout this project.

Many more people have given expert commentary on specific chapters, provided specific figures, advised on us end-of- chapter problems, or provided particular insights for either the first or second edition:

Chapter 1

Bill Gelbart (University of California, Los Angeles), Shura Grosberg (New York University), Randy Hampton (University of California, San Diego), Sanjoy Mahajan (Olin College), Ron Milo (Weizmann Institute of Science), Michael Rubinstein (University of North Carolina, Chapel Hill).

Chapter 2

John A. G. Briggs (European Molecular Biology Laboratory), James Boedicker (California Institute of Technology), James Brody (University of California, Irvine), Titus Brown (Michigan State University), Ian Chin-Sang (Queen's University), Avigdor Eldar (Tel Aviv University), Scott Fraser (California Institute of Technology), CT Lim (National University of Singapore), Dianne Newman (California Institute of Technology), Yitzhak Rabin (Bar-Ilan University), Manfred Radmacher (University of Bremen), Michael Rubinstein (University of North Carolina, Chapel Hill), Steve Small (New York University), Linda Song (Harvard University), Dave Tirrell (California Institute of Technology), Jon Widom (Northwestern University).

Chapter 3

Tom Cech (University of Colorado), Andreas Matouschek (Northwestern University), Yitzhak Rabin (Bar-Ilan University), Michael Reddy (University of Wisconsin, Milwaukee), Nitzan Rosenfeld (Rosetta Genomics), Michael Rubinstein (University of North Carolina, Chapel Hill), Antoine van Oijen (Rijksuniversiteit Groningen), Jon Widom (Northwestern University).

Chapter 4

Elaine Bearer (Brown University), Paul Jardine (University of Minnesota, Twin Cities), Michael Reddy (University of Wisconsin, Milwaukee), Michael Rubinstein (University of North Carolina, Chapel Hill).

Chapter 5

James Boedicker (California Institute of Technology), Ken Dill (Stony Brook University), Randy Hampton (University of California, San Diego), Rick James (University of Minnesota, Twin Cities), Heun Jin Lee (California Institute of Technology), Bill Klug (University of California, Los Angeles), Steve Quake (Stanford University), Elio Schaechter (San Diego State University).

Chapter 6

Ken Dill (Stony Brook University), Dan Herschlag (Stanford University), Terry Hwa (University of California, San Diego), Arbel Tadmor (California Institute of Technology).

Chapter 7

Gary Ackers (Washington University in St. Louis), Olaf Andersen (Cornell University), Ken Dill (Stony Brook University), Henry Lester (California Institute of Technology).

Chapter 8

Ken Dill (Stony Brook University), Shura Grosberg (New York University), Michael Rubinstein (University of North Carolina, Chapel Hill), Jeremy Schmit (Kansas State University), Andy Spakowitz (Stanford University), Paul Wiggins (University of Washington).

Chapter 9

Mike Hagan (Brandeis University), Thomas Record (University of Wisconsin, Madison), Bob Schleif (Johns Hopkins University), Pete von Hippel (University of Oregon).

Chapter 10

Zev Bryant (Stanford University), Carlos Bustamante (University of California, Berkeley), Hans-Günther Döbereiner (University of Bremen), Paul Forscher (Yale University), Ben Freund (Brown University), Bill Gelbart (University of California, Los Angeles), Paul Grayson (California Institute of Technology), Mandar Inamdar (Indian Institute of Technology, Bombay), Bill Klug (University of California, Los Angeles), Joy Rimchala (Massachusetts Institute of Technology), Doug Smith (University of California, San Diego), Megan Valentine (University of California, Santa Barbara), Jon Widom (Northwestern University), Paul Wiggins (University of Washington).

Chapter 11

Ashustosh Agrawal (University of Houston), Patricia Bassereau (Institut Curie), Hans-Günther Döbereiner (University of Bremen), Evan Evans (University of British Columbia), Dan Fletcher (University of California, Berkeley), Terry Frey (San Diego State University), Christoph Haselwandter (University of Southern California), KC Huang (Stanford University), Sarah Keller (University of Washington, Seattle), Bill Klug (University of California, Los Angeles), Carmen Mannella (State University of New York, Albany), Eva Schmid (University of California, Berkeley), Pierre Sens (ESPCI, Paris), Sergei Sukharev (University of Maryland, College Park), Tristan Ursell (Stanford University), Paul Wiggins (University of Washington).

Chapter 12

Howard Berg (Harvard University), Justin Bois (University of California, Los Angeles), Zev Bryant (Stanford University), Ray Goldstein (University of Cambridge), Jean-François Joanny (Institut Curie), Sanjoy Mahajan (Olin College), Tom Powers (Brown University), Todd Squires (University of California, Santa Barbara), Howard Stone (Harvard University).

Chapter 13

Howard Berg (Harvard University), Ariane Briegel (California Institute of Technology), Dan Gillespie, Jean-François Joanny (Institut Curie), Martin Linden (Stockholm University), Jennifer Lippincott-Schwartz (National Institutes of Health), Ralf Metzler (Technical University of Munich), Frosso Seitaridou (Emory University), Pierre Sens (ESPCI, Paris), Dave Wu (California Institute of Technology).

Chapter 14

Jean-François Joanny (Institut Curie), Randy Kamien (University of Pennsylvania), Martin Linden (Stockholm University), Ralf Metzler (Technical University of Munich), Pierre Sens (ESPCI, Paris), Arbel Tadmor (California Institute of Technology).

Chapter 15

Anders Carlsson (Washington University in St. Louis), Marileen Dogterom (Institute for Atomic and Molecular Physics), Dan Fletcher (University of California, Berkeley), Dan Herschlag (Stanford University), Jean-François Joanny (Institut Curie), Tom Pollard (Yale University), Dimitrios Vavylonis (Lehigh University).

Chapter 16

Bill Gelbart (University of California, Los Angeles), Jean-François Joanny (Institut Curie), Tolya Kolomeisky (Rice University), Martin Linden (Stockholm University), Jens Michaelis (Ludwig-Maximilians University), George Oster (University of California, Berkeley), Megan Valentine (University of California, Santa Barbara), Jianhua Xing (Virginia Polytechnic Institute and State University).

Chapter 17

Olaf Andersen (Cornell University), Chris Gandhi (California Institute of Technology), Jean-François Joanny (Institut Curie), Stephanie Johnson (California Institute of Technology), Rod MacKinnon (Rockefeller University), Chris Miller (Brandeis University), Paul Miller (Brandeis University), Phil Nelson (University of Pennsylvania).

Chapter 18

Maja Bialecka-Fornal (California Institute of Technology), Bill Bialek (Princeton University), David Chandler (University of California, Berkeley), Anna Damjanovic (Johns Hopkins University), Govindjee (University of Illinois, Urbana-Champaign), Harry Gray (California Institute of Technology), Heun Jin Lee (California Institute of Technology), Rudy Marcus (California Institute of Technology), Tom Miller (California Institute of Technology), Ron Milo (Weizmann Institute of Science), Jose Onuchic (Rice University), Nipam Patel (University of California, Berkeley), Mark Ratner (Northwestern University), Mattias Rydenfelt (California Institute of Technology), Dave Savage (University of California, Berkeley), Klaus Schulten (University of Illinois, Urbana-Champaign), Kurt Warncke (Emory University), Jay Winkler (California Institute of Technology).

Chapter 19

James Boedicker (California Institute of Technology), Robert Brewster (California Institute of Technology), Titus Brown (Michigan State University), Nick Buchler (Duke University), Eric Davidson (California Institute of Technology), Avigdor Eldar (Tel Aviv University), Michael Elowitz (California Institute of Technology), Robert Endres (Imperial College London), Daniel Fisher (Stanford University), Scott Fraser (California Institute of Technology), Uli Gerland (Ludwig-Maximilians University), Ido Golding (Baylor College of Medicine), Mikko Haataja (Princeton University), Terry Hwa (University of California, San Diego), Daniel Jones (California Institute of Technology), Tom Kuhlman (University of Illinois, Urbana-Champaign), Wendell Lim (University of California, San Francisco), Chris Myers (Cornell University), Bob Schleif (Johns Hopkins University), Vivek Shenoy (Brown University), Steve Small (New York University), Peter Swain (McGill University), David Van Valen (California Institute of Technology), Ned Wingreen (Princeton University), Sunney Xie (Harvard University).

Chapter 20

Justin Bois (University of California, Los Angeles), Thomas Gregor (Princeton University), KC Huang (Stanford University), Frank Julicher (Max Planck Institute of Complex Systems, Dresden), Karsten Kruse (University of Saarlandes), Andy Oates (Max Planck Institute of Molecular Cell Biology and Genetics, Dresden), Jordi Garcia Ojalvo (Polytechnic University of Catalonia), George Oster (University of California, Berkeley), Andrew Rutenberg (Dalhousie University), David Sprinzak (Tel Aviv University), Carolina Tropini (Stanford University).

Chapter 21

Ralf Bundschuh (The Ohio State University), Uli Gerland (Ludwig-Maximilians University), Daniel Jones (California Institute of Technology), Justin Kinney (Cold Spring Harbor Laboratory), Chris Myers (Cornell University), Eric Peterson (California Institute of Technology), Frank Pugh (Pennsylvania State University), Jody Puglisi (Stanford University), Oliver Rando (University of Massachusetts Medical School), Tony Redondo (Los Alamos National Laboratory), Eran Segal (Weizmann Institute of Science), Boris Shraiman (University of California, Santa Barbara) Peter Swain (University of Edinburgh), Jon Widom (Northwestern University), Chris Wiggins (Columbia University).

Contents

PART 4 THE MEANING OF LIFE

Contents in Detail

Chapter 10 Beam Theory: Architecture for Cells and Skeletons 383

Chapter 11 Biological Membranes: Life in Two Dimensions 427

PART 3 LIFE IN MOTION 481

Chapter 12 The Mathematics of Water 483

Chapter 13 A Statistical View of Biological Dynamics 509

Chapter 17 Biological Electricity and the Hodgkin–Huxley Model 681

Chapter 18 Light and Life 717

Chapter 21 Sequences, Specificity, and Evolution 951

Chapter 22 Whither Physical Biology? 1023

Special Sections

There are five classes of special sections indicated with icons and colored bars throughout the text. They perform order of magnitude estimates, explore biological problems using computation, examine the experimental underpinnings of topics, and elaborate on mathematical details.

COMPUTATIONAL EXPLORATION

ESTIMATE

EXPERIMENTS

MATH

TRICKS

Map of the Maps

Part 1: Map of Alfred Russel Wallace's voyage with the black lines denoting Wallace's travel route and the red lines indicating chains of volcanoes. From *The Malay Archipelago* (1869) by Alfred Russel Wallace.

Chapter 1: Map of the world according to Eratosthenes (220 B.C.E.). Erastosthenes is known for, among many other things, his measurement of the circumference of the Earth, and is considered one of the founders of the subject of geography. From *Report on the Scientific Results of the Voyage of the H.M.S. Challenger During the Years 1872–76*, prepared under the superintendence of C. Wyville Thompson and John Murray (1895).

Chapter 2: Population density in Los Angeles County, as determined in the 2000 census. Darker colors represent denser populations (up to 100,000 people per square mile). From the United States Census Bureau.

Chapter 3: Sedimentary rock layers in the Grand Canyon. Geology and cross section by Peter J. Conley, artwork by Dick Beasley. From the United States National Park Service (1985).

Chapter 4: *Carta marina*, a map of Scandinavia, by Olaus Magnus. A translation of the Latin caption reads: *A Marine map and Description of the Northern Lands and of their Marvels, most carefully drawn up at Venice in the year 1539 through the generous assistance of the Most Honourable Lord Hieronymo Quirino*. This detail shows the sea monsters in the ocean between Norway and Iceland.

Part 2: Tourist map of Père Lachaise cemetery, Paris, France.

Chapter 5: Airplane routes around the nearly spherical Earth. Courtesy of OpenFlights.com.

Chapter 6: Josiah Willard Gibbs articulated the variational principle that shows how to find the equilibrium state of a system by maximizing the entropy. Gibbs spent his entire career in New Haven, Connecticut at Yale University. This 1886 map shows the university buildings during Gibbs' time. Source: Yale University Map Collection. Courtesy of the Yale University Map Collection.

Chapter 7: County map of Virginia and West Virginia, drawn by Samuel Augustus Mitchell Jr. in 1864, after the American Civil War.

Chapter 8: Aerial view of the hedge maze at Longleat Safari and Adventure Park, near Warminster, United Kingdom. Courtesy of Atlaspix/Alamy.

Chapter 9: Topographic map of the Great Salt Lake (Utah, United States) and surrounding region. From the United States Geological Survey (1970).

Chapter 10: Blueprint diagram of the Golden Gate Bridge, San Francisco, California, United States. Courtesy of EngineeringArtwork.com

Chapter 11: Digital elevation map of Mount Cotopaxi in the Andes Mountains, near Quito, Ecuador. Blue and green correspond to the lowest elevations in the image, while beige, orange, red, and white represent increasing elevations. Courtesy of the NASA Earth Observatory (2000).

Part 3: Migration tracks of the sooty shearwater, a small seabird, tracked with geolocating tags from two breeding colonies in New Zealand. Breeding season is shown in blue, northward migration in yellow, and wintering season and southward migration in orange. Over about 260 days, an individual animal travels about 64,000 km in a figure-8 pattern across the entire Pacific Ocean. From S. A. Shaffer et al., "Migratory shearwaters integrate oceanic resources across the Pacific Ocean in an endless summer," *Proceedings of the National Academy of Sciences USA*, **103**: 12799–12802, 2006.

Chapter 12: Worldwide distribution of ocean currents (warm in red, cold in green). Arrows indicate the direction of drift; the number of strokes on the arrow shafts denote the magnitude of the drift per hour. Sea ice is shown in purple. Prepared by the American Geographical Society for the United States Department of State in 1943.

Chapter 13: Temperature map of the sun's corona, recorded by the Extreme Ultraviolet Imaging Telescope at the Solar and Heliospheric Observatory on June 21, 2001. Courtesy of ESA/NASA.

Chapter 14: John Snow's map of the 1854 cholera outbreak in the Soho neighborhood of London. By interviewing residents of the neighborhood where nearly 500 people died of cholera in a ten-day period, Snow found that nearly all of the deaths occurred in homes close to the water pump in Broad Street, which he hypothesized was the source of the epidemic. Reproduced from *On the Mode of Communication of Cholera, 2nd Edition*, John Snow (1855).

Chapter 15: Positron emission tomography (PET scan) map of a healthy human brain, showing the rate of glucose utilization in various parts of the right hemisphere. Warmer colors indicate faster glucose uptake. Courtesy of Alzheimer's Disease Education and Referral Center, a service of the National Institute on Aging (United States National Institutes of Health).

Chapter 16: High speed train routes of France, mapped as a transit diagram. Courtesy of Cameron Booth.

Chapter 17: Nile River delta at night, as photographed by the crew in Expedition 25 on the International Space Station on October 28, 2010. Courtesy of Image Science & Analysis Laboratory, Johnson Space Center, Earth Observatory, NASA/GSFC SeaWiFS Project.

Chapter 18: Single-celled photosynthetic organisms such as the coccolithophore *Emiliana huxleyi* can form gigantic oceanic blooms visible from space. In this April 1998 image, the Aleutian Islands and the state of Alaska are visible next to the Bering Sea that harbors the algal bloom. Courtesy of NASA/GSFC SeaWiFS Project.

Part 4: A map of the infant universe, revealed by seven years of data from the Wilkinson Microwave Anisotropy Probe (WMAP). The image reveals 13.7 billion year old temperature fluctuations (the range of ±200 microKelvin is shown as color differences) that correspond to the seeds that grew to become the galaxies. Courtesy of NASA/WMAP Science Team.

Chapter 19: Map of the Internet, as of September, 1998, created by Bill Cheswick. *Courtesy of* Lumeta Corporation 2000–2011. Published in Wired Magazine, December 1998 (issue 6.12).

Chapter 20: The Sloan Great Wall measured by J. Richard Gott and Mario Juric shows a wall of galaxies spanning 1.37 billion light years. It stands in the Guinness Book of Records as the largest structure in the universe. Courtesy of Michael Blanton and the Sloan Digital Sky Survey Collaboration, www.sdss.org.

Chapter 21: This map shows the patterns of human migration as inferred from modern geographical distributions of marker sequences in the Y chromosome (blue), indicating patrilineal inheritance, and in the mitochondrial DNA (orange), indicating matrilineal inheritance. Courtesy of National Geographic Maps, Atlas of the Human Journey.

Chapter 22: "The Lands Beyond" drawn by Jules Feiffer for The Phantom Tollbooth (1961) by Norton Juster. Courtesy of Knopf Books for Young Readers, a division of Random House, Inc.

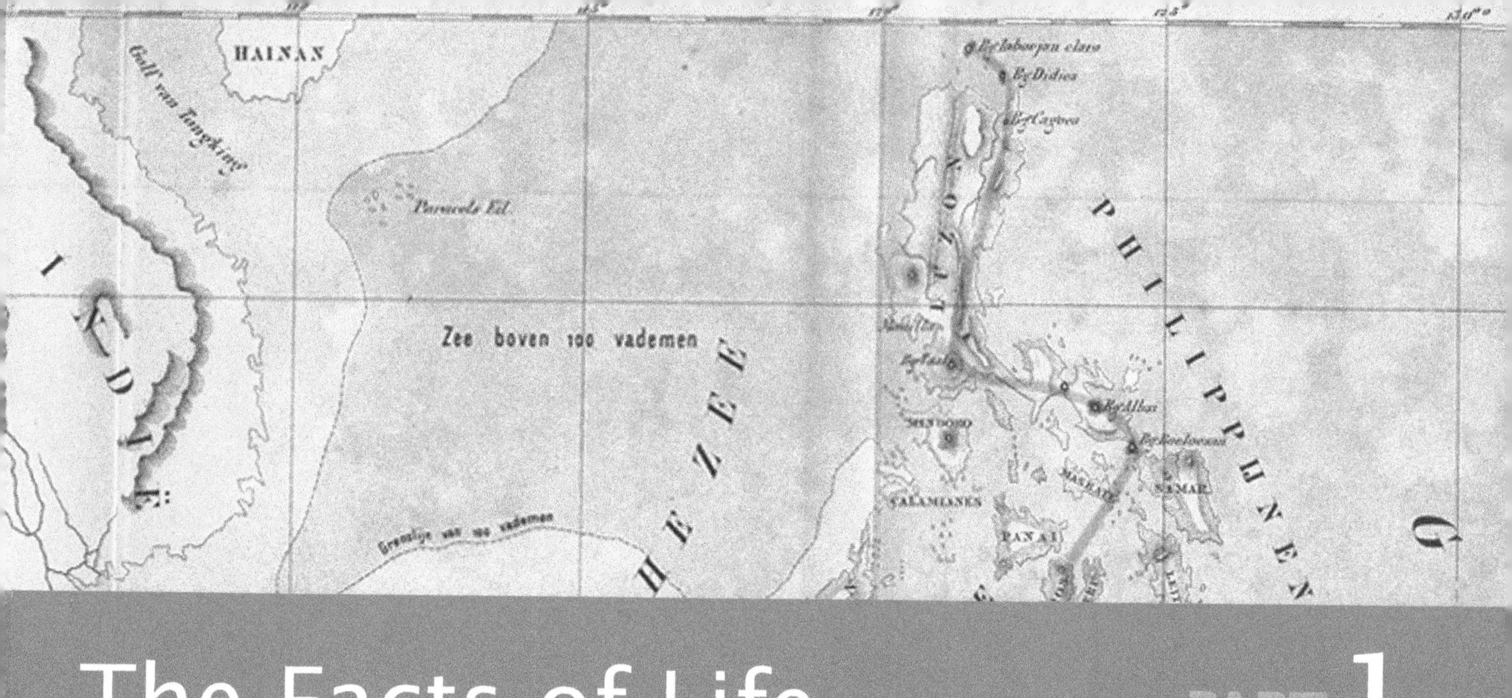

The Facts of Life

PART 1

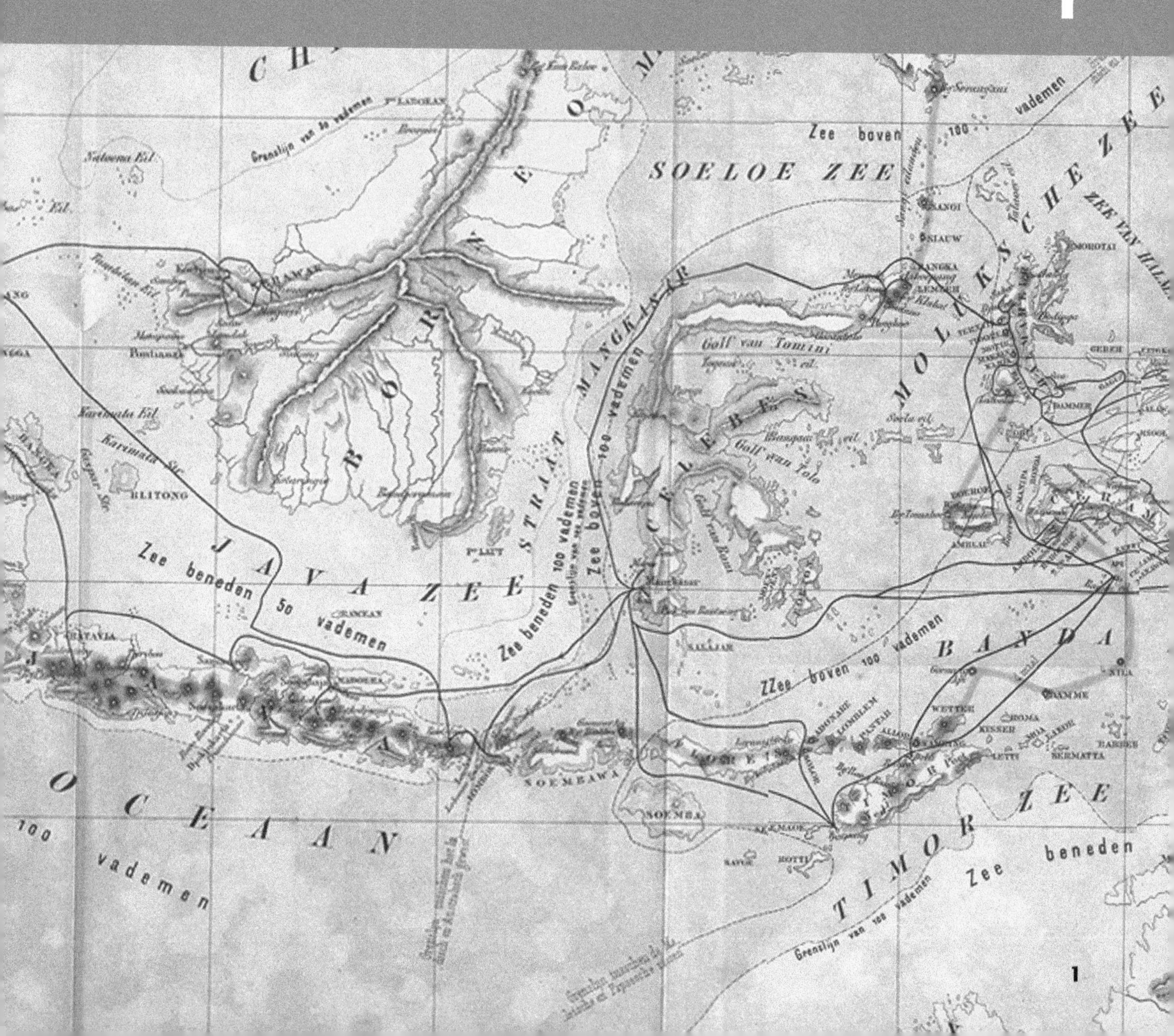

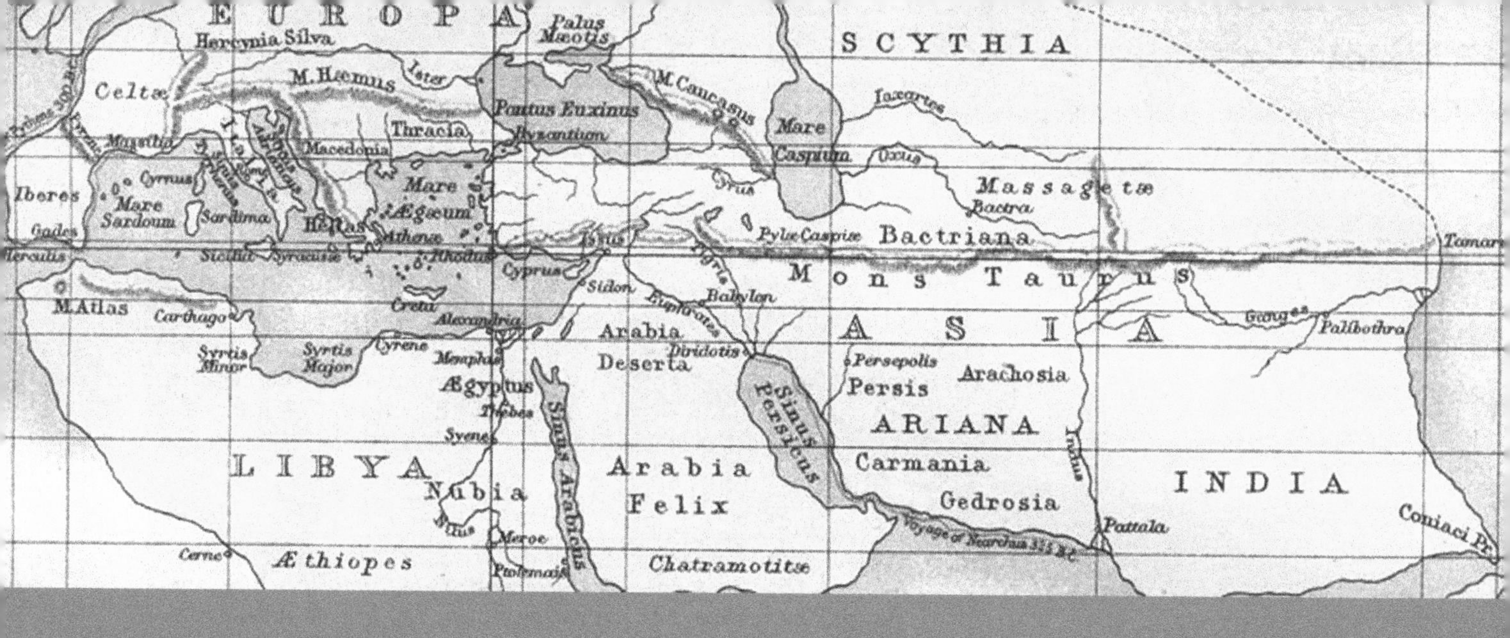

Why: Biology by the Numbers 1

Overview: In which the notion of biological numeracy is examined

> To simplify is the greatest form of sophistication.
>
> **Attributed to Leonardo Da Vinci**

Charles Darwin once said "All observation should be for or against some view if it is to be of any service," meaning that data are most meaningful against the backdrop of some conceptual or theoretical framework (Darwin, 1903, p. 194). The goal of this chapter is to develop a sense of the kinds of quantitative data that are being obtained in all fields of biology and the models that must be put forth to greet this data. Like with any useful map, we will argue that any good model has to overlook some of the full complexity and detail of a given biological problem in order to generate an abstraction that is simple enough to be easily grasped by the human mind, as an aid to developing intuition and insight. At the same time, it is critical that useful models make meaningful predictions, so they must include at least some of the realistic details of the biological system. The art of model building lies in striking the proper balance between too little detail and too much. Part of our emphasis on model building centers on the role of having a feeling for the numbers: sizes, shapes, times, and energies associated with biological processes. Here we introduce the style of making numerical estimates that will be used throughout the book.

1.1 Biological Cartography

In one of the great stories in scientific history, Eratosthenes used the shadow of the noon-day sun at two different positions to work out a first, and surprisingly accurate, measurement of the radius of the Earth. Measurements like those of Eratosthenes made it possible to construct rudimentary maps of the world as it was then known to the Greeks with results such as that decorating the opening to this chapter. Such maps provide a distorted but recognizable view of the world.

In the chapters that follow, our aim is to similarly show how a few fundamental measurements and calculations suffice to provide the rough contours of what is clearly a much more nuanced biological landscape. The idea of such maps is to reveal enough detail to provide a conceptual framework that helps us navigate such biological landscapes, and serve as a starting point to orient ourselves for more in-depth exploration. In constructing such maps, we walk a perilous ridge between descriptions that are so distorted as to be uninformative and those that are so detailed that it is no longer correct to even think of them as maps. These ideas find their literary embodiment in one of Borges' shortest short stories, which serves as the battle cry for the chapters that follow.

> *On Exactitude in Science . . .*
> In that Empire, the Art of Cartography attained such Perfection that the map of a single Province occupied the entirety of a City, and the map of the Empire, the entirety of a Province. In time, those Unconscionable Maps no longer satisfied, and the Cartographers Guilds struck a Map of the Empire whose size was that of the Empire, and which coincided point for point with it. The following Generations, who were not so fond of the Study of Cartography as their Forebears had been, saw that that vast Map was Useless, and not without some Pitilessness was it, that they delivered it up to the Inclemencies of Sun and Winters. In the Deserts of the West, still today, there are Tattered Ruins of that Map, inhabited by Animals and Beggars; in all the Land there is no other Relic of the Disciplines of Geography. Suàrez Miranda: Viajes de varones prudentes, Libro Cuarto, cap. XLV, Lérida, 1658.
>
> Jorge Luis Borges and Adolfo Bioy Casares

1.2 Physical Biology of the Cell

With increasing regularity, the experimental methods used to query living systems and the data they produce are quantitative. For example, measurements of gene expression in living cells report how much of a given gene product there is at a given place at a given time. Measurements on signaling pathways examine the extent of a cellular response such as growth, DNA replication, or actin polymerization, as a function of the concentration of some upstream signaling molecule. Other measurements report the tendency of DNA to wrap into compact structures known as nucleosomes as a function of the underlying sequence. The list of examples goes on and on and is revealed by the many graphs depicting quantitative measurements on biological problems that grace the current research literature.

The overarching theme of the developing field of physical biology of the cell is that quantitative data demand quantitative models and, conversely, that quantitative models need to provide experimentally testable quantitative predictions about biological phenomena. What this means precisely is that in those cases where a biological problem has resulted in a quantitative measurement of how a particular biological output parameter varies as some input is changed (either because it is manipulated by the experimenter or because it varies as part of a natural biological process), the interpretation of that biological data should also be quantitative. In the chapters that follow, our aim is to provide a series of case studies involving some of the key players and processes in molecular and cell biology that demonstrate how physical model building can respond to this new era of quantitative data and sharpen the interplay between theory and experiment.

Model Building Requires a Substrate of Biological Facts and Physical (or Chemical) Principles

One of the first steps in building models of quantitative experiments like those described above is trying to decide which features of the problem are of central importance and which are not. However, even before this, we need to know facts about our system of interest. Throughout this book, we will assert a series of "facts" that fall into several different categories of certainty. We are most comfortable with statements describing observations that you as an individual observer can readily confirm, such as the statement that cells contain protein molecules. A second layer of facts with which we are nearly as comfortable are those arrived at and repeatedly confirmed by decades of experimentation using many different, independent experimental techniques. One fact of this kind is the statement that protein molecules within cells are synthesized by a large piece of machinery called the ribosome. A third category of "facts" that we will take care to mark as speculative are those that stand as compelling explanations for biological observations, but that may rest on unproven or controversial assumptions. In this category, we would include the commonly accepted proposition that modern-day mitochondria are the descendants of what was once a free-living bacterium that entered into a symbiotic relationship with an ancestral eukaryotic cell. In these cases, we will generally explain the observation and the interpretation that underlies our modeling approach. One of the most basic classes of facts that we will need throughout the remainder of the book concerns the nature of the molecules that make up cells and organisms. Because this will set up the common molecular language for the rest of our discussions, we will describe these molecules at some length in the next section.

Complementary to these biological observations, we embrace the proposition that biological entities cannot violate the laws of physics and chemistry. Many fundamental physical and chemical principles can be directly applied to understanding the behavior of biological systems, for example, the physical principle that the average energy of a molecule increases with increasing environmental temperature, and the chemical principle that oil and water do not mix. Along with biological facts described above, such principles will serve as the basis for the models to be described throughout the book.

1.3 The Stuff of Life

Scientists and other curious humans observe the natural world around them and try to make sense of its constituent parts and the ways that they interact. The particular subset of the natural world called life has always held special fascination. Any 3-year-old child knows that a rock is not alive but that a puppy, a tree, and indeed the child himself are alive. Yet the 3-year-old would be hard pressed to produce a rigorous definition of life and, indeed, scientists who have thought about this for decades can only approximate a description of this fascinating property. No single feature can reliably discriminate life from non-life, but all living systems do share certain central properties. Living systems grow, consume energy from their environment, reproduce to form offspring that are rather similar to themselves, and die.

In the quest to understand the nature of life, scientists over hundreds of years have investigated the properties of living organisms

Figure 1.1: Atomic-level structural representation of members of each of the major classes of macromolecules, all drawn at the same scale. Nitrogen is colored in blue, oxygen in red, phosphorus in yellow, carbon in gray, and hydrogen in white. (A) Atomic structure of a small fragment of the nucleic acid DNA in the B form, (B) atomic structure of the oxygen-carrying protein hemoglobin (PDB 1hho), (C) phosphatidylcholine lipid molecule from a cell membrane, and (D) branched complex carbohydrate (M41 capsular polysaccharide) from the surface of the bacterium *Escherichia coli* (PDB 1cap). (Illustrations courtesy of D. Goodsell.)

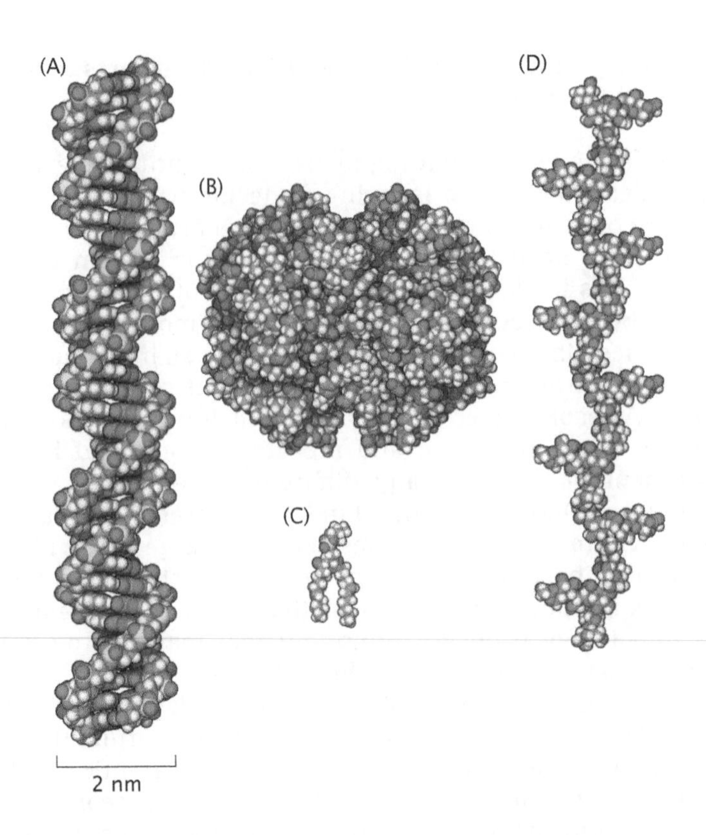

and their constituent materials with an overarching goal of trying to understand why a puppy is different from a rock, even though they are both made of atoms that obey the same physical laws. Although this quest is still ongoing, some useful generalizations about the material nature of life are now widely accepted. For example, one observed feature of many of the molecules that make up living organisms is that they tend to be relatively large and structurally complex, and are hence called macromolecules. Living organisms also contain a large number of small, simple molecules that are critical to their function, ranging from water and metal ions to sugars such as glucose. Although these small molecules are important to life processes and, indeed, are used by the cell as building blocks to make the macromolecules, a distinction is usually drawn between small molecules that can arise through nonliving chemical processes and characteristic biological macromolecules that are found nowhere but in living organisms. Despite the diversity of macromolecules found in living organisms, the kinds of atoms found within them are surprisingly restricted, consisting mainly of carbon (C), oxygen (O), nitrogen (N), and hydrogen (H), with a smattering of sulfur (S) and phosphorus (P). It is largely the ways in which this humble suite of atoms are connected to each other that create the special properties of the macromolecules found in living systems. Of course, many other kinds of atoms and ions are critical for the biochemical processes of life, but they are generally not covalently incorporated in macromolecular structures.

Organisms Are Constructed from Four Great Classes of Macromolecules

If we take any organism, ranging from a small soil bacterium to a giant redwood tree or a massive whale, and separate out its constituent macromolecules, we will find a small number of distinct classes, chief among these being proteins, carbohydrates, lipids, and nucleic acids. Figure 1.3 shows representative examples of each of these classes. Broadly speaking, these different classes of macromolecules perform different functions in living organisms. For example, proteins form structural elements within cells and are largely responsible for catalyzing specific chemical reactions necessary to life. Lipids form membrane barriers that separate the cell from the outside world and subdivide the cell into interior compartments called organelles. Carbohydrates are used for energy storage, for creating specific surface properties on the outside of cell membranes, and, sometimes, for building rigid structural units such as cell walls. Nucleic acids perform a critical function as the memory and operating instructions that enable cells to generate all the other kinds of macromolecules and to replicate themselves. Within each of these broad categories there is tremendous diversity. A single organism may have tens of thousands of structurally distinct proteins within it. Similarly, the membranes of cells can comprise hundreds of different lipid species.

Why are these particular classes of macromolecules the stuff of life? A complete understanding of the answer to this question would require knowledge of the specific chemical and physical conditions on the early Earth that gave rise to the first living cells. However, by studying living organisms on Earth today, it is seen that these molecules are wonderfully suited for the perpetuation of life for several reasons. First, each of these classes of molecules can be assembled by the cell from a small number of simpler subunit or precursor molecules. It is the combinatorial assembly of these simple subunits that gives rise to the tremendous structural diversity mentioned above. What this means is that a cell needs only a relatively restricted repertoire of chemical reactions to be able to synthesize these sets of subunits from the food in its environment. The fact that most cells on Earth make a living by consuming other living organisms is probably both a cause and a consequence of the fact that they all share very similar suites of these fundamental building blocks.

Nucleic Acids and Proteins Are Polymer Languages with Different Alphabets

The rest of this book will be largely devoted to exploring the special properties of biological macromolecules and the ways that they are used by cells, with the large-scale goal of understanding how the special properties of life may emerge from fundamental physical and chemical interactions. We will begin by discussing one specific and fundamentally important aspect of macromolecular behavior that is necessary for life as we know it. In particular, a great triumph of the study of molecular biology in the middle part of the twentieth century was the profound realization that the sequence of nucleic acid subunits in the cell's DNA is directly responsible for determining the sequence of amino acids in that same cell's proteins (that is, the discovery of the genetic code). The relationship between a DNA sequence and a protein sequence has been described by Francis Crick as comprising the "two great polymer languages" of cells.

As shown in Figure 1.2, both nucleic acids and proteins are built up from a limited alphabet of units. The nucleic acids are built up from an alphabet of four letters that each represent a chemically distinct "nucleotide" (A – adenine, G – guanine, T – thymine, and C – cytosine, for DNA). The chemical structure of these four nucleotides is shown in Figure 1.3. The protein language is constructed from an alphabet of 20

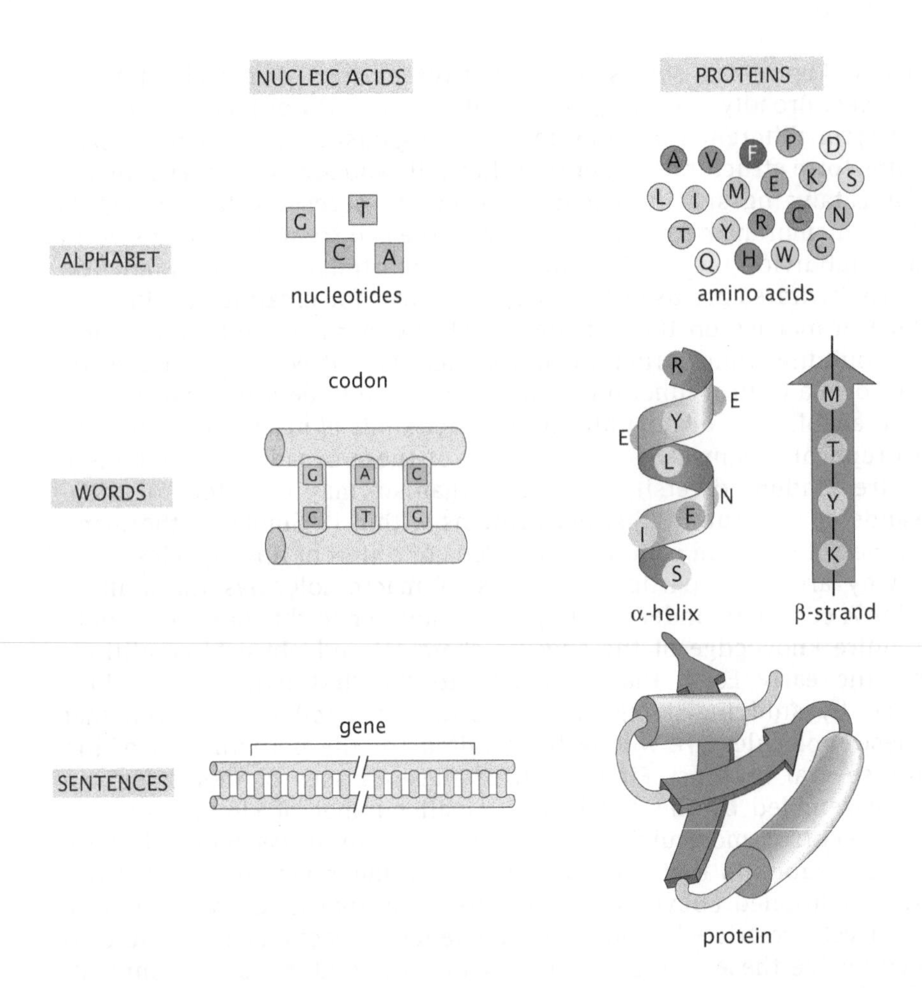

Figure 1.2: Illustration of Crick's "two great polymer languages." The left-hand column shows how nucleic acids can be thought of in terms of letters (nucleotides), words (codons), and sentences (genes). The right-hand column illustrates a similar idea for proteins, where the letters correspond to amino acids, the words to elements of secondary structure such as α helices and β strands, and the sentences to fully folded functional proteins.

distinct amino acids. How can just four nucleotide bases be translated into protein sequences containing 20 different amino acids? The key realization is that the DNA letters are effectively read as "words," called codons, made up of three sequential nucleotides, and each possible combination of three nucleotides encodes one amino acid (with some built-in redundancy in the code and a few exceptions such as the codon for "stop"). In the protein language, groups of amino acid "letters" also form "words," which give rise to the fundamental units of protein secondary structure, namely α helices and β strands, which we will discuss below. Finally, in the DNA language, sentences are formed by collections of words (that is, collections of codons) and correspond to genes, where we note that a given gene codes for a corresponding protein. An example of a complete sentence in the protein language is a fully folded, compact, biologically active enzyme that is capable of catalyzing a specific biochemical reaction within a living cell. Much of the rest of the book will be built around developing models of a host of biological processes in cells that appeal to the macromolecular underpinnings described here.

The sequences associated with nucleic acids and proteins are linked mechanistically through the ribosome, which takes nucleic acid sequences (in the form of messenger RNA (mRNA)) and converts them into amino acid sequences (in the form of proteins). These two polymer languages are linked informationally in the form of the genetic code. In particular, each three-letter collection from the nucleic acid language has a corresponding letter that it codes for in the protein language. The determination of this genetic code is one of the great chapters in the history of molecular biology and the results of that

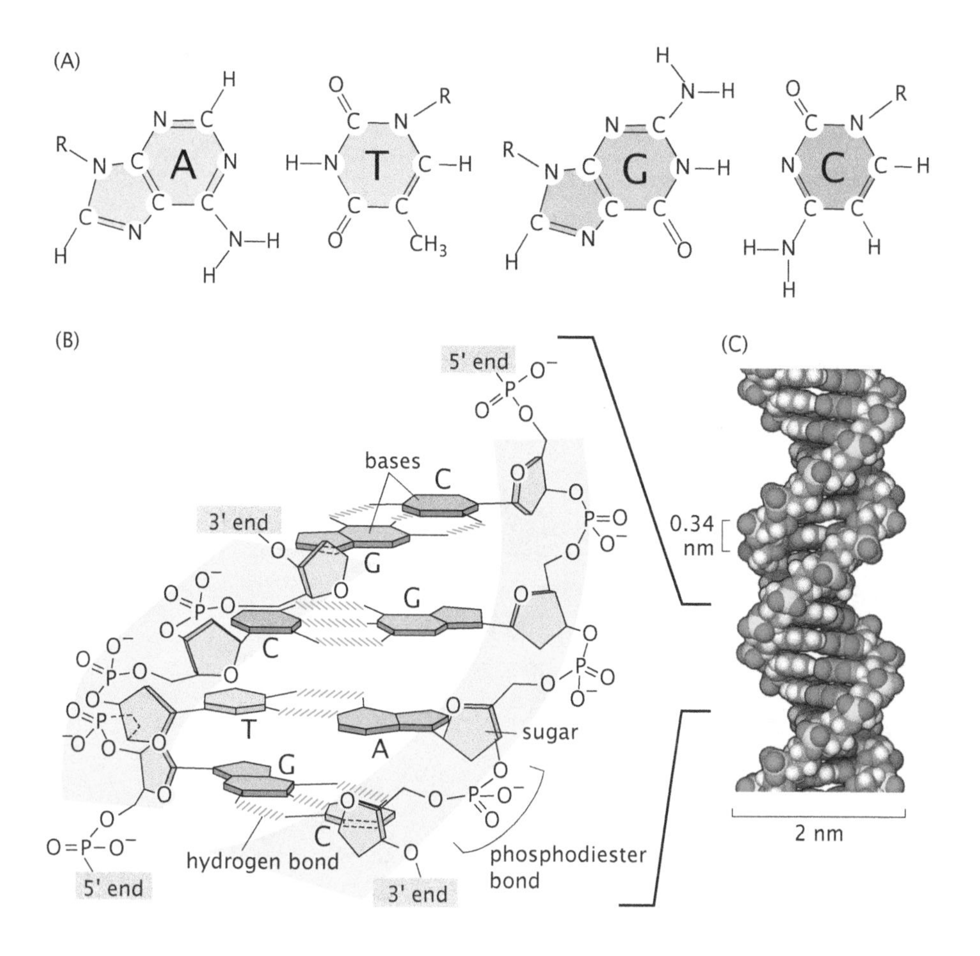

Figure 1.3: The chemical structure of nucleotides and DNA. (A) In DNA, the four distinct subunits or bases are abbreviated: A (adenine), T (thymine), G (guanine), and C (cytosine). In these diagrams, carbon is represented by C, oxygen by O, nitrogen by N and hydrogen by H. The letter R indicates attachment to a larger chemical group (the rest of the molecule); for nucleotides, the R group consists of the pentose sugar deoxyribose attached to phosphate. A single line connecting two atoms indicates a single covalent bond and a double line indicates a double covalent bond. The two large bases, A and G, are called purines and the two small bases, C and T, are called pyrimidines. (B) Illustration of how bases are assembled to form DNA, a double helix with two "backbones" made of the deoxyribose and phosphate groups. The four bases are able to form stable hydrogen bonds uniquely with one partner such that A pairs only with T and G pairs only with C. The structural complementarity of the bases enables the faithful copying of the nucleotide sequence when DNA is replicated or when RNA is transcribed. (C) Space-filling atomic model approximating the structure of DNA. The spacing between neighboring base pairs is 0.34 nm. (Adapted from B. Alberts et al., Molecular Biology of the Cell, 5th ed. Garland Science, 2008.)

quest are shown in Figure 1.4, which shows the information content recognized by the universal translating machine (the ribosome) as it converts the message contained in mRNA into proteins.

1.4 Model Building in Biology

1.4.1 Models as Idealizations

In trying to understand the nature of life, the approach that we take in this book is the same as that followed by scientists for generations. We will approach each system with the goal of gaining some useful insight into its behavior by abstracting and simplifying the highly complex materials of living organisms so that we can apply simple, analytical models that have some predictive value. By definition, we cannot retain a complete atomic description of each macromolecule. Instead, we aim to select only the relevant properties of the macromolecule that speak to the particular aspect of that molecule's behavior that we are trying to address. Therefore, it would be inaccurate to say that we have in mind "a" simple model for a complex macromolecule such as DNA. Instead, we will use a suite of simple models that can be thought of as projections of the complex, multifaceted reality of the DNA molecule into a variety of simpler DNA maps. In the same way, any useful map of a territory emphasizes only the features that will be most relevant to the person who will be using the map. A classic example of the extraction and elegant presentation of only the most useful features is Harry Beck's famous map of the London Underground system, which shows the

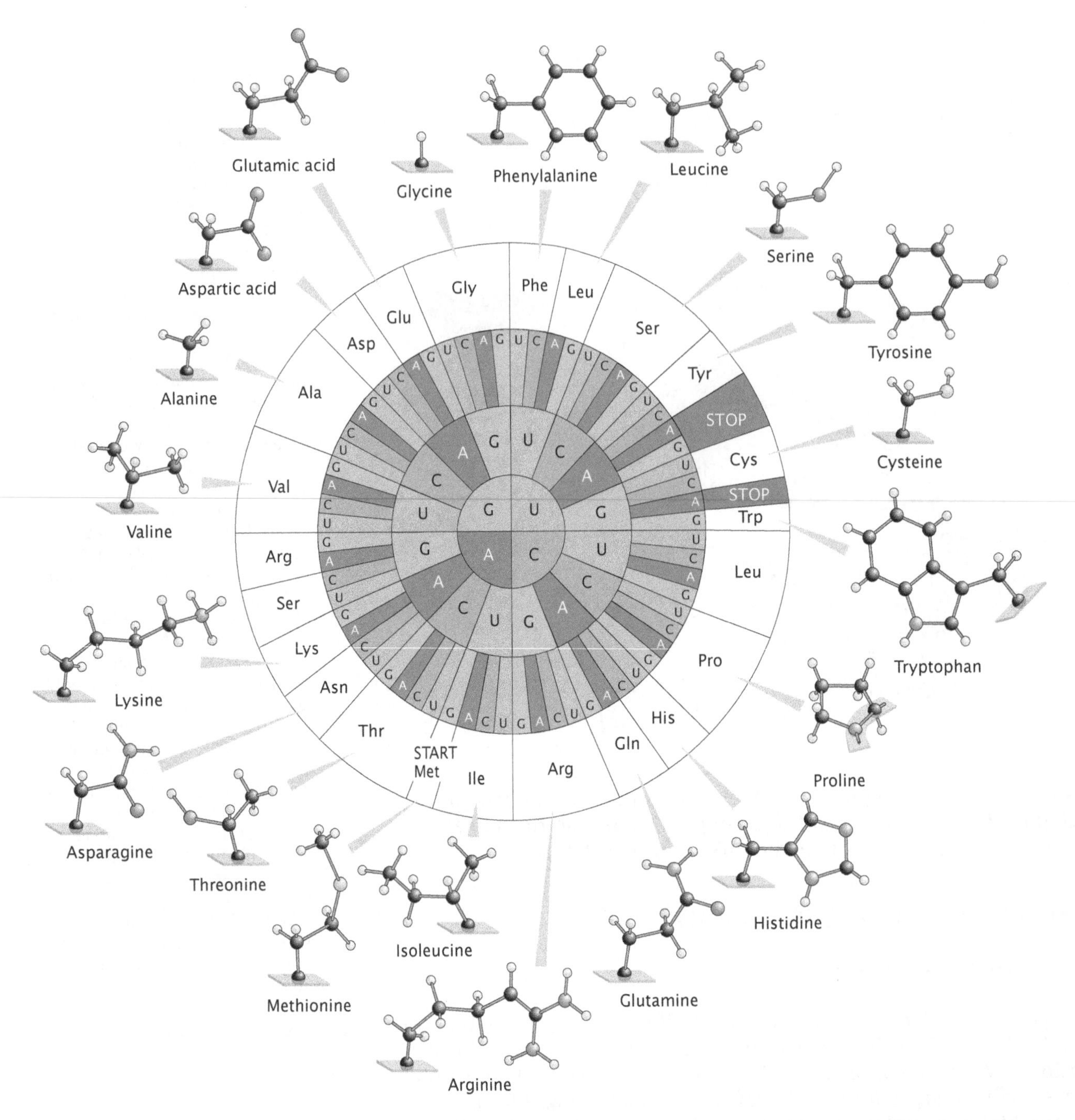

Figure 1.4: Genetic code. In this schematic representation, the first nucleotide in a coding triplet is shown at the center of the ring, the second nucleotide in the middle colored ring, and the third nucleotide in the outer colored ring. In this representation of the genetic code, the four bases are adenine (A), cytosine (C), guanine (G), and uracil (U). Uracil is structurally very similar to thymine (T), and is used instead of thymine in messenger RNA. The amino acids corresponding to each group of triplets are illustrated with their names (outer ring) and atomic structures. Two amino acids, tryptophan and methionine, are encoded by only a single triplet, whereas others, including serine, leucine, and arginine, are encoded by up to six. Three codons do not code for any amino acid and are recognized as stop signals. The unique codon for methionine, AUG, is typically used to initiate protein synthesis.

stations and connecting lines with an idealized geometry that does not accurately represent the complex and rather frustrating topography of London's streets. Although certain features of London's geography are deliberately ignored or misrepresented, such as the actual physical distances between stations, the London Underground map shows the user exactly enough information about the connectivity of the transport system to go anywhere he wishes. In this section, we will

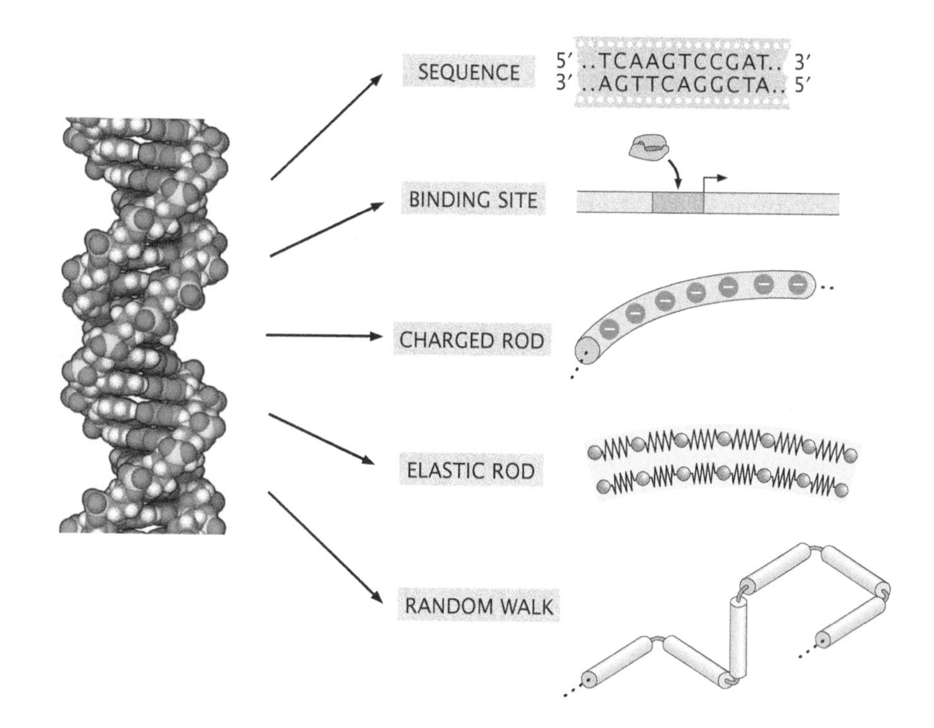

Figure 1.5: Idealizations of DNA. DNA can be thought of as a sequence of base pairs, as a series of binding sites, as a charged rod, as an elastic rod, or as a freely jointed polymer arranged in a random walk, depending upon the problem of interest.

summarize the kinds of physical properties of biological materials that we may wish to consider in isolation. In Section 1.5, we will list the kinds of fundamental physical models that will be applied throughout the book.

Biological Stuff Can Be Idealized Using Many Different Physical Models

Often, the essence of building a useful and enlightening model is figuring out what features of the problem can be set aside or ignored. Throughout the book, we will come to view molecules and cells from many different angles. An example of this diversity of outlook is shown in Figure 1.5. The nature of the simplified space, and therefore the nature of the projected DNA model, depends upon the specific question that we are asking. Figure 1.5 shows five different projections for the meaning of DNA that we will use repeatedly throughout the book.

DNA is one of the most important molecules, both because it is the carrier of genetic information and because of its iconic role in the history of molecular biology. There is a polarity to each strand of this helical polymeric molecule; that is, the two ends of the strand are structurally and chemically distinct, and are called the 5′ and 3′ ends. In addition, there is heterogeneity along their length, since the chemical identity of the bases from one nucleotide to the next can be different. It is often the informational content of the DNA molecule that garners attention. Indeed, the now-routine sequencing of whole genomes of organisms ranging from bacteria to humans permits us to forget the molecular details of the DNA molecule and to focus just on the sequence of A's, T's, G's, and C's that determines its information content. This representation will be most useful, for example, when we are trying to discern the history of mutational events that have given rise to diversification of species or when we are trying to calculate the information content of genomes. The second representation shown in Figure 1.5 abstracts variations in DNA sequence in large blocks where one particular stretch of sequence (shown in the figure as the more darkly colored block) represents a portion of sequence that has particularly strong affinity for a protein binding partner, such as the RNA polymerase enzyme that copies the sequence encoded by the DNA into a molecule of mRNA. Both these first two representations draw attention to dissimilarities in detailed chemical structure within the DNA molecule. However, sometimes we will consider DNA simply as a physical entity within the cell where the specific nature of its sequence is not relevant. Instead, other properties of the molecule will be emphasized, such as its net distribution of charge, bending elasticity, or behavior as a long polymer chain subject to thermal undulations. Schematics of models emphasizing these physical properties of DNA are also shown in Figure 1.5.

Proteins are subject to many different idealizations as well. Just as nucleic acids can be thought of as sequences of nucleotides, we can think of proteins in terms of the linear sequence of amino acids of which they are made. The second great polymer language gives rise to proteins, which are considered to provide most of the important biological functions in a cell, including the ability to catalyze specific chemical reactions. The detailed atomic structure of proteins is substantially more complex than that of DNA, both because there are 20 rather than 4 fundamental subunits and because the amino acid chain can form a wider variety of compact three-dimensional structures than can the DNA double helix. Even so, we will extract abstract representations of proteins emphasizing only one feature or another as shown in Figure 1.6 in much the same way we did for DNA.

A protein can be represented as a simple linear sequence of amino acids. In cases where this representation contains too much detail, we will elect to simplify further by grouping together all hydrophobic amino acids with a common designation H and all polar amino acids with a common designation P. In these simple models, the physical differences between hydrophobic and polar amino acids (that is, those that are like oil versus those that are like water) affect the nature of their interactions with water molecules as well as with each other and thereby dictate the ways that the amino acid chain can fold up to make the functional protein. Similarly, we can think of the three-dimensional folded structure of proteins at different levels of abstraction depending on the question under consideration. A common simplification for protein structure is to imagine the protein as a series of cylinders (α-helices) and ribbons (β-strands) that are connected to one another in a defined way. This representation ignores much of the chemical complexity and atomic-level structure of the protein but is useful for considering the basic structural elements. An even more extreme representation that we will find to be useful for thinking about protein folding is a class of lattice models of compact polymers in which the amino acids are only permitted to occupy a regular array of positions in space.

Fundamentally, the amino acid sequence and folded structure of the protein serve a biological purpose expressed in protein function or activity, and we will also develop classes of models that refer only to the activity of the molecule. For example, the vast majority of proteins in the cell are capable of forming specific binding interactions with one another or with other biomolecules; these binding events are usually important to the protein's biological function. When modeling these kinds of interactions, we will envision proteins as "receptors" carrying binding sites that may be occupied or unoccupied by a binding partner, called a "ligand," and our description will make no further

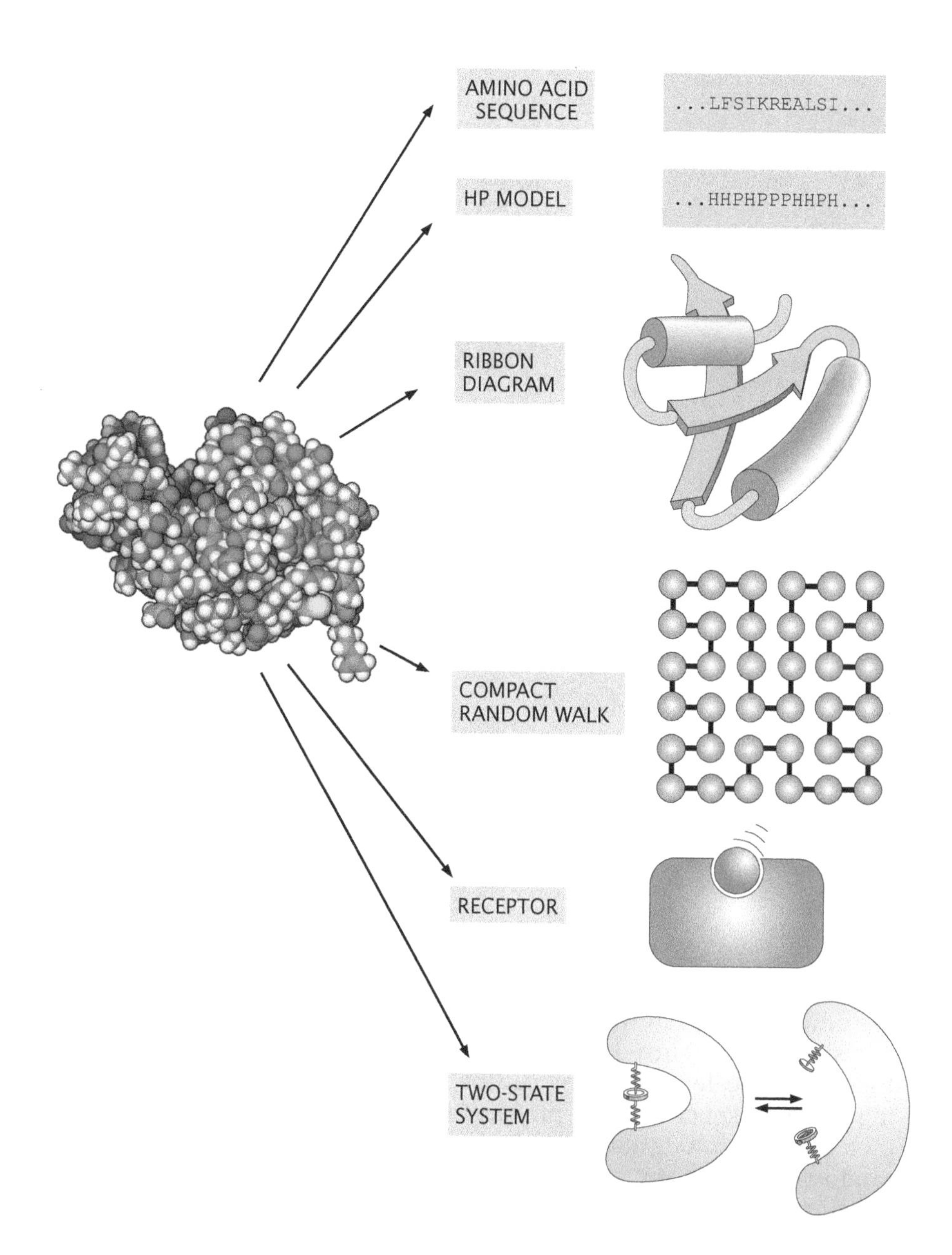

Figure 1.6: Idealizations of protein. Proteins can be thought of as a particular sequence of amino acids, as a simplified sequence reporting only the hydrophobic (H, oil-like) or polar (P, water-like) chemical character of the amino acids, as a collection of connected ribbons and cylinders, as a compact polymer on a lattice, as a binding platform for ligands, or as a two-state system capable of interconverting between different functional forms.

reference to the protein's complex internal structure. For some models, we will focus on changes in activity of proteins, considering them as simple two-state systems that can interconvert between two different functional forms. For example, an enzyme that catalyzes a particular biochemical reaction may exist in an "active" or "inactive" state, depending on the presence of ligands or other modifications to its structure.

By performing these simplifications and abstractions, we are not intending to deny the extraordinary complexity of these molecules and are fully cognizant that amino acid sequence, folding and compaction properties, and conformational changes between states dictate binding affinity, and it is an artificial simplification to consider these properties in isolation from one another. Nonetheless, it is generally useful to consider one single level of description at a time when trying to gain intuition from simple estimates and models, which will serve as our primary emphasis.

We will extend this approach beyond individual macromolecules to large-scale assemblies of macromolecules such as cell membranes

Figure 1.7: Idealization of membranes. A membrane can be modeled as an elastic object that deforms in response to force, as a random surface fluctuating as a result of collisions with the molecules in the surrounding medium, as an electrical circuit element, or as a barrier with selective permeability.

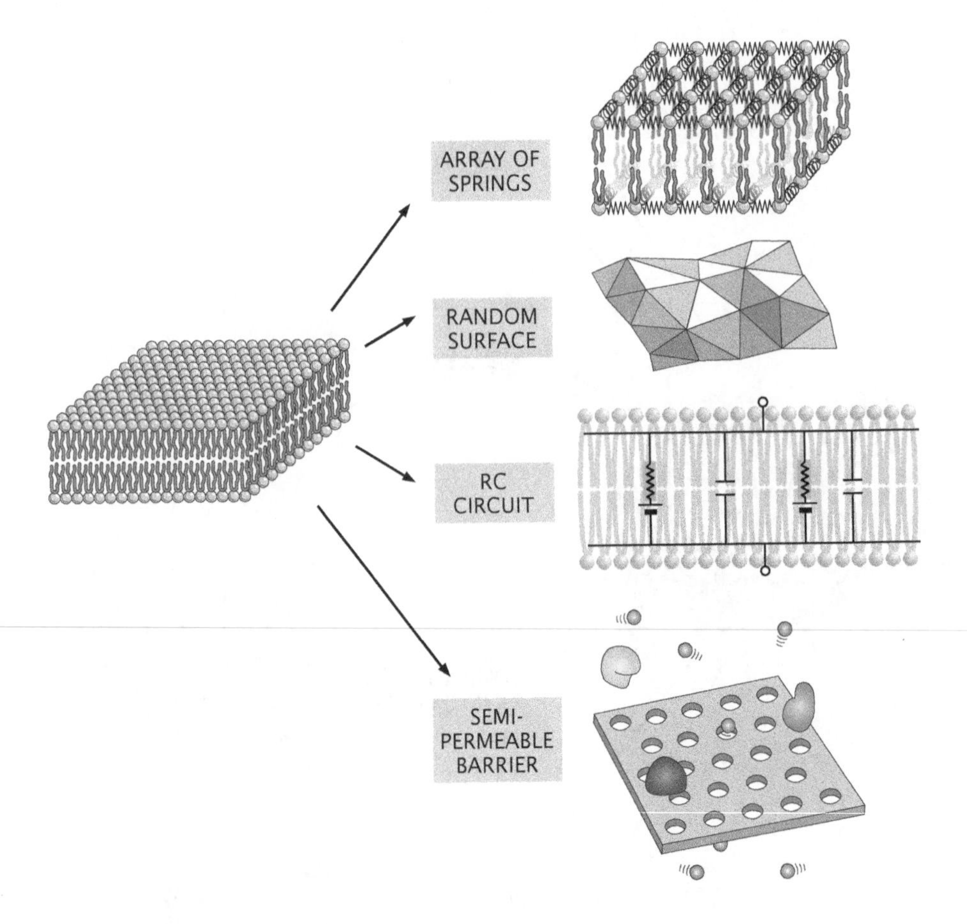

made up of millions of individual lipid and protein molecules. Some of the different ways in which we will think about biological membranes are shown in Figure 1.7. Just as with individual macromolecules, these large-scale assemblies can be usefully characterized by focusing on one aspect of their physical behavior at a time. For example, membranes can be thought of as bendable, springy elastic sheets, as random surfaces, or even as arrays of electrical elements such as resistors and capacitors. In their interactions with other molecules, membranes often serve as selective barriers that will enable some molecular species to cross while blocking others. Depending upon the particular question being asked, we will exploit different idealizations of the membrane.

Audaciously, we will bring these same idealization techniques to bear even on living cells, as shown in Figure 1.8. Although a living cell is exceedingly complex, we will nonetheless choose to extract one physical property of the cell at a time so that we can gain insight from our simple approach. For example, the bacterium *Escherichia coli*, a beloved experimental organism, can for some purposes be thought of as an object carrying an array of protein receptors on its surface. For other purposes, for example when we think of how a bacterium swims through water, we will think of it as an elastohydrodynamic object, that is, a mechanical object that can bend and can interact with the flow of water. The physical properties of the path it follows through water can be further abstracted by considering the cell's large-scale motion in the context of a biased random walk, ignoring the hydrodynamic details that actually enable that motion. Finally, because much biological research focuses on how cells alter the expression of their polymer languages in response to changing conditions, we

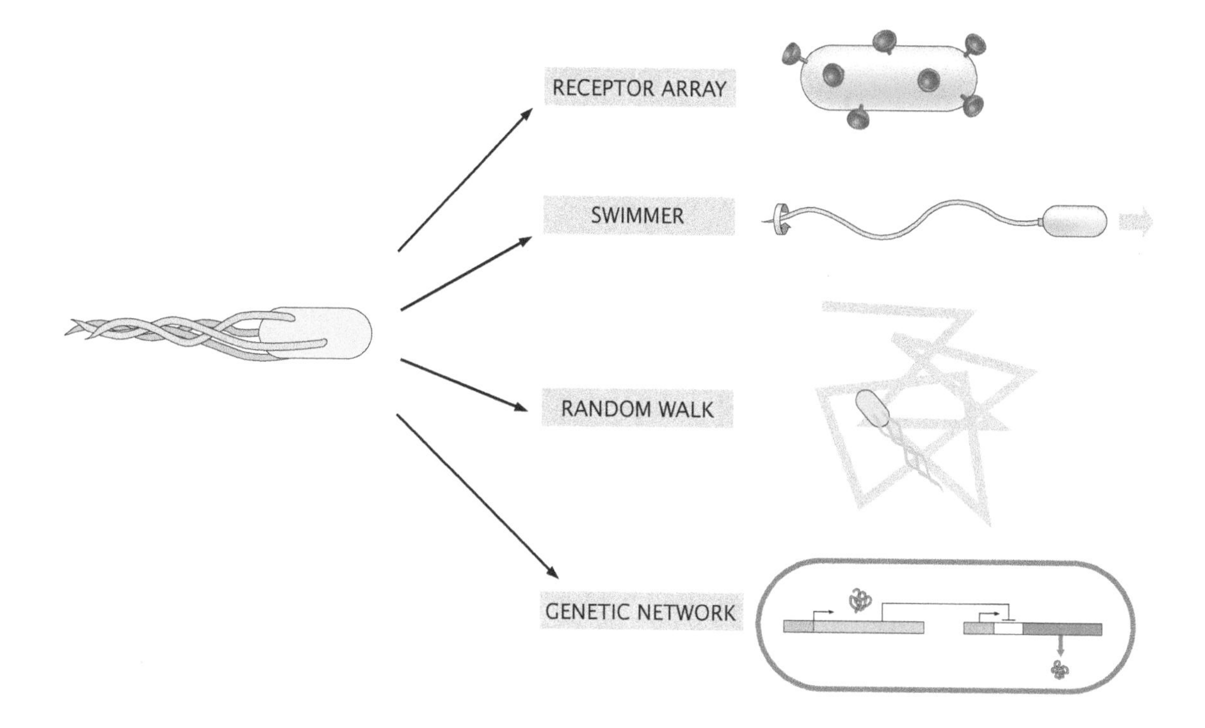

Figure 1.8: *E. coli* idealization. The cell can be modeled as an array of receptors for a ligand of interest, as an elastohydrodynamic object, as a biased random walker moving through water, or as an information processing device characterized by a series of genetic networks.

will frequently find it useful to consider the cell as an information processing device, for example as a network governing the flow of information in the form of gene expression. These descriptions are not mutually exclusive, nor are any of them comprehensive. They are again deliberate abstractions made for practical reasons.

Although so far we have emphasized the utility of abstract projections centering on a single physical property for descriptions of biological entities, the same approach is actually extremely useful in consideration of other systems such as solutions of charged ions in water. Various abstractions of the watery medium of life are shown in Figure 1.9. Sometimes we will pretend that a solution is a regular lattice. For example, if we are interested in ligands in solution and their tendency to bind to a receptor, we adopt a picture of the solution in which the ligands are only permitted to occupy specific discrete positions. Although this is an extreme approximation, it is nonetheless immensely valuable for calculations involving chemical equilibrium and remarkably gives accurate quantitative predictions. As the watery medium interacts with the living creatures it contains, sometimes its hydrodynamic flow properties (or viscous properties) are of most interest. When considering rates of chemical reactions taking place in water, we will find it useful to think of water as the seat of diffusive fluctuations. The macromolecules that exist in the watery medium of cells also interact with specific properties of water. For example, some of the chemical groups in proteins are polar or "hydrophilic" (water-loving), meaning that they are able to form hydrogen bonds with water, while other groups are oil-like and repel water. In the folded protein, the oil-like portions tend to cluster on the inside. In contrast, charged molecules on the surface can interact with the dielectric character of water molecules, which are neutral overall but do exhibit a slight charge separation.

Each of these representations will resurface multiple times throughout the book in different contexts related to different biological questions. Although none of them can give a complete understanding

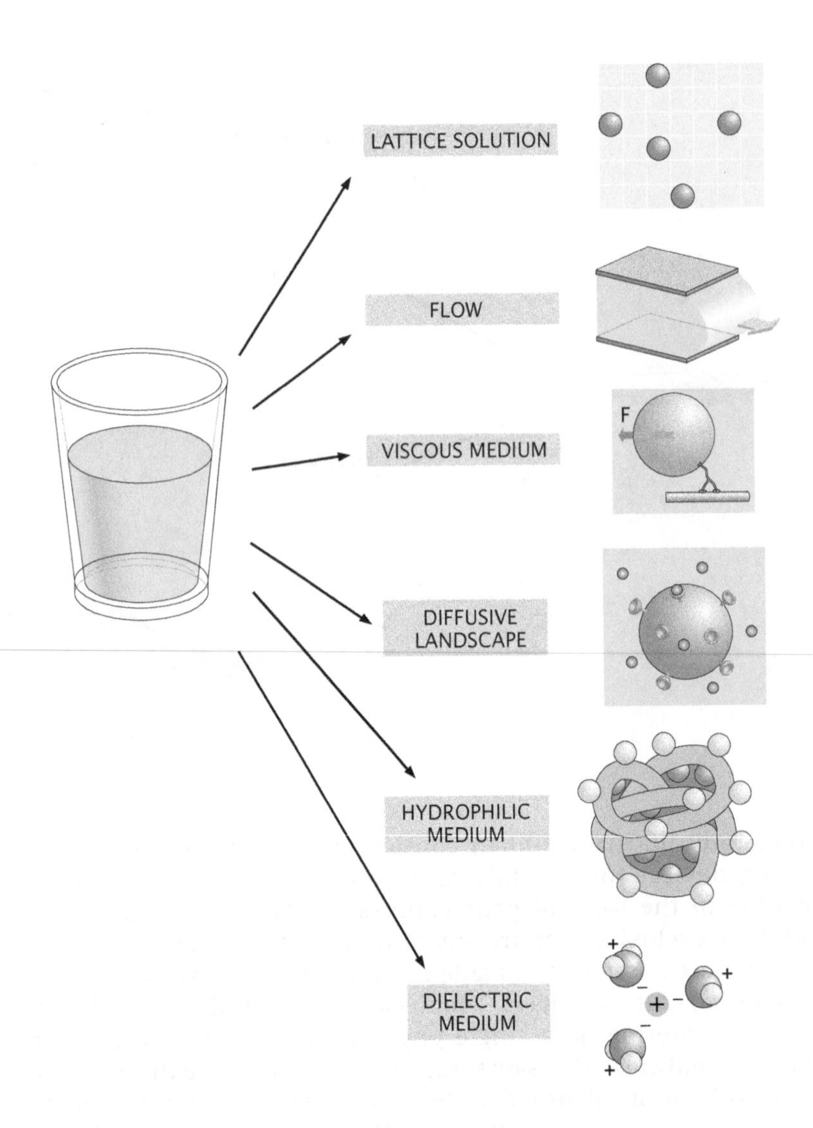

Figure 1.9: Idealization of a solution. A solution can be represented as a lattice of discrete positions where solutes might exist, as a fluid with a mean flow, as a viscous medium that exerts a drag force on objects moving through it, as a fluctuating environment that induces random motion of macromolecules and ligands, as a hydrophilic medium that readily dissolves polar molecules but repels oil-like molecules, or as a dielectric medium that is a poor electrical conductor but can support electrostatic fields.

of the behavior of living cells, each of them serves to provide quantitative insight into some aspect of life and, taken as a whole, they can begin to project a more realistic image of cells than any single abstraction can do alone.

1.4.2 Cartoons and Models

Biological Cartoons Select Those Features of the Problem Thought to Be Essential

We have argued that the art of model building ultimately reflects a tasteful separation of that which is essential for understanding a given phenomenon from that which is not. As is true of all sciences, the history of experimental progress in biology has always involved researchers constructing explicit conceptual models to help them make sense of their data. In some cases, the theoretical framework developed for biological systems is of precisely the mathematical character that will be presented throughout the book, such as that illustrated in the use of thermodynamics and statistical mechanics to understand biochemical reactions or in the development of the theory of the action potential. More commonly, much of the important modeling of biological systems has been in the form of visual schematics that illustrate the most essential features in a given biological process and how they interact. In fact, the act of drawing the type of cartoon found in molecular biology textbooks or research papers reflects an important type of conceptual model building and either implicitly or explicitly reflects choices about which features of the problem are really important and which can be ignored.

As an example of the model building that takes place in constructing biological cartoons, consider the structure of one of the most beautiful and intriguing eukaryotic organelles, namely, the mitochondrion. Figures 1.10(A) and (B) show a planar section of a mitochondrion as obtained using electron microscopy and part of a three-dimensional structure resulting from cryo-electron microscopy. Figure 1.10(C) shows two cartoons that describe the structure of this important organelle. This pair of drawings emphasizes the central importance of the membranes which bound the organelle and which separate the interior into different, relatively isolated compartments. Though the three-dimensional model represents more detail regarding the precise geometry of the membranes, the essential *conceptual* elements are the same, namely, (a) the mitochondria are closed, membrane-bound organelles and (b) the inner membrane has a complex folded structure. Furthermore, the folds of the inner membrane greatly increase the total amount of surface area of this membrane and therefore increase the area available for the ATP-synthesizing machinery that lies within this membrane, as will be discussed below.

The other cartoons in this figure also represent the mitochondrion, but in very different ways, focusing on specific conceptual elements involved in different aspects of mitochondrial function and behavior. For example, Figure 1.10(D) reveals the surprising existence of mitochondrial DNA, which is a vestige of the microbial origins of this important organelle. Interestingly, these organelles contain their very own DNA genomes, completely distinct from the genome found in the cell's nucleus. As indicated in the diagram, the genome is a circular piece of DNA (like the genome found in bacterial cells, but very different from eukaryotic chromosomes, which are linear pieces of DNA). Each mitochondrion contains several copies of the genome, and they are distributed between the two daughters when the mitochondrion divides.

The diagram provided in Figure 1.10(E) shows a highly magnified segment of the mitochondrial inner membrane, illustrating the proteins that cross the membrane and that are involved in the generation of ATP. A central function of mitochondria in eukaryotic cells is to convert one form of chemical energy, high-energy electrons in the compound NADH, into a different form of chemical energy, a phosphodiester bond in ATP (these different forms of chemical energy and their significance will be discussed in more detail in Chapter 5). The process of performing this conversion requires the cooperative action of several different large protein complexes that span the inner mitochondrial membrane. Sequential transfer of electrons from one protein complex to another results in the transport of hydrogen ions from the mitochondrial matrix across the inner membrane into the intermembrane space, establishing a gradient with an excess of hydrogen ions on the outside. In the final step, the hydrogen ions are allowed to travel back down their concentration gradient into the mitochondrial matrix, by passing through a remarkable protein machine called ATP synthase, which we will revisit several times throughout the rest of the book. ATP synthase catalyzes the construction of a molecule of ATP by combining a molecule of ADP and an

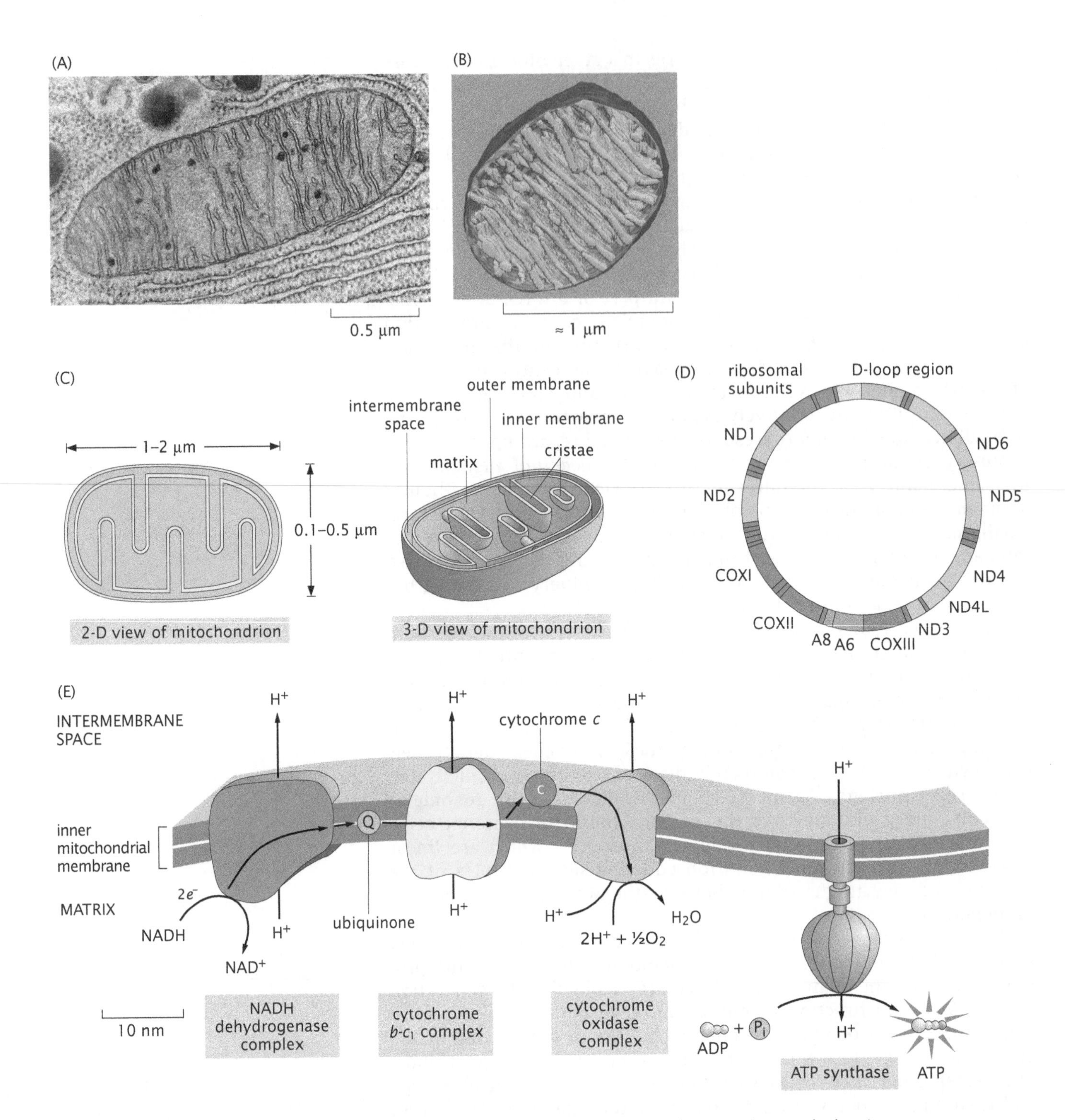

Figure 1.10: Several different ways of illustrating mitochondria. (A) Thin-section electron micrograph showing a mitochondrion found in a cell within the pancreas of a bat. (B) Cryo-electron microscopy reconstruction of the structure of a mitochondrion, shown cut in half to reveal the internal structure. (C) Diagrams showing the arrangement of membranes dividing the mitochondrion into distinct compartments. While the outer membrane forms a smooth capsule, the inner membrane is convoluted to form a series of cristae. Distinct sets of proteins are found in the matrix (inside the inner membrane) and in the intermembrane space (between the inner and outer membranes). (D) Schematic of mitochondrial DNA. (E) Schematic illustration of the major proteins involved in the electron transport chain and in ATP synthesis in the inner membrane of the mitochondrion. The overall purpose of the complex chemical reactions carried out by this series of proteins is to catalyze the creation of the important energy carrier molecule ATP. (A, from D. W. Fawcett, The Cell, Its Organelles and Inclusions: An Atlas of Fine Structure, W. B. Saunders, 1966; B, courtesy of T. Frey; D, adapted from S. DiMauro and E. A. Schon, *N. Engl. J. Med.* 348:2656, 2003; E, adapted from B. Alberts et al., Molecular Biology of the Cell, 5th ed., Garland Science, 2008.)

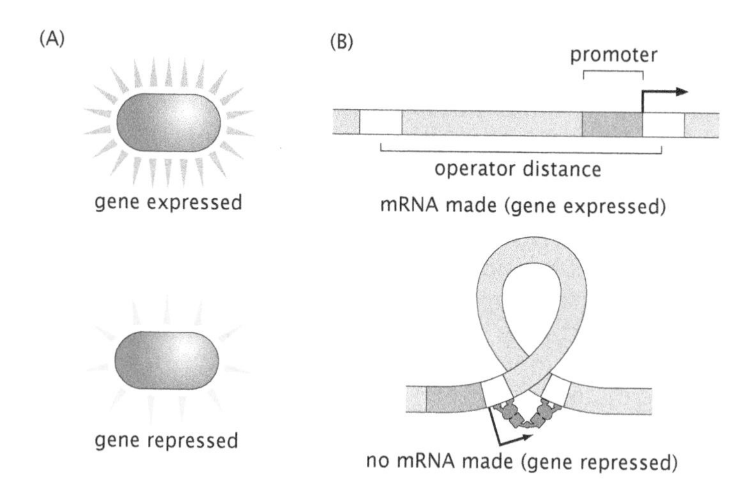

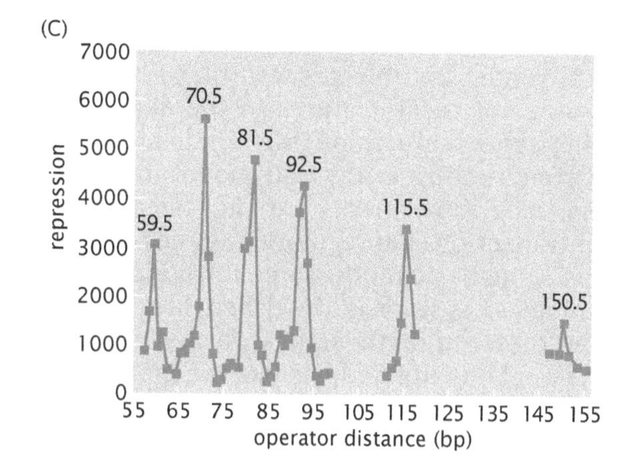

Figure 1.11: Qualitative and quantitative illustrations of gene repression. (A) Bacteria containing a gene whose expression level is regulated by the Lac repressor protein may exist in different states. When repression is low, the gene is expressed, and large amounts of the gene product are present. If the gene being regulated encodes a fluorescent protein, the bacteria will glow brightly when the gene is expressed. In contrast, when the gene is repressed, relatively little fluorescent protein will be produced, and the bacteria will glow dimly. (B) Schematic showing events at the level of the DNA during gene expression and repression. During repression, the Lac repressor protein binds to two distinct operator sites on the DNA (white boxes) and forces the DNA to form a loop. This prevents binding of the protein machine that copies the DNA into mRNA. When no mRNA is produced, the gene is repressed. (C) Quantitative measurements of the strength of gene repression as a function of the exact distance (expressed in base pairs) between the two operator sites. (C, data from J. Müller et al., *J. Mol. Biol.* 257:21, 1996.)

inorganic phosphate ion. The precise molecular details of this process are fascinating and complex; our intention here is merely to point out that this kind of cartoon represents an abstraction of the mitochondrion that emphasizes a completely different set of components and concepts than the membrane-focused cartoons shown above.

Each of these different kinds of cartoon is a valid and informative model of certain aspects of the mitochondrion, but they serve essentially nonoverlapping conceptual purposes. Furthermore, each cartoon represents the culmination of thousands of separate experiments covering decades of hard-won, detailed knowledge; experts have carefully sifted the raw data and made difficult decisions about which details are important and which are dispensable in considering each of these aspects of mitochondrial reality. Throughout the rest of this book, we will take advantage of the hard work of conceptualization that has already gone into constructing these kinds of diagrams. Our usual goal will be to take these existing biological models one step further, by rendering them into a mathematical form.

Quantitative Models Can Be Built by Mathematicizing the Cartoons

One of the key mantras of the book is that the emergence of quantitative data in biology generates a demand that the cartoons of molecular biology be mathematicized. In particular, in some cases only by constructing a mathematical model of a given biological problem can our theoretical description of the problem be put on the same footing as the data itself, that is, quantitatively. A concrete example that will arise in Chapter 19 is the measurement of gene expression as a function of the distance between two binding sites on DNA as shown in Figure 1.11. The mechanistic basis of these data is that a certain protein (Lac repressor) binds simultaneously at two sites on the DNA and forms a loop of the DNA between the two binding sites. When the Lac repressor protein is bound to the DNA, it prevents the binding of the protein machine that copies the DNA into RNA. The strength of this effect (repression) can be measured quantitatively. The experimental data show that the amount of repression depends in a complex way on the precise distance between the two binding sites.

Our argument is that a cartoon-level model of this process, while instructive, provides no quantitative basis for responding to these data. For example, why are there clear periodic peaks in the data? Why does the height of the peaks first increase and then decrease? In Chapter 19, using the framework of statistical mechanics, we will derive a mathematical model for this process that attempts to predict the shapes and sizes of these features in the data while at the same time highlighting some features of DNA mechanics that remain unclear. Further, by constructing a physical model in mathematical terms of this process, new experiments are suggested that sharpen our understanding of transcriptional regulation. In nearly every chapter, we will repeat this same basic motif: some intriguing quantitative observation on a biological system is described first in terms of a cartoon, which is then recast in mathematical form. Once the mathematical version of the model is in hand, we use the model to examine previous data and to suggest new experiments.

As a result of examples like this (and many others to be seen later), one of the central arguments of the present chapter is that there are many cartoons in molecular biology and biochemistry that have served as conceptual models of a wide variety of phenomena and that reflect detailed understanding of these systems. To make these models quantitatively predictive, they need to be cast in mathematical form; this in keeping with our battle cry that quantitative data demand quantitative models. As a result, one way of viewing the chapters that follow is as an attempt to show how the cartoons of molecular biology and biochemistry have been and can be "translated into a mathematical form."

1.5 Quantitative Models and the Power of Idealization

Quantitative models of the natural world are at the very heart of physics and in the remainder of this book we will illustrate how pervasive these ideas have become in biology as well. This approach is characterized by a rich interplay between theory and experiment in which the consequences of a given model are explored experimentally. In return, novel experimental results beckon for the development of new theories. To illustrate the way in which particular key models are recycled again and again in surprisingly varied contexts, we consider one of the most fundamental physical models in all of science, namely, the simple spring (also known as the harmonic oscillator).

1.5.1 On the Springiness of Stuff

The concept of springiness arises from the confluence of the very important mathematical idea of a Taylor series (explained in detail in "The Math Behind the Models" on p. 215) and an allied physical idea known as Hooke's law (the same Hooke who ushered in the use of microscopy in biology and coined the term "cell"). These ideas will be developed in detail in Chapter 5 and for now we content ourselves with the conceptual framework. The fundamental mathematical idea shared by all "springs" is that the potential energy for almost any system subjected to small displacements from equilibrium is well approximated by a quadratic function of the displacement. Mathematically, we can write this as

$$\text{energy} = \tfrac{1}{2}kx^2. \tag{1.1}$$

What this equation states is that the potential energy increases as the square of the displacement x away from the equilibrium position. The "stiffness" k is a measure of how costly it is to move away from equilibrium and reflects the material properties of the spring itself.

Another way of thinking about springs is that they are characterized by a restoring force that is proportional to how far the spring has been displaced from its equilibrium position. Mathematically, this idea is embodied in the equation

$$F = -kx, \tag{1.2}$$

where F is the restoring force, x is the displacement of the spring from its equilibrium position, and k is the so-called spring constant (the stiffness). The minus sign signals that the force is a restoring one towards $x = 0$, which designates the equilibrium position. This result is known as Hooke's law. When stated in this way, this kind of physical model gives the impression of an abstract example of masses on frictionless tables with pulleys and springs. But empirically, scientists have found that precisely this mathematical model arises naturally in many different practical contexts, as shown in Figure 1.12.

In biology, the simple spring comes cloaked in many different disguises, as we will see throughout the book. From a technological perspective, many of the key single-molecule techniques used to study macromolecules and their assemblies invoke this description. Both optical tweezers and the atomic force microscope can be mapped precisely and unequivocally onto spring problems. However, as shown in Figure 1.12, there are many other unexpected examples in which the problem can be recast in a way that is mathematically equivalent to the simple spring problem introduced above. Simple spring models have been invoked when considering the bending of DNA, the beating of the flagellum on a swimming sperm, and even the fluctuations of membranes at the cell surface. As these are all mechanical processes, it is fairly straightforward to see how the idea of a spring can apply to them. But the real power of the simple spring model will be revealed when we see how it can be directly applied to nonmechanical problems. For example, when we discuss biochemical reactions in a cell, we will imagine molecules moving on an "energy landscape" where they accumulate in "potential wells." The mathematics of simple springs will help us understand the rates that govern these biochemical reactions. The ideas of harmonic oscillators and potential wells will also illuminate our exploration of protein conformational changes. Even more abstractly, we will see that changes in

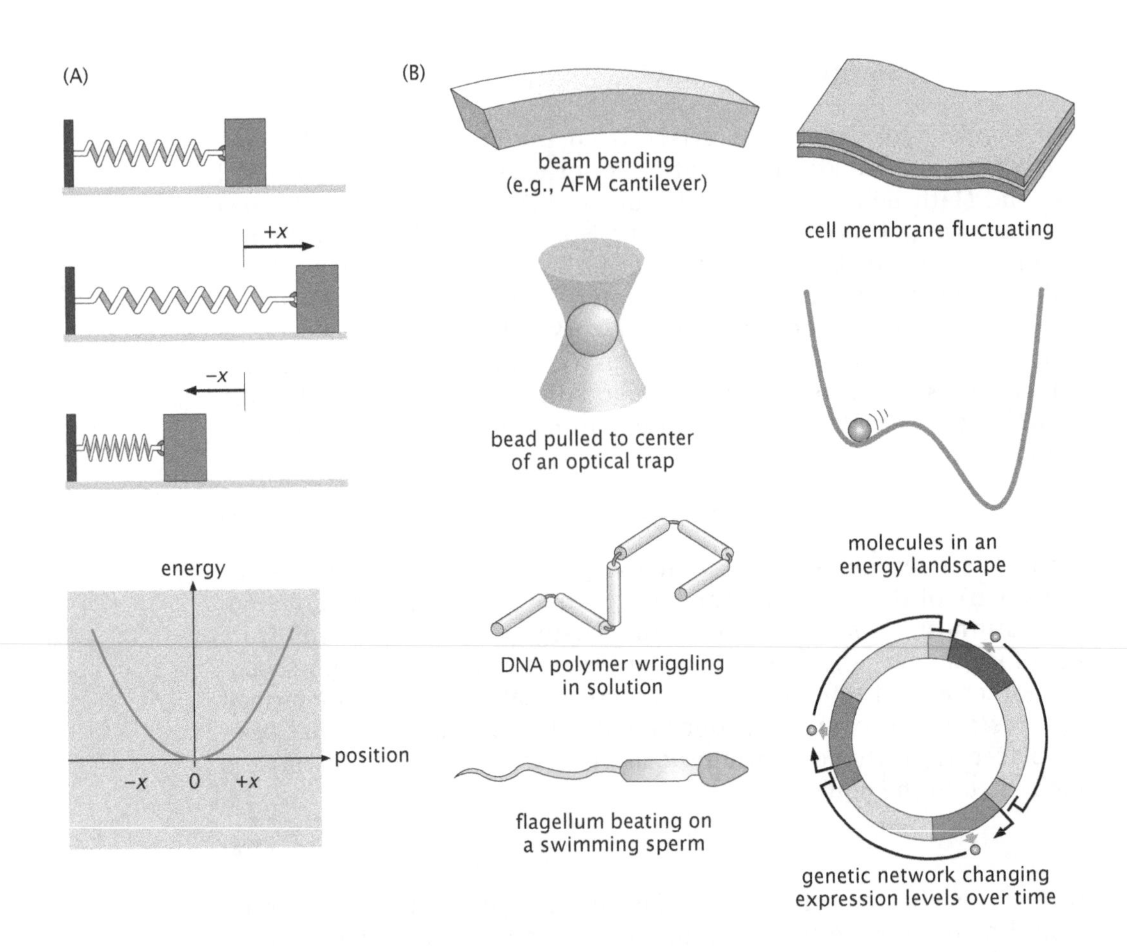

Figure 1.12: Gallery showing some of the different ways in which the idea of a spring is used in biological model building. (A) Classical physics representation of the harmonic oscillator. A simple spring is attached at one end to an immovable wall, and at the other end to a block-shaped mass that can slide on a frictionless table. The position of the mass changes as the spring extends and compresses. The graph at the bottom shows the energy of the spring as a function of the mass position. When the spring is relaxed at its resting length, the energy is at a minimum. (B) Multiple manifestations of the simple harmonic oscillator relevant to biological systems.

gene expression over time for certain interesting kinds of genetic networks can also be surprisingly well described using the mathematics of simple harmonic oscillators.

1.5.2 The Toolbox of Fundamental Physical Models

The overall strategy of this book will be to apply this kind of fundamental physical thinking to the widest possible range of biological problems. Through careful analysis of the quantitative assumptions that go into the analysis of biological data, it has become clear that a mere handful of fundamental physical models are often sufficient to provide a rigorous framework for interpretation of many kinds of quantitative biological data.

Beyond the harmonic oscillator introduced above, we believe that there are fewer than 10 other key physical concepts that require mastery. A scientist who understands and appreciates the applications of these concepts in biology is afforded a powerful framework for zeroing in on the most important, unexplained open questions. It is not our contention that this family of broad reaching physical models is sufficient to explain biological phenomena, far from it. Rather, we contend that any observations that can be completely quantitatively accounted for within the framework of one of these models require no further speculation of unidentified mechanism, while data that cannot be embraced within this physical framework cry out for more investigation and exploration.

Our choice of this set of concepts is purely utilitarian. We are not attempting to write a book surveying all of physics, or surveying all of biology. Modern, cutting-edge research at the interdisciplinary interface between biology and physics demands that scientists trained in one discipline develop a working knowledge of the other. Students currently training in these fields have the opportunity to explore both in detail and to understand their complementarity and interconnectedness. In this book, we have attempted to address all three audiences: physicists curious about biology, biologists curious about physics, and broad-minded learners ready to delve into both. Our hope is that these ideas will provide a functional foundation of some of the key physical concepts most directly relevant to biological inquiry. At the same time, we hope to provide enough examples of specific biological phenomena to allow a physical scientist to identify further important unsolved problems in biology.

The bulk of this book comprises detailed dissections and case studies of these fundamental concepts and concludes with chapters showing applications of several of these concepts simultaneously. As noted above, many individual chapters each focus on a single class of physical model. The key broad physical foundations are as follows:

- simple harmonic oscillator (Chapter 5)
- ideal gas and ideal solution models (Chapter 6)
- two-level systems and the Ising model (Chapter 7)
- random walks, entropy, and macromolecular structure (Chapter 8)
- Poisson–Boltzmann model of charges in solution (Chapter 9)
- elastic theory of one-dimensional rods and two-dimensional sheets (Chapters 10 and 11)
- Newtonian fluid model and the Navier–Stokes equation (Chapter 12)
- diffusion and random walks (Chapter 13)
- rate equation models of chemical kinetics (Chapter 15)

1.5.3 The Unifying Ideas of Biology

One of the most pleasing and powerful aspects of science is its ability to link together seemingly unrelated phenomena. Few examples from the physical sciences are more potent at making this point than the surprising unification of the age-old phenomenon of magnetism known from lodestones and the early days of compasses and navigation to the diverse manifestations of electricity such as the lightning that attends summer thunderstorms to the many and varied appearances of light such as rainbows or the apparent bending of a spoon in a glass of water. All of these phenomena and many more were brought under the same roof in the form of the modern theory of electromagnetism.

The toolbox of fundamental physical ideas introduced above focused on the way in which certain key ideas from physics have served to buckle together a huge number of observations in a way that is not just descriptive, but, more importantly, is predictive. But such unifying concepts are certainly not unique to the physical sciences. Biology has far-reaching ideas of its own. Just as it is true that a mere handful of fundamental physical models are often sufficient to provide a rigorous framework for interpretation of many kinds of quantitative biological data, there is a similar handful of fundamental biological ideas that serve as an overarching foundation for all that we think about in biology. Indeed, many earlier authors (see the references at the end of the chapter) have provided various interesting perspectives on what one might call the "great ideas of biology."

On the second page of the nearly 1400 pages of the *Feynman Lectures on Physics*, Feynman says "If, in some cataclysm, all of scientific knowledge were to be destroyed, and only one sentence passed on to the next generation of creatures, what statement would contain the most information in the fewest words?" Feynman chooses the idea that "all things are made of atoms". However, a survey of scientists would not be unanimous on this point, and some fraction of them would come down in favor of something along the lines of "all living organisms are related through descent from a common ancestor through the process of evolution by natural selection." Though perhaps it has become hackneyed through overuse, the words of Dobzhansky remain both timely and true: "Nothing in biology makes sense except in the light of evolution." Indeed, there is a strong argument that evolution is biology's greatest idea and will leave its indelible stamp on nearly everything we say in the chapters that follow.

While Darwin and Wallace labored to find an understanding of the process of evolution, Mendel was hard at work in establishing the early ideas on the underlying mechanisms of inheritance, resulting in discoveries that founded another one of biology's uniquely great ideas, namely, genetics. One of the most remarkable adventures of the human mind has been the quest to relate Mendel's abstract particles of inheritance to the molecules of the cell, with the successful insight being the realization that DNA is the carrier of genetic information. History is often characterized by various ages and it is no exaggeration to say that we are now living in the DNA Age.

Another of the key ideas that is now largely taken for granted, but is a unifying concept all the same and was hard won in the making, is the cell theory, the insight that all living organisms are made up of cells and that individual cells are themselves the quantum unit of living matter. The biochemical processes that take place within cells also reveal a remarkable unity, whether we think of the common metabolic pathways shared by so many organisms or the fact that the molecular machinery that implements our shared genetic heritage is based upon a few common macromolecular assemblies such as polymerases and the ribosome.

The key broad biological foundations that will serve as a point of departure for much of what follows in the remainder of the book are as follows:

- the theory of evolution (Chapters 3, 4, 18, 20, and 21, and many others less explicitly)
- genetics and the nature of inheritance (Chapters 2–5 and 21)
- the cell theory (Chapters 2, 3, 11, and 14)
- the unity of biochemistry (Chapters 2, 3, 5, 6, and 15)

One of the features we find especially intriguing about physical biology is that this perspective yields a different view of the proximity of different biological phenomena. As illustrated in Figure 1.13, the way we think about problems depends greatly upon both the kinds of questions we are asking and on how we go about answering them. The development of a frog and a fly both raise questions that a developmental biologist would recognize as related: what determines the timing and synchronization of the developmental process,

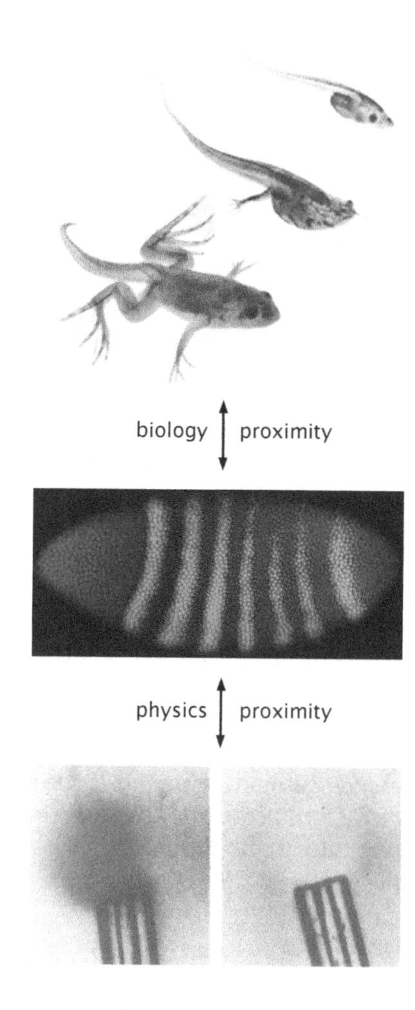

Figure 1.13: Proximity of ideas in cell biology and in physical biology. During the early development of the fruit fly *Drosophila* (middle), the eventual fate of the cells is determined by the patterned expression of a group of regulatory proteins (colored stripes). In a traditional organization of biological topics, the process of cell fate determination in *Drosophila* development might be discussed in the context of development of other animals such as the frog (top). However, the critical physical concepts that enable cell specification in the fly embryo involve processes such as diffusion and precise measurement of chemical concentrations processes that are also central to very different biological phenomena that have nothing to do with animal development. One of these phenomena is illustrated at the bottom, where on the left a group of bacteria is seen swarming toward the tip of a glass micropipette that is leaking a delicious amino acid (serine) into the medium. When the serine is switched for a toxin (phenol), as shown on the right, the bacteria quickly reverse course and swim away. In both the developing embryo and the swimming bacterium, cells measure the concentrations in their environment and make decisions as a result of similar physical measurements. (Top, courtesy of R. J. Denver, photo by David Bay; middle, adapted from E. Poustelnikova et al., *Bioinformatics* 20:2212, 2004; bottom, adapted from B. A. Rubik et al., *Proc. Natl Acad. Sci. USA* 75:2820, 1978.)

how is polarity established, etc.? On the other hand, pattern formation in the fly embryo and bacterial chemotaxis are conceptually linked when viewed as questions about the limits of chemical gradient detection. In the chapters that follow, we seek these kinds of unconventional statements of proximity with the ambition of using proximity in the physical biology sense to shed light on a host of different problems.

1.5.4 Mathematical Toolkit

One of the most surprising aspects of science is the way in which the same ideas show up again and again. For example, the random walk model does duty not only as a way to think about the diffusive motions of molecules in solution such as described in Figure 1.13 but also as a tool for describing the conformations of polymers. Thus far, we have argued for such unity and pervasiveness in the context of physics and biology. But as will be seen throughout the book, certain overarching mathematical ideas reveal themselves again and again as well. Indeed, this observation has been canonized in much the same way as Dhobzhansky's remark about evolution, in the title of another famous essay, this time by Eugene Wigner and entitled "The Unreasonable Effectiveness of Mathematics in the Natural Sciences."

One example of this unreasonable effectiveness has to do with our ability to assign numbers to our degree of belief when faced with questions where there are many outcomes, such as whether a receptor will be bound by a ligand or not. Probability theory will serve as one of the mathematical cornerstones of what we have to say in coming chapters. In particular, we will repeatedly and often appeal to four key distributions: the binomial distribution, the Gaussian distribution, the Poisson distribution, and the exponential distribution, each of which will help us sharpen the way we think about many different biological questions.

However, we will need more than just the powerful calculus of uncertainty provided by the modern theory of probability and statistics. Ideas from the differential and integral calculus are also foundational in much of what we will say, especially through our use of differential equations. Often, we will want to know how the quantity of some particular species, such as the number of microtubules, varies in time. With our calculus toolkit, one of our favorite tools will concern the mathematics of superlatives, how do we find the largest or smallest values of some function or functional? These ideas will be explored in detail in Section 5.3.1 (p. 209).

Sometimes, the push for a strictly analytical treatment of the mathematics that arises in thinking about our models will be either too tedious or demanding. In such cases, we will turn to the amplification in personal mathematical power that is available to us all now as a result of computers. Indeed, the exploitation of such computation will serve as one of the centerpieces through the remainder of the book in the form of "Computational Explorations" in which we use Matlab to numerically explore our models.

1.5.5 The Role of Estimates

In this era of high-speed computation, the temptation to confront new problems in their full mathematical complexity is seductive. Nevertheless, the core philosophy of the present book advocates a spirit that is embodied in a story, perhaps apocryphal, of the demise of Archimedes. Plutarch tells us that Archimedes was deeply engaged in a calculation, which he was performing with a stick in the sand, at the time when the Roman army invaded Syracuse. When confronted by a Roman soldier and asked to identify himself, Archimedes told the soldier to wait until he was done with his calculation. The soldier replied by impaling Archimedes, bringing the life of one of the great scientists of history to an early and unjust end. Our reason for recounting this story is as a reminder of the mathematical character of many of the models and estimates that we will report on in this book (but hopefully with less dire personal consequences). In particular, we like to imagine that a model is of the proper level of sophistication to help build intuition and insight if, indeed, its consequences can be queried by calculations carried out with a stick in the sand, without recourse to sophisticated computers or detailed references.

Throughout the book, we will repeatedly make use of estimates to develop a feeling for the numbers associated with biological structures and biological processes. This philosophy follows directly from the biological tradition articulated by the pioneering geneticist Barbara McClintock, who emphasized the necessity of developing a "feeling for the organism," in her case the crop plant maize. We believe that a quantitative intuition is a critical component of having a feeling for the organism. Indeed, we think of model building as proceeding through a series of increasingly sophisticated steps starting with simple estimates to develop a feel for the relevant magnitudes, followed by the development of toy models that can (usually) be solved analytically, followed in turn by models with an even higher degree of realism that might require numerical resolution.

The estimates that we advocate are all meant to be sufficiently simple that they can be done on the back of an envelope or written with a stick in the sand on a desert island. In keeping with this purpose, we will rarely use more than one significant digit in our estimates. One useful rule of thumb that we will adopt is the geometric mean rule (see *Guesstimation* by Weinstein and Adam for a more complete description). The idea is that when making an estimate where we are uncertain of the magnitudes, begin by attempting to guess a lower and upper bound for the magnitude of interest. With these two guesses in hand, multiply them together and take the square root of the result as your best guess for that poorly known quantity. This strategy is illustrated in Figure 1.14. What this scheme really amounts to is averaging the exponents rather than the quantities themselves.

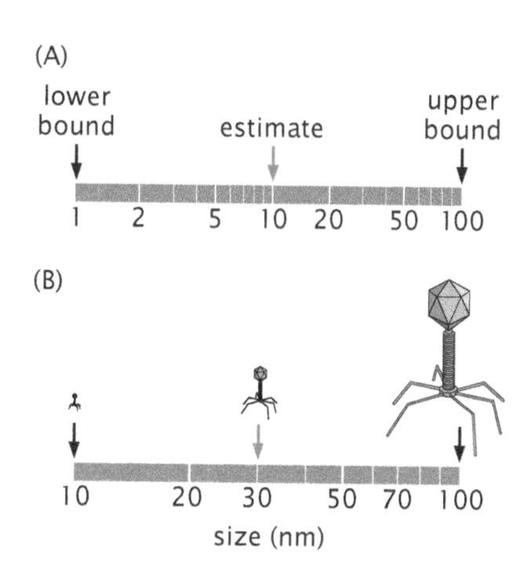

Figure 1.14: Geometric mean rule for estimates. (A) For cases in which we don't know the relevant magnitude, a powerful scheme is to guess an upper and lower bound and to take their geometric mean $\sqrt{x_{lower}x_{upper}}$ as our best guess. (B) Application of the geometric mean rule to guess the size of bacteriophages that are larger than 10 nm and smaller than 100 nm. The geometric mean guess is for a typical size of roughly 30 nm.

Naturally, this introduces somewhat low precision, for example, an estimate may yield an answer of 100 molecules in a system when the precisely measured value is actually 42, but has the benefit of greater simplicity and clarity. In general we will consider an estimate to be successful if the result is within an order of magnitude of the measured value. Indeed, as shown in Figure 1.15, we will often resort to simple arithmetic rules in which we are intentionally numerically imprecise in the name of rapidly getting to an order-of-magnitude result. In cases where our estimates differ by more than an order of magnitude, it will be clear that we have made an incorrect assumption and we will use this as an opportunity to learn something more about the problem of interest.

The precept that animates much of the book is that we are closest to understanding a phenomenon, not when we unleash the full weight of a powerful numerical simulation of the problem at hand, but rather when we are engaged in developing a feel for the scaling of the quantities of interest and their associated numerical values.

In many cases, one of the most powerful tools that can be unleashed in order to understand the relevant numerical magnitudes in a given problem is known as *dimensional analysis*. Indeed, in many examples throughout the book, whether when talking about fluid flow using the Navier–Stokes equations or the nature of the genetic switch, we will write our equations in dimensionless form. One of the reasons that these dimensionless forms are so powerful is that they permit us to write the equations describing the phenomenon of interest in the "natural variables" of the problem. For example, when thinking about a circuit with a resistor and a capacitor, the natural time unit is given by $\tau = RC$, the product of the resistance and the capacitance. In thinking about the establishment of morphogen gradients during anterior–posterior patterning in flies, the fundamental length scale is given by $\lambda = \sqrt{D\tau}$, where D is the diffusion constant of the morphogen and τ is its degradation time. Further, when we are trying to estimate some quantity of interest, a powerful trick is to simply use the relevant constants describing that problem to construct a quantity with the dimensions of the quantity we are trying to estimate.

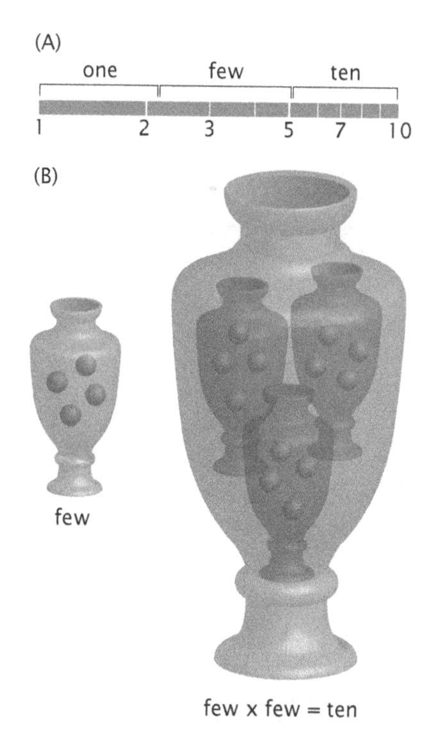

Figure 1.15: Few times one rule for multiplication. (A) Scheme for assigning numbers either to "one," "few," or "ten (concept from "Street Fighting Mathematics" by Sanjoy Mahajan)." (B) Demonstration that few $\times$ few $\approx$ 10.

In the morphogen gradient example, if we assume that the length scale λ over which the pattern changes significantly is determined by the diffusion constant of the morphogen and its mean lifetime, then the dimensions of these two quantities determine the formula for the length scale λ up to a numerical constant. To be specific, the diffusion constant has dimensions of $\text{Length}^2/\text{Time}$, while the mean lifetime has dimensions of Time. The only way these two quantities can be combined into one that has dimensions of Length is by multiplying them and then taking a square root, leading to the formula for λ given above. The particular power of dimensional analysis comes from comparing the estimate it yields with the measured value of the relevant quantity. In the morphogen example, a large discrepancy of the measured λ and the one estimated from the formula obtained by dimensional analysis would indicate that our determination of the morphogen diffusion constant or its lifetime is wrong. If more careful measurements of these two parameters still lead to a discrepancy, then we are led to an even more exciting conclusion, namely, that there is something fundamentally missing in our understanding of how the morphogen gradient is established, something that would introduce yet another parameter that determines λ.

Taylor and the Cover of Life *Magazine* One of the greatest stories of the power and subtlety of estimation and, particularly, of dimensional analysis is associated with G. I. Taylor. During the early days of the atomic bomb, the energetic yield of the first explosion was classified as a military secret. This did not keep the people who make such decisions from releasing several time lapse photos of the explosion that showed the radius of the fireball as a function of time. On this meager information, shown on the cover of *Life* magazine, Taylor was able to use the principles of fluid mechanics to *estimate* the yield of the explosion, an estimate that was exceedingly close to the secret value of the actual yield.

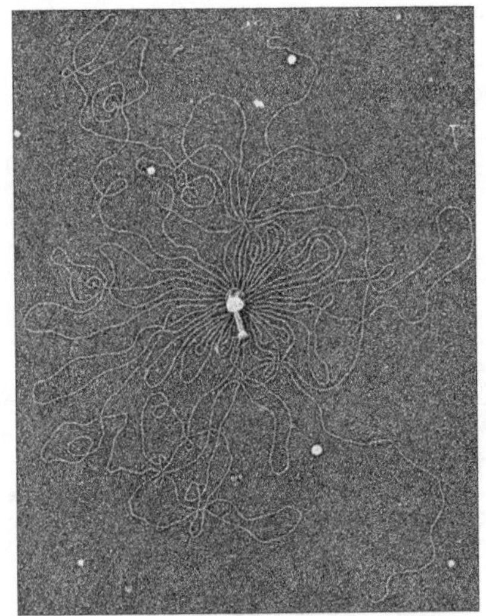

Figure 1.16: Electron microscopy image of a bacteriophage genome that has escaped its capsid. Simple arguments from polymer physics can be used to estimate the genomic size of the DNA by examining the physical size of the randomly spread DNA. We will perform these kinds of calculations in Chapter 8. (Adapted from G. Stent, Molecular Biology of Bacterial Viruses. W. H. Freeman, 1963.)

Interestingly, this famous episode from the cultural history of the physical sciences could have happened just as well in biology. Pictures like that shown in Figure 1.16 were the thrilling outcomes of the early days of the use of electron microscopy in biology. Here too, by observing the physical region occupied by the DNA molecule from the ruptured bacterial virus (bacteriophage), it is possible to make a highly simplified estimate of the *genomic* size (that is, the length in base pairs) of the bacteriophage's DNA.

Why do numbers matter? Why is it useful to know the number of copies of a given molecule that is controlling the expression of a gene of interest within an *E. coli* cell? What merit is there in estimating the rate of actin polymerization at the leading edge of a motile cell? What is the value in knowing the time it takes a given protein to diffuse a millimeter? First, simple estimates serve as a reality check to help us see whether or not our impressions of how a system works are reasonable at all. Second, a sense of the numbers can tell us what kinds of physical constraints are in play. For example, by knowing the size of a bacterium and its swimming speed, we can construct a characteristic number (the Reynolds number) that tells us about the importance of acceleration in the dynamical problem of swimming. Similarly, a knowledge of the number of the specific DNA-binding proteins known as transcription factors that control the expression of a gene of interest can tell us about the importance of fluctuations in the gene expression process.

As will be seen in Chapters 3 and 19, biological oscillators are one of the centerpieces of the regulatory architecture of almost all living organisms. Oscillatory behavior is seen in examples ranging from the clocks that control the cell cycle in developing embryos to the circadian rhythms that make jet lag the unwanted but constant companion of the weary traveler. As we explore mathematical models of regulatory biology, we will see that minimal oscillators can be constructed from genetic networks involving just two proteins, namely, an activator and a repressor. But what we will also see in contemplating these important networks is that for many of the possible parameters we might choose for the protein production and degradation rates, they don't yield oscillatory behavior at all, but rather stable and boring solutions in which the protein concentrations relax after some transient period into stationary values. Hence, even a clear picture of the wiring of some network cannot tell us what that network does in the normal life of a cell and calls for a quantitative examination of how that network behaves as a function of its governing parameters.

1.5.6 On Being Wrong

When performing simple quantitative estimates, or when attempting to write simple mathematical models to describe some aspect of a biological process, we are easily tempted to worry about whether our estimates and models are "right" or "wrong." There are at least four different ways in which a model might justly be called "wrong," but we believe that committing each of these kinds of wrongness can be a useful and worthwhile experience.

One category of wrong models might fail because they use an inappropriate type of fundamental physical model to abstract an aspect of a particular biological problem. This kind of wrongness might be illustrated by the question: how many hours can fit in the back of a pickup truck? The investigator will quickly learn that the predictions of such a model will simply have no relevance to the actual experiment or biological phenomenon being considered. It is inevitable that any person learning to conceptualize the kinds of complicated issues we explore in biology will make some of these kinds of mistakes at first. With time and experience, these errors will become less frequent.

A second category of wrong models might fail because they do not include enough detail about the system. For example, when we try to predict how fast *E. coli* bacterial cells can swim through water, we might be able to write an equation that correctly predicts the order of magnitude of the average speed of the cells, but fails to provide any basis for predicting how the speed might fluctuate over time. This kind of error, we believe, is not so much wrongness as incompleteness. As we have argued, it is actually extremely important that physical models be incomplete in order to be simple enough to help in building intuition. A model that is nearly as complex as the system itself may be able to make very accurate predictions, but will not be useful for promoting understanding.

A third category of wrong models includes those where inappropriate assumptions are made about the relative importance of different physical considerations. Although these models may be mathematically correct, they will fail to make accurate predictions for the behavior of cells in the real world. For example, if the swimming *E. coli* cell stops rotating its flagellum, then its future trajectory will be governed by a combination of inertia, viscous drag, and Brownian motion. A modeler whose primary experience lies in hydrodynamics for large objects such as submarines or fish may be inclined to ignore the contribution of Brownian motion, but for the bacterium this is the primary determinant of its behavior. Familiarity with the biological system and the opportunity to directly compare model predictions with experimental data for iteratively refined generations of models will usually allow this problem to be overcome.

The fourth category of wrong models is by far our favorite. These are the wrong models that drive breakthrough experiments. When a person who is familiar with a particular biological system writes down a quantitative model for that system incorporating the correct basic physical models and uses the correct order-of-magnitude estimates for relevant parameters nevertheless finds that it makes dramatically incorrect predictions, this is almost always a golden opportunity to learn something fundamentally new about the system. Science advances when people notice that something does not quite make sense, that something is puzzling, that something seems to be missing. It is our hope that readers of this book will develop sufficient confidence in their ability to create appropriate models that when they find a model that seems to be wrong in this way they will seize the opportunity to make a new scientific discovery.

1.5.7 Rules of Thumb: Biology by the Numbers

A mastery of any field requires knowledge of a few key facts. For the physicist who uses quantum mechanics, this means knowing the ground state energy of the hydrogen atom, the charge on an electron, Planck's constant, and several other key physical parameters. In physical biology, there are similarly a small set of numbers that are worth carrying around as the basis of simple estimates of the kind one might carry out with a stick in the sand. In that spirit, the key numbers given in Table 1.1 are those that we find useful when thinking about physical biology of the cell. In the chapters that follow, these key numbers will permit us to form the habit of making quantitative estimates as a preliminary step when confronting new problems.

Throughout the remainder of the book, we will repeatedly make reference to certain numerical rules of thumb such as that the volume of an *E. coli* cell is roughly $1\,\mu m^3$ or that the natural energy scale of physical biology is $k_B T$ (that is, Boltzmann's constant multiplied by room temperature in kelvins ≈ 4 pN nm (piconewton nanometers)). It is convenient to express this natural energy scale in terms of piconewtons and nanometers when we are considering the forces generated by protein molecular motors moving along DNA, for example, but this natural energy scale can also be expressed in different units depending on the type of calculation we are doing. When considering processes involving charge and electricity, it is convenient to recall that $k_B T$ is approximately equal to 25 meV (millielectron volts). When considering biochemical reactions, it is more useful to express $k_B T$ in biochemical units, as 0.6 kcal/mol (kilocalories per mole) or 2.5 kJ/mol (kilojoules per mole).

The quantitative rules of thumb in Table 1.1 will serve as the basis of our rough numerical estimates that could be carried out using a stick in the sand without reference to books, papers, or tables of data. Where do these numbers come from? Each comes from the results of a long series of experimental measurements of many different kinds. Right now we will simply assert their values, but later in the book we will provide numerous examples showing how these kinds of measurements have been made and also noting that in many cases, there remains uncertainty about the precise values of the parameters that emerge from such measurements. How precisely must we know these numbers? As stated earlier, our approach is to be generally satisfied

Table 1.1: Rules of thumb for biological estimates.

	Quantity of interest	Symbol	Rule of thumb
E. coli			
	Cell volume	$V_{E.\,coli}$	$\approx 1\ \mu m^3$
	Cell mass	$m_{E.\,coli}$	≈ 1 pg
	Cell cycle time	$t_{E.\,coli}$	≈ 3000 s
	Cell surface area	$A_{E.\,coli}$	$\approx 6\ \mu m^2$
	Macromolecule concentration in cytoplasm	$c_{E.\,coli}^{macromol}$	≈ 300 mg/mL
	Genome length	$N_{bp}^{E.\,coli}$	$\approx 5 \times 10^6$ bp
	Swimming speed	$v_{E.\,coli}$	$\approx 20\ \mu m/s$
Yeast			
	Volume of cell	V_{yeast}	$\approx 60\ \mu m^3$
	Mass of cell	m_{yeast}	≈ 60 pg
	Diameter of cell	d_{yeast}	$\approx 5\ \mu m$
	Cell cycle time	t_{yeast}	≈ 200 min
	Genome length	N_{bp}^{yeast}	$\approx 10^7$ bp
Organelles			
	Diameter of nucleus	$d_{nucleus}$	$\approx 5\ \mu m$
	Length of mitochondrion	l_{mito}	$\approx 2\ \mu m$
	Diameter of transport vesicles	$d_{vesicle}$	≈ 50 nm
Water			
	Volume of molecule	V_{H_2O}	$\approx 10^{-2}\ nm^3$
	Density of water	ρ	$1\ g/cm^3$
	Viscosity of water	η	≈ 1 centipoise (10^{-2} g/(cm s))
	Hydrophobic embedding energy	$\approx E_{hydr}$	2500 cal/(mol nm^2)
DNA			
	Length per base pair	l_{bp}	$\approx 1/3$ nm
	Volume per base pair	V_{bp}	$\approx 1\ nm^3$
	Charge density	λ_{DNA}	2 e/0.34 nm
	Persistence length	ξ_p	50 nm
Amino acids and proteins			
	Radius of "average" protein	$r_{protein}$	≈ 2 nm
	Volume of "average" protein	$V_{protein}$	$\approx 25\ nm^3$
	Mass of "average" amino acid	M_{aa}	≈ 100 Da
	Mass of "average" protein	$M_{protein}$	$\approx 30{,}000$ Da
	Protein concentration in cytoplasm	$c_{protein}$	≈ 150 mg/mL
	Characteristic force of protein motor	F_{motor}	≈ 5 pN
	Characteristic speed of protein motor	v_{motor}	≈ 200 nm/s
	Diffusion constant of "average" protein in cytoplasm	$D_{protein}$	$\approx 10\ \mu m^2/s$
Lipid bilayers			
	Thickness of lipid bilayer	d	≈ 5 nm
	Area per molecule	A_{lipid}	$\approx \frac{1}{2}\ nm^2$
	Mass of lipid molecule	m_{lipid}	≈ 800 Da

in those cases where our numerical estimates are of the right order of magnitude. To that end, we make certain gross simplifications. For example, when thinking about proteins and the amino acids that make them up, we will often adopt a minimalistic perspective in which the mass of amino acids is *approximated* to be 100 Da (daltons, or grams per mole). In actuality, real amino acids have masses that range over a wider set of values from 75 to 204 Da. Likewise, the rest of the numbers in Table 1.1 represent only rough average values to use as a starting point for quantitative estimates. As we progress through the examples in the rest of the book, these values will become increasingly familiar. However, our book does not make a systematic attempt at the important task of collecting values for a wide range of biologically important numbers. For this, the interested reader is encouraged to visit the "BioNumbers" website to search out what is known about the quantitative status of a given biological parameter and to be on the lookout for those cases in which we cite such BioNumbers by their BioNumber ID (BNID).

1.6 Summary and Conclusions

When applying physical and quantitative thinking to biological problems, two different skills are critically important. First, the biological system must be described in terms of fundamental physical models that can help to shed light on the behavior of the system. Fewer than 10 truly basic physical models are able to provide broad explanatory frameworks for many problems of interest in biology. Second, appropriate quantitative estimates to give order-of-magnitude predictions of sizes, numbers, time scales, and energies involved in a biological process can serve as a reality check that our conception of the system is appropriate and can help predict what kinds of experimental measurements are likely to provide useful information. Armed with these skills, an investigator can begin to systematize the large amounts of detailed quantitative data generated by biological experimentation and determine where new scientific breakthroughs might be sought.

1.7 Further Reading

Alberts, B, Johnson, A, Lewis, J, et al. (2008) Molecular Biology of the Cell, 5th ed., Garland Science. This book provides a discussion of much of the biological backdrop for our book.

Schaechter, M, Ingraham, JL, & Neidhardt, FC (2006) Microbe, ASM Press. This book is full of insights into the workings of microscopic organisms.

Stryer, L (1995) Biochemistry, 5th ed., W. H. Freeman. We like to imagine our readers with this book at their side. Stryer is full of interesting insights presented in a clear fashion.

Schleif, R (1993) Genetics and Molecular Biology, 2nd ed., The Johns Hopkins University Press. Schleif's book discusses many of the same topics found throughout our book from a deeper biological perspective, but with a useful and interesting quantitative spin.

Vogel, S (2003) Comparative Biomechanics—Life's Physical World, Princeton University Press. Though largely aimed at describing problems at a different scale than those described in our book, Vogel's book (as well as his many others) presents an amazing merger of biology and physical reasoning.

Nurse, P (2003) The Great Ideas of Biology, *Clinical Medicine* **3**, 560. This article describes Nurse's views on some of the greatest ideas in biology.

Bonner, JT (2002) The Ideas of Biology, Dover Publications. This book espouses Bonner's view on the overarching themes of modern biology.

The spirit of physical biology we are interested in developing here has already been described in a number of other books. The following are some of our favorites:

Dill, K, & Bromberg, S (2011) Molecular Driving Forces, 2nd ed., Garland Science. Despite a deceptive absence of difficult mathematics, their book is full of insights, subtlety, and tasteful modeling.

Boal, D (2012) Mechanics of the Cell, Cambridge University Press. Boal's book illustrates the way in which physical reasoning can be brought to bear on problems of biological significance. We have found that many of the exercises at the back of each chapter are the jumping off point for the analysis of fascinating biological problems.

Howard, J (2001) Mechanics of Motor Proteins and the Cytoskeleton, Sinauer Associates. Howard's book is a treasure trove of information and concepts.

Nelson, P (2004) Biological Physics: Energy, Information, Life, W. H. Freeman. This book discusses a wide range of topics at the interface between physics and biology.

Benedek, GB, & Villars, FMH (2000) Physics with Illustrative Examples from Medicine and Biology, 2nd ed., Springer-Verlag. This three-volume series should be more widely known. It is full of useful insights into everything ranging from the analysis of the Luria–Delbrück experiment to the action potential.

Sneppen, K, & Zocchi, G (2005) Physics in Molecular Biology, Cambridge University Press. This book presents a series of vignettes that illustrate how physical and mathematical reasoning can provide insights into biological problems.

Jackson, MB (2006) Molecular and Cellular Biophysics, Cambridge University Press. Jackson's book covers many themes in common with our own book and is especially strong on ion channels and electrical properties of cells.

Waigh, TA (2007) Applied Biophysics, Wiley. This interesting book covers many of the same topics covered here.

Harte, J (1988) Consider a Spherical Cow, University Science Books. This book does a beautiful job of showing how order of magnitude estimates can be used to examine complex problems. Those that are intrigued by this book might consider dipping into his Consider a Cylindrical Cow (University Science Books, 2001) as well.

Weinstein, L, & Adam, JA (2008) Guesstimation: Solving the World's Problems on the Back of a Cocktail Napkin, Princeton University Press. This wonderful little book shows by example how to do the kind of estimates advocated here and also is the basis for the rule captured in Figure 1.14.

Goldreich, P, Mahajan, S, & Phinney, S, Order-of-Magnitude Physics: Understanding the World with Dimensional Analysis, Educated Guesswork, and White Lies, online book, available at www.inference.phy.cam.ac.uk/sanjoy/oom/. This excellent work embodies the science of estimation so helpful to biology as well as physics.

Cohen, J (2004) Mathematics is biology's next microscope, only better; Biology is Mathematics' next physics, only better, *PLoS Biology* **2**, e439. This article argues for the way that quantitative reasoning will enrich biology and for the ways that biology will enrich mathematics.

Mahajan, S (2010) Street-Fighting Mathematics, The MIT Press. Much of our mathematics will be of the street-fighting kind and there is nowhere better to get a feeling for this than from Mahajan's book.

1.8 References

Bray, D (2001) Reasoning for results, *Nature* **412**, 863. Source of Darwin's quote at the beginning of the chapter.

Darwin, F (ed.) (1903) More Letters of Charles Darwin: A Record of His Work in a Series of Hitherto Unpublished Letters, Volume I, John Murray.

DiMauro, S, & Schon, EA (2003) Mitochondrial respiratory-chain diseases, *N. Engl. J. Med.* **348**, 2656.

Fawcett, DW (1996) The Cell, Its Organelles and Inclusions: An Atlas of Fine Structure, W. B. Saunders. Fawcett's atlas is full of wonderful electron microscopy images of cells.

Milo, R, Jorgensen, P, Moran, U, et al. (2010) BioNumbers—the database of key numbers in molecular and cell biology, *Nucl. Acids Res.* **38**, D750.

Müller, J, Oehler, S, & Müller-Hill, B (1996) Repression of *lac* promoter as a function of distance, phase and quality of an auxiliary *lac* operator, *J. Mol. Biol.* **257**, 21.

Poustelnikova, E, Pisarev, A, Blagov, M, et al. (2004) A database for management of gene expression data *in situ*, *Bioinformatics* **20**, 2212.

Rubik, BA, & Koshland, DE, Jr. (1978) Potentiation, desensitization, and inversion of response in bacterial sensing of chemical stimuli, *Proc. Natl. Acad. Sci.*, **75**, 2820.

Stent, G (1963) Molecular Biology of Bacterial Viruses, W. H. Freeman.

What and Where: Construction Plans for Cells and Organisms

2

Overview: In which we consider the size of cells and the nature of their contents

Cells come in a dazzling variety of shapes and sizes. Even so, their molecular inventories share many common features, reflecting the underlying biochemical unity of life. In this chapter, we introduce the bacterium *Escherichia coli* (commonly abbreviated as *E. coli*) as our biological standard ruler. This cell serves as the basis for a first examination of the inventory of cells and will permit us to get a sense of the size of cells and the nature of their contents. Using simple estimates, we will take stock of the genome size, numbers of lipids and proteins, and the ribosome content of bacteria. With the understanding revealed by *E. coli* in hand, we then take a powers-of-10 journey down and up from the scale of individual cells. Our downward journey will examine organelles within cells, macromolecular assemblies ranging from ribosomes to viruses, and then the macromolecules that are the engines of cellular life. Our upward journey from the scale of individual cells will examine the biological structures resulting from multicellularity, this time with an emphasis on how cells act together in contexts ranging from bacterial biofilms to the networks of neurons in the brain.

> "By the help of Microscopes there is nothing so small as to escape our inquiry; hence there is a new visible World discovered."
>
> **Robert Hooke**

2.1 An Ode to *E. coli*

Scientific observers of the natural world have been intrigued by the processes of life for many thousands of years, as evidenced by early written records from Aristotle, for example. Early thinkers wondered about the nature of life and its "indivisible" units in much the same way that they mused about the fundamental units of matter. Just as physical scientists arrived at a consensus that the fundamental unit of matter is the atom (at least for chemical transactions), likewise, observers of living organisms have agreed that the indivisible unit of life is the cell. Nothing smaller than a cell can be shown to be alive in a sense that is generally agreed upon. At the same time, there are no currently known reasons to attribute some higher "living" status to multicellular organisms, in contrast to single-celled organisms.

Cells are able to consume energy from their environments and use that energy to create ordered structures. They can also harness energy and materials from the environment to create new cells. A standard definition of life merges the features of metabolism (that is, consumption and use of energy from the environment) and replication (generating offspring that resemble the original organism). Stated simply, the cell is the smallest unit of replication (though viruses are also replicative units, but depend upon their infected host cell to provide much of the machinery making this replication possible).

The recognition that the cell is the fundamental unit of biological organization originated in the seventeenth century with the microscopic observations of Hooke and van Leeuwenhoek. This idea was put forth as the modern cell theory by Schwann, Schleiden, and Virchow in the mid-nineteenth century and was confirmed unequivocally by Pasteur shortly thereafter and repeatedly in the time since. Biologists agree that all forms of life share cells as the basis of their organization. It is also generally agreed that all living organisms on Earth shared a common ancestor several billion years ago (or, more accurately, a community of common ancestors) that would be recognized as cellular by any modern biologist.

In terms of understanding the basic rules governing metabolism, replication, and life more generally, one cell type should be as good as any other as the basis of experimental investigations of these mechanisms. For practical reasons, however, biologists have focused on a few particular types of cell to try to illuminate these general issues. Among these, the human intestinal inhabitant *E. coli* stands unchallenged as the most useful and important representative of the living world in the biologist's laboratory.

Several properties of *E. coli* have contributed to its great utility and have made it a source of repeated discoveries. First, it is easy to isolate because it is present in great abundance in human fecal matter. Unlike most other bacteria that populate the human colon, *E. coli* is able to grow well in the presence of oxygen. In the laboratory, it replicates rapidly and can easily adjust to changes in its environment, including changes in nutrients. Second, it is nearly routine to deliberately manipulate these cells (using insults such as radiation, chemicals, or genetic tricks, for example) to produce mutants. Mutant organisms are those that differ from their parents and from other members of their species found in the wild because of specific changes in DNA sequence that give rise to biologically significant differences. For example, *E. coli* is normally able to synthesize purines to make DNA and RNA from sugar as a nutrient source. However, particular mutants of *E. coli* with enzymatic deficiencies in these pathways have lost the ability to make their own purines and become reliant on being fed precursors for these molecules. A more familiar and frightening example of the consequence of mutations is the way in which bacteria acquire antibiotic resistance. Throughout the book, we will be using specific examples of biological phenomena to illustrate general physical principles that are relevant to life. Often, we will have recourse to *E. coli* because of particular experiments that have been performed on this organism. Further, even when we speak of experiments on other cells or organisms, often *E. coli* will be behind the scenes coloring our thinking.

2.1.1 The Bacterial Standard Ruler

The Bacterium *E. coli* Will Serve as Our Standard Ruler

Throughout this book, we will discuss many different cells, which all share with *E. coli* the fundamental biological directive to convert energy from the environment into structural order and to perpetuate their species. On Earth, it is observed that there are certain minimal requirements for the perpetuation of cellular life. These are not necessarily absolute physical requirements, but in the competitive environment of our planet, all surviving cells share these features in common. These include a DNA-based genome, mechanisms to transcribe DNA into RNA, and, subsequently, translation mechanisms using ribosomes to convert information in RNA sequences into protein sequence and protein structure. Within those individual cells, there are many substructures with interesting functions. Larger than the cell, there are also structures of biological interest that arise because of cooperative interactions among many cells or even among different species. In this chapter, we will begin with the cell as the fundamental unit of biological organization, using *E. coli* as the standard reference and standard ruler. We will then look at smaller structures within cells and, finally, larger multicellular structures, zooming in and out from our fundamental cell reference frame.

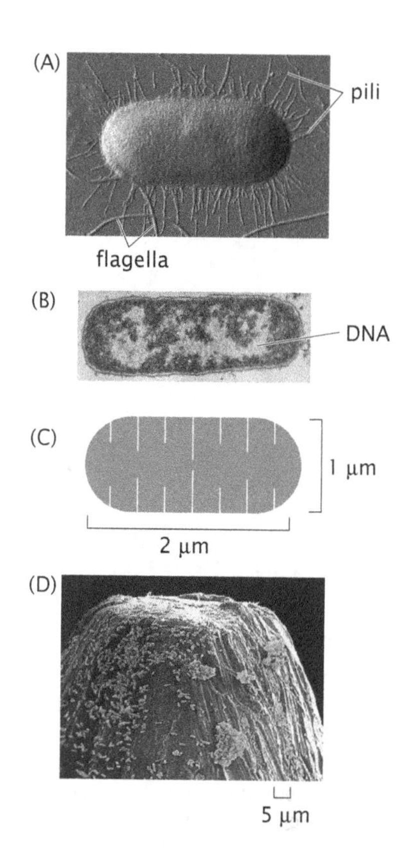

Figure 2.1: *E. coli* as a standard ruler for characterizing spatial scales. (A) Atomic-force microscopy (AFM) image of an *E. coli* cell, (B) electron micrograph of a sectioned *E. coli* bacterium, and (C) the *E. coli* ruler. (D) Bacteria on the head of a pin giving an impression of how our standard ruler compares with the dimensions of everyday objects. (A, courtesy of Ang Li; D, courtesy of Tony Brain.)

Figure 2.1 shows several experimental pictures of an *E. coli* cell and its schematization into our standard ruler. The AFM image and the electron micrograph in Figures 2.1(A) and (B) show that these bacteria have a rod-like morphology with a typical length between 1 and 2 μm and a diameter between 0.5 and 1 μm. The reader is invited to use light microscopy images of these bacteria and to determine the size of an *E. coli* cell for him or herself in the "Computational Exploration" at the end of this section.

The length unit of 1 μm, or micrometer, is so useful for the discussion of cell biology that it has a nickname, the "micron." To put this standard ruler in perspective, we note that, with its characteristic length scale of 1 μm, it would take roughly 50 such cells lined up end to end in order to measure out the width of a human hair. On the other hand, we would need to divide the cell into roughly 500 slices of equal width in order to measure out the diameter of a DNA molecule. Using these insights, the question of how many bacteria can dance on the head of a pin is answered unequivocally in Figure 2.1(D).

Note that the average size of these cells depends on the nutrients with which they are provided, with those growing in richer media having a larger mass. An extremely elegant experiment that explores the connection between growth rate and mass is shown in Figure 2.2. Our reference growth condition throughout the book will be a chemically defined solution referred to by microbiologists as "minimal medium" that is a mixture of salts along with glucose as the sole carbon source. In the laboratory, bacteria are often grown in "rich media," which are poorly defined but nutrient-rich mixtures of extracts from organic materials such as yeast cultures or cow brains. Although microorganisms can grow very rapidly in rich media, they are rarely used for biochemical studies because their exact contents are not known. For consistency, we will therefore refer primarily to experimental results for bacterial growth in minimal media.

Because of its central role as the quantitative standard in the remainder of the book, it is useful to further characterize the geometry of *E. coli*. One example in which we will need a better sense of the geometry of cells and their internal compartments is in the context of reconciling experiments performed *in vitro* (that is, in test tubes) and *in vivo* (that is, in living cells). Results from experiments done *in vitro* are based upon the free concentrations of different molecular species. On the other hand, in *in vivo* situations, we might know the number of copies of a given molecule such as a transcription factor inside the cell (transcription factors are proteins that regulate expression of genes by binding to DNA). To reconcile these two pictures, we will need the cellular volume to make the translation between molecular counts and concentrations. Similarly, when examining the distribution of membrane proteins on the cell surface, we will need a sense of the cell surface area to estimate the mean spacing between these proteins, which will tell us about the extent of interaction between them. For most cases of interest in this book, it suffices to attribute a volume $V_{E.\ coli} \approx 1\ \mu m^3 = 1$ fL to *E. coli* and an area of roughly $A_{E.\ coli} \approx 6\ \mu m^2$ (see the problems for examples of how to work out these numbers from known cellular dimensions).

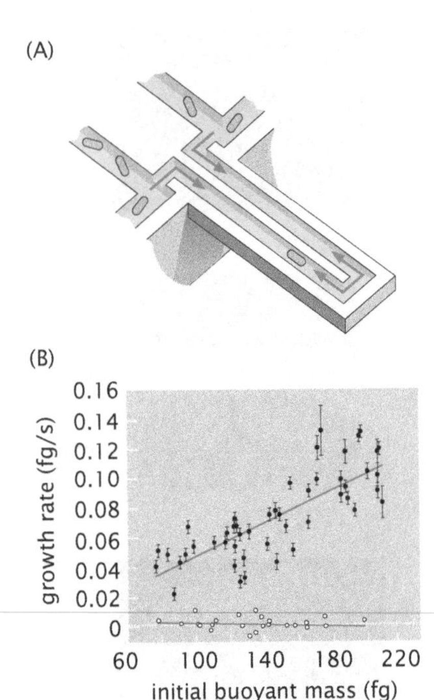

Figure 2.2: *E. coli* mass and growth rate as measured using a cantilever assay. (A) Schematic of the geometry of the device used to characterize the buoyant mass of bacterial cells by dynamically trapping them within the hollow cantilever. The hollowed out cantilever oscillates with a slightly different frequency when there is a cell present in its interior. (B) Relation between the buoyant mass at the start of the experiment and the growth rate. *E. coli* K12 cells grown at 37 °C. Filled circles correspond to cells that are growing normally and open circles correspond to fixed cells. (Figures adapted from M. Godin et al., *Nat. Meth.* 7: 387, 2010.)

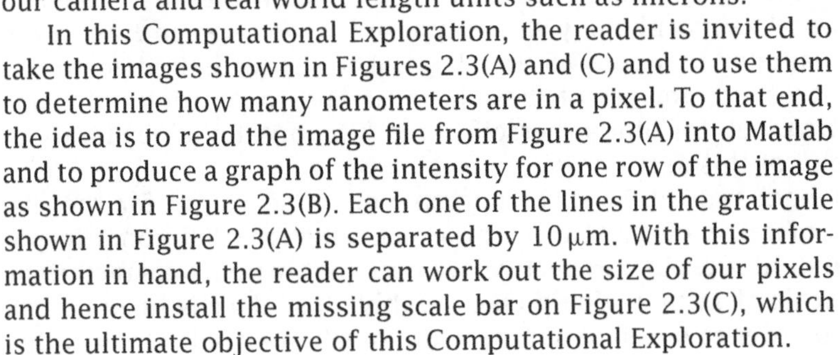

COMPUTATIONAL EXPLORATION

Computational Exploration: Sizing Up *E. coli* As already revealed in Figure 2.1, the simplest way for us to figure out how big bacteria are is to stick them under a microscope and to look at them. Of course, to do this in reality requires some way to relate the cells we see magnified in our microscope and their actual size. In modern terms, what this really means is figuring out the relationship between the size of the pixels produced by our camera and real world length units such as microns.

In this Computational Exploration, the reader is invited to take the images shown in Figures 2.3(A) and (C) and to use them to determine how many nanometers are in a pixel. To that end, the idea is to read the image file from Figure 2.3(A) into Matlab and to produce a graph of the intensity for one row of the image as shown in Figure 2.3(B). Each one of the lines in the graticule shown in Figure 2.3(A) is separated by 10 μm. With this information in hand, the reader can work out the size of our pixels and hence install the missing scale bar on Figure 2.3(C), which is the ultimate objective of this Computational Exploration.

2.1.2 Taking the Molecular Census

In the remainder of this section, we will proceed through a variety of estimates to try to get a grip on the number of molecules of different kinds that are in an *E. coli* cell. Why should we care about these numbers? First, a realistic physical picture of any biological phenomenon demands a quantitative understanding of the individual particles involved (for biological phenomena, this usually means molecules) and the spatial dimensions over which they have the freedom to act. One of the most immediate outcomes of our cellular census will be the realization of just how crowded the cellular interior really is, a subject explored in detail in Chapter 14. Our census will paint a very different picture of the cellular interior as the seat of biochemical reactions than is suggested by the dilute and homogeneous environment of the biochemical test tube. Indeed, we will see that the mean spacing between protein molecules within a typical cell is less than 10 nm. As we will see below this distance is comparable to the size of the proteins themselves!

Taking the molecular census is also important because we will use our molecular counts in Chapter 3 to estimate the rates of macromolecular synthesis during the cell cycle. How fast is a genome replicated? What is the average rate of protein synthesis during the cell cycle and, given what we know about ribosomes, how do they maintain this rate of synthesis? A prerequisite to beginning to answer these questions is the macromolecular census itself.

Ultimately, to understand many experiments in biology, it is important to realize that most experimentation is comparative. That is, we compare "normal" behavior to perturbed behavior to see if some measurable property has increased or decreased. To make these statements meaningful, we need first to understand the quantitative baseline relative to which such increases and decreases are compared. There is another sense in which numbers of molecules are particularly meaningful which will be explored in detail in subsequent chapters that has to do with whether we can describe a cell as having "a lot" or "a few" copies of some specific molecule. If a cell has a lot of some particular molecule, then it is appropriate to describe the concentration of that molecule as the basis for predicting cellular function. However, when a cell has only a few copies of a particular molecule, then we need to consider the influence of random chance (or stochasticity) on its function. In many cases, cells have an interesting intermediate number of molecules where it is not immediately clear which perspective is appropriate. However, knowing the absolute numbers always gives us a reality check for subsequent assumptions and approximations for modeling biological processes.

Because of these considerations, much effort among biological scientists has been focused on the development of quantitative techniques for measuring the molecular census of living cells (both bacteria and eukaryotes). In this chapter, we will rely primarily on order-of-magnitude estimates based on simple assumptions. These estimates are validated by comparison with measurements. In subsequent chapters, these estimates will be refined through explicit model building and direct comparison with quantitative experiments.

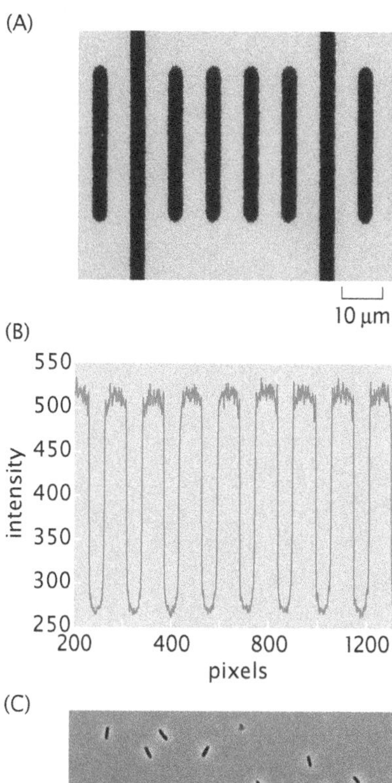

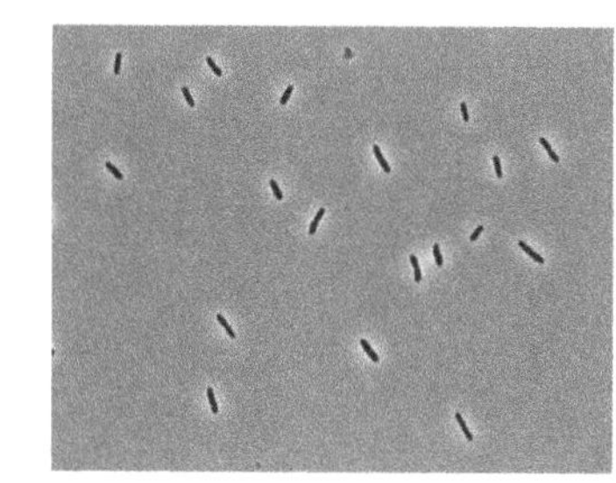

Figure 2.3: Sizing up *E. coli*. (A) Image of a graticule at 100× magnification. (B) Matlab plot of the intensity for a horizontal cut through the image. (C) Phase contrast image of a field of view with several *E. coli* cells taken at the same magnification as in (A).

Estimate: Sizing Up *E. coli* As already noted in the previous chapter, cells are made up of an array of different macromolecules as well as small molecules and ions. To estimate the number of proteins in an *E. coli* cell, we begin by noting that, with its 1 fL volume, the mass of such a cell is roughly 1 pg, where we have assumed that the density of the cell is that of water, which is 1 g/mL, though clearly Figure 2.2 shows that this assumption is not true. Measurements reveal that the dry weight of the cell is roughly 30% of its total and half of that mass is protein. As a result, the total protein mass within the cell is roughly 0.15 pg. We can also estimate the number of carbon atoms in a bacterium by considering the chemical composition of the macromolecules of the cell. This implies that roughly half the dry mass comes from the carbon content of these cells, a figure that reveals of the order of 10^{10} carbon atoms per cell. Two of the key sources that have served as a jumping off point for these estimates are Pedersen et al. (1978) and Zimmerman and Trach (1991), who describe the result of a molecular census of a bacterium.

As a first step toward revealing the extent of crowding within a bacterium, we can estimate the number of proteins by assuming an average protein of 300 amino acids with each amino acid having a characteristic mass of 100 Da. These assumptions are further examined in the problems at the end of the chapter. Using these rules of thumb, we find that the mean protein has a mass of 30,000 Da. Using the conversion factor that $1\,\text{Da} \approx 1.6 \times 10^{-24}\,\text{g}$, we have that our typical protein has a mass of $5 \times 10^{-20}\,\text{g}$. The number of proteins per *E. coli* cell is estimated as

$$N_{\text{protein}} = \frac{\text{total protein mass}}{\text{mass per protein}} \approx \frac{15 \times 10^{-14}\,\text{g}}{5 \times 10^{-20}\,\text{g}} \approx 3 \times 10^6. \quad (2.1)$$

If we invoke the rough estimate that one-third of the proteins coded for in a typical genome correspond to membrane proteins, this implies that the number of cytoplasmic proteins is of the order of 2×10^6 and the number of membrane proteins is 10^6, although we note that not all of these membrane-associated proteins are strictly transmembrane proteins.

Another interesting use of this estimate is to get a rough impression of the number of ribosomes—the cellular machines that synthesize proteins. We can estimate the total number of ribosomes by first estimating the total mass of the ribosomes in the cell and then dividing by the mass per ribosome. To be concrete, we need one other fact, which is that roughly 20% of the protein complement of a cell is ribosomal protein. If we assume that all of this protein is tied up in assembled ribosomes, then we can estimate the number of ribosomes by noting that (a) the mass of an individual ribosome is roughly 2.5 MDa and (b) an individual ribosome is roughly one-third by mass protein and two-thirds by mass RNA, facts that can be directly confirmed by the reader by inspecting the structural biology of ribosomes. As a result, we have

$$\begin{aligned} N_{\text{ribosome}} &= \frac{0.2 \times 0.15 \times 10^{-12}\,\text{g}}{830{,}000\,\text{Da}} \times \frac{1\,\text{Da}}{1.6 \times 10^{-24}\,\text{g}} \\ &\approx 20{,}000 \text{ ribosomes}. \end{aligned} \quad (2.2)$$

The numerator of the first fraction has 0.2 as the fraction of protein that is ribosomal, 0.15 as the fraction of the total cell mass that is protein, and 1 pg as the cell mass. Our estimate for that part of the ribosomal mass that is protein is 830,000 Da. The size of a ribosome is roughly 20 nm (in "diameter") and hence the total volume taken up by these 20,000 ribosomes is roughly $10^8\,\text{nm}^3$. This is 10% of the total cell volume.

Idealizing an *E. coli* cell as a cube, sphere, or spherocylinder yields (see the problems) that the surface area of such cells is $A_{E.\,coli} \approx 6\,\mu\text{m}^2$. This number may be used in turn to estimate the number of lipid molecules associated with the inner and outer membranes of these cells as

$$N_{\text{lipid}} \approx \frac{4 \times 0.5 \times A_{E.\,coli}}{A_{\text{lipid}}} \approx \frac{4 \times 0.5 \times (6 \times 10^6\,\text{nm}^2)}{0.5\,\text{nm}^2} \approx 2 \times 10^7, \quad (2.3)$$

where the factor of 4 comes from the fact that the inner and outer membranes are each *bilayers*, implying that the lipids effectively cover the cell surface area four times. A lipid bilayer consists of two sheets of lipids with their tails pointing toward each other. The factor of 0.5 is based on the crude estimate that roughly half of the surface area is covered by membrane proteins rather than lipids themselves. We have made the similarly crude estimate that the area per lipid is 0.5 nm^2. The measured number of lipids is of the order of 2×10^7 as well.

In terms of sheer numbers, water molecules are by far the majority constituent of the cellular interior. One of the reasons this fact is intriguing is that during the process of cell division, a bacterium such as *E. coli* has to take on a very large number of new water molecules each second. The estimate we obtain here will be used to examine this transport problem in the next chapter. To estimate the number of water molecules, we exploit the fact that roughly 70% of the cellular mass (or volume) is water. As a result, the total mass of water is 0.7 pg. We can find the approximate number of water molecules as

$$N_{H_2O} \approx \frac{0.7 \times 10^{-12}\,\text{g}}{18\ \text{g/mol}} \times 6 \times 10^{23}\ \text{molecules/mol}$$
$$\approx 2 \times 10^{10}\ \text{water molecules.} \tag{2.4}$$

It is also of interest to gain an impression of the content of inorganic ions in a typical bacterial cell. To that end, we assume that a typical concentration of positively charged ions such as K^+ is 100 mM, resulting in the estimate

$$N_{\text{ions}} \approx \frac{(100 \times 10^{-3}\ \text{mol}) \times (6 \times 10^{23}\ \text{molecules/mol})}{10^{15}\ \mu\text{m}^3} \times 1\ \mu\text{m}^3$$
$$= 6 \times 10^7. \tag{2.5}$$

Here we use the fact that $1\,\text{L} = 10^{15}\ \mu\text{m}^3$. This result could have been obtained even more easily by noting yet another simple rule of thumb, namely, that one molecule per *E. coli* cell corresponds roughly to a concentration of 2 nM.

The outcome of our attempt to size up *E. coli* is illustrated schematically in summary form in Figure 2.4. A more complete census of an *E. coli* bacterium can be found in Neidhardt et al. (1990). The outcome of experimental investigations of the molecular census of an *E. coli* cell is summarized (for the purposes of comparing with our estimates) in Table 2.1.

How is the census of a cell taken experimentally? This is a question we will return to a number of different times, but will give a first answer here. For the case of *E. coli*, one important tool has been the use of gels like that shown in Figure 2.5. Such experiments work by breaking open cells and keeping only their protein components. The complex protein mixture is then separated into individual molecular species using a polyacrylamide gel matrix. First, the protein mixture is distributed through a tube-shaped polyacrylamide gel that has been polymerized to contain a stable pH gradient, and then an electric field is applied across the gel. The net charge on each protein depends on the pH and on the number and type of charged (protonatable) amino acid side chains that it contains. For example, the carboxylic acid group on aspartate will be negatively charged at high pH, but

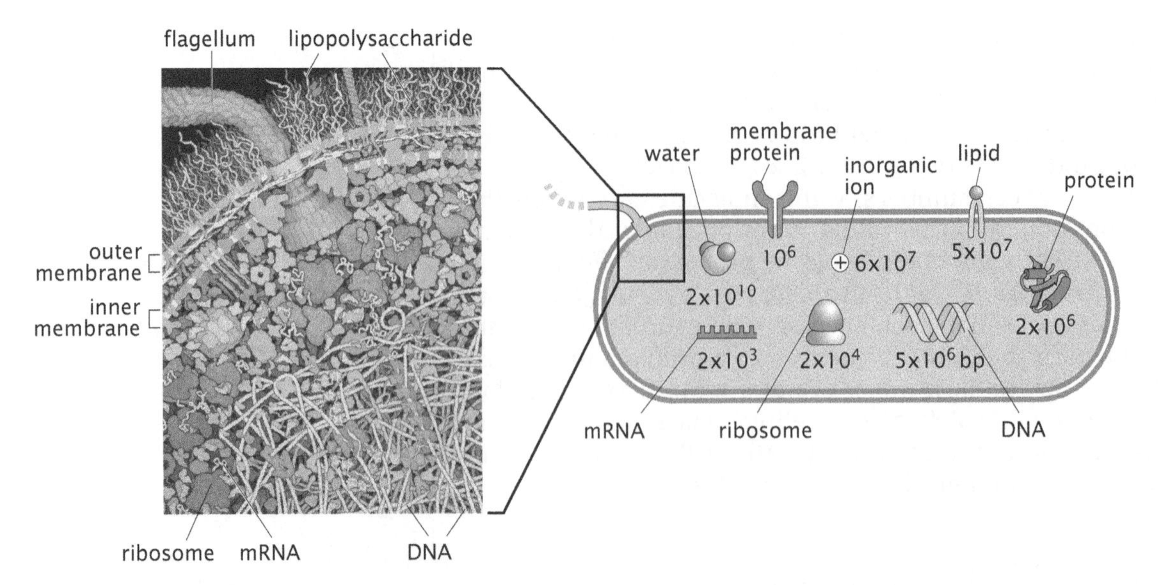

Figure 2.4: Molecular contents of the bacterium *E. coli*. The illustration on the left shows the crowded cytoplasm of the bacterial cell. The cartoon on the right shows an order-of-magnitude molecular census of the *E. coli* bacterium with the approximate number of different molecules in *E. coli*. (Illustration of the cellular interior courtesy of D. Goodsell.)

Table 2.1: Observed macromolecular census of an *E. coli* cell. (Data from F. C. Neidhardt et al., Physiology of the Bacterial Cell, Sinauer Associates, 1990 and M. Schaechter et al., Microbe, ASM Press, 2006.)

Substance	% of total dry weight	Number of molecules
Macromolecules		
Protein	55.0	2.4×10^6
RNA	20.4	
23S RNA	10.6	19,000
16S RNA	5.5	19,000
5S RNA	0.4	19,000
Transfer RNA (4S)	2.9	200,000
Messenger RNA	0.8	1,400
Phospholipid	9.1	22×10^6
Lipopolysaccharide (outer membrane)	3.4	1.2×10^6
DNA	3.1	2
Murein (cell wall)	2.5	1
Glycogen (sugar storage)	2.5	4,360
Total macromolecules	**96.1**	
Small molecules		
Metabolites, building blocks, etc.	2.9	
Inorganic ions	1.0	
Total small molecules	**3.9**	

will pick up a hydrogen ion and will be neutral at low pH. Conversely, the amine group on lysine will be neutral at high pH, but will pick up a hydrogen ion and will be positively charged at low pH. The pH where a protein's charge is net neutral is called its "isoelectric point."

When a protein finds itself in a region of the gel where the pH is below its isoelectric point, there will be an excess of positive charge associated with the protein, and it will move toward the cathode when the electric field is applied. When a protein finds itself in a region of the gel where the pH is above its isoelectric point, it will have a net negative charge, and will therefore move toward the anode. When all the proteins in the mixture have moved to the location of the pH where each is neutral, then all movement stops. At this point, "isoelectric focusing" is complete, and all of the proteins are arrayed along the tube-shaped gel in positions according to their net charge. Next, a charged detergent is added that binds to all proteins so the total number of detergent molecules associated with an individual protein is roughly proportional to the protein's overall size. The detergent-soaked isolectric focusing gel tube is placed at one end of a flat, square gel that also contains detergent, and an electric field is applied in a direction perpendicular to the first field. Because the net charge on the detergent molecules is much larger than the original net charge of the protein, the rate of migration in the second direction through the gel is determined by the protein size.

After the procedure described above, the individual protein species in the original mixture have been resolved into a series of spots on the gel, with large, negatively charged proteins at the upper left-hand corner and small, positively charged proteins at the lower right-hand corner for the gel shown in Figure 2.5. The proteins can then be stained with a nonspecific dye so that their locations within the gel can be directly observed. The intensity of the spots on such a gel can then be used as a basis for quantifying each species. The identity of the protein that congregates in each spot can be determined by physically cutting each spot out of the gel, eluting the protein, and determining its size and amino acid content using mass spectrometry. Similar tricks are used to characterize the amount of RNA and lipids, for example, resulting in a total census like that shown in Table 2.1.

More recently, several other methods have been brought to bear on the molecular census of cells. As shown in Figure 2.6, two of these methods are the use of mass spectrometry and fluorescence microscopy. In mass spectrometry, fragments of the many proteins contained within the cell are run through the mass spectrometer and their absolute abundances determined. The results of this technique applied to *E. coli* are shown in Figure 2.6(A). Alternatively, in fluorescence microscopy, a library of cells is created where each strain in this library expresses a particular protein from the *E. coli* proteome fused to a fluorescent protein. By calibrating the fluorescence corresponding to an individual fluorescent protein, one can measure the amount

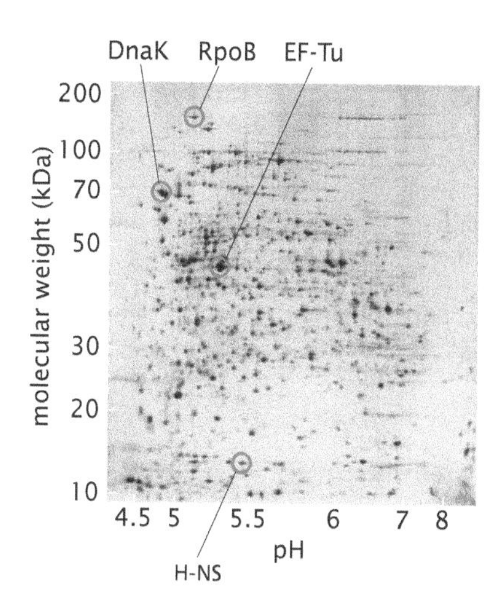

Figure 2.5: Protein census of the cell. Measurement of protein census of *E. coli* using two-dimensional polyacrylamide gel electrophoresis. Each spot represents an individual protein with a unique size and charge distribution. The spots arising from several well-known bacterial proteins are labeled. (Copyright Swiss Institute of Bioinformatics, Geneva, Switzerland.)

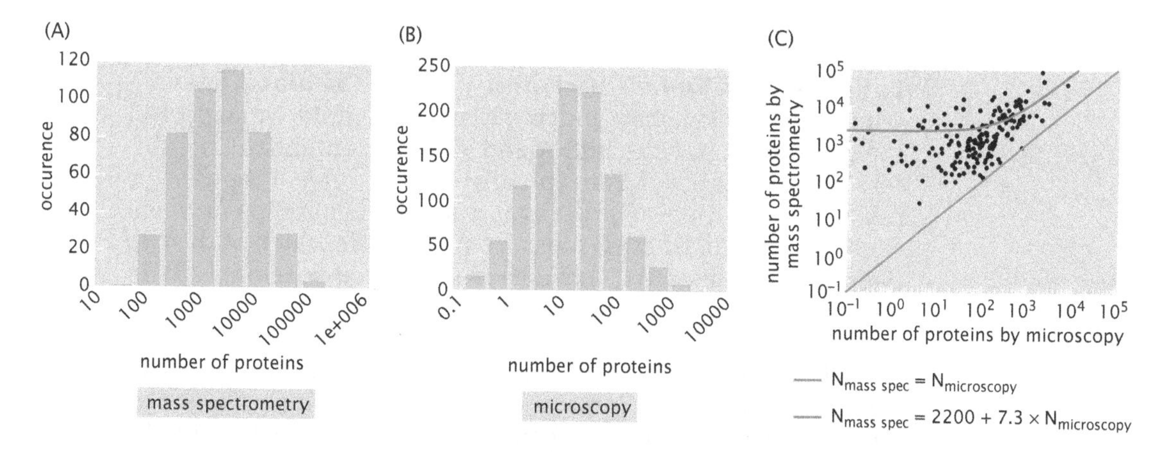

Figure 2.6: Protein census of *E. coli* using several techniques. A histogram of protein number in *E. coli* is shown from results using (A) mass spectrometry and (B) fluorescence microscopy with protein fusions to fluorescence proteins. (C) Comparison of mass spectrometry and fluorescence methods for the same collection of proteins showing a discrepancy between the two techniques as revealed by the fact that the best fit is not given by a line of slope one. (A, adapted from P. Lu et al., *Nat. Biotechnol.* 25:117, 2007; B, adapted from Y. Taniguchi et al., *Science* 329:533, 2010.)

Figure 2.7: Schematic of the random partitioning of molecules during the process of cell division. (A) When the cell divides, each of the molecules chooses a daughter cell via a coin flip. (B) Probability distribution for the number of heads for different choices of the total number of coin flips. (C) The width of the distribution as a function of the number of coin flips.

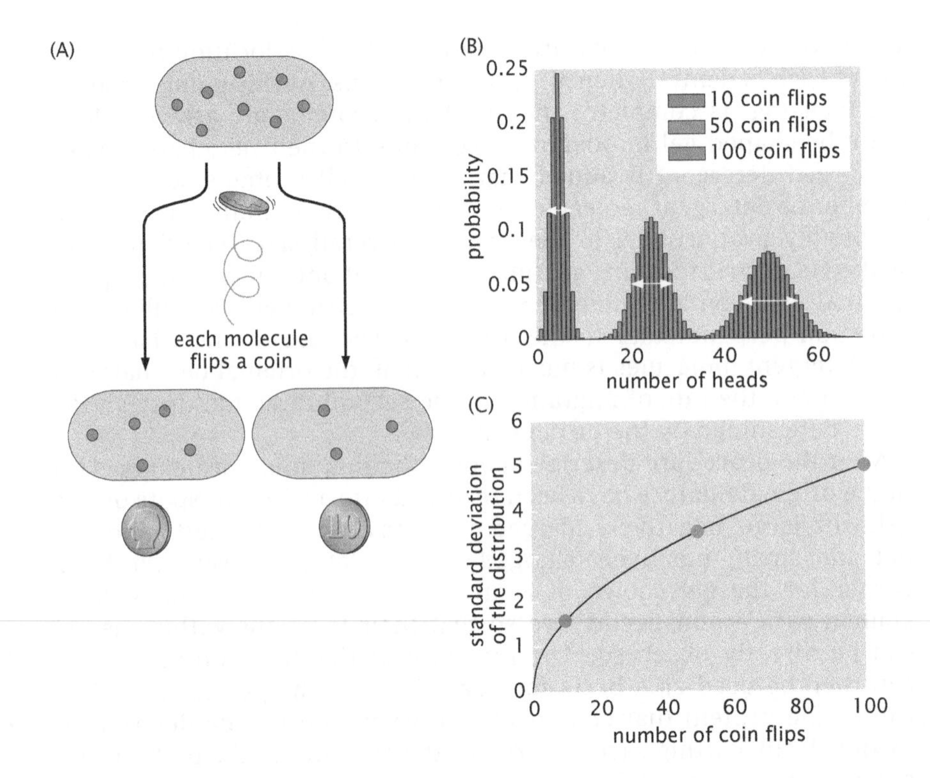

of protein fusion in each strain of the library by comparing the total fluorescence in the cell with the single-molecule standard. Such a measurement for *E. coli* results in the histogram showed in Figure 2.6(B). It is important to point out, however, that these two methods are not in agreement, as shown in Figure 2.6(C). It is seen that, for a given protein, fluorescence microscopy tends to undercount proteins with respect to mass spectrometry (or mass spectrometry tends to overcount with respect to fluorescence microscopy). There are various possible sources for this discrepancy, ranging from systematic errors in the two experimental techniques to the fact that the experiments were done under different growth conditions leading to very different cell cycle times and cell sizes. One of the key features revealed by fluorescence measurements is the importance of cell-to-cell variability, which we can estimate using simple ideas from probability theory.

Estimate: Cell-To-Cell Variability in the Cellular Census

When cells divide, how much variability should we expect in the partitioning of the mRNA and protein contents of the different daughter cells? Conceptually, we can think of the passive molecular partitioning process as a series of coin flips in which each molecule of interest flips a coin to decide which of the two daughters it will go to. This idea is illustrated schematically in Figure 2.7.

To make a simple estimate of the variability, we turn to one of the most important probability distributions in all of science, namely, the binomial distribution. Despite the apparently contrived nature of coin flips, they are precisely the statistical experiment we need when thinking about cellular concentrations of molecules that are passively partitioned during cell division. The idea of the binomial distribution is that we carry out N trials of our coin flip process with the outcomes being heads and tails. However, a more biologically relevant way to think of these outcomes is that with each cell division, the macromolecule of interest partitions either to daughter 1 or to daughter 2, with the probability of going to daughter 1 given by p and the probability of going to daughter 2 given by $q = 1 - p$. If we are dealing with a fair coin, or if the sizes of the two daughters are equal and there are no active segregation mechanisms in play, then both outcomes are equally likely and have probability $p = q = 1/2$.

To be precise, the probability of having n_1 of our N molecules partition to daughter cell 1 is given by

$$p(n_1, N) = \frac{N!}{n_1!(N - n_1)!} p^{n_1} q^{N-n_1}. \quad (2.6)$$

The factor $N!/[n_1!(N - n_1)!]$ reflects the fact that there are many different ways of flipping n_1 heads (h) and $N - n_1$ tails (t) and is given by the famed binomial coefficients (see Figure 2.8). For example, if $N = 3$ and $n_1 = 2$, then our three trials could turn out in three ways as *thh*, *hth*, and *hht*, exactly what we would have found by constructing $3!/(2!1!)$. Note also that throughout this book we will see that, in certain limits, the binomial distribution can be approximated by the Gaussian distribution (large-N case) or the Poisson distribution ($p \ll 1$ case).

If the number of copies of our mRNA or protein of interest is 10, then on division each daughter will get 5 of these molecules on average. More formally, this can be stated as

$$\langle n_1 \rangle = Np. \quad (2.7)$$

For the case of a fair coin (i.e., $p = 1/2$), this reduces to precisely the result we expect. This intuitive result can be appreciated quantitatively in the distributions shown in Figure 2.7(B). Here, the average number of heads obtained for a number of coin flips is given by the center of the corresponding distribution. The width of these distributions then gives a sense for how often there will be a result that deviates from the mean. As a result, to assess the cell-to-cell variability, we need to find a way to measure the width of the distribution. One way to characterize this width is by the standard deviation, which is computed from the function given in Equation 2.6 as follows:

$$\langle n_1^2 \rangle - \langle n_1 \rangle^2 = Npq, \quad (2.8)$$

a result the reader is invited to work out in the problems at the end of the chapter.

With these results in hand, we can now turn to quantifying the fluctuations that arise during partitioning. There is no one way to characterize the fluctuations, but a very convenient one is to construct the ratio

$$\frac{\sqrt{\langle n_1^2 \rangle - \langle n_1 \rangle^2}}{\langle n_1 \rangle} = \frac{1}{\sqrt{N}} \quad (2.9)$$

This result tells us that when the number of molecules is small (i.e., <100), the cell-to-cell variation is itself a significant fraction of the mean itself.

Such claims are not a mere academic curiosity and have been the subject of careful experimental investigation. On multiple occasions in the remainder of this book, we will examine partitioning data (for mRNA, for proteins, for carboxysomes,

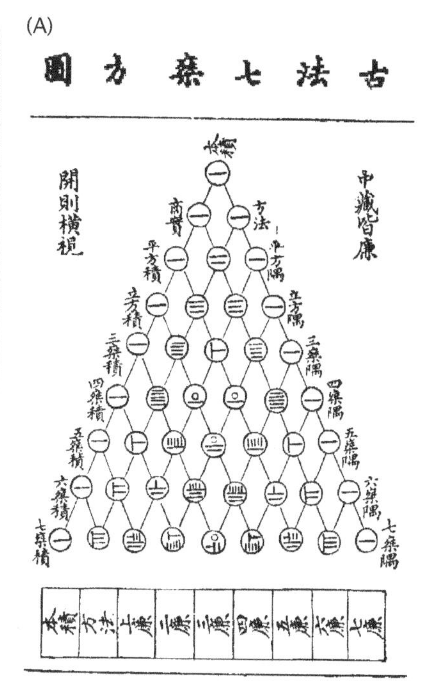

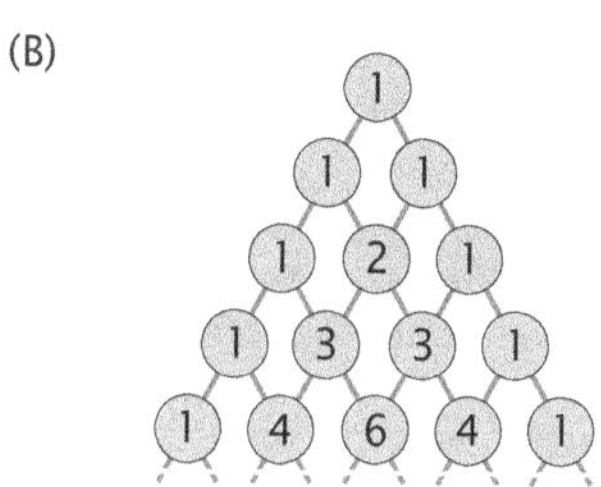

Figure 2.8: Binomial coefficients. (A) Though commonly known as Pascal's triangle, Chinese mathematicians had also earlier appreciated the symmetry of this collection of numbers. (B) The binomial coefficients that arise when raising $p + q$ to the power n, where n characterizes which row we are talking about, with the top "row" denoted by $n = 0$.

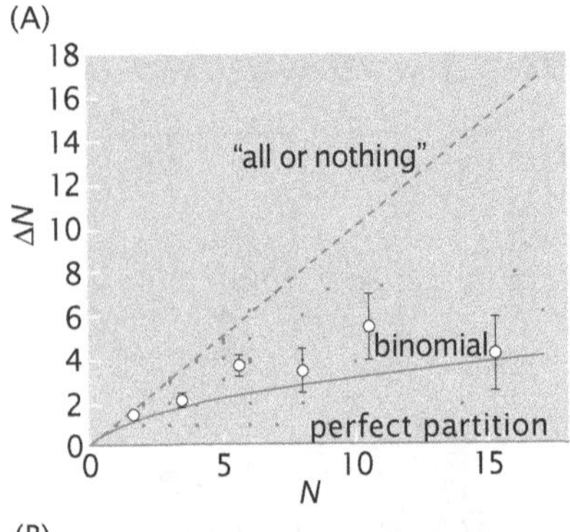

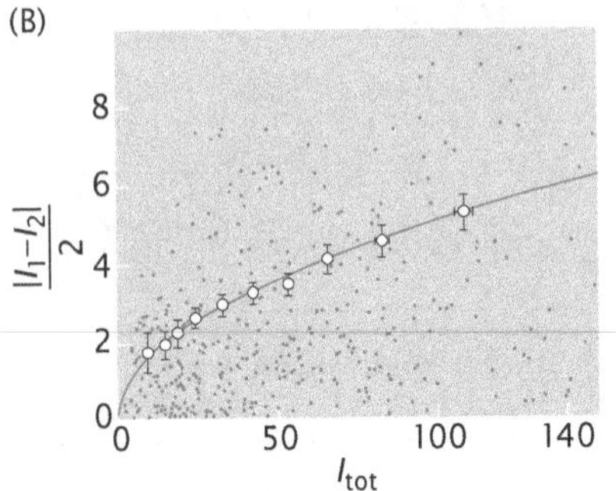

Figure 2.9: Binomial partitioning of mRNA and proteins in *E. coli* during cell division. (A) Difference ΔN in the number of mRNAs between the two daughters given that the mother has N mRNAs. The curves show three possible partitioning mechanisms involving "all or nothing" in which one daughter takes all of the mRNAs, binomial partitioning, and "perfect partitioning" in which each daughter gets exactly half of the proteins from the mother cell. (B) Difference in the fluorescence intensity of the two daughter cells for a fluorescent fusion to a repressor protein as a function of the fluorescence intensity of the mother cell. The line corresponds to binomial partitioning model. (A, adapted from I. Golding et al., *Cell* 123:1025, 2005; B, adapted from N. Rosenfeld et al., *Science* 307:1962, 2005.)

etc.) in which cell lineages are followed and the partitioning of the macromolecules or complexes of interest are measured directly. An example of such data for the case of mRNA and proteins is shown in Figure 2.9. As a result of the ability to fluorescently label individual mRNAs as they are produced, it is possible to characterize the number of such mRNAs in the mother cell before division and then to measure how they partition during the division process. Using fusions of fluorescent proteins to repressor molecules, similar measurements were taken in the context of proteins as shown in Figure 2.9(B). In both of these cases, the essential idea is to count the number of molecules in the mother cell and then to count the number in the two daughters and to see how different they are. Similar data is shown in Figure 18.7 (p. 724) for the case of the organelles in cyanobacteria known as carboxysomes, which contain the all-important enzyme Rubisco so critical to carbon fixation during the process of photosynthesis.

COMPUTATIONAL EXPLORATION

Computational Exploration: Counting mRNA and Proteins by Dilution As will be highlighted throughout this book, many of our key results depend critically upon the number of molecular actors involved in a given process. As shown in Figure 18.7 (p. 724), for example, fluorescence microscopy has made it possible to track the dynamics and localization of key cellular structures such as the carboxysomes. One of the main conclusions of that study was that the partitioning of the carboxysomes is not a random process since the *statistics* of the partitioning of these organelles between the two daughter cells does not follow the binomial distribution, which would be characteristic of random partitioning. The goal of this Computational Exploration is to explore the statistics of random partitioning from the perspective of simulations, with reference not only to the carboxysome example, but also to clever new ways of taking the molecular census. Using these simulations, we will show how the fluctuations in the partitioning of a fluorescent protein between daughter cells is related to the cells' intensity as schematized in Figure 2.10. Data from this type of experiment are shown in Figure 19.15 (p. 816).

The advent of fluorescent protein fusions now makes it possible to assess the relative quantities of a given protein of interest on the basis of the fluorescence of its fusion partner. However, for the purposes of the strict comparison between theory and experiment that we are often interested in, it is necessary to have a calibration that permits us to convert between the arbitrary fluorescence units as measured in a microscope and the more interesting molecular counts necessary for plugging into our various formulae or that we use in dissecting some molecular mechanism. To effect this calibration, we exploit the idea of a strict linear relation between the observed intensity and the number of fusion proteins as indicated by the relation

$$I_{tot} = \alpha N_{tot}, \tag{2.10}$$

where α is the unknown calibration factor linking fluorescence and the number of fluorescent molecules (N_{tot}).

As indicated in Figure 2.10, a series of recent experiments all using the same fluctuation method make it possible to

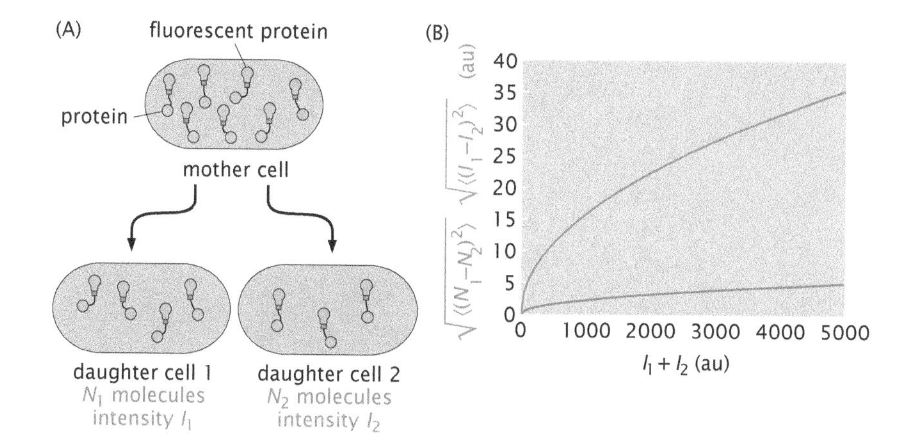

Figure 2.10: Intensity of the daughter cells as a result of binomial partitioning. (A) The mother cell starts out with a certain number of fluorescent proteins (represented by light bulbs) and upon division, these proteins are partitioned according to the binomial distribution. The number of fluorescent proteins and their corresponding total fluorescent intensity are related by the calibration factor α. (B) Plot of the intensity difference between the two daughters as a function of the intensity of the mother cell and the corresponding difference in partitioning of fluorescent molecules as a function of the total number of molecules in the mother cell.

actually determine N_{tot} by determining the unknown calibration factor. The idea of the experiment is to exploit the natural fluctuations resulting from imperfect partitioning of the fluorescent proteins between the two daughter cells. The key point is that for the case in which the partitioning between the two daughters is random (i.e., analogous to getting heads or tails in a coin flip), the size of those fluctuations depends in turn upon the total number of proteins being segregated between the two daughters. The larger the number of proteins, the larger the fluctuations. In particular, as shown in the problems at the end of the chapter, we can derive the simple and elegant result that the average difference in intensity between the two daughter cells is given by

$$\langle (I_1 - I_2)^2 \rangle = \alpha I_{\text{tot}}, \tag{2.11}$$

where I_1 and I_2 are the intensities of daughters 1 and 2, respectively, and I_{tot} is the total fluorescence intensity of the mother cell ($I_{\text{tot}} = I_1 + I_2$).

The simple analytic result of Equation 2.11 can perhaps be best appreciated through an appeal to a stochastic simulation. The idea is to "simulate" the dilution of fluorescent proteins during repeated cell divisions where all the mother cells are presumed to start with N copies of the fluorescent protein and the partitioning of those proteins to the two daughters is determined strictly by a simple binary random process (i.e., heads or tails of a coin flip of an honest coin). Conceptually, the algorithm for performing this simulation can be summarized as follows:

1. Choose the number of fluorescent molecules N in the mother cell and compute the intensity $I_{\text{tot}} = \alpha N$.
2. Generate N random numbers between 0 and 1. If a given random number is between 0 and 0.5, assign the protein to daughter 1 and otherwise assign the protein to daughter 2.
3. Compute the "intensity" of the daughters by multiplying N_1 and N_2 by α.
4. Perform the same operation for 100 cells (for example).
5. Choose a new N and repeat the process.

By considering the same division process again and again, for different choices of N, we can generate a curve that captures

Figure 2.11: Simulation results of fluorescent protein dilution and counting experiment. For a given total number of proteins, N, in the mother cell, we simulate 100 different outcomes of partitioning by cellular division and plot their corresponding partitioning error (green dots). For each value of N, we can calculate the root mean square in partitioning (red dots) and fit them to the expected functional form given by the square root of Equation 2.11.

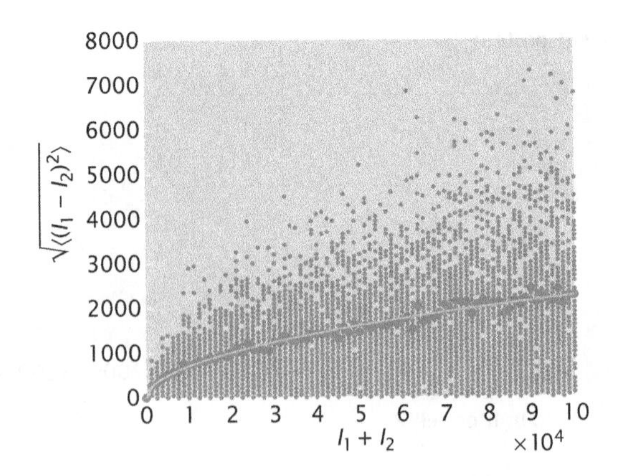

the statistics set forth above. In particular, for each and every N, 100 separate "divisions" were simulated and the results of each such division process are shown as the data points. For a given value of N, we can calculate the root mean square error in partitioning and its corresponding intensity, given by $\sqrt{\langle(I_1 - I_2)^2\rangle}$. To see an example of the outcome of such a simulation, examine Figure 2.11.

Examples of Matlab code that could be used to perform this 'Computational Exploration' can be found on the book's, website.

The Cellular Interior Is Highly Crowded, with Mean Spacings Between Molecules That Are Comparable to Molecular Dimensions

One of the most intriguing implications of our census of the molecular parts of a bacterium is the extent to which the cellular interior is crowded. Because of experiments and associated estimates on the contents of *E. coli*, it is now possible to construct illustrations to depict the cellular interior in a way that respects the molecular census. The crowded environs of the interior of such a cell are shown in Figure 2.4. This figure gives a view of the crowding associated with any *in vivo* process. In Chapter 14, we will see that this crowding effect will force us to call in question our simplest models of chemical potentials, the properties of water, and the nature of diffusion. The generic conclusion is that the mean spacing of proteins and their assemblies is comparable to the dimensions of these macromolecules themselves. The cell is a very crowded place!

The quantitative significance of Figure 2.4 can be further appreciated by converting these numbers into concentrations. To do so, we recall that the volume of an *E. coli* cell is roughly 1 fL. The rule of thumb that emerges from this analysis is that a concentration of 2 nM implies roughly one molecule per bacterium. A concentration of 2 μM implies roughly 1000 copies of that molecule per cell. Concentration in terms of our standard ruler is shown in Figure 2.12. This figure shows the number of copies of the molecule of interest in such a cell as a function of the concentration.

We can use these concentrations directly to compute the mean spacing between molecules. That is, given a certain concentration, there is a corresponding average distance between the molecules. Having a sense of this distance can serve as a guide to thinking about the likelihood of diffusive encounters and reactions between various molecular constituents. If we imagine the molecules at a given concentration arranged on a cubic lattice of points, then the mean spacing between

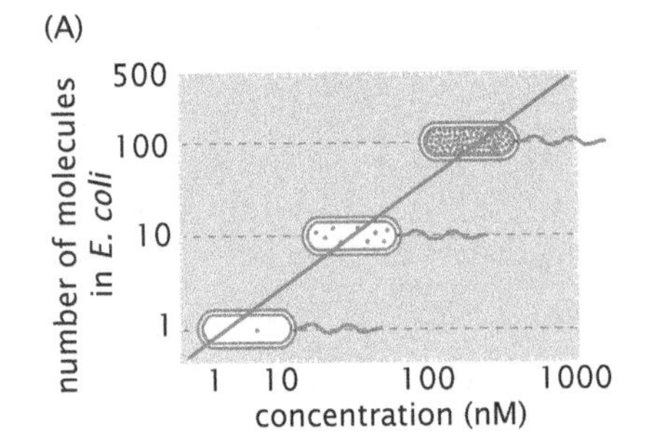

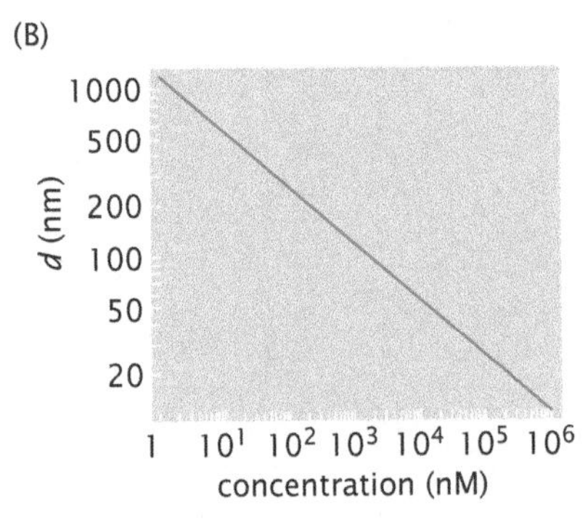

Figure 2.12: Physical interpretation of concentration. (A) Concentration in *E. coli* units: number of copies of a given molecule in a volume the size of an *E. coli* cell as a function of the concentration. (B) Concentration expressed in units of typical distance d between neighboring molecules measured in nanometers.

those points is given by

$$d = c^{-1/3}, \tag{2.12}$$

where c is the concentration of interest (measured in units of number of molecules per unit volume). Larger concentrations imply smaller intermolecular spacings. This idea is formalized in Figure 2.12(B), which shows the relation between the mean spacing measured in nanometers and the concentration in nanomolar units.

2.1.3 Looking Inside Cells

With our reference bacterium in mind, the remainder of the chapter focuses on the various structures that make up cells and organisms. To talk about these structures, it is helpful to have a sense of how we know what we know about them. Further, model building requires facts. To that end, we periodically take stock of the experimental basis for our models. For this chapter, the "Experiments Behind the Facts" focuses on how biological structures are explored and measured.

EXPERIMENTS

Experiments Behind the Facts: Probing Biological Structure To size up cells and their organelles, we need to extract "typical" structural parameters from a variety of experimental studies. Though we leave a description of the design and setup of such experiments to more specialized texts, the goal is to provide at least enough detail so that the reader can see where some of the key structural facts that we will use throughout the book originated. We emphasize two broad categories of experiments: (i) those in which some form of radiation interacts with the structure of interest and (ii) those in which forces are applied to the structure of interest.

Figure 2.13 shows three distinct experimental strategies that feed into our estimates, all of which reveal different facets of biological structure. One of the mainstays of structural analysis is light microscopy. Figure 2.13(A) shows a schematic of the way in which light can excite fluorescence in samples that have some distribution of fluorescent molecules within them. In particular, this example shows a schematic of a microtubule that has some distribution of fluorophores along its length. Incident photons of one wavelength are absorbed by the fluorophores and this excitation leads them to emit light of a different wavelength, which is then detected. As a result of selective labeling of only the microtubules with fluorophores, it is only these structures that are observed when the cell is examined in the microscope. These experiments permit a determination of the size of various structures of interest, how many of them there are, and where they are localized. By calibrating the intensity from single fluorophores, it has become possible to take a single-molecule census for many of the important proteins in cells. For an example of this strategy, see Wu and Pollard (2005). A totally different window on the structure of the cell and its components is provided by tools such as the atomic-force microscope (AFM). As will be explained in Chapter 10, the AFM is a cantilever beam with a sharp tip on its end. The tip is brought very close to the surface where the structure of interest is present and is then scanned in the

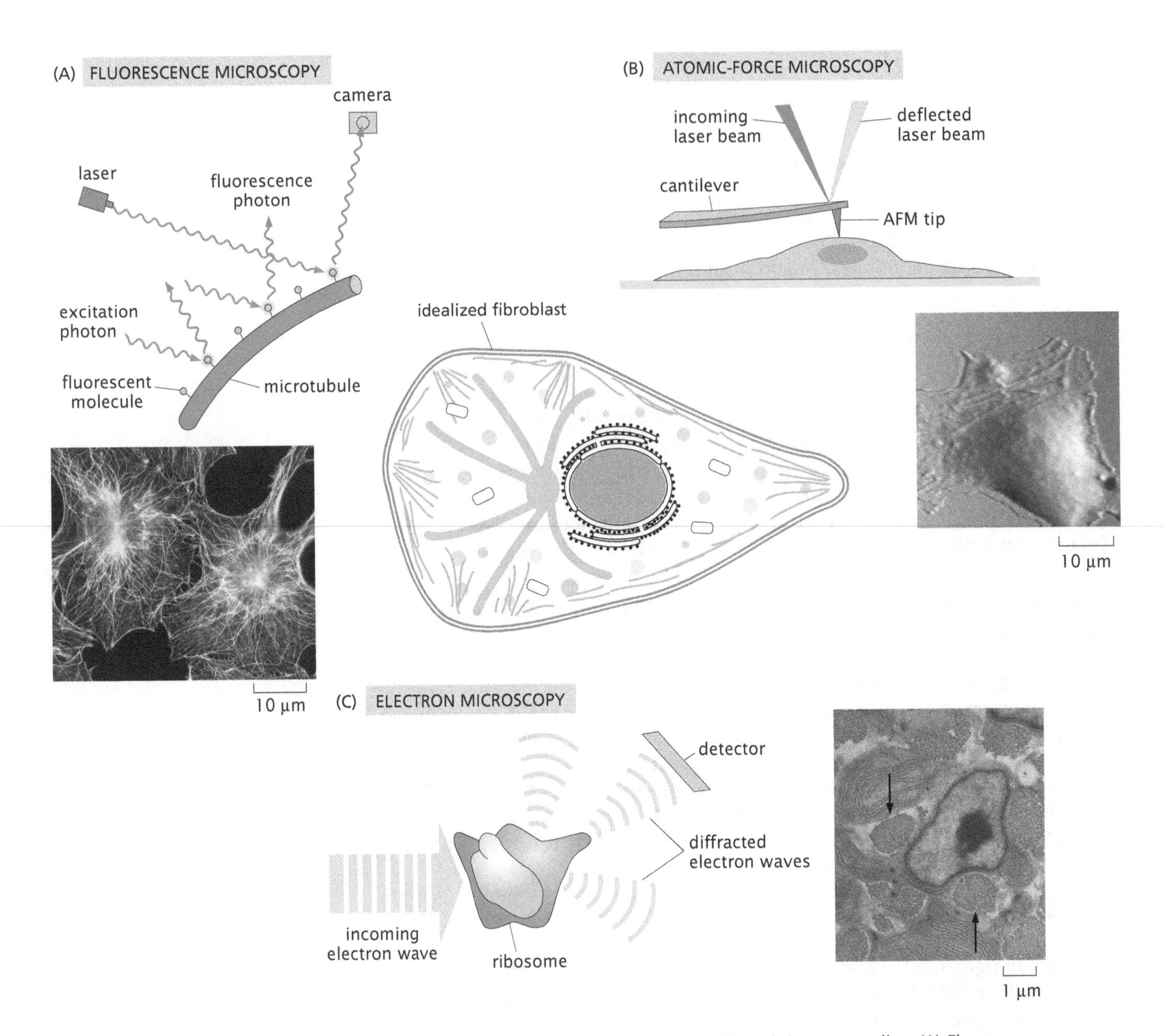

Figure 2.13: Experimental techniques that have revealed the structure of both cells and their organelles. (A) Fluorescence microscopy and the associated image of a fibroblast with labeled microtubules (yellow) and DNA (green). (B) Atomic-force microscopy schematic and the associated image of the surface topography of a fibroblast. (C) Electron microscopy schematic and image of cross-section through a fibroblast in an animal tissue. Arrows indicate bundles of collagen fibers. (A, courtesy of Torsten Wittman; B, adapted from M. Radmacher, *Meth. Cell Biol.* 83:347, 2007; C, adapted from D. E. Birk and R. L. Trelstad, *J. Cell Biol.* 103:231, 1986.)

plane of the sample. One way to operate the instrument is to move the cantilever up and down so that the force applied on the tip remains constant. Effectively, this demands a continual adjustment of the height as a function of the x–y position of the tip. The nonuniform pattern of cantilever displacements can be used to map out the topography of the structure of interest. Figure 2.13(B) shows a schematic of an AFM scanning a typical fibroblast cell as well as a corresponding image of the cell.

Figure 2.13(C) gives a schematic of the way in which X-rays or electrons are scattered off a biological sample. The schematic shows an incident plane wave of radiation that interacts with the biological specimen and results in the emergence of radiation with the same wavelength but a new propagation direction. Each point within the sample can be thought of as a source of radiation, and the observed intensity at the detector reflects the interference from all of these different sources. By observing the pattern of intensity, it is possible to deduce something about the structure that did the scattering. This same basic idea is applicable to a wide variety of radiation sources, including X-rays, neutrons, and electrons.

An important variation on the theme of measuring the scattered intensity from irradiated samples is cryo-electron tomography. This technique is one of the centerpieces of structural biology and is built around uniting electron microscopy with sample preparation techniques that rapidly freeze the sample. The use of tomographic methods has made it possible to go beyond the planar sections seen in conventional electron microscopy images. The basis of the technique is indicated schematically in Figure 2.14, and relies on rotating the sample over a wide range of orientations and then assembling a corresponding three-dimensional reconstruction on the basis of the entirety of these images. This technique has already revolutionized our understanding of particular organelles and is now being used to image entire cells.

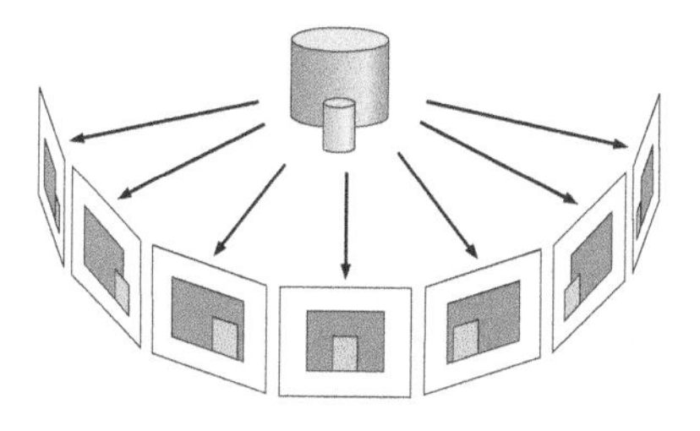

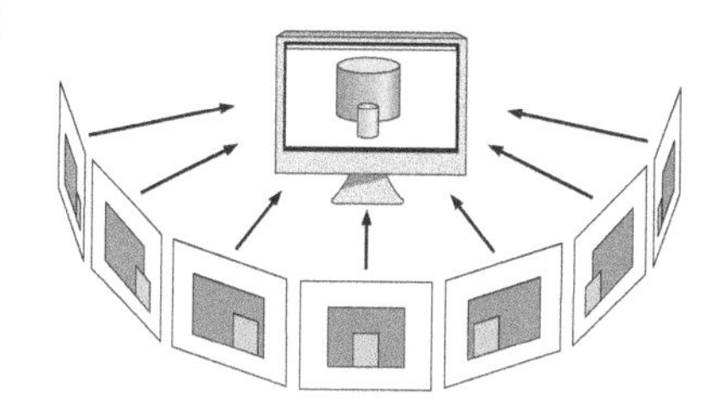

Figure 2.14: Schematic of tomographic reconstruction. (A) The three-dimensional sample is rotated in the electron microscope and imaged repeatedly to generate a series of two-dimensional images representing the pattern of radiation scattered at slightly different orientations. (B) Information from the image series is recombined computationally to generate a three-dimensional rendering of the original object.

2.1.4 Where Does *E. coli* Fit?

Biological Structures Exist Over a Huge Range of Scales

The spatial scales associated with biological structures run from the nanometer scale of individual molecules all the way to the scale of the Earth itself. Where does *E. coli* fit into this hierachy of structures? Figure 2.15 shows the different structures that can be seen as we scale in and out from an *E. coli* cell. A roughly 10-fold increase in magnification relative to an individual bacterium reveals the viruses that attack bacteria. These viruses, known as bacteriophages, have a characteristic scale of roughly 100 nm. They are made up of a protein shell (the capsid) that is filled with the viral genome. Continuing our downward descent using yet higher magnification, we see the ordered packing of the viral genome within its capsid. These structures are intriguing because they involve the ordered arrangement of more than 10 μm of DNA in a capsid that is less than 100 nm across. Another rough factor-of-10 increase in resolution reveals the structure of the DNA molecule itself, with a characteristic cross-sectional radius of roughly 1 nm and a length of 3.4 nm per helical repeat.

A similar scaling out strategy reveals new classes of structures. As shown in Figure 2.15, a 10-fold increase in spatial scale brings us to the realm of eukaryotic cells in general, and specifically, to the scale of the epithelial cells that line the human intestine. We use this example because bacteria such as *E. coli* are central players as part of our intestinal ecosystem. Another 10-fold increase in spatial scale reveals one of the most important inventions of evolution, namely, multicellularity. In this case, the cartoon depicts the formation of planar sheets of epithelial cells. These planar sheets are themselves the building blocks of yet higher-order structures such as tissues and organs. Scaling out to larger scales would bring us to multicellular organisms and the structures they build.

The remainder of the chapter takes stock of the structures at each of these scales and provides a feeling for the molecular building blocks

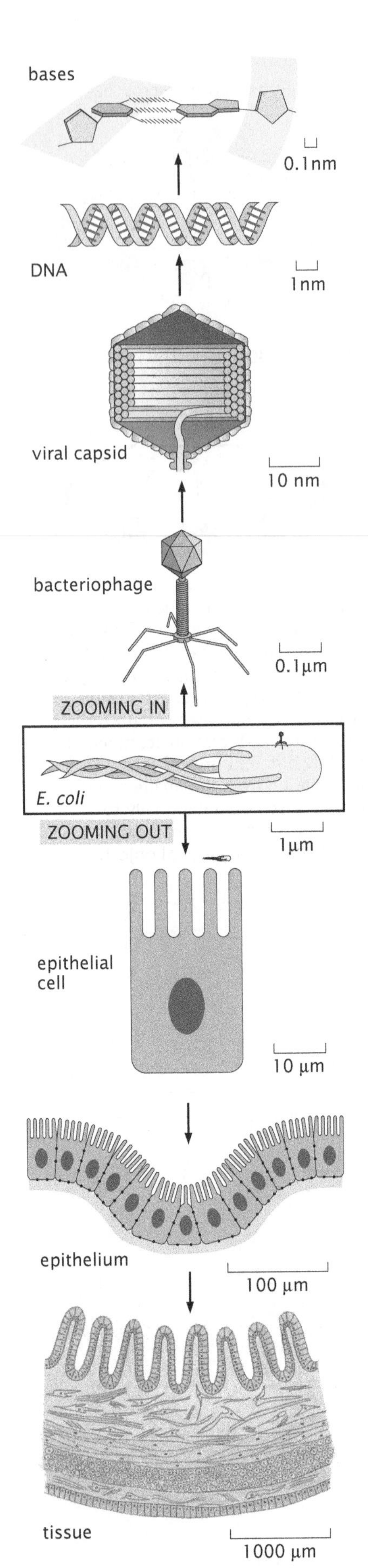

that make up these different structures. Our strategy will be to build upon our cell-centered view and to first descend in length scale from that of cells to the molecules of which they are made. Once this structural descent is complete, we will embark on an analysis of biological structure in which we zoom out from the scale of individual cells to collections of cells.

2.2 Cells and Structures within Them

2.2.1 Cells: A Rogue's Gallery

All living organisms are based on cells as the indivisible unit of biological organization. However, within this general rule there is tremendous diversity among living cells. Several billion years ago, our last common ancestor gave rise to three different lineages of cells now commonly called Bacteria, Archaea, and Eukarya, a classification suggested by similarities and differences in ribosomal RNA sequences. Every living organism on Earth is a member of one of these groups. Most bacteria and archaea are small (3 μm or less) and extremely diverse in their preferred habitats and associated lifestyles ranging from geothermal vents at the bottom of the ocean to permafrost in Antarctica. Bacteria and Archaea look similar to one another and it has only been within the last few decades that molecular analysis has revealed that they are completely distinct lineages that are no more closely related to each other than the two are to Eukarya.

The organisms that we most often encounter in our everyday life and can see with the naked eye are members of Eukarya (individuals are called eukaryotes). These include all animals, all plants ranging from trees to moss, and also all fungi, such as mushrooms and mold. Thus far, we have focused on *E. coli* as a representative cell, although we must acknowledge that *E. coli*, as a member of the bacterial group, is in some ways very different from a eukaryotic or archaeal cell. The traditional definition of a eukaryotic cell is one that contains its DNA genome within a membrane-bound nucleus. Most bacteria and archaea lack this feature and also lack other elaborate intracellular membrane-bound structures such as the endoplasmic reticulum and the Golgi apparatus that are characteristic of the larger and more complex eukaryotic cells.

Cells Come in a Wide Variety of Shapes and Sizes and with a Huge Range of Functions

Cells come in such a wide variety of shapes, sizes, and lifestyles that choosing one representative cell type to tell their structural story is misleading. In Figure 2.16, we show a rogue's gallery illustrating a small segment of the variety of cell sizes and shapes found in the eukaryotic group, all referenced to the *E. coli* standard ruler. This gallery is by no means complete. There is much more variety than we can illustrate, but this covers a reasonable range of eukaryotic cell

Figure 2.15: Powers-of-10 representation of biological length scales. The hierarchy of scales is built around the *E. coli* standard ruler. Starting with *E. coli*, Section 2.2 considers a succession of 10-fold increases in resolution as are shown in the figure. Section 2.3 zooms out from the scale of an *E. coli* cell.

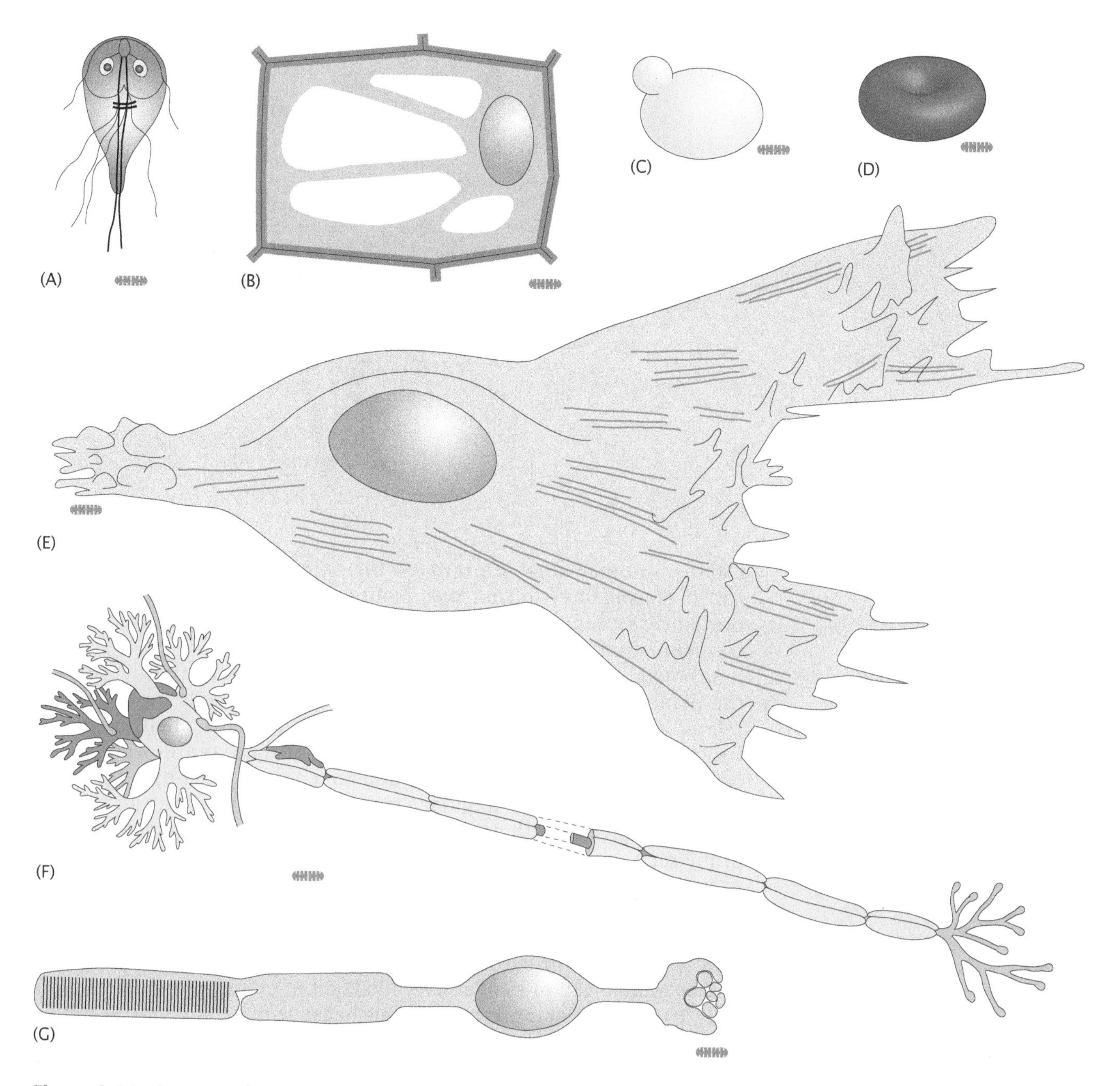

Figure 2.16: Cartoons of several different types of cells all referenced to the standard *E. coli* ruler. (A) The protist *Giardia lamblia*, (B) a plant cell, (C) a budding yeast cell, (D) a red blood cell, (E) a fibroblast cell, (F) a eukaryotic nerve cell, and (G) a retinal rod cell.

types that have been well studied by biologists. In this figure, we have chosen a variety of examples that represent experimental bias among biologists where more than half of the examples are human cells and the others represent the rest of the eukaryotic group.

The vast majority of eukaryotes are members of a group called protists, as shown in Figure 2.17. This poorly defined group encompasses all eukaryotes that are neither plants nor animals nor fungi. Protists are extremely diverse in their appearance and lifestyles, but they are all small (ranging from 0.002 mm to 2 mm). Some examples of protists are marine plankton such as *Emiliana huxleyi*, soil amoebae such as *Dictyostelium discoideum*, and the lovely creature *Paramecium* seen in any sample of pond water and familiar from many high-school biology

Figure 2.17: Protist diversity. This figure illustrates the morphological diversity of free-living protists. The various organisms are drawn to scale relative to the head of a pin. (Adapted from B. J. Finlay, *Science* 296:1061, 2002.)

classes. Another notable protist is the pathogen that causes malaria, called *Plasmodium falciparum*. Figure 2.16(A) shows the intriguing protist *Giardia lamblia*, a parasite known to hikers as a source of water contamination that causes diarrhea.

Although protists constitute the vast majority of eukaryotic cells on the planet, biologists are often inclined to study cells more closely related to or directly useful to us. This includes the plant kingdom, which is obviously important as a source of food and flowers. Plant cells like that shown in Figure 2.16(B) are characterized by a rigid cell wall, often giving them angular structures like that shown in the figure. The typical length scale associated with these cells is often tens of microns. One of the distinctive features of plant cells is their large vacuoles within the intracellular space that hold water and contribute to the mechanical properties of plant stems. These large vacuoles are very distinct from animal cells, where most of the intracellular space is filled with cytoplasm. Consequently, in comparing a plant and an animal cell of similar overall size, the plant cell will have typically 10-fold less cytoplasmic volume because most of its intracellular space is filled with vacuoles. Hydrostatic forces matter much more to plants than animals. For example, a wilting flower can be revived simply by application of water since this allows the vacuoles to fill and stiffen the plant stem.

Among the eukaryotes, the group most closely related to the animals (as proved by ribosomal RNA similarity and many other lines of evidence) is, surprisingly, the fungi. The representative fungus shown in Figure 2.16 is the budding yeast *Saccharomyces cerevisiae* (which we will refer to as *S. cerevisiae*). *S. cerevisiae* was domesticated by humans several thousand years ago and continues to serve as a treasured microbial friend that makes our bread rise and provides alcohol in our fermented beverages such as wine. Just as *E. coli* often serves as a key model prokaryotic system, the yeast cell often serves as the model single-celled eukaryotic organism. Besides the fact that humans are fond of *S. cerevisiae* for its own intrinsic properties, it is also useful to biologists as a representative fungus. Of all the other organisms on Earth, fungi are closest to animals in terms of evolutionary descent and similarity of protein functions. Although there are no single-celled animals, there are some single-celled fungi, including

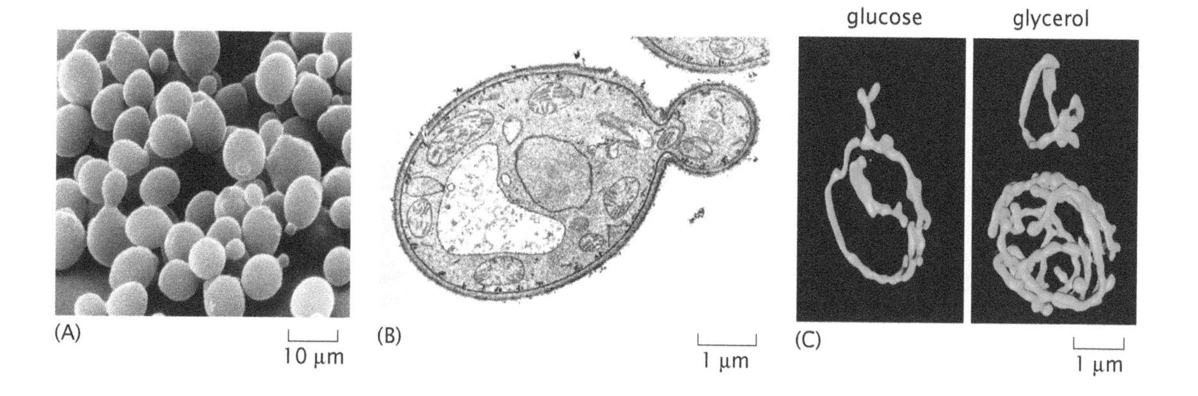

S. cerevisiae. Therefore, many laboratory experiments relying on rapid replication of single cells are most easily performed on this organism and its relatives *Candida albicans* and *Schizosaccharomyces pombe* (a "fission" yeast that divides in the middle). Figure 2.18 shows a scanning electron microscope image of a yeast cell engaged in budding. As this image shows, the geometry of yeast is relatively simple compared with many other eukaryotic cells and it is also a fairly small member of this group, with a characteristic diameter of roughly 5 μm. Nonetheless, it possesses all the important structural hallmarks of the eukaryotes, including, in particular, a membrane-bound nucleus, separating the DNA genome from the cytoplasmic machinery that performs most metabolic functions.

Figure 2.18: Microscopy images of yeast and their organelles. (A) Scanning electron micrograph of the yeast *Candida albicans* revealing the overall size scale of budding yeast. (B) Electron microscopy image of a cross-section through a budding *Candida albicans* yeast cell. (C) Confocal microscopy images of the mitochondria of *S. cerevisiae* under different growth conditions. (A, courtesy of Ira Herskowitz and Eric Schabtach; B, adapted from G. M. Walker, Yeast, Physiology and Biotechnology, John Wiley, 1998; C, adapted from A. Egner et al., *Proc. Natl Acad. Sci. USA* 99:3370, 2002.)

Earlier, we estimated the molecular census of an *E. coli* cell. It will now be informative to compare those estimates with the corresponding model eukaryotic cell that will continue to serve as a comparative basis for all our eukaryotic estimates.

Estimate: Sizing Up Yeast The volume of a yeast cell can be computed in *E. coli* volume units, $V_{E.\ coli}$. In particular, if we recall that $V_{E.\ coli} \approx 1.0\,\mu\text{m}^3 = 1.0$ fL and think of yeast as a sphere of diameter 5 μm, then we have the relation $V_{\text{yeast}} \approx 60\,V_{E.\ coli}$; that is, roughly 60 *E. coli* cells would fit inside of a yeast cell. This result is consistent with experimental observations of the yeast cell size distribution such as shown in Figure 2.19. The surface area of a yeast cell can be estimated using a radius of $r_{\text{yeast}} \approx 2.5\,\mu\text{m}$, which yields $A_{\text{yeast}} \approx 80\,\mu\text{m}^2$. If we treat the yeast nucleus as a sphere with a diameter of roughly 2.0 μm, its volume is roughly 4 μm^3. Within this nucleus is housed the 1.2×10^7 bp of the yeast genome, which is divided among 16 chromosomes. The DNA in yeast is packed into higher-order structures mediated by protein assemblies known as histones. In particular, the DNA is wrapped around a series of cylindrical cores made up of eight such histone proteins each, with roughly 150 bp wrapped around each histone octamer, and approximately a 50 bp spacer between. As a result, we can estimate the number of nucleosomes (the histone–DNA complex) as

$$N_{\text{nucleosome}} \approx \frac{12 \times 10^6\ \text{bp}}{200\ \text{bp/nucleosome}} \approx 60{,}000. \tag{2.13}$$

Experimentally, the measured number appears to be closer to 80,000, with a mean spacing between nucleosomes of the order

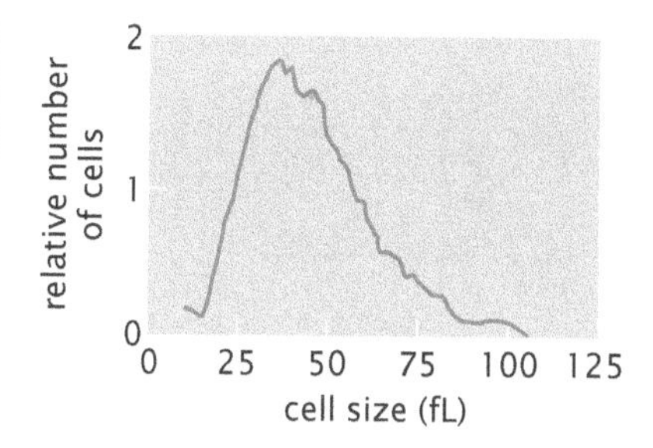

Figure 2.19: Yeast cell size distribution. Distribution of cell volumes measured for wild-type yeast cells. (Adapted from P. Jorgensen et al., *Science* 297:395, 2002.)

of 170 bp. The total volume taken up by the histones is roughly $230\,\text{nm}^3$ per histone (thinking of each histone octamer as a cylindrical disk of radius 3.5 nm and height 6 nm), resulting in a total volume of $14 \times 10^6\,\text{nm}^3$ taken up by the histones. The volume taken up by the genome itself is comparable at $1.2 \times 10^7\,\text{nm}^3$, where we have used the rule of thumb that the volume per base pair is $1\,\text{nm}^3$. The packing fraction (defined as the ratio of the volume taken up by the genome to the volume of the nucleus) associated with the yeast genomic DNA can be estimated by evaluating the ratio

$$\rho_{\text{pack}} \approx \frac{(1.2 \times 10^7\,\text{bp}) \times (1\,\text{nm}^3/\text{bp})}{4 \times 10^9\,\text{nm}^3} \approx 3 \times 10^{-3}. \tag{2.14}$$

Note that we have used the fact that the yeast genome is 1.2×10^7 bp in length and is packed in the nucleus, which has a volume $\approx 4\,\mu\text{m}^3$.

These geometric estimates may be used to make corresponding molecular estimates, such as the number of lipids and proteins in a typical yeast cell. The number of proteins can be estimated in several ways—perhaps the simplest is just to assume that the fractional occupancy of yeast cytoplasm is identical to that of *E. coli*, with the result that there will be 60 times as many proteins in yeast as in *E. coli* based strictly on scaling up the cytoplasmic volume. This simple estimate is obtained by *assuming* that the composition of the yeast interior is more or less the same as that of an *E. coli* cell. This strategy results in

$$N_{\text{protein}}^{\text{yeast}} \approx 60 \times N_{\text{protein}}^{E.\ coli} \approx 2 \times 10^8. \tag{2.15}$$

The number of lipids associated with the plasma membrane of the yeast cell can be obtained as

$$N_{\text{lipid}} \approx \frac{2 \times 0.5 \times A_{\text{yeast}}}{A_{\text{lipid}}} \approx \frac{2 \times 0.5 \times (80 \times 10^6\,\text{nm}^2)}{0.5\,\text{nm}^2} \approx 2 \times 10^8, \tag{2.16}$$

where the factor of 0.5 is based on the idea that roughly half of the surface area is covered by membrane proteins rather than lipids themselves and the factor of 2 accounts for the fact that the membrane is a bilayer. This should be contrasted with the situation in *E. coli*, which has a double membrane surrounding the cytoplasm.

Another interesting estimate suggested by Figure 2.18(C) is associated with the organellar content of these cells. In particular, this figure shows the mitochondria of yeast that are being grown in two different media. These pictures suggest several interesting questions such as what fraction of the cellular volume is occupied by mitochondria and what is the surface area tied up with the mitochondrial outer membranes? Using the image of the cells grown on glucose in the left of Figure 2.18(C), we see that the mitochondria have a reticular form made up of several long tubes with a radius of roughly $r = 100$ nm. These tubes are nearly circular like the inner tubes of a bicycle with a radius of curvature of approximately $R = 1\ \mu\text{m}$, meaning that each such tube has a length of roughly $L = 2\pi R \approx 6\ \mu\text{m}$, for a total tube length of 12 μm given that there appear to be two such tubes. Hence, the volume taken up by the mitochondria is $V = \pi r^2(2L) \approx 0.5\ \mu\text{m}^3$, with an area of $A = 2\pi r(2L) \approx 6\ \mu\text{m}^2$, meaning that about 1% of the volume of the cells grown on glucose is taken up by their mitochondria and an area is tied up in the mitochondrial outer membrane that is roughly 10% of the total plasma membrane area. In other cell types, the fraction of the area taken up by the mitochondria can be much larger, as will be seen in Chapter 11 (p. 427).

Our estimates are brought into sharpest focus when they are juxtaposed with actual measurements. The census of yeast cells has been performed in several distinct and fascinating ways. For a series of recent studies, the key idea is to generate thousands of different yeast strains, a strain library, each of which has a tag on a different one of the yeast gene products. Each yeast protein can then be specifically recognized by antibodies in the specific strain where it has been tagged. A second scheme is to construct protein fusions in which the protein of interest is attached to a fluorescent protein such as the green fluorescent protein (GFP). Then, by querying each and every yeast strain in the library either by examining the extent of antibody binding or fluorescence, it is possible to count up the numbers of each type of protein. Figure 2.20(B) shows a histogram of the number of proteins that occur with a given protein copy number in yeast. Although the average protein appears to be present in the cell with a few thousand copies, this histogram shows that some proteins are present with fewer than 50 copies and others with more than a million copies. Further, similarly quantifying mRNA as shown in Figure 2.20(C) reveals that many genes, including essential genes, are expressed with an average of less than one molecule of RNA per cell. By adding up the total number of proteins on the basis of this census, we estimate there are 50×10^6 proteins in a yeast cell, somewhat less than suggested by our crude estimate given above. Now that we have completed our first introduction to the important yeast as a representative eukaryote, we return to our tour of cell types in Figure 2.16.

Cells from Humans Have a Huge Diversity of Structure and Function

The remainder of the cells in Figure 2.16 are all human cells and show another interesting aspect of cellular diversity. To a first approximation, every cell in the human body contains the same DNA genome. And yet, individual human cells differ significantly with respect to their sizes (with sizes varying from roughly 5 μm for red blood cells to 1 m for the largest neurons), shapes, and functions. For example, rod cells in the retina are specialized to detect incoming light and transmit that information to the neural system so that we can see. Red blood cells are primarily specialized as carriers of oxygen and, in fact, are dramatically different from almost all other cells in having dispensed with their nucleus as part of their developmental process. As we will discuss extensively throughout the book, other cells have other unique classes of specialization.

Figure 2.16(D) is a schematic of the structure of a red blood cell. Note that the shapes of these cells are decidedly not spherical, raising interesting questions about the mechanisms of cell-shape maintenance. Despite their characteristic size of order 5 μm, these

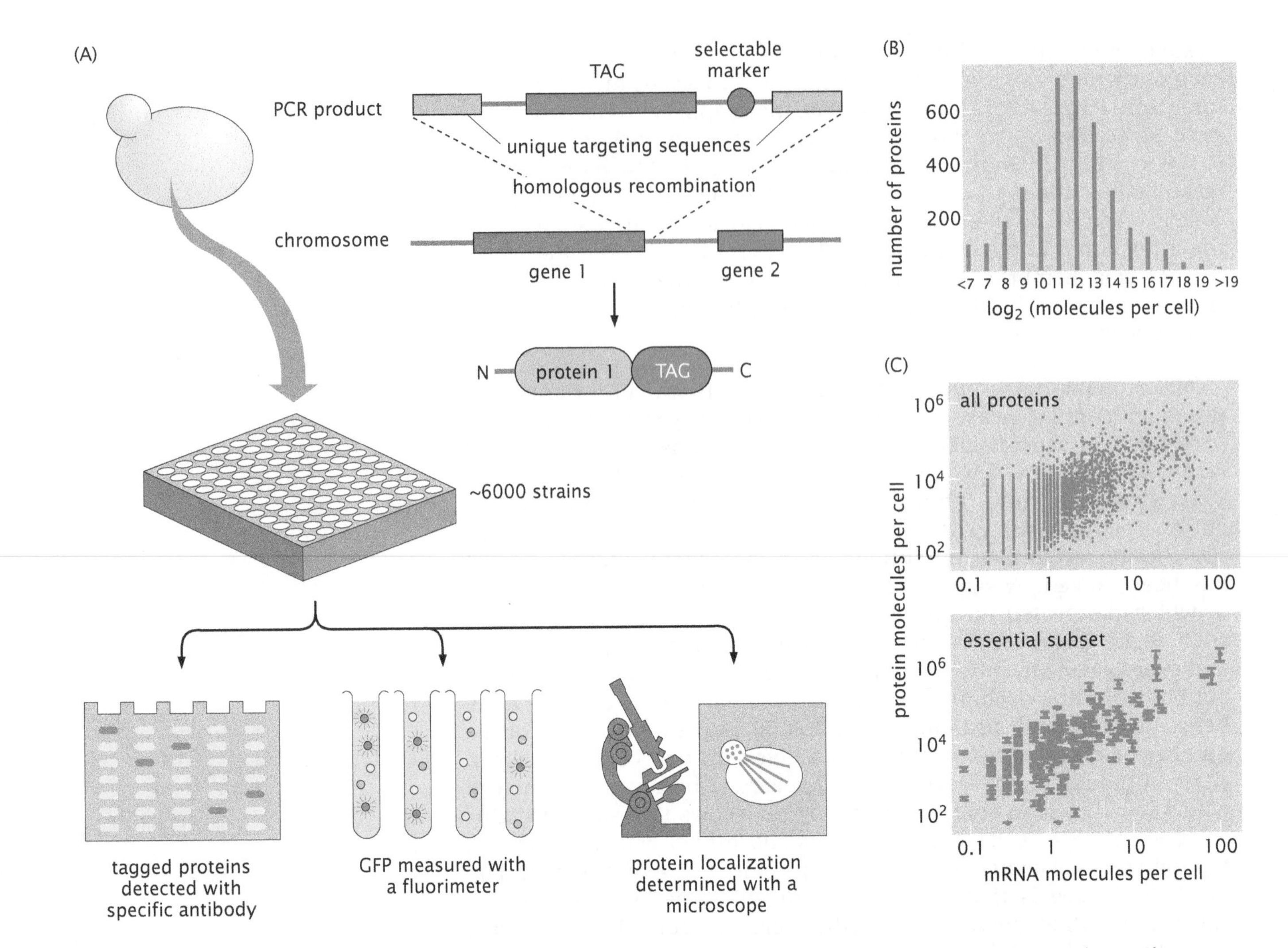

Figure 2.20: Protein copy numbers in yeast. (A) Schematic of constructs used to measure the protein census and several different methods for quantifying protein in labeled cells. The "tag" attached to each gene may encode a fluorescent protein or a site for antibody recognition, depending on the experiment. (B) Result of antibody detection of various proteins in yeast showing the number of proteins that have a given copy number. The number of copies of the protein per cell is expressed in powers of 2 as 2^N. (C) Abundances of various proteins as a function of their associated mRNA copy numbers. The bottom plot shows this result for essential soluble proteins. (A–C, adapted from S. Ghaemmaghami et al. *Nature* 425:737, 2003.)

cells easily pass through capillaries with less than half their diameter as shown in Figure 2.21, implying that their shape is altered significantly as part of their normal life cycle. While in capillaries (either artificial or *in vivo*), the red blood cell is severely deformed to pass through the narrow passage. In their role as the transport vessels for oxygen-rich hemoglobin, these cells will serve as an inspiration for our discussion of the statistical mechanics of cooperative binding. Red blood cells are a target of one of the most common infectious diseases suffered by humans, malaria, caused by the invasion of a protozoan. Malaria-infected red blood cells are much stiffer than normal cells and cannot deform to enter small capillaries. Consequently, people suffering from malaria experience severe pain and damage to tissues because of the inability of their red blood cells to enter those tissues and deliver oxygen.

One of the most commonly studied eukaryotic cells from multicellular organisms is the fibroblast, which is shown schematically in Figure 2.16(E) and in an AFM image in Figure 2.22. These cells will serve as a centerpiece for much of what we will have to say about "typical" eukaryotic cells in the remainder of the book. Fibroblasts are associated with animal connective tissue and are notable for secreting the macromolecules of the extracellular matrix.

Cells in multicellular organisms can be even more exotic. For example, nerve cells (Figure 2.16F) and rod cells (Figure 2.16G) reveal a great deal more complexity than the examples highlighted above. In these cases, the cell shape is intimately related to its function. In the case of nerve cells, their sinewy appearance is tied to the fact that the various branches (also called "processes") known as dendrites and axons convey electrical signals that permit communication between distant parts of an animal's nervous system. Despite having nuclei with typical eukaryotic dimensions, the cells themselves can extend processes with characteristic lengths of up to tens of centimeters. The structural complexity of rod cells is tied to their primary function of light detection in the retina of the eye. These cells are highly specialized to perform transduction of light energy into chemical energy that can be used to communicate with other cells in the body and, in particular, with brain cells that permit us to be conscious of perceiving images. Rod cells accomplish this task using large stacks of membranes which are the antennas participating in light detection. Figure 2.16 only scratches the surface of the range of cellular size and shape, but at least conveys an impression of cell sizes relative to our standard ruler.

2.2.2 The Cellular Interior: Organelles

As we descend from the scale of the cell itself, a host of new structures known as organelles come into view. The presence of these membrane-bound organelles is one of the defining characteristics that distinguishes eukaryotes from bacteria and archaea. Figure 2.23 shows a schematic of a eukaryotic cell and an associated electron microscopy image revealing some of the key organelles. These organelles serve as the specialized apparatus of cell function in capacities ranging from genome management (the nucleus) to energy generation (mitochondria and chloroplasts) to protein synthesis and modification (endoplasmic reticulum and Golgi apparatus) and beyond. The compartments that are bounded by organellar membranes can have completely different protein and ion compositions. In addition, the membranes of each of these different membrane systems are characterized by distinct lipid and protein compositions.

A characteristic feature of many organelles is that they are compartmentalized structures that are separated from the rest of the cell by membranes. The nucleus is one of the most familiar examples since it is often easily visible using standard light microscopy. If we use

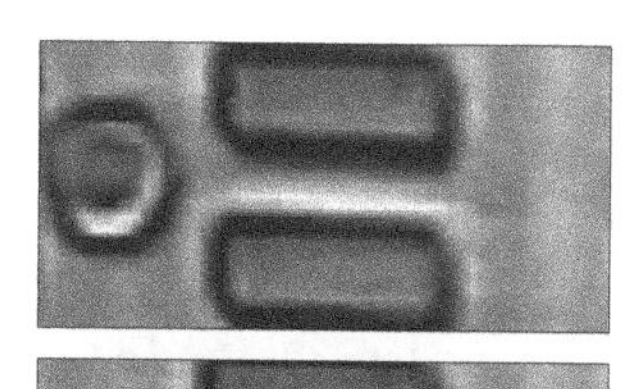

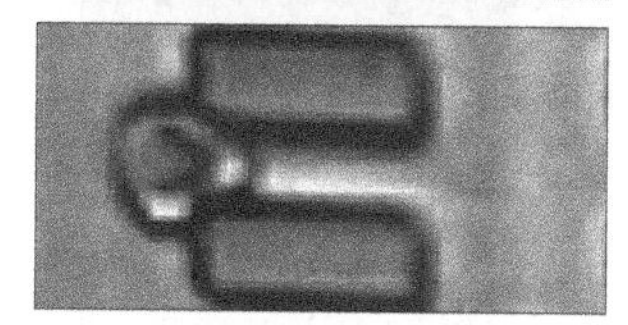
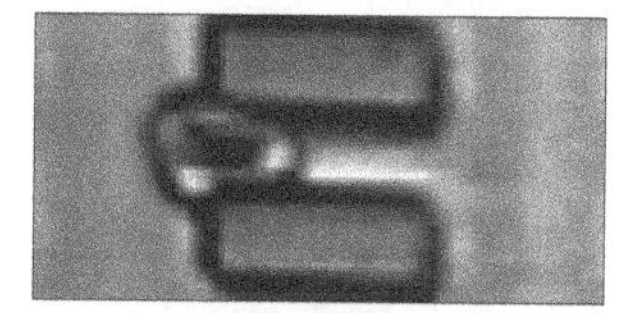
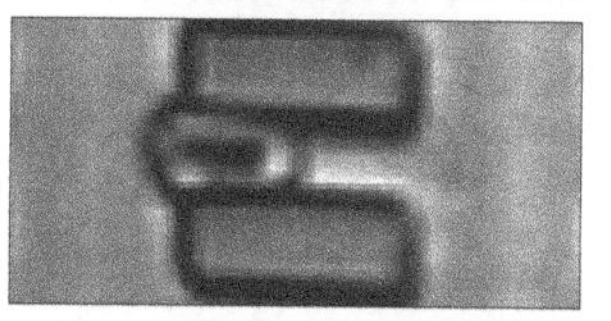
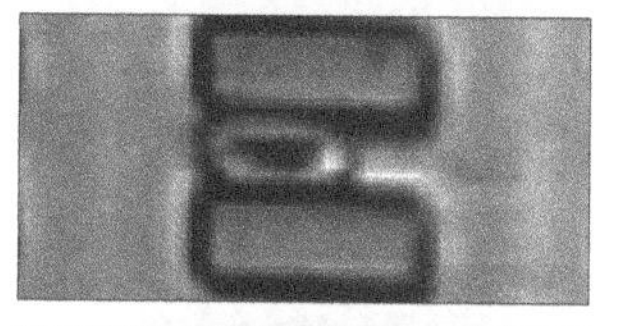
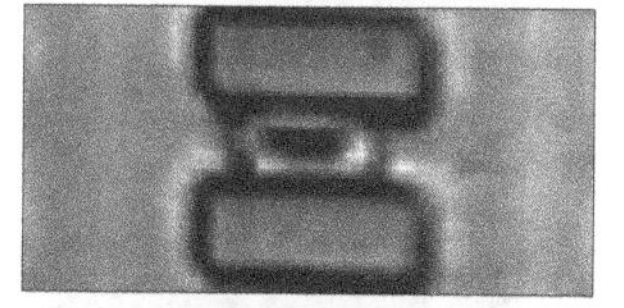
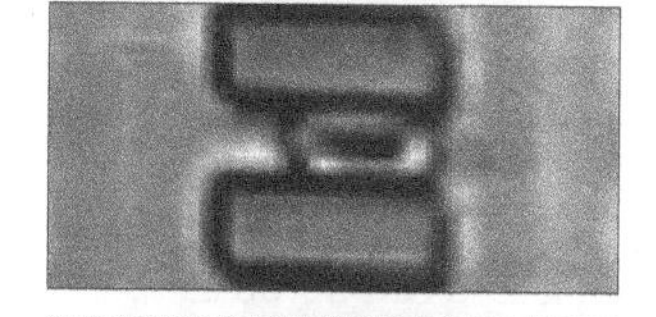
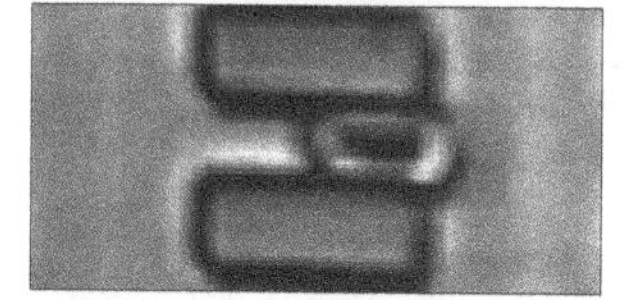
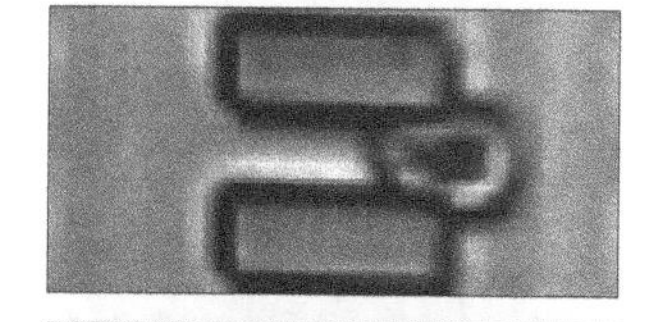
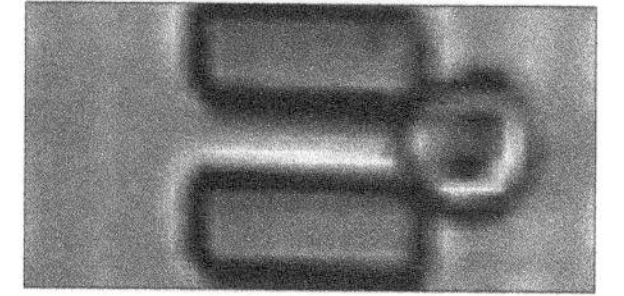

10 μm

Figure 2.21: Deformability of red blood cells. To measure the deformability of human red blood cells, an array of blocks was fabricated in silicon, each block was 4 μm × 4 μm × 12 μm. The blocks were spaced by 4 μm in one direction and 13 μm in the other. A glass coverslip covered the top of this array of blocks. A dilute suspension of red blood cells in a saline buffer was introduced to the system. A slight pressure applied at one end of the array of blocks provided bulk liquid flow, from left to right in the figure. This liquid flow carried the red blood cells through the narrow passages. Video microscopy captured the results. The figure shows consecutive video fields with the total elapsed time just over one-third of a second. (Courtesy of J. Brody.)

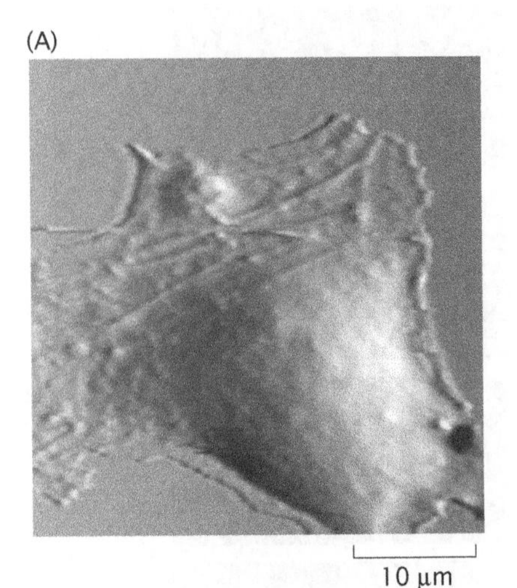

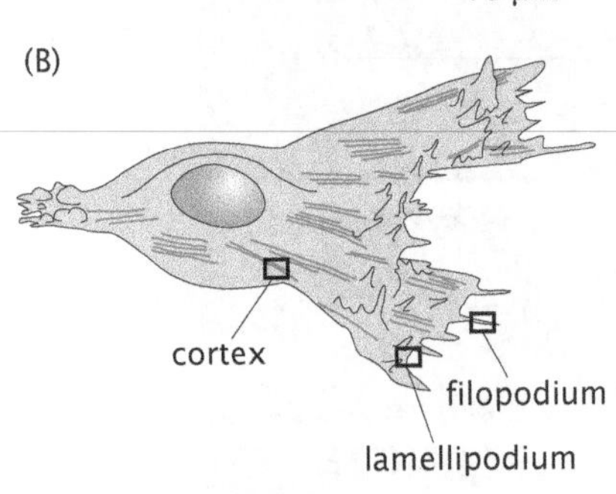

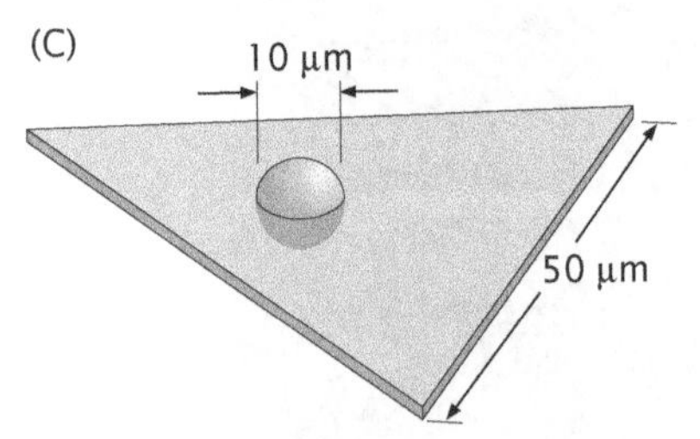

Figure 2.22: Structure of a fibroblast. (A) Atomic-force microscopy image of a fibroblast, (B) cartoon of the external morphology of a fibroblast, and (C) characteristic dimensions of a "typical" fibroblast. (A, adapted from M. Radmacher, *Meth. Cell Biol.* 83:347, 2007.)

the fibroblast as an example, then the cell itself has dimensions of roughly 50 μm, while the nucleus has a characteristic linear dimension of roughly 10 μm, as shown schematically in Figure 2.22. From a functional perspective, the nucleus is much more complex than simply serving as a storehouse for the genetic material. Chromosomes are organized within the nucleus, forming specific domains as will be discussed in more detail in Chapter 8. Transcription as well as several kinds of RNA processing occur in the nucleus. There is a flux of molecules such as transcription factors moving in and completed RNA molecules moving out through elaborate gateways in the nuclear membrane known as nuclear pores. Portions of the genome involved in synthesis of ribosomal RNA are clustered together, forming striking spots that can be seen in the light microscope and are called nucleoli.

Moving outward from the nucleus, the next membranous organelle we encounter is often the endoplasmic reticulum. Indeed, the membrane of the endoplasmic reticulum is contiguous with the membrane of the nuclear envelope. In some cells such as the pancreatic cell shown in Figure 2.24, the endoplasmic reticulum takes up the bulk of the cell interior. This elaborate organelle is the site of lipid synthesis and also the site of synthesis of proteins that are destined to be secreted or incorporated into membranes. It is clear from images such as those in Figures 2.24 and 2.25 that the endoplasmic reticulum can assume different geometries in different cell types and under different conditions. How much total membrane area is taken up by the endoplasmic reticulum? How strongly does the specific membrane morphology affect the total size of the organelle?

Estimate: Membrane Area of the Endoplasmic Reticulum

One of the most compelling structural features of the endoplasmic reticulum is its enormous surface area. To estimate the area associated with the endoplasmic reticulum, we take our cue from Figure 2.24, which suggests that we think of the endoplasmic reticulum as a series of concentric spheres centered about the nucleus. We follow Fawcett (1966), who characterizes the endoplasmic reticulum as forming "lamellar systems of flat cavities, rather uniformly spaced and parallel to one another" as shown in Figure 2.24.

An estimate can be made by adding up the areas from each of the concentric spheres making up our model endoplasmic reticulum. This can be done by simply noticing that the volume enclosed by the endoplasmic reticulum can be written as

$$V_{\mathrm{ER}} = \sum_i A_i d, \quad (2.17)$$

where A_i is the area of the ith concentric sphere and d is the distance between adjacent cisternae. Since two membranes bound each cisterna, the total area of the endoplasmic reticulum membrane is $A_{\mathrm{ER}} = 2 \times \sum_i A_i$. In our model, the total volume of the endoplasmic reticulum can be written as the difference between the volume taken up by the outermost sphere and the volume of the innermost concentric sphere (which is the same as the volume of the nucleus). This results in

$$V_{\mathrm{ER}} = \frac{4\pi}{3} R_{\mathrm{out}}^3 - \frac{4\pi}{3} R_{\mathrm{nucleus}}^3. \quad (2.18)$$

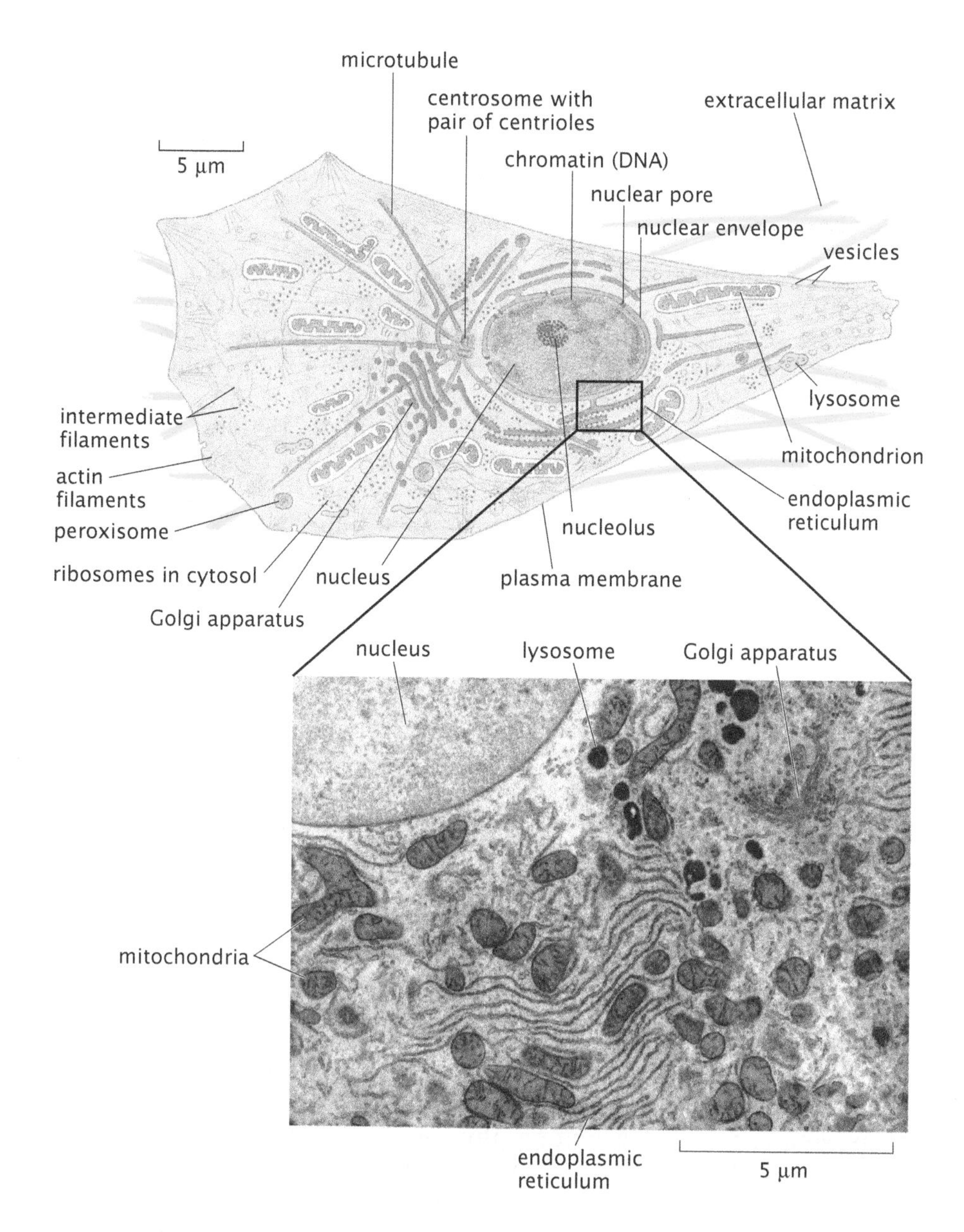

Figure 2.23: Eukaryotic cell and its organelles. The schematic shows a eukaryotic cell and a variety of membrane-bound organelles. A thin-section electron microscopy image shows a portion of a rat liver cell approximately equivalent to the boxed area on the schematic. A portion of the nucleus can be seen in the upper left-hand corner. The most prominent organelles visible in the image are mitochondria, lysosomes, the rough endoplasmic reticulum, and the Golgi apparatus. (Eukaryotic cell from Alberts et al., Molecular Biology of the Cell, 5th ed., New York, Garland Science, 2008; electron micrograph from D. W. Fawcett, The Cell, Its Organelles and Inclusions: An Atlas of Fine Structure. W. B. Saunders, 1966.)

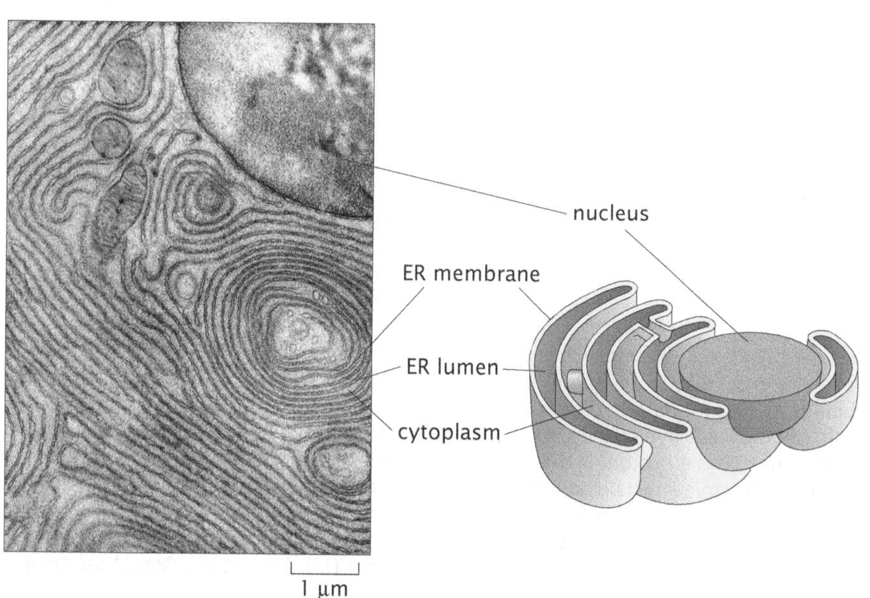

Figure 2.24: Electron micrograph and associated schematic of the endoplasmic reticulum (ER). The left-hand panel shows a thin-section electron micrograph of an acinar cell from the pancreas of a bat. The nucleus is visible at the upper right and the dense and elaborate endoplasmic reticulum structure is strikingly evident. The right-hand panel shows a schematic diagram of a model for the three-dimensional structure of the endoplasmic reticulum in this cell. Notice that the size of the lumen in the endoplasmic reticulum in the schematic is exaggerated for ease of interpretation. (Electron micrograph from D. W. Fawcett, The Cell, Its Organelles and Inclusions: An Atlas of Fine Structure. W. B. Saunders, 1966.)

Combining the two ways of computing the volume of the endoplasmic reticulum, Equations 2.17 and 2.18, we arrive at an expression for the endoplasmic reticulum area,

$$A_{\mathrm{ER}} = \frac{8\pi}{3d}\left(R_{\mathrm{out}}^3 - R_{\mathrm{nucleus}}^3\right). \tag{2.19}$$

Using the values $R_{\mathrm{nucleus}} = 5\,\mu\mathrm{m}$, $R_{\mathrm{out}} = 10\,\mu\mathrm{m}$, and $d = 0.05\,\mu\mathrm{m}$, we get, at an estimate, $A_{\mathrm{ER}} = 15 \times 10^4\,\mu\mathrm{m}^2$. This result should be contrasted with a crude estimate for the outer surface area of a fibroblast, which can be obtained by using the dimensions in Figure 2.22(C) and which yields an area of $10^4\,\mu\mathrm{m}^2$ for the cell membrane itself. To estimate the area of the endoplasmic reticulum when it is in reticular form, we describe its structure as interpenetrating cylinders of diameter $d \approx 10\,\mathrm{nm}$ separated by a distance $a \approx 60\,\mathrm{nm}$, as shown in Figure 2.25. The completion of the estimate is left to the problems, but results in a comparable membrane area to the calculation here assuming concentric spheres.

The other major organelles found in most cells and visible in Figure 2.23 include the Golgi apparatus, mitochondria, and lysosomes. The Golgi apparatus, similar to the endoplasmic reticulum, is largely involved in processing and trafficking of membrane-bound and secreted proteins (for another view, see Figure 11.38 on p. 463). The Golgi apparatus is typically seen as a pancake-like stack of flattened compartments, each of which contains a distinct set of enzymes. As proteins are processed for secretion, for example by addition and remodeling of attached sugars, they appear to pass in an orderly fashion through each element in the Golgi stack. The mitochondria are particularly striking organelles with a smooth outer surface housing an elaborately folded system of internal membrane structures (for a more detailed view, see Figure 11.39 on p. 464). The mitochondria are the primary site of ATP synthesis for cells growing in the presence of oxygen, and their fascinating physiology and structure have been well studied. We will return to the topic of mitochondrial structure in Chapter 11 and discuss the workings of the tiny machine responsible for ATP synthesis in Chapter 16. Lysosomes serve a major role in the degradation of cellular components. In some specialized cells such as macrophages, lysosomes also serve as the compartment where bacterial invaders can be degraded. These membrane-bound organelles are filled with acids, proteases, and other degradative enzymes. Their shapes are polymorphous; resting lysosomes are simple and nearly spherical, whereas lysosomes actively involved in degradation of cellular components or of objects taken in from the outside may be much larger and complicated in shape.

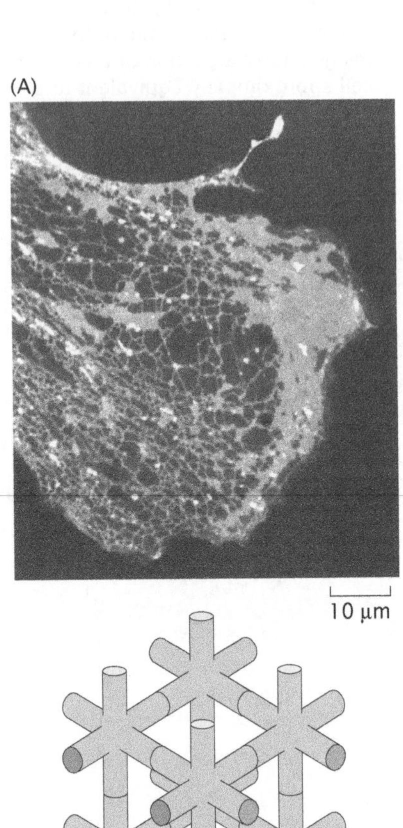

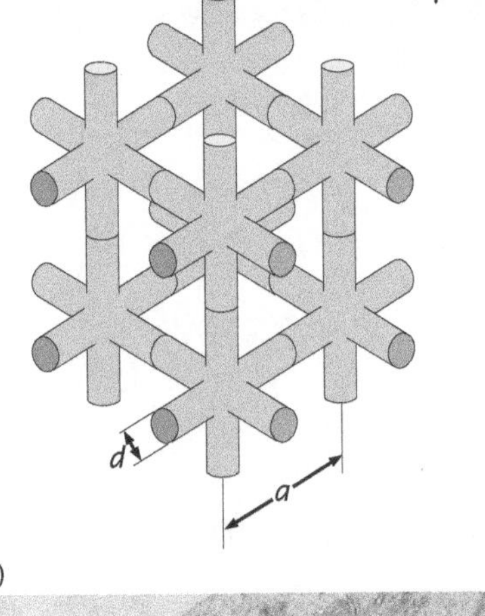

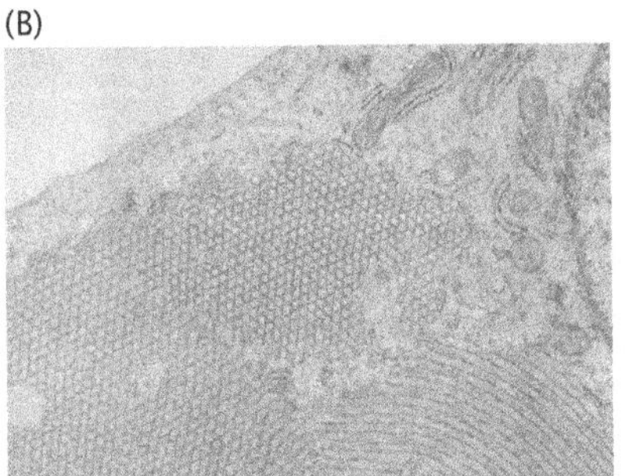

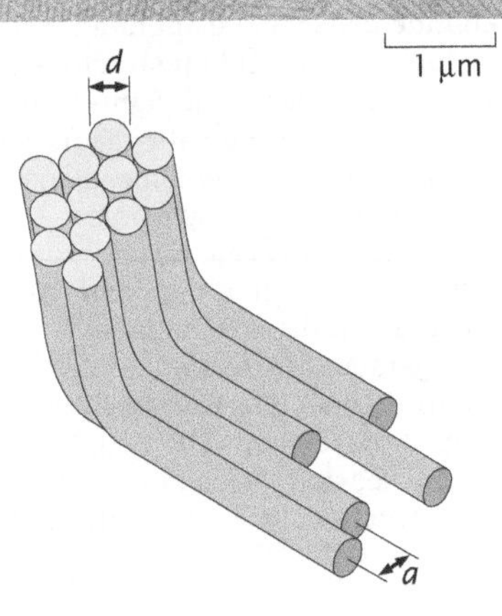

Figure 2.25: Variable morphology of the endoplasmic reticulum. (A) In most cultured cells, the endoplasmic reticulum is a combination of a web-like reticular network of tubules and larger flattened cisternae. In this image, a cultured fibroblast was stained with a fluorescent dye called DiOC6 that specifically labels endoplasmic reticulum membrane. On the bottom is a schematic of an idealized three-dimensional reticular network. (B) Some specialized cells and those treated with drugs that regulate the synthesis of lipids reorganize their endoplasmic reticulum to form tightly-packed, nearly crystalline arrays that resemble piles of pipes. (A, adapted from M. Terasaki et al., *J. Cell. Biol.* 103:1557, 1986; B, adapted from D. J. Chin et al., *Proc. Natl Acad. Sci. USA* 79:1185, 1982.)

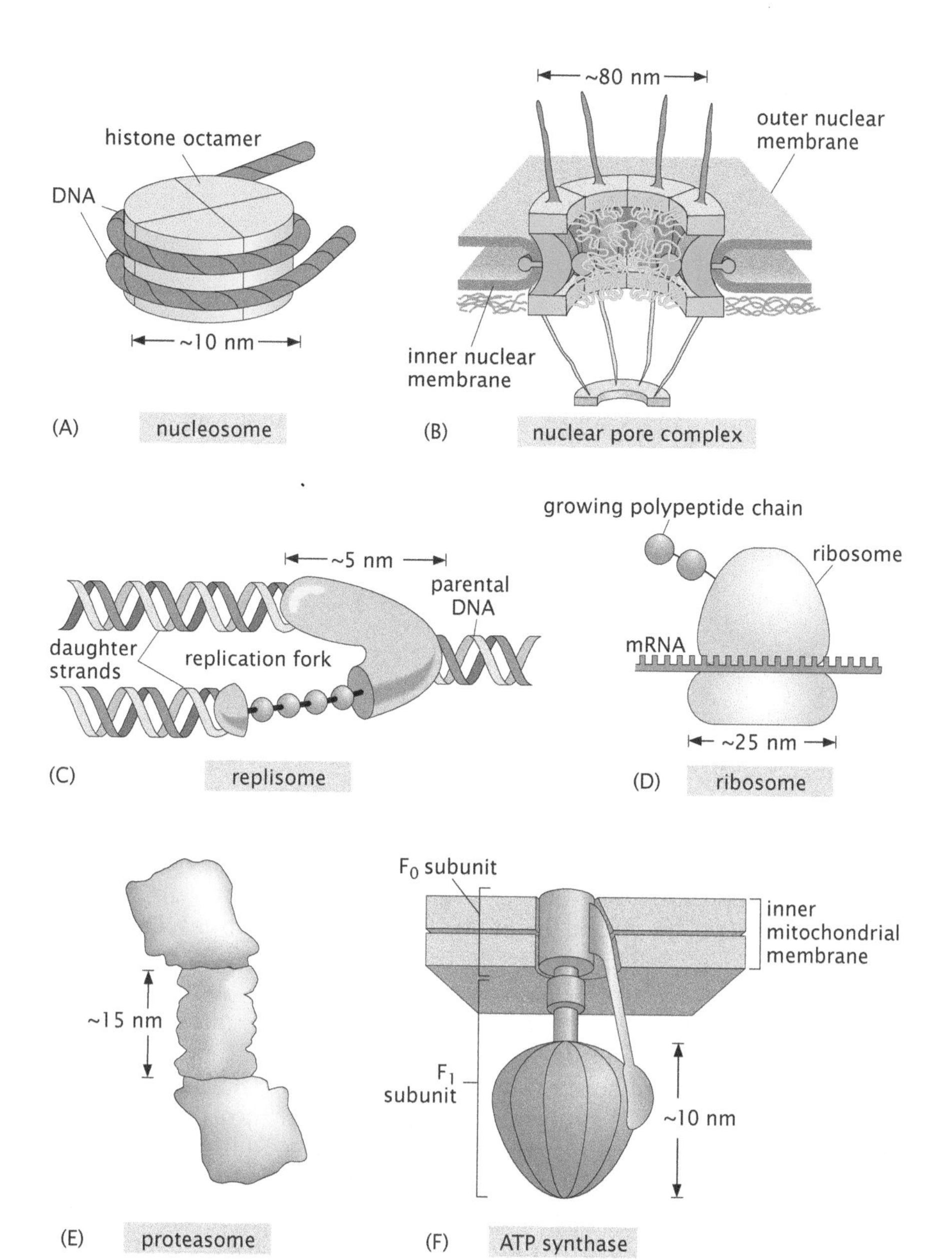

Figure 2.26: The macromolecular assemblies of the cell. (A) The nucleosome is a complex of eight protein units and DNA. (B) The nuclear pore complex mediates the transport of material in and out of the nucleus. (C) The replisome is a complex of proteins that mediate the copying of the genetic material. (D) The ribosome reads mRNA and synthesizes the corresponding polypeptide chain. (E) The proteasome degrades proteins. (F) ATP synthase is a large complex that synthesizes new ATP molecules.

These common organelles are only a few of those that can be found in eukaryotic cells. Some specialized cells have remarkable and highly specialized organelles that can be found nowhere else, such as the stacks of photoreceptive membranes found in the rod cells of the visual system as indicated schematically in Figure 2.16(G) (p. 53). The common theme is that all organelles are specialized subcompartments of the cell that perform a particular subset of cellular tasks and represent a smaller, discrete layer of organization one step down from the whole cell.

2.2.3 Macromolecular Assemblies: The Whole is Greater than the Sum of the Parts

Macromolecules Come Together to Form Assemblies

Proteins, nucleic acids, sugars, and lipids often work as a team. Indeed, as will become clear throughout the remainder of this book,

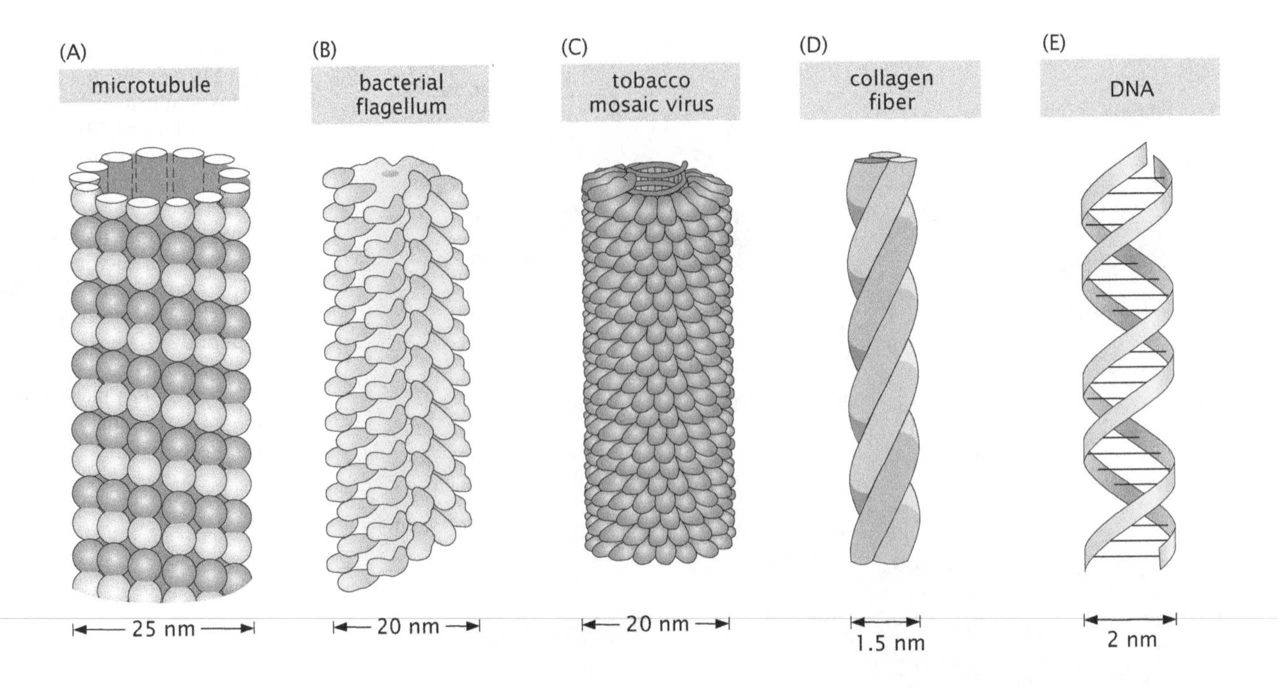

Figure 2.27: Helical motifs of molecular assemblies. Macromolecular assemblies have a variety of different helical structures, some formed from individual monomeric units (such as (A)–(C)), others resulting from coils of proteins (D), and yet others made up of paired nucleotides (E).

these macromolecules often come together to make assemblies. We think of yet another factor-of-10 magnification relative to the previous section, and with this increase of magnification we see assemblies such as those shown in cartoon form in Figure 2.26. The genetic material in the eukaryotic nucleus is organized into chromatin fibers, which themselves are built up of protein–DNA assemblies known as nucleosomes (Figure 2.26A). Traffic of molecules such as completed RNA molecules out of the nucleus is mediated by the elaborate nuclear pore complex (Figure 2.26B). The replication complex that copies DNA before cell division is similarly a collection of molecules, which has been dubbed the replisome (Figure 2.26C). When the genetic message is exported to the cytoplasm for translation into proteins, the ribosome (an assembly of proteins and nucleic acids) serves as the universal translating machine that converts the nucleic acid message from the RNA into the protein product written in the amino acid alphabet (Figure 2.26D). When proteins have been targeted for degradation, they are sent to another macromolecular assembly known as the proteasome (Figure 2.26E). The production of ATP in mitochondria is similarly mediated by a macromolecular complex known as ATP synthase (Figure 2.26F) . The key idea of this subsection is to show that there is a very important level of structure in cells that is built around complexes of individual macromolecules and with a characteristic length scale of 10 nm.

Helical Motifs Are Seen Repeatedly in Molecular Assemblies

A second class of macromolecular assemblies, characterized not by function but rather by structure, is the wide variety of helical macromolecular complexes. Several representative examples are depicted in Figure 2.27. In Figure 2.27(A), we show the geometric structure of microtubules. As will be described in more detail later, these structures are built up of individual protein units called tubulin. A second example shown in Figure 2.27(B) is the bacterial flagellum of *E. coli.* Here too, the same basic structural idea is repeated, with the helical geometry built up from individual protein units, in this case

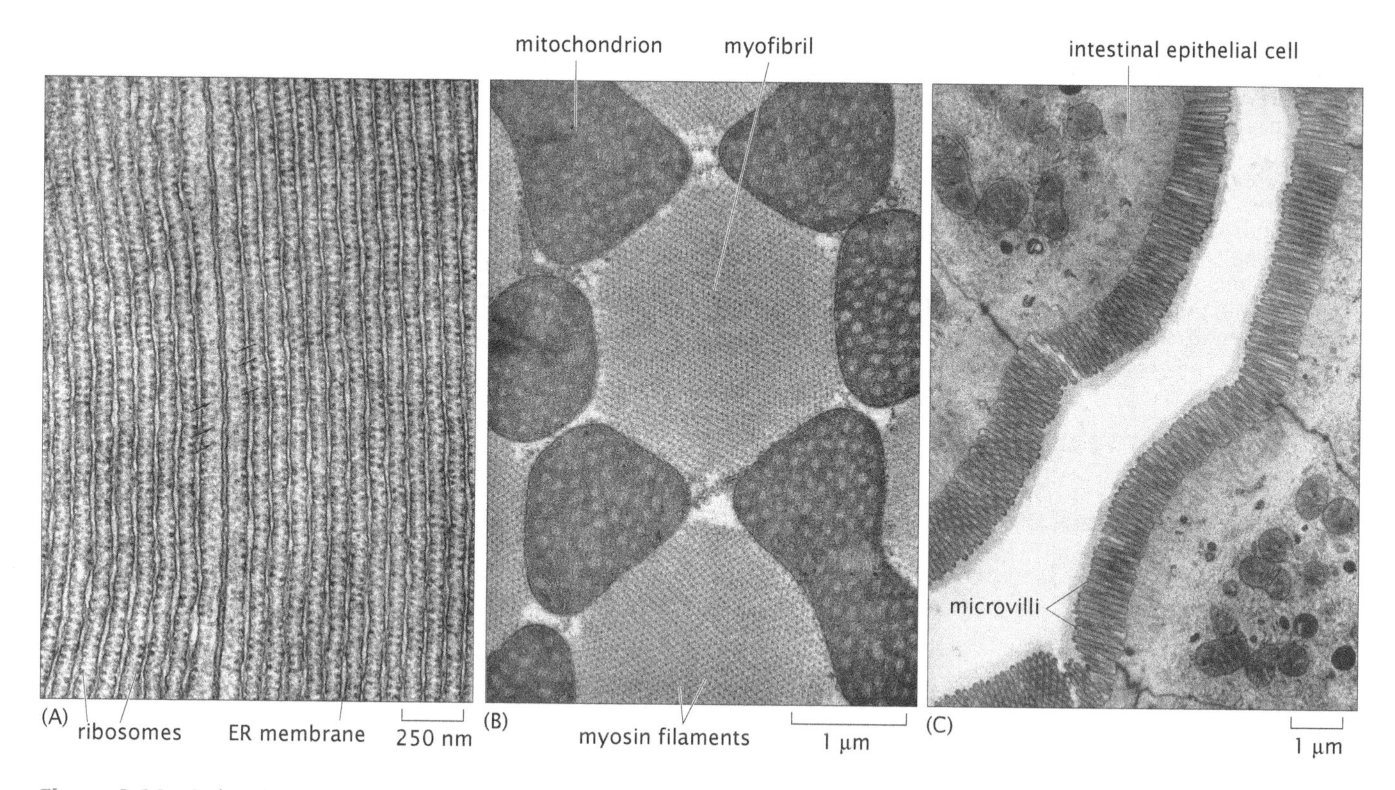

Figure 2.28: Ordered macromolecular assemblies. Collage of examples of macromolecules organized into superstructures. (A) Ribosomes on the endoplasmic reticulum ("rough ER"), (B) myofibrils in the flight muscle, and (C) microvilli at the epithelial surface. (A–C, adapted from D. W. Fawcett, The Cell, Its Organelles and inclusions: An Atlas of Fine Structure. W. B. Saunders, 1966.)

flagellin. The third example given in the figure is that of a filamentous virus, with tobacco mosaic virus (TMV) chosen as one of the most well-studied viruses.

The helical assemblies described above are characterized by individual protein units that come together to form helical filaments. An alternative and equally remarkable class of filaments are those in which α-helices (chains of amino acids forming protein subunits with a precise, helical geometry) wind around each other to form superhelices. The particular case study that will be of most interest in subsequent discussions is that of collagen, which serves as one of the key components in the extracellular matrix of connective tissues and is one of the major protein products of the fibroblast cells introduced earlier in the chapter (see Figure 2.16 on p. 53). The final example is the iconic figure of the DNA molecule itself.

Macromolecular Assemblies Are Arranged in Superstructures

Assemblies of macromolecules can interact with each other to create striking instances of cellular hardware with a size comparable to organelles themselves. Figure 2.28 depicts several examples. Figure 2.28(A) shows the way in which ribosomes are organized on the endoplasmic reticulum with a characteristic spacing that is comparable to the size of the ribosomes ($\approx$20 nm). A second stunning example is the organization of myofibrils in muscles as shown in Figure 2.28(B). This figure shows the juxtaposition of the myofibrils and mitochondria. The myofibrils themselves are an ordered arrangement of actin filaments and myosin motors, as will be discussed in more detail in

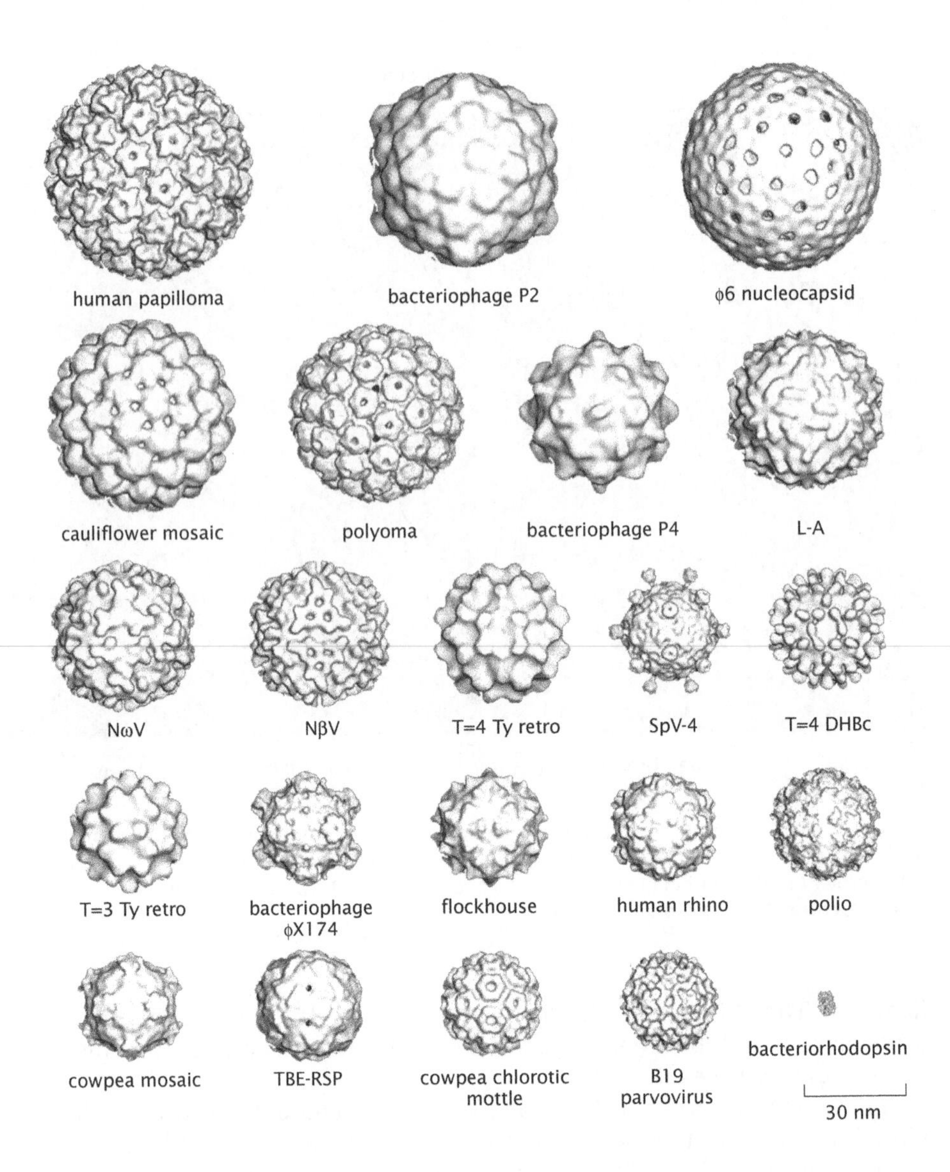

Figure 2.29: Structures of viral capsids. The regularity of the structure of viruses has enabled detailed, atomic-level analysis of their construction patterns. This gallery shows a variety of the different geometries explored by the class of nearly spherical viruses. For size comparison, a large protein, bacteriorhodopsin, is shown in the bottom right. (Adapted from T. S. Baker et al., *Microbiol. Mol. Biol. Rev.* 63:862, 1999.)

Chapter 16. The last example, shown in Figure 2.28(C), is of the protrusions of microvilli at the surface of an epithelial cell. These microvilli are the result of collections of parallel actin filaments. The list of examples of orchestration of collections of macromolecules could go on and on and should serve as a reminder of the many different levels of structural organization found in cells.

2.2.4 Viruses as Assemblies

Viruses are one of the most impressive and beautiful examples of macromolecular assembly. These assemblies are a collection of proteins and nucleic acids (though many viruses have lipid envelopes as well) that form highly ordered and symmetrical objects with characteristic sizes of tens to hundreds of nanometers. The architecture of viruses is usually a protein shell where the so-called capsid is made up of a repetitive packing of the same protein subunits over and over to form an icosahedron. Within the capsid, the virus packs its genetic material, which can be either single-stranded or double-stranded DNA or RNA depending upon the type of virus. Figure 2.29 is a gallery of the capsids of a number of different viruses. Different viruses have different elaborations on this basic structure and can include lipid coats, surface receptors, and internal molecular machines such as polymerases and proteases. One of the most amazing features of viruses is that by hijacking the machinery of the host cell, the viral genome commands the construction of its own inventory of parts within the host cell and then, in the crowded environment of that host, assembles into infectious agents prepared to repeat the life cycle elsewhere.

Human immunodeficiency virus (HIV) is one of the viruses that has garnered the most attention in recent years. Figure 2.30 shows cryo-electron microscopy images of mature HIV virions and gives a sense of both their overall size (roughly 130 nm) and their internal structure. In particular, note the presence of an internal capsid shaped like an ice-cream cone. This internal structure houses the roughly 10 kilobase (kb) RNA viral genome. As with our analysis of the inventory of a cell considered earlier in the chapter, part of developing a "feeling for the organism" is to get a sense of the types and numbers of the different molecules that make up that organism. In the case of HIV, these numbers are interesting for many reasons, including that they say something about the "investment" that the infected cell has to make in order to construct new virions.

For our census of an HIV virion, we need to examine the assembly of the virus. In particular, one of the key products of its roughly 10 kb genome is a polyprotein, a single polypeptide chain containing what will ultimately be distinct proteins making up the capsid, matrix, and nucleocapsid, known as Gag and shown schematically in Figure 2.31. The formation of the *immature* virus occurs through the association of

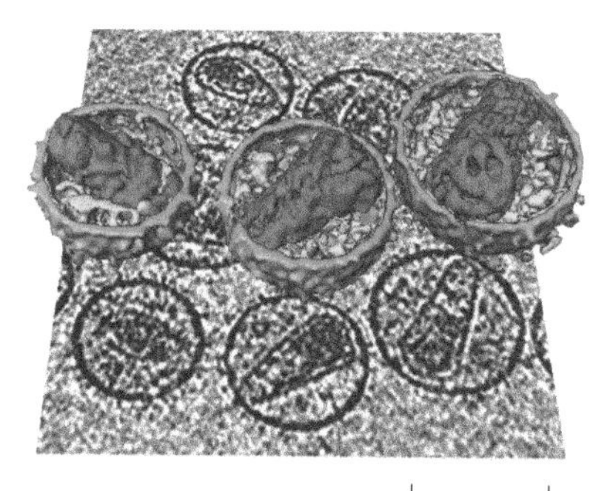

Figure 2.30: Structure of HIV viruses. The planar image shows a single frame from an electron microscopy tilt series. The three-dimensional images show reconstructions of the mature viruses featuring the ice-cream-cone-shaped capsid on the interior. (Adapted from J. A. G. Briggs et al., *Structure* 14:15, 2006.)

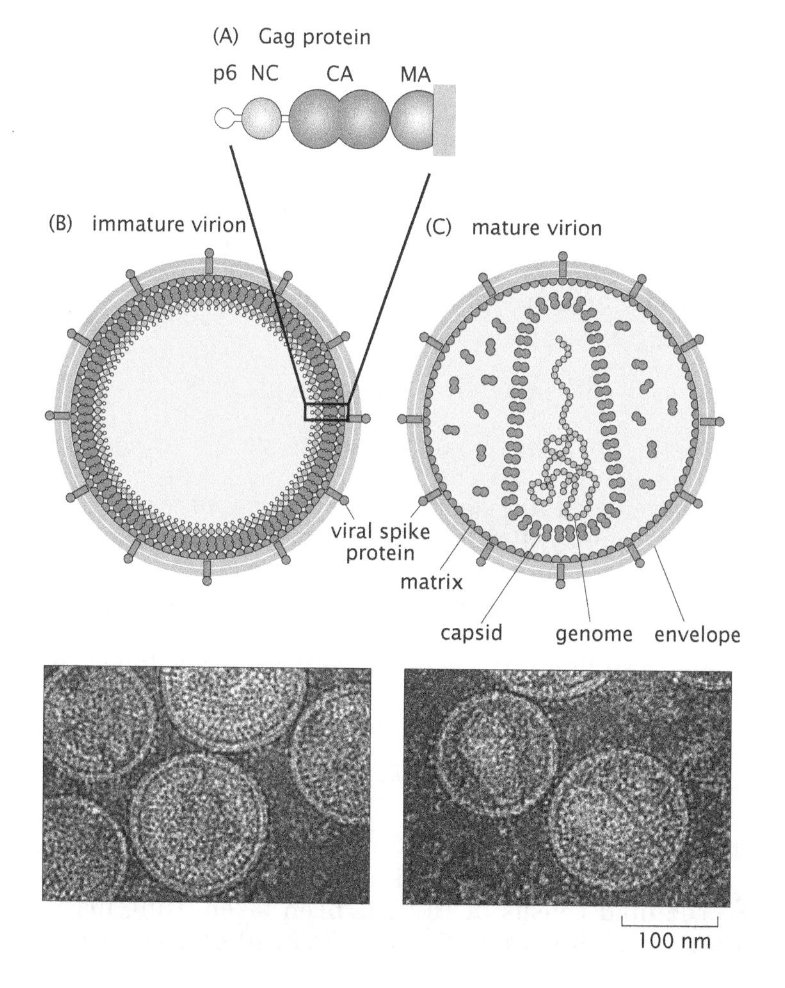

Figure 2.31: HIV architecture. (A) Schematic of the Gag polyprotein, a 41,000 Da architectural building block. (B) Immature virions showing the lipid bilayer coat and the uncut Gag shell on the interior. (C) Mature virions in which the Gag protein has been cut by proteases and the separate components have assumed their architectural roles in the virus. The associated electron microscopy images show actual data for each of the cartoons. (Adapted from J. A. G. Briggs et al., *Nat. Struct. Mol. Biol.* 11:672, 2004.)

the N-terminal ends of these Gag proteins with the lipid bilayer of the host cell, with the C-termini pointing radially inward like the spokes of a three-dimensional wheel. (N-terminus and C-terminus refer to the two structurally distinct ends of the polypeptide chain. During protein synthesis, translation starts at the N-terminus and finishes at the C-terminus.) As more of these proteins associate on the cell surface, the nascent virus begins to form a bud on the cell surface, ultimately resulting in spherical structures like those shown in Figure 2.31. During the process of viral maturation, a viral protease (an enzyme that cuts proteins) clips the Gag protein into its component pieces known as matrix (MA), capsid (CA), nucleocapsid (NC), and p6. The matrix forms a shell of proteins just inside of the lipid bilayer coat. The capsid proteins form the ice-cream-cone-shaped object that houses the genetic material and the nucleocapsid protein is complexed with the viral RNA.

ESTIMATE

Estimate: Sizing Up HIV Unlike many of their more ordered viral counterparts, HIV virions have the intriguing feature that the structure from one to the next is not exactly the same. Indeed, they come in both different shapes and different sizes. As a result, our attempt to "size up" HIV will be built around some representative numbers for these viruses, but the reader is cautioned to think of a statistical distribution of sizes and shapes. As shown in the cryo-electron microscopy picture of Figure 2.30, the diameter of the virion is between 120 nm and 150 nm and we take a "canonical" diameter of 130 nm.

We begin with the immature virion. To find the number of Gag proteins within a given virion, we resort to simple geometrical reasoning. Since the radius of the overall virion is roughly 65 nm, and the outer 5 nm of that radius is associated with the lipid bilayer, we imagine a sphere of radius 60 nm that is decorated on the inside with the inward facing spokes of the Gag proteins. If we think of each such Gag protein as a cylinder of radius 2 nm, this means they take up an area $A_{\text{Gag}} \approx 4\pi \text{nm}^2$. Using this, we can find the number of such Gag proteins as

$$N_{\text{Gag}} = \frac{\text{surface area of virion}}{\text{area per Gag protein}} \approx \frac{4\pi(60\,\text{nm})^2}{4\pi\,\text{nm}^2} \approx 3500. \tag{2.20}$$

The total mass of these Gag proteins is roughly

$$M_{\text{Gag}} \approx 3500 \times 40{,}000\ \text{Da} \approx 150\ \text{MDa}, \tag{2.21}$$

where we have used the fact that the mass of each Gag polyprotein is roughly 40 kDa. This estimate for the number of Gag proteins is of precisely the same magnitude as those that have emerged from cryo-electron microscopy observations (see Briggs et al., 2004).

The number of lipids associated with the HIV envelope can be estimated similarly as

$$N_{\text{lipids}} \approx \frac{2 \times 4\pi(65\,\text{nm})^2}{1/2\,\text{nm}^2} \approx 200{,}000\ \text{lipids}, \tag{2.22}$$

where the factor of 2 accounts for the fact that the lipids form a bilayer, and we have used a typical area per lipid of $0.5\,\text{nm}^2$. The lipid census of HIV has been taken using mass spectrometry, which permits the measurement of each of the different types of lipids forming the viral envelope (see Brügger et al., 2006). Interestingly, the diversity of lipids in the HIV envelope is enormous, with the lipid composition of the viral envelope distinct from that of the host cell membrane. The measured total number of different lipids is roughly 300,000. Further analysis of the parts list of HIV is left to the problems at the end of the chapter.

Ultimately, viruses are one of the most interesting cases of the power of macromolecular assembly. These intriguing machines occupy a fuzzy zone at the interface between the living and the nonliving.

2.2.5 The Molecular Architecture of Cells: From Protein Data Bank (PDB) Files to Ribbon Diagrams

If we continue with another factor of 10 in our powers-of-10 descent, we find the individual macromolecules of the cell. In particular, this increase in spatial resolution reveals four broad categories of macromolecules: lipids, carbohydrates, nucleic acids, and proteins. As was shown in Chapter 1 (p. 3), these four classes of molecules make up the stuff of life and have central status in building cells both architecturally and functionally. Though often these molecules are highly anisotropic (for example, a DNA molecule is usually many orders of magnitude longer than it is wide), their characteristic scale is between 1 and 10 nm. For example, as shown earlier in the chapter, a "typical" protein has a size of several nanometers. Lipids are more anisotropic, with lengths of 2–3 nm and cross-sectional areas of roughly 0.5 nm^2.

The goal of this section is to provide several different views of the molecules of life and how they fit into the structural hierarchy described throughout the chapter.

Macromolecular Structure Is Characterized Fundamentally by Atomic Coordinates

The conjunction of X-ray crystallography, nuclear magnetic resonance, and cryo-electron microscopy has revealed the atomic-level structures of a dazzling array of macromolecules of central importance to the function of cells. The list of such structures includes molecular motors, ion channels, DNA-binding proteins, viral capsid proteins, and various nucleic acid structures. The determination of new structures is literally a daily experience. Indeed, as will be asked of the reader in the problems at the end of the chapter, a visit to websites such as the Protein Data Bank or VIPER reveals just how many molecular and macromolecular structures are now known.

Though the word *structure* can mean different things to different people (indeed, that is one of the primary messages of this chapter and Chapter 8), most self-identified structural biologists would claim that the determination of structure ultimately refers to a list of atomic coordinates for the various atoms making up the structure of interest. As an example, Figure 1.3 (p. 6) introduced detailed atomic portraits of nucleic acids, proteins, lipids, and sugars. In such descriptions, the structural characterization of the system amounts to a set of coordinates

$$\mathbf{r}_i = x_i\mathbf{i} + y_i\mathbf{j} + z_i\mathbf{k}, \tag{2.23}$$

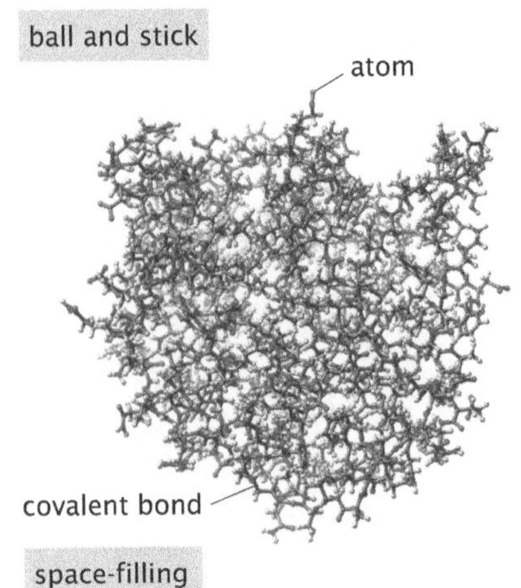

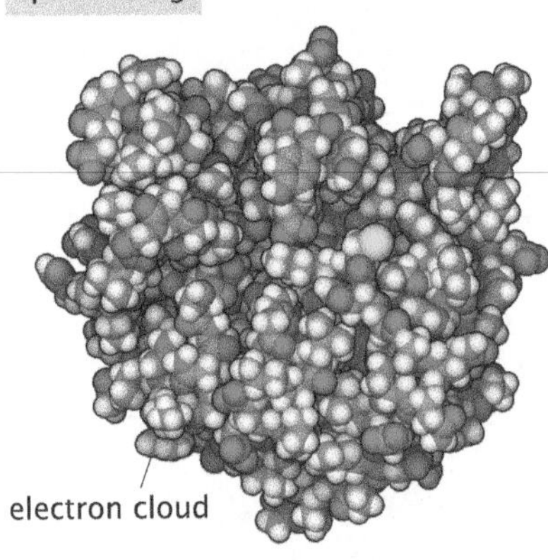

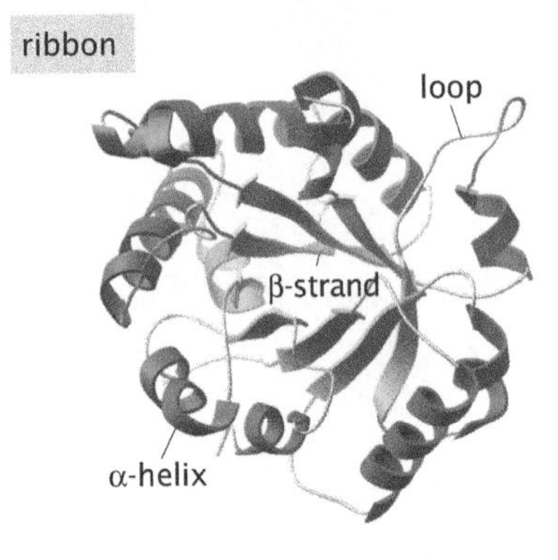

Figure 2.32: Three representations of triose phosphate isomerase. This enzyme is one of the enzymes in the glycolysis pathway (PDB 3tim). (Courtesy of D. Goodsell.)

where, having chosen some origin of coordinates, the coordinates of the ith atom in the structure are given by (x_i, y_i, z_i). That is, we have some origin of Cartesian coordinates and every atomic position is an address on this three-dimensional grid.

Because the macromolecules of the cell are subject to incessant jiggling due to collisions with each other and the surrounding water, a static picture of structure is incomplete. The structural snapshots embodied in atomic coordinates for a given structure miss the fact that each and every atom is engaged in a constant thermal dance. Hence, the coordinates of Equation 2.23 are really of the form

$$\mathbf{r}_i(t) = x_i(t)\mathbf{i} + y_i(t)\mathbf{j} + z_i(t)\mathbf{k}, \tag{2.24}$$

where the t reminds us that the coordinates depend upon time and what is measured in experiments might be best represented as $\langle \mathbf{r}_i(t) \rangle_{\text{time}}$, where the brackets $\langle\ \rangle_{\text{time}}$ signify an average over time.

An example of an atomic-level representation of one of the key proteins of the glycolysis pathway is shown in Figure 2.32. We choose this example because glycolysis will arise repeatedly throughout the book (see p. 191 for example) as a canonical metabolic pathway. The figure also shows several alternative schemes for capturing these structures such as using ribbon diagrams which highlight the ways in which the different amino acids come together to form elements of secondary structure such as α-helices and β-sheets.

Chemical Groups Allow Us to Classify Parts of the Structure of Macromolecules

When thinking about the structures of the macromolecules of the cell, one of the most important ways to give those structures functional meaning (as opposed to just a collection of coordinates) is through reference to the chemical groups that make them up. For example, the structure of the protein shown in Figures 1.3 and 2.32 is not just an arbitrary arrangement of carbons, nitrogens, oxygens, and hydrogens. Rather, this structure reflects the fact that the protein is made up of a linear sequence of amino acids that each have their own distinct identity as shown in Figure 2.33. The physical and chemical properties of these amino acids dictate the folded shape of the protein as well as how it functions.

Amino acids are but one example of a broader class of sub-nanometer-scale structural building blocks known as "chemical groups." These chemical groups occur with great frequency in different macromolecules and, like the amino acids, each has its own unique chemical identity. Figure 2.34 shows a variety of chemical groups that are of interest in biochemistry and molecular biology. These are all biologically important chemical functional groups that can be attached to a carbon atom as shown in Figure 2.34 and are all found in protein structures. The methyl and phenyl groups contain only carbon and hydrogen and are hence hydrophobic (unable to form hydrogen bonds with water). To the right of these are shown two chemically similar groups, alcohol and thiol, consisting of oxygen or sulfur plus a single hydrogen. The key feature of these two groups is that they are highly reactive and can participate in chemical reactions forming new covalent bonds. Amino acids containing these functional groups (serine, threonine, tyrosine, and cysteine) are frequently important active site residues in enzyme-catalyzed reactions. The next row starts with a nitrogen-containing amino group, which is

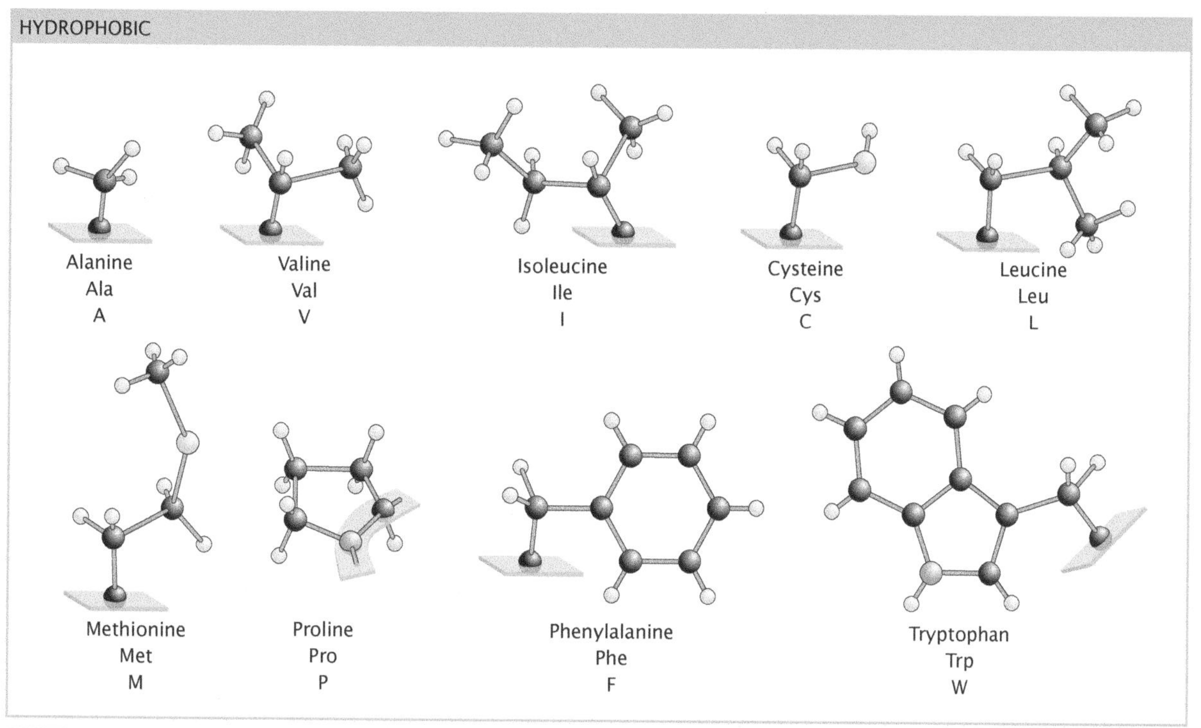

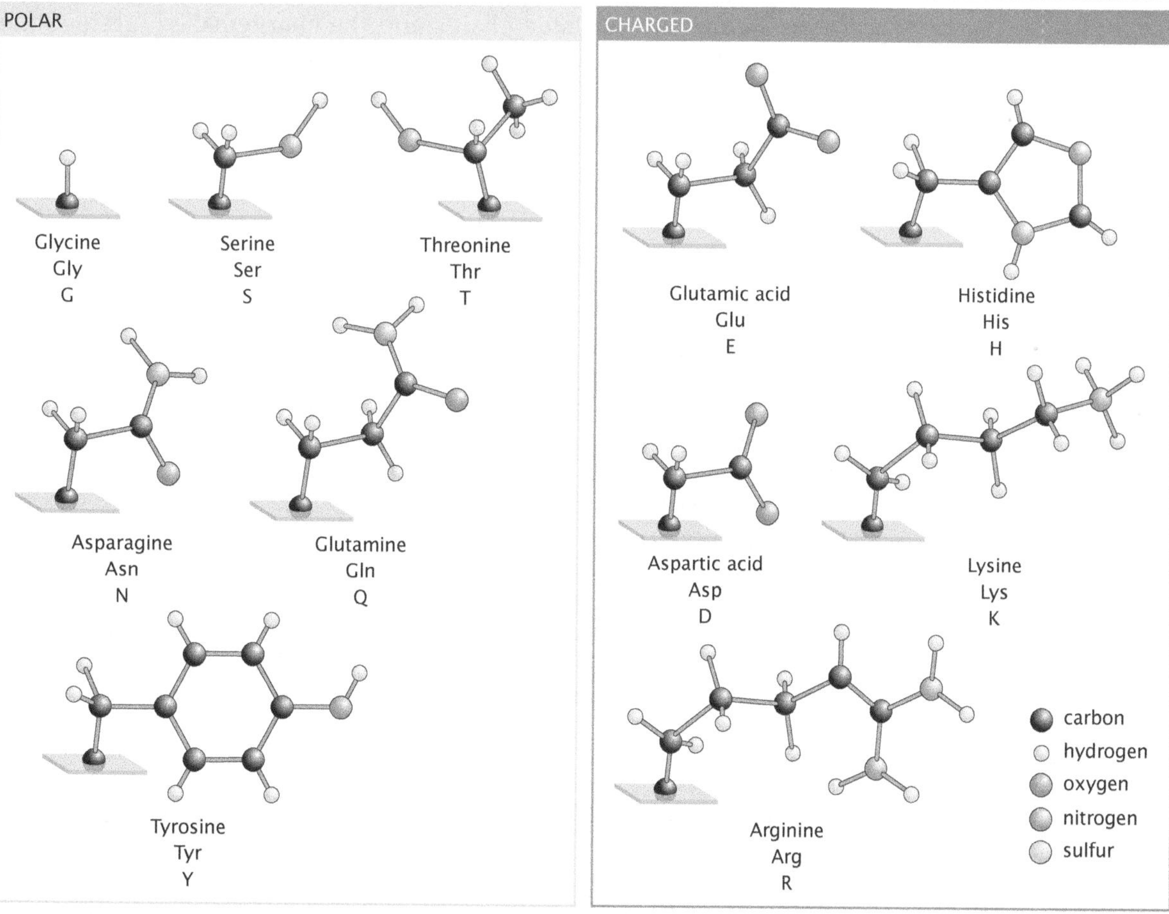

Figure 2.33: Amino acid side chains. The amino acids are represented here in ball-and-stick form, where a black ball indicates a carbon atom, a small white ball a hydrogen atom, a red ball an oxygen atom, a blue ball a nitrogen atom, and a yellow ball a sulfur atom. Only the side chains are shown. The peptide backbone of the protein to which these side chains are attached is indicated by an orange tile. The amino acids are subdivided based upon their physical properties. The group shown at the top are hydrophobic and tend to be found on the interior of proteins. Those at the bottom are able to form hydrogen bonds with water and are often found on protein surfaces.

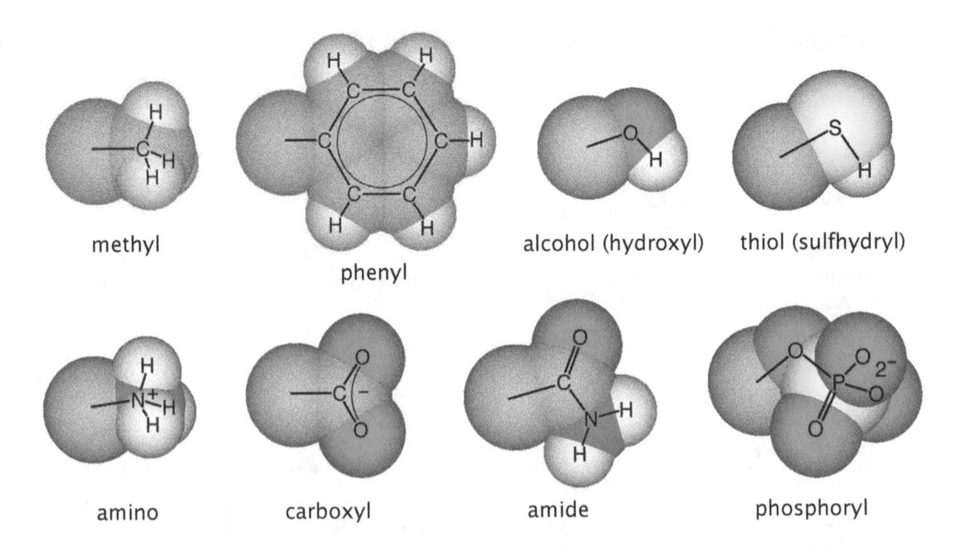

Figure 2.34: Chemical groups. These are some of the most common groups found in organic molecules such as proteins. (Courtesy of D. Goodsell.)

usually positively charged at neutral pH, and a carboxylic acid, which is usually negatively charged. All amino acids in monomeric form have both of these groups. In a protein polymer, there is a free amino group at the N-terminus of the protein and a free carboxylic acid group at the C-terminus of the protein. Several amino acids also contain these groups as part of their side chains and the charge-based interactions are frequently responsible for chemical specificity in molecular recognition as well as some kinds of catalysis. An amide group is shown next. This group is not generally charged but is able to participate in a variety of hydrogen bonds. The last group shown is a phosphoryl group, which is not part of any amino acid that is incorporated by the ribosome in a polypeptide chain during translation. However, phosphoryl groups are frequently added to proteins as a post-translational modification and perform extremely important regulatory functions.

Nucleic acids can similarly be thought of from the point of view of chemical groups. In Figure 1.3 (p. 9), we showed the way in which individual groups can be seen as the building blocks of DNA structures such as that shown in Figure 1.3(A). In particular, we note that the backbone of the double helix is built up of sugars (represented as pentagons) and phosphates. Similarly, the nitrogenous bases that mediate the pairing between the complementary strands of the backbone are represented diagrammatically via hexagons and pentagons, with hydrogen bonds depicted as shown in the figure.

This brief description of the individual molecular units of the machines of the cell brings us to the end of our powers-of-10 descent that examined the structures of the cell. Our plan now is to zoom out from the scale of individual cells to examine the structures they form together.

2.3 Telescoping Up in Scale: Cells Don't Go It Alone

Our powers-of-10 journey has thus far shown us the way in which cells are built from structural units going down from organelles to macromolecular assemblies to individual macromolecules to chemical groups, atoms, and ions. Equally interesting hierarchies of structures are revealed as we reduce the resolution of our imaginary camera and zoom out from the scale of individual cells. What we see once we

Figure 2.35: Representative examples of different communities of cells. (A) These stromatolites living in Shark Bay (Western Australia) are layered structures built by cyanobacteria. Stromatolites dominate the fossil record between 1 and 2 billion years ago, but are rare today. (B) This scanning electron micrograph showing a bacterial biofilm growing on an implanted medical device illustrates the morphological variety of bacterial species that can cooperate to form an interdependent community. (C) The social amoeba *Dictyostelium discoideum* forms fruiting bodies. The picture shows a fruiting body with spores—the tall stalk with a bulb at the top is a collection of amoebae. (D) A close-up photograph of the head of the robber fly, *Holocephala fusca*, shows the ordered structure of the compound eyes found in most insects. (A, Courtesy of Bellarmine University Department of Biology; B, Reproduced with permission of the ASM Microbe Library; C, Courtesy of Pauline Schaap; D, Courtesy of Thomas Shahan.)

begin to zoom out from the scale of single cells is the emergence of communities in which cells do not act independently.

2.3.1 Multicellularity as One of Evolution's Great Inventions

Life has been marked by several different evolutionary events that each wrought a wholesale change in the way that cells operate. One important category of such events is the acquisition of the ability of cells to communicate and cooperate with one another to form multicellular communities with common goals. This has happened many times throughout all branches of life and has culminated in extremely large organisms such as redwood trees and giraffes. In this section, we explore the ways in which cell–cell communication and cooperation have given rise to new classes of biological structures. Figure 2.35 shows a variety of different examples of cellular communities, some of which form the substance of the remainder of the chapter.

Bacteria Interact to Form Colonies such as Biofilms

The oldest known cellular communities recorded in the fossil record are essentially gigantic bacterial colonies called stromatolites. Although most stromatolites were outcompeted in their ecological niches by subsequent fancier forms of multicellular life, a few can still be found today taking essentially the same form as their 2-billion-year-old fossil brethren. Bacterial colonies of this type can still be found today in Australia, such as those shown in Figure 2.35(A). These communities have a characteristic size of about a meter and reflect collections of bacterial cells held together by an extracellular matrix secreted by these cells.

Many interesting kinds of bacterial communities consist of more than one species. Indeed, through a sophisticated system of signaling,

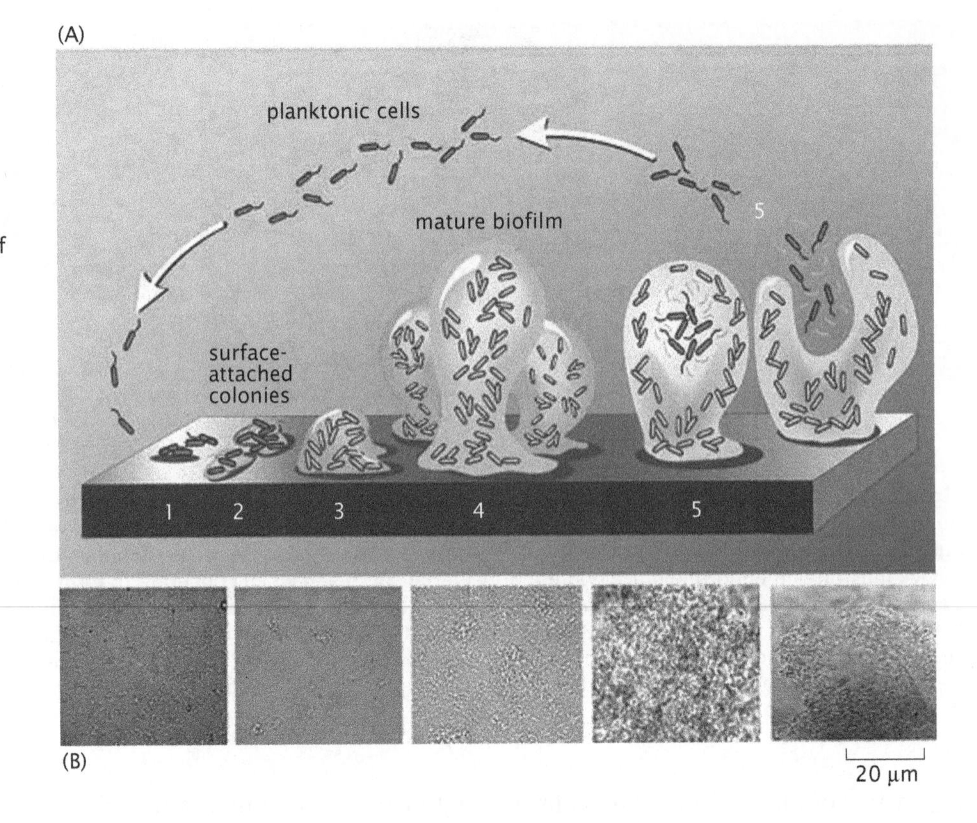

Figure 2.36: Schematic of the formation of a biofilm by bacteria. (A) The various stages in the formation of the biofilm are (1) attachment to surface, (2) secretion of extracellular polymeric substance (EPS), (3) early development, (4) maturation, and (5) shedding of cells from the biofilm. (B) The microscopy images show biofilms in various stages of the film formation process. (Adapted from P. Stoodley et al., *Annu. Rev. Microbiol.* 56:187, 2002. Original figure created by Peg Dirckx, MSU-CBE.)

detection, and organization, bacteria form colonies of all kinds ranging from biofilms to the ecosystems within animal guts. Bacterial biofilms are familiar to us all as the basis of the dentist's warning to floss our teeth every night. These communities are functionally as well as structurally interdependent. Other biofilms are noted for their destructive force when they attach to the surfaces of materials. Figure 2.35(B) shows a biofilm that grew on a silicon rubber voice prosthesis that had been implanted in a patient for about three months.

Structurally, a biofilm is formed as shown schematically in Figure 2.36. The key building blocks of such structures are a population of bacteria, a surface onto which these cells may adhere, and an aqueous environment. The formation of a biofilm results in a population of bacterial cells that are attached to a surface and enclosed in a polymeric matrix built up of molecules produced by these very same bacteria. The early stages of biofilm formation involve the adhesion of the bacteria to a surface, followed by changes in the characteristics of these bacteria such as the loss of flagella and micro-colony formation mediated by pili-twitching motility. At a larger scale, these changes at the cellular level are attended by the formation of colonies of cells and differentiation of the colonies into structures that are embedded in extracellular polysaccharides. Though there are a variety of different morphologies that are adopted by such films, roughly speaking, these biofilms are relatively porous structures (presumably to provide a conduit for import and export of nutrients and waste, respectively) that typically take on mushroom-like structures such as indicated schematically in Figure 2.36. These films have a relative proportion of something like 85% of the mass taken up by extracellular matrix, while the remaining 15% is taken up by cells themselves. A typical thickness for such films ranges from 10 to 50 μm.

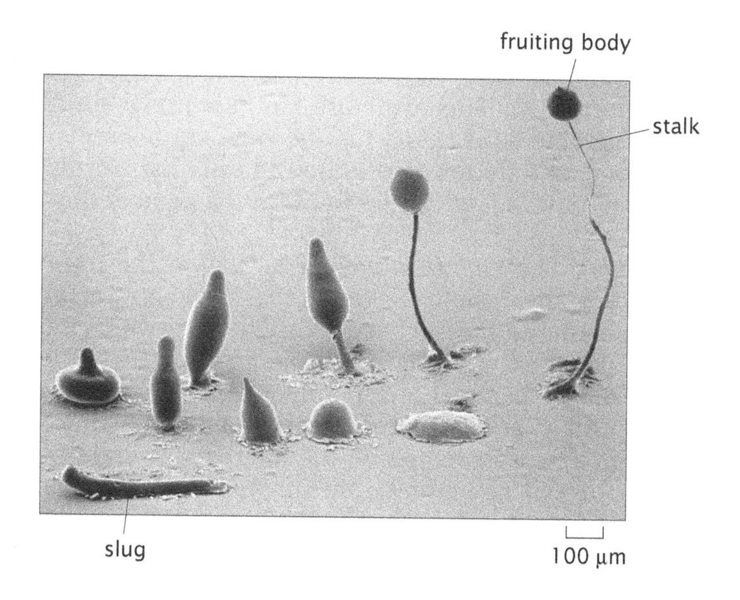

Figure 2.37: Formation of a multicellular structure during starvation. The social amoeba *Dictyostelium discoideum* responds to starvation by forming a structure made up of tens of thousands of cells in which individual cells suffer different fates. Cells near the top of the structure form spores, which are resurrected once conditions are favorable. The figure shows the developmental stages that take place on the way to making fruiting bodies, starting in the bottom right and proceeding clockwise. (Copyright, M. J. Grimson and R. L. Blanton, Biological Sciences Electron Microscopy Laboratory, Texas Tech University.)

Teaming Up in a Crisis: Lifestyle of *Dictyostelium discoideum*

Although bacteria can form communities, eukaryotes have clearly raised this to a high art. One particularly fascinating example that may give clues as to the origin of eukaryotic multicellularity is the cellular slime mold *Dictyostelium discoideum*, as shown in Figure 2.35(C). This small, soil-dwelling amoeba pursues a solitary life when times are good but seeks the comfort of its fellows during times of starvation. *Dictyostelium* is usually content to wander around as an individual with a characteristic size between 5 and 10 μm. However, when deprived of their bacterial diet, these cells undertake a radical change in lifestyle, which involves both interaction and differentiation through a series of fascinating intermediate steps as shown in Figure 2.37. A group of *Dictyostelium* amoebae in a soil sample that find themselves faced with starvation send chemical signals to one another resulting in the coalescence of thousands of separate amoebae to form a slug that looks like a small nematode worm. These cells appear to be poised on the brink between unicellular and multicellular lifestyles and can readily convert between them. Ultimately, as shown in the figure, the slug stops moving and begins to form a stalk. At the tip of the stalk is a nearly spherical bulb that contains many thousands of spores, essentially cells in a state of suspended animation. When environmental conditions are appropriate for individual amoeba to thrive, the spores undergo the process of germination, with each spore becoming a functional, individual amoeba.

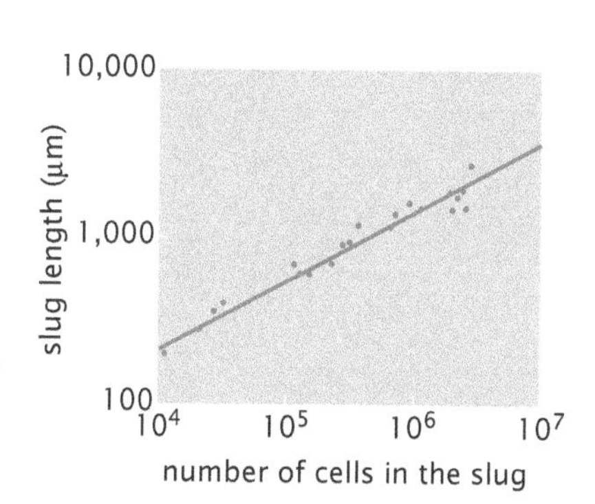

Figure 2.38: Slug size in *Dictyostelium discoideum*. The plot shows a relation between the size of the *Dictyostelium* slug and the corresponding number of cells. Points represent different individuals and the line shows the overall trend. (Adapted from Bonner, www.dictybase.org.)

ESTIMATE

Estimate: Sizing Up the Slug and the Fruiting Body The relation between the number of cells in a slug and its size is shown in Figure 2.38, where we see that these slugs can range in length between several hundred and several thousand microns with the number of cells making up the slug being between tens of thousands and several million. The next visible stage in the development is the sprouting of a stalk with a bulb at the top known as a fruiting body. The stalk is of the order of a millimeter in length, while the fruiting body itself is several hundred microns across. This fruiting body is composed of thousands of spores, that are effectively

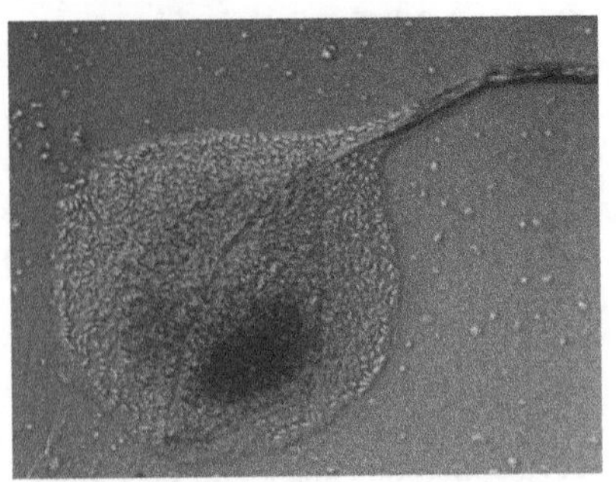

Figure 2.39: Microscopy image of a fruiting body. The fruiting body has been squished on a microscope slide, revealing both the size of the spores and their numbers.

in a state of suspended animation. An example of such a fruiting body that has been squished on a microscope slide is shown in Figure 2.39. This structure has functional consequences. In particular, those cells that become spores remain poised to respond to a better day, while the cells that formed the stalk have effectively ended their lives for the good of those that survive.

An immediate question of interest concerning the multicellular fruiting bodies shown in Figure 2.37 is how many cells conspire to make up such structures. An estimate of the number of cells in a fruiting body can be constructed by examining the nearly hemispherical collection of cells shown in Figure 2.39. Note that we treat the fruiting body as a hemisphere since, once it is squashed on the microscope slide, its shape is flattened and resembles a hemisphere more closely than a sphere. The diameter of this hemisphere is roughly 200 μm. Our rough estimate for the number of spores in the fruiting body is obtained by evaluating the ratio

$$\text{number of cells} = \frac{V_{\text{body}}}{V_{\text{cell}}}, \tag{2.25}$$

where we assume that the entirety of the fruiting body volume is made up of cells. If we assume that the cell is 4 μm in diameter, this yields

$$\text{number of cells} = \frac{\frac{2}{3}\pi(100\,\mu\text{m})^3}{\frac{4}{3}\pi(2\,\mu\text{m})^3} \approx 6 \times 10^4 \text{ cells}. \tag{2.26}$$

Note that the size of the ball of cells in a fruiting body can vary dramatically from the 200 μm scale shown here to several times larger, and, as a result, the estimate for the number of cells in a fruiting body can vary. Also, note that a factor-of-2 error in our estimate of the size of an individual cell will translate into a factor-of-8 error in our count of the number of cells in the slug or fully formed fruiting body.

Multicellular Organisms Have Many Distinct Communities of Cells

The three branches of life that have most notably exploited the potential of the multicellular lifestyle are animals, plants, and fungi. While bacterial and protist colonial organisms rarely form communities with characteristic dimensions of more than a few millimeters (with the exception of stromatolites), individuals in these three groups routinely grow to more than a meter in size. Their enormous size and corresponding complexity can be attributed to at least three factors: (i) production of extracellular matrix material that can provide structural support for large communities of cells, (ii) a predilection towards cellular specialization such that many copies of cells with the same genomic content can develop to perform distinct functions, and (iii) highly sophisticated mechanisms for the cells to communicate with one another within the organism. We emphasize that these traits are not unique to animals, plants, and fungi, but they are used more extensively there than elsewhere. A beautiful illustration of these principles is seen in the eye of a fly as shown in Figure 2.35(D). The eye is made up of hundreds of small units called ommatidia, each of which contains a group of eight photoreceptor cells, support cells, and a cornea. During development of the eye, these cells signal to one another to establish their identities and relative positions to create a stereotyped structure that is repeated many times. The overarching theme of the remainder of this chapter is the exploration of how cells come together to form higher-order structures and how these structures fit into the overall hierarchy of structures formed by living organisms.

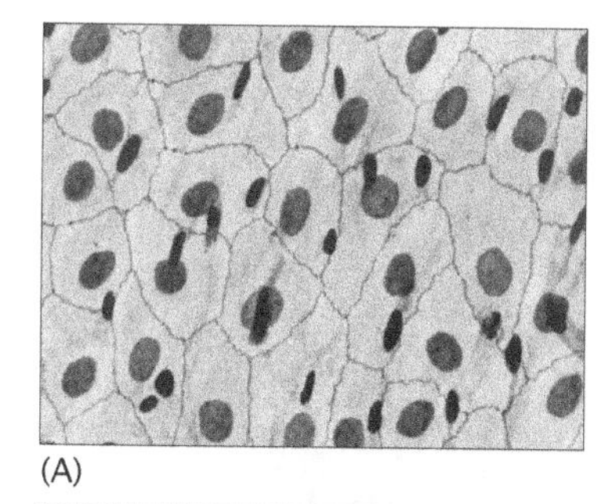

(A)

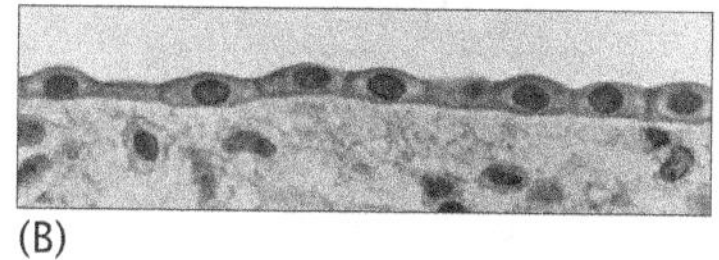

(B)

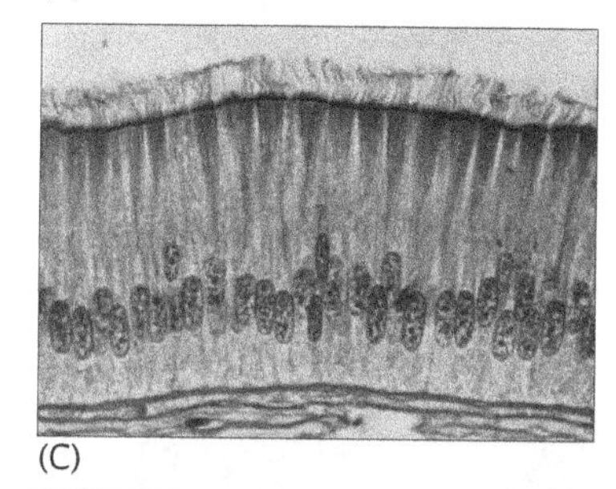

(C)

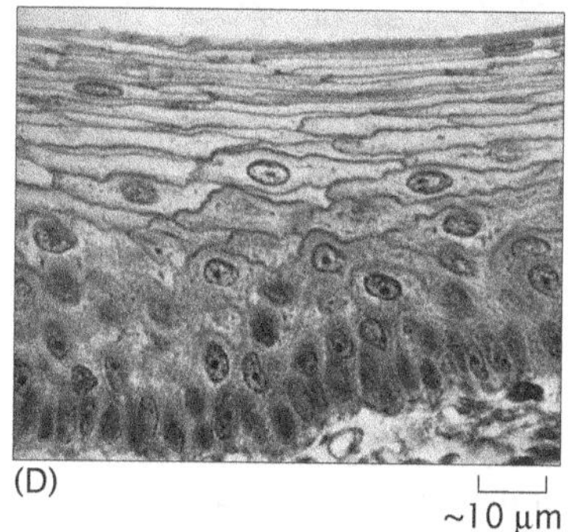

(D)

~10 μm

Figure 2.40: Shapes and architecture of epithelial cells. Epithelia are tissues formed by continuous sheets of cells that make tight contacts with one another. (A) Viewed from above, a simple epithelial sheet resembles a tiled mosaic. The dark ovals are the cell nuclei stained with silver. (B) Viewed from the side, simple epithelia such as this from the dog kidney may form a single layer of flattened cells above a loose, fibrous connective tissue. (C) In some specialized epithelia, the cells may extend upwards forming elongated columns and develop functional specializations at the top surface such as the beating cilia shown here. This particular ciliated columnar epithelium is from the alimentary tract of a freshwater mussel. (D) In other epithelia such as skin or the kitten's gum shown here, epithelial cells may form multiple layers. (A–D, adapted from D. W. Fawcett, The Cell, Its Organelles and Inclusions: An Atlas of Fine Structure. W. B. Saunders, 1966.)

2.3.2 Cellular Structures from Tissues to Nerve Networks

Multicellular structures are as diverse as cells themselves. Often, the nature of these structures is a reflection of their underlying function. For example, the role of epithelia as barriers dictates their tightly packed, planar geometries. By way of contrast, the informational role of the network of neurons dictates an entirely different type of multicellular structure.

One Class of Multicellular Structures is the Epithelial Sheets

Epithelial sheets form part of the structural backdrop in organs ranging from the skin to the bladder. Functionally, the cells in these structures have roles such as serving as a barrier to transport of molecules, providing an interface at which molecules can be absorbed into cells, and serving as the seat of certain secretions. Several different views of these structures are shown in Figure 2.40.

The morphology of epithelial sheets is diverse in several ways. First, the morphology of the individual cells can be different (isotropic versus anisotropic). In addition, the assemblies of cells themselves have different shapes. For example, the different structures can be broadly classified into those that are a monolayer sheet (simple epithelium) and those that are a multilayer (stratified epithelium). Within these two broad classes of structures, the cells themselves have different morphologies. The cells making up a given epithelial sheet can be flat, pancake-like cells, denoted as *squamous* epithelium. If the cells making up the epithelial sheet have no preferred orientation, they are referred to as *cuboidal*, while those that are elongated perpendicular to the extracellular support matrix are known as *columnar* epithelia. Epithelial sheets have as one of their functions (as do lipid bilayers) the segregation of different media, which can have highly different ionic concentrations, pH, macromolecular concentrations, and so on.

Tissues Are Collections of Cells and Extracellular Matrix

We have seen that cells can interact to form complexes. An even more intriguing example of a multicellular structure is provided by tissues such as that shown in Figure 2.41. These connective tissues are built up from a diverse array of cells and materials they secrete. Beneath the epithelial surface, fibroblasts construct extracellular matrix. This connective tissue is built up of three main components: cells such as fibroblasts and macrophages, connective fibers, and a supporting substance made up of glycosaminoglycans (GAGs). What is especially appealing and intriguing about these tissues is the orchestration, in both space and time, leading to the positioning of cells and fibers. The fibroblasts, which were featured earlier in this chapter, serve as factories for the proteins that make up the extracellular matrix. In particular, they synthesize proteins such as collagen and elastin that, when secreted, assemble to form fibrous structures that can support mechanical loads. The medium within which these fibers (and the cells) are embedded is made up of the third key component of the connective tissue, namely, the hydrated gel of GAGs.

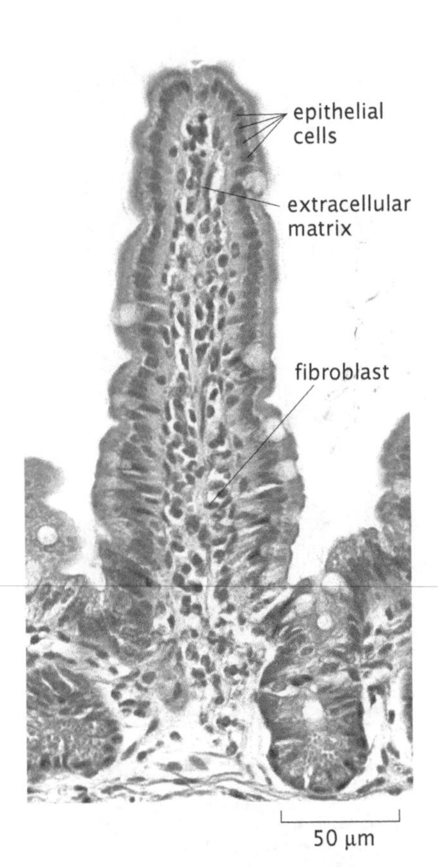

Figure 2.41: Tissue organization. This thin section of a villus in the human small intestine was stained with hematoxylin and eosin, a mixture of dyes that stains DNA and other negatively charged macromolecules blue or dark purple, and stains most proteins pink. In this image, a finger-shaped intestinal villus sticks up into the lumen of the intestine, covered with a single continuous monolayer of epithelial cells, whose function is to absorb nutrients. The supporting structure underneath the epithelial layer is composed of extracellular matrix secreted by fibroblasts. Numerous immune system cells also patrol this territory. (Adapted from L. G. van der Flier and H. Clevers, *Annu. Rev. Physiol.* 71:241, 2009.)

Nerve Cells Form Complex, Multicellular Complexes

A totally different example of structural organization involving multiple cells is illustrated by collections of neurons. Neurons are the specialized cells in animals that we associate with thinking and feeling. These cells allow for the transmission of information over long distances in the form of electrical signals. Neurons are constructed with many input terminals known as dendrites and a single output path known as the axon. The fascinating structural feature of these cells is that they assemble into complex networks that are densely connected in patterns where dendrites from a given cell reach out to many others, from which they take various inputs. An example of a collection of fluorescently labeled neurons is shown in Figure 2.42. Note that the branches (dendrites and axons) that reach out from the various cells have lengths far in excess of the 10 μm scale characteristic of typical eukaryotes. Indeed, axons of some neurons can have lengths of centimeters and more.

One fascinating example of neuronal contact is offered by the so-called neuromuscular junction as shown in Figure 2.43. These junctions are the point of contact between motor neurons, which convey the marching orders for a given muscle, and the muscle fiber itself. As is seen in the figure, the axon from a given motor neuron makes contacts with various muscle fibers. As will be described in more detail in Chapter 17, when an electrical signal (action potential) arrives at the contact point known as a synapse, chemical neurotransmitters are released into the space between the nerve and muscle. These neurotransmitters result in the opening of ligand-gated ion channels in the muscle, which results in a change in the electrical state of the muscle and leads to its contraction. These contrasting examples provide a small window into the diversity of cell–cell contacts.

2.3.3 Multicellular Organisms

The highest level in the structural hierarchy to be entertained here is that of individual multicellular organisms. The diversity of multicellular organisms is legendary, ranging from roses to hummingbirds, from Venus flytraps to the giant squid. What is especially remarkable about this diversity is contained in the simple statement of the cell theory: each and every one of these organisms is a collection of cells and their products. And, each and every one of these organisms is the result of a long history of evolution resulting in specialization and diversification of the various cells that make up that organism.

Cells Differentiate During Development Leading to Entire Organisms

The fruit fly *Drosophila melanogaster* has had a long and rich history as one of the key "model" organisms of biology and is a useful starting point for thinking about the size of organisms. As shown in

Figure 2.44, the mature fly has a size of roughly 3 mm that can be thought of morphologically as being built up of 14 segments: 3 segments making up the head, 3 segments making up the thorax and 8 segments making up the abdomen. We will further explore both the life cycle and the biological significance of *Drosophila* in Chapters 3 (p. 87) and 4 (p. 137), respectively, and for now content ourselves with examining the spatial scales associated with both the mature fly (Figure 2.44) and the embryos from which they are derived.

Drosophila has attained its legendary status in part because of the way it has revealed so many different ideas about genetics and embryonic development. One of the most well-studied features of the development of *Drosophila* is the way in which it lays down its anterior–posterior architecture during early development. This body plan is dictated by different cells adopting different patterns of gene expression depending upon where they are within the embryo. The pattern of expression of the *even-skipped* gene is shown in Figure 2.45. The gene *even-skipped* (*eve*) is expressed in seven stripes corresponding to 7 of the 14 *Drosophila* segments. One of the most remarkable features of the pattern of gene expression illustrated in the figure is its spatial sharpness, with the discrimination between the relevant gene being "on" or "off" taking place on a length scale comparable to the cell size itself. A fundamental challenge in the study of embryonic development has been the quest to understand the connection between patterns of gene expression in the embryo and the generation of macroscopic form.

Figure 2.42: Illustration of a complex network of cells formed by neurons. Using multiple fluorophores simultaneously to create a BrainBow, the structure of the nerves is elucidated. (Adapted from J. W. Lichtman et al., *Nat. Rev. Neurosci.* 9:417, 2008.)

Estimate: Sizing Up Stripes in *Drosophila* Embryos The sharp patterns in gene expression exhibited by the pair-rule genes such as the *even-skipped* gene shown in Figure 2.45 result from spatial gradients of DNA-binding proteins known as transcription factors that either increase (activation) or decrease (repression) the level of gene expression. To get a feeling for the scales associated with the gradients in transcription factors that dictate developmental decisions and the features they engender, we idealize the *Drosophila* embryo as a spherocylinder. We characterize the geometry by two parameters, namely, the length of the cylindrical region, L, and its radius R,

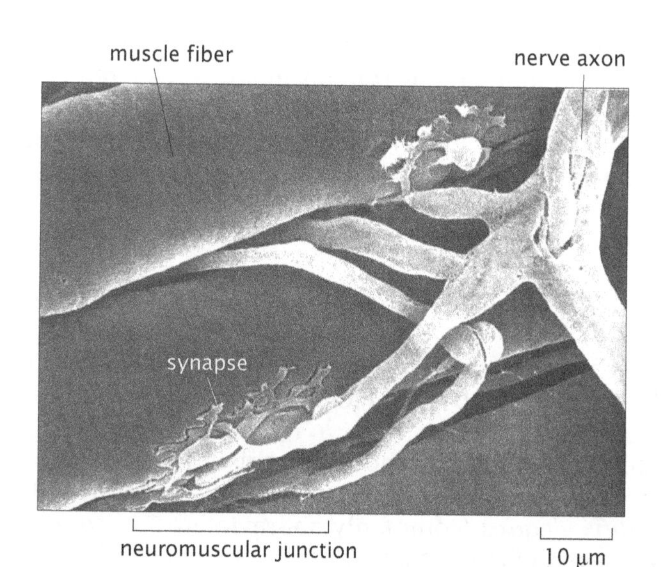

Figure 2.43: Neuromuscular junction. The axon from a nerve cell makes contact with various skeletal muscle fibers. Each muscle fiber is a giant multinucleated cell formed by the fusion of hundreds or thousands of precursor cells. Neurotransmitters secreted by the nerve cell at the synapse initiate contraction of the muscle fibers. (Adapted from D. W. Fawcett and R. P. Jensh, Bloom and Fawcett's Concise Histology. Hodder Arnold, 1997.)

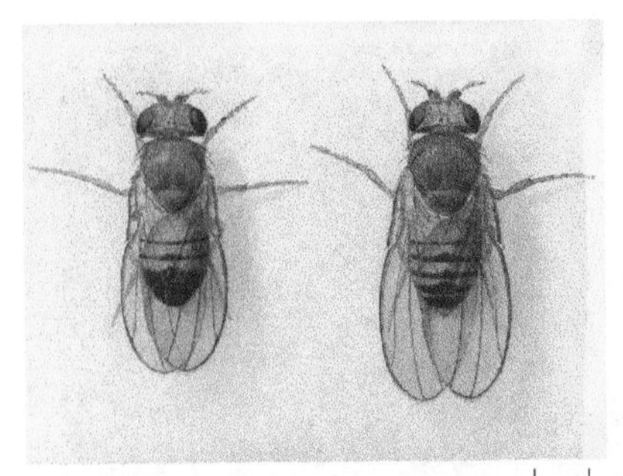

Figure 2.44: The male (left) and female (right) adult wild-type *Drosophila melanogaster*. After the 1934 drawing by Edith M. Wallace. (Courtesy of the Archives, California Institute of Technology.)

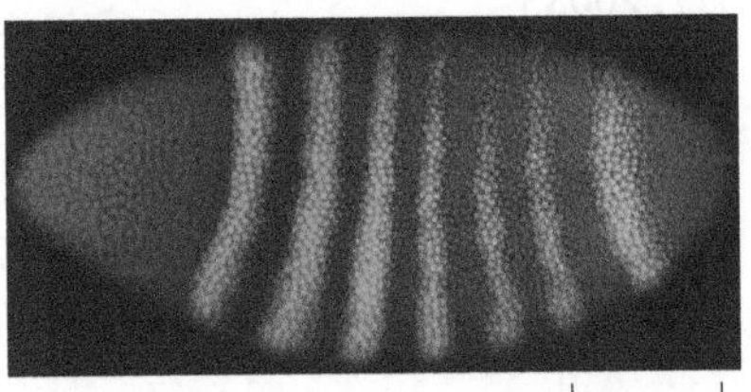

Figure 2.45: Pattern of gene expression in the *Drosophila* embryo. This early-stage fly embryo has undergone 13 rounds of cell division. The protein products of three different genes have been labeled using immunofluorescence: Bicoid in blue, Even-skipped in green, and Caudal in red. The small dots are individual nuclei. (Embryo be10 from the FlyEx database: http://urchin.spbcas.ru/flyex)

where we use approximate values of 500 μm for the length of the cylindrical region and 100 μm for the radius. In the embryo, many of the nuclei are on the surface and, as a result, our estimates will depend upon having a rough estimate of their areal density. The area of the embryo in the simple geometrical model used here is given by

$$A = 4\pi R^2 + 2\pi RL. \tag{2.27}$$

We consider the embryo after 13 divisions and before gastrulation, which brings the precursors of the muscles and internal organs to the interior of the embryo. One interesting fact about fruit fly development is that these initial divisions occur as nuclear divisions without the formation of cellular membranes which would give rise to cells. Also, note that there are not 2^{13} nuclei, since, during the preceding divisions, not all of the nuclei went to the surface of the embryo, and those that are left behind in the interior of the embryo do not keep pace with those on the surface during subsequent divisions. Instead, at this stage in development, there are on the order of $N = 6000$ nuclei on the surface. See Zalokar and Erk (1976) for details of the nuclear census during embryonic development. The areal density (that is, the number per unit area) is given by

$$\sigma = \frac{N}{4\pi R^2 + 2\pi RL}. \tag{2.28}$$

Using the numbers described above leads to an areal density of 0.01 cells/μm². As seen in Figure 2.45, the stripe patterns associated with the *Drosophila* embryo are very sharp and reflect cells making decisions at a very localized level.

Inspection of Figure 2.45 reveals that the stripes are roughly 20 μm wide. As a result, we can estimate the total number of nuclei participating in each stripe as

$$n = \sigma 2\pi R l_{\text{stripe}} \approx 0.01\ \text{cells}/\mu\text{m}^2 \times 2 \times 3.14 \times 100\ \mu\text{m} \times 20\ \mu\text{m} \approx 125\ \text{nuclei}. \tag{2.29}$$

Note that the average area per nucleus is given by $1/\sigma \approx 100\ \mu\text{m}^2$, suggesting that the radii of these nuclei is roughly 6 μm. Our main purpose in carrying out this exercise is to demonstrate the length scale of the structures that are put down during embryonic development. In this case, what we have seen is that out of the roughly 6000 nuclei that characterize the *Drosophila* embryo at the time of gastrulation, groups of roughly 120 nuclei have begun to follow distinct pathways as a result of differential patterns of gene expression that foreshadow the heterogeneous anatomical features of the mature fly.

The Cells of the Nematode Worm, *Caenorhabditis Elegans*, Have Been Charted, Yielding a Cell-by-Cell Picture of the Organism

A more recently popularized model organism for studying the genetics and development of multicellular animals is the nematode worm *Caenorhabditis elegans* (commonly abbreviated *C. elegans*) shown in Figure 2.46. Several factors that make this worm particularly attractive

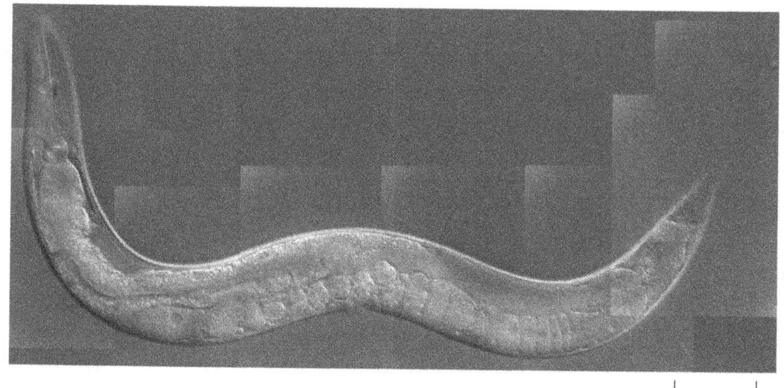

Figure 2.46: *C. elegans.* This differential interference contrast image of a single adult worm was assembled from a series of high-resolution micrographs. The worm's head is at the top left-hand corner and its tail is at the right. Its gut is visible as a long tube going down the animal's body axis. Its egg cells are also visible as giant ovals towards the bottom of the body. (Courtesy of I. D. Chin-Sang.)

in its capacity as a model multicellular eukaryote are that (i) its complete genome has been determined, (ii) the identity of each and every one of its 959 cells has been determined (see Figure 2.47), and (iii) the worms are transparent, permitting a variety of structural investigations. Amazingly, *all* of the cells of this organism have had their lineages traced from the single cell that is present at the moment of fertilization. What this means precisely is that all of the roughly 1000 cells making up this organism can be assigned a lineage of the kind cell A begat cell B which begat cell C, and so on.

As shown in Figure 2.46, these worms are roughly 1 mm in length and 0.05 mm across. Like *Drosophila,* they too have been subjected to a vast array of different analyses, including, for example, how their behavior is driven by the sensation of touch. One of the most remarkable outcomes of the series of experiments leading to the lineage tree for *C. elegans* was the determination of the connectivity of the 302

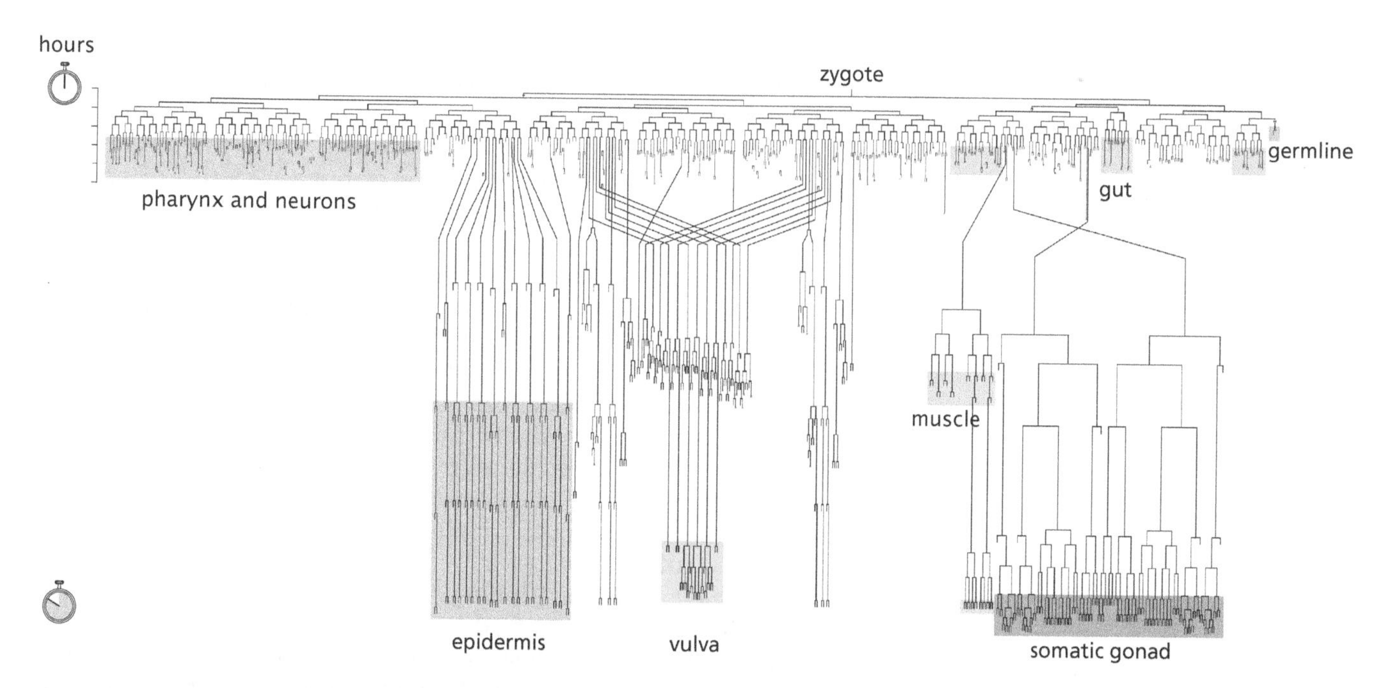

Figure 2.47: Cell lineages in *C. elegans.* The developmental pattern of every cell in the worm is identical from one animal to another. Therefore, it has been possible for developmental biologists to determine the family tree of every cell present in the entire animal by patient direct observation. In this schematic representation, the cell divisions that occur during embryogenesis are shown in the band across the top. The later cell divisions of the epidermis, vulva, and somatic gonad all take place after the animal has hatched.

neurons present in this nematode. By using serial thin sections from electron microscopy, it was possible to map out the roughly 7000 neuronal connections in the nervous system of this tiny organism. The various nerve cells are typically less than 5 μm across.

ESTIMATE

Estimate: Sizing Up *C. elegans* As an estimate of the cellular content of a "simple" animal, *C. elegans* provides a way of estimating the characteristic size of eukaryotic cells. As suggested by Figure 2.46, these small worms can be thought of as cylinders of length 1 mm and with a width of 0.1 mm. The total volume of such a worm is computed simply as roughly $7 \times 10^6\ \mu m^3$. If we use the fact that there are roughly 1000 cells in the organism, this tells us that the characteristic volume of the cells is $7000\ \mu m^3$, implying a typical radius for these cells of roughly 10 μm. A reasonable estimate for the volume of eukaryotic cells is somewhere between 2000 and $10{,}000\ \mu m^3$.

Many different species of nematodes share a common body plan involving a small number of well-defined cells; however, they may vary greatly in size. One of the largest nematodes, *Ascaris*, is a common parasite of pigs and humans. It has been estimated that up to 1 billion people on the planet carry *Ascaris* in their intestine. These worms closely resemble *C. elegans* except that they may be up to 15 cm in length and their cells are correspondingly larger. This kind of observation is not unusual throughout the animal kingdom. Many species have close relatives that differ enormously in size. The reasons and mechanisms that determine the overall size of multicellular organisms remain poorly understood.

Higher-Level Structures Exist as Colonies of Organisms

Organisms do not exist in isolation. Every organism on the planet is part of a larger ecological web that features both cooperation and competition with other individuals and species ultimately ranging up to the interconnected ecosystem of planet Earth. Coral reefs represent a vivid example of the interdependence of huge varieties of species living together in close quarters. An equally vivid and diverse community, though one frequently less appreciated for its intrinsic beauty, is found closer to home within our own intestines, which are inhabited by a teeming variety of bacteria. It has been estimated that the human body harbors nearly 10 times more microbial cells than it does human cells, and at least 100 times more bacterial genes than human genes. Thus, we should think of ourselves not really as individuals but rather as complex ecosystems of which the human cells form only a small part. Thus, our story of the hierarchy of structures that make up the living world began with the bacterium *E. coli* and will end there now. It is an irony that we as humans, it might be argued, have been colonized by our standard ruler *E. coli*.

Though we have examined biological structures over a wide range of spatial scales, our powers-of-10 journey still falls far short of being comprehensive. Ultimately, what we learn from this is that the handiwork of evolution has resulted in biological structures ranging from the nanometer scale of molecular machines all the way to the scale of the planet itself. Trying to understand what physical and biological factors drive the formation of these structures will animate much of the remainder of this book.

2.4 Summary and Conclusions

Biological structures range from the scale of nanometers (individual proteins) to hundreds of meters (redwood trees) and beyond (ecological communities). In this chapter, we have explored the sizes and numbers of biological entities, starting with the unit of life, the individual cell, and working our way down and up. We have found that sometimes biological objects show up in very large numbers of identical copies such that averages, for example of the concentration of ribosomes in the *E. coli* cytoplasm, are reasonable approximations. On the other hand, sometimes biological things show up in very few copies such that the exact number can make a big difference in the behavior of the system. We have found that cells are crowded, that there really is a world in a grain of sand (or in a biological cell). We have also explored some of the ways that biological units such as proteins or cells may interact with one another to form larger and more complex entities. Having developed an intuitive feeling for size and scale will better enable us to realistically envision the biological processes and problems described throughout the remainder of the book.

2.5 Problems

Key to the problem categories: • Model refinements and derivations • Estimates • Computational simulations

• 2.1 Revisiting the *E. coli* mass

We made a simple estimate of the mass of an *E. coli* cell by assuming that such cells have the same density of water. However, a more reasonable estimate is that the density of the macromolecules of the cell is 1.3 times that of water (BNID 101502, 104272). As a result, the estimate of the mass of an *E. coli* cell is off by a bit. Using that two-thirds of the mass is water and that the remaining one-third is macromolecular, compute the percentage error made by treating the macromolecular density as the same as that of water.

• 2.2 A feeling for the numbers: the chemical composition of a cell by pure thought

Make an estimate of the composition of carbon, hydrogen, oxygen, and nitrogen in the dry mass of a bacterium. Using knowledge of the size and mass of a bacterium, the fraction of that mass that is "dry mass" (that is, $\approx 30\%$) and the chemical constituents of a cell, figure out the approximate small integers (<10) for the composition $C_mH_nO_pN_q$, that is, find m, n, p, and q.

• 2.3 A feeling for the numbers: microbes as the unseen majority

(a) Use Figure 2.1 to justify the assumption that a typical bacterial cell (that is, *E. coli*) has a surface area of $6\,\mu m^2$ and a volume of $1\,\mu m^3$. Also, express this volume in femtoliters. Make a corresponding estimate of the mass of such a bacterium.

(b) Roughly 2–3 kg of bacteria are harbored in your large intestine. Make an estimate of the total number of bacteria inhabiting your intestine. Estimate the total number of human cells in your body and compare the two figures.

(c) The claim is made (see Whitman et al., 1998) that in the top 200 m of the world's oceans, there are roughly 10^{28} prokaryotes. Work out the total volume taken up by these cells in m^3 and km^3. Compute their mean spacing. How many such cells are there per milliliter of ocean water?

• 2.4 A feeling for the numbers: molecular volumes and masses

(a) Estimate the volumes of the various amino acids in units of nm^3.

(b) Estimate the mass of a "typical" amino acid in daltons. Justify your estimate by explaining how many of each type of atom you chose. Compare your estimate with the actual mass of several key amino acids such as glycine, proline, arginine, and tryptophan.

(c) On the basis of your result for part (b), deduce a rule of thumb for converting the mass of a protein (reported in kDa) into a corresponding number of residues. Apply this rule of thumb to myosin, G-actin, hemoglobin, and hexokinase and compare your results with the actual number of residues in each of these proteins. Relevant data for this problem are provided on the book's website.

• 2.5 Minimal media and *E. coli*

Minimal growth medium for bacteria such as *E. coli* includes various salts with characteristic concentrations in the mM range and a carbon source. The carbon source is typically glucose and it is used at 0.5% (a concentration of 0.5 g/100 mL). For nitrogen, minimal medium contains ammonium chloride (NH_4Cl) with a concentration of 0.1 g/100 mL.

(a) Make an estimate of the number of carbon atoms it takes to make up the macromolecular contents of a bacterium such as *E. coli*. Similarly, make an estimate of the number of nitrogens it takes to make up the macromolecular contents of a bacterium? What about phosphate?

(b) How many cells can be grown in a 5 mL culture using minimal medium before the medium exhausts the carbon? How many cells can be grown in a 5 mL culture using minimal medium before the medium exhausts the nitrogen? Note that this estimate will be flawed because it neglects the *energy* cost of synthesizing the macromolecules of the cell. These shortcomings will be addressed in Chapter 5.

• 2.6 Atomic-level representations of biological molecules

(a) Obtain coordinates for several of the following molecules: ATP, phosphatidylcholine, B-DNA, G-actin, the lambda repressor/DNA complex or Lac repressor/DNA complex, hemoglobin, myoglobin, HIV gp120, green fluorescent protein (GFP), and RNA polymerase. You can find the coordinates on the book's website or by searching in the Protein Data Bank and various other Internet resources.

(b) Download a structural viewing code such as VMD (University of Illinois), Rasmol (University of Massachusetts), or DeepView (Swiss Institute of Bioinformatics) and create a plot of each of the molecules you downloaded above. Experiment with the orientation of the molecule and the different representations shown in Figure 2.32.

(c) By looking at phosphatidylcholine, justify (or improve upon) the value of the area per lipid ($0.5\,\text{nm}^2$) used in the chapter.

(d) Phosphoglycerate kinase is a key enzyme in the glycolysis pathway. One intriguing feature of such enzymes is their enormity in comparison with the sizes of the molecules upon which they act (their "substrate"). This statement is made clear in Figure 5.5 (p. 195). Obtain the coordinates for both phosphoglycerate kinase and glucose and examine the relative size of these molecules. The coordinates are provided on the book's website.

• 2.7 Coin flips and partitioning of fluorescent proteins

In the estimate on cell-to-cell variability in the chapter, we learned that the standard deviation in the number of molecules partitioned to one of the daughter cells upon cell division is given by

$$\langle n_1^2 \rangle - \langle n_1 \rangle^2 = Npq. \tag{2.30}$$

(a) Derive this result.

(b) Derive the simple and elegant result that the average difference in intensity between the two daughter cells is given by

$$\langle (I_1 - I_2)^2 \rangle = \alpha I_{\text{tot}}, \tag{2.31}$$

where I_1 and I_2 are the intensities of daughters 1 and 2, respectively, and I_{tot} is the total fluorescence intensity of the mother cell and assuming that there is a linear relation between intensity and number of fluorophores of the form $I = \alpha N$.

• 2.8 HIV estimates

(a) Estimate the total mass of an HIV virion by comparing its volume with that of an *E. coli* cell and assuming they have the same density.

(b) The HIV maturation process involves proteolytic clipping of the Gag polyprotein so that the capsid protein CA can form the shell surrounding the RNA genome and nucleocapsid NC can complex with the RNA itself. Using Figures 2.30 and 2.31 to obtain the capsid dimensions, estimate the number of CA proteins that are used to make the capsid and compare your result with the total number of Gag proteins.

• 2.9 Areas and volumes of organelles

(a) Calculate the average volume and surface area of mitochondria in yeast based on the confocal microscopy image of Figure 2.18(C).

(b) Estimate the area of the endoplasmic reticulum when it is in reticular form using a model for its structure of interpenetrating cylinders of diameter $d \approx 10\,\text{nm}$ separated by a distance $a \approx 60\,\text{nm}$, as shown in Figure 2.25.

• 2.10 An open-ended "feeling for the numbers": the cell

Using the figures of cells and their organelles provided on the book website, carry out estimates of the following:

(a) The number of nuclear pores in the nucleus of a pancreatic acinar cell.

(b) The spacing between mitochondrial lamellae and the relative area of the inner and outer mitochondrial membranes.

(c) The spacing and areal density of ribosomes in the rough endoplasmic reticulum of a pancreatic acinar cell.

(d) The DNA density in the head of a sperm.

(e) The volume available in the cytoplasm of a leukocyte.

2.6 Further Reading

Goodsell, D (1998) The Machinery of Life, Springer-Verlag. This book illustrates the crowded interior of the cell and the various macromolecules of life.

Fawcett, DW (1966) The Cell, Its Organelles and Inclusions: An Atlas of Fine Structure, W. B. Saunders. We imagine our reader comfortably seated with a copy of Fawcett right at his or her side. Fawcett's electron microscopy images are stunning.

Fawcett, DW, & Jensh, RP (2002) Bloom and Fawcett: Concise Histology, 2nd ed., Hodder Arnold. This book shows some of the beautiful diversity of both cells and their organelles.

Gilbert, SF (2010) Developmental Biology, 9th ed., Sinauer Associates. Gilbert's book is a source for learning about the architecture of a host of different organisms during early development. Chapter 6 is especially relevant for the discussion of this chapter.

Wolpert, L, Tickle, C, Lawrence, P, et al. (2011) Principles of Development, 4th ed., Oxford University Press. Wolpert's book is full of useful cartoons and schematics that illustrate many of the key ideas of developmental biology.

2.7 References

Alberts, B, Johnson, A, Lewis, J, et al. (2008) Molecular Biology of the Cell, 5th ed., Garland Science.

Baker, TS, Olson, NH, & Fuller, SD (1999) Adding the third dimension to virus life cycles: three-dimensional reconstruction of icosahedral viruses from cryo-electron micrographs, *Microbiol. Mol. Biol. Rev.* **63**, 862.

Birk, DE, & Trelstad, RL (1986) Extracellular compartments in tendon morphogenesis: collagen fibril, bundle, and macroaggregate formation, *J. Cell Biol.*, **103**, 231.

Brecht, M, Fee, MS, Garaschuk, O, et al. (2004) Novel approaches to monitor and manipulate single neurons *in vivo*, *J. Neurosci.* **24**, 9223.

Briggs, JAG, Simon, MN, Gross, I, et al. (2004) The stoichiometry of Gag protein in HIV-1, *Nat. Struct. Mol. Biol.* **11**, 672.

Briggs, JAG, Grünewald, K, Glass, B, et al. (2006) The mechanism of HIV-1 core assembly: Insights from three-dimensional reconstruction of authentic virions, *Structure* **14**, 15.

Brügger, B, Glass, B, Haberkant, P, et al. (2006) The HIV lipidome: a raft with an unusual composition, *Proc. Natl Acad. Sci. USA* **103**, 2641.

Chin, DJ, Luskey, KL, Anderson, RGW, et al. (1982) Appearance of crystalloid endoplasmic reticulum in compactin-resistant Chinese hamster cells with a 500-fold increase in 3-hydroxy-3-methylglutaryl-coenzyme A reductase, *Proc. Natl Acad. Sci. USA* **79**, 1185.

Cross, PC, & Mercer, KL (1993) Cell and Tissue Ultrastructure, W. H. Freeman. A great pictorial source for anyone interested in the beauty of cells and their organelles.

Egner, A, Jakobs, S, & Hell, SW (2002) Fast 100-nm resolution three-dimensional microscope reveals structural plasticity of mitochondria in live yeast, *Proc. Natl. Acad. Sci.*, **99**, 3370.

Fawcett, DW (1986) Bloom and Fawcett—A Textbook of Histology, 11th ed., W. B. Saunders.

Finlay, BJ (2002) Global dispersal of free-living microbial eukaryote species, *Science* **296**, 1061.

Ghaemmaghami, S, Huh, W-K, Bower, K, et al. (2003) Global analysis of protein expression in yeast, *Nature* **425**, 737.

Godin, M, Delgado, FF, Son, S, et al. (2010) Using buoyant mass to measure the growth of single cells, *Nat. Methods*, **7**, 387.

Golding, I, Paulsson, J, Zawilski, SM, & Cox, EC (2005) Real-time kinetics of gene activity, *Cell* **123**, 1025.

Jorgensen, P, Nishikawa, JL, Breitkreutz, BJ, & Tyers M (2002) Systematic identification of pathways that couple cell growth and division in yeast, *Science*, **297**, 395.

Lichtman, JW, Livet, J, & Sanes, JR (2008) A technicolour approach to the connectome, *Nat. Rev. Neurosci.* **9**, 417.

Lu, P, Vogel, C, Wang, R, et al. (2007) Absolute protein expression profiling estimates the relative contributions of transcriptional and translational regulation, *Nat. Biotechnol.*, **25**, 117.

Neidhardt, FC, Ingraham, JL, & Schaechter, M (1990) Physiology of the Bacterial Cell, Sinauer Associates. This outstanding book is full of interesting facts and insights concerning bacteria.

Pedersen, S, Bloch, PL, Reeh, S, & Neidhardt, FC (1978) Patterns of protein synthesis in *E. coli*: A catalog of the amount of 140 individual proteins at different growth rates, *Cell* **14**, 179.

Poustelnikova, E, Pisarev, A, Blagov, M, et al. (2004) A database for management of gene expression data *in situ*, *Bioinformatics* **20**, 2212.

Radmacher, M (2007) Studying the mechanics of cellular processes by atomic force microscopy, *Methods Cell. Biol.*, **83**, 347.

Rosenfeld, N, Young, JW, Alon, U, Swain, PS, & Elowitz, MB (2005) Gene regulation at the single-cell level, *Science* **307**, 1962.

Schaechter, M, Ingraham, JL, & Neidhardt, FC (2006) Microbe, ASM Press.

Schulze, E, & Kirschner, M (1987) Dynamic and stable populations of microtubules in cells, *J. Cell Biol.* **104**, 277.

Stoodley, P, Sauer, K, Davies, DG, & Costerton, JW (2002) Biofilms as complex differentiated communities, *Annu. Rev. Microbiol.* **56**, 187.

Taniguchi, Y, Choi, PJ, Li, GW, et al. (2010) Quantifying *E. coli* Proteome and Transcriptome with Single-Molecule Sensitivity in Single Cells, *Science*, **329**, 533.

Terasaki, M, Chen, LB, & Fujiwara, K (1986) Microtubules and the endoplasmic reticulum are highly interdependent structures, *J. Cell. Biol.* **103**, 1557.

van der Flier, LG, & Clevers, H (2009) Stem cells, self-renewal, and differentiation in the intestinal epithelium, *Annu. Rev. Physiol.*, **71**, 241.

Visser, W, Van Sponsen, EA, Nanninga, N, et al. (1995) Effects of growth conditions on mitochondrial morphology in *Saccharomyces cerevisiae*, *Antonie van Leeuwenhoek* **67**, 243.

Walker, GM (1998) Yeast, Physiology and Biotechnology, Wiley.

Whitman, WB, Coleman, DC, & Wiebe, WJ (1998) Prokaryotes: the unseen majority, *Proc. Natl Acad. Sci. USA* **95**, 12, 6578.

Wu, J-Q, & Pollard, TD (2005) Counting cytokinesis proteins globally and locally in fission yeast, *Science* **310**, 310. This paper illustrates the strategy of using fluorescence microscopy for taking the census of a cell.

Zalokar, M, & Erk, I (1976) Division and migration of nuclei during early embryogenesis of *Drosophila melanogaster*, *J. Microscopie Biol. Cell.* **25**, 97. This paper makes a census of the number of cells in the *Drosophila* embryo at various stages in development.

Zimmerman, SB, & Trach, SO (1991) Estimation of macromolecule concentrations and excluded volume effects for the cytoplasm of *Escherichia coli*, *J. Mol. Biol.* **222**, 599. Discussion of crowding in the bacterial cell.

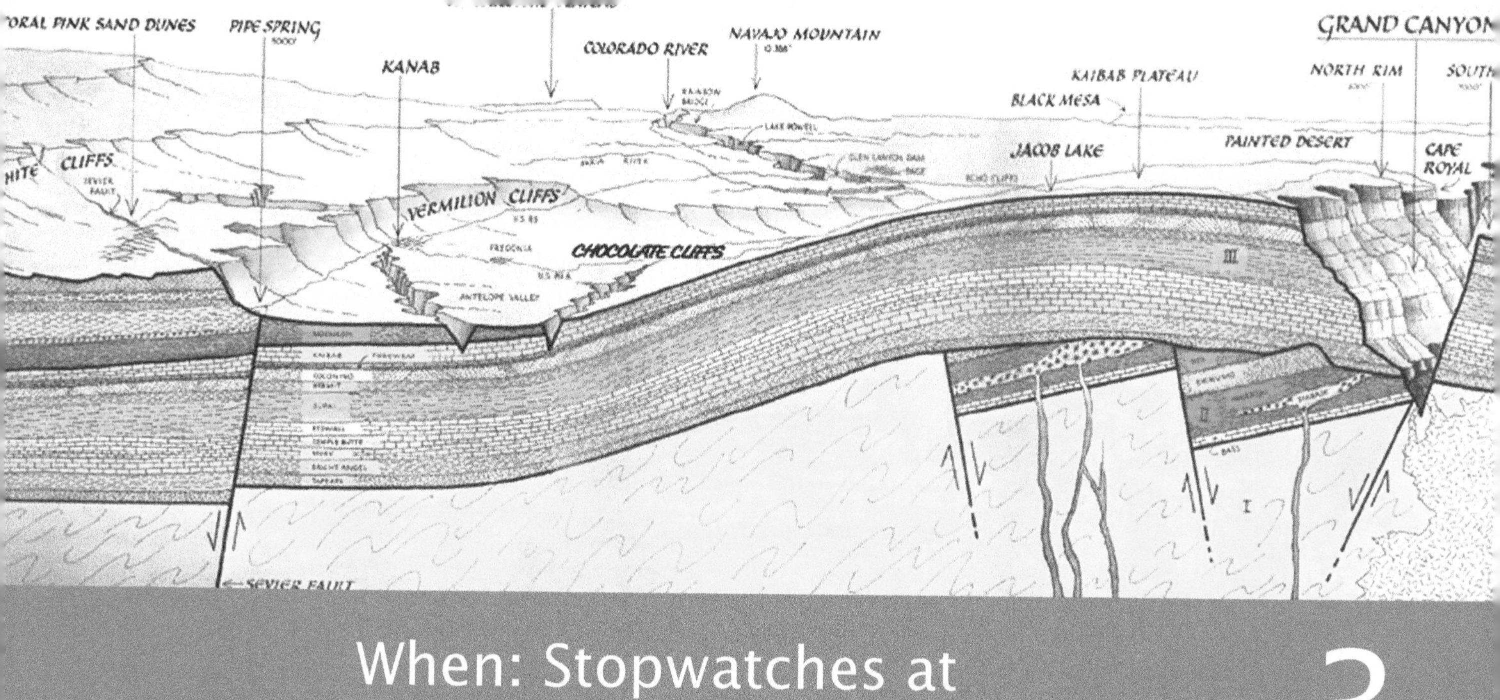

When: Stopwatches at Many Scales

3

Overview: In which various stopwatches are used to measure the rate of biological processes

> "Some day people may look back on the isotope as being as important to medicine as the microscope."
>
> **Archibald Hill**

Just as biological structures exist over a wide range of spatial scales, biological processes take place over time scales ranging from less than a microsecond to the age of the Earth itself. Using the cell cycle of *E. coli* as a standard stopwatch, this chapter develops a feeling for the rates at which different biological processes occur and illustrates the great diversity of these processes, mirroring the structural diversity seen in the previous chapter. With this "feeling for the numbers" in hand, we then explore several different views of the passage of biological time. In particular, we show how sometimes cells manage time by stringing together processes in succession while at other times they manipulate time (using enzymes to speed up reactions, for example) to alter the intrinsic rates of processes.

3.1 The Hierarchy of Temporal Scales

One of the defining features of living systems is that they are dynamic. The time scales associated with biological processes run from the nanosecond (and faster) scale of enzyme action to the more than 10^9 years that cover the evolutionary history of life and the Earth itself, as indicated in the map that opens this chapter. The inexorable march of biological time is revealed over many orders of magnitude in time scale, as illustrated in Figure 3.1. If we watch biological systems unfold with different stopwatches in hand, the resulting phenomena will be different: at very fast time scales, we will see the molecular dance of different biochemical species as they interact and change identity; at much slower scales, we see the unfolding of the lives of individual cells. If we slow down our stopwatch even more, what we see are the trajectories of entire species. To some extent, there is a coupling between the temporal scales described in this chapter and

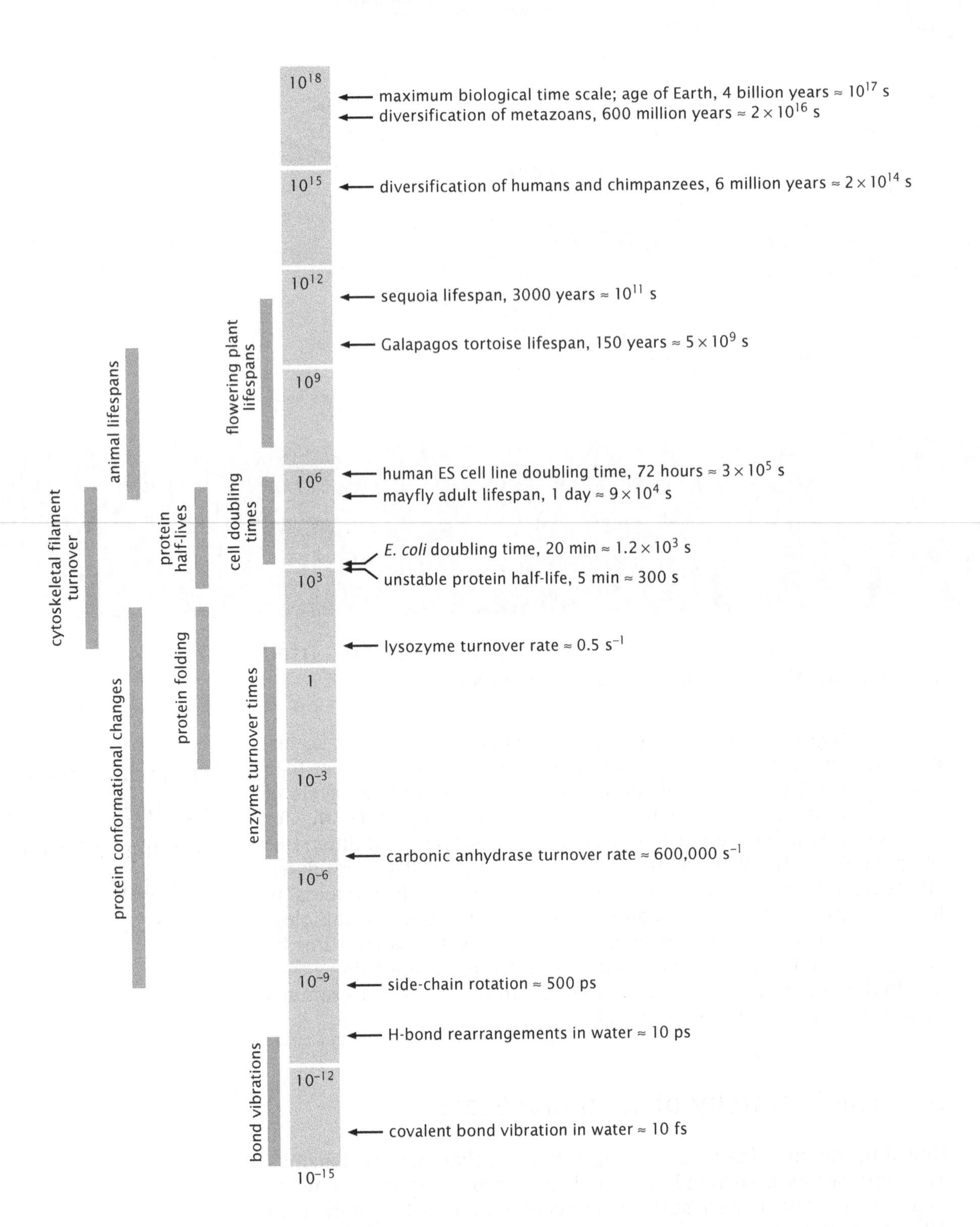

Figure 3.1: Gallery of biological time scales. Logarithmic scale showing the range of time scales associated with various biological processes. The time scale is in seconds.

the spatial scales described in the previous chapter; small things such as individual molecules tend to operate at fast rates, and large things such as elephants tend to move around at slow rates.

This chapter describes the time scales of biological phenomena from a number of different perspectives. In this section, we develop a feeling for biological time scales by examining the range of different time scales seen in molecular, cell, developmental, and evolutionary biology. This discussion is extended by describing the experimental basis for what we know about time scales in biology. As in Chapter 2, we once again invoke *E. coli* as our reference system, this time by using the cell cycle of our "reference cell" as the standard stopwatch. The remainder of the chapter is built around viewing time in biology from three distinct perspectives. In Section 3.2, we show how the time scale of certain biological processes is dictated by how long it takes some particular procedure (such as replication) to occur. We will refer to this as *procedural time*. Section 3.3 explores time from a different angle. In this case, we consider a broad class of biological processes whose timing is of the "socks before shoes" variety. That is, processes are linked in a sequential string, and in order for one process to begin, another must have finished. We will refer to this kind of time keeping as *relative time*. Finally, Section 3.4 reveals a third way of viewing time in biological processes, as a commodity to be manipulated. In a process we will call *manipulated time*, we show how cells and organisms find ways to either speed up or slow down key processes such as replication and metabolism.

3.1.1 The Pageant of Biological Processes

Biological Processes Are Characterized by a Huge Diversity of Time Scales

A range of different processes associated with individual organisms, and their associated time scales, is shown in Figure 3.2 (we leave a discussion of evolutionary processes for the next section). Broadly speaking, this figure shows a loose powers-of-10 representation of different biological processes in the same spirit as Figure 2.15 (p. 52) showed a powers-of-10 representation of spatial scales. Although this serves as a useful starting point, as we will see later in the chapter, an *absolute* measure of time in seconds or minutes is sometimes not the most useful way to think about the passage of time within cells. For example, embryonic development for humans takes drastically longer than for chickens, but the relative timing of common events can be meaningfully compared. For the moment, our discussion of Figure 3.2 is designed to give a feeling for the numbers; how long do various key biological processes actually take in absolute terms as measured in seconds, minutes and hours?

We begin (Figure 3.2A,B) with some of the processes associated with the development of the fruit fly *Drosophila melanogaster*. *Drosophila* has been one of the key workhorses of developmental biology, and much that we know about embryonic development was teased out of watching the processes that take place over the roughly 10 days between fertilization of the egg and the emergence of a fully functioning adult fly. If we increase our temporal resolution by a factor of 10, we see the processes in the development of the fly embryo. Over the first 10 hours or so after fertilization, as shown in Figure 3.2(B), a single cell is turned into an organized collection of thousands of cells with particular spatial positions and functions. One of the most dramatic parts of embryonic development is the process of gastrulation when the future gut forms as a result of a series of folding events in the embryo. This process is indicated schematically in Figure 3.2(B).

Individual cells have a natural developmental cycle as well. The *cell cycle* refers to the set of processes whereby a single cell, through the process of cell division, becomes two daughter cells. The time scales associated with the cell cycle of a bacterium such as *E. coli* are shown in Figure 3.2(C), with a characteristic scale of several thousand seconds. The lives of individual cells are fascinating and complex. If we

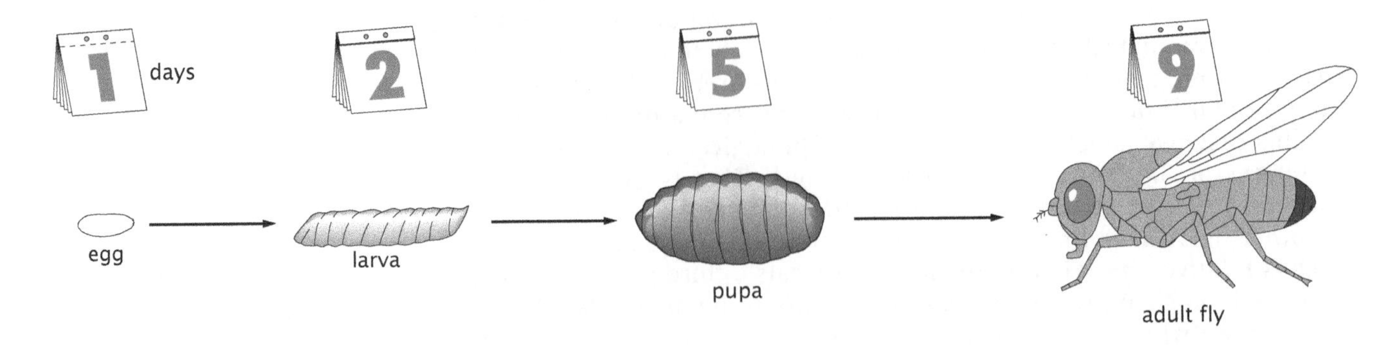

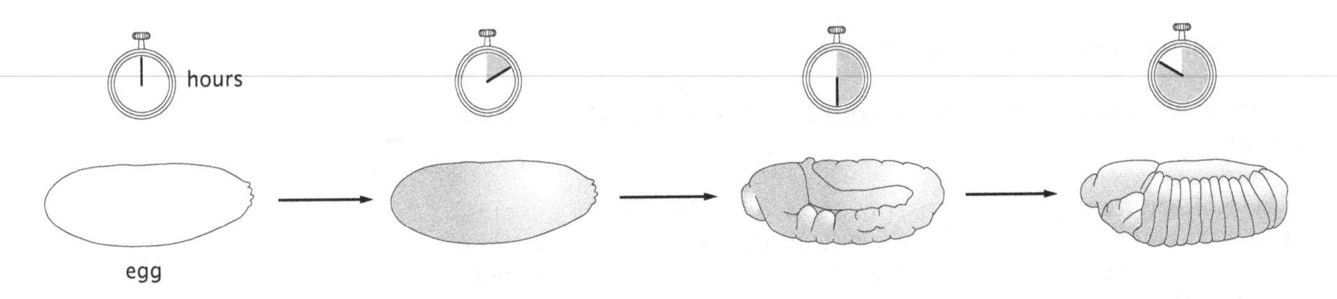

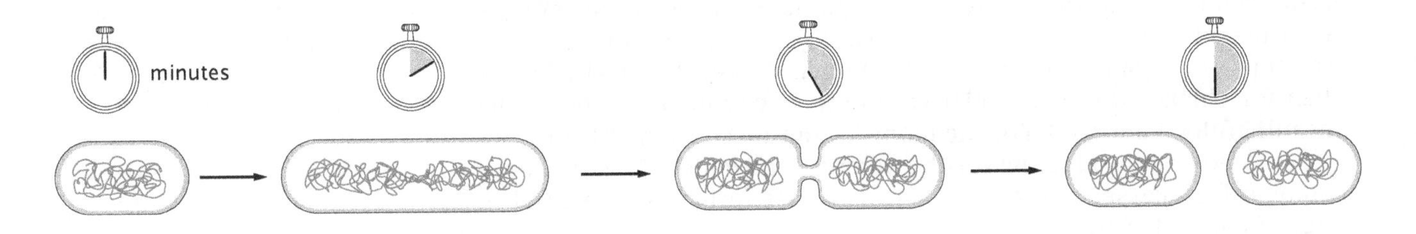

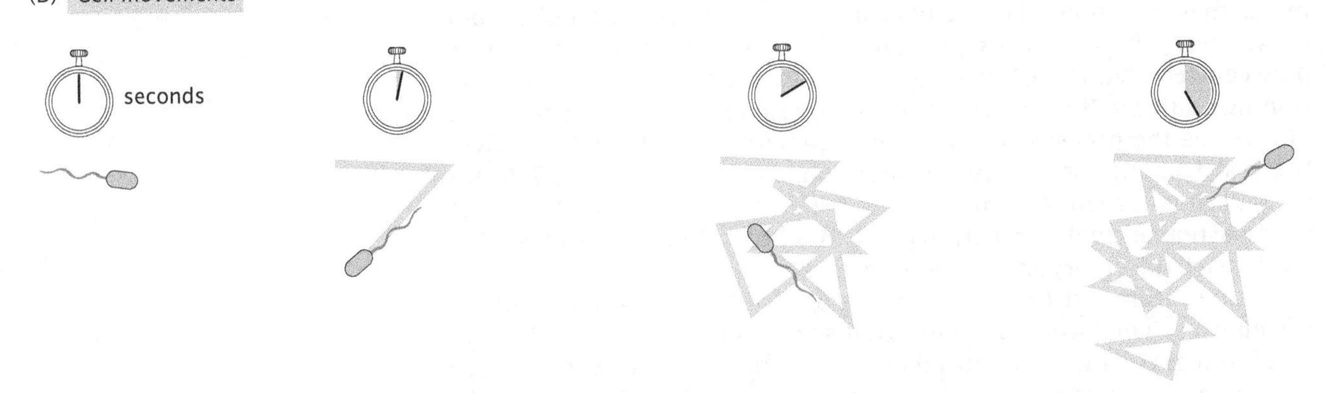

Figure 3.2: Hierarchy of biological time scales. Cartoon showing range of time scales associated with different biological processes. (A) Development of *Drosophila*. (B) Early development of *Drosophila* embryo. (C) Bacterial cell division. (D) Cell movements. (E) Protein synthesis. (F) DNA transcription. (G) Gating of ion channels. (H) Enzyme catalysis.

(E) Protein synthesis

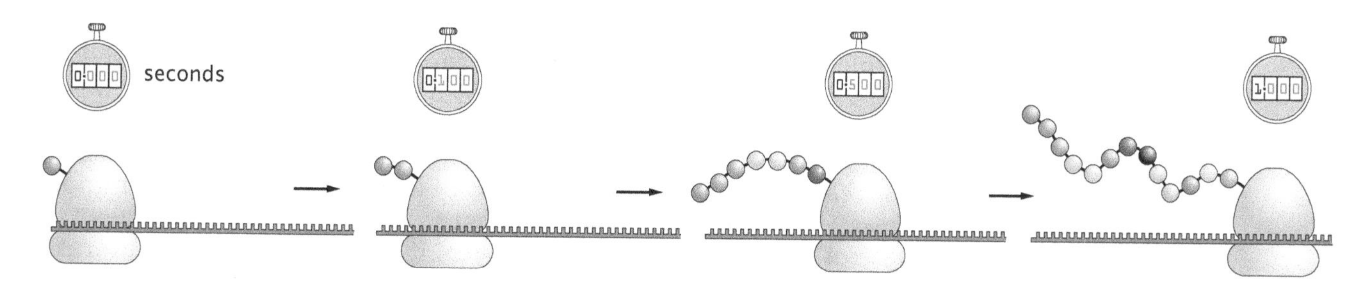

(F) Transcription

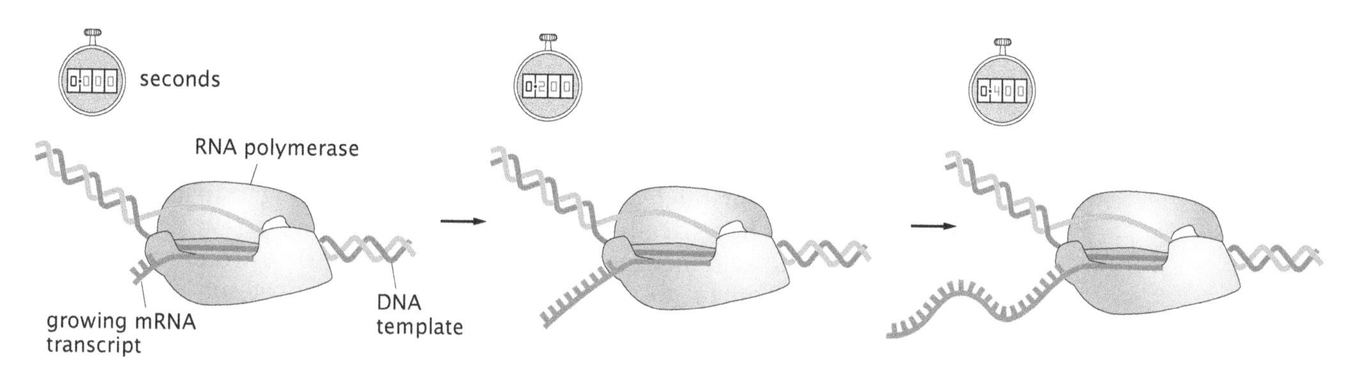

(G) Gating of ion channels

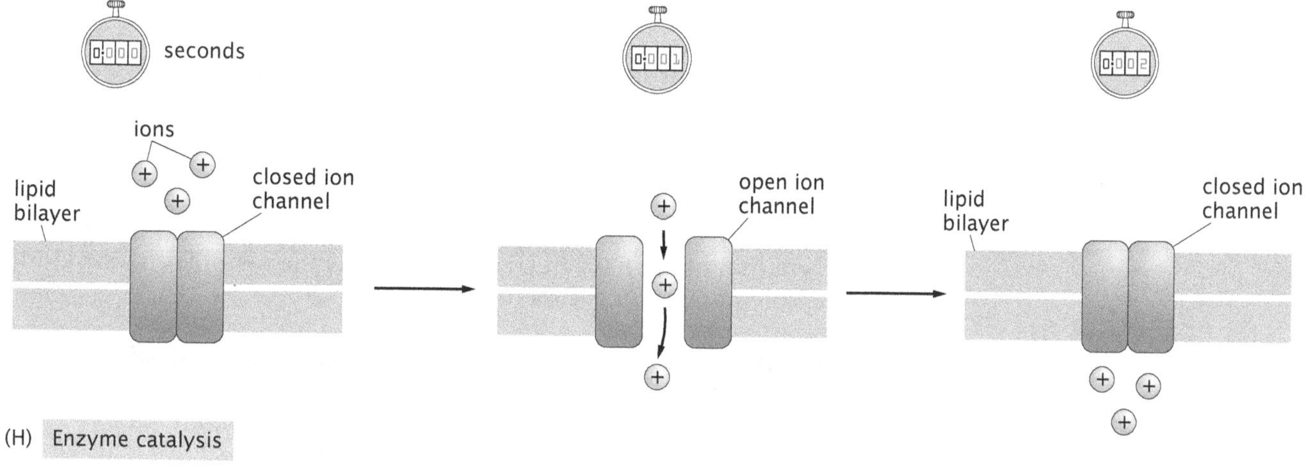

(H) Enzyme catalysis

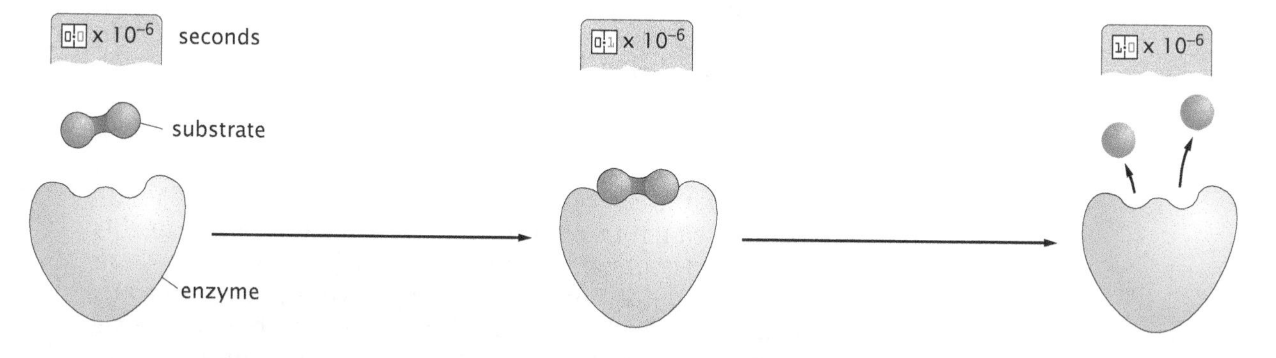

Figure 3.2: Continued.

were to dissect the activities of an individual cell as it goes about its business between cell divisions, we would find a host of processes taking place over a range of different time scales. If we stare down a microscope at a swimming bacterium for several seconds, we will notice episodes of directed motion, punctuated by rapid directional changes. Figure 3.2(D) shows the time scales over which an individual bacterium such as *E. coli* exercises its random excursion during movement. If our stopwatch runs a factor of 10 faster, we are now operating at the scale of deciseconds, a scale that characterizes the rate of amino acid incorporation during protein synthesis, a process represented in Figure 3.2(E). Macromolecular synthesis is one of the most important sets of processes that any cell must undertake to make a new cell. Another key part of the macromolecular synthesis required for cell division is the process of transcription, which is the intermediate step connecting the genetic material as contained in DNA and the readout of that message in the form of proteins. Transcription refers to the synthesis of mRNA molecules as faithful copies of the nucleotide sequence in the DNA, a polymerization process catalyzed by the enzyme RNA polymerase. The incorporation by RNA polymerase of nucleotides onto the mRNA during transcription, as depicted schematically in Figure 3.2(F), happens a few times faster than amino acid incorporation by ribosomes during protein synthesis.

In the moment-to-moment life of the cell, proteins do most of the work. Many proteins are able to operate at time scales much faster than the relatively stately machinery carrying out the central dogma operations. For example, a great number of biological processes are dictated by the passage of ions across ion channels, with a characteristic time scale of milliseconds as shown in Figure 3.2(G). A factor-of-1000 speed-up of our stopwatch brings us to the world of enzyme kinetics at the microsecond time scale (Figure 3.2H) and faster. It is important to note that these time scales merely represent a general rule of thumb. For example, turnover rates for individual enzymes may range from $0.5\,s^{-1}$ to $600{,}000\,s^{-1}$.

Before proceeding, one of the questions we wish to consider is how the time scales depicted in Figure 3.2 are actually known. As with much of our story, the stopwatches associated with each of the cartoons in that figure have been determined as the results of many kinds of complementary experiments.

EXPERIMENTS

Experiments Behind the Facts: Measurements of Biological Time Broadly speaking, the experiments that characterize the dynamics of cells and the molecules that populate them are ultimately based on tracking transformations. We can divide these experiments into four broad categories that can be applied across all levels of spatial scale from molecular to ecological. These methods are summarized in Figure 3.3.

1. *Direct observation* The first and most obvious way to characterize time in a biological process is simply to observe the process unfold and to record the absolute time at which transformation occurs. An example of this strategy is shown in Figure 3.3, which shows the motion of a neutrophil. Similarly, looking down a microscope at a mammalian cell in tissue culture, it is possible to observe many of the steps in the cell cycle (see Figure 3.21 on p. 116) unfolding over real time, including condensation of the chromosomes, alignment of the chromosomes

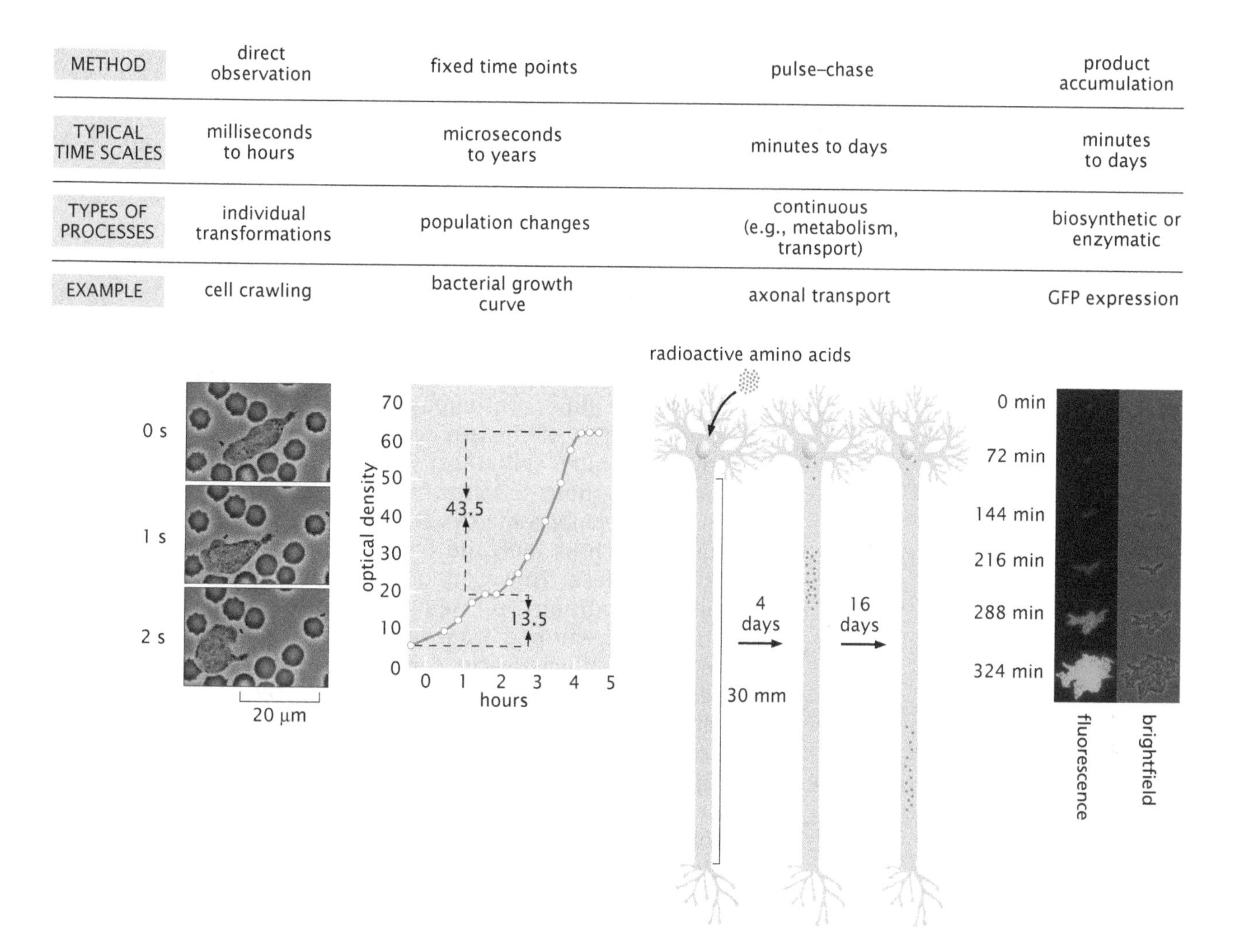

Figure 3.3: Experiments to measure the timing of biological processes. The figure summarizes four strategies for measuring biological rates. For *direct observation*, the example shows three frames from a video sequence taken by David Rogers in the 1950s of a single white blood cell (neutrophil) pursuing a bacterium through a forest of red blood cells. The movement of the cell is sufficiently fast that it can be directly observed by the human eye. For *fixed time points*, the experiment shown is a classic performed by Monod, who tracked the growth of *E. coli* in a single culture when two different nutrient sugars were mixed together. The bacteria initially consumed all of the available glucose and then their growth rate slowed as they switched over into a new metabolic mode enabling them to use lactose. For *pulse–chase*, labeling proteins at their point of synthesis in a neuron cell body with a pulse of radioactive amino acids followed by a chase of unlabeled amino acids was used to measure the rate of continuous axonal transport. *Product accumulation* is illustrated by the expression of GFP under a regulated promoter of interest in a bacterial cell. The rate of gene transcription off of such promoter can be inferred by measuring the amount of GFP present as a function of time. (Growth curve adapted from B. Müller-Hill, The *lac* Operon: A Short History of a Genetic Paradigm, Walter de Gruyter, 1996; neuronal transport adapted from B. Droz and C. P. Leblond, *Science* 137:1047, 1962; fluorescent image series adapted from N. Rosenfeld et al., *Science* 307:1962, 2005.)

through the action of the mitotic spindle, their segregation into daughter nuclei, and finally cytokinesis when the cell is pinched into two fully formed daughter cells. Although this is easy to do for processes that take minutes to hours and occur over spatial scales that can be observed with a light microscope or the unaided human eye, it is extremely difficult to measure time simply by observation for events that are very fast, very slow, very small, or very large. Over the past few decades, there have been vast experimental improvements in direct or near-direct observation of single molecules such that this naturalistic approach to "observing a lot just by watching" can be applied all the way down to the molecular scale. We will see many examples of this approach throughout the book.

2. *Fixed time points* When events of interest cannot be directly observed, there are other ways to probe their duration. Rather than continuously observing an individual over time, one can draw individuals from a population at given time intervals and examine their properties at this series of fixed time points. For example, a bacterial population in a liquid culture started from a single cell will grow exponentially, then plateau, and eventually die off over a period of several days. Rather than staring at the tube continuously for several days, the essential kinetics of this process can be measured simply by examining cell density at some fixed interval such as every hour as shown in Figure 3.3. Similarly, the events of embryonic development for useful model organisms such as flies and frogs unfold over a period of days to weeks. However, under a given set of environmental conditions, the sequence and timing of these events are stereotyped from one individual to another. Therefore, the investigator can accurately describe the sequence of events in frog development by examining one dish of embryos an hour after fertilization, a second dish of embryos two hours after fertilization, etc. This is useful when the methods used to examine the embryos result in their death, for example, fixing them and staining for a particular protein of interest or preparing them for electron microscopy examination. At a much smaller spatial scale and faster time, the method of stopped-flow kinetics enables investigators to follow enzymatic events by mixing together an enzyme and its substrate and then squirting the mixture into a denaturing acid bath after fixed intervals of time. These methods are all more indirect than direct observation, but in many cases are technically easier, and different kinds of complementary information can be gleaned by comparing both for a single process.

3. *Pulse–chase* Many biological processes operate in a continuous fashion. For example, bacteria constantly take in sugar from their medium for energy and to generate the molecular building blocks to synthesize new constituents. The process of glycolysis converts a molecule of glucose into two molecules of pyruvate. Because glucose is continuously taken up and pyruvate is continuously generated, it is extremely difficult to measure how long the conversion process takes. The set of methods used to tackle these kinds of problems are generally called pulse–chase experiments. In this particular example, a bacterial cell may be fed glucose tagged with radioactive carbon for a very brief period of time, for example 1 minute. This is followed by feeding with nonradioactive glucose. Cells are then removed from the bacterial culture at various time intervals and their metabolites are examined to see which contain the

radioactive carbon. Over time, the amount of labeled glucose will decrease and the major radioactive species will pass through a series of intermediates until finally most of the radioactive carbon will be found in pyruvate. Thus, a pulse–chase experiment can be used to determine the order of intermediates in a metabolic pathway and also the amount of time it takes for the cell to perform each transformation. A classic example of this strategy to examine transport in neurons is shown in Figure 3.3. Essentially the same method is used by naturalists examining dispersion times of birds and other animals by tagging individuals with a band or radio-transmitter and releasing them back into their natural population to see where they end up and when.

4. *Product accumulation* The final type of experiment used to determine biological rates is exemplified by an assay with a purified enzyme where a colorless substrate is converted into a colored product over time. By measuring the concentration of the colored product as a function of time, the investigator can extrapolate the average turnover rate given the known concentration of enzymes in the test tube. Similar experiments where observation of the accumulation of a product can be used as a surrogate for rate measurements can also be performed in living cells. A particularly useful example is expressing GFP (green fluorescent protein) downstream of a promoter of interest as shown in Figure 3.3. When the promoter is induced (that is, for example, by exposing the cells to some molecule that turns the gene of interest on) GFP begins to accumulate and the amount of fluorescence can be directly measured and converted into numbers of GFP molecules. Because GFP is remarkably stable, its accumulation can often represent a more accurate reporter for promoter activity than the promoter's natural product, which may be subject to other layers of regulation including rapid degradation.

3.1.2 The Evolutionary Stopwatch

The general rule that all biological processes are dynamic and undergo change over time applies to molecules, cells, organisms, and species. The evolutionary clock started more than 3 billion years ago with the appearance of the first cellular life forms on Earth. It is generally accepted that there were complex life-like processes occurring prior to the emergence of the first recognizable cells, though we cannot learn very much about what they were like either from the fossil record or from comparative studies among organisms living today.

All of the astonishing diversity of cellular life currently existing on the planet, ranging from archaea living in geothermal vents deep in the ocean to giant squid to redwood trees to the yeast that make beer, was descended from a universal common ancestor, probably a population of cells rather than an individual. This last universal common ancestor (LUCA) would have been clearly recognizable as cellular: it contained DNA as a genetic material, it transcribed its DNA into mRNA, and it translated mRNA into proteins using ribosomes. It

(A)

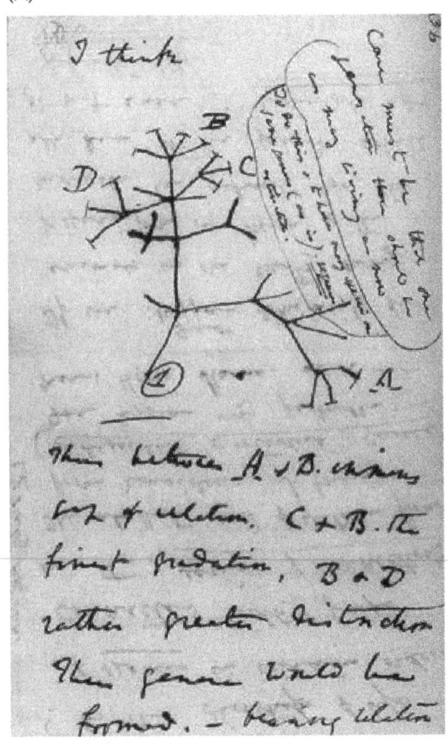

(B)

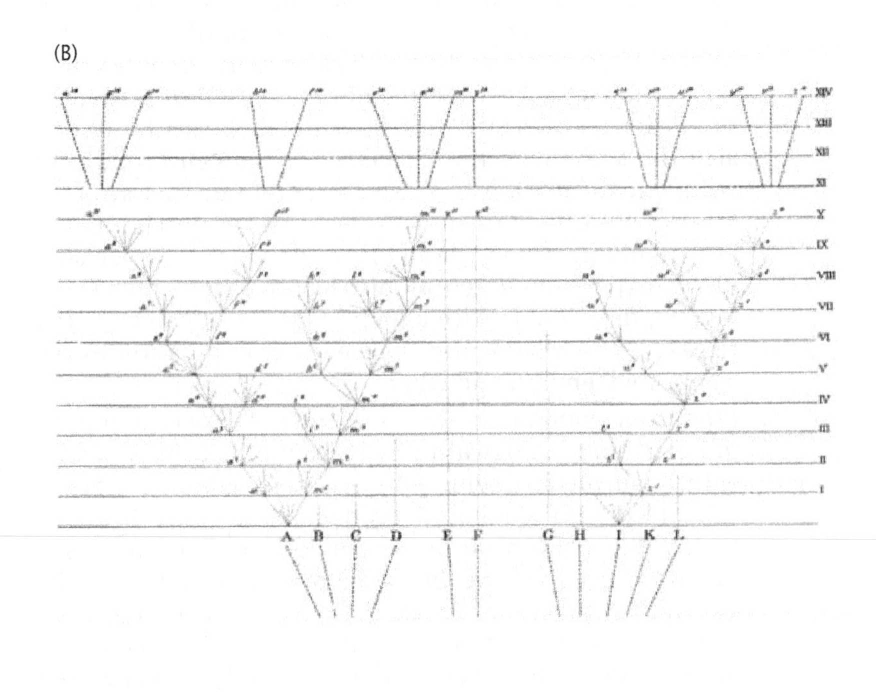

Figure 3.4: Two versions of Darwin's phylogenetic tree. (A) In his notebooks, Darwin drew the first version of what we now recognize as a common schematic demonstrating the relatedness of organisms. He introduced this speculative sketch with the words "I think," as his theory was beginning to take form. (B) In the final published version of *On the Origin of Species*, the tree had assumed more detail, showing the passage of time and explicitly indicating that most species have gone extinct. (Adapted from C. Darwin, *On the Origin of Species*, John Murray, 1859. Courtesy of The American Museum of Natural History.)

also processed sugar to make energy through the process of glycolysis and contained a rudimentary cytoskeleton consisting of an actin-like molecule and a tubulin-like molecule. We can attribute all of these features to LUCA because they are universally shared among all existing branches of cellular life. However, the demonstrable differences between redwood trees and giant squids accumulated slowly over evolutionary time as individual cellular populations became genetically isolated from one another and underwent change and divergence to fill different ecological niches. As the planet Earth is constantly being reshaped and remodeled by the uncounted legion of organisms that inhabit it, environmental niches are always unstable and can be changed by geological processes, global and local climate alterations, or the actions of competing organisms.

We can fruitfully think of evolution as the process of change in the genetic information carried by a population of related organisms. Sometimes, a single lineage can be seen as altering over time as its environment changes. More commonly, a single population will subdivide into populations that will become isolated and suffer different fates. Some will die off, some will remain similar to the ancestral population, and some will undergo significant biochemical, morphological, or behavioral alterations over time until they are ultimately recognized as new species. These basic ideas were beautifully articulated by Charles Darwin in *On the Origin of Species* and illustrated by the single figure in that book reproduced here as Figure 3.4.

How long does this evolutionary process take and how can we measure the passage of evolutionary time? It is unsatisfying to rely on

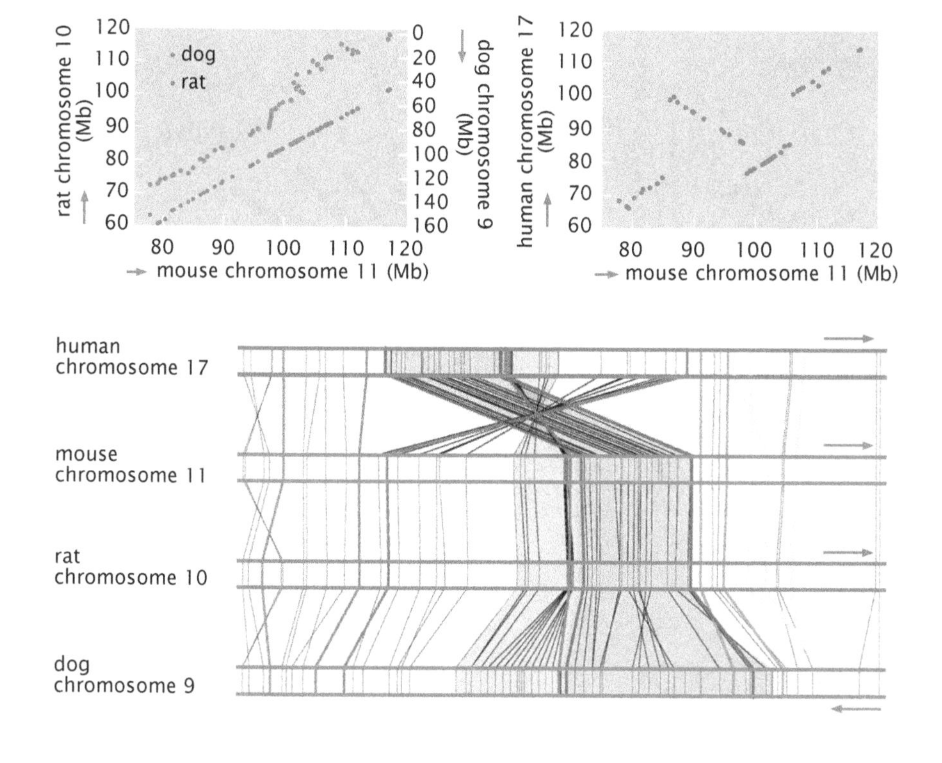

Figure 3.5: Inferring evolutionary relatedness by chromosome alignment. Equivalent regions of four chromosomes from human, mouse, rat, and dog were compared to find the location of homologous genes. The graphs at the top show the position of each gene in the rat, dog, and human sequences as a function of their positions on the mouse sequence. Because little change has occurred in chromosomal structure between the mouse and the rat, the points representing the locations of homologous genes form a nearly perfectly straight line. On the equivalent chromosomal segment from the dog, the genes are again mostly in the same order, but the spacing between them has changed substantially. The numbering convention for positions along the dog chromosome runs in the opposite direction compared with the mouse and rat chromosomes, as indicated by the arrows. Comparing the human with the mouse, a large inversion can be detected. The same data are shown in a different form in the chart at the bottom. Each vertical line on the chromosome represents a particular gene and the diagonal lines between the chromosomes link up homologs between human and mouse, mouse and rat, and rat and dog. The gray region shows a portion of the chromosome that can be followed through all the species; it is inverted between human and mouse, very similar in position and size between mouse and rat, and similar in orientation but increased in size between rat and dog because of changes in the spacing between the genes. (Adapted from G. Andelfinger et al., *Genomics* 83:1053, 2004.)

the extrapolation of mutation rates measured in artificial laboratory experiments to the evolution of species over time in the real world. Real-world conditions are much less stable or controlled than laboratory conditions, and furthermore the time scales of greatest interest for studying the evolution of species are much longer than can be achieved in the laboratory by even the most patient experimentalist. Traditionally, our understanding of evolutionary alterations depended upon two kinds of observations, namely, comparisons among currently living species and examination of the fossil record. Information about the age of particular fossils can be inferred from identification of the geological strata in which they are found, and also by examining the proportions of different radioisotopes that decay at a regular rate and thereby provide information about when the rock was formed.

Comparison of living species to ascertain the degree of relatedness was carried out for many hundreds of years before the modern theory of evolution was first described. It is immediately obvious that some organisms are more closely related than others. For example, horses and donkeys are clearly more similar to each other than either is to a dog, but horses and dogs are more similar to each other than either is to a squid. These obvious morphological differences have been the basis of the science of systematics going back to Linnaeus in the eighteenth century. In the modern era of molecular genetics, we can more easily ascribe a universal metric for genetic similarity among organisms based on similarities and differences in DNA sequence. As a population evolves over time, its DNA complement will change by several mechanisms. First, small-scale mutations or large-scale rearrangements of its genome may occur (an illustration of the consequences of this kind of rearrangement is shown in Figure 3.5). Second, the population may acquire new genes or even entire groups of genes by horizontal transfer from other organisms (horizontal transfer refers to cells passing genetic material to cells other than their own descendants). And third, the population may simply lose large chunks

Figure 3.6: Universal phylogenetic tree. This diagram shows the similarity among ribosomal RNA sequences for representative organisms from all major branches of life on Earth. The total length of the lines connecting any two organisms indicates the magnitude of the difference in their ribosomal RNA sequences.

of DNA. Thus different organisms contain different complements of genes as well as sequence differences between homologous copies of the same gene. The term "homologous" refers to descent from a common ancestor. For example, ribosomal RNAs are homologous in all cells. In Chapter 21, we will give some examples of ribosomal RNA sequences and show how they can be used to build a universal phylogenetic tree (see Figure 21.32 on p. 1009). One example of a tree based on ribosomal RNA sequences that attempts to demonstrate the relatedness among all branches of existing life is shown in Figure 3.6.

Phylogenetic trees established by molecular methods tend to be in excellent agreement with analogous trees of similarity based on morphological or biochemical criteria as have been established by botanists, zoologists, and microbiologists over the past several hundred years. We will examine statistical methods for constructing such trees in Chapter 21.

What does any of this have to do with the determination of evolutionary time? In the laboratory, we can observe that certain types of changes in DNA sequences within a population happen frequently (for example, single-point mutations changing a C to a T), while others happen more rarely (cross-over events reversing the order of all the genes within a segment of a chromosome). We can even measure the time constants that characterize such events. If we assume that these kinds of mutational events happen with the same frequency in wild populations as they do in the laboratory, then we can estimate divergence times for organismal populations based on calculating how long on average it would take to achieve the observed number of sequence alterations given known rates of sequence alteration events. In a few cases, these time estimates can be anchored by reference to the fossil record. In reality, inferring evolutionary time from sequence similarity is fraught with peril because not all sequence alterations are equally likely to be randomly incorporated into the genetic heritage of a population of organisms. Some mutations will prove to be unfavorable for a given organism's lifestyle and individuals carrying those mutations will be eliminated from the population by natural selection. Other mutations will prove to be advantageous and organisms carrying those mutations will quickly outcompete other members of their species. These selection effects can make the sequence-determined evolutionary clock appear to run too slow or too fast. Biologists face challenges similar to those faced by astronomers. In the astronomical setting, continual refinements in cosmological distance scales based on various types of standard candle (light sources of known absolute intensity) have led to increasingly refined measurements of astronomical distance. Similarly, biologists have a number of different standard stopwatches that can be used to calibrate the flow of evolutionary time.

3.1.3 The Cell Cycle and the Standard Clock

The *E. coli* Cell Cycle Will Serve as Our Standard Stopwatch

In Figure 2.1 (p. 37) we used the size of an *E. coli* cell as our standard measuring stick. Similarly, we now invoke the time scale of the *E. coli* cell cycle as our standard stopwatch. The goal of Figure 3.2 was to illustrate the variety of different processes that occur in cell biology and the time scales over which they are operative. As with our discussion of structural hierarchies, we once again use the trick of invoking *E. coli* as our reference, this time with the several thousand seconds of its cell cycle as our reference time scale.

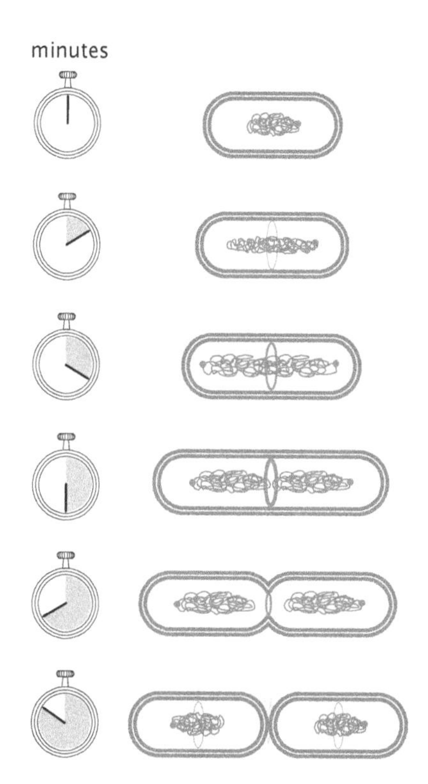

Figure 3.7: Schematic of an idealized bacterial cell cycle. A newborn cell shown at the top has a single chromosome with a single origin of replication marked by the green dot. The cell cycle initiates with the duplication of the origin, and DNA replication then proceeds in an orderly fashion around the circular chromosome. At the same time, a group of cell division proteins beginning with the tubulin analog FtsZ form a ring at the center of the cell that will dictate the future site of septum formation. As DNA replication proceeds and the cell elongates, the two origins become separated from each other, with one traveling the entire length of the cell to take up residence at the opposite pole. As the septum begins to close down, the two chromosomal masses are physically separated into the two daughter cells, where the cycle can begin anew.

As shown in Figure 3.2(C), the bacterial cell cycle will be defined as the time between the "birth" of a given cell resulting from division of a parental cell to the time of its own subsequent division. This cell cycle is characterized structurally by the segregation of the duplicated bacterial chromosome into two separate clumps and the construction of a new portion of the cell wall, or septum, that separates the original cell into two daughters. These processes are illustrated in more detail in Figure 3.7. Because *E. coli* is a roughly cylindrical cell that maintains a nearly constant cross-sectional area as it grows longer, the total cell volume can be easily estimated simply from measuring the length, and this also provides a guide as to the point in the cell cycle. As cell division proceeds, *E. coli* doubles in length and hence also doubles in volume. The time scale associated with the binary fission process of interest here is of the order of an hour (to within a factor of 2), though division can take place in under 20 minutes under optimal growth conditions.

In the previous chapter, we argued that having a proper molecular inventory of a cell is a prerequisite to building models of many problems of biological interest. Here we argue that a similar "feeling for the numbers" is needed concerning biological time scales. How long does it take for an *E. coli* cell to copy its genome and is this rate consistent with the speed of the molecular machine (DNA polymerase) that does this copying? On what time scale do newly formed proteins in neurons reach the ends of their axons and can this be explained by diffusion? Often, the time scale associated with a given process will provide a clue about what physical mechanisms are in play. In addition, one of our biggest concerns in coming chapters will be to figure out under what conditions we are justified in using ideas from equilibrium physics (as opposed to nonequilibrium physics). The answer to this question will be determined by whether or not there is a separation of time scales, and the only way we can know that is by having a feeling for what time scales are operative in a given problem. To that end, we begin by taking stock of the processes that an *E. coli* cell must undergo to copy itself.

For estimates in this book, we will choose a standard for bacterial growth in a minimal defined medium with glucose as the sole carbon source. As mentioned previously, the rate of cell division can vary more than 10-fold depending upon nutrient availability and temperature, so we must define the terms under which we will proceed with our estimates. The choice of minimal medium with glucose at 37°C is a practical one since many quantitative experiments have been performed under these conditions. With sufficient aeration, *E. coli* in this medium typically double in 40–50 minutes, and we will use 3000 s as our canonical cell cycle time. In general, time scales for biological processes are much more variable than spatial scales, although it is true that rapidly growing *E. coli* are slightly larger than slowly growing *E. coli*. The difference in size for cells growing under different conditions may be an order of magnitude less than the difference in cycle time.

COMPUTATIONAL EXPLORATION

Computational Exploration: Timing *E. coli* We noted above that the division time for *E. coli* has a characteristic time scale of a few thousand seconds. In this Computational Exploration, we make a crude estimate of the mean division time by using a time lapse video of dividing bacteria like that shown in Figure 3.8. In such a movie, a microscope with a 100× objective is set up to take a snapshot roughly every 10 minutes, though here we only show images from every 30 minutes. When viewed one after the other, these movies reveal that the cells undergo a succession of growth and division processes.

The idea in this exercise is to write a Matlab script that examines the images and determines the total area of the cells as a function of time. The phase contrast microscopy used to make this video is an imaging technique that makes it possible to clearly visualize objects like *E. coli* that would otherwise be nearly transparent when illuminated directly by white light. The first step is to find the cells in some automated fashion, a process known as segmentation, though in the simple algorithm described here we fall short of the identification of individual cells. There are many different techniques for carrying out the segmentation process. One of the simplest ideas is to use thresholding, in which all intensities below (or above) a certain threshold are assigned a value of zero (or one). One of the problems with this approach is that it is indiscriminate in the sense that objects other than cells that have different intensities than the background will be detected by the thresholding. To eliminate these false positives, we could

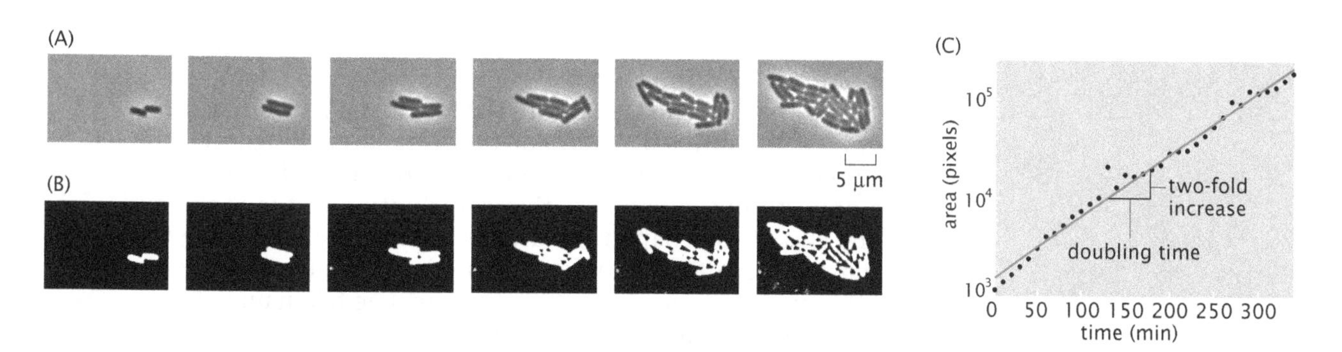

Figure 3.8: Time lapse movie of dividing *E. coli* cells. (A) Phase contrast snapshots taken 30 minutes apart showing dividing cells. (B) Thresholded versions of the same images shown in (A). (C) Plot of the area of the cells as a function of time in units of pixels. A division time of roughly 45 minutes can be read off the graph directly, which jibes with the fact that these cells were grown on agar pads with minimal media. The full set of images for this movie needed for the Computational Exploration can be found on the book's website.

go further, for example by accepting only those objects whose area falls between some upper and lower bound and with an aspect ratio that is appropriate.

For the reader who has not had much experience with image analysis, the unsophisticated treatment recommended here is intended to develop intuition in segmentation techniques such as thresholding. To be concrete, as shown in Figure 3.8(A), we have a series of images of dividing cells. We suggest that the reader start with the first image and read it into Matlab. Then, rescale the values in the image from 0 to 1. Make a plot of this new rescaled image. The next step is to set all pixels lower than some threshold value (for example, 0.4) to the value of 1 and all pixels with intensity higher than this value to 0. This algorithm is shown in schematic form in Figure 3.9. The outcome of this procedure is a series of new images like those shown in Figure 3.8(B). The reader can then compare different choices of the threshold used to see how this choice works both for this image and others from the image sequence. With a satisfactory choice of threshold, compute the area of the "cells" from each image by summing over the values on all of the pixels in the image and make a plot of this area as shown in Figure 3.8(C).

If we assume that the cells are doubling steadily, this implies that the number of cells as a function of time can be written as $N(t) = N_0 e^{kt}$. On the assumption that all cells have the same area, by multiplying both sides by the area per cell A_0, we can rewrite this equation as $A(t) = A_0 e^{kt}$. The time constant for doubling can be determined by noting that $2N_0 = N_0 e^{kt_{\text{double}}}$, which results in $t_{\text{double}} = (1/k) \ln 2$. The rate constant, in turn, can be read off of a graph like that shown in Figure 3.8(C).

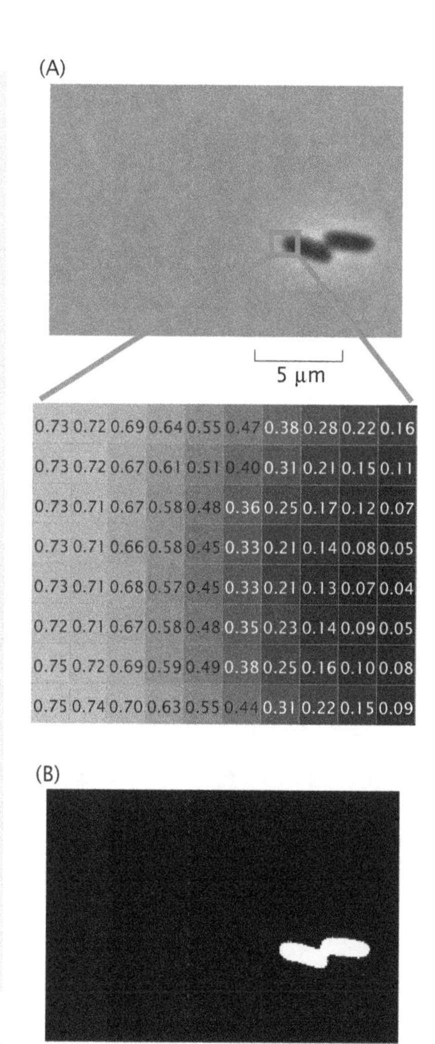

Figure 3.9: Segmentation of cells using thresholding. (A) A phase contrast image of part of a field of dividing cells. The figure below zooms in on a small region of the image and shows the intensity values of the image for all pixels within that region. (B) The result of segmenting the image with a threshold value of 0.4.

Estimate: Timing *E. coli* In Section 2.1.2 (p. 38), we sized up *E. coli* by giving a series of rough estimates for its parts list. We now borrow those estimates to gain an impression of the rates of various processes in the *E. coli* cell cycle. The simple idea behind these estimates is to take the total quantity of material that must be used to make a new cell and to divide by the time (≈3000 s) of the cell cycle. When *E. coli* is grown on minimal medium with glucose as the sole carbon source, six atoms of carbon are added to the cellular inventory for each molecule of glucose taken up. In the previous chapter, we estimated that the number of carbon atoms it takes to double the material in a cell so that it can divide in two (just the construction material) is of the order of 10^{10}. For this estimate, we ignored the material released as waste products and the reader will have the opportunity to estimate this contribution in the problems at the end of the chapter. We are also deliberately ignoring the glucose molecules that must be consumed to generate energy for the synthesis reactions—this topic will be taken up in Chapter 5. At this point, we can estimate the rate of sugar uptake required simply to deliver the 10^{10} carbon molecules necessary for building the material of the new cell: 10^{10} carbons must be captured over 3000 s, with 6 carbons per glucose molecule, giving an average rate of roughly 5×10^5 glucose molecules every second.

Of course, having the carbon present is merely a prerequisite for the macromolecular synthesis required to make a new cell. One of the most important processes in the cell cycle is replication. Given that the complete *E. coli* genome is about 5×10^6 bp in size, we can estimate the required rate of replication as

$$\frac{dN_{\text{bp}}}{dt} \approx \frac{N_{\text{bp}}}{\tau_{\text{cell}}} \approx \frac{5 \times 10^6\,\text{bp}}{3000\,\text{s}} \approx 2000\,\text{bp/s}. \tag{3.1}$$

Note that we have rounded to the nearest thousand.

Similarly, the rate of protein synthesis can be estimated by recalling from the previous chapter that the total number of proteins in *E. coli* is roughly 3×10^6, implying a protein synthesis rate of

$$\frac{dN_{\text{protein}}}{dt} \approx \frac{N_{\text{protein}}}{\tau_{\text{cell}}} \approx \frac{3 \times 10^6\ \text{proteins}}{3000\,\text{s}} \approx 1000\,\text{proteins/s}. \tag{3.2}$$

A similar estimate can be performed for the rate of lipid synthesis, resulting in

$$\frac{dN_{\text{lipid}}}{dt} \approx \frac{N_{\text{lipid}}}{\tau_{\text{cell}}} \approx \frac{5 \times 10^7\ \text{lipids}}{3000\,\text{s}} \approx 20{,}000\,\text{lipids/s}. \tag{3.3}$$

Yet another intriguing aspect of the mass budget associated with the cell cycle is the control of the water content within the cell. Recalling our estimate from the previous chapter that an *E. coli* cell has roughly 2×10^{10} water molecules results in the estimate that the rate of water uptake during the cell cycle is

$$\frac{dN_{\text{H}_2\text{O}}}{dt} \approx \frac{N_{\text{H}_2\text{O}}}{\tau_{\text{cell}}} \approx \frac{2 \times 10^{10}\ \text{waters}}{3000\,\text{s}} \approx 7 \times 10^6\,\text{waters/s}. \tag{3.4}$$

This rate of water uptake can be considered slightly differently by working out the average mass flux across the cell membrane. The flux is defined as the amount of mass crossing the membrane per unit area and per unit time, and in this instance is given by

$$j_{\text{water}} \approx \frac{dN_{\text{H}_2\text{O}}/dt}{A_{E.coli}} \approx \frac{7 \times 10^6\ \text{waters/s}}{6 \times 10^6\,\text{nm}^2} \approx 1\ \text{water/(nm}^2\text{s)}, \tag{3.5}$$

though we also note that this mass transport is mediated primarily by proteins that are distributed throughout the membrane.

Each of these estimates tells us something about the nature of the machinery that mediates the processes of the cell. In the remaining sections, these estimates will serve as our jumping-off point for estimating the rate at which individual molecular machines carry out the processes of synthesis and transport needed to support metabolism and the cell cycle.

COMPUTATIONAL EXPLORATION

Computational Exploration: Growth Curves and the Logistic Equation One of the consequences of all of this busy molecular synthesis and cell division is that cells grow and divide, grow and divide, over and over again. After less than 24 hours, a 5 mL culture that starts out with just one cell will be populated with roughly 5×10^9 cells. Of course, such growth cannot go on unchecked indefinitely. In the remainder of the book, we will repeatedly have occasion to ask for a detailed accounting of the trajectories of some quantity that is changing over time, such as the number of cells in our culture, $N(t)$ described above. In many cases, we will be able to resort to an analytic treatment of the differential equations that describe these trajectories. However, there will also be times where the differential equation of interest is beyond our means to attack analytically and we will have to resort to numerical methods. To see how the numerical approach to these problems unfolds, we consider bacterial growth.

The study of growth curves was once referred to by Jacques Monod as "the basic method of microbiology." We have already seen an example of such a growth curve in Figure 3.3, which shows data from one of the original Monod experiments that resulted in the first understanding of transcriptional regulation. This fervor for the study of bacterial growth was articulated even more forcefully and amusingly by Frederick Neidhardt in his article aptly entitled "Bacterial growth: constant obsession with dN/dt" where he reports that "One of life's inevitable disappointments—one felt often by scientists and artists, but not only by them—comes from expecting others to share the particularities of one's own sense of awe and wonder. This truth came home to me recently when I picked up Michael Guillens fine book 'Five Equations That Changed the World' and discovered that my equation—the one that shaped my scientific career—was not considered one of the five."

The simplest version of Neidhardt's life-changing equation is

$$\frac{dN}{dt} = rN. \tag{3.6}$$

This equation has far-reaching consequences since it shows how some quantity grows over time and has the familiar solution

$$N(t) = N_0 e^{rt} \tag{3.7}$$

that signals exponential growth. With the inclusion of a minus sign in front of the right-hand side, this same equation

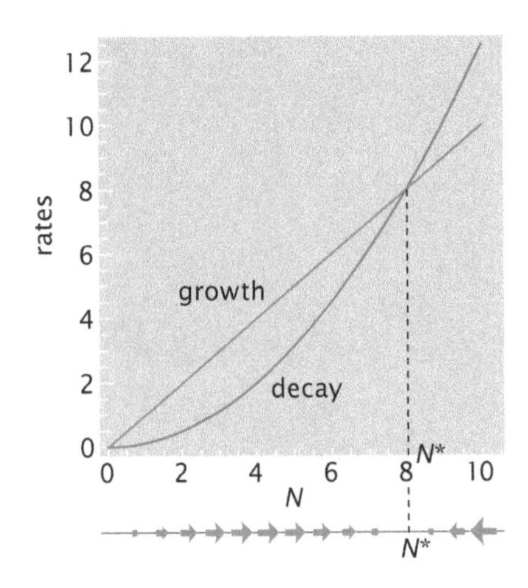

Figure 3.10: Rates as a function of number of cells. The blue curve shows how the growth rate depends upon the number of cells. The red curve shows the negative value of the decay rate. The vectors under the figure show the net growth rate. The length and size of these vectors is proportional to the rate and their direction signifies whether there is net growth or decay in the population. N^* is the fixed point of this simple dynamical system.

captures problems ranging from radioactive decay to the degradation of messenger RNAs in the cellular interior.

Of course, as noted above, despite the best efforts of humans with both population and spending, such growth cannot go unchecked. One of the simplest models that respects the fact that such growth cannot be sustained is the so-called logistic equation given by

$$\frac{dN}{dt} = rN\left(1 - \frac{N}{K}\right). \tag{3.8}$$

In this equation, K represents the maximum population that can be sustained by the environment. When N is very small compared with K, the equation shows nearly the same exponential behavior as the simple growth described by Equation 3.6. However, as N gets larger, the growth rate slows down, until when $N = K$ (that is, the population has reached the maximum carrying capacity of the environment), dN/dt falls to 0. An alternative way of looking at Equation 3.8 is to think of the cellular birth and death processes independently. This can be depicted graphically as shown in Figure 3.10. In particular, note that there is a growth rate characterized by the term rN and a death rate characterized by $-rN^2/K$. Each of these rates depends on N with different functional forms as shown in Figure 3.10. Further, the fixed point of the system is revealed by the vectors plotted beneath the graph, which show a simple one-dimensional phase portrait that tells us whether the *net* growth will be positive or negative as a function of the population size. These phase portraits allow us to run our dictum of the necessity of mathematicizing the cartoons in reverse by making it possible to cartoonize the mathematics in a clear and simple way that reveals the key features of the dynamical system. We will use such phase portraits to cartoonize the mathematics throughout this book.

Though the logistic equation given in Equation 3.8 is amenable to analytic resolution with the solution given by

$$N(t) = \frac{KN_0 e^{rt}}{K + N_0\left(e^{rt} - 1\right)}, \tag{3.9}$$

we will use it as the basis for our illustration of the simplest numerical method for integrating such equations.

Both of the differential equations written above for the time evolution of our growing bacterial culture are special cases of the more general form

$$\frac{dN}{dt} = f(N, t). \tag{3.10}$$

In this case, the function $f(N, t)$ is the growth rate and tells us how the rate depends upon the current population size. To integrate this equation, we note that we can rewrite it in terms of the discrete approximation to the derivative as

$$\frac{N(t + \Delta t) - N(t)}{\Delta t} \approx f(N, t). \tag{3.11}$$

Using this definition, we can rearrange the terms to find

$$N(t + \Delta t) = N(t) + f(N, t)\Delta t. \tag{3.12}$$

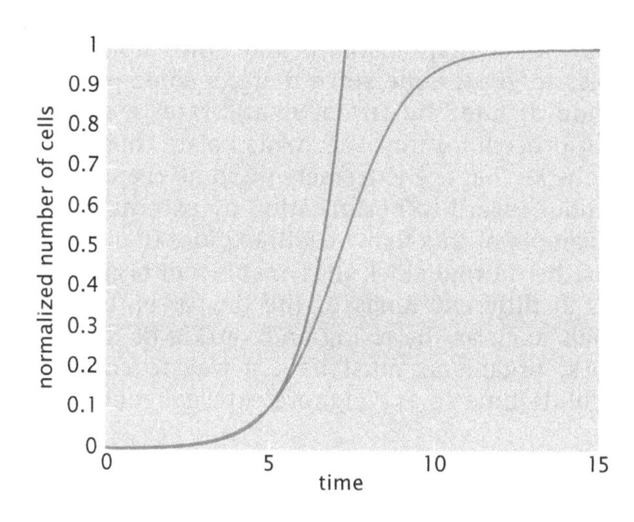

Figure 3.11: Integration of differential equations for bacterial growth. Using the simple algorithm described in the Computational Exploration, we integrate both the equation for exponential growth and the logistic equation "by hand." The red line shows the result of integration of Equation 3.6 for the case in which $r = 1$. The blue line shows the result of integration of Equation 3.8 for the case in which $r = 1$ and $K = 1$.

This formula can be easily implemented in a number of different ways within Matlab, but perhaps the easiest scheme is simply to construct a loop that repeatedly takes the current population and increments it to produce the population size a time step Δt later. The result of integrating both of the equations given above using this integration scheme is shown in Figure 3.11.

The Matlab code included on the book's website provides two versions of this exercise. One of these versions illustrates the numerical integration "by hand" described above and the other shows how to use the built-in function within Matlab for integrating differential equations. Generally, rather than writing the integration routine itself, it is often easier to exploit such built-in functions.

3.1.4 Three Views of Time in Biology

Modern humans have built much of the activity of our societies around an obsession with absolute time. This obsession is revealed by the propensity for events to occur at a certain time of day, for example, class starts at 9 a.m., or scheduling our activity by measured blocks of time, for example, you must practice the piano for half an hour. It is not clear, however, that other organisms relate to time in this manner. In the remainder of this chapter, we will discuss three different views of time that seem to be important to life, and we will term them *procedural time*, *relative time*, and *manipulated time*.

In the previous chapter, we explored the question of why biological things are a certain size, and the ultimate reason is the finite extent of the atoms that make up biological molecules. Here we are trying to understand why biological processes take a certain amount of time, a difficult task. For the most part, the size of things does not strongly depend on environment and external conditions, but the time scale of processes often does. For example, bacteria growing in leftover potato salad will replicate rapidly when the salad is left on a picnic table in full sun but much more slowly in a refrigerator. The fundamental reason for the difference in replication rates as a function of temperature can be attributed to the slowing or the acceleration of the many individual enzymatic steps that must take place for the cell to double in size and divide. In this sort of context, it appears that organisms pay attention to *procedural time* (an idea to be fleshed out in Section 3.2) rather than absolute time: they do something for as long as it takes to get it done since there is some procedure such as DNA replication dictated by an enzymatic rate. A particularly interesting class of procedural time mechanisms are those that organisms use to build clocks that are extremely good at keeping track of absolute time without regard to perturbation by external conditions. One fascinating example of this that we will explore in more detail later in the chapter is the diurnal clock that enables an organism to perform different acts at different times of the day, even in the absence of external signals such as the rising and setting of the sun. For these clocks to work, organisms must have a way to convert procedural time into absolute time so as to ignore external conditions, including temperature.

Although calculating procedural time for a process of interest can often put a lower limit on how fast that process can occur, cells often seem to put as much effort into making sure that processes occur in the correct order as in making sure that they occur quickly. In the context of cell division, for example, it would be problematic for a cell to try to segregate its chromosomes into the two daughters until the process of DNA replication is complete. The result would be that at least one daughter would lack the full genetic complement of the mother cell. We will refer to processes where one must be complete before another can start under the category of *relative time* (that is, before or after rather than how long). This topic will be explored in Section 3.3.

Third, and perhaps most interestingly, it appears that living organisms are rarely content to accept time as it is. In some cases, they seem to be impatient, demanding that their life processes occur more quickly than permitted by the underlying chemical and physical mechanisms. Rate acceleration by enzyme catalysis is a prime example. In other cases, they seem to delay the intrinsic proceeding of events, freezing time in "suspended animation" as in formation of bacterial spores that can survive for hundreds or thousands of years, only to be reanimated when conditions become favorable. In Section 3.4, we will argue that these processes are examples of what we will refer to as *manipulated time.*

3.2 Procedural Time

The underlying idea of measurements of procedural time is simply that the chemical and physical transformations characteristic of life do not happen instantaneously. Complex processes can be thought of as being built up from many small steps, each of which takes a finite amount of time. For many biological processes that are intrinsically repetitive, such as the replication of DNA or the synthesis of proteins, the same step is used over and over again, namely, addition of single nucleotides to a growing daughter strand or addition of single amino acids to a growing polypeptide chain. In this section on procedural time, we will begin by making some estimates about these processes of the central dogma as an example of the general issues of computing procedural time for multistep biological processes. Then we will move on to the interesting special examples of clocks and oscillators where procedural times are calibrated so that cell cycles and diurnal cycles can follow the constant ticking of a reliable clock.

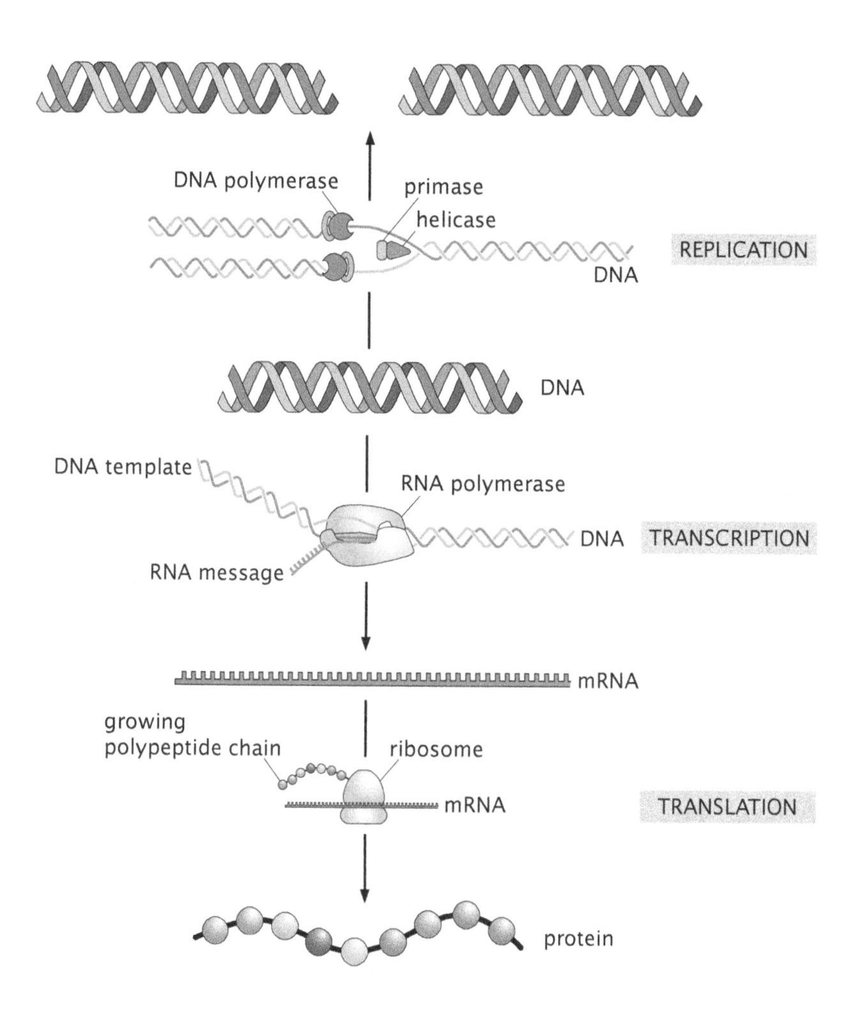

Figure 3.12: The processes of the central dogma. DNA is replicated to make a second copy of the genome. Transcription refers to the process in which RNA polymerase makes an mRNA molecule. Translation refers to the synthesis of a polypeptide chain whose sequence is dictated by the arrangement of nucleotides on mRNA.

3.2.1 The Machines (or Processes) of the Central Dogma

The Central Dogma Describes the Processes Whereby the Genetic Information Is Expressed Chemically

One of the most important classes of processes in cellular life is that associated with the so-called central dogma of molecular biology. The suite of processes associated with the central dogma are those related to the polymerization of the polymer chains that make up the nucleic acids and proteins that are at the heart of cellular life. The fundamental processes of replication, transcription, and translation and their linkages are shown in Figure 3.12. The basic message of this "dogma" in its least sophisticated form is that DNA leads to RNA, which leads to proteins. From the standpoint of cellular timing, the processes of the central dogma will serve as a prime example of procedural time. A typical circular bacterial genome, for example, is replicated by just two DNA polymerase complexes that take off in opposite directions from the origin of replication and each travels roughly half-way around the bacterial genome to meet on the opposite side. The time to replicate a bacterial genome is governed by the rate at which these polymerase motors travel to copy the roughly 5×10^6 bp of the bacterial genome. Similarly, the time to synthesize a new protein is governed by the rate of incorporation of amino acids by the ribosome.

The Processes of the Central Dogma Are Carried Out by Sophisticated Molecular Machines

One of the primary processes shown in Figure 3.12 is the copying of the genome, also known as replication. DNA replication must take place before a cell divides. As shown in Figure 3.12, the process of DNA replication is mediated by a macromolecular complex (the replisome), which has a variety of intricate parts such as the enzyme DNA polymerase, which incorporates new nucleotides onto the nascent DNA molecule, and helicases and primases that pry apart the DNA at the replication fork and prime the polymerization reaction, respectively.

The DNA molecule serves as a template in two different capacities. As described above, a given DNA molecule serves as a template for its own replication (DNA $\rightarrow$ DNA). However, in its second capacity as the carrier of the genetic material, a DNA molecule must also dictate the synthesis of proteins (the expression of its genes). The first stage in this process of gene expression is the synthesis of an mRNA molecule (DNA $\rightarrow$ RNA) with a nucleotide sequence complementary to the DNA strand from which it was copied, which will serve as the template for protein synthesis. This transcription process is carried out by a molecular machine called RNA polymerase that is shown schematically in Figure 3.12. In eukaryotes, transcription takes place in the nucleus while subsequent protein synthesis takes place in the cytoplasm, so there must be an intermediate step of mRNA export.

Once the messenger molecule (mRNA) has been synthesized, the translation process can begin in earnest (RNA $\rightarrow$ protein). As already described in Section 2.2.3 (p. 63), translation is mediated by one of the most fascinating macromolecular assemblies, namely, the ribosome. The ribosome is the apparatus that speaks both of the two great polymer languages and, in particular, forms a string of amino acids (a polypeptide chain) that are dictated by the codons (represented by three letters) on the mRNA molecule. The structure of the ribosome is indicated in cartoon form in Figure 3.12. As might be expected for a bilingual machine, the ribosome contains structural components of both RNA and protein. The two halves of the ribosome clamp an mRNA and then the ribosome moves processively down the length of the mRNA. As the ribosome moves along, successive triplets of nucleotides are brought into registry with active sites in the ribosomal machinery that align special RNA molecules (tRNA), charged with various amino acids, to recognize the complementary triplet codon. Subsequently, the ribosome catalyzes transfer of the correct amino acid from the tRNA onto a growing polypeptide chain and releases the now empty tRNA. This set of processes was indicated schematically in Figure 3.2(E). As shown in Figure 3.13, the nascent mRNA molecules in bacteria are immediately engaged by ribosomes so that protein translation can occur before transcription is even finished.

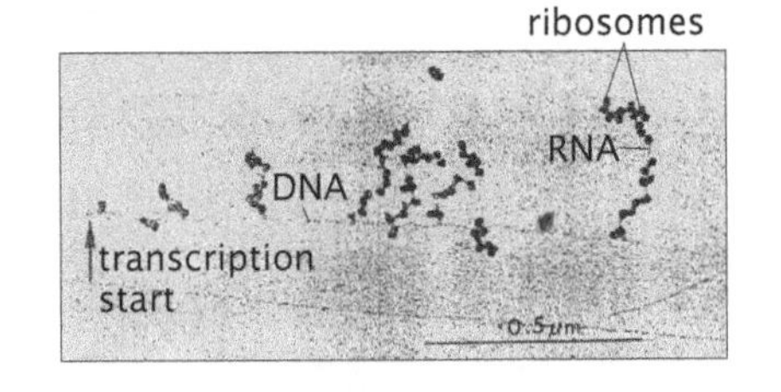

Figure 3.13: Electron microscopy image of simultaneous transcription and translation. The image shows bacterial DNA and its associated mRNA transcripts, each of which is occupied by ribosomes. (Adapted from O. L. Miller et al., *Science* 169:392, 1970.)

The timing of all three of these processes is dictated by the intrinsic rate at which these machines carry out their polymerization reactions. All of them can be thought of in the same framework as repetitions of N essentially identical reactions, each of which takes an average time Δt to perform (though clearly there are substantial fluctuations). We will now estimate total times for each of the three central processes of the central dogma.

Estimate: Timing the Machines of the Central Dogma

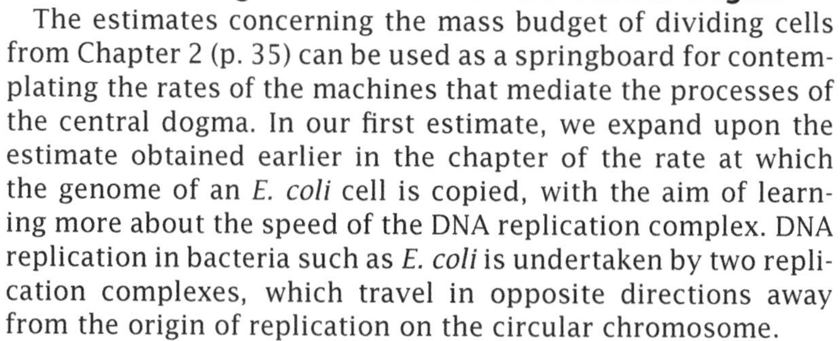

The estimates concerning the mass budget of dividing cells from Chapter 2 (p. 35) can be used as a springboard for contemplating the rates of the machines that mediate the processes of the central dogma. In our first estimate, we expand upon the estimate obtained earlier in the chapter of the rate at which the genome of an *E. coli* cell is copied, with the aim of learning more about the speed of the DNA replication complex. DNA replication in bacteria such as *E. coli* is undertaken by two replication complexes, which travel in opposite directions away from the origin of replication on the circular chromosome.

Given that the complete *E. coli* genome is about 5×10^6 bp in size and it is copied in the 3000 s of the cell cycle, we have already found that the rate of DNA synthesis is roughly 2000 bp/s, or 1000 bp/s per DNA replication complex (replisome) since there are two of these complexes moving along the DNA simultaneously. Biochemical studies have found rates for the DNA polymerase complex in the 250–1000 bp/s range. As we have mentioned, *E. coli* are actually capable of dividing in much less than 3000 s, in fact, in as little as 1000 s, although their DNA replication machinery cannot proceed any faster than this absolute speed limit. How do they achieve this? For now, we will leave this as an open mystery and will return to the question in the final section of the chapter on manipulating time.

For a bacterial cell, transcription involves the synthesis of mRNA molecules with a typical length of roughly 1000 bases. Our reasoning is that the typical protein has a length of 300 amino acids, with 3 bases needed to specify each such amino acid. Both bulk and single-molecule studies have revealed that a characteristic transcription rate is tens of nucleotides per second. Using 40 nucleotides/s, we estimate that the time to make a typical transcript is roughly 25 s.

Yet another process of great importance in the central dogma is protein synthesis by ribosomes. Recall from our estimates in the previous chapter that the number of proteins in a "typical" bacterial cell like *E. coli* is of the order of 3×10^6. This suggests, in turn, that there are of the order of 9×10^8 amino acids per *E. coli* cell, which are incorporated into new proteins over the roughly 3000 s of the cell cycle. We have made the assumption that each protein has 300 amino acids. This implies that the mean rate of amino acid incorporation per second is given by

$$\frac{dN_{aa}}{dt} \approx \frac{9 \times 10^8 \text{ amino acids}}{3000\,\text{s}} \approx 3 \times 10^5 \text{ amino acids/s.} \qquad (3.13)$$

The number of ribosomes at work on synthesizing these new proteins is roughly 20,000, which implies that the rate per ribosome is 15 amino acids/s, while the measured value is 25 amino acids incorporated per second. These numbers also imply that the mean time to synthesize a typical protein is roughly 20 s.

One of our conclusions is that the rate of protein synthesis by the ribosome is slower than the rate of mRNA synthesis by RNA polymerase. However, as shown in Figure 3.13, multiple ribosomes can simultaneously translate a single mRNA by

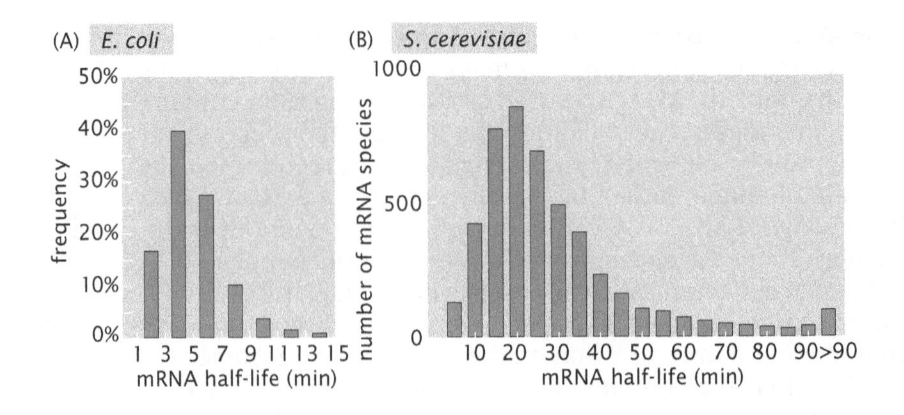

Figure 3.14: mRNA lifetimes. (A) Measurements of the lifetimes of thousands of different mRNA transcripts in *E. coli* using microarrays. The mean lifetime is slightly in excess of 5 minutes. (B) Measurements of the lifetimes of thousands of different mRNA transcripts in the yeast *S. cerevisiae* using microarrays. (A, adapted from J. A. Bernstein et al., *Proc. Natl Acad. Sci. USA* 99:9697, 2002; B, adapted from Y. Wang et al., *Proc. Natl Acad. Sci. USA* 99:5860, 2002.)

proceeding linearly in an orderly fashion and indeed, multiple RNA transcripts may exist in different degrees of completion, being transcribed from the same genetic locus. Thus, when considering the net rates of processes in cells, the number of molecular players is clearly as important as the intrinsic rate.

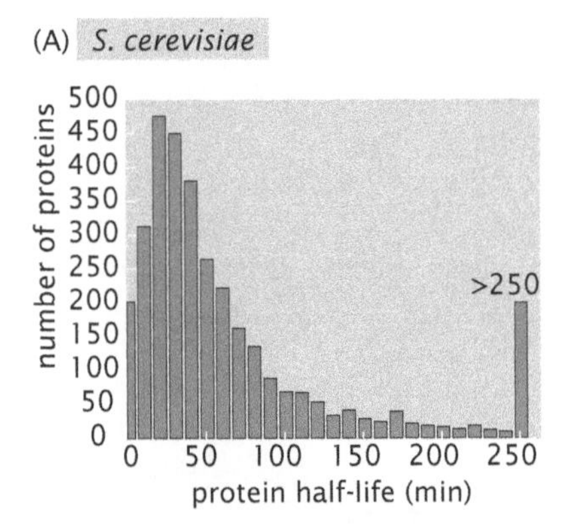

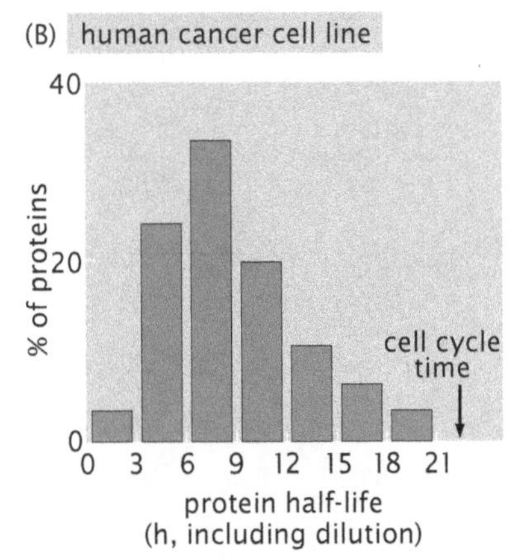

Figure 3.15: Protein lifetimes. Measurements of the lifetimes of thousands of different proteins. (A) Distribution of lifetimes of proteins in the yeast *S. cerevisiae*. (B) Distribution of lifetimes of proteins in human cells. (A, adapted from A. Belle et al., *Proc. Natl Acad. Sci. USA* 103:13004, 2006; B, adapted from E. Eden et al., *Science* 331:764, 2011.)

The estimates we have constructed for the rates at which the processes of the central dogma play are not the entire story of the overall census of mRNAs and their protein partners. Both mRNA transcripts and their downstream protein products are subjected to decay processes that take them out of circulation. To see this more explicitly, consider the results presented in Figure 3.14, which reflect measurements on literally thousands (that is, >2000) of different mRNA transcripts in *E. coli* and yeast. In both of the studies presented here, the factors determining mRNA lifetime were more subtle than some of the favored hypotheses such as a relation to secondary structure in the mRNA or the abundance of the gene products. One of the fascinating outcomes of the yeast study was the finding that mRNAs whose protein products are part of the same macromolecular assembly tend to have similar lifetimes.

Generally, the decay times for proteins are longer than those for mRNAs, as shown in Figure 3.15. Indeed, the half-lives are generally sufficiently long that the dominant effect in bacteria is protein elimination through cell division rather than by active degradation. That is, when cells divide, the average number of proteins per cell is reduced by a factor of two. In the case of eukaryotes like those shown in Figure 3.15, the lifetimes are between tens and hundreds of minutes. Degradation dynamics is explored further in the problems at the end of the chapter.

3.2.2 Clocks and Oscillators

In the context of the central dogma, we have described measurements of procedural time for processes that essentially happen once and run to completion, such as the synthesis of a protein molecule. However, many cellular processes run in repeated regular cycles. These cyclic or oscillatory processes frequently represent control systems where procedural times of some subprocess can be used to set the oscillation period. Two widely studied examples are the oscillators used to drive the cell division cycle and the mechanisms governing behavioral switches between daytime and nighttime, which will be explored in detail below. These daily clocks are called circadian or diurnal oscillators. Other everyday oscillators run the beating of our hearts and the pattern of our breathing.

Developing Embryos Divide on a Regular Schedule Dictated by an Internal Clock

One of the best understood examples of an oscillatory clock used by cells is seen in the early embryonic cell cycle of many animals. The best-studied example is the South African clawed frog, *Xenopus laevis* (abbreviated *X. laevis*). After the giant egg (1 mm) is fertilized, a cell division cycle proceeds roughly every 20 minutes until the egg has been cleaved into approximately 4000 similarly sized cells. The regularity and synchrony of these cell divisions reflects an underlying oscillatory clock based on a clever manipulation of procedural time. The clock starts each cell division with the synthesis of a protein called cyclin. Cyclin is made from a relatively rare mRNA. As a result, the protein accumulates slowly. The biological function of cyclin is to activate a protein kinase, an enzyme that covalently attaches phosphate groups to amino acid side chains on other proteins. This process, known as phosphorylation, is one of the key ways that protein activity can be controlled after translation. Essentially, the protein is inactive in the absence of its phosphate group. Kinase activation cannot begin until the cyclin protein has accumulated to a certain threshold level. After the kinase is activated, one of its targets is an enzyme, which in turn catalyzes the destruction of the cyclin protein. All cyclin in the cell is quickly destroyed with a half-life of 90 s, resetting the clock to its zero position.

The regularity of this oscillatory clock depends upon several measurements of procedural time. First, accumulation of the cyclin protein to its threshold level depends upon the rate of ribosomal synthesis of that protein. Second, activation of the cyclin-dependent protein kinase kicks off a second procedural time measurement, which reflects the length of time required by the kinase to encounter and phosphorylate its enzymatic substrates. Third, the degradation of the cyclin protein also requires a fixed, but brief, amount of time. The sum of these three procedural times gives the total period of the clock. The outcome of these molecular events in terms of molecular concentrations is illustrated in Figure 3.16. Just as for all the examples of procedural time described above, the amount of absolute time in seconds, minutes, or hours may change depending upon external conditions such as temperature.

This cyclin-driven cell-cycle oscillator is one example of a very general category of two-component oscillatory systems found throughout biology. A simplified idealized representation of such an oscillator is shown in Figure 3.17(A), while a more accurate representation of the real cell-cycle control system from the yeast *S. cerevisae* is shown in Figure 3.17(B).

Diurnal Clocks Allow Cells and Organisms to Be on Time Everyday

A second example of the use of procedural time to build a clock is when cells arrange a series of molecular processes in such a way that they can measure an absolute time. Unlike the cell-cycle clock, it is critical that the diurnal clock does not change its period when the temperature changes, such as during the change of seasons. Many

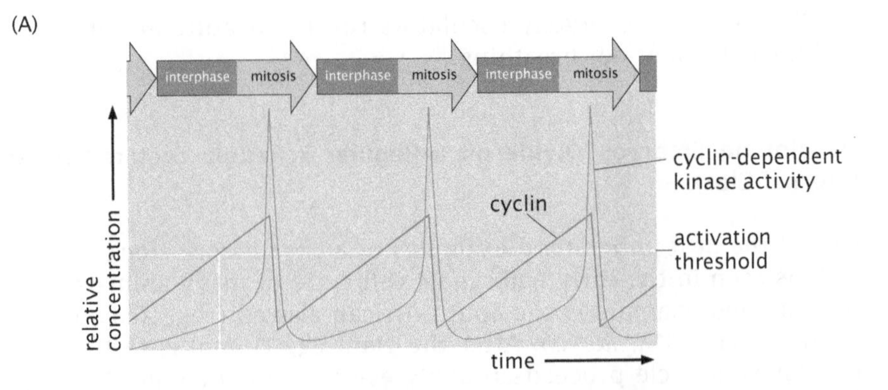

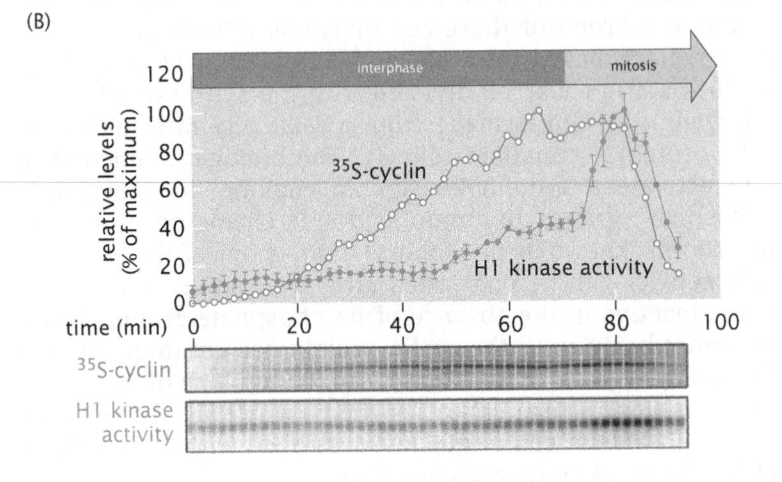

Figure 3.16: The oscillatory cell-cycle clock. This diagram shows the procedural events that underlie the regular oscillations of the cell-cycle clock in the *X. laevis* embryo. (A) Cyclin protein concentration rises slowly over time until it reaches a threshold at which point it activates cyclin-dependent kinase. Cyclin-dependent kinase activity increases sharply at this threshold and in turn activates enzymes involved in cyclin protein degradation. Once the degradation machinery is turned on, cyclin protein levels quickly fall back to zero. Cyclin-dependent kinase activity also falls and the degradation machinery inactivates. This oscillatory cycle is repeated many times. (B) Actual data for a single cycle of the oscillation. (Adapted from J. R. Pomerening et al., *Cell* 122:565, 2005.)

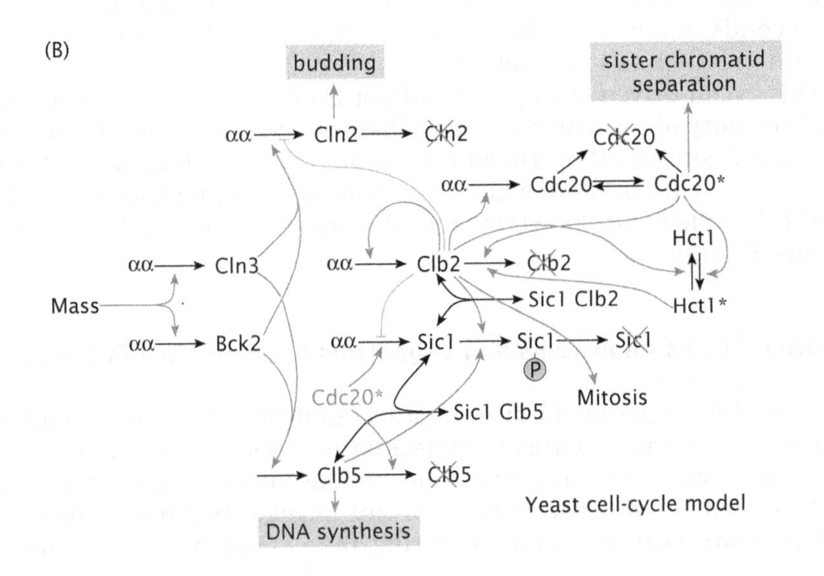

Figure 3.17: Logic diagrams for the construction of cell-cycle oscillators. (A) The minimal oscillator requires only two components. The first component activates the second component, for example by catalyzing its synthesis. The second component inhibits the first, for example by catalyzing its degradation. (B) A biochemically realistic representation of the cell-cycle oscillator in yeast is outrageously more complicated. This is because the real oscillator must work under a wide variety of conditions, be insensitive to fluctuations in the concentrations or activities of its components, and be subject to multiple kinds of regulatory inputs. (Adapted from N. W. Ingolia and A. W. Murray, *Curr. Biol.* 14:R771, 2004.)

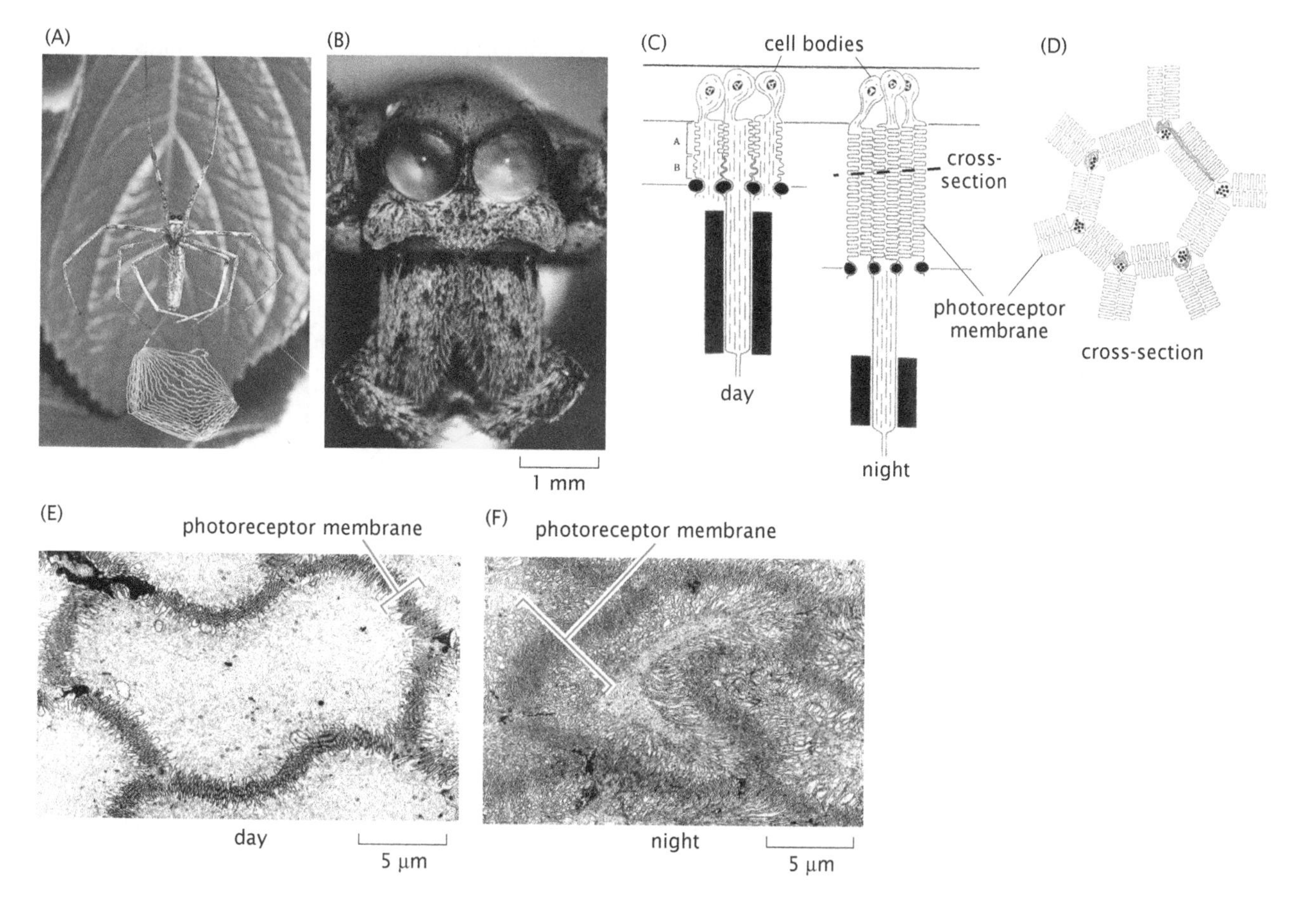

Figure 3.18: An extreme example of a structural change driven by the diurnal clock. (A) The net-casting spider, *Deinopis subrufa*, is a nocturnal hunter with an unusual strategy. It spins a small net, which it holds with its legs and tosses to entangle unwary passing prey. (B) In order to see the prey and know when to toss its net, the spider must have excellent night vision. Two of its eight eyes are extremely enlarged and exquisitely light-sensitive. (C) The light sensitivity of the spider's eyes changes by a factor of approximately 1000 between daytime and nighttime. During the day, the photoreceptor cell processes are short and fairly disorganized. At night, the total amount of membrane containing light sensors increases both by lengthening of the cells and by the construction of convoluted membrane folds, all packed with photoreceptor molecules. (D) In cross-section, the photosensitive membranes of neighboring cells abut each other, forming a regular tile-like pattern. (E) A cross-section through the photoreceptor cells of a spider sacrificed during the day shows relatively modest thickening of the boundary membranes. (F) An equivalent section taken from a spider sacrificed at night shows a vast increase in the number and size of the membrane folds. At dawn, these membranes will all be degraded, only to be resynthesized the following dusk. (A, courtesy of Mike Gray, Australian Museum; B, courtesy of The Smith-Kettlewell Eye Research Institute; C–F, adapted from A. D. Blest, *Proc. R. Soc. Lond.* B200:463, 1978.)

organisms perform some specific task at the same time everyday. A spectacular example is shown in Figure 3.18, where an animal alters the light sensitivity of its eyes in anticipation of sundown. While we might imagine that these kind of daily changes are triggered by, for example, the intensity of sunlight, it has been demonstrated for many organisms that they continue to perform their diurnal cycle even when kept in the dark. Direct observation of these cycles over long periods of time in cyanobacteria has demonstrated that they can operate with tight precision over weeks without any external cues about absolute time.

Different organisms use information about the time of day for vastly different purposes. Nevertheless, as illustrated in Figure 3.19, the molecular circuitry governing their circadian rhythms conserves certain common features. Generally, these systems include positive elements that activate transcription of so-called clock genes that drive rhythmic biological outputs as well as promoting the expression of

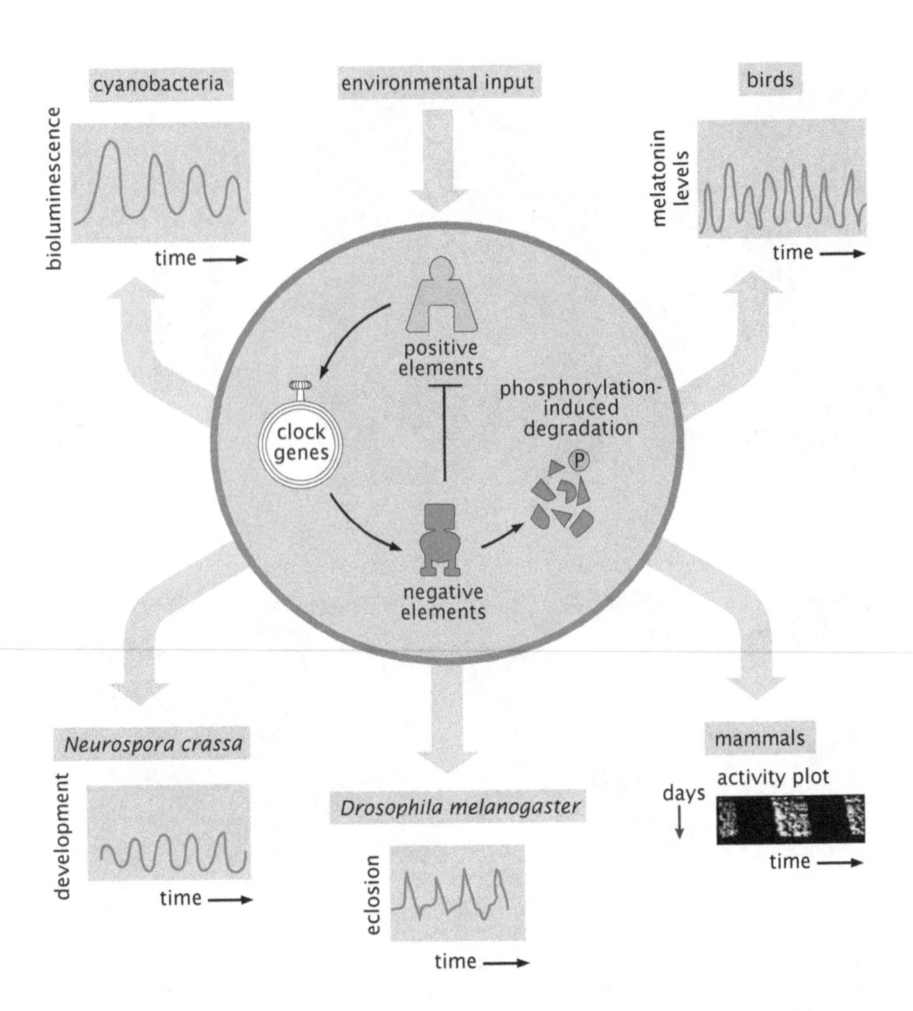

Figure 3.19: Schematic showing generic features of circadian clocks. Circadian clock mechanisms are autonomously driven oscillators that can be modulated by external inputs. Different organisms ranging from cyanobacteria to fungi to insects, birds, and mammals use their circadian timers to regulate different kinds of biological outputs and also use very different kinds of protein components in the internal circuitry. (Adapted from D. Bell-Pedersen et al., *Nat. Rev. Genet.* 6:544, 2005.)

negative elements that inhibit the activities of the positive elements. Phosphorylation of the negative elements leads to their degradation, allowing them to restart the cycle. Although the circadian oscillators are capable of continuing to measure time in constant light or constant darkness, they can nevertheless accept inputs from environmental signals such as the sun to reset their phase. Humans commonly experience the inefficiency of the phase-resetting mechanisms as the phenomenon of jetlag.

The circadian oscillator known to function with the fewest components is the one from the photosynthetic cyanobacterium *Synechococcus elongatus.* This organism's clock requires just three proteins, called KaiA, KaiB, and KaiC, and remarkably it appears that neither gene transcription nor protein degradation is necessary for this clock to function. A purified mixture of just these three proteins together with ATP is capable of sustaining an oscillatory cycle of KaiC protein phosphorylation over periods of at least several days. KaiC is able to catalyze both its own phosphorylation and its own dephosphorylation. KaiA enhances KaiC autophosphorylation and KaiB inhibits the effect of KaiA. Figure 3.20 shows the data supporting this remarkable finding.

3.3 Relative Time

The examples in the previous section on procedural time have emphasized the ways that cells set and measure the time that it takes to accomplish specific tasks. Some processes are rapid and others are

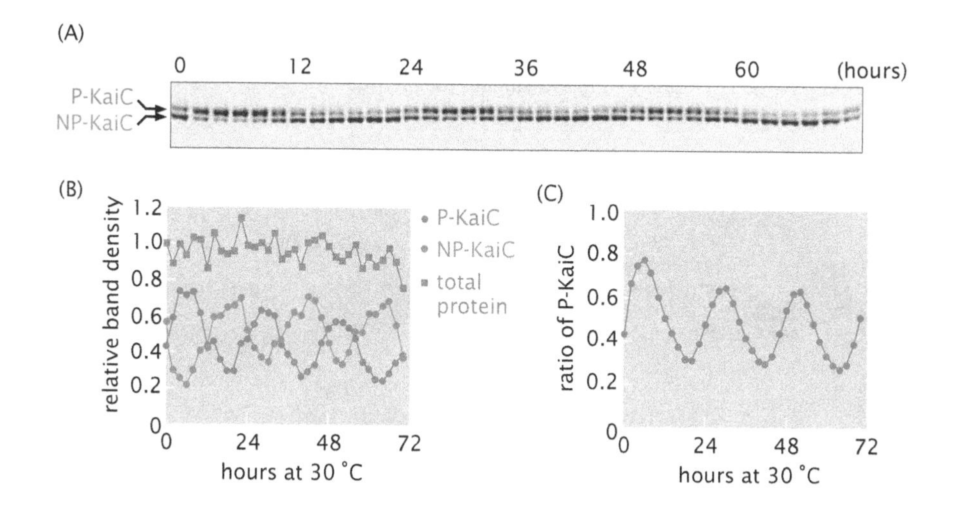

Figure 3.20: Reconstitution of the circadian oscillator. (A) In a mixture of purified KaiC protein together with KaiA, KaiB, and ATP, the molecular weight of KaiC can be seen to shift up and down slightly over a 24 h period. The upper band on this gel shows the phosphorylated form of KaiC (P-KaiC) protein and the lower band is the nonphosphorylated form (NP-KaiC). (B) Quantitation of the density of these two bands over time reveals that their concentration oscillates in a reciprocal manner such that the total amount of KaiC protein remains roughly constant. (C) The ratio of the amount of phosphorylated KaiC to total KaiC oscillates with a regular period of slightly under 24 h. (Adapted from M. Nakajima et al., *Science* 308:414, 2005.)

slow because of intrinsic features or environmental circumstances. In the well-regulated life of the cell, it is frequently important that fast and slow processes not be permitted to run independently of one another, but instead be linked in a logical sequence that depends upon the cell's needs. In this context, we now turn to what we will call *relative time*, which includes the governing mechanisms that ensure that related processes can be strung together in a "socks before shoes" fashion in which event A must be completed before event B can begin. Event C dutifully awaits the completion of event B before it begins, and so on.

3.3.1 Checkpoints and the Cell Cycle

In our initial discussion of the eukaryotic cell division cycle in the context of clocks and oscillators, we used the example of early embryonic divisions in the frog *X. laevis* and asserted that the underlying driver was a simple two-component oscillator. Once past the earliest stages of embryonic development, the cell cycle becomes much more complex and, in particular, becomes sensitive to feedback control from the cell's environment. The points in the cell cycle that are subject to interruption by external signals are referred to as checkpoints. These checkpoints ensure, for example, that chromosomes are not segregated until the DNA replication process is complete.

The Eukaryotic Cell Cycle Consists of Four Phases Involving Molecular Synthesis and Organization

Figure 3.21 shows the key features of the eukaryotic cell cycle, with an emphasis on the regulatory checkpoints that ensure that all processes will occur in the correct order. There is no single universal time scale for the eukaryotic cell cycle, which can vary greatly from one cell type to the next. In the human body, some cells in the intestinal lining can divide in as little as 10–12 hours, while others such as some tissue stem cells have cell cycles measured in days or weeks. The eukaryotic cell cycle is usually described in terms of four stages denoted as G_1, S, G_2, and M, where the M phase includes the most recognized features, namely, nuclear division (mitosis) and cell division (cytokinesis), the two G (gap) phases are periods of growth, and the S (synthesis) phase is the period during which the nuclear DNA is replicated. Together, the phases other than the M phase constitute

Figure 3.21: The eukaryotic cell cycle. (A) This cartoon shows some of the key elements of the process of cell division, including the four phases G_1, S, G_2, and M, as well as some of the most important checkpoints. Cells that stop proliferating can exit the cell cycle at G_1 and enter a resting phase, called G_0. Most fully differentiated cells in the adult human body are in G_0. Under particular circumstances, non-dividing cells in G_0 can reenter the cell cycle at G_1. (B) Time course of cell mass and DNA content during the cell cycle. Cell mass can increase continuously, while DNA content increases only during S phase. (A, adapted from T. D. Pollard and W. C. Earnshaw, Cell Biology, W. B. Saunders, 2007; B, adapted from A. Murray and T. Hunt, The Cell Cycle, Oxford University Press, 1993.)

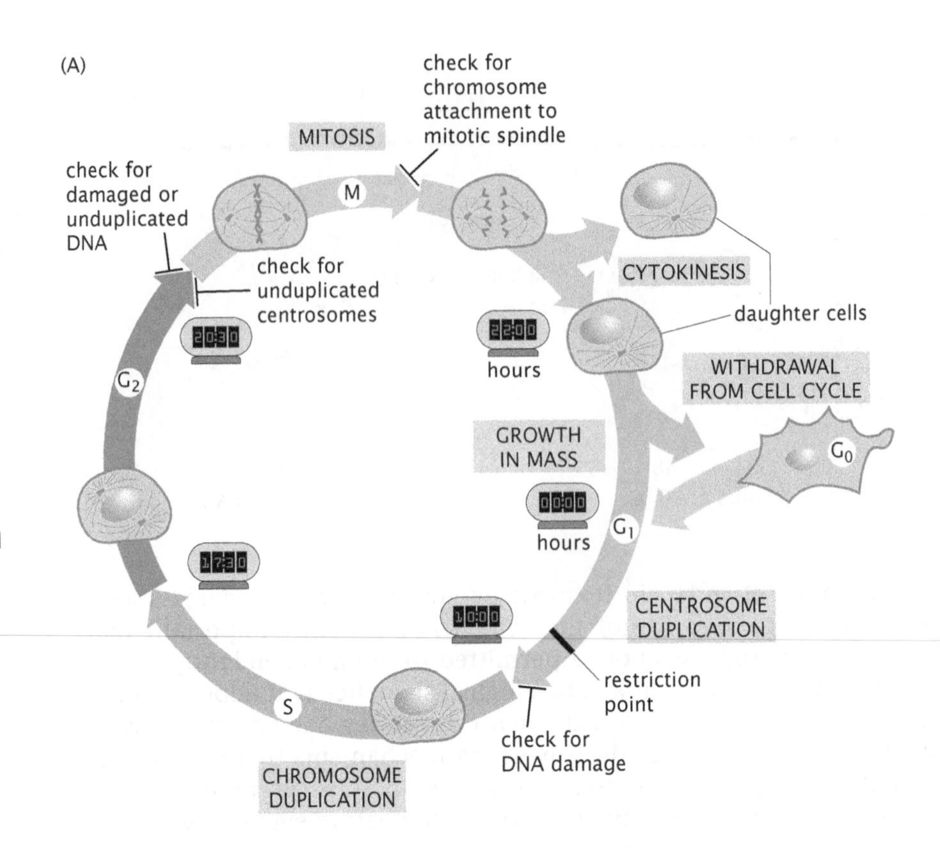

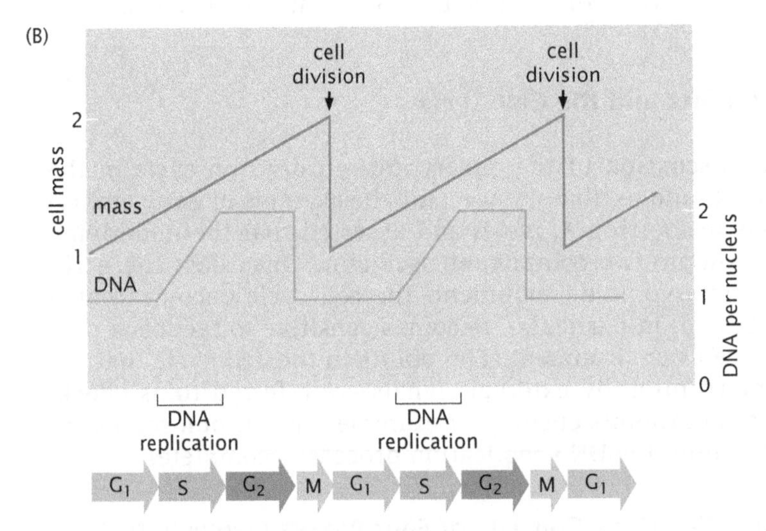

interphase. During interphase, the mass content of the cell increases, as does its size.

If we use a cultured animal cell such as a fibroblast as our standard, G_1 is roughly 10 hours long, is characterized by a significant increase in cell mass, and culminates in a checkpoint to insure sufficient cell size and appropriate environmental conditions to pass to the next stage. Throughout most of G_1, the cell has only a single centrosome, forming the core of the microtubule organizing center. Near the end of G_1, the centrosome is duplicated so that the two daughter centrosomes can eventually form the two poles of the microtubule-rich mitotic spindle. At this point, the cell examines itself for DNA damage. Any signs of damage such as double-strand breaks will trigger a checkpoint control that prevents the cell from initiating DNA replication until the damage is completely repaired. This checkpoint also insures a critical aspect of the regulation of relative time for the cell's replication, specifically, that it must have grown to approximately twice its prior size before it is allowed to begin to divide. If this checkpoint is successfully passed, then DNA replication can begin. In the S phase, the eukaryotic DNA is replicated over a period of about 6 hours.

Following the S phase and a shorter gap phase called G_2, another checkpoint verifies that every chromosome has been completely replicated and the centrosome has been properly duplicated before the initiation of the assembly of the mitotic spindle, the microtubule-based apparatus that physically separates the chromosomes into the two daughter cells. This enforcement of relative time is particularly critical because if a cell were to try to segregate the chromosomes before replication was complete, then at least one of the daughters would inherit an incomplete copy of the genome. After passing this checkpoint, the M phase begins. This relatively brief period of the cell cycle (of the order of 1 hour) involves most of the spectacular events of cell division that can be directly observed in a light microscope. Within the M phase, again it is critical that events occur in the proper order. The bipolar mitotic spindle built from microtubules forms symmetric attachments to each pair of replicated sister chromosomes. When they have all been attached to the spindle, the chromosomes all suddenly and simultaneously release their sisters and are pulled to opposite poles. A spindle-assembly checkpoint insures that every chromosome is properly attached before segregation begins. The molecular mechanisms governing the enforcement of relative time in the cell cycle involve protein phosphorylation and degradation events as well as gene transcription. In order to delve deeper into the principles governing the measurement and enforcement of relative time, we will now turn to a different example where gene transcription is the principal site of regulation.

3.3.2 Measuring Relative Time

There is great regulatory complexity involved in orchestrating the cell cycle. That is, the time ordering of the expression of different genes follows a complex program, with certain parts clearly following a progression in which some processes must await others before beginning. By measuring the pattern of gene expression, it becomes possible to explore the relative timing of events in the cell cycle. To get a better idea of how this might work, we need to examine how networks of genes are coupled together.

Genetic Networks Are Collections of Genes Whose Expression Is Interrelated

Sets of coupled genes are shown schematically in Figure 3.22. For simplicity, this diagram illustrates how the product of one gene can alter the expression of some other gene. Perhaps the simplest regulatory motif is direct negative control in which a specific protein binds to the promoter region on DNA of a particular gene and physically blocks binding of RNA polymerase and subsequent transcription. This protein is itself the result of some other gene, which can in turn be subject to control by yet other proteins (or perhaps the output of the gene that it controls). The second broad class of regulatory motifs is referred to as activation and results when a regulatory protein (a transcription

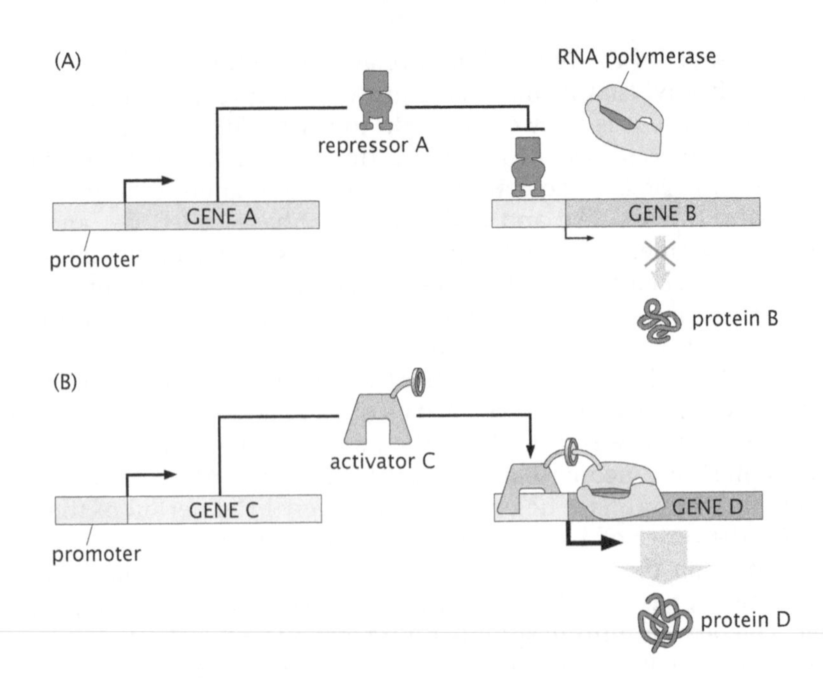

Figure 3.22: Simple networks of interacting genes. (A) Gene A encodes a repressor protein, which binds to the promoter region of gene B and prevents it from being transcribed by RNA polymerase. Only very small amounts of protein B are expressed. (B) Gene C encodes an activator protein, which helps to recruit RNA polymerase to the promoter of gene D. Large amounts of protein D are expressed.

factor called an activator) binds in the vicinity of the promoter and "recruits" RNA polymerase to its promoter.

One way to measure the extent of gene expression is to use a DNA microarray. The idea behind this technique is that a surface is decorated in an ordered x–y array with fragments of DNA, and the sequence of each spot on this array is different. To take a census of the current mRNA contents of a cell (which gives a snapshot of the current expression level of all genes), the cell is broken open and the mRNA contents are hybridized (that is, they bind to their complementary strand via base pairing) to the DNA on the microarray. DNA molecules on the surface that are complementary to RNA molecules in the cell lysate will hybridize with their complementary fragments, and those molecules that bind are labeled with fluorescent tags so that the fluorescence intensity can be used as a readout of the number of bound molecules. There is a bit more subtlety in the procedure than we have described, since really the mRNA is turned into DNA first, but we focus on the concept of the measurement rather than its practical implementation. The intensity of the spots on the microarray report the extent to which each gene of interest was expressed. By repeating this measurement again and again at different time points, it is possible to profile the state of gene expression for a host of interesting genes at different times in the cell cycle. These measurements yield a map of the *relative* timing of different genes.

One of the key model systems for examining the bacterial cell cycle is *Caulobacter crescentus*. In this system, DNA microarray analysis was used to examine the timing of various processes during the course of the cell cycle. In a beautiful set of experiments, roughly 20% of the *Caulobacter* genome was implicated in cell-cycle control as a result of time-varying mRNA concentrations that were slaved to the cell cycle itself. The idea of the experiment was to break open synchronized cells every 15 minutes and to harvest their mRNA. Then, by using a DNA microarray to find out which genes were being expressed at that moment, it was possible to put together a profile of which genes were expressed when. The outcome of this experiment is shown in

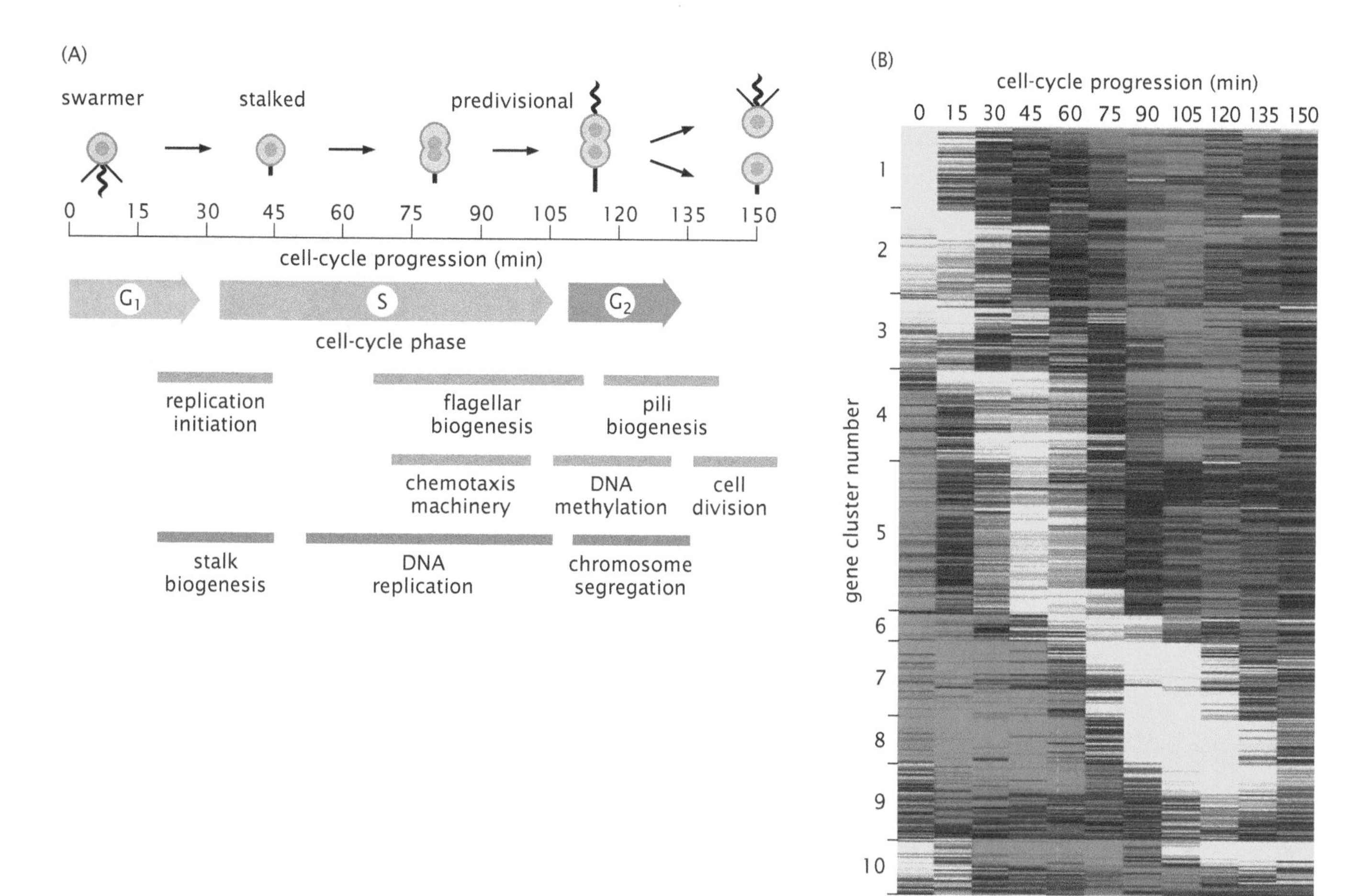

Figure 3.23: Gene expression during the cell cycle for *Caulobacter crescentus*. (A) The 150 minute cell cycle of *Caulobacter* is shown, highlighting some of the key morphological and metabolic events that take place during cell division. (B) The microarray data show how different batteries of related genes are expressed in a precise order. The genes are organized by time of peak expression. Each row corresponds to the time history of expression for a given gene, with time running from 0 to 150 minutes from left to right. For each gene, yellow indicates the time with the highest level of gene expression and blue indicates the time with the lowest level of gene expression. From top to bottom, the genes are organized into different clusters associated with different processes such as DNA replication and chromosome segregation. (Adapted from M. T. Laub et al., *Science* 290:2144, 2000.)

Figure 3.23. What these experiments revealed is the *relative* timing of different events over the roughly 150 minutes of the *Caulobacter* cell cycle.

The Formation of the Bacterial Flagellum Is Intricately Organized in Space and Time

A higher-resolution look at the relative timing of cellular events is offered by the macromolecular synthesis of one of the key organelles for cell motility, the bacterial flagellum. Figure 3.24 shows the various gene products (FlgK, MotB, etc.) that are involved in the formation of the bacterial flagellum. Essentially, each of these products corresponds to one of the protein building blocks associated with flagellar construction. Once the flagella are assembled, the cell propels itself around by spinning them. The dynamical question posed in the experiment is the extent to which the expression of the genes associated with these different building blocks is orchestrated in time.

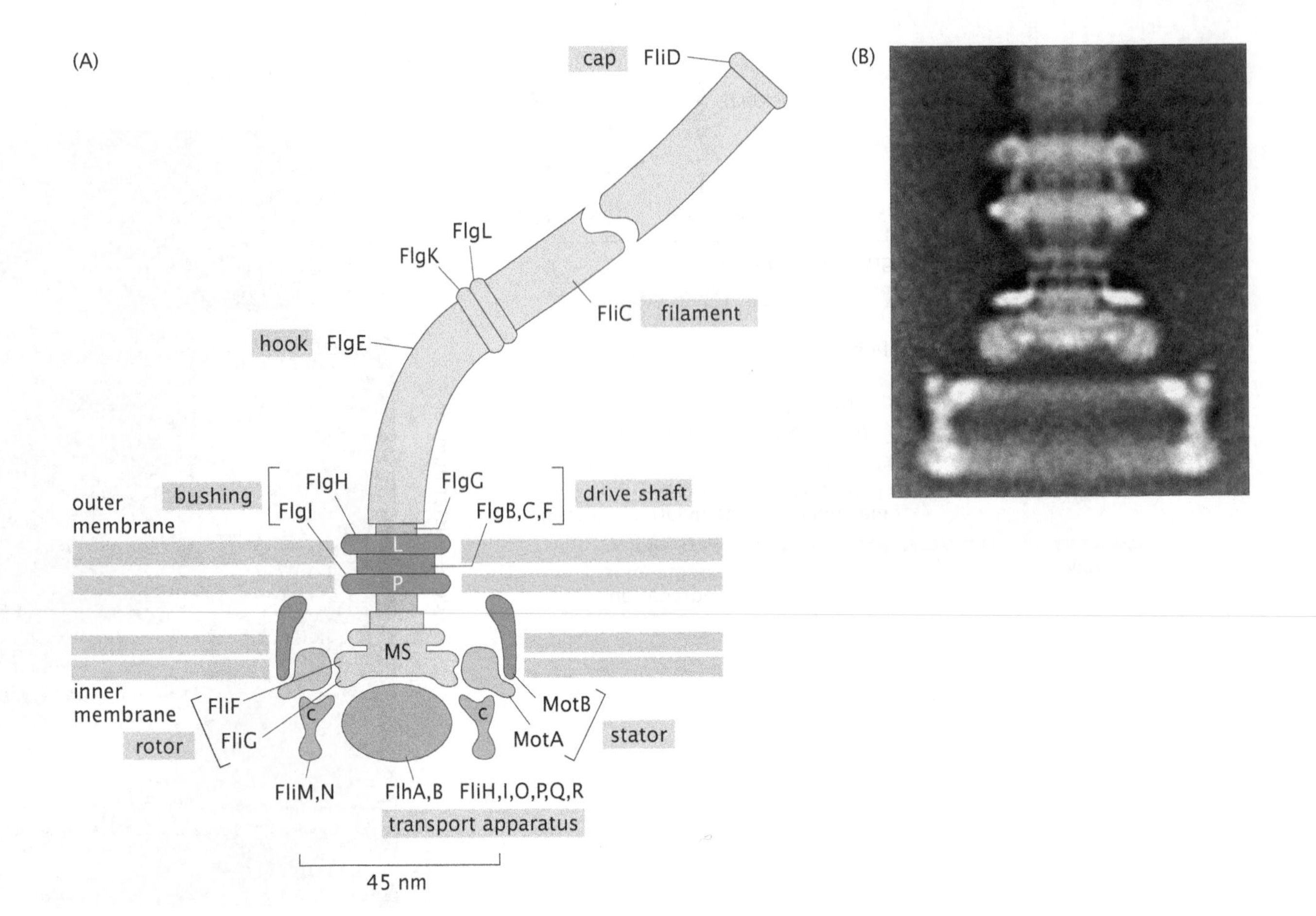

Figure 3.24: Molecular architecture of the bacterial flagellum. (A) The diagram shows the membrane-bound parts of the flagellar apparatus as well as the flagellum itself. The labels refer to the various gene products involved in the assembly of the flagellum. (B) The large ring at the bottom is the C ring, found on the cytoplasmic face. The smaller ring immediately above it is the MS ring, which spans the inner membrane, and the L and P rings (which span the outer membrane) are seen near the top. (Adapted from H. C. Berg, *Phys. Today* 53:24, 2000.)

The idea of the experiment is to induce the growth of flagella in starved *E. coli* cells (which lack flagella) and to use a reporter gene, namely, a gene leading to the expression of GFP, to report on when each of the different genes associated with the flagellar pathway is being expressed. This experiment permits us to peer directly into the dynamics of assembly of the bacterial flagellum, which reveals a sequence of events that are locked into succession in exactly the sort of way that we argued is characteristic of relative time. Figure 3.25 shows the results of this experiment. To deduce a time scale from this figure, we consider the band of expression and note that the roughly 15 genes are turned on sequentially over a period of roughly 180 minutes starting about 5 hours after the beginning of the experiment. This implies an approximate delay time between each product of roughly 12 minutes. Note that the entire experiment was run over a period of roughly 10 hours, though induction of new flagella (the part of the experiment of interest here) begins after three or four rounds of cell division have occurred.

3.3.3 Killing the Cell: The Life Cycles of Viruses

Cells are not the only biological entities that care about relative timing. Once viruses have infected a host cell, they are like a ticking time bomb with an ever-shortening fuse of early, middle, and late genes. Once these genes have been expressed and their products assembled, hundreds of new viruses emerge from the infected (and now defunct) cell to repeat the process elsewhere.

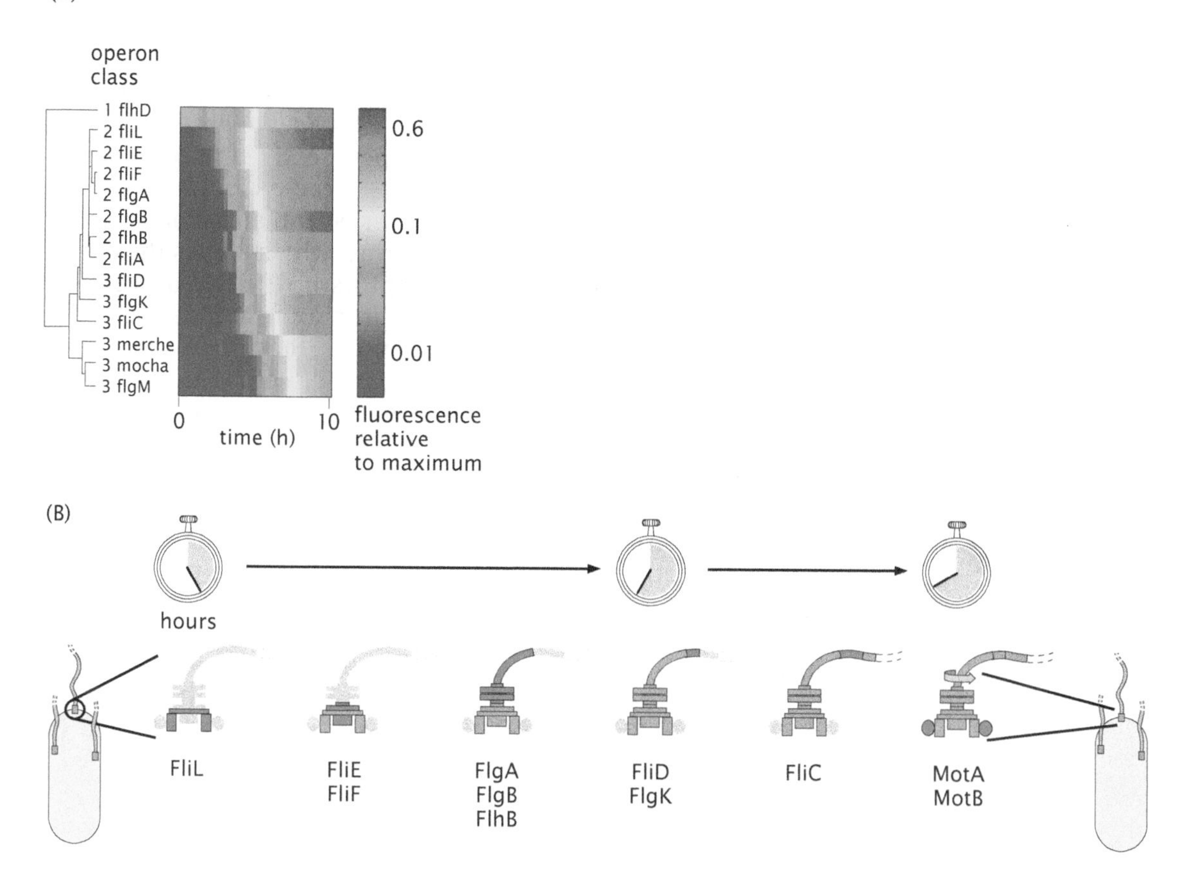

Figure 3.25: Timing of gene expression during the process of formation of the bacterial flagellum. (A) The extent of gene expression (as measured by fluorescence intensity) of each of the gene products as a function of time. (B) The cartoons show the timing of synthesis of different parts of the flagellum. The stopwatch times are chosen by assuming that in total the process takes roughly 3 hours and that flagellar synthesis begins roughly 5 hours after the beginning of the experiment. (Adapted from S. Kalir et al., *Science* 292:2080, 2001.)

Viral Life Cycles Include a Series of Self-Assembly Processes

We have already described the cell cycle as a master process characterized by an enormous number of subprocesses. A more manageable example of an entire "life cycle" is that of viruses, which illustrates the intricate relative timing of biological processes. An example of the viral life cycle of a bacteriophage (introduced in the last chapter as a class of viruses that attack bacterial cells) is shown in Figure 3.26, which depicts the key components in the life history of the virus. The key stages in this life cycle are captured in kinetic verbs such as infect, transcribe, translate, assemble, package, and lyse. Infection is the process of entry of the viral DNA into the host cell. Transcription and translation refer to the hijacking of the cellular machinery so as to produce viral building blocks (both nucleic acid and proteins). Assembly is the coming together of these building blocks to form the viral capsid. Packaging, in turn, is the part of the life cycle when the viral genome is enclosed within the capsid. Finally, lysis refers to the dissolution of the host cell and the emergence of a new generation of phage that go out and repeat their life cycle elsewhere.

As illustrated by the cartoon in Figure 3.26 and, in particular, by our use of the stopwatch motifs, the time between the arrival of the virus at the bacterial surface and the destruction of that very same membrane during the lysis phase when the newly formed viruses are released seems very short at 30 minutes. Indeed, one of our goals in the chapters that follow will be to come to terms with the 30 minute characteristic time scale of the viral life cycle and the various processes that make it up. On the other hand, though the absolute units (30 minutes) are interesting, it is important to emphasize that this set of processes is locked together sequentially in a progression of relative timing events (as with the synthesis of the bacterial flagellar apparatus).

Because of their stunning structures and rich lifestyles, we now examine a second class of viruses, namely, RNA animal viruses such as HIV already introduced from a structural perspective in Section 2.2.4 (p. 66). As shown in Figure 3.27, the infection process for these viruses is quite distinct from that in bacterial viruses. In particular, we note the presence of a membrane coat on the virus, which allows the entire virus to attach to membrane-bound receptors on the victim cell. As a result of this interaction between the virus and the host cell, the virus is swallowed up by the cell that is under attack in a process of membrane fusion. Once the virus has entered the embattled cell, the genetic material (RNA) is released and reverse transcriptase creates a DNA molecule encoding the viral genome using the viral RNA as a template, which is then delivered to the host nucleus and incorporated into its genome. After the genetic material has been delivered to the nucleus, a variety of synthesis processes are undertaken that result in copies of the viral RNA as well as fascinating polyproteins (the Gag proteins described in Chapter 2), which are exported to the plasma membrane, where they undergo an intricate process of self-assembly at the membrane of the infected cell. Once the newly formed virus is exported, it undergoes a maturation process resulting in new, fully infectious, viral particles. Each of these processes is locked in succession in a pageant of relatively timed events.

3.3.4 The Process of Development

As already demonstrated in Section 2.3.3 (p. 78), one of the most compelling, mysterious, and visually pleasing processes in biology is the development of multicellular organisms. Like the cell cycle of individual cells, development depends upon a fixed relative ordering of events. Perhaps the most studied organism from the developmental perspective is the fruit fly *Drosophila melanogaster*. The process of *Drosophila* development has already been schematized in Figures 3.2(A) and (B). Embryonic development represents the disciplined outcome of an encounter between an egg and its partner sperm. In the hours that follow this encounter for the fruit fly, the nascent larva undergoes a series of nuclear divisions and migrations as shown in Figure 3.28. In particular, as the nuclei divide to the 10th generation (512 nuclei), they also start to collect near the surface of the developing larva, forming the syncytial blastoderm, an ovoid object with nearly all of the cells localized to the surface. At the 13th generation, the individual nuclei are enclosed by their own membranes to form the cellular blastoderm. The structural picture by the end of this process is a collection of roughly 5000 cells that occupy the surface of an ovoid object (roughly), which is 500 μm in length and roughly 200 μm in cross-sectional diameter.

Accompanying these latter stages of the developmental pathway is the beginning of a cellular dance in which orchestrated cell movements known as gastrulation lead to the visible emergence of the macroscopic structures associated with the nascent embryo. Snapshots from this process are shown in Figure 3.29 with a time scale associated with the process of gastrulation of the order of hours. We have already proclaimed the importance and beauty of the temporal organization of gene expression associated with the cell cycle. We now add to that compliment by noting that during the development of the

viral attachment

50 nm

bacteriophage

bacterium

host cell DNA

minutes

2 μm

DNA injection

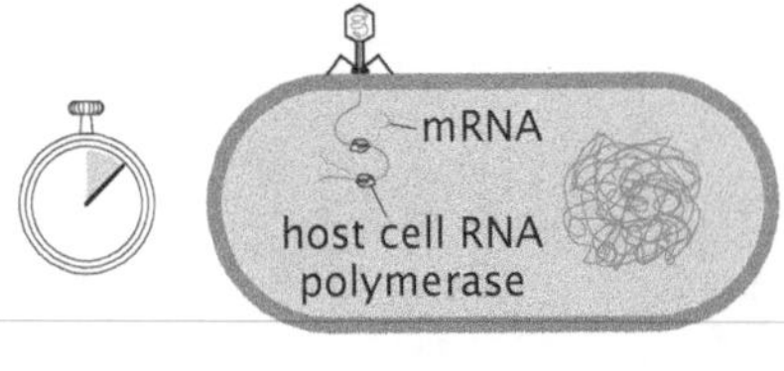

macromolecular synthesis and self-assembly

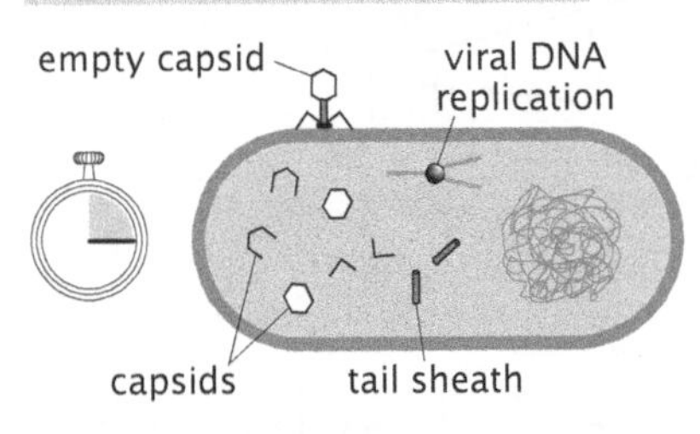

DNA packaging

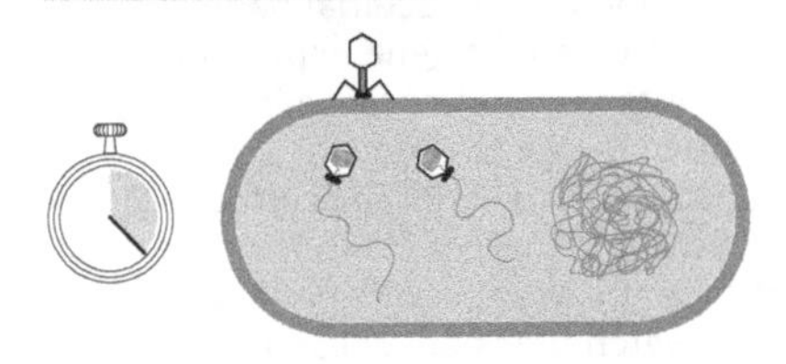

assembly completion and lysis

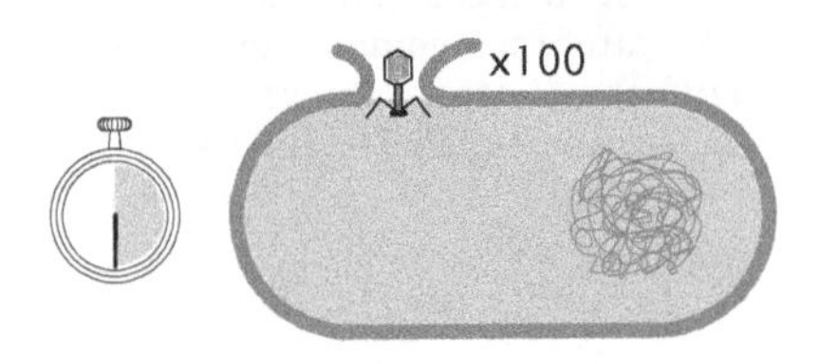

Figure 3.26: Timing the life cycle of a bacteriophage. The cartoon shows stages in the life cycle of a bacteriophage and roughly how long after infection these processes occur. Note that the head and the tail follow distinct assembly pathways and join only after they have separately assembled.

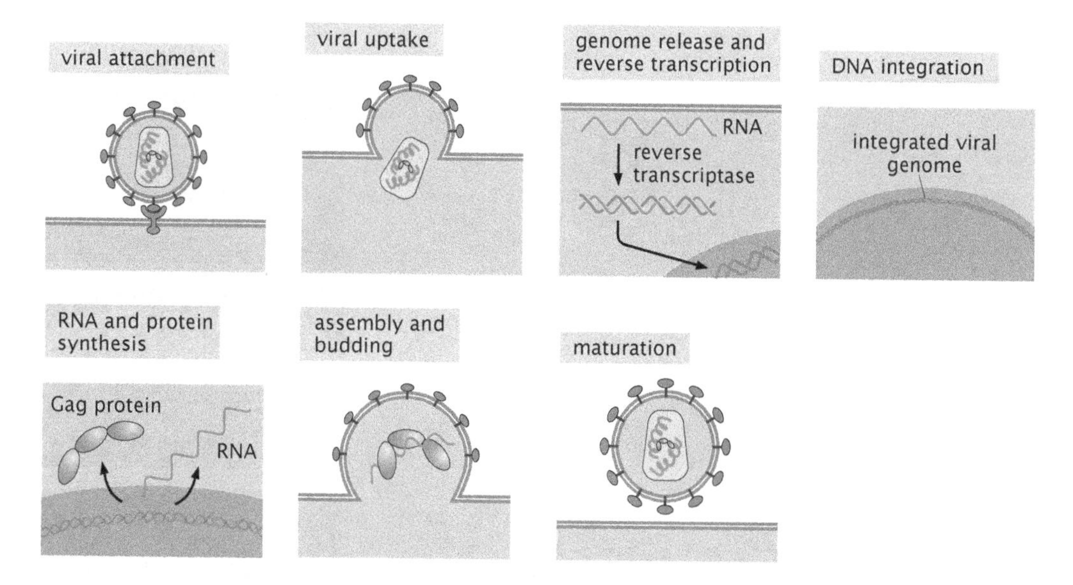

Figure 3.27: Stages in the life cycle of HIV. The timing of the events in the life cycle of HIV can be explored in Reddy and Yin (1999).

Drosophila body plan, there is an ordered spatial pattern of expression of genes with colorful names such as *hunchback* and *giant*, which determine the spatial arrangement of different cells.

These developmental processes make their appearance here because they too serve as an example of relative time. In particular, an example of the time ordering is the cascade of genes associated with the segmentation of the fly body plan into its anterior and posterior parts as already introduced in Section 2.3.3 (p. 78). The long axis of the *Drosophila* embryo is subject to increasing structural refinement as a result of a cascade of genes known collectively as segmentation genes. This collection of genes acts in a cascade, which is a code word for precisely the kind of sequential processes that are behind *relative time* as introduced in this section. The first set of genes in the cascade are known as the *gap* genes. These genes divide the embryo into three regions: the anterior, middle, and posterior. The *gap* genes have as protein products transcription factors that control the next set of genes in the cascade, which are known as *pair-rule* genes. The *pair-rule* genes begin to form the identifiable set of seven stripes of cells. Finally, the *segment polarity* genes are expressed in 14 stripes.

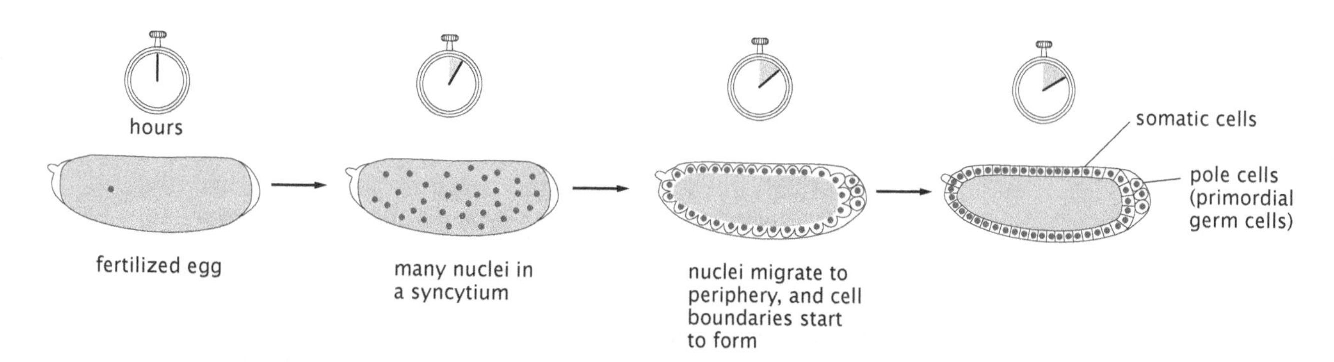

Figure 3.28: Early development of the *Drosophila* embryo. After fertilization, the single nucleus undergoes a series of eight rapid divisions producing 256 nuclei that all reside in a common cytoplasm. At this point, the nuclei begin to migrate toward the surface of the embryo while continuing to divide. After reaching the surface and undergoing a few more rounds of division, cell boundaries form by invaginations of the plasma membrane. At this early stage, the cells that are destined to give rise to sperm or eggs segregate themselves and cluster at the pole of the embryo. All of these events happen within roughly 2 hours.

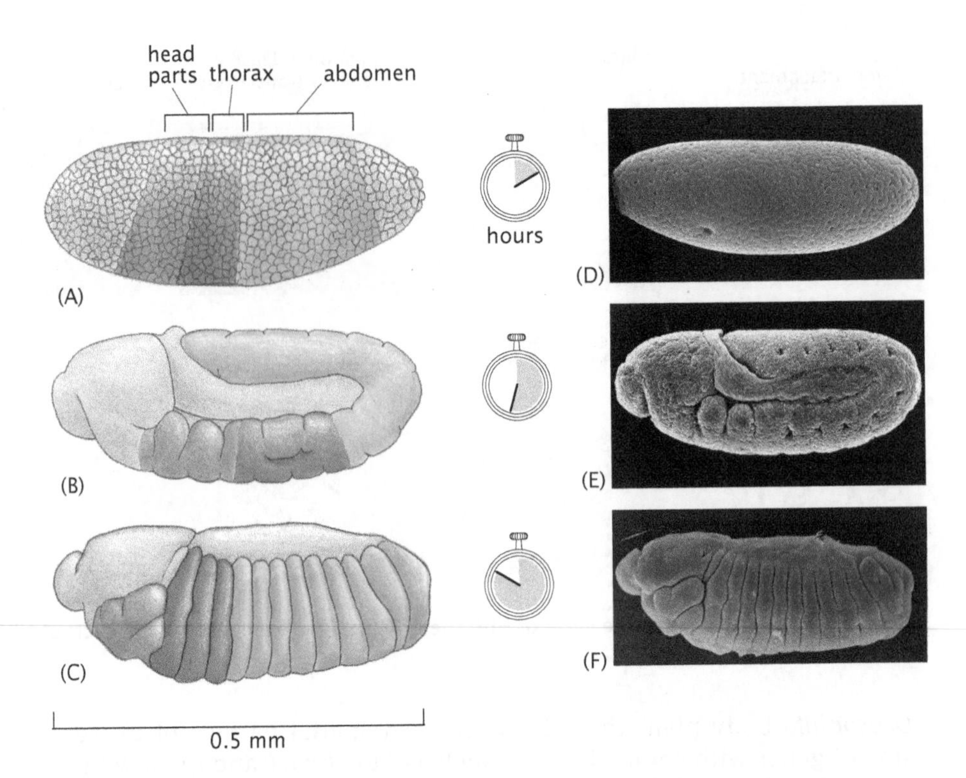

Figure 3.29: Early pattern formation in the *Drosophila* embryo. (A)–(C) show schematics of the shape of the *Drosophila* embryo at 2, 6, and 10 hours after fertilization, respectively. At 2 hours, no obvious structures have yet begun to form, but the eventual fates of the cells that will form different parts of the animal's body have already been determined. By 6 hours, the embryo has undergone gastrulation and the body axis of the embryo has lengthened and curled back on itself to fit in the egg shell. By 10 hours, the body axis has contracted and the separate segments of the animal's body plan have become clearly visible. (D), (E) and (F) show scanning electron micrographs of embryos at each of these stages. Remarkably, all of this pattern formation takes place without any growth. (A, adapted from B. Alberts et al., Molecular Biology of the Cell, 5th ed., Garland Science, 2008. D and E, courtesy of F. R. Turner and A. P. Mahowald, *Dev. Biol.* 50:95–108, 1976; F, from J. P. Petschek, N. Perrimon and A. P. Mahowald, *Dev. Biol.* 119:175–189, 1987. All with permission from Academic Press.)

ESTIMATE

Estimate: Timing Development A simple estimate of the number of cells associated with a developing organism can be obtained by assuming perfect synchrony from one generation to the next,

$$\text{number of cells} \approx 2^N, \tag{3.14}$$

where N is the number of generations. Further, if we assume that the cell cycle is characterized by a time τ_{cc}, then the number of cells as a function of time can be written simply as

$$\text{number of cells} \approx 2^{t/\tau_{\text{cc}}}. \tag{3.15}$$

Interestingly, in the early stages of *Drosophila* development, since it is only nuclear division (and hence largely DNA replication) that is taking place, the mean doubling time is 8 minutes. Thus after 100 minutes, roughly 10 generations worth of nuclear division will have occurred, with the formation of the approximately 1000 cells that form the syncytial blastoderm.

3.4 Manipulated Time

Sometimes, the cell is not satisfied with the time scales offered by the intrinsic physical rates of processes and has to find a way to beat these speed limits. For example, in some cases, the bare rates of biochemical reactions are prohibitively slow relative to characteristic cellular time scales and, as a result, cells have tied their fate to enzymatic manipulation of the intrinsic rates. In a similar vein, diffusion as a means of intracellular transport is ineffective over large distances. In this case, cells have active transport mechanisms involving molecular motors and cytoskeletal filaments that can overcome the diffusive speed limit, or alternatively, the cell might resort to reduced-dimensionality diffusion to beat the diffusive speed limit. There are even more tricky ways in which cells manipulate time, such as in the case of beating the bacterial replication limit. These examples and others will serve as the basis of our discussion of the way cells manipulate time.

3.4.1 Chemical Kinetics and Enzyme Turnover

Some chemical reactions proceed much more slowly than would be necessary for them to be biologically useful. For example, the hydrolysis of the peptide bonds that make up proteins would take times measured in years in the absence of proteases, which are the enzymes that cleave these bonds. Triose phosphate isomerase, the enzyme in the glycolysis pathway featured in Figure 2.32 (p. 70), is responsible for a factor-of-10^9 speed-up in the glycolytic reaction it catalyzes. What these numbers show is that even if a given reaction is favorable in terms of free energy, the energy barrier to that reaction can make it prohibitively slow. As a result, cells have found ways to manipulate the timing of reactions using enzymes as catalysts. Indeed, the whole of biochemistry is in some ways a long tale of catalyzed reactions, many of which take place on time scales much shorter than milliseconds, whereas, in the absence of these enzymes, they might not take place in a year! The individual players in the drama of glycolysis such as hexokinase, phosphofructokinase, triose phosphate isomerase, and pyruvate kinase reveal their identity as enzymes with the ending *ase* in their names. Enzymes are usually denoted by the ending *ase* and are classified according to the reactions they catalyze.

The basis of enzyme action is depicted in Figure 3.30. For concreteness, we consider an isomerization reaction where a molecule starts out in some high-energy state A and we interest ourselves in the transitions to the lower-energy state B. For the molecule schematized in the figure, the energy associated with the conformation has an electrostatic contribution and an "internal" contribution. As seen in Figure 3.30(A), the internal contribution has two minima, while the electrostatic contribution to the free energy is a monotonically decreasing function. In the presence of the enzyme, the height of the barrier is suppressed because the charges on the molecule of interest have an especially favorable interaction with the enzyme at the transition state.

The key point about the reaction rate is that, as shown in the figure, it depends upon the energy barrier separating the two states according to

$$\Gamma_{A\to B} \propto e^{-G_{\text{barrier}}/k_{\text{B}}T}, \qquad (3.16)$$

where $\Gamma_{A\to B}$ is the transition rate with units of s^{-1}, G_{barrier} is the size of the energy barrier and $k_{\text{B}}T$ is the thermal energy scale with k_{B} known

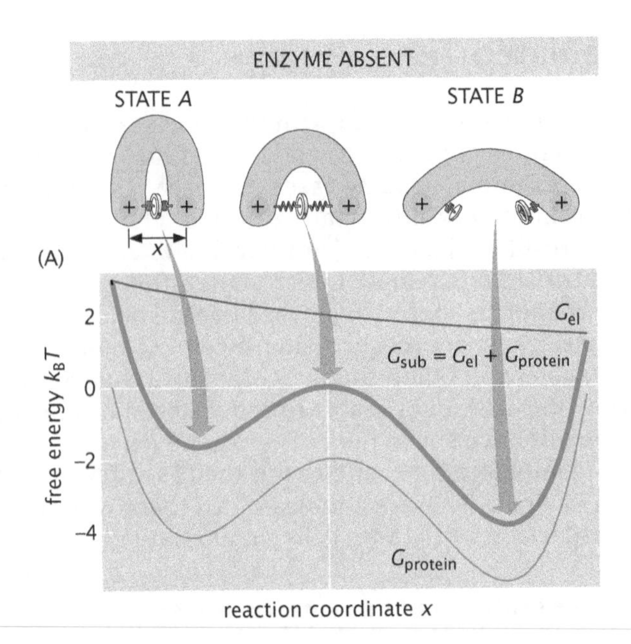

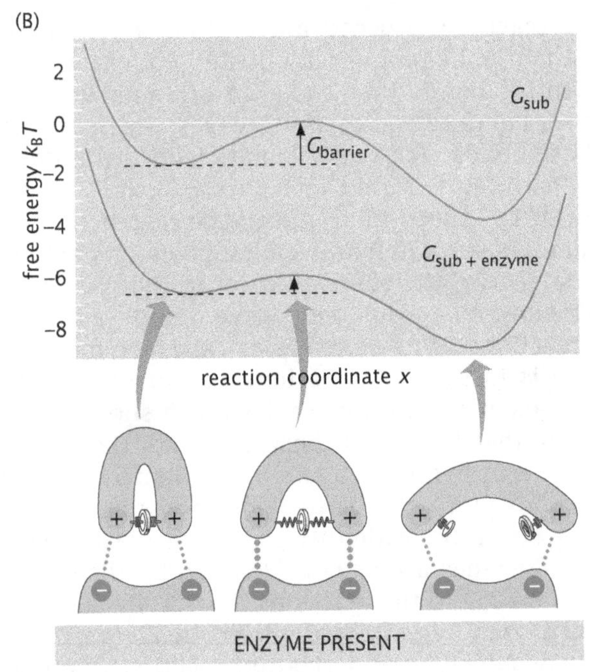

Figure 3.30: Enzymes and biochemical rates. (A) The electrostatic (G_{el}) and internal ($G_{protein}$) contributions to the overall free energy. The sum of these two contributions gives the total conformational free energy G_{sub} (in the absence of the enzyme). (B) A simple one-dimensional representation of the energy landscape for a biochemical reaction in the absence and presence of an enzyme to catalyze the reaction. The presence of the enzyme lowers the energy barrier through favorable charge interactions at the transition state and increases the rate of the reaction.

as Boltzmann's constant. As will be shown in Chapters 5 and 6, k_BT sets the scale of thermally accessible fluctuations in a system. Even though the energy of state *B* might be substantially lower than that of state *A*, the transitions can be exceedingly slow because of large barrier heights (that is, $G_{barrier} \gg k_BT$). The presence of an enzyme does not alter the end states or their relative energies, but it suppresses the barrier between the two states.

3.4.2 Beating the Diffusive Speed Limit

A second example of the way cells manipulate time is the way in which they perform transport and trafficking. Organelles, proteins, nucleic acids, etc. are often produced in one part of the cell only to be transported to another part where they are needed. For example, the mRNA molecules produced in the nucleus need to make their way to the ribosomes, which are found in the cytoplasm on the endoplasmic reticulum. One physical process that can move material around is passive diffusion.

Diffusion Is the Random Motion of Microscopic Particles in Solution

Ions, molecules, macromolecular assemblies, and even organelles wander around aimlessly as a result of diffusion. Diffusion refers to the random motions suffered by microscopic particles in solution, and is sometimes referred to as Brownian motion in honor of the systematic investigations made by Robert Brown in the 1820s. Brown noticed the random jiggling of pollen particles suspended in solution, even for systems that are ostensibly in equilibrium and have no energy source. Indeed, so determined was he to find out whether or not this was some effect intrinsic to living organisms, he even examined exotic suspensions using materials such as the dust from the Sphinx and found the jiggling there too. The effects of Brownian motion are palpable for particles in solution that are micron size and smaller, exactly the length scales that matter to cells. Diffusion results from the fact that in the cell (and for microscopic particles in solution), deterministic forces are on a nearly equal footing with thermal forces, an idea to be fleshed out in Section 5.1.1 (p. 189) and illustrated graphically in Figure 5.1 (p. 190). Thermal forces result from the random collisions between particles that can be attributed to the underlying jiggling of atoms and molecules. This fascinating topic will dominate the discussion of Chapter 13.

Estimate: The Thermal Energy Scale One way to quantify the relative importance of the energy scale of a given process and thermal energies is by measuring the energy of interest in $k_B T$ units. At room temperature, typical thermal energies are of the order of $k_B T$, with a value of

$$k_B T = 1.38 \times 10^{-23}\,\text{J/K} \times 300\,\text{K} \approx 4.1 \times 10^{-21}\,\text{J} = 4.1\,\text{pN nm}. \tag{3.17}$$

One way to see the importance of this energy scale is revealed by Equation 3.16 (and will also be revealed by the Boltzmann distribution (to be described in Chapter 6), which says that the probability of a state with energy E_i is proportional to $e^{-E_i/k_B T}$). These expressions show that when the energy is comparable to $k_B T$, barriers will be small (and probabilities of microstates high). The numerical value ($k_B T \approx 4\,\text{pN nm}$) is especially telling since many of the key molecular motors relevant to biology act with piconewton forces over nanometer distances, implying a competition between deterministic and thermal forces. This discussion tells us that for many problems of biological interest, thermal forces are on a nearly equal footing with deterministic forces arising from specific force generation.

Diffusion Times Depend upon the Length Scale

One simple and important biological example of diffusion is the motion of proteins bound to DNA, which can be described as one-dimensional diffusion along the DNA molecule. For example, it is thought that regulatory proteins (transcription factors) that control gene expression find their specific binding sites on DNA partly through this kind of one-dimensional diffusive motion. Another example is provided by the arrival of ligands at their specific receptors. The basic picture is that of molecules being battered about and every now and then ending up by chance in the same place at the same time. To get a feeling for the numbers, it is convenient to consider one of the key equations that presides over the subject of diffusion, namely,

$$t_{\text{diffusion}} \approx x^2/D, \tag{3.18}$$

where D is the diffusion constant. This equation tells us that the time scale for a diffusing particle to travel a distance x scales as the square of that distance.

Estimate: Moving Proteins from Here to There For molecules and assemblies that move passively within the cell, the time scale can be estimated using Equation 3.18. For a protein with a 5 nm diameter, the diffusion constant in water is roughly $100\,\mu\text{m}^2/\text{s}$; this estimate can be obtained from the Stokes–Einstein equation (to be discussed in more detail in Chapter 13 in Equation 13.62 on p. 531), which gives the diffusion constant of a sphere of radius R moving through a fluid of viscosity η at temperature T as $D = k_B T/6\pi\eta R$. The time scale for such a typical protein to diffuse a distance of our standard ruler (that is, across an *E. coli*) is

$$t_{E.coli} \approx \frac{L^2_{E.coli}}{D} \approx \frac{1\,\mu\text{m}^2}{100\,\mu\text{m}^2/\text{s}} \approx 0.01\text{ s}. \tag{3.19}$$

This should be contrasted with the time scale required for diffusion to transport molecules from one extremity of a neuron to the other as shown in Figure 3.3. In particular, the diffusion time for the squid giant axon, which has a length of the order of 10 cm, is $t_{\text{diffusion}} \approx 10^8$ s! The key conclusion to take away from such an estimate is the impossibly long time scales associated with diffusion over large distances. Nature's solution to this conundrum is to exploit *active* transport mechanisms in which ATP is consumed in order for motor molecules to carry out directed motion.

ESTIMATE

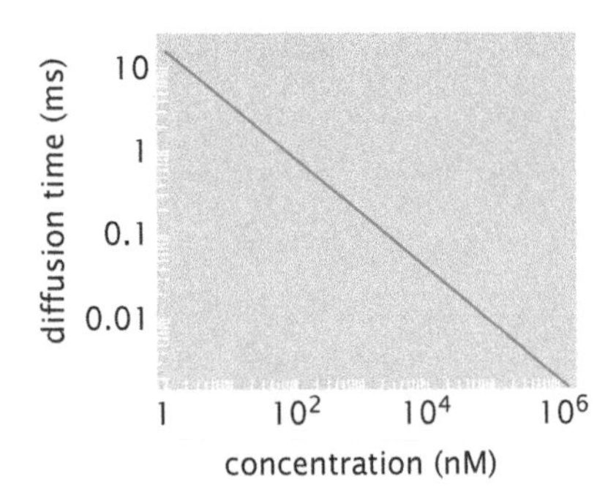

Figure 3.31: Concentration measured in diffusion times. The characteristic time scale for diffusion-mediated molecular encounters depends upon the concentration of the diffusing species.

Another way of coming to terms with diffusion is suggested by thinking about concentration. In Figure 2.12 (p. 48), we described physical interpretations of concentration in terms of number of molecules per *E. coli* cell and in terms of mean molecular spacings. We can also think of concentrations in terms of diffusion times by using Equation 3.18. In particular, this equation can be used to estimate a characteristic time scale for reactants to find each other as a function of the concentration as shown in Figure 3.31.

Diffusive Transport at the Synaptic Junction Is the Dynamical Mechanism for Neuronal Communication

One of the classic episodes in the history of biology centered around the battle between Golgi, Ramon y Cajal, and their followers as to the nature of nerve transmission. At the heart of the controversy over soup versus sparks was the question of whether nerves are reticula (the view of Golgi) or rather built up of individual nerve cells, with communication between adjacent cells carried out by molecular transport. This debate was largely settled in favor of the "soup" view.

Estimate: Diffusion at the Synaptic Cleft The ideas introduced above can help us understand the dynamics of neurotransmitters at synapses. An example of the geometry of such a synapse is shown in Figure 3.32. Using exactly the same ideas as developed in the previous section, we can work out the time scale for neurotransmitters released at one side of the synapse to reach the receptors on the neighboring cell.

To be concrete, consider the diffusion of acetylcholine across a synaptic cleft with a size of roughly 20 nm. Given a diffusion constant for acetylcholine of $\approx 100\ \mu\text{m}^2/\text{s}$, the time for these molecules to diffuse across the cleft is

$$t = L^2/D \approx \frac{400\ \text{nm}^2}{10^8\ \text{nm}^2/\text{s}} \approx 4\ \mu\text{s}. \tag{3.20}$$

(A)

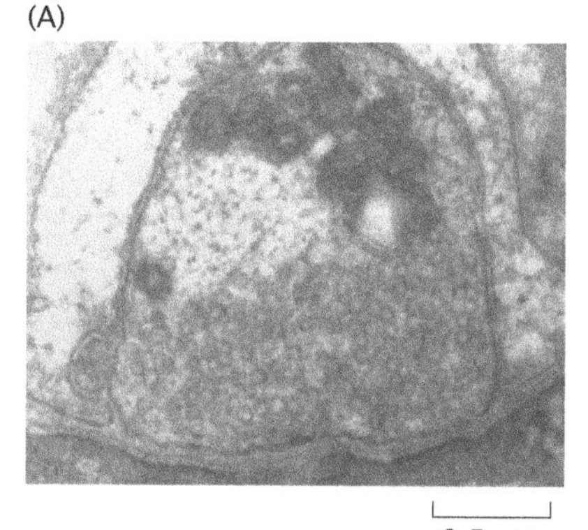

(B)

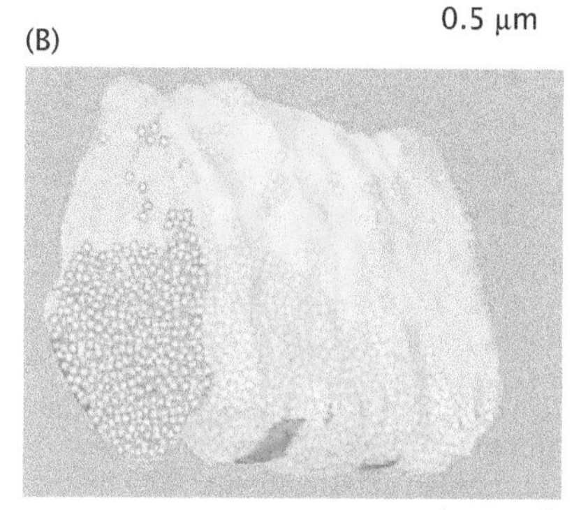

(C)

Figure 3.32: Geometry of the synapse. (A) Electron micrograph image of a nerve terminal. (B) Reconstruction of the synaptic geometry for a nerve terminal segment with a length of roughly 2 μm. The many vesicles that are responsible for the active response of the synapse are shown in red. (C) Schematic of the diffusive transport across a synaptic cleft. (A and B, adapted from S. O. Rizzoli and W. J. Betz, *Nat. Rev. Neurosci.*, 6:57, 2005.)

Molecular Motors Move Cargo over Large Distances in a Directed Way

In many instances, particularly for movement over long distances, diffusion is too slow to be of any use for intracellular transport. To beat the diffusive speed limit, cells manipulate time with a sophisticated array of molecular machines (usually proteins) that result in directed transport. These processes are collectively powered by the consumption of some energy source (usually ATP). Broadly construed, the subject of active transport allows us to classify a wide variety of molecules as *molecular motors*, the main subject of Chapter 16 (p. 623). We have already seen the existence of such motors in a number of different contexts, with both DNA polymerase and RNA polymerase, which were introduced earlier in the chapter, satisfying the definition of active transport because they use chemical energy to move along their DNA substrate in a directed way.

Concretely, the class of motors of interest here are those that mediate transport of molecules from one place in the cell to another in eukaryotes. Often, such transport takes place as vesicular traffic, with the cargo enclosed in a vesicle (a flexible spherical shell made up of lipid molecules in the form of a lipid bilayer) that is in turn attached to some molecular motor. These molecular motors travel in a directed fashion on the cytoskeletal network that traverses the cell. For example, traffic on microtubules runs in both directions as a result of two translational motors, kinesin and dynein. Molecular-motor-mediated transport on actin filaments is shown in cartoon form in Figure 3.33. Note that this cartoon gives a rough idea of the relative proportions of the motors and the actin filaments on which they move and conveys the overall structural features, such as two heads, of the motors themselves. In addition, Figure 3.33(B) shows a time trace of the position of a fluorescently labeled myosin motor, which illustrates the discrete steps of the motor, also permitting a measurement of the mean velocity.

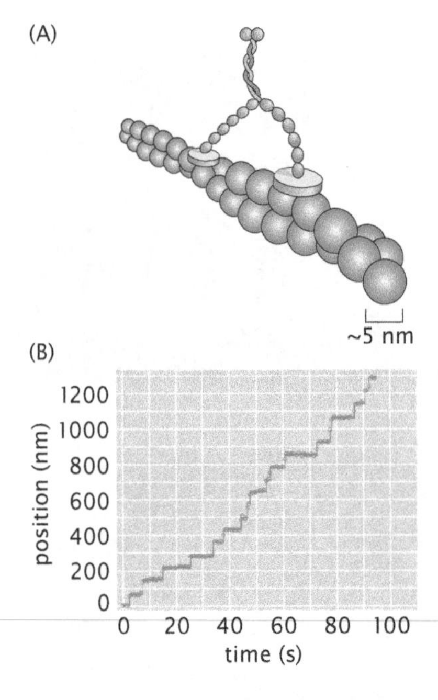

Figure 3.33: Motion of myosin on an actin filament. (A) Schematic of the motor on an actin filament. Note that the step size is determined by the periodicity of the filament. (B) Position as a function of time for the motor myosin V as measured using single-molecule techniques. (Adapted from A. Yildiz et al., *Science* 300:2061, 2003.)

Estimate: Moving Proteins from Here to There, Part 2 We have already noted that biological motility is based in large measure on diffusion. On the other hand, there are a host of processes that cannot wait the time required for diffusion. In particular, recall that our estimate for the diffusion time for a typical protein to traverse an axon was a whopping 10^8 s, or roughly 3 years. For comparison, we can estimate the transport time for kinesin moving on a microtubule over the same distance. As the typical speed of kinesin in a living cell is $1\,\mu\text{m/s}$, the time for it to transport a protein over a distance of 10 cm is 10^5 s, or just over a day.

To see these ideas play out more concretely, we can return to Figure 3.3 (p. 93). The classic experiment highlighted there traces the time evolution of radioactively labeled proteins in a neuron. The figure shows that the radio-labeled proteins travel roughly 18 mm in 12 days, which translates into a mean speed of roughly 20 nm/s. Observed axonal transport speeds for single motors are a factor of 10 or more larger, but we can learn something from this as well. In particular, motors are not perfectly "processive"—that is, they fall off of their cytoskeletal tracks occasionally and this has the effect of reducing their mean speed. Observed motor velocities are reported, on the other hand, on the basis of tracking individual motors during one of their processive trajectories.

Membrane-Bound Proteins Transport Molecules from One Side of a Membrane to the Other

Another way in which cells manipulate transport rates is by selectively and transiently altering the permeability of cell membranes through protein channels and pumps. Many ionic species are effectively unable to permeate (at least on short time scales) biological membranes. What this means is that concentration gradients can be maintained across these membranes until these protein channels open, which then permits a flow of ions down their concentration gradient. In fancier cases, such as indicated schematically in Figure 3.34, ions can be pumped up a concentration gradient through mechanisms involving ATP hydrolysis. In Figure 3.2(G) we showed the process of ion transport across a membrane with a characteristic time of microseconds.

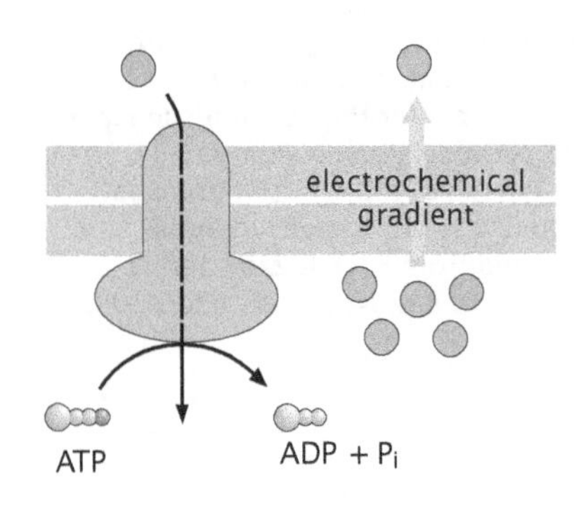

Figure 3.34: Active transport across membranes. Molecular pumps consume energy in the form of ATP hydrolysis and use the liberated free energy to pump molecules against their concentration gradient.

Estimate: Ion Transport Rates in Ion Channels An ion channel embedded in the cell membrane can be thought of as a tube with a diameter of approximately $d = 0.5$ nm (size of hydrated ion) and a length $l = 5$ nm (width of the lipid bilayer). With these numbers in hand, and a typical value of the diffusion constant for small ions (e.g. sodium), $D \approx 2000\,\mu\text{m}^2/\text{s}$, we can estimate the flux of ions through the channel, assuming that their motion is purely diffusive.

To make this estimate, we invoke an approximate version of Fick's law (to be described in detail in Chapter 8) that says that the flux (number of molecules crossing unit area per unit time) is proportional to the difference in concentration and inversely proportional to the distance between the two "reservoirs." Mathematically, this can be written as

$$J_{\text{ion}} \approx D\frac{\Delta c}{l}, \tag{3.21}$$

where Δc is the difference in ion concentration across the cell membrane. For typical mammalian cells the concentration difference for sodium or potassium is $\Delta c \approx 100\,\text{mM}$ (or $\approx 6 \times 10^{-2}$ molecules/nm^3) and the distance across the membrane is $l \approx 5\,\text{nm}$, resulting in

$$\begin{aligned} J_{\text{ion}} &\approx 2 \times 10^9 \frac{\text{nm}^2}{\text{s}} \times \frac{6 \times 10^{-2}\ \text{molecules/nm}^3}{5\,\text{nm}} \\ &\approx 2 \times 10^7\,\text{nm}^{-2}\,\text{s}^{-1}, \end{aligned} \tag{3.22}$$

where we have used a diffusion constant of $D = 2 \times 10^9\,\text{nm}^2/\text{s}$ appropriate for an ion. Given the cross-sectional area of a typical channel $A_{\text{channel}} = d^2\pi/4 \approx 0.2\,\text{nm}^2$, the number of ions traversing the membrane per second is estimated to be

$$\frac{\text{d}N_{\text{ion}}}{\text{d}t} = J_{\text{ion}}A_{\text{channel}} \approx 2 \times 10^7\,\text{nm}^{-2}\,\text{s}^{-1} \times 0.2\,\text{nm}^2 = 4 \times 10^6\,\text{s}^{-1}. \tag{3.23}$$

This estimate does remarkably well at giving a sense of the time scales associated with ion transport across ion channels.

Enzymes, molecular motors, and ion channels (and pumps) are all ways in which the cell uses proteins to circumvent the intrinsic rates of different physical or chemical processes.

3.4.3 Beating the Replication Limit

The most fundamental process of cellular life is to form two new cells. A minimal requirement for this to take place is that an individual cell must duplicate its genetic information. Replication of the genetic material proceeds through the action of DNA polymerase, an enyzme that copies the DNA sequence information from one DNA strand into a complementary strand. Like all biochemical reactions, this requires a certain amount of time, which we estimated earlier in the chapter. Why should any cell accept this speed limit on its primary directive of replication? When we calculated the replication time for the *E. coli* chromosome, we concluded that two replication forks operating at top speed would be sufficient to replicate the chromosome in approximately the 3000 s division time that we stipulated for a bacterium growing in a minimally defined medium with glucose as the sole carbon source under a continuous supply of oxygen. While this set of conditions is in many ways convenient for the human experimentalist, it is by no means ideal for the bacterium. If instead of only supplying glucose we add a rich soup of amino acids, *E. coli* will grow much faster, with a doubling time of order 20 minutes (1200 s).

How can the bacterium double more quickly than its chromosome can replicate? For *E. coli* and other fast-growing bacteria, the answer is a simple and elegant manipulation of the procedural time limit imposed by the DNA replication apparatus. The bacterium begins replicating its chromosome a second time before the first replication is complete. In a rapidly growing *E. coli*, there may be between four and eight copies of the chromosomal DNA close to the replication origin, even though there may be only one copy of the chromosome close to the replication terminus. In other words, the bacterium has started replicating its daughter's, grandaughter's or even great grandaughter's chromosome before its own replication is complete. The newborn *E. coli* cell is thus essentially already pregnant with partially replicated chromosomes preparing for the next one or two generations.

As we have noted above, the genome size for bacteria tends to be substantially smaller than the genome size for eukaryotic cells. Nonetheless, eukaryotic cells are still capable of replicating at a remarkably fast rate. For example, early embryonic cells of the South African clawed frog *X. laevis* can divide every 30 minutes. Despite the fact that its genome (3100 Mbp) is more than 600-fold larger than the genome of *E. coli*, two mechanical changes enable rapid replication of the *Xenopus* genome (and the genome of other eukaryotes). First, the genome is subdivided into multiple linear chromosomes rather than a single circular chromosome. Second, and more importantly, replication is initiated simultaneously from many different origins sprinkled throughout the chromosome as opposed to the single origin of bacterial chromosomes. This parallel processing for the copying of genomic information enables the task to be completed more rapidly than would be dictated by the procedural time limit imposed by a single molecule of DNA polymerase.

3.4.4 Eggs and Spores: Planning for the Next Generation

We have been considering the processes of cell growth and division as though they are tightly coupled with one another. In some cases, however, organisms may separate the processes of growth and division so that they occur over different spans of time. The most dramatic example of this is in the growth of giant egg cells followed by extremely rapid division of early embryos after fertilization. For example, in the frog *X. laevis*, each individual egg cell is enormous—up to 1 mm across—and grows gradually within the body of the female over a period of 3 months. Following fertilization, cell division occurs without growth so that a tadpole hatches after 36 hours that has the same mass as the egg from which it is derived.

Even for organisms where cell growth normally is coupled to cell division, there are several mechanisms whereby cells may choose to postpone either growth or division if conditions are unfavorable. For example, many cells ranging from bacteria through fungi to protozoans such as *Dictyostelium* are capable of creating spores. Spores are nearly metabolically inert and serve as a storage form for the genomic information of the species that can survive periods of drought or low nutrient availability. This is a mechanism by which an organism can effectively exist in suspended animation, waiting for however much time is needed to pass until conditions become favorable again. When fortune finally favors the spore, it can germinate, releasing a rapidly growing cell. The maximum survival time of spores is unknown. However, there are *Bacillus* spores that were put into storage by Louis Pasteur in the late nineteenth century that appear to be fully viable today and it is generally accepted that some spores may be able to survive for thousands of years. Some controversial reports even suggest that viable bacterial spores can be recovered from insect bodies trapped in amber for a few million years. The seeds of flowering plants perform a similar role, though they are typically not as robust as spores.

Although animals do not form true spores, several do have forms that permit long-term survival under starvation conditions. The most familiar examples are hibernation of large animals such as bears, which can survive an entire winter season without eating. Smaller animals perform similar tricks. These include dauer (German for "enduring") form larvae of several worms. The most impressive example of "suspended animation" among animals is presented by the tardigrade or water bear, a particularly adorable segmented metazoan that rarely grows larger than 1 mm. The tardigrade normally lives in water, but when it is dried out, it slows its metabolism and alters its body shape, extruding almost all of its water, to form a dried out form called a tun. Tardigrade tuns can be scattered by wind and can survive extreme highs and lows of temperature and pressure. When the tun falls into a favorable environment like a pond, the animal will reanimate. Each of these examples shows how organisms have evolved mechanisms that are completely indifferent to the absolute passage of time.

3.5 Summary and Conclusions

Because life processes are associated with constant change, it is important to understand how long these processes take. In the case of the diurnal clock, organisms are able to measure the passage of time with great regularity to determine their daily behaviors. For most other kinds of biological processes, times are not absolute. In this chapter, we explored several different views of time in biological systems starting with the straightforward assignment of procedural time as measured by the amount of time it takes to complete some process. In complex biological systems, processes that occur at intrinsically different rates may be linked together such that one must be completed before another can begin. Examples of this kind of measurement of relative time are found in the regulation of the cell cycle, the assembly of complex structures such as the bacterial flagellum, etc. Finally, we briefly explored some of the ways that organisms manipulate biological processes to proceed faster or slower than the normal intrinsic rates. Armed with these varying views of time, we will use time as a dimension in our estimates and modeling throughout the remainder of the book. Time will particularly take center stage in Part 3, Life in Motion, where dynamic processes will be revealed in all their glory.

3.6 Problems

Key to the problem categories: • Model refinements and derivations • Estimates

• 3.1 Growth and the logistic equation

In the chapter, we described the logistic equation as a simple toy model for constrained growth of populations. In this problem, the goal is to work out the dynamics in more detail.

(a) Rewrite the equation in dimensionless form and explain what units this means time is measured in.

(b) Find the value of N at which the growth rate is maximized.

(c) Find the maximum growth rate.

(d) Use these results to make a one-dimensional phase portrait like that shown in Figure 3.10.

• 3.2 Protein synthesis and degradation in the *E. coli* cell cycle

Improve the estimates for the protein synthesis rates of *E. coli* during a cell cycle from Sections 3.1.3 and 3.2.1 by including the effect of protein degradation. For simplicity, assume that all proteins are degraded at the same rate with a half-life of 60 minutes and work out the number of ribosomes needed to produce the protein content of a new bacterium given that part of the synthesis is required for the replacement of degraded proteins. Compare with the results from the estimate in Section 2.1.2 (p. 38).

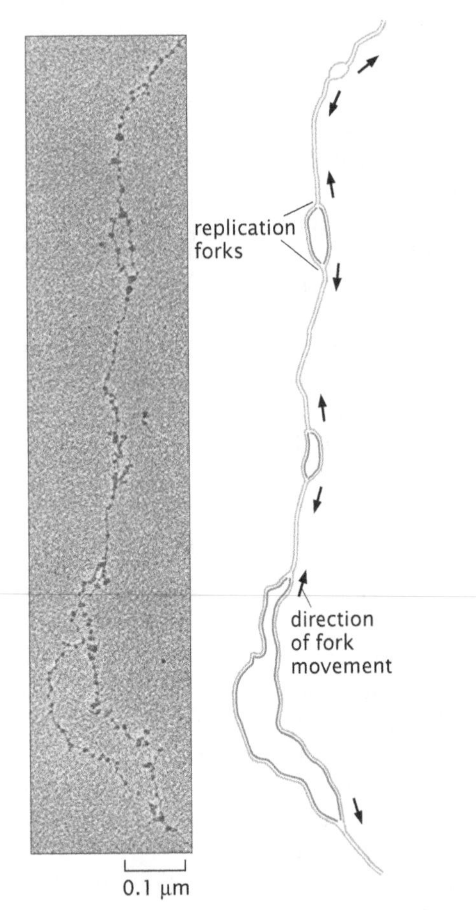

Figure 3.35: Replication forks in *D. melanogaster*. Replication forks move away in both directions from replication origins. (Electron micrograph courtesy of Victoria Foe. Adapted from B. Alberts et al., Molecular Biology of the Cell, 5th ed. Garland Science, 2008.)

• 3.3 DNA replication rates

Assuming that Figure 3.35 is a representative sample of the replication process:

(a) Estimate the fraction of the total fly genome shown in the micrograph. Note that the fly genome is about 1.8×10^8 nucleotide pairs in size.

(b) Estimate the number of DNA polymerase molecules in a eukaryotic cell like this one from the fly *D. melanogaster*.

(c) There are eight forks in the micrograph. Estimate the lengths of the DNA strands between replication forks 4 and 5, counting up from the bottom of the figure. If a replication fork moves at a speed of roughly 40 bp/s, how long will it take for forks 4 and 5 to collide?

(d) Given the mean spacing of the bubbles, estimate how long it will take to replicate the entire fly genome.

• 3.4 RNA polymerase and ribosomes

(a) If RNA polymerase subunits β and β′ together constitute approximately 0.5% of the total mass of protein in an *E. coli* cell, how many RNA polymerase molecules are there per cell, assuming each β and β′ subunit within the cell is found in a complete RNA polymerase molecule? The subunits have a mass of 150 kDa each. (Adapted from Problem 4.1 of Schleif, 1993.)

(b) Rifampin is an antibiotic used to treat *Mycobacterium* infections such as tuberculosis. It inhibits the initiation of transcription, but not the elongation of RNA transcripts. The time evolution of an *E. coli* ribosomal RNA (rRNA) operon after addition of rifampin is shown in Figures 3.36(A)–(C). An operon is a collection of genes transcribed as a single unit. Use the figure to estimate the rate of transcript elongation. Use the beginning of the "Christmas-tree" morphology on the left of Figure 3.36(A) as the starting point for transcription.

(c) Using the calculated elongation rate, estimate the frequency of initiation off of the rRNA operon. These genes are among the most transcribed in *E. coli*.

(d) As we saw in the chapter, a typical *E. coli* cell with a division time of 3000 s contains roughly 20,000 ribosomes. Assuming there is no ribosome degradation, how many RNA polymerase molecules must be synthesizing rRNA at any instant? What percentage of the RNA polymerase molecules in *E. coli* are involved in transcribing rRNA genes?

• 3.5 Cell cycle and number of ribosomes

Read the paper "Ribosome content and the rate of growth of *Salmonella typhimurium*" by R.E. Ecker and M. Schaechter (*Biochim. Biophys. Acta* **76**, 275, 1963).

(a) The authors show that their data can be fit by a simple assumption that the rate of production of soluble protein, P, is proportional to the number of ribosomes. But this leaves out the need for the ribosomes to also make ribosomal protein, in an amount, R. The simplest assumptions are that the total protein production rate, dA/dt, is proportional to the number of ribosomes, and the total amount of non-ribosomal protein (also knows as soluble protein) needed per cell is independent of growth rate. Given these assumptions, how would the ratio R/A, with A the total protein content, depend on the growth rate? Does this give a maximum growth rate? If so, what is it?

(b) With a 3000 s division time for *E. coli*, about 25% of its protein is ribosomal. Note that the microbe *Salmonella typhimurium* considered here is very similar to *E. coli*. Using these numbers and your results from above, what fraction of the protein would be ribosomal for the highest growth rate studied in the paper? How does this compare to their measured ratio of ribosomes to soluble protein, R/P at these growth rates? How does the predicted R/P at high growth rates change if you now assume that dP/dt is proportional to R as they did in the paper?

(c) Another factor that needs to be taken into account is the decay of proteins. If all proteins decayed at the same rate, γ, how would this modify your results from (a)? How does the predicted functional from of R/A versus growth rate change? Explain why the data rule out γ being too large and hence infer a rough lower bound for the lifetime, $1/\gamma$, of "average" proteins.

(Problem courtesy of Daniel Fisher.)

• 3.6 The bleach-chase method and protein degradation

Protein production is tempered both by active degradation and by dilution due to cell division. A clever recent method (see Eden et al. 2011) makes it possible to measure net degradation rates by taking two populations of the same cells that have a fluorescent protein fused to the protein of interest. In one of those populations, the fluorescent proteins are photobleached and in the other they are not. Then, the fluorescence in the two populations is monitored over time as the photobleached population has its fluorescence replenished by protein production. By monitoring the time dependence in the difference between these two populations, the degradation rate constant can be determined explicitly.

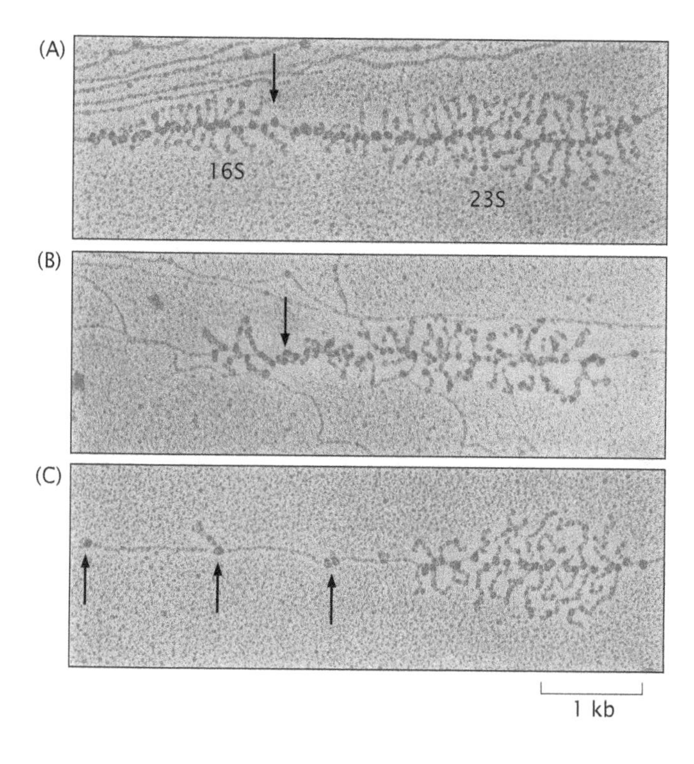

Figure 3.36: Effect of rifampin on transcription initiation. Electron micrographs of *E. coli* rRNA operons: (A) before adding rifampin, (B) 40 s after addition of rifampin, and (C) 70 s after exposure. No new transcripts have been initiated, but those already initiated are carrying on elongation. In (A) and (B) the arrow is used as a fiducial marker and signifies the site where RNaseIII cleaves the nascent RNA molecule producing 16S and 23S ribosomal subunits. RNA polymerase molecules that have not been affected by the antibiotic are marked by the arrows in (C). (Adapted from L. S. Gotta et al., *J. Bacteriol.* 20:6647, 1991.)

(a) At a certain instant in time, we photobleach one of the two populations so that their fluorescent intensity is now reduced relative to its initial value and relative to the value in the unphotobleached population. The number of fluorescent proteins N_f in the unphotobleached population varies in time according to the simple dynamical equation

$$\frac{dN_f}{dt} = \beta - \alpha N_f, \tag{3.24}$$

which acknowledges a rate of protein production β and a degradation rate α. Explain what this equation means and what it implies about the steady-state value of the number of fluorescent proteins per cell.

(b) A similar equation describes the dynamics of the unphotobleached molecules in the cells that have been subjected to photobleaching, with the number that that are unphotobleached given by N_u and described by the dynamical equation

$$\frac{dN_u}{dt} = \beta - \alpha N_u. \tag{3.25}$$

On the other hand, the number of photobleached proteins is subject to a different dynamical evolution described by the equation

$$\frac{dN_p}{dt} = -\alpha N_p, \tag{3.26}$$

since all that happens to them over time is that they degrade. Explain why there are two populations of proteins within the photobleached cells and why these are the right equations.

(c) In the paper, the authors then tell us to evaluate the difference in the number of fluorescent proteins in the two populations. A critical assumption then is that

$$\frac{dN_f}{dt} = \frac{dN_u}{dt} + \frac{dN_p}{dt} \tag{3.27}$$

The point is that, over time after photobleaching, the photobleached cells will become more fluorescent again as new fluorescent proteins are synthesized.

Plot the difference between the intensity of the cells that were not disturbed by photobleaching and those that were. In particular, we have

$$\frac{d(N_f - N_u)}{dt} = -\alpha(N_f - N_u). \tag{3.28}$$

Note that this quantity is experimentally accessible since it calls on us to measure the level of fluorescence in the two populations and to examine the difference between them. Integrate this equation and show how the result can be used to determine the constant α that characterizes the dynamics of protein decay.

• 3.7 The sugar budget in minimal medium

In rapidly dividing bacteria, the cell can divide in times as short as 1200 s. Make a careful estimate of the number of sugars (glucose) needed to provide the carbon for constructing the macromolecules of the cell during one cell cycle of a bacterium. Use this result to work out the number of carbon atoms that need to be taken into the cell each second to sustain this growth rate.

• 3.8 Metabolic rates

Assume that 1 kg of bacteria burn oxygen at a rate of 0.006 mol/s. This oxygen enters the bacterium by diffusion through its surface at a rate given by $\Phi = 4\pi DRc_0$, where $D = 2\,\mu m^2/ms$ is the diffusion constant for oxygen in water, $c_0 = 0.2\,mol/m^3$ is the oxygen concentration, and R is the radius of the typical bacterium, which we assume to be spherical.

(a) Show that the amount of oxygen that diffuses into the bacterium is greater than the amount used by the bacterium in metabolism. For simplicity, assume that the bacterium is a sphere.

(b) What conditions does (a) impose on the radius R for the bacterial cell? Compare it with the size of *E. coli*.

• 3.9 Evolutionary time scales

To get a sense of evolutionary time scales, it is useful to draw an analogy with human time scales. If the age of life on Earth is analogous to one year, what would, proportionally, various other time scales be analogous to? Specifically, the time since the origin of animal life? Of primates? Of the genus *Homo*? Of civilization? Since the extinction of dinosaurs? A human life span? (Problem courtesy of Daniel Fisher.)

3.7 Further Reading

Murray, A, & Hunt, T (1993) The Cell Cycle, Oxford University Press. This book has a useful discussion of the cell cycle.

Morgan, DO (2007) The Cell Cycle: Principles of Control, New Science Press. Morgan's book is full of interesting insights into the cell cycle.

Bier, E (2000) The Coiled Spring: How Life Begins, Cold Spring Harbor Laboratory Press. This is the best introduction to developmental biology of which we are aware. This book is a wonderful example of the seductive powers of development.

Carroll, SB (2005) Endless Forms Most Beautiful, W. W. Norton. One of us (RP) read this book twice in the first few weeks after it hit the shelves. From the perspective of the present chapter, this book illustrates the connection between developmental and evolutionary time scales.

Neidhardt, F (1999) Bacterial growth: constant obsession with *dN/dt*, *J. Bacteriol.* **181**, 7405. Bacterial growth curves are one of the simplest and most enlightening tools for peering into the inner workings of cells. Neidhardt's ode to growth curves is both entertaining and educational.

Dressler, D, & Potter, H (1991) Discovering Enzymes, W. H. Freeman. This book is full of fascinating insights into enzymes.

3.8 References

Alberts, B, Johnson, A, Lewis, J, et al. (2008) Molecular Biology of the Cell, 5th ed., Garland Science.

Andelfinger, G, Hitte, C, Etter, L, et al. (2004) Detailed four-way comparative mapping and gene order analysis of the canine *ctvm* locus reveals evolutionary chromosome rearrangements, *Genomics* **83**, 1053.

Bell-Pedersen, D, Cassone, VM, Earnest, DJ, et al. (2005) Circadian rhythms from multiple oscillators: lessons from diverse organisms, *Nat. Rev. Genet.* **6**, 544.

Belle, A, Tanay, A, Bitincka, L, et al. (2006), Quantification of protein half-lives in the budding yeast proteome, *Proc. Natl Acad. Sci. USA* **103**, 13004.

Berg, HC (2000) Motile behavior of bacteria, *Phys. Today* **53**(1), 24.

Blest, AD (1978) The rapid synthesis and destruction of photoreceptor membrane by a dinopid spider: a daily cycle, *Proc. R. Soc. Lond.* **B200**, 463.

Darwin, C (1859) On the Origin of Species, John Murray.

Droz, B, & Leblond, CP (1962) Migration of proteins along the axons of the sciatic nerve, *Science* **137**, 1047.

Eden, E, Geva-Zatorsky, N, Issaeva I, et al. (2011) Proteome half-life dynamics in living human cells, *Science* **331**, 764.

Gotta, LS, Miller, OL, Jr, & French, SL (1991) rRNA transcription rate in *Escherichia coli*, *J. Bacteriol.* **20**, 6647.

Ingolia, NW, & Murray, AW (2004) The ups and downs of modeling the cell cycle, *Curr. Biol.* **14**, R771.

Kalir, S, McClure, J, Pabbaraju, K, et al. (2001) Ordering genes in a flagella pathway by analysis of expression kinetics from living bacteria, *Science* **292**, 2080.

Laub, MT, McAdams, HH, Feldblyum, T, et al. (2000) Global analysis of the genetic network controlling a bacterial cell cycle, *Science* **290**, 2144.

Miller, OL, Jr, Hamkalo, BA, & Thomas, CA, Jr (1970) Visualization of bacterial genes in action, *Science* **169**, 392.

Müller-Hill, B (1996) The *lac* Operon: A Short History of a Genetic Paradigm, Walter de Gruyter. A very stimulating and interesting book to read.

Nakajima, M, Imai, K, Ito, H, et al. (2005) Reconstitution of circadian oscillation of cyanobacterial KaiC phosphorylation *in vitro*, *Science* **308**, 414.

Pollard, TD, & Earnshaw, WC (2007) Cell Biology, 2nd ed., W. B. Saunders.

Pomerening, JR, Kim, SY, & Ferrell, JE (2005) Systems-level dissection of the cell-cycle oscillator: bypassing positive feedback produces damped oscillations, *Cell* **122**, 565.

Reddy, B, & Yin, J (1999) Quantitative intracellular kinetics of HIV type 1, *AIDS Res. Hum. Retroviruses* **15**, 273.

Rizzoli, SO, & Betz, WJ (2005), Synaptic vesicle pools, *Nat. Rev. Neurosci.* **6**, 57.

Rosenfeld, N, Young, JW, Alon, U, et al. (2005) Gene regulation at the single-cell level, *Science* **307**, 1962.

Schleif, R (1993) Genetics and Molecular Biology, The Johns Hopkins University Press.

Yildiz, A, Forkey, JN, McKinney, SA, et al. (2003) Myosin V walks hand-over-hand: single fluorophore imaging with 1.5-nm localization, *Science* **300**, 2061.

Who: "Bless the Little Beasties" 4

Overview: In which key model systems are introduced

> To see a World in a Grain of Sand, And a Heaven in a Wild Flower, Hold Infinity in the Palm of your hand, And Eternity in an hour.
>
> **William Blake,** *Auguries of Innocence*

Most organisms on the planet share some fundamental similarities, as epitomized by the near universality of the genetic code. For exploring these kinds of processes, the choice of any cell as a subject should be as good as any other. Yet, for various reasons, often certain organisms become the beloved model systems that propel biological investigation forward. At the same time, many organisms exhibit fantastic specializations such that they may be considered experts in particular types of biological processes. For studying these specialized processes, such as rapid electrical conductance down the squid giant axon, it is almost always best to go to the expert. Many kinds of biological experimentation make other demands on their subjects. The subjects must be willing to grow in laboratory conditions, they must be amenable to genetic or biochemical analysis, and they must not be terribly hazardous to human researchers. These competing demands have resulted in the choice among biological researchers of a certain subset of living organisms that are widely used as model systems. In this chapter, we will explore a brief history of the contributions of several of the most noted model organisms who will reappear time and again throughout the remainder of the book.

4.1 Choosing a Grain of Sand

The preceding two chapters have celebrated the extraordinary diversity of living organisms on Earth. We have emphasized the fascinating and useful fact that all organisms alive today are part of a common lineage and therefore share many features, underlying their extraordinary diversity. From this perspective, it should be possible to undertake the exploration of biological science using any living organism as a "model system" for study. In practice though, a handful of organisms have been most widely used and proved to be most informative. In Chapter 2, we introduced *E. coli* as the ultimate experimental workhorse, and certainly for many processes represented in the bacterial domain of life *E. coli* has provided most of our understanding. There are many things that *E. coli* cannot do, such as the kind of sexual reproduction characteristic of multicellular eukaryotes or differentiation into multiple cell types to form a multicellular organism. From each of the different branches of the great tree of life, one or a few organisms have been selected as models that are experimentally tractable and that provide a means of introduction for the curious biologist to the less accessible organisms within their branch. The examples that we have chosen to discuss in this chapter will reappear throughout the book. However, a quick perusal of the biological literature reveals many other model organisms that are used with comparable frequency to those discussed here, and this chapter should be considered introductory rather than comprehensive.

Modern Genetics Began with the Use of Peas as a Model System

In the late nineteenth century at about the same time that Darwin and Wallace were first elucidating the interconnectedness of all living organisms and Louis Pasteur was articulating the biochemical unity of life, a humble Austrian monk, Gregor Mendel, was establishing a new science of genetics. He did this not by focusing on the broad sweep of biology as had these other contemporaneous giants, but rather by focusing in great detail on a single organism, the pea plant *Pisum sativum*. Mendel noticed that variations in biological form and function occurred not just among different species, but also among individuals of the same species such as the pea. Furthermore, plants with a particular trait such as unusually wrinkly peas seemed particularly likely to give rise to offspring that shared that same trait. Through a series of carefully controlled pea breeding experiments, Mendel established the statistical rules governing inheritance and provided compelling and concrete evidence for the idea that large-scale traits of organisms were passed from parent to offspring in the form of "irreducible units" of inheritance that passed unaltered.

Peas were Mendel's model organism of choice not merely because of their central role in the monastery kitchen garden, but also because they have a short life cycle, a wide variety of easily measured heritable traits, and ease of large-scale stable cultivation. These same traits have been sought by subsequent generations of biologists in the choice of many other model systems that we will explore, ranging from bacteriophage to bacteria to yeast to fruit flies. The choice of a model system is, of course, only one step towards exploring the mechanistic basis of complex biological phenomena. The organism must lend itself to certain kinds of experimentation as detailed below.

4.1.1 Biochemistry and Genetics

What makes an organism a useful model organism? Throughout the twentieth century, a major focus of biological research was on understanding the underlying molecular basis for complex life processes. This meant that a useful model organism must in one way or another facilitate the process of making connections between events at the molecular scale and events at the level of whole organisms. Two very general complementary approaches encompass most of this work: genetics and biochemistry. Genetics is the study of heritable variation and takes advantage of the fact that living organisms give rise to offspring that are typically similar but not identical to themselves. Biochemistry demands the isolation of a component of a living organism that retains some interesting function. This takes advantage of the fact that the structure of organisms is divisible.

The goal of both biochemistry and genetics is to identify the molecules that are key players in important biological processes. But the approach of these two disciplines to molecular identification is nearly opposite. In the field of genetics, important molecules are identified within the context of an enormously complex living organism. By removing one component at a time and then asking whether the organism can still perform the process of interest, this procedure will determine whether a particular gene product is *necessary* for the process under investigation. In the complementary approach of biochemistry, the complex organism is subdivided into simpler and simpler subsets or fractions and each fraction is then tested to determine whether it can perform the process of interest. In many classic cases, a single polypeptide chain, laboriously purified from an organism, can perform an important biological process completely on its own in a test tube. This kind of analysis can be used to determine whether a particular molecule or set of molecules is *sufficient* to perform the process of interest. In the modern era of molecular analysis, we consider a process to be "understood" in terms of its molecular players if we can identify all molecules that are necessary and sufficient.

EXPERIMENTS

Experiments Behind the Facts: Genetics The practice of genetics involves identification and isolation of genes responsible for heritable variations within a species. In effect, this is applied Darwinism in action since it takes advantage of the two pillars of the theory of evolution, namely, (i) heritable variation and (ii) selection. One of the most familiar long-term genetics projects undertaken by humans has been the creation of hundreds of distinct dog breeds from a common wolf-like ancestor. In the laboratory, the first step in a genetics experiment is usually the isolation of a mutant that has either lost or gained some property of interest to the experimenter. For example, isolation of a fruit fly mutant with white instead of the normal red eyes can be the first step towards characterizing the pathway responsible for the generation of red eye color.

The earliest geneticists such as Mendel took advantage of naturally occurring mutants and chose to cross only those individuals carrying the trait of interest. Impatient modern biologists will usually start by deliberately generating a large collection of random mutants and then sorting through them to find those with interesting properties.

Kinds of mutagenesis Generation of mutants implies deliberate damage to DNA. In the mid-twentieth century, this was most commonly done with nonspecific chemical or radiological insults. X-rays or ultraviolet radiation or nonspecific chemical agents such as nitrosoguanidine that react with DNA bases induce random DNA damage that often the cell cannot accurately repair. These agents can cause alterations in single base pairs or small insertions or deletions that may alter the reading frame of a gene. An alternative approach that renders the subsequent steps of locating the site of the mutation much easier is to induce mutagenesis by introducing a transposon, a small, self-replicating fragment of DNA that will insert itself into the genome of the organism and typically disrupt the expression of genes at the site of insertion. In some cases, a biologist may wish to target the mutations more narrowly in the pool and perhaps explore only sequence variations within a particular gene. For these narrow purposes, many techniques of site-directed mutagenesis have been developed that take advantage of recombinant DNA and polymerase chain reaction (PCR) technology specifically to alter DNA in a designed, nonrandom way.

Selections and screens With the collection of mutants in hand, the next step is to identify the individuals with the desired phenotype. An organism's phenotype consists of its observable properties; in contrast, its genotype is the underlying heritable information that encodes for the observable properties. The easiest way to identify mutants with the desired phenotype is to design an experiment that will kill all of the organisms that lack the desired trait. For example, bacterial mutants resistant to a certain antibiotic can be readily isolated by simply pouring the antibiotic into the bacterial culture and isolating the survivors. This kind of mutant hunt is called a *selection.*

Because it is rare that such a convenient life or death situation can be established, it is more common that geneticists will design *screens,* methods of sorting mutant populations to enrich or identify the mutants of interest. For *Drosophila,* a screen may be as simple as a visual examination of the mutant pool. This approach is sufficient for isolating white-eyed flies or wrinkled peas. More typically, the organisms must be challenged or assayed in some way. For example, identification of *E. coli* mutants that are unable to grow on lactose as a carbon source requires replica plating as shown in Figure 4.1. Clever biochemists, wary of replica plating, have worked out complementary methods by designing chemicals that serve as reporters for functional biological pathways. For example, normal *E. coli* are able to convert the colorless chemical nicknamed X-gal (whose real name is 5-bromo-4-chloro-3-indolyl-β-D-galactopyranoside) into a blue product if and only if their lactose utilization pathway is intact. When grown on plates containing X-gal, mutant colonies stand out by virtue of their white (clear) color.

Conditional mutants In studying processes that are essential to life (for example, the ability to replicate DNA, the ability to generate ATP, etc.) it is obvious that there must be a catch. Genetics can only study heritable variation in individuals that are able to survive and reproduce. However, these essential life processes are certainly of interest to biological scientists. The indispensable trick that has been developed to allow geneticists to study these essential processes is the generation of "conditional mutants," that is, genes whose mutant phenotype is expressed under only some conditions and not others. A classic example is the isolation of temperature-sensitive mutants in microbial species. Normal *E. coli* can replicate over a wide range of temperatures from below 10 °C to over 45 °C. Bacteria carrying a mutation in DNA polymerase that alters protein stability such that the polymerase is functional at a low temperature

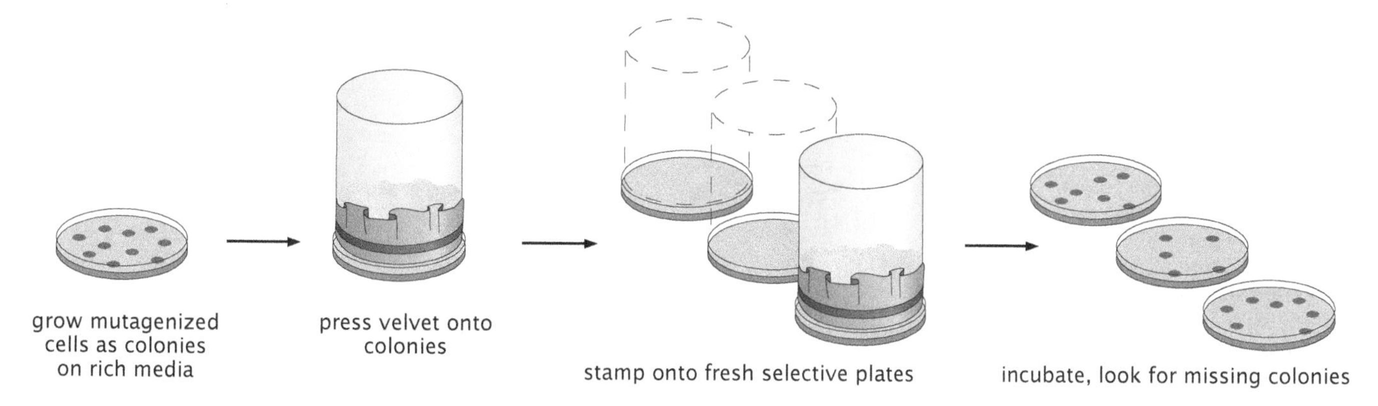

Figure 4.1: Replica plating method for isolating bacterial mutants. In these experiments, an entire distribution of colonies is lifted off one plate and put onto other plates that have some different nutrients. By comparing the original and secondary plates, it is possible to find colonies that did *not* grow on the secondary plates and identify them as mutants in some characteristic of interest.

(30 °C) but unstable at a high temperature (40 °C) can be propagated at low temperature even though they carry a lethal mutation in an essential product. Modern implementations of this approach involve construction of altered genes whose products are only expressed when the organisms are given an artificial chemical signal.

Experiments Behind the Facts: Biochemistry The practice of biochemistry is complementary to that of genetics and involves an application of reductionism, the scientific philosophy that complex processes can be broken down into simpler constituents. One of the earliest applications of biochemistry followed up the brilliant insight that the desirable ability of yeast to convert sugar into alcohol does not require the presence of living yeast cells, but rather only the presence of a few of its critical proteins that catalyze this chemical transformation. The central philosophy of biochemistry can be summarized as "grind and find" as exemplified by putting an organism into a blender and sorting out the molecules of the resulting organismal smoothie to find the one responsible for the process of interest.

Assay development The most important feature of a biochemical project is the development of an appropriate assay, a testing mechanism to determine whether a given test tube contains the molecule or molecules responsible for a process of interest. For example, we mentioned above the chromogenic substrate X-gal, which turns blue when acted upon by an enzyme in the *E. coli* pathway for lactose utilization. Besides being used on plates to assay the phenotypes of intact cells, X-gal can also be added to a test tube containing a purified protein and the development of blue color can be measured quantitatively. This is a simple example. Most biochemical assay development requires substantially more effort and ingenuity.

Fractionation With an assay in hand, the next necessary step is to subdivide the components of the organism. Numerous techniques have been developed for separating molecular constituents on the basis of their chemical or physical properties. Among the most powerful of these are the many varieties of chromatography. The basic idea is to pass a complex mixture of

Figure 4.2: Biochemical columns. The column filters different molecular species by virtue of the different speeds that these species have as they traverse the column. (A) In size-exclusion chromatography, the column is filled with a matrix of small beads that have many pores of varying sizes. When a mixture of molecules of different sizes is poured in at the top of the column, the largest molecules will flow through fastest because they cannot enter the small pits and holes in the beads. The smallest molecules will move slowest. (B) In affinity chromatography, the matrix is chemically conjugated to a molecule that will bind to the protein of interest. When a complex mixture is poured into this kind of column, most of the proteins will flow through immediately and only those that can bind to the ligand in the column will be retained. To elute the protein of interest from the column, a soluble form of the same ligand can be added. This will displace the protein from the column matrix and allow it to be collected as a fraction at the bottom.

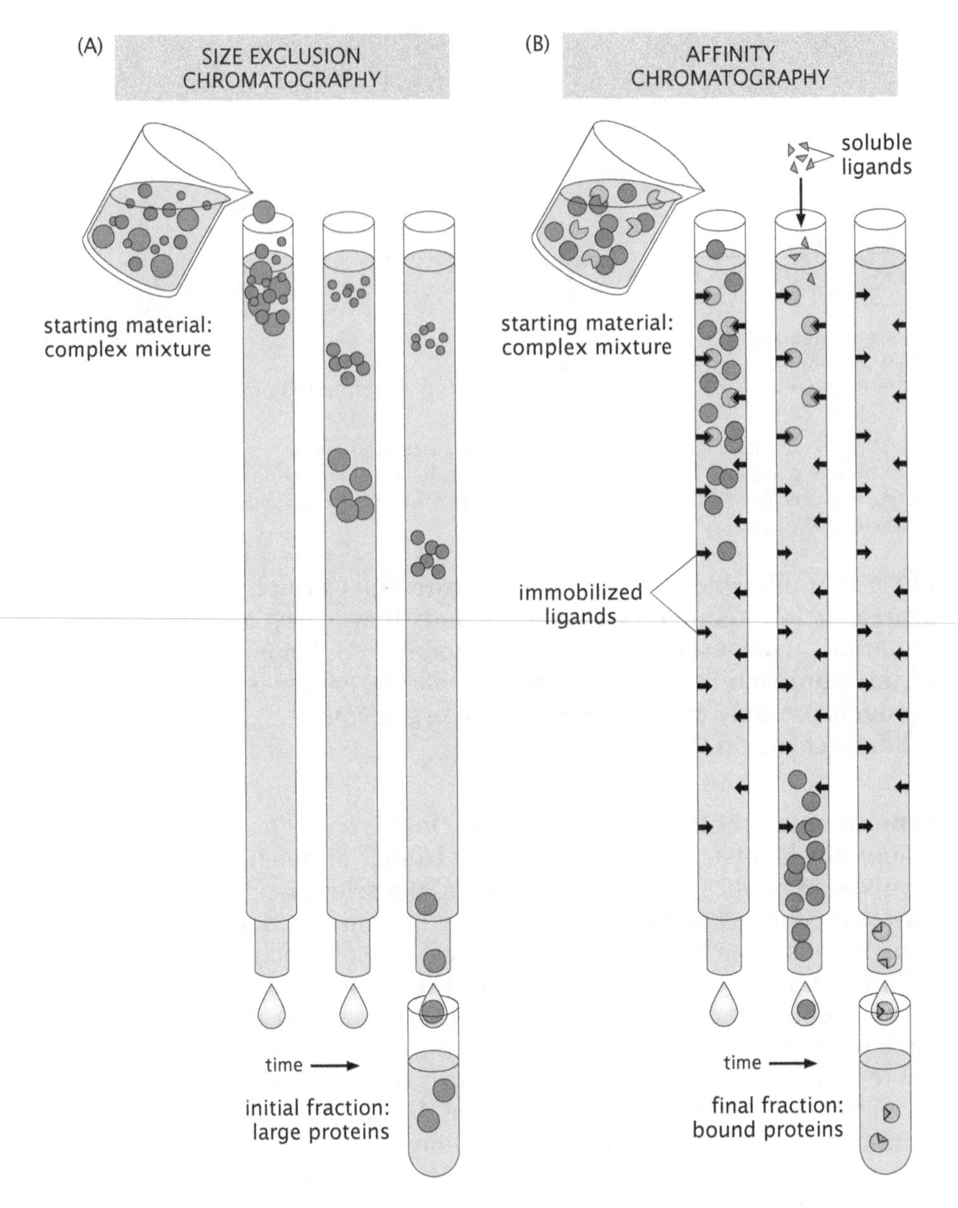

components through a matrix of material where some components are more free to move than others. A useful geometry is the biochemical column shown in Figure 4.2, in which a small volume of material is added to the top and fractions are collected from the bottom. Molecules that interact only weakly with the material of the column pass through quickly and can be collected in early fractions. Molecules whose transit time is delayed appear at the bottom, only later. Different kinds of columns can separate different molecules based on their physical size, net charge in solution, ability to bind to particular molecular targets, etc. By using a series of fractionation steps operating on different principles, it is frequently possible to isolate a single molecular species or complex that can perform the desired function.

Armed with a sense of two of the central experimental pillars of biology, we now explore some of the hall-of-fame model systems that have led to the emergence of modern biology and that can serve as testbeds for physical biology as well.

4.2 Hemoglobin as a Model Protein

This section explores the question of how the study of the structure and function of the hemoglobin molecule has served as one of the dominant themes in biochemistry for more than a century. As noted in the previous section, genetics and biochemistry have served as two of the driving forces behind the emergence of modern biology. One of the most significant goals of biochemistry has been the dissection of the structure and function of all of the molecules of life. This quest has been spectacularly successful, and can be illustrated through the case study of hemoglobin, a molecule we will appeal to repeatedly to illustrate different ideas in physical biology.

4.2.1 Hemoglobin, Receptor–Ligand Binding, and the Other Bohr

The Binding of Oxygen to Hemoglobin Has Served as a Model System for Ligand–Receptor Interactions More Generally

The function of macromolecules in the living world emerges from interaction. That is, it is only in their relations with one another (Lac repressor and DNA, hemoglobin and oxygen, lipids and water, etc.) that the molecules of life begin to be animated with the first shadowy resemblance of their role in the living realm. Figure 4.3 shows some examples of receptor–ligand linkages that will be revisited throughout the book. Klotz (1997), himself paraphrasing Paul Ehrlich, noted "a substance is not effective unless it is linked to another." Probably the most studied example of such linkage is that of the hemoglobin (Hb) molecule in its partnership with oxygen. Hemoglobin is present in red blood cells and is actively responsible for transporting oxygen to cells. Target tissues such as muscle also contain oxygen-binding proteins. The most abundant of these is myoglobin. Whereas hemoglobin has four binding sites for oxygen molecules, myoglobin has only one. As we will see below, this structural difference between the two proteins causes very interesting functional differences. This cooperative relationship (ligand (O_2) and receptor (Hb)) is but one example of the nearly endless variety of such relationships that occur in biological chemistry. From a physical model-building viewpoint, the key questions that will occupy our attention center on estimating the fraction of receptors that will be bound by ligands as a function of ligand concentration and external conditions such as pH and temperature.

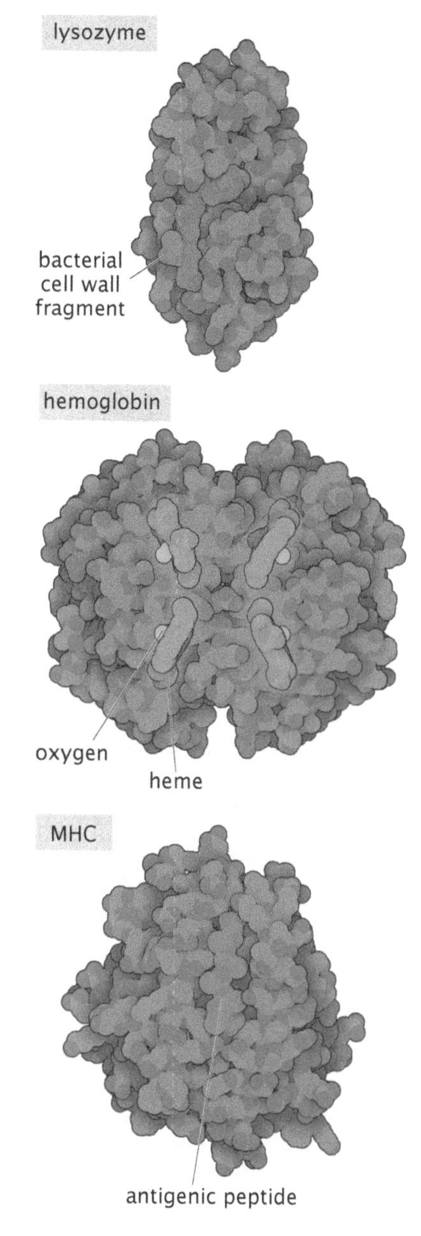

Figure 4.3: Examples of ligand–receptor interactions. From top to bottom the proteins are lysozyme (PDB 148l), hemoglobin (PDB 1hho), and major histocompatability complex protein (MHC, PDB 1hsa). Lysozyme binds molecules making up the bacterial cell wall resulting in its degradation, hemoglobin binds oxygen, and, in its role in the immune system, MHC presents antigens at the cell surface. (Courtesy of D. Goodsell.)

Estimate: Hemoglobin by the Numbers There are hundreds of millions of hemoglobin molecules in each of the red blood cells that populate our bodies. To get a feeling for the numbers, we examine the results of a complete blood count (CBC) test at the doctor. Such tests report a number of interesting features about our blood. For an example of the outcome of such a test, see Table 4.1 in the problems at the end of the chapter. One of the outcomes of such a test is the observation that typically there are roughly 5×10^6 red blood cells per microliter of blood and that there is roughly 15 g of hemoglobin for every deciliter of blood (the so-called mean corpuscular hemoglobin (MCH)). If we estimate that there are roughly 5 L of blood in a typical adult human, this amounts to roughly 25×10^{12} red blood cells, containing in total some 750 g of hemoglobin. Given that the mass of a hemoglobin molecule is roughly 64,000 Da, we arrive at the conclusion that there are of the order of 7×10^{21} hemoglobin molecules in the blood of the average adult. Finally, these observations from an everyday blood test thus lead to the conclusion that in each red blood cell there are of the order of 3×10^{8} hemoglobin molecules. This large number of molecules trapped inside a membrane results in an osmotic pressure, as we will calculate in Chapter 14. Estimates based on the quantities measured in a blood test are explored further in the problems at the end of the chapter.

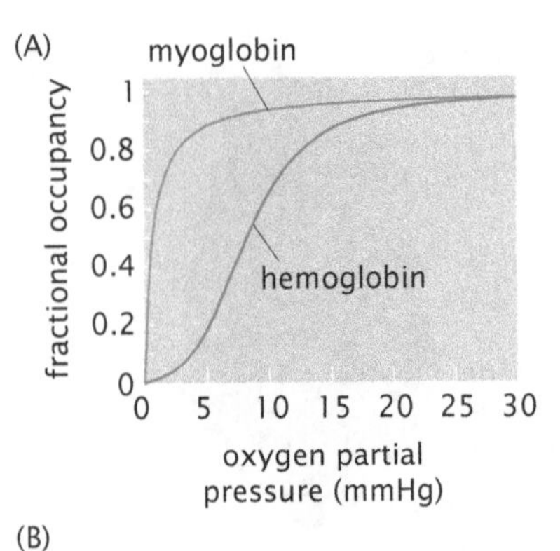

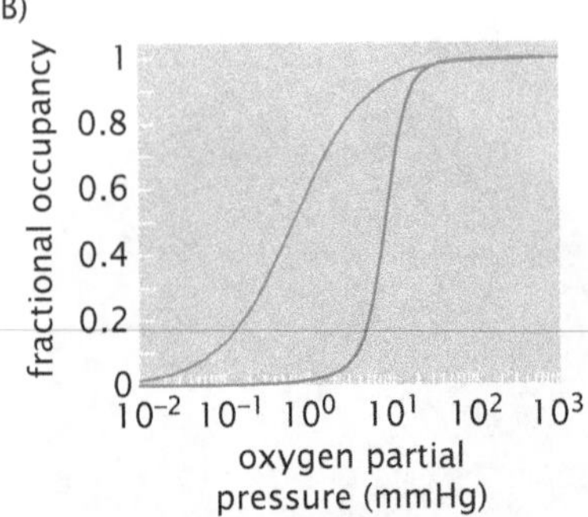

Figure 4.4: Binding curves for oxygen uptake by myoglobin and hemoglobin. The fractional occupancy refers to the fraction of available oxygen-binding sites that are occupied in equilibrium as a function of the oxygen concentration as dictated by the pressure. (A) Binding curves on a linear scale. (B) Binding curves on a log–linear plot. The curves were generated based on experimental data. (Data for hemoglobin from K. Imai, *Biophys. Chem.* 37:197, 1990; data for myoglobin from A. Rossi-Fanelli and E. Antonini, *Arch. Biochem. Biophys.* 77:478, 1958.)

Quantitative Analysis of Hemoglobin Is Based upon Measuring the Fractional Occupancy of the Oxygen-Binding Sites as a Function of Oxygen Pressure

One of the key questions that one might ask about the binding relationship between a ligand and its receptor is what fraction of the receptor-binding sites is occupied as a function of the concentration of ligand molecules. This question is of physiological interest in the setting of hemoglobin because it tells us something about the uptake and delivery of oxygen as a function of the environment within which the hemoglobin finds itself. If a hemoglobin molecule is in an oxygen-starved environment, it has a higher propensity to drop bound oxygen molecules. These insights are captured quantitatively in binding curves like those in Figure 4.4, which show the fractional occupancy of sites in both hemoglobin and myoglobin as a function of the oxygen pressure. What we note in such curves is that when the concentration of oxygen is very high, essentially all of the available sites on hemoglobin (or myoglobin) are occupied by O_2 molecules. By way of contrast, in those environments such as active muscle where the concentration of oxygen is "low" (as defined precisely by the binding curve itself), the hemoglobin molecule has largely delivered its cargo.

Interestingly, much of the work that ushered in the quantitative approach to hemoglobin structure and function was carried out by Christian Bohr, a Danish physiologist, perhaps more familiar to some of our readers as the father of Niels Bohr. Indeed, there is an effect associated with hemoglobin known as the Bohr effect. This effect refers to data like those shown in Figure 4.5, where there are shifts in the oxygen-binding curve as a function of the pH. The Bohr effect is of interest to athletes since it reveals the fact that in a more acidic environment (like starved muscles), the competition of H^+ to bind to hemoglobin makes the release of O_2 even more likely than at normal pH. From the point of view of the present book, the beauty of measurements like those shown in the figure is that they were instrumental in driving research aimed at quantitative modeling of protein structure and function and serve as a compelling example of our battle cry that quantitative data demand quantitative models.

4.2.2 Hemoglobin and the Origins of Structural Biology

In parallel with efforts to dissect the functional characteristics of hemoglobin, there was a strong push to understand the structure of this important molecule. The idea of connecting structure and function is such a central tenet of current thinking that it is perhaps difficult to realize how much needs to be known before one can embark upon the attempt to effect this linkage.

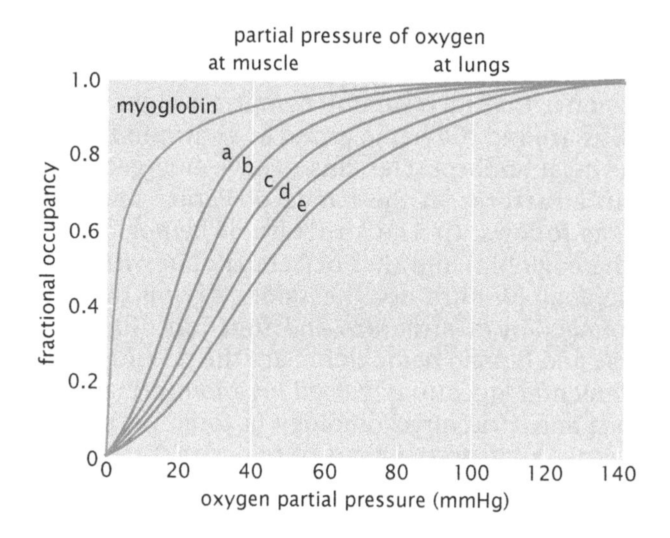

Figure 4.5: Binding curves for oxygen uptake by hemoglobin as a function of pH revealing the Bohr effect. The hemoglobin binding curves are shown for five values of the pH: (a) 7.5, (b) 7.4, (c) 7.2, (d) 7.0, and (e) 6.8. The vertical lines indicate the partial pressures experienced in muscle and in the lungs. (Adapted from R. E. Dickerson and I. Geis, Hemoglobin: Structure, Function, Evolution and Pathology. Benjamin/Cummings, 1983.)

The Study of the Mass of Hemoglobin Was Central in the Development of Centrifugation

One of the first "structural" questions that was tackled (with great difficulty) is what is the mass of the hemoglobin molecule? It is easy to take the answer to such questions for granted. Historically, the proper response to this question was tied to the development of one of the most important tools in the biochemist's and biologist's modern arsenal, the centrifuge. Svedberg's development of the centrifuge represents one of the key technological breakthroughs that permitted the advance of biological science, with particular implications for the debate over the nature of proteins and the constitution of hemoglobin (see Tanford and Reynolds, 2001—referenced in Further Reading at the end of the chapter). In particular, as a result of centrifugation studies, it was concluded that hemoglobin is a tetramer made up of four units each with a mass of roughly 16,000 Da.

Structural Biology Has Its Roots in the Determination of the Structure of Hemoglobin

The Protein Data Bank has obscured the extreme difficulty underlying the determination of protein structure. With a simple mouse click, the structures of thousands of different proteins are available for anyone to see. However, we are the beneficiaries of more than 30 years of dogged effort on the part of the founders of structural biology. With reference to the structure of hemoglobin, Max Perutz is the prime hero. One of the most inspiring features of Perutz's battle to obtain the structure of hemoglobin is the length of that battle. From the time (1937) at which he set himself the task of determining this structure to the time of publication of the first structural reports was a period of nearly 25 years during which a great number of technical obstacles were overcome.

The outcome of the structural studies of Perutz (and others) was the recognition that hemoglobin is a tetramer composed of four subunits, two α-chains and two β-chains. The focal point of each subunit is a heme group, which is made up of an iron atom at the center of a porphyrin ring. As noted above, each of these subunits has a molecular mass of roughly 16,000 Da and an amino acid sequence of roughly 145 amino acids (the sequence can be seen in Figure 21.23 on p. 995).

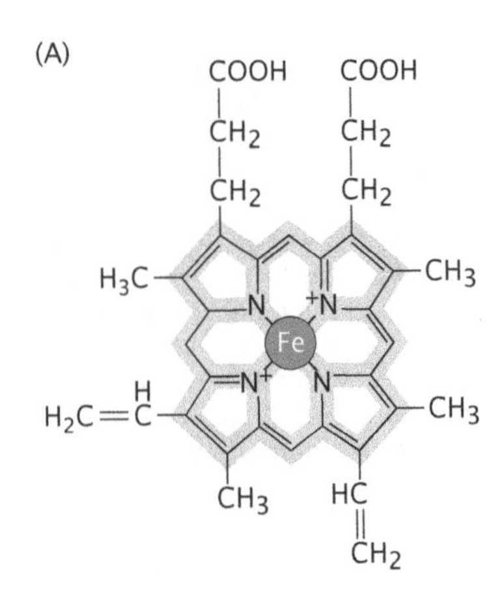

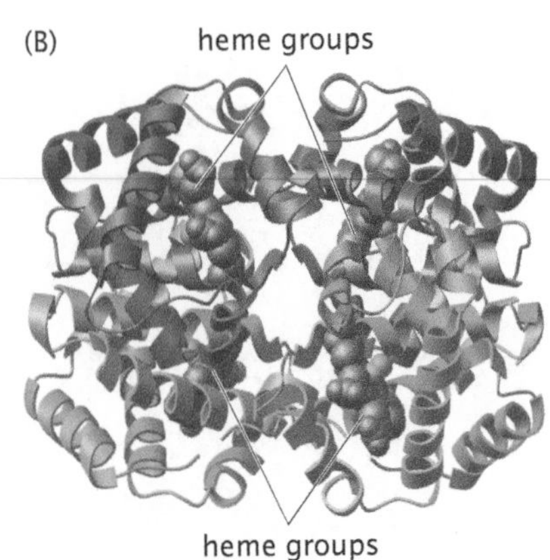

Figure 4.6: Structure of hemoglobin. (A) The heme group includes a porphyrin ring (light red line) that can bind an iron ion. The iron can bind to oxygen or carbon monoxide molecules and is responsible for the red color of blood. (B) The protein hemoglobin consists of four separate polypeptide chains shown in different colors, two α-chains and two β-chains (PDB 1hho). Each one carries a heme group, so the overall protein can bind up to four molecules of oxygen.

The heme groups are the focal point of a functional discussion of hemoglobin because they are the site of O_2 binding. One useful structural description is that each subunit comprises eight α-helical regions that form a box around the heme group as shown in Figure 4.6.

Some of the most important lessons to have emerged from the analysis of protein structures in the hands of Perutz and others can be summarized as follows. (i) The structure of hemoglobin with bound oxygen (oxyhemoglobin) and that of hemoglobin with no bound oxygen (deoxyhemoglobin) are not the same, paving the way for ideas about the connection of structure and function. (ii) The similarities between horse and human hemoglobin are the primitive roots of some of the fundamental questions raised in bioinformatics concerning both sequence and structure homology (a topic we take up in Chapter 21), and strike right to the heart of how structural biologists have contributed to our further understanding of evolution.

4.2.3 Hemoglobin and Molecular Models of Disease

Another fascinating episode in the history of hemoglobin as a model system is the quest to understand sickle-cell disease (sometimes called sickle-cell anemia). This disease is characterized morphologically by severe alterations in the structure of the red blood cell and is manifested physiologically as trapping of the red blood cells in capillaries, cutting off oxygen to some tissues and causing severe pain. Figure 2.21 (p. 59) shows the passage of healthy red blood cells through narrow man-made capillaries. The proposal made by Pauling was that sickle-cell disease is a "molecular disease" resulting from a defect in hemoglobin. The histological manifestation of this defect is the banana-shaped cells characteristic of sickle-cell disease rather than the more familiar biconcave disc shape characteristic of healthy red blood cells (shown in Figure 5.10 on p. 204).

The particular fault with the hemoglobin molecule in the sickle-cell case is the replacement of a charged glutamic acid residue with a hydrophobic valine at one specific location on the β-chains in the hemoglobin tetramer. The formation of the peculiar shapes of the red blood cells in those with the disease is a result of the fact that the hemoglobin molecules aggregate to form fibers. These fibers push on the cell membrane and deform it. Our estimate at the beginning of this section led us to the conclusion that there are 3×10^8 hemoglobin molecules per red blood cell. To get an idea of the mean spacing of these molecules, we can use the formula $d = (V/N)^{1/3}$ (this formula appeared as Equation 2.12 on p. 49), which, taking a volume of 30 μm^3 per red blood cell, leads us to a center-to-center distance between hemoglobin molecules of less than 5 nm, to be compared with a typical protein radius of the order of 2–3 nm. What we learn from these numbers will be revealed as a recurring theme throughout the book (indeed, all of Chapter 14 will discuss these effects), namely, that cells are very crowded. With regard to hemoglobin in red blood cells, the crowding is so high that apparently replacing a charged residue with a hydrophobic residue suffices to drive crystallization of the hemoglobin molecules.

4.2.4 The Rise of Allostery and Cooperativity

One of the related roles in which hemoglobin has served as a model system of paramount importance is in the context of the physical origins of the curves shown in Figure 4.4. In particular, as will be shown in Chapter 7, the most naive model that one would be tempted to apply to the problem of oxygen binding to a protein does a remarkable job of reproducing the features shown in the *myoglobin* binding curve. However, the sigmoidal shape of the hemoglobin binding curve is a different matter altogether and helped give rise to two key notions, namely, cooperativity and allostery. As noted by Brunori (1999), "I suspect that no other protein will ever be studied in comparable detail: many facets of the biology, chemistry and physics of proteins were addressed initially by researchers working on hemoglobin." As we will show in Chapter 7, cooperativity refers to the fact that binding of ligands on different sites on the same molecule is not independent. For hemoglobin, binding of an oxygen molecule at one of the four binding sites provided by the heme groups effectively increases the likelihood that a second oxygen molecule will bind at a second site, relative to completely deoxygenated hemoglobin. A new molecule of oxygen is even more likely to bind to hemoglobin where two of the oxygen-binding sites are occupied, and still more likely to bind to hemoglobin where three sites are already occupied. Consequently, it is usually the case that a solution of hemoglobin with a limiting amount of oxygen present will contain many individual hemoglobin molecules that have no bound oxygen at all, and some hemoglobin molecules that have four bound oxygens. The idea that the binding of a ligand at one site on a protein can affect the properties of the protein at a distant site is generally referred to as allostery. For hemoglobin, the binding of an oxygen at one site allosterically increases its affinity for oxygen at the other three sites. For many enzymes, the binding of an allosteric activator or inhibitor at one site will affect the enzymatic activity of the enzyme at a distant site. This idea, which has been served immensely by studies on hemoglobin, has turned out to be one of the central ideas concerning the biological uses of binding.

In addition to the notion of cooperativity, studies of hemoglobin have provided the basis for the development of the idea that proteins can exist in multiple conformational states. This is a theme we will develop in detail in Chapter 7 and explains some of the added complexity that proteins adopt as a result of post-translational modifications such as phosphorylation. For enzymes, it is frequently useful to imagine that the protein can exist in several distinct structural states and that it is constantly switching back and forth among these states. One of these structural states can be thought of as the active or functional state, and environmental changes such as tuning the pH or binding of ligands to allosteric sites on these molecules can alter the fraction of time that the molecule spends in each state. The idea that ligand binding can induce conformational change is one that saw concrete experimental realization in the context of hemoglobin and inspired a classic model known as the Monod–Wyman–Changeux (MWC) model, a two-state model of allostery that had as one of its proving grounds the analysis of oxygen-binding curves in hemoglobin. This model will be described in Chapter 7.

4.3 Bacteriophages and Molecular Biology

One of the most enticing classes of biological organisms that has had flagship status as a model system consists of the bacterial viruses known as bacteriophages. The present section highlights a few representative examples of the way in which bacteriophages have figured centrally (and still do) in the explication of biological phenomena.

4.3.1 Bacteriophages and the Origins of Molecular Biology

Bacteriophages Have Sometimes Been Called the "Hydrogen Atoms of Biology"

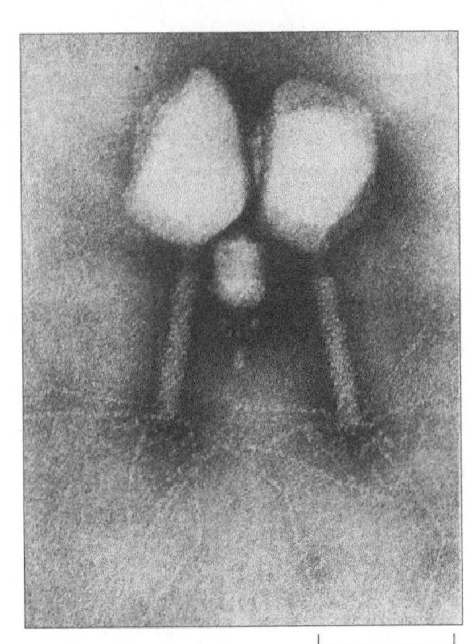

Figure 4.7: Electron micrograph of the T2 (large) and ϕ29 (small) bacteriophages. (Adapted from S. Grimes et al., *Adv. Virus Res.* 58:255, 2002.)

The choice of model organisms often reflects experimental convenience. Bacteriophages and their bacterial hosts were so experimentally convenient that together they served to launch modern molecular biology. The use of bacterial viruses as model organisms resulted from a broad variety of such conveniences, including a 20 minute life cycle and the fact that huge quantities (in excess of 10^{10}) of such viruses could be assayed simultaneously. Further, bacteriophages have relatively compact genomes (usually much less than 1 Mbp) and are constructed from a limited parts list. Because of this simplicity, Max Delbrück referred to bacteriophages as the "hydrogen atom of biology," a reference to the fact that in physics the hydrogen atom is the simplest of atoms and that lessons from this simple case are instructive in building intuition about more complex atoms as well. The basic architectural element of the bacteriophage is the presence of a protein shell (the capsid) that is filled with genetic material. Several examples of bacteriophages are shown in Figure 4.7, which has been dubbed the phage family portrait, though the diversity of bacterial viruses goes well beyond the two examples shown here. This particular electron micrograph shows the morphology and relative size of the T2 and ϕ29 phages.

Experiments on Phages and Their Bacterial Hosts Demonstrated That Natural Selection Is Operative in Microscopic Organisms

The unity of life is an idea that we now take for granted. Nevertheless, less than 150 years ago, it was not clear that microbes were even "cells." As will be discussed in more detail in Chapter 21, only in the latter half of the twentieth century were the Archaea discovered. It has always been a challenge to try to figure out the relations between the world of microscopic living organisms on the one hand and the macroscopic living organisms of everyday experience such as giraffes and redwood trees on the other.

One of the hard-won insights into our understanding of microbes and the viruses that infect them had to do with the applicability of the ideas of evolution to microorganisms. The question was: are microbes subject to variation and selection in the same way as larger organisms such as peas and fruit flies? The answer to this question came from a famous experiment by Delbrück and Luria known as the "fluctuation test" that relied on bacteriophages.

Their reasoning was based on the observation that infection of a culture of bacteria with bacteriophages led to some subset of cells that were resistant. Their aim was to determine whether this resistance arose as a result of spontaneous mutations or was somehow induced by the phages themselves. If the resistance was induced, they reasoned that if they diluted some original culture into several different cultures and infected them with phages, roughly the same number of resistant colonies should be found per culture. On the other hand, if

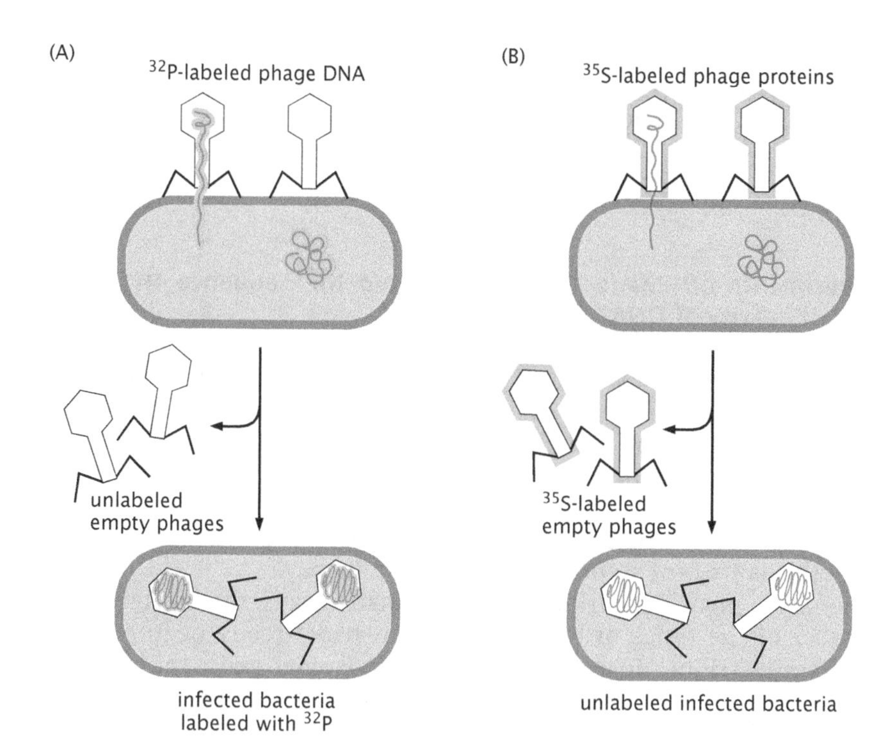

Figure 4.8: The Hershey–Chase experiment. (A) For the case in which the nucleic acid of the phage is labeled, radioactive phosphorus is found in the infected bacteria. (B) For the case in which the protein of the phage is labeled, no radioactivity is found in the infected bacteria.

the resistance arose spontaneously, there should be a large variation in the number of resistant colonies from one culture to the next. The latter eventuality is what was observed, demonstrating the important role of spontaneous mutations as the raw material for evolution, even in the case of microscopic organisms.

The Hershey–Chase Experiment Both Confirmed the Nature of Genetic Material and Elucidated One of the Mechanisms of Viral DNA Entry into Cells

One of the classic experiments in the history of molecular biology that exploited bacteriophages is the so-called Hershey–Chase experiment shown in Figure 4.8. The idea of the experiment is to build phage particles with packaged DNA that is radioactively labeled with the phosphorus isotope ^{32}P and proteins associated with the capsid containing the sulfur isotope ^{35}S. These labeled phages are used to infect bacteria and then the phages and the bacteria are separated by vigorous stirring in a blender. By querying the bacteria to find out which radioactive species they contain, it is possible to determine whether protein or DNA is the "genetic material" of the viruses themselves.

The Hershey–Chase experiment was a direct and convincing experimental confirmation that DNA (and not proteins) carries the genetic material. However, in addition to this deep-rooted insight with implications for all of biology, several more specialized insights emerged from this experiment as well. First, this experiment provides insights into the nature of viral infection, at least in the case of bacteriophages with the viral DNA delivered to the bacterial interior while the protein shell remains outside. In Chapters 10 and 16, we will explore how physical reasoning can be used to understand viral packing and ejection, respectively. A second powerful set of insights embodied in this experiment concern the nature of viral self-assembly, in particular, and biological self-assembly, more generally. Specifically, this experiment showed that the synthetic machinery (for both DNA and proteins) of the host cell had to be hijacked to do the bidding of the virus; namely, it had to make copies of the viral DNA and express the proteins required for the assembly of virulent new phage particles.

Experiments on Phage T4 Demonstrated the Sequence Hypothesis of Collinearity of DNA and Proteins

Much of what we now "know" and take for granted was hard won in the knowing. Despite the ease with which every high-school science student can rattle off the connection between the two great polymer languages, the collinearity of DNA and protein, itself an article of belief of the central dogma, was at the start nothing but hypothesis. It was the work of Seymour Benzer on phage T4 that gave this hypothesis its first steps towards being a theory.

Benzer's phage experiment was in many ways the bacteriophage partner of the types of experiments that had already been perfected by Thomas Hunt Morgan and his chromosomal geographers whose work on fruit flies will be discussed later in this chapter. Benzer used mutants of a particular gene known as *rII* of bacteriophage T4. At this point in time, the idea that genes were physical entities arranged in a linear fashion along chromosomal DNA had already been established by the work of Morgan, but most people tended to think of individual genes as being point objects that could be either normal or mutant. In Benzer's experiment, he examined what could happen when two different mutant copies of a gene were allowed to recombine with one another. He found that recombination events could glue the healthy parts of the sequences back together, resulting in a wild-type phage. The conclusion to be drawn from the recovery of wild-type phages from mutants was that in some cases mutations could exist at different locations within a single gene. Benzer could map out the geography of the *rII* gene by measuring the frequency which with wild-type T4 phages were restored between multiple pairwise combinations of distinct mutations. The more likely the restoration, the more distant the mutations within the *rII* gene.

Most fundamentally, Benzer showed experimentally that individual genes are endowed with physical extent (that is, some region of base pairs along the DNA double helix) exactly as was demanded by the model of DNA structure proposed by Watson and Crick. There was a long gap of roughly 20 years between Benzer's gene mapping experiments and the explicit sequencing of DNA, again involving a bacteriophage, though this time the phage ϕX174.

The Triplet Nature of the Genetic Code and DNA Sequencing Were Carried Out on Phage Systems

Brenner and Crick, inspired by Benzer's mapping experiments, saw that these same methods could provide an answer to the fundamental coding problem: is the genetic code built up of triplet codons? The triplet codon hypothesis was predicated on the idea that a given gene is read in groups of three letters starting from some unique point. Though seemingly arcane, it was also suspected that the chemical acridine yellow caused mutations in DNA by either adding or deleting a single base. The realization of Crick and Brenner was that if

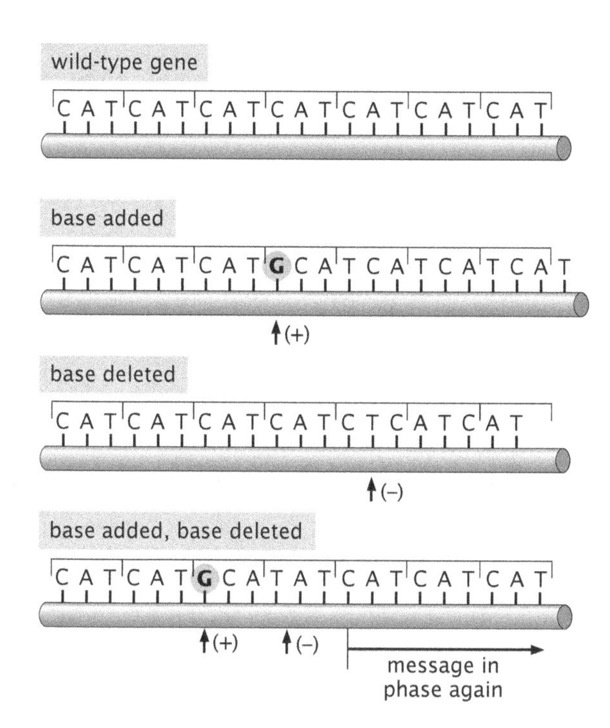

Figure 4.9: Cartoon showing the restoration of phase in DNA sequences through collections of insertion and deletions. (Adapted from H. F. Judson, The Eighth Day of Creation. Cold Spring Harbor Laboratory Press, 1996.)

the triplet hypothesis was correct, then a collection of three insertions or deletions (or alternatively an insertion and a deletion) should restore nonfunctional genes to functional status since they would restore the reading of the gene to the correct, original phase as shown in Figure 4.9. As a result, by mixing and matching insertion and deletion mutants, it was possible to show that the genetic code was functionally based on collections of triplets.

Phages Were Instrumental in Elucidating the Existence of mRNA

Though it is easy for any reader of a textbook to quote with authority the various biological roles of RNA, it is quite another matter to demonstrate such assertions experimentally, especially for the first time. Another of the classic experiments that ushered in molecular biology on the heels of phage research and that served to strengthen the experimental backdrop to the central dogma concerned the functional role of what is now known as mRNA (messenger RNA).

One of the early pieces of evidence in favor of the idea that mRNA served as an intermediate in protein synthesis came from the observation that in cells infected by a phage, there was an unstable RNA fraction (as opposed to the relatively stable RNA that makes up the ribosome) with a similar base composition as the phage DNA itself. In an experiment carried out by Brenner, Jacob, and Meselson, T4 phage infection was used to determine whether this second class of RNA was or was not related to the ribosome structure. The idea was that if the RNA synthesized after viral infection were exclusively related to ribosome structure, then this RNA should only be found in ribosomes synthesized after infection. Alternatively, if the RNA was a messenger, then it should be found in association with old and new ribosomes alike. By employing growth media in which different isotopes of nitrogen and carbon were used, heavy and light ribosomes would be synthesized and could be separated by density gradient centrifugation. The result of the experiment was the observation that the viral-derived RNA was associated with the heavy ribosomes (old) and

Figure 4.10: Lifestyles of bacterial viruses. Lysogenic and lytic pathways for bacteriophage lambda. In the lysogenic pathway, the viral genome is incorporated into the bacterial genome to form a "prophage." In the lytic pathway, new viruses are assembled to completion and are able to repeat their deeds elsewhere after being released from the lysed cell. (Adapted from B. Alberts et al., Molecular Biology of the Cell, 5th ed. Garland Science, 2008.)

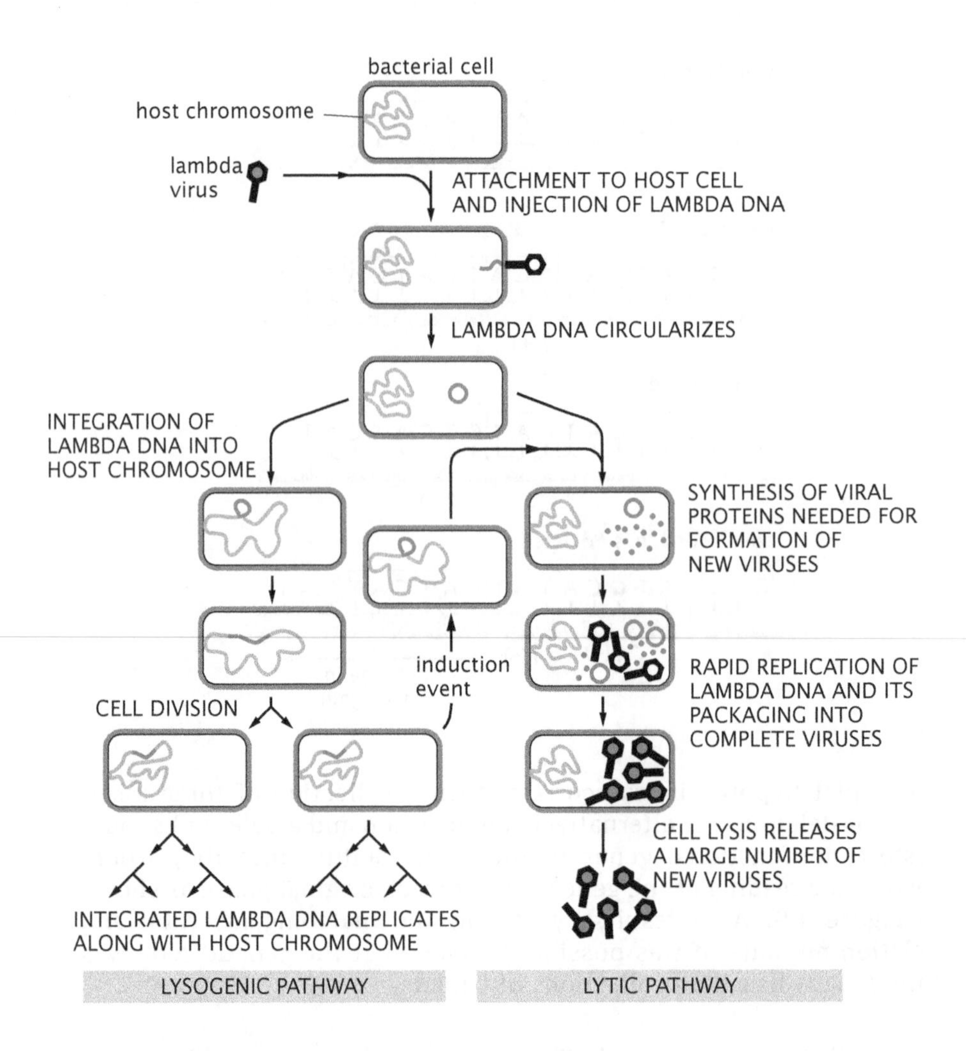

hence served in an informational rather than structural role. For the details of this fascinating episode, see Echols (2001—referenced in Further Reading at the end of the chapter).

General Ideas about Gene Regulation Were Learned from the Study of Viruses as a Model System

We have already described the viral life cycle of certain bacteriophages in rough terms in Figure 3.26 (p. 122). However, one of the subtleties left out of that description is the observation that in certain instances, after infection, the viral genome can be peacefully incorporated into the host cell genome and can lie dormant for generations in a process known as lysogeny. These two post-infection pathways are shown in Figure 4.10. As a result of observations on the phage switch, it was wondered how the "decision" is made either to follow the lytic pathway, culminating in destruction of the host cell or, alternatively, to follow the lysogenic pathway with the result that the viral genome lies dormant. The pursuit of the answer to this question, in conjunction with studies of the *lac* operon (to be discussed in Section 4.4.2), resulted in profound insights into the origins of biological feedback and control and the emergence of the modern picture of gene regulation.

The key idea to emerge from experiments on the lifestyle choices of bacteriophages (and the metabolic choices of their bacterial hosts) is the idea that there are genes whose primary purpose is to control

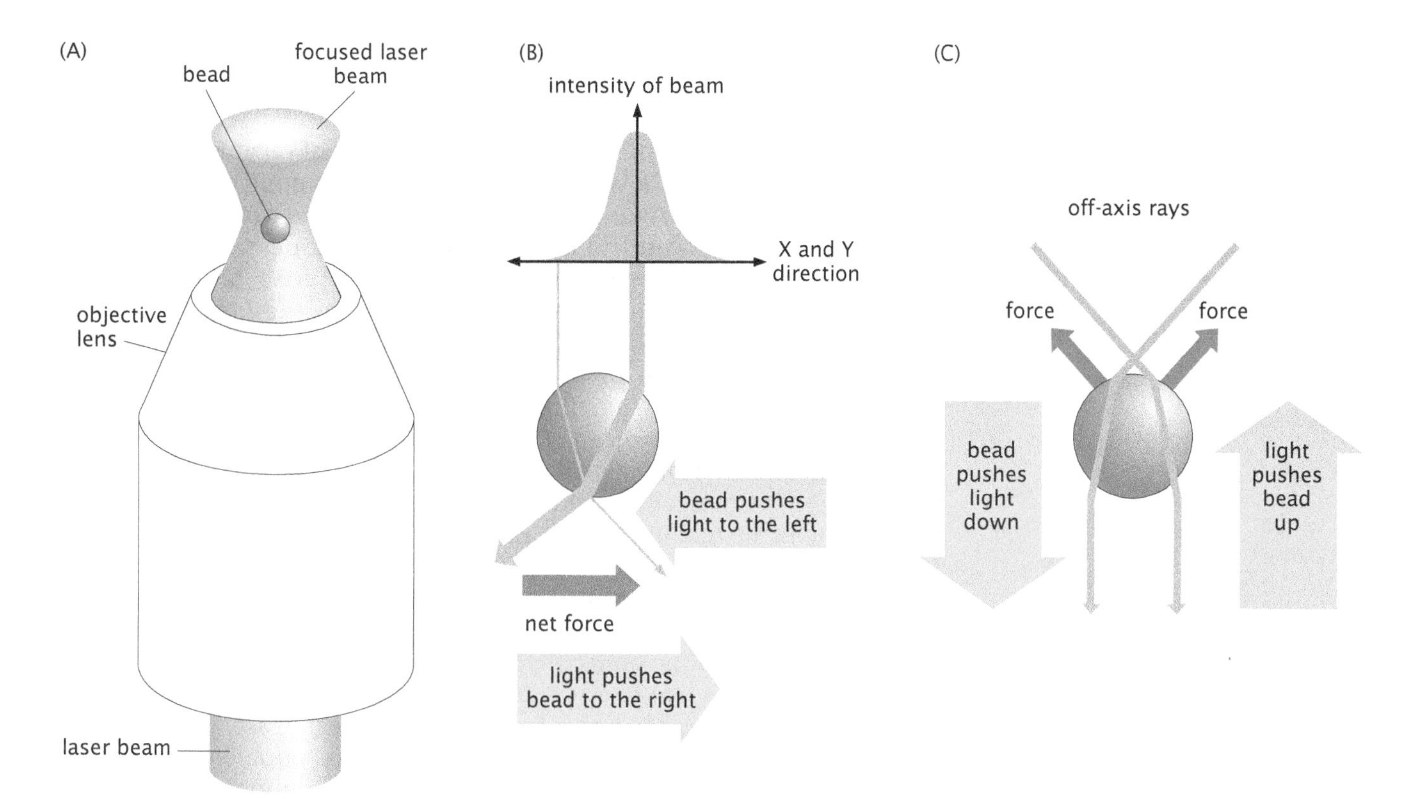

Figure 4.11: Schematic of optical tweezers. (A) A simple optical trap can be made just by focusing a laser beam through a high-numerical-aperture microscope objective. A glass or plastic bead of the order of one micron in size will be held at the focal point of the laser light. (B) In cross-section, the focused light has an approximately Gaussian intensity profile. If the bead drifts slightly to the left, the imbalanced force of the refracted light will push it back toward the center of the beam. (C) Similarly, when the bead drifts down with respect to the focal point of the laser light, the unbalanced force from the refracted light will push it back up.

other genes. Using physical biology to dissect these processes will serve as one of the threads of the entire book and will be the centerpiece of Chapter 19. In the bacteriophage setting, the examination of the decision of lysis versus lysogeny led to the emergence of the idea of genetic switches, controlled at the molecular level by a host of molecular assistants (known as transcription factors) that turn genes either off (repressors) or on (activators). Regulatory decisions are clearly one of the great themes of modern biology and the emergence of this field owes much to work on bacteriophages and their bacterial hosts.

4.3.2 Bacteriophages and Modern Biophysics

As with many of the systems described in this chapter, just when it seems that a particular model system has exhausted its usefulness, like a phoenix from the ashes, these model systems reemerge in some new context providing a model basis for some new class of phenomena. Use of bacterial viruses has seen a renaissance as a result of progress on two disparate fronts, namely, (i) the successes of structural biologists in crystallizing and determining the structures of both individual parts of viruses and entire viruses (illustrated in Figure 2.29 on p. 66) and (ii) the emergence of the discipline of single-molecule biophysics, which has permitted the investigation of processes attending the viral life cycle.

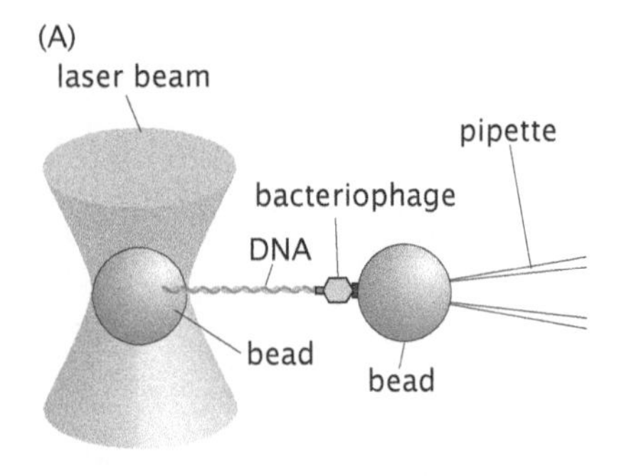

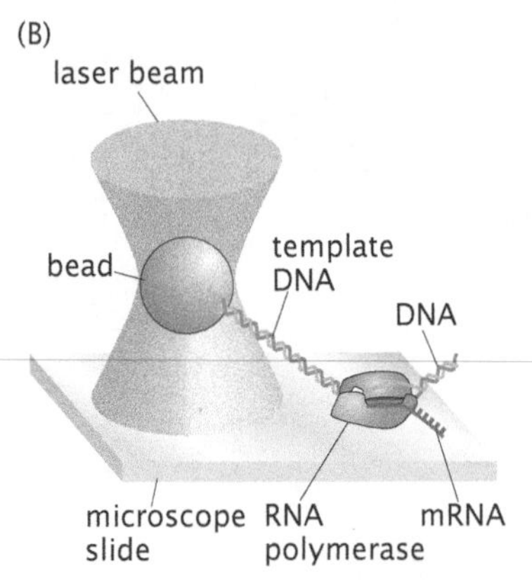

Figure 4.12: Cartoon showing two single-molecule experiments based on bacteriophages. (A) Single-molecule analysis of DNA packaging in bacteriophage $\phi 29$. The molecular motor in the phage pulls the viral genome into the capsid against the force produced by the optical trap. (B) Single-molecule studies of transcription using RNA polymerase from bacteriophage T7. During the process of *in vitro* transcription, the DNA molecule is pulled into the immobilized RNA polymerase. Since the bead is held in the optical trap, the forces produced by RNA polymerase can be measured.

Many Single-Molecule Studies of Molecular Motors Have Been Performed on Motors from Bacteriophages

One of the most exciting experimental advances of recent decades has been the advent of a host of single-molecule techniques for querying the properties of individual molecular actors in the drama of the cell. Some of the most beautiful applications of these ideas have involved individual bacteriophages and their particular gene products. One class of single-molecule experiments centers on the idea of holding a particular molecule (or set of molecules) during some process such as the packing of viral DNA into the capsid and examining how the applied force alters the ensuing dynamics.

The use of optical tweezers such as shown in Figure 4.11 permits the application of piconewton forces over nanometer length scales to a variety of molecular machines. Though there are many different single-molecule techniques available, for the moment we consider just the use of light to apply forces. The basic idea of the optical tweezers method is relatively simple: radiation pressure from laser light is used to apply forces to a micron-scale glass bead. The trick is to attach this bead to the molecule of interest such that the motion of the bead provides a surrogate for the dynamics of the macromolecule of interest. These linkages between bead and molecule are implemented using a molecular analog of Velcro. One very popular and strong Velcro-like linkage exploits the ability of the protein streptavidin to bind very specifically to the small vitamin molecule biotin. This interaction is widely used in the laboratory because biotin can be easily chemically coupled to nearly any macromolecule of interest. By applying known forces to the bead, the mechanochemistry of the attached macromolecule can be studied as a function of the applied load.

Using optical tweezers, a number of brilliant *in vitro* experiments have been performed on biological processes relevant to the bacteriophage life cycle. One such experiment involves the measurement of the forces that are operative when the portal motor is in the process of packing the viral DNA into the capsid. A schematic of the experiment is shown in Figure 4.12(A). This class of experiments reveals the build up of a resistive force as more and more DNA is packaged into the capsid. We will use simple ideas from electrostatics and elasticity to compute these forces in Chapter 10. A second class of experiments of the single-molecule variety that has been similarly revealing and founded upon the use of bacteriophage constructs involves the measurement of the rate of transcription by T7 RNA polymerase as shown in Figure 4.12(B). As will be seen in the remainder of this book (especially Chapter 16), these experiments provide a direct window onto the dynamics of the molecules that make cells work.

4.4 A Tale of Two Cells: *E. coli* As a Model System

4.4.1 Bacteria and Molecular Biology

We have already seen the important role played by bacteriophages as model systems in the creation of the field of molecular biology. There is an intimate partnership between these phages and the bacteria they infect and perhaps most notably the bacterium *E. coli*. As stated eloquently in the 1980 Nobel Lecture of Paul Berg (Berg, 1993), "Until a few years ago, much of what was known about the molecular details of

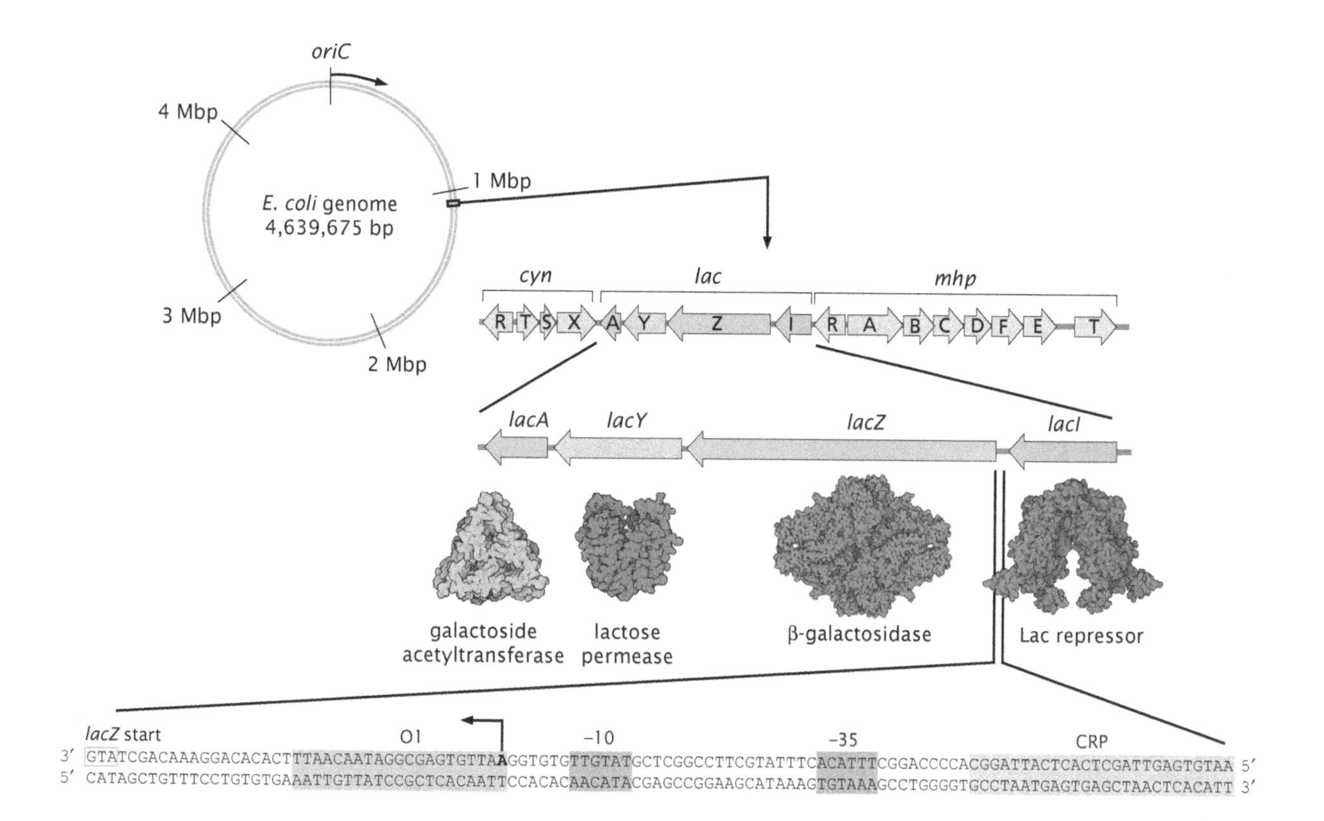

Figure 4.13: Circular map of the *E. coli* genome with a higher-resolution view of the region near the *lac* operon. At the top, the entire genome is represented as a circle. The origin of replication is referred to as *oriC* and base pairs are numbered starting from 1 in the direction of the arrow. The *lac* operon is located about one-quarter of the way around the circle, nestled between the *cyn* and *mhp* operons, whose functions are cyanate utilization and degradation of 3-(3-hydroxyphenyl)propionic acid, respectively. The protein-encoding genes in the *lac* operon are oriented such that they are all transcribed toward the origin. Each of the four genes *lacI* (pdb 1LBH, 1EFA), *lacZ* (pdb 1BGL), *lacY* (pdb 1PV7), and *lacA* (pdb 1KRV) encodes a distinct protein involved in lactose utilization. The proteins are not all drawn to the same scale since β-galactosidase, for example, is so large. The DNA sequence around the beginning of the *lacZ* gene is shown at higher magnification. The start of transcription is indicated by the arrow. The codon for the first amino acid in the sequence (methionine) is boxed. The two binding sites for RNA polymerase are indicated with red shaded boxes at the −10 and −35 positions counting from the transcriptional start site. Two important regulatory sites are shown as shaded boxes; O1 is a binding site for the Lac repressor protein (encoded by the *lacI* gene) and CRP is a binding site for another regulatory protein. (Structural illustrations from D. Goodsell.)

gene structure, organization and function had been learned in studies with prokaryote microorganisms and the viruses that inhabit them, particularly, the bacterium *Escherichia coli* and the T and lambdoid bacteriophages." The goal of this section is to explore the various ways in which *E. coli* has earned this status and to show how many of these discoveries now provide the impetus for physical biology.

The structure of *E. coli* was described in Chapter 2 and some of the key processes that attend its life cycle were described in Chapter 3. These rich structures and processes are dictated by the 4.6 Mbp genome illustrated in Figure 4.13. If we invoke our rule of thumb that each gene encodes for a protein with a length of 300 amino acids, then this amounts to roughly 1000 bp/gene. Using this rule of thumb, we estimate in turn that the *E. coli* genome codes for roughly 4600 distinct proteins. Genome sequencing has revealed that the actual number of genes is 4435, of which 4131 encode proteins, 168 encode structural RNA, and the remainder appear to be nonfunctional. While this is a relatively long "parts list," it is much simpler than the parts list for a jumbo jet and gives us hope that the functioning of the

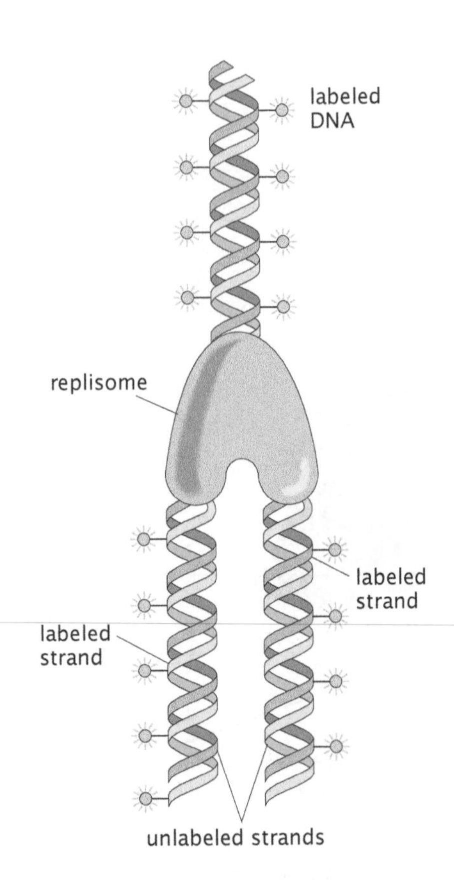

Figure 4.14: The Meselson–Stahl experiment and the hypothesis of semi-conservative replication. When DNA labeled on both strands with a heavy nitrogen isotope is copied by the replisome, each of the two daughter molecules ends up with one heavy strand and one light strand. Note that the schematic intentionally ignores subtleties about the distinction between leading and lagging strands during replication.

E. coli cell may be understandable in terms of the functions of its constituent parts.

4.4.2 *E. coli* and the Central Dogma

The *central dogma* of molecular biology (introduced in Chapter 3 and illustrated in Figure 3.12 on p. 107) specified the mechanism whereby the information content of genomes is turned into the proteins that implement cellular functions. At the outset, many of the ideas associated with the central dogma were speculative and required the first generation of molecular biologists to design and implement a series of clever experiments to work out the processes that connect the structure and information content of DNA to its downstream consequences. Much of that work was carried out using the bacterium *E. coli* and its associated viruses.

The Hypothesis of Conservative Replication Has Falsifiable Consequences

From the moment of its inception, thinking on the structure of DNA was intimately tied to ideas about how the genetic material was faithfully passed on from one generation to the next. One of the hypotheses associated with the development of the central dogma was that of semiconservative replication. That is, when a particular DNA molecule is copied, the two daughter molecules each have a single strand from the parent molecule. This model is clearly experimentally falsifiable and led to one of the most famed experiments in the history of molecular biology, the Meselson–Stahl experiment. Once again (like with the Hershey–Chase experiment), the concept of the experiment was built around the idea of using different isotopes to track the evolution of the parent and daughter molecules. From the perspective of mass conservation, the key idea is that replication involves doubling the total number of nucleotides relative to the parent molecule. One scenario to imagine is that somehow the parent molecule is copied but that the daughter molecule is built up entirely of new nucleotides. Alternatively, in the hypothesis of semiconservative replication, each of the daughter molecules has one strand from the parent and one comprising new nucleotides.

The idea of the Meselson–Stahl experiment was to use nucleotides with the nitrogen isotope ^{15}N, rather than the ordinary ^{14}N isotope. By growing *E. coli* cells in a medium that made available only heavy nitrogen, a culture of cells was grown whose DNA consisted strictly of ^{15}N. By then growing the cells for one generation in a medium in which only ^{14}N is provided, the daughter strands would bear the mark of this change in nitrogen isotopes, which would show up as a difference in mass. The semiconservative and conservative replication hypotheses give rise to different predictions for the observed masses. An example of the distinction between parent and daughter strands is shown in Figure 4.14. To separate the different molecules, the samples were centrifuged in a gradient of CsCl (cesium chloride), with the effect that each DNA molecule found the appropriate equilibrium depth where the buoyant force just matched the downward-acting effective force of gravitation (effective because the force is really due to the centrifugation). What Meselson and Stahl found is that the newly synthesized DNA had a mass intermediate between the ^{15}N and ^{14}N versions of the molecule, supporting the hypothesis of semiconservative replication.

One of the key insights to emerge from the experiment of Meselson and Stahl was the demonstration that DNA replication is a templated polymerization process. That is, the parent strand serves as a template for the daughter and one-half of the parent molecule is donated to each daughter cell. Indeed, from the standpoint of the physics or chemistry of polymerization, this is one of the resounding differences between biological polymers and ordinary synthetic polymers like polyethylene—biological polymers are built from a template and hence, in principle, every copy (as long as no mistakes are made) should have precisely the same sequence as the parent molecule.

Extracts from *E. coli* Were Used to Perform *In Vitro* Synthesis of DNA, mRNA, and Proteins

The understanding of DNA replication on the basis of experiments involving *E. coli* did not stop with Meselson and Stahl. One line of research of particular importance used extracts from *E. coli* cells in order to carry out *in vitro* synthesis of new DNA molecules. In a similar vein, another episode in the history of molecular biology using *E. coli* as the relevant experimental organism centered on the isolation of the enzyme responsible for creating mRNA. The genetic tradition launched by Mendel was consummated in one of the greatest discoveries of modern biology, namely, the cracking of the genetic code, which was also carried out using extracts of *E. coli*, resulting in the understanding embodied in Figure 1.4 (p. 10).

4.4.3 The *lac* Operon as the "Hydrogen Atom" of Genetic Circuits

Gene Regulation in *E. coli* Serves as a Model for Genetic Circuits in General

We have already described the study of lytic control in bacteriophages as one of the two key episodes in the development of the modern view of genetic control. The other such episode, developed concurrently in the Paris laboratories of Jacob and Monod, concerned the genetic decisions made in *E. coli* associated with metabolism of the sugar lactose. The remarkable idea that emerged from these studies is that there are certain genes whose task it is to control other genes. Prior to the advent of the operon concept to be described below, the notion of a gene was generally tied to what might be called a "structural gene," responsible for the construction of molecules such as hemoglobin and DNA polymerase and many others. However, the operon concept called for a new type of gene whose role was to control the expression of some other gene.

The ways in which such control can be effected are wide and varied. One important type of genetic control is known as transcriptional regulation and refers to molecular interventions that control whether or not an mRNA molecule for a gene of interest is synthesized. Negative control (or repression) is mediated by molecules known as repressors that can bind the DNA and inhibit the ability of RNA polymerase to bind and make mRNA. Positive control (or activation) refers to the case in which a gene is generically dormant unless some molecule induces transcription by "recruiting" RNA polymerase to the relevant promoter. Activation is mediated by molecules known as activators that increase the probability of RNA polymerase binding. Note that

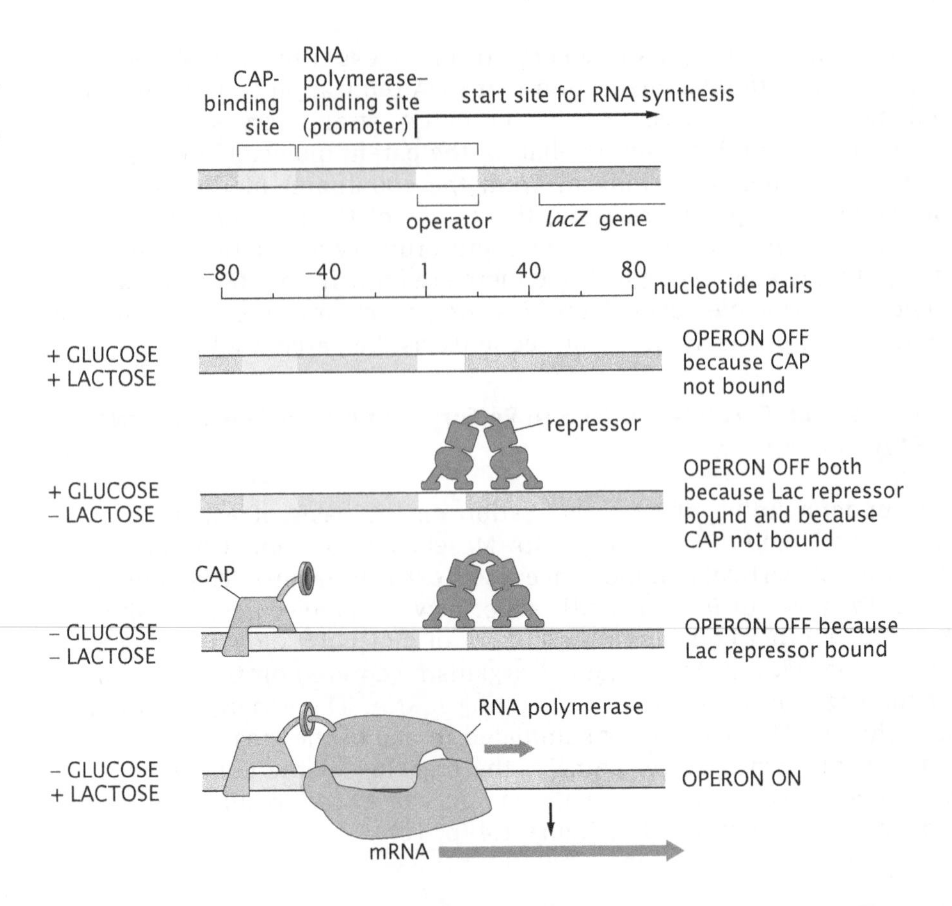

Figure 4.15: Schematic of the regulatory region of the *lac* operon. This figure shows the genography of the *lac* operon in the form of the function and products of different parts of this part of the *E. coli* genome. CAP is the catabolite activator protein, which is an activator of the *lac* operon in the absence of glucose. Combinatorial control by CAP and Lac repressor ensures that the genes of the operon are expressed only when lactose is present and glucose is absent.

there is a wide variety of other mechanisms that regulate gene expression after the polymerase is already bound, though the recruitment mechanism described here will garner the most attention throughout the book.

The *lac* Operon Is a Genetic Network That Controls the Production of the Enzymes Responsible for Digesting the Sugar Lactose

The classic example of negative control is that provided by the *lac* operon shown in Figure 4.15, one of the many profound and lasting insights to have emerged from the study of bacteria in general, and *E. coli* in particular. The work on this operon was so convincing and paradigmatic that it took over a decade for the first example of positive genetic control to be widely accepted. The chemical reactions that are tied to this genetic circuit involve the digestion of the sugar lactose in *E. coli* cells that have been deprived of other sugars and that find themselves in a medium rich in lactose. An example of a growth curve in which cells make a decision to turn on the genes responsible for metabolism of a sugar source other than glucose is shown in Figure 3.3 (p. 93), where we see that once the cells have exhausted the glucose, there is a waiting period during which the genes necessary to use lactose are turned on. The enzyme responsible for this digestion is called β-galactosidase (shown in Figure 4.13) and is substantively present only when lactose is the only choice of sugar available.

The *lac* operon refers to the region in the *E. coli* genome shown in Figure 4.13 that includes the promoter site for the binding of RNA polymerase upstream from the structural genes of the *lac* operon, the operator site where Lac repressor binds and inhibits transcription by

RNA polymerase and then the genes encoding β-galactosidase (*lacZ*), the lactose permease (*lacY*), which builds membrane pores through which lactose can be brought into the cell, and transacetylase (*lacA*), an enzyme that transfers an acetyl group from acetyl CoA to the sugar. (CoA, or coenzyme A, is a small carrier molecule that can be attached to acetyl or acyl groups and is often used as an intermediate in metabolic reactions.) The *lac* operon is shown in Figures 4.13 and 4.15, which give a feel for the number of base pairs involved, the appearance of the relevant gene products, and the relative positions of the various regions in the operon. The region of control is associated with the promoter (with the arrow signifying the start of RNA synthesis) and the operator (the repressor binding site). The promoter is the region where RNA polymerase binds prior to transcription. This region partially overlaps the operator region, which is the site where Lac repressor binds to prevent transcription. The product of the *lacI* gene is the Lac repressor molecule itself. In addition, there is an activator known as CAP (catabolite activator protein) that binds in the vicinity of the promoter and effects positive control. For the present discussion, the most important reason for bringing all of this up is the observation that studies on this seemingly obscure feature of *E. coli* served as a model system for the construction of the modern theory of gene regulation, which has implications for topics as seemingly distant as the emergence of the molecular understanding of development and the development of animal body plans.

4.4.4 Signaling and Motility: The Case of Bacterial Chemotaxis

E. coli Has Served as a Model System for the Analysis of Cell Motility

A typical bacterium such as *E. coli* gets around. Not only that, but just as these cells express metabolic preferences, their motility reveals that they can express preferences about their environments as well. In particular, if some favorable chemical species (sugars, dipeptides, and other "chemoattractants") is pipetted locally into a solution, the bacteria will actually bias their motion so as to swim towards these chemoattractants. Similarly, if repellants are pipetted into the medium, the bacteria will swim away from them. This process of chemotaxis is one of the great stories of biology and is not just some obscure practice of bacteria. Indeed, chemotaxis is part of the social decision making of *Dictyostelium* (see p. 75) and is a key ingredient of the workings of our immune systems as neutrophils hunt down foreign invaders as shown in Figure 3.3 (p. 93).

The study of bacterial chemotaxis in *E. coli* has resulted in the emergence of a beautiful story that relates molecular signaling and cell motility as shown in Figure 4.16. Using the tools of genetics and biochemistry, it has been possible to identify the molecular machinery responsible for chemotactic decision making, with many of the lessons emerging from this study relevant to other organisms as well. The study of chemotaxis and the resulting bacterial motility represents a profound meeting point of subjects, including low-Reynolds-number hydrodynamics, cell signaling, and the study of polymorphism of protein complexes such as the bacterial flagellum (also relevant in the setting of viral self-assembly).

As can be seen in an optical microscope, a motile bacterium finds its way across the field of view very quickly. The mechanistic underpinnings of this motion are tied to the rotation of the bacterial

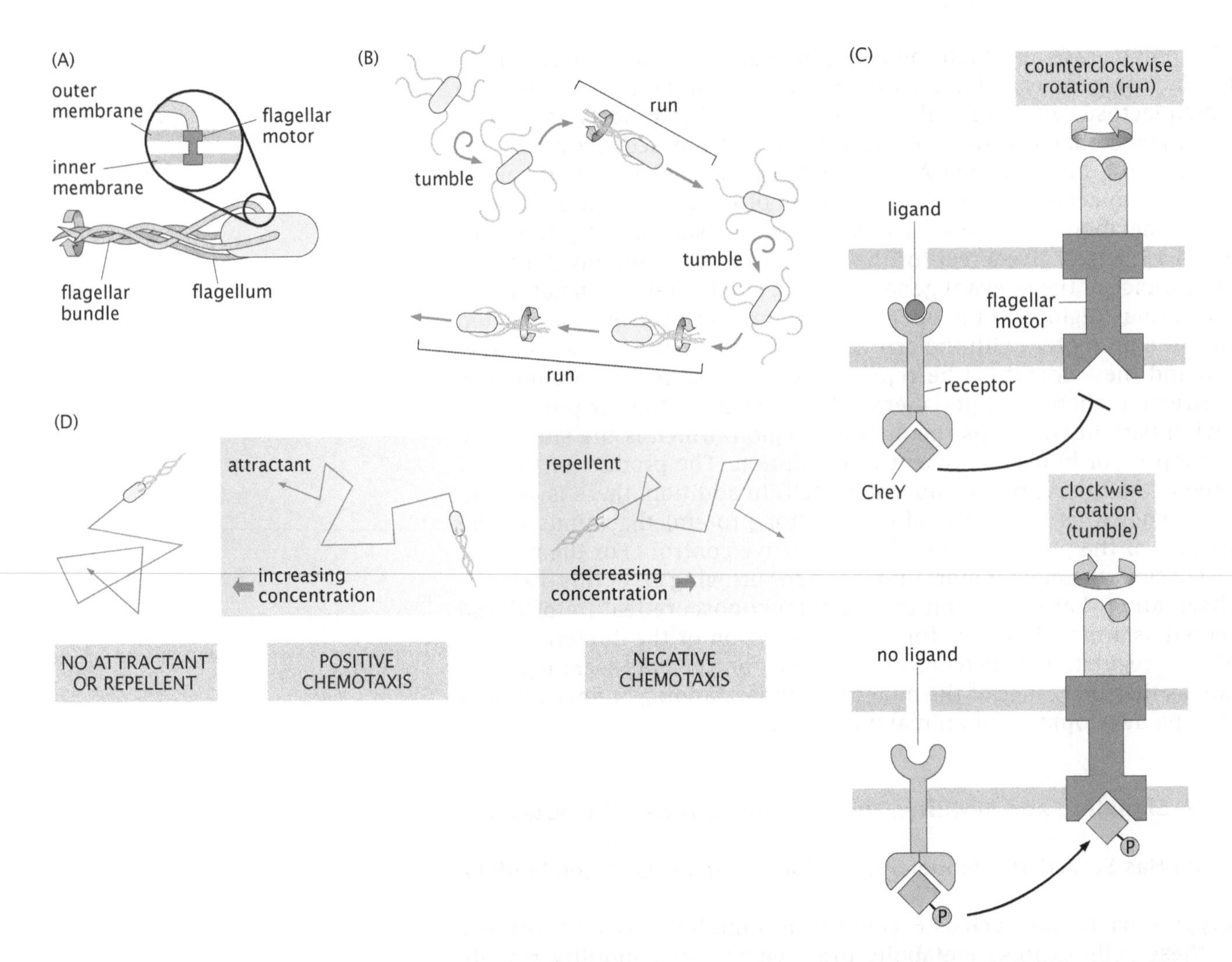

Figure 4.16: Mechanism of bacterial chemotaxis. (A) Many bacteria, including *E. coli*, are equipped with flagella. The helical flagellar filaments are turned by a motor embedded in the cell's wall and membranes. (B) When all of the flagellar motors on an individual cell are rotating counterclockwise, the filaments bundle together and the bacterium moves efficiently in a straight line called a run. When the motors reverse direction and spin clockwise, the flagellar bundle flies apart and the bacterium rotates in an apparently random manner called a tumble. Over long distances, bacterial trajectories appear as run segments arranged at random angles to one another at junctures where a tumble event occurred. (C) The switching of the motor from counterclockwise to clockwise rotation, and hence the switching of the behavior from running to tumbling, can be controlled by the presence of small molecules in the bacterium's environment. These molecules bind to receptors that then induce phosphorylation of a signaling protein, which in turn binds to and affects the mechanics of the motor. (D) Tuning of swimming behavior by small molecules can cause the bacterium to swim either toward desirable food sources, such as sugars or amino acids, or away from noxious toxins.

flagellum. A typical *E. coli* has on the order of 5 such flagella with a characteristic length of the order of 10 μm. One of the most remarkable classes of molecular motors is those that exercise rotary motion. The first such motor to be characterized in detail is that associated with the rotary motion of the *E. coli* flagellum and shown in Figure 3.24 (p. 120). This tiny rotary motor is capable of spinning the flagellum at roughly 6000 rpm using energy stored in an ion gradient across the bacterial cell membrane. The remarkable mechanical feature of this motor is that it can reverse its direction, rotating either clockwise or counterclockwise at comparable speeds. It is the likelihood of switching between counterclockwise and clockwise motion that regulates bacterial trajectories in response to chemoattractants. When all of the motors on a particular bacterium are rotating in a counterclockwise direction, the five or so helical flagella line up into a well-ordered bundle and their spinning cooperates to propel the bacterium forward in a straight line at a rapid speed ($\approx$10–30 μm/s). When the motors reverse direction, the bundle is disrupted and the individual flagella fly apart. This disruption in the straight swimming of the bacterium is referred to as a tumble, because when the motor switches back to counterclockwise rotation and the bacterium begins swimming in a straight line again, the bacterium will be pointing in a new direction in space. During chemotaxis, bacteria swim towards attractive chemicals and away from repellant molecules simply by changing the frequency of tumbling. In Chapter 19, we will describe the molecular underpinnings of this behavior as well as physical models that have been put forth to explain the phenomenon.

All told, *E. coli* has been a remarkable model organism that has yielded a long list of biological discoveries.

4.5 Yeast: From Biochemistry to the Cell Cycle

Yeast Has Served as a Model System Leading to Insights in Contexts Ranging from Vitalism to the Functioning of Enzymes to Eukaryotic Gene Regulation

Several technical features of *E. coli* made it a nearly ideal organism for early geneticists and biochemists. First, it grows quickly. Second, it can grow on plates in colonial form such that all the descendants of a single individual cell can easily be gathered together. Third, it exhibits a variety of alternative metabolic lifestyles: for example, the ability to grow with and without oxygen, the ability to grow on different carbon sources, and the ability to either synthesize its own amino acids or take them up from the surrounding medium as environmental conditions permit. This flexibility facilitated isolation of bacterial mutants that were able to grow in one condition but not another, resulting in the identification of the enzymes involved in most major metabolic pathways. However, *E. coli* is a bacterium and is in many ways profoundly different from eukaryotic organisms such as ourselves.

To elucidate processes specific to eukaryotic organisms, another model system must be adopted that ideally would share many of the desirable features of *E. coli*. Such an organism is *Saccharomyces cerevisiae*, the brewer's or baker's yeast. Humans domesticated this wild fungus that normally grows on the skin of fruit some thousands of years ago. It has long played a critical role in the economy of human societies as it is necessary for making leavened bread and creating a variety of delicious intoxicating beverages, including beer and wine. Because of the economic importance of having reliable and reproducible brewing and baking processes, the inner secrets of *S. cerevisiae* were of intense interest to early microbiologists (see Barnett (2003) in Further Reading at the end of the chapter).

Although wild *S. cerevisiae* typically grows in unruly, stringy cellular groups, strains of domesticated *S. cerevisiae* were chosen by microbiologists to form neat, round colonies when grown on agar plates, looking much like colonies of *E. coli* but smelling substantially better. *S. cerevisiae* also grows remarkably fast for a eukaryotic cell, doubling in less than 90 minutes under favorable conditions. Also like *E. coli*, yeast can pursue a variety of metabolic lifestyle choices, and so it is similarly easy to isolate auxotrophic mutants (that is, mutants unable to synthesize an essential amino acid or other metabolite). However, *S. cerevisiae* is a full-fledged card-carrying eukaryote. It has a membrane-enclosed nucleus and a full complement of eukaryotic membrane-bound organelles, including an endoplasmic reticulum, Golgi apparatus, and mitochondria, all of which are missing in *E. coli*. Even better, *S. cerevisiae* has sex the way that most other eukaryotes do, blending together the complete genomes of two parent cells that are then recombined in the offspring. Intriguingly, yeast can grow in either haploid or diploid forms, that is, either with one copy of every chromosome or with two copies of every chromosome where each of the two copies is derived from different parents. Thus, genetic and molecular dissection of eukaryotic-specific behaviors has been largely carried out using yeast. Furthermore, yeast (being a fungus) is much more closely related to animals than any bacterium, or indeed most other eukaryotic microorganisms. It therefore shares with us many specific pathways, including certain kinds of signaling pathways that are not found in *E. coli*. Below, we summarize some particular case studies where yeast has provided initial critical insights into processes that are important to many or all eukaryotic cells.

4.5.1 Yeast and the Rise of Biochemistry

Yeast served as a central player in the debate over vitalism and in launching the field of biochemistry. Many of the greatest chemists of the eighteenth century participated in the heated and vitriolic debate over the nature of fermentation. Fermentation was particularly important in France, at that time the premier wine-making country in the world. Prior to his decapitation by guillotine during the French Revolution, the great French chemist Antoine Lavoisier was the first to measure accurately the conversion of sugar into alcohol and carbon dioxide. Over the next several decades as the resolution of optical microscopes improved, it became possible actually to see the yeast cells responsible for fermentation and to observe their process of growth and division, establishing that fermentation was a function of a plant-like living organism and not a purely chemical process. Although it eventually did become accepted that living yeast cells were normally responsible for fermentation, many chemists held to the complementary view that life itself was not essential for fermentation to occur.

In an interesting irony of history, both sides in this heated debate were correct: as part of their normal life cycle, yeast cells convert sugar into alcohol. At the same time, the proteins (enzymes) that carry out this process within yeast can be extracted and purified to perform precisely the same biochemical conversion. Careful experiments by Eduard Buchner and others revealed that cell-free extracts from yeast could catalyze the fermentation of sugar, resulting in the realization that there are specific chemical substances that catalyze the key chemical reactions of the biological realm. Insights on the biochemistry of yeast were foundational in unearthing the features of the glycolysis pathway shown in Figure 5.2 (p. 191) and discussed in Chapter 5.

4.5.2 Dissecting the Cell Cycle

As yeast proceeds through its cell division cycle, it undergoes a series of morphological changes such that it is immediately apparent to the eye of a microscopist where a particular individual yeast cell is in its

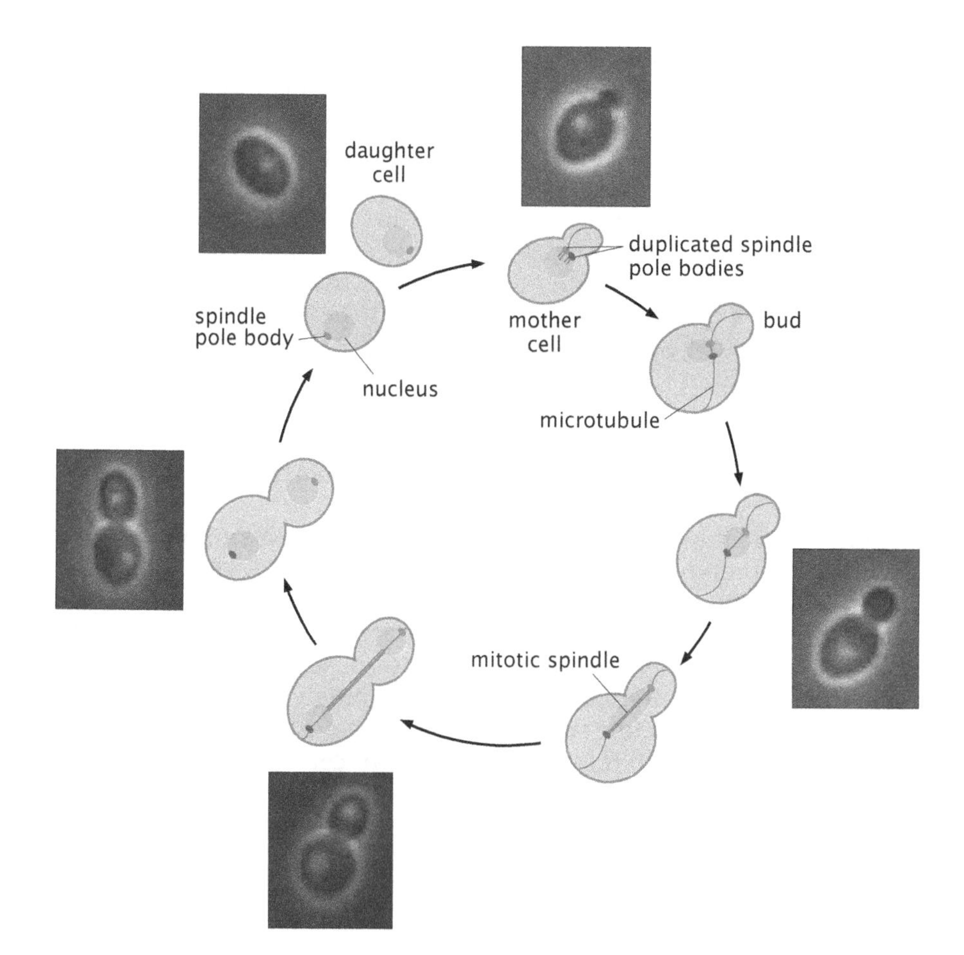

Figure 4.17: Cell cycle of yeast. Diagram of the process of yeast budding and division. As a single yeast cell progresses through the cell cycle, the future daughter cell, or bud, grows in size until it is large enough to receive one copy of the replicated genome. Then the bridge between mother and daughter cells is severed and both mother and daughter can go on to bud. The microscopy images show the yeast *S. cerevisiae* at different stages in the cell cycle. (Images from R. Menssen et al., *Curr. Biol.* 11:345, 2001.)

cell-cycle progression, as shown in Figure 4.17. It is easier to detect progression through the yeast cell cycle by visual means in *S. cerevisiae* than most other cell types because of its unusual propensity to grow by budding rather than by binary fission. The yeast cell cycle starts with an ovoid mother cell that then begins to grow a small bud from a single point on the surface. As the cell progresses through the cell cycle, the bud grows while the mother cell remains nearly the same size. After the chromosomes have been duplicated, the mitotic spindle delivers one set of chromosomes into the bud which eventually pinches off as a daughter cell. The mother cell can then go on to produce more daughters and the daughter can go on to have daughters of her own.

The correspondence between cell morphology and position in the cell cycle made it straightforward for investigators to isolate a large series of mutants involved in cell-cycle progression. Because cell division is a process essential to life, it was necessary to isolate conditional mutants (conditional mutants were defined in the Experiments Behind the Facts section on p. 140 at the beginning of the chapter). The initial strategy was to identify a large number of mutant yeast strains that were able to grow at a normal temperature (30 °C) but unable to grow at an elevated temperature (37 °C) where wild-type yeast can still flourish. Next, these temperature-sensitive growth mutants were visually evaluated to determine whether all the cells in a dying population at the high temperature appeared to be stuck at the same point in the cell cycle. Each of these mutant strains (called Cdc mutants for "cell division cycle") lacked a critical protein involved in cell-cycle progression such that a cell at any stage in the division cycle when moved to an elevated temperature would continue to progress until it reached the point where the missing protein was essential.

Many of the genes identified in this manner as critical to cell-cycle progression in *S. cerevisiae* have since proved to be critical for cell-cycle progression in most other eukaryotes as well, including in humans and frogs. As described in Chapter 3, biochemical experiments focused on understanding the cell-cycle oscillator in frog embryos were also used to identify proteins critical in cell-cycle progression. It is a satisfying demonstration of the complementarity of biochemistry and genetics as well as of the unity of life that both approaches pointed towards many of the same candidate molecules.

4.5.3 Deciding Which Way Is Up: Yeast and Polarity

The major benefits of yeast as an experimental organism, namely, its rapid and stereotypical growth pattern and the ease with which conditional mutants can be isolated, have been repeatedly exploited in many different kinds of studies to elucidate the basic processes of eukaryotic cell biology, analogous to those described above for the cell cycle. Another interesting and general biological process, critically informed by experiments on yeast, is the establishment of cell polarity. Polarity simply refers to the ability of cells to localize subcellular components in a nonuniform way in response to either external or internal spatial signals. Establishment of polarity is a feature of essentially all eukaryotic cells and many bacterial cells. It is particularly evident in cell types where spatial differentiation between one end of the cell and the other is critical for biological function. Some examples include motile cells that must distinguish front from back, epithelial cells forming barriers in animal tissues that must distinguish inside from outside, and neurons, which typically receive electrical signals from neighboring cells at one end and transmit them at the other end. As is clear from the images of budding yeast progressing through its cell cycle shown in Figure 4.17, the yeast cell targets growth in a polarized fashion such that the bud grows while the mother cell stays about the same size.

A series of clever genetic screens has enabled the identification of the protein machinery in yeast responsible for the establishment of large-scale cell polarity and for targeted growth. The master regulator determining cell polarity was actually first identified in the Cdc screen described above. Cells with a conditional mutation in the protein Cdc42 did not arrest at a particular point in the cell cycle, unlike the others described, but instead continued to grow uniformly and isotropically, becoming larger and larger, without ever choosing a bud site. Subsequent studies have shown that the Cdc42 protein is the primary signal that eventually directs construction of large-scale internal polarized structures, including actin cables, which run from one end of the cell to the other, that provide the underlying scaffolding for all kinds of directed transport into the growing bud. In many kinds of animal cells, the homologs of Cdc42 have also been found to play important instructional roles in establishing cytoskeletal polarity.

The same polarization machinery is used for a different purpose when yeast cells feel the urge to mate with one another. Haploid yeast cells of either of two mating types called *a* and α secrete chemical pheromones that can be detected by a nearby member of the opposite sex. Because yeast cells are unable to crawl towards one another for consummation, they instead simply grow in the direction of the

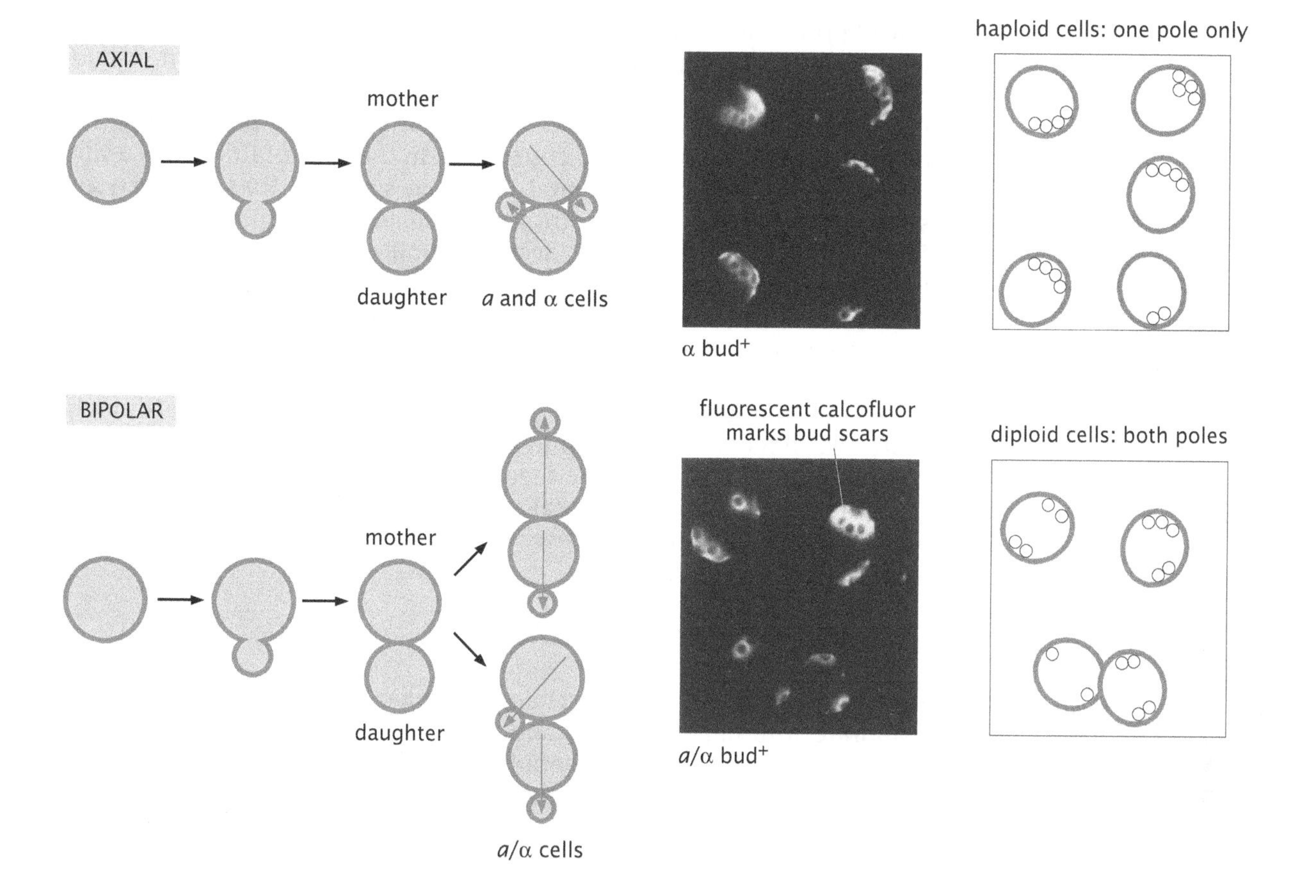

Figure 4.18: Bud site selection in growing yeast. Haploid cells (*a* cells or α cells) typically form a new bud immediately adjacent to their most recent site of cell division. This can be seen in two ways. First, looking at the pattern of small microcolonies (at the four-cell stage) shows every cell in contact with every other cell. Second, examining the position of bud scars on the cell surface (irreversible cell wall deformations that report on sites of previous cell division) shows that all bud scars are found in a single clump on one end of the cell. In contrast, diploid cells grow in a bipolar pattern where bud site choice alternates between the two ends of the cell. This results in microcolonies that are more extended with less cell–cell contact. Bud scars are found in two distinct clumps on opposite ends of the cell. (Adapted from J. Chant and I. Herskowitz, *Cell* 65:1203, 1991.)

pheromone signal until they make contact. Under this circumstance, the localization of Cdc42 to a single patch on the surface is directed by signals from the pheromone receptor in the yeast membrane. This is an example of the coupling between external spatial information, in this case provided by the pheromone secreted by a nearby amorous mating partner, and the internal polarization machinery of the cell. Similar hierarchical signaling events using many of the same proteins have since been shown to be involved in the polarization of motile animal cells such as neutrophils during chemotaxis and in the polarized growth of nerve cell extensions, all in the presence of external cues.

In the absence of an external signal, how does the growing yeast cell choose where to put its Cdc42 protein and, therefore, where to put its single bud? It has long been observed that yeast cells bud in regular patterns that are determined by their haploid or diploid state as shown in Figure 4.18. Haploid cells typically start a new bud site immediately adjacent to the site from which they last budded, such that the cells in a microcolony tend to all pile up in a small clump. In contrast, diploid cells alternate their bud site selection between their two poles such that microcolonies grow in a more extended form. A visual screen was used to find yeast mutants that formed an apparently random pattern of buds on their surface. These mutants were fully viable and exhibited no significant growth defect in the laboratory. Their only deficiency was in their ability to tell which end is which.

In all the cases described above, the yeast cell grows in one and only one direction, although in the case of diploid cells growing in a bipolar fashion, this cell clearly has the capability of growing from either end. How does the cell know that it has formed a localized patch of

(A)

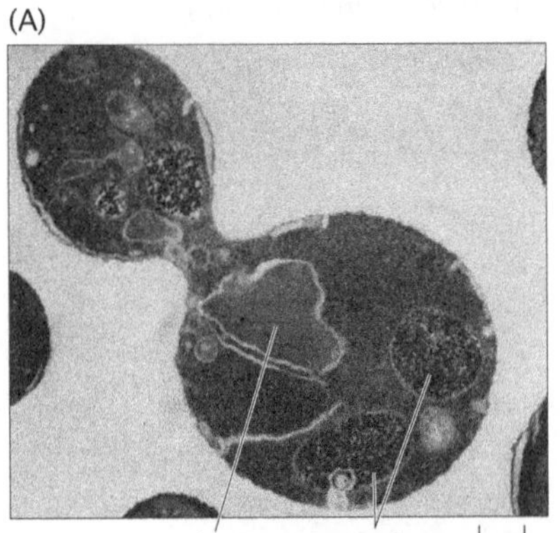

(B)

nucleus accumulated vesicles 0.5 μm

Figure 4.19: Accumulation of membrane transport vesicles in secretion-deficient yeast cells. (A) A conditional mutant in the secretory pathway gene grown at the permissive temperature showing normal internal membrane morphology. (B) Mutant grown at the restrictive temperature that has accumulated large numbers of internal transport vesicles. (Adapted from P. Novick and R. Schekman, *Proc. Natl Acad. Sci. USA* 76:1858, 1979.)

Cdc42 protein at exactly one site? Why do yeast almost never make the mistake of initiating two different buds simultaneously, even in bud site selection mutants that cannot tell one location from another? This is a specific instance of a more general problem that cells frequently have of needing to be able to count to 1. In the case of Cdc42 accumulation, the bulk of experimental evidence suggests that the unity of site selection depends on both positive feedback loops and negative feedback loops. Initial slight accumulation of Cdc42 at one location is positively reinforced as activated Cdc42 tends to self-associate and then growth of polarized actin structures directed by Cdc42 directs yet more Cdc42 to the same site. At the same time, negative feedback loops, which are not so well understood, seem to actively suppress Cdc42 accumulation everywhere else on the cell. Again, although molecular details vary from one species to another, it seems that the combination of local positive feedback and long-range negative feedback is used in many biological systems to enforce selection of one, and only one, location for subsequent polarization.

4.5.4 Dissecting Membrane Traffic

In contrast to bacteria and archaea, all eukaryotes have extensive and elaborate internal, membrane-bound organelles which perform distinct functions in cellular metabolism and behavior. These membrane-bound organelles, such as the endoplasmic reticulum and Golgi apparatus, are carefully positioned within the cell and specific subsets of their contents are trafficked from one organelle to another and also between organelles and the cell surface. In *S. cerevisiae*, directed membrane traffic is responsible for delivering newly synthesized cell wall material to sites of polarized growth as described above. Membrane traffic is also necessary for proper delivery of enzymes to their appropriate subcellular compartments, such as delivery of degradative proteases to the yeast vacuole and delivery of enzymes involved in nutrient digestion to the cell surface.

As the simplest and most genetically tractable eukaryote to exhibit all of these kinds of membrane trafficking, yeast was the target of mutational abuse designed to uncover critical genes involved in both the mechanics and signaling of membrane transport. Examples of such abused cells are shown in Figure 4.19. A critical observation was that yeast unable to deliver secretory organelles properly to the cell surface would become denser over time as protein synthesis continued without concomitant cell growth because the material responsible for cell-wall extension was no longer transported to the site of cell-wall synthesis. Furthermore, such mutants have defects in expressing secreted enzymes used for yeast metabolism, such as the enzyme invertase that clips sucrose into glucose and fructose during fermentation. To isolate mutants involved in this process, researchers first performed an enrichment by spinning yeast cells through a density gradient made of commercial floor polish. Extremely dense cells were isolated from the gradient, rinsed clean of floor polish, and then plated on a medium that would change color in the presence of invertase or other cell-surface enzymes. Several dozen different mutant strains were isolated that had various defects in processes associated with cell wall and enzyme secretion. These ranged from some cells unable to translocate proteins from the cytoplasm into the endoplasmic reticulum to others deficient in vesicle transport from the endoplasmic reticulum to the Golgi apparatus, to others deficient only in the final fusion event delivering secretory vesicle contents to the cell surface. Members of this last class tended to accumulate very large numbers of frustrated vesicles immediately subjacent to what should have been the site of cell growth during bud formation. In parallel, biochemical experiments performed on animal cells specialized for secretion, including pancreatic cells and brain cells, identified proteins necessary for different mechanical steps of membrane fusion and vesicle budding. Nearly all of the proteins found by these biochemical methods in animal cells turned out to be homologs of genes identified in the yeast screen.

4.5.5 Genomics and Proteomics

All of the specific cell biological case studies described above have illustrated how the excellence of yeast as an organism for both biochemical and genetic experimentation has led to fundamental insights relevant to all eukaryotic cells. After the invention of recombinant DNA technology in the 1970s, classical genetics was complemented by reverse genetics. Instead of isolating mutants based on phenotype and then trying to identify the molecular lesions involved, researchers were now able to start with a particular molecule of interest and ask what phenotype would result when that molecule was disrupted. For example, for a particular protein, isolated through some biochemical means, site-specific disruption of the gene encoding that protein would generate an engineered mutant with that specific deficiency and no other, which could then be examined for phenotypic alteration. This process is called reverse genetics because it starts with known molecules and asks for phenotypes, in contrast to classical genetics, which starts with phenotypes and looks for molecules. It will come as no surprise that yeast was also one of the first eukaryotic organisms in which reverse genetics was fruitfully applied. Replication of engineered plasmids in yeast was first reported in 1978 and gene replacement by homologous recombination just a few years later in 1983. Thus, the full toolkit for site-specific DNA manipulation, originally developed in *E. coli*, was also realized in yeast. Along the same lines, yeast was the first eukaryotic cell whose genome was fully sequenced and assembled in 1996.

The sequencing of the yeast genome opened the door for a new class of genome-wide investigations that celebrated the now known and finite boundaries of yeast genetic material. With this knowledge in hand, it was possible to ask truly comprehensive questions. Rather than examining changes in transcription levels for a single gene at a time, it became conceivable to measure transcription levels for every gene in the genome. Similarly, rather than performing stepwise classical or reverse genetic analysis, one mutant at a time, it became conceivable to ask how a disruption in any of the yeast genes might affect a particular process. These large-scale, grandiose imaginings were made realizable by the relatively modest size of the yeast genome (only about 6000 genes) and its fairly simple genetic structure (very few introns interrupting gene sequences).

Over the years since the completion of the yeast genome sequence, a breathtaking set of experimental resources have been developed and made available by the community of yeast researchers. One early example was the production of complementary DNA (cDNA) microarrays that enabled the measurement of the level of every mRNA in the cell simultaneously under a variety of experimental conditions.

(cDNA is the DNA synthesized from an RNA template using the enzyme reverse transcriptase.) Massive databases collate and correlate experimental information about mRNA levels measured in this way for thousands of different individual experiments. An even more ambitious resource has been the construction of a comprehensive library of targeted yeast gene disruption mutants where, one by one, each gene in the genome has been deleted and replaced with a molecular bar code, a specific DNA sequence that identifies that mutant strain from amongst all of the others. These tagged deletion strains either can be ordered and used one by one for standard reverse genetic analysis or else can be pooled and assayed simultaneously by following their barcodes to find all of the targeted disruptions that affect a particular process of interest. This family of techniques collectively called *genomics* represents the evolution of Mendel's field of genetics into the modern era where our access to complete genome sequence information enables the possibility of performing studies that are comprehensive and complete in a way not previously imaginable. These large-scale projects in yeast paved the way for similarly ambitious genome-scale projects in many other model organisms, including humans. In many of these projects involving technique development, research in the yeast community led the way.

As we stated at the beginning of this chapter, molecular understanding of biological processes requires complementary information from experiments based in genetics and from experiments based in biochemistry. The biochemical equivalent of genomic-scale experimentation is a growing field called *proteomics*. Again, because of the relatively small number and limited complexity of protein products encoded in the yeast genome, it has been possible to perform several different kinds of comprehensive analyses of protein behavior, localization, and functional interaction. For example, in a project of similar technical scope to the generation of the deletion collection described above, every identified yeast gene has been replaced one at a time with a modified gene fused to the gene encoding GFP as indicated in Figure 2.20 (p. 58). Of the 6000 or so strains constructed in this manner, about 4000 actually expressed green fluorescence under standard laboratory growth conditions. This made it possible to catalog in a systematic way the subcellular localization of every single one of these thousands of proteins in living cells. Besides the subcellular location of a protein, much important information about its function and biological role can often be gleaned by examining its binding partners.

In a similar proteomic-scale approach, every protein-encoding gene in the yeast genome was replaced with an engineered version fused to an affinity purification tag, a specific protein sequence that is easily fished out of a complex mixture using biochemical techniques. Each tagged protein isolated in this manner could then be examined for binding partners that tended to purify along with it. The study of binding in biology will be one of the threads that runs through the entire book, with our first serious encounter with this subject to begin in Chapter 6. This is one of several kinds of large data sets that have been used to build comprehensive maps of the network of protein interactions in yeast. A small portion of this network is illustrated in Figure 4.20. The affinity purification approach will find only a subset of relevant interactions, specifically, those that are stable enough to survive disruption of the yeast cell and passage over a purification column.

Other kinds of experiments, which may now also be implemented on a genome-wide or proteome-wide scale, can reveal transient

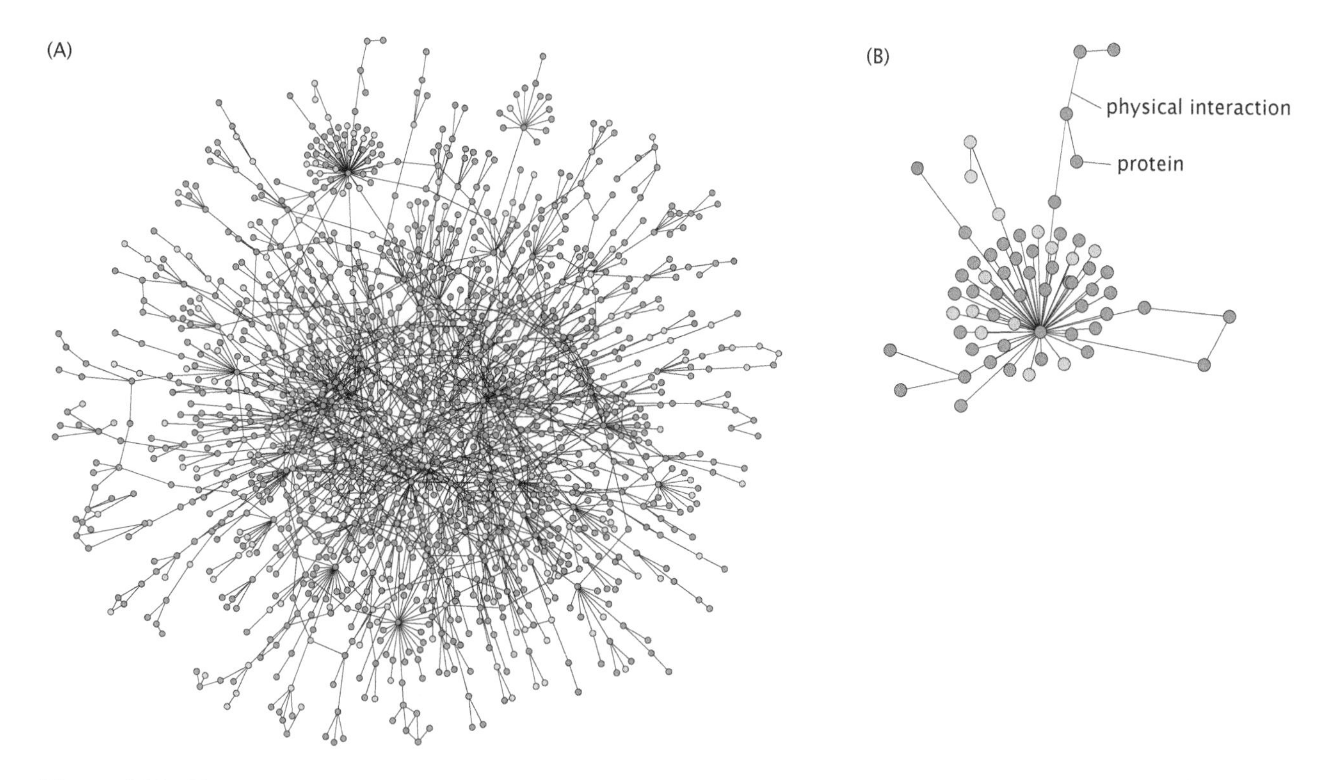

Figure 4.20: The yeast protein–protein interaction network. (A) In this diagram, each dot (or node) represents a single protein. Lines connecting the nodes indicate that those two proteins have been shown to interact with one another, biochemically or genetically. This kind of analysis reveals that all protein products in yeast participate in a web of interactions such that no protein is an island unto itself. (B) A magnified view of part of the network from (A). (Adapted from H. Jeong et al., *Nature* 411:41, 2001)

interactions or functional interactions that may be missed when searching simply for stable complexes. One commonly used approach to find interacting proteins is referred to as a two-hybrid screen. In the original implementation, a protein of choice called "bait" was fused to a partially functional protein fragment that would be able to regulate the expression of a reporter gene if and only if the bait protein was bound to a target protein carrying the other half of the functional tag. Taking a single bait protein and challenging it with a library of tagged potential partners can rapidly lead to identification of proteins that bind to the bait in living cells without the need for any biochemical purification. The rapidly growing databases that connect this kind of biochemical information with information from other large-scale screens showing genetic or functional interactions can instantly provide a researcher with a molecular-level context to understand the likely consequences of an alteration in any gene of interest.

Why has the rapid development of all these new genomic and proteomic techniques happened so quickly in yeast as compared with other model organisms? As we have described, part of the answer has to do with the properties of yeast itself. However, a perhaps equally important part of the answer comes from the nature of the yeast research community. Thousands of different individual laboratory groups use yeast as a primary or secondary model organism and this community of researchers has recognized the extraordinary power of creating and sharing large-scale resources. At present, both the concrete resources such as yeast strains and plasmids and the intangible resources such as the large-scale quantitative data sets generated by these techniques are freely and efficiently shared among all the members of the research community. This amplifies the power of any individual group, rendering the yeast research community a large-scale analog of the genetic and proteomic networks that we have just described.

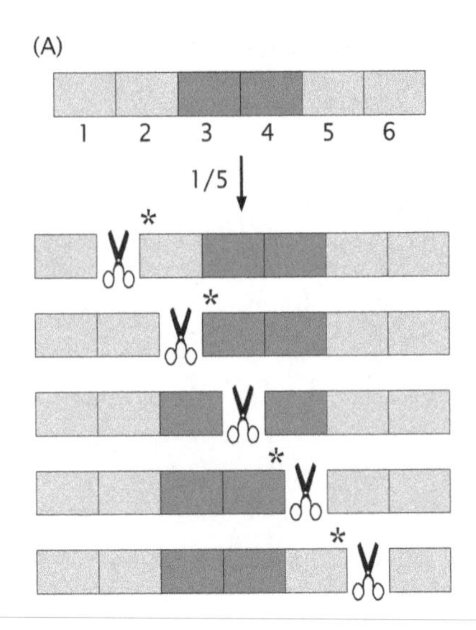

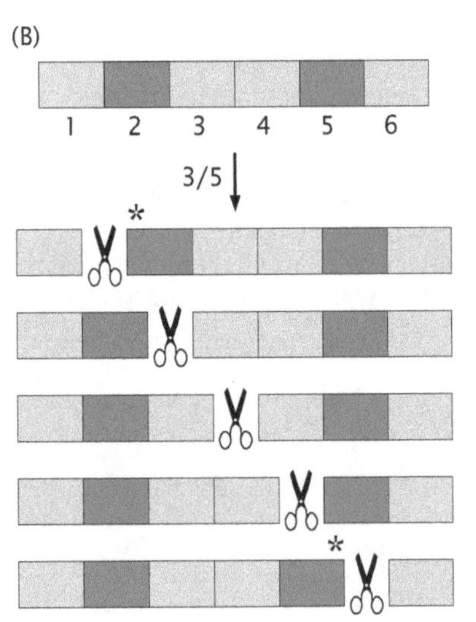

Figure 4.21: Concept of mutation correlation and physical proximity on the gene illustrated by labeling a string in two distinct points and then making random cuts of the string. The probability that the two labels will remain on the same part of the string *after* the cut depends upon their physical proximity *before* the cut. (A) Two mutations (red boxes) that are close to one another are likely to remain on the same part of the string (four times out of five). (B) Two mutations that are further apart are more likely to be separated (remaining together only two times out of five).

4.6 Flies and Modern Biology

4.6.1 Flies and the Rise of Modern Genetics

Drosophila melanogaster Has Served as a Model System for Studies Ranging from Genetics to Development to the Functioning of the Brain and Even Behavior

One of the recurring themes in the history of molecular biology is the effort that surrounds the discovery of a concrete realization of a concept that is already known abstractly. One such example is the realization that the abstract concept of a gene (a unit of inheritance) is physically manifested in chromosomes, a discovery resulting from the work of Thomas Hunt Morgan and his group on the fruit fly *D. melanogaster*. The experiments of Morgan and his students concerned the study of correlations in mutations. Morgan and his gene hunter students were, like Mendel, concerned with the nature of inheritance and mutation. The significant point of departure for their experiments was a single fruit fly with white eyes rather than the red eyes of the wild type.

The beautiful and powerful idea that ties together the various experiments from Morgan, Sturtevant, and co-workers is that correlations in inheritance give a measure of physical proximity of genes on the chromosome, as illustrated schematically in Figure 4.21. The simple idea that permitted them to map the location of particular genes on a chromosome of *Drosophila* was that during the cross-over process those genes that are physically closer on the chromosome have a greater chance of being inherited together in the offspring. Crossover is the process whereby genes are scrambled as the fruit fly is making eggs, through cutting and ligation of genomic DNA. For example, starting from a mother who carries mutations for both white eyes and miniature wings (a double mutation), the offspring can carry both mutations, one or the other, or neither. Morgan and co-workers observed that offspring of mothers carrying both white eyes and miniature wings were less likely to coinherit both mutations than offspring of mothers carrying mutations for white eyes and yellow body color. They interpreted this to mean that the gene encoding yellow bodies is physically closer on the chromosome to the gene encoding white eyes than is the gene encoding for miniature wings and therefore less likely to be separated during the process of recombination.

Another key observation enabling the evolution of the idea of a gene from an abstract unit of inheritance to a physical segment of DNA on a chromosome was facilitated by a peculiar habit of cells in the *Drosophila* salivary gland. These giant cells replicate multiple copies of their chromosomes without separating them. The multiple copies of replicated DNA remain aligned next to one another, forming a giant polytene chromosome that can be easily seen with a light microscope as shown in Figure 4.22. Characteristic dark and light banding patterns, visible on the polytene chromosomes, represent a barcode identifying certain stretches of the chromosome. The

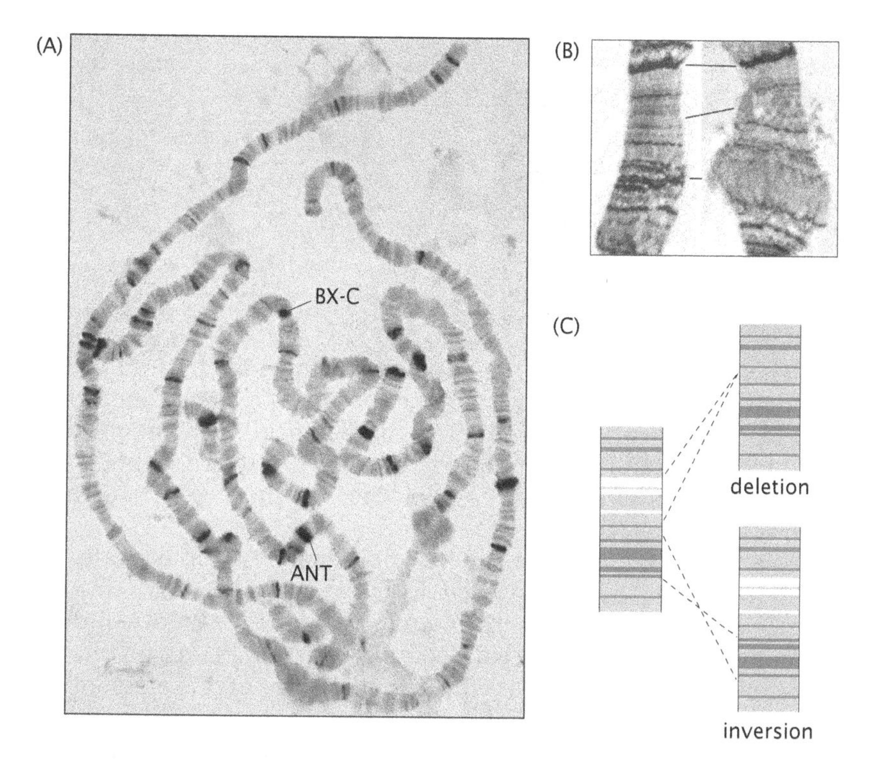

Figure 4.22: Polytene chromosomes and genetic cartography. (A) A light microscope image of a giant polytene chromosome from a *Drosophila* salivary gland cell. The positions of two genetic loci, *Antennapedia* (ANT) and *bithorax* (BX-C), are indicated. (B) A magnified view of a small portion of one of the chromosomes shows the effects of transcription on polytene chromosome structure. The images on the left and right show the same chromosomal region, before and after a specific phase of developmental transcription is initiated by the hormone ecdysone. The portion of the polytene chromosome bundle encoding a gene that is being actively transcribed decondenses to form a characteristic "puff." The identities of these developmentally regulated genes can be correlated both with their positions in the genetic map and with their physical positions on the polytene chromosome. (C) Schematic showing how a physical deletion or inversion of a portion of the chromosome is reflected in both the appearance of the polytene chromosome and the apparent order of genes on the chromosome as determined by genetic mapping. (A, adapted from B. Alberts et al., Molecular Biology of the Cell, 5th ed., Garland Science, 2008; B, adapted from I. F. Zhimulev et al., *Int. Rev. Cytol.* 241:203, 2004.)

number of distinct kinds of chromosomes visible in the microscope was intriguingly the same as the number of linkage groups (sets of genes that are generally transmitted together) identified by genetic analysis, suggesting a direct correspondence between those visible chromosomes and the linked units of inheritance. It was found that large physical rearrangements such as inversions or deletions that visibly change the banding pattern on the polytene chromosomes were correlated with genetic inversions or deletions describing changes in the linkage of particular groups of genes. This firmly established the concept that genes were physical entities aligned in order on chromosomal DNA. It is not an exaggeration to say that Morgan and his gene hunters, by studying the geography of genes on chromosomes, launched the scientific program that culminated in the sequencing of entire genomes and the biological mapmaking that is the hallmark of modern genetics.

4.6.2 How the Fly Got His Stripes

A second key role for the fruit fly has been in its role as a model system for the study of development. Unlike bacteriophages, *E. coli*, or yeast, *Drosophila* are multicellular animals, much like ourselves, with heads, eyes, and legs. Several famous *Drosophila* mutations illustrated the idea that single genetic changes could result in complete respecification of complex body parts made up of thousands of cells. For example, the mutations *eyeless* and *wingless*, as their names imply, result in the development of flies that completely fail to form a critical body part. Even more amazing are mutations such as *Antennapedia* and *bithorax* that result in transformation of one body part into another as shown in Figure 4.23. The heads of flies carrying *Antennapedia* mutations grow legs where their antennae ought to be. Flies with the *bithorax* mutation develop a second set of full-size wings directly behind their normal single pair. These kinds of observations

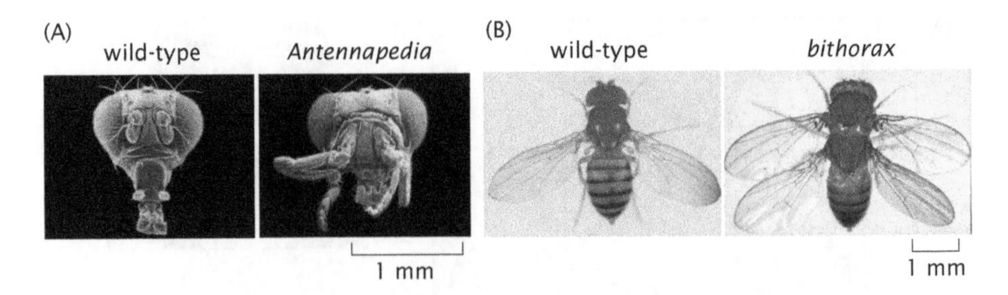

Figure 4.23: *Drosophila* mutations that transform one body part into another. (A) Scanning electron microscopy of fly heads showing the normal head with normal antennae and an *Antennapedia* mutant. (B) Fly carrying *bithorax* alongside a normal fly. (A, Courtesy of F. R. Turner.; B, courtesy of E. B. Lewis.)

established the idea that large-scale developmental patterns can be controlled by master regulatory genes. These are single genes that initiate a cascade of events leading to specification of many different cell types in carefully regulated spatial order to create body parts such as eyes, legs, and wings. We will discuss these genes further in Chapter 21.

More generally, *Drosophila* genetics and analysis of physically deficient mutant flies have provided a great deal of insight into our understanding of how cells communicate with one another during development to form large-scale complex patterns. One of the most visually striking examples is the specification of a series of segments in the fly embryo along the head-to-tail axis that will eventually become distinct segments of the adult fly. These patterns were shown in Figure 2.45 (p. 80) and discussed in that section. Specification of the segments requires several different kinds of cell–cell communication that take advantage of physical processes such as diffusion to set up steady-state distributions of signaling molecules and also uses interesting kinds of positive and negative feedback to refine positional information. In Chapter 19, we will return to this example and explore the elegant molecular and cellular interactions involved in this kind of pattern formation.

Another kind of developmental event that has been well studied in *Drosophila* and is also relevant to other kinds of animal development involves the situation where a field of identical cells becomes specified to establish distinct cellular identities. A dramatic and well-studied example is the specification of photoreceptor cells in the developing *Drosophila* eye (see Figure 2.35(D) on p. 73 for eyes in another species of fly). The *Drosophila* eye, like our eye, contains a series of light-sensitive cells that respond to different segments of the visible light spectrum enabling the animal to see in color. At early stages of development, each of the *Drosophila* photoreceptor cell precursors has the potential to become any photoreceptor cell type. As development proceeds, the cells communicate with their neighbors in such a way as to enforce the final outcome that each cluster contains one and only one copy of each cell type. This local cell–cell communication is mediated by proteins expressed on the surface of the cells that allow each cell to keep track of what its neighbor is doing and to direct the neighbors not to follow the same developmental program as the cell is undergoing itself.

Drosophila has been a model system that, because of its facility as a subject for genetic analysis, has provided some of the first molecular clues for understanding each of these kinds of developmental processes. Information gleaned from experiments on *Drosophila*, both the identities of specific key molecules and the paradigms for pattern formation events, has been key to understanding developmental processes in animals such as mice and humans which are not so genetically tractable.

4.7 Of Mice and Men

Just as yeast proved more useful for studies of eukaryotic biology than *E. coli* and *Drosophila* more useful than yeast for issues having to do with multicellularity and animal development, likewise, some biological issues of paramount interest to humans, such as development of cancer and mammalian-specific elaborations of the central nervous system (such as a highly specific sense of smell), can best be studied only in a mammalian model system. Many different kinds of mammals have been used as laboratory research subjects, ranging from horses to dogs to rodents. In general, rodents have been most widely used because of their small size and fairly rapid generation time. Rats have been particularly widely used for psychological and physiological experimentation, giving rise to the vernacular term "lab rat." Guinea pigs have also put in heroic work as laboratory subjects, giving rise to another vernacular term, being treated like a "guinea pig." But of all the rodents, and indeed, all the mammals, that have been used in biomedical research, mice (*Mus musculus*) are by far the most important.

Mice have probably been kept as pets by humans since we began storing grain and it has long been noted that occasional mutants would arise with coloring or behavior different from normal. By the eighteenth century, mouse fanciers in Japan were deliberately breeding strangely colored mice and in the nineteenth century, mouse breeding as a hobby began to catch on in Europe. The life span of mice is short (typically about 2 years) and they become sexually mature at about 6–8 weeks, facilitating genetic breeding experiments. Although it is relatively easy to isolate mutant strains of mice, for a long time it was extremely difficult to identify the molecular lesions responsible for the change in color or behavior. Advances in cloning and genotyping techniques over the past 50 years have now made it routine to identify the genetic location of a particular mutation. More importantly, it is now actually possible to deliberately genetically engineer mice in much the same way we can genetically engineer *E. coli*, yeast, or *Drosophila*. Specific genes can be deleted by site-specific recombination and altered genes can be introduced and expressed. For example, GFP can be expressed in mice either specifically in a tissue where gene expression is controlled by a known promoter or more generally, in all cells. Three such mice are shown in Figure 4.24.

One extremely interesting property of mammalian development that is not shared by flies is that every cell at a particular early stage is totipotent, that is, it can turn into any type of cell present in the adult

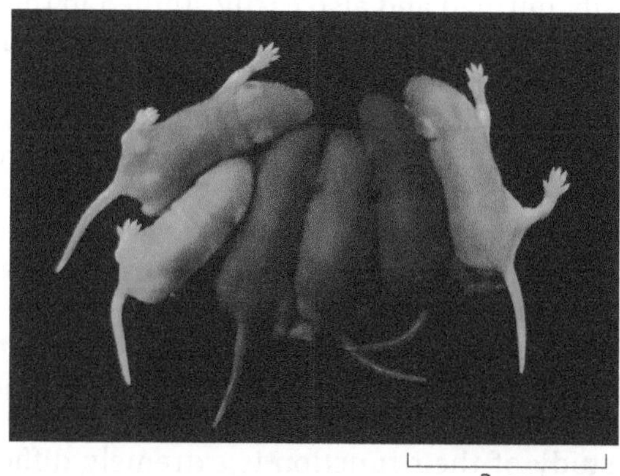

Figure 4.24: Genetic manipulation of the mouse. Mouse embryonic stem cells were engineered in culture so that they could express the green fluorescent protein at all stages of development. These engineered cells were then introduced into a normal mouse embryo and the resulting chimeric mouse bred to generate animals who expressed GFP in every cell in the body. Here, three GFP-expressing mouse pups are shown under fluorescent illumination along with three of their normal littermates. (Courtesy of Ralph Brinster, University of Pennsylvania.)

organism. A common manifestation of this totipotency is seen in identical twins; when the cells of the early embryo accidentally split into two groups, each group is fully able to form a complete human being. Embryonic cells with this property can be isolated and genetically engineered. Manipulation of these so-called embryonic stem cells is a critical tool in the study of mammalian developmental biology. Embryonic stem cells were first isolated from human embryos in 1998 and research on their properties and potential medical use is proceeding rapidly in many parts of the world.

4.8 The Case for Exotica

Most of the model systems discussed in this chapter are laboratory workhorses that have been used by very large numbers of laboratories to make a wide range of distinct discoveries. However, there are many cases where a single specialization in an organism has proved to be of extraordinary utility in illuminating general principles in biological function. Many of these organisms cannot be cultured in the laboratory and their scientific claim to fame may be based on a single finding. However, biology would be profoundly different if any of these beasties had not been discovered.

4.8.1 Specialists and Experts

Already we have frequently fallen into the trap of describing conditions within "the cell" as if all cells are identical or similar. Obviously, this is a drastic oversimplification. Even within a single organism such as the human body, different cells that contain the same genome are highly specialized for different purposes. Experimental biologists have long taken advantage of the diversity and specialization of different cells within organisms and different organisms within ecosystems to examine the mechanism of a particular process of interest that may be employed by very many cells but is brought to a fine art in only a few.

For example, the cytoskeletal proteins actin and myosin are involved in the production of contractile force in all eukaryotic cells. However, the skeletal muscle cells of vertebrates contain extraordinarily high concentrations of both of these proteins arranged into regular arrays that are able to produce a large amount of force and contract over long distances to move animal limbs. Thus, biochemists interested in the mechanism of force generation by actin and myosin have long focused on skeletal muscle cells as cellular specialists where these proteins are easily purified and also easily studied in their native habitats. Some animals go further and modify their already specialized cells in particularly interesting ways that are also useful for identification of important proteins involved in general processes. The electric fish *Torpedo* uses a highly specialized organ derived from muscle to generate an electric charge that can shock and stun its prey. In normal muscle cells, the electrical signaling directing their contraction occurs over a very small part of their surface area where a motor neuron makes contact with a small field of receptors called acetylcholine receptors (we will describe the mechanisms of this electrical signaling in Chapter 17). In a typical skeletal muscle cell, this field of receptors covers less than 0.1% of the cell surface, so purification of the receptors and study of their function is extremely difficult. However, in the *Torpedo* electric organ, the electric receptor domains have been greatly expanded, covering more than 10% of the cell surface. Therefore, researchers wishing to purify and examine the acetylcholine receptors generally do so from wild-caught *Torpedo* fish.

Another example where the very large size of individual cells facilitated investigation of general cell processes is in the salivary glands of *Drosophila*. As discussed in the previous section, in these large cells, the chromosomes are duplicated many times without cell division and the four chromosomes form aligned bundles. Striped patterns on these chromosome bundles reflecting underlying differences in chromatin organization are readily visible at the level of light microscopy. This special property of these cells was of great use to early *Drosophila* geneticists in establishing the physical realization of the abstract idea of the gene.

Important specialists have also been identified outside of the animal kingdom. One that has proved particularly fruitful in the study of molecular biology has been the small teardrop-shaped ciliate *Tetrahymena*. This organism first gained celebrity status with the identification of the biosynthetic machinery that makes telomeres, specialized segments at the end of linear chromosomes that prevent information being lost due to structural difficulties of DNA replication at the ends. In human cells, telomeres represent a very small fraction of the total DNA, with only one telomere at each of the ends of our 46 chromosomes. *Tetrahymena*, however, undergoes an unusual developmental process where important segments of its chromosomes are copied many times and stored in a separate specialized nucleus, resulting in many thousands of telomeres per cell. As with the examples discussed above, this specialization facilitated biochemical identification of the enzyme complex responsible for telomere formation, called telomerase, subsequently found to be well conserved among all eukaryotes. *Tetrahymena* is also famous for being the first cell type where an RNA molecule was proved to have catalytic function. Catalytic RNAs, called ribozymes, have also since been found throughout all domains of life. This ribozyme is able to catalyze a biochemical cleavage reaction in a manner that would normally be part of the bailiwick of proteins. Structural and chemical studies of the nature of this catalysis have helped reveal principles of how biochemical catalysis works in a more general sense.

4.8.2 The Squid Giant Axon and Biological Electricity

Repeatedly in the history of biology, we will find other examples where an important molecule or biological process was easy to identify or study from an exotic cell from an exotic creature. A beautiful example that we will revisit in Chapter 17 is the giant axon from the squid *Loligo*, which was used to great effect to understand the mechanisms of electrical signaling in neurons and also the mechanisms of long-distance intracellular transport. The squid giant axon operates under the same principle as all other animal nerve cells, but, as its name suggests, it is enormous, with a size of up to 1 mm in diameter. One key challenge in studying the electrical properties of cells is how to interface our macroscopic measurement apparatus and the cell membrane. This question translates into the technical challenge of connecting electrodes to nerve cells. Because of its huge size, the squid giant axon is ideal to manipulate in the laboratory, to impale with electrodes and to observe under the microscope. As a result, the squid giant axon allowed scientists to meet this challenge such that the nerve cells could be subjected to direct electrical query through the use of electrodes. A schematic of the squid and its axon is shown in Figure 4.25. Subsequent development of much smaller and finer electrodes enabled the study of electrical signaling in much smaller neurons and revealed that the principles elucidated in the squid are general.

Much of the business of cells is mediated by the presence of charged species both within cells and in the extracellular medium. The study of the control of spatial position and temporal evolution of charge has relevance to a huge variety of processes and most notably to the functioning of nerve cells. The present discussion will show the way in which studies on the giant axons of squids served as the basis of our current understanding of action potentials.

There Is a Steady-State Potential Difference Across the Membrane of Nerve Cells

Cell membranes mediate the compartmentalization of different chemical species (ranging from simple ions to proteins) within cells. One consequence of such compartmentalization is the possibility of a mismatch in the charge state from one side of the membrane to the other, with the consequence that an electrical potential is set up across that membrane. Such potentials are known as membrane potentials. To get a sense for the types of ionic mismatches encountered in real cells, we note that the concentration of K^+ ions within the cell is roughly 20 times greater than outside the cell. A crude estimate of the corresponding membrane potential for such concentrations can be obtained by using the Nernst equation (to be discussed in detail in Chapter 17), which tells us that

$$V = \frac{k_B T}{q} \ln \frac{C_{out}}{C_{in}}, \tag{4.1}$$

where q is the charge on the species of interest, C_{out} is the extracellular concentration of these charges, and C_{in} is their concentration within the cell. Using the concentration ratio suggested above, the Nernst equation suggests a membrane potential of the order of -75 mV, indicating that the potential inside the cell is 75 mV lower than the potential outside.

Nerve Cells Propagate Electrical Signals and Use Them to Communicate with Each Other

There are a number of different mechanisms cells use to convey information. One key mechanism involves the presence of chemical messengers that are released as the basis of cell signaling. A second key mechanism of information transmission is the use of electrical signals and, in particular, the so-called action potential. The physical picture associated with the action potential is that of a traveling pulse of membrane potential difference that propagates along nerve cells at speeds of up to 100 m/s. The successive arrival of an electrical pulse along the axon of a nerve cell is shown schematically in Figure 4.25. The idea is that the local value of the membrane potential at some point along the axon is disturbed from its equilibrium value as a result of a transient change in the permeability of the membrane due to the opening of ion channels and a concomitant flow of ions across that membrane. This spatially localized change in potential has the effect of disturbing the ion concentration and potential

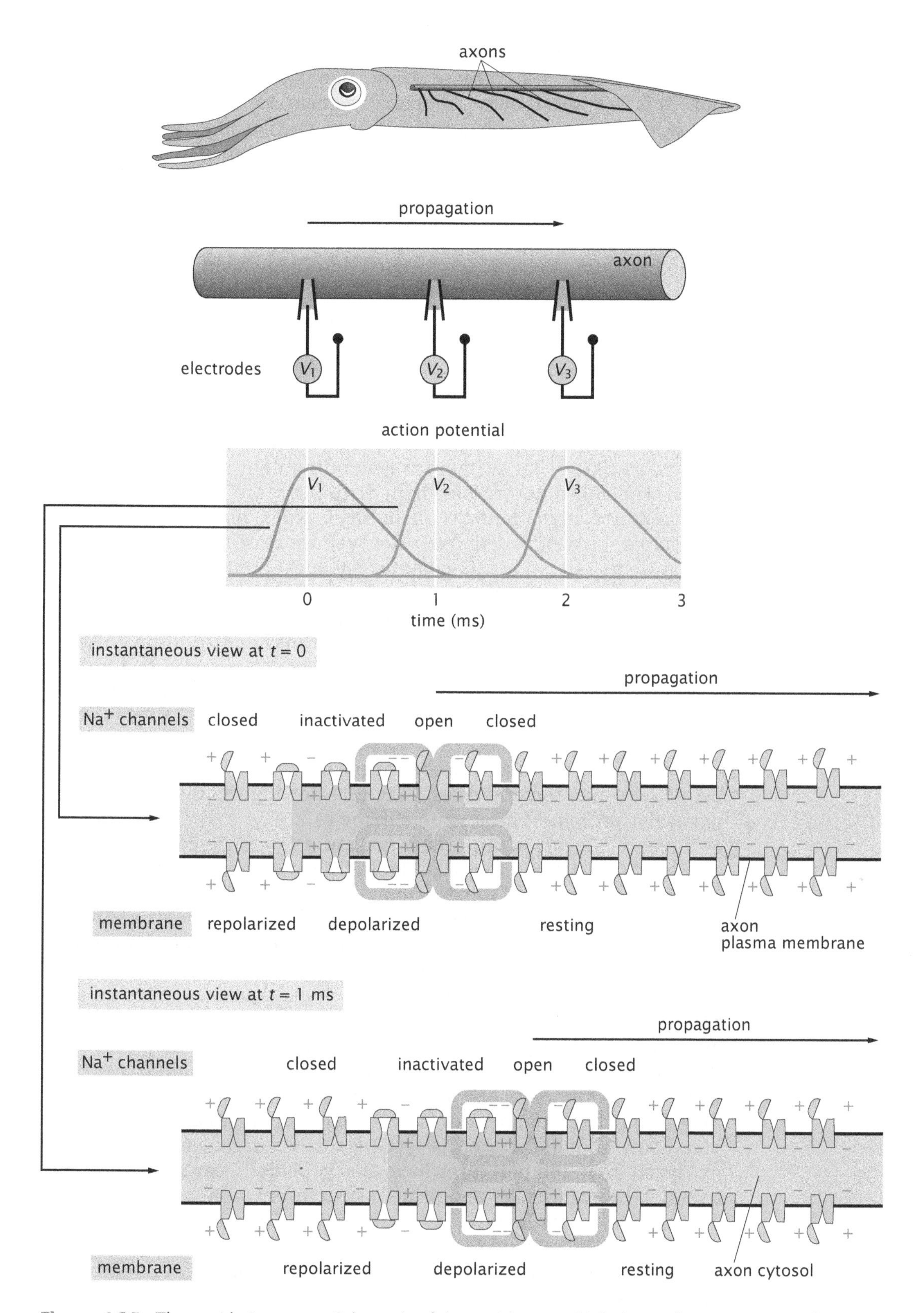

Figure 4.25: The squid giant axon. Schematic of the squid axon which shows the axon as part of the squid's anatomy, illustrates the propagation of an action potential in abstract electrical terms, and shows how that action potential is mediated by the presence of ion channels. (Adapted from B. Alberts et al., Molecular Biology of the Cell, 5th ed. Garland Science, 2008.)

in its vicinity, which unleashes transient changes in permeability in those regions with a corresponding change in the membrane potential. This cascade moves sequentially down the axon and, as noted above, can serve as the basis of signaling such as is evidenced by the contraction of muscles in response to external stimuli.

4.8.3 Exotica Toolkit

In the preceding section, we focused on organisms that specialize in producing a molecule or performing a function because the molecule or function is widespread. However, in many cases, the real strength of exotica lies in their ability to do something unique. Clever scientists have repeatedly exploited the very unusual properties of some organisms to generate reagents that can serve as a critical piece of the biologist's toolkit. The most familiar example is the green fluorescent protein (GFP). This protein is produced by the jellyfish *Aequoria victoria* as part of its system for generating light signals deep in the ocean. The ability to produce light or to fluoresce under illumination is shared by many organisms, including fireflies, many bacteria, fungi, and corals, as well as jellyfish. However, most of these light production systems rely on highly complex chemical groups that are created through elaborate biosynthetic pathways involving many enzymes. The very special feature of GFP that has made it so useful is that this protein is able to fluoresce on its own without any additional chemical groups. Thus, when a gene encoding GFP is expressed in a foreign organism from a bacterium to a tobacco plant to a rabbit, the protein is still able to fluoresce. Repeatedly throughout this book, we will see examples of the uses of GFP, sometimes as a reporter for transcription, sometimes as a probe for measuring physical properties of cytoplasm, and sometimes as a tag for showing the distribution of a particular protein of interest in a living cell.

Arguably the most important reagent ever to be identified from an exotic organism is the heat-stable DNA polymerase from *Thermus aquaticus* that enabled the development of the polymerase chain reaction (PCR). This revolutionary technique is used in all fields of biology ranging from field identification of organisms to cloning of useful products for biotechnology to identification of forensic evidence at crime scenes. PCR amplification of rare DNA sequences relies on thermal cycling where the temperature of the sample is repeatedly raised and lowered. DNA polymerase from any normal organism would be denatured by the rise in temperature. However, *T. aquaticus* prefers to live in geothermal hot springs and would die of cold at normal mammalian body temperatures. Its DNA polymerase can therefore survive many rounds of thermal cycling. Other organisms that live at extremely high temperatures have also provided heat-stable proteins useful in biotechnology and protein structural studies.

One of the major features of organisms on Earth is the large amount of energy that they devote to interacting with other organisms, largely to kill and eat them. Some predatory organisms have developed great speed or strength to catch their prey. But consider the plight of the cone snail. This family of slow-moving mollusks aspires to a predatory lifestyle, but speed and strength are not part of their repertoire. Instead, they have developed a remarkable witch's brew of potent neurotoxins that can paralyze a wide range of prey organisms. A similar strategy of hunting by chemical warfare has been developed by many snakes, spiders, and scorpions. Still other organisms create neurotoxins for defensive reasons, including sponges, frogs, and pufferfish. Why are all these toxins useful rather than just exotic? Neurotoxins work by binding with extraordinarily high affinity to receptors in their target animals. Inhibition of these receptors at low concentrations of neurotoxin must be sufficient to paralyze or kill the target animal. Therefore, their receptors are almost always proteins critically important in neuronal signaling or muscle contraction. Because of the strong binding affinity of neurotoxins for their receptors, they can be used biochemically to purify and identify their receptors. Specific toxins are also extremely useful for studying the sequence of events involved in a particular form of signaling, since they will typically permit upstream events to occur but will block those that are downstream.

This is by no means an exhaustive list of the important contributions that have been made to biology by nonstandard organisms. The point we wish to emphasize is that the study of life processes is greatly enriched by examining the ways in which organisms are different from each other as well as by examining the ways in which they are similar.

4.9 Summary and Conclusions

This chapter has introduced many of the key model molecules and organisms that will be returned to as the basis for the development of physical biology in the remainder of the book. Our argument is that once the biology of a particular organism is well in hand, it becomes possible to shift the emphasis to the kind of rich interplay between quantitative theory and experiment that is the hallmark of physical biology to be explored throughout the book. Some of the key model systems that we will return to are hemoglobin, bacteriophages, *E. coli*, the yeast *S. cerevisiae*, and the fruit fly *D. melanogaster*. In addition to the celebrity status of these key organisms, we have also argued that there is an important role for the study of exotica. In many ways, the pleasure of biology is to explore the differences among organisms and to understand the ways in which particular specializations have arisen from the evolutionary canvas. In the remainder of the book, we will explore these themes again and again.

4.10 Problems

Key to the problem categories: • Estimates • Data interpretation

• 4.1 Structure of hemoglobin and myoglobin

(a) As in Problem 2.6, obtain the atomic coordinates for hemoglobin and myoglobin. Measure their dimensions, identify the different subunits and the heme groups.

(b) Expand the analysis of hemoglobin on p. 143 by calculating the mean spacing between hemoglobin molecules inside a red blood cell. How does this spacing compare with the size of a hemoglobin molecule?

(c) Typical results for a complete blood count (CBC) are shown in Table 4.1. Assume that an adult has roughly 5 L of blood in his or her body. Based on these values, estimate:
(i) the number of red blood cells;
(ii) the percentage in volume they represent in the blood;
(iii) their mean spacing;
(iv) the total amount of hemoglobin in the blood;
(v) the number of hemoglobin molecules per cell;
(vi) the number of white blood cells in the blood.

• 4.2 The number of phages on Earth

(a) An estimate for the number of phages on Earth can be obtained using data from Bergh et al. (1989). By taking water samples from lakes and oceans and counting the various phages, one arrives at counts of the order of 10^6–10^7 phages/mL. Using reasonable assumptions about the amount of water on Earth, make an estimate of the number of phages.

(b) Work out the mass of all of the phage particles on the Earth using your result from (a).

(Problem suggested by Sherwood Casjens.)

Table 4.1: Typical values from a CBC. (Adapted from R. W. Maxwell, Maxwell Quick Medical Reference, Maxwell Publishing, 2002.)

Test	Value
Red blood cell count (RBC)	Men: $\approx(4.3–5.7) \times 10^6$ cells/μL Women: $\approx(3.8–5.1) \times 10^6$ cells/μL
Hematocrit (HCT)	Men: ≈39–49% Women: ≈35–45%
Hemoglobin (HGB)	Men: ≈13.5–17.5 g/dL Women: ≈12.0–16.0 g/dL
Mean corpuscular hemoglobin (MCH)	≈26–34 pg/cell
MCH concentration (MCHC)	≈31–37%
Mean corpuscular volume (MCV)	≈80–100 fL
White blood cell count (WBC)	$\approx(4.5–11) \times 10^3$ cells/μL
Differential (% of WBC):	
Neutrophils	≈57–67
Lymphocytes	≈23–33
Monocytes	≈3–7
Eosinophils	≈1–3
Basophils	≈0–1
Platelets	$\approx(150–450) \times 10^3$ cell/μL

Table 4.2: Fraction of crossovers of six sex-linked factors in *Drosophila*. (Adapted from A. H. Sturtevant, *J. Exp. Zool.* 14:43, 1913.)

Factors	Fraction of crossovers
BR	115/324
B(C,O)	214/21736
(C,O)P	471/1584
(C,O)R	2062/6116
(C,O)M	406/898
PR	17/573
PM	109/458
BP	1464/4551
BM	260/693

• 4.3 Genetics by the numbers

Three successive nucleotides along a DNA molecule, called a *codon*, encode one amino acid. The weight of an *E. coli* DNA molecule is about 3.1×10^9 daltons. The average weight of a nucleotide pair is 660 daltons, and each nucleotide pair contributes 0.34 nm to the overall DNA length.

(a) What fraction of the mass of an *E. coli* cell corresponds to its genomic DNA?

(b) Calculate the length of the *E. coli* DNA. Comment on how it compares with the size of a single *E. coli* cell.

(c) Assume that the average protein in *E. coli* is a chain of 300 amino acids. What is the maximum number of proteins that can be coded by the *E. coli* DNA?

(d) On an alien world, the genetic code consists of two base pairs per codon. There are still four different bases. How many different amino acids can be encoded?

• 4.4 Mutation correlation and physical proximity on the gene

In Section 4.6.1, we briefly described Sturtevant's analysis of mutant flies that culminated in the generation of the first chromosome map. In Table 4.2, we show the crossover data associated with the different mutations that he used to draw the map. A crossover refers to a chromosomal rearrangement in which parts of two chromosomes exchange DNA. An illustration of the process is shown in Figure 4.26. The six factors looked at by Sturtevant are B, C, O, P, R, and M. Flies recessive in B, the black factor, have a yellow body color. Factors C and O are completely linked, they always go together and flies recessive in both of these factors have white eyes. A fly recessive in factor P has vermilion eyes instead of the ordinary red eyes. Finally, flies recessive in R have rudimentary wings and those recessive in M have miniature wings. For example, the fraction of flies that presented a crossover of the B and P factors is denoted

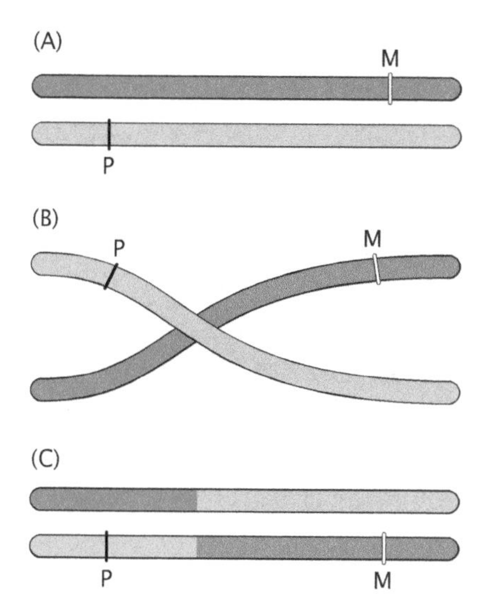

Figure 4.26: Crossing over of chromosomes. (A) Chromosomes before crossing over showing two loci labeled P and M. (B) Illustration of the crossing-over event. (C) Chromosomes after crossover.

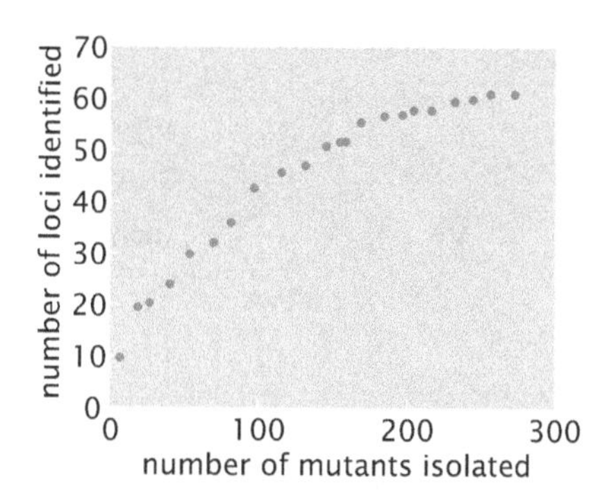

Figure 4.27: Saturation of a mutant library. Number of different identified loci as a function of the number of mutants isolated. (Adapted from C. Nusslein-Volhard et al., *Roux's Arch. Dev. Biol.* 193:267, 1984.)

as BP. Assume that the frequency of recombination is proportional to the distance between loci on the chromosome.

Reproduce Sturtevant's conclusions by drawing your own map using the first seven data points from Table 4.2.

Keep in mind that shorter "distances" are more reliable than longer ones because the latter are more prone to double crossings. Are distances additive? For example, can you predict the distance between B and P from looking at the distances B(C,O) and (C,O)P? What is the interpretation of the two last data points from Table 4.2?

• 4.5 Saturation of mutants in libraries

In a set of classic experiments, the second chromosome of *D. melanogaster* was mutagenized and the effects of these mutations characterized based on their phenotype in embryonic development. The experimenters found 272 mutants with phenotypes visibly different from wild-type embryos. However, when they determined the location of the mutations using the method outlined in Figure 4.21 and worked out in Problem 4.4, they discovered that these mutations only mapped to 61 different positions or loci on that chromosome. Figure 4.27 shows how, as more mutants were scored, ever more mutants corresponded to previously identified loci. Using a model that assumes a uniform probability of mutation in any locus, calculate the number of new loci found as a function of the number of mutants isolated. Explain the saturation effect and plot your results against the data. Provide a judgment on whether it is useful to continue searching for loci. (*Hint*: Start by writing down the probability that a specific locus has not been mapped after scoring the first M mutants). Relevant data for this problem are provided on the book's website.

• 4.6 Split genes and the discovery of introns

In a classic experiment, it was discovered that eukaryotic genomic DNA codes for proteins in a noncontinuous fashion, culminating in the idea of introns, regions in a given gene that are actually spliced out of the mRNA molecule before translation. The concept of the experiment was to hybridize the unspliced DNA with a DNA mimic of the spliced mRNA. Using Figure 4.28, estimate the sizes of the spliced regions.

• 4.7 Open reading frames in random DNA

In this problem, we will compute the probabilities of finding specific DNA sequences in a perfectly random genome, for which we assume that the four different nucleotides appear randomly and with equal probability.

(a) From the genetic code shown in Figure 1.4, compute the probability that a randomly chosen sequence of three nucleotides will correspond to a stop codon. Similarly, what is the probability of a randomly chosen sequence corresponding to a start codon?

(b) A reading frame refers to one of three possible ways that a sequence of DNA can be divided into consecutive triplets of nucleotides. An open reading frame (ORF) is a reading frame that contains a start codon and does not contain a stop codon for at least some minimal length of codons. Derive a formula for the probability of an ORF having a length of N codons (not including the stop codon).

(c) The genome of *E. coli* is approximately 5×10^6 bp long and is circular. Again assuming a that the genome is a random configuration of base pairs, how many ORFs of length 1000 bp (a typical protein size) would be expected by chance? Note that there are six possible reading frames.

(Problem courtesy of Sharad Ramanathan)

4.11 Further reading

Davis, RH (2003) The Microbial Models of Molecular Biology: From Genes to Genomes, Oxford University Press. This book examines the role of microbial model systems.

Johnson, GB (1996) How Scientists Think – Twenty-one Experiments That Have Shaped Our Understanding of Genetics and Molecular Biology, Wm C. Brown Publishers. This book makes for fascinating reading and describes many of the classic experiments discussed in this chapter in more detail.

Tanford, C, & Reynolds, J (2001) Nature's Robots: A History of Proteins, Oxford University Press. This book gives an

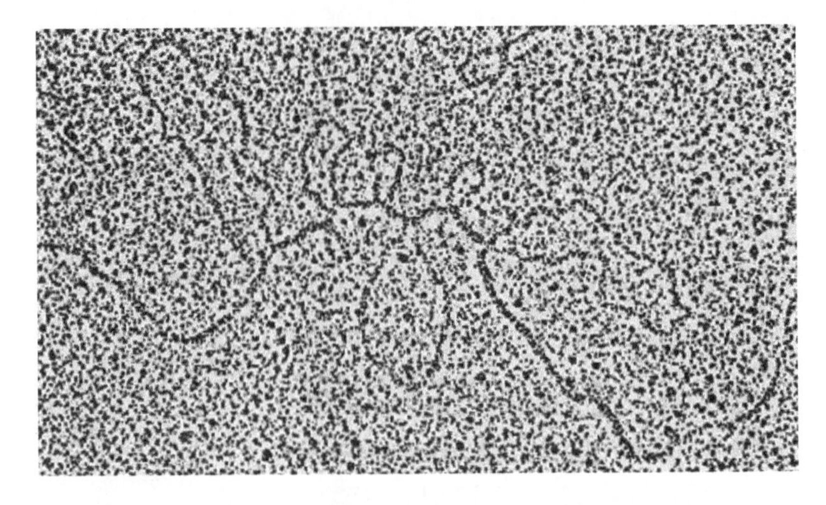

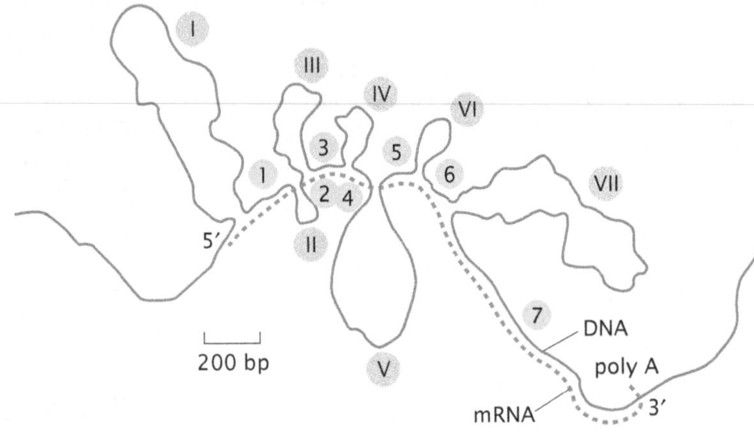

Figure 4.28: Experiment by Chambon to demonstrate the existence of introns. Unspliced DNA is hybridized to spliced mRNA. The loops correspond to regions of the DNA that have been removed in the mRNA. (Adapted from P. Chambon, *Sci. Am.* 244(5):60, 1981.)

informative account of the history of ideas associated with the determination of the nature of proteins. Within the context of the present chapter, the reader will find that hemoglobin enters this historical stage with great regularity.

Dickerson, RE (2005) Present at the Flood: How Structural Molecular Biology Came About, Sinauer Associates. This wonderful book is full of commentaries and reprints from the revolution in structural biology. For the purposes of the present chapter, it is Dickerson's story of "How to Solve a Protein Structure" with particular emphasis on the globins that touches our story.

Berg, H (2000) Motile Behavior of Bacteria, *Phys. Today*, January. Berg provides a masterful, though brief, account of the nature of motility in *E. coli*, describing many of the topics developed in the present chapter. In addition, Berg has written two wonderful books, *E. coli* in Motion (2003, Springer) and Random Walks in Biology (1993, Princeton University Press), both of which explain many fascinating features of bacteria, including chemotaxis.

Echols, H (2001) Operators and Promoters: The Story of Molecular Biology and Its Creators, University of California Press. Echols' book is a profoundly interesting and unique description of the history of much of molecular biology. In the context of the present chapter, it has a number of useful insights into the use of phages and *E. coli* as model systems, and describes topics such as gene regulation masterfully.

Judson, HF (1996) The Eighth Day of Creation: The Makers of the Revolution in Biology, Cold Spring Harbor Laboratory Press. Judson's book, like that of Echols, recounts the history of the development of molecular biology and describes many of the experiments and advances discussed in the present chapter. This book is fascinating.

Barnett, JA (2003) Beginnings of microbiology and biochemistry: the contribution of yeast research, *Microbiology* **149**, 557. Barnett has written a series of articles on the history of yeast in the emergence of biochemistry and microbiology. This article gives a flavor of the more complete set of articles, which we highly recommend.

Ghaemmaghami, S, Huh, W-K, Bower, K, et al. (2003) Global analysis of protein expression in yeast, *Nature* **425**, 737. This paper gives a flavor of the genome-wide yeast experiments that permit the study of nearly the full complement of yeast proteins.

Kornberg, A (1991) For the Love of Enzymes: The Odyssey of a Biochemist, Harvard University Press. This book describes Kornberg's work on DNA replication and much more and includes the charming story (see p. 106) which gives our chapter its title.

Weiner, J (1999) Time, Love, Memory: A Great Biologist and His Quest for the Origins of Behavior, Alfred A. Knopf. In the present chapter, we described the way in which *D. melanogaster* has repeatedly served as a model system for characterizing both genetics and development in eukaryotic systems. Weiner's book provides a thoughtful and interesting side-by-side biography of *Drosophila* and Seymour Benzer, who used this organism as the basis of his studies of the biology of behavior.

4.12 References

Alberts, B, Johnson, A, Lewis, J, et al. (2008) Molecular Biology of the Cell, 5th ed., Garland Science.

Berg, P (1993) Dissections and reconstructions of genes and chromosomes. In Nobel Lectures, Chemistry 1971–1980, World Scientific.

Bergh, O, Borsheim, KY, Bratbak, G, & Heldal, M (1989) High abundance of viruses found in aquatic environments, *Nature* **340**, 467.

Brunori, M (1999) Hemoglobin is an honorary enzyme, *Trends Biochem. Sci.* **24**, 158.

Chambon, P (1981) Split genes, *Scientific American* **244**(5), 60.

Chant, J, & Herskowitz, I (1991) Genetic control of bud site selection in yeast by a set of gene products that constitute a morphogenetic pathway, *Cell* **65**, 1203.

Dickerson, RE, & Geis, I (1983) Hemoglobin: Structure, Function, Evolution and Pathology, The Benjamin/Cummings Publishing Company, Inc. This lovely book is a testament to the status of hemoglobin as a model system. Chapter 2 gives an excellent discussion of the key role of hemoglobin as a model system with relevance to physiology, structural biology and the biophysics of allostery and cooperativity.

Grimes, S, Jardine, PJ, & Anderson, D (2002) Bacteriophage phi29 DNA packagings, *Adv. Virus Res.* **58**, 255.

Imai, K (1990) Precision determination and Adair scheme analysis of oxygen equilibrium curves of concentrated hemoglobin solution. A strict examination of Adair constant evaluation methods, *Biophys. Chem.* **37**, 197.

Jeong, H, Mason, SP, Barabasi, AL, & Oltvai, ZN (2001) Lethality and centrality in protein networks, *Nature* **411**, 41.

Klotz, IM (1997) Ligand–Receptor Energetics – A Guide for the Perplexed, John Wiley and Sons.

Maxwell, RW (2002) Maxwell Quick Medical Reference, 5th ed., Maxwell Publishing Company.

Menssen R, Neutzner A, & Seufert, W (2001) Asymmetric Spindle pole localization of yeast Cdc15 Kinase links mitotic exit and Cytokinesis, *Curr. Biol.* **11**(5), 345.

Novick, P, & Schekman, R (1979) Secretion and cell-surface growth are blocked in a temperature sensitive mutant of *Saccharomyces cerevisiae*, *Proc. Natl Acad. Sci. USA* **76**, 1858.

Nusslein-Volhard, C, Wieschaus, E, & Kluding, H (1984) Mutations affecting the pattern of the larval cuticle in *Drosophila melanogaster*. 1. Zygotic loci on the 2nd chromosome, *Roux's Arch. Dev. Biol.* **193**, 267.

Rossi-Fanelli, A, & Antonini, E (1958) Studies on the oxygen and carbon monoxide equilibria of human myoglobin, *Arch. Biochem. Biophys.* **77**, 478.

Sturtevant, AH (1913) The linear arrangement of six sex-linked factors in *Drosophila* as shown by their mode of association, *J. Exp. Zool.* **14**, 43.

Zhimulev, IF, Belyaeva, ES, Semeshin, VF, et al. (2004) Polytene chromosomes: 70 years of genetic research, *Int. Rev. Cytol.* **241**, 203.

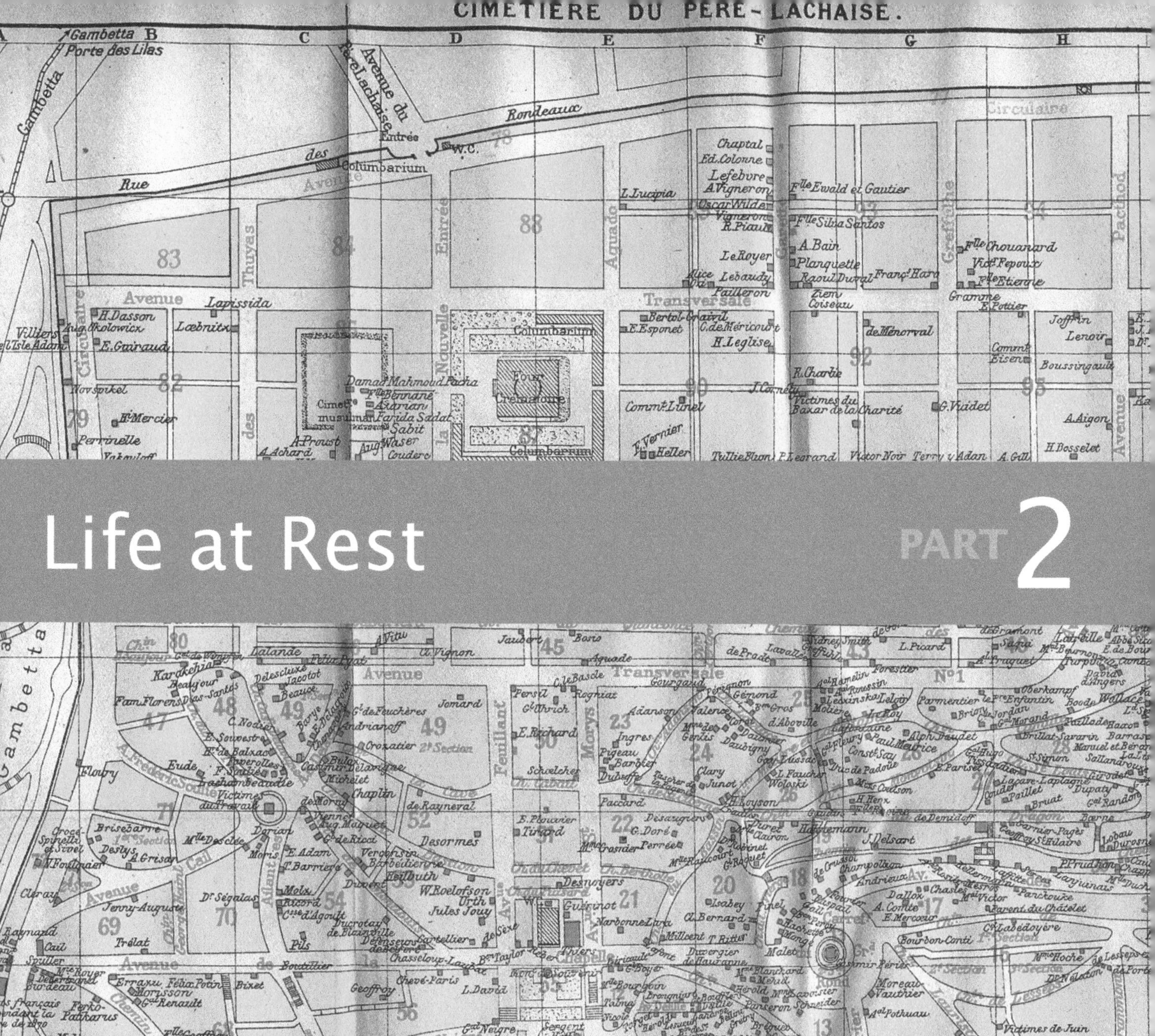

Life at Rest

PART 2

Mechanical and Chemical Equilibrium in the Living Cell

5

Overview: In which we examine how cells manage energy and how scientists compute energy transformations

> "An equation is worth a thousand words."
>
> **A physical biologist**

Chemical transformations that consume and liberate energy are one of the hallmarks of living systems. Living cells follow the same principles of conservation of matter and energy as do all other physical systems, though they also operate under an additional set of constraints imposed by their evolutionary history. In this chapter, we first summarize how cells manipulate and store chemical energy in ways that can be used to perform material transformations, such as macromolecular synthesis, mechanical work, such as muscle contraction, or even production of light energy, as in a firefly's abdomen. In order to develop the mathematical tools necessary to model these kinds of biological transformations, we exploit the useful simplification that many chemical and mechanical systems can be treated as if they are close to an equilibrium state. As we will see, many real-world situations can be surprisingly well modeled using equilibrium assumptions. This perspective alone is enough to provide useful insight into biological phenomena as fundamental and diverse as protein folding, binding reactions, and the formation of lipid bilayers. This useful oversimplification will form the substance of the next several chapters.

5.1 Energy and the Life of Cells

Much of the business of cellular life involves transformations of energy. Most familiar organisms make their living by eating other living or freshly dead things, thereby consuming energy-carrying organic molecules that have been generated and shepherded by other living organisms. This material transfer process makes all forms of life on Earth an interconnected and interdependent web where a key mode of communication is energy transfer. Humans consume food made up largely of fats, proteins, and carbohydrates, initially synthesized in other organisms, mostly plants, animals, and fungi. We use the molecules we consume not only to create material but also to fuel the energy-requiring processes of daily life including muscle contraction, heat generation, and brain activity.

Ultimately, it would not be possible for life to survive merely by recycling or exchanging energy among organisms—there must be an outside energy source. For most ecosystems on Earth, the ultimate energy source is sunlight which is harvested by various cells in plants, in many unicellular eukaryotes and in many kinds of bacteria. See Morton (2007) in "Further Reading" for a fascinating discussion of life and light. The light gathered by these cells serves not only their own energy needs, but eventually provides the energy for the remainder of the interdependent web of life. These cells exploit the energy of sunlight using an array of specialized molecules, with one of the outcomes being to transfer ions across membranes. The ion gradients store energy in a battery-like form that can then be coupled to enzymes that can, for example, convert CO_2 (and H_2O) from the air into sugar and indirectly into all other biomolecules, including proteins, lipids, and nucleic acids.

At least four key kinds of energy are relevant in biological systems: chemical energy, mechanical energy, electromagnetic energy, and thermal energy. Each of these forms of energy may be converted by living organisms into each of the others, with the interesting and important exception that thermal energy is generally a dead end since harnessing thermal (random) motions to carry out useful work is prohibited (the second law of thermodynamics). For example, electromagnetic energy in the form of a photon from the sun can be harnessed by a cell during photosynthesis to generate chemical energy that may be used for metabolic processes. Conversely, fireflies and other bioluminescent organisms convert chemical energy into photons. Energy gathered by organisms from their environments can be stored for later use, primarily in chemical form. Energy-storing molecules are used by all organisms for a host of important cellular processes. For example, ATP is used to pump molecules across membranes, to create the specialized polymeric apparatus of cellular motility, and to power the motors that allow our muscles to contract.

Throughout the book, we will invoke a wide range of different models to explore how these energy transformations take place. The main goal of this chapter is to demonstrate how the principles of energy minimization and free energy minimization can be used to predict the direction of transformations occurring in living systems. In considering energy minimization calculations, we will often make an explicit or implicit assumption that the system is operating close to equilibrium, such that any small excursion of the system will typically result in it returning to its original state. How do we reconcile the mathematically convenient equilibrium assumption with the real-world observation that biological systems are constantly dynamic and changing? The key insight is that different processes occur at different time scales, and so we can frequently isolate some small part of a biological process occurring at a relatively rapid time scale and pretend that it is at equilibrium with respect to its effects on processes that occur more slowly. In Chapters 15 and 19, we will explore this "quasiequilibrium" condition in the context of a few specific biological problems.

The next several chapters will build up the tools of equilibrium thermodynamics and statistical mechanics for treating equilibrium problems. Before embarking on our journey through the biological uses of statistical mechanics, we begin by taking stock of the interplay of thermal and deterministic forces (and energies) in biology and we examine the chemical basis for biological energy storage.

5.1.1 The Interplay of Deterministic and Thermal Forces

One of the important characteristics of the cellular interior that makes it so different from the world of our everyday experience is the fact that thermal and deterministic forces are on an equal footing. By thermal forces, we refer to the forces exerted on macromolecular structures as a result of the incessant jiggling of all of the molecules (such as water) that surround them. When we are considering the transformations that a biological system can undergo, it is useful to picture the range of available possibilities in terms of an energy landscape. For example, a protein may exist in a large number of possible conformations, but some will be energetically preferred over others. In any given biological system, the shape of the peaks and valleys on the free energy landscape can be changed by an energy input. For example, mechanical stretching of a membrane containing an ion channel will tend to make the open conformation of the channel more favorable relative to the closed conformation. At the same time, the rates at which molecules explore the energy landscape tend to be primarily determined by thermal forces. These different effects can be quantitatively related to one another through the use of common units.

Thermal Jostling of Particles Must Be Accounted for in Biological Systems

Perhaps the most famed example of thermal effects is that of Brownian motion (the microscopic basis of diffusion), already introduced in Chapter 3 (p. 127). Observation of small particles ($\approx 1\ \mu m$), fluorescently labeled molecules, and even macromolecules within cells reveals the fact that they suffer excursions that are, to all appearances, completely random. This jostling is a reflection of the fact that in addition to whatever deterministic forces might be applied to the particle or molecule of interest (such as electrostatic interactions, attachment to springs, etc.), they are also subjected to forces due to constant collisions with the molecules that make up the surrounding medium, which in turn constantly collide with one another.

To get a sense of the relative importance of thermal and deterministic forces, we need a numerical measure of the contribution of thermal effects. One way to compare the thermal and deterministic scales is through ratios of the form E_{det}/k_BT, where E_{det} represents the scale of deterministic energies in the problem of interest. For example, we might ask for the energy scale associated with breaking a hydrogen bond, or the energetic cost of bending a DNA molecule. In Chapter 6, we will show that the probability of a given "microstate" of a system is proportional to the Boltzmann factor, e^{-E_{det}/k_BT}, revealing the quantitative interplay of thermal and deterministic energies. The natural energy unit for a single molecule inside a cell is set by the thermal energy scale at room temperature, namely,

$$k_BT = 4.1\ \text{pN nm}. \tag{5.1}$$

We can see that this energy scale will be of central importance to the life and times of macromolecules such as sugars, lipids, proteins, and

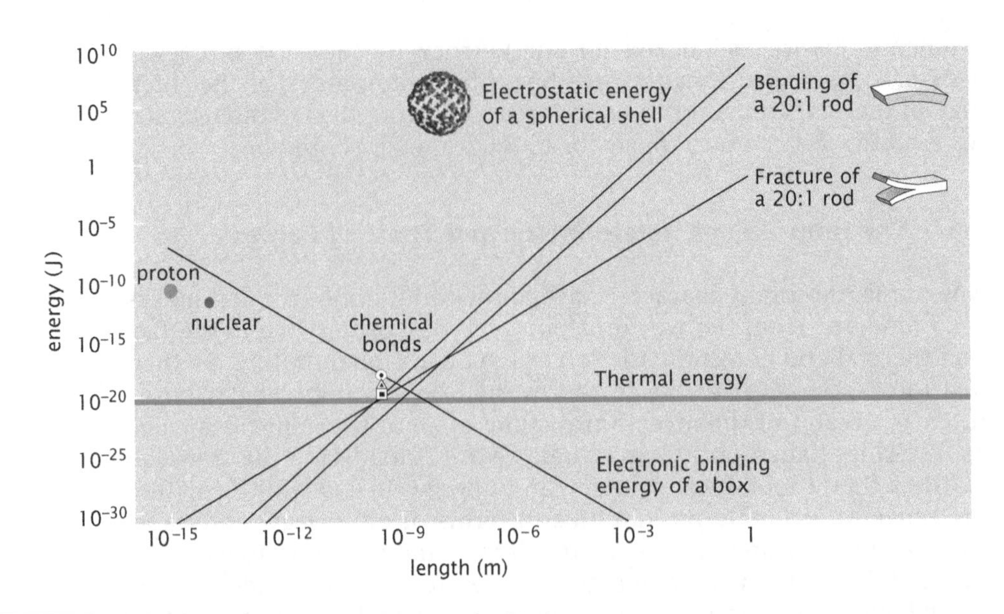

Figure 5.1: Energy as a function of length scale for a number of different energetic mechanisms. The graph shows how thermal, chemical, mechanical, and electrostatic energies associated with an object scale with the size of the object. As the characteristic object size approaches that of biological macromolecules, all of the energy scales converge to a single regime. The horizontal line shows the thermal energy scale. The bending energy is estimated by considering an elastic rod with an aspect ratio of 20:1 that is bent into a semicircular arc. The electrostatic energy is estimated for a model spherical protein with polar residues on its surface and for which all of the polar residues are stripped of a single charge (see Chapter 9). Chemical energy as a function of length, or binding energy, is estimated approximately by considering the effects of confining a free electron in a box of that length scale. For comparison, measured binding energies are shown for three chemical bonds (hydrogen bonds, phosphate groups in ATP, and covalent bonds). On this log–log scale they all appear very similar to one another at the point of convergence. (Adapted from R. Phillips and S. Quake, *Phys. Today* 59(5):38, 2006.)

nucleic acids because the energy delivered by ATP hydrolysis is tens of $k_B T$ and many of the motors that perform the functions of the cell exert piconewton forces over nanometer length scales. For other kinds of biological transformations, it is sometimes more useful to consider the thermal energy scale in different units. For example, for biochemical reactions, $k_B T = 0.6\,\text{kcal/mol}$ or $2.5\,\text{kJ/mol}$, and when considering thermal motions of charge, we will use $k_B T = 25\,\text{meV}$.

These ideas on the relative importance of thermal and deterministic forces are made more concrete in Figure 5.1. The horizontal line in the figure corresponds to the thermal energy scale, represented here as $k_B T = 4.1 \times 10^{-21}$ J. The other lines illustrate the energy cost associated with particular deterministic scenarios such as stripping a fraction of the charge off spheres of different size, bending rods of different sizes, and confining electrons within boxes of different sizes, which conveys a feeling for the energy scale of binding (which is also captured explicitly as the energy of hydrogen and van der Waals bonding in the figure). The key point of the figure is to note that at the *nanometer* scale (precisely the scale of the macromolecules of the cell), thermal energies and the deterministic energies of properties like charge rearrangement, bonding, and molecular rearrangement are comparable, unlike at the familiar centimeter or meter scales, where deterministic forces predominate.

Because each of these forms of energy is of comparable scale and effectively interchangeable at the molecular level, a living organism that needs to generate motion, heat, electricity, and biomolecular synthesis is expert at energetic interconversions. For the most part, energy used by living cells is derived from chemical energy in food and used to generate all the other forms. An interesting exception is photosynthesis, where electromagnetic energy is first converted into chemical energy and then into everything else. To develop a feeling for the numbers, we will now consider the molecular basis for generation and storage of chemical energy in cells.

5.1.2 Constructing the Cell: Managing the Mass and Energy Budget of the Cell

In Chapter 2 (p. 35), we estimated the number of all the different kinds of macromolecules in a cell and worked out the number of glucose

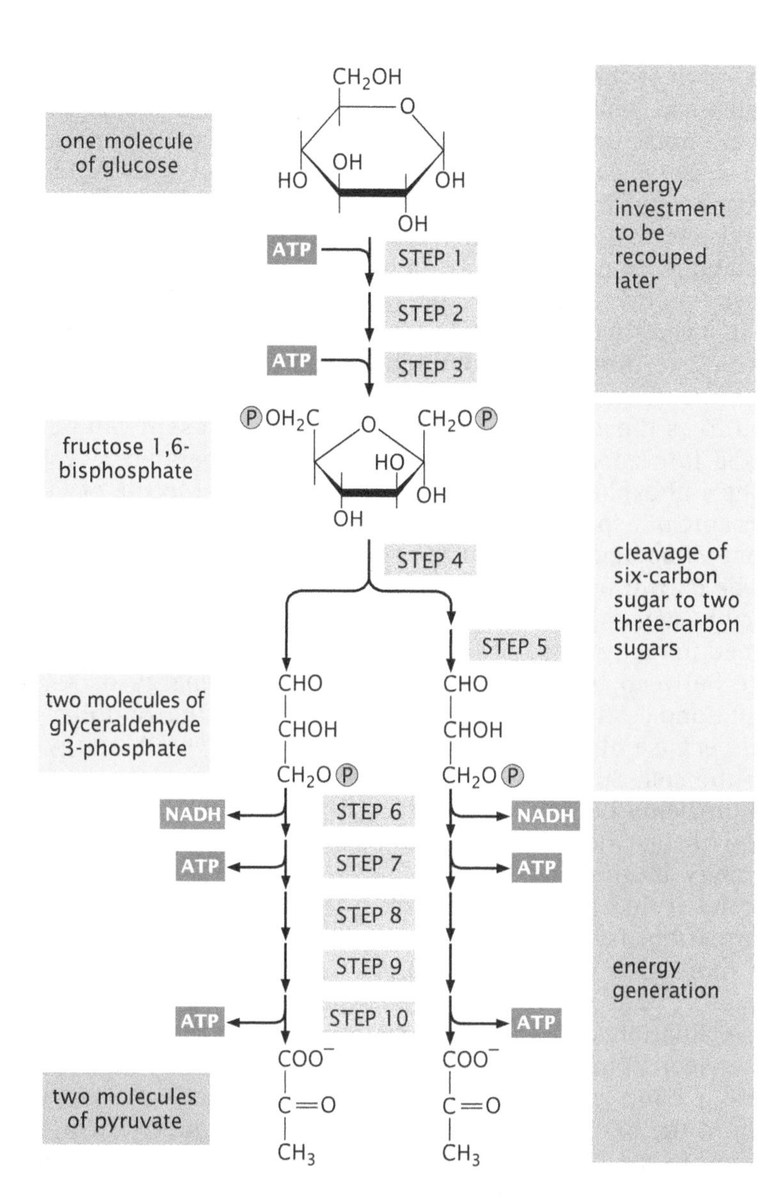

Figure 5.2: A schematic outlining the overall organization of the glycolytic pathway. The outcome of the 10 steps of glycolysis is the conversion of a single molecule of glucose into two molecules of pyruvate and the concomitant net production of two molecules of ATP and two of NADH. (Adapted from B. Alberts et al., Molecular Biology of the Cell, 5th ed. Garland Science, 2008.)

molecules required to build these constituents if glucose is the sole carbon source. What we neglected was the significant metabolic work that must be performed to transform the carbon atoms of glucose into the carbon atoms of amino acids, nucleotides, fatty acids, etc. Metabolism is the general term used to refer to cellular transformations of one molecule into another. The specific transformation of glucose into the amino acid lysine, for example, requires the ordered action of many enzymes, several of which consume energy while performing the necessary transformations. In living cells, energy is stored and transferred in several forms, most commonly in the form of a high-energy chemical bond on the molecule adenosine triphosphate (ATP; we will discuss cellular energy in more detail below). The ultimate source of the energy used to synthesize ATP in fact comes from metabolic breakdown of glucose in a pathway known as glycolysis and illustrated schematically in Figure 5.2. For *E. coli* growing in the presence of oxygen, a single molecule of glucose can be metabolically broken down to form up to 30 molecules of ATP from ADP since, in this case, the pyruvate emerging from the glycolysis pathway can be used to fuel further energy-producing reactions. This process results in carbon dioxide as a waste product. One interesting question is, what fraction of the glucose taken up by the cell is used to make new molecules and what fraction is used to provide the energy to make those new molecules? In this section, we will estimate the energy budget of a single *E. coli* proceeding through one round of its cell cycle.

In order to perform this estimate, we need to understand the nature of energy storage in cells, the typical amounts of energy required for metabolic transformations, and the ways in which cells allocate their energy and material resources. In most cells, energy is stored in a variety of forms that can be interconverted with very high efficiency. The three most commonly used are ATP, NADH (NADPH), and transmembrane H^+ gradients as shown in Figure 5.3. ATP is often referred to as the energy currency of the cell because it can be easily converted into goods and services. The energy liberated by hydrolysis of the γ-phosphate bond on ATP to generate a molecule of ADP and an inorganic phosphate ion P_i (PO_4^{2-}) is approximately 20 $k_B T$ (though the exact value depends upon the concentrations of ATP, ADP, and P_i, as will be shown in Section 6.4.4 on p. 274). ATP is a useful energy currency because this amount of energy is comparable to the energy consumed in many kinds of biochemical transformations and is intermediate between thermal energy ($k_B T$) and the energy of a typical covalent bond (150 $k_B T$). ATP can be considered as the 20 dollar bill of the cell because of its intermediate value in the overall energy economy of the cell. Spending money in large chunks such as 100 dollar bills is unwieldy because they are hard to break. On the other hand, paying with just dollar bills is a nuisance because it takes many of them to buy anything useful.

A second form of chemical energy that is important for metabolic transformations is carried in the form of easily transferrable electrons on the molecules NADH and NADPH. Many metabolic transformations require that the level of oxidation of an organic molecule be altered. Oxidation and reduction reactions involve the transfer of electrons between compounds. A compound is oxidized when electrons are removed and reduced when electrons are added. For organisms growing in the oxygen-rich environment of the modern Earth, oxidation reactions are usually spontaneous. However, reduction reactions require energy input. For reductive biosynthesis, a pair of electrons is usually donated by NADPH, creating an oxidized form of this carrier molecule $NADP^+$. Hence, NADPH gives up its hydride ion (H^-) in the same way that ATP gives up P_i, in both cases liberating energy for doing useful biochemical work.

Another use of reducing energy in cells is to establish H^+ gradients across membranes. This is an example of the third major form of biological energy storage. The electrical consequences of charge separation by transmembrane ion gradients will be the focus of Chapter 17. Ion gradients are easily interconvertible with either ATP energy or NADH energy. NADH can donate its high-energy electrons to electron carrier molecules in the plasma membrane of bacterial cells or in the inner mitochondrial membrane of eukaryotic cells that ultimately liberate H^+ ions on the opposite side, generating a gradient. The energy stored in this kind of ion gradient can be converted into ATP through the action of the enzyme F_1-F_0 ATP synthase. Energy release is effected here by letting ions flow across the membrane through transmembrane proteins, such as ATP synthase. This is analogous to the way a hydroelectric plant uses the kinetic energy of water supplied by gravity. Here the role of water is played by H^+ ions. However, instead of gravity, electrostatic and entropic forces drive the flow of ions and the ATP synthase plays the role of the turbine.

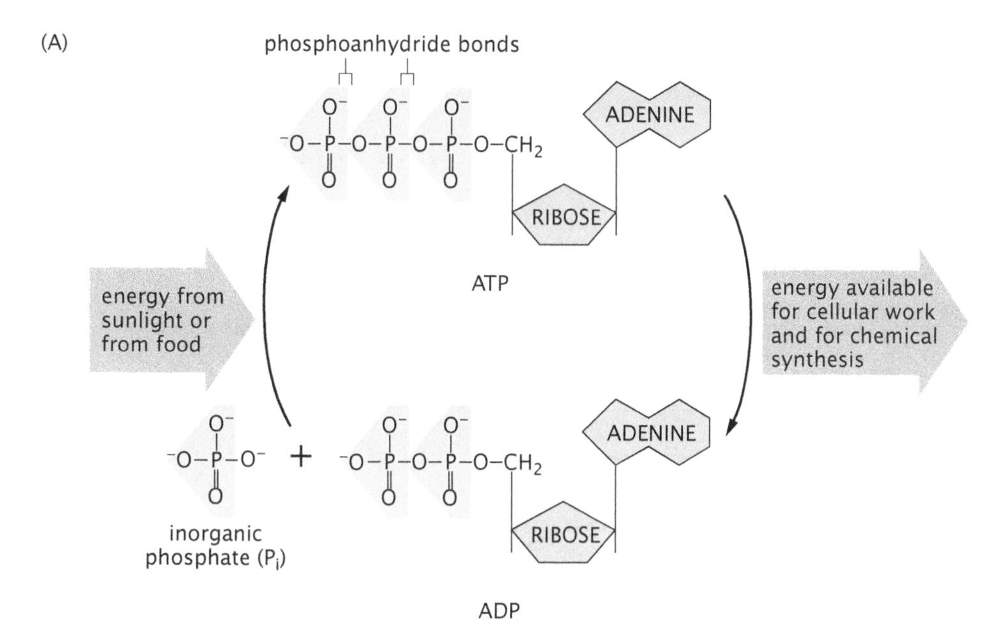

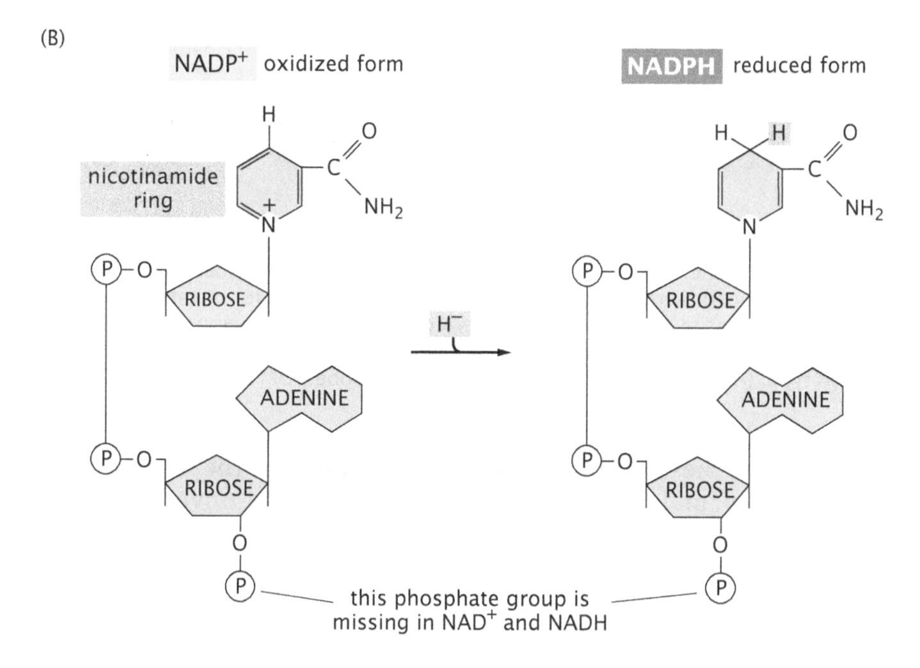

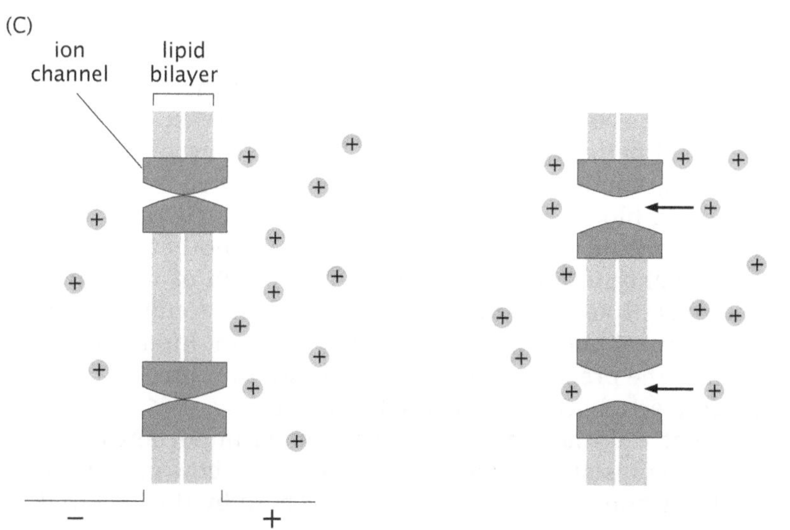

Figure 5.3: Three important forms of biological energy. (A) Energy for chemical synthesis and for force generation is stored in the form of ATP, which can be converted to ADP + P_i releasing roughly 20 k_BT of useful energy. ADP + P_i can then be converted back to ATP. While many enzymes use ATP itself, others use guanosine triphosphate (GTP), uridine triphosphate (UTP), or cytosine triphosphate (CTP), but the energies are equivalent. (B) Reducing potential is carried in the form of transferrable high-energy electrons on NADH (or the very similar molecule NADPH). Two electrons can be transferred from NADPH to reduce an oxidized organic compound, liberating one hydrogen ion (H^+) and the oxidized form of the carrier molecule $NADP^+$. In this case, the energy liberated by oxidation of one mole of NADH can be used to synthesize roughly two or three moles of ATP. (C) Transmembrane ion gradients, particularly in the form of H^+ gradients, are also used to store energy. The H^+ gradient across the membrane yields a negative potential on the left and a positive potential on the right. When ion channels open, the ions can flow down their electrochemical gradient. (A–C, adapted from B. Alberts et al., Molecular Biology of the Cell, 5th ed. Garland Science, 2008.)

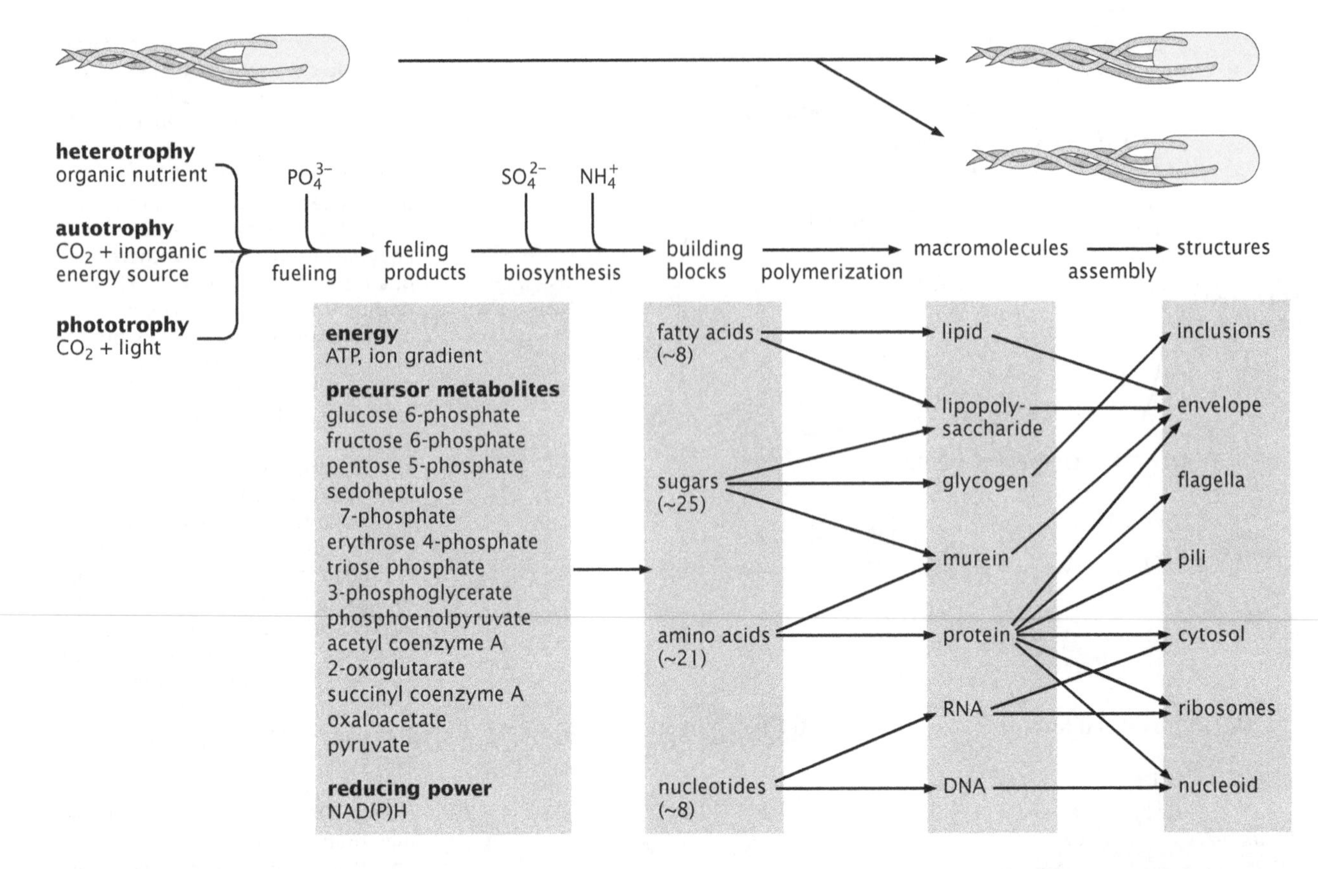

Figure 5.4: Energy and mass costs to make a new bacterial cell. This diagram illustrates the flow of materials and energy required for bacterial duplication. Nutrients are taken from the environment, either organic molecules provided by other organisms or carbon dioxide and light in the case of photosynthetic bacteria. Together with a few inorganic ions such as phosphate, sulfate, and ammonium, the carbon sources consumed by the bacterium are converted into precursor metabolites and then into the fatty acids, sugars, amino acids, and nucleotides that are used to build macromolecules. The macromolecules are further assembled into large-scale structures of the cell. The numbers shown in the "building blocks" column correspond to the rough number of molecular building blocks of each type. (Adapted from M. Schaechter et al., Microbe. ASM Press, 2006.)

How do all of these forms of energy contribute to the synthesis of the biological molecules that make up the cell? Conversion of a carbon source such as glucose into the carbon skeletons of any of the other necessary organic molecules (amino acids, nucleotides, phospholipids, etc.) proceeds through an intricate series of step-wise chemical transformations where metabolic enzymes catalyze the rearrangement of atoms within a substrate molecule, the cleavage of covalent chemical bonds in the substrate, and the formation of new covalent bonds. Figure 5.4 gives a schematic of the chain of reactions connecting the food source (for example, glucose) to the final product, namely, two cells. The product of one biochemical reaction in this kind of biochemical pathway goes on to be the substrate in the next reaction. In the diagram, this chain of reactions is denoted by "fueling products," "building blocks," "macromolecules," and "structures." All organic molecules within the cell are linked to one another through an intricate and highly interconnected network or web of metabolic reactions. The overall architecture of the network is dominated by the existence of critical nodes represented by important intermediate molecules (such as the precursor metabolites shown in Figure 5.4) that can in turn be converted into various final products.

A highly schematized diagram of one of the most important metabolic networks (glycolysis) of *E. coli* and other cells is shown in Figure 5.2. After the six-carbon molecule glucose is taken up by the bacterial

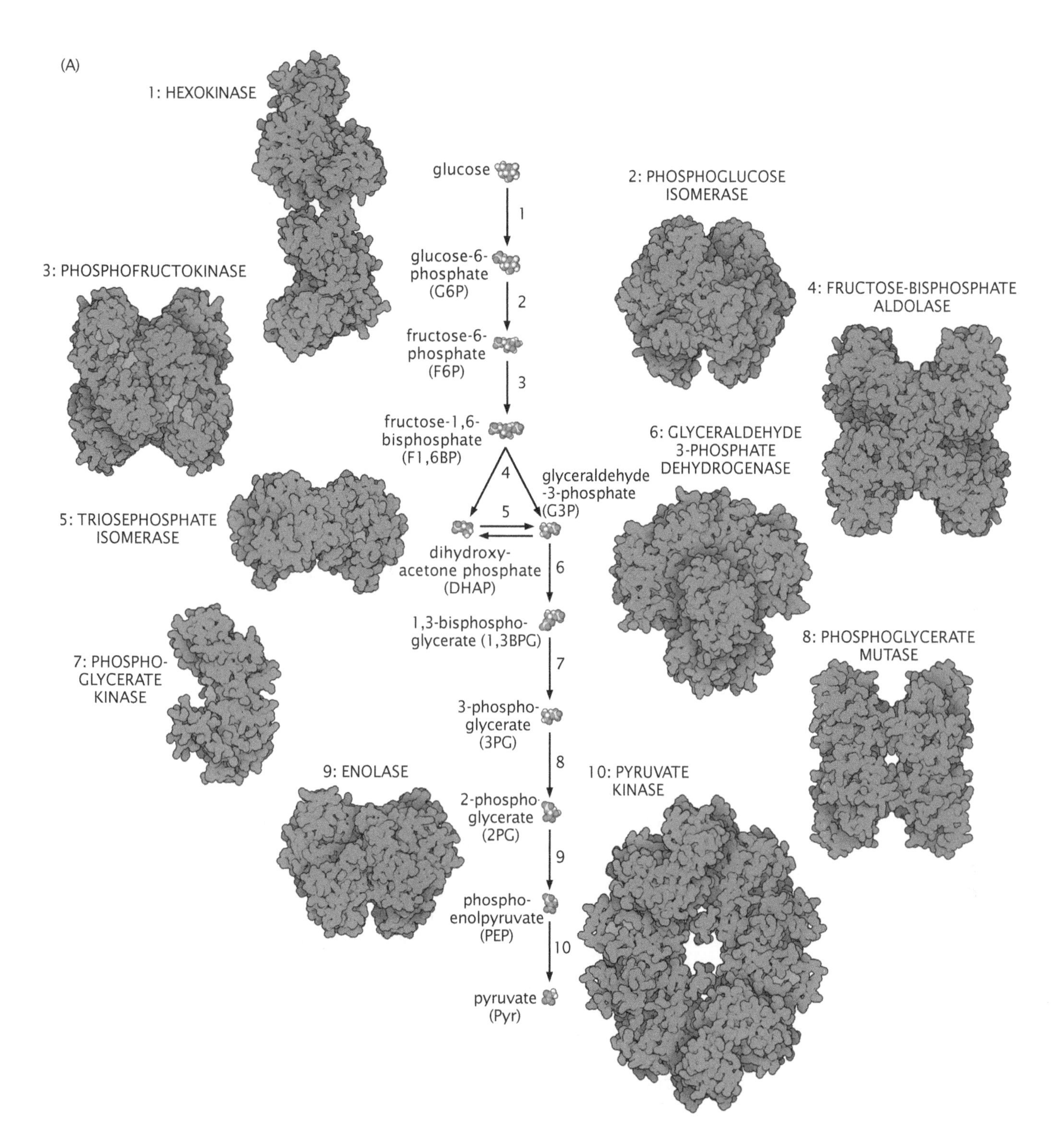

Figure 5.5: The molecules of the glycolytic pathway and the energy landscape for their transformations. (A) By a series of 10 chemical steps, one molecule of glucose is converted into two molecules of pyruvate. Each step is catalyzed by a specific enzyme, all of which are shown here as space-filling models. The enzymes are substantially larger than the small-molecule substrates on which they act. (B) The downward energetic progression of the glycolytic pathway is illustrated graphically, where each horizontal bar represents the relative energy level of one of the glycolytic intermediates. The approximate concentration of each intermediate in micromolar units (μM) is shown in parentheses. Overall, the transformation of glucose to pyruvate is extremely energetically favorable. Some of the energy liberated during each of these transformation steps is captured by the high-energy carrier molecules, ATP and NADH. Three of the steps in glycolysis have such large negative energy changes associated with them that they are considered irreversible: phosphorylation of glucose to glucose 6-phosphate, phosphorylation of fructose 6-phosphate to fructose 1,6-bisphosphate, and conversion of phosphoenolpyruvate to pyruvate with the concomitant synthesis of ATP. Many of the other steps take place with little net energy change. (A, courtesy of D. Goodsell, for PDB accession codes see website; B, adapted from C. K. Mathews et al., Biochemistry, Addison-Wesley Longman, 2000)

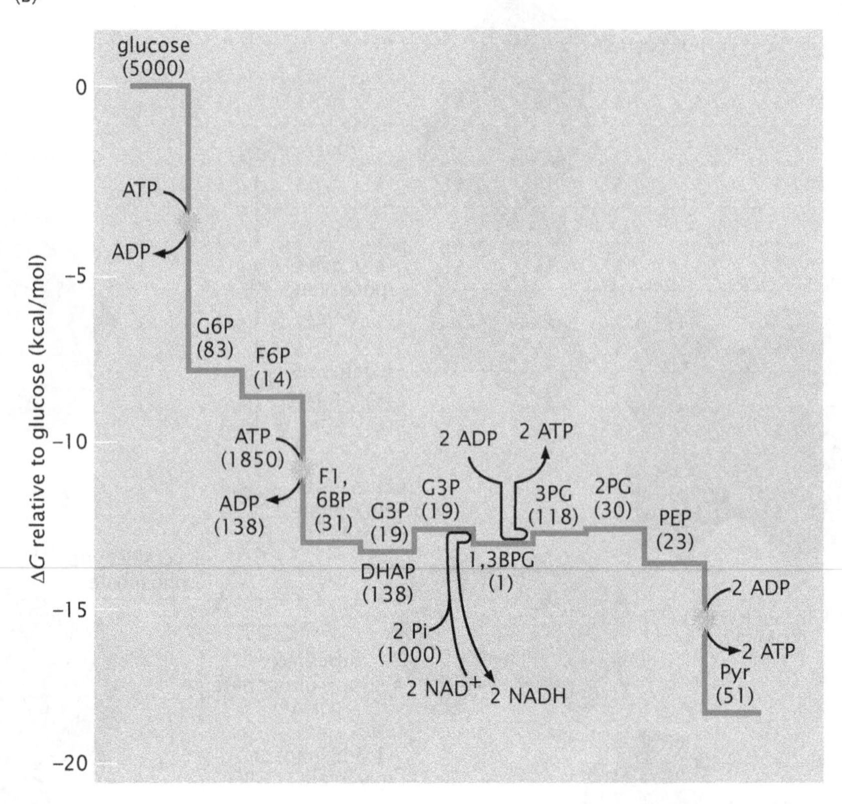

Figure 5.5: Continued.

cell, it is broken down by the process of glycolysis to form two copies of the three-carbon molecule pyruvate. This overall set of transformations takes place through 10 distinct chemical steps as shown at the molecular level in Figure 5.5. Pyruvate, in turn, can be used to synthesize a variety of amino acids or fatty acids. As glucose is broken down to form pyruvate, some of the chemical energy stored in its covalent bonds is used to synthesize ATP and NADH. These high-energy carrier molecules can then donate their energy to drive forward biosynthetic reactions that are not intrinsically energetically favorable.

A useful way to envision the energetic transformations during glycolysis is to picture each molecular species as having a characteristic energy and, as glucose goes through its series of transformations, the molecule travels up and down on a hilly energy landscape. This idea is related to the treatment of enzymes that we introduced in Figure 3.30 (p. 126). There we showed that the bent form of the substrate molecule resides at a slightly higher energy level than the straight form. For real molecules such as the intermediates in the glycolytic pathway, the energy of a particular species depends not only on its molecular structure but also on its concentration in the cell. We will consider the exact definition of these molecular energies in Chapter 6.

This general framework has prepared us to estimate the amount of glucose needed to provide energy for a single round of cell division compared with the amount required to provide structural building blocks. Accurate estimation of this number requires a detailed examination of each of the biosynthetic pathways in *E. coli* and tabulation of the energy consumed or liberated at every enzymatic step. Such calculations have been undertaken by brave biochemists. For our purposes, we will instead attempt a cruder but much simpler scheme in which we posit "typical" costs and gains associated with each class of molecule as counted in Chapter 2 (proteins, nucleic acids, etc.).

ESTIMATE

Estimate: The Energy Budget Required to Build a Cell Our estimates of the inventory of a cell given in Chapter 2 (p. 38) provided a feeling for the numbers of each kind of macromolecule needed to make a new cell. If the cell has glucose as its sole carbon source, the carbons in the sugar need to be taken apart and reassembled as useful building blocks such as amino acids and nucleotides that make the construction of the macromolecules of the cell possible.

The concept of the estimate we undertake here is represented by the analogy of considering the cost of constructing a building. Overall costs can be subdivided into the costs of the physical construction materials themselves and the cost of the labor required to put them together. In the cell, both the construction material (in the form of organic molecules) and the energy source ultimately are derived from nutrients taken up by the cell. As with our earlier estimates regarding *E. coli* in Chapters 2 and 3, we will consider cells growing in a medium where glucose is its sole source of carbon and biosynthetic energy. In Chapter 2, we estimated that the total number of carbon atoms required to construct a new *E. coli* cell is approximately 10^{10}. At 6 carbon atoms per glucose molecule, this means the cell must take on roughly 2×10^9 glucose molecules simply to provide the raw construction materials for doubling its mass so that it can divide. How many additional glucose molecules must the cell take up and convert in order to have the biosynthetic energy required to refashion all those carbon skeletons into cellular material?

In *E. coli*, there are seven major classes of macromolecular components whose synthesis we must consider: protein, DNA, RNA, phospholipid, lipopolysaccharide, peptidoglycan, and glycogen. Because each of these kinds of components involves its own elaborate biosynthesis pathways, we must consider them separately. Rather than going through all of them, we will start with the illustrative example of proteins, briefly discuss DNA and RNA, and then assert the final outcome of the energy budget calculation.

For biosynthesis of proteins when glucose is the sole carbon source, the glucose carbon skeletons must first be converted into amino acids, and then those amino acids must be polymerized to form new proteins. As can be easily appreciated by a glance at Figure 2.33 (p. 71), amino acids vary significantly in their structure and some are more complicated to synthesize than others. Over the past 100 years, the metabolic pathways for synthesis of each of these amino acids have been determined and the responsible enzymes identified using methods such as the pulse–chase method (see Figure 3.3 on p. 93) and generation of auxotrophic mutants, which need to be fed precursor molecules to survive. All of the amino acid synthetic pathways are connected directly or indirectly to the glycolytic pathway shown in Figure 5.5. Indeed, all metabolites in the cell are connected to all others through the elaborate metabolic web, which can be graphically represented in a summary diagram that covers most of a wall and that resembles the

Tokyo subway map but is substantially more intricate. Alanine, for example, can be synthesized from pyruvate in a single step, by a single enzyme. Tryptophan, in contrast, requires the coordinated action of 12 enzymes. The net synthesis cost for making each of the amino acids is summarized in Table 5.1. For purposes of calculating the energy budget, we must also take into account the fact that some amino acids are much more abundant than others. For example, glycine is approximately 10-fold more abundant than tryptophan. By multiplying the energetic cost of making each amino acid by its relative abundance in the cell, we can estimate that the average energetic cost to synthesize an amino acid is roughly 1.2 ATP equivalents for cells growing aerobically and 4.7 ATP equivalents for cells growing anaerobically.

After the amino acids have been synthesized, they must be strung together to make proteins. This painstaking assembly work requires a large input energy to the tune of roughly 4 ATP equivalents for each amino acid, including the cost to attach the amino acids to tRNAs and to power the movement of the ribosome. As a result, the cost of adding each amino acid is 5.2 ATP equivalents corresponding to the 1.2 ATP equivalents

Table 5.1: Amino acid abundance and cost of synthesis for making the amino acids under both aerobic and anaerobic growth conditions. "Glucose equivalents" refers to the number of glucose molecules that must be used to generate the carbon skeletons of each amino acid (for example, one mole of alanine, an amino acid containing three carbons, can be synthesized from one half mole of glucose, a sugar containing six carbons). "ATP equivalents" refers to the approximate amount of biosynthetic energy required to synthesize the amino acid from glucose as a starting material. A negative value indicates that synthesis of the amino acid from glucose is favorable so energy is generated rather than consumed. These numbers are not absolute; they depend on several assumptions about metabolic energetics and pathway utilization. We have assumed that the energetic value of one molecule of NADH or NADPH is equivalent to two ATP molecules. We have not accounted for the biosynthetic cost of sulfate, ammonium, or single carbon units. (Data from F. C. Neidhardt et al., Physiology of the Bacterial Cell, Sinauer Associates, 1990; M. Schaechter et al., Microbe, ASM Press, 2006; and EcoCyc, Encyclopedia of *Escherichia coli* K-12 Genes and Metabolism, www.ecocyc.org.)

Amino acid	Abundance (molecules per cell)	Glucose equivalents	ATP equivalents (aerobic)	ATP equivalents (anaerobic)
Alanine (A)	2.9×10^8	0.5	−1	1
Arginine (R)	1.7×10^8	0.5	5	13
Asparagine (N)	1.4×10^8	0.5	3	5
Aspartate (D)	1.4×10^8	0.5	0	2
Cysteine (C)	5.2×10^7	0.5	11	15
Glutamate (E)	1.5×10^8	0.5	−7	−1
Glutamine (Q)	1.5×10^8	0.5	−6	0
Glycine (G)	3.5×10^8	0.5	−2	2
Histidine (H)	5.4×10^7	1	1	7
Isoleucine (I)	1.7×10^8	1	7	11
Leucine (L)	2.6×10^8	1.5	−9	1
Lysine (K)	2.0×10^8	1	5	9
Methionine (M)	8.8×10^7	1	21	23
Phenylalanine (F)	1.1×10^8	2	−6	2
Proline (P)	1.3×10^8	0.5	−2	4
Serine (S)	1.2×10^8	0.5	−2	2
Threonine (T)	1.5×10^8	0.5	6	8
Tryptophan (W)	3.3×10^7	2.5	−7	7
Tyrosine (Y)	7.9×10^7	2	−8	2
Valine (V)	2.4×10^8	1	−2	2

it costs to make the average amino acid and the 4 ATP equivalents it takes to add the amino acid onto the peptide chain. Multiplying this by the total number of amino acids that need to be strung together to make a cell, we find

$$\text{protein energy cost} \approx 5.2\,\text{ATP} \times 300 \times 3 \times 10^6$$
$$\approx 4.5 \times 10^9\,\text{ATP equivalents.} \quad (5.2)$$

We have taken 300 as the number of amino acids in the "average" protein, and used an approximate number of 3×10^6 proteins per bacterium.

It is possible to perform similar calculations for each of the other six classes of macromolecules. Interestingly, while we found for proteins that synthesis of the amino acid precursors is relatively energetically inexpensive and assembly into proteins relatively costly, the situation is the opposite for DNA and RNA. Here, the energy required to synthesize a nucleotide triphosphate precursor is large, of the order of 10–20 ATP equivalents depending upon growth conditions, but the additional cost required to assemble the polymers is small. Whereas amino acid synthesis consumed less than a quarter of the total energy required to make proteins, nucleotide synthesis requires nearly 90% of the energy needed to make nucleic acids.

Table 5.2 summarizes the biosynthetic cost for each of the major classes of macromolecule. As noted above, the exact numbers vary depending upon growth conditions, but these serve as a reasonable estimate for our standard *E. coli* growing under standard conditions. Recall that an *E. coli* cell must take up roughly 2×10^9 glucose molecules for building materials to double its mass. Growing with maximum efficiency under aerobic conditions, a single molecule of glucose can generate up to 30 molecules of ATP, with carbon dioxide as the waste product. Comparing the total amount of biosynthetic energy required by adding up all of the components in Table 5.2, about 10^{10} ATP equivalents are required, or about 3×10^8 molecules of glucose. Thus, it requires about one-fifth as much glucose just to pay for labor as it does to provide the actual building materials for constructing a new cell. Under less efficient growth conditions, the cost of biosynthesis can actually exceed the cost of materials by as much as 10-fold. Furthermore, in the estimates above, we have ignored the fact that macromolecules are constantly degraded and replenished inside the cell. This will surely increase the overall energy budget, but will not affect our estimates by more than an order of magnitude.

Table 5.2: Biosynthetic cost in ATP equivalents to synthesize the macromolecules of a single *E. coli* cell.

Class	Biosynthetic cost (aerobic) – ATP equivalent
Protein	4.5×10^9
DNA	3.5×10^8
RNA	1.6×10^9
Phospholipid	3.2×10^9
Lipopolysaccharide	3.8×10^8
Peptidoglycan	1.7×10^8
Glycogen	3.1×10^7

We have seen that much of the useful energy available to cells is stored as the energy of phosphate bonds. This energy can be released in a variety of different ways for processes such as those associated with the central dogma, for cell motility, and for setting up ion gradients across cells. One of the tools we will use to examine these transformations of energy is the theory of mechanical and chemical equilibrium as embodied in the laws of thermodynamics and statistical mechanics. That is, the calculus of equilibrium to be set up in this and the following chapter is an abstract tool that permits us to predict the direction and extent of important energy transactions in the cell. With a sense of the energy scales associated with important biological transformations now in hand, we turn to an analysis of the mathematical tools used to characterize these transformations.

5.2 Biological Systems as Minimizers

In the previous section, we considered the energetics of macromolecular synthesis, only one of many activities undertaken by busy cells. In a more general sense, our consideration of the role of energy and energy transformations in the processes of life can also be applied to many other kinds of problems beyond biosynthesis. What determines the shape of a red blood cell? Given a particular oxygen partial pressure in the lungs, what is the fractional binding occupancy of the hemoglobin within red blood cells? How much force is required to package the DNA within the capsid of a bacteriophage? What fraction of Lac repressor molecules in an *E. coli* cell are bound to DNA? Each of these questions is ultimately about energy transactions and can be couched in the form of a minimization problem in which we seek the least value of some function. For example, as will be shown in Chapter 11, the question of the shape of red blood cells will be formulated mathematically as the problem of minimizing the free energy of the membrane and associated architectural filaments that bound the cell. Similarly, our discussion of chemical equilibrium and equilibrium constants for problems ranging from the occupancy of hemoglobin by oxygen to the binding of Lac repressor to DNA will be founded upon equality of chemical potentials, which is a simple consequence of minimizing the free energy. As we will see, questions in both mechanical and chemical equilibrium can be stated in the language of minimization principles.

In the remainder of this chapter, we commence our efforts to develop mathematical models of biological energy transformations viewed through the prism of minimization principles. A complementary view of transformations will be developed in Chapter 15, when we explicitly consider rates and dynamics. The key point of the present discussion is how our understanding of the equilibrium configurations of systems ranging from DNA–protein complexes to bones under stress can be built around the idea of minimizing an appropriate energy quantity. The quest to develop this intuition will lead us to the mathematics of the calculus of variations and will culminate in the elucidation of Gibbs' calculus of equilibrium in the form of the principle of minimum free energy.

5.2.1 Equilibrium Models for Out of Equilibrium Systems

Given that living organisms are quintessential examples of systems that are out of equilibrium, it is natural to ask to what extent the tools of equilibrium physics are of any use in biology. Perhaps surprisingly, in fact there is a wealth of examples where the use of equilibrium ideas is well justified.

Equilibrium Models Can Be Used for Nonequilibrium Problems if Certain Processes Happen Much Faster Than Others

The decision of whether an equilibrium description is appropriate for a given problem often comes down to a question of time scales. As the simplest example, we examine the validity of treating a cell as though it is in mechanical equilibrium. Mechanical equilibrium is characterized by the absence of any unbalanced forces in a system. However, a more nuanced description of mechanical equilibrium appropriate for some biological problems is the idea that all of the forces in the system are balanced on the time scales at which the biological process is taking place. For example, as a cell crawls across a surface, the cytoskeleton is pushing on parts of the plasma membrane. In some cases, the response of the membrane can be thought of as so fast on the time scale of the underlying cytoskeletal dynamics that at every instant the membrane has equilibrated mechanically with respect to the forces produced by the cytoskeleton.

Similar arguments apply in the case of chemical equilibrium. For concreteness, consider the reaction,

$$\mathrm{A} \underset{k_-}{\overset{k_+}{\rightleftharpoons}} \mathrm{B} \overset{r}{\longrightarrow} \mathrm{C} \tag{5.3}$$

where we have assumed for simplicity that the backwards reaction from C to B has a negligible rate (this approximation is useful for thinking about processes such as transcription and translation). The basic argument being made verbally in this section and mathematically later (in Chapter 15) is that if the rates associated with the conversion between A and B are sufficiently fast in comparison with the rate at which B is depleted as a result of conversion into C, then we can think of the reaction $\mathrm{A} \rightleftharpoons \mathrm{B}$ as being in equilibrium. The concrete signature of this rapid preequilibrium is that the amounts of A and B occur in a fixed ratio determined by the ratio of the forward and backward rates for the reaction.

The outcome of this kind of analysis is shown in Figure 5.6. The key point of the calculation is embodied in the fact that after an initial transient period, the ratio [A]/[B] (we use the notation [A] to mean "concentration of A") is constant for all subsequent times even though the absolute number of A and B molecules is decreasing over time. This fixed ratio is the equilibrium constant for the reaction $\mathrm{A} \rightleftharpoons \mathrm{B}$. If the rate of conversion to the product C is too fast, the rapid preequilibrium condition is no longer satisfied, yielding a situation like that shown in Figure 5.6(B).

As will become clear in subsequent chapters, there are many cases in which the *numbers* associated with various kinetic processes justify the use of equilibrium arguments like those to be developed in this chapter. The mindset that justifies this approach is one of time scales; namely, when the rate constants for some initial reaction in a series of reactions are fast (in a way that can be evaluated mathematically), then that reaction can be treated as an equilibrium reaction.

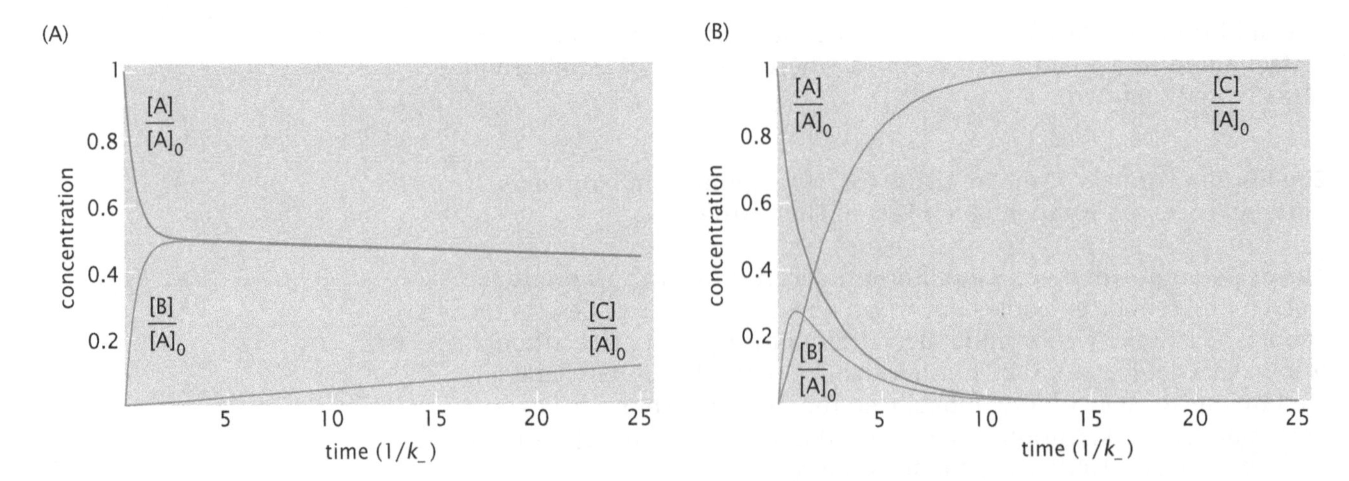

Figure 5.6: Rapid approach to equilibrium of a subprocess. Plot of the time-dependence of the concentrations in the reaction $A \rightleftharpoons B \rightarrow C$ of $A(t)$, $B(t)$ and $C(t)$ described by Equation 5.3. (A) For the case in which the rate for converting B to C is slow in comparison with the rates for the reaction between A and B, after an initial transient period, A and B reach their equilibrium values relative to each other for the remainder of the process. (B) Plot showing the case in which there is no rapid preequilibrium. Time is shown in nondimensional units by expressing it in units of the inverse of the rate k_{-}.

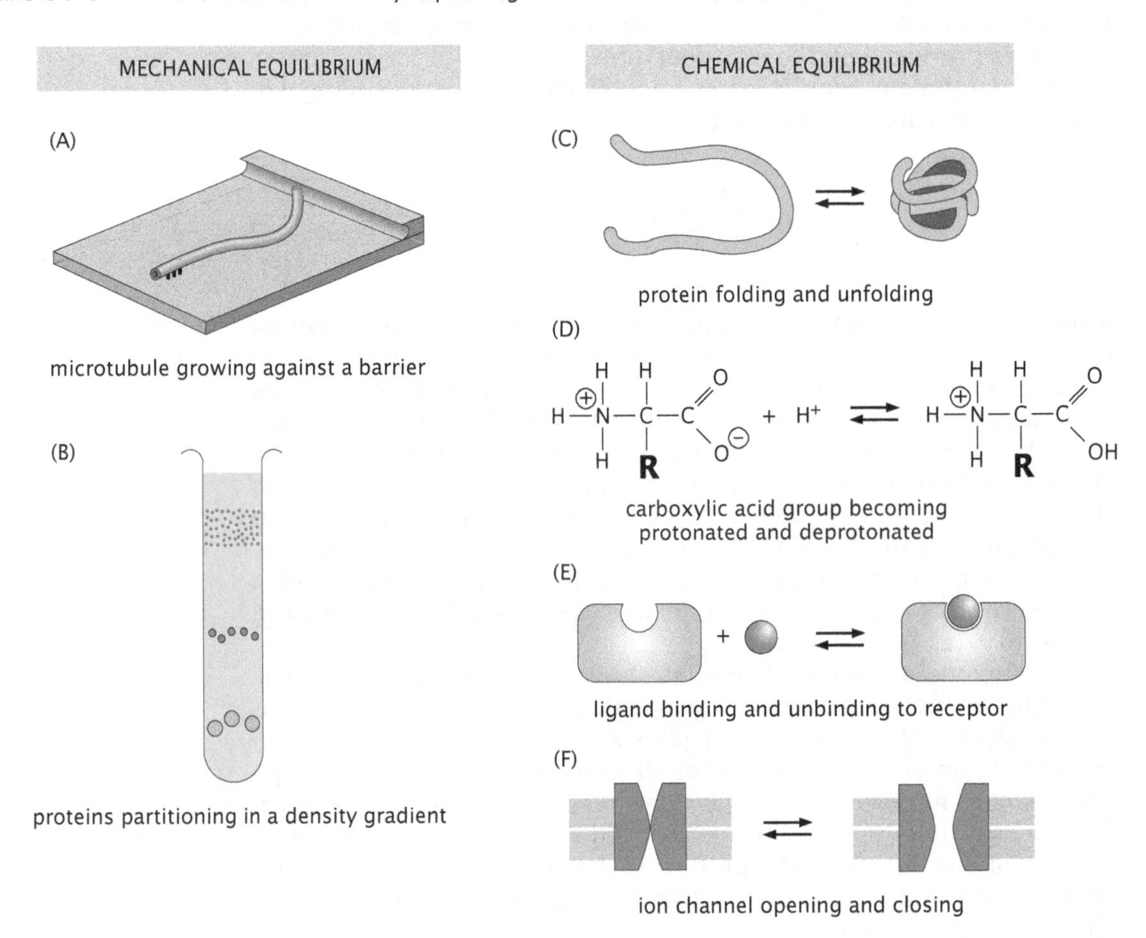

Figure 5.7: Proteins in equilibrium. Schematic showing many examples of the way in which proteins are approximated as being in equilibrium.

5.2.2 Proteins in "Equilibrium"

To set the stage concretely for some of the ways in which we will invoke equilibrium models to think about problems of biological interest, Figure 5.7 shows some examples where we treat proteins from an equilibrium perspective. In some instances (Figures 5.7A,B), our analysis can be built strictly around the notion of mechanical equilibrium. The examples of chemical equilibrium begin with the claim that it is useful to think of the folded state of a protein as a free-energy minimizer. The next example of protein properties from an equilibrium perspective is the treatment of the way in which the charge state of a protein depends upon the pH of the solution. Here the idea is that the charge state of the protein reflects a competition between the entropy gained by permitting charges to wander in solution and the corresponding energy cost associated with removing those charges from their protein host. Yet another example that will arise repeatedly throughout the book is the treatment of binding, where in the case of a protein we can think of it as being complexed with some ligand of interest. Here too, the basic picture is one of an interplay between the entropy associated with free ligands and the energetic gain they garner as a result of being bound to their protein host. A final example where at times it is convenient to think of a protein as being in equilibrium is when that protein coexists in an active and an inactive form and where the relative probability of these states is dictated by some external influence. For example, as will be discussed in Chapter 7, phosphorylation of a protein can shift it from an inactive to an active state. A second example, also to be examined in Chapter 7, is the gating of ion channels. Here too, channel gating can sometimes be treated as an equilibrium problem where some tuning parameter such as the external tension in the membrane or an applied voltage can alter the probability that the channel is open.

Protein Structures are Free-Energy Minimizers

As a result of the sequencing of an ever-increasing number of genomes, the challenge to assign meaning to that genomic information has also increased. In particular, with the genetic sequence in hand, what can be said about the structure and function of the various proteins coded for in these genomes? Assuming that a particular gene within a genome has been identified, the question can be posed differently. We have already seen in Figure 1.4 (p. 10) that the languages of nucleic acids and proteins are related by the universal genetic code that tells us how to translate the DNA sequence into a corresponding amino acid *sequence*. However, once the relevant amino acid sequence has been determined, we still do not know the structure implied by that given primary sequence.

A first step in solving this problem corresponds to answering the question: of all of the possible ways that that particular chain of amino acids can fold up, which has the lowest free energy? From an intuitive perspective, we already possess heuristic ideas for thinking about protein folding as illustrated in Figure 5.8. In particular, certain amino acid side chains can happily participate in the hydrogen bonding network of the surrounding aqueous solution, while those residues with hydrophobic side chains are sequestered from the surrounding solution. From a quantitative perspective, these structural preferences have a corresponding free-energy benefit.

A second way in which proteins are conveniently viewed from the equilibrium perspective has to do with their charge state. As the pH of the solution is varied, the charge on different amino acid residues in a particular polypeptide chain will vary. An example from the amino acid glycine is shown in Figure 5.9. We can think about the liberation of charge in solution as a result of the competition between the energetic favorability of keeping unlike charges near to each other and the entropic benefit of letting the charges stray from their protein

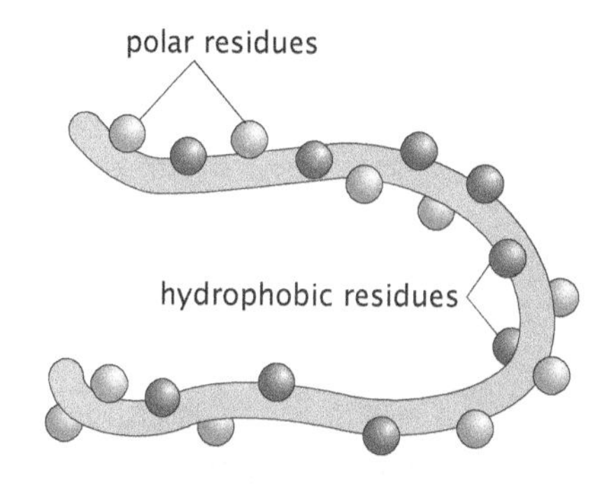

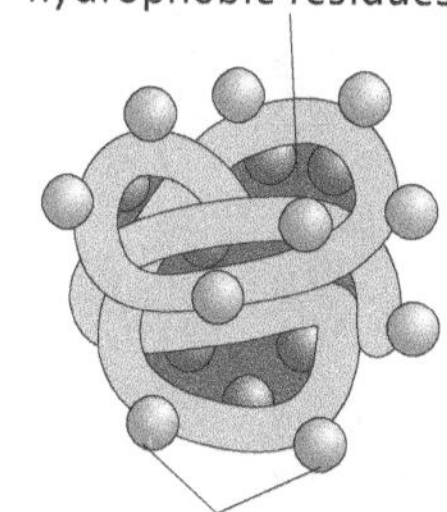

Figure 5.8: Schematic of the way in which protein folding sequesters hydrophobic amino acids while leaving their polar counterparts in contact with the surrounding solution.

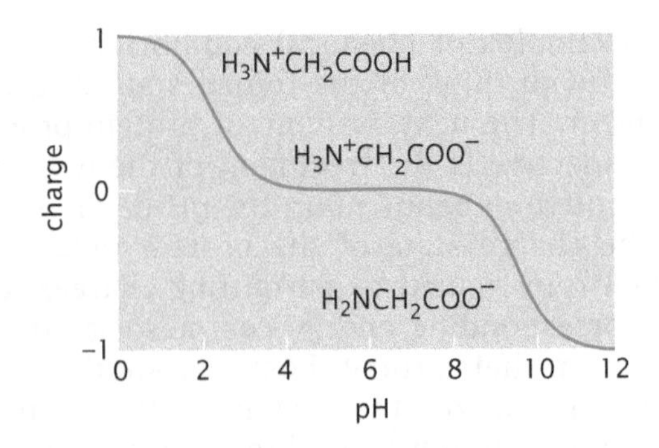

Figure 5.9: Titration curve showing the charge state of the amino acid glycine as a function of the pH of the solution. Low pH corresponds to a high concentration of H^+ ligands, resulting in saturation of the glycines. (Adapted from K. Dill and S. Bromberg, Molecular Driving Forces, 2nd ed. Garland Science, 2011.)

host. The reason for bringing up these protein examples is to highlight the way in which equilibrium ideas are often a starting point for the analysis of important biological problems.

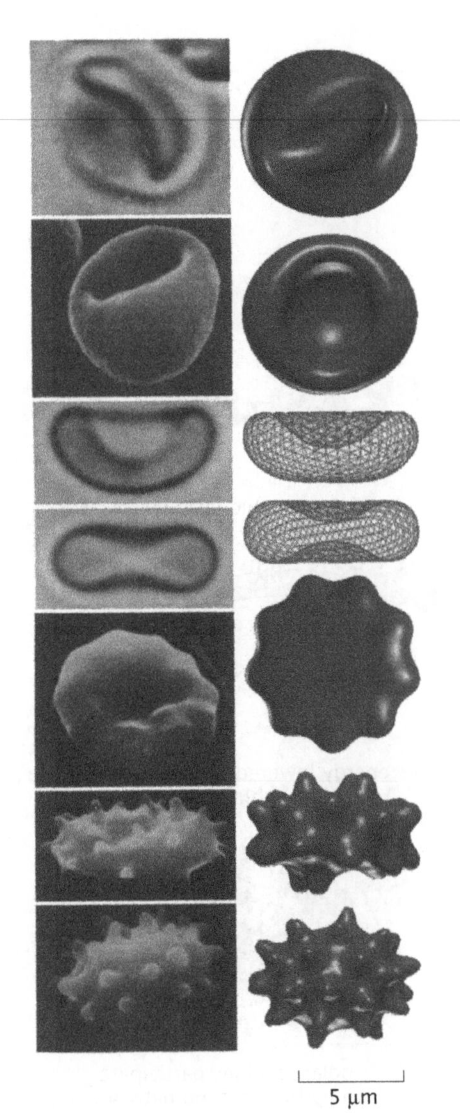

Figure 5.10: Red blood cell shapes. The left-hand column shows shapes of red blood cells as observed experimentally and the right-hand column shows calculations of the shapes. (Adapted from G. Lim et al., *Proc. Natl Acad. Sci. USA* 99:16766, 2002.)

5.2.3 Cells in "Equilibrium"

We have seen that there are many circumstances in which the molecules and macromolecular assemblies of the cell can be viewed from an equilibrium perspective. At the larger scales representative of cells themselves, there are many cases in which we can consider some particular part of the cell (such as the membrane) as being in local mechanical or chemical equilibrium. One example of this kind of thinking is that of the equilibrium shapes of red blood cells. As shown in Figure 5.10, the shapes of such cells have been precisely characterized experimentally and can similarly be calculated.

As will be introduced in the remainder of this chapter and driven home as a key part of the rest of the book, in problems of free-energy minimization there are two key steps: first, the selection of a class of competitors, and, second, the determination of the free energy associated with each such competitor. In the setting of red blood cells of interest here, the class of competitors is the set of all shapes satisfying two geometric constraints, namely, that the overall area of the red blood cell surface be the same from one shape to the next and also that the volume enclosed by that area be the same. Figure 5.10 shows in the right-hand panel the shapes that have the lowest free energy for different choices of a control parameter that is the difference in area between the two leaflets of the membrane.

5.2.4 Mechanical Equilibrium from a Minimization Perspective

As argued above, there are a variety of different biologically interesting examples that, when examined in physical terms, amount to problems in minimization. One class of problems that can be thought of in this way centers on mechanical equilibrium.

The Mechanical Equilibrium State is Obtained by Minimizing the Potential Energy

One way to think about the mechanics of bodies at rest is Newton's first law of motion: namely, that if a body is in equilibrium, then there are no unbalanced forces on that body. Stated mathematically, the condition of translational equilibrium is

$$\sum_i \mathbf{F}_i = 0, \tag{5.4}$$

where $\mathbf{F}_i$ is the ith force acting on the body. The use of the bold-face letter in writing the forces $\mathbf{F}_i$ reflects the fact that the force is a vector quantity. For example, if we consider the hook shown in Figure 5.11, there is a force acting on that hook due to the spring and a second force due to the hanging weight and these forces balance each other at equilibrium. Their force vectors have the same length and point in opposite directions. However, it is not always most convenient or enlightening to consider equilibrium problems in the vectorial language of forces. The alternative that will often be favored throughout the remainder of the book is the equivalent formulation of the problem of mechanical equilibrium as one of minimization. The principle of minimum potential energy asserts that the mechanical state of equilibrium is the one (out of all of the possible alternatives) that has the lowest potential energy.

To write the equilibrium of the system shown in Figure 5.11 in terms of energy, we can write the potential energy as the sum of two terms, one of which captures the energy of the stretched spring while the other describes the "loading device," namely, the lowering of the weight. Given these concepts, the potential energy can be written as

$$U(x) = \underbrace{\tfrac{1}{2}k(x - x_0)^2}_{\text{PE of spring}} - \underbrace{mg(x - x_0)}_{\text{PE of weight}}, \tag{5.5}$$

where x_0 is the length of the spring when it is unstretched and will also serve as our zero point for the potential energy of the hanging weight. We use the label "PE" to denote potential energy. These two terms are shown in Figure 5.11(B) and we see that their sum has a minimum (that is, the equilibrium point). To find the point x_{eq} at which the minimum occurs, we note that at x_{eq}, the slope of the function $U(x)$ is zero—this condition corresponds to the mathematical statement $dU/dx = 0$, as will be shown in more detail below. Minimization of the potential energy in this case corresponds physically to finding that choice of the displacement x_{eq} that leads to the lowest energy and is determined by the condition

$$\frac{dU}{dx} = k(x_{eq} - x_0) - mg = 0. \tag{5.6}$$

This result can be rewritten as

$$x_{eq} = x_0 + \frac{mg}{k}, \tag{5.7}$$

which tells us the size of the excursion made by the spring away from its equilibrium position. The result jibes with our intuition in the sense that heavier weights (mg large) lead to larger excursions and a stiffer spring (k large) results in a smaller excursion.

The idea of the potential energy of the loading device (analogous to the weight) introduced above is pervasive and will be used repeatedly in the book. As shown in Figure 5.12, we will think about the energy associated with deforming cantilevers, such as in the atomic-force microscope, polymers, and membranes. In all of these cases, when we write down the total energy (or free energy) of the system, we will have to account for the way in which the deformation of our system of interest (that is, the polymer or membrane) leads to an attendant change in the energy of the loading device, as depicted here by the lowering of a weight.

As noted as early as Figure 1.12 (p. 22), "springs" show up in a surprising variety of circumstances. One example that we are particularly

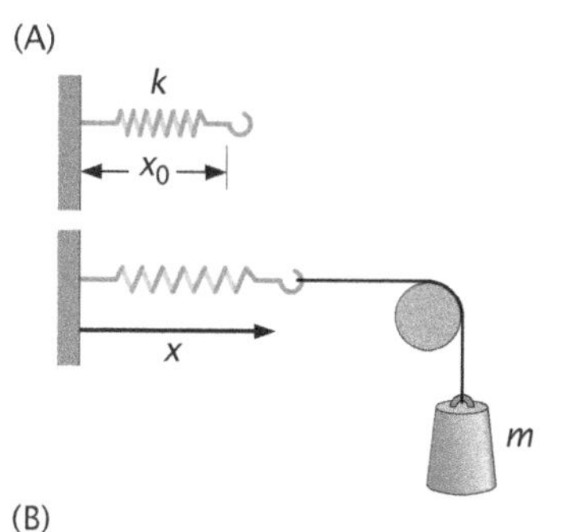

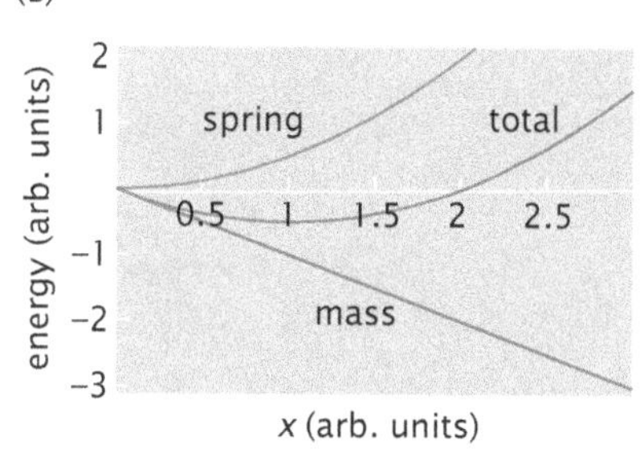

Figure 5.11: Mechanical equilibrium as potential-energy minimization. (A) Schematic showing how the mechanical equilibrium of a system can be thought of from the point of view of minimization of the potential energy. (B) Potential energy of the spring and the weight and their sum as a function of the displacement.

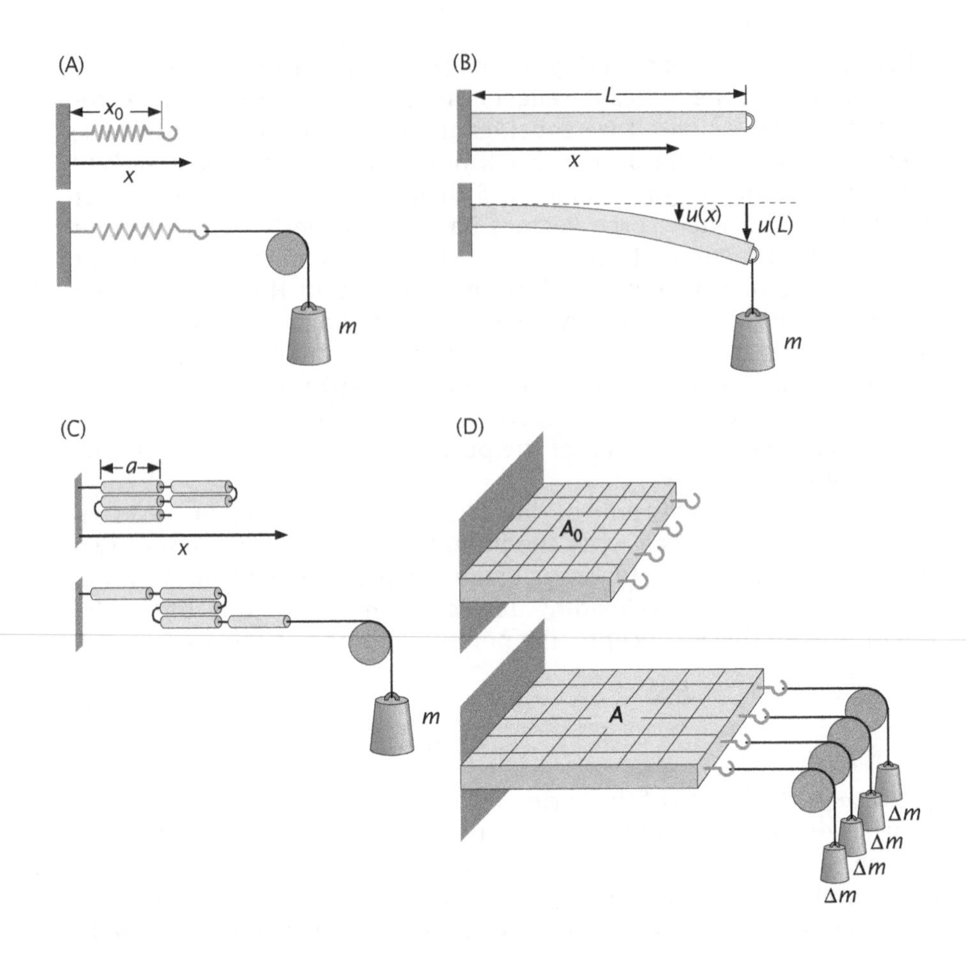

Figure 5.12: Mechanics of loading devices. (A) Mass–spring system, (B) beam under the action of an applied force, (C) polymer chain subjected to a load, and (D) membrane subjected to an applied tension.

fond of is the use of laser light to make a spring in the form of optical tweezers (introduced in Figure 4.11 on p. 153), such as shown in Figure 5.13. In particular, we consider the case of a bead in an optical trap that is subject to a load due to a piece of tethered DNA. We ask what displacement the bead suffers in the trap as a result of the applied load. We can write down the potential energy function in the form

$$U(x) = \frac{1}{2}k_{\text{trap}}x^2 - Fx, \tag{5.8}$$

where we have assumed that the optical trap can be treated as a spring with stiffness k_{trap} and that the applied force is characterized by a magnitude F. If we now seek the energy-minimizing choice of x, obtained through solving $\mathrm{d}U/\mathrm{d}x = 0$, we find that the equilibrium displacement in this case is given by

$$x_{\text{eq}} = \frac{F}{k_{\text{trap}}}. \tag{5.9}$$

The characteristic scales for an experiment like this are forces between 1 and 100 piconewtons. This kind of experiment permits the measurement of a variety of interesting single-molecule properties, such as the force–extension characteristics of macromolecules (DNA, RNA, proteins, etc.) and the force–velocity characteristics of molecular motors. Several examples of the uses of optical trapping were introduced in Figure 4.12 (p. 154). An example of force–extension data for DNA obtained by using single-molecule methods (in this case magnetic tweezers rather than optical tweezers) is shown in Figure 5.14. This experiment allows a range of different forces to be applied to DNA and the corresponding elongation of the DNA molecule to be examined. At low forces, the extension increases linearly with force, while at high forces the extension saturates since the molecule has been stretched to its full contour length. As will be seen in Chapter 8, these experiments can be compared directly with our theoretical understanding of DNA mechanics.

These examples of mechanical equilibrium provide an introduction to the key precept of the present chapter, which is the idea that equilibrium structures are minimizers of potential energy (at zero temperature) or free energy (at finite temperature). To perform such a minimization, we need to write the energy or free energy in terms of some set of variables that characterize the geometric state (that is, the structure) of the system. Once we have written the energy or free energy in terms of the parameters characterizing the system, then our task is reduced to the mathematics of determining which out of all of the various structural competitors leads to the lowest value of the potential energy or free energy. For the simple mass–spring systems introduced in this section, the minimization required the evaluation of a derivative. However, more generally we must address the mathematical question: given a function, out of all of the possible competitors, how do we find the one that minimizes the value of that function?

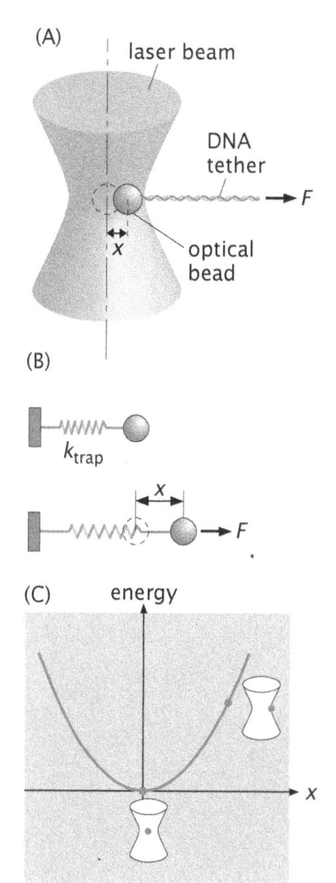

Figure 5.13: Representation of an optical trap as a mass–spring system. (A) Schematic showing how force moves the bead away from the center of the trap. (B) Replacement of the optical problem with a corresponding effective spring. (C) Energy of the bead in the trap as a function of its position in the trap. Note that the energy is only quadratic for sufficiently small displacements.

Computational Exploration: Determining the Spring Constant of an Optical Trap As noted above, a bead trapped by an optical trap undergoes Brownian motion in an approximately quadratic potential energy landscape induced by the laser light. In this Computational Exploration, we use a series of images from such an experiment to determine the stiffness of the trap. With a knowledge of this stiffness, we can then determine the magnitude of force exerted by the trapped bead on some tethered protein such as a molecular motor or the force necessary to extend DNA to a certain length by measuring the size of the force-induced excursion of the bead due to this pulling. Some typical geometries for such an experiment are shown in Figure 4.12 (p. 154) and 5.13.

For the purposes of this exercise, we propose a crude algorithm based upon simple thresholding in which we turn the pixels associated with the bead into 1's and all other pixels into 0's. We then ask Matlab to tell us the position of the centroid of the bead and then find its mean excursion squared. With that

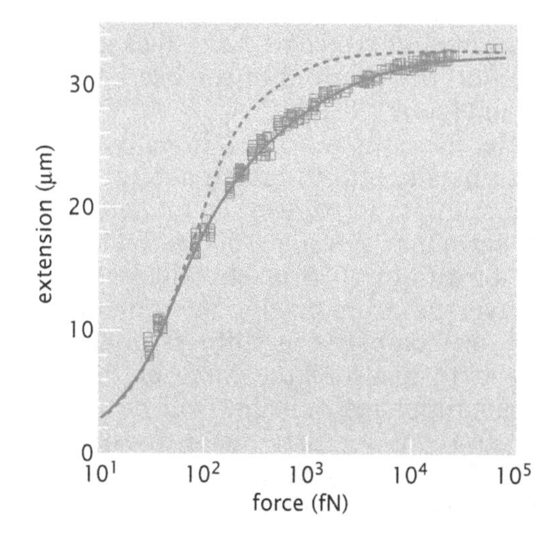

Figure 5.14: Force–extension curve for double-stranded DNA. Data for force versus extension for double-stranded DNA from λ-phage (with the DNA molecule here resulting from linking two such molecules for a total length of 97 kb) illustrating the distinction between the freely jointed chain model (dashed line) and the worm-like chain model (solid line). The freely jointed chain model will be discussed in detail in Chapter 8 and the worm-like chain model will be discussed in Chapter 10. (Adapted from C. Bustamante et al., *Science* 265:1599, 1994.)

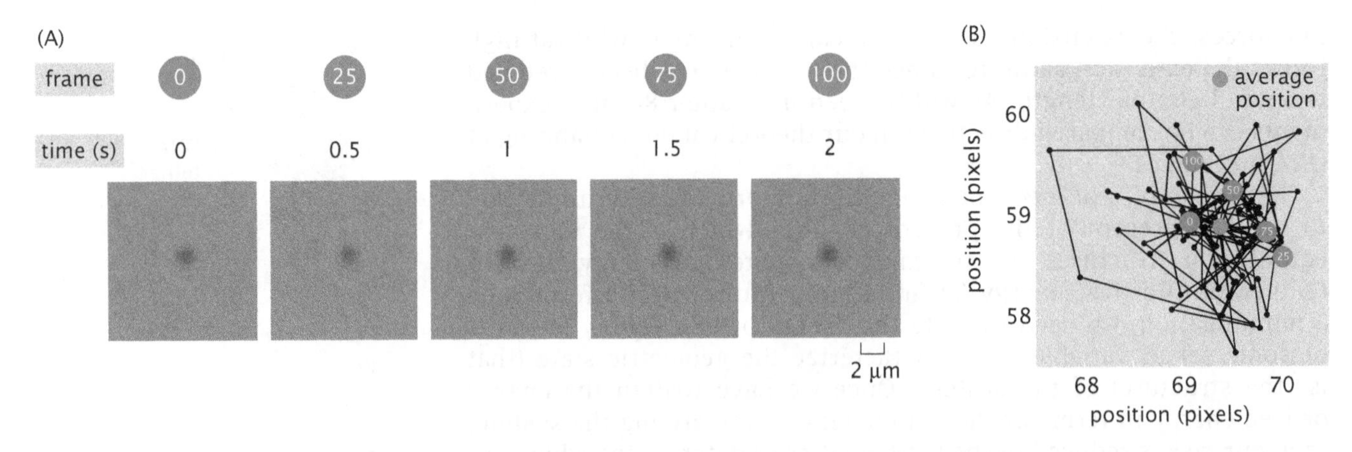

Figure 5.15: Determination of the spring constant of an optical trap. (A) Sequence of images of the trapped bead showing its position at various times. The frame capture rate is 50 frames/s. (B) Centroid position as a function of time for the full trajectory. The positions of the bead during the frames shown in part (A) are indicated with numbered red dots.

quantity in hand, we exploit the expression that gives the trap stiffness in terms of these excursions as

$$k_{\text{trap}} = \frac{k_B T}{\langle x^2 \rangle}. \quad (5.10)$$

The derivation of this equation is left as an exercise for the reader at the end of Chapter 6 since it involves the use of statistical mechanics.

To be concrete, Figure 5.15 shows the logical progression of ideas for determining the trap stiffness. The sequence of images in Figure 5.15(A) shows a trapped bead in a very simple homemade optical trap using a 10 mW laser at 632.8 nm. Images were acquired at a rate of 50 frames per second. As seen in the images, the bead appears darker than the surrounding area. This is very similar to the ideas developed when we segmented bacteria in Figure 3.9 (p. 101), where the bacteria appeared darker as well. We rescale all of the pixel values from 0 to 1 such that the pixels associated with the bead have values close to 0. The idea of the most naive Matlab code we can use to track the bead is to simply threshold the image by asking it to convert all pixels below some threshold to 1's and all pixels above that threshold to 0's so that in the thresholded image the beads now appear white. Once we have the series of thresholded images, we find the centroid of the bead, which can be plotted as shown in Figure 5.15(B).

To complete the exercise, we need to convert the length units, which are in pixels, into the more natural micron units that will give us a stiffness in pN/μm. To that end, we use the fact that for this particular setup, a 100 μm feature on a calibration slide has a width of 1028 pixels. This result serves as the basis for a conversion from pixel to micron units.

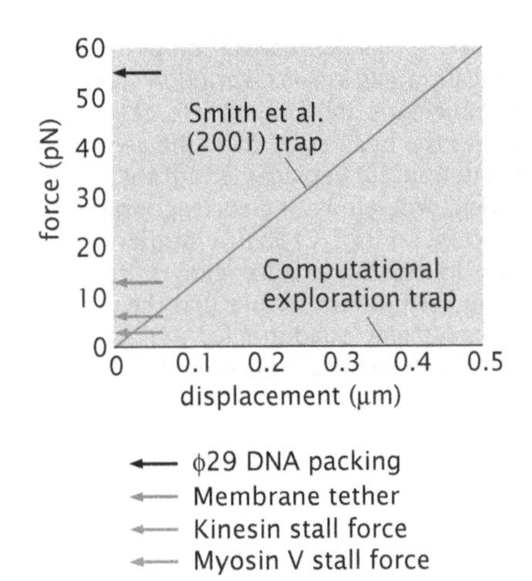

Figure 5.16: Force versus displacement for optical traps. The force as a function of displacement for the very weak optical trap from this Computational Exploration and for a much stronger optical trap that was used to do the experiments that will be introduced in Figure 10.18 (p. 403). The measured stalling forces for several molecular motors are also shown. The optical trap used to construct the trap resulting in the data of Figure 5.15 exploited a laser that is much weaker than a "research grade" optical trap, but is excellent for classroom exercises.

In Figure 5.16, we compare the stiffness obtained for this trap with that used to measure the force of the DNA packaging motor in bacteriophage φ29 that will be introduced in Figure 10.18 (p. 403). We see that the stiffness of our trap, though very useful for pedagogical purposes, is much smaller than that used for the actual single-molecule experiment. This is further revealed by the typical forces exerted by several motors overlaid with the stiffness for both traps in Figure 5.16.

Finally, this crude method of finding the stiffness of the trap based on the position of the bead determined by thresholding is just one possible strategy. An alternative approach for obtaining the stiffness of the trap is by computing the power spectrum of the bead's position.

5.3 The Mathematics of Superlatives

The search for extrema is a mathematical embodiment of the human instinct for superlatives. In casual conversation, rarely an hour goes by without injecting words such as "best" and "worst" into our speech. Our technologies similarly reflect the pressure to make things faster, smaller, lighter, safer, etc. The development of modern mathematics included tools for finding functions that could be characterized by superlatives such as biggest and smallest. The present section is a mathematical excursion that aims to show how to replace the verbal and intuitive case-by-case discriminations with precise mathematical tools that permit us to search over what amounts to an infinite set of competitors. The reason such a mathematical interlude is necessary is that the study of equilibrium demands that we minimize functions such as the potential energy or the free energy and, as a result, we need the mathematics that permits us to effect such minimizations.

5.3.1 The Mathematization of Judgement: Functions and Functionals

The translation from everyday language, where superlatives are characterized by words such as "best," "fastest," etc. to the mathematical form of these same concepts requires the introduction of a scheme for attaching numbers to the degree of "bestness."

As a concrete example in the mathematization of superlatives, we consider the bending of a beam as shown in Figure 5.17. This particular example will arise repeatedly throughout the remainder of the book in many disguises. For example, when we think about the geometry of deformed DNA, the buckling of microtubules under force, and the use of cantilevers as tools for applying force to macromolecules, in each case we will write the energy of the system in terms of the geometry of these bent beams and will seek the configuration that leads to the lowest energy cost. The question we are interested in answering is: what choice of the displacement function $u(x)$ leads to the lowest value of the potential energy of the beam and the loading device? Note that in this case, the potential energy depends upon the specification of an entire function, $E_{\text{tot}}[u(x)]$, where we have introduced the square bracket notation [...] to call attention to the fact that the energy depends upon a function rather than a finite set of parameters. An alternative that sometimes comes in handy is to discretize the geometry of the beam as shown in Figure 5.17(B). In this case, we treat the beam as a series of discrete masses where now there is a set $(u_1, u_2, \ldots, u_N)$ of displacements that determine the potential energy. In this case, the energy is a *function* of the unknowns $(u_1, u_2, \ldots, u_N)$ and can be written as $E_{\text{tot}}(u_1, u_2, \ldots, u_N)$.

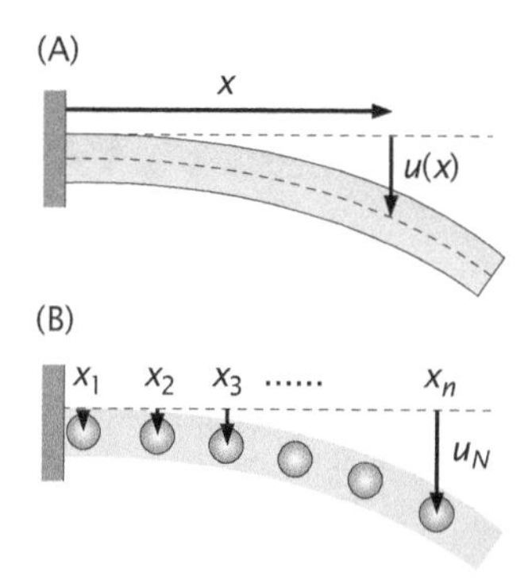

Figure 5.17: Two different representations of the geometry of a beam subjected to a load on its end. (A) Continuous representation of the beam geometry. (B) Discretization of the geometry of the beam.

Functionals Deliver a Number for Every Function They Are Given

When we write the energy in the form $E_{\text{tot}}(u_1, u_2, \ldots, u_N)$, we are on familiar mathematical turf. A discrete set of parameters u_1, u_2, etc. suffices to describe the geometry of the system and the energy is a *function* of these geometric parameters. In writing the energy in the form $E_{\text{tot}}[u(x)]$, we have implicitly introduced a new mathematical idea (a functional), since in this case it takes a function $u(x)$ to characterize the geometry of the deformed beam and the energy depends upon the function. An energy functional assigns an energy to each configuration, where the configuration itself is characterized by an entire function.

To be concrete in our thinking, Figure 5.18 shows several examples in which the free energy depends upon the disposition of the system as characterized by a function. Figure 5.18(A) shows different

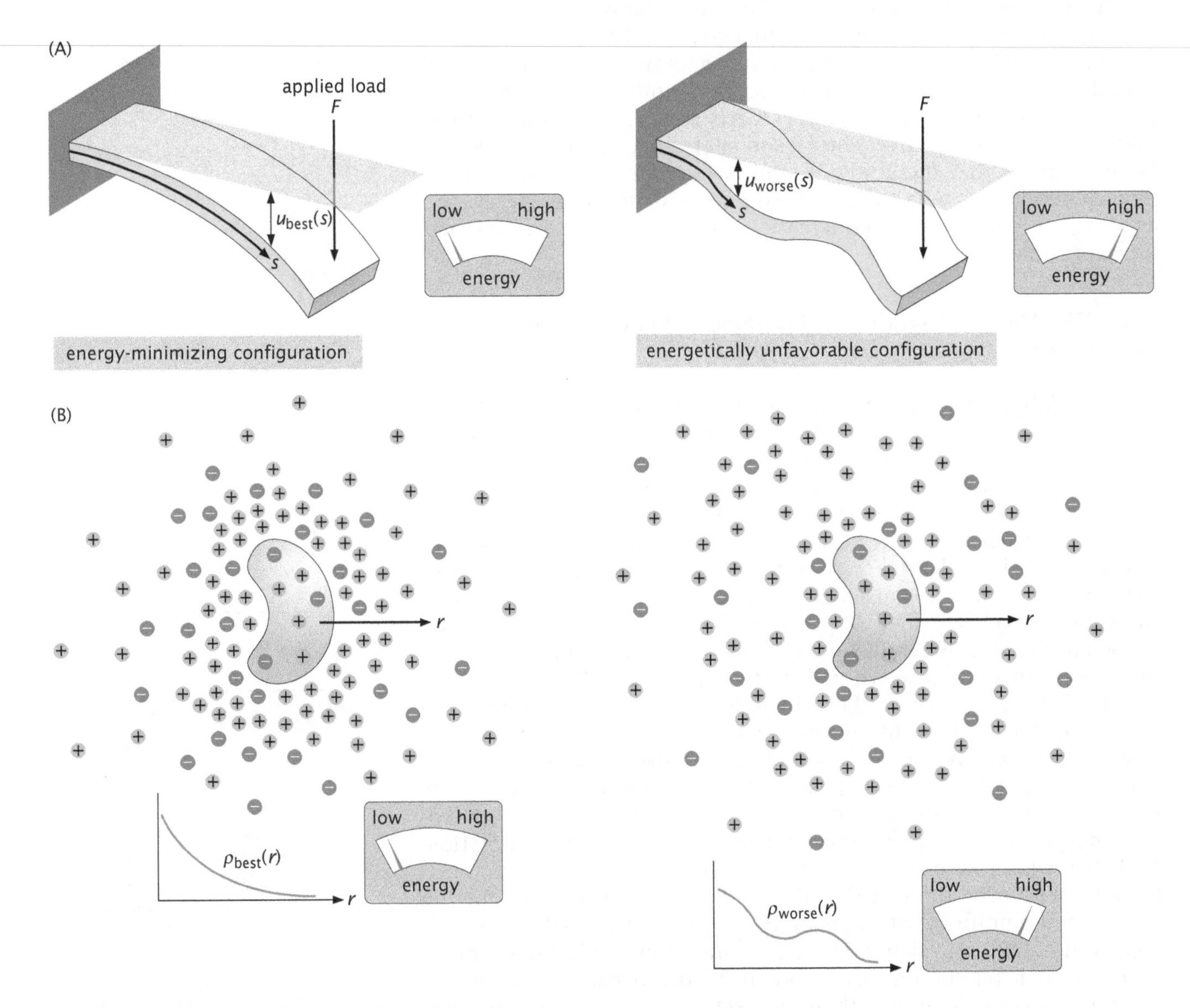

Figure 5.18: Examples of two functionals. (A) Energy as a function of the shape, $u(x)$, of a beam. Different shapes have different strain energies. (B) Free energy as a function of the distribution of ions in solution in the vicinity of a protein. The density of ions is characterized by the function $\rho(\mathbf{r})$. Note that we plot the density as a function of the distance from the origin, r.

structures for a deformed beam. Each deformed configuration of the beam is described by a *different* function $u(x)$. Further, each such $u(x)$ corresponds to a different energy. The figure shows the energy-minimizing configuration as well as a particularly bad guess (that is, high energy) for the deformed geometry. The energy meter icon shows how the energy of the latter configuration is higher than that of the energy-minimizing structure. A more subtle example to be taken up again in Chapter 9 concerns the distribution of ions around a protein. In this case, the unknown function is $\rho(\mathbf{r})$, the density of ions as a function of position in space. Here too, it is possible to write down a free-energy functional that delivers a free energy for each and every guess we might make for the density of ions. Figure 5.18(B) shows both the free-energy minimizing distribution of ions as well as a less-than-optimal distribution of ions, and the energy meter reports their respective overall free energies.

The overarching theme of this section is the idea of a cost function or functional. The key point is that we want to compute some quantity that we are interested in minimizing. In many cases, this "cost function" is the energy or free energy. If that cost function depends upon the disposition of a finite set of parameters such as the $(u_1, u_2, \ldots, u_N)$ that characterized our beam represented discretely in Figure 5.17, then indeed the cost function is a function. On the other hand, if it takes an entire function such as $u(x)$ to characterize the state of the beam, then we have a cost functional since we have to specify a function in order to determine the energy or free energy.

5.3.2 The Calculus of Superlatives

The previous discussion showed the way in which we can cast our ideas about the extent to which some quality or quantity of interest is best or worst, biggest or smallest, and so on. This led us naturally to the idea of functions and functionals. Now that we are able to say how good or bad, big or small a particular quantity is as a function of some control parameters (or control function), we pose the question of how to discriminate amongst all the competitors to find the winner. We begin by discussing the implementation of these ideas in the context of ordinary calculus. The more general question of finding the extreme values of functionals is treated in the appendix at the end of the chapter.

Finding the Maximum and Minimum Values of a Function Requires That We Find Where the Slope of the Function Equals Zero

As our first foray into the question of how to cast our search for superlatives in mathematical terms, we recall a few ideas from the ordinary calculus of maxima and minima. We consider a function $f(u_1, u_2, \ldots, u_N) = f(\{u_i\})$, which depends upon the N variables $\{u_i\}$, where we have introduced the notation $\{\}$ to indicate a set of objects. We imagine the variables $\{u_i\}$ are allowed to range over some set of values, and we ask the question: what choice of the values $\{u_i\}$ renders the function $f(\{u_i\})$ maximum or minimum? To be concrete, we remind the reader of the discussion surrounding Figure 5.11 on how mechanical equilibrium can be thought off as the minimization of potential energy. In that case, our minimization problem involved one parameter, the displacement x.

To find the maxima and minima of functions, we find those values of the function for which the slope is zero as embodied in

$$\frac{\partial f}{\partial u_i} = 0 \quad (i = 1, 2, \ldots N). \tag{5.11}$$

That is, our problem amounts to solving the N equations in N unknowns given by Equation 5.11. The notation $\partial f/\partial u_i$ refers to the partial derivative and is explained in The Math Behind the Models below. We have said nothing about how we might go about solving such equations, but the prescription for obtaining them is now clear. Though we will not have the space to go into the subtlety of solving such equations for a generic nonlinear problem, we refer the reader to the entertaining cautionary tales of Acton (1990) in the Further Reading at the end of the Chapter.

The Math Behind the Models: the Partial Derivative

MATH

Throughout the book, it will be of interest to find out how functions vary as we change a variable. Often, however, we will be interested in functions that depend upon more than one variable simultaneously. For example, in minimization problems, often the energy (or free energy) depends upon more than one parameter: the free energy can depend upon both the volume of the system and the number of particles. Another important example is a function $f(x, t)$ that depends upon both position (x) and time (t) simultaneously. For example, we might like to know the deflection of a beam characterized by the function $u(x, t)$ that tells us how much deflection there is a distance x along the beam at a time t. Alternatively, we might interest ourselves in the concentration of some molecule, $c(\mathbf{r}, t)$, at every position in space. In this case, the function depends upon four variables since the vector $\mathbf{r}$ is really (x, y, z).

In these cases, the notion of a derivative is more subtle because we have to say with respect to what variable. The mathematical tool that arises in this case is the partial derivative. The idea is illustrated in Figure 5.19. The derivative of ordinary calculus tells how a function changes as a result of a small excursion. The partial derivative generalizes that idea by telling us how a function changes when we make an excursion in *one* of the variables in the function while leaving the others constant. An intuitive example from everyday experience is illustrated by walking in a mountain pass. The shape of a mountain pass is like a saddle. In particular, walking in one direction leads us down whereas walking in a perpendicular direction leads us up the peaks that bound that mountain pass. In these two cases, the partial derivatives actually have different signs since in one case the curve is sloping downward and in the other it is sloping upward.

If we think of the height of the local topography of a mountain as $f(u_1, u_2)$, where u_1 and u_2 correspond to two perpendicular axes, then the partial derivative tells us how the function changes when we walk along these two directions. These ideas are represented mathematically through

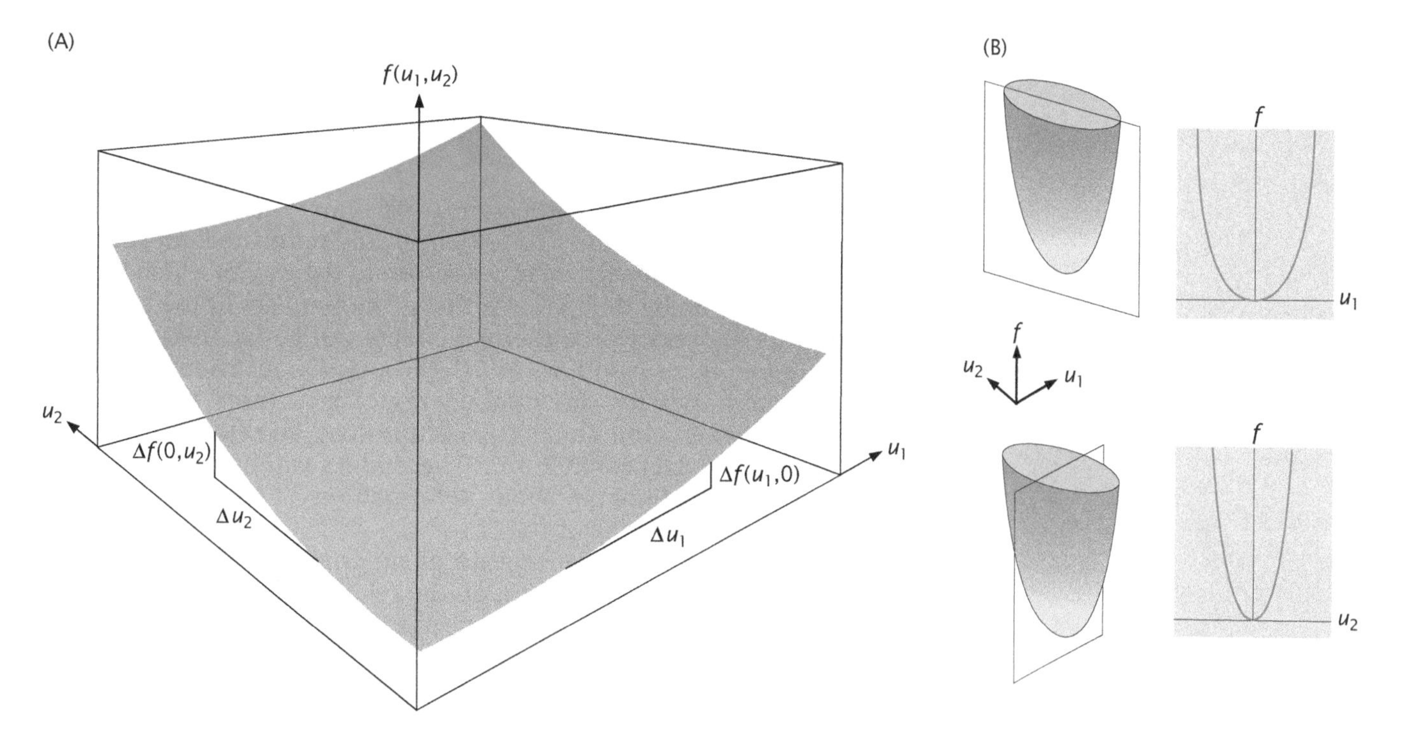

Figure 5.19: Illustration of the concept of a partial derivative. (A) The plot shows a function $f(u_1, u_2)$ that depends upon the variables u_1 and u_2. If u_2 is held fixed, the surface is reduced to a curve and the partial derivative is nothing more than the ordinary derivative familiar from calculus, but on this particular curve. (B) Planar cuts through the function $f(u_1, u_2)$.

the definitions

$$\frac{\partial f(u_1, u_2)}{\partial u_1} = \lim_{\Delta u_1 \to 0} \frac{f(u_1 + \Delta u_1, u_2) - f(u_1, u_2)}{\Delta u_1}, \tag{5.12}$$

and

$$\frac{\partial f(u_1, u_2)}{\partial u_2} = \lim_{\Delta u_2 \to 0} \frac{f(u_1, u_2 + \Delta u_2) - f(u_1, u_2)}{\Delta u_2}. \tag{5.13}$$

For the sake of concreteness in finding minima, we consider the simple example of quadratic functions like those shown in Figure 5.20. The two-dimensional example has the functional form

$$f(u_1, u_2) = \tfrac{1}{2}(A_{11}u_1^2 + A_{22}u_2^2 + 2A_{12}u_1u_2). \tag{5.14}$$

If we now implement the injunction of Equation 5.11 (that is, $\partial f/\partial u_1 = 0$ and $\partial f/\partial u_2 = 0$), we find

$$\begin{aligned} A_{11}u_1 + A_{12}u_2 &= 0, \\ A_{21}u_1 + A_{22}u_2 &= 0, \end{aligned} \tag{5.15}$$

a pair of coupled, linear equations for the minimizing values of u_1 and u_2. Assuming the equations have a unique solution (which is true if the determinant of the matrix A made up of the coefficients A_{11}, A_{22}, A_{21}, and A_{12} is nonzero), $u_1 = u_2 = 0$ is clearly such a solution, indicating that the function f has a minimum (or maximum) at $(0, 0)$.

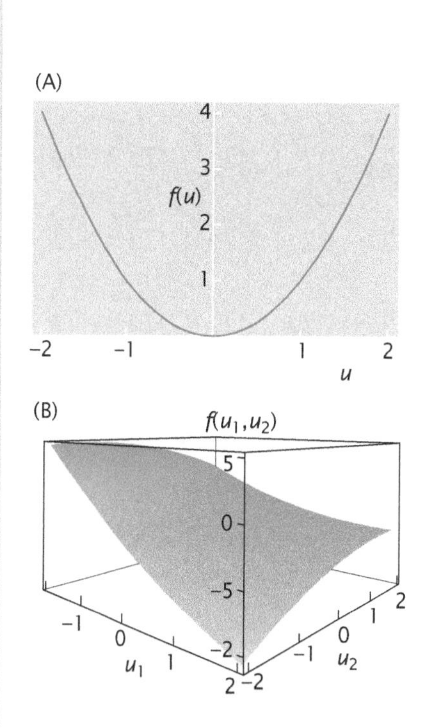

Figure 5.20: Quadratic energy functions. (A) Case of $f(u) = \frac{1}{2}Au^2$. (B) Case of $f(u_1, u_2) = \frac{1}{2}(A_{11}u_1^2 + A_{22}u_2^2 + 2A_{12}u_1u_2)$.

5.4 Configurational Energy

In Mechanical Problems, Potential Energy Determines the Equilibrium Structure

Our brief foray into the mathematical machinery used to find minimizers leaves us poised now to ask physically motivated questions of biological interest. In particular, we return to the way in which biological structures can be thought of either as minimizers of the potential energy (this in cases where thermal effects can be ignored) or of the free energy. In this section, we attack the strictly mechanical question of what determines the potential energy of structures and ask how the potential-energy-minimizing structure may be selected from the class of all structural competitors. These ideas will be used in subsequent chapters in thinking about deformations of DNA, cytoskeletal filaments, and membranes.

In order to apply the mathematics of superlatives, we must first be able to pass energetic judgement on the relative goodness or badness of a given structure. In particular, to pass this judgement, we require an energy function (or functional) that delivers an energy for each and every value of the structural parameters for the structure of interest. The nature of such energy functions forms the backdrop for much of the history of physics.

One class of energy functions with deep significance are those that posit a quadratic dependence of the energy on the departure from equilibrium. The motivation for this class of energy function is the idea that, regardless of the detailed features of a given energy landscape, near equilibria any such function can be treated as a quadratic function of the variables that describe the excursion from equilibrium. Concretely, if we consider the one-dimensional case where the potential energy is of the form $U(x)$ as shown in Figure 5.21 and there is a point of equilibrium at x_{eq}, then we may expand the function $U(x)$ in a Taylor series, keeping terms only up to quadratic order. The idea of the Taylor series is pervasive and is explained in The Math Behind the Models after this section. The Taylor series for our potential is of the form

$$U(x) = U(x_{\text{eq}} + \Delta x) \approx U(x_{\text{eq}}) + \left.\frac{\mathrm{d}U}{\mathrm{d}x}\right|_{\text{eq}} \Delta x + \frac{1}{2}\left.\frac{\mathrm{d}^2U}{\mathrm{d}x^2}\right|_{\text{eq}} \Delta x^2, \tag{5.16}$$

where we have introduced the notation Δx to characterize the excursion about the equilibrium point. In this one-dimensional case, Δx can be thought of as the distance traveled away from the equilibrium point x_{eq}. This situation is shown in Figure 5.21. Equilibrium demands that $\mathrm{d}U/\mathrm{d}x|_{\text{eq}} = 0$ since at the equilibrium point there are no unbalanced forces, and hence we are left with

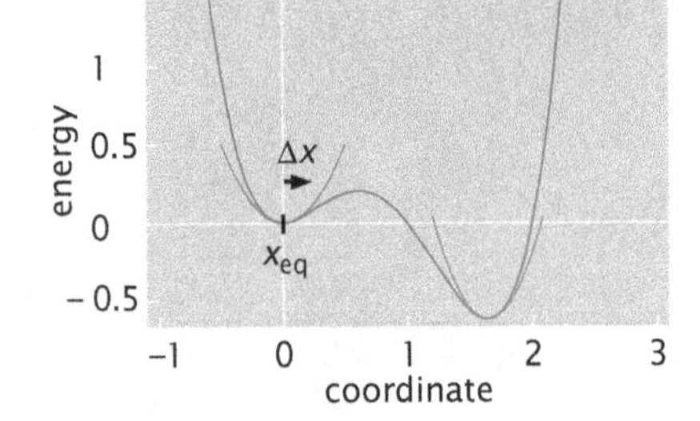

Figure 5.21: Potential energy as a function of the coordinate x. A quadratic representation of that energy landscape is shown in the vicinity of two equilibrium points.

$$U(x_{\text{eq}} + \Delta x) \approx U(x_{\text{eq}}) + \frac{1}{2}\left.\frac{\mathrm{d}^2U}{\mathrm{d}x^2}\right|_{\text{eq}} \Delta x^2, \tag{5.17}$$

which is of the form $U(x) = \frac{1}{2}kx^2$, where the "stiffness" of the "spring" holding the system at equilibrium is given by $k = \mathrm{d}^2U/\mathrm{d}x^2$. This same idea can be generalized to higher dimensions, in which excursions are permissible in multiple directions (for example, on a mountain top, we can choose to walk off in two orthogonal directions and the energy cost of doing so is quadratic in the excursion variables).

One of the most powerful incarnations of the idea developed above is provided by the theory of elasticity, which teaches us how to write down the energy of a continuous body, such as a rod or a membrane, as a quadratic function of the strain, which measures the amount of deformation. We take up the elastic energy of deformation (and Hooke's law) in the next section.

MATH

The Math Behind the Models: The Beauty of the Taylor Expansion A very important tool invoked in the mathematical analysis of physical models is the use of the so-called Taylor expansion. Series expansions of this kind will be one of our primary mathematical tools in the remainder of this book. The idea is very simple and amounts to replacing a function $f(x)$ in some neighborhood with a simple polynomial. As will be seen repeatedly throughout this book, the virtue of these approximations is that they allow us often to replace intractable nonlinear expressions with simple algebraic surrogates that we can handle analytically and that give an intuitive sense of the mathematics.

The idea of the Taylor expansion is embodied in the simple formula

$$f(x) = a_0 + a_1 x + a_2 x^2 + \cdots . \tag{5.18}$$

Most of the time, we will only keep terms up to second order, and as a result the Taylor series algorithm reduces to the question: what three coefficients a_0, a_1, and a_2 should we use to best approximate the function $f(x)$?

For concreteness, let us consider the case in which we are interested in the behavior of the function $f(x)$ near $x = 0$. If we set $x = 0$ on both sides of Equation 5.18, we see that $a_0 = f(0)$. But we already know the function $f(x)$, so all we have to do is find its value at $x = 0$ to obtain the first coefficient. Next, let us take the derivative of both sides of Equation 5.18 with respect to x. We are left with the equation

$$f'(x) = a_1 + 2a_2 x + \cdots . \tag{5.19}$$

Once again, if we set $x = 0$, we are left with $a_1 = f'(0)$. We can continue to play the same game, this time evaluating the second derivative, with the result

$$f''(x) = 2a_2 + \cdots , \tag{5.20}$$

which leads to $a_2 = \frac{1}{2} f''(0)$. This same basic analysis can be carried on indefinitely if one is interested in higher-order terms. Most of the time we will be content with the expression

$$f(x) \approx f(0) + f'(0)x + \tfrac{1}{2} f''(0)x^2 . \tag{5.21}$$

The symbol $\approx$ refers to the fact that in the neighborhood of the point x, the left- and right-hand sides of this equation are *approximately* equal. The conclusion of this little analysis is that if we want to find a simple quadratic surrogate for our function of interest, all we need to know is the value of the function and its first two derivatives at the point around which

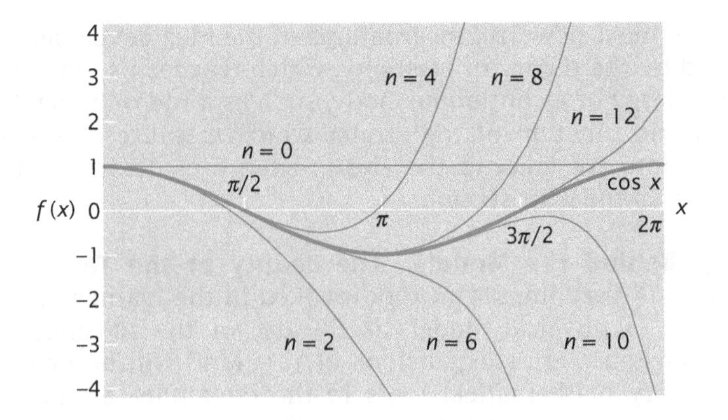

Figure 5.22: Comparison of the function $\cos x$ and its Taylor expansion. The curves are labeled by the order of the highest term kept in the Taylor series. For example, $n = 2$ means that the series goes to quadratic order, etc. The cosine function we are approximating is shown in bold for comparison with the approximate expressions.

we are expanding. An example of this kind of analysis for the case of cos x is shown in Figure 5.22. In particular, using the rules given above, the Taylor series for this function is given by

$$\cos x \approx 1 - \frac{x^2}{2!} + \frac{x^4}{4!} - \frac{x^6}{6!} + \frac{x^8}{8!} - \frac{x^{10}}{10!} + \cdots. \tag{5.22}$$

Figure 5.22 compares the function $\cos x$ with various approximations based upon the Taylor series. We see that as more terms are included, the approximation is good for a wider range of values of x. Of course, there are mathematical subtleties that arise when considering a generic function, such as the question of convergence of the Taylor series. For example the function $1/(1-x)$ has the Taylor series, $1 + x + x^2 + x^3 + \cdots$, which is finite only for values of x such that $-1 < x < 1$.

5.4.1 Hooke's Law: Actin to Lipids

There is a Linear Relation Between Force and Extension of a Beam

To see how these ideas about small departures from equilibrium can be applied to continuous bodies of biological significance such as DNA, cytoskeletal filaments, and membranes, we begin by examining the elasticity of a stretched rod. This subject will be taken up in detail in Chapter 10 and our aim here is to present the conceptual underpinnings of the ideas of elasticity theory. Consider a beam of undeformed length L that is stretched by an amount ΔL as shown in Figure 5.23. The geometric state of deformed objects is most naturally captured in terms of a quantity known as the strain and defined in the current setting as

$$\varepsilon = \frac{\Delta L}{L}. \tag{5.23}$$

The central idea captured by the notion of strain is that adjacent points of the material suffer *different* displacements. In our current example, the displacement of a given point depends upon how far it is from the origin. The result of such relative displacements is that the bonds in the material are stretched as depicted schematically in Figure 5.23. Though our thought experiment considers the case of extension, one can just as easily consider the case of compression, in which case $\Delta L < 0$. Note that for simplicity we ignore the small displacements perpendicular to the direction of stretch known as the Poisson effect.

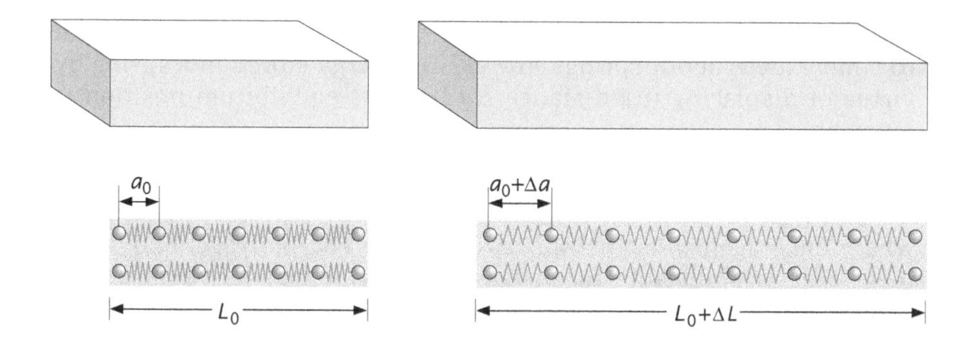

Figure 5.23: Energetics of beam stretching. The continuum description of the beam as a deformable solid can be interpreted in terms of the stretching of the individual atomic bonds.

To garner an idea of the mechanical interpretation of these deformations, Figure 5.23 suggests that we can think of the overall macroscopic deformation as imposing the stretching of a huge set of microscopic springs that correspond to the bonds between the atoms making up that beam. We recall that the relation between force and stretch for a spring is given by

$$F = -k\Delta a, \tag{5.24}$$

where k is the spring constant, Δa is the extension of the spring, and F is the force it engenders. Macroscopically, this same idea is written as

$$\frac{F}{A} = E\frac{\Delta L}{L}, \tag{5.25}$$

where F is the applied force, A is the cross-sectional area of the beam, and E is a material property known as the Young modulus, which reflects the stiffness of the beam. Note that the Young modulus has units of force/area or energy/volume, since the strain is dimensionless. The quantity F/A is known as the stress and has dimensions of force per unit area.

For the simple model of a beam composed of many microscopic springs shown in Figure 5.23, where two nearby springs are separated by distance a_0, Equation 5.25 can be derived from Equation 5.24. In particular, if a force F is applied to the beam, it is balanced by all the springs in the cross-section of the beam, which each stretch by the same amount. Springs in "parallel" (like resistors) share the load. Since the number of such springs is A/a_0^2, where a_0^2 is area taken up by an individual spring, each spring will be extended by $\Delta a = (F/k)/(A/a_0^2)$. This result is a reflection of the fact that n equivalent springs in parallel will each suffer a displacement $(F/k)/n$. Since all the springs that make up the beam have the same extension, the net extension of the beam will be $\Delta L = (L/a_0)\Delta a$, where L/a_0 is the number of springs along the length of the beam, each contributing amount Δa to ΔL. If we substitute for Δa in the last equation the expression derived in the previous one, we arrive at Equation 5.25, with $E = k/a_0$. In this simple model of an elastic solid, the Young modulus is the ratio of the spring constant associated with a bond between two atoms, divided by the typical distance between them.

The Energy to Deform an Elastic Material is a Quadratic Function of the Strain

As yet, we have presented Hooke's law as a statement about the forces that result from deforming elastic materials. However, in many circumstances, it is more useful to characterize the elastic properties of a deformable material through reference to its energy. If we refer back to simple ideas about springs, the elastic energy stored in a spring by virtue of displacing it a distance Δa from its equilibrium position is given by

$$E_{\text{strain}} = \tfrac{1}{2}k(\Delta a)^2, \tag{5.26}$$

where once again k is the spring constant. The more general statement that is applicable to an elastic material that has suffered an extensional strain like that shown in Figure 5.23 can be obtained by a divide-and-conquer strategy in which the material is divided up into a bunch of little volume elements. In each such volume element, we compute the strain energy density and multiply by the volume of that element to obtain the energy for that little chunk of material. By summing (in fact, integrating) over all of the material elements in the material, we find the total strain energy as

$$E_{\text{strain}} = \frac{EA}{2}\int_0^L \left(\frac{\Delta L}{L}\right)^2 \mathrm{d}x, \tag{5.27}$$

where A is the cross-sectional area of the beam. This equation can be derived for the simple model of a beam shown in Figure 5.23, in the same way that Equation 5.25 was derived above, by adding up the elastic energies of all the microscopic springs that make up the beam.

If we consider the more general case in which the relative stretch is a function of position along the axis of the beam, the energy associated with deformation is given by

$$E_{\text{strain}} = \frac{EA}{2}\int_0^L \left(\frac{\mathrm{d}u(x)}{\mathrm{d}x}\right)^2 \mathrm{d}x. \tag{5.28}$$

The key point of all of this is the existence of an energy function that penalizes *relative* changes in length of adjacent material points in a body.

In the remainder of the book, we will appeal to elastic arguments like those described above. The virtue of these elastic arguments is that they will permit us to probe the energy cost of processes such as DNA packing (in nucleosomes and viruses), the buckling of microtubules at high force, and the deformation of lipid bilayers in the neighborhood of ion channels, to name but a few examples. Two examples are illustrated in Figure 5.24, where we see that analysis of elements of the cytoskeleton can be couched in the language of elasticity theory. In addition, the figure also foreshadows our examination of lipid bilayer membranes in Chapter 11. In both cases, the basic idea is the same, namely, that there is a quadratic energy cost associated with small excursions of the system about its equilibrium configuration.

So far, we have argued that in many instances it is convenient to represent mechanical equilibrium as the condition of minimum potential energy. Of course, to carry out such a minimization, we must first have a way of assigning potential energy to different configurations. We have seen that for systems *near* equilibrium, the energy cost can be written as a quadratic function of the excursions about that equilibrium. These quadratic energies emerge both when characterizing the elastic response of materials treated as a continuous medium and when carrying out an atom-by-atom reckoning of the energy of configuration. However, there is often more to the delicate balance that determines structures than their potential energy alone. Thermal forces also make their presence known and we take up the apparatus

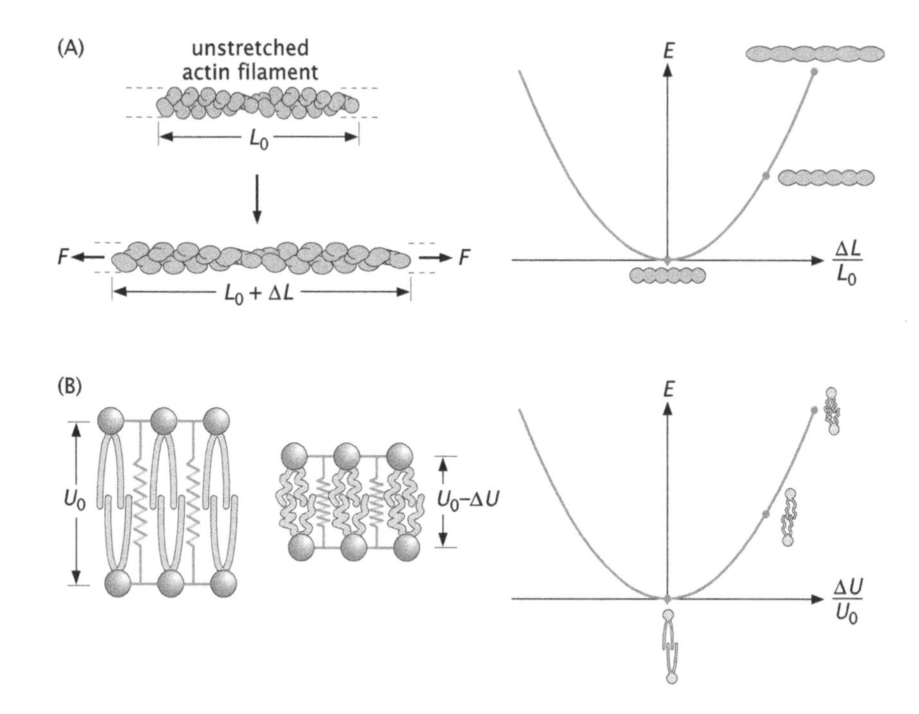

Figure 5.24: Deformation of macromolecular assemblies and the corresponding elastic energy cost associated with these deformations. (A) Schematic of F-actin stretching in response to a force applied along the filament axis. The energy curve shows a quadratic cost to either elongate or shrink the filament relative to its equilibrium length. (B) Schematic of deformation in which the thickness of the lipid bilayer is changed relative to its equilibrium value. The energy curve shows the elastic energy cost to change the thickness of a lipid bilayer from its equilibrium thickness.

to handle this part of the free-energy budget in the remainder of the chapter.

5.5 Structures as Free-Energy Minimizers

In the previous section, we have seen how mechanics can provide insights into the equilibrium configurations of systems at different scales. However, our arguments were incomplete because we neglected the role of thermal fluctuations in dictating equilibria. The aim of the present section is to explore the extension of our discussion of equilibrium to supplement energy minimization with the often conflicting demand of maximizing the entropy.

Though we will derive the result in its full glory later in the chapter, for the moment we examine the notion of free energy qualitatively. The concept of the free energy is embodied in the equation

$$\text{free energy} = \text{energy} - \text{temperature} \times \text{entropy}, \tag{5.29}$$

where the entropy (as will be shown below) is a measure of the number of different ways of rearranging the system. The fundamental argument of the remainder of the chapter and one of the foundational tools of the rest of the book is the idea that *the equilibrium state of a system is that choice out of all states available to the system that minimizes the free energy.*

The Entropy is a Measure of the Microscopic Degeneracy of a Macroscopic State

From a mathematical perspective, the ideas introduced above about thermal forces are codified in the notion of the entropy. Though we are coached in thinking about energy from our earliest exposures to science, in fact, there is a much more intuitive state variable that provides deep insight into the factors determining the equilibrium states

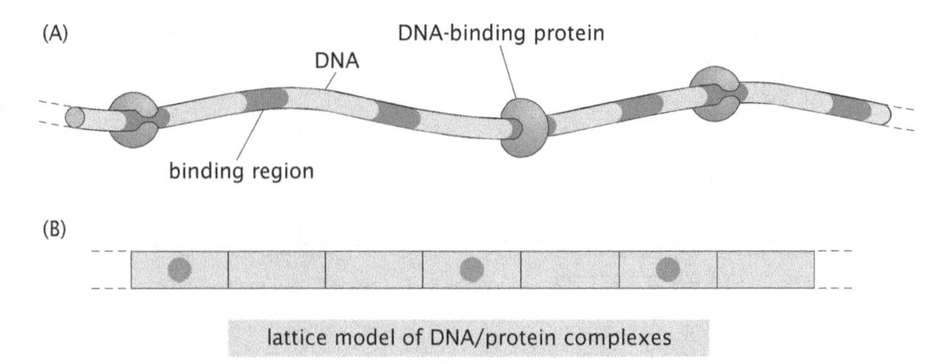

Figure 5.25: Possible arrangements of proteins on a DNA molecule. (A) The cartoon schematizes a DNA molecule on which there is a series of binding sites, which are shaded dark orange. The DNA-binding proteins can occupy any of these sites. (B) The lattice model represents a further idealization in which we imagine the DNA molecule as a series of boxes into which we can put the DNA-binding proteins.

of complex systems such as a solution with a number of interacting species. In particular, the entropy of a closed system provides a measure of the number of different microscopic ways that we can realize a given observed macroscopic state, and can be written as

$$S = k_B \ln W, \tag{5.30}$$

where W is the number of microstates compatible with the macrostate of interest and k_B is the Boltzmann constant. In light of this definition, we see that when minimizing the free energy, the energetic terms tend to favor lower energy while the entropy contribution favors the macrostates that can be realized in the most ways.

As a concrete example relevant to our attempt to quantitatively unravel gene expression (see Chapters 6 and 19), we consider the role of entropy in the context of DNA-binding proteins. In particular, the entropy in this case reveals the number of distinct ways that we can arrange the bound proteins (nonspecifically) along the entire DNA molecule, as shown in Figure 5.25. We imagine that our DNA molecule has a total of N binding sites, N_p of which are occupied by the protein of interest. Further, we assume that the binding energies when the proteins are bound nonspecifically are the same regardless of which nonspecific sites are occupied (though, in reality, even the energetics of nonspecific binding varies from site to site). As stated above, the entropy is a measure of the number of distinct ways of realizing a given macroscopic situation, in this case characterized by the number of possible binding sites and the number of binding proteins, and is given in most general terms as

$$S = k_B \ln W(N_p; N), \tag{5.31}$$

where S is the entropy and $W(N_p; N)$ is the multiplicity factor, which reflects the number of ways of rearranging the N_p proteins on the N binding sites. This definition of the entropy results from key consistency conditions such that the entropy of a composite system should be additive (although the total number of microstates for such a composite system is multiplicative).

For our example of DNA-binding proteins, we note that we have N choices as to where we lay down the first of the N_p proteins. Once this protein has been put down, we only have $N-1$ remaining sites where we might elect to put down the second protein. The third protein may now be put down on the DNA in any one of the remaining $N-2$ binding sites. Hence, the total number of ways of laying down our N_p proteins is $N \times (N-1) \times (N-2) \times \cdots \times (N-N_p+1)$. However, we have ignored the fact that these are not distinct configurations since we have no way to distinguish the case in which the first protein landed on site 10 and the second protein on site 15 and vice versa. As a result, we have overcounted and must divide by the number of rearrangements of those N_p proteins on the occupied sites, which, following the same argument as above, is $N_p \times (N_p - 1) \times \cdots \times 1$. This product of all integers from 1 to N_p is N_p-factorial, which is denoted $N_p!$. We are now in a position to write the total number of *microscopic* arrangements as

$$W(N_p; N) = \frac{N \times (N-1) \times (N-2) \times \cdots \times (N - N_p + 1)}{N_p \times (N_p - 1) \times \cdots \times 1}. \quad (5.32)$$

If we now multiply top and bottom of the equation by $(N - N_p)!$, it results in the more pleasingly symmetric form

$$W(N_p; N) = \frac{N!}{N_p!(N - N_p)!}. \quad (5.33)$$

For a DNA-binding protein such as Lac repressor, there are roughly 10 copies of this protein bound on the roughly 5×10^6 DNA-binding sites within the *E. coli* genome. The formula above then tells us that there are roughly 3×10^{60} distinct arrangements of the Lac repressor bound to the *E. coli* genome.

Now that the counting has been effected, we are prepared to invoke Boltzmann's equation for the entropy given in Equation 5.31. To compute this entropy, we need to evaluate

$$S = k_B \ln \frac{N!}{N_p!(N - N_p)!}. \quad (5.34)$$

One of the key approximations needed in cases like this is known as the Stirling approximation, which in its simplest form can be written as

$$\ln N! \approx N \ln N - N. \quad (5.35)$$

The origins of this approximation are taken up in The Math Behind the Models below and in the problems at the end of the chapter. In the context of our DNA–protein problem, if we invoke the Stirling approximation, we find

$$S = -k_B N[c \ln c + (1 - c) \ln (1 - c)], \quad (5.36)$$

where we have introduced the more convenient concentration variable, $c = N_p/N$. The entropy as a function of concentration is shown in Figure 5.26. The key insight to emerge from this expression is the way in which the number of different ways of arranging the two species of interest depends upon their relative numbers. We see that the entropy is maximal when half of the sites are occupied—this situation reflects the fact that this concentration permits the most distinct arrangements.

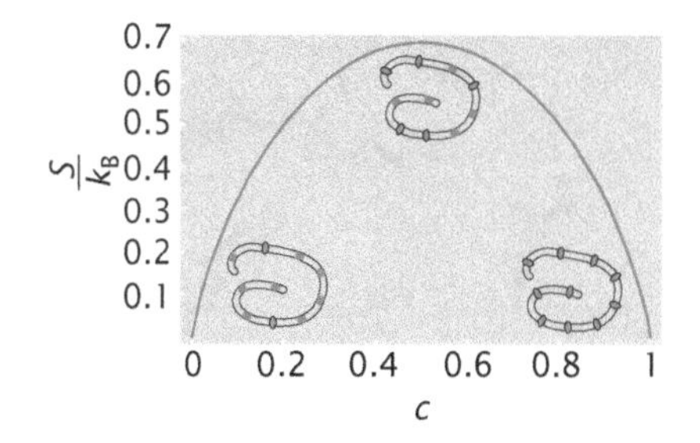

Figure 5.26: Entropy as a function of concentration of DNA binding proteins. The schematics show a DNA molecule with binding sites labeled in red and with DNA binding proteins as green ovals on the binding sites. In going from left to right, the fraction of sites occupied by proteins increases.

The Math Behind the Models: The Stirling Approximation The Stirling approximation arises as a result of the need to evaluate expressions of the form ln $N!$. The simplest heuristic argument to derive the result is based on the observation that

$$\ln N! = \ln\,[N(N-1)(N-2)\times\cdots\times 1]. \tag{5.37}$$

On the other hand, by virtue of the property of logarithms that $\ln AB = \ln A + \ln B$, we can rewrite Equation 5.37 as

$$\ln N! = \sum_{n=1}^{N} \ln n. \tag{5.38}$$

We can now replace this sum with the approximate integral

$$\sum_{n=1}^{N} \ln n \approx \int_{1}^{N} \ln x\, dx = N \ln N - N. \tag{5.39}$$

In the problems at the end of the chapter, this approximation is treated more carefully.

MATH

5.5.1 Entropy and Hydrophobicity

To gain a little more practice in the use of the entropy idea, we consider a toy model of one of the most important molecular driving forces in biological systems, namely, the hydrophobic effect. The qualitative idea is that when a hydrophobic molecule is placed in water, it prevents the water molecules in its vicinity from participating in some of the hydrogen bonds that they would have in the hydrophobic molecule's absence. An example of the highly idealized hydrogen bonding network in water is illustrated in Figure 5.27.

Hydrophobicity Results from Depriving Water Molecules of Some of Their Configurational Entropy

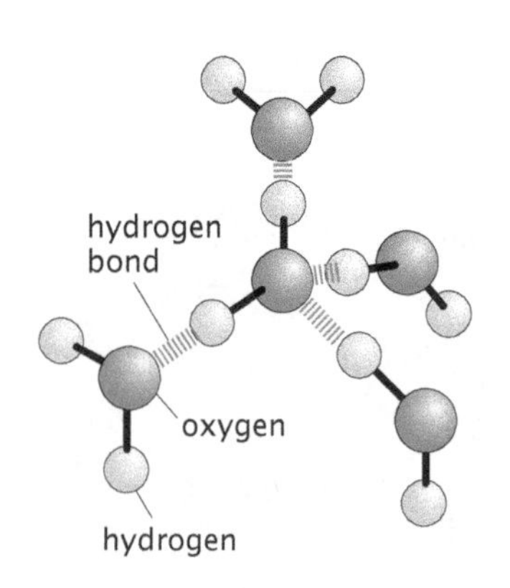

Figure 5.27: The hydrogen bonding network in water. Water molecules participate in hydrogen bonding (illustrated by the striped lines joining adjacent water molecules). A given water molecule can be idealized as having neighbors arranged in a tetrahedral structure.

The objective of the present section is to make an estimate of the magnitude of these hydrophobic effects. The basic thrust of the argument will be to describe how nonpolar molecules in solution deprive water molecules of the capacity to engage in hydrogen bonding and thereby take away part of their orientational entropy. This simple model borrows from a model originally formulated by Pauling (1935) to capture the entropy of ice. With this mechanism in hand, we then carry out numerical estimates of the size of this effect.

The structural idea suggested by Figure 5.27 is that the oxygen atoms of neighboring water molecules form a tetrahedral network. As further suggested by Figure 5.27, these water molecules form a dynamic network of hydrogen bonds, where each oxygen, on average, makes hydrogen bonds with two of the four water molecules surrounding it. A useful conceptual framework for thinking about hydrophobicity is that when nonpolar molecules are placed in solution, the water molecules that neighbor the nonpolar molecule of interest have a restricted set of choices for effecting such hydrogen bonding. We can coarse grain the continuum of possible orientations available to a water molecule to the six distinct orientations shown in Figure 5.28. As a result, it is possible to estimate the entropic

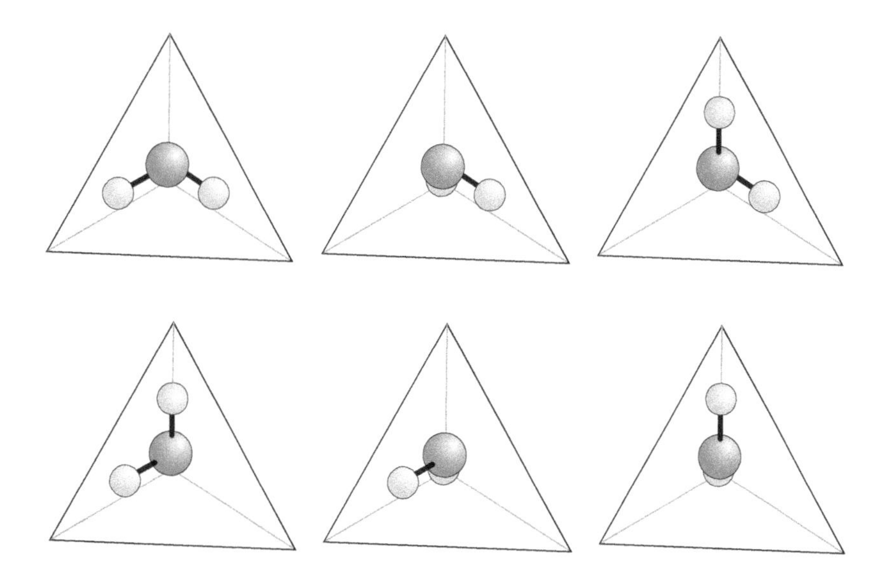

Figure 5.28: Orientations of water molecules in a tetrahedral network. Each image shows a different arrangement of the water molecule that permits the formation of hydrogen bonds with neighboring water molecules. The hydrogen bonds are in the directions of the vertices that are *not* occupied by hydrogens in the figure (Adapted from K. Dill and S. Bromberg, Molecular Driving Forces, 2nd ed., Garland Science, 2011.)

disadvantage associated with the presence of nonpolar molecules (see Dill and Bromberg (2011) for a clear description of this effect).

The six orientations that a water molecule can assume derive from the six ways of choosing to point the hydrogen atoms associated with the water molecule of interest towards the vertices of a tetrahedron. If one of the four water molecules in its immediate vicinity is replaced by a nonpolar molecule, then the number of available orientations drops to three since one of the possible hydrogen bonding partners is now gone. For example, if we assume that the neighboring water molecule in the direction of the lower right-hand vertex of Figure 5.28 is removed, this means that hydrogen bonds can no longer be formed with the oxygen on the water shown in the figure and the hydrogen atoms on the missing water molecule. As a result, the three configurations in the bottom of Figure 5.28 are now forbidden. This simple model predicts that the presence of the nonpolar molecule deprives each neighboring water molecule of half of its possible orientations as a participant in the hydrogen bonding network. The entropy change of each such water molecule is given by

$$\Delta S_{\text{hydrophobic}} = \underbrace{k_B \ln 3}_{\text{constrained } H_2O} - \underbrace{k_B \ln 6}_{\text{unconstrained } H_2O} = -k_B \ln 2. \quad (5.40)$$

Thus far, we have determined the entropy loss per water molecule. To make our estimate useful, we now need to estimate the number of water molecules that are impacted by the presence of the nonpolar (that is, hydrophobic) molecule of interest.

We can obtain a quantitative description of the hydrophobic cost to place a hydrophobic molecule in water as

$$\Delta G_{\text{hydrophobic}}(n) = nk_B T \ln 2, \quad (5.41)$$

where n is the number of water molecules adjacent to the nonpolar molecule of interest. Here we have accounted only for the entropic contribution to the free energy cost, which is given by $-T\Delta S_{\text{hydrophobic}}$. One particularly useful way of characterizing our result is to say that the presence of hydrophobic molecules incurs some free-energy cost per unit area ($\gamma_{\text{hydrophobic}}$) and hence that the free-energy cost to embed a given hydrophobic molecule in water is obtained as $\Delta G_{\text{hydrophobic}} = \gamma A$, where A is the effective area of the interface between the hydrophobic molecule and the surrounding water. As said above, a more convenient representation of this result is to assign a free energy per unit area, which can be obtained by determining the area per water molecule that is contributed. Using the simple estimate that 10 water molecules cover an area of approximately 1 nm^2 and that $\ln 2 \approx 0.7$, we can see that the interfacial free energy required to submerse a hydrophobic object in water is roughly 7 k_BT/nm^2. For a small molecule such as oxygen (O_2) that has an approximate surface area between 0.1 and 0.2 nm^2, the energetic cost of putting this molecule in water is roughly 1 k_BT. As a result, oxygen can be readily dissolved in water, even though it is nonpolar and cannot form hydrogen bonds. However, larger hydrophobic molecules comparable in size to a protein or even a sugar molecule would require significant free-energy input to dissolve in water.

The hydrophobic effect is responsible for the everyday observation that oil and water do not mix. The free-energy cost resulting from this simple model for putting an individual hydrocarbon molecule such as octane into a watery environment is of the order of 15 k_BT. Each addition of a new molecule of octane to water costs the same amount of free energy additively. However, if the octane molecules clump together, the total surface area of the clump may be much less than the sum of their individual surface areas. In water at room temperature, where individual molecules can jiggle around rapidly, it usually takes no more than a few seconds for the molecules to sort themselves out such that the interfacial surface is minimized.

Amino Acids Can Be Classified According to Their Hydrophobicity

The energies associated with the hydrophobic effect are extremely important at both the molecular and cellular scales in dictating the formation of structures. For example, consider a protein that contains a variety of amino acid side chains, some of which are hydrophilic (able to form hydrogen bonds with water) and others of which are hydrophobic. From the argument outlined above, it is clear that there must be a free-energy cost for the hydrophobic side chains to exist in an aqueous environment. As a first approximation, the folding of proteins into defined three-dimensional structures can be thought of as an application of the principle of the separation of oil and water. The protein is made up of an elastic backbone from which dangle a mixture of hydrophobic and hydrophilic amino acid side chains. As described above, the entropic demands of the system will tend to force the hydrophobic side chains to gather together in a sequestered internal oil droplet at the heart of the protein. Hydrophilic amino acid side chains will tend to remain on the protein surface where they can form hydrogen bonds with water. This concept was illustrated in Figure 5.8 (p. 203).

To make an accurate quantitative model describing the role of the hydrophobic effect in protein folding, we would have to know the relative free-energy cost for water exposure for each of the 20 amino acids. However, as a useful simplified strategy for building intuition, we will start by simply dividing the amino acids into two broad groups, one that includes all hydrophobic residues (H) and the other that includes all hydrophilic or polar residues (P). As we will explore in more detail in Chapter 8, this drastically oversimplified model provides useful estimates for many aspects of structure.

As a result of the arguments given above, we can rank the various hydrophobic amino acids most simply through reference to the effective area that they present to the surrounding water. Within this framework, the hydrophobic cost of exposing such a residue is of the form

$$\Delta G_{\text{hydrophobic}} \approx \underbrace{\gamma_{\text{hydrophobic}}}_{\text{cost/area}} \underbrace{A_{\text{hydrophobic}}}_{\text{hydrophobic area}} . \tag{5.42}$$

The detailed implementation of this strategy is left to the reader in the problems at the end of the chapter.

When in Water, Hydrocarbon Tails on Lipids Have an Entropy Cost

These same ideas can also be used to give an approximate description of the free energy associated with lipids when they are isolated in solution. Lipid molecules are characterized by polar head groups that are attached to long, fatty acid tails that are hydrophobic. The simple and useful idea in this case is to consider each such tail as though it presents a cylinder of hydrophobic material and to assign a free-energy cost to isolated lipids given by the product of the hydrophobic free-energy cost per unit area computed above and the area presented by the "cylinder" from the lipid tails. The free-energy cost associated with isolated lipids leads to the key driving force resulting in the formation of lipid bilayers.

5.5.2 Gibbs and the Calculus of Equilibrium

We have already observed that the principle of minimum potential energy presides over questions of the mechanical equilibrium configurations of systems at zero temperature. We now enter into a discussion of a principle that plays precisely the same role for systems in equilibrium at finite temperature. Once again, we will see that the equilibrium edict can be couched in variational language (that is, as a minimization problem). In particular, our discussion will culminate with the statement that out of all competing states of a system, the equilibrium state minimizes the relevant free energy. However, before discussing free-energy minimization, we reflect on the even more fundamental embodiment of the second law of thermodynamics, namely, that for a closed system in equilibrium the entropy is a *maximum*.

Thermal and Chemical Equilibrium are Obtained by Maximizing the Entropy

The study of thermal and chemical equilibrium is presided over by the second law of thermodynamics. In words, this law can be stated as the assertion that the macroscopic equilibrium state of an *isolated* system is that state that can occur in the largest number of microscopic ways (that is, that maximizes the entropy). Stated differently, when faced with the question of choosing from the space of all macroscopic competitors, choose that one that has the most microscopic representatives. This is the governing principle of all of thermodynamics in the same sense as all of mechanics derives from Newton's second law, $F = ma$.

How the injunction of entropy maximization plays out in real but simple circumstances is illustrated in an isolated system like that

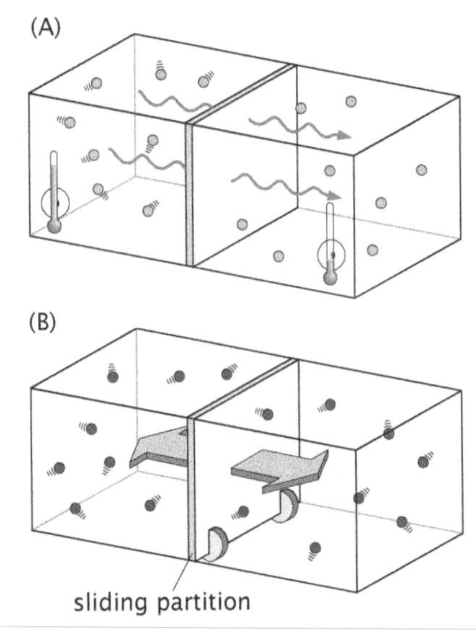

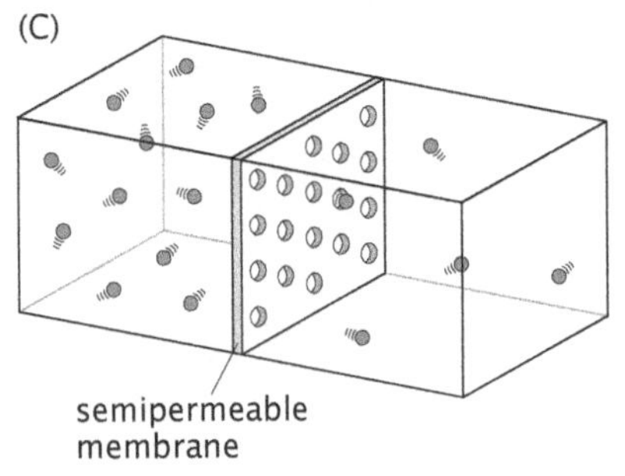

Figure 5.29: Schematic representation of an isolated system with two subcompartments and a barrier between these two compartments that permits transfer of (A) energy, (B) volume, and (C) particles.

shown in Figure 5.29, which has an internal partition. Our use of the word "isolated" refers to the fact that the contents of the container are entirely indifferent to anything and everything that we do outside—hence, we are unable to communicate with that system by doing work on it, by heating it, or by applying any sort of fields such as magnetic or electric fields. The thought experiment of interest here involves the idea of spontaneously removing the constraint implied by the internal partition. For example, as shown in Figure 5.29(A), if we permit the flow of energy between the two compartments, there will be a transfer of energy until the entropy is maximized (which corresponds to equality of temperature). As shown in Figure 5.29(B), if the brakes that hold the partition fixed are released, this partition is free to slide until the overall entropy of the system has reached a maximum. Depending upon which side has the greater pressure, the partition will roll either to the left or to the right until the pressures on the two sides are equal, resulting in the state of maximum entropy. Finally, the case of most biological interest is that shown in Figure 5.29(C), in which the internal partition is rendered permeable to the flow of particles. In this case, the particles will flow across the partition until the entropy is maximized, a state that we will show later corresponds to equality of chemical potentials in the two regions.

To show that the idea of entropy maximization leads to consequences that are consistent with our physical intuition, we reason quantitatively about the isolated system with a partition shown in Figure 5.29. We claimed that upon removal of the constraints represented by the partition, there would be a redistribution of energy (heat will flow), an adjustment in the volume (the partition will roll in one direction or the other), and a redistribution of particles (particles will diffuse) until the entropy of the closed system reaches a maximum. Mathematically, we can see this by examining $S_{tot} = S_1(E_1, V_1, N_1) + S_2(E_2, V_2, N_2)$, the total entropy, which is an additive function of the entropy on the two sides of the partition. Note that because our system is isolated, there are constraints of the form $E_{tot} = E_1 + E_2$, $V_{tot} = V_1 + V_2$, and $N_{tot} = N_1 + N_2$, where E_{tot} is the total energy of the closed system, V_{tot} is the total volume, and N_{tot} is the total number of particles. As noted above, when the conditions implied by the initial constraints are relaxed (for example, the brake is released and the partition can roll), there will be a spontaneous change in the state of the system until the system entropy is maximized. Mathematically, for the case in which the partition permits exchange of energy, the entropy maximization takes the form

$$dS = \left(\frac{\partial S_1}{\partial E_1}\right) dE_1 + \left(\frac{\partial S_2}{\partial E_2}\right) dE_2 = 0. \tag{5.43}$$

If we now invoke the fact that the total energy is conserved, we have $dE_2 = -dE_1$, which when substituted into Equation 5.43 yields

$$\left(\frac{\partial S_1}{\partial E_1} - \frac{\partial S_2}{\partial E_2}\right) dE_1 = 0. \tag{5.44}$$

To see our derivation through to the end, we now introduce the thermodynamic definition of temperature, $(\partial S/\partial E)_{V,N} = 1/T$, which reveals that our result is equivalent to the statement that $T_1 = T_2$. That is, when the partition permits energy transfer, heat will flow until the temperatures on the two sides are equal. Note that we cannot derive every result from thermodynamics here and encourage readers unfamiliar with thermodynamic identities to consult the Further Reading at the end of the chapter.

The argument goes in precisely the same way when we consider the case where the brakes are removed and the partition is permitted to slide. In this case it is the volume that will be adjusted in such a way as to maximize the system entropy. In particular, the condition of entropy maximization is

$$dS = \left(\frac{\partial S_1}{\partial V_1}\right) dV_1 + \left(\frac{\partial S_2}{\partial V_2}\right) dV_2 = 0. \quad (5.45)$$

Once again, we exploit the constraint, which tells us that $dV_2 = -dV_1$, resulting in

$$\left(\frac{\partial S_1}{\partial V_1} - \frac{\partial S_2}{\partial V_2}\right) dV_1 = 0. \quad (5.46)$$

At this point, we use the thermodynamic identity that $p/T = (\partial S/\partial V)_{E,N}$, resulting in the observation that entropy maximization corresponds to equality of pressure.

The case that is probably of greatest biological interest is that in which the partition permits the flow of particles. In this case, entropy maximization corresponds to the statement

$$dS = \left(\frac{\partial S_1}{\partial N_1}\right) dN_1 + \left(\frac{\partial S_2}{\partial N_2}\right) dN_2 = 0. \quad (5.47)$$

Exploiting the constraint that the overall number of particles is fixed, we have that $dN_1 = -dN_2$, resulting in

$$\left(\frac{\partial S_1}{\partial N_1} - \frac{\partial S_2}{\partial N_2}\right) dN_1 = 0. \quad (5.48)$$

Here we use the result that the chemical potential is defined in terms of the entropy change as $\mu/T = -(\partial S/\partial N)_{E,V}$, and hence entropy maximization implies equality of chemical potentials on both sides of the partition.

The key point of these arguments has been to highlight the variational description of thermodynamic equilibrium. That is, the privileged equilibrium state of a system can be found by maximizing the entropy. The driving forces implied by entropy maximization have a variety of interesting consequences, which we take up presently. Though many problems of interest will require a more sophisticated implementation of the second law in the form of the principle of minimum free energy, there are a number of problems where entropy maximization can be used directly. Important examples include the notion of an entropic spring, which describes the force–extension characteristics of molecules like DNA, the notion of depletion forces between macromolecular assemblies in solution, and the origins of osmotic pressure. In anticipation of the role of entropy maximization in coming sections, it is of interest here to show how order can arise from entropy maximization.

5.5.3 Departure from Equilibrium and Fluxes

So far, our discussion has emphasized the nature of the terminal privileged states that are dictated by the second law of thermodynamics.

And yet, it is clear that one of the defining features of living organisms is that they are not in equilibrium! We can go farther than our previous analysis by thinking about what happens when we start in a situation in which the system is not yet at equilibrium. In these cases, we know that the system will evolve in such a way that $\Delta S > 0$. If we consider the case in which our two subsystems permit the flow of particles, then the change in entropy can be written as

$$\left(\frac{\partial S_1}{\partial N_1} - \frac{\partial S_2}{\partial N_2}\right) \mathrm{d}N_1 \geq 0, \tag{5.49}$$

implying that mass will flow in such a way to guarantee that the expression on the left is positive. Similar arguments can be made for the volume and the energy change during a spontaneous process. We can think of the term in parentheses in the inequality 5.49 as constituting the "driving force" for mass transfer.

Often, the relation between driving force and transfer of quantities such as energy and particles is captured in simple linear relations of the form

$$\text{flux} = \text{transport coefficient} \times \text{driving force}. \tag{5.50}$$

Flux is defined as the number of units of the quantity of interest (energy or particle number, for example) crossing unit area per unit time. We will return to these kinds of descriptions repeatedly throughout the book.

5.5.4 Structure as a Competition

Thus far, we have asserted that equilibrium structures reflect energy minima (zero temperature) and entropy maxima (finite temperature). However, the case of greatest interest for biological model building is associated with a variational middle ground between the strictly mechanical ambition of minimizing energy and the statistical ambition of maximizing entropy. In particular, both the *in vitro* assays of solution biochemistry and the *in vivo* chemical action of cellular life reflect a more subtle situation in which the system of interest can exchange energy or matter (or both) with the surroundings. The variational injunction in these cases is to minimize the free energy, which can be thought of intuitively as teasing out the competition between maximizing multiplicity and minimizing energy. The variational principle that is equipped to permit the playing out of this competition is the principle of minimum free energy introduced in words earlier in the chapter.

Free Energy Minimization Can Be Thought of as an Alternative Formulation of Entropy Maximization

As developed above, Gibbs' calculus of equilibrium asserts that when contemplating *isolated* systems, our best guess as to the equilibrium state is that macroscopic state that can happen in the most ways microscopically. This injunction is translated into mathematical terms by virtue of the introduction of the entropy, which we are asked to maximize in order to find equilibria. On the other hand, there are a number of problems of interest for which the system may not be thought of as isolated. Indeed, most interesting systems (like

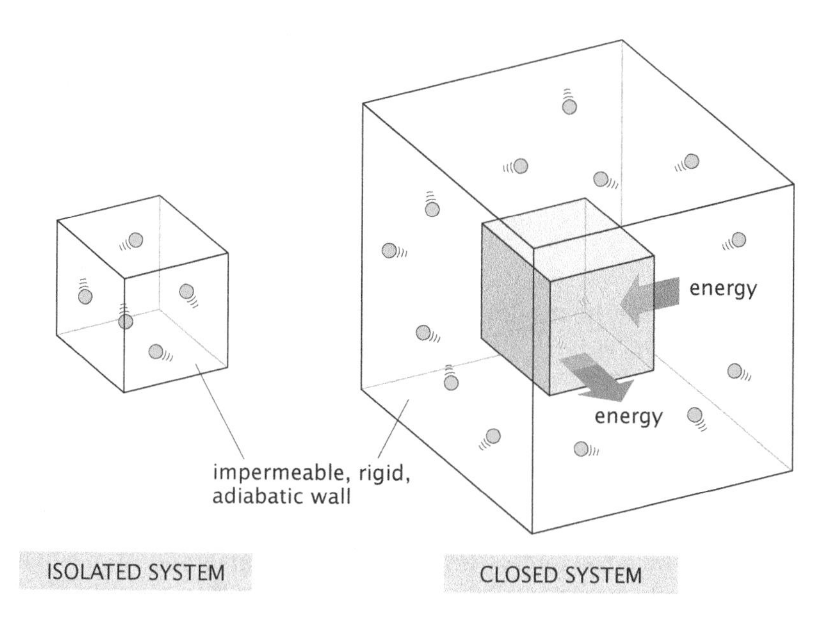

Figure 5.30: Isolated and closed systems. The isolated system is unable to exchange energy and matter with the rest of the world. In the closed system, there is an exchange of energy between the system and the surrounding reservoir.

macromolecules in the cell, macromolecular assemblies such as RNA polymerase or the ribosome, or indeed cells themselves) are in contact with an external medium with which they can exchange energy and matter. Whether we think of a cell in the ocean or DNA polymerase in a thermal cycler used for doing the polymerase chain reaction, our system of interest is in contact with the rest of the world. Interestingly, the problem of maximizing the entropy of a system plus the reservoir with which it is in contact is equivalent to minimizing the free energy of just the system itself. The beauty of this principle is that it allows us to dismiss the huge potential complexity engendered by the fact that our system is in contact with a reservoir and to consider only those degrees of freedom that describe the system itself.

To see how this discussion goes, consider Figure 5.30, which shows examples of both isolated and closed systems. Note that a closed system is characterized by the ability to exchange energy with its environment. An open system is free to exchange both energy and matter with its environment. The theoretical trick that allows us to exploit the principle of entropy maximization in the context of these systems is to turn a closed system into an isolated system by putting our original system in contact with a reservoir. From this perspective, our overall system consists of the original system and its associated reservoir and its equilibrium is now dictated by entropy maximization. For example, in the *in vitro* situation of solution biochemistry, a test tube in contact with a water bath is an example of a closed system that can exchange energy with its environment. Similarly, within the confines of a cell, we can think of a particular site on a DNA molecule as the system of interest, and the reservoir as the DNA-binding proteins in the cytoplasm or on other nonspecific sites on the DNA.

To see the analysis of this problem through to the end, we now maximize the entropy of our composite system made up of the original system and reservoir. An alternative way of thinking of our requirement that the entropy be maximized is to say that during any spontaneous process that follows the removal of a constraint, the entropy will increase. We may write this statement as

$$\mathrm{d}S_{\mathrm{tot}} = \mathrm{d}S_{\mathrm{r}} + \mathrm{d}S_{\mathrm{s}} \geq 0. \qquad (5.51)$$

where S_s refers to the entropy of our system and S_r to the entropy of the reservoir. We may borrow from our knowledge of the first law of thermodynamics, which permits us to write the change in energy of the reservoir as

$$dE_r = T\,dS_r - p\,dV_r, \tag{5.52}$$

where $T\,dS_r$ is the heat added to the reservoir and $p\,dV_r$ is the work done by the reservoir. If we substitute this relation into our entropy inequality, we have

$$dS_s + \frac{dE_r}{T} + \frac{p}{T}\,dV_r \geq 0. \tag{5.53}$$

Like before, since the energy and volume are conserved in our overall isolated system, we have $dE_r = -dE_s$ and $dV_r = -dV_s$, resulting in

$$dS_{tot} = dS_s - \frac{1}{T}\,dE_s - \frac{p}{T}\,dV_s \geq 0. \tag{5.54}$$

Finally, if we multiply both sides of our equation by $-T$, resulting in a change in direction of the inequality, we are left with

$$dG = d(E_s + pV_s - TS_s) \leq 0, \tag{5.55}$$

where we have introduced the Gibbs free energy $G = E - TS + pV$ and have shown that in a spontaneous process, the free energy will be reduced or, what is the same, that the Gibbs free energy will take its minimum value in the equilibrium state. We remind the reader that the statement that the Gibbs free energy is minimized in equilibrium is founded upon the more fundamental statement of entropy maximization for the entire system that is built up from the subsystem of interest and the reservoir. Further, note that once we elect to invoke the Gibbs free energy, there is no reference to the coordinates associated with the reservoir, other than to say that our system of interest is kept at fixed temperature T and pressure p by virtue of its contact with the reservoir.

5.5.5 An Ode to ΔG

The Free Energy Reflects a Competition Between Energy and Entropy

The interesting physical insight tied to the free energy is that equilibrium structures reflect a competition between energetic and entropic influences. If the temperature goes to zero, we recover the minimization of the energy E, which is the principle that presides over questions of strictly *mechanical* equilibrium. On the other hand, as the temperature rises, the entropic term makes itself heard with increasing forcefulness until at sufficiently high temperatures, it dominates the decision concerning the equilibrium state.

To be specific, this battle between energy and entropy is well illustrated by the examples presented in Figure 5.7 (p. 202). For example, in the context of the folding of a protein into its native state, the energetic component of the problem has to do with the formation of various native contacts between amino acids that result in a net energy lowering, while the entropic part of the free-energy budget has to do with the number of alternative conformations available to the protein when it is *not* folded. At high enough temperatures, the entropic imperative must be obeyed and the protein denatures. A similar competition is seen in the context of the charges on proteins, where the energy of electrostatic interactions tends to keep charges localized to their molecular hosts while the entropic part of the free-energy budget prefers to see these charges delocalized. As a final case study in deconstructing the free energy in the context of the examples of Figure 5.7, consider the case of binding of ligands to a protein. In this case, the ligands are afforded an entropic advantage if they wander around in solution. On the other hand, binding to their molecular hosts confers an energetic advantage and the interplay between these competing demands is the province of free-energy minimization.

In the coming chapters, we will invoke this idea of a competition between energetic and entropic factors repeatedly in contexts ranging from protein folding to the distribution of ions around a ribosome to the deflection of biofunctionalized cantilevers in the presence of particular antigens. In each case, our arguments will be formulated first in terms of an energetic term that tends to pull the system in one direction—for example, the Coulomb attraction between ions in solution and some macromolecule will tend to localize those ions near the macromolecule. The second competing term will reflect the will of the entropic term, which will favor the spreading out of the ions around the macromolecule. Such arguments ultimately serve as the concrete outcome of the present chapter, which has argued for the idea that the question of equilibrium structures can be seen as one of minimization of the relevant potential. As such, the free energy presides over all questions requiring that we determine the equilibrium state. Note that there are different free energies that are most convenient (Helmholtz free energy, Gibbs free energy) and, for simplicity, we will ignore these subtleties and always use the symbol G for the free energy in the remainder of the book.

5.6 Summary and Conclusions

Much of the busy activity of cellular life involves transformations of matter and energy. In this chapter, we showed that the study of these physical and chemical transformations can be couched in the language of finding the maximum (entropy) or minimum (potential energy) value of some physical quantity. In particular, we argued that Newtonian statics can be reformulated through the idea that we seek the lowest value of the potential energy of the system that is consistent with whatever constraints are imposed. Though these mechanical principles are important, we saw that the principle of minimum free energy is of even greater biological importance. In particular, we have argued that isolated systems containing many particles (and thus which can be realized by astronomical numbers of microscopically distinct but macroscopically identical states) can be thought of as satisfying a different variational imperative, namely, the maximization of their multiplicity. The calculus of equilibrium for many-particle systems is founded on the idea that macroscopic states are those which can be realized in the largest number of ways microscopically. Coming chapters will show how this simple idea can be used to understand a huge variety of different biological phenomena.

5.7 Appendix: The Euler–Lagrange Equations, Finding the Superlative

Finding the Extrema of Functionals Is Carried Out Using the Calculus of Variations

This chapter centered on our ability to find the maxima and minima of energies, entropies, and free energies. On the other hand, as we saw in Figure 5.18, often the energy and free energy dictated by our biological problems will be functionals rather than functions. This raises the mathematical question: given an infinite set of competitor functions, how can we find that function that minimizes the energy or free-energy functional? The aim of the present discussion is to generalize the earlier discussion based on ordinary calculus and to examine the functional analog of finding the extremum of a function. For a deeper discussion of these issues, we refer the reader to both Lanczos (1986) and Gelfand and Fomin (2000).

The Euler–Lagrange Equations Let Us Minimize Functionals by Solving Differential Equations

In many cases, the type of functional minimization described above can be written in a very specialized form, namely, as the search for that function that leads to an extremum for an integral. In this case, we are asked to minimize a functional of the form

$$E[u(s)] = \int_{a_1}^{a_2} f(u(s), u'(s))\, ds. \tag{5.56}$$

As with the calculation of extrema of functions, the defining condition is that, for any "small excursion" about the extrema, there should be no first-order change in the value of the functional. To make this idea more concrete, we consider excursions of the form $\eta(s)$, and demand that

$$\frac{\delta E[u(s)]}{\delta u(s)} = \lim_{\varepsilon \to 0} \frac{E[u(s) + \varepsilon\eta(s)] - E[u(s)]}{\varepsilon} = 0. \tag{5.57}$$

If we reason by analogy with ordinary calculus, what this expression tells us is that if we have found the minimizing function $u(s)$, then any small excursion (namely, $\varepsilon\eta(s)$) about that minimum will lead to no change in the functional. The notation $\delta E[u(s)]/\delta u(s)$ reminds us that we are taking the "functional derivative" as opposed to the derivative of ordinary calculus. The class of admissible excursions, $\eta(s)$ is further specified by boundary conditions imposed on the competitor functions. For example, for the case when the values of the competitor functions are fixed at the boundaries, the admissible excursions must satisfy $\eta(a_1) = \eta(a_2) = 0$. The condition expressed in Equation 5.57 may now be written as

$$\frac{\delta E}{\delta u(s)} = \lim_{\varepsilon \to 0} \frac{1}{\varepsilon} \left[\int_{a_1}^{a_2} f(u(s) + \varepsilon\eta(s), u'(s) + \varepsilon\eta'(s))\, ds - \int_{a_1}^{a_2} f(u(s), u'(s))\, ds \right]. \tag{5.58}$$

We now consider a Taylor series expansion of the integrand $f(u(s) + \varepsilon\eta(s), u'(s) + \varepsilon\eta'(s))$, and, in particular, we consider such an expansion to first order, resulting in

$$f(u(s)+\varepsilon\eta(s), u'(s)+\varepsilon\eta'(s)) \approx f(u(s), u'(s)) + \varepsilon\frac{\partial f}{\partial u'}\eta'(s) + \varepsilon\frac{\partial f}{\partial u}\eta(s). \quad (5.59)$$

As a result, we may now write

$$\frac{\delta E}{\delta u(s)} = \int_{a_1}^{a_2}\left[\frac{\partial f}{\partial u'}\eta'(s) + \frac{\partial f}{\partial u}\eta(s)\right] \mathrm{d}s. \quad (5.60)$$

Until now, our results have shown us how to reexpress our original problem in terms of the rate of change of the function $f(u(s), u'(s))$. At this point, we have two essential steps that remain. First, we rearrange Equation 5.60 by exploiting a single integration by parts. In particular, we note

$$\int_{a_1}^{a_2}\frac{\partial f}{\partial u'}\eta'(s)\,\mathrm{d}s = \eta(s)\frac{\partial f}{\partial u'}\bigg|_{a_1}^{a_2} - \int_{a_1}^{a_2}\frac{\mathrm{d}}{\mathrm{d}s}\frac{\partial f}{\partial u'}\eta(s)\,\mathrm{d}s, \quad (5.61)$$

where the first term on the right-hand side of the equation is zero because $\eta(a_1) = \eta(a_2) = 0$ as dictated by boundary conditions. As a result of these manipulations, we may rewrite Equation 5.60 as

$$\frac{\delta E}{\delta u(s)} = \int_{a_1}^{a_2}\left[-\frac{\mathrm{d}}{\mathrm{d}s}\left(\frac{\partial f}{\partial u'}\right) + \frac{\partial f}{\partial u}\right]\eta(s)\,\mathrm{d}s. \quad (5.62)$$

We now carry out the second step, which is to acknowledge that the condition for an extremum really corresponds to the statement $\delta E/\delta u(s) = 0$. In light of this condition and as a result of the fact that $\eta(s)$ is arbitrary, we are left with

$$\frac{\mathrm{d}}{\mathrm{d}s}\left(\frac{\partial f}{\partial u'}\right) - \frac{\partial f}{\partial u} = 0. \quad (5.63)$$

What we have learned is that the problem of minimizing a functional is equivalent to that of solving a differential equation. In particular, the differential equation associated with variational problems like that given above is the so-called Euler–Lagrange equation of the problem of interest. The beauty of this result is that it has replaced a problem that we do not know how to do, namely, minimizing a functional, with another that we might know how to do, namely, solving a differential equation. The minimization of a functional to solve problems of interest will show up again in Chapter 9 when we think about the charge distribution around a protein and in Chapter 11 when we work out the deformation in a lipid bilayer membrane induced by membrane proteins such as ion channels.

5.8 Problems

Key to the problem categories: • Model refinements and derivations • Estimates • Model construction • Database mining

• 5.1 Energy cost of macromolecular synthesis

In this problem, you will determine the biosynthetic cost of generating the amino acid serine from glucose used as food. The biosynthetic costs of all the amino acids are given in Table 5.1, and now you will get a first-hand sense of how to obtain those values. Visit the website www.ecocyc.org and find the metabolic pathways for serine synthesis and glycolysis. The starting molecule of the serine synthesis pathway is an intermediate in glycolysis. How many molecules of glucose must be taken up to provide the carbon skeleton used to make serine? Draw the biochemical pathway starting with glucose and ending in serine, labeling the energy-requiring and energy-generating steps. How many molecules of ATP are consumed and created along the way? How many reducing equivalents of NADH and NADPH are consumed or created along the way? In order to answer this question completely, you may need to consider a few other pathways as well. For example, it is likely that the displayed glycolysis pathway does not actually start with glucose. To get glucose 6-phosphate from glucose requires 1 ATP. One of the steps in the serine synthesis pathway is coupled to a conversion of L-glutamate to 2-ketoglutarate. You will have to look up the "glutamate biosynthesis III" pathway to determine the cost of regenerating L-glutamate from 2-ketoglutarate. Assuming that each NADH or NADPH is equivalent to 2 ATP, which is a reasonable conversion factor for bacteria, what is the net energy cost to synthesize one molecule of serine in units of ATP and units of $k_B T$?

• 5.2 The sugar budget revisited

In Chapter 3, we worked out the rate of sugar uptake to provide the construction materials for a dividing bacterium. However, as shown in this chapter, sugar molecules also provide the *energy* needed to perform macromolecular synthesis. Amend the estimate of Chapter 3 to include the fact that sugar supplies construction materials and the energy needed to assemble them. You might find it useful to look at the macromolecular energy costs revealed in Table 5.2. How many sugars are needed to provide the energy and construction materials for making a new cell? Make an estimate for the average rate of sugar uptake for a dividing bacterium in light of this amendment to our earlier estimates.

• 5.3 A feeling for the numbers: covalent bonds

(a) Based on a typical bond energy of 150 $k_B T$ and a typical bond length of 0.15 nm, use dimensional analysis to estimate the frequency of vibration of covalent bonds.

(b) Assume that the Lennard–Jones potential given by

$$V(r) = \frac{a}{r^{12}} - \frac{b}{r^6}$$

describes a covalent bond (though real covalent bonds are more appropriately described by alternatives such as the Morse potential that are not as convenient analytically). Using the typical bond energy as the depth of the potential and the typical bond length as its equilibrium position, find the parameters a and b. Do a Taylor expansion around this equilibrium position to determine the effective spring constant and the resulting typical frequency of vibration.

(c) Based on your results from (a) and (b), estimate the time step required to do a classical mechanical simulation of protein dynamics.

• 5.4 Taylor expansions

(a) Repeat the derivation of the Taylor expansion for $\cos x$ given in the chapter, but now expand around the point $x = \pi/2$.

(b) Do a Taylor expansion of the function e^x around zero. Generate a plot of the fractional error as a function of x for different orders of the approximation and a comparison of the function and the different-order expansions such as was shown in Figure 5.22.

(c) A one-dimensional potential-energy landscape is given by the equation $f(x) = x^4 - 2x^2 - 1$. Find the two minima and do a Taylor expansion around one of them to second order. Show the original function and the approximation on the same plot in the style of Figure 5.21.

• 5.5 A feeling for the numbers: comparing multiplicities

Boltzmann's equation for the entropy (Equation 5.30) tells us that the entropy difference between a gas and a liquid is given by

$$S_{\text{gas}} - S_{\text{liquid}} = k_B \ln \frac{W_{\text{gas}}}{W_{\text{liquid}}}. \tag{5.64}$$

From the macroscopic definition of entropy as $dS = dQ/T$, we can make an estimate of the ratios of multiplicities by noting that boiling of water takes place at fixed T at 373 K.

(a) Consider a cubic centimeter of water and use the result that the heat needed to boil water (the latent heat of vaporization) is given by $Q_{\text{vaporization}} = 40.66$ kJ/mol (at 100 °C) to estimate the ratio of multiplicities of water and water vapor for this number of molecules. Write your result as 10 to some power. If we think of multiplicities in terms of an ideal gas at fixed T, then

$$\frac{W_1}{W_2} = \left(\frac{V_1}{V_2}\right)^N. \tag{5.65}$$

What volume change would one need to account for the liquid/vapor multiplicity ratio? Does this make sense?

(b) In the chapter, we discussed the Stirling approximation and the fact that our results are incredibly tolerant of error. Let us pursue that in more detail. We have found that the typical types of multiplicities for a system like a gas are of the order of $W \approx e^{10^{25}}$. Now, let us say we are off by a factor of 10^{1000} in our estimate of the multiplicities, namely, $W = 10^{1000} e^{10^{25}}$. Show that the difference in our evaluation of the entropy is utterly negligible whether we use the first or second of these results for the multiplicity. This is the error tolerance that permits us to use the Stirling approximation so casually! (This problem was adapted from Ralph Baierlein, Thermal Physics, Cambridge University Press, 1999.)

• 5.6 Stirling approximation revisited

The Stirling approximation is useful in a variety of different settings. The goal of the present problem is to work through a more sophisticated treatment of this approximation than the simple heuristic argument given in the chapter. Our task is to find useful representations of $n!$ since terms of the form $\ln n!$ arise often in reasoning about entropy.

(a) Begin by showing that

$$n! = \int_0^\infty x^n e^{-x}\, dx. \tag{5.66}$$

To demonstrate this, use repeated integration by parts. In particular, demonstrate the recurrence relation

$$\int_0^\infty x^n e^{-x}\, dx = n \int_0^\infty x^{n-1} e^{-x}\, dx, \tag{5.67}$$

and then argue that repeated application of this relation leads to the desired result.

(b) Make plots of the integrand $x^n e^{-x}$ for various values of n and observe the peak width and height of this integrand. We are interested now in finding the value of x for which this function is a maximum. The idea is that we will then expand about that maximum. To carry out this step, consider $\ln(x^n e^{-x})$ and find its maximum—argue why it is acceptable to use the logarithm of the original function as a surrogate for the function itself, that is, show that the maxima of both the function and its logarithm are at the same x. Also, argue why it might be a good idea to use the logarithm of the integrand rather than the integrand itself as the basis of our analysis. Call the value of x for which this function is maximized x_0. Now expand the logarithm about x_0. In particular, examine

$$\ln[(x_0+\delta)^n e^{-(x_0+\delta)}] = n\ln(x_0+\delta) - (x_0+\delta) \quad (5.68)$$

and expand to second order in δ. Exponentiate your result and you should now have an approximation to the original integrand that is good in the neighborhood of x_0. Plug this back into the integral (be careful with limits of integration) and, by showing that it is acceptable to send the lower limit of integration to $-\infty$, show that

$$n! \approx n^n e^{-n} \int_{-\infty}^{\infty} e^{-\delta^2/2n}\, d\delta. \quad (5.69)$$

Evaluate the integral and show that in this approximation

$$n! = n^n e^{-n}(2\pi n)^{1/2}. \quad (5.70)$$

Also, take the logarithm of this result and make an argument as to why most of the time we can get away with dropping the $(2\pi n)^{1/2}$ term.

5.7 Counting and diffusion

In this chapter, we began practising with counting arguments. One of the ways we will use counting arguments is in thinking about diffusive trajectories.

Consider eight particles, four of which are black and four white. Four particles can fit left of a permeable membrane and four can fit right of the membrane. Imagine that due to random motion of the particles every arrangement of the eight particles is equally likely. Some possible arrangements are BBBB|WWWW, BBBW|BWWW, WBWB|WBWB; the membrane position is denoted by |.

(a) How many different arrangements are there?

(b) Calculate the probability of having all four black particles on the left of the permeable membrane. What is the probability of having one white particle and three black particles on the left of the membrane. Finally, calculate the probability that two white and two black particles are left of the membrane. Compare these three probabilities. Which arrangement is most likely?

(c) Imagine that, in one time instant, a random particle from the left-hand side exchanges places with a random particle on the right-hand side. Starting with three black particles and one white particle on the left of the membrane, compute the probability that after one time instant there are four black particles on the left. What is the probability that there are two black and two white particles on the left, after that same time instant? Which is the more likely scenario of the two?

(Adapted from Example 2.3 from K. Dill and S. Bromberg, Molecular Driving Forces, 2nd ed. Garland Science, 2011.)

5.8 Molecular driving forces

In Section 5.5.2, we showed that entropy maximization leads to our intuitive ideas about equilibrium. However, that discussion can be extended to reveal the direction of spontaneous processes. In particular, during any spontaneous process, we know that the entropy will increase. Use this fact in the form of the statement that $(\mu_2 - \mu_1)\, dN_1 \geq 0$ to deduce the role of differences in chemical potential as a "driving force" for mass transport. If $\mu_2 > \mu_1$, in which direction will particles flow? Make analogous arguments for the flow of energy and changes in volume.

5.9 Further Reading

Morton, O (2007) Eating the Sun: How Plants Power the Planet, Fourth Estate. This excellent book describes the story of how our modern understanding of photosynthesis was developed.

Neidhardt, FC, Ingraham, JL, & Schaechter, M (1990) Physiology of the Bacterial Cell, Sinauer Associates. Chapter 5 on "Biosynthesis and Fueling" is particularly relevant for the present chapter.

Gottschalk, G (1986) Bacterial Metabolism, Springer-Verlag. This book is full of interesting insights into the census and energy budget of bacterial cells.

Price, NC, Dwek, RA, Ratcliffe, RG, & Wormald, MR (2001) Principles and Problems in Physical Chemistry for Biochemists, 3rd ed., Oxford University Press. This excellent book describes many of the important chemical reactions of biology from a thermodynamic perspective.

Lemons, DS (1997) Perfect Form: Variational Principles, Methods, and Applications in Elementry Physics, Princeton University Press. Lemons' book is a pedagogical delight and offers a variety of interesting, yet simple, insights into how mechanics can be couched in the language of minimization.

Nahin, PJ (2004) When Least is Best: How Mathematicians Discovered Many Clever Ways to Make Things as Small (or as Large) as Possible, Princeton University Press. Nahin gives a sense of the wide range of problems that can be formulated as questions of minimization.

Acton, FS (1990) Numerical Methods That Work, The Mathematical Association of America. Acton is thoughtful and very amusing and describes some of the pitfalls of numerical mathematics. See also his book Real Computing Made Real: Preventing Errors in Scientific and Engineering Calculations (Princeton University Press, 1996) for more fun and insights.

Kittel, C, & Kroemer, H (1980) Thermal Physics, 2nd ed., W. H. Freeman. This book will provide background on thermodynamics and statistical mechanics for the interested reader.

Callen, HB (1985) Thermodynamics and an Introduction to Thermostatistics, 2nd ed., Wiley. Callen champions the idea of maximum entropy as the basis for finding equilibrium states. Our treatment mirrors his Chapter 2 on "The Conditions of Equilibrium."

Tanford, C (1962) Contribution of Hydrophobic Interactions to the Stability of the Globular Conformation of Proteins, *J. Am. Chem. Soc.* **84**, 4240, and Tanford, C (1991) The Hydrophobic Effect: Formation of Micelles and Biological Membranes, Krieger Publishing Company. Tanford has a nice touch in describing hydrophobicity.

5.10 References

Alberts, B, Johnson, A, Lewis, J, et al. (2008) Molecular Biology of the Cell, 5th ed., Garland Science.

Bustamante, C, Marko, JF, Siggia, ED, & Smith, S (1994) Entropic elasticity of λ-phage DNA, *Science* **265**, 1599.

Dill, K, & Bromberg, S (2011) Molecular Driving Forces, 2nd ed., Garland Science.

Gelfand, IM, & Fomin, SV (2000) Calculus of Variations (translated and edited by R Silverman), Dover Publications. This is an unabridged reprint of the work originally published by Prentice-Hall in 1963.

Lanczos, C (1986) The Variational Principles of Mechanics, Dover Publications. This is an unabridged and unaltered reprint of the 4th edition of the work originally published by the University of Toronto Press in 1970.

Lim, G, Wortis, M, & Mukhopadhyay, R (2002) Stomatocyte–discocyte–echinocyte sequence of the human red blood cell: Evidence for the bilayer-couple hypothesis from membrane mechanics, *Proc. Natl Acad. Sci. USA* **99**, 16766.

Mathews, CK, van Holde, KE, & Ahern, KG (2000) Biochemistry, 3rd ed., Addison Wesley Longman. Their Figure 13.6 gives an energy profile of the glycolysis pathway.

Neidhardt, FC, Ingraham, JL, & Schaechter, M (1990) Physiology of the Bacterial Cell: A Molecular Approach, Sinauer Associates.

Pauling, L (1935) The structure and entropy of ice and of other crystals with some randomness of atomic arrangement, *J. Am. Chem. Soc.* **57**, 2680. Pauling's paper clearly outlines the structural assumptions behind his ice model, which can also be used to estimate the hydrophobic cost of placing molecules in water.

Phillips, R, & Quake, S (2006) The biological frontiers of physics, *Phys. Today* **59**(5), 38.

Schaechter, M, Ingraham, JL, & Neidhardt, FC (2006) Microbe, ASM Press.

Smith, DE, Tans, SJ, Smith, SB, et al. (2001) The bacteriophage straight ϕ29 portal motor can package DNA against a large internal force, *Nature* **413**, 748.

Smith, SB, Cui, Y, & Bustamante, C (1996) Overstretching B-DNA: the elastic response of individual double-stranded and single-stranded DNA molecules, *Science* **271**, 795.

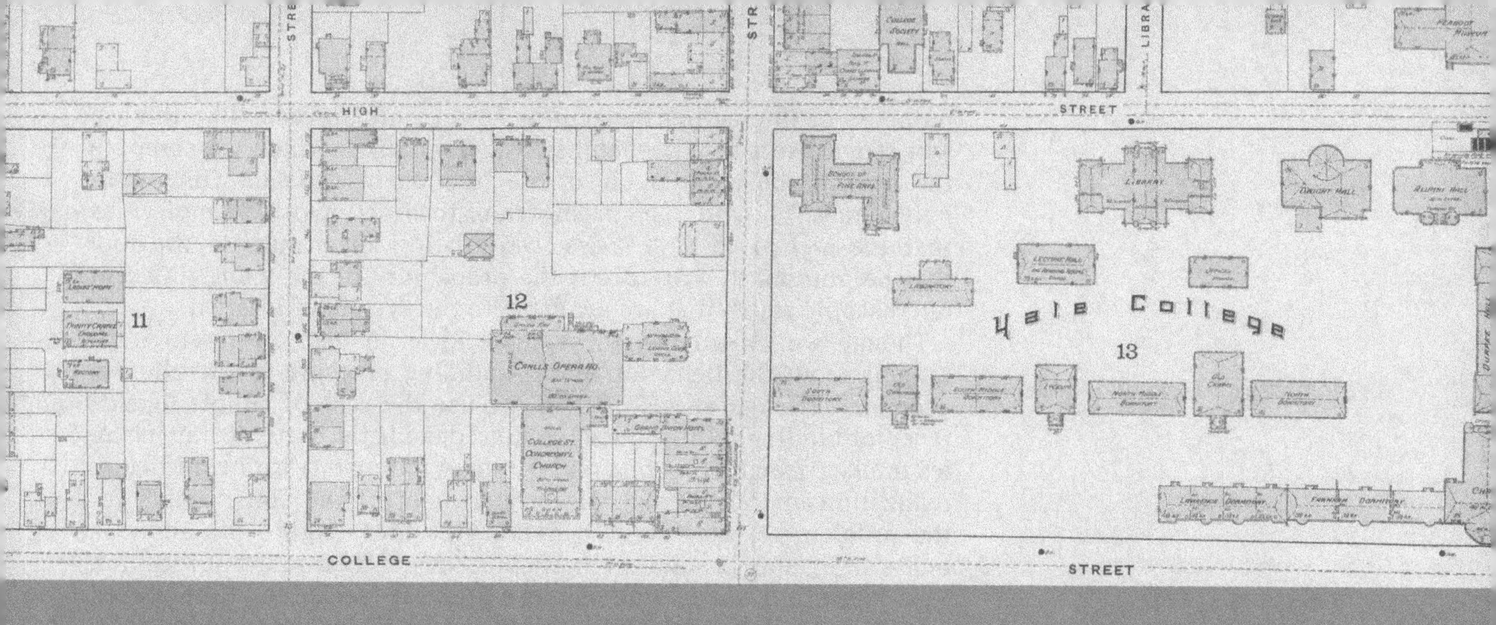

6 Entropy Rules!

Overview: In which we develop the tools of statistical mechanics and use them to study a variety of biological problems

In the previous chapter, we argued that despite the fact that cells are a profound example of systems that are out of equilibrium, in many instances we can choose some subset of a cell and proceed as though it is in mechanical or chemical equilibrium. That equilibrium, in turn, can often be seen as a reflection of the competing demands of lowering the energy and increasing the entropy. To go beyond an abstract statement that equilibrium states are free-energy minimizers, we need a way to explicitly compute the probability of all of the different microscopic arrangements of the system of interest. To that end, this chapter introduces the fundamental result of statistical mechanics, the Boltzmann distribution, which tells us how the probability of a given microstate of a system depends upon its energy. Once we have this distribution in hand, we use it to study a number of important biological problems, including ligand–receptor interactions, RNA polymerase binding to promoter sites on DNA, the origins of osmotic pressure, the law of mass action, and how the energy delivered by ATP hydrolysis depends upon the concentrations of ATP, ADP, and inorganic phosphate (P_i).

> "The fascination of a growing science lies in the work of the pioneers at the very borderland of the unknown, but to reach this frontier one must pass over well travelled roads; of these one of the safest and surest is the broad highway of thermodynamics."
>
> **G. N. Lewis and M. Randall**

6.1 The Analytical Engine of Statistical Mechanics

The goal of this section is to develop the formalism of statistical mechanics in a way that can be used for biological applications. Whether we think of the enzymes, nucleotides, and DNA that make up a PCR in solution, or the binding of RNA polymerase to its DNA target site (the promoter) in a living cell, these situations involve large numbers of interacting molecules. As a result, in cases like this we will often resort to a probabilistic description in which we compute the *average* behavior such as the fraction of cells in which RNA polymerase is bound to its promoter, or the fraction of DNA molecules that are melted in a PCR assay. Statistical mechanics is the tool that permits us to write down the probability of the many different microscopic states that are available to the system of interest.

Though we will invoke these same ideas in many different ways ranging from the behavior of DNA-binding proteins to the allowed geometric configurations of DNA, we use the case of simple ligand–receptor binding to introduce the first idea for the statistical mechanics toolkit, namely, that of a microstate. A microstate is one particular realization of the microscopic arrangement of the constituents for the problem of interest. To define this idea, Figure 6.1 shows a model of ligands in solution in the presence of their relevant receptor. This figure introduces one of our statistical mechanics workhorses, namely, a lattice model of the solution in which we imagine the solution as a series of tiny boxes amongst which we can distribute the ligands. When we use the term microstates in this setting we are referring to all of the different ways of arranging these L ligands in solution and on the receptor. The particular cases in the figure show several different microstates in which all of the ligands are free in solution and one microstate in which the receptor is occupied. The total number of microstates corresponding to the unoccupied receptor is given by

$$\text{number of microstates} = \frac{\Omega!}{L!(\Omega - L)!}, \tag{6.1}$$

where Ω is the number of tiny boxes available in the system, L is the number of ligands, and the symbol $\Omega!$ ("Ω-factorial") is shorthand for $\Omega(\Omega-1)(\Omega-2)\cdots\times 1$. The mathematics of this kind of counting argument is explained in more detail in the The Math Behind the Models below.

A second example of the idea of a microstate is illustrated in Figure 6.2. If we were to look down a microscope at fluorescently labeled DNA molecules such as those from bacteriophage λ with a genome length of 48 kb and a contour length of roughly 16 μm, what we would see is a fluctuating blob of fluorescence. If we had very high spatial and temporal resolution, what we would observe is the DNA molecule passing through a series of distinct conformations. The shapes of the molecule (that is, the microstates) can be characterized by using the distance along the molecule (labeled s) as a label to characterize a point on the molecule and the function $\mathbf{r}(s)$ to characterize the positions of all such points. In other words, the function $\mathbf{r}(s)$ characterizes the instantaneous microstate of the system.

It should be evident from the above two examples that there is no absolute definition of a microstate for a biological system of interest. In the DNA example, one might alternatively describe the configuration of the molecule by listing the instantaneous coordinates of all the atoms, in which case the microstates are all such lists of coordinates. Therefore the microstates one chooses to describe a system of interest depend on the particular question one is looking to answer. The description of DNA configurations in terms of $\mathbf{r}(s)$ is appropriate when analyzing single-molecule pulling experiments at low forces, but might be inadequate if we want to investigate structural changes at the base-pair level due to large forces.

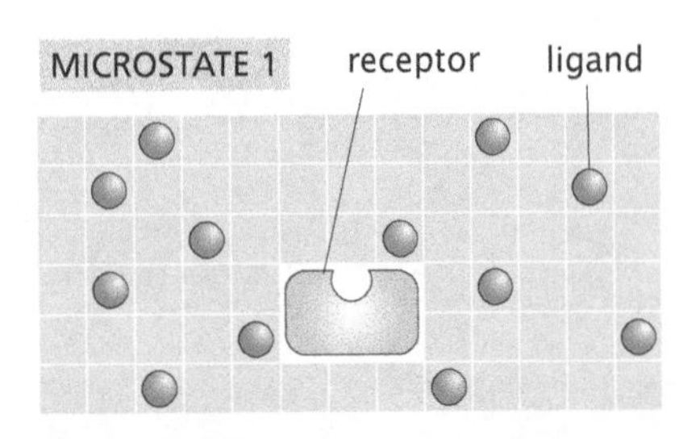

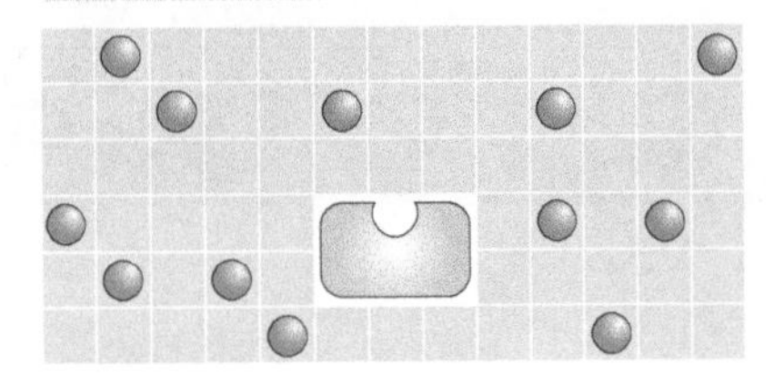

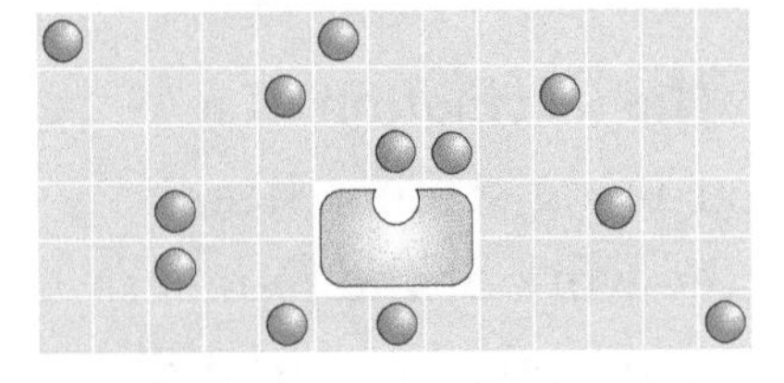

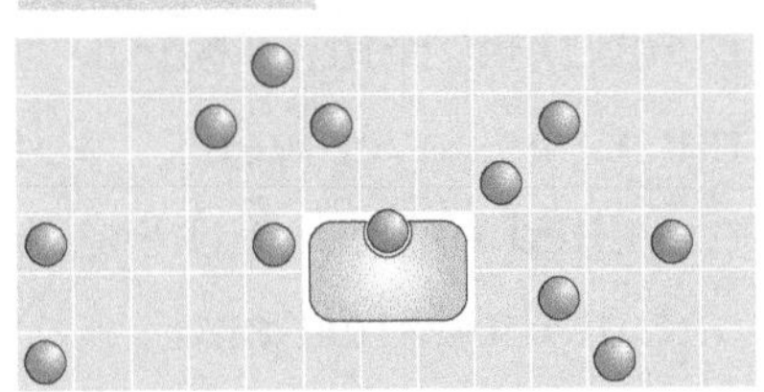

etc.

Figure 6.1: Simple model of ligand–receptor binding. The solution is treated using a "lattice model" in which the positions that can be occupied by ligands are dictated by a discrete set of lattice sites. The microstates of the system correspond to the different ways of arranging the L ligands among the different lattice sites. The first three microstates correspond to an unoccupied receptor and microstate 4 has the receptor occupied, leaving $L-1$ ligands in solution.

(A)

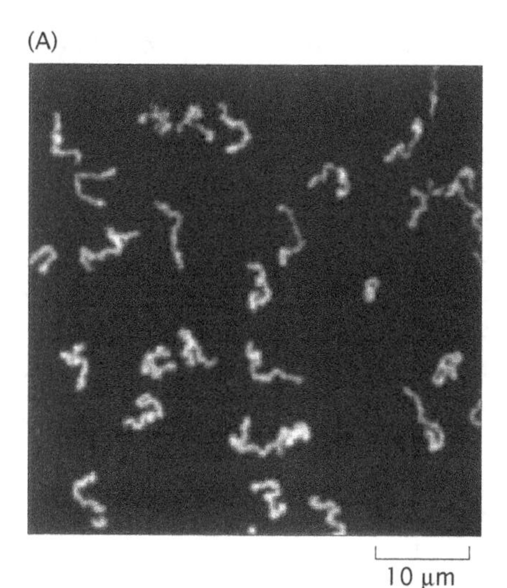

(B)

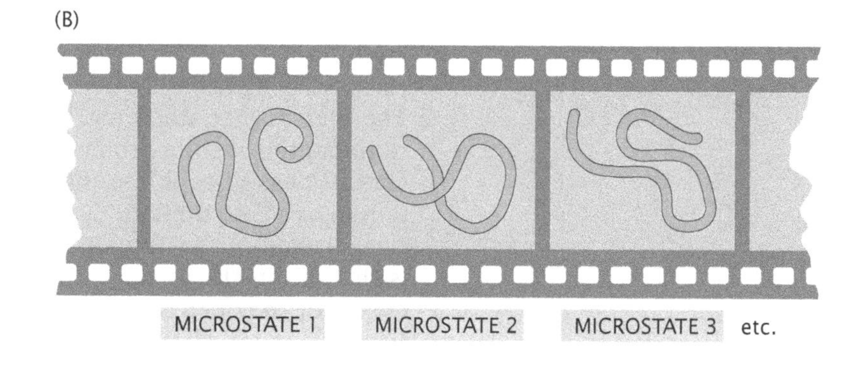

Figure 6.2: Microstates of DNA in solution. DNA molecules (and other polymers) jiggle around in solution exploring different conformational states. Each one of these conformations represents a different microstate. (A) Fluorescence microscopy image of λ-DNA confined to a surface. (B) Film strip showing conformations of a single DNA molecule at different instants. (A, adapted from B. Maier and J. O. Rädler, *Phys. Rev. Lett.* 82:1911, 1999.)

MATH

The Math Behind the Models: Counting Arrangements of Particles One of the mathematical tools we will use repeatedly throughout the book is counting up the number of ways of arranging L indistinguishable objects in Ω boxes without permitting any box to have more than one object. When we put the first object in a box, we have Ω choices for where it may be placed. Once that first object has been placed, there are only $\Omega - 1$ boxes left to put one of the remaining $L - 1$ objects. This same basic argument is repeated L times with the result that we have $\Omega(\Omega-1)(\Omega-2) \times \cdots \times [\Omega-(L-1)]$ ways of distributing the L objects. This can be written more simply as

$$\text{number of arrangements} = \frac{\Omega!}{(\Omega - L)!}. \qquad (6.2)$$

However, so far, we have neglected the fact that the L objects are indistinguishable. For example, if we have L shiny black billiard balls, we cannot distinguish the "state" in which one of the balls is in box 5 and the second in box 12 from the opposite case. Another way to say this is that if we were to take a picture of the occupancy of the different boxes and then rearranged the objects while using the same subset of boxes, the photographs would reveal no difference. To cure this overcounting, we have to divide out the $L!$ ways of arranging the L objects in the boxes they occupy. As a result, the actual number of distinct "states" for these L objects in Ω boxes is

$$\text{number of states} = \frac{\Omega!}{L!(\Omega - L)!}. \qquad (6.3)$$

This important equation will be used again and again throughout the book. It will allow us to examine the competition between translational entropy and binding energy in a number of different biologically relevant situations.

The Probability of Different Microstates Is Determined by Their Energy

The idea of statistical mechanics is to deliver a probability distribution that tells us the probability of all of the different microstates. One simple class of problem that illustrates this idea is the two-state system to be taken up in detail in Chapter 7 in which the system has only *two* allowed microstates. A classic example of importance in biology is that of ion channels, which in the simplest case can be in either the closed or the open state. Figure 6.3 shows the current as a function of time for an ion channel that illustrates this two-state behavior. When the channel is open, there is current flowing through the channel and when the channel is closed, there is no current. The probability of the closed state is reflected by the fraction of the total time that the channel is closed. Similarly, the probability of the open state is given by the fraction of the total time that the channel spends open.

The key distinguishing feature of different microstates is their energy E_i, where the subscript i is a label for the ith microstate. For example, as shown in Figure 6.3, the energy of the closed state is $\varepsilon_{\text{closed}}$ and the energy of the open state is $\varepsilon_{\text{open}}$. What statistical mechanics tells us precisely is that the probability of finding a given microstate with energy E_i is

$$p(E_i) = \frac{1}{Z} \mathrm{e}^{-E_i/k_B T}, \tag{6.4}$$

where $1/Z$ is a constant set by the requirement that when we sum over all probabilities the result is 1 (that is, the probability distribution is normalized). The quantity Z has legendary status in statistical mechanics and is known as the partition function. The probability distribution given in Equation 6.4 is known as the Boltzmann distribution and has the same central role in statistical mechanics as does the equation $F = ma$ in mechanics. We will derive the Boltzmann distribution in Section 6.1.3. The normalization constant $1/Z$ can be determined by requiring that $\sum_{i=1}^{N} p(E_i) = 1$, which implies

$$Z = \sum_{i=1}^{N} \mathrm{e}^{-E_i/k_B T}. \tag{6.5}$$

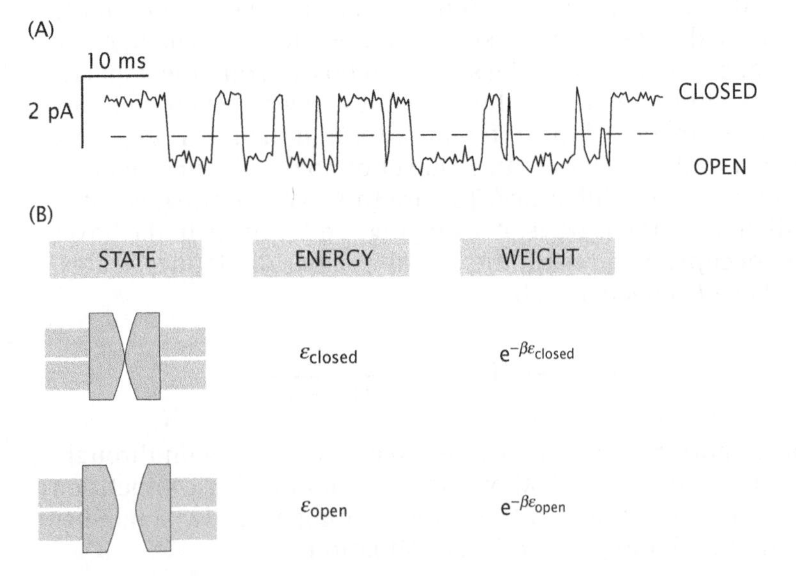

Figure 6.3: Probability of microstates from the Boltzmann distribution. Microstates available to a model ion channel that has two states, namely, closed and open. (A) Example of current as a function of time for an ion channel. (B) States and weights for the closed and open states of the channel as dictated by the Boltzmann distribution. (A, adapted from B. U. Keller et al., *J. Gen. Physiol.* 88:1, 1986.)

The reason for the importance placed on knowing the probability distribution for the different microstates is that once we have it, we can compute key observables such as the average current flowing through the channel or the average energy given by

$$\langle E\rangle = \sum_{i=1}^{N} E_i p(E_i), \tag{6.6}$$

where we have introduced the notation $\langle\cdots\rangle$ to indicate averages. What this equation tells us is that to find the average energy of the system, we sum over all of the energies of the possible microscopic outcomes, each weighted appropriately by its probability $p(E_i)$. One class of averages that will come to centerstage throughout the book is that concerned with the ligand occupancy, where we will compute the probability that a given receptor is bound as a function of the ligand concentration.

The Tricks Behind the Math: Differentiation with Respect to a Parameter The partition function serves as the analytical engine of statistical mechanics and permits us to directly calculate quantities such as the free energy ($G = -k_B T \ln Z$) and the average energy. To see the simple connection between the partition function and the average energy, we invoke a useful mathematical observation. Our interest is in the average energy, which can be written as

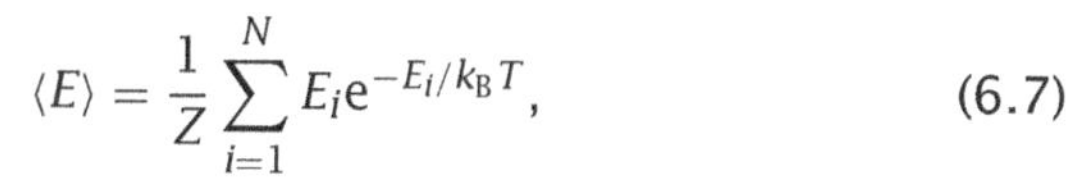

$$\langle E\rangle = \frac{1}{Z}\sum_{i=1}^{N} E_i e^{-E_i/k_B T}, \tag{6.7}$$

a result we obtain by substituting Equation 6.4 into Equation 6.6. Note that by virtue of the definition of the partition function as $Z = \sum_{i=1}^{N} e^{-E_i/k_B T}$, we can write the average energy as

$$\langle E\rangle = -\frac{1}{Z}\frac{\partial}{\partial\beta}Z, \tag{6.8}$$

where we have introduced the notation $\beta = 1/k_B T$. The point is that when we differentiate $e^{-\beta E_i}$ with respect to β, the result is $-E_i e^{-\beta E_i}$, exactly the quantity we need to compute the average. We can go even further by using the identity $d[\ln f(x)]/dx = (1/f)(df/dx)$. This permits us to rewrite Equation 6.8 as

$$\langle E\rangle = -\frac{\partial}{\partial\beta}\ln Z. \tag{6.9}$$

This very important trick will be used repeatedly for computing key observables of biological interest such as the probability that an ion channel is open and the average number of ligands bound to a receptor (such as the number of oxygen molecules bound to hemoglobin).

6.1.1 A First Look at Ligand–Receptor Binding

One of the most powerful uses to which we will put the tools developed in this chapter is to the broad class of binding interactions of interest in biology. Figure 6.1 introduced an example of the kinds of problems we will encounter and for which we can use the Boltzmann distribution and the partition function. Examples of this kind of

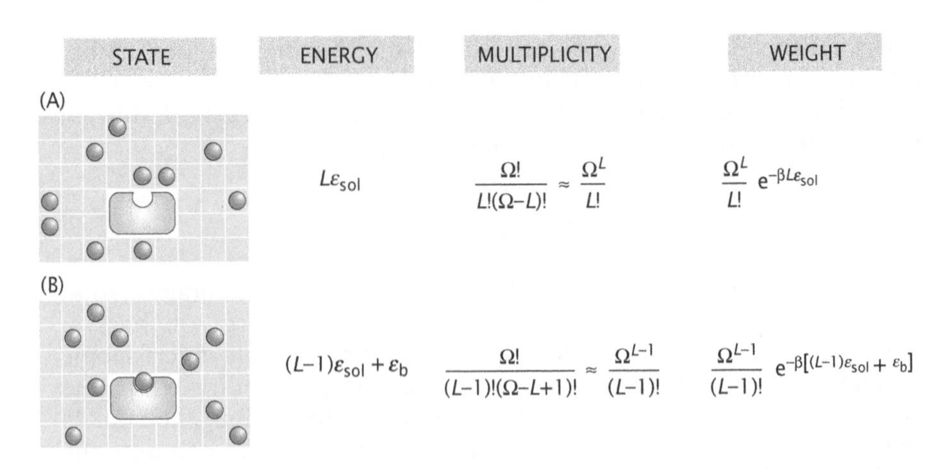

Figure 6.4: States and weights diagram for ligand–receptor binding. The cartoons show a lattice model of the solution for the case in which there are L ligands. In (A), the receptor is unoccupied. In (B), the receptor is occupied by a ligand and the remaining $L-1$ ligands are free in solution. A given state has a weight dictated by its Boltzmann factor. The multiplicity refers to the number of different microstates that share that same Boltzmann factor (for example, all of the states with no ligand bound to the receptor have the same Boltzmann factor).

binding include the binding of oxygen to hemoglobin, the binding of transcription factors to DNA, and the binding of acetylcholine to the nicotinic acetylcholine receptor. To examine the physics of Figure 6.1, imagine there are L ligand molecules in the box characterized by Ω lattice sites as well as a single receptor with one binding site as shown. For simplicity, we ignore any configurational degrees of freedom associated with the receptor itself. Our ambition is to compute the probability that a receptor is occupied by a ligand (p_{bound}) as a function of the number (or concentration) of ligands.

To see the logic of this calculation more clearly, Figure 6.4 shows the states available to this system, as well as their energies, their multiplicities, and overall statistical weights. The key point is that there are only two classes of states, namely, (i) all of those states for which there is no ligand bound to the receptor and (ii) all of those states for which one of the ligands is bound to the receptor. The neat feature of this situation is that although there are many realizations of each class of state, the Boltzmann factor is the same for each realization of these classes of state as shown in Figure 6.4.

To compute the probability that a ligand is bound, we need to construct a ratio in which the numerator is the weight of all states in which one ligand is bound to the receptor and the denominator is the sum over all states. This idea is represented graphically in Figure 6.5. What the figure shows is that there are a host of different states in which the receptor is occupied: first, there are L different ligands that can bind to the receptor; second, the $L-1$ ligands that remain behind in solution can be distributed amongst the Ω lattice sites in many different ways. In particular, we have

$$\text{weight when receptor occupied} = \underbrace{e^{-\beta\varepsilon_b}}_{\text{bound ligand}} \times \underbrace{\sum_{\text{solution}} e^{-\beta(L-1)\varepsilon_{\text{sol}}}}_{\text{free ligands}}, \tag{6.10}$$

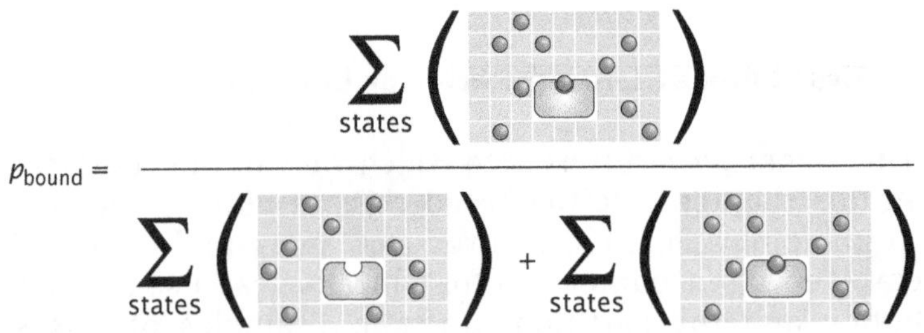

states

$p_{\text{bound}} =$

states + states

Figure 6.5: Probability of receptor occupancy. The figure shows how the probability of receptor occupancy can be written as a ratio of the weights of the favorable outcomes and the weights of *all* outcomes. The notation in the numerator instructs us to sum over the Boltzmann factors for all microstates of the system in which the receptor is occupied.

where we have introduced ε_b as the binding energy for the ligand and receptor and ε_{sol} as the energy for a ligand in solution. The summation $\sum_{solution}$ is an instruction to sum over all of the ways of arranging the $L-1$ ligands on the Ω lattice sites in solution, with each of those states assigned the weight $e^{-\beta(L-1)\varepsilon_{sol}}$. Since the Boltzmann factor is the same for each of these states, what this sum amounts to is finding the number of arrangements of the $L-1$ ligands amongst the Ω lattice sites, which yields

$$\sum_{\text{solution}} e^{-\beta(L-1)\varepsilon_{sol}} = \frac{\Omega!}{(L-1)![\Omega-(L-1)]!} e^{-\beta(L-1)\varepsilon_{sol}}. \tag{6.11}$$

The denominator of the expression shown in Figure 6.5 is the partition function itself, since it represents the sum over *all* possible arrangements of the system (both those with the receptor occupied and not) and is given by

$$Z(L,\Omega) = \underbrace{\sum_{\text{solution}} e^{-\beta L\varepsilon_{sol}}}_{\text{none bound}} + \underbrace{e^{-\beta\varepsilon_b} \sum_{\text{solution}} e^{-\beta(L-1)\varepsilon_{sol}}}_{\text{ligand bound}}. \tag{6.12}$$

We have already evaluated the second term in the sum culminating in Equation 6.11. To complete our evaluation of the partition function, we have to evaluate the sum $\sum_{solution} e^{-\beta L\varepsilon_{sol}}$ over all of the ways of arranging the L ligands on the Ω lattice sites, with the result

$$\sum_{\text{solution}} e^{-\beta L\varepsilon_{sol}} = e^{-\beta L\varepsilon_{sol}} \frac{\Omega!}{L!(\Omega-L)!}. \tag{6.13}$$

In light of these results, the partition function can be written as

$$Z(L,\Omega) = e^{-\beta L\varepsilon_{sol}} \frac{\Omega!}{L!(\Omega-L)!} + e^{-\beta\varepsilon_b} e^{-\beta(L-1)\varepsilon_{sol}} \frac{\Omega!}{(L-1)![\Omega-(L-1)]!}. \tag{6.14}$$

We can now simplify this result by using the approximation that

$$\frac{\Omega!}{(\Omega-L)!} \approx \Omega^L, \tag{6.15}$$

which is justified as long as $\Omega \gg L$. The approximation amounts to taking the largest term in the sum that would result from resolving the parentheses in Equation 6.15. To see why this is a good approximation, consider the case when $\Omega = 10^6$ and $L = 10$, resulting in

$$\frac{10^6!}{(10^6-10)!} = 10^6 \times (10^6-1) \times (10^6-2) \times \cdots \times (10^6-9) \approx (10^6)^{10}. \tag{6.16}$$

The error made by effecting this approximation can be seen by multiplying out all the terms in parentheses in Equation 6.16 and keeping the terms of order $(10^6)^9$. We find that this next term has the value $45 \times (10^6)^9$, which is roughly four orders of magnitude smaller than the leading term, demonstrating the legitimacy of the approximation.

With these results in hand, we can now write p_{bound} as

$$p_{bound} = \frac{e^{-\beta\varepsilon_b} \frac{\Omega^{L-1}}{(L-1)!} e^{-\beta(L-1)\varepsilon_{sol}}}{\frac{\Omega^L}{L!} e^{-\beta L\varepsilon_{sol}} + e^{-\beta\varepsilon_b} \frac{\Omega^{L-1}}{(L-1)!} e^{-\beta(L-1)\varepsilon_{sol}}}. \tag{6.17}$$

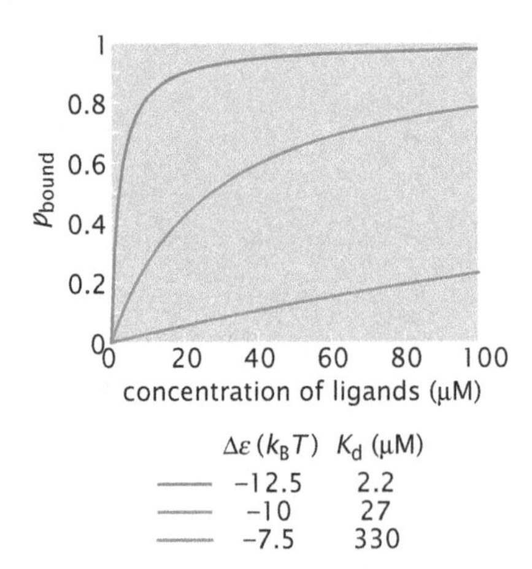

Figure 6.6: Average occupancy as a function of ligand concentration. The figure shows the average number of ligands bound as a function of the number of ligands in solution. The plot shows curves for three choices of $\Delta\varepsilon$: -7.5, -10, and $-12.5\ k_BT$, and a standard state $c_0 = 0.6$ M. The binding energies are also translated into the language of equilibrium dissociation constants.

This result can be simplified by multiplying the top and bottom by $(L!/\Omega^L)e^{\beta L\varepsilon_{sol}}$, resulting in

$$p_{bound} = \frac{(L/\Omega)e^{-\beta\Delta\varepsilon}}{1 + (L/\Omega)e^{-\beta\Delta\varepsilon}}, \tag{6.18}$$

where we have defined $\Delta\varepsilon = \varepsilon_b - \varepsilon_{sol}$. The overall volume of the box is V_{box} and this permits us to rewrite our results using concentration variables. In particular, this can be written in terms of ligand concentration $c = L/(\Omega V_{box})$, where ΩV_{box} is the total volume considered. We introduce $c_0 = 1/V_{box}$, a "reference" concentration (effectively, an arbitrary "standard state") corresponding to having all sites in the lattice occupied. This results in

$$p_{bound} = \frac{(c/c_0)e^{-\beta\Delta\varepsilon}}{1 + (c/c_0)e^{-\beta\Delta\varepsilon}}. \tag{6.19}$$

This classic result goes under many different names depending upon the field (such as the Langmuir adsorption isotherm or a Hill function with Hill coefficient $n=1$). Regardless of names, this expression will be our point of departure for thinking about all binding problems. Though many problems of biological interest exhibit binding curves that are "sharper" than this one (that is, they exhibit cooperativity—to be discussed in detail in Chapter 7), ultimately, even those curves are measured against the standard result derived here.

To make a simple estimate of the parameters appearing in Equation 6.19, we choose the size of the elementary boxes in our lattice model to be 1 nm^3, which corresponds to $c_0 \approx 1.6$ M. This is comparable to the standard state of 1 M used in many biochemistry textbooks. Given this choice of standard state, we can plot p_{bound} as a function of the concentration of ligands for different choices of the binding energy with characteristic binding energies ranging from -7.5 to $-12.5\ k_BT$. The result is plotted in Figure 6.6. As said before, many problems in statistical mechanics can be seen as the playing out of a competition between energetic and entropic contributions to the overall free energy. In this case, the interesting concentration of ligand corresponds to that choice of L for which the two terms in the denominator of Equation 6.17 are approximately equal. Equality of these two terms roughly amounts to the statement that the entropy lost in stealing one of the ligands from solution to bind it to the receptor is just made up for by the energetic gain ($\Delta\varepsilon$) associated with binding the ligand to the receptor. Notice also that this concentration corresponds to having half occupancy ($p_{bound} = 0.5$). At low concentrations, the entropic term is dominant, while at high enough concentrations, the energetic term dominates.

6.1.2 The Statistical Mechanics of Gene Expression: RNA Polymerase and the Promoter

An exciting application of the ideas on ligand–receptor binding developed above is to the problem of gene regulation. Cells make "decisions" all the time. One of the key manifestations of cellular decision making is the expression of different genes at different places at different times and to different extents. In Section 3.2.1, we introduced the central dogma. However, our treatment of replication, transcription, and translation was barren because it failed to acknowledge all of the possible cellular interventions that can occur during each of these processes. Figure 6.7 shows a more complete view of the processes attending the central dogma, with special reference to the control that is exercised over transcription and translation. One of the most important examples of genetic control is that of transcriptional regulation, which refers to the decision making about whether or not an mRNA molecule will be made. In particular, there are molecules known as transcription factors that serve to either summon or forbid RNA polymerase from binding in the vicinity of a given promoter. The role of transcription factors will be examined more precisely in Chapter 19 and for now we content ourselves with exploring the problem of RNA polymerase binding.

A Simple Model of Gene Expression Is to Consider the Probability of RNA Polymerase Binding at the Promoter

We begin with the cellular decision of whether or not a messenger RNA molecule for a given gene of interest will be made. That is, we begin by mapping the problem of gene expression onto the "simpler" subproblem of transcriptional regulation. However, we go farther yet. In particular, we replace the question of how much mRNA has been produced with the even simpler question of whether or not the promoter associated with the gene of interest is occupied by RNA polymerase. Figure 6.8 is a simplified view of the geography of the promoter region of a given gene. In particular, we view the promoter as a landing pad for the relevant RNA polymerase on the DNA. Transcription refers to the process that begins once the polymerase has escaped the promoter and starts to move along the gene of interest, resulting in an mRNA molecule (the transcript). A beautiful microscopy image of this process was shown in Figure 3.13 (p. 108). In our simple picture of transcription, we replace the complexity of the entire pathway linking DNA and its protein product with a simpler equilibrium question of whether or not the promoter for the gene of interest is occupied.

Of course, in embarking on such a calculation, we must at some point argue for why this strategy is physically reasonable. In particular, is it reasonable to suppose that the question of probability of RNA polymerase binding at the promoter is a surrogate for the question of how much mRNA there is at a given instant? Further, to what extent are we justified in pretending that the system is in equilibrium? Recall that in Chapter 5, we began with an analysis of the conditions the rates must satisfy in order for the equilibrium picture to be appropriate. In Chapter 19, we will examine the *rates* of the various processes associated with transcription, which will provide a standardized metric for deciding when the equilibrium assumption is legitimate and when it is not. For now, we will consider the equilibrium assumption valid and we will explore its implications.

Most Cellular RNA Polymerase Molecules Are Bound to DNA

There are an enormous number of sites (both other promoters and nonspecific sites) to which RNA polymerase molecules can bind other than to the promoter of interest. In addition, some of those

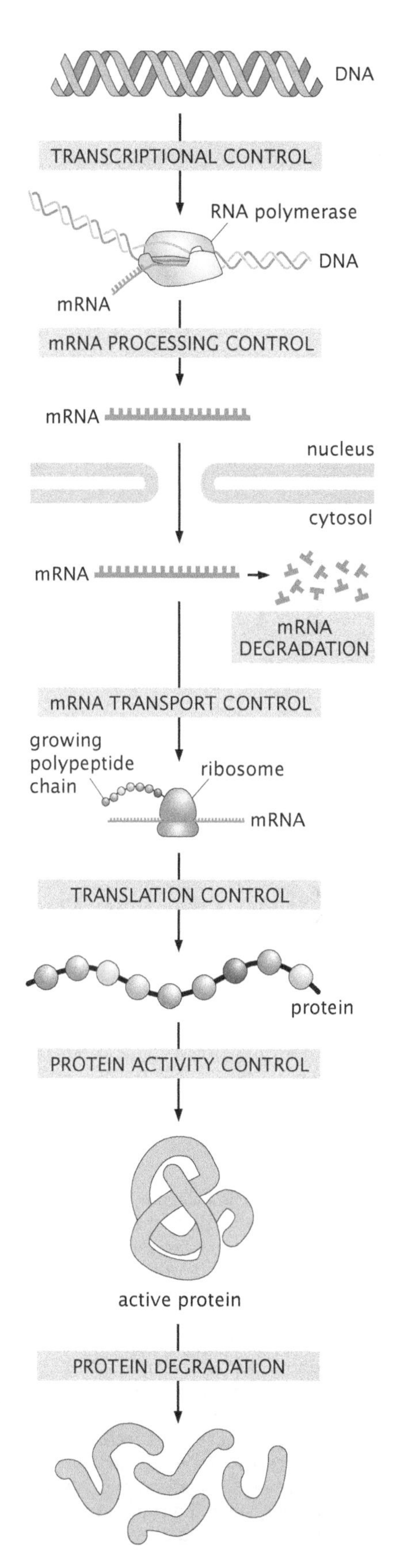

Figure 6.7: Genetic control in the central dogma. Schematic of the ways in which control can be exercised over expression of a given gene. At each step in the central dogma, there are control mechanisms that say when and where this step can take place. For example, transcriptional control refers to whether or not the mRNA associated with a given gene is synthesized.

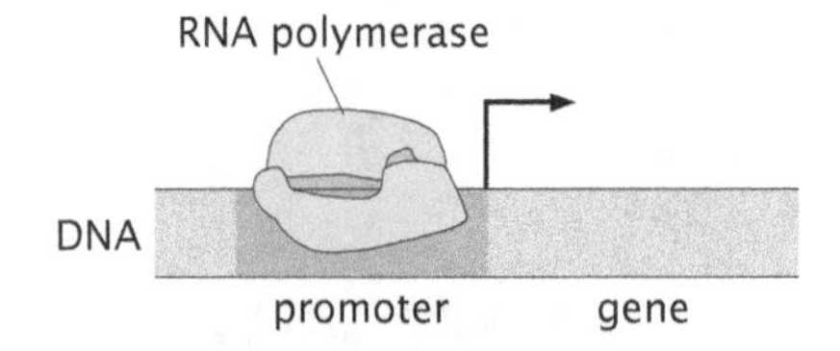

Figure 6.8: Geography of the promoter region. The promoter is the site where the RNA polymerase binds just before the transcription start site (denoted by the arrow) for the gene of interest.

polymerase molecules might not be bound to DNA at all, but rather could be diffusing freely in solution. However, a number of lines of experimental evidence such as the examination of the RNA polymerase content of DNA-free mini-cells (bacteria that lack a copy of the genome as a result of its mother cell dividing near one of its poles) reveal that the vast majority of the thousand or so RNA polymerase molecules present in *E. coli* are actually bound to DNA. We make the simplifying assumption that all RNA polymerase molecules are bound to DNA (either to the promoter of interest or nonspecifically) and examine the competition between the nonspecific sites and the promoter. In particular, we examine how the presence of these competitor sites determines the probability of bare RNA polymerase (in the absence of any regulatory proteins) binding.

In order to assess the competition between nonspecific sites and the promoter itself, we must first examine the partition function for the P RNA polymerase molecules when they are distributed amongst the nonspecific sites as shown in Figure 6.9. This is precisely the same problem considered earlier in the chapter when looking at ligand–receptor binding. The simplest model for RNA polymerase binding argues that the DNA can be viewed as N_{NS} distinct boxes (N_{NS} is the number of *nonspecific* sites on the DNA) where we need to place P RNA polymerase molecules, only allowing one such molecule per site. This results in the partial partition function

$$Z_{NS}(P, N_{NS}) = \underbrace{\frac{N_{NS}!}{P!(N_{NS}-P)!}}_{\text{multiplicity}} \times \underbrace{e^{-\beta P \varepsilon_{pd}^{NS}}}_{\text{Boltzmann factor}} . \tag{6.20}$$

We will use the notation ε_{pd}^{S} to characterize the binding energy of RNA polymerase to the specific promoter site of interest and ε_{pd}^{NS} to characterize the binding energy for nonspecific sites as shown in Figure 6.10. (A simplification made by the model is that a continuum of different binding energies is replaced by one representative value.) We are now poised to write down the *total* partition function for this problem in which the promoter must do battle with all of the nonspecific competitor sites. There are two sets of configurations to sum over, namely, all of those states in which the promoter is unoccupied, and all of those states for which the promoter is occupied leaving only $P-1$

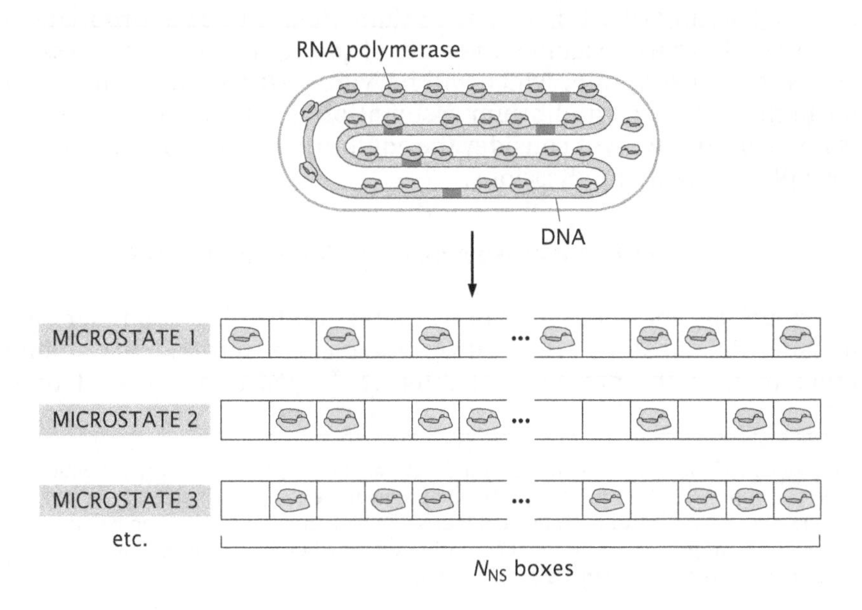

Figure 6.9: The RNA polymerase reservoir. Schematic of the ways in which RNA polymerase can be distributed throughout a bacterial cell. These molecules can be bound at any of the N_{NS} nonspecific sites on the DNA, at promoters, or be distributed throughout the cytoplasm. Each microstate corresponds to a different arrangement of the polymerase molecules on these nonspecific binding sites. We consider a model in which RNA polymerase is bound exclusively on the DNA, since the cytoplasmic contribution is negligible.

RNA polymerase molecules to be distributed amongst the nonspecific competitor sites. These different states with their corresponding statistical weights are shown in Figure 6.11.

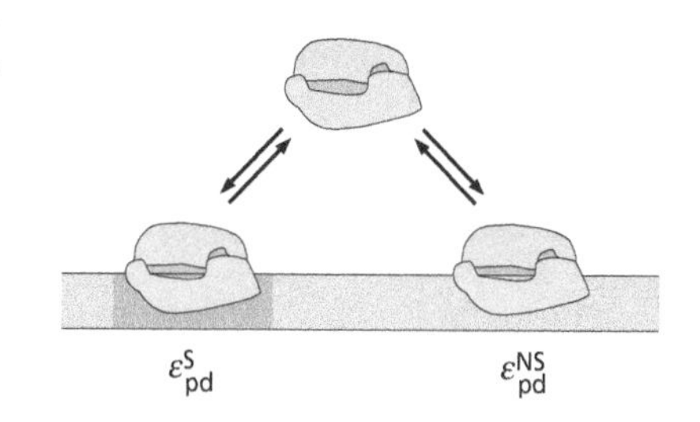

Figure 6.10: Polymerase binding energies. Illustration of the difference in binding energy for RNA polymerase when it is bound specifically (ε^{S}_{pd}) and nonspecifically (ε^{NS}_{pd}).

The Binding Probability of RNA Polymerase to Its Promoter Is a Simple Function of the Number of Polymerase Molecules and the Binding Energy

Given the states and weights as shown in Figure 6.11, we can write the *total* partition function as

$$Z(P, N_{NS}) = \underbrace{Z_{NS}(P, N_{NS})}_{\text{no RNAP on promoter}} + \underbrace{Z_{NS}(P-1, N_{NS})e^{-\beta\varepsilon^{S}_{pd}}}_{\text{one RNAP on promoter}}. \tag{6.21}$$

To find the probability that RNA polymerase (RNAP) is bound to the promoter of interest, we find the ratio of the weights of the configurations for which the RNA polymerase is bound to its promoter to the weights associated with all configurations, resulting in

$$p_{bound} = \frac{\dfrac{N_{NS}!}{(P-1)![N_{NS}-(P-1)]!}e^{-\beta\varepsilon^{S}_{pd}}e^{-\beta(P-1)\varepsilon^{NS}_{pd}}}{\dfrac{N_{NS}!}{P!(N_{NS}-P)!}e^{-\beta P\varepsilon^{NS}_{pd}} + \dfrac{N_{NS}!}{(P-1)![N_{NS}-(P-1)]!}e^{-\beta\varepsilon^{S}_{pd}}e^{-\beta(P-1)\varepsilon^{NS}_{pd}}}. \tag{6.22}$$

This result is shown in icon form in Figure 6.12. At this point, we invoke the approximation $N_{NS}!/(N_{NS}-P)! \approx (N_{NS})^P$, which holds if $P \ll N_{NS}$ (this approximation was already introduced in Equation 6.15 (p. 243)). In light of this approximation, if we multiply the top and bottom of Equation 6.22 by $[P!/(N_{NS})^P]e^{\beta P\varepsilon^{NS}_{pd}}$, we can write our final expression for p_{bound} as

$$p_{bound} = \frac{\dfrac{P}{N_{NS}}e^{-\beta\Delta\varepsilon_{pd}}}{1+\dfrac{P}{N_{NS}}e^{-\beta\Delta\varepsilon_{pd}}} = \frac{1}{1+\dfrac{N_{NS}}{P}e^{\beta\Delta\varepsilon_{pd}}}, \tag{6.23}$$

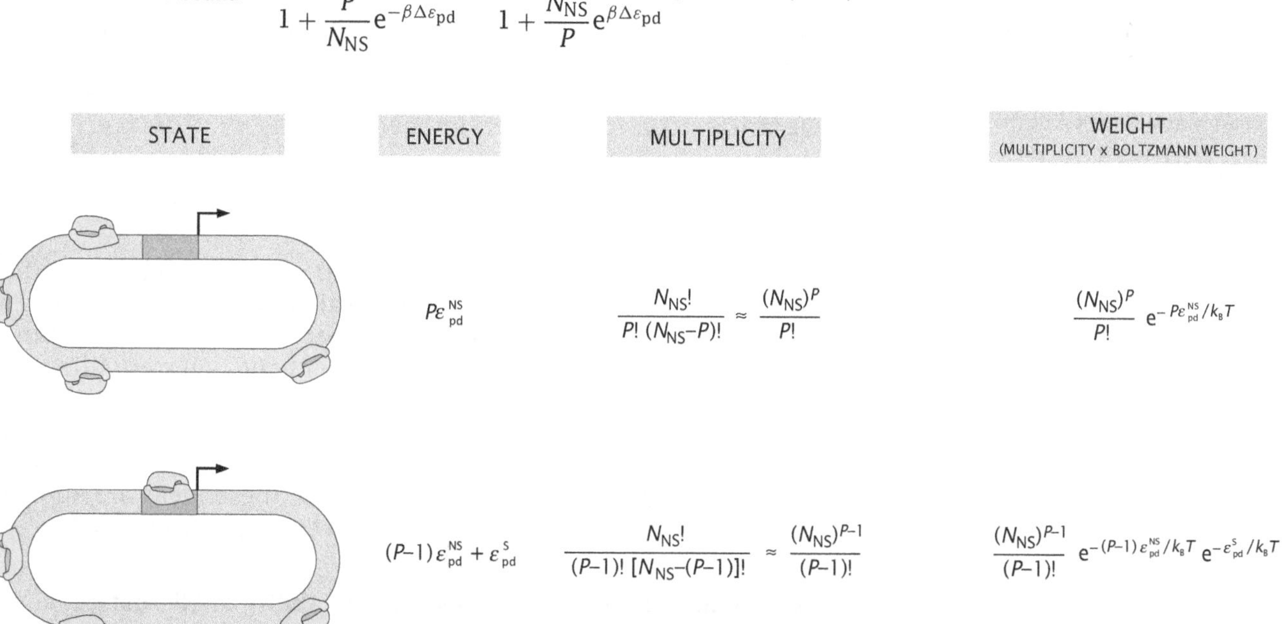

Figure 6.11: States and weights for promoter occupancy. The states are divided into two categories: those in which one of the RNA polymerase molecules is on the promoter and those for which the promoter is unoccupied.

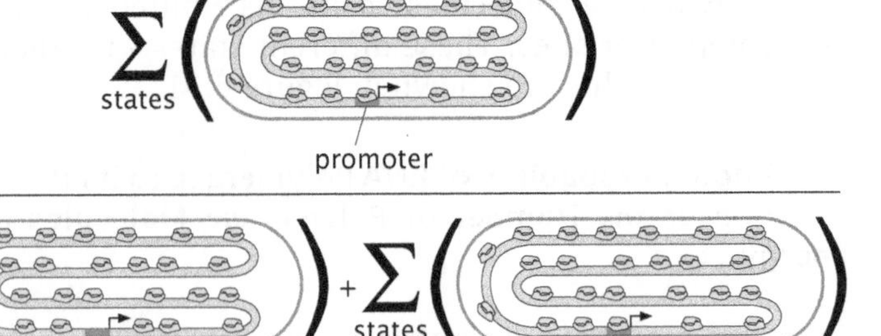

Figure 6.12: Promoter occupancy as a ratio. The probability of promoter occupancy by RNA polymerase is the ratio of the weights of all states with the promoter occupied to the weights of all possible states of the polymerase on the DNA.

where $\Delta\varepsilon_{pd} = \varepsilon^{S}_{pd} - \varepsilon^{NS}_{pd}$. This formula can be reinterpreted in terms of a model that has only two states, namely, promoter occupied by the RNA polymerase and promoter unoccupied; see Figure 6.11. The unoccupied state of the promoter has weight 1, while the occupied state has weight $(P/N_{NS})e^{-\beta\Delta\varepsilon_{pd}}$.

Once again, it is the energy difference $\Delta\varepsilon$ that matters rather than the absolute value of any of the particular binding energies. Furthermore, the difference between a strong promoter and a weak promoter can be considered as equivalent to a difference in $\Delta\varepsilon$. Equation 6.23 will serve as the standard against which all other results involving activators and repressors (which are two important classes of transcription factors) will be judged throughout the remainder of the book and especially in Chapter 19. For now, we can use this result to consider some case studies in RNA polymerase binding.

Figure 6.13 shows the probability that RNA polymerase will be bound to its promoter in this simple model as a function of the number of RNA polymerase molecules in our hypothetical bacterial cell. Note that in order to make explicit graphs like those shown in the figure, we have to make an estimate of the effective binding free energy. In this particular case, the two graphs correspond to two different choices of promoter strength. How to compute these binding energies from measurements of equilibrium constants will be explained in Section 6.4.1. In the first instance, we consider a promoter in *E. coli* for which we estimate $\Delta\varepsilon_{pd} = -2.9\ k_BT$. The second case corresponds to the promoter associated with bacteriophage T7 with a promoter strength characterized by $\Delta\varepsilon_{pd} = -8.1\ k_BT$. Both these estimates can be made based on *in vitro* measurements.

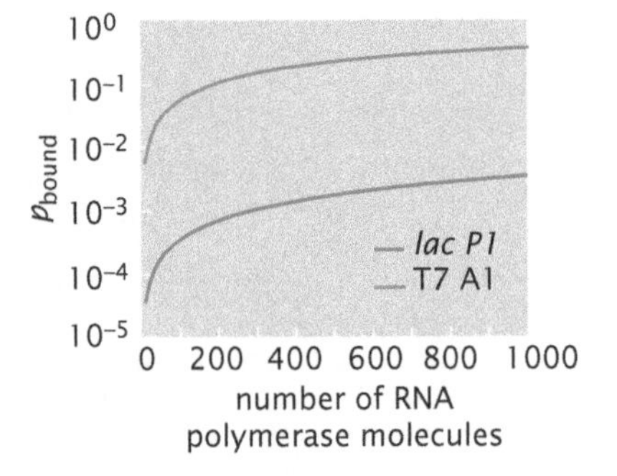

Figure 6.13: Probability of promoter occupancy as a function of the number of RNA polymerase molecules. p_{bound} is computed using values for the specific and nonspecific binding of RNA polymerase obtained *in vitro* and corresponding to the *lac* promoter, and the A1 promoter from the phage T7.

6.1.3 Classic Derivation of the Boltzmann Distribution

As we have shown, from an axiomatic perspective, the Boltzmann distribution is all we need to know to begin calculating biologically interesting quantities such as the mean occupancy of a receptor as a function of the ligand concentration. On the other hand, before continuing, we examine several ways of seeing where the Boltzmann distribution comes from. To that end, we examine three different views of its origins. For those readers who are prepared to use the Boltzmann distribution without probing where it comes from, these ideas are applied to various problems of biological interest starting in Section 6.2.

The Boltzmann Distribution Gives the Probability of Microstates for a System in Contact with a Thermal Reservoir

The setup we consider for our derivation of the probability of microstates for systems in contact with a thermal reservoir is shown in Figure 6.14. The idea is that we have a box that is separated from the rest of the world by rigid, impermeable, and adiabatic walls. As a result, the total energy and the total number of particles within the box are constant. Within the box, we now consider two regions, namely, one that is our *system* of interest and the other that is the reservoir. We are interested in how the system and the reservoir share their energy.

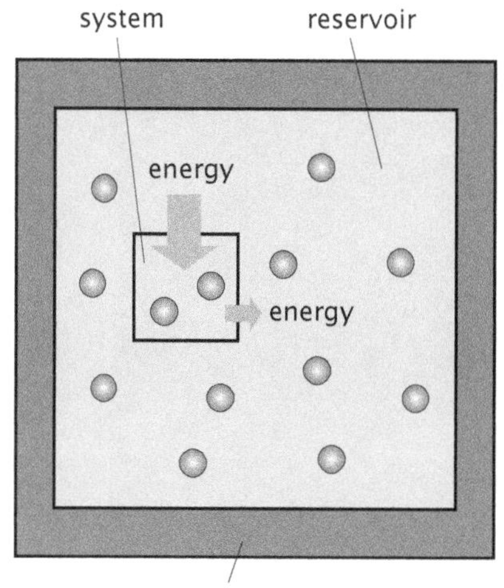

Figure 6.14: System in contact with a heat bath (thermal reservoir). The system and its reservoir are completely isolated from the rest of the world by walls that are adiabatic (forbid the flow of heat out or in), rigid (forbid the change of volume), and impermeable (forbid the flow of matter). Energy can flow across the wall separating the system from the reservoir, and as a result, the energy of the system (and reservoir) fluctuates.

The total energy is $E_{\text{tot}} = E_{\text{r}} + E_{\text{s}}$, where the subscripts "r" and "s" signify reservoir and system, respectively. Our fundamental assertion is that the probability of finding a given microstate of the system with energy E_{s} is proportional to the number of states available to the *reservoir* when the system is in this state. The ratio of the probability of the system having energy $E_{\text{s}}^{(1)}$ to that of the system having energy $E_{\text{s}}^{(2)}$ is then

$$\frac{p(E_{\text{s}}^{(1)})}{p(E_{\text{s}}^{(2)})} = \frac{W_{\text{r}}(E_{\text{tot}} - E_{\text{s}}^{(1)})}{W_{\text{r}}(E_{\text{tot}} - E_{\text{s}}^{(2)})}, \tag{6.24}$$

where $W_{\text{r}}(E_{\text{tot}} - E_{\text{s}}^{(i)})$ is the number of states available to the *reservoir*, when the system is in the particular state $E_{\text{s}}^{(i)}$. The logic is that we assert that the system is in *one* particular microstate that is characterized by its energy E_{s}. When the system is assigned this energy, the reservoir has available a particular number of states, $W_{\text{r}}(E_{\text{tot}} - E_{\text{s}})$, that depends upon how much energy, $E_{\text{tot}} - E_{\text{s}}$, it has. Though the equations may seem cumbersome, in fact, it is the underlying conceptual idea that is the subtle (and beautiful) part of the argument. The point is that the total number of states available to the universe of system plus reservoir when the system is in a particular microstate with energy $E_{\text{s}}^{(1)}$ is given by

$$W_{\text{tot}}(E_{\text{tot}} - E_{\text{s}}^{(1)}) = \underbrace{1}_{\text{states of system}} \times \underbrace{W_{\text{r}}(E_{\text{tot}} - E_{\text{s}}^{(1)})}_{\text{states of reservoir}}. \tag{6.25}$$

This is because we have asserted that the system itself is in one particular microstate that has energy $E_{\text{s}}^{(1)}$. Though there may be other microstates with the same energy, we have selected one particular microstate of that energy and ask for its probability.

We can rewrite Equation 6.24 as

$$\frac{W_{\text{r}}(E_{\text{tot}} - E_{\text{s}}^{(1)})}{W_{\text{r}}(E_{\text{tot}} - E_{\text{s}}^{(2)})} = \frac{e^{S_{\text{r}}(E_{\text{tot}} - E_{\text{s}}^{(1)})/k_{\text{B}}}}{e^{S_{\text{r}}(E_{\text{tot}} - E_{\text{s}}^{(2)})/k_{\text{B}}}}, \tag{6.26}$$

where we have invoked the familiar Boltzmann equation for the entropy, namely $S = k_{\text{B}} \ln W$, which can be rewritten as $W = e^{S/k_{\text{B}}}$. To complete the derivation, we now expand the entropy as

$$S_{\text{r}}(E_{\text{tot}} - E_{\text{s}}) \approx S_{\text{r}}(E_{\text{tot}}) - \frac{\partial S_{\text{r}}}{\partial E} E_{\text{s}}, \tag{6.27}$$

where we have only kept terms that are first order in the system energy. Finally, if we recall the thermodynamic identity $(\partial S/\partial E)_{V,N} = 1/T$, we can write our result as

$$\frac{p(E_{\text{s}}^{(1)})}{p(E_{\text{s}}^{(2)})} = \frac{e^{-E_{\text{s}}^{(1)}/k_{\text{B}}T}}{e^{-E_{\text{s}}^{(2)}/k_{\text{B}}T}}. \tag{6.28}$$

The resulting probability for finding the system in state i with energy $E_{\text{s}}^{(i)}$ is

$$p(E_{\text{s}}^{(i)}) = \frac{e^{-\beta E_{\text{s}}^{(i)}}}{Z}, \tag{6.29}$$

where Z is the partition function that makes the probabilities all add up to 1. This is precisely the Boltzmann distribution introduced earlier.

MATH

The Math Behind the Models: One Person's Macrostate Is Another's Microstate The Boltzmann formula is the foundation of statistical mechanics. It relates the probability of a microstate to its energy, $p_i = \mathrm{e}^{-\beta\varepsilon_i}/Z$. The partition function is defined as $Z = \sum_i \mathrm{e}^{-\beta\varepsilon_i}$, where the sum extends over all microstates of the system.

In practice, we are often interested in the likelihood that the system will adopt a conformation that is characterized by some macroscopic parameter, call it X. Typically, X is a variable that we can measure or otherwise control. For example, for a DNA molecule inside a cell, an interesting quantity, which can be measured using fluorescent markers, is the distance R between two sites on the DNA chain. Multiple measurements of R can be used to construct a histogram of R-values, which when normalized is the probability distribution $p(R)$.

In general, the probability of the macrostate X is given by the sum of probabilities of all the microstates of the system that adopt the specified value X,

$$p(X) = \sum_{i_X} p_i = \sum_{i_X} \frac{1}{Z} \mathrm{e}^{-\beta\varepsilon_i}. \tag{6.30}$$

For the DNA example, the sum in the above equation would run over only those microstates that have the prescribed distance between the two labeled sites on the polymer.

Using the basic relation between the partition function and the free energy, $G = -k_{\mathrm{B}}T \ln Z$, we can express the probability of the macrostate X as

$$p(X) = \frac{1}{Z} \mathrm{e}^{-\beta G(X)}, \tag{6.31}$$

where

$$G(X) = -k_{\mathrm{B}}T \ln\left(\sum_{i_X} \mathrm{e}^{-\beta\varepsilon_i}\right) \tag{6.32}$$

is the free energy of the macrostate X. Note that the formula for $p(X)$ is identical to the Boltzmann formula for the probability of a microstate, with the energy of the microstate replaced by the free energy of the macrostate. Similarly, when writing down the states and weights for the macrostates X, the energy is replaced by the free energy, as shown in Figure 6.15. In this sense, one person's macrostate is truly another person's microstate.

(A)

	STATE	WEIGHT
X_1	1	$\mathrm{e}^{-\beta\varepsilon_1}$
	2	$\mathrm{e}^{-\beta\varepsilon_2}$
	3	$\mathrm{e}^{-\beta\varepsilon_3}$
	⋮	⋮
	n_1	$\mathrm{e}^{-\beta\varepsilon_{n_1}}$
X_2	n_1+1	$\mathrm{e}^{-\beta\varepsilon_{n_1+1}}$
	⋮	⋮
	n_2	$\mathrm{e}^{-\beta\varepsilon_{n_2}}$
X_3	n_2+1	$\mathrm{e}^{-\beta\varepsilon_{n_2+1}}$
	⋮	⋮
	n_3	$\mathrm{e}^{-\beta\varepsilon_{n_3}}$

(B)

STATE	WEIGHT
X_1	$\mathrm{e}^{-\beta G(X_1)}$
X_2	$\mathrm{e}^{-\beta G(X_2)}$
X_3	$\mathrm{e}^{-\beta G(X_3)}$
⋮	⋮

Figure 6.15: Ambiguity in the definition of microstates and macrostates. (A) The states and weights associated with all the microstates of the system, grouped according to the value of a macroscopic parameter X. (B) States and weights for macrostates characterized by the parameter X. The free energies appearing in the weights are computed by doing a restricted sum over all the microstates contributing to the specified value of X.

6.1.4 Boltzmann Distribution by Counting

Different Ways of Partitioning Energy Among Particles Have Different Degeneracies

The Boltzmann distribution is so central to statistical reasoning about complex, many-body systems that it is worthwhile to develop different ways of viewing its underpinnings. Our second approach is to imagine a total energy E to be shared among n particles, and we ask for the probability that the energy of one particular particle will be E_i. The essential intuition is that as the energy donated to the particle of interest is increased, we reduce the total number of ways of distributing the remaining energy among the $n-1$ other particles, thus reducing the likelihood of our state of interest. (This argument is really equivalent to that of the previous subsection since in that case as the system takes more energy it reduces the number of states available to the reservoir.)

Consider a "universe" consisting of n particles. These particles have energies E_i ($i = 1, 2, 3, \ldots, n$) which are integer multiples of an energy unit ε (0, ε, 2ε, etc.). We assume that the total energy of all the particles in our "universe" is constant and given by $N\varepsilon$. Mathematically, we represent this injunction as $\sum_{i=1}^{n} E_i = N\varepsilon$. Furthermore, we assume that any partitioning of the energy of the "universe" among the n particles is equally likely. With the stage set this way, we ask: what is the probability $p(E)$, that a chosen particle has energy $E = m\varepsilon$?

The calculation of the probability $p(E)$ can be broken up into two steps. First, we compute $W(N, n)$, the number of ways that N units of energy can be partitioned among n particles. With this quantity in hand, we can compute the desired probability as

$$p(E) = p(m) = \frac{W(N-m, n-1)}{W(N, n)}. \tag{6.33}$$

The numerator is the number of ways of distributing $N-m$ units of energy over $n-1$ particles, which is what is left over in our "universe" once we have assigned m units of energy to the particle of interest. The denominator is the total number of ways it is possible to share the energy $N\varepsilon$ among all n particles.

The counting problem that leads to $W(N, n)$ can be reformulated in the following way. We graphically represent the N units of energy as N batteries. To assign energies to particles, we draw $n+1$ dividers, shown in Figure 6.16 as gray bars; the number of batteries between two consecutive dividers then indicates the energy of one particular particle. We carry out the bookkeeping by making use of a binary notation where each divider is a 0 and each battery is a 1, so that a binary word of zeroes and ones labels a particular partitioning of the energy.

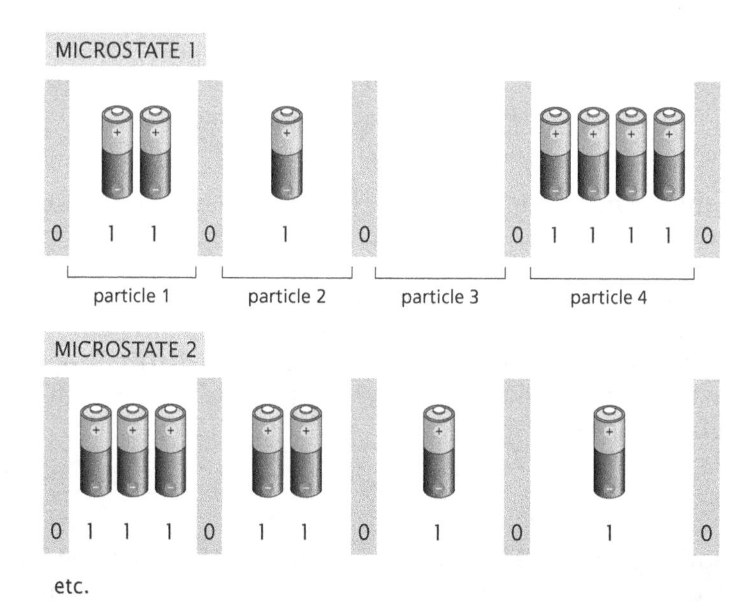

Figure 6.16: Graphical representation of the problem of counting the number of ways of partitioning N energy units among n particles. Each battery corresponds to a single unit of energy, ε, and the counting problem reduces to finding how many ways these energy units can be distributed amongst the four particles. This particular case shows two microstates corresponding to an energy of 7ε distributed among 4 particles, hence $N = 7$ and $n = 4$.

In the binary representation, the number of ways of partitioning the total energy $N\varepsilon$ is simply the number of distinct words made from the $n+1$ zeroes and the N ones. Since each word starts and ends with a zero (see Figure 6.16), we are left with $n-1$ zeroes and N ones to play with, and therefore the number of such words is

$$W(N, n) = \frac{(N+n-1)!}{N!(n-1)!}. \tag{6.34}$$

The logic of this expression is as follows: the numerator is the number of permutations of the N batteries and $n-1$ internal dividers, assuming that they are distinguishable from each other. The denominator is there to account for the fact that all the batteries (N of them) are indistinguishable and similarly the internal dividers ($n-1$ of them) are indistinguishable.

The formula for the probability that a chosen particle has m units of energy is obtained by using Equation 6.34 for the numerator and the denominator in Equation 6.33, resulting in the expression

$$p(m) = (n-1)\frac{N!}{(N-m)!}\frac{[N+n-1-(m+1)]!}{(N+n-1)!}. \tag{6.35}$$

To gain further intuition about the result we have just obtained, we make use of the approximate formula already used several times in the chapter, namely,

$$\frac{N!}{(N-i)!} \approx N^i, \tag{6.36}$$

which holds for $N \gg i$. We use this to approximate the two fractions appearing in Equation 6.35, to obtain the formula

$$p(m) = \frac{n-1}{N+n-1}\left(\frac{N+n-1}{N}\right)^{-m}. \tag{6.37}$$

If we make the further simplifying assumption that both the number of particles n and the total number of energy units N are much greater than 1, and that $N \gg n$, we arrive at

$$p(m) = \frac{n}{N}\mathrm{e}^{-m(n/N)}, \tag{6.38}$$

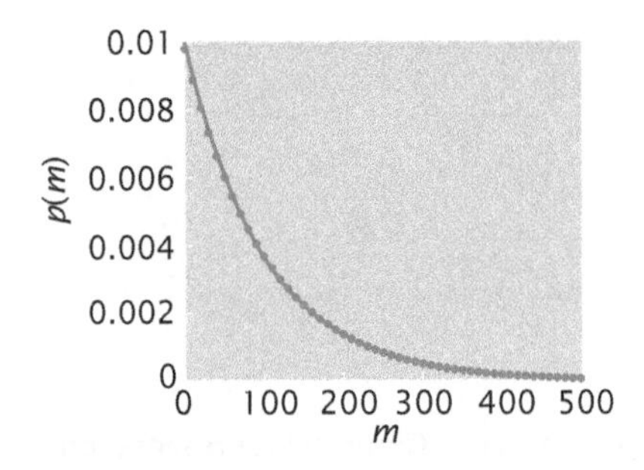

Figure 6.17: Probability of a system having energy E. The plot shows both the exact (dots: Equation 6.35) and approximate (solid curve: Equation 6.39) results for the probability of a microstate as a function of its energy. The number of energy units is $N = 10{,}000$ and this energy is shared among $n = 100$ particles.

where the exponential factor follows from the identity

$$\left(\frac{N+n-1}{N}\right)^{-m} = \mathrm{e}^{-m\ln[(N+n-1)/N]}$$

and the approximation $\ln(1+n/N) \approx n/N$, which is accurate for $N \gg n$ and results from doing a Taylor expansion of the logarithm (see The Math Behind the Models on p. 215).

The final approximate formula tells us that the probability of a chosen particle having an energy $E = m\varepsilon$ is an exponentially decreasing function of the energy (that is, like the Boltzmann distribution). Moreover, since $N\varepsilon/n$ is the average energy per particle ($\langle E\rangle$), we can rewrite the final formula as

$$p(E) = \frac{\varepsilon}{\langle E\rangle}\mathrm{e}^{-E/\langle E\rangle}. \tag{6.39}$$

This is nothing but the Boltzmann distribution. The exact and approximate results of Equations 6.35 and 6.39 are compared in Figure 6.17.

6.1.5 Boltzmann Distribution by Guessing

We have seen that the Boltzmann distribution can be derived on the basis of counting arguments as a reflection of the number of ways of partitioning energy between a system and a reservoir. A completely different way of deriving the Boltzmann distribution is on the basis of information theory and effectively amounts to making a *best* guess about the probability distribution given some limited knowledge about the system such as the average energy.

Maximizing the Entropy Corresponds to Making a Best Guess When Faced with Limited Information

The starting point for our argument is the idea that there are many possible probability distributions that one might assign to a given set of microstates. Two extreme examples are: (i) we could say that each and every microstate has the same probability, and (ii) we could say that one of the microstates has probability 1 and all of the other microstates have probability 0. Information theory provides a selection principle that tells us, out of all of the possible competing assignments of the probability distribution, which one is the least biased.

Qualitatively, the starting point for our argument is the Shannon entropy given by

$$S(p_1, p_2, \ldots, p_N) = S(\{p_i\}) = -\sum_{i=1}^{N} p_i \ln p_i, \tag{6.40}$$

where p_i is the probability of the ith microstate (or outcome). We will also find it convenient at times to refer to the Shannon entropy as the missing information, $\mathrm{MI}(p_1, p_2, \ldots, p_N)$. We can think of the Shannon entropy as a probe that reports the state of ignorance implied by a particular probability distribution. Note that in the limit where we know nothing other than that there are N states (that is, $p_i = 1/N$), the Shannon entropy takes the value $S = \ln N$, the maximum value that it can attain for N states. Alternatively, in the case where we are certain of the disposition of the system (for example, $p_4 = 1$ and all the other p_i are zero), the Shannon entropy is zero. In the language of missing information, what this means is that in the first case, we "learn" the most when we are told which of the N outcomes is actually realized. By way of contrast, in the latter case where $p_4 = 1$, we have no missing information and it is not informative at all to be told that the outcome is state 4.

To use the Shannon entropy as the basis of a selection principle, we implement the principle of maximum entropy, which states that out of all of the possible probability distributions, we should choose that distribution which has the maximum Shannon entropy. The rationale for this choice is that it corresponds to the distribution that introduces the least bias—that is, it is maximally noncommittal with respect to the missing information. The achievement of Shannon was the insight that in problems of inference where one assigns probability distributions based on some limited information, there is a *best* choice (in the variational sense introduced in Chapter 5).

To see how this all works, consider a situation where we know that there are N outcomes, but nothing more. Intuitively, the best guess seems to be that we have no reason to favor one outcome over the others and hence we should assign equal probabilities to all outcomes. The edict of entropy maximization tells us to maximize the Shannon entropy subject to whatever constraints we might have. In this case, the only relevant constraint is that the probability distribution is normalized, namely, $\sum_{i=1}^{N} p_i = 1$. To that end, we must maximize

$$S' = -\sum_i p_i \ln p_i - \lambda\left(\sum_i p_i - 1\right), \tag{6.41}$$

where the first term is the Shannon entropy and the second term introduces a so-called Lagrange multiplier λ (see The Math Behind the Models section below) to enforce the constraint of normalization. Our procedure is to find that set of probabilities p_i which maximize this augmented entropy function. Note that maximization of S' with respect to the Lagrange multiplier yields the normalization constraint

$$\frac{\partial S'}{\partial \lambda} = 0 = -\left(\sum_i p_i - 1\right). \tag{6.42}$$

Differentiating with respect to the unknown probability p_i, we have

$$\frac{\partial S'}{\partial p_i} = 0 = -\ln p_i - 1 - \lambda. \tag{6.43}$$

The solution to this equation is

$$p_i = \mathrm{e}^{-1-\lambda}. \tag{6.44}$$

Even before determining the Lagrange multiplier λ, we note that the probabilities of all outcomes do not depend on i and are therefore equally likely, consistent with our intuition.

To continue with the development, we now determine the Lagrange multiplier λ by imposing the constraint $\sum_i p_i = 1$. If we imagine there are N possible outcomes, this constraint becomes

$$\sum_{i=1}^{N} \mathrm{e}^{-1-\lambda} = 1, \tag{6.45}$$

or, what is the same thing,

$$\mathrm{e}^{-1-\lambda} = \frac{1}{N}. \tag{6.46}$$

What this implies then is that the probabilities are given by

$$p_i = \frac{1}{N}, \tag{6.47}$$

meaning that each of the N possible outcomes is equally likely. Using information theory in this case is much like using a sledgehammer to crack open a peanut. On the other hand, the mathematical essence of using entropy maximization was all contained in this simple example, and there are other examples where guessing the right answer without some analysis is not so easy.

The Math Behind the Models: The Method of Lagrange Multipliers Minimization and maximization problems involving a number of different parameters carry their own special challenges when the parameters are not free to vary independently. To be concrete, think of the shortest distance between Los Angeles and Paris. The correct answer depends upon whether or not there are constraints. In the constraint-free case, the

MATH

shortest distance between these cities is a straight line passing through the Earth. On the other hand, for the practical problem of interest to humans, the better answer is a great circle. The great circle solution is the minimizer in the constrained set of solutions which are curves on the surface of a sphere (that is, the Earth).

To make the mathematics more transparent, consider the maximization of the function $A = xy$ (the area of a rectangle) subject to the constraint that the perimeter is fixed with value $P = 2x + 2y$. The point is that we are not free to maximize $A(x, y)$ by searching over all x and y. Instead, our choices of x and y must respect the condition $P - 2x - 2y = 0$. In this case, the easiest approach is to solve the constraint equation for y, resulting in $y = (P - 2x)/2$, and substitute into the maximization problem, where now we only have to maximize with respect to x. This is a perfectly reasonable strategy. On the other hand, this kind of strategy is of limited use when the constraint equation(s) involve many variables that are related in some ugly, nonlinear way.

The method of Lagrange multipliers permits us to circumvent the (sometimes) challenging step of eliminating variables. In particular, if we introduce the augmented function

$$A_a(x, y) = A(x, y) - \lambda(P - 2x - 2y), \tag{6.48}$$

then we are free to solve $\partial A_a/\partial x = 0$, $\partial A_a/\partial y = 0$, and $\partial A_a/\partial \lambda = 0$. The result is that we have taken a problem in which we are maximizing a function of two variables (but only over a subset of allowed values of x and y) with a maximization over three variables (x, y, and λ), without any limitation on the allowed values of x and y. If we carry out this strategy for this problem we find that $x = y = P/4$, which tells us that the rectangle giving the largest enclosed area at fixed perimeter is a square!

Entropy Maximization Can Be Used as a Tool for Statistical Inference

The use of entropy maximization for probability assignments has a rich history and one of the more amusing and instructive examples is the application of these ideas to assigning probabilities when faced with a dishonest die. The setup is this: we repeatedly throw a die and after a huge number of trials report the average value given by $\langle i \rangle = (1/N)\sum_{n=1}^{N} i_n$, where we have introduced the notation i_n to signify the value on the die after the nth roll. The question then becomes: what is the best probability assignment that we can make faced with this limited but very important knowledge?

For concreteness, we consider the case in which the experiment yields $\langle i \rangle = 4.5$, distinct from the case of an honest die, which would have yielded $\langle i \rangle = 3.5$. With only this information in hand, our task is to make a best guess as to the probabilities (p_i) of each of the different faces. In terms of these probabilities, the average can be written as

$$\langle i \rangle = \sum_{i=1}^{6} p_i i, \tag{6.49}$$

where p_i is the probability of obtaining a face with i spots. For an honest die, where the probability of obtaining any number is the same (that is, $p_i = 1/6$ for all i), we have

$$\langle i \rangle = \frac{1}{6}\sum_{i=1}^{6} i = 3.5. \qquad (6.50)$$

We are faced with the more interesting case where this average implies that the probabilities of finding the different faces are different.

There are two pieces of information that we know about the probability distribution governing our dishonest die, namely, (i) that the sum of the probabilities has to be 1 and (ii) that the average value over many rolls of the die is 4.5. This knowledge will yield a probability distribution with lower Shannon entropy than the uniform case in which $p_i = 1/6$. These two constraints mean that we are not free to minimize over the probabilities of all six outcomes as though they are independent. For example, we can eliminate p_6 by invoking the normalization constraint as

$$p_6 = 1 - p_1 - p_2 - p_3 - p_4 - p_5. \qquad (6.51)$$

In addition, we could use the constraint on the average value to eliminate p_5, for example. Were we to follow this analysis through to its conclusion, the next step would be to maximize the entropy over the four remaining p_i. On the other hand, the whole point of the Lagrange multiplier idea is that it permits us to forego the explicit elimination of variables due to the constraints.

For the problem of the dishonest die, we can include our constraints naturally by writing the augmented Shannon entropy as

$$S_{\text{constrained}} = -\sum_{i=1}^{N} p_i \ln p_i - \lambda\underbrace{\left(\sum_i p_i - 1\right)}_{\text{normalization}} - \mu\underbrace{\left(\sum_i i p_i - \langle i \rangle\right)}_{\text{average number}}, \qquad (6.52)$$

where λ and μ are the Lagrange multipliers corresponding to each of the constraints. We want to maximize this expression with respect to p_i, which yields

$$\frac{\partial S_{\text{constrained}}}{\partial p_i} = 0 = -\ln p_i - 1 - \lambda - \mu i. \qquad (6.53)$$

This result can be solved for p_i, resulting in the probability assignment

$$p_i = \mathrm{e}^{-1-\lambda}\mathrm{e}^{-\mu i}. \qquad (6.54)$$

By imposing the normalization condition $\sum_i p_i = 1$ we find

$$\sum_{i=1}^{6} \mathrm{e}^{-1-\lambda}\mathrm{e}^{-\mu i} = 1. \qquad (6.55)$$

This result implies in turn that

$$\mathrm{e}^{-1-\lambda} = \frac{1}{\sum_{i=1}^{6} \mathrm{e}^{-\mu i}}, \qquad (6.56)$$

which permits us to write the full distribution as

$$p_i = \frac{\mathrm{e}^{-\mu i}}{\sum_{i=1}^{6} \mathrm{e}^{-\mu i}}. \qquad (6.57)$$

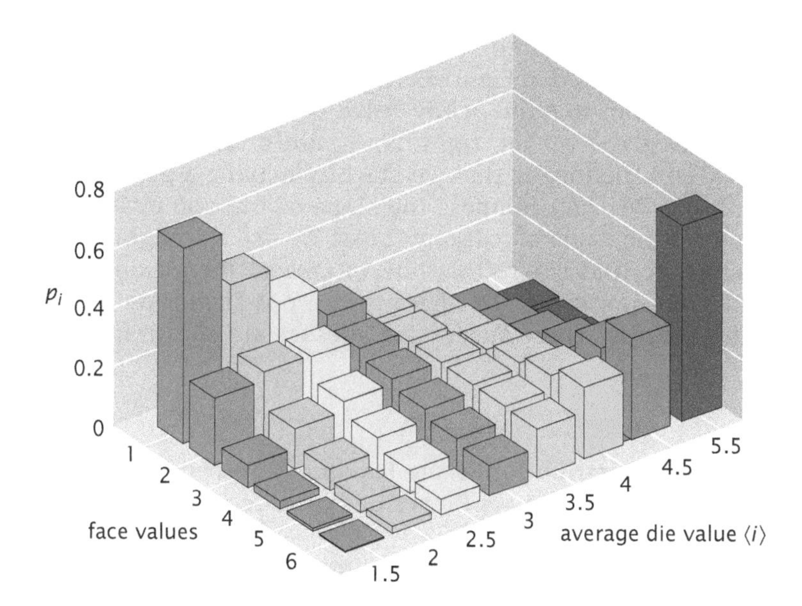

Figure 6.18: Probability distribution for a dishonest die. The plot shows the probability for getting each of the six faces for various choices of $\langle i \rangle$.

At this point, the remaining unknown is the Lagrange multiplier μ. To determine this constraint, we begin by defining $x = e^{-\mu}$, which permits us to rewrite p_i as

$$p_i = \frac{x^i}{x + x^2 + x^3 + x^4 + x^5 + x^6}. \tag{6.58}$$

If we substitute this expression for the probability into the constraint given in Equation 6.49, we find

$$\langle i \rangle = \frac{x + 2x^2 + 3x^3 + 4x^4 + 5x^5 + 6x^6}{x + x^2 + x^3 + x^4 + x^5 + x^6}. \tag{6.59}$$

This general equation can be solved for x given a particular $\langle i \rangle$ as shown in the problems at the end of the chapter. Figure 6.18 shows the resulting probability distribution for several choices of $\langle i \rangle$.

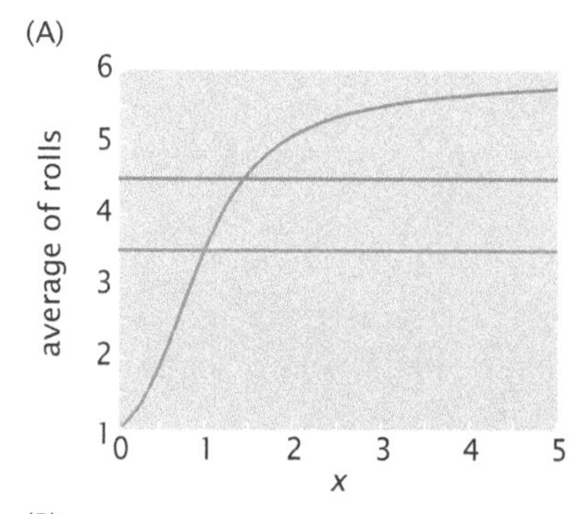

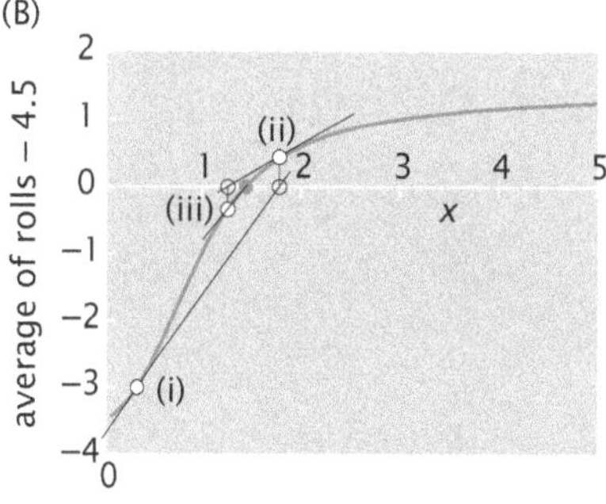

Figure 6.19: Methods for finding roots. (A) Graphical scheme for determining the Lagrange multipliers for the problem of the dishonest die. The horizontal lines correspond to $\langle i \rangle = 3.5$ (green) and $\langle i \rangle = 4.5$ (red). (B) Newton method for determining the Lagrange multipliers. (i), (ii), and (iii) represent successive guesses for the solution using Newton's method.

COMPUTATIONAL EXPLORATION

Computational Exploration: Numerical Root Finding

Equation 6.59 requires us to find the value of x given a knowledge of the average value on the faces of the die denoted $\langle i \rangle$. Often, we are faced with the problem of finding roots to equations such as this of a highly nonlinear nature where simple analytic approaches are not feasible. In this Computational Exploration, we will examine several different ways to find the roots of such a problem and the reader is urged to follow our steps for him or herself.

Perhaps the most conceptually simple approach to this problem is shown in Figure 6.19(A), where we plot the two sides of Equation 6.59 as a function of x and simply look for the points of intersection of the two curves. In this case, plotting $\langle i \rangle$ as a function of x amounts to nothing more than drawing a horizontal line, since it is a constant. As seen in the figure, we examine several cases. We start with the case in which $\langle i \rangle = 3.5$ and note immediately that this intersects with the curve for the right-hand side of Equation 6.59 for the value of $x = 1$. For the more interesting case in which $\langle i \rangle = 4.5$, clearly $x \neq 1$ and the actual value should be determined by the reader.

The second scheme for finding our unknown Lagrange multiplier is illustrated in Figure 6.19(B). Newton's method provides an algorithm to numerically solve the equation $f(x) = 0$, where $f(x)$ is given by the right-hand side of Equation 6.59 minus $\langle i \rangle$. Here the idea is that we start by making a guess for the solution x. We then compute the slope of $f(x)$ and evaluate that slope for the value of x at our guess. We use a straight line of this slope to find where this line intersects the x-axis and use this as our new guess for where the zero is found. We then repeat the process. Three successive guesses are shown in the figure.

The Boltzmann Distribution is the Maximum Entropy Distribution in Which the Average Energy is Prescribed as a Constraint

So far, our adventures in entropy maximization have centered on toy problems with no physical or biological significance. On the other hand, they have served as a warm-up exercise for thinking about the Boltzmann distribution itself. Indeed, aside from a change in notation, the derivation of the Boltzmann distribution is essentially a rehashing of the problem of the dishonest die. The basic idea is this: we are faced with a system in contact with a heat bath—for example, imagine a small test tube immersed in a water bath held at some constant temperature. In this case, the system in the test tube can be in any of a number of different energy states and we are after the probability p_i associated with a state whose energy is E_i. The effect of the surrounding water bath is to fix the average energy of our test tube.

The average energy constraint is physically imposed as a result of the fact that the system is in contact with a reservoir at fixed temperature and is written as $\langle E \rangle = \sum_i E_i p_i$. Once again, we consider an augmented Shannon entropy function, which involves two Lagrange multipliers, one tied to each constraint, and given by

$$S' = -\underbrace{\sum_i p_i \ln p_i}_{\text{Shannon entropy}} \underbrace{-\lambda\left(\sum_i p_i - 1\right)}_{\text{normalization}} \underbrace{-\beta\left(\sum_i p_i E_i - \langle E \rangle\right)}_{\text{average energy}}. \tag{6.60}$$

Note that once again the first constraint imposes normalization of the distribution. Maximizing this augmented function leads to the conclusion that $p_i = \mathrm{e}^{-1-\lambda-\beta E_i}$. Imposition of the normalization requirement, $\sum_i p_i = 1$, implies that

$$\mathrm{e}^{-1-\lambda} = \frac{1}{\sum_i \mathrm{e}^{-\beta E_i}}. \tag{6.61}$$

We identify the partition function as

$$Z = \sum_i \mathrm{e}^{-\beta E_i}. \tag{6.62}$$

As a result of the analysis given above, we now see that the probabilities are given by

$$p_i = \frac{\mathrm{e}^{-\beta E_i}}{\sum_i \mathrm{e}^{-\beta E_i}}. \tag{6.63}$$

The analysis given above has so far shown us how to obtain the Lagrange multiplier λ. A deeper question concerns the Lagrange multiplier β. It will be shown in the next subsection, using the physics of the ideal gas, that this constraint is related to the temperature through $\beta = 1/k_B T$. Formally, the Lagrange multiplier β is determined by recourse to the original constraint $\langle E \rangle = \sum_i E_i p_i$. To see how this constraint plays out concretely, we now turn to the example of a "gas" of noninteracting particles.

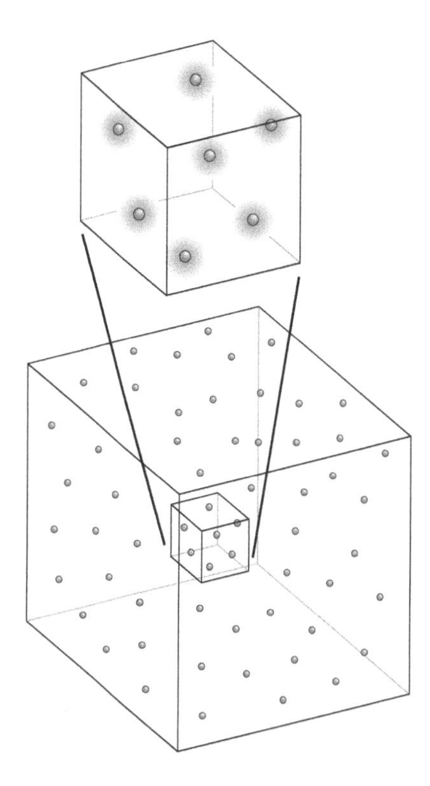

Figure 6.20: Interactions in the ideal solution (and gas) model. Schematic of a small volume element of molecules in the dilute limit described by the ideal gas and ideal solution models. The shaded region around each molecule illustrates the range of their interactions with each other.

6.2 On Being Ideal

We have argued that entropy will serve as one of the central actors in our description of the physics of biological systems. Though the generic description of entropy as a measure of the number of available microstates given in Equation 5.31 (p. 220) is useful in the abstract, we will most often appeal to a concrete description of entropy that is inspired by the ideal gas, but is applicable more widely to any set of noninteracting molecules or particles that explore space independently. We go beyond the abstraction of the previous chapter by showing the power of models in which we make the approximation not only that the system of interest is in equilibrium, but that the constituents of that system do not interact with each other. The idea behind these models is schematized in Figure 6.20. What this figure depicts is the idea that the range of interaction between the molecules of interest is small in comparison with their mean spacing. As a result, they are largely free to explore space independently. This picture will inspire our treatment of gases and solutions.

As a result of making simplifying assumptions like those described above, we can compute the entropy exactly in a form that is useful in a number of different biological contexts. In coming chapters, the methods we develop here will be used to examine the binding of oxygen to hemoglobin (Chapter 7), the force–extension properties of DNA and proteins (Chapter 8), the unwrapping of nucleosomes (Chapter 10), and the regulation of gene expression (Chapter 19).

6.2.1 Average Energy of a Molecule in a Gas

The Ideal Gas Entropy Reflects the Freedom to Rearrange Molecular Positions and Velocities

In the previous section, we showed how the probability distribution for a system with average energy $\langle E \rangle$ could be guessed by using the principle of maximum entropy. However, to finish that calculation, we need to determine the meaning and significance of the Lagrange multiplier β. For a gas of noninteracting molecules, we can focus on just one of the molecules since each and every molecule will be described by a probability distribution of the same form. Abstractly, if we have a system described by N independent variables $(x_1, x_2, \ldots, x_N)$, then the probability distribution $P(x_1, x_2, \ldots, x_N)$ factorizes into the form $P(x_1, x_2, \ldots, x_N) = p(x_1) \times p(x_2) \times \cdots \times p(x_N)$. What this means is that finding the probability distribution for one of the variables is as good as finding the entire probability distribution.

To apply this idea to a gas of noninteracting molecules, we take our cue from Figure 6.20 and imagine N molecules confined to a box of volume V. The microscopic state of the system is characterized by the spatial coordinates and momenta of each molecule. On the other hand, by invoking the fact that the molecules are noninteracting, we can ignore the configurational part and just compute $P(p_x)$, the probability density that a given molecule will have a momentum in the range p_x to $p_x + \mathrm{d}p_x$. We can ignore the configuration part of the probability distribution because the probability of being in any little volume element is the same—the spatial distribution of molecules is uniform. In order to use the Boltzmann distribution, we have to assign an energy to each "microstate." In this case, we have simplified the set of microstates to consider the x-component of momentum of one molecule with the knowledge that all components of momentum for all molecules are described by the same probability. The contribution to the kinetic energy of the system due to this component of the molecule's motion is

$$E = \frac{p_x^2}{2m}, \tag{6.64}$$

where we use the fact that the kinetic energy is $\mathrm{KE} = \frac{1}{2}mv_x^2$ and the momentum is given by $p_x = mv_x$.

Recall from the previous section that we found that the probability distribution for a system with average energy $\langle E \rangle$ is given by Equation 6.63 and is proportional to $\mathrm{e}^{-\beta E}$, where E is the energy of the state of interest. To apply this result to the ideal gas, from Equation 6.63, we can write down the probability distribution directly as

$$P(p_x) = \frac{\mathrm{e}^{-\beta\left(p_x^2/2m\right)}}{\sum_{\text{states}} \mathrm{e}^{-\beta\left(p_x^2/2m\right)}}. \tag{6.65}$$

To determine the constraint β, we now need to probe in a bit more detail what we really mean by the notion of $\sum_{\text{states}}$. The interpretation of $\sum_{\text{states}}$ is that we need to sum over all of the possible values of the momentum p_x for our molecule of interest. Since p_x is a continuous variable that can take on any value, the summation is replaced by the integral,

$$\sum_{\text{states}} \longrightarrow \int_{-\infty}^{\infty} \mathrm{d}p_x. \tag{6.66}$$

To proceed, we invoke the result that

$$\int_{-\infty}^{\infty} \mathrm{e}^{-\beta p_x^2/2m}\, \mathrm{d}p_x = \sqrt{\frac{2m\pi}{\beta}}, \tag{6.67}$$

one of the most important integrals in mathematics and its applications to physics, which is discussed in more detail in The Math Behind the Models below.

To determine the Lagrange multiplier β, we need to invoke the constraint on the average energy. In particular, we use the constraint

$$\langle E \rangle = \tfrac{1}{2}k_{\mathrm{B}}T. \tag{6.68}$$

This constraint is a restatement of the law of equipartition of energy, which says that the average energy associated with every degree of freedom is $\frac{1}{2}k_{\mathrm{B}}T$ and is one of the principal results of statistical mechanics (see Dill and Bromberg, 2011, in the Further Reading at the end of the chapter). With this result in hand, we can now work out the parameter β appearing in the Boltzmann distribution from the average energy

$$\langle E\rangle = \frac{\int_{-\infty}^{\infty} \frac{p_x^2}{2m} e^{-\beta(p_x^2/2m)}\, dp_x}{\sqrt{2m\pi/\beta}}. \tag{6.69}$$

If we make the definition $\alpha = \beta/2m$, this can be rewritten as

$$\langle E\rangle = \frac{(1/\beta)\int_{-\infty}^{\infty} \alpha p_x^2 e^{-\alpha p_x^2}\, dp_x}{\sqrt{\pi/\alpha}}. \tag{6.70}$$

The integral can now be evaluated by resorting to the trick of differentiation with respect to a parameter introduced in The Tricks Behind the Math on p. 241. In particular, we can rewrite the average energy as

$$\langle E\rangle = \frac{\alpha^{3/2}}{\beta\sqrt{\pi}}\left(-\frac{\partial}{\partial\alpha}\right)\int_{-\infty}^{\infty} e^{-\alpha p_x^2}\, dp_x. \tag{6.71}$$

If we now carry out the relevant integration and associated differentiation, we are left with the important conclusion that our Lagrange multiplier β is given by

$$\beta = \frac{1}{k_B T}. \tag{6.72}$$

With this result, we have come full circle to show that by using the maximum entropy approach to guess the relevant distribution, this is equivalent to the Boltzmann distribution.

The Math Behind the Models: The Gaussian Integral Our calculations on the ideal gas demanded the evaluation of the integral

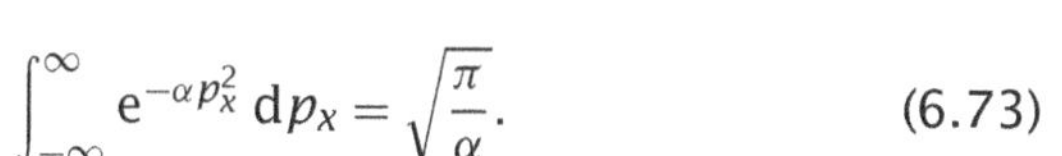

$$\int_{-\infty}^{\infty} e^{-\alpha p_x^2}\, dp_x = \sqrt{\frac{\pi}{\alpha}}. \tag{6.73}$$

To see this, we resort to a beautiful trick that argues that evaluating such a nice integral once is just not enough. Hence, we write

$$\int_{-\infty}^{\infty} e^{-\alpha x^2} dx \times \int_{-\infty}^{\infty} e^{-\alpha y^2} dy = \int_0^{2\pi} d\theta \int_0^{\infty} r\, dr\, e^{-\alpha r^2}. \tag{6.74}$$

The key observation inherent in writing things this way is that we are integrating $e^{-\alpha(x^2+y^2)}$ over the entire plane and that this integral can also be written in polar coordinates. The θ-integration can be done immediately and yields 2π. As a result, we are left with

$$\int_{-\infty}^{\infty} e^{-\alpha x^2} dx \times \int_{-\infty}^{\infty} e^{-\alpha y^2} dy = 2\pi \int_0^{\infty} r\, dr\, e^{-\alpha r^2}. \tag{6.75}$$

If we make the substitution $u = -\alpha r^2$, the integral on the right-hand side reduces to π/α, which tells us then that

$$\left(\int_{-\infty}^{\infty} e^{-\alpha x^2} dx\right)^2 = \frac{\pi}{\alpha}, \tag{6.76}$$

which demonstrates the result quoted at the beginning.

6.2.2 Free Energy of Dilute Solutions

One of the central ideas about entropy in this book is that there are a few basic implementations of Boltzmann's assertion that $S = k_B \ln W$ that arise over and over in carrying out statistical mechanical reasoning. The ideal gas illustrates the way in which kinetic energy may be shared among a bunch of different molecules. The other crowning example in which we can simply calculate the entropy is embodied in the formula

$$W(N, \Omega) = \frac{\Omega!}{N!(\Omega - N)!}, \tag{6.77}$$

which instructs us about the configurational entropy associated with rearranging N objects amongst Ω boxes. How can this simple formula help us think about the free energy of solutions? In Section 5.5.2 (p. 225), we showed that two subsystems are in chemical equilibrium when their chemical potentials are equal. If we are to consider an aqueous solution with some dilute population of molecules, or alternatively, if we are to think about the interactions of gene regulatory proteins and DNA in solution, how are we to write the chemical potentials of the various species of interest?

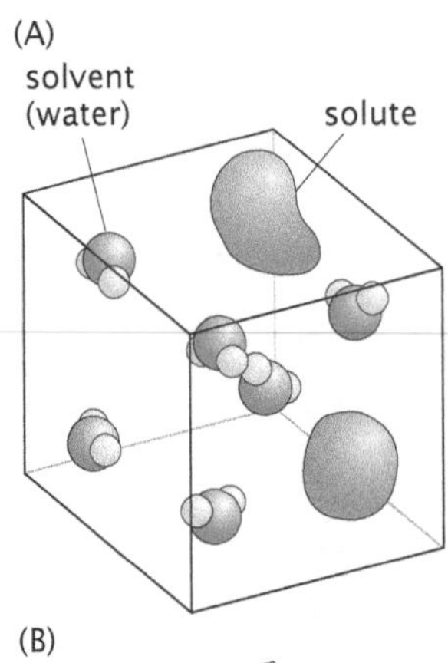

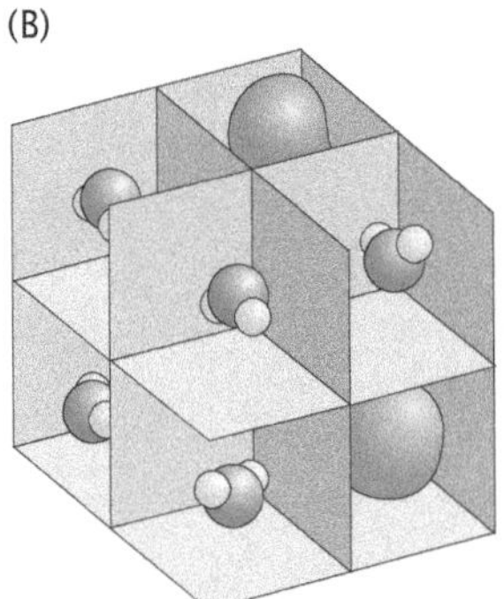

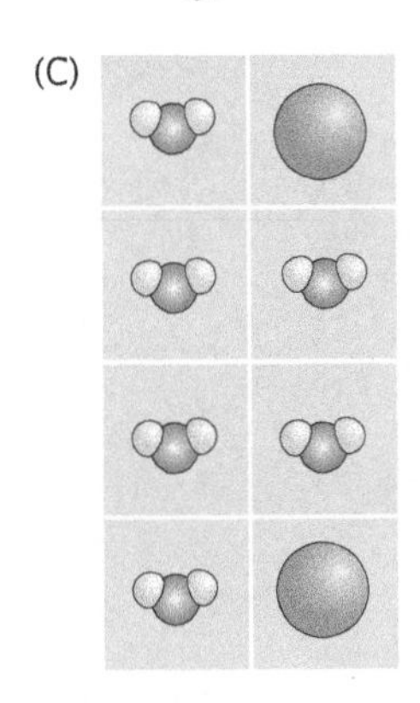

Figure 6.21: Idealization of a dilute solution as a system on a lattice. (A) Cartoon of the actual solution in which the species of interest is dilute and wanders around freely in solution. (B) Three-dimensional and (C) two-dimensional lattice idealization of the situation in (A) in which the molecules of interest are restricted to visit sites on a lattice.

The Chemical Potential of a Dilute Solution Is a Simple Logarithmic Function of the Concentration

To make progress on the question of how to write the chemical potentials for dilute solutions like those described above, we appeal to lattice models. Lattice models have been used to great advantage as a discretization trick that allows for the performance of combinatoric arguments in a way that is analytically tractable. For our present purposes, we imagine our system (water + solute) as being built up as a series of lattice sites as shown in Figure 6.21. These lattice models were introduced in Section 6.1.1 (p. 241) and illustrated in Figure 6.1 (p. 238). By restricting the set of allowed positions for the molecules of interest to the sites of a lattice, we have a countable set of distinct states that can be enumerated explicitly.

We write the number of water molecules as N_{H_2O}, while the number of solute molecules is given by N_s. Our objective is to write the total free energy of this system and then to obtain the solute chemical potential through the relation

$$\mu_{\text{solute}} = \left(\frac{\partial G_{\text{tot}}}{\partial N_s}\right)_{T,p}. \tag{6.78}$$

Intuitively, the chemical potential really tells us the free energy cost associated with changing the number of solute molecules in solution by 1 as

$$\mu_{\text{solute}} = G_{\text{tot}}(N_s + 1) - G_{\text{tot}}(N_s). \tag{6.79}$$

The reader is invited to explore this more deeply as well as the connection to the partition function in the problems at the end of the chapter. We argue that the free energy is given by

$$G_{\text{tot}} = \underbrace{N_{H_2O}\mu^0_{H_2O}}_{\text{water free energy}} + \underbrace{N_s\varepsilon_s}_{\text{solute energy}} - \underbrace{TS_{\text{mix}}}_{\text{mixing entropy}}. \tag{6.80}$$

This equation exploits the fact that the Gibbs free energy of the system of water + solute molecules is a sum of the free energy of water in the absence of solute, a contribution due to the presence of each solute molecule, and a mixing entropy term due to the possibility of rearranging the solute molecules. Note that the assumption that our solute molecules are dilute has snuck into the calculation by virtue of the fact that we write the energy of the solute molecules as a contribution per solute molecule multiplied by the total number of such molecules. This implicitly assumes no interaction between the various solute molecules and ascribes a single quantity ε_s as the cost to surround each such molecule with molecules of water.

To make progress, we now need a model for the configurational entropy, and it is here that we appeal to our lattice model with a total number of sites given by $N_{H_2O} + N_s$. We exploit our old friend, $S = k_B \ln W$, where in this case

$$W(N_{H_2O}, N_s) = \frac{(N_{H_2O} + N_s)!}{N_{H_2O}! N_s!}. \tag{6.81}$$

This quantity tells us the number of ways of distributing the N_{H_2O} and N_s molecules on the sites of the lattice. To compute the entropy, we compute the logarithm of $W(N_{H_2O}, N_s)$. As usual, we exploit the Stirling approximation (see p. 222), which results in

$$S_{mix} = -k_B \left(N_{H_2O} \ln \frac{N_{H_2O}}{N_{H_2O} + N_s} + N_s \ln \frac{N_s}{N_{H_2O} + N_s} \right). \tag{6.82}$$

Now we may exploit the smallness of the ratio N_s/N_{H_2O} (that is, there are many fewer solute molecules than there are water molecules), with the result that

$$S_{mix} \approx -k_B \left[N_{H_2O} \ln \left(1 - \frac{N_s}{N_{H_2O}} \right) + N_s \ln \frac{N_s}{N_{H_2O}} \right], \tag{6.83}$$

which, after Taylor expansion of the logarithm (that is, $\ln(1+\varepsilon) \approx \varepsilon$, see The Math Behind the Models on p. 215), results in

$$S_{mix} \approx -k_B \left(N_s \ln \frac{N_s}{N_{H_2O}} - N_s \right), \tag{6.84}$$

and hence

$$G_{tot}(T, p, N_{H_2O}, N_s) = N_{H_2O} \mu^0_{H_2O}(T, p) + N_s \varepsilon_s(T, p) + k_B T \left(N_s \ln \frac{N_s}{N_{H_2O}} - N_s \right). \tag{6.85}$$

With this result in hand, we can now find the solute chemical potential by effecting the derivative $\mu_s = \partial G / \partial N_s$, which is

$$\mu_s = \varepsilon_s + k_B T \ln \frac{c}{c_0}. \tag{6.86}$$

In this expression, we have introduced the symbols $c = N_s/V_{box}$ and $c_0 = N_{H_2O}/V_{box}$, which permit us to replace number variables by concentration variables. The key lesson we draw from this analysis is that the chemical potential for noninteracting systems, whether the ideal gas or the dilute solution considered here, is of a particularly simple form that will see duty in a range of applications throughout the remainder of the book and is written generically as

$$\mu_i = \mu_{i0} + k_B T \ln \frac{c_i}{c_{i0}}, \tag{6.87}$$

where the subscript zero refers to some convenient reference state. The convention in chemistry is usually to take the state where each of the molecular components is at 1 M concentration.

6.2.3 Osmotic Pressure as an Entropic Spring

Osmotic Pressure Arises from Entropic Effects

Our derivation of the chemical potential has led us to consider the lives of individual particles in solutions. A cell living in an aqueous environment will exchange material with the solution by taking nutrients in from the outside and excreting waste products. As we will further explore, the rates of these processes is strongly dependent upon the chemical potential of each of these components both inside and outside of the cell. Indeed, the second law of thermodynamics tells us that during any spontaneous process,

$$(\mu_1 - \mu_2)\,\mathrm{d}N_1 \leq 0. \tag{6.88}$$

Hence if $\mu_1 > \mu_2$, this implies that $\mathrm{d}N_1 < 0$ (that is, the number of particles in subsystem 1 will decrease as a result of the "driving force" $\mu_1 - \mu_2$ towards mass transport). For a reminder, see Section 5.5.2 (p. 225).

We begin by considering a fundamental problem faced by every cell. As we described in Chapter 2, the inside of the cell is jam-packed with a huge range of closely spaced molecules and macromolecular assemblies. For a bacterium living in a puddle of water, the exterior environment has components in solution at much lower concentrations. As was discussed in Chapter 5, this difference in concentration can provide the driving force for mass transport, determining whether a particular component will move into or out of the cell.

Because of the enormous concentration of internal components, the cell itself is under constant stress from the tendency of water to move from the outside in and the tendency of nearly all cellular components to move from the inside out. This difference between the inside and the outside leads to a mechanical force called osmotic pressure. In an extreme case, a vulnerable cell dropped into distilled water will simply swell and explode. Indeed, this has been frequently used in the laboratory to induce cells and viruses to release their DNA. Considering the fundamental and inescapable physical nature of this osmotic pressure effect, why do humans not explode when we dive into swimming pools? Humans, like most other living organisms, expend a lot of energy on coping with the dangerous effects of the osmotic pressure difference. One common mechanism used by many bacteria and plant cells is to build a rigid cell wall outside of the plasma membrane that physically prevents cell swelling. In addition, many cells use energy to actively pump water out and to modulate the concentrations of ions and other small solutes. We will discuss these channels and transporters further in Chapter 11. It is remarkable how rapid and robust the cell's response to sudden osmotic changes may be. This is particularly clear for organisms that are able to thrive in multiple different environments. An *E. coli* cell living in the lower intestine of a human is surrounded by an environment with high concentrations of other cells and nutrient-rich organic molecules. The same bacterium may suddenly be released into the freshwater environment of the toilet bowl and maintain viability by rapidly adjusting to this osmotic shock. In this section, we will explore a simple mathematical formulation of the idea of osmotic pressure.

An immediate and important biological application of our analysis of dilute solutions is to the problem of osmotic pressure. The basic idea behind the emergence of osmotic pressure is that the presence of a semipermeable membrane can result in the rearrangement of those molecules to which the membrane is permeable and the setting up of an attendant pressure. To see this, consider the model geometry shown in Figure 6.22, which shows a container with an internal membrane that is permeable to water but not to the solute molecules of interest. As a result of the permeability of the membrane to water, the equilibrium state is characterized by equality of chemical potentials for the water molecules on both sides of the container. (Revisit Section 5.5.2 (p. 225) to see how entropy maximization implies equality of chemical potentials.)

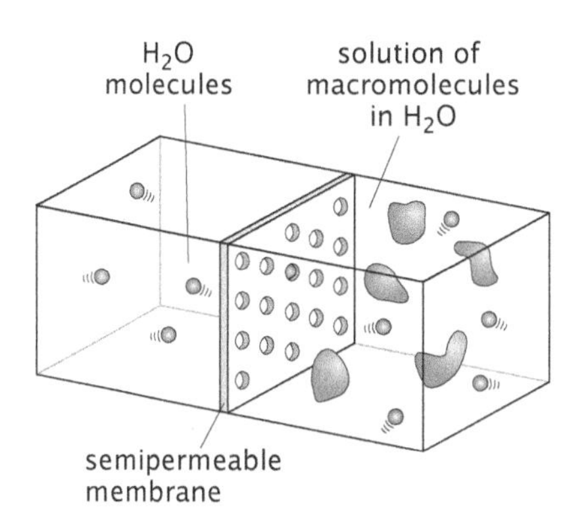

Figure 6.22: Schematic of a container with a semipermeable membrane with molecules in solution in one side of the container and pure solvent in the other.

The chemical potential of the water molecules on the side of the membrane with the dilute concentration of solute molecules can be derived by appealing to Equation 6.85 and in particular by evaluating $\mu_{H_2O}(T,p) = \partial G_{tot}/\partial N_{H_2O}$. Note that here we make explicit the dependence of the chemical potential on the pressure and the temperature. The resulting expression is

$$\mu_{H_2O}(T,p) = \mu^0_{H_2O}(T,p) - \frac{N_s}{N_{H_2O}} k_B T. \tag{6.89}$$

The equilibrium between the two sides may now be expressed via the equation

$$\underbrace{\mu^0_{H_2O}(T,p_1)}_{\text{solute-free side}} = \underbrace{\mu^0_{H_2O}(T,p_2) - \frac{N_s}{N_{H_2O}} k_B T}_{\text{side with solutes}}. \tag{6.90}$$

Note that we have already asserted that there will be a pressure difference between the two sides by introducing the notation p_1 for the pressure on the pure water side of the container and p_2 for the pressure on the side containing the solute molecules. We now expand the chemical potential on the right-hand side around p_1 as

$$\mu^0_{H_2O}(T,p_2) \approx \mu^0_{H_2O}(T,p_1) + \left(\frac{\partial \mu^0_{H_2O}}{\partial p}\right)(p_2 - p_1). \tag{6.91}$$

As a result of the thermodynamic relation $\partial\mu/\partial p = v$, where v is the volume per molecule, this equation can be transformed into

$$p_2 - p_1 = \frac{N_s}{V} k_B T, \tag{6.92}$$

where $V = N_{H_2O} v$ is, to a very good approximation, equal to the total volume on the solute side of the semipermeable membrane. This relation is known as the van't Hoff formula and it gives the osmotic pressure as a function of the concentration of the solute.

Viruses, Membrane-Bound Organelles, and Cells Are Subject to Osmotic Pressure

Osmotic pressure arises in many different contexts. From a biological point of view, the existence of osmotic pressure can give rise to mechanical insults that cells and viruses must find ways to endure. In particular, cell membranes and viral capsids are permeable to certain molecules and not others, giving rise to the possibility of osmotic shock.

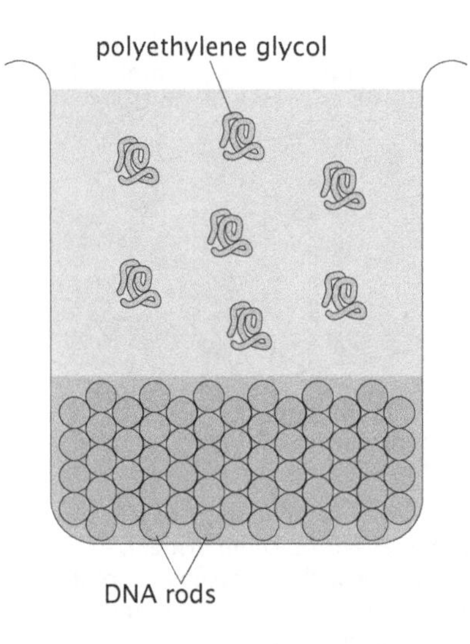

Figure 6.23: Experiment to measure DNA interstrand interactions. The experiment uses an osmolyte such as polyethylene glycol to apply an osmotic pressure to hexagonally ordered DNA and X-rays to measure the interstrand spacing. The DNA is arranged into a lattice with the double-helix molecules pointing out of the page in the schematic.

Estimate: Osmotic Pressure in a Cell To estimate the osmotic pressure faced by a bacterium such as *E. coli*, we return to the molecular census of Chapter 2. In particular, the characteristic concentration of inorganic ions in cells is 100 mM. Using the rough rule of thumb that a concentration of 1 nM corresponds to 0.6 molecules/fL, and using the fact that the volume of such a cell is roughly 1 fL, the number of inorganic ions in a bacterium is roughly 6×10^7. To compute the osmotic pressure using Equation 6.92, we need the concentration in units of number/nm^3, which is

$$c \approx \frac{6 \times 10^7 \text{ molecules}}{10^9 \text{ nm}^3} \approx 0.06 \frac{\text{molecules}}{\text{nm}^3}. \tag{6.93}$$

If we are now to invoke the van't Hoff formula of Equation 6.92, this translates into a pressure of order

$$p \approx 0.06 \frac{\text{molecules}}{\text{nm}^3} \times 4 \text{ pN nm} \approx 0.24 \text{ pN/nm}^2. \tag{6.94}$$

This result can be translated into units of atmospheric pressure using the conversion $1 \text{ pN/nm}^2 \approx 10$ atm, which yields the estimate that the osmotic pressure of a cell corresponds to several atmospheres of pressure.

ESTIMATE

Osmotic Forces Have Been Used to Measure the Interstrand Interactions of DNA

In addition to the concrete biological significance for the osmotic pressure described above, osmotic pressure has been used ingeniously as a tool to measure the forces between macromolecules. One example of this type of experiment is the use of osmotic pressure to measure the interaction between adjacent strands of DNA. Such experiments are important in contexts ranging from the packing of DNA in viruses to the *in vitro* condensation of DNA by condensing agents such as spermine, spermidine, and putrescine. A schematic of the types of experiments we have in mind is shown in Figure 6.23. In these experiments, double-helical DNA condenses in the presence of a polymer such as polyethylene glycol. The condensed structure forms an ordered lattice with hexagonal symmetry.

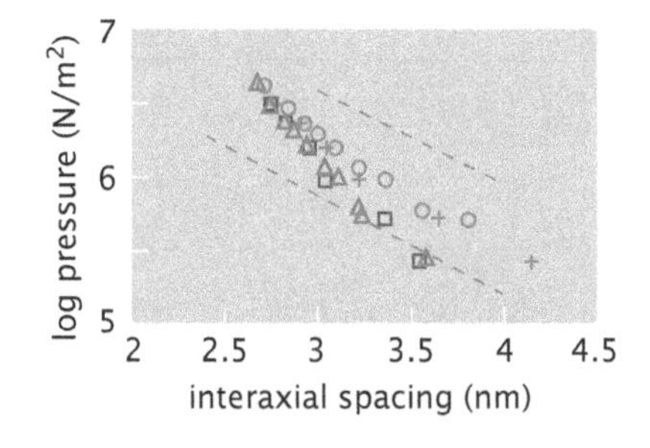

Figure 6.24: Equation of state relating pressure and intermolecular spacing for DNA. Different concentrations of the osmolyte of interest were used to generate different osmotic pressures resulting in different interstrand spacings. The symbols represent different salt conditions. (Adapted from D. C. Rau et al., *Proc. Natl Acad. Sci. USA* 81:2621, 1984.)

In the experiment shown in Figure 6.23, polyethylene glycol is too large to fit between the DNA strands, and, as such, exerts an osmotic pressure on the condensed DNA. The spacing between parallel strands of DNA was measured using X-rays. By making repeated measurements with different concentrations of osmolyte (and hence different osmotic pressures), it is possible to determine the relation between the osmotic pressure and the DNA spacing. An example of the data that emerge from such experiments is shown in Figure 6.24, which shows the relation between the osmotic pressure and the interstrand spacing between DNA strands. According to these experiments, the dependence of the pressure associated with the interactions in hexagonally packed DNA on the spacing between the strands can be described by the functional form

$$P(d_s) = F_0 e^{-d_s/c}, \tag{6.95}$$

where d_s is the interstrand spacing, F_0 is a constant whose magnitude depends on the type and strength of the ionic solution, and $c \approx 0.27$ nm is a decay length that is approximately constant over a wide range of salt conditions.

6.3 The Calculus of Equilibrium Applied: Law of Mass Action

One of the crowning achievements of statistical mechanics and the application with arguably the broadest biological reach is the analysis of chemical equilibrium. In biology, essentially every dynamic life process requires some sort of chemical transformation, either two separate entities coming together to form a complex, a previous existing complex separating into constitutent parts, or an actual conversion of one type of molecule into another. Up until now, we have only needed to keep track of the numbers and energies of molecules in our system. But if we are going to attack real biochemical problems, we also need formal ways to keep track of their identities.

In standard biochemical language, molecular numbers are usually formalized as concentrations and the tendency of one molecular species to convert into another is summarized by an equilibrium constant. In the simplest case, we can consider two molecules that come together to make a complex according to the reaction $A + B \rightleftharpoons AB$. The bi-directional arrows indicate that the complex can also dissociate to reform the original parts. Our biochemical intuition tells us that if we start with a test tube containing high concentrations of A and B, but no AB complex, then AB complex will be formed. Conversely, if we begin with a tube containing exclusively a high concentration of the AB complex, but no A or B, then some of the complex will dissociate. After sufficient time has passed, we should not be able to distinguish between the two test tubes. That is, their final equilibrium state does not depend upon the initial conditions. However, the initial direction of the reaction in each case can be reliably predicted from our intuitive expectation that a final equilibrium state will eventually be attained. Throughout biology, we repeatedly use this intuition to predict future occurrences in metabolic reactions, in responses of cells to hormonal signals, in the dynamics of gene expression, and in practically every other biological process that one can imagine. In this section, we will use the concepts of the Boltzmann distribution developed thus far to recreate the chemical laws of equilibrium from the point of view of individual molecules and energies. Understanding the connection between these two different formalisms for describing the same real process is a central achievement in connecting statistical mechanics to the functions of life.

6.3.1 Law of Mass Action and Equilibrium Constants

Equilibrium Constants are Determined by Entropy Maximization

As with all equilibrium analyses, the basic rules are the same as those laid down in Chapter 5. In particular, we seek that configuration which minimizes the relevant free energy with respect to whatever types of excursion our system is capable of making. In the case of chemical equilibrium, we are interested in changes in the number of reactant and product molecules. That is, we consider a system that permits molecules to neither enter nor leave. However, these molecules are permitted to interact among themselves with the effect that the number of molecules of each type *can* change. For simplicity, we consider the reaction

$$A + B \rightleftharpoons AB. \tag{6.96}$$

To characterize the configurational state of this system, we introduce the number of molecules of each species, N_A, N_B, and N_{AB}. Given this notation, the excursions the system may exercise correspond to changes in these numbers, ΔN_A, ΔN_B, and ΔN_{AB}. The statement of equilibrium, at fixed pressure and temperature, then becomes

$$dG = \left(\frac{\partial G}{\partial N_A}\right)_{T,p} dN_A + \left(\frac{\partial G}{\partial N_B}\right)_{T,p} dN_B + \left(\frac{\partial G}{\partial N_{AB}}\right)_{T,p} dN_{AB} = 0. \tag{6.97}$$

To turn this statement into a concrete result requires the imposition of correct stoichiometric reasoning. That is, we need to account for the fact that when the numbers of A and B molecules decrease by 1, there is a corresponding increase by 1 in the number of AB molecules. To that end, we introduce stoichiometric coefficients ν_i that count the change in the number of particles of the ith type during the reaction. For the reaction considered above we have $\nu_A = \nu_B = -1$ and $\nu_{AB} = 1$, which signifies that the creation of a new AB molecule requires a concomitant reduction in the number of individual A and B molecules.

If we recall that the derivatives featured in Equation 6.97 are actually the chemical potentials of the individual species (that is, $\partial G/\partial N_i = \mu_i$), then, at equilibrium, for a fixed value of T and P we have

$$dG = 0 \Longrightarrow \mu_A \, dN_A + \mu_B \, dN_B + \mu_{AB} \, dN_{AB} = 0, \tag{6.98}$$

which in the more general case of N species can be written as

$$dG = 0 \Longrightarrow \sum_{i=1}^{N} \mu_i \, dN_i = 0. \tag{6.99}$$

If we exploit the stoichiometric coefficients as defined above through the statement $dN_i = \nu_i \, dN$, then Equation 6.97 takes the general but abstract form

$$\sum_i \nu_i \mu_i = 0. \tag{6.100}$$

So far, what we have written makes no specific assumptions about the system of interest—a particular system requires us to make a statement about the precise functional form of the chemical potentials. To make progress, we invoke the specific form of the chemical potential derived earlier in the chapter, namely, $\mu_i = \mu_i^0 + k_B T \ln(c_i/c_{i0})$, where c_{i0} is a standard state concentration, which is generally defined as 1 M in the case of a solution. With that result in hand, Equation 6.100 can be recast in a highly useful form that is an instance of the law of mass action and will lead immediately to the notion of an equilibrium constant for the reaction of interest.

The logic of our derivation of the law of mass action amounts to first substituting Equation 6.87 into Equation 6.100. This results in

$$\sum_{i=1}^{N} \nu_i \mu_{i0} = -k_B T \sum_{i=1}^{N} \ln\left(\frac{c_i}{c_{i0}}\right)^{\nu_i}. \tag{6.101}$$

If we now invoke the rule that $\ln x_1 + \ln x_2 + \cdots + \ln x_N = \ln(x_1 x_2 \cdots x_N)$, we can rewrite this result as

$$-\frac{1}{k_B T}\sum_{i=1}^{N} \nu_i \mu_{i0} = \ln\left[\prod_{i=1}^{N}\left(\frac{c_i}{c_{i0}}\right)^{\nu_i}\right], \tag{6.102}$$

where we have used the notation $x_1 x_2 \cdots x_N = \prod_{i=1}^{N} x_i$. If we now exponentiate this result, noting that $e^{\ln x} = x$, we find

$$\prod_{i=1}^{N}\left(\frac{c_i}{c_{i0}}\right)^{\nu_i} = e^{-(1/k_B T)\sum_{i=1}^{N}\nu_i\mu_{i0}}. \tag{6.103}$$

This can be simplified by passing the c_{i0} factors to the right-hand side, resulting in

$$\prod_{i=1}^{N} c_i^{\nu_i} = \underbrace{\left(\prod_{i=1}^{N} c_{i0}^{\nu_i}\right) e^{-(1/k_B T)\sum_{i=1}^{N}\nu_i\mu_{i0}}}_{K_{eq}(T)}, \tag{6.104}$$

a result known as the law of mass action. This equation tells us the relation between the concentrations of reactants and products at a given temperature and pressure. So important is the right-hand side of this equation that it has been distinguished as the equilibrium constant K_{eq},

$$K_{eq} = \left(\prod_i c_{i0}^{\nu_i}\right) e^{-\sum_i \mu_{i0}\nu_i/k_B T}, \tag{6.105}$$

where $K_{eq} = 1/K_d$, with K_d the so-called dissociation constant. Often, the equilibrium (or dissociation) constant is treated as an empirical parameter that can be measured experimentally at a particular temperature and pressure. On the other hand, our statistical mechanical treatment of the problem shows that there is a corresponding microscopic interpretation as well. In the case of the example that was introduced in Equation 6.96, this yields the ratio

$$\prod_{i=1}^{N} c_i^{\nu_i} = c_A^{-1} c_B^{-1} c_{AB}^{1} = \frac{c_{AB}}{c_A c_B} = K_{eq}. \tag{6.106}$$

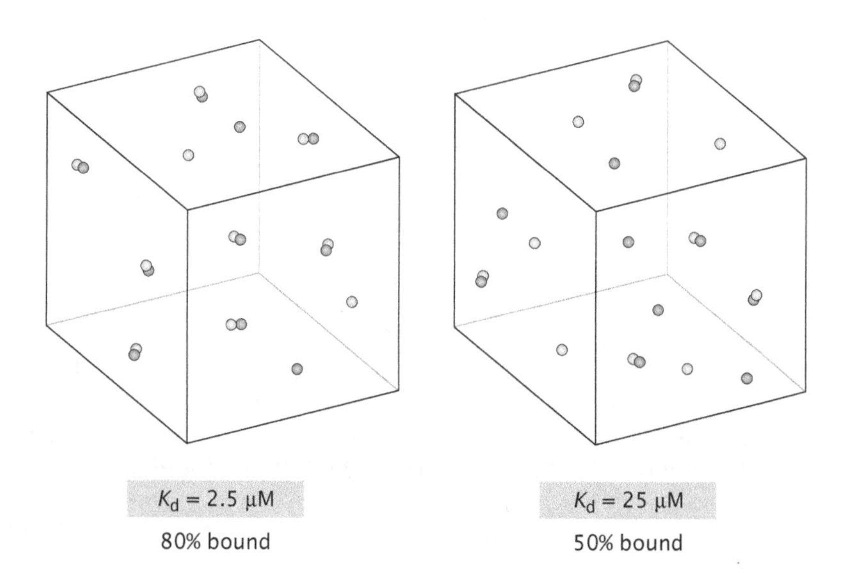

Figure 6.25: The dissociation constant. The relative amounts of reactant and product are determined by the dissociation constant. In these two figures, the numbers of reactants and products are different because of different values of K_d. The total concentration of dark and light molecules is 50 μM.

A graphical representation of the idea embodied in the law of mass action is shown in Figure 6.25. The idea of the figure is to show the relative number of reactant and product molecules for different choices of the dissociation constant.

These results are admittedly abstract, both notationally and conceptually. As a result, to make the entire picture more concrete, we now turn to some specific examples that will illustrate the meaning of our various notational conventions and that begin to highlight their biological relevance.

6.4 Applications of the Calculus of Equilibrium

6.4.1 A Second Look at Ligand–Receptor Binding

In Section 6.1.1 (p. 241), we examined the problem of ligand–receptor binding from the point of view of a microscopic model of the statistical mechanics of binding. That same problem can be addressed from the point of view of equilibrium constants and the law of mass action, and it is enlightening to examine the relation between the two points of view. The reaction of interest is characterized by the stoichiometric equation

$$\mathrm{L} + \mathrm{R} \rightleftharpoons \mathrm{LR}. \tag{6.107}$$

This reaction is described by a dissociation constant given by the law of mass action as

$$K_{\mathrm{d}} = \frac{[\mathrm{L}][\mathrm{R}]}{[\mathrm{LR}]}. \tag{6.108}$$

It is convenient to rearrange this expression in terms of the concentration of ligand–receptor complexes as

$$[\mathrm{LR}] = \frac{[\mathrm{L}][\mathrm{R}]}{K_{\mathrm{d}}}. \tag{6.109}$$

As before, our interest is in the probability that the receptor will be occupied by a ligand. In terms of the concentration of free receptors and ligand–receptor complexes, p_{bound} can be written as

$$p_{\mathrm{bound}} = \frac{[\mathrm{LR}]}{[\mathrm{R}] + [\mathrm{LR}]}. \tag{6.110}$$

This amounts to the statement that we can use the fraction of receptors that are bound by ligands as a surrogate for the probability that a *single* receptor will be occupied by a ligand. We are now poised to write p_{bound} itself by invoking Equation 6.109, with the result that

$$p_{\mathrm{bound}} = \frac{[\mathrm{L}]/K_{\mathrm{d}}}{1 + ([\mathrm{L}]/K_{\mathrm{d}})}. \tag{6.111}$$

What we see is that K_{d} is naturally interpreted as that concentration at which the receptor has a probability of 1/2 of being occupied. Furthermore, we can compare this equation for p_{bound} with the one derived earlier in the chapter (Equation 6.19 on p. 244) in the context of a lattice model. We see that the dissociation constant can be expressed in terms of the microscopic parameters of the lattice model as

$$K_{\mathrm{d}} = \frac{1}{v} e^{\beta \Delta \varepsilon}, \tag{6.112}$$

where v is the volume taken up by a single lattice site and $\Delta\varepsilon$ is the binding energy. This formula allows us to go back and forth between the statistical mechanics language of binding energies and the chemical language of dissociation constants. In the problems at the end of the chapter, we explore the limitations of this approach. Notice that our expression for p_{bound} depends on the concentration of free ligands and not on their total concentration. This is an important subtlety since in an experiment we pipette in some total concentration and the free concentration is determined by the molecular properties of the interaction. As a result, when determining dissociation constants, it is often convenient to work at concentrations where the ligands are in significant excess to the receptors. In this case, the free ligand concentration is nearly equal to the total ligand concentration. The reader can explore this issue further in the problems.

Note that in each of our treatments of the ligand–receptor binding problem, we have determined the quantity p_{bound}, which depends upon the concentration of the ligands. The receptor can only be in one of two states, namely, bound or unbound. One interesting way to characterize this binding probability is through the use of the "missing information" as provided by the Shannon entropy (Equation 6.40, p. 253). Using this definition and noting that the ligand–receptor problem is an effective two-state system, we can write the missing information as

$$\text{MI}(l) = -p_{\text{bound}} \ln p_{\text{bound}} - p_{\text{unbound}} \ln p_{\text{unbound}}. \tag{6.113}$$

If we introduce the notation $l = [L]/K_{\text{d}}$ and recall that $p_{\text{bound}} = l/(1+l)$ and $p_{\text{unbound}} = 1/(1+l)$, we can rewrite the missing information as

$$\text{MI}(l) = -\frac{l}{1+l} \ln \frac{l}{1+l} - \frac{1}{1+l} \ln \frac{1}{1+l}. \tag{6.114}$$

The first term corresponds to the bound state and the second to the unbound state. This can be simplified to the form

$$\text{MI}(l) = -\frac{l}{1+l} \ln l + \ln(1+l) \tag{6.115}$$

and is plotted in Figure 6.26. The most immediate conclusion from this analysis is that the missing information is maximum when

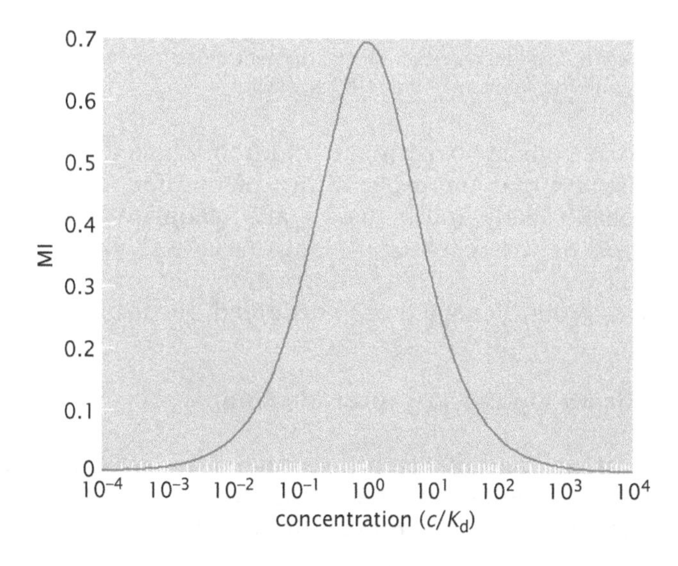

Figure 6.26: The missing information for the ligand–receptor problem.

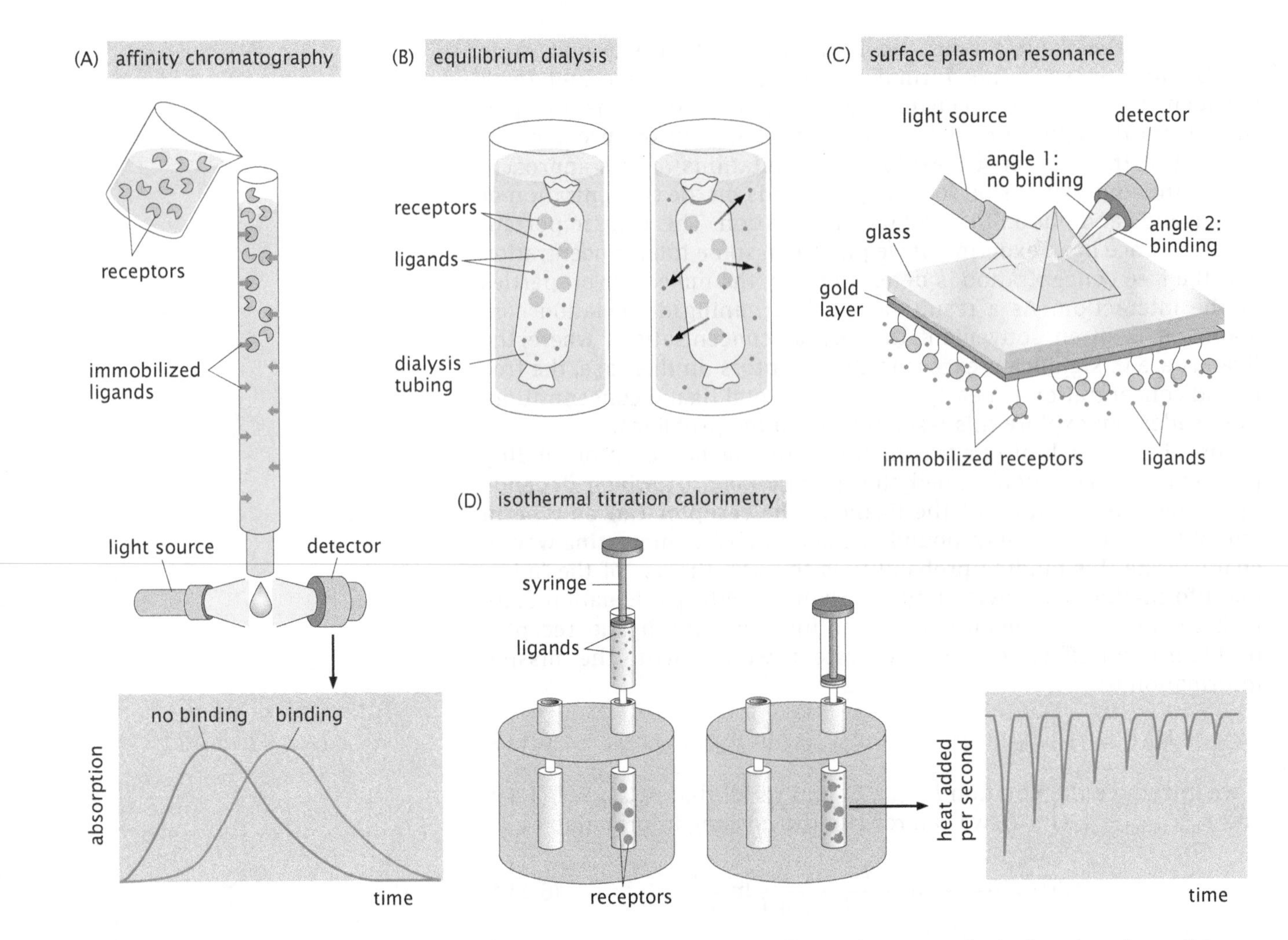

Figure 6.27: Different methods for measuring equilibrium binding. (A) The affinity of a receptor for its ligand can be easily measured using a version of affinity chromatography (see Figure 4.2 on p. 142). When a solution containing the receptor is passed over a column where the ligand is immobilized on the matrix, the receptor's progress down the column will be slowed due to repeated rounds of association and dissociation with the ligand. The strength of the binding can be observed by measuring the time it takes for the protein to emerge from the end of the column. (B) In cases where the receptor and the ligand are very different in size such that the ligand is able to pass freely through a dialysis membrane while the receptor is not, the equilibrium binding affinity can be measured by determining the concentration of ligand inside and outside of the dialysis bag after a long equilibration period. (C) Other methods for measuring binding affinity can be performed using smaller amounts of material. One commonly used method, surface plasmon resonance, uses receptors immobilized to a surface. The instrument uses the intensity and angle of reflected light, which are functions of the amount of mass attached to the surface. When a soluble ligand is added and binds to the surface, the mass changes. (D) The physical consequences of binding are exploited in measurements using isothermal titration calorimetry. A chamber containing the receptors is brought into thermal equilibrium with a reference cell and ligands are then injected in small increments. The instrument measures the amount of heat that must be added or taken away to keep the temperature the same as in the reference cell.

$[L] = K_d$. This is consistent with our intuition, since it is precisely at this concentration that the probabilities of the bound and unbound states are equally likely and hence we are maximally ignorant about the disposition of the receptor. At both large and small concentrations, there is very little missing information, because in those cases the receptor is either occupied or unoccupied, respectively.

6.4.2 Measuring Ligand–Receptor Binding

Binding curves are one of the primary windows on the function of many molecules of biological interest. As already shown in Figure 6.6, what such curves show is the probability that the receptor of interest will have a bound ligand as a function of the concentration of ligands. There are a variety of elegant methods for measuring binding curves, several of which are highlighted in Figure 6.27. Often, these methods take advantage of the fact that a binding event has the effect of changing some physical property. For example, surface plasmon resonance exploits a change in the dielectric properties of a surface layer of receptors when they are bound to ligands. Isothermal titration calorimetry exploits the energy released (or taken up) during a binding reaction. In coming chapters, we will have occasion to refer to binding curves in a number of different settings and the reader should always think back to how such curves are measured.

Several examples of the kind of data that arise from binding measurements are shown in Figure 6.28 as well as fits to those data using the simple binding curve given in Equation 6.111. Stated simply, these curves illustrate the fractional occupancy of myoglobin by oxygen as the concentration of oxygen (as measured by the pressure) is increased, the binding of HIV viral proteins to cell surface receptors, and the binding of the transcription factor NtrC to its target DNA. The reasonable accord between the measured and computed binding curves also indicates that the simple (noncooperative) binding characterized by Equation 6.111 describes these binding processes.

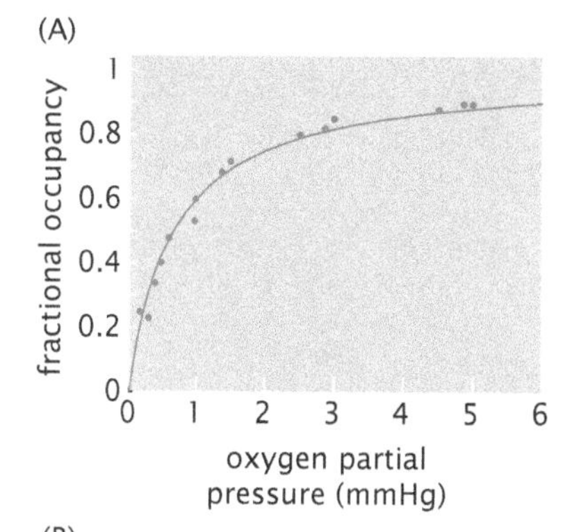

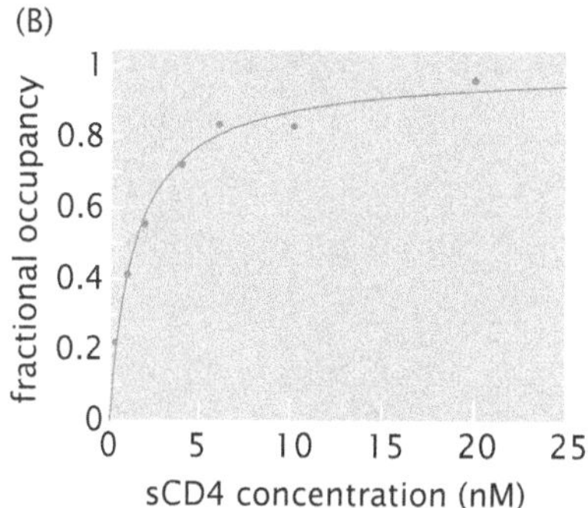

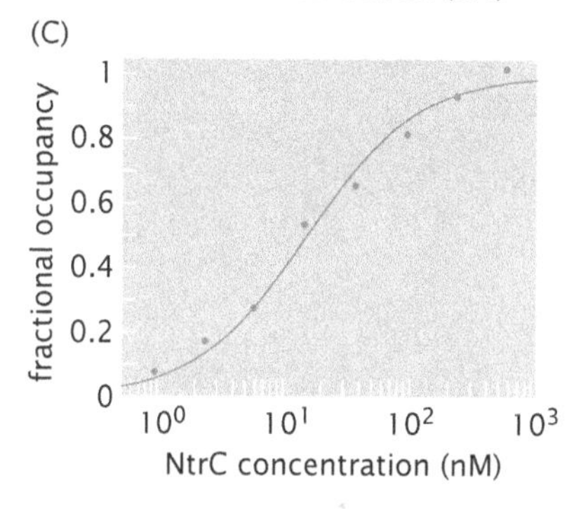

Figure 6.28: Examples of ligand–receptor binding. (A) The binding of oxygen to myoglobin as a function of the oxygen partial pressure. The points correspond to the measured occupancy of myoglobin as a function of the oxygen partial pressure and the curve is a fit to Equations 6.19 and 6.111. The fit yields $\Delta\varepsilon \approx -7.04 k_B T$ using a standard state $c_0 = 760$ mmHg $= 1$ atm, which also corresponds to a dissociation constant $K_d = 0.666$ mmHg. (B) Binding of HIV protein gp120 to a soluble form of the receptor CD4, giving $\Delta\varepsilon \approx -19.84 k_B T$ or $K_d = 1.4578$ nM with $c_0 = 0.6$ M. (C) Binding of NtrC to DNA, giving $\Delta\varepsilon = -17.47 k_B T$ or $K_d = 15.5$ nM with $c_0 = 0.6$ M. Note that the horizontal axis is logarithmic rather than linear, so the standard binding curve appears sigmoidal in shape. (A, data from A. Rossi-Fanelli and E. Antonini, *Arch. Biochem. Biophys.* 77:478, 1958; B, data from D. W. Brighty et al., *Proc. Natl Acad. Sci. USA* 88:7802, 1991; C, data from V. Weiss et al., *Proc. Natl Acad. Sci. USA* 89:5088, 1992.)

6.4.3 Beyond Simple Ligand–Receptor Binding: The Hill Function

The mathematical treatment of binding reactions is usually built upon the foundations laid in the previous subsection, using the assumption that a single ligand binds to a single receptor in an independent reaction. The familiar consequence of this assumption is a concentration-dependent binding curve such as is shown in Figure 6.6 (p. 244). At low concentrations of ligand, the amount of occupied receptor is a nearly linear function of ligand concentration. At high concentrations of ligand, nearly all the receptors are occupied and the binding curve saturates. However, there are many classes of biological reactions for which this kind of simple concentration dependence would be inadequate. For cases when it is important that a molecule or a cell effectively be either "on" or "off" as a function of the concentration of the signal received, it would be preferable for the binding curve to be switch-like (more like a step function).

For example, cells in an adult human are constantly receiving signals to either survive or die (commit apoptosis). There is no in-between state. The decision to die is irrevocable. Imagining a concentration-dependent pro-apoptotic signal in the context of the normal binding curve, it is not clear where on this continuum the commitment to death should lie. For this and many other biological processes where an all-or-none response is required, it is important that cells be able to turn an essentially analog signal into a digital output. One common mechanism used to generate switch-like responses is the construction of receptors where multiple ligands are able to bind simultaneously, subject to the constraint that the binding of one ligand significantly enhances the affinity for the binding of subsequent ligands. We have already seen one example of the consequences of cooperative binding in the case of hemoglobin in Chapter 4.

One way to see how such switch-like behavior (dubbed *cooperativity*) emerges is by appealing to the reaction

$$\mathrm{L} + \mathrm{L} + \mathrm{R} \rightleftharpoons \mathrm{L_2R}. \qquad (6.116)$$

In this case, the dissociation constant can be written as

$$K_\mathrm{d}^2 = \frac{[\mathrm{L}]^2[\mathrm{R}]}{[\mathrm{L_2R}]}. \tag{6.117}$$

Since we are interested in the probability that the receptor will have bound ligand, we make an argument like that given in Equation 6.110, resulting in

$$p_\mathrm{bound} = \frac{[\mathrm{L_2R}]}{[\mathrm{R}]+[\mathrm{L_2R}]}. \tag{6.118}$$

Using the definition of the dissociation constant, this can be rewritten as

$$p_\mathrm{bound} = \frac{([\mathrm{L}]/K_\mathrm{d})^2}{1+([\mathrm{L}]/K_\mathrm{d})^2}. \tag{6.119}$$

This is a simple instance of a *Hill function* with Hill coefficient $n = 2$. More generally, Hill functions are of the form

$$p_\mathrm{bound} = \frac{([\mathrm{L}]/K_\mathrm{d})^n}{1+([\mathrm{L}]/K_\mathrm{d})^n}, \tag{6.120}$$

and a commonly adopted strategy is to fit binding data to such curves directly without reference to the underlying origins of a given Hill coefficient. In Figure 6.29(A), we show the cases $n = 1, 2,$ and 4 to illustrate the way in which more switch-like behavior emerges with increasing Hill coefficient.

Our derivation here has made the simplifying assumption that the concentration of receptor bound to a single ligand (that is, [LR]) is negligible, that is, the binding is either all or nothing. In reality, of course, this is not strictly true and therefore the measured value of the Hill coefficient for a particular binding reaction may not be an integer. For example, the theoretical Hill coefficient for the binding of oxygen to hemoglobin if it were entirely cooperative would be 4 since there are four binding sites, but the measured value is approximately 3.0.

Just as we earlier examined the missing information as offered by the Shannon entropy for the problem of simple ligand–receptor binding, it is also of interest to characterize the missing information for binding described by a Hill function. For binding in the case of a Hill function, the system can once again exist in two states, bound and unbound, and their probabilities can be used to compute the missing information as

$$\mathrm{MI}(l) = -\frac{l^n}{1+l^n}\ln\frac{l^n}{1+l^n} - \frac{1}{1+l^n}\ln\frac{1}{1+l^n}, \tag{6.121}$$

where once again we define $l = [\mathrm{L}]/K_\mathrm{d}$. The missing information as a function of this dimensionless concentration is shown in Figure 6.29(B), where it is seen that as the cooperativity increases, the width of the missing information decreases since only over a very narrow concentration range are we uncertain as to whether ligand is bound or not.

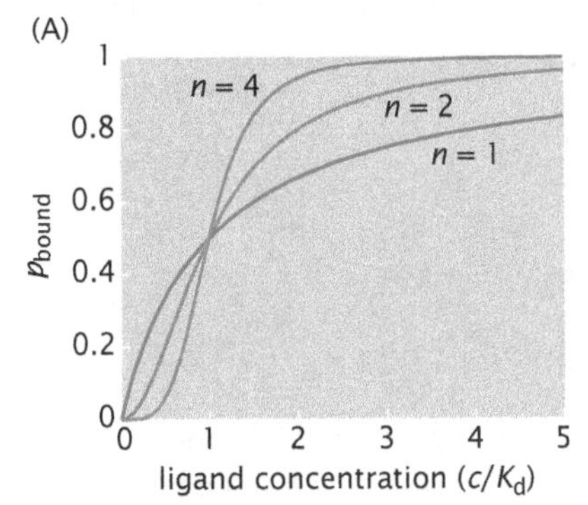

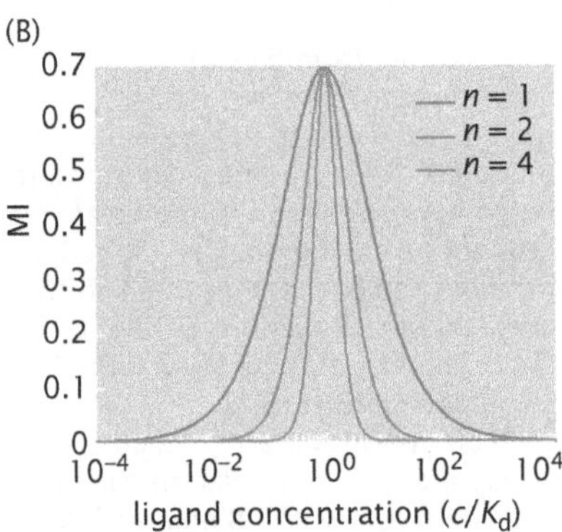

Figure 6.29: The Hill function and binding. (A) Family of binding curves with different Hill coefficients. The graph compares binding curves with different choices of Hill coefficient labeled by $n = 1, 2, 4$. (B) The missing information as a function of concentration for the three different Hill functions shown in (A). The missing information takes its maximum value of ln 2 when $c = K_\mathrm{d}$, corresponding to equal probability of the bound and unbound states.

6.4.4 ATP Power

As shown in Section 5.1.2 (p. 190), ATP drives many of the chemical transformations of the cell. Further, we argued that the energy released upon ATP hydrolysis was of just the right magnitude to mediate these transformations, in much the same way a 20 dollar bill is a convenient amount of spending money. To more precisely characterize the energy that is made available upon hydrolysis, we need to consider the way in which the free energy of hydrolysis depends upon the concentrations of reactants and products. We consider the free energy released when one ATP molecule is hydrolyzed to give an ADP molecule and inorganic phosphate. As discussed previously, if the concentrations of these three molecular species were to take on their equilibrium values, no free-energy release would accompany the reaction. In this case, equilibrium truly does mean death. To avoid this predicament, the cell maintains a nonequilibrium concentration of ATP, ADP, and inorganic phosphate.

The Energy Released in ATP Hydrolysis Depends Upon the Concentrations of Reactants and Products

One of the key questions that we might wish to address in light of this discussion is precisely how the free energy available from ATP hydrolysis depends upon the concentrations of both reactants and products. To make this estimate, we begin by using the results from Section 6.3.1 to write the change in free energy upon hydrolysis as

$$dG = \sum_i \mu_i \, dN_i = \sum_i \mu_i \nu_i \, dN. \tag{6.122}$$

If we use our standard expression for the chemical potential (Equation 6.87 on p. 263) the energy released per reaction, ΔG, corresponds to $dN = 1$, giving

$$\Delta G = \sum_i \left[\mu_{i0} + k_B T \ln\left(\frac{c_i}{c_{i0}} \right) \right] \nu_i. \tag{6.123}$$

If we use this equation twice, once as above and once for a reference concentration (ΔG_{ref}), and take the difference in these two free-energy differences, we find that these two free-energy differences are related by

$$\Delta G = \Delta G_{\text{ref}} + k_B T \ln\left(\frac{\prod_{i=1}^{N} c_i^{\nu_i}}{\prod_{i=1}^{N} c_{i,\text{ref}}^{\nu_i}} \right). \tag{6.124}$$

A convenient choice of reference state is the equilibrium state. Namely, if the concentrations of the various molecular species taking part in the reaction are at their equilibrium values, then $\Delta G_{\text{ref}} = 0$. Furthermore, as shown in the previous section, in equilibrium, $\prod_{i=1}^{N} c_{i,\text{eq}}^{\nu_i} = K_{\text{eq}}$, and therefore

$$\Delta G = k_B T \ln\left(\frac{\prod_{i=1}^{N} c_i^{\nu_i}}{K_{\text{eq}}} \right). \tag{6.125}$$

Note that this equation can in turn be rewritten as

$$\Delta G = \Delta G_0 + k_B T \ln\left[\prod_{i=1}^{N} \left(\frac{c_i}{1\,\text{M}} \right)^{\nu_i} \right]. \tag{6.126}$$

where $\Delta G_0 = -k_B T \ln K_{eq}$ is the so-called standard state free energy, assuming that the equilibrium constant is also expressed in molar units.

For the particular case of ATP hydrolysis, described by the reaction, $ATP \rightleftharpoons ADP + P_i$, Equation 6.126 becomes

$$\Delta G = \underbrace{\Delta G_0}_{\text{standard state}} + \underbrace{k_B T \ln \frac{[ADP][P_i]}{[ATP]}}_{\text{concentration dependence}}, \tag{6.127}$$

where all concentrations are measured relative to the 1 M standard state. Using the measured standard state free energy $\Delta G_0 \approx -12.5\, k_B T$/molecule, and the typical values for the *in vivo* concentrations found in *E. coli* (see Bionumbers for representative values), which are roughly $[ATP] = 5 \times 10^{-3}$, $[ADP] = 0.5 \times 10^{-3}$, and $[P_i] = 10 \times 10^{-3}$, we compute $\Delta G \approx -20 k_B T$. In most cells, the concentrations of ATP, ADP, and P_i are roughly in the millimolar range, so we take $-20\, k_B T$ as our rule of thumb for the free energy released in ATP hydrolysis *in vivo*.

6.5 Summary and Conclusions

Statistical mechanics is one of the most powerful and far-reaching tools of all of physics. To the extent that the use of equilibrium ideas makes sense for biological problems, statistical mechanics is a powerful part of the set of tools that life scientists bring to their quantitative work as well. The starting point for the application of statistical mechanics is the Boltzmann distribution. Speaking of this distribution, Feynman (1972) noted: "This fundamental law is the summit of statistical mechanics, and the entire subject is either the slide-down from this summit, as the principle is applied to various cases, or the climb-up to where the fundamental law is derived." In this chapter, we started with the climb up to the summit by introducing the notion of a microstate and by showing the meaning and derivation of the Boltzmann distribution. With this distribution in hand, we then turned to the application of these ideas to ligand–receptor binding, the promoter occupancy and cellular decision making, the origins and significance of osmotic pressure, and the calculus of equilibrium in the form of the law of mass action. With these tools in hand, we now turn to a special class of problems that can be considered using statistical mechanics, namely, the two-state problem.

6.6 Problems

Key to the problem categories: • Model refinements and derivations • Estimates • Data interpretation • Model construction

• 6.1 Disulfide bond formation

A protein, shown schematically in Figure 6.30, may contain several cysteines (represented as light colored balls), which may pair together to form disulfide bonds. For six cysteines, three disulfide bonds can form; the figure shows the pairing arrangement 1–6, 2–5, 3–4. How many different disulfide pairing arrangements are possible? Derive the general formula for the number of different pairing arrangements when there are n cysteines (and n is even).

(Adapted from problem 1.11 of K. Dill and S. Bromberg, Molecular Driving Forces, 2nd ed. Garland Science, 2011.)

• 6.2 Statistical mechanics of an optical trap.

In the Computational Exploration in Chapter 5 (p. 207) we described the use of laser light to trap micron-sized beads. The dynamics of such beads can be

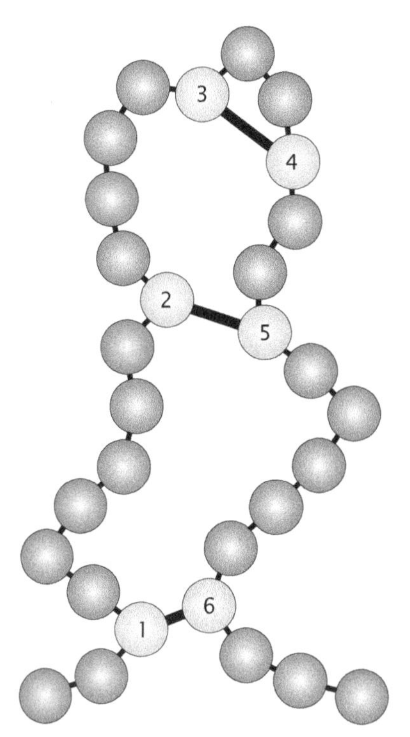

Figure 6.30: Simplified representation of protein structure. Each amino acid is represented as a shiny ball. Cysteine residues are called out via labels. (Adapted from K. Dill and S. Bromberg, Molecular Driving Forces, 2nd ed. Garland Science, 2011.)

thought of as the Brownian motion of a particle in a quadratic energy well. Compute the mean-squared excursion $\langle x^2 \rangle$ of such a bead in a one-dimensional quadratic well with a potential-energy profile $U(x) = \frac{1}{2}kx^2$ and show that we can then determine the trap stiffness as

$$k = \frac{k_B T}{\langle x^2 \rangle}. \tag{6.128}$$

6.3 Polymerase binding to the promoter revisited

The probability of promoter occupancy can be computed using both statistical mechanics and thermodynamics (that is, using equilibrium constants). These two perspectives were already exploited for simple ligand–receptor binding in Sections 6.1.1 and 6.4.1.

(a) Write an expression for the probability of finding RNA polymerase bound to the promoter as a function of the equilibrium constants for specific and nonspecific binding.

(b) *In vitro*, the dissociation constant of RNA polymerase binding to nonspecific DNA is approximately $10\,\mu$M and the dissociation constants of RNA polymerase to the *lac P1* and T7A1 promoters are 550 nM and 3 nM, respectively. Use these constants and the results from (a) to estimate the *in vivo* binding energies of RNA polymerase to *lac P1* and T7A1 promoters.

6.4 Free versus bound ligand

The expression found in Equation 6.111 gave the probability of binding of a ligand to a receptor as a function of the free concentration of ligand, [L]. However, quantities easier to tune experimentally are the total concentrations of ligand, $[\mathrm{L}]_{\mathrm{tot}}$, and receptor, $[\mathrm{R}]_{\mathrm{tot}}$.

(a) Obtain the probability of binding as a function of the total concentration of ligand and receptor.

(b) Examine the limit where $[\mathrm{L}]_{\mathrm{tot}} \gg [\mathrm{R}]_{\mathrm{tot}}$. Notice that this is precisely the limit used in Section 6.1.1 to calculate the probability of binding using a lattice model. Comment on the relation between the statistical mechanics and thermodynamic identifications made in Section 6.4.1 and in Problem 6.3.

(c) Maeda et al. (2000) measured binding curves of different σ subunits to RNA polymerase core enzyme. Fit the data from their Figure 1(A) (available on the book's website) to the model you derived in (a), assuming a total concentration of receptor (σ^{70}) of 0.4 nM. Compare that with the results of fitting to the expression derived in (b).

6.5 Distinguishable ligands

Derive the probability that a receptor is occupied by a ligand using a model that treats the L ligands in solution as distinguishable particles. Show that the expression is the same as obtained in the text (Equation 6.19), where the ligands were treated as indistinguishable.

6.6 Derivation of the Boltzmann distribution

Derive the Boltzmann distribution using the counting argument from Section 6.1.4. Modify the argument given there such that every particle has at least an energy of ε.

6.7 Binding polynomials and polymerization

Many cellular processes involve polymerization, where a bunch of monomers bind together to form a polymer. Examples include transcription, translation, construction of the cytoskeleton, etc. Here we consider a simple model of polymerization where each monomer is added to the growing chain with the same equilibrium binding constant K.

This situation is described by the following set of chemical equations:

$$\begin{aligned} X_1 + X_1 &\overset{K}{\rightleftharpoons} X_2 \\ X_1 + X_2 &\overset{K}{\rightleftharpoons} X_3 \\ &\vdots \\ X_1 + X_{n-1} &\overset{K}{\rightleftharpoons} X_n \\ &\vdots \end{aligned} \tag{6.129}$$

The symbol X_n denotes a polymer n monomers in size. In equilibrium, there will be polymers of all different sizes. We use this model to compute the average polymer size, and how it depends on the concentration of monomers.

(a) Find an expression for the probability that a polymer is n monomers in length in terms of K and $x = [X_1]$, the concentration of free monomers. Use this result to get an expression for the average polymer size $\langle n \rangle$.

(b) Show that the average polymer size can be written as

$$\langle n \rangle = \frac{d \ln Q}{dKx},$$

where $Q = 1 + Kx + (Kx)^2 + (Kx)^3 + \cdots$ is the *binding polynomial*.

(c) Plot $\langle n \rangle$ as a function of Kx. Show that $\langle n \rangle$ diverges as Kx approaches 1 from below. What is the physical interpretation of this divergence? (In reality, there is no such thing as a polymer that is infinite in size.)

(Adapted from Problem 28.4 of K. Dill and S. Bromberg, Molecular Driving Forces, 2nd ed. Garland Science, 2011.)

6.8 Lattice model of the chemical potential

Intuitively, the chemical potential really tells us the free energy cost associated with changing the number of solute molecules in solution by 1 as

$$\mu_{\text{solute}} = G_{\text{tot}}(N_{\text{s}} + 1) - G_{\text{tot}}(N_{\text{s}}). \tag{6.130}$$

Using the lattice model of a solution, compute the free energy and obtain an expression for the chemical potential.

6.9 Osmotic pressure of a cell

In Section 6.2.3, we derived the van't Hoff formula for the osmotic pressure and performed an estimate of the osmotic pressure experienced by a bacterium as a result of its impermeability to inorganic ions. Examine the contribution to the osmotic pressure of a bacterium coming from the presence of proteins within the cell.

How does this compare with the contribution to the osmotic pressure from inorganic ions described in the section? See Table 2.1 and Figure 2.4 for the relevant data.

6.10 The die problem revisited

Work out the probability distribution for a dishonest die that has average values $\langle i \rangle = 2.5$, $\langle i \rangle = 3.5$, and $\langle i \rangle = 4.5$. Report the values of the Lagrange multipliers in each case and make a plot (a bar plot) of the probability distribution for all three of these cases.

6.11 The missing information revisited

For the single-receptor binding problem, show that the missing information for ligand concentrations fK_{d} and K_{d}/f are the same. This explains the symmetry of the curve shown in Figure 6.26.

6.12 Simple estimate of first passage times

The rate of a molecule passing over a free-energy barrier of height B is proportional to the Boltzmann factor $e^{-B/k_{\text{B}}T}$, with T the (absolute) temperature and k_{B} the Boltzmann constant. The exponential dependence on B can be understood heuristically as follows: Thermal fluctuations typically give the molecule an energy of about $k_{\text{B}}T$. To go up in energy by another factor of $k_{\text{B}}T$ is somewhat more unlikely than to go down. Consider the barrier of height B as being a staircase of energy steps each of height $k_{\text{B}}T$. Next suppose that the probability that the next step of the particle is up is q and the probability that the next step is down is given by $1 - q$ with $q < 1/2$. Make a crude estimate of the probability p of getting to the top of the barrier in "one try." You will need to say what you mean by "one try." Compare your answer with the Boltzmann factor for $B \gg k_{\text{B}}T$. (Problem courtesy of Daniel Fisher.)

6.7 Further Reading

Ackers, GK, Johnson, AD, & Shea, MA (1982) Quantitative model for gene regulation by lambda phage repressor, *Proc. Natl Acad. Sci. USA* **79**, 1129. This paper introduced the thermodynamic models of gene regulation with reference to the genetic switch in phage lambda.

Dill, K, & Bromberg, S (2011) Molecular Driving Forces, 2nd ed., Garland Science. This book includes an interesting discussion of the information theoretic approach to statistical mechanics and justifies many of the results we have *assumed*. Their treatment of binding is a great source.

Schroeder, DV (2000) An Introduction to Thermal Physics, Addison–Wesley Longman. Our treatment of the free energy of dilute solutions and the osmotic pressure follows that given by Schroeder.

Kittel, C, & Kroemer, H (1980) Thermal Physics, 2nd ed., W. H. Freeman. An excellent development of the tools of statistical mechanics.

Baierlein, R (1999) Thermal Physics, Cambridge University Press. Our treatment of the law of mass action follows the excellent treatment in Baierlein's Chapter 11.

Jaynes, ET (1989) Papers on Probability and Statistics, edited by R. D. Rosenkranz, Kluwer Academic. Jaynes' unconventional approach to statistical mechanics inspires passionately positive and negative reactions. We recommend this series of papers if for no other reason than the fun of hearing a clear and thoughtful case favoring the use of information theory in statistical mechanics.

Buchler, NE, Gerland, U, & Hwa, T (2003) On schemes of combinatorial transcription logic, *Proc. Natl Acad. Sci. USA* **100**, 5136. This paper presents a general statistical mechanical formulation of transcription factors and their interactions with DNA.

Bintu, L, Buchler, NE, Garcia, HG, et al. (2005) Transcriptional regulation by the numbers 1: Models, *Curr. Opin. Gen. Dev.* **15**, 116. This paper gives a pedagogical description of the use of "thermodynamic models" for examining transcriptional regulation.

6.8 References

Brighty, DW, Rosenberg, M, Chen, IS, & Ivey-Hoyle, M (1991) Envelope proteins from clinical isolates of human immunodeficiency virus type 1 that are refractory to neutralization by soluble CD4 possess high affinity for the CD4 receptor, *Proc. Natl Acad. Sci. USA* **88**, 7802.

Feynman, RP (1972) Statistical Mechanics, W.A. Benjamin.

Keller, BU, Hartshorne, RP, Talvenheimo, JA, Catterall, WA, & Montal, M (1986) Sodium channels in planar lipid bilayers: channel gating kinetics of purified sodium channels modified by batrachotoxin, *J. Gen. Physiol.* **88**, 1.

Maeda, H, Fujita, N, & Ishihama, A (2000) Competition among seven *Escherichia coli* σ subunits: relative binding affinities to the core RNA polymerase, *Nucleic Acids. Res.* **28**, 3497.

Maier, B, & Rädler, JO (1999) Conformation and self-diffusion of single DNA molecules confined to two dimensions, *Phys. Rev. Lett.* **82**, 1911.

Rau, DC, Lee, B, & Parsegian, VA (1984) Measurement of the repulsive force between polyelectrolyte molecules in ionic solution: hydration forces between parallel DNA double helices, *Proc. Natl Acad. Sci. USA* **81**, 2621.

Rossi-Fanelli, A, & Antonini, E (1958) Studies on the oxygen and carbon monoxide equilibria of human myoglobin, *Arch. Biochem. Biophys.* **77**, 478.

Weiss, V, Claverie-Martin, F, & Magasanik, B (1992) Phosphorylation of nitrogen regulator I of *Escherichia coli* induces strong cooperative binding to DNA essential for activation of transcription, *Proc. Natl Acad. Sci. USA* **89**, 5088.

Two-State Systems: From Ion Channels to Cooperative Binding

7

Overview: In which we apply statistical mechanics to macromolecules that can be represented as two-state systems

> There are two models that a biological statistical mechanician is interested in, the one they are working on and the MWC model.
>
> **A Statistical Mechanician**

In the last few chapters, we have developed the tools of statistical mechanics. The goal of this chapter is to use those tools to examine a very special set of biological problems where the *state* of the system is described by simple two-state variables we label σ. For example, ion channels can be either closed ($\sigma = 0$) or open ($\sigma = 1$) and proteins that accept binding partners have their binding site either vacant ($\sigma = 0$) or occupied ($\sigma = 1$). This important class of biological problems can be treated in a unified way using statistical mechanics and permits us to investigate a range of interesting topics, including ion channel gating, phosphorylation, and ligand–receptor binding and cooperativity, with special reference to the case of hemoglobin. In addition, this chapter adds a new tool from statistical mechanics to our arsenal in the form of the so-called Gibbs distribution. This distribution is especially useful for problems in which the system of interest is in contact with a particle reservoir.

7.1 Macromolecules with Multiple States

7.1.1 The Internal State Variable Idea

Proteins and nucleic acids often exist in multiple states. Molecular motors adopt a range of different conformations depending upon the presence or absence of bound nucleotides such as ATP as well as the presence or absence of the cytoskeletal tracks that they move on. Many proteins have an active and inactive form, and the relative probabilities of these different states depend, for example, upon whether or not the enzyme has been phosphorylated or whether or not an allosteric effector is bound. In addition, macromolecules bind both other substrates and each other, and the different states of binding can be thought of as different "states" of the molecules of interest.

The goal of this chapter is to attack this wide class of problems characterized by several distinct states (bound and unbound, phosphorylated and unphosphorylated, etc.) by invoking powerful tools from the statistical mechanics of two-level systems. This formalism will permit us to assign an internal state variable σ to characterize the state of the macromolecule of interest (ion channel open or closed, ligand bound or not bound, protein phosphorylated or not, etc.). The reason this is important is that with such internal state variables in hand, we will be able to write the energy of the system in a simple analytic form that can be used in the statistical mechanics setting. This is significant, in turn, because it affords us the chance to compute quantities of interest such as the probability that an ion channel is open or closed, the average state of occupancy of a receptor, whether a protein is active or not, and the average number of melted base pairs for a molecule of DNA. Each of these observable quantities dictates important biological processes.

The State of a Protein or Nucleic Acid Can Be Characterized Mathematically Using a State Variable

Figure 7.1 shows a gallery of different examples of the use of internal state variables to characterize the state of macromolecules. The critical feature of these molecules that makes this approximation possible is the fact that the molecule adopts a countable set of distinct states that can be labeled by integers. In the two-state case, we find it convenient to label the two states by 0 and 1, though these choices are arbitrary. The examples shown in this gallery will serve as the backdrop for the analysis carried out throughout this and remaining chapters.

One simple example of the way in which state variables can be used to describe the internal states of macromolecules is for ion channels. Indeed, the measured current as a function of time for a channel shown in Figure 7.2 illustrates the way in which the channel switches back and forth between two discrete and well-defined states. In this

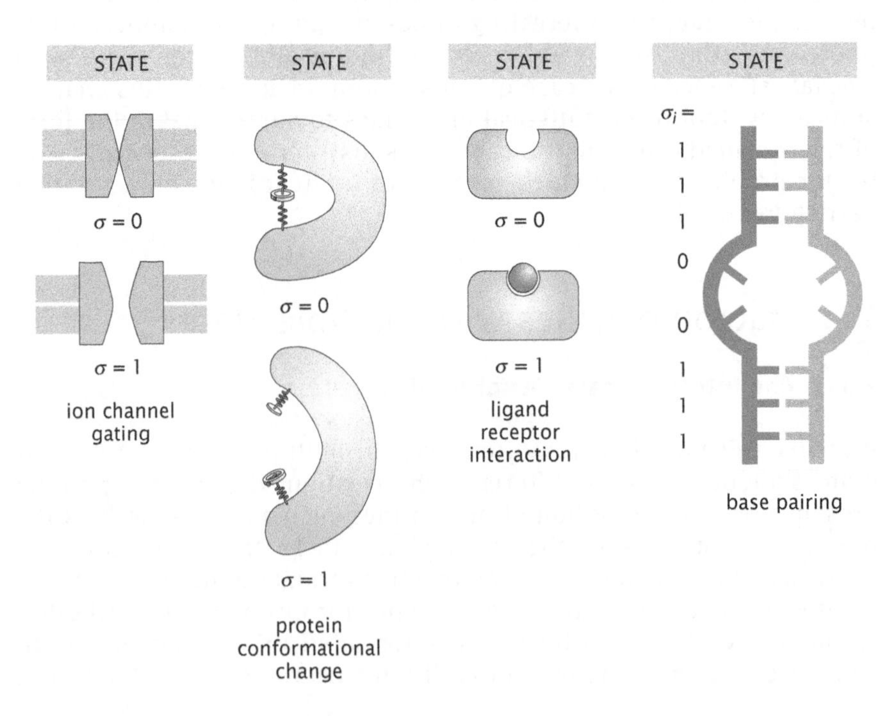

Figure 7.1: Examples of the internal state variable description of macromolecules. From left to right, the figures show an ion channel, a protein with an active and an inactive state, a receptor with its ligand partner, and a partially melted strand of dsDNA (dsDNA refers to double-stranded DNA, while ssDNA refers to single-stranded DNA). In each case, the state of the system can be described by one or several variables, each of which can only take the values 0 or 1.

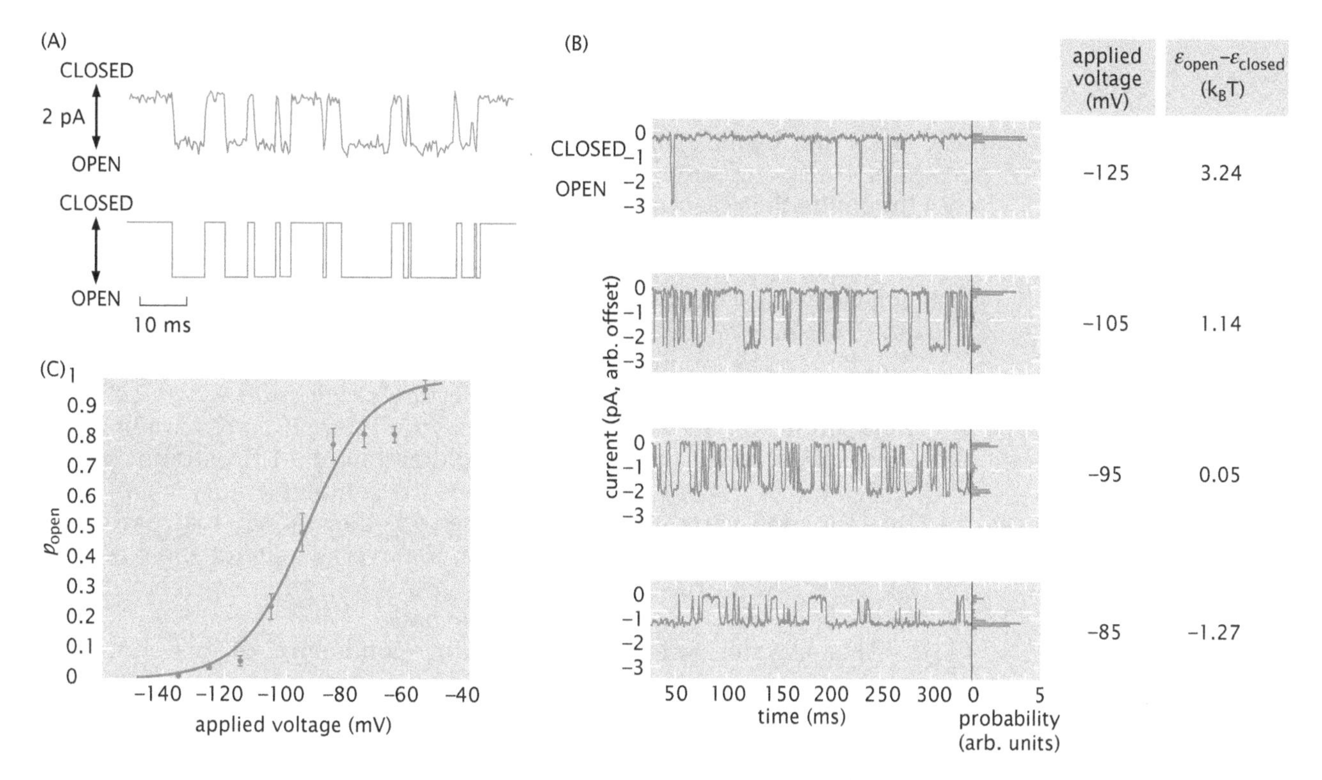

Figure 7.2: Experimental data showing measured current trajectories for a sodium ion channel subjected to different voltages. The fact that the transition times are fast in comparison with the dwell times favors the use of a two-state approximation. (A) Current trace and its two-state idealization. (B) A single channel incorporated into a pure lipid bilayer is subjected to different voltages and the resulting current is measured. Tuning the driving force for gating only changes the relative dwell times and not the existence of the states themselves. For different ion channels, the driving force might be (i) membrane tension, (ii) voltage, (iii) ligand binding, and (iv) phosphorylation. In this case, changing the voltage leads to a change in the fraction of time the channel spends open vs closed. (C) The open probability as a function of the voltage. (Adapted from B. U. Keller et al., *J. Gen. Physiol.* 88:1, 1986.)

case, we invoke $\sigma = 0$ to characterize the closed state of the channel and $\sigma = 1$ to characterize the open state. The dynamical trajectory of the channel can then be written as $\sigma(t)$ and if we imagine discretizing time into segments Δt, the gating trajectory amounts to a string of zeroes and ones switching back and forth over time. Using the state variable σ, the open probability is given by the average value of σ itself as $\langle\sigma\rangle$. In particular, for the case in which $\langle\sigma\rangle = 1/2$, the same amount of time is spent in the open and in the closed state. An interesting feature of the energy $E(\sigma)$ is that it depends upon the external driving force such as the applied tension or the voltage. The reader is invited to explore how to naively threshold noisy data like that shown in the figure in order to obtain the open probability in the Computational Exploration at the end of this section.

A second example of the two-state idea is provided by protein phosphorylation. In Figure 6.7 (p. 245), we showed the many cellular interventions that regulate the various steps in the central dogma. For example, activators and repressors alter the probability of the process of transcription. One of the most intriguing ways in which cells exert control over the current stockpile of proteins is through post-translational modifications. By adding (phosphorylation) or taking away (dephosphorylation) a phosphate group from certain residues such as serine and histidine, for example, a protein can be taken from an active to an inactive state. For further details, see Walsh (2006) in the Further Reading at the end of the chapter. Like with the ion channel example, the protein state as a result of post-translational modification can be described in terms of two-state internal state variables.

Another simple example of this kind of description is provided by the familiar problem of ligand–receptor binding, where the *internal* state of the receptor is captured by $\sigma = 0$ if no ligand is bound and $\sigma = 1$ if a ligand is bound. We can write the energy of the system as $E(\sigma) = \sigma\varepsilon_b + (1 - \sigma)\varepsilon_{sol}$, which yields an energy ε_b when the ligand is bound to the receptor and an energy ε_{sol} when the ligand is in solution. As will be shown in Section 7.2.4, this same kind of description can be especially powerful for problems such as O_2 binding in hemoglobin, where we will introduce site occupancy variables σ_1, σ_2, σ_3, and σ_4 to describe the state of occupancy of the four sites available for O_2 binding in the hemoglobin molecule. For a reminder on the structure of hemoglobin and its binding properties, see Section 4.2 (p. 143).

A subtle way in which one can use the internal state variable idea is to characterize the extent of hybridization of a DNA strand. In this case, for every base pair along the two complementary chains, we have a state variable which captures whether or not that particular base pair is hybridized. In this case, the overall state of the molecule is described by the vector $(\sigma_1, \sigma_2, \ldots, \sigma_N) = \{\sigma_i\}$, where σ_i refers to the state of hybridization of the ith base pair.

This chapter reflects an intriguing confluence of two powerful threads from entirely different fields. On the one hand, in the biological setting, the idea of macromolecules that can exist in several distinct states is ubiquitous. As a result, exploring the biological consequences of multiple states is natural and important. On the other hand, from the standpoint of statistical mechanics, there are few problems that have garnered as much attention as two-state systems where some state variable adopts one of two different values. The interested reader is invited to look up some of the vast literature on the "Ising model" to see the wide implications of this model in the physics context. In the remainder of the chapter, we will show how the biological and physical consequences of two-state systems can be merged into a coherent and quantitative picture of a variety of different problems. One cautionary note is that for the present chapter, we continue with the assumption that ideas from *equilibrium* physics are often useful to characterize living matter. For many problems, we can go a long way with this picture, but it is important to bear in mind when it has been assumed.

COMPUTATIONAL EXPLORATION

Computational Exploration: Determining Ion Channel Open Probability by Thresholding Figure 7.2(C) shows the probability that an ion channel is open as a function of the applied voltage. But how are such open probabilities determined from the ion channel current measurements resulting in time traces like those shown in Figure 7.2(B)? In this Computational Exploration, we take these current traces for the sodium ion channel, threshold them to find out what fraction of the time the channel spends in the open state, and then use that information to determine the open probability. We start by obtaining the time traces from the book website and loading them into Matlab, a subset of which are shown in Figure 7.2(B). As with other examples throughout the book, the data is obtained by digitizing the results from the original research literature. As a result the time resolution from these data sets does not really reflect the time resolution of their experimental setup and hence this analysis should be viewed as a pedagogical exercise rather than a "research grade" result.

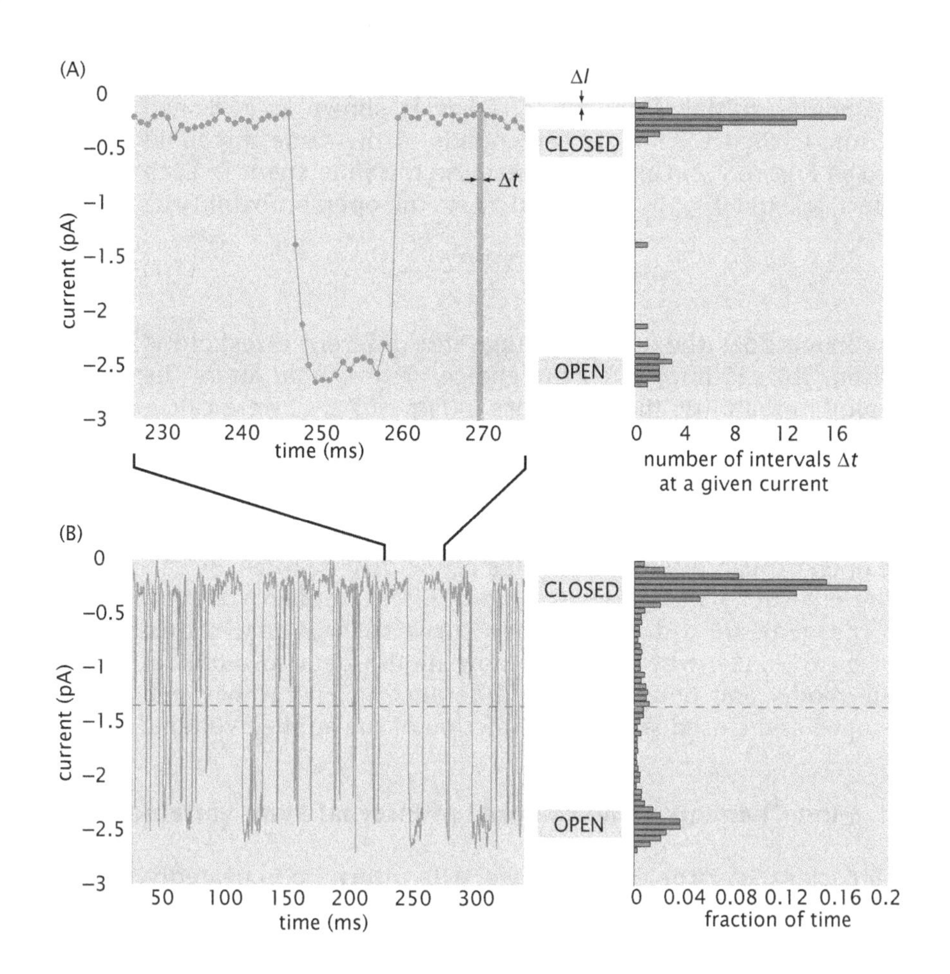

Figure 7.3: Determining ion channel open probability. (A) Binning strategy to construct current histograms. On the left, the current as a function of time is shown for a sodium ion channel subjected to a voltage difference of -105 mV. This image shows a small part of the full trajectory. Time is sliced into segments of duration Δt and the current is similarly binned into fragments of width ΔI. (B) Current as a function of time of an sodium ion channel subjected to a voltage difference of -105 mV and histogram for the fraction of time each current value was detected. The dashed red line represents the threshold used in the computational exploration.

We begin by plotting the current passing through the channel as a function of time when the voltage across the channel is set at -105 mV. The resulting current trace is shown in Figure 7.3(A), where the trace clearly shows at least two distinct states. This is revealed by the fact that the current spends most of its time either around -0.25 pA or around -2.5 pA. An alternative way to reveal what fraction of the total observation time that the channel spends in one or the other state is by appealing to a histogram like that shown in the right panel of Figure 7.3(A). To construct such a histogram, we divide time into short intervals of length Δt and then ask for the current value during each such interval and increment a counter for the current of interest each time a current within some small range of that value is observed. Although Figure 7.3(B) tells us that there are two main states, there also appears to be some statistical weight associated with an intermediate value. This could be a real feature of the data or an artifact resulting from time averaging and can only be revealed as such by increasing the time resolution of the experiment.

As noted above, the probability of being in a particular state can be computed by evaluating the fraction of time spent in that state. If we integrate the histogram over all bins corresponding to a state of interest, we obtain the total fraction of time spent in that particular state. For example, we can define a threshold in the current. For all currents below this threshold, we will consider the channel as open. On the other hand, if the current is above the threshold we will take it to be closed. One way to estimate a value for the threshold is to choose a midpoint between the maximum and minimum currents. The threshold resulting from this strategy is shown as a dashed red line in Figure 7.3 and corresponds to the same threshold used in Figure 7.2. Once we determine the time spent in each state T_{open} and T_{closed}, we can calculate the open probability as

$$p_{\text{open}} = \frac{T_{\text{open}}}{T_{\text{open}} + T_{\text{closed}}}. \tag{7.1}$$

How much does the answer change if a different threshold is chosen? To see how much the choice of threshold alters the looping probability, the error bars in Figure 7.2(C) were calculated by changing the threshold by 25% and determining the corresponding change in the open channel probability.

With this algorithm in hand, we can go through the time traces corresponding to the different voltages and determine the open probability for each value of the driving force. By carrying out such calculations for the different applied voltages, we can generate a dose–response curve such as that shown in Figure 7.2(C), where we can now fit the data to our two-state model and determine the difference in energy between the open and closed states as a function of the applied voltage.

7.1.2 Ion Channels as an Example of Internal State Variables

One class of problems that we will return to repeatedly is the structure and function of ion channels. Ion channels are transmembrane proteins that mediate the transport of ions in and out of cells. Our workhorse simplification will be a two-state model of ion channels in which the channel is imagined to exist in two conformations, namely, the closed and open states. Though the choices of 0 and 1 for our state variables are arbitrary, they are very useful. In particular, this choice has the feature that the open probability of the channel can be obtained directly by computing $\langle\sigma\rangle$, the average value of σ. A schematic of the geometry of a channel described by the σ variables is shown in Figure 7.4. Of course, what makes ion channels so remarkable is that the probability of being in these different states can be tuned by various external agents such as the binding of ligands, the application of a voltage, or the application of tension in the surrounding membrane.

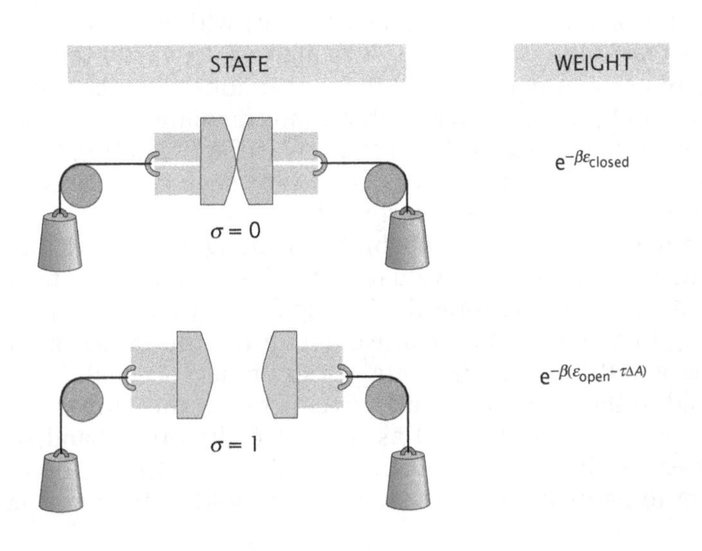

Figure 7.4: States and weights for a two-state ion channel. The ion channel is assumed to exist in one of two states characterized by $\sigma = 0$ (closed) and $\sigma = 1$ (open). When the channel is closed it has an energy $\varepsilon_{\text{closed}}$ and when it is open it has an energy $\varepsilon_{\text{open}}$. In addition, in the open state, the energy of the loading device is reduced, resulting in the additional term $\tau\Delta A$, where τ is the tension and ΔA is the area change upon gating. The energy of the loading device was introduced in Figure 5.12 (p. 206) and will be developed more fully in Figure 7.6 and the associated discussion.

In this first serious encounter with ion channels, we want to show how the open probability can be tuned by the action of external driving forces. In particular, our aim is to compute the open probability p_{open}, which, in terms of our internal state variable σ, can be written as $\langle\sigma\rangle$. When $\langle\sigma\rangle \approx 0$, this means that the probability of finding the channel open is low. Similarly, when $\langle\sigma\rangle \approx 1$, this means that it is almost certain that we will find the channel open. Of course, to evaluate these probabilities, we need to invoke the Boltzmann distribution, which tells us that the probability of finding a state with energy E is $p(E) = e^{-\beta E}/Z$, where Z is the partition function introduced in Section 6.1 (p. 237). To evaluate this probability, we need the energy as a function of the internal state variable, $E(\sigma)$. For the case in which there is no external driving force, this energy can be written as

$$E(\sigma) = \sigma\varepsilon_{\text{open}} + (1-\sigma)\varepsilon_{\text{closed}}, \tag{7.2}$$

where we have introduced the energies $\varepsilon_{\text{open}}$ and $\varepsilon_{\text{closed}}$ for the open and closed states, respectively.

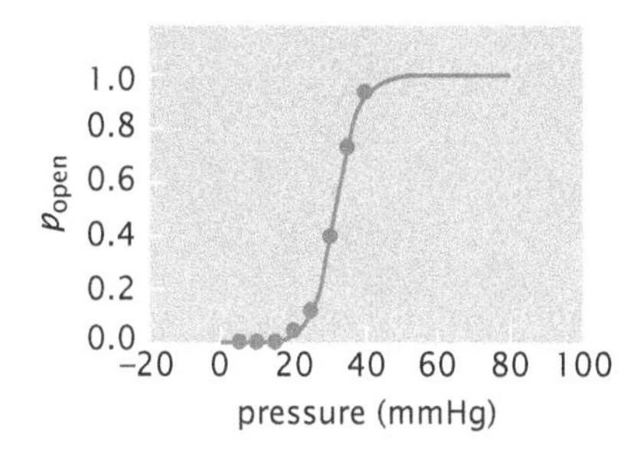

Figure 7.5: Ion channel open probability. The graph shows the open probability as a function of the applied tension for the mechanosensitive channel MscL. The measurement is obtained by applying a suction pressure on a pipette to a membrane patch containing channels. The x-axis in the graph shows the applied suction pressure rather than the membrane tension. (Adapted from E. Perozo et al., *Nat. Struct. Biol.* 9:696, 2002.)

The Open Probability $\langle\sigma\rangle$ of an Ion Channel Can Be Computed Using Statistical Mechanics

The interesting case for ion channel dynamics arises when the energy balance between the open and closed states is altered by the presence of some driving force. For concreteness, we consider the case of mechanosensitive channels where the driving force is the tension in the membrane. Mechanosensitive channels in bacteria have been hypothesized to serve as safety valves to protect against membrane rupture due to osmotic imbalance. In particular, when a bacterial cell is subjected to osmotic shock (see Section 6.2.3 on p. 264), the resulting flow of water across the cell membrane results in membrane tension. These channels reply by opening. An example of how the open probability depends upon the applied tension is shown in Figure 7.5.

Though we will consider this case in much more detail in Chapter 11, at this point we introduce the energy as a function of the applied tension τ, which is given by

$$E(\sigma) = \sigma\varepsilon_{\text{open}} + (1-\sigma)\varepsilon_{\text{closed}} - \sigma\tau\Delta A. \tag{7.3}$$

The term $\sigma\tau\Delta A$ favors the open state and reflects the fact that the energy of the *loading* device is lowered in the open state. We have already introduced the idea of accounting for the energy of loading devices in Figure 5.12 (p. 206), and this idea is of broad significance in a number of different situations. In particular, there are many circumstances in which we will want to know the free energy for systems that include some external load. In these cases, we must consider the free energy of our biological system of interest as well as that of the "loading device" and we find the use of lowering weights a convenient pedagogical fiction to emphasize the need to include the loading device in the overall free energy budget.

To see better how the external tension couples to the energy, Figure 7.6 shows a scheme for using weights to load the channel. In this case, we imagine that the total area of the membrane is constant and, as a result, when the channel opens, because the radius gets larger, the weights are lowered by some amount, which reduces their potential energy. The greater the weights, the larger the change in

(A)

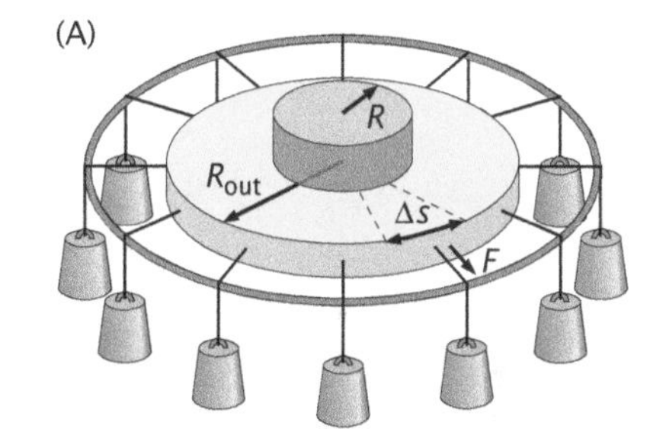

(B)

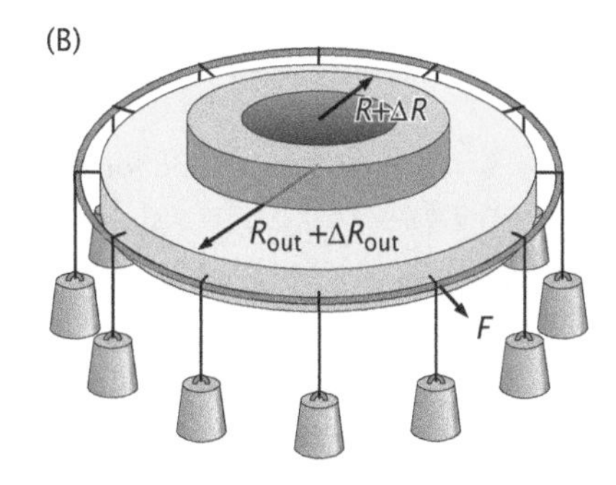

Figure 7.6: Schematic of how channel opening results in a relaxation in the loading device. For simplicity, we represent the loading device as a set of weights attached to the membrane far from the channel. (A) Channel in closed configuration. (B) When the channel opens, the weights are lowered, resulting in a reduction in the potential energy of the loading device.

potential energy. This fiction of using weights (we will use a similar scheme to examine the force–extension properties of DNA in Chapter 8) is a simple representation of externally applied forces. If we imagine a finite membrane as shown in Figure 7.6, when the channel opens, the outer radius will change as $\Delta R_{\rm out} \approx (R/R_{\rm out})\Delta R$, a constraint that follows immediately from insisting that the total area of the membrane is fixed. We are interested in evaluating the change in potential energy of the loading device (that is, the dropping of the weights in our simple cartoon scenario) as a result of channel opening. To do so, we compute the work associated with the forces F, which are most conveniently parameterized through a force/length (the tension τ) acting through the distance $\Delta R_{\rm out}$. To find the extent to which the weights drop, we use $2\pi R \Delta R = 2\pi R_{\rm out} \Delta R_{\rm out}$, where $\Delta R_{\rm out}$ is the displacement of the weights at the periphery. This results in

$$\Delta G_{\rm tension} = -\underbrace{\tau \Delta s}_{\text{force on arc}} \times \underbrace{\frac{R}{R_{\rm out}}\Delta R}_{\text{displacement of patch}} \times \underbrace{\frac{2\pi R_{\rm out}}{\Delta s}}_{\text{no. of ``patches''}}. \tag{7.4}$$

We have introduced the variable Δs for the tiny increment of arc length such that $\tau = F/\Delta s$. Given these definitions, we see that the energy change of the loading device is given by

$$\Delta G_{\rm tension} = -\tau 2\pi R \Delta R. \tag{7.5}$$

Note that this is precisely the expression we invoked in Equation 7.3 if we note that the change in area is $\Delta A = 2\pi R \Delta R$.

A more general way of representing the effect of an external driving force on channel open probability is shown in Figure 7.7. This figure depicts a schematic representation of the entire energy landscape (as opposed to just the energy of the closed and open states) for a channel as a function of the radius for several different choices of the external tension. The diagram shows the case of external tension, but a similar landscape would be found as a function of voltage or ligand binding. Notice that when there is no tension, the closed state has lower energy than the open state. With increasing tension, the energy of the open state is lowered with respect to that of the closed state. The key point is to see how the presence of a driving force can shift the overall balance between the energies of the closed and open states.

To compute the open probability of the channel using the statistical mechanics tools introduced in Chapter 6, we need to evaluate the partition function

$$Z = \sum_{\sigma=0}^{1} {\rm e}^{-\beta E(\sigma)}. \tag{7.6}$$

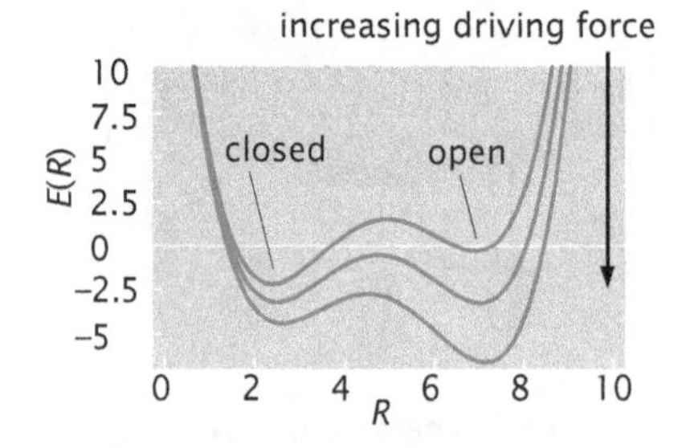

Figure 7.7: Energy landscape for an ion channel. Schematic of the free energy as a function of the channel radius for a model two-state ion channel. The different curves correspond to different values of the external parameter that drives gating (such as membrane tension, ligand concentration, or applied voltage).

Using the energy given in Equation 7.3, we find

$$Z = {\rm e}^{-\beta \varepsilon_{\rm closed}} + {\rm e}^{-\beta(\varepsilon_{\rm open} - \tau \Delta A)}. \tag{7.7}$$

This permits us to write the open probability at once as

$$p_{\rm open} = \frac{{\rm e}^{-\beta(\varepsilon_{\rm open} - \tau \Delta A)}}{{\rm e}^{-\beta(\varepsilon_{\rm open} - \tau \Delta A)} + {\rm e}^{-\beta \varepsilon_{\rm closed}}}. \tag{7.8}$$

It is important to note that an entirely equivalent (and probably more elegant) way of evaluating the open probability is

$$\langle \sigma \rangle = \sum_{\sigma=0}^{1} \sigma p(\sigma) = p(1) = p_{\rm open}. \tag{7.9}$$

The open probability as a function of the applied tension is shown in Figure 7.8, which demonstrates a higher probability of the channel being open as the tension increases.

To understand how a particular channel is going to behave under a driving force, we need to know two things. First, we need to understand the intrinsic preference for each of its two states. Second, we need to understand how the external driving force alters the relative energies of these different states. With these two quantitative measurements in hand, the engine of statistical mechanics allows us to compute the behavior of the channel under a range of driving forces. This section has served as an introduction to the idea of state variables using ion channels as a convenient example. We will return to ion channels in earnest in Chapters 11 and 17. We now take up the subject of ligand–receptor binding from the point of view of state variables.

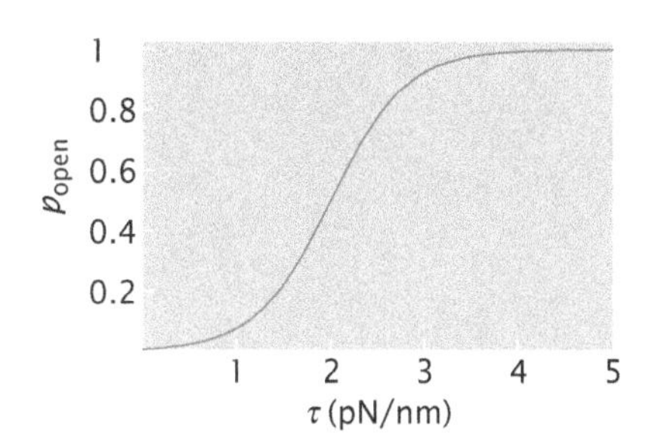

Figure 7.8: Ion channel open probability as a function of driving force for gating. The plot shows $p_{open} = \langle \sigma \rangle$ as a function of the applied tension τ. The parameters used in the plot for a model mechanosensitive channel are $\Delta\varepsilon = -5\,k_BT$ and $\Delta A = 10\,\text{nm}^2$. The plot would have the same functional form if the driving force for gating were voltage rather than tension.

7.2 State Variable Description of Binding

In Sections 6.1.1 (p. 241) and 6.4.1 (p. 270), we have already encountered the problem of receptor–ligand binding, once from the perspective of the Boltzmann (canonical) distribution, and once from the perspective of the law of mass action and equilibrium constants. In this section, we tackle this problem yet a third time, but this time with an eye to introducing the tools from statistical mechanics (Gibbs distribution and the grand partition function) that will permit us to generalize our discussion of binding to the case of ligands with multiple cooperative receptors and to cases for which there are multiple competing ligands. Indeed, the tools developed here will carry us through the remainder of the book in a diverse set of applications.

In Chapter 6, we showed that the Boltzmann distribution presides over questions about the probabilities of different microstates for a system in contact with a heat bath. The idea there was that by putting our system in contact with a huge energy reservoir, it is free to exchange energy with its environment and hence occupy any of its different energy microstates. For many of the binding problems of interest throughout the remainder of the book, an important generalization is to think of our system as being in contact not only with a heat bath, but with a particle reservoir as well, as shown in Figure 7.9. Just as the temperature of the reservoir controls the average energy of our system, the chemical potential of the particle reservoir dictates the average number of particles in our system. The driving force leading to ligand binding is the concentration in the particle reservoir. As we will see in later sections, this powerful idea can be used to characterize chemical reactions of all kinds.

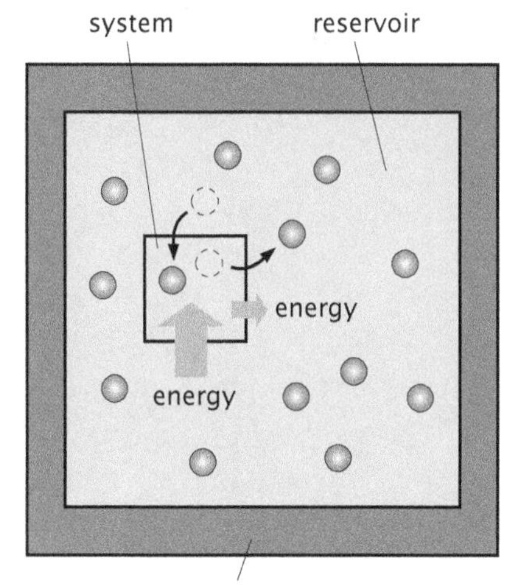

Figure 7.9: System in contact with a particle and energy reservoir. The system and reservoir are isolated from the rest of the world by adiabatic, rigid, and impermeable walls. However, the system and reservoir themselves can exchange both energy and particles.

7.2.1 The Gibbs Distribution: Contact with a Particle Reservoir

The Gibbs Distribution Gives the Probability of Microstates for a System in Contact with a Thermal and Particle Reservoir

In Chapter 6, we showed how the Boltzmann distribution could be derived both using counting arguments and on the basis of information theory. Another popular class of derivation examines the coupling between the system and the reservoir as we will do here. These three distinct classes of derivations for the key distributions of statistical mechanics (both Boltzmann and Gibbs can be derived in all of these different ways) should give the reader a better intuition for how these distributions work and what they mean.

The setup we consider for our derivation of the probability of microstates for systems in contact with a hybrid thermal–particle reservoir is shown in Figure 7.9. The idea is that we have a box that closes off our little world of interest from everything else—it is effectively isolated. As a result, the total energy and the total number of particles within the box are constant. Within the box, we now consider two regions, one of which is our *system* of interest and the other of which is the reservoir. Probably the most familiar and intuitive example of this idea is an open container sitting outside in the atmosphere. The presence of the container will not make any difference to either the thermal or particle-number state of the reservoir. On the other hand, the contents of the container will certainly come to equilibrium with the atmosphere.

The total number of particles in the box is $N_{\text{tot}} = N_{\text{r}} + N_{\text{s}}$, where the subscript "r" refers to the reservoir and the subscript "s" refers to the system. Similarly, the total energy is $E_{\text{tot}} = E_{\text{r}} + E_{\text{s}}$. Our fundamental assertion is that the probability of finding a given state of the system (characterized by an energy E_{s} and number of particles N_{s}) is proportional to the number of states available to the *reservoir* when the system is in this state. That is,

$$\frac{p(E_{\text{s}}^{(1)}, N_{\text{s}}^{(1)})}{p(E_{\text{s}}^{(2)}, N_{\text{s}}^{(2)})} = \frac{W_{\text{r}}(E_{\text{tot}} - E_{\text{s}}^{(1)}, N_{\text{tot}} - N_{\text{s}}^{(1)})}{W_{\text{r}}(E_{\text{tot}} - E_{\text{s}}^{(2)}, N_{\text{tot}} - N_{\text{s}}^{(2)})}, \tag{7.10}$$

where $W_{\text{r}}(E_{\text{tot}} - E_{\text{s}}^{(1)}, N_{\text{tot}} - N_{\text{s}}^{(1)})$ is the number of states available to the *reservoir* when the system is in the particular state $(E_{\text{s}}^{(1)}, N_{\text{s}}^{(1)})$. The total number of states available to the universe of system plus reservoir when the system is in the particular state $(E_{\text{s}}^{(1)}, N_{\text{s}}^{(1)})$ is given by

$$W_{\text{tot}}(E_{\text{tot}} - E_{\text{s}}^{(1)}, N_{\text{tot}} - N_{\text{s}}^{(1)}) = \underbrace{1}_{\text{states of system}} \times \underbrace{W_{\text{r}}(E_{\text{tot}} - E_{\text{s}}^{(1)}, N_{\text{tot}} - N_{\text{s}}^{(1)})}_{\text{states of reservoir}} \tag{7.11}$$

because we have asserted that the system itself is in one particular microstate that has energy $E_{\text{s}}^{(1)}$ and particle number $N_{\text{s}}^{(1)}$. Though there may be other microstates with the same energy and particle number, we have selected one particular microstate of that energy and particle number and ask for its probability.

Using the fact that the entropy is defined as $S = k_{\text{B}} \ln W$, which can be rewritten as $W = \text{e}^{S/k_{\text{B}}}$, we can rewrite Equation 7.10 as

$$\frac{W_{\text{r}}(E_{\text{tot}} - E_{\text{s}}^{(1)}, N_{\text{tot}} - N_{\text{s}}^{(1)})}{W_{\text{r}}(E_{\text{tot}} - E_{\text{s}}^{(2)}, N_{\text{tot}} - N_{\text{s}}^{(2)})} = \frac{\text{e}^{S_{\text{r}}(E_{\text{tot}} - E_{\text{s}}^{(1)}, N_{\text{tot}} - N_{\text{s}}^{(1)})/k_{\text{B}}}}{\text{e}^{S_{\text{r}}(E_{\text{tot}} - E_{\text{s}}^{(2)}, N_{\text{tot}} - N_{\text{s}}^{(2)})/k_{\text{B}}}}. \tag{7.12}$$

To complete the derivation, we Taylor-expand the entropy as

$$S_{\text{r}}(E_{\text{tot}} - E_{\text{s}}, N_{\text{tot}} - N_{\text{s}}) \approx S_{\text{r}}(E_{\text{tot}}, N_{\text{tot}}) - \frac{\partial S_{\text{r}}}{\partial E} E_{\text{s}} - \frac{\partial S_{\text{r}}}{\partial N} N_{\text{s}}, \tag{7.13}$$

where we have only kept terms to first order in E_{s} and N_{s}. Finally, if we recall the thermodynamic identities $(\partial S/\partial E)_{V,N} = 1/T$ and $(\partial S/\partial N)_{E,V} = -\mu/T$, we can write our result as

$$\frac{p(E_{\text{s}}^{(1)}, N_{\text{s}}^{(1)})}{p(E_{\text{s}}^{(2)}, N_{\text{s}}^{(2)})} = \frac{\text{e}^{-(E_{\text{s}}^{(1)} - \mu N_{\text{s}}^{(1)})/k_{\text{B}}T}}{\text{e}^{-(E_{\text{s}}^{(2)} - \mu N_{\text{s}}^{(2)})/k_{\text{B}}T}}, \tag{7.14}$$

which tells us that the probability of a given state is proportional to the Gibbs factor (that is, $p(E_s^{(1)}, N_s^{(1)}) \propto e^{-(E_s^{(1)}-\mu N_s^{(1)})/k_BT}$). As a result, the probability of finding the system in state i with energy $E_s^{(i)}$ and particle number $N_s^{(i)}$ is

$$p(E_s^{(i)}, N_s^{(i)}) = \frac{e^{-\beta(E_s^{(i)}-\mu N_s^{(i)})}}{\mathcal{Z}}, \tag{7.15}$$

a result known as the Gibbs distribution, where we have defined the grand partition function as

$$\mathcal{Z} = \sum_i e^{-\beta(E_s^{(i)}-N_s^{(i)}\mu)}. \tag{7.16}$$

This equation instructs us to sum over all of the possible microstates (labeled by i).

In Section 6.1, we argued that the partition function serves as the analytical engine of statistical mechanics and illustrated that claim by showing how the average energy $\langle E\rangle$ can be computed as a derivative of the partition function (see Equation 6.9 on p. 241). The grand partition function permits us to go further and to compute the average number of particles in our system as

$$\langle N\rangle = \frac{1}{\beta}\frac{\partial}{\partial\mu}\ln\mathcal{Z}. \tag{7.17}$$

To see this, we note that the average particle number can be written as

$$\langle N\rangle = \frac{1}{\mathcal{Z}}\sum_i N_i e^{-\beta(E_i-N_i\mu)}, \tag{7.18}$$

where we have dropped the cumbersome notation involving subscript "s" to signify that we are referring to the "system." Indeed, from now on, whenever we invoke the Gibbs distribution, we will tacitly assume that the terms E_i and N_i refer exclusively to our system, which is in contact with a hybrid thermal–particle reservoir. What this perspective offers from the point of view of our microscopic models is that the chemical potential serves as a shorthand for the explicit treatment of the reservoir.

7.2.2 Simple Ligand–Receptor Binding Revisited

To apply the Gibbs distribution, we revisit the problem of ligand–receptor binding introduced in Sections 6.1.1 (p. 241) and 6.4.1 (p. 270). In this case, we imagine a slightly different setup where all of our attention is focused on a single receptor that is in contact with the surrounding heat bath and particle reservoir. The beauty of treating the problem in this way is that it will provide a much simpler treatment of the particles in the reservoir than was offered by the Boltzmann distribution. Further, this approach will permit us to generalize easily to more complicated cases such as hemoglobin in which there are multiple binding sites. In this case, the receptor can be in one of two states, bound or unbound, with σ serving as an indicator of the state of binding, with $\sigma = 0$ corresponding to the unbound state and $\sigma = 1$ to the bound state. The energy in this case is $E = \varepsilon_b\sigma$, where $\varepsilon_b < 0$, revealing a favorable interaction between ligand and receptor. The particular choices of σ are arbitrary and had we made different choices, then we would have had to make a corresponding change in

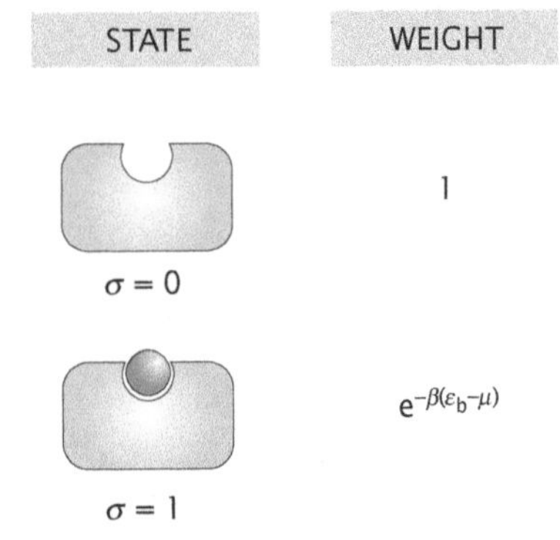

Figure 7.10: States and weights for ligand–receptor binding. The schematic shows the states of the receptor and their corresponding statistical weights as computed using the Gibbs distribution.

the energy. The virtue of this particular choice of the σ's is that $\langle\sigma\rangle$ reports the average number of bound ligands. The states and weights for our ligand–receptor problem are shown in Figure 7.10.

Using the grand canonical distribution as the basis of our evaluation of $\langle N\rangle$, we need to evaluate

$$\mathcal{Z} = \sum_{\text{states}} e^{-\beta(E_{\text{state}} - N_{\text{state}}\mu)}, \tag{7.19}$$

where we have switched notation to sum over "states" instead of the nondescript index i. The variable $\beta = 1/k_BT$ reflects the contact of the system with a thermal reservoir and the presence of μ reflects contact with a particle reservoir. The sum over states is very simple since there are only two states to consider, namely, (i) the state where the receptor is not occupied, which is characterized by $\sigma = 0$, and (ii) the state where the receptor is occupied, which is characterized by $\sigma = 1$. As a result, we write

$$\mathcal{Z} = \sum_{\sigma=0}^{1} e^{-\beta(\varepsilon_b\sigma - \mu\sigma)}. \tag{7.20}$$

The resulting sum is of the form

$$\mathcal{Z} = 1 + e^{-\beta(\varepsilon_b - \mu)}. \tag{7.21}$$

The average number of ligands bound is equal to the normalized weight of the occupied state, and is given by

$$\langle N\rangle = \frac{e^{-\beta(\varepsilon_b-\mu)}}{1 + e^{-\beta(\varepsilon_b-\mu)}}. \tag{7.22}$$

This result can also be computed by taking the derivative as described in Equation 7.17.

For the case of a receptor that can bind only one ligand, the average number of bound ligands, $\langle N\rangle$, is equivalent to the quantity p_{bound} introduced in Chapter 6. As a result, the calculation done here should yield precisely the same result we found in Equation 6.19 (p. 244). To see that equivalence, we recall that the chemical potential of an ideal solution can be written as $\mu = \mu_0 + k_BT\ln(c/c_0)$ as shown in Section 6.2.2 (p. 262). If we substitute this expression for the chemical potential into Equation 7.22, we find

$$\langle N\rangle = \frac{(c/c_0)e^{-\beta\Delta\varepsilon}}{1 + (c/c_0)e^{-\beta\Delta\varepsilon}}, \tag{7.23}$$

where we have introduced the notation $\Delta\varepsilon = \varepsilon_b - \mu_0$. Here $\Delta\varepsilon$ is the energy difference upon taking the ligand from solution and placing it on the receptor. This result is equivalent to that found in Chapter 6. We have revisited the problem of ligand–receptor binding for two reasons. First, the present treatment gives us a chance to see the idea of our internal state variables σ in action. Second, this example also served as our maiden example of the use of the Gibbs distribution which will be used again to describe O_2 binding in hemoglobin, the equilibrium accessibility of nucleosomes, and other problems as well.

7.2.3 Phosphorylation as an Example of Two Internal State Variables

The idea of a two-state system is extremely powerful in biology and applies to many cases beyond those already mentioned. One application where these ideas can be useful even when applied in a nontraditional manner is in understanding the control of protein activities by covalent post-translational modifications. It is very common in biological systems that the activity of an enzyme must be rapidly altered depending upon changes in environmental conditions. An example that we mentioned previously is the case of the enzymes controlling the cell cycle (see Section 3.2.2 on p. 110).

Regulation of protein activity by covalent attachment of phosphate groups is one of the most important regulatory modes in all of biology. It is estimated that the human genome encodes more than 500 kinase enzymes that attach phosphate groups to specific substrates. Several features of phosphorylation render it a particularly effective regulatory mechanism. The substrates for protein phosphorylation are the target protein itself and ATP, which is found abundantly in all living cells. The kinase enzymes catalyze the transfer of the terminal phosphate group from ATP to an appropriate nucleophilic chemical group on the protein, releasing ADP as a product as shown in Figure 7.11(A). In eukaryotic cells, the amino acid side chains that are most commonly phosphorylated are those containing hydroxyl groups: serine, threonine, and tyrosine. In bacteria, by way of contrast, histidine and aspartate are commonly phosphorylated. The attachment of a phosphate group which carries two negative charges causes a dramatic change in the local charge distribution on the surface of the protein. This alteration can have drastic, large-scale effects on overall protein structure and on the protein's ability to associate with binding partners or its intrinsic state of activity as indicated schematically in Figure 7.11(B). Furthermore, this dramatic alteration is readily reversible through the action of a separate set of enzymes called phosphatases. Phosphatases use water to hydrolyze the bonds connecting the phosphate to the protein's amino acid side chain releasing free inorganic phosphate and the original unmodified protein as products as also shown in Figure 7.11(A).

For the large number of enzymes and other proteins whose activity is regulated in this way, the quantitative analysis that we have so far presented for simple two-state systems can be readily adapted to make a simple model that quantifies the fraction of activated proteins and how it depends on the state of phosphorylation.

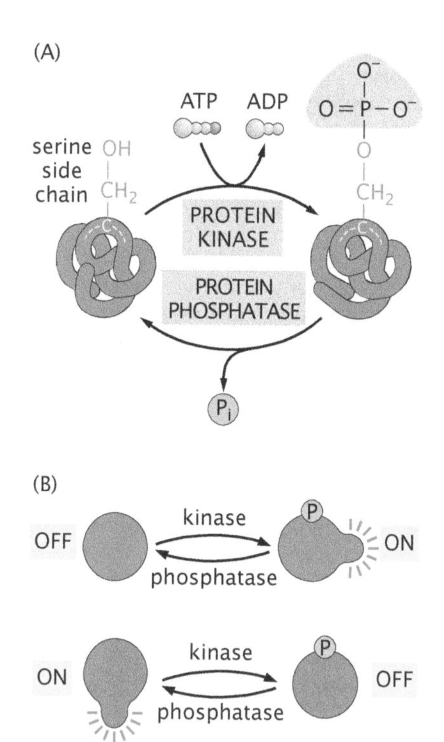

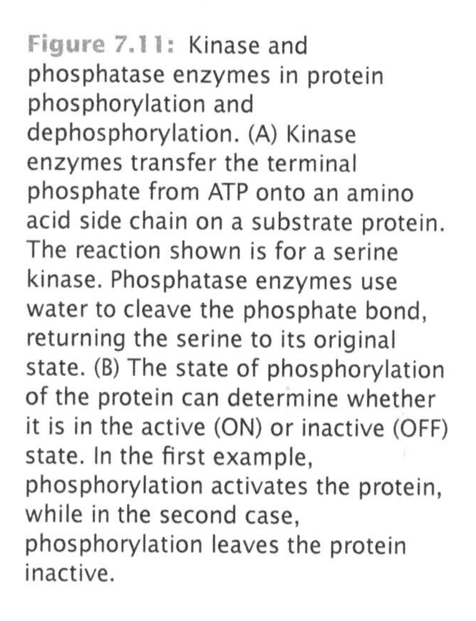

Figure 7.11: Kinase and phosphatase enzymes in protein phosphorylation and dephosphorylation. (A) Kinase enzymes transfer the terminal phosphate from ATP onto an amino acid side chain on a substrate protein. The reaction shown is for a serine kinase. Phosphatase enzymes use water to cleave the phosphate bond, returning the serine to its original state. (B) The state of phosphorylation of the protein can determine whether it is in the active (ON) or inactive (OFF) state. In the first example, phosphorylation activates the protein, while in the second case, phosphorylation leaves the protein inactive.

Phosphorylation Can Change the Energy Balance Between Active and Inactive States

To describe the "structural" state of the protein, which signifies whether it is active or not, we introduce the state variable σ_S, such that $\sigma_S = 0$ implies that the protein is inactive and $\sigma_S = 1$ signifies that the protein is active. However, in addition to characterizing the activity of the protein, we need a second variable that characterizes the state of phosphorylation of the protein which we label σ_P, where $\sigma_P = 0$ signifies the unphosphorylated state and $\sigma_P = 1$ the phosphorylated state. This toy model of phosphorylation foreshadows the powerful Monod–Wyman–Changeux (MWC) model to be explored in more detail later in the chapter.

As shown in Figure 7.12, the state of phosphorylation can alter the relative energies of the active and inactive states. This has the effect of shifting the equilibrium from the inactive to the active state or vice versa. To gain intuition about this we consider a toy model of phosphorylation whereby the addition of a phosphate group introduces

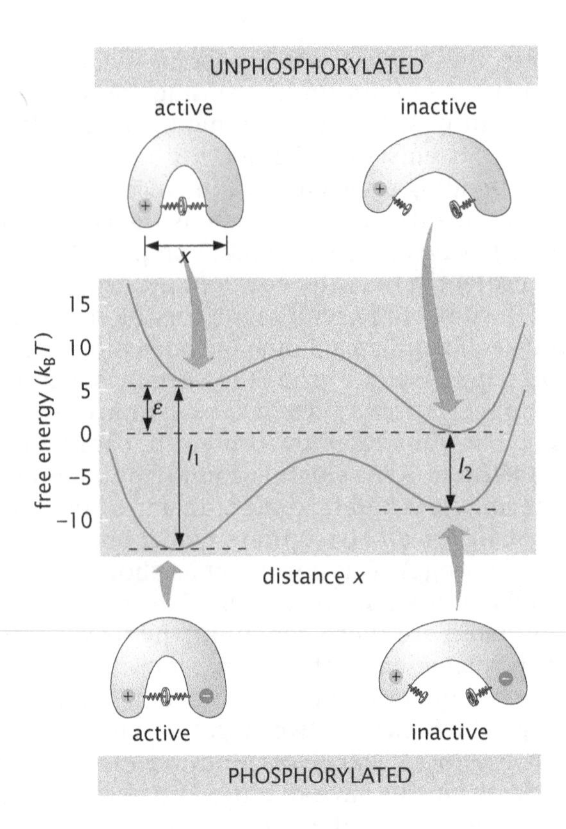

Figure 7.12: Energy landscape for a protein in its phosphorylated and unphosphorylated states. An idealized protein can exist in two conformations where only the bent form is enzymatically active. For the native protein, the inactive form is in a lower energy state, and therefore in equilibrium most molecules will be inactive rather than active. Protein phosphorylation adds negative charge to the surface of the protein. Due to favorable electrostatic interactions with positive charges elsewhere on the protein, phosphorylation stabilizes both the active and inactive conformations. However, the active conformation is stabilized to a greater degree than the inactive conformation. Consequently, at equilibrium, most of the phosphorylated molecules will be in the active form.

a favorable electrostatic interaction that lowers the active state free energy with respect to the inactive state free energy.

The states and weights for the active and the inactive state are summarized in Figure 7.13. The interaction energies appearing in the states and weights figure, I_1 and I_2, correspond to the electrostatic energy between the two charges in the active and the inactive states, respectively. Using the σ variables introduced above, we can write the free energy of the protein as

$$G(\sigma_P, \sigma_S) = (1 - \sigma_P)[(1 - \sigma_S)0 + \sigma_S \varepsilon] + \sigma_P[(1 - \sigma_S)(-I_2) + \sigma_S(\varepsilon - I_1)], \quad (7.24)$$

which simplifies to

$$G(\sigma_P, \sigma_S) = \varepsilon \sigma_S - I_2 \sigma_P + (I_2 - I_1)\sigma_S \sigma_P. \quad (7.25)$$

From the states and weights shown in Figure 7.13, we find that the probability of the enzyme being in the active state, if it is not phosphorylated, is

$$p_{\text{active}} = \frac{e^{-\beta G(\sigma_S=1,\sigma_P=0)}}{\sum_{\sigma_S=0,1} e^{-\beta G(\sigma_S,\sigma_P=0)}} = \frac{e^{-\beta\varepsilon}}{e^{-\beta\varepsilon} + 1}, \quad (7.26)$$

while in the phosphorylated state,

$$p^*_{\text{active}} = \frac{e^{-\beta G(\sigma_S=1,\sigma_P=1)}}{\sum_{\sigma_S=0,1} e^{-\beta G(\sigma_S,\sigma_P=1)}} = \frac{e^{-\beta(\varepsilon - I_1)}}{e^{-\beta(\varepsilon - I_1)} + e^{\beta I_2}}. \quad (7.27)$$

The change in activity due to phosphorylation, which we characterize by the ratio $p^*_{\text{active}}/p_{\text{active}}$, is then

$$\frac{p^*_{\text{active}}}{p_{\text{active}}} = \frac{1 + e^{\beta\varepsilon}}{1 + e^{\beta(\varepsilon + I_2 - I_1)}}. \quad (7.28)$$

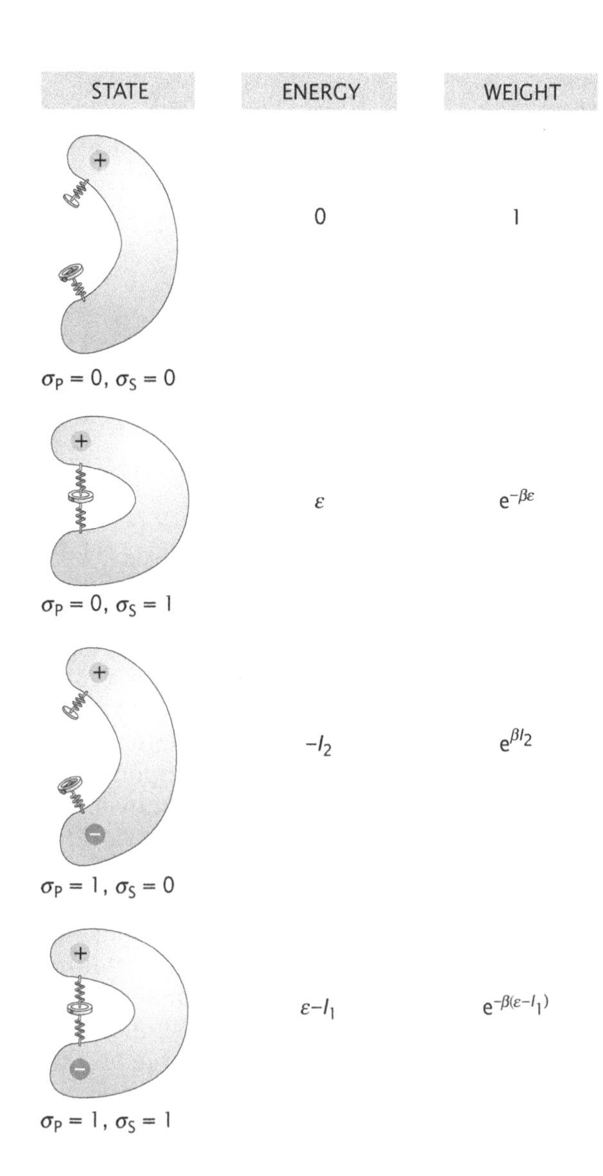

Figure 7.13: States and weights for phosphorylation that follow from the free-energy landscape shown in Figure 7.12.

In the toy model described by the free-energy landscape in Figure 7.12, we have $\varepsilon \approx 5\,k_BT$ and $I_2 - I_1 \approx -10\,k_BT$, which gives a $p^*_{\text{active}}/p_{\text{active}} \approx 150$ increase in activity upon phosphorylation. For enzymes encountered in cells, the increase in activity upon phosphorylation spans a wide range, from factors of 2 or so, to 1000.

Two-Component Systems Exemplify the Use of Phosphorylation in Signal Transduction

A widespread and biologically important use of the ability of protein phosphorylation to drastically alter protein activity is found in the two-component signal transduction systems. These systems are extremely common among bacteria, but are also found in half of archaea and in a few eukaryotes. The purpose of signal transduction systems is to translate environmental information into a macromolecular language that can be understood by the cell. A very simple example of this that we have encountered repeatedly is the Lac repressor, which is a protein that is able to bind to DNA and influence the ability of RNA polymerase to transcribe a set of linked genes. The critical feature of the Lac repressor that enables it to perform

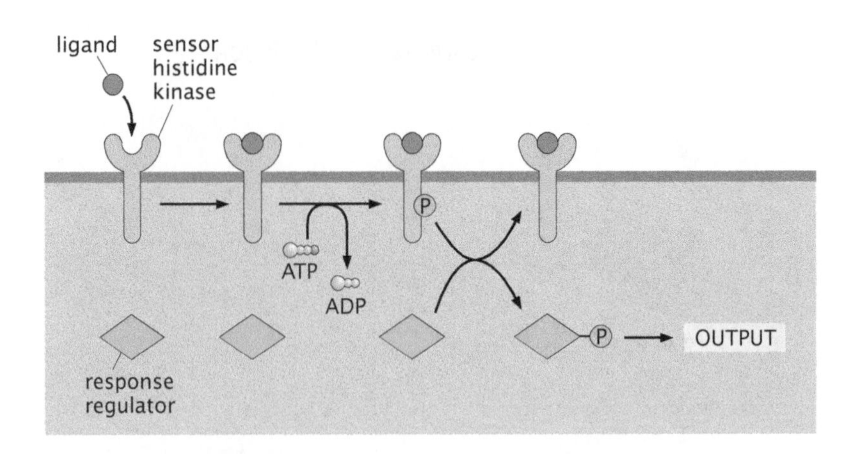

Figure 7.14: Schematic showing reactions of the sensor histidine kinase and response regulator in a typical bacterial two-component system.

signal transduction is that it can also bind a sugar related to lactose, and the affinity with which it binds DNA is modulated by this other binding event.

In the case of most two-component signal transduction systems, the environmental signal that the cell is monitoring is found outside the cytoplasm. In these situations, it is not practical to have the small-molecule binding function and the DNA regulatory function coexist in the same polypeptide chain. For two-component systems, the small-molecule detection is typically carried out by a transmembrane protein, a sensor histidine kinase, and the transcriptional regulatory function is carried out by a second protein called a response regulator. These two partners communicate with one another by transfer of a phosphate from the histidine kinase to the response regulator. The molecular players in a typical two-component system are shown in Figure 7.14. For example, binding of a small molecule to the external domain of a sensor can activate an autokinase activity, such that a phosphate group is attached to a histidine residue on the cytoplasmic domain of that same protein. Subsequently, the phosphate is transferred to an aspartate residue on the partner response regulator. If this response regulator is a transcription factor, its affinity for DNA may be modulated by as much as 1000-fold.

As many different features of the environment are important to the well-being of a bacterial cell, many organisms encode dozens or even hundreds of two-component systems, each tuned to recognize a different condition and induce a different cellular response. Besides binding interesting small molecules, some sensor histidine kinases are able to detect environmental properties such as temperature and osmolarity. Among the response regulators, most influence transcription but others activate various kinds of enzymatic activities within the cell. The diversity of two-component systems that have been identified within the *E. coli* genome is illustrated in Figure 7.15.

The proliferation of such large numbers of structurally related histidine kinases and response regulators raises an interesting problem for the cell. How is it possible to ensure that a given signal will produce only the desired specific cellular response without crosstalk to other pathways? Part of the answer is illustrated in Figure 7.16. The isolated cytoplasmic histidine kinase domains of most sensor proteins exhibit a marked preference for transfer of the phosphate to their own specific partner response regulators, ignoring a vast array of structurally similar but inappropriate targets.

In this way, the cell can quickly alter its transcriptional program in response to an environmental change. Because the phosphorylated

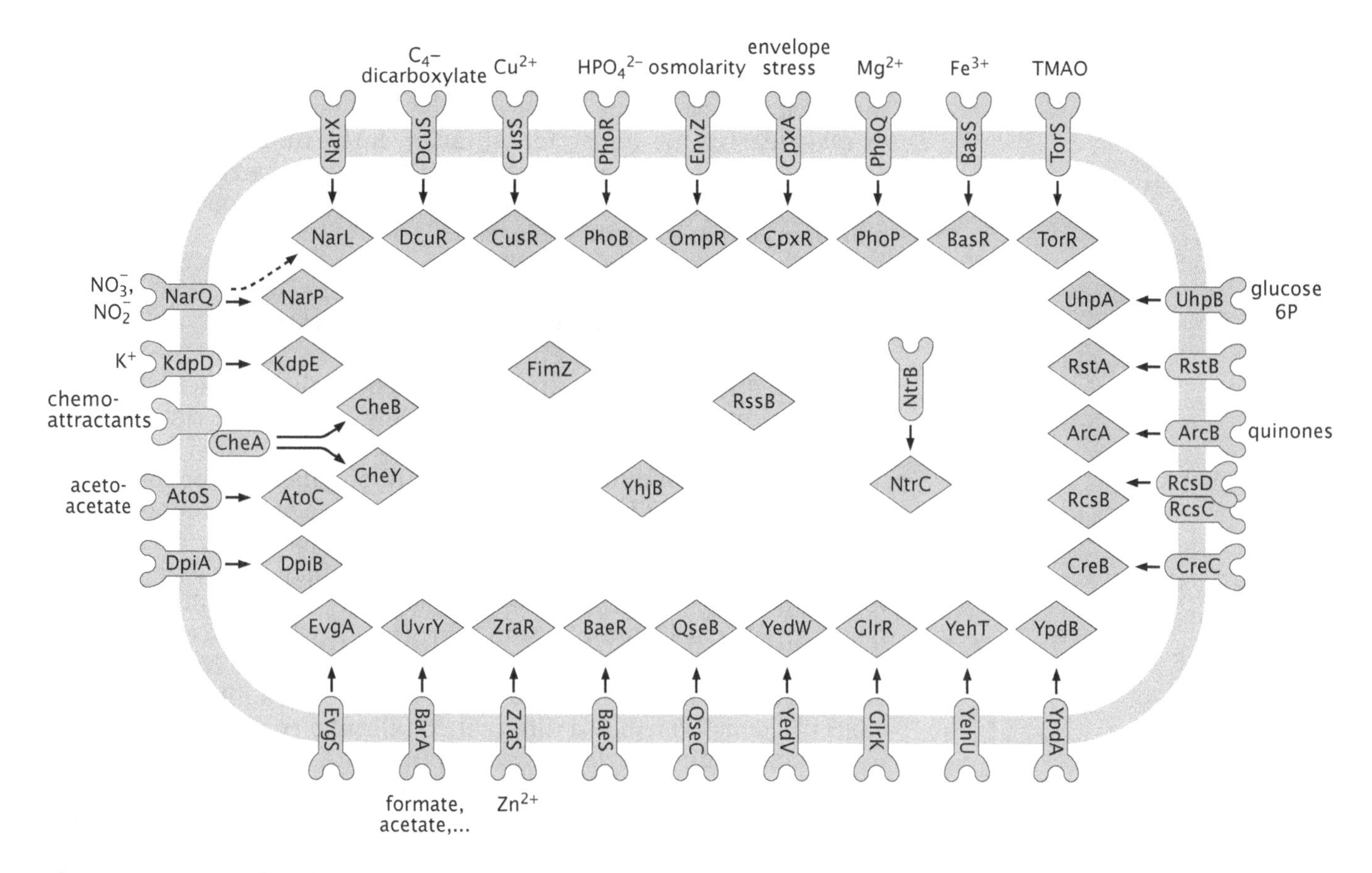

Figure 7.15: Sensor histidine kinases and response regulators found in *E. coli*. The cognate pairs of histidine kinases (green) and response regulators (blue) have been identified by a combination of genetic and biochemical studies. (Courtesy of Mark Goulian and Michael Laub.)

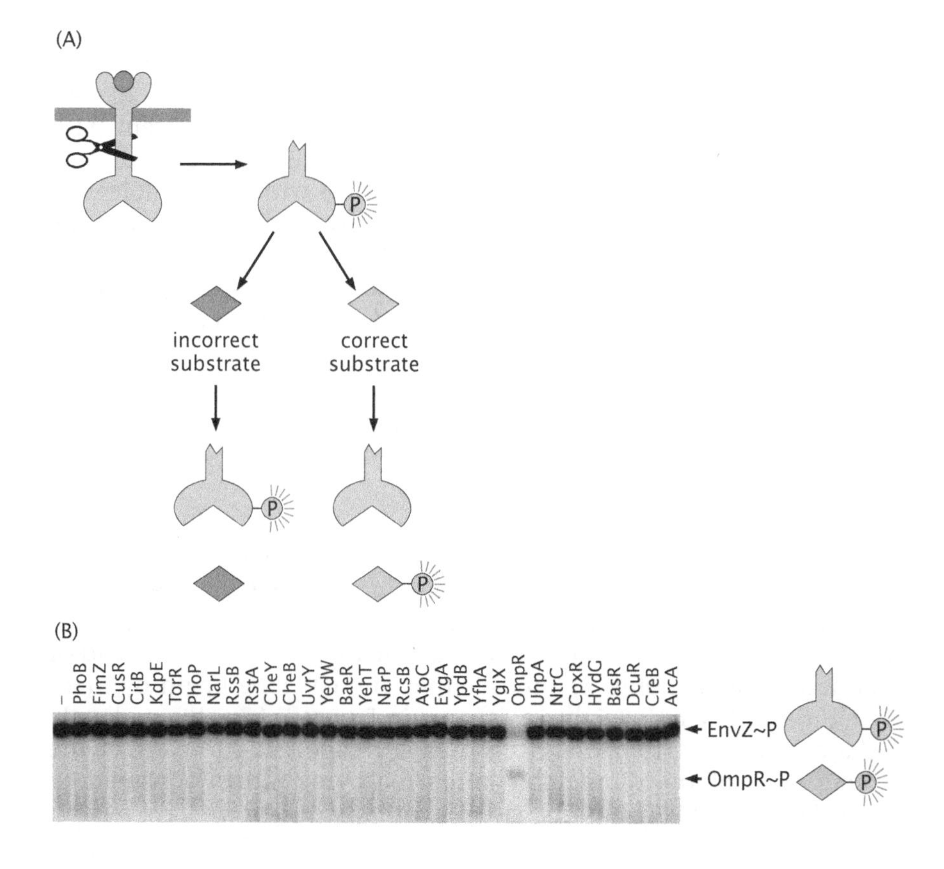

Figure 7.16: Sensor histidine kinases and response regulators found in *E. coli*. (A) Experiment demonstrating specificity of a histidine kinase for its substrate. The cytoplasmic, enzymatically active portion of the kinase can be purified and tested for its ability to transfer a radioactive phosphate to a purified candidate substrate. (B) The osmolarity-sensing histidine kinase EnvZ is able only to transfer phosphate to OmpR, out of 32 possible substrates tested. (Adapted from J. M. Skerker et al., *PLoS Biol.* 3:1770, 2005.)

amino acid can also be rapidly dephosphorylated by a protein phosphatase enzyme, the signal can be just as quickly switched off, and then switched back on again if need be, without the need to degrade or resynthesize the transcription factor. Interestingly, for most two-component systems, the phosphatase activity is carried out by the same sensor histidine kinase protein that was responsible for the phosphorylation in the first place.

7.2.4 Hemoglobin as a Case Study in Cooperativity

In Section 4.2 (p. 143), we argued that hemoglobin has served as the classic example of ligand–receptor binding. One of the rich features offered by hemoglobin above and beyond the results for simple ligand–receptor binding we have already obtained in Sections 6.1.1 (p. 241), 6.4.1 (p. 270), and 7.2.2 is the existence of cooperativity. Cooperativity refers to the fact that the binding energy for a given ligand depends upon the number of ligands that are already bound to the receptor. Intuitively, the cooperativity idea results from the fact that when a ligand binds to a protein, it will induce some conformational change. As a result, when the next ligand binds, it finds an altered protein interface and hence experiences a different binding energy (characterized by a different equilibrium constant). This effect is reflected in binding data as shown in Figure 4.4 (p. 144). From the point of view of statistical mechanics, we will interpret cooperativity as an interaction energy—that is, the energies of the various ligand binding reactions are not simply additive.

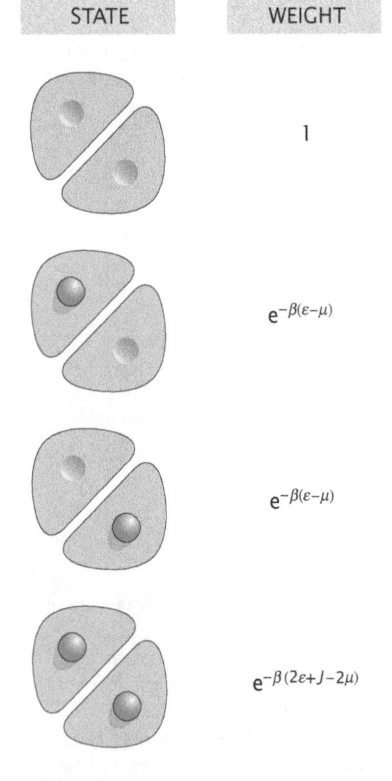

Figure 7.17: States and weights diagram for a toy model of dimoglobin. Each state of occupancy is characterized by a pair (σ_1, σ_2) denoting whether the first and second sites are occupied by an oxygen molecule. The weights show the Gibbs factors for each of the different states.

The Binding Affinity of Oxygen for Hemoglobin Depends upon Whether or Not Other Oxygens Are Already Bound

In keeping with the overarching theme of the chapter, our treatment of ligand–receptor binding in the classic case of hemoglobin can be couched in the language of two-state occupation variables. In particular, for hemoglobin, we describe the state of the system with the vector $(\sigma_1, \sigma_2, \sigma_3, \sigma_4)$, where σ_i takes the values 0 (unbound) or 1 (bound) characterizing the occupancy of site i within the molecule. Figure 4.6 (p. 146) showed the structure of hemoglobin revealing the four binding sites for oxygen molecules. One of the main goals of a model like this is to address questions such as the average number of bound oxygen molecules as a function of the oxygen concentration (or partial pressure).

A Toy Model of a Dimeric Hemoglobin (Dimoglobin) Illustrate the Idea of Cooperativity

In order to make analytic progress in revealing the precise nature of cooperativity, we examine a toy model that reflects some of the full complexity of binding in hemoglobin. In particular, we imagine a fictitious dimoglobin molecule that has two O_2-binding sites. (Indeed, some clams actually do have a dimeric hemoglobin instead of a tetrameric hemoglobin like most other animals.) We begin by identifying the states and weights as shown in Figure 7.17. This molecule is characterized by four distinct states corresponding to each of the binding sites of the dimoglobin molecule being either occupied or empty. For example, if binding site 1 is occupied then we have $\sigma_1 = 1$,

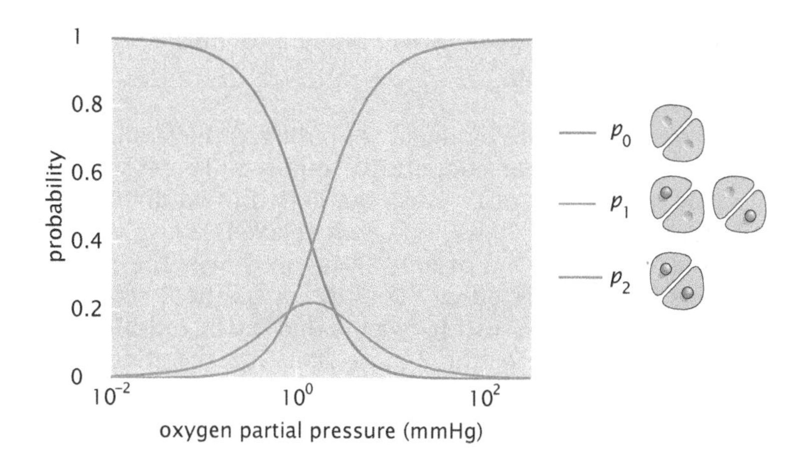

Figure 7.18: Probabilities of oxygen binding to dimoglobin. The plot shows the probability of finding no oxygen molecules bound to dimoglobin (p_0), that of finding one molecule bound (p_1), and that of finding two molecules bound (p_2). The parameters used are $\Delta\varepsilon = -5\ k_BT$, $J = -2.5\ k_BT$, and $c_0 = 760$ mmHg.

and if it is unoccupied then $\sigma_1 = 0$. The energy of the system can be written as

$$E = \varepsilon(\sigma_1 + \sigma_2) + J\sigma_1\sigma_2, \tag{7.29}$$

where ε is the energy associated with an oxygen being bound to one of the two sites. The parameter J is a measure of the cooperativity and implies that when both sites are occupied, the energy is not just the sum of the individual binding energies.

The grand partition function is obtained by summing over the four states shown in Figure 7.17 and is given by

$$\mathcal{Z} = \underbrace{1}_{\text{unoccupied}} + \underbrace{e^{-\beta(\varepsilon-\mu)} + e^{-\beta(\varepsilon-\mu)}}_{\text{single occupancy}} + \underbrace{e^{-\beta(2\varepsilon+J-2\mu)}}_{\text{both sites occupied}}. \tag{7.30}$$

With the partition function in hand, we can compute the probabilities of each of the distinct classes of states: unoccupied, single occupancy, double occupancy. In Figure 7.18, we plot these probabilities as a function of the oxygen partial pressure.

Using Equation 7.17, we can find the average occupancy as a function of the ligand chemical potential as

$$\langle N\rangle = \frac{2e^{-\beta(\varepsilon-\mu)} + 2e^{-\beta(2\varepsilon+J-2\mu)}}{1 + e^{-\beta(\varepsilon-\mu)} + e^{-\beta(\varepsilon-\mu)} + e^{-\beta(2\varepsilon+J-2\mu)}}. \tag{7.31}$$

This simple result now permits us to write the occupancy in terms of the concentration of oxygen by remembering that $\mu = \mu_0 + k_BT\ln(c/c_0)$ (this was shown in Section 6.2.2 on p. 262), and it is given by

$$\langle N\rangle = \frac{2(c/c_0)e^{-\beta\Delta\varepsilon} + 2(c/c_0)^2e^{-\beta(2\Delta\varepsilon+J)}}{1 + 2(c/c_0)e^{-\beta\Delta\varepsilon} + (c/c_0)^2e^{-\beta(2\Delta\varepsilon+J)}}, \tag{7.32}$$

where we define $\Delta\varepsilon = \varepsilon - \mu_0$. This result is shown in Figure 7.19. To further probe the nature of cooperativity, a useful exercise is to examine the occupancy in the case where the interaction term J is zero. In this case, we find the average occupancy is given by the sum of two independent single-site occupancies as

$$\langle N\rangle = 2\frac{(c/c_0)e^{-\beta\Delta\varepsilon}}{1 + (c/c_0)e^{-\beta\Delta\varepsilon}}. \tag{7.33}$$

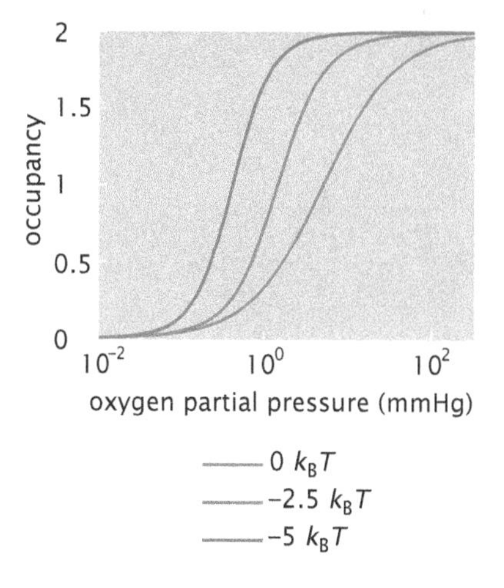

Figure 7.19: Average number of oxygen molecules bound to dimoglobin as a function of oxygen concentration. The parameters used to make the plots are $\Delta\varepsilon = \varepsilon_b - \mu_0 = -5\ k_BT$, $c_0 = 760$ mmHg and $J = 0$ (no cooperativity), $J = -2.5\ k_BT$ and $J = -5\ k_BT$.

The Monod–Wyman–Changeux (MWC) Model Provides a Simple Example of Cooperative Binding

One of the classic two-state models for binding is the Monod–Wyman–Changeux (MWC) model for cooperative binding. The essence of the model is that the protein of interest can exist in two distinct conformational states known as "tense" (T) and "relaxed" (R). In the absence of ligand, the T state of the protein is favored over the R state. We represent this unfavorable energy cost to access the R state with the energy ε. However, the interesting twist is that the ligand-binding reaction has a higher affinity for the R state. This has the effect that with increasing ligand concentration, the balance will be tipped toward the R state, despite the cost, ε, of accessing that state. We label the binding energies ε_T and ε_R, which signify the favorable energy upon binding to the molecule when it is in the T and R states, respectively. If we maintain our use of σ_1 and σ_2 to characterize the state of ligand occupancy of our toy model of dimoglobin and, in addition, introduce the variable σ_m to indicate whether the molecule is in the T ($\sigma_m = 0$) or R ($\sigma_m = 1$) state, then the energy of our system can be written as

$$E = (1 - \sigma_m)\varepsilon_T \sum_{i=1}^{2} \sigma_i + \sigma_m \left(\varepsilon + \varepsilon_R \sum_{i=1}^{2} \sigma_i \right). \tag{7.34}$$

In order to find the occupancy (that is, $\langle N \rangle$) of the dimoglobin, we need to compute the grand partition function. As usual, it is illuminating to depict the various allowed states and their corresponding statistical weights as shown in Figure 7.20. There are a total of eight distinct states and we can sum over all of them to obtain the grand partition function

$$\mathcal{Z} = \underbrace{1 + 2e^{-\beta(\varepsilon_T - \mu)} + e^{-\beta(2\varepsilon_T - 2\mu)}}_{\text{T terms}} + \underbrace{e^{-\beta\varepsilon}(1 + 2e^{-\beta(\varepsilon_R - \mu)} + e^{-\beta(2\varepsilon_R - 2\mu)})}_{\text{R terms}}. \tag{7.35}$$

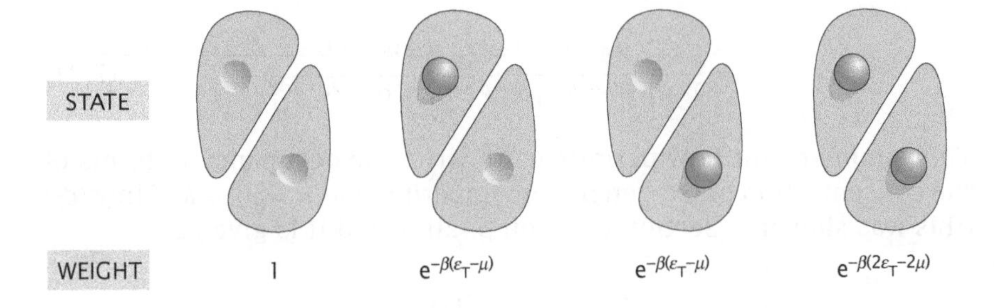

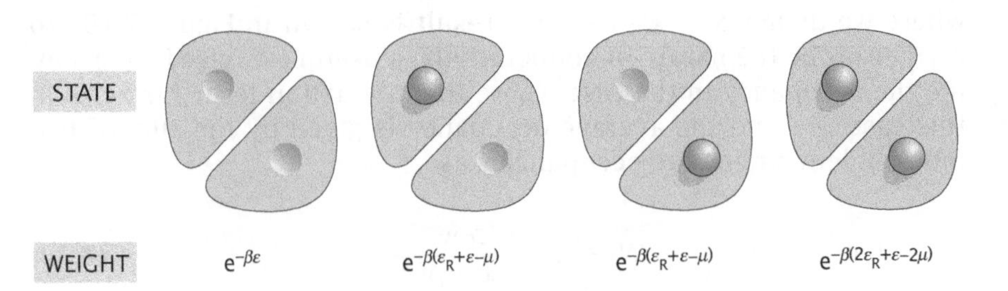

Figure 7.20: States and weights for the MWC model. The upper row shows the occupancies for the T state of the molecule and the lower row shows the occupancies for the R state of the molecule. The set of allowed states amounts to permitting 0, 1, or 2 O_2 molecules to bind to either the T or the R state of the molecule.

As usual, we can find the average occupancy by evaluating

$$\langle N \rangle = k_B T \frac{\partial}{\partial \mu}(\ln \mathcal{Z}),$$

with the result

$$\langle N \rangle = \frac{2}{\mathcal{Z}}[x + x^2 + e^{-\beta\varepsilon}(y + y^2)], \tag{7.36}$$

where we have defined $x = (c/c_0)e^{-\beta(\varepsilon_T - \mu_0)}$ and $y = (c/c_0)e^{-\beta(\varepsilon_R - \mu_0)}$. The average number of bound ligands as a function of the concentration of oxygen is shown in Figure 7.21.

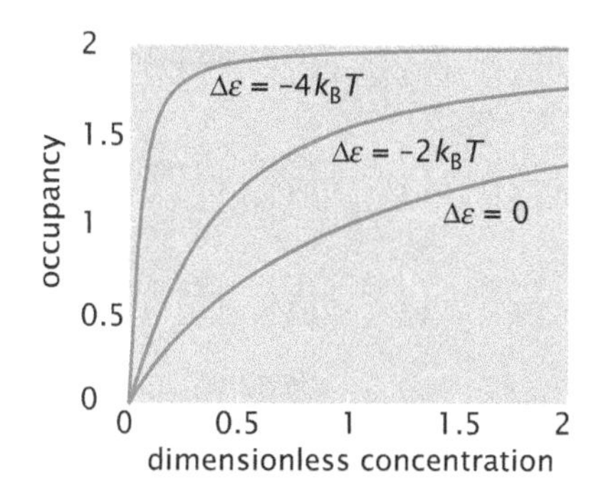

Figure 7.21: Average number of bound receptors in dimoglobin for the MWC model. The dimensionless concentration is written as $x = (c/c_0)e^{-\beta(\varepsilon_T - \mu_0)}$, where ε_T is the binding free energy of the ligand in the tense state. $\Delta\varepsilon$ is the difference between the binding energy in the relaxed and tense states. For the plots shown here, $\varepsilon = 2\,k_B T$.

Using the hypothetical molecule dimoglobin, we have examined oxygen binding from several different perspectives. The first model we introduced is mechanistically more detailed. On the other hand, in many practical cases involving real proteins the coupling energies cannot be easily measured. In such circumstances, the MWC approximation allows quantitative treatments of cooperative protein behavior using only two states and a few parameters. This can be particularly useful when there are many different ligands interacting with the same protein, each of which can affect its overall enzymatic activity.

Statistical Models of the Occupancy of Hemoglobin Can Be Written Using Occupation Variables

Using the simple occupation variable formalism introduced above, we can now examine a hierarchy of models that have been set forth in the attempt to understand cooperative oxygen binding in hemoglobin. In each of these cases, the occupation variable language permits a simple statement of the degree of oxygen binding. Further, the energy of the system itself both with and without cooperativity may be easily written in this language. In particular, we now characterize the binding state of the hemoglobin molecule with four state variables, $\{\sigma_1, \sigma_2, \sigma_3, \sigma_4\}$. Each of these variables can take the value 0 or 1, with $\sigma_\alpha = 0$ corresponding to site α unoccupied and $\sigma_\alpha = 1$ corresponding to site α occupied by an oxygen. Figure 7.22 shows a series of models of hemoglobin binding that account for the cooperativity with different degrees of sophistication.

There is a Logical Progression of Increasingly Complex Binding Models for Hemoglobin

Noncooperative Model We begin with the simplest model, in which the binding on the different sites is independent. In this case, the energy of the system is given by

$$E = \varepsilon \sum_{\alpha=1}^{4} \sigma_\alpha, \tag{7.37}$$

where ε is the energy associated with an oxygen molecule binding to one of the sites on hemoglobin. As usual, the injunction of statistical mechanics is to use the energy in order to compute the partition function. In this case, the grand partition function corresponds to summing over the 16 possible states of the system corresponding to all of the choices of the σ_i. More concretely, the grand partition function is written as

$$\mathcal{Z} = \sum_{\sigma_1=0}^{1} \sum_{\sigma_2=0}^{1} \sum_{\sigma_3=0}^{1} \sum_{\sigma_4=0}^{1} e^{-\beta(\varepsilon-\mu)\sum_{\alpha=1}^{4}\sigma_\alpha}. \tag{7.38}$$

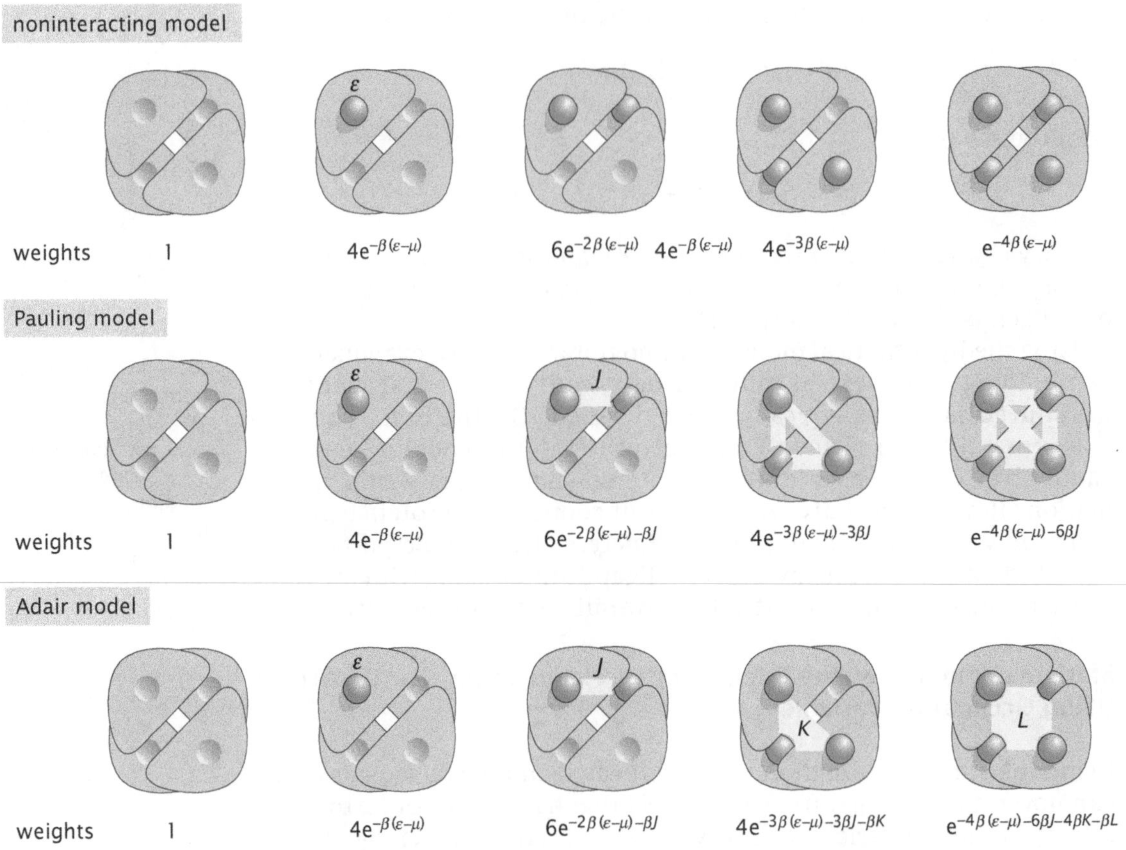

noninteracting model

weights 1 $4e^{-\beta(\varepsilon-\mu)}$ $6e^{-2\beta(\varepsilon-\mu)}$ $4e^{-\beta(\varepsilon-\mu)}$ $4e^{-3\beta(\varepsilon-\mu)}$ $e^{-4\beta(\varepsilon-\mu)}$

Pauling model

weights 1 $4e^{-\beta(\varepsilon-\mu)}$ $6e^{-2\beta(\varepsilon-\mu)-\beta J}$ $4e^{-3\beta(\varepsilon-\mu)-3\beta J}$ $e^{-4\beta(\varepsilon-\mu)-6\beta J}$

Adair model

weights 1 $4e^{-\beta(\varepsilon-\mu)}$ $6e^{-2\beta(\varepsilon-\mu)-\beta J}$ $4e^{-3\beta(\varepsilon-\mu)-3\beta J-\beta K}$ $e^{-4\beta(\varepsilon-\mu)-6\beta J-4\beta K-\beta L}$

Figure 7.22: Hierarchy of models that can be used to characterize the cooperativity in oxygen binding to hemoglobin. In each case, four state variables $\{\sigma_1, \sigma_2, \sigma_3, \sigma_4\}$ are used to characterize whether the four sites are occupied by oxygen or not. The difference from one model to the next is how the energy depends upon the four state variables. These differences are reflected in the statistical weights for the different states.

By noting that each of the terms is independent, this may be simplified to the form

$$\mathcal{Z} = \sum_{\sigma_1=0}^{1} e^{-\beta(\varepsilon-\mu)\sigma_1} \sum_{\sigma_2=0}^{1} e^{-\beta(\varepsilon-\mu)\sigma_2} \sum_{\sigma_3=0}^{1} e^{-\beta(\varepsilon-\mu)\sigma_3} \sum_{\sigma_4=0}^{1} e^{-\beta(\varepsilon-\mu)\sigma_4}. \quad (7.39)$$

These sums are each evaluated to $1 + e^{-\beta(\varepsilon-\mu)}$ and, as a result, the total partition function is of the form

$$\mathcal{Z} = (1 + e^{-\beta(\varepsilon-\mu)})^4. \quad (7.40)$$

To find the occupancy of a given hemoglobin molecule, we resort to the usual trick introduced in Equation 7.17. For the result given in Equation 7.40, this yields

$$\langle N \rangle = \frac{4e^{-\beta(\varepsilon-\mu)}}{1 + e^{-\beta(\varepsilon-\mu)}}. \quad (7.41)$$

We can rewrite this result in terms of the oxygen concentration by using our simple model of the chemical potential, namely, $\mu = \mu_0 + k_B T \ln(c/c_0)$, with the result that the occupancy is given by

$$\langle N \rangle = 4\frac{(c/c_0)e^{-\beta(\varepsilon-\mu_0)}}{1 + (c/c_0)e^{-\beta(\varepsilon-\mu_0)}}. \quad (7.42)$$

Note that this result is just four times the result we would obtain for a single binding site—ligand binding to the different sites is completely independent. If we compare this model with observed oxygen binding curves as shown in Figure 7.23, we see that the noncooperative binding model is completely inconsistent with the data.

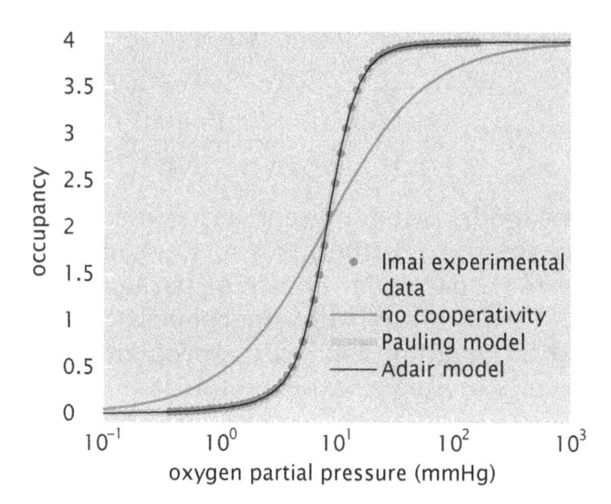

Figure 7.23: Hemoglobin binding. Comparison of the oxygen binding isotherms for different models of hemoglobin using the two-level system description. (Data from K. Imai, *Biophys. Chem.* 37:1, 1990.)

Pauling Model The next model in this hierarchy of models that can be written in the language of the two-state system is the so-called Pauling model. In this case, there is no change to the way in which we characterize the system using the set of variables $\{\sigma_1, \sigma_2, \sigma_3, \sigma_4\}$. What changes from one example to the next is our choice of energy function. In the case of the Pauling model, the physical content of the cooperativity arises because it is assumed that there is a pairwise interaction between oxygens on different sites. If we think of the four binding sites as the vertices of a tetrahedron, there are six interactions corresponding to the six edges of the tetrahedron. If we label the four vertices 1, 2, 3, and 4, these pairwise interactions are between 1 and 2, 1 and 3, etc. and there are a total of six distinct such interactions.

Within this model, the energy of the system is written in the form

$$E = \varepsilon \sum_{\alpha=1}^{4} \sigma_\alpha + \frac{J}{2} {\sum_{(\alpha,\gamma)}}' \sigma_\alpha \sigma_\gamma, \tag{7.43}$$

where the sums over α and γ run from 1 to 4, the prime $\sum'$ instructs us *not* to include terms in the sum when $\alpha = \gamma$, and J is divided by 2 to account for the presence of terms like $\sigma_1\sigma_2$ and $\sigma_2\sigma_1$ which both occur in the sum. Whenever two different sites are occupied, there is a corresponding term in the energy with a contribution J. The partition function corresponding to this energy is given by

$$\mathcal{Z} = \sum_{\sigma_1=0}^{1} \sum_{\sigma_2=0}^{1} \sum_{\sigma_3=0}^{1} \sum_{\sigma_4=0}^{1} e^{-\beta(\varepsilon-\mu)\sum_{\alpha=1}^{4}\sigma_\alpha - \beta(J/2)\sum'_{\alpha,\gamma}\sigma_\alpha\sigma_\gamma}, \tag{7.44}$$

which once again corresponds to summing over all 16 states of occupancy of the hemoglobin molecule by its partner oxygens. As before, the partition function can be evaluated analytically and is given by

$$\mathcal{Z} = \underbrace{1}_{\text{0 bound}} + \underbrace{4e^{-\beta(\varepsilon-\mu)}}_{\text{1 bound}} + \underbrace{6e^{-2\beta(\varepsilon-\mu)-\beta J}}_{\text{2 bound}} + \underbrace{4e^{-3\beta(\varepsilon-\mu)-3\beta J}}_{\text{3 bound}} + \underbrace{e^{-4\beta(\varepsilon-\mu)-6\beta J}}_{\text{4 bound}}. \tag{7.45}$$

Once the partition function is in hand, computing the average occupancy is a matter of computing a single derivative of the form given in Equation 7.17, resulting in

$$\langle N \rangle = \frac{4e^{-\beta(\varepsilon-\mu)} + 12e^{-\beta(\varepsilon-\mu)-\beta J} + 12e^{-3\beta(\varepsilon-\mu)-3\beta J} + 4e^{-4\beta(\varepsilon-\mu)-6\beta J}}{1 + 4e^{-\beta(\varepsilon-\mu)} + 6e^{-2\beta(\varepsilon-\mu)-\beta J} + 4e^{-3\beta(\varepsilon-\mu)-3\beta J} + e^{-4\beta(\varepsilon-\mu)-6\beta J}}. \tag{7.46}$$

If we adopt the notation $j = \mathrm{e}^{-\beta J}$ and $x = (c/c_0)\mathrm{e}^{-\beta(\varepsilon-\mu_0)}$, then we are left with

$$\langle N\rangle = \frac{4x + 12x^2 j + 12x^3 j^3 + 4x^4 j^6}{1 + 4x + 6x^2 j + 4x^3 j^3 + x^4 j^6}. \tag{7.47}$$

The beauty of this model is that it is entirely minimalistic and involves only two free parameters. Further, this model reduces to the noncooperative model for a particular choice of parameters. In particular, if $J = 0$ (that is, $j = 1$), this abolishes the cooperativity and we restore our earlier result of Equation 7.42. In the problems at the end of the chapter, the reader is invited to examine the data in Figure 7.23 using this model.

Adair Model The next level in the hierarchy of models that we examine is the so-called Adair model, which goes beyond the Pauling model in accounting for three- and four-body interactions. What this means concretely is that if three sites are occupied by oxygens, there is an energy that is different than the sum of all of the pair interactions. The reason for the proliferation of parameters (four parameters in the Adair model, in comparison with two in the Pauling model) is to account for the richness of the binding data, which can include competitive binding by other ligands and mutants of the hemoglobin protein. The reader is referred back to Figure 7.22 to get a sense for the types of interactions included in the Adair model. The energy in the Adair model is written as

$$E = \varepsilon\sum_{\alpha=1}^{4}\sigma_\alpha + \frac{J}{2}\sum_{\alpha,\gamma}{}'\sigma_\alpha\sigma_\gamma + \frac{K}{3!}\sum_{\alpha,\beta,\gamma}{}'\sigma_\alpha\sigma_\beta\sigma_\gamma + \frac{L}{4!}\sum_{\alpha,\beta,\gamma,\delta}{}'\sigma_\alpha\sigma_\beta\sigma_\gamma\sigma_\delta, \tag{7.48}$$

where the parameters K and L capture the energy of the three- and four-body interactions, respectively. Note that the sums for the terms involving the parameters K and L are only over those cases where the σ's refer to different binding sites as indicated by the prime on the summation sign.

The grand partition function for this model has the same basic structure as we found in the previous cases. In particular, we are invited to sum over the 16 distinct binding configurations of the molecule, with each one assigned the appropriate energy. This results in the somewhat daunting expression

$$\mathcal{Z} = \sum_{\sigma_1=0}^{1}\sum_{\sigma_2=0}^{1}\sum_{\sigma_3=0}^{1}\sum_{\sigma_4=0}^{1}\exp\left[-\beta(\varepsilon-\mu)\sum_{\alpha=1}^{4}\sigma_\alpha - \frac{J}{2}\sum_{\alpha,\beta}{}'\sigma_\alpha\sigma_\beta - \frac{K}{3!}\sum_{\alpha,\beta,\gamma}{}'\sigma_\alpha\sigma_\beta\sigma_\gamma - \frac{L}{4!}\sum_{\alpha,\beta,\gamma,\delta}{}'\sigma_\alpha\sigma_\beta\sigma_\gamma\sigma_\delta\right]. \tag{7.49}$$

On the other hand, this expression is not nearly as bad as it looks, and the partition function can be evaluated, resulting in

$$\mathcal{Z} = \underbrace{1}_{\text{0 bound}} + \underbrace{4\mathrm{e}^{-\beta(\varepsilon-\mu)}}_{\text{1 bound}} + \underbrace{6\mathrm{e}^{-2\beta(\varepsilon-\mu)-\beta J}}_{\text{2 bound}} + \underbrace{4\mathrm{e}^{-3\beta(\varepsilon-\mu)-3\beta J-\beta K}}_{\text{3 bound}} + \underbrace{\mathrm{e}^{-4\beta(\varepsilon-\mu)-6\beta J-4\beta K-\beta L}}_{\text{4 bound}}. \tag{7.50}$$

As before, the occupancy is obtained by evaluating the derivative with respect to the chemical potential using Equation 7.17, and this results in

$$\langle N \rangle = \frac{4x + 12x^2 j + 12x^3 j^3 k + 4x^4 j^6 k^4 l}{1 + 4x + 6x^2 j + 4x^3 j^3 k + x^4 j^6 k^4 l}, \tag{7.51}$$

where we have introduced the notation $k = e^{-\beta K}$ and $l = e^{-\beta L}$. In the absence of the interaction terms (that is, $j = k = l = 1$), we once again recover the result in Equation 7.42.

Another way of viewing the results of this section is shown in Figure 7.24. These plots illustrate the probability of the various allowed states of the system as a function of the concentration of oxygen. In particular, we plot the probability of finding no oxygen bound (p_0), one oxygen bound (p_1), and so on. By comparing Figures 7.24(A) and (B), we see that in the case of cooperative binding, the intermediate states are effectively eliminated, with the dominant states being either unoccupied or saturated. The reader is invited to examine this result in more detail in the problems at the end of the chapter.

We have considered a hierarchy of models for the binding of oxygen in hemoglobin. As shown in the analysis, all of these models can be written in terms of the two-state occupation variables $\{\sigma_i\}$ and their differences correspond to the different ways in which they handle cooperativity (which we conveniently model as interactions between the different binding sites). One of the key outputs of these models is the binding curves, which show how the occupancy depends upon the concentration of oxygen. We compare these different models in Figure 7.23, where it is seen that the Pauling and Adair models have introduced cooperativity. The parameters we use to obtain these curves come from measurements on the equilibrium constants for hemoglobin binding in which the binding curves are *fit* to the functional form

$$\langle N \rangle = \frac{4K_1 x + 12K_1 K_2 x^2 + 12K_1 K_2 K_3 x^3 + 4K_1 K_2 K_3 K_4 x^4}{1 + 4K_1 x + 6K_1 K_2 x^2 + 4K_1 K_2 K_3 x^3 + K_1 K_2 K_3 K_4 x^4}, \tag{7.52}$$

with the parameters $K_1 = 1.51 \times 10^{-2}\text{mmHg}^{-1}$, $K_2 = 1.52 \times 10^{-2}$ mmHg^{-1}, $K_3 = 3.47 \times 10^{-1}\,\text{mmHg}^{-1}$, and $K_4 = 3.2\,\text{mmHg}^{-1}$. "Cooperativity" is one of the key facts of biochemical interaction.

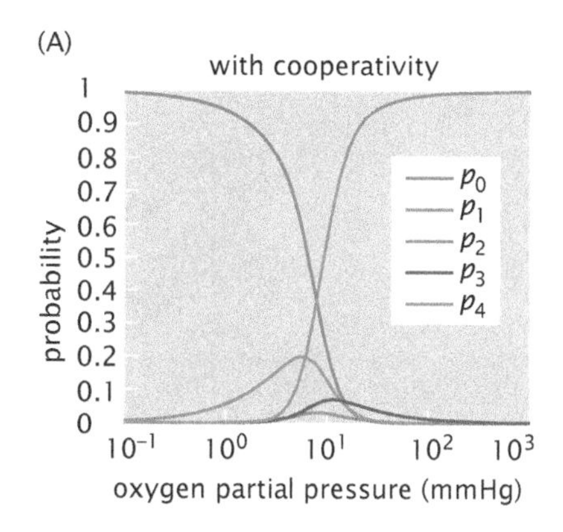

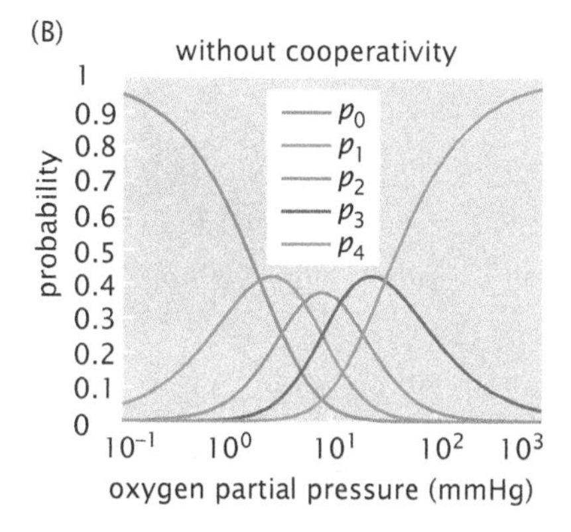

Figure 7.24: Probability of various states in hemoglobin binding. The plot shows the probabilities of the states p_0, p_1, p_2, p_3, and p_4, where p_n refers to the probability of the state with n oxygen molecules bound to the hemoglobin. (A) Adair model treatment of the probabilities of the different states (parameters shown after Equation 7.52). (B) Plot for the case in which there is no cooperativity in the model ($K_d = 8.07\,\text{mmHg}$).

7.3 Ion Channels Revisited: Ligand-Gated Channels and the MWC Model

Earlier in the chapter, we began our discussion of two-state systems by appealing to the example of ion channels. We noted that there are many different kinds of driving forces that can tip the balance between the closed and open states. Ligand-gated channels are ion channels whose opening and closing is regulated by binding of ligands to the protein that makes up the channel. One of the best studied examples is the nicotinic acetylcholine receptor, which plays a role in the neuromuscular junction. It has two binding sites for acetylcholine and the equilibrium between the open and closed state of the channel is shifted toward the open state by the binding of acetylcholine, as shown in Figure 7.25.

To study the opening of the channel as a function of acetylcholine concentration we make use of a statistical mechanics model of a channel, which is analogous to the Monod–Wyman–Changeux (MWC) model of dimoglobin discussed in Section 7.2.4. In the dimoglobin case, the protein was in the T or R state and the ligand had a higher affinity for

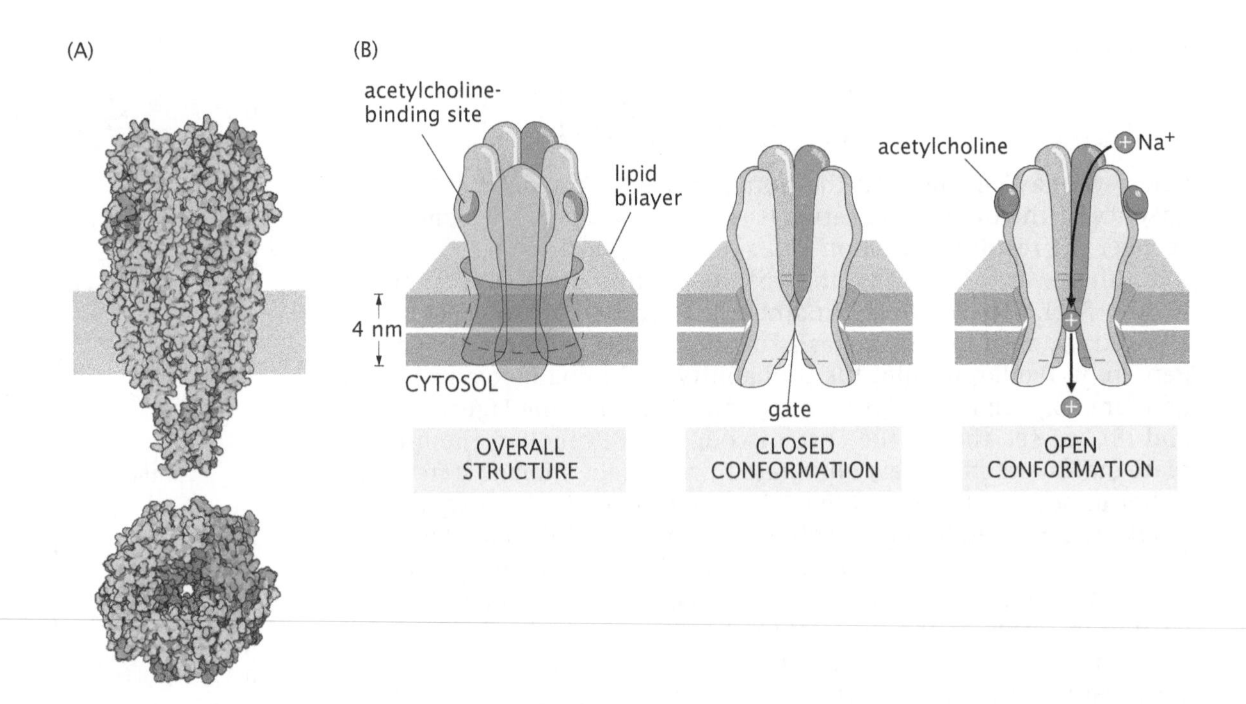

Figure 7.25: Nicotinic acetylcholine receptor. (A) Structure of the receptor. (B) Binding of acetylcholine leads to the channel opening (A, courtesy of D. Goodsell).

the two binding sites in the R state, which itself was higher in energy than the T state. Then, by changing the concentration of the ligands, one could shift the equilibrium from the T to the R state. Similarly, for the ligand-gated channel, we consider the open state of the channel to have a higher energy than the closed channel, so that in the absence of acetylcholine the channel is closed. By having the acetylcholine have a higher affinity for each of the two binding sites when the channel is in the open state as opposed to the closed state, we once again have a model in which there is a rich interplay between binding and conformational state just as we already saw in the case of hemoglobin. To compute the probability that the channel will be open we make use of the states and weights diagram shown in Figure 7.26.

The probability of the channel being open is the sum of the weights of the open states (w_{open}) divided by the sum of all the weights ($w_{\text{open}} + w_{\text{closed}}$), which produces the formula

$$p_{\text{open}} = \frac{e^{-\beta\varepsilon}\left(1 + e^{-\beta(\varepsilon_b^{\text{open}}-\mu)}\right)^2}{e^{-\beta\varepsilon}\left(1 + e^{-\beta(\varepsilon_b^{\text{open}}-\mu)}\right)^2 + \left(1 + e^{-\beta(\varepsilon_b^{\text{closed}}-\mu)}\right)^2}. \tag{7.53}$$

The chemical potential of acetylcholine molecules in solution is related to their concentration, $e^{\beta\mu} = (c/c_0)e^{\beta\mu_0}$ (see Section 6.2.2), and therefore we can rewrite Equation 7.53 as

$$p_{\text{open}} = \frac{e^{-\beta\varepsilon}\left(1 + \frac{c}{K_d^{\text{open}}}\right)^2}{e^{-\beta\varepsilon}\left(1 + \frac{c}{K_d^{\text{open}}}\right)^2 + \left(1 + \frac{c}{K_d^{\text{closed}}}\right)^2}. \tag{7.54}$$

To further simplify the formula, we have introduced the dissociation constants $K_d^{\text{open}} = c_0 e^{\beta(\varepsilon_b^{\text{open}}-\mu_0)}$ and $K_d^{\text{closed}} = c_0 e^{\beta(\varepsilon_b^{\text{closed}}-\mu_0)}$, for the acetylcholine molecule binding to the channel in the open and closed states, respectively.

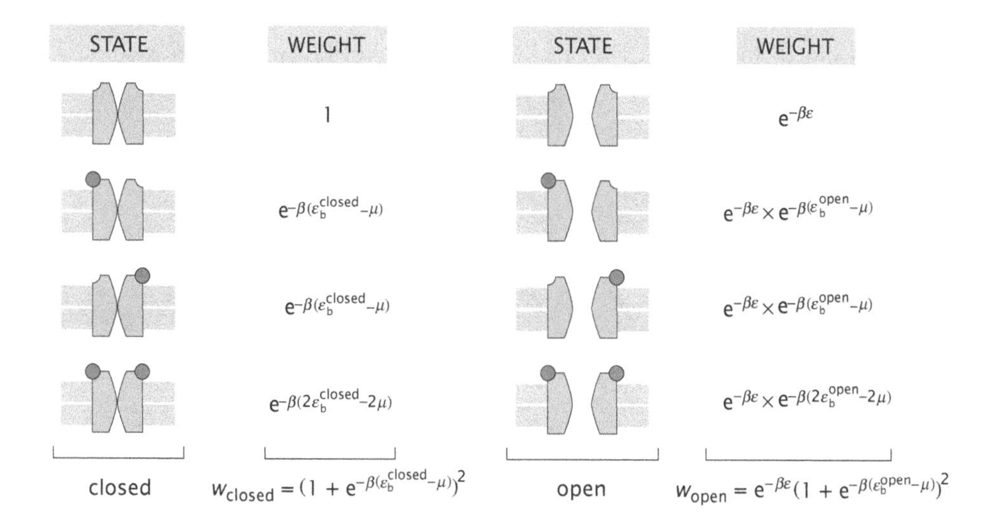

Figure 7.26: States and weights for the MWC model of the ligand-gated ion channel.

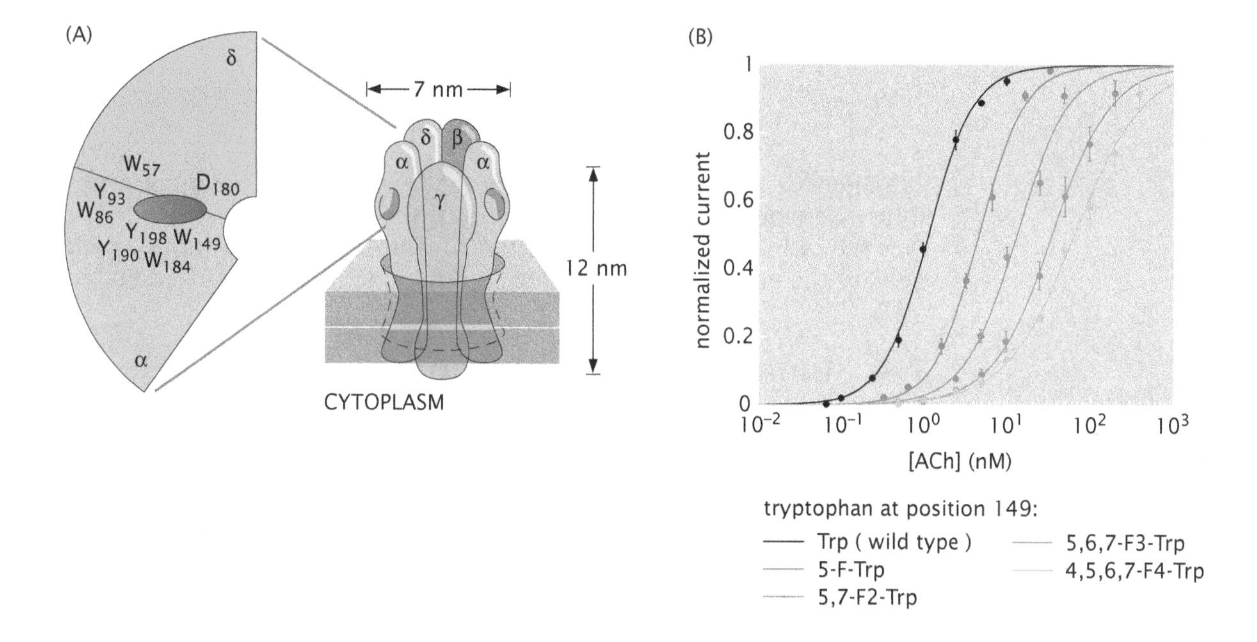

Figure 7.27: Probability of open channel as a function of acetylcholine concentration, for different mutant channels. (A) Structural description of the channel, where W stands for tryptophan, Y for tyrosine, and D for aspartic acid. The numbers indicate the position of the amino acid on the peptide chain. (B) Open probability of channel as a function of acetylcholine concentration, for different mutant channels. In these channels, modified tryptophans were used to replace the tryptophan at position 149. (Adapted from W. Zhong et al., *Proc. Natl Acad. Sci. USA* 95:12088, 1998).

We can use the formula for p_{open} to comment on experiments that make use of channel recordings to measure the fraction of time that the ligand-gated channel spends in the open state as a function of the concentration of acetylcholine. In Figure 7.27 we show data for one such experiment where different mutant channels were investigated. The data shows two trends: the opening probability increases with increasing acetylcholine concentration and the concentration at which the probability reaches half-maximum is different for different mutant channels. The slopes in the middle portions of the graph, associated with the degree of cooperativity, also seem to show a decreasing trend.

The MWC model of the ligand gated channel is determined by three parameters, the energy difference between the open and closed states, ε, and the two acetylcholine dissociation constants, one for the open and one for the closed state of the receptor. If we measure the concentration of acetylcholine in units of K_d^{closed}, that is, $\tilde{c} = c/K_d^{closed}$, then there are two parameters left, namely, the ratio of the two dissociation

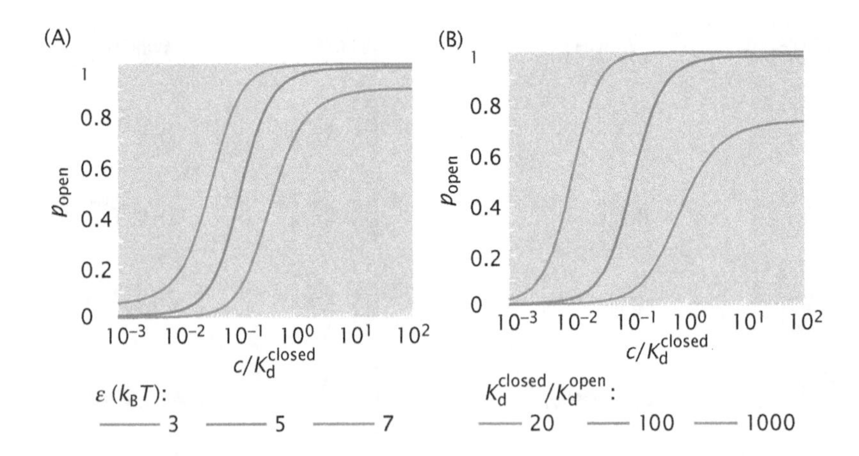

Figure 7.28: Channel open probability. (A) Open probability as a function of the normalized ligand concentration for different values of the energy difference between being open and closed, ε, and $K_d^{closed}/K_d^{open} = 100$. (B) Open probability as a function of ligand concentration for different values of the ratio of the dissociation constant in the closed and open states, K_d^{closed}/K_d^{open}, and $\varepsilon = 5k_BT$.

constants and the Boltzmann factor associated with the energy difference between the closed and open state of the channel. This results in an expression for the opening probability of the form

$$p_{open} = \frac{e^{-\beta\varepsilon}\left(1 + \tilde{c}\frac{K_d^{closed}}{K_d^{open}}\right)^2}{e^{-\beta\varepsilon}\left(1 + \tilde{c}\frac{K_d^{closed}}{K_d^{open}}\right)^2 + (1+\tilde{c})^2}. \quad (7.55)$$

In Figure 7.28, we show two plots that explore how the open probability depends upon the parameters in the model, one in which we vary the ratio of the two dissociation constants and the other in which we vary the energy difference between the open and the closed state.

7.4 Summary and Conclusions

Many important biological problems involve macromolecules that exist in one of two states. For example, the simplest picture of ion channels considers them as either "open" or "closed." Similarly, receptors can either have a ligand bound to a particular binding site or not. In this chapter, we examined the idea of two-state variables as a way to capture the essence of problems of this type in a way that is amenable to analysis using statistical mechanics. Further, we saw that in many cases, these two-state problems involve "cooperativity" which reflects the interdependence of different binding sites within the same molecule. These same ideas will arise repeatedly throughout the remainder of the book.

7.5 Problems

Key to the problem categories: • Model refinements and derivations • Model construction

• 7.1 Loading in mechanosensitive channels

Figure 7.6 shows the coupling between an external load and a mechanosensitive channel. Show that when the channel radius changes by ΔR, area conservation implies $\Delta R_{out} = (R/R_{out})\Delta R$ and use this to fill in all the missing steps leading to Equation 7.5. Use the resulting energy to generate your own version of Figure 7.8.

• 7.2 Average occupancy from statistical mechanics

Derive Equation 7.17. Use this equation to derive the average occupancy of a receptor with a single binding site as a function of the ligand concentration.

7.3 Dimoglobin revisited

(a) Use the canonical distribution (as in Section 6.1.1 on p. 241) to redo the problem of dimoglobin binding. For simplicity, imagine a box with N O_2 molecules that can be distributed among Ω sites. This simple lattice model of the solution is intended to account for the configurational entropy available to the O_2 molecules when they are in solution. This disposition of the system is shown in Figure 7.29. Use the energy given in Equation 7.29 when constructing the partition function.

(b) Figure 7.18 shows the probabilities of the various states available to dimoglobin in its interactions with its oxygen binding partners. Write expressions for the probabilities p_0, p_1, and p_2 corresponding to occupancy 0, 1, and 2, respectively. Using the parameters shown in the caption to Figure 7.18, reproduce the plot.

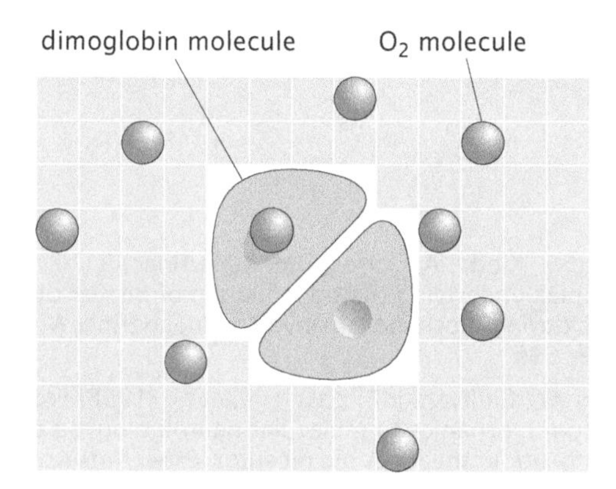

Figure 7.29: Schematic of the binding assay of interest in which the oxygen molecules can be either bound to dimoglobin or in solution.

7.4 State probabilities in the MWC model

Plot p_0, p_1, and p_2, the probabilities of different states of occupancy for both the T and R states for the MWC model of dimoglobin. Use the same parameters to generate your plot that were used to generate Figure 7.21.

7.5 Carbon monoxide and hemoglobin

Carbon monoxide is a deadly gas that binds hemoglobin roughly 240 times as tightly as oxygen does (this means that CO has 1/240 the dissociation constant of O_2, or $240K_{CO} = K_{O_2}$, where $K_{O_2} = 26$ mmHg).

(a) When both CO and O_2 are present, use the Hill equation introduced in Section 6.4.3 to calculate the probability that hemoglobin will be saturated with oxygen. Similarly, compute the probability that hemoglobin will be saturated with CO. Calculate the partial pressure of oxygen using the fact that atmospheric oxygen constitutes roughly 21% of air and assume a partial pressure of CO of 2 mmHg. Hemoglobin binding to carbon monoxide has a Hill coefficient of 1.4 and hemoglobin binding to oxygen has a Hill coefficient of 3.0.

(b) Plot the probability of O_2 binding to hemoglobin as a function of the partial pressure of CO assuming the oxygen partial pressure remains constant.

(c) Show that CO and O_2 will have an equal probability of binding when the condition

$$\left(\frac{[O_2]}{K_{O_2}}\right)^{n_{O_2}} \left(\frac{K_{CO}}{[CO]}\right)^{n_{CO}} = 1$$

is satisfied and work out the partial pressure of CO at which this occurs.

7.6 Toy model of protein folding

A four-residue protein can take on the four different conformations shown in Figure 7.30. Three conformations are open and have energy ε ($\varepsilon > 0$) and one is compact, and has energy zero.

(a) At temperature T, what is the probability, p_o, of finding the molecule in an open conformation? What is the probability, p_c, that it is compact?

(b) What happens to the probability p_c, calculated in (a), in the limit of very large and very low temperatures.

(c) What is the average energy of the molecule at temperature T?

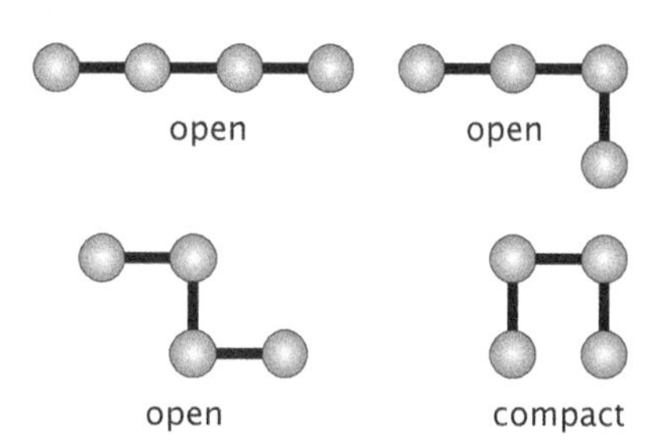

Figure 7.30: Toy model of protein folding showing four different conformations. (Adapted from K. Dill and S. Bromberg, Molecular Driving Forces, 2nd ed. Garland Science, 2011.)

7.7 Chemical potentials and channel open probabilities

An alternative way to think of the probability of gating of membrane-bound channels is to think of the membrane as consisting of two species of channel, closed and open, at concentrations c_{closed} and c_{open}, respectively. These two species are subject to constant interconversion characterized by an equilibrium in which their respective chemical potentials are equal. By setting the chemical potentials for these two species equal, work out an expression for the open probability.

7.8 Energy landscapes in the two-state model

Draw an energy landscape such as shown in Figure 7.7 for the voltage-gated sodium channel presented in Figure 7.2. In particular, show how the landscape changes as a function of the applied voltages shown in Figure 7.2(C). Explain your reasoning carefully.

7.6 Further Reading

Andersen, O, & Koeppe, R (1992) Molecular determinants of channel function, *Physiol. Rev.* **72**, S89. This interesting review gives a broad and well-organized discussion of many of the issues that arise in thinking about ion channel gating. Section VII on gating describes two-state channel models like those in this chapter.

Walsh, CT (2006) Posttranslational Modification of Proteins—Expanding Nature's Inventory, Roberts and Company Publishers. Besides the provocative subtitle, this book is full of interesting insights into the phenomenology of posttranslational modification.

Two excellent treatments of the Gibbs distribution are those in Kittel, C, & Kroemer, H (1980) Thermal Physics, 2nd ed., W. H. Freeman and in Schroeder, DV (2000) An Introduction to Thermal Physics, Addison-Wesley Longman.

Dill, K, & Bromberg, S (2011) Molecular Driving Forces, 2nd ed., Garland Science. This book has an excellent discussion of binding and cooperativity.

Goldbeter, A, & Koshland, DE (1981) An amplified sensitivity arising from covalent modification in biological systems, *Proc. Natl Acad. Sci. USA* **78**, 6840. This illustrates sensitivity of response of proteins subject to covalent modifications such as phosphorylation.

Graham, I, & Duke, T (2005) The logical repertoire of ligand-binding proteins, *Phys. Biol.* **2**, 159. Another interesting discussion of how proteins can exist in multiple states.

Mello, BA, & Tu, Y (2005) An allosteric model for heterogeneous receptor complexes: Understanding bacterial chemotaxis responses to multiple stimuli, *Proc. Natl Acad. Sci. USA* **102**, 17354. This paper is an interesting example of the use of state variables like those described here and the MWC model as it applies to bacterial chemotaxis.

7.7 References

Imai, K (1990) Precision determination and Adair scheme analysis of oxygen equilibrium curves of concentrated hemoglobin solution. A strict examination of Adair constant evaluation methods. *Biophys. Chem.* **37**, 1.

Keller, BU, Hartshorne, RP, Talvenheimo, JA, Catterall, WA, & Montal, M (1986) Sodium channels in planar lipid bilayers: Channel gating kinetics of purified sodium channels modified by batrachotoxin, *J. Gen. Physiol.* **88**, 1.

Perozo, E, Kloda, A, Cortes, DM, & Martinac, B (2002) Physical principles underlying the transduction of bilayer deformation forces during mechanosensitive channel gating, *Nat. Struct. Biol.* **9**, 696.

Zhong, W., Gallivan, J. P., Zhang, Y., et al. (1998) From *ab initio* quantum mechanics to molecular neurobiology: a cation-pi binding site in the nicotinic receptor, *Proc. Natl Acad. Sci. USA* **95**, 12088.

Random Walks and the Structure of Macromolecules

8

Overview: In which we think of macromolecules as random walks

There are many different ways of characterizing biological structures. A useful alternative to the deterministic description of structure in terms of well-defined atomic coordinates is the use of statistical descriptions. For example, the arrangement of a large DNA molecule within the cell is often best characterized statistically in terms of average quantities such as the mean size and position. The goal of this chapter is to examine one of the most powerful ideas in all of science, namely, the random walk, and to show its utility in characterizing biological macromolecules such as DNA. We will show how these ideas culminate in a probability distribution for the end-to-end distance of polymers and how this distribution can be used to compute the "structure" of DNA in cells as well as to understand single-molecule experiments in which molecules of DNA (or proteins) are pulled on and the subsequent deformation is monitored as a function of the applied force. In addition, we will show how these same ideas may be tailored to thinking about proteins.

> "I only went out for a walk and finally concluded to stay out till sundown, for going out, I found, was really going in."
>
> **John Muir**

8.1 What Is a Structure: PDB or R_G?

The study of structure is often a prerequisite to tackling the more interesting question of the functional dynamics of a particular macromolecule or macromolecular assembly. Indeed, this notion of the relation between structure and function has been elevated to the status of the true central dogma of molecular biology, namely, "sequence determines structure determines function" (Petsko and Ringe, 2004), which calls for uncovering the relation between sequence and consequence. The idea of structure is hierarchical and subtle, with the relevant detail that is needed to uncover function often occurring at totally disparate spatial scales. For example, in thinking about nucleosome positioning, an atomic-level description of the state of methylation of the DNA might be required, whereas in thinking about cell division, a much coarser description of DNA is likely more useful.

The key message of the present chapter is that there is much to be gained in some circumstances by abandoning the deterministic, PDB mentality described in earlier chapters for a *statistical* description in which we attempt only to characterize certain average properties of the structure. We will argue that this type of thinking permits immediate and potent contact with a range of experiments.

8.1.1 Deterministic versus Statistical Descriptions of Structure

PDB Files Reflect a Deterministic Description of Macromolecular Structure

The notion of structure is complex and ambiguous. In the context of crystals, we can think of structure at the level of the monotonous regular packing of the atoms into the unit cells of which the crystal is built. This thinking applies even to crystals of nucleic acids, proteins, or complexes such as ribosomes, viruses, and RNA polymerase. Indeed, it is precisely this regularity that makes it possible to deposit huge PDB files containing atomic coordinates on databases such as the PDB and VIPER. In this world view, a structure is the set $(\mathbf{r}_1, \mathbf{r}_2, \ldots, \mathbf{r}_N)$, where $\mathbf{r}_i$ is the vector position $\mathbf{r}_i = (x_i, y_i, z_i)$ of the ith atom in this N-atom molecule. However, the structural descriptions that emerge from X-ray crystallography provide a deceptively static picture that can only be viewed as a starting point for thinking about the functional dynamics of macromolecules and their complexes in the crowded innards of a cell.

Statistical Descriptions of Structure Emphasize Average Size and Shape Rather Than Atomic Coordinates

In the context of polymeric systems such as DNA, the notion of structure brings us immediately to the question of the relative importance of universality (for example, how size scales with the number of monomers) and specificity in macromolecules. In particular, there are certain things that we might wish to say about the structure of polymeric systems that are indifferent to the precise chemical details of these systems. For example, when a DNA molecule is ejected from a bacteriophage into a bacterial cell, all that we may really care to say about the disposition of that molecule is how much space it takes up and where within the cell it does so. Similarly, in describing the geometric character of a bacterial genome, it may suffice to provide a description of structure only at the level of characterizing a blob of a given size and shape. Indeed, these considerations bring us immediately to the examination of statistical measures of structure. As hinted at in the title of this section, one statistical measure of structure is provided by the radius of gyration, R_G, which, roughly speaking, gives a measure of the size of a polymer blob. In the remainder of the chapter, we show the calculable consequences of statistical descriptions of structure.

8.2 Macromolecules as Random Walks

Random Walk Models of Macromolecules View Them as Rigid Segments Connected by Hinges

One way to characterize the geometric disposition of a macromolecule such as DNA is through the *deterministic* function $\mathbf{r}(s)$. This function

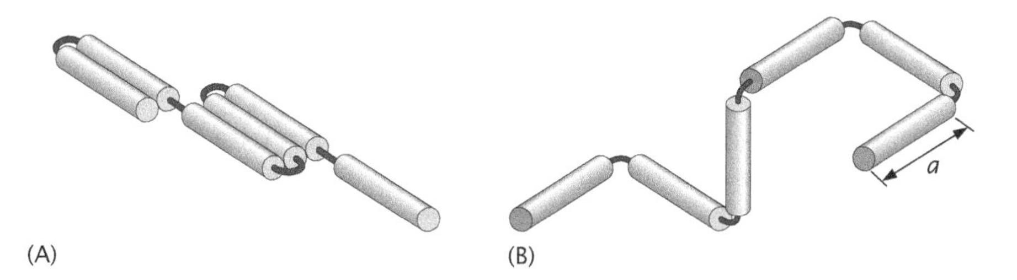

Figure 8.1: Random walk model of a polymer. Schematic representation of (A) a one-dimensional random walk and (B) a three-dimensional random walk as an arrangement of linked segments of length a.

tells us the position $\mathbf{r}$ of that part of the polymer which is a distance s along its contour. An alternative we will explore here is to discretize the polymer into a series of segments, each of length a, and to treat each such segment as though it is rigid. The various segments that make up the macromolecular chain are then imagined to be connected by flexible links that permit the adjacent segments to point in various directions. The one- and three-dimensional versions of this idea are shown in Figure 8.1. In the one-dimensional case, the segments are at $\pm 180^\circ$ with respect to each other. We draw them as nonoverlapping for clarity. For the three-dimensional case, we illustrate the situation in which the links are restricted to 90° angles, though there are many instances in which we will consider links that can rotate in arbitrary directions (the so-called freely jointed chain model).

Figure 8.2 shows an example of the correspondence between the real structures of these molecules and their idealization in terms of the lattice model of the random walk. In particular, Figure 8.2 shows a conformation of DNA on a surface. Using the discretization advocated above, we show how this same structure can be approximated using a series of rigid rods (the Kuhn segments) connected by flexible hinges. We will argue that this level of description can be useful in settings ranging from estimating the entropic cost of confining DNA to a bacterial cell to the stretching of DNA by laser tweezers.

(A)

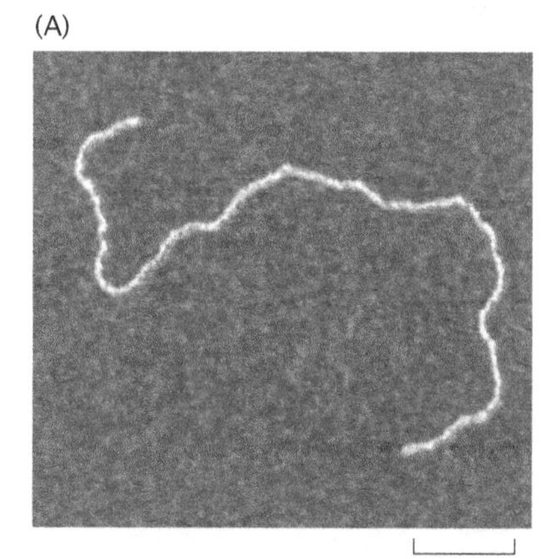

100 nm

(B)

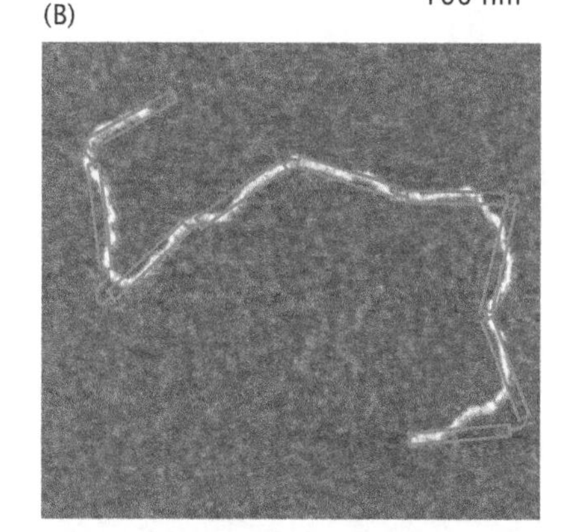

Figure 8.2: DNA as a random walk. (A) Structure of DNA on a surface as seen experimentally using atomic-force microscopy. (B) Representation of the DNA on a surface as a random walk. (Adapted from P. A. Wiggins et al., *Nat. Nanotech.* 1:37, 2006.)

8.2.1 A Mathematical Stupor

In Random Walk Models of Polymers, Every Macromolecular Configuration Is Equally Probable

In this section, we work our way up by degrees to some of the full beauty and depth of the random walk model. The aim of the analysis is to obtain a probability distribution for each and every macromolecular configuration and to use these probabilities to compute properties of the macromolecule that can be observed experimentally, such as the mean size of the macromolecule and the free energy required to deform that molecule. Our starting point will be an analysis of the random walk in one dimension, with our discussion being guided by the ways in which we will later generalize these ideas and apply them in what might at first be considered unexpected settings.

We begin by imagining a single random walker confined to a one-dimensional lattice with lattice parameter a as already shown in Figure 8.1(A). The life history of this walker is built up as a sequence of left and right steps, with each step constituting a single segment in the polymer. In addition, for now, we postulate that the probabilities of right and left steps are given as $p_r = p_\ell = 1/2$. The trajectory of the walker is built up by assuming that at each step the walker starts anew with no concern for the orientation of the previous segment. We note that for a chain with N segments, this implies that there are a total of 2^N different permissible macromolecular configurations, each with probability $1/2^N$.

The Mean Size of a Random Walk Macromolecule Scales as the Square Root of the Number of Segments, $\sqrt{N}$

Given the spectrum of possible configurations and their corresponding probabilities, one of the most immediate questions we can pose concerns the mean distance of the walker from its point of departure as a function of the number of segments in the chain. In the context of biology, this question is tied to problems such as the cyclization of DNA, the likelihood that a tethered ligand and receptor will find each other, and the gross structure of plasmids and chromosomal DNA in cells. To find the end-to-end distance for the molecule of interest, we can use both simple arguments as well as brute force calculation, and we will take up each of these options in turn. The simple argument notes that the expected value of the walker's distance from the origin, R, after N steps can be obtained as

$$\langle R \rangle = \left\langle \sum_{i=1}^{N} x_i \right\rangle, \tag{8.1}$$

where $x_i = \pm a$ is the displacement suffered by the walker during the ith step and where we have introduced the bracket notation $\langle \cdots \rangle$ to signify an average. Recall that, to obtain such an average, we sum over all possible configurations with each configuration weighted by its probability (in this case, the probabilities are all equal). This result may be simplified by noting that the averaging operation represented by the brackets $\langle \cdots \rangle$ on the right-hand side of the equation can be taken into the summation symbol (that is, the average of a sum is the sum of the averages) and through the recognition that $\langle x_i \rangle = 0$. Indeed, this leaves us with the conclusion that the mean displacement of the walker is identically zero.

A more useful measure of the walker's departure from the origin is to examine

$$\langle R^2 \rangle = \left\langle \sum_{i=1}^{N} \sum_{j=1}^{N} x_i x_j \right\rangle. \tag{8.2}$$

This is the variance of the probability distribution of R, while $\sqrt{\langle R^2 \rangle}$ is the standard deviation. Its significance is that the probability of finding our random walker within one standard deviation of the mean is close to 70%. In other words, the standard deviation is the measure of the typical excursion of the random walker after N steps, and therefore serves as a good surrogate for the typical size of the related polymer.

In order to make progress on Equation 8.2, we break up the sum into two parts as

$$\langle R^2 \rangle = \sum_{i=1}^{N} \langle x_i^2 \rangle + \sum_{i \neq j=1}^{N} \langle x_i x_j \rangle. \tag{8.3}$$

Note that each and every step is independent of all steps that precede and follow it. This implies that the second term on the right-hand side is zero. In addition, and since $x_i = \pm a$, we note that $\langle x_i^2 \rangle = a^2$, with the result that

$$\langle R^2 \rangle = Na^2. \tag{8.4}$$

Thus, we have learned that the walker's departure from the origin is characterized statistically by the assertion that $\sqrt{\langle R^2 \rangle} = a\sqrt{N}$, meaning that the distance from the origin grows as the square root of the number of segments in the chain.

The Probability of a Given Macromolecular State Depends Upon Its Microscopic Degeneracy

In addition to the simple argument spelled out above, it is also possible to carry out a brute force analysis of this problem using the conventional machinery of probability theory. We consider this an important alternative to the analysis given above, since it highlights the fact that there are many microscopic configurations that correspond to a given macroscopic configuration. In particular, in the case in which the walker makes a total of N steps, we pose the question: what is the probability that n_{r} of those steps will be to the right (and hence $n_{\ell} = N - n_{\mathrm{r}}$ to the left)? Since the probability of each right or left step is given by $p_{\mathrm{r}} = p_{\ell} = 1/2$, the probability of a *particular* sequence of N left and right steps is given by $(1/2)^N$. On the other hand, we must remember that there are many ways of realizing n_{r} right steps and n_{ℓ} left steps out of a total of N steps. In particular, there are

$$W(n_{\mathrm{r}}; N) = \frac{N!}{n_{\mathrm{r}}!(N - n_{\mathrm{r}})!}, \tag{8.5}$$

distinct ways of achieving this outcome. This kind of counting result was derived in The Math Behind the Models on p. 239. A particular example of this thinking to the case $N = 3$ is shown in Figure 8.3, where we see that there is one configuration in which all three segments are right-pointing, one configuration in which all three

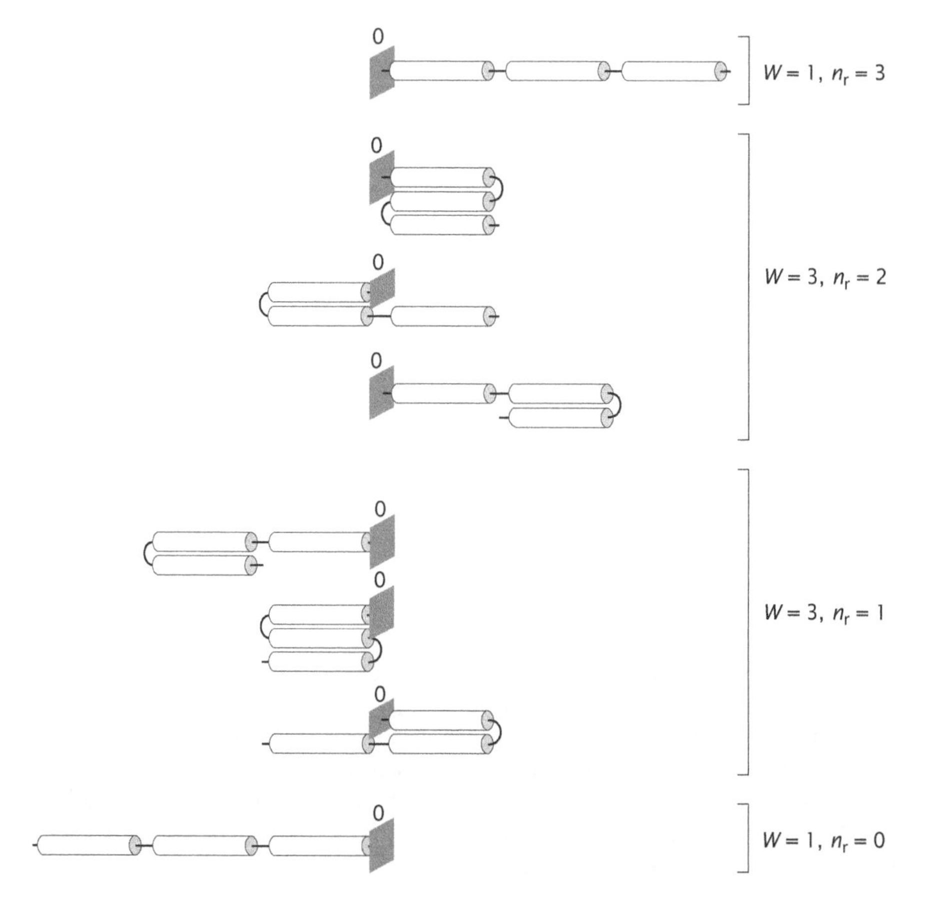

Figure 8.3: Random walk configurations. The schematic shows all of the allowed conformations of a polymer made up of three segments ($2^3 = 8$ conformations) and their corresponding degeneracies.

segments are left-pointing, and three configurations each for the cases in which $n_r = 2, n_\ell = 1$, and $n_r = 1, n_\ell = 2$.

We have now enumerated the microscopic degeneracies of each macroscopic configuration (characterized by a given end-to-end distance). As a result, we are poised to write down the probability of an overall departure n_r from the origin: this is given by

$$p(n_r; N) = \frac{N!}{n_r!(N-n_r)!}\left(\frac{1}{2}\right)^N. \tag{8.6}$$

With this probability distribution in hand, we can now evaluate any average characterizing the geometric disposition of the chain by summing over all of the configurations.

To develop facility in the use of this probability distribution, we begin by confirming that it is normalized. To do so, we ask for the outcome of the sum

$$\sum_{n_r=0}^{N} p(n_r; N) = \sum_{n_r=0}^{N} \frac{N!}{n_r!(N-n_r)!}\left(\frac{1}{2}\right)^N. \tag{8.7}$$

To evaluate this sum, we recall the binomial theorem, which tells us

$$(x+y)^N = \sum_{n_r=0}^{N} \frac{N!}{n_r!(N-n_r)!} x^{n_r} y^{N-n_r}. \tag{8.8}$$

For the case in which $x = y = 1$, we see that this implies

$$\sum_{n_r=0}^{N} \frac{N!}{n_r!(N-n_r)!} = 2^N. \tag{8.9}$$

Plugging this result back into Equation 8.7 demonstrates that the probability distribution is indeed normalized (that is, $\sum_{n_r=0}^{N} p(n_r; N) = 1$).

Entropy Determines the Elastic Properties of Polymer Chains

The probability distribution for n_r can be used to deduce a more telling quantity, namely, the probability distribution for the end-to-end distance, $R = (n_r - n_\ell)a$. If we use the condition $n_r + n_\ell = N$ to solve for n_ℓ and substitute this into $R = (n_r - n_\ell)a$, it follows that $n_r = (N + R/a)/2$, and Equation 8.6 can be rewritten as

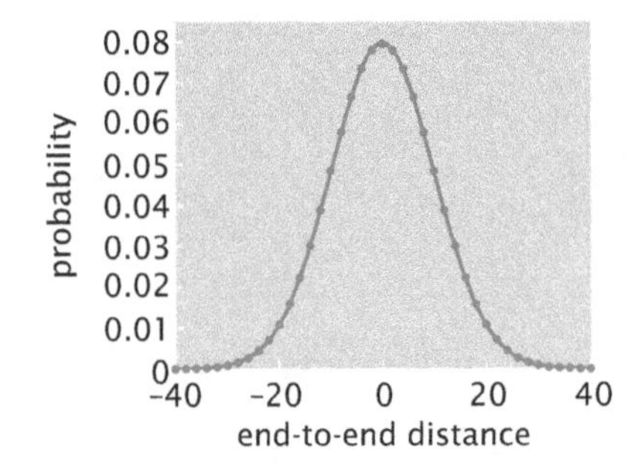

Figure 8.4: End-to-end probability distribution for a one-dimensional "macromolecule" with 100 segments. The figure shows a comparison of the binomial distribution (dots) given in Equation 8.10 and the approximate Gaussian distribution (curve) given in Equation 8.16.

$$p(R; N) = \frac{N!}{\left(\frac{N}{2}+\frac{R}{2a}\right)!\left(\frac{N}{2}-\frac{R}{2a}\right)!}\left(\frac{1}{2}\right)^N, \tag{8.10}$$

to give the probability distribution of the end-to-end distance. This distribution is plotted in Figure 8.4. For large N, this probability distribution is sharply peaked at $R = 0$. Next we show that it takes on the form of a Gaussian distribution for $R \ll Na$. This calculation involves two math methods we have discussed previously, namely, the Stirling approximation (p. 222 and the problems at the end of Chapter 5), $\ln n! \approx n \ln n - n + \frac{1}{2}\ln(2\pi n)$ for $n \gg 1$, and the Taylor expansion (p. 215), $\ln(1+x) \approx x - x^2/2$ for $x \ll 1$. Note that here we take the first three terms in the Stirling approximation, and keep terms up to x^2 in the Taylor expansion, in anticipation of the fact that the leading term in $\ln p(R; N)$ is of the order of R^2.

We begin by taking the logarithm of the probability distribution for R shown in Equation 8.10 and then we apply the Stirling approximation to each of the three factorials, resulting in

$$\ln p(R;N) = \underbrace{N\ln N - N + \frac{1}{2}\ln(2\pi N)}_{\ln N!}$$

$$-\underbrace{\left\{\left(\frac{N}{2}+\frac{R}{2a}\right)\ln\left(\frac{N}{2}+\frac{R}{2a}\right)-\left(\frac{N}{2}+\frac{R}{2a}\right)+\frac{1}{2}\ln\left[2\pi\left(\frac{N}{2}+\frac{R}{2a}\right)\right]\right\}}_{\ln[(N/2)+(R/2a)]!}$$

$$-\underbrace{\left\{\left(\frac{N}{2}-\frac{R}{2a}\right)\ln\left(\frac{N}{2}-\frac{R}{2a}\right)-\left(\frac{N}{2}-\frac{R}{2a}\right)+\frac{1}{2}\ln\left[2\pi\left(\frac{N}{2}-\frac{R}{2a}\right)\right]\right\}}_{\ln[(N/2)-(R/2a)]!}$$

$$-N\ln 2 \qquad (8.11)$$

In the next step, we rewrite the logarithms,

$$\ln\left(\frac{N}{2}\pm\frac{R}{2a}\right) = \ln\left[\frac{N}{2}\left(1\pm\frac{R}{Na}\right)\right] = \ln\frac{N}{2} + \ln\left(1\pm\frac{R}{Na}\right), \qquad (8.12)$$

where we have used the rule about logarithms that $\ln(AB) = \ln(A) + \ln(B)$. We can now make use of the Taylor expansion,

$$\ln\left(1\pm\frac{R}{Na}\right) \approx \pm\frac{R}{Na} - \frac{1}{2}\left(\pm\frac{R}{Na}\right)^2, \qquad (8.13)$$

which we substitute repeatedly in Equation 8.11. After a bit of algebra (which is left as an exercise for the reader), we arrive at the formula

$$\ln p(R;N) = \ln 2 - \frac{1}{2}\ln(2\pi N) - \frac{R^2}{2Na^2}. \qquad (8.14)$$

If we now exponentiate both sides of this equation, we find the coveted Gaussian distribution,

$$p(R;N) = \frac{2}{\sqrt{2\pi N}}\mathrm{e}^{-R^2/2Na^2}. \qquad (8.15)$$

Note that the derived approximate formula is a probability for values of R that come in multiples of $2a$, since R is either always an even or odd multiple of $2a$, depending on whether N is even or odd. To turn this into a probability distribution function, $P(R;N)$, such that $P(R;N)\,\mathrm{d}R$ is the probability that R falls within an interval of length $\mathrm{d}R$, all that remains is to divide out the result in Equation 8.15 by the density of integer R values per unit length, which is $1/2a$. This yields the result for the probability distribution function for the end-to-end distance of a freely jointed chain,

$$P(R;N) = \frac{1}{\sqrt{2\pi Na^2}}\mathrm{e}^{-R^2/2Na^2}, \qquad (8.16)$$

which we will make use of repeatedly throughout the book.

The result derived above is a special case of the so-called central limit theorem, which is arguably the most important result of probability theory. In a nutshell, it states that the probability distribution of $x_1 + x_2 + \cdots + x_N$, which is a sum of identically distributed independent random variables, is Gaussian in the limit of large N, as long as the mean and variance of each individual x_i are finite. Since the individual displacements of the random walker satisfy this condition, it immediately follows that for a large number of steps N, the total displacement R is Gaussian-distributed, with mean $\langle \mathbf{R} \rangle = 0$ and variance $\langle \mathbf{R}^2 \rangle = Na^2$. Note that this holds regardless of whether the walk is executed in one, two, or three dimensions, and independent of the allowed angles between subsequent steps of the walk, as long as each step is taken independently of the previous one.

We leave it as a homework problem to show that the Gaussian distribution of R for a one-dimensional walk given in Equation 8.16 indeed has the required mean and variance. Here we make use of this result to derive the large-N distribution for the end-to-end distance of a three-dimensional random walk. Since the mean is zero, the distribution is of the form

$$P(\mathbf{R}; N) = \mathcal{N} \mathrm{e}^{-\kappa R^2}, \tag{8.17}$$

where the parameters $\mathcal{N}$ and κ are to be determined from the two identities

$$\begin{aligned} &\int_{-\infty}^{+\infty}\int_{-\infty}^{+\infty}\int_{-\infty}^{+\infty} P(\mathbf{R}, N)\, \mathrm{d}^3 R = 1 \qquad &&\text{(normalization)},\\ &\int_{-\infty}^{+\infty}\int_{-\infty}^{+\infty}\int_{-\infty}^{+\infty} R^2 P(\mathbf{R}, N)\, \mathrm{d}^3 R = Na^2 \qquad &&\text{(variance)}. \end{aligned} \tag{8.18}$$

Since both integrands are functions of R^2, we can transform the volume integral in both cases to an integral over spherical shells of radius R to obtain

$$\begin{aligned} &\int_{0}^{+\infty} P(\mathbf{R}, N) 4\pi R^2\, \mathrm{d}R = 1 \qquad &&\text{(normalization)},\\ &\int_{0}^{+\infty} R^2 P(\mathbf{R}, N) 4\pi R^2\, \mathrm{d}R = Na^2 \qquad &&\text{(variance)}. \end{aligned} \tag{8.19}$$

To compute the integrals in the above equations, we make use of the Gaussian integral formulas

$$\begin{aligned} &\int_{0}^{+\infty} 4\pi \mathcal{N} R^2 \mathrm{e}^{-\kappa R^2}\, \mathrm{d}R = 4\pi \mathcal{N} \frac{1}{4}\sqrt{\frac{\pi}{\kappa^3}} = 1,\\ &\int_{0}^{+\infty} 4\pi \mathcal{N} R^4 \mathrm{e}^{-\kappa R^2}\, \mathrm{d}R = 4\pi \mathcal{N} \frac{3}{8}\sqrt{\frac{\pi}{\kappa^5}} = Na^2. \end{aligned} \tag{8.20}$$

To compute κ, we can divide the second of Equations 8.20 by the first to give

$$\kappa = \frac{3}{2Na^2}. \tag{8.21}$$

Substituting this result into the integral in the first of Equations 8.20 gives us

$$\mathcal{N} = \left(\frac{\kappa}{\pi}\right)^{3/2} = \left(\frac{3}{2\pi Na^2}\right)^{3/2}, \tag{8.22}$$

the normalization constant. Putting this all together, we obtain the end-to-end distribution for a three-dimensional random walk with N Kuhn segments of length a:

$$P(\mathbf{R}; N) = \left(\frac{3}{2\pi Na^2}\right)^{3/2} \mathrm{e}^{-3R^2/2Na^2}. \tag{8.23}$$

Note that $P(\mathbf{R}; N)$ has units of inverse volume, or concentration, and has an intuitive interpretation as the concentration of one end of the random-walk polymer at position $\mathbf{R}$ in the vicinity of the other end. Furthermore, $P(\mathbf{R}; N)$ is sharply peaked at $\mathbf{R} = 0$, and this property underlies the elasticity of polymer chains. Namely, if you imagine stretching a polymer (say, the *E. coli* DNA) so that R is nonzero, then upon release it will quickly find itself in the $R \approx 0$ state solely by virtue of this being a much more likely state. Note that this is not the result of a physical force, such as the electric force, which is ultimately responsible for the elastic properties of crystals, but purely a result of statistics. As such, it is, like the case of the pressure of an ideal gas, another example of an entropic force.

Estimate: End-to-End Probability for the *E. coli* genome

One interesting application of these ideas that will be explored more throughout the chapter is to the structure of chromosomal DNA. The circular DNA associated with an *E. coli* cell is roughly 5 million base pairs long. An open DNA chain of the same size can be modeled as a random walk of roughly $N = 15{,}000$ steps since the Kuhn length (the length of the "rigid" segments in the chain model) for bare DNA is roughly 300 bp. The probability that the end-to-end distance is zero for a one-dimensional walk of this many steps can be estimated from Equation 8.15 and is 7×10^{-3}. The probability that $R = 500a$ is 2×10^{-6}, while for $R = 1000a$ the probability drops all the way down to 2×10^{-17}. As discussed above, this overwhelming probability that R is close to zero is responsible for the elastic response of polymer chains due to an applied load.

The Persistence Length Is a Measure of the Length Scale Over Which a Polymer Remains Roughly Straight

With the random walk model in hand, we can describe the structure of long polymers, whose contour length L is much larger than the persistence length ξ_p, which is the length over which the polymer is essentially straight. In particular, the persistence length is the scale over which the tangent–tangent correlation function decays along the chain. Figure 8.5 conveys this idea in the case of DNA by illustrating the length scale over which the genomic DNA of a bacterium meanders.

To see this idea more clearly, we imagine a polymer as a curve in three-dimensional space. At each point along that curve, we can draw a tangent vector that points along the polymer at that point. As a result of thermal fluctuations, the polymer meanders in space and the persistence length is the length scale over which "memory" of the tangent vector is lost. From a mathematical perspective, we can write the tangent–tangent correlation function as $\langle \mathbf{t}(s) \cdot \mathbf{t}(u) \rangle$, where $\mathbf{t}(s)$ is the tangent vector evaluated at a distance s along the polymer and the notation $\langle \cdots \rangle$ is an instruction to average over all the configurations. The persistence length determines the scale over which correlations in tangent vectors decay through the equation

$$\langle \mathbf{t}(s) \cdot \mathbf{t}(u) \rangle = e^{-|s-u|/\xi_p}. \tag{8.24}$$

In Chapter 10, we derive this equation in the context of a model where the polymer is thought of as a long and thin elastic beam. Furthermore, we note that Equation 8.24 is not universally valid. For example,

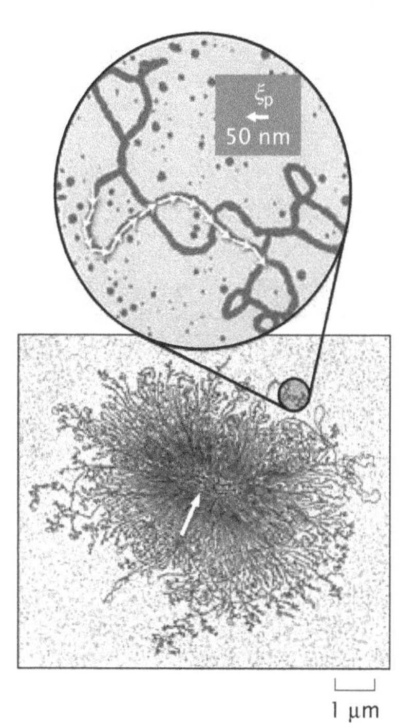

Figure 8.5: Illustration of the spatial extent of a bacterial genome that has escaped the bacterial cell. The expanded region in the figure shows a small segment of the DNA and has a series of arrows on the DNA, each of which has a length equal to the persistence length in order to give a sense of the scale over which the DNA is stiff. (Adapted from an original by Ruth Kavenoff.)

if the tangents are kept fixed and equal at the ends of the polymer, say by laser tweezers, then $\langle \mathbf{t}(0) \cdot \mathbf{t}(s) \rangle$ will decay at first, but as s approaches the contour length of the polymer, L, it will necessarily increase, since $\mathbf{t}(0) \cdot \mathbf{t}(L) = 1$. Other constraints on the polymer, such as confinement by the cell wall, will also lead to deviations from Equation 8.24. Still, for small enough separations $|s - u|$, the exponential law is expected to hold.

A good example of a long flexible polymer is provided by genomic DNA of viruses such as λ-phage, with a contour length of 16.6 μm. This should be compared with the persistence length $\xi_p \approx 50$ nm of DNA at room temperature and solvent conditions typical of the cellular environment. Since the persistence length is the length over which the tangent vectors to the polymer backbone become uncorrelated, we can think of the polymer as consisting of $N \sim L/\xi_p$ connected links that take random orientations with respect to each other. This is the logic that gives rise to the *freely jointed chain* model (essentially the random walk picture undertaken in the previous section). As already described, in the freely jointed chain model, polymer conformations are random walks of N steps. The length of the step is the *Kuhn length*, which is roughly equal to the persistence length. As promised in the earlier discussion, we now establish the relation between the persistence length and the Kuhn length invoked in the random walk model. To make a more precise determination of the Kuhn length, we calculate the mean-squared end-to-end distance of an elastic beam undergoing thermal fluctuations, and compare it with the same quantity obtained for the freely jointed chain. The end-to-end vector $\mathbf{R}$ of a beam can be expressed in terms of the tangent vector $\mathbf{t}(s)$:

$$\mathbf{R} = \int_0^L \mathrm{d}s \, \mathbf{t}(s). \tag{8.25}$$

As a result, we can write

$$\langle \mathbf{R}^2 \rangle = \left\langle \int_0^L \mathrm{d}s \, \mathbf{t}(s) \cdot \int_0^L \mathrm{d}u \, \mathbf{t}(u) \right\rangle, \tag{8.26}$$

where $\langle \cdots \rangle$ is an average over all polymer configurations. Using the average of the tangent–tangent correlation function, Equation 8.24, we find

$$\langle \mathbf{R}^2 \rangle = 2 \int_0^L \mathrm{d}s \int_s^L \mathrm{d}u \, \mathrm{e}^{-(u-s)/\xi_p}. \tag{8.27}$$

The above result is obtained by splitting up the integration over the $L \times L$ box in s–u space into integrals over the two triangles, one with $s < u$ and the other with $s > u$, which give equal contributions (thus the factor of 2). In the limit $L \gg \xi_p$ we are considering here, we have

$$\langle \mathbf{R}^2 \rangle \approx 2 \int_0^L \mathrm{d}s \int_0^\infty \mathrm{d}x \, \mathrm{e}^{-x/\xi_p} = 2L\xi_p. \tag{8.28}$$

To obtain this result, we made a change of variables $x = u - s$ in the second integral and then replaced the upper bound of integration $L - s$ by ∞, which is justified in the $L \gg \xi_p$ limit. Comparing the above formula with the result that follows from the random walk model, Equation 8.4, $\langle \mathbf{R}^2 \rangle = aL$, we see that Kuhn length a is twice the persistence length, $a = 2\xi_p$. In rewriting the random walk result, we made use of the relation between the length of the walk and the number of Kuhn segments, $L = Na$. We are now prepared to make estimates of the physical size of genomes in solution.

8.2.2 How Big Is a Genome?

A simple estimate of the size of a polymer in solution can be obtained using the end-to-end distance,

$$\sqrt{\langle R^2 \rangle} = \sqrt{2L\xi_p}. \tag{8.29}$$

The radius of gyration is perhaps a more precise measure of the polymer size and is defined through the expression

$$\langle R_G^2 \rangle = \frac{1}{N}\sum_{i=1}^{N}\langle(\mathbf{R}_i - \mathbf{R}_{CM})^2\rangle. \tag{8.30}$$

Roughly speaking, it measures the average distance between the monomers and the center of mass of the polymer. The center of mass is defined as

$$\mathbf{R}_{CM} = \frac{1}{N}\sum_{i=1}^{N}\mathbf{R}_i. \tag{8.31}$$

With this definition of the radius of gyration in hand, a simple relation between radius of gyration, contour length L, and persistence length ξ_p can be written as (proven by the reader in the problems at the end of the chapter)

$$\sqrt{\langle R_G^2 \rangle} = \sqrt{\frac{L\xi_p}{3}}. \tag{8.32}$$

We may write this result in an alternative form in terms of the number of base pairs in the genome of interest by noting that $L \approx 0.34\,\text{nm} \times N_{bp}$, and hence

$$\sqrt{\langle R_G^2 \rangle} \approx \tfrac{1}{3}\sqrt{N_{bp}\xi_p}\ \text{nm}. \tag{8.33}$$

This relation between the radius of gyration of DNA in solution and the number of base pairs is plotted in Figure 8.6.

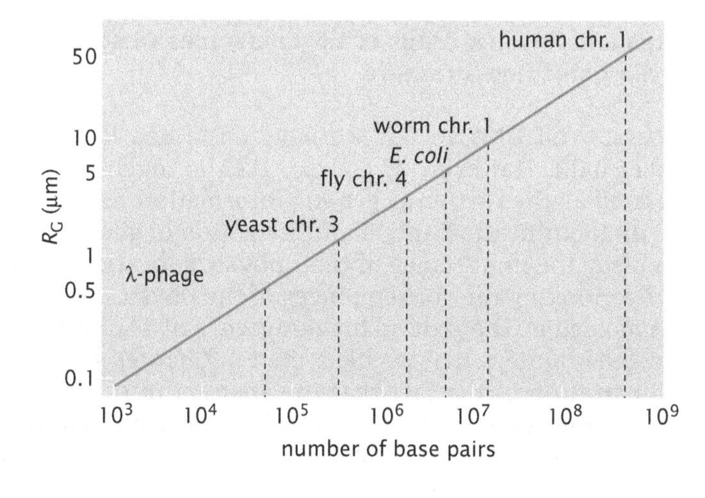

Figure 8.6: Size of genomic DNA in solution. Plot of the average size of a DNA molecule in solution as a function of the number of base pairs using the random walk model. The labels correspond to particular chromosomes from viruses, bacteria, yeast, flies, worms, and humans.

ESTIMATE

Estimate: The Size of Viral and Bacterial Genomes One application of ideas like those described above in the setting of biological electron microscopy is to images of viruses and cells that have ruptured and are thus surrounded by the DNA debris from their genome. We already mentioned in conjunction with Figure 1.16 (p. 28) that the appearance of DNA in electron microscopy images can be used as the basis of an estimate of genome length. A second example is shown in Figure 8.5, where it is seen that the DNA adopts a configuration in solution that is much larger than the configuration it has when packed inside of the virus or bacterium. To develop intuition for what is seen in such images, we exploit Equation 8.32 to formulate an estimate of the size of the DNA. Consider Figure 1.16, which shows bacteriophage T2. As seen in the figure, the viral genome has leaked from what is apparently a ruptured capsid and we will assume that this DNA in solution has adopted an equilibrium configuration. The genomes of T2 and T4 are very similar, with a genome length of roughly 150 kb. Recalling that the persistence length is $\xi_p \approx 50\,\text{nm}$, Equation 8.33 tells us that the mean size of the DNA seen in Figure 1.16 is $\sqrt{\langle R_G^2 \rangle} = (1/3)\sqrt{150 \times 10^3 \times 50}\,\text{nm} \approx 0.9\,\mu\text{m}$. This result is comparable to though larger than the length scale of the exploded DNA seen in Figure 1.16. Given the crudeness of the model and, probably more importantly, the fact that the DNA seems to be constrained via links to the capsid itself, this analysis provides a satisfactory first approximation to the structures seen in electron microscopy.

These same arguments can be invoked again to coach our intuition concerning the size of the DNA cloud surrounding a bacterium that has lost its DNA. In this case, the genome length is substantially larger than that of the T2 phage, namely, $N_{bp} \approx 4.6 \times 10^6$ base pairs. Once again invoking Equation 8.33 tells us that the mean size of the DNA seen in Figure 8.5 is $\sqrt{\langle R_G^2 \rangle} \approx 5\,\mu\text{m}$. As with the phage calculation, the random walk calculation should be seen as an overestimate, since the bacterial genome is circular and the DNA is clearly forced to return to the bacterium repeatedly, inhibiting the structure from adopting a fully expanded configuration.

8.2.3 The Geography of Chromosomes

Genetic Maps and Physical Maps of Chromosomes Describe Different Aspects of Chromosome Structure

In our discussion of DNA so far, we have described it as a featureless polymer chain. However, of course, DNA is much better known and appreciated as the carrier of genetic information. Classical genetics focused on identification and characterization of genes as abstract entities, ignoring the importance of their physical location on chromosomes and overlooking the consequences of the physical nature of the carrier DNA molecule. The ground breaking work of Thomas Hunt Morgan and his gene hunters that we described in Chapter 4 was an early and vivid illustration of the fact that the abstract informational entities known as genes exist with concrete physical relationships to one another. As we have learned more about the regulation and activity of genes, it has become more and more clear that the physical location and dynamic properties of the DNA molecule that carries them are critical components of their biological activity. For example, Morgan's mapping strategy relied on measuring the frequency of recombination between two or more genes. The physical process of recombination requires that two homologous DNA molecules be mobile within a nucleus such that they can physically encounter one another with a measurable frequency. Recombinations do not seem to occur in all nuclei. In the fruit fly, chromosomes are able to recombine in meiosis during oogenesis in the female germline, but not during spermatogenesis in the male germline. Why is it that sometimes DNA segments are able to physically encounter one another and sometimes they are not? What determines the probability of such encounters? These issues in polymer conformations set physical limits on genetic events ranging from transformation and transduction in bacterial cells to the generation of diverse antibodies in the immune system of mammals.

Different Structural Models of Chromatin Are Characterized by the Linear Packing Density of DNA

One of the themes that we will keep revisiting is the question of DNA packing. In eukaryotic cells, DNA is condensed into chromatin fibers. The basic unit of chromatin is the nucleosome. How nucleosomes are packaged into chromatin depends on whether the cell is dividing or not. During interphase, the cell is actively transcribing genes, and the chromosomes are not as condensed as during mitosis when the two copies of the complete genome need to be equally divided among the two daughter cells.

One measure of the degree of DNA packaging into chromosomes is the linear density of chromatin, ν, which specifies the number of base pairs of DNA in a nanometer of chromatin fiber. For the 30 nm fiber, shown in Figure 8.7(A), $\nu \approx 100\,\text{bp/nm}$, while for the 10 nm fiber the packing density is about an order of magnitude smaller. A simple estimate of ν can be made based on the micrograph in Figure 8.7(B), which shows individual nucleosomes along the 10 nm fiber. We see that there are on average two nucleosomes for every 50 nm of fiber. We assume there are 200 bp per nucleosome (150 bp wound around the histones plus 50 bp of linker DNA), and therefore $\nu \approx 2 \times 200\,\text{bp}/50\,\text{nm} = 8\,\text{bp/nm}$. For comparison, for metaphase chromosomes, $\nu \approx 30{,}000\,\text{bp/nm}$.

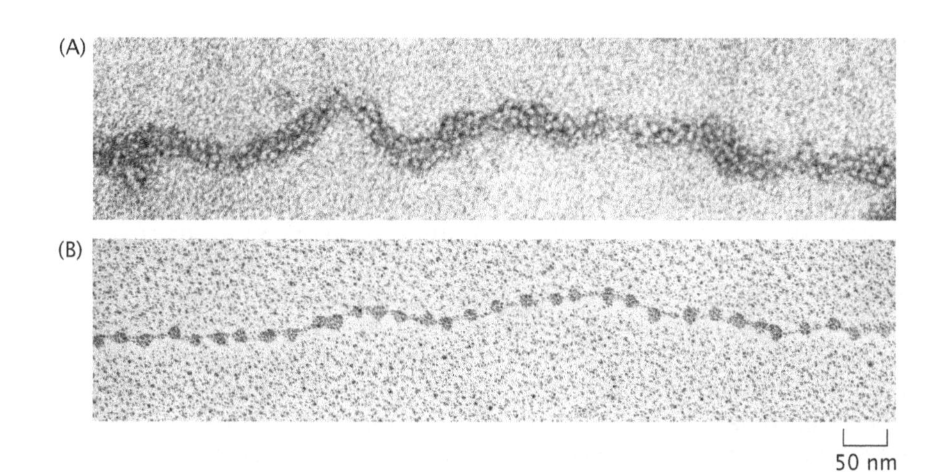

Figure 8.7: Electron microscopy images of chromatin. (A) Chromatin extracted from an interphase nucleus appears as a 30 nm thick fiber. (B) Stretching out a part of the chromatin reveals the "beads-on-a-string" structure of the 10 nm fiber, where each bead is an individual nucleosome. (A, courtesy of Barbara A. Hamkalo; B, courtesy of Victoria Foe.)

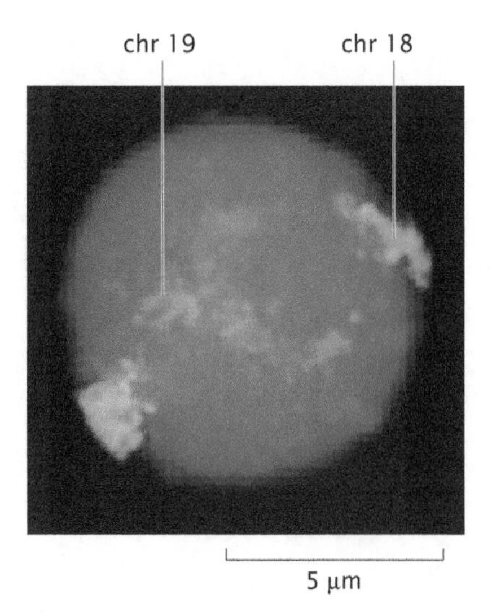

Figure 8.8: Fluorescently stained chromosomes 18 (green) and 19 (red) in the nucleus of a human cell. The two copies of chromosome 18 typically assume positions near the periphery of the nucleus, while the two copies of chromosome 19 are located closer to the center. (From J. A. Croft et al., *J. Cell Biol.* 145:1119, 1999.)

Spatial Organization of Chromosomes Shows Elements of Both Randomness and Order

It used to be believed that interphase chromosomes were randomly distributed within the cell nucleus resembling a bowl of spaghetti. Contrary to this view, there is mounting evidence from experiments with fluorescently tagged chromosomes that the spatial organization of genes in the cell is ordered, as depicted in Figure 8.8. These experiments have put forward the notion of chromosome territories, whereby individual chromosomes and particular genetic loci are always found in the same region of the nucleus. The existence of chromosome territories raises a number of questions about how gene expression and pairing interactions of genes (such as during recombination) are orchestrated in space and time.

The observation that interphase chromosomes are segregated would not be surprising if we were dealing with a polymer system that was very dilute, but in a dense situation free polymers in solution will interpenetrate each other. Simple estimates can be made for the density of chromatin within the nucleus, and they typically lead to the conclusion that the expected equilibrium state of chromosomes should be that of a dense polymer system. The fact that segregation is nonetheless observed points to the existence of mechanisms beyond polymer chain entropy and confinement that affect the spatial distribution of chromosomes. We will examine chromosome tethering as one such mechanism. Tethering scenarios posit that chromosomes have particular physical locations because they are held there by tethering molecules. Possible tethering scenarios are shown in Figure 8.9.

Estimate: Chromosome Packing in the Yeast Nucleus

Using polymer physics, here we examine the question of whether chromosomes in yeast are more likely to resemble spaghetti mixed in a bowl or segregated blobs not unlike meatballs. The yeast cell has 16 chromosomes in its nucleus. The diameter of the interphase nucleus is about $2\,\mu\text{m}$. The chromosome size varies between 230 kb and 1500 kb, with a total genome size of 12 Mb. This gives a mean density of $c = 12\,\text{Mb}/[(4\pi/3) \times (1\,\mu\text{m})^3] \approx 3\,\text{Mb}/\mu\text{m}^3$. We now compare this density with the density of a typical yeast chromosome released from the confines of the cell nucleus. If we adopt the random walk model of a polymer to describe chromatin free in solution, this density can be estimated as $c^* = N_{\text{bp}}/(4\pi R_G^3/3)$, where N_{bp} is the chromosome size in base pairs and R_G is the radius of gyration of the polymer. If we take an average size of a yeast chromosome to be $12\,\text{Mb}/16 = 750\,\text{kb}$ and a packing density of 8 bp/nm, the length of this polymer is $750\,\text{kb}/(8\,\text{bp/nm}) = 94\,\mu\text{m}$. Using the *in vitro* measured value of the persistence length for a 10 nm fiber, $\xi_p = 30\,\text{nm}$, the estimate for the radius of gyration is $R_G = 0.97\,\mu\text{m}$. This then leads to a density for a "free" chromosome $c^* = 750\,\text{kb}/[(4\pi/3) \times (0.97\,\mu\text{m})^3] \approx 200\,\text{kb}/\mu\text{m}^3$, which is about 10 times smaller than the density of chromosomes in the nucleus. The same qualitative conclusion is reached assuming a 30 nm fiber model for the chromosomes. Namely, using a packing density of 100 bp/nm and the reported persistence length of 200 nm, an average chromosome has a density of

$c^* \approx 500\,\mathrm{kb}/\mu\mathrm{m}^3$. This indicates that the chromosomes in the yeast nucleus should typically be found in an entangled melt-like configuration since there is not enough room for them to adopt their preferred configurations without overlap. The fact that yeast chromosomes are segregated, with each chromosome taking up a well-defined region of the nucleus, indicates the need for a specific mechanism for segregation, such as tethering to the nuclear periphery, as shown in Figure 8.9.

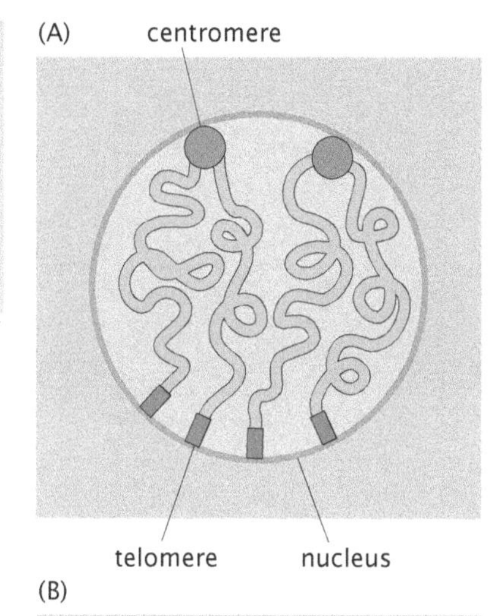

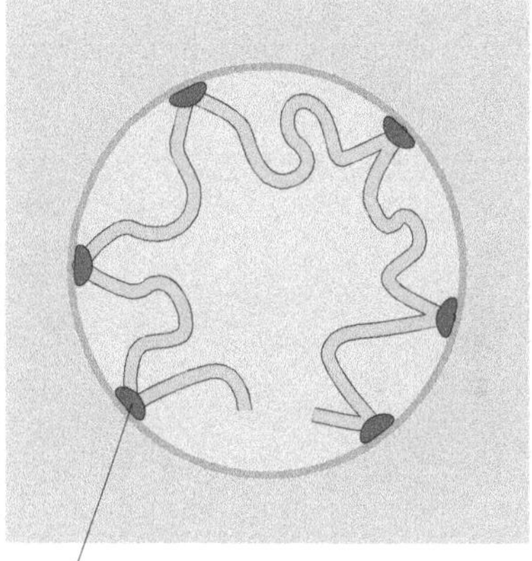

Figure 8.9: Cartoon representation of possible tethering scenarios of interphase chromosomes. (A) Tethering at the centromere and the two telomeres at the nuclear periphery and (B) tethering at intermediate locations. (Adapted from W. F. Marshall, *Curr. Biol.* 12:R185, 2002.)

Chromosomes Are Tethered at Different Locations

One experimental trick that has made it possible to examine chromosome geography is the use of repeated DNA-binding sites that are the target of particular fluorescently labeled proteins. Conceptually, the experiment can be designed by having two distinct sets of DNA-binding sites that are separated by a known *genomic* distance. Then, by measuring the *physical* distance between these binding sites in space as revealed by where the colored spots appear in a fluorescence image, it is possible to map out the spatial distribution of different sites on the genome.

Experiments that utilize fluorescence *in situ* hybridization, or *lacO* arrays inserted into the chromosomes and labeled with GFP-fused Lac repressors, can yield detailed information about the distribution of distances between chromosomal loci. Note that our use of the word "distance" depends upon context; in some cases we will be referring to the scalar distance between two points and in other cases to the displacement vector connecting them. We will pass freely back and forth between these two cases, and their relation is explored in the problems at the end of the chapter. In the absence of tethering (or if there is a single tether present) a random walk model of chromatin predicts a Gaussian distribution of distances $\mathbf{r}$ between the two fluorescent markers,

$$P(\mathbf{r}) = \left(\frac{3}{2\pi Na^2}\right)^{3/2} \mathrm{e}^{-3\mathbf{r}^2/2Na^2}. \tag{8.34}$$

Here $a = 2\xi_p$ is the Kuhn or segment length of the polymer and N is the total number of Kuhn segments between the two markers.

The simplest tethering configuration that leads to a distance distribution different than that described above is one with two tethers, as shown in Figure 8.10. One tether is assumed to coincide with the location of one of the two fluorescent markers, and the other tether is at a position $\mathbf{R}$ between the two markers. This configuration of markers and tethers leads to a displaced Gaussian distribution of distances

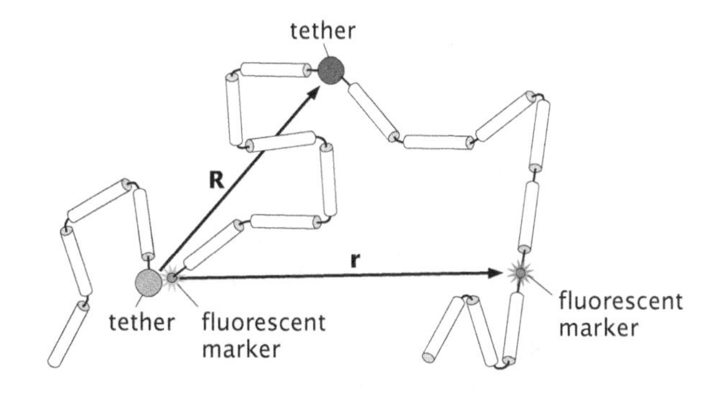

Figure 8.10: Simple configuration of a tethered chromosome. The two tethers are at fixed locations in space, and the second tether is at position **R** with respect to the first. The distribution of distances between the two fluorescent markers, one being at the same position on the chromosome as the tether, is a displaced Gaussian.

r between the markers,

$$P(\mathbf{r}) = \left(\frac{3}{2\pi N' a^2}\right)^{3/2} e^{-3(\mathbf{r}-\mathbf{R})^2/2N'a^2}, \tag{8.35}$$

where N' is now the number of Kuhn segments between the second tether and the second marker. This formula follows simply from Equation 8.34 when applied to the distribution of distances $\mathbf{r} - \mathbf{R}$ between the second tether and the second marker. It is interesting to note that mathematical properties of Gaussian distributions, like the one that says that a convolution of two Gaussian distributions is a Gaussian distribution, dictate that *any* tethering configuration will result in a displaced Gaussian distribution of distances.

The implicit assumption we have made in writing Equations 8.34 and 8.35 is that chromosome configurations can be described by random walks. In light of the dense packing of chromosomes in cells, this might seem like an overly zealous use of a simple physical model. However, as we demonstrate using several examples later in this section, this model captures key features of experimental data on chromosomes and, more importantly, it makes falsifiable predictions suggesting new directions for experimentation. As a result, this model is a good starting point for quantitative investigations of chromosome geography. This idea is further bolstered by the Flory theorem, which states that for dense polymer systems, such as chromosomes confined to cells, distributions of distances between monomers are described by random walk statistics.

The contour length of the chromosome between the two tagged loci, Na, can be expressed in terms of the genomic distance between the two fluorescent markers as $Na = N_{\mathrm{bp}}/\nu$, where ν is the linear packing density of DNA in chromatin. For example, two genomic loci $N_{\mathrm{bp}} = 100\,\mathrm{kb}$ apart would be separated by a 30 nm fiber, which is $100\,\mathrm{kb}/(100\,\mathrm{bp/nm}) = 1\,\mu\mathrm{m}$ in contour length. Assuming that the chromatin structure is that of a 10 nm fiber the contour distance along the fiber between the loci would be 10 times as large given the 10 times smaller packing density.

The end-to-end distribution function for a random walk polymer is determined by a single parameter Na^2, the mean end-to-end distance squared. Since the contour length $Na = N_{\mathrm{bp}}/\nu$, the mean end-to-end distance squared can also be written as $\langle R^2 \rangle = N_{\mathrm{bp}} a/\nu$. Therefore the material parameter that characterizes the random walk model of chromosomes is the ratio of the Kuhn length and the packing density. This parameter can be determined from measurements of the average distance squared between two labeled regions of the chromosome as a function of their genomic distance. The results of such a measurement on human chromosome 4 are shown in Figure 8.11. The fit to the data yields an estimate of $a/\nu = 2\,\mathrm{nm}^2/\mathrm{bp}$, which is nothing but the initial slope of the linear portion of the data. The fact that the data level off at large genomic distance can be attributed to the effect of chromosome confinement within the cell nucleus. Below we analyze this confining effect using a random walk model for chromosome configurations in the bacterium *Vibrio cholerae*.

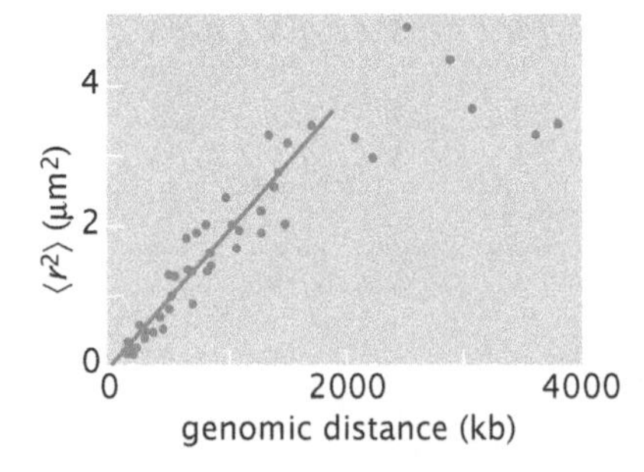

Figure 8.11: Physical distance between two fluorescently labeled loci on human chromosome 4 as a function of the genomic distance. The physical distance is measured in terms of the average squared distance between the two labels (dots). The curve corresponds to a linear fit as discussed in the text. (Adapted from G. van den Engh et al., *Science* 257:1410, 1992.)

With a measurement of the chromatin material parameter a/ν in hand, we can compute the expected probability distribution of distances between fluorescently tagged loci on the chromosome. Typically, due to random orientations of cells in the microscope, experiments with tagged chromosomes only yield information about the magnitude r of the distance vector $\mathbf{r}$ between the two marked spots on the chromosome. Probability distributions for this quantity follow from Equations 8.34 and 8.35 by integrating out the angular variables θ and ϕ associated with the vector $\mathbf{r}$. This procedure yields

$$P(r) = \left(\frac{3}{2\pi Na^2}\right)^{3/2} 4\pi r^2 e^{-3r^2/2Na^2}, \tag{8.36}$$

for the free-polymer case, and

$$P(r) = \left(\frac{3}{2\pi N'a^2}\right)^{1/2} \frac{r}{R}\left(e^{-3(r-R)^2/2N'a^2} - e^{-3(r+R)^2/2N'a^2}\right) \tag{8.37}$$

when the polymer is tethered. Note that tethering gives a different functional form for the distribution of distances. This provides us with a mathematical tool with which to detect tethering of chromosomes in cells.

Measurement of the distribution of distances between tagged regions on yeast chromosome III suggests that this difference in distributions can be observed *in vivo*. In Figure 8.12, we show the distance distribution measured between two fluorescent tags, one placed near the so-called HML region of chromosome III of budding yeast and the other on the spindle pole body, which is at a fixed location on the nuclear periphery and essentially marks the location of the centromere. The measured distribution is poorly fitted by the free-polymer formula, Equation 8.36, while the tethered-polymer formula, Equation 8.37 does the job well.

The fit to the tethered-polymer distribution yields two quantities that characterize the model, namely, the mean-squared distance between the tether and the fluorescent marker at HML, $N'a^2 = 0.5\,\mu\text{m}^2$, and $R \approx 0.9\,\mu\text{m}$, the distance from the spindle pole body to the tethering point. Note that in order to compute the genomic location of the putative tethering point, we need the quantity a/v that characterizes chromatin structure. For that, measurements like those leading to Figure 8.11 for human chromosome 4 are needed.

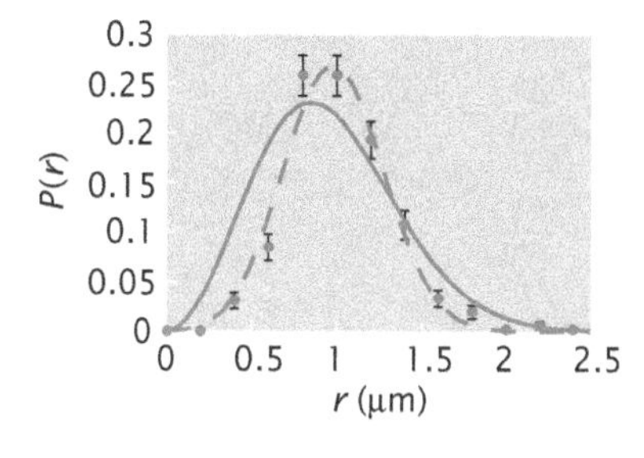

Figure 8.12: Statistics of yeast chromosome III. Distribution of distances between two fluorescent tags placed in proximity of the centromere and the HML region on yeast chromosome III. These two regions are separated by approximately 100 kb in genomic distance. The full line is a fit to the free-polymer distance distribution, Equation 8.36, while the dashed line is a fit to the tethered-polymer formula, Equation 8.37. (Courtesy of S. Gordon-Messer, J. Haber, and D. Bressan.)

Chromosome Territories Have Been Observed in Bacterial Cells

Bacterial chromosomes used to be thought of as unstructured and random. This view has been seriously challenged by experiments that utilize fluorescent markers placed at different genomic locations, as shown in Figure 8.13. In this experiment, 112 different mutants of *Caulobacter crescentus* were created with fluorescent tags placed at 112 different locations covering the length of its circular chromosome. Measurements of the average position of the marker along the length of the cell revealed a linear relationship between the genomic distance from the origin of replication and the physical distance away from the pole of the bacterium. This is not to be expected assuming a simple model of the 4 Mb circular chromosome as a polymer loop confined to the cell.

Estimate: Chromosome Organization in *C. crescentus*
Another measure of the organization of the chromosome in *C. crescentus* is provided by the width of the distribution of positions of the marked regions. As shown in Figure 8.13, the standard deviation of the position is independent of genomic distance from the origin of replication, and is approximately $0.2\,\mu\text{m}$ (cell length $L \approx 2\,\mu\text{m}$). We can rationalize this measurement within a simple model where the chromosome is partitioned into loops. This can be effected by proteins that make contact between different locations on the chromosome (H-NS

ESTIMATE

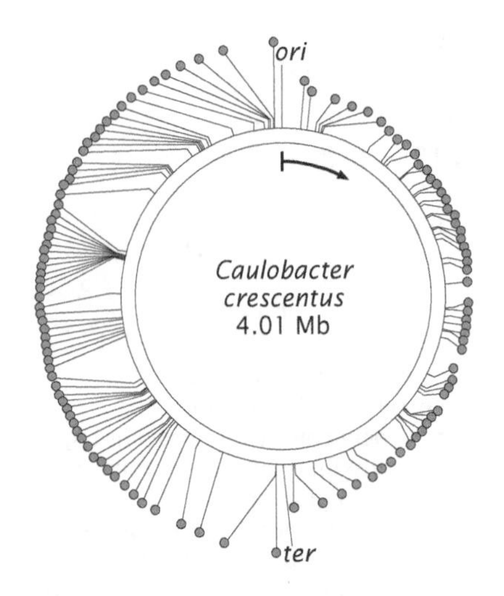

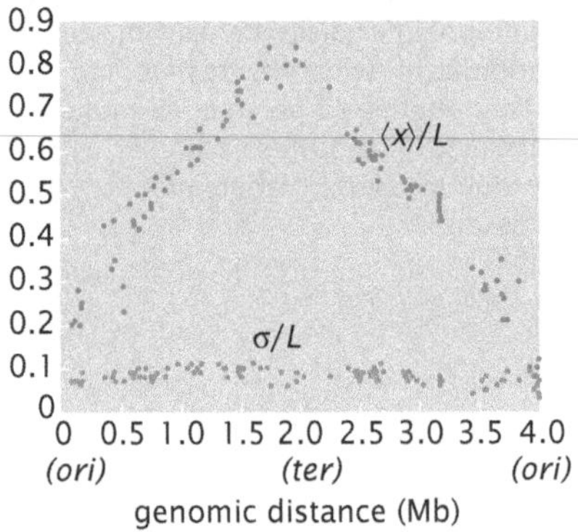

Figure 8.13: Chromosome geography in *C. crescentus.* Average positions (x/L) and the standard deviation ($\Delta x/L$) of the position along the long axis of the cell are plotted for 112 different fluorescently tagged locations along the chromosome of *C. crescentus.* The mean locations of the fluorescent tags and their corresponding standard deviations are shown on the diagram. (Adapted from P. H. Viollier et al., *Proc. Natl Acad. Sci. USA* 101:9257, 2004.)

is one example). To estimate the size of a loop, we assume that the observed dispersion of the position is due to the random walk nature of the loop. Since the mean of the square of the end-to-end distance, $r^2 = x^2 + y^2 + z^2$, is Na^2, the mean of x^2 is three times less (assuming a spherically symmetric distribution), or $Na^2/3$. Using the relation between genomic distance and the mean distance squared, $Na^2 = N_{\text{bp}}a/\nu$, and assuming that the chromosome has the same Kuhn length ($a = 100\,\text{nm}$) and packing density ($\nu = 3\,\text{bp/nm}$) as naked DNA, we arrive at an estimate $(0.2\,\mu\text{m})^2 = Na^2/3 = N_{\text{bp}}/3(100/3)\,\text{nm}^2/\text{bp}$, $N_{\text{bp}} \approx 4\,\text{kb}$, which means that the loop should be 8 kb or less. (A more careful analysis would take into account the closed nature of a loop, yielding an estimate that is higher by a factor of 2.) This correlates well with other measurements of topological domains in bacterial chromosomes, which find them to be roughly 10 kb in size.

Chromosome Territories in *Vibrio cholerae* Can Be Explored Using Models of Polymer Confinement and Tethering

Another experiment placed fluorescent markers close to each of the two origins of replication on the two chromosomes of the bacterium *V. cholerae.* This bacterium has two chromosomes, roughly 3 Mb and 1 Mb in size. In this case, the position of the fluorescent marker along the length of the cell (x) and perpendicular to it (y) were both measured. The distribution of x and y are shown in Figure 8.14 for the origin of replication for the larger of the two chromosomes. For comparison, the length of the cell is about $3.2\,\mu\text{m}$, while its diameter is roughly $0.8\,\mu\text{m}$.

The width of the distribution of x positions is roughly half a micron, which is considerably less than the length of the cell. The distribution is centered around $x_0 = 0.6\,\mu\text{m}$, consistent with a tether located at this position in the cell, and is well described by a Gaussian, as expected for a random walk polymer that is unaffected by the presence of cell walls. By fitting the Gaussian distribution for the end-to-end distance of a simple one-dimensional random walk polymer, Equation 8.16,

$$P(x) = \sqrt{\frac{1}{2\pi Na^2}}\,e^{-(x-x_0)^2/2Na^2}, \tag{8.38}$$

we extract the parameter $Na^2 = 0.16\,\mu\text{m}^2$. Assuming once again the Kuhn length of bare DNA, $a = 0.1\,\mu\text{m}$, we conclude that the number of Kuhn segments between the fluorescent marker and the tethering point at $x_0 = 0.6\,\mu\text{m}$ is $N = 16$. Taking $\nu = 3\,\text{bp/nm}$, this gives a genomic distance of $16 \times 0.1\,\mu\text{m} \times 3\,\text{bp/nm} = 4.8\,\text{kb}$ to the tether. Therefore, the simple one-dimensional model of the chromosome is consistent with the existence of a tether at the genomic position roughly 5 kb away from the location of the fluorescent marker. This kind of distribution may also be consistent with other mechanisms for determining the positions of chromosomal loci in living cells, including the existence of multiple tethers.

The distribution of positions along the y-direction is spread over the width of the cell and is centered at zero. The latter is a consequence of the experimental procedure used to collect distance data from cells whose orientation along the azimuthal direction was

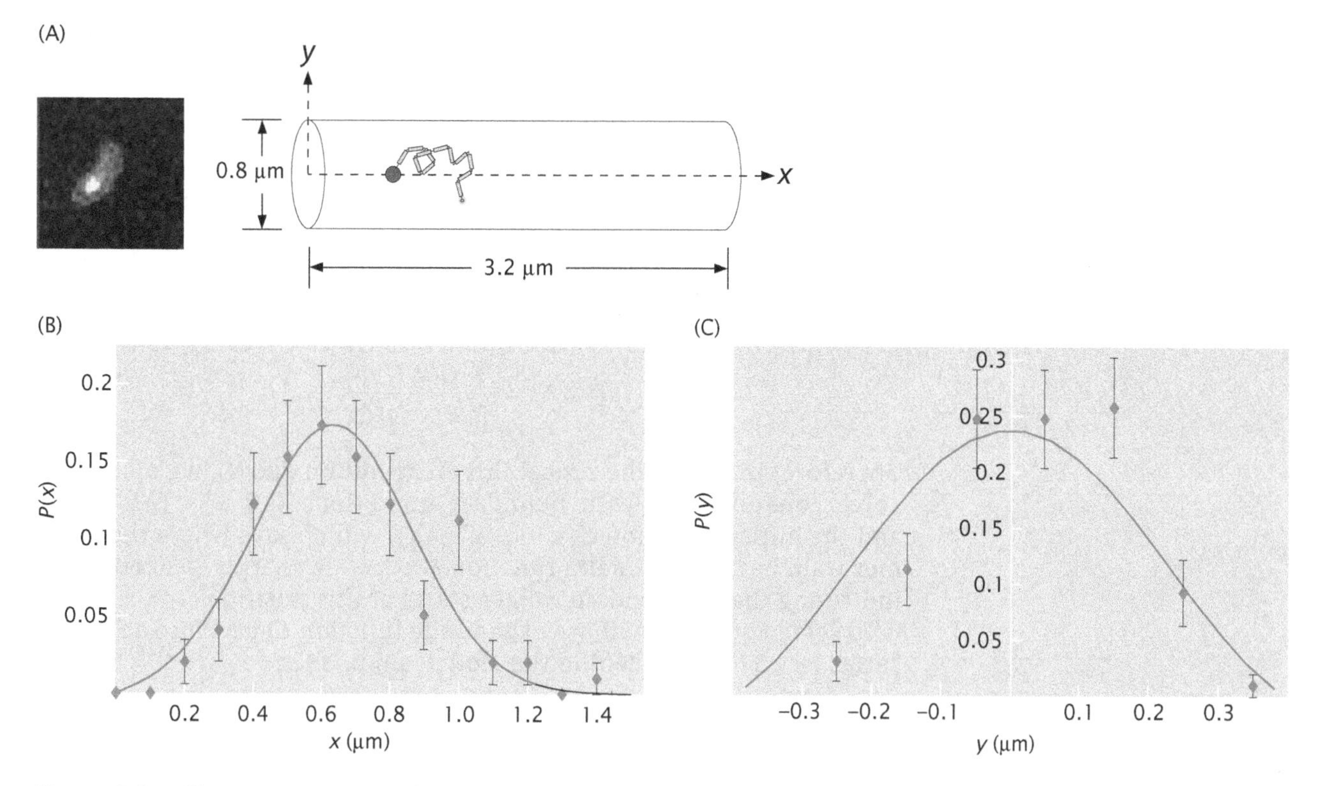

Figure 8.14: Chromosome position distributions *in vivo*. (A) The position of the fluorescently tagged origin of replication on the larger of the two *V. cholerae* chromosomes is measured along the long axis of the cell (x-direction) and perpendicular to it (y-direction). The cell can be modeled as a cylinder, while the distribution of x- and y-positions can be predicted using a model of a chromosome as a confined and tethered random walk polymer. (B, C) Measured distance distribution functions and comparison with theory. $P(x)$ is the Gaussian distribution characteristic of a free random walk polymer, while $P(y)$ is non-Gaussian due to effects of confinement by the cell walls. Calculation of $P(y)$ for a random walk polymer confined to a cylinder is left as a problem at the end of the chapter. (Courtesy of A. Fiebig, J. Schmit, and J. Theriot, unpublished.)

random. Furthermore, the distribution is not Gaussian, indicative of confinement by the cell walls. To develop quantitative intuition about confinement, we develop a model of a one-dimensional polymer made up of N segments, each of length a, tethered at position x_0 and confined to a cell of size L as shown in Figure 8.15. Our goal is to calculate the distribution of the end-to-end distance, $P(x; N)$.

To compute $P(x; N)$, we once again make use of the mapping to the random walk model in which polymer configurations are identified with trajectories of a random walker that has taken N steps starting at position x_0. As we are only interested in those random walks that stay within the cell, we impose absorbing boundary conditions. In other words, we demand that $P(x; N)$ vanishes for $x = 0$ and $x = L$ for any choice of N. This guarantees that any walk that crosses the boundary of the cell is excluded from the ensemble of allowed walks. The fraction of random walks that start at $x = x_0$ and end up at x without

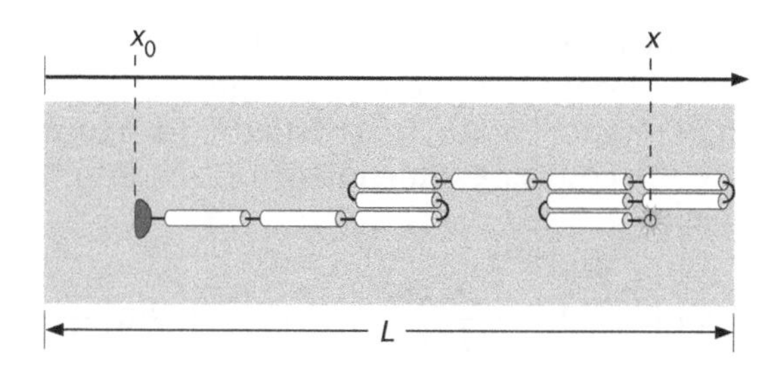

Figure 8.15: Simplified one-dimensional model of a chromosome confined to a cell of size L and tethered at position x_0. The model makes a prediction for the distribution of distances to the fluorescent marker, $P(x)$.

leaving the cell is then $G(x; N)$. This quantity satisfies the diffusion equation,

$$\frac{\partial G(x;N)}{\partial N} = \frac{a^2}{2}\frac{\partial^2 G(x;N)}{\partial x^2}. \tag{8.39}$$

This connection between random walks and diffusion leading to the above equation is explored in Chapter 13 (see the discussion on p. 518) and in the problems at the end of this chapter.

The probability that a walk that stays in the cell also ends up at position x is then

$$P(x;N) = \frac{G(x;N)}{\int_0^L G(x;N)\,\mathrm{d}x}. \tag{8.40}$$

Therefore, to obtain the probability distribution $P(x; N)$, we must first solve Equation 8.39 with boundary conditions $G(0; N) = G(L; N) = 0$ and the initial condition $G(x; 0) = \delta(x - x_0)$, which says where the polymer walk begins. The delta function $\delta(x - x_0)$ is sharply peaked at x_0, indicating that the random walker starts at this position.

To solve Equation 8.39, we expand the function $G(x; N)$ into a Fourier series (see The Math Behind the Models on p. 332),

$$G(x;N) = \sum_{n=1}^{\infty} A_n(N)\sin\left(\frac{n\pi}{L}x\right). \tag{8.41}$$

Note that every term in the sum satisfies the absorbing boundary condition. We still need to satisfy the initial condition and the differential equation itself.

The initial condition states

$$\delta(x - x_0) = \sum_{n=1}^{\infty} A_n(0)\sin\left(\frac{n\pi}{L}x\right), \tag{8.42}$$

and it needs to be solved for the constants $A_n(0)$. To do this, we multiply both sides by $\sin(m\pi x/L)$ and integrate the equation from 0 to L. The left-hand side gives $\sin(m\pi x_0/L)$, while the right-hand side is

$$\sum_{n=1}^{\infty} A_n(0)\int_0^L \sin\left(\frac{n\pi}{L}x\right)\sin\left(\frac{m\pi}{L}x\right)\mathrm{d}x = A_m(0)\frac{L}{2}, \tag{8.43}$$

where we have used the orthogonality property of sine functions given by

$$\int_0^L \sin\left(\frac{n\pi}{L}x\right)\sin\left(\frac{m\pi}{L}x\right)\mathrm{d}x = \delta_{n,m}\frac{L}{2}. \tag{8.44}$$

Putting the results of integrating the left- and right-hand sides of Equation 8.42 together, we find

$$A_m(0) = \frac{2}{L}\sin\left(\frac{m\pi}{L}x_0\right). \tag{8.45}$$

Now we turn to the differential equation itself. The question at hand is what should we choose for the coefficients $A_n(N)$ so that the diffusion equation, Equation 8.39, is satisfied? To figure this out, we simply substitute the Fourier expansion of $G(x; N)$ into the differential equation. This yields

$$\sum_{n=1}^{\infty}\frac{\mathrm{d}A_n(N)}{\mathrm{d}N}\sin\left(\frac{n\pi}{L}x\right) = -\frac{a^2}{2}\sum_{n=1}^{\infty} A_n(N)\left(\frac{n\pi}{L}\right)^2\sin\left(\frac{n\pi}{L}x\right). \tag{8.46}$$

Now we once again use the trick of multiplying both sides of this equation by $\sin(m\pi x/L)$ and integrating from 0 to L. Employing the orthogonality property this time yields a differential equation for the coefficient $A_m(N)$ given by

$$\frac{\mathrm{d}A_m(N)}{\mathrm{d}N} = -\frac{a^2}{2}\left(\frac{m\pi}{L}\right)^2 A_m(N). \tag{8.47}$$

The solution to this equation is an exponential function,

$$A_m(N) = A_m(0)\exp\left[-\left(\frac{m\pi}{L}\right)^2 \frac{a^2}{2}N\right], \tag{8.48}$$

where the coefficient $A_m(0)$ was determined above (Equation 8.45) from the initial condition.

Finally, the solution to Equation 8.39 that satisfies the initial condition that all walkers start at x_0 and the absorbing boundary conditions at the cell boundaries is

$$G(x;N) = \sum_{n=1}^{\infty} \frac{2}{L}\sin\left(\frac{n\pi}{L}x_0\right)\sin\left(\frac{n\pi}{L}x\right)\exp\left[-\left(\frac{n\pi}{L}\right)^2\frac{a^2}{2}N\right]. \tag{8.49}$$

To turn this quantity into the probability distribution for the end-to-end distance of a polymer confined in a cell, we make use of Equation 8.40, to yield

$$P(x;N) = \frac{1}{L}\frac{\sum_{n=1}^{\infty}\sin\left(\frac{n\pi}{L}x_0\right)\sin\left(\frac{n\pi}{L}x\right)\exp\left[-\left(\frac{n\pi}{L}\right)^2\frac{a^2}{2}N\right]}{\sum_{n=1}^{\infty}\sin\left(\frac{n\pi}{L}x_0\right)\frac{1}{n\pi}[1-\cos(n\pi)]\exp\left[-\left(\frac{n\pi}{L}\right)^2\frac{a^2}{2}N\right]}. \tag{8.50}$$

This probability distribution is plotted in Figure 8.16(A) for DNA ($a = 100$ nm) confined to a cell 2 μm in length, for DNA lengths ranging from 0.5 μm to 10 μm. Note that, for the shortest chain, the confining cell has no effect and the end-to-end distance distribution is a simple Gaussian function like that given in Equation 8.38. For the intermediate chain length, $Na = 2$ μm, the effect of the cell is to skew the distribution owing to the fact that the tethering point, $x_0 = 0.75$ μm, was chosen closer to the left cell boundary. Finally, for very long DNA lengths, the distribution is once again symmetric, with all memory of the tethering point lost.

The confined random walk model provides us with the quantitative intuition that allows us to conclude that the observed distribution of average positions of markers along the *C. crescentus* chromosome shown in Figure 8.13 is inconsistent with a model of a polymer confined to the cell interior and tethered only at the poles of the bacterium. This tethering configuration would lead to the average position for most markers (except ones close to the origin and terminus of replication, which are thought to be colocalized with the poles) being at the midpoint of the cell. Therefore, further constraints need to be imposed on the chromosome to establish the observed chromosome geography.

In Figure 8.16(B), we once again plot the end-to-end distance distribution using Equation 8.50, but this time for a DNA molecule that has a length of $Na = 1$ μm ($a = 100$ nm), tethered at the center of the confining box, for box lengths ranging from 1 μm to 3 μm. We note

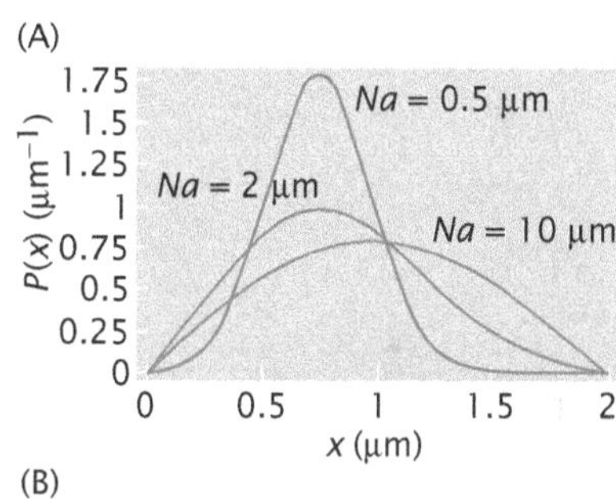

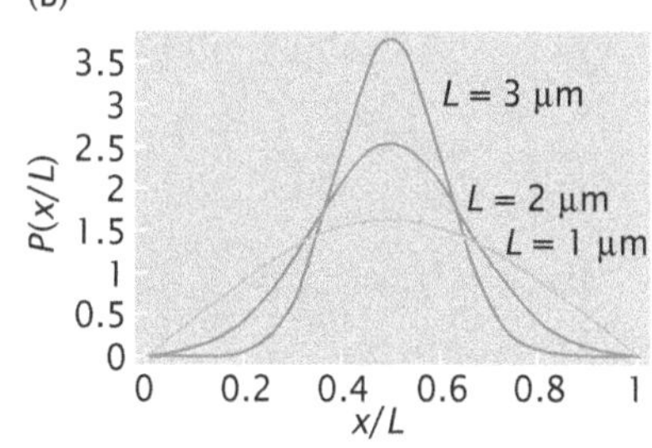

Figure 8.16: Distributions for confined polymers. (A) The distribution of distances to the fluorescent marker for the one-dimensional chromosome model for different contour lengths of the chromatin fiber between the tethering point (at $x_0 = 0.75$ μm) and the fluorescent marker. The cell size is $L = 2$ μm, and the packing density and Kuhn length are those of bare DNA. (B) Same as in (A), for a 1 μm long chromatin fiber confined to cells of different size and tethered in the middle of the cell.

that the effect of confinement sets in rather rapidly: there is little evidence for it in the largest box size, while for the smallest one the distribution is practically that of a very long polymer confined to a small box. This provides an explanation of the difference in the observed distance distributions in the x- and y-directions for the fluorescent markers placed on the *V. cholerae* chromosome as shown in Figure 8.14(C). We can check this assertion quantitatively by fitting the measured x-distribution to the derived formula. This gives two parameters, namely, the position of the assumed tether, x_0, and the size of the chain characterized by the quantity Na^2. With the quantity Na^2 in hand and assuming the y-position of the tether to be at $y = 0$ (this has little effect given the strong confinement in the y-direction, which, as remarked above, erases the effect of the tether position), we can simply plot the expected y-distribution and ask whether it matches the data. This comparison is shown in Figure 8.14, where we model the cell as a cylinder and take into account the fact that the experimentally measured y-position of the fluorescent marker is the projection of its radial distance onto the plane of the coverslip, on which the cells rest. The details of this calculation are left as a homework exercise.

The Math Behind the Models: Expanding in Sines and Cosines Throughout the book, we are often invited to consider functions that are defined on the interval between 0 and L. A useful property of such functions that we employ over and over again is that they can be expanded into a Fourier series given by

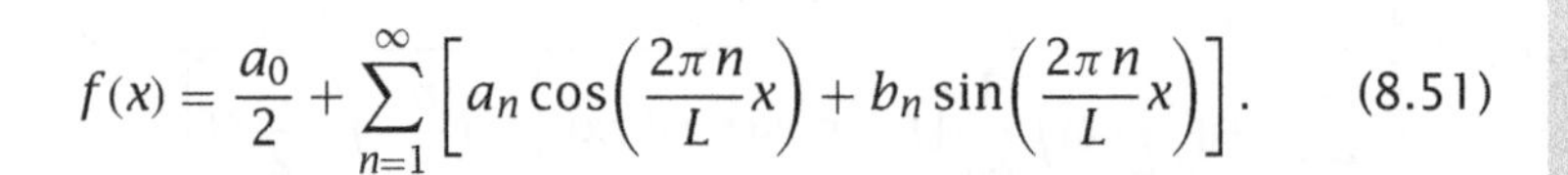

$$f(x) = \frac{a_0}{2} + \sum_{n=1}^{\infty} \left[a_n \cos\left(\frac{2\pi n}{L}x\right) + b_n \sin\left(\frac{2\pi n}{L}x\right) \right]. \tag{8.51}$$

Here a_n and b_n are Fourier coefficients, numbers that need to be computed for a given function f and that encode the special features of the function of interest. The above equality is true for all points on the interval, with the possible exception of $x = 0$ and $x = L$. Since all the functions appearing in the sum on the right-hand side take on the same value at 0 and L, we would have to conclude that $f(0) = f(L)$ is also true. If this is not the case, it can be shown that the Fourier series representation of $f(x)$ takes on the value $[f(0) + f(L)]/2$ at the boundaries of the interval.

Computing the Fourier coefficients relies on the orthogonality property of sine and cosine functions. In particular, the integral of the product of two such functions is nonzero only in the case in which both functions are sines, or both are cosines, and they have the same period; the period of $\sin(2\pi n/L)$ is L/n. We can restate this mathematically as

$$\begin{aligned} &\int_0^L \sin\left(\frac{2\pi n}{L}x\right) \cos\left(\frac{2\pi m}{L}x\right) \mathrm{d}x = 0, \\ &\int_0^L \sin\left(\frac{2\pi n}{L}x\right) \sin\left(\frac{2\pi m}{L}x\right) \mathrm{d}x = \delta_{n,m}\frac{L}{2}, \\ &\int_0^L \cos\left(\frac{2\pi n}{L}x\right) \cos\left(\frac{2\pi m}{L}x\right) \mathrm{d}x = \delta_{n,m}\frac{L}{2}, \end{aligned} \tag{8.52}$$

where the Kronecker symbol $\delta_{n,m}$ is 1 for $n = m$ and zero otherwise. With these identities in hand, we can compute the Fourier coefficients of the function $f(x)$ by multiplying it with sines and cosines with different periods, and integrating over the interval between 0 and L. Looking at the right-hand side of Equation 8.51 and taking into account the orthogonality identities above, we see that the only surviving term on the right-hand side is the sine or cosine term with the same period. Therefore, we have the following identities:

$$\begin{aligned} \int_0^L f(x)\,dx &= \frac{a_0}{2}L, \\ \int_0^L f(x)\cos\left(\frac{2\pi n}{L}x\right)dx &= a_n\frac{L}{2}, \\ \int_0^L f(x)\sin\left(\frac{2\pi n}{L}x\right)dx &= b_n\frac{L}{2} \end{aligned} \tag{8.53}$$

from which we can compute the Fourier coefficients

$$\begin{aligned} a_0 &= \frac{2}{L}\int_0^L f(x)\,dx, \\ a_n &= \frac{2}{L}\int_0^L f(x)\cos\left(\frac{2\pi n}{L}x\right)dx, \\ b_n &= \frac{2}{L}\int_0^L f(x)\sin\left(\frac{2\pi n}{L}x\right)dx. \end{aligned} \tag{8.54}$$

To illustrate the procedure of expanding a function into a Fourier series, consider the simple example given by the function $f(x)$ that is equal to 1 for $0 < x < L/2$ and equal to zero for $L/2 < x < L$ (that is, a square wave). Fourier coefficients are computed using Equation 8.54, and we find $a_0 = 1, a_n = 0$ for any other value of n, $b_n = 0$ for n even, and $b_n = 2/(\pi n)$ for n odd. How the function $f(x)$ emerges from the Fourier series as more and more terms are kept in the sum is shown in Figure 8.17.

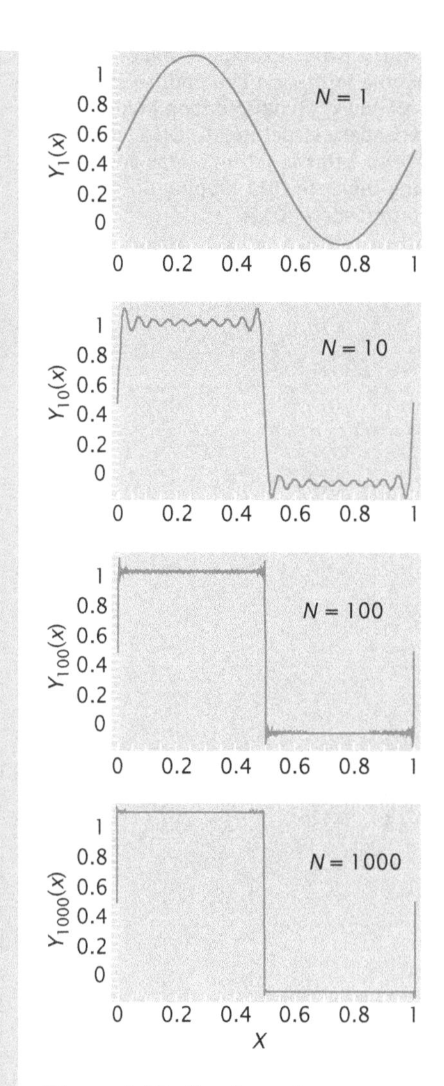

Figure 8.17: Fourier series representation of a square wave. Different graphs correspond to the Fourier series representation of the square wave function where the first N terms have been retained in the sum on the right-hand side of Equation 8.51.

8.2.4 DNA Looping: From Chromosomes to Gene Regulation

The organization of genomes occurs at many different scales. A shorter-scale phenomenon of widespread significance is the formation of loops of various kinds in both genomic DNA and RNAs. Figure 8.18 shows how nucleic acids form "loops" in a wide variety of different settings. For example, as illustrated in Figure 8.18(A), melting of DNA results in bubbles of single-stranded fragments, and the meandering of the single-stranded fragments can be evaluated as a problem in random walks. Similar ideas are relevant in evaluating the propensity of RNA to form hairpin loops, which are an important element of RNA secondary structure. Another favorite example involves the formation of DNA loops by transcription factors as part of the process of gene regulation. Yet another example, shown in Figure 8.18(D), involves genetic recombination in which distant parts of chromosomal DNA find one another as a precursor to the recombination event itself. These events are important in situations ranging from mating type switching in yeast to V(D)J recombination in B cells, to the stochastic decision making that attends olfactory receptor selection.

Figure 8.18: Examples of looping: (A) bubble formation in a double- stranded DNA helix, (B) hairpin loop in RNA secondary structure, (C) DNA looping due to a transcription factor, and (D) long-distance DNA looping of chromosomal DNA.

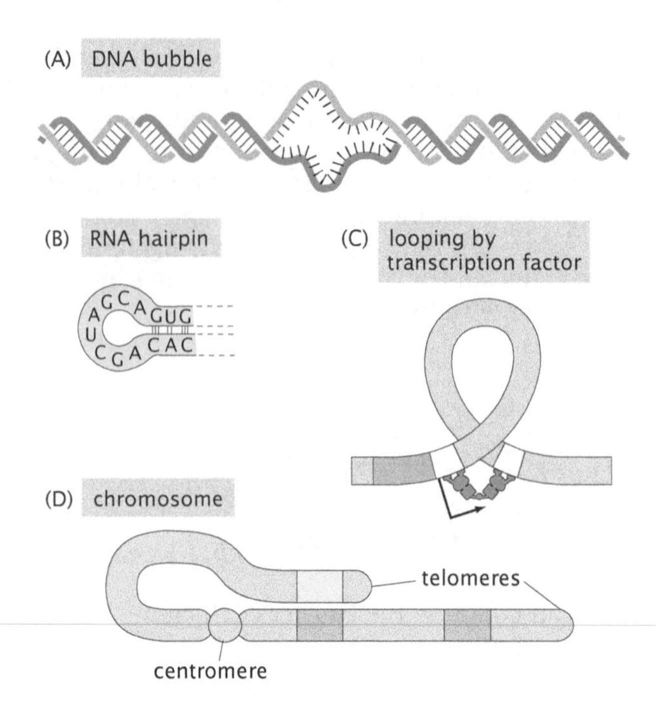

The Lac Repressor Molecule Acts Mechanistically by Forming a Sequestered Loop in DNA

In Figure 4.13 (p. 155) and Section 4.4.3 (p. 157), we introduced the *lac* operon as a particularly notable example of gene regulation. One part of the *lac* operon story is how the genes of this operon are repressed by the Lac repressor protein as shown in Figure 8.19. Thus far, our description of the Lac repressor has been largely schematic without particular reference to the mechanical actions responsible for repression. The actual story of the action of the Lac repressor is more complicated than that illustrated in Figure 4.15 (p. 158). In fact, there are several other operator sites (O2 and O3) in addition to the primary operator site (O1) described there where the repressor can bind resulting in a DNA loop like that shown in Figure 8.19. The effectiveness of repression is highest when the Lac repressor tetramer (built up from four copies of the product of the *lacI* gene) binds to two operators simultaneously. The results of an experiment in which the loop length was varied is shown in Figure 1.11 (see p. 19).

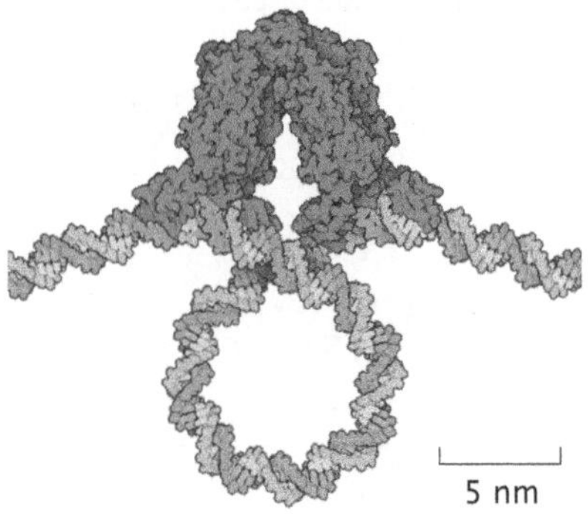

Figure 8.19: Model for DNA loop formation by the Lac repressor. The interface between the protein and the DNA was determined by X-ray crystallography (PDB 1lbh, 1efa), but the overall position and shape of the DNA in the loop is an artist's rendition. (Courtesy of D. Goodsell.)

Looping of Large DNA Fragments Is Dictated by the Difficulty of Distant Ends Finding Each Other

In order for a protein molecule such as the Lac repressor to spontaneously form a loop in the DNA, the DNA and protein must together suffer a fluctuation that brings all of the pieces into physical proximity. As will be shown in Chapter 10, for the DNA to bend in this way costs elastic energy. However, there is also a contribution to the free energy of looping from entropy, since when the DNA is looped there are fewer conformations available to the system and hence a reduction in the entropy.

As a warm-up exercise to evaluate the entropic cost of loop formation, we consider a one-dimensional model and examine the fraction of conformations which close on themselves. The probability, $p_\circ$, of loop formation is the probability that the one-dimensional random walker returns to the origin. Using Equation 8.10 for $R=0$, we conclude

$$p_\circ = \frac{\text{number of looped configs}}{\text{total number of configs}} = \frac{\dfrac{N!}{(N/2)!(N/2)!}}{2^N}, \tag{8.55}$$

where N is the number of Kuhn segments. Here we are interested in the long-chain limit, which corresponds to $N \gg 1$. This is also the limit in which the random walk model can be applied to DNA conformations, as discussed previously. To further simplify Equation 8.55, we make use of our trusty Stirling formula (p. 222), $N! \approx (N/\mathrm{e})^N \sqrt{2\pi N}$, which holds for $N \gg 1$ and implies

$$p_\circ \approx \sqrt{\frac{2}{\pi N}}. \tag{8.56}$$

The interesting prediction of the model is that the cyclization probability of long DNA strands will decay with polymer length to the power $-1/2$.

This result for the probability that the two ends will be within some small distance of each other can also be obtained using the Gaussian approximation to the end-to-end distribution derived in Section 8.2.1. To use the continuous distribution, we need the probability that the two ends of the chain are within some critical distance of one another, namely, $\delta \ll \sqrt{Na^2}$. In this case, the end-to-end distribution of Equation 8.16 can be approximated by

$$P(R; N) \approx \frac{1}{\sqrt{2\pi Na^2}}, \tag{8.57}$$

where we have made the substitution $\mathrm{e}^{-R^2/2Na^2} \approx 1$, valid for $-\delta < R < \delta$. The cyclization probability is obtained by integrating over all the distances of near contact in the form

$$p_\circ = \int_{-\delta}^{\delta} \frac{1}{\sqrt{2\pi Na^2}}\, \mathrm{d}R = \sqrt{\frac{2}{\pi N}}\frac{\delta}{a}, \tag{8.58}$$

which, as expected, is the same as Equation 8.56 for $\delta = a$.

Unlike the scaling of the polymer size with its length, which we found to be independent of the dimensionality of space, the effect of dimensionality on cyclization is quite significant. In particular, the cyclization probability has a different form depending upon whether we evaluate this quantity for one-, two-, or three-dimensional random walks. To see this, consider the three-dimensional random walk of N steps. The probability of returning to the origin can be written as the ratio of the number of walks that return to the origin to the total number of walks in much the same way as we did above (the precise details of this calculation in the discrete language is left to the problems at the end of the chapter). However, a more immediate route to the result can be obtained by exploiting the continuous distribution.

Consider the end-to-end distribution of a three-dimensional random walk. In particular, the probability that the two ends of the chain are at distance δ or smaller is given by the integral

$$p_\circ = \int_0^{\delta} 4\pi R^2 P(R; N)\, \mathrm{d}R = \int_0^{\delta} 4\pi R^2 \left(\frac{3}{2\pi Na^2}\right)^{3/2} \mathrm{e}^{-3R^2/2Na^2}\, \mathrm{d}R. \tag{8.59}$$

Since we are interested in cyclization, we can assume that the distance δ is much smaller than the polymer size, $N^{1/2}a$. In this case, the exponential function in the integrand can be approximated by 1, and the resulting integral is

$$p_\circ = \int_0^\delta 4\pi R^2 \left(\frac{3}{2\pi Na^2}\right)^{3/2} \mathrm{d}R = \left(\frac{6}{\pi N^3}\right)^{1/2} \left(\frac{\delta}{a}\right)^3. \quad (8.60)$$

The main conclusion that follows from this calculation is that the cyclization probability decays as the number of Kuhn segments of the chain to the power −3/2. In Section 10.3 (p. 394), we will finish these arguments by showing how to link the entropic and energetic description of DNA looping. These ideas will then be applied to compute the probability of gene expression in Section 19.2.5 (p. 822).

Chromosome Conformation Capture Reveals the Geometry of Packing of Entire Genomes in Cells

Thus far, our treatment of chromosome geography has made use of fluorescently labeled genetic loci to address the question of the spatial organization of chromosomes. The key weakness of this approach is that it provides information about the position and dynamics of one, or at most a few, genetic loci at a time. The spatial organization of the whole genome can be built up by repeatedly examining one genetic locus after another, as was done in the experiments on *C. crescentus* described in Figure 8.13, but this is time-consuming and even impractical for very large eukaryotic genomes. Interestingly, an alternative has been developed that appeals to ideas about looping like those described above.

One method that overcomes the locus-by-locus problem is chromosome conformation capture (3C), illustrated in Figure 8.20(A). This method is used to obtain genome-wide information about the spatial proximity of any two genetic loci. In particular, for any two genetic loci on the genome, it provides the frequency of close contacts. These close contacts are detected by first crosslinking the entire genome. The concept is that two sites on the DNA are crosslinked when they are in close proximity and connected to each other via DNA-bound proteins. The DNA is then digested, leaving the cross-linked fragments bound to proteins intact. Now these pieces of DNA are ligated together and the proteins removed, thus creating new DNA strands that contain sequence from the two genetic loci that were in close proximity in the cell before fixation. In the original 3C method, these sequences were then detected using the polymerase chain reaction with primers that correspond to the individual loci being probed for their proximity, while in more recent implementations of the method this has been superseded by directly sequencing the resulting fragments.

The 3C method and its generalizations and extensions have been used extensively to study the relation between different functions of the chromosome (transcription, recombination) and its spatial organization. It has also provided quantitative tests of polymer models of chromosomes, as shown in Figures 8.20(B) and (C). In the previous section we calculated the probability of cyclization for a random walk polymer and found that in three-dimensions it scales with the polymer length to the power −3/2. This is indeed the result found from applying 3C to yeast chromosome III, as shown in Figure 8.20(B). On the other hand, in human cells the measurements of contact frequency as a function of genomic distance yield the exponent −1 inconsistent with the expectation provided by a simple random walk model of the chromosome. Interestingly, the so-called crumpled globule model, predicts this kind of scaling.

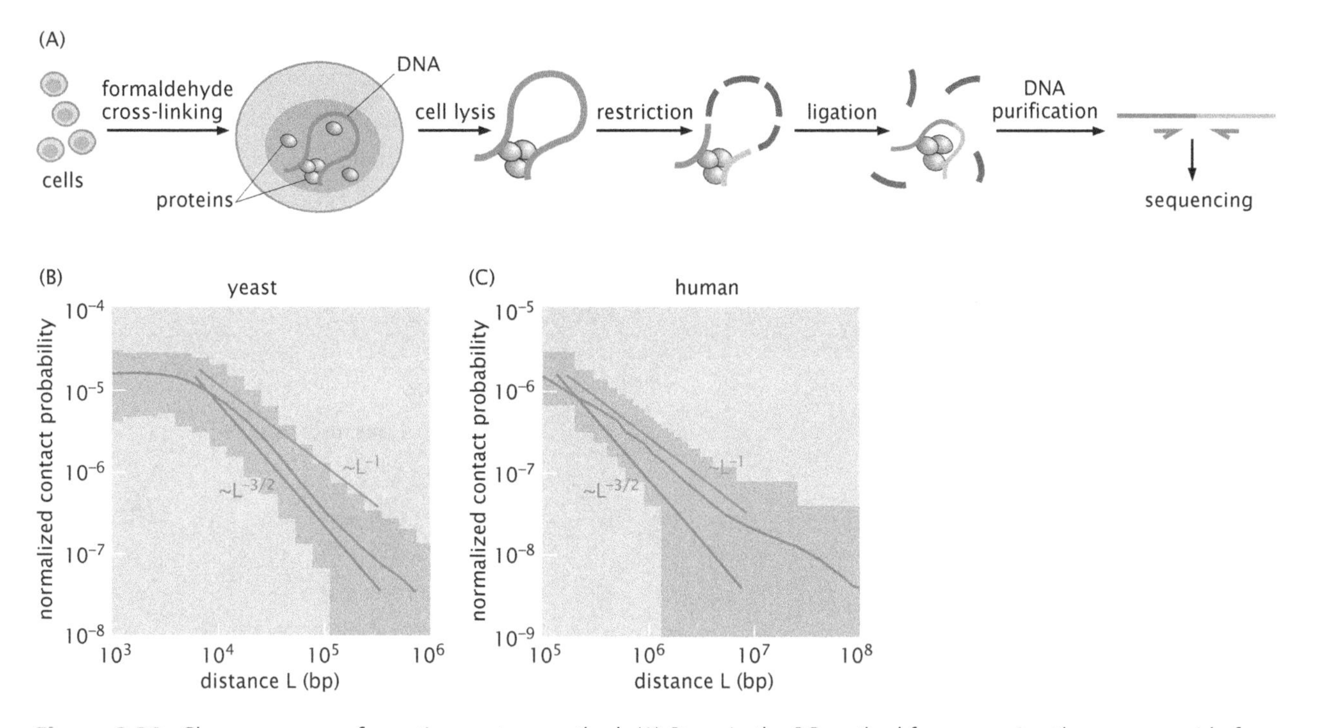

Figure 8.20: Chromosome conformation capture method. (A) Steps in the 3C method for measuring the genome-wide frequency of close contacts between genetic loci. Frequency of intra-chromosomal contacts between genetic loci as a function of the genomic distance between them in (B) yeast and (C) human. The measured frequency in yeast is consistent with a random walk model of the yeast chromosomes (red line), while the data from human cells is not. The human data is consistent with the crumpled globule model. (B, Courtesy of Geoffrey Fudenberg and Leonid Mirny; data in (B) Courtesy of Job Dekker and Jon-Matthew Belton, unpublished; C, adapted from E. Lieberman-Aiden, et al., *Science* 326:289, 2009.)

In crumpled globule models, the chromosome is assumed to take on a non-equilibrium configuration of a random walk polymer obtained by collapsing small segments of the polymer into globules, which then form a beads-on-strings structure, with the globules playing the role of monomers. This structure is then further collapsed into a new set of super-globules that form yet again a beads-on-strings structure, with this hierarchical packing carried on *ad infinitum*. This process leads to a self-similar fractal structure with interesting properties such as the absence of knots and other topological obstacles to the chromosome easily folding and unfolding. It has been speculated that this is a structural feature that can be useful in making genetic loci easily accessible to protein machines involved in transcription and recombination.

8.3 The New World of Single-Molecule Mechanics

Models such as the random walk model described here have extraordinary reach. Yet another interesting application of these ideas is to the development of single-molecule techniques for measuring the response of macromolecules to external forcing.

Single-Molecule Measurement Techniques Lead to Force Spectroscopy

There are a number of different ways of applying forces to individual macromolecules. Several of these techniques are represented in schematic form in Figure 8.21. One such technique, shown in Figure 8.21(A), involves the use of micron-sized cantilevers that are

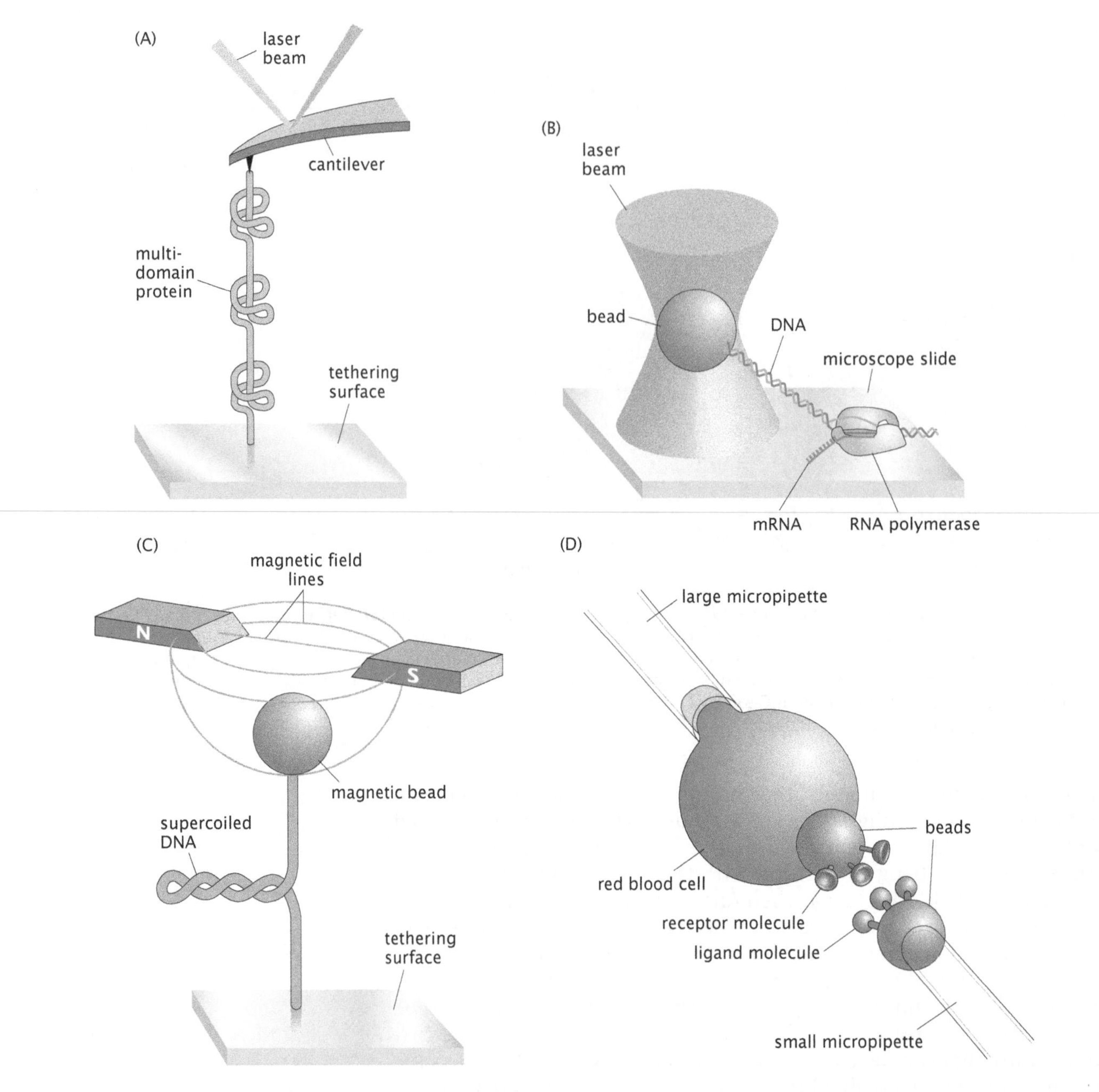

Figure 8.21: Schematic showing a variety of single-molecule techniques. (A) Single-molecule atomic-force microscopy being used to stretch a multidomain protein. (B) Optical tweezers being used to measure the rate of transcription. (C) Magnetic tweezers being used to measure the torsional properties of DNA. (D) Pipette-based force apparatus being used to measure ligand–receptor adhesion forces.

attached to a macromolecule that is, in turn, tethered to a surface. Through control of the height of the surface to which the molecule is tethered, for example, the cantilever will suffer a deflection, which can be measured using reflected laser light. A second example, shown in Figure 8.21(B), is optical tweezers, which permit the application of forces of order 1–50 pN on macromolecules of interest. In this case, the key idea is that by attaching a macromolecule to a micron-sized bead, it is possible to pull on the bead (and hence the molecule) by shining laser light on the bead and using the resulting radiation pressure to manipulate the bead. The same concept is similarly played out in the context of the magnetic tweezers shown in Figure 8.21(C), where the bead is manipulated by magnetic fields rather than laser light. One of the interesting variations on the forcing scheme provided by the magnetic tweezer is the opportunity to apply torsional forces, which examine the response of molecules to twist. The final example, shown in Figure 8.21(D), is the use of a pipette-controlled force apparatus in which the strengths of ligand–receptor interactions as well as the mechanical response of lipid bilayer vesicles can be examined. Our main point in this discussion is to alert the reader to the emergence of single-molecule techniques that complement the tools of traditional solution biochemistry and permit the measurement of not only the average properties of the various macromolecules of biological interest, but also the fluctuations about this average response.

8.3.1 Force–Extension Curves: A New Spectroscopy

Different Macromolecules Have Different Force Signatures When Subjected to Loading

The techniques introduced above permit the explicit measurement of the force–extension characteristics of a range of different molecules. Figure 8.22 shows the force–extension properties of several characteristic examples ranging from DNA to proteins. In particular, Figure 8.22(A) shows the force–extension characteristics of a single DNA molecule subjected to loading (a similar example was shown in Figure 5.13, p. 207). Note that the same characteristic force–extension signature will be found for a given DNA molecule regardless of which of the various techniques is used to measure it, and further that this curve provides a unique fingerprint, which serves as the basis of *force spectroscopy* of macromolecules. Figure 8.22(B) shows a plot of the force–extension properties of a particular RNA molecule. Note that the character of the secondary structure associated with a given RNA molecule is translated, in turn, into the character of the force–extension curve, illustrating the idea that the force–extension curve provides a spectroscopic fingerprint of different macromolecules. Figure 8.22(C) shows yet a third example of the intriguing diversity of force–extension curves associated with different macromolecules, this time revealing how the multidomain protein titin unfolds in the presence of force. One immediate statement that can be made in this

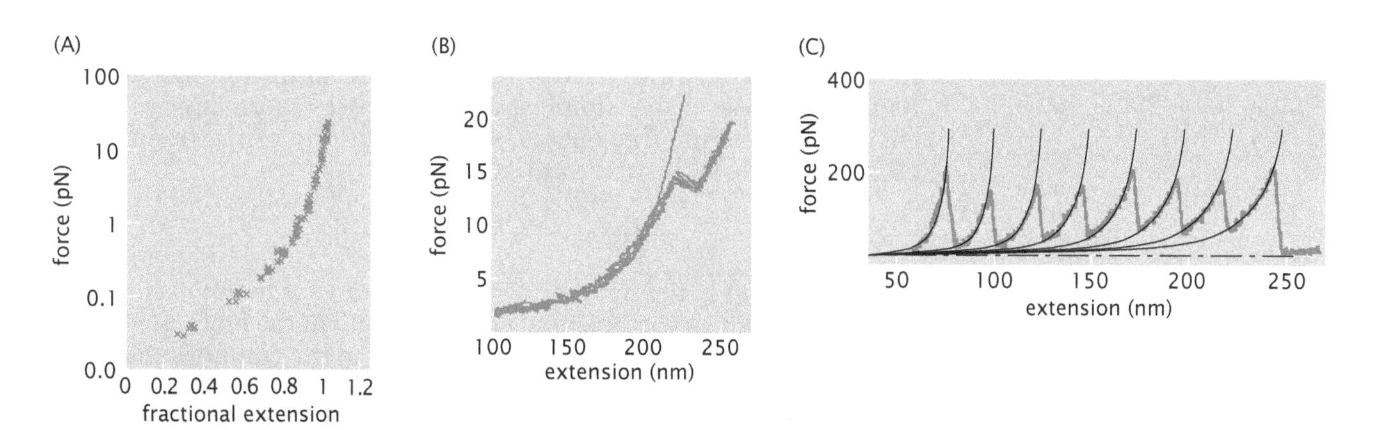

Figure 8.22: Force–extension curves for a variety of different macromolecules: (A) dsDNA, (B) RNA, and (C) protein made of repeats of Ig module 27 of the I band of human cardiac titin. The measured curves illustrate the sense in which single-molecule experiments serve as the basis of force spectroscopy. Nonmonotonic features seen in the plots correspond to changes in structure due to an applied force. (A, adapted from C. Bustamante et al., *Curr. Opin. Struct. Biol.* 10:279, 2000; B, adapted from J. Liphardt et al., *Science* 292:733, 2001; C, adapted from M. Carrion-Vazquez et al., *Proc. Natl Acad. Sci. USA* 96:3694, 1999.)

example is that the number of load drops in the curve corresponds to the number of unfolded domains in the protein. We emphasize that these three examples are but a tiny representation of the broad class of measurements that have been made on polysaccharides, lipids, proteins, and nucleic acids as well as their assemblies.

8.3.2 Random Walk Models for Force–Extension Curves

Given that different macromolecules exhibit different force–extension signatures, it is of interest to see if we can compute some characteristics of these curves using what we know about random walks. Indeed, the calculation of these force–extension curves gives us the opportunity to further explore entropic forces.

The Low-Force Regime in Force–Extension Curves Can Be Understood Using the Random Walk Model

One of the simplest models that can be written to capture the relation between force and extension in polymers is based on a strictly entropic interpretation of the free energy. In particular, by remembering that as the chain molecule is stretched to lengths approaching its overall contour length, the overall number of configurations available to the molecule goes down, and with it so too does the entropy. This reduction in entropy corresponds to an increase in the free energy. If the pulling experiment is done sufficiently slowly, we can think of the force as being given by

$$\text{force} = -\frac{\partial G}{\partial L}, \tag{8.61}$$

where G is the free energy and L is the length.

We begin with a one-dimensional rendition of the freely jointed chain model. We imagine a polymer of overall length $L_{\text{tot}} = Na$, where N is the number of monomers and a is the length of each monomeric segment. The basic thrust of our argument will be to construct the free energy $G(L)$ as a function of the length $L = (n_{\text{r}} - n_{\ell})a$, from which the force necessary to arrive at that extension is given by Equation 8.61. As before, we use the notation n_{r} and n_{ℓ} to signify how many of the total links are right-pointing (n_{r}) and how many are left-pointing (n_{ℓ}). In order to proceed, we need an explicit formula for the free energy. As noted above, in this simplest of models, we ignore any enthalpic contributions to the free energy, with the entirety of the free energy of the molecule taking the form

$$G(L) = -k_{\text{B}}T \ln W(L; L_{\text{tot}}), \tag{8.62}$$

where $W(L; L_{\text{tot}})$ is the number of configurations of the molecule that have length L given that the total contour length of the molecule is L_{tot}.

As shown in Figure 8.23, we are interested in the equilibrium of our random walk representation of the polymer when it is subjected to external forcing such as can be provided by an optical tweezers setup. A particularly transparent way to imagine this problem is to think of weights being dangled from the ends of the polymer as shown in Figure 8.24 (this idea of representing the energy of the loading device via weights was introduced in Figure 5.12 on p. 206). In this case, the free energy of Equation 8.62 must be supplemented with a term of the

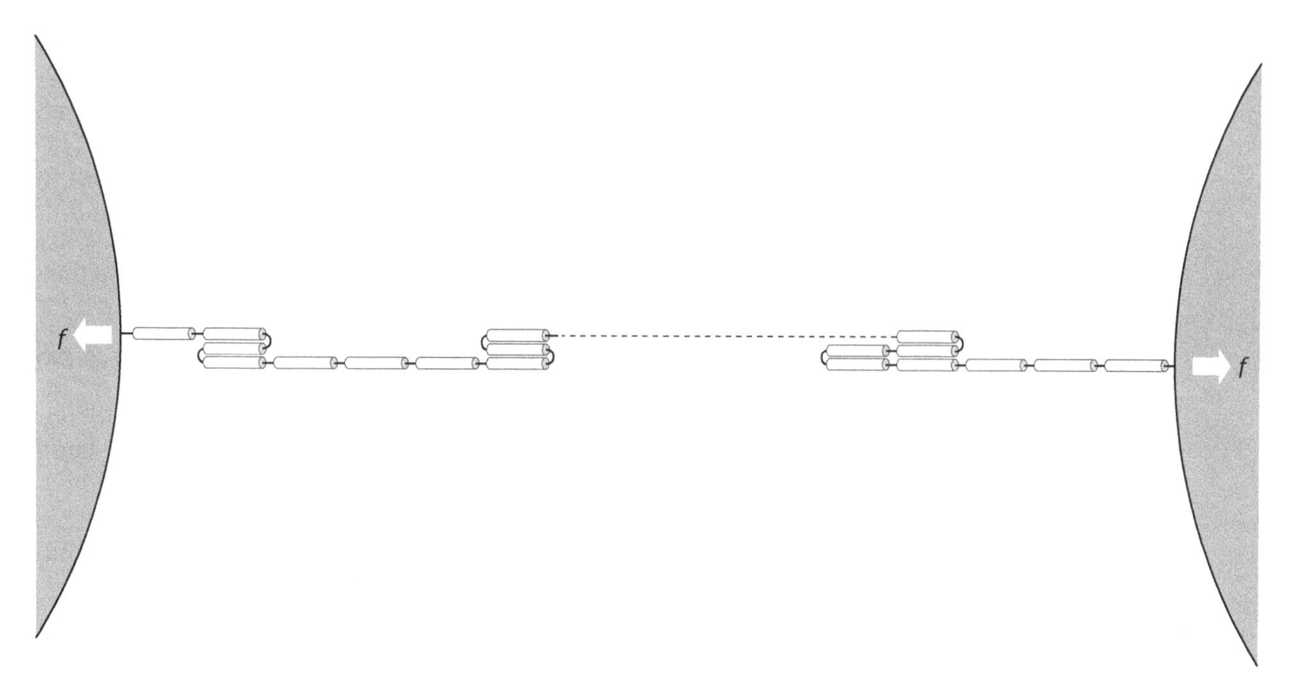

Figure 8.23: Model polymer subjected to loading. Schematic of a model one-dimensional polymer subjected to external forcing by optical tweezers.

form $U_{\text{weights}} = -mgL$. What this term says physically is that the more the molecule is stretched, the lower the weights will dangle, with the result that their potential energy is decreased.

Putting together this term and the contribution from Equation 8.62, we have for the total free energy of the system

$$G(L) = \underbrace{-mgL}_{\text{contribution from weights}} - \underbrace{k_B T \ln W(L; L_{\text{tot}})}_{\text{entropic contribution of polymer conformations}}$$

To make further progress with this result, and, in particular, to obtain the free-energy minimizing length as a function of the applied force, we must first find a concrete expression for $W(L; L_{\text{tot}})$. To that end, we note that this reduces to nothing more than the combinatoric question of how many different ways there are of arranging N arrows, n_R of which are right-pointing and $n_L = N - n_R$ of which are left-pointing. The result is

$$W(n_R; N) = \frac{N!}{n_R!(N - n_R)!}, \tag{8.63}$$

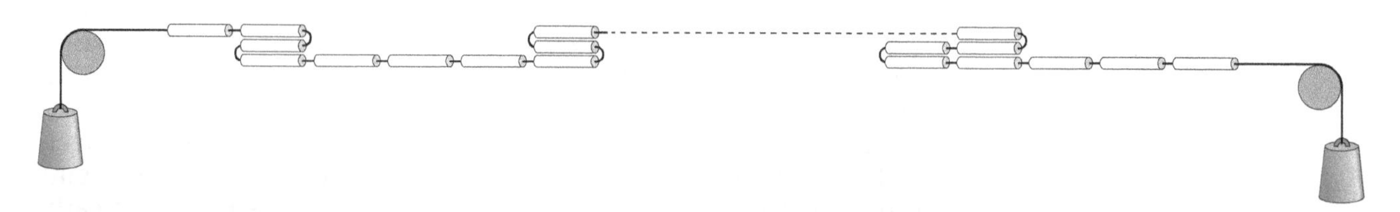

Figure 8.24: Polymer subject to an external load. Schematic of a model one-dimensional polymer subjected to external forcing through the attachment of weights on the ends. This scenario is a pedagogical device to illustrate how to include the forcing in the overall free energy budget.

where we have found it convenient to replace our reference to L and L_{tot} with a reference to the number of right pointing arrows and the total number of such arrows with the recognition that they are related by $L = (n_R - n_L)a$ and $L_{tot} = Na$.

Given the free energy, our task now is to minimize it with respect to length (or n_R). To that end, we first invoke the Stirling approximation (p. 222), which, we remind the reader, allows us to replace $\ln N!$ by $N \ln N - N$. In light of this approximation, the overall free energy may be written as

$$G(n_R) = -2mgn_R a + k_B T[n_R \ln n_R + (N - n_R) \ln(N - n_R)]. \tag{8.64}$$

Note that we have neglected all constant terms since they will not contribute during the minimization. Differentiation of this expression with respect to n_R results in

$$\frac{\partial G}{\partial n_R} = -2mga + k_B T \ln n_R - k_B T \ln(N - n_R) = 0. \tag{8.65}$$

Solving this equation for the quantity n_R/n_L, we obtain

$$\frac{n_R}{n_L} = \mathrm{e}^{2mga/k_B T}, \tag{8.66}$$

which we use to obtain a simple relation for the extension

$$z = \frac{\langle L \rangle}{L_{tot}} = \frac{n_R - n_L}{n_R + n_L} = \tanh\left(\frac{mga}{k_B T}\right). \tag{8.67}$$

The construct of using weights to load the molecule was a convenient pedagogical device to provide a concrete mechanism for seeing how the energy of the loading device can be included in the free-energy budget. More generally, the two ends are subjected to a force f with the result that $z = \tanh(fa/k_B T)$. This force–extension relation is shown in Figure 8.25. To gain further insight into the quantitative aspects of the model, we consider the limiting case of a small force, that is, $fa \ll k_B T$. For a dsDNA molecule in physiological conditions ($a \approx 100\,\text{nm}$), this corresponds to $f \ll 40\,\text{fN}$, while for the much more flexible ssDNA ($a \approx 1.5\,\text{nm}$), the small-force regime is obtained for $f \ll 3\,\text{pN}$. In the small-force limit, the force–extension curve is linear (as shown in the problems at the end of the chapter),

$$\langle L \rangle = \frac{L_{tot} a}{k_B T} f, \tag{8.68}$$

that is, in this regime, the polymer behaves like an ideal Hookean spring with a stiffness constant $k = k_B T / L_{tot} a$. The fact that the stiffness of this spring is proportional to the temperature reveals its true entropic nature. For λ-phage dsDNA, whose contour length is $L_{tot} = 16.6\,\mu\text{m}$, the effective spring constant is $k \approx 2.3\,\text{fN}/\mu\text{m}$, while for the same length ssDNA, the stiffness is given by $k \approx 160\,\text{fN}/\mu\text{m}$. Note that the larger flexibility of ssDNA, as evidenced by its smaller persistence length, leads to a larger value for the effective spring stiffness.

Thus far, our model of the macromolecule has been highly idealized in that we have imagined that each monomer can only point in one of two directions. Though that model is instructive, clearly it is of interest to expand our horizons to the more physically realistic three-dimensional case. The generalization of our freely jointed chain analysis to three dimensions holds no particular surprises. The

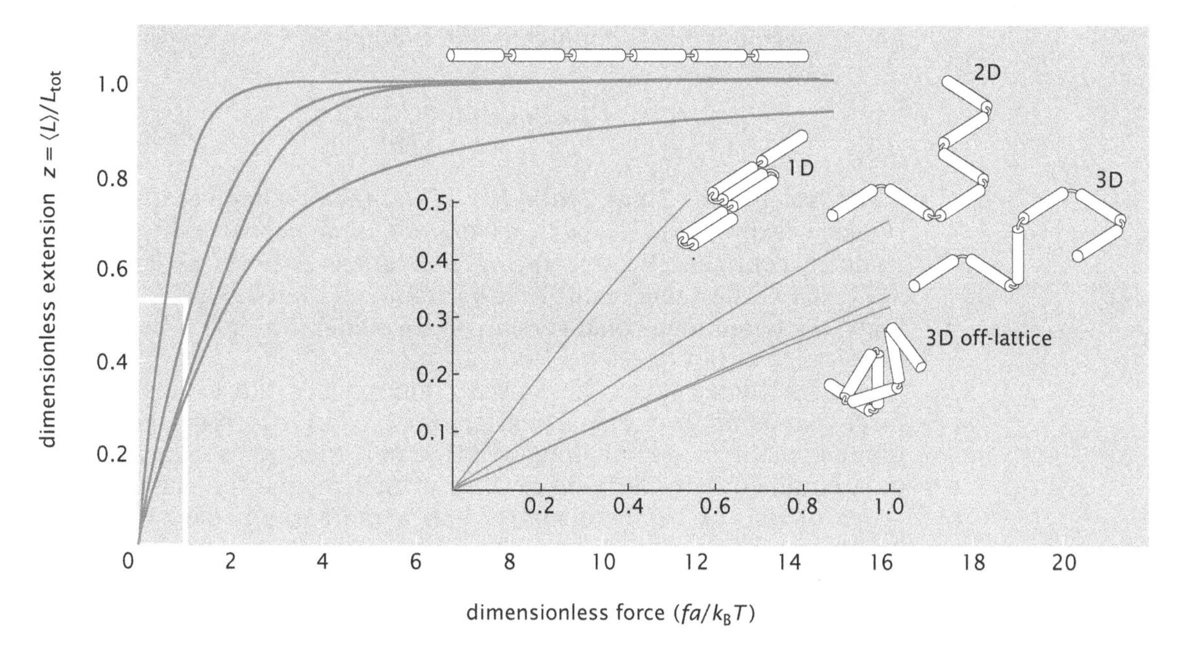

Figure 8.25: Relation between force and extension as obtained using the freely jointed chain model. Results for one, two, and three dimensions are shown and the three-dimensional case is shown both for the version in which the monomers can only point in the Cartesian directions and for the case in which they can point in any direction. The region of the graph shown in the white box is expanded in the inset. The curves are related to their corresponding model by the cartoon showing the random walk chain. A comparison of this model with the data was shown in Figure 5.14, p. 207.

fundamental idea is that now, instead of constraining the monomers that make up the molecule of interest to point only right or left, we give them full three-dimensional motion. The simplest variant of this model is to permit each monomer to point in one of six directions (that is, $\mathbf{e}_1$, $-\mathbf{e}_1$, $\mathbf{e}_2$, $-\mathbf{e}_2$, $\mathbf{e}_3$, and $-\mathbf{e}_3$). We quote the result for this model, namely,

$$z = \frac{\langle L \rangle}{L_{\text{tot}}} = \frac{2\sinh(\beta fa)}{4 + 2\cosh(\beta fa)}, \tag{8.69}$$

and leave the details as an exercise for the reader.

The more interesting case which we work out in greater detail is that in which each monomer can point in *any* direction. In this case, rather than writing out the free energy explicitly, we compute the partition function and use it to deduce the relevant averages, such as the average length at a given applied force. As each link in the chain is independently fluctuating, the partition function for $N = L_{\text{tot}}/a$ links is $Z_N = Z_1^N$, with

$$Z_1 = \int_0^{2\pi} \mathrm{d}\phi \int_0^{\pi} \mathrm{e}^{fa\cos\theta/k_BT} \sin\theta \, \mathrm{d}\theta. \tag{8.70}$$

This equation instructs us to compute the Boltzmann factor for every orientation of the monomer, characterized by the angles ϕ and θ, and then sum (integrate) over all possible values of the two angles. This integral over the unit sphere can be evaluated with the change of variables $x = \cos\theta$, to give

$$Z_1 = 4\pi \frac{k_BT}{fa} \sinh\left(\frac{fa}{k_BT}\right). \tag{8.71}$$

Now the free energy $G(f) = -k_BT \ln Z_N$ is a function of the applied force f and we differentiate it with respect to f to obtain an expression for its thermodynamic conjugate, the average polymer length,

$$\langle L \rangle = -\frac{\partial G}{\partial f} = Na\left[\coth\left(\frac{fa}{k_B T}\right) - \frac{k_B T}{fa}\right]. \tag{8.72}$$

The small-force limit, $fa/k_B T \ll 1$, in this case gives the same Hookean expression, $f = k\langle L \rangle$, as the one-dimensional freely jointed chain, except the effective spring constant is 3 times as large, $k = 3k_B T/L_{tot}a$. The same result follows from Equation 8.69. Not surprisingly, the two-dimensional version of the model, whether it be defined on a lattice or not, gives $k = 2k_B T/L_{tot}a$.

At large forces when the polymer approaches full extension, the force–extension formula, Equation 8.72, derived from the freely jointed chain model no longer adequately describes experimental data obtained by pulling on dsDNA. In that regime the elastic properties of dsDNA begin to matter and a more sophisticated model, which incorporates bending stiffness, describes the experimental data much better. This so-called worm-like chain model is taken up in Chapter 10.

8.4 Proteins as Random Walks

One of the key ideas driving research in structural biology, which seeks to describe protein structure in atomic detail, is that the function of a protein follows from its structure. So far, we have shown how the random walk model can be applied to nucleic acids. Proteins are polymers comprising amino acids. Therefore, a natural question to ask is what, if any, aspects of protein structure can be understood from simple coarse-grained models of polymers, such as the various random walks introduced in this chapter?

Globular proteins in their native state form compact structures that are quite different from the open configurations implied by the random walk model. Therefore, we might be tempted to conclude that the random walk model has no business commenting on proteins. Instead, we consider a modification of the random walk model we have employed so far by explicitly accounting for the compact nature of proteins.

The compact random walk model we employ in this section is defined on a lattice, meaning that the random walker, whose trajectories represent polymer configurations, jumps from one lattice site to the next. Usually when representing the polymer by a random walk on a lattice, the sites not occupied by the monomers (or, equivalently, those sites not visited by the random walker) are thought of as representing the solvent molecules. Simple random walks described in the previous sections are open structures with the monomer sites typically surrounded by solvent sites. As remarked above, this is inadequate for describing protein conformations, which are compact with solvent typically making contact only with amino acids at the surface of the protein. To mimic this property of proteins, we invoke compact random walks (also referred to as Hamiltonian walks), which are self-avoiding random walks that visit every site of the lattice, usually taken to be cubic, as depicted in Figure 8.26. By virtue of covering all the lattice sites by monomers, all the solvent sites are pushed to the surface. These compact random walks are a very coarse-grained model of proteins and, as with all coarse-grained models, one is limited in scope and precision of the questions that the model is equipped to address.

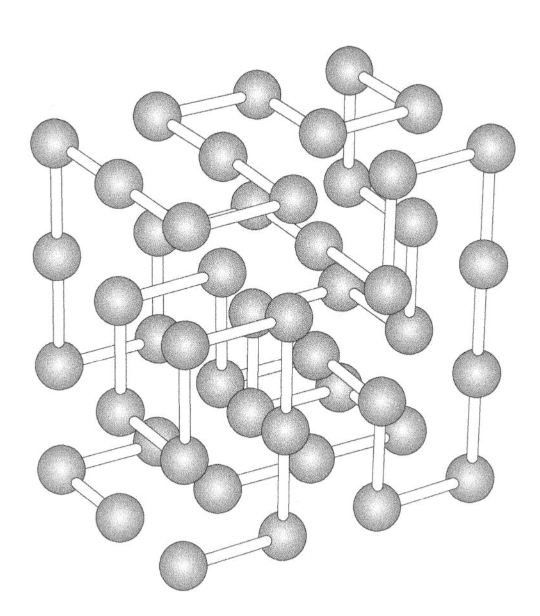

Figure 8.26: Compact polymer configuration on a 4 × 4 × 3 cubic lattice. Each ball represents an amino acid. (Adapted from P. D. Thomas and K. A. Dill, *Proc. Natl Acad. Sci. USA* 93:11628, 2006.)

The rewards, on the other hand, come in the form of simplicity and generality of the answers obtained. Furthermore, as with any good model, compact random walks reveal new questions and sharpen old ones about the structure of naturally occurring proteins.

8.4.1 Compact Random Walks and the Size of Proteins

The Compact Nature of Proteins Leads to an Estimate of Their Size

Possibly the simplest property of a globular protein is its size, as measured by its linear dimensions, or, more precisely, its radius of gyration. Examination of representative proteins from the PDB reveals a systematic dependence of the protein's size on its mass. In particular, for globular proteins, the radius of gyration scales roughly with the cube root of the mass. The relation between the physical size of proteins and their sequence size is shown in Figure 8.27.

The observed scaling is a simple consequence of the compact nature of proteins, and is thus also a property that is captured by compact random walks. Since a compact random walk completely fills the lattice (see Figure 8.26), its linear size will scale with the linear dimension of the lattice or with the cube root of the number of lattice sites, given that we have in mind a three-dimensional lattice. If we associate a single residue with each site, and take these to be of roughly equal mass, we arrive at the scaling law observed for many real proteins. Compactness implies that all the space occupied by proteins is filled, with no holes present. Therefore, the volume occupied by the protein, which necessarily scales as the cube of its linear dimension, is proportional to the mass. For proteins in the unfolded state, the structures are better described as random walks. The size of a random walk polymer, unlike compact polymers, scales as the 1/2 power of the mass. If one were to examine random self-avoiding walks (random walks with the additional constraint of no self-intersections), an argument due to Flory predicts scaling of the linear size with mass to the 3/5 power, indicating a structure that is even more expanded than that of a simple random walk.

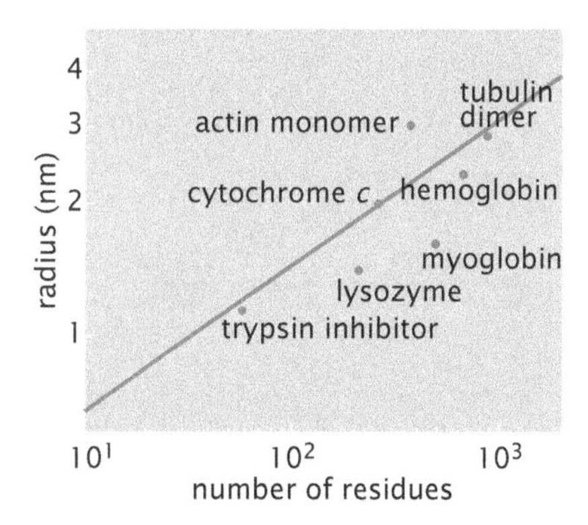

Figure 8.27: Scaling of protein size as a function of the number of amino acid residues. The line has a slope of 1/3, corresponding to a space-filling packing.

8.4.2 Hydrophobic and Polar Residues: The HP Model

One of the challenges brought in on the heels of the successes of the great genomic sequencing initiatives is that of figuring out the structural and functional implications of these vast libraries of genes. One step in unraveling the meaning of all of these genomic data is to figure out how to go from a particular protein sequence to the corresponding structure. The problem is that when confronted with some new gene sequence, one would like to be able to state what proteins are implied by the various sequences and what structures these proteins have. Like for the analysis of protein–ligand binding in Chapter 7, here too we will find that the use of internal state variables to characterize the amino acid identity of a given residue is extremely powerful.

The process by which a chain of amino acids assumes the specific three-dimensional native structure of a protein is often not understood in enough detail to allow a prediction of the structure based on the known sequence. The complexity of the problem is illustrated in part by the observation that the number of possible three-dimensional conformations of a protein is so large that a random search in structure space would never uncover the native state. Though nature is clever enough to wiggle its way out of this problem, sometimes we are not. Even if we are to model structures using a highly simplified and contracted scheme in which a given structure is viewed as a random walk on a cubic lattice as introduced above, the number of structures for a 100-monomer chain is 6^{100}, or 6.5×10^{77}. The way we obtain this estimate is based on the idea that the link connecting every successive set of residues can point in one of the six directions along the three Cartesian axes. If we imagine doing a random search among these structures at a (very optimistic) rate of one structure per femtosecond (10^{-15} s), it would take roughly 2×10^{55} years to complete the search. This is about 10^{45} times the age of the Universe!

The above estimate tells us that the folding of a protein into its native structure is most certainly not a random process. The hydrophobic interaction between amino acid residues and the water molecules that surround them leads to a collapse of the chain as was illustrated in Figure 5.8 (p. 203). As a result, the hydrophobic residues are sequestered to the interior of the protein, while the surface is populated by polar residues. Thus hydrophobicity is one force that can steer the protein to a folded state avoiding a random search of configuration space. Indeed, the spirit of the class of models introduced here is that collapse induced by hydrophobic effects drives the formation of secondary structure as opposed to an alternative view in which the formation of the hydrogen bonds that define secondary structure leads to collapse.

The HP Model Divides Amino Acids into Two Classes: Hydrophobic and Polar

The idea that the hydrophobic force plays a prominent role in protein folding has led to coarse-grained models of proteins where the 20 naturally occurring amino acids are replaced with a two-letter alphabet that identifies each amino acid as being hydrophobic (H) or polar (P). This leads to a drastic reduction of the complexity of the sequence space as the number of possible sequences for a 100-mer goes down from $20^{100} \approx 10^{130}$ to $2^{100} \approx 10^{30}$. To implement such a model, we need to decide how to partition the 20 amino acids into the two

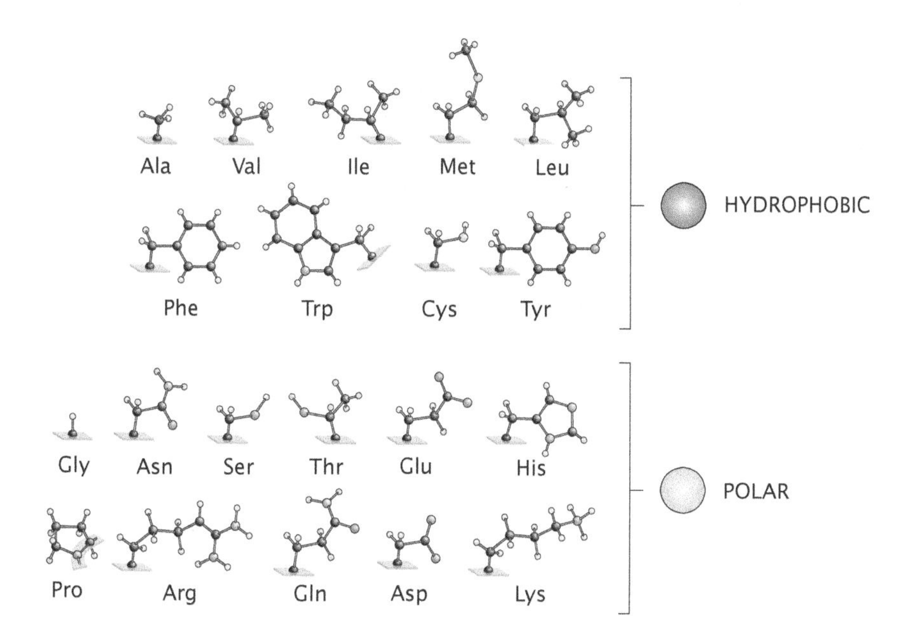

Figure 8.28: Mapping of the amino acids onto an HP alphabet. The 20 amino acids are coarsely separated into two categories, namely, hydrophobic (H) or polar (P).

categories H and P. An example of such a partitioning is shown in Figure 8.28. Indeed, as shown in Figure 8.29, there is a hierarchy of possible classifications of the amino acids based on various properties for grouping them.

In the remainder of the book, we will use the HP model introduced here as the basis of a variety of different discussions. Our reasoning is that classifying amino acids according to just these two broad categories allows us to take otherwise analytically intractable problems and render them tractable. For example, in Section 21.5.1 (p. 1011), we will consider an HP model of translation and kinetic proofreading

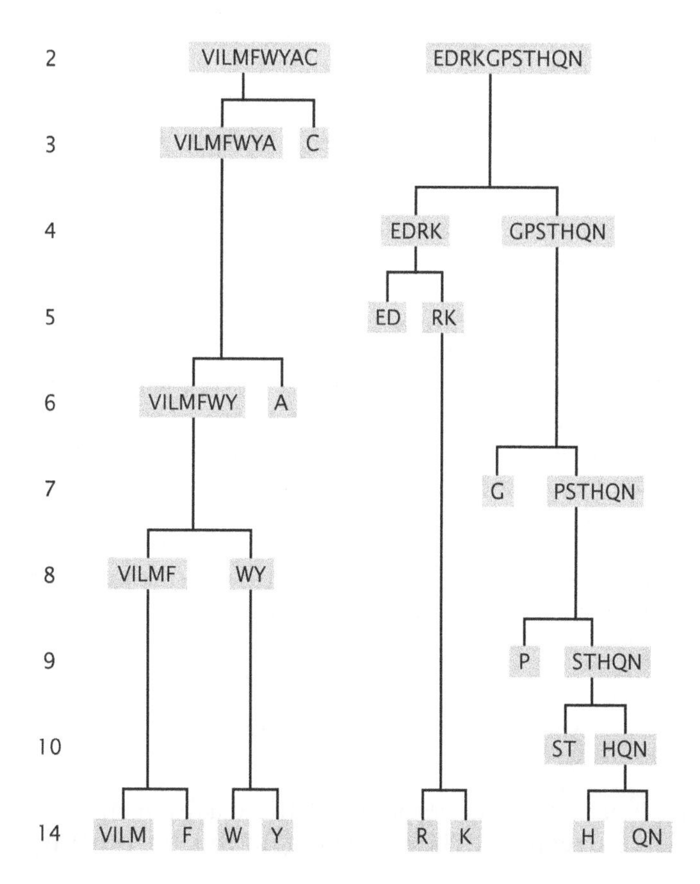

Figure 8.29: Hierarchy of amino acid classifications. Groupings of amino acids into "classes" with similar properties. At the top of the figure, the amino acids are grouped into two categories, hydrophobic (H) on the left and polar (P) on the right. At each level, the number of distinct classes is shown by the integer on the left. (Adapted from P. D. Thomas and K. A. Dill, *Proc. Natl Acad. Sci. USA* 93:11628, 1996.)

featuring only two species of tRNA. This simplification will allow us to carry out the analysis completely. Similarly, the entirety of Chapter 21 on bioinformatics will be based on sequence alignments using only the HP alphabet. Though we compromise on biological realism, our sense is that the pedagogical payoff is worth it.

8.4.3 HP Models of Protein Folding

The protein folding problem of finding the native structure given the amino acid sequence of a protein is one of a class of problems concerning the relationship between the amino acid sequence space and the space of three-dimensional protein structures. Just as introducing a two-letter alphabet greatly reduces the sequence space, constraining the space of structures to compact random walks on a lattice makes the exploration of structure space more tractable. In particular the number of compact polymer structures on a $3 \times 3 \times 3$ lattice, often used in numerical studies, is 103,346, while the number of possible sequences is $2^{27} = 134,217,728$.

To gain intuition about lattice HP models, we will investigate a toy model that consists of six monomers on a 2×3 lattice. The number of possible *sequences* is $2^6 = 64$. We consider three possible structures shown in Figure 8.30(A). Beside the list of sequences and structures, the other ingredient of the model is the hydrophobic energy, which measures the extent to which the H monomers make energetically unfavorable contacts with the solvent and with P monomers. (Solvent molecules are the lattice sites on the outside surface of the six-mer.) A simple model of this interaction is to assign a free-energy penalty ε for every contact an H monomer makes with either a solvent molecule or a P monomer. These unfavorable contacts are shown as dashed lines in Figure 8.30(B). A more refined model might distinguish the interaction energy associated with an H–solvent and an H–P contact.

The protein folding problem within this toy model can be formulated in the following way: given an HP sequence, which of the possible structures minimizes the hydrophobic interaction energy? Then the lowest-energy state is identified as the native state of the protein. To shed more light on this question, we consider the example of two model sequences, HPHPHP and PHPPHP. The energies for each of these two sequences in each of the three possible compact structures are given in Figure 8.30(B). We see that the first sequence has the same energy regardless of the compact structure the six-mer assumes. This implies that, independent of temperature, the probability of finding the polymer in any of the three compact structures is 1/3. Such a sequence is not protein-like in the sense that it does not have a unique, low-energy, native structure.

On the other hand the sequence PHPPHP has a unique native structure, the Π-shaped structure shown in Figure 8.30(B). The probability of finding the chain in the native structure is proportional to the Boltzmann factor associated with its energy,

$$p_{\text{fold}} = \frac{e^{-2\beta\varepsilon}}{e^{-2\beta\varepsilon} + 2e^{-4\beta\varepsilon}}, \tag{8.73}$$

where the denominator is nothing but the partition function for the three possible structures. The probability of this toy protein adopting

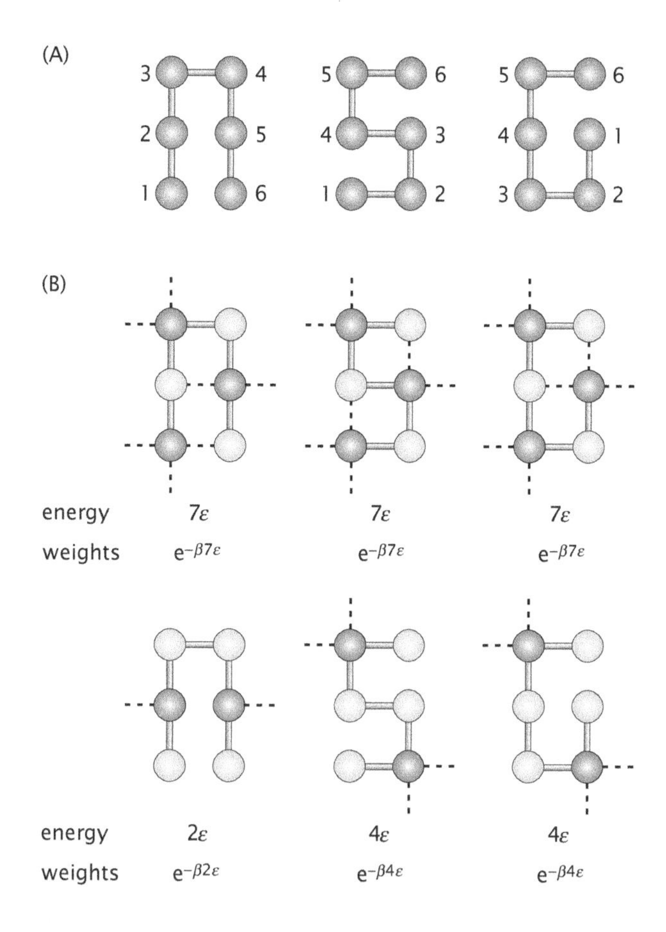

Figure 8.30: HP model of protein folding. (A) Compact structures of an HP six-mer on a 2 × 3 lattice, unrelated by symmetries. (B) The hydrophobic energy of an HP six-mer in a particular compact structure depends on its sequence. The energy function assigns a cost ε for every contact, represented here by a dashed line, between an H monomer (red) and either a P monomer (gray) or a solvent molecule. The sequence in the top panel, HPHPHP, has the same energy in all three compact structures, while PHPPHP has one structure as its unique lowest energy state, which is characteristic of protein-like sequences.

the folded state as a function of temperature is shown in Figure 8.31. Note the sigmoidal character of the plot, which is characteristic of many real proteins.

Another interesting question we can pose in the context of this toy model of folding is: what sequences are protein-like? Such questions are practically impossible to address in more realistic models of proteins given the astronomically large (literally!) number of sequences and structures. The hope is that by asking these types of questions in simple lattice models, one might uncover patterns that are also present in real proteins.

In the context of our toy model, we can address this question systematically as there are only 64 sequences to go through. For every sequence, we would need to determine its energy in each one of the

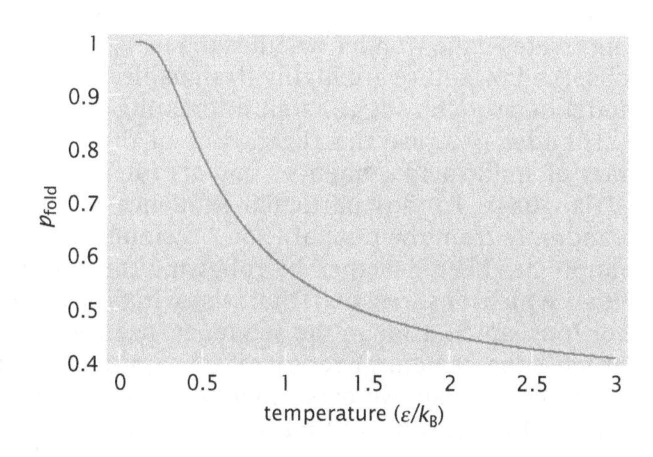

Figure 8.31: Probability of finding the PHPPHP polymer in its native state.

Figure 8.32: Protein-like sequences fold into a unique compact structure. The number of protein-like sequences varies from compact structure to compact structure. The structures with a particularly large number of protein-like sequences associated with them are highly *designable*. Hydrophobic residues are shown in red and polar residues in gray.

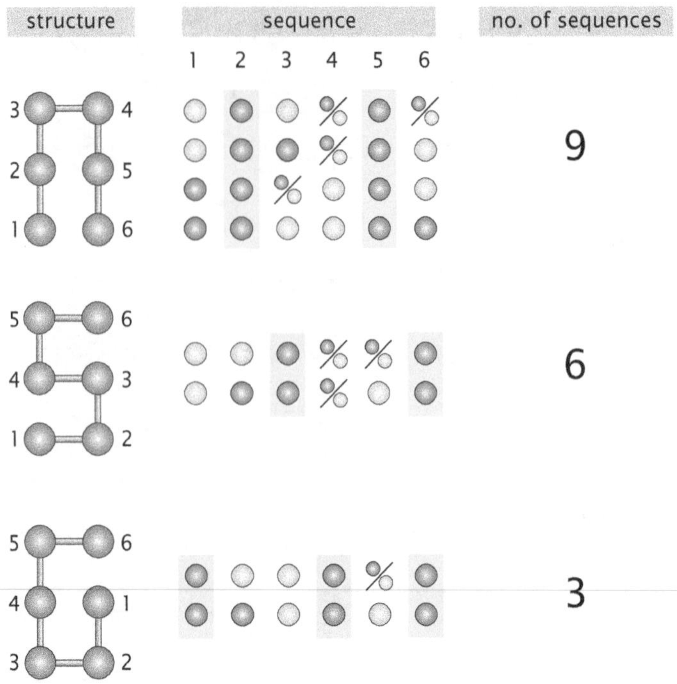

three possible compact structures, in order to identify the protein-like sequences with a unique lowest-energy structure.

Instead of going through all the sequences, a simple solution presents itself if we notice that a necessary condition for a sequence to have a unique native structure is for there to be at least one HH contact. This is the case for the PHPPHP sequence in Figure 8.30(B). Then we can construct, for each of the three possible compact structures, all the sequences that have that particular structure as its unique native state. One strategy is to begin by choosing two residues that are in contact in the chosen structure and not in any other; for example, this is the case for residues 2 and 5 in the Π structure. We make both these residues an H and then we assign an H or a P to all the other residues so that no *favorable* contacts are made in any of the other compact structures. The outcome of implementing this algorithm is shown in Figure 8.32.

An interesting feature of this model is that it predicts the Π structure to be the most designable one. Out of the 64 sequences, 9 have this particular structure as their unique ground state. The least designable structure has only three sequences that fold into it. This little calculation motivates us to wonder whether the more complex protein structures observed in Nature are highly designable or not.

The HP model of proteins suggests an interesting strategy for protein design. The idea is to use the degeneracy of the genetic code to create a library of amino acid sequences that are identical when translated into HP language. For any particular sequence, the amino acids are chosen randomly from the pool of H or P residues. For example, a four-helix bundle has been designed by following the pattern HPPHHPPHPPHHPPH..., which ensures that there is a hydrophobic residue every three or four amino acids in the sequence; see Figure 8.33. This is consistent with the structural repeat of 3.6 amino acids per turn of an α-helix. It has been shown experimentally that these sequences not only properly fold into helices but also have enzymatic activity.

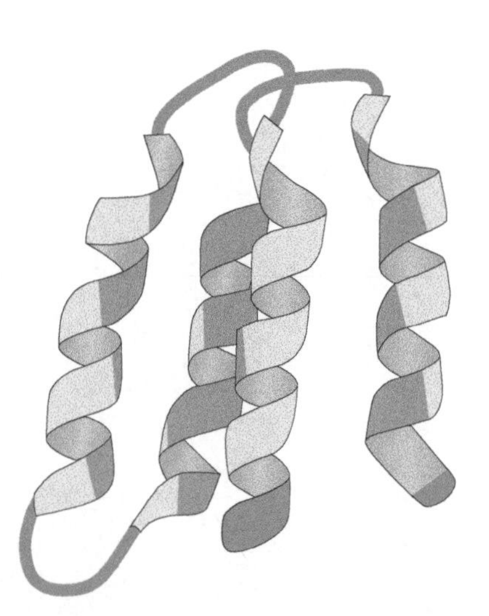

Figure 8.33: The four-helix bundle designed by using an HP sequence strategy. The HP sequence is chosen to conform to the 3.6 amino acids per turn of an α-helix. The hydrophobic residues (green) are sequestered in the interior of the bundle, while the polar ones (yellow) are on the surface facing the solvent. (Adapted from M. H. Hecht et al., *Protein Sci.* 13:1711, 2004.)

Identical design principles have been used to make β-sheets which can aggregate into structures akin to amyloid fibers.

8.5 Summary and Conclusions

The random walk model is useful in many different scientific settings. One powerful application of these ideas is to the structure and properties of polymers, including many of the "giant molecules" of life. In this chapter, we have shown how simple ideas from the physics of random walks can be used to explore the size and distribution of DNA, the force–extension properties of polymers and the emergence of entropic elasticity, and as a toy model that captures some of the features of protein folding.

8.6 Problems

Key to the problem categories: • Model refinements and derivations • Data interpretation • Database mining

• 8.1 One-dimensional random walk

(a) Perform all the steps that lead to Equation 8.14.

(b) Show that the Gaussian distribution of R for a one-dimensional random walk given in Equation 8.16 indeed has the required mean and variance.

(c) Use the Gaussian chain model to compute the entropy of a polymer chain as a function of its extension, compute the force–extension curve, and deduce the spring constant of the polymer.

• 8.2 Radius of gyration

Prove the relation between the contour length and the radius of gyration given by Equation 8.32.

• 8.3 Mean departure from the origin

Compute the mean departure of a one-dimensional random walker from its starting point. In particular, use the fact that the mean excursion can be written as $\langle R \rangle = (\langle n_r \rangle - \langle n_\ell \rangle)a$ and that the probability distribution for n_r right steps out of a total of N steps is given by the binomial distribution.

• 8.4 Binomial and Gaussian distributions

Investigate by plotting how as the number of segments N of a polymer chain is increased, the binomial end-to-end distribution becomes a Gaussian distribution. Compare the different distributions in a way analogous to Figure 8.4. Also, investigate the fractional error made by approximating the binomial distribution with the Gaussian. What conclusions do you draw?

• 8.5 Polymer configurations and the diffusion equation

Equation 8.39 characterizes the probability distribution for random walkers. Derive this equation by using the fact that the probability that the walker will be at position x at step N implies that the walker was either at $x - a$ or $x + a$ at step $N - 1$. In particular, write an equation for $p(x, N)$ in terms of $p(x \pm a, N - 1)$ and by Taylor expanding $p(x \pm a, N - 1) \approx p(x, N) +$ derivative terms.

• 8.6 Random walk in a cylinder

Use a generalization of Equation 8.39 to three dimensions to solve for the probability distribution for the end-to-end distance for a polymer tethered along the axis of a cylinder with radius R and length L at position x_0. Use your results to compare with the data shown in Figure 8.14.

• 8.7 Chromosome tethering

(a) Starting with Equation 8.34, derive Equation 8.36 for the probability density $P(r)$ for a random walk polymer in three dimensions, where r is the distance between its ends. What are the units of $P(r)$?

(b) Starting with Equation 8.34, derive Equation 8.37 for the end-to-end distance distribution in the case of a random walk polymer tethered at distance R away from one of its ends; see Figure 8.10.

• 8.8 Three-dimensional random walk and polymer cyclization

Calculate the cyclization probability of a discrete random walk in one, two, and three dimensions.

• 8.9 Freely jointed chain in three dimensions

(a) Derive Equation 8.69, use the result to derive the relation between force and extension, and make a plot of the resulting function.

(b) In the small-force limit, the force–extension curve is linear; that is, in this regime, the polymer behaves like an ideal Hookean spring with a stiffness constant $k \propto k_B T / L_{tot} a$. Demonstrate this claim and deduce the numerical factors that replace the proportionality with a strict equality.

• 8.10 Force-induced unfolding of multidomain proteins can be modeled using the random walk model

We can generalize the discussion of force–extension curves to the case of multidomain proteins. The data relevant to the particular case of the muscle protein titin has already been shown in Figure 8.22(C). The idea of the analysis we will bring to bear on this problem is shown in Figure 8.34, where it is seen that the overall contour length of the chain increases in a systematic and calculable way as a function of

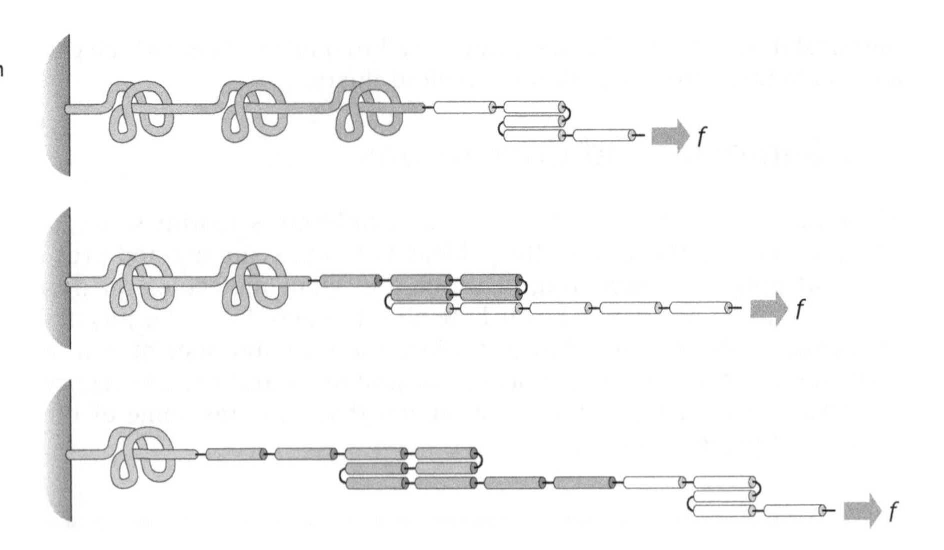

Figure 8.34: Random walk model for elastic parts of the titin force–extension curve.

the number of domains that have unfolded. Use a model like that suggested in Figure 8.34 to compare with the data shown in Figure 8.22(C). Your fits to the data should focus only on the rising parts of the force–expansion curve. The relevant data can be found on the book's website.

• 8.11 Transition between B- and S-form DNA

DNA subjected to a stretching force exceeding 60 pN undergoes a structural transition from the usual B form to the so-called S form ("S" for stretch). Here we examine a simple model of this transition based on the freely jointed chain model of DNA and compare it with experimental data.

(a) Consider the freely jointed chain model in one dimension. Each link of the polymer points in the $+x$- or the $-x$-direction. There is a force f in the $+x$-direction applied at one of the ends (see Figure 8.35A). To account for the B-to-S transition, we assume that links are of length b (B state) or a (S state), with $a > b$. Furthermore, there is an energy penalty ε of transforming the link from a B state to an S state. (This is the energy, presumably, for unstacking the base pairs.) Write down the expressions for the total energy and the Boltzmann factor for each of the four states of a single link.

(b) Compute the average end-to-end distance for one link. The average end-to-end distance for a chain of N links is N times as large.

(c) Plot the average end-to-end distance normalized by Nb (that is, the relative extension) as a function of force using the numbers appropriate for DNA, namely, $b = 100$ nm and, $a = 190$ nm. To estimate ε, take the *energy per base pair* for transforming B-DNA to S-DNA to be 5 k_BT (the length of one base pair is approximately 1/3 nm for B DNA). How does your plot compare with Figure 8.35(B)?

Relevant data for this problem are provided on the book's website. (Problem and solution suggested by Andy Spakowitz.)

• 8.12 Scaling of protein size

The scaling of a polymer's size as a function of the number of monomers is one of the central results to emerge from simple lattice models of polymers. The goal of this problem

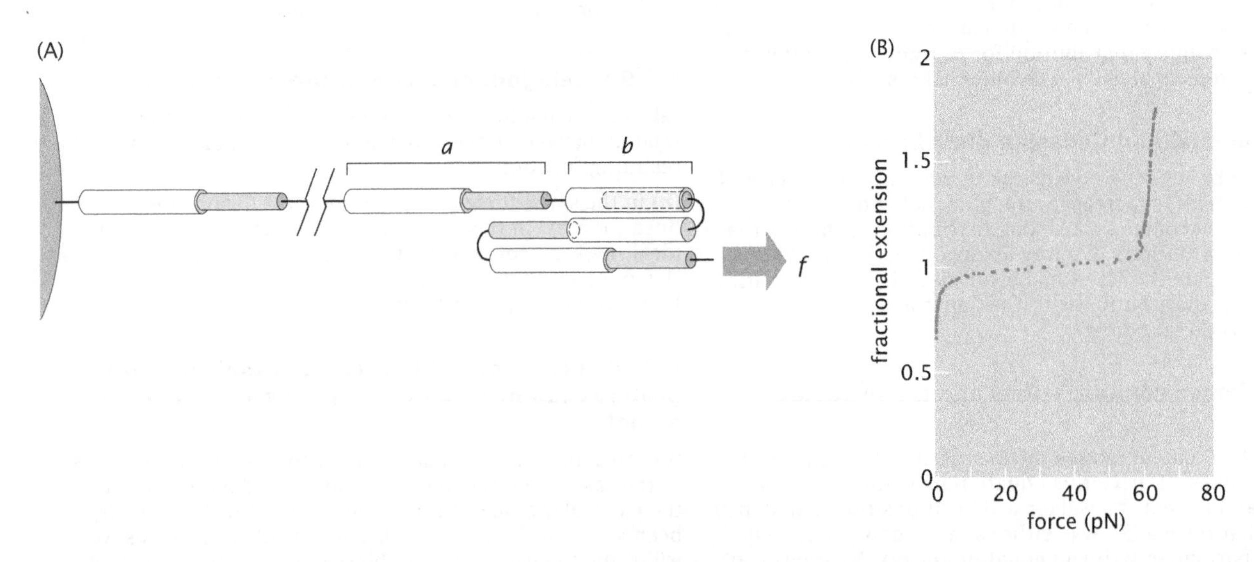

Figure 8.35: (A) Random-walk model for B to S DNA transition. The B DNA links have length b and the S DNA links have length a. (B) Force–extension curve for dsDNA. (B, adapted from C. Bustamante et al., *Nature* 421:423, 2003.)

is to investigate the extent to which such arguments are in fact appropriate for biological polymers, and, in particular, proteins. To that end, use the PDB in order to download the coordinates for a variety of globular proteins, including myoglobin, hemoglobin, bovine pancreatic trypsin inhibitor (BPTI), lysozyme, cytochrome *c*, G-actin, and tubulin. In each case, compute the radius of gyration and then make a plot that shows the radius of gyration for each of these proteins as a function of the number of residues in the protein. The goal of the problem is to reproduce Figure 8.27.

8.7 Further Reading

Grosberg, AY, & Khokhlov, AR (1997) Giant Molecules: Here, There, and Everywhere . . . , Academic Press. This book is a thoughtful discussion of polymer physics that is pleasing to novices and professionals alike. Interested readers should also see their more advanced Statistical Physics of Macromolecules (American Institute of Physics, 1994).

Benedek, GB, & Villars, FMH (2000) Physics With Illustrative Examples from Medicine and Biology: Statistical Physics, 2nd ed., Springer-Verlag. We have referred to the series by Benedek and Villars throughout the book—as always, a great source of interesting material.

Berg, H (1993) Random Walks in Biology, Princeton University Press. Berg's book is an enlightening classic. The discussion on random walks pertains to diffusion, but the understanding garnered in that setting can be used to think about polymers.

Doi, M (1995) Introduction to Polymer Physics, Oxford University Press, and Doi, M, & Edwards, SF (1986) The Theory of Polymer Dynamics, Oxford University Press. These books give the interested reader insights into the statistical physics of polymers.

Chandrasekhar, S (1943) Stochastic problems in physics and astronomy, *Rev. Mod. Phys.* **15**, 1 (1943). Chandrasekhar's amazing article is a compendium of elegant and useful results pertaining to random walks and more general ideas on stochastic processes.

Nelson, P (2004) Biological Physics: Energy, Information, Life, W. H. Freeman. Nelson's treatment of the elasticity and force–extension properties of DNA is excellent.

Poland, D, & Scheraga, HA (1970) Theory of Helix–Coil Transitions in Biopolymers: Statistical Mechanical Theory of Order–Disorder Transitions in Biological Macromolecules, Academic Press. This book illustrates the use of simple lattice models like that we used for DNA melting applied also to problems in protein folding.

De Gennes, PG (1979) Scaling Concepts in Polymer Physics, Cornell University Press. One of the great classics in the field of polymer physics. De Gennes' approach to intuitive models and simple arguments should inspire the next generation of physical biologists.

Fiebig, A, Keren, K, & Theriot, JA (2006) Fine-scale time-lapse analysis of the biphasic, dynamic behaviour of the two *Vibrio cholerae* chromosomes, *Mol. Microbiol.* **60**, 1164. An interesting example of the experimental study of chromosome geography.

Dill, KA, Bromberg, S, Yue, K, et al. (1995) Principles of protein folding—A perspective from simple exact models, *Protein Sci.* **4**, 561 and Dill, K (1999) Polymer principles and protein folding, *Protein Sci.* **8**, 1166. These articles give many interesting insights into the use of lattice models and reduced alphabet amino acid repertoires to examine protein folding.

Hecht, MH, Das, A, Go, A, Bradley, LH, & Wei, Y (2004) *De novo* proteins from designed combinatorial libraries, *Protein Sci.* **13**, 1711. This very interesting review describes the use of the HP model in carrying out protein design.

Rippe, K (2001) Making contacts on a nucleic acid polymer, *Trends Biochem. Sci.* **26**, 733. This article demonstrates some of the interesting ways that polymer physics can be used to study biological problems pertaining to chromosome structure and organization.

8.8 References

Alberts, B, Johnson, A, Lewis, J, et al. (2008) Molecular Biology of the Cell, 5th ed., Garland Science.

Bustamante, C, Smith, SB, Liphardt, J, & Smith, D (2000) Single-molecule studies of DNA mechanics, *Curr. Opin. Struct. Biol.* **10**, 279.

Bustamante, C, Bryant, Z, & Smith, SB (2003) Ten years of tension: single-molecule DNA mechanics, *Nature* **421**, 423.

Carrion-Vazquez, M, Oberhauser, AF, Fowler, SB, et al. (1999) Mechanical and chemical unfolding of a single protein: A comparison, *Proc. Natl Acad. Sci. USA* **96**, 3694.

Liphardt, J, Onoa, B, Smith, SB, Tinoco, I, Jr, & Bustamante, C (2001) Reversible unfolding of single RNA molecules by mechanical force, *Science* **292**, 733.

Marshall, WF (2002) Order and disorder in the nucleus, *Curr. Biol.* **12**, R185.

Petsko, GA, & Ringe, D (2004) Protein Structure and Function, New Science Press.

Thomas, PD, & Dill, KA (1996) An iterative method for extracting energy-like quantities from protein structures, *Proc. Natl Acad. Sci. USA* **93**, 11628.

van den Engh, G, Sachs, R, & Trask, BJ (1992) Estimating genomic distance from DNA sequence location in cell nuclei by a random walk model, *Science* **257**, 1410.

Viollier, PH, Thanbichler, M, McGrath, PT, et al. (2004) Rapid and sequential movement of individual chromosomal loci to specific subcellular locations during bacterial DNA replication, *Proc. Natl Acad. Sci. USA* **101**, 9257.

Wiggins, PA, van der Heijden, T, Moreno-Herrero, F, et al. (2006) High flexibility of DNA on short length scales probed by atomic force microscopy, *Nat. Nanotech.* **1**, 137.

Electrostatics for Salty Solutions

9

Overview: In which the behavior of salty solutions is described using electrostatics and statistical mechanics

The macromolecules of life inhabit a watery medium. The charge state of DNA and proteins is largely controlled by the ionic character of the surrounding solution. Lipid molecules form ordered bilayers as a result of the unfavorable interactions of their tails with water. In this chapter, we explore the electrostatics of charges in water and how the energetics of such charges plays a role in dictating the binding and assembly reactions that are so central to biology. The culmination of this discussion is the Poisson–Boltzmann theory, which unites simple ideas from electrostatics with the central features of statistical mechanics. These ideas will be explored particularly through case studies in viral packing and assembly.

> “Hic rhodus, hic salta.”
>
> **Aesop's Fables**

9.1 Water as Life's Aether

All life on Earth depends on water. Denial of water to cells and organisms rapidly leads to death. In Chapter 12, we will discuss some of the special mechanical properties of water that cells and organisms experience. A second and even more important aspect of water as the medium for life is its unusual ability to interact with charged molecules ranging from individual ions such as K^+ and Na^+ to enormous charged macromolecules such as DNA. Charge interactions play a very special role in the function of biological molecules because they can act over long distances. We have frequently discussed binding reactions between various macromolecules as a result of the binding energy between those molecules. However, biological function usually requires specificity. That is, a macromolecule will bind to some very specific partners but will not bind to others of similar shape and size. In many cases, binding specificity arises because of complementary patterns of charge distribution on the surface of the two macromolecules involved in the interaction.

One of the special properties of the H_2O molecule is that it can dissociate into H^+ and OH^-. For pure water, on average 1 out of every 10^7 molecules will undergo this separation. However, the likelihood of separation can be influenced by the presence of other molecules that have the tendency to bind either H^+ or OH^-. When in water solution, molecules with this property are said to alter the solution's pH. pH is defined via $\mathrm{pH} = -\log_{10}[\mathrm{H}^+]$ in a water-based solution. Because proteins contain many amino acid side chains that can either accept or donate H^+ ions to the water solution, their overall structure is strongly dependent on pH. Cells frequently take advantage of the pH dependence of protein structure as a regulatory mechanism and therefore feature many pathways for modulating local and global pH. For example, in the human body, blood pH is very tightly regulated at a normal value of around 7.3 and an increase or decrease in blood pH by a single unit can cause death. However, some compartments within cells may have very different pH's. For example, lysosomal compartments involved in protein degradation usually maintain a pH below 4. Cells manipulate pH by actively transporting H^+ ions across membranes. Other effects of charge transport across membranes will be discussed in Chapter 17.

ESTIMATE

Estimate: The Eighth Continent It has been said that more is known about our nearby celestial neighbors than about our own mysterious oceans. These giant liquid reservoirs have been the scene for some of life's greatest evolutionary stories ranging from the tiny world of microbes and their viruses to the majestic whales that are the largest animals on the planet. Indeed, the subject of biogeography has made much of the biodiversity of the different continents, though the Earth's great oceans harbor some of the weirdest organisms of all, making them perhaps the greatest "continent" of them all.

The backdrop for life in the ocean is a salty environment characterized by roughly 1 mole of salt for every kilogram of water. Given that there are 18 g of water per mole and 1000 g of water in each such kilogram, this means that the number of moles of water is

$$\text{moles of water in a kilogram} = \frac{1000\ \text{g/L}}{18\ \text{g/mol}} \approx 55.5\ \text{M}. \qquad (9.1)$$

What this means in turn is that the mole fraction of salt in the ocean is

$$\text{mole fraction of salt} = 1/55.5 \approx 0.018. \qquad (9.2)$$

More extreme examples such as Mono Lake in California and the Great Salt Lake in Utah are characterized by concentrations of salt between 3 and 8 times higher than that found in the ocean, with these high concentrations placing special demands on the living occupants of these lakes.

In Section 6.2.3, we already saw the significance of these kinds of salt concentrations for osmotic pressure and we may use the relation $\Delta\Pi = ck_BT$ to derive a simple rule of thumb that a concentration difference of 100 mM corresponds to a pressure difference of

$$\frac{100 \times 10^{-3} \times 6 \times 10^{23}}{1000\ \text{cm}^3} \times \frac{10^6\ \text{cm}^3}{1\ \text{m}^3} \times 4 \times 10^{-21}\ \text{J}$$
$$\approx 24 \times 10^4 \frac{\text{N}}{\text{m}^2} = 2.4\ \text{atm}. \tag{9.3}$$

When seen in this way, it is clear that the salt content of life's watery medium can lead to large osmotic pressures.

The utility of pH regulation of protein structure is vividly illustrated during the invasion of host cells by influenza virus. Influenza virus is surrounded by a membrane containing many copies of a protein called hemagglutinin (HA) that is capable of mediating membrane fusion. When the virus is outside a host cell, for example when it is being scattered in the sneeze droplets of an infected person, HA is in an inactive conformation. Only after being taken up by a cell into a compartment with low pH does the HA protein change into a form that can fuse with the membrane as shown in Figure 9.1. In this way, the virus takes advantage of the host cell's preference to regulate pH within certain cellular compartments as a clue to let the virus know that it has been taken up by an unsuspecting cell and should initiate its program of replication and destruction.

The present chapter examines water, animated by the view that water is the medium of biochemical life. The reader is reminded of Figure 2.4 (p. 42), in which we carried out the single-molecule census of a typical bacterium. From our present perspective, what is particularly important about the numbers quoted there is that they reveal the enormous quantity of both water molecules and their attendant ions. Using the rough estimate that 70% by mass of a typical cell is water, we found that there are roughly 2×10^{10} water molecules in each bacterium. Similarly, estimates of the concentrations (mM) of ions such as K^+ and Na^+ tell us that there are roughly 10^7 ions populating a typical bacterium.

In addition to the interesting mechanical questions that can be raised about biological fluids in motion to be taken up in Chapter 12, it is also of interest to examine the way in which water serves as a kind of aether for biochemical action. One of our key points in undertaking

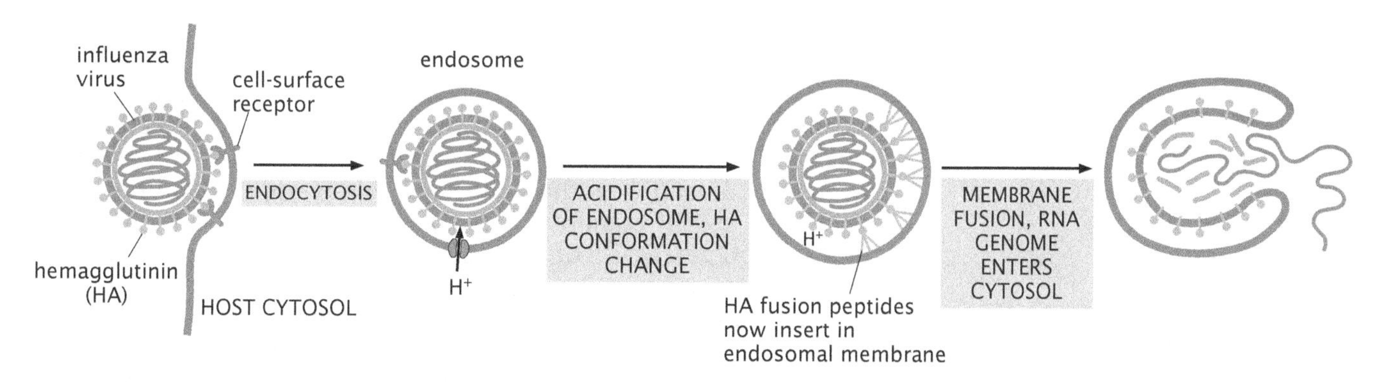

Figure 9.1: Influenza virus uptake. This diagram shows the pH-dependent process by which the nucleocapsid of the influenza virus is delivered into the cytoplasm of a mammalian host cell. The virus binds to receptors on the cell surface and is taken up by receptor-mediated endocytosis. After endocytosis, specialized proton pumps are added to the endosomal membrane, transporting H^+ ions from the cytoplasm, consequently causing a drop in pH. The pH decrease causes a dramatic conformational change in the influenza envelope protein, HA. The extended, low-pH form of the protein catalyzes fusion of the viral membrane with the endosomal membrane, releasing the virus contents into the cytoplasm of the cell. (Adapted from B. Alberts et al., Molecular Biology of the Cell, 5th ed., Garland Science, 2008.)

an examination of the role of water as the seat of biochemical action will be a recognition of the central importance of charges that both are free in solution and can be liberated from their macromolecular hosts. In particular, we will examine how these issues are central to questions such as the salt dependence of DNA–protein interactions and the assembly of viruses.

9.2 The Chemistry of Water

9.2.1 pH and the Equilibrium Constant

Dissociation of Water Molecules Reflects a Competition Between the Energetics of Binding and the Entropy of Charge Liberation

To examine pH from a quantitative perspective requires that we consider the interplay of energetic and entropic effects in dictating charge separation. In particular, the competition between energy minimization (bound water molecules) and entropy maximization (ionic dissociation of water) sets the stage for many biological reactions and is a compelling and enlightening example of the uses of the law of mass action developed in Section 6.3 (p. 267). The reaction of interest can be written as

$$H_2O \rightleftharpoons H^+ + OH^-, \tag{9.4}$$

with the idea being that we are interested in finding out what fraction of the water molecules in a sample of water are in a dissociated state. For the purposes of notational simplicity, we replace the ν_i favored in the general derivation of Equation 6.104 (p. 269) with ν_{H_2O}, ν_{H^+}, and ν_{OH^-}, with the further realization that for this reaction we have $\nu_{H_2O} = -1$, $\nu_{H^+} = 1$, and $\nu_{OH^-} = 1$. Using these definitions, the law of mass action (see Section 6.3 on p. 267) for the problem of dissociation becomes

$$\frac{[H^+][OH^-]}{[H_2O]} = \frac{[H^+]_0[OH^-]_0}{[H_2O]_0} \exp\left(-\frac{\mu^0_{H^+} + \mu^0_{OH^-} - \mu^0_{H_2O}}{k_B T}\right). \tag{9.5}$$

What this equation tells us may be understood as follows. The left-hand side tells us the ratio of the concentrations of products to reactants at a given temperature. The right-hand side is the dissociation constant and tells us the balance between reactants and products in terms of a few key quantities. First, when we write $[A]_0$, we refer to the concentration of species A in some *standard state*. Ultimately, the choice of standard state is arbitrary and is usually made for reasons of convenience. In addition, the right-hand side features quantities such as μ^0_i, which are the standard state chemical potentials.

For the concrete case of dissociation of water being considered here, more can be said. We begin by assuming that the presence of H^+ ions results strictly from dissociation. This means in turn that $[H^+] = [OH^-]$. A further simplification arises from the realization that, for all practical purposes, $[H_2O] = [H_2O]_0$. This claim results from the fact that the reaction in Equation 9.4 leads to so little product that it barely perturbs the initial concentration of water, $[H_2O]_0 = 55\,M$. As a result, we can assume $[H_2O] = [H_2O]_0$ in Equation 9.5. This means that the denominators on both sides of the equation cancel. Given that $[H^+]_0 = [OH^-]_0 = 1\,M$ and using the fact that the measured energy change in the reaction is $\mu^0_{H^+} + \mu^0_{OH^-} - \mu^0_{H_2O} \approx 79.9\,kJ/mol$, we find

$[H^+][OH^-] = [H^+]^2 = 1.0 \times 10^{-14}\,M^2$, and hence $[H^+] = 1.0 \times 10^{-7}\,M$. Recalling that the definition of pH is

$$pH = -\log_{10}[H^+], \tag{9.6}$$

where $[H^+]$ is in molar units, we see that we have recovered pH = 7 for water under standard conditions. To gain further intuition about pH, in Figure 9.2 we plot the average distance between H^+ ions as a function of the pH as another way of viewing the significance of the pH concept.

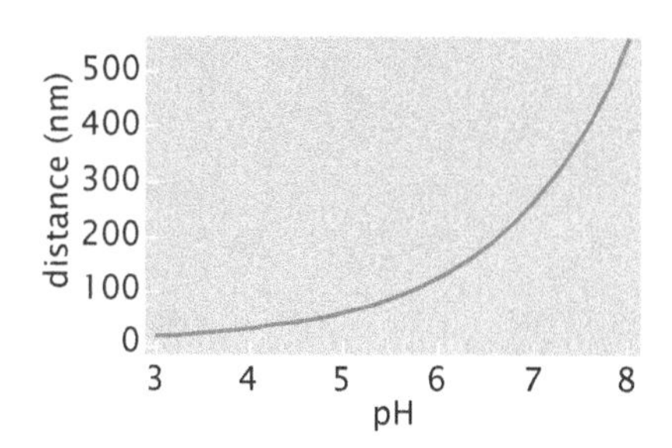

Figure 9.2: Distance between H^+ ions as a function of pH.

9.2.2 The Charge on DNA and Proteins

The Charge State of Biopolymers Depends upon the pH of the Solution

We have already noted that the charge state of macromolecules is a critical factor in determining both their structure and function. To investigate the way that this charge state is tuned, we consider the generic reaction

$$HM \rightleftharpoons H^+ + M^-, \tag{9.7}$$

where M is the macromolecule of interest. From our earlier discussion of the law of mass action (see Section 6.3 on p. 267), we note that there is a dissociation constant for this reaction given by

$$K_d = \frac{[H^+][M^-]}{[HM]}. \tag{9.8}$$

A useful measure of the tendency of a particular molecule to undergo the dissociation reaction is known as the pK and is defined by

$$pK = -\log_{10} K_d. \tag{9.9}$$

If we take the logarithm of both sides of Equation 9.8, we are left with

$$-\log_{10} K_d = -\log_{10}[H^+] - \log_{10}[M^-] + \log_{10}[MH]. \tag{9.10}$$

Recalling the definitions of both pK and pH we see that this may be rewritten as

$$pH = pK + \log_{10}\frac{[M^-]}{[MH]}, \tag{9.11}$$

a result known as the Henderson–Hasselbalch equation. The physical significance of this result may be seen by setting $[M^-] = [MH]$. In this case, we see that pK corresponds to that pH at which half of [MH] has dissociated. Hence, when pK values are quoted for a given molecule, this tells us that at a pH with the same numerical value, half of the molecules will have suffered the dissociation reaction.

To see these ideas in action, we begin by considering the case of DNA, which has $pK \approx 1.0$. This tells us that at normal pH, the phosphates on the DNA backbone are essentially fully dissociated, resulting in two electronic charges for every base pair, or a linear charge density of $\lambda = 2e/0.34\,\text{nm}$. DNA is a strong acid and readily surrenders the charges on its phosphates.

Different Amino Acids Have Different Charge States

A more compelling application of these ideas is to the charge state of proteins since different side groups have different dissociation tendencies, resulting in the fact that at different pHs, different side groups will be in different states of dissociation (we already saw this in the titration curve in Figure 5.9 on p. 204). In Chapter 10, we will discuss the spontaneous self-assembly of the nucleosome as DNA segments wrap onto histone octamers. We will see that the lysine and arginine residues on the histones present positive charges that interact favorably with the negative charges on the DNA and make it possible to overcome the steep elastic cost of bending the DNA to fit on the histones. More generally, the interplay of protein charge and pH is one of the primary ways that the repertoire of protein responses is extended.

9.2.3 Salt and Binding

Many events in the lives of biological molecules involve specific recognition of one molecule by another. Some examples we have considered (see Chapter 6) include RNA polymerase binding to DNA specifically at promoter sites and receptors binding their specific ligands. The surface distribution of charges on these macromolecules plays a critical role in determining the specificity of these recognition events as well as their strength. The importance of electrostatics in macromolecular binding is easily observed in the laboratory because nearly all such binding interactions are strongly dependent on the concentration of ions in the solution in which they are measured.

One set of examples of salt effects on binding are those related to DNA–protein interactions, and especially the way in which regulatory proteins (transcription factors) bind to DNA. In Section 6.1.2 (p. 244) we examined the binding of RNA polymerase to promoters. We also noted that under most circumstances, the decision to create an mRNA transcript at a gene of interest is dictated by the binding of repressors and activators, which often work by inhibiting or enhancing the binding of polymerase. The case study that will run as a thread through the entire book is that of the *lac* operon, introduced in Figures 4.13 (p. 155) and 4.15 (p. 158). An example of how the affinity of a transcription factor (that is, Lac repressor) for DNA is tuned by the concentration of salt is shown in Figure 9.3.

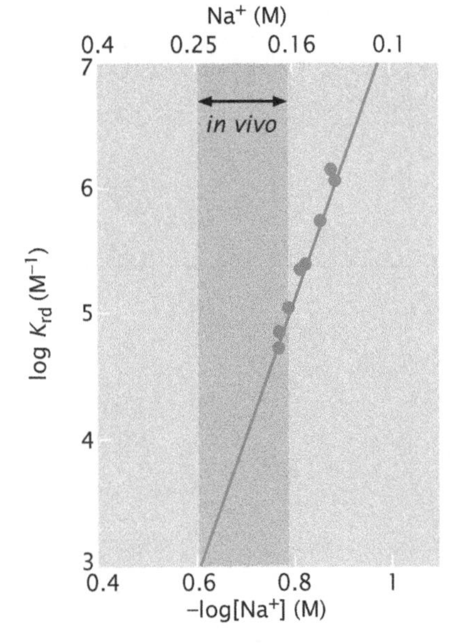

Figure 9.3: Salt dependence of protein–DNA binding. Equilibrium constant for Lac repressor binding to nonspecific DNA as a function of Na^+ concentration. The range of *in vivo* salt concentrations is shaded. (Adapted from Y. Kao-Huang et al., *Proc. Natl Acad. Sci. USA* 74:4228, 1977.)

There are several intuitive ways to begin to think about these results. One of the ideas to be explored in the remainder of the chapter is that the ions in salty solutions assemble in a screening cloud around the macromolecules that have the potential to bind. When this binding reaction occurs, there is a release of the ions in these screening clouds, which results in an entropy increase. A second feature that can be attributed to such binding interactions is that, with increasing ion concentration, the ions can actually interact with the receptor in a way that competes with the ligands. Toy models that explore these effects are presented in the problems at the end of the chapter, though we note that there are many subtleties that these toy models do not even begin to acknowledge.

9.3 Electrostatics for Salty Solutions

As shown above, the use of measured equilibrium (or dissociation) constants can take us a long way toward understanding the charge state of macromolecules in solution. On the other hand, to really examine the microscopic underpinnings of the equilibrium constant, we need to be able to assess electrostatic potentials. Further, a host of problems, such as the energetics of viral DNA packing, the molecular origins of nucleosomal stability, and the nature of membrane potentials, require us to have certain key facts about electrostatics in hand. To that end, the present section provides a primer on electrostatics.

9.3.1 An Electrostatics Primer

We have seen that charge in solution can be moved around. The extent to which charges leave their macromolecular hosts is dictated by a competition between energy and entropy. The energy of interaction between positive and negative charges favors reducing the separation between them. On the other hand, the entropic advantage of letting charges wander in solution drives the system towards charge liberation. To see how that competition works we need an explicit formula for the interaction energy between charges. The first physical fact that most of us learn about charged objects is that like charges repel and unlike charges attract. To go further with this qualitative observation, we need a way to assign energies to different configurations of charges. This question falls within the province of the broad subject of electrostatics, which we will describe in some detail throughout the remainder of the chapter.

In this mini-review of electrostatics, the main concept that we develop is that of the electric potential and its relation to a distribution of charges. We begin by recalling the force between two charged particles,

$$F = \frac{1}{4\pi\varepsilon_0 D} \frac{q_1 q_2}{r^2}, \tag{9.12}$$

where we have assumed that the particles are much smaller than the distance separating them (so-called "point charges"). In the formula for the force, the constants are given by $1/4\pi\varepsilon_0 = 9 \times 10^9\,\mathrm{N\,m^2/C^2}$ and D is the dielectric constant ($D = 80$ for water), q_1 and q_2 are the charges measured in coulombs, while r is the distance between the two charged particles. If the charges are of the same sign, the force F is repulsive, while if they are oppositely charged, they will attract. This is the celebrated Coulomb law. Strictly speaking, this formula relates vector quantities. For our purposes, we note that the force is along the line joining the two charges.

In air and other nonpolar media, the dielectric constant (D appearing in Equation 9.12) is small (≈ 1), while in polar solvents like water, D is large (≈ 80). We can understand this based on the schematic shown in Figure 9.4, where the electrostatic force is diminished between two point charges due to the intervening molecules, which have permanent dipoles. The origin of these permanent dipoles is illustrated in Figure 9.5 and results from molecular asymmetries. The way that

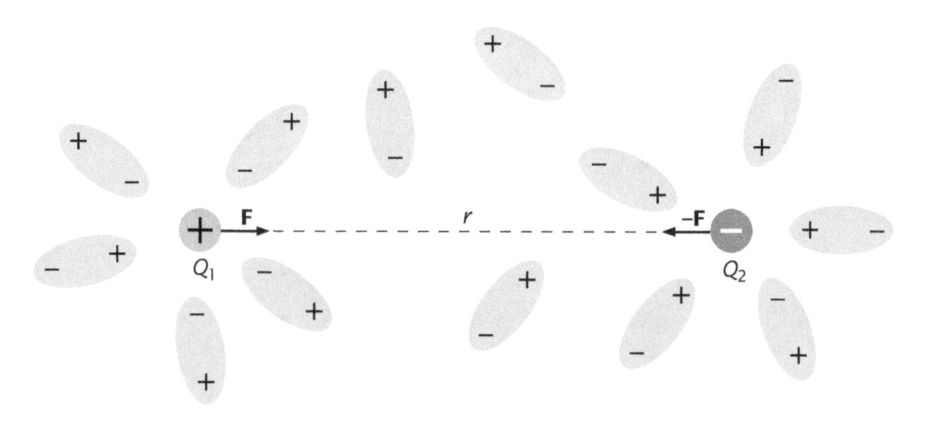

Figure 9.4: Illustration of polarization of molecules by an electric field. A positive and negative charge in solution interact with the molecules in their vicinity. If those neighboring molecules have a permanent dipole, they will be oriented so that their charges are favorably aligned.

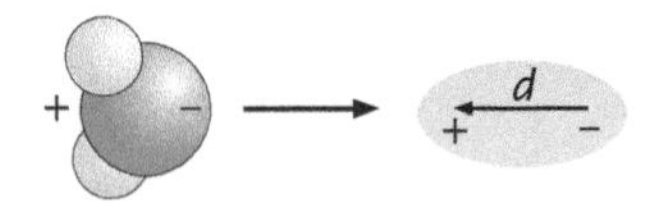

Figure 9.5: The dipole moment of water. Asymmetries in the distribution of charge in the water molecule lead to a permanent dipole moment that can be represented by thinking of two opposite charges separated by a distance d.

screening is accomplished through the intervention of these permanent dipoles is by having the negative charge of the dipoles oriented toward the positive point charge and the positive charges of the dipoles oriented toward the negative point charge as illustrated in Figure 9.4. As will be shown later, in the presence of free charges, such as salts, this screening effect is even more extreme.

A Charge Distribution Produces an Electric Field Throughout Space

Instead of considering the forces between charges, we can introduce the concept of an *electric field* **E**, so that the force acting on a test charge q_{test} placed in the field is $q_{\text{test}}\mathbf{E}$. Note that this change in perspective places the emphasis on the space surrounding charges (that is, on the fields) rather than on the charges themselves. For the case of a point charge Q at the origin, from Equation 9.12 we derive the expression

$$\mathbf{E}(\mathbf{r}) = \frac{1}{4\pi\varepsilon_0 D}\frac{Q}{r^2}\,\mathbf{e}_r, \tag{9.13}$$

for the electric field at position $\mathbf{r}$; $\mathbf{e}_r$ is a unit vector along the radial direction $\mathbf{r}$. In the case of a charge distribution, where charges q_i are placed at specified positions $\mathbf{r}_i$, the electric field at position $\mathbf{r}$ is obtained by summing contributions from each of the individual charges,

$$\mathbf{E}_{\text{tot}}(\mathbf{r}) = \sum_i \mathbf{E}_i = \sum_i \mathbf{E}(\mathbf{r} - \mathbf{r}_i). \tag{9.14}$$

The fact that the total field is simply obtained by adding the fields that would have been produced by individual charges assuming all other charges were absent is the *principle of superposition.*

The concept of a charge distribution is very useful for describing charged macromolecules such as proteins and DNA in solution. Instead of a discrete distribution like that described above, which is specified by charges q_i and their positions $\mathbf{r}_i$, it is often useful to introduce a continuous charge distribution specified by a density $\rho(\mathbf{r})$. This idea is represented in Figure 9.6. Given the charge density, the charge ΔQ within a small volume ΔV located at position $\mathbf{r}$ is $\rho(\mathbf{r})\Delta V$. To compute the field due to a charge distribution specified by a charge density, we proceed in the same fashion as in the discrete case, using the principle of superposition. The only difference is an operational one whereby we substitute the sum over discrete charges in Equation 9.14 with an integral over volume elements dV. Various examples of the principle of superposition for both discrete and continuous charge distributions are shown in Figure 9.7.

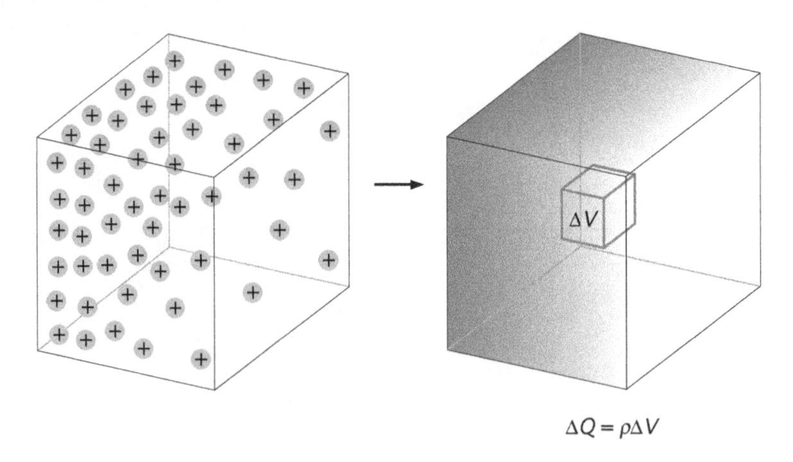

Figure 9.6: Representation of a discrete charge distribution as a continuous charge distribution. The schematic on the left shows a discrete arrangement of charges. In the continuum limit shown on the right, a small volume element ΔV is considered and we count up the number of charges ΔQ in that volume.

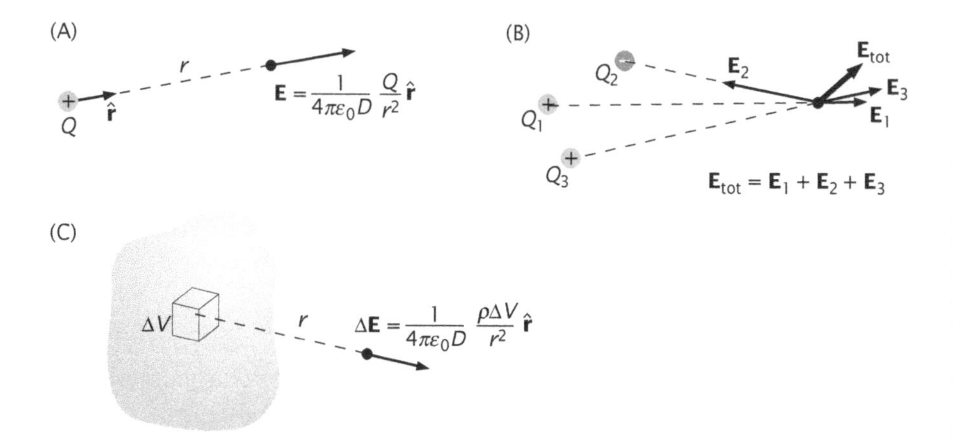

Figure 9.7: Electric field of a charge distribution. (A) For a single charge, the field lines emanating from the charge are directed radially outward (positive charge) or inward (negative charge). (B) For the case of a discrete set of charges, the electric field at a given point is obtained by computing the vector sum of the electric fields due to the individual charges. (C) For a continuous charge distribution, the electric field at some point is obtained by adding up (by integration) the contributions to the field due to every little volume element in the charged body.

The Flux of the Electric Field Measures the Density of Electric Field Lines

Often, a powerful alternative to the integral formulation for calculating fields is to use so-called *local* descriptions that involve differential equations relating the field and charge density at a point. For electrostatics, we now develop a local relation that connects the electric field at position $\mathbf{r}$ to the charge density at the same position. To uncover this relation, we consider the *flux* of the electric field. If we think of the lines of electric field distributed through space, we define the flux as the number of lines per unit area, where the area element is perpendicular to the field direction. Lines that penetrate the surface and point inwards give a negative contribution to the flux, while the outward pointing ones give a positive contribution. For a closed surface, we can evaluate the total flux through that surface by visiting each little area element of the surface and summing the flux through each such element. Mathematically, this operation amounts to an integration in which we break the surface up into small area elements ΔA_i with surface normals $\mathbf{n}_i$, and for each element we compute $\mathbf{E}(\mathbf{r}_i) \cdot \mathbf{n}_i \Delta A_i$. Note that these surface normals $\mathbf{n}_i$ are vectors and we will adopt the convention that they are pointing outward away from the surface. The flux Φ is then the sum of all such terms from all the surface elements i.

For example, the flux of the electric field due to a point charge is calculated by integrating the formula for the field in Equation 9.13 over the surface of a sphere. In this case, the flux is given by

$$\Phi = \int_{\partial V} \mathbf{E}(\mathbf{r}) \cdot \mathbf{n} \mathrm{d}A = \int_{\partial V} \frac{1}{4\pi\varepsilon_0 D}\frac{Q}{r^2}\mathbf{e}_r \cdot \mathbf{e}_r \,\mathrm{d}A. \quad (9.15)$$

The resulting integral can be evaluated simply, resulting in

$$\Phi = \frac{1}{4\pi\varepsilon_0 D}\frac{Q}{r^2}4\pi r^2 = \frac{Q}{D\varepsilon_0}. \quad (9.16)$$

Note that the flux does not depend on the size of the sphere. This is in line with the intuitive definition of the flux as the number of field lines piercing the sphere. If we take into account the fact that field lines start and end only at charges, then the number of field lines through the sphere will not depend on how big it is. Moreover, this number will not change if we deform the sphere to an arbitrary shape. Examples of the flux are shown in Figure 9.8. If we use the principle of superposition, we can conclude that the relation between the flux of the electric field through a closed surface, and the total charge within

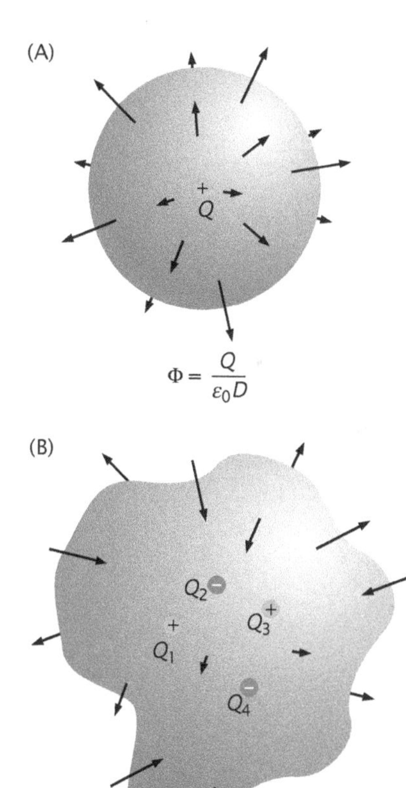

Figure 9.8: Electric flux and Gauss' law. (A) Flux of the field of a point charge through a spherical surface. (B) Generalization of the case shown in (A) to an arbitrary charge distribution and an arbitrary closed surface.

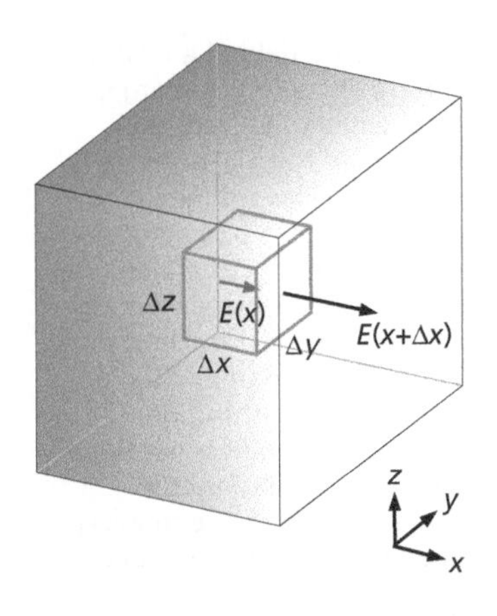

Figure 9.9: Electric flux emanating from a box. A small volume element with edge lengths Δx, Δy, and Δz is used to compute the flux. The flux on the faces perpendicular to the x-direction is given by $E_x(x+\Delta x)\Delta y \Delta z - E_x(x)\Delta y \Delta z$, where the negative sign results from the fact that the electric field at x is pointing inward.

it, Q_{tot}, is given by Equation 9.16 with Q replaced by Q_{tot}. This is a statement of Gauss' law.

This same result applies just as well to a spherical charge distribution that has a charge Q spread uniformly throughout its volume. In this case, we evaluate the flux through a surface with radius larger than that of the ball of charge itself. The total flux of the electric field through a sphere of radius r is $4\pi r^2 E(r)$, while the enclosed charge is Q. From Equation 9.16, the field is

$$E(r) = \frac{1}{4\pi\varepsilon_0 D}\frac{Q}{r^2}$$

which is the same as for a point charge with effective charge Q placed at the center of the ball.

The point of these results is that they illustrate the key message of Gauss' law, namely, that the net flux through a given closed surface is given by

$$\Phi = \frac{Q_{tot}}{D\varepsilon_0}. \tag{9.17}$$

Though this equation may seem like an abstraction, its power comes from the fact that this expression can be used to deduce a key governing equation for electrostatics. To see that, consider a charge distribution where the charge density only depends on the coordinate x while it is uniform in y and z, as depicted in Figure 9.9. This will produce a nonuniform field in the x-direction. If we consider a small rectilinear volume element with sides Δx, Δy, and Δz, the charge contained within it will be $\Delta Q = \rho(x)\Delta x \Delta y \Delta z$. The flux of the electric field through the outer surface of this volume element is

$$\Phi = E_x(x+\Delta x)\Delta y \Delta z - E_x(x)\Delta y \Delta z, \tag{9.18}$$

where the contributions to the flux from the surface elements perpendicular to the y- and z-directions cancel as the field is uniform in these directions. The subscript x in E_x refers to the fact that we are dealing with the x-component of the electric field. If we Taylor-expand using $E_x(x+\Delta x) \approx E_x(x) + (\mathrm{d}E_x(x)/\mathrm{d}x)\Delta x$, we find

$$\Phi = \frac{\mathrm{d}E_x(x)}{\mathrm{d}x}\Delta x \Delta y \Delta z. \tag{9.19}$$

Substituting this value of the flux in Equation 9.17 and using $\Delta Q = \rho(x)\Delta x \Delta y \Delta z$, we arrive at the relation

$$\frac{\mathrm{d}E_x(x)}{\mathrm{d}x} = \frac{\rho(x)}{D\varepsilon_0}, \tag{9.20}$$

which is yet another version of Gauss' law. The case of an arbitrary charge distribution can be handled using a similar strategy, resulting in

$$\frac{\partial E_x(x,y,z)}{\partial x} + \frac{\partial E_y(x,y,z)}{\partial y} + \frac{\partial E_z(x,y,z)}{\partial z} = \frac{\rho(x,y,z)}{D\varepsilon_0}. \tag{9.21}$$

This result is the most general form of Gauss' law and, as promised, provides a local relation between the electric field and the charge density. The reader is asked to derive this result in the problems.

The Electrostatic Potential Is an Alternative Basis for Describing the Electrical State of a System

So far we have described the electric field produced by a charge distribution in terms of the vector field $\mathbf{E}(\mathbf{r})$. What this means is that to specify the field we need to specify three numbers, the three components of the vector **E**, at every point in space. A simpler description of the field is obtained in terms of the electric potential $V(\mathbf{r})$, which is a scalar field. What is remarkable is that this description based upon specifying a single number (the electric potential $V(\mathbf{r})$) at every point in space is completely equivalent to providing the electric field *vector* at every point. To come to grips with this, we must understand the relation between the three components of the electric field and the electric potential.

The electric potential at position **r** is defined as the work per unit charge done in bringing a test charge from a point at infinity to the position **r**. The process of transporting the charge is assumed to be very slow so that at any given instant the force applied on the charge is exactly cancelled by the force due to the electric field. To get a better understanding of this definition, it is best to work out an example. Consider the case of an electric field due to a charge distribution that varies along the x-direction only, as was introduced above. From the symmetries of this charge distribution, we see that the electric field points along the x-direction. Since the electric force on a unit test charge in this case is $E_x(x)\mathbf{e}_x$, where $\mathbf{e}_x$ is the unit vector along the x direction, and the displacements when bringing the charge from infinity are $\Delta x(-\mathbf{e}_x)$, the electric potential at position x is

$$V(x) = \int_x^{\infty} \{[-E_x(x')\mathbf{e}_x] \cdot [-\mathbf{e}_x \mathrm{d}x']\} = \int_x^{\infty} E_x(x')\,\mathrm{d}x'. \tag{9.22}$$

The first minus sign is in the integral since the applied force on the charge is equal and opposite to the force of the electric field. We have also made use of the expression for the work of a force **F** over a displacement element $\Delta\mathbf{x}$, $\Delta W = \mathbf{F}\cdot\Delta\mathbf{x}$. Finally, differentiating the left- and the right-hand sides of Equation 9.22, we arrive at a relation between the electric field and the potential

$$E_x(x) = -\frac{\mathrm{d}V(x)}{\mathrm{d}x}. \tag{9.23}$$

These ideas are depicted graphically in Figure 9.10. The idea is to compute the work done (that is, the change in energy) upon moving the test charge between two neighboring points. In Figure 9.10(A), we show the energy cost to move a test charge between the points 0 and X in a *uniform* field. Using the definition of the electric potential given in Equation 9.22, we have the result

$$V(x) - V(0) = -\int_0^X E_x(x')\,\mathrm{d}x'. \tag{9.24}$$

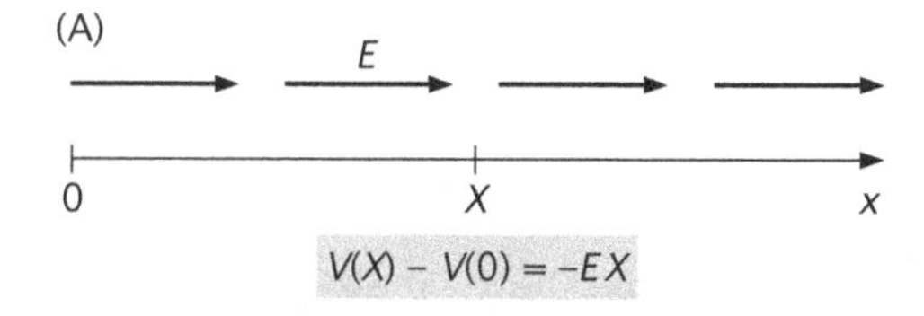

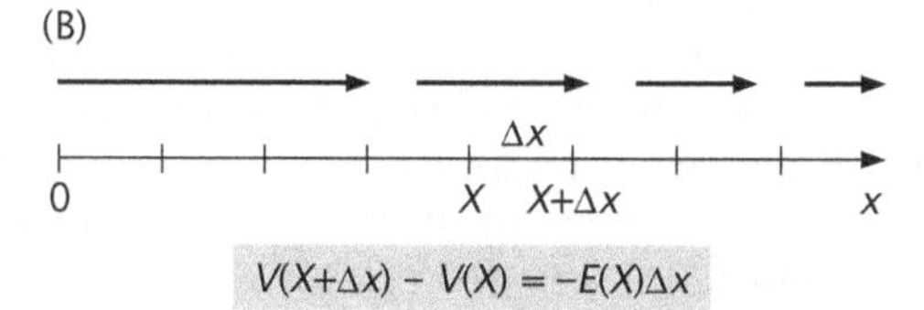

Figure 9.10: Local relation between the electric field and the electrostatic potential. (A) For the case of a uniform electric field, the work done in moving a test charge between X and the origin is given by $-EX$. (B) The change in energy for moving a test charge between X and $X + \Delta x$ in a nonuniform field is given by $-E(X)\Delta x$.

For the special case of a uniform field, the integral results in $V(x) - V(0) = -E_x x$. More importantly, if we are interested in the difference in potential between two neighboring points, we have

$$V(x + \Delta x) - V(x) = -\int_x^{x+\Delta x} E_x(x')\,\mathrm{d}x' \approx -E_x(x)\Delta x. \tag{9.25}$$

This last result is equivalent to Equation 9.23.

In a more general setting, when the charge distribution has variation in all three directions, this relation generalizes to

$$\left(E_x(x, y, z), E_y(x, y, z), E_z(x, y, z)\right) = -\left(\frac{\partial V(x, y, z)}{\partial x}, \frac{\partial V(x, y, z)}{\partial y}, \frac{\partial V(x, y, z)}{\partial z}\right). \tag{9.26}$$

In words, Equation 9.26 states that, given the electric potential, the electric field is a vector pointing in the direction of most rapid decrease of the potential, with a magnitude equal to the rate of change of the potential along this direction. Finally, substituting the relation between V and $\mathbf{E}$ into Gauss' law as given by Equation 9.21, we arrive at the Poisson equation

$$\frac{\partial^2 V(x, y, z)}{\partial x^2} + \frac{\partial^2 V(x, y, z)}{\partial y^2} + \frac{\partial^2 V(x, y, z)}{\partial z^2} = -\frac{\rho(x, y, z)}{D\varepsilon_0}, \tag{9.27}$$

which relates the electric potential to the charge density. This governing differential equation is the jumping off point for examining the fields and energies due to charges around macromolecules.

The Math Behind the Models: The Gradient Operator and Vector Calculus In writing Equations 9.26 and 9.27, the notation has become cumbersome and brings us to one of the most powerful tools in the arsenal of mathematical physics, the so-called vector calculus. In particular, given a scalar field such as $V(\mathbf{r})$, we can compute an associated vector (known as the gradient vector) of the form

$$\mathbf{E}(\mathbf{r}) = -\nabla V(\mathbf{r}), \tag{9.28}$$

where we define the gradient operator (often called just "grad") as

$$\nabla = \left(\frac{\partial}{\partial x}, \frac{\partial}{\partial y}, \frac{\partial}{\partial z}\right). \tag{9.29}$$

When operating on a scalar function such as $V(\mathbf{r})$, the gradient operator returns a vector. Similarly, we can rewrite Equation 9.27 as

$$\nabla^2 V(\mathbf{r}) = -\frac{\rho(\mathbf{r})}{D\varepsilon_0}. \tag{9.30}$$

The operator ∇^2 is often denoted as the "Laplacian" of the scalar function of interest.

Though for our purposes ∇ and ∇^2 will serve as a convenient shorthand for writing expressions involving various partial derivatives, these operators have a deeper mathematical and physical significance that is sometimes obscured by writing out the full expressions as we have done in Equations 9.26 and 9.27. The reader is invited to work through this beautiful topic in the book by Schey (1996).

MATH

There Is an Energy Cost Associated With Assembling a Collection of Charges

Cells invest a great deal of their energy budget in moving charges around. To understand better the energetic implications of charge management, we introduce the energy associated with a charge distribution. In particular, we consider a charge distribution and we ask the question: what is the work done in bringing isolated charges from far away, where they do not interact, to form such a distribution? The electric energy is equal to this work. A positive value of the electrical energy means that another form of energy has to be expended in order to assemble the charge distribution. For example, consider a charge distribution consisting of two point charges separated by a distance r. The work in bringing the second charge to the first, and therefore the electric energy, is $q_2 V_2$, where V_2 is the electric potential at the position occupied by the second charge due to the field of the first charge. We could compute this energy also by considering the work needed to bring charge 1 into the vicinity of charge 2. In that case, the energy would be expressed as $q_1 V_1$. We can therefore write the energy in a symmetric form as $\frac{1}{2}(q_1 V_1 + q_2 V_2)$. For a general charge distribution, we have

$$U_{\mathrm{el}} = \underbrace{\frac{1}{2}\sum_i q_i V_i}_{\text{discrete}} = \underbrace{\frac{1}{2}\int V(\mathbf{r})\rho(\mathbf{r})\,\mathrm{d}^3\mathbf{r}}_{\text{continuous}}, \tag{9.31}$$

for a discrete and a continuous charge distribution, respectively.

As an example we compute the electrical energy of a ball of radius R with a charge Q uniformly distributed throughout its volume as shown in Figure 9.11. To do this calculation, we consider the work in assembling this ball by breaking it up into the work needed to put together spherical shells of thickness $\mathrm{d}r$ and charge $\mathrm{d}q$ to form the ball. If the charge of the assembled shells is q, the increase in electric energy when another shell is added is

$$\mathrm{d}U_{\mathrm{el}} = V(r)\mathrm{d}q. \tag{9.32}$$

This equation says that when adding layers in the onion, adding the layer at radius r costs an energy dictated by the potential of the charge already there, $V(r)$, times the amount of charge added in that shell, $\mathrm{d}q$. The potential of the charge already there once the ball has been built up to a radius r is gotten by remembering that the electric field outside a charged ball is the same as for a point charge placed at its center. Using the definition of the potential as the work of the electric field, we find

$$V(r) = \int_r^\infty E(r')\,\mathrm{d}r' = \int_r^\infty \frac{1}{4\pi\varepsilon_0 D}\frac{q}{r'^2}\,\mathrm{d}r' = \frac{1}{4\pi\varepsilon_0 D}\frac{q}{r}. \tag{9.33}$$

To proceed, we need to figure out how much charge is inside of the sphere of radius r, which can be found through use of the charge density $\rho = Q/\frac{4}{3}\pi R^3$ to compute the charges $q = \rho\frac{4}{3}\pi r^3$ and $\mathrm{d}q = \rho 4\pi r^2\mathrm{d}r$. With these results in hand, we can write the energy to add a shell of charge as

$$\mathrm{d}U_{\mathrm{el}} = V(r)\mathrm{d}q = \frac{1}{4\pi\varepsilon_0 D}\frac{\rho\frac{4}{3}\pi r^3}{r}\rho 4\pi r^2\mathrm{d}r. \tag{9.34}$$

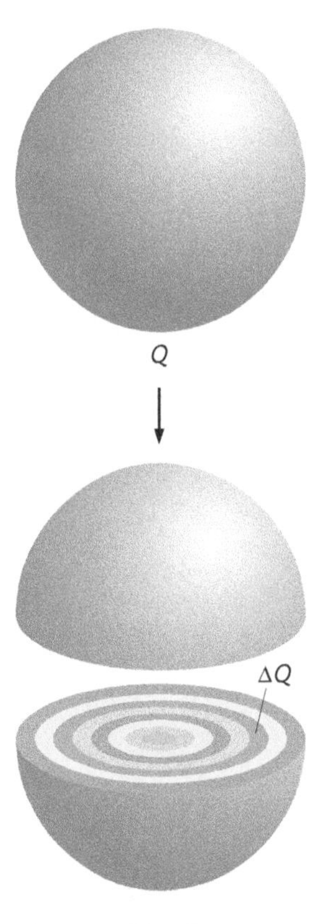

Figure 9.11: Electric energy of a charged ball. The energy to assemble a spherical charge distribution is obtained by summing up the energy cost associated with adding each new spherical shell of charge like the layers in an onion.

Summing up over all shells, which amounts to integrating Equation 9.34 over r, we find

$$U_{\text{el}} = \int_0^R \frac{1}{4\pi\varepsilon_0 D} \frac{16\pi^2}{3} \rho^2 r^4 \, \mathrm{d}r = \frac{1}{4\pi\varepsilon_0 D} \frac{3}{5} \frac{Q^2}{R}. \tag{9.35}$$

ESTIMATE

Estimate: DNA Condensation in Bacteriophage ϕ29? DNA inside a virus is contained in the volume provided by the viral capsid. Examples of such capsids were given in Figure 2.29 (p. 66). Since DNA is a strong acid, it is highly charged in solution. Therefore, energy needs to be expended in order to bring all the charge carried by the DNA into close proximity within the capsid. For the bacteriophage ϕ29, this energy is provided in the form of mechanical work done by the portal motor (a protein machine that translocates DNA), which in turn is fueled by ATP hydrolysis. Measurements of the work done by these motors are shown in Figure 10.19 (p. 404). These measurements reveal that upon packaging of the DNA there is an internal force build-up. The work against this internal force can be estimated from the measurements by evaluating the area under the force–length curve and is about

$$W_{\text{int}} \approx \frac{1}{2} \times 7000\,\text{nm} \times 60\,\text{pN} \approx 200{,}000\,\text{pN nm}.$$

This estimate uses the fact that length of the ϕ29 DNA is roughly 7 μm and the maximum force is 60 pN (for this rough estimate, we assume a simple linear rise of the force with the amount of DNA packaged).

The total charge carried by the ϕ29 genome, which is roughly 20,000 base pairs long, is

$$Q = \frac{2e}{\text{bp}} \times 20{,}000\,\text{bp}$$

where $e = 1.6 \times 10^{-19}$ C is the charge of an electron. Assuming that all this charge is spread uniformly throughout the viral capsid, which we approximate as a sphere of radius $R \approx 20$ nm, the work needed to condense this charge is given by Equation 9.35, which yields the estimate

$$W_{\text{charge}} \approx \frac{9 \times 10^9}{80} \frac{3}{5} \frac{(40{,}000 \times 1.6 \times 10^{-19})^2}{20 \times 10^{-9}}\,\text{J} \approx 10^8\,\text{pN nm}.$$

This estimate of the charging energy is 2000 times greater than the measured work against the internal force! This apparent violation of energy conservation is resolved once we realize that the electrical forces between the charges on the DNA are *screened* by the presence of counterions, making them effectively much weaker than Equation 9.35 would lead us to believe. Examining the nature of screening and its biological implications is the primary mission of the remainder of this chapter.

The Energy to Liberate Ions from Molecules Can Be Comparable to the Thermal Energy

There is a great deal of energy locked up in electrostatic interactions. Indeed, the existence of charge neutrality can be attributed to the steep cost of liberating charges. However, part of the story that is often left untold is that this steep cost can be seriously reduced once the charges in question are in water. A particularly revealing exercise in this regard can be found in Nelson (2004), where a simple estimate is made as to the costs associated with separating positive and negative charges when they are in vacuum as opposed to when they are in water. The intriguing outcome of this estimate is that when in water, charges are largely free to wander. The case studies of greatest interest often involve the separation of an OH^- and an H^+.

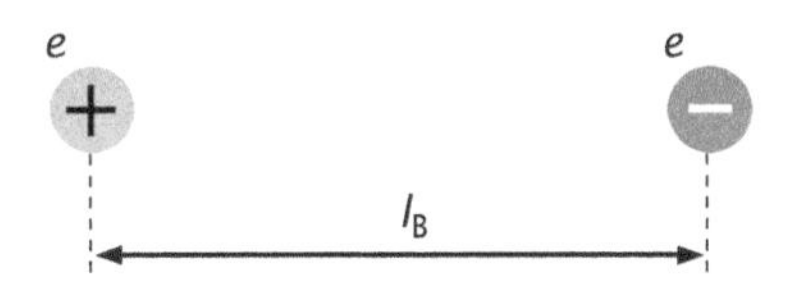

Figure 9.12: Illustration of the Bjerrum length. Two equal and opposite charges in solution can drift apart by a distance l_B with an energetic cost of k_BT.

A more intuitive and precise rendition of these ideas is found by noting that a balance can be struck between the electrostatic energy associated with stripping a charge away from a molecule in solution and the thermal energy scale. Stated differently, the tendency toward ionization can be thought of as a competition between the energetic tendency of binding and the entropic gain associated with permitting an ion to wander. We are interested in balancing the electrostatic interaction energy (as obtained from Coulomb's law) and the thermal energy. The idea is illustrated in Figure 9.12, which shows two equal and opposite charges that have been separated by a distance (the so-called Bjerrum length) such that their electrostatic interaction energy is equal to the thermal energy k_BT.

To estimate the Bjerrum length in terms of known quantities, we balance the electrostatic and thermal energies as

$$\frac{e^2}{4\pi\varepsilon_0 D l_B} = k_B T, \tag{9.36}$$

resulting in the determination of a fundamental length, denoted the Bjerrum length and given by

$$l_B = \frac{e^2}{4\pi\varepsilon_0 D k_B T}, \tag{9.37}$$

where we use $e^2/4\pi\varepsilon_0 \approx 2.3 \times 10^{-28}$ J m, resulting in $l_B \approx 0.7$ nm for water with its dielectric constant $D \approx 80$.

9.3.2 The Charged Life of a Protein

To gain quantitative intuition about the charging of proteins in solution we consider a simple model of a globular protein as a ball of radius R composed of amino acids, which are represented as tightly packed smaller balls whose radius is r.

A particularly revealing way to look at this problem (Nelson 2004) is to imagine a protein like that shown in Figure 9.13 in which all of the hydrophobic residues are sequestered within the protein and the surface is *assumed* to be composed strictly of polar residues, each of which is assumed to be able to surrender a single charge. The question we then pose is related to the self-energy of a charged sphere as captured by Equation 9.35. Here we ask how that self-energy depends upon the size of the protein, which is characterized by its radius R. The difference between this calculation and the previous one is that here, rather than thinking of the charge being added like layers in an onion, all of the charge is concentrated on the shell of radius R. The plan of our calculation is to work out the number of polar residues per protein as a function of the radius R and then to use that number to compute the self-energy as

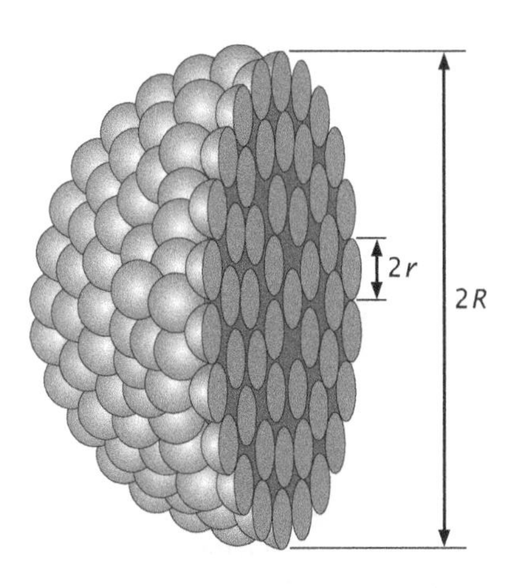

Figure 9.13: Schematic of the analysis of the energy cost associated with removing charge from the surface of a protein. The protein has a radius R and the individual residues have a mean radius r. In this simplified model, it is assumed that only the amino acids on the protein surface are polar and permit their charges to be stripped away.

$$U = \frac{1}{4\pi\varepsilon_0 D}\frac{Q_{\mathrm{tot}}^2}{2R}, \tag{9.38}$$

where the electrical energy of a charged shell shown here differs by numerical factors from the electrical energy of a charged ball (this different result is derived in the problems at the end of the chapter). The number of residues is given by

$$N = \frac{4\pi R^2}{\pi r^2}, \tag{9.39}$$

where we make the simplifying assumption that the radii of all of the polar residues are the same and given by r. If we now make the allied assertion that the total charge is $Q_{\text{tot}} = Ne$, then we can write the energy as

$$U = \frac{e^2}{4\pi\varepsilon_0 D}\frac{8R^3}{r^4}. \tag{9.40}$$

We can rewrite this in terms of the Bjerrum length by noting that

$$\frac{e^2}{4\pi\varepsilon_0 D} = l_{\text{B}} k_{\text{B}} T, \tag{9.41}$$

such that we can rewrite the energy as

$$U = k_{\text{B}} T l_{\text{B}} \frac{8R^3}{r^4}. \tag{9.42}$$

If we now use $l_{\text{B}} \approx 0.7\,\text{nm}$ and $r \approx 0.5\,\text{nm}$, then the energy of the protein as a function of its size is given by

$$U = k_{\text{B}} T \frac{8 \times 0.7}{(0.5)^4} R^3, \tag{9.43}$$

where the energy is measured in $k_{\text{B}}T$ units and the radius of the protein is measured in nanometers. The resulting energy as a function of protein radius is shown in Figure 9.14. This graph gives a qualitative feeling for the way in which nanometer-sized objects can tolerate charge loss, while their larger counterparts cannot, an idea already introduced in Figure 5.1 (p. 190).

9.3.3 The Notion of Screening: Electrostatics in Salty Solutions

Ions in Solution Are Spatially Arranged to Shield Charged Molecules Such as DNA

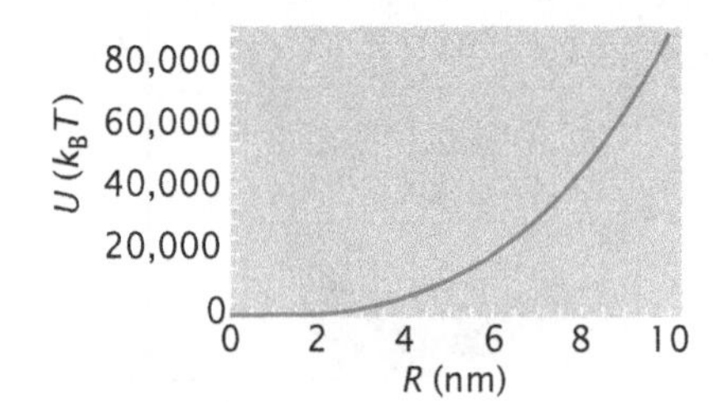

Figure 9.14: Energy cost to charge a protein. The graph shows the energy (in $k_{\text{B}}T$ units) of a charged sphere in which the surface is charged with a density dictated by the density of polar residues. The graph shows how this charging energy depends upon the radius of the model protein.

The ions that make up the salty solutions that serve as the biochemical aether for the great macromolecules of life are mobile. As a result, the distribution of ions around a given protein or DNA molecule depends upon the shape of that molecule and the concentration of these surrounding ions. As an introduction to the qualitative features of the behavior of charges in solution, Figure 9.15 shows a fragment of a DNA molecule that is highly negatively charged and its complement of counterions in the surrounding solution. The goal of the theory to be highlighted below is to develop a precise reckoning of the distribution of such ions around the DNA molecule and to find how the presence of such molecules alters the electric potential set up by the molecule.

A charged macromolecule such as a protein or nucleic acid will acquire a screening cloud in a salty solution such that the total charge on the macromolecule is neutralized as shown in Figure 9.16(A). The

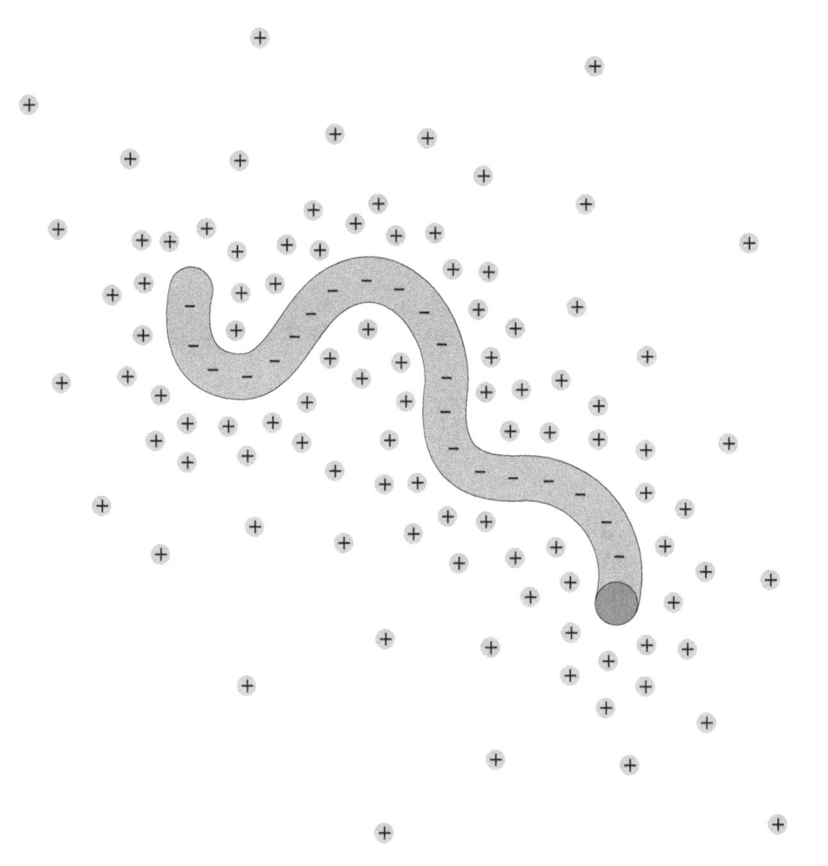

Figure 9.15: DNA in an ionic solution. The schematic shows the large negative charge density on the DNA molecule and the positive counterions in the surrounding solution. Note that the negative ions in solution required for charge neutrality have been omitted for clarity.

distance over which the cloud extends is determined by a competition between the energy and entropy of the ions. To be concrete, we assume that the macromolecule is negatively charged, with total charge $-Q$. In this case, positive counterions are drawn into the vicinity of the macromolecule, thus forming a screening cloud. These counterions adopt this configuration since by doing so they lower their electrostatic energy. At the same time, the local increase of the concentration of counterions in the screening cloud is characterized by an entropic penalty since they are effectively more confined. Similarly, negative ions are distributed in a way that balances the energy gain and the entropy cost of excluding them from the vicinity of the negatively charged macromolecule.

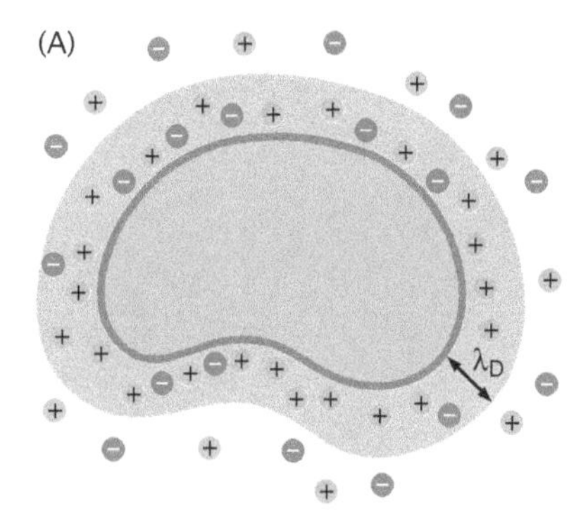

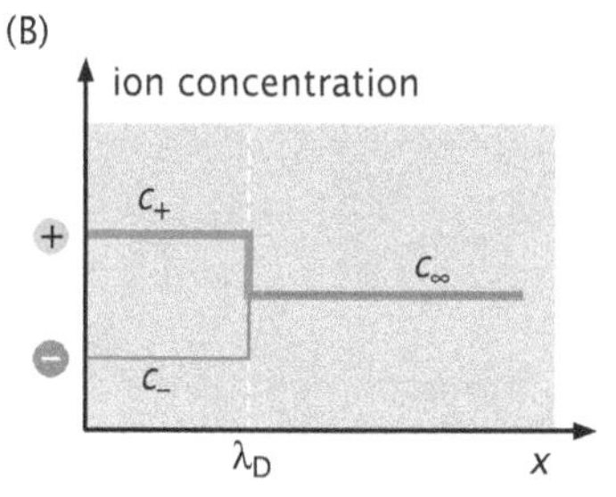

Figure 9.16: Charge density in salty solutions. (A) The screening cloud surrounding a charged surface. The macromolecule is assumed to have a uniform, negative charge density and in this simple model a uniform screening cloud is set up within a distance λ_D of the macromolecule. (B) Simple model for the structure of the screening cloud.

The Size of the Screening Cloud Is Determined by a Balance of Energy and Entropy of the Surrounding Ions

To estimate the width of the screening cloud, we assume that the concentration of counterions is uniform in the cloud, and given by

$$c_+ = c_\infty + \frac{\Delta c}{2}, \tag{9.44}$$

where c_∞ is the concentration of positive ions beyond the screening cloud as shown in Figure 9.16(B). Similarly, the concentration of negative ions is reduced in the screening cloud by $\frac{1}{2}\Delta c$. The quantity Δc is defined as $\Delta c = c_+ - c_-$ and measures how the far-field concentration c_∞ is perturbed by the macromolecule within the screening cloud. In particular, Δc characterizes the depletion of negative charge in the screening cloud and the augmentation of positive charge. The net charge of the screening cloud is equal to $e\Delta c A\lambda_D$, where A is the surface area of the charged macromolecule and λ_D is the width of the screening cloud, also known as the Debye length. The idea of this simplified model is that over a region of thickness λ_D, the ion distribution is uniform but distinct from the far-field value. The total charge of the macromolecule and its associated screening cloud is zero, implying that $Q = e\Delta c A \lambda_D$, which can be solved for Δc yielding

$$\Delta c = \frac{Q}{e}\frac{1}{A\lambda_D}, \tag{9.45}$$

where the factor of Q/e is the number of charges and permits us to compute the concentration.

We approximate the electric field in the screening cloud by

$$E_{\text{cloud}} = \frac{Q}{D\varepsilon_0 A}, \tag{9.46}$$

which assumes that all the screening charge is at distance λ_D from the surface of the macromolecule. The electric field associated with the charge distribution shown in Figure 9.16(B) is considered in the problems at the end of the chapter. The average electrostatic potential within the screening cloud is therefore

$$V_{\text{cloud}} = -\frac{1}{2}E_{\text{cloud}}\lambda_D = -\frac{1}{2D\varepsilon_0}\frac{Q\lambda_D}{A}. \tag{9.47}$$

In equilibrium, the chemical potential for the positive counterions is everywhere the same, in keeping with the derivation of Section 5.5.2 (p. 225). To exploit this idea, we use the fact that the chemical potential beyond the screening cloud, where the concentration of ions is c_∞ and the electrical potential is zero, is $\mu = \mu_0 + k_B T \ln(c_\infty/c_0)$. On the other hand, the chemical potential of an ion that finds itself in the screening cloud is

$$\mu = \mu_0 + k_B T \ln\left(\frac{c_\infty + \frac{1}{2}\Delta c}{c_0}\right) + eV_{\text{cloud}}, \tag{9.48}$$

where the last term in the equation captures the energy of the charges in the cloud. Equating the chemical potentials for the counterions inside and outside the screening cloud and substituting Equation 9.45 for Δc and Equation 9.47 for V_{cloud} leads to

$$k_B T \ln\left(1 + \frac{Q}{2eA\lambda_D c_\infty}\right) - \frac{1}{2D\varepsilon_0}\frac{Qe\lambda_D}{A} = 0. \tag{9.49}$$

If we assume that the change in concentration of the positive ions in the screening cloud is small compared with c_∞, we can expand the logarithm ($\ln(1+x) \approx x$, see the discussion of Taylor series on p. 215) in the above equation. Further, if we recall that the Bjerrum length is given by $l_B = e^2/4\pi\varepsilon_0 D k_B T$, we find

$$\lambda_D = \sqrt{\frac{1}{4\pi l_B c_\infty}}, \tag{9.50}$$

which is within a factor of $\sqrt{2}$ of the exact result obtained for the Debye screening length from the so-called Debye–Hückel theory to be described in more detail below.

We have used a toy model to show that a charged macromolecule acquires a counterion cloud in a salty solution. This cloud has a finite extent given by the Debye length. For distances greater than the Debye length, the charged macromolecule and its associated screening cloud are effectively neutral. This raises the question of the nature of the electrostatic interaction between two charged macromolecules. The picture of the screening cloud we have developed so far would suggest that this interaction is short-ranged. Namely, we expect there to be no interaction unless the two macromolecules are brought in sufficiently close proximity that their screening clouds overlap. At these short distances, the interaction is a combination of entropic and energetic effects associated with changing the screening cloud configuration.

Estimate: DNA Condensation in Bacteriophage $\phi29$ Redux

ESTIMATE

The physical picture associated with screening can serve as the basis for making a new estimate of the energy expended in packing DNA into a viral capsid that goes beyond the estimate given earlier in the chapter (p. 368). To make a rough estimate, we consider the screening cloud to be made up of positively charged counterions only. Furthermore we assume that the charge on the DNA is perfectly screened so that no electrical interaction between these charges remains. The free-energy cost for packing DNA is then solely the result of decreasing the entropy of the charges that make up the cloud of counterions as they are transferred to the confines of the capsid.

Given the assumptions stated above, for $\phi29$ with its roughly 20,000 base pairs of DNA, the number of positive charges in the screening cloud is approximately 40,000. The size of the screening cloud is given by the Debye length, which under physiological conditions with a characteristic salt concentration of 100 mM results in a screening length $\lambda_D \approx 1$ nm.

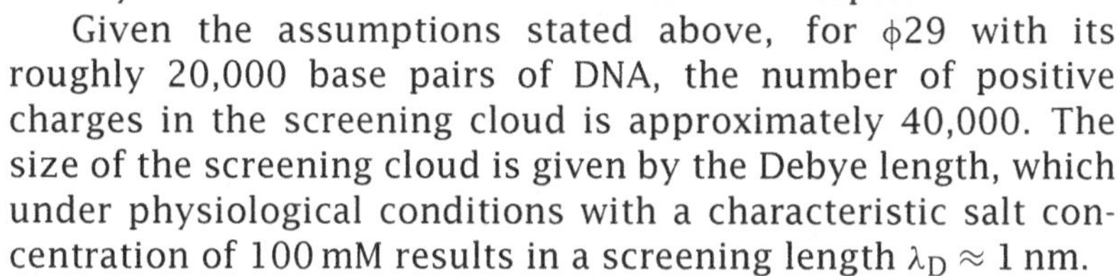

When the DNA is packed into the viral capsid, the counterions enter the capsid as well to neutralize the DNA. The energy cost associated with this process corresponds to the pV work for squeezing the counterions into a smaller volume than they occupied when the DNA was outside the capsid as is shown in Figure 9.17, and is given by

$$W_{\text{charge}} = Nk_BT \ln \frac{V_{\text{cloud}}}{V_{\text{capsid}}}. \qquad (9.51)$$

In this equation, $N = 40{,}000$ is the number of counterions, V_{cloud} is the volume of the screening cloud, while V_{capsid} is the volume available to the counterions in the capsid.

The volume of the screening cloud is obtained by thinking of the DNA as a long cylindrical rod of radius R_{DNA} surrounded by a cloud of thickness λ_D. This results in

$$V_{\text{cloud}} = L\pi \left[(R_{\text{DNA}} + \lambda_D)^2 - R_{\text{DNA}}^2 \right] \approx 64{,}000\,\text{nm}^3 \qquad (9.52)$$

where we have taken $L = 20{,}000 \times 0.34\,\text{nm} = 6800\,\text{nm}$ for the length of the DNA, $R_{\text{DNA}} = 1$ nm is the DNA radius, and $\lambda_D = 1$ nm is the Debye length.

The volume available to the counterions in the capsid is the difference between the volume of the capsid and that of the DNA,

$$V_{\text{capsid}} = \frac{4\pi}{3} R_{\text{capsid}}^3 - L\pi R_{\text{DNA}}^2 \approx 12{,}000\,\text{nm}^3, \qquad (9.53)$$

where we have taken $R_{\text{capsid}} = 20$ nm for the radius of the $\phi29$ viral capsid.

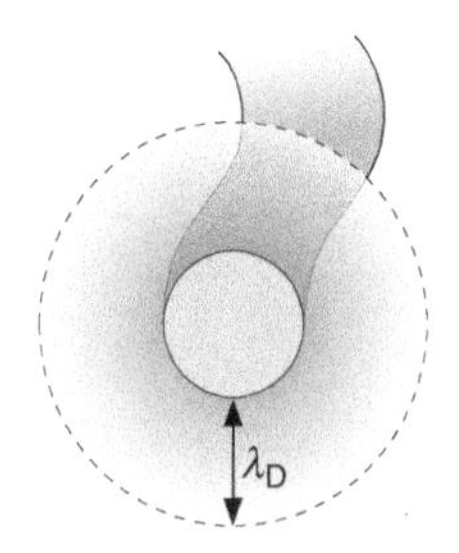

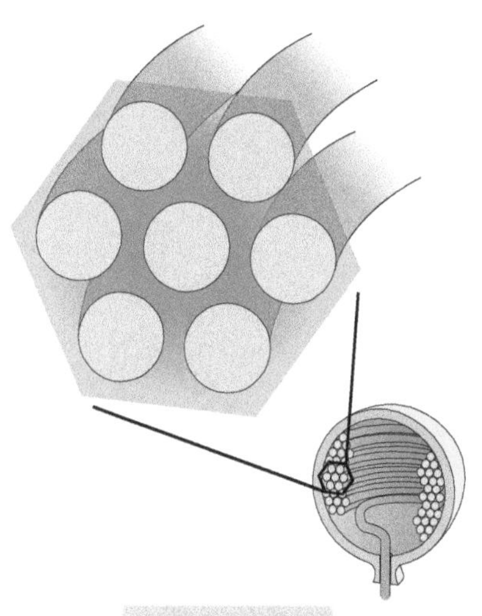

Figure 9.17: Estimate of the energy to put DNA inside a viral capsid. Before being stuffed in the capsid, the counterions surrounding the DNA are in their equilibrated form. Once the DNA is in the capsid, there is no longer sufficient free volume for these counterions and they are squeezed into a reduced volume.

Substituting these estimates for the volume of the screening cloud and the free volume of the capsid into Equation 9.51, we arrive at an estimate $W_{\text{charge}} \approx 65{,}000\, k_B T$, which is roughly a factor of 3 larger than the work measured in single-molecule experiments. Given the crudeness of our approach, this is a satisfactory agreement, especially when we compare it with the estimate we obtained by completely ignoring the screening of the DNA charge by counterions. One factor that is missing in our estimate is the lowering of the electrostatic energy for the counterions in the capsid since they will now be on average closer to the DNA than they were in the screening cloud around the DNA free in solution.

9.3.4 The Poisson–Boltzmann Equation

Thus far, our treatment of the effects of screening has been extremely approximate and really amounts to a series of estimates. The more formal theory of charge rearrangements in solution is based upon the Poisson–Boltzmann equation. This approach merges two of the most important results of theoretical physics, namely, the Poisson equation (Equation 9.27, p. 366), which relates the electric potential to the charge density, and the Boltzmann distribution (Equation 6.4, p. 240), which provides the probability of different microstates (such as the different configurations that can be adopted by ions in solution). The key conceptual idea behind this approach is that the ions in solution are equilibrated with whatever molecules they are screening. That is, the probability of finding an ion at a given point in the solution is dictated by the Boltzmann weight.

The Distribution of Screening Ions Can Be Found by Minimizing the Free Energy

We consider the distribution of ions around a charged object such as a macromolecule in solution containing positive and negative ions. To simplify matters, we consider ions that have a charge $\pm ze$. If the macromolecule of interest is negatively charged, we expect that close to it there will be an increase in the density of positive ions, which are attracted, and a decrease in the density of negative ions, which are repelled. Far from the macromolecule, electrical neutrality of the solution demands that the number densities for the two ion species are the same, $c_+ = c_- = c_\infty$. More generally, in the case of many different ionic species, the neutrality condition can be written as $\sum_i z_i c_i = 0$, where z_i is the valency of the ith species.

To address the ionic charge density quantitatively, we first consider a very simple scenario where the charged object in solution is a membrane of negative charge with charge per unit area given by σ as shown in Figure 9.18. The units of σ are coulombs/m^2 (C/m^2). In this case, as long we stay close to the membrane so that it is effectively infinite in extent, the number densities for positive and negative ions will vary only in the direction perpendicular to the plane of the membrane, which we denote by x; $x = 0$ is the position of the charged surface. The number densities for the ions in solution are then given by the Boltzmann formula

$$\begin{aligned} c_+(x) &= c_\infty \mathrm{e}^{-zeV(x)/k_B T} \\ c_-(x) &= c_\infty \mathrm{e}^{+zeV(x)/k_B T} \end{aligned} \tag{9.54}$$

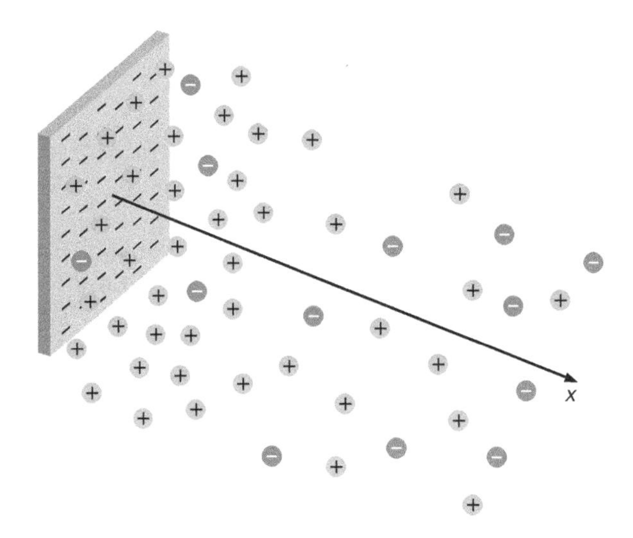

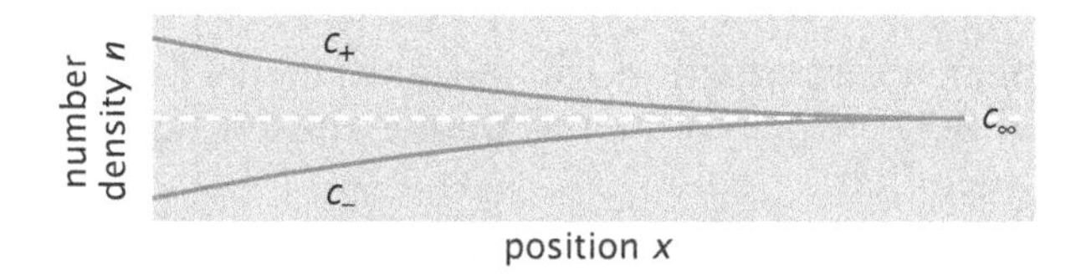

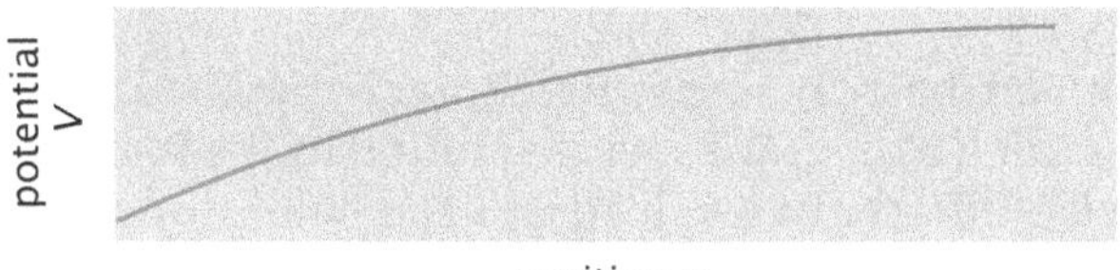

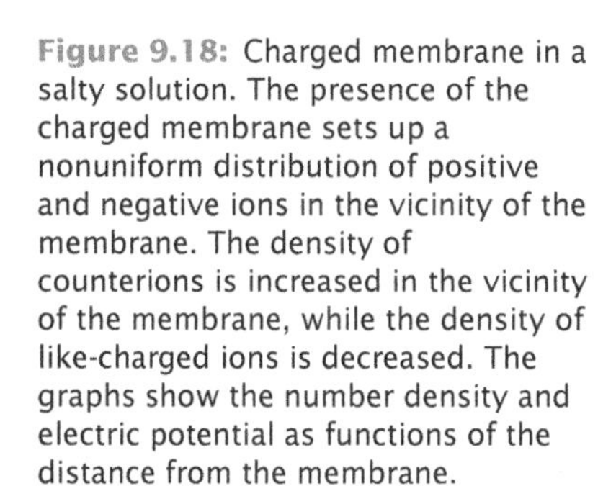

Figure 9.18: Charged membrane in a salty solution. The presence of the charged membrane sets up a nonuniform distribution of positive and negative ions in the vicinity of the membrane. The density of counterions is increased in the vicinity of the membrane, while the density of like-charged ions is decreased. The graphs show the number density and electric potential as functions of the distance from the membrane.

where $\pm zeV(x)$ is the electric energy of positive and negative ions in the field described by the potential $V(x)$. The potential, like the number densities, also varies only along the direction perpendicular to the charged membrane. In order to compute the distribution of ions in solution, we must first find the electric potential along the x-direction. The electric potential itself is determined by the distribution of ion charges and we therefore seek a second relation between the two. It is provided by the Poisson equation derived in Section 9.3.1 (p. 361).

The total charge density at position x is obtained by summing over the contributions from both the negative and positive charges and is given by

$$\rho(x) = zec_+(x) - zec_-(x). \tag{9.55}$$

This charge density is related to the electric potential at the same position by the Poisson equation and is given by

$$\frac{\mathrm{d}^2 V(x)}{\mathrm{d}x^2} = -\frac{\rho(x)}{D\varepsilon_0}. \tag{9.56}$$

Substituting Equation 9.54 into Equation 9.55 and the resulting expression for $\rho(x)$ into the Poisson equation, we arrive at

$$\frac{\mathrm{d}^2 V(x)}{\mathrm{d}x^2} = \frac{zec_\infty}{D\varepsilon_0}\left(\mathrm{e}^{zeV(x)/k_\mathrm{B}T} - \mathrm{e}^{-zeV(x)/k_\mathrm{B}T}\right). \tag{9.57}$$

This nonlinear differential equation for the electric potential in a salty solution is the Poisson–Boltzmann equation. Its solution, with boundary conditions set by the presence of the charged membrane, provides us with $V(x)$, which can then be substituted into Equation 9.54 to obtain the charge distribution itself.

The Screening Charge Decays Exponentially Around Macromolecules in Solution

For situations in which the electric potential is "small," the Poisson–Boltzmann equation can be simplified by approximating the exponentials $e^{\pm zeV(x)/k_B T}$ appearing in Equation 9.57 with their Taylor expansions, $1 \pm zeV(x)/k_B T$ (a procedure known as *linearization*) to obtain the Debye–Hückel equation. This is a good approximation in the limit when the potential is much smaller than $k_B T/e = 25\,\text{mV}$. To get a sense of the size of this limiting potential, we can compare it with the potential at the surface of a charged protein in water. For a charge of $q = 1e$ and taking the protein diameter to be $R = 4\,\text{nm}$, we estimate the potential at the surface to be roughly $(1/4\pi\varepsilon_0 D)(q/R) \approx 4\,\text{mV}$. In the presence of counterions, this is certainly an overestimate, as will be shown later.

With the linear approximation for the Boltzmann factors in hand, from Equation 9.57 we arrive at the Debye–Hückel equation for our simple planar geometry, namely,

$$\frac{d^2 V(x)}{dx^2} = \frac{2z^2 e^2 c_\infty}{D\varepsilon_0 k_B T} V(x). \tag{9.58}$$

This equation can be solved by elementary methods for the planar geometry described above. We seek a solution in the form of a function that when twice differentiated leads to the same function. This is a property of the exponential function. Indeed, we can check that

$$V(x) = Ae^{-x/\lambda_D} + Be^{x/\lambda_D} \tag{9.59}$$

is the solution by substituting for $V(x)$ in Equation 9.58. Here A and B are integration constants and

$$\lambda_D = \sqrt{\frac{D\varepsilon_0 k_B T}{2z^2 e^2 c_\infty}} \tag{9.60}$$

is the Debye screening length introduced approximately in the previous section.

The constants A and B are fixed by boundary conditions. The first condition states that the potential far from the charged membrane is zero, which implies that $B = 0$. We conclude that the electric potential falls off exponentially with a characteristic distance given by the Debye screening length λ_D. To determine the remaining integration constant, we can make use of the fact that the electric field at the surface of the membrane is

$$E_x(0) = \frac{\sigma}{\varepsilon_0 D}, \tag{9.61}$$

a result once again implied by Gauss' law. Using the relation between the field and the potential, $E_x(x) = -dV(x)/dx$, and Equation 9.59 for $V(x)$, we find $A = \sigma\lambda_D/D\varepsilon_0$, and

$$V(x) = \frac{\sigma\lambda_D}{D\varepsilon_0} e^{-x/\lambda_D}. \tag{9.62}$$

Substituting this solution into the equation relating the potential to the charge density away from the charged membrane, Equation 9.56, we find

$$\rho(x) = -\frac{\sigma}{\lambda_D} e^{-x/\lambda_D}. \tag{9.63}$$

Similarly, the number density for positive and negative ions is exponential in the distance from the charged membrane. Assuming a negative surface charge for the membrane, the density of positive ions is maximum at the surface and then it drops off exponentially until it reaches c_∞. For negative ions, the situation is reversed.

As already shown with the simple toy model of screening in the previous discussion and confirmed by this more precise calculation, the key idea of the Debye length is that beyond this length the potential is essentially zero and the charge distribution is uniform. For a charged protein in a salt solution with charge density $c_\infty = 200\,\text{mM}$, typical for potassium ions in the cell interior, the Debye screening length is roughly 0.7 nm. This means that beyond this distance the charge on the protein will not be "felt" by other charges. This signals the importance of the three-dimensional shapes of proteins in matters electrical, since for two proteins to experience electric interactions, their charged residues have to be able to come in close proximity of each other. This will be allowed geometrically if their shapes are complementary to each other. Of course, all of these musings on the distributions of charges near proteins implicitly *assume* that a continuum approach for the distribution of a discrete set of charges makes sense at these scales.

9.3.5 Viruses as Charged Spheres

So far, we have seen how screening works for the abstract case of a planar interface. That geometry is relevant to the charge state near biological membranes. A second class of important problems concerns the charge state around proteins and macromolecular assemblies such as ribosomes and viral capsids. In this case, a useful idealization of these structures is a charged sphere. Here we make use of the Debye–Hückel model to compute the electric energy of a charged sphere in a salty solution. We use this result to estimate the electrostatic energy associated with the assembly of a viral capsid. In particular, we address the experimentally relevant question of the salt dependence of the equilibrium constant for this process.

The energy cost for assembling a spherical shell of charge Q and radius R is $\frac{1}{2}QV(R)$, where $V(R)$ is the potential on the surface of the sphere, as shown in Equation 9.31. For a charge distribution that is spherically symmetric, we can compute the electric potential by making use of Coulomb's law in integral form, Equation 9.16. The flux of the electric field through a sphere of radius $r > R$ is

$$\Phi(r) = E_r(r) 4\pi r^2. \tag{9.64}$$

The total charge contained within the sphere can be computed by making use of the charge density ρ:

$$q(r) = \int_0^r \rho(r') 4\pi r'^2 \, dr'. \tag{9.65}$$

Taking the relation between the flux and the charge, $\Phi(r) = q(r)/D\varepsilon_0$, and differentiating both sides with respect to r, we arrive at the Poisson equation for the spherically symmetric case,

$$\frac{d}{dr}\left[-r^2 \frac{dV(r)}{dr}\right] = \frac{\rho(r)}{D\varepsilon_0} r^2, \tag{9.66}$$

where we have also made use of the relation between the electric field and the potential, $E_r(r) = -\mathrm{d}V(r)/\mathrm{d}r$. This result may be simplified to the form

$$\frac{1}{r}\frac{\mathrm{d}^2[rV(r)]}{\mathrm{d}r^2} = -\frac{\rho(r)}{D\varepsilon_0}. \tag{9.67}$$

If we substitute the linearized version of the Boltzmann equation for the charge density,

$$\rho(r) \approx -\frac{2z^2e^2c_\infty}{k_\mathrm{B}T}V(r), \tag{9.68}$$

into Equation 9.66, we arrive at the Debye–Hückel equation for the case of spherical symmetry,

$$\frac{\mathrm{d}^2[rV(r)]}{\mathrm{d}r^2} = \frac{1}{\lambda_\mathrm{D}^2}[rV(r)], \tag{9.69}$$

where λ_D is the Debye screening length (Equation 9.60).

Note that the equation for $rV(r)$ is the same as the one we derived for $V(x)$ in the case of planar symmetry. As a result, we can write the solution to this equation without further ado by noting that $rV(r)$ is an exponentially decaying function, and hence

$$V(r) = \frac{A}{r}\mathrm{e}^{-r/\lambda_\mathrm{D}}. \tag{9.70}$$

To determine the integration constant A, we should make use of the fact that we seek the potential in the presence of a sphere of charge Q and radius R. Since the electric field at the surface of the sphere is

$$E_r(R) = \frac{Q}{4\pi\varepsilon_0 DR^2}, \tag{9.71}$$

we can determine A from the relation $E_r(r) = -\mathrm{d}V(r)/\mathrm{d}r$, implying

$$A = \frac{Q\mathrm{e}^{R/\lambda_\mathrm{D}}}{4\pi\varepsilon_0 D(1+R/\lambda_\mathrm{D})}. \tag{9.72}$$

Substituting this formula for the constant A into Equation 9.70, we arrive at an expression for the potential at the surface of a charged sphere,

$$V(R) = \frac{1}{4\pi\varepsilon_0 D}\frac{Q}{R}\frac{\lambda_\mathrm{D}}{R+\lambda_\mathrm{D}}. \tag{9.73}$$

The electrostatic energy of the spherical shell is now simply

$$U(R) = \frac{1}{2}QV(R) = \frac{1}{2}k_\mathrm{B}T\left(\frac{Q}{e}\right)^2\frac{l_\mathrm{B}\lambda_\mathrm{D}}{R(R+\lambda_\mathrm{D})}, \tag{9.74}$$

where we have used the Bjerrum length, defined as $l_\mathrm{B} = e^2/(4\pi\varepsilon_0 Dk_\mathrm{B}T)$.

An interesting application of the formula for the electrostatic energy of a charged sphere in a salty solution is as the basis for developing intuition about the nature of viral assembly. As discussed in Section 2.2.4 (p. 66), viral capsids consist of hundreds of repeating protein units arranged into a thin shell, typically with icosahedral symmetry. The protein building blocks making up the capsid are held together by contact forces, such as the hydrophobic force, that arise between the surfaces of the protein units that are in contact.

(A)

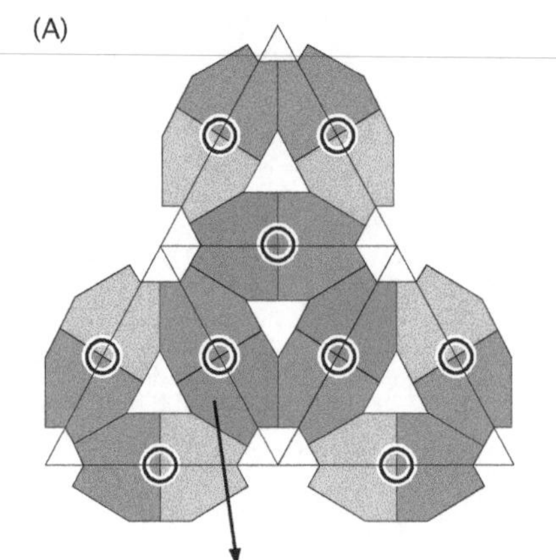

(B)

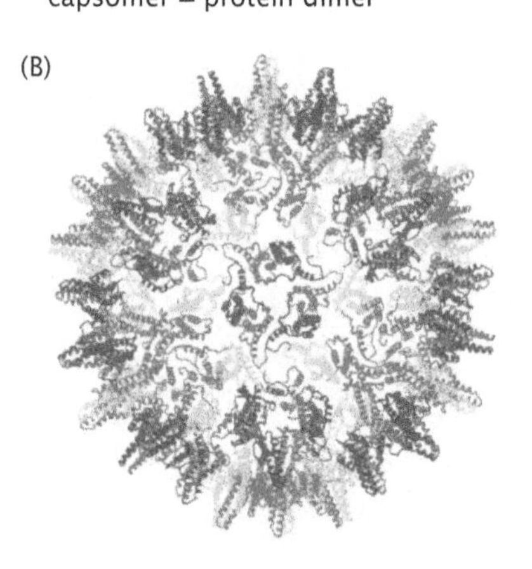

Figure 9.19: Viral capsid structure. (A) Close-up view of the contacts between adjacent capsomers making up one icosahedral face. Each capsomer is formed from two protein units to make a dimer (represented by two different colors for each capsomer). The circle in the center of each represents α-helices emerging perpendicular to the structure (seen as spikes in (B)).
(B) Three-dimensional view of the human hepatitis B virus. (Adapted from S. A. Wynne et al., *Mol. Cell* 3:771, 1999.)

These protein building blocks are charged in solution. Therefore, an important contribution to the energy budget of capsid assembly is the electrostatic cost for bringing like charged capsomers in close proximity to form the capsid. This is described in cartoon form in Figure 9.19, which shows the human hepatitis B virus, which is built up of 120 dimer capsomers.

A simple model that accounts for the electrostatic and contact (typically hydrophobic in origin) forces can be written as

$$\Delta G_{\text{capsid}} = \Delta G_{\text{contact}} + \frac{1}{2} k_B T N^2 z^2 \frac{l_B \lambda_D}{R^2}, \qquad (9.75)$$

where the electrostatic piece is obtained from Equation 9.74 assuming that N capsomers make up the capsid and that each carries a charge ze (in other words, the total charge is $Q = Nze$). Also, since at 100 mM monovalent salt, the Debye length ($\lambda_D = 1$ nm) is at least an order of magnitude smaller than the radius of a typical virus, we make the approximation $R + \lambda_D \approx R$. This expression for the free energy of assembly is interesting because the second term depends sensitively on the salt concentration through the screening length.

This model can be tested against experiments that measure the equilibrium constant K_{capsid} of capsid assembly at different salt concentrations. The equilibrium constant is related to the free energy by the usual relation, $\ln K_{\text{capsid}} = -\Delta G_{\text{capsid}}/k_B T$. Results of experiments on the thermodynamics of assembly of hepatitis B virus are shown in Figure 9.20. The equilibrium constant is determined by measuring the fraction of capsids in solution as a function of the concentration of capsomers, which in this particular experiment was done using size exclusion chromatography. We explore these measurements in more detail in the problems at the end of the chapter.

The data in Figure 9.20(A) show a linear increase in $\ln K_{\text{capsid}}$ as a function of temperature. This is to be expected given the small range of temperatures over which assembly of the virus is observed. Furthermore, we see that the slope of the curves is independent of salt concentration, indicating that the temperature dependence of ΔG_{capsid} comes primarily from the contact interactions. The model (Equation 9.75) then predicts that the salt-dependent shift of the $\ln K_{\text{capsid}}(T)$ curve, observed in the data plots, is dictated by the electrostatic part of ΔG_{capsid}. Since the salt concentration c enters this part of the free energy through the Debye screening length $\lambda_D = 1/\sqrt{8\pi l_B c}$, the model predicts a shift linear in $c^{-1/2}$. This expectation is borne out by the data, as shown in Figure 9.20(B). Therefore the data are in quantitative agreement with a simple model of capsid assembly, whereby the equilibrium between capsomers and assembled capsids is established as a competition between favorable contact interactions and unfavorable electrostatic interactions between the capsomers in a capsid.

(A)

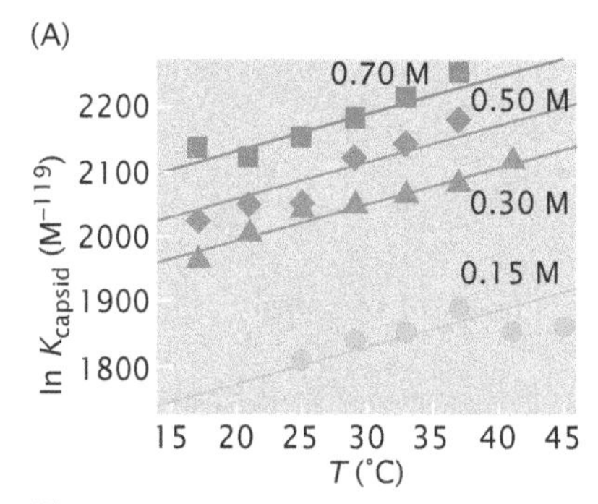

(B)

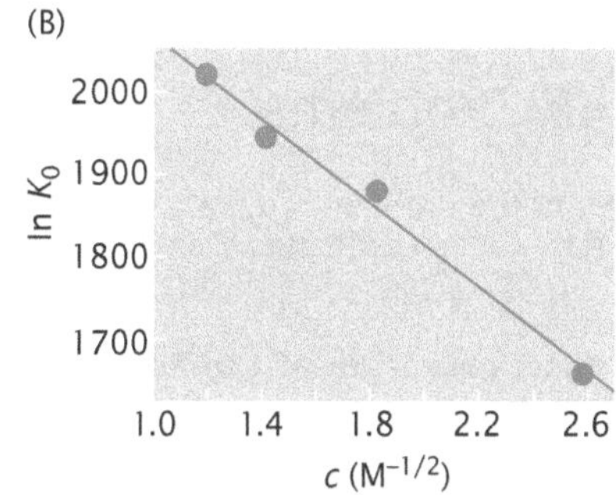

Figure 9.20: Equilibrium constants for assembly of the hepatitis B virus. (A) The equilibrium constant is plotted at various temperatures and concentrations of NaCl. The dimensionless equilibrium constant was determined from the measured mole fractions of completed capsids and capsomers in solution, using size-exclusion chromatography. (B) Salt dependence of the equilibrium constant for viral assembly. The salt-dependent shift of the equilibrium constant, which is clearly seen is quantified by evaluating $\ln K_{\text{capsid}}$ at $T_0 = 0\,°\text{C}$ by linear extrapolation, follows the $c^{-1/2}$ scaling obtained from a model of a virus capsid as a charged spherical shell in salty water. (Adapted from P. Ceres and A. Zlotnick, *Biochemistry* 41:11525, 2002 and W. K. Kegel and P. van der Schoot, *Biophys. J.* 86:3905, 2004).

9.4 Summary and Conclusion

Whether in the sterile setting of the biochemist's test tube or the crowded and heterogeneous environment of the cell, charged macromolecules are surrounded by a nonuniform distribution of "screening" ions. The presence of these ions has many consequences for cellular life, as evidenced by the host of pH-dependent processes in cells and the strong salt dependence of many important binding reactions. The key theoretical tool that is used to think about these problems involves a merger of two important results from electrostatics and statistical mechanics, namely, the Poisson equation, which relates charge density to electric potential, and the Boltzmann distribution, which assigns probability to microstates. In general, the resulting model involves a nonlinear partial differential equation for the electric potential. However, for "small" potentials, this equation can be approximated by a linear equation resulting in the Debye screening length. For simple planar and spherical geometries, these equations can be solved all the way to the end, and they reveal an exponential decay of the screening charge and potential.

9.5 Problems

Key to the problem categories: • Model refinements and derivations • Estimates • Model construction

• 9.1 Derivation of the Poisson equation

(a) In Equation 9.21, we asserted a relation between the divergence of the electric field and the charge density. Imitate the one-dimensional derivation culminating in Equation 9.20 to deduce Equation 9.21. To do this, explicitly compute the flux through a small cubical volume element like that shown in Figure 9.9.

(b) By using the fact that the electric field can be written as $\mathbf{E}(\mathbf{r}) = -\nabla V(\mathbf{r})$, write a partial differential equation relating the potential and the charge density.

• 9.2 pH and protons in *E. coli*

(a) The pH of the *E. coli* cytosol is about 7.6–7.8. How many free protons per cell does this equate to?

(b) The pH of the periplasmic space is about 7.0. Estimate the number of protons in the periplasmic space.

(Problem courtesy of Andrew Chisholm)

• 9.3 Gauss' law in action

(a) Deduce the electric field of a uniformly charged sphere of radius R and total charge Q. Obtain the field both within and on the exterior of the sphere.

(b) Deduce the electric field of a uniformly charged spherical shell of radius R and total charge Q.

(c) In Section 9.3.3, we considered the screening of the charge on a protein by charges in solution. In that toy model, we approximated the field in the screening cloud by assuming that all the screening charges are at a fixed distance from the protein surface, which led to the formula

$$E = \frac{Q}{D\varepsilon_0 A} \tag{9.76}$$

Use Gauss' law for two uniformly and oppositely charged parallel plates separated by distance d, with surface charge density $\pm Q/A$, to show that this is the field in between the plates.

• 9.4 Energy of a charged spherical shell

Deduce the energy of a charged shell as given by Equation 9.38.

• 9.5 A simple model for viral capsid assembly

In the chapter, we considered the equilibrium constant for the assembly of a viral capsid, and its salt dependence. In experiments, this equilibrium constant is usually determined by measuring the relative amounts of free capsomers and completed capsids using size exclusion chromatography, for example. Furthermore, in the data analysis leading up to the equilibrium constant, one typically assumes that partially formed capsids are present in negligible amounts in solution, and can be ignored. Here we investigate this assumption in the context of a simple version of an equilibrium model for the assembly of icosahedral viral capsids described by Zlotnick (1994). Our model for a capsid is a dodecahedron that assembles from 12 identical pentagonal subunits. When subunits associate, they make favorable contacts along edges with an energy $\Delta\varepsilon < 0$ per edge, and pay a translational entropy penalty due to the *loss* of translational degrees of freedom once the subunits are part of the larger assembly.

We assume that assembly occurs through binding of subunits in such a way that the only allowed species are those consistent with all or part of the final dodecahedron product. We seek to evaluate the equilibrium amount of free capsomers, complete dodecahedrons, and partially assembled structures. The energy for a structure of size n is given by

$$\varepsilon_n = \sum_{m=1}^{n} f_m \Delta\varepsilon, \tag{9.77}$$

where f_n is the number of additional contacts created when a capsomer binds to a structure of size $n-1$ to form a structure of size n. To simplify matters further, we consider the following form for the number of contacts

$$f_n = \begin{cases} 1 & (n=2), \\ 2 & (3 \leq n \leq 7), \\ 3 & (8 \leq n \leq 10), \\ 4 & (n=11), \\ 5 & (n=12). \end{cases} \tag{9.78}$$

Note that within this model, $f_1 = 0$, meaning that individual capsomers set the zero of energy.

To describe the state of the solution containing N_{tot} capsomers, we make use of the volume fractions ϕ_n, $n = 1, 2, 3, \ldots, 12$, which are defined as $\phi_n = N_n v_n / V$, where N_n is the number of partially formed capsids made of n capsomers, v_n is the volume of each of these structures, while V is the volume of the solution. The goal of the problem is to compute ϕ_n for different values of n, as a function of $\Delta\varepsilon$ and $\phi_{\mathrm{T}} = N_{\mathrm{T}} v_1 / V$, the total volume fraction for all the capsomers in solution.

(a) Using a lattice model for solution as we have done throughout the book, show that the total free energy of the capsomer solution is

$$G_T = \sum_{n=1}^{12} \left\{ N_n \varepsilon_n + \frac{V}{v_n} k_B T [\phi_n \ln \phi_n + (1-\phi_n)\ln(1-\phi_n)] \right\}.$$

Now show that by minimizing this free energy with respect to N_n, with the constraint that the total number of capsomers is constant (use a Lagrange multiplier to enforce this constraint) and equal to N_T, the volume fraction of intermediates of size n is given by

$$\phi_n = (\phi_1)^n \, e^{-\varepsilon_n / k_B T}. \tag{9.79}$$

To get this result, you will need to use the fact that $\varepsilon_1 = 0$. Finally, show that the constraint on the state variable N_n can be rewritten in terms of the volume fractions as

$$\phi_T = \sum_{n=1}^{12} \phi_n.$$

(b) Assume that the only species with significant volume fractions are those of sizes $n = 1$ and $n = 12$. Show that in this case the critical value, ϕ_C, of ϕ_T for which half of all capsomers are in complete capsids is given by

$$\ln(\phi_C/2) = \frac{\varepsilon_{12}}{11 k_B T}. \tag{9.80}$$

(c) Carry out a numerical solution for ϕ_n, $n = 1, 2, \ldots, 12$, as a function of ϕ_T and $\Delta\varepsilon$. Plot ϕ_n as a function of n for $\phi_T = \phi_C$ and $\Delta\varepsilon = -1, -5$, and $-10\ k_B T$. How are the capsomers distributed among the 12 different structures in each of these cases? What happens to the fraction of capsomers in complete capsids as the total volume fraction is varied from below to above ϕ_C, in the case $\Delta\varepsilon = -5\ k_B T$? (Problem courtesy of Mike Hagan.)

• 9.6 Debye length

In Section 9.3.3, we computed the Debye length from a toy model of the screening cloud. The idea was to find the width of the cloud that minimizes its free energy. The free energy accounted for the electrostatic energy and the entropy of the charges in the cloud. The electrostatic energy was estimated by assuming a uniform electric field throughout the cloud, as though all the counterions were located at distance λ_D from the negatively charged macromolecule. Here we reconsider this estimate by computing the electric field and the electrostatic energy associated with a uniform charge distribution, as shown in Figure 9.16.

(a) Compute the electric field inside the screening cloud, $E(x)$, a distance x from the surface of the negatively charged macromolecule, using Gauss' law. Assume that the field in the cloud is along the x-direction, perpendicular to the surface of the macromolecule.

(b) Compute the electrostatic energy associated with the screening cloud using the relation

$$U = \frac{D\varepsilon_0}{2} \int_{\text{cloud}} E^2 \, dV,$$

where the integral runs over the volume of the cloud.

(c) Compute the entropy cost associated with the screening cloud by taking into account the contribution of all the positive and negative ions in the cloud. Then, compute the Debye length as the cloud width that minimizes the cloud free energy. How does this value compare with the one obtained in the chapter?

• 9.7 Poisson–Boltzmann revisited

(a) Using the linearized Poisson–Boltzmann equation, calculate the positive and negative charge concentrations, $c_+(x)$ and $c_-(x)$, and the electric potential $V(x)$ at a distance x from a charged plane in a salty solution. If the charge per unit area on the plane is $\sigma = e/a$, derive a condition on the area a that makes the linear approximation valid. Assume a salt concentration of $c_\infty = 100$ mM.

(b) Plot the electric potential and the concentrations of positive and negative charges as functions of the distance from the charged plane, assuming that the charge on the plane is one electron per 100 nm^2, and $c_\infty = 100$ mM.

(c) Add up all the charges in solution and show that they exactly compensate for those on the charged plane.

• 9.8 Potential near a protein

Consider a protein sphere with a radius of 1.8 nm, and charge $Q = -10e$, in an aqueous solution of $c_\infty = 0.05$ M NaCl at 25 °C. Consider the small ions as point charges and use the linear approximation to the Poisson–Boltzmann equation.

(a) Fill in the steps leading to Equations 9.70 and 9.72, and derive the expression for the potential $V(r)$ of a charged sphere in a salty solution.

(b) What is the surface potential of the protein in units $k_B T/e$?

(c) What is the concentration of Na^+ ions and of Cl^- ions at the surface of the protein?

(d) What is the concentration of Na^+ and Cl^- ions at a distance of 0.3 nm from the protein surface?

(Adapted from Problem 23.2 of K. Dill and S. Bromberg, Molecular Driving Forces, 2nd ed. Garland Science, 2011.)

• 9.9 Charging energy of proteins in salty water

In the toy model of a protein described in Section 9.3.2, we assumed that a protein can be thought of as a charged sphere in water. Here, we consider the effect of salt on its electrical energy.

(a) Compute the energy of a charged spherical protein of radius R, in water, and in the presence of monovalent salt at concentration c_∞. Assume that the charged residues are uniformly distributed on the surface of the protein.

(b) Redo the calculation leading to the plot in Figure 9.14. Plot the electrical energy of the protein as a function of its radius for different salt concentrations, ranging between 1 mM and 100 mM. What conclusion do you draw about the effect of salt on the charged state of a protein?

• 9.10 Binding to a membrane

Consider a phospholipid bilayer membrane consisting of a mixture of 90% uncharged lipid and 10% singly charged acid lipid. Assume 0.68 nm^2 surface area per lipid head group, and assume further that the charged lipids are uniformly distributed and immobile. The membrane is in contact with an aqueous solution of NaCl at 25 °C. The salt concentration is $c_\infty = 100$ mM.

(a) Calculate the areal charge density (that is, charge per unit surface area) of the membrane.

(b) Calculate the surface potential of the membrane.

What is the electrostatic energy (in $k_B T$ units) of binding to the membrane of a trivalent positive ion such as spermidine (a biologically active polyamine) assuming that:

(c) Binding occurs at the membrane surface?

(d) Owing to steric factors, the charges of the bound spermidine stay in the water 0.5 nm distant from the membrane surface?

(Adapted from Problems 23.6 and 23.7 of K. Dill and S. Bromberg, Molecular Driving Forces, 2nd ed. Garland Science, 2011.)

• 9.11 Toy model of salt-dependent ligand–receptor binding

In Section 9.2.3, we discussed experiments that demonstrate the effect of salts on binding of proteins to DNA. Here we explore this effect in the context of two models, both of a simple ligand–receptor system, where we take into account the charge state and the related electrostatic energy of the relevant proteins. We assume that the charge of the receptor is $+ze$ and the ligand has charge $-ze$, where e is the elementary charge of the electron.

For both models, the state in which the ligand is bound to the receptor is the same and corresponds to a situation where the charges on the two cancel each other out. The difference comes in the unbound state.

(a) Consider a model in which the unbound ligand and the receptor each have z tightly bound monovalent counterions that neutralize their charge. Compute the probability that the ligand is bound to the receptor and the dissociation constant. How does the dissociation constant scale with the concentration of counterions?

(b) Next, consider a model in which ligand and receptor in the unbound state are surrounded by a screening cloud of counterions, whose width is given by the Debye screening length λ_D. What is the scaling of the dissociation constant with the concentration of counterions in this case?

(c) Compare the predictions of the two models with the data shown in Figure 9.3. Which model is favored by the data?

• 9.12 Membrane pores

A neutral protein "carrier" may help an ion to transfer into and cross a lipid membrane.

(a) What is the electrostatic free-energy change when a monovalent ion is transferred from water at 25 °C to a hydrocarbon solvent with dielectric constant $D = 2$? The radius of the ion is 0.2 nm.

(b) Comment on the ability of the ions to diffuse through lipid bilayers. How has nature solved this problem? How do ions manage to get across?

(Adapted from Problem 22.11 of K. Dill and S. Bromberg, Molecular Driving Forces, 2nd ed. Garland Science, 2011.)

9.6 Further Reading

Dill, K, & Bromberg, S (2011) Molecular Driving Forces, 2nd ed., Garland Science. Their discussions of charges in solution are highly enlightening.

Bradley, T (2009) Animal Osmoregulation, Oxford University Press. Extremely interesting discussion of how organisms manage life in salty environments.

Record, MT, Jr, Zhang, W, & Anderson, CF (1998) Analysis of effects of salts and uncharged solutes on protein and nucleic acid equilibria and processes: a practical guide to recognizing and interpreting polyelectrolyte effects, Hofmeister effects, and osmotic effects of salts, *Adv. Protein Chem.* **51**, 281. A thorough discussion of the relation between salt and binding.

Feynman, RP, Leighton, RB, & Sands, M (1965) The Feynman Lectures on Physics, Addison-Wesley. Volume II of these lectures gives a beautiful description of electricity and magnetism.

Daune, M (1999) Molecular Biophysics—Structures in Motion, Oxford University Press. Part IV, entitled "Biopolymers as polyelectrolytes," has a detailed discussion of many of the topics covered here.

Purcell, EM (1984) Electricity and Magnetism, 2nd ed., McGraw-Hill. Purcell's book is a classic introduction to electricity and magnetism.

Schey, HM (1996) Div, Grad, Curl and All That: An Informal Text on Vector Calculus, 4th ed., W. W. Norton and Company. This book provides a fun and thorough introduction to the tools of vector calculus.

9.7 References

Alberts, B, Johnson, A, Lewis, J, et al. (2008) Molecular Biology of the Cell, 5th ed., Garland Science.

Jensen, DE & von Hippel, PH (1976) DNA "melting" proteins. I. Effects of bovine pancreatic ribonuclease binding on the conformation and stability of DNA, *J. Biol. Chem.* **251**, 7198.

Kao-Huang, Y, Revzin, A, Butler, AP, et al. (1977) Nonspecific DNA binding of genome-regulating proteins as a biological control mechanism: Measurement of DNA-bound *Escherichia coli* lac repressor *in vivo*, *Proc. Natl Acad. Sci. USA* **74**, 4228.

Kegel, WK, & van der Schoot, P (2004) Competing hydrophobic and screened-coulomb interactions in hepatitis B virus capsid assembly, *Biophys. J.* **86**, 3905.

Nelson, P (2004) Biological Physics: Energy, Information, Life, W. H. Freeman.

Wynne, SA, Crowther, RA, & Leslie, AGW (1999) The crystal structure of the human hepatitis B virus capsid, *Mol. Cell* **3**, 771.

Zlotnick, A (1994) To build a virus capsid. An equilibrium model of the self-assembly of polyhedral protein complexes, *J. Mol. Biol.* **241**, 59.

Beam Theory: Architecture for Cells and Skeletons

10

Overview: In which we apply the theory of elastic rods to a diverse array of biological filaments

> "I demolish my bridges behind me - then there is no choice but forward."
>
> **Fridtjof Nansen**

Many of the macromolecules of living organisms are filamentous. Not only are they a striking visual feature within cells, but their structure is intimately tied to the ways in which such molecules are used in cells. The representation of geometric structures as networks of one-dimensional elements is a perspective of great power and applicability. That part of mechanics which has grown up around this approximation is known traditionally as "beam theory." Historically, the study of the structural mechanics of beams has been equally rewarding whether applied to the flying buttresses of Notre Dame or to the wings of the jumbo jet that carried the tourist to Paris. The theme of the present chapter is the observation that the biological world has put one-dimensional beams to similar architectural uses. The physical ideas introduced here all amount to arrangements of springs (that is, Hooke's law) and permit us to examine a range of seemingly unrelated problems such as DNA bending during transcriptional regulation, packaging of DNA in viruses and in the eukaryotic nucleus, and the properties of the cytoskeleton.

10.1 Beams Are Everywhere: From Flagella to the Cytoskeleton

One-Dimensional Structural Elements Are the Basis of Much of Macromolecular and Cellular Architecture

When we reflect on structures such as trees, animal skeletons, the cytoskeletal networks of cells, or the molecules that make up Crick's "two great polymer languages," these structures are all based upon elements with one dimension that is much larger than the other two.

Geometric structures with this property are known as beams (or rods) and have interesting mechanical features that affect their biological function. Figure 10.1 shows a variety of different examples of structures that can be viewed mechanically as one-dimensional beams. Figure 10.1(A) is a reminder of the role of beams as structural elements in conventional architecture. A much smaller example that has been the basis of impressive single-molecule studies of biological molecules is the atomic-force microscope based upon cantilevers like that shown in Figure 10.1(B).

In this chapter, we focus on the interesting biological examples of structural elements that can be thought of as elastic beams. In particular, Figures 10.1(C–G) reveal a number of examples of this kind. Figure 10.1(C) illustrates the network of one-dimensional filaments that make up the cytoskeleton of a typical eukaryotic cell. Another beautiful example is that of the hair cells of the mammalian inner ear as shown in Figure 10.1(D). These hair cells have slender protrusions, tens of microns in length, known as stereocilia, which vibrate in response to sound. Figure 10.1(E) suggests that the flagella that propel cells ranging from *E. coli* to eukaryotic sperm can be viewed as elastic rods. Flagella like that shown in the figure can be as much as 10 times longer than the cell bodies that they are used to propel, while their cross-section can be orders of magnitude smaller. Figures 10.1(F) and (G) show the way in which individual polymers (microtubules and DNA, respectively) have geometries that are very long in comparison with their width.

In the sections to follow, we develop the quantitative framework for thinking about the structure and energetics of biological structures that can be approximated as one-dimensional beams. Two of the most important case studies involve the deformations that DNA

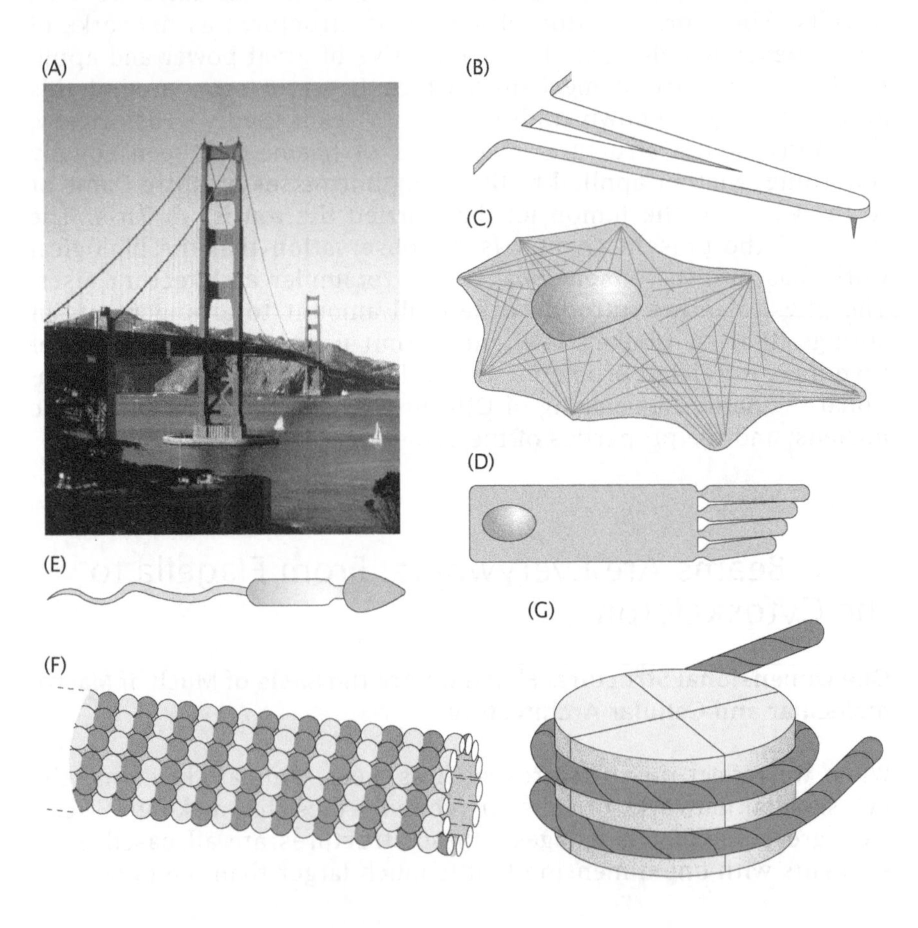

Figure 10.1: Diverse examples of the way in which structures can be interpreted using beam theory. (A) A bridge as a collection of beams and cables. (B) A small-scale cantilever used in an atomic-force microscope. (C) The cytoskeleton in a eukaryotic cell. (D) Stereocilia on an inner ear hair cell. (E) The flagellum of a sperm. (F) An individual microtubule. (G) A representation of DNA as an elastic rod in the context of the nucleosome.

is subjected to both in the gene regulatory setting as a result of interactions with DNA-binding proteins and in the DNA-packaging context. DNA packaging, whether in viruses, prokaryotes, or eukaryotes, involves compaction of huge DNA molecules into confined spaces with an attendant energy cost associated with elastic bending. A second class of case studies will involve the mechanics of the cytoskeleton when considered as a collection of elastic beams. One of the interesting observations offered by physical biology is that problems that may seem very distantly related when viewed from the biological perspective are very close when viewed from the physical biology perspective. In this case, the study of elastic deformations of beams will lead us to speak of transcriptional regulation, DNA packaging, and cytoskeletal mechanics using one common language.

10.2 Geometry and Energetics of Beam Deformation

10.2.1 Stretch, Bend, and Twist

Biological filaments are characterized by one dimension (the length) that is much greater than their transverse dimensions. For example, in the case of the bacterial flagellum, the structure has a length in excess of microns with a diameter that is measured in only tens of nanometers. As was illustrated in Figure 2.27 (p. 64), the same can be said of tobacco mosaic virus, actin, and microtubules, which similarly have a characteristic length of order microns with cross-sectional dimensions measured in nanometers. Because of this geometric asymmetry, it is possible to invoke key simplifying assumptions that permit us to write the energy of deformations of these asymmetric structures very simply.

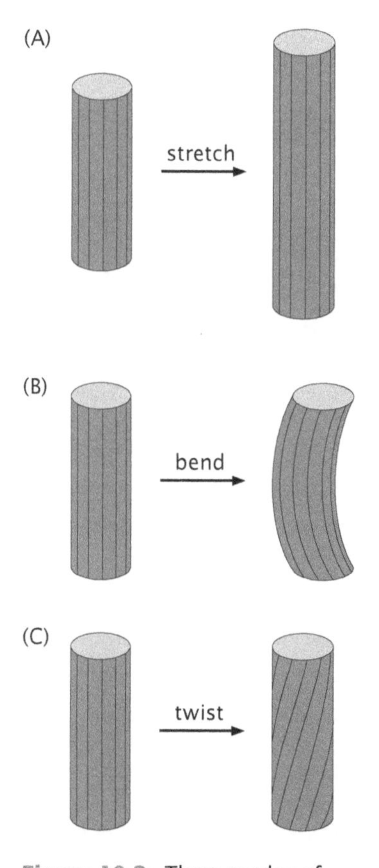

Figure 10.2: Three modes of deformation of a beam: (A) stretching of a beam, (B) bending of a beam, and (C) twisting of a beam.

Beam Deformations Result in Stretching, Bending, and Twisting

We begin with a qualitative discussion of the geometric character of the three key independent modes of deformation to which a beam may be subjected. In particular, the deformation of beams can be described in terms of extension, bending, and torsion as shown in Figure 10.2. Extensional deformations have already been discussed in Section 5.4.1 (p. 216) and correspond to simple extension (or compression) of the beam along its long axis from length L_0 to length $L_0 + \Delta L$. In this chapter, we complement the earlier discussion by thinking about two other modes of deformation that are also important in the mechanics of biological systems. As seen in Figure 10.2(B), the bending of a beam can, at its simplest, be thought of as a deformation that takes a straight beam and bends it into an arc of a circle. The other key mode of deformation corresponds to twisting the beam about its long axis as shown in Figure 10.2(C).

A Bent Beam Can Be Analyzed as a Collection of Stretched Beams

The state of deformation that will animate the majority of our discussion in this chapter is that of bending. Bending is of interest in many of the examples already revealed in Figure 10.1. For example, when determining the free energy associated with the wrapping of DNA around the histone octamer as shown in Figure 10.1(G), the energy

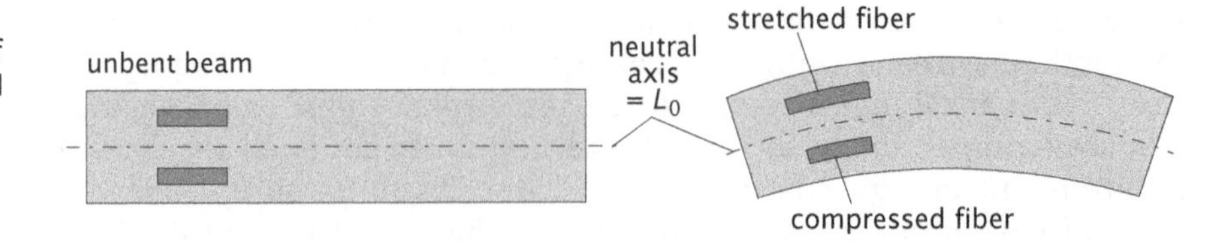

Figure 10.3: Stretching of material elements in a bent beam. The length of the dotted line representing the neutral axis is unchanged after bending. Material elements above the neutral axis are stretched and those below it are compressed.

of bending the DNA serves as one of the two key ingredients in the overall free energy and competes with the DNA–protein interaction energy to determine the overall free energy of nucleosome formation.

We begin by examining the nature of the geometric assumptions that are made about the state of deformation of a bent beam. Indeed, the first thing we will say about the geometry of a bent beam is that there is a certain axis, known as the neutral axis, that is neither stretched nor compressed during the act of beam bending, as shown in Figure 10.3. The key observation embodied in the figure is that material elements above the neutral axis are stretched, while those below the neutral axis are compressed.

The geometric characterization of a beam subjected to some complicated (that is, nonuniform) state of bending as shown in Figure 10.4 is analyzed by recourse to a divide-and-conquer strategy that will be taken up repeatedly throughout the book. The idea of this strategy is that we divide the beam up into a collection of small segments,

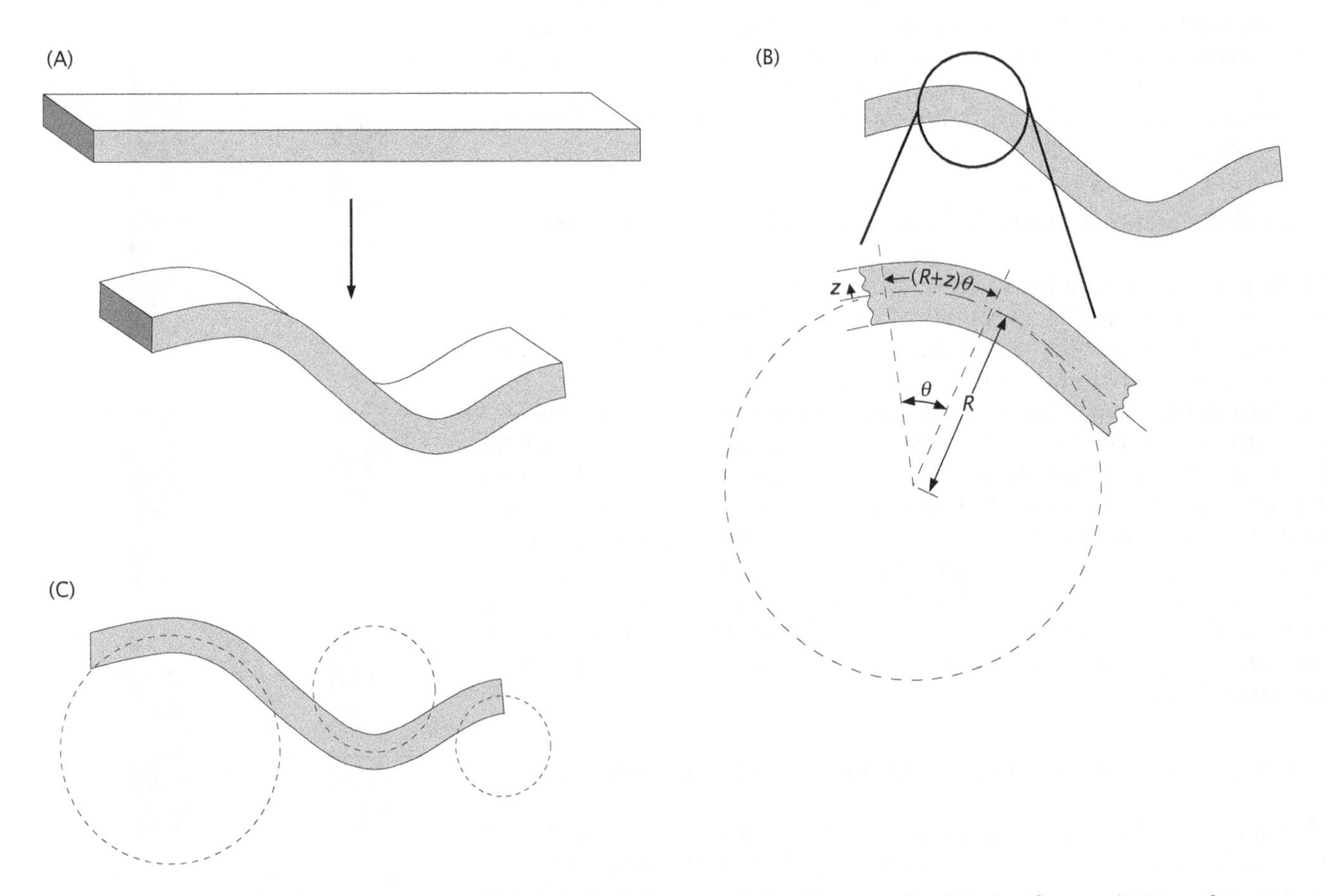

Figure 10.4: Curvature of a beam. The schematic shows the way in which a beam can be thought of as a collection of segments, each of which is bent locally into an arc of a circle. (A) Nonuniform bending of a beam. (B) Determination of curvature at a given point along the beam. (C) Illustration of different curvature at different points along the beam.

each of which can be thought of as being a part of an arc of a circle. We will argue below that for beams bent into circular arcs, it is a straightforward exercise to compute both the curvature ($1/R$) and the resulting deformation energy. Consequently, when trying to evaluate the bending energy of the entire beam, we just add up the contributions from each and every segment (by mathematical integration). The implementation of this geometric idea is illustrated in Figure 10.4. The figure invites us to ignore the dimension perpendicular to the plane of the page since by symmetry all beam elements along this direction are equivalent. The second key point illustrated in the figure is that, regardless of the complexity of the bending, we may think of the beam as a collection of circular arcs all glued to each other. That is, each cross-section along the length of the beam can be fitted onto a circle with a particular radius, this radius corresponding to the *local* radius of curvature of the beam itself.

Once the radius of curvature is in hand, we are prepared to compute the state of strain within the beam. As was shown in Section 5.4.1 (p. 216), given the strain for a beam that has been stretched, we can use Hooke's law to compute the strain energy of bending. The strain (for the simple cases of interest here) is a measure of the fractional extension ($\Delta L/L_0$) of a particular little element of the material. In particular, we note that the extent of compression or extension is a linear function of the perpendicular distance of a given material point from the neutral axis (that is, $\Delta L(z) = \alpha z$). On geometric grounds, we may determine the unknown constant α through an examination of Figure 10.4, where we note that all parts of the beam subtend the angle θ as shown in the figure. As a result, the fiber at distance z from the neutral axis is of overall length $L(z) = (R + z)\theta$, which may be rewritten as $L(z) = (R + z)L_0/R$, where we have used the fact that $\theta = L_0/R$. As a result, we can write the change in length of the fiber at distance z from the neutral axis as

$$\Delta L(z) = L(z) - L_0 = (R + z)\frac{L_0}{R} - L_0 = \frac{zL_0}{R}. \qquad (10.1)$$

Continuing along these same lines, we conclude that the extensional strain $\varepsilon(z)$ at a distance z from the neutral axis is given by

$$\varepsilon(z) = \frac{\Delta L(z)}{L_0} = \frac{z}{R}. \qquad (10.2)$$

Recall that z is measured from the neutral axis and hence material elements above this axis in Figure 10.4 are stretched ($\varepsilon > 0$) while those below this axis are compressed ($\varepsilon < 0$).

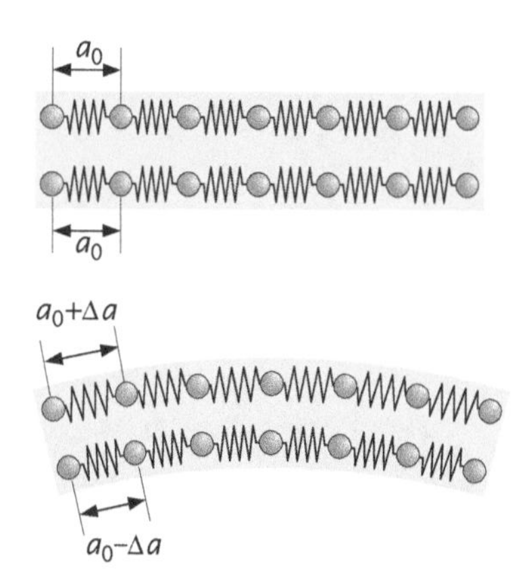

Figure 10.5: Microscopic interpretation of the energetics of beam bending. When a beam is subjected to bending, the bonds above the neutral axis are stretched while those below the neutral axis are compressed.

The Energy Cost to Deform a Beam Is a Quadratic Function of the Strain

Given the state of strain within a beam described above, it is now possible to assess the energy stored in the beam by virtue of such deformation. The beautiful idea is to use what we already know about the energetics of stretching beams to work out the energy cost to *bend* a beam. The principle is illustrated in Figure 10.5, which is a microscopic impression of the idea that the bonds between atoms above the neutral axis are stretched while those below the neutral axis are compressed. Though the detailed forces between atoms in macromolecules are complicated, in the case of small deformations the energetics of bond bending and stretching is simply quadratic.

The energy associated with the extension of a beam can be *estimated* easily by ignoring the Poisson effect (essentially the effect that if we pull in one direction, the material contracts in the others) and noting that the strain energy density (that is, energy per unit volume) is given by

$$W(\varepsilon) = \frac{1}{2}E\varepsilon^2 = \frac{1}{2}E\left(\frac{\Delta L}{L_0}\right)^2, \tag{10.3}$$

where we remind the reader that E is the Young modulus of the material of interest (see Section 5.4.1 on p. 216 for a refresher). Given this estimate for the strain energy density, reckoning the total stored elastic energy in the little volume of deformed material requires multiplying the energy density by the volume of the little volume element.

A more precise treatment of the energy of deformation accounts for the fact that the state of strain depends upon where we are within the deformed beam. In particular, we invoke Equation 10.2, which tells us how the strain depends upon the perpendicular distance from the neutral axis. From an energetic point of view, this means that we have to add up the contributions from each little material element separately. In light of the z dependence of the strain itself, the total strain energy stored in the beam is written as

$$E_{\text{bend}} = L_0 \int_{\partial\Omega} \mathrm{d}A \frac{E}{2R^2} z^2, \tag{10.4}$$

where the integration is over the cross-sectional area $\partial\Omega$ perpendicular to the beam axis. The strain energy integral is shown schematically in Figure 10.6.

More generally, our result may be written as

$$E_{\text{bend}} = \frac{EIL}{2R^2}, \tag{10.5}$$

where we have introduced the geometric moment

$$I = \int_{\partial\Omega} z^2 \,\mathrm{d}A, \tag{10.6}$$

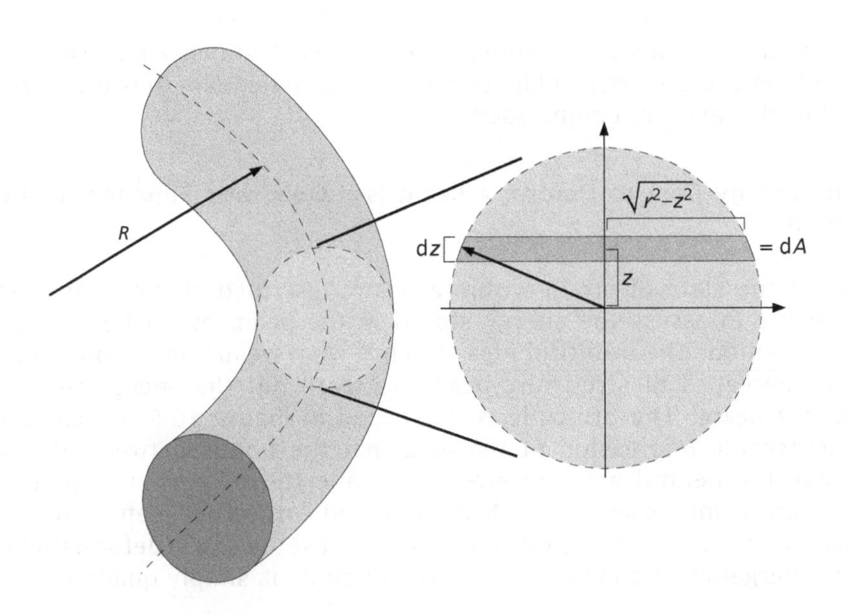

Figure 10.6: Computation of the bending energy. The bending energy is obtained by integrating over the cross-sectional area of the beam.

where once again the area element is perpendicular to the beam axis, and the quantity z is a measure of the perpendicular distance from the neutral axis. Equation 10.5 is one of the more important equations to be considered in this book, since it tell us the bending energy associated with bending a beam into a circular arc of length L with radius of curvature R. We may take that result even further through the recognition that for a circular loop with radius of curvature R, its length is $L = 2\pi R$ and the bending energy is

$$E_{\text{loop}} = \frac{\pi EI}{R}. \tag{10.7}$$

The energetics of simple circular loops will be applied in numerous places throughout the remainder of the chapter and especially in our consideration of transcriptional regulation and DNA packaging.

As yet, our analysis has been for the case in which the entire beam of length L is uniformly deformed into a single arc of a circle with radius R. More generally, we will be interested in states of deformation in which the local curvature differs from one point to the next. In this case, we invoke a *locality* assumption in which we pretend that the energy of a given material particle is that of the beam described above and for which the curvature is constant. The total energy is gotten by adding up the energy material particle by material particle and is of the form

$$E_{\text{bend}} = \frac{K_{\text{eff}}}{2}\int_0^L \mathrm{d}s\,\frac{1}{R(s)^2}, \tag{10.8}$$

where we have introduced the flexural rigidity $K_{\text{eff}} = EI$, which embodies both material parameters (that is, the Young modulus E) and the geometric shape (through the geometric moment I). This equation is the mathematical embodiment of the cartoon shown in Figure 10.4, where we show how each point on the beam can be thought of locally as part of a circle. There are several other convenient ways of writing Equation 10.8 that will arise in the remainder of the book. First, we note that $\kappa(s) = 1/R(s)$, where we have defined the curvature $\kappa(s)$. This implies that we can rewrite the equation as

$$E_{\text{bend}} = \frac{K_{\text{eff}}}{2}\int_0^L \left|\frac{\mathrm{d}\mathbf{t}}{\mathrm{d}s}\right|^2 \mathrm{d}s, \tag{10.9}$$

where we use the fact that the curvature can itself be written as the derivative of the tangent vector.

10.2.2 Beam Theory and the Persistence Length: Stiffness is Relative

Thermal Fluctuations Tend to Randomize the Orientation of Biological Polymers

One intriguing theme that we will return to repeatedly throughout the book (already introduced in Section 5.1.1 on p. 189) is the idea of a competition between thermal effects and deterministic forces. For example, in Figure 9.12 (p. 369) we introduced the Bjerrum length as the length scale over which charges can wander from a protein without offending the Coulomb interaction too significantly. Here we examine another example of this same kind of argument in which a biological polymer such as DNA is kicked around into various different orientations as a result of interactions with the surrounding fluid. In this case, it is the elastic forces rather than the Coulomb forces that set the length scale over which such fluctuations are tolerated. In particular, the intuition corresponding to this idea is that if the polymer is too short, it will be indifferent to thermal fluctuations. By way of contrast, for very long polymers, the orientation at one extremity is completely indifferent to that at the other.

The Persistence Length Is the Length Over Which a Polymer Is Roughly Rigid

The competition described above between thermal fluctuations and the energetic cost associated with beam bending is succinctly captured in the emergence of a single length scale, namely, the persistence length. The persistence length is a measure of the competition between the entropic parts of the free energy, which tend to randomize the orientation of the polymer, and the energetic cost of bending. Throughout physical biology, there are a variety of different length scales that arise that, at the deepest level, reflect the interplay between energy and entropy. Generally, length scales that reflect the competition between thermal and deterministic energies can be estimated by equating the deterministic energy cost for the particular mechanism of interest to $k_B T$, the thermal energy. When elastic energies are competing with thermal fluctuations, the relevant comparison is

$$k_B T \approx \frac{EIL}{2R^2}, \tag{10.10}$$

where L is the length of the fragment of interest. Roughly speaking, the persistence length ξ_p is that length of polymer for which the radius of curvature is equal to the length of polymer itself. Hence, if we set L and R both equal to ξ_p, we see that our estimate for the persistence length is given by

$$\xi_p \approx \frac{EI}{2k_B T}. \tag{10.11}$$

The Persistence Length Characterizes the Correlations in the Tangent Vectors at Different Positions Along the Polymer

An alternative view of the persistence length is to think of biological polymers from the standpoint of the geometry of space curves. There are a number of different ways of characterizing the mathematics of space curves. One way is via correlation functions, which measure the extent to which the geometry of one part of the polymer is correlated with some other part.

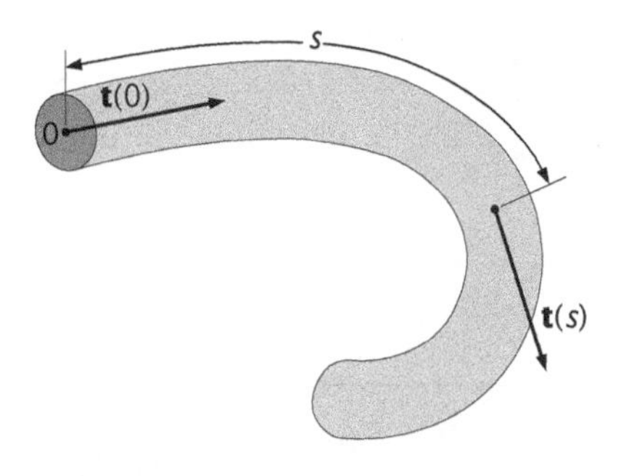

Figure 10.7: Tangent vectors on a fluctuating polymer chain. The parameter s measures the distance along the polymer. The tangent vector at contour length s is represented as $\mathbf{t}(s)$.

To be concrete, consider a space curve in parametric representation given by the vector $\mathbf{r}(s)$, where s is the arclength parameter. The function $\mathbf{r}(s)$ is a rule that assigns a position vector for every value of the arclength s. Such a configuration is shown in Figure 10.7. The question we pose concerning a given space curve is the nature of the tangent–tangent correlation function defined as $g(s) = \langle \mathbf{t}(\tau + s) \cdot \mathbf{t}(\tau) \rangle$. Our notation is built around the unit tangent vector, $\mathbf{t}(s)$, which is defined as the unit vector tangent to the curve at arclength s. The tangent–tangent correlation function as defined above is thus far a purely mathematical notion and provides a measure of the relation between the tangents at arclength τ and arclength $\tau + s$. When the correlation function is 1, the tangent vectors are parallel.

This idea was made explicit in Figure 8.5 (p. 319), where we showed the genome of a bacterium that had been released from the cell. In the inset to that figure, we labeled the persistence length with small white arrows, each with a length equal to that of the persistence length of DNA. These arrows reveal that on the scale of tens of nanometers, the tangents are correlated while on the scale of microns they are not.

The Persistence Length Is Obtained by Averaging Over All Configurations of the Polymer

Qualitatively, we see that the actual numerical value of the persistence length depends upon the stiffness of the filament. Microtubules and DNA do not have the same persistence length, with DNA characterized by a persistence length of roughly 50 nm while microtubules have a persistence length of the order of 6 mm. To compute the persistence length, we average over all configurations of the polymer, with each such configuration assigned a Boltzmann weight that depends upon the bending energy. Before we plunge into the mathematics of the tangent–tangent correlation function

$$g(s) = \langle \mathbf{t}(s) \cdot \mathbf{t}(0) \rangle, \tag{10.12}$$

it is useful to think about its limits. First, since the tangent vector is of unit length, $g(0) = 1$. On the other hand, for s much larger than the persistence length, we expect the two tangent vectors to be independent and $g(s) \to 0$. A simple function with these properties is an exponential function

$$g(s) = e^{-s/\xi_p} \tag{10.13}$$

where ξ_p is the persistence length. The mathematical proof that $g(s)$ is indeed exponential is somewhat subtle. It is based on the property that two adjacent segments of an elastic beam are buffeted by independent thermal forces and hence the tangent–tangent correlation function over both segments is the product of the individual ones, that is, $g(s_1 + s_2) = g(s_1)g(s_2)$; s_1 and s_2 are the lengths of the two segments (further details can be found in Nelson 2004).

We seek a relation between the persistence length, as defined by Equation 10.13, and the flexural rigidity $K_{eff} = EI$ of an elastic beam. To this end, we take a short beam of length $s \ll \xi_p$ and compute $g(s)$. For a short beam, thermal forces can only bend the beam slightly and the shape at any instant can be approximated by an arc of a circle of radius R. The energy of such a configuration is given by Equation 10.5, which can also be written as

$$E_{bend} = \frac{EI}{2s}\theta^2, \tag{10.14}$$

with $\theta = s/R$. Taking the tangent of one end of the beam to point along the z-direction, the tangent–tangent correlation function becomes

$$g(s) = \langle \cos\theta(s) \rangle, \tag{10.15}$$

with $\theta(s)$ the angle between the tangent vector at the other end and the z-direction. Given our assumption of small deflection angles ($\theta(s) \ll 1$) appropriate for a short beam, the cosine function can be expanded into a Taylor series (see The Math Behind the Models on p. 215) to yield a simplified expression

$$g(s) = \left\langle 1 - \frac{\theta^2(s)}{2} \right\rangle. \tag{10.16}$$

The thermal average in the above equation is computed by summing over all possible orientations of the tangent vector at s, which in three dimensions traces out the surface of a unit sphere. Therefore,

$$\left\langle \theta^2(s) \right\rangle = \frac{1}{Z} \int_0^{2\pi} \mathrm{d}\phi \int_0^{\pi} \mathrm{d}\theta \sin\theta \; \theta^2 \mathrm{e}^{-(EI/2k_B Ts)\theta^2}, \tag{10.17}$$

where

$$Z = \int_0^{2\pi} \mathrm{d}\phi \int_0^{\pi} \mathrm{d}\theta \sin\theta \; \mathrm{e}^{-(EI/2k_B Ts)\theta^2}, \tag{10.18}$$

is the normalization factor (partition function). To simplify calculations we note that the integrand in Equation 10.17 differs from the integrand for Z by the factor θ^2, which we can pull down from the Boltzmann factor in Equation 10.18 by differentiating with respect to E as introduced in Section 6.9 (p. 241). Then we can write

$$\left\langle \theta^2(s) \right\rangle = \frac{1}{Z}\left(-\frac{2k_B Ts}{I} \frac{\partial Z}{\partial E} \right) = -\frac{2k_B Ts}{I} \frac{\partial \ln Z}{\partial E}, \tag{10.19}$$

and all that remains to be calculated is the integral in Equation 10.18.

We compute the integral in question by making the substitution $\sin\theta \approx \theta$, valid for small angles, and by making use of a change of variables $u = (EI/2k_B Ts)\theta^2$,

$$Z = \frac{2\pi k_B Ts}{EI} \int_0^{\infty} \mathrm{d}u \, \mathrm{e}^{-u} = \frac{2\pi k_B Ts}{EI}. \tag{10.20}$$

Note that the upper integration bound for the variable u tends to infinity in the limit when s is much smaller than the persistence length. Using the result $\partial \ln Z/\partial E = -1/E$ in Equation 10.19, and substituting the value for $\langle \theta^2(s) \rangle$ obtained in this way in Equation 10.16, we arrive at the result

$$g(s) = 1 - \frac{k_B T}{EI} s. \tag{10.21}$$

Comparing this with Equation 10.13 in the $s \ll \xi_p$ limit, we conclude

$$\xi_p = \frac{EI}{k_B T}. \tag{10.22}$$

Note that the scaling is precisely that found in Equation 10.11, which was based on a simple comparison of the thermal energy and the bending energy of an elastic beam bent into a circular arc of 1 radian. This result also allows us to rewrite the flexural rigidity as $EI = \xi_p k_B T$.

10.2.3 Elasticity and Entropy: The Worm-Like Chain

The Worm-Like Chain Model Accounts for Both the Elastic Energy and Entropy of Polymer Chains

So far in the book, we have used two different physical models to characterize the structure and free energy of polymers. In Chapter 8, we examined the behavior of polymers as random walks. These ideas permitted us to consider experiments such as force–extension measurements that probe the entropic forces tending to keep polymers collapsed. In this chapter, we have taken a complementary view that considers the energetic cost of deforming polymers. Evidently, both the bending energy and entropic contributions to the overall free-energy budget are important. To that end, a class of models known as worm-like chain models has been set forth in which one evaluates the partition function associated with different polymer configurations, where each such configuration is appropriately weighted by its corresponding elastic energy cost. The concept of these models is shown in Figure 10.8.

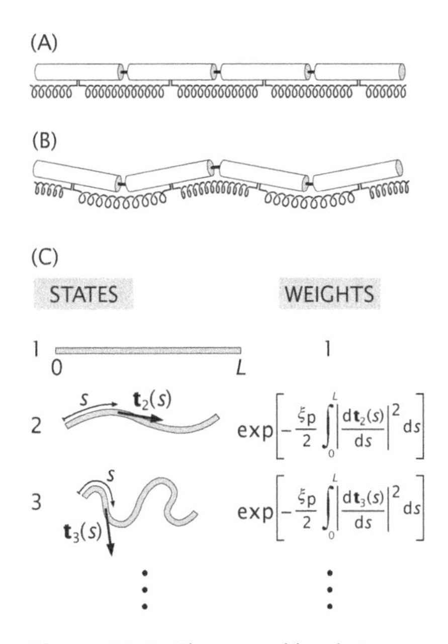

Figure 10.8: The worm-like chain concept. As in Chapter 8, chains are represented by cylinders connected by flexible links. (A) The undeformed configuration showing that the springs are unstretched, but the links are deprived of entropy because there is only one such possible arrangement of the segments. (B) A deformed configuration showing that there is an energetic cost to bend the chain, but there are more configurations of the system. (C) States and weights for the worm-like chain model in the absence of applied force. There are an infinite number of different states corresponding to all the different space curves that can be made with the polymer, but to find the energy of each such configuration, we simply compute the bending energy of the conformation of interest.

The competition between chain entropy and bending energy is captured by the partition function, which for the worm-like chain model reads

$$Z = \int \mathcal{D}\mathbf{t}(s) \exp\left(-\frac{\xi_p}{2}\int_0^L \left|\frac{d\mathbf{t}}{ds}\right|^2 ds\right). \quad (10.23)$$

The mathematics expressed by the above equation translates into a simple algorithm:

(1) draw a curve of length L representing a possible DNA configuration;
(2) evaluate its bending energy

$$E_{\text{bend}} = \frac{\xi_p k_B T}{2}\int_0^L \left|\frac{d\mathbf{t}}{ds}\right|^2 ds$$

and the corresponding Boltzmann factor $e^{-E_{\text{bend}}/k_B T}$; and
(3) repeat (1) and (2) for all possible curves and sum all the Boltzmann factors to obtain Z.

The sum over curves is the celebrated Feynman path integral, written here as $\int \mathcal{D}\mathbf{t}(s)$, where $\mathbf{t}(s)$ is the tangent vector of the curve, parameterized by the arclength s. For the present purposes, Equation 10.23 should just be thought of as a fancy, shorthand notation for the partition function that reminds us that there are an infinite number of different ways of configuring the polymer.

In order to compute the force–extension curve of the worm-like chain model, we consider a force F applied to one end of the chain in the z-direction. The energy of the worm-like chain then acquires an additional $-F\int_0^L t_z\, ds$ term, where t_z is the z-component of the tangent vector $\mathbf{t}$. This is equivalent to the energy of the freely jointed chain model in the presence of an applied force, discussed previously, and amounts to accounting for the lowering of weights when the polymer is elongated as shown schematically in Figure 8.24 (p. 341). For fixed F, the chain extension $z = \int_0^L t_z\, ds$ averaged over all configurations is

$$\langle z\rangle = \frac{1}{Z(f)}\int \mathcal{D}\mathbf{t}(s)\; z \exp\left(-\frac{\xi_p}{2}\int_0^L \left|\frac{d\mathbf{t}}{ds}\right|^2 ds + f\int_0^L t_z\, ds\right), \quad (10.24)$$

where $Z(f)$ is the partition function in the presence of the applied force $F = fk_B T$. Note that the reduced force f has units of inverse length. Equation 10.24 can be rewritten as

$$\langle z\rangle = \frac{d\ln Z(f)}{df}, \quad (10.25)$$

owing to the fact that z is the thermodynamic conjugate of the force. The problem of computing the force–extension curve therefore reduces to computing the partition function $Z(f)$. This is a rather difficult mathematical problem, since one has to make sense of summing over all curves. Fortunately, this problem greatly simplifies in the low- and high-force limits. In Section 10.7 (the appendix at the end of the chapter), we examine both the low- and high-force limits, which permit a determination of the force–extension properties of the worm-like chain model in these limits.

The outcome of these calculations can be summarized in an interpolation formula

$$f\xi_p \approx \frac{z}{L} + \frac{1}{4(1 - z/L)^2} - \frac{1}{4} \tag{10.26}$$

that permits an evaluation of the extensions suffered by a worm-like chain over the entire range of applied forces. The high-force limit in this formula is very different from the freely jointed chain result for the force–extension curve (see Section 8.3.2 on p. 340), which predicts a $1/f\xi_p$ approach to full extension $z = L$. The slower approach to full extension in this case is a consequence of the finite bending rigidity at scales below the Kuhn length in the worm-like chain model.

10.3 The Mechanics of Transcriptional Regulation: DNA Looping Redux

The physical properties of DNA as a deformable polymer complement its well-known properties as the storage medium for genetic information. Indeed, there are many cases where the interplay between the informational and physical properties of the DNA molecule have intriguing biological consequences. One example of great importance is that of gene regulation. There are a variety of examples in which transcriptional regulation is carried out by the binding of proteins, which have the effect of forming loops in the DNA. An example of this effect was introduced in Figure 1.11 (p. 19) and the entropic contribution to the free energy of looping was discussed in Section 8.2.4 (p. 333). The aim of this section is to take stock of how the ideas on beam bending developed in this chapter can be applied to examine DNA deformations in transcriptional regulation.

10.3.1 The *lac* Operon and Other Looping Systems

As was already emphasized in Section 4.4.3 (p. 157), one of the most celebrated examples of gene regulation is the *lac* operon. This genetic network oversees the metabolism of the sugar lactose in *E. coli*. Historically, this system served as one of the focal points resulting in the emergence of a picture of how genes are controlled. In the present setting, we consider repression of transcription in the *lac* operon, which is mediated by the formation of loops. The argument of this section is that such loops can be considered from the perspective of the elastic theory of beams. As a brief reminder, we recall that the *lac* operon is subject to both positive and negative control. In particular, RNA polymerase is "recruited" to the promoter by an activator protein known as CAP. Negative control of transcription is dictated by the presence of the DNA-binding protein Lac repressor.

Transcriptional Regulation Can Be Effected by DNA Looping

Lac repressor, as shown in Figure 8.19 (p. 334), is a tetrameric protein that binds two distinct DNA sites (one of which is in the vicinity of the promoter) and loops the intervening DNA. In particular, there are a total of three specific binding sites for Lac repressor, denoted O1, O2, and O3, with O2 being 401 base pairs downstream of O1 and O3 being 92 base pairs upstream of O1. Full repression by the repressor molecule demands that the tetrameric repressor molecule bind O1, whose center is located 11 base pairs downstream from the RNA polymerase transcription start site, and one of the auxiliary operators simultaneously, thus forming a loop with the intervening DNA. Depending upon which operators are linked, the looped region can either be 401 or 92 base pairs in length. Note that the 92 base pair loop appears to involve substantial bending of a DNA fragment that is notably smaller than the persistence length.

In addition to the example of the *lac* operon, there are a host of other intriguing prokaryotic and eukaryotic examples of regulatory regions involving DNA looping. In prokaryotes, another important example is the *ara* operon, which is associated with metabolism of the sugar arabinose. More generally in eukaryotes, there are a wide variety of examples of *cis*-regulatory regions that control developmental processes that also involve DNA looping. To give an impression of the complexity of *cis*-regulatory regions associated with eukaryotic promoters, we consider one of the well-characterized regulatory networks associated with sea urchin development. Figure 10.9 shows the regulatory region associated with the *cyIIIa* gene in sea urchin development as well as electron micrographs of the looped segments of DNA resulting from the binding of particular transcription factors.

10.3.2 Energetics of DNA Looping

In Section 8.2.4 (p. 333) we examined the entropic consequences of DNA looping. That discussion now must be complemented by an evaluation of the energetic consequences of looping. We begin by estimating the amount of elastic energy that has to be paid in order for the repressor molecule to effect its binding. From the perspective of elasticity, a simple estimate for the energetics of DNA looping can be obtained on the assumption that the loops are circular. In particular, once we accept this geometric picture, the energetics of a loop of radius R is given by $E_{\rm loop} = \xi_{\rm p}\pi k_{\rm B}T/R$, as was deduced in Equation 10.7. A more useful way of writing this is in terms of the number of base pairs involved in the loop. In particular, if we recall that the length of a given fragment of DNA is given by $L \approx 0.34 N_{\rm bp}$ nm, then we can write the loop energy more directly as

$$\frac{E_{\rm loop}}{k_{\rm B}T} = \frac{2\pi^2}{N_{\rm bp}}\left(\frac{\xi_{\rm p}}{\delta}\right) \approx \frac{3000}{N_{\rm bp}}, \tag{10.27}$$

where we have introduced the parameter $\delta \approx 0.34$ nm, which is the length of DNA per base pair, and to obtain the numerical expression we have used a persistence length $\xi_{\rm p} = 50$ nm. This expression for the loop formation energy is plotted in Figure 10.10.

Figure 10.9: Gene regulation and DNA looping. (A) Schematic of the DNA *cis*-regulatory region associated with the *cyIIIa* cytoskeletal actin gene. (B) Electron micrograph of looped DNA. (C) Schematic showing which regions of the DNA shown in (A) join together to form a loop. (D) Electron micrograph of a second looped configuration. (E) Schematic showing which regions of the DNA shown in (A) join together to form the loop. (Adapted from R. W. Zeller et al., *Proc. Natl Acad. Sci. USA* 92:2989, 1995.)

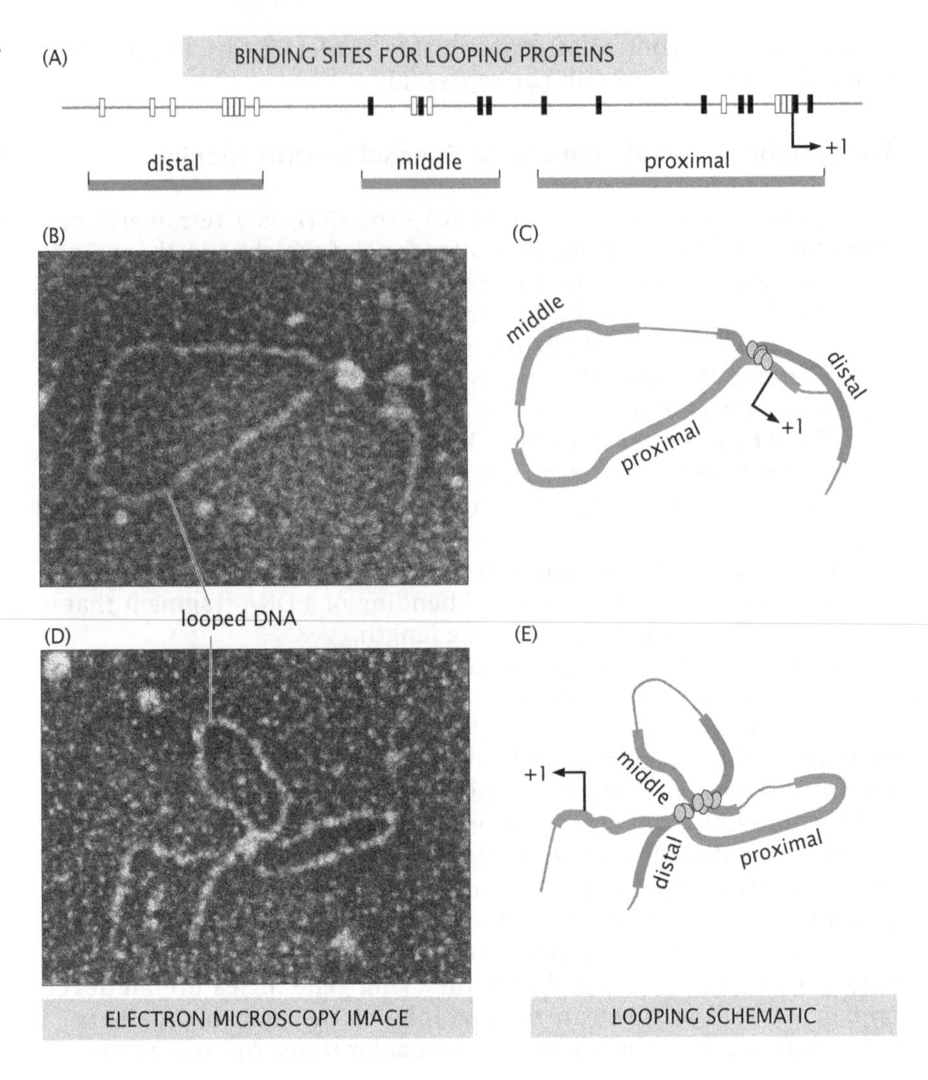

10.3.3 Putting It All Together: The J-Factor

The simple model of the energy to form a DNA circle can be married with the estimate of the looping entropy described in Section 8.2.4 (p. 333) in order to compute the overall free energy of looping. The idea is that for long fragments, the free-energy cost of looping is dominated by the entropic cost of depriving the DNA of configurations as a result of forming the loop. On the other hand, at very short DNA lengths, the energetic cost of bending the DNA dominates the free energy. We are now ready to examine the interplay of these different effects and to compute an expression for the probability of looping that respects both the energetic and entropic contributions to the free energy.

Equation 10.27 reveals the elastic energy cost to form a circular loop of DNA and can be rewritten as

$$\Delta E_{\text{loop}} \approx 3000\, k_{\text{B}}T/N_{\text{bp}}, \tag{10.28}$$

where only the bending energy is taken into account. As shown in Chapter 8, in the simplest random walk model of a long polymer chain, the probability of loop formation, $p_\circ$, is proportional to $N_{\text{bp}}^{-3/2}$ (see Equation 8.60 on p. 336). This implies that the entropy loss due to

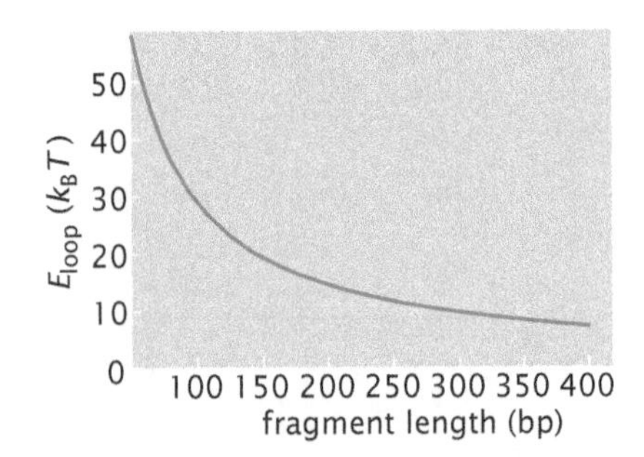

Figure 10.10: Energy of DNA loops as a function of the number of base pairs in the DNA. The calculation assumes that the loops are perfect circles.

loop formation is

$$\Delta S_{\text{loop}} = k_B \ln p_\circ = k_B \left(-\frac{3}{2} \ln N_{\text{bp}} + \text{const} \right), \qquad (10.29)$$

where the constant term does not depend on the length of the chain. This follows because $\Delta S_{\text{loop}} = S_{\text{loop}} - S_{\text{total}} = k_B \ln(W_{\text{loop}}/W_{\text{total}})$, and $W_{\text{loop}}/W_{\text{total}}$ is nothing more than $p_\circ$, the probability to form a loop. Our results for the looping energy and entropy can be assembled to form an approximate free energy for loop formation of the form

$$\Delta G_{\text{loop}} = \Delta E_{\text{loop}} - T\Delta S_{\text{loop}} \approx k_B T \left(\frac{3000}{N_{\text{bp}}} + \frac{3}{2} \ln N_{\text{bp}} + \text{const} \right). \qquad (10.30)$$

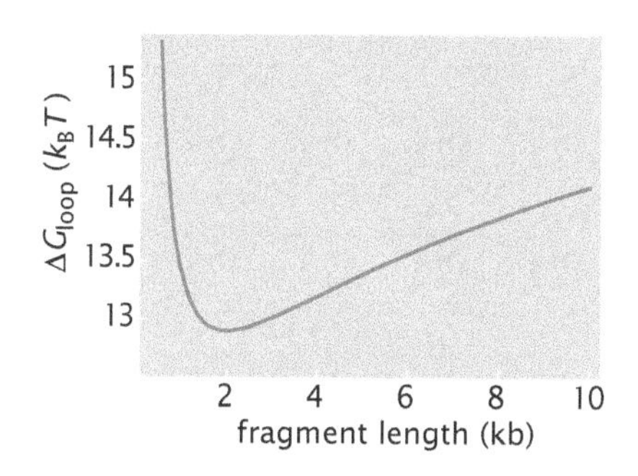

Figure 10.11: Looping free energy. The free energy to form a circular loop as a function of the number of base pairs in the loop. This result is based upon a toy model in which the energy is computed using a simple elastic model of rod bending and an entropy based upon the random walk model.

This free energy is plotted in Figure 10.11. One notable feature of this curve is that there is an optimal DNA length at which the looping free energy is at its minimum, slightly under 2000 bp.

The simple estimates described above are amenable to experimental verification through *in vitro* studies of DNA cyclization (see Figure 10.12). In particular, in a series of now-classic experiments, the probability of looping could be assessed explicitly and compared with the results of the simple linear elastic model described above. The idea of such experiments is to trap spontaneous thermal fluctuations of the DNA fragments into the looped configuration when the two ends of the linear DNA fragment are in close proximity. To do this, the DNA of interest is prepared with complementary sticky ends. When this DNA is examined at a given concentration, two key outcomes are possible: either the DNA will form circles or it will form dimers. An enzyme called ligase that sews together fragments of DNA that have complementary sticky ends is used to lock the DNA into one or the other of these conformations and the relative concentrations of these two species are measured. The probability of loop formation directly measured in these cyclization experiments has a maximum closer to 500 bp. The quantitative discrepancy is not surprising, since our approximate model fails at intermediate chain

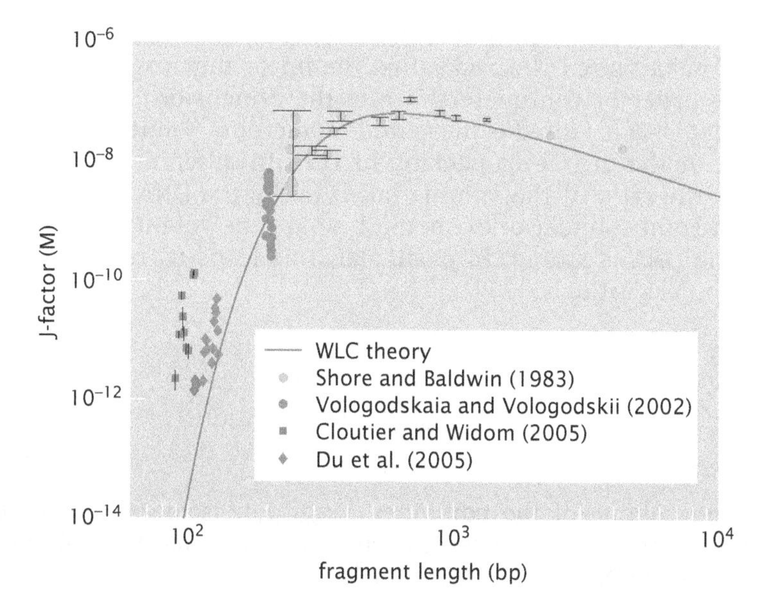

Figure 10.12: J-factor as a function of DNA fragment length. The curve shows the result of using the worm-like chain model to describe the cyclization. This model is more quantitatively realistic than the toy model considered in this chapter, since rather than only considering the elastic energy of circular conformations at small DNA segment lengths, it considers all shapes that close upon themselves and their corresponding entropies. (Data adapted from T. E. Cloutier and J. Widom, *Proc. Natl Acad. Sci. USA* 102:3645, 2005; Q. Du et al., *Proc. Natl Acad. Sci. USA* 102:5397, 2005; D. Shore and R. L. Baldwin, *J. Mol. Biol.* 170:957, 1983; M. Vologodskaia and A. Vologodskii, *J. Mol. Biol.* 317:205, 2002.)

lengths when entropic and energetic contributions to the free energy of looping are comparable.

In previous chapters, we have shown how the binding of proteins to their substrates can be characterized both in terms of binding energies and in terms of dissociation constants. Similarly, there are two different languages that can be used to describe DNA mechanics and probabilities of DNA adopting particular configurations. Cyclization experiments are often reported using a quantity known as the J-factor, which has units of concentration. This magnitude can be interpreted as the effective concentration of one end of the DNA molecule in the vicinity of the other. The J-factor is proportional to $e^{-\beta \Delta G_{\text{loop}}}$, where ΔG_{loop} is the looping free energy and the proportionality factor, which has units of concentration, depends on the microscopic model employed. The reader is invited to consider this in detail in Problem 10.4.

The results of a variety of different measurements of the J-factor are shown in Figure 10.12. The trends are consistent with the model described here, namely, the long length limit is entropy-dominated, while the short-length limit is dictated by the energy of bending.

10.4 DNA Packing: From Viruses to Eukaryotes

The Packing of DNA in Viruses and Cells Requires Enormous Volume Compaction

One of the many remarkable features of DNA is its length relative to the linear dimensions of the volumes within which it is usually confined. For example, in the case of a bacteriophage, in excess of 10 μm of DNA is packed inside a capsid that is roughly 50 nm across. To put this in terms of everyday dimensions, this is like putting 500 m of cable (with a diameter of the order of 10 cm) from the Golden Gate Bridge into the back of a delivery truck. There are several ways to view the extent to which a particular genome is compacted. The simplest measure, though misleading, is, as was suggested above, to compare the linear dimensions of the DNA of interest with those of the region within which that DNA is confined. For example, in the case of "typical" eukaryotic DNA packaging, the linear dimension of the DNA is of the order of centimeters, while the dimension of the nucleus where that DNA is stored is measured in microns. A much more useful measure of the degree of packing of DNA in different settings is to consider the ratio of the volume taken up by the DNA (when viewed as a long solid cylinder of diameter 2 nm) to the volume of the region where the DNA is stored. In particular, this leads us to consider the dimensionless ratio

$$\nu = \frac{\Omega_{\text{genome}}}{\Omega_{\text{container}}} \approx \frac{N_{\text{bp}}}{\Omega_{\text{container}}}, \tag{10.31}$$

where the volume of the container, $\Omega_{\text{container}}$, is measured in cubic nanometers and we have used the rule of thumb that the volume per base pair for dsDNA is roughly 1 nm^3.

Estimate: The DNA Packing Compaction Ratio. Equation 10.31 permits us to examine the degree of relative compaction of the DNA in the different examples shown in Figure 10.13. As a concrete example of a bacteriophage, we consider the famed lambda phage. The capsid of lambda phage can be thought of as a sphere of radius 27 nm that holds a 48,500 bp genome. Note that there is a slight subtlety due to the thickness of the capsid itself which we will ignore. The packing ratio for lambda is given by

$$\nu_{\text{lambda}} = \frac{N_{\text{bp}}\,\text{nm}^3}{\frac{4}{3}\pi R^3\,\text{nm}^3} \approx \frac{5 \times 10^4}{4 \times 27^3} \approx 0.6. \tag{10.32}$$

This result reveals that the DNA inside a bacteriophage is packed to nearly crystalline densities, a fact that has far-reaching consequences for bacteriophage infectivity and lifestyle. In particular, as will be shown in the remainder of this section, as a result of these high packing densities, there is a large free-energy cost to packing the genomic DNA, which requires the services of an ATP motor. It has also been speculated that these large free energies of packing are responsible for assisting the translocation of the genetic material during the infection process itself.

The second example of interest is that of the bacterial nucleoid. In this case, while apparently less structured than in the viral context, the DNA is still highly localized. We idealize the nucleoid as a sphere of radius 0.25 μm. Given the genome length of *E. coli* of 5×10^6 bp, this implies a packing ratio

$$\nu_{\text{bacterium}} \approx \frac{5 \times 10^6\,\text{nm}^3}{4 \times (250\,\text{nm})^3} \approx 0.1. \tag{10.33}$$

In the case of a eukaryotic sperm cell, the DNA is highly compacted, as was already revealed in Figure 10.13. In particular, the sperm head (which we treat as a sphere) is roughly 5 μm in diameter, resulting in a compaction ratio

$$\nu_{\text{sperm}} \approx \frac{10^9\,\text{nm}^3}{4 \times (2500\,\text{nm})^3} \approx 0.02. \tag{10.34}$$

ESTIMATE

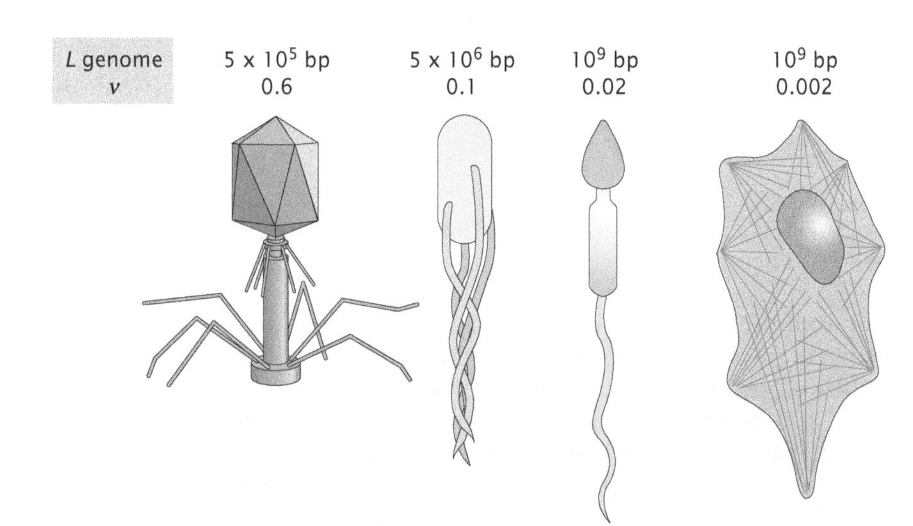

Figure 10.13: Schematics of some of the different scenarios associated with DNA packaging in lambda phage, an *E. coli* cell, a human sperm cell, and a human fibroblast. In each case, the genome length and the packing ratio ν (see Equation 10.31) are shown.

The final case we will consider is that of the eukaryotic nucleus. In this case, we can think of a genome with an approximate length of 10^9 bp housed in a nucleus with a radius of roughly 5 μm. These dimensions imply a packing ratio

$$\nu_{\text{nucleus}} \approx \frac{10^9\ \text{nm}^3}{4 \times (5000\ \text{nm})^3} \approx 0.002. \qquad (10.35)$$

The problem of DNA packaging is intriguing not only on the grounds of sheer geometric crowding, but also because of the recognition that the regions within which DNA is packaged (such as in a viral capsid) have linear dimensions that are comparable to the persistence length of the DNA. The claim that animates the remainder of our discussion of DNA packaging is that there is an elastic energy cost to be paid to effect such packaging, which complements the electrostatic contributions introduced in Chapter 9. Two examples to be examined presently both concern the packaging of genomic data, on the one hand, in the context of bacterial viruses and, on the other, the structure of eukaryotic chromosomes. In both cases, the smallness of the packaging region in comparison with the size of the persistence length indicates the presence of elastic (and other) free-energy penalties to pay. Indeed, the packaging of viral DNA exacts such a high cost that there is a molecular motor that carries out the necessary work to compress the viral DNA.

10.4.1 The Problem of Viral DNA Packing

In rapid succession, a number of new experimental insights into the way DNA in viruses is packaged and ejected have come into focus. As the basis for this discussion, we remind the reader of the character of the phage life cycle already shown in Figure 3.26 (p. 122). For a bacteriophage such as lambda phage, the virus attaches to the host cell receptor (the outer membrane protein LamB in the case of the *E. coli* host of lambda phage) and ejects its ~50 kb genome in roughly a minute. Over the next 10–20 minutes, the viral genome is replicated and the proteins coded for by that genome are synthesized. Once a sufficient concentration of proteins is available, the capsids spontaneously assemble, followed by a fascinating packaging process in which the portal motor consumes ATP and pushes the DNA into the capsid. A quantitative analysis of molecular motors will be carried out in Chapter 16. Upon completion of packaging, the capsid assembly is finished with the addition of the tails, and the mature virions are released from the ravaged bacterium to repeat their evil deeds elsewhere.

Structural Biologists Have Determined the Structure of Many Parts in the Viral Parts List

Structural biologists have begun to determine the structures of many of the molecular participants in the viral life cycle. For example, the structure of part of the portal motor that packs the DNA in the capsid has been determined, as has the structure of the membrane-puncturing device that leads to the delivery of the T4 viral genome. More important for the present discussion, there has also been a great deal of progress in determining the structures of the viral capsids themselves as well as the DNA packaged within them. Figure 2.29 (p. 66) shows a famous gallery revealing the structures of a host of different viral capsids. One of the most notable features of these capsids is their impressive icosahedral symmetry. Another important feature is the observation that, to within a factor of 2 or so, viral capsids are all roughly the same size. Just as there are databanks that serve as a repository for protein structures, there is a databank of viral structures known as VIPER that provides atomic-level structures of a range of different capsids in various states during their maturation.

A second key structural question concerns the geometry of the packaged genome itself. Cryo-electron microscopy experiments as well as X-ray scattering experiments have revealed that the DNA within viral capsids is highly ordered. In particular, these experiments reveal a definite spacing between adjacent strands of the packed DNA. An example of such results is shown in Figure 10.14 for phage T7, while Figure 10.15 is a gallery of such structures showing both the beautiful symmetry of the phage capsids and their packaged DNA. As a result of these kinds of experiments, a number of structural hypotheses for the packaged DNA have been advanced, with one of the most favored structural models being that the DNA is packed in concentric arrangements within viral capsids as shown in Figure 10.16.

In addition to data obtained using electron microscopy, X-ray diffraction experiments have also shed light on the structure of the packed DNA. For example, measurements have been made on the spacing between the DNA segments as a function of the fractional filling of the capsid as shown in Figure 10.17. These data can be interpreted using a simple geometric model of the packaging. As shown in Figure 10.16, if the spacing between DNA strands is uniform and equal to d_s, then the volume taken up by the genome is approximately Ld_s^2, with L being the length of the genome packed. Since the total volume taken up by the genome is roughly equal to the capsid volume, that is, $V_{\text{capsid}} \approx Ld_s^2$, it follows that the spacing between DNA strands $d_s \approx \sqrt{V_{\text{capsid}}/L}$. The predicted scaling, $d_s \sim L^{-1/2}$, seems to hold true for the experimental data, as shown in Figure 10.17, albeit over a fairly small range in genome length.

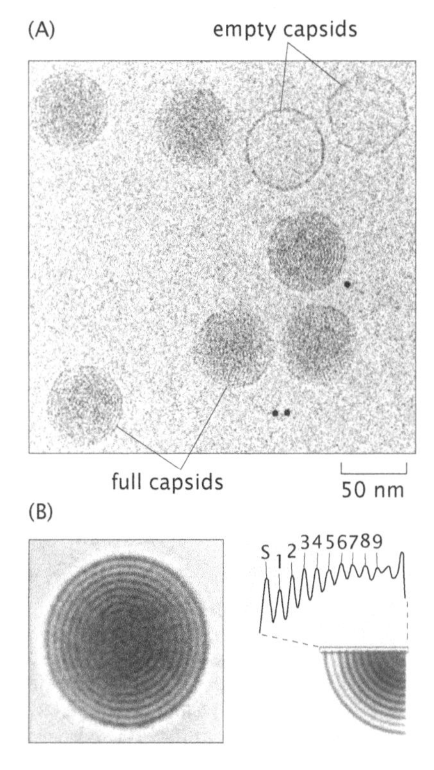

Figure 10.14: Cryo-electron microscopy images of packaged DNA. (A) Image showing several T7 capsids with their complement of packed DNA. (B) High-resolution view of a single capsid, which demonstrates the rings of ordered DNA within the capsid. On the right, the series of peaks show the alternating density as a function of radius. The numbering corresponds to the different shells of DNA, while the label "S" refers to the protein shell of the capsid itself. (Adapted from M. E. Cerritelli et al., *Cell* 91:271, 1997.)

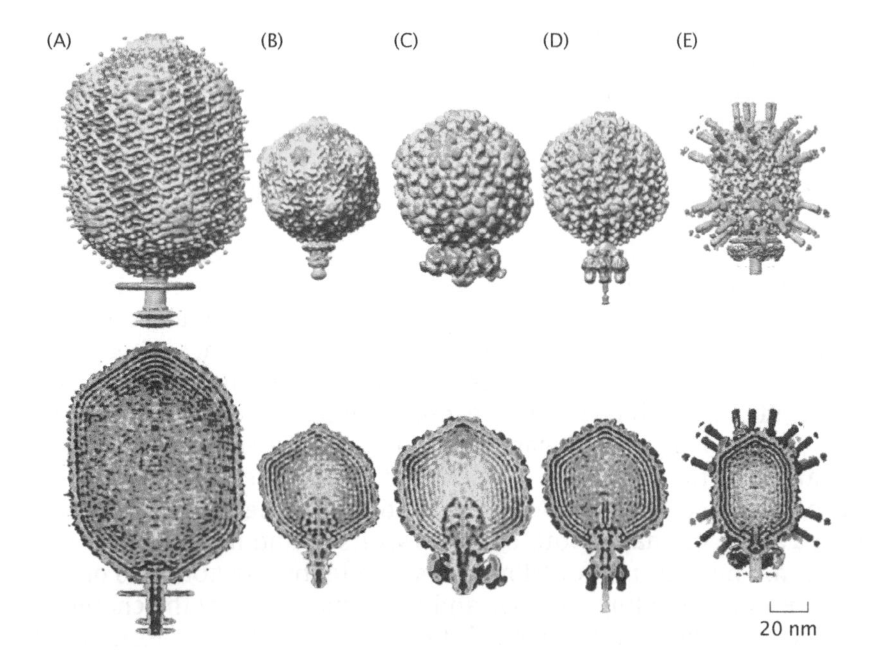

Figure 10.15: Structure of bacteriophage capsids and their packaged DNA. Cryo-electron microscopy images of the capsid structure (top row) and packaged DNA (bottom row) for bacteriophages (A) T4, (B) T7, (C) epsilon 15, (D) P22, and (E) ϕ29. (Adapted from J. E. Johnson and W. Chiu, *Curr. Opin. Struct. Biol.* 17:237, 2007.)

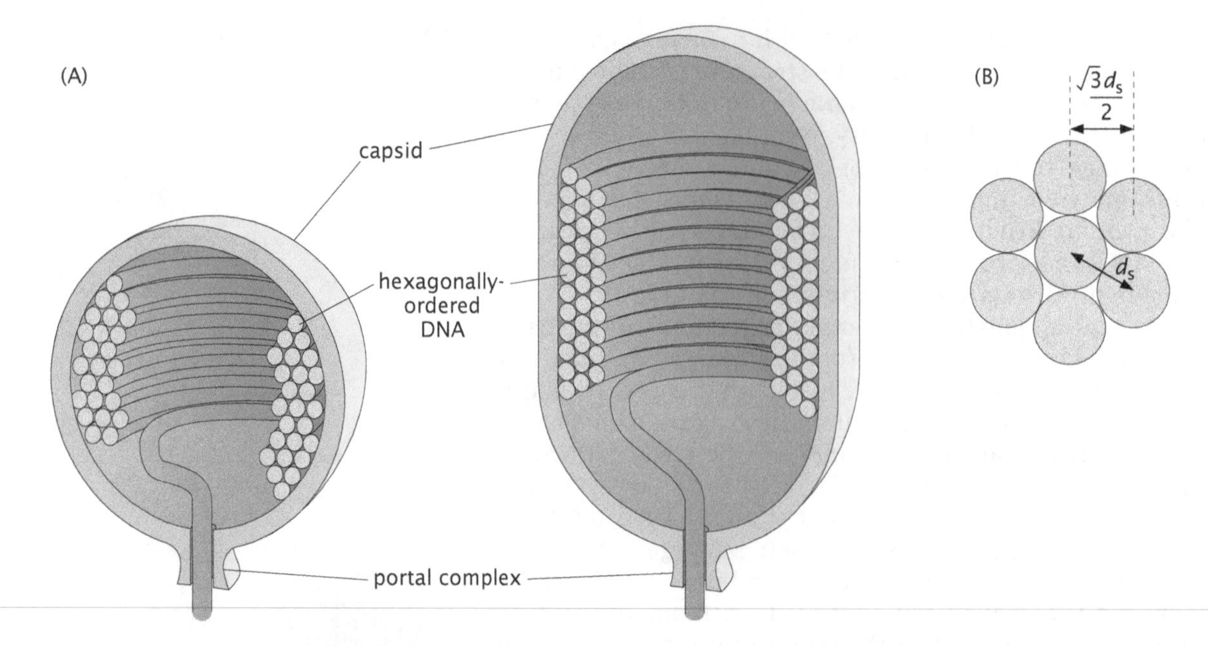

Figure 10.16: Two simplified models used to illustrate the calculation of free energy associated with packaged DNA. (A) The spherical capsid on the left is a simplified version of T7 or lambda phage. The elongated capsid on the right is a simplified version of the ϕ29 capsid. (B) Close up of cross-sections of DNA packed inside the capsids. Note that a given segment of DNA is surrounded by six neighboring segments in a hexagonal arrangement.

The Packing of DNA in Viruses Results in a Free-Energy Penalty

These structural insights have been complemented by single-molecule experiments on the DNA packaging process as well as *in vitro* studies of DNA ejection. The single-molecule-packaging experiments measure the force exerted by the packaging motor during the process of viral packaging itself, as shown schematically in Figure 10.18. These techniques make it possible to grab the viral DNA while it is being packaged into the viral capsid and to measure the force exerted by the portal motor as it overcomes the increasing stored energy of the packed DNA. The kind of data that emerge from these measurements is shown in Figure 10.19, which illustrates both the packaging rate and the forces that build up during packaging as a function of the percentage of the genome that is actually inside of the capsid. All told, the combination of structural insights like those described above in conjunction with biochemical and single-molecule experiments has led to a picture of the viral life cycle that can be examined quantitatively.

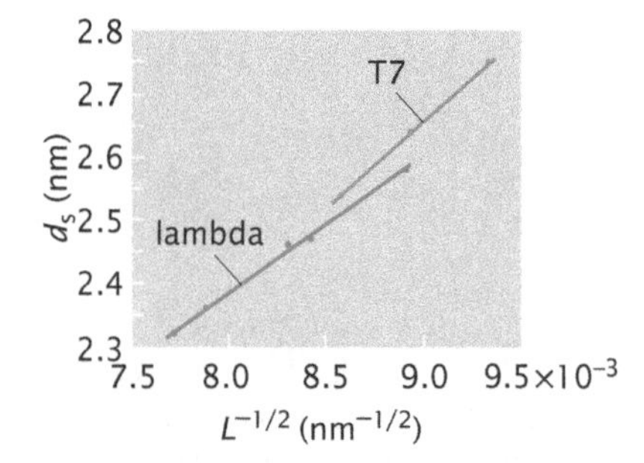

Figure 10.17: DNA spacing as a function of the packaged length. Data on the ordered packaging of DNA in two distinct bacteriophages (T7 and lambda) have been measured using X-rays and cryo-electron microscopy. (Adapted from P. K. Purohit et al., *Biophys. J.* 88:851, 2005.)

The aim of this section is to take stock of the mechanical forces that come into play during viral packaging and to reckon these forces explicitly in closed form in a simple model of DNA elasticity and electrostatic repulsions. These forces are then compared with those measured in the fascinating single-molecule experiments on DNA packaging mentioned above. The energetics of viral packaging is characterized by a number of different factors, including (i) the entropic spring effect, which favors a spread-out configuration for DNA in solution; (ii) the energetics of elastic bending, which results from inducing curvature in the DNA with a scale that is considerably smaller than the persistence length of ~50 nm; and (iii) those factors related to the presence of charge both on the DNA itself and in the surrounding solution. The entropic contribution is smaller by a factor of 10 or more relative to the bending energies and those mediated by the charges on the DNA and the surrounding solution. Our strategy is to examine the elastic and charge-related interaction energies separately and then to assemble them to obtain a complete estimate of the energetics of viral DNA packaging.

In particular, we seek a free energy of packaged DNA of the form

$$G_{\mathrm{tot}}(d_s, L) = G_{\mathrm{bend}}(d_s, L) + G_{\mathrm{charge}}(d_s, L), \tag{10.36}$$

where d_s is the spacing of DNA in the capsid (defined in Figure 10.16) and L is the length of DNA that is packed. With this quantity in hand we can compute the equilibrium spacing for the DNA as a function of the amount of genome packed, by setting $\partial G_{\mathrm{tot}}/\partial d_s = 0$ and solving for d_s. Substituting the function $d_s(L)$ into G_{tot}, we can compute the force that resists packaging as $F = -\mathrm{d}G_{\mathrm{tot}}(d_s(L), L)/\mathrm{d}L$. In doing this, we make the implicit assumption that the packaging process is quasistatic and the DNA remains in equilibrium. As already discussed in detail in Chapter 6, the use of equilibrium assumptions is a matter of the various time scales that are in play. In this case, the rate of packaging and the DNA relaxation time are the key variables of interest. Since roughly 100 bp are packaged each second, this leaves sufficient time for the DNA within the capsid to rearrange itself.

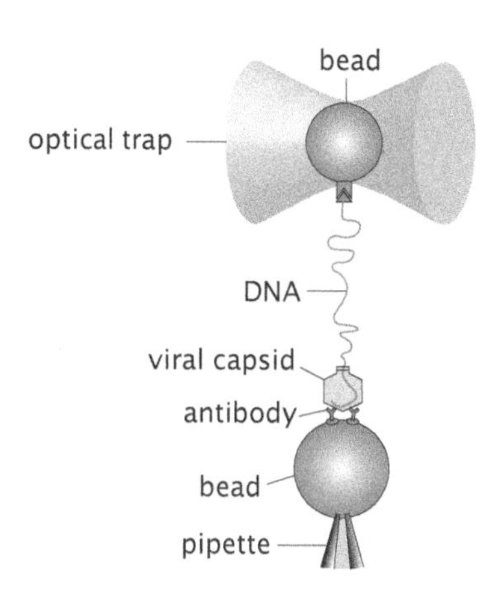

Figure 10.18: Optical tweezers measurement of the forces during DNA packaging. The viral capsid is attached to one bead using antibodies and the viral genome is attached to a second bead. This second bead is held in an optical trap and the forces are monitored as the DNA is reeled into the capsid by the ATP-consuming portal motor. (Adapted from D. E. Smith et al., *Nature* 413:748, 2001.)

A Simple Model of DNA Packing in Viruses Uses the Elastic Energy of Circular Hoops

The elastic estimate that we obtain for the forces associated with viral packaging is predicated upon the most naive usage of the linear elastic theory of beams developed earlier in the chapter. In particular, we neglect the accumulation of stored elastic energy as a result of twist, for which little is known in the context of these viral DNA packing problems, and concentrate instead only on the contribution of the bending to the stored energy. Within this approximation, the elastic contribution to the free energy can be obtained by using Equation 10.7.

Though we imagine the viral DNA to be packed in the form of a helix as shown in schematic form in Figure 10.16(A), from the perspective of our elastic energy functional, the geometry may be thought of as a stacking of hoops of radius R. The key point is that although the actual radius of curvature is given by $R_c = R(1 + p^2/4\pi R^2)$, where p is the helical pitch, for the geometries of interest here, $p \approx 2$ nm while $R \approx 20$ nm and hence the parameter $p^2/4\pi R^2 \ll 1$ and can be neglected without compromising on the key features of the analysis. In light of this approximation and using the fact that for circular hoops $\kappa = 1/R$, the elastic energy can be written as

$$G_{\mathrm{bend}} = \pi \xi_p k_B T \sum_i \frac{N(R_i)}{R_i}, \tag{10.37}$$

where $N(R_i)$ is the number of hoops that are packed at the radius R_i. The presence of this term reflects the fact that, due to the shape of the capsid, as the radius becomes smaller the DNA can pack higher up into the capsid, thus increasing the number of allowed hoops.

To make analytic progress with the expression for the stored elastic energy given above, we convert it into an integral of the form

$$G_{\mathrm{bend}} = \frac{2\pi \xi_p k_B T}{\sqrt{3} d_s} \int_R^{R_{\mathrm{out}}} \frac{N(R')}{R'} \, \mathrm{d}R'. \tag{10.38}$$

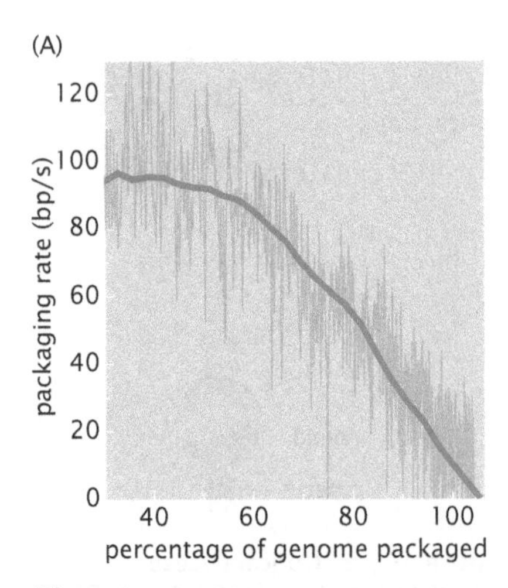

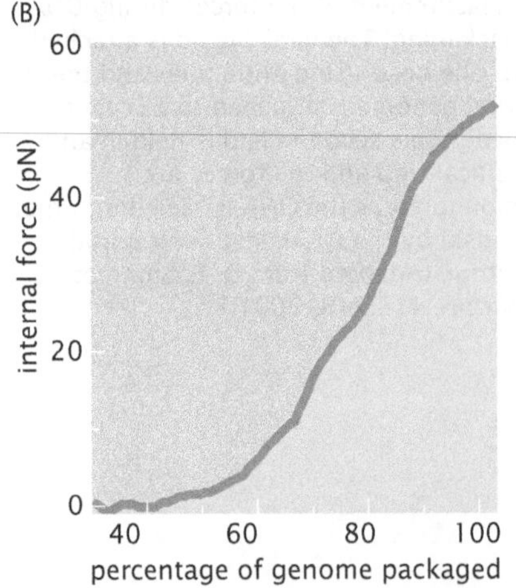

Figure 10.19: Data on DNA packaging in bacteriophage ϕ29. (A) Rate of DNA packaging and (B) force resisting further packaging as a function of the amount of genome present in the capsid. (Adapted from D. E. Smith et al., *Nature* 413:748, 2001.)

The summation $\sum_i$ has been replaced by an integral $\int 2\,\mathrm{d}R'/\sqrt{3}d_\mathrm{s}$, where $(\sqrt{3}/2)d_\mathrm{s}$ is the center-to-center distance between adjacent strands of the DNA along the radial direction; see Figure 10.16(B). Replacing a discrete sum by an integral would in general lead to some error, but this error diminishes as the number of hoops becomes large. This calculation will yield the elastic energy in terms of the radius of the packed DNA. It is necessary to convert it to an expression in terms of the total length packed for purposes of comparison with experimental results. The length packed is generally given as

$$L = \frac{2}{\sqrt{3}d_\mathrm{s}} \int_R^{R_\mathrm{out}} 2\pi R' N(R')\,\mathrm{d}R'. \tag{10.39}$$

To illustrate how to explicitly reckon the elastic energy, we first consider this energy for the simplest (and perhaps unrealistic) case of a cylindrical capsid with DNA in the inverse spool configuration. Such a geometry may be obtained, for instance, by neglecting the conical caps of an icosahedral capsid. In particular, this cylinder is characterized by the geometric parameters z (the height) and R_out (the radius). In this case, $N(R) = z/d_\mathrm{s}$ since the height of the cylinder is z and the spacing between two adjacent DNA strands is d_s, and the corresponding elastic energy gotten by using Equation 10.38 is given by

$$G_\mathrm{bend}(R) = \frac{2\pi \xi_\mathrm{p} k_\mathrm{B} T z}{\sqrt{3}d_\mathrm{s}^2} \ln\left(\frac{R_\mathrm{out}}{R}\right). \tag{10.40}$$

It follows from Equation 10.39 that the packed length is given by

$$L(R) = \frac{2\pi z}{\sqrt{3}d_\mathrm{s}^2}(R_\mathrm{out}^2 - R^2) \tag{10.41}$$

or, equivalently, $R = R_\mathrm{out}\sqrt{1 - \sqrt{3}d_\mathrm{s}^2 L/2\pi z R_\mathrm{out}^2}$.

The elastic energy may now be rewritten as a function of the packed length of DNA,

$$G_\mathrm{bend} = -\frac{\pi \xi_\mathrm{p} k_\mathrm{B} T z}{\sqrt{3}d_\mathrm{s}^2} \ln\left(1 - \frac{\sqrt{3}d_\mathrm{s}^2 L}{2\pi z R_\mathrm{out}^2}\right). \tag{10.42}$$

This result may be used in turn to compute the force associated with the accumulation of elastic energy. In particular, differentiating the energy obtained above with respect to the length of packed DNA, we find

$$f(L) = -\frac{\mathrm{d}G_\mathrm{bend}}{\mathrm{d}L} = -\frac{\xi_\mathrm{p} k_\mathrm{B} T/2R_\mathrm{out}^2}{1 - \sqrt{3}d_\mathrm{s}^2 L/2\pi z R_\mathrm{out}^2}. \tag{10.43}$$

Though the algebra is messier for other capsid shapes, these same basic steps may be imitated in each case to obtain the elastic contribution to the energy. Performing these calculations is left to the problems at the end of the chapter.

DNA Self-Interactions Are also Important in Establishing the Free Energy Associated with DNA Packing in Viruses

Simple as it is, our model thus far really does not do justice to all of the competing energies in the problem of DNA packaging. In particular, the elastic contributions must be supplemented by energetic terms related to the presence of charges both on the DNA and on the counterions in the surrounding solution. Under physiological conditions, these charge interactions manifest themselves as a repulsion between nearby segments of DNA. Rather than attempting a "first principles" treatment of this interaction, we appeal instead to experiments that attempt to measure that interaction directly. The idea of the experiments was introduced in Chapter 6 and involves the use of osmotic pressure to squeeze an ordered arrangement of DNA as shown in Figure 6.23 (p. 266) and to simultaneously measure the spacing between adjacent strands using X-rays. The osmotic pressure is built up by the presence of polyethylene glycol in solution. What emerges from such experiments is a relation between the osmotic pressure and the spacing between adjacent strands as shown in Figure 6.24 (p. 266). To this piece of empirical evidence we add the assumption that parallel strands interact through a pair potential $v(d_s)$ per unit length and that interactions are limited only to the first nearest neighbors.

Once the pressure has been determined, it is possible to compute the interaction energy, in turn, and from that energy to obtain the pair potential itself. To that end, consider N parallel strands of length l, each packed in a hexagonal array with a spacing d_s. The total energy stored in these strands by virtue of their interactions is

$$G_{\text{charge}} = 3Nlv(d_s), \tag{10.44}$$

where the factor 3 appears because each strand interacts with 6 nearest neighbors (ignoring surface effects) and we multiply by 1/2 to take care of double counting. The total volume of the assemblage is

$$V = N\frac{\sqrt{3}}{2}d_s^2 l. \tag{10.45}$$

The pressure is $p(d_s) = -\mathrm{d}G_{\text{charge}}/\mathrm{d}V$, where $\mathrm{d}V = Nl\sqrt{3}d_s\,\mathrm{d}d_s$. As a result, we may write the force in terms of the pressure as

$$f(d_s) = -\frac{\mathrm{d}v(d_s)}{\mathrm{d}d_s} = \frac{1}{\sqrt{3}}p(d_s)d_s, \tag{10.46}$$

where $f(d_s)$ is the force per unit length of the strands. We now substitute $p(d_s) = F_0\mathrm{e}^{-d_s/c}$ (the result of a fit to the data of Figure 6.24 on p. 266) and solve the resulting differential equation that determines the pair potential, remembering that $v(\infty) = 0$. This procedure leads to the potential

$$v(d_s) = \frac{1}{\sqrt{3}}F_0(c^2 + cd_s)\mathrm{e}^{-d_s/c}. \tag{10.47}$$

With the interaction potential in hand, we can write the interaction contribution to the free energy as

$$G_{\text{charge}} = \sqrt{3}F_0(c^2 + cd_s)L\mathrm{e}^{-d_s/c}, \tag{10.48}$$

where $L = Nl$ is the total length of the strands. An interesting fact about the interaction energy is that it can be controlled by the experimenter to a certain extent by changing the ionic strength of the buffer solution, a topic that was discussed in detail in Chapter 9. This, in turn, our simple model would predict, will change the pressure with which DNA is packaged inside the viral capsid, possibly affecting the ability of the virus to inject its DNA into the host cell.

DNA Packing in Viruses Is a Competition Between Elastic and Interaction Energies

Our calculations of the elastic and interaction terms can now be assembled together to provide an expression for the total free energy associated with the packaged DNA. These two terms result in a competition that sets the length scale of the hoop spacing in the capsid. In particular, the interaction terms favor keeping the strands as far apart as possible. On the other hand, keeping the strands too far apart means that the loops will have to adopt configurations with small radii of curvature. Using the cylindrical capsid as a concrete example, the total free energy can be written as

$$G_{\text{tot}} = G_{\text{bend}} + G_{\text{charge}} = -\frac{\pi \xi_{\text{p}} k_{\text{B}} T z}{\sqrt{3} d_{\text{s}}^2} \ln\left(1 - \frac{\sqrt{3} d_{\text{s}}^2 L}{2\pi z R_{\text{out}}^2}\right) + \sqrt{3} F_0 (c^2 + c d_{\text{s}}) L e^{-d_{\text{s}}/c}. \tag{10.49}$$

Note that these equations reflect the simplifying assumption that the bending energy and the energy associated with the charges are independent, though, in fact, the persistence length (and hence bending energy) does depend upon the charges that are present. In order to compute the forces associated with the packaged DNA, we now minimize this free energy, at fixed L, to find the optimal spacing of the DNA at that particular packaged length. Similarly, we can find the resistive force that is set up as a result of the packaged DNA by evaluating $F(L) = -\partial G_{\text{tot}}/\partial L$, while using the optimal d_{s} at each L. The results of this model are shown in Figure 10.20; the results for ϕ29 should be compared with Figure 10.19(B). The reader is invited to make that comparison explicitly in the problems at the end of the chapter. More important is that we can now use the same model on a number of different bacteriophages and predict what the resistive force that builds up during packaging would be if the same laser-tweezers experiment were performed on the new phage; see Figure 10.20. Note that for T7 we predict a maximum force in excess of 100 pN, suggesting either a more powerful motor than the one in ϕ29, or that in the buffer conditions used in the ϕ29 experiment this packaging reaction would not go to completion. Either way, the model suggests that measuring the force build-up during packaging in T7 would yield further insight into this fascinating biological process.

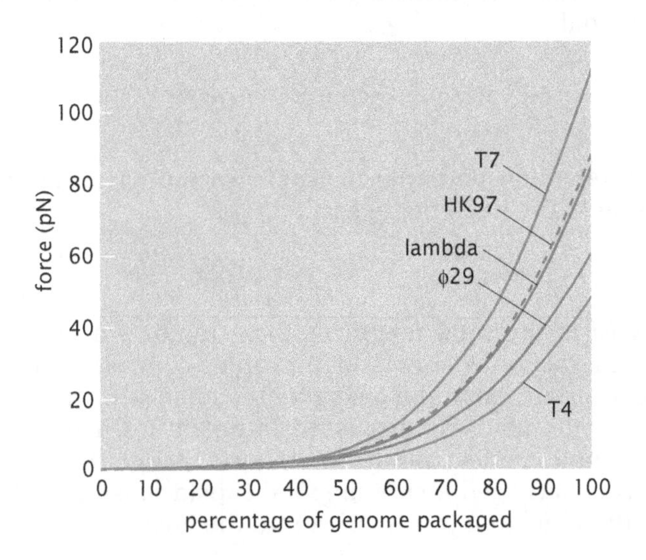

Figure 10.20: Packing forces for different viruses. Different bacteriophages have different capsid sizes and genome lengths, resulting in different overall packaging energies. These curves show the predicted forces for a number of different bacteriophages, (T7, HK97, lambda, and ϕ29) for the case in which the solution conditions are the same as those used in the optical-tweezers experiment on phage ϕ29. (Adapted from P. K. Purohit et al., *Biophys. J.* 88:851, 2005.)

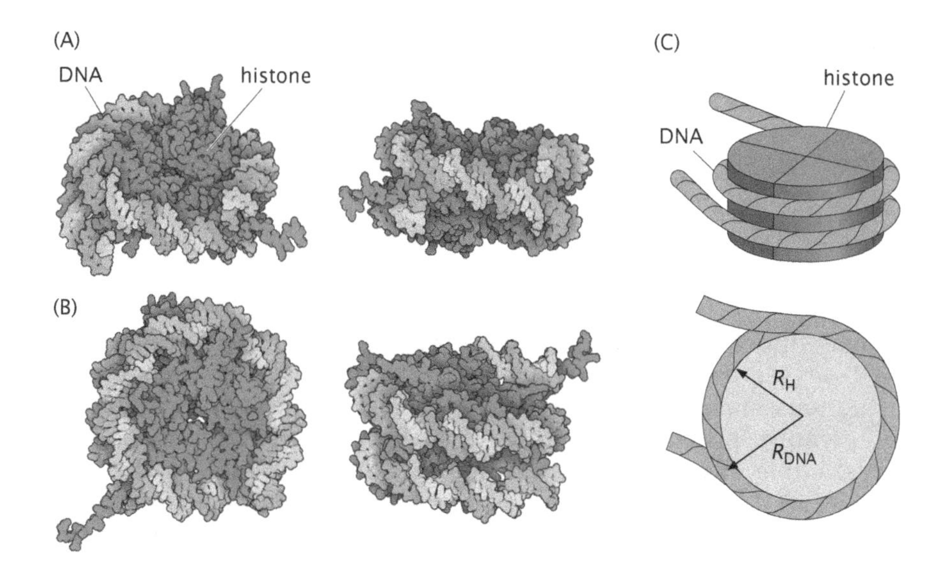

Figure 10.21: Structure of the nucleosome. (A) Atomic-level depiction of the nucleosome core particle revealing both the histone octamer and the encircling DNA. (B) Alternative atomic-level view of the nucleosome. (C) Cartoon representation of the nucleosome core particle where the histone and DNA radii (R_H and R_{DNA}, respectively) are shown. (A, B, courtesy of D. Goodsell.)

10.4.2 Constructing the Nucleosome

The second key example of DNA packaging that we aim to examine in quantitative detail from the point of view of beam theory is the eukaryotic nucleus. DNA packaged in the eukaryotic nucleus is arranged in a hierarchical structure that at the smallest scales involves the wrapping of DNA around a protein assembly known as the histone octamer to form the nucleosome already introduced in Figure 2.26 (p. 63). At the next scale up, individual nucleosomes are arranged to make the 30 nm chromatin fiber in which the nucleosomes form an orderly array like beads on string and are further condensed into a helical coil. In this discussion, we consider the smallest link in the chain of structures, namely, the nucleosome core particle. To shed light on this structure with a level of detail in excess of that given in Figure 2.26, Figure 10.21 shows an atomic-level depiction of the wrapping of DNA around the histone octamer. The relevant orders of magnitude to bear in mind are that the histone octamer is roughly 7 nm in diameter and 6 nm high, and will be treated presently as a tiny cylinder. The DNA wraps around this cylindrical particle roughly 1.75 times with 147 bp.

Estimate: Sizing Up Nucleosomes To get a rough impression of the packaging of eukaryotic DNA, we note that there are 147 bp assigned to each histone, with intervening linker regions of roughly 50 bp. As a result, we make the simple estimate that there are roughly 200 bp per histone. For a genome with 3×10^9 bp, this implies that the number of histone octamers is roughly

$$N_{\text{octamer}} = \frac{3 \times 10^9 \text{ bp}}{200 \text{ bp}} \approx 10^7. \tag{10.50}$$

This estimate should be seen as an underestimate because some fraction of the histones will probably be free in solution, for example.

It is also of interest to estimate the fraction of the nuclear volume that is taken up by histones. Given that we think of the histone octamer as a cylinder of diameter roughly 7 nm and height 6 nm, each such histone has a volume of

$$\Omega_{\text{histone}} = \pi r^2 h \approx 225 \text{ nm}^3. \quad (10.51)$$

Hence, the total volume taken up by the 10^7 histones is $\sim 2.3 \times 10^9 \text{ nm}^3$, which should be compared with the roughly $5 \times 10^{11} \text{ nm}^3$ of volume available within the nucleus of $5 \,\mu\text{m}$ radius.

Eukaryotic DNA is packed in a hierarchical structure, with much of the DNA sequestered within nucleosomes. This fact is biologically mysterious since many of the most essential transactions involving DNA require proteins to bind to specific target sites on the DNA. For example, we have already discussed the observation that the process of transcription is mediated by batteries of transcription factors (activators and repressors) that bind the DNA and modulate transcription. As a result, it is of great interest to understand the accessibility of nucleosomal DNA. How easily can target sites be accessed as a function of their distance from the unwrapped ends of the nucleosomal DNA? To examine this question, we begin by considering the free energy of nucleosome assembly, followed by a statistical mechanics treatment of the equilibrium accessibility of nucleosomal DNA.

Nucleosome Formation Involves Both Elastic Deformation and Interactions Between Histones and DNA

One of the goals of this section is to examine the free-energy balance associated with the assembly of the nucleosome core particle. We recall that the persistence length for DNA is of the order of 50 nm. By way of contrast, note that the radius of curvature associated with the DNA wrapped around the histone octamer is roughly 4.5 nm. Like with the viral packaging example considered in the previous section, there are clearly elastic consequences to such highly deformed DNA fragments. In addition to the cost of elastic deformation, there are other energy contributions to consider as well. In particular, we need to consider the favorable electrostatic interactions between the DNA and the histone octamer since the DNA itself is negatively charged and the histone octamer surface is covered with lysine and arginine residues that present a compensating positive charge. Within this simple picture, the free energy of formation for a nucleosome can be written as

$$G_{\text{nucleosome}} = G_{\text{bend}} + G_{\text{DNA–histone}}. \quad (10.52)$$

To construct a preliminary estimate of the deformation energy associated with the formation of the nucleosome core particle, we treat the DNA as a featureless rod subject to a uniform state of deformation. In light of Equation 10.8, the energy stored in each turn of the DNA by virtue of its deformation is given by

$$G_{\text{bend}} = \frac{\pi \xi_p k_B T}{R_{\text{DNA}}}, \quad (10.53)$$

where R_{DNA} is the radius of curvature of the wrapped DNA.

To proceed with our estimate, we need to compute the curvature associated with the 147 bp segment when it is wrapped around the histone octamer. As shown in Figure 10.21, the DNA wraps around the histone octamer in a helical pattern (we are not speaking here of the DNA double helix itself) with a small pitch. For simplicity of calculation, we use an approximate expression for the curvature in which the helical pitch is neglected. In addition, we assume that the DNA is wrapped fully around the histone octamer two times, resulting in a total bending free energy cost of $G_{\text{bend}} = 2\pi\xi_{\text{p}} k_{\text{B}}T/R_{\text{DNA}} \approx 70\, k_{\text{B}}T$.

The second key contribution to the free energy of formation of the nucleosome is dictated by the interactions between the DNA and the positively charged residues on the histones. As a simplest model, we characterize this interaction energy via an adhesive energy γ_{ad}, which has units of energy/length. In light of this model, the adhesive contribution to the total free energy is

$$G_{\text{DNA–histone}} = 4\pi R_{\text{DNA}}\gamma_{\text{ad}}, \tag{10.54}$$

where we have assumed that the DNA is twice wrapped fully around the histone octamer. In the next section, we show how this adhesive energy can be deduced from experimental data on the equilibrium accessibility of nucleosomes.

10.4.3 Equilibrium Accessibility of Nucleosomal DNA

As noted above, DNA molecules in eukaryotic cells wind around histone octamers to form the nucleosome. In this state, the DNA is not directly accessible to regulatory proteins since their binding sites are occluded by the nucleosome. We extend the discussion of nucleosome assembly given above to investigate the statistical mechanics of binding of regulatory proteins to DNA in the nucleosome. Key experiments on nucleosome accessibility were performed *in vitro* by assessing the susceptibility of particular sites on the DNA to cleavage by restriction enzymes as a function of the distance of these sites from the unwrapped ends of the nucleosomal DNA. These restriction enzymes are proteins that cleave DNA at specific recognition sites and served as a convenient readout for assessing nucleosome accessibility. Effectively, these experiments provide a position-dependent equilibrium constant that depends upon the distance of the site of interest from the unwrapped ends of the nucleosomal DNA.

The Equilibrium Accessibility of Sites within the Nucleosome Depends upon How Far They Are from the Unwrapped Ends

The model put forward to interpret these results envisions the binding of a DNA-binding protein (for the experiment in question, restriction enzymes were used as the protein of interest) to its target site as a two-step process: first the DNA unwraps from the histones simply as a result of thermal fluctuations, and then the restriction enzyme binds to its specific site, which is no longer occluded by the nucleosome. This process is shown schematically in Figure 10.22. In Figure 10.23 we present the results for the equilibrium constant of DNA unwrapping from the nucleosome. The simple observation is that this equilibrium constant decreases the further away the binding site is from the unwrapped ends of the nucleosome. To address these data in a quantitative way, we turn once again to statistical mechanics and

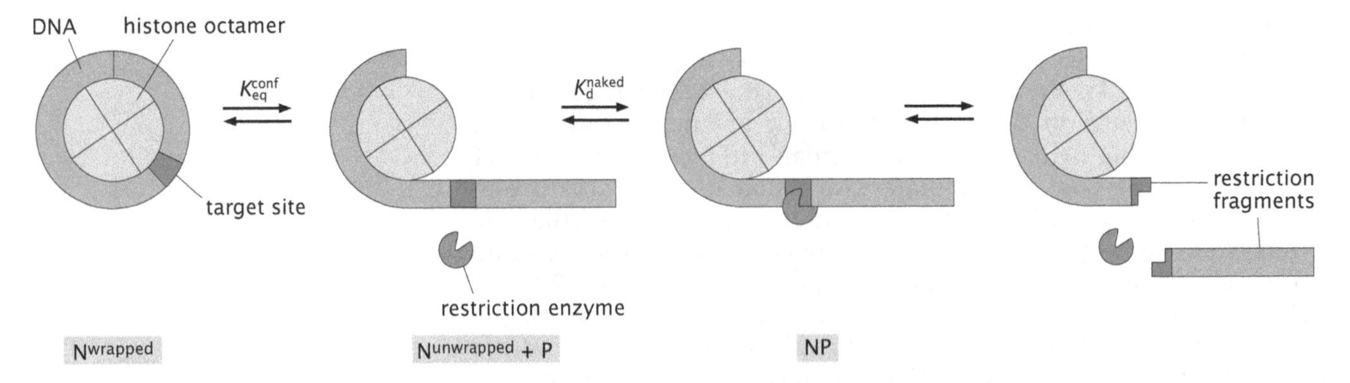

Figure 10.22: Experiment to measure equilibrium accessibility of nucleosomes. Nucleosomal DNA is prepared with a binding site for a restriction enzyme. A wrapped nucleosome, $N^{wrapped}$ can transiently unwrap ($N^{unwrapped}$) and interact with a restriction enzyme, P, upon exposure of its target binding site forming the NP complex. A measurement is made of the probability of restriction digestion as a function of the distance of the target site from the unwrapped ends of the nucleosomal DNA.

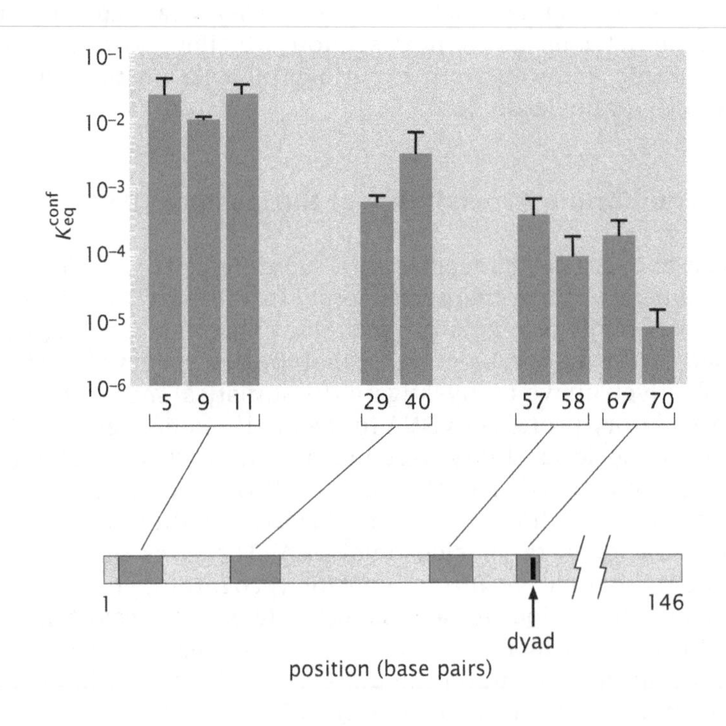

Figure 10.23: Equilibrium constant for site exposure as a function of the distance of the site from the unwrapped end of the nucleosomal DNA. The approximate positions of the binding sites are shown. The dyad is the center of symmetry of the DNA molecule when it is wrapped around the histone octamer. (Adapted from K. J. Polach and J. Widom, *J. Mol. Biol.* 254:130, 1995.)

compute the probability that an enzyme is bound to the target site as a function of the target site location.

The simple model we consider has two degrees of freedom as shown in Figure 10.24. One of the relevant degrees of freedom is the length of DNA unwound from the nucleosome, x, which is continuous and takes values between 0 and $L \approx 25$ nm; this length is half of the total DNA length wound in a typical nucleosome. The other degree of freedom is discrete, and it simply keeps track of whether a restriction enzyme is bound or not to its specific site. The binding site for the enzyme we take to be located at position x_{re}, which is somewhere along the DNA length between 0 and L. The quantity we wish to compute is the probability p_{bound} that the restriction enzyme is bound to its site, as a function of the restriction site location.

The probability that an enzyme is bound is readily computed in the grand canonical ensemble introduced in Section 7.2.1 (p. 289). The

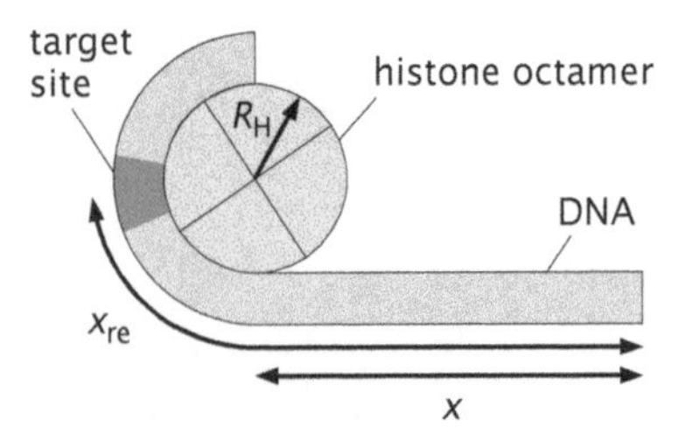

Figure 10.24: Geometry of site accessibility of the nucleosome. The coordinate x refers to how much the DNA is unwrapped and x_{re} refers to the minimum distance that the DNA needs to be unwrapped in order to access the target site.

grand partition function is

$$\mathcal{Z} = \underbrace{\int_0^L e^{\frac{1}{k_BT}(\gamma_{ad}-\gamma_{bend})(L-x)} \frac{dx}{a}}_{\text{no enzyme bound}} + \underbrace{e^{\frac{\mu}{k_BT}} e^{\frac{\varepsilon_{bind}}{k_BT}} \int_{x_{re}}^L e^{\frac{1}{k_BT}(\gamma_{ad}-\gamma_{bend})(L-x)} \frac{dx}{a}}_{\text{enzyme bound}}. \tag{10.55}$$

The first term in $\mathcal{Z}$ is a sum over all possible values for the length x of DNA unwound from the nucleosome, with no restriction enzyme bound. The sum over discrete base pairs is for convenience replaced by an integral over the DNA length, in which case a is the distance between consecutive base pairs; this guarantees that the number of states, $\int_0^L dx/a$, is equal to the number of base pairs of DNA in the nucleosome. The statistical weight for a given x is determined by the energy per unit length for DNA binding to histones ($-\gamma_{ad}$) and the energy per unit length associated with bending the DNA (γ_{bend}).

The second term accounts for unwound states of the DNA when there is a restriction enzyme bound to its specific binding site. For this binding to occur, the DNA must unwind at least by an amount given by the position of the binding site x_{re}, which sets the lower bound of the integral. If the DNA is unwound by less than this amount, the binding site remains inaccessible. The weight associated with a given x now has two additional factors, one that takes into account the chemical potential of the restriction enzymes in solution, μ, and the other arising from the binding energy of the enzyme to its specific site on the DNA, $-\varepsilon_{bind}$.

Now, the probability that the enzyme is bound to its restriction site is the ratio of the second term in the partition function in Equation 10.55 to the total partition function,

$$p_{bound} = \frac{\frac{[P]}{[P]_0} e^{\frac{\varepsilon_{bind}}{k_BT}} \left(e^{\frac{1}{k_BT}(\gamma_{ad}-\gamma_{bend})(L-x_{re})} - 1 \right)}{\frac{[P]}{[P]_0} e^{\frac{\varepsilon_{bind}}{k_BT}} \left(e^{\frac{1}{k_BT}(\gamma_{ad}-\gamma_{bend})(L-x_{re})} - 1 \right) + \left(e^{\frac{1}{k_BT}(\gamma_{ad}-\gamma_{bend})L} - 1 \right)}, \tag{10.56}$$

where we have made use of the ideal-solution result for the chemical potential, $e^{\mu/k_BT} = [P]/[P]_0$. Here [P] is the concentration of proteins (restriction enzymes) in solution, while $[P]_0$ is the concentration at which the chemical potential is zero.

To make contact with the experimental results, we should translate the above expression into the language of dissociation constants. The probability that a restriction enzyme is bound to DNA can be expressed in terms of the equilibrium dissociation constant

$$K_d = \frac{[P][N]}{[NP]} \tag{10.57}$$

for the reaction $P + N \rightleftarrows NP$. Here P denotes free restriction enzymes, N are nucleosomes free of restriction enzymes, and NP are nucleosome–protein complexes.

The probability that a restriction enzyme is bound can now be expressed in terms of the ratio of the concentration of restriction enzyme–nucleosome complexes and the total concentration of all nucleosomes,

$$p_{bound} = \frac{[NP]}{[NP] + [N]}. \tag{10.58}$$

Using Equation 10.57, this leads to the expression

$$p_{\text{bound}} = \frac{[\text{P}]/K_{\text{d}}}{1 + [\text{P}]/K_{\text{d}}}. \tag{10.59}$$

Finally, comparing this equation with Equation 10.56, we deduce

$$K_{\text{d}} = \frac{[\text{P}]_0}{\text{e}^{\frac{\varepsilon_{\text{bind}}}{k_{\text{B}}T}}} \frac{\text{e}^{\frac{1}{k_{\text{B}}T}(\gamma_{\text{ad}} - \gamma_{\text{bend}})L} - 1}{\text{e}^{\frac{1}{k_{\text{B}}T}(\gamma_{\text{ad}} - \gamma_{\text{bend}})(L - x_{\text{re}})} - 1} \tag{10.60}$$

for the equilibrium dissociation constant.

We now wish to connect our model with the quantity measured experimentally, namely the equilibrium constant for nucleosomal accessibility, $K_{\text{eq}}^{\text{conf}}$. It is defined as the ratio of the concentration of nucleosomes whose DNA is sufficiently unwrapped so as to allow the restriction enzymes to bind to the concentration of nucleosomes whose DNA is wrapped such that no DNA binding is allowed. We call these concentrations $[\text{N}^{\text{unwrapped}}]$ and $[\text{N}^{\text{wrapped}}]$, respectively as shown in Figure 10.22. Our claim is that $K_{\text{eq}}^{\text{conf}}$ is embedded as a part of the total K_{d}. In order to see this, we multiply and divide K_{d} by $[\text{N}^{\text{unwrapped}}]$, resulting in

$$K_{\text{d}} = \frac{[\text{N}][\text{P}]}{[\text{NP}]} \cdot \frac{[\text{N}^{\text{unwrapped}}]}{[\text{N}^{\text{unwrapped}}]}. \tag{10.61}$$

Note that [NP] is the same as $[\text{N}^{\text{unwrapped}}\text{P}]$, since in order to form the nucleosome–restriction-enzyme complex, we need the nucleosomes to be sufficiently unwrapped. With this in mind, we can express the dissociation constant as

$$K_{\text{d}} = \underbrace{\frac{[\text{N}^{\text{unwrapped}}][\text{P}]}{[\text{N}^{\text{unwrapped}}\text{P}]}}_{K_{\text{d}}^{\text{naked}}} \frac{[\text{N}]}{[\text{N}^{\text{unwrapped}}]}, \tag{10.62}$$

where $K_{\text{d}}^{\text{naked}} = [\text{P}_0]/e^{\varepsilon_{\text{bind}}/k_{\text{B}}T}$ is the dissociation constant associated with naked DNA, where the equilibrium between restriction enzymes bound to the DNA and those in solution is established in the absence of histones. The last term in Equation 10.62 can be written in terms of the wrapped and unwrapped nucleosome species. In order to do that, we remind ourselves that $[\text{N}] = [\text{N}^{\text{unwrapped}}] + [\text{N}^{\text{wrapped}}]$, leading to

$$\frac{[\text{N}]}{[\text{N}^{\text{unwrapped}}]} = \frac{[\text{N}^{\text{unwrapped}}] + [\text{N}^{\text{wrapped}}]}{[\text{N}^{\text{unwrapped}}]} = 1 + \underbrace{\frac{[\text{N}^{\text{wrapped}}]}{[\text{N}^{\text{unwrapped}}]}}_{\left(K_{\text{eq}}^{\text{conf}}\right)^{-1}}, \tag{10.63}$$

which results in a dissociation constant as a function of $K_{\text{d}}^{\text{naked}}$ and $K_{\text{eq}}^{\text{conf}}$,

$$K_{\text{d}} = K_{\text{d}}^{\text{naked}} \left(1 + \frac{1}{K_{\text{eq}}^{\text{conf}}}\right). \tag{10.64}$$

Comparing this with Equation 10.60, we arrive at the expression

$$K_{\text{eq}}^{\text{conf}} = \frac{\text{e}^{\frac{1}{k_{\text{B}}T}(\gamma_{\text{ad}} - \gamma_{\text{bend}})L} - \text{e}^{\frac{1}{k_{\text{B}}T}(\gamma_{\text{ad}} - \gamma_{\text{bend}})x_{\text{re}}}}{\text{e}^{\frac{1}{k_{\text{B}}T}(\gamma_{\text{ad}} - \gamma_{\text{bend}})L}\left(\text{e}^{\frac{1}{k_{\text{B}}T}(\gamma_{\text{ad}} - \gamma_{\text{bend}})x_{\text{re}}} - 1\right)}, \tag{10.65}$$

which depends only on the position of the restriction binding site along the DNA sequence, and not on the strength of the binding site, $\varepsilon_{\mathrm{bind}}$.

We are now in the position to compare the theoretical results with the measurements discussed earlier. Figure 10.25 shows both the experimental data points and a best fit to these points using the adhesive energy as a free parameter. We conclude that $K_{\mathrm{eq}}^{\mathrm{conf}}$ decreases by four orders of magnitude as the position of the binding site shifts from the end of the wound DNA to the dyad at $x = 70$ bp ≈ 25 nm, in good agreement with the measurements of Polach and Widom (1995). (Note that the sharp increase at $x_{\mathrm{re}} = 0$ seen in Figure 10.25 is an artifact of the continuum approximation assumed by the model; it leads to an infinite $K_{\mathrm{eq}}^{\mathrm{conf}}$ when $x_{\mathrm{re}} \to 0$. Deriving the discrete model that remedies this pathology of the continuous model is left as a problem for the reader at the end of the chapter.) Although these experiments address the unwrapping of short DNA segments from histones *in vitro*, it is likely that similar energetic considerations apply to DNA wrapping and unwrapping for chromosomes in the nuclei of living eukaryotic cells.

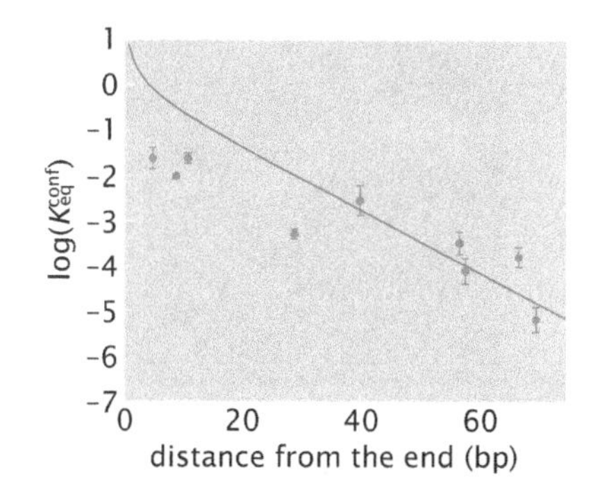

Figure 10.25: Equilibrium constant for DNA accessibility. The data points result from measurements described in this section. The curve is the result of the model worked out in Section 10.4. We define the parameter $\Delta\gamma = \gamma_{\mathrm{ad}} - \gamma_{\mathrm{bend}}$. Using a persistence length of 50 nm implies $\gamma_{\mathrm{bend}} = 1.2\ k_BT$/nm. (Data adapted from K. J. Polach and J. Widom, *J. Mol. Biol.* 254:130, 1995.)

10.5 The Cytoskeleton and Beam Theory

The mechanical behavior of cells is largely determined by the material properties and architectural arrangements of structural elements that tend to look like beams or networks of beams, collectively known as the cytoskeleton. A particularly dramatic illustration of the centrality of these filamentous structural elements in determining cell shape and mechanics is afforded by the neuron. In the late nineteenth century, the Spanish anatomist Santiago Ramón y Cajal keenly observed neurons and then drew spectacular pictures to reflect his observations, as shown in Figure 10.26. The most striking feature of this cell is its enormously elaborate and yet tightly controlled morphology. No simple physical process such as surface tension or bending can account for this kind of elaborate structure; it must be carefully constructed from within the cell.

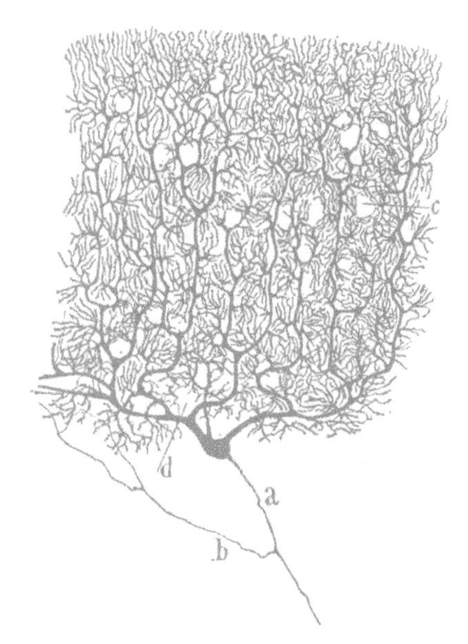

Figure 10.26: Drawing of a neuron. This figure shows one of the many impressive drawings from Ramón y Cajal of a neuron. In particular, this figure shows a human Purkinje cell and illustrates the morphological complexity dictated by cytoskeletal proteins. (Adapted from S. Ramón y Cajal, Histology of the Nervous System. Oxford University Press, 1995.)

Eukaryotic Cells Are Threaded by Networks of Filaments

If we look inside a cell, we can see that it is filled with filamentous structures that appear to provide the structural support for these elaborate and complex shapes. These filaments are typically made of protein, but it is striking that they are extremely large, much larger than any individual polypeptide chain could actually be. In fact, these filaments are made of many identical copies of protein subunits that assemble in helical aggregates, as already shown schematically in Figure 2.27 (p. 64). Thousands or even tens of thousands of subunits come together to form a single filament. If we experimentally disrupt the association of these subunits, for example using drugs or mutations, then cell shape is horrifically compromised. Quantitatively, how can we think about the ways in which the mechanical properties of these filaments are responsible for the structure and organization of the cell? From a mechanical perspective, the most striking feature of these filaments is their aspect ratio: their length is orders of magnitude larger than their cross-sectional dimension, an observation we also made of DNA. As a result, we can use a similar

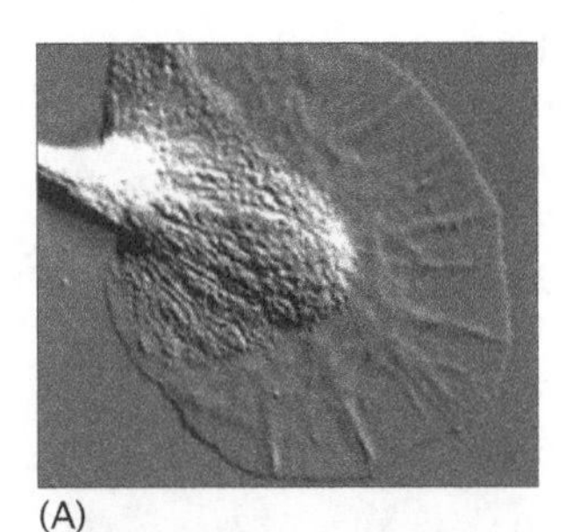

(A)

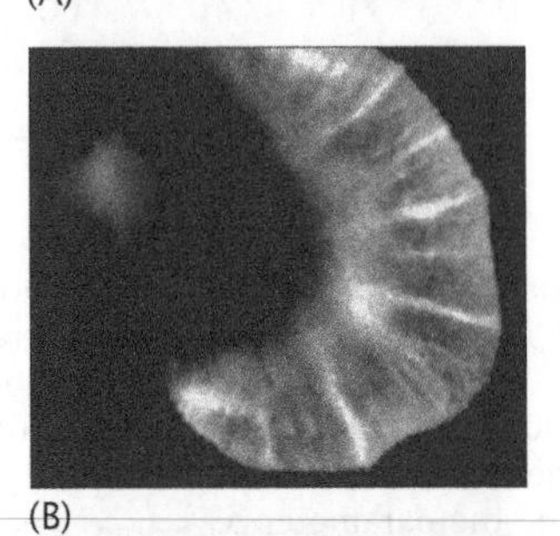

(B)

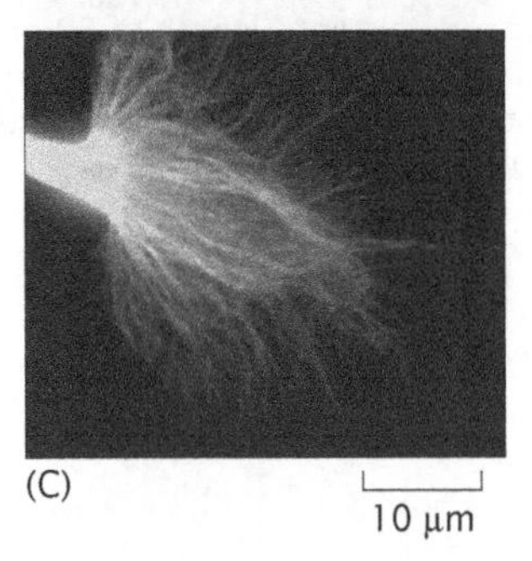

(C)

Figure 10.27: Cytoskeletal filaments in a neural growth cone. (A) Differential interference contrast image of an *Aplysia* (sea slug) neuron growth cone. (B) Fluorescence image of actin in the growth cone. (C) Fluorescence image of microtubules in the growth cone. The distribution of intermediate filaments (neurofilaments) is similar to the distribution of microtubules shown here. (Courtesy of P. Forscher.)

mechanical treatment to describe the properties of these filaments as structural elements.

10.5.1 The Cellular Interior: A Structural Perspective

If we take the neuron shown in Figure 10.26, and separate out the molecularly distinct filament types that it contains, we will find that three predominate. One group of these filaments are microtubules, which tend to be straight filaments about 25 nm in diameter. We will also find actin filaments, which are roughly 8 nm in diameter, and finally, intermediate filaments, so-called because they are intermediate in size between actin and microtubules. These different filaments have distinct locations within the neuronal cell as depicted in Figure 10.27, which shows actin and microtubules within the growth cone found at the tip of a developing neuron. The various filaments also have different mechanical properties and perform distinct biological functions.

Microtubules Microtubules have been implicated in a variety of key cellular processes, serving as a conduit for the motion of molecular motors such as kinesin and dynein as well as presiding over the organization of cell division. Several layers in the structural hierarchy of microtubule organization are shown in Figure 10.28. These filaments are made up of individual tubulin subunits with a characteristic scale of about 8 nm. These subunits bind to each other in a head-to-tail fashion, forming strings called protofilaments. A test tube containing purified tubulin subunits will spontaneously assemble to form microtubules under appropriate conditions. These microtubules are hollow cylindrical structures in which a cross-section contains 13 tubulin subunits arranged in a ring. Looking along the side of a microtubule, we can see that each individual subunit forms two kinds of protein–protein contacts with its neighbors. Around the ring, these subunits have lateral interactions, while along their length, they form head-to-tail contacts such that all individual subunits are pointing in the same direction along the microtubule lattice. Within cells, microtubules are frequently found organized into bundles. For example, microtubules in the mitotic spindle are responsible for separating duplicated chromosomes during cell division. Each chromosome is anchored to a bundle of approximately 30 microtubules that are all oriented in the same direction and cross-linked together. As we will discuss below, bundling of cytoskeletal filaments alters their mechanical properties in ways that are important for cell function.

Actin Filaments Like microtubules, actin filaments are helical assemblies of globular subunits. Although the individual subunits are comparable in size (5 nm for actin versus 8 nm for microtubules), the filaments are smaller because they are only 2 protofilaments across in cross-section rather than 13, as shown in Figure 10.29. Inside cells, actin filaments are also frequently found in bundles. Actin serves a wide variety of structural and functional roles in cells, with some of its most familiar activities related to motility. As a result of their ability to hydrolyze ATP, these filaments can perform useful work resulting in motile functions such as the creation of protrusions in motile cells (see Figure 15.2 on p. 576) and the motion of bacterial pathogens (see Figure 15.3 on p. 577). Together with

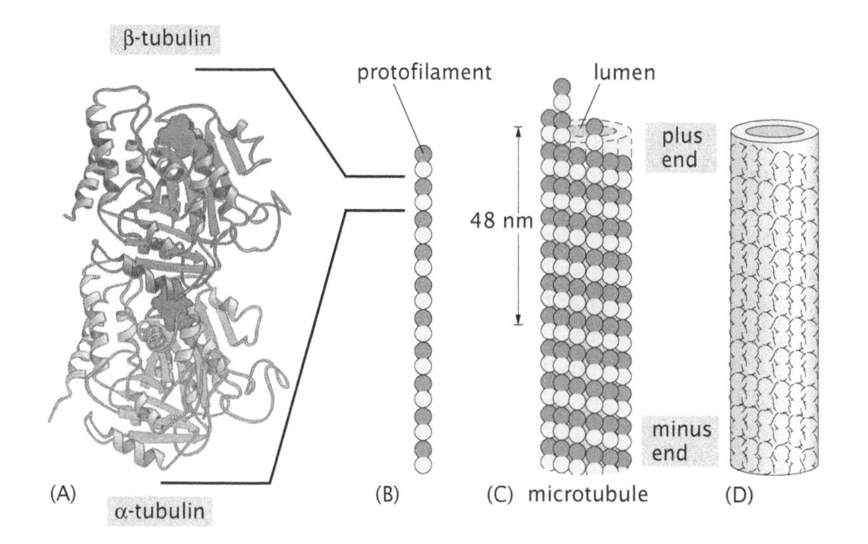

Figure 10.28: Structure of a microtubule. (A) Ribbon diagram depiction of a tubulin subunit, consisting of a dimer of the two proteins α-tubulin and β-tubulin. Bound GTP molecules are shown in red. (B) Individual tubulin subunits assemble in a head-to-tail fashion. A single string of subunits is called a protofilament (such structures are unstable on their own). (C) Microtubules are hollow cylinders of 13 protofilaments. Because of the intrinsic polarity of the tubulin subunit, the microtubule itself is also polarized, with all β-tubulin proteins exposed at the plus end and all α-tubulin proteins exposed at the minus end. (D) For idealized representations, we will treat the microtubule as a uniform hollow cylinder with a diameter of 25 nm. (Adapted from B. Alberts et al., Molecular Biology of the Cell, 5th ed. Garland Science, 2008.)

myosin motors, actin filaments are also required for muscle contraction. The development of the complex network of actin filaments seen in cells is tightly controlled in both space and time by an array of actin-related proteins, as will be discussed in more detail in Chapter 15. One of the outcomes of the activity of these actin-related proteins is the ability to form higher-order structures such as branched networks.

Intermediate Filaments The subunits of intermediate filaments are very different from either actin or tubulin subunits in that they are elongated proteins rather than globular. Assembly of intermediate filaments involves lateral bundling and twisting of these subunits into coiled-coil structures. Their functional role is, at least in part, to provide mechanical resistance to stretching of the cells in tissues. These coiled-coil dimers join to form antiparallel tetramers that are,

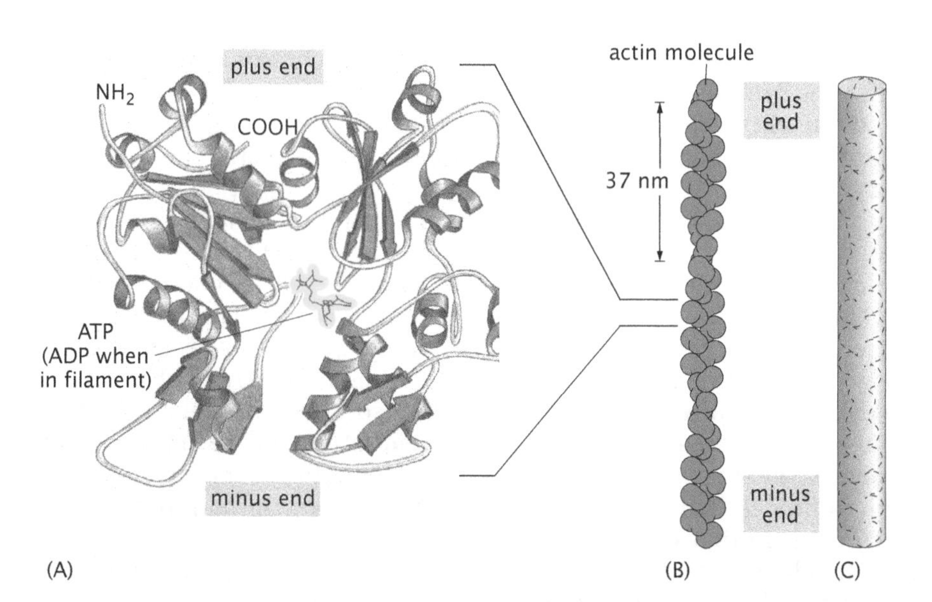

Figure 10.29: Structure of actin. (A) Ribbon diagram depiction of a single G-actin monomer. Each actin monomer is a globular protein 5.4 nm in diameter that can bind a single nucleotide (ATP or ADP). (B) F-actin filaments are assembled from two protofilaments that wrap around each other with a 37 nm repeat. All actin subunits within the filament are oriented in the same direction. Because the filament consists of two parallel protofilaments, the addition of a single subunit extends the overall filament length by 2.7 nm. (C) For idealized treatments of actin filaments, we will regard them as uniform cylinders with a diameter of 8 nm. (Adapted from B. Alberts et al., Molecular Biology of the Cell, 5th ed. Garland Science, 2008.)

in turn, wound together in a rope-like structure that lends them their mechanical strength.

Prokaryotic Cells Have Proteins Analogous to the Eukaryotic Cytoskeleton

The assembly of large helical filaments from small protein subunits appears to be a universally conserved feature of cellular organization. In particular, actin and tubulin are found in nearly all cells on the planet, including bacteria and archaea, as well as eukaryotes. In bacteria, for historical reasons, their names are different (for example, the tubulin homolog in bacteria and archaea is called FtsZ and various actin homologs go by names such as MreB and ParM; often, bacterial proteins are named based on the genetic screen in which they were discovered rather than on their structure) but it is clear from protein crystal structures that the proteins are extraordinarily similar. The protein structure of these organizational molecules has been assiduously conserved over three billion years of evolution. In all cells, cytoskeletal filaments contribute to the determination of cell shape and the mechanical processes associated with cell division. Their exact roles vary substantially from cell to cell. However, the basic theme that they provide large-scale organization by self-assembly is universal.

10.5.2 Stiffness of Cytoskeletal Filaments

The Cytoskeleton Can Be Viewed as a Collection of Elastic Beams

We have asserted that these cytoskeletal filaments are responsible for the organization and mechanical properties of cells. This implies that they can provide mechanical support against applied loads and overcome competing forces that arise from a variety of mechanisms such as membrane elasticity, binding reactions, surface tension, and so on. A prerequisite to deciphering the role of architectural elements is to characterize their response as individual mechanical elements. In particular, as was already shown for DNA, a critical measure of beam stiffness is captured by the persistence length or flexural rigidity.

For cytoskeletal elements, three different factors determine the effective stiffness of structures within the cell. The first, of course, is the intrinsic flexibility of the filaments themselves. The second is the presence or absence of filament binding proteins that modulate filament stiffness, and the third is larger-scale organization of filaments into bundles, networks, or gels. Multiple techniques have been developed to measure stiffness at each of these scales. At the level of single filaments, thermal motions are sufficient to induce filament bends over a spatial scale that can be directly observed in a light microscope. Therefore, the most straightforward and arguably the least invasive way to measure filament stiffness has been by direct observation of filaments either fluorescently labeled so that individual filaments can be visualized or, in the case of microtubules, using enhanced contrast microscopy, which permits visualization even of unlabeled microtubules. Other measurements use mechanical manipulation of filaments so that their deformation can be observed under a specific applied load. These kinds of measurements may use either calibrated microneedles or beads held in optical traps to deliver known forces to the ends of captured filaments. A dramatic example of this approach is

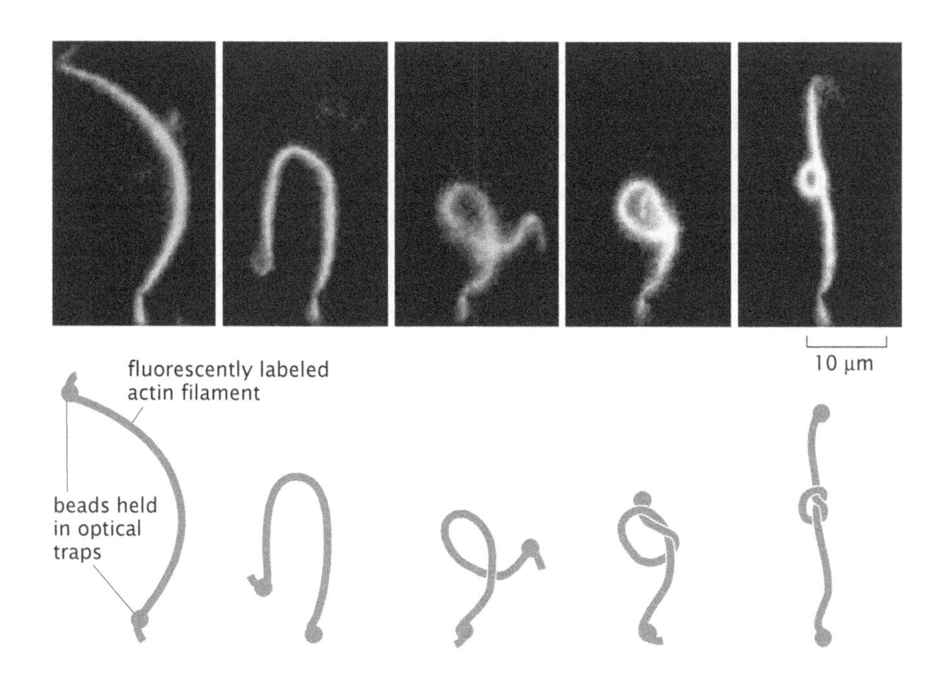

Figure 10.30: Actin tied in a knot. The flexibility of the actin filament is vividly illustrated in this experiment where two beads are attached to opposite ends of a filament and manipulated using optical traps to tie a tiny knot. (Adapted from Y. Arai et al., *Nature* 399:446, 1999.)

shown in Figure 10.30, where optical tweezers are used to tie a single actin filament into a knot.

Cells build mechanical structures by combining individual cytoskeletal filaments in ways that alter their large-scale mechanical properties. This is most easily envisioned in the common situation where cytoskeletal filaments are bundled together to create stronger elements that resist bending. Figure 10.31 shows bundles of actin filaments forming protrusions of the cell membrane on the surface of human intestinal epithelial cells. The function of these protrusions is to vastly increase the surface area through which nutrients can be absorbed. As we will explore in detail in Chapter 11, bending the cell plasma membrane, particularly in such a tortuous manner, requires energy input. Looked at from another perspective, the membrane exerts a force on these bundles of actin filaments. The persistence of

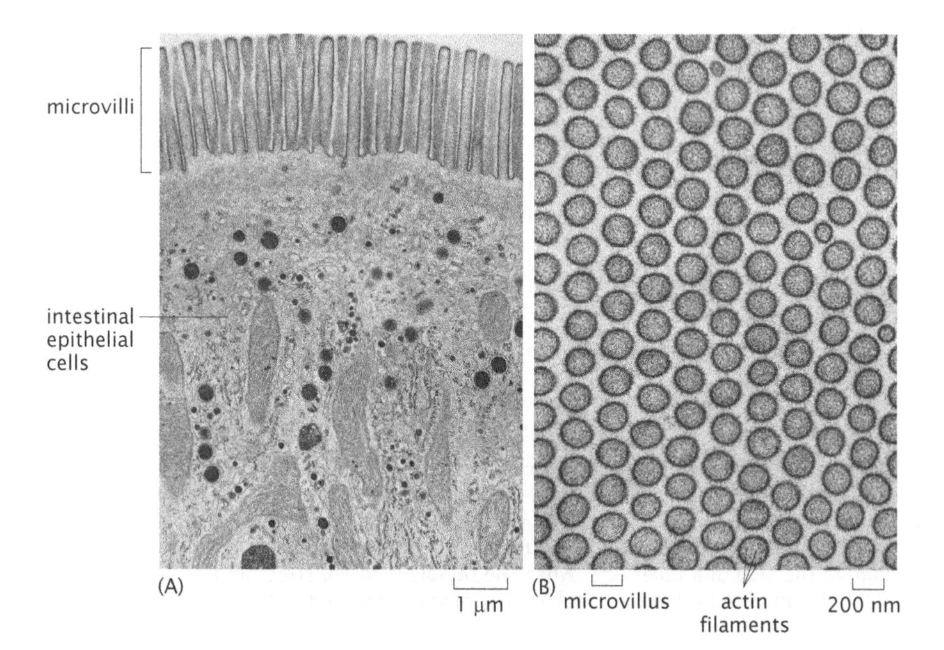

Figure 10.31: Intestinal brush border. Two electron micrographs show the organization of actin filaments and cell membrane at the absorptive surface of human intestinal epithelial cells. (A) Section of intestinal epithelial cells seen from the side with the absorptive microvilli at the top. (B) Glancing cross-section across the top of the cell showing the close packing and organization of the microvilli with their internal bundles of actin filaments. (Adapted from D. W. Fawcett, The Cell, Its Organelles and Inclusions: An Atlas of Fine Structure. W. B. Saunders, 1966.)

the bundle structures at the surface of epithelial cells demonstrates that the force exerted by the membrane is not sufficient to bend or buckle these filament bundles. In the next section, we will make an estimate showing that a single filament with length comparable to a microvillus cannot support the forces applied by the membrane and will buckle. Therefore, it seems reasonable to conclude that the biological function of bundle formation in this case is to increase the stiffness of the composite filament beyond that of single filaments. In a human, a typical microvillus includes roughly 30 actin filaments packed together in a hexagonal lattice.

A very different and even more elegant use of the enhanced stiffness of actin-filament bundles is found in the vertebrate inner ear hair cell, the cell type responsible for converting motions of air into perceived sound. Hair cells within the inner ear lie adjacent to a tectorial membrane as shown in Figure 10.32, which forms one wall of a fluid-filled chamber that vibrates in response to sound waves hitting the ear drum. At the top surface of the hair cells where they contact the tectorial membrane are a series of rigid projections called stereocilia that, like microvilli, are constructed of a bundle of actin filaments closely wrapped by the cell plasma membrane. Unlike microvilli, which are all approximately the same length, the stereocilia within a bundle on an individual hair cell present a graded series of lengths resembling a staircase. Movements of the tectorial membrane cause the longest of these stereocilia to lean relative to the surface of the cell. Because the individual stereocilia are so stiff, the bundles do not bend. So instead, as the stereocilia lean over, the shorter ones slide relative to the longer ones. This leads to the opening of an ion channel that converts the mechanical signal to an electrical signal that can be perceived by the brain. The extreme stiffness of the stereocilia is a necessary element of this mechanical signal transduction event and could not be achieved using a single cytoskeletal filament.

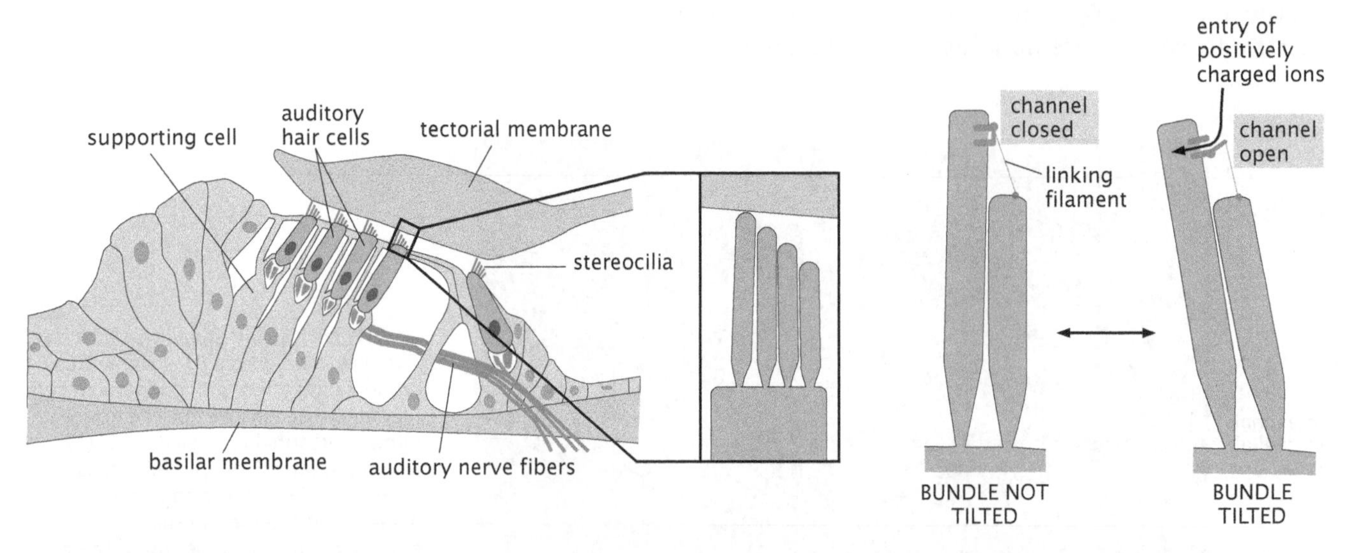

Figure 10.32: Beam mechanics in hearing. Sound is detected and converted into an electrical signal in the nervous system by hair cells. The upper surface of hair cells has a projecting bundle of stereocilia that make contact with a structure in the inner ear called the tectorial membrane. The tectorial membrane vibrates at specific points in response to sounds. Vibration of the membrane pushes on the stereocilia bundle. Neighboring stereocilia within the bundle are linked to one another by elastic filaments that connect the top of one stereocilium to the side of a taller one. When the bundle is tilted, stretching of the linking filaments leads to opening of ion channels. (Adapted from B. Alberts et al., Molecular Biology of the Cell, 5th ed. Garland Science, 2008.)

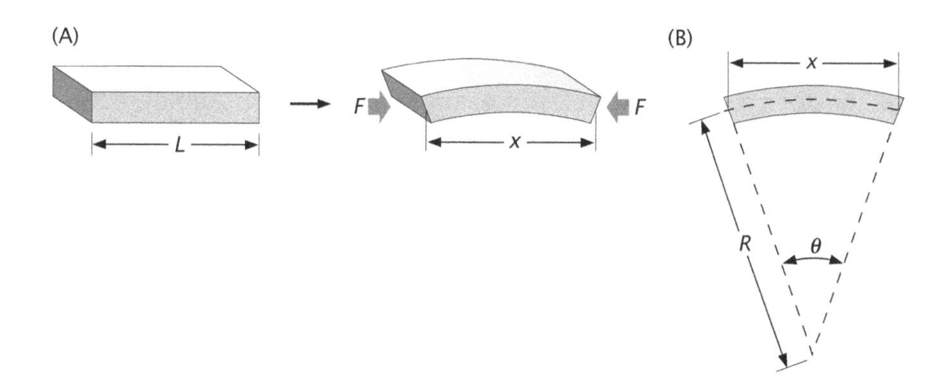

Figure 10.33: Schematic of the buckling process. (A) A beam of length L is loaded at the ends with compressive forces. At sufficiently large forces, the beam undergoes buckling. (B) Geometric parameters used to characterize the buckling process.

10.5.3 Cytoskeletal Buckling

A Beam Subject to a Large Enough Force Will Buckle

Our previous analysis of the deformation of beams has been restricted to those cases in which the material is subject to deformations that can be considered "small." Of course, from an experimental perspective, such restrictions are entirely artificial and we are free to push and pull on beams with forces that engender deformations that exceed those characterized by the theory of elasticity. One of the ways in which materials can deform other than the simple deformations considered thus far is buckling. An example of the type of process we have in mind is shown in Figure 10.33. In the discussion to follow, we show how a simple estimate of the buckling force can be determined by examining the energetics required to deform the beam into either a semicircular arc or a sinusoidal profile.

Cytoskeletal filaments can exert forces during their polymerization process. One especially intriguing class of processes is that in which the leading edge of polymerizing actin filaments is responsible for deforming the membrane of a cell, such as in the filopodia of motile cells on surfaces. An intriguing question that arises in this context is how much force such filaments can take before they suffer a buckling deformation. Indeed, in a series of *in vitro* experiments, polymerizing microtubules were grown on a surface into a groove that would stall their progress and induce buckling. A schematic of such an experiment is shown in Figure 10.34. We can begin to interpret this experiment with simple ideas on elastic buckling.

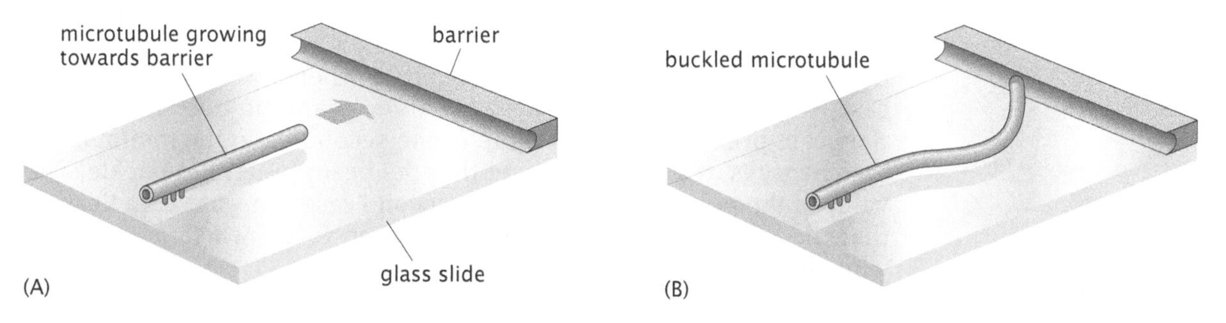

Figure 10.34: Microtubule buckling during polymerization. (A) At the start of the experiment, a microtubule is tightly bound to a glass slide, oriented such that its plus end is growing towards a barrier wall with a small overhang. With the addition of tubulin subunits, the microtubule grows freely until it contacts the wall. (B) Small thermal fluctuations in the position of the tip of the growing microtubule are thought to allow sufficient space for insertion of tubulin subunits occasionally, even when the microtubule tip is in contact with the wall. The force built up at this interface causes the microtubule to bend and then to buckle. (Adapted from M. Dogterom and B. Yurke, *Science* 278:856, 1997.)

10.5.4 Estimate of the Buckling Force

We consider a long, thin, elastic beam such as a microtubule with a force F applied at the two ends as shown in Figure 10.33. Experience tells us that if we increase the force applied to the beam, it will buckle at some point. Here we investigate this phenomenon in the context of beam theory. To provide an estimate of the critical load for buckling, we once again adopt the simple geometry of a circular arc already used earlier in the chapter. In particular, we assume a simple shape for the buckled beam, an arc of a circle of radius R and length L; $\theta = L/R$ is the angle made by the arc. We are interested in determining the critical force for which $\theta \neq 0$ is the energy-minimizing value. That is, as the beam buckles, the energy associated with the applied force decreases while the bending energy increases. If the bending energy increase dominates, the beam will remain straight ($\theta = 0$); this will always be the case at sufficiently small forces. At very high forces, the opposite will be true and the beam will buckle. Making use of the circular geometry for the buckled beam, we can easily estimate the value of the critical force, F_{crit}.

For the present situation, we compute the total energy as a sum of the beam bending energy and the the energy associated with the applied force, resulting in

$$E_{\text{tot}} = \underbrace{\frac{\xi_{\text{p}} k_{\text{B}} T}{2} \frac{L}{R^2}}_{\text{elastic energy}} - \underbrace{F(L - x)}_{\text{loading device}} , \tag{10.66}$$

where

$$x = 2R \sin \frac{\theta}{2} \tag{10.67}$$

is the distance between the two ends of the beam. As elsewhere in the book, the loading device term captures the energy relaxation associated with the device applying the force F. In terms of the angle θ, the total energy of the buckled beam is

$$\frac{E_{\text{tot}}}{k_{\text{B}} T} = \frac{\xi_{\text{p}}}{L} \frac{\theta^2}{2} - \frac{FL}{k_{\text{B}} T} \left(1 - \frac{2}{\theta} \sin \frac{\theta}{2}\right). \tag{10.68}$$

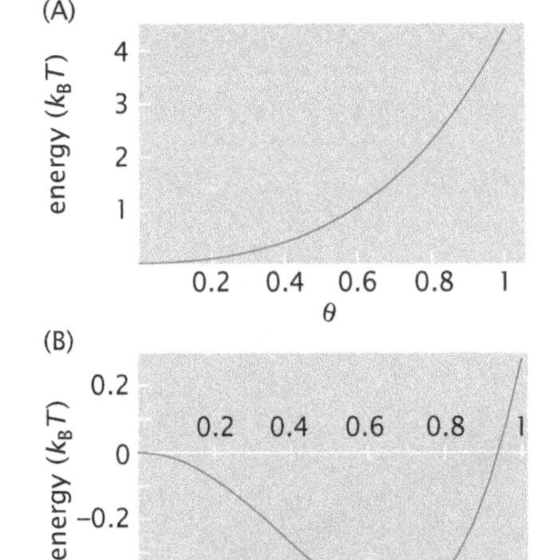

Figure 10.35: Energetics of beam buckling. Energy of a buckled beam as a function of the angle θ for (A) $F < F_{\text{crit}}$ and (B) $F > F_{\text{crit}}$. The parameters used in this plot are $L = 20\ \mu\text{m}$ and $\xi_{\text{p}} k_{\text{B}} T = 30\ \text{pN}\ \mu\text{m}^2$, which correspond to the numbers appropriate to the experiment indicated schematically in Figure 10.34.

Our strategy is to search for a buckling instability by seeing if at large enough forces the lowest-energy state corresponds to a nonzero value of θ. To see this, we plot the energy as a function of the angle θ for different values of the force. In Figure 10.35, we show two plots, one of the energy for $F < F_{\text{crit}}$ and the other when $F > F_{\text{crit}}$. In the first case, the minimum of the energy is at $\theta = 0$ and no buckling occurs, the beam prefers to stay straight. In the second case, the minimum of the energy is at $\theta \neq 0$, indicating that the beam prefers to buckle. The value of the preferred energy in the second case is negative, indicating that the energy associated with the applied force, which is negative, wins over the positive bending energy.

Beam Buckling Occurs at Smaller Forces for Longer Beams

We can take this result further and obtain an estimate for the critical value of the force. In particular, by expanding Equation 10.68 in small θ, which is legitimate if the applied force is close to its critical value, we arrive at

$$\frac{E_{\text{tot}}}{k_B T} = \frac{\xi_p}{L}\frac{\theta^2}{2} - \frac{FL}{k_B T}\frac{\theta^2}{24}. \tag{10.69}$$

From this expression, we see that for $F < F_{\text{crit}}$, where

$$F_{\text{crit}} = 12\frac{k_B T \xi_p}{L^2}, \tag{10.70}$$

the straight-beam solution with zero energy is the energy-minimizing solution. As soon as the force rises above F_{crit} this is no longer the case and the beam buckles. More to the point, for $F > F_{\text{crit}}$, the sign of the energy for small θ goes from positive to negative, indicating that the solution at $\theta = 0$ goes from being stable to unstable. Note that the critical force for buckling scales as L^{-2}, meaning that longer beams can be buckled more easily than shorter beams. For the experiment indicated schematically in Figure 10.34, the length of the microtubules is roughly 20 μm and they are characterized by a flexural rigidity $\xi_p k_B T = 30$ pN μm^2, corresponding to a critical force of 0.9 pN.

The more sophisticated approach to this problem would be to relax the assumption of a circular shape for the buckled beam and allow any shape. Then the minimizing procedure would have to be performed over all shapes, which, mathematically speaking, corresponds to a variational problem. Such an approach would yield a sinusoidal shape as the energy minimizer and the expression for the critical force would differ from our Equation 10.70 only by a numerical factor, with π^2 replacing our 12.

10.6 Summary and Conclusions

Although DNA, microtubules, and large bones such as femurs are very different in nearly every way in their biological function, their size scale, and their material nature, nonetheless, many aspects of their mechanical behavior in biological systems can be quantitatively described using the same simple set of linear elasticity models. Indeed, this demonstrates the assertion that problems that are quite distinct in a biological sense can be seen as quite similar when viewed from the physical biology perspective. This chapter developed the relevant tools for using linear elasticity to model elastic beams and illustrated how these ideas can be used for a host of distinct and concrete problems. The consequences of elasticity can be found in such unexpected places as the efficiency of gene expression and the development of technology to create cantilever-based biosensors that can serve as artificial noses.

10.7 Appendix: The Mathematics of the Worm-Like Chain

In Section 10.2.3, we introduced the worm-like chain model as a way of uniting the chain statistics introduced in Chapter 8 and the elasticity of beam bending introduced in this chapter. However, evaluation of the force–extension properties of this model is considerably more complicated than the freely jointed chain model introduced in Section 8.3.2 (p. 340). We now examine the worm-like chain model in a more detailed fashion than presented in the chapter.

First, we examine the low-force limit defined by $f\xi_p \ll 1$; for dsDNA this becomes $F \ll k_B T/\xi_p = 0.08$ pN, a small force indeed. In this limit, the partition function can be expanded in powers of $f\xi_p$,

$$Z(f) = \int \mathcal{D}\mathbf{t}(s) \left\{ \exp\left(-\frac{\xi_p}{2} \int_0^L \left| \frac{d\mathbf{t}}{ds} \right|^2 ds \right) \left[1 + f \int_0^L t_z(s)\, ds \right. \right.$$

$$\left. \left. + \frac{f^2}{2} \int_0^L t_z(s)\, ds \int_0^L t_z(u)\, du + O\left((f\xi_p)^3 \right) \right] \right\}, \tag{10.71}$$

and we can safely retain only the first three terms in the expansion. The approximate expression that we get in this way can be rewritten as

$$Z(f) = Z(0) \left[1 + f \int_0^L \langle t_z(s) \rangle_0\, ds + \frac{f^2}{2} \int_0^L \int_0^L ds\, du \langle t_z(s) t_z(u) \rangle_0 \right]. \tag{10.72}$$

Here $\langle \cdots \rangle_0$ is the average evaluated using the zero-force partition function, Equation 10.23. The second term does not contribute, since the average extension in the zero-force case is zero. The third term in the expansion gives a non-vanishing contribution, which can be evaluated with the help of the tangent–tangent correlation function $\langle \mathbf{t}(s) \cdot \mathbf{t}(u) \rangle_0 = e^{-|s-u|/\xi_p}$. Recall that the tangent–tangent correlation function was defined in Section 10.2.2. Since $\mathbf{t}(s) \cdot \mathbf{t}(u) = t_x(s)t_x(u) + t_y(s)t_y(u) + t_z(s)t_z(u)$, and since the worm-like chain energy is invariant under rotations, $\langle t_z(s)t_z(u) \rangle_0 = \frac{1}{3} \langle \mathbf{t}(s) \cdot \mathbf{t}(u) \rangle_0$ follows. Substituting this into Equation 10.72, and using $\int_0^L \int_0^L ds\, du\, e^{-|s-u|/\xi_p} = 2L\xi_p$, which holds in the $L \gg \xi_p$ limit, gives

$$Z(f) = Z(0) \left(1 + \frac{f^2 L \xi_p}{3} \right). \tag{10.73}$$

Finally, making use of the relation given in Equation 10.25 we arrive at

$$\frac{\langle z \rangle}{L} = \frac{2 f \xi_p}{3}, \tag{10.74}$$

which is the same as for the freely jointed chain as long as we take the length of the Kuhn segment to be twice the persistence length.

Next, we consider the limit of very large forces ($f\xi_p \gg 1$), which, as remarked earlier, for dsDNA corresponds to stretching forces in excess of 0.08 pN. In this limit, chain configurations that contribute appreciably to the partition function $Z(f)$ have their tangent vectors pointing roughly in the direction of the force. In other words, the components of the tangent vector in the x- and y-directions can be considered small, and

$$\mathbf{t} \approx \left(t_x, t_y, 1 - \tfrac{1}{2}(t_x^2 + t_y^2) \right). \tag{10.75}$$

This approximate expression for the tangent vector turns the formula for the energy into a quadratic form in t_x and t_y given by

$$E_{tot} = \frac{\xi_p k_B T}{2} \int_0^L ds \left[\left(\frac{dt_x}{ds} \right)^2 + \left(\frac{dt_y}{ds} \right)^2 \right] + \frac{f k_B T}{2} \int_0^L ds \left(t_x^2 + t_y^2 \right) - f k_B T L. \tag{10.76}$$

For quadratic energies such as this, the path integral for the partition function turns into a Gaussian integral, which can be evaluated explicitly.

Using Equation 10.75, the average extension in the high-force limit can be written as

$$\langle z \rangle = L - \frac{1}{2}\int_0^L \mathrm{d}s \left\langle t_x^2 + t_y^2 \right\rangle, \qquad (10.77)$$

where $\langle \cdots \rangle$ is the average with respect to $Z(f)$. To compute $\left\langle t_x^2 + t_y^2 \right\rangle$, we first express the energy functional in terms of the Fourier components of the tangent vector,

$$t_\alpha(s) = \sum_w \mathrm{e}^{iws} t_\alpha(w) \quad (\alpha = x, y), \qquad (10.78)$$

where the frequencies are defined by $w = 2\pi j/L$ with j an integer. In Fourier space, the energy takes on the form of the potential energy of a collection of harmonic oscillators, two for each value of the frequency w and given by

$$E_{\text{tot}} = \frac{Lk_\mathrm{B}T}{2} \sum_w \left(\xi_\mathrm{p} w^2 + f\right) \left(|t_x(w)|^2 + |t_y(w)|^2\right). \qquad (10.79)$$

This observation allows us to compute the average $|t_\alpha(w)|^2$ without explicitly computing the path integral. Namely, we remind ourselves of the equipartition theorem of equilibrium statistical mechanics, which states that the average energy for every quadratic degree of freedom is $k_\mathrm{B}T/2$. Therefore,

$$\left\langle \frac{Lk_\mathrm{B}T}{2} \left(\xi_\mathrm{p} w^2 + f\right) |t_\alpha(w)|^2 \right\rangle = \frac{k_\mathrm{B}T}{2} \quad (\alpha = x, y), \qquad (10.80)$$

and

$$\frac{\langle z \rangle}{L} = 1 - \frac{1}{L} \sum_w \frac{1}{\xi_\mathrm{p} w^2 + f} \qquad (10.81)$$

follows immediately from Equation 10.77 if we replace the integral in that equation by a sum over Fourier modes. Now the sum in Equation 10.81 can be evaluated by taking a continuum limit $\sum_w \to (L/2\pi)\int_{-\infty}^{+\infty} \mathrm{d}w$; the remaining integral over w gives

$$\frac{\langle z \rangle}{L} = 1 - \frac{1}{2\sqrt{f\xi_\mathrm{p}}}. \qquad (10.82)$$

The results obtained here in the low-force and the high-force limit can be used to construct an approximate formula for the force–extension relationship at all forces. If we rewrite Equation 10.74 as

$$f\xi_\mathrm{p} = \frac{3}{2}\frac{\langle z \rangle}{L} \qquad (10.83)$$

and Equation 10.82 as

$$f\xi_\mathrm{p} = \frac{1}{4\left(1 - \langle z \rangle/L\right)^2}, \qquad (10.84)$$

then these two limiting forms can be combined to obtain Equation 10.26. This interpolation formula has the correct behavior at low

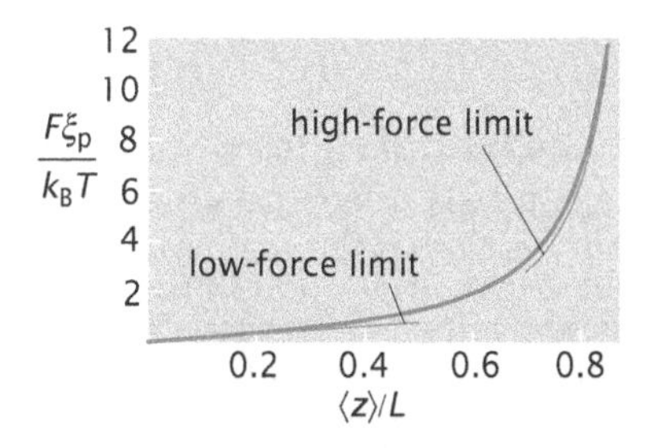

Figure 10.36: Force–extension curve for a worm-like chain. The red line is the interpolation formula, Equation 10.26, while the two blue lines are the limiting results obtained in the low- and high-force limits.

force and at high force, as shown in Figure 10.36, and it deviates at most by 10% from the exact result, which can be obtained numerically using a number of different schemes.

10.8 Problems

Key to the problem categories: • Model refinements and derivations • Data interpretation • Model construction

• 10.1 Persistence length and Fourier analysis

In the chapter, we computed the tangent–tangent correlation function for a polymer, which we modeled as an elastic beam undergoing thermal fluctuations. The calculation was carried out in the limit of small fluctuations and it led to an expression for the persistence length in terms of the flexural rigidity of the polymer. Here we reexamine this problem for a two-dimensional polymer, but without the assumption of small fluctuations.

(a) For a polymer confined to a plane, the tangent vector $\mathbf{t}(s)$ can be written in terms of the polar angle $\theta(s)$ as $\mathbf{t}(s) = (\cos\theta(s), \sin\theta(s))$. Rewrite the beam bending energy, Equation 10.9, in terms of the polar angle $\theta(s)$.

(b) Expand the polar angle $\theta(s)$ into a Fourier series, taking into account the boundary conditions $\theta(0) = 0$ and $d\theta/ds = 0$ for $s = L$. The first boundary condition comes about by choosing the orientation of the polymer so that the tangent vector at $s = 0$ is always along the x-axis. Convince yourself that the second is a consequence of there being no force acting on the end of the polymer.

(c) Rewrite the bending energy in terms of the Fourier amplitudes $\tilde{\theta}_n$, introduced in (b), and show that it takes on the form equivalent to that of many independent harmonic oscillators. Use equipartition to compute the thermal average of each of the Fourier amplitudes.

(d) Make use of the identity $\langle \cos X \rangle = e^{-X^2/2}$, which holds for a Gaussian distributed random variable X, to obtain the equation for the tangent–tangent correlation function, $\langle \mathbf{t}(s) \cdot \mathbf{t}(0) \rangle = e^{-\theta(s)^2/2}$. Then compute $\langle \theta(s)^2 \rangle$ by using the Fourier series representation of $\theta(s)$ and the average values of the Fourier amplitudes $\tilde{\theta}_n$ obtained in (c). Convince yourself either by plotting or Fourier analysis that on the interval $0 < s < L$, $\langle \theta(s)^2 \rangle = s/\xi_p$.

(e) How does the persistence length in two dimensions compare with the value obtained in three dimensions? Explain why the tangent–tangent correlation function decays slower in two dimensions. What is the situation in one dimension?

• 10.2 Flexural rigidity of biopolymers

(a) Recall from p. 389 that when treating macromolecules as elastic beams it is the combination $K_{eff} = EI$, the flexural rigidity, that dictates the stiffness of that molecule. The flexural rigidity is a product of an energetic (E, Young modulus) and geometric (I, areal moment of inertia) factor. Reproduce the argument given in the chapter that culminated in Equation 10.8 for the bending energy of a beam and show that the flexural rigidity enters as claimed above.

(b) Using what you know about the geometry of DNA, actin filaments, and microtubules, determine the areal moment of inertia I for each of these molecules. Be careful and remember that microtubules are hollow. Make sure that you comment on the various simplifications that you are making when you replace the macromolecule by some simple geometry.

(c) Given that the elastic modulus of actin is 2.3 GPa, take as your working hypothesis that E is universal for the macromolecules of interest here and has a value 2 GPa. In light of this choice of modulus, compute the stress needed to stretch both actin and DNA with a strain of 1%. Convert this result into a pulling force in piconewtons.

(d) Using the results from (b) and (c), compute the persistence lengths of all three of these molecules.

(e) Given that the measured persistence length of DNA is 50 nm, and using the areal moment of inertia you computed in (b), compute the Young modulus of DNA. How well does it agree with our 2 GPa rule of thumb from above?

• 10.3 Twisting DNA

For small torques exerted on a beam of elastic material, there is a linear relationship between the torque and the torsional strain, $\tau = C d\theta/dz$. This is analogous to the bend elasticity worked out in the chapter. Single-molecule experiments using magnetic or optical tweezers allow measurements of the twist elasticity of DNA molecules by systematically winding up the tethered DNA molecule and examining the build up of torque as indicated schematically in Figure 10.37.

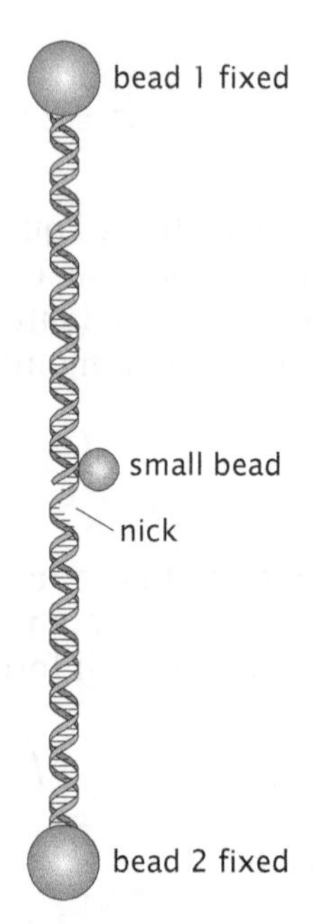

Figure 10.37: DNA twist elasticity experiment. A DNA molecule is constrained rotationally by two beads at both ends. A third bead is used to monitor the unwinding when a nick is introduced. (Adapted from Z. Bryant et al., *Nature* 424:338, 2003.)

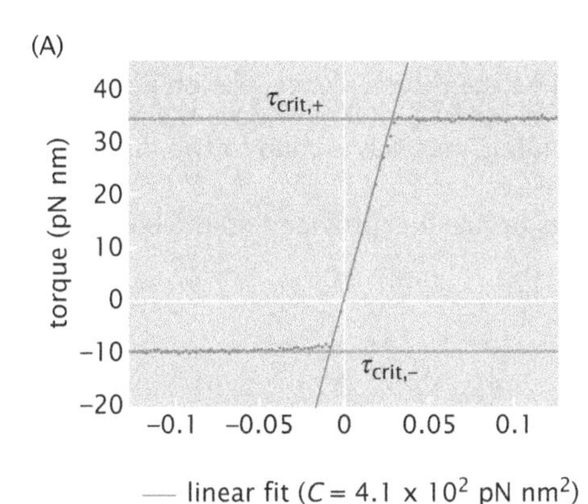

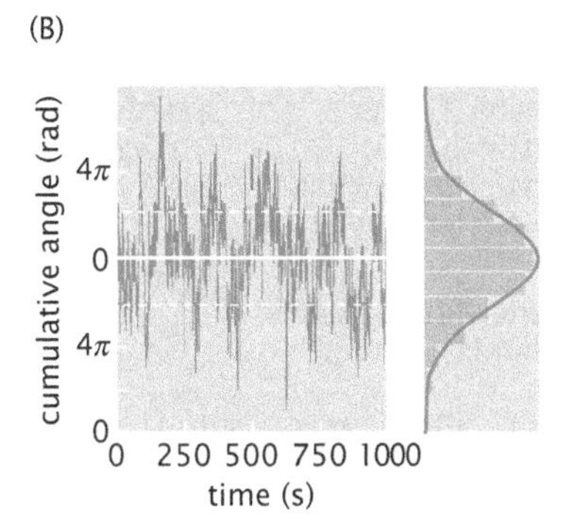

Figure 10.38: Experimental data from DNA twisting experiment. (A) Experimental results for the relation between the torque and the angular deflection of the DNA molecule. (B) Distribution of angular fluctuations in the bead position. (Adapted from Z. Bryant et al., *Nature* 424:338, 2003.)

(a) Use the data shown in Figure 10.38(A) to estimate the twist modulus.

(b) Estimate the torque needed to wind a 14.8 kb DNA molecule by 10 rotations and by 100 rotations.

(c) Write an expression for the energy stored in the twisted DNA molecule by virtue of its twist deformation. Using this expression and the data given in Figure 10.38(A), estimate the energy per base pair stored in the DNA molecule of 14.8 kb in length after the molecule has been subjected to 50 complete revolutions.

(d) An alternative way to measure the twist modulus (as with the bending modulus) is to probe the thermal fluctuations of the molecule. Figure 10.38(B) shows the angular excursion of the molecule (as reported by the motion of a bead stuck to the DNA in the middle of the molecule as shown in Figure 10.37) due to thermal fluctuations. Derive an expression linking the twist fluctuations to the stiffness and use the data to estimate the twist modulus. In particular, find the probability distribution $p(\theta)$ by computing the partition function. Show that $\langle(\Delta\theta)^2\rangle = k_B TL/C$ and use this result and the data to find C.

• 10.4 The J-factor connection

Derive the relation between the J-factor and the free energy of cyclization described in Section 10.3.3. In particular, construct a lattice model of the cyclization process by imagining a box of Ω lattice sites each with volume v. We consider three species of DNA: monomers with sticky ends that are complementary to each other; dimers, which reflect two monomers sticking together; and DNA circles in which the two ends on the same molecule have stuck together; The number of molecules of each species is N_1, N_2, and N_c, respectively.

(a) Use the lattice model to write down the free energy of this assembly and attribute an energy ε_b to the binding of complementary ends, ε_{loop} to the looped configurations, and ε_{sol} as the energy associated with DNA–solvent interactions when the length of the DNA corresponds to one monomer. Further, *assume* that the solvent energy for a dimer is $2\varepsilon_{sol}$ and for a looped configuration is identical to that of a monomer. Use a Lagrange multiplier μ to impose the constraint that $N_{tot} = N_1 + 2N_2 + N_c$.

(b) Minimize the free energy with respect to N_1, N_2, and N_c to find expressions for the concentrations of the three species. Then use the fact that J is the concentration of monomers at which $N_2 = N_c$ to solve for the unknown Lagrange multiplier μ and to obtain an expression for the J-factor in terms of the looping free energy.

• 10.5 Packing free energy for a spherical virus

Repeat the calculations of the energy of DNA packing for a spherical capsid. Contrast this result with that obtained in the chapter where it was assumed that the capsid is a cylinder. Use the experimental data provided on the book's website and compare the model with the data shown in Figure 10.19(B).

Relevant data for this problem is provided on the book's website.

• 10.6 Entropy of packing ϕ29

(a) Estimate the total work that the portal motor of ϕ29 needs to perform in order to overcome the entropy loss of the DNA after it has been packaged. (*Hint*: To count states, assume that the conformations of the DNA out of the capsid can be modeled by a freely jointed chain with 90° angles between the segments.)

(b) What is the total work of the motor when packaging the ϕ29 DNA, as measured by Smith et al. (2001)? What conclusion do you draw from comparing this value with the one obtained in (a)?

Relevant data for this problem is provided on the book's website.

• 10.7 Nucleosomes in a box

Repeat the derivation of nucleosome accessibility using the canonical distribution. This means that you should imagine a nucleosome in a box with L DNA-binding proteins and then work out the probability that the nucleosome will be occupied by one of those proteins as a function of their concentration and as a function of the position of the binding site on the DNA.

• 10.8 Nucleosome formation and assembly

(a) Repeat the derivation given in the chapter for the nucleosome formation energy, but now assuming that there is a discrete number of contacts ($N = 14$) between the DNA and the histone octamer.

(b) Use the discrete model to calculate the equilibrium accessibility of binding sites wrapped around nucleosomes. Apply this model to the data by Polach and Widom (1995) and fit the adhesive energy per contact $\gamma_{discrete}$.

(c) Reproduce Figure 10.25 and compare your results for the equilibrium accessibility versus burial depth from (a) and (b) with the continuum model.

(d) Look at some of the binding affinities of different DNA sequences to histones reported by Lowary and Widom

(1998). Once again, assume that the electrostatic interaction between the histone and the different DNA molecules does not vary, that is, it is not sequence-dependent. This is equivalent to saying that the difference between each sequence lies in its flexibility, in its bending energy. What would one expect the difference in their persistence lengths to be?

(e) Model the case of having two binding sites for the same DNA-binding protein on a DNA molecule that is wrapped around a histone octamer. How does the equilibrium accessibility depend on the protein concentration and the relative position of the binding site? How does the problem change if the two binding sites correspond to two different DNA-binding proteins?

Relevant data for this problem is provided on the book's website.

10.9 Further Reading

Calladine, CR, Drew, HR, Luisi, BF, & Travers, AA (2004) Understanding DNA: The Molecule and How It Works, 3rd ed., Academic Press. This book gives an in-depth description of many topics we have only touched on in this chapter.

Gordon, JE (1978) Structures or Why Things Don't Fall Down, Penguin Books. A beautiful description of the relevance of beam theory to the world of macroscopic structures is to be found in Chapter 11, which is entitled "The advantage of being a beam."

Alberts, B, Johnson, A, Lewis, J, et al. (2008). Molecular Biology of the Cell, 5th ed., Garland Science. We will make reference to Alberts et al. repeatedly in this book for a compelling reason: their book is a treasure trove. Chapter 16 on the cytoskeleton is a useful starting point.

Amos, LA, & Amos, WB (1991) Molecules of the Cytoskeleton, The Guilford Press. An interesting source on the cytoskeleton.

Bray, D (2001) Cell Movements: From Molecules to Motility, Garland Science. Bray's unique book is full of great stuff ranging from the nature of the cytoskeleton to the nature of bacterial motion.

Boal, D (2012) Mechanics of the Cell, 2nd ed., Cambridge University Press. Boal's treatment of the elastic theory of beams and rods is excellent.

Marko, JF, & Siggia, E (1995) Stretching DNA, *Macromolecules* **28**, 209. A classic reference on force–extension in the worm-like chain model.

Riemer, SC, & Bloomfield, VA (1978) Packaging of DNA in bacteriophage heads: some considerations on energetics, *Biopolymers* **17**, 785. A beautiful paper describing the free-energy contribution to DNA packaging in bacteriophages.

Kindt, J, Tzlil, S, Ben-Shaul, A, & Gelbart, WM (2001) DNA packaging and ejection forces in bacteriophage, *Proc. Natl Acad. Sci. USA* **98**, 13671.

Fuller, DN, Rickgauer, JP, Jardine, PJ, et al. (2007) Ionic effects on viral DNA packaging and portal motor function in bacteriophage ϕ29, *Proc. Natl Acad. Sci. USA* **104**, 11245. This paper raises questions about the ability of the simple model presented in the chapter to account for measurements of the packaging forces during viral packaging.

Schiessel, H (2003) The physics of chromatin, *J. Phys.: Condens. Matter* **15**, R699. This excellent article describes the structure and energetics of nucleosomes.

Dogterom, M, & Yurke, B (1997) Measurement of the force–velocity relation for growing microtubules, *Science* **278**, 856. This paper illustrates how buckling of microtubules as shown in Figure 10.34 is used to measure the relation between force and velocity for growing microtubules.

10.10 References

Alberts, B, Johnson, A, Lewis, J, et al. (2008) Molecular Biology of the Cell, 5th ed., Garland Science.

Arai, Y, Yasuda, R, Akashi, K-I, et al. (1999) Tying a molecular knot with optical tweezers, *Nature* **399**, 446.

Bryant, Z, Stone, MD, Gore, J, et al. (2003) Structural transitions and elasticity from torque measurements on DNA, *Nature* **424**, 338.

Cerritelli, ME, Cheng, N, Rosenberg, AH, et al. (1997) Encapsidated conformation of bacteriophage T7 DNA, *Cell* **91**, 271.

Cloutier, TE, & Widom, J (2005) DNA twisting flexibility and the formation of sharply looped protein–DNA complexes, *Proc. Natl Acad. Sci. USA* **102**, 3645.

Du, Q, Smith, C, Shiffeldrim, N, et al. (2005) Cyclization of short DNA fragments and bending fluctuations of the double helix, *Proc. Natl Acad. Sci. USA* **102**, 5397.

Fawcett, DW (1966) The Cell, Its Organelles and Inclusions: An Atlas of Fine Structure, W. B. Saunders.

Johnson, JE, & Chiu, W (2007) DNA packaging and delivery machines in tailed bacteriophages, *Curr. Opin. Struct. Biol.* **17**, 237.

Lowary, PT, & Widom, J (1998) New DNA sequence rules for high affinity binding to histone octamer and sequence-directed nucleosome positioning, *J. Mol. Biol.* **276**, 19.

Nelson, P (2004) Biological Physics: Energy, Information, Life, W. H. Freeman.

Polach, KJ, & Widom, J (1995) Mechanism of protein access to specific DNA sequences in chromatin: a dynamic equilibrium model for gene regulation, *J. Mol. Biol.* **254**, 130.

Purohit, PK, Inamdar, MM, Grayson, PD, et al. (2005) Forces during bacteriophage DNA packaging and ejection, *Biophys. J.* **88**, 851.

Ramón y Cajal, S (1995) Histology of the Nervous System (translated by Neely Swanson and Larry W. Swanson), Oxford University Press.

Shore, D, & Baldwin, RL (1983) Energetics of DNA twisting. I. Relation between twist and cyclization probability, *J. Mol. Biol.* **170**, 957.

Smith, DE, Tans, SJ, Smith, SB, et al. (2001) The bacteriophage ϕ29 portal motor can package DNA against a large internal force, *Nature* **413**, 748.

Vologodskaia, M, & Vologodskii, A (2002) Contribution of the intrinsic curvature to measured DNA persistence length, *J. Mol. Biol.* **317**, 205.

Zeller, RW, Griffith, JD, Moore, JG, et al. (1995) A multimerizing transcription factor of sea urchin embryos capable of looping DNA, *Proc. Natl Acad. Sci. USA* **92**, 2989.

Biological Membranes: Life in Two Dimensions

11

Overview: In which biological membranes are viewed as elastic media

> "I well remember how some of the orthodox chemists of that time (around 1920) ridiculed the pictures drawn by Langmuir and Harkins, of the molecules of long chain compounds standing on end in surfaces, with their water-soluble ends in the water and their hydrocarbon ends away from it ..."
>
> N. K. Adam

One of the hallmark features of cellular life is the presence of membranes that separate the cell from the rest of the world and its compartments from each other. These membranes are characterized by a host of interesting chemical and physical properties, including mechanical properties that demand an energy cost for bending membranes away from their equilibrium configurations. This chapter further develops our thinking on elasticity by extending the one-dimensional treatment of elasticity used in Chapter 10 to the two-dimensional setting of biological membranes. As a result, like many of our chapters, this chapter merges an extremely important area of cell biology, namely, the study of membranes and the proteins that occupy them, with a correspondingly important area of physics, namely, the elastic theory of membranes. As with any good merger, the whole is greater than the sum of the parts.

11.1 The Nature of Biological Membranes

11.1.1 Cells and Membranes

Cells and Their Organelles Are Bound by Complex Membranes

Membranes are a critical part of all forms of life on Earth because they separate the cellular contents from the external world, maintaining vast differences in chemical composition between a cell and its surroundings. At the same time, membranes must enable the passage of critical nutrients into the cell and the passage of waste products out. They must also be physically flexible to enable cells to grow or otherwise change shape and they must have malleable topologies, for example, such that a single cell can divide into two parts each of which has a completely closed, contiguous membrane. Membranes in biological systems have balanced these multiple demands by exploiting the special physical properties of the molecules that make them up.

The primary constituent of any biological membrane must be a molecule with two physically separated subdomains, one part of which is hydrophilic (able to form hydrogen bonds with water) and the other part of which is hydrophobic. The most common membrane constituent is a lipid molecule as illustrated in Figure 11.1(A). The nearly cylindrical shape of most membrane lipids makes the *bilayer* the most

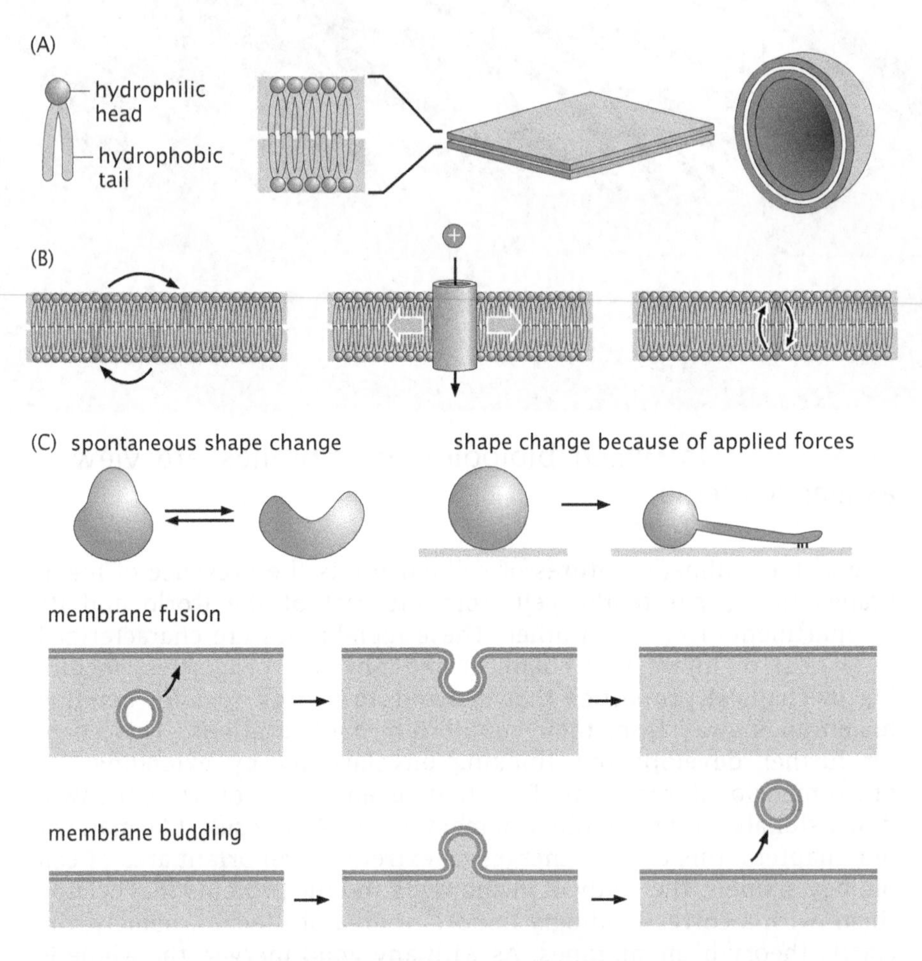

Figure 11.1: Structure and properties of biological membranes. (A) A major constituent of membranes is a lipid molecule, shown here diagrammatically as having a hydrophilic (water-loving) head group and a hydrophobic (water-fearing) tail. One spontaneous arrangement of these molecules consists of a bilayer, two molecules thick, where all the head groups are arranged on the two surfaces and the tails span the space in between. Membrane bilayers of this kind can be extremely large. In order to avoid edge effects, it is common for bilayers to form closed structures such as the sphere shown here. (B) Within the lipid bilayer, individual molecules can move in several ways. Lipid molecules may spontaneously diffuse laterally within each of the two leaflets within the bilayer. Protein molecules embedded in the bilayer may also diffuse laterally, and, in addition, some interesting membrane proteins may undergo conformational changes that open selective channels, for example, to allow the passage of charged ions across the hydrophobic barrier. At a very slow spontaneous rate, individual lipids may flip over from one leaflet to the other. This process of flipping may be sped up by certain membrane proteins called flippases. (C) Membrane shape changes associated with biological function. All membranes undergo spontaneous shape changes and fluctuations due to thermal energy. Application of external forces, for example by molecular motor proteins, can further deform equilibrium membrane shapes. A common geometry is the extrusion of a tube. Membranes may also undergo fusion and fission (or budding) to change their topology.

common geometrical organization for spontaneous self-assembly of lipid molecules in water. When lipid molecules are present in water at a sufficiently high concentration, they will spontaneously organize into structures with the hydrophobic tails sequestered such as in the bilayered plasma membrane that surrounds living cells.

From a mechanical perspective, the defining feature of these membranes is their extremely tiny aspect ratio, with their lateral dimensions typically orders of magnitude larger than their thickness of only 5 nm. As a result, the elastic models that we will use to examine membrane deformation are two-dimensional and emphasize the stretching and bending of the planar sheet. Despite the extreme thinness of these membranes, the hydrophobic region between the pairs of head groups nevertheless forms a formidable barrier against diffusion of any hydrophilic molecule. Passage of a hydrophilic molecule, such as an ion or an amino acid nutrient, requires the activity of particular proteins embedded in the bilayer that can serve as selective channels or pumps. The hydrophobic barrier also makes it an extremely slow process for a single lipid molecule to flip from one leaflet of the bilayer to the other since this would require its hydrophilic head group to pass through the hydrophobic region. In contrast, lateral movements of lipids within a single leaflet can be very fast, essentially diffusive. These processes are illustrated in Figure 11.1(B).

Zooming outward from this molecular focus, many of the interesting biologically related activities of membranes can be approximated by considering the membrane as a two-dimensional, featureless, elastic sheet. Closed forms made by these elastic sheets may change shape either spontaneously or due to the application of forces either from the inside or the outside. Even more interestingly, for many biological processes, it is necessary that separate membrane structures either fuse together to form a new whole membrane or separate through the process of fission. For cells, these processes are equivalent to mating and replication, the spice of life.

Figure 11.2 shows different manifestations of the basic membrane bilayer organization in different parts of both eukaryotic and prokaryotic cells. As we will see below, cell membranes are incredibly complex. Their molecular components include hundreds of different types of lipids, associated proteins that may span the entire membrane or only associate with one side, carbohydrate modifications (strings of sugars) on both protein and lipid components, and often association with other structures such as cell walls or ribosomes.

Electron Microscopy Provides a Window on Cellular Membrane Structures

One of the ways that we know about cell membranes is through electron microscopy. The presence of membranes is revealed through a sharp contrast between the membrane itself and the surrounding bulk of the cell. Several electron microscopy images of various membrane examples are shown in Figure 11.3. For example, Figure 11.3(A) shows one image from a cryo-electron microscopy tilt series of the bacterium *C. crescentus*, already introduced in Figure 3.23 (p. 119). The inner and outer membranes of the bacterium are clearly visible in the image. A corresponding image of the cellular boundary for a eukaryotic cell is depicted in Figure 11.3(B), which shows the complex and dense arrangement of microvilli on the surface of an epithelial cell. Many of the most amazing membrane structures found in cells

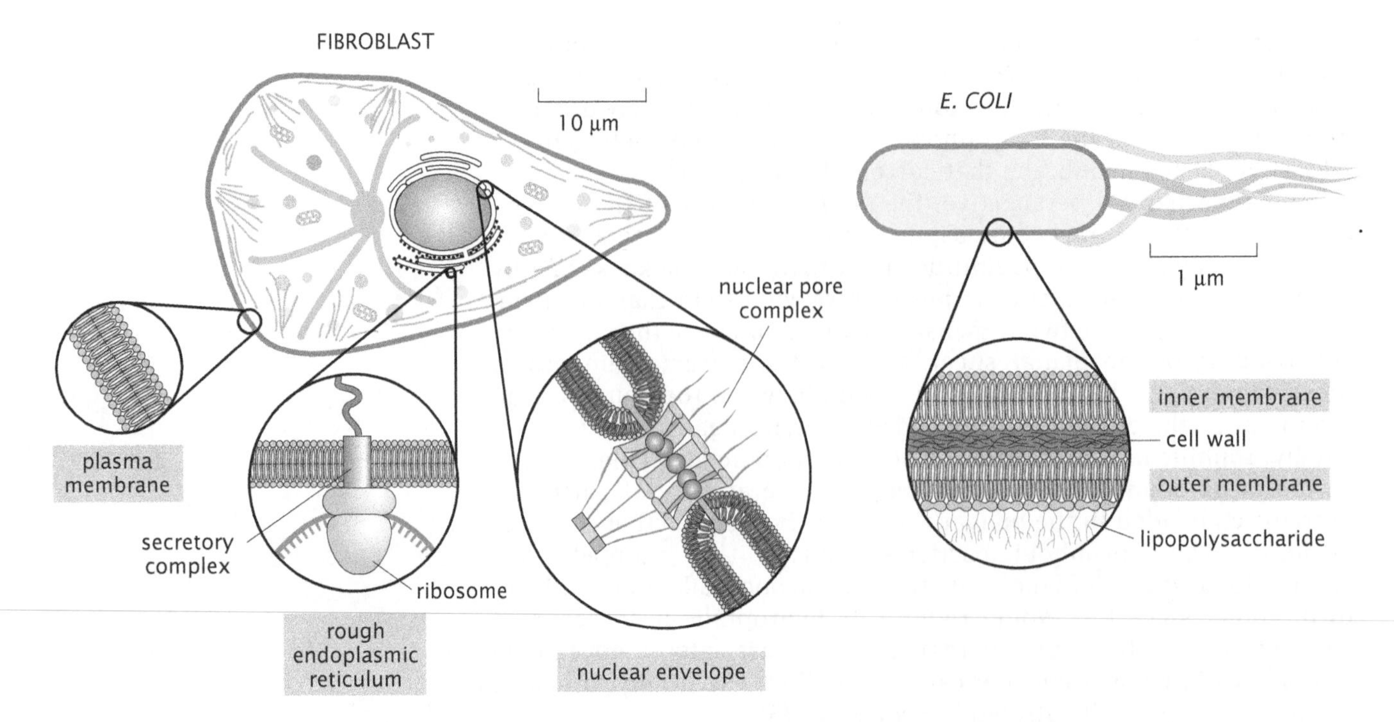

Figure 11.2: Key examples of membranes in biological systems. Eukaryotic cells, such as this fibroblast, are rife with many specialized membranes. The plasma membrane is a single phospholipid bilayer riddled with membrane proteins. The rough endoplasmic reticulum, also a single bilayer, is the site of synthesis of membrane-bound and secreted proteins. The ribosomes synthesizing these proteins are intimately associated with a transport apparatus in the endoplasmic reticulum membrane. The nuclear envelope consists of two phospholipid bilayers with a thin space between them. This nuclear envelope is perforated by nuclear pores that permit transport of materials from the cytoplasm to the nucleus and back. Bacterial cells rarely have internal membranous organelles, but may have very complex external membranes. For *E. coli*, the cell envelope consists of two bilayers—an inner membrane and an outer membrane—separated by a rigid cell wall. The outer leaflet of the outer membrane is largely composed of an unusual molecule called lipopolysaccharide, rather than of phospholipids.

are associated with their organelles as illustrated in Figures 11.3(C) and (D), which show the layered membrane structure in a rod cell and in a mitochondrion with surrounding endoplasmic reticulum, respectively.

The starting point for thinking about membrane organization is that its shape is dictated by the physical properties of the two layers of phospholipids that make up the lipid bilayer. This lipid bilayer is two lipid molecules thick, riddled with a dazzling array of membrane proteins. Figure 11.4 shows several generations of models for cell membrane structure. The fluid mosaic model of Singer and Nicolson (1972) envisioned a lipid bilayer as a two-dimensional fluid in which embedded membrane proteins were able to easily move laterally in the plane of the membrane, but could not move out of the plane. Later versions of this model acknowledged the fact that there is a great deal of structural heterogeneity within the lipid bilayer. For example, membranes containing multiple types of lipids that tend to mix nonideally can have a complex organization in which structurally compatible lipids assemble into microdomains. Along similar lines, membrane proteins can generate local order in the lipids that surround them and lipid domains can strongly influence protein organization. In living cells, the membrane does not exist in isolated two-dimensional splendor—long branched chains of carbohydrates protrude into the third dimension and structural elements within the cell such as the cytoskeleton interact extensively with membrane components to shape the membrane surface.

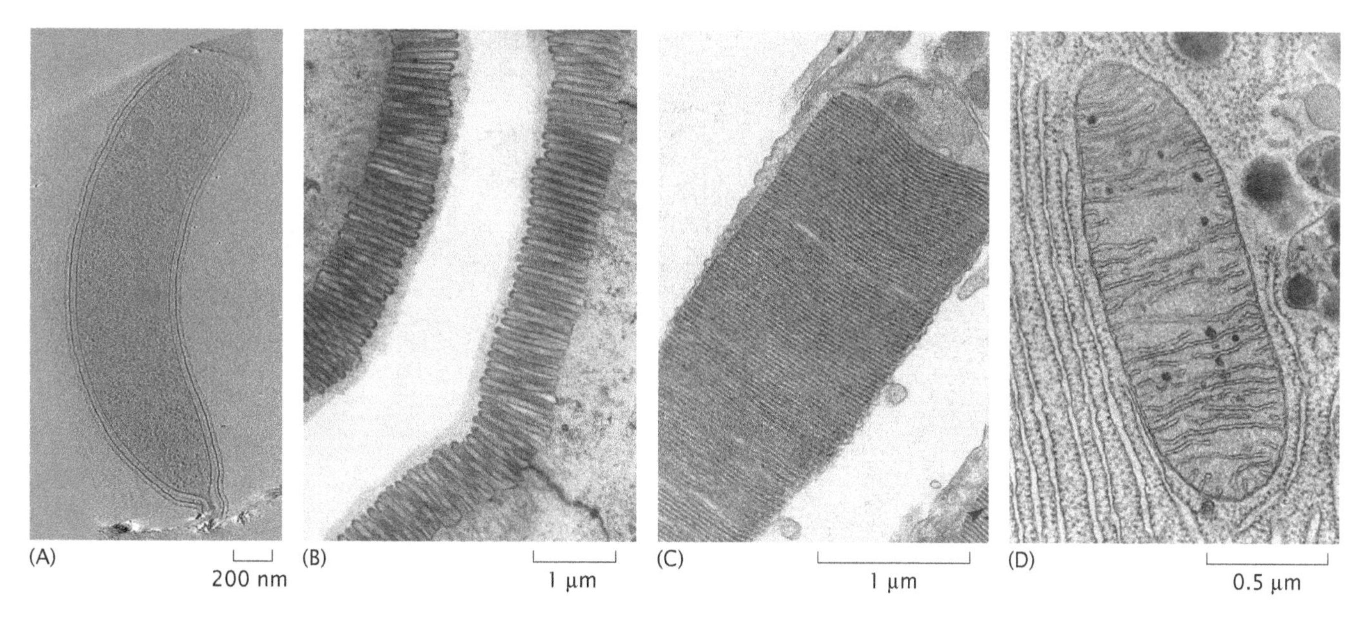

Figure 11.3: Electron microscopy images of a variety of membranes. (A) Cryo-electron microscopy image of *C. crescentus*. The several layers of membrane and cell wall can be easily seen. (B) Apical surface of intestinal epithelial cells showing the dense membrane folds around the microvilli. The sugar chains extending outwards from the surface of the membrane can also be seen as a fuzzy layer above the microvilli. (C) Stacks of membranes packed with photoreceptors in the outer segment of a rod cell. (D) Thin section of a mitochondrion surrounded by rough endoplasmic reticulum from the pancreas of a bat. (A, courtesy of G. Jensen; B, D, adapted from D. W. Fawcett, The Cell, Its Organelles and Inclusions: An Atlas of Fine Structure. W. B. Saunders, 1966; C, adapted from P. C. Cross and K. L. Mercer, Cell and Tissue Ultrastructure. W. H. Freeman, 1993.)

11.1.2 The Chemistry and Shape of Lipids

Membranes Are Built from a Variety of Molecules That Have an Ambivalent Relationship with Water

Lipids are often referred to as one of the four basic building blocks of the living world (along with proteins, nucleic acids, and sugars). Figure 11.5 shows the chemical and geometrical makeup of five distinct classes of macromolecule associated with cell membranes in eukaryotes, bacteria, and archaea. All of these disparate molecular structures share an elongated hydrophobic domain, usually made up of fatty acid tails, associated with a hydrophilic headgroup. The hydrophilic headgroups are easily incorporated in the hydrogen-bonding network of the surrounding water. In contrast, the hydrophobic tails are generally obliged to be sequestered from contact with water, and typically form bilayers. The conflicting relationships between different regions of lipid molecules and water have resulted in the designation of such molecules as *amphipathic*, which signifies their love–hate relationship with water. There is a great diversity of molecular structures for both the hydrophilic and hydrophobic domains, but this common overall organization allows different lipids to happily assemble together to form mixed membranes. For the ball-and-stick models of representative lipids shown in Figure 11.5, the brightly colored atoms represent polar, water-loving groups, and the large gray domains represent the hydrophobic portions.

The most common lipid type found in most cell membranes is the glycerol-based phospholipid, one of which is shown in Figure 11.5(A). Glycerol is a small molecule with a chain of three carbons and three associated hydroxyl groups. In order to make a phospholipid, the cell attaches two long hydrophobic tails to two of these three hydroxyl

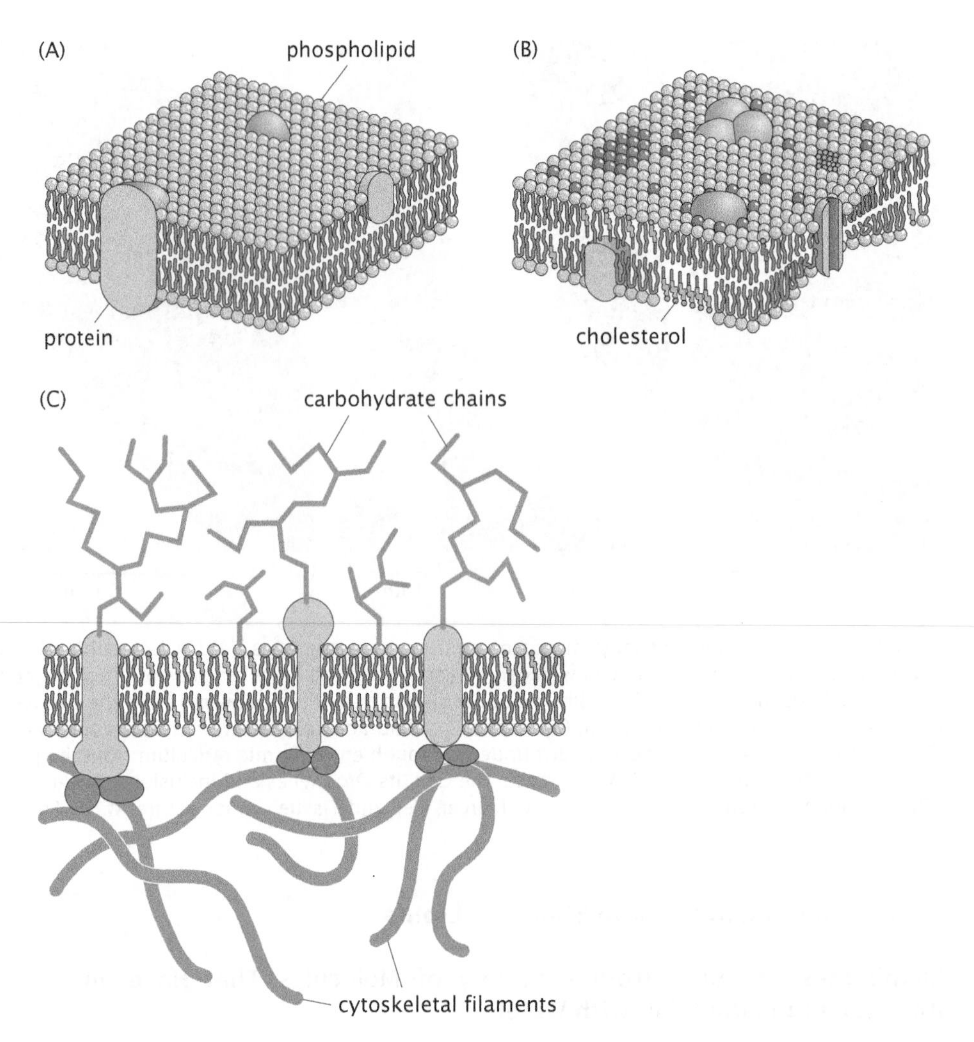

Figure 11.4: Schematic models of our understanding of cell membranes. (A) In the early fluid mosaic hypothesis, the membrane was envisioned as a two-dimensional fluid of phospholipids studded with occasional proteins. (B) Further refinements of this model have acknowledged multiple kinds of heterogeneity. Phospholipids with different chemical character and cholesterol can self-associate to form subdomains. Proteins also may tend to self-associate within the plane of the membrane and, furthermore, can distort the membrane by locally altering its thickness or composition. (C) Beyond the plane of the phospholipid bilayer, complex carbohydrate chains attached to both lipids and proteins extend outward and many proteins and lipids are attached to structural elements within the cell, further restricting their movement. (Adapted from O. G. Mouritsen, Life—As a Matter of Fat. Springer-Verlag, 2005.)

groups, and a hydrophilic head group to the third. In eukaryotes and bacteria, the hydrophobic tails are usually fatty acids, which are long linear hydrocarbon chains with a single carboxylic acid group at one end. The carboxylic acid can form an ester linkage to the glycerol hydroxyl group, a type of chemical bond that is easily formed and cleaved. The hydrocarbon portion is a repeating series of units of the form $(CH_2)_n$, with n typically ranging from 10 to 18. In addition, the shapes of these tails are determined by the number of carbon–carbon double bonds, which induce kinks in the tail. Many archaea use slightly different hydrophobic tails, which are based on branched aliphatic molecules called isoprenoids, connected to the glycerol backbone by ether linkages instead of ester linkages. The hydrophilic group starts with a phosphate group attached directly to the third glycerol hydroxyl. If the head group consists of the phosphate alone, then the lipid is called "phosphatidic acid"; the chemical structure is drawn in Figure 11.6(A) (on the left). Cells may further elaborate the head by attaching additional molecules to the phosphate such as ethanolamine, choline, or serine. Some of the resulting structures are also shown in Figure 11.6(A). Another option is to form a dimer of two phosphatidic acid lipids, resulting in a large amphipathic molecule with two phosphates, two glycerol backbones, and four fatty acid tails. One such molecule, cardiolipin, is diagrammed alongside the other glycerol-based phospholipids in Figure 11.6(A).

Many eukaryotic cells (and a few bacteria) augment their repertoires of glycerol-based phospholipids with sphingolipids. At first glance,

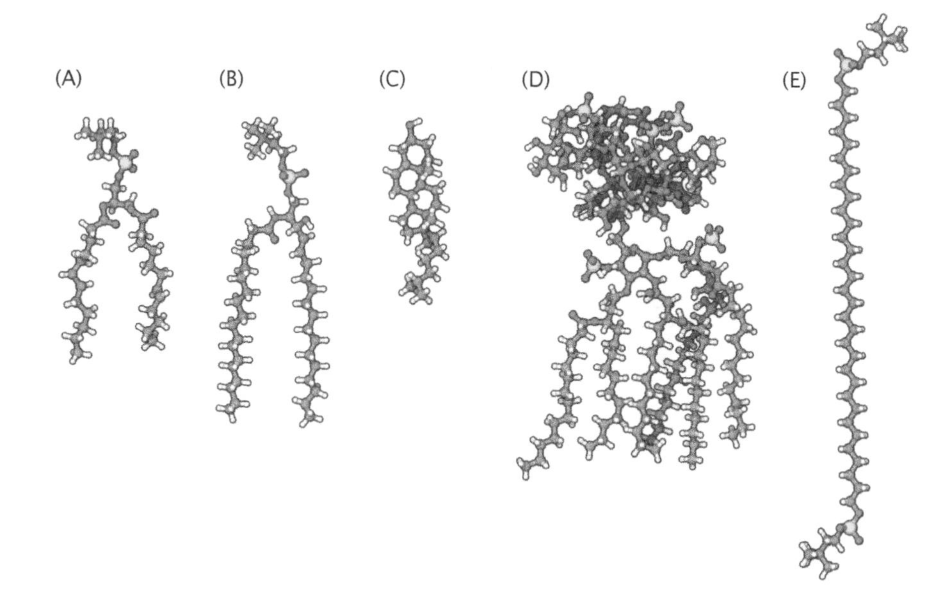

Figure 11.5: Examples of key molecules of the cell membrane. The atomic structures of (A) a phospholipid, (B) a sphingolipid, (C) cholesterol, (D) lipopolysaccharide (PDB 1z2r), and (E) a bolalipid (from archaea). (Courtesy of D. Goodsell.)

the sphingolipid shown in Figure 11.5(B) looks very similar to the glycerol-based phospholipid, and yet it contains no glycerol. Instead, this molecule is built starting from a molecule called sphingosine, which already has one long hydrocarbon chain and the cell need only attach one other chain plus the headgroup. The chemical formula for a particular sphingolipid, sphingomyelin, is shown in Figure 11.6(A), second from the right. Because the overall size and shape of sphingolipids are very similar to those of phospholipids, they can easily coexist in the same bilayers, and their lateral mobilities within the fluid mosaic bilayer are comparable.

Cholesterol is dramatically different in size and structure from the phospholipids and sphingolipids (see Figure 11.5C). The hydrophobic portion consists of a rigid planar array of four rings with an attached hydrocarbon tail, and the headgroup is just a tiny hydroxyl. Cholesterol and related sterol lipids are found in many eukaryotic cells, but not in bacteria or archaea. Because of the large planar ring structure embedded within the hydrophobic portion of the molecule, the addition of cholesterol to a phospholipid bilayer will disrupt the packing order of the other molecules. Because of this, cholesterol tends to increase the fluidity of phospholipid bilayers at low temperatures.

The relative amounts of cholesterol versus phospholipids can vary dramatically among different eukaryotic species, among different cell types within a multicellular organism, and even among different organellar lipids within a single cell type. As shown in Figure 11.6(B), the plasma membrane of a rat hepatocyte (liver cell) is rich in cholesterol and sphingomyelin but has only relatively modest amounts of the glycerol-based phospholipids phosphatidylcholine and phosphatidylethanolamine, while in the nuclear membrane the glycerol-based phospholipids dominate.

One special type of lipid that is found only in (some) bacterial cells is the lipopolysaccharide (LPS) shown in Figure 11.5(D). This monstrously large molecule is found only in the outer leaflet of the outer membrane on the surface of so-called Gram-negative bacterial cells such as *E. coli*, which have the cell wall sandwiched between their inner and outer membranes. LPS molecules are built starting with

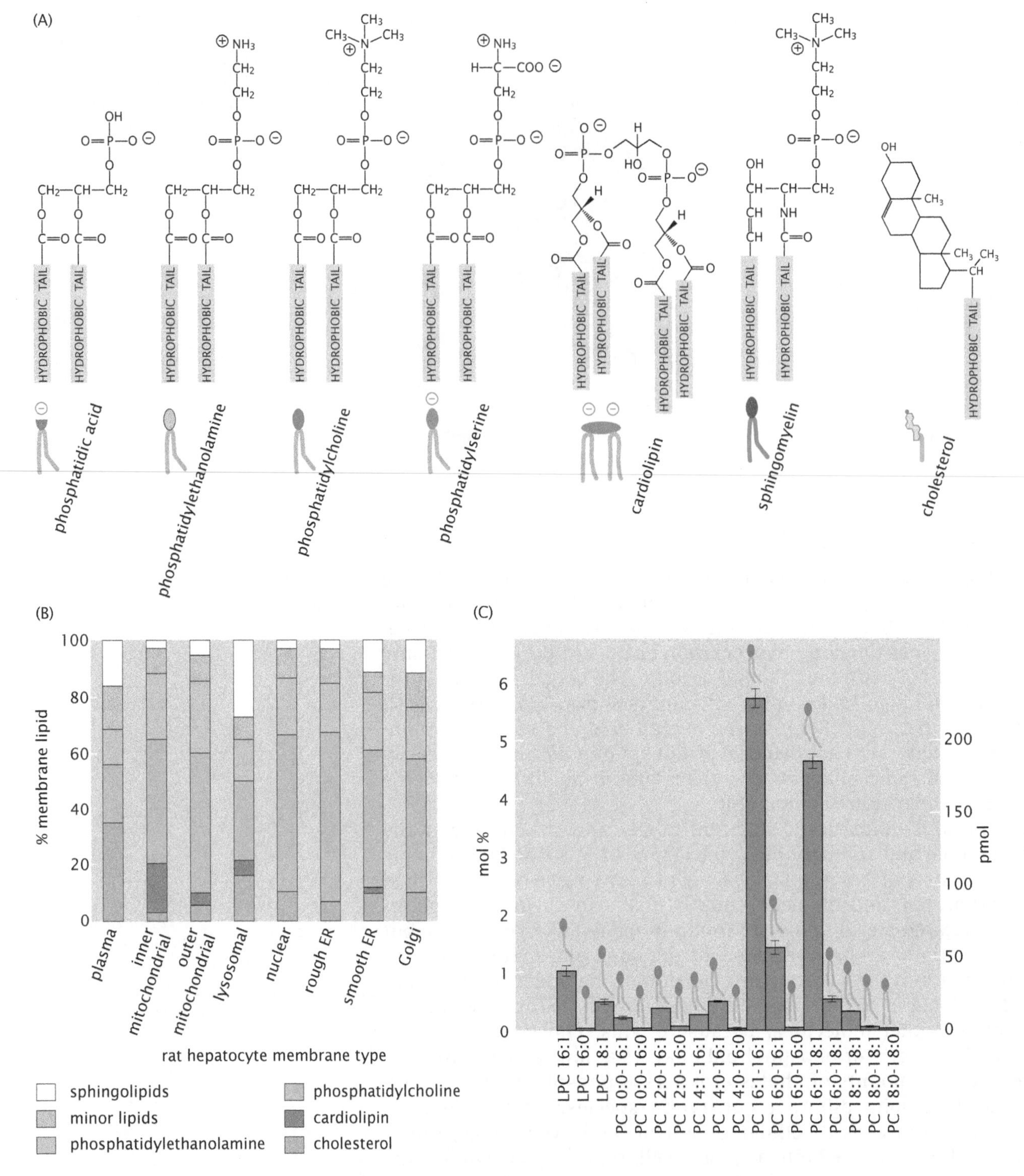

Figure 11.6: Lipid diversity in eukaryotic cells. (A) Chemical structures of various lipids commonly found in animal cells. The phospholipids phosphatidic acid and phosphatidylserine are both negatively charged. Phosphatidylethanolamine and phosphatidylcholine are zwitterionic molecules that stably maintain one negative and one positive charge, and are therefore net neutral. Cardiolipin resembles two phospholipids joined at the head, and carries two negative charges. Sphingomyelin and cholesterol are non-glycerol-based amphipathic molecules that are both found commonly in eukaryotic cells but rarely in bacterial or archaeal cells. The cartoons show the relative shapes of the different molecules. (B) Relative concentrations of different lipid types in various membrane compartments of a rat hepatocyte (liver cell). (ER = endoplasmic reticulum). (C) Relative amounts of 18 different molecular variants of phosphatidylcholine, as measured for normal yeast cells by mass spectrometry. (LPC = lysophosphatidylcholine, or one-tailed phosphatidylcholine). The molecules are distinguished by the number of carbons in their tails and the presence or absence of double-bonds; for example, 16:1 represents a 16-carbon-long tail with one double bond. The cartoons represent the relative shapes of the distinct species. The presence of a double bond causes a kink in the hydrocarbon tail. (B, adapted from D. L. Nelson and M. M. Cox, Lehninger Principles of Biochemistry, 3rd ed., Worth Publishers, 2000; C, adapted from C. S. Ejsing et al., *Proc. Natl Acad. Sci. USA* 106:2136, 2009.)

a pair of linked sugars to which a total of six hydrocarbon chains are attached. The two sugars are also linked typically to two phosphate groups rendering the molecule negatively charged. The charged headgroup is usually further elaborated by linkage to an enormous complex carbohydrate chain, which can be elaborately branched. The exact composition and organization of this complex carbohydrate chain varies among different bacterial species and even among strains of the same species. Because LPS is made only by bacterial cells (not by human cells) and can serve as a specific identification card for different kinds of bacteria, LPS is the major antigen used by the human immune system to recognize and target pathogenic bacterial invaders.

The third major kingdom of life, the Archaea, features a particularly intriguing kind of lipid, sometimes called a bolalipid, shown in Figure 11.5(E). These special lipids are most common in archaea that live in particularly harsh environments, including those with high salt or high temperature. Here the hydrocarbon chains are about twice as long as those found in phospholipids and are capped by two headgroups, one at either end. Rather than forming bilayers, these bolalipids actually form monolayers that are less susceptible than bilayers to disruption by chemical and thermal insults.

The above discussion makes it clear that there are many different kinds of lipids, but fails to convey the dramatic dimensions of this diversity. For each phospholipid type with a given head group, phosphatidylcholine for example, further diversity can exist within a cell based on the combinatorial variety of different possible tails. Recent experiments using mass spectrometry to enumerate and measure the concentrations of all the lipid species in cells (an approach sometimes called *lipidomics*) has begun to reveal the real complexity within the cell. Figure 11.6(C) shows the various flavors of phosphatidylcholine in normally growing yeast cells whose masses can be distinguished. These include 18 different tail molecules, where the phosphatidylcholine may have one or two tails, either tail may vary in length from 10 to 18 carbons, and each of the one or two tails may or may not have a double bond. The relative abundance of these slightly different molecular species varies over a hundred-fold range, with the most abundant species (phosphatidylcholine with two 16-carbon chains where each chain has a single double bond) represents about 5% of the total lipid. Indeed, this mass-based measurement still underestimates the actual structural diversity, as two 16-carbon tails with a single double bond may have the double bond at different locations (for example, between the 4th and 5th carbons from the end, or between the 8th and 9th carbons from the end); although these molecules may have slightly different behaviors in the context of a lipid bilayer, their masses will be identical. Even so, the yeast study identifying these 18 versions of phosphatidylcholine found a total of over 250 measurably different lipid molecules. Other lipid species may be present at levels too low to detect by this method. The identities and relative amounts of all the lipids may change when cells are grown under different conditions, or respond to signals in their environment. A comprehensive and exhaustive nomenclature has been developed for distinguishing all the possible varieties of lipid molecule structure in a systematic way; the interested reader is invited to consult Fahy et al. (2005). In general, our appreciation of the significance of lipid diversity in cells lags far behind our understanding of the significance of protein diversity, but it seems likely that cells find both kinds of diversity to be important and useful.

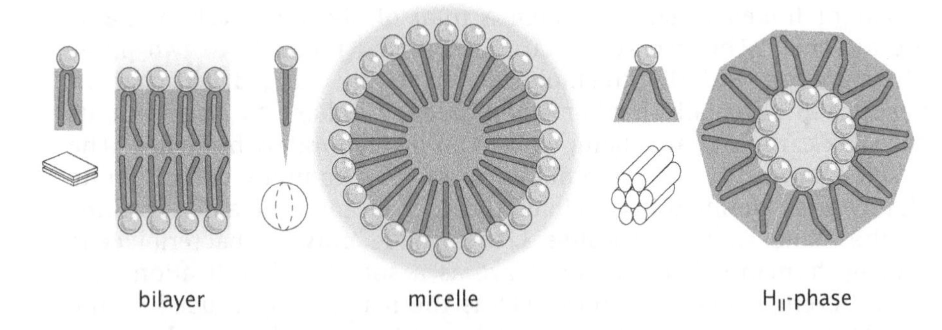

Figure 11.7: Geometrical shape of lipids. This figure shows a coarse-grained representation of lipid geometries that is useful in developing intuition for the spontaneous curvature induced by different lipid types. The small insets show the kinds of three-dimensional geometries adopted by these lipids.

The Shapes of Lipid Molecules Can Induce Spontaneous Curvature on Membranes

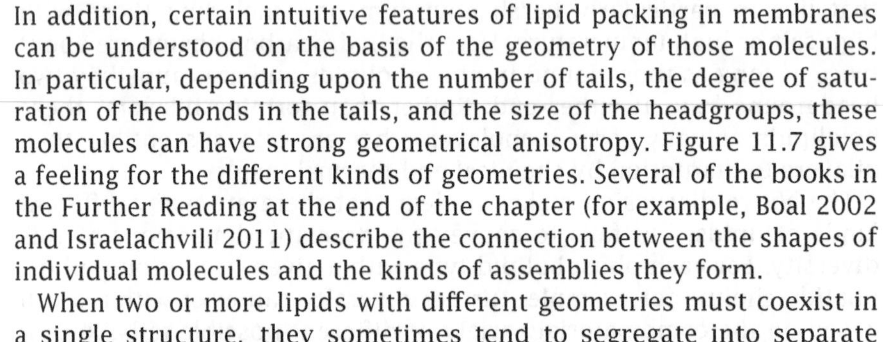

Thus far, we have largely focused on the chemical properties of lipids. In addition, certain intuitive features of lipid packing in membranes can be understood on the basis of the geometry of those molecules. In particular, depending upon the number of tails, the degree of saturation of the bonds in the tails, and the size of the headgroups, these molecules can have strong geometrical anisotropy. Figure 11.7 gives a feeling for the different kinds of geometries. Several of the books in the Further Reading at the end of the chapter (for example, Boal 2002 and Israelachvili 2011) describe the connection between the shapes of individual molecules and the kinds of assemblies they form.

When two or more lipids with different geometries must coexist in a single structure, they sometimes tend to segregate into separate domains. Under some conditions, this spontaneous segregation can induce the generation of complex vesicle geometries where different domains have different curvatures as shown in Figure 11.8. Inside cells, this kind of lipid heterogeneity together with the influence of membrane-associated proteins as discussed below can contribute to formation of elaborate intracellular membrane systems.

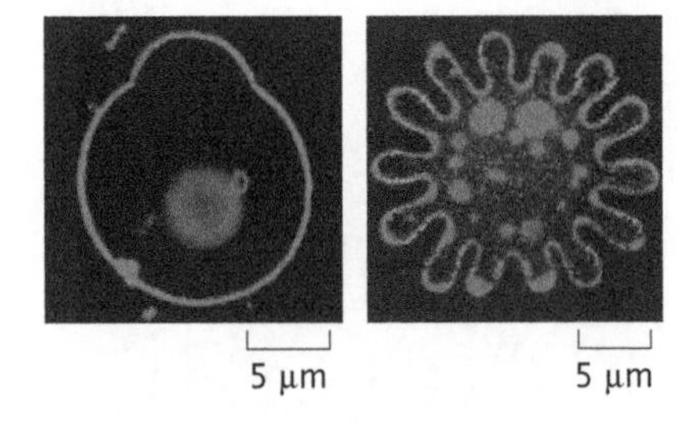

Figure 11.8: Structures of multicomponent vesicles. Vesicles made of more than one species of lipid can give rise to structures with complex geometries. Different lipid species are labeled with different fluorescent dyes, shown here in red and blue. The lipids with distinct physical properties tend to spontaneously segregate into domains. On the left, a vesicle at low temperature (25 °C) exhibits two large domains. The line tension caused by the mismatch at the boundary between the two domains (discussed later in the chapter) causes a deformation of the vesicle, such that one domain (blue) adopts a higher curvature than the other (red). On the right, a similar vesicle held at a higher temperature (50 °C) adopts a much more complicated shape. The individual domains are smaller, and the overall shape again separates regions with high curvature (red) from regions with low curvature (blue). (Adapted from T. Baumgart et al., *Nature* 425:821, 2003).

11.1.3 The Liveliness of Membranes

The major biological function of membranes is to separate cells and organelles from their surroundings, but at the same time, cells and organelles must communicate and exchange material with the external world. The critical functional modifications to biological membranes that enable this exchange are generally mediated by proteins, which make up a large fraction of the mass of the membrane (see the estimate below to get a feeling for the numbers). From a general perspective, the structures of many of these proteins can be thought of as having three distinct parts: an intracellular domain, a transmembrane domain, and an extracellular domain. Figure 11.9 shows a gallery of some representative examples of such proteins whose functions are detailed in the following paragraphs.

Estimate: Sizing Up Membrane Heterogeneity In Chapter 2, we estimated our way to a census of a bacterium like *E. coli* that included an estimate of the protein complement of the membranes. One way to state the importance of membrane proteins is through the estimate that roughly one-third of the genes in a typical genome encode membrane proteins. In the case of *E. coli*, this led us to the estimate that there are 10^6 such proteins per cell. Since a bacterium like *E. coli* has two membrane systems, we can naively imagine that there

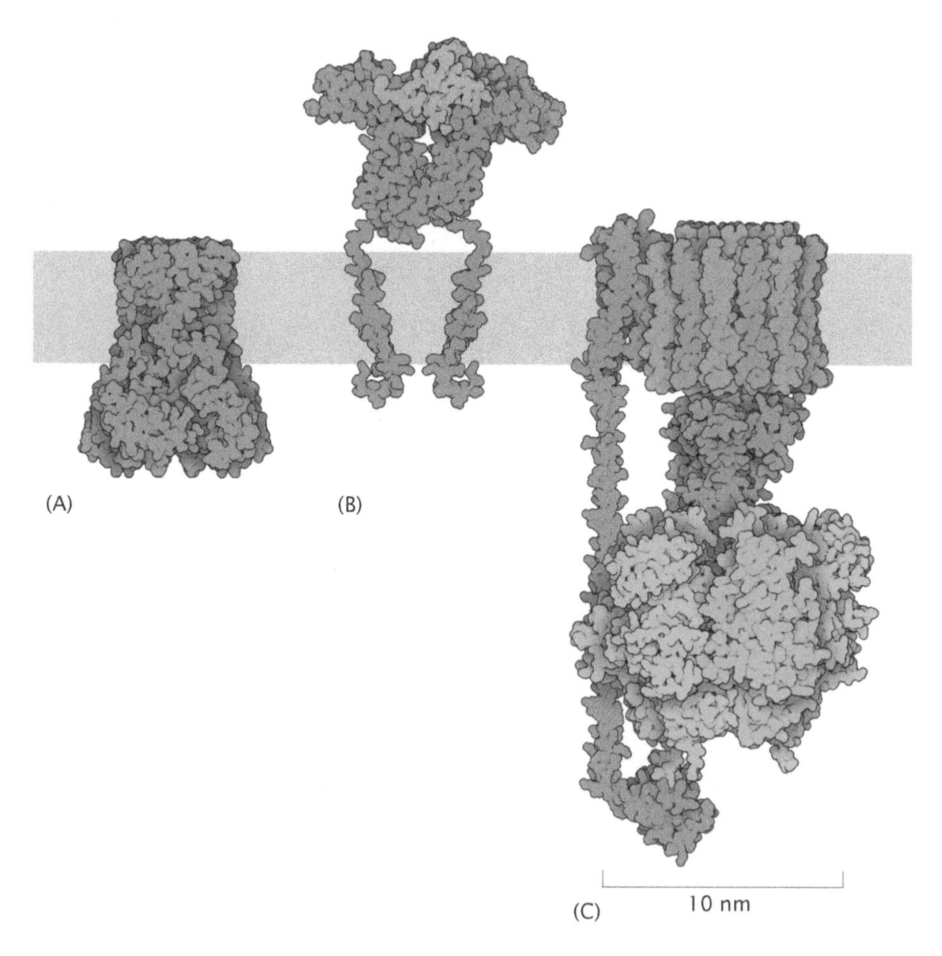

Figure 11.9: Examples of various membrane proteins that are responsible for the flow of information and material across membranes. (A) Potassium channel (PDB 1p7b), (B) growth hormone receptor (PDB 3hhr, 2ks1), and (C) ATP synthase (PDB 1c17, 1e79, 1l2p, 2a7u). (Courtesy of D. Goodsell.)

are roughly 500,000 membrane proteins in each membrane. Given that the area of these membranes is roughly $6\,\mu m^2$, this implies an area per protein of roughly $12\,nm^2$. If we assume that these membrane proteins are closely packed in a square array, this, in turn, implies a mean spacing between proteins of roughly 3.5 nm.

In the case of mitochondrial membranes, it has been estimated that roughly 70% of the mass of these membrane systems comes from their protein complement. This idea is presented in a striking visual form in Figure 11.10, which shows electron microscopy images of mitochondrial membranes revealing their membrane proteins. A useful compendium of protein to phospholipid mass ratios can be found in Mitra et al. (2004).

Membrane Proteins Shuttle Mass Across Membranes

In one generation, a typical cell must double in size in preparation for division into two daughters. Obviously, it is necessary for a cell to import from its surroundings the organic and inorganic molecules that it uses to maintain itself and to build a new daughter cell. These include sugars, amino acids, ions, and even water. Indeed, these are the raw materials that we described in our estimates of the rates of different processes in Chapter 3. The lipid bilayer has different intrinsic permeabilities to these different components, as illustrated

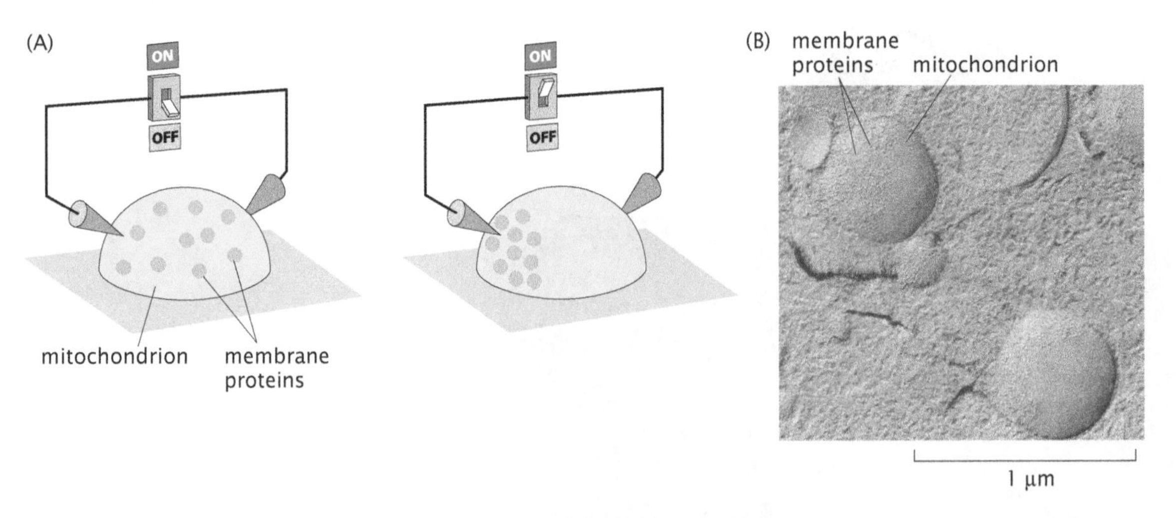

Figure 11.10: An experiment demonstrating the mobility of proteins in the mitochondrial membrane. (A) Isolated mitochondrial membranes were exposed to an electric field. When the field is off, the proteins are uniformly distributed on the mitochondrial surface. When the field is turned on, the membrane proteins migrate preferentially to one pole. If the field is turned off again, they return to their uniformly distributed state in several seconds. (B) Freeze fracture electron microscopy shows the membrane proteins as small bumps collected on the left half of the two membrane regions shown here. (B, Adapted from A. E. Sowers and C. R. Hackenbrock, *Proc. Natl Acad. Sci. USA* 78:6246, 1981.)

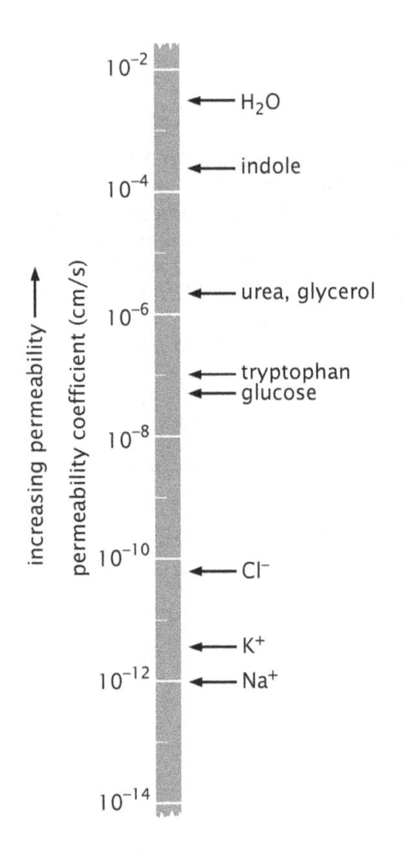

Figure 11.11: Range of measured membrane permeabilities for different molecular species. Membranes are most permeable to water and least permeable to ions. (Adapted from R. N. Robertson, The Lively Membranes. Cambridge. University Press, 1983.)

in Figure 11.11. The permeability is defined through

$$\text{flux} = P(c_{\text{in}} - c_{\text{out}}), \tag{11.1}$$

where the flux is the number of molecules crossing a unit area per unit time. The permeability coefficient P has units of length/time.

For example, water has a fairly high permeability coefficient and, as a result, water concentrations or pressures can equalize across a lipid bilayer spontaneously. However, the permeability of glucose is four orders of magnitude lower, and the permeability of potassium ions is four orders of magnitude lower still. Therefore, one of the most biologically important classes of membrane proteins are the transporters that facilitate the movement of these molecules across membranes. Some of these transporter proteins act as channels that selectively increase the permeability for a molecule to move down its concentration gradient. In many cases, these channels are gated to open only when a certain signal is received in the form of a ligand binding or a change in transmembrane voltage, though other kinds of driving forces can gate channels as well. An example of this is the potassium channel shown in Figure 11.9(A). An even more interesting type of transporter uses cellular energy to actively transport a small molecule up its concentration gradient. These active transporters (or pumps) are used by cells to accumulate nutrients such as sugars and amino acids from the environment, and play a critical role in establishing large concentration gradients of ions such as K^+, Na^+, and H^+. A particularly important class of these proteins are those that pump ions across membranes using the energy resulting from the absorption of a photon of light. This is the initial event in photosynthesis, and is ultimately the energy source for almost the entire web of life on the surface of planet Earth. Cells can use ion concentration gradients established by transporters and pumps in many ways, for example, to drive yet more transport processes, to generate ATP, and in cells such as neurons to generate electrical signals.

Membrane Proteins Communicate Information Across Membranes

Frequently the cell needs to react to changes in its environment without necessarily importing material. An example of this kind of event centers on the cellular decision made by a neuron as it extends the processes that become its axon and dendrites. We saw the cytoskeletal arrangements giving rise to these morphological changes in Figure 10.27 (p. 414). Certain signaling molecules known as growth factors are secreted by target cells such as muscle cells that invite the attentions of a neuron. These growth factors bind to receptors (see Figure 11.9B) on the neuron and initiate a cascade of events within the cell that leads to assembly of cytoskeletal structures that push out new appendages (axons and dendrites). For the signal to be communicated, it is not necessary that the growth factor itself enter the neuron. Instead, its binding to the receptor causes a conformational change in the transmembrane receptor protein that can result in the activity of an enzyme associated with the domain of the receptor that resides on the inside of the cell. Sometimes the relevant enzyme is actually a part of the same receptor protein that crosses the membrane and binds to the growth factor, and sometimes the enzyme is a separate protein that can detect the conformational change in the receptor that occurs after ligand binding.

Specialized Membrane Proteins Generate ATP

As already described in a number of places throughout the text, ATP is the central energy currency of the cell. As a result of the staggering amount of ATP turned over (see the estimate on p. 575) and hence new ATP molecules synthesized from ADP and inorganic phosphate P_i, it is of great interest to learn more about the machine responsible for ATP synthesis. Indeed, your authors have heard it said that ATP synthase is the "second most important molecule in biology" (see the structure in Figure 11.9C). This fabulous molecular machine converts a transmembrane H^+ gradient into ATP by coupling the transport of H^+ ions down their concentration gradient to a forced rotation of a protein turbine driving conformational changes in three separate enzyme active sites that grab an ADP and an inorganic phosphate P_i and smash them together, resulting in a molecule of ATP. The machine can also run backwards, consuming ATP to generate a transmembrane H^+ gradient. We will revisit the mechanism of this machine in Chapter 16. A functionally similar membrane-bound molecular machine is the flagellar motor apparatus. This machine converts the chemical energy stored in an H^+ gradient into the mechanical energy needed for cellular movement.

Membrane Proteins Can Be Reconstituted in Vesicles

A very powerful technique for the analysis of membrane proteins is to reconstitute them in artificial phospholipid vesicles where they are isolated from the cellular environment. One example we find particularly elegant involves the simultaneous reconstitution of bacteriorhodopsin (a membrane protein that pumps ions in response to light) and ATP synthase, shown in Figure 11.12. What makes this example so provocative is the fact that when reconstituted simultaneously into the same vesicles, these proteins have a symbiotic relationship in the sense that the light-driven H^+ pumping exercised by

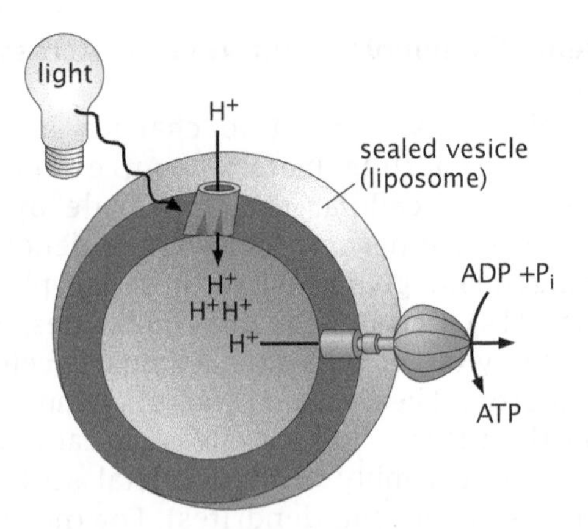

Figure 11.12: Reconstituted vesicle with both bacteriorhodopsin and ATP synthase present. (Adapted from B. Alberts et al., Molecular Biology of the Cell, 5th ed., Garland Science, 2008.)

bacteriorhodopsin serves to create precisely the H^+ gradient needed to drive the ATP synthase. This tiny manmade chemical plant converts light energy into ATP chemical energy like a minimalist green plant.

11.2 On the Springiness of Membranes

Whether we are interested in the shapes of organelles or the membrane deformation around a membrane protein, one of the first quantitative questions that we must answer is how the energy depends upon membrane shape. Our approach will be to develop the simplest elasticity theory that captures the energy of the various kinds of membrane deformation. As we saw in Chapter 10, materials are often conveniently described without reference to the underlying discreteness of matter by using continuum elasticity theories, and our treatment of membranes will be developed in that spirit. In Chapter 10, we congratulated ourselves on the ability to take advantage of the reduced dimensionality of elastic beams (which feature one dimension that is much larger than the other two). This leads to an elastic theory that is much more tractable than its full three-dimensional counterpart. Similarly, for membranes, we will use the fact that the thickness is much smaller than the lateral extent of membranes in order to construct a reduced dimensionality surrogate for the full three-dimensional elasticity.

11.2.1 An Interlude on Membrane Geometry

In order to assign energies to different membrane configurations, we need some mathematical tools to distinguish one membrane configuration from another. In Figure 10.2 (p. 385), we introduced three distinct classes of beam deformation of interest for characterizing the *geometry* of deformation for beams: stretch, bend, and twist. Here we set up similar mathematical and physical tools that will permit us to consider the geometry of deformation of membranes. Our first task is to describe the geometry of the membrane surface itself.

Figure 11.13 illustrates four classes of membrane deformation that will be particularly useful in our efforts to model biological membranes and the proteins that act within them. For example, our analysis of micropipette aspiration experiments will demand that we consider membrane stretching. When thinking about the way in which molecular motors might create tubules in the endoplasmic reticulum, we will be interested in the bending imposed on the membrane. Our discussion of mechanosensitive ion channels will center on the way in

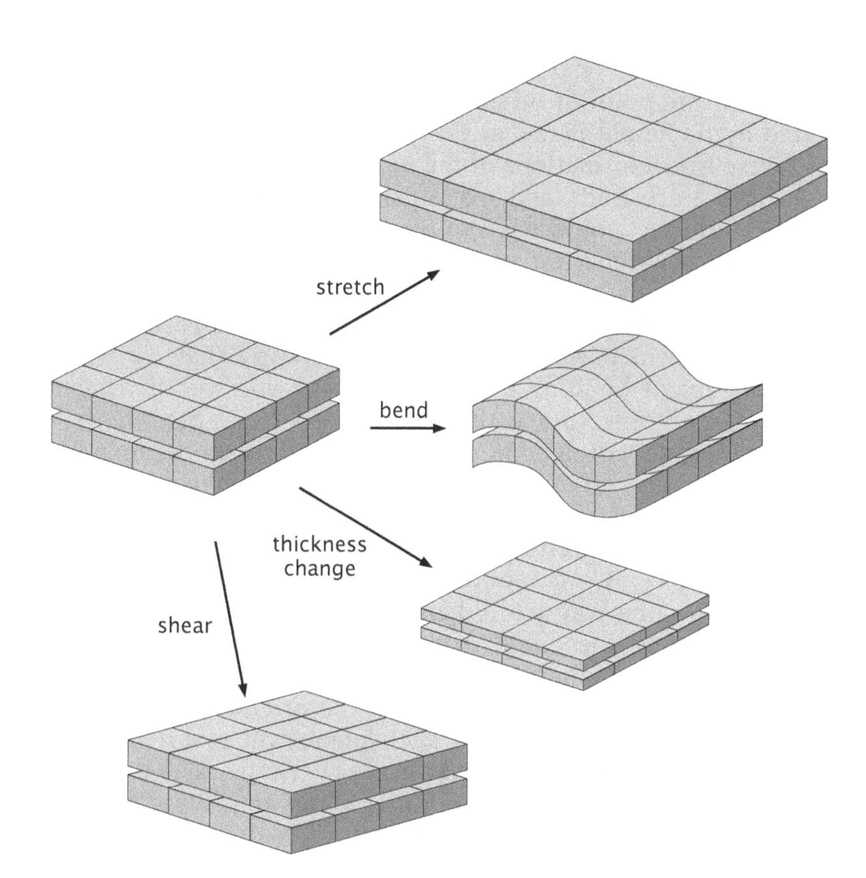

Figure 11.13: The geometry of membrane deformation. From top to bottom we illustrate stretching of a membrane, bending of a membrane, thickness deformation of a membrane, and shearing of a membrane.

which proteins can influence the thickness of the surrounding bilayer. Finally, to understand the various shapes of red blood cells, we will have to consider shear deformations of the cell membrane and its associated spectrin network. To get a sense geometrically for how such deformations work, we will repeatedly appeal to the square patch of membrane shown in Figure 11.13. There are many subtleties layered on top of the treatment here, but a full treatment of this rich topic would take us too far afield, and we content ourselves with the pictorial representations shown here.

Membrane Stretching Geometry Can Be Described by a Simple Area Function

The top image in Figure 11.13 illustrates the first class of deformations we will consider, namely, when the area of the patch of membrane is increased by an amount Δa. Just as the parameter ΔL was introduced in Section 5.4.1 (p. 216) to characterize the homogeneous stretching of a beam, the parameter Δa will provide a simple way to characterize the change in the area of a membrane. To be explicit about the fact that the amount of stretch could in principle vary at different points on the membrane, we introduce a function $\Delta a(x, y)$ that tells us how the area of the patch of membrane at position (x, y) is changed upon deformation.

Membrane Bending Geometry Can Be Described by a Simple Height Function, $h(x, y)$

To consider bending deformations, we treat surfaces as shown in Figure 11.14. We lay down an x–y grid on the reference plane and we

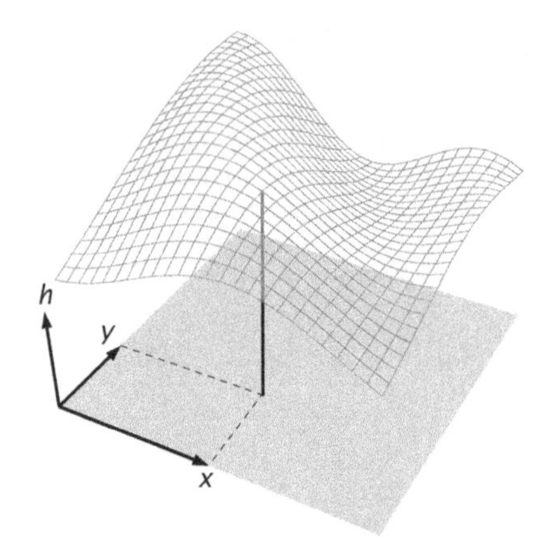

Figure 11.14: The height function $h(x, y)$. The surface of the membrane is characterized by a height at each point (x, y). This height function tells us how the membrane is disturbed locally from its preferred flat reference state.

use the variable h to characterize the height of the membrane above that plane at the point of interest. The geometry of the membrane is captured by its height $h(x, y)$ at every point in the plane. Note that in cases where the deformations of the membrane are sufficiently severe (that is, there are folds and overlaps), this simple description will not suffice and we would have to work using an intrinsic treatment of the geometry without reference to the planar reference coordinates described here.

Once we have the height function in hand, we can then compute the curvature, which we will see is the key way that we will capture the extent of bending deformations. As with our treatment of beams, we are going to see that the energetics of bending a lipid bilayer membrane will depend upon the curvature of the membrane. To explore the idea of membrane curvature, we take the approach shown in Figure 11.15. We can cut through our surface with a plane, and in so doing, the intersection of the surface with that plane results in a curve. We compute the curvature of that space curve in exactly the same way we did in Chapter 10 (see Figure 10.4 on p. 386) by finding the circle that best fits the curve at the point of interest. However, there is a problem with this story. The value we get for the curvature depends upon the orientation of the plane we use to cut the surface. Each such plane will result in a different curve and a correspondingly different curvature. The way around this impasse is a beautiful theorem that states that there is one particular choice of two orthogonal planes for which the curvature will take two extreme values, one high and one low. These are the so-called principal curvatures. This theorem guarantees that it takes two numbers to capture the curvature of a surface, as opposed to the

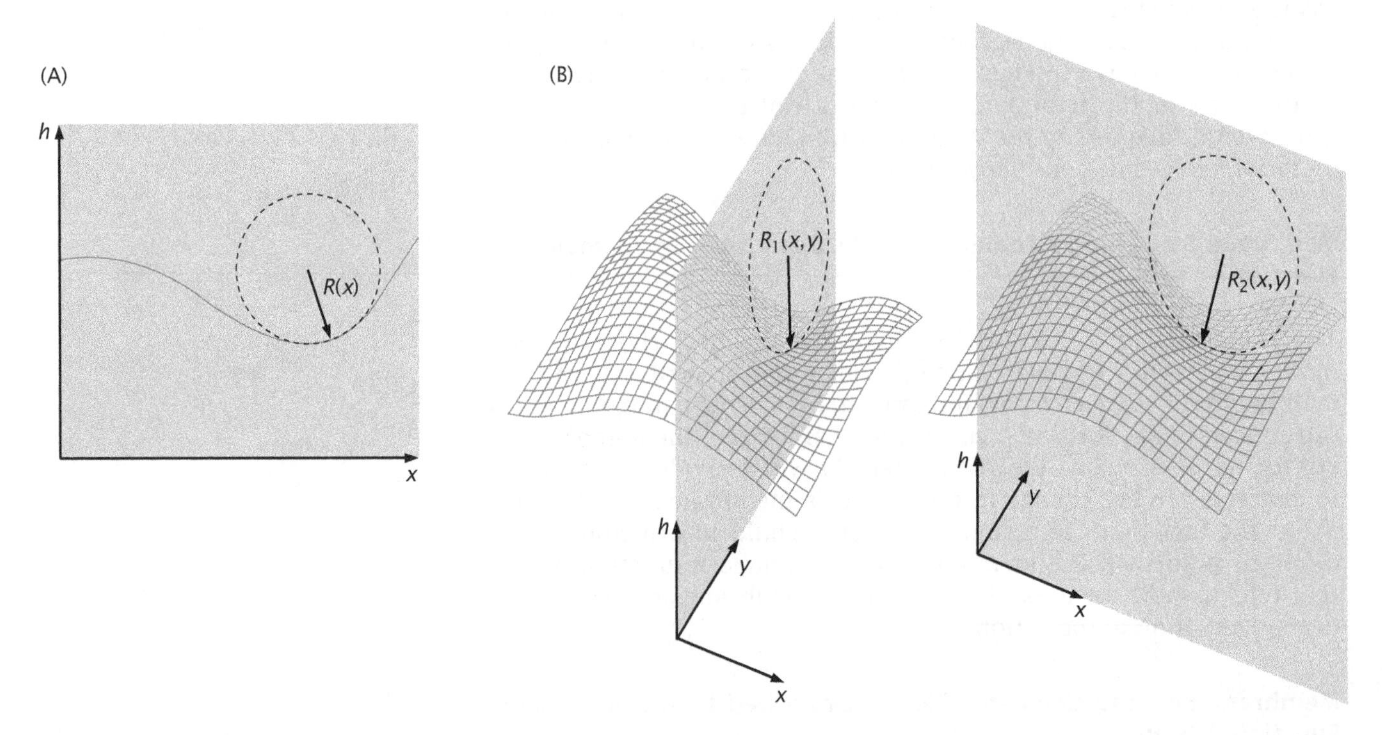

Figure 11.15: The curvature of space curves and surfaces. (A) The curvature of a curve is found by making the best fit of a circle to the point at which we are computing the curvature. (B) The curvature of a surface is obtained by finding the best circle along two orthogonal directions on the surface. This figure shows the intersection between a surface and a plane parallel to the y-axis and a second intersection between the surface and a plane parallel to the x-axis.

single number $(1/R)$ that it takes to capture the curvature of a curve (or a beam).

From a practical point of view, the easiest way to *compute* the curvature is to visit the point of interest on the surface where we want to find the curvature, to construct the tangent plane at that point with Cartesian coordinates that are defined locally as shown in Figure 11.16, and then to expand the height in powers of x_1 and x_2, where we have adopted the notation $x_1 = x$ and $x_2 = y$, which is more convenient for evaluating sums. The resulting quadratic function, $h(x_1, x_2) = \sum_{i,j=1}^{2} \kappa_{ij} x_i x_j$, has coefficients that are directly related to the curvature. In particular, we evaluate the curvature κ by computing the matrix

$$\kappa = \begin{pmatrix} \kappa_{11} & \kappa_{12} \\ \kappa_{21} & \kappa_{22} \end{pmatrix}, \tag{11.2}$$

where we have introduced the notation

$$\kappa_{ij} = \frac{\partial^2 h}{\partial x_i \partial x_j}. \tag{11.3}$$

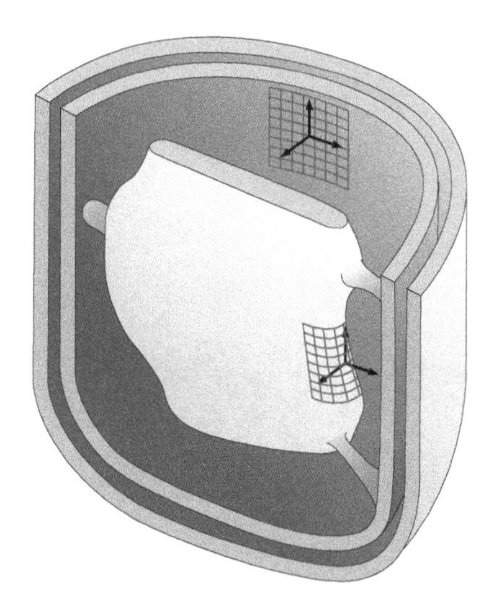

Figure 11.16: Assignment of local coordinate patches to the surface of a membrane. Here we have taken a complex membrane in the mitochondrion and illustrated two local fits to that surface.

The eigenvalues of this matrix are the two principal curvatures. Eigenvalues and eigenvectors are discussed in detail in The Math Behind the Models on p. 595. Finding the eigenvalues and eigenvectors for this problem corresponds to rotating the two orthogonal planes shown in Figure 11.15(B) and finding the choice of coordinates for which the curvature matrix is diagonal. To see these definitions in action, consider two simple but important cases. First, for a sphere, the two principal curvatures are identical and are both equal to $1/R$, where R is the radius of the sphere. The second case to consider is a cylinder of radius r. In this case, the principal curvatures are 0 and $1/r$.

One other way to think about our analysis is to consider the geometrical interpretation of the Taylor expansion of the height function $h(x, y)$ shown in Figure 11.14 around the point (x_0, y_0). That is, we write the height in the neighborhood of the point (x_0, y_0) as

$$h(x, y) = h(x_0, y_0) + \frac{\partial h}{\partial x}\Delta x + \frac{\partial h}{\partial y}\Delta y + \frac{1}{2}\left(\frac{\partial^2 h}{\partial x^2}\Delta x^2 + 2\frac{\partial^2 h}{\partial x \partial y}\Delta x \Delta y + \frac{\partial^2 h}{\partial y^2}\Delta y^2\right). \tag{11.4}$$

Here we have kept terms up to second order in Δx and Δy, and the partial derivatives are all computed at (x_0, y_0). If we consider the formula above as a mapping of a small $\Delta x \times \Delta y$ square patch of membrane around the point (x_0, y_0) to a small membrane patch on the surface, we can give each term a simple geometrical interpretation as shown in Figure 11.17. The constant term corresponds to a simple shift of the patch in the z-direction by an amount $h(x_0, y_0)$. The first-order terms represent a rotation of the small patch so that it is a tangent to the surface, while the second-derivative terms correspond to the bending of the small membrane patch so that it has the right curvature.

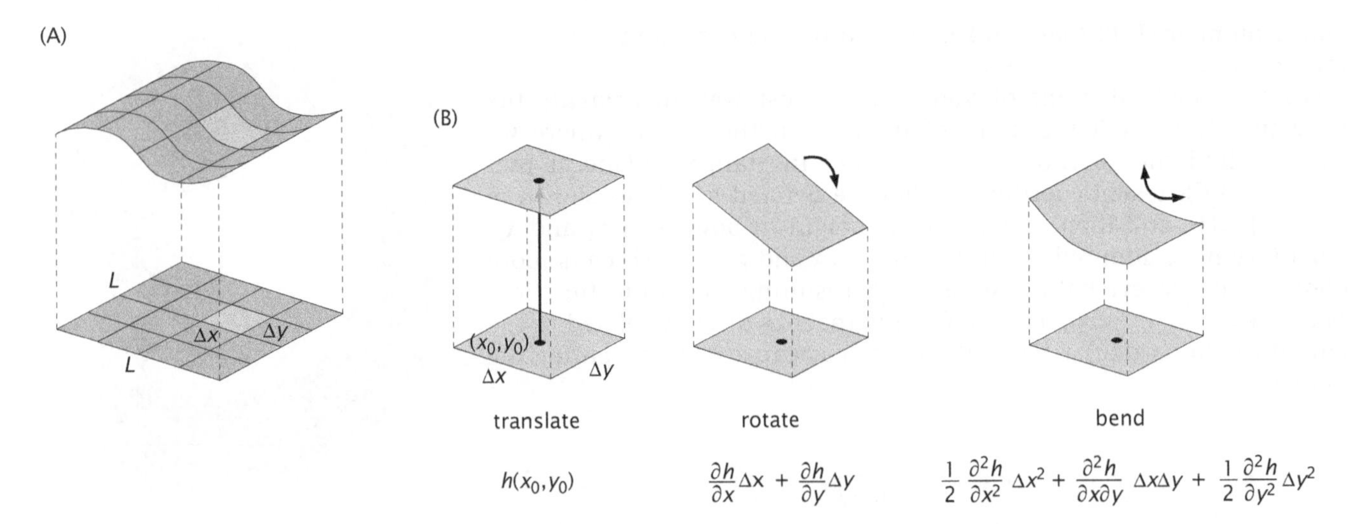

Figure 11.17: Geometrical interpretation of the Taylor expansion of the height function. (A) Representation of membrane as a series of small patches. (B) The terms in the Taylor expansion of the height function take a flat patch of the reference plane and map it onto a patch of the deformed membrane.

One of the limitations of the approach we have described is that a generic surface can have such a complex structure that it might not be describable in terms of the height function $h(x, y)$. For example, a sufficiently wiggly surface can actually have overhangs, which means that there is more than one height corresponding to a given point (x, y). Our antidote to this problem is to divide the surface up into a bunch of little coordinate patches and on each such patch to erect a local frame (x, y, z). An example of this strategy for a mitochondrion-like geometry is shown in Figure 11.16.

Membrane Compression Geometry Can Be Described by a Simple Thickness Function, $w(x,y)$

As will be seen later in Section 11.6.2, a membrane protein might have some part that is hydrophobic and that aligns with the surrounding membrane. This has the effect of altering the membrane thickness. As a result, another key way in which membranes can change is that their thickness can be altered relative to their equilibrium value w_0, where w_0 refers to the equilibrium half-width of the bilayer. This idea is captured in the field $w(x, y)$, which reports the thickness at point (x, y).

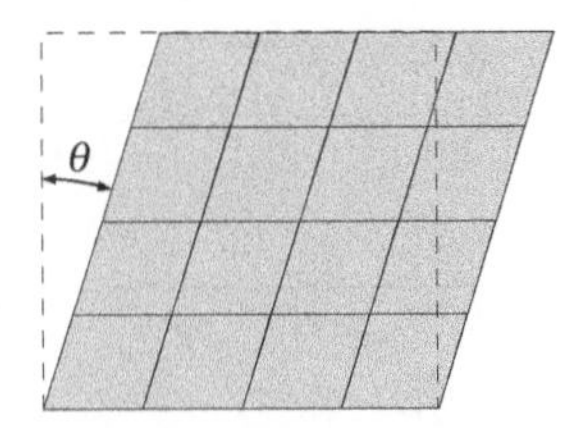

Figure 11.18: Shear deformation. The sheared patch of membrane (solid line) is characterized by the angle θ, with respect to the undeformed patch (dashed line).

Membrane Shearing Can Be Described by an Angle Variable, θ

Pure lipid bilayer membranes are fluid and therefore cannot support a shear deformation. However, cell membranes are often attached to an elastic network, such as the one formed by spectrin in the case of red blood cells, which provides them with shear rigidity. To describe a shear deformation, we consider a small square patch of the undeformed membrane, as showed in Figure 11.18. In the deformed membrane, the square becomes a rhombus of equal area, characterized by the angle θ, which we use to specify the state of shear deformation.

11.2.2 Free Energy of Membrane Deformation

With a description of the deformed membrane geometry in hand, our next step is to write down the free energy. In particular, we are interested in writing down a free energy that is a functional of the shape, G[shape]. Recall from the discussion in Chapter 5 that a functional is a rule that gives us a number every time we hand it a function. In this case, the function we hand to our free-energy functional is the shape (as characterized by geometric functions such as $\Delta a(x, y)$, $h(x, y)$, and $w(x, y)$) and the resulting number is the free energy. Different problems will force us to focus on different classes of deformation. By way of introduction, we now examine the energetic consequences of area change, bending, and thickness variation.

There Is a Free-Energy Penalty Associated with Changing the Area of a Lipid Bilayer

One class of deformation to which we can subject a membrane is stretching. There is a corresponding energy cost. A schematic of this kind of membrane stretching that appeals to an analogy based on stretching springs is shown in Figure 11.19. As shown in the figure, this area stretch term penalizes changes in the area of the membrane relative to some reference value and can be written as

$$G_{\text{stretch}} = \frac{K_a}{2} \int \left(\frac{\Delta a}{a_0} \right)^2 \mathrm{d}a, \tag{11.5}$$

where Δa is the change in area and a_0 is the reference area. The fact that we have written this free energy as an integral means that the area change might be different at every point within the membrane. As usual, we adopt a "divide and conquer" approach in which we visit each little membrane area, compute its deformation energy, and then add up the resulting energies by summing over every membrane patch.

In the case when the areal strain, $\Delta a/a$, is uniform over the surface of the membrane, the above formula simplifies to

$$G_{\text{stretch}} = \frac{K_a}{2} \frac{\Delta a^2}{a_0}, \tag{11.6}$$

where a_0 is the reference area. The parameter K_a is the area-stretch modulus and has characteristic values in the range 55–70 $k_B T/\text{nm}^2$ (often reported in units of mN/m, ranging from 230 to 290 mN/m).

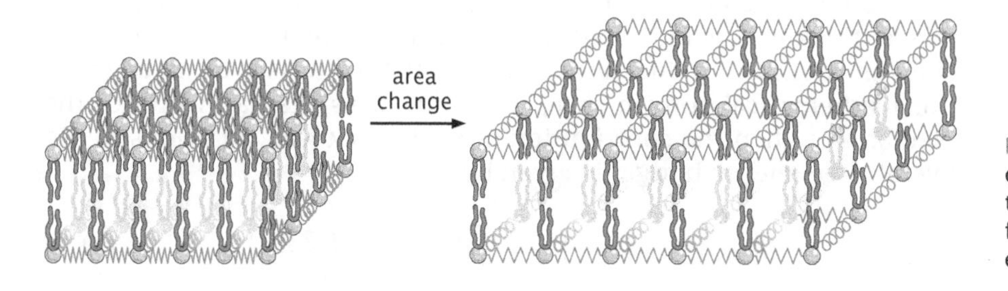

Figure 11.19: Schematic of the energy cost for stretching a membrane. When the area of the membrane is changed from its equilibrium value, there is an energetic cost.

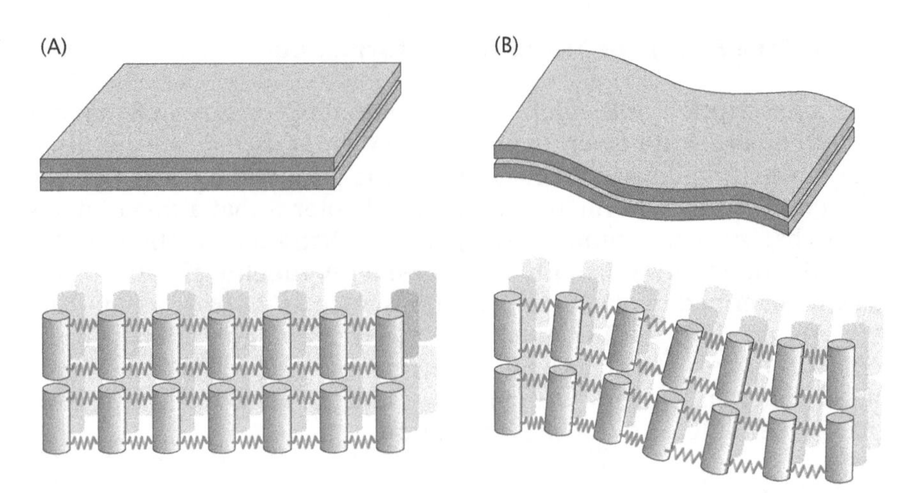

Figure 11.20: Schematic of the energy penalty associated with average bending of the bilayer. (A) The undeformed state of the membrane. (B) A membrane that has suffered a bending deformation resulting in relative tilt of neighboring lipid molecules.

There Is a Free-Energy Penalty Associated with Bending a Lipid Bilayer

When a patch of lipid bilayer is bent away from its flat zero-energy state (assuming there is no spontaneous curvature), there is a rearrangement of both the headgroups and the tails of the lipids within the bilayer. Loosely speaking, we can think of these rearrangements as being equivalent to stretching and compressing springs as shown in Figure 11.20. More precisely, the free energy of bending can be written in terms of the curvature as

$$G_{\text{bend}}[h(x,y)] = \frac{K_{\text{b}}}{2}\int \mathrm{d}a\,[\kappa_1(x,y)+\kappa_2(x,y)]^2, \qquad (11.7)$$

a model sometimes known as the Helfrich–Canham–Evans free energy. We have introduced the notations κ_1 and κ_2 to represent the principal curvatures of the surface at the point of interest and they may be thought of as the outcome of diagonalizing the matrix κ_{ij} that appears in Equation 11.3. The mean curvature is defined as $\bar{\kappa} = (\kappa_1 + \kappa_2)/2$. What this equation really instructs us to do is to visit every point on the surface, find its curvature, and compute $\bar{\kappa}^2$ at that point, and then to add up the energy over all points on the membrane.

Note that the bending free energy introduced in Equation 11.7 involves a new material parameter, K_{b}, the bending rigidity. Since the units of κ are 1/length (as can be seen from the definition of the curvature as the second derivative, length/length^2) and da has units of length^2, and since the overall unit of the expression is an energy, we see that K_{b} has units of energy with typical values in the range 10–20 $k_{\text{B}}T$. We will describe how the bending rigidity is measured in Section 11.3.1.

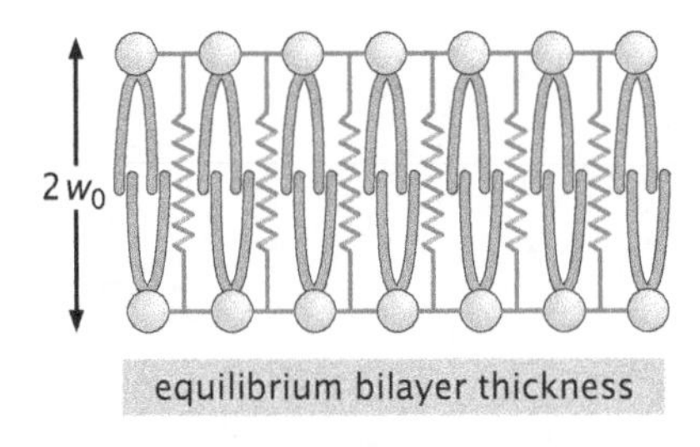

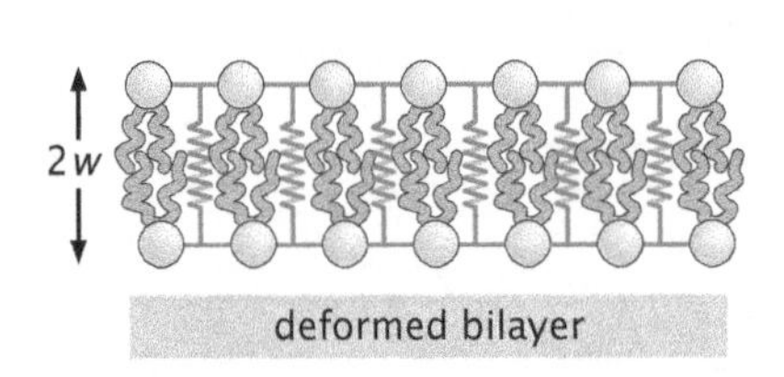

Figure 11.21: Energy penalty for bilayer thickness changes. Springs are used to illustrate the idea that there is an energy cost to change the thickness of a lipid bilayer from its equilibrium value, $2w_0$.

There Is a Free-Energy Penalty for Changing the Thickness of a Lipid Bilayer

Yet another type of membrane springiness results from changing the bilayer thickness as indicated schematically in Figure 11.21. If we consider an equilibrium thickness $2w_0$ and the thickness is changed to $2w$ (for mathematical convenience, we define w as the half-width of the membrane), then the contribution of such thickness variations to the overall free energy budget is given by

$$G_{\text{thickness}}[w(x,y)] = \frac{K_{\text{t}}}{2}\int \mathrm{d}a\left[\frac{w(x,y)-w_0}{w_0}\right]^2, \qquad (11.8)$$

where K_t is a stiffness that assigns an energy cost to thickness variation. It has units of energy/area and a typical value of roughly $60\,k_BT/\text{nm}^2$. As an example of the kind of problem that will demand this free energy, the thickness of the bilayer in the vicinity of a membrane protein varies continuously, adopting its equilibrium thickness $2w_0$ only sufficiently far from the protein. To account for the spatial variation of the thickness around the protein, we have to compute the free-energy cost, membrane patch by membrane patch, and add them all up.

This section has described the geometry of some of the deformations that biological membranes suffer, and how we can write the free-energy cost such deformations imply. Equations 11.5, 11.7, and 11.8 will serve as the basis for our analysis of a host of problems involving membrane deformation in the remainder of the chapter. Even though we will not consider explicit applications of this idea, we should mention that shearing of cell membranes is also accompanied by an energy cost that is quadratic in the shear deformation θ. This contribution to the total free energy is necessary when considering the equilibrium shapes of red blood cells, for example (see Figure 5.10 on p. 204).

There Is an Energy Cost Associated with the Gaussian Curvature

From our experience with macroscopic materials, we know that given a one-dimensional beam, there are geometrical transformations that do not cost energy. For example, we can translate the rod horizontally, having no effect on its energy. We can rotate the rod (about its center of mass) without a change in energy. But, regardless of orientation, it requires force to bend the rod, and hence there is an energy penalty for this deformation. While this simple notion of bending has a pleasingly direct analog in a membrane through the mean curvature energy developed above, there are additional energies that are not present in the one-dimensional case.

Recall that the curvature of the membrane was described by the matrix given in Equation 11.2, which has two eigenvalues called the principal curvatures. If we were to translate or rotate the surface, it would not affect these curvatures and hence they are said to be "invariant" under translation and rotation. Similarly, we know that it should not matter in which direction the membrane is bent. If the membrane has positive or negative curvature, both should have identical energy costs. This led us to the fact that it is the mean curvature squared that dictates the energy cost. We can ask whether there is another way of formulating an energy using the two principal curvatures that is distinct from the mean curvature, but is still both quadratic and invariant under rotation and translation. Without going into the details, we note that there is such a contribution to the total bending energy resulting from the Gaussian curvature of the surface. This is the product of the two principal curvatures, as opposed to their sum as in the case of mean curvature. This contribution to the membrane deformation energy has its own bending modulus, κ_G.

The Gaussian curvature has the possibility of being biologically interesting since it relates directly to the topology (that is, number of handles) of a surface. If we consider a closed surface, like a vesicle, no matter what deformations the vesicle suffers, the contribution from Gaussian curvature remains fixed and hence can be ignored. It is only when the topology of the surface changes (like going from a sphere to a donut) that the Gaussian curvature comes into play. This means that in the complicated morphologies of the endoplasmic reticulum and mitochondria, Gaussian curvature may be a significant portion of the energy budget when membrane holes, tubes, and tori are frequently created and destroyed. The second very interesting feature is that, unlike the mean curvature modulus, the sign of κ_G is not necessarily positive. In fact, for some kinds of artificial membranes, it has been found to be of the order of the mean curvature modulus, but negative. Membranes with different signs of κ_G prefer different generic membrane shapes. A positive modulus indicates that the membrane is most likely to adopt the shape of a saddle point; such shapes are found in the fusion and fission stalks of vesicles, shown in Figure 11.1(C), and in the crista junctions of mitochondria, for example. If the modulus is negative, the Gaussian curvature prefers a tightly wrapped morphology. Unfortunately, because the Gaussian curvature is topologically invariant, it is rather difficult to measure the sign or magnitude of its modulus in real membranes and it will generally not enter the remainder of our discussion.

One of the main shortcomings of our use of the ideas on membrane elasticity discussed in this section is that we have emphasized mechanics rather than statistical mechanics. Concretely, this means we have ignored the complicated thermal dance of membranes. For many problems, an examination of membrane biophysics as a problem in elasticity is an enlightening first step. However, the role of thermal fluctuations in dictating membrane response is an important part of the total story and the interested reader is invited to consult the Further Reading at the end of the chapter.

11.3 Structure, Energetics, and Function of Vesicles

11.3.1 Measuring Membrane Stiffness

Membrane Elastic Properties Can Be Measured by Stretching Vesicles

The free energies introduced in Equations 11.5, 11.7, and 11.8 all involve material parameters that characterize the resistance of membranes to different classes of deformation. Micropipette aspiration experiments have been used to measure several of these parameters. These experiments subject lipid bilayer vesicles (closed spherical shells made of lipid bilayers) to known deformations. The key features of the experiment are shown in Figure 11.22. These experiments exploit a small pipette (with diameter $\approx 10\,\mu m$ or smaller), which serves as a probe in a solution containing the vesicles (or cells) under investigation. This pipette is connected to a reservoir that permits the application of a known pressure difference Δp between the micropipette and the surrounding medium. As a result, it is possible to grab vesicles (as shown in Figure 11.22) and then to apply a suction pressure that has the effect of deforming the vesicle. Using video microscopy, one can obtain images of the deformed vesicle like those shown in Figure 11.23. The key measurables in this experiment are geometrical, namely, the radius of the vesicle R_V, the length of the aspirated vesicle tether l, and the radius of its hemispherical cap R_1, which is roughly equal to half the diameter d of the micropipette.

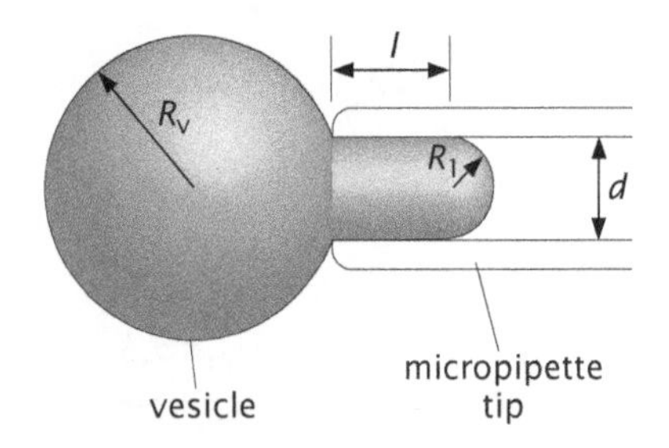

Figure 11.22: Pipette aspiration technique. A micropipette is used to grab a lipid bilayer vesicle. Suction pressure is used to pull part of the vesicle into the pipette. The geometry of the aspirated vesicle in conjunction with the known pipette pressure difference permit a determination of the material constants defining the mechanical response of the lipid bilayer.

Interestingly, on the basis of the known pressure difference and these measured geometric parameters, it is possible to determine the tension in the membrane as well as the material properties such as K_b and K_a, the bending and the area stretch moduli, respectively. The pressure difference between the interior of the vesicle and the surrounding medium, Δp_{out}, is given by the Laplace–Young law,

$$\Delta p_{out} = \frac{2\tau}{R_v}, \tag{11.9}$$

where τ is the tension in the membrane. This relation is familiar from the study of soap bubbles and tells us that to maintain a sphere of radius R_v requires a pressure difference Δp. The detailed analysis, including a derivation of the Laplace–Young relation, is left to the problems at the end of the chapter. Similarly, the pressure difference between the inside of the vesicle and the inside of the micropipette can be written as

$$\Delta p_{in} = \frac{2\tau}{R_1}. \tag{11.10}$$

Since $\Delta p_{in} - \Delta p_{out}$ is equal to the pressure difference Δp, which is controlled experimentally, we can take the difference of last two equations to obtain an expression for the tension in terms of the measured geometry as

$$\tau = \frac{\Delta p}{2} \frac{R_1}{1 - (R_1/R_v)}. \tag{11.11}$$

As a result, for each Δp, we can measure the geometry and determine the tension. With geometry and tension in hand, we are prepared to determine the moduli.

For example, to measure the area stretch modulus K_a, one can make use of the relation

$$\tau = K_a \frac{\Delta a}{a_0} \tag{11.12}$$

between the areal strain $\Delta a/a_0$ and the tension τ; here Δa is the change in membrane area and a_0 is the area of the reference state. In writing this relation, we have assumed that the tension is large enough to have "ironed out" all the wrinkles in the membrane due to thermal fluctuations. This entropic effect is the analog of the force–extension properties of polymers explored in Section 8.3.2 (p. 340). At low tensions (like with a polymer at low forces) the membrane is full of spontaneous undulations. With increasing tension, these fluctuations can literally be stretched out of the membrane. In the discussion here, we are assuming that the applied tension simply results in an overall increase of the average distance between the lipid molecules. It should be noted that measurements of the areal stress–strain relation in the entropic regime can also be utilized and they lead to the determination of the bending modulus K_b. The full treatment of this experiment, including the determination of the bending term, is explored in the problems at the end of the chapter.

The areal strain can be determined from the images shown in Figure 11.23 by measuring the total area of the membrane as that of the surface of revolution obtained by rotating the image around the symmetry axis along the centerline of the pipette. A simple approximation to the area can be obtained by thinking of the deformed vesicle

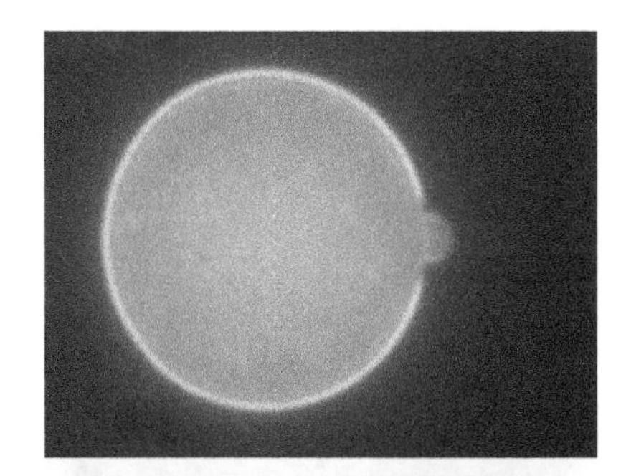

Figure 11.23: Images from a micropipette aspiration experiment. This series of fluorescence images shows how increasing hydrostatic pressure deforms a giant unilamellar vesicle. The membrane has been doped with lipids whose headgroups are labeled with rhodamine (a small fluorescent molecule). In each frame, the area, and hence the tension, is increased. From top to bottom, the tensions are 1.1 mN/m, 3.2 mN/m, and 7.4 mN/m. (Courtesy of T. Ursell.)

as a large sphere with a small cylinder ending in a hemispherical cap attached to it (this is the part in the pipette). The area of the capped cylinder is

$$\Delta a = 2\pi R_1 l + 2\pi R_1^2, \tag{11.13}$$

while the reference area is $a_0 = 4\pi R_v^2$, corresponding to a spherical vesicle of radius R_v (note that we have made the further simplifying assumption that the vesicle radius remains unchanged in the course of the experiment). These two formulas combined give the following approximate expression for the areal strain:

$$\frac{\Delta a}{a_0} = \frac{R_1^2(1 + l/R_1)}{2R_v^2}. \tag{11.14}$$

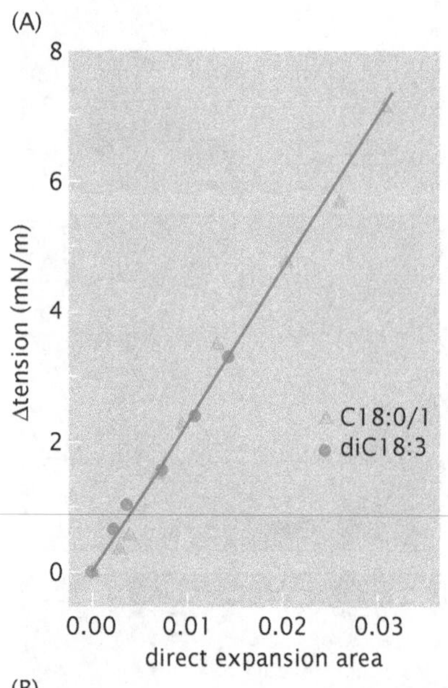

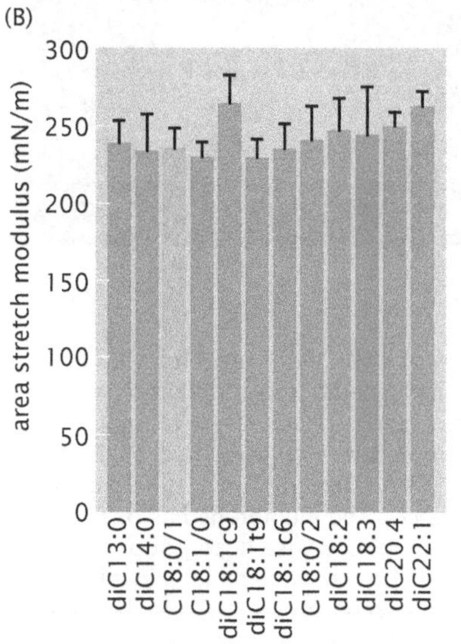

Figure 11.24: Experimental data from a micropipette aspiration experiment. (A) A series of tensions were applied by controlling the pipette pressure, and the associated tension and area were measured for each such pressure. The slope of the graph is the area stretch modulus. (B) Measured values of the area stretch modulus for different lipids. Lipid tail structure is denoted by the number of carbons and the number of unsaturated bonds on the carbon chain; for instance C18:0/1 has two 18 carbon chains, one of which has an unsaturated bond, likewise diC18:3 has two identical (the "di" prefix) 18 carbon chains, each with three unsaturated bonds. (Adapted from W. Rawicz et al., *Biophys. J.* 79:328, 2000.)

Therefore, to determine the area stretch modulus K_a, we need to plot the right-hand side of Equation 11.11 against the right-hand side of Equation 11.14, and the resulting slope is K_a. Examples of the results of this kind of analysis are shown in Figure 11.24.

One of the difficulties of the approach described here if we are concerned with real cell membranes is that these measurements are usually performed in the pristine setting of pure, single-component, lipid bilayer vesicles. In contrast, because cell membranes are littered with membrane proteins and lipid compositional heterogeneity and form connections with both intracellular and extracellular structural elements, it remains a challenge to really know how to characterize the energetics of a complex cell membrane.

11.3.2 Membrane Pulling

In the experiment described in the section above, the geometry of the micropipette constrains the shape and size of the membrane tubule that is pulled out by suction. For most kinds of membrane tubule formation inside cells, the nature of the force is very different and yet the final outcome is still frequently a long, cylindrical tubule extending from the membrane reservoir. In living cells, there are at least two organelles characterized by tubular, reticular structure: the endoplasmic reticulum and the trans-Golgi network. Many kinds of transport intermediates that carry proteins from one distinct membrane compartment to another are also tubular or reticular in structure. Careful observations tracking the dynamics of these membranous organelles together with microtubules inside cells have led to the hypothesis that the formation of cell biological tubules often requires membrane-attached motor proteins working in gangs. Video sequences showing *in vivo* tubule formation and a speculative model for how the motor proteins might be acting are shown in Figure 11.25. This leads us to wonder, for a membrane vesicle with given mechanical properties, if a pulling force is applied at a single point, can we predict whether the vesicle will simply be elongated into an ellipsoid or will it generate an extruded tubule? What are the magnitudes of the forces that must be involved for this to be the true *in vivo* mechanism of endoplasmic reticulum structure formation?

Experiments show that a point force applied to a lipid membrane will indeed result in a long thin tether. Forces that produce tethers are applied by either laser tweezers or molecular motors. To gain intuition about the forces and energies involved, we assume a simplified model of a spherical vesicle of radius R with a tether of length L and radius r

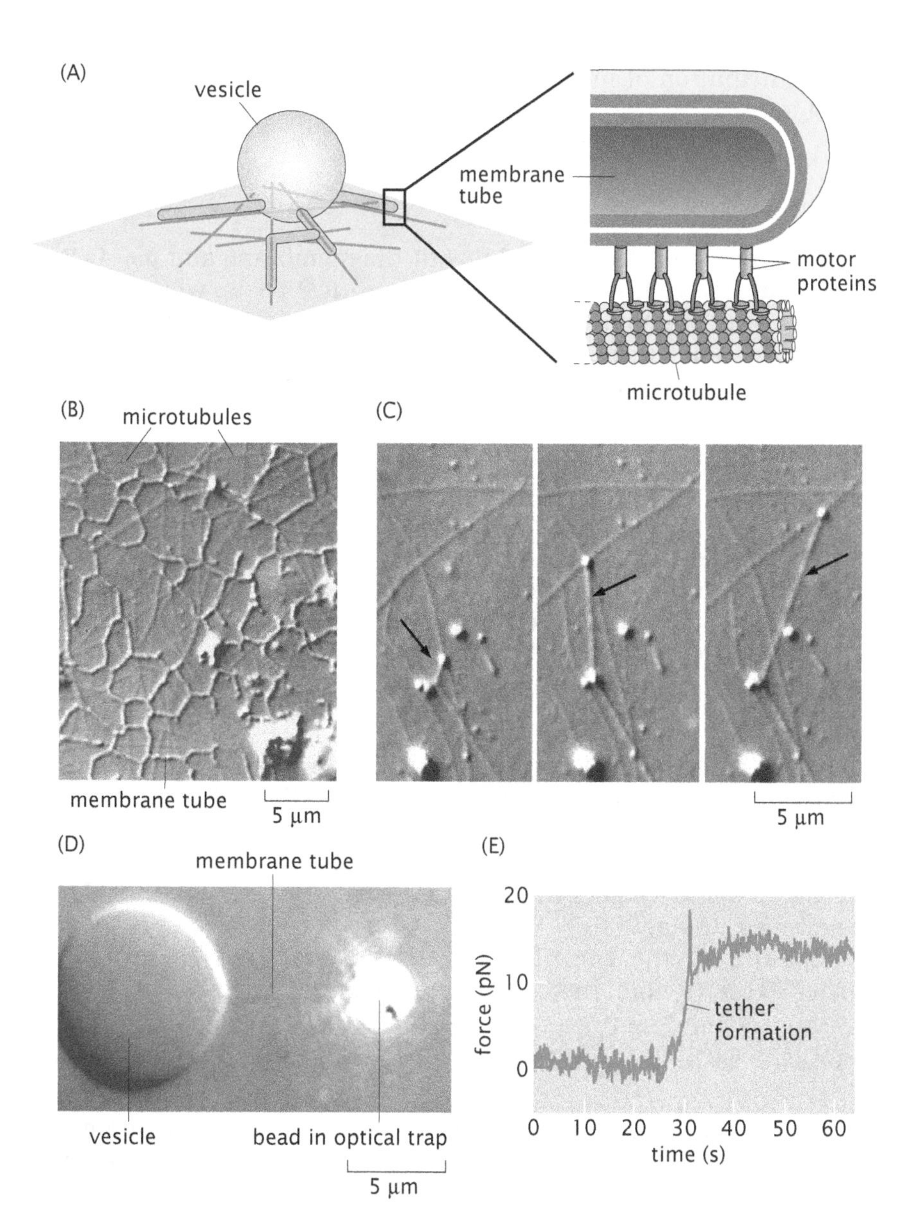

Figure 11.25: Force-induced tether formation. (A) Long thin membrane tubes can be pulled out from a nearly spherical vesicle by the action of groups of motor proteins. (B) In an *in vitro* experiment combining microtubules with rat liver endoplasmic reticulum membranes and *Xenopus* egg cytosolic proteins, an elaborate web of endoplasmic reticulum membrane forms on the microtubules, as can be readily seen by differential interference contrast microscopy. (C) Over a period of 11 s, a single tubule can be seen to be pulled out (black arrows) and switch tracks from one microtubule to another. (D) In order to measure the force associated with tether formation, a bead in an optical trap can be stuck to a vesicle membrane and then pulled to extend the membrane tube. (E) A time trace for an experiment like that depicted in (D) shows that the force on the bead increases suddenly when the tether is formed, and then remains fairly constant as it is extended. Typical forces associated with tether formation are of the order of 10 pN. (A, D, E, adapted from G. Koster et al., *Proc. Natl Acad. Sci. USA* 100:15583, 2003; B, C, adapted from V. Allan and R. Vale, *J. Cell Sci.* 107:1885, 1994.)

pulled out by a force f, as illustrated in Figure 11.26. Experiments with beads typically measure the force to be of the order of 10 pN, for a variety of membranes, including those from endoplasmic reticulum and the Golgi apparatus.

In order to compute the force needed to pull out a tether, we begin by writing down the free energy for the vesicle + tether system. The free energy has a contribution from membrane bending,

$$G_{\text{bend}} = \underbrace{8\pi K_b}_{\text{spherical part}} + \underbrace{\pi K_b \frac{L}{r}}_{\text{cylindrical part}} + \underbrace{4\pi K_b}_{\text{hemispherical end cap}} = 12\pi K_b + \pi K_b \frac{L}{r}. \tag{11.15}$$

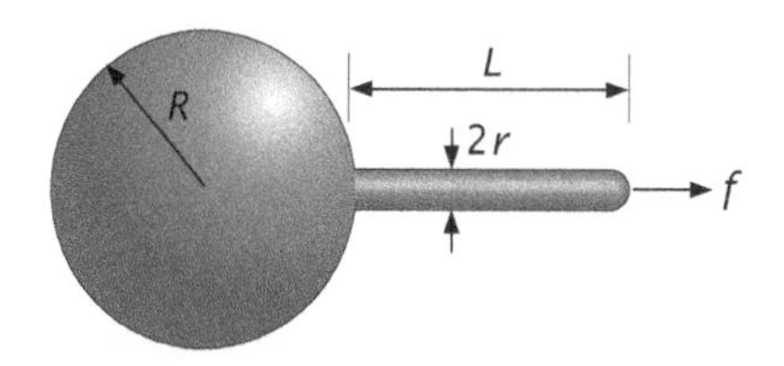

Figure 11.26: Model of force-induced tether formation. To account for the tether formation due to an applied force, we consider a simplified model of a spherical vesicle of radius R. The tether, formed by the application of force f, is a capped cylinder of length L and radius r.

The expressions for the sphere and the cylinder are obtained by taking the sum of the principal curvatures squared, which is $4/R^2$ for the sphere and $1/r^2$ for the cylinder, and multiplying it with the corresponding area.

The contribution of membrane stretching to the overall free-energy budget is

$$G_{\text{stretch}} = \frac{K_{\text{a}}}{2} \frac{(a - a_0)^2}{a_0}. \tag{11.16}$$

Here K_{a} is the area stretch modulus of the membrane and $a = 4\pi R^2 + 2\pi rL$ is the area of the spherical and cylindrical parts, while a_0 is the reference area. We ignore the area of the hemispherical cap at the end of the tether since throughout we assume $r \ll R, L$.

There are two other terms in the total free energy of the vesicle–tether system. One corresponds to the work against the pressure difference Δp between the inside and the outside of the vesicle,

$$G_{pV} = -\Delta p \left(\frac{4}{3} \pi R^3 + r^2 \pi L \right), \tag{11.17}$$

where the term in parentheses is the total volume (again, we ignore the contribution from the hemispherical cap). The other term is the work of the load applied at the end of the tether,

$$G_{\text{load}} = -fL. \tag{11.18}$$

The minus signs in the last two equations remind us that the excess internal pressure Δp and the load f have the effect of increasing the volume and the tether length, respectively. As usual, if the reader finds it convenient, the load term can be visualized as reflecting the lowering of some weights when the tether is elongated (as introduced in Figure 5.12 on p. 206).

Combining all the contributions to the total free energy of the vesicle + tether system discussed, we arrive at

$$G_{\text{tot}} = 12\pi K_{\text{b}} + \pi K_{\text{b}} \frac{L}{r} + G_{\text{stretch}} - \Delta p \left(\frac{4}{3} \pi R^3 + \pi r^2 L \right) - fL. \tag{11.19}$$

The equilibrium shape of the vesicle + tether is one that minimizes G_{tot} with respect to the three parameters r, R, and L, which completely specify the shape in the constrained set of geometries we consider. In other words, in equilibrium, the partial derivatives of G_{tot} with respect to each of these three variables are zero, which leads to the following three equations:

$$\frac{\partial G_{\text{tot}}}{\partial R} = 0 \Rightarrow 8\pi \tau R - 4\pi \Delta p R^2 = 0, \tag{11.20}$$

$$\frac{\partial G_{\text{tot}}}{\partial r} = 0 \Rightarrow -\pi K_{\text{b}} \frac{L}{r^2} + 2\pi \tau L - 2\pi \Delta p r L = 0, \tag{11.21}$$

$$\frac{\partial G_{\text{tot}}}{\partial L} = 0 \Rightarrow \pi K_{\text{b}} \frac{1}{r} + 2\pi \tau r - \pi \Delta p r^2 - f = 0. \tag{11.22}$$

To obtain these equations, we have made use of the identity

$$\frac{\partial G_{\text{stretch}}}{\partial R} = K_{\text{a}} \frac{a - a_0}{a_0} \frac{\partial a}{\partial R} = \tau 8\pi R, \tag{11.23}$$

where in the second step we have used the linear relation between the membrane tension and the areal strain, $\tau = K_{\text{a}}(a - a_0)/a_0$. A completely analogous procedure leads to the relations $\partial G_{\text{stretch}}/\partial r = \tau 2\pi L$ and $\partial G_{\text{stretch}}/\partial L = \tau 2\pi r$.

Equation 11.20 yields the Laplace–Young relation between the membrane tension and the pressure difference across the membrane,

$\Delta p = 2\tau/R$. If we substitute this result into Equation 11.21, we find that the Δp term is smaller than the τ term by a factor r/R, and can be safely ignored. Then solving this equation for r, the radius of the tether, yields

$$r = \sqrt{\frac{K_b}{2\tau}}. \tag{11.24}$$

Using the Laplace–Young relation in Equation 11.22, again leads us to conclude that the Δp term can be safely ignored compared with the τ term. Replacing the above result for r in the remaining terms of Equation 11.22 leads to

$$f = 2\pi\sqrt{2K_b\tau}, \tag{11.25}$$

which relates the force applied by the load to the tension and the bending modulus of the membrane.

Given that the measured forces for pulling tethers from the endoplasmic reticulum and Golgi are of the order of 10 pN, we can use the above formulas to make rough estimates of the tension in these membranes, as well as the expected tether radius. Rewriting Equation 11.25 as $\tau = f^2/(4\pi^2)(2K_b)$, and recalling that a typical bending modulus has a magnitude of $K_b \approx 20\ k_BT$ ($k_BT = 4$ pN nm), we find $\tau \approx 0.015$ pN/nm (also 0.015 mN/m), which is about three orders of magnitude smaller than the typical tension at which membranes rupture. Similarly, from Equation 11.24, we estimate the tether radius to be $r \approx 50$ nm, which is close to what is observed in experiments on endoplasmic reticulum and Golgi membranes such as those shown in Figure 11.25.

11.3.3 Vesicles in Cells

Vesicles Are Used for a Variety of Cellular Transport Processes

If we broadly define vesicles as small, roughly spherical compartments bounded by a single lipid bilayer membrane, there are three general uses for such vesicles in cells. In particular, these vesicles are used for taking things up from the outside (endocytosis), delivering things to the outside (exocytosis), and moving things from compartment to compartment within the cell (intracellular transport). Figure 11.27 shows examples of each of these three classes.

The first example of uptake is specialized for cells taking up small particles from their environment. A classic example is cells taking up small cholesterol-containing particles called LDL from the bloodstream. Cholesterol is a necessary component for animal cell membranes, as shown in Figure 11.6. Typically about half the cholesterol in our cell membranes we synthesize and half comes from our diet and must, as a result, be internalized by cells. Cholesterol consumed in the diet is packaged by the liver into LDL particles, which are a complex formed of a protein and a phospholipid monolayer surrounding a fatty droplet of cholesterol-containing compounds. The length scale of these particles is of the order of 20 nm. For another cell to use this particle to make membranes, the particle has to be taken up and disassembled. The large size precludes entry into the cell through protein transporters. Instead, LDL is bound by a receptor on the surface of the cell. These bound receptors then cluster together because the internal or cytoplasmic domain of the transmembrane receptor can be bound indirectly by a molecule called

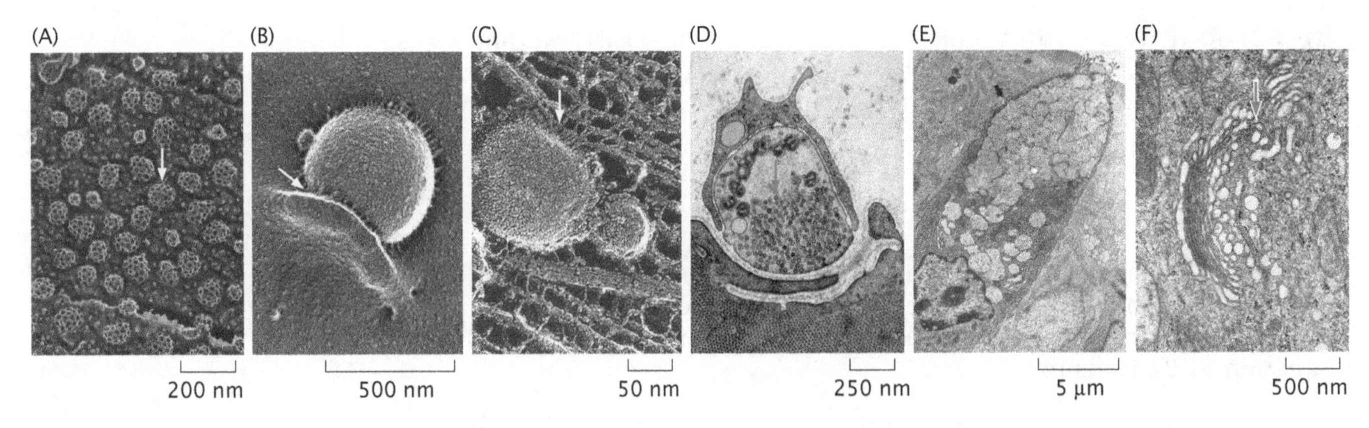

Figure 11.27: Gallery of examples of vesicles in cells. (A) Clathrin-coated pits (see white arrow) forming on the membrane of a cell performing endocytosis. (B) Phagocytosis of a large particle involving wrapping of the membrane at the point shown by the arrow. (C) A transport vesicle (see arrow) within the axon of a neuron tightly linked to microtubules. (D) Thin section through a synapse of a junction between a muscle cell and a nerve cell packed full of synaptic vesicles (arrow). (E) Goblet cell in the intestine dumping giant mucus-containing vesicles (highlighted by arrow) out into the lumen. (F) Thin section of the Golgi apparatus showing small transport vesicles (see arrow) that mediate transport among the stacks. (A, B, D, courtesy of J. Heuser; C, adapted from N. Hirokawa, *Science* 279:519, 1998; E, adapted from P. C. Cross and K. L. Mercer, Cell and Tissue Ultrastructure. W. H. Freeman, 1993; F, adapted from D. W. Fawcett, A Textbook of Histology. W. B. Saunders, 1986.)

clathrin, which self-assembles to form curved cage-like structures. As more bound LDL receptors cluster in this clathrin-coated pit, assembly of the clathrin coat deforms and bends the membrane to form a tear-drop-shaped invagination that eventually pinches off to make a coated, spherical vesicle. Figure 11.27(A) shows a beautiful array of clathrin-coated pits in various states of assembly.

The second example, shown in Figure 11.27(B), is a similar uptake process specialized for particles that are too large to be taken up by the clathrin-based mechanism. An example is when an immune system cell such as a macrophage or a neutrophil engulfs an invading disease-causing bacterium in order to destroy it. This process requires a combination of adhesion of the macrophage membrane to the bacterium and active energy-requiring assembly of cytoskeletal elements including actin and myosin to protrude the membrane so that it engulfs the bacterium, forming a giant vesicle bounded by a single lipid bilayer.

The next two examples, shown in Figures 11.27(C) and (D), show the opposite kind of process, whereby a cell delivers components to the outside rather than taking them up. A spectacular example occurs at the junction between a motor nerve cell and the muscle cell that it tells to contract as shown in Figure 11.28. The terminus of the nerve cell is packed with small vesicles called synaptic vesicles lined up right next to the place where it makes contact with the muscle cell. When an electrical signal (the so-called action potential to be described in Chapter 17) reaches the nerve terminus, these synaptic vesicles very quickly fuse to the plasma membrane, dumping their contents into the gap between the two cells. These synaptic vesicle contents are primarily a molecule called acetylcholine, which binds to receptors on the muscle cell, opening ion channels that depolarize the muscle cell membrane and trigger it to contract.

Just as we saw above with endocytosis, the mechanisms of exocytosis are specialized for small rather than large export events. While synaptic vesicles are very small (40–60 nm diameter), other exocytic vesicles called granules can be very large (several microns across), as shown in Figure 11.27(E). Some of these are found in goblet cells in the

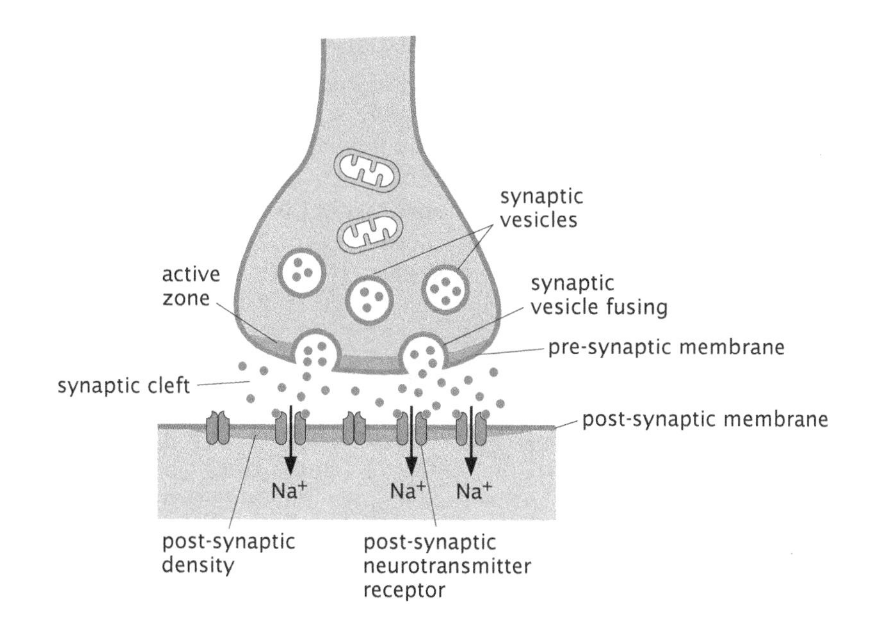

Figure 11.28: Schematic illustration of a synapse. A neuron extends its axon to a target cell where the axon terminates in a specialized pre-synaptic structure that is filled with small vesicles containing neurotransmitter molecules. On the membrane of the target cell, a specialized post-synaptic domain is packed with a high density of neurotransmitter receptors. When an action potential reaches the synapse, the synaptic vesicles fuse, emptying their contents into the cleft between the two cells.

lining of the intestine whose function is to secrete mucus that coats the intestine and protects it against a variety of insults.

The next example, Figure 11.27(F), shows cases where vesicles are used to transport materials from one region of the cell to another. In some cases, such as vesicular transport in the Golgi apparatus, vesicles move over very short distances (of the order of tens of nanometers) from stack to stack within the Golgi. In other cases, such as in a nerve axon, vesicles may be transported over very large distances of up to a meter. What all of these examples demonstrate is that vesicle management by cells is a key facet of their overall behavior.

There Is a Fixed Free-Energy Cost Associated with Spherical Vesicles of All Sizes

One question of immediate importance in thinking about the vesicle budget of cells is how much free energy it costs to make them. An estimate of this free-energy cost can be made by recourse to the bending energy introduced in Equation 11.7. Recall that the idea embodied in that equation is that we should visit each little patch of area on the membrane of interest, compute the mean curvature, and then compute the energy. With their spherical geometries, vesicles are probably the simplest of all shapes to consider and have an energy of the form

$$G_{\text{vesicle}} = \frac{K_{\text{b}}}{2} \int_{\partial\Omega} \left(\frac{2}{R}\right)^2 \mathrm{d}a, \tag{11.26}$$

where $\partial\Omega$ indicates that we should integrate over the whole membrane area. Since the curvature is constant everywhere on the sphere, we can do the integration immediately, yielding

$$G_{\text{vesicle}} = \frac{K_{\text{b}}}{2} \left(\frac{2}{R}\right)^2 4\pi R^2 = 8\pi K_{\text{b}}. \tag{11.27}$$

This result is very instructive for several reasons. First, the energy to make a vesicle (when viewed from the perspective of the bending energy) is independent of the vesicle size. That is, there is a fixed energy cost for all vesicles. The second lesson to emerge from this result is numerical and amounts to the observation that since $K_b \approx 10$–$20\,k_BT$, the free energy to form vesicles is in the range 250–500 k_BT. As a result, it is clear that cells must have some active mechanism to overcome the large energy penalty to make vesicles. One way in which this intrinsic energy cost can be overcome is by incorporating proteins that favor the curved state. A second mechanism involves distributions of curvature-inducing lipids. Before examining the mechanism of vesicle formation in cells, we first take stock of the vesicle budget in cells.

Estimate: Vesicle Counts and Energies in Cells A census of the vesicles of eukaryotic cells is intriguing for a number of reasons, including that it can help us size up how much energy is invested in vesicular traffic. To get a sense of the cargo-carrying capacity of these vesicles, we consider synaptic vesicles with a typical diameter of 60 nm. This corresponds to a volume per vesicle of roughly 100,000 nm^3. Synaptic vesicles carry the neurotransmitter acetylcholine, which has a molecular weight of roughly 450 Da, suggesting a volume per molecule of roughly 1 nm^3. As a result, if we think of these vesicles as being densely packed with this neurotransmitter, each such vesicle has a cargo of roughly 100,000 acetylcholine molecules. The actual number is smaller, and in the region of many thousands (our estimate assumes that the entire vesicle volume is occupied by acetylcholine and is an overestimate).

ESTIMATE

A rule of thumb is that for a cell such as a fibroblast, the membrane flux associated with vesicles is such that the amount of membrane area taken up by vesicles in 1 hour is equivalent to the entire plasma membrane area. To get a sense for what this means in terms of number of vesicles and their rate of production, we must first estimate the membrane area of a fibroblast. Figure 2.22 (p. 60) depicts a highly schematized fibroblast and leads to a rough estimate of an area $a_{\text{fibroblast}} \approx 2000\,\mu m^2$. If we think of each vesicle as having a diameter of roughly 100 nm (probably an overestimate), the area of each such vesicle is roughly $a_{\text{vesicle}} \approx 30{,}000\,nm^2$. To bud off an entire plasma membrane's worth of vesicles would hence require roughly 60,000 vesicles. The claim that this takes place over the course of an hour suggests a rate of vesicle production of about 1000 vesicles per minute. If we recall our energy estimate and assume that each such vesicle costs 500 k_BT worth of energy (equivalent to roughly 25 ATP molecules), then the total energy equivalent associated with producing all of these vesicles is roughly 25,000 ATP/minute.

Vesicle Formation Is Assisted by Budding Proteins

As was shown in Figure 11.27 and was demanded by our energy estimate that it costs hundreds of k_BT to make a vesicle, there are a variety of proteins that specialize in vesicle formation. Figure 11.29 gives a schematic view of the secretory pathways associated with eukaryotic cells and shows the way in which vesicles are shed from the membranes of the endoplasmic reticulum, the Golgi apparatus, and the plasma membrane. In particular, the protein complexes known as COPI and COPII mediate vesicle transport between the endoplasmic reticulum and the Golgi apparatus, with COPI overseeing transport

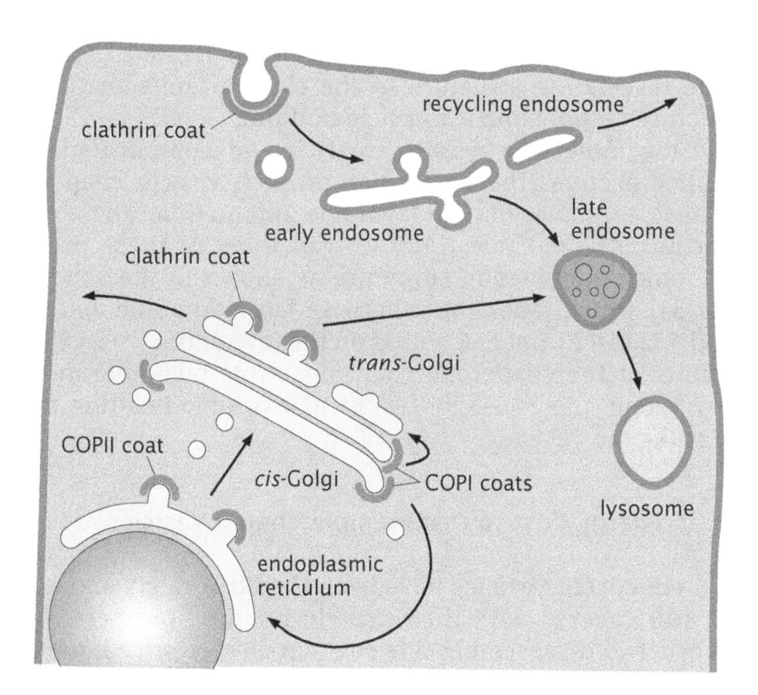

Figure 11.29: Protein coats that mediate vesicle trafficking. Vesicles from the endoplasmic reticulum are formed with a coat of COPII protein and then trafficked to the *cis* face of the Golgi apparatus. Vesicles moving from stack to stack in the Golgi apparatus are formed by a different protein, COPI. At the *trans* face of the Golgi, vesicles bud with a clathrin coat. Clathrin is also involved in endocytosis at the plasma membrane. (Adapted from T. Kirchhausen, *Nat. Rev. Mol. Cell Biol.* 1:187, 2000.)

from the Golgi to the endoplasmic reticulum and within the Golgi cisternae, while COPII leads from the endoplasmic reticulum to the Golgi.

Clathrin mediates the uptake of material at the plasma membrane, which then passes to the early endosome and also permits vesicular transport from the Golgi to endosomes. Clathrin is also involved in

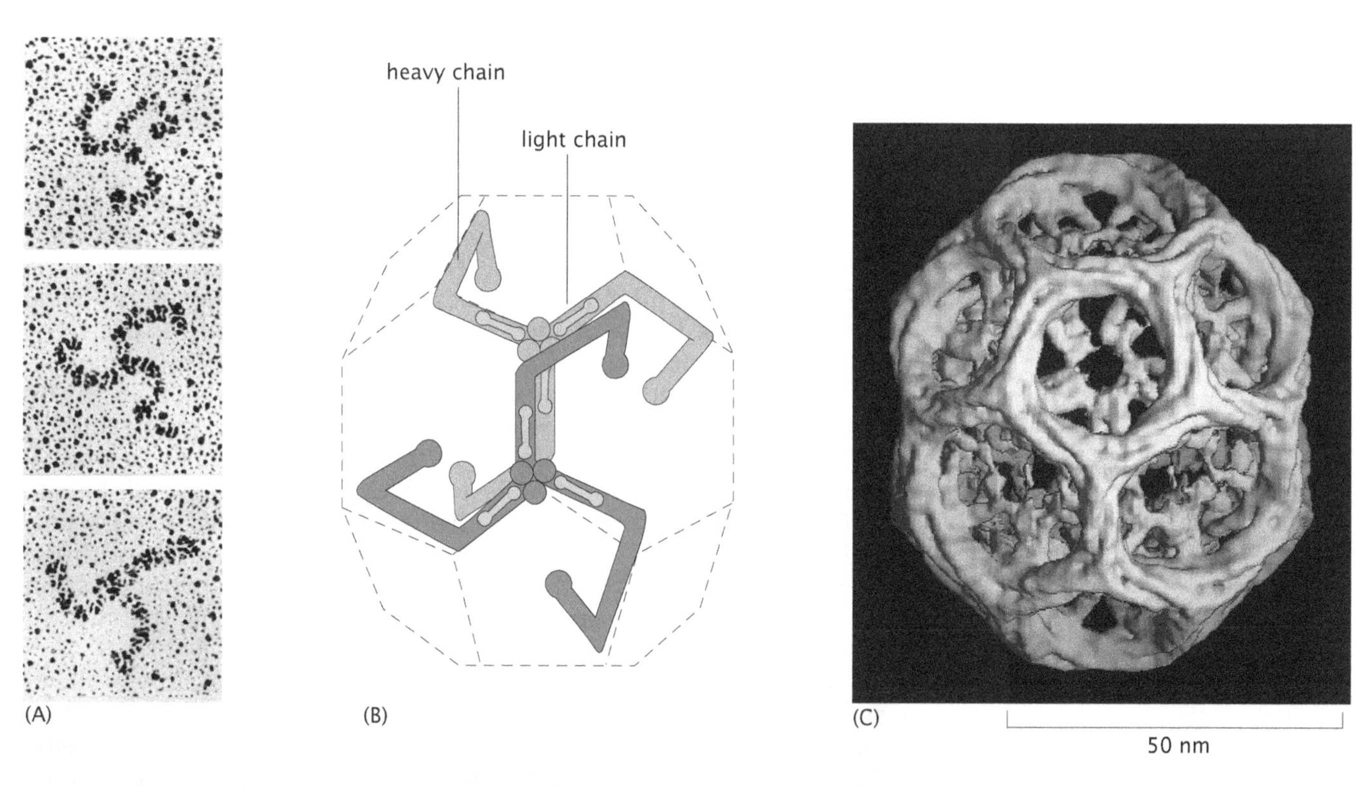

Figure 11.30: Structure of clathrin. (A) Individual, isolated clathrin molecules appear in the electron microscope as three-legged structures known as triskelions. (B) As shown in the schematic, triskelions can associate with one another to form a regular lattice structure. (C) Completely closed cages of clathrin, as shown in this cryo-electron microscopy reconstruction, form around budding vesicles and contribute energy to the budding process. (A, adapted from E. Ungewickell and D. Branton, *Nature* 289:420, 1981; B, adapted from I. S. Nathke et al., *Cell* 68:899, 1992; C, courtesy of B. M. F. Pearse, from C. J. Smith et al., *EMBO J.* 17:4943, 1998.)

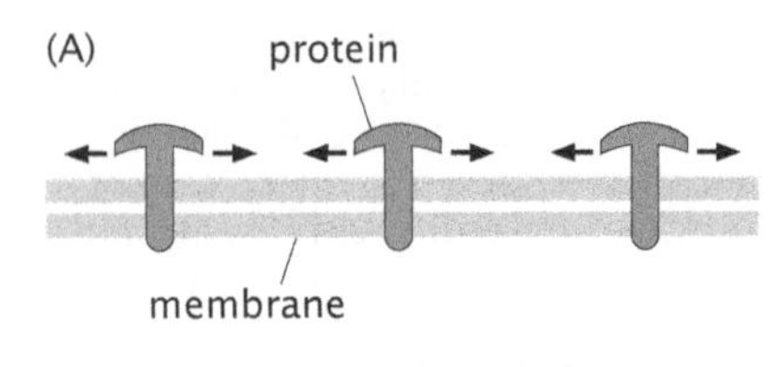

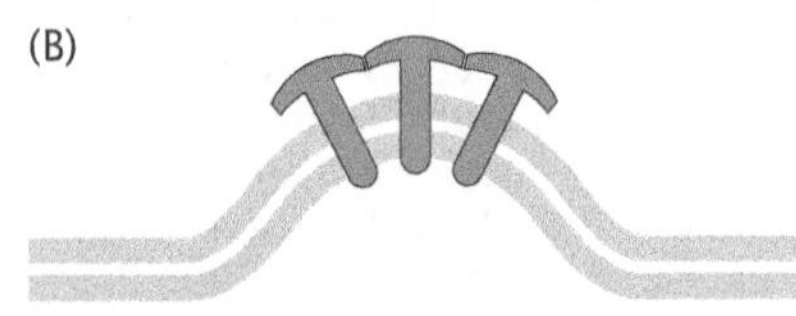

Figure 11.31: Protein-induced membrane curvature. (A) The most energetically favorable shape for a small patch of membrane will typically be a flat plane in the absence of spontaneous-curvature-inducing lipids. Membrane-embedded proteins can move laterally within this plane. (B) If some of the proteins have an affinity for one another in a curved conformation, the favorable energy of protein–protein binding can outweigh the unfavorable energy of membrane bending.

carrying cargo from the Golgi to the plasma membrane. Figure 11.30 shows the fascinating structure of the clathrin units that make the cage structure responsible for vesicle budding.

Though the molecular details are rich and complicated, from a physical perspective, the basic idea of such vesicle chaperones is that because of their mutual favorable interaction, these proteins self-associate. At the same time, we can think of these proteins as carrying some spontaneous curvature as shown in the schematic of Figure 11.31, which takes the otherwise flat membrane and bends it into a bud. We can estimate a bound on the interaction energy between the subunits of the chaperone coat, since this favorable interaction has to overcome the 250–500 k_BT of unfavorable bending energy it takes to make the vesicle.

There Is an Energy Cost to Disassemble Coated Vesicles

When protein-coated vesicles have been shed from a given membrane, they are still covered with their protein budding partners. Interestingly, in order to disassemble this coat, an energy price must be paid in the form of ATP hydrolysis. This follows from the fact that the formation of the protein coat surrounding the vesicle was a downhill reaction energetically. If we adopt a minimal model in which it is assumed that the binding energy to form the vesicle is exactly compensated by the bending energy (that is, 500 k_BT), then the number of ATP equivalents (20 k_BT) it will take to remove the coat is of order 25.

11.4 Fusion and Fission

Within a living eukaryotic cell, formation of vesicles is rarely an end in itself. Instead, vesicles are usually intermediate steps in a complex membrane traffic system that moves proteins and other cellular components from one location to another and enables complex morphological specialization of different domains of the cell, such as the asymmetry of membrane structures between two sides of epithelial cells facing the inside of the gut and the bloodstream, respectively. For these biological purposes, the formation of vesicles by membrane budding and the addition of vesicle material to other membranes by vesicle fusion must be carefully regulated, efficient, and rapid. It appears to be a common solution to this problem for cells to employ sets of proteins called SNAREs to catalyze the process of fusion, as illustrated in Figure 11.32. Conversely, vesicle budding also normally involves a set of specialized proteins that cooperate with the coat proteins such as clathrin to ease the path of membrane fission for a budding vesicle and dynamin, which pinches off the nascent vesicles.

Both the flat membrane and the sphere are perfectly happy of themselves. As a result, to make them fuse or separate, there are protein machines whose job is to lower the free energy of the intermediate. GTP hydrolysis is involved in making these processes occur. Though the real details of the mechanisms are unknown, all these things share the appearance that the proteins function to lower the barrier to fusion or fission by manipulating the geometry of the membrane system. In the future, it will be an interesting challenge to see if the effects of these proteins can be modeled convincingly.

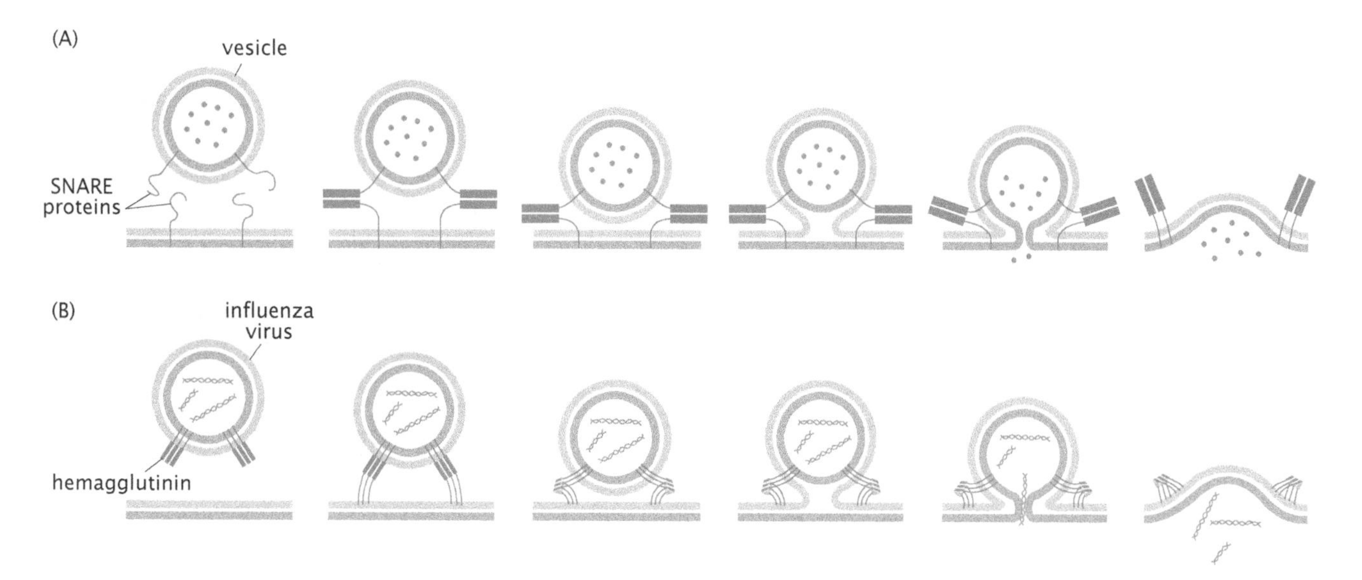

Figure 11.32: Vesicle fusion catalyzed by proteins. (A) During normal membrane trafficking in eukaryotic cells, specialized SNARE proteins help transport vesicles find their appropriate targets and fuse. Complementary SNARE proteins in the membranes of the vesicle and the target form tightly bundled helices with one another, pulling the two members physically close together. Fusion is thought to proceed in several steps, first with fusion of the outer leaflet, forming an unstable intermediate, and then fusion of the inner leaflets, forming a leaky pore, and eventually a completely fused new membrane. (B) Viruses may use analogous processes to fuse with cell membranes. For example, the influenza virus is initially taken up into host cell endosomes and then fusion of the viral membrane with the endosomal membrane ejects the virus genome into the cytoplasm of the host cell. Fusion is mediated by the viral hemagglutinin protein, which undergoes a pH-dependent conformational change as the endosome is acidified.

11.4.1 Pinching Vesicles: The Story of Dynamin

Our study of trafficking can shed yet further light on the energetics associated with membrane rearrangements in cells. As shown in Figure 11.33, the last step prior to the existence of a detached (but coated) vesicle is that the tube of membrane that links the vesicle to the membrane must be severed. The protein dynamin, a GTPase that polymerizes on tubules, is thought to pinch off the last connection between the vesicle and the membrane from which it has been produced.

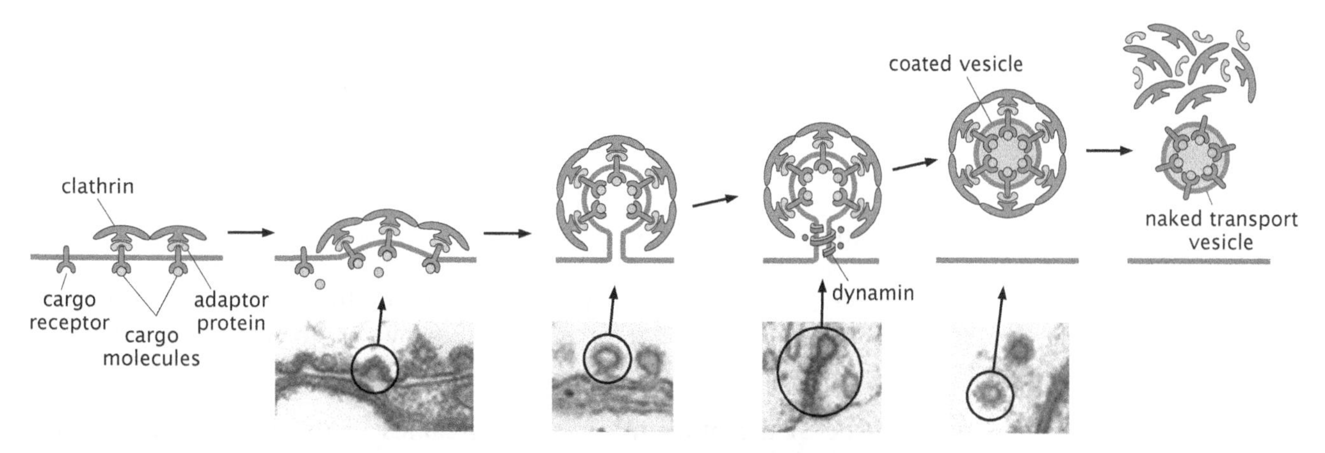

Figure 11.33: Membrane scission by accessory proteins. The electron microscopy images show clathrin-coated vesicles in various stages of the process of vesicle creation. The schematics above show the hypothesized stages in the process of membrane scission and subsequent formation of a vesicle filled with cargo. (Top, adapted from B. Alberts et al., Molecular Biology of the Cell, 5th ed., Garland Science, 2008; bottom, adapted from V. I. Slepnev and P. De Camilli, *Nat. Rev. Neurosci.* 1:161, 2000).

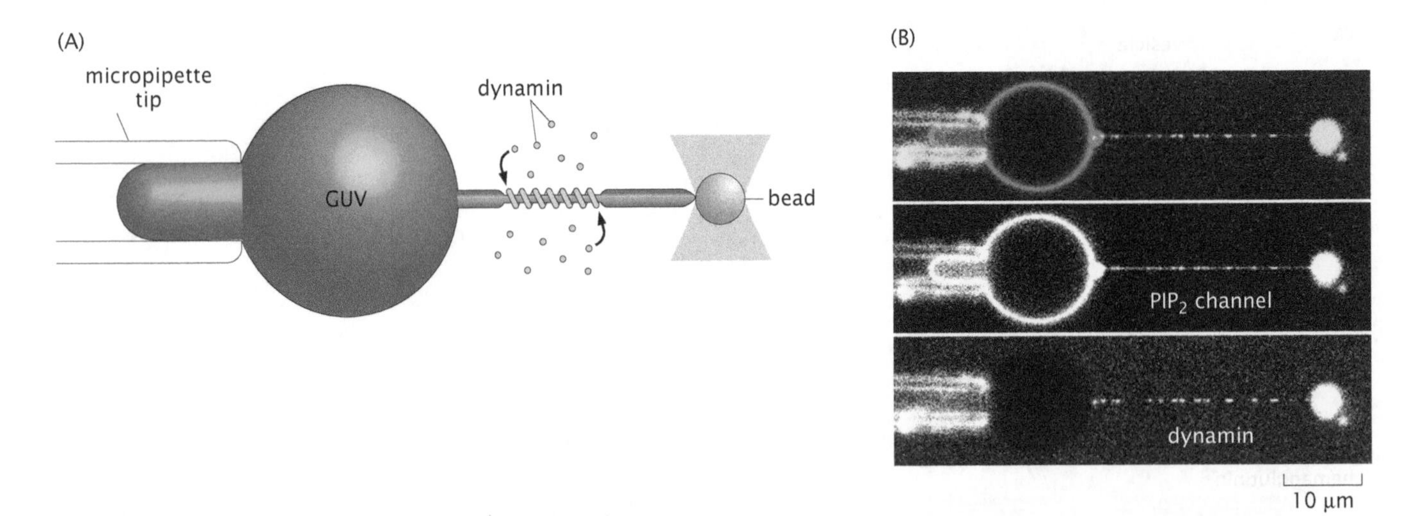

Figure 11.34: Dynamin and tethers. (A) Schematic of the experiment in which a tether is pulled from a vesicle and dynamin (green) is permitted to polymerize on the tether. (B) Image of the experiment using confocal microscopy to examine several fluorescence channels simultaneously. A subset of the lipids in the vesicle are labeled in red (GloPIP2) and the dynamin is labeled with Alexa 488 (green). (Adapted from A. Roux et al., *Proc. Natl Acad. Sci. USA* 107:4141, 2010.)

To learn more about the details of processes like those described above, a series of elegant *in vitro* experiments have been performed based precisely on the physical ideas presented earlier in this chapter. Indeed, one of our objectives in presenting this example is to show the rich interplay between simple theoretical ideas and clever *in vitro* biophysical experiments that can, in turn, shed light on a topic that is absolutely central to cell biology. As shown in Figure 11.34, a tubule can be pulled from a pipette-bound vesicle as discussed earlier in the chapter (p. 450). By exploiting the relation between force and radius and the bending modulus and tension already worked out there (that is, Equations 11.24 and 11.25 on p. 453), two key insights have been garnered. First, as a result of the polymerization of the dynamin on the tubule, the pulling force exerted by the optical trap is relaxed since the polymer of dynamin itself exerts a force that prevents the tubule from retracting. This makes it possible to measure how much force is imposed on tubules by the presence of dynamin. The second insight we will explore that emerges from this experiment has to do with how the propensity of the dynamin to polymerize depends upon the preexisting curvature of the tubule.

To explore the polymerization force, a tubule is pulled as already indicated schematically in Figure 11.34(A). With an equilibrated tubule, dynamin is then injected into the experimental chamber as shown in Figure 11.35 and the subsequent force is monitored as reflected by the position of the bead in the trap. The logic of how bead position in the trap is used to obtain force was already illustrated in Figure 5.13 (p. 207) and the associated text and it is precisely these kinds of ideas that are in play to make the force measurements here. While measuring the force, simultaneous fluorescence observations are performed to determine the stages of dynamin polymerization on the tubule.

One of the other interesting outcomes of this experimental setup is the ability to measure the extent to which dynamin will nucleate on the tubules as a function of the radius of the tether of interest. As shown in Figure 11.36, by controlling the tension τ in the membrane through application of pipette pressure, it is possible to tune

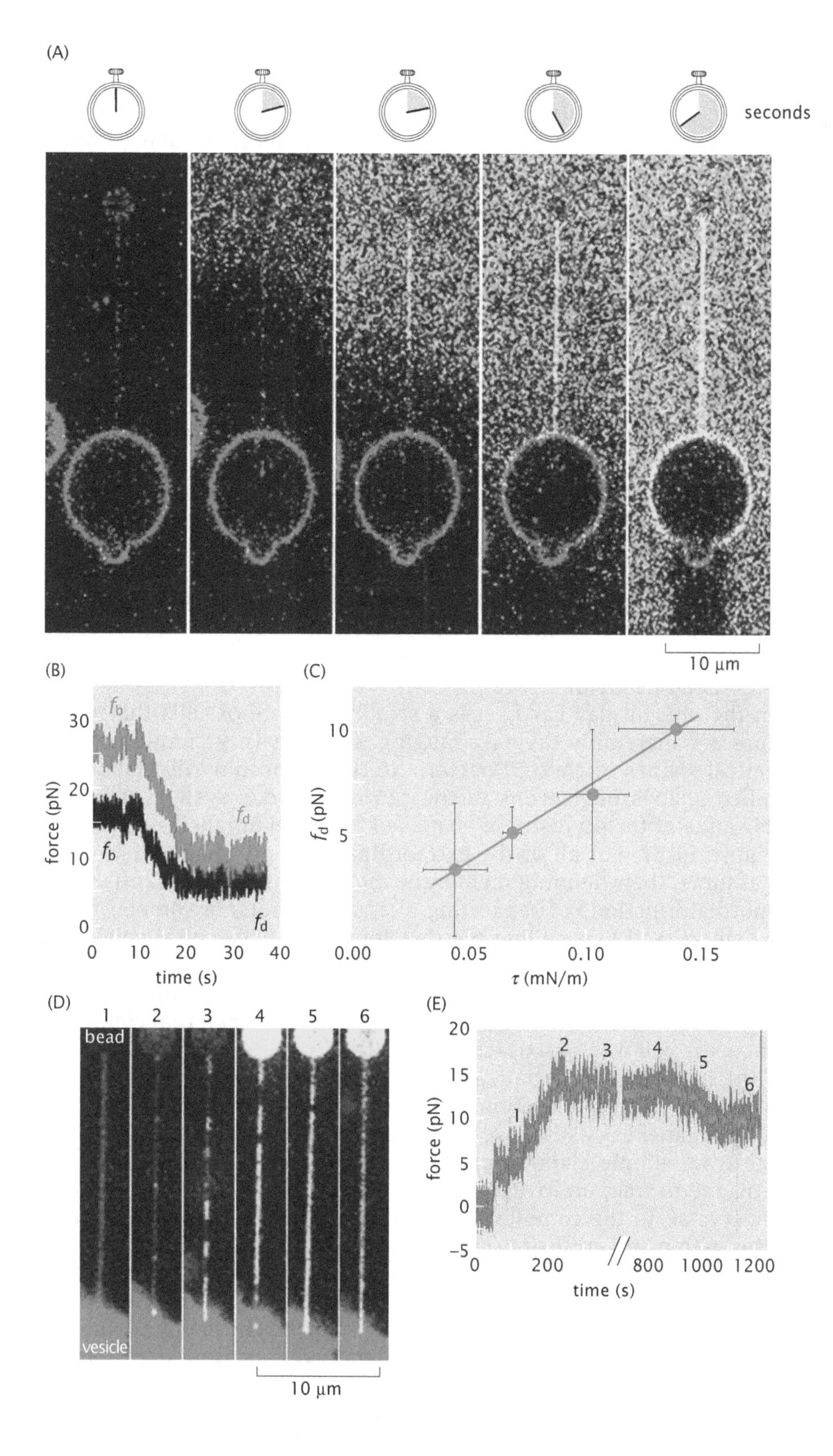

Figure 11.35: Measurement of the force due to dynamin polymerization. (A) Confocal images showing the tubule at various time instants after the injection of fluorescently labeled dynamin (green). The time is measured in seconds and the amount of dynamin used was 12 μM. (B) The force required to hold the tubule is measured using an optical trap. The graph shows two force histories for different vesicles after the injection of dynamin into the chamber. f_b refers to the force to hold the tube before dynamin is present and f_d refers to the tube-holding force after injection of dynamin. (C) Dependence of force on membrane tension. (D) Nucleation and growth of dynamin polymers on the tubule. The concentration used in this case is 440 nM. (E) Force profile for series of images shown in (D) and labeled with different numbers. (Adapted from A. Roux et al., *Proc. Natl Acad. Sci. USA* 107:4141, 2010.)

the radius of the tubule according to the equation

$$r = \frac{f}{4\pi\tau}. \tag{11.28}$$

By simultaneously monitoring the tubules in the fluorescent channel corresponding to the labeled dynamin, one can determine when the tubules begin to be occupied by nascent dynamin filaments. The outcome of a variety of such experiments is shown in Figure 11.36(C). All told, these experiments reveal the power of biophysical experiments to precisely measure the properties of individual molecular players such as dynamin.

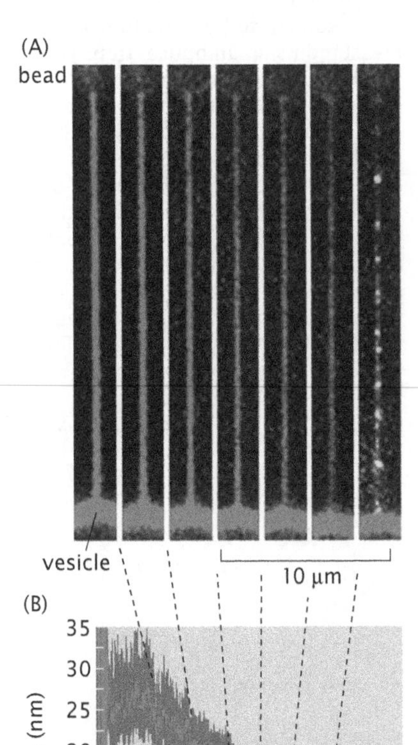

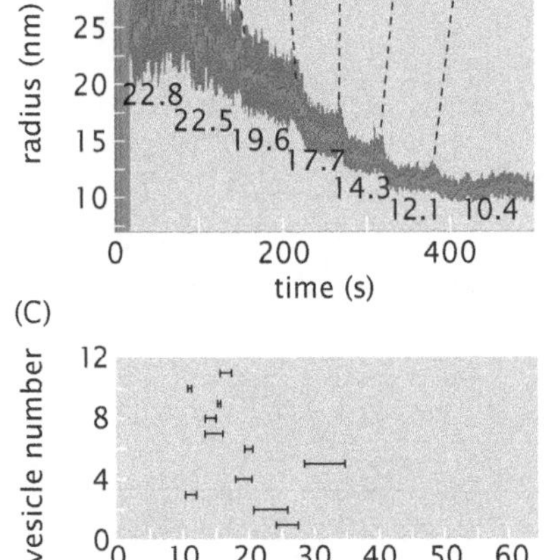

Figure 11.36: Dependence of dynamin polymerization on tubule radius. (A) Images of the tubule as a function of time. The red fluorescence channel corresponds to the membrane and the green fluorescence channel to dynamin. Polymerization of the dynamin occurs only in the last of the images and corresponds to the smallest of the radii shown. (B) Reduction of the tubule radius as a result of stepwise changes in the pipette pressure. (C) Critical radii for dynamin polymerization measured using this experimental technique for 11 different vesicles. (Adapted from A. Roux et al., *Proc. Natl Acad. Sci. USA* 107:4141, 2010.)

11.5 Membranes and Shape

In the early days of cellular life on Earth, the primary function of the membrane was probably fairly simple. It provided at least two important functions. By separating inside from outside with a selectively permeable barrier, the membrane was critical in maintaining concentration gradients of certain molecules and ions. Second, by separating self from non-self, the membrane was critical in establishing the identity of individual cell lineages and thereby providing a framework of inheritance on which competition and natural selection could act. As soon as these lineages were established early in evolutionary history, cells began to compete with one another for resources and also began to specialize. For many kinds of cell specialization, size, shape, and motion are critical and can confer selective advantage. At this point, the function of the membrane moved beyond being a simple barrier to being a structural element contributing to many aspects of cell function and behavior.

On the present-day Earth, it is a striking feature of cells that their shapes are incredibly diverse, ranging from the very simple, nearly spherical shapes of coccoid bacteria to the extraordinarily elaborate, ramified shapes of neurons in the central nervous system of mammals. Some of the diversity of shapes of cells and organelles is shown in Figure 11.37. For all of the extraordinarily many shapes that cells can assume, their defining membrane must maintain its essential and primordial functions of separating inside from outside and self from non-self, as well as enabling the development of structural complexity. For modern cells, the elaboration of shape depends not only upon the mechanical properties of the bilayer, but also on its interactions with internal elements such as the cytoskeleton and external elements such as the cell wall or extracellular matrix.

With increasing specialization and diversification of cellular metabolism during evolution, a variety of intracellular membrane-bound organelles also arose. Intracellular membranous organelles range from simple elaborations of the boundary membrane such as the magnetosome, an invagination of the membrane housing a magnetic crystal, to the complex, multicompartment secretory apparatus of the mammary epithelial cells that produce milk. The shapes of intracellular membranous organelles are nearly as diverse as the shapes of cells themselves ranging from simple, nearly spherical vesicles to elaborate tubular networks such as those found in the endoplasmic reticulum and stacks of flattened pancake-like cisternae as in the Golgi apparatus.

What gives rise to this broad variety of cell and organelle shapes and to what extent are membrane deformations a part of the answer to this question?

11.5.1 The Shapes of Organelles

Many membrane-bound organelles have shapes that are not even approximately spherical. For example, Figure 11.38 (and Figures 2.18

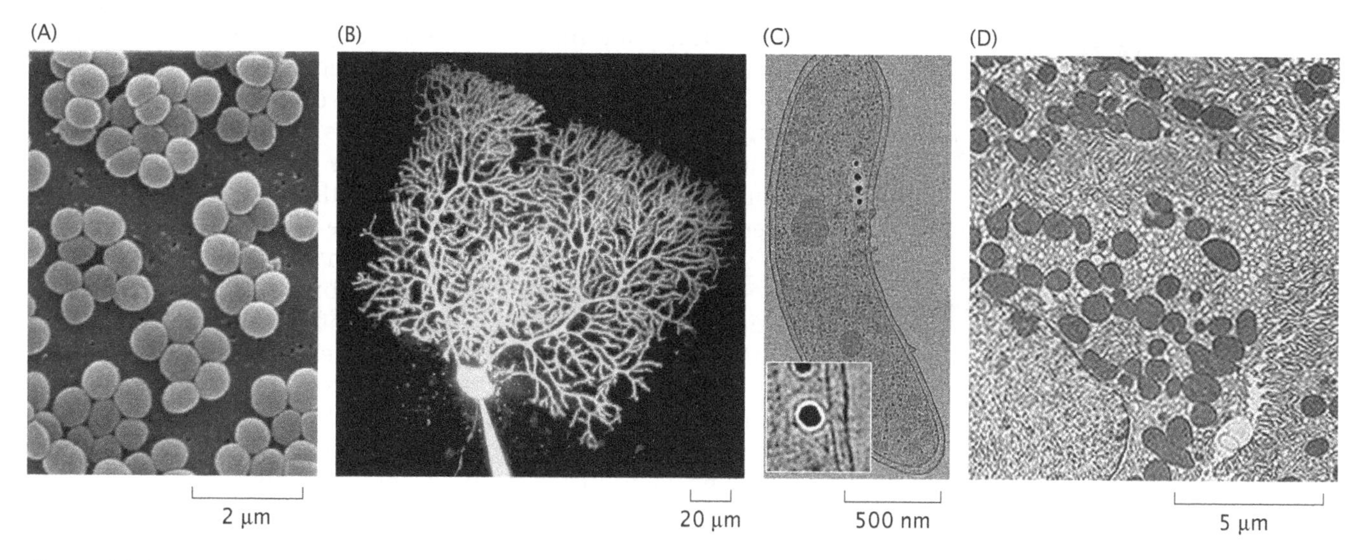

Figure 11.37: Structural variation in the complexity of shapes of cells and organelles. (A) *Staphylococcus aureus* is a nearly spherical bacterium, one of the simplest imaginable cell shapes. (B) A Purkinje cell in a mammalian cerebellum forms an elaborate branched tree-like structure as revealed here by injecting the cell body with a fluorescent molecule. (C) Magnetosomes are simple, nearly spherical organelles found in magnetotactic bacteria. Within the organelle membrane, which is contiguous with the plasma membrane, is housed a crystal of magnetite as can be seen in the inset. (D) Mammalian cells specialized for secretion, such as the parietal cell from the stomach shown in this micrograph, may have extremely complex membrane-bound organelles. (A, adapted from Public Health Image Library—image 6486; B, adapted from S.-H. Wang et al., *Nat. Neurosci.* 3:1266, 2000; C, adapted from A. Komeili et al., *Science* 311:242, 2006; D, adapted from P. C. Cross and K. L. Mercer, Cell and Tissue Ultrastructure, W. H. Freeman, 1993.)

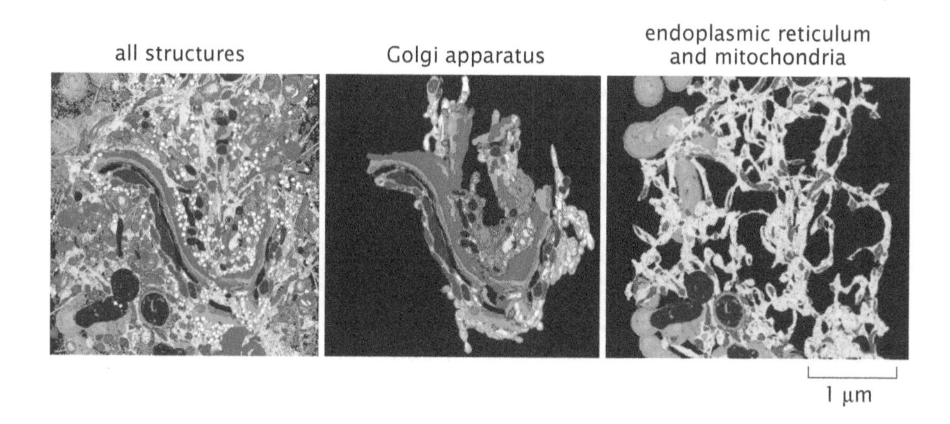

Figure 11.38: Varieties of organelle shapes within a single cell. The extent of all membranous structures in a cultured pancreatic cell was laboriously reconstructed by tracing their outlines on serial thin section electron microscopy images. All the organelles within a single small segment of the cell are shown, with different compartments represented by different colors. Separate views of the Golgi apparatus, endoplasmic reticulum, and mitochondria within this region are separated out so that their elegant and varied shapes can be appreciated. (Adapted from B. Marsh et al., *Proc. Natl Acad. Sci. USA* 98:2399, 2001.)

(p. 55), 2.23 (p. 61), 2.24 (p. 61), and 2.25 (p. 62)) shows the structures adopted by the mitochondria, endoplasmic reticulum, and the Golgi apparatus. One of the most visually striking features of these organelles is the convoluted and layered structures adopted by their membranes. Interestingly, despite these complex shapes, from one cell to the next, the same stereotyped structures appear again and again. Clearly, one of the most intriguing questions at the interface between cell biology and membrane biophysics is what drives and maintains these structures?

The Surface Area of Membranes Due to Pleating Is So Large That Organelles Can Have Far More Area than the Plasma Membrane

One useful way to gain intuition about the nature of organellar membranes is to examine how much area is tied up in them. As shown in Figure 11.39, the mitochondria are built up of two distinct membrane

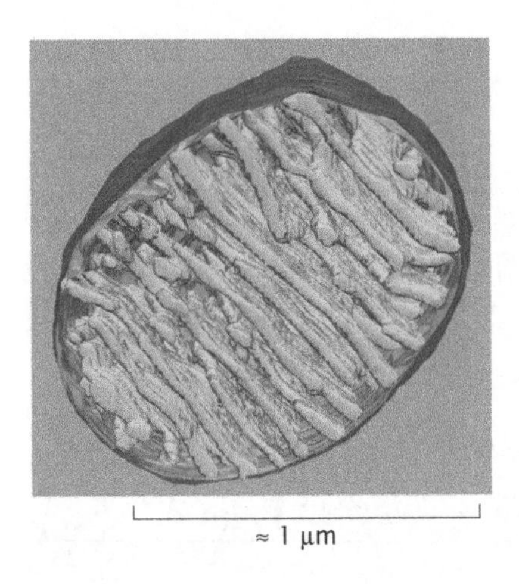

Figure 11.39: Cryo-electron microscopy reconstruction of the structure of a mitochondrion, shown cut in half to reveal the internal structure. (Courtesy of T. Frey.)

systems. An outer bounding membrane, roughly the shape of a bacterium, encloses two distinct regions. The second membrane system (the inner mitochondrial membrane) has a complex topology, often involving many parallel layers known as cristae. Interestingly, as indicated in schematic form in Figure 11.40, the endoplasmic reticulum may be thought of roughly as a structure with inverted properties to those presented by mitochondria in the sense that it too involves two membrane systems, but with the outer one harboring the larger area in this case. In particular, the nuclear membrane can be thought of as the inner surface of the enclosed region of the endoplasmic reticulum. In this case, there are protrusions emerging from the nucleus forming a second membrane system, and the net area of this second membrane is much larger than that of the nuclear membrane. The relative areas of these different membrane systems can be evaluated using mitochondria as a case study as shown below.

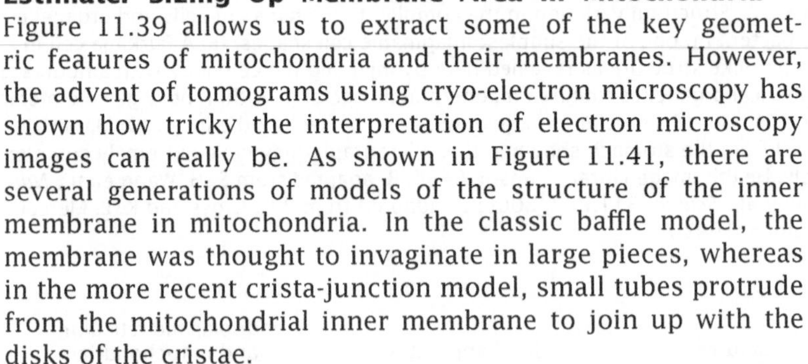

Estimate: Sizing Up Membrane Area in Mitochondria
Figure 11.39 allows us to extract some of the key geometric features of mitochondria and their membranes. However, the advent of tomograms using cryo-electron microscopy has shown how tricky the interpretation of electron microscopy images can really be. As shown in Figure 11.41, there are several generations of models of the structure of the inner membrane in mitochondria. In the classic baffle model, the membrane was thought to invaginate in large pieces, whereas in the more recent crista-junction model, small tubes protrude from the mitochondrial inner membrane to join up with the disks of the cristae.

To characterize the geometry of mitochondria, we assume that an "average" mitochondrion is a spherocylinder with length $1\,\mu\text{m}$ and diameter $0.8\,\mu\text{m}$. The volume of such a structure is

$$\Omega_{\text{mito}} = \tfrac{4}{3}\pi r^3 + \pi r^2 h \approx 4 \times (0.4\,\mu\text{m})^3 + 3 \times (0.4\,\mu\text{m})^2 \times 0.2\,\mu\text{m} \approx 0.4\,\mu\text{m}^3. \quad (11.29)$$

An estimation that reflects more directly on mitochondrial function is to examine the area of the inner and outer membranes of the mitochondrion as well as the volume enclosed in the different spaces within. The area of the outer membrane can be estimated as

$$A^{\text{out}}_{\text{mito}} = 4\pi r^2 + 2\pi r h \approx 4 \times 3 \times (0.4\,\mu\text{m})^2 + 2 \times 3 \times 0.4\,\mu\text{m} \times 0.2\,\mu\text{m} \approx 2.5\,\mu\text{m}^2, \quad (11.30)$$

where we have followed our usual rules for rounding and taken $\pi \approx 3$. The area of the inner membrane can be estimated by assuming that the cristae are a series of disks that are packed along the long axis of the mitochondrion. As a result, the total area associated with the inner membrane is estimated to be

$$A^{\text{inner}}_{\text{mito}} = A^{\text{outer}}_{\text{mito}} + 2N_{\text{disk}}A_{\text{disk}} \approx 2.5\,\mu\text{m}^2 + 2 \times 15 \times 3 \times (0.4\,\mu\text{m})^2 \approx 20\,\mu\text{m}^2, \quad (11.31)$$

where the factor of 2 multiplying the number of disks accounts for the fact that the cristae are hollow pancakes and we have

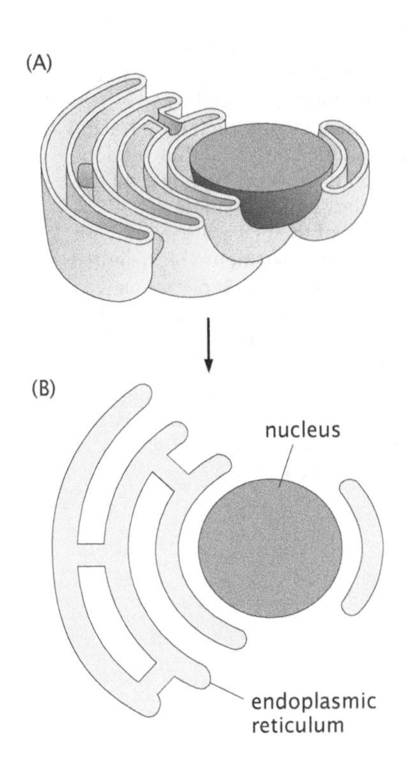

Figure 11.40: Schematic of the structure of the endoplasmic reticulum. (A) Three-dimensional schematic of a cut through a eukaryotic cell showing the endoplasmic reticulum and the nucleus. (B) A two-dimensional model of the membrane system formed by the endoplasmic reticulum.

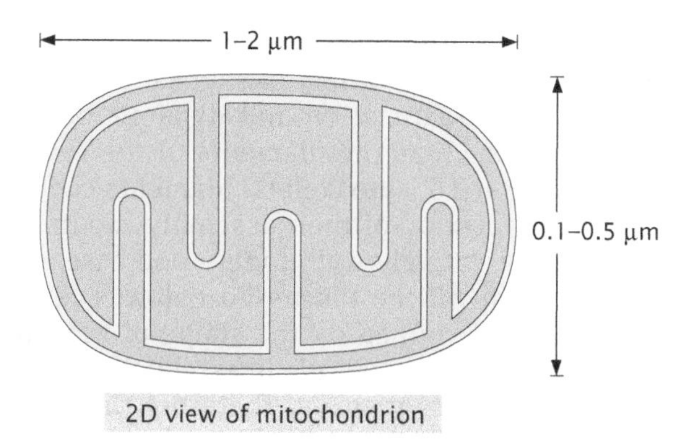

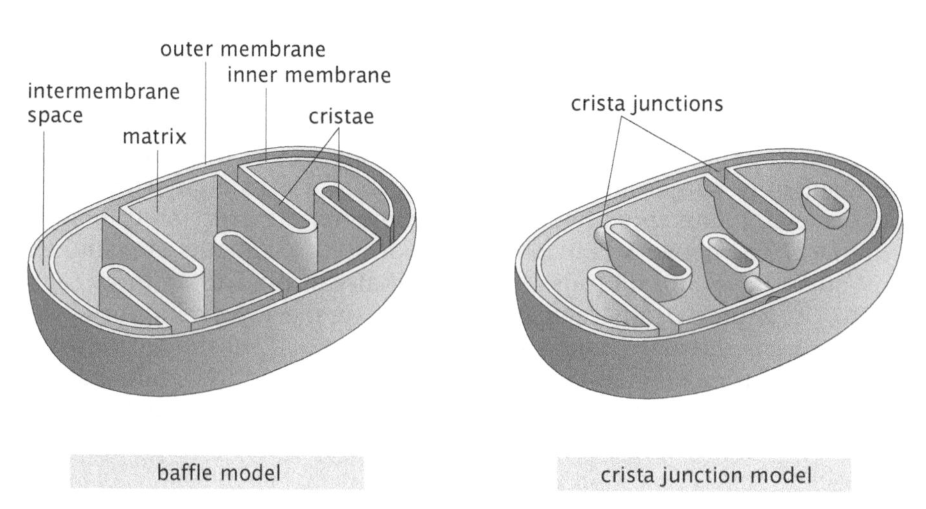

Figure 11.41: Cartoons of competing models of mitochondria. Planar cuts through the two models are consistent with 2D images from electron microscopy. The crista-junction model emerged as a result of careful studies using cryo-electron microscopy.

taken $N_{\text{disk}} = 15$. This number should be compared with the area of the entire plasma membrane itself, which for a fibroblast we saw in Figure 2.22 (p. 60) is roughly $2000\,\mu\text{m}^2$. What we see is that a single mitochondrion by itself already carries a respectable area in comparison with the total membrane area of the cell. Hence, when we add up the membrane area associated with all organelles, it is in excess of the area devoted to the plasma membrane.

11.5.2 The Shapes of Cells

The special elastic properties of membranes are necessary for determining a wide variety of shapes in different biological contexts including organelles. However, at the level of the entire cell, membrane properties alone are not sufficient to determine cell shape. Cells are full of stiffening agents to support the phospholipid membrane in some particular shape. These stiffening agents are generally elastic beams such as those described in the previous chapter that may be assembled or cross-linked into large-scale networks with remarkable elastic strength. These structural supports may be found inside or outside the cell membrane, or both. The cell wall of plants, for example, is constructed of a dense, cross-linked meshwork of cellulose fibers that is sufficiently strong to allow a Redwood tree to reach a mature height greater than 100 m and sufficiently tough to survive the death of the cells themselves and even the death of the tree to be used by humans for building furniture and houses. Likewise, fungi, bacteria, and archaea often possess stiff external cell walls.

Animal cells, on the other hand, never have external cell walls and yet they come in a fantastic variety of shapes. Generally, these shapes arise because of the organization and dynamic behavior of internal structural elements such as the filaments of the cytoskeleton. As we will discuss in Chapter 16, cytoskeletal filaments can assemble to generate force and the location of their assembly can be regulated by the cell. A "simple" and relatively well-understood case is the mammalian red blood cell, which will be discussed below. Even for cells whose shape is primarily determined by the shape of the external cell wall, the cytoskeleton plays a critical role in establishing the distribution of cell wall biosynthetic enzymes and therefore the overall shape. This is thought to be true for organisms as diverse as plants and bacteria. There also may be information communicated in reverse such that the distribution of cell wall elements influences the internal structure of the cytoskeleton.

The characteristic shape and mechanical properties of red blood cells are not solely determined by the properties of the lipid bilayer. Immediately subjacent to the membrane is a regular meshwork of cytoskeletal filaments attached to transmembrane proteins to hold it in place. The network is made up of short actin filaments crosslinked by spectrin, a long fibrous molecule, that can bind to actin at each end. The geometrically regular spectrin–actin meshwork supports and strengthens the membrane and, indeed, it is most fruitful to consider these structural elements acting as a unit. Because of the contribution of the supporting cytoskeletal network, the membrane no longer acts as a fluid and can support shear stresses.

The Equilibrium Shapes of Red Blood Cells Can Be Found by Minimizing the Free Energy

The simplest model of the free energy required to explore the phase diagram of red blood cell shapes uses only the bending-energy term introduced earlier in the chapter, but with the added constraint that we consider only those competitors for which both the total surface area of the membrane and the volume enclosed by that surface are fixed. However, this scheme is unable to account for the entire sequence of shapes in the series stomatocyte–discocyte–echinocyte evidenced in Figure 5.10 (p. 204). In order to account for the richness and diversity of red blood cell shapes, it is necessary to expand the simple bending-energy model in two key ways. First, a term must be added to the free energy that accounts for the free-energy cost associated with having the area of the inner and outer leaflets differ from some equilibrium value ΔA_0. The second key extension of the model is induced by the presence of the underlying cytoskeletal network. As a result of the presence of the spectrin network, elastic terms must be introduced that account explicitly for the cost to shear and stretch the cell. With these extra features in hand, the model can predict a progression of shapes as shown in Figure 5.10 (p. 204). From top to bottom in that figure, the parameter being tuned is the difference in area between the inner and outer leaflets of the membrane. In the top image, the inner leaflet is roughly 1% larger than the outer leaflet, and in the bottom image, the outer leaflet has roughly 2% greater area than the inner leaflet.

This same model has also been used as the basis of theoretical investigations of the deformation of red blood cells as they pass through capillaries or artificial channels as shown in Figure 2.21 (p. 59). The key insight to emerge from these calculations is the ability of such a simple model to account for the observed red blood cell shapes both under different solution conditions and in the presence of forces that deform the cells. Although red blood cells have relatively simple shapes, the expanded model including the two elements of area difference on the inside versus outside and the attachment to the cytoskeleton supporting shear stress can serve as a first step toward exploring shape determination for much more complex cells such as neurons.

11.6 The Active Membrane

11.6.1 Mechanosensitive Ion Channels and Membrane Elasticity

Mechanosensitive Ion Channels Respond to Membrane Tension

Although we have acknowledged that proteins comprise a major fraction of membrane components, we have not yet explicitly examined how embedded proteins might influence membrane mechanics and how membrane mechanics might in turn affect protein function. One intriguing biological application of the ideas developed thus far in this chapter is to bacterial mechanosensitive ion channels already introduced in Section 7.1.2 (p. 286). Mechanosensitive channels in bacteria appear to serve as safety valves to protect against membrane rupture due to osmotic imbalance. In particular, when a bacterial cell is subjected to osmotic shock, the resulting flow of water across the cell membrane results in a rapid change in membrane tension. These channels reply by opening, saving the cell from bursting.

We will emphasize the mechanosensitive channel of large conductance (MscL) as a model system that permits an examination of the way in which gating is tied to membrane deformations and external tension. One of the features that makes this an especially appealing model problem is the existence of both structural and functional data. As shown in Figure 11.42, the structure of this channel is known in the closed state and plausible hypotheses have been advanced for its structure in the open state as well. Similarly, patch clamp experiments

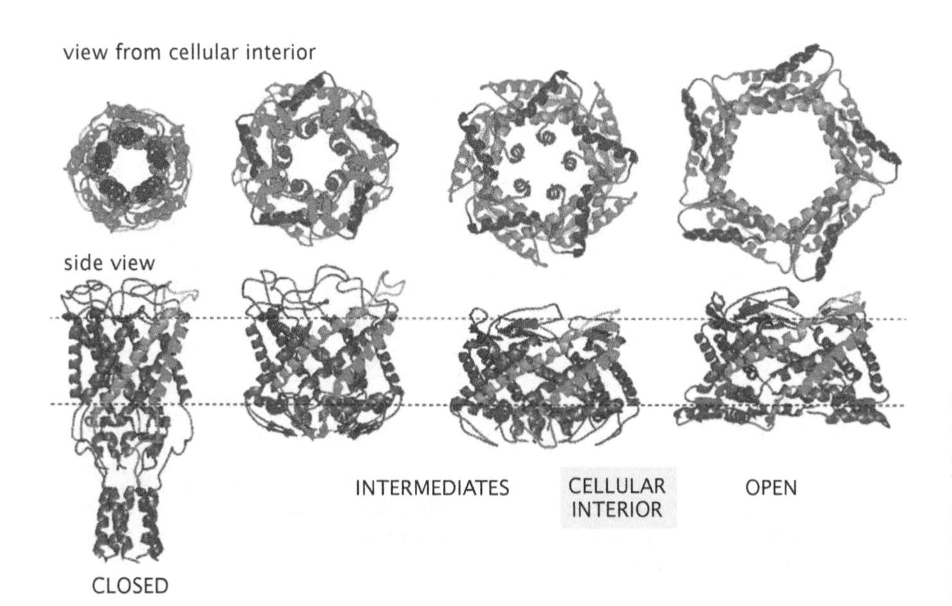

Figure 11.42: Schematic of MscL. The figure shows the closed state as determined using X-ray crystallography. The intermediate states and the open states are models based upon various other kinds of data. (Adapted from S. Sukharev et al., *Biophys. J.* 81:917, 2001.)

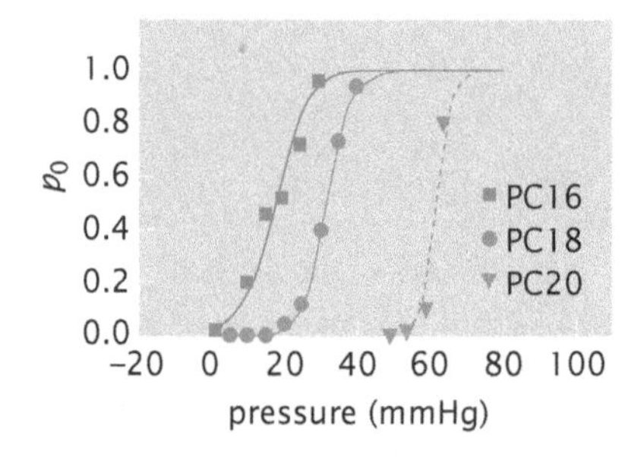

Figure 11.43: Ion channel open probability. The graph shows the open probability as a function of the pressure for the mechanosensitive channel, MscL. Different curves correspond to different length tails for the lipids in the surrounding membrane. The particular cases are tails with 16, 18, and 20 carbon atoms in their backbone. (Adapted from E. Perozo et al., *Nat. Struct. Biol.* 9:696, 2002.)

reveal the relation between the applied tension and the current flowing through the channel. One of the key observables to emerge from these experiments is the open probability p_{open} as a function of the applied pressure as shown in Figure 11.43. This figure goes beyond the version shown in Figure 7.5 (p. 287) by illustrating the intriguing way in which the open probability depends upon the lengths of the lipid tails of the membrane in which the channel lives. The fact that the open probability depends upon lipid-tail length provides a clue as to the importance of membrane deformation in dictating part of the free-energy budget associated with channel gating. As a result, we now consider how membrane proteins deform the lipid bilayer that surrounds them.

11.6.2 Elastic Deformations of Membranes Produced by Proteins

Proteins Induce Elastic Deformations in the Surrounding Membrane

As argued at the beginning of this chapter, the liveliness of membranes owes much to the activity of their proteins. The hypothesis we explore now is that the energetics of the surrounding membrane contributes to the functioning of a wide class of membrane proteins. Our examples of choice are the mechanosensitive proteins because of the availability of quantitative data such as those shown in Figure 11.43. This class of ion channels is gated by the presence of tension in the surrounding membrane. However, more generally, any membrane protein that undergoes some conformational change that alters the shape that it presents to the surrounding membrane will deform that membrane. The interesting idea explored in this section is that this membrane deformation feeds back to the protein and can alter its conformational preference.

To see how membrane deformation might couple to protein function, we begin by considering a membrane protein like that shown schematically in Figure 11.44(A). For mathematical convenience, we consider the one-dimensional geometry shown in Figure 11.44(B), where we define the functions $h_+(x)$ and $h_-(x)$ that characterize the height of the upper and lower leaflets of the lipid bilayer. The basis for this figure is the idea that, like their lipid partners, membrane

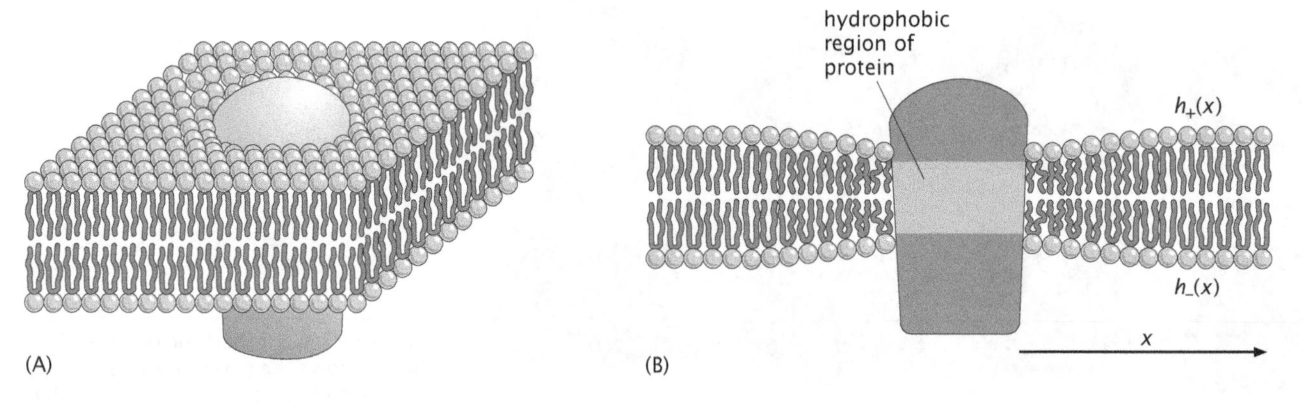

Figure 11.44: Protein-induced membrane deformation. (A) Protein in a membrane. (B) Schematic showing the nature of the deformations in the vicinity of a membrane protein. The heights of the upper and lower leaflets of the bilayer are defined by the two fields, $h_+(x)$ and $h_-(x)$. The lipids near the protein are deformed as a result of a hydrophobic matching to the hydrophobic patches of the protein. The lipid schematic ignores the fact that the lipids are fluctuating, resulting in many different conformations.

proteins present hydrophobic patches that have an energetic imperative to be sequestered from water and one way they can satisfy this need is by gluing to the tails of the surrounding lipids. However, the lengths of the lipid tails dictate an equilibrium thickness for the membrane. Further, the hydrophobic thicknesses of the bilayer and the protein might not be the same. This results in an energetically costly transition region where the lipids are deformed to have a length intermediate between their far-field equilibrium value and that of the protein. Obviously, to really characterize the deformations induced by membrane proteins, we need to perform a full two-dimensional calculation which is left to the problems at the end of the chapter. Further, this highly idealized picture ignores the fact that the lipids are engaged in a constant thermal dance and are jiggling around rather than statically distributed around the protein.

To capture the geometry of protein-induced membrane deformation, it is convenient to define two related sets of fields, namely,

$$h(x) = \frac{h_+(x) + h_-(x)}{2} \tag{11.32}$$

and

$$w(x) = \frac{h_+(x) - h_-(x)}{2}. \tag{11.33}$$

We can think of $h(x)$ as defining the midplane of the lipid bilayer, while $w(x)$ defines the half-width of the bilayer at position x. In light of these definitions, we can rewrite the leaflet variables as $h_+(x) = h(x) + w(x)$ and $h_-(x) = h(x) - w(x)$.

Protein-Induced Membrane Bending Has an Associated Free-Energy Cost

Figure 11.44 shows that there can be bending deformations caused by membrane proteins. In Section 11.2.2, we showed how to think about the energetics of membrane bending, and we now adapt that treatment to the problem of membrane proteins. From a microscopic perspective, this term may be viewed as the penalty incurred for locally changing the orientation of the lipid molecules in the vicinity of the membrane protein. In terms of the fields $h_+(x)$ and $h_-(x)$, this term may be written as

$$G_{\text{bend}}[h_+(x), h_-(x)] = \frac{K_{\text{b}}^{(1)}}{2} \int \mathrm{d}x \left[\left(\frac{\partial^2 h_+(x)}{\partial x^2} \right)^2 + \left(\frac{\partial^2 h_-(x)}{\partial x^2} \right)^2 \right], \tag{11.34}$$

where $K_{\text{b}}^{(1)} = K_{\text{b}}/2$ is the bending modulus for an individual leaflet, and is half that of the membrane as a whole. The bending contribution to the free energy is more conveniently written in terms of the fields $h(x)$ and $w(x)$ as

$$G_{\text{bend}}[h(x), u(x)] = \frac{K_{\text{b}}}{2} \int \mathrm{d}x \left[\left(\frac{\partial^2 h(x)}{\partial x^2} \right)^2 + \left(\frac{\partial^2 w(x)}{\partial x^2} \right)^2 \right]. \tag{11.35}$$

This equation tells us that both the midplane and thickness variations have associated bending energies.

11.6.3 One-Dimensional Solution for MscL

Membrane Deformations Can Be Obtained by Minimizing the Membrane Free Energy

Our strategy in examining the connection between tension and channel gating is to use a simple model of membrane elasticity to probe the energetic consequences of channel opening. In particular the presence of a channel in the membrane results in a deformation of the surrounding bilayer as already described above. The additional feature associated with gating is that the amount of membrane deformed by virtue of the presence of the channel depends upon the channel conformation. To capture the energetics of channel gating, we resort to a one-dimensional model as shown in Figure 11.45. Our plan is to compute the energy per unit length associated with deforming the membrane due to a *linear* inclusion. Once we have that energy, then, as shown in the figure, we will wrap that line into a circle, attributing an energy per unit length to the result derived using the one-dimensional model. The mathematically correct way of treating the channel inclusion is to consider the full two-dimensional geometry of the problem using cylindrical coordinates. In the axisymmetric situation assumed here, this too will lead to a one-dimensional model for the energy of membrane deformation, but the resulting equations for the minimizing deformation profile would no longer be as simple as the one derived below. While mathematically rigorous, the cylindrical coordinate approach does not add anything to our understanding of the role of membrane mechanics in the gating of MscL.

For simplicity, we focus on a model of a mechanosensitive protein that only induces thickness variation and no midplane bending. In this case, for the one-dimensional problem, the energy depends upon the deformation $u(x) = w(x) - w_0$ as

$$G_{\rm h}[u(x)] = \underbrace{\frac{K_{\rm b}}{2}\int_R^\infty \left(\frac{{\rm d}^2 u}{{\rm d}x^2}\right)^2 {\rm d}x}_{\text{bending energy}} + \underbrace{\frac{K_{\rm t}}{2w_0^2}\int_R^\infty u(x)^2\,{\rm d}x}_{\text{hydrophobic mismatch}}. \qquad (11.36)$$

We introduce the notation $G_{\rm h}$ with subscript "h" to signify that the free energy arises from the *hydrophobic mismatch* between the membrane protein and surrounding lipids. In order to compute the energy associated with membrane deformation, we must find the profile $u(x)$ that minimizes this free energy subject to the boundary conditions. When solving a partial differential equation like that derived from

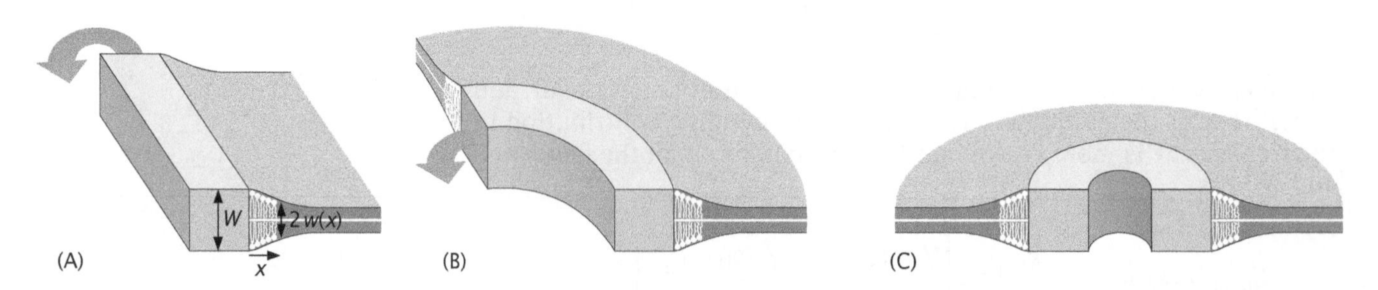

Figure 11.45: Deformation energy due to an ion channel. Schematic showing how to take a one-dimensional elastic problem for an "ion channel" and to turn it into a two-dimensional membrane deformation problem for a cylindrical (approximately) membrane protein such as MscL. (A) Geometry of the one-dimensional problem. (B) Wrapping the one-dimensional solution to create a deformed annulus. (C) The final deformed configuration showing the annulus of deformed material.

Equation 11.36, we have to say something about how the field $u(x)$ behaves at the boundaries—namely, at the channel itself (for $x = R$) and also very far away from the channel (the far field as $x \to \infty$). One of the conditions is the hydrophobic "gluing" condition, which is represented mathematically as

$$u(R) = \tfrac{1}{2}W - w_0, \tag{11.37}$$

and insists that the lipid bilayer at the protein–lipid interface has a half-width given by the size of the hydrophobic patch on the channel, W. A second condition is present on the slope of the field $u(x)$, and is given by

$$u'(R) = 0, \tag{11.38}$$

which says that in this simple model of a cylindrical protein, the upper and lower leaflets of the lipid bilayer glue onto the protein symmetrically. The other two conditions say that the membrane is undisturbed very far from the protein, and are given by

$$u(\infty) = 0 \tag{11.39}$$

and

$$u'(\infty) = 0. \tag{11.40}$$

As was shown in the appendix of Chapter 5, the minimization of energy functionals like that of Equation 11.36 is an exercise in the calculus of variations. Recall that the idea of such calculations is analogous to minimizing an ordinary function by setting its derivative equal to zero. Here, we set the functional derivative equal to zero, and the outcome is a partial differential equation. To compute the functional derivative, we let the function $u(x)$ undergo an excursion $\eta(x)$, $u(x) \to u(x) + \varepsilon\eta(x)$, with the boundary conditions on u above ensuring that η vanishes at the boundaries. In this case, taking the functional equivalent of the derivative to minimize the free-energy functional amounts to

$$\frac{\delta G_\mathrm{h}}{\delta u(x)} = 0 = K_\mathrm{b} \int_R^\infty u_{xx}\eta_{xx}\,\mathrm{d}x + \frac{K_\mathrm{t}}{w_0^2} \int_R^\infty u\eta\,\mathrm{d}x, \tag{11.41}$$

where we use the notation $u_{xx}(x) = \partial^2 u/\partial x^2$. To proceed, we begin by integrating by parts on the term involving η_{xx}, resulting in

$$\frac{\delta G_\mathrm{h}}{\delta u(x)} = K_\mathrm{b} u_{xx}\eta_x|_R^\infty - K_\mathrm{b} \int_R^\infty u_{xxx}\eta_x\,\mathrm{d}x + \frac{K_\mathrm{t}}{w_0^2} \int_R^\infty u\eta\,\mathrm{d}x. \tag{11.42}$$

The first term above is zero because u_x is fixed at the boundaries, which requires that η_x be zero at these boundaries. We can integrate by parts a second time on the term involving u_{xxx} to find

$$\frac{\delta G_\mathrm{h}}{\delta u(x)} = -K_\mathrm{b} u_{xxx}\eta|_R^\infty + K_\mathrm{b} \int_R^\infty u_{xxxx}\eta\,\mathrm{d}x + \frac{K_\mathrm{t}}{w_0^2} \int_R^\infty u\eta\,\mathrm{d}x = 0, \tag{11.43}$$

with the boundary term once again vanishing because $u(x)$ is fixed both at R and very far away from the protein, and hence $\eta(R) = \eta(\infty) = 0$. This yields the equilibrium equation for this system, namely,

$$K_\mathrm{b}\frac{\mathrm{d}^4 u}{\mathrm{d}x^4} + \frac{K_\mathrm{t}}{w_0^2}u = 0. \tag{11.44}$$

The solution to this differential equation minimizes the free energy G_h, and hence gives us the "best" profile for the deformed membrane around the protein.

The Membrane Surrounding a Channel Protein Produces a Line Tension

The equilibrium equation resulting from minimization of the free-energy functional is a linear differential equation with constant coefficients. A useful approach to solving equations of this form is often by recourse to the trial solution $u = e^{\Lambda x}$. Substitution of this solution (and its derivatives) into Equation 11.44 results in the quartic equation

$$\Lambda^4 + \frac{K_t}{K_b w_0^2} = 0. \tag{11.45}$$

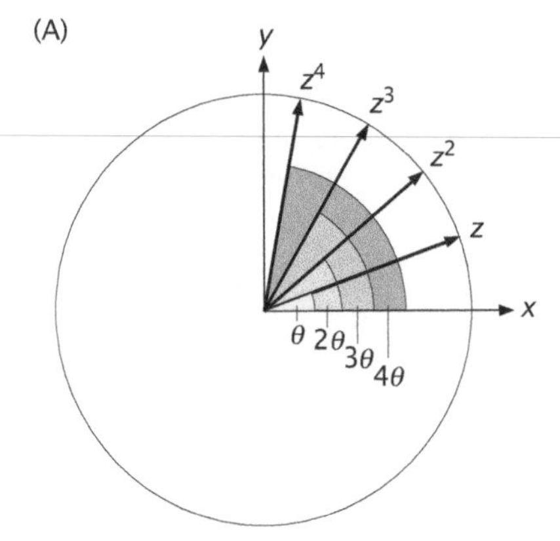

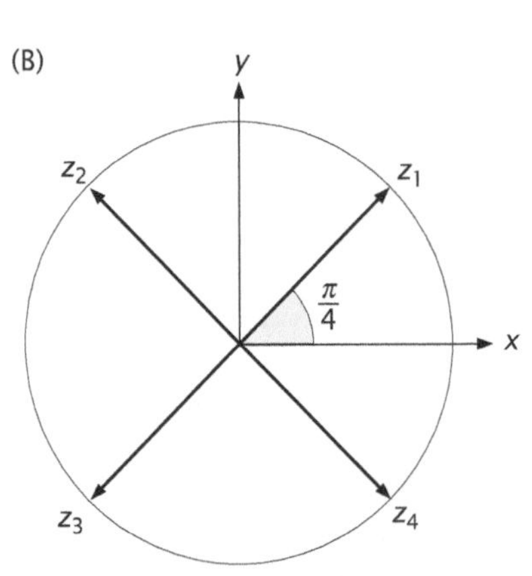

Figure 11.46: Fourth roots of −1. (A) Complex numbers are represented as vectors. For a complex number z whose modulus is 1 (length of corresponding vector is 1), the complex numbers obtained by raising it to the second, third, and fourth powers are vectors also of length 1 and they are rotated versions of the starting vector. The angles they make with the horizontal axis are two, three, and four times the angle θ the original vector makes with the same axis. (B) There are four complex numbers that when raised to the fourth power yield −1. They are at angles $\pi/4$ (indicated in the diagram as z_1), $3\pi/4$ (z_2), $5\pi/4$ (z_3), and $7\pi/4$ (z_4) with respect to the horizontal axis.

Just as there are two solutions to a quadratic equation of the form $x^2 + 1 = 0$, the quartic equation found here has four distinct solutions. Formally, this means the solution is

$$\Lambda = \sqrt[4]{\frac{K_t}{K_b w_0^2}} \times \sqrt[4]{-1}. \tag{11.46}$$

The four distinct solutions to this equation correspond to the four distinct fourth roots of −1 and are of the form

$$\Lambda_1 = \sqrt[4]{\frac{K_t}{K_b w_0^2}}\, e^{i\pi/4} = \sqrt[4]{\frac{K_t}{K_b w_0^2}} \left(\frac{\sqrt{2}}{2} + i\frac{\sqrt{2}}{2}\right) \tag{11.47}$$

$$\Lambda_2 = \sqrt[4]{\frac{K_t}{K_b w_0^2}}\, e^{i3\pi/4} = \sqrt[4]{\frac{K_t}{K_b w_0^2}} \left(-\frac{\sqrt{2}}{2} + i\frac{\sqrt{2}}{2}\right) \tag{11.48}$$

$$\Lambda_3 = \sqrt[4]{\frac{K_t}{K_b w_0^2}}\, e^{i5\pi/4} = \sqrt[4]{\frac{K_t}{K_b w_0^2}} \left(-\frac{\sqrt{2}}{2} - i\frac{\sqrt{2}}{2}\right) \tag{11.49}$$

$$\Lambda_4 = \sqrt[4]{\frac{K_t}{K_b w_0^2}}\, e^{i7\pi/4} = \sqrt[4]{\frac{K_t}{K_b w_0^2}} \left(\frac{\sqrt{2}}{2} - i\frac{\sqrt{2}}{2}\right) \tag{11.50}$$

To explore the mathematics of this solution, see The Math Behind the Models section below.

The Math Behind the Models: Fourth Roots of −1 The starting point for our analysis is Euler's formula $e^{i\theta} = \cos\theta + i\sin\theta$. This result tells us in turn that $e^{i\pi} = -1$ and more generally that $e^{i(2n+1)\pi} = -1$, where n is an integer. As a result, the solutions for $\sqrt[4]{-1}$ can be written as $e^{i[(2n+1)\pi/4]}$, for example, since $(e^{i[(2n+1)\pi/4]})^4 = e^{i(2n+1)\pi} = -1$. As shown in Figure 11.46, there are four unique roots, corresponding to the angles $\pi/4$, $3\pi/4$, $5\pi/4$, and $7\pi/4$.

As a result of the solutions given above, we now have four solutions to the equilibrium equation. However, since $u(x) \to 0$ as $x \to \infty$, we only keep Λ_2 and Λ_3, since these two solutions (as seen from their real parts) decay for large x while the other two solutions grow with increasing distance from the protein. Realistically, our physical intuition tells us that membrane far away from the protein must be less affected than membrane close by, so we can simply eliminate these possible solutions. This physical statement is equivalent to the mathematical boundary conditions described in Equations 11.39 and 11.40. What we learn from this analysis is that the displacement profile has the general form

$$u(x) = C_2 e^{\Lambda_2 x} + C_3 e^{\Lambda_3 x}, \tag{11.51}$$

which decays as a function of distance from the channel on a length scale λ^{-1}, where

$$\lambda = \sqrt[4]{\frac{K_t}{K_b w_0^2}}. \tag{11.52}$$

In order to determine the unknown constants in the solution, we apply the boundary conditions at the edge of the channel. These conditions were given in Equations 11.37 and 11.38, which state that the lipid bilayer is hydrophobically matched with the protein. Using these conditions, we have

$$C_2 e^{\Lambda_2 R} + C_3 e^{\Lambda_3 R} = U, \tag{11.53}$$

$$\Lambda_2 C_2 e^{\Lambda_2 R} + \Lambda_3 C_3 e^{\Lambda_3 R} = 0, \tag{11.54}$$

where we have introduced a new variable $U = W/2 - w_0$. These two equations depend upon the two unknowns C_2 and C_3 and when solved (see the problems at the end of the chapter) yield the solution

$$u(x) = U e^{-(\sqrt{2}/2)\lambda(x-R)} \left\{ \cos\left[\frac{\sqrt{2}}{2}\lambda(x-R)\right] + \sin\left[\frac{\sqrt{2}}{2}\lambda(x-R)\right] \right\}. \tag{11.55}$$

With this solution in hand, we can also compute the second derivative of the displacement profile,

$$u''(x) = U\lambda^2 e^{-(\sqrt{2}/2)\lambda(x-R)} \left\{ \sin\left[\frac{\sqrt{2}}{2}\lambda(x-R)\right] - \cos\left[\frac{\sqrt{2}}{2}\lambda(x-R)\right] \right\}, \tag{11.56}$$

which permits us to compute the energy given in Equation 11.36.

In particular, if we now substitute the displacement profile and its derivatives back into Equation 11.36, this results in

$$G_h = \frac{K_b}{2} \int_R^\infty dx\, U^2 \lambda^4 e^{-\sqrt{2}\lambda(x-R)} \left\{ 1 - 2\sin\left[\frac{\sqrt{2}}{2}\lambda(x-R)\right] \cos\left[\frac{\sqrt{2}}{2}\lambda(x-R)\right] \right\}$$
$$+ \frac{K_t}{2w_0^2} \int_R^\infty dx\, U^2 e^{-\sqrt{2}\lambda(x-R)} \left\{ 1 + 2\sin\left[\frac{\sqrt{2}}{2}\lambda(x-R)\right] \cos\left[\frac{\sqrt{2}}{2}\lambda(x-R)\right] \right\}. \tag{11.57}$$

At first, this equation looks horrible. But recall the definition of λ given in Equation 11.52 and note that the first integral uses the constant λ^4 in the integrand. If we bring this constant outside of the integral and substitute in its definition from Equation 11.52, we find that the constant coefficients for both integrals actually end up being the same. Then we can cancel out the second terms in the two integrals, and the remaining integrals are elementary, resulting in

$$G_h = \frac{K_t U^2}{\sqrt{2}\lambda w_0^2}. \tag{11.58}$$

Note that what we have computed is the energy per unit length associated with the deformation of the membrane surrounding the "channel." We are now prepared to exploit the mathematical idea embodied in Figure 11.45: namely, we make the assumption that this energy per unit length can even be used for the circular deformation field around a cylindrically symmetric ion channel, resulting in the total free energy

$$G_{\text{MscL}} = G_{\text{h}} + G_{\text{tension}} = G_0 + \underbrace{\frac{K_{\text{t}} U^2}{\sqrt{2} w_0^2 \lambda}}_{\text{energy/length}} \underbrace{2\pi R}_{\text{circumference}} - \underbrace{\tau \pi R^2}_{\text{loading device}} . \tag{11.59}$$

The outcome of our long calculation turns out to be a startlingly simple and physically intuitive result. In particular, the membrane deformation term depends upon the channel radius linearly while the load term scales quadratically with radius according to

$$G_{\text{MscL}} = G_0 + \tfrac{1}{2} \mathcal{K} U^2 2\pi R - \tau \pi R^2, \tag{11.60}$$

where we have introduced the compact notation

$$\mathcal{K} = \sqrt{2} \sqrt[4]{\frac{K_{\text{t}}^3 K_{\text{b}}}{w_0^6}}, \tag{11.61}$$

for the "spring constant" induced by membrane deformation. The significance of this result for the gating of mechanosensitive channels is shown in Figure 11.47. What we note is that at low tension, the closed state has the lower free energy. With increasing tension, the minimum at the open radius decreases in energy and ultimately supersedes the closed state. The picture, then, is of a competition between the unfavorable energy of channel gating (due to a larger annulus of deformed lipid bilayer) and the favorable energy associated with the tension.

Indeed, the critical tension can be computed by searching for that tension at which the free energies of the open and closed states are equal, namely, $G_{\text{MscL}}(R_{\text{C}}) = G_{\text{MscL}}(R_{\text{O}})$, which results in

$$\tfrac{1}{2} \mathcal{K} U^2 2\pi R_{\text{C}} - \tau \pi R_{\text{C}}^2 = \tfrac{1}{2} \mathcal{K} U^2 2\pi R_{\text{O}} - \tau \pi R_{\text{O}}^2. \tag{11.62}$$

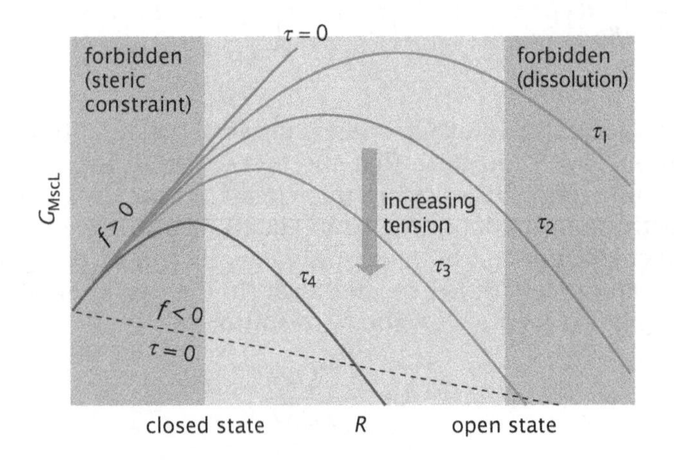

Figure 11.47: Schematic of the free energy of the membrane as a function of the channel radius for different values of the applied tension, τ. The parameter f is the effective line tension given by $f = K_{\text{t}} U^2 / \sqrt{2} w_0^2 \lambda$. Note that there is a small-R cutoff that results from the fact that there is a minimum channel radius corresponding to the closed state. Similarly, there is a large-R cutoff corresponding to dissolution of the channel into isolated α-helices—a physical event to which we attribute a huge energy. (Adapted from P. A. Wiggins and R. Phillips, *Proc. Natl Acad. Sci. USA* 101:4071, 2004.)

Table 11.1: Comparison of theory and experiment for the mechanosensitive channel, MscL. Experiments measure the pressure difference at which the open probability is 1/2. Comparison between theory and experiment requires knowledge of the pipette radius to convert pressure into tension. n refers to the number of carbons in the lipid chains. (Data taken from E. Perozo et al., *Nat. Struct. Biol.* 9:696, 2002.)

	Theory		Experiment	
n	τ_{crit} (k_BT/nm^2)	$\Delta G(\tau=0)$ (k_BT)	$P_{1/2}$ (mmHg)	$\Delta G(\tau=0)$ (k_BT)
16	0.23	5	24 ± 2	4
18	0.52	11.5	42 ± 5	9.4
20	0.93	20.4	72 ± 8	23.5

This equation can now be solved for the critical tension as

$$\tau_{crit} = \mathcal{K}U^2 \frac{1}{R_c + R_0}. \tag{11.63}$$

In addition, the model allows us to probe the free-energy difference at zero tension, which is

$$\Delta G(\tau = 0) = G_{MscL}^{\tau=0}(R_0) - G_{MscL}^{\tau=0}(R_c) = \pi \mathcal{K} U^2 (R_0 - R_c). \tag{11.64}$$

These results can be compared with the experimental analysis of this problem by appealing to measured values of the various model parameters such as $R_c \approx 2.3$ nm, $R_0 = 3.5$ nm, $U = W/2 - w_n = w_{n=12} - w_n$, and $\mathcal{K} = 20\, k_BT/\text{nm}^3$. Using reasonable choices for w_n, the half-width of the undeformed lipid bilayer, for different sized lipids, we obtain $U = 0.065(12 - n)$ nm. The theoretical predictions of the model obtained from Equations 11.63 and 11.64 are compared with results of experiments in Table 11.1.

Our discussion of the mechanosensitive channel has laid the groundwork for a more general discussion in Chapter 17 of the important role of ion channels. The present discussion has used simple ideas about membrane elasticity as a window on channel function and our later discussions will shift the emphasis to simple ideas from electrostatics.

11.7 Summary and Conclusions

Membranes play a central role in cell biology, separating self from non-self, inside from outside, and distinct specialized compartments within a cell from one another. The study of membranes requires understanding the convergence of a number of different physical principles, ranging from the self-organizing power of the hydrophobic effect to the peculiar properties of two-dimensional fluids to the energetics of membrane bending. Many of the properties of cellular membranes are determined not by their phospholipid constituents but by their many and varied proteins. Some membrane-associated proteins catalyze transport of ions and other molecules across the hydrophobic barrier, while others govern the budding, fusion, and tubule formation processes that are characteristic of the life of membranes inside a cell. The elegant and mutually beneficial interactions between phospholipids and proteins conspire to make cellular membranes lively, but well ordered.

11.8 Problems

Key to the problem categories: • Model refinements and derivations • Estimates

• 11.1 Membrane protein census

(a) Table 1 of Mitra et al. (2004) reports the mass ratio of proteins and phospholipids in the membranes of various cells and organelles. Use the asserted 2.0 mg of protein for every 1.0 mg of phospholipid in the *E. coli* membrane to compute the areal density of membrane proteins and their mean spacing. Make a corresponding estimate for the membrane of the endoplasmic reticulum using the fact that the mass ratio in this case is 2.6. Explain all of your assumptions in making the estimate.

(b) Dupuy and Engelman (2008) report that the area fraction associated with membrane proteins in the red blood cell membrane is roughly 23%, while the lipids themselves take up roughly 77% of the membrane area. Use these numbers to estimate the number of membrane proteins in the red blood cell membrane and their mean spacing. Explain all the assumptions you make in constructing the estimate.

• 11.2 Glucose and membrane permeability

(a) Compute the time scale for molecules to empty out of a "spherical bacterium" of radius 1 μm due to the permeability of the membrane to various molecular species. In particular, use the relation $j = -P\Delta c$ between the flux of molecules j across the membrane and its permeability P. Make a plot of the time scale as a function of the parameter P using the data in Figure 11.11, and then estimate the time scale for glucose. The permeability for phosphorylated glucose is thought to be much lower due to its charge, which makes traversing the membrane much less favorable. How much longer will the glucose take to leak out if the permeability is 100-fold smaller in the phosphorylated than unphosphorylated state, for example?

(b) To go beyond the simple estimate of (a), write a differential equation for the rate of change of concentration of molecules assuming that the initial concentration within the cell is c_{in} and the concentration outside of the cell is $c_{out} = 0$. Solve the differential equation and find the time scale for the molecules to exit the cell. How does this compare with the simple estimate you made in (a)?

• 11.3 Mathematics of curvature

Consider the function $h(x_1, x_2) = x_1^2 + x_1 x_2 - 2x_2^2$, which we assume describes the shape of a deformed lipid bilayer membrane. As shown in Figure 11.14, x_1 and x_2 are the coordinates of the reference plane below the membrane.

(a) Make a plot of the height as a function of x_1 and x_2.

(b) Compute the principal radii of curvature as functions of x_1 and x_2.

(c) Compute the bending free energy for the piece of membrane corresponding to the square $0 \leq x_1 \leq 1$ and $0 \leq x_2 \leq 1$ in the reference plane.

• 11.4 Membrane area change

The height function used in the chapter can be used to compute the change in area of a membrane when it suffers some deformation. Consider the height function $h(x, y)$ defined over an $L \times L$ square in the x–y plane. The function $h(x, y)$ tells us the deformed state of the $L \times L$ square. To compute the change in area, we divide the square into small $\Delta x \times \Delta y$ squares and consider how each of them is deformed on the surface $h(x, y)$ (see Figure 11.17A). The function $h(x, y)$ maps the small square into a tilted parallelogram.

(a) Show that the two sides of the parallelogram are given by the vectors

$$\Delta x\left(\mathbf{e}_x + \frac{\partial h}{\partial x}\mathbf{e}_z\right) \quad \text{and} \quad \Delta y\left(\mathbf{e}_y + \frac{\partial h}{\partial y}\mathbf{e}_z\right), \tag{11.65}$$

where $\mathbf{e}_x$, $\mathbf{e}_y$, and $\mathbf{e}_z$ are unit vectors.

(b) The magnitude of the vector product of two vectors is equal to the area subsumed by them, that is, $\mathbf{a} \times \mathbf{b} = ab\sin\theta$, where θ is the angle between vectors $\mathbf{a}$ and $\mathbf{b}$. Using this fact, show that the area of the parallelogram is

$$a' = \sqrt{1 + \left(\frac{\partial h}{\partial x}\right)^2 + \left(\frac{\partial h}{\partial y}\right)^2}\,\Delta x \Delta y. \tag{11.66}$$

(c) Now assume that the partial derivatives appearing in the above equation are small and expand the square-root function using $\sqrt{1 + x^2} \approx 1 + \frac{1}{2}x^2$, to demonstrate the useful formula for the area change,

$$\Delta a = a' - a = \frac{1}{2}\left[\left(\frac{\partial h}{\partial x}\right)^2 + \left(\frac{\partial h}{\partial y}\right)^2\right]\Delta x \Delta y. \tag{11.67}$$

By integrating this expression over the entire domain, we can find the total area change associated with the surface. Divide both sides by $a = \Delta x \times \Delta y$ to find a formula for the relative change of area $\Delta a/a$.

• 11.5 Laplace–Young relation

(a) To derive the Laplace–Young relation (Equation 11.9), imagine a spherical object with internal pressure p_{in} and external pressure p_{out}. This pressure difference leads to an outward-pointing normal force at all points on the membrane. Perform the thought experiment shown in Figure 11.48 in which the spherical shell is cut at the equator and the tension in the membrane acts downward to counterbalance the upward force due to the pressure difference. Derive the Laplace–Young relation by balancing the forces acting in the positive z-direction due to the pressure and in the negative z-direction due to the surface tension. Note that the surface tension can be interpreted as a force per unit length acting downward at the equator. Ignore any contribution due to bending energy and focus strictly on the role of the surface tension.

(b) Use energy arguments (rather than force balance arguments) to derive the same result. To do so, consider a small change in the radius of the sphere. As a result of this change in radius, we will incur a free-energy change associated with pV work and a part having to do with γA work. By insisting that the change in free energy for such an excursion be zero, obtain the Laplace–Young relation.

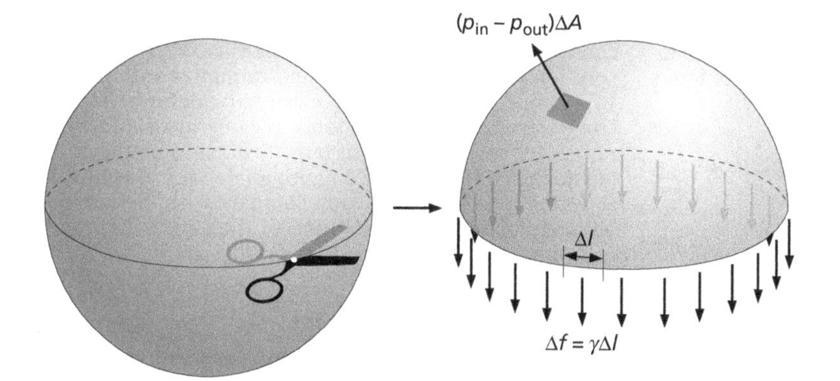

Figure 11.48: Balance of forces in a membrane. The vesicle is cut at the equator and the forces due to the surface tension are replaced with a set of downward acting forces.

11.6 Membrane deformation and adhesion in pipettes

The patch clamp technique is widely used in electrophysiology to study the gating properties of ion channels. A glass micropipette with an open tip is brought into contact with a cell or a vesicle containing the ion channel of interest. As a result of the strong adhesion between the glass and the lipid membrane, the membrane gets pulled into the pipette as shown in Figure 11.49.

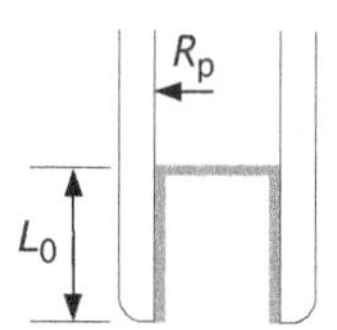

Figure 11.49: Membrane patch and pipette. Schematic showing the membrane patch shape (grey) at zero applied pressure. The pipette has a radius of R_p and as a result of adhesion, the membrane is pulled up the pipette by a distance L_0.

Although the glass-membrane adhesion is crucial to avoiding spurious leakage currents not resulting from the channels themselves, this adhesion can also have negative side effects. For example, as the membrane adheres to the pipette, it induces a tension that is different from the resting tension in the membrane patch, and this could lead to errors in the estimates of gating tension of mechanosensitive channels, for example. In this problem, we use our understanding of elasticity of membranes to estimate the size of this adhesion-induced tension.

(a) Consider the geometry shown in Figure 11.49, where the membrane patch has a cylindrical adhered domain and a circular free domain. The elastic energy associated with the patch comes from two contributions, namely, the stretch energy and the glass–bilayer adhesion energy. The stretch energy results from the fact that the membrane is stretched when it is pulled into the pipette. This contribution to the free energy has a quadratic dependence on the areal strain $\phi = (A - A_0)/A_0$, and is given by

$$G_{\text{stretch}} = \frac{1}{2} K_A \phi^2 A_0 , \tag{11.68}$$

where A_0 is the unstretched area of the membrane patch and $K_A \approx 50\, k_BT/\text{nm}^2$ is the stretch modulus.

The adhesion energy is proportional to the area of the contact domain A_{adh} and is given by $G_{\text{adh}} = -\gamma A_{\text{adh}}$, where γ is the adhesion energy (that is, the energy per unit area). Using these facts and the geometry shown in Figure 11.49, obtain the expression for the total elastic energy of the membrane patch in terms of the areal strain ϕ.

(b) Minimize the elastic energy with respect to ϕ to obtain the expression for the equilibrium areal strain. Compute the tension τ in the membrane patch in equilibrium using the relation $\tau = K_A \phi$. For $\gamma \sim 0.5\, k_BT/\text{nm}^2$, a typical value for glass–bilayer interaction, estimate the equilibrium areal strain and the tension in the patch.

(c) Assuming that the unstretched area of the patch is $A_0 = 2 \times \pi R_p^2$, what is the length of membrane tube L that lines the sides of the pipette? How does this length compare with the length computed from simple geometrical considerations, namely, by assuming that the membrane patch does not get stretched when it is drawn into the pipette.

(d) Show that membrane bending does not contribute significantly to the energy budget of the membrane patch. Do this by computing the bending energy of cylindrical membrane material within the pipette and compare the resulting energy with the stretch energy and adhesion energy you already obtained.

11.7 Bending modulus and the pipette aspiration experiment

The pipette aspiration experiment described in the chapter can be used to measure the bending modulus K_b as well as the area stretch modulus. Lipid bilayer membranes are constantly jostled about by thermal fluctuations. Even though a flat membrane is the lowest-energy state, fluctuations will cause the membrane to spontaneously bend. The goal of this problem is to use equilibrium statistical mechanics to predict the nature of bending fluctuations and to use this understanding as the basis of experimental measurement of the bending modulus. (Note: This problem is challenging and the reader is asked to consult the hints on the book's website to learn more of our Fourier transform conventions, how to handle the relevant delta functions, the subtleties associated with the limits of integration, etc.).

(a) Write the total free energy of the membrane as an integral over the area of the membrane. Your result should have a contribution from membrane bending and a contribution from membrane tension. Write your result using the function $h(\mathbf{r})$ to characterize the height of the membrane at position $\mathbf{r}$.

(b) The free energy can be rewritten using a decomposition of the membrane profile into Fourier modes. Our Fourier transform convention is

$$h(\mathbf{r}) = \frac{A}{(2\pi)^2} \int \tilde{h}(\mathbf{q}) e^{-i\mathbf{q}\cdot\mathbf{r}}\, d^2\mathbf{q}, \tag{11.69}$$

where $A = L^2$ is the area of the patch of membrane of interest. Plug this version of $h(\mathbf{r})$ into the total energy you derived above (that is, bending energy and the energy related to tension) to convert this energy in real space to an energy in q-space. Note that the height field in q-space looks like a sum of harmonic oscillators.

(c) Use the equipartition theorem in the form $\langle E(\mathbf{q}) \rangle = k_B T/2$, where $E(\mathbf{q})$ is the energy of the qth mode and the free energy can now be written as

$$F[\tilde{h}(\mathbf{q})] = \frac{A}{(2\pi)^2} \int E(\mathbf{q})\, d^2\mathbf{q}. \tag{11.70}$$

Use this result to solve for $\langle |\tilde{h}(\mathbf{q})|^2 \rangle$, which will be used in the remainder of the problem.

(d) We now have all the pieces in place to compute the relation between tension and area and thereby the bending modulus. The difference between the actual area and the projected area is

$$A_{act} - A = \frac{1}{2} \int [\nabla h(\mathbf{r})]^2\, d^2\mathbf{r}. \tag{11.71}$$

This result can be rewritten in Fourier space as

$$\langle A_{act} - A \rangle = \frac{1}{2} \frac{A^2}{(2\pi)^2} \int_{\pi/\sqrt{A}}^{\pi/\sqrt{a_0}} q^2 \langle |\tilde{h}(\mathbf{q})|^2 \rangle 2\pi q\, dq. \tag{11.72}$$

Work out the resulting integral, which relates the areal strain to the bending modulus, membrane size, temperature, and *tension*. The limits of integration are set by the overall size of the membrane (characterized by the area A) and the spacing between lipids (a_0 is the area per lipid), respectively. This result can be directly applied to micropipette experiments to measure the bending modulus. In particular, using characteristic values for the parameters appearing in the problem suggested by Boal (2002) such as $\tau \approx 10^{-4}$ J/m^2, $K_b \approx 10^{-19}$ J, and $a_0 \approx 10^{-20}$ m^2, show that

$$\frac{\Delta A}{A} \approx \frac{k_B T}{8\pi K_b} \ln\left(\frac{A\tau}{K_b \pi^2}\right), \tag{11.73}$$

and describe how this result can be used to measure the bending modulus. (For an explicit comparison with data, see Rawicz et al. (2000). For further details on the analysis, see Helfrich and Servuss (1984). An excellent account of the entire story covered by this problem can be found in Chapter 6 of Boal (2002).)

11.8 Variational approach to deformation induced by MscL

In the chapter, we explicitly minimized Equation 11.36 by solving a partial differential equation for the unknown deformation field $u(x)$. As an alternative, consider the powerful technique of adopting a variational solution. As a first try, consider the function $u(x) = Ae^{-x/x_0}$ for the deformation induced by the membrane protein and plug this "trial solution" into Equation 11.36. Determine the constant A by insisting that the hydrophobic mismatch at the protein–lipid interface is u_0. Perform the integrations required to obtain the free energy and then minimize with respect to the parameter x_0 to find the lowest-energy solution consistent with this functional form for the mismatch profile. Use the resulting solution to compute G_h and to obtain the critical tension. Compare your result with the "exact" result found in the chapter. One difficulty with the trial solution adopted above is that it does not respect the condition that $u'(x) = 0$ at the protein–lipid interface. A more appropriate trial solution is of the form

$$u(x) = u_0 \left(1 + \frac{x}{x_0}\right) e^{-x/x_0}. \tag{11.74}$$

Verify that this trial solution satisfies the boundary conditions at the lipid–protein interface. Plug this trial solution into the free energy and minimize with respect to x_0 and find the free-energy-minimizing deformation profile. Use this result to plot the total free energy as a function of radius for several different tensions including the critical tension (that is, produce a figure analogous to Figure 11.47). How does the critical tension compare with the value obtained using the full solution of the partial differential equation?

11.9 Dynamin on tubules

In the chapter, we considered an experiment in which a tubule was pulled from a giant unilamellar vesicle and then dynamin was added into the solution. The dependence of nucleation on the radius was inferred by appealing to the force measured in the trap as shown in Figure 11.36. Here we rederive this force.

(a) Write a free energy for the tubule in terms of the bending energy and the tension.

(b) Find the force due to the bending energy and the surface tension by evaluating $f = -\partial G_{tubule}/\partial L$.

(c) Now consider the effect of the optical trap with stiffness k_{trap} on the free energy of the tubule/bead system. Assume that the equilibrium position of the bead corresponds to a tubule of length L_0 and write down the expression for the total free energy that includes the free energy of the tubule and the energy of the bead in the trap. Find the equilibrium value of the tubule length L^*, and show that it is a result of balancing the tubule force and the force applied by the optical trap.

11.10 Two-dimensional treatment of MscL

Consider a two-dimensional generalization of Equation 11.36.

(a) Show that the free energy associated with membrane deformation may be written as

$$G_{hydrophobic} = \frac{K_b}{2} \int (\nabla^2 u)^2\, d^2r + \frac{K_t}{2} \int \left(\frac{u}{a}\right)^2 d^2r \tag{11.75}$$

To this add the change in free energy due to the change in tension, namely,

$$G_{tension} = G_0 - \tau\pi R^2, \tag{11.76}$$

where G_0 is a constant that sets the reference value for the energy of the loading device. Explain why the free-energy expression for the composite system of channel and loading device can now be written as

$$G_{MscL} = G_0 - \tau\pi R^2 + \frac{K_b}{2} \int (\nabla^2 u)^2\, d^2r + \frac{K_t}{2} \int \left(\frac{u}{a}\right)^2 d^2r. \tag{11.77}$$

To compute this free energy explicitly, we need to know the profile of the field $u(\mathbf{r})$ around the membrane.

(b) Find the Euler–Lagrange equation for this free-energy functional.

(c) Solve the partial differential equation for the deformation profile due to the membrane protein and imitate the various steps leading up to Equation 11.58, but now done for the full two-dimensional deformation profile.

(d) Compare this solution with that of the one-dimensional approximation given in the chapter.

11.9 Further Reading

Robertson, RN (1983) The Lively Membranes, Cambridge University Press. Though this lovely book is by now more than 20 years old, it is replete with beautiful and instructive diagrams and excellent discussions concerning the biological significance, chemical makeup, and physical properties of membranes.

Tanford, C (2004) Ben Franklin Stilled the Waves: An Informal History of Pouring Oil on Water with Reflections on the Ups and Downs of Scientific Life in General, Oxford University Press. A very interesting account of our understanding of lipids and the structures they form. Tanford is opinionated, but it is fun to hear people's honest thoughts rather than neutral droning.

Israelachvili, J (2011) Intermolecular and Surface Forces, 3rd ed., Academic Press. This book has a useful discussion on the relation of lipid molecule shape and macroscopic structure.

Mouritsen, OG (2005) Life—As a Matter of Fat, Springer-Verlag. A broad discussion of the role of lipids in living matter.

Engelman, DM (2005) Membranes are more mosaic than fluid, *Nature* **438**, 578. This review gives a status report on the current understanding of the heterogeneous nature of real cell membranes.

Boal, D (2012) Mechanics of the Cell, 2nd ed., Cambridge University Press. Boal's treatment of equilibrium shapes is very instructive. This book provides an excellent discussion of the important topic of thermal fluctuations in membranes that we have neglected.

Helfrich, W, & Servuss, R-M (1984) Undulations, steric interaction and cohesion of fluid membranes, *Nuovo Cim.* **3D**, 137. This paper illustrates the relation between thermal fluctuations and the bending rigidity of a lipid bilayer membrane.

Munn, EA (1974) The Structure of Mitochondria, Academic Press. This book is a beautiful compendium of the diversity found in mitochondrial structures.

Frey, TG, & Mannella, CA (2004) The internal structure of mitochondria, *Trends Biochem. Sci.* **25**, 319. This inspiring article is full of beautiful pictures and insights into mitochondrial shape and function.

Dobereiner, H-G (2000) Properties of giant vesicles, *Curr. Opin. Coll. Interf. Sci.* **5**, 256. Discussion of lipid bilayer mechanics and membrane shape.

Seifert, U (1997) Configurations of fluid membranes and vesicles, *Adv. Phys.* **46**, 13. An excellent source for a detailed description of the physics of lipid bilayers.

Lim, G, Wortis, M, & Mukhopadhyay, R (2002) Stomatocyte–discocyte–echinocyte sequence of the human red blood cell: Evidence for the bilayer-couple hypothesis from membrane mechanics, *Proc. Natl Acad. Sci. USA* **99**, 16766 and Noguchi, H, & Gompper, G (2005) Shape transitions of fluid vesicles and red blood cells in capillary flows, *Proc. Natl Acad. Sci. USA* **102**, 14159. These papers describe red blood cell shapes.

Upadhyaya, A, & Sheetz, MP (2004) Tension in tubulovesicular networks of Golgi and endoplasmic reticulum membranes, *Biophys. J.* **86**, 2923. Optical tweezers used to measure membrane tension.

Smith, A, Sackmann, E, & Seifert, U (2004) Pulling tethers from adhered vesicles, *Phys. Rev. Lett.* **92**, 208101. This interesting little article has a very nice treatment of the tether pulling problem.

Sprong, H, van der Sluijs, P, & van Meer, G (2001) How proteins move lipids and proteins move lipids, *Nat. Rev. Mol. Cell Biol.* **2**, 504. This article explores the connection between lipids and proteins in cell membranes.

Andersen, OS, & Koeppe, RE (2007) Bilayer thickness and membrane protein function: an energetic perspective, *Annu. Rev. Biophys. Biomol. Struct.* **36**, 107. This article explores the coupling between membrane deformation and membrane protein function. The membrane protein gramicidin provides a powerful case study in the coupling between membranes and membrane proteins and this article explores that connection.

Sukharev, S, & Corey, DP (2004) Mechanosensitive channels: multiplicity of families and gating paradigms, *Sci. STKE* **3**, 219. A useful review on mechanosensation.

Gillespie, PG, & Walker, RG (2001) Molecular basis of mechanosensory transduction, *Nature* **413**, 194. An extremely interesting review describing different case studies in mechanosensation.

Hamill, OP (ed.) (2007) Mechanosensitive Ion Channels, Part A, Academic Press. This book is full of enlightening articles on mechanosensation. See especially the article by Markin and Sachs on "Thermodynamics of mechanosensitivity."

Fain, GL (2003) Sensory Transduction, Sinauer Associates. This book describes various kinds of sensation and the role played by ion channels.

Takamori, S, Holt, M, Stenius, K, et al. (2006) Molecular anatomy of a trafficking organelle, *Cell* **127**, 831. This article reports on a quantitative description of the molecular participants in trafficking organelles.

Dupuy, AD, & Engleman, DM (2008) Protein area occupancy at the center of the red blood cell membrane, *Proc. Natl Acad. Sci. USA* **105**, 2848. This article describes a measurement of the relative area occupancy.

11.10 References

Alberts, B, Johnson, A, Lewis, J, et al. (2008) Molecular Biology of the Cell, 5th ed., Garland Science.

Allan, V, & Vale, R (1994) Movement of membrane tubules along microtubules *in vitro*: evidence for specialised sites of motor attachment, *J. Cell Sci.* **107**, 1885.

Baumgart, T, Hess, ST, & Webb, WW (2003) Imaging coexisting fluid domains in biomembrane models coupling curvature and tension, *Nature* **425**, 821.

Cross, PC, & Mercer, KL (1993) Cell and Tissue Ultrastructure: A Functional Perspective, 2nd ed., W. H. Freeman.

Fahy, E, Subramaniam, S, Brown, HA, et al. (2005) A comprehensive classification system for lipids, *J. Lipids Res.* **46**, 839.

Fawcett, DW (1966) The Cell, Its Organelles and Inclusions: An Atlas of Fine Structure, W. B. Saunders.

Fawcett, DW (1986) A Textbook of Histology, W. B. Saunders.

Hirokawa, N (1998) Kinesin and dynein superfamily proteins and the mechanism of organelle transport, *Science* **279**, 519.

Kirchhausen, T (2000) Three ways to make a vesicle, *Nat. Rev. Mol. Cell Biol.* **1**, 187.

Komeili, A, Li, Z, Newman, DK, & Jensen, GJ (2006) Magnetosomes are cell membrane invaginations organized by the actin-like protein MamK, *Science* **311**, 242.

Koster, G, VanDuijn, M, Hofs, B, & Dogterom, M (2003) Membrane tube formation from giant vesicles by dynamic association of motor proteins, *Proc. Natl Acad. Sci. USA* **100**, 15583.

Marsh, B, Mastronarde, DN, Buttle, KF, Howell, KE, & McIntosh, JR (2001) Organellar relationships in the Golgi region of the pancreatic beta cell line, HIT-T15, visualized by high resolution electron tomography, *Proc. Natl Acad. Sci. USA* **98**, 2399.

Mitra, K, Ubarretxena-Belandia, I, Taguchi, T, Warren, G, & Engelman, DM (2004) Modulation of the bilayer thickness of exocytic pathway membranes by membrane proteins rather than cholesterol, *Proc. Natl Acad. Sci. USA* **101**, 4083. Table 1 of this paper gives a useful list of protein to phospholipid mass ratios for a number of different membranes.

Nathke, IS, Heuser, J, Lupas, A, et al. (1992) Folding and trimerization of clathrin subunits at the triskelion hub, *Cell* **68**, 899.

Nelson, DL, & Cox, MM (2008) Lehninger Principles of Biochemistry, 5th ed., W. H. Freeman.

Perozo, E, Kloda, A, Cortes, DM, & Martinac, B (2002) Physical principles underlying the transduction of bilayer deformation forces during mechanosensitive channel gating, *Nat. Struct. Biol.* **9**, 696.

Rawicz, W, Olbrich, KC, McIntosh, T, Needham, D, & Evans, E (2000) Effect of chain length and unsaturation on elasticity of lipid bilayers, *Biophys. J.* **79**, 328.

Roux, A, Koster, G, Lenz, M, et al. (2010) Membrane curvature controls dynamin polymerization, *Proc Nat Acad Sci USA* **107**, 4141.

Singer, SJ, & Nicolson, GL (1972) The fluid mosaic model of the structure of cell membranes, *Science* **175**, 720.

Slepnev, VI & De Camilli, P (2000) Accessory factors in clathrin-dependent synaptic vesicle endocytosis, *Nat Rev Neurosci* **1**, 161.

Smith, CJ, Grigorieff, N, & Pearse, BM (1998) Clathrin coats at 21 Å resolution: a cellular assembly designed to recycle multiple membrane receptors, *EMBO J.* **17**, 4943.

Sowers, AE, & Hackenbrock, CR (1981) Rate of lateral diffusion of intramembrane particles: measurement by electrophoretic displacement and rerandomization, *Proc. Natl Acad. Sci. USA* **78**, 6246.

Sukharev, S, Durell, SR, & Guy, HR (2001) Structural models of the MscL gating mechanism, *Biophys. J.* **81**, 917.

Ungewickell, E, & Branton, D (1981) Assembly units of clathrin coats, *Nature* **289**, 420.

Wang, S-H, Denk, W, & Häusser, M (2000) Coincidence detection in single dendritic spines mediated by calcium release, *Nat. Neurosci.* **3**, 1266.

Wiggins, PA, & Phillips, R (2004) Analytic models for mechanotransduction: gating a mechanosensitive channel, *Proc. Natl Acad. Sci. USA* **101**, 4071.

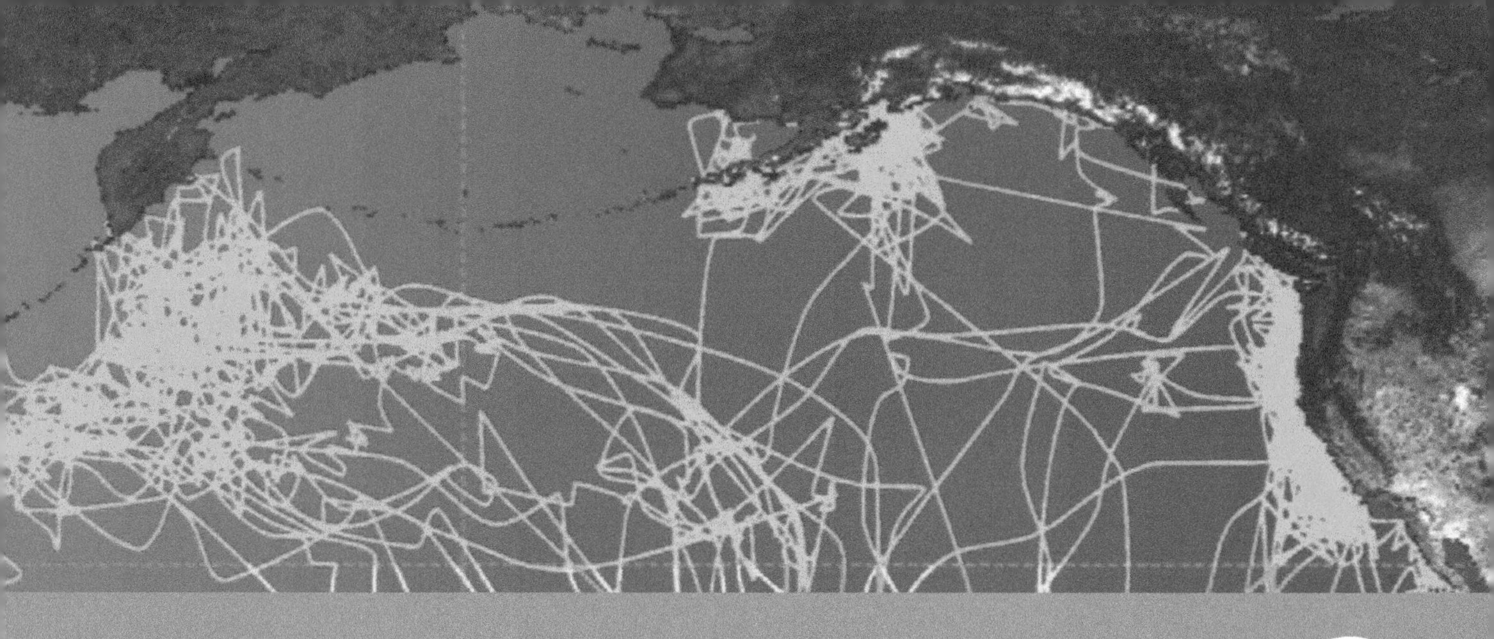

Life in Motion

PART 3

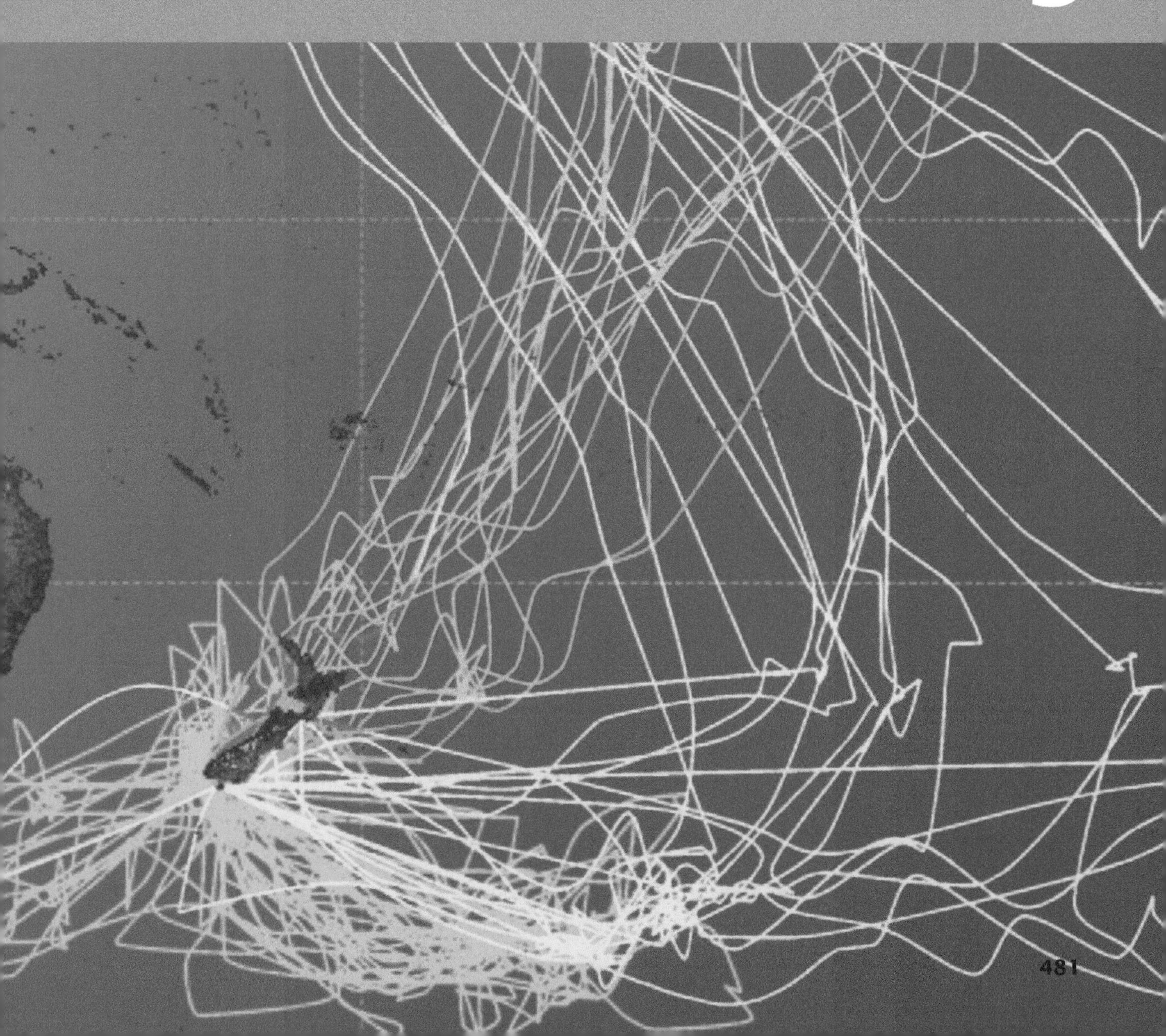

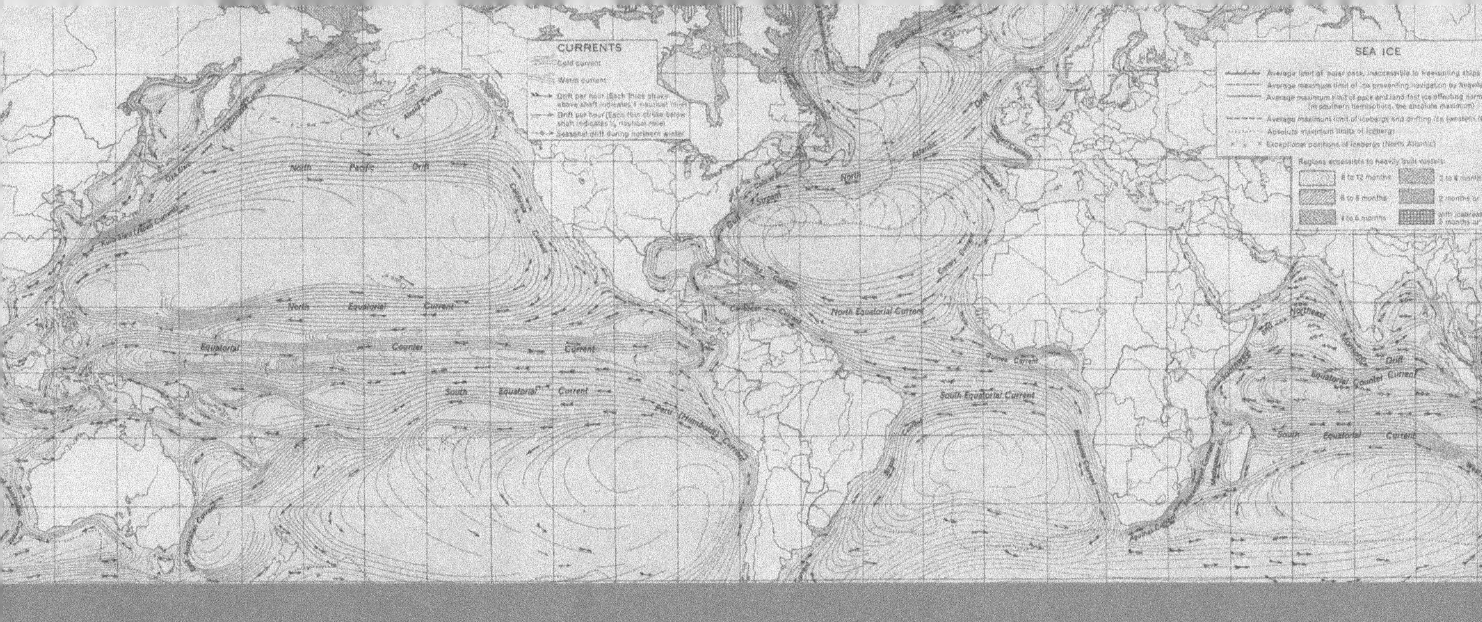

The Mathematics of Water 12

Overview: In which life's watery medium is studied as a dynamical object

> "There are these two young fish swimming along and they happen to meet an older fish swimming the other way, who nods at them and says 'Morning, boys. How's the water?' And the two young fish swim on for a bit, and then eventually one of them looks over at the other and goes 'What the hell is water?'"
>
> **David Foster Wallace**

The preceding six chapters have all shared a common assumption that has been both their strength and their weakness. That assumption is that approximate physical descriptions of biological systems can invoke the concept of equilibrium. We have taken pains to delimit the conditions under which this assumption is valid or reasonable. But now it is time to acknowledge the reality that living systems generally exist far from equilibrium and that *dynamic* descriptors such as time, rates, and trajectories are frequently the natural language of physical biology. This chapter undertakes an analysis of the dynamics of life's watery medium. In particular, we will develop the physical and mathematical tools to make estimates of fluid dynamics both for phenomena within cells and for environmental interactions such as swimming.

12.1 Putting Water in Its Place

In a Darwinian environment where organisms compete for limited resources, the ability to consume, grow, and reproduce more rapidly than one's rivals may lead to a selectable advantage. In this chapter, we will begin our exploration of biological dynamics with a celebration of water, the medium of life on Earth. Oceans cover 70% of the Earth's surface and humans diving into these oceans are greeted by a rich diversity of living forms. But this panoply of life is nothing in comparison with the teeming masses that are revealed upon glancing through a microscope at a drop of pond water. The many organisms that actively swim through water, chasing prey, avoiding predators, and seeking more favorable environments for life are practical masters at managing fluid dynamics. No vessel yet built by man can move through water with the energetic efficiency of a blue whale, anchovy, or *Paramecium*. Even organisms that have left the ocean and carry

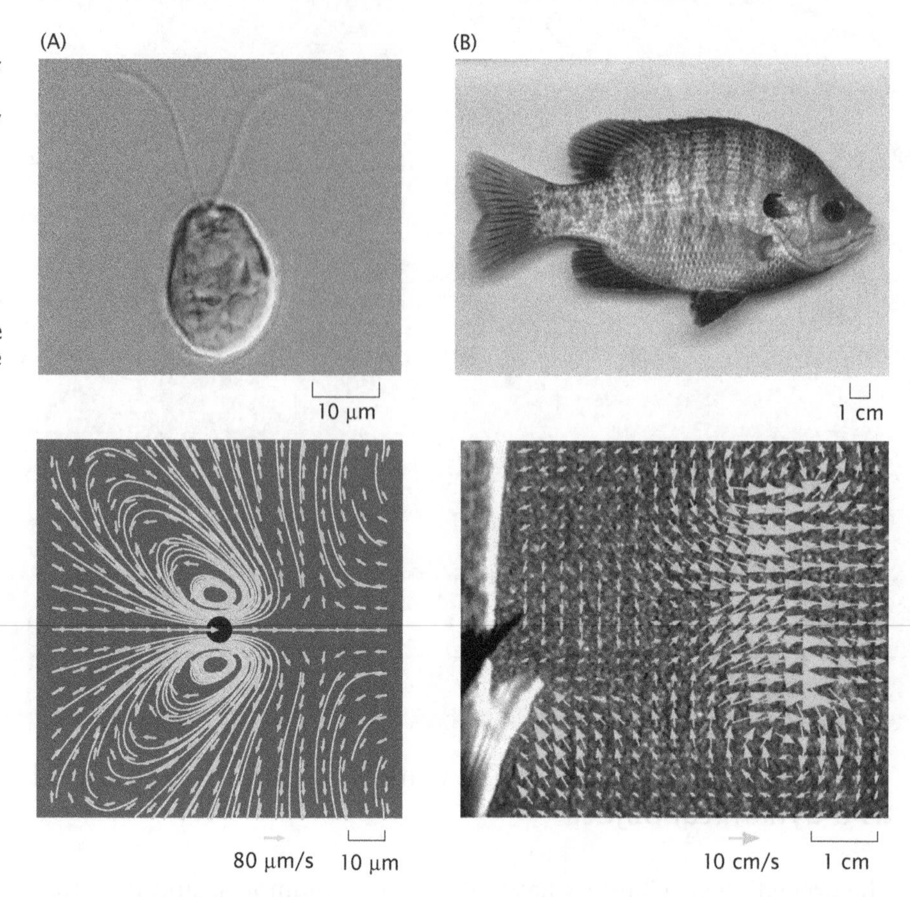

Figure 12.1: Swimming across scales. (A) The single-cell alga *Chlamydomonas reinhardtii* swims with the help of its flagella at about 100 μm/s. The velocity field of the fluid around it can be visualized using fluorescent beads as tracer particles. (B) Bluegill sunfish swimming at a speed of about 10 cm/s produce a flow field in the surrounding water, which can be visualized in a similar fashion, using fluorescent tracer particles. In this chapter, we explore the physics of fluids in motion and describe the key differences between flows at length and velocity scales associated with microorganisms and fish. (A (top), courtesy of Brian Piasecki, Lawrence University; A (bottom), adapted from J. S. Guasto et al., *Phys. Rev. Lett.* 105:168102, 2010; B (top), courtesy of the Ohio Department of Natural Resources, Division of Wildlife; B (bottom), adapted from E. G. Drucker and G. V. Lauder, *J. Exp. Biol.* 204:431, 2001)

their own water internally require exquisite control over the water's movement as in blood circulation or the rising sap in plants.

Water plays an important part in the living world on scales ranging from a single cell to groups of organisms swimming in schools, as illustrated in Figure 12.1. Its physical and chemical properties are of direct relevance to the structure and function of biomolecules. Therefore, intuition about water, and fluids in general, is of considerable importance when exploring the living world. In this chapter, we investigate the properties of fluids with regards to their motion by developing the model of the Newtonian fluid.

12.2 Hydrodynamics of Water and Other Fluids

12.2.1 Water as a Continuum

Though Fluids Are Composed of Molecules It Is Possible to Treat Them as a Continuous Medium

In order to formulate a mathematical description of a fluid such as water, we must first decide on the proper "coordinates." In this regard, a fluid can be thought of as a "box of molecules" as shown in Figure 12.2. The water molecules are thermally agitated and interact with each other via forces such as the van der Waals force. In this case, the description of a fluid would entail providing a full list of position and momentum coordinates of all the molecules. This is the worldview of molecular dynamics, which we will largely avoid in our search

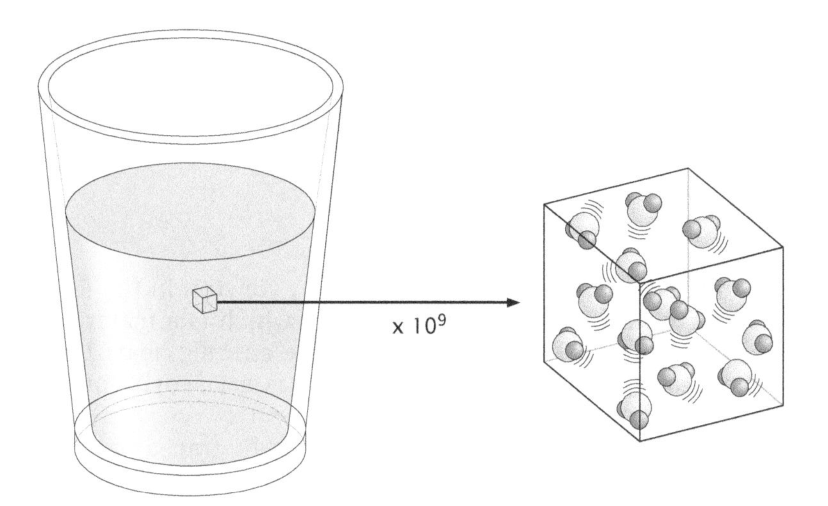

Figure 12.2: Representations of water. The continuous description is a smeared-out representation of an underlying discrete set of molecules.

for reduced descriptions, though this is not to deny that much insight has been garnered from all-atom descriptions.

The view of fluids as continuous media is complementary to that provided by molecular mechanics. In the continuum view (like with our treatment of the elastic theory of beams and membranes), the fluid is thought of as a featureless blob of density ρ. Its state of motion is described by a velocity field $\mathbf{v}(\mathbf{r}, t)$, which can be thought of as a set of arrows stuck to every point in the flow, each denoting the direction and magnitude of the fluid's velocity at the location $\mathbf{r}$ and at the time instant t. This is shown in Figure 12.3. The idea of a mechanical description of the fluid is to find a way to compute the forces on each such volume element and to use these forces to deduce the governing equations of motion for each such fluid element.

12.2.2 What Can Newton Tell Us?

Gradients in Fluid Velocity Lead to Shear Forces

We have studied solids and their response to applied forces in Section 5.4.1 (p. 216) and Chapter 10. Now we turn to a comparable discussion for fluids. One of the most notable distinctions between fluids and solids is that fluids flow and solids do not (or, more precisely, when solids flow, they do so very slowly). The mechanical engineer would say that while to strain a solid one must apply a force, no such force is needed if we are dealing with a fluid. In fact, the response of a fluid to an applied force is flow, or a rate of change of strain. To illustrate this, consider placing a spoon in a jar of honey. If it is slowly retracted from the jar, there will be very little force resisting the motion of the spoon. On the other hand, if the spoon is pulled out quickly, the whole jar might come along for the ride. Central to this phenomenon is the notion of viscosity, which, like the Young modulus for solids, is a material parameter that describes the mechanical properties of fluids.

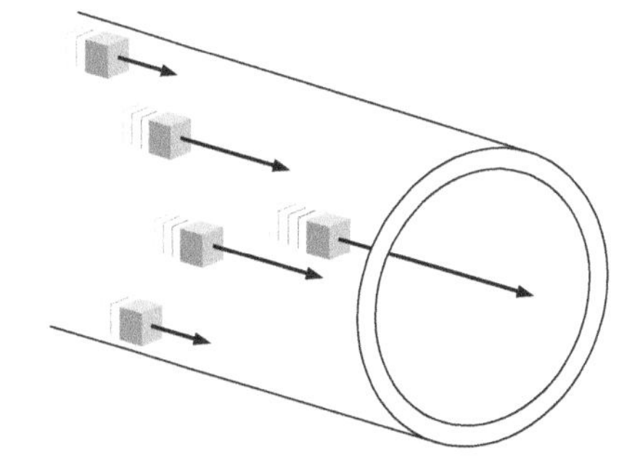

Figure 12.3: The velocity field. Schematic illustrating the replacement of a discrete (molecular) description of a fluid with a continuum (field) description in which each little material volume element is assigned a velocity vector that characterizes the velocity at that point in the fluid.

A first stab at a definition of viscosity can be fashioned from observations that follow from a simple experiment. Consider two large parallel plates, each of area A, separated by a small distance d, with fluid between them, as shown in Figure 12.4. Now imagine we pull on one plate with force F while the other plate is kept fixed. For simple fluids, termed "Newtonian," what we will observe is that the plate

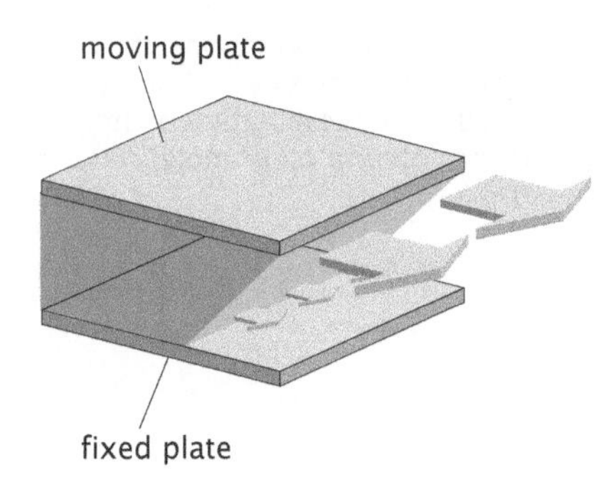

Figure 12.4: Measurement of fluid viscosity. A fluid of unknown viscosity is injected between two flat plates. Because of the no-slip condition at the solid–fluid interface on the two plates, which states that the fluid at the plates has to move with the same velocity as the plates themselves, movement of the upper plate at a constant velocity will impart a linear velocity gradient on the fluid. The viscosity can be measured by determining the amount of force required to move the upper plate at a particular constant velocity.

we are pulling on will respond by moving with a constant velocity v, which is proportional to the applied force. In fact, doing experiments with different-sized plates and various separations between the plates will yield a relation of the form

$$\frac{F}{A} = \eta \frac{v}{d}. \tag{12.1}$$

Indeed, this is the formula first proposed by Newton in the *Principia*. The constant η is the viscosity of the fluid, which is a material property of the fluid, just like the density or the specific heat. The units of viscosity are pascal seconds and water has a viscosity of roughly 1 mPa s.

The microscopic origins of the viscosity can be traced back to thermal diffusion of the molecules of the liquid. If a velocity gradient is set up in a fluid, then diffusion of molecules from regions of space characterized by a smaller fluid velocity to ones where the velocity is higher, and vice versa, will result in transfer of momentum and consequently a force. An appealing analogy can be drawn with cars on a two-lane highway. If one lane is moving much faster than the other, there will be a tendency of the slower cars to move into the faster lane, leading to a slowing down of this lane. The same thing happens in the world of molecules except that "lane switches" are not acts of impatient drivers but the result of random thermal motion of the molecules. In what follows, we will develop the laws of fluid dynamics in mathematical form, based on the very simple relation shown in Equation 12.1. Though the model of the Newtonian fluid will serve as our conceptual framework, many of the fluids relevant to biology exhibit more complex material behavior known as non-Newtonian. These materials are familiar from cases such as egg whites or mucus and their response to an applied force can exhibit both fluid-like and solid-like characteristics.

12.2.3 $F = ma$ for Fluids

To develop the equations of motion of a fluid, we concentrate on the motion of a small fluid element and apply to it Newton's second law, $F = ma$. The size of the fluid element is chosen to be small compared with the scale over which the fluid velocity changes appreciably.

We proceed by deriving a formula for the acceleration of a fluid element. We observe a fluid element at position $\mathbf{r} = (x, y, z)$, at time t, as it moves over a time interval Δt. To simplify the math, we will assume that the fluid velocity vector is in the x-direction and that it varies only along this direction. The initial velocity of the fluid element is $\mathbf{v}(x, t) = v(x, t)\mathbf{e}_x$, where $\mathbf{e}_x$ is a unit vector pointing in the x-direction. After the prescribed time interval, the fluid element will be at position $x + \Delta x$, and its velocity will be $v(x + \Delta x, t + \Delta t)$ as shown in Figure 12.5. Taylor-expanding the final velocity to first order in Δx and Δt leads to

$$v(x + \Delta x, t + \Delta t) = v(x, t) + \frac{\partial v}{\partial x}\Delta x + \frac{\partial v}{\partial t}\Delta t. \tag{12.2}$$

Using the fact that the fluid element moves by $\Delta x = v(x, t)\Delta t$ in time Δt, from the above equation we can obtain the change in velocity of the fluid element,

$$\Delta v(x, t) = \left(\frac{\partial v}{\partial t} + v\frac{\partial v}{\partial x}\right)\Delta t. \tag{12.3}$$

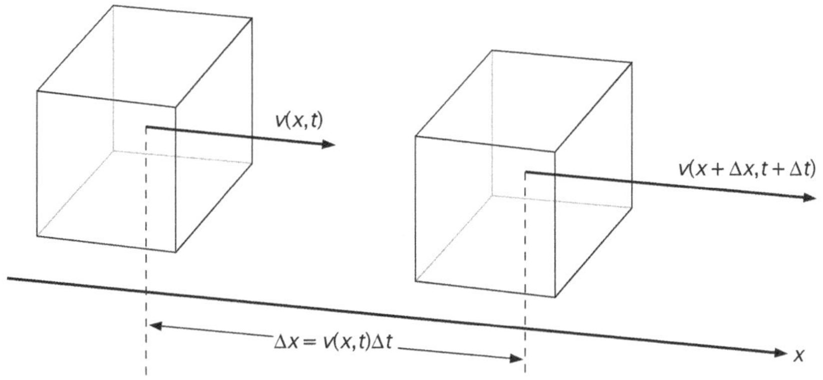

Figure 12.5: Schematic illustrating the calculation of the acceleration of a small parcel of fluid.

Since the acceleration is given by $a = \Delta v/\Delta t$, this leads to an acceleration of the form

$$a(x,t) = \frac{\partial v(x,t)}{\partial t} + v(x,t)\frac{\partial v(x,t)}{\partial x}. \tag{12.4}$$

Note that the rate of change of the velocity of a fluid element consists of two parts, namely, one that accounts for the explicit time dependence of the velocity field, and the so-called convective term, which accounts for the spatial variation of the velocity field. To better grasp the convective term, imagine a tube that is bent into a circular arc and filled with water flowing at a constant rate. The velocity field at different points in the flow does not change with time. Nonetheless, individual fluid elements necessarily have centripetal acceleration, which results in uniform circular motion. This acceleration follows from the change in direction of the velocity as the fluid element traverses the arc of a circle.

For an arbitrary velocity field, the same reasoning employed above leads to a general formula for the acceleration of the fluid element, which can be written as $\mathbf{a} = \mathrm{D}\mathbf{v}/\mathrm{D}t$. The "material time derivative" $\mathrm{D}/\mathrm{D}t$ evaluates the time rate of change of a quantity for a given material particle carried with the flow. The formula we derived for the acceleration, Equation 12.4, in the more general setting gives

$$\frac{\mathrm{D}}{\mathrm{D}t} = \frac{\partial}{\partial t} + \mathbf{v}\cdot\nabla, \tag{12.5}$$

where the first term accounts for the explicit time dependence and the second arises in response to the convective terms. Here we have made use of the gradient operator ∇ discussed in The Math Behind the Models on p. 366. The dot product between the velocity and the gradient operator signifies a spatial derivative in the direction of the velocity.

This notation represents a novel feature that arises in the continuum setting as a result of the fact that there are two possible origins for the time dependence of field variables. First, as noted above, there is the conventional explicit time dependence, which merely reflects the fact that the field variable may change with time. The second source of time dependence is a convective term and is tied to the fact that as a result of the motion of the continuum, the material particle can be dragged into a region of space where the field is different. As a concrete realization of this idea, we might imagine a tiny ship at sail in a stream, armed with a thermometer that measures the temperature of the water in its vicinity. If the tiny ship is docked at one position, but the temperature changes over time, the thermometer will report a temperature change. On the other hand, if the temperature depends upon position but not time, if the tiny ship sails some distance down the stream, there will be a different temperature measured by virtue of the position dependence of the temperature.

These ideas can be represented mathematically as follows. The temperature T can vary with time, even if the fluid is at rest, by heating, which will change the temperature of the fluid and result in a nonzero $\mathrm{D}T/\mathrm{D}t$ because of the $\partial T/\partial t$ term. The opposite extreme is when the temperature of the water is varying along the downstream direction, but is uniform in time. In this instance, the temperature associated with a given material particle can change by virtue of the fact that the material particle enters a series of spatial regions with different temperatures. In particular, if we imagine a steady flow in the x-direction with velocity v_x, then $\mathrm{D}T/\mathrm{D}t$ will be nonzero because of the term $v_x\partial T/\partial x$.

Now that we have the geometry of fluid motion in hand, we turn to the calculation of the forces acting on the fluid element. For simplicity, we take the fluid element to be a box of dimensions $\Delta x \times \Delta y \times \Delta z$ whose faces are perpendicular to the x-, y-, and z-axes. For a fluid in flow, the force due to interaction with other fluid elements can be split into a pressure force and a viscous force. The first is a consequence of pressure variations in the fluid, while the second is the frictional force that arises due to gradients in the flow velocity. Both contributions are illustrated in Figure 12.6.

The pressure field $p(x, y, z, t)$ exerts opposing forces on parallel faces of the fluid element as shown in Figure 12.6(A). We examine the force on the two faces of the fluid element that are perpendicular to the x-axis. The face at position x experiences a force $p(x, y, z, t)\Delta y\Delta z$ in the positive x-direction, while the force on the face at $x + \Delta x$ is $p(x + \Delta x, y, z, t)\Delta y\Delta z$ in the negative x-direction. The total pressure force in the positive x-direction can be obtained by adding these two forces as

$$\delta F_x^{\mathrm{p}} = p(x, y, z, t)\Delta y\Delta z - p(x + \Delta x, y, z, t)\Delta y\Delta z. \tag{12.6}$$

By rewriting the second term using the Taylor expansion $p(x + \Delta x, y, z, t) \approx p(x, y, z, t) + (\partial p/\partial x)\Delta x$, we find

$$\delta F_x^{\mathrm{p}} = -\frac{\partial p}{\partial x}\Delta x\Delta y\Delta z. \tag{12.7}$$

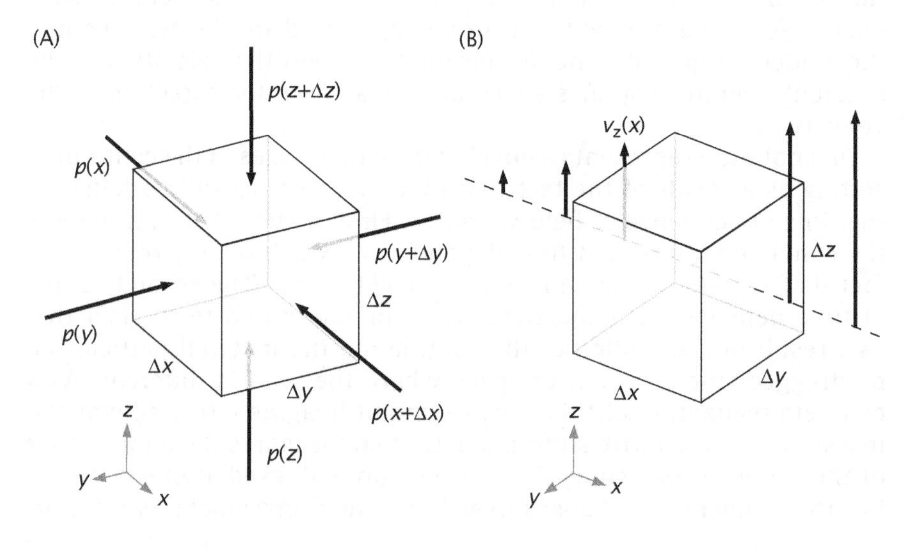

Figure 12.6: Pressure and viscous forces acting on a small volume element of fluid. (A) Pressure acts on the faces of the volume element perpendicular to each face, with the force pointing into the fluid element. (B) Spatial gradients of the fluid velocity lead to shear forces on the faces of the fluid element. If the fluid velocity points in the z-direction, and it spatially varies along the x-axis, then shear forces will be acting in the z-direction on the faces of the fluid element that are perpendicular to the x-axis.

The y- and z-components are obtained in an analogous fashion, giving a pressure force that can be written as

$$\delta \mathbf{F}^{\mathrm{p}} = -\nabla p \, \Delta x \Delta y \Delta z, \tag{12.8}$$

where $\nabla p = (\partial p/\partial x, \partial p/\partial y, \partial p/\partial z)$.

To evaluate the viscous force on the fluid element considered above, we consider a simple shear flow given by the velocity field $\mathbf{v} = (0, 0, v_z(x))$, shown in Figure 12.6(B). We make use of Newton's formula for the viscous force, Equation 12.1, to determine the force acting on each of the faces of the fluid element. In particular, Newton's formula tells us that the viscous force on each of the faces of the fluid element is proportional to the rate of change of the fluid velocity in the direction perpendicular to that face, and is also proportional to the area of the face. Given that the velocity only changes along the x-direction, we conclude that viscous forces will be nonzero only on the two faces perpendicular to x, that is, those in the y–z plane, whose area is $\Delta y \Delta z$.

The viscous force on the face at position x is $-\eta[\partial v_z(x)/\partial x]\Delta y \Delta z \mathbf{e}_z$, while the force on the face at position $x + \Delta x$ is $\eta[\partial v_z(x + \Delta x)/\partial x]\Delta y \Delta z \mathbf{e}_z$, where $\mathbf{e}_z$ is a unit vector along the z-direction. The reason the first term has a negative sign is that the fluid velocity at x is larger just within the volume element than outside. As a result, the viscous force is acting to reduce the velocity of the volume element. By way of contrast, on the face at $x + \Delta x$, the fluid velocity outside the volume element is larger than within it and the viscous force is therefore acting to increase the velocity of the fluid element, hence the positive sign of the force. The resulting viscous force on the volume element of fluid is obtained by adding up the contributions on the two faces. In particular, we have

$$\delta F_z^{\mathrm{v}} = -\eta \frac{\partial v_z(x)}{\partial x} \Delta y \Delta z \mathbf{e}_z + \eta \frac{\partial v_z(x + \Delta x)}{\partial x} \Delta y \Delta z \mathbf{e}_z, \tag{12.9}$$

as the total force. By Taylor-expanding the second term to first order in Δx, we find

$$\delta F_z^{\mathrm{v}} = \eta \frac{\partial^2 v_z}{\partial x^2} \Delta x \Delta y \Delta z. \tag{12.10}$$

The formula for the viscous force is easily generalized to a fully three-dimensional situation with the velocity field a function of all three coordinates. Instead of listing all possible cases, it is best to proceed by observing that our simple derivation leads us to conclude that the total viscous force is proportional to the spatial second derivatives of the fluid velocity. Since the force is a vector, and taking the simplest assumption of an isotropic fluid, we should consider all the ways of forming a vector quantity out of two ∇ operators (which provide the two derivatives) and the velocity field $\mathbf{v}$. The two independent combinations that can be formed are $\nabla^2 \mathbf{v}$ and $\nabla(\nabla \cdot \mathbf{v})$. For an incompressible fluid, which is the only situation we will consider, $\nabla \cdot \mathbf{v} = 0$, and the formula for the viscous force is

$$\delta \mathbf{F}^{\mathrm{v}} = \eta \nabla^2 \mathbf{v} \Delta x \Delta y \Delta z. \tag{12.11}$$

In an incompressible flow, the volume of a fluid element, like the one shown in Figure 12.7, does not change over time. If we start with an element whose dimensions are $\Delta x \times \Delta y \times \Delta z$, then an instant Δt later

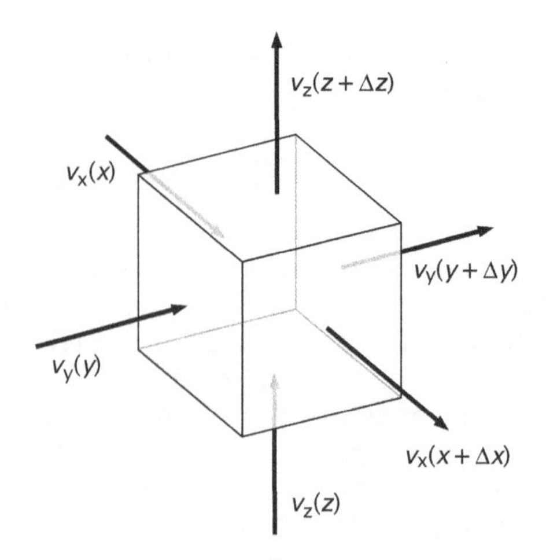

Figure 12.7: Schematic showing the velocity field on a small material volume element.

the size of the element in the x-direction is

$$\Delta x' = \Delta x\left(1 + \frac{\partial v_x}{\partial x}\Delta t\right), \tag{12.12}$$

with equivalent formulas holding for $\Delta y'$ and $\Delta z'$. Equation 12.12 is derived by considering the difference of the displacements of the two faces of the fluid element along the x-axis. Namely, in time Δt, the left face displaces by $v_x(x, y, z)\Delta t$, while for the right face we have $v_x(x + \Delta x, y, z)\Delta t$ for the displacement. The difference of the two gives the increase in the fluid element length in the x-direction. With this in hand, we can calculate the change in volume of the fluid element to first order in Δt, resulting in

$$\Delta x'\Delta y'\Delta z' = \Delta x\Delta y\Delta z\left[1 + \left(\frac{\partial v_x}{\partial x} + \frac{\partial v_y}{\partial y} + \frac{\partial v_z}{\partial z}\right)\Delta t\right]. \tag{12.13}$$

Demanding that the volume of the fluid element does not change is then equivalent to setting the second term to zero, or $\nabla \cdot \mathbf{v} = 0$, known as the incompressibility condition.

12.2.4 The Newtonian Fluid and the Navier–Stokes Equations

We are now finally in a position to write down Newton's second law for a fluid element, namely,

$$\Delta m\mathbf{a} = \delta\mathbf{F}^{\mathrm{p}} + \delta\mathbf{F}^{\mathrm{v}}, \tag{12.14}$$

where $\Delta m = \rho\Delta x\Delta y\Delta z$. Using the full vector version of Equation 12.4 for the acceleration of the fluid element, Equation 12.8 for the force due to pressure, and Equation 12.11 for the viscous force, we find

$$\frac{\partial \mathbf{v}}{\partial t} + (\mathbf{v}\cdot\nabla)\mathbf{v} = -\frac{1}{\rho}\nabla p + \nu\nabla^2\mathbf{v}. \tag{12.15}$$

This is the celebrated Navier–Stokes equation, where ρ is the density of the fluid and $\nu = \eta/\rho$ is the kinematic viscosity. In fact, this single vector equation is shorthand notation for three separate equations in the components of the velocity, (v_x, v_y, v_z). Presumably all the complexities associated with flows that can be captured by the simple Newtonian fluid constitutive model and that we observe in nature are encoded in this single equation. The fact that it is nonlinear makes it analytically intractable for all but the simplest flows.

An important example to be considered later in the chapter is that of low Reynolds number flow, in which the acceleration of fluid elements can be ignored rendering the equation linear.

The Velocity of Fluids at Surfaces Is Zero

To solve the Navier–Stokes equation, which mathematically speaking is a partial differential equation, we must specify boundary conditions that characterize the fluid velocity at surfaces, for example. Fluid flows are typically considered as bounded by solid walls, so the velocity of the fluid at the surface of the wall must be taken into account. The boundary condition that seems to work most of the time is the no-slip condition, which states that at the solid boundary the fluid is at rest with respect to the wall. Evidence for this is seen in the dust that collects on fan blades and remains even after the fan is turned on. Even more unequivocally, since fluid cannot pass through surfaces, the normal component of the velocity must surely vanish.

12.3 The River Within: Fluid Dynamics of Blood

Within the human body, fluid dynamics is vividly illustrated by the process of blood circulation. As the heart pumps to circulate blood throughout the body, blood travels through a system of tubes (the blood vessels) ranging in size from roughly 1 cm to as small as 2 μm. Figure 2.21 (p. 59) illustrated that red blood cells with sizes in excess of 5 μm can squeeze through these very small capillaries.

The simplest model of blood flowing through a vessel is that of a Newtonian fluid (a vast oversimplification), characterized by viscosity η and density ρ, steadily flowing through a pipe of circular cross-section of diameter d, due to a pressure difference Δp over the length l of the pipe as shown in Figure 12.8(A). At the walls of the pipe, the fluid velocity is zero and the symmetry of the model dictates that inside the pipe the fluid velocity only varies with the distance from the central axis of the pipe, r. In other words, we can write the fluid velocity vector as

$$\mathbf{v} = v(r)\mathbf{e}_z, \tag{12.16}$$

where $\mathbf{e}_z$ is a unit vector in the direction of fluid flow, which we identify with the z-axis.

To compute the fluid velocity profile $v(r)$, we make use of the Navier–Stokes equation, which in this situation dramatically simplifies. First, in a steady-state situation, which is the one we are considering, $\partial \mathbf{v}/\partial t = 0$. Second, the nonlinear term also vanishes, since there is no spatial variation of the velocity field in the direction of the flow (the z-direction). Therefore, the left-hand side of Equation 12.15 is

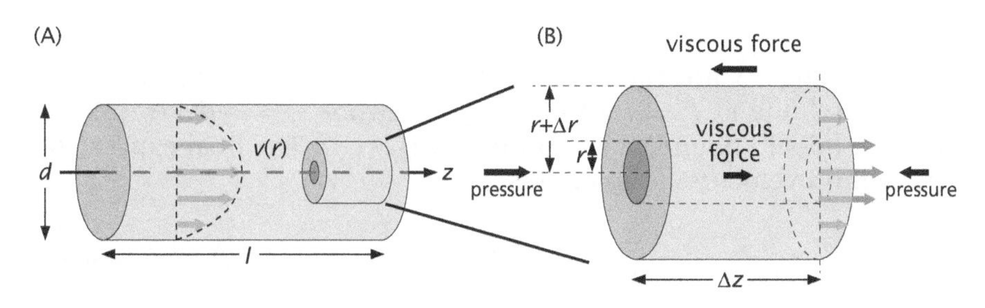

Figure 12.8: Fluid dynamics of pipe flow. (A) Steady-state flow through a pipe of diameter d and length l has a velocity profile that varies along the radial direction. (B) For steady flow through a pipe, the viscous forces acting on every fluid element are balanced by the pressure forces. The fluid element here is taken to be a hollow cylinder of thickness Δr.

identically zero, and the Navier–Stokes equation in this situation simply describes, for every fluid element, a balance between the pressure force and the viscosity-dependent frictional force.

Rather than writing the Navier–Stokes equations in cylindrical coordinates explicitly, we derive the relevant equation by considering a small fluid element in the shape of a hollow cylinder of length Δz. The inner and outer radii of the cylinder are r and $r+\Delta r$, respectively, as shown in Figure 12.8(B). The forces acting on the cylinder are the opposing pressure forces,

$$\delta F_z^{\mathrm{p}} = [p(z) - p(z+\Delta z)] 2\pi r \Delta r = -\frac{\mathrm{d}p}{\mathrm{d}z} 2\pi r \Delta r \Delta z, \tag{12.17}$$

and the opposing viscous forces,

$$\begin{aligned} \delta F_z^{\mathrm{v}} &= \eta v'(r+\Delta r) 2\pi (r+\Delta r)\Delta z - \eta v'(r) 2\pi r \Delta z \\ &= \eta v' 2\pi \Delta r \Delta z + \eta \frac{\mathrm{d}v'}{\mathrm{d}r} 2\pi r \Delta r \Delta z, \end{aligned} \tag{12.18}$$

obtained from Equation 12.1. These viscous forces act on the inner and the outer surface of the cylindrical fluid element, as shown in Figure 12.8(B). The area of the inner surface is $2\pi r \Delta z$, while $2\pi(r+\Delta r)\Delta z$ is the area of the outer surface. The symbol v' in the above equation is shorthand for $\mathrm{d}v/\mathrm{d}r$, and we Taylor-expanded the terms $p(z+\Delta z)$ and $v'(r+\Delta r)$ to first order in Δz and Δr to obtain the final formula.

The force balance on the fluid element can be written as $\delta F_z^{\mathrm{p}} + \delta F_z^{\mathrm{v}} = 0$, and if we make use of Equations 12.17 and 12.18, we arrive at a differential equation for $v(r)$ of the form

$$\frac{1}{\eta}\frac{\mathrm{d}p}{\mathrm{d}z} = \frac{1}{r}\frac{\mathrm{d}v}{\mathrm{d}r} + \frac{\mathrm{d}^2 v}{\mathrm{d}r^2} = \frac{1}{r}\frac{\mathrm{d}}{\mathrm{d}r}\left(r\frac{\mathrm{d}v}{\mathrm{d}r}\right). \tag{12.19}$$

Those well versed in vector calculus will recognize the right-hand side of Equation 12.19 to be nothing but the Laplacian of the velocity field $v(r)\mathbf{e}_z$ in cylindrical coordinates, as is dictated by the Navier–Stokes equation in vectorial form and given by Equation 12.15.

To solve this equation, we can first integrate both sides with respect to z along the length of the pipe, to obtain

$$-\frac{1}{\eta}\Delta p = \frac{1}{r}\frac{\mathrm{d}}{\mathrm{d}r}\left(r\frac{\mathrm{d}v}{\mathrm{d}r}\right) l, \tag{12.20}$$

where $\Delta p = p(0) - p(l)$ is the pressure drop down the pipe.

The solution to the above differential equation is obtained by integrating twice and is given by

$$v(r) = -\frac{\Delta p}{\eta l}\left(\frac{r^2}{4} - C_1 \ln r + C_2\right), \tag{12.21}$$

where C_1 and C_2 are the two integration constants. Since the fluid velocity is finite at $r = 0$, we have $C_1 = 0$, while $C_2 = -d^2/16$ follows from the no-slip boundary condition, $v(d/2) = 0$. The final expression for the fluid velocity in the pipe is

$$v(r) = \frac{\Delta p}{4\eta l}\left(\frac{d^2}{4} - r^2\right). \tag{12.22}$$

From this expression, other useful formulas can be determined, such as those for the average fluid velocity across the cross-section of the pipe,

$$\overline{v} = \frac{\int_0^{d/2} v(r) 2\pi r\, dr}{\pi d^2/4} = \frac{\Delta p d^2}{32\eta l}, \tag{12.23}$$

and for the volumetric flow rate $Q = \overline{v}\pi d^2/4$, which upon substitution of $\overline{v}$ yields

$$Q = \frac{\pi \Delta p d^4}{128\eta l}. \tag{12.24}$$

The formula for the flow rate is a classic result in fluid dynamics, and it is usually attributed to Poiseuille. It predicts that the flow rate scales as the fourth power of the diameter of the pipe, for a constant pressure difference. This tells us that small changes in the width of a blood vessel, say due to plaque accumulation, can lead to a significant decrease in the rate of blood flow.

Estimate: Blood Flow Through Capillaries Capillaries of different animals are all roughly equal in size, about $d = 5\,\mu\text{m}$ in diameter. We can make a rough estimate of blood flow through a capillary from the measured pressure difference and the size. The pressure difference across a capillary is about $\Delta p \approx 20\,\text{mmHg} \approx 3000\,\text{Pa}$ over the length of a capillary, $l \approx 1\,\text{cm}$. Using the equation for the average flow velocity through a pipe,

$$v = \frac{\Delta p d^2}{32\eta l}, \tag{12.25}$$

we estimate the rate of blood flow to be $v \approx 0.02\,\text{cm/s}$, while the measured flow is roughly 0.05 cm/s. Note that in the above estimate we used the viscosity of water $\eta = 10^{-3}\,\text{Pa s}$; the fact that the estimated flow is slightly lower than measured seems to imply that the viscosity of blood in capillaries is lower than that of water. In fact, the viscosity of blood is roughly three times that of water, which would make the discrepancy between our estimate and the observed velocity even larger. This is not too surprising given the fact that the size of a red blood cell is roughly equal to the diameter of a capillary. A more faithful treatment requires taking into account the non-Newtonian character of the fluid.

12.3.1 Boats in the River: Leukocyte Rolling and Adhesion

Our previous discussion of the fluid dynamics of blood again made the approximation that blood is a simple fluid. However, as is easily seen in a low-powered microscope, blood is actually a dense suspension of many cells that have a wide variety of appearances, sizes, and functions in the body. By far the most abundant are the red blood cells already discussed (see Section 4.2 on p. 143) as vessels for hemoglobin to carry oxygen around the body. Another major cellular component of blood comprises the so-called white blood cells that are agents of the immune system. The white blood cells use blood circulation as a rapidly moving superhighway to enable them to reach any

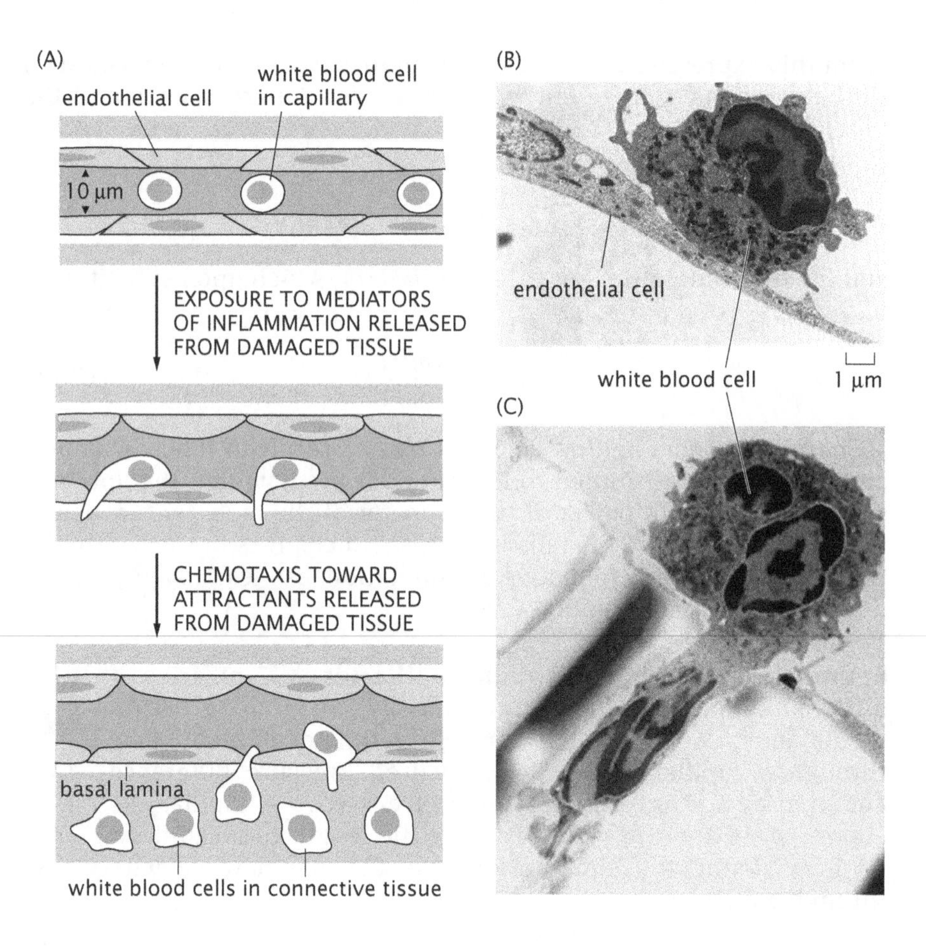

Figure 12.9: Invasion of a leukocyte into tissue. (A) Schematic of how leukocytes are recruited to damaged tissue. (B) Electron micrograph of a leukocyte adhering to an endothelial cell. (C) Electron micrograph of a leukocyte caught in the act of extravasation, that is, crossing the endothelial cell barrier, in tissue culture. (A, adapted from B. Alberts et al., Molecular Biology of the Cell, 5th ed. Garland Science, 2008; B, Courtesy of *The Irish Scientist*; C, Courtesy of Clare O'Connor and Jill Mackarel.)

part of the body very quickly to respond to an infection. The process of invasion of infected tissues by these white blood cells (or leukocytes) involves several different phases that illustrate the complexity and the regulable interactions between the fluid dynamics of the blood and the walls of the blood vessels.

In a healthy human with no infections, white blood cells are passive travelers in the current of blood like the red blood cells. However, in response to injury or infection, cells in a damaged tissue secrete molecules that signal to the cells forming the lining of the blood vessel to alter their adhesive surface properties and also to the circulating leukocytes to alter their own adhesive properties in a complementary way. This process is shown in Figure 12.9. As the leukocytes and endothelial cells that make up the vessel walls initially slightly increase their binding affinity for one another, the leukocytes undergo a striking behavior called rolling. Specific molecular binding events occur between the leukocytes and the endothelial surfaces that are not sufficiently strong to hold the leukocytes stuck to the wall against the powerful current of the flowing blood. As leukocytes balance the viscous force driving their motion and the adhesive surface, the stable dynamics involves rolling of the cells, with attachment and detachment occurring at a rate resulting in rolling without detachment. In response to stronger distress signals from damaged tissue, the strength of these contacts is further enhanced such that the spherical rolling leukocytes will plaster themselves against the endothelial wall. After sticking, the leukocyte actually exits the blood vessel and enters the damaged tissue, where it fights infection. Different leukocytes use different weapons in the fight against microbial invaders.

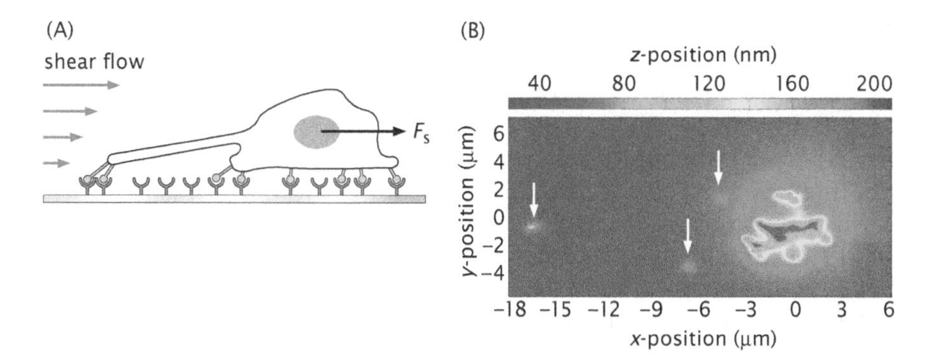

Figure 12.10: Mechanics of leukocyte rolling. (A) Schematic of the forces involved in leukocyte rolling. The shear flow in the blood vessel leads to a forward shear force on the leukocyte, which is balanced by the bonds formed between ligands on the cell surface (light green) and the P-selectin receptors (green) on the surface of the endothelium. (B) Total internal reflection microscopy image of the surface of the leukocyte cell rolling on a P-selectin covered slide showing points (indicated by white arrows) at which filopodia, protrusions of the cell membrane, are making contact with the slide surface. The color code indicates the height of the cell surface above the slide. (Adapted from P. Sundd et al., *Nat. Meth.* 7:821, 2010.)

Estimate: Mechanics of leukocyte rolling Leukocytes in blood vessels experience forces due to fluid flow around them, and due to binding of ligand molecules on their surface to receptor molecules on the surface of endothelial cells. The result of these forces is leukocyte rolling, which occurs at about $10\,\mu\text{m/s}$, which is considerably smaller than the few hundred microns per second typical of average blood flow velocity in capillaries.

The typical shear stress experienced by leukocytes due to blood flow is $\sigma = 1\,\text{Pa}$ and it leads to a shear force $F_s \approx \sigma 4\pi r^2$, where r is the radius of a leukocyte, which we approximate by a sphere. The size of a leukocyte can be estimated from Figure 12.9(B), $r \approx 5\,\mu\text{m}$, leading to an estimate $F_s \approx 300\,\text{pN}$. In order for the leukocytes to roll at a constant speed, the drag force must be balanced by the force provided by its adhesion to the endothelial surface. Adhesive forces are a result of the formation of transient bonds between the selectin receptors on the surface of the endothelium and their ligands (various glycoproteins) on the surface of the leukocyte; see Figure 12.10(A). Given that the typical rupture force for a ligand–receptor pair is $100\,\text{pN}$, multiple bonds have to be formed at the same time in order to provide the necessary counterbalance to the shear force exerted by the moving fluid. In fact, recent experiments, shown in Figure 12.10(B), have uncovered three to four long tethers (filopodia) that extend behind the rolling leukocyte cell. Each filopodium forms multiple bonds with the selectin covered surface, as shown schematically in Figure 12.10(A). The forces that arise from the stretching of these bonds provide a counterbalance to the force exerted on the leukocyte by the flow of blood.

12.4 The Low Reynolds Number World

12.4.1 Stokes Flow: Consider a Spherical Bacterium

Interestingly, for many of the problems of interest in biology, the equations of fluid mechanics can be considerably simplified because the dynamics is like that familiar from the world of everyday experience when a marble is dropped in a jar of honey. Insights into "life at low Reynolds number" were developed in a classic article of the same title by Edward Purcell, which is recommended to all of our readers. To illustrate the peculiar properties of "low Reynolds number" flows, an experiment has been underway in Australia since 1927. This experiment, shown in Figure 12.11, examines the "flow" of a highly viscous fluid under the action of gravity. So far, in the roughly 80 years since its inception, there have been eight drops of fluid. The reason this experiment is important for developing biological intuition is that this intensely viscous environment should conjure up images of the world of a bacterium, one that is characterized by severe viscous drag, as will be illustrated more concretely below when we compute the stopping distances of bacteria.

Figure 12.11: The 100-year experiment. Illustration of a "fluid" of everyday proportions that would present the same viscous properties to a fish as does water to an *E. coli* cell. The black goo in the middle can be seen well along the way to the next drop, which is dangling precariously above the container, which is filled with the results of the previous drops. (Courtesy of Professor J. S. Mainstone.)

The nonlinear term in the Navier–Stokes equation can be neglected for flows dominated by viscosity. To see this, consider the flow due to a rigid body of linear size L moving through the fluid at speed U. The inertial terms appearing on the left-hand side of Equation 12.15 can be approximated in magnitude by $\rho U^2/L$, while the magnitude of the viscous term is given by $\eta U/L^2$. The idea of these estimates is to approximate the magnitude of the derivatives of the fluid velocity by U/L and U/L^2, for the first and second derivatives, respectively. The ratio of the magnitudes of the inertial and the viscous terms is the dimensionless Reynolds number,

$$Re = \frac{\rho LU}{\eta} = \frac{LU}{\nu}. \tag{12.26}$$

The significance of the Reynolds number is that in the case $Re \ll 1$, the Navier–Stokes equation simplifies to the Stokes equation,

$$\nabla p = \eta \nabla^2 \mathbf{v}. \tag{12.27}$$

The Stokes equation is linear, and when supplemented with the incompressibility condition $\nabla \cdot \mathbf{v} = 0$, it leads to important analytic solutions such as the flow field around a sphere moving at a uniform velocity $\mathbf{v}$.

An alternative scheme for examining the meaning of the Reynolds number as a dimensionless characterization of the relative importance of viscous and inertial force is to make a simple comparison of kinetic energy and viscous dissipation. If we consider a fluid parcel of size a traveling at velocity u, then the kinetic energy (KE) of this fluid parcel is its mass, ρa^3, times the square of its velocity, or

$$\text{KE} \sim \rho a^3 u^2. \tag{12.28}$$

This number should be contrasted with the energy dissipated by viscous work done as the fluid parcel moves a distance comparable to its size. This work is given by

$$W \sim \underbrace{\eta \frac{u}{a}}_{\text{viscous stress}} \times \underbrace{a^2}_{\text{area}} \times \underbrace{a}_{\text{distance traveled}}, \tag{12.29}$$

where the rate of change of the fluid velocity, which produces the viscous stress, is once again estimated as u/a. The ratio of the kinetic energy and the viscous work reproduces Equation 12.26, and we conclude that in the low-Re world the kinetic energy of a fluid element is rapidly dissipated by friction. The same conclusion follows if we estimate the time it would take to dissipate the kinetic energy of the fluid element and compare it with the time it would take it to move a distance comparable to its size. The first is given by the ratio of the kinetic energy ($\rho a^3 u^2$) and the power of the friction force ($\eta a u^2$),

$$\tau_{\text{viscous}} = \rho a^2/\eta, \tag{12.30}$$

while the second is

$$\tau_{\text{inertial}} = a/u. \qquad (12.31)$$

The ratio of these two time scales once again produces Equation 12.26.

An important solution of the Stokes flow equation given by Equation 12.27 is for the case of a sphere of radius R moving through a fluid with a constant velocity $\mathbf{v}$ and far from any walls. The different kinds of flows that develop in this case as a function of the Reynolds number are shown in Figure 12.12. The fundamental solution for a sphere in a flow will serve as the basis for developing intuition about many of the key problems in biological fluid dynamics that will concern us here. An alternative way of viewing this same problem is to set up a coordinate system that moves along with the sphere. In this case, we need to solve Equation 12.27 assuming that the velocity of the fluid vanishes at the boundary of the sphere (the no-slip boundary condition) and that it equals $-\mathbf{v}$ very far from the ball. Instead of going through the mathematics that ultimately leads to the celebrated Stokes formula for the drag force on a sphere, which can be found in most fluid dynamics books, we present an intuitive argument based on dimensional analysis.

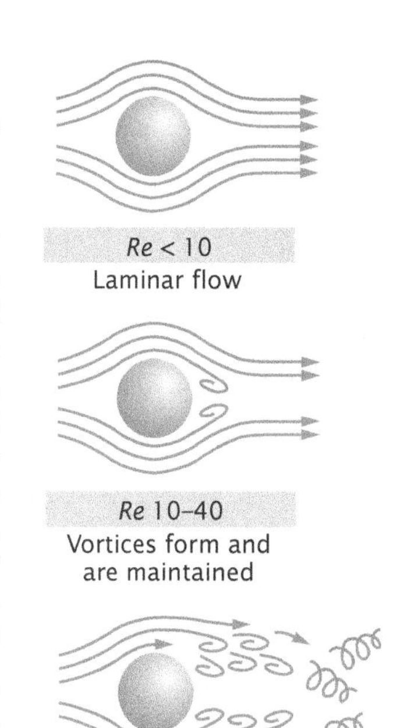

Figure 12.12: Flow around a spherical object as a function of the fluid velocity. With increasing Reynolds number, the flow transitions from laminar to turbulent. (Adapted from R. P. Feynman et al., The Feynman Lectures on Physics. Addison-Wesley, 1965.)

To calculate the force exerted on the sphere by the fluid, we need to integrate the viscous stress and the pressure over the surface of the sphere. As described in the derivation of the Navier–Stokes equation, the viscous stress is proportional to the spatial derivative of the velocity, which we can approximate with v/R, using simple dimensional arguments. The same holds true for the pressure, as is seen directly from Equation 12.27. As the area of the sphere is $4\pi R^2$, we can evaluate the "integral" over the entire surface by multiplying the characteristic force scale $(\eta v/R)$ by this area, resulting in the Stokes force $F_S = C\eta R v$, with C a dimensionless constant. Note that this equation also follows from the linearity of the Stokes equation (pressure and fluid velocity are linearly related), which implies $F_S \sim v$, where the coefficient of proportionality has units N s/m. The only way to construct such a constant from the remaining dimensionful parameters in the problem, namely, the viscosity η and the radius of the ball R, is via the combination ηR. The precise formula for the Stokes drag,

$$F_S = 6\pi\eta R v, \qquad (12.32)$$

obtained by doing the math and solving the differential equation, only adds the value of the constant $C = 6\pi$ to the story.

This mode of thinking also allows us to calculate the drag on a moving cylinder in low Reynolds number conditions. This calculation relates to the problem of bacterial propulsion by rotating flagella as well as to the many important examples of cilia. For a cylinder of length L and radius r moving at velocity $\mathbf{v}$, we expect fluid velocity gradients of size v/r. Since the surface area of the cylinder is $2\pi rL$, we estimate that the viscous drag is of the form $C_{\text{cyl}}\eta v L$, with C_{cyl} being of the order of unity. This simple argument can be repeated for a solid body of arbitrary shape, leading to the rule of thumb that the largest dimension of the body is the one that sets the scale of the frictional drag force.

There is a detail missing in the above argument, which will prove important when we take on the issue of propulsion of bacteria by twirling of their flagella. In particular, with a cylinder, there are two inequivalent directions, one along its axis and the other perpendicular to it, and the drag depends on whether the motion is along one or the

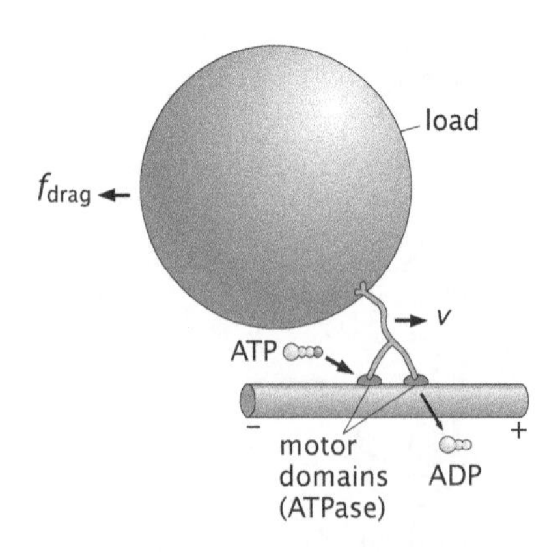

Figure 12.13: Viscous forces during optical tweezer experiments. A molecular motor with nanometer dimensions is attached to a bead hundreds of nanometers in size. As the motor moves along its cytoskeletal track, there is a small viscous drag force on the bead. This figure is not drawn to scale.

other direction. In particular, the coefficient C_{cyl} is twice as large in the direction perpendicular to the long axis than when the motion is along it. Furthermore, dimensional analysis for the cylinder case is not as powerful as for the sphere. For the cylinder, there is a dimensionless parameter L/r, and therefore dimensional analysis alone allows for C_{cyl} to be an arbitrary function of it. In fact, detailed calculations based on the Stokes flow equations have shown that C_{cyl} has a weak, logarithmic dependence on L/r.

12.4.2 Stokes Drag in Single-Molecule Experiments

Single-molecule experiments allow us to spy individually on the molecules of the cell. We have already seen in Chapter 8 how these tools can be used to explore the mechanics of DNA. Another important application of these techniques has been to the study of the *dynamics* of a host of machines of the cell. To interpret these experiments, we have to understand what artifacts (if any) are introduced by the beads and filaments we attach to these macromolecules.

Stokes Drag Is Irrelevant for Optical Tweezers Experiments

Single-molecule experiments with optical tweezers are used to study molecular motors and how the interplay of chemistry and mechanics gives them the functionality of cargo-carrying transport machines in the cell. Such experiments, as shown in Figure 3.33 (p. 130), permit the measurement of the velocity of molecular motors such as myosin as a function of the applied load. From the standpoint of the present chapter, one question that comes to mind is revealed by Figure 12.13, which shows the mismatch between the size of the motor (measured in nanometers) and the size of the bead (with micron-scale dimensions) that is tethered to the motor and is used to control it. In particular, as the motor pulls the bead along, how large are the drag forces imposed by the presence of the bead and how do these forces compare with those due to the motor itself?

To estimate the magnitude of this force, we invoke Equation 12.32, where R is the radius of the bead and v is its velocity. If we use numbers appropriate for the motion of a molecular motor, we have

$$F_S \approx 6 \times \pi \times \underbrace{10^{-3}\,\mathrm{N\,s/m^2}}_{\text{viscosity}} \times \underbrace{500 \times 10^{-9}\,\mathrm{m}}_{\text{bead radius}} \times \underbrace{10^{-6}\,\mathrm{m/s}}_{\text{motor speed}} \approx 10^{-2}\,\mathrm{pN}.$$

This simple estimate illustrates that the drag force due to the presence of the bead is nearly three orders of magnitude smaller than the force exerted by the motor itself (roughly 5 pN). As a result, despite the fact that the motor with its tethered bead might evoke the image of Atlas carrying the Earth, the extra force that the motor needs to supply to overcome viscous drag on the bead is negligible.

Similar conclusions regarding the relevance of the drag force can be reached in experiments of the kind illustrated in Figure 12.14. These remarkable single-molecule experiments have been used to demonstrate unequivocally that the ATP synthase is a rotary motor. The idea of the experiments is to attach a fluorescently labeled actin filament to the ATP synthase. The reasoning is that if the motor rotates, the actin filament will be spun around and this rotation, given the size of the actin filament, can be easily observed under a microscope. Such

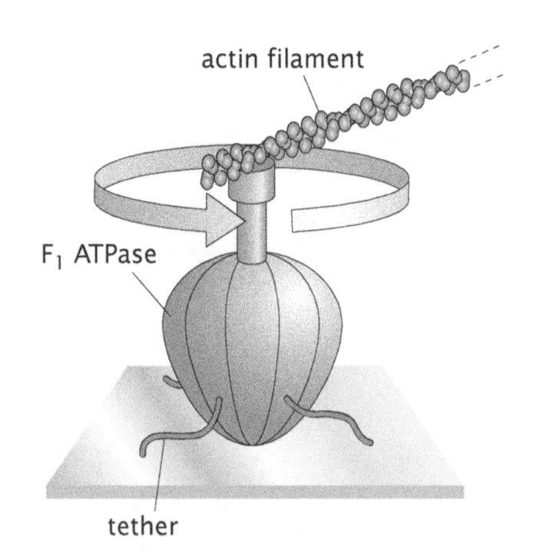

Figure 12.14: Cartoon of single-molecule experiment in which an actin filament is attached to F_1 ATPase. The rotation of the motor translates to the observable rotation of the filament, allowing for a detailed study of the mechanics of this important enzyme involved in ATP synthesis.

experiments have provided a window on the mechano-chemistry of these fascinating machines and demonstrate the existence of pauses.

12.4.3 Dissipative Time Scales and the Reynolds Number

We can gain more intuition about the low-*Re* world by invoking a mechanical analogy for motion dominated by viscous forces. Consider a damped harmonic oscillator, described by the equation

$$m\ddot{x} + \gamma\dot{x} + kx = 0, \tag{12.33}$$

where m is the mass, γ the damping coefficient, and k the spring constant. The symbols $\ddot{x}$ and $\dot{x}$ are shorthand for the acceleration and velocity, that is, the second and first derivatives of x in time. We are interested in the motion of the oscillator in the viscous limit. Our idea is to examine the relative importance of the inertial and drag terms in this simple case. For the Navier–Stokes equation, this is the limit in which the viscous time scale is shorter than any other time scale in the problem. The viscous time scale is the time for the kinetic energy to be dissipated by the viscous forces. We can see this as the time scale over which kinetic energy is dissipated by the drag term. This can be written as

$$\frac{m\dot{x}^2}{\tau_V} \approx \gamma\dot{x}^2, \tag{12.34}$$

where the right-hand side is the rate of dissipation, given by the product of the viscous force ($\gamma\dot{x}$) and the velocity ($\dot{x}$). As a result, we can write the viscous time scale as

$$\tau_V = \frac{m\dot{x}^2}{\gamma\dot{x}^2} = \frac{m}{\gamma}. \tag{12.35}$$

The claim is that in the viscous limit the inertial term ($m\ddot{x}$) can be dropped in Equation 12.33, as long as we are interested in the position of the oscillator at times greater than the viscous time, m/γ. This then leads to a simpler equation for $x(t)$,

$$\gamma\dot{x} + kx = 0, \tag{12.36}$$

whose solution is readily obtained,

$$x(t) = x_0 e^{-(k/\gamma)t}, \tag{12.37}$$

where x_0 is the initial position of the oscillator.

A direct check on the above claim can be performed by solving Equation 12.33, assuming an initial position (x_0) and velocity (v_0) for the oscillator. The solution to the differential equation is

$$x(t) = C_+ e^{\lambda_+ t} + C_- e^{\lambda_- t}, \tag{12.38}$$

with

$$\lambda_\pm = \frac{-\gamma \pm \gamma\sqrt{1 - 4km/\gamma^2}}{2m}, \tag{12.39}$$

and the constants $C_\pm = (\lambda_\mp x_0 - v_0)/(\lambda_\mp - \lambda_\pm)$. In the viscous limit, the period of the oscillator, $\sqrt{m/k}$, is much larger than the viscous time scale m/γ, and consequently $km/\gamma^2 \ll 1$. This strong inequality permits us to Taylor-expand the square root in Equation 12.39. Expanding

Equation 12.39 to lowest order in km/γ^2, and substituting $\lambda_\pm$ into Equation 12.38 leads to

$$x(t) = C_+ e^{-(k/\gamma)t} + C_- e^{-(\gamma/m)t}. \quad (12.40)$$

At times t much longer than the viscous time scale, this equation reduces to Equation 12.37, which was obtained by ignoring the inertial term in the damped harmonic oscillator equation. It is interesting to note that dependence on the initial velocity also drops out of the problem, just as Equation 12.37 would have us believe, as long as $m/\gamma \ll x_0/v_0$ is also satisfied. This additional assumption states that the viscous time scale is much smaller than the time it would take the particle to traverse a distance x_0 if it were moving at a constant velocity v_0.

The point of this exercise was to illustrate the relative importance of inertial and drag terms in the case of a system with a single degree of freedom. The lessons learned here can be used to better understand the significance of the Reynolds number in the fluid-mechanical setting.

12.4.4 Fish Gotta Swim, Birds Gotta Fly, and Bacteria Gotta Swim Too

The low Reynolds number regime is especially pertinent for the swimming of microbes such as *E. coli.* To see how the Reynolds number measures up for bacteria in comparison with macroscopic swimmers like fish, we resort to order-of-magnitude estimates. For a fish of size 10 cm swimming through water at a speed of 1 m/s, the viscous time scale given by Equation 12.30 is $\tau_v = 10^4$ s, which can be estimated using the kinematic viscosity of water at room temperature, which is $\eta/\rho = 10^{-6}\,\mathrm{m^2/s}$. The inertial time scale, Equation 12.31, is $\tau_i = 0.1$ s. This leads to a large Reynolds number of 10^5, which implies that inertial effects dominate viscous ones. On the other hand, for an *E. coli* bacterium, which is roughly 1 μm in size, swimming through water at 10 μm/s leads to $\tau_v = 1\,\mu$s and $\tau_i = 0.1$ s, and a Reynolds number of 10^{-5}. In this case, the kinetic energy is dissipated practically instantaneously, as the bacterium will move over a very short distance (of the order of $\tau_v u \approx 0.01$ nm) before it comes to a stop.

From the example above, we see that there is a difference of 10 orders of magnitude in Reynolds number describing swimming of macroscopic objects, such as fish, and microscopic ones, such as *E. coli.* Consequently, as we will discuss in more detail shortly, the mechanical world inhabited by a bacterium is a very strange place, one where Aristotle reigns supreme (in the sense that the dynamics in this world is one in which *velocity* and force are linearly related). To gain more intuition about this large difference in Reynolds numbers, we can ask how viscous should a substance be so that a fish swimming through that medium had a Reynolds number like that of *E. coli* in water? The answer is that the kinematic viscosity should be 10 orders of magnitude larger than that of water, or $10^4\,\mathrm{m^2/s}$. This is roughly the value for pitch at room temperature, exactly like the experiment revealed in Figure 12.11.

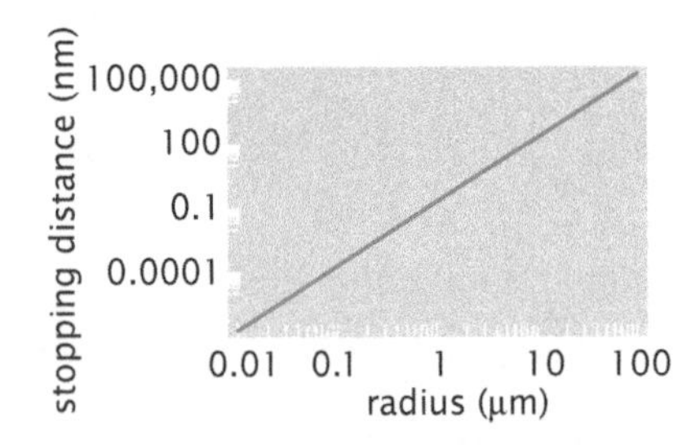

Figure 12.15: Stopping distance for a sphere with initial velocity equal to its diameter per second. The y-axis reports the distance over which a spherical particle will stop if it is moving with an initial velocity corresponding to traveling its diameter in a second.

A second way to illustrate this peculiar world of low Reynolds number dynamics is gained by considering the stopping distances for objects of different sizes. The idea is that we consider a spherical object with some initial velocity equal to the sphere's diameter per

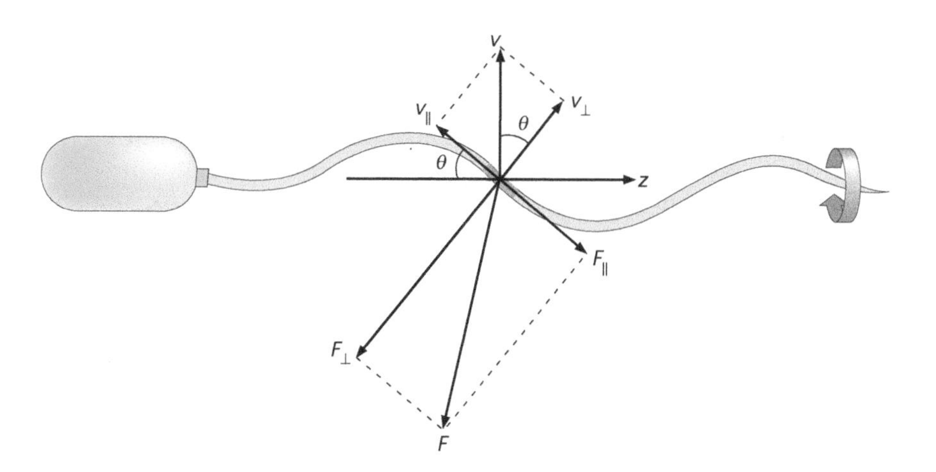

Figure 12.16: Force diagram for a swimming bacterium. The schematic shows how the velocity at a given point on the flagellum can be resolved into components that are parallel and perpendicular to the flagellum. The parallel and perpendicular components of the drag force acting on the small element (red) of the flagellum are also shown. Note that the sum of the two force components, the total force, is not parallel to the velocity, owing to the difference in drag coefficients for motion parallel and perpendicular to the long axis of the rod-like element.

second. We then ask how long it will take for drag forces to bring that sphere to a stop given this initial velocity. The result of such a calculation is shown in Figure 12.15. For the case of a sphere with a radius of 1 μm, the stopping distance is less than a nanometer!

An important consequence of the low Reynolds number world bacteria inhabit is the way in which they propel themselves in search of nutrients. For example, *E. coli* moves by rotating its flagella, which have the form of thin helical filaments, as already shown in Figure 4.16 (p. 160). The propulsive force that this generates relies on the difference in the drag coefficients between parallel and perpendicular motion of a rod in a fluid. To obtain an estimate of the propulsive force and the velocity it generates, we take the flagellum to be a helical filament whose length is $L = 10\,\mu\text{m}$ and that rotates around its long axis at a frequency of $f = 100\,\text{Hz}$. The pitch and diameter of the helix are $P = 2\,\mu\text{m}$ and $D = 0.5\,\mu\text{m}$. If we take a small segment of the flagellum of length l, it will be at an angle θ with respect to the z-direction and moving at a linear velocity $v = \pi Df$. The geometry we have in mind is shown in Figure 12.16. The angle is related to the pitch and diameter of the helix, $\tan\theta = \pi D/P$. The forces parallel and perpendicular to the flagellum segment are

$$F_{\|} = \gamma_{\|} v \sin\theta \quad \text{and} \quad F_{\perp} = \gamma_{\perp} v \cos\theta, \tag{12.41}$$

where $\gamma_{\|} \approx 2\pi\eta l$ and $\gamma_{\perp} \approx 4\pi\eta l$ are the drag coefficients for a thin rod moving in a direction parallel and perpendicular, respectively, to its long axis. Note that we have neglected logarithmic corrections to these drag coefficients. The key point is the fact that the drag coefficient is different in the two directions and this leads to a nonzero propulsive force.

The propulsive force

$$F_{\text{p}} = 2\pi\eta L v \cos\theta \sin\theta \tag{12.42}$$

is the sum of the z-components of the parallel and perpendicular forces, over all segments, L/l in number. If the bacterium is moving at a speed V, then the drag force $2\pi\eta LV$ is balanced by the propulsive force F_{p}, leading to an expression for V, namely,

$$V = v\cos\theta \sin\theta = \pi Df \sin\theta \cos\theta. \tag{12.43}$$

Using the numbers for *E. coli* given above, we estimate $V \approx 70\,\mu\text{m/s}$, while the measured speed is roughly $30\,\mu\text{m/s}$.

Reciprocal Deformation of the Swimmer's Body Does Not Lead to Net Motion at Low Reynolds Number

Twirling of a helical filament is one of many strategies used by life forms that live at low Reynolds number to move around. Interestingly, some swimming strategies that are quite effective for humans and other large swimmers fail miserably at low Reynolds number. Take for example a simple strategy that is similar to that used by a scallop. The "swimmer" is made of two plates connected by a hinge, and the plates come together and then swing apart at different speeds. Even though this kind of motion will lead to the propulsion of the swimmer at large Reynolds numbers, a low Reynolds number swimmer that adopts this strategy will be rewarded by fatigue without forward progress. To show this, we exploit the property of low Reynolds number flows that the forces on moving objects are proportional to their velocity.

The angle between the plates, $\theta(t)$, varies in time in such a way that at the beginning and the end of the swimming stroke the angle is the same, $\theta(0) = \theta(T)$. The force that the fluid exerts on the "scallop," by linearity, is proportional to the angular velocity $d\theta/dt$. On the other hand, the linear velocity of the scallop v is proportional to the force, and therefore $v = \text{const}\ d\theta/dt$. Integrating both sides over the period of the scallop motion (from 0 to T) gives $x(T) - x(0) = \text{const}[\theta(T) - \theta(0)] = 0$, in other words the net displacement is zero. We see that this is always going to be the case when the motion of the swimmer is reciprocal, meaning that the swimming cycle can be divided into two "strokes" that are time inverses one of the other. A simple generalization of the "scallop," which now has three plates and two hinges, can produce net motion as long as the order in which the plates fold back up is the same as the order in which they swing open (thus avoiding the trap of reciprocal motion); see Figure 12.17. Exactly which way a swimmer like this will move turns out to be a rather subtle question, and the answer even depends on the amplitude of the plates' motion.

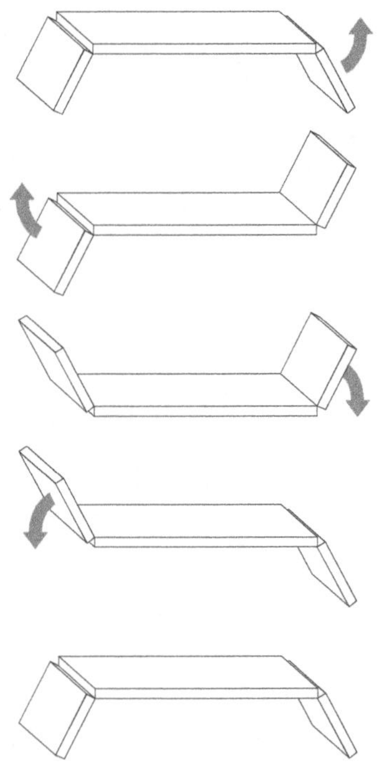

Figure 12.17: Purcell's simple swimmer. Movement of the plates as indicated by the arrows leads to an overall translation, swimming, of the object. It has been suggested that this model might explain the mechanism by which small helical bacteria known as *Spiroplasma* swim. (Adapted from E. M. Purcell, *Am. J. Phys.* 45:3, 1977.)

12.4.5 Centrifugation and Sedimentation: Spin It Down

A different class of problems in the mechanics of fluids is relevant to one of the most familiar tools of the biochemist and the molecular biologist, namely, the centrifuge. Often, it is of interest for the purposes of biochemical purification to separate macromolecules or their complexes from the surrounding solution. Centrifugation is a particularly effective way of achieving this, and it provides a simple example of diffusion in the presence of drift and invokes the Stokes formula once again.

The process of centrifugation involves spinning a sample consisting of large molecules in solution at rates that can be as high as 100,000 rpm (corresponding to accelerations of roughly $10^6\,g$). The goal is to separate the macromolecules by size, which is achieved as follows: the process of spinning the sample leads to an apparent centrifugal force that acts on every molecule. The force is directed away from the rotation axis and its magnitude is $m\omega^2 r$, where m is the mass of the molecule, ω is the rotation rate, and r is the molecule's distance from the rotation axis. If we assume that the size of the sample is much smaller than its distance from the axis of rotation, then we can say that the centrifugal force per unit mass is, to a good approximation, constant. This is like the more familiar example of gravity, which bestows the same acceleration of $9.8\,\mathrm{m/s^2}$ upon all free falling bodies near the surface of the Earth. This analogy also reminds us that the mass of the molecule that enters the formula for the centrifugal force should, following Archimedes, be corrected by the mass of the displaced solvent, $m \to m - \rho V$, where ρ is the density of the solvent and V is the volume taken up by the molecule.

This centrifugal force imparts a drift velocity to the molecules. In the friction-dominated world of macromolecules in solution, the frictional force is proportional to the drift velocity. This frictional force will exactly match the centrifugal force, leading to the expression

$$v_{\mathrm{drift}} = \frac{m g_{\mathrm{c}}}{\gamma}, \tag{12.44}$$

where g_{c} is the centrifugal force per unit mass (assumed constant) and γ is the friction coefficient. For a spherical particle of radius R moving in a fluid with viscosity η, the friction coefficient is given by the Stokes formula, $\gamma = 6\pi\eta R$.

The quantity m/γ is independent of the rotation speed of the centrifuge and is often quoted for macromolecules in units of svedbergs; 1 svedberg is 10^{-13} s, which roughly corresponds to a globular protein 1 nm in radius, moving in water. This estimate follows from taking the typical density of proteins to be $1.35\ \mathrm{g/cm^3}$ and using Equation 12.45 below.

The drift velocity as given by Equation 12.44 is a quadratic function of size. If we compare particles that are all roughly spheres of radius R made of the same material with density ρ_{p}, then

$$v_{\mathrm{drift}} = \frac{(\rho_{\mathrm{p}} - \rho)\frac{4}{3}\pi R^3 g_{\mathrm{c}}}{6\pi\eta R} = \frac{2(\rho_{\mathrm{p}} - \rho)}{9\eta} R^2 g_{\mathrm{c}}. \tag{12.45}$$

This strong dependence of the drift velocity on particle size allows for effective separation of particles using centrifugation.

In a spinning sample, the centrifugal force is not the only one acting on the macromolecules in solution. There is also the ubiquitous random thermal force due to constant collisions of the macromolecules with the solvent molecules. The thermal forces lead to diffusion of the macromolecules. This has a negative effect on our ability to separate molecules.

Consider an initial distribution of molecules such that they form a very narrow band at the top of the centrifuge tube. This is the case in rate zonal centrifugation, where the sample is deposited on top of a density gradient of sucrose or some other substance in water; see Figure 12.18(A). The positions of the molecules along the tube are labeled by x and the band is initially centered around $x = 0$. To simplify the analysis, we assume that at time zero the centrifugal force is turned on. The molecules will begin to drift while at the same time their distribution will spread due to diffusion. The increase in the width of the band will be given by the diffusive law (see p. 524), $\Delta x = \sqrt{2Dt}$, while the position of the center of the band will be given by $x_{\mathrm{c}} = v_{\mathrm{drift}} t$. Note that the width of the band increases only as the square root of time while its position goes linearly with time. This distinction is what allows for separation of molecules.

To investigate the conditions under which two species of macromolecules will be separated, we consider the simple situation where initially both molecules are distributed in a very narrow band at $x = 0$.

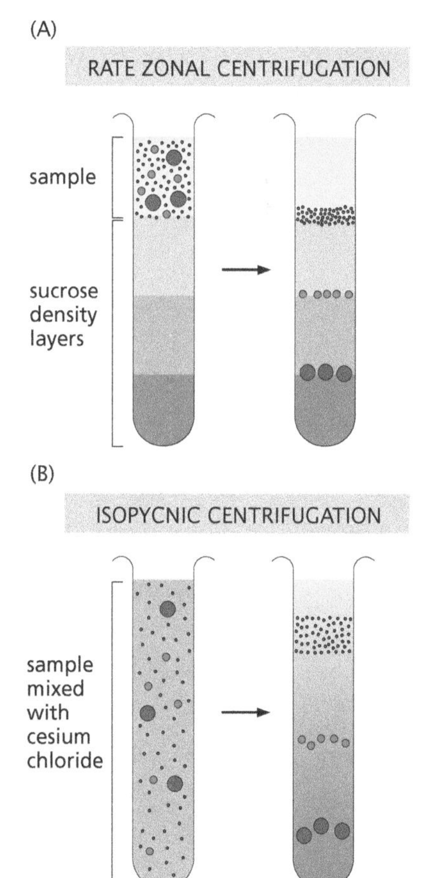

Figure 12.18: Fluid mechanics of centrifugation. (A) In rate zonal centrifugation, the sample is initially prepared in such a way that the macromolecules of interest are in a narrow zone at the top of the centrifuge tube. (B) In isopycnic centrifugation, the macromolecules are distributed throughout the volume of the tube. The density gradient covers the range of all macromolecules that are to be separated from solution.

Given that the two species are characterized by different drift velocities and different diffusion constants, separation will take place when the distance between the centers of the bands of molecular species 1 and 2 is greater than the sum of the widths of the two bands. In other words, we require

$$|v_{\text{drift}\,1} - v_{\text{drift}\,2}|t > (\sqrt{2D_1} + \sqrt{2D_2})\sqrt{t}, \tag{12.46}$$

which is always true, assuming we can wait long enough, that is, for times $t > t_{\text{sep}}$, where the separation time is

$$t_{\text{sep}} = \left(\frac{\sqrt{2D_1} + \sqrt{2D_2}}{|v_{\text{drift}\,1} - v_{\text{drift}\,2}|} \right)^2. \tag{12.47}$$

Given that the separation occurs in a test tube of finite length, the separation time has to be small enough so that the separation occurs before the bands of macromolecules drift all the way to the end of the test tube. This places a constraint on the design parameters of the centrifuge if it is going to be successful in separating the two particular molecular species. If the greater of the two drift velocities is $v_{\text{drift}\,1}$, then the condition for the separation to occur in a tube of length L is $v_{\text{drift}\,1} t_{\text{sep}} < L$. Using the formulas for the drift velocity and the separation time, we can rewrite this last condition as a condition on the centrifugal force (which is in turn a condition on the rotational speed),

$$g_c > \frac{1}{L} \frac{m_1}{\gamma_1} \left(\frac{\sqrt{2D_1} + \sqrt{2D_2}}{m_1/\gamma_1 - m_2/\gamma_2} \right)^2. \tag{12.48}$$

Instead of waiting for a prescribed time so as to achieve separation, the "isopycnic" method relies on a density gradient that covers the range of all the macromolecules in the centrifuge tube. The schematic of this method is shown in Figure 12.18(B). In the isopycnic case, a particular macromolecule will drift either down the tube or up the tube, depending on whether it is more or less dense than the solvent at that particular position. When the molecule reaches the position at which its density is matched by that of the solvent, drift will stop altogether. This situation corresponds to Equation 12.45, which states that $v_{\text{drift}} = 0$ when $\rho_p = \rho$. Therefore the final outcome, after spinning the sample for a long enough time, will be that all the macromolecules will be separated into bands that sit at the position where the densities are matched.

12.5 Summary and Conclusions

Looking down a microscope at swimming bacteria is an exciting and deceptive experience. As they swim about, these tiny cells give the false impression that their relation to their watery medium is the same as ours, but this is far from the truth. This chapter has shown how Newton's second law of motion (**F** = *m***a**) can be applied straightforwardly to fluid elements and how the resulting equations of motion can be used to develop intuition about the lives of cells such as red blood cells as they travel through our bodies and bacteria as they swim about.

12.6 Problems

Key to the problem categories: • Model refinements and derivations, • Estimates, • Model construction

• 12.1 Bacterial foraging

Bacteria use swimming to seek out food. Imagine that the bacterium is in a region of low food concentration. For the bacterium to profit from swimming to a region with more food, it has to reach there before diffusion of food molecules makes the concentrations in the two regions the same. Here we find the smallest distance that a bacterium needs to swim so it can outrun diffusion.

(a) Make a plot in which you sketch the distance traveled by a bacterium swimming at a constant velocity v as a function of time t, and the distance over which a food molecule will diffuse in that same time. Indicate on the plot the smallest time and the smallest distance that the bacterium needs to swim to outrun diffusion.

(b) Make a numerical estimate for these minimum times and distances for an *E. coli* swimming at a speed of $30\,\mu m/s$. The diffusion constant of a typical food molecule is roughly $500\,\mu m^2/s$.

(c) Estimate the number of ATP molecules the bacterium must consume (hydrolyze) per second in order to travel at this speed, assuming that all of the energy usage goes into overcoming fluid drag. The amount of energy released from one ATP molecule is approximately $20\,k_BT$. Note that the bacterial flagellar motor is actually powered by a proton gradient and this estimate focuses on the ATP equivalents associated with overcoming fluid drag.

• 12.2 Pressure and the doctor's office

In this problem, we examine the hydrostatic pressure in a fluid at rest.

(a) Consider a small fluid element $\Delta x \times \Delta y \times \Delta z$ at rest. Write down the balance of forces on the fluid element due to the fluid pressure p and the gravitational pull of the Earth. Show that this leads to the differential equation

$$\nabla p = \rho \mathbf{g}, \tag{12.49}$$

where $\mathbf{g}$ is the acceleration due to gravity.

(b) Solve the differential equation derived in (a) assuming a uniform fluid density. Show that the pressure in the fluid is given by $p(z) = p_0 - \rho g z$, where the z-axis is in the direction opposite to $\mathbf{g}$.

(c) Estimate the atmospheric pressure. (*Hint:* Look up the density of air and make a reasonable guess for the height of the atmosphere.)

(d) If you raise your arm above your head as the doctor is measuring your blood pressure, how much will the measurement change compared with when you keep your arm level with the heart. Try it next time you are at the doctor's office!

• 12.3 Coccolithophores, buoyancy, and blooms

(a) Use the Internet to search for "Emiliana huxleyi bloom" and enjoy the images from space of the ocean blooms produced by these coccolithophores. One of our favorite pictures can be found on the book's website.

(b) By examining a scanning electron micrograph of *Emiliana huxleyi*, estimate its size and mass.

(c) Assuming that the density of a hypothetical cell is 1.3 times that of water, work out its average height from the bottom of a beaker filled with air and filled with water at room temperature. Plot the average height as a function of the cell radius.

(d) Use the result from (c) to deduce a length scale at which the effects of gravity and thermal forces are comparable. Consider the gravitational energy released when a spherical particle drops over a distance equal to its radius, and compare it with the thermal energy.

• 12.4 Experiment at high Reynolds number

(a) At high Reynolds number, the drag force on a moving object becomes largely independent of the viscosity of the fluid. Given that the drag force in this case is a function of the fluid density ρ, the object size R, and its speed v, use dimensional analysis to obtain a formula for the drag force.

(b) From a sheet of paper, cut out two circles, one with a radius twice as large as the other, and using scissors and some Scotch tape turn them into two cones as shown in Figure 12.19(A). Stand on a chair and drop the two cones

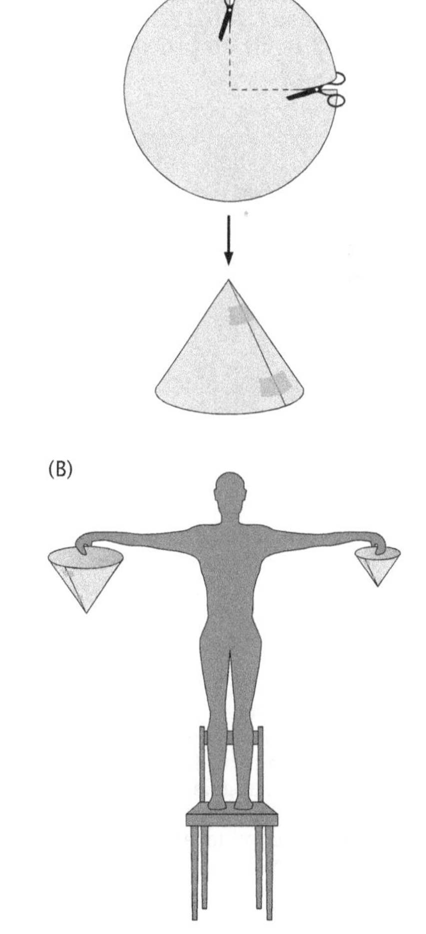

Figure 12.19: Measuring drag by hand.

from the same height, pointy end down, as shown in Figure 12.19(B). You will observe that after a short period, both cones reach a terminal speed. Which cone is moving faster? Explain your observation based on the formula derived in (a). (*Hint:* Derive a formula for the terminal speed as a function of the cone size by balancing the drag force and the force of gravity.)

(Adapted from a problem courtesy of S. Mahajan.)

• 12.5 DNA twist experiment

A molecule of DNA (length L) is attached at both ends to beads via the two strands of the double helix, so it is torsionally constrained. Assume that you have twisted one bead relative to the other by a total angle ϕ, and you keep the molecule under tension to prevent it from supercoiling. Now you attach a small bead of radius R to both the strands and break one of the strands below the tiny bead, so that the molecule can unwind by rotating around the single bond swivel in the unbroken strand as shown in Figure 10.38. To understand this experiment, we need to consider the drag forces on the rotating bead.

(a) Show that the drag coefficient for a sphere of radius R rotating at angular velocity ω is given by $K\eta R^3$, where K is a numerical factor. For rotational motion, the drag coefficient relates the angular velocity to the frictional torque. Use the approach that led to the Stokes formula in the chapter. The idea is to say that the size of the viscous stresses is $v/R = \omega$, and the area over which they act is R^2. Combine these two to get the force, and then the torque.

(b) When DNA is highly twisted, the torque as a function of twist angle ϕ is constant at a value of roughly $\tau = 33$ pN nm. Use the data shown in Figure 12.20 to determine the numerical factor K in the drag coefficient, in multiples of π.

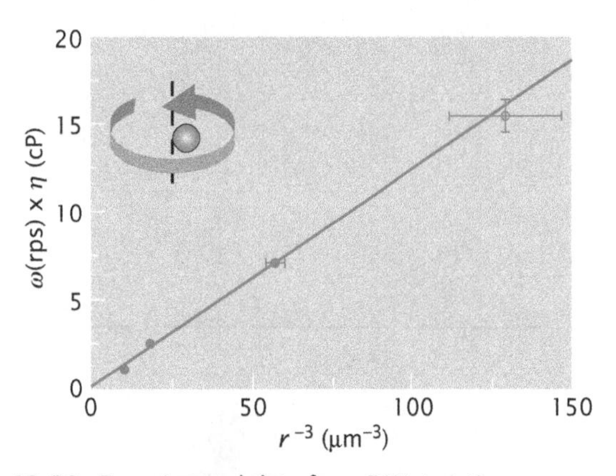

Figure 12.20: Experimental data from DNA twisting experiment. Rotation rate as a function of bead size. (Adapted from Z. Bryant et al., *Nature* 424:338, 2003.)

(c) Write an expression for the angular velocity of the small bead in terms of the viscosity η, the radius of the bead R, the constant of torsional stiffness C, and the length and the number of whole extra turns in the DNA molecule, N, where $N = \phi/2\pi$. Estimate the angular velocity of an $R = 400$ nm bead if the length of DNA is 14.8 kb and it has been twisted by 50 extra turns. The torsional stiffness of DNA is approximately 400 pN nm^2. Express your answer in revolutions per second.

(d) Use the results obtained in (a) and (b) to determine the rotational drag coefficient of a sphere rotating around its center of mass. Use the fact that the rotational motion of the bead in the experiment can be decomposed into translational motion of the center of mass and rotation around the center of mass.

(Adapted from a problem courtesy of C. Bustamante.)

• 12.6 Life at low *Re*

(a) *E.coli* swims at about 20 μm/s by rotating a bundle of helical flagella. If the motors were to turn 10 times faster than normal, what would their swimming speed be? If their fluid environment were made 10 times more viscous, but the motors were to turn at the same rate, what would the swimming speed be? How does the power output of the motor change in these two hypothetical situations?

(b) Two micron-sized spheres, one made of silver and the other gold, sediment (that is, fall under gravity) in a viscous fluid. The silver sphere has twice the radius of the gold one. Which sediments faster?

(c) The left ventricle of the human heart expels about 50 cm^3 of blood per heartbeat. Assuming a pulse rate of 1 heartbeat per second and a diameter of the aorta of about 2 cm, what is the mean velocity of blood in the aorta? What is the Reynolds number?

(d) What is the Reynolds number of a swimming bacterium? A tadpole? A blue whale?

(Adapted from a problem courtesy of H. C. Berg and D. Nelson.)

• 12.7 Protein centrifugation

Proteins and other macromolecules can be separated by size using centrifugation. The idea is to spin a sample containing proteins of different size in solution. The spinning produces a centrifugal force per unit mass g_c, which leads to diffusion with a drift velocity that depends on the protein size. We assume that a protein in the sample can be approximated as a ball of radius R.

(a) Following the discussion in the chapter, fill in the steps leading up to the formula for the drift velocity of the protein as a function of its radius (Equation 12.45),

$$v_{\text{drift}} = \frac{2(\rho_{\text{protein}} - \rho_{\text{solvent}})g_c R^2}{9\eta}, \qquad (12.50)$$

where ρ_{protein} and ρ_{solvent} are the densities of the protein and the solvent and η is the solvent viscosity.

(b) Estimate the drift velocity for hemoglobin in water in an ultracentrifuge with $g_c \approx 10^5 g$, where $g \approx 10$ m/s^2 is the acceleration of freely falling objects in Earth's gravitational field. Assume a typical protein density of 1.2 g/cm^3.

(c) We would like to separate two similar proteins, having the same density, $\rho' = 1.35$ g/cm^3. They have diameters of 4 nm and 5 nm, respectively. The two protein species start out mixed together in a thin layer at the top of a 1 cm long centrifuge tube. How large should the centrifuge acceleration g_c be so that the two proteins are separated before they drift to the end of the tube?

12.7 Further Reading

Purcell, EM (1977) Life at low Reynolds number, *Am. J. Phys.* **45**, 3. The classic of the field of low Reynolds number hydrodynamics.

Bray, D (2001) Cell Movements: From Molecules to Motility, 2nd ed., Garland Science. A rich source for studying the diversity of cellular motion.

Berg, H (1993) Random Walks in Biology, Princeton University Press. A must read for anyone interested in physical biology.

Berg, H (2004) *E. coli* in Motion, Springer–Verlag. Another Berg classic!

Denny, M (1993) Air and Water: The Biology and Physics of Life's Media, Princeton University Press. Denny's book is full of interesting discussion about the way that living organisms confront air and water.

Vogel, S (1994) Life in Moving Fluids: The Physical Biology of Flow, 2nd ed., Princeton University Press. The emphasis of Vogel's book is more on macroscopic organisms than on microbes. Enlightening and entertaining.

Van Dyke, M (1982) Album of Fluid Motion, Parabolic Press. A classic collection of images of fluid motion.

Guyon, E, Hulin, JP, Petit, L, & Mitescu, CD (2001) Physical Hydrodynamics, Oxford University Press. This excellent book provides a deep introduction to fluid mechanics.

Warhaft, Z (1997) Introduction to Thermal-Fluid Engineering: The Engine and the Atmosphere, Cambridge University Press. Though this book emphasizes entirely different scales than the discussion here, it is an elegant and stimulating discussion of many of the same ideas, with a similar emphasis on order-of-magnitude estimates.

Hinch, EJ (1988) Hydrodynamics at low Reynolds numbers: a brief and elementary introduction, in Disorder and Mixing, edited by E Guyon, J-P Nadal, and Y Pomeau, Kluwer Academic Publishers.

Happel, J, & Brenner, H (1983) Low Reynolds Number Hydrodynamics, Martinus Nijhoff Publishers. A serious discussion of low Reynolds number phenomena for those interested in the mathematical details.

Witten, TA, & Pincus, PA (2004) Structured Fluids: Polymers, Colloids, Surfactants, Oxford University Press. A clear description of many important topics not covered in our discussion, but relevant to understanding the physical behavior of biological materials.

Chen, S, & Springer, TA (1999) An automatic braking system that stabilizes leukocyte rolling by an increase in selectin bond number with shear, *J. Cell Biol.* **144**, 185. This paper describes physical features of leukocyte rolling.

Chen, S, & Springer, TA (2001) Selectin receptor–ligand bonds: formation limited by shear rate and dissociation governed by the Bell model, *Proc. Natl Acad. Sci.* USA **98**, 950. This paper gives a flavor of the physical issues associated with leukocyte rolling.

Bird, RB, & Curtiss, CF (1984) Fascinating polymeric liquids, *Phys. Today* January, 36. A fun introduction to the world of non-Newtonian fluids.

12.8 References

Alberts, B, Johnson, A, Lewis, J, et al. (2008) Molecular Biology of the Cell, 5th ed., Garland Science.

Bryant, Z, Stone, MD, Gore, J, et al. (2003) Structural transitions and elasticity from torque measurements on DNA, *Nature* **424**, 338.

Drucker, EG & Lauder, GV (2001) Wake dynamics and fluid forces of turning maneuvers in sunfish, *J. Exp. Biol.* **204**, 431.

Feynman, RP, Leighton, RB, & Sands, M (1965) The Feynman Lectures on Physics, Addison-Wesley. Chapter 41 of Volume II on the "Flow of wet water" describes many important ideas from fluid mechanics.

Guasto, JS, Johnson, KA & Gollub, JP (2010) Oscillatory flows induced by microorganisms swimming in two dimensions, *Phys. Rev. Lett.* **105**, 168102.

Sundd, P, Edgar Gutierrez, E, Pospieszalska, MK, et al. (2010) Quantitative dynamic footprinting microscopy reveals mechanisms of neutrophil rolling, *Nat. Meth.* **7**, 821.

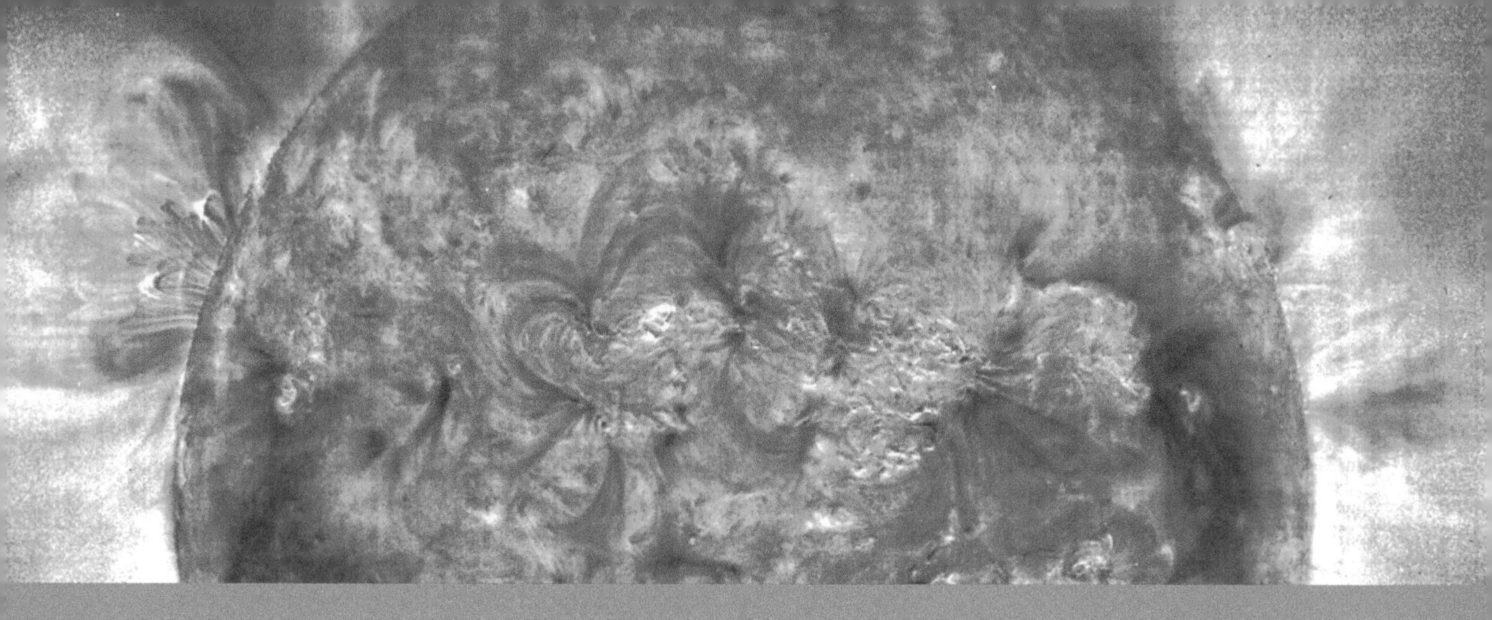

A Statistical View of Biological Dynamics

13

Overview: In which the random walk model is applied to the motion of macromolecules

Dynamics in cells comes in a number of different forms. One of the most important classes of dynamical process is diffusion, the random jiggling of individual molecules in solution. When many such molecules are diffusing simultaneously, the overall concentration field undergoes changes in space and time that give the appearance of ordered and directed movements of molecules down concentration gradients. The goal of this chapter is to illustrate the important role of diffusion in living systems, to compare and contrast microscopic and continuum descriptions of diffusion, and to apply these ideas to important problems such as how the method of fluorescence recovery after photobleaching (FRAP) works and how receptors mediate signaling.

"To do science is to search for repeated patterns, not simply to accumulate facts..."

Robert MacArthur

13.1 Diffusion in the Cell

Living systems are subject to incessant and tireless change. In Chapter 12, we eased gently into the treatment of biological dynamics by considering directed movements in water, though we relied heavily on the simplifying assumption that water can be treated as a continuous fluid rather than as a collection of interacting molecules. We are now ready to take the next step and consider the individual movements of discrete particles in water, ranging from molecules to organelles to viruses and the cells they attack. Over the next four chapters, we will develop this theme of biological motions, starting with the simplest case applying to nonliving and living systems alike, namely, Brownian or diffusive motion. What makes such processes especially intriguing is that despite the stochastic microscopic underpinnings, huge numbers of diffusing molecules over a large number of time steps can give the appearance of purposeful dynamics of particles down a concentration gradient.

As discussed in Section 3.4.2 (p. 126), Brownian motion is an inevitable outcome of the thermal jiggling of water molecules and does not indicate the activities of a living system. However, diffusive motion is always present at molecular length scales, and biological systems must tolerate, exploit, or inhibit Brownian motion in order to perform directed dynamic processes. A familiar example of the physical limits put on organisms by the process of diffusion is something you experience with every breath you take. Human metabolism demands a constant high concentration of oxygen supplied to mitochondria throughout the body. Much smaller organisms that are oxygen-dependent can rely simply on diffusion of oxygen as a delivery mechanism, but this is only efficient over distances of the order of tens of microns.

In order to grow to sizes exceeding 1 m, humans and other large animals have developed elaborate mechanisms to circulate oxygen and effectively enable its delivery to all tissues. In Chapter 7, we examined hemoglobin as a protein specialized for the sole purpose of carrying oxygen to parts of the body far from the lungs. Oxygen inhaled in air can diffuse through lung tissue over an effective distance of roughly $100\,\mu m$ that is set not only by the free diffusion of oxygen, but also by its rate of consumption by cells in the tissue. In the lung, a fine network of capillaries surrounds each air sac and diffusion is sufficient for oxygen to travel from inhaled air to the hemoglobin-filled blood in the capillaries. Rapid fluid circulation, driven by your beating heart, carries the oxygen around the body much more rapidly than would be possible by simple diffusion. Reaching the tissues in the capillaries, oxygen molecules are again able to diffuse on a scale of $100\,\mu m$. This sets the constant spacing of the finest branches of capillaries for all mammals from mice to blue whales.

In this chapter, we will make simple estimates about the distances over which passive transport (that is, diffusion) is effective and derive and apply the mathematical formalism of diffusion.

13.1.1 Active versus Passive Transport

The cell is teeming with motion. One of the first questions that one might ask about all of this bustling is to what extent it is random and to what extent active and directed. As we discussed in detail in Chapter 5, the interplay between thermal and deterministic forces is one of the hallmarks of cellular dynamics. On the thermal side of the ledger, one of the dominant effects of the thermal forces is the very existence of diffusive motion itself. An example of the diffusion of various ion species after the opening of an ion channel is shown in Figure 13.1.

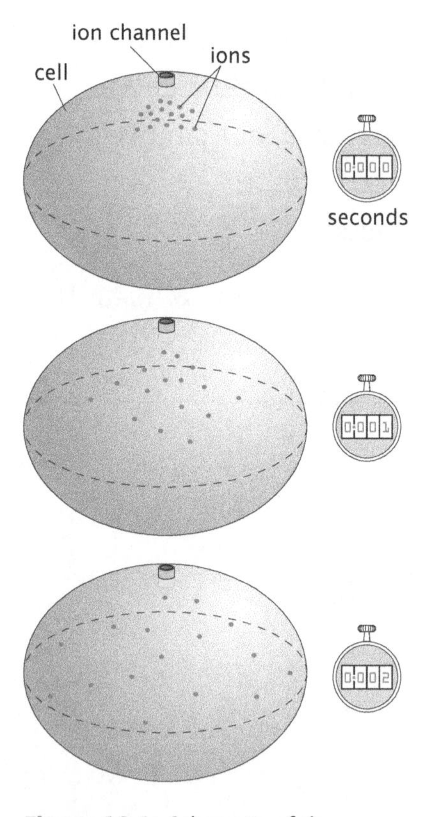

Figure 13.1: Schematic of the diffusion of ions just after opening of an ion channel. The three snapshots give a qualitative illustration of the distribution of ions as a function of time.

A second example, which contrasts the nature of diffusive and active transport, is shown in Figure 13.2. This figure shows the motion of an RNA polymerase molecule during "free" diffusive transport within the cell as well as when it is engaged in active motion during transcription. The hallmark of directed motion is the existence of some energy source. The key point of the figure is to contrast diffusive and directed motion. Ideas like those to be developed in this chapter are useful not only for the macromolecules of the cell, but for cells themselves. As shown in Figure 13.3, a swimming bacterium such as *E. coli* when viewed at low resolution over long times looks as though it too is

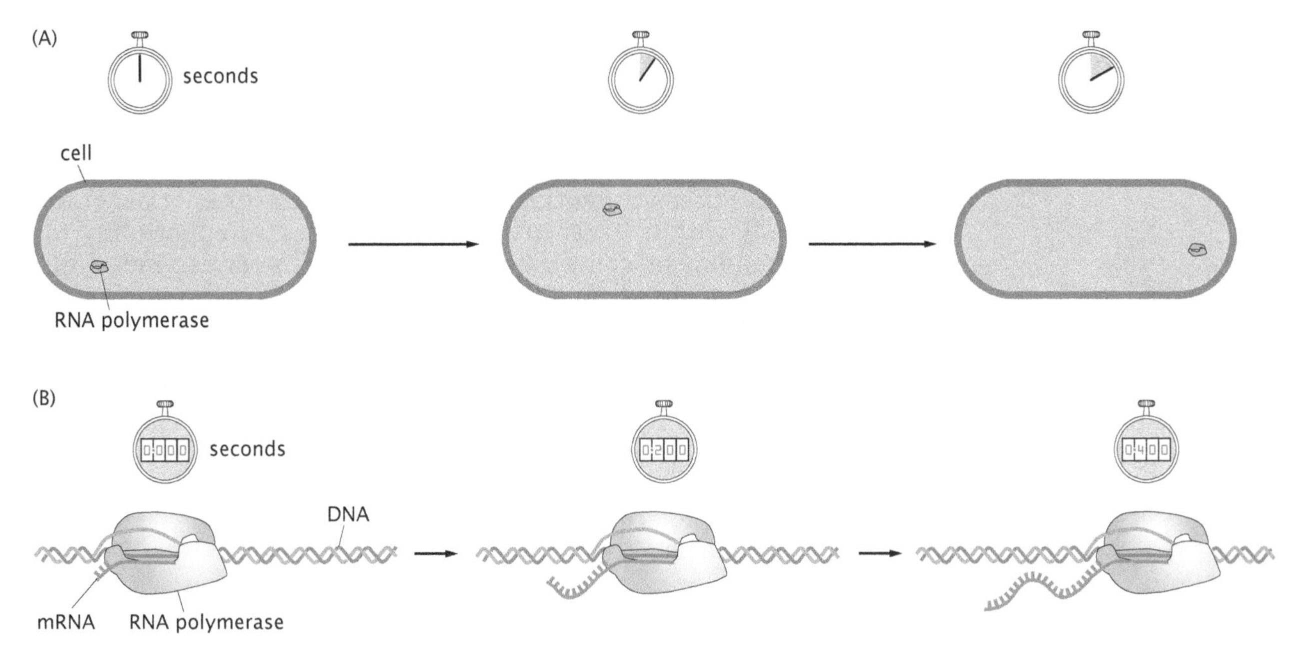

Figure 13.2: Comparison of diffusive and directed motions of RNA polymerase. (A) Free RNA polymerase molecule diffusing in a bacterial cell. (B) One-dimensional motion of RNA polymerase along DNA characteristic of active transport.

undergoing a random walk characteristic of diffusive motions. On the other hand, at short time scales, it is noticed that the cell undergoes a series of runs and tumbles as was introduced in our discussion of bacterial chemotaxis in Section 4.4.4 (p. 159). The language of random walks will be useful for thinking about a variety of different examples.

13.1.2 Biological Distances Measured in Diffusion Times

One of the simplest and most far-reaching results that will emerge from the present chapter is the derivation of simple estimates for the time it takes for diffusion to transport molecules to different distances. In particular, we will show that the typical time it takes for a particle to diffuse a distance L is given by $t \approx L^2/D$, where D is the diffusion constant of the particle. The diffusion constant has units of length^2/time, and it depends on the size of the particle, the temperature, and the viscosity of the surrounding fluid. We will discuss this in detail later in this chapter. For the moment, we examine the numerical consequences of this simple, but important result.

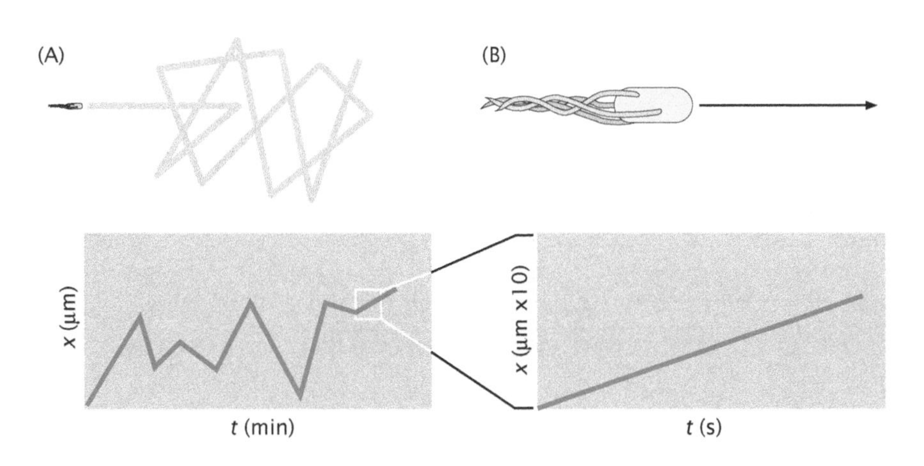

Figure 13.3: Patterns of *E. coli* swimming at different scales. (A) At low magnification, the swimming movement of a single bacterium appears to be a random walk, that is, a series of steps oriented at random angles. Plotting the *x*-component of the bacterial position vector as a function of time will show a chaotic series of back-and-forth movements. (B) At higher magnification, it is clear that each step of this random walk is made up of very straight, regular movements.

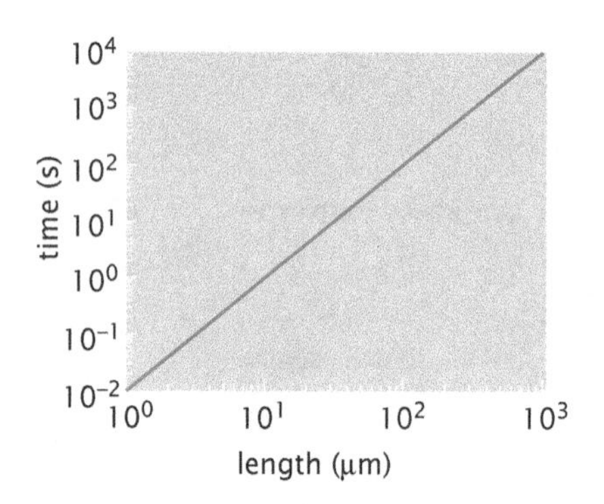

Figure 13.4: Diffusion time as a function of the length traveled for a typical value of the diffusion coefficient ($D = 100\,\mu m^2/s$) of a protein in water. The straight line on the log–log plot has a slope of 2 since time and distance are related by $t = x^2/D$.

The Time It Takes a Diffusing Molecule to Travel a Distance *L* Grows as the Square of the Distance

Unlike in the case of ballistic motion with constant velocity, where the time to travel a distance L grows linearly with the length scale of interest, diffusive dynamics implies that the time scale grows quadratically with distance. This result provides an opportunity for intuition-building by converting distances into the corresponding diffusion time as shown in Figure 13.4. In this figure, we plot the diffusion time as a function of the distance for $D = 100\,\mu m^2/s$, a characteristic diffusion coefficient for a typical globular protein in water at room temperature. Note that the time scale associated with diffusion over a distance $L \approx 10^6\,\mu m$ (1 m) is 10^{10} s (≈300 years)! This should make it clear that transport in cells and organisms requires mechanisms other than diffusion. We will come back to this point when thinking about molecular motors in Chapter 16.

Diffusion Is Not Effective Over Large Cellular Distances

Though diffusion is clearly a part of the overall machinery associated with cellular dynamics, as shown in Figure 13.5, there are also instances where diffusive motion is too slow to link different regions of the cell. One of the most dramatic examples of this is illustrated by nerve cells, as was already introduced in Figure 3.3 (p. 93). In Figure 13.5(A), we show the motion of a molecule by random jiggling along an axon. Using results like those shown in Figure 13.4, it is seen that for diffusion to be effective over the meter length scale implies a time scale that is absurdly long. On the other hand, as will be seen more explicitly in coming chapters, if the molecule of interest is transported in a directed fashion by virtue of molecular motors,

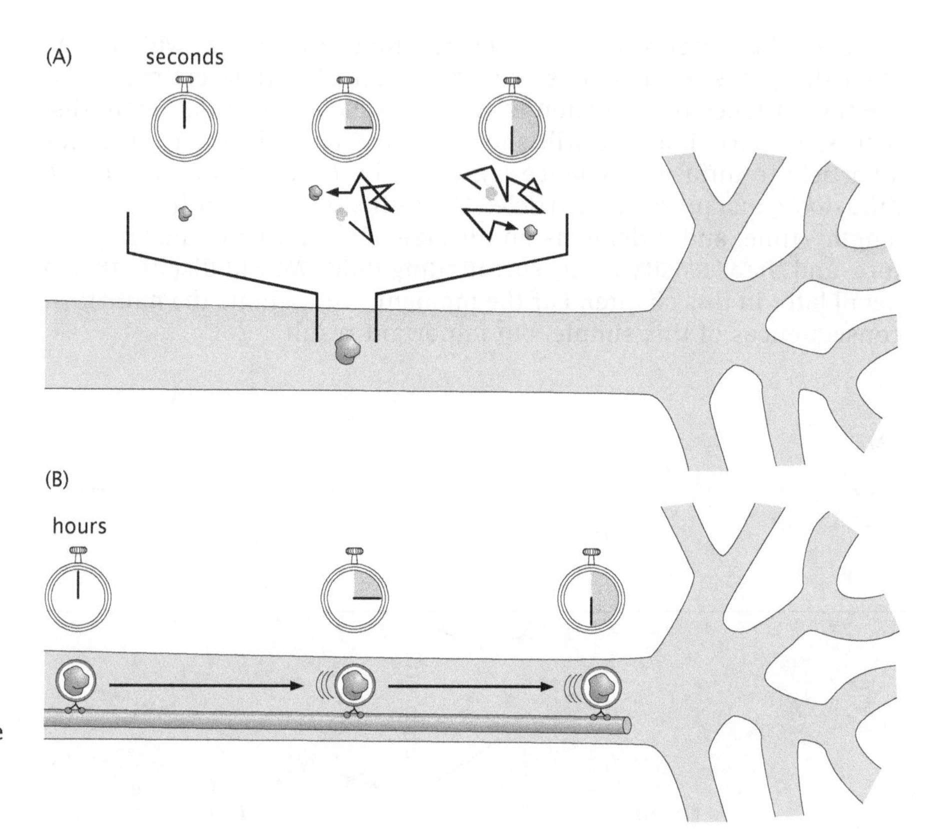

Figure 13.5: Transport within a nerve cell. (A) Passive transport of a molecule by diffusion. (B) Active transport of a molecule through directed motion of a molecular motor.

the time scale of the resulting motion is shortened by many orders of magnitude.

To better understand diffusive dynamics (especially in living cells), we need a sense of what experimental techniques can be used to probe such dynamics. In the Experiments Behind the Facts section below, we examine several especially useful techniques.

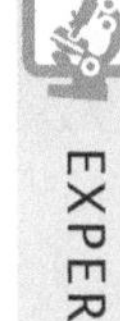

Experiments Behind the Facts: Measuring Diffusive Dynamics There are a number of different techniques that permit the investigation of diffusive dynamics within cells. Here, we consider three examples of some of the ingenious techniques that have been introduced to measure diffusion: FRAP, single-particle tracking, and fluorescence correlation spectroscopy (FCS).

FRAP takes advantage of the annoying feature of fluorescently labeled molecules that when they are exposed to too much light they no longer fluoresce, since the fluorophores can only emit a limited number of photons. In this instance, this weakness is turned into a strength by virtue of the fact that it can be used to measure diffusive dynamics within cells. The technique is illustrated schematically in Figure 13.6. In particular, a laser is focused on a certain spot in the cell with characteristic dimensions of a micron or larger. After the laser pulse, other fluorescently labeled molecules from elsewhere within the cell diffuse back into the space that had previously been photobleached by the laser light. By watching the time course of the recovery of fluorescence, it is possible to extract features of the diffusive dynamics.

An example of the appearance of cells during the FRAP experiment is shown in Figure 13.7. In the series of snapshots, the bleached region shows increasing fluorescence over time as new molecules from outside the bleached region diffuse into that region in a process known as recovery. Section 13.2.3 (p. 525) explores the mathematical foundations of this technique and describes the insights it provides.

A more direct technique for monitoring diffusive dynamics is through explicit particle tracking in which individual trajectories are monitored. This technique is as old as the subject of Brownian motion itself, and was used as the basis of measuring atomic dimensions (and Avogadro's number) in the classic experiments of Perrin (1990). The notion of trajectory

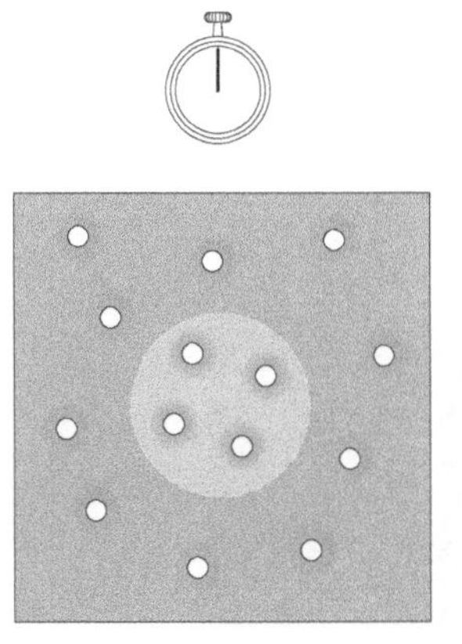

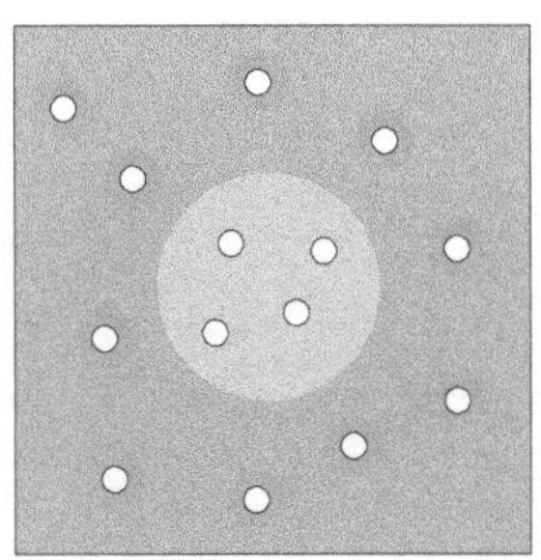

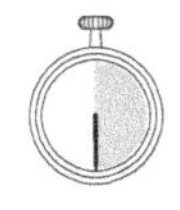

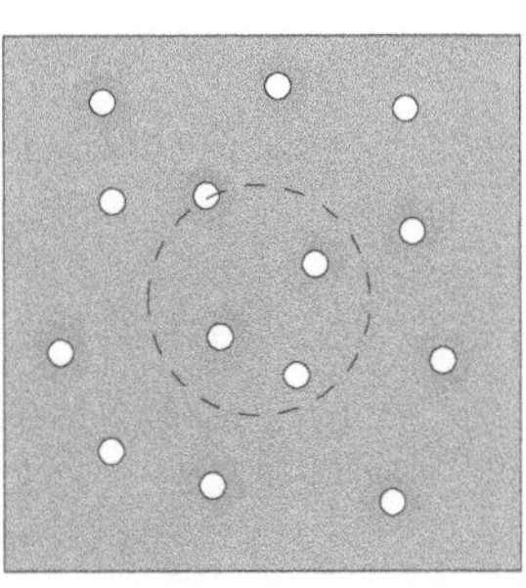

Figure 13.6: Schematic of the FRAP experiment. A particular region of the cell is photobleached, effectively destroying the fluorescent molecules in that region (as shown in the second frame). Recovery of fluorescence in the photobleached region results from fluorescent molecules from elsewhere in the sample diffusing into the photobleached region.

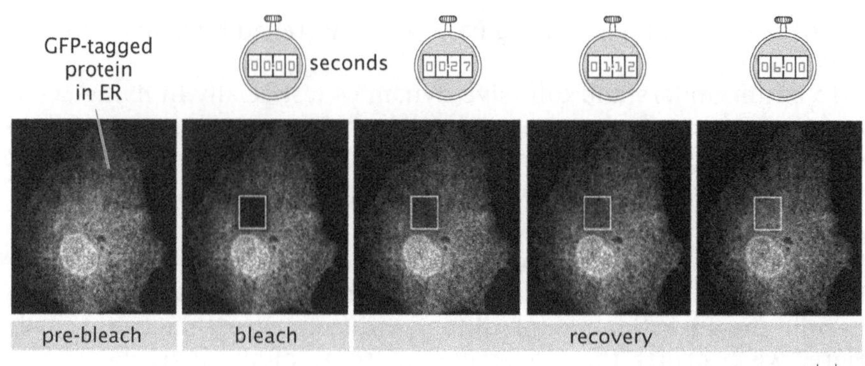

Figure 13.7: FRAP experiment showing recovery of a GFP-labeled protein confined to the membrane of the endoplasmic reticulum. The boxed region is photobleached at time instant, $t = 0$. In subsequent frames, fluorescent molecules from elsewhere in the cell diffuse into the bleached region. (Adapted from J. Ellenberg et al., *J. Cell Biol.* 138:1193, 1997.)

mapping is of widespread interest and has been used on problems ranging from the motions of bacteria to the wandering of individual proteins along DNA. In this case, the idea usually involves video microscopy in which images of the moving species of interest are captured at a fixed interval and the corresponding trajectory is constructed. In the bacterial setting, the nature of the trajectories of motile cells reveals that they have preferences for moving in the direction of certain nutrients, the phenomenon of chemotaxis to be taken up in detail in Chapter 19.

Another technique of great utility for probing diffusive dynamics within cells is FCS. The idea of FCS is to measure the fluorescence intensity in a small region of the cell as a function of time as shown in Figure 13.8. The intensity fluctuates as the fluorescent molecules enter and leave the region under observation. By analyzing the temporal fluctuations of the intensity through the use of time-dependent correlation functions, the diffusion constant and other characteristics of the molecular motion can be uncovered.

13.1.3 Random Walk Redux

The present chapter is a continuation of the story already begun in Chapter 8. One of the key themes of the book is the idea that certain models have superstar status because of their ability to shed light on a range of different problems. Here we take a second look at the random walk model, a model with such hall of fame status that it gets double billing. Recall that in Chapter 8 we invoked the random walk model as a scheme for examining the structure of long-chain molecules. In this

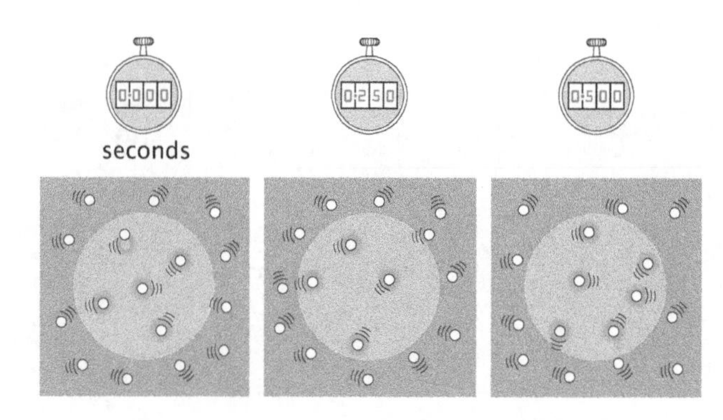

Figure 13.8: Schematic of the FCS technique. The intensity of fluorescent light coming from illuminated molecules is measured as a function of time. The intensity varies in time as a result of molecules diffusing in and out of the observation region, shown in pale green.

chapter, we examine a second powerful role for random walk models, namely, as the basis for considering the problem of diffusion. One of the interesting contrasting perspectives that will be seen to emerge when comparing the results of Chapter 8 and the present chapter is that in the earlier case, we invoked the random walk model as an *equilibrium* model. In this chapter, it is doing double time as a model of nonequilibrium dynamics.

The logic of the remainder of the chapter will be to develop the formalism of diffusive dynamics from two distinct perspectives. First, we will think in macroscopic terms by smearing out the collection of diffusing molecules into a concentration field and then by writing macroscopic evolution equations for the changes in the concentration over space and time. The second perspective will be strictly microscopic and will consider the individual hopping events of single particles as they wander aimlessly through the volume of interest (usually the cell itself). Interestingly, these two views will be reconciled as it will be seen that together all of the randomly wandering molecules give the macroscopic appearance of directed motion driven by concentration gradients. Once we have these diffusive models in hand, we will turn to a range of interesting biological applications of these ideas.

In Section 5.5.2 (p. 225), we introduced the molecular driving forces that drive a system towards equilibrium. Our argument was that if we remove some internal constraint on a system (such as those shown in Figure 5.29 on p. 226), the system will change until it reaches some terminal privileged state known as the equilibrium state. Prior to reaching the equilibrium state, there is an imbalance in temperature or pressure or chemical potential. For the case of mass transport of interest here, we focus on the the fact that for a system that is not in equilibrium with respect to mass transport, the chemical potential is not equal across the system and this serves as a driving force for diffusive motion.

13.2 Concentration Fields and Diffusive Dynamics

Our first foray into the world of diffusive dynamics will be founded upon the idea of macroscopic concentrations and fluxes. The notion of a concentration has already been used a number of times throughout the book and will continue to serve as one of the key conceptual tools for much of what happens in the remainder of the book as well. As a reminder, Figure 13.9 shows that the concentration field tells us the average number of molecules per unit volume. More precisely, the concept is that we divide space up into a bunch of small boxes (such as shown in Figure 13.9), with the boxes large enough to include many molecules, but small enough so that the density is nearly uniform over the scale of the box. We use the notation $c(\mathbf{r}, t)$ to signify the concentration in a box centered at position $\mathbf{r}$ in three-dimensional space (with units of number of particles per unit volume) and $c(x, t)$ to signify the concentration field in one-dimensional problems (with units of number of particles per unit length).

With the idea of a concentration in hand, we can consider the origins of diffusive dynamics. In particular, we begin by noting that in this macroscopic worldview, diffusive dynamics is the result of concentration gradients. What we mean precisely by the term "gradient"

Figure 13.9: Schematic illustrating the definition of a concentration field. The system is divided up into small boxes of volume ΔV. The overall concentration field is changing sufficiently slowly that in each small box the "concentration" is constant.

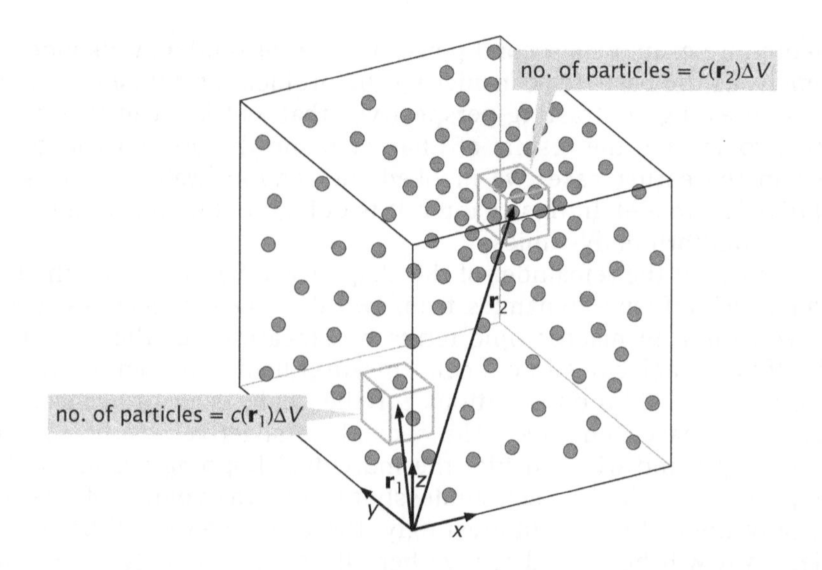

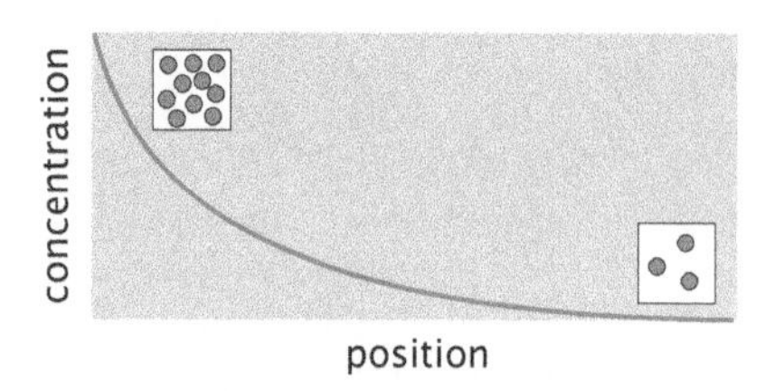

Figure 13.10: Example of a concentration profile. The plot shows the variation of the concentration with distance along some spatial domain, illustrating the idea of a position-dependent concentration.

is a spatial variation in the concentration field. Figure 13.10 shows a simple concentration profile where on the left-hand side of the domain of interest, the concentration of the molecule of interest is high, while on the right-hand side of the domain of interest, the concentration is low.

The other key quantity of interest for our macroscopic description of diffusion is the flux. Flux can be thought of conceptually as shown in Figure 13.11, where it is seen that we identify a plane with some area A and then count the net number of molecules that cross that area per unit time. That is the component of the flux vector in that direction. In its full generality, the idea is more subtle than this, since in three dimensions, the flux is actually a vector whose components give the flux across planes that are perpendicular to the x-, y-, and z-directions. The goal of our thinking is to determine what amounts to an "equation of motion" that tells how the concentration field changes in both space and time.

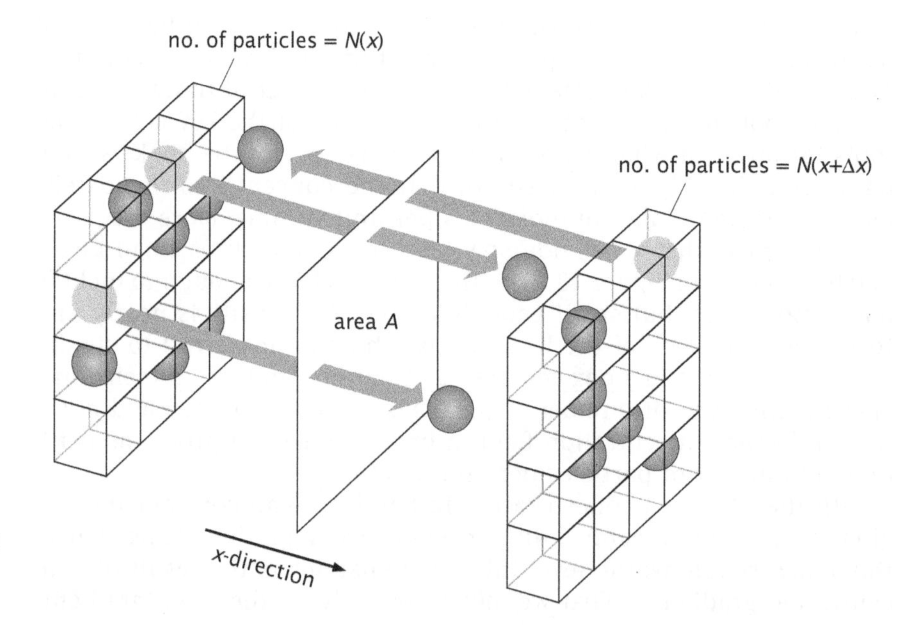

Figure 13.11: Schematic of the flux concept. In one dimension, space is discretized into a series of planes separated by a distance Δx. Particles can hop left or right, and the flux across the plane between two adjacent planes is computed by counting the net number of particles crossing unit area per unit time.

Fick's Law Tells Us How Mass Transport Currents Arise as a Result of Concentration Gradients

As a first cut, we treat this problem on strictly phenomenological grounds. Later in the chapter, we will show how this phenomenological law can be derived from microscopic considerations. For the time being, we restrict our attention to one-dimensional concentration profiles so that the resulting mathematics is simplified. Fick's first law is the assertion that the flux is linearly related to the concentration gradient, namely,

$$j = -D\frac{\partial c}{\partial x}, \tag{13.1}$$

where j is a current density per unit time, which can be thought of as the number of particles crossing unit area per unit time and where D is the diffusion coefficient. (For a brief review of partial derivatives we refer the reader to The Math Behind the Models on p. 212.) The minus sign in Fick's law guarantees that the particle flux is in the right direction. For example, if we consider the profile shown in Figure 13.10, the concentration profile decreases with increasing x ($\partial c/\partial x < 0$). On the other hand, it is clear that molecules flow from the region of higher concentration to lower concentration, down the concentration gradient, in the positive x-direction. The units of D can be determined by examining the units of all of the other quantities in Fick's law. Note that it is conventional notation to characterize the units of a quantity D as $[D]$ and the reader is asked to bear this in mind since the same notation is used to specify concentrations. Exploiting this scheme for Fick's law, we have

$$[j] = \frac{1}{\text{length}^2 \times \text{time}}, \tag{13.2}$$

which signifies number per unit area per unit time. The units of the right-hand side of the equation are

$$\left[\frac{\partial c}{\partial x}\right] = \frac{\text{number of particles/length}^3}{\text{length}} = \frac{\text{number of particles}}{\text{length}^4}. \tag{13.3}$$

By rearranging our equation, we are left with the units of the diffusion coefficient, which are length^2/time. Note that the units of the diffusion coefficient are independent of the dimensionality of space. Typical values for the diffusion constant are shown in Table 13.1.

Our goal is to assess the rate at which the concentration in a small region changes over time. For concreteness, consider the box shown in Figure 13.12, where we have specialized to the case in which the flux is only in the x-direction. This means that particles are flowing in and out on the two faces of the cube that are perpendicular to the x-direction.

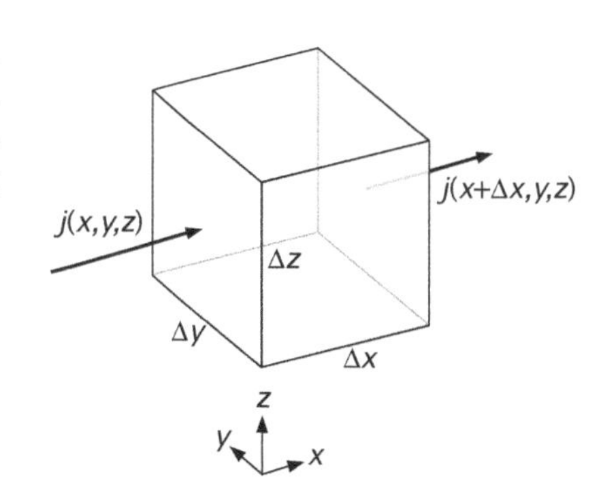

Figure 13.12: Mass transport out of a small volume element. The fluxes across the faces of the box change the number of particles in that volume.

Table 13.1: Table of diffusion coefficients for different molecules. (Data for GFP from M. B. Elowitz et al., *J. Bacteriol.* 181:197, 1999 and yeast data from W. F. Marshall et al., *Curr. Biol.* 7:930, 1997.)

Molecule	Diffusion coefficient
Potassium ion in water	$\approx 2000\,\mu m^2/s$
GFP in *E.coli* cytoplasm	$\approx 7\,\mu m^2/s$
DNA in yeast nucleus	$5 \times 10^{-4}\,\mu m^2/s$

The basic strategy is to assess how many particles enter or leave at the face at position x and similarly across the face at position $x + \Delta x$. We define $N_{\text{box}}(x, y, z, t)$ as the number of particles in the box at time t and note that this can be computed as $N_{\text{box}}(x, y, z, t) = c(x, y, z, t)\Delta x \Delta y \Delta z$. Since mass is conserved (that is, we are not yet thinking about the case where there are reactions that can alter the number of particles of a given species), the change in $N_{\text{box}}(x, y, z, t)$ can only arise from the fluxes across the faces of the box.

The Diffusion Equation Results from Fick's Law and Conservation of Mass

First, note that the change in the number of particles, N_{box}, per unit time is the change in concentration per unit time times the volume of the box, and can be written as

$$\frac{\partial N_{\text{box}}}{\partial t} = \frac{\partial c}{\partial t}\Delta x \Delta y \Delta z. \tag{13.4}$$

By mass conservation, this result has to be equal to the number of particles going into the box per unit time, $j(x, y, z)\Delta y \Delta z$, *minus* the number of particles going out of the box per unit time, $j(x + \Delta x, y, z)$ $\Delta y \Delta z$, and is reckoned as

$$\frac{\partial c}{\partial t}\Delta x \Delta y \Delta z = j(x, y, z)\Delta y \Delta z - j(x + \Delta x, y, z)\Delta y \Delta z. \tag{13.5}$$

We can then Taylor-expand $j(x + \Delta x, y, z)$ to first order in Δx (see the discussion of Taylor expansions on p. 215) to give

$$\frac{\partial c}{\partial t}\Delta x \Delta y \Delta z \approx j(x, y, z)\Delta y \Delta z - \left[j(x, y, z) + \frac{\partial j}{\partial x}\Delta x\right]\Delta y \Delta z. \tag{13.6}$$

If we now collect terms, the local statement of conservation of mass can be written as

$$\frac{\partial c}{\partial t} = -\frac{\partial j}{\partial x}. \tag{13.7}$$

Note that the significance of this equation is that it is a statement about the relation between the flux and concentration in every little neighborhood of the volume of interest.

By combining the statement of mass conservation (Equation 13.7) and the relation between flux and concentration gradient (Equation 13.1), we can generate a very useful relation, namely,

$$\frac{\partial c}{\partial t} = D\frac{\partial^2 c}{\partial x^2}, \tag{13.8}$$

which is the classic law of diffusion in one dimension. Note that to derive this particular form of the diffusion equation, we had to assume that D is independent of concentration. This single equation embodies two key ideas, namely, (i) mass conservation and (ii) a material law relating flux and concentration. Note that the first of these ideas is independent of material particulars, while Fick's law need not be satisfied in all circumstances since flux might depend on concentration in a more complicated, nonlinear fashion.

13.2.1 Diffusion by Summing Over Microtrajectories

The diffusion equation derived in the previous section can be obtained completely differently from a microscopic perspective. The key idea in

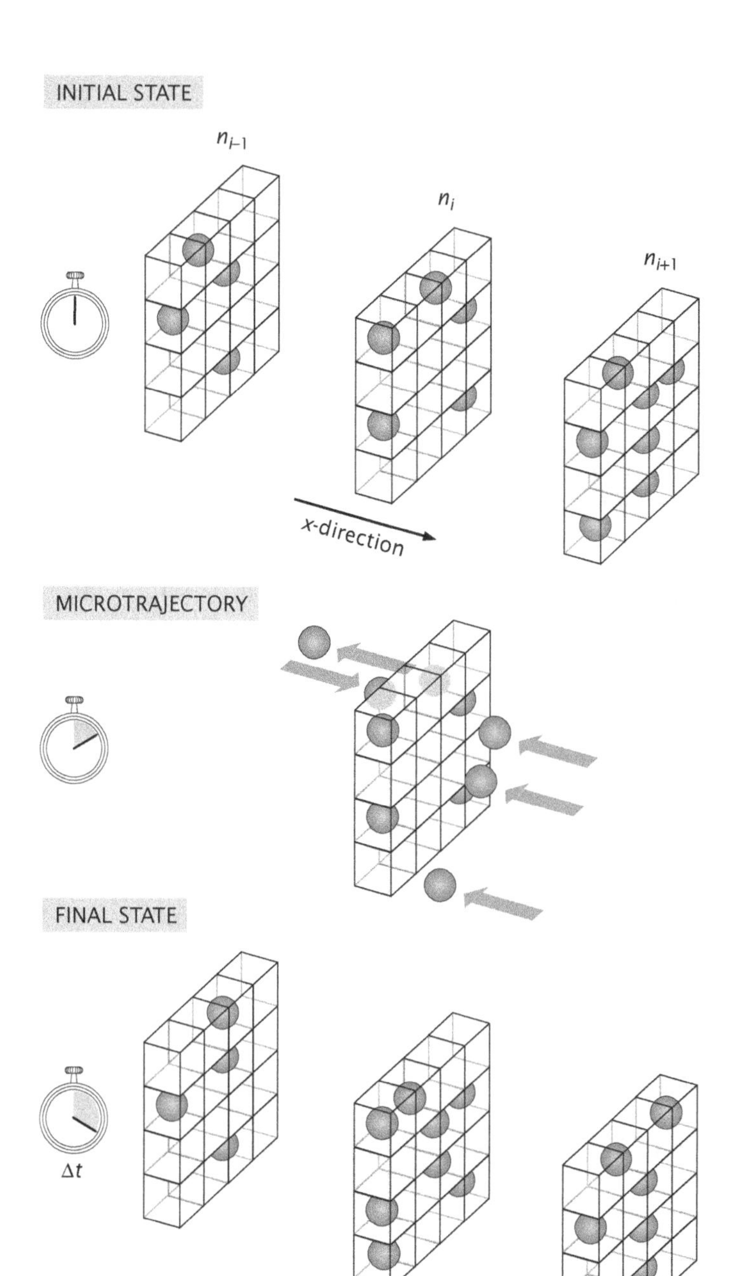

Figure 13.13: Schematic of a one-dimensional array of random walkers. Space is discretized into a set of planes.

this case is to consider the motions of individual diffusing molecules (or particles) and to sum over all of the possible microscopic trajectories of the system. The overall macroscopic response emerges as the average over all of these underlying microtrajectories. An example of one particular microtrajectory for a one-dimensional diffusion problem is shown in Figure 13.13.

Particles or fluorescently labeled molecules observed in a microscope are seen to undergo random jiggling, with each particle suffering a different trajectory. We now place these random trajectories front and center and elaborate on a quantitative treatment of diffusion that parallels the states-and-weights approach to computing equilibrium probabilities already used throughout the book. In later chapters, we will make use of this "trajectories-and-weights" approach to random dynamics of diffusing particles, molecular motors, polymerization motors, etc. Here we illustrate this procedure on the

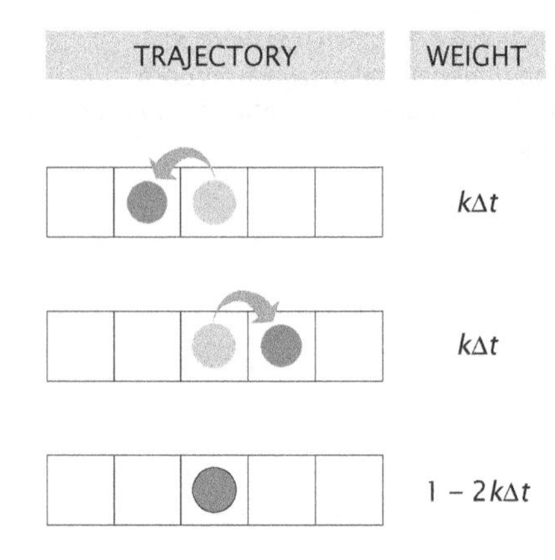

Figure 13.14: Trajectories and weights for simple diffusion. A given diffusing particle can do one of three things at every time step: jump left, jump right, or stay put. Each of these microtrajectories has an associated statistical weight.

simple diffusion process. The key idea is to describe a random trajectory by the probability density for finding a particle at a particular position at a given instant in time, $p(\mathbf{x}, t)$. In particular, the probability of finding the particle in a box of width Δx centered at point x at time t is given by $p(x, t)\Delta x$. To simplify the math, we specialize to one-dimensional motion, and discretize space and time. In this case, particle trajectories can be compactly denoted as long lists of integers that specify the position of the particle in units of a at different instants of time, measured in units of Δt. To derive the governing equation for the probability $p(x, t)$, we only have to specify the weights of all realizations of microtrajectories that can occur over time Δt.

Microtrajectories and their corresponding weights are shown in Figure 13.14. The diffusing particle, over time Δt, stays put, or jumps to the left or right a distance a, where we imagine that the particles can only occupy lattice sites on a lattice with spacing a. The probability of making a jump in either direction is $k\Delta t$, while the probability of staying put is $1 - 2k\Delta t$, insuring that the probabilities for all three possible outcomes add up to 1. We can use this model to compute a number of quantities associated with the particle trajectories. We begin by computing the mean and the variance of the particle displacement over time t. In time t, the particle makes a total of $N = t/\Delta t$ steps, each accompanied by a displacement Δx_i, $i = 1, 2, \ldots, N$. The total displacement, $\Delta x_{\text{tot}} = \Delta x_1 + \Delta x_2 + \cdots + \Delta x_N$, is a sum of independent identically distributed random variables. Therefore, as shown in The Tricks Behind the Math at the end of this section, the mean and the variance of Δx_{tot} are simply N times the mean and variance of Δx, the displacement for one time step. These are readily calculated from Figure 13.14 by summing over microtrajectories. We obtain the mean by summing over the three microtrajectories that can occur during a given time step as

$$\langle \Delta x \rangle = a \times k\Delta t + (-a) \times k\Delta t + (0) \times (1 - 2k\Delta t) = 0. \tag{13.9}$$

We can compute the variance as the average of the square of displacement once again by summing over all of the eventualities at a given instant as

$$\langle \Delta x^2 \rangle = a^2 \times k\Delta t + (-a)^2 \times k\Delta t + (0)^2 \times (1 - 2k\Delta t) = 2a^2 k\Delta t. \tag{13.10}$$

The variance of the total displacement is $N = t/\Delta t$ times greater, resulting in

$$\left\langle \Delta x_{\text{tot}}^2 \right\rangle = 2(a^2 k)\, t, \tag{13.11}$$

which is the result for diffusive spreading if we identify $a^2 k$ with the diffusion constant D.

The trajectories-and-weights approach can also be used to derive the governing equation for $p(x, t)$, the probability density that the particle is at position x at time t. The idea is to sum over all the microtrajectories starting at time instant t that result in the particle being at position x at time $t + \Delta t$. For this to happen, the particle needs to be at position x (if it is to stay put on the next time step), $x - a$ (if it is to jump to the right at the next time step), or $x + a$ (if it is to jump to the left at the next time step) at time t, and the associated probabilities are $p(x, t)$, $p(x - a, t)$, and $p(x + a, t)$, respectively. Using the probabilities in Figure 13.14, we can write $p(x, t + \Delta t)$ as a sum over microtrajectories,

$$p(x, t+\Delta t) = \underbrace{(1-2k\Delta t)\times p(x,t)}_{\text{stay put}} + \underbrace{k\Delta t \times p(x-a,t)}_{\text{jump right}} + \underbrace{k\Delta t \times p(x+a,t)}_{\text{jump left}}, \tag{13.12}$$

which leads to a discrete differential (or difference) equation for $p(x, t)$. Also, in writing the above equation, we have used the so-called Markov property of the process, namely the fact that the probability of a microtrajectory at time t is independent of the previous history of the particle; this is what allows us to express the probability of each outcome as a product of probabilities. To arrive at the more familiar, continuous diffusion equation, we once again make use of the Taylor expansion

$$\begin{aligned} p(x, t+\Delta t) &\approx p(x,t) + \Delta t \frac{\partial p(x,t)}{\partial t}, \\ p(x \pm a, t) &\approx p(x,t) \pm a\frac{\partial p(x,t)}{\partial x} + \frac{a^2}{2}\frac{\partial^2 p(x,t)}{\partial x^2}. \end{aligned} \tag{13.13}$$

Substituting these formulas into Equation 13.12 gives

$$\frac{\partial p(x,t)}{\partial t} = (a^2 k)\frac{\partial^2 p(x,t)}{\partial x^2}. \tag{13.14}$$

This is the diffusion equation derived in the previous section from Fick's law, with $D = a^2 k$, the same identification we made above.

Like with many fundamental results, there are multiple ways of deriving the diffusion equation. It is instructive to examine yet another way of deriving this equation, which is another way of summing over all of the microscopic trajectories available to the system at every instant. The approach adopted here is that taken by Einstein in one of his classic 1905 papers. We imagine that time is sliced up into intervals of length Δt and that, at every time step, particles can either jump or stay put. Einstein starts by writing the concentration at position x and time $t + \Delta t$ as the following integral:

$$\underbrace{c(x, t+\Delta t)}_{\substack{\text{concentration} \\ \text{at } x \text{ now}}} = \int_{-\infty}^{+\infty} \underbrace{c(x+\Delta, t)}_{\substack{\text{concentration at} \\ x+\Delta \text{ earlier}}} \underbrace{\phi(\Delta)}_{\substack{\text{probability of a} \\ \text{jump of length } \Delta}} \, \mathrm{d}\Delta. \tag{13.15}$$

What this integral says precisely is that to find the concentration at position x at time $t + \Delta t$, we need to sum over all of the possible ways that particles could have gotten there. In particular, at time t, the particle could have been at position $x + \Delta$ and then jumped to position x during the time step. (Note we follow Einstein's notation precisely, so the reader is warned that what Einstein calls Δ is our Δx in our earlier derivation of the diffusion equation.) The microtrajectory that we described above can be true for any choice of Δ. This means that in order to obtain the concentration at x, we have to sum over all of the possible jumping events with each one weighted by $\phi(\Delta)$, the probability of jumping a distance Δ. Effectively, Einstein considers the possibility that particles can jump *any* distance, whereas in our earlier derivation we permitted jumps only of size a. Einstein makes two further assumptions. First, he posits a symmetry in the jump probabilities of the form

$$\phi(\Delta) = \phi(-\Delta), \tag{13.16}$$

which states that the probability of jumping a certain distance to the right is the same as the probability of jumping that same distance to the left, that is, that there is no bias in the chosen direction. If we included a bias, we would get a driven diffusion equation, which we will encounter in the context of molecular motors. The other key feature of the distribution $\phi(\Delta)$ is

$$\int_{-\infty}^{+\infty} \phi(\Delta)\,\mathrm{d}\Delta = 1, \tag{13.17}$$

which guarantees that the molecules do *something* at every time step. We now make a familiar refrain by Taylor-expanding both terms appearing in Equation 13.15, which results in

$$c(x, t+\Delta t) \approx c(x, t) + \frac{\partial c}{\partial t}\Delta t \tag{13.18}$$

and

$$c(x+\Delta, t) \approx c(x, t) + \frac{\partial c}{\partial x}\Delta + \frac{1}{2}\frac{\partial^2 c}{\partial x^2}\Delta^2. \tag{13.19}$$

If we substitute these results into Equation 13.15, we find

$$\begin{aligned} c(x, t) + \frac{\partial c}{\partial t}\Delta t \approx{}& c(x, t)\int_{-\infty}^{+\infty} \phi(\Delta)\,\mathrm{d}\Delta + \frac{\partial c}{\partial x}\int_{-\infty}^{+\infty} \Delta\phi(\Delta)\,\mathrm{d}\Delta \\ &+ \frac{1}{2}\frac{\partial^2 c}{\partial x^2}\int_{-\infty}^{+\infty} \Delta^2\phi(\Delta)\,\mathrm{d}\Delta. \end{aligned} \tag{13.20}$$

The right-hand side of this equation can be examined term by term. The integral in the first term is 1 by Equation 13.17. As a result, we have a term of the form $c(x, t)$ on both sides of the equation that will cancel out of the final result. The second term is zero because we are integrating an odd function, Δ, times an even function, $\phi(\Delta)$. If we define the integral in the last term as

$$D \equiv \frac{1}{2\Delta t}\int_{-\infty}^{+\infty} \Delta^2\phi(\Delta)\,\mathrm{d}\Delta, \tag{13.21}$$

we can write Equation 13.20 as

$$\frac{\partial c}{\partial t} = D\frac{\partial^2 c}{\partial x^2}, \tag{13.22}$$

which is precisely the same result for the one-dimensional diffusion equation that we obtained earlier.

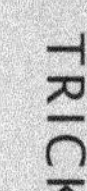

TRICKS

The Tricks Behind the Math: Averaging Sums of Random Variables Independent identically distributed random variables $\sigma_1, \sigma_2, \ldots, \sigma_N$ all have the same probability distribution, $P(\sigma)$, and their joint probability distribution factorizes,

$$P_{\text{joint}}(\sigma_1, \sigma_2, \ldots, \sigma_N) = P(\sigma_1) \times P(\sigma_2) \times \cdots \times P(\sigma_N). \tag{13.23}$$

The factorization property simply means that the probability that one of the random variables takes on a particular value is independent of all the other random variables in the bunch. If the variables σ_i take on two values, say 1 and 0, this mathematical construct could be used, for example, to describe N noninteracting ion channels, with 0 and 1 corresponding to a channel being closed or open. Beyond this example, there are many more that we will encounter, so we take a brief interlude here to derive two useful identities for the sum of independent identically distributed random variables.

We begin by showing that the average value of the sum, $\sigma_1 + \sigma_2 + \cdots + \sigma_N$, is equal to N times the average of one of the random variables (since they are identical, it doesn't matter which one we choose). We start by writing the average of the sum using the joint probability distribution,

$$\left\langle \sum_{i=1}^{N} \sigma_i \right\rangle = \sum_{\sigma_1} \sum_{\sigma_2} \cdots \sum_{\sigma_N} \left[\sum_{i=1}^{N} \sigma_i P_{\text{joint}}(\sigma_1, \sigma_2, \ldots, \sigma_N) \right]. \quad (13.24)$$

Then, we make use of the factorization property given in Equation 13.23, and the above equation can be written as

$$\begin{aligned} \left\langle \sum_{i=1}^{N} \sigma_i \right\rangle &= \sum_{\sigma_1} \sigma_1 P(\sigma_1) \sum_{\sigma_2} P(\sigma_2) \cdots \sum_{\sigma_N} P(\sigma_N) \\ &+ \sum_{\sigma_1} P(\sigma_1) \sum_{\sigma_2} \sigma_2 P(\sigma_2) \cdots \sum_{\sigma_N} P(\sigma_N) \\ &+ \cdots \sum_{\sigma_1} P(\sigma_1) \sum_{\sigma_2} P(\sigma_2) \cdots \sum_{\sigma_N} \sigma_N P(\sigma_N). \end{aligned} \quad (13.25)$$

Finally, using the fact that all the probabilities must add up to 1 ($\sum_\sigma P(\sigma) = 1$) and the fact that all the random variables are identical, we arrive at the desired result, namely,

$$\left\langle \sum_{i=1}^{N} \sigma_i \right\rangle = N \sum_{\sigma} \sigma P(\sigma) = N \langle \sigma \rangle . \quad (13.26)$$

Next, we compute the variance of the sum of N independent identically distributed random variables. The variance is the average of the square of the difference between the random variable and its mean,

$$\operatorname{var}\left(\sum_{i=1}^{N} \sigma_i \right) = \left\langle \left(\sum_{i=1}^{N} \sigma_i - \left\langle \sum_{i=1}^{N} \sigma_i \right\rangle \right)^2 \right\rangle . \quad (13.27)$$

Using the average computed above, and expanding the square, we can simplify the above equation to read

$$\operatorname{var}\left(\sum_{i=1}^{N} \sigma_i \right) = \left\langle \left(\sum_{i=1}^{N} \sigma_i \right)^2 \right\rangle - N^2 \langle \sigma \rangle^2 . \quad (13.28)$$

Writing the square in the above equation as a product of two equal terms, and being mindful to use different summation variables i and j in the two sums, we arrive at

$$\operatorname{var}\left(\sum_{i=1}^{N} \sigma_i \right) = \left\langle \sum_{i,j=1}^{N} \sigma_i \sigma_j \right\rangle - N^2 \langle \sigma \rangle^2 . \quad (13.29)$$

Now, to compute $\left\langle \sum_{i,j=1}^{N} \sigma_i \sigma_j \right\rangle$, we break up the double sum into two pieces, one with N terms where $i = j$, and the other with the remaining $N^2 - N$ terms where $i \neq j$:

$$\left\langle \sum_{i,j=1}^{N} \sigma_i \sigma_j \right\rangle = \left\langle \sum_{i=1}^{N} \sigma_i^2 \right\rangle + \left\langle \sum_{i \neq j; i,j=1}^{N} \sigma_i \sigma_j \right\rangle . \tag{13.30}$$

Since all the σ_i are independent, for $i \neq j$ we have

$$\langle \sigma_i \sigma_j \rangle = \langle \sigma_i \rangle \langle \sigma_j \rangle = \langle \sigma \rangle^2 .$$

Putting all this back into Equation 13.29, we arrive at the result

$$\begin{aligned} \text{var}\left(\sum_{i=1}^{N} \sigma_i \right) &= N \langle \sigma^2 \rangle + (N^2 - N) \langle \sigma \rangle^2 - N^2 \langle \sigma \rangle^2 \\ &= N \left(\langle \sigma^2 \rangle - \langle \sigma \rangle^2 \right) = N \, \text{var}(\sigma). \end{aligned} \tag{13.31}$$

In other words, the variance of the sum of N independent identically distributed random variables is equal to N times the variance of one.

13.2.2 Solutions and Properties of the Diffusion Equation

Concentration Profiles Broaden Over Time in a Very Precise Way

Now that we have the diffusion equation in hand, it is of great interest to examine its biological consequences. One of the most useful tools corresponds to knowing how to solve this equation for a spike of concentration at the origin at time $t = 0$. In particular, if at time $t = 0$ we start with N molecules in an infinitesimally small region around $x = 0$, the concentration profile will evolve in the following way:

$$c(x,t) = \frac{N}{\sqrt{4\pi Dt}} \mathrm{e}^{-x^2/4Dt} . \tag{13.32}$$

The solution itself is left to the problems at the end of the chapter. Further, by dividing by N, this equation can then be interpreted as giving the probability density for finding a particle between x and $x + \mathrm{d}x$. The solution quoted above is often denoted as the Green's function of the diffusion equation and its evolution can be seen in Figure 13.15. This equation for the concentration tells us that the profile has the form of a Gaussian. The width of the Gaussian is $\sqrt{2Dt}$ and hence it increases as the square root of the time. One of the most beautiful features of a solution like this is that once it is known, by exploiting the linearity of the diffusion equation itself, we are then free to write the solution for an arbitrary initial distribution of diffusing molecules. This idea will be taken up in the problems at the end of the chapter.

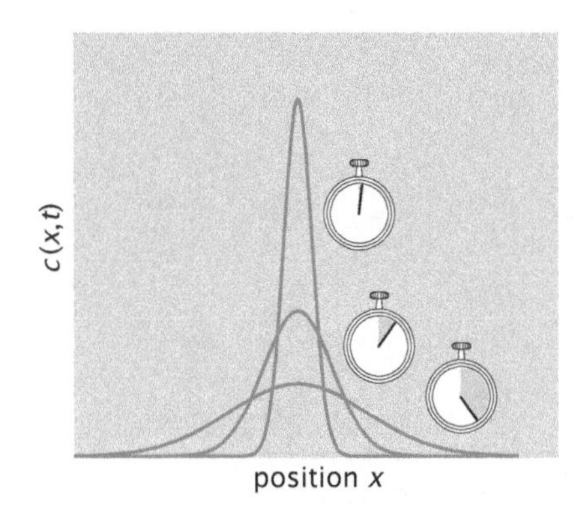

Figure 13.15: Time evolution of the concentration field. The plot shows the solution for the diffusion equation at different times for an *initial* concentration profile that is a spike at $x = 0$.

Note in Figure 13.15 that the mean position of the concentration distribution does not change with time. This corresponds to the absence of a drift term, though drift terms will form a centerpiece of our discussion in Chapter 16 when we turn to the dynamics of molecular motors. On the other hand, even in the absence of drift, the diffusive dynamics is rich and interesting. One of the most interesting quantities to feature is the width of the distribution, $\langle x^2 \rangle$, which broadens over time. Since the distribution is Gaussian, we can essentially read off the dynamics of the width, but we take this opportunity to compute it explicitly since it is instructive both physically and mathematically. To compute this broadening, we need to evaluate $\langle x^2 \rangle$ as

$$\langle x^2 \rangle = \frac{\int_{-\infty}^{+\infty} x^2 \frac{N}{\sqrt{4\pi Dt}} e^{-x^2/4Dt}\, dx}{N} = \frac{1}{\sqrt{4\pi Dt}} \int_{-\infty}^{+\infty} x^2 e^{-x^2/4Dt}\, dx, \quad (13.33)$$

where we have made use of the probability distribution for finding a particle at position x at time t, which is related to the concentration distribution, Equation 13.32, by $c(x,t)/N$. Using the trick introduced in The Tricks Behind the Math below, we can evaluate this integral straightaway to find

$$\langle x^2 \rangle = \frac{1}{\sqrt{4\pi Dt}} \int_{-\infty}^{+\infty} x^2 e^{-x^2/4Dt}\, dx = \frac{1}{\sqrt{4\pi Dt}} \frac{\sqrt{\pi}}{2} (4Dt)^{3/2} = 2Dt. \quad (13.34)$$

Note that this is the key result that we have already invoked a number of times (for example, see Section 13.1.2 on p. 511) throughout the book as the basis of intuition about diffusive processes. In particular, this is the result that we have argued reveals how diffusion times scale with the square of the distance over which diffusion must act.

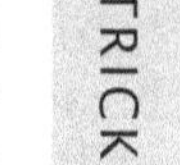

TRICKS

The Tricks Behind the Math: Doing Integrals by Differentiating With Respect to a Parameter Sometimes the knowledge of one integral in terms of a parameter appearing in the integrand can be used to compute a number of related integrals obtained by differentiating with respect to the parameter in question. Indeed, we have already invoked this trick in our discussion of the partition function (p. 241). In the case of the Gaussian integral (discussed on p. 261),

$$\int_{-\infty}^{\infty} e^{-\alpha x^2}\, dx = \sqrt{\frac{\pi}{\alpha}}, \quad (13.35)$$

we can differentiate both sides of the equation with respect to the parameter α to obtain

$$\int_{-\infty}^{+\infty} x^2 e^{-\alpha x^2}\, dx = -\frac{\partial}{\partial \alpha} \int_{-\infty}^{+\infty} e^{-\alpha x^2}\, dx = -\frac{\partial}{\partial \alpha} \sqrt{\frac{\pi}{\alpha}} = \frac{\sqrt{\pi}}{2\alpha^{3/2}}. \quad (13.36)$$

If we differentiate with respect to α a second time, this yields $\langle x^4 \rangle$. Repeated differentiation with respect to α allows us to determine all of the even moments.

13.2.3 FRAP and FCS

One of the merits of the techniques described earlier in the chapter (p. 513) for measuring diffusion is that they can be readily applied to living cells. The diffusive behavior of a molecule in the environment of a cell depends upon the physical structure of the molecule itself and also on the structure of its environment as well as its interactions with other molecules. In Chapter 14 we will explore some specific cases of how information can be gleaned about cell structure by examining deviations from the diffusive behavior expected in dilute solutions. Here we start with a simple experimental case examining the motion of tracer molecules such as GFP within living cells where the tracer molecule does not form any specific binding

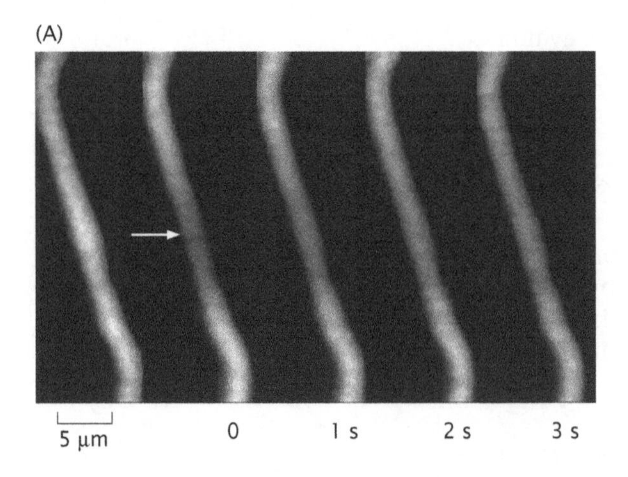

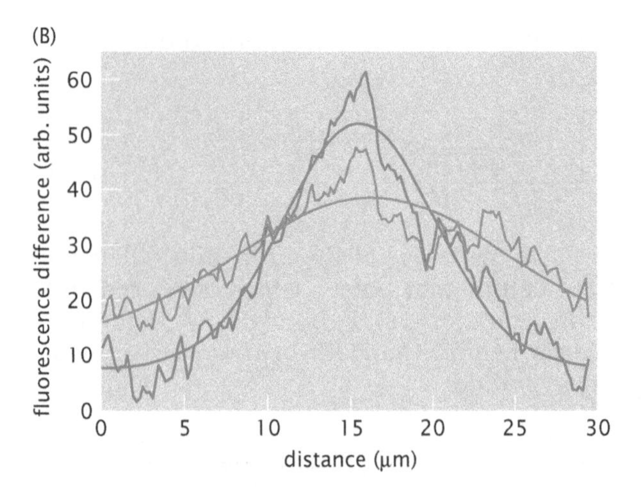

Figure 13.16: FRAP data from a bacterium labeled with GFP. (A) The images show an elongated bacterium with the pre-bleach image on the left and subsequent images taken at different times (reported in seconds) after photobleaching. The arrow shows the photobleached region. (B) The curves show the difference in the intensity profiles before and after photobleaching, along the long axis of the bacterium. The red curve is measured right after photobleaching while the blue curve was obtained from an image acquired 4 s later. (Adapted from C. W. Mullineaux et al., *J. Bacteriol.* 188:3442, 2006.)

reactions with cellular constituents. An example of this sort of experiment is shown in Figure 13.16 for an experiment in which elongated *E. coli* cells are photobleached and the resulting fluorescence intensity is measured over time.

To calculate the expected time evolution of GFP following photobleaching, consider a one-dimensional *E. coli* such as might be found in the gut of a spherical cow (see Figure 22.2 on p. 1030). We leave the more realistic two-dimensional problem to the end of the chapter, though we note that the key features of the problem are already revealed in the one-dimensional case. We consider a one-dimensional model of a FRAP experiment. The fluorescent molecules diffuse in a box of length $2L$, which for convenience we place between $-L$ and L along the x-axis. The initial concentration is equal to c_0 on the intervals $-L < x < -a$ and $a < x < L$ and is zero on the interval $-a < x < a$ as is shown in Figure 13.17. In other words, we imagine that we start with a uniform concentration c_0 of fluorescent molecules in the width-$2L$ box and then we photobleach all the molecules in a smaller box of size $2a$ by exposing them to intense laser light. If we were to look under a microscope, we would observe the recovery of fluorescence as the nonbleached molecules made their way into the box of size $2a$. Based on the speed of fluorescence recovery, the diffusion constant of the fluorescent molecules can be measured.

We can use the simple one-dimensional model to gain quantitative insight into the recovery process. To compute the recovery curves, we first solve the diffusion equation

$$\frac{\partial c}{\partial t} = D\frac{\partial^2 c}{\partial x^2} \tag{13.37}$$

for the concentration of fluorescent molecules $c(x, t)$, with the initial concentration after photobleaching given by

$$c(x, 0) = \begin{cases} c_0 & \text{for } -L < x < -a, \\ 0 & \text{for } -a < x < a, \\ c_0 & \text{for } a < x < L. \end{cases} \tag{13.38}$$

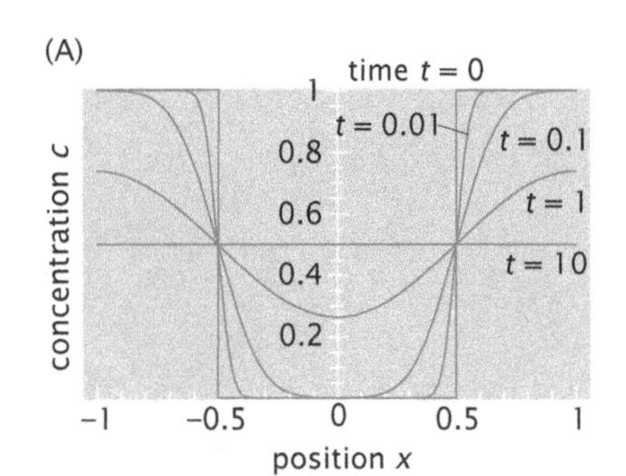

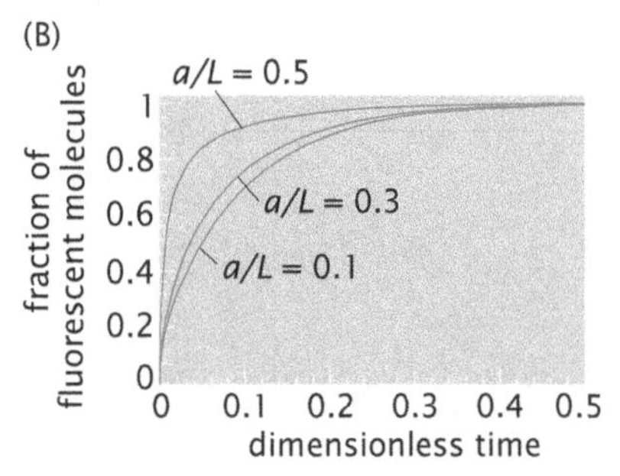

Figure 13.17: One-dimensional model of FRAP. (A) Concentration profile for different times after photobleaching. The bleached region is half the size of the confining region, $2L$. (B) Fluorescence recovery as a function of time for different sizes of bleached regions. Recovery is fastest when the bleached region is half the size of the confining region. In both graphs, time is measured in units of L^2/D and length in units of L.

We also impose the boundary condition $\partial c/\partial x = 0$ for $x = \pm L$, which says that the flux of fluorescent molecules vanishes at the boundaries of the one-dimensional cell (no material flows in or out). This mimics the real-life situation with fluorescent proteins confined to the volume of the cell, to the cell membrane, or to some other subcellular structure.

To solve the diffusion equation with the prescribed initial and boundary conditions, we begin by expanding the concentration profile $c(x, t)$ in terms of cosine functions (introduced in The Math Behind the Models on p. 332),

$$c(x, t) = A_0(t) + \sum_{n=1}^{\infty} A_n(t) \cos\left(\frac{x}{L} n\pi\right). \tag{13.39}$$

This expansion guarantees that the boundary conditions are met, namely each of the functions $A_n(t)\cos(xn\pi/L)$ has vanishing first derivatives at $x = \pm L$. Furthermore, since the initial concentration profile takes the same values for positive and negative x, it is readily expanded in cosine functions. The solution of the diffusion equation now boils down to finding the functions $A_n(t)$ such that both Equation 13.37 and the initial condition, Equation 13.38, are satisfied.

To proceed, we substitute the series expansion of $c(x, t)$ into the diffusion equation. This yields

$$\frac{\partial A_0}{\partial t} + \sum_{n=1}^{\infty} \frac{\partial A_n(t)}{\partial t} \cos\left(\frac{x}{L} n\pi\right) = D \sum_{n=1}^{\infty} \left[-A_n(t) \frac{n^2\pi^2}{L^2}\right] \cos\left(\frac{x}{L} n\pi\right), \tag{13.40}$$

which, due to the orthogonality property of the cosine functions for different n (see Equation 13.44 below), turns into a set of independent differential equations,

$$\begin{aligned} \frac{\partial A_0}{\partial t} &= 0, \\ \frac{\partial A_n}{\partial t} &= -\frac{Dn^2\pi^2}{L^2} A_n(t) \quad (n \geq 1). \end{aligned} \tag{13.41}$$

The solution to each one of these (infinite in number) equations is an exponential function

$$A_n(t) = A_n(0) \mathrm{e}^{-(Dn^2\pi^2/L^2)t}, \tag{13.42}$$

which when substituted into Equation 13.39 gives

$$c(x, t) = A_0(0) + \sum_{n=1}^{\infty} A_n(0) \mathrm{e}^{-(Dn^2\pi^2/L^2)t} \cos\left(\frac{x}{L} n\pi\right). \tag{13.43}$$

The final piece of the puzzle is the determination of the constants $A_n(0)$.

To compute the initial amplitudes of the cosine functions, we once again resort to the orthogonality property of these functions,

$$\int_{-L}^{L} \cos\left(\frac{x}{L} n\pi\right) \cos\left(\frac{x}{L} m\pi\right) \mathrm{d}x = L\delta_{n,m}. \tag{13.44}$$

In particular, we multiply both sides of Equation 13.43 by $\cos(n\pi x/L)$ for different values of n, and then integrate over x, which provides us with the equations

$$A_0(0) = \frac{1}{2L}\int_{-L}^{L} c(x,0)\,dx,$$
$$A_n(0) = \frac{1}{L}\int_{-L}^{L} c(x,0)\cos\left(\frac{x}{L}n\pi\right)dx \quad (n \geq 1) \tag{13.45}$$

for the initial amplitudes. Substituting the initial concentration profile, $c(x,0)$, into these equations, and performing the integrals, we arrive at

$$A_0(0) = c_0\frac{L-a}{L},$$
$$A_n(0) = -2c_0\frac{\sin(n\pi a/L)}{n\pi} \quad (n \geq 1). \tag{13.46}$$

Putting these results back into the derived formula for $c(x,t)$, Equation 13.43 gives us the solution for the concentration profile as a function of time,

$$c(x,t) = c_0\left[1 - \frac{a}{L} - \sum_{n=1}^{\infty}\frac{2\sin(n\pi a/L)}{n\pi}e^{-(Dn^2\pi^2/L^2)t}\cos\left(\frac{x}{L}n\pi\right)\right], \tag{13.47}$$

which is plotted as a function of x for different times (and setting $a = L/2$) in Figure 13.17(A). Note that at long times, such that t is much greater than L^2/D, which is the diffusion time for a box of length L, the concentration profile tends to a constant value equal to $c_\infty = c_0(1 - a/L)$. This can be understood in a very simple way. Namely, at long times, we expect diffusion to make the concentration profile uniform over the $2L$ interval. Then, the fact that the number of fluorescent molecules does not change in time leads to the equation

$$c_\infty(2L) = c_0[2(L-a)], \tag{13.48}$$

which gives the computed value of the concentration at long times.

Given the concentration profile as a function of time, we are now in the position to compute a FRAP recovery curve within our simple one-dimensional model. We ask how many fluorescent molecules there are in the bleached region as a function of time. In our simple model, the bleached region is a box that spans from $-a$ to a on the x-axis. We already know that at $t = 0$, the number of fluorescent molecules in the bleached region is $N_f = 0$, while at times much longer than the diffusion time, this number tends to $c_\infty(2a)$. For intermediate times, we need to compute

$$N_f(t) = \int_{-a}^{a} c(x,t)\,dx. \tag{13.49}$$

Substituting our result for the concentration profile given in Equation 13.47 into the integral leads to an expression for the recovery curve,

$$N_f(t) = 2c_0a\left(1 - \frac{a}{L}\right)$$
$$\times\left[1 - \frac{1}{(a/L)(1-a/L)}\sum_{n=1}^{\infty}\frac{2}{n^2\pi^2}\sin^2\left(\frac{n\pi a}{L}\right)e^{-(Dn^2\pi^2/L^2)t}\right]. \tag{13.50}$$

Note that at very long times, N_f approaches $c_\infty(2a) = 2c_0a(1 - a/L)$, as expected.

In Figure 13.17(B), we plot a FRAP recovery curve normalized by $2c_0a(1 - a/L)$, the total number of fluorescent molecules in the bleached region in the long-time limit. The model makes an interesting prediction that the recovery is fastest when the size of the bleached region is equal to half the size of the confining region. Furthermore, the recovery curves are identical for bleached regions of fractional size a/L and $1 - a/L$, which follows directly from Equation 13.50. In particular, the right-hand side of this equation is invariant under the exchange $a/L \leftrightarrow 1 - a/L$.

13.2.4 Drunks on a Hill: The Smoluchowski Equation

Thus far, our treatment of diffusion has been based upon those problems in which there are no external forces acting on the particle of interest. On the other hand, there are a number of diffusive processes in which the diffusing species is subjected to a force. For example, we can imagine ion diffusion in the presence of an electric field. In the trajectories-and-weights treatments of diffusion, we assumed that the probabilities of jumping in any of the allowed directions are equal. This is not the case if an external applied force biases the motion of the particle in some particular direction. In this case, we expect the rates to be asymmetrical, since a jump in the direction of the force will be more probable than a jump against the direction of the force. To see the effect of this asymmetry, we can repeat the analysis that led to the derivation of the diffusion equation, but now with the force-induced asymmetry in jump rates.

First we compute the mean and the variance of the particle displacement after time t. Once again, both the mean and variance of the total displacement are $N = t/\Delta t$ times greater than the mean and variance of the displacement Δx resulting from a single time step. These in turn are readily computed from the trajectories and weights as shown in Figure 13.18, resulting in

$$\begin{aligned} \langle \Delta x \rangle &= a \times k_+\Delta t + (-a) \times k_-\Delta t = a(k_+ - k_-)\Delta t, \\ \text{var}(\Delta x) &= a^2 \times k_+\Delta t + (a)^2 \times k_-\Delta t - \langle \Delta x \rangle^2 = a^2(k_+ + k_-)\Delta t \end{aligned} \tag{13.51}$$

where in obtaining the final result for the variance we have dropped the $\langle \Delta x \rangle^2$ term on account of it being much smaller than the first one; this is because the term that is omitted is quadratic in Δt, or, more precisely, because $k_\pm \Delta t \ll 1$.

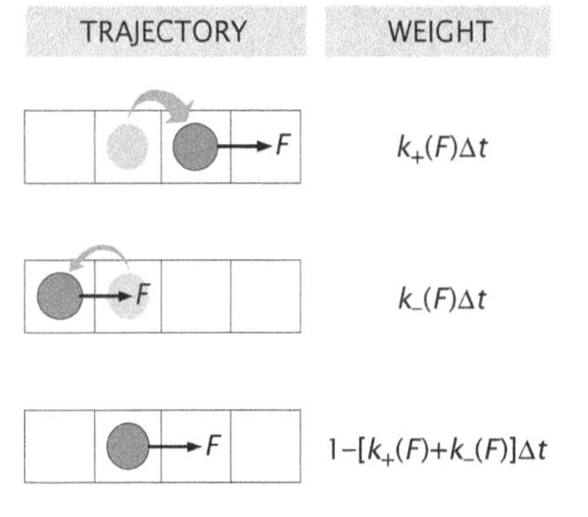

Figure 13.18: Trajectories and weights for driven diffusion. The probabilities of jumping to the left and right are unequal since jumping in the direction of the force is more likely than jumping in the opposite direction to the applied force.

We see that the variance of the displacement is the same as for unbiased diffusion, with the diffusion constant now being given by $D = (k_+ + k_-)a^2/2$. On the other hand, the mean is now nonzero, and it increases linearly with time. Therefore, the overall motion of the particle can be described as diffusion with drift, with a drift velocity

$$v = \frac{\langle \Delta x \rangle}{\Delta t} = a(k_+ - k_-). \tag{13.52}$$

Note that for a particle moving through a fluid, and in the limit of low Reynolds number, the drift velocity is related to the applied force on the particle, $F = \gamma v$, with γ the friction coefficient. This idea was already discussed in Section 12.4.1 (p. 495).

Just as we did for the unbiased diffusion case, we can use the trajectories-and-weights approach to derive the governing equation for $p(x, t)$, the probability density of finding a particle at position x at time t. Again, we consider all the microtrajectories that in time Δt end up at position x. There are three such trajectories, with the particle being initially at x and staying put, the particle starting at $x - a$ and jumping to the right, and, finally, the particle starting at $x + a$ and jumping to the left. The probability $p(x, t + \Delta t)$ is given by the sum over these trajectories,

$$p(x, t + \Delta t) = [1 - (k_+ + k_-)\Delta t] \times p(x, t) + k_+\Delta t \times p(x - a, t) + k_-\Delta t \times p(x + a, t). \quad (13.53)$$

We can turn this difference equation into the more familiar continuum form by Taylor-expanding $p(x, t + \Delta t)$ and $p(x \pm a, t)$, as was done in Equation 13.13. We arrive at

$$\frac{\partial p(x, t)}{\partial t} = -v\frac{\partial p(x, t)}{\partial x} + D\frac{\partial^2 p(x, t)}{\partial x^2}, \quad (13.54)$$

with the drift velocity $v = a(k_+ - k_-)$ and diffusion coefficient $D = (k_+ + k_-)a^2/2$, as stated above.

13.2.5 The Einstein Relation

The microscopic derivation of diffusion with drift given above can be complemented by a macroscopic derivation. As we will see below, this leads to an important relation between diffusion and friction first derived by Einstein. We consider a generalization of Fick's law to account for the bias that will arise in the presence of driving forces. In particular, a force F exerted on a particle results in a drift velocity $v = F/\gamma$, where γ is the friction coefficient. For a spherical particle of radius a moving through a fluid of viscosity η, $\gamma = 6\pi\eta a$. (See Equation 12.32 and the accompanying discussion on p. 497.) The presence of a net drift of a collection of particles all moving with the same mean velocity v, illustrated in Figure 13.19, results in a flux of the form

$$J_F = \frac{(v\Delta t)c}{\Delta t} = \frac{F}{\gamma}c. \quad (13.55)$$

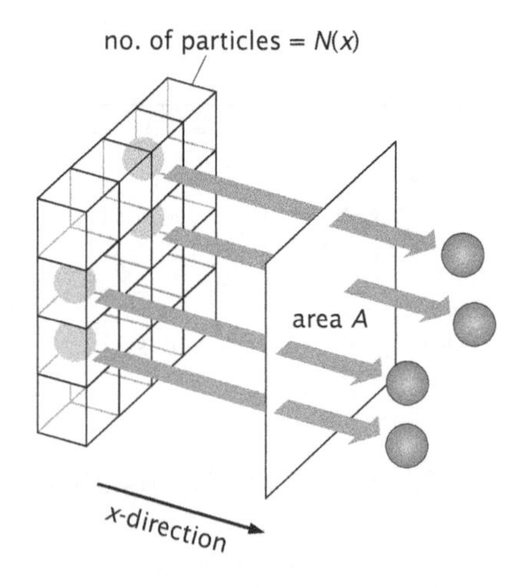

Figure 13.19: Flux due to an applied force. All the particles are moving with a drift velocity v to the right, resulting in $\Delta N = cvA\Delta t$ particles crossing the surface of area A in time Δt. The flux through the surface is $\Delta N/A\Delta t = cv$.

We are now in the position to write the total flux as a sum of those parts due to random hopping and those parts due to the applied force. In particular, the total flux takes the form

$$J(x) = -D\frac{\mathrm{d}c}{\mathrm{d}x} + \frac{F}{\gamma}c. \quad (13.56)$$

Note that in equilibrium, the net flux vanishes, $J(x) = 0$, resulting in

$$D\frac{\mathrm{d}c}{\mathrm{d}x} = \frac{F}{\gamma}c. \quad (13.57)$$

This differential equation describes the way in which a nonuniform concentration distribution can be set up by the presence of a force and will be used, for example, to characterize the distribution of ions near a membrane. To explore the consequences of this result, we must first solve this simple linear differential equation, which we can do using the separation-of-variables technique. Indeed, separation of variables results in

$$\gamma D\frac{\mathrm{d}c}{c} = F\,\mathrm{d}x. \quad (13.58)$$

It is convenient at this point to restrict our attention to those forces that can be derived from a potential energy as $F = -\mathrm{d}U/\mathrm{d}x$. As a result, we have

$$\gamma D \frac{\mathrm{d}c}{c} = -\mathrm{d}U. \tag{13.59}$$

This equation can be integrated to yield

$$\frac{c(x)}{c(0)} = \frac{\mathrm{e}^{-U(x)/\gamma D}}{\mathrm{e}^{-U(0)/\gamma D}}. \tag{13.60}$$

This is exactly what we would expect from the Boltzmann distribution, provided we set

$$D = k_\mathrm{B} T/\gamma, \tag{13.61}$$

which is precisely the Einstein relation!

The Einstein relation is remarkable. It relates two quantities, diffusion and friction, that at first glance might seem worlds apart. The diffusion constant captures the random motion of a microscopic particle through a fluid due to thermal agitation by the surrounding molecules. The friction coefficient, on the other hand, talks about the macroscopic effect of resistance to motion through a fluid experienced by objects large and small. Still, since both of these effects depend on the interaction of the particle with the molecules of the surrounding fluid, one might expect that there would be a relation between the two.

Beyond its power as a conceptual tool, the Einstein relation also provides a very useful formula for estimating the diffusion constant of microscopic particles moving through a fluid, such as ions, proteins, and plastic beads used in single-molecule experiments. If we approximate the particle as a sphere with a radius a, the Stokes formula discussed in Chapter 12 (see Equation 12.32, p. 497) provides the relation between the particle size and its friction coefficient, $\gamma = 6\pi\eta a$. Here η is the viscosity of the fluid. Combining this equation with the Einstein formula, Equation 13.61, leads to the celebrated Stokes–Einstein relation

$$D = \frac{k_\mathrm{B} T}{6\pi\eta a}. \tag{13.62}$$

The case of diffusion with drift also has important consequences out of equilibrium. Like before, mass conservation tells us that

$$\frac{\partial c}{\partial t} = -\frac{\partial J}{\partial x}. \tag{13.63}$$

On the other hand, since the flux has an extra term, the resulting governing equation itself has a new term and is given by

$$\frac{\partial c}{\partial t} = D\frac{\partial^2 c}{\partial x^2} - \frac{F}{\gamma}\frac{\partial c}{\partial x}, \tag{13.64}$$

effectively the same result already shown in Equation 13.54. As will be shown in the remainder of the book, this governing equation is important for describing a range of processes involving both external forcing and diffusion simultaneously.

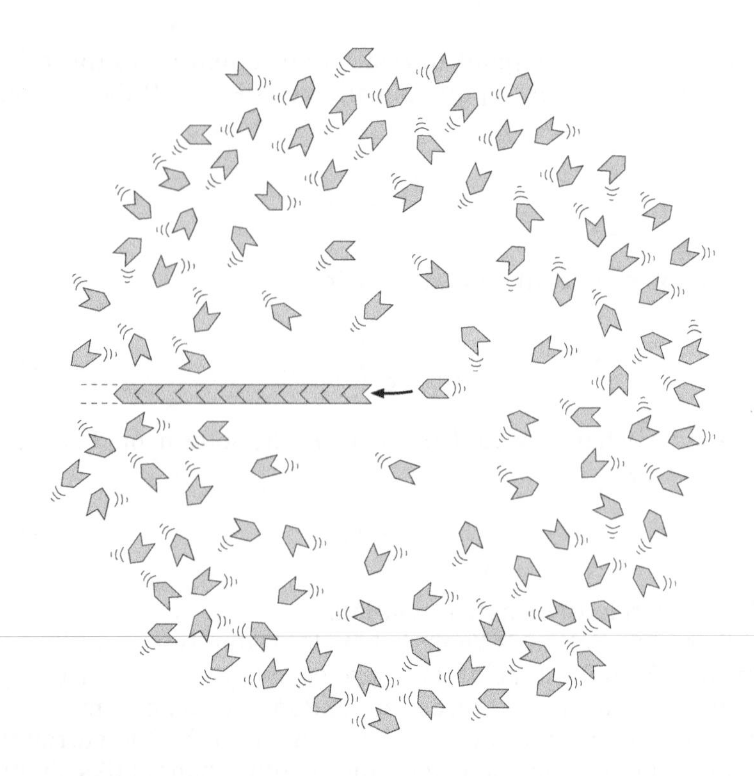

Figure 13.20: Monomer diffusion to the tip of a growing cytoskeletal filament. The figure illustrates that there is a depletion of monomers in the immediate vicinity of the growing tip. The schematic of the concentration profile around the filament ignores any disturbance to the distribution of monomers caused by the filament itself.

13.3 Diffusion to Capture

Another interesting class of problems associated with diffusion that show up in a number of different settings are those in which we are interested in the rate at which some diffusing species arrives at a given region of space. For example, the polymerization of cytoskeletal filaments such as actin requires the arrival of unencumbered actin monomers at the tip of the growing filament, as indicated schematically in Figure 13.20.

Similarly, a host of signaling processes depend upon the arrival of some mobile species of interest at the cell surface, where the species attaches to some receptor and induces the resulting signal cascade. A generic representation of chemoreceptors on the cell surface is shown in Figure 13.21, though the idealization of uniform receptors shown in the schematic is a vast oversimplification (for example, see Kentner and Sourijk, 2006). The mathematical problem we are interested in solving in this case is illustrated schematically in Figure 13.22, in which we idealize a cell as a sphere and imagine a uniform distribution of receptor molecules that decorate the cell surface. In particular, we consider the case where there is a radial distribution of molecules centered around the cell and with a far-field concentration c_0. The question we then pose is what the rate is at which signaling molecules find their way to the surface receptors.

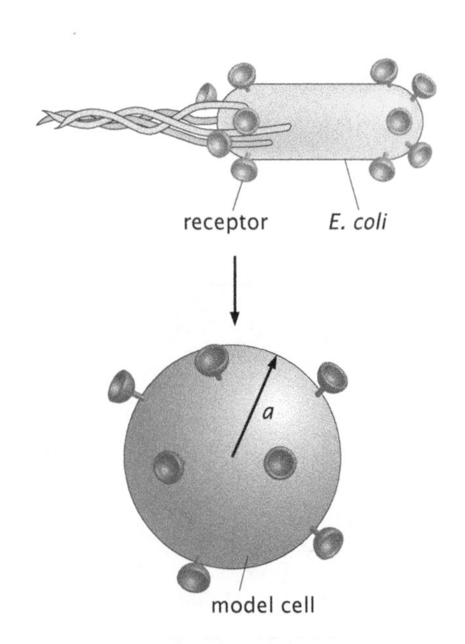

Figure 13.21: Cartoon representing the chemoreception process. The cell surface is decorated with receptors that tend to be clustered at the cell poles. For the purposes of simple analytical calculation, we idealize a bacterium as a sphere with a uniform density of receptors.

13.3.1 Modeling the Cell Signaling Problem

We now make a concrete and mathematical description of the diffusion to capture process in the setting of the highly idealized geometry shown in Figure 13.22. As we have already mentioned, the goal of our calculation will be to compute the number of signaling molecules that bind to the receptors per unit time and represented mathematically as dn/dt. We assume the cell is a sphere of radius a that has M

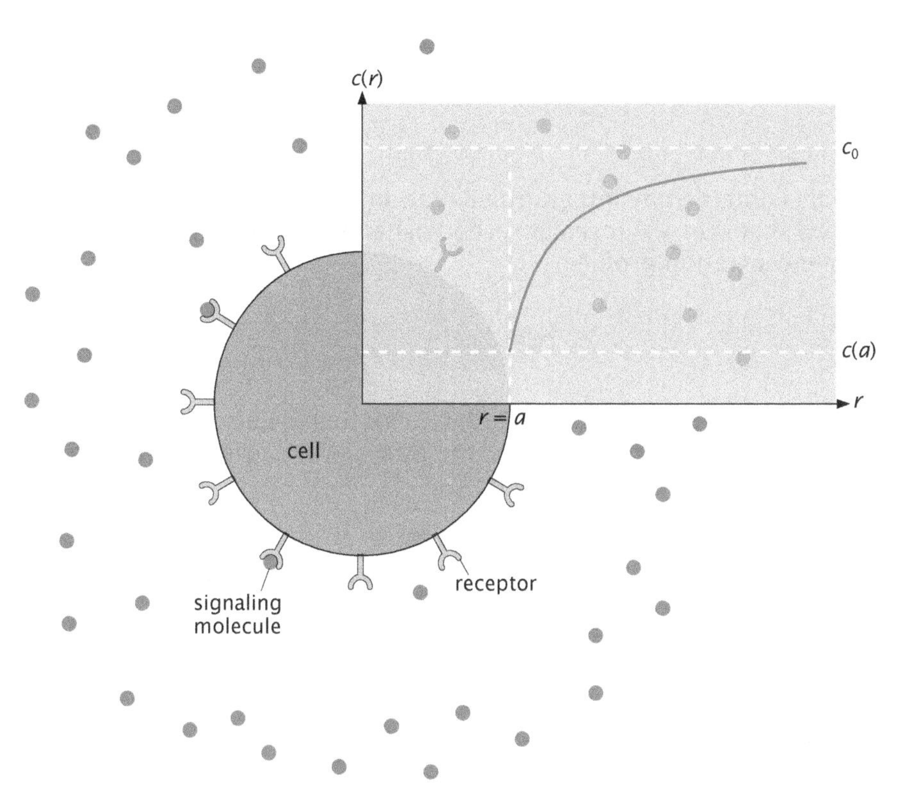

Figure 13.22: Concentration profile in the neighborhood of a spherical cell. The concentration profile $c(r)$ is spherically symmetric and characterizes the concentration of ligands as a function of distance from the cell surface.

receptors on its surface. Further, we assume that the concentration profile has spherical symmetry ($c(\mathbf{r}) = c(r)$) and that there is a far-field concentration of signaling molecules $c(\infty) = c_0$.

Perfect Receptors Result in a Rate of Uptake $4\pi Dc_0a$

In order to solve the problem using the *diffusion equation*, we interest ourselves in the steady-state condition characterized by $\partial c/\partial t = 0$. In this case, the diffusion equation given in Equation 13.8 reduces to

$$D\nabla^2 c = 0. \tag{13.65}$$

For the special case of spherical symmetry, the Laplacian can be written only in terms of the radial variables and results in a simplification of the diffusion equation to the form

$$\nabla^2 c = \frac{1}{r^2}\frac{\partial}{\partial r}\left(r^2\frac{\partial c}{\partial r}\right) = 0. \tag{13.66}$$

Since if the derivative of something is zero this implies that the something itself is a constant, we may rewrite this equation as

$$r^2\frac{\partial c}{\partial r} = A, \tag{13.67}$$

where A is a constant to be determined by the two boundary conditions $c(a) = 0$ and $c(\infty) = c_0$. The condition $c(a) = 0$ amounts to asserting that the receptors on the cell surface are perfect absorbers. That is, all molecules that arrive at the surface are swallowed up by the receptors. Our strategy unfolds as follows. First, we use the diffusion equation to determine the concentration profile $c(r)$. This can be worked out explicitly by recognizing that the solution to

Equation 13.67 is given by

$$c(r) = -\frac{A}{r} + B, \tag{13.68}$$

which is obtained by integrating Equation 13.67. By imposing the conditions that $c(a) = 0$ (perfect receptors) and $c(\infty) = c_0$, we can rewrite the concentration profile as

$$c(r) = c_0\left(1 - \frac{a}{r}\right). \tag{13.69}$$

Once the concentration is in hand, we can then use Fick's law to relate it to a mass flux. In particular, the flux at the surface of the cell is $j(a) = -D\partial c/\partial r|_a$, which is given by

$$j(a) = -\frac{Dc_0}{a}. \tag{13.70}$$

If we take this flux and multiply by the area of the sphere, what results is the total number of particles arriving per unit time, given by

$$\frac{\mathrm{d}n}{\mathrm{d}t} = -j(a)4\pi a^2 = 4\pi Dac_0. \tag{13.71}$$

This simple result, namely, $\mathrm{d}n/\mathrm{d}t = 4\pi Dac_0$, is one of the most useful interpretative tools for examining the rates of reactions ranging from receptor-mediated signaling to polymerization at the tip of a growing cytoskeletal filament. In particular, it is the basis of our ability to distinguish diffusion-limited and reaction-rate-limited reactions.

A Distribution of Receptors Is Almost as Good as a Perfectly Absorbing Sphere

In essence, the previous calculation assumed that no matter how fast diffusion delivers fresh ligands to the surface-bound receptors, they are prepared to take up those ligands. What happens in the case where the rate of uptake at the receptors is not fast enough to keep up with diffusion? To explore this, we assume that the cell has a finite rate of signaling molecule absorption, k_{on}. Note that we will recover our previous result in the limit that k_{on} is sufficiently large. The number of molecules adsorbed per unit time will be given by the rate equation

$$\frac{\mathrm{d}n}{\mathrm{d}t} = Mk_{\mathrm{on}}c(a), \tag{13.72}$$

where M is the number of surface-bound receptors. The essence of the argument we pursue is that in steady state a concentration profile will be established that guarantees that the diffusive flux is just large enough to feed the receptor absorption process. Note that, using mass conservation, the flux across an imaginary sphere at any radius is given by $-j(r)4\pi r^2$. The minus sign is present because the current density is defined to be positive when it points outward. If we use Fick's law together with Equation 13.72, we find

$$-j(r)4\pi r^2 = D\frac{\partial c}{\partial r}4\pi r^2 = Mk_{\mathrm{on}}c(a), \tag{13.73}$$

which can be written in integrated form as

$$\int_{c(a)}^{c(r)} \mathrm{d}c = \int_a^r \frac{Mk_{\mathrm{on}}c(a)}{4\pi Dr^2}\,\mathrm{d}r. \tag{13.74}$$

After integrating both sides, we obtain

$$c(r) - c(a) = \frac{Mk_{\mathrm{on}}c(a)}{4\pi Da} - \frac{Mk_{\mathrm{on}}c(a)}{4\pi Dr}. \tag{13.75}$$

Finally, if we use the condition $c(\infty) = c_0$, the concentration at the cell's surface can be written as

$$c(a) = \frac{c_0}{1 + (Mk_{\mathrm{on}}/4\pi Da)}. \tag{13.76}$$

There are two interesting limits to this expression and each one implies something different about $c(a)$. The first limit of interest corresponds to

$$\frac{Mk_{\mathrm{on}}}{4\pi Da} \gg 1 \Rightarrow c(a) = 0. \tag{13.77}$$

Note that we have recovered the limit of perfect absorbers considered earlier in this section. This result implies that the receptors will be able to absorb signaling molecules as fast as diffusive processes can deliver them. The other limit corresponds to

$$\frac{Mk_{\mathrm{on}}}{4\pi Da} \ll 1 \Rightarrow c(a) = c_0, \tag{13.78}$$

and implies that the on rate is so slow that the background concentration is not depleted at all.

These results teach an important biological lesson. What we learn is that decorating a cell surface with too many receptors adds nothing further to the ability of that surface to take on board further ligands. To state this issue more precisely, we ask how many receptors we need before we have a situation almost as good as a fully adsorbing surface. Plugging in the result from Equation 13.76 and assuming we have M receptors on the cell's surface, we find

$$\frac{\mathrm{d}n}{\mathrm{d}t} = Mk_{\mathrm{on}}c(a) = M\frac{k_{\mathrm{on}}c_0}{1 + (Mk_{\mathrm{on}}/4\pi Da)}. \tag{13.79}$$

This result can be used to ask how many receptors we need on the surface in order to get an absorption rate that is half that of the diffusive speed limit given by $4\pi Dac_0$, for example. This condition amounts to solving the equation

$$\frac{\mathrm{d}n}{\mathrm{d}t} = \frac{4\pi Dac_0}{2} = \frac{(Mk_{\mathrm{on}}/4\pi Da)4\pi Dac_0}{1 + (Mk_{\mathrm{on}}/4\pi Da)}, \tag{13.80}$$

for the parameter M. More precisely, this amounts to solving the equation

$$\frac{1}{2} = \frac{\beta}{1+\beta}, \tag{13.81}$$

where $\beta = Mk_{on}/4\pi Da$. The solution to the equation is $\beta = 1$, which implies

$$M = \frac{4\pi Da}{k_{on}}. \tag{13.82}$$

To see what this solution really means, we now resort to particular numerical examples. We consider a "typical" eukaryotic cell, which we idealize as a sphere of radius $a \approx 10\,\mu m$. Further, we assume a diffusion coefficient for the ligand species of $100\,\mu m^2/s$ and a $k_{on} \approx 10\,\mu M^{-1}\,s^{-1}$, which is the typical on rate for actin or ParM (a bacterial analog of actin) polymerization. Before we proceed, we have to turn the concentration in k_{on} into useful units, namely,

$$1\,\mu M = 6 \times 10^{23} \times 10^{-6}\,L^{-1} = 600/\mu m^3. \tag{13.83}$$

Using Equation 13.82, we can now compute $M \approx 10^5$ receptors.

It is interesting to consider whether or not 10^5 is a large number of receptors to have on the surface of a cell of radius $10\,\mu m$. One measure of how crowded this distribution of receptors is can be garnered by computing the mean spacing d and the fraction of the surface that will be covered by these receptors. The surface area of the sphere (that is, the cell) is $A_{sphere} = 4\pi(10\,\mu m)^2 \approx 12 \times 100\,\mu m^2$. The mean spacing can then be estimated as

$$d^2 = \frac{12 \times 100\,\mu m^2}{10^5} = 12 \times 10^3\,nm^2 \approx 100\,nm \times 100\,nm, \tag{13.84}$$

which says that the average spacing between receptors will be 100 nm. We can also ask what fraction of the area of the membrane is actually taken up by these receptors. If we consider a receptor with an area $b = 10\,nm^2 = 10^{-5}\,\mu m^2$, the fraction of the membrane area taken up by receptors is

$$\text{covered membrane fraction} = \frac{10^5 \times 10^{-5}\,\mu m^2}{1200\,\mu m^2} = \frac{1}{1200}. \tag{13.85}$$

Interestingly, these simple estimates demonstrate that even a relatively sparse distribution of membrane-bound receptors can rival a perfectly absorbing sphere. Further, this estimate also reveals that many different species of receptor can decorate the cell surface simultaneously while leaving room for the others and with all receptors operating nearly as perfect absorbers.

Real Receptors Are Not Always Uniformly Distributed

The argument above indicates that the spatial distribution of the receptors on the surface of the bacterium should not matter at all if the only thing that is important to the cell is the probability of receptor occupancy. The actual distribution of chemotaxis receptors on the *E. coli* surface can be visualized experimentally as shown in Figure 13.23. In contrast to the expectation from our simple treatment, the bacterium clusters the receptors together in geometric highly ordered arrays found at the two poles. Similar arrangements of chemotaxis receptor clusters have been found in at least a dozen widely divergent bacterial species. This observation indicates that the bacteria must care about some feature of the spatial distribution of their receptors that we have overlooked in our argument

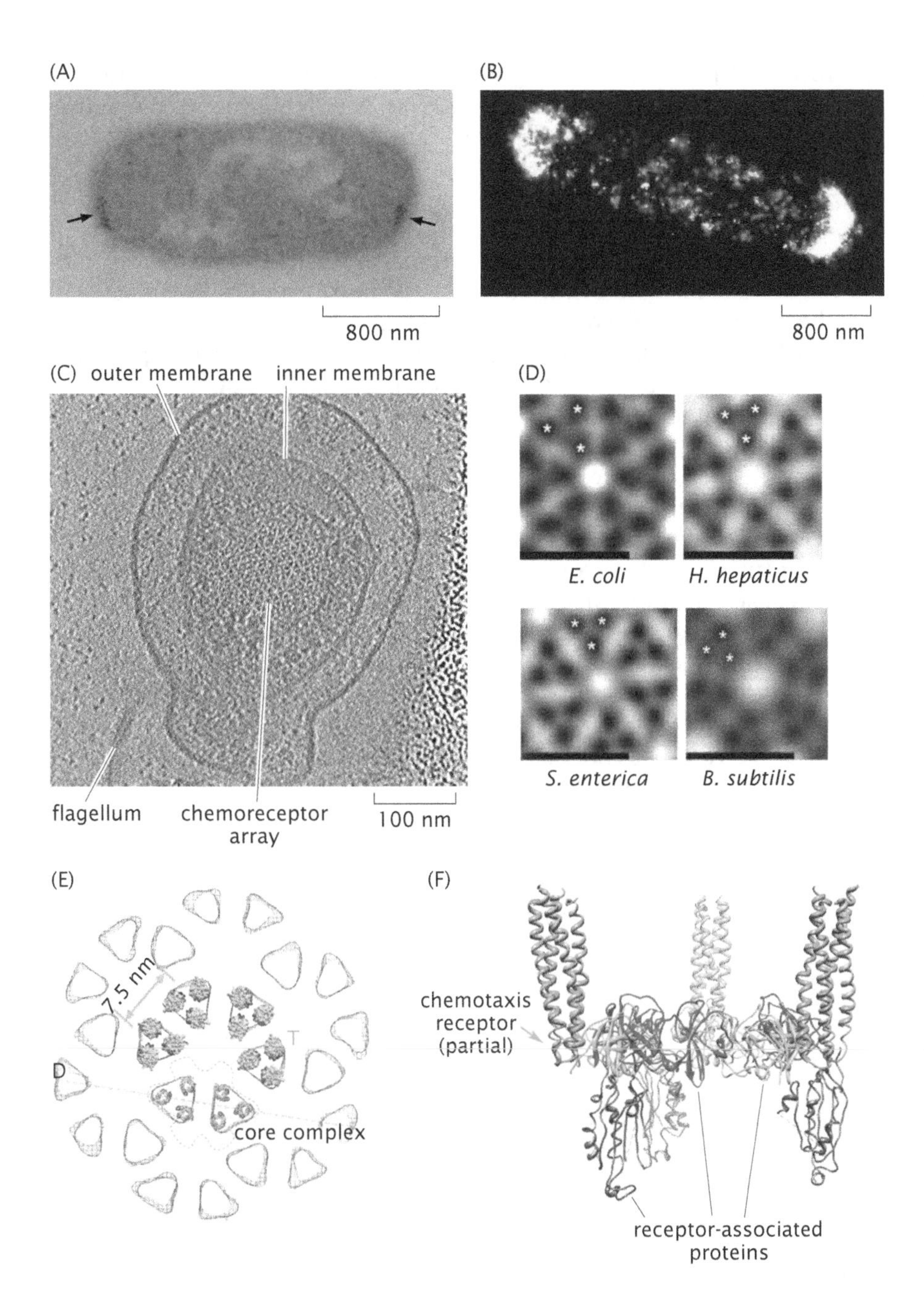

Figure 13.23: Spatial organization of chemotaxis receptor clusters. (A) Localization of chemotaxis receptors in *E. coli* as shown by immunogold labeling in a thin-section electron micrograph. Two clusters are apparent, one at each pole of the cell, indicated by arrows. (B) High-resolution fluorescence image of *E. coli* chemotaxis receptors. Here it is clear that there are many small clusters along the sides of the bacterium as well as the two major ones at the poles. (C) Cryo-electron microscopy showing a tomographic slice through the pole of *Salmonella enterica* (a pathogen closely related to *E. coli*). At this resolution, it becomes apparent that the receptors are packed into regular hexagonal arrays. (D) Averaged tomograms from four different bacteria show similar receptor packing. Each asterisk indicates a single receptor dimer. These group together to form triangular "trimers of dimers" which then pack hexagonally. (E) Using data from cryoeletron microscopy images (gray), and the crystal structures of the receptor dimers (cyan), the positions and orientations of the individual proteins within the larger array can be estimated, as shown here. (F) The regular spacing of the receptor dimers (magenta) is established and maintained by groups of associated proteins (blue and gray) that bind on the cytoplasmic side. This crystal structure shows the arrangement of the receptor-associated proteins within a single trimer of dimers. (A, adapted form J. R. Maddock and L. Shapiro, Science 259:1717, 1993; B, adapted from D. Greenfield et al., PLoS Biol. 7:e1000137, 2009; C, D and F adapted from A. Briegel et al., *Proc. Natl. Acad. Sci. USA* 109:3766, 2012; E adapted from J. Liu et al., *Proc. Natl Acad. Sci. USA* 109:E1481, 2012.)

on the diffusion-to-capture problem. Other experiments suggest that the precise tuning of the bacterial chemotaxis response to combinations of different chemoattractants requires complex communication among different classes of receptors and their downstream signaling molecules. This large-scale signal integration appears to be facilitated by the regular arrangement of all the components. We will revisit this problem in Chapter 19.

13.3.2 A "Universal" Rate for Diffusion-Limited Chemical Reactions

The ideas about diffusion to capture introduced in the previous section have a very interesting application to the analysis of the rates of chemical reactions. We focus on a simple bimolecular reaction of the form $A + B \rightleftharpoons AB$. We imagine that the overall reaction rate is a combination of two distinct factors. First, the overall rate of reaction clearly depends upon how often reactants arrive in each others vicinity as a result of diffusion. However, proximity is not necessarily a guarantee of of reactivity. Once A and B have found each other, it may take many tries for them to join up to form AB. The argument we make here is that these two effects can result in limiting scenarios known as "diffusion-limited reaction" and "rate-limited reaction." For the moment, we examine the case in which once the reactants are nearby, they react. The point of the exercise is to compute the universal speed limit for such diffusion-limited reactions.

One of the schemes we will return to for determining the rates of biological processes is to bound the rate of a given process on the assumption that its rate is dictated entirely by diffusion. For example, we can estimate the time it takes for capsid proteins to form a viral capsid by working out the diffusion time for individual proteins to be captured by the growing capsid. Of course, in reality many of the processes of interest involve several steps, including that (i) the relevant molecular participants find each other (both in terms of spatial position and orientation) and (ii) once they do find each other, they still need to carry out the interaction of interest, which itself has some rate dictated by an energy barrier. The idea advanced here is that the diffusive rate provides a maximum rate speed limit. Examples of interest include the reaction $H^+ + OH^- \rightarrow H_2O$, the binding of oxygen to hemoglobin, the binding of *lac* repressor to DNA, and the assembly of individual monomeric units to form viral capsids. The outcome of the most naive diffusive arguments (and in the case where the two diffusing particles are treated as being of the same size) is that the rate constant for molecular association is given by

$$k_{\text{diffusive}} = \frac{8 k_B T}{3\eta}, \tag{13.86}$$

where η is the viscosity of the medium. This is simply obtained from Equation 13.71 by replacing D with $2D$ and a with $2a$, to account for the mutual diffusion of two particles with equal diffusion constants, $D = k_B T/6\pi\eta a$, and the fact that they need to be within distance $2a$ to interact. If we use the numbers for water at room temperature to determine the viscosity ($\eta \approx 10^{-3}$ Pa s), this results in the value $k_{\text{diffusive}} \approx 7 \times 10^9\ \text{M}^{-1}\,\text{s}^{-1}$, a value that is helpful for estimating rates for point-like particles, but overestimates the rates associated with protein–ligand interactions, a shortcoming that can be amended by acknowledging that a ligand can only interact correctly with its receptor in certain precise orientations. In this case, the rotational diffusion of the ligand and the protein receptor must also be taken into account.

13.4 Summary and Conclusions

Because so many of life's processes occur at the molecular scale, the thermally driven diffusion of molecules is a major influence governing how rapidly and at what location biochemical reactions can occur. In this chapter we have examined several of the interesting dynamic consequences of the diffusive behavior of molecules. Diffusion as a transport mechanism is efficient over short, but not long, distances. As a mechanism for delivering ligands to receptors, the dynamics of diffusion generates a built-in limit such that sparsely distributed receptors on a cell surface are nearly as efficient at receiving signals as a surface completely covered with receptors would be. We have also considered the consequences of allowing other processes to influence diffusive behaviors such as in the case of an external applied force or in the context of diffusion to capture. A remarkable number of cases in biological dynamics are well approximated by one of these simple scenarios. Reconciling the diffusive behavior of individual molecules with the macroscopic evolution of concentration gradients reveals one of the many fascinating ways that apparently directed behaviors at a macroscopic scale can arise from individually random and uncoupled behaviors of molecules. Diffusion is a fact of life at molecular scales; all molecular processes must either exploit diffusion or overcome it.

13.5 Problems

Key to the problem categories: • Model refinements and derivations • Estimates • Data interpretation • Model construction

• 13.1 Biological distances and diffusion times

Generate a series of plots like that shown in Figure 13.4 for all three choices of diffusion constant shown in Table 13.1. Justify these choices of diffusion coefficients by using the Stokes–Einstein relation $D = k_B T/6\pi\eta a$.

• 13.2 Diffusion from a point source

The idea in this problem is to derive the solution to the one-dimensional diffusion equation for a point source, given by Equation 13.32. The tools we invoke in this problem may seem heavy-handed on the first try, but illustrate a bevy of important ideas from mathematical physics.

(a) Take the Fourier transform of the diffusion equation by transforming in the spatial variables to obtain a new differential equation for $\tilde{c}(k, t)$.

(b) Solve the resulting differential equation for $\tilde{c}(k, t)$. Then compute the inverse Fourier transform to arrive at the solution in real space, $c(x, t)$.

(c) Show that the solution for an arbitrary initial concentration distribution $c(x, t = 0)$ can be written as an integral over the solution for a point source. In particular, consider an initial concentration profile of the form $c(x, 0) = c_0$ for $x < 0$ and $c(x, 0) = 0$ for $x > 0$ and find the resulting diffusive profile.

(d) Formally derive the relation $\langle x^2 \rangle = 2Dt$.

• 13.3 Diffusive flux distribution.

In the chapter, we argued that diffusion is the result of nothing more than molecules flipping coins. In one dimension, this leads to a simple and beautiful "flux distribution" function. In this problem, consider two adjacent planes, one of which has N_1 molecules and the other of which has N_2 molecules, and derive the flux distribution function.

• 13.4 FRAP of one-dimensional *E. coli* revisited

Figure 13.16(A) shows different snapshots of an *E. coli* cell after it has been subjected to photobleaching. Use the solution for the FRAP problem of a one-dimensional bacterium (that is, Equation 13.47) to produce a plot of the difference between the initial concentration (that is, before photobleaching) and the concentration at time t as shown in Figure 13.16(B). Make a series of plots for different time points using a diffusion constant for GFP in *E. coli* of $D = 7\,\mu\text{m}^2/\text{s}$. The ambitious reader is encouraged to use a more realistic treatment of the $t = 0$ concentration profile than the highly simplified uniform hole worked out in the chapter. Relevant data for this problem is provided on the book's website.

• 13.5 Two-dimensional FRAP analysis

The goal of this problem is to generalize the one-dimensional treatment of FRAP given in the chapter. Consider a cell as a planar circle of radius R uniformly covered with freely diffusing fluorescent proteins. Imagine that the laser photobleaches a hole of radius a in the middle of the cell. Note that we ignore the presence of the nucleus. Work out the concentration of fluorescent proteins in the cell as a function of position and time in analogy with the one-dimensional treatment of the problem done in the chapter. Compute the number of molecules in the hole after photobleaching as a function of time.

• 13.6 Rotational diffusion

(a) Consider a sphere of radius R in water. Due to random collisions with the water molecules, the sphere will rotationally diffuse. The diffusion law in this case is analogous to the one obtained for translational motion,

$$\left\langle \Delta\theta^2 \right\rangle = 2D_r t. \tag{13.87}$$

What are the units of the rotational diffusion coefficient D_r? Write down the formula for D_r using the Einstein relation and the rotational friction coefficient obtained in Problem 12.5, and convince yourself that the units are correct.

(b) Estimate how long it takes for an *E. coli* to diffuse over an angle equal to 1 radian. What is the distance traveled by the bacterium during that time?

• 13.7 Diffusion to capture and the diffusive speed limit

In the chapter, we solved the problem of diffusion to capture using physical arguments to bypass explicitly solving the diffusion equation. In this problem, we do the math.

(a) Write the diffusion equation for the perfect-absorber case in spherical coordinates. Use the method of separation of variables and reproduce the solution given in the chapter.

(b) Use the flux to compute the number of molecules absorbed per unit time and find the corresponding k_{on} implied by this solution. Plug in reasonable numbers to compute the diffusive speed limit for the case of oxygen binding to hemoglobin.

• 13.8 Chemoreceptor clustering

There is strong evidence that chemoreceptors in *E. coli* tend to cluster near one pole (see Kentner and Sourjik (2006) and Figure 13.23). One hypothesis about the role of such clustering is that it might increase the ability of a bacterium to better detect molecules in its environment. Determine if this is the most efficient strategy for counting (absorbing) molecules of chemoattractant. Approximate *E. coli* as a sphere $a = 1\ \mu\text{m}$ in radius and neglect its motion. Then compare the diffusive current to $N = 1000$ receptors (absorbing patches of radius $s = 1$ nm) scattered over the surface of the cell with the diffusive current to the same receptors incorporated into a single patch with the same total area. Make use of the result that the diffusive current onto a sphere of radius a with N absorbing patches of radius s spread uniformly over its surface is

$$I = \frac{4\pi D a c_\infty}{1 + \pi a / Ns}, \tag{13.88}$$

where D is the diffusion constant of the molecules, while c_∞ is their concentration far from the cell. (Adapted from a problem courtesy of H. C. Berg.)

13.6 Further Reading

Bray, D (2001) Cell Movements: From Molecules to Motility, 2nd ed., Garland Science. Bray's book is a beautiful description of a host of dynamical processes associated with cells.

Berg, H (1993) Random Walks in Biology, Princeton University Press. Berg's book is a classic, but not in the sense of Mark Twain who quipped that a classic is a book that is talked about by all and read by none. Berg's book is widely read and deservedly so.

Berg, H (2003) *E. coli* in Motion, Springer-Verlag. Another Berg classic!

Benedek, GB, & Villars, FMH (2000) Physics with Illustrative Examples from Medicine and Biology: Statistical Physics, Springer-Verlag. Benedek and Villars have made their way to our Further Reading list in many chapters because they have interesting things to say on many topics.

Perrin, J (1990) Atoms, Ox Bow Press (originally published in 1913). This book is full of interesting insights to reward the curious reader.

Einstein, A (1956) Investigations on the Theory of the Brownian Movement, Dover Publications (originally published in 1926; the papers in this collection were published in 1905–1906). Einstein's treatment of diffusion still serves as a fine introduction.

Chandrasekhar, S (1943) Stochastic problems in physics and astronomy, *Rev. Mod. Phys.* **15**, 1. Chandrasekhar's amazing article is a compendium of elegant and useful results pertaining to random walks and more general ideas on stochastic processes.

Axelrod, D, Koppel, DE, Schlessinger, J, Elson, E, & Webb, W (1976) Mobility measurement by analysis of fluorescence photobleaching recovery kinetics, *Biophys. J.* **16**, 1055. This paper describes the theoretical underpinnings of the use of photobleaching as a tool to study dynamics within cells.

Lippincott-Schwartz, J, Snapp, E, & Kenworthy, A (2001) Studying protein dynamics in living cells, *Nat. Rev. Mol. Cell Biol.* **2**, 444. An excellent account of the use of techniques such as FRAP for studying dynamics within cells.

Verkman, AS (2002) Solute and macromolecule diffusion in cellular aqueous compartments, *Trends Biochem. Sci.* **27**, 27. This article describes the use of photobleaching as a tool to study dynamics in living cells.

Berg, HC, & Purcell, EM (1977) Physics of chemoreception, *Biophys. J.* **20**, 193. One of the classic papers in physical biology, illustrating how ideas about diffusion to capture can be used to think about cell signaling.

Kentner, D, & Sourjik, V (2006) Spatial organization of the bacterial chemotaxis system, *Curr. Opin. Microbiol.* **9**, 619. This paper demonstrates that the idea of uniformly distributed receptors is not appropriate for the chemotaxis system and that the receptors are localized to the poles.

13.7 References

Briegel, A, Ortega, DR, Tocheva, EI, et al. (2009) Universal architecture of bacterial chemoreceptor arrays, *Proc. Natl Acad. Sci. USA* **106**, 17181.

Ellenberg, J, Siggia, ED, Moreira, JE et al. (1997) Nuclear membrane dynamics and reassembly in living cells: targeting of an inner nuclear membrane protein in interphase and mitosis, *J. Cell Biol.* **138**, 1193.

Elowitz, MB, Surrettee, MG, Wolf, P-E, Stock, JB, & Leibler, S (1999) Protein mobility in the cytoplasm of *Escherichia coli*, *J. Bacteriol.* **181**, 197.

Greenfield, D, McEvoy, AL, Crooks, GE, et al. (2009) Self-organization of the *Escherichia coli* chemotaxis network imaged with super-resolution light microscopy, *PLoS Biol.* **7**, e1000137.

Liu, J, Hua, B, Moradoa, DR, et al. (2012) Molecular architecture of chemoreceptor arrays revealed by cryoelectron tomography of Escherichia coli minicells, *Proc. Natl. Acad. Sci. USA* **109**, E1481.

Maddock, JR, & Shapiro, K (1993) Polar location of the chemoreceptor complex in the *Escherichia coli* cell, *Science* **259**, 1717.

Marshall, WF, Straight, A, Marko, JF, et al. (1997) Interphase chromosomes undergo constrained diffusional motion in living cells, *Curr. Biol.* **7**, 930.

Mullineaux, CW, Nenninger, A, Ray, N, & Robinson, C (2006) Diffusion of green fluorescent protein in three cell environments in *Escherichia coli*, *J. Bacteriol.* **188**, 3442.

Life in Crowded and Disordered Environments 14

Overview: In which we reexamine the picture of dilute solutions and account for cellular crowding

> "Nobody goes there anymore. It's too crowded."
>
> **Yogi Berra**

The cellular interior is so crowded that the distance between neighboring proteins is comparable to protein size itself. Similarly, the cell membrane is richly inhabited with large numbers of proteins of different types in a lipid background that is highly varied. This means that the material environment of cells is far different from the conditions found in most biochemical experiments and featured in many of the models described so far. In particular, we have ignored the effects of this crowding by exploiting simple ideas about binding that make use of the "dilute-solution" limit. Similarly, our treatment of diffusive kinetics has been built around a random-walk picture in which molecules are free to wander unencumbered by interactions with neighboring walkers. This chapter examines how crowding alters both equilibria and kinetics and examines some simple toy models of these effects.

14.1 Crowding, Linkage, and Entanglement

Much of the substance of our quantitative story of cells has thus far centered on different ways of viewing binding and diffusion problems. Binding and diffusion are central to the biochemical reactions that are the engine of cellular life. However, many of the assumptions that are hidden behind our use of these concepts seem to be at odds with the way cells really work. Two of the key assumptions that have been behind the scenes in much of what we have done are (i) the assumption of *ideality* in which it is supposed that the molecules of interest are sufficiently dilute that they do not interact either through direct contact or even through longer range potentials and (ii) the assumption of *homogeneity*, which posits that the environment at one place is just like it is somewhere else. The present chapter examines how to relax these two assumptions and to ask what new insights into biological structure and function then emerge.

Cellular life is colored by various broad classes of phenomena that are at odds with the assumptions of ideality and homogeneity. One of the simplest facts of cellular life that contradicts the ideality assumptions favored so far in the book is the intense crowding of the cellular interior. A mean spacing of less than 10 nm between proteins is a conservative estimate of the extent of crowding, a fact that we will show has implications both for equilibria and for kinetics. This mean spacing should be contrasted with the roughly 100 nm spacing between molecules that would be found in the "high" concentration of an *in vitro* experiment taking place at millimolar concentrations.

Examination of cells using either light microscopy or an electron microscope reveals a wide variety of organized structures that demonstrate that the homogeneity assumptions we have invoked so far are often inappropriate. What is revealed is a heterogeneous environment characterized by *linked* polymer networks such as the actin network that crisscrosses the leading edge of motile cells or the peptidoglycan network that confers bacterial shape. In addition, the presence of organelles such as are found in eukaryotes reveals even more subtle and intriguing features such as different lipid compositions for adjacent organelles, for example, the endoplasmic reticulum and the Golgi apparatus.

This chapter probably raises more questions than it answers, but it is built around two broad thrusts, namely, (i) how we should amend our treatment of binding and interaction to account for crowding effects on biochemical equilibria, and (ii) how the diffusive dynamics seen in cells are different than that found in dilute solutions.

14.1.1 The Cell Is Crowded

The ability to conduct a quantitative census of cells experimentally as was described in Chapter 2 has transformed our understanding of living matter. Experiments like those shown in Figures 2.5 (p. 43) and 2.20 (p. 58) make it possible to count up the number of copies of many of the different proteins that make cells function. In addition, such experiments have served as the inspiration for a new generation of iconic illustrations, such as Figure 2.4 (p. 42), the aim of which is "to help imagine biological molecules in their proper context: packed into living cells" (Goodsell 1998). In biochemical reactions in the laboratory, much care is taken to insure that features such as the concentration and the charge state (as reflected in the pH and the ionic strength) mimic those of living cells. On the other hand, many of these experiments ignore the need to "correct for extract dilution with molecular crowding," a dictum elevated to the status of a biochemical commandment by Arthur Kornberg (Kornberg 2000).

To get a sense of the crowded nature of the cellular interior, we return to our estimates of Section 2.1.2 (p. 38), where we took stock of the census of a bacterium. Hidden within those calculations is a simple way to view the extent of crowding within the cellular interior. The simplest estimate is to imagine a bacterium as a cube with edge length of 1000 nm (that is, 1 μm). As shown in Figure 2.4 (p. 42), we estimated that there are roughly several million proteins within an *E. coli* cell. For simplicity, imagine that there are 10^6 such proteins in the cytoplasm of our bacterium. This means that in our hypothetical cubic bacterium, we can line them up 100 on a side in all three directions. This means that the mean center-to-center spacing is 10 nm. To simplify our estimate further, now imagine that our proteins are all spheres with a radius of 2 nm. What this means is that the "solution" between proteins is slightly more than 5 nm thick. The picture this estimate paints of the cellular interior is one in which the diluteness and homogeneity of ideal-solution theory is drastically overthrown. However, crowding is more diverse and subtle than the mere fact that there are lots of proteins squashed together. One way in which this crowding is more subtle is in the form of the many networks of filamentous molecules that crisscross cells, a topic we explore presently.

14.1.2 Macromolecular Networks: The Cytoskeleton and Beyond

As will be shown below, crowding of spheres is one instructive way to understand excluded-volume effects and the depletion forces they engender. However, these ideas fall short as we try to more closely approximate structures and conditions within living organisms. This is because many of the components responsible for structural organization and mechanical properties of living cells and organisms are better approximated by elastic beams rather than spheres. In Chapter 10, we considered the properties of individual filaments, but now in our attempt to effect a more realistic treatment of living conditions we must consider what happens when large numbers of filaments are densely crowded together. This happens not only within cells, but also outside cells in the extracellular matrix. Indeed, the implausibly tall size of animals and green plants can only be maintained against gravity because of the remarkable properties of filamentous structures of the extracellular matrix.

Aligned filament structures are commonly seen in organisms. Some examples are shown in Figure 14.1. We have already seen how parallel bundles of actin filaments and microtubules are used in building cell surface projections that resist buckling (see Chapter 10), such as the microvilli seen in Figure 14.1(A). Bundles of filaments are also found within cells contributing to intracellular traffic. A dramatic example of this is the aligned bundles of microtubules and intermediate filaments in the axon of a neuron as shown in Figure 14.1(B). Aligned filaments of extracellular matrix shown in Figure 14.1(C) give our tendons and ligaments their remarkable resistance against shearing. The alignment of collagen in tendons can be easily seen even at the level of light microscopy. Similar alignment of cellulose fibers in the plant extracellular matrix allows, as shown in Figure 14.1(D), trees to grow against gravity to be over a hundred meters tall.

Although formation of aligned filament bundles can greatly enhance the strength and mechanical properties beyond that of individual filaments, a shortcoming of this organizational motif is that it is extremely anisotropic. In many cases, cells must be able to resist mechanical insults coming from all directions, not just those conveniently aligned with the bundle axis. To build mechanically strong three-dimensional structures, cells often crosslink their filaments into networks where the angle of intersection between neighboring filaments may be quite large. Again, network organizations are found throughout nature, made of many types of filaments. In Figure 14.1(E), we see the branched network of actin filaments at the leading edge of a crawling cell. Extracellular connective tissue such as the basement

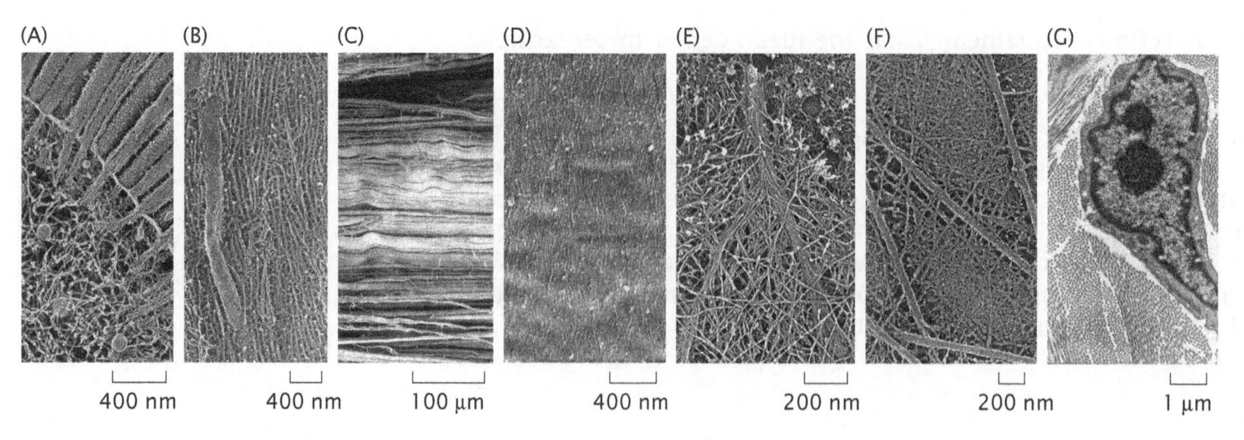

Figure 14.1: Gallery of filamentous network architectures found in cells and extracellular matrix. (A) The upper edge of an intestinal epithelial cell reveals tightly packed bundles of actin in the microvilli and a less organized but still dense actin meshwork just below the membrane. (B) Within the axon of a nerve cell, aligned and tightly packed bundles of neurofilaments and microtubules allow the axon to grow very long and serve as tracks for transporting organelles such as mitochondria. (C) In animal ligaments such as this one from a rat, aligned collagen bundles resist shear stresses. (D) The cell wall of plants is made of aligned bundles of cellulose for organisms ranging from this tiny weed *Arabidopsis* to the mighty Redwood. (E) At the leading edge of a moving cell, actin filaments in webs and bundles grow to push the membrane forward. (F) Animal tissues made of epithelial cells usually include a basement membrane lining underneath the cells to improve the mechanical strength of the tissue. The basement membrane is not a phospholipid membrane but rather a felt-like mat of collagen, proteoglycans, and other filamentous molecules. (G) Cells such as the fibroblast shown here navigate through densely packed extracellular matrix. In this thin section, collagen fibrils have been sliced in several different orientations, revealing their regular packing. (A, F, courtesy of J. Heuser; B, adapted from N. Hirokawa, *J. Cell Biol.* 94:129, 1982; C, adapted from P. P. Provenzano et al., *J. Orthop. Res.* 20:975, 2002; D, adapted from D. H. Burk and Z. H. Ye, *Plant Cell* 14:2145, 2002; E, adapted from T. M. Svitkina et al., *J. Cell Biol.* 160:409, 2003; G, adapted from P. C. Cross and K. L. Mercer, Cell and Tissue Ultrastructure. W. H. Freeman, 1993.)

membrane shown in Figure 14.1(F) and the collagen-rich tissue shown in Figure 14.1(G) can also be arranged to resist shear and stretching forces at multiple orientations. The rigid cell walls of bacterial cells also follow this structural theme where long fibrils made of chains of sugars are crosslinked by small peptides, creating a three-dimensional structure that resists external stresses.

Cells not only allow filament bundling and superstructural organization to happen but encourage it with crosslinking proteins that tie filaments together. Spontaneous alignment alone will give a mixed orientation of polar filaments, but in many of the examples seen in cells the filaments are all pointing in the same direction. This cannot arise from entropy alone, but rather emerges from the action of local nucleating sites that make sure everything grows in the same direction or from the exploitation of motor proteins to sort out filaments based on their orientations. This will be further discussed in Chapter 16. A higher-resolution image of filamentous organization at the leading edge of a motile cell is shown in Figure 14.2. One of the intriguing features of this organization is that it varies as a function of distance from the leading edge itself. In Chapter 15, we will explore how the time-dependent growth of actin filaments can lead to this organization.

14.1.3 Crowding on Membranes

In addition to the important role of crowding in the three-dimensional setting of the cellular interior, crowding is also a fact of life in the cell membrane. The existence of such crowding in cell membranes was depicted in Figure 11.10 (p. 438), which illustrates how membrane

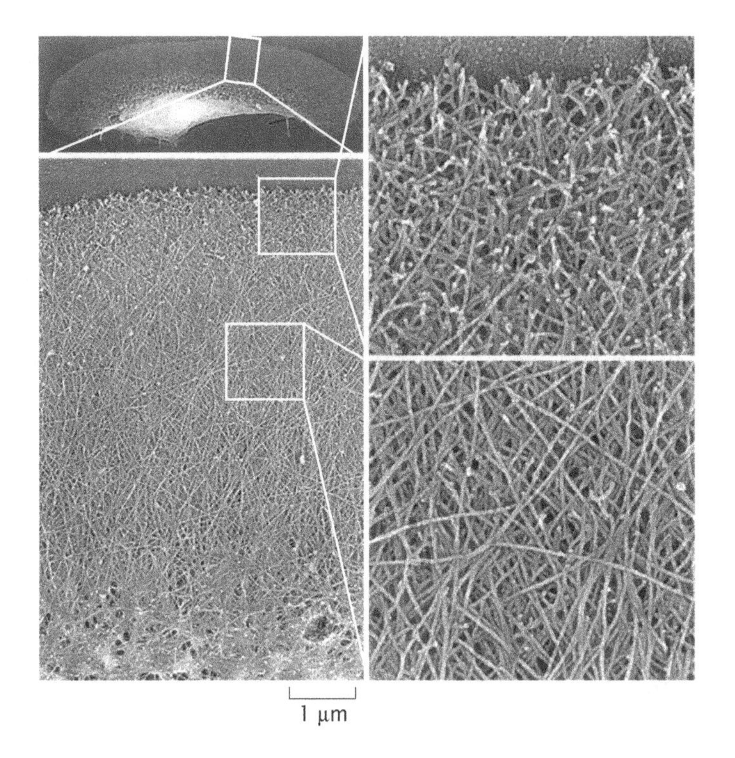

Figure 14.2: Organization of actin filaments at the leading edge of a crawling cell. This fish skin keratocyte was moving upward at the time that it was fixed. Its membrane was stripped off and the filamentous structures imaged after being coated with a very thin layer of platinum. The large white blob towards the bottom of the cell shows the location of the nucleus and the membranous organelles. The area within the white box is blown up below. Essentially all of the filaments are actin. Two regions are shown at still higher magnification on the right. At the top, near the front edge of the cell, the actin filaments are short and frequently branched. Further back, individual actin filaments are longer and overlaid at nearly random angles. (Adapted from T. M. Svitkina et al., *J. Cell Biol.* 139:397, 1997.)

proteins in mitochondria can be manipulated by the application of an electric field. The result of the application of a field is to segregate the membrane proteins to one side of the membrane, as shown in the electron micrograph of freeze-fractured membranes. This image and others demonstrate the large fraction of membrane devoted to membrane proteins. Indeed, in the case of the mitochondria, more than 50% of the membrane mass is donated by proteins. Interesting measurements of the relative mass of phospholipids and membrane proteins are reported in Mitra et al. (2004).

Just as simple estimates on protein concentrations in the cytoplasm reveal a mean spacing comparable to protein size, similar estimates can be carried out for the cell membrane as well. To see this, we recall that the membrane area for a bacterium like *E. coli* is roughly $6\,\mu m^2$ and that both the inner and outer membranes are home to roughly 500,000 proteins. This tells us that the area per protein is roughly $10\,nm^2$, or that the typical distance between two proteins is of the order of 3 nm. The cell membrane is tightly packed indeed, with a crude estimate being that roughly half of the membrane area is occupied by proteins.

14.1.4 Consequences of Crowding

So far, we have examined the structural features of the crowded environment of cells. We have seen that this crowded structure that serves as the backdrop for the bustling biochemical metropolis of the cell is quite different from the dilute and homogeneous environments of solution biochemistry. What are the consequences of this crowding for cells? We explore two broad classes of consequence arising from this crowding, namely, how equilibrium binding is modified by crowding and how diffusive processes are altered as a result of the tight packing of molecules within cells.

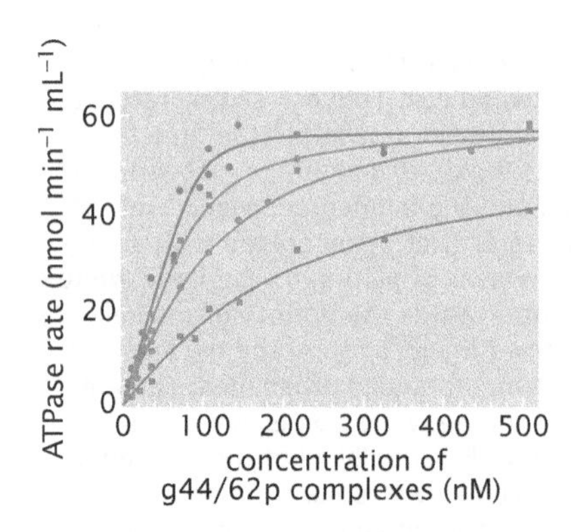

Figure 14.3: ATPase rate associated with T4 DNA replication. The different curves measure the ATPase rate as a function of the g44/62p (clamp loader) protein concentration as measured using different concentrations of polyethylene glycol as a crowding agent. The concentrations of polyethylene glycol going from the bottom to the top curve are 0, 2.5, 5, and 7.5 weight percent. (Adapted from T. C. Jarvis et al., *J. Biol. Chem.* 265:15160, 1990.)

Crowding Alters Biochemical Equilibria

The first broad category of effects that can be blamed on crowding are the modifications that take place in equilibria. For example, crowding can alter the equilibrium state of a system relative to the dilute limit of solution biochemistry. In Chapter 6, we illustrated how statistical mechanics can be used to compute binding curves that tell us the occupancy of a receptor, for example, as a function of the concentration of ligands. However, such results are modified in various ways by the presence of "passive" crowding agents. What this means concretely is that the binding probability for some substrate as a function of the substrate concentration will be enhanced relative to the case in which there are no crowding agents. A second intriguing outcome of the existence of crowding is the onset of new types of entropic forces that have nothing to do with van der Waals forces or charges. These entropic forces are known alternatively as "excluded-volume forces" or "depletion interactions" and can have the counterintuitive effect of apparently introducing "order."

An example of how binding may be altered by crowding effects is shown in Figure 14.3. In this case, the measurement examines accessory protein complexes associated with DNA replication in a bacteriophage. In particular, these data show the effect of crowding molecules on the likelihood of binding of gene products 45 (sliding clamp) and 44/62 (clamp loader) which then themselves bind to the DNA polymerase and enhance its activity. These four molecules together suffice to produce leading-strand synthesis in DNA replication. The crowding agent used in this case is polyethylene glycol with an average mass of 12 kDa. The rate of ATP hydrolysis by the clamp loader (g44/62p) is used as a readout of the extent of binding of g45p to g44/62p, and this is what is plotted in the figure for the case in which the concentration of g44/62p is titrated at a fixed concentration of g45p. As a control, it was demonstrated that changing the concentration of crowding molecules did not by itself change the catalytic activity of the g44/62p. The data presented in this figure is but one example of a widespread phenomenon in which the addition of inert crowding molecules shifts biochemical equilibria.

Crowding Alters the Kinetics within Cells

Not only does crowding alter equilibrium properties, it also can significantly impact a variety of dynamical processes within cells. A simple starting point is the question of how diffusion is altered in cells relative to in dilute solutions. One way to examine this question is to use FRAP as introduced in Section 13.1.2 (p. 511). The idea is to fluorescently label some macromolecule of interest within the cell and to measure the diffusive dynamics of this protein after some region within the cell has been photobleached. An example of such an experiment is shown in Figure 13.7 (p. 514). The outcome of FRAP experiments is a measure of the slowing down of diffusion relative to its dilute-solution values as shown in Figure 14.4, where the relative diffusion coefficient in the cell interior with respect to water is plotted as a function of the size of the diffusing particle. These curves show a fourfold decrease in the diffusion constant, with an even stronger effect for larger molecules.

A second interesting way in which diffusive dynamics is altered is revealed by examining the dynamics of membrane proteins at the cell surface. Video microscopy in which individual membrane proteins

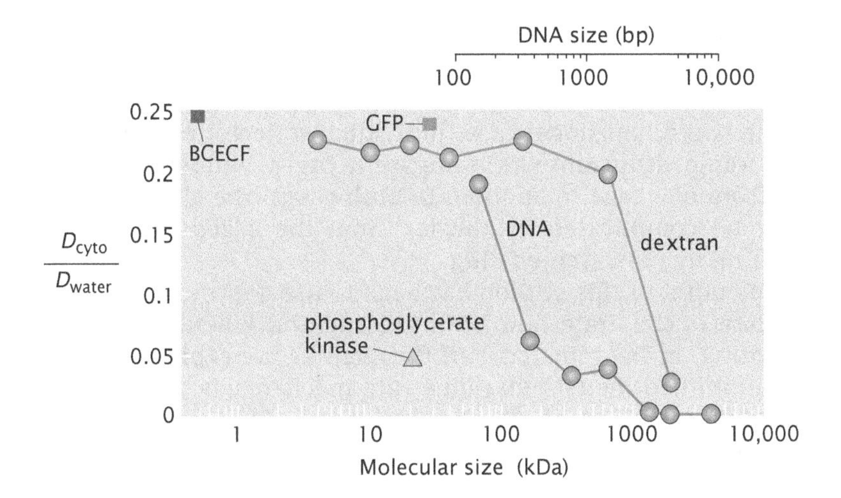

Figure 14.4: Diffusion constants in cells. The plot shows the ratio of the measured cellular diffusion constant to that for the same molecule in water for several different molecules, including a series of DNA molecules of different sizes and BCECF, a small fluorescent molecule. (Adapted from A. S. Verkman, *Trends Biochem. Sci.* 27:27, 2002.)

are followed reveals that the diffusive trajectories suffered by these molecules are quite distinct from the traditional two-dimensional random walk that would be expected of free diffusion. In particular, what is observed are several different categories of phenomena. In one class of motions, illustrated in Figure 14.5, episodes of free diffusion are punctuated by transient association with other membrane proteins. A related class of diffusive motions reveals free diffusion within what appear to be two-dimensional cages, followed by escape to some adjacent cage, followed by more localized diffusion. The same

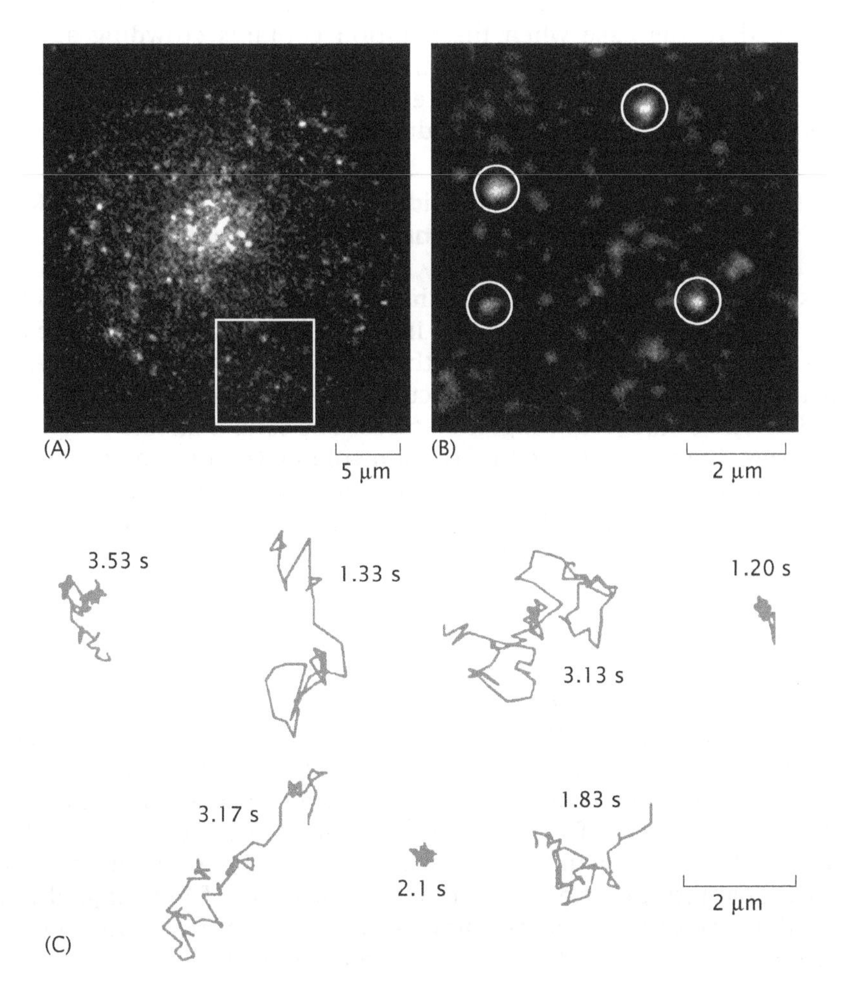

Figure 14.5: Single-molecule measurements for diffusion of membrane-associated proteins. (A) Cells were transfected with a construct encoding GFP fused to the membrane protein Lck. In the fluorescence microscope, the cells appear to be covered with randomly distributed spots. (B) In a magnified view of the region bounded by the box in (A), individual molecules can be clearly seen (circles). Their movements can be tracked over time by video microscopy. (C) A series of tracks measured for individual molecules ranging over total times of about 1–3.5 s show very heterogeneous individual behavior. Some molecules appear to be trapped and nearly stationary, while others travel long distances. (Adapted from A. D. Douglass and R. Vale, *Cell* 121:937, 2005.)

phenomenon has been observed in three-dimensional diffusion both for labeled RNAs inside living cells and for various kinds of tracer particles in three-dimensional filamentous networks constructed *in vitro*. As in the two-dimensional case, individual trajectories are characterized by rapid diffusion within apparent cages, interspersed by rare jumps from one cage to another. In both cases, the observed dynamics may be complicated by the fact that the cages themselves are undergoing dynamic remodeling.

The key point of this section has been to use a few examples of real-world data to illustrate how both binding and kinetics are modified by crowding. In the remainder of the chapter, we explore a variety of highly simplified models that illustrate mechanisms where crowding agents have the effect of altering equilibrium binding curves, inducing entropic forces between particles that are indifferent to each other in dilute solution, and slowing down diffusion. It is important to bear in mind that the models introduced here have as their ambition to illustrate a minimal treatment of the role of crowding. More faithful models of crowding phenomena require more sophisticated treatments.

14.2 Equilibria in Crowded Environments

14.2.1 Crowding and Binding

In Chapter 6, we considered the statistical mechanics of binding of a ligand to a protein using a lattice model. This approach can be extended to the case when the solution contains crowding agents, molecules that do not interact with the ligand or the protein but simply take up space. We consider the effect of crowding molecules on the probability that the receptor's binding site is occupied by the ligand.

Lattice Models of Solution Provide a Simple Picture of the Role of Crowding in Biochemical Equilibria

To simplify matters, we assume that the reaction volume is divided into Ω cells of volume v as shown in Figure 14.6. Each of the elementary cells in the lattice model of the solution is empty (which really means occupied by a solvent molecule), occupied by a ligand, or occupied by one of the "crowding" molecules. The total numbers of ligands and crowding molecules in the reaction volume Ωv are L and C, respectively, as shown in Figure 14.6. If we consider the situation where in solution there are only ligand and crowding molecules present, then in this simple lattice model of the solution, the partition function is

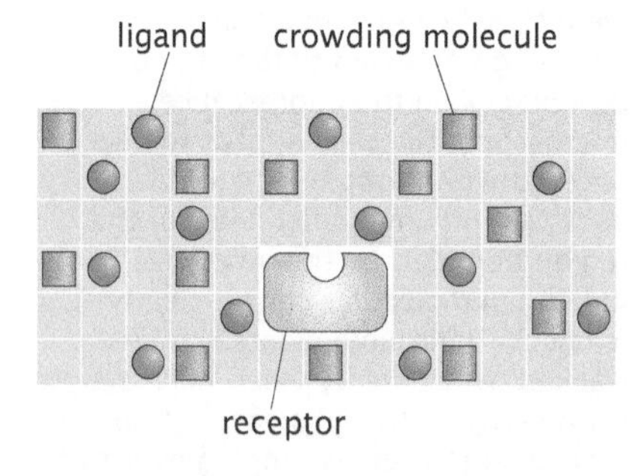

Figure 14.6: Lattice model of crowding and its effect on ligand–protein binding. The reaction volume is occupied by solvent molecules, crowding molecules, and ligands. A single ligand can bind to the protein.

$$Z_{\text{sol}}(L, C) = \frac{\Omega!}{L!C!(\Omega - L - C)!} e^{-\beta L \varepsilon_L^{\text{sol}}} e^{-\beta C \varepsilon_C^{\text{sol}}}, \tag{14.1}$$

where $\varepsilon_L^{\text{sol}}$ and $\varepsilon_C^{\text{sol}}$ are the energies of the ligand and crowding molecules in solution. The combinatorial factor in Equation 14.1 simply counts the number of ways of distributing L ligands and C crowding molecules among the Ω cells that make up the reaction volume.

Given this model of the solution, we now ask about the probability that the receptor in solution will be bound by a ligand and how this probability depends upon the concentration of both ligand and crowding molecules. The receptor can be in one of two states: either it has the ligand bound, or not. The weights of these two states are

$Z_{sol}(L-1, C)e^{-\beta\varepsilon_L^b}$ and $Z_{sol}(L, C)$, respectively, where ε_L^b is the energy of the ligand bound to the protein. The probability that the ligand will be bound to the protein is therefore

$$p_{bound} = \frac{Z_{sol}(L-1, C)e^{-\beta\varepsilon_L^b}}{Z_{sol}(L-1, C)e^{-\beta\varepsilon_L^b} + Z_{sol}(L, C)}. \quad (14.2)$$

If we divide the numerator and denominator of the above equation by $Z_{sol}(L-1, C)e^{-\beta\varepsilon_L^b}$ and make the additional assumption that $\Omega - L - C \gg 1$, the equation takes on the simple form

$$p_{bound} = \frac{1}{1 + \frac{\Omega - L - C}{L} e^{\beta\Delta\varepsilon_L}}, \quad (14.3)$$

where $\Delta\varepsilon_L \equiv \varepsilon_L^b - \varepsilon_L^{sol}$ is the change in the energy of the ligand in going from solution to the receptor.

From Equation 14.3, we conclude that the crowding molecules will have an appreciable effect on the probability that a ligand is bound to the protein only in the limit when C is comparable to Ω (we assume that the number of ligands is small, $L \ll \Omega$). Since the effect of C is effectively to reduce the volume in which the ligands can distribute themselves, we see that an increase in C leads to an increase in p_{bound}. This is illustrated in Figure 14.7, where p_{bound} is plotted for a number of different concentrations of the crowding molecules.

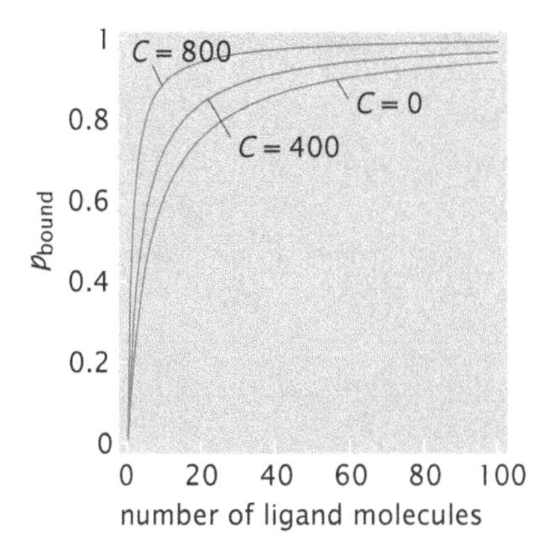

Figure 14.7: Probability of a protein binding site being occupied by a ligand for a number of different concentrations of the crowding molecules. The reaction volume is $\Omega = 1000$ and $\Delta\varepsilon_L = -5\ k_BT$.

A more sophisticated lattice model treatment of crowding and binding allows us to reflect concretely on the data already introduced in Figure 14.3. The theoretical curves shown in that figure are fits to binding curves like those used throughout the book and they yield the dissociation constant as a function of polyethylene glycol concentration. The resulting dissociation constants are plotted in Figure 14.8. Qualitatively, what these data show is that as the concentration of crowding agent is increased, the dissociation constant decreases, meaning that the binding reaction is more favorable. Our interest is in exploiting simple models to give an intuitive sense of the origins of such data.

To model the polyethylene glycol dependence of the dissociation constant, we make use of the lattice model of crowding described above. The new twist to the story is that the depletant molecules (polyethylene glycol) are considerably smaller (12 kDa) than the protein complex (gene product 44/62 with a mass of 164 kDa) whose binding is being affected. To account for the size difference, in the lattice model of the solutions, we assume that the protein takes up r boxes, where a box can accommodate a single polyethylene glycol molecule. To further simplify the combinatorics, we assume that the reaction volume is broken up into Ω large boxes, each consisting of r smaller boxes, as shown in Figure 14.9. The proteins take up the larger boxes, while the polyethylene glycol takes up unoccupied smaller boxes. A more detailed calculation would allow the protein to take up any available square region r boxes in size.

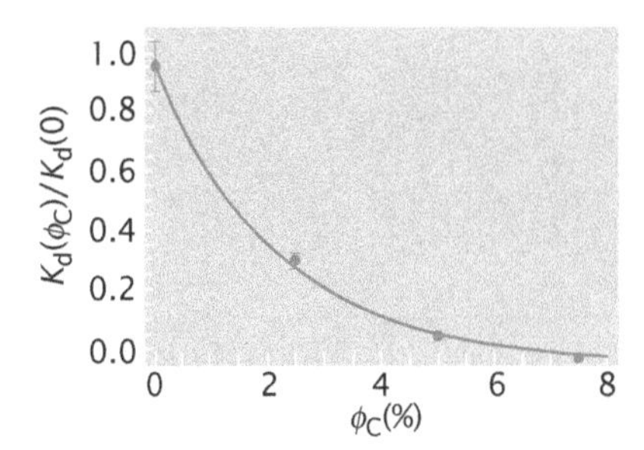

Figure 14.8: Dissociation constant as a function of crowding. Comparison of theory (full line) and experiment (filled circles) for binding in the presence of crowding agents. Measured values of the dissociation constant for g44/62p and g45p components of T4 DNA replication complex as a function of polyethylene glycol concentration measured in percent volume fraction. The theoretical curve was obtained by fitting Equation 14.6 for the effective size ratio r of the protein components to polyethylene glycol 12000 (i.e., with a molecular mass of 12 kDa) molecules. (Adapted from T. C. Jarvis et al., *J. Biol. Chem.* 265:15160, 1990.)

For this situation, the partition function of the solution of L proteins and C crowding molecules is given by

$$Z_{sol}(L, C) = \frac{\Omega!}{(\Omega - L)!L!} \frac{(r\Omega - rL)!}{(r\Omega - rL - C)!C!} e^{-\beta L\varepsilon_L^{sol}} e^{-\beta C\varepsilon_C^{sol}} \quad (14.4)$$

which reduces to Equation 14.1 when $r = 1$. The probability of the protein being bound to its receptor can now be computed just as we

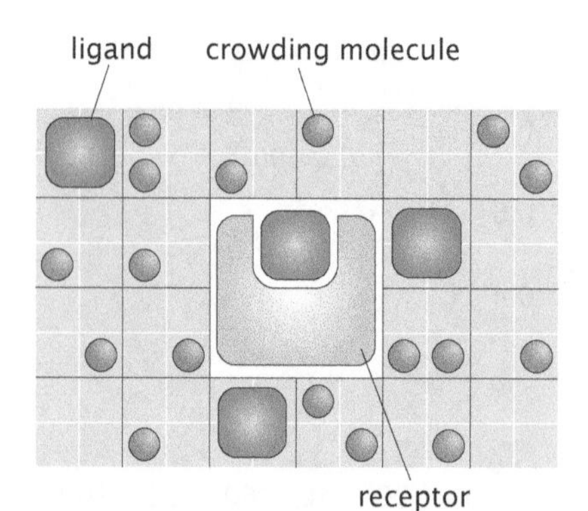

Figure 14.9: Lattice model for large ligands. This lattice model describes binding in the presence of crowding agents where the size of the crowder is different from that of the ligands. This is represented by using different-size boxes for the crowder and the ligand.

did for the case when the crowding molecules and the ligands were assumed to be of the same size. In particular, we construct a ratio in which the numerator is the statistical weight of the bound states and the denominator sums over the statistical weights of all states, just as we did in Equation 14.3. This results in a probability of binding of the form

$$p_{\text{bound}} = \frac{1}{1 + \frac{Z_{\text{sol}}(L, C)}{Z_{\text{sol}}(L-1, C)} e^{\beta \Delta \varepsilon_L}} = \frac{1}{1 + \frac{\Omega}{L}(1 - \phi_C)^r e^{\beta \Delta \varepsilon_L}} \tag{14.5}$$

where $\phi_C = C/r\Omega$ is the volume fraction of the crowding molecules in solution. To obtain the last equality, we assumed that $L \ll \Omega$, that is, the volume fraction of the binding proteins is much less than 1, and we repeatedly made use of the formula $(N+r)!/N! \approx N^r$, which holds for $N \gg r$.

By comparing Equation 14.5 with Equation 6.111 (p. 270), we can read off the volume-fraction-dependent dissociation constant, and we find that

$$\frac{K_d(\phi_C)}{K_d(\phi_C = 0)} = (1 - \phi_C)^r. \tag{14.6}$$

In Figure 14.8, we compare this formula with the data for the g44/62p protein complex from the T4 bacteriophage. The theoretical curve in the figure results from a best fit to the data using the parameter r, which is a measure of the relative "size" of the ligands and the crowder. From *a priori* knowledge of the masses of both the ligands and the crowders, a first guess is that $r \approx 15$, while a best fit yields the value $r = 45$.

The treatment given here only scratches the surface of the way in which crowding agents alter biochemical equilibria. The model introduced here is a caricature of crowding effects and neglects a variety of important features, one of which is indicated schematically in Figure 14.10. More generally, as written, the models presented in this section ignore the real "excluded-volume" effect in which particles effectively take up more space than their physical volume, and the amount of space they occupy depends upon their overall concentration! The implications of this idea for particle interactions will be taken up in Section 14.2.3.

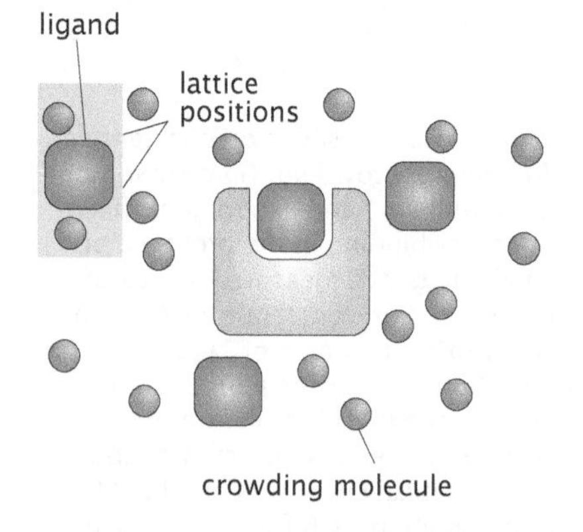

Figure 14.10: Limitations of the lattice model of crowding. This figure shows an allowed configuration for a ligand that is artificially *forbidden* in the lattice model.

14.2.2 Osmotic Pressures in Crowded Solutions

Osmotic Pressure Reveals Crowding Effects

One interesting outcome of the presence of proteins and ions in the cellular interior is that they induce an osmotic pressure. Experiments on concentrated solutions of proteins such as hemoglobin can shed light on this phenomenon. An example of such data that reveals the nonlinear dependence of the osmotic pressure on the protein concentration is shown in Figure 14.11. We can think about these data using the lattice model of a crowded solution introduced above. At low concentrations, the pressure depends linearly on the amount of hemoglobin present as described by the van't Hoff formula discussed in Section 6.2.3 (p. 264). As the hemoglobin solution becomes concentrated, the interactions between the molecules become important and lead to deviations from the simple law.

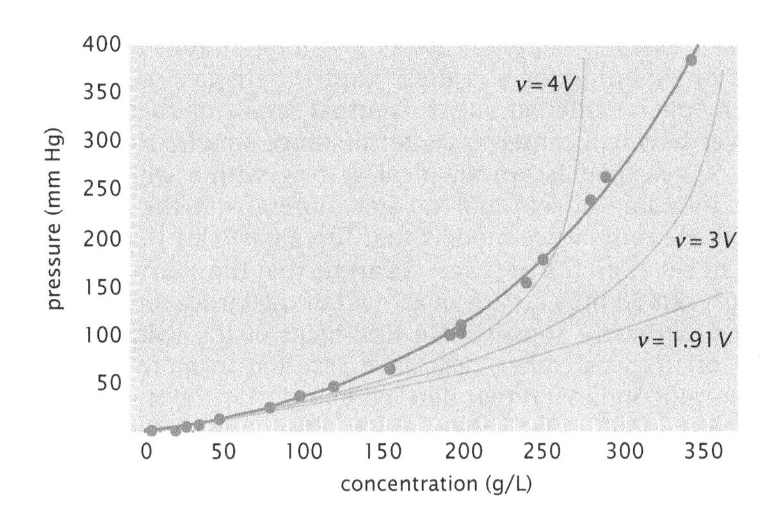

Figure 14.11: Osmotic pressure of a concentrated solution of hemoglobin at 0 °C. The filled circles are the experimental data points. The light red lines are predictions of the lattice gas, while the full red line is the pressure of a gas of hard spheres as described by Equation 14.11, with each sphere having a volume V corresponding to a diameter of 5.8 nm. The labels on the lines indicate the volume of a single box in the lattice model given in Equation 14.10. (Data taken from P. D. Ross and A. P. Minton, *J. Mol. Biol.* 112:437, 1977.)

For the case of a concentrated solution of hemoglobin, we employ a lattice model that consists of two species of molecules: filled boxes correspond to hemoglobin molecules and empty boxes represent the solvent. The partition function for this model is

$$Z_{\mathrm{sol}}(H, \Omega) = \frac{\Omega!}{H!(\Omega - H)!} \mathrm{e}^{-\beta H \varepsilon_{\mathrm{H}}^{\mathrm{sol}}}, \tag{14.7}$$

This is nothing but Equation 14.1, where now H, the number of hemoglobin molecules, takes the place of the crowding agent C, and $L = 0$.

The osmotic pressure p in the lattice model can be computed by considering the change in free energy, $G = -k_{\mathrm{B}}T \ln Z_{\mathrm{sol}}$, of the system when its total volume decreases by a single cell of volume v. Setting the work equal to the change in free energy,

$$pv = G(\Omega - 1) - G(\Omega) = -k_{\mathrm{B}}T \ln \frac{Z_{\mathrm{sol}}(H, \Omega - 1)}{Z_{\mathrm{sol}}(H, \Omega)}, \tag{14.8}$$

leads to an expression for the pressure

$$p = \frac{k_{\mathrm{B}}T}{v} \ln \frac{\Omega}{\Omega - H}, \tag{14.9}$$

where we have made use of Equation 14.7. We can rewrite this equation in terms of the concentration of the hemoglobin molecules, $[H] = H/\Omega v$, as

$$p = -\frac{k_{\mathrm{B}}T}{v} \ln(1 - [H]v). \tag{14.10}$$

In Figure 14.11, we compare the pressure predicted by the lattice model with experimental data. Note that the only free parameter in the model is the volume of a single box on the lattice, namely the parameter v. To make the comparison concrete, we convert the experimentally determined concentration, which has units g/L, to a molar concentration. To do so, we divide by the mass of a single hemoglobin molecule, $m_{\mathrm{hemoglobin}} \approx 64\,\mathrm{kDa}$. We see that the lattice model reproduces the nonlinearity and the right scale of the pressure when we use v that is comparable to the size of a hemoglobin molecule as determined from X-ray scattering.

Despite the favorable qualitative features of the lattice model, it still does not give an entirely satisfactory description of the observed osmotic pressure. Happily, there is a long tradition in statistical physics of exploring the so-called "hard-sphere gas," a model system in which spheres interact only by mutual repulsion that forbids them from ever having a center-to-center distance smaller than twice their radius. These models are an ideal setting within which to explore excluded-volume effects and do not suffer from the artificial constraint present in lattice models that force particles to occupy only a restricted set of points in space. We argue that the shortcomings in the model presented thus far are an artifact of the lattice model that result from the extremely approximate treatment of the volume exclusion. Studies of the hard-sphere gas have resulted in more sophisticated partition functions than that derived using a lattice model and result in the pressure of a hard-sphere gas as a function of its concentration of the form

$$p = k_B T[H](1 + x + 0.625x^2 + 0.287x^3 + 0.110x^4). \qquad (14.11)$$

This formula was first obtained by Boltzmann in 1899. The variable x is defined as $x = 4V[H]$, where V is the volume of a single hard sphere. The result of using the hard-sphere analysis is shown in Figure 14.11 and gives a much better treatment of the pressure, as the reader will demonstrate in the problems at the end of the chapter. The simple hard-sphere gas description of a concentrated hemoglobin solution gives a quantitative explanation of the observed osmotic pressure.

14.2.3 Depletion Forces: Order from Disorder

The Close Approach of Large Particles Excludes Smaller Particles Between Them, Resulting in an Entropic Force

One of the most intriguing and counterintuitive results of crowding is the way in which entropy can induce forces and structural ordering. A beautiful effect that arises on strictly entropic grounds is that of depletion forces, illustrated in Figure 14.12. The idea is that in a solution consisting of some large molecules (or particles) in the presence of some much smaller particles, as the large particles approach one another with a distance comparable to the size of the smaller particles, they will exclude the smaller particles from the volume between them. As they get even closer, the volume available to the smaller particles will *increase*. This has the effect of inducing an entropic force of attraction between the large particles. Macromolecules and macromolecular complexes in the cell are typically separated by distances that are comparable to their size. As a result, this kind of force should enter into the discussion of how macromolecules in the cell interact.

To see how depletion forces arise, we begin with the case of a two-dimensional system of large and small disks in the presence of a surface as shown in Figure 14.13. The large disk has radius R while the smaller disks have a radius r. In this idealized two-dimensional geometry, our interest is in computing the total area available to the small disks as a function of the distance z between the large disk and the surface. When the large disk is far from the surface, the total excluded area is the sum of the areas excluded by the large disk and the surface. As the large disk gets closer to the surface, there is a decrease in the area that is inaccessible to the small disks (shown as the shaded fragment of a circle) and this *increases* their entropy. The strategy of our calculations is then to compute the entropy of the small particles as a function of the distance between the large disk and the surface (or

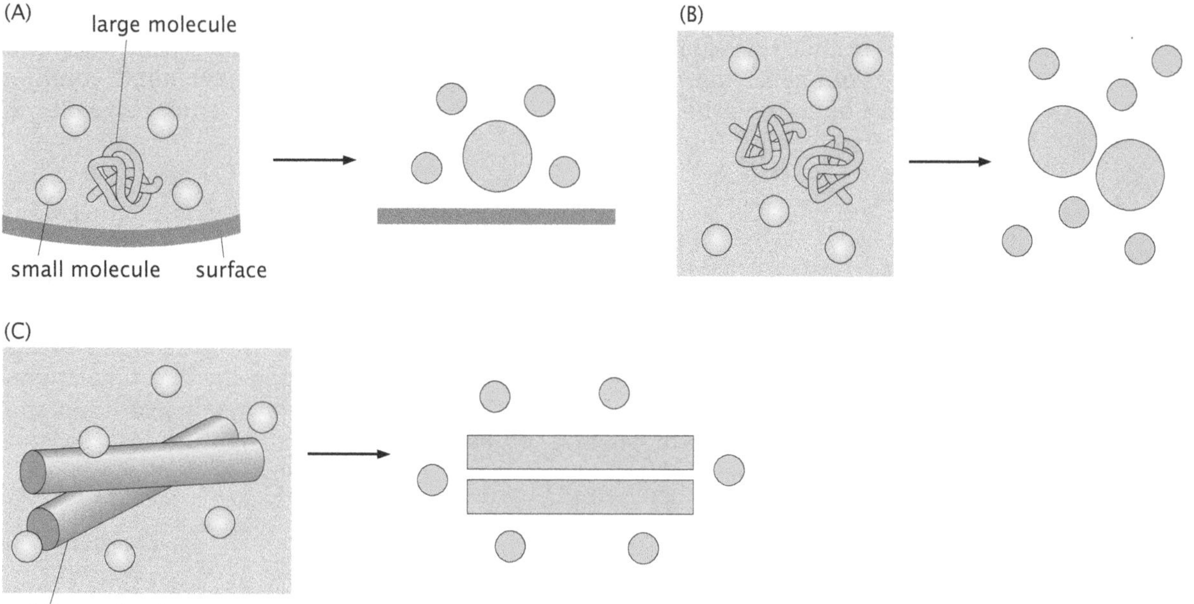

Figure 14.12: Schematic of the forces resulting from volume exclusion. In each of the schematics we show a two-dimensional idealization of the configuration of interest. (A) Large particle near a surface, (B) two large particles in solution, and (C) two rod-like molecules in solution. As the two large objects come closer together, the total volume they exclude for the small disks decreases, resulting in an effective attractive force between the large particle and the surface, the two large particles, and the two rod-like molecules, respectively.

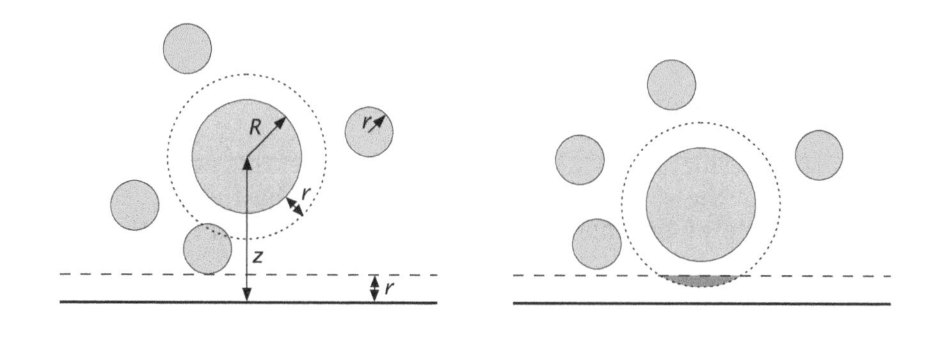

Figure 14.13: Geometry of excluded-volume interactions. A large disk of radius R in the presence of a surface and multiple small disks of radius r. The dotted lines depict the volume excluded to the small disks. As the large disk approaches the surface, the total excluded volume for the small disks decreases due to the overlap of the volumes excluded by the large disk and the surface (depicted by dashed lines). This overlapping excluded volume is shown as the gray shaded region in the right-hand figure. Because of the increase in the area available to the disks, there is an increase in their entropy and an effective attraction between the large disk and the surface.

other interesting examples). This entropy will permit us to compute the free energy itself and, once we have this free energy, we can compute $F = -\partial G/\partial z$ to obtain the effective depletion force between the disk and the surface. This calculation itself is given in the problems at the end of the chapter. At this point, we show how to consider these problems in general terms and then use the general formalism for a particular case study.

In the previous section, we considered the crowding effect of one molecular species on the binding of another. We can use the same lattice model to build intuition about excluded-volume interactions. Formally, we consider a solution filled with two species, namely, two large molecules in the presence of a large number of smaller crowding agents. We consider the situation in which there are no conventional forces (van der Waals, electrostatics, etc.) between the various molecules. Rather, the forces that arise do so strictly on the basis of the entropy changes incurred by the close proximity of the two large molecules. In this situation, the free energy of the small molecules depends on the positions and orientations (if the molecules have some structure to them) of the large ones. That is, the large molecules reduce the volume of space available to the small ones from V_{box}, the volume of the box, to $V_{box} - V_{ex}$, where V_{ex} is the excluded volume. Concretely, the free-energy change induced by this excluded volume is given by

$$G_{ex} = -Nk_BT\ln\left(\frac{V_{box} - V_{ex}}{v}\right) + Nk_BT\ln\left(\frac{V_{box}}{v}\right), \tag{14.12}$$

where N is the number of small molecules and v is a constant with units of volume (for example, for a lattice model, this would be the volume of a unit cell). This effect was introduced in Figure 14.13 and is shown more precisely in Figure 14.14. Combining the two logarithms and assuming that the excluded volume V_{ex} is much less than the overall volume of the box, we can use the approximation $\ln(1 + x) \approx x$ to obtain a simpler expression for this free energy as

$$G_{ex} = Nk_BT\frac{V_{ex}}{V_{box}}. \tag{14.13}$$

Note that the parameter V_{ex} depends upon the distance between the two large molecules and hence that there is a distance-dependent force. In particular, once the excluded-volume regions of the two large particles begin to overlap, this reduces V_{ex} and increases the entropy of the small particles. In addition to its simplicity, Equation 14.13 also has an appealing physical interpretation. We can identify Nk_BT/V_{box} with the ideal-gas (osmotic) pressure of the small particles in the box, and therefore the excluded-volume free energy is equal to the pressure–volume work done on the small-molecule gas in reducing the volume it occupies by V_{ex}.

To see how the calculations we have done thus far lead to depletion forces, consider a small change in the distance between the two large molecules. If this leads to a change in the free energy of the small molecules via Equation 14.13, then we can interpret this as a depletion force between the two large molecules mediated by the small ones. The magnitude of this depletion force is simply the absolute value

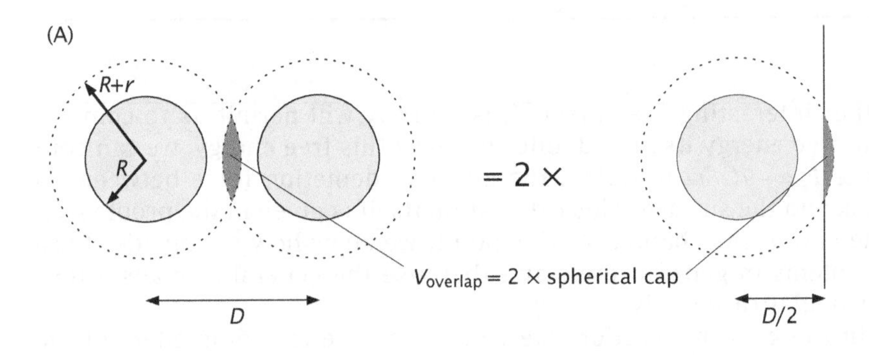

Figure 14.14: Depletion interaction between two spheres. (A) The depletion force is nonzero when two spheres come within a distance $D < 2(R + r)$, and is determined by the overlap volume, which is twice the volume of a spherical cap. (B) The volume of a spherical cap is equal to the difference between the volume of a spherical cone and a cone.

of the derivative of G_{ex} with respect to the distance. The sign of the derivative indicates the direction of the force: if it is negative, G_{ex} decreases with distance and the depletion force is repulsive, while a positive derivative indicates an attractive force.

To build quantitative intuition about the depletion interaction, we consider the case of two large spherical particles in a sea of smaller spherical particles. As discussed in the previous paragraphs, the calculation of the depletion force boils down to determining the excluded volume V_{ex}. We first examine the case of two spheres of radius R surrounded by smaller molecules of radius r.

The volume excluded by a single large sphere is a sphere of radius $R+r$. That is, the centers of the small spheres are unable to occupy any point within a distance of $R+r$ of the centers of the large spheres. If two spheres are present, then the total volume excluded for occupancy by the small spheres is twice as large unless the distance between the spheres is such that their excluded volumes overlap. From the discussion above, it should be clear that only in this case, when the distance between the two spheres, D, is less than $2(R+r)$, will there be an attractive depletion force between them. In this case, the total excluded volume from both spheres is *less* than twice the excluded volume of each sphere individually and is given by

$$V_{ex} = 2 \times \frac{4\pi}{3}(R+r)^3 - V_{overlap}. \tag{14.14}$$

The overlap volume, shown in Figure 14.14(A), consists of two spherical caps whose volume can be computed by subtracting the volume of the cone and that of the spherical cone, as shown in Figure 14.14(B).

The volume of the spherical cone is given by the integral

$$V_{spherical\,cone} = \int_0^{2\pi} d\phi \int_0^{\theta} \sin\theta\, d\theta \int_0^{R+r} r^2\, dr = \frac{2\pi}{3}(R+r)^3(1-\cos\theta). \tag{14.15}$$

Since θ is also the angle subtended by the cone, we can replace its cosine by $\cos\theta = (D/2)/(R+r)$ to give

$$V_{spherical\,cone} = \frac{2\pi}{3}(R+r)^2(R+r-D/2). \tag{14.16}$$

The volume of the cone is 1/3 of the volume of a cylinder with the same radius and height. The radius of the cone in Figure 14.14(B) is $\sqrt{(R+r)^2-(D/2)^2}$ and its height is $D/2$; therefore,

$$V_{cone} = \frac{\pi}{3}(D/2)\left[(R+r)^2-(D/2)^2\right]. \tag{14.17}$$

Finally we obtain the overlap volume as twice the difference between the volume of the spherical cone and the cone,

$$V_{overlap} = \frac{2\pi}{3}(R+r-D/2)^2(2R+2r+D/2). \tag{14.18}$$

To estimate the depletion force between the two large spheres, we are left with taking the derivative of the free energy G_{ex} in Equation 14.13 with respect to the distance D, which yields

$$F_{depletion} = -\frac{\partial G_{ex}}{\partial D} = -p\pi\left[(R+r)^2 - \frac{D^2}{4}\right]. \tag{14.19}$$

Here $p = nk_BT$ is the osmotic pressure of the small molecules ($n = N/V_{box}$ is the concentration), and the distance between the two large spheres satisfies $2R < D < 2(R + r)$; for larger distances, the overlap volume and the force are both zero. To get a feeling for the numbers, we take $R \gg r$ and $D \approx 2R + r$, in which case $F_{\text{depletion}} \approx \pi n k_B T r R$. For a bead whose radius is $R = 1\,\mu\text{m}$ surrounded by small molecules with $r = 2\,\text{nm}$ at a concentration $n = 1\,\text{mM}$, the size of the depletion force is $F_{\text{depletion}} \approx 15\,\text{pN}$.

Depletion forces have been explored experimentally as shown in Figure 14.15. In the case shown in the figure, two beads of $1.25\,\mu\text{m}$ diameter were confined to move along the line joining their centers. This confinement to one dimension was effected using an optical trapping system in which the laser is scanned so as to make a linear region in which the beads are trapped, but at constant potential. The depleting agent in the experiment is λ-phage DNA, $16\,\mu\text{m}$ in length, which at concentrations shown in the figure forms spherical globules with a radius of roughly $r \approx 500\,\text{nm}$. Polymer entropy in this case prevents overlap of different DNA molecules, so they effectively behave as hard spheres.

The experiment measures the distance between the beads using light microscopy. Repeated measurements of the distance lead to a determination of the probability distribution of distances, $p(D)$. Since in equilibrium $p(D)$ is proportional to the Boltzmann factor $e^{-\beta G_{\text{ex}}(D)}$, the logarithm of the measured distribution yields the free energy $G_{\text{ex}}(D)$. This is the quantity plotted in Figure 14.15, for different DNA concentrations. (The free energy is determined up to a constant, which in the experiment was chosen so that $G_{\text{ex}}(D)$ goes to zero at large D.)

The lines in Figure 14.15 are fits to the formula $G_{\text{ex}}(D) = pV_{\text{overlap}}$, where V_{overlap} is given by Equation 14.18. The fitting parameters used were the effective radius of the DNA molecules (r in Equation 14.18) and the osmotic pressure p. The effective radius of DNA was found to be concentration-independent and its value consistent with independent measurements of the same quantity. The osmotic pressure, $p/k_B T$, was found to be proportional to the concentration of DNA, n, as van't Hoff's formula says, but with a coefficient of 0.5 instead of 1. The origins of this discrepancy are unclear and could be a source of further inquiry.

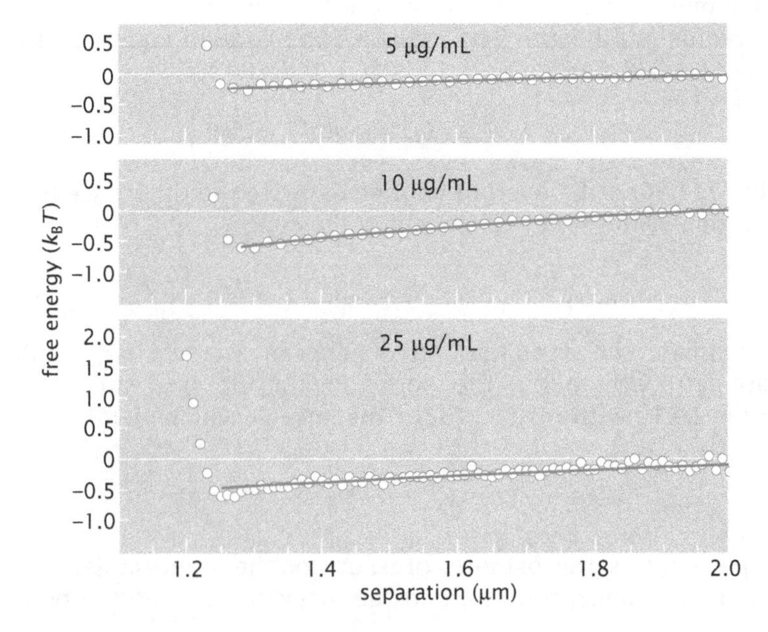

Figure 14.15: Measured free energies due to excluded volume. The free energy of interaction of two $1.25\,\mu\text{m}$ diameter beads as a function of the concentration of the depleting agent (labeled in each panel). The depleting agent is monodisperse DNA from λ-phage with a radius of gyration of approximately 500 nm as measured using light scattering. (Adapted from A. G. Yodh et al., *Phil. Trans. R. Soc. Lond.* A359:921, 2001.)

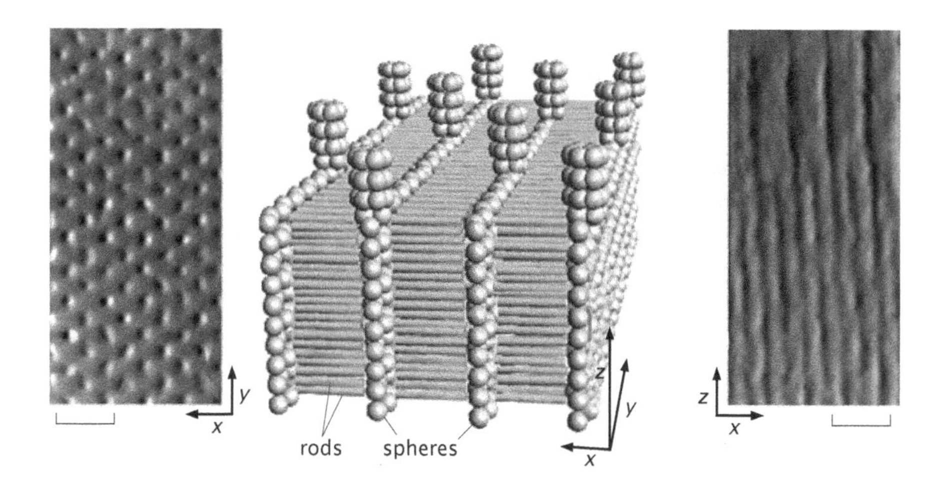

Figure 14.16: Entropic ordering of mixtures of hard rods and spheres. A solution of hard rods and spheres has a rich phase diagram that depends on the volume fractions of the two components, and it includes layered structures such as the one shown here. The left-hand panel is a micrograph along the z-axis, while the one on the right was taken in the y-direction. The schematic in the middle depicts the proposed layered structure. The scale bars are 3 μm. (Adapted from M. Adams et al., *Nature* 393:349, 1998.)

Depletion Forces Can Induce Entropic Ordering!

So far we have shown how excluded-volume effects can lead to short-range attractive depletion forces between two macromolecules. Beside the effect on bimolecular interactions, depletion forces can also produce ordered structures of surprising complexity. This is clearly seen in multicomponent systems such as the one made of filamentous viruses and small spheres. From a geometrical point of view, we think of the viruses as hard rods that interact with the nearby spheres to produce the layered structures shown in Figure 14.16. The surprising thing about these structures is that entropy alone leads to microphase separation, in which layers of balls are interspersed with layers of rods. Macrophase separation, in which the spheres and balls take up residence in different parts of the reaction volume, is the expected outcome, but experiments have shown that there are regions of phase space (determined by volume fractions of rods and spheres) where layered structures are preferred.

In light of these findings on model systems of rods and spheres, it is intriguing to consider whether depletion forces might be contributing to the organization of macromolecules and macromolecular complexes inside cells. As more quantitative data are obtained on the spatial arrangement of macromolecular complexes within cells, these questions might very well come to the forefront of physical biology.

14.2.4 Excluded Volume and Polymers

Excluded Volume Leads to an Effective Repulsion Between Molecules

We showed above that the presence of small molecules in the solution can lead to an effective attraction between two larger molecules, by considering the volume from which the large molecules exclude the smaller ones. Here we take up the issue of the mutual exclusion of a collection of N macromolecules confined to the interior of a cell, and each occupying a volume v. The excluded-volume interaction refers to the effect that two molecules are not allowed to occupy the same location in the cell.

We adopt the lattice model used in Section 14.2.2, where N macromolecules are distributed among Ω boxes each of volume v. The excluded-volume interaction manifests itself in the property that every box can be occupied by at most one macromolecule. To obtain an estimate of the effective repulsion between molecules due to the excluded-volume interaction, we compute the free-energy difference between the state where excluded volume is enforced and one in which it is not.

For the situation when the excluded volume is enforced, the partition function is simply the number of ways of choosing N boxes from the total number of boxes Ω in which to place the macromolecules. This is given by

$$Z_{\mathrm{ex}}(N)=\frac{\Omega!}{N!(\Omega-N)!}. \tag{14.20}$$

On the other hand, if we do not enforce the excluded volume condition, the partition function is

$$Z_{\mathrm{nex}}(N)=\frac{(N+\Omega-1)!}{\Omega!N!}, \tag{14.21}$$

which is the number of ways of partitioning N molecules among Ω boxes. This is the same counting problem as the one encountered on p. 252 (see Equation 6.34) where we considered the problem of partitioning energy units among identical particles, which eventually led to the Boltzmann distribution.

Using the canonical relation between free energy and the partition function, $G=-k_{\mathrm{B}}T\ln Z$, we can compute the free-energy difference between the two states as

$$\Delta G_{\mathrm{ex}}=G_{\mathrm{ex}}-G_{\mathrm{nex}}=-k_{\mathrm{B}}T\ln\frac{Z_{\mathrm{ex}}}{Z_{\mathrm{nex}}}. \tag{14.22}$$

To make further progress we make use of the Stirling approximation, $n!\approx(n/\mathrm{e})^n$, which is valid for $n\gg 1$ (described in "The Math Behind the Models" on p. 222 and in the problems at the end of Chapter 5). Within this approximation, and assuming $\Omega\gg N$,

$$\frac{\Omega!}{(\Omega-N)!}=\left(\frac{\Omega}{e}\right)^N \tag{14.23}$$

and the ratio of partition functions appearing in the above formula can be expressed as:

$$\frac{Z_{\mathrm{ex}}}{Z_{\mathrm{nex}}}=\frac{\Omega!}{(\Omega-N)!}\frac{(\Omega-1)!}{(\Omega+N-1)!}=\left(\frac{\Omega}{\Omega+N}\right)^N. \tag{14.24}$$

Finally, if we plug this expression for the ratio of partition functions into eqn 14.22 for the free-energy difference due to the excluded-volume effect, we find

$$\Delta G_{\mathrm{ex}}=Nk_{\mathrm{B}}T\ln\left(1+\frac{N}{\Omega}\right)\approx k_{\mathrm{B}}T\frac{N^2}{\Omega}, \tag{14.25}$$

where we have used the Taylor expansion for the logarithm ($\ln(1+x)\approx x$). Note that we can interpret the last formula by saying that the excluded-volume interaction raises the free energy of the system of macromolecules by $k_{\mathrm{B}}T\phi$ per molecule, where $\phi=N/\Omega$ is the volume fraction occupied by the macromolecules in the cell.

Self-avoidance Between the Monomers of a Polymer Leads to Polymer Swelling

This chapter has argued that crowding reveals itself in many different ways. These effects can be observed experimentally, and also force us to reconsider many of the powerful theoretical tools developed so far in the book. In Chapter 8, we developed the random walk model of polymers and we have shown its utility in describing macromolecules such as DNA. One feature of the random walk model that is disturbing, however, is that it permits polymer configurations in which its monomers occupy the same position in space and the chain is allowed to self-intersect. Clearly these conformations are unphysical and one would be tempted to discard the random walk model on these grounds. Still we should remember that the random walk model is probabilistic in nature, and therefore, if the offending states are almost never realized (that is, they have a vanishingly small probability), the model will provide reliable results for the average properties of flexible macromolecules. For example, if we are dealing with DNA whose length is only a few persistence lengths, then the energetic cost of making bends that would lead to self-intersections is too great and self-intersection will never happen. This observation begs the question: how long must a DNA molecule be before the self-avoidance effect starts rearing its head? For example, we have repeatedly used the formula $\sqrt{Na^2}$ for the size of a macromolecule such as DNA, which was derived based on the random walk model with no self-avoidance. We will demonstrate shortly that taking self-avoidance into account produces a very different result.

To determine the effect of self-avoidance on average polymer size, we make use of an estimate suggested by Flory. The idea is to consider the competing effects of polymer chain entropy, which has the tendency to make the chain compact, and that of self-avoidance, which tends to swell up the chain. To account for these two effects, we start by writing down an approximate expression for the free energy of the polymer chain as a function of its size R as

$$G(R) = -TS_{\text{rw}}(R) + G_{\text{ex}}(R), \tag{14.26}$$

where $S_{\text{rw}}(R)$ is the random walk entropy for a chain of size R and $G_{\text{ex}}(R)$ is the excluded-volume interaction discussed above.

To estimate the entropy of the polymer chain, we make use of the end-to-end distribution for a random walk, $P(R;N)$, given by Equation 8.23 on p. 318. In this case, we can write that result as

$$S_{\text{rw}}(R) = k_B \ln P(R;N) + \text{const} = -k_B \frac{3R^2}{2Na^2} + \text{const}, \tag{14.27}$$

where the constant term does not depend on the size (R) of the polymer chain.

For the excluded-volume interaction (G_{ex}) we approximate the polymer chain with a gas of hard cylinders of length a ($\approx$100 nm is the Kuhn length for DNA) and diameter d ($\approx$2 nm for DNA). Unlike the example treated previously, where we considered the excluded-volume interaction for a gas of hard spheres, here we have to take into account that the volume that one cylinder excludes for the center of mass of another depends on their mutual orientation. As shown in Figure 14.17, for fixed orientation, the excluded volume is estimated to be $v = 2da^2 \sin\theta$, where θ is the angle between the long axes of the cylinders. (The exact result for v obtained by Onsager reduces to

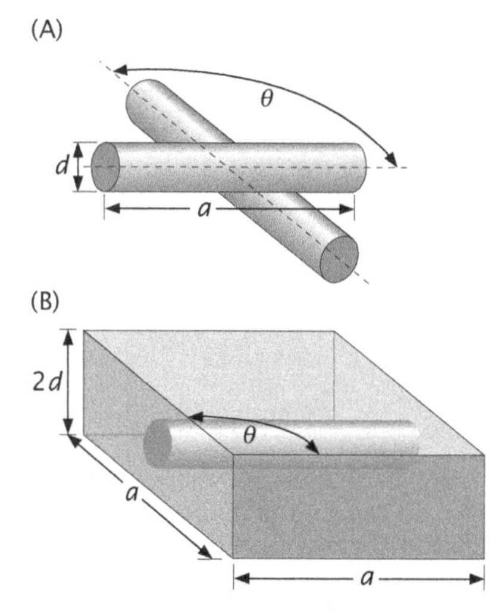

Figure 14.17: The excluded volume for two cylinders. (A) The volume that one cylinder excludes for another depends on their mutual orientation characterized by the angle θ between their long axes. (B) For the situation when the length of the cylinder is much greater than its diameter, the excluded volume is well approximated by the shaded parallelepiped.

this formula for the case $a \gg d$.) Averaging of $\sin\theta$ over all possible orientations gives

$$\langle \sin\theta \rangle = \frac{1}{4\pi}\int_0^{2\pi} d\phi \int_0^{\pi} \sin^2\theta \, d\theta = \frac{\pi}{4}, \tag{14.28}$$

to yield a final estimate for the excluded volume $\pi a^2 d/2$. With this result in hand, we can adapt Equation 14.25 to this situation. In order to use Equation 14.25 in the form $G_{\text{ex}} = k_B TN\phi$, we need to determine the volume fraction ϕ. Here N is the number of hard cylinders (that is, the number of Kuhn segments that make up the polymer chain) and permits us to write the total volume fraction occupied by the cylinders making up the polymer chain as

$$\phi(R) = N\frac{\frac{1}{2}\pi a^2 d}{\frac{4}{3}\pi R^3} = N\frac{3a^2 d}{8R^3}. \tag{14.29}$$

Equation 14.25 tells us how to construct the contribution of the excluded volume to the free energy and results in

$$G_{\text{ex}}(R) = k_B TN^2 \frac{3a^2 d}{8R^3}. \tag{14.30}$$

Putting the excluded-volume interaction term and the entropy term together, we arrive at the Flory estimate for the free energy of a polymer chain of size R, namely,

$$G_{\text{Flory}}(R) = k_B T\frac{3R^2}{2Na^2} + k_B TN^2\frac{3a^2 d}{8R^3}, \tag{14.31}$$

where we have dropped the unimportant constant term from the entropy estimate. To obtain the size of the polymer chain at equilibrium, all that remains is to determine that value of R that minimizes the Flory free energy. This is accomplished by taking the derivative of Equation 14.31 and setting it to zero, which yields

$$R_{\text{Flory}} = \left(\tfrac{3}{8}a^4 d\right)^{1/5} N^{3/5}. \tag{14.32}$$

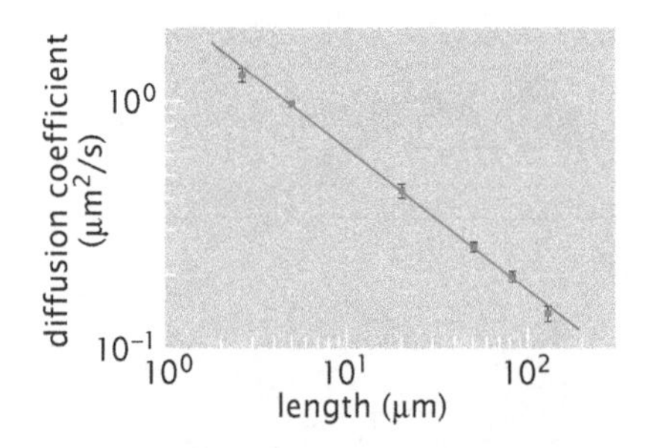

Figure 14.18: Measurements of the diffusion constant of fluorescently labeled DNA as a function of its length. Data is plotted on a log–log plot and the best fit line corresponds to a scaling relationship $D \sim L^{-0.57}$. The diffusion constant is related to the polymer size R by the Einstein relation, $D = k_B T/6\pi\eta R$. Assuming the Flory result for the size of a self-avoiding polymer $R \sim L^{0.6}$ (Equation 14.32) leads to $D \sim L^{-0.6}$, which is in good agreement with experiment. (Figure adapted from R. M. Robertson et al., *Proc. Natl Acad. Sci. USA* 103:7310, 2006.)

Most notably, the size of the polymer with self-avoidance scales with the number of segments N to the power 3/5. The fact that this number is greater than 1/2, the exponent associated with simple random walks, attests to the swelling of the chain induced by the excluded-volume interaction. Experiments on fluorescently labeled DNA are consistent with this prediction, as illustrated in Figure 14.18.

The Flory estimate of the free energy of a self-avoiding polymer can also be used to address the question raised earlier, namely, how long must a polymer be before excluded-volume interactions lead to observable deviations from simple random walk behavior? To arrive at such a criterion, we can compare the sizes of the excluded-volume interaction term, $G_{\text{ex}}(R)$, and the entropy term, $S_{\text{rw}}(R)$, in G_{Flory}, assuming that the size of the polymer is given by the random walk result $R = \sqrt{Na^2}$. In this case,

$$G_{\text{ex}} \approx k_B T\frac{d}{a}N^{1/2} \tag{14.33}$$

and

$$S_{\text{rw}} \approx k_{\text{B}}, \tag{14.34}$$

follow from Equations 14.30 and 14.27, respectively, where we have ignored the numerical constants (the Flory estimate does not get these right anyway). We see that for N large enough, the excluded-volume interaction term will always dominate. More precisely, taking the parameter values appropriate for DNA, $d = 2\,\text{nm}$ and $a = 100\,\text{nm}$, we can conclude that for $N \ll (a/d)^2 = 2500$, the excluded-volume interaction term will be negligible compared with the random walk entropy term, and it can be safely ignored. This many Kuhn segments of DNA corresponds to a molecule that is $L = Na = 2500 \times 100\,\text{nm}$, or $250\,\mu\text{m}$, in length. This conclusion seems to contradict experimental results shown in Figure 14.18, which show scaling characteristic of a self-avoiding polymer extending down to DNA lengths of only a few microns. A possible resolution is that in solution DNA is effectively thicker than what its structure would suggest, due to its highly charged nature. Namely, experiments at different salt concentrations and theory that takes into account DNA electrostatics have lead to proposals for an effective thicknesses for DNA of the order of 20 nm, ten times its bare diameter. This would reduce the estimate for how long DNA needs to be before excluded-volume interactions begin to matter down to $N \ll 25$ Kuhn segments, or $L \ll 2.5\,\mu\text{m}$, consistent with Figure 14.18.

14.2.5 Case Study in Crowding: How to Make a Helix

Helices are among the most common of molecular structures. Proteins gone both good and bad form helices. Alpha helices are among the most ubiquitous structural motifs in all of biology. In addition, we have told the story of cytoskeletal filaments and flagella several times and they too exemplify the importance of helical motifs in the biological setting. Proteins are also known to aggregate into helical structures mistakenly, with potentially devastating consequences.

There have been several intriguing theoretical studies of this propensity for helix formation. Early studies focused on how local rules could lead to the formation of helices. As shown in Figure 14.19, helix formation was front and center in Linus Pauling's thinking about the structures formed by proteins. An allied set of extremely interesting observations was made by Crane, who noticed that a collection of identical but intrinsically unsymmetrical objects will naturally form helices if they are decorated by the same pattern of Velcro and anti-Velcro. Figure 14.19(D) shows several examples that illustrate this simple principle.

Another way of thinking about helix formation has been suggested, as shown in Figure 14.20. In this case, the mechanism builds upon ideas introduced in Section 14.2.3, in which we described order induced by entropy, with special reference to excluded-volume interactions. In this case, the words are intuitive, though the geometrical exercise that goes with those words is more tricky.

As shown in Figures 14.20(A) and (B), we consider a polymer of diameter $2t$ surrounded by a collection of "crowders" of radius r. Because of the finite size of the crowders, they are forbidden from entering the region of thickness r around the cylinder. If the polymer folds into a helical conformation, this reduces the excluded volume since there is an overlap of different regions of this excluded volume. That

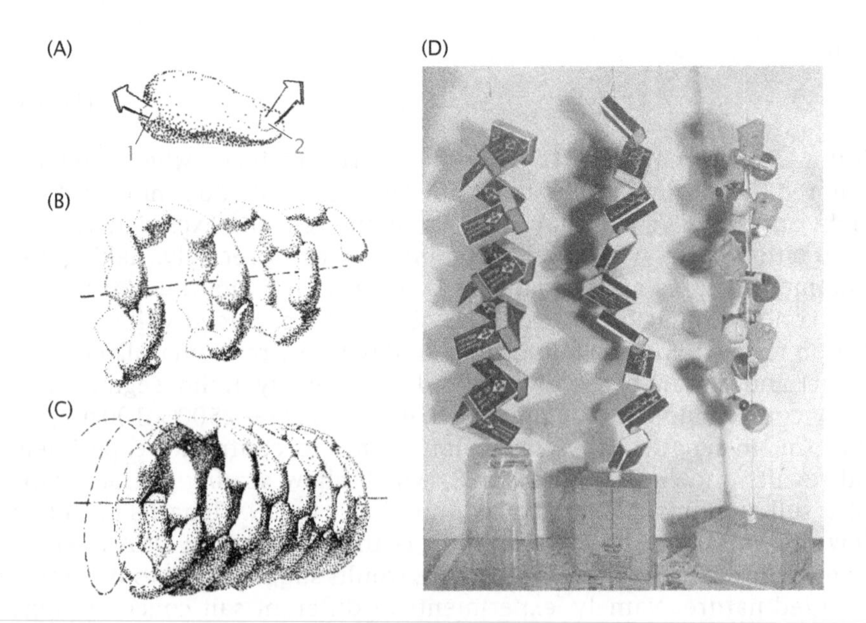

Figure 14.19: Helical structures generated by local rules. (A) The compact protein shown here has an asymmetric surface, covered with a non-repeating pattern of positive and negative charges and hydrophobic patches of various sizes. If you take two copies of this asymmetric protein, and consider all the potential ways that they could be glued together, there will be some pair of surface patches whose interaction is slightly more energetically favorable than any other possible pair of interactions. (B) Now taking many copies of this protein, and allowing patch 1 (see part A) on the first protein to bind to patch 2 on the second protein, then patch 1 on the second protein to bind to patch 2 on the third protein, etc., the proteins can be assembled into a regular helix. (C) Spaces in between the turns of the helix can be filled in by interdigitating another helix of the same geometry. Now, the neighboring intertwined helices are so close together that additional binding interactions between protein subunits in neighboring helices can contribute to the energetic stability of the polymer, even if they are not quite as energetically favorable as the original interaction between patch 1 and patch 2. (D) This concept can apply to any object with a patterned surface, including matchboxes and wooden balls as well as proteins. These helical models were built in 1950 by the physicist H. Richard Crane, at the University of Michigan. (A–C, adapted from L. Pauling, *Faraday Soc. Discuss.* 13:170, 1953; D, adapted from H. R. Crane, *The Scientific Monthly* 70(6):376, 1950.)

reduction is taken advantage of by the crowders, which now have more room in which to jiggle around, with a concomitant increase in their entropy. Hence, the entropic contribution to the overall free-energy budget favors helix formation.

One of the beautiful outcomes of this simple argument is that the shape of the helix (that is, its pitch P and radius R) will depend upon the size of the crowders. This relationship is shown in Figure 14.20(C), which plots the ratio of pitch to radius P/R as a function of the dimensionless crowder radius r/t. For small crowders, this ratio saturates to a value of $P/R \approx 2.5$, as shown in the figure, which should be compared with the value seen in proteins of $P/R = 2$. Though this mechanism is not offered as the final word on the origin of helical assemblies, it illustrates yet another way in which simple physical principles can induce recognizable structural patterns.

Of course, if the polymer itself has some intrinsic rigidity, then a price must be paid to effect this bending. In this case, there is a competition between the entropic benefit of collapse into the helical state and the elastic penalty that attends it. Figure 14.20(D) shows the regimes in which either the entropy wins the competition and the helix state is favored or the elastic deformation energy wins and the stretched polymer conformation is favored.

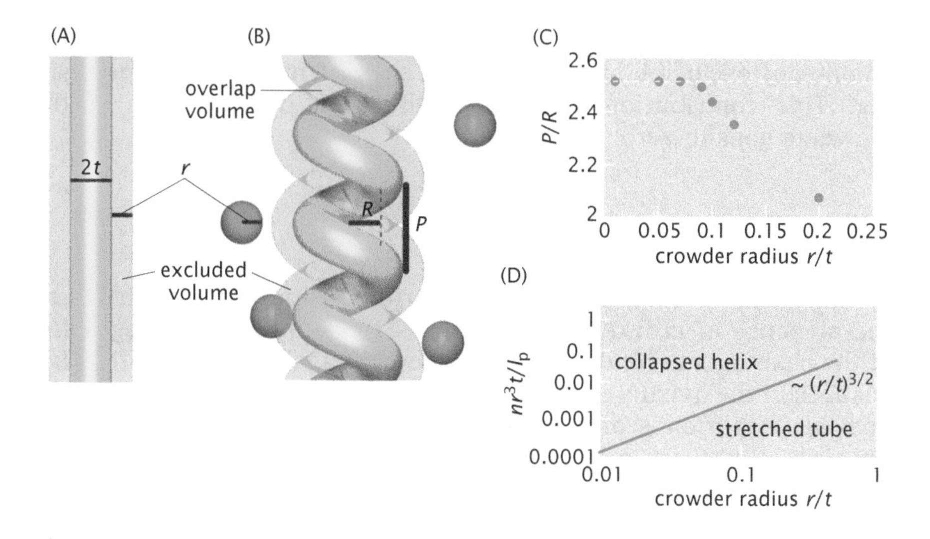

Figure 14.20: Excluded-volume interactions and helix formation. (A) A polymer of radius t is surrounded by a cylindrical shell of thickness r from which "crowder" particles are excluded. (B) When the polymer adopts a helical configuration, the excluded-volume region is decreased, thus *increasing* the entropy of the crowding particles. (C) Ratio of pitch to radius of the helix as a function of the size of the crowding particles. (D) Balance of elastic energy and crowder entropy. The line shows the value of the free energy (dimensionless) at which the collapsed helical state and the straight polymer have the same free energy as a function of the crowder radius. n is the concentration of crowders and l_p is the persistence length of the polymer. (Adapted from Y. Snir and R. D. Kamien, *Science* 307: 1067, 2005.)

14.2.6 Crowding at Membranes

Precisely these same ideas about the entropy of crowding have implications for membranes as well. One of the simplest cases to consider is highlighted schematically in Figure 14.21, which shows an ion channel that changes its radius upon gating. For the simple fixed two-dimensional box geometry shown in the figure, the key point is that when the channel opens, the rest of the proteins have less area to jiggle around in and hence their entropy is decreased, resulting in an entropic driving force favoring the closed state.

To see this explicitly, consider the translational entropy of a single crowder. This can be computed as the logarithm of the area available to the center of mass of our single crowder, resulting in a free energy

$$g_{\text{crowd}}(R) = -k_B T \ln\left[\frac{L^2 - \pi(R+R_p)^2 - A_{\text{edge}}}{A_{\text{lattice}}}\right], \qquad (14.35)$$

where R_p is the radius of the crowder, R is the radius of the channel, and A_{edge} is the band of thickness R_p around the edge of the box from which the crowder center of mass is excluded. This expression can be simplified by appealing to a Taylor expansion in the small parameter $[\pi(R+R_p)^2 + A_{\text{edge}}]/L^2$. Using this expansion the difference in free energy between the open ($R = R_o$) and closed ($R = R_c$) states due to crowding is given as

$$\Delta G_{\text{crowd}} = N\Delta g_{\text{crowd}} \approx c_A k_B T[\pi(R_o^2 - R_c^2) + 2\pi R_p(R_o - R_c)], \qquad (14.36)$$

with $c_A = N/L^2$ denoting the crowder concentration. The intuitive interpretation of these terms can be seen by rewriting the area and circumference change upon gating, respectively, as $\Delta A = \pi(R_o^2 - R_c^2)$ and $\Delta C = 2\pi(R_o - R_c)$. We can then divide the entropic crowding tension into surface and line tension contributions of the form

$$\Delta G_{\text{crowd}} \approx -\sigma_{\text{crowd}}\Delta A + \tau_{\text{crowd}}\Delta C. \qquad (14.37)$$

This "ideal gas" approximation ignores the overlapping exclusion zones of the crowders themselves, which are important at high areal densities. Note that the surface tension is given by $-\sigma_{\text{crowd}} = c_A k_B T$

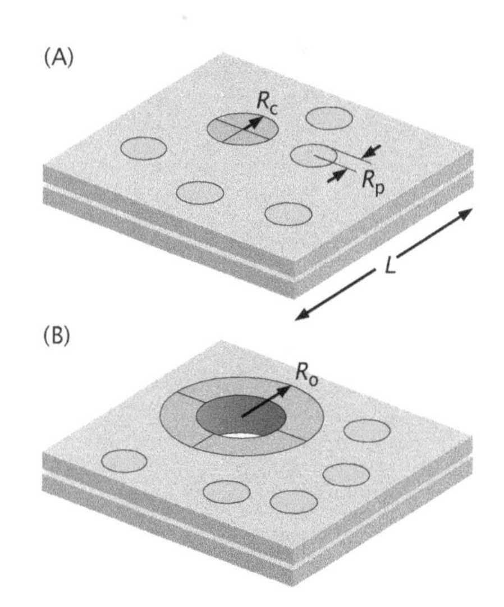

Figure 14.21: Excluded-volume forces and channel gating. (A) Area available to crowding molecules when the channel is closed. (B) Reduced area available to crowding molecules when the channel is open. The net result is an entropic tension which favors the closed state.

and the line tension $\tau_{crowd} = c_A R_p k_B T$ originates in the fact that the annulus of exclusion around the ion channel changes size upon gating. This contribution vanishes in the limit that the size of the crowders goes to zero.

14.3 Crowded Dynamics

So far, we have examined how equilibrium properties are altered by the presence of crowding. However, as mentioned at the beginning of the chapter, a second way in which crowding is revealed is through changes in the dynamics within cells. In light of the observation of the possible importance of macromolecular crowding in producing new physical effects, we now return to some of the questions presented earlier in the chapter concerning the nature of diffusion, but now from the point of view of the nature of the dynamics when the diffusive medium is dense and complex.

14.3.1 Crowding and Reaction Rates

Enzymatic Reactions in Cells Can Proceed Faster than the Diffusion Limit Using Substrate Channeling

In dilute *in vitro* biochemical experiments, the reaction rate for an enzymatic conversion of a substrate into a product depends upon the concentration of the enzyme, the concentration of the substrate, and the intrinsic turnover rate of the enzyme. It is generally accepted that the concentration dependence of these rates describes the time it takes for an enzyme diffusing freely in solution to randomly encounter its substrate. At higher concentrations of either, these random collisions become more frequent.

Inside cells, enzymes are rarely free to diffuse as they would in dilute solution since proteins tend to form large complexes or associate with membranes or cytoskeletal elements. One example of this effect can be seen in Figure 14.4, where it is shown that the glycolytic enzyme phosphoglycerate kinase has an effective diffusion coefficient in cells that is roughly 5 times lower than that of GFP despite the fact that the two molecules are nearly the same size. We might expect then that reaction rates inside cells would be largely dominated by substrate concentration. However, in many critical metabolic pathways, such as the Krebs cycle operating inside mitochondria, the concentration of small-molecule substrates, such as the intermediate oxaloacetate, is much too low to account for the overall flux through the metabolic pathway.

How can a cell drive an enzymatic transformation more rapidly than the physical law of diffusion should allow? One common solution, which may prove to be nearly universal, is that cells simply do not allow substrate molecules that are intermediates in metabolic pathways to diffuse from the active site of one enzyme in the pathway to the next. Instead, specific, though low-affinity, protein–protein interactions among all the enzymes in the pathway are used to assemble a giant multienzyme complex in which the substrate can travel directly from the active site of one enzyme to another without ever freely diffusing. This behavior is known as substrate channeling.

Because of this effect, it is extremely challenging to predict the actual rate of a biochemical transformation *in vivo*, even if the

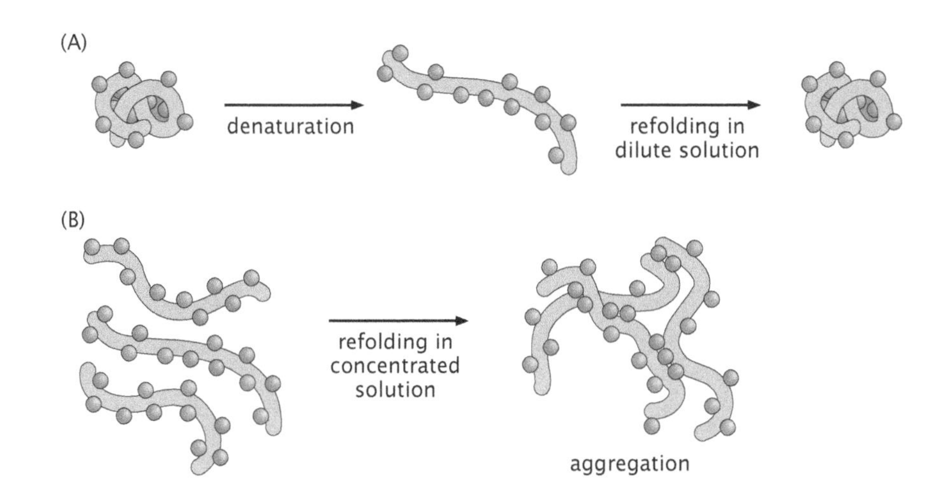

Figure 14.22: Protein folding and aggregation. A protein folded in its native state sequesters hydrophobic domains on the inside to hide the hydrophobic core. Denaturation disrupts the native structure, exposing these hydrophobic patches. (A) When the protein is allowed to refold in very dilute solution, the hydrophobic patches within a single molecule self-associate to reform the native hydrophobic core. (B) At high concentration, the hydrophobic patch of one protein molecule may associate with the hydrophobic patch of another, triggering protein aggregation rather than native refolding. Hydrophobic residues are shown in red, while hydrophilic residues are shown in blue.

concentrations of both substrate and enzyme are known and the turnover rate of the enzyme has been accurately measured.

Protein Folding Is Facilitated by Chaperones

Another case where dilute *in vitro* biochemical experiments fail to accurately represent the complexities of protein behavior inside cells is in the study of protein folding. Many small proteins of relatively simple structure can be purified and denatured with harsh chemical agents such as urea or guanidinium chloride. When the denaturing agents are removed, the proteins refold *in vitro* to their original native structure. These kinds of experiments are successful only when the protein concentration is several orders of magnitude lower than the actual concentrations of protein inside of cells. In more crowded solutions, denatured proteins tend to aggregate by intermolecular association of their hydrophobic patches or domains, preventing proper intramolecular association of these domains to form the protein's hydrophobic core as shown in Figure 14.22.

How do cells prevent aggregation of proteins as they are synthesized from ribosomes in the highly crowded cytoplasmic environment? Specialized proteins called chaperones facilitate protein folding both by increasing its rate and by preventing aggregation of partially folded protein intermediates. These chaperones come in two flavors. Chambered chaperones such as GroEL in bacteria and TRiC in eukaryotic cells actually form a tiny private room in which an individual polypeptide chain is free to fold with no danger of random collision with the hydrophobic patches of others. These chambered chaperones consume ATP in the process of opening and closing the door to the room. The second class of chaperone, exemplified by small heat-shock proteins such as HSP70, tend not to require ATP for their action. These bind to the hydrophobic domains of nascent proteins as they emerge from the ribosome and prevent their aggregation until the entire protein domain has been translated and is ready to fold.

14.3.2 Diffusion in Crowded Environments

As was illustrated in Figures 14.4 and 14.5, diffusion in crowded environments is more subtle than its dilute-solution counterpart. Theoretical responses to this challenging problem are usually all built

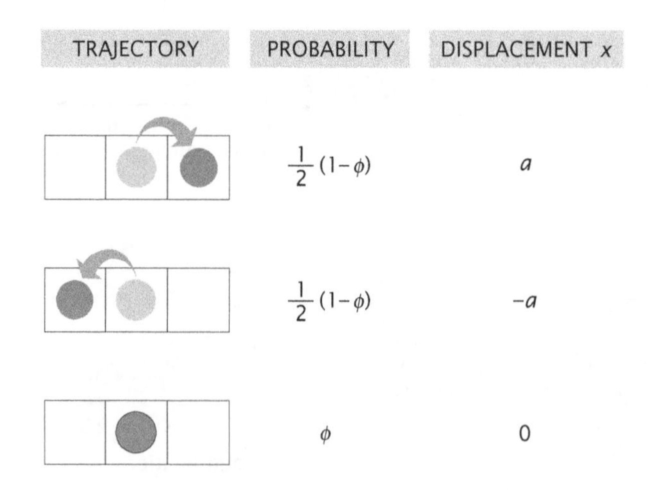

Figure 14.23: Trajectories and weights for a one-dimensional random walk that takes into account the effect of crowding.

around the same fundamental and intuitive idea: for a particle to hop to a new position, that new position cannot already be occupied. A simple random walk model can be used to illustrate the effect of crowding on molecular diffusion, though we note that, like with the treatments earlier in the chapter, this is only a modest attempt to come to terms with the problem. To make the model concrete, assume that the volume fraction occupied by the molecules is ϕ and that no two molecules can occupy the same site. In this case, the probability that a chosen site is occupied by a molecule is ϕ. Further, for simplicity, we consider only a one-dimensional walker.

The random walk now proceeds in the usual way: at every time instant τ, the particle makes a jump to the left or to the right with equal probability. The jump is successful only if there is no particle at the new location. Therefore, the particle jumps to the right, or to the left, with probability

$$p_{\text{right}} = p_{\text{left}} = \tfrac{1}{2} \times (1 - \phi), \tag{14.38}$$

where $\frac{1}{2}$ is the probability for attempting the jump, while $1 - \phi$ is the probability that the new site is unoccupied, thus allowing for a successful completion of the jump. If the neighboring site is occupied, the attempted jump will be unsuccessful and the particle will stay put. The probability of that outcome is equal to the probability that a neighboring site is occupied, which results in

$$p_{\text{stay}} = \phi, \tag{14.39}$$

which is also equal to $1 - p_{\text{jump}}$. These three possible outcomes are illustrated in Figure 14.23.

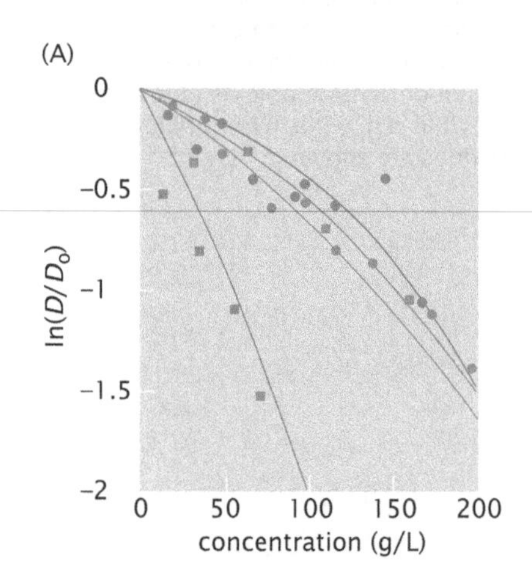

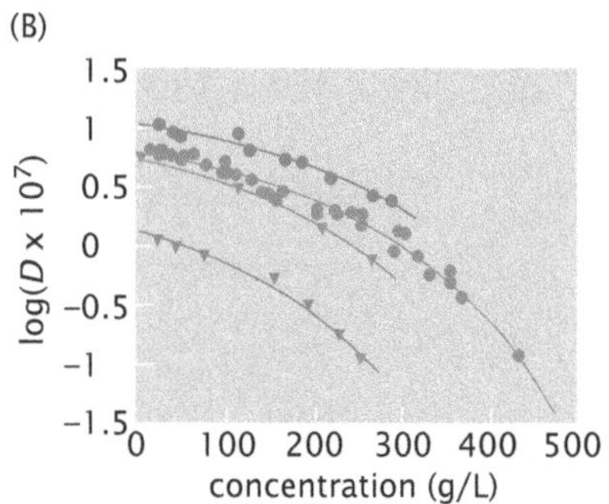

Figure 14.24: Diffusion and crowding. (A) Tracer diffusion as a function of protein concentration measured relative to the diffusion in the absence of protein. FITC–aldolase diffusing in a background of aldolase (red circles), BSA (blue circles), ovalbumin (green squares), and ribonuclease (purple squares). (B) Self-diffusion of globular proteins as a function of the protein concentration: myoglobin (red circles), hemoglobin (blue circles), ovalbumin (green triangles), invertebrate hemoglobin (purple triangles). Diffusion constants are given in units of cm^2/s. (A, adapted from N. Muramatsu and A. P. Minton, *Proc. Natl Acad. Sci. USA* 85:2984, 1988; B, adapted from S. B. Zimmerman and A. P. Minton, *Annu. Rev. Biophys. Biomol. Struct.* 22:27, 1993.)

To compute the diffusion constant associated with this random walk, we evaluate the mean-square displacement $\langle x^2 \rangle$ as a function of time t. For $t = \tau$, the trajectory consists of a single step and

$$\langle x^2 \rangle(\tau) = a^2 \times p_{\text{right}} + a^2 \times p_{\text{left}} + 0 \times p_{\text{stay}} = a^2(1 - \phi). \tag{14.40}$$

After time t, the molecule has made $N = t/\tau$ steps and the mean-square displacement is N times larger than that in Equation 14.40,

$$\langle x^2 \rangle(t) = \frac{t}{\tau} \times \langle x^2 \rangle(\tau) = \frac{a^2}{\tau}(1 - \phi)t. \tag{14.41}$$

As we did in Chapter 13, we read off the diffusion constant from Equation 14.41, and

$$D = D_0(1 - \phi), \tag{14.42}$$

where $D_0 = a^2/2\tau$ is the result we obtained for the diffusion constant of a random walker when no other molecules are present.

We see that the effect of crowding is to reduce the diffusion constant by an amount proportional to the volume fraction occupied by the molecules. This will in turn affect the diffusion-limited on rate. Several examples of measurements on diffusion both of tracer molecules and of self-diffusion are shown in Figure 14.24. Qualitatively, it is seen that the model results are consistent with the trends revealed by the data. On the other hand, the precise functional form yielded by the model is based upon a coarse, "mean field" description and a more sophisticated treatment based on the excluded volume of hard spheres provides a semiquantitative explanation of the data (for details, see Muramatsu and Minton 1988).

14.4 Summary and Conclusions

This chapter has explored one of the exciting frontiers at the interface between cell biology and physical theory, namely, how the properties of living matter differ from conventional solutions as a result of their extreme crowding. We have examined two broad classes of consequences, namely, (i) how equilibrium reactions are altered and (ii) how the dynamics of diffusion and biochemical rates are altered. Though this chapter appears in the part of the book entitled "Life in Motion," the alert reader will notice that the majority of our calculations have centered on equilibrium phenomena. This is a reflection of our inability to put together a compelling set of simple models that respond to the interesting dynamical data on crowding.

14.5 Problems

Key to the problem categories: • Model refinements and derivations • Estimates • Data interpretation • Model construction

• 14.1 A feeling for the numbers: comparing *in vitro* and *in vivo* concentrations

In the chapter, we argued that the mean spacing between molecules in an *in vitro* biochemical experiment is roughly 100 nm at μM concentrations while in the cell the spacings are a factor of 10 smaller. Justify these statements with simple estimates. The biochemical "standard state" is often taken as 1 M. Work out the mean spacing between molecules at this concentration.

• 14.2 Effective concentrations and activity

The effect of crowding on the chemical potential of a molecular species in solution can be captured by the equation $\mu = \mu_0 + k_B T \ln(c\gamma/c_0)$, where the subscripts zero are for a reference state. The effective concentration is given by $c\gamma$, where c is the actual concentration that is present in the solution and γ is called the activity coefficient. The simple lattice model of proteins in solution used repeatedly in the chapter implies a corresponding model for the activity coefficient. Work out this activity coefficient and compare your formula with the experimental results shown in Figure 14.25.

• 14.3 Osmotic pressure of hemoglobin

Use the approximate formula for the pressure of a gas of hard spheres, Equation 14.11, to extract an effective hard-sphere radius for hemoglobin from the data given in Figure 14.11. How does this effective radius compare with the dimensions of the molecule obtained by X-ray scattering?

Relevant data for this problem are provided on the book's website.

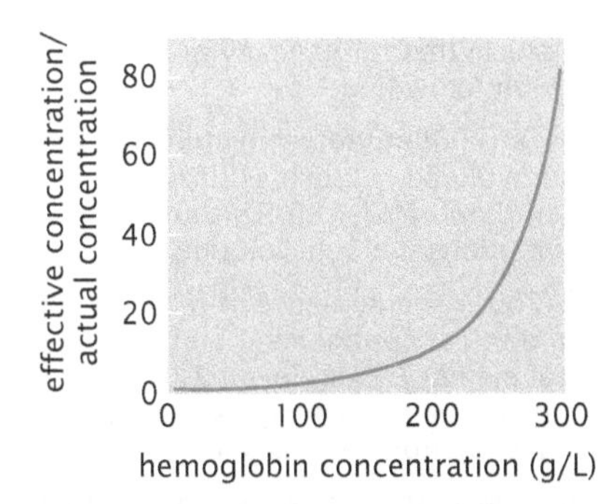

Figure 14.25: Crowding and the activity coefficient. Effect of hemoglobin concentration on its effective concentration. (Adapted from A. P. Minton, *J. Biol. Chem.* 276:10577, 2001.)

14.4 Depletion force between a sphere and a surface

Compute the depletion force between a sphere of radius R and a planar surface by carrying out the calculation indicated schematically in Figure 14.13. The radius of the small spheres is r.

14.5 Excluded-volume interactions

In the chapter, we worked out a general statement of the free energy for two large objects in solution and in the presence of small depletant molecules, which was based upon the osmotic pressure of the molecules and the volume excluded by the objects when at distance D. Repeat that derivation leading up to the formula

$$G(D) = \Pi_0 V_{ex}(D), \quad (14.43)$$

where $\Pi_0 = Nk_BT/V$ is the osmotic pressure of the depletant molecules. Estimate the force between two beads whose radius is 1 μm in a concentrated protein solution, when the concentration is equal to the value characteristic of an *E. coli* cell. Use $r = 3$ nm for the typical radius of a protein.

14.6 Self-avoidance in flatland

Repeat the Flory calculation from the chapter (Section 14.2.4) for DNA confined to a two-dimensional surface.

(a) Find the scaling of the size of the polymer as a function of its length, incorporating self-avoidance.

(b) Estimate the DNA length for which self-avoidance becomes important. How does this compare with the length of genomic DNA from a λ-phage?

14.7 Diffusion and crowding

In this problem, we extend the one-dimensional model of diffusion in the presence of crowding molecules to account for the difference in size between a tracer particle (considered to be present at low concentration) and the crowders. This situation is relevant for the data shown in Figure 14.24(A). The tracer particles are assumed to be undergoing random walk motion on the larger tracer lattice with lattice constant b, while the crowders are hopping between adjacent sites of the smaller lattice, with lattice constant a (see Figure 14.26). The two lattices are introduced to account for the difference in size between the two molecular species.

(a) Calculate the diffusion coefficient by considering the possible trajectories of the tracer particles and their probabilities. Note that the tracer can hop to an adjacent site of the tracer lattice only if there are no crowders present. Express your answer in terms of the diffusion coefficient D_0 of the tracer particles in the absence of crowders, the volume fraction of the crowders ϕ, and the ratio of the tracer and crowder sizes $r = b/a$.

(b) Plot $\ln(D/D_0)$ as a function of the volume fraction for different values of r. How well does this model explain the data shown in Figure 14.24(A)? To make this comparison, you will need to estimate the sizes of the molecules used in the experiment from their molecular masses and a typical protein density that is 1.3 times that of water. The data are provided on the book's website.

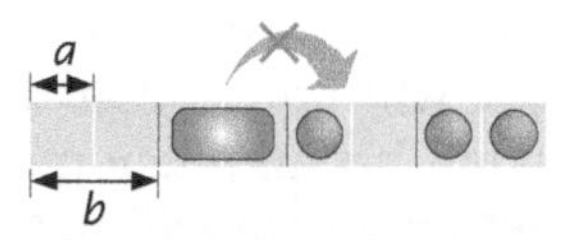

Figure 14.26: Lattice model of tracer particles of size b diffusing in the presence of crowding molecules of size a. The tracer particle can hop to the neighboring tracer site only if there are no crowding molecules present in the $r = b/a$ adjacent crowding molecule sites. The fraction of sites occupied by crowding molecules is ϕ.

14.6 Further Reading

Goodsell, DS (1996) Our Molecular Nature: The Body's Motors, Machines and Messages, Springer-Verlag. Goodsell's illustrations provide a compelling, visual demonstration of the extent of crowding in cells.

Zimmerman, SB, & Minton, AP (1993) Macromolecular crowding: biochemical, biophysical and physiological consequences, *Annu. Rev. Biophys. Biomol. Struct.* **22**, 27. This is a classic article that raises many of the important issues associated with "crowding."

Luby-Phelps, K (2000) Cytoarchitecture and physical properties of cytoplasm: volume, viscosity, diffusion, intracellular surface area, *Int. Rev. Cytol.* **192**, 189. This article spells out the key hidden assumptions in dilute-solution biochemistry.

Minton, AP (2001) The influence of macromolecular crowding and macromolecular confinement on biochemical reactions in physiological media, *J. Biol. Chem.* **276**, 10577. This paper lays out many of the interesting issues that arise as a result of the crowded environment in cells.

Ellis, RJ (2001) Macromolecular crowding: obvious but underappreciated, *Trends Biochem. Sci.* **26**, 597. A very useful introduction to the subject of crowding and its relevance to both equilibrium phenomena and kinetics.

Bray, D (2001) Cell Movements: From Molecules to Motility, 2nd ed., Garland Science.

Asakura, S, & Oosawa, F (1958) Interaction between particles suspended in solutions of macromolecules, *J. Polymer Sci.* **33**, 183. A beautiful example of theoretical reasoning to predict new phenomena. The authors note that "experimental proof of the reality of force derived here has not yet been obtained," a reminder that theoretical ideas can be useful even without the so-called "supporting data."

De Gennes, PG (1979) Scaling Concepts in Polymer Physics, Cornell University Press. De Gennes gives a clear derivation of the role of excluded-volume effects in polymers.

Lindén, M, Sens, P, & Phillips, R (2012) Entropic tension in crowded membranes, *PLoS Comput. Biol.* **8**, e1002431. This paper shows how the entropic tension due to membrane protein crowding affects ion channel gating.

14.7 References

Adams, M, Dogic, Z, Keller, SL, & Fraden, S (1998) Entropically driven microphase transitions in mixtures of colloidal rods and spheres, *Nature* **393**, 349.

Burk, DH, & Ye, ZH (2002) Alteration of oriented deposition of cellulose microfibrils by mutation of a katanin-like microtubule-severing protein, *Plant Cell* **14**, 2145.

Crane, HR (1950) Principles and problems of biological growth, *The Scientific Monthly* **70**, 376.

Cross, PC, & Mercer, KL (1993) Cell and Tissue Ultrastructure, 2nd ed., W. H. Freeman.

Douglass, AD, & Vale, R (2005) Single-molecule microscopy reveals plasma membrane microdomains created by protein–protein networks that exclude or trap signaling molecules in T cells, *Cell* **121**, 937.

Goodsell, DS (2009) The Machinery of Life, 2nd ed., Springer-Verlag.

Hirokawa, N (1982) Cross-linker system between neurofilaments, microtubules, and membranous organelles in frog axons revealed by the quick-freeze, deep-etching method, *J Cell Biol.* **94**, 129.

Jarvis, TC, Ring, DM, Daube, SS, & von Hippel, PH (1990) Macromolecular crowding: thermodynamic consequences for protein–protein interactions within the T4 DNA replication complex, *J. Biol. Chem.* **265**, 15160.

Kornberg, A (2000) Ten commandments: lessons from the enzymology of DNA replication, *J. Bacteriol.* **182**, 3613.

Meroueh, SO, Bencze, KZ, Hesek, D, et al. (2006) Three-dimensional structure of the bacterial cell wall peptidoglycan, *Proc. Natl Acad. Sci. USA* **103**, 4404.

Mitra, K, Ubarretxena-Belandia, I, Taguchi, T, Warren, G, & Engelman, DM (2004) Modulation of the bilayer thickness of exocytic pathway membranes by membrane proteins rather than cholesterol, *Proc. Natl Acad. Sci. USA* **101**, 4083. Table 1 of this paper gives a useful list of protein to phospholipid mass ratios for a number of different membranes.

Muramatsu, N, & Minton, AP (1988) Tracer diffusion of globular proteins in concentrated protein solutions, *Proc. Natl Acad. Sci. USA* **85**, 2984.

Pauling, L (1953) Aggregation of globular proteins, *Faraday Soc. Discuss.* **13**, 170.

Provenzano, PP, Hayashi, K, Kunz, DN, Markel, MD, & Vanderby, R (2002) Healing of subfailure ligament injury: comparison between immature and mature ligaments in a rat model, *J. Orthop. Res.* **20**, 975.

Ross, PD, & Minton, AP (1977) Analysis of non-ideal behavior in concentrated hemoglobin solutions, *J. Mol. Biol.* **112**, 437.

Robertson, RM, Laib, S, & Smith, DE (2006) Diffusion of isolated DNA molecules: dependence on length and topology, *Proc. Natl Acad. Sci. USA* **103**, 7310.

Snir, Y, & Kamien, RD (2005) Entropically driven helix formation, *Science* **307**, 1067.

Svitkina, TM, Bulanova, EA, Chaga, OY, et al. (2003) Mechanism of filopodia initiation by reorganization of a dendritic network, *J. Cell Biol.* **160**, 409.

Svitkina, TM, Verkhovsky, AB, McQuade, KM, & Borisy, GG (1997) Analysis of the actin–myosin II system in fish epidermal keratocytes: mechanism of cell body translocation, *J. Cell Biol.* **139**, 397.

Verkman, AS (2002) Solute and macromolecule diffusion in cellular aqueous compartments, *Trends Biochem. Sci.* **27**, 27.

Yodh, AG, Lin, K-H, Crocker, JC, et al. (2001) Entropically driven self-assembly and interaction in suspension, *Phil. Trans. R. Soc. Lond.* **A359**, 921.

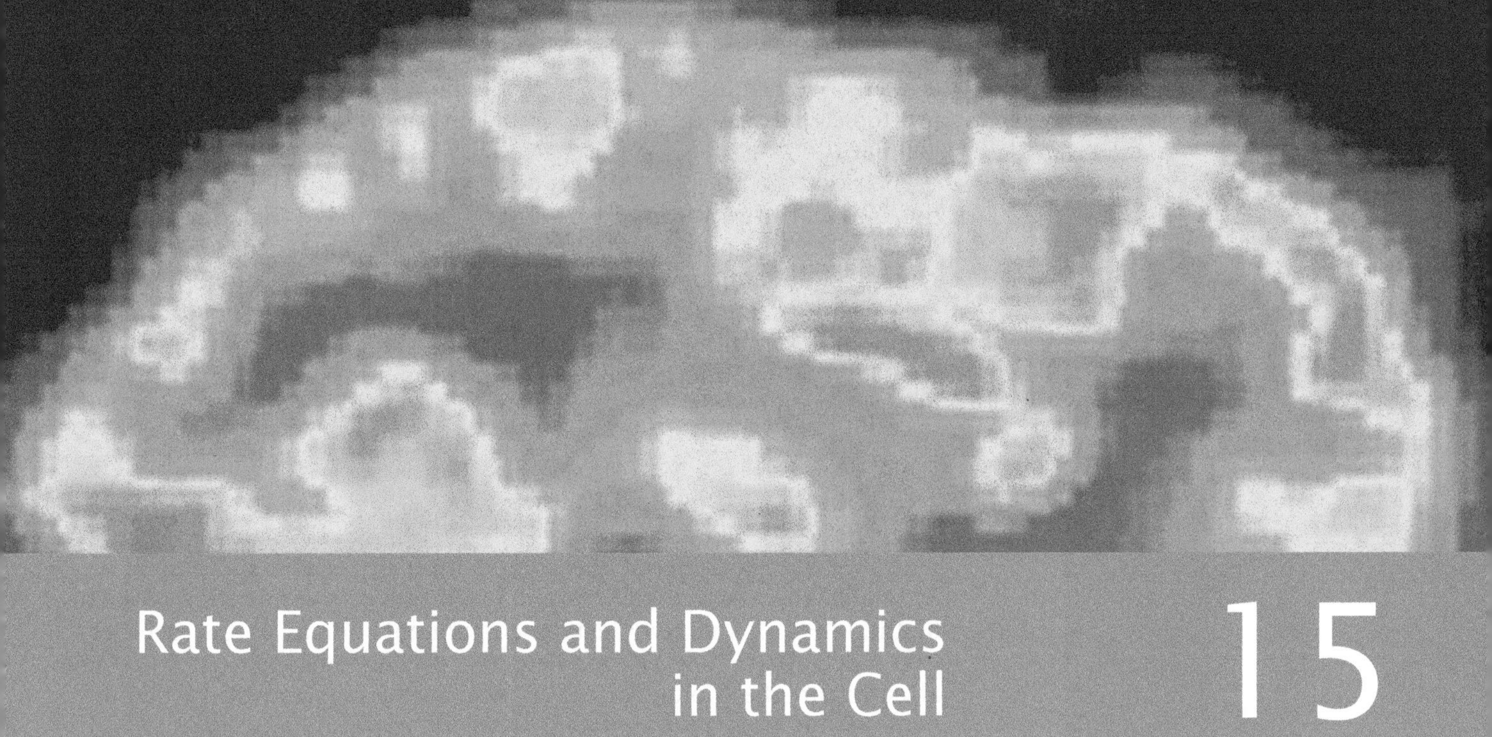

Rate Equations and Dynamics in the Cell

15

Overview: In which dynamic trajectories of molecules are described using rate equations

Much of our discussion of the interplay of the molecules and macromolecular complexes that fill the cell, both large and small, has been founded upon the use of arguments about chemical potentials and equilibrium constants. In living cells, however, circumstances are constantly changing. Ion channels are constantly opening and closing, enzymes are constantly catalyzing biochemical transformations, and the elements of the cytoskeleton are growing and shrinking. In this chapter, we introduce deterministic and statistical approaches for handling these kinds of dynamical processes within cells. In particular, this chapter introduces the use of rate equations, which provide a description of the time variation of populations of different chemical species. We illustrate these ideas with special reference to the dynamics of ion channels, enzyme kinetics, and cytoskeletal assembly.

"Mathematics is biology's next microscope, only better; biology is mathematics' next physics, only better."

Joel Cohen

15.1 Biological Statistical Dynamics: A First Look

A cell is a bustling metropolis of chemical reactions that are linked together in complex networks of reactants and products. Whether we think about the chemical reactions that make up the metabolic pathways that guarantee the energy solvency of living cells such as are shown in Figure 5.2 (p. 191) or about the synthetic pathways that construct the structural components of such cells, ultimately, all of these pathways are built up of chemical reactions. If we measure the chemical activity of living cells, we see that the rates and identities of these reactions are constantly changing to reflect the dynamics of metabolism, cell division, and motility, to name a few examples. As a result, we are faced with the challenge of putting together modeling tools that are up to the task of describing the rich dynamics of cellular life.

Thus far in this part of the book, we have considered the fluid dynamics of life's watery medium and the diffusive dynamics of random walks in sparse and crowded environments. To truly tackle biological dynamics, we must also consider transformations in which a molecule changes its identity rather than just its position. We begin by examining a few case studies in cellular dynamics and then derive the mathematical toolbox to address these problems quantitatively.

15.1.1 Cells as Chemical Factories

When we first considered the composition of cells in Chapter 2, we estimated the average cell contents in terms of the numbers of each of the kinds of macromolecular component. For a review, see Section 2.1.2 on p. 38. In deriving the average values, we paid no attention to the biological reality that every cell and every component within it is in a constant state of material flux. As cells go about their daily business of consuming, excreting, growing, dividing, moving around, etc., their compositions are constantly changing. Even for a cell under fairly constant conditions, which is not actively changing in size, its individual constituents are constantly being synthesized and degraded in a steady-state manner. When conditions do change, living organisms rapidly adapt by altering their complement of proteins, RNAs, lipids, other macromolecules, and even small metabolites.

Examples of the variation in cells over time at the level of the genes that are expressed were shown in Figures 3.23 (p. 119) and 3.25 (p. 121). It is technically more difficult to measure changes in protein, lipid, or metabolite levels than it is to measure changes in RNA levels, but many such experiments have indeed shown that the compositions can vary to a surprising degree over a very short time. In order to quantitatively understand the nature of these changes, we must be able to describe the processes by which one kind of chemical species is transformed into another. The general theme of transformation is embodied over many different time scales and size scales in living cells. Rapid transformations can include processes such as protein phosphorylation and steps in metabolic pathways catalyzed by enzymes. Slower transformations may include the construction of new large-scale assemblies such as the bundles of stereocilia at the top of an inner ear hair cell during differentiation. Because cells usually contain many copies of each individual kind of molecule, it is often useful to think of transformations at a population scale as changes in concentrations. Accordingly, several of our derivations will consider rate problems as descriptions of changes in concentrations over time. However, ultimately, the transformations so conveniently characterized in terms of concentrations are nothing but time-varying reactions of molecules. One of the main goals of the chapter will be to integrate the molecular-level and population-level views of rate equations.

In order to give a feeling for the numbers with respect to rates of biological transformations, we consider an everyday example, literally the amount of energy that is used by the human body every day.

Estimate: Rate of ATP Synthesis in Humans To get a sense of the cellular ATP budget, consider the ATP equivalent of the average daily human diet of 2000 kcal. If we use an approximate figure of 12 kcal/mol as the energy liberated by the hydrolysis of ATP and further assume that half of the energy input in the form of our diet is turned into ATP, the number of moles of ATP synthesized each day is equal to (1000 kcal/day)/(12 kcal/mol) = 80 mol/day. Given that the molecular weight of ATP is roughly 500 g/mol, this implies a daily turnover of more than 40 kg of ATP! Obviously a human body does not at any moment have 40 kg of ATP; this mass reflects the constant turnover accompanying metabolic processes. A total of 40 kg of ATP is synthesized as part of this busy metabolic enterprise.

15.1.2 Dynamics of the Cytoskeleton

In addition to the constant flux of small metabolites of the cell such as ATP, there is also a surprising degree of turnover among the cell's structural elements. We have already considered the filaments of the cytoskeleton in a number of contexts, for example, as macromolecular assemblies in Chapter 2, as beams that can bend and buckle in Chapter 10, and as one of the key elements that make the cell so crowded in Chapter 14. However, we have neglected to explore the construction of these filaments and their extraordinarily dynamic nature.

All cytoskeletal polymers within living cells are constantly growing and shrinking by addition and loss of protein subunits at the same time as they are serving as construction beams and tracks for molecular motors. The rates of the process of filament assembly are

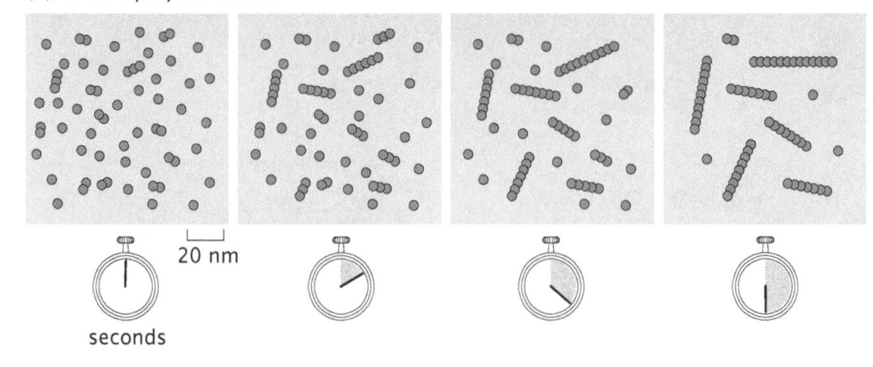

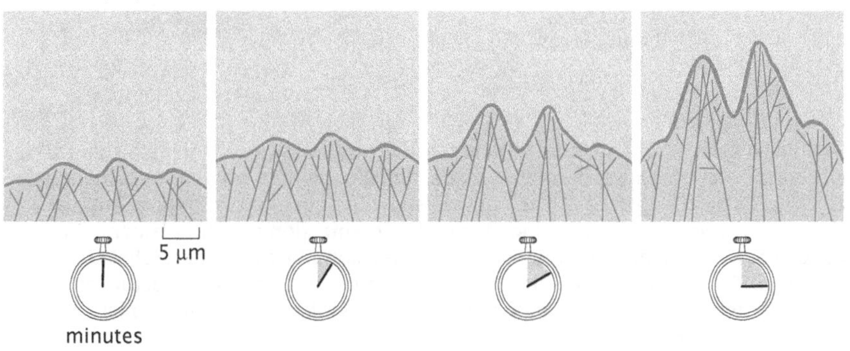

Figure 15.1: Snapshots in the life history of cytoskeletal filaments. These schematics aim to give a sense of the spatial and temporal dynamics of cytoskeletal assembly. (A) *In vitro* assay revealing the growth of cytoskeletal filaments assembled from monomeric subunits. (B) *In vivo* time series showing the dynamics of actin filaments growing at the leading edge of a crawling cell.

illustrated in Figure 15.1 for actin filament assembly both *in vitro* and *in vivo. In vitro,* a solution of purified actin monomers can assemble into filaments over a time scale of a few minutes. In cells, this same kind of assembly over comparable time scales can be harnessed to push forward the leading edge of a crawling cell. The microscopic processes that attend polymerization are more complex than those shown here. For example, subunits can dissociate from the filaments as well as assemble, and in addition there are ATP hydrolysis reactions that take place within the polymer. Furthermore, the rate at which monomers are added to the growing filament depends upon the concentration of actin monomers and also on the regulatory influences of many other proteins in the cell.

The rates associated with cytoskeletal dynamics can be estimated by observing the motions of living cells. Figure 15.2 shows a fish skin cell called a keratocyte migrating on a surface. Note that actin filaments are polymerizing at the leading edge of the cell and pushing the membrane forward with the cell moving at an average rate of 0.2 μm/s. Interestingly, this same fundamental process of orchestrated cytoskeletal assembly can be hijacked by infectious bacteria such as *Listeria monocytogenes* and used as the basis of their own motility within the cytoplasm of the infected human host cell as shown in Figure 15.3. In this case, the bacterium manipulates the host cell cytoskeletal self-assembly process to form a comet tail made up of actin filaments that pushes the bacterium along at rates ranging between 0.05 and 1.4 μm/s, depending upon the host cell type. As shown in Figure 15.3(C), the key components of this system can

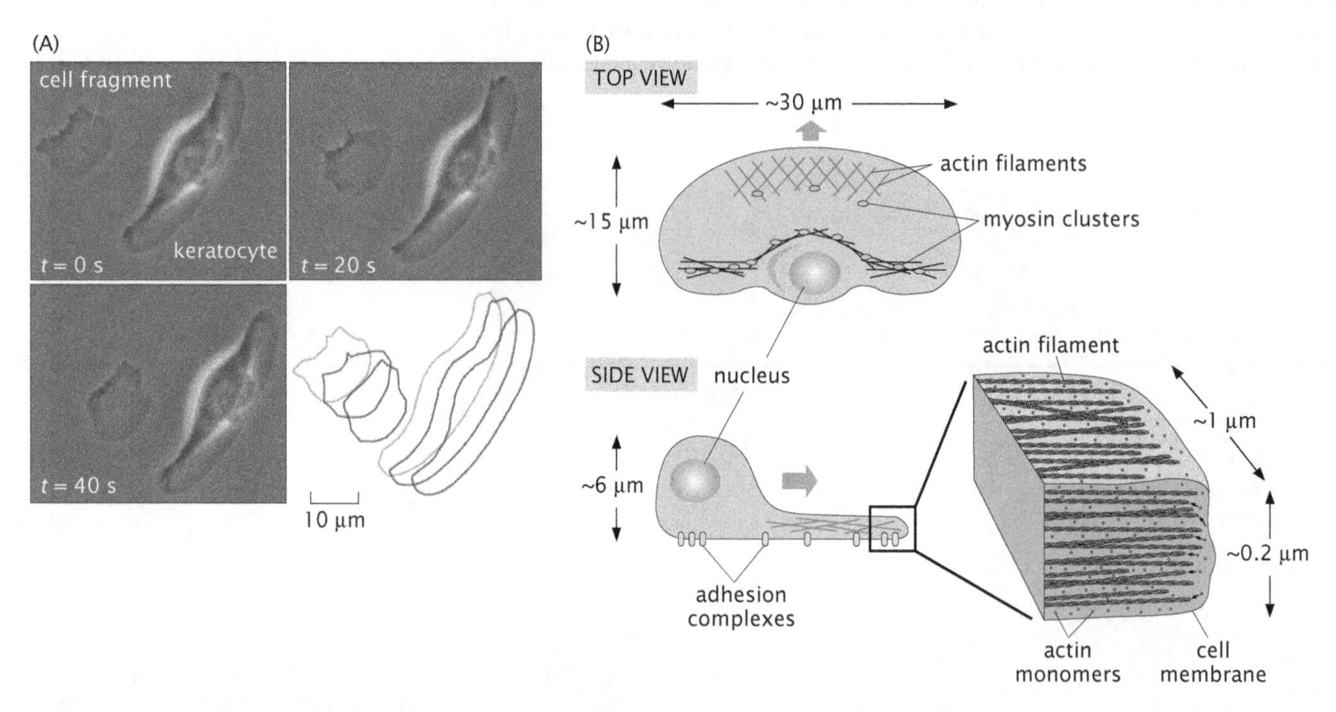

Figure 15.2: Actin-based crawling motility of epithelial cells. (A) This series of time-lapse images shows a single fish skin cell moving across a glass cover slip. Behind the cell an actin-rich fragment of another cell's lamellipodium is crawling autonomously without a nucleus or cell body. The frames were captured at 20 s intervals. The outlines show the positions of the cell and cell fragment at these time intervals. Note that they glide forward without changing shape. (B) Organization of actin filaments in the keratocyte. Seen from the top as in the microscope images, the lamellipodium is a large extension filled with a crosslinked network of actin filaments (see also Figure 14.2 on p. 547). Seen from the side, the lamellipodium is a very flat structure that drags the rounded cell body and nucleus behind it. A schematic illustration of the leading edge indicates the approximate density of actin filaments in this structure. (A, courtesy of G. Allen, K. Keren, and J. Theriot.)

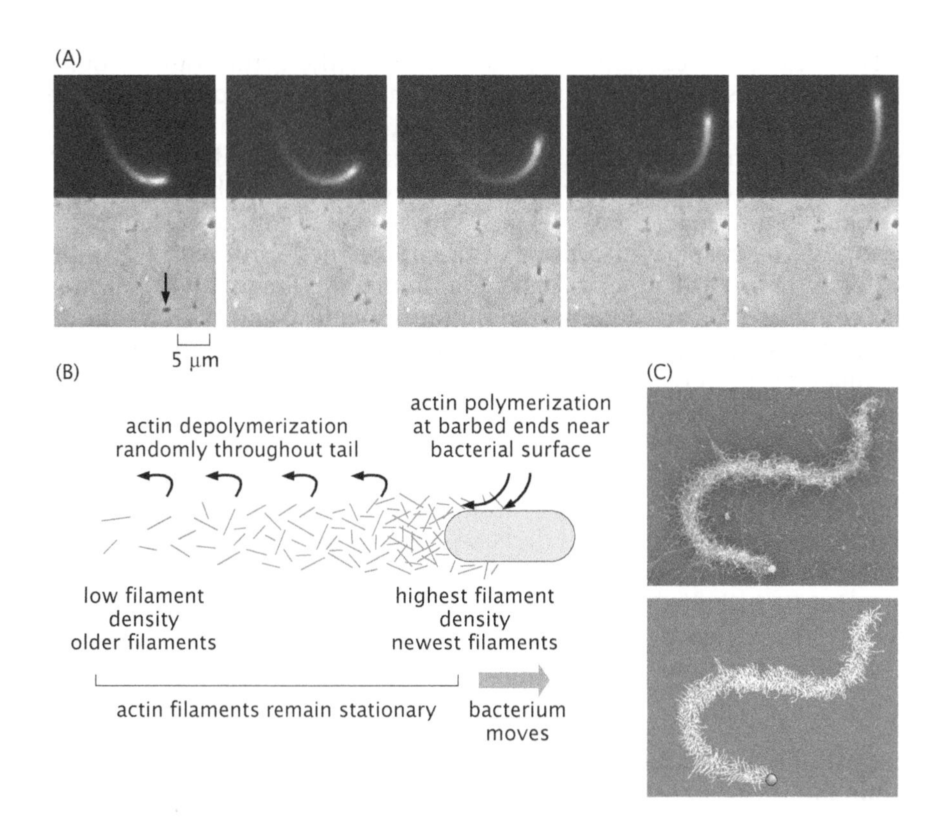

Figure 15.3: Actin-polymerization-driven movement of *L. monocytogenes*. (A) This series of time-lapse images shows the movement of a bacterium in a cytoplasmic extract. Phase-contrast images of the bacterium are shown on the bottom in each frame (the bacterium is indicated with an arrow) and fluorescently labeled actin is shown on the top frame. The frames were collected at 30 s intervals. (B) Schematic diagram showing the dynamics of actin filaments in the comet tail. (C) Electron micrograph of plastic bead driven by actin polymerization with schematic shown below. (A, courtesy of D. Fung and J. Theriot; C, courtesy of L. Cameron, T. Svitkina, J. Theriot, and G. Borisy.)

be abstracted from their cellular context and induced to drive the motions of synthetic beads as well.

Estimate: The Rate of Actin Polymerization We can use the observed rate of cell migration in the examples considered in Figures 15.2 and 15.3 to estimate the mean rate of polymerization of actin filaments. Because the motion of both the keratocyte and *Listeria* reflect the incorporation of monomers on linear actin filaments, we can make a simple estimate of the rate of polymerization through a knowledge of the speed of the cell and the size of the individual monomers that make up the actin filament. The mean velocity of a *Listeria* bacterium in a typical epithelial host cell is comparable to the speed of keratocyte migration, about 0.2 μm/s. For each G-actin subunit added to the growing filament, it increases in length by approximately 3 nm (see Figure 10.29 on p. 415). If we assume that there is a perfect linear relation between the growth of individual filaments and the macroscopic motion of the cell (obviously, this is a gross oversimplification, but nevertheless provides a useful bound), then we estimate that the mean incorporation rate is

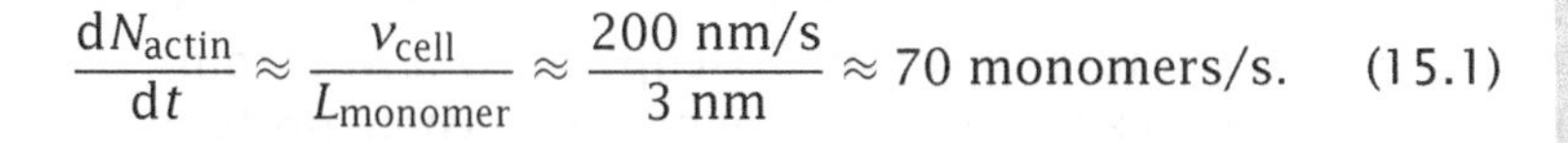

$$\frac{\mathrm{d}N_{\mathrm{actin}}}{\mathrm{d}t} \approx \frac{v_{\mathrm{cell}}}{L_{\mathrm{monomer}}} \approx \frac{200\ \mathrm{nm/s}}{3\ \mathrm{nm}} \approx 70\ \mathrm{monomers/s}. \qquad (15.1)$$

Note that this rate of polymerization is very characteristic of cellular polymerization and will serve as a useful rule of thumb in subsequent discussions.

Three notes of caution are in order. First, the actual rate of polymerization is strongly dependent upon the concentration of available subunits, as we will explore in more detail later in the chapter. Second, actin polymerization is not the rate-determining step for movement of either the keratocyte or *Listeria*. Other forces such as adhesion actually limit their speeds. Third, living cells employ a legion of cytoskeleton-associated proteins to regulate the location and dynamics of actin and the other cytoskeletal filaments. Frequently, the rates of cellular events are determined by the activation and localization of these accessory proteins, rather than actin itself. Nevertheless, this estimate accurately demonstrates that cytoskeletal dynamics must be considered on time scales in which individual events may take place in only tens of milliseconds.

Besides the assembly of the actin structures at the leading edge during cell motility, another dramatic example of regulated actin dynamics is seen during cell division, when a belt of actin and myosin filaments quickly assembles around the center of a cell and contracts to pinch it in half. In the fission yeast *Schizosaccharomyces pombe*, about 30 accessory factors have been identified that modulate the assembly of actin filaments in the cleavage ring.

EXPERIMENTS

Experiments Behind the Facts: Taking the Molecular Census As stated above, rates for cytoskeletal reactions depend upon the concentrations of the proteins involved. Many proteins are not uniformly distributed throughout the cell, and we need to know the local concentration rather than the global average concentration of the protein of interest. For these purposes, merely knowing the number of molecules per cell is not enough, we must also know their spatial distribution. The combination of fluorescence microscopy with other quantitative techniques permits this kind of measurement.

An example of a census of actin-related proteins is shown in Figure 15.4. It is first necessary to determine the total concentration of each protein and then to characterize the distribution of the protein within individual cells. For a large number of proteins believed to play roles in cell division, *S. pombe* strains were generated in which exactly one protein of interest was fused to a yellow fluorescent protein, YFP. In order to determine the number of molecules of each tagged protein present in the cells, the average protein concentration can be measured using a quantitative technique such as Western blotting. In a Western blot, the proteins from a large population of cells are all run out together on a gel and then probed with an antibody. A specific antibody against YFP recognizes all of the tagged proteins with equal affinity, and the Western blot signal can be calibrated against a known dilution of purified YFP. Dividing this by the number of cells loaded on a gel gives a measurement of the average concentration of each tagged protein per cell. Having calibrated the actual number of molecules per cell using the Western blot, the measured fluorescence intensity (observed either by fluorescence microscopy or by flow cytometry) can be related to an absolute scale of cytoplasmic protein concentration. Finally, local concentrations can be determined from the fluorescence microscopy images. Proteins involved in building the cytokinetic furrow (plane of cell division) are concentrated up to 100-fold on the furrow relative to their global abundance.

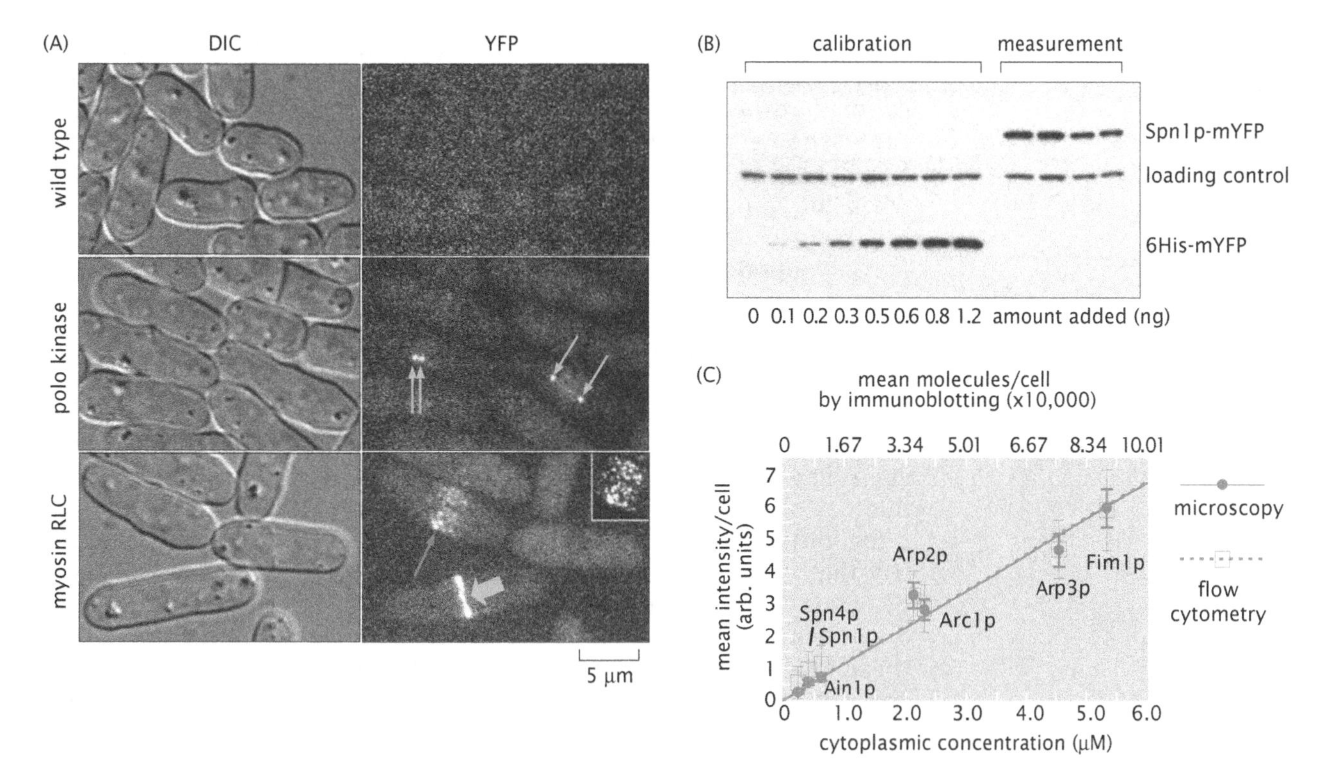

Figure 15.4: Calibration and quantitation of cell-division-associated proteins in the fission yeast *S. pombe.* (A) Wild-type yeast cells, clearly visible as individuals by differential interface contrast microscopy, have no fluorescent signal in the YFP channel. Polo kinase is not visible in most cells, but, in two cells caught in the process of chromosome segregation, the kinase is seen concentrated in pairs of dots (yellow arrows). The myosin regulatory light chain accumulates in a narrow ring as seen in the cell at the bottom of the image (blue arrow), immediately before the cell divides. At a slightly earlier stage in the cell cycle, myosin regulatory light chain begins to accumulate near the presumptive division site in a loose band of dots (red arrow). (B) The average concentration of each protein fusion can be measured using calibrated Western blots. On the left-hand half of the gel, known amounts of purified YFP protein are loaded together with a control protein present in the same amount in every lane. On the right-hand part of the gel, four replicate measurements are made for a single YFP-tagged protein. The amount of the tagged protein is determined by comparing the intensity of the bands with the calibration curve. (C) The number of molecules per cell determined by immunoblotting is linearly related to the average fluorescence intensity per cell measured by flow cytometry or fluorescence microscopy. (A–C, adapted from J.-Q. Wu and T. D. Pollard, *Science* 310:310, 2005.)

15.2 A Chemical Picture of Biological Dynamics

We have given a few qualitative descriptions of cases where understanding rates of transformation is critical to understanding biological processes, with examples ranging from the energy budget of ATP in the cell to the dynamics of the cytoskeleton. We now turn to building the quantitative toolkit that will allow us to treat these cases rigorously by tracking the concentrations of the different molecular players over time. A quantitative theory of such reactions must ultimately answer three key questions about all of the molecular participants: how many molecules are there, where are they, and when are they there? To that end, we will first focus on the general ideas associated with the rate equation paradigm, and then provide several different case studies, including the dynamics of ion channel gating and the time evolution of cytoskeletal filaments.

15.2.1 The Rate Equation Paradigm

The starting point for our discussion is the idea of a chemical concentration. Recall that the concentration tells us the number of molecules of a given species per unit volume (see Figure 13.9 on p. 516). The conventional reason for using a concentration rather than explicitly calling out the *number* of molecules is that the concentration is applicable whether we talk about a drop of water or an entire lake, though, as will be seen when discussing the Gillespie algorithm in Chapter 19, when the number of molecular players is very small these ideas can break down. As we will see below, concentrations (and not absolute particle numbers) often dictate the rates of chemical reactions.

Chemical Concentrations Vary in Both Space and Time

For the purposes of the present discussion, we now imagine that the concentration of the ith species varies with both position and time such that we think of the concentration as $c_i(\mathbf{r}, t)$, which tells us the number of molecules of type i per unit volume at position $\mathbf{r}$ at time t. Exploiting a definition of concentration as a "field variable" (a quantity that varies in space) anticipates the fact that the chemical state of a test tube or a cell can develop spatial nonuniformity (that is, dependence on $\mathbf{r}$) and time variation. The assumption that makes these ideas tolerable is that the spatial variation of the concentration changes over distances that are large compared with the mean spacing of the molecules themselves. That is, every little box within the overall volume behaves like a box of uniform concentration.

We note that in the context of a living cell, there may be reasons to doubt the validity of the concentration idea itself since (i) the number of molecules can be exceedingly small and (ii) such molecules can be localized to membranes or particular organelles. Thus it is important to bear two things in mind: first, local concentration rather than global concentration is the appropriate parameter for cases where molecules are confined to particular organelles or regions, and, second, for those cases where actual molecular numbers are small, it will be important to consider the stochastic behaviors of individual trajectories rather than global averages. With these provisos, we forge ahead.

Rate Equations Describe the Time Evolution of Concentrations

We begin with the simplest case, in which we assume that the concentration at one point in space is identical to that at another. As a result, we drop the label $\mathbf{r}$ in our description of the concentration and focus only on $c_i(t)$, which changes over the course of time as a result of the reactions that link the various reactants, $\{c_j\}$. We introduce the notation $\{c_j\}$ as shorthand for the set $(c_1, c_2, c_3, \ldots, c_n)$ where each subscript refers to a different species. The fundamental postulate of the rate equation paradigm is that we can write the time evolution of the concentration in the form of a differential equation as

$$\frac{\mathrm{d}c_i(t)}{\mathrm{d}t} = f(\{c_j\}; \{k_i\}), \tag{15.2}$$

where $f(\{c_j\}; \{k_i\}) = f(c_1, c_2, \ldots, c_n; k_1, k_2, \ldots, k_m)$. These various concentrations $c_1, c_2, \ldots, c_n$ are for all of the species implicated in the reactions of interest and the parameters k_i are "rate constants" that dictate how fast the various reactions go. Equation 15.2 says that the concentration of the ith species changes in time. How much it will change depends upon the concentrations of all of the various other species that couple to c_i. To make these ideas precise, we now consider explicit examples of different types of reaction that take $f(\{c_j\}; \{k_i\})$ from abstract to concrete form.

15.2.2 All Good Things Must End

Macromolecular Decay Can Be Described by a Simple, First-Order Differential Equation

One of the key physical processes that must be considered when trying to endow the function $f(\{c_j\})$ with real content is the idea that macromolecules decay and are degraded. For example, if we consider the mRNA within a cell, these molecules are generally short-lived in comparison with the average division time of the cell. Similarly, proteins in the cell have a characteristic tendency to decay. The actual lifetime of an individual protein may vary from one kind of protein to another and may be regulated within the cell.

For these decay reactions, just as for radioactive decay, the material does not vanish but rather is transformed into something else. Degraded mRNA is broken down into individual nucleotides and degraded proteins are broken down into individual amino acids. For the purposes of our discussion on rates of decay reactions, we will begin by ignoring these complexities and instead consider the concentration of only one species at a time.

Our principal biological example in this section will be the chemical reaction that lies at the heart of photosynthesis and also vision. A small organic molecule called retinal, illustrated in Figure 15.5, can exist in two slightly different conformational forms. As found in its natural environment associated with a protein in the membrane of photosynthetic archaea, the lower-energy form of retinal, or ground state, is referred to as "all-*trans*-retinal" because of the arrangement of the double bonds in its long tail. A second form, 13-*cis*-retinal, exists in a slightly higher-energy state. To go from the all-*trans* form to the 13-*cis* form, one of the carbon–carbon double bonds must undergo a 180° rotation introducing a kink in the middle of the molecule's tail. In nature, retinal and related molecules can undergo this conformational change (also called an isomerization reaction) when they collide with a photon carrying the appropriate energy. Absorption of the energy from the photon enables retinal to overcome the significant energy barrier for rotation of this double bond as illustrated in Figure 15.5(B). Given sufficient time, the slightly higher-energy 13-*cis*-retinal will decay back to the lower-energy form of all-*trans*-retinal.

This tiny conformational change is exploited by cells in several remarkable ways. In photosynthetic archaea, retinal is embedded at the heart of a large transmembrane protein called bacteriorhodopsin. The photon-induced kinking of the retinal is amplified by the protein to generate a large-scale conformational change that eventually results in the transport of a single hydrogen ion from the inside of the cell to the outside of the cell. Thus, the energy of the photon has been converted into an electrochemical transmembrane gradient, which can be used by the archeon to generate ATP, to spin its flagella, and for any other purpose it desires. In the human retina, protein homologs of bacteriorhodopsin are found in light-sensitive rod and cone cells. Here a slightly different isomerization of retinal (from the 11-*cis* form to the all-*trans* form) initiates a signal transduction cascade that triggers neurons to communicate the change in state to the visual cortex of the brain, where we perceive retinal isomerization as light.

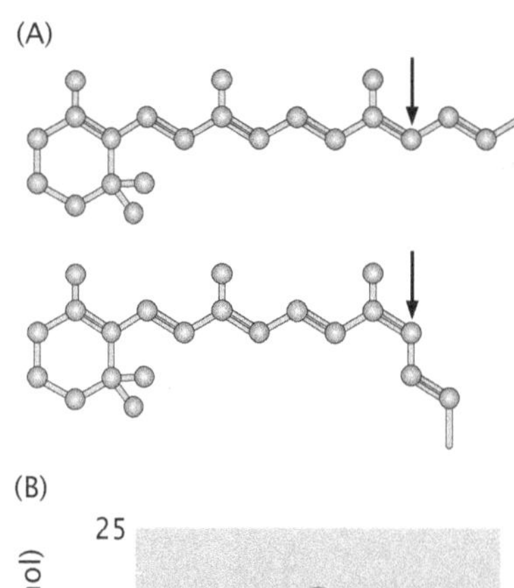

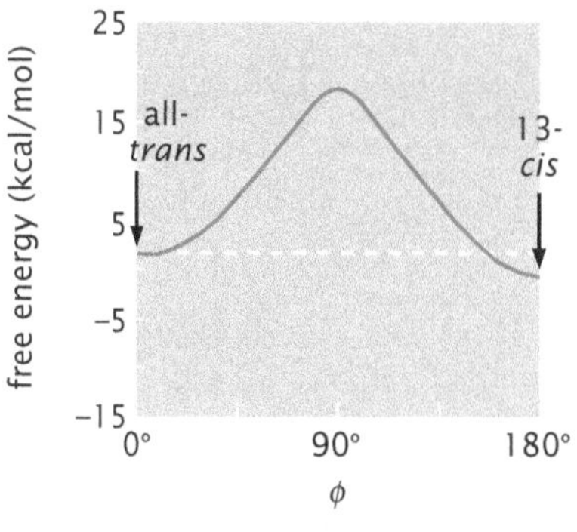

Figure 15.5: The conformational change of retinal. (A) The all-*trans* form of retinal, shown at the top, is its lowest-energy state. The 13-*cis* form, shown at the bottom, is formed by rotation around the atom indicated with the arrow. This rotation is commonly caused by absorption of a photon. (B) The calculated energy landscape for retinal as a function of the degree of rotation around the carbon atom reveals a large energy barrier between the *cis* and *trans* forms. (B, adapted from A. Hermone and K. Kuczera, *Biochemistry* 37:2843, 1998.)

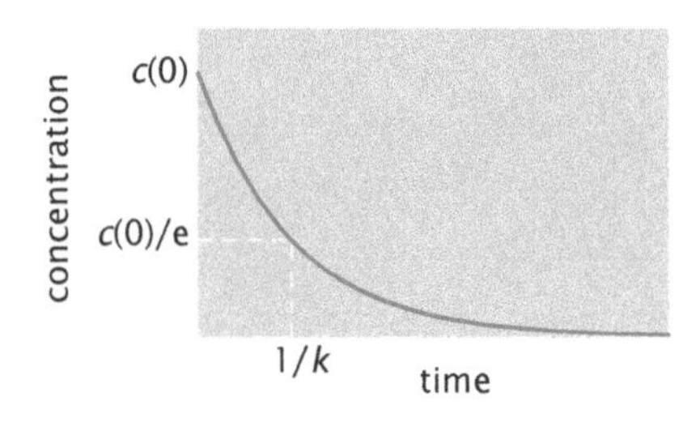

Figure 15.6: The number of molecules as a function of time during a decay process. The time scale $1/k$ sets the time it takes before the concentration has decayed to $1/e$ of its initial value.

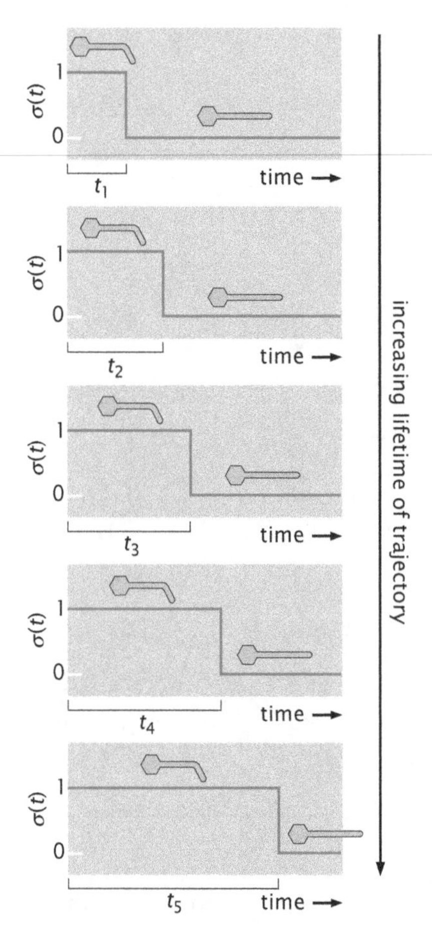

Figure 15.7: Schematic showing the class of microscopic trajectories for a system undergoing simple isomerization/decay dynamics. The two states of the system are labeled by 1 and 0, and hence a given trajectory is characterized by some waiting time in the 1 state followed by a fast decay to the 0 state. The transition itself is assumed to be very fast relative to the waiting time. The set of trajectories shown here provides a few representative examples. The waiting time t_i is actually a continuous variable, so there is a continuum of allowed trajectories.

Decay reactions like the conversion of 13-*cis*-retinal into all-*trans*-retinal can be captured very simply using equations of the form of Equation 15.2. In particular, for the case in which we are interested in the simple decay of a given reactant, we have $f(c;k) = -kc$, which says that the rate of decay of the reactant is proportional to how much reactant there is around. The parameter that characterizes that decay rate for a given molecular species is k and the evolution equation itself is

$$\frac{dc(t)}{dt} = -kc(t). \tag{15.3}$$

Note that the constant k that parameterizes the rate of decay has units of 1/time.

This linear differential equation can be solved by recourse to the method of separation of variables (see Equation 13.58 on p. 530) and results in a solution of the form $c(t) = c(0)e^{-kt}$, yielding a time-dependent concentration profile of the form

$$c(t) = c_0 e^{-t/\tau}, \tag{15.4}$$

where c_0 is the initial concentration (that is, $c(0) = c_0$) and we have defined a characteristic time $\tau = 1/k$. The time evolution in the concentration of molecules that suffer such decay is shown in Figure 15.6. This simple calculation shows that if we start out with some initial concentration c_0, this concentration will steadily decline over time with an average exponential profile like that shown in the figure and symbolized in Equation 15.4. A cautionary note is in order: the use of rate equations like those advocated here focuses on the average values of quantities of interest. This automatically deprives us of deeper contact with fluctuational quantities. This is one reason we now turn to considering these processes from the single-molecule perspective.

15.2.3 A Single-Molecule View of Degradation: Statistical Mechanics Over Trajectories

Molecules Fall Apart with a Characteristic Lifetime

As noted above, the simplest mathematical description of decay dynamics is founded upon introducing a characteristic decay constant. It is of interest to examine this decay constant more deeply from both a phenomenological and a microscopic perspective. We adopt the view that the molecule can undergo any of an infinite number of different microscopic trajectories. The idea is that when the system has not yet decayed it is in a state labeled with a 1 and when it has decayed it is labeled with a 0. Here we are generalizing the equilibrium ideas of Chapter 7 to the dynamical case where we consider the transitions between the two states over time. An example of this class of trajectories is shown in Figure 15.7.

Mathematically, we label these trajectories $\sigma_i(t)$, where the trajectory is defined as

$$\sigma_i(t) = \begin{cases} 1 & \text{if } t < t_i, \\ 0 & \text{otherwise,} \end{cases} \tag{15.5}$$

where the time t_i is the waiting time associated with the trajectory of interest. The role of the constant k (or alternatively, of $\tau = 1/k$) is to characterize the *average* lifetime of the macromolecule of interest. Because the rate constant k has units of inverse time (for example, s^{-1}) as it describes the probability that something will happen per unit time, its reciprocal τ simply has units of time and can be thought of intuitively as dictating the average lifetime of the molecule. That is, a given macromolecule, such as a mRNA, can decay after a waiting time of one second or one minute or one week. However, when we consider a huge collection of such molecules, it makes sense to speak of an average lifetime associated with these molecules, and that is what the decay constant captures.

We can use this trajectory picture with our "trajectories-and-weights" approach introduced in Figure 13.14 (p. 520). We discretize time into N intervals of length Δt (so the total time $t = N\Delta t$) and then find the probability of a given trajectory by multiplying the probabilities of what happens at each time step over the entire trajectory. For example, if the molecule survives for N time steps and decays on the $(N+1)$th time step, then the probability of that process is

$$p(t)\Delta t = \underbrace{(1 - k\Delta t) \times (1 - k\Delta t) \times \cdots \times (1 - k\Delta t)}_{N \text{ time steps}} \times k\Delta t, \tag{15.6}$$

where $k\Delta t$ is the probability of a decay during any time step and $1 - k\Delta t$ is the probability of no decay at a given time step. The function $p(t)$ is a probability density and requires that we compute $p(t)\Delta t$ to compute the probability that the decay occurs between time t and $t + \Delta t$. If we use the fact that $\Delta t = t/N$, this result can be rewritten as

$$p(t)\Delta t = (1 - k\Delta t)^N k\Delta t = \left(1 - \frac{kt}{N}\right)^N k\Delta t \approx k e^{-kt}\Delta t, \tag{15.7}$$

where we have used the identity that $e^{-x} = \lim_{N\to\infty}(1 - x/N)^N$. This result tells us that once we have set the mean lifetime ($\tau = 1/k$), the probability of decay in the time interval between t and $t + \Delta t$ is

$$p(t)\Delta t = \frac{1}{\tau} e^{-t/\tau}\Delta t. \tag{15.8}$$

Decay Processes Can Be Described with Two-State Trajectories

Another way of illustrating the trajectory perspective for decay processes is to exploit the maximum-entropy approach introduced in Section 6.1.5 (p. 253). Recall that each decay process is characterized by a trajectory corresponding to how long the system is in state A before decaying to state B. It is convenient to characterize the molecular state with the discrete index 1 for state A and 0 for state B. Our aim here is to use the information-theoretic approach to statistical mechanics introduced in Section 6.1.5 (p. 253) to obtain the probability distribution on these trajectories. Just as our discussion in Chapter 6 showed several ways to obtain the Boltzmann distribution, our present discussion illustrates several distinct ways of deriving the waiting time distribution, $p(t)$.

We are interested in the probability $p(t)$, which signifies the probability of a microtrajectory with lifetime t, precisely the same quantity considered in Equation 15.8. The only information about the overall process that we invoke in our analysis is that the average lifetime is τ. In light of this constraint, we can write the constrained Shannon entropy functional as

$$S = -\underbrace{\int_0^\infty p(t)\ln p(t)\,\mathrm{d}t}_{\text{Shannon entropy}} - \underbrace{\lambda\left(\int_0^\infty p(t)\,\mathrm{d}t - 1\right)}_{\text{normalization constraint}}$$

$$-\underbrace{\mu\left(\int_0^\infty tp(t)\,\mathrm{d}t - \tau\right)}_{\text{average lifetime constraint}}. \tag{15.9}$$

More concretely, the constraints captured with the Lagrange multipliers above are

$$\int_0^\infty p(t)\,\mathrm{d}t = 1, \tag{15.10}$$

which specifies that the probabilities sum to 1, and

$$\int_0^\infty tp(t)\,\mathrm{d}t = \tau, \tag{15.11}$$

which guarantees that the average lifetime is τ. Refer to The Math Behind the Models on p. 254 for a reminder on the Lagrange multiplier concept. As usual, we are now faced with the prospect of maximizing this constrained Shannon entropy by equating its derivative to zero and solving for $p(t)$, resulting in

$$-\ln p(t) - \lambda - 1 - \mu t = 0. \tag{15.12}$$

This results in an expression for the probability of a trajectory with lifetime t of

$$p(t) = \mathrm{e}^{-1-\lambda}\mathrm{e}^{-\mu t}. \tag{15.13}$$

Now, we have to impose the constraints in order to determine the Lagrange multipliers λ and μ. The normalization constraint results in the condition

$$\int_0^\infty p(t)\,\mathrm{d}t = \mathrm{e}^{-1-\lambda}\underbrace{\int_0^\infty \mathrm{e}^{-\mu t}\,\mathrm{d}t}_{=1/\mu} = 1. \tag{15.14}$$

As a result of this analysis, we have

$$\mathrm{e}^{-1-\lambda} = \mu. \tag{15.15}$$

Given that we have the normalization of the probability distribution in hand, which leads us to

$$p(t) = \mu\mathrm{e}^{-\mu t}, \tag{15.16}$$

we now determine the second Lagrange multiplier by exploiting the average lifetime $\tau = \int_0^\infty tp(t)\,\mathrm{d}t$. In light of our determination of the first Lagrange multiplier, this condition results in

$$\tau = \mu\int_0^\infty t\mathrm{e}^{-\mu t}\,\mathrm{d}t. \tag{15.17}$$

We can now perform this integration by recourse to the simple trick of differentiating with respect to μ to bring the factor t into the integrand (introduced in The Tricks Behind the Math on p. 525), resulting in

$$\tau = \mu\left(-\frac{\mathrm{d}}{\mathrm{d}\mu}\right)\int_0^\infty \mathrm{e}^{-\mu t}\,\mathrm{d}t = \mu\left(-\frac{\mathrm{d}}{\mathrm{d}\mu}\right)\frac{1}{\mu} = \frac{1}{\mu}. \tag{15.18}$$

Hence, our second Lagrange multiplier μ is now determined as $\mu = 1/\tau$. Substituting this result into Equation 15.16 yields $p(t) = (1/\tau)\mathrm{e}^{-t/\tau}$, leading in turn to the result that the probability that the decay will occur between times t and $t + \Delta t$ is given by

$$p(t)\Delta t = \frac{1}{\tau}\mathrm{e}^{-t/\tau}\Delta t. \tag{15.19}$$

We will come back to this in the context of molecular motors when analyzing the different lifetimes of a motor's internal states (see Section 16.2.3 on p. 647).

Decay of One Species Corresponds to Growth in the Number of a Second Species

The reactions we have been thinking about are described by the simple chemical formula

$$\mathrm{A} \rightarrow \mathrm{B}. \tag{15.20}$$

This kind of process is depicted schematically in Figure 15.8 for the example of retinal. Thus far, we have considered only the fate of species A, but of course in the real reaction the number of molecules of the product species B also changes with time and the time-dependent equations describing the fates of A and B are inextricably linked to one another. In this case, for every decrease in the number of A molecules, we have a corresponding increase in the number of B molecules, which implies the constraint

$$c_\mathrm{A}(t) + c_\mathrm{B}(t) = c_0. \tag{15.21}$$

Indeed, the dynamics of these two species as embodied in Equation 15.21 is nothing more than an expression of mass conservation. The condition of mass conservation implies that

$$\frac{\mathrm{d}c_\mathrm{A}}{\mathrm{d}t} = -\frac{\mathrm{d}c_\mathrm{B}}{\mathrm{d}t} = -kc_\mathrm{A}. \tag{15.22}$$

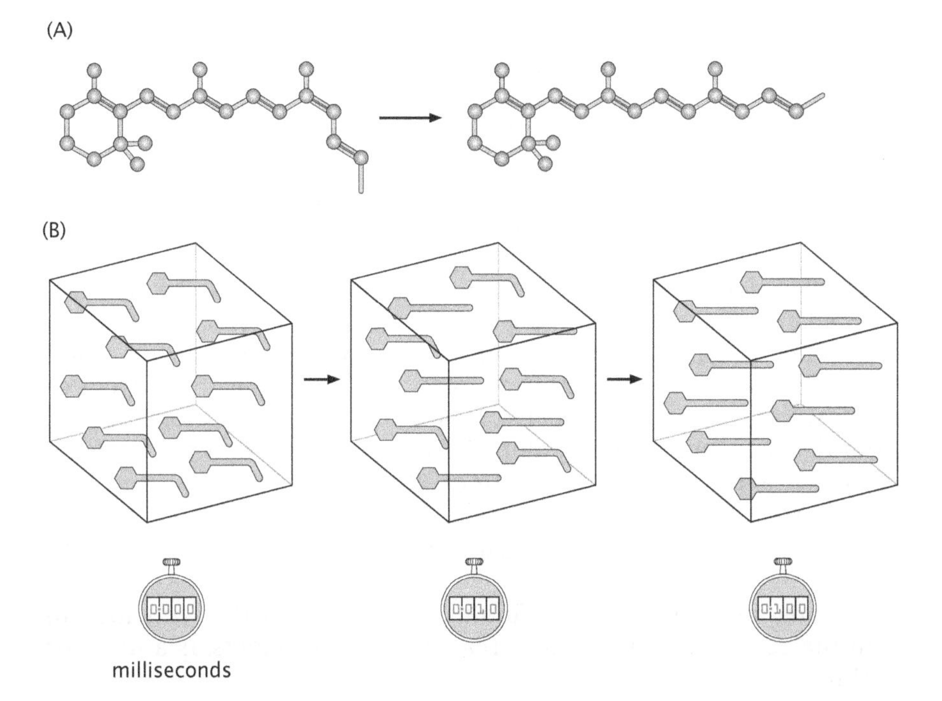

Figure 15.8: Different views of the isomerization process. (A) Schematic of an isomerization process where species A is decaying into species B. In this case, we use the two forms of retinal described earlier in the chapter to characterize the process. (B) Schematic of the change in the populations of the two species over time.

If we consider the situation where initially all of the molecules are A molecules, then we have $c_A(0) = c_0$ and $c_B(0) = 0$. From our earlier solution of the decay problem, it is now straightforward to write down the time evolution of both species. Using the mass conservation condition given in Equation 15.21, we find

$$\frac{dc_B}{dt} = kc_A = k[c(0) - c_B] \Rightarrow c_B(t) = c_0(1 - e^{-kt}). \tag{15.23}$$

The time evolution of both populations under these initial conditions is indicated schematically in Figure 15.8, and the reader is asked to flesh out this solution and make the corresponding plots in the problems at the end of the chapter.

The same ideas introduced above can be used just as well for examining the reversible reaction

$$A \rightleftharpoons B. \tag{15.24}$$

In this case, the rate equations of interest can be written as

$$\frac{dc_A}{dt} = -k_+ c_A + k_- c_B \tag{15.25}$$

and

$$\frac{dc_B}{dt} = -\frac{dc_A}{dt}, \tag{15.26}$$

with the constraint of mass conservation of the form given in Equation 15.21. Equation 15.25 says that the change in concentration of species A comes from two sources, namely, (i) the decay of A into B with rate constant k_+ and (ii) the decay of B into A with rate constant k_-. In this case, the long-time behavior is nonzero concentrations of both A and B, with their ratio dictated by the ratio k_+/k_-. The details of this analysis are left to the reader in the problems.

15.2.4 Bimolecular Reactions

Chemical Reactions Can Increase the Concentration of a Given Species

Decay and isomerization are only a limited subset of the repertoire of interesting biochemical transactions. Indeed, these reactions are unusual in that there is only one molecule involved; this one molecule may change its state, but does not explicitly interact with other molecules. In general, the more interesting cases are those in which two different reactants come together to form some third product. In these cases, the function $f(\{c_j\})$ must have contributions coming from the presence of interactions between the different reactants. In particular, we now consider reactions of the form

$$A + B \rightleftharpoons AB. \tag{15.27}$$

As before, our interest is in finding a description that provides us with the concentrations of both reactants and products as a function of time.

Intuitively, we can argue that the contribution of the association reaction to the overall concentration of the products can be written as

$$\frac{\mathrm{d}c_{\mathrm{AB}}(t)}{\mathrm{d}t} = k_{\mathrm{AB}}c_{\mathrm{A}}c_{\mathrm{B}}, \tag{15.28}$$

where k_{AB} is a constant that reflects the rate of association of A and B molecules. Conceptually, this equation says that A and B molecules come together to form AB molecules and the rate at which they do so is proportional to the probability that A's and B's are at the same place (that is, their concentrations). Note that bimolecular reactions lead to a new feature in our description. In particular, we note that the dynamical equations of different species are coupled. In our earlier treatment of decay reactions, the equation for $c_{\mathrm{A}}(t)$ made no appeal to the concentrations of other species. In the present case, changes in the concentration of one species are accompanied by a concomitant change in the concentrations of other species. Mathematically, this amounts to the fact that the differential equations for the system are *coupled* (in more than the simple way revealed in Equation 15.22).

To see how this works more concretely, we return to one of our workhorse problems, namely, ligand–receptor binding, and now explicitly consider how the system changes over time. We consider a ligand and receptor whose concentrations are [L] and [R], which are either free in solution or bound in the LR complex, whose concentration is [LR]. To model the reaction

$$\mathrm{L} + \mathrm{R} \rightleftharpoons \mathrm{LR}, \tag{15.29}$$

we consider a lattice model as shown in Figure 15.9, in which there are a total of Ω distinct "lattice" sites, L ligands, and R receptors. We treat the solution as an ideal solution of ligand and receptor where the probability of occupancy of an elementary box by ligand is L/Ω and the probability of occupancy of a box by receptor is R/Ω. Furthermore we assume that for L and R to form a complex, they must be in the same box. Given these assumptions, in time Δt, the change in

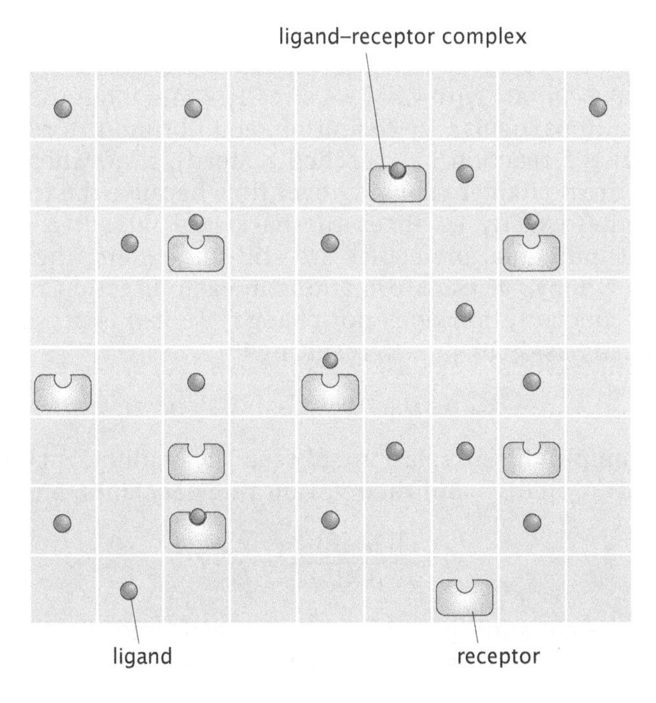

Figure 15.9: Lattice model of ligand–receptor binding dynamics. The system is divided up into Ω distinct boxes. There are L ligands and R receptors in solution and the binding reaction between them to form the complex LR can occur when a ligand and receptor are found at the same lattice site. When a ligand and receptor are found in the same box, the probability that they will react to form LR in time Δt is given by $k'_{\mathrm{on}}\Delta t$.

the number of ligand–receptor complexes due to binding of ligand to receptor can be written as

$$\Delta N_{LR} = \underbrace{-(k_{off}\Delta t)N_{LR}}_{\text{decay term}} + \left(\underbrace{\Omega}_{\text{no. of boxes}} \times \underbrace{\frac{N_L}{\Omega}\frac{N_R}{\Omega}}_{\text{box occupancy prob.}} \times (k'_{on}\Delta t) \right), \tag{15.30}$$

where the first term is the decay process discussed earlier (see Section 15.2.2 on p. 581), while the second term represents the rate at which LR is produced and is obtained by working out the rate per elementary box times the total number of such boxes; k'_{on} is the rate at which ligand–receptor pairs that are colocalized in the same elementary box transform to a ligand–receptor complex.

This result can be recast in the more familiar differential form by dividing the left- and right-hand sides by Δt and then by the volume Ωv, where v is the volume of each elementary box in our lattice model. In particular, we have

$$\frac{d}{dt}\left(\frac{N_{LR}}{\Omega v}\right) = -k_{off}\left(\frac{N_{LR}}{\Omega v}\right) + \frac{\Omega}{\Omega v}\frac{N_L}{\Omega v}\frac{N_R}{\Omega v}v^2 k'_{on}, \tag{15.31}$$

where in the second term on the right we have divided and multiplied by v twice in order to convert to concentration variables such as $[L] = N_L/\Omega v$, $[R] = N_R/\Omega v$, and $[LR] = N_{LR}/\Omega v$. These manipulations transform Equation 15.30 into

$$\frac{d[LR]}{dt} = -k_{off}[LR] + k_{on}[L][R], \tag{15.32}$$

where the bimolecular on-rate is related to the lattice model rate constant by $k_{on} = vk'_{on}$, and has units of $M^{-1}\,s^{-1}$. This heuristic derivation shows how to link simple lattice models with macroscopic rate equations defined in terms of concentration variables.

Equilibrium Constants Have a Dynamical Interpretation in Terms of Reaction Rates

The rate equation formalism described above provides a useful opportunity to make contact with what we already know about the equilibria of reactions. In particular, by definition, equilibrium is a reflection of the fact that the reaction has reached a steady state where the concentrations are no longer changing over time because the forward flux in the reaction exactly balances the backward flux. It is important to note that individual molecules are still undergoing the reactions, but the net number of transformations in each direction is equal, so the overall concentration does not change. We can express this idea mathematically as $d[LR]/dt = 0$, resulting in

$$-k_{off}[LR]_{eq} + k_{on}[L]_{eq}[R]_{eq} = 0. \tag{15.33}$$

This equation provides a relation between the equilibrium concentrations ($[L]_{eq}, [R]_{eq}, [LR]_{eq}$) and the relevant rate constants, namely,

$$K_d = \frac{[L]_{eq}[R]_{eq}}{[LR]_{eq}} = \frac{k_{off}}{k_{on}}. \tag{15.34}$$

Using our rate equation picture, we have recovered the law of mass action derived using statistical mechanics in Section 6.3 (p. 267) where we demonstrated that K_d is the concentration at which $p_{bound} = 1/2$. This equation demonstrates the connection between the equilibrium concentrations and the associated kinetic rate constants. However, as indicated schematically in Figure 15.10, the rate picture allows us to say much more than what the terminal, privileged (equilibrium) state will be. The dynamical picture captured by the rate equations permits us to examine the variation in the concentrations over time for different choices of their initial values (that is, at time $t = 0$).

15.2.5 Dynamics of Ion Channels as a Case Study

An interesting application of the ideas developed so far is the dynamics of ion channels. In Chapter 7, we discussed ion channels as an example of a real biological system that can be fruitfully viewed using the two-state paradigm because in the simplest picture, they exist only in open or closed states (see Figure 7.2, p. 283). There we considered channel behavior only from an equilibrium perspective and now we are ready to turn to the question of rates. For example, when the membrane of a nerve cell is depolarized, how long does it take for the voltage-gated calcium channels to open, and how does the time-dependent behavior of the ion distribution along the whole membrane depend on the behavior of the individual channels themselves? Our goal is to write a rigorous description that relates the trajectories of individual channels, which are intrinsically binary, to the

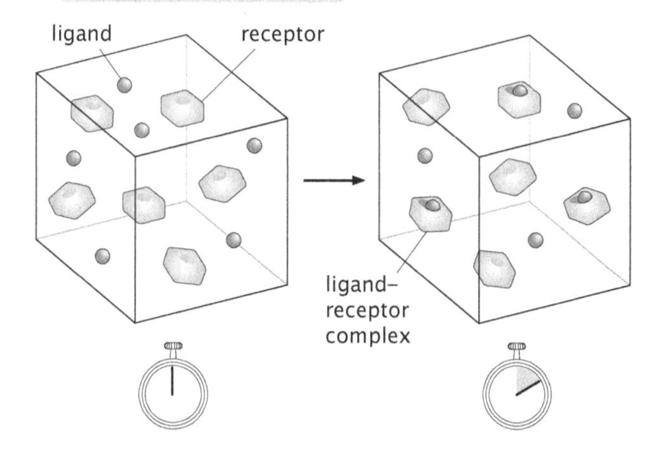

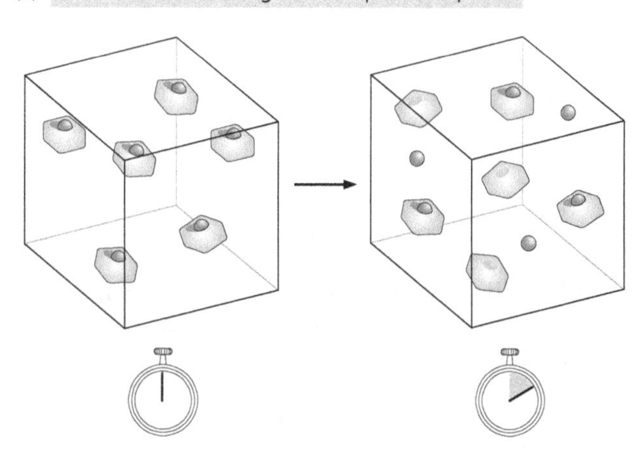

Figure 15.10: Direction of reaction. (A) Time evolution of the system in the case in which the initial concentrations of [L] and [R] are in excess of their equilibrium concentrations. (B) Time evolution of the system in which the initial concentration of [LR] is in excess of its equilibrium concentration.

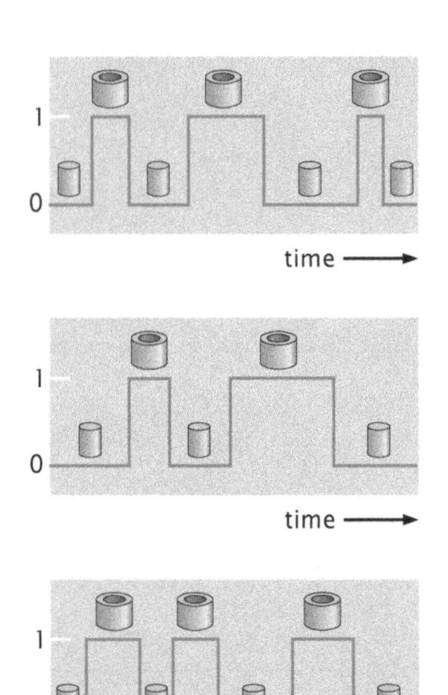

Figure 15.11: Ion channel trajectories. Three representative examples of the time evolution of a two-state ion channel that switches back and forth between the closed and open states. For simplicity, we assume that the switching itself is *instantaneous* in comparison with the time spent in either of the two states. The icons represent the open and closed states.

overall behavior of a large-scale system that may have an analog-like response. In addition, we explore a probabilistic description of how the behavior of identical channel molecules may differ over time. We will model the channel as existing in only two states, open and closed, with only a single relevant transformation reaction converting one state into the other. This will be a warm-up problem for more complex systems later in the chapter that involve more molecular species and more complex transformations.

Rate Equations for Ion Channels Characterize the Time Evolution of the Open and Closed Probability

Examples of the microtrajectories available to individual channels are shown in schematic form in Figure 15.11. Our goal is to write a kinetic description that allows us to classify and analyze different trajectories. One quantity we might like to use to characterize ion channel dynamics is the probability that, if we start in the open state at time $t = 0$, it will still be open a time τ later.

To exploit our rate equation paradigm for the problem of ion channel dynamics, we appeal to Figure 15.12. The idea is that we have a patch of membrane that is occupied by a total of N distinct two-state channels. Our goal is to write a rate equation that characterizes the time evolution of the probability of being in the open state and the way we will do this is to compute $N_O(t)/N$ (where $N_O(t)$ is the number of open channels at time t), which will effectively determine that probability. We adopt a strategy in which time is discretized into steps of length Δt and, at each time step, the channel can undergo a transition from its current state or it can remain in the same state. The "reaction" of interest is of the form

$$\mathrm{O} \underset{k_-}{\overset{k_+}{\rightleftharpoons}} \mathrm{C}, \tag{15.35}$$

where O signifies the open state, C signifies the closed state and k_+ and k_- are the rate constants that determine the probability of a change of state during a given time step. The change in the number of open channels in a given time step can be written as

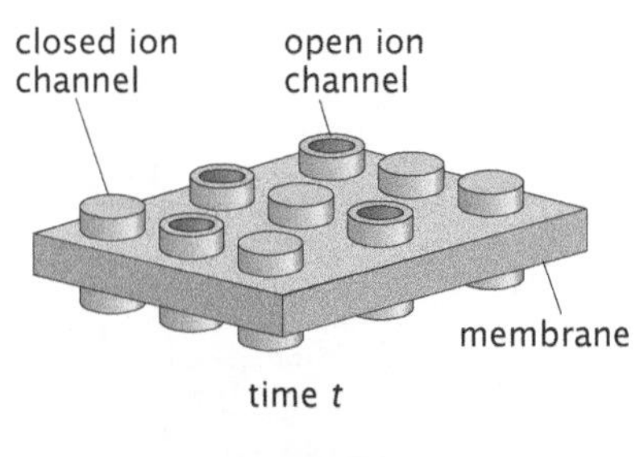

$$\Delta N_O = \underbrace{-k_+ N_O \Delta t}_{\mathrm{O}\to\mathrm{C}} + \underbrace{k_- N_C \Delta t}_{\mathrm{C}\to\mathrm{O}}. \tag{15.36}$$

If we divide both sides by Δt, we find the rate of change in the number of open channels as

$$\frac{\Delta N_O}{\Delta t} = -k_+ N_O + k_- N_C, \tag{15.37}$$

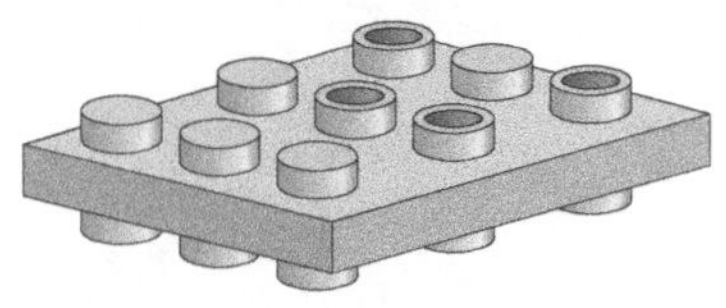

Figure 15.12: Channel gating kinetics. A patch of membrane with a collection of ion channels, some of which are open, some of which are closed. At every time step, channels can either switch their state or stay in the same state.

which can be further simplified by dividing this equation by N itself and using $p_O = N_O/N$ and $p_C = N_C/N$. If we examine the limit as $\Delta t \to 0$, this results in the differential equation

$$\frac{dp_O}{dt} = -k_+ p_O + k_- p_C. \tag{15.38}$$

If we now exploit the fact that $p_O + p_C = 1$, which amounts to the statement that the channels are either open or closed (that is, it is a two-state system), this may be rewritten as

$$\frac{dp_O}{dt} = -k_+ p_O + k_-(1 - p_O). \tag{15.39}$$

This equation is more transparent if written in the form

$$\frac{dp_O}{dt} = -(k_+ + k_-)p_O + k_-. \tag{15.40}$$

In particular, we can solve this equation by resorting to separation of variables and by adding a constant term, resulting in

$$p_O(t) = \frac{k_-}{k_+ + k_-} + Ae^{-(k_+ + k_-)t}. \tag{15.41}$$

The specific solution cancels off the constant term in the differential equation and the exponential term captures the time dependence. If we now exploit the initial condition that $p_O(0) = 1$ (that is, all the channels are open at $t = 0$), we can determine the constant A, resulting in

$$p_O(t) = \frac{k_-}{k_+ + k_-} + \frac{k_+}{k_+ + k_-}e^{-(k_+ + k_-)t}. \tag{15.42}$$

The dynamics of the channel is usefully characterized using this probability, which we will show below can also be thought of as providing the *correlation function*. In particular, this solution tells us how to view trajectories like those given in Figure 15.11 with particular reference to how long the channel stays in the closed and open states between switching events. Consider a single-molecule experiment in which we observe a channel switching back and forth between the closed and open states (experimental data of this kind are shown in Figure 7.2 on p. 283). Imagine we start our stopwatch when the protein is in the open state, and we record the state of the protein in time, assigning 0 to the closed state and 1 to the open state. This will result in a random telegraph signal $\sigma(t)$ like that shown schematically in Figure 15.11. Our task is to compute the correlation function $\langle\sigma(0)\sigma(t)\rangle$. The average can be computed by repeating the experiment over and over again; the figure shows the hypothetical outcomes of three such experiments. Alternatively, we could watch a single channel over a long time period and compute a time average. In the case of a stationary process (one that looks the same whenever we view it), the two procedures give the same result. For a stationary process, the random graph you get by shifting $\sigma(t)$ is just as good as the original, that is, it is drawn from the same distribution.

For a *given* process $\sigma(t)$, the quantity $\sigma(0)\sigma(t)$ is either 1 or 0. The average $\langle\sigma(0)\sigma(t)\rangle$ is therefore equal to the probability of $\sigma(t)$ being 1, *conditioned on* $\sigma(0)$ being equal to 1. But this is precisely the probability $p_O(t)$ that we computed above. Experimentally, one way to evaluate these probabilities is to examine channel trajectories like those shown in Figure 7.2 (p. 283) and to make a histogram of the number of times each waiting time (in the interval t and $t + \Delta t$, where Δt is the bin size) appears. Experimental data of this variety for a sodium channel are shown in Figure 15.13.

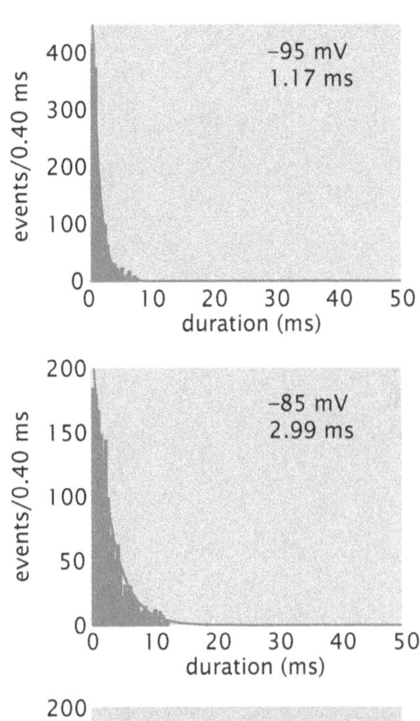

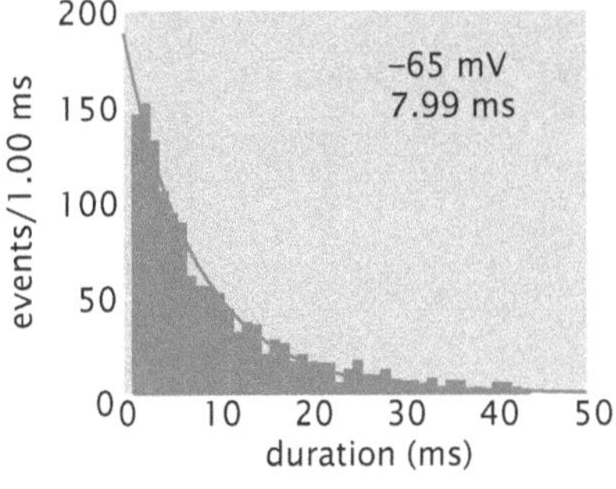

Figure 15.13: Measured channel gating kinetics. Each histogram shows the number of events observed for the open-state lifetime for a voltage-gated sodium channel. The frequencies shown on the y-axis are given as number of events per bin size (measured in milliseconds). Note that the bin size is larger for the bottom graph. The three histograms correspond to different membrane voltages as indicated in the figure. The average lifetime τ is shown for each voltage. (Adapted from B. U. Keller et al., *J. Gen. Physiol.* 88:1, 1986.)

15.2.6 Rapid Equilibrium

Having flexed our muscles with relatively simple two-state ion channels, we are now ready to turn to more complex biological systems that involve multiple transformations. One useful exercise we are now prepared for is to return to the discussion of Section 5.2.1 (p. 200). The goal of that discussion was to assess the circumstances under which we can get away with approximating a nonequilibrium process using equilibrium ideas. Despite the fact that the physics of life is typically portrayed as strictly the province of nonequilibrium thinking, there are many instances in which equilibrium ideas are quite useful. Further, we argued that the question of whether or not equilibrium ideas are appropriate is ultimately a question of time scales. The goal of the present discussion is to use rate equations that explicitly account for the changes in concentration with time so that we can determine whether equilibrium assumptions are appropriate for a particular experiment.

Here we illustrate this idea using a toy model of three chemical species interconverting between one another, with two fast and one slow step characterized by the reaction scheme

$$\mathrm{A} \underset{k_-}{\overset{k_+}{\rightleftharpoons}} \mathrm{B} \overset{r}{\rightarrow} \mathrm{C}. \tag{15.43}$$

We wish to compute the concentration of the three species as a function of time, starting from a state where we have species A only. The fast steps have rate constants $k_\pm \gg r$. Also, we simplify the kinetic scheme by assuming that the $\mathrm{B} \rightarrow \mathrm{C}$ reaction is irreversible. (A more careful treatment would keep the $\mathrm{C} \rightarrow \mathrm{B}$ rate and then analyze the solutions by assuming it to be much smaller than all the other ones in the problem.) Our goal is to show that in this explicitly dynamical situation, equilibrium is established rapidly, and thereafter maintained between the fast steps of the reaction.

Based on the kinetic scheme outlined in Equation 15.43, the rate equations for the three chemical species are

$$\begin{aligned} \frac{\mathrm{d}A}{\mathrm{d}t} &= -k_+A + k_-B, \\ \frac{\mathrm{d}B}{\mathrm{d}t} &= k_+A - (k_- + r)B, \\ \frac{\mathrm{d}C}{\mathrm{d}t} &= rB. \end{aligned} \tag{15.44}$$

These can be written in dimensionless form by dividing both sides of the equations by k_-, which effectively corresponds to measuring time in units of $1/k_-$. We introduce the dimensionless time $\tau = k_-t$. It is unfortunate that the conventional notation for dimensionless time τ uses the same symbol as the conventional notation for average lifetime introduced earlier in the chapter. Throughout the remainder of this section, τ will be used to refer to dimensionless time. By additionally introducing the dimensionless rate constants $k = k_+/k_-$ (this is also the equilibrium constant for the $\mathrm{A} \rightleftharpoons \mathrm{B}$ reaction) and $\varepsilon = r/k_-$, we can rewrite the rate equations as

$$\begin{aligned} \frac{\mathrm{d}A}{\mathrm{d}\tau} &= -kA + B, \\ \frac{\mathrm{d}B}{\mathrm{d}\tau} &= kA - (1 + \varepsilon)B, \\ \frac{\mathrm{d}C}{\mathrm{d}\tau} &= \varepsilon B. \end{aligned} \tag{15.45}$$

Since the first two equations are decoupled from the third, we solve those using the matrix method, with initial condition $(A(0), B(0)) = (1, 0)$. In matrix form, the above rate equations can be written as

$$\frac{d}{d\tau}\begin{pmatrix}A\\B\end{pmatrix} = \begin{pmatrix}-k & 1\\ k & -(1+\varepsilon)\end{pmatrix}\begin{pmatrix}A\\B\end{pmatrix}. \tag{15.46}$$

The general solution to this matrix equation is

$$\begin{pmatrix}A\\B\end{pmatrix}(\tau) = a_1\begin{pmatrix}A_1\\B_1\end{pmatrix}e^{\omega_1\tau} + a_2\begin{pmatrix}A_2\\B_2\end{pmatrix}e^{\omega_2\tau}, \tag{15.47}$$

where $\begin{pmatrix}A_{1,2}\\B_{1,2}\end{pmatrix}$ are the eigenvectors of the 2×2 matrix in Equation 15.46, while $\omega_{1,2}$ are the corresponding eigenvalues; $a_{1,2}$ are the integration constants which we will determine from the initial conditions. The Math Behind the Models below summarizes how matrices can be manipulated to find the eigenvalues and eigenvectors.

The eigenvalues of our matrix are solutions to the equation

$$\det\begin{pmatrix}-k-\omega & 1\\ k & -(1+\varepsilon)-\omega\end{pmatrix} = 0, \tag{15.48}$$

which, once the determinant has been resolved, becomes a quadratic equation

$$\omega^2 + (1+k+\varepsilon)\omega + k\varepsilon = 0 \tag{15.49}$$

in the unknown eigenvalues ω. The two solutions of the quadratic equation,

$$\omega_{1,2} = -\tfrac{1}{2}(1+k+\varepsilon) \pm \tfrac{1}{2}\sqrt{(1+k+\varepsilon)^2 - 4k\varepsilon} \tag{15.50}$$

are the eigenvalues that dictate the time evolution of the various concentrations.

To simplify the arithmetic, let us make good use of the assumptions we made at the beginning, namely, that k is of the order of 1 and that ε is much smaller than 1. In this case, we can Taylor-expand Equation 15.50 in powers of ε using the rule that $\sqrt{1+x}\approx 1+x/2$ (see The Math Behind the Models on p. 215) to obtain approximate values of the eigenvalues,

$$\omega_1 = -\frac{k}{k+1}\,\varepsilon, \quad \omega_2 = -(k+1) \tag{15.51}$$

to leading order in ε. The lowest-order term in the first eigenvalue is of the order of ε^1, while the lowest-order term in the second eigenvalue is of the order of ε^0. It is interesting to note that we obtain a slow ($\omega_1 \ll 1$) and a fast ($\omega_2 \approx 1$) rate, as expected. In particular, the fast rate should correspond to the rate at which equilibrium is established between species A and B, while we expect the slow rate to describe the overall decay of A and B giving rise to chemical species C.

In order to complete the calculation, we still need to compute the eigenvectors corresponding to the eigenvalues ω_1 and ω_2. These are obtained by solving the following 2×2 system of linear equations:

$$\begin{pmatrix}-k-\omega_{1,2} & 1\\ k & -(1+\varepsilon)-\omega_{1,2}\end{pmatrix}\begin{pmatrix}A_{1,2}\\B_{1,2}\end{pmatrix} = 0, \tag{15.52}$$

here written in matrix form. To leading order in ε the eigenvectors are

$$\begin{pmatrix}A_1\\B_1\end{pmatrix} = \begin{pmatrix}1\\k\end{pmatrix}, \quad \begin{pmatrix}A_2\\B_2\end{pmatrix} = \begin{pmatrix}1\\-1\end{pmatrix} \tag{15.53}$$

and the general solution of the differential equation, Equation 15.46 is

$$\begin{pmatrix} A \\ B \end{pmatrix}(\tau) = a_1 \begin{pmatrix} 1 \\ k \end{pmatrix} \mathrm{e}^{-[k\varepsilon/(k+1)]\tau} + a_2 \begin{pmatrix} 1 \\ -1 \end{pmatrix} \mathrm{e}^{-(k+1)\tau}. \tag{15.54}$$

Finally, to obtain the integration constants $a_{1,2}$, we make use of the initial condition, $\begin{pmatrix} A \\ B \end{pmatrix}(0) = \begin{pmatrix} 1 \\ 0 \end{pmatrix}$, which, when substituted into Equation 15.54, gives

$$\begin{aligned} a_1 + a_2 &= 1, \\ ka_1 - a_2 &= 0. \end{aligned} \tag{15.55}$$

The solution to this set of linear equations is $a_1 = 1/(k+1)$ and $a_2 = k/(k+1)$, and we can now write the full solution of the rate equations to lowest order in ε for species A and B as

$$\begin{pmatrix} A \\ B \end{pmatrix}(\tau) = \begin{pmatrix} 1/(k+1) \\ k/(k+1) \end{pmatrix} \mathrm{e}^{-[k\varepsilon/(k+1)]\tau} + \begin{pmatrix} k/(k+1) \\ -k/(k+1) \end{pmatrix} \mathrm{e}^{-(k+1)\tau}. \tag{15.56}$$

This solution confirms our intuition. In Figure 15.14, we show the exact solution of the rate equations, for parameters $k = 1$ and $\varepsilon = 0.01$. As anticipated, we observe a fast decay away from the initial state (all A) until the concentrations of A and B are in a fixed ratio (that is, the equilibrium constant), which is followed by a slow decay during which the species A and B are in equilibrium with each other, but their overall quantity is decaying. In particular, for times such that $\tau \gg 1/(k+1)$, we can ignore the second term in Equation 15.56 and

$$\begin{pmatrix} A \\ B \end{pmatrix}(\tau) = \begin{pmatrix} 1/(k+1) \\ k/(k+1) \end{pmatrix} \mathrm{e}^{-[k\varepsilon/(k+1)]\tau}. \tag{15.57}$$

This equation describes a slow (rate $\approx \varepsilon$) overall decay of species A and B, while their relative concentrations are related by the equilibrium constant, $B(t)/A(t) = k$. What this exercise illustrates is the way in which a dynamical process involving multiple reactions can have some subset of reactions that behave as though they are in instantaneous equilibrium at all times. The classic example of this kind of thinking is its application to reactions catalyzed by enzymes, generally referred to as Michaelis–Menten kinetics.

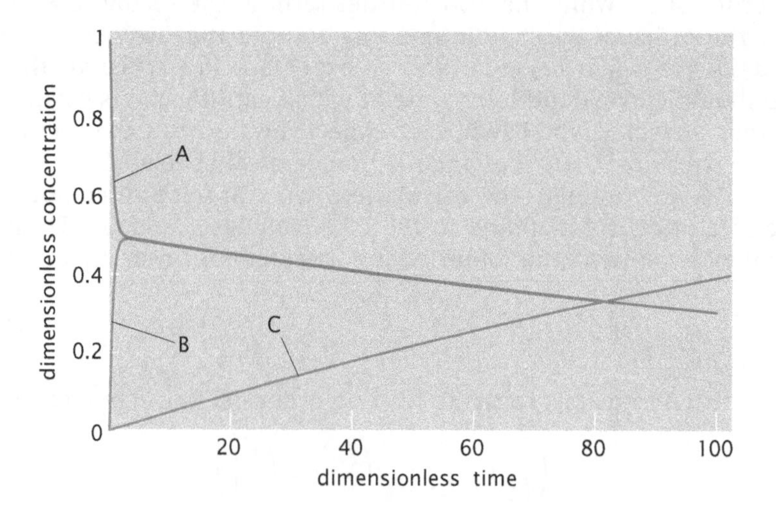

Figure 15.14: The time variation of the concentrations of chemical species A and B, for $k = 1$ and $\varepsilon = 0.01$. Initially, there is one unit of concentration of A and none of B. After a fast decay of A and rise of B, the two settle into an equilibrium situation where A = B. The accumulation of product C is also shown in the figure.

The Math Behind the Models: Eigenvalues and Eigenvectors We are all familiar with equations of the form $ax + b = 0$, where a and b are parameters and x is an unknown. The temptation is to say that the solution of this equation is $x = -b/a$. This is certainly true if a is not zero. If, on the other hand, $a = 0$, then either there is no solution for $b \neq 0$ or any number x is a solution for $b = 0$. These might seem like mathematical niceties, but when a is promoted to a matrix and $b = 0$, the question of when the above equation has nonzero solutions is at the heart of a number of interesting problems, and in particular the one we are about to face, namely, finding the eigenvectors and eigenvalues of a matrix.

We concern ourselves with square matrices, so that the number of rows and the number of columns are the same. We further specialize to 2×2 matrices so as to simplify the algebra, but all our conclusions will apply to matrices of arbitrary size.

The action of a matrix on a vector is illustrated in Figure 15.15. Multiplying a matrix and a vector produces a new vector, which is in general related to the original one by rotating it, that is, changing the direction it is pointing in, and changing its length. A vector for which the action of the matrix leaves the direction it is pointing in unchanged is called an *eigenvector*. The new vector will still typically have a new length, and the ratio of the new to the old length is the *eigenvalue* associated with that particular eigenvector. In order to find the eigenvectors of a matrix, one strategy would be to try all vectors and see which ones remain in the same direction upon action of the matrix. This is clearly not practical, since there are too many vectors to try, an infinite number in fact.

To find an eigenvector of a matrix

$$\mathbf{M} = \begin{pmatrix} m_{11} & m_{12} \\ m_{21} & m_{22} \end{pmatrix}, \tag{15.58}$$

we need to find the solution to the equation $\mathbf{Me} = \lambda\mathbf{e}$, which says that the vector $\mathbf{e}$ is the eigenvector of the matrix with eigenvalue λ. Using the rules of matrix multiplication, we can expand this matrix equation into a set of two equations with two unknowns,

$$\begin{aligned} m_{11}e_1 + m_{12}e_2 &= \lambda e_1, \\ m_{21}e_1 + m_{22}e_2 &= \lambda e_2. \end{aligned} \tag{15.59}$$

This can be rewritten as

$$\begin{aligned} (m_{11} - \lambda)e_1 + m_{12}e_2 &= 0, \\ m_{21}e_1 + (m_{22} - \lambda)e_2 &= 0, \end{aligned} \tag{15.60}$$

so it takes on the form discussed above, $(\mathbf{M} - \lambda\mathbf{I})\mathbf{e} = 0$, where we have rewritten Equations 15.60 in matrix form using the identity matrix $\mathbf{I} = \begin{pmatrix} 1 & 0 \\ 0 & 1 \end{pmatrix}$. To solve this system of equations, we first express e_2 in terms of e_1 from the first of the two equations to obtain

$$e_2 = -\frac{(m_{11} - \lambda)}{m_{12}}\, e_1, \tag{15.61}$$

MATH

(A)

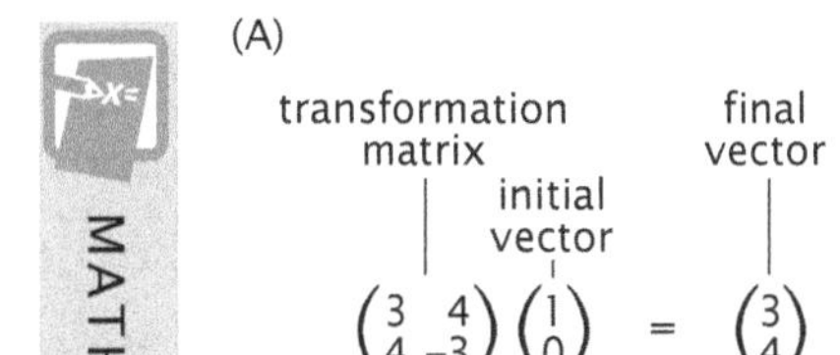

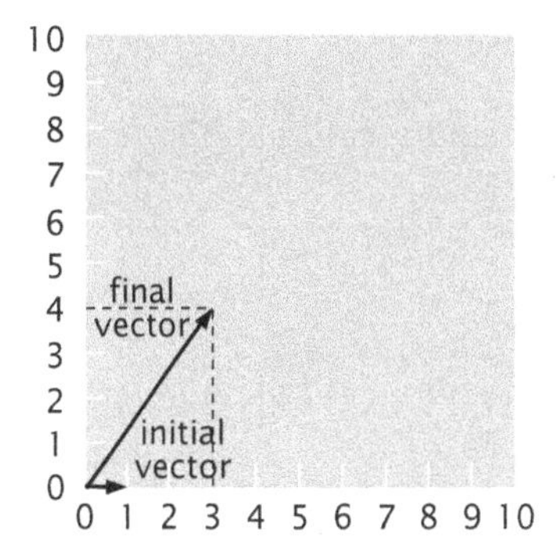

(B)

$$\begin{pmatrix} 3 & 4 \\ 4 & -3 \end{pmatrix}\begin{pmatrix} 2 \\ 1 \end{pmatrix} = \begin{pmatrix} 10 \\ 5 \end{pmatrix} = 5\begin{pmatrix} 2 \\ 1 \end{pmatrix}$$

eigenvalue

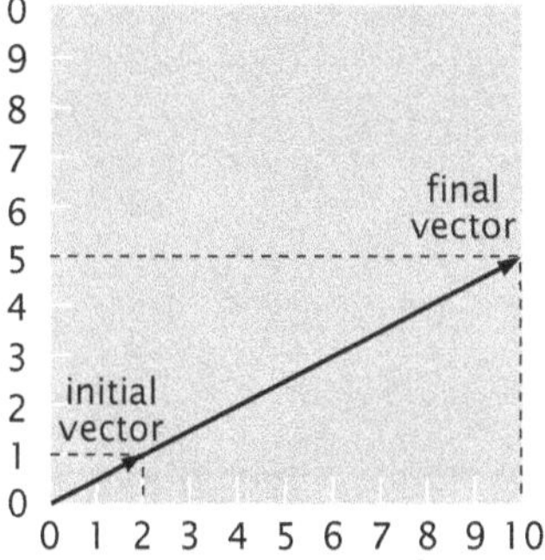

Figure 15.15: Eigenvectors and eigenvalues of a matrix. (A) A matrix acting on an arbitrary vector changes both its direction and length. (B) Eigenvectors when acted on by a matrix only change their length. The factor by which the length increases is the eigenvalue.

where we have assumed that $m_{12} \neq 0$. In the case $m_{12} = 0$, we can repeat the same procedure as below by solving the second equation in Equations 15.60 for e_1. Finally, if both m_{12} and m_{21} are zero, then the matrix is diagonal and $\begin{pmatrix}1\\0\end{pmatrix}$ and $\begin{pmatrix}0\\1\end{pmatrix}$ are its eigenvectors, with m_{11} and m_{22} being the respective eigenvalues.

Next we substitute Equation 15.61 into the second of Equations 15.60, and we arrive at

$$\frac{m_{21}m_{12} - (m_{11} - \lambda)(m_{22} - \lambda)}{m_{12}} e_1 = 0. \tag{15.62}$$

Since we are looking for a nonzero solution to this equation, the coefficient in front of e_1 must be zero. In other words,

$$(m_{11} - \lambda)(m_{22} - \lambda) - m_{12}m_{21} = 0 \tag{15.63}$$

needs to hold. This is a quadratic equation whose two solutions λ_1 and λ_2 are the eigenvalues. To get the eigenvectors, we can set $e_1 = 1$ and get e_2 from Equation 15.61 using one or the other eigenvalue.

It is important to note that the quadratic equation leading to the eigenvalues can also be rewritten in matrix form,

$$\det(\mathbf{M} - \lambda\mathbf{I}) = 0, \tag{15.64}$$

where the operation det is the determinant of the matrix. In this form, the equation for the eigenvalues works for matrices of arbitrary size. For a 2×2 matrix, the determinant is the difference between the product of the matrix elements along the two diagonals.

15.2.7 Michaelis–Menten and Enzyme Kinetics

One of the most powerful tools for analyzing biochemical kinetics that exploits rapid pre-equilibrium assumptions like those described in the previous section is the model of Michaelis–Menten kinetics. This simple kinetic scheme will arise in our thinking about cytoskeletal polymerization and molecular motors, and has also been applied to many different kinds of enzymes involved in a wide variety of biological processes ranging from central metabolism to signaling during embryonic development. The Michaelis–Menten framework for thinking about enzyme kinetics offers the power of a unifying approximate description for the huge variety of biological catalysts that use many different detailed mechanisms to perform a wide range of transformations on a plethora of substrates. Within this unifying framework, every enzyme can be approximately characterized by just two parameters, namely, one rate constant describing the maximum turnover rate at which the enzyme can operate if its substrate is present at saturating concentrations, and a second constant describing the magnitude of the substrate concentration required for the enzyme to operate at half its maximal rate. Although the Michaelis–Menten scheme is clearly oversimplified for dealing with the true kinetic complexity of biological enzyme mechanisms, its great simplicity and broad applicability have earned it a central place of honor in the history of biochemistry. Many biochemistry reference books include large tables comparing the activity of different enzymes as characterized by the two quantitative parameters derived from this scheme.

The essential idea in the Michaelis–Menten scheme is that the reaction catalyzed by a given enzyme is composed of two separable steps. The first step (assumed to be readily reversible) involves the formation of an intermediate complex of enzyme and substrate. Once this first step has taken place, the complex can undergo a second reaction resulting in the formation of the product. There are a host of cases in which this kind of thinking is appropriate: (i) promoter occupancy by RNA polymerase is often much faster than escape to form transcripts; (ii) binding of tRNA in the ribosome can be much faster than peptide bond formation; (iii) ATP binding to an enzyme can be much faster than the hydrolysis step; etc.

This class of reactions can be written as

$$\mathrm{E}+\mathrm{S} \underset{k_-}{\overset{k_+}{\rightleftharpoons}} \mathrm{ES} \xrightarrow{r} \mathrm{E}+\mathrm{P}. \quad (15.65)$$

We assume that the final step is irreversible and characterized by the rate constant r. This equation is similar in structure to Equation 15.43, with the important addition of the enzyme E as a catalyst. Note that the enzyme is regenerated after the reaction is complete so that it can go on to catalyze the same reaction for a new substrate molecule.

Given this reaction scheme, we can write the rate equations using the rules described earlier in the chapter, resulting in

$$\begin{aligned}
\frac{\mathrm{d[E]}}{\mathrm{d}t} &= -k_+[\mathrm{E}][\mathrm{S}] + k_-[\mathrm{ES}] + r[\mathrm{ES}],\\
\frac{\mathrm{d[S]}}{\mathrm{d}t} &= -k_+[\mathrm{E}][\mathrm{S}] + k_-[\mathrm{ES}],\\
\frac{\mathrm{d[ES]}}{\mathrm{d}t} &= k_+[\mathrm{E}][\mathrm{S}] - (k_- + r)[\mathrm{ES}],\\
\frac{\mathrm{d[P]}}{\mathrm{d}t} &= r[\mathrm{ES}].
\end{aligned} \quad (15.66)$$

One simple and immediate approach to these equations is to solve them numerically. That is, by starting with some set of initial conditions (that is, $[\mathrm{E}]_0$, $[\mathrm{S}]_0$, $[\mathrm{ES}]_0$, and $[\mathrm{P}]_0$, where we use the subscript "0" to signify initial values), we can perform an integration of these dynamical equations to find the resulting time history for the concentrations of all four species. A choice of initial concentrations that illustrates the essence of the dynamics is $[\mathrm{E}]_0 = [\mathrm{S}]_0 = (k_- + r)/k_+$ and $[\mathrm{ES}]_0 = [\mathrm{P}]_0 = 0$. In this case, we start out with no intermediate or product. The time evolution of the system in this case is shown in Figure 15.16. Note that the concentration of enzyme initially decreases, but ultimately starts to climb again once the available substrate starts to become too depleted. The reader is invited to explore this analysis in detail in the problems at the end of the chapter.

Of course, often it is of great interest to have an approximate analytic description of a given model for the purposes of developing intuition. The dynamical equations given above can be solved approximately by recourse to certain simplifying assumptions. One powerful scheme is to assume that the intermediate is in a steady state corresponding to $\mathrm{d[ES]}/\mathrm{d}t = 0$. This approximation is reasonable during the relatively early part of the reaction time series where the rate of change of product is approximately linear with time, that is, each enzyme molecule is working at a nearly constant rate. If we make this

(A)

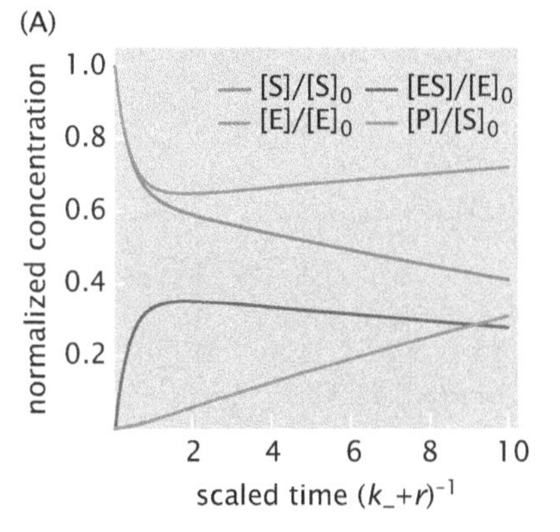

(B)

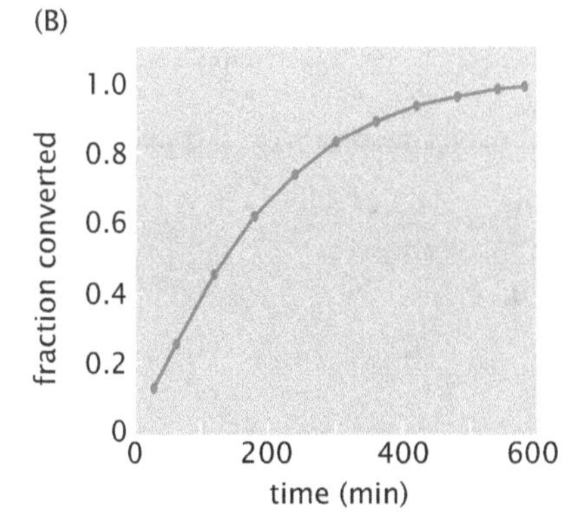

Figure 15.16: Concentrations of reactants and products in an enzyme catalyzed reaction. (A) The graph shows the time evolution of [E], [S], [ES], and [P]. Time is defined in units of $1/(k_- + r)$ and $r/(r + k_-) = 0.1$. (B) Experimental data showing the amount of product as a function of time for the hydrolysis of sucrose as catalyzed by the yeast enzyme invertase. At early time points, the function showing product over time is nearly linear as illustrated in (A). At later time points (after about 100 min) the rate of reaction slows down because the substrate is being depleted. (Data adapted from A. J. Brown, *J. Chem. Soc. Trans.* 81:373, 1902.)

assumption, then we can obtain the concentration [ES] in terms of [E] and [S] as

$$\frac{[E][S]}{[ES]} = \frac{k_- + r}{k_+} = K_m, \quad (15.67)$$

where we have introduced a new constant, K_m, usually called the Michaelis constant. The value K_m can be intuitively understood as the concentration of substrate where the reaction is proceeding at half the maximum possible rate. If $r \ll k_-$, then K_m is approximately equal to k_-/k_+, that is, the dissociation constant K_d for the binding of the substrate to its enzyme.

Our interest is in the rate of product formation under steady-state conditions. We can rewrite the equation for $d[P]/dt$ as

$$\frac{d[P]}{dt} = r\frac{[E][S]}{K_m}. \quad (15.68)$$

This equation can be conveniently rewritten by virtue of the realization that the *maximum* rate of reaction corresponds to $V_{max} = r[E_{tot}]$, which assumes that the entirety of the stockpile of enzyme is bound to substrate and ready to form product. This will be the case when the concentration of substrate is extremely high relative to K_m. Using this definition, we can now rewrite Equation 15.68 as

$$\frac{d[P]}{dt} = V_{max}\frac{[E][S]/K_m}{[E_{tot}]}. \quad (15.69)$$

If we now invoke the fact that $[E_{tot}] = [E] + [ES]$ and use $[ES] = [E][S]/K_m$, we obtain the Michaelis–Menten result,

$$\frac{d[P]}{dt} = V_{max}\frac{[S]/K_m}{1 + ([S]/K_m)}, \quad (15.70)$$

which permits us to compute the rate of reaction as a function of substrate concentration.

The most important conclusion of this analysis for our purposes is that the *rate* of the reaction is going to depend on the concentration of substrate in a very systematic way that is recognized as having the same functional form as the binding curves introduced in Chapter 6. The functional form is plotted in Figure 15.17(A), where it can be seen that K_m corresponds to the substrate concentration needed to reach half maximum velocity. The excellent fit of the curve to experimental data for ATP hydrolysis by the motor protein myosin is shown in Figure 15.17(B).

Although this framework has proved valuable as a basis for thinking about how reaction rates depend upon substrate concentration for almost 100 years, for many real enzymes this analysis is too simplified to be quantitatively faithful to the data, particularly when enzyme rates are measured over a very large range of substrate concentrations. In many cases, the simple Michaelis–Menten kinetics is not applicable, because the enzyme–substrate complex must proceed through multiple intermediate steps prior to releasing product. Further complications may arise when an enzyme uses two or more substrate molecules, which may bind to the active site cooperatively, and when enyzmes are allosterically activated by regulators that bind at locations other than the active site. In addition, experimental factors often complicate the analysis. For example, product molecules often act as inhibitors for their enzymes so the effective enzyme concentration decreases over time. The reader is invited to explore these

(A)

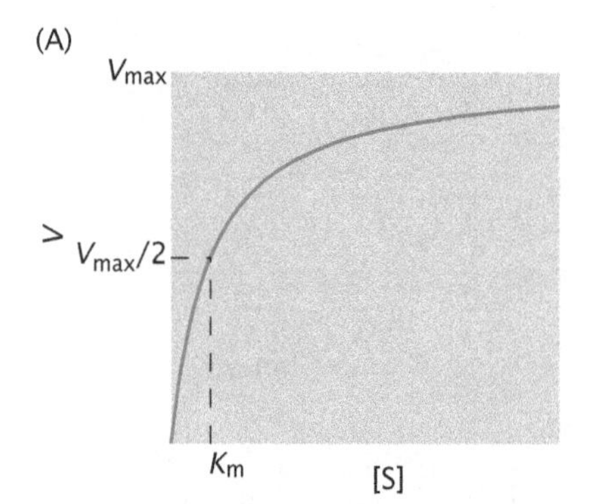

(B)

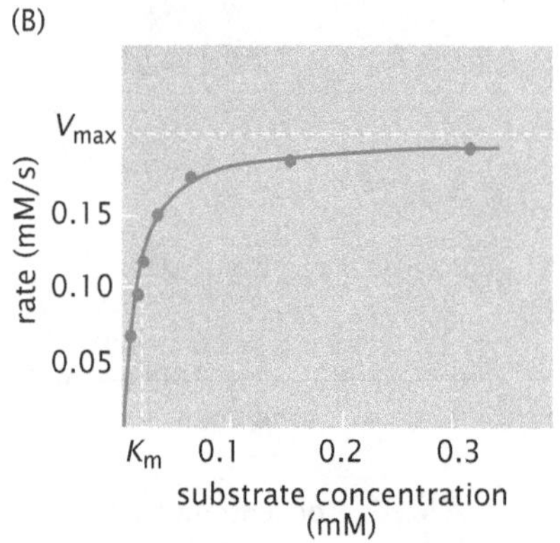

Figure 15.17: Michaelis–Menten kinetics. (A) The general form of the Michaelis–Menten equation showing the enzyme rate as a function of substrate concentration can be captured by just two parameters, namely, the maximal rate for saturating substrate concentrations (V_{max}) and the substrate concentration at which the rate is half of the maximum (K_m). (B) Experimental data showing the rate of ATP hydrolysis by rabbit skeletal muscle myosin as a function of ATP concentration. (B, adapted from L. Ouellette et al., *Arch. Biochem. Biophys.* 39:37, 1952.)

questions further both in the problems and in the Further Reading at the end of the chapter.

15.3 The Cytoskeleton Is Always Under Construction

So far, this chapter has emphasized rate equations as a conceptual framework for treating problems involving the time evolution of reactants and products in chemical reactions. These ideas can be used in a powerful way to explore the dynamics of the cytoskeleton. As a result, we review some of the key examples of cytoskeletal dynamics, and then illustrate how to use the rate equation paradigm to consider these examples.

15.3.1 The Eukaryotic Cytoskeleton

The Cytoskeleton Is a Dynamical Structure That Is Always Under Construction

The eukaryotic cytoskeleton in mammalian cells is built around three main molecular actors, namely, actin, microtubules, and intermediate filaments as already introduced in Section 10.5.1 (p. 414). One of the most intriguing features of these biological polymers is the fact that they exhibit rich dynamical behavior whether examined *in vitro* or within cells. Indeed, there is a wide variety of cellular processes that depend upon this rich behavior of cytoskeletal filaments. For example, Figure 15.18 shows the way in which microtubules play a role in segregating chromosomes. Similarly, in Figures 15.2 and 15.3, we saw how actin assembly mediates cell motility. The main lesson of these observations is that the cytoskeleton is a constantly changing network that is subject to precise control that dictates when and where filaments will nucleate, attach, branch, and grow.

Though spectacular cellular events like those shown in Figure 15.18 reveal the intricacy of cytoskeletal dynamics, to begin our thinking on model building of such dynamics we begin with the more tractable *in vitro* situation shown in Figure 15.19. The main idea of the experiments that such cartoons depict is that actin monomers in solution spontaneously aggregate to form filaments. Such experiments were indicated schematically in Figure 15.1(A). One way to perform these measurements is to attach a small molecule covalently to a specific site on actin that is poorly fluorescent when the actin is in its soluble monomeric form but whose fluorescence is strongly increased once the actin monomers are part of a filament. By monitoring the fluorescence as a function of time, it is possible to measure the rate at which filaments nucleate and grow. Data of this kind are shown in Figure 15.19(C).

As shown in Figure 15.19(A), if actin monomers are placed in solution, after some initial lag time they will nucleate filaments, which will then grow until an equilibrium is reached where the rates of detachment and attachment are equal. By way of contrast, as shown in Figure 15.19(B), if the solution is seeded with preexisting nuclei, the growth phase starts immediately and is characterized by the same growth rate. These experiments provide a first clue as to how the cell might control when and where filaments are present by tuning their nucleation. Indeed, hundreds of different kinds of actin-associated

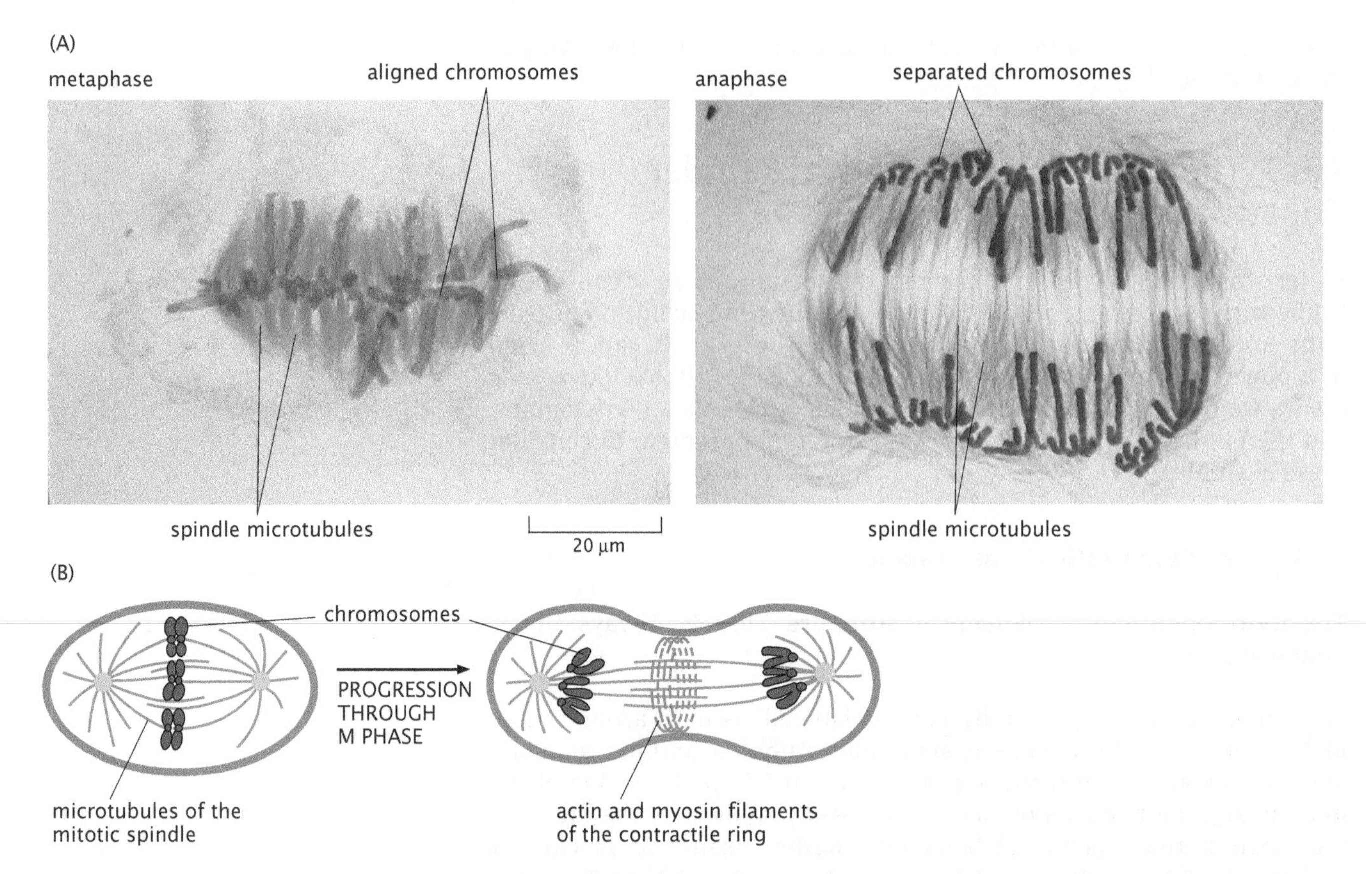

Figure 15.18: Chromosome segregation by microtubules. (A) Micrographs of the transition from metaphase (where the duplicated sister chromosomes are all aligned) to anaphase (where the chromosomes have been separated into two groups) in endosperm cells from a lily. The tubulin is labeled by gold-labeled antibodies. (B) Cartoon showing the analogous process in an animal cell, where microtubules build a spindle to separate the chromosomes and an actin filament ring forms in the middle of the mother cell to split it into two daughters. (A, courtesy of A. Bajer; B, adapted from B. Alberts et al., Molecular Biology of the Cell, 5th ed. Garland Science, 2008.)

proteins inside living cells are able to modulate every aspect of actin filament dynamics, including the nucleation rate, elongation rate, dissociation rate, ATP exchange rate, large-scale architecture, etc. The concentrations of some of these actin-modulating proteins have been determined by experiments such as those shown in Figure 15.4. Actual actin polymerization data showing the effects of such proteins are provided in Figure 15.19(C).

15.3.2 The Curious Case of the Bacterial Cytoskeleton

Historically, it was thought that one of the defining distinctions between bacteria and eukaryotes was that bacteria lacked a cytoskeleton. However, this picture has now changed dramatically, with tubulin and actin homologs having been identified in bacteria and possibly an intermediate filament-like protein as well. One example is that of ParM, a bacterial homolog of eukaryotic actin, which is responsible for segregating plasmids upon cell division. In particular, ParM was discovered on a plasmid carrying antibiotic-resistance genes that were associated with an outbreak of epidemic typhoid in Mexico and the southwestern USA. ParM and its molecular partners are responsible for making sure that in those bacteria that carry the plasmid both daughter cells receive a copy. A schematic of the mechanism of ParM is shown in Figure 15.20.

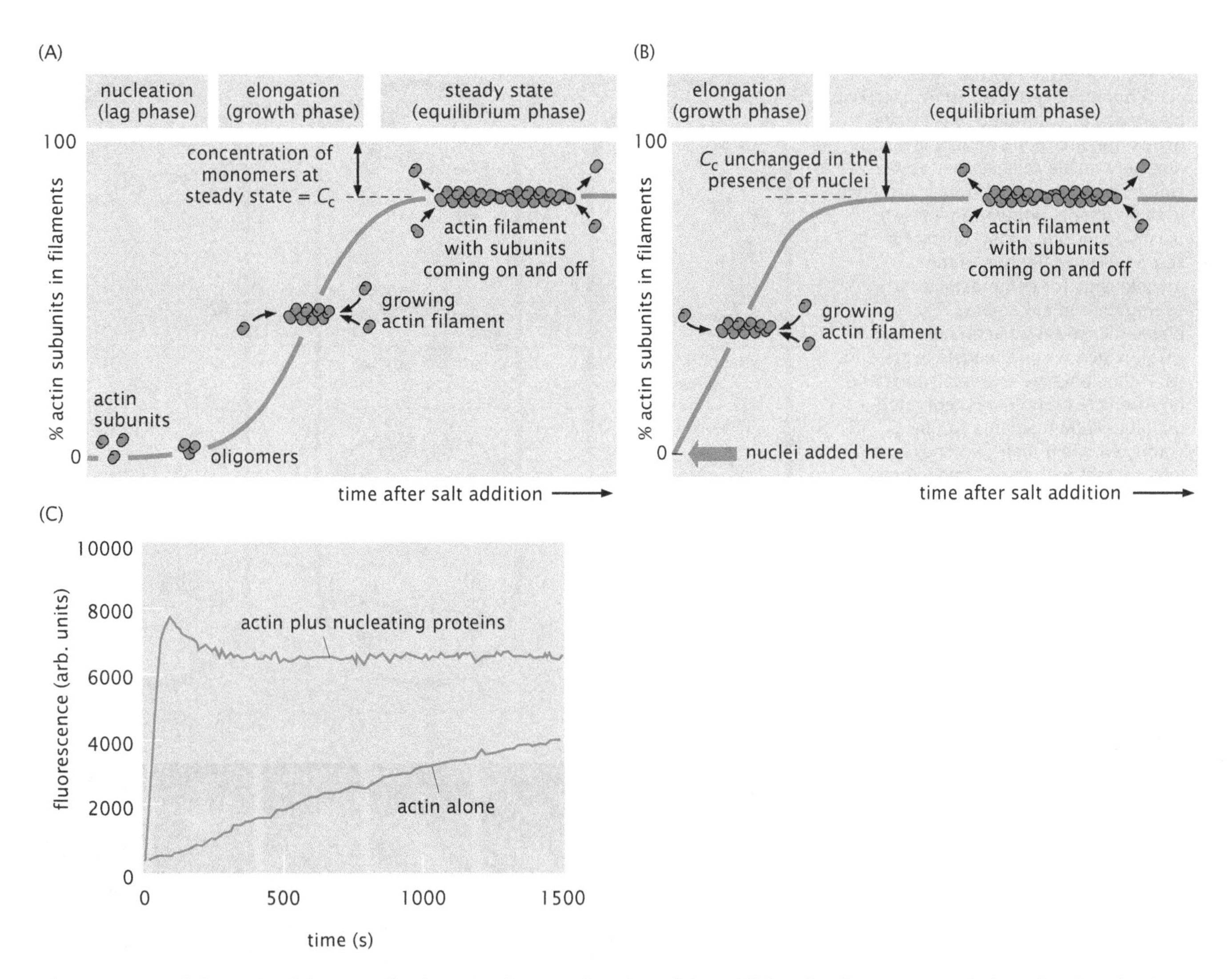

Figure 15.19: Schematic of the rate of polymerization as a function of time. (A) Starting from a pure solution of actin subunits, the assembly of filaments is triggered by the addition of salt. The limiting step in filament assembly is the random collision of at least three subunits to form a nucleus, which can then be rapidly elongated. Because nucleus formation is slow compared with filament elongation, a measurement of the total amount of actin in filament form as a function of time will show an initial lag phase. When polymerization is complete, there remains a small amount of actin in solution; this concentration of leftover monomers is called the critical concentration. (B) If stabilized nuclei are added at the beginning of the polymerization reaction, elongation will start immediately and there is no lag phase. The critical concentration is not affected. (C) Experimental growth curves for actin as reported by fluorescence with and without the activity of proteins (Arp2/3 and ActA) that nucleate filaments. The slow-growth case corresponds to no Arp2/3. (A, B, adapted from B. Alberts et al., Molecular Biology of the Cell, 5th ed. Garland Science, 2008; C, adapted from M. D. Welch et al., *Science* 281:105, 1998.)

ParM is an intriguing example since in some ways it can be thought of as a simple version of a mitotic spindle. In Figure 15.18, we showed a picture of a mitotic spindle in action where it is seen how microtubules separate the chromosomes and an actin filament ring forms in the middle of the mother cell. Figure 15.20 elaborates that analogy by depicting the way in which ParM carries out similar functions in the prokaryotic setting. The basic idea is that the ParM filaments grow and shrink spontaneously until both ends successfully encounter one of the two copies of the antibiotic-resistance plasmid. When both plasmids are attached to the different extremities, the filaments polymerize without shrinkage, effectively pushing their plasmid cargo to opposite poles of the dividing cell. Images from a video microscopy sequence of plasmid segregation are shown in Figure 15.20(B).

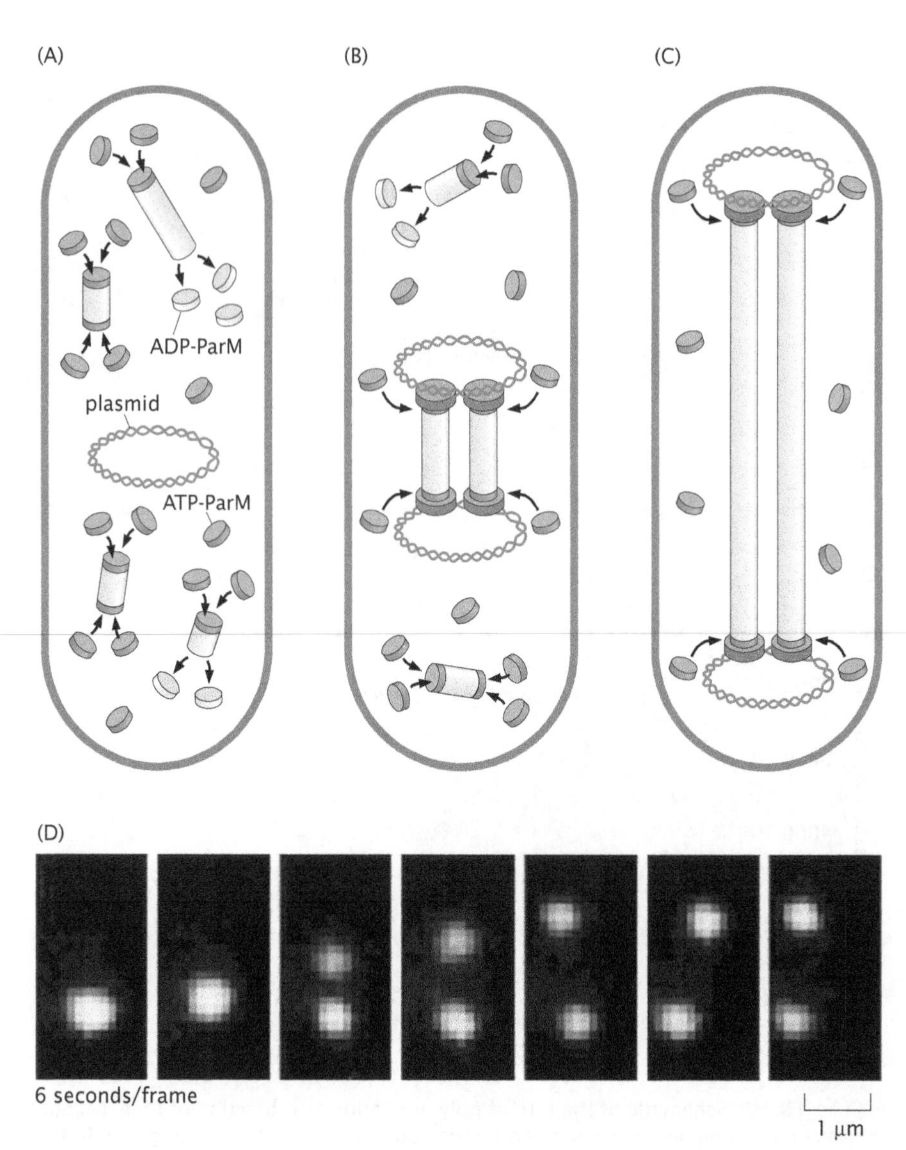

Figure 15.20: A model for plasmid segregation driven by the bacterial actin homolog ParM. (A) Some bacterial plasmids carry antibiotic resistance genes that are found at very low copy numbers inside cells, as few as 1–2 per cell. In order to maintain their presence in the bacterial population, these plasmids require a mechanism for segregation analogous to the mechanisms for chromosome segregation in eukaryotes. The R1 plasmid encodes an actin homolog ParM, which is continuously expressed. ParM filaments are extremely unstable. (B) After the plasmid has replicated, unstable ParM filaments can be stabilized when their two ends come into contact with ParR protein (gray) which in turn binds to specific sites on the two plasmids. The filaments can continue to elongate at the site of DNA binding. (C) The insertional elongation of ParM filaments pushes the two copies of the plasmid to opposite poles of the cell. (D) Time-lapse sequence showing segregation of two fluorescently tagged plasmids in a living *E. coli* cell. (A–C, adapted from E. C. Garner et al., *Science* 306:1021, 2004; D, adapted from C. S. Campbell and R. D. Mullins, *J. Cell Biol.* 179:1059, 2007.)

Other examples of bacterial cytoskeletal homologs are MreB, an actin homolog that is related to genome position and cell shape in some bacteria, and FtsZ, which is a tubulin homolog and one of the main players in the formation of the ring that dictates the plane of cell division. One of the peculiar features of these prokaryotic cytoskeletal proteins is that there is a role reversal relative to their eukaryotic counterparts, with actin homologs undertaking tasks reserved for microtubules in the eukaryotic setting and vice versa. The study of the cytoskeleton in both prokaryotes and eukaryotes provides an interesting opportunity to put the rate equation ideas developed in this chapter to work.

15.4 Simple Models of Cytoskeletal Polymerization

Huge classes of cellular processes can ultimately be thought of in terms of the addition of monomers to the ends of growing chains: transcription and translation, cytoskeletal assembly, polymerization of filamentous viruses such as tobacco mosaic virus, etc. The simplest picture of such polymerization reactions amounts to assuming

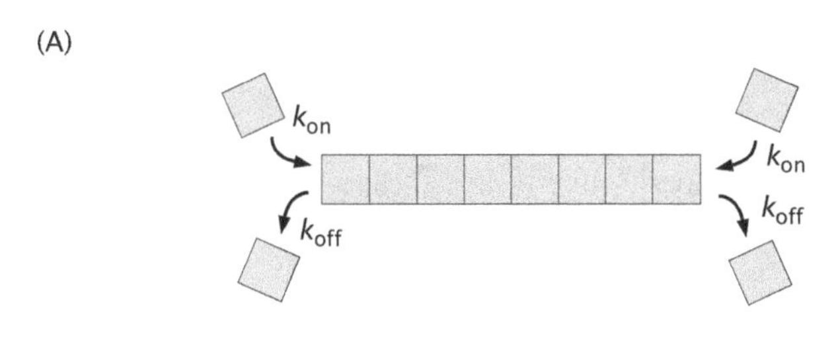

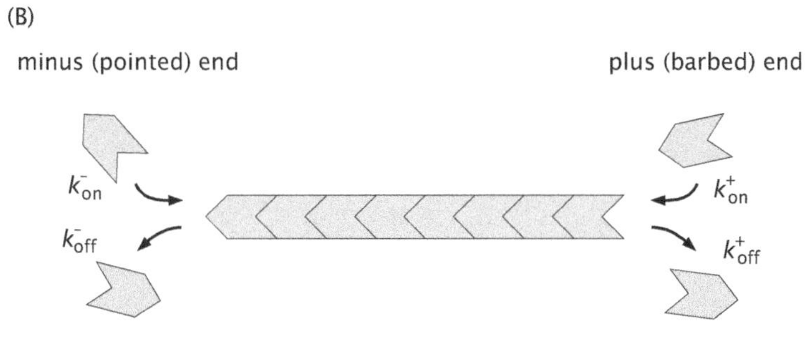

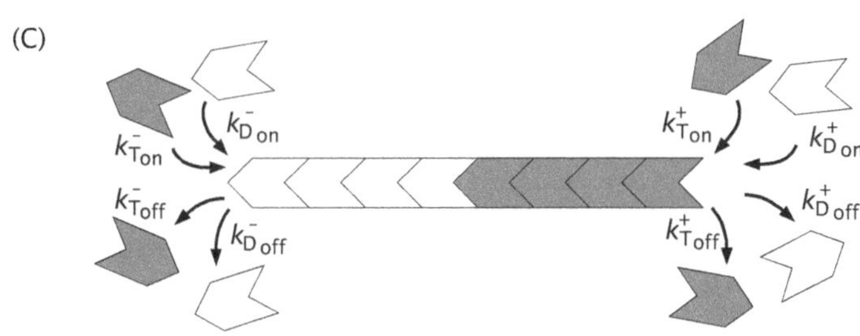

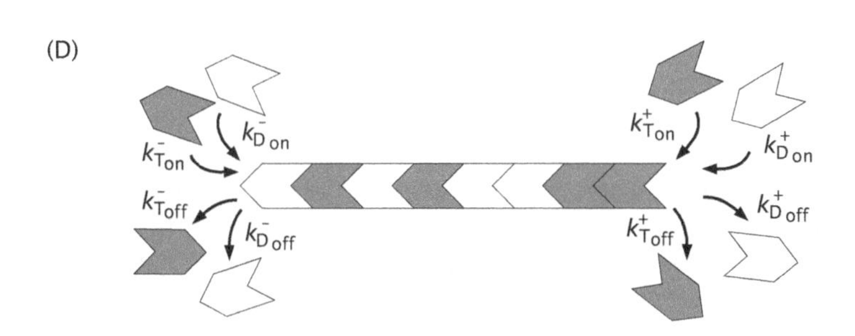

Figure 15.21: Hierarchy of models of cytoskeletal polymerization, which include features such as asymmetric rates and hydrolysis.
(A) Polymerization with equal rates on both ends. (B) Polymerization with unequal rates on the two ends due to structural asymmetry of the filament. The two structurally distinct ends are often called the plus and minus ends or, for actin, the barbed and pointed ends. (C) Polymerization with unequal rates and vectorial hydrolysis reaction of nucleotides on incorporated monomers. (D) Polymerization with hydrolysis that can take place on any monomer within the filament. For each case, the rate constants are illustrated. Different rate constants may apply for on rates and off rates, plus and minus ends of filaments, and ATP versus ADP-containing subunits.

that there is a characteristic rate of addition of monomers to the end of a growing polymer and that there is a characteristic rate at which monomers fall off the end of such polymers. These ideas lead naturally to various classes of dynamic models.

The Dynamics of Polymerization Can Involve Many Distinct Physical and Chemical Effects

For concreteness, we use actin as our first case study. Our plan is to consider a hierarchy of models of increasing complexity as shown in Figure 15.21. The key quantities that we will reason about are the probability distribution $P_n(t)$, which delivers the probability of having a filament of length n at time t, and $\langle L \rangle$, which characterizes the average length of the polymers. The simplest level of approximation for determining these quantities posits polymerization processes in which the on and off rates at the two ends of the growing polymer are the same and there are no associated nucleotide hydrolysis reactions.

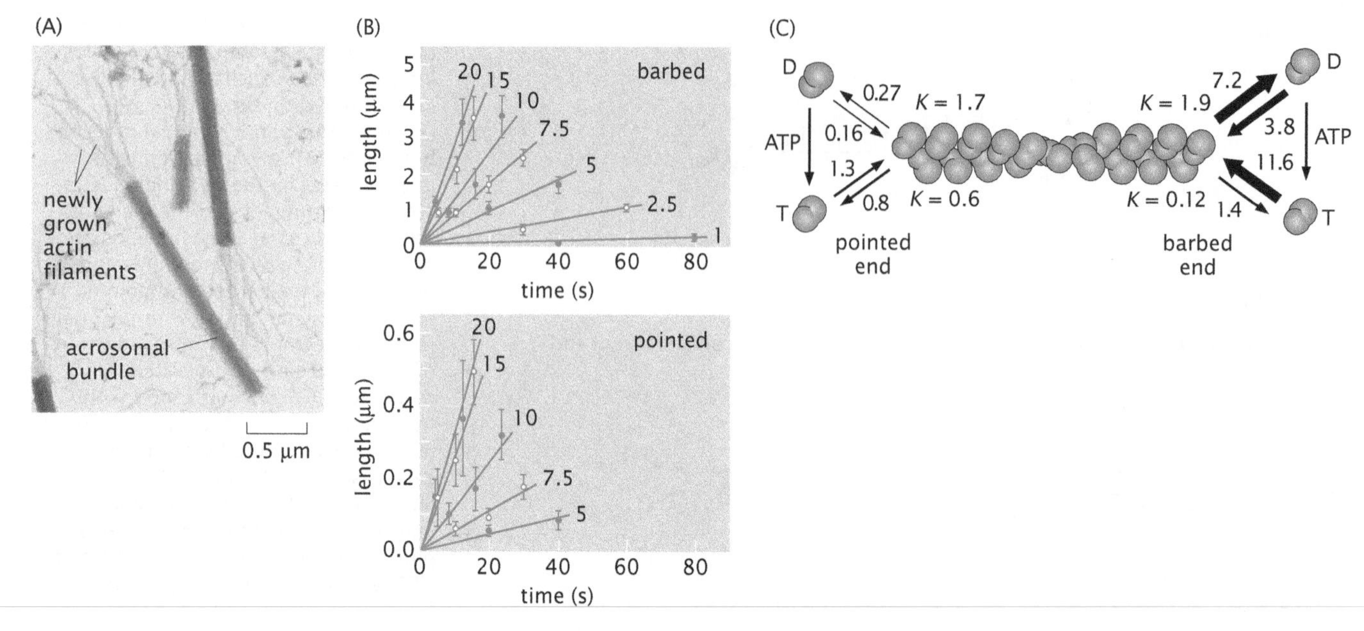

Figure 15.22: Rates of actin polymerization. (A) Electron micrograph showing polarized growth of actin filaments from the ends of horseshoe crab sperm acrosomal bundles. These bundles consist of a parallel array of stabilized actin filaments. When monomeric actin is added, it will form new filaments at different rates on the barbed and on the pointed ends of the acrosome. Here tufts of long filaments can be seen growing from the barbed ends of several acrosomal bundles. The fact that the longer filaments grown from the acrosomes correspond to those growing at the barbed ends can be confirmed by binding isolated myosin motor heads to the filaments and examining them by electron microscopy. The myosin heads bind at an angle so that a pair resembles an arrowhead, giving the two structurally distinct ends of the actin filament their traditional "barbed" and "pointed" names. (B) Length as a function of time for actin growth from barbed and pointed ends of acrosomal bundles. The actin monomer concentration in μM is shown next to each time course. Note the difference in scales; barbed end growth is approximately 10 times faster than pointed end growth. (C) Cartoon of G-actin polymerization. On rates are given in $(\mu M\ s)^{-1}$, off rates in s^{-1}, and dissociation constants in μM. (A, courtesy of M. Footer; B, C, adapted from T. D. Pollard, *J. Cell Biol.* 103:2747, 1986.)

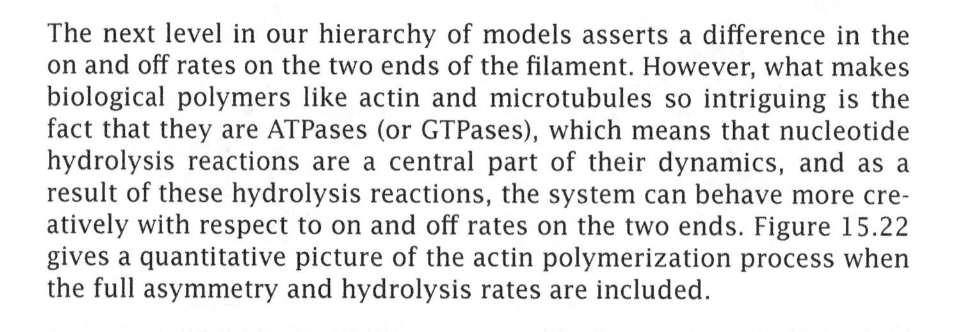

The next level in our hierarchy of models asserts a difference in the on and off rates on the two ends of the filament. However, what makes biological polymers like actin and microtubules so intriguing is the fact that they are ATPases (or GTPases), which means that nucleotide hydrolysis reactions are a central part of their dynamics, and as a result of these hydrolysis reactions, the system can behave more creatively with respect to on and off rates on the two ends. Figure 15.22 gives a quantitative picture of the actin polymerization process when the full asymmetry and hydrolysis rates are included.

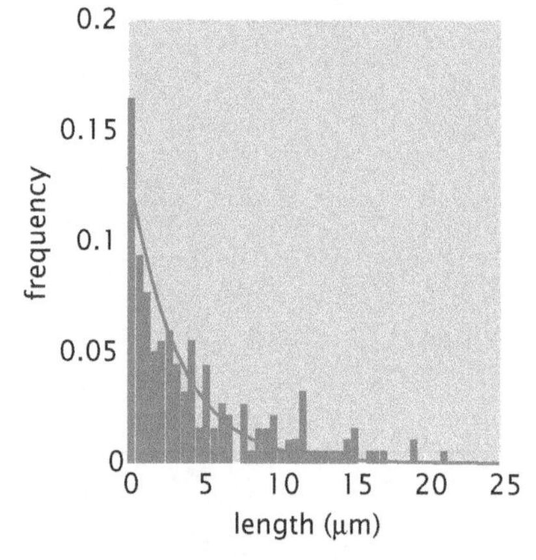

Figure 15.23: Distribution of filament lengths for actin filaments. (Adapted from S. Burlacu et al., *Am. J. Physiol.* 262:C569, 1992.)

15.4.1 The Equilibrium Polymer

Equilibrium Models of Cytoskeletal Filaments Describe the Distribution of Polymer Lengths for Simple Polymers

The simplest model that one might try in thinking about biological polymers is an equilibrium model in which the dynamics results from statistical fluctuations at the two ends. Such models provide a useful description of the formation of biological filaments such as bacterial flagella, filamentous viruses, and amyloid protein fibers. At the same time, such models lack the nuance to capture some of the most intriguing dynamical features of cytoskeletal polymers (actin, microtubules, and their bacterial homologs) such as treadmilling and dynamic instability. One of the first classes of data that such a model can confront is the distribution of filament lengths such as is shown in Figure 15.23. In particular, this figure shows a histogram of the distribution of filament lengths and effectively counts up how many filaments of each length are found.

Our starting point for the equilibrium analysis is the assertion that the probability distribution for the different lengths is stationary, which implies $dP_n/dt = 0$. In this case, the equilibrium is characterized by

$$P_n + P_1 \overset{K_d}{\rightleftharpoons} P_{n+1}, \tag{15.71}$$

which can also be written in a more useful form as

$$K_d = \frac{[P_n][P_1]}{[P_{n+1}]}. \tag{15.72}$$

In principle, K_d should be a function of n since the reaction of two monomers to form a dimer is not necessarily the same as the reaction of a 1000-mer and a monomer to form a 1001-mer. For simplicity, we make the extra assumption that K_d does not depend upon the number of monomers. If we consider the case when $n = 1$ in Equation 15.72, we find

$$K_d = \frac{[P_1][P_1]}{[P_2]} = \frac{[P_1]^2}{[P_2]} \Rightarrow [P_2] = \frac{[P_1]^2}{K_d}. \tag{15.73}$$

We can adopt the same strategy again by taking $n = 2$ and use the result from the previous iteration to find

$$K_d = \frac{[P_2][P_1]}{[P_3]} = \frac{[P_1]}{[P_3]}\frac{[P_1]^2}{K_d} \Rightarrow [P_3] = \frac{[P_1]^3}{K_d^2}. \tag{15.74}$$

By inspection, we can now write the general result as

$$[P_n] = \frac{[P_1]^n}{K_d^{n-1}} = K_d\left(\frac{[P_1]}{K_d}\right)^n. \tag{15.75}$$

This result now permits us to examine the equilibrium distribution of lengths of simple polymers. In particular, it provides a basis for estimating the number of 2-mers, 3-mers and so on. We can rewrite Equation 15.75 in a friendlier way as

$$[P_n] = K_d e^{n\ln([P_1]/K_d)} = K_d e^{-\alpha n}, \tag{15.76}$$

where we define $\alpha = -\ln([P_1]/K_d)$. For the case in which $[P_1] < K_d$, we see immediately that the distribution is a decreasing function of the length as is revealed in Figure 15.23.

Given the distribution of polymer lengths in this simple equilibrium model, we are now poised to compute the average length of these filaments. In particular, the calculation of interest is

$$\langle n \rangle = \frac{\int_0^\infty n e^{-\alpha n}\, dn}{\int_0^\infty e^{-\alpha n}\, dn} = \frac{1}{\alpha} = \frac{1}{\ln(K_d/[P_1])}. \tag{15.77}$$

Using the value of K_d determined above, we can plot the average length as a function of the monomer concentration as shown in Figure 15.24. The average length is a very sensitive function of the monomer concentration. As a result, if the cell were to exploit an equilibrium mechanism of this form, it would necessarily involve very

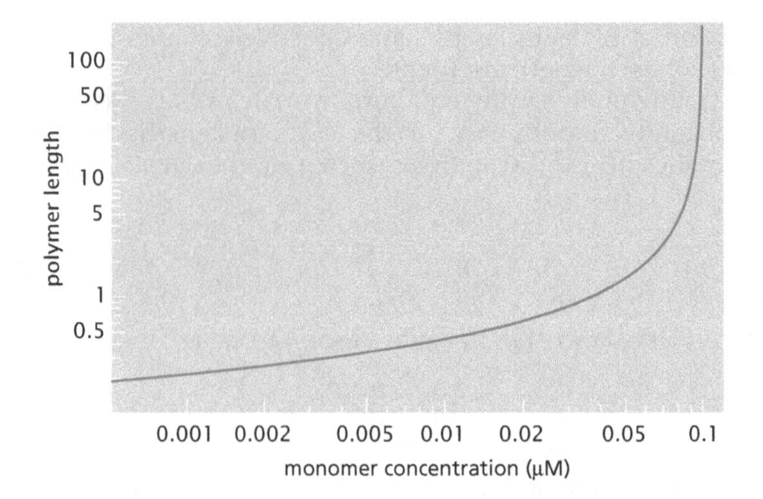

Figure 15.24: Average length of the filaments as a function of the monomer concentration for the equilibrium model.

fine tuning of monomer concentrations. In reality, some polymers that polymerize using this mechanism have other means for determining their preferred length. For example, filamentous viruses such as the tobacco mosaic virus (TMV) use their nucleic acid genome as a template to decide how long to grow. To get a better feeling for the implications of this model, we can also examine the nature of the fluctuations it predicts.

Estimate: Equilibrium Polymers? In order to see what this simple equilibrium model of biological polymers has to say, we require an estimate of K_d. One way that we can obtain an estimate of K_d is as the ratio k_{off}/k_{on}. By appealing to the known rates for actin and ignoring the fact that these rates depend upon whether the monomers are bound to ATP or ADP, we have $k_{on} \approx 10\,\mu M^{-1}\,s^{-1}$ and $k_{off} \approx 1\,s^{-1}$. Using these numbers, we find $K_d = k_{off}/k_{on} \approx 0.1\,\mu M$. For self-assembling polymers, K_d is also often referred to as the critical concentration, c^*. Starting with a solution of monomers, if the concentration is above c^*, then filaments will grow, but below c^* they will shrink. As usual, we find it useful to convert concentrations to mean spacings. In this case, this concentration corresponds to a mean spacing of roughly 200 nm. Recall that the size of the individual monomers is of the order of 5 nm, and the mean spacing between macromolecules in a typical cell is less than 10 nm. The fact that monomers can find each other and rapidly form filaments over such large intermolecular spacings emphasizes the importance of rapid diffusive motion.

An Equilibrium Polymer Fluctuates in Time

The fact that a system is in equilibrium does *not* mean that it is static. As already illustrated in the case of Brownian motion, systems that are in equilibrium undergo diffusive fluctuations. We can borrow this kind of thinking to examine how the length of an equilibrium polymer will vary in time as a result of the growth and shrinkage of filaments due to their natural fluctuations. In particular, we ask the question: how long do we have to wait for the length to change by 1 μm due to these fluctuations? The model we examine is shown schematically in Figure 15.25.

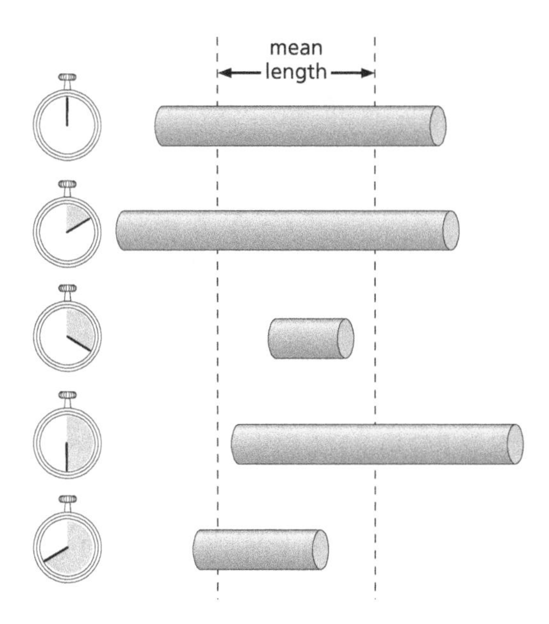

Figure 15.25: Cartoon of a filament growing and shrinking by fluctuations. At $t = 0$, the length of the filament is slightly larger than the mean length. Over time, the two ends are subject to fluctuations as monomers are added and removed from both ends.

To examine the dynamics of an equilibrium filament, we discretize time into steps of length Δt. At every instant, as shown in Figure 15.26, there are three possible fates at both ends of the filament: the end can remain unchanged, the end can lose a monomer, or the end can take on another monomer. With the trajectories and weights in hand, we can work out the average dynamical excursion of the system by summing over all microtrajectories of the system with each one appropriately weighted by the statistical weight. Within this framework, the probability of growth by an amount a (the length of a monomer) is given by

$$P(a) = k_{\mathrm{on}} c_0 \Delta t, \qquad (15.78)$$

where c_0 is the concentration of monomers. The probability of shrinkage by a length a is given by

$$P(-a) = k_{\mathrm{off}} \Delta t. \qquad (15.79)$$

Finally, the probability of nothing happening is

$$P(0) = (1 - k_{\mathrm{on}} c_0 \Delta t - k_{\mathrm{off}} \Delta t). \qquad (15.80)$$

Recall that in equilibrium the critical concentration is given by $c^* = k_{\mathrm{off}}/k_{\mathrm{on}}$, which implies that $P(a) = P(-a)$. As a result, the average

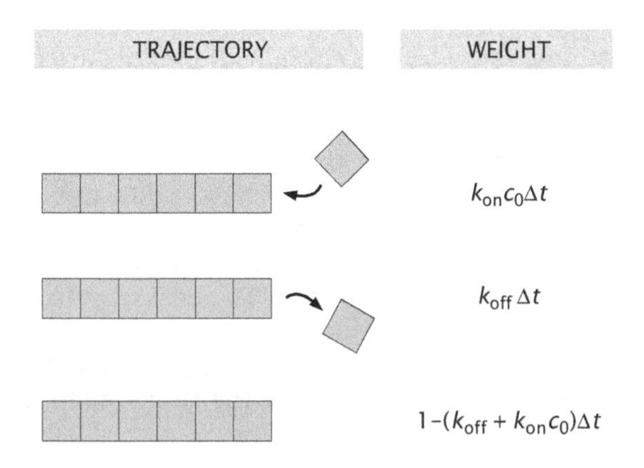

Figure 15.26: Trajectories and weights for an equilibrium polymer. The schematic shows the three processes that are available to a polymer at every time step.

change in length in one time step is

$$\underbrace{\langle x \rangle}_{\text{in } \Delta t} = aP(a) - aP(-a) = 0. \qquad (15.81)$$

This result is not surprising, since, effectively, the tip of the filament may be thought of as a random walker that has the same probability of going to the right or to the left. A more telling measure of the fluctuations is provided by the mean-square displacement, given by

$$\langle x^2 \rangle = a^2 P(a) + a^2 P(-a) = a^2 k_{\text{on}} c_0 \Delta t + a^2 k_{\text{on}} c_0 \Delta t = 2a^2 k_{\text{on}} c_0 \Delta t. \qquad (15.82)$$

We are interested in finding out how many time steps it will take to accumulate a total excursion that is typical of the length change seen in cytoskeletal filaments, such as 1 μm. Since each instant is independent of the previous instant (that is, there is no memory), the accumulated variance over N steps is N times the variance associated with one step. In particular, we have

$$\langle L^2 \rangle = 2Na^2 k_{\text{off}} \Delta t, \qquad (15.83)$$

which can be rewritten by noting that $N = t/\Delta t$, resulting in

$$\langle L^2 \rangle = 2a^2 k_{\text{off}} t. \qquad (15.84)$$

The significance of this result is best served by considering some real-world numbers. Our interest is in the time we have to wait to see an excursion of length L, which is given by

$$\sqrt{\langle L^2 \rangle} = \sqrt{2a^2 k_{\text{off}} t}. \qquad (15.85)$$

For the special case in which we query the time needed for an excursion of length 1 μm, we have

$$t = \frac{\langle L^2 \rangle}{2a^2 k_{\text{off}}}. \qquad (15.86)$$

Using the various approximate parameters described above, this yields

$$t = \frac{1\ \mu\text{m}^2}{2\left(\underbrace{\frac{4}{1000}\ \mu\text{m}}_{a}\right)^2 1\ \text{s}^{-1}} \approx 9\ \text{hours}. \qquad (15.87)$$

The time scales generated by the equilibrium model are clearly inconsistent with the observed time scales of cytoskeletal dynamics. The easiest way to note that is to watch a time-lapse video of a motile or dividing cell and to observe that these processes take place on minute rather than hour time scales. These rapid dynamics require nucleotide hydrolysis by actin and microtubules and also the activity of myriad other cytoskeleton-associated proteins. The fact that the cytoskeleton is *not* in equilibrium is necessary to give the cell morphological flexibility over rapid time scales. In Section 15.4.3, we will return to the issue of nucleotide hydrolysis by cytoskeletal filaments. In preparation, we first consider the time-dependent behavior of simple nonhydrolyzing polymers.

15.4.2 Rate Equation Description of Cytoskeletal Polymerization

Polymerization Reactions Can Be Described by Rate Equations

Cytoskeletal polymerization consists of three very distinctive phases as shown in Figure 15.19(A). If we start with a situation where we only have monomers free in solution, there will be a *lag phase* during which these monomers assemble into small oligomers. After this nucleation phase and once these nucleation seeds are assembled, the *growth phase* begins. Filament elongation will continue until a *steady-state phase* is reached. Here there will still be an interchange of monomers between the solution and the filaments, but the net length change will be zero. In those cases where nuclei are added directly to the solution, the lag phase can be bypassed, resulting in growth like that shown in Figure 15.19(B).

As an alternative to the equilibrium model described above and in response to experiments like those indicated schematically in Figure 15.19, we consider the simplest case of a system that is seeded with M "nuclei" that can serve to nucleate growing filaments. Our plan is to write down a rate equation that examines the number of monomers, $n(t)$, associated with the "average" filament at each instant in time. Since our rate equation picture does not consider fluctuations, we note that the dynamical equation describing the number of monomers, n, in each filament is identical for each filament and is given by

$$\frac{\mathrm{d}n}{\mathrm{d}t} = \underbrace{k_{\mathrm{on}}\left(c_0 - \frac{Mn(t)}{V}\right)}_{\text{growth}} - \underbrace{k_{\mathrm{off}}}_{\text{shrinkage}}, \tag{15.88}$$

where c_0 is the initial concentration of monomers in the volume V and M is the number of nuclei that seed the growth. This equation says that there will be an addition of monomers with rate k_{on}, which is proportional to the monomer concentration. In addition to the growth, there is also the potential of shrinkage of the filaments as a result of the k_{off} term. As the monomer concentration varies, there is a competition between these two terms. The significance of the factor $c_0 - Mn/V$ is that it captures the instantaneous concentration. Our picture is of a box of fixed volume V that has some initial number of monomers c_0V available for polymerization. As the filament growth process proceeds, the number of available monomers is reduced since ever more monomers are now tied up in filaments (the number of monomers in the filaments is Mn) and are hence unavailable for further polymerization. Note that in this simple model, the total number of nuclei, M, is fixed.

The solution to this dynamical equation is found by noting that we have a first-order linear differential equation that can be rewritten as

$$\frac{\mathrm{d}n}{\mathrm{d}t} + \frac{k_{\mathrm{on}}Mn}{V} = k_{\mathrm{on}}c_0 - k_{\mathrm{off}}. \tag{15.89}$$

When written in this form, we see that it can be solved by first guessing a "particular" solution, in this case

$$n_{\mathrm{particular}}(t) = \frac{V}{Mk_{\mathrm{on}}}(k_{\mathrm{on}}c_0 - k_{\mathrm{off}}). \tag{15.90}$$

To this particular solution should then be added a solution to the equation for the case in which the right-hand side of Equation 15.89

has been set to zero. The solution in this case is

$$n_{\text{homogeneous}}(t) = A\mathrm{e}^{-k_{\text{on}}Mt/V}, \tag{15.91}$$

where the constant A is an arbitrary constant that is fixed by initial conditions. One particular case of interest is the one in which the initial size of the nuclei is zero (of course, this contradicts the idea that a critical nucleus is needed to start the polymerization process, but this choice is mathematically convenient and gives rise to the same key features). Using $n(0) = 0$, we can fix the constant A as

$$A = -\frac{V}{Mk_{\text{on}}}(k_{\text{on}}c_0 - k_{\text{off}}). \tag{15.92}$$

As a result of these manipulations, the solution to our problem is given by

$$n(t) = \frac{V}{Mk_{\text{on}}}(k_{\text{on}}c_0 - k_{\text{off}})(1 - \mathrm{e}^{-k_{\text{on}}Mt/V}). \tag{15.93}$$

In order to compute the length of the growing filament, we note that the length is given by

$$L(t) = an(t), \tag{15.94}$$

where a is the length per monomer. This result is plotted in Figure 15.27, which exhibits two key regimes. First, in the short-time limit, defined as times much shorter than the time

$$\tau = \frac{V}{Mk_{\text{on}}}, \tag{15.95}$$

the filaments grow linearly with time. This can be seen by considering the behavior at short times, which is revealed by expanding $\mathrm{e}^{-k_{\text{on}}Mt/V}$ in a Taylor series (that is, using $\mathrm{e}^x \approx 1 + x$; see The Math Behind the Models on p. 215). The second key regime is the long-time limit for which the growth saturates and the length of each filament is constant.

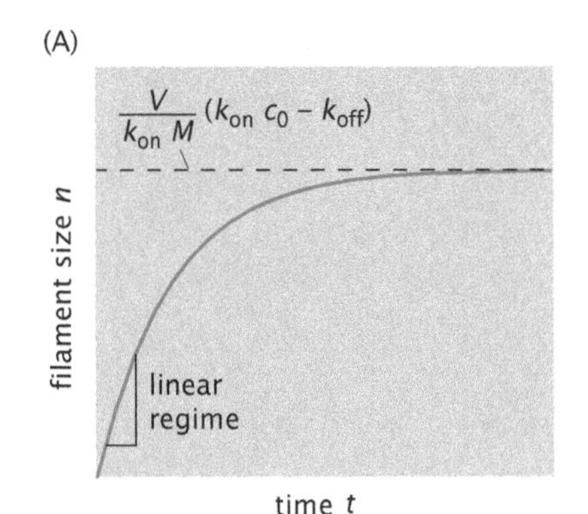

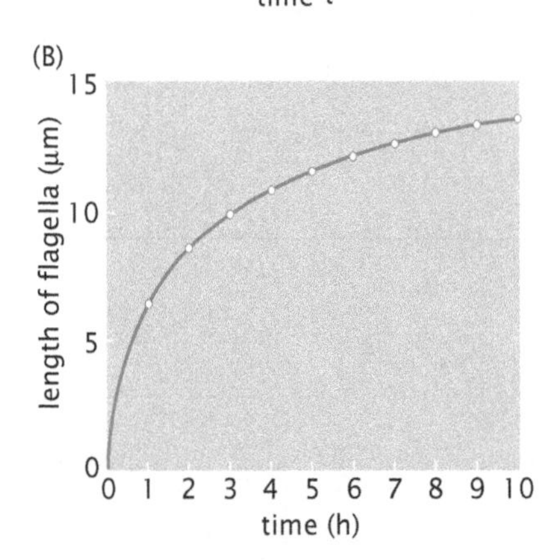

Figure 15.27: Average filament size versus time. (A) Length of the filaments as a function of time assuming M initial nuclei. During the initial stages of growth, the growth rate is linear, while at long times the system reaches a steady state. (B) Data showing length of filaments versus time for assembly of bacterial flagellin. (B, adapted from T. Iino, *J. Supramol. Struct.* 2:372, 1974.)

The Time Evolution of the Probability Distribution $P_n(t)$ Can Be Written Using a Rate Equation

A more sophisticated version of the model we developed above asks for the time evolution of the probability distribution $P_n(t)$, which reflects the probability of finding filaments of length n as a function of time. In particular, $P_n(t)$ is the probability that a given filament is an n-mer, and is given as

$$P_n(t) = \frac{N_n}{\sum_{n=1}^{\infty} N_n}, \tag{15.96}$$

where N_n is the number of filaments containing n monomers. Figure 15.28 shows how one takes a distribution of filaments and uses a histogram to determine experimentally the probabilities of the various polymer lengths.

The rate equation that captures the change in the number of n-mers can be written symbolically as

$$\frac{\mathrm{d}P_n}{\mathrm{d}t} = \text{rate of monomer addition to } n\text{-mer} - \text{rate of monomer removal.} \tag{15.97}$$

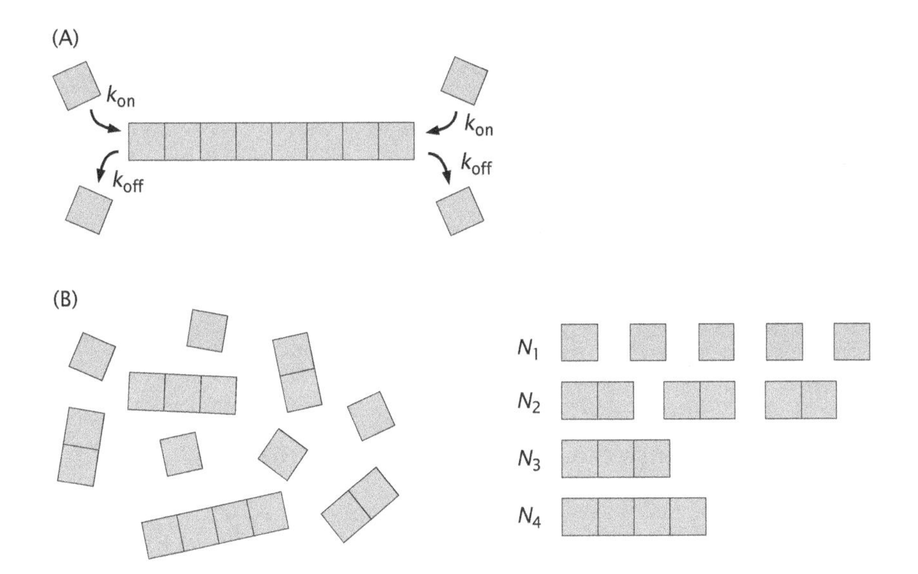

Figure 15.28: Illustration of the idea of the probability distribution $P_n(t)$. (A) The reactions that can change the current length of the filament. (B) A distribution of filaments of various lengths. We count up the number of each species N_1, N_2, etc.

To proceed, we make the simplifying assumption that the only way a given n-mer can change its length is by either the addition or the loss of a single monomer. For example, we reject the process in which a 5-mer is added to a 4-mer resulting in a 9-mer. In addition, we begin our analysis by assuming that both ends of the growing filament are the same. This assumption will be suppressed later in favor of the experimentally relevant situation in which cytoskeletal filaments have polarity.

The reaction we are proposing is

$$\mathrm{P}_n + \mathrm{P}_1 \underset{k_{\mathrm{off}}}{\overset{k_{\mathrm{on}}}{\rightleftharpoons}} \mathrm{P}_{n+1}. \tag{15.98}$$

Our goal is to write a rate equation for the dynamics of this reaction. However, the equations are coupled. That is, the equation for the evolution of P_n depends upon the evolution of both P_{n-1} and P_{n+1}. To find the time evolution of P_n, we have to consider the "adjacent" reaction given by

$$\mathrm{P}_{n-1} + \mathrm{P}_1 \underset{k_{\mathrm{off}}}{\overset{k_{\mathrm{on}}}{\rightleftharpoons}} \mathrm{P}_n, \tag{15.99}$$

which will account for the "decay" of a polymer of length n into one of length $n-1$. The time evolution of the probability distribution is governed by four distinct classes of process and is captured mathematically as

$$\frac{\mathrm{d}P_n}{\mathrm{d}t} = \underbrace{k_{\mathrm{on}}P_{n-1}P_1}_{\text{addition to } P_{n-1}} + \underbrace{k_{\mathrm{off}}P_{n+1}}_{\text{removal from } P_{n+1}} - \underbrace{k_{\mathrm{on}}P_nP_1}_{\text{addition to } P_n} - \underbrace{k_{\mathrm{off}}P_n}_{\text{removal from } P_n}. \tag{15.100}$$

One question of immediate interest that emerges from a model of this kind is: what is the average length of a polymer as a function of time whose growth is described by Equation 15.100? If a is the length of a monomer, we can write the total average length of the filament as

$$\langle L \rangle = \sum_{n=1}^{\infty} na P_n. \tag{15.101}$$

By taking the time derivative of this expression, we can then write an equation for the time evolution of the average length as

$$\frac{\mathrm{d}\langle L\rangle}{\mathrm{d}t} = \sum_{n=1}^{\infty} na\frac{\mathrm{d}P_n}{\mathrm{d}t}. \tag{15.102}$$

We can now substitute our expressions for the time derivatives themselves from the original rate equation (Equation 15.100), with the result

$$\frac{\mathrm{d}\langle L\rangle}{\mathrm{d}t} = \sum_{n=1}^{\infty} na(k_{\mathrm{on}}P_{n-1}P_1 + k_{\mathrm{off}}P_{n+1} - k_{\mathrm{on}}P_nP_1 - k_{\mathrm{off}}P_n). \tag{15.103}$$

We may now rearrange the right-hand side, resulting in

$$\frac{\mathrm{d}\langle L\rangle}{\mathrm{d}t} = \sum_{n=1}^{\infty} ank_{\mathrm{on}}P_1(P_{n-1} - P_n) + \sum_{n=1}^{\infty} ank_{\mathrm{off}}(P_{n+1} - P_n). \tag{15.104}$$

The next step in our argument is to factor out P_1 and to simplify the functional form of the expressions using the identity

$$\sum_{n=1}^{\infty} nP_{n-1} = 2P_1 + 3P_2 + \cdots = \sum_{n=1}^{\infty}(n+1)P_n. \tag{15.105}$$

Using this identity several times in our original expression results in

$$\begin{aligned} ak_{\mathrm{on}}P_1\sum_{n=1}^{\infty} n(P_{n-1} - P_n) &= ak_{\mathrm{on}}P_1\sum_{n=1}^{\infty}[(n+1) - n]P_n \\ &= ak_{\mathrm{on}}P_1\sum_{n=1}^{\infty} P_n = ak_{\mathrm{on}}P_1. \end{aligned} \tag{15.106}$$

The net result of these manipulations is that we recover precisely the same expression determined earlier for the mean length, namely,

$$\frac{\mathrm{d}\langle L\rangle}{\mathrm{d}t} = (k_{\mathrm{on}}P_1 - k_{\mathrm{off}})a. \tag{15.107}$$

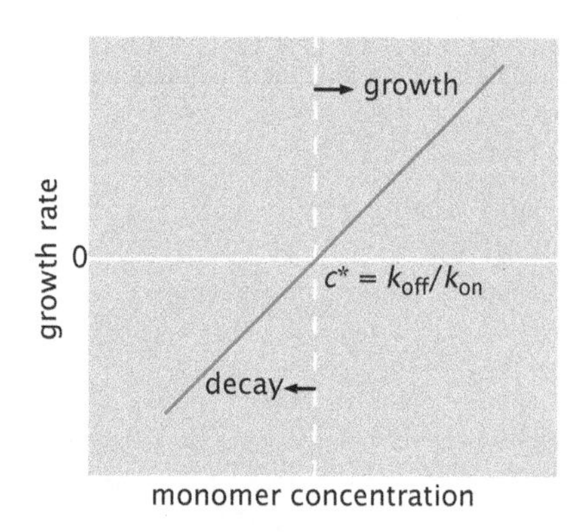

Figure 15.29: Growth rate for filaments. The graph shows the growth rate as a function of the monomer concentration. Below a critical concentration, the filaments shrink. For concentrations greater than the critical concentration, the filaments grow.

One useful way to visualize the significance of this result is to plot the rate of growth as a function of monomer concentration as is shown in Figure 15.29. Note that, depending upon the concentration, the filaments will either grow or shrink. In particular, for concentrations smaller than $k_{\mathrm{off}}/k_{\mathrm{on}}$, the filaments will shrink, whereas for larger values of monomer concentration, the filaments will grow. In the special case where P_1 is equal to the critical concentration $c^* = k_{\mathrm{off}}/k_{\mathrm{on}}$, the average length will not change, because $\mathrm{d}\langle L\rangle/\mathrm{d}t = 0$.

Rates of Addition and Removal of Monomers Are Often Different on the Two Ends of Cytoskeletal Filaments

As was already hinted at in Figure 15.21, there is a hierarchy of models of biological polymerization reactions of increasing physical realism. The next refinement of the models we have considered thus far, which brings us one step closer to the behavior of real cytoskeletal filaments, is to consider the case in which the rates on the two ends of the growing filament are different. A schematic depiction of this situation is shown in Figure 15.30 and reflects the idea that this difference in rates

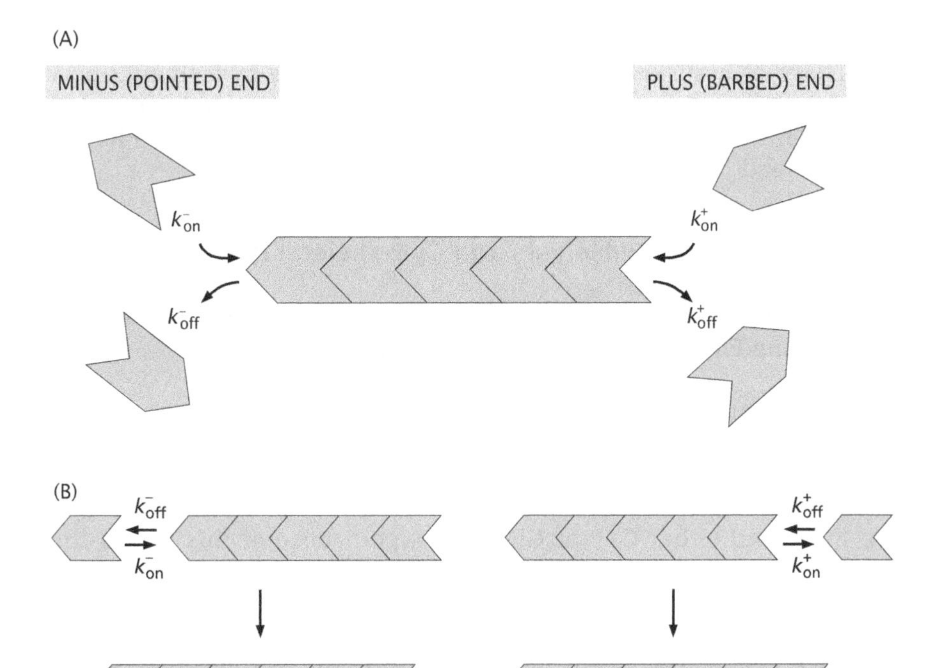

Figure 15.30: Model of polymerization where each end has its own rates. (A) Illustration of the different rate constants on the barbed and pointed ends. (B) Addition of a single monomer to the growing filament.

on the two ends is a result of structural asymmetry on the two ends. Stated differently, cytoskeletal polymers such as actin are characterized by a polarity with the distinct ends being labeled as "plus" and "minus" (for historical reasons, the plus end on actin is also called the barbed end and the minus end is called the pointed end). The plus end has a faster growth rate than the minus end, and also a faster shrinkage rate.

One intriguing feature of this model is that the ratios of the rates on the two ends must equal each other. The simplest way to see this is to note that the molecular interfaces at the two ends are identical and hence they imply the same free energy associated with the contacts. Inspection of the schematic in Figure 15.30(A) reveals that although the monomers are themselves asymmetric, once two of them have joined together, the common interface they share is indifferent to whether the monomer was added on the plus or minus end. This condition is represented mathematically as

$$\Delta G^+ = \Delta G^- \Rightarrow \frac{k_{off}^+}{k_{on}^+} = \frac{k_{off}^-}{k_{on}^-} = \frac{1}{V}e^{\Delta G/k_BT}. \tag{15.108}$$

The rate equations for this situation are identical in form on the two ends, though characterized by different rates, and are given by

$$\frac{dn_+}{dt} = k_{on}^+ c_0 - k_{off}^+ \tag{15.109}$$

and

$$\frac{dn_-}{dt} = k_{on}^- c_0 - k_{off}^-. \tag{15.110}$$

The growth rate as a function of monomer concentration is shown in Figure 15.31 and reveals that when one end grows the other suffers the same fate; this will always be the case when the concentration of monomers in solution is above c^*. This model is a reasonably accurate description of the behavior of simple polymers, such as bacterial

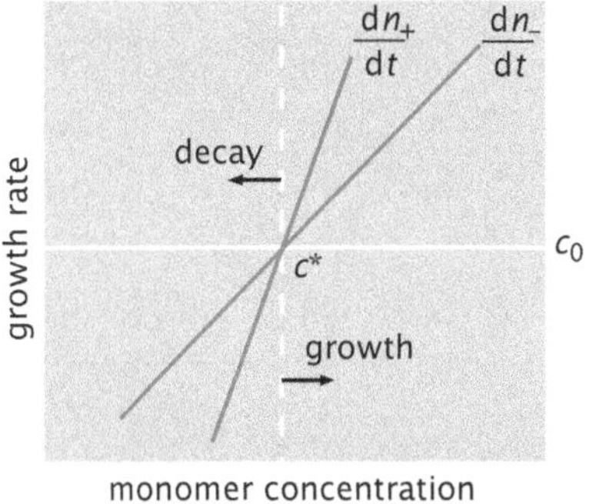

Figure 15.31: Growth rate for the case where the rates for each end of the filament are different. Above the critical concentration c^*, both ends grow, and below the critical concentration, both ends shrink.

flagellin. However, it falls short of being able to describe the entire repertoire of cytoskeletal dynamics for nucleotide-hydrolyzing polymers, such as actin and microtubules. For these remarkable polymers, it is possible for one end to grow while the other is shrinking, a feature that can lead to the phenomenon of treadmilling.

15.4.3 Nucleotide Hydrolysis and Cytoskeletal Polymerization

ATP Hydrolysis Sculpts the Molecular Interface, Resulting in Distinct Rates at the Ends of Cytoskeletal Filaments

The full beauty and complexity of cytoskeletal polymerization is tied to the fact that the protein monomers interact with nucleotides and, indeed, can catalyze nucleotide hydrolysis. For example, actin monomers can bind either ATP or ADP, while in the case of tubulin, they bind either GTP or GDP. The significance of this fact is that these cytoskeletal polymers behave as ATPases (or as GTPases) and, by virtue of the coupling of polymerization to ATP (or GTP) hydrolysis, they can undergo conformational changes that alter the molecular interface between adjacent monomers and hence adjust the rates of binding and unbinding. The full complexity of the polymerization process was shown in Figure 15.21. For our present purposes, the key point of these observations is that there are different on and off rates for ATP-bound and ADP-bound monomers at each end of the structurally polarized actin filaments. This effect is represented schematically in Figures 15.21(C) and 15.21(D).

The rate equations describing this situation can be substantially simplified if we assume that on the barbed ends, only ATP-bound monomers are coming on and off, while on the pointed ends, only ADP-bound monomers are exchanged between the filament and the reservoir. In this situation, the rate equations have an identical form to those presented in Equations 15.109 and 15.110. The difference is that in this case the restriction on the ratio of the rates embodied in Equations 15.108 is no longer present. The resulting growth rates on the two ends as a function of concentration are shown in Figure 15.32.

These growth rates have a novel property relative to the other models in the hierarchy of models we have considered, namely, they can undergo the process of treadmilling. In treadmilling, one end grows while the other end shrinks. As shown in Figure 15.32, for a certain critical concentration c_{TM}, the barbed end can be growing at the same rate that the pointed end is shrinking, a condition represented mathematically as

$$\frac{\mathrm{d}n_+}{\mathrm{d}t} = -\frac{\mathrm{d}n_-}{\mathrm{d}t}. \tag{15.111}$$

This condition can be written directly in terms of the rates as

$$k_{\mathrm{on}}^{+} c_{\mathrm{TM}} - k_{\mathrm{off}}^{+} = -k_{\mathrm{on}}^{-} c_{\mathrm{TM}} + k_{\mathrm{off}}^{-}. \tag{15.112}$$

Solving this equation for c_{TM} results in the condition

$$c_{\mathrm{TM}} = \frac{k_{\mathrm{off}}^{+} + k_{\mathrm{off}}^{-}}{k_{\mathrm{on}}^{+} + k_{\mathrm{on}}^{-}}, \tag{15.113}$$

which tells us how the treadmilling concentration depends upon the on and off rates on the two ends of the model polymer.

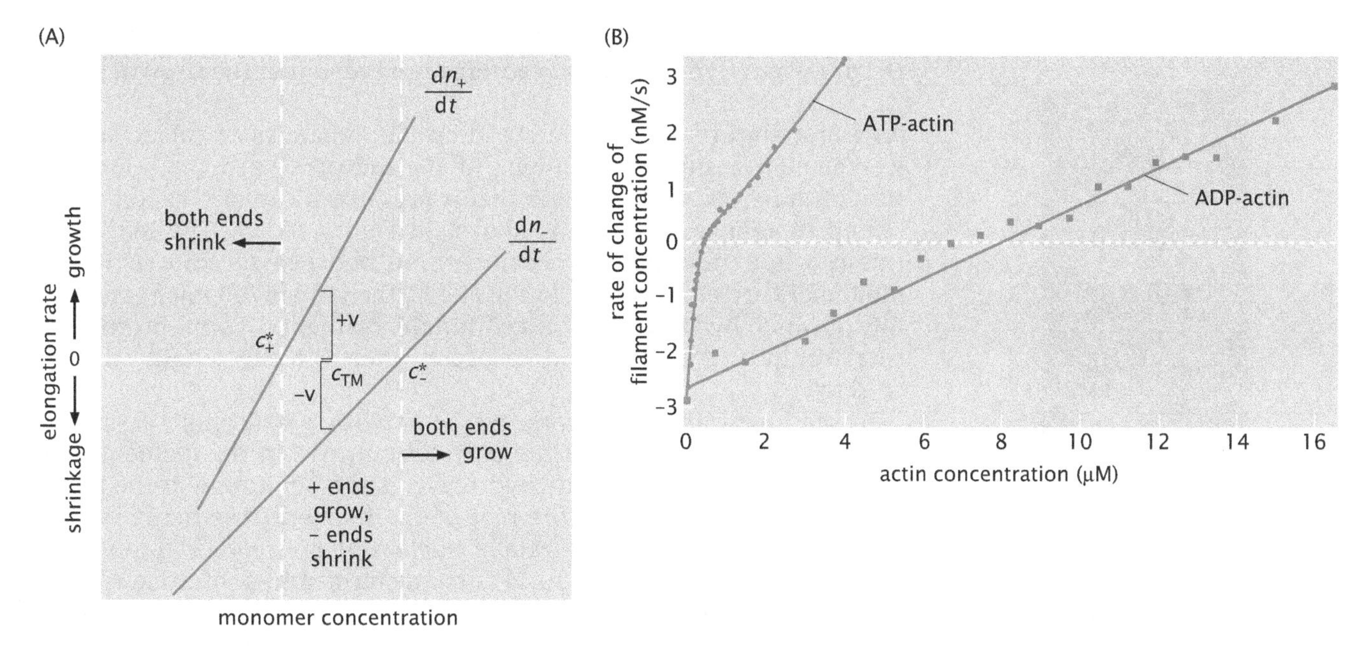

Figure 15.32: Growth rates on the two ends as a function of concentration for the case in which there is nucleotide hydrolysis. (A) The two curves show the growth rates on the two ends of a microtubule as a function of the concentration of monomers. The upper curve shows the faster growth rate on the plus end. At the critical concentration, c_{TM}, the growth rate on the plus end is equal to the shrinkage rate on the minus end, resulting in treadmilling. (B) Growth rate as a function of actin concentration for ATP–actin (left) and ADP–actin (right). ADP–actin behaves as predicted by the model described in the text, displaying a critical concentration at about 8 μM. For ATP–actin, the growth rate above the critical concentration is a linear function of actin concentration as expected; however, the curve becomes nonlinear at lower concentrations. This is because ATP hydrolysis is accelerated by polymerization so the disassembling species is actually ADP–actin. In other words, the ATP cap (see Figure 15.35 below) has vanished on the shrinking filaments. (B, adapted from M. F. Carlier et al., *J. Biol. Chem.* 259:9983, 1984.)

These ideas serve as the basis for thinking about the observed dynamical behavior of microtubules. In particular, under the right circumstances, it is possible to observe microtubules that are simultaneously growing on one end while shrinking at the other, with the result that particular points within the microtubule lattice are fixed with respect to the absolute coordinate system of the cell. An example of this behavior is shown in Figure 15.33.

15.4.4 Dynamic Instability: A Toy Model of the Cap

One of the most intriguing features of microtubule dynamics is the existence of an effect known as "dynamic instability." The basic observation is that an individual microtubule can undergo repeated cycles of growth and shrinkage, even when its chemical environment remains essentially constant. Similarly, a population of microtubules in the same test tube or cell will typically include some individuals that are growing and some individuals that are shrinking at any given instant in time. If the length of a single microtubule is measured as a function of time, the resulting graph is a series of sawtooths such as are shown in Figure 15.34. That is, the dynamics of microtubule growth is punctuated by occasional catastrophes, where the filament shrinks, and rescues, where it begins to grow again. Dynamic instability has also been observed in the bacterial actin-like filament ParM. It has been argued that dynamic instability provides a mechanism for repeated attempts for cytoskeletal filaments to explore the large cytoplasmic space and find their targets during chromosome segregation and other spatially complex processes.

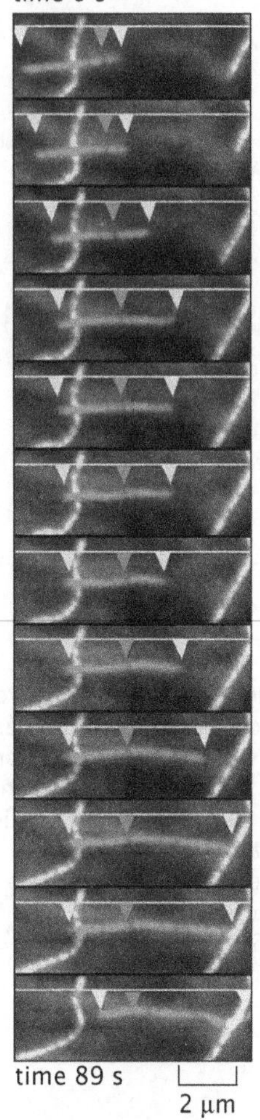

Figure 15.33: Microtubule treadmilling. Individual microtubules have been observed to undergo treadmilling inside living cells. The fluorescently labeled microtubule looks as though it is sliding from left to right. However, as indicated by the middle arrow, which points to a fixed point within the microtubule, in fact one end (the minus end, on the left) is shortening and the other (the plus end, on the right) is elongating. (Adapted from C. M. Waterman-Storer and E. D. Salmon, *J. Cell Biol.* 139:417, 1997.)

A Toy Model of Dynamic Instability Assumes That Catastrophe Occurs When Hydrolyzed Nucleotides Are Present at the Growth Front

As a first step towards understanding the dynamics of catastrophes, we consider a simple toy model of the unhydrolyzed cap. Our picture of the growing filament is that it presents a cap of GTP- (or ATP-) bound monomers such as is shown in Figure 15.35. The length of this cap is determined by a competition between the rate at which monomers are being added to the end of the growing filament and the rate at which hydrolysis occurs within the filament. Within this model, catastrophe is hypothesized to occur when the cap length shrinks to zero.

To see how this might work, we begin with a deterministic model in which it is assumed that the only subunit within the filament that is able to perform nucleotide hydrolysis is the one found at the interface between the cap and the rest of the filament. Furthermore, for this vulnerable subunit, the rate of nucleotide hydrolysis (conversion of GTP to GDP or of ATP to ADP) is characterized by a lifetime τ. This deterministic picture implies that the position of the interface moves along the filament with a characteristic velocity a/τ. Also note that the pointed ends have rate constants characterized by ADP-bound filament ends while the growing barbed ends are characterized by the ATP rate constants. A more sophisticated and physically realistic scenario, where hydrolysis can happen at any point along the cap, would provide a more complete picture of the cap dynamics.

Within this simple model, the onset of catastrophe occurs when and if the hydrolysis "front" catches up with the growing tip on the barbed end. One physical mechanism that could lead to this effect would be a slowing down of the growing tip due to depletion of the reservoir of available monomers. Recall that our picture is that the leading edge of the filament has a velocity v_{tip} and that the hydrolysis takes place at a rate $1/\tau$. Within this model, the critical condition for a catastrophe is that the velocity of the hydrolysis front equals that of the growing tip and is given by

$$\underbrace{\frac{dx_{\text{tip}}}{dt}}_{\text{growth rate of leading edge}} = \underbrace{\frac{a}{\tau}}_{\text{speed of hydrolysis front}} . \quad (15.114)$$

This equation says that in the time it takes for the next monomer to be added to the growing tip, the monomer at the tip will have time to hydrolyze its nucleotide triphosphate, where we have made the simplifying assumption that the cap is only one monomer in width. The critical condition written mathematically above really sets the bound, since it is only when the hydrolysis front is moving faster than the growing tip that we are guaranteed that the front will catch up with the tip and induce a catastrophe.

The physical picture of the catastrophe, then, is that the catastrophe starts once the hydrolysis front catches the leading edge, because at that point the GTP (or ATP) growth rates are superseded by their GDP (or ADP) counterparts. Said differently, the off rate dominates for the GDP monomers, whereas the on rate dominates for the GTP monomers. This model predicts a very specific dependence of the catastrophe rate on the concentration of monomers available for growth. To see this, we turn to the mathematical implementation of the model.

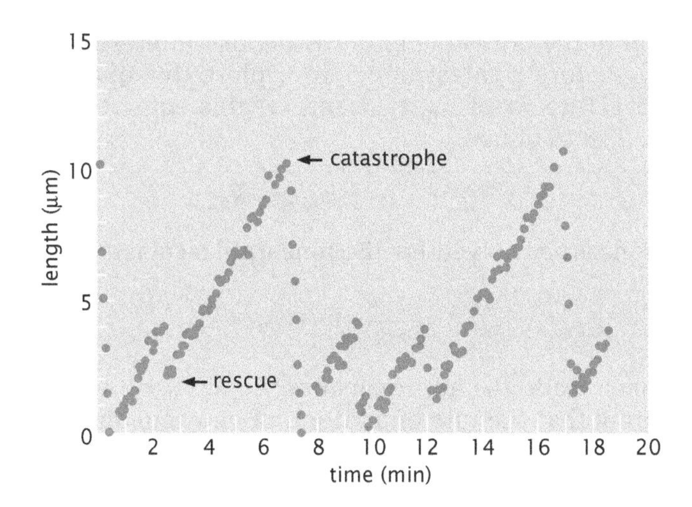

Figure 15.34: Natural history of the length variation for a single microtubule over time. A 10 μm long microtubule was imaged in the microscope and its length measured roughly 10 times per minute. Almost immediately upon the beginning of this observation, the microtubule underwent a catastrophe and shrank rapidly to near zero length. It then underwent a rescue and began to grow again. Over the subsequent 20 minutes, the cycle repeated 5 more times, with the lengths of the growing and shrinking phases varying randomly. Microtubule growth took place at a nearly constant rate for all cycles, and shrinkage also took place at a constant, though faster, rate. (Adapted from D. K. Fygenson et al., *Phys. Rev.* E50:1579, 1994.)

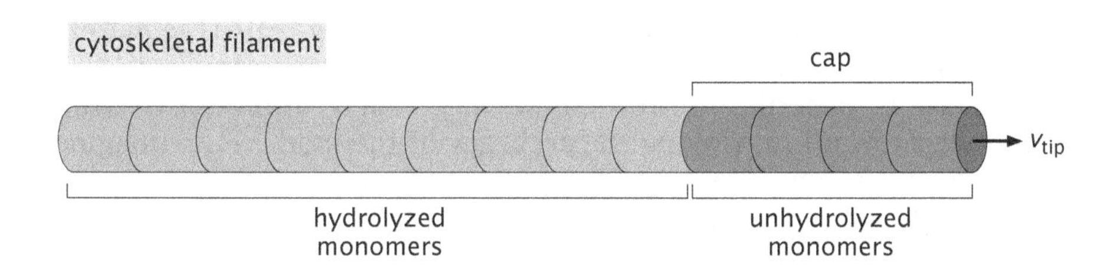

Figure 15.35: A filament presenting a cap of GTP- (or ATP-) bound monomers. The monomers in the cap have not yet performed nucleotide hydrolysis.

The growth of the tip is essentially described by Equation 15.88, with the observation that $dx_{tip}/dt = a dn/dt$, and results in

$$\frac{dx_{tip}}{dt} = a\left[k_{on}\left(c_0 - \frac{x_{tip}M}{Va}\right) - k_{off}\right], \tag{15.115}$$

where M is the number of growing filaments. Note that x_{tip}/a tells us the number of monomers per filament, which when multiplied by M gives the total number of monomers tied up in filaments. The solution to this equation proceeds exactly as did the solution to Equation 15.88 and is plotted in Figure 15.36, and given by

$$x_{tip}(t) = \frac{aV}{Mk_{on}}(k_{on}c_0 - k_{off})(1 - e^{-Mk_{on}t/V}). \tag{15.116}$$

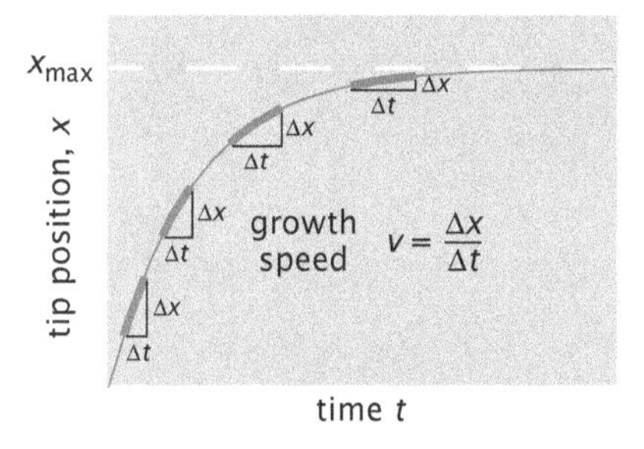

Figure 15.36: Position of the tip of a growing filament as a function of time. The instantaneous velocity of the tip (that is, the growth rate) is given as the slope of this curve. As a result of monomer depletion, the tip growth rate slows down over time.

In this toy model, a catastrophe occurs when the speed of the tip drops just below the hydrolysis speed. This happens at a certain critical length of the cap x_{tip}^{crit} where, because of depletion of the reservoir of soluble monomers, the tip growth rate has slowed sufficiently to match the hydrolysis speed. This is characterized by the condition

$$a\left[k_{on}\left(c_0 - \frac{x_{tip}M}{Va}\right) - k_{off}\right] = \frac{a}{\tau}. \tag{15.117}$$

This equation can be solved for the critical length, which is given by

$$x_{tip}^{crit} = \frac{Va}{Mk_{on}}\left(k_{on}c_0 - k_{off} - \frac{1}{\tau}\right). \tag{15.118}$$

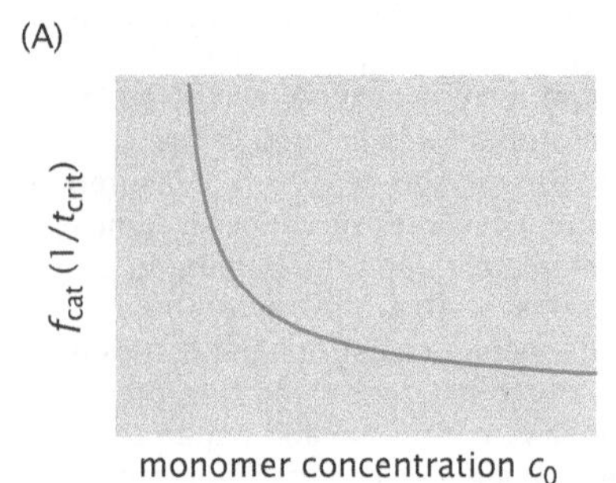

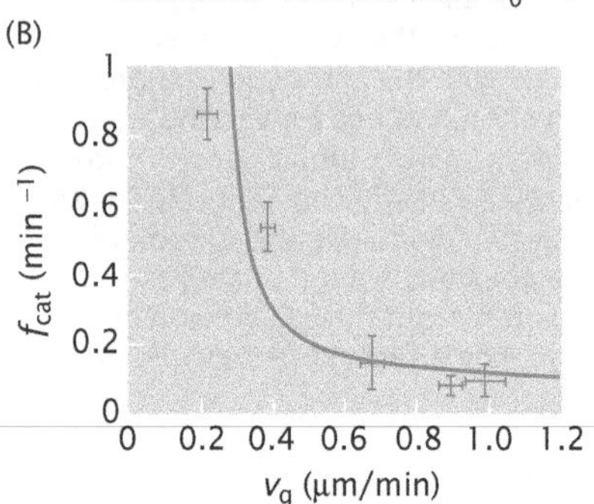

Figure 15.37: Catastrophe rate for filaments undergoing dynamic instability. (A) Catastrophe rate as a function of the monomer concentration as predicted by the toy model of cap dynamics. (B) Measured catastrophe rate as a function of the growth velocity (which is linearly related to the monomer concentration). (B, adapted from H. Flyvbjerg et al., *Phys. Rev.* E54:5538, 1996.)

In addition to the critical length, we are also interested in the waiting time t_{crit} before a catastrophe. To explore this question, we use Equation 15.116 to write $x_{tip}(t_{crit})$ and set this equal to the result in Equation 15.118 to obtain

$$e^{Mk_{on}t_{crit}/V} = \tau(k_{on}c_0 - k_{off}). \tag{15.119}$$

This equation can be solved for the time until a catastrophe as

$$t_{crit} \simeq \frac{V}{Mk_{on}} \ln(\tau k_{on} c_0), \tag{15.120}$$

where we have made the approximation $k_{off} \ll k_{on}c_0$, which is equivalent to stating that the rate of addition of new monomers to the cap will be much higher than the rate at which monomers are removed. We can now plot the catastrophe rate $1/t_{crit}$ (Figure 15.37A).

One of the shortcomings of this model is that it predicts that once the monomers are depleted, all microtubules will suffer catastrophes at essentially the same time. However, this prediction is clearly at odds with experimental observations on microtubule dynamics, where some microtubules can be seen to shrink (that is, undergo catastrophe) while others are growing nearby inside a living cell. This merely demonstrates that our toy model, while a useful starting point, is inadequate to describe the full complexity of microtubule behavior. A first step toward improving the accuracy of the model is to imagine that hydrolysis can occur not only at the hydrolyzed–nonhydrolyzed interface shown in Figure 15.35, but also at other positions within the cap. That is, by chance, the hydrolysis event can occur very near to the end of the cap and a hydrolysis "front" can propagate away from this hydrolysis site in both the plus and minus directions. This conceptual scheme is illustrated in Figure 15.21(D). The results of this more sophisticated model for the catastrophe rate are shown in Figure 15.37(B). Here, instead of plotting the rate of catastrophes as a function of the monomer concentration, the experimental results are shown as a function of the growth velocity since this velocity is itself a linear function of the concentration. Though an analysis of this model is beyond the scope of this chapter, the reader is encouraged to examine Flyvbjerg et al. (1996) to see how the details play out. Another consideration that can lead to different behaviors for two individual microtubules in the same chemical environment is the fact that the average velocities associated with both tip growth and nucleotide hydrolysis are subject to the kinds of fluctuations illustrated in Figure 15.25. Thus, a particular microtubule may lose its cap simply by bad luck even when overall conditions favor net polymerization. This stochastic consideration almost certainly applies to microtubule dynamics in living cells.

15.5 Summary and Conclusions

Cells are a bustling environment in which huge numbers of chemical reactions are taking place constantly. As a result, an important part of the modeling toolkit for thinking about the behavior of cells are tools that model reaction dynamics. The main goal of this chapter has been the development of one particularly important paradigm for modeling chemical reactions, namely, the theory of rate equations. These equations characterize the time variation of concentrations of reactants and products in a chemical reaction of interest.

15.6 Problems

Key to the problem categories: • Model refinements and derivations • Estimates • Data interpretation • Model construction

• 15.1 Census of cytoskeletal proteins in fission yeast

Consider the measurements shown in Figure 15.4 where global and local concentrations of various cytoskeletal proteins in fission yeast cells were determined using fluorescence microscopy. Here we make use of the measured concentrations, reported in Table 1 of Wu and Pollard (2005), to make estimates for the fission yeast cytoskeleton.

(a) Estimate the volume of a fission yeast cell. Next, given the concentration of actin monomers, capping proteins, and formins, work out the mean spacing between these proteins. Then, use your estimated volume and the measured concentrations to estimate the number of copies of each of these proteins in the "typical" fission yeast cell.

(b) About half of the actin monomers in the yeast cell are in filamentous form at any given time. Given the dissociation constant for capping proteins, $K_d = 1$ nM, estimate the average length of actin filaments, assuming one capping protein per filament. Compare your estimate with the typical value, which is of the order of 100 monomers per filament.

• 15.2 Dynamics of isomerization reactions

In Equation 15.22, we described the kinetics of an isomerization reaction. Here we do the math and explore a couple of examples of this reaction.

(a) Work out the solution for the concentrations of both species and make plots of $c_A(t)$, $c_B(t)$, and their sum. Assume that initially there are only molecules of species A present, at concentration c_0.

(b) Apply the results of (a) to the decay of 13-*cis*-retinal to all-*trans*-retinal, as was illustrated in Figure 15.8. The half-life of this reaction is $\tau = 2$ s.

(c) At a very different time scale, these same ideas apply to radiometric dating. A celebrated example is the decay of potassium to argon with a half-life of 1.26 billion years. Plot the amount of argon as a function of time, assuming initially only potassium is present.

• 15.3 Rate equations for interconversion

Consider the reversible reaction

$$A \rightleftharpoons B. \tag{15.121}$$

In this case, the rate equations of interest can be written as

$$\frac{dc_A}{dt} = -k_+ c_A + k_- c_B \tag{15.122}$$

and

$$\frac{dc_B}{dt} = -\frac{dc_A}{dt}, \tag{15.123}$$

with the constraint of mass conservation of the form given in Equation 15.21. In this case, the long-time behavior is nonzero concentrations of both A and B.

(a) Solve these equations for $c_A(t)$ and $c_B(t)$ assuming that only the molecular species A is present at time $t = 0$. Make plots of the solutions and demonstrate that the long-time behavior is dictated by the ratio k_+/k_-.

(b) Using the current recording for a single sodium ion channel shown in Figure 7.2(A) (p. 283), estimate the opening and the closing rate of the channel. Use the result from (a) to plot the probability that the channel is open and the probability it is closed as a function of time. Assume the channel to be closed initially.

• 15.4 Lattice model for bimolecular reactions

Fill in all of the details in the derivation of Equation 15.32 for the bimolecular reaction rate, given in Section 15.2.4.

• 15.5 Michaelis–Menten versus exact kinetics

Write the full set of rate equations for the reaction described by the Michaelis–Menten kinetic model. Introduce dimensionless variables by measuring time in units of $(k_- + r)^{-1}$ and concentrations in units of $K_m = (k_- + r)/k_+$ and define $\varepsilon = r/(r + k_-)$.

(a) Solve these equations numerically and reproduce Figure 15.16(A).

(b) Make a plot of $d[P]/dt$ versus [S] using both the Michaelis–Menten form and the exact solution for the following choices of parameters: (i) $[S]_0 = 100, [E]_0 = 1, \varepsilon = 0.1$ and (ii) $[S]_0 = 1, [E]_0 = 1, \varepsilon = 1$. What conclusion do you draw about the validity of the Michaelis–Menten form?

• 15.6 Microtubule dynamics and dynamic instability

In this problem, we consider a phenomenological model for steady state microtubule dynamics that was introduced by Dogterom and Leibler (1993). Note that there is a more interesting class of models that include GTP hydrolysis explicitly (see Flyvbjerg et al. 1996). The goal of such models is to respond to data such as those in Figure 15.34. This figure shows a record of the length of a single microtubule as a function of time and reveals the series of "catastrophes" and "rescues" as the polymer changes its length.

(a) Deduce the following equations for the probability distributions $p_+(n, t)$ and $p_-(n, t)$, which give the probability of finding a microtubule:

$$\frac{\partial p_+}{\partial t} = -f_{+-}p_+ + f_{-+}p_- - v_+\frac{\partial p_+}{\partial z}, \tag{15.124}$$

$$\frac{\partial p_-}{\partial t} = +f_{+-}p_+ - f_{-+}p_- + v_-\frac{\partial p_-}{\partial z}. \tag{15.125}$$

Write a master equation for $p_+(n, t)$ and $p_-(n, t)$ by noting that there are four processes that can change the probability at each instant. Consider the + case: (i) the $n-1$ polymer can grow and become an n polymer, characterized by a rate v_+; (ii) the n polymer can grow and become an $n+1$ polymer, also characterized by a rate v_+; (iii) the $n+$ polymer can switch from growing to shrinking with a rate f_{+-}; and (iv) the $n-$ polymer can switch from shrinking to

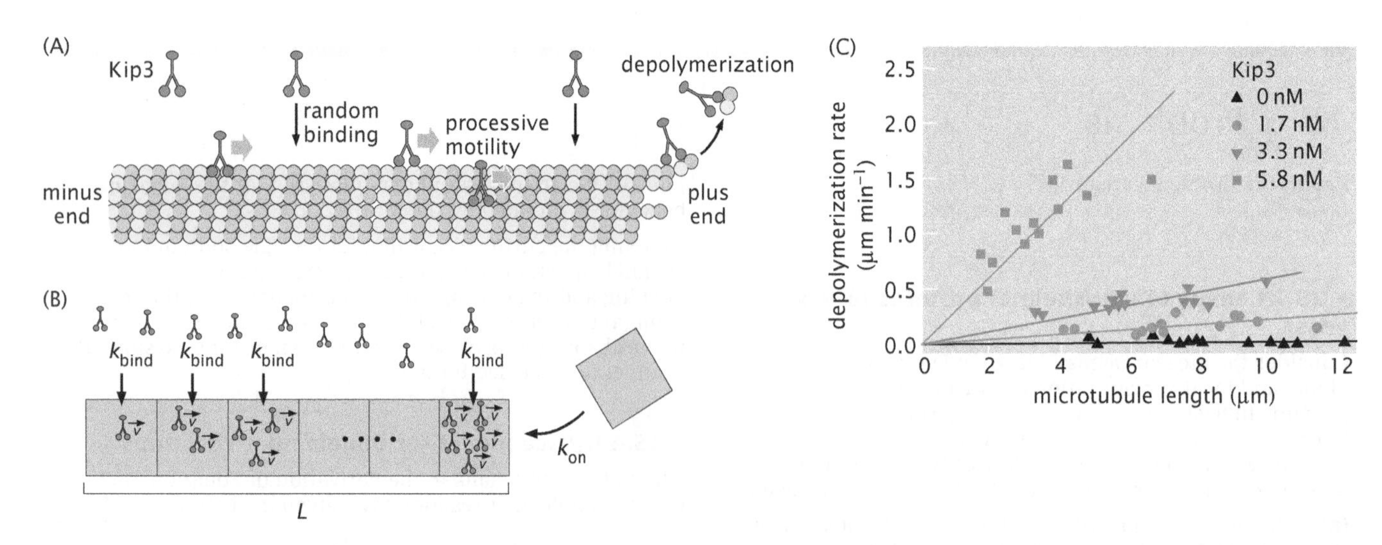

Figure 15.38: Antenna mechanism of microtubule length control. (A) Schematic of antenna mechanism of microtubule length control. (B) Toy model of the antenna mechanism. Kip3 molecules free in solution associate with microtubule monomers depicted as blocks, and then proceed toward the plus end moving at a constant speed v. Once at the end, they detach from the microtubule, taking a block/monomer with them. This Kip3-dependent depolymerization is counteracted by monomer addition at the plus end. (C) Data showing the length-dependent depolymerization rate for different concentrations of Kip3. (A, adapted from V. Varga et al., *Cell* 138:1174, 2009; C, adapted from V. Varga et al., *Nat. Cell Biol.* 8:957, 2006.)

growing with a rate f_{-+}. If you account for all four of these possibilities, you will have the correct master equation. Use a Taylor expansion on factors like $p_+(n-1,t) - p_+(n,t)$ to obtain Equations 15.124 and 15.125.

(b) Solve these equations for $p_\pm(n)$ in the steady state (that is, $\partial p_\pm(n,t)/\partial t = 0$). Show that in the steady state, the probabilities $p_\pm(n)$ decay exponentially with constant

$$\sigma = \frac{v_+ f_{-+} - v_- f_{+-}}{v_+ v_-}. \quad (15.126)$$

(c) Use Figure 15.34 to estimate the parameters v_+, v_-, f_{+-}, f_{-+}, and then find the average length of the polymers that is *predicted* by this simple model. To find the average length, you will need to sum over all lengths with their appropriate probability. The slopes of the growth and decay regions tell you about the on and off rates, and the durations of the growth and decay periods tell you something about the parameters f_{+-} and f_{-+}. Note that by fitting the dynamical data, you are deducing/predicting something about the distribution of lengths.

• 15.7 Antenna model for microtubule length control

In the chapter, we examined the equilibrium polymer and found that it is characterized by a broad, exponential distribution of lengths. Contrary to this observation, cytoskeleton filaments in cells often have well-defined lengths that appear to be properly adjusted to their cellular function. How length control is achieved is an interesting problem, since the building blocks of cytoskeletal filaments and their interactions take place at the nanometer scale, yet they assemble into structures whose length is on the micron scale. It has been hypothesized that the cell needs to set up some sort of mechanism by which length is measured. One such mechanism was proposed for microtubule length control whereby the kinesin Kip3 walks toward the plus end of a microtubule and once there leads to depolymerization of the terminal tubulin dimers as depicted in Figure 15.38(A).

In this problem, we examine a toy model of this mechanism and show that it leads to a steady-state length for microtubules. The model assumes that Kip3 molecules in solution bind to the filament monomers represented as blocks in Figure 15.38(B) with rate k_{bind}, and then proceed to step toward the plus end of the filament with rate v. Only at the end of the filament do the Kip3 molecules fall off taking a monomer block along with them. This depolymerization process competes with monomer attachment at the plus end with rate k_{on}.

(a) Show that the model predicts a steady-state occupation of the filament by Kip3 molecules such that the number of Kip3 molecules per monomer increases linearly with position toward the plus end. This is observed in experiments. For example, see Figure 3(b) in Varga et al. (2006).

(b) Compute the rate of monomer detachment at the plus end in steady state assuming that it is equal to the flux of Kip3 molecules into the terminal monomer. Compare your result with the data shown in Figure 15.38(C) and estimate the rate k_{bind} for different Kip3 concentrations from the plots. Explore the dependence of k_{bind} on Kip3 concentration. What do you conclude?

(c) Write down the equation for $L(t)$, the filament length as a function of time, by considering polymerization at the plus end and the length-dependent depolymerization, which we derived in the previous part. Determine the steady-state length of the microtubule at 3 nM Kip3 concentration typical of a cell, assuming a typical attachment rate of $k_{on} = 1\ \mu m/min$. (Problem courtesy of Justin Bois.)

15.7 Further Reading

Segel, LA (1984) Modeling Dynamic Phenomena in Molecular and Cellular Biology, Cambridge University Press. Segel's book provides a wonderful discussion of many of the topics covered in the present chapter.

Michaelis, L, & Menten, ML (1913) Kinetics of invertase action, *Z. Biochem.* **49**, 333. This classic paper outlines the theoretical framework used by biochemists to study the kinetics of enzymatic catalysis.

Crane, HR (1950) Principles and problems of biological growth, *Sci. Monthly* **70**, 376. This fascinating article shows how the assembly of objects having no particular symmetry favors helical geometries.

Abraham, VC, Krishnamurthi, V, Taylor, DL, & Lanni, F (1999) The actin-based nanomachine at the leading edge of migrating cells, *Biophys. J.* **77**, 1721. This paper, like that of Wu and Pollard (2005), presents a census of actin, in this case at the leading edge of motile cells.

Oosawa, F, & Asakura, S (1975) Thermodynamics of the Polymerization of Protein, Academic Press. This classic book is full of interesting pictures and discussion of the polymerization problem.

Amos, LA, & Amos, WB (1991) Molecules of the Cytoskeleton, The Guilford Press. This book is full of interesting insights into the cytoskeleton.

Hill, TL, & Kirschner, MW (1982) Bioenergetics and kinetics of microtubules and actin filament assembly–disassembly, *Int. Rev. Cytol.* **78**, 1. This comprehensive paper lays out the theoretical basis for kinetics and thermodynamics of protein polymerization for every possible case: with and without structural polarity, with and without nucleotide hydrolysis, in the presence of capping proteins and in the presence of loads or barriers.

Flyvbjerg, H, Holy, TE, & Leibler, S (1996) Microtubule dynamics: caps, catastrophes, and coupled hydrolysis, *Phys. Rev.* **E54**, 5538. This interesting paper shows how to go beyond the simple model of the cap described in the chapter to include stochastic dissolution of the cap by hydrolysis.

Caplow, M, & Shanks, J (1996) Evidence that a single monolayer tubulin–GTP cap is both necessary and sufficient to stabilize microtubules, *Mol. Biol. Cell* **7**, 663. This paper describes experiments on the nature of the GTP cap in microtubules.

15.8 References

Alberts, B, Johnson, A, Lewis, J, et al. (2008) Molecular Biology of the Cell, 5th ed., Garland Science.

Brown, AJ (1902) Enzyme action, *J. Chem. Soc. Trans.* **81**, 373.

Burlacu, S, Janmey, PA, & Borejdo, J (1992) Distribution of actin filament lengths measured by fluorescence microscopy, *Am. J. Physiol.* **262**, C569.

Campbell, CS, & Mullins, RD (2007) *In vivo* visualization of type II plasmid segregation: bacterial actin filaments pushing plasmids, *J. Cell Biol.* **179**, 1059.

Carlier, MF, Pantaloni, D, & Korn, ED (1984) Evidence for an ATP cap at the ends of actin filaments and its regulation of the F-actin steady state, *J Biol. Chem.* **259**, 9983.

Dogterom, M, & Leibler, S (1993) Physical aspects of the growth and regulation of microtubule structures, *Phys. Rev. Lett.* **70**, 1347.

Fygenson, DK, Braun, E, & Libchaber, A (1994) Phase diagram of microtubules, *Phys. Rev.* **E50**, 1579.

Garner, EC, Campbell, CS, & Mullins, RD (2004) Dynamic instability in a DNA-segregating prokaryotic actin homolog, *Science* **306**, 1021.

Hermone, A, & Kuczera, K (1998) Free-energy simulations of the retinal *cis* $\rightarrow$ *trans* isomerization in bacteriorhodopsin, *Biochemistry* **37**, 2843.

Iino, T (1974) Assembly of *Salmonella* flagellin *in vitro* and *in vivo*, *J. Supramol. Struct.* **2**, 372.

Keller, BU, Hartshorne, RP, Talvenheimo, JA, Catterall, WA, & Montal, M (1986) Sodium channels in planar lipid bilayers. Channel gating kinetics of purified sodium channels modified by batrachotoxin, *J. Gen. Physiol.* **88**, 1.

Ouellete, L, Laidler, KJ, & Morales, MF (1952) Molecular kinetics of muscle adenosine triphosphatase, *Arch. Biochem. Biophys.* **39**, 37.

Pollard, TD (1986) Rate constants for the reactions of ATP– and ADP–actin with the ends of actin filaments, *Cell Biol.* **103**, 2747.

Varga, V, Helenius, J, Tanaka, K, et al. (2006) Yeast kinesin-8 depolymerizes microtubules in a length-dependent manner, *Nat. Cell Biol.* **8**, 957.

Varga, V, Leduc, C, Bormuth, V, Diez, S, & Howard, J (2009) Kinesin-8 motors act cooperatively to mediate length-dependent microtubule depolymerization, *Cell* **138**, 1174.

Waterman-Storer, CM, & Salmon, ED (1997) Actomyosin-based retrograde flow of microtubules in the lamella of migrating epithelial cells influences microtubule dynamic instability and turnover and is associated with microtubule breakage and treadmilling, *J. Cell Biol.* **139**, 417.

Welch, MD, Rosenblatt, J, Skoble, J, Portnoy, DA, & Mitchison, TJ (1998) Interaction of human Arp2/3 complex and the *Listeria monocytogenes* ActA protein in actin filament nucleation, *Science* **281**, 105.

Wu, J-Q, & Pollard, TD (2005) Counting cytokinesis proteins globally and locally in fission yeast, *Science* **310**, 310. This paper illustrates the strategy of using fluorescence microscopy for taking the census of a cell.

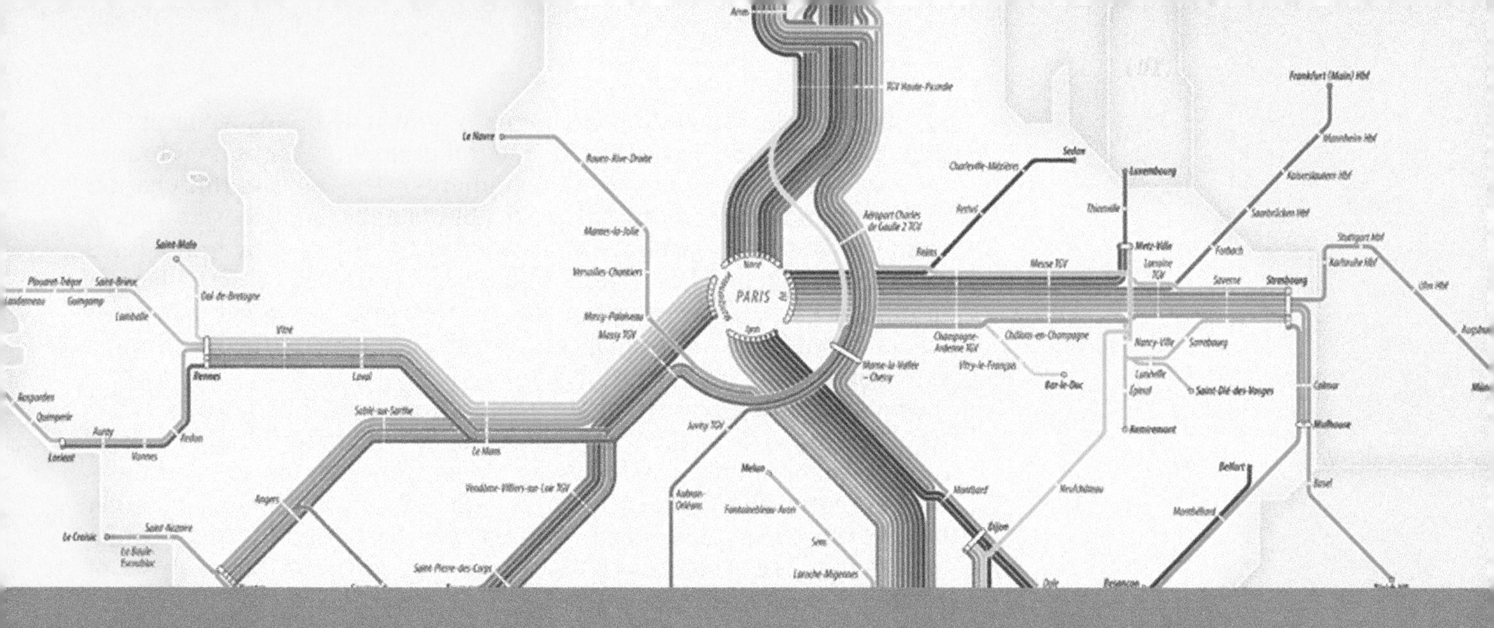

Dynamics of Molecular Motors 16

Overview: In which the dynamics of molecular motors are studied using rate equations

> Like a flash of lightning and in an instant the truth was revealed. I drew with a stick on the sand the diagrams of my motor.
>
> **Nikola Tesla**

The question of cellular dynamics has arisen throughout the book in a number of different contexts. One of the conclusions we have drawn in our analysis is that active transport plays an important role in mediating cellular life. Molecular motors are key tools used by the cell to perform active transport and to maintain its nonequilibrium character. In particular, we interest ourselves in molecular machines that can perform work, usually by the consumption of ATP. This chemomechanical coupling will be the centerpiece of the present chapter. We begin by considering several different classes of molecular motors and the cellular processes they mediate. A vast array of different classes of experiments ranging from structure determination to single-molecule biophysics has resulted in a range of quantitative data on motors that we examine systematically. We then use rate equations to characterize the dynamics of motors.

16.1 The Dynamics of Molecular Motors: Life in the Noisy Lane

Directed and purposeful movement is one of the properties we most closely associate with living organisms. Even organisms that appear immotile to the naked eye, such as green plants, are very busy on the cellular level, exhibiting rapid directed movements of organelles such as chloroplasts and segregation of chromosomes during cell division. The past several chapters have largely focused on diffusion and random walks as modes of motion, but these are clearly insufficient to explain directed movements such as chromosome segregation or muscle contraction. Nonrandom movements cost energy! They demand mechanisms that convert chemical energy into mechanical energy (work). In Chapter 5, we described the forms of chemical energy storage in cells, including high-energy covalent bonds such as the phosphoanhydride bonds in ATP and ion gradients across membranes (or, more generally, concentration gradients). The focus of this chapter will be on molecular motors that are able to use one of these forms of chemical energy to generate a mechanical force acting over a defined distance in a defined direction. In our exploration of these molecular motors, we will blend many of the physical and biological principles encountered in previous chapters, ranging from beam theory through rate equations.

One fascinating feature of the motors found in cells is their structural and functional diversity. Nevertheless, many different kinds of motors share similar fundamental physical mechanisms, so analysis of one motor can frequently shed light on the functioning of other motors that are evolutionarily unrelated and are used for distinct biological purposes. For convenience, we will separate our analysis of motors into four broad classes:

(i) *Translational motors.* These are motors that move in a one-dimensional fashion, by stepping along a linear "track" as a substrate. Motors in this class include myosin, which causes muscle contraction by walking along actin filaments, and helicases, which move along DNA and use energy to unwind the double helix.

(ii) *Rotary motors.* These remarkable motors are usually embedded in the cell membrane and generate torque by rotation of mechanical elements. One of the best-understood rotary motors in biology is the bacterial flagellar motor, shown in Figure 3.24 (p. 120).

(iii) *Polymerization motors.* As we began to explore in the preceding chapter, energy can be released during the process of subunit polymerization and depolymerization. Both actin assembly and microtubule assembly are harnessed by cells to generate force directly, as well as to provide linear tracks for other force-generating translational motors.

(iv) *Translocation motors.* These motors involve threading a structure such as DNA or an unfolded protein through a hole and then pushing it or pulling it through the hole, frequently (though not always) across a membrane. The importing of proteins into specific cellular organelles, such as the mitochondrion, frequently requires the action of translocation motors.

This scheme for dividing motors into four classes will simplify our modeling efforts later in the chapter, but this list is neither comprehensive nor mutually exclusive. Some important motors combine features of more than one class. For example, RNA polymerase moves along its DNA substrate like a translational motor, but uses energy derived from polymerization of nucleotide subunits. The motor that packages bacteriophage DNA into the capsid is thought to combine features of both rotary and translocation motors. Also, there are several kinds of force-generating systems known to operate in specialized cells that do not fit easily within this framework. A famous example is the sperm of the horseshoe crab, which carries a coiled pre-stressed spring that can uncoil rapidly when triggered to pierce the jelly coat of the horseshoe crab egg during fertilization. We will not attempt to explore the mechanisms of such exotic motors here, although they must obey the same physical principles as the more familiar motors that we will use as our examples.

Our plan in this chapter is to examine the structure and function of these different classes of motor and then to show how simple rate equations can be used to understand many aspects of their operation. The study of biological motors has been a fertile area for the fusion of biology and physics in research, because understanding their function demands an integration of biochemistry with mechanics. Characterization of the properties of a molecular motor such as kinesin, a translational motor that moves along microtubules, requires measuring the biochemical rate constants associated with its hydrolysis of ATP, but also physical properties such as the speed with which it moves and the amount of force that it can generate. Because of the small size of most biological motor proteins (typically a few nanometers) and the fact that they operate under conditions where thermal motions are significant, measurement of physical motor properties such as speed and force has been technically challenging. As we will see, numerous clever and elegant measurement techniques that exploit a wide range of physics concepts have been developed for this purpose, largely by teams of biologists and physicists working together.

Within a unified physical framework where molecular motors are treated as tiny machines that convert chemical energy into mechanical work, the characteristics of different motors can be compared and the mechanisms of their inner gearboxes deduced. As an example, Figure 16.1 shows the behavior of three different motors—kinesin, RNA polymerase, and the phage packaging motor—when they are subjected to increasing loads. Just like human-manufactured motors, molecular motors tend to slow down when they are forced to do more work, but the three different motors slow down in distinct ways. The biological function of motors depends on the way that they convert the energy released from a chemical reaction to the conformational change that causes movement and generates force. Conversely, external forces applied to moving motors can affect their biochemical reaction rates. Therefore, the motor's speed depends on the applied force in a way that must reflect its mechanochemical energy conversion mechanism. As shown in the graph, RNA polymerase tends to move at top speed as its load increases until it reaches a threshold where its speed slows suddenly and it stalls. In contrast, the phage packaging motor slows down very gradually with increasing force, and kinesin exhibits a complex force–velocity relationship somewhere in between. Later in the chapter, we will see how different physical models for the coupling between mechanics and biochemistry inside a molecular motor can predict different shapes for these characteristic force–velocity curves. Before proceeding to the modeling, we will begin with a brief tour of the biological roles and characteristics of the four major motor classes.

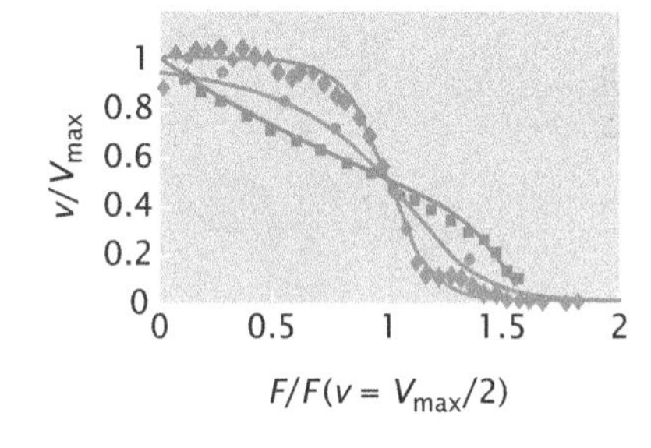

Figure 16.1: Effects of load force on motor velocity for kinesin, RNA polymerase, and the bacteriophage packaging motor. For all three motors, the speed has been normalized to the maximum speed (v/V_{max}) and the force has been normalized to the force where the speed is decreased to half its maximum ($F(v = V_{max}/2)$). The different shapes of the force–velocity curves imply distinct mechanisms. (Adapted from C. Bustamante et al., *Annu. Rev. Biochem.* 73:705, 2004.)

16.1.1 Translational Motors: Beating the Diffusive Speed Limit

One of the most important and familiar classes of molecular motors are those associated with the cytoskeleton, specifically with microtubules and actin filaments. As we have seen throughout the book, these motors are responsible for motile processes as diverse as the directed transport of vesicles in individual cells to the running, swimming, and flying of animals on the basis of muscle contractions. These translational cytoskeletal motors can be classified into three protein families—myosin, kinesin, and dynein—one member of each

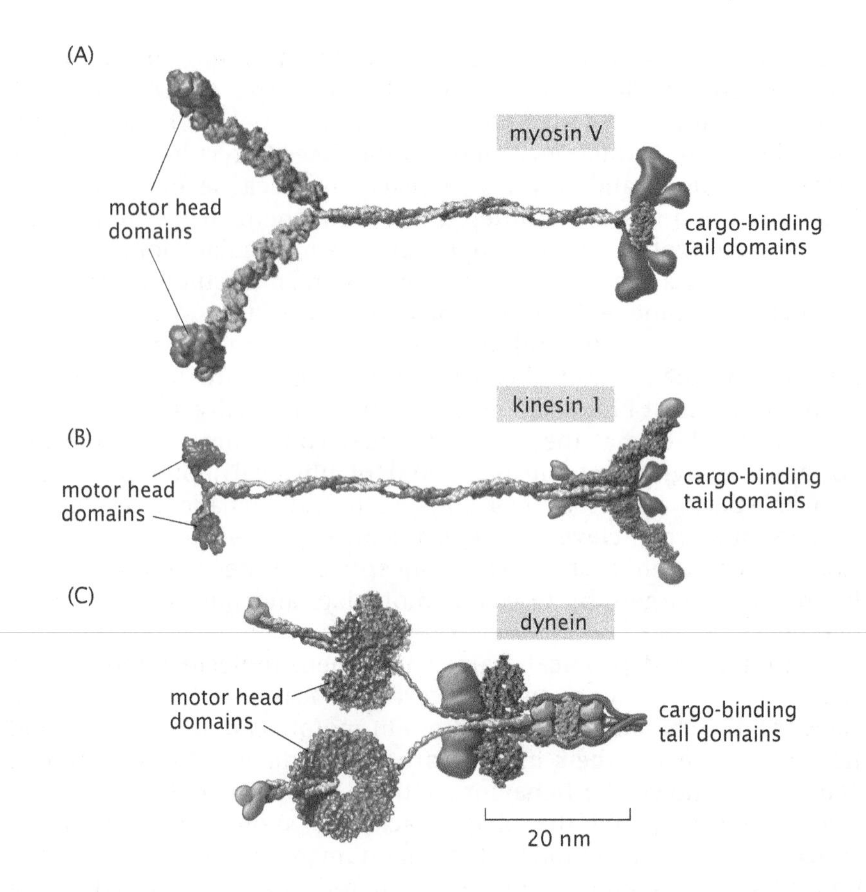

Figure 16.2: Key classes of translational motor. (A) A myosin V molecule is one of about 20 different types of myosins that move on actin filaments. (B) Kinesin 1 is also a member of a large family of related molecular motor proteins, but these move on microtubules rather than on actin. Although myosins and kinesins have different substrates, the detailed structures of their motor heads are quite similar and they are thought to be derived from a single common molecular ancestor. (C) Cytoplasmic dynein represents a different class of microtubule-based motors that appears to be unrelated to kinesin or myosin. (Adapted from R. D. Vale, *Cell* 112:467, 2003.)

being shown in Figure 16.2. Because both microtubules and actin filaments are constructed from asymmetric subunits that self-assemble by binding to each other in a head-to-tail orientation, the tracks have a distinct structural polarity, and each type of motor is able to move in one direction. Members of the dynein protein family, for example, move along microtubules toward the minus end, while most members of the kinesin family move in the opposite direction, toward the plus end (although there are a few special types of kinesin that are minus-end-directed). Myosin family members all move along actin filaments, mostly toward the barbed end (equivalent to the microtubule plus end), but with a few exceptions that move toward the pointed end.

Each of these motors moves along its substrate track by taking a series of discrete steps of fixed size. Each mechanical step is tightly coupled to a single biochemical cycle of ATP binding, hydrolysis, and product release. The net speed of a translation motor, then, is the product of its step size and the number of steps it can take per unit time. Another important feature of translation motors is their processivity, that is, the number of steps that a single motor can take along a single filament before it falls off. Some translation motors, such as RNA polymerase, are extremely processive, so that an entire mRNA molecule that may include thousands of nucleotide bases can be synthesized by one polymerase in a single run. Other motors are much less processive. As we will see below, the myosin motor involved in muscle contraction typically takes only one or two steps along its actin filament before it falls off; this low processivity is important for its normal biological function. The amount of force that can be exerted by a single motor taking a single step is ultimately limited by the amount of chemical energy liberated by hydrolysis of a single molecule of ATP.

ESTIMATE

Estimate: Force Exerted During a Single Motor Step The kinesin motor moves 8 nm per ATP hydrolysis event. As a result, we can estimate the force exerted by the motor as

$$F_{\mathrm{max}} = \frac{\text{free energy of ATP hydrolysis}}{\text{step size}} \approx \frac{20\, k_{\mathrm{B}}T}{8\,\mathrm{nm}} \approx 10\,\mathrm{pN},$$

where we have used our usual rule of thumb that the thermal energy scale is $k_{\mathrm{B}}T \approx 4\,\mathrm{pN\,nm}$. Of course, this is an over-estimate because we have assumed that the entirety of the free energy offered up by ATP hydrolysis can be converted into mechanical work; real motors may work with lower efficiency.

Each of the three major cytoskeletal motor families contains numerous distinct members that are specialized for different functions in the cell, for example different kinesin family members bind to different cargoes and move along microtubules at different speeds. Overall, the three classes of motor share similar domain organization, as illustrated in Figure 16.2. One unifying structural theme is the existence of an ATP-binding domain, called the motor head domain, in the parts of the motors that are able to bind to their target filaments. This domain catalyzes the hydrolysis of ATP and undergoes a conformational change that is guided and amplified into a large-scale conformational change of the whole molecule, allowing it to make a step along its linear track. In order for movement to be productive, the cycle of ATP hydrolysis must be coupled to a second cycle of binding and unbinding to the track. The three sample motors shown in Figure 16.2 all have two heads, but some motors have just one, or even three. At the opposite end of the molecule, the tail domain is able to bind to cargo. Within each protein family, the motor head domains are evolutionarily more conserved than the cargo-binding tail domains. This makes sense, because all the motors of a given class walk along the same type of track, but may carry different kinds of cargo.

A schematic of the diversity of trafficking action associated with these motors is shown in Figure 16.3. A concrete example of this kind of trafficking is seen in neurons, where microtubule-based motors carry vesicles, proteins, and mRNA from their sites of synthesis in the cell body to the sites of synapse formation at the far end of the

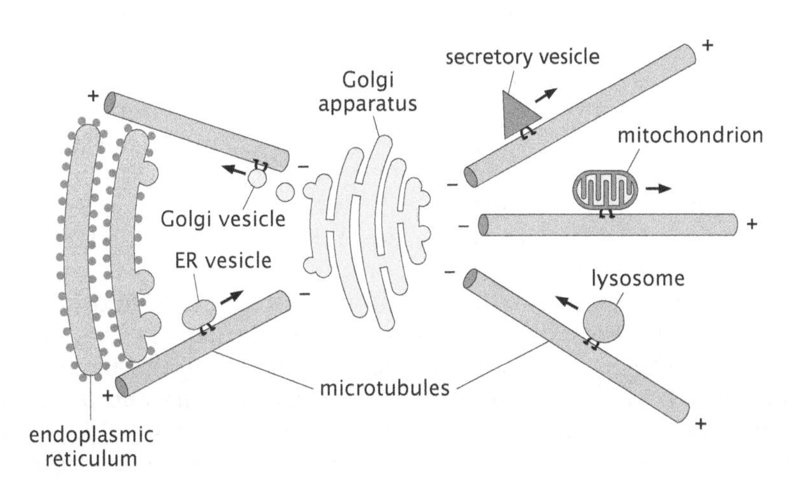

Figure 16.3: Directional transport of membrane-bound organelles. This schematic shows a few of the different types of microtubule-based organelle transport that must coexist in a typical eukaryotic cell. Different members of the kinesin and dynein motor protein families bind to distinct organelles or transport vesicles and mediate their movement through the cell. (Adapted from N. Hirokawa, *Science* 279:519, 1998.)

axon. In a giraffe, this distance for the longest neurons in the body can be several meters, and the microtubule-based motors carry material at the rate of about 20–40 cm/day.

Yet another intriguing example of the role of translational motors in cells is in the process of cell division. Throughout the book, we have made much of the story of genome management, both in terms of the physical demands associated with having the genetic material in the right place at the right time as well as with the informational demands tied to making sure that genes are expressed when they need to be. A compelling part of that story is chromosome segregation into two daughter cells after DNA replication during the cell cycle. In this setting, translational microtubule-based motors are partially responsible for the separation of the chromosomes by the mitotic spindle during cell division, where over a dozen distinct kinesin family members collude with one another and with a cytoplasmic form of dynein to generate relative translation of the microtubules within the spindle and to position and move the chromosomes. This phenomenon was shown in Figure 15.18 (p. 600) and we remind the reader of it here as a shining example of the orchestrated activities of motors and their cytoskeletal partners.

The Motion of Eukaryotic Cilia and Flagella Is Driven by Translational Motors

Another example of the action of translational motors is provided by active filamentous structures in cells, which embody a subtle coupling between beam-bending dynamics and translational, force-generating motors. In particular, a wide range of processes important to eukaryotic organisms are mediated by flagella and cilia, which are cell-surface projections containing bundles of microtubules crosslinked by special forms of dynein motor proteins, surrounded by the cell plasma membrane. The motion of cilia pervades the living world, with examples ranging from the surfaces of embryos to the linings of various mammalian tissues to the swimming of unicellular organisms such as *Paramecium*. One familiar example is provided by the mucociliary escalator in the lungs. As a result of waves of ciliary motion, harmful materials are transported up and out of the lungs so that they can be spat out or harmlessly swallowed. Paralysis of the cilia due to superviscous mucus in cystic fibrosis patients almost universally causes early death due to the inability to clear bacterial infections. Likewise, dynein paralysis due to dynein mutations in Kartagener's syndrome tends to cause early death from lung infections. The flagella in swimming cells such as sperm cells are structurally closely related to beating cilia; both are based upon beautiful structures known as axonemes. (It is important to remember that eukaryotic flagella, such as those on sperm cells, and bacterial flagella are completely different kinds of structures. The fact that they share the same name is an unfortunate historical accident.) A schematic of the microtubule and motor-based architecture of the axoneme is shown in Figure 16.4.

The way in which translational motors are harnessed to drive oscillations of structures like cilia and eukaryotic flagella is shown in Figure 16.5. The concept is that adjacent filaments in the axoneme are connected to each other both by translational motors and by linking devices that prevent much relative motion of these parallel filaments. As a result, when the motors make a stepping motion, they produce a

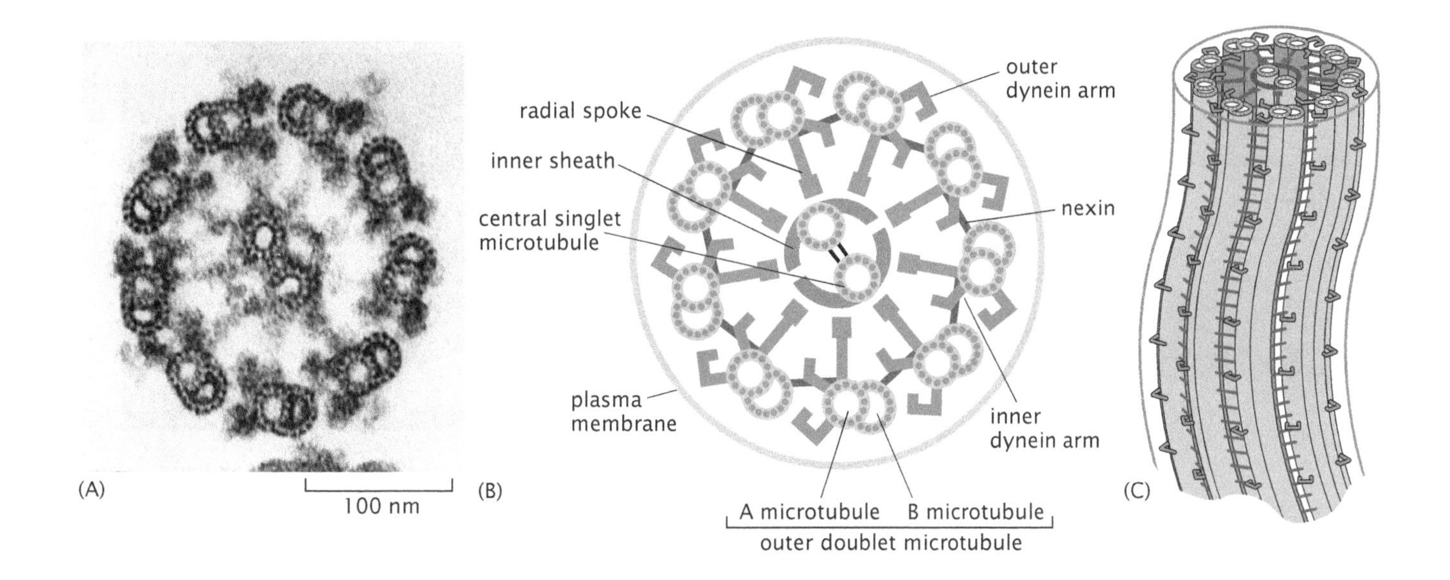

Figure 16.4: Structure of the axoneme. (A) Thin-section electron micrograph of a cross-section of the flagellum of the microscopic algae *Chlamydomonas*. (B) Diagram of the flagellar parts. The inner dynein arms and outer dynein arms are motor proteins that make the flagellum beat by sliding the microtubules relative to one another. (C) A side view of the flagellum. (A, courtesy of Lewis Tilney.)

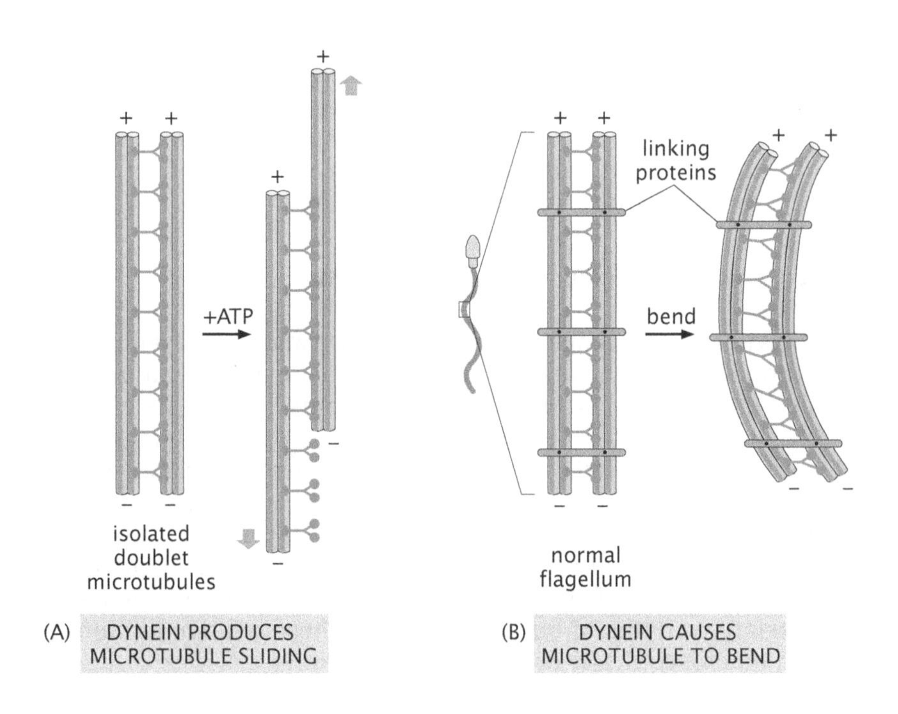

Figure 16.5: Bending of flagella and cilia due to translational motors. (A) When adjacent filaments are *not* tethered together, stepping of the motors will result in sliding of the adjacent filaments. (B) When the adjacent filaments are tethered together, stepping of the motors will result in deformation of the filaments. (Adapted from B. Alberts et al., Molecular Biology of the Cell, 5th ed. Garland Science, 2008.)

compressive force that induces bending. In Figure 16.5(A), we see that if two adjacent filaments are not linked, then the motion of the molecular motors will induce sliding. On the other hand, if the filaments are tied together, these motors will generate bending, as is shown in Figure 16.5(B). The energetics of filament bending can be modeled using beam theory, as discussed in Chapter 10. Spatial and temporal coordination of dynein motor stepping along the length of the cilium or flagellum gives rise to regular beat patterns that propagate down the structure. For example, a snapshot of the flagellum of a swimming

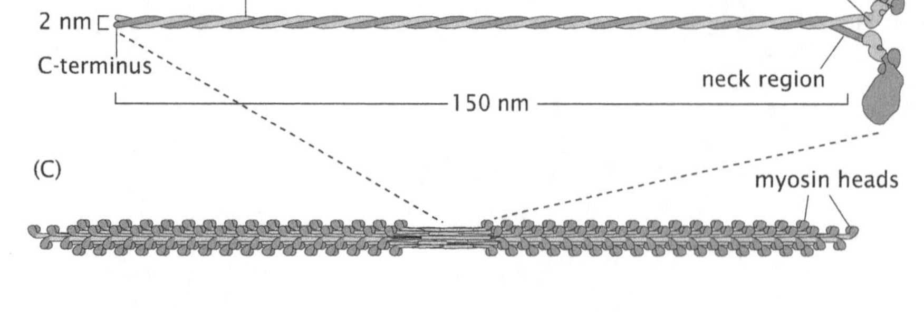

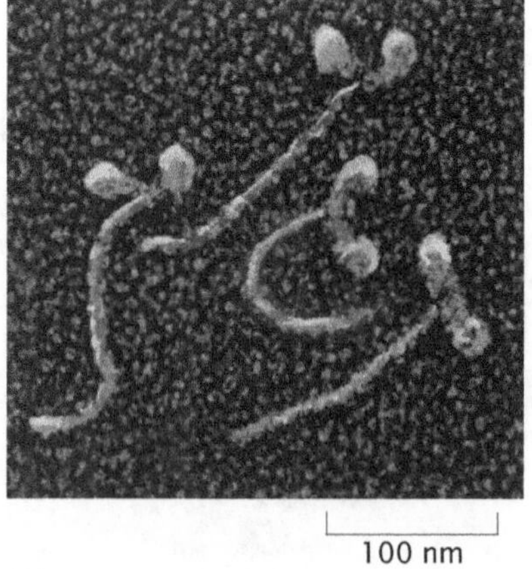

Figure 16.6: Structure of muscle myosin II. (A) Myosin II from skeletal muscle is a hexamer consisting of two extremely large heavy chains and four much smaller light chains. The heavy chains include a long coiled-coil domain at the C-terminus and the actin-binding, force-generating motor head at the N-terminus. (B) Platinum replica imaging of individual myosin molecules reveals the beautiful regularity of their structure. (C) Several hundred individual myosin II hexamers can self-assemble to form a thick filament. In this cylindrical bundle, the myosin molecules in the left half are all pointing toward the left, and those in the right half are all pointing toward the right. This antiparallel orientation is critical for muscle contraction. (A, C, adapted from B. Alberts et al., Molecular Biology of the Cell, 5th ed. Garland Science, 2008; B, courtesy of J. Heuser.)

sperm will show the flagellum bent in an elegant, near-sinusoidal curve. A moment later, the sine wave will appear to have shifted down the flagellum. The propagation of this bending wave drives the sperm head forward through its low Reynolds number environment with an efficiency and speed that depend in part on the viscosity of the medium, as we explored in Chapter 12. The ability of human sperm to swim to find an egg to fertilize thus depends on the convergence of several important principles in physical biology.

Muscle Contraction Is Mediated by Myosin Motors

Though kinesin will usually serve as our canonical example of a translational motor, the action of translational cytoskeletal motors is probably most renowned in the context of muscles. In this case, it is the translational motor myosin that occupies centerstage as shown in Figure 16.6. The strikingly regular structure of striated skeletal muscle is built up from many repeated units of individual contractile elements known as sarcomeres. As shown in Figure 16.7, sarcomeres are made up of regularly spaced and precisely oriented filaments and motors. Collections of myosin molecules make thick filaments that induce relative sliding of bundles of actin. These relative motions are revealed macroscopically as muscle contractions.

Estimate: Myosin and Muscle Forces Molecular motors are molecules that consume some form of chemical energy in order to deliver mechanical work. For the translational motors of interest in the present section, the kind of information we seek in characterizing a given motor includes how fast the motor goes, how much force it can apply, and how processive it is (that is, how many steps does it take before falling off its filament).

A fascinating feature of muscles is that they reflect the action of many motors acting simultaneously. Given the

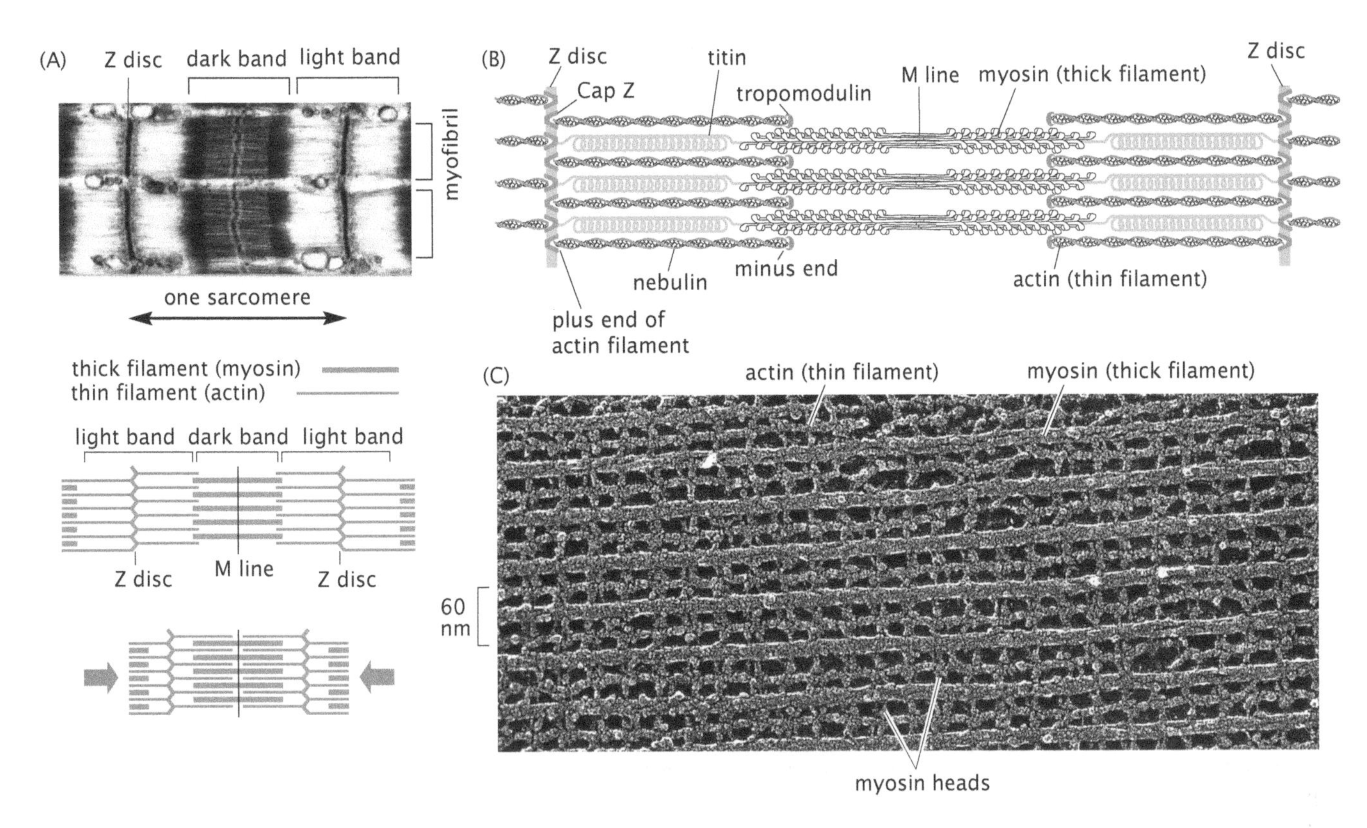

Figure 16.7: The structure of muscle. (A) Thin-section electron micrograph showing the organization of a single sarcomere when a muscle is stretched. The dark band in the middle represents the location of the aligned myosin thick filaments and the light bands on the sides show the position of actin. The diagrams below show the change in sarcomere length during muscle contraction. (B) The regular structure of the sarcomere depends on the precise arrangement and alignment of many structural proteins. The long proteins titin and nebulin help to set the length of the actin thin filaments and determine the overall length of the sarcomere. The Z disc serves as an anchor point for the actin thin filaments, and ensures that they are all oriented in the same direction, so that the myosin heads walking toward the actin filament plus ends will cause the sarcomere to shorten. (C) A quick-freeze deep-etch electron micrograph shows the extremely regular spacing of thick myosin filaments alternating with thin actin filaments and the myosin heads bridging the gap between them. (A, courtesy of Roger Craig; B, adapted from B. Alberts et al., Molecular Biology of the Cell, 5th ed. Garland Science, 2008; C, courtesy of J. Heuser.)

properties of individual motors, how much force might we estimate can be applied by an array of motors such as are found in a muscle? We can develop an estimate of this force by examining the structure and function of muscles. Figure 16.7(B) shows a cartoon of a muscle cell made up of muscle fibers or myofibrils. The myofibrils are themselves composed of contractile units called sarcomeres. Myosin molecules are arranged in a cylindrically symmetric structure called the thick filament, and exert forces on the outer actin filaments. We can estimate the net force per myosin molecule by appealing to a simple picture in which muscles are thought of as arrays of springs in series and parallel as shown in Figure 16.8. The number of myosins in a cross-section of muscle is roughly

$$N_{\text{myosin}} \approx \frac{\text{cross-sectional area of muscle}}{\text{cross-sectional area of thick filament}} \times N_{\text{myosin/thick filament}} \approx \frac{\pi(3\,\text{cm})^2}{\pi(60\,\text{nm})^2} \times 300 \approx 10^{14}.$$

If we assume that the force scale associated with a muscle like the biceps in the upper arm is 100 N (corresponding to

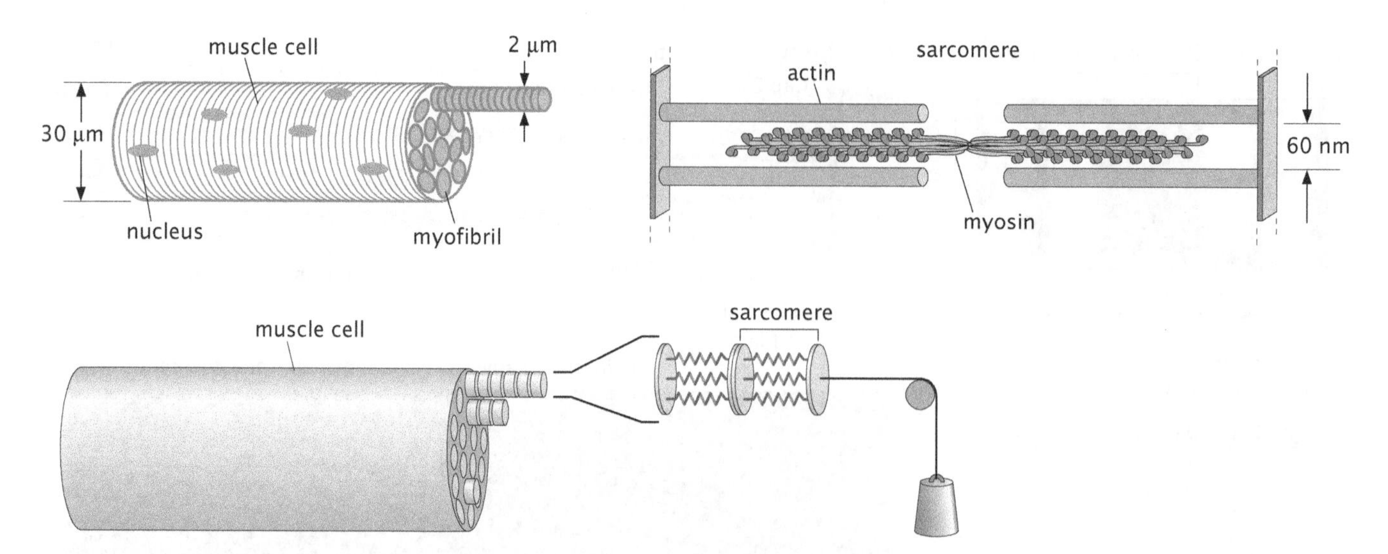

Figure 16.8: Idealization of muscle. The diagram at the upper left shows a single skeletal muscle cell, with multiple nuclei. Each muscle cell contains many myofibrils (shown as red cylinders with a diameter of 2 μm) in a parallel bundle, forming a cell with an overall diameter of typically 10–100 μm. The myofibrils have regular striations that correspond to the Z discs at the end of each sarcomere. Within a single myofibril, the sarcomeres all contract in series. A normal skeletal muscle will contain many muscle cells arranged in parallel; most skeletal muscles in humans have an overall thickness of a few millimeters to a few centimeters. The diagram at the upper right shows the approximate size of one myosin thick filament in a sarcomere. The diagram at the bottom shows an idealized mechanical conception of a muscle cell, where each Z disc is treated as a rigid flat disc connected by several springs, and each spring represents a myosin thick filament. The contraction of the muscle "springs" generates force, schematized as the lifting of the weight over the pulley.

lifting a 10 kg mass), and that this force is partitioned equally amongst the myosins, then this results in an estimate of the force generated by a single myosin head of

$$F_{\text{myosin}} \approx \frac{10\,\text{kg} \times 10\,\text{m/s}^2}{10^{14}} = 1\,\text{pN}.$$

Despite the inexactness of several of the assumptions in this estimate, the final number of 1 pN per myosin head is of the same order of magnitude as the actual force revealed by measurements that have been made on single myosin motor proteins *in vitro*.

Translational motors of the kind introduced here serve in a huge variety of different capacities and reflect the divergence and specialization that is the hallmark of evolution. Much of what we know about these motors is the result of a variety of beautiful experiments, some of which we recount now.

Experiments Behind the Facts: Measuring Motor Action

EXPERIMENTS

Our understanding of the structure, function, and relatedness of different motors comes from decades of effort using a host of different techniques. Although both measurements in living cells and measurements on isolated proteins *in vitro* have contributed significantly to our understanding of motor protein action, here we will explore only *in vitro* experiments.

One of the most dramatic assays is the so-called gliding motility assay in which motors such as myosin are fixed to a microscope slide and fluorescently labeled actin filaments are then added to the system. As shown in Figure 16.9, these filaments bind to the motors and, after the addition of ATP, are

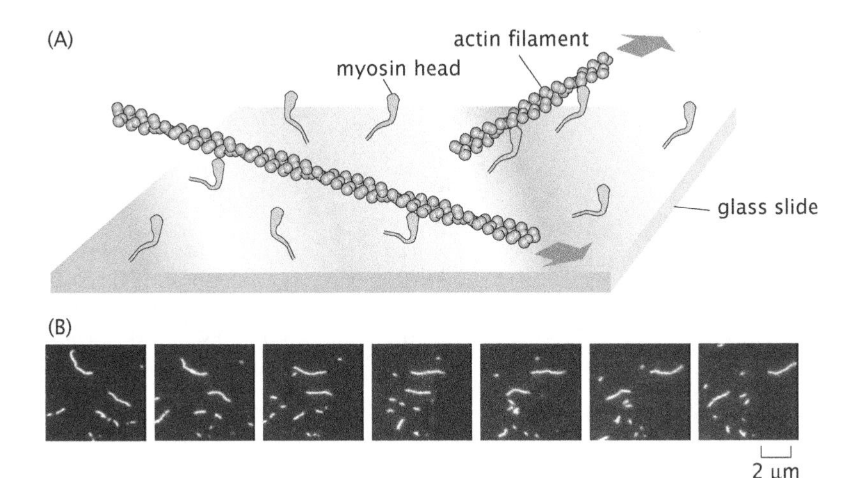

Figure 16.9: Gliding motility assay with myosin. (A) Individual motors are attached to the surface of a microscope slide and fluorescently labeled actin filaments are observed as they glide across the surface after the addition of ATP. (B) Frames from a video sequence spaced at 30 s intervals show the progress of individual filaments in many different directions. (A, adapted from B. Alberts et al., Molecular Biology of the Cell, 5th ed. Garland Science, 2008; B, courtesy of J. A. Spudich.)

then translated relative to the stationary motors. In this kind of gliding filament assay, a large number of different motor heads may bind simultaneously to the same filament and work cooperatively. This is a good imitation of *in vivo* situations such as that found for myosin II in the muscle sarcomere.

Another option for watching the movement of purified molecular motor proteins is to attach them to a bead that can be imaged with a conventional light microscope. As a result, the bead serves as a passive reporter for the actual motion of the motor itself. This geometry more closely imitates the *in vivo* situation where kinesin, for example, performs vesicle transport. Movements of the large (micron-scale) bead can be tracked with nanometer-scale precision, reflecting the behavior of the motor proteins. Figure 16.10 shows an example of this kind of measurement, where the beads and the microtubules can be directly observed using DIC microscopy. Attaching a bead to a motor also offers another advantage. The bead can be trapped using optical tweezers, as illustrated in Figure 4.12 (p. 154), and used to apply force to the walking motor, in any direction chosen by the investigator. This is one of the most powerful techniques used to measure the response of motors to applied forces, generating data such as that shown in Figure 16.1.

Both the filament gliding assay and the bead-based assay require that the motor proteins be immobilized: in the former on the microscope cover slip and in the latter on the bead. An alternative approach is to observe the movement of individual molecules by labeling them with a small fluorescent tag (Figure 16.11) or to watch them using atomic-force microscopy (Figure 16.12). For example, as shown in Figure 16.11, fluorescent molecules can be attached to the heads of myosin V molecules. As a result of the actual walking of the motor, the fluorophore is observed to undergo distinct steps of a precise length. Note that, from a quantitative perspective, one of the key outcomes of single-molecule experiments like these is that they provide a window on the stochastic nature of motor dynamics. For example, the time between steps of the motor is different for each step. In Section 16.2.4, we will calculate the expected distribution of these waiting times for simple models of motor function.

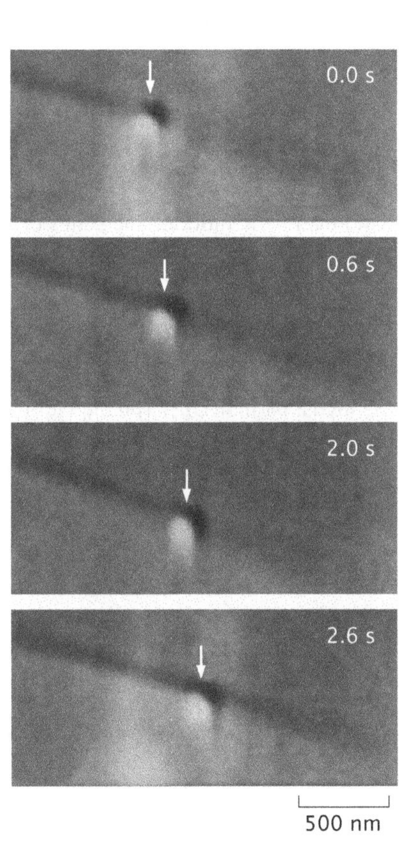

Figure 16.10: Kinesin-driven movement. A glass bead coated with kinesin motors was brought in contact with a microtubule using an optical trap. Both the microtubule and the bead can be seen using DIC microscopy and the optical trap is visible as a slightly shiny spot around the bead. When the trap is shut off, the bead begins to move down the microtubule processively over several seconds. (Adapted from S. M. Block et al., *Nature* 348:348, 1990.)

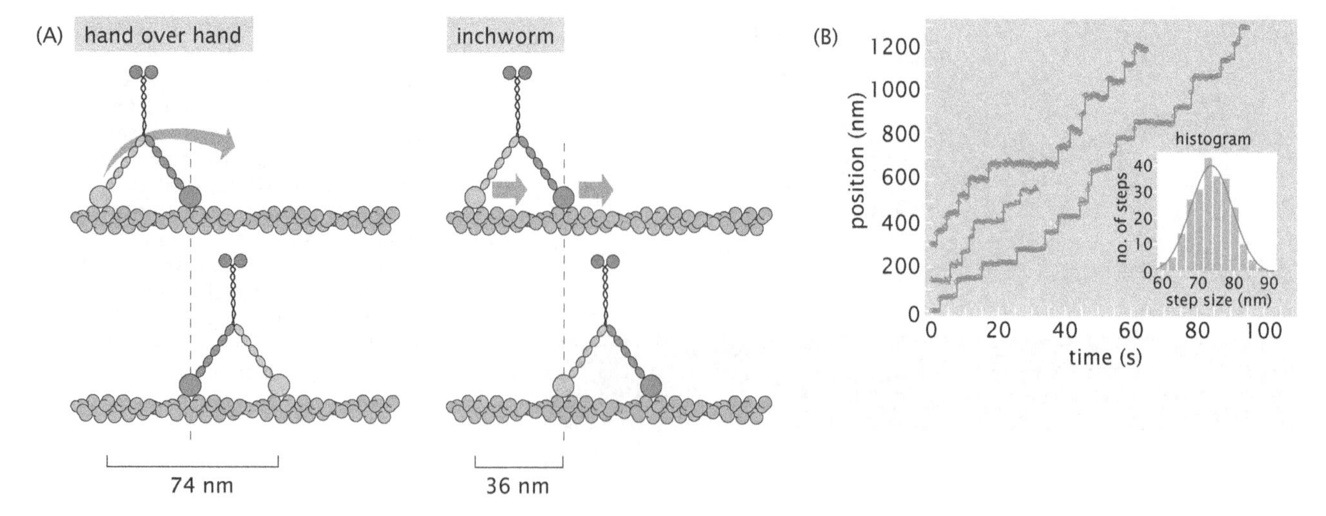

Figure 16.11: Single-molecule dynamics of myosin V stepping. (A) Cartoons showing the differences between hand-over-hand and inchworm models for motion. (B) Position as a function of time for myosin V molecules labeled with a fluorophore on one of the arms of the molecule. Three different traces from different molecules are shown. The average step size is slightly over 70 nm, consistent with a hand-over-hand stepping mechanism but inconsistent with an inchworm mechanism. (Adapted from A. Yildiz et al., *Science* 300:2061, 2003.)

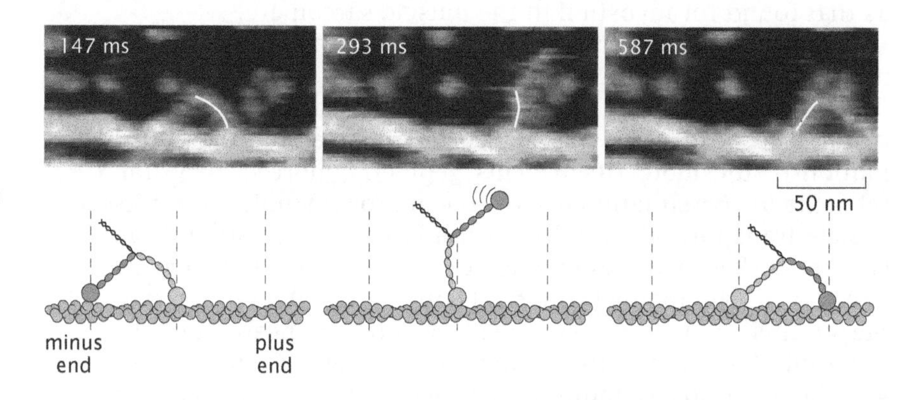

Figure 16.12: Single myosin V walking along an actin filament. Using high-speed microscopy, the movement of a single myosin V on an actin filament can be visualized revealing hand-over-hand motion. The series of images shown at the top are taken from a single high-speed AFM movie, where the actin filament is visible as a thick band across the bottom with its characteristic helical pitch. The white lines highlight one of the two heads of the myosin V molecule. The highlighted head corresponds to the yellow head in the diagrams below. (Adapted from N. Kodera et al., *Nature* 468:72, 2010.)

One of the applications of these particular experiments is their ability to distinguish different hypothesized mechanisms of walking such as inchworm or hand-over-hand motions. As shown in Figure 16.11, the motor has been labeled by placing a single fluorophore on one of the motor heads. In the case where the fluorophore is situated on the end of one of the arms of the motor, individual steps will be revealed as motions of the fluorophore with a step size of $\sim$70 nm. This measurement is most consistent with a model where the two heads of myosin V take turns taking steps.

The size of the characteristic steps depends on the structure of the motor itself. An interesting demonstration of this is shown in Figure 16.13. In this experiment, the motor head domain of myosin V was progressively truncated to make it shorter and shorter. Because the conformational change of myosin that enables it to take a step involves rotation of a long, rigid "lever arm," the shortened proteins take shorter steps.

16.1.2 Rotary Motors

Evolution has generated not only an exquisite variety of translational motors, but also motors whose motions are rotational. Two of the

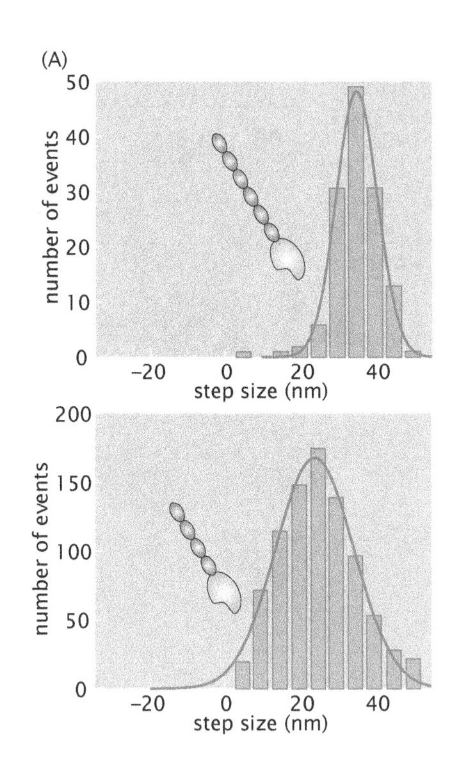

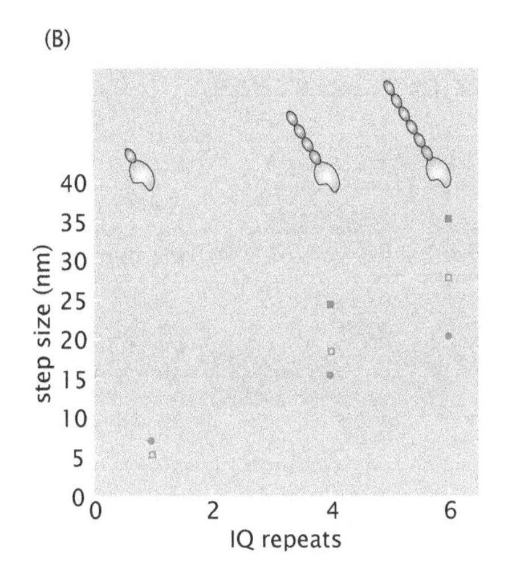

Figure 16.13: Motor step size depends on motor structure. (A) Motor heads of myosin V were purified, either in their normal state (top) or after the rotating lever arm had been shortened by protein engineering (bottom). The characteristic step size for the truncated protein is shorter than for the full-length protein. (B) For myosin V, the lever arm is built from a series of rigid subdomains called IQ repeats, so arms of different lengths can be easily constructed by including different numbers of repeats. Over a range from 1 to 6, the step size increases linearly with the number of repeats. The three different symbols show three different types of protein constructs; all follow the linear rule. (Adapted from T. J. Purcell et al., *Proc. Natl Acad. Sci. USA* 99:14159, 2002.)

most widely studied examples of rotary motors are the bacterial flagellar motor and ATP synthase, both of which were introduced as membrane proteins in Chapter 11. Schematics of both of these motors are shown in Figure 16.14.

The bacterial flagellar motor is embedded in the cell membrane of bacterial cells and is attached to the long filamentous flagellum. A bacterial cell may have either only a single polar flagellum, like the cholera-causing pathogen *Vibrio cholerae*, or it may have several flagella distributed over the surface, like our old friend *E. coli*, or even more complicated arrangements. When the motor rotates, it induces a rotary motion in the flagellum that propels the bacterium along in its highly viscous environs (as measured by the low Reynolds number already described in Chapter 12). Interestingly, the bacterial flagellar motor uses an ion gradient (rather than ATP) as the basis of its mechanical cycle, as shown in Figure 16.14. In particular, the motor is driven by a flow of hydrogen ions due to a concentration gradient between the inside of the cell and the space between the two bacterial membranes (a few exotic flagellar motors use sodium ions rather than hydrogen ions). The resulting motion has a rate in excess of 100 rotations per second. One particularly astonishing feature of this motor is that it can reverse its direction of rotation without reversing the direction of ion flow.

ATP synthase is one of the central powerhouses of living cells, found in the inner membrane of bacterial cells and also in the mitochondria of eukaryotic cells. It is an amazing molecular machine that is constructed of two different rotary motors connected to a common drive shaft. The F_0 motor of ATP synthase is similar to the flagellar rotary motor, in that it uses the energy stored in the transmembrane gradient of hydrogen ions to rotate. The F_1 motor uses ATP hydrolysis to rotate in the opposite direction. Under normal circumstances, when the transmembrane electrochemical gradient is strong, the F_0 motor generates more torque than the F_1 motor, and so the F_0 motor forces the F_1 motor to rotate in reverse, and thereby synthesize ATP

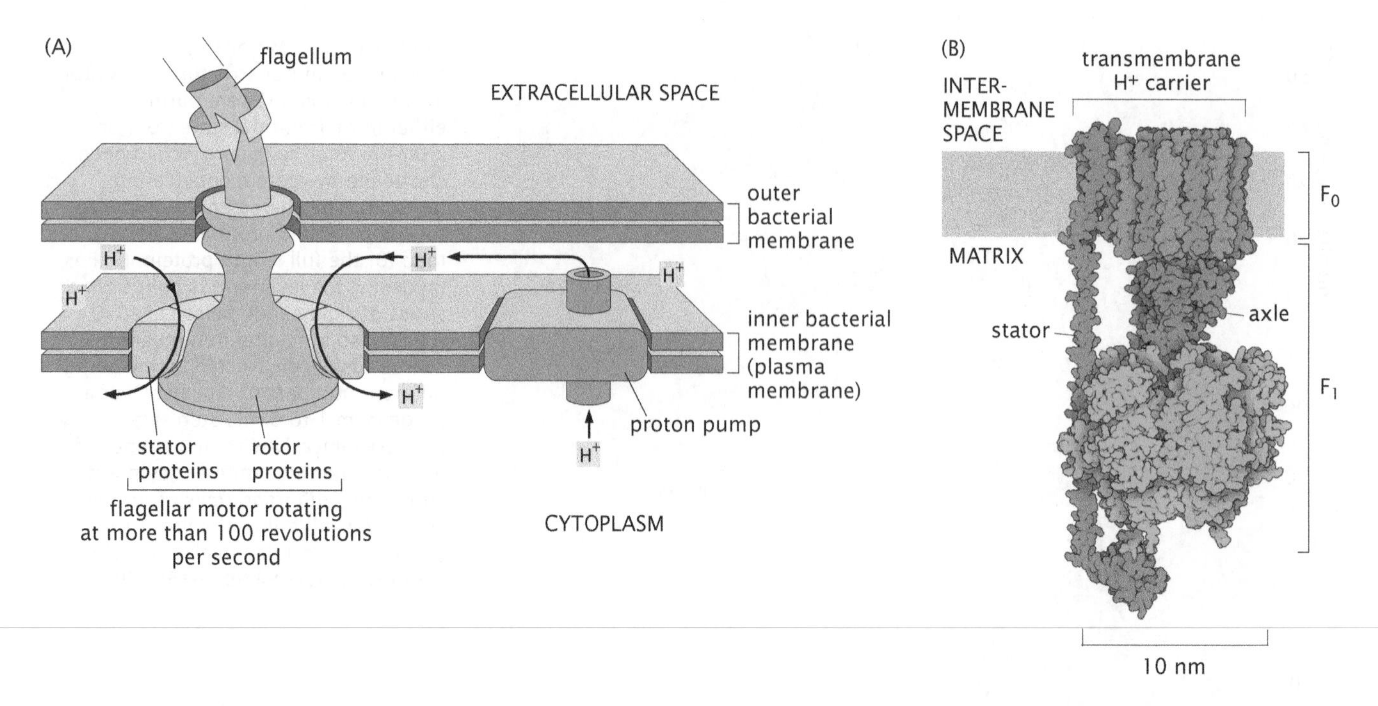

Figure 16.14: Examples of rotary motors. (A) The bacterial flagellum is like a tiny propeller driven by the gradient of hydrogen ions across the bacterial inner membrane. Continuous operation of the motor requires that the gradient be replenished by ATP-consuming proton pumps. The flagellum itself is an extremely long (10 μm) helical filament attached at its base to the motor apparatus. The motor is embedded in the bacterial inner membrane and anchored to the cell wall with a shaft passing through the outer membrane. The motor is capable of rotating in either direction at speeds up to 100 Hz. (B) ATP synthase is a rotary motor that uses the transmembrane electrical potential of the hydrogen ion gradient between the matrix and the intermembrane space of the mitochondrion in order to drive a mechanical rotation that in turn drives the chemical synthesis of ATP from ADP and inorganic phosphate. (A, adapted from B. Alberts et al., Molecular Biology of the Cell, 5th ed. Garland Science, 2008; B, courtesy of David Goodsell.)

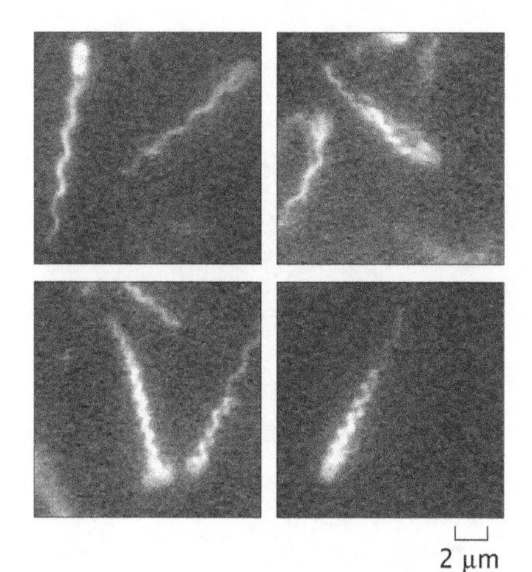

Figure 16.15: Flagellar movement in bacteria. Live bacterial cells were labeled on their surface using a fluorescent molecule revealing the helical shape of the flagella. For the different individuals shown, helical pitch and amplitude vary significantly. In a few cases, multiple helical forms can be seen attached to the same bacterium resulting in frayed bundles. (Adapted from L. Turner et al., *J. Bacteriol.* 182:2793, 2000.)

from ADP plus inorganic phosphate. However, if the transmembrane electrochemical gradient is weak, the balance can tilt in the other direction, and the F_1 motor will generate more torque than F_0. Under these circumstances, the coupled motor uses ATP hydrolysis to pump hydrogen ions out of the cell.

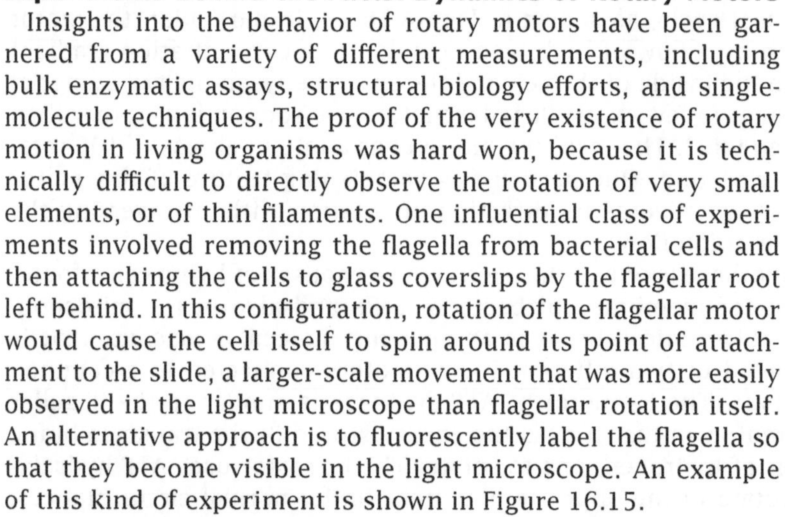

Experiments Behind the Facts: Dynamics of Rotary Motors

Insights into the behavior of rotary motors have been garnered from a variety of different measurements, including bulk enzymatic assays, structural biology efforts, and single-molecule techniques. The proof of the very existence of rotary motion in living organisms was hard won, because it is technically difficult to directly observe the rotation of very small elements, or of thin filaments. One influential class of experiments involved removing the flagella from bacterial cells and then attaching the cells to glass coverslips by the flagellar root left behind. In this configuration, rotation of the flagellar motor would cause the cell itself to spin around its point of attachment to the slide, a larger-scale movement that was more easily observed in the light microscope than flagellar rotation itself. An alternative approach is to fluorescently label the flagella so that they become visible in the light microscope. An example of this kind of experiment is shown in Figure 16.15.

Insights into the behavior of rotary motors can also be gleaned from *in vitro* measurements on single motors (as opposed to *in vivo* measurements). One of the most famed

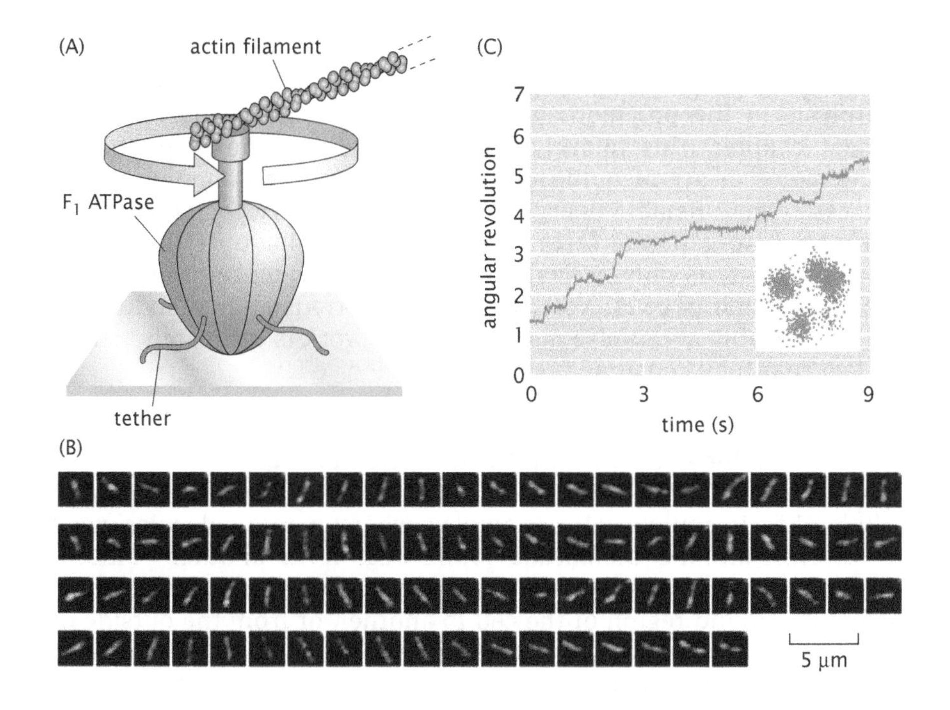

Figure 16.16: Single-molecule observation of a rotary motor using actin filaments to reveal the motor rotation. (A) The F_1 portion of ATP synthase is tethered to a glass slide. The top of the rotating shaft is attached to a fluorescently labeled actin filament. (B) As the F_1 shaft turns, the actin filament swings around. The time interval between images is 133 ms. (C) At low ATP concentrations, it is clear that the rotation occurs in three evenly spaced angular substeps. The graph shows the angular revolution for a single actin filament over a period of a few seconds and the inset shows the positions of the filament end over a longer movie. (A, B, adapted from H. Noji et al., *Nature* 386:299, 1997; C, adapted from R. Yasuda et al., *Cell* 93:1117, 1998.)

experiments of single-molecule biophysics involves the direct observation of the rotary motion of individual F_1 motors. This rotation has been measured by attaching a fluorescently labeled actin filament to the drive shaft of the motor and watching the rotation of the actin filament when the motor is provided with excess ATP (note that this is the opposite direction from its normal function in cells, where F_0 forces F_1 to run in reverse and to synthesize ATP rather than to hydrolyze it). A schematic diagram and typical data from this clever experiment are shown in Figure 16.16. These measurements revealed that the F_1 motor rotates in distinct steps of 120°, tightly coupled to its ATP hydrolysis activity. The quantized nature of the steps and the tight coupling between mechanical movement and ATP hydrolysis are strongly reminiscent of the translation motors discussed in the preceding section.

16.1.3 Polymerization Motors: Pushing by Growing

The coupling of hydrolysis to force generation can take place in unexpected ways. In addition to the translational and rotary motors described above, cells have other mechanisms such as the use of polymerization of cytoskeletal filaments as a means of force generation. We will refer to these cases as polymerization motors. Several examples of polymerization motors have been shown in the previous chapter. Figure 15.3 (p. 577) illustrates one of our favorite examples, namely, the way in which some bacterial pathogens such as *Listeria* hijack the host cytoskeleton to drive their motility. Figure 15.20 (p. 602) is a schematic of the way that polymerization motors are thought to segregate antibiotic resistance plasmids in dividing bacteria.

In Chapter 15, we described some of the features of these cytoskeletal filaments and how rate equations can be used to describe their dynamics. However, one of the most intriguing features of these filaments that was not elaborated on there is their ability to apply forces and to do mechanical work. Recall that the culmination of our discussion of the cytoskeleton was the analysis of the role of nucleotide hydrolysis in the polymerization process. For the purposes of the present chapter, this hydrolysis will be seen to contribute to the maintenance of a nonequlibrium state where soluble monomers are in excess; this unstable energetic state is necessary for polymerization motors to generate force. These examples illustrate the way in which filamentous polymerization has the consequence of mediating directed motion in cells. Note that, like with translational and rotary motors, the ability of these machines to do work is ultimately tied to the fact that they consume fuel such as ATP or GTP.

16.1.4 Translocation Motors: Pushing by Pulling

Cellular life is replete with examples where macromolecules need to travel from one membrane-bound region to another. Indeed, nucleic acids, proteins, sugars, and even lipids have to go from one membrane-bound region of the cell to another, or from the outside of a cell to the inside. We use the term "translocation motor" to refer to the broad class of molecular machines whose job is to mediate such transport. When viewed most broadly, this category includes the nuclear pore complex, mitochondrial import and export proteins, the proteasome, DNA packing motors in bacteriophage, ion channels, transporters, the Sec complexes, and even flippases that take lipid molecules from one side of a lipid bilayer to the other and maintain highly asymmetric lipid distributions in the face of entropy.

Examples of two of the speculated possible mechanisms of one of these motors are shown in Figure 16.17. The mitochondrion has

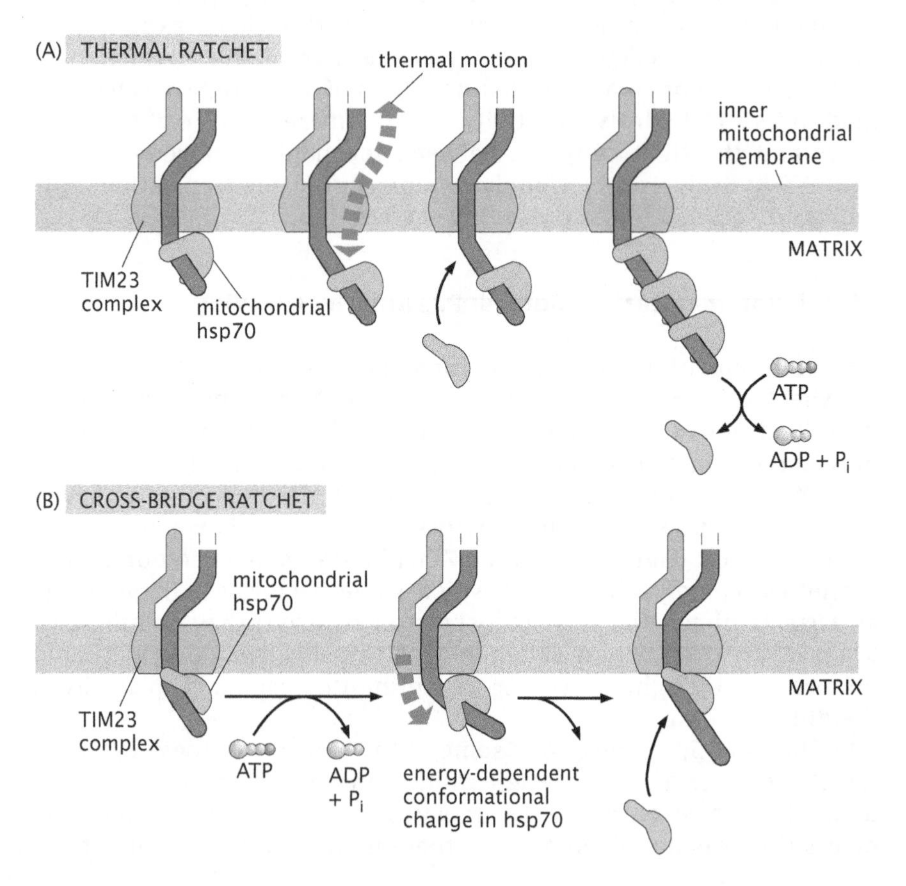

Figure 16.17: Thermal ratchet model and cross-bridge ratchet model of protein import into the mitochondria. (A) Thermal ratchet model of protein translocation. Thermal motion of the polymer in and out of the pore is biased by the presence of binding proteins on only one side of the barrier. (B) Cross-bridge ratchet model of translocation. Binding proteins on one side of the barrier may also use energy-dependent conformational changes to further insure that the cargo polymer moves in only one direction. (Adapted from B. Alberts et al., Molecular Biology of the Cell, 5th ed. Garland Science, 2008.)

served as one of the centerpiece examples throughout the book and offers us an opportunity to consider translocation motors as well. TIM and TOM are two important complexes responsible for the import and export of proteins in mitochondria. One of the mechanisms that has been proposed for membrane transport by TIM, for example, is that the unfolded peptide will be recognized in a way that permits free diffusion of the peptide into the mitochondria. As shown in Figure 16.17(A), binding partners of the translocating peptide start binding to the peptide that has already made it through the channel, generating a ratchet effect. An alternative mechanism envisions a power stroke, where a molecule binds to the part of the peptide that has already diffused in and pulls it in by an ATP-driven conformational change (cross-bridge ratchet). This is depicted in Figure 16.17(B).

To drive home the significance of this kind of motor action, we consider several different examples. As shown in Chapter 10, bacteriophage DNA is highly pressurized in the capsid, contributing to the overall driving force behind the infection process. However, this pressure effect is apparently insufficient by itself to mediate full translocation of the viral genome. The mechanism of translocation in the case of phage T7 is thought to be that RNA polymerase binds to a promoter located near the end that is first delivered into the bacterium and starts transcribing. The subsequent energy-dependent translocation of RNA polymerase along the DNA template contributes to pulling the viral genome into the host cell. This is shown schematically in Figure 16.18.

Yet another example of this same basic theme is the proteasome. The proteasome serves as the cell's protein garbage disposal, taking unwanted or misfolded polypeptides and digesting them. The protease active sites of the proteasome are sequestered on the inside of a large cylindrical complex, so that the protease activity does not accidentally degrade nonspecific targets in the cell. Delivery of polypeptide chains into the proteasome for degradation requires recognition of a specific protein tag by a cap complex that sits on both ends of the cylindrical proteasome. This cap complex uses energy derived from ATP hydrolysis to unfold the targeted polypeptide and thread it into the degrading maw of the proteasome. Once in the core of the proteasome, the polypeptides are cleaved into short peptides, and the amino acids can be recycled as building blocks for new proteins. Although protein degradation itself is not an energy-requiring process, the active unfolding of the polypeptide as it is fed into the proteasome does require the consumption of ATP.

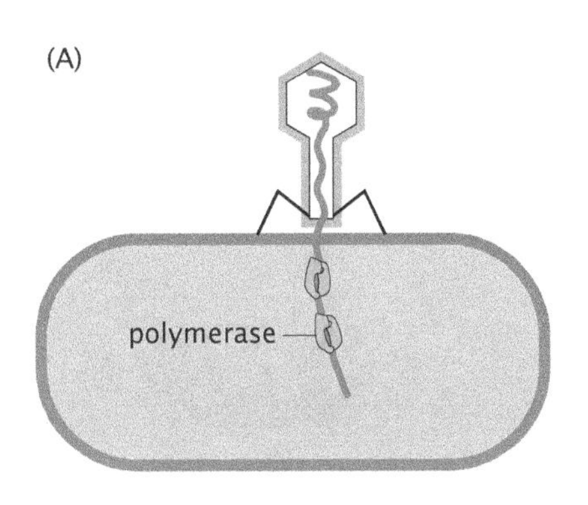

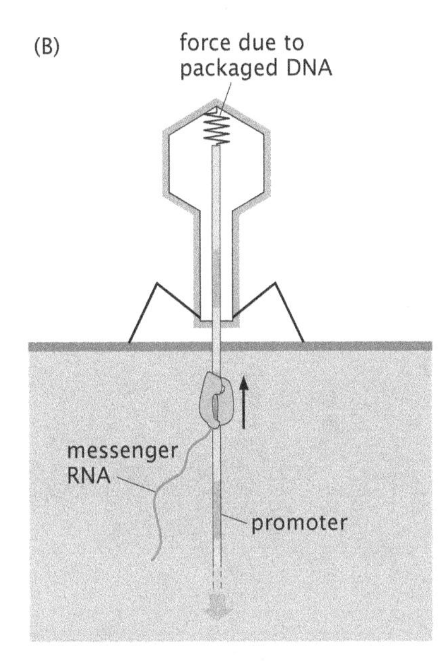

Figure 16.18: Translocation of phage DNA. (A) Illustration of the driving force on phage DNA. The packed DNA produces a driving force. In addition, the RNA polymerase can begin transcription and help to pull the DNA out. (B) Close up view of the effect of polymerase transcription on DNA translocation.

16.2 Rectified Brownian Motion and Molecular Motors

The previous section gave a quick tour of some of the key classes of molecular motors. Now our goal is to see what kind of theoretical framework can be built for the various observations that have been made on these motors. Though there is much that one might ask about motors, we will limit the discussion to a few key questions: What is the mean velocity of a motor and how does it depend upon the applied load? How much force can a motor exert before it will stall? How does the velocity depend upon the concentration of ATP or some other fuel? How different is the trajectory of a given motor from one experiment to the next?

One of the interesting challenges that will dictate the kinds of models we consider is shown in Figure 16.19. When viewed at very high spatial and temporal resolution, we will see that a molecular motor moves stochastically. There will be pauses between steps and sometimes the motor will lurch backward rather than forward. In constrast, if we look over longer time scales, the motor will appear to move steadily forward with a characteristic mean velocity. Different classes of models capture different key features of the motor's dynamics, operating over different time and length scales.

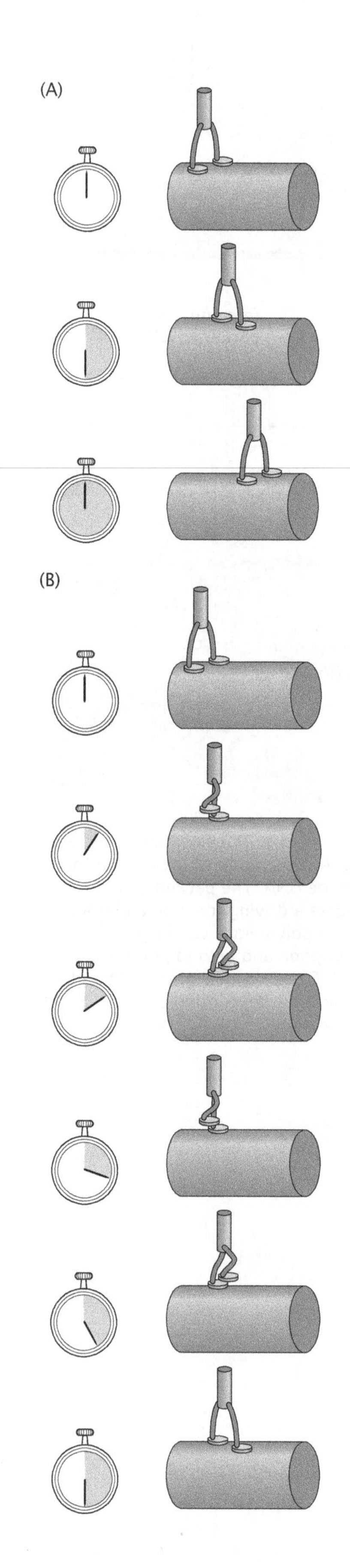

16.2.1 The Random Walk Yet Again

Molecular Motors Can Be Thought of as Random Walkers

Throughout the book, we have seen the appeal of the broad class of random walk models for characterizing a host of biological processes. The present section shows how driven random walk models arise naturally as a scheme for characterizing motor dynamics. We will begin by using cytoskeletal translational motors to develop these ideas, but emphasize that the conceptual treatment here is very general and can apply to rotational, polymerization, and translocation motors as well as translation motors. The basic idea is to consider the range of possible conformational states of a motor, the ways that the energy levels of the conformational states are influenced by the biochemical mechanism of energy utilization and external influences such as applied forces, and the ways that an individual motor can change from one state to another. One of the ways we will describe motor dynamics is shown in Figure 16.20. For a translational motor moving along some periodic track such as an actin filament or a microtubule, this class of models imagines n slots along the track that are each the same. Within each of these slots, the motor can be in any one of P distinct states. That is, motors are characterized by a *state space* in which the motor can occupy a set of geometric positions, and at each such position it can occupy a set of internal structural states. This perspective is particularly transparent for a simple translational motor such as kinesin, where the geometric positions correspond to different positions of the motor molecule along the microtubule, and the internal state refers to both conformational states of the motor as well as its binding to other molecules such as ATP, ADP, and inorganic phosphate P_i.

The simplifying assumption that the motor position along the track can be discretized into equal-sized boxes is a reflection of the observation that real motors generally move in quantized steps of a characteristic size. We have already seen this for myosin V, where each head moves about 74 nm in each cycle of ATP hydrolysis (Figure 16.11) and for F_1, which rotates in 120° steps (Figure 16.16).

To model the dynamics of the motor over this set of discrete states, we introduce the probability $p_m(n, t)$, which is the probability that the motor is at position na along its polymer track and has internal state m at time t. The length a is defined as the distance between successive periodic positions of the track along which the translational motor moves as shown in Figure 16.20. We claim that knowledge of

Figure 16.19: Motion of motors at different time scales. (A) If we watch the motor with low temporal resolution, it will appear to move steadily at some mean velocity. (B) If we watch the motor at high temporal (and spatial) resolution, we will see that its position fluctuates.

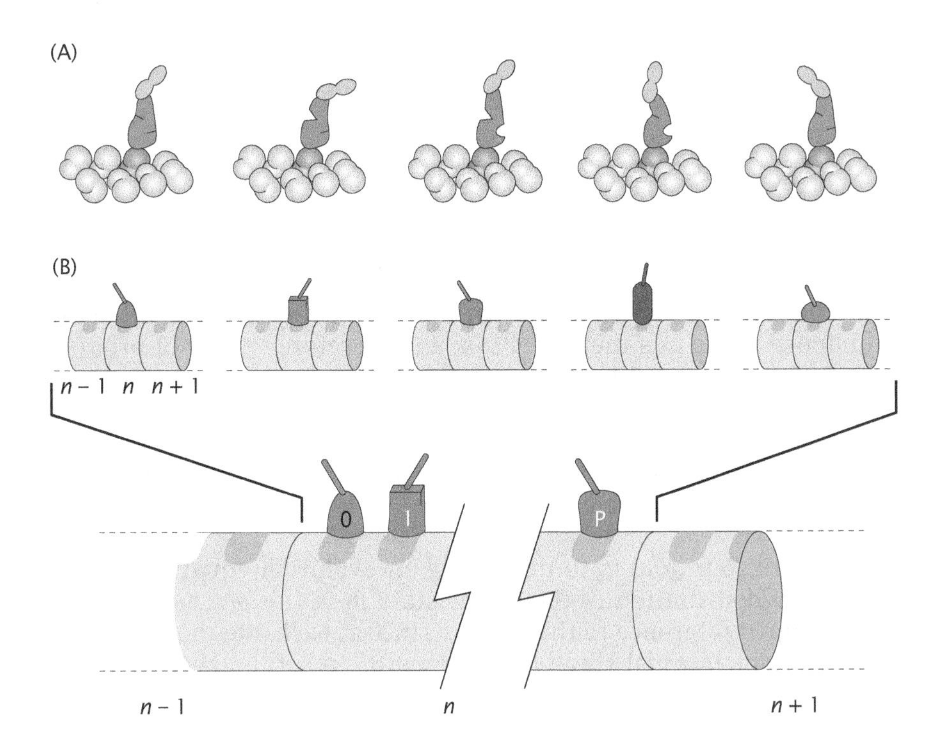

Figure 16.20: Schematic of position–state models. (A) An individual molecular motor, such as the myosin II head bound to an actin filament shown here, can exist in a wide variety of distinct conformational states. For myosin, the differences among the conformational states include the position of the lever arm, which can point either to the right or to the left, as well as the internal structure of the motor head itself, which may be different depending on whether it is bound to ATP, ADP plus inorganic phosphate, ADP alone, or no nucleotide at all, as well as on whether it is being stretched or compressed by the action of an external force or biological load. (B) A complete description of the state of an individual motor molecule must encompass its position with respect to the filament track as well as its internal conformation. The track along which the translational motor moves is divided into a set of boxes that are labeled by the parameter n. Here, each of five different protein conformations is schematized at the same position on the filament, position n. If the motor steps forward or backward, it will find itself at position $n+1$ or position $n-1$, where again it can assume any of several internal conformations. To generalize this treatment, we imagine that the motor can assume any of $P+1$ distinct states at any position along the filament as shown in the bottom schematic. The description of a particular motor at any instant in time can be summarized by just two parameters, namely, its position ($n-1$, n, $n+1$, etc.) and its internal state (0, 1, 2, etc.).

the function $p_m(n, t)$ permits the calculation of quantities of interest such as the mean velocity as a function of applied load. In practice, experimenters usually perform this operation in reverse; speed is directly measured under a variety of externally imposed load forces, and the underlying kinetic rate constants are determined by fitting kinetic models to the data.

16.2.2 The One-State Model

As a first example, we adopt the simplest possible model of the motion of a motor as depicted in Figure 16.21. In this case, we assume that the motor has no internal states and simply hops from one site to the next with forward rate $k_+(F)$ and backward rate $k_-(F)$ under the action of the applied force F. The probability per unit time of the motor moving forward by one site is $k_+(F)$, while the probability per unit time of the motor moving backward by one site is $k_-(F)$. Note that there is an explicit dependence of these rates on the applied force, which we assume to be applied in the backward direction. We begin with this oversimplified case as a way to explore the position–state treatment, but recognize that it cannot be directly applied to any real motor, because motors must couple energy utilization (for example, in the form of ATP hydrolysis or ion transport down an electrochemical gradient) to a mechanical conformational change, indicating that in reality they must exist in at least two states in order to do useful work. For the simplicity of the mathematical derivation, we have chosen to acknowledge the importance of energy utilization by allowing the forward rate constant $k_+(F)$ and the backward rate constant $k_-(F)$ to be different, even when the applied force F is zero. For a truly one-state motor, it would not be able to distinguish its position along the filament or distinguish a forward step from a backward step, because it would remain in the same single state regardless of its location, so there would be no way for these two rates to differ. Such a molecule

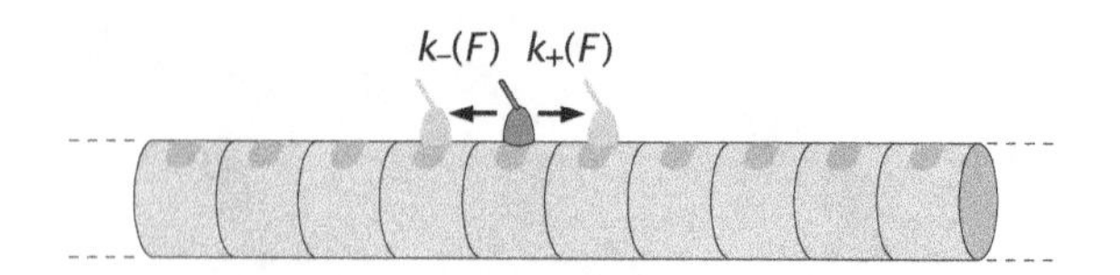

Figure 16.21: Schematic of a one-state motor. The motor can only be in one state in each box. The rate constants characterize the probability that the motor will jump left or right per unit time.

would be able to move in a random walk along its filament, but could not impose a bias in either direction, and therefore could not produce useful work. So in reality, our treatment here is of a "quasi-one-state" motor, which does have different rate constants for taking a forward step versus a backward step, but we will set aside for the moment the internal complexities of the motor that permit these differences (we will subsequently derive a two-state model that treats the differences explicitly).

At present, our goal is to determine an evolution equation for the probability distribution of the motor state $p_1(n, t) = p(n, t)$, where we have dropped reference to the label m since at each site the motor can only be in one internal state. In other words, $p(n, t)$ is the probability of finding a motor at position n at time t, where for convenience we take the time to be discretized in steps Δt.

The Dynamics of a Molecular Motor Can Be Written Using a Master Equation

At this point, we argue that the probability $p(n, t + \Delta t)$ can be gotten by summing over all of the processes that take place during time Δt, that start at time t and that have as their outcome the motor ending up at site n at time $t + \Delta t$. That is, we sum over all of the individual microtrajectories available to the system with the constraint that the final state is the prescribed one. For example, the site at $n + 1$ could be occupied at time t and the motor could hop backwards, resulting in it being at site n in time $t + \Delta t$. As we have done throughout the book, one convenient scheme for examining dynamical processes of this sort is through "trajectories and weights" diagrams such as that shown in Figure 16.22.

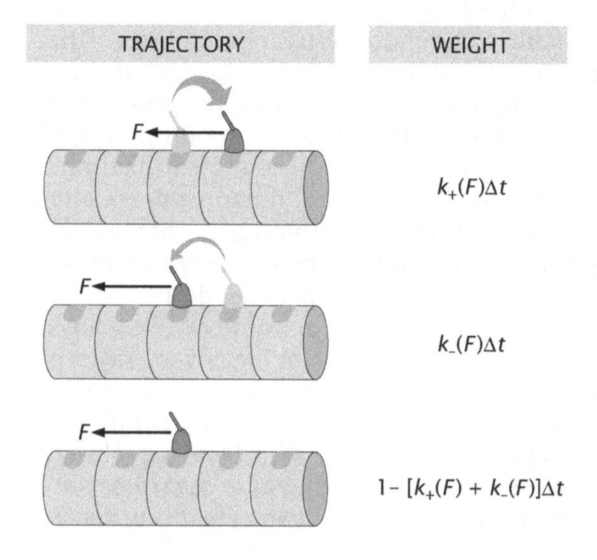

Figure 16.22: Trajectories and weights for molecular motors. The diagram shows all of the trajectories for a given motor during a time step Δt. The motor can jump forward (to the right, against the applied force) or backward (to the left, in the same direction as the applied force), or it can stay put.

To determine the evolution equation, we sum over all microtrajectories and the result can be written as

$$p(n,t+\Delta t) = \underbrace{k_+\Delta t p(n-1,t)}_{\text{jump from site to left}} + \underbrace{k_-\Delta t p(n+1,t)}_{\text{jump from site to right}} + \underbrace{(1-k_-\Delta t - k_+\Delta t)p(n,t)}_{\text{stay put}}. \tag{16.1}$$

This "master equation" adds up the probabilities of all the trajectories that lead to a given site being occupied as a result of the microscopic steps of the motor. Further, this equation can be recast in a more useful form by bringing $p(n, t)$ to the left-hand side and dividing through by Δt. This results in

$$\frac{p(n,t+\Delta t)-p(n,t)}{\Delta t} = k_+[p(n-1,t)-p(n,t)] + k_-[p(n+1,t)-p(n,t)]. \tag{16.2}$$

The next key point in the analysis is to think of the probability distribution as a continuous function of the position x along the filament, where the probability $p(x, t)$ for finding the motor at position x is equal to $p(n, t)$, when $x = na$. For the probability $p(x, t)$, we can use a Taylor expansion (see The Math Behind the Models on p. 215)

$$p(x \pm a, t) \approx p(x,t) \pm \frac{\partial p}{\partial x}a + \frac{1}{2}\frac{\partial^2 p}{\partial x^2}a^2, \tag{16.3}$$

where we ignore higher-order terms. Substituting this result back into Equation 16.2 leads to the differential equation

$$\frac{\partial p}{\partial t} = -(k_+ - k_-)\frac{\partial p}{\partial x}a + \frac{1}{2}(k_+ + k_-)\frac{\partial^2 p}{\partial x^2}a^2, \tag{16.4}$$

where we have also used the approximation

$$\frac{p(x,t+\Delta t)-p(x,t)}{\Delta t} \approx \frac{\partial p(x,t)}{\partial t}. \tag{16.5}$$

Note that this equation is of precisely the form already described in Chapter 13, namely,

$$\frac{\partial p}{\partial t} = -V\frac{\partial p}{\partial x} + D\frac{\partial^2 p}{\partial x^2}, \tag{16.6}$$

where we have made the definitions

$$V = a[k_+(F) - k_-(F)] \tag{16.7}$$

and

$$D = \frac{a^2}{2}[k_+(F) + k_-(F)]. \tag{16.8}$$

What we have recovered (not surprisingly) is the equation for diffusion in the presence of a force. What we have learned is that the probability distribution $p(x, t)$ describing a one-state motor can be characterized by Equation 16.6, which is a driven or biased diffusion equation also known as the Smoluchowski equation. The physical essence of this equation is that the motors move with an average velocity V. However, if we start a collection of motors on parallel filaments all at the same time (like a collection of sprinters in a race), we will find that, over time, they spread out in a way characterized by the diffusion constant D. To see this explicitly, we can solve the equation.

The Driven Diffusion Equation Can Be Transformed into an Ordinary Diffusion Equation

One way to solve the equation is to perform a change of variables that turns it into the conventional diffusion equation, without a driven term, for which we already know the solution as was shown in Chapter 13 (see Equation 13.32 on p. 524). The relevant change of variables is given by

$$\bar{t} = t \tag{16.9}$$

and

$$\bar{x} = x - Vt. \tag{16.10}$$

The fun of a transformation like this is that it amounts to shifting to a frame of reference that is moving along at the mean velocity. The derivatives appearing in the driven diffusion equation are now determined as follows. First, the time derivative is given by

$$\frac{\partial p}{\partial t} = \frac{\partial p}{\partial \bar{t}}\frac{\partial \bar{t}}{\partial t} + \frac{\partial p}{\partial \bar{x}}\frac{\partial \bar{x}}{\partial t} = \frac{\partial p}{\partial \bar{t}} - V\frac{\partial p}{\partial \bar{x}}. \tag{16.11}$$

We can perform a similar exercise with the spatial derivatives, resulting in

$$\frac{\partial p}{\partial x} = \frac{\partial p}{\partial \bar{x}} \tag{16.12}$$

and

$$\frac{\partial^2 p}{\partial x^2} = \frac{\partial^2 p}{\partial \bar{x}^2}. \tag{16.13}$$

Using these transformations, we see that Equation 16.6 takes the form

$$\frac{\partial p}{\partial \bar{t}} = D\frac{\partial^2 p}{\partial \bar{x}^2}, \tag{16.14}$$

precisely the familiar diffusion equation already discussed in detail in Chapter 13.

The solution in the case of initial conditions where the motor is localized at $\bar{x} = 0$ at $t = 0$ is given by

$$p(\bar{x}, \bar{t}) = \frac{1}{\sqrt{4\pi D\bar{t}}}\mathrm{e}^{-\bar{x}^2/4D\bar{t}}. \tag{16.15}$$

This can be recast in terms of the original variables as

$$p(x, t) = \frac{1}{\sqrt{4\pi Dt}}\mathrm{e}^{-(x-Vt)^2/4Dt}. \tag{16.16}$$

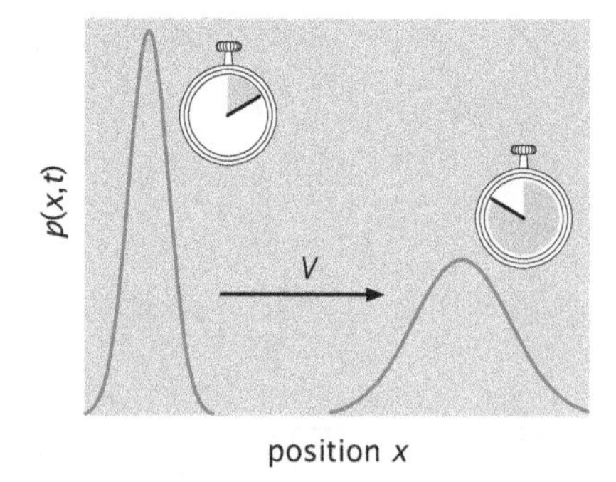

Figure 16.23: Solution to the driven diffusion equation. As time passes, the center of mass of the probability distribution moves to the right with mean velocity V. As time passes, the distribution becomes wider.

The evolution of this probability distribution is presented in Figure 16.23, where it is seen that the probability distribution broadens due to diffusion as it propagates along with a mean velocity V.

Our discussion thus far has centered on the overall dynamics of the one-state motor without reference to the actual values adopted by parameters such as $k_+(F)$ and $k_-(F)$. As was noted above, one of the primary physical aspects of the model is the presence of the asymmetric jump rates k_+ and k_-. In particular, these jump rates are related to the force acting on the motor, shown schematically in Figure 16.24.

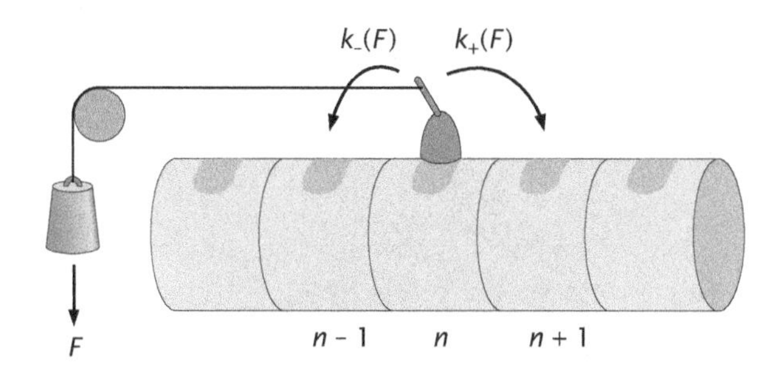

Figure 16.24: The effect of force on jump rate. A molecular motor subjected to an applied load will have its bare forward and backward jump rates modified.

In equilibrium, we know that the ratio of the rates has to satisfy a special relation that is dictated by the characteristics of the energy landscape such as is shown in Figure 16.25. By equilibrium, we mean that there is no net flux from one state, where the motor can be found with probability p_i, to the other, where the occupancy of the motor has probability p_j. For neighboring sites n and $n+1$, this condition can be written as

$$k_+ p_n = k_- p_{n+1}. \tag{16.17}$$

Equilibrium probabilities are given by the Boltzmann formula $p_n = \mathrm{e}^{-\beta G_n}/Z$, where G_n is the free energy of the motor when on the nth site along the filament. Using this definition in the equilibrium condition above leads to the relation between the forward and backward rates,

$$\frac{k_+}{k_-} = \mathrm{e}^{-\beta \Delta G}, \tag{16.18}$$

where the free-energy change is given by $\Delta G = G_{n+1} - G_n$.

In the presence of force applied in the backward direction, the free energy of the nth site is raised above the no-force value by the work of the motor against the applied force Fna, that is, $G_n \to G_n + Fna$. This is depicted in Figure 16.24 in our usual way by using a pulley and mass to represent the applied force on the motor. Now applying the same reasoning as above for the no-force case, we arrive at

$$\frac{k_+(F)}{k_-(F)} = \mathrm{e}^{-\beta(\Delta G + Fa)}. \tag{16.19}$$

The significance of this expression is described in Figure 16.25, where it is seen that the effect of the force is to tilt the energy landscape and thereby change the barrier height and the allied rates.

This idea can be tested experimentally by using an optical trap to apply force to a single motor and pulling the load either forward or backward relative to the motor's preferred direction. Data from one such experiment are shown in Figure 16.26. In this case, pulling forward on myosin V increases its stepping speed by nearly 10-fold compared with pulling backward with the same force.

To say anything more about how this class of model behaves with force, we actually have to make a precise statement about how the forward and backward rates depend individually on the force. To that end, we examine several different case studies in which the applied force is present in different parts of the rates. As our first example, we assume that all the dependence of the force is in k_+.

This means that

$$k_+(F) = k_- \mathrm{e}^{-\beta(\Delta G + Fa)}, \tag{16.20}$$

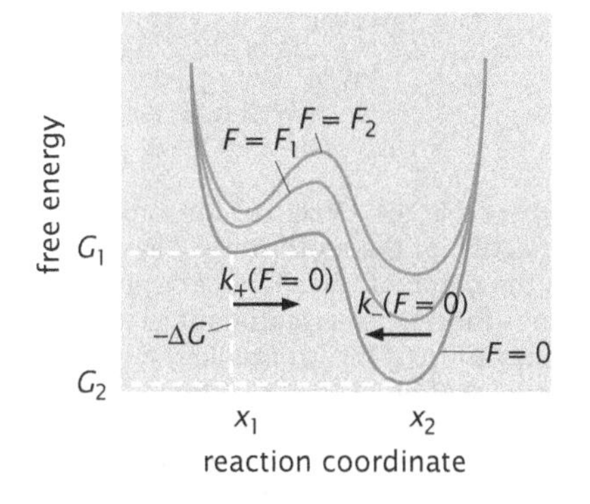

Figure 16.25: Energy landscape of a one-state model in the absence of any force (red) and when the application of forces pulling backward on the motor tilts the landscape (blue and green).

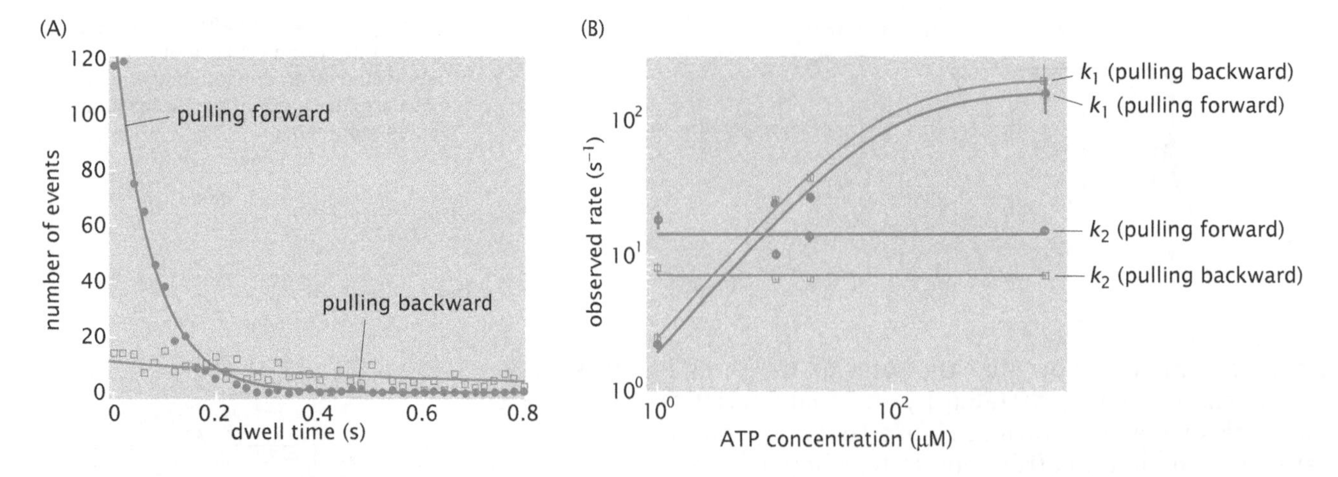

Figure 16.26: Effects of forward and backward forces on stepping rates for myosin V. (A) The average time between steps is a reflection of the motor rate. When a forward pulling force is applied to myosin V, individual steps take place in rapid succession, so the graph of "dwell times" is dominated by very brief pauses between steps (filled circles). In contrast, the histogram of dwell times for the same motor subjected to a backward-pulling force shows brief pauses and many more long ones compared to the behavior for a forward pulling force. (B) Over a range of ATP concentrations, the dwell time graphs for myosin V are best fit by a model that invokes two different internal states and two rate constants (see Section 16.2.4). The rate constant for one of the state transitions, k_1, is dependent on ATP concentration but does not change with applied force. The second rate constant, k_2, is independent of ATP concentration, but strongly force-dependent, increasing significantly when forward force is applied rather than backward force. This result shows that the force dependence of motor activity can be at least partially uncoupled from the mechanism of energy utilization. (Adapted from T. J. Purcell et al., *Proc. Natl Acad. Sci. USA* 102:13873, 2005.)

which can be plugged into Equation 16.7 to obtain

$$V(F) = ak_-(\mathrm{e}^{-\beta(\Delta G + Fa)} - 1). \tag{16.21}$$

This result is shown in Figure 16.27(A). A second case study is built around the idea that k_+ is independent of the force, assigning all the force dependence to k_- and resulting in

$$k_-(F) = k_+\mathrm{e}^{\beta(\Delta G + Fa)}, \tag{16.22}$$

which implies in turn that

$$V(F) = ak_+(1 - \mathrm{e}^{\beta(\Delta G + Fa)}). \tag{16.23}$$

This result is shown in Figure 16.27(B). In both cases, if the force is large enough, the motor will start moving backward. Backward stepping under large applied forces has been directly observed for both myosin V and kinesin. Data from an experiment measuring velocity as a function of applied force for myosin V are shown in Figure 16.27(C), revealing a trend more consistent with the version of the model in which the force primarily influences the magnitude of k_-. However, the reader should bear in mind that this simple one-state model is primarily aimed to illustrate a style of analysis rather than to convey molecular realism consistent with translational motors such as myosin V and kinesin.

In addition to the force dependence of the velocity, this model can help us begin to think about another important experimental parameter called the randomness. If we introduce a characteristic time $\tau = a/V$, we can ask what is the diffusive excursion, $\Delta x^2 = 2D\tau = 2Da/V$, over that time. The randomness is defined as

$$\text{randomness} = r \equiv \frac{\Delta x^2}{a^2} = \frac{2D}{Va} = \frac{k_+(F) + k_-(F)}{k_+(F) - k_-(F)}, \tag{16.24}$$

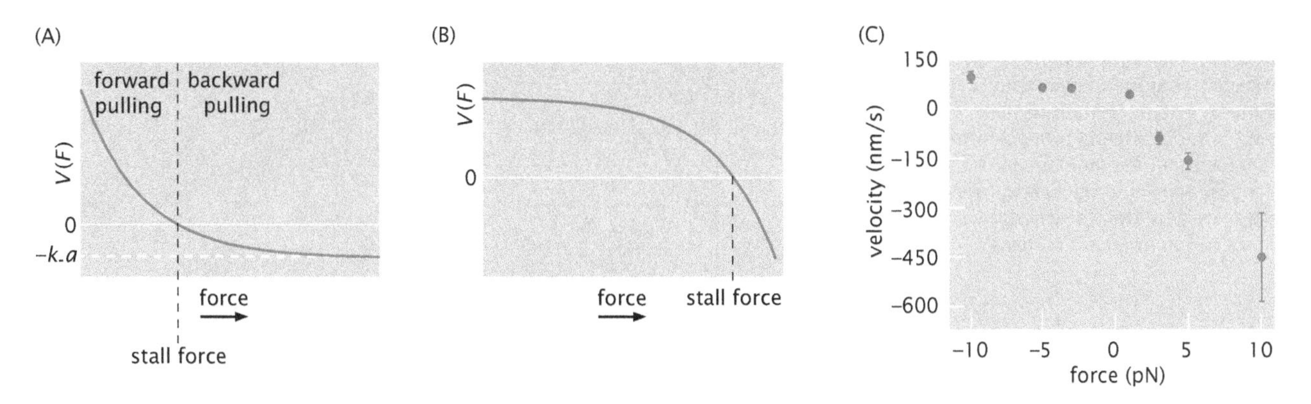

Figure 16.27: Molecular motor velocity as a function of applied force. (A) Theoretical result assuming force dependence is in the forward rate. (B) Theoretical result assuming force dependence is in the backward rate. (C) Data for myosin V, where negative forces represent forward pulling and positive forces represent backward pulling. The shape of the curve most closely resembles the curve in (B), which would be interpreted in the simple one-state model as indicating that the effects of force on this motor result in changes in the backward rate, not the forward rate. (C, adapted from J. C. Gebhardt et al., *Proc. Natl Acad. Sci. USA* 103:8680, 2006.)

where the actual final formula relating the randomness to the rate constants is only valid for the one-state motor we are considering in this model. Note that if $k_+ = k_0[\mathrm{ATP}]$, then the randomness will approach 1 as the ATP concentration becomes large. It is more difficult to measure the randomness parameter accurately than it is to measure speed, but nevertheless some data are available. Figure 16.28 shows one such set of measurements for kinesin. In this case, the experimental results are more subtle than can be captured by the simple one-state model, but are fit reasonably well by a more realistic two-state model, which we will develop below. The problems at the end of the chapter present another opportunity to explore randomness within the one-state model.

16.2.3 Motor Stepping from a Free-Energy Perspective

So far, we have made use of a master equation approach whereby the motor motion is described as a random walk along the filament.

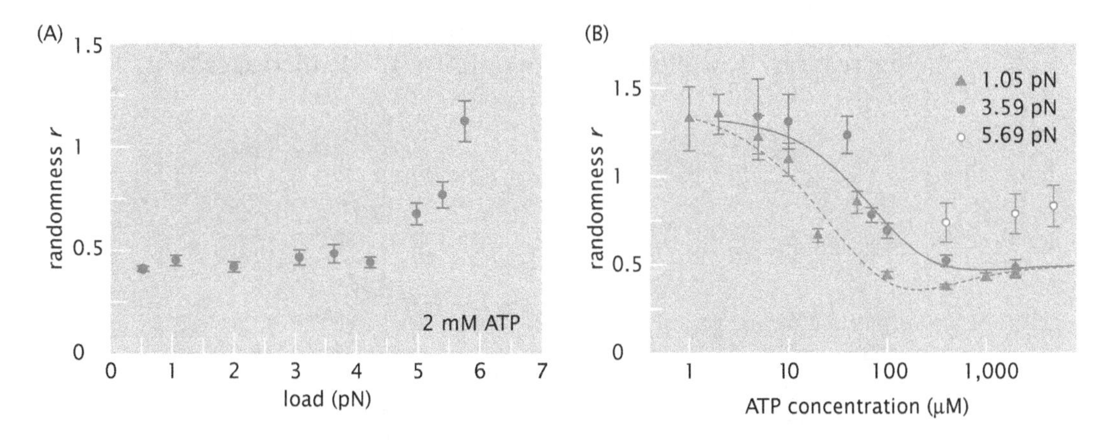

Figure 16.28: Randomness parameter measured for kinesin under a variety of conditions. (A) At high (saturating) concentrations of ATP and low loading force, the randomness is generally low, indicating that the motor moves at a fairly constant speed. When the load force is very high, approaching the stall force, randomness increases. (B) Randomness also increases at very low levels of ATP (that is, when the binding rate of ATP may become rate-limiting), and there is a complex relationship between force and randomness at low ATP concentrations, again with higher forces causing higher randomness, or more variability in motor speed. Although these data cannot be fit by a simple one-state model like that described in the text, models invoking more internal states do a better job of accounting for the measurements. (Adapted from K. Visscher et al., *Nature* 400:184, 1999.)

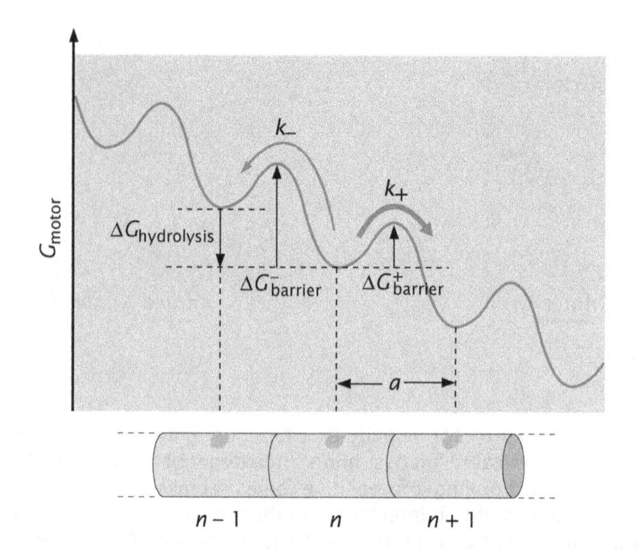

Figure 16.29: The free energy of a motor moving along a filament. n labels the discrete positions of the motor on the filament. The overall tilt of the free-energy surface leads to motion of the motor with an average speed to the right. The forward and backward rates are determined by the free-energy barriers between adjacent states.

The rates k_+ and k_- describe the probability of the motor stepping a distance a to the right, or to the left, in unit time. An alternative view is provided by the model that depicts the molecular motor as diffusing on a free-energy landscape, as shown in Figure 16.29. In this case, the rate k_+ can be computed as the inverse average time for the motor to diffuse over the barrier from the nth to the $(n+1)$th site on the filament, while for k_- the diffusion from $n+1$ to n is the one to consider. Here we make the simplifying assumption that the transition rates are given by the height of the barrier separating the two states via the Arrhenius relation, $k = \Gamma e^{-\beta \Delta G_{\text{barrier}}}$, where Γ is the frequency of attempts to go over the barrier and $\Delta G_{\text{barrier}}$ is the height of the barrier on the free-energy landscape.

As in our analysis of the effect of force on motor speed, we can make use of equilibrium arguments to constrain the space of allowed models for the stepping rates k_+ and k_- and how they depend on the concentration of ATP in the toy model of a one-state motor. To begin, we compute the change in free energy associated with a single step of the motor. We assume that each step is accompanied by hydrolysis of a single ATP molecule, which in the lattice model shown in Figure 16.30 is represented by having the number of ATP molecules in solution decrease by 1, while the number of ADP and P_i increases by 1.

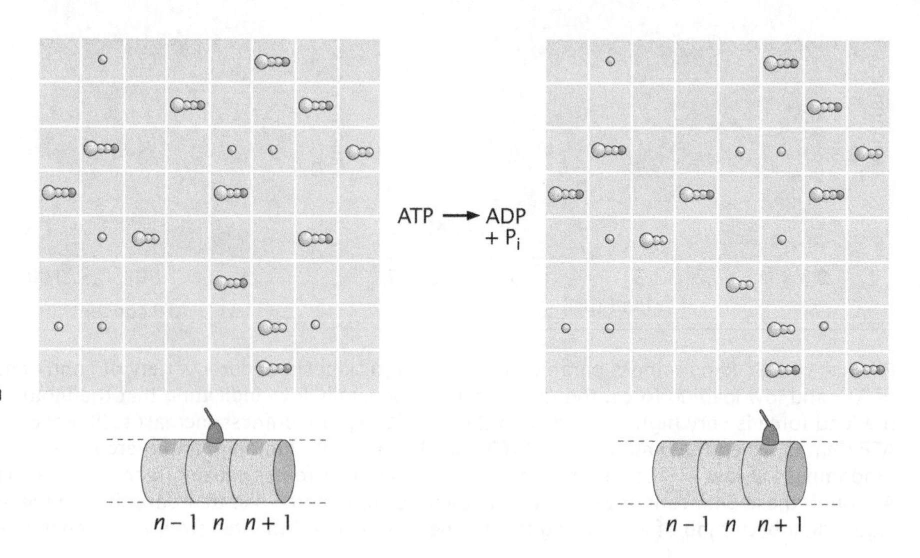

Figure 16.30: Lattice model of motor motility coupled to ATP hydrolysis. Each step of the motor is accompanied by a single hydrolysis event, which decreases the number of ATP molecules by 1 and increases the number of ADP and P_i molecules by 1.

If the state that corresponds to the motor being at site n has A ATP molecules in solution, D ADP molecules, and P inorganic phosphates, the statistical weight of that state is

$$w_n = \frac{\Omega!}{A!D!P!(\Omega - A - D - P)!} \mathrm{e}^{-\beta A \varepsilon_{\text{bond}}}, \tag{16.25}$$

where the combinatorial term accounts for all the ways of rearranging the molecules in solution among the Ω boxes, and the Boltzmann factor corresponds to all the bonds between the inorganic phosphate and the ADP present in each of the ATP molecules. Upon hydrolysis, the motor moves to the $n+1$ site and one ATP molecule breaks down into an ADP and a P_i. Therefore, the statistical weight of this state is

$$w_{n+1} = \frac{\Omega!}{(A-1)!(D+1)!(P+1)![\Omega - (A-1) - (D+1) - (P+1)]!} \mathrm{e}^{-\beta (A-1) \varepsilon_{\text{bond}}}. \tag{16.26}$$

The free-energy difference between the state $n+1$ and the state n, which is the free energy of hydrolysis $\Delta G_{\text{hydrolysis}}$, is related to the ratio of the statistical weights for the nth and the $(n+1)$th state,

$$\mathrm{e}^{-\beta \Delta G_{\text{hydrolysis}}} = \frac{w_{n+1}}{w_n} = \frac{A}{DP} \Omega \mathrm{e}^{\beta \varepsilon_{\text{bond}}}. \tag{16.27}$$

In the last equality, we have made the simplifying assumption that the number of boxes in the lattice model is much greater than the number of molecules of the three species, which are in turn all much bigger than 1. Also note that we have rederived from a lattice model perspective the formula for the free energy of ATP hydrolysis (see Equation 6.127 on p. 276)

$$\Delta G_{\text{hydrolysis}} = \Delta G_0 + k_B T \ln\left(\frac{[D][P]}{[A]} v\right), \tag{16.28}$$

where the concentrations of the different molecular species are given by the number divided by the volume of the solution, Ωv, with v the volume of a single box. The "standard-state free energy" $\Delta G_0 = -\varepsilon_{\text{bond}}$ in this case corresponds to the energy released upon breaking a single chemical bond in an ATP molecule. The standard state for the lattice model is one where all the boxes are occupied by the three molecular species, that is, when $[A] = [D] = [P] = 1/v$. (We choose this as the standard state for convenience; biochemists traditionally and arbitrarily use a standard state of 1 M concentration for all chemical species.) Clearly, in this state, there is no entropy change when a single ATP breaks up into ADP and P_i since there is no freedom for placing these molecules into different boxes, and the free energy of hydrolysis is just the bond-breaking energy.

The free energy of hydrolysis we have computed above corresponds to the difference in free energy of successive minima on the free-energy landscape on which our motor diffuses and therefore it is equal to the overall slope of the diagram in Figure 16.29. We see that changes in the concentration of ATP, ADP, or P_i will lead to a change in the slope of the free-energy landscape as required by Equation 16.28. Hence, within this highly simplified model, the motor velocity will inherit this concentration dependence.

To obtain a formula for how the motor speed depends on ATP concentration, equilibrium considerations are not enough. In the simple model described by the free-energy landscape shown in Figure 16.29,

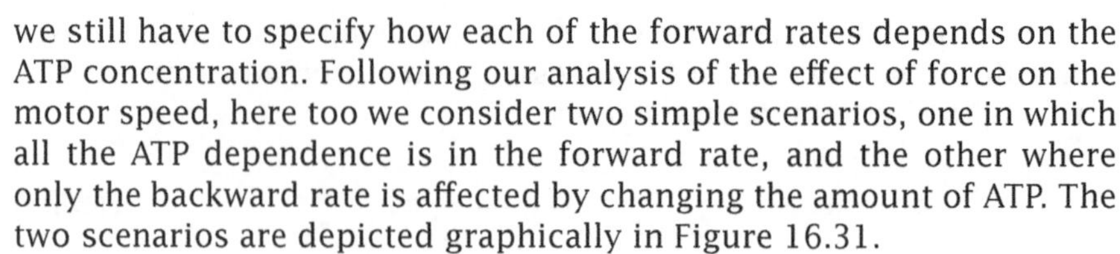
we still have to specify how each of the forward rates depends on the ATP concentration. Following our analysis of the effect of force on the motor speed, here too we consider two simple scenarios, one in which all the ATP dependence is in the forward rate, and the other where only the backward rate is affected by changing the amount of ATP. The two scenarios are depicted graphically in Figure 16.31.

For the case when only the forward rate is ATP-dependent, we can use the Arrhenius formula to obtain

$$\begin{aligned} k_+ &= \Gamma_+ \mathrm{e}^{-\beta(\Delta G^-_{\text{barrier}} + \Delta G_{\text{hydrolysis}})} \\ k_- &= \Gamma_- \mathrm{e}^{-\beta \Delta G^-_{\text{barrier}}}. \end{aligned} \tag{16.29}$$

Since the motor speed is $V = a(k_+ - k_-)$, using Equation 16.28 in the above equation for k_+ yields

$$V = a(k^0_+[A]v - k_-), \tag{16.30}$$

where the rate $k^0_+ = \Gamma_+ \mathrm{e}^{-\beta(\Delta G^-_{\text{barrier}} + \Delta G_0)}/[D][P]v^2$ is independent of ATP, and is the forward rate for the standard-state concentration of ATP, $[A] = 1/v$. The prediction is that the speed of the motor increases linearly with ATP concentration. This is clearly contradicted by experiments in the large-ATP-concentration limit, where the motor speed saturates. In fact, the model itself becomes inconsistent for large enough ATP concentrations. Raising the ATP concentration has the effect of decreasing the barrier for stepping from the nth site to the $(n+1)$th site. When the barrier is gone, that is, when $\Delta G^-_{\text{barrier}} + \Delta G_{\text{hydrolysis}} = 0$, the Arrhenius form for the transition rate is no longer valid. At best the present model can be used to give an idea of how the motor behaves at concentrations of ATP that are well below saturation.

The case of the backward rate having all the ATP dependence can be analyzed in an analogous way. In this case, the rates have the form

$$\begin{aligned} k_+ &= \Gamma_+ \mathrm{e}^{-\beta \Delta G^+_{\text{barrier}}}, \\ k_- &= \Gamma_- \mathrm{e}^{-\beta\left(\Delta G^+_{\text{barrier}} - \Delta G_{\text{hydrolysis}}\right)}, \end{aligned} \tag{16.31}$$

and the speed of the motor is described by

$$V = a\left(k_+ - \frac{k^0_-}{[A]v}\right), \tag{16.32}$$

where $k^0_- = \Gamma_- \mathrm{e}^{-\beta(\Delta G^+_{\text{barrier}} - \Delta G_0)}[D][P]v^2$. This model does predict saturation of the motor speed at large ATP concentrations, but, as in the case of the previous model for high ATP concentrations, at low enough ATP concentrations the barrier for a backward step vanishes, rendering the Arrhenius formula meaningless.

The predictions for the dependence of motor speed as a function of ATP concentration for the two models described above are shown in Figure 16.32. They lack the nuance to respond to the full range of experimental data. Still, the one-state motor provides a good starting point for building intuition about the nature of motor dynamics. To actually make convincing contact with experimental data on motors like kinesin, in the next section we begin to analyze a two-state motor.

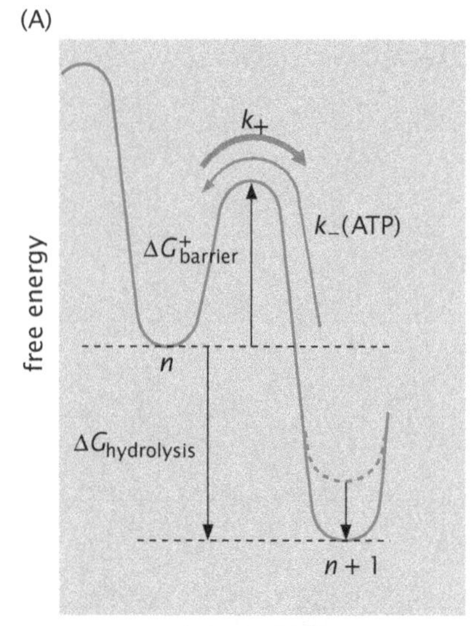

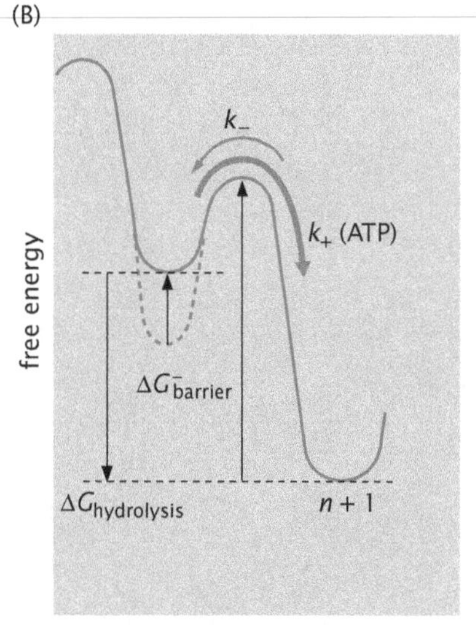

Figure 16.31: Possible scenarios for how the rates of motors depend on the ATP concentration. (A) The backward rate is the ATP-dependent step, while the forward rate is not affected by the concentration of ATP. (B) The forward step is the only one that is ATP-dependent. The dashed red curve represents the free-energy surface for a smaller concentration of ATP and the vertical black arrow directly underneath indicates the change upon increasing the amount of ATP in solution.

16.2.4 The Two-State Model

The most immediate generalization of the model presented in the previous section is to consider the case in which there are two internal states associated with each position. A schematic description of this model is shown in Figure 16.33. We adopt the indices 0 and 1 to characterize the internal states. An interesting feature of the two-state model, which is different from the one-state picture, is that some of the allowed transitions are not associated with geometric displacements along the filament, but rather refer to internal transitions such as those involving hydrolysis. For real molecular motors, the internal conformational transitions associated with ATP hydrolysis are coupled to a protein conformational change that generates a single step along the substrate and also to a change in the affinity of the motor head for its substrate. The coupled cycles of binding, ATP hydrolysis, stepping, and unbinding generate directed movement. In the one-state model, we were forced to make the nonphysical assumption that the forward and backward rate constants were different from one another even though the internal state of the motor was unchanged regardless of the stepping direction; we are now prepared to render the model more realistic (and, as we will see, surprisingly powerful in its ability to make quantitative predictions) by incorporating the second state.

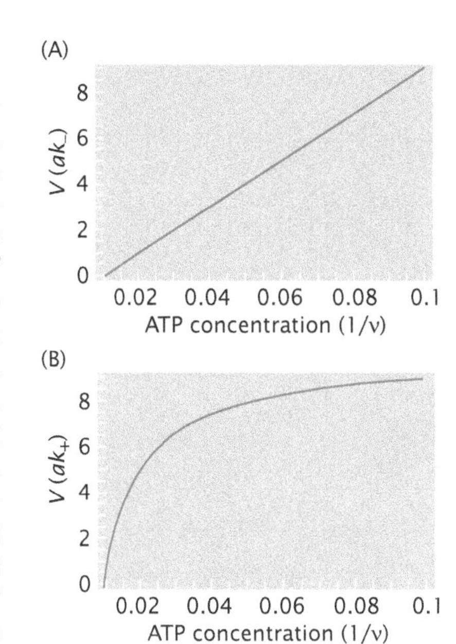

Figure 16.32: Speed of a molecular motor as a function of the ATP concentration as predicted by the one-state model. (A) All the dependence in ATP concentration is in the forward stepping rate. (B) The backward step is ATP-dependent while the forward step is not.

The Dynamics of a Two-State Motor Is Described by Two Coupled Rate Equations

As with our analysis of the one-state motor, our ambition is to write a set of rate equations that describe the time evolution of the probability distribution $p_i(n, t)$, where the label i refers to the internal state (either 0 or 1) and n refers to the position on the linear track. For

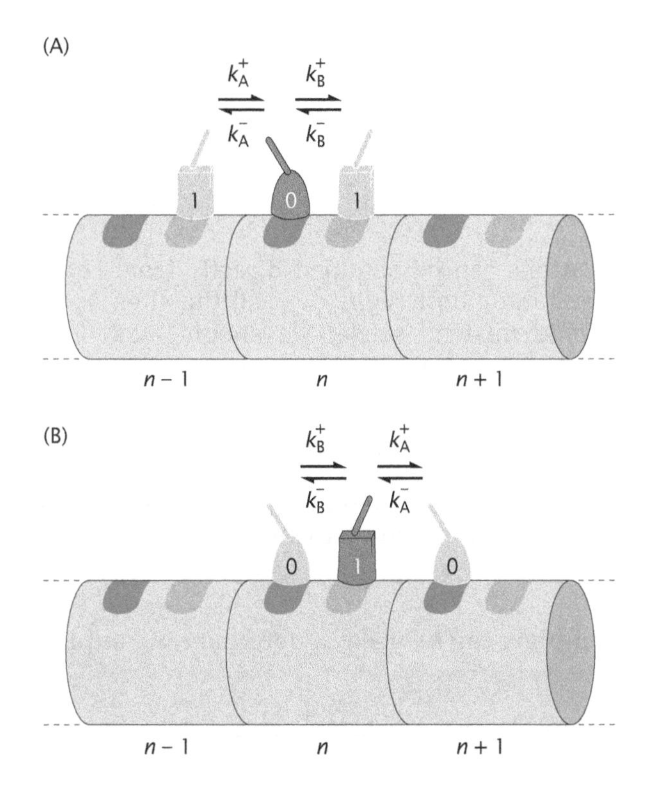

Figure 16.33: Two-state motor model. (A) The rates for the transitions that can occur to change the occupancy of internal state 0. The dark icon indicates the current state of the motor head and the two light icons indicate the two possible states in the next time step. In the two-state model, the motor head is constrained to convert from internal state 0 to internal state 1. This can occur either while the motor remains stationary with respect to the filament (with rate constant k_B^+) or while the motor takes a single step backwards (with rate constant k_A^-). (B) The rates for the transitions that can occur to change the occupancy of internal state 1. The motor head can convert to internal state 0, either while remaining in place, or while taking a single step forward.

example, if we interest ourselves in $p_0(n, t)$, we need to consider several different processes that result in the motor being at time t in state 0 at site n. In particular, the motor can come into this state from state 1 at position $n-1$ with a rate k_A^+. In addition, we need to consider the processes in which the motor arrives in this state from state 1 at position n with a rate k_B^-. Finally, there is the uneventful situation where the motor stays put and remains in state 0 at position n during the time interval from t to $t+\Delta t$. This last microtrajectory occurs with probability $1-(k_A^- + k_B^+)\Delta t$.

Following a strategy similar to that we used in the case of the one-state motor, we can write the time evolution of $p_0(n, t)$ as

$$\frac{\mathrm{d}p_0(n,t)}{\mathrm{d}t} = k_A^+ p_1(n-1,t) + k_B^- p_1(n,t) - k_A^- p_0(n,t) - k_B^+ p_0(n,t). \quad (16.33)$$

Similar reasoning applies to the equation for the time evolution of $p_1(n, t)$, which can be written as

$$\frac{\mathrm{d}p_1(n,t)}{\mathrm{d}t} = k_A^- p_0(n+1,t) + k_B^+ p_0(n,t) - k_A^+ p_1(n,t) - k_B^- p_1(n,t). \quad (16.34)$$

We are interested in the position of the motor as a function of time, which we now derive using certain general observations about the problem.

Though we do not want to enter into the details of how to fully solve models such as the two-state model presented above and we leave this as a homework problem, there is a clever and simple scheme for deducing the velocity in such a model. The trick is to introduce the probabilities P_0 and P_1, which are the probabilities that the motor will be at any site n, but in state 0 or state 1, respectively. That is, we ignore the question of which site the motor occupies and only concern ourselves with its internal state. As a result, we have effectively mapped our problem onto a two-state problem. We can immediately write down the rate equations that capture the dynamics of this two-state system as

$$\frac{\mathrm{d}P_0}{\mathrm{d}t} = k_A^+ P_1 + k_B^- P_1 - k_A^- P_0 - k_B^+ P_0 \quad (16.35)$$

and

$$\frac{\mathrm{d}P_1}{\mathrm{d}t} = k_A^- P_0 + k_B^+ P_0 - k_A^+ P_1 - k_B^- P_1. \quad (16.36)$$

These two equations can be obtained directly from Equations 16.33 and 16.34 by summing both sides over all the sites $n = 1$ to ∞. (For this to make mathematical sense, we should make the additional assumption that the motor is moving along a microtubule that is infinitely long.) In steady state, we see that

$$(k_A^+ + k_B^-)P_1 = (k_A^- + k_B^+)P_0. \quad (16.37)$$

A second condition on our probabilities is that the motor be in either state 0 or state 1, which, stated mathematically, is

$$P_0 + P_1 = 1. \quad (16.38)$$

These two conditions can be used to determine P_0 and P_1 themselves as

$$P_0 = \frac{k_A^+ + k_B^-}{k_A^- + k_B^- + k_A^+ + k_B^+} \quad (16.39)$$

and

$$P_1 = \frac{k_A^- + k_B^+}{k_A^- + k_B^- + k_A^+ + k_B^+}. \qquad (16.40)$$

Given these probabilities, we can now compute the motor velocity by examining the motions implied by each of the possible $0 \to 1$ and $1 \to 0$ transitions. For concreteness, we assume that in going from state 0 to state 1 at a given site, the distance traveled by the motor is δ and in going from state 1 at one site to state 0 at the next site, the distance moved by the motor is $a - \delta$, where a is the periodicity of the cytoskeletal filament on which the motor moves.

The basic picture is that during each time interval Δt, there are four possible transitions the system can make: a step to the right or the left of length δ can be made by the transition between the 0 and 1 states on the same site, or, alternatively, a step of length $a - \delta$ can be made to the right or to the left by transitions between states 1 and 0 on different sites of the filament. The net result is that the average velocity is given by

$$V = \delta(P_0 k_B^+ - P_1 k_B^-) + (a - \delta)(P_1 k_A^+ - P_0 k_A^-). \qquad (16.41)$$

If we now use the results for P_0 and P_1 given in Equations 16.39 and 16.40, the velocity may be rewritten simply as

$$\langle V \rangle = a \frac{k_A^+ k_B^+ - k_A^- k_B^-}{k_A^- + k_B^- + k_A^+ + k_B^+}. \qquad (16.42)$$

An interesting outcome is that the average velocity does not depend on the change in position of the motor (characterized by δ) as it transitions between the two internal states on the same site.

Note that we can rewrite the motor velocity in much the same way that we did for the one-state motor, $V = a(k_+ - k_-)$, but with the definitions

$$k_+ = \frac{k_A^+ k_B^+}{k_A^+ + k_B^+ + k_A^- + k_B^-} \qquad (16.43)$$

and

$$k_- = \frac{k_A^- k_B^-}{k_A^+ + k_B^+ + k_A^- + k_B^-}. \qquad (16.44)$$

In order to obtain the diffusion constant of the motor, a more detailed analysis of Equations 16.33 and 16.34 is needed. This is left to the reader as a problem at the end of the chapter.

Models with two or more internal states are more adept at responding to experimental data than the one-state model introduced earlier. An example of the kind of data that have been experimentally measured for kinesin is shown in Figure 16.34. In these experiments, the speed of kinesin-driven movement was carefully measured over a range of ATP concentrations and load forces. The speed of the motor increases when more ATP is present, approaching a constant rate at very high concentrations of ATP. This is reminiscent of the turnover rates of enzymes behaving with Michaelis–Menten kinetics (see Figure 16.54 on p. 678). For kinesin, the motor speed can be assumed to be proportional to the ATPase enzyme turnover rate, because individual motor steps are uniform in size and are tightly

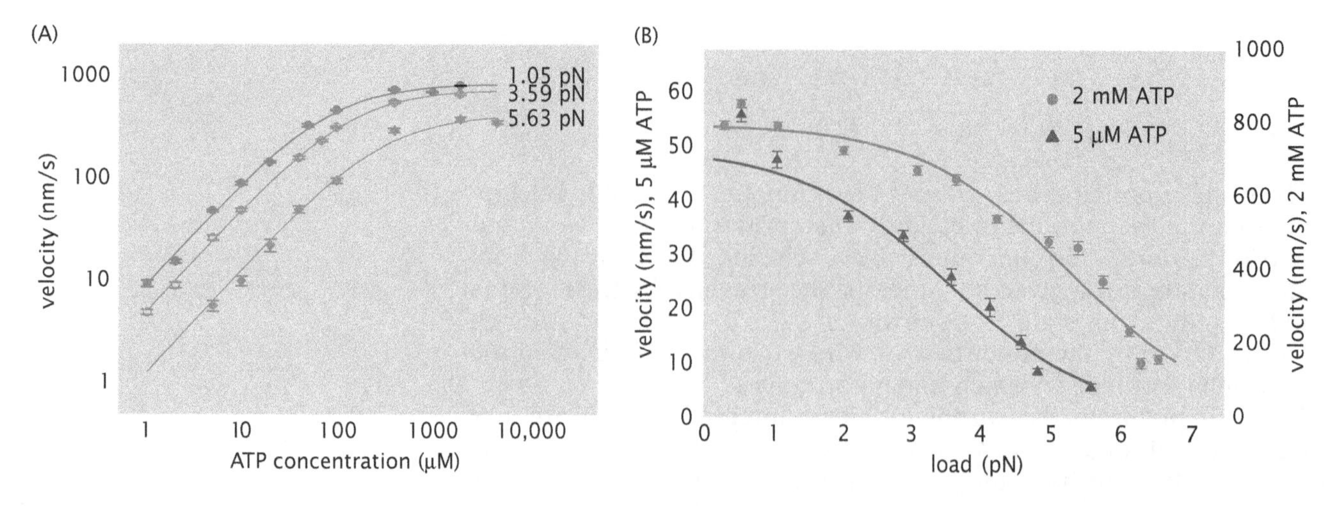

Figure 16.34: Speed of kinesin motors varying with ATP concentration and applied force. (A) Kinesin speed increases with increasing ATP at low concentrations and saturates at high concentrations. (B) Load dependence of motor velocity for several ATP concentrations. (Adapted from K. Visscher et al., *Nature* 400:184, 1999 and M. J. Schnitzer et al., *Nat. Cell Biol.* 2:718, 2000.)

coupled to ATP hydrolysis. Performing the same experiment under different applied loads reveals that increasing the load not only decreases the maximum speed that can be achieved by the motor, but also increases the apparent binding constant of ATP for the motor (that is, the ATP concentration at which the speed is one-half of its maximum value).

Internal States Reveal Themselves in the Form of the Waiting Time Distribution

The two-state motor involves internal states that are characteristic of any general treatment of motor dynamics. Some experiments have suggested the existence of multiple substates for movements of real molecular motors. Even in cases where we are not able to resolve the internal states in a direct manner, there are still some indirect consequences of their existence. One of the most immediate ways in which these internal states reveal themselves is through the waiting time distribution, which provides a measure of the pauses between forward steps of the motor. To show how these internal states alter the waiting time distribution, we will consider the case in which the ith substate has a characteristic lifetime τ_i. This implies that the composite process resulting in net translation of the motor is going to be related to some combination of these times.

Our maximum-entropy analysis from Chapter 15 showed that the probability density function for waiting times given some average waiting time $\langle t \rangle$ is of the form

$$p(t) = \frac{1}{\langle t \rangle} \mathrm{e}^{-t/\langle t \rangle}. \tag{16.45}$$

Recall that this distribution emerges as a result of the constraint

$$\langle t \rangle = \int_0^\infty t\, p(t)\, \mathrm{d}t. \tag{16.46}$$

In order to determine the waiting time distribution for a composite process built up of two subprocesses both of which are characterized by exponential waiting times such as in Equation 16.45, the net waiting time distribution is obtained as

$$p(t) = \int_0^t p_{\text{A}}(\tau) p_{\text{B}}(t - \tau)\, \text{d}\tau. \tag{16.47}$$

$p_{\text{A}}(\tau)$ is the probability that the first step occurs after time τ and $p_{\text{B}}(t - \tau)$ is the probability that the second step occurs after time $t - \tau$. The probability of a total waiting time for the composite process is $p(t)$. The integral written in Equation 16.47 represents a sum over the set of *all* allowed microtrajectories. For example, if $t = 5$ s, then the integral is a sum over all those intermediate steps whose waiting times add up to a total of 5 s. When $\tau = 2$ s for process A, this implies that process B will take 3 s. Examples of different allowed composite microtrajectories are shown in Figure 16.35.

We already know the probabilities for the separate subprocesses, which can be written in the form

$$p_{\text{A}}(t) = \tau_{\text{A}}^{-1} \text{e}^{-t/\tau_{\text{A}}} \tag{16.48}$$

and

$$p_{\text{B}}(t) = \tau_{\text{B}}^{-1} \text{e}^{-t/\tau_{\text{B}}}. \tag{16.49}$$

TRAJECTORY 1

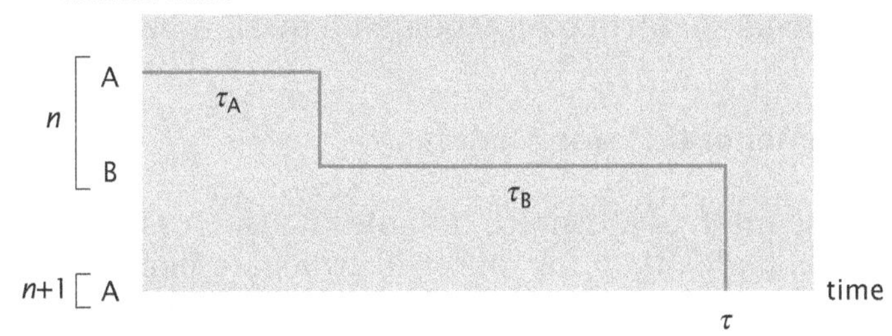

TRAJECTORY 2

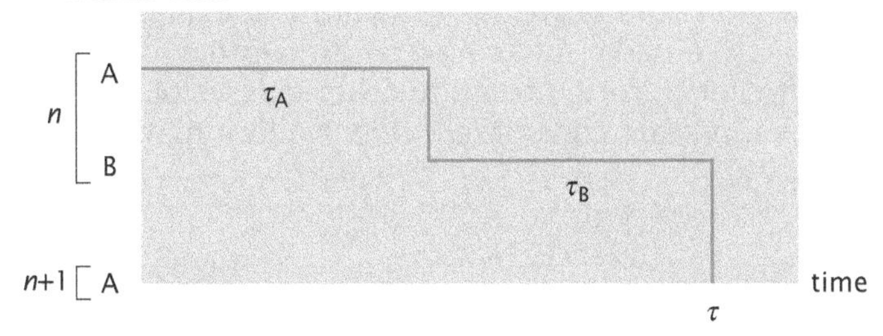

TRAJECTORY 3

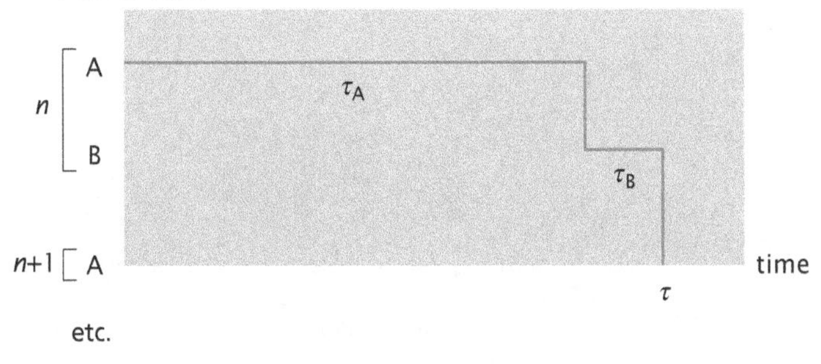

Figure 16.35: Trajectories for a two-state system. The total time to make the two transitions is τ. There are many different ways that the system can make the two transitions during this time τ. There is a waiting time τ_{A} to make the first step and τ_{B} to make the second step.

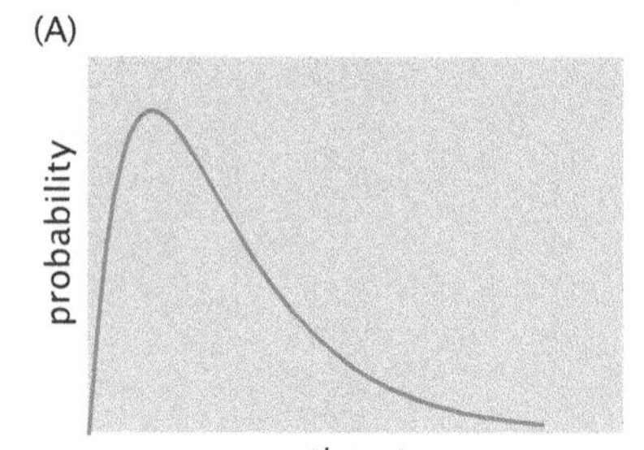

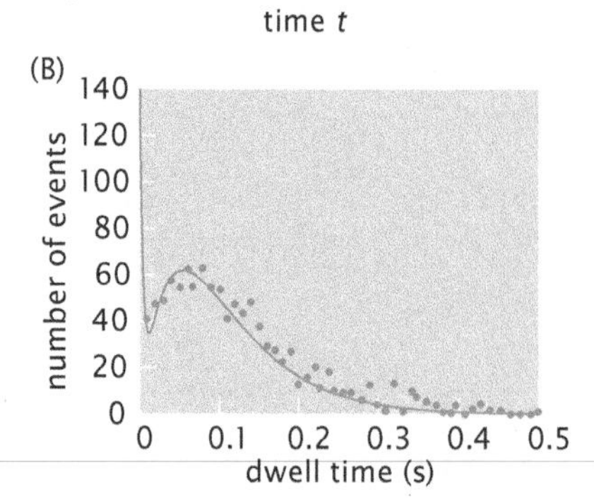

Figure 16.36: Probability distribution for waiting times for a two-state motor. (A) Theoretical result, as stated in Equation 16.51. (B) Data on dwell time distribution for a single-headed version of myosin V, at 10 μM ATP. (B adapted from T. J. Purcell et al., *Proc. Natl Acad. Sci. USA* 102:13873, 2005 and J. C. Liao et al., *Proc. Natl Acad. Sci. USA* 104:3171, 2007.)

As a result, Equation 16.47 can be written explicitly as

$$p(t) = \int_0^t e^{-\tau/\tau_A} e^{-(t-\tau)/\tau_B} d\tau \frac{1}{\tau_A \tau_B}. \tag{16.50}$$

This integral can be evaluated (this is assigned as a problem at the end of the chapter) and results in

$$p(t) = \frac{1}{\tau_B - \tau_A}(e^{-t/\tau_B} - e^{-t/\tau_A}). \tag{16.51}$$

Figure 16.36 shows the functional form of this class of waiting time distribution. In practice, observed dwell times for various molecular motors fit well to this kind of scheme.

Because the rates of internal conformational changes for motors in position–state models depend on nucleotide hydrolysis state and on applied force, we expect that dwell time distributions should also depend on these factors. Figure 16.37 shows average dwell times for myosin V as a function of force for two different concentrations of ATP. At high ATP, dwell times are initially very short, but increase with load as the motor is forced to do more work. At very low concentrations of ATP, the behavior of the motor is dominated by the state while it is waiting for a nucleotide to bind, and is less sharply dependent on applied force. The conclusion of this analysis is that the overall distribution of dwell times provides a window on the composite processes that make up the overall motor dynamics and a filter on the different classes of models set forth to greet data on dwell times.

16.2.5 More General Motor Models

In general, the internal dynamics of molecular motors can be considerably more complex than the one- and two-state motors considered thus far. The *P*-state, *N*-position class of models introduced in Figure 16.20 is probably still a useful framework for thinking about these motors. For example, in order to fit all the data available for kinesin, it appears that a model based upon four internal states is most realistic. The complexity of the mathematics and the proliferation of parameters make these models largely beyond the scope of this book. However, before turning to other classes of motors besides the translational motors considered here, we first make some general observations.

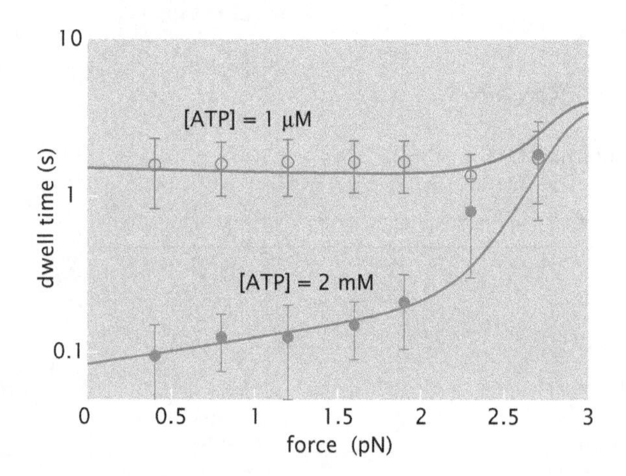

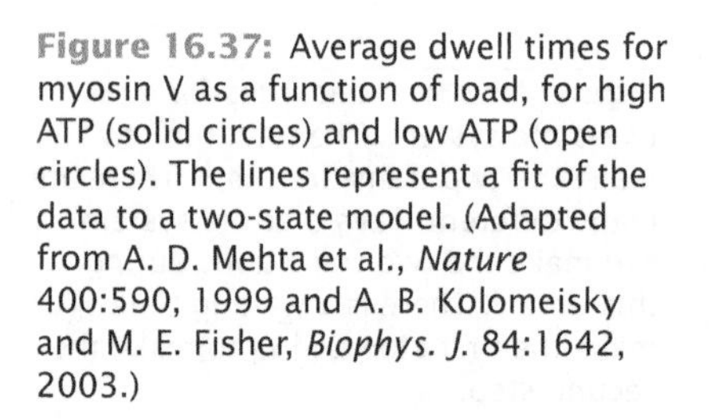
Figure 16.37: Average dwell times for myosin V as a function of load, for high ATP (solid circles) and low ATP (open circles). The lines represent a fit of the data to a two-state model. (Adapted from A. D. Mehta et al., *Nature* 400:590, 1999 and A. B. Kolomeisky and M. E. Fisher, *Biophys. J.* 84:1642, 2003.)

As an example of the proliferation of internal states, consider the particular case of myosin shown in Figure 16.38. This figure schematizes the various subprocesses present in the cycle of a myosin molecule as it moves along actin. In the first stage, a myosin head is bound to an actin filament (attached), from which it unbinds upon ATP binding, allowing it to move along the filament (released). ATP is hydrolyzed (the resulting ADP and phosphate remain bound) producing a larger conformational change, which causes the head to be displaced around 5 nm along the filament (cocked). The phosphate is released in order to bind the head to the new position on the

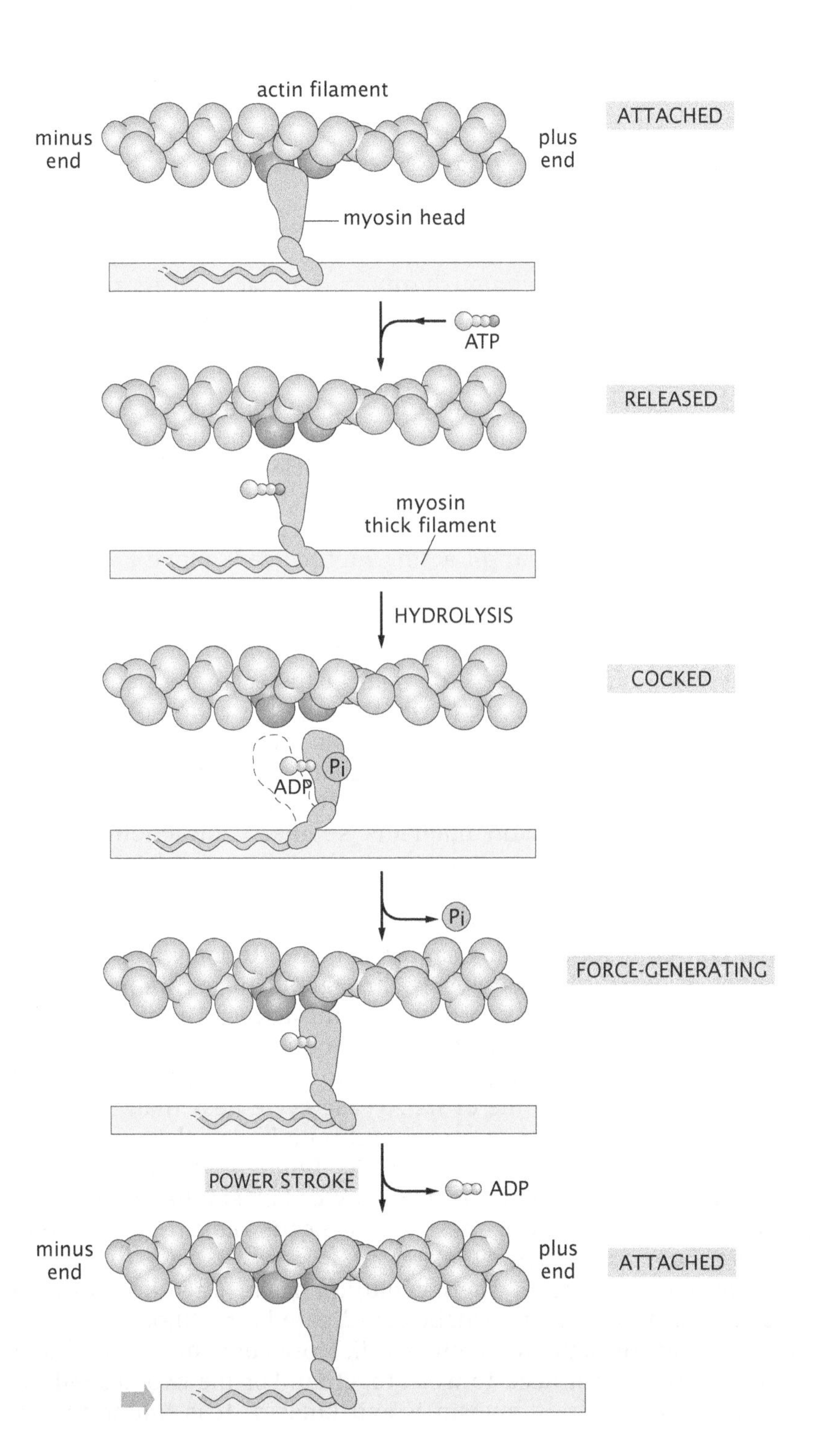

Figure 16.38: Model of myosin walking on an actin filament in a muscle. Actin thin filaments and myosin thick filaments are arranged in a regular array in muscle such that individual myosin heads protruding from the thick filaments are conveniently positioned to step along the actin filament tracks. At the beginning of a single-step cycle, the myosin motor head domain is attached to the actin filament. Binding of ATP to the myosin releases it. ATP hydrolysis is coupled to a conformational change such that the head is poised over the next subunit on the actin filament. When phosphate is released from the myosin head, it is able to bind in this new position. Release of ADP from the myosin head is coupled to the force-generating power stroke. (Adapted from B. Alberts et al., Molecular Biology of the Cell, 5th ed. Garland Science, 2008.)

filament and the power stroke starts, during which the ADP molecule is released (force-generating), returning to the initial stage. The argument is that each of the internal states exhibited by myosin can be represented in the framework introduced above, but now with further internal states per geometric position.

There are a number of different experimental dials that can be controlled to elicit different motor responses. Two of the most interesting ways to perturb the dynamics of molecular motors like myosin and kinesin are to change the ATP concentration and to apply forces to the motor, as we have seen in several cases above. Another particularly interesting kind of perturbation is to link several individual motors so that they are forced to work together. We now turn to this fascinating topic.

16.2.6 Coordination of Motor Protein Activity

Thus far, we have focused on the remarkable mechanical properties of individual motor molecules. Nevertheless, in cells, it is rare for a single molecule to accomplish anything notable on its own. More typically, the action of many individual motors must be coordinated to achieve a larger-scale goal. An impressive example of this is seen in muscle fibers, where the 10^{14} heads (as estimated in Section 16.1) must all cooperate and generate force nearly simultaneously without interfering with one another. These concerted motor motions generate the macroscopic shortening of the muscle that enables us to run, jump, and swim. How is this coordination achieved?

There are three requirements for motor coordination during muscle contraction: first, that all acting motor heads move in the same direction; second, that they all generate force at the same time; and third, that they not interfere with one another. In muscle, the first requirement is achieved by the exquisite geometrical control exercised by cells during the development of the sarcomere as shown in Figure 16.7. The second is achieved through the action of a group of actin-binding proteins called tropomyosin and the troponins, whose choreographed response to calcium signals is shown in Figure 16.39. The long tropomyosin protein physically prevents binding of the myosin heads to the actin filaments so that contraction can only occur in response to a positive signal, the influx of calcium into the cell. Because calcium ions flood the entire cell extremely rapidly as a consequence of the rapid propagation of membrane depolarization (which will be discussed in Chapter 17), the tropomyosin block can be removed nearly simultaneously throughout the entire muscle cell, thus enabling large-scale coordinated contraction. The third requirement is alleviated by the fact that an individual myosin head in a skeletal muscle remains tightly bound to the actin filament for only a short fraction of the time of its ATPase cycle (less than 5%). Thus, one head generates a power stroke and rapidly detaches so as not to interfere with the action of other heads. Tellingly, this feature is not true for motor proteins such as myosin V or kinesin that have to work processively on their own; these motor heads remain tightly bound for more than 50% of their ATPase cycle.

Muscle is a special case, where evolution has resulted in a system in which motor head behavior can be coordinated for millions of individual molecules throughout an entire cell. More commonly, it is smaller teams of motors that need to work together. For the two-headed processive motors such as myosin V and kinesin, it is important that

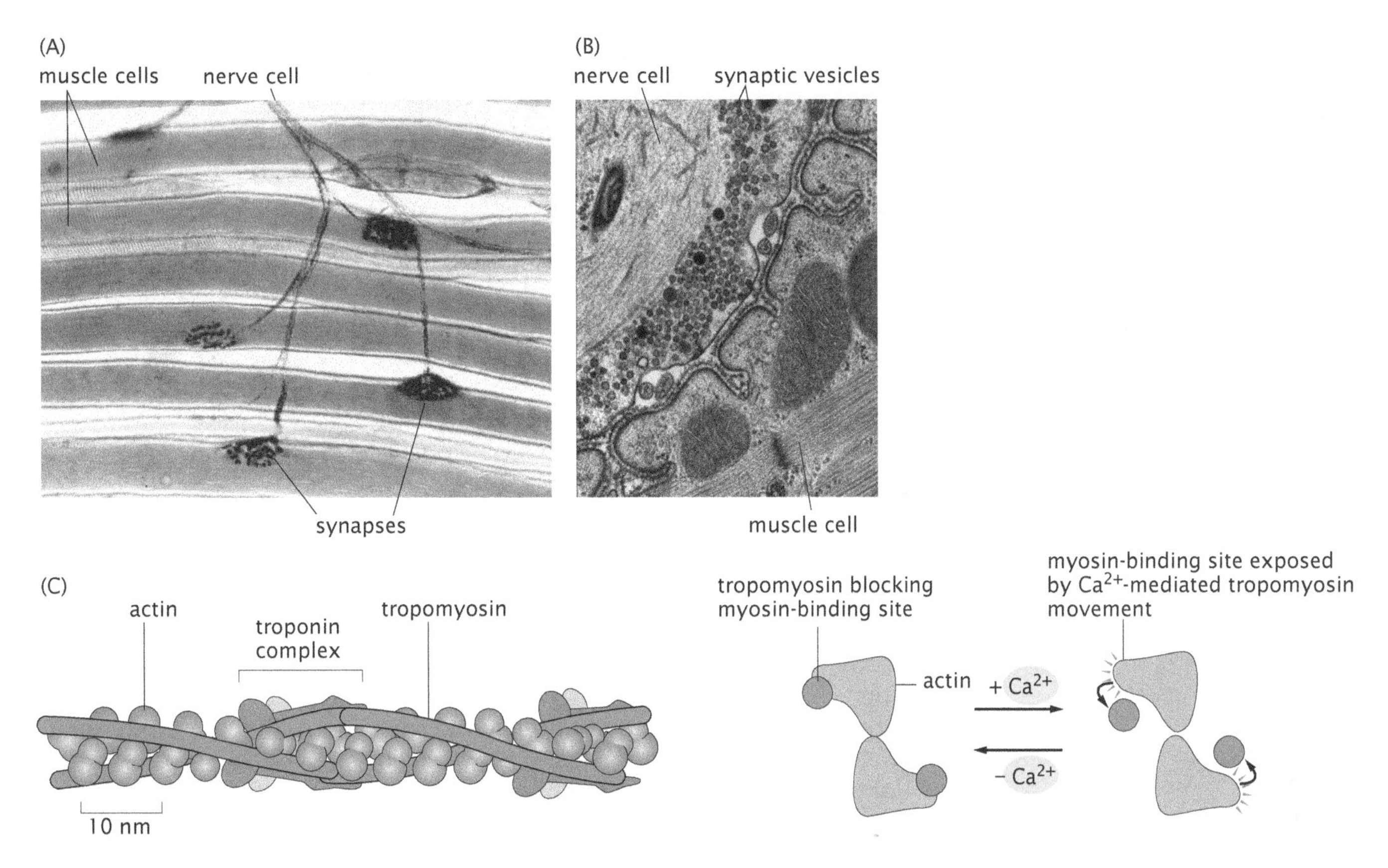

Figure 16.39: Schematic of coordination in muscle contraction. (A) A bundle of muscle fibers is innervated by a single motor neuron that forms communication synapses with all the fibers in the bundle. (B) At each of these synapses, a differentiated portion of the nerve terminal is stuffed with acetylcholine-containing vesicles lined up across the intercellular gap from their target muscle cell. After the neuron releases its neurotransmitters, acetylcholine binding to receptors on the muscle cell triggers a nearly instantaneous influx of calcium ions throughout the entire cytoplasm of the giant muscle cell. (C) In the resting muscle cell, myosin heads are prevented from binding to the actin filaments because the actin is coated with a series of copies of a long, skinny protein called tropomyosin, which is also associated with several small calcium-binding proteins called troponins. When calcium ions bind to troponin, a conformational change causes tropomyosin to shift its position on the actin filament, revealing the myosin-binding sites, which can then be simultaneously attacked by the myosin motor heads poised nearby. (A, courtesy of T. Caceci; B, courtesy of J. Heuser; C, adapted from B. Alberts et al., Molecular Biology of the Cell, 5th ed. Garland Science, 2008.)

the rear head not release from the filament until the front head is firmly bound. Otherwise, the motor and its attached cargo would be at risk of drifting away due to thermal motion, which is significant at molecular distance scales. Thus, the two heads must somehow communicate through their linked domains to influence one another's cycles of ATP hydrolysis and filament binding. The mechanisms of this form of molecular communication are currently under investigation. One interesting approach to this problem is illustrated in Figure 16.40. In this experiment, protein engineering was used to create a series of artifically oligomerized kinesin motor heads, linked by a series of rigid rod-shaped protein domains and elastic spring-like protein domains. While individual kinesin heads were able to generate microtubule gliding at reasonable rates, linked pairs and triplets were able to cooperate to generate progressively faster motion. Cooperation was manifested even by these artificially engineered protein constructs, where the molecular structure of the links between the motor heads bears no resemblance to the links found in any real molecular motors. This result provides hope that the mechanisms of cooperation between linked heads can perhaps be understood in purely mechanical terms.

Frequently, it is necessary for multiple copies of different kinds of motors to work together, for example in the intracellular transport of

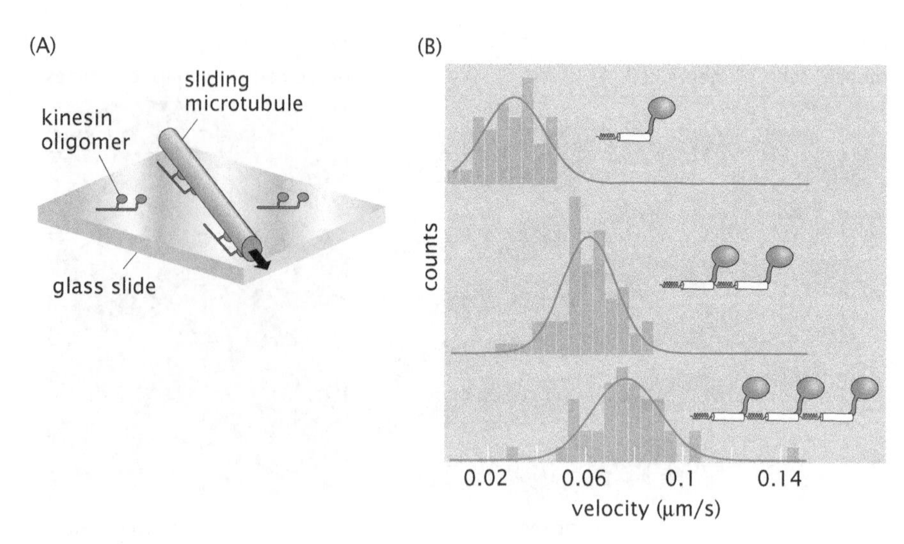

Figure 16.40: Cooperation among linked motors. (A) Schematic of the experiment, showing immobilized kinesin oligomers moving microtubules in a gliding assay. (B) Speed measurement histograms for individual microtubules being moved on slides coated with single kinesin heads (top), linked pairs of kinesin heads (middle), or linked triplets of kinesin heads (bottom). Although the individual molecular motor head units are identical in each case, coupling multiple heads together enables them to generate faster gliding speeds. (Adapted from M. R. Diehl et al., *Science* 311:1468, 2006.)

membrane-bound organelles and vesicles. One example of this kind of transport is shown in Figure 16.41. These cells, melanocytes, contain many small vesicles filled with dark pigment called melanosomes. In some fish and amphibians, the cells can change colors in response to hormonal signals, communicating when the animal is frightened, angry, or sexually aroused. This is accomplished by virtue of the fact that each melanosome is coated with three different types of motor protein: a dynein motor that moves the melanosomes inwards towards the center of the cell along microtubule tracks, a kinesin motor that moves in the opposite direction along the same track, and a myosin V motor that moves along actin filaments at the cell periphery. Strikingly, when a cell receives a hormonal signal, all the melanosomes in the cell switch directions simultaneously. This is because the hormonal signals indirectly induce phosphorylation of proteins that regulate the motors' activity such that either the inward motor or the outward motor predominates. The examination of coordinated motor activities is one of the exciting current frontiers in motor research.

16.2.7 Rotary Motors

Throughout the preceding section, we have developed the idea of position–state models for translational motors, imagining that each energy-requiring step is coupled to the movement of a motor along a linear track. This same formalism can easily be generalized to the second major biological class of motors, the rotary motors, by imagining the track bent into a closed circle. As illustrated in Figure 16.16, rotary motors also move in a series of discrete steps, with each rotational step tightly coupled to an energy-releasing reaction. These are typically either the hydrolysis of ATP or the transport of an ion across a membrane down an electrochemical gradient.

Important mechanical quantities introduced to describe translational motors all have direct analogs in the rotational motor setting. For example, linear speed in the translational context becomes angular velocity in rotary motors, and the force generated becomes torque. Interesting questions to consider then are: how is the angular velocity of a rotary motor related to the applied torque, and how does it depend on the available energy (that is, ATP concentration or electrochemical potential)?

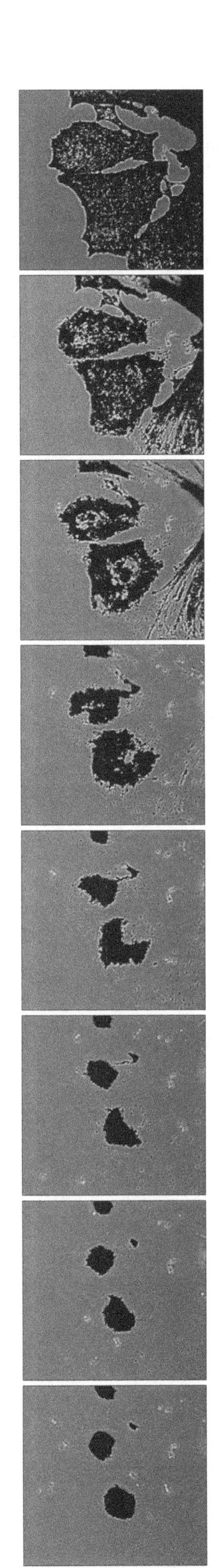

Figure 16.41: Transport of pigment granules. Pigment cells cultured from the skin of a black tetra reorganize their pigment granules after stimulation with adrenaline. Frames shown from a video sequence are separated by approximately 2-minute intervals. The aggregation of the pigment granules in the center of the cell makes the fish appear to change to a lighter color. (Adapted from V. I. Rodionov and G. G. Borisy, *Nature* 386:170, 1997.)

As shown in Figure 16.42, rotary motors are typically large multiprotein machines with a series of similar subunits arranged in a ring. In the example illustrated in this figure, the transport of a single Na^+ ion across the membrane is accompanied by a conformational change in the motor that is equivalent to a single step of the moving ring (called the rotor) with respect to a complementary immobile unit (called the stator).

The motor is driven by thermal fluctuations that result in rotational diffusion of the rotor. Diffusion of the rotor is rectified, leading to directed motion, by electrostatic interactions between the charges on the rotor, the stator, and the Na^+ ions. Specifically, the rotor charge is captured by the stator charge, which is of opposite sign. The captured rotor charge can then diffuse away from the stator charge until it finds itself in the input channel (Figure 16.42), where it is exposed to a large concentration of Na^+ ions on the periplasmic side. The rotor charge can then capture a Na^+ ion from the periplasm, rendering it partially neutral, which allows it to diffuse away from the stator charge. Once it leaves the stator, the Na^+ ion is released on the low-concentration cytoplasmic side. The driving force leading to the rotation of the motor is thus the free-energy difference experienced by the Na^+ ion as it travels from the periplasmic to the cytoplasmic side.

This type of mechanism applies to torque generation by both the bacterial flagellar rotor and the F_0 subunit of ATP synthase. In the case of ATP synthase, we have an interesting opportunity to compare torque generation by ion transport (for F_0) and torque generation by ATP hydrolysis (for F_1).

ESTIMATE

Estimate: Competition in the ATP Synthase One of the remarkable features of the ATP synthase is that it can rotate in either direction depending upon the magnitude of the ion gradient across the membrane and upon the concentrations of ATP, ADP, and P_i. In this estimate, our interest is in examining the magnitude of the torque that can be applied in each of the two directions. In particular, the F_0 unit can be induced to rotate as a result of an ion flux across the membrane, with a 30° rotation resulting from each ion transported. By way of contrast, the F_1 subunit rotates as a result of ATP hydrolysis, with a 120° rotation for each ATP hydrolyzed.

We can estimate the torque produced by the motor in both of these modes of operation by assuming that the entire free-energy change associated with ion transport (F_0) or ATP hydrolysis (F_1) can be converted into useful work. The torque is equivalent to the tilt of the free-energy landscape on which the motor diffuses, and as such serves as the driving force for directed rotational motion, analogously to the way that applied force tilts the free-energy landscape for a linear translation motor. For F_0, we can write

$$\Delta G_{tot} = \Delta G_{pot} + \Delta G_{conc}, \quad (16.52)$$

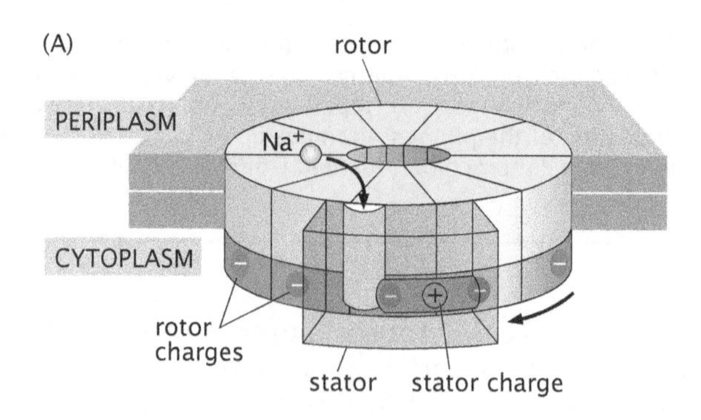

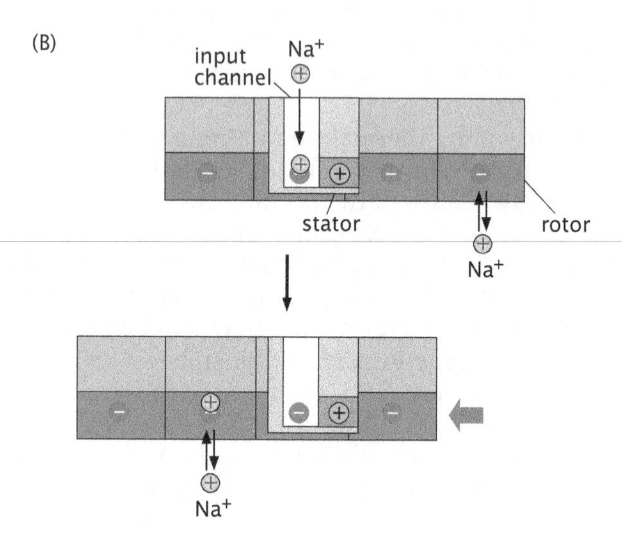

Figure 16.42: Operation of a Na^+ ion transporter. (A) Diagram of the F_0 motor from the bacterium *Propionigenium modestum*. While the F_0 of most bacteria transports H^+ ions, this interesting example uses Na^+ ions instead. The ions move down their electrochemical gradient from the outside of the cell to the inside. The rotor, shown in green, is a ring that spans the membrane. The rotor is made up of a series of protein subunits that each have a critical negative charge on the side of the rotor that is exposed to the cytoplasm. The rotor is surrounded by a series of stator elements, one of which is shown here in brown. Na^+ ions enter an aqueous channel between the rotor and stator from the periplasmic side, but cannot immediately escape into the cytoplasm because the channel is blocked at the bottom. (B) A single step of the rotation is shown schematically. When the rotor is positioned relative to the stator such that one of the rotor's negative charges is positioned near the bottom of the channel, the entering Na^+ ion can associate with the rotor's negative charge and neutralize it. At this point, the rotor can turn one step clockwise (shown here as moving to the left). The neutralizing Na^+ ion is now exposed to the cytoplasm, where it can freely dissociate. At the same time, the next rotor charge is brought into register at the bottom of the channel. A fixed positive charge in the stator element prevents the rotor from turning backward. (Adapted from C. Bustamante et al., *Acc. Chem. Res.* 34:412, 2001.)

where ΔG_{pot} is the free-energy change associated with moving an ion across the transmembrane electrical potential and ΔG_{conc} is the free-energy change associated with the change in entropy corresponding to taking the charge at a concentration c_{out} and moving it to a region with concentration c_{in}. The contribution to the free energy resulting from moving the ion across the transmembrane potential is given by

$$\Delta G_{\text{pot}} = e\Delta V. \tag{16.53}$$

For a typical membrane potential difference of 90 mV, this results in an energy release of roughly 4 k_BT. The entropic contribution to the free-energy release associated with taking ions at the concentration outside the cell and installing them inside the cell is given by

$$\Delta G_{\text{conc}} = k_BT \ln \frac{c_{\text{out}}}{c_{\text{in}}}. \tag{16.54}$$

If we assume perfect free-energy conversion, this means that the entirety of the free-energy difference due to ion transport is available to do work and results in a torque

$$\tau_{F_0} \approx \frac{4\ k_BT}{\pi/6} + \frac{k_BT}{\pi/6} \ln \frac{c_{\text{out}}}{c_{\text{in}}} \approx 30 + 20\Delta\text{pH pN nm}, \tag{16.55}$$

where ΔpH is the difference in pH between the inside and the outside of the cell.

As noted above, the motor can run in reverse if the concentration of ions is sufficiently low and the concentration of

ATP is sufficiently high. When running in reverse, the maximum torque that can be generated by F_1 can be estimated as

$$\tau_{F_1} \approx \frac{\Delta G_0}{2\pi/3} + \frac{k_B T}{2\pi/3} \ln \frac{[ADP][P_i]}{[ATP]} \approx 40 \text{ pN nm}, \tag{16.56}$$

where in the last equation we have made use of the typical value, $20\, k_B T$, for the free energy released during a hydrolysis event of a single ATP molecule under physiological conditions. Comparing the two equations for the torque, we see that by tuning ΔpH, one can reverse the direction of the motor. The cross-over point occurs at approximately $\Delta \text{pH} \approx 0.5$. Healthy *E. coli* normally maintain ΔpH around 0.75, meaning that normally the torque generated by F_0 is stronger and ATP is synthesized. ΔpH can vary from about 2.0 to about −0.25 depending on the pH of the external medium. When the external medium is very alkaline (that is, pH > 8.0), ATP synthase runs backwards.

16.3 Polymerization and Translocation as Motor Action

16.3.1 The Polymerization Ratchet

Until now, our quantitative discussion of motors has primarily centered on translational motors that move along cytoskeletal tracks. Interestingly, the cytoskeletal filaments that serve as the scaffolding for translational motors themselves exhibit motor action. Indeed, cytoskeletal polymerization is one of the cell's main mechanisms for applying forces in particular places. An interesting model for force-generating polymerization is the polymerization ratchet. The basic idea is that the addition of monomers onto the end of a growing filament can result in a pushing force on some resisting barrier, by virtue of fluctuations either in the position of the barrier or in the position of the filament itself. For concreteness, we can think schematically of growing actin filaments pushing against a cell membrane at the leading edge of a crawling cell.

We begin by thinking of a cytoskeletal filament that has grown to a length such that there is no room between it and the resistive barrier. This kind of experiment was illustrated in Figure 10.34 (p. 419). Once the cytoskeletal filament has encountered the barrier, it appears that there is no room for the next monomer to come in and attach. It would seem that the polymerization process would grind to a halt. However, if the barrier jiggles and a new monomer sneaks in now and then when there is room for it to fit, the net result will be an ever-increasing displacement of the membrane. Similarly, the filament itself will fluctuate and might bend away from the barrier so that a new monomer has room to bind. Regardless of the precise mechanism, the fluctuation is "frozen" in, leading to the generation of a deforming force that acts on the membrane. The net result is that a growing actin filament will do work against the elastic forces that would like to keep the barrier undeformed.

Based on the ratchet picture described above and using thermodynamic reasoning, we can estimate the maximum force that the growing filament can exert. In the absence of a force, the free-energy landscape for monomer addition looks like the schematic shown in Figure 16.43.

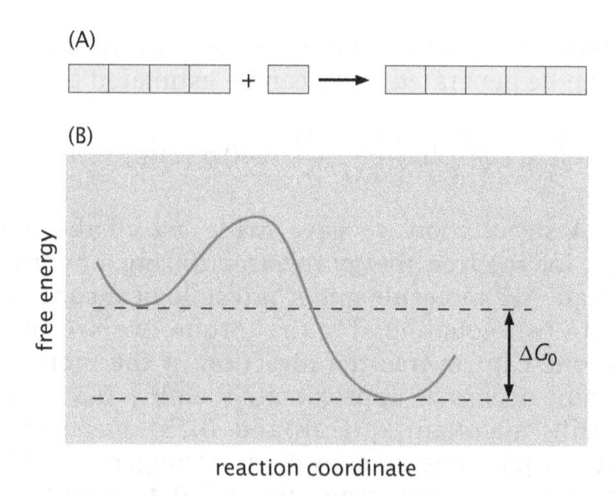

Figure 16.43: Free-energy picture of equilibrium polymerization. The addition of each monomer to the growing filament leads to a decrease in the net free energy of the system. Cells can harness the free-energy decrease associated with protein polymerization to do useful mechanical work.

In the presence of a force, the addition of a new monomer has an additional free-energy cost because of the work $F\delta$ that it has to do in the presence of the barrier. This is simply the work that the filament does against the applied force over a distance equal to the length of the monomer. In terms of a free-energy diagram, as shown in Figure 16.44, the effect of the force is to raise the free energy of the on state compared with the off state by an amount $F\delta$. This change in the free-energy difference between the two states leads to an increase in the dissociation constant for the polymerization reaction,

$$K_{\mathrm{d}}(F) = K_{\mathrm{d}}(0)\mathrm{e}^{F\delta/k_{\mathrm{B}}T}. \tag{16.57}$$

As discussed in Chapter 15, the dissociation constant is equal to the monomer concentration m^* at which the average filament length is not changing in time. Therefore the maximum force that can be generated by the growth of a filament when the monomer concentration is m is given by the relation

$$m = K_{\mathrm{d}}(0)\mathrm{e}^{F_{\mathrm{max}}\delta/k_{\mathrm{B}}T}. \tag{16.58}$$

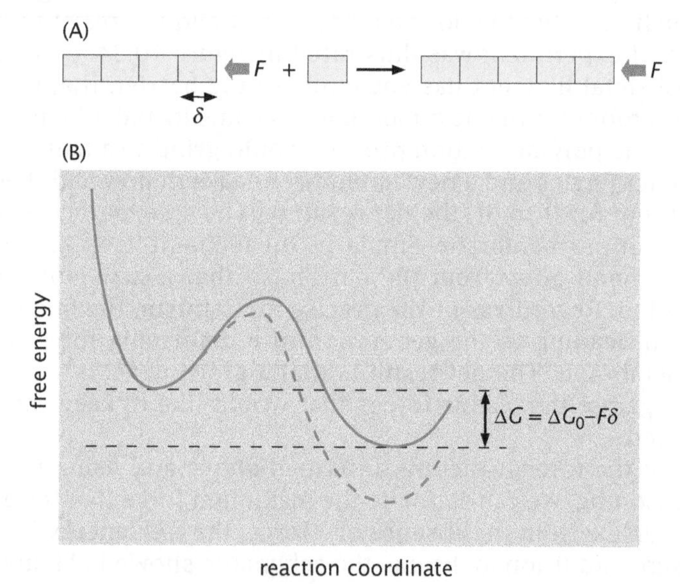

Figure 16.44: Polymerization against a force. The presence of an applied force tilts the free-energy landscape and makes the addition of the next monomer less favorable than it would be in the absence of a force.

Solving this equation for the maximum force, we arrive at

$$F_{\text{max}} = \frac{k_B T}{\delta} \ln \frac{m}{m^*}, \tag{16.59}$$

where $m^* = K_d(0)$ is the critical concentration in the absence of force.

Using this simple formula, we can obtain estimates for the maximum force exerted by polymerization of actin and microtubules. Inside living cells, the concentration of actin monomers is estimated to be $\approx 20\,\mu\text{M}$, that is, $m = 100m^*$, and the length δ that the filament grows by addition of a single monomer is given by

$$\delta \approx \frac{5.4 \text{ nm}}{2 \text{ protofilaments}} = 2.7 \text{ nm}; \tag{16.60}$$

we get the maximum force as

$$F_{\text{max}} = \frac{4 \text{ pN nm}}{2.7 \text{ nm}} \times \ln 100 \approx 7 \text{ pN}. \tag{16.61}$$

For microtubules, for the sake of an estimate, we take excess tubulin concentration $m = 100m^*$ to be of the same order as the excess actin concentration, with a length change contributed by addition of a single monomer given by

$$\delta = \frac{8 \text{ nm}}{13 \text{ protofilaments}} \approx 0.6 \text{ nm}. \tag{16.62}$$

Hence, the critical force is given by

$$F_{\text{max}} = \frac{4 \text{ pN nm}}{0.6 \text{ nm}} \times \ln 100 \approx 30 \text{ pN}. \tag{16.63}$$

This result suggests that, all other things being equal, polymerization motors that take smaller unit steps are able to generate larger forces. In real cells, it is sometimes observed that individual growing or shrinking microtubules may exert biologically significant forces, while actin filaments almost always work in groups.

One limiting factor that is not taken into account by the previous estimate is that the maximum force exerted by a polymerizing filament might not be attained if the filament buckles. While the polymerization force generated by a single filament is constant regardless of filament length, longer filaments are more easily buckled than shorter filaments. As we discussed in Chapter 10 and derived in Equation 10.69 (p. 421), we can estimate the buckling force as

$$F_{\text{buckle}} \approx \frac{k_{\text{flex}}}{L^2}, \tag{16.64}$$

where the flexural rigidity of the filament is the parameter that characterizes its resistance to bending, and it is related to its persistence length via $k_{\text{flex}} = \xi_p k_B T$. From this formula, we see that k_{flex} has units of energy × length, and therefore we must divide by length^2 to obtain a force.

Since the buckling force is inversely proportional to the length squared, it is only very short filaments that will be able to sustain a load equal to F_{max} without buckling. The critical length is determined by setting the two forces equal. For an actin filament with a persistence length of 15 μm, setting $F_{\text{buckle}} = F_{\text{max}}$ and using Equation 16.64 yields

$$\frac{4 \text{ pN nm} \times 15\,\mu\text{m}}{L_{\text{max}}^2} = 7 \text{ pN} \tag{16.65}$$

and $L_{max} \approx 0.1\ \mu m$ is the maximum length that can withstand buckling. Arrays of growing actin filaments that need to exert a force inside cells are usually linked together by crosslinking proteins, forming bundles or networks where the average length of a free filament end is much shorter than this limit, as shown in Figures 14.2 (p. 547) and 15.3 (p. 577). We can make the analogous estimate for microtubules, for which $F_{max} \approx 30$ pN, and which are considerably stiffer than F-actin, with a persistence length of 3 mm, to obtain $L_{max} \approx 0.6\ \mu m$.

The Polymerization Ratchet Is Based on a Polymerization Reaction That Is Maintained Out of Equilibrium

The thermodynamic arguments put forward above give upper bounds on the size of the force that a growing filament can exert on a load. In order to judge the usefulness of this estimate, as well as gain quantitative intuition about the time scales involved in filament polymerization against a load, we require a dynamical model that can provide an estimate for the length of time that it will take to add each monomer to a growing filament. At a molecular scale, it is likely that thermal motion of some element in the system will be sufficient to open up a monomer-sized space between the end of the filament and the barrier load. The general class of kinetic models that assume thermal motion as a rate-limiting step is commonly referred to as the thermal ratchet or Brownian ratchet model.

A schematic of one form of the Brownian ratchet is shown in Figure 16.45. A polymer made up of monomers of length δ is growing near a load object, such as the cell membrane. Each time thermal fluctuations cause the gap between the polymer and the load to be greater than δ, a monomer can attach itself to the growing tip, thereby effectively pushing the load forward. One of the key quantities we wish to compute is the velocity with which the load will move.

In the absence of a force on the load object, we can obtain an estimate of its velocity by making two simplifying assumptions, namely, (i) once monomers attach they do not fall off and (ii) monomers attach immediately upon appearance of a gap that is larger than the monomer size δ. The time for the load to diffuse over a distance δ can be estimated from its diffusion constant, $t_D = \delta^2/2D$. Using the simplifying assumptions given above, we can estimate the velocity on the grounds that after this time the load will have displaced permanently by an amount δ. Therefore its average ("ideal") velocity is $v_{id} = \delta/t_D$ and

$$v_{id} = \frac{2D}{\delta}. \tag{16.66}$$

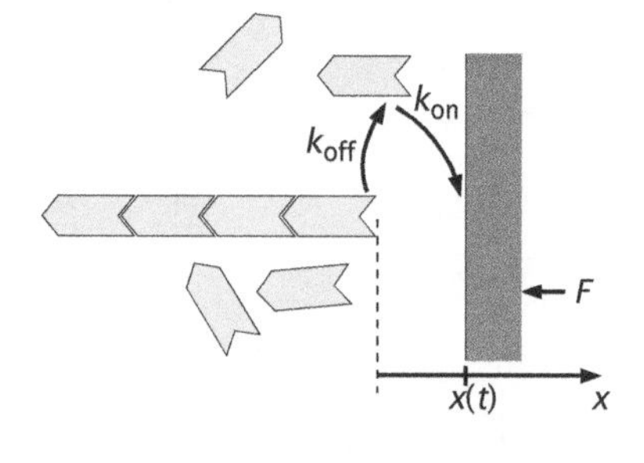

Figure 16.45: Polymerization Brownian ratchet. The growing filament induces a displacement of the obstacle, which pushes back with a force F. In this particular model, the barrier is able to move back and forth along the x-axis due to thermal fluctuations, occasionally opening up a large enough space for the insertion of a new monomer at the growing tip.

The result is that the ideal velocity of the tip must be proportional to the diffusion constant of the load object.

This idea has been tested experimentally, by using different sizes of bacteria or of polystyrene beads coated with bacterial proteins to generate polymerization-driven actin comet tails such as those shown in Figure 15.3 (p. 577). Contrary to the predictions of this very simple model, the speed depends only weakly on the size of the load object, and is not proportional to its diffusion coefficient. An alternative model in which the fluctuations of the growing filament itself are responsible for opening up monomer-sized gaps has been proposed to account for this discrepancy, and makes more accurate predictions of how speed depends on varying experimental conditions. A further complication of real systems compared with this idealized treatment is the influence of other forces on speed, including adhesive forces connecting a subset of the growing actin filaments to the load object. Nevertheless, regardless of details, the basic idea of ratcheting thermal fluctuations of some mechanical element of the system (including the load and the filament tips) by maintaining a nonequilibrium concentration of monomers is a good starting point for building a quantitative model of the polymerization motor.

We can generalize the kinetic model in a way that is applicable to any of these particular implementations of the idea of the Brownian ratchet by calculating the speed in terms of the probability per unit time that a monomer will be added to the growing filament, regardless of the precise physical mechanism that opens up the space. Here, we need to consider two slightly different kinetic scenarios, illustrated in Figure 16.46. In the first regime, termed monomer-addition-limited, the rate-limiting step for elongation of a filament is simply the time it takes for a diffusing monomer to attach to the filament tip. In the second regime, termed gap-opening-limited, the opening of a monomer-sized gap between the filament and the load is rare, but as soon as it opens, a monomer is incorporated into a filament. In either case, the probability that a monomer attaches to a filament in time interval Δt can be written as the probability that an attachment is allowed times the probability that an attachment will occur *assuming* it is allowed, $\Delta p(\text{on}) = p(\text{allowed}) \times k_{on} m \Delta t$. The probability that an attachment is allowed is equal to the probability that the load position from the edge of the growing filament is greater than δ. In the monomer-addition-limited regime, the time between attachments is long enough so that the load can sample all possible positions. Therefore, the probability that an attachment is allowed is equal to the *equilibrium* probability that the load position is greater than δ, $p_{eq}(x > \delta)$.

Now we can write the average polymerization rate as

$$v = \delta[k_{on} m p_{eq}(x > \delta) - k_{off}]. \tag{16.67}$$

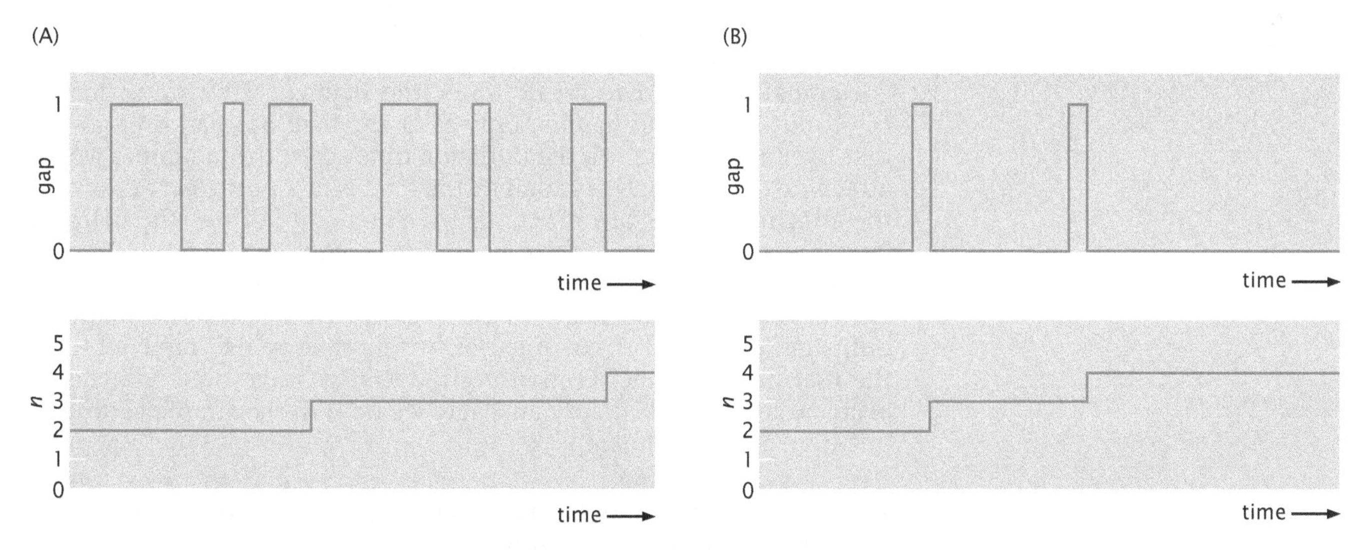

Figure 16.46: Trajectories for the gap between the load and the filament juxtaposed with polymer length (bottom, n) for (A) the monomer-addition-limited and (B) the gap-opening-limited regimes of the simple Brownian ratchet. When the elongation reaction is monomer-addition-limited, the load is frequently in a position favorable for monomer addition (gap = 1), but monomers attach to the filament end comparatively rarely. When the elongation reaction is gap-opening-limited, the load is rarely in a position favorable for monomer addition, but monomer addition occurs every time it is permitted. Note that when a monomer is added, the gap closes (gap = 0).

The equilibrium probability that the gap between the polymer and the load is greater than x is given by

$$p_{\mathrm{eq}}(x > \delta) = \int_{\delta}^{\infty} p_{\mathrm{eq}}(x)\,\mathrm{d}x, \tag{16.68}$$

where

$$p_{\mathrm{eq}}(x) = \frac{F}{k_{\mathrm{B}}T}\mathrm{e}^{-Fx/k_{\mathrm{B}}T} \tag{16.69}$$

is the equilibrium probability of finding the load at position x away from the growing tip of the filament. Doing the integration in Equation 16.68 and substituting the result into Equation 16.67 leads to an expression for the force-dependent polymerization rate in the monomer-addition-limited regime,

$$v = \delta[k_{\mathrm{on}} m \mathrm{e}^{-F\delta/k_{\mathrm{B}}T} - k_{\mathrm{off}}]. \tag{16.70}$$

An implicit assumption we have made all along is that the off rate is not affected by the force. One experimentally testable prediction of the model is the exponential fall-off of the rate on applied force with a characteristic force of $k_{\mathrm{B}}T/\delta$. For microtubules, the characteristic force is approximately 7 pN, while the experiments find a corresponding value of roughly 2 pN, which is reasonably good agreement considering the extreme simplifications incorporated in the model. Also, note that the model predicts the maximum force equal to that predicted earlier from equilibrium considerations. At this value of the force, the filament switches from growth to shrinkage.

To make further progress, we now switch our attention to the assumption that the monomer-addition time is much less than the gap-opening time. In this case, a movie of the growing filament would reveal rapid fluctuations of the filament between states with length n and $n-1$ with the filament in the n state most of the time (this assumes that $k_{\mathrm{on}} m > k_{\mathrm{off}}$), as shown in Figure 16.46. (One might observe shorter filament lengths as well, but these would require two consecutive events with a monomer falling off the filament, which is considerably less likely than an off event followed by an on event.)

Eventually the movie would show the filament growing to length $n+1$, but this would happen only after the load has had time to diffuse over a distance δ. On the diffusion time scale, the monomer would attach instantaneously, as soon as the load is at a position to allow for the attachment of a new monomer. At this point in time, the filament length would begin to fluctuate between lengths $n+1$ and n. We conclude that in this gap-opening-limited regime, the average filament growth rate would be $1/\tau_1$, where τ_1 is the average time it takes the diffusing load to reach the position δ; this time is also referred to as the first passage time. Computing the first passage time is a classic problem in diffusion theory and here we do it in the simple setting of one-dimensional diffusion.

The Polymerization Ratchet Force–Velocity Can Be Obtained by Solving a Driven Diffusion Equation

As a warm-up exercise, we compute the first passage time in the absence of a force. We consider the diffusion equation for the steady-state probability $p(x)$ of finding a particle at position x within an interval of width δ as shown in Figure 16.47. We assume that when

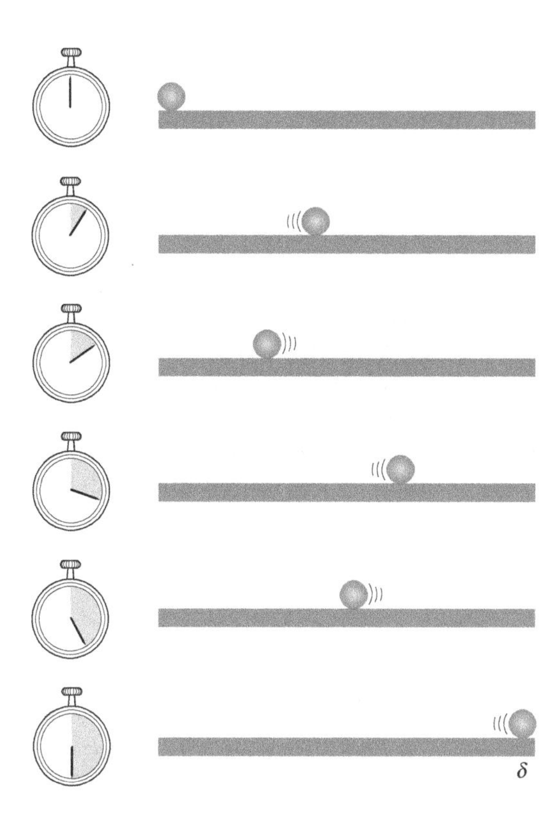

Figure 16.47: Mean first passage time. In the one-dimensional version of this problem, a particle starts at the origin and diffuses along a line. The time it takes for the particle to reach the position δ for the first time is the so-called first passage time. The average of this time over many trajectories is the mean first passage time.

a particle crosses the $x = \delta$ boundary, it disappears and then immediately reappears at $x = 0$. This condition insures that the probability of finding the particle anywhere in the interval $(0, \delta)$ is equal to 1. Furthermore, with this boundary condition the mean rate at which the particle reaches the boundary at $x = \delta$ starting at $x = 0$ is nothing but the steady-state diffusion current $j_0 = -D\partial p/\partial x$.

The steady-state diffusion equation states that

$$\frac{\partial^2 p}{\partial x^2} = 0, \tag{16.71}$$

which means that the probability is a linear function of position, that is, $p(x) = Ax + B$. To compute the coefficients A and B, we make use of two conditions, namely, first we have the normalization condition,

$$\int_0^\delta p(x)\,\mathrm{d}x = B\delta + A\frac{\delta^2}{2} = 1, \tag{16.72}$$

while the second condition insures that the probability vanishes at $x = \delta$. This can be written as

$$p(\delta) = A\delta + B = 0; \tag{16.73}$$

this is the condition that the particle will vanish when it reaches this boundary. From these two equations, we find $A = -2/\delta^2$ and $B = 2/\delta$. The current is therefore

$$j_0 = -DA = \frac{2D}{\delta^2}, \tag{16.74}$$

while the first passage time is the inverse of this quantity, $\tau_1 = \delta^2/2D$. The filament polymerization rate is then finally

$$v = \frac{\delta}{\tau_1} = \frac{2D}{\delta}. \tag{16.75}$$

Now we can repeat the calculation in the presence of a force acting on the diffusing particle (in the case of the polymerization ratchet, this would be the load). In this case, the current also has a contribution from the drift produced by the applied force,

$$j_0 = -D\frac{\partial p}{\partial x} - \frac{F}{\gamma}p, \tag{16.76}$$

where γ is the friction coefficient that relates the force to the velocity. The negative sign in front of the drift term is due to the fact that the force is taken to point in the negative x-direction. The friction coefficient is related to the diffusion constant by the Einstein relation, $\gamma D = k_B T$. All of these ideas for treating driven diffusion were developed in Sections 13.2.4 (p. 529) and 13.2.5 (p. 530).

In steady state, the current is constant and we can solve the differential equation for $p(x)$ in terms of the unknown current j_0. The solution is the sum of the general solution of the homogeneous equation and a particular solution to the inhomogeneous equation,

$$p(x) = \underbrace{Ae^{-Fx/k_B T}}_{\text{homogeneous}} - \underbrace{j_0\frac{\gamma}{F}}_{\text{inhomogeneous}}. \tag{16.77}$$

Since once again, $p(\delta) = 0$ and $\int_0^\delta p(x)\,dx = 1$ we get two equations for the two unknown constants, A and j_0, namely,

$$Ae^{-F\delta/k_B T} - j_0\frac{\gamma}{F} = 0,$$

$$-A\frac{k_B T}{F}(e^{-F\delta/k_B T} - 1) - j_0\frac{\gamma}{F}\delta = 1.$$

These two equations yield a steady-state current of the form

$$j_0 = \frac{1}{(k_B T\gamma/F^2)\left(e^{F\delta/k_B T} - 1\right) - \gamma\delta/F} \tag{16.78}$$

which, like in the no-force case, leads to an expression for the polymerization rate of the form

$$v = \delta j_0 = \frac{D}{\delta}\frac{(F\delta/k_B T)^2}{e^{F\delta/k_B T} - 1 - F\delta/k_B T}. \tag{16.79}$$

It is comforting to find that in the limit $F\delta \ll k_B T$, this expression reduces to the one obtained in the zero-force limit. This makes intuitive sense since in this case the thermal energy is so large that the diffusing "particle" does not "know" about the potential.

In the other limiting case, when $F\delta \gg k_B T$, the formula for the polymerization rate reduces to $v = [F^2 D\delta/(k_B T)^2]e^{-F\delta/k_B T}$. This corresponds to a first passage time for the "particle" that is of the form $\tau_1 = \delta/v = [(k_B T/F)^2/D]e^{F\delta/k_B T}$. The prefactor is the time to diffuse a distance $k_B T/F$, while the exponential Arrhenius factor accounts for the probability that the particle finds itself with energy $F\delta$ while in thermal equilibrium at temperature T.

16.3.2 Force Generation by Growth

Polymerization Forces Can Be Measured Directly

Several techniques have been developed enabling direct measurement of forces generated by polymerizing filaments. The first direct measurement of force generated by polymerization of a single microtubule used the elastic beam properties of the microtubule itself as a transduction device to measure force. We have already discussed this experiment in Figure 10.34 (p. 419), but there we focused on calculation of the force needed to bend a microtubule. In actuality, the purpose of this experiment was not to measure the bending force, which was known through other means, but rather to measure the polymerization force generated at the interface between the growing microtubule and the wall, using the bending only as a readout. This classic experiment demonstrated that an individual microtubule can generate several piconewtons of actual pushing force while growing in this constrained geometry, as shown in Figure 16.48. This magnitude is similar to the force generated by a single kinesin or dynein motor as discussed earlier in this chapter.

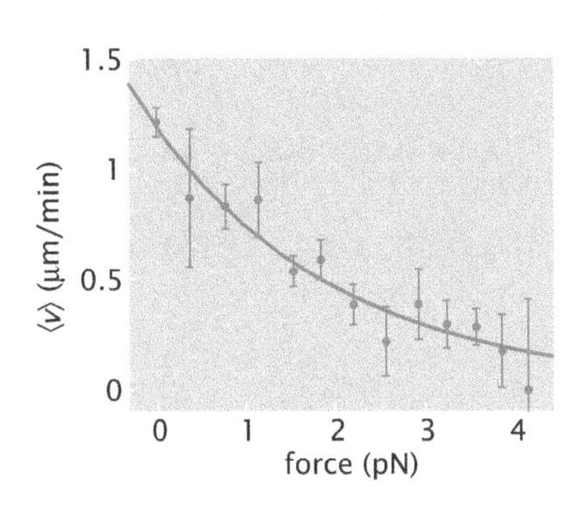

Figure 16.48: Average polymerization velocity for growing microtubules pushing up against a rigid barrier, as a function of applied force. In this experiment, the force was applied by the elastic bending of the growing microtubule itself. As for other molecular motors, the speed of the polymerization motor decreases as the force increases. (Adapted from M. Dogterom and B. Yurke, *Science* 278:856, 1997.)

Some complications arise from using the growing microtubule itself as the device to measure force generation. In particular, if the stiffness of the microtubule changes as it grows, a proposition for which there is experimental support, this may lead to an incorrect deduction of the force. It is therefore useful to have an independent way of measuring the polymerization force. This has been achieved using an optical trap as shown in Figure 16.49. As we have seen, optical traps are well suited for measuring forces in the piconewton range, with step sizes of several nanometers. These optical-trap-based measurements are in excellent agreement with the measurements based on microtubule bending. This demonstrates that growth-dependent changes in microtubule flexibility are insufficient to significantly alter the interpretation of these measurements.

In vitro, it is possible to measure forces generated by single motors, whether translational or polymerization motors. However, in cells, it is rare for any kind of motor to operate in isolation. Most biological force generation involves arrays of cooperating force-generating elements. Optical traps are not strong enough to measure the forces provided by a network of actin filaments. For this kind of measurement, the cantilever-based microscopies described in Chapter 10

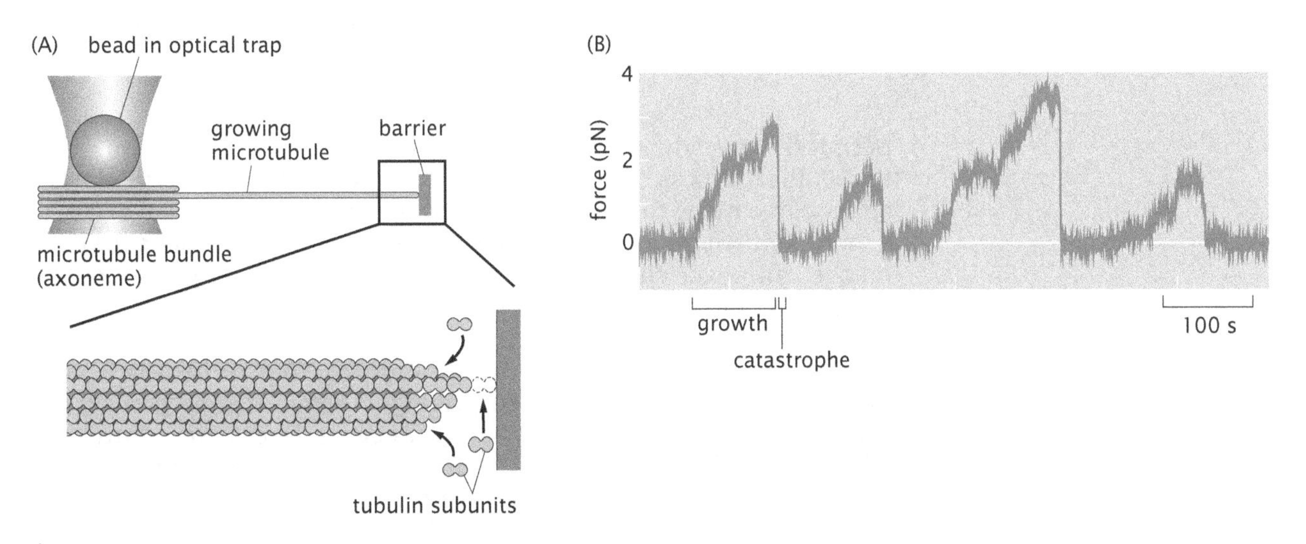

Figure 16.49: Optical-trap measurement of the force of microtubule polymerization. (A) A bundle of microtubules is attached to a bead, which is then held in an optical trap. The end of the trapped bundle is brought into close proximity of a rigid, nanofabricated wall. Addition of soluble tubulin dimers permits growth at the interface between the bundle and the wall, pushing the bead backward in the optical trap. (B) An individual trajectory measured for a single experiment reveals phases of microtubule growth followed by force-induced catastrophe. (Adapted from J. W. Kerssemakers et al., *Nature* 442:709, 2006.)

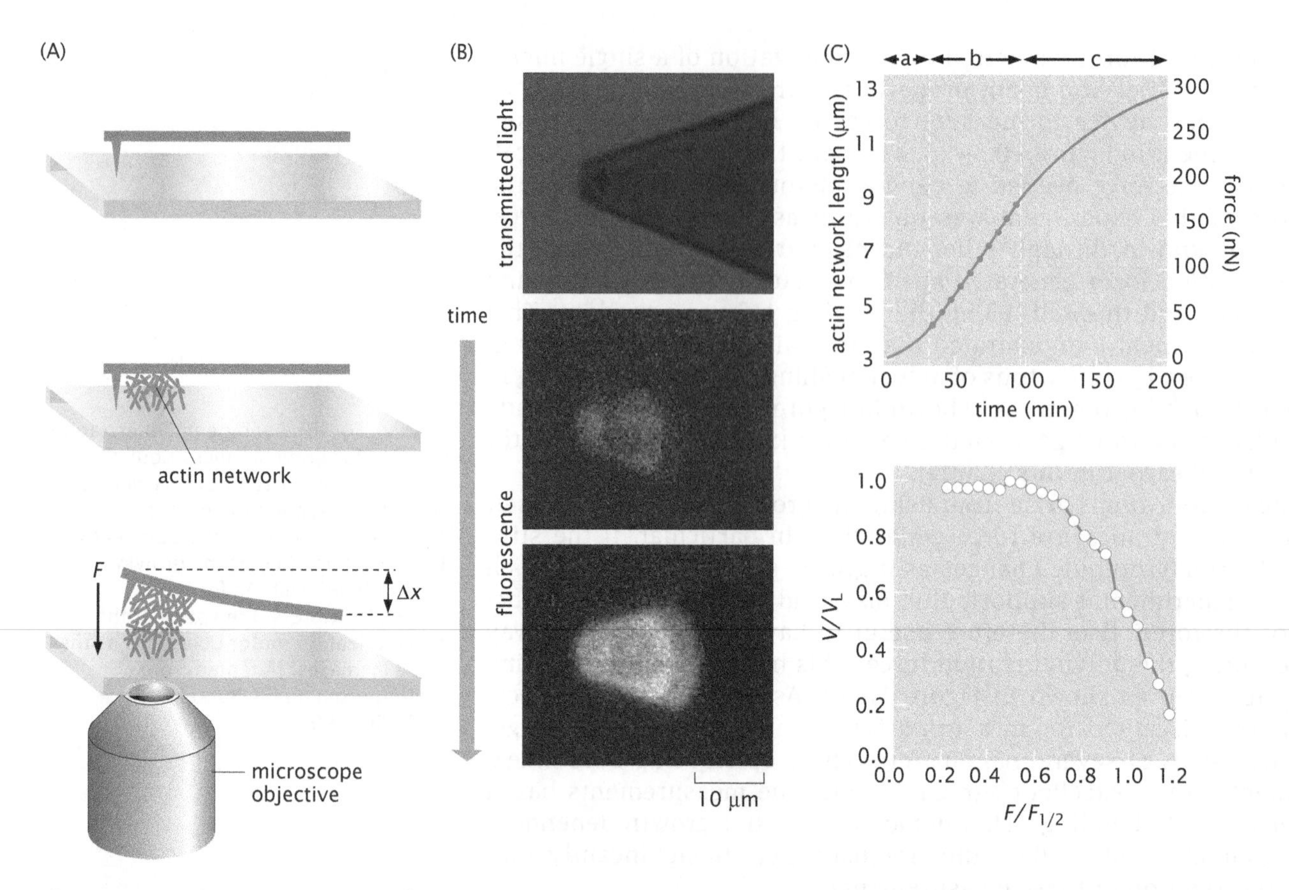

Figure 16.50: AFM measurement of polymerization forces. (A) Schematic of the experiment. A cantilever coated with a protein that promotes actin nucleation is brought close to a glass surface. When a solution is added containing actin monomers and crosslinking proteins, network growth is initiated on the cantilever. Continued network growth deflects the cantilever upwards. (B) Images of the cantilever photographed through the glass slide showing the accumulation of fluorescently labeled actin as the experiment progresses. (C) The top graph shows the total length of the actin network as a function of time. As the gel grows, the cantilever is deflected, and the amount of applied force increases. This deflection occurs in three phases. In the initial phase (a), the actin is just beginning to assemble and the deflection accelerates over time. During the intermediate phase (b), the velocity of network growth remains nearly constant; the dots show a straight-line fit to this portion. Finally, in the terminal phase (c), the growth velocity slows as the total force approaches the stall force. The bottom graph shows the same data (from phases b and c) presented as a force–velocity curve for this experiment. Here the growth velocity shown on the vertical axis has been normalized to the maximum velocity V_L, and the force shown on the horizontal axis has been normalized to the the force required to decrease the velocity to half its maximum value, $F_{1/2}$. The flat plateau on the left part of the graph demonstrates that network growth velocity remains constant even as force increases. (Adapted from S. H. Parekh et al., *Nat. Cell Biol.* 7:1219, 2005.)

permit the investigation of much larger forces. As we showed in Chapter 10, the stiffness of a cantilever is a simple function of its thickness, so the force range can be tuned. Figure 16.50 shows a cantilever-based experiment where growth of an actin network was triggered to occur on the cantilever surface held in close proximity to a glass surface. Growth of the network deflecting the cantilever can be used to measure both force and growth velocity. This experiment demonstrated that arrays of actin filaments do not act together to generate force in a simple additive manner. Surprisingly, network growth occurs at a nearly constant rate over a large range of forces, in contrast to all the individual molecular motors discussed so far, which exhibit a monotonically decreasing force–velocity curve.

Polymerization Forces Are Used to Center Cellular Structures

The fact that cytoskeletal filament growth can produce pushing forces can be used by a cell to do much more than simply move. In particular,

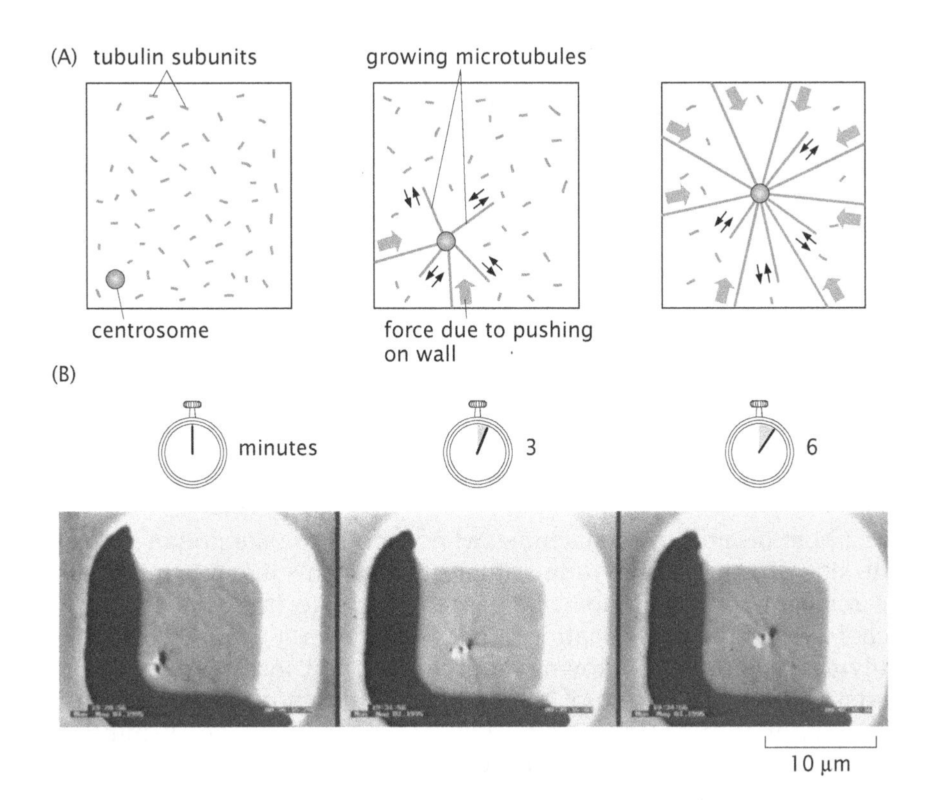

Figure 16.51: Self-centered centrosomes. (A) Diagrams showing time sequence of a self-centering experiment. Initially, a centrosome is added to a microfabricated square well along with soluble tubulin subunits. As the centrosome nucleates growth of microtubules, individual microtubules grow and shrink in a process of dynamic instability. When a microtubule tip contacts the wall of the chamber, it can continue to grow, resulting in a pushing force on the centrosome. Eventually, as microtubules grow longer and push against the wall in all directions, the centrosome finds a stable position of the geometrical center of the well. (B) Frames from a video sequence using differential interference contrast (DIC) microscopy show this process over a period of several minutes. (Adapted from T. E. Holy et al., *Proc. Natl Acad. Sci. USA* 94:6228, 1997.)

this phenomenon is exploited by cells to set up a universal coordinate system whereby they are able to position their organelles at precise geographical locations in the enormous cellular volume. Specifically, if microtubules within a cell are labeled fluorescently, it can be seen that they typically emanate in a star-like array from a single point known as the centrosome. The centrosome is a tiny object, less than 0.5 μm across, that nonetheless can position itself near the middle of a cell with typical sizes of tens of microns. How does the centrosome find the cell center? One possible mechanism can be demonstrated by a clever experiment illustrated in Figure 16.51 that involves isolating a centrosome and dropping it into a nanofabricated hole with dimensions comparable to those of a cell. The centrosome on its own diffuses around aimlessly. However, if tubulin is added so that microtubules can grow from the centrosome, the centrosome quickly (within minutes) zooms in to the geometric center of the hole, regardless of the hole shape. If the centrosome is grabbed with a laser trap and displaced from its central location, it will gently but insistently return to the center. The mechanism relies on the pushing forces at the tips of all of the microtubules when they run into the walls. Only when the centrosome is at the geometrical center of the enclosure do all of the forces cancel out. In living cells of the fission yeast *Schizosaccharomyces pombe*, this mechanism has been directly observed centering the nucleus halfway down this rod-shaped cell by virtue of microtubules pushing against the two opposite ends.

16.3.3 The Translocation Ratchet

Another fascinating kind of motor action introduced at the beginning of this chapter is associated with translocation. We argued that because of the division of cells into different compartments, there

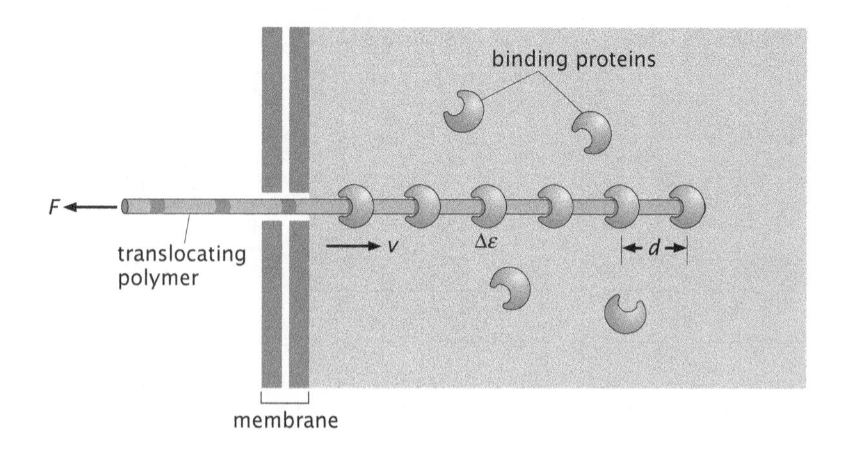

Figure 16.52: Translocation ratchet. Motion of a polymer across a membrane is ratcheted by the presence of binding proteins that prevent the back motion of the polymer.

are a host of molecular machines whose job is to take polymers from one side of a membrane to the other (mechanisms like this might also be relevant for macromolecular assemblies like the proteasome). A schematic of the translocation process is shown in Figure 16.52. The polymer of interest is moving from left to right in the presence of a resistive force. By virtue of the action of binding proteins, the diffusive motion is ratcheted. The goal of this subsection is to examine the relation between the force and the velocity.

Protein Binding Can Speed Up Translocation through a Ratcheting Mechanism

A rough quantitative feel for the speed-up due to the presence of binding proteins and their associated ratchet action can be obtained by comparing the diffusion time in the absence and presence of the binding proteins. If we consider a polymer of length $L = nd$, then the time scale for diffusive motion over the entire length is

$$t_{\text{diffusion}} \approx \frac{n^2 d^2}{D}, \tag{16.80}$$

where d is the mean spacing between binding sites. The diffusive motion of a biopolymer traversing such a membrane will be sped up as a result of the binding of molecules to the polymer at a progression of sites once they have crossed the membrane. In this case, the translocation time takes the form

$$t_{\text{translocate}} \approx n\frac{d^2}{D}. \tag{16.81}$$

This expression can be argued for simply by noting that d^2/D is the time for a single binding site to diffuse over a distance d. Assuming that this exposes the binding site to a binding protein, which then binds and prevents any backward motion of the polymer past this position, the total time for translocation is the time for all n binding sites to complete this process. That is, the speed-up results from the fact that it is faster to make n diffusive trajectories of overall length d than it is to make a single diffusive trajectory to diffuse an overall distance nd.

In the case of a bacteriophage injecting its DNA into a bacterial cell, the binding of RNA polymerases can serve to produce a ratchet effect (at the same time, of course, the RNA polymerase also performs the critical service of beginning to express the genes of the bacteriophage). For DNA with a characteristic length of 10 μm, typical of a genome of a bacterial virus, and with roughly one polymerase binding site per micron, the speed-up due to polymerase binding will be $n = L/d = 10$. This will in fact be an upper bound given the fact that all binding is reversible and once in a while an RNA polymerase will detach from the DNA, spoiling the ratcheting action. Because RNA polymerase is also a translational motor, it can probably also contribute to injection of the DNA genome by active energy-dependent movement. As we have previously discussed in Chapter 10, the differential pressure between the bacteriophage capsid and the host cell cytoplasm can also contribute to the injection of DNA. However, other biological systems, including the import of newly translated proteins into the lumen of the endoplasmic reticulum and the delivery of cytoplasmically expressed proteins into the various compartments of the mitochondria, appear to use mechanisms that can be schematized as translocation ratchets even though the relevant binding proteins are not themselves translational motors and there is no significant pressure differential.

To make these estimates more precise, we model the polymer as a rod diffusing through an opening (channel) in the membrane, the translocation pore. The polymer contains binding sites for proteins that cannot pass through the translocation pore, which are a distance d apart. We make the further simplifying assumption that the proteins are only present on one side of the membrane in the cellular compartment to which the polymer in question is translocating. We also assume that the binding of proteins is irreversible; a more realistic model would take into account the on and off rates for protein binding. The translocation process then proceeds in the following manner. The translocating polymer, buffeted by the solution, undergoes diffusion through the translocation pore. By pure chance, a protein-binding site will find itself on the side where the binding proteins are present. If a protein then binds, the polymer will be prevented from diffusing back and it will have translocated by distance d. We can use this basic picture to calculate the average flux of protein-binding sites through the translocation pore.

To address the question of translocation velocity, we focus on the protein-binding site that has *last* emerged from the pore. To describe the stochastic motion of this binding site, we introduce the quantity $p(x, t)\,\mathrm{d}x$, which is the probability of finding the last binding site to have crossed the pore at position $(x, x + \mathrm{d}x)$ at time t. The translocation problem formulated in this way now maps to the Smoluchowski equation for the variable $p(x, t)$, the probability density. That is, the flux of protein binding sites is given by

$$J_x = \underbrace{-\frac{DF}{k_B T} p}_{\text{drift}} - \underbrace{D\frac{\partial p}{\partial x}}_{\text{diffusion}}, \tag{16.82}$$

where we have included a drift term to account for the possible presence of a force acting on the polymer during translocation; for $F > 0$, this is a force that opposes translocation. In the example of DNA translocation during viral infection, a negative force is supplied during translocation by the host cell's RNA polymerases, which help the translocation process along, speeding up the cell's demise.

The Translocation Time Can Be Estimated by Solving a Driven Diffusion Equation

The differential equation that prescribes the evolution of the binding-site probability is once again the Smoluchowski equation,

$$\frac{\partial p}{\partial t} + \frac{\partial J_x}{\partial x} = 0, \tag{16.83}$$

which in this case is the statement of the conservation of the number of binding sites, and is, upon substituting Equation 16.82, the diffusion equation for $p(x, t)$. "Well this is strange," a careful reader might complain, "we've arrived at the diffusion equation once again, so where is all the information about proteins, their binding sites, etc." The answer is that this is what the boundary conditions are for! The boundary condition appropriate to our problem is

$$p(d, t) = 0, \tag{16.84}$$

which states that the probability of the protein-binding site being at position $x = d$ is identically zero. This follows from the fact that we are following the motion of the last binding site to have emerged from the pore. As soon as this site finds itself at position d, a new binding site emerges from the pore and binds a protein, and is now the new "last" site to emerge. The old "last" site effectively disappears at this point.

To calculate the steady-state current ($J = \text{const}$), we solve $J_x = J$ with the boundary condition from Equation 16.84, essentially exactly as we did in the previous section for the polymerization motor. The solution to the homogeneous part of the resulting differential equation is

$$p_{\text{hom}}(x) = C\mathrm{e}^{-Fx/k_B T}, \tag{16.85}$$

while

$$p_{\text{part}}(x) = -J\frac{k_B T}{DF} \tag{16.86}$$

is a particular solution; the sum of the two is the general solution, with C a constant that can be determined from the boundary condition. After a bit of algebra, we arrive at

$$p(x) = J\frac{k_B T}{DF}(\mathrm{e}^{F(d-x)/k_B T} - 1), \tag{16.87}$$

which gives us the probability of finding the last protein-binding site to have crossed the pore a distance x away from the pore, *in the steady state* (thus the lack of any time dependence).

At this point, we should remind ourselves of the original goal we set out to achieve. We are really interested in the translocation velocity for the polymer in the steady state v_{tr}, which is simply related to the current J: $v_{\text{tr}} = Jd$. In order to calculate J, we make use of the fact that the probability of finding the last protein site anywhere on the interval $(0, d)$ is 1, that is,

$$\int_0^d p(x)\,\mathrm{d}x = J\frac{d^2}{D}\left(\frac{k_B T}{Fd}\right)\left(\mathrm{e}^{Fd/k_B T} - \frac{Fd}{k_B T} - 1\right) = 1. \tag{16.88}$$

Finally, solving the above equation for the steady-state current leads to a formula for the translocation velocity as a function of the force acting on the polymer,

$$v_{tr} = \frac{D}{d} \frac{w^2}{e^w - w - 1} \tag{16.89}$$

where $w = Fd/k_BT$ is the dimensionless force. Note that in the small-load-force limit, we arrive at a simple expression, $v_{tr} = 2D/d$, which follows from a simple argument: the time for the protein-binding site to diffuse over a distance d unassisted by an applied force is $d^2/2D$; once this occurs, a new protein is bound and the polymer has translocated by distance d. Therefore, the translocation velocity is given by $d/(d^2/2D) = 2D/d$, the same as above.

In a more realistic model of translocation, the irreversibility of the protein binding should be relaxed. Instead, one has on and off rates for the binding of these proteins. This will enter the model described above, again through the boundary condition for $p(x)$ in the equation for the steady-state current. The main qualitative difference that this imparts on the predicted translocation velocity dependence on the force is the emergence of a finite stall force at which v_{tr} goes to zero. In the model worked out above, Equation 16.89 leads to an infinite stall force, which is an unphysical feature.

16.4 Summary and Conclusions

Directed forces in biological systems are powered by molecular motors devoted to the conversion of chemical energy into mechanical energy. Although their detailed mechanisms and biological functions are impressively diverse, we have shown in this chapter that many aspects of motor function can be fruitfully considered by breaking motor activity down into a series of small discrete steps. Transitions between steps can be considered using either discrete or continuous formalisms that acknowledge differences in energy for different internal motor states and also the influence of diffusive and stochastic events. Many mechanical properties of motors, including speed, force or torque, power, and variability, can be thus considered in a unifying theoretical framework.

16.5 Problems

Key to the problem categories: • Model refinements and derivations • Estimates • Data interpretation

• 16.1 Randomness in the one-state model

An alternative definition for the randomness to that given in the chapter is

$$r = \lim_{t\to\infty} \frac{\langle x(t)^2\rangle - \langle x(t)\rangle^2}{a\langle x(t)\rangle}. \tag{16.90}$$

(a) Using the definition of randomness given above and the probability distribution for the one-state motor in the continuum limit, Equation 16.16, work out an explicit expression for the randomness.

(b) Equation 16.24 provides an expression for the randomness in terms of the force-dependent rate constants of one state of the model. Show that randomness in this case is insensitive to whether the force dependence is in the forward or the backward rate. Make a plot of the randomness as a function of force. How does your result compare with the experimental curve shown in Figure 16.28(A)?

• 16.2 Stepping of myosin V

Single-molecule experiments have been performed on myosin V where a fluorescent marker was placed at different locations on the light-chain domain and individual steps were recorded. It was found that the average step size is about 37 nm; see Figure 16.53.

(a) If the dye is placed on the light-chain domain at a distance x along the direction of motion from the midpoint between the two heads of the motor, what step size do you expect to observe? What value of x explains the data shown in Figure 16.53? What do you suppose is the origin of the peak at approximately 74 nm in the step-size histogram?

Figure 16.53: Single-molecule measurements of steps taken by myosin V along actin filaments. The histogram in the inset is of all the measured step sizes, which are measured with 1.5 nm precision. (Adapted from A. Yildiz et al., *Science* 300:2061, 2003.)

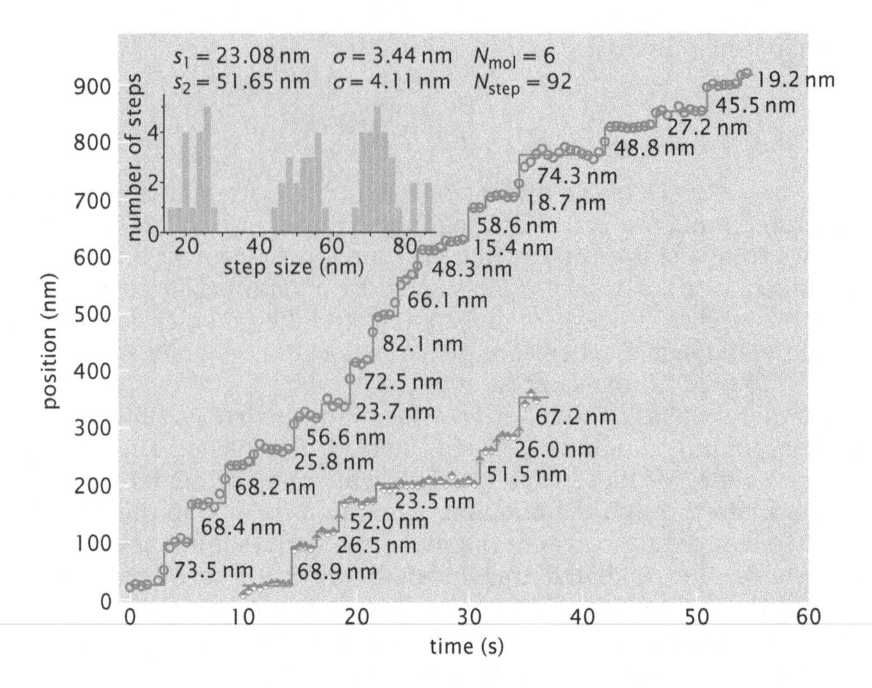

(b) Assume that the stepping rate is k. This is the probability per unit time that the motor will make a step. Calculate the waiting time distribution between the two steps observed in the experiment if the fluorescent marker is placed at position x found in (a). What is the expected distribution, assuming a hand-over-hand stepping mechanism if the marker is placed very close to one of the heads (that is, $x \approx 18$ nm)? Use your calculated distributions to rationalize and fit the data (from Yildiz et al., 2003) provided on the book's website. Does your analysis support the hand-over-hand mechanism? What value of k do you obtain?

• 16.3 Kinesin as an ATP-hydrolyzing enzyme

As described in the chapter, careful measurements have been performed that examine the dependence of motor velocity on ATP concentration. Under certain conditions, the hydrolysis reaction performed by a molecular motor can be described using the Michaelis–Menten model introduced in Section 15.2.7 (p. 596). In the particular case of kinesin, its stepping is strongly coupled to its ATPase activity, which translates into relatively constant step sizes. Finally, its high processivity allows for a clear definition of a speed, since kinesin takes many steps before falling off the microtubule.

Relate the reaction speed (the rate of ATP hydrolysis) to the maximum stepping speed of kinesin and determine its dependence on ATP concentration. Fit your model to the data by Schnitzer and Block (1997) shown in Figure 16.54 and provided on the book's website. Then work out what change in substrate concentration is needed to increase the reaction rate from $0.1 v_{max}$ to $0.9 v_{max}$.

• 16.4 Kinetics of two-state motors

In the chapter, in order to obtain the velocity of a two-state motor, we made use of a trick to circumvent solving the master equation directly. Here we take up this task and in the process also derive an expression for the diffusion constant.

(a) Consider a trial solution of the system of equations for $p_0(n, t)$ and $p_1(n, t)$, given by Equations 16.33 and 16.34, that is of the form

$$\begin{pmatrix} p_0(n,t) \\ p_1(n,t) \end{pmatrix} = e^{i(Kn-\omega t)} \begin{pmatrix} A \\ B \end{pmatrix}. \tag{16.91}$$

Find a relation between K and ω that guarantees the existence of a solution of this form. This is the so-called dispersion relation.

(b) By substituting the trial solution $e^{i[(K/a)-\omega t]}$ into the differential equation for diffusion with drift (Equation 13.54, p. 530), show that the dispersion relation in this case is

$$\omega = v\frac{K}{a} - iD\frac{K^2}{a^2}. \tag{16.92}$$

where v is the drift velocity and D is the diffusion constant.

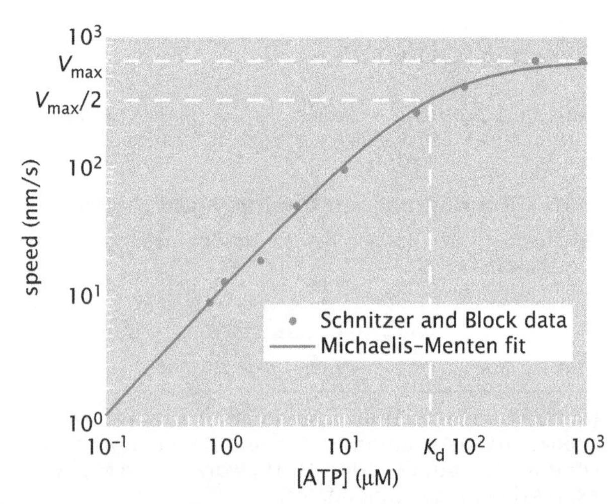

Figure 16.54: ATP hydrolysis by kinesin. Speed of a kinesin motor as a function of ATP concentration and fit of the data to a Michaelis–Menten model. (Adapted from M. J. Schnitzer and S. M. Block, *Nature* 388:386, 1997.)

(c) Demonstrate that in the limit $K \ll 1$, the dispersion relation for the two-state motor is the same as that for diffusion with drift. To do this, Taylor-expand $\omega(K)$ in K and solve the equation for ω obtained in (a) order by order in K, which amounts to computing the coefficients in the Taylor expansion. Compare your result with the dispersion relation for dispersion with drift and read off the diffusion coefficient for the motor and its speed. Check that the formula for the speed matches the one obtained in the chapter.

- **16.5 Polymerization ratchet estimates**

In the chapter, we analyzed the polymerization ratchet in the two limiting cases of diffusion-limited and reaction-limited polymerization. By comparing the time for a load, a polystyrene sphere 1 μm in diameter, to diffuse a distance given by the actin monomer size, and the average time for an actin monomer to be added to the growing end of the filament, find the condition for the free actin monomer concentration that is necessary for the polymerization in the presence of the load to be reaction-limited. Compare this concentration with the critical concentration for actin filament growth.

16.6 Further Reading

Vogel, S (2001) Prime Mover: A Natural History of Muscle, W. W. Norton. Like all of Vogel's books, this one makes for fascinating reading and is pertinent to part of the story developed in this chapter.

Howard, J (2001) Mechanics of Motor Proteins and the Cytoskeleton, Sinauer Associates. Howard's book is the first source that we go to when trying to learn more about motors.

Duke, T (2002) Modelling motor proteins, in Physics of Bio-Molecules and Cells, edited by H Flyvbjerg, F Jülicher, P Ormos & F David, EDP Sciences/Springer-Verlag. An instructive discussion of motor proteins.

Nelson, P (2004) Biological Physics: Energy, Information, Life, W. H. Freeman. Very interesting discussion of motor proteins.

Fisher, ME, & Kolomeisky, AB (1999) The force exerted by a molecular motor, *Proc. Natl Acad. Sci. USA* **96**, 6597.

Bustamante, C, Macosko, JC, & Wuite, GJL (2000) Grabbing the cat by the tail: manipulating molecules one by one, *Nat. Mol. Cell Biol.* **1**, 131. This paper has an excellent table showing the strengths and limitations of various single molecule methods.

Fisher, ME, & Kolomeisky, AB (2001) Simple mechanochemistry describes the dynamics of kinesin molecules, *Proc. Natl Acad. Sci. USA* **98**, 7748. This interesting paper examines how simple models like those described in the chapter can respond to single-molecule data on motor dynamics.

Thomas, N, Imafuku, Y, & Tawada, K (2001) Molecular motors: thermodynamics and the random walk, *Proc. R. Soc. Lond.* **B268**, 2113.

Mahadevan, L, & Matsudaira, P (2000) Motility powered by supramolecular springs and ratchets, *Science* **288**, 95. This interesting review summarizes several exotic kinds of force generation that we do not address.

Jülicher, F, Ajdari, A, & Prost, J (1997) Modeling molecular motors, *Rev. Mod. Phys.* **69**, 1269. A comprehensive review with many useful and general insights into motor function.

Simon, SM, Peskin, CS, & Oster, GF (1992) What drives the translocation of proteins? *Proc. Natl Acad. Sci. USA* **89**, 3770. An important and thought-provoking paper on the translocation ratchet.

Theriot, JA (2000) The polymerization motor, *Traffic* **1**, 19. This review summarizes the theoretical basis for force generation by filament assembly.

Peskin, CS, Odell, GM, & Oster, GF (1993) Cellular motions and thermal fluctuations: the Brownian ratchet, *Biophys J.* **65**, 316. This beautiful paper describes the physics of motors like those described in the tail end of our chapter.

Zandi, R, Reguera, D, Rudnick, J, & Gelbart, WM (2003) What drives the translocation of stiff chains? *Proc. Natl Acad. Sci. USA* **100**, 8649. More recent insights into the nature of translocation ratchets.

16.7 References

Alberts, B, Johnson, A, Lewis, J, et al. (2008) Molecular Biology of the Cell, 5th ed. Garland Science.

Block, SM, Goldstein, LS, & Schnapp, BJ (1990) Bead movement by single kinesin molecules studied with optical tweezers, *Nature* **348**, 348.

Bustamante, C, Keller, D, & Oster, G (2001) The physics of molecular motors, *Acc. Chem. Res.* **34**, 412.

Bustamante, C, Chemla, YR, Forde, NR, & Ishaky, D (2004) Mechanical processes in biochemistry, *Annu. Rev. Biochem.* **73**, 705.

Diehl, MR, Zhang, K, Lee, HJ, & Tirrell, DA (2006) Engineering cooperativity in biomotor–protein assemblies, *Science* **311**, 1468.

Dogterom, M, & Yurke, B (1997) Measurement of the force–velocity relation for growing microtubules, *Science* **278**, 856.

Gebhardt, JC, Clemen, AE, Jaud, J, & Rief, M (2006), Myosin-V is a mechanical ratchet, *Proc. Natl Acad. Sci. USA* **103**, 8680.

Hirokawa, N (1998) Kinesin and dynein superfamily proteins and the mechanism of organelle transport, *Science* **279**, 519.

Holy, TE, Dogterom, M, Yurke, B, & Leibler, S (1997) Assembly and positioning of microtubule asters in microfabricated chambers, *Proc. Natl Acad. Sci. USA* **94**, 6228.

Kerssemakers, JW, Munteaunu, EL, Laan, L, et al. (2006) Assembly dynamics of microtubules at molecular resolution, *Nature* **442**, 709.

Kodera, N, Yamamoto, D, Ishikawa, R, & Ando, T (2010) Video imaging of walking myosin V by high-speed atomic force microscopy, *Nature* **468**, 72.

Kolomeisky, AB, & Fisher, ME (2003) A simple kinetic model describes the processivity of myosin-V, *Biophys J.* **84**, 1642.

Liao, JC, Spudich, JA, Parker, D, & Delp, SL, (2007). Extending the absorbing boundary method to fit dwell-time distributions of molecular motors with complex kinetic pathways, *Proc. Natl Acad. Sci. USA* **104**, 3171.

Mehta, AD, Rock, RS, Spudich, JA, Mooseker, MS, & Cheney, RE (1999) Myosin-V is a processive actin-based motor, *Nature* **400**, 590.

Noji, H, Yasuda, R, Yoshida, M, & Kinoshita, K, Jr. (1997) Direct observation of the rotation of F1-ATPase, *Nature* **386**, 299.

Parekh, SH, Chaudhuri, O, Theriot, JA, & Fletcher, DA (2005) Loading history determines the velocity of actin-network growth, *Nat. Cell Biol.* **7**, 1219.

Purcell, TJ, Morris, C, Spudich, JA, & Sweeney, HL (2002) Role of the lever arm in the processive stepping of myosin V, *Proc. Natl Acad. Sci. USA* **99**, 14159.

Purcell, TJ, Sweeney, HL, & Spudich, JA (2005) A force-dependent state controls the coordination of processive myosin V, *Proc. Natl Acad. Sci. USA* **102**, 13873.

Rodionov, DI, & Borisy, GG (1997) Self-centering activity of cytoplasm, *Nature* **386**, 170.

Schnitzer, MJ, & Block, SM (1997) Kinesin hydrolyses one ATP per 8-nm step, *Nature* **388**, 386.

Schnitzer, MJ, Visscher, K, & Block, SM (2000) Force production by single kinesin motors, *Nat. Cell Biol.* **2**, 718.

Turner, L, Ryo, WS, & Berg, HC (2000) Real-time imaging of fluorescent flagellar filaments, *J. Bacteriol.* **182**, 2793.

Vale, RD (2003) The molecular motor toolbox for intracellular transport, *Cell* **112**, 467.

Visscher, K, Schnitzer, MJ, & Block, SM (1999) Single kinesin molecules studied with a molecular force clamp, *Nature* **400**, 184.

Yasuda, R, Noji, H, Kinosita, K, Jr., & Yoshida, M (1998) F1-ATPase is a highly efficient molecular motor that rotates with discrete 120 degree steps, *Cell* **93**, 1117.

Yildiz, A, Forkey, JN, McKinney, SA, et al. (2003) Myosin V walks hand-over-hand: single fluorophore imaging with 1.5-nm localization, *Science* **300**, 2061.

Biological Electricity and the Hodgkin–Huxley Model 17

Overview: In which the propagation of nerve impulses is examined as a problem in biological electricity

> "Action is eloquence."
>
> **William Shakespeare, *Coriolanus***

In this chapter, we will focus on the electrical consequences of charge separation across membranes. We will explore both how cells establish charge separation and how they use it. Arguably, the most spectacular application of biological electricity by cells is the action potential in nerve cells, a mechanism used to rapidly propagate information from the nerve cell body to the tip of its axon, traveling at a rate of 10–100 m/s. This rapid transmission of information occurs 9–10 orders of magnitude faster than it takes molecules to diffuse over the length of a typical axon, and about 7 orders of magnitude faster than motor-driven transport can communicate molecular information between two points. Similar electrical coupling along the membrane of a muscle cell enables a giant myofibril, which may be several millimeters long, to contract nearly simultaneously along its entire length. In animals, the rapid electrical transmission of information in the nerve and muscle systems allows us to run from predators and to catch our prey.

17.1 The Role of Electricity in Cells

In Chapter 9, we discussed the movement of ions in aqueous solution and found that although charges can be separated over distances such as the Bjerrum length (see Figure 9.12 on p. 369), this distance is small (<1 nm) and, given the rapid motion of ions in water, any transient charge imbalance will quickly return to an equilibrium state by the flow of ions. However, as we know from our macroscopic experience with batteries, separation of charge is a powerful means of storing energy. Cells take advantage of charge separation as a means of energy (and information) storage by exploiting their membranes and the protein machines (channels and pumps) in them, to prevent

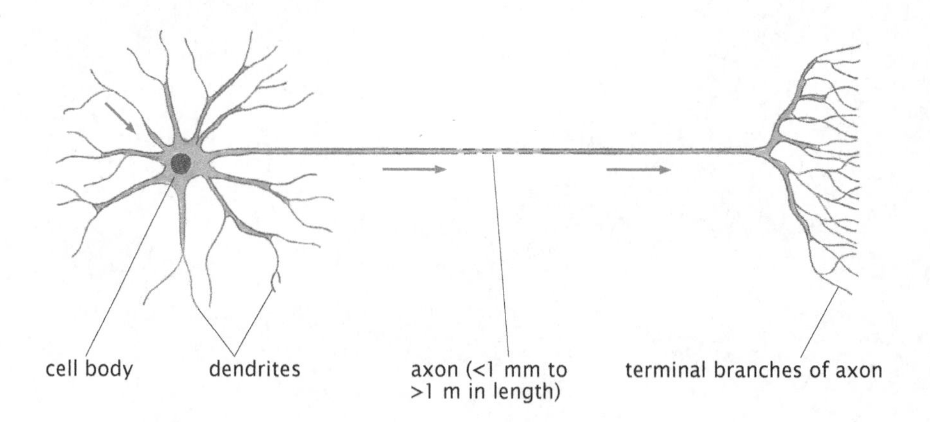

Figure 17.1: Schematic of a nerve cell. Signals are propagated along the axon through the action potential.

charge imbalances from equilibrating. As we discussed in Chapter 11, phospholipid bilayers represent an insulating barrier that is essentially impermeable to aqueous ions unless specific protein channels are activated.

Beyond preserving the integrity of the cell, the membrane also serves an important function in electrical signaling. By maintaining a potential difference between the cellular interior and its exterior, the membrane provides the physical basis for this form of signaling in which the steady-state value of the membrane potential can be disrupted due to a chemical, electrical, or mechanical stimulus and this disturbance can propagate along the cell membrane. One of the most celebrated examples of this phenomenon is the action potential seen in nerve cells such as that shown schematically in Figure 17.1 (and shown in a fluorescence image in Figure 2.42 on p. 79). The similarity between this biological electricity and inorganic electricity was vividly illustrated in the 1780s by Galvani, who showed that applying electric shocks to the leg muscle of a recently deceased frog would make it twitch.

While the focus of this chapter is electrical signaling in cells, it is worth noting that both electrical and chemical connections are required for the functioning of the elaborate networks formed by neurons. Roughly speaking, electrical transmission conveys information within a single cell and chemical signals bridge the gap from one cell to another. Electrical signals can be translated into chemical signals, and vice versa, as we saw in Chapter 11 (Figure 11.28 on p. 455). The arrival of an electrical signal from the neuron cell body triggers the process of vesicle fusion at the synaptic site. Fusion of synaptic vesicles at the synaptic site releases a neurotransmitter that causes membrane depolarization and electrical signaling in the target cell.

17.2 The Charge State of the Cell

17.2.1 The Electrical Status of Cells and Their Membranes

In our everyday macroscopic encounters with electricity, we are used to thinking of electrons traveling down metallic wires with a flow of negative charge. However, movement of ions, including those with positive charge, can also produce electrical currents and potential differences. Indeed, the earliest batteries invented by Volta and Daniell in the late eighteenth and early nineteenth centuries relied on current carried by ions such as Zn^{2+}, Cu^{2+}, and Ag^{+} ions. In cells, in a few cases, electron transport is used directly to produce charge separation, as in photosynthesis in chloroplasts and electron transport in mitochondria. More commonly though, cells create and manipulate transmembrane gradients of positive ions, the most important being Na^{+}, K^{+}, and Ca^{2+} ions. As discussed in Chapter 9, many of the macromolecules in the cell, particularly nucleic acids, are intrinsically negatively charged at physiological pH, so it is not surprising that small positive ions dominate the ionic landscape of the cell's electrical state. Among small negative ions, Cl^{-} ions are the only monatomic ions whose distribution is manipulated and exploited by cells. Phosphate and sulfate ions are also important in many biochemical reactions, but play little role in setting up the charge state of the cell.

In order to generate electric potentials and currents, ions alone are not enough. It is necessary that the cell create mechanisms for separating and concentrating ions and permitting them to pass only through predefined conduits. The cell membrane provides these important functions as a combination of a thin but very effective insulator in the form of the phospholipid bilayer littered with many different transmembrane proteins that act as ion channels and ion pumps. The permeability of the membrane itself was illustrated in Figure 11.11 (p. 438). The combined electrochemical properties of the lipid bilayer, channels, and pumps provide a patch of membrane with electrical properties equivalent to a set of resistors (which are voltage-dependent), batteries (whose voltage is set by the ion concentration difference), and a capacitor, all connected in parallel. By selectively opening and closing transmembrane ion channels, which can specifically conduct only a subset of ions (usually only a single type of ion), the cell can tune its membrane electrical potential. This process is at the heart of electrical signaling in excitable cells.

17.2.2 Electrochemical Equilibrium and the Nernst Equation

In the absence of a cell membrane, cellular components would diffuse away and cellular processes, including the all-important membrane potentials necessary for maintaining life, would cease.

Ion Concentration Differences Across Membranes Lead to Potential Differences

To understand the emergence of a membrane potential, we step back and consider a simplified situation that illustrates the energetic and entropic competition that plays out when there are concentration gradients of charged species. To study this process of electrochemical equilibrium, we consider an isolated system with an internal barrier separating the system into two regions as shown in Figure 17.2. On one side, we have a charge-neutral solution of K^{+} and Cl^{-} ions, for example, at concentration c_1, and on the other side, we have a charge-neutral solution of the same ions, but at concentration c_2. If we now remove the constraint of impermeability for the K^{+} ions, these ions will start flowing from the region of high concentration to low concentration. As already highlighted in Chapter 13, the diffusion of these ions down their concentration gradient can be thought of as a response to the driving force tending to maximize the entropy of

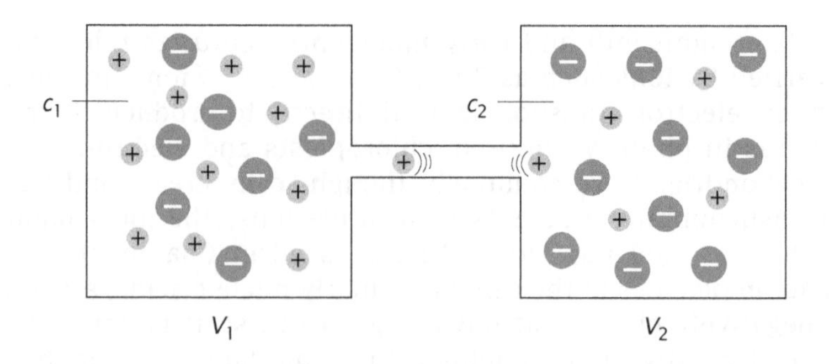

Figure 17.2: Charge separation in a model system. The box is separated into two equal domains by a barrier that is only permeable to positive ions. The solutions on both sides were initially charge-neutral but at different concentrations. Because only the positively charged ions are permitted to move across the barrier down their concentration gradient, a concentration difference between positive ions in the two boxes has developed, leading to a charge separation and a potential difference. The potential difference in equilibrium is the Nernst potential.

the system. As these ions flow down their concentration gradient, there will be an ever-increasing charge imbalance and a concomitant *increase* in the electric energy of the system. Equilibrium is reached when these two effects balance. This occurs at a particular value of the potential difference between the two regions, which depends on the ratio of the *final* concentrations of K^+ ions. For numbers typical of cellular processes, the change of concentration that accompanies the transfer of ions across the cell membrane is many orders of magnitude smaller than the concentrations themselves, which we can therefore treat as constant.

To apply these ideas mathematically, we recall from Chapter 9 that the electrical energy of an ion of unit charge e, when placed in a region of potential V_1, is eV_1. For an ion whose valence is z, the corresponding charge is ze and the energy is zeV_1. Consider two small regions in our salt solution (small so we can consider the electrical potential roughly uniform). In one, the potential is V_1 and the concentration of positive ions is c_1, while in the other we have V_2 and c_2 for the potential and the concentration. In equilibrium, the probability of finding a positive ion in region 1 or 2 is

$$p_{1,2} = \frac{1}{Z} e^{-zeV_{1,2}/k_B T}, \tag{17.1}$$

resulting from application of the Boltzmann distribution to this problem in which the energies are electrical. The subscript 1, 2 on p and V means the subscript should be either 1 or 2; there is a distinct equation for each box. The ratio of the two probabilities is equal to the ratio of the number of ions in region 1 and the number in region 2, which in turn are proportional to the ion concentrations in the two regions. We can use this proportionality to write the ratio of concentrations as

$$\frac{c_1}{c_2} = \frac{p_1}{p_2} = \frac{e^{-zeV_1/k_B T}}{e^{-zeV_2/k_B T}}. \tag{17.2}$$

If we take the logarithm of both sides of this equation, we are left with the celebrated Nernst equation, which relates the difference of electrical potentials to the ratio of concentrations of ions and is given by

$$V_2 - V_1 = \frac{k_B T}{ze} \ln \frac{c_1}{c_2}. \tag{17.3}$$

The Nernst equation will be our starting point for trying to understand the way cells manage charge in order to set up membrane potentials.

As seen in Equation 17.3, the fundamental energy scale of these charge transactions is set by the ratio $k_B T/e$. If we substitute the relevant numbers, we find $k_B T/e \approx 25\,\text{mV}$. An alternative way of stating

Table 17.1: Ion concentrations and the Nernst potential for small ions within the cell. The numbers are typical of mammalian skeletal muscle cells, which have a resting potential of $V_{mem} = -90$ mV. (Data adapted from B. Hille, Ion Channels of Excitable Membranes. Sinauer Associates, 2001.)

Ion species	Intracellular concentration (mM)	Extracellular concentration (mM)	Nernst potential (mV)
K^+	155	4	−98
Na^+	12	145	67
Ca^{2+}	10^{-4}	1.5	130
Cl^-	4	120	−90

this is that the thermal energy scale corresponds to a 25 mV potential difference. The fact that the voltage across the cell membrane of neurons and other excitable cells is in the 10–100 mV range shows that thermal effects play an important role since the thermal energy scale and the charge energy scale are comparable.

Table 17.1 shows the intracellular and extracellular concentrations of the four major ions responsible for the cell's electrical state, as measured for a skeletal muscle cell in the rest state. The excess positive charge inside the cell is primarily balanced out by the negative charge on key macromolecules such as nucleic acids and proteins. All of these charges, both inside and outside the cell, conspire to produce a potential drop across the cell membrane. This resting potential for the skeletal muscle cell is about −90 mV, indicating that the interior of the cell is at a potential that is 90 mV lower than that of the exterior. For different cell types and under different environmental conditions, the voltage drop across the cell membrane ranges from −10 to −95 mV, and the ionic imbalance varies as well.

This table also shows the Nernst potential for each of these ionic species given the measured intracellular and extracellular ion concentrations. The balance between the two opposing influences, entropic and electrostatic, can be assessed by comparing the membrane potential with the Nernst potential. Given the concentrations inside (c_{in}) and outside (c_{out}) the cell, the Nernst potential is given by $V_{Nernst} \approx (27\,\text{mV}/z) \times \ln(c_{out}/c_{in})$. To get this expression, we have made use of Equation 17.3 with $e = 1.6 \times 10^{-19}$ C and $k_B T = 4.3 \times 10^{-21}$ J (here the temperature is taken to be 37 °C, appropriate for the muscle cell, resulting in 27 mV rather than 25 mV as the prefactor in the Nernst equation). In equilibrium, the Nernst potentials of all the diffusing ionic species are the same and equal to the membrane potential; this situation is referred to as Donnan equilibrium. As indicated in the table, the ionic species are *not* in a state of equilibrium, because, in the cases of Na^+ and Ca^{2+} ions, the Nernst potentials do not even have the same sign as the observed membrane potential.

17.3 Membrane Permeability: Pumps and Channels

A Nonequilibrium Charge Distribution Is Set Up Between the Cell Interior and the External World

Conditions within a cell are considerably more complex than our idealized description above would indicate. Although we can certainly consider a cell membrane as a semipermeable barrier separating two salt reservoirs at different concentrations, we must keep track of many different species of ions. Our description must also account for the fact that membrane permeability to specific ions can be regulated by the cell and altered rapidly and deliberately.

As noted above, electrically excitable cells are far from equilibrium in several interesting ways. First, all four ion species maintain at least a 10-fold difference in concentration across the cell membrane, with the extreme being Ca^{2+} ions, which maintain a 10,000-fold difference. From our discussion about the Nernst potential, we might speculate that these concentration differences could be balanced by an appropriate electrical potential drop across the cell membrane. Quantitatively, this would amount to the Nernst potential of every ionic species being equal to the potential drop across the cell membrane. However, the actual transmembrane potential is close to the Nernst potential for only two of the four major ions, K^+ and Cl^-. For Na^+ and Ca^{2+} ions, the Nernst potential is actually of the "wrong" sign, indicating that the high relative concentration of sodium and calcium outside the cell is maintained against both an unfavorable chemical gradient and an unfavorable electrical potential. Next, we see that there is something special about calcium, both because its intracellular concentration is vanishingly small and because the ratio between extracellular and intracellular concentrations is anomalously high (although the cytoplasmic concentration of Ca^{2+} ions is low, there is a large reservoir of Ca^{2+} ions within the cell sequestered inside the endoplasmic reticulum). Finally, it seems curious that the cell selectively separates K^+ and Na^+ ions, with K^+ predominating inside the cell and Na^+ predominating outside the cell, because the chemical nature of these two ions is very similar and for most chemical reactions they are functionally interchangeable.

Our challenge now is to understand both how the cell sets up and maintains this striking nonequilibrium condition and how these chemical and electrical imbalances are exploited for cellular function. This requires detailed understanding of two important classes of transmembrane proteins that primarily concern themselves with ions. In particular, this intriguing, dynamic charge state is mediated by voltage-dependent, ion-selective channels and energy-consuming ion pumps. As will become clear in our discussion below, it is critical for their biological function that these channels and pumps be choosy about which ions they permit to pass. Some proteins are more selective than others, with a few passing only a single ionic species, while others pass several species that share the same valence and sign of charge, and some are fairly promiscuous in their gatekeeping. There are estimated to be several hundred distinct ion channels encoded in the human genome. The differential expression and regulation of these many selective channels gives rise to the elaborate electrochemistry of human existence.

Signals in Cells Are Often Mediated by the Presence of Electrical Spikes Called Action Potentials

Life is dynamic. One manifestation of this dynamism is the fact that the membrane potential can be rapidly changed. As shown in Figure 4.25 (p. 177), the time dependence of the local (that is, at a given point on the membrane) membrane potential results in a profile that can be thought of as a traveling pulse. These traveling pulses are known as action potentials and have a variety of interesting features gleaned from experiments on the squid giant axon and other nerve cells. The prototypical excitable behavior of these cells has a characteristic voltage profile like that shown in Figure 17.3. These plots show the membrane potential as a function of time at one particular point on the membrane. What we see is that the membrane is completely depolarized (even changing the sign of the membrane potential) and that this excitation lasts for a time measured in milliseconds and then decays back to its equilibrium value. If we were to make a similar measurement somewhere further down an axon, for example, after some delay time set by the velocity of propagation of the action potential, we would see this same stereotyped pulse pass by that point as well.

One of the most intriguing features of the action potential is the observation that the creation of a pulse requires the stimulating

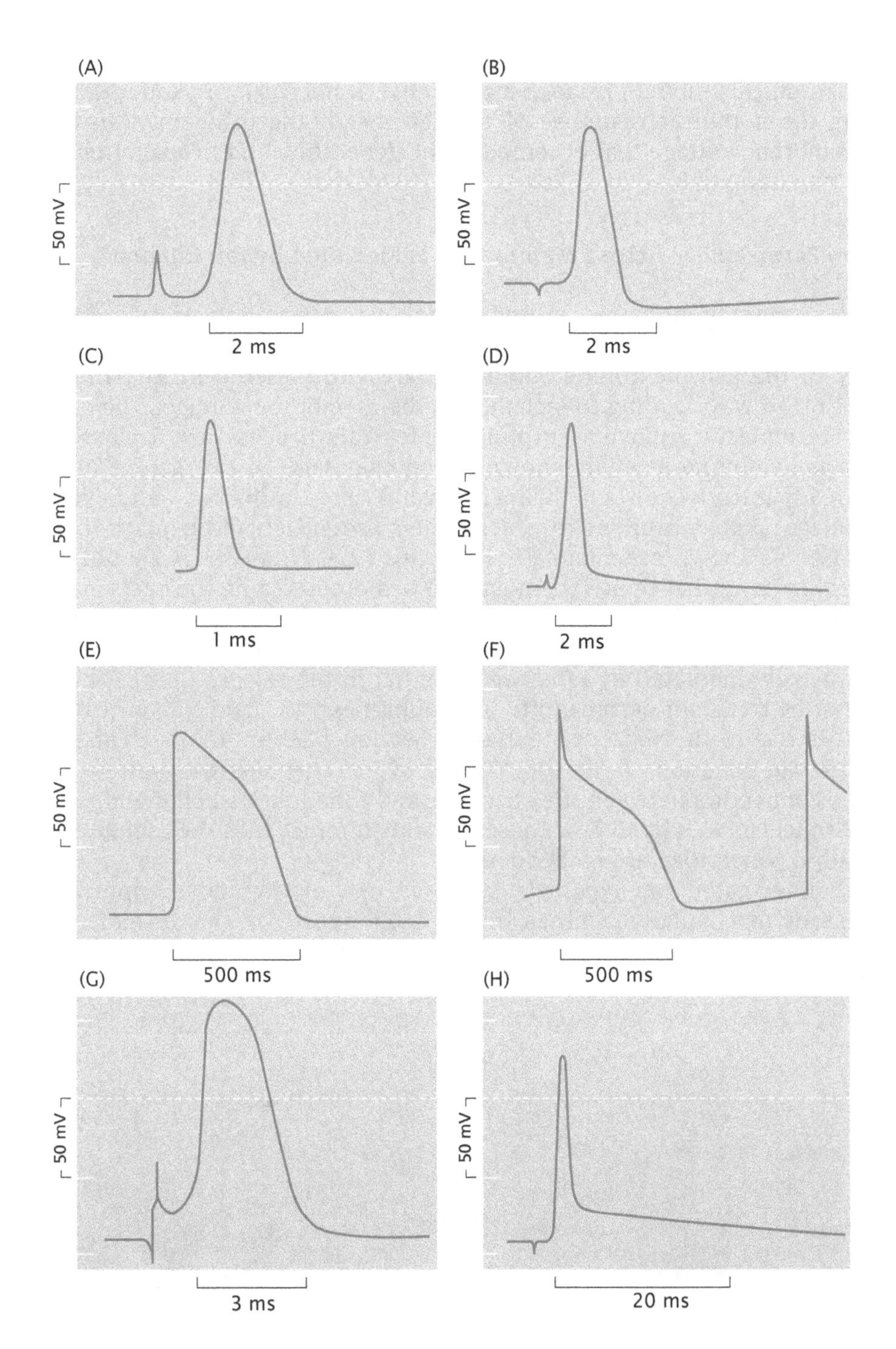

Figure 17.3: The many shapes of action potentials. This gallery shows examples of action potentials recorded from different cell types in different animals. The horizontal dashed white line is zero potential. The top row compares the action potentials in the squid giant axon as recorded (A) in the animal and (B) after isolation by dissection. (C) This action potential was measured in the axon of a cell from a cat's peripheral nervous system. (D) This similar potential was recorded from the cell body of a neuron from a cat's spinal cord. (E) This action potential was recorded from a muscle fiber rather than from a neuron, this one in the heart of a frog. (F) This action potential was recorded from a regulatory neuron in a sheep's heart. (G) Potential generated in the remarkable specialized muscle tissue that forms the electric organ of a fish that can shock its prey. (H) Action potential recorded from a more typical muscle from a frog's thigh. The amplitudes of all of these action potentials are comparable and are between 90 and 150 mV; however, their timing duration, as shown in milliseconds, is extremely variable. (Adapted from R. D. Keynes and D. J. Aidley, Nerve and Muscle. Cambridge University Press, 2001.)

voltage difference to exceed a threshold value. That is, if the local stimulus leads to a transient voltage difference that is less than this critical value, there will be no resulting pulse. By way of contrast, once this threshold value has been exceeded, the magnitude of the action potential takes a constant value that is indifferent to the particulars of the stimulation. Once the action potential has been generated, it propagates down an axon essentially without attenuation at speeds between 10 and 100 m/s.

17.3.1 Ion Channels and Membrane Permeability

From a model-building perspective like that advocated here, the study of action potentials resulted in one of the most successful models at the interface of biology, chemistry, and physics in the history of science, namely, the Hodgkin–Huxley model. As will be seen later in the chapter, this model provides a predictive framework for understanding the nonlinear response of the cell membrane to changes in the membrane voltage that accompany the generation and propagation of action potentials.

Ion Permeability Across Membranes Is Mediated by Ion Channels

An intriguing outcome of the Hodgkin–Huxley analysis of action potentials was the prediction of a mechanism whereby ion permeability of the cell membrane could be transiently altered in a spatially localized way, leading to a change in the membrane voltage. The concrete molecular players responsible for this process are a class of transmembrane proteins known as ion channels. Of course, we have already introduced such channels several times in the book and every college biology student can recite their properties and explain their action as shown schematically in Figure 17.4. However, it should be kept in mind that their very existence was predicted on the basis of an analysis of electrical signaling in cells long before any of this molecular information had been gathered. Recalling that biological electricity is usually mediated by ions, we note that in the case of interest here it is the transient permeability of membranes to K^+ and Na^+ ions that gives rise to the action potential. In Section 11.6 (p. 467) we considered the question of the functioning of ion channels with an eye to the connection between the structure and function of such membrane channels as well as to how these proteins interact with the membrane within which they are embedded.

Voltage-gated ion channels serve as one of the most important classes of cellular machines in the cell's quest for charge management. We have seen throughout the book that ion channels can be thought of as having at least two key roles: (i) they are able to detect

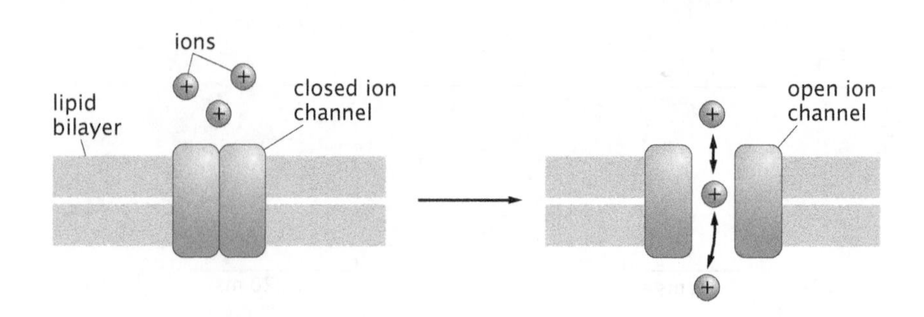

Figure 17.4: Ion channels and transient membrane permeability. When the ion channel is closed, the concentration gradient is maintained across the cell membrane. When the channel opens, ions flow across the membrane until an equilibrium distribution is established. The equilibrium distribution of ions depends upon the electric potential across the membrane.

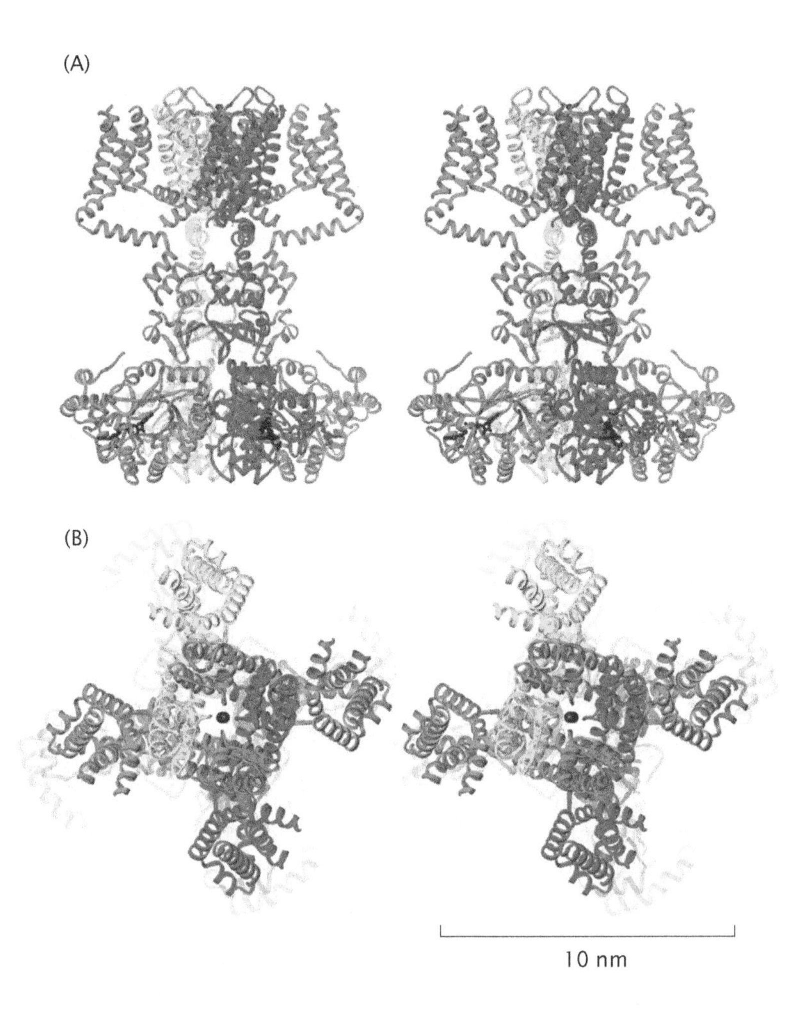

Figure 17.5: Structure of a voltage-gated ion channel. This figure shows stereo ribbon diagram views of the mammalian Kv1.2 structure. (A) Side view of the channel with the extracellular solution above and the interior of the cell below. (B) Top view of the structure. Each subunit is shown in a different color. (Adapted from S. B. Long et al., *Science* 309:897, 2005.)

some stimulus such as an increase in the concentration of a particular ligand, the tension in a membrane, or a particular voltage, and (ii) as a result of detecting the relevant stimulus, these channels open and permit (often selectively) the flow of ions through the channel. One of the important ways that our understanding of voltage-gated ion channels has evolved is as a result of a number of successful structural studies of these channels. Remarkably, under some circumstances, these highly irregular, membrane-bound proteins can be crystallized, enabling them to be analyzed using X-ray crystallography. An example of the structure of a mammalian voltage-gated channel is shown in Figure 17.5.

As experimental insights into both the structure and function of these channels have increased, so too has the effort to model their behavior. In Chapter 7, we saw how two-state models can be used to give a qualitative feeling for the mechanisms and energetics of ion-channel gating. In the case described there, we examined how membrane tension couples to channel gating. We now use a similar two-state analysis of voltage gating.

A Simple Two-State Model Can Describe Many of the Features of Voltage Gating of Ion Channels

The patch clamp measurements shown in Figure 7.2 (p. 283), which measure the current as a function of applied voltage, show how

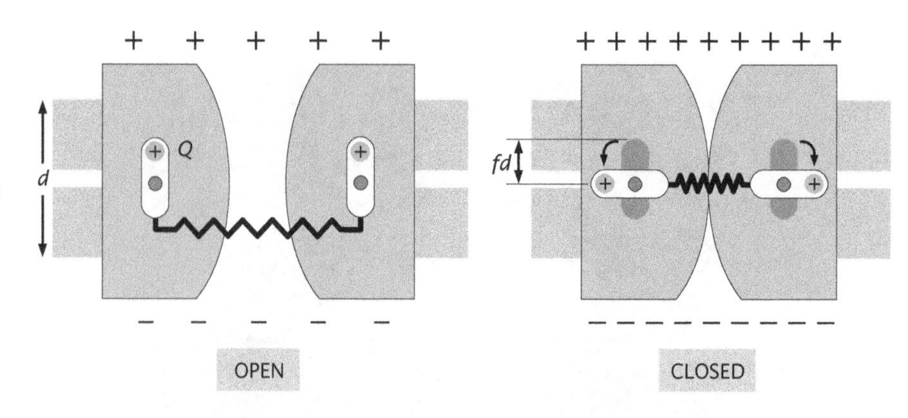

Figure 17.6: Two-state model of ion-channel voltage gating. In the open state, the "spring", representing the conformational degrees of freedom of the channel, is in the relaxed state and the gating charge, associated with the voltage-sensor part of the channel, is in the energetically costly state. In the closed state, the spring is in a compressed, high-energy state, but the gating charge has shifted favorably, by a fraction f of the membrane thickness d, with respect to the applied field. Note that we assume the total gating charge (the sum of the charges at both ends of the spring) to be Q.

voltage-gated channels respond to changes in the membrane voltage. Interestingly, as the membrane voltage is made less negative, the channels spend more and more time in the open state. What this tells us is that the preferred state of these channels is *open* and the presence of the applied voltage (in the form of the membrane potential) is what keeps them closed. A simple two-state model, sketched in Figure 17.6, provides an intuitive but quantitative description of voltage gating. Of course, the model is *not* intended to describe the actual structural details of real channels. Rather, it is intended to illustrate the way in which competing physical effects can drive gating. The essence of the model is a competition between the protein conformational degrees of freedom, represented in the toy model by a nonlinear spring, and the gating charge, which has an interaction energy with the applied field imposed by the membrane potential. Qualitatively, when the channel is open, the protein has lower energy (the spring is in the undeformed state) and the energy of the charge is higher than it would be in the closed state. When the channel is closed, the electric energy dominates and the protein energy is less favorable (that is, the spring is deformed).

As was derived in Chapter 7, the probability of the open state is given by

$$p_{\text{open}} = \frac{e^{-\beta\Delta\varepsilon}}{1 + e^{-\beta\Delta\varepsilon}} \tag{17.4}$$

where $\Delta\varepsilon$ is the energy difference between the open and the closed state. For the simple model shown in Figure 17.6, the energy difference upon gating is given by

$$\Delta\varepsilon = \Delta\varepsilon_{\text{conf}} - Q f V_{\text{mem}}. \tag{17.5}$$

$\Delta\varepsilon_{\text{conf}}$ is the difference in energy between the open and closed states that is associated with the protein conformational degrees of freedom (here captured by the spring). The second term accounts for the contribution due to motion of the charges in the potential. The part of the energy arising from the spring is negative and favors the open state. On the other hand, the electric potential contribution to $\Delta\varepsilon$ is positive (recall that V_{mem} is negative), implying that the electric field across the membrane stabilizes the closed state. The second term introduces the quantity f, shown in Figure 17.6, which is the fraction of the membrane thickness d that the gating charge Q moves by. In the uniform electric field $E = V_{\text{mem}}/d$ that is present in the membrane, $(fd)E = fV_{\text{mem}}$ corresponds to the drop in electrical potential experienced by the gating charge when it shifts from the open to the closed position. Depolarization of the cell membrane results in V_{mem} being less negative and it leads to a smaller potential drop for the gating charge in the closed state. This eventually favors the open state because the charge can assume the less favorable position with a lower energy penalty while at the same time *relaxing* the protein degrees of freedom represented by the spring. The midpoint value of the membrane potential, at which $p_{open} = 1/2$, corresponds to $\Delta\varepsilon = 0$ and is given by

$$V^* = \frac{\Delta\varepsilon_{conf}}{Qf}. \tag{17.6}$$

In terms of the membrane potential, and using Equation 17.5, the probability for the channel to be open can be written as

$$p_{open} = \frac{1}{1 + e^{\beta q(V^* - V_{mem})}}, \tag{17.7}$$

where we have defined $q = Qf$, the effective gating charge. The graph of p_{open} as a function of the membrane voltage has the measured sigmoidal shape, as seen in Figure 17.7. This will turn out to be one of the key ingredients in our model of the action potential, which we discuss toward the end of the chapter.

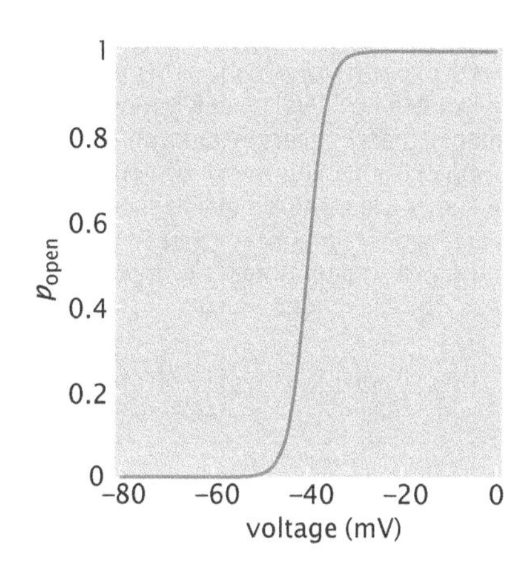

Figure 17.7: Probability of the open state of a voltage-gated channel in the simple two-state model; p_{open} is plotted as a function of the applied voltage. The parameters used are $V^* = -40$ mV and $q = 12e$, corresponding to the experimentally observed 12 charges that move across the membrane upon channel opening.

17.3.2 Maintaining a Nonequilibrium Charge State

Ions Are Pumped Across the Cell Membrane Against an Electrochemical Gradient

What are the biological consequences of transient permeability of the cell membrane to particular classes of ions? Entry of calcium ions into the cell is an important and dramatic signaling event that causes major changes in cell behavior, ranging from neuronal signaling and muscle contraction to fertilization and immunological activation. In these events, it is a particular kind of calcium channel whose permeability is regulated by transmembrane voltage that serves to translate an electrical signal into a chemical signal. The entry of calcium ions into the cell causes alterations in the conformation and activity of a breathtaking range of effector proteins that go on to direct the biochemical events associated with myosin activation in the muscle cell, fusion of vesicles with the plasma membrane, etc. We can fruitfully consider both the establishment and exploitation of transmembrane ionic and electric potentials as the gears in the heavy-lifting machinery that allows biological use to be made of calcium.

When a calcium channel opens, positively charged Ca^{2+} ions flow into the cell. Both the concentration gradient of calcium across the cell membrane and the electric field drive the flow. The concentration gradient is maintained by ATP-consuming ion pumps. One of these pumps actively transports Ca^{2+} ions out of the cell and the other exchanges sodium for potassium. Just as water pumps can keep cellars dry when heavy rain falls, the cell makes use of membrane-bound proteins that actively pump these ions out of the cell.

Excitable cells propagate electrical signals by manipulating the transmembrane voltage. The transmembrane voltage is largely established by the imbalances in ion concentrations of sodium and potassium, which are much more abundant than calcium. To see how this works, we need to consider the action of two proteins in parallel. First,

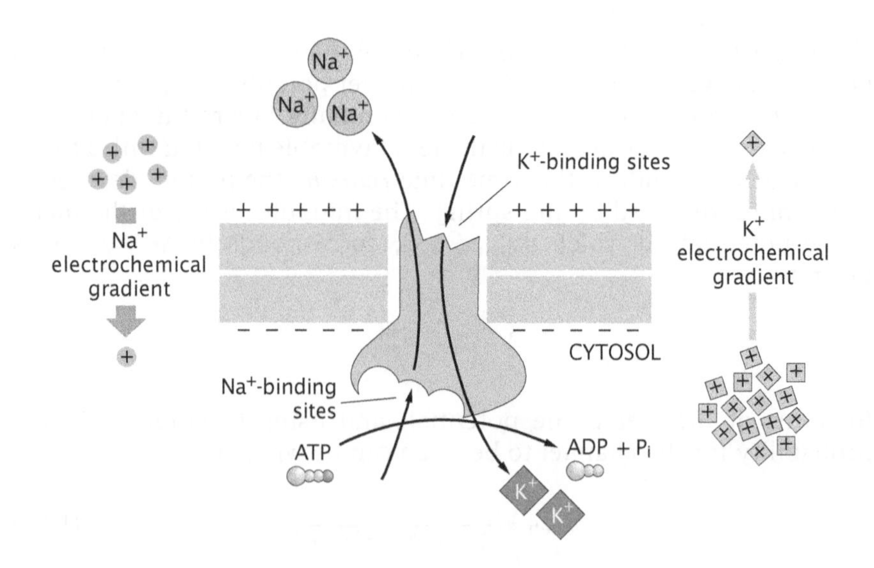

Figure 17.8: Operation of the sodium–potassium pump. This pump transports both Na^+ and K^+ ions against their concentration gradients through the consumption of ATP; three Na^+ ions are pumped out of the cell for every two K^+ ions that are pumped in. The figure schematizes the "binding sites" for these ions.

we need an ATP-consuming pump that functions similarly to the calcium pump, but instead of transporting calcium out, it ejects Na^+ ions from the cell and simultaneously adds K^+ ions into the cell. In particular, for every stroke of the pump, two K^+ ions are pumped into the cell and three Na^+ ions are pumped out, as shown in Figure 17.8. The operation of the sodium–potassium pump generates steep concentration gradients for both sodium and potassium ions across the membrane, but the electrical potential is nearly neutral. (To be precise, the electrical potential would be slightly negative, as the pump actually exchanges only two potassium ions for three sodium ions, leaving the cell with a slight deficit of positive charge on the inside.)

The second membrane protein that is critical for setting up the membrane resting potential is the potassium "leak" channel, which is selectively permeable to K^+ but not Na^+ ions. As shown in Figure 17.9, the opening of the leak channel in the presence of an active sodium–potassium pump can have a dramatic effect on the membrane potential. When the leak channels are open, the K^+ ions are moving down their concentration gradient, which tends to make the inside of the cell relatively more negative. The loss of the positive charge from the inside of the cell decreases the overall electrical driving potential for potassium ions to leave. That is, although the ions are moving with their concentration gradient, they are moving against their electrical potential. Eventually, the electrical and chemical driving potentials balance each other out, leaving the cell at a net negative resting potential (typically around −40 to −70 mV) that can approach the Nernst potential for potassium (see Table 17.1). At this point, opening a sodium-specific channel can allow sodium ions to enter the cell, in the direction favored by both their concentration gradient and the membrane potential (negative inside, attracting positively charged sodium ions). Regulated opening of sodium channels can cause the normally negative resting potential to flip to positive (up to around +30 mV), approaching the Nernst potential of sodium. In general, the identity of the open channels with maximum ion flux will determine the potential of the cell, where the ion with the greatest permeability will drive the resting potential of the membrane toward its own preferred Nernst potential.

The establishment of a transmembrane resting potential is an example of how a cell can convert chemical energy (in the form of ATP) into electrical energy. The cartoon shown in Figure 17.8 reveals that there are three Na^+-binding and two K^+-binding sites, leading to a balance of three Na^+ ions transported for every two K^+ ions. The energy necessary to accomplish this transfer comes from the hydrolysis of one ATP molecule, which is accompanied by the release of roughly 20 k_BT worth of free energy. In order to get a sense of whether the chemical-to-electrical energy conversion process is efficient or inefficient, we can compare the amount of free energy released in hydrolysis with the work that the pump needs to do against the electrochemical gradients for potassium and sodium.

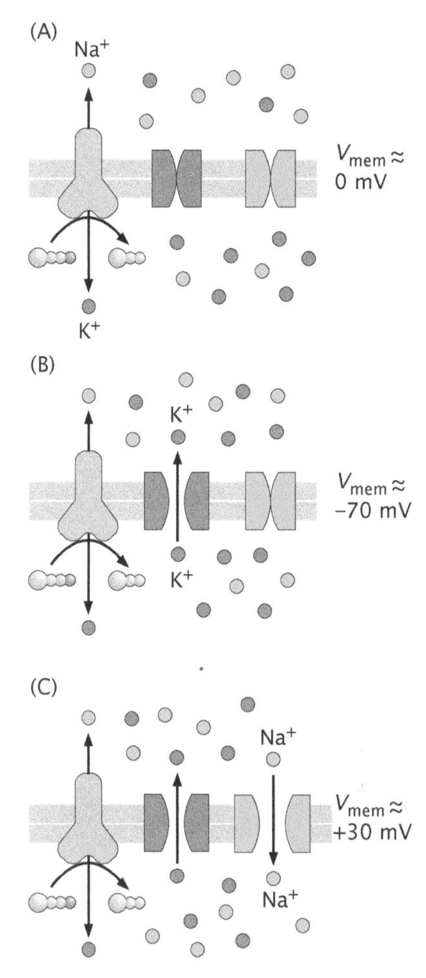

Figure 17.9: Establishment of transmembrane electrical potential differences. (A) The sodium–potassium pump (green) uses the energy of ATP hydrolysis to pump sodium ions out of the cell and pump potassium ions in. In a membrane where the pump is active but most ion channels are closed, both Na^+ and K^+ maintain steep concentration gradients. (B) Under normal circumstances, potassium-specific "leak channels" (purple) are in an open state, allowing potassium ions to exit the cell. The exit of K^+ ions renders the membrane potential negative on the inside. (C) Opening a sodium channel (orange) (for example, a voltage-gated sodium channel triggered by the propagation of an action potential) will allow Na^+ to enter, inverting the membrane potential so that it is now positive on the inside.

ESTIMATE

Estimate: Charge Pumping at Membranes The free-energy cost for transporting ions across the cell membrane is partly due to the entropy cost of transferring them against the concentration gradient and partly due to the motion of the charges in an electrical potential. To compute the "chemical" contribution (that is, the reduction in entropy), we recall from Chapter 6 that the chemical potential is the free energy per ion (or atom or molecule) and for ideal solutions is given by $\mu = \mu_0 + k_BT \ln(c/c_0)$. Hence, we can estimate the free-energy gain when we take three Na^+ ions from one side of the membrane to the other using the chemical potential difference, resulting in $W_{en} = 3k_BT \ln(c_{out}^{Na}/c_{in}^{Na}) \approx 7.5\, k_BT$, where we have used the ion concentrations given in Table 17.1. The total work for transporting the Na^+ ions consists of an additional contribution due to the electric potential and it is given by $W_{el} = -3eV_{mem} \approx 10.4\, k_BT$, which is positive as the inside of the cell is at a lower electrical potential than the outside. Putting these two contributions together, we find a total work of $17.9\, k_BT$ associated with pumping Na^+ ions. Similarly, we can determine the work that the pump does in transporting the two K^+ ions. In this case, the work against the electrical field is negative, indicating that the positive ions are being transported to the inside of the cell, which is at a lower potential than the outside. For potassium, $W_{el} = 2eV_{mem} \approx -6.9\, k_BT$, $W_{en} = 2\, k_BT \ln(c_{in}^{K}/c_{out}^{K}) \approx 7.3\, k_BT$, giving a total work of $0.4\, k_BT$. We note that this is about 40 times less than the work needed to transport the three Na^+ ions, indicative of the fact that potassium is close to being in Donnan equilibrium while sodium is not. The total work done by the sodium–potassium pump in transporting the five ions during one cycle is $W_{tot} \approx 18.3\, k_BT$, which is only slightly less than the free energy of ATP hydrolysis, of about 20 k_BT (p. 274), suggesting that the pump operates at high efficiency.

17.4 The Action Potential

17.4.1 Membrane Depolarization: The Membrane as a Bistable Switch

Cells that carry voltage-gated sodium or calcium channels are termed excitable cells and they include neurons, muscle cells, spermatazoa and eggs, and many cells involved in hormonal signaling. Like all cells, excitable cells maintain a membrane potential. The magnitude and contribution of different ions vary somewhat from cell to cell, but, generically, the potential is negative in the cell interior. As we will discuss in detail below, this state of affairs depends upon active pumps and the fact that the membrane is more permeable to potassium and chloride than it is to sodium and calcium. Because of this, any signaling event within the cell that results in a selective change in ion permeability through the membrane may result in a (transient) change in transmembrane potential away from its resting value. When channels open, allowing passage of sodium or calcium, the transmembrane potential becomes less negative and in extreme cases may even switch over to being positive. This positive change in potential is called depolarization. Conversely, if potassium channels are opened, then the potential will hew even more closely to the equilibrium potential of potassium resulting in hyperpolarization.

Particular fun happens in cells where selective ion channels can be opened not just by ligand binding, but also in response to a change in membrane voltage, as described in the previous section. In these cases, a local and transient depolarization of the membrane can be amplified and propagated to travel across the entire length of the cell. This process is illustrated in Figure 4.25 (p. 177), where a local depolarization of the cell membrane propagates along the axon due to the action of sodium channels. Sodium channels take center stage as they are responsible for the propagation of a local depolarization of the cell membrane from one patch of the axon membrane to the next. Before the action potential arrives, the sodium channels are closed. They open transiently due to depolarization of the adjacent membrane patch where the action potential has already arrived. In response to the channels opening, the membrane potential of this patch then depolarizes by jumping to the the Nernst potential of sodium, and the stage is set for depolarization of the next membrane patch. The membrane returns to the resting state by the inactivation of sodium channels, which prevents further sodium ions from crossing the membrane, followed by the action of the sodium pumps, which restore the original sodium ion concentration. We see that in order to formulate a quantitative model of the action potential, the workings of ion channels and pumps have to be understood in some detail, and we turn to that task later in this section.

Coordinated Muscle Contraction Depends Upon Membrane Depolarization

Consider the example of a skeletal muscle cell responsible for your ability to run away from a tiger. Figure 2.43 (p. 79) shows an example of the neuromuscular junction where the command to contract a muscle is executed. When the muscle cell receives a signal from the nervous system to contract, it is important that all of the myosin motors in the muscle work at the same time to give large-scale, coordinated contraction. The electrical properties of the muscle cell membrane are responsible for this spectacularly fast coordination of the contractile event. This communication of the signal received from the neuron is transferred down the muscle cell by a traveling wave of membrane depolarization. During depolarization, the electrical potential across the membrane goes from roughly $-90\,\mathrm{mV}$ to $+60\,\mathrm{mV}$. This charge reversal reflects a collusion between several different classes of ion channel and can propagate quickly along the entire membrane of the cell. This collaboration between different classes of ion channel is shown in Figure 17.10.

The first event in depolarization leading to muscle contraction is the binding of the neurotransmitter acetylcholine to receptors on the

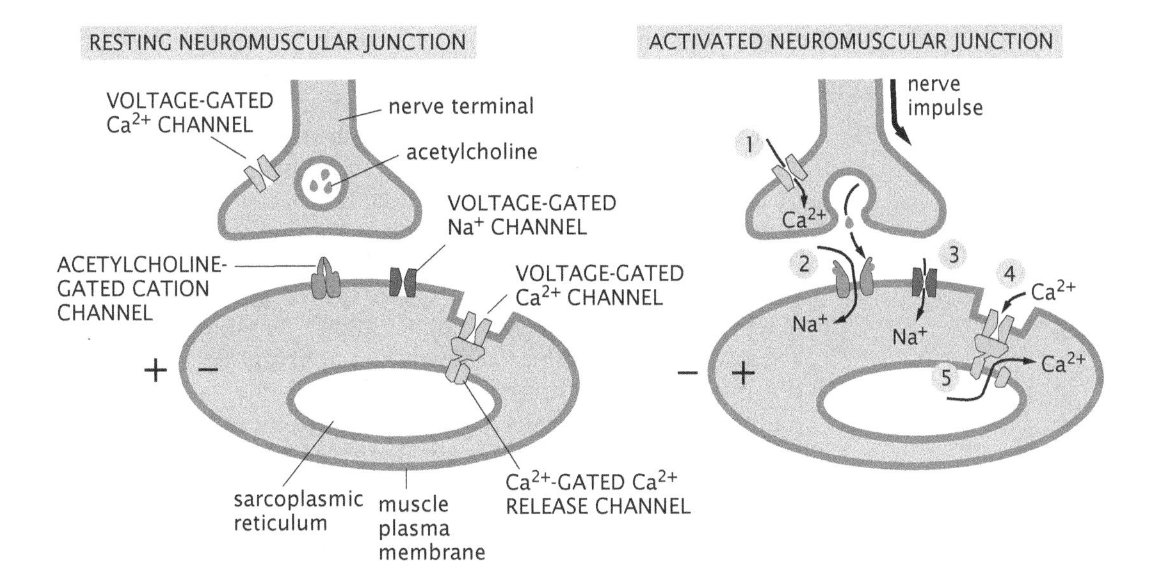

Figure 17.10: Collaboration among ion channels in muscle cell contraction. Contraction of the muscle cell is initiated by binding of acetylcholine to acetylcholine receptors. This opens channels that allow Na^+ ions to enter the cell, causing membrane depolarization. In neighboring patches of membrane, this depolarization opens voltage-dependent sodium channels, which in turn cause depolarization, which in turn opens more voltage-dependent sodium channels, etc. This self-propagating depolarization travels as a wave along the cell membrane of a muscle cell. At the same time, voltage-dependent calcium channels are opened, raising the intracellular calcium concentration above its resting level of 100 nM. Structurally, the plasma membrane forms periodic deep invaginations, the transverse tubules, that come into close contact with tubules of the sarcoplasmic reticulum which itself is filled with a high density of Ca^{2+} ions trapped in the lumen. Ca^{2+} ions entering the cell through voltage-dependent calcium channels near these points of contact can bind to and open calcium-dependent calcium channels that reside in the sarcoplasmic reticulum membrane. This influx of calcium in turn opens more calcium-dependent calcium channels, resulting in a massive dumping of Ca^{2+} ions into the myofibril's cytoplasm. This large concentration of calcium enables binding of myosin to actin and muscle contraction. (Adapted from B. Alberts et al., Molecular Biology of the Cell, 5th ed. Garland Science, 2008.)

muscle cell. The neurotransmitter is released at the terminus of the motor neuron after another electrical cellular event (an action potential) to be discussed below. Acetylcholine binds to a class of receptors on the muscle cell that do double duty as both acetylcholine-binding proteins and as channels that are highly permeable to both sodium and potassium. When these channels open due to ligand binding (see Chapter 7), there is little net flow of K^+ ions, because potassium is already close to electrochemical equilibrium. However, Na^+ ions take advantage of the open channel to rush into the cell, driven by entropic and electrical forces due to concentration and charge imbalances. As we saw in Table 17.1, the Nernst potential for Na^+ ions given the observed concentration difference across the membrane is roughly 65 mV. As a result, with the opening of the acetylcholine receptor channels, the local membrane voltage near the patch of open channels rises so much that it reverses sign and becomes positive inside. Now a second type of channel comes into play, namely, the voltage-gated sodium channel (voltage-gating data for this channel were shown in Figure 7.2 on p. 283). The membrane depolarization due to the local opening of acetylcholine receptor channels then drives a conformational change in these channels, which are much more selective for sodium than potassium. As these channels open, the local permeability ratio between sodium and potassium changes from the resting state of 0.02 (in favor of potassium) to nearly 20 (in favor of sodium). This results in the local transmembrane potential jumping from being roughly equal to the equilibrium potassium potential to the Nernst potential for sodium, since sodium permeability now dominates the system. This local change in potential serves as a driving force to gate nearby channels, which then transiently permit the flow of Na^+ ions in their neighborhood, resulting in a propagation of the depolarization across the entire cell at a rate of about a meter per second.

How does this simultaneous membrane depolarization across the cell membrane lead to simultaneous activation of contractile activity of all the myosin heads in the muscle cells? As we saw in Chapter 16, skeletal muscle myosin heads cannot actually bind to actin until Ca^{2+} ions have bound to troponin, displacing it from its position masking the myosin-binding sites (this was shown in Figure 16.39 on p. 659). It will come as no surprise that voltage-dependent calcium channels are partly responsible. However, considering the very large volume of the muscle cell compared with the small surface area of the plasma membrane, the cell uses a second trick to increase the total amount of calcium released following an electrical signal. The myofibrils in the muscle cell are enmeshed in a web of membrane, a specialized form of endoplasmic reticulum called the sarcoplasmic reticulum, that contains a high intracellular store of Ca^{2+} ions. At many places across the muscle cell, the plasma membrane forms deep invaginations that come into close apposition with the web of the sarcoplasmic reticulum. At these points of contact, calcium entering the cell across the plasma membrane triggers the opening of yet another specialized class of ion channel, the calcium-dependent calcium channel, which allows Ca^{2+} ions to flood the cytoplasm from this intermembrane store. Calcium-dependent calcium release is autocatalytic in the sense that release of calcium from a local region of the sarcoplasmic reticulum membrane will activate nearby calcium-dependent calcium channels.

As you continue to run from the tiger, the muscle cell can contract repeatedly. Thus, there must be some mechanisms for restoring the resting state after a depolarization event has occurred. There are two important components. First, voltage-gated sodium channels remain open only for a brief time. We will examine the consequences of channel closing and inactivation below when we discuss action potentials in nerve cells. Second, ATP-driven pumps restoring the nonequilibrium resting distribution of sodium and calcium operate continuously and restore the electrical and chemical imbalance, characteristic of the resting state.

The importance of gated ion channels to the functioning of animal neuromuscular systems is dramatically illustrated by the fact that these channels are the molecular targets of many of the most poisonous venoms and toxins known to humans. For example, the adorable Japanese pufferfish, which can kill an unwary diner if inexpertly prepared for sushi, carries in its liver and ovaries a large concentration of a toxin called tetrodotoxin that binds to and blocks the voltage-gated sodium channel, resulting in muscular paralysis and death by suffocation. The venom of the banded krait (a snake found in India and southeast Asia) contains α-bungarotoxin, which prevents the acetylcholine receptor channel from opening.

A Patch of Cell Membrane Can Be Modeled as an Electrical Circuit

So far, our discussion of the function of voltage-gated channels has been largely qualitative. Our goal now is to produce a quantitative description of the electrical properties of excitable membranes. In Chapter 11 we considered the cell membrane as an elastic sheet. But, as shown in Figure 1.7 (p. 14), there are many different idealizations of membranes that are useful, and one of them is as an electric circuit. In particular, in the context of biological electricity, the appropriate picture is that of a collection of resistors, batteries, and capacitors, as shown in Figure 17.11. The key point is that the presence of the lipid bilayer membrane confers a capacitance to the composite system of membrane and channels, while the presence of ion channels makes the membrane behave as a series of resistors in parallel characterized by conductances that are different for each channel type. The battery elements in Figure 17.11 denote the presence of Nernst potentials that arise due to the concentration difference of ions across the cell membrane. The Nernst potential for each ionic species can be represented as a battery because it will have to be overcome or act in concert with the electrical potential across the membrane.

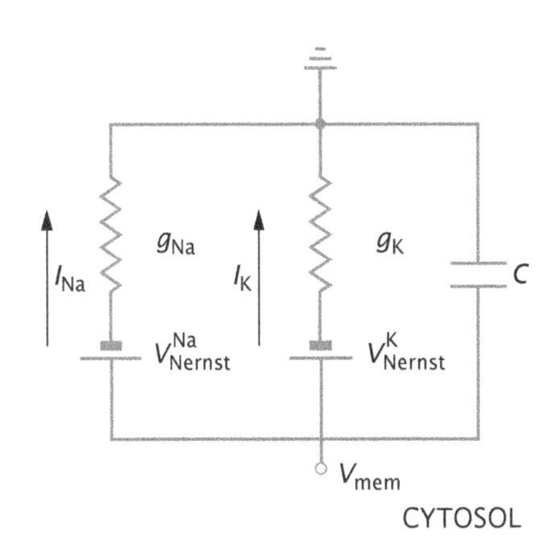

Figure 17.11: Mapping of a membrane onto an equivalent electrical circuit. The lipid bilayer yields the behavior of a capacitor of capacitance C, while the presence of channels can be modeled as resistors whose conductances (inverse resistances) are g_{Na} and g_K for the sodium and potassium channels, respectively. The batteries represent the Nernst potentials for each of the individual ionic species, which account for the ionic currents due to the concentration gradient across the membrane.

Capacitance is a measure of the ability of a circuit element to store charge. As shown in Chapter 9, a local disruption of charge neutrality is permitted near surfaces, and how this relates to a membrane is illustrated in Figure 17.12. In particular, in this setting, the capacitance is defined as the ratio of the excess charge on either side of the membrane and the membrane potential, $C_{patch} = Q_{patch}/V_{mem}$. The capacitance of a patch of the cell membrane can be approximated by thinking of it as a parallel-plate capacitor. The charge on the capacitor plates is $\pm\sigma A_{patch}$, where σ is the excess charge per unit area of membrane and A_{patch} is the area. The electric field inside a parallel-plate capacitor is uniform and equal to $\sigma/\varepsilon_0 D$, where D is the dielectric constant of the material between the plates, as discussed in Chapter 9. Therefore the potential drop across the membrane is $V_{mem} = \sigma d/\varepsilon_0 D$, where d is the thickness of the membrane, or the distance between the plates of the parallel-plate capacitor. Dividing the charge by the membrane voltage leads to the formula $C_{patch} = \varepsilon_0 D A_{patch}/d$ for the capacitance of a patch of membrane. Since the cell membrane has a thickness of $d \approx 5$ nm and a dielectric constant $D_{mem} = 2$, its capacitance is predicted to be $C = C_{patch}/A_{patch} \approx 0.4\,\mu\text{F/cm}^2$. The typical measured value for the capacitance per unit area in cell membranes is $C = 1\,\mu\text{F/cm}^2$.

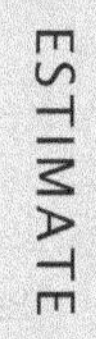

ESTIMATE

Estimate: Charge Transfer During Depolarization The depolarization of the cell membrane of a skeletal muscle cell is accompanied by a transfer of charge down the electrochemical gradient. Given that the membrane potential changes by roughly $\Delta V_{mem} = 100$ mV, and the fact that the capacitance per unit area is $C = 1\,\mu\text{F/cm}^2$, the net charge transfer, in units of electron charge $e = 1.6 \times 10^{-19}$ C, is

$$\Delta Q_{patch} \approx \Delta V_{mem} C \times 2\pi r l = 10^{10}\, e/\text{cm} \times l. \tag{17.8}$$

Here we have approximated the muscle cell as a cylinder of radius $r = 25\,\mu$m and length l. We compare this number with the typical number of ions in the cell, which can be approximated by taking the concentration of ions and multiplying it by the cell volume. In particular, we approximate the charge content inside the cell as

$$Q_{in} \approx e c_{in} \times \pi r^2 l = 10^{15}\, e/\text{cm} \times l, \tag{17.9}$$

where we use $c_{in} \approx 100$ mM as a typical ion concentration. The ratio of the charge transferred during the depolarization process to the total charge already inside the cell is

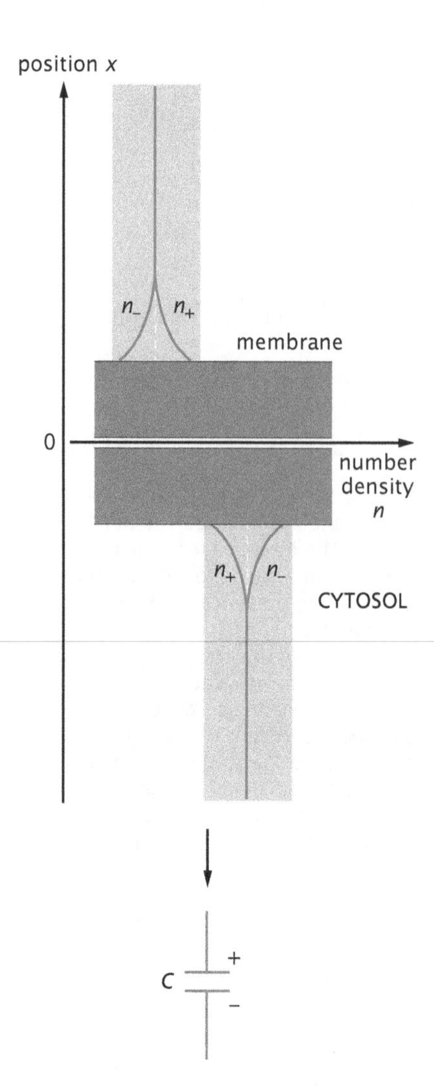

Figure 17.12: Membrane capacitance. There is an excess negative charge inside the cell (cytosol side) and an excess positive charge on the outside, leading to a voltage drop across the membrane, just like in a parallel-plate capacitor. The charge imbalance is present in a thin layer next to the cell membrane, whose width is of the order of the Debye length. The net charge at the surface of the membrane and the membrane voltage are related to each other by the membrane capacitance $C = Q/V_{mem}$.

given by

$$\frac{\Delta Q_{patch}}{Q_{in}} \approx 10^{-5} \quad (17.10)$$

and does not depend on the cell length l. The important conclusion to come out of this estimate is that the charge transferred during the process of depolarization is a very small perturbation to the concentrations of ions inside the cell.

The Difference Between the Membrane Potential and the Nernst Potential Leads to an Ionic Current Across the Cell Membrane

The idea of the membrane as a resistor and a battery in series can be understood using the general relation between current and driving force. A chemical potential difference signals that the system is not in equilibrium and it serves as a driving force for a flux of particles. This was one of the key ideas introduced in Section 5.5.2 (p. 225). We can go beyond the treatment given there by arguing that the fluxes that restore the system to equilibrium are most simply modeled as being linearly related to the driving forces themselves (this is the essence of Fick's law for diffusion introduced in Chapter 13).

In the case of interest here, the difference in chemical potential between the two sides of the cell membrane for any particular ion species is given by

$$\Delta\mu = \underbrace{\left(\mu_0 + k_B T \ln\frac{c_{in}}{c_0} + zeV_{in}\right)}_{\text{chemical potential inside}} - \underbrace{\left(\mu_0 + k_B T \ln\frac{c_{out}}{c_0} + zeV_{out}\right)}_{\text{chemical potential outside}} \quad (17.11)$$

and leads to a net flux of these ions across the membrane. Just as in Chapter 13, where the flux of diffusing molecules was proportional to the difference in concentration, here too the ionic current I (per unit area of membrane) will be proportional to $\Delta\mu$ and is given by

$$I = g\left(\frac{k_B T}{ze}\ln\frac{c_{in}}{c_{out}} + V_{in} - V_{out}\right), \quad (17.12)$$

where the coefficient of proportionality, g, is the conductance per unit area. Since current flows from higher to lower chemical potential, a positive current indicates flow of positive ions out of the cell.

Using the defining relation for the Nernst potential, namely, $V_{Nernst} = (k_B T/ze)\ln(c_{out}/c_{in})$, and the relation $V_{mem} = V_{in} - V_{out}$, for the membrane potential, we can rewrite Equation 17.12 as

$$I = g(V_{mem} - V_{Nernst}). \quad (17.13)$$

This linear relation between the current and the voltage is known as Ohm's law. Note that the Nernst potential plays the role of a battery in series with a resistor of conductance g, as depicted in Figure 17.11. Furthermore, since the total current through the membrane has contributions from all the ionic species that can pass through the membrane (through channels), the equivalent circuit diagram consists of resistor and battery elements for each ionic species connected in parallel.

Voltage-Gated Channels Result in a Nonlinear Current–Voltage Relation for the Cell Membrane

Channel gating has functional consequences for the cell. In particular, when the potential of the cell changes in a way that permits channel gating, the membrane permeability is transiently altered as a result of open channels. While open, these channels permit the flow of current with a characteristic response that is provided by the so-called *I–V* curve: a quantitative relation between the current and the voltage. Figure 17.13 shows the current-voltage (*I–V*) curves for the Na^+ and K^+ ions through the membrane of a squid axon. The simple model for the ionic current across the membrane presented in Equation 17.13 predicts a linear relationship between the membrane potential and the current. The experimentally measured *I–V* curve is roughly linear for the potassium current, while it is highly nonlinear for the sodium current (though this nonlinearity can be roughly interpreted as interpolating between two linear conductance regimes as shown below). One way to think about this is to assert that the ionic conductance g, as defined by Equation 17.13, is *voltage-dependent*.

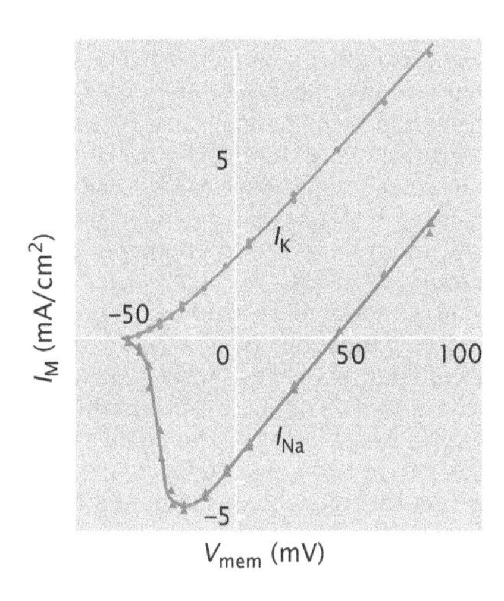

Figure 17.13: Measured *I–V* curves for sodium and potassium. The circles are the data for the current associated with K^+ ions and the triangles correspond to the Na^+ ions. In this experiment, the voltage was stepped from a negative "holding" potential to the the value shown along the *x*-axis. (Adapted from B. Hille, Ion Channels of Excitable Membranes. Sinauer Associates, 2001.)

To understand the voltage dependence of the sodium conductance in the cell membrane, we need to consider the molecular parts list of the cell membrane. Conduction of ions through the cell membrane is intimately tied to the gating properties of the ion channels themselves. If G_1 is the conductance of a single ion channel in the open state, then the conductance of a membrane patch will be $N_{open}G_1$, where N_{open} is the number of open channels in the patch of interest. The number of open channels can be computed from the probability that a particular channel is open, p_{open}. In particular, the number of open channels is given as $N_{open} = N_{patch}p_{open}$, where N_{patch} is the number of channels, open or closed, in the patch of membrane. The fact that an ion channel can be either open or closed is clearly observed in current recordings from a single channel, as shown in Figure 7.2 (p. 283), which shows single-channel observations on a sodium channel subjected to different voltages. The conductance of a membrane patch will therefore be proportional to the probability that a channel is open, $g_{patch} = N_{patch}p_{open}G_1$, as shown in Figure 17.14.

Because the overall electrical properties of the membrane depend upon the nonlinear gating properties of channels, it is useful to examine simple models of gating and to explore their consequences for membrane conductance. The fact that the *I–V* curves are not linear is a macroscopic manifestation of channel gating. To demonstrate this explicitly, we plot theoretical *I–V* curves for the Na^+ and K^+ ions

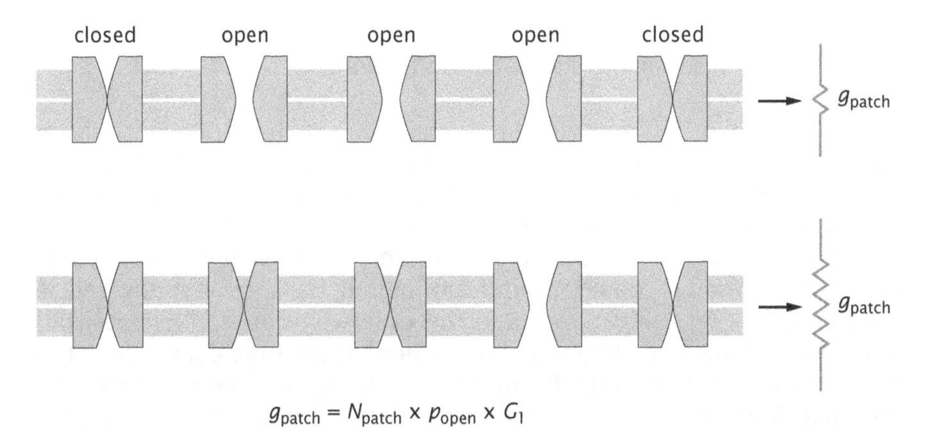

Figure 17.14: Conductance of a membrane patch. The conductance of a membrane patch depends upon the number of channels in the patch, N_{patch}, the probability that a given channel will be open, p_{open}, and the conductance of individual channels, G_1. Because the open probability is a nonlinear function of the applied voltage, so too is the overall conductance.

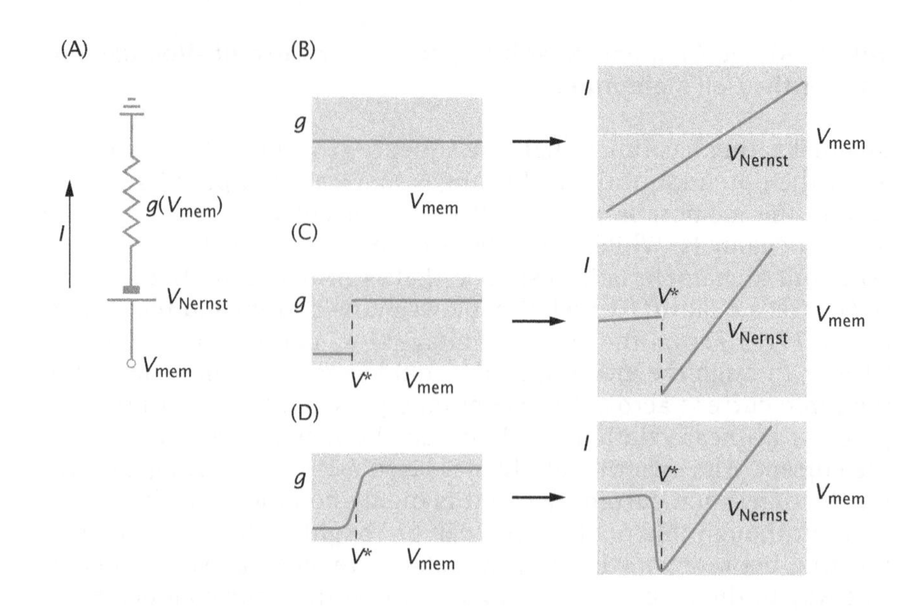

Figure 17.15: I–V curves deduced using simple arguments. (A) Circuit representing a patch of membrane containing ion channels as a resistor in series with a battery. (B) Constant-conductance model. The current in this case is given by Ohm's law, though the zero-current condition occurs when $V_{mem} = V_{Nernst}$. (C) Simple model in which the conductance is a step function: at $V_{mem} = V^*$, the conductance switches from a low value to a higher value. Each conductance regime has a linear relation between current and voltage. (D) Sigmoidal model for p_{open}. In this case, the I–V relation has an interpolating region between small and large V_{mem}.

that follow from the simple two-state voltage-gating model that led to Equation 17.7. The two main features of these curves are that they intersect the voltage axis at the Nernst potential and that their slope has a jump at V^* (the potential at which $p_{open} = 1/2$). In the first case study shown in Figure 17.15, we consider the simplest model of Ohmic conduction, which *assumes* that the channels are always open. In this case, the current is a linear function of the potential, as shown in the figure. The more realistic models acknowledge that with increasing V_{mem} the probability of channel gating will change, resulting in a concomitant change in the overall membrane conductance. This leads to two distinct regimes: at low voltage, the probability that the channel is open is low and the conductance has a small value, while at large voltage, the channels are open and the conductance has a large value. In both of these regimes, we assume a linear relation between voltage and current; it is an inheritance from p_{open} that gives the I–V curves their distinctive features.

Even though individual ion channels are either open or not (as already shown in Figure 7.2 on p. 283 and reiterated in Figure 17.16), the behavior of many channels together in a patch of membrane gives rise to conductance like that shown in Figure 17.13. In particular, even though individual channels will themselves either be open or closed in an all-or-none fashion, if we imagine a membrane patch with many such channels, the macroscopic behavior is such that it corresponds to summing over the current traces for all the individual channels.

A Patch of Membrane Acts as a Bistable Switch

With the notion of voltage-gated channels in hand, we are fully equipped to tackle the problem of biological membranes as electrically excitable media. First, we consider a patch of membrane, which we model as a resistor plus a battery element for every ionic species, and a capacitor, all connected in parallel, as already shown in Figure 17.11. We now consider what happens if the inside of the cell is made slightly less negative, as is the case when an action potential propagates. This corresponds to a net flow of positive charge into the cell, which raises the membrane potential. The charge per unit area ΔQ that flows across the membrane in time Δt can then be calculated

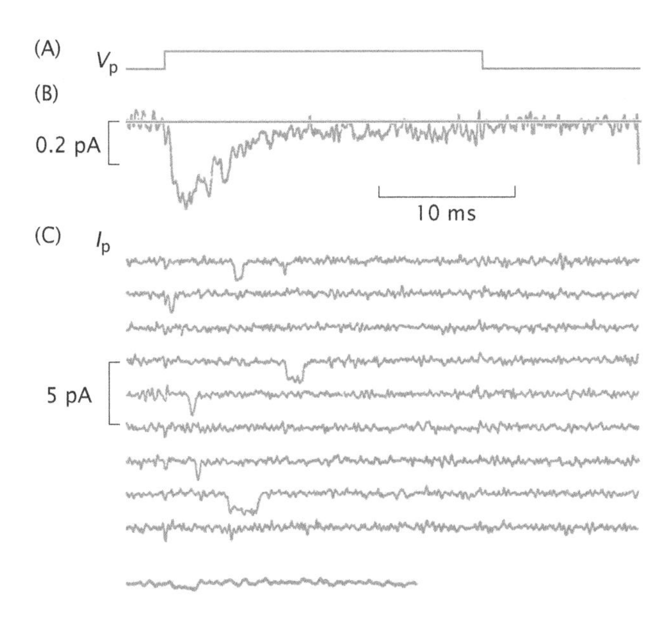

Figure 17.16: Superposition of currents from individual channels. This figure shows current traces from individual Na^+ channels in cultured muscle cells of rats. (A) A 10 mV depolarizing voltage step V_p is applied. (B) Summing the currents from individual channels over 300 trials results in the macroscopic current. (C) The individual traces show the currents through individual channels. Six of the nine traces show gating events which are revealed as downward excursions in the trajectories. (Adapted from E. R. Kandel et al., eds, Principles of Neural Science, 4th ed. McGraw-Hill, 2000.)

in two ways: from the total current across the membrane (*into* the cell),

$$\Delta Q = -(I_K + I_{Na})\Delta t, \tag{17.14}$$

or, using the capacitance per unit area of the membrane, from the change in the membrane potential,

$$\Delta Q = C\Delta V_{mem}. \tag{17.15}$$

Putting these two equations together, and using Equation 17.13, we arrive at a differential equation for the voltage across a patch of membrane,

$$C\frac{dV_{mem}}{dt} = g_K(V^K_{Nernst} - V_{mem}) + g_{Na}(V^{Na}_{Nernst} - V_{mem}). \tag{17.16}$$

Here we have in mind a small enough patch so that the voltage drop across it can be considered constant.

The equation for the membrane voltage derived above shows bistability. That is, there are two steady-state solutions for V_{mem}. We can see this by setting the left-hand side of Equation 17.16 to zero, which leads to an expression for the steady-state membrane potential,

$$V_{mem} = \frac{g_K V^K_{Nernst} + g_{Na} V^{Na}_{Nernst}}{g_K + g_{Na}}. \tag{17.17}$$

This equation has a very intuitive interpretation. To see this, consider the two cases when $g_K \gg g_{Na}$ and $g_{Na} \gg g_K$. In the former case, Equation 17.17 yields $V_{mem} \approx V^K_{Nernst}$. In the case when $g_{Na} \gg g_K$, Equation 17.17 yields $V_{mem} \approx V^{Na}_{Nernst}$. In other words, the ion species with the highest conductance sets the membrane potential approximately equal to its Nernst potential. The same conclusion follows from the phase portrait of Equation 17.16, which is shown in graphical form in Figure 17.17. The goal of our analysis is to figure out what steady-state voltages are available to the composite system of membrane and ion channels and to illustrate that there are two "fixed points" corresponding to the potassium and sodium Nernst potentials.

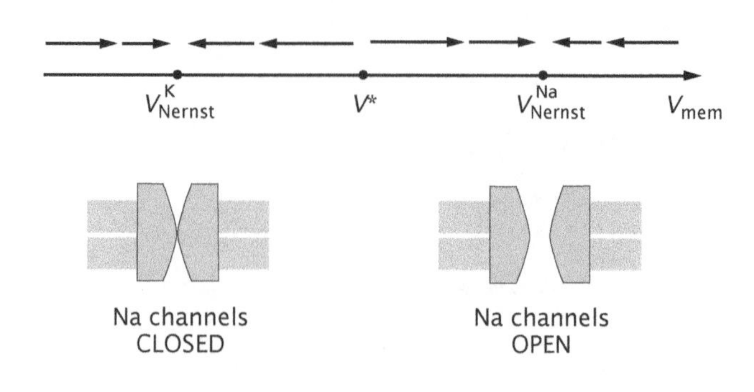

Figure 17.17: Dynamics of the membrane potential. Using Equation 17.16, we can compute dV_{mem}/dt, which depends upon V_{mem}, $V^{\mathrm{K}}_{\mathrm{Nernst}}$, and $V^{\mathrm{Na}}_{\mathrm{Nernst}}$. Depending upon the relative values of these three parameters, dV_{mem}/dt will be negative, zero, or positive, with the potential at which the sodium channels are gated, V^*, dictating this change of sign. The sign and magnitude of dV_{mem}/dt is plotted as a function of V_{mem} and reveals two "fixed points" corresponding to $V^{\mathrm{K}}_{\mathrm{Nernst}}$ and $V^{\mathrm{Na}}_{\mathrm{Nernst}}$. When the sodium channels are closed, the potassium Nernst potential dominates, and when they are open, the sodium Nernst potential dominates. Therefore, the membrane is a dynamical system that behaves as a bistable switch.

To simplify the analysis, we assume that the conductance of the potassium channel is voltage-independent. In reality, potassium channels are also voltage-gated, as we have already seen. Still, the assumption is not completely devoid of merit, since, upon depolarization of the membrane, the opening of potassium channels occurs in about 5 ms, while for sodium channels the time to open is roughly 1 ms. Using the formula for p_{open}, obtained from the two-state model of voltage gating, the voltage dependence of the conductance of sodium channels is

$$g_{\mathrm{Na}} = g^{\mathrm{open}}_{\mathrm{Na}} \frac{1}{1 + \mathrm{e}^{\beta q (V^* - V_{\mathrm{mem}})}}. \qquad (17.18)$$

We denote the conductance when all channels in the membrane patch are open as $g^{\mathrm{open}}_{\mathrm{Na}}$. With this expression in hand, we can analyze Equation 17.16 for bistability.

It follows from Equation 17.18 that for $V_{\mathrm{mem}} < V^*$, the sodium conductance will be small, and we assume that it is considerably smaller than the potassium conductance. Then, it follows from Equation 17.17 that the membrane voltage will be close in value to the Nernst potential of potassium. In the case when the initial membrane voltage is above V^*, the sodium channels open up and the sodium conductance dominates potassium. In this case, the membrane voltage tends to the Nernst potential of sodium, which is positive. The existence of the two stable fixed points, $V_{\mathrm{mem}} = V^{\mathrm{K}}_{\mathrm{Nernst}}$ and $V_{\mathrm{mem}} = V^{\mathrm{Na}}_{\mathrm{Nernst}}$, reveals the bistable behavior illustrated schematically in Figure 17.17. The existence of this bistability provides an opportunity for switching behavior of excitable membranes, which is the basis for action potential propagation.

The Dynamics of Voltage Relaxation Can Be Modeled Using an *RC* Circuit

As discussed above, a patch of membrane will act as a bistable switch due to the voltage dependence of its ability to conduct Na^+ ions. The question then arises, if we excite the membrane patch by raising the membrane voltage above its resting value, but not so much so as to flip the switch, how quickly will V_{mem} relax back to its resting value?

To answer this question, we can make use of Equation 17.16 and solve it assuming $g_{\mathrm{Na}} \ll g_{\mathrm{K}}$, which implies that the steady-state membrane potential is approximately equal to $V^{\mathrm{K}}_{\mathrm{Nernst}}$. The initial, $t = 0$, value of the membrane potential is taken to be $V_{\mathrm{mem}}(0) = V^{\mathrm{K}}_{\mathrm{Nernst}} + \Delta V$, where ΔV is such that $V_{\mathrm{mem}}(0) < V^*$. The solution to Equation 17.16 with this prescribed initial condition is

$$V_{\mathrm{mem}}(t) = V^{\mathrm{K}}_{\mathrm{Nernst}} + \Delta V \mathrm{e}^{-(g_{\mathrm{K}}/C)t}. \qquad (17.19)$$

In words, the membrane potential after a subthreshold stimulus ΔV decays back to the resting value exponentially fast, with a time constant

$$\tau = \frac{C}{g}, \tag{17.20}$$

where g is the conductance of the membrane patch, which we assumed above to be given approximately by g_K. For a squid giant axon in the resting state, the values for the conductance and capacitance of a membrane patch are $5\,\Omega^{-1}\,\mathrm{m}^{-2}$ and $10^{-2}\,\mathrm{F/m}^2$, respectively, yielding an estimate of the time constant $\tau \approx 2$ ms.

17.4.2 The Cable Equation

So far we have considered a patch of membrane where the membrane voltage only varied in the direction perpendicular to the patch. In experiments, this is achieved by placing a long thin wire along the axon axis, which is then held at constant voltage (the so-called voltage clamp). When considering the propagation of an action potential, on the other hand, the membrane voltage varies as a function of position along the axon. At any given time, the membrane voltage, roughly speaking, is at its resting value everywhere except at the position of the action potential spike, where the membrane is depolarized. To gain some quantitative intuition about the spatial variation of the membrane voltage, we consider a situation where one end of a cylindrical axon is held at a fixed voltage V_0 and we ask how the membrane voltage will change as a function of position along the axon. The setup is shown in Figure 17.18(A). To examine this situation quantitatively, we consider the membrane as a series of resistors, batteries, and capacitors connected in a circuit as shown in Figure 17.18(B). To simplify our discussion here, we describe the membrane patch with only one type of resistor and battery, which can be thought of as the equivalent resistance and Nernst potential for the total current (sodium plus potassium) across the membrane.

To obtain an equation that describes the spatial variation of the membrane voltage, we note that the change of voltage, $V(x + \Delta x) - V(x)$, is due to the longitudinal current flowing inside the axon, $i(x)$. If ΔR_{int} is the electrical resistance of an axon of length Δx for current propagating along the long axis of the axon, then Ohm's law states

$$V(x + \Delta x) - V(x) = -i(x)\Delta R_{\mathrm{int}}. \tag{17.21}$$

The longitudinal current $i(x)$ changes with distance x along the axon because of losses due to currents in the radial direction, $i_r(x)$, perpendicular to the cell membrane. Looking at the vertex in the circuit marked $V(x)$ in Figure 17.18, there is a current $i(x - \Delta x)$ coming in, and two currents, $i(x)$ and $i_r(x)$, going out. Conservation of current says that the amount of current flowing in must equal the amount flowing out. This gives us another relation between $i(x)$ and $V(x)$,

$$i(x - \Delta x) - i(x) = i_r(x) = \Delta g_{\mathrm{patch}}[V(x) - V_{\mathrm{Nernst}}]. \tag{17.22}$$

We define $\Delta g_{\mathrm{patch}}$ as the conductance of the tiny patch of membrane associated with the small piece of the axon of length Δx. Taylor-expanding the left-hand sides of the above two equations in Δx leads to a pair of differential equations for the membrane voltage and the

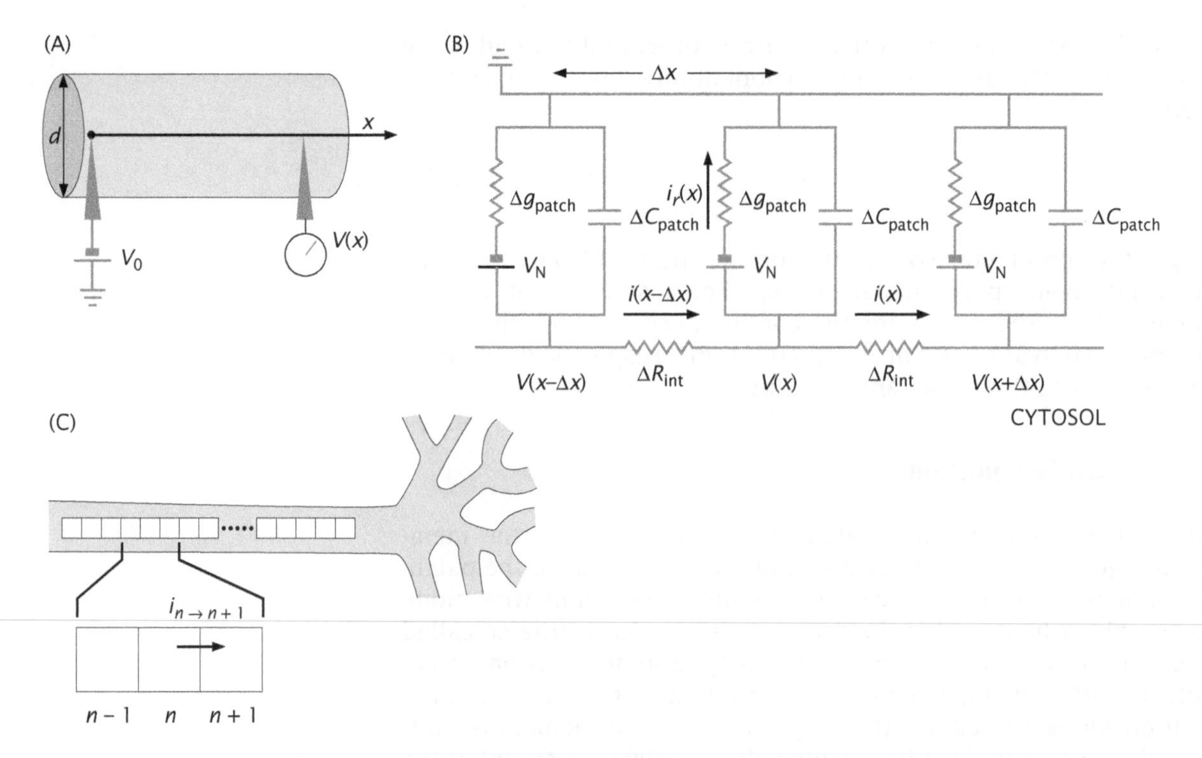

Figure 17.18: Spatial variation of voltage during an action potential. (A) An axon of diameter d is held at fixed voltage V_0 at one of its ends; the membrane voltage $V(x)$ is measured a distance x away from the end. (B) The spatial dependence of the membrane voltage is modeled by a lattice of circuit elements, where Δg is the equivalent membrane conductance, V_N is the equivalent Nernst potential, and C is the membrane capacitance. The lattice spacing is Δx. Each axon element is characterized by a membrane voltage $V(x)$ and a radial current flowing out of the cell, $i_r(x)$. In the steady-state situation that we consider here, the current through the capacitors is zero. The current $i(x)$ is the longitudinal current flowing from one axon element to the next. These currents encounter an internal resistance ΔR_{int} due to the cytoplasm. (C) Discrete model for the propagation of an electrical impulse along the axon. Each box corresponds to a circuit element connected to adjacent elements as shown in (B).

longitudinal current,

$$\frac{\mathrm{d}V(x)}{\mathrm{d}x} = -\frac{\Delta R_{int}}{\Delta x} i(x), \tag{17.23}$$

$$\frac{\mathrm{d}i(x)}{\mathrm{d}x} = -\frac{\Delta g}{\Delta x}[V(x) - V_{Nernst}]. \tag{17.24}$$

Before attempting to solve these equations, we consider the meaning of the conductance Δg_{patch} and the resistance ΔR_{int}. The conductance of a patch is proportional to its area and can be written as $\Delta g_{patch} = g\pi d\Delta x$, where d is the axon diameter and g is the conductance per unit area. The longitudinal resistance, on the other hand, can be written as $\Delta R_{int} = \rho\Delta x/(\pi d^2/4)$, where ρ is the resistivity of the intracellular medium (we will similarly use $\Delta C_{patch} = C\pi d\Delta x$ in the next section). For a squid giant axon, we take $d \approx 0.5$ mm, $\rho \approx 0.3\,\Omega$ m, and $g \approx 5\,\Omega^{-1}\,\mathrm{m}^{-2}$.

In order to solve the equations for the membrane voltage and current, we begin by taking the derivative of both sides of Equation 17.23 and substituting for $\mathrm{d}i(x)/\mathrm{d}x$ from Equation 17.24. This gives a single second-order equation for the membrane voltage,

$$\frac{\mathrm{d}^2V(x)}{\mathrm{d}x^2} = \frac{1}{\lambda^2}[V(x) - V_{Nernst}], \tag{17.25}$$

where

$$\lambda = \sqrt{\frac{d}{4\rho g}} \tag{17.26}$$

has units of length. The meaning of the length λ is revealed by the solution to the above equation for the membrane potential, with the boundary condition $V(0) = V_0$, namely,

$$V(x) = (V_0 - V_{\text{Nernst}})e^{-x/\lambda} + V_{\text{Nernst}}. \tag{17.27}$$

Therefore, λ is the characteristic distance over which the voltage perturbation (raising of the voltage with respect to V_{Nernst}) diminishes along the long axis of the axon. For the squid giant axon, this distance is $\lambda \approx 9$ mm. Still, this result falls short of the full repertoire of dynamical electrical behavior seen in cells and showcased in Figure 17.3 (p. 687). To consider action potentials in their full glory, we need to merge an understanding of the spatial and temporal variations in the membrane potential.

17.4.3 Depolarization Waves

To make further quantitative progress on the question of propagation of the membrane depolarization, we combine our analysis of a single patch of membrane as a bistable switch with the cable equation, which accounts for the electrical communication between different patches along the axon. The current conservation that led to Equation 17.22 is now modified to include separately the sodium and potassium currents, as well as the capacitive current given by the change in charge on the capacitor (membrane) due to the change in membrane voltage as described by Equation 17.15. Putting all of this together results in

$$i(x - \Delta x, t) - i(x, t) = \Delta g^{\text{Na}}_{\text{patch}}[V(x, t) - V^{\text{Na}}_{\text{Nernst}}] + \Delta g^{\text{K}}_{\text{patch}}[V(x, t) - V^{\text{K}}_{\text{Nernst}}] + \Delta C_{\text{patch}}\frac{\partial V(x, t)}{\partial t}, \tag{17.28}$$

which is an equation for the current $i(x)$ flowing through the cell as depicted in Figure 17.18(B).

Waves of Membrane Depolarization Rely on Sodium Channels Switching into the Open State

As illustrated in Figure 17.18, the current $i(x, t)$ is the axial current at time t between the membrane patch at position x and the patch at position $x + \Delta x$, while $V(x, t)$ is the membrane potential at position x at time t. Each patch is a cylinder Δx in length and of diameter d. The most important new feature is that now we explicitly account for the voltage dependence of the sodium conductance as described by Equation 17.18. We also, once again, make the simplifying assumption that the potassium conductance does not change with voltage, so as to mimic the fact discussed earlier that the potassium channels open on a longer time scale compared with the sodium channels upon depolarization of the membrane. The conductances per unit area that we use in this model of the action potential propagation are plotted in Figure 17.19. In particular, we model the conductance (per unit area) for Na^+ ions as

$$g_{\text{Na}} = \left(\frac{100}{1 + e^{\beta q(V^* - V)}} + \frac{1}{5}\right)\Omega^{-1}\,\text{m}^{-2}, \tag{17.29}$$

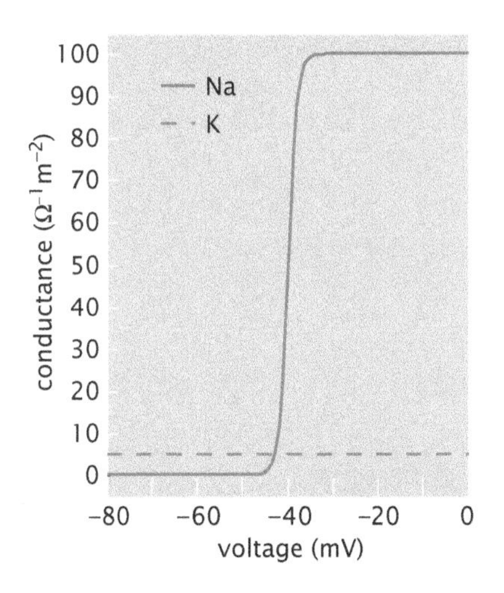

Figure 17.19: Channel conductances. Sodium and potassium conductance of a membrane patch of unit area used in the numerical solutions to the time-dependent cable equation.

with $\beta q = 0.5\,\text{mV}^{-1}$, while the potassium conductance is constant and given by $g_K \approx 5\,\Omega^{-1}\,\text{m}^{-2}$. This form for the sodium conductance is dictated by the experimental observation that in the resting state $g_{Na} \approx (1/25) g_K$, while at the peak of the action potential, $g_{Na} \approx 20 g_K$.

To arrive at the governing equation for the membrane potential, we follow the same steps as those leading to Equation 17.25. In particular, we combine the current conservation equation, Equation 17.28, with the equation describing the voltage drop from one patch to the next due to the axial current, Equation 17.21, to obtain the time-dependent cable equation

$$\lambda^2 \frac{\partial^2 V(x,t)}{\partial x^2} - \tau \frac{\partial V(x,t)}{\partial t} = \frac{g_{Na}(V(x,t))}{g_K}[V(x,t) - V^{Na}_{Nernst}] + [V(x,t) - V^{K}_{Nernst}], \tag{17.30}$$

where the characteristic length scale λ is given by Equation 17.26 while the characteristic time τ is defined in Equation 17.20 (but with g replaced with g_K in both equations).

The left-hand side of Equation 17.30 evokes the diffusion equation for the concentration of random walkers that we discussed in Chapter 13. In fact, it is not exactly the same as the diffusion equation, since the right-hand side is a rather complicated nonlinear function of the voltage. Still, we can gain some insight about its solutions if we rely on our intuition about the diffusion equation as well as the bistable nature of a membrane patch.

To make quantitative progress, we assume that initially a current is injected into the axon at position $x = 0$ over a small region of width 1 mm, thus changing the membrane voltage in that region of the axon. If the membrane voltage in this small region around $x = 0$ is below the threshold for the opening of the sodium channels, then these channels will remain closed and the membrane voltage will relax to its resting value, which is close to the Nernst potential of potassium. When the membrane voltage is close to its resting value, the right-hand side of Equation 17.30 is small and the dynamics of the voltage is essentially diffusive. In other words, the relaxation of the voltage pulse is equivalent to the relaxation to equilibrium of a spatially localized spike in the concentration of diffusing particles. This is shown in Figure 17.20(A).

When the initial voltage spike is above the threshold, the sodium channels open and the membrane voltage in the small region around $x = 0$ tends to the Nernst potential of sodium. At the same time, this perturbation propagates into regions away from $x = 0$, which leads to the opening of sodium channels in those regions as well, driving the membrane voltage there to the Nernst potential of sodium. This leads to the establishment of a propagating front, as shown in Figure 17.20(B). The scale of the front propagation speed is necessarily set by the ratio λ/τ, since these are the only two parameters in Equation 17.30 that have units of length and time. Figure 17.21 shows the results of numerical determination of the front propagation speed obtained from Equation 17.30 as a function of axon diameter d. Since $v \approx \lambda/\tau$ and since Equation 17.26 tells us that $\lambda \sim \sqrt{d}$, we expect the velocity to scale with axon diameter as $v \sim \sqrt{d}$, which is confirmed by numerical solution of the time-dependent cable equation. Biologically, the implication is that larger-diameter nerve axons can transmit information more quickly. In the squid, the giant axon coordinates an escape response enabling the animal to flee from a predator. It is clear why the speed of this response should be important. The reader

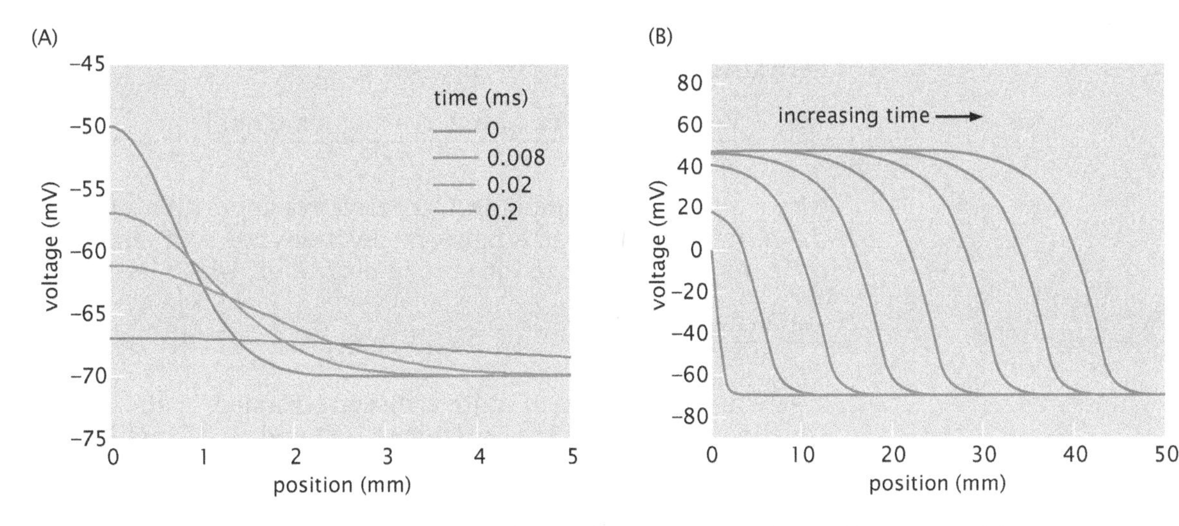

Figure 17.20: Numerical solutions to the time-dependent cable equation, Equation 17.30, using $\lambda = 9.1$ mm, $\tau = 2$ ms, $V^{\mathrm{Na}}_{\mathrm{Nernst}} = 54$ mV, $V^{\mathrm{K}}_{\mathrm{Nernst}} = -75$ mV, and $V^* = -40$ mV. The initial membrane voltage is a Gaussian spike above the resting membrane potential, $V(x, 0) = \Delta V e^{-(x/\sigma)^2} + V_{\mathrm{mem}}$, of amplitude ΔV and width $\sigma = 1$ mm. The resting potential V_{mem} is related to the Nernst potentials of Na^+ and K^+ by Equation 17.17. (A) Solution when the initial voltage spike is subthreshold, showing a decaying signal. (B) When the initial voltage spike is above the threshold, a propagating front solution is obtained. The snapshots in the propagating front are shown at 200 ms intervals.

is invited to explore numerical solutions of the cable equation in the Computational Exploration below and in the problems at the end of the chapter.

Depolarization waves similar to these are responsible for the coordinated calcium influx that enables all of the sarcomeres in the muscle cell to contract simultaneously. Like action potentials, which have the form of traveling spikes, these waves appear as an all-or-nothing response to the stimulating current, and their shape and propagation speed are independent of the size of the stimulus. The missing ingredient that leads to spikes is the inactivation of sodium channels.

Computational Exploration: Numerical Solution of the Cable Equation In the Computational Exploration "Growth Curves and the Logistic Equation" in Section 3.1.3 (p. 99), we showed how we can numerically integrate the logistic equation. In that case, we only had to integrate the equation with respect to time. The present problem generalizes these earlier ideas to a partial differential equation in which we have to solve for an unknown function (namely, the current in the axon) over both space and time.

COMPUTATIONAL EXPLORATION

In order to solve Equation 17.30 numerically, we need to discretize the current in both space and time. As in the case of the logistic equation done earlier (Section 3.1.3, p. 99), we divide time into small intervals of length Δt. Additionally, we divide the axon into small segments of length Δx as shown in Figure 17.18(C). With such a discretization in hand, we then need to find approximate representations of the derivatives that appear in the partial differential equation. Let's start by looking at the time derivative in Equation 17.30. Using the definition of the derivative, we can discretize this term as follows:

$$\frac{\partial V(x,t)}{\partial t} \approx \frac{V(x,t+\Delta t) - V(x,t)}{\Delta t}. \tag{17.31}$$

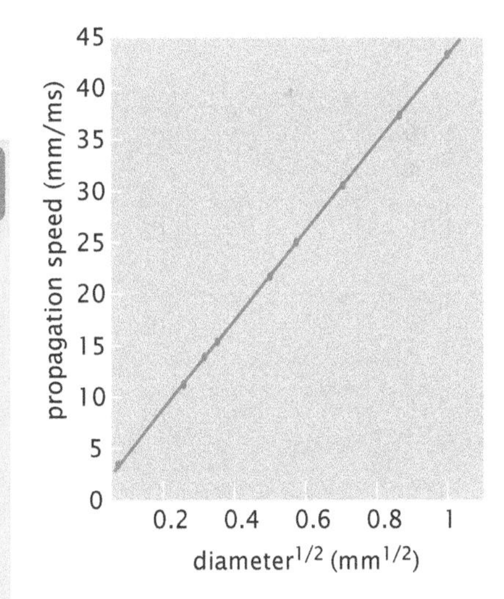

Figure 17.21: Propagation speed of depolarization wave. The propagation speed scales with the square root of the axon diameter, as predicted from the dimensional analysis of Equation 17.30. The data points show the propagation speed for different diameter values calculated numerically using the cable equation. The parameters used are the same as those used to produce Figure 17.20.

For the second derivative in x, we use the definition of the derivative several times in succession, as

$$\frac{\partial^2 V(x,t)}{\partial x^2} \approx \frac{\partial}{\partial x}\left[\frac{V(x+\Delta x/2,t) - V(x-\Delta x/2,t)}{\Delta x}\right], \tag{17.32}$$

where we have calculated the first derivative at position x using the difference in voltage between positions $x+\Delta x/2$ and $x-\Delta x/2$. We now discretize the second derivative, which leads to

$$\frac{\partial^2 V(x,t)}{\partial x^2} \approx \frac{1}{\Delta x^2}\left[V(x+\Delta x,t) - 2V(x,t) + V(x-\Delta x,t)\right]. \tag{17.33}$$

We are now ready to write a discrete version of the cable equation. By substituting Equations 17.31 and 17.33 into Equation 17.30, we can calculate the voltage at time $t+\Delta t$ at position x as a function of the voltages at time t at that same position and the adjacent "boxes" located at $x-\Delta x$ and $x+\Delta x$, namely

$$\begin{aligned} V(x,t+\Delta t) = V(x,t) + \frac{1}{\tau}\Bigg\{ & \frac{\lambda^2}{\Delta x^2}\left[V(x+\Delta x,t) - 2V(x,t)\right. \\ & \left. + V(x-\Delta x,t)\right] - \frac{g_{\mathrm{Na}}(V(x,t))}{g_{\mathrm{K}}}\left[V(x,t) - V^{\mathrm{Na}}_{\mathrm{Nernst}}\right] \\ & - \left[V(x,t) - V^{\mathrm{K}}_{\mathrm{Nernst}}\right]\Bigg\}, \end{aligned} \tag{17.34}$$

where the conductance of sodium (per unit area) is given by Equation 17.29 and the sodium and potassium conductances are also shown graphically in Figure 17.19. The voltage drop between adjacent discrete elements of our imaginary axon leads to a current flow, which in turn leads to a change in the voltage at the position of interest.

When solving partial differential equations like this, we need to provide both initial conditions and boundary conditions. These boundary conditions tell us about the disposition of the current at the edge of the domain. If solving this equation analytically, we could, for example, assume an infinite neuron. This strategy of assuming an infinite domain of integration is not practical in the context of numerical integration. A simple way of dealing with this is to set up periodic boundary conditions. This means that we will join both ends of our neuron. Specifically, if we think of a neuron of length $2L$, the boundary condition at $x=-L$ is implemented as follows:

$$\begin{aligned} V(-L,t+\Delta t) = V(-L,t) + \frac{1}{\tau}\Bigg\{ & \frac{\lambda^2}{\Delta x^2}\left[V(-L+\Delta x,t) - 2V(-L,t)\right. \\ & \left. + V(L,t)\right] - \frac{g_{\mathrm{Na}}(V(-L,t))}{g_{\mathrm{K}}}\left[V(-L,t) - V^{\mathrm{Na}}_{\mathrm{Nernst}}\right] \\ & - \left[V(-L,t) - V^{\mathrm{K}}_{\mathrm{Nernst}}\right]\Bigg\}, \end{aligned} \tag{17.35}$$

where we have imposed our periodic boundary condition through the assignment

$$V(-L-\Delta x,t) = V(L,t). \tag{17.36}$$

Similarly, the evolution of the voltage at $x = L$ is given by

$$V(L, t+\Delta t) = V(L, t) + \frac{1}{\tau}\left\{\frac{\lambda^2}{\Delta x^2}\left[V(-L, t) - 2V(L, t) + V(L - \Delta x, t)\right] - \frac{g_{\mathrm{Na}}(V(L, t))}{g_{\mathrm{K}}}\left[V(L, t) - V_{\mathrm{Nernst}}^{\mathrm{Na}}\right] - \left[V(L, t) - V_{\mathrm{Nernst}}^{\mathrm{K}}\right]\right\}. \tag{17.37}$$

Here, we have used the condition

$$V(L + \Delta x, t) = V(-L, t). \tag{17.38}$$

Finally, before proceeding to the numerical solution of the equation, we need to set up an initial condition. To be concrete, we imagine an initial pulse captured by a Gaussian of width $\sigma = 1$ mm and height V_0 given by

$$V(x, t = 0) = (V_0 - V_{\mathrm{mem}})\,\mathrm{e}^{-(x/\sigma)^2} + V_{\mathrm{mem}}. \tag{17.39}$$

Notice that the initial voltage is then V_0 at $x = 0$ and V_{mem} far away from that position, where V_{mem} is given by Equation 17.17. Our goal is to see how this pulse evolves over time.

We are now almost ready to proceed with the numerical integration. Before we move forward though, we must be careful with our choices of Δt and Δx. Our instinct might be that by choosing Δt and Δx to be ever smaller will increase the accuracy of our integration, but at the cost of longer computation time. However, the choice of these two parameters is more nuanced. As shown in Figure 17.20, we expect a propagating voltage front for certain initial conditions. From the text, we know that the front should propagate at a speed $v \approx \lambda/\tau$. In order to follow the propagating front, we want to solve our equation at a resolution faster than that speed, namely,

$$v < \frac{\Delta x}{\Delta t}, \tag{17.40}$$

which can also be written as

$$\frac{v\Delta t}{\Delta x} < 1. \tag{17.41}$$

However, the situation is even more subtle, because Equation 17.30 has diffusive terms as well. As a result, we do not just want to integrate on a time scale faster than the propagation of the front, we also want to integrate on a time scale faster than the diffusion process, which is given by $D = \lambda^2/\tau$. This condition can be written as

$$D = \frac{\lambda^2}{\tau} < \frac{\Delta x^2}{\Delta t} \tag{17.42}$$

or

$$D\frac{\Delta t}{\Delta x^2} < 1. \tag{17.43}$$

These kinds of conditions are known as Courant–Friedrichs–Lewy conditions. We see from here that, for example, as we make Δx small, we will have to also make Δt small in order to keep their ratio within the allowed range. For the conditions shown in Equation 17.41, for example, how much smaller than one the ratio $v\Delta t/\Delta x$ needs to be varies with the actual equation being considered. To address this, one has to resort to tools such as linear stability analysis, but for our purposes it suffices to alert the reader to the possible pitfalls.

In order to be able to obtain the numerical solution in a reasonable time, we want to keep Δt relatively large. A way to achieve this while still respecting the condition given by Equation 17.41 for the cable equation considered here is to decrease the speed v and diffusion constant D, which can be done by decreasing λ, as can be seen from Equation 17.30. For example, some reasonable parameters are $\lambda = 0.18$ mm, $\Delta t = 5 \times 10^{-4}$ ms, and $\Delta x = 0.01$ mm, such that $v\Delta t/\Delta x \approx 5 \times 10^{-3}$ and $D\Delta t/\Delta x^2 \approx 0.08$.

One extra subtlety to keep in mind when performing such a numerical integration has to do with the amount of data that is going to be produced. If we integrate over, for example, 20 ms, this means that we would be saving $20\,\text{ms}/(5 \times 10^{-4}\ \text{ms}) = 4 \times 10^4$ time points. Large datasets from long integration times can easily overflow the memory that Matlab or any similar software can handle. Because the action happens on a time scale that is longer than Δt, we can choose, for example, to save only one of every hundred data points for later use such as generating plots like those shown in Figure 17.20 .

Taking into account all the considerations described above, we can generate time traces such as those shown in Figure 17.20(A) for a value of V_0 below the threshold and Figure 17.20(B) for an initial voltage amplitude that is above the threshold. Interestingly, as shown in the problems at the end of the chapter, functions such as `pdepe` in Matlab can automatically decide on values of Δx and Δt in an adaptive way. Namely, it can change their values at different time points of the simulation. Still, this computational exploration should give the reader a sense for the subtleties involved in solving such partial differential equations, which will perhaps be overlooked if functions such as `pdepe` are used blindly.

17.4.4 Spikes

Our discussion so far of membrane depolarization and the rapid traveling waves of both depolarization and rapid calcium influx can apply to all excitable cells. Neurons, however, are the fanciest and most beloved of excitable cells because they go beyond these blunt global types of electrical events to generate a nearly digital code of signaling that conveys information about signal strength and duration encoded in pulse trains.

The analog-to-digital signal conversion undertaken by neurons in their electrical activity can enable our brains to communicate directly with computers and vice versa since they use a fundamentally similar bit-based informational mechanism. The engineering of such interfaces has enabled paralyzed people to move computer cursors with their thoughts. The specific mechanism by which a neuron takes an analog input and converts it into a digital yes–no output is to integrate numerous small signaling events as they converge at the cell body,

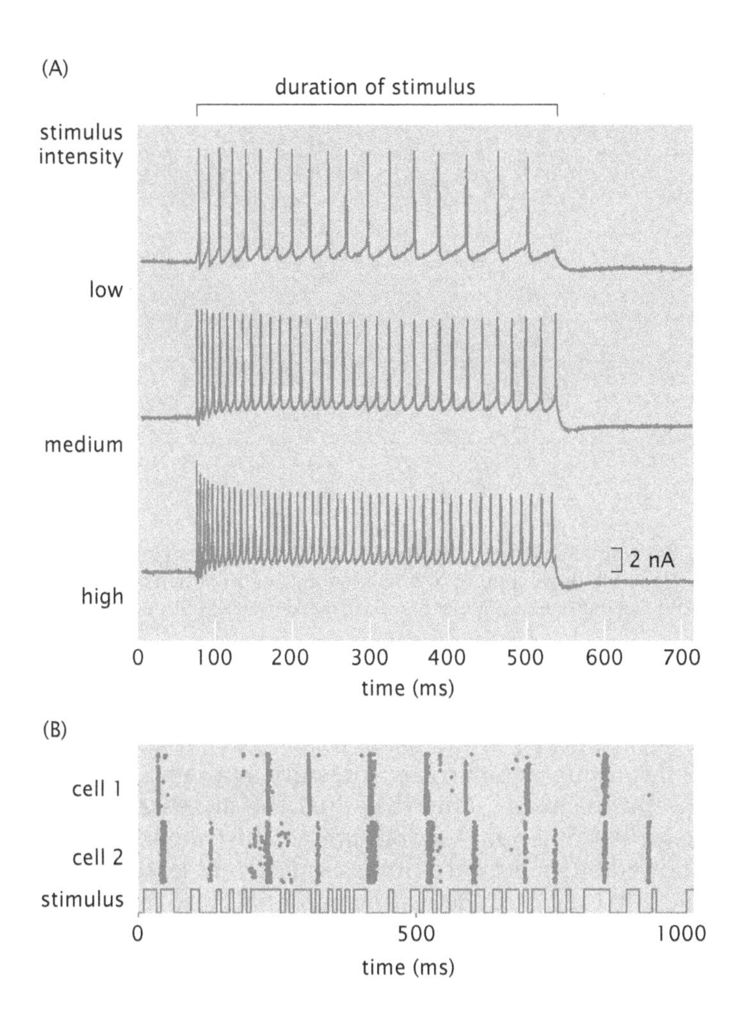

Figure 17.22: Information-encoding in spike patterns. (A) Stimulating neurons with current from an electrode typically results in the initiation of a series of action potentials as shown here in this series of experiments done using a preparation from the brain of a cat. When the strength of the stimulating current is increased, the action potentials maintain their stereotyped amplitude but appear more frequently. (B) More complex information can apparently be encoded in the pattern as well as the frequency of the spikes. This diagram shows the firing of neurons in part of the visual system of an anesthetized cat. In these graphs, a single spike is shown with a red dot. The visual stimulus was a flickering pattern as shown at the bottom. When a single cell (cell 1) is recorded as the animal is shown the pattern 128 times, the cell responds with nearly the same firing pattern, as shown at the top. Remarkably, when an equivalent cell from a different animal (cell 2) is recorded while being exposed to the same stimulus, its characteristic response pattern shares many of the same features. (A, adapted from X. J. Zhan et al., *J. Neurophysiol.* 81:2360, 1999; B, adapted from P. Reinagel and R. C. Reid, *J. Neurosci.* 22:6837, 2002.)

resulting in local membrane polarization or depolarization. A typical nerve cell receives many input signals all over its dendrites, but has a single output point, the axon, which carries a digital bit in the special, stereotyped traveling depolarization event (that is, the action potential). As we can see from the trace recording from a single neuron in Figure 17.3 (p. 687), such an action potential represents a spike of membrane depolarization that travels down the axon to the terminus at a finite speed and is always the same shape and size, resulting in a blip at the output end. Neurons receiving high levels of excitatory input will fire trains of spikes in rapid succession like those shown in Figure 17.22. The same neuron receiving a lower level of excitatory signals or receiving a combination of excitatory and inhibitory signals will fire spikes that are spaced more widely apart. Yet, each individual spike is essentially indistinguishable.

The stereotyped shape, size, and propagation speed of these spikes relies on the details of time-dependent behaviors of the critical channels discussed earlier in the chapter. In particular, the voltage-gated sodium channels open within less than half a millisecond after depolarization, but they open only transiently and close again with a time constant on the order of 2 ms. After closing, they become "inactivated" and will not reopen for a period of some tens of milliseconds, even if depolarization of the membrane is maintained. In contrast, voltage-gated potassium channels open in response to depolarization with a much slower time constant (of the order of several milliseconds) and

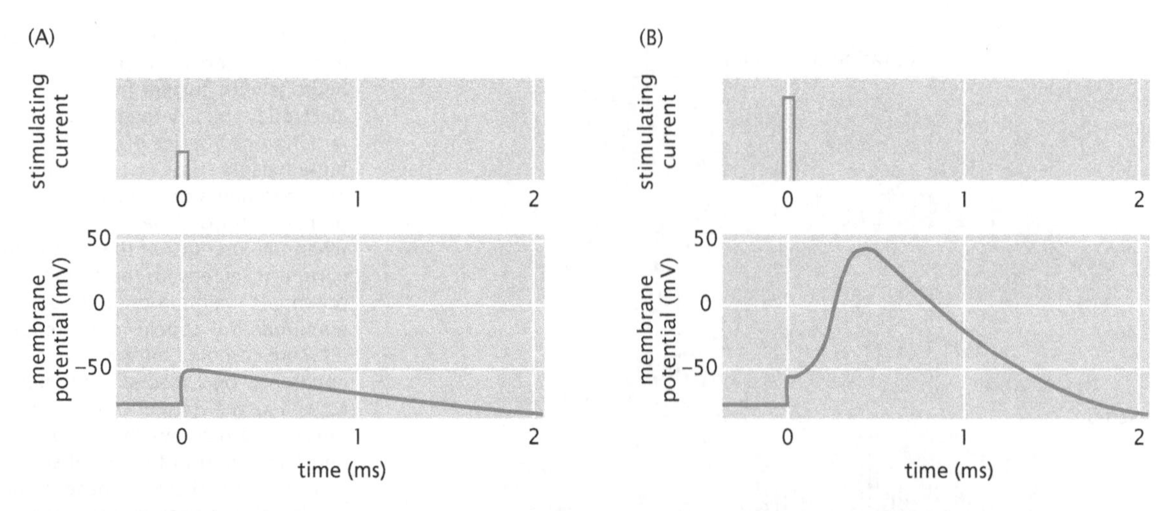

Figure 17.23: Membrane potential for different stimuli. (A) A small injection of current results in a transient change in the membrane potential. (B) A larger injection of current results in the development of a full-fledged action potential with the stereotyped response shown in Figure 17.3. (Adapted from B. Alberts et al., Molecular Biology of the Cell, 5th ed. Garland Science, 2008.)

remain open as long as depolarization is maintained. Thus, for a single patch of membrane in the axon, as long as the input stimulus is above the threshold, depolarization results in a rapid inward sodium current in approximately 1 ms that flips the membrane potential to nearly the Nernst potential for sodium, as illustrated in Figure 17.23. This is followed by an outward potassium current lasting over several milliseconds that restores the resting potential close to the potassium-determined level. The particular feature of sodium channels that they are inactivated after depolarization results in a unidirectional propagation of the action potential down the axon.

17.4.5 Hodgkin–Huxley and Membrane Transport

The action potential is an all-or-nothing event. If a charge stimulus applied to the cell membrane raises the membrane voltage above a threshold ($V^* \approx -40\,\text{mV}$), the membrane potential shoots up to a value roughly given by the Nernst potential of sodium, $V_{\text{Nernst}}^{\text{Na}} \approx 50\,\text{mV}$. This sharp increase of the membrane potential of a patch of membrane raises the membrane potential of a nearby patch above threshold, leading to its depolarization, which in turn depolarizes the next patch, and so on, leading to an action potential.

The denouement of the work of the entire chapter is the determination of the mathematical governing equation for the propagation of an action potential. Our work until now has prepared us for these ideas, since we have focused on both the molecular behavior of the ion channels of excitable cells and the coarse-grained treatment of these cells as lattices of circuit elements. The molecular treatment of action potential propagation was introduced in Figure 4.25 (p. 177) and gives a hint of how electrical signals can be propagated over long distances at constant speed and without dissipation.

Inactivation of Sodium Channels Leads to Propagating Spikes

To turn the cartoon shown in Figure 4.25 into a mathematical model, we add inactivation of sodium channels to the time-dependent cable equation given in Equation 17.30. The idea of inactivation is that, after opening, the sodium channels can go into a nonconducting state that is not the closed state. Only after a transition to the closed state (which is also nonconducting) can the channel be opened again by applying a membrane voltage above the threshold.

To describe inactivation quantitatively, we use the tools developed in Chapter 15. We denote the probability that the sodium channel is in the closed state by $p_C(t)$, while $p_O(t)$ and $p_I(t)$ are the probabilities for the channel to be in the open and the inactive states, respectively. Given these definitions, we can write the master equations describing the switching from closed to open and from open to inactive states as

$$\begin{aligned} \frac{\mathrm{d}p_C}{\mathrm{d}t} &= -k_{\text{open}} p_C, \\ \frac{\mathrm{d}p_O}{\mathrm{d}t} &= k_{\text{open}} p_C - k_{\text{inactive}} p_O, \\ \frac{\mathrm{d}p_I}{\mathrm{d}t} &= k_{\text{inactive}} p_O. \end{aligned} \tag{17.44}$$

Here we have made a number of simplifying assumptions, such as there being no transition from open back to closed or inactive to closed. This allows us to study the propagation of a single spike. In order to account for the voltage-dependent opening of the sodium channel, we assume that the opening rate is proportional to the equilibrium probability for the channel to be open, which is described by Equation 17.7. We obtain

$$k_{\text{open}} = k_{\text{open}}^{\max} \frac{1}{1 + e^{\beta q [V^* - V(x,t)]}}, \tag{17.45}$$

while the inactivation rate is a constant.

As in the time-dependent cable equation that we used to study the propagation of the depolarization front, here we also assume that the potassium conductance stays constant in time and as a function of membrane voltage. The sodium conductance depends on whether the sodium channels are open, closed, or inactivated, and we model it by the equation

$$g_{\text{Na}} = g_{\text{Na}}^{\text{open}} p_O + g_{\text{Na}}^{\text{closed}} p_C. \tag{17.46}$$

The conductance in the open state is taken to be much larger than the conductance of potassium (roughly 20-fold, for the squid giant axon), while the conductance in the closed state is about 25 times smaller than g_K.

Numerical solution of the time-dependent cable equation, Equation 17.30, where now the sodium conductance is both time- and voltage-dependent and is described by Equations 17.44–17.46, is shown in Figure 17.24. We see that with the addition of inactivation, the propagating solutions are spikes. These spikes are generated by an initial voltage depolarization above a threshold. Like the propagating depolarization fronts, these spikes have a shape that is constant in time and their speed is set by the physical properties of the axon in the same way as for the depolarization front described above.

In detail, the actual model written by Hodgkin and Huxley was different than the model we have sketched here. Their model involved similar conceptual elements but a different mathematical implementation. The reader is encouraged to read the classic original paper by Hodgkin and Huxley, listed in the Further Reading at the end of the chapter, for the initial formulation of the breakthrough model.

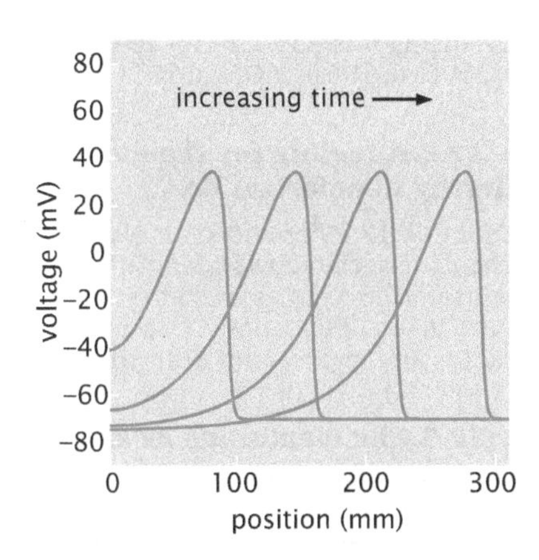

Figure 17.24: Numerical solution of the time-dependent cable equation in the presence of inactivation of the sodium channels. The propagating signal is in the form of a spike, similar to the action potentials that propagate along nerve cells. The parameters used to solve the equation, and the initial membrane voltage, are the same as those used for Figure 17.20(B). Furthermore, all the sodium channels are initially closed, and $k_{\text{open}}^{\max} = 20\ \text{ms}^{-1}$, $k_{\text{inactive}} = 2\ \text{ms}^{-1}$, $g_{\text{Na}}^{\text{open}} = 100\ \Omega^{-1}\ \text{m}^{-2}$ and $g_{\text{Na}}^{\text{closed}} = 0.2\ \Omega^{-1}\ \text{m}^{-2}$.

17.5 Summary and Conclusions

As with many other interesting physical phenomena, cells exploit electricity in important and subtle ways. In this chapter, we have shown how the composite system of cell membranes and their attendant ion channels can be modeled as an array of resistors, capacitors, and batteries. One of the most interesting outcomes of this analysis is the fact that excitable membranes behave as bistable switches. We have shown how the behavior of membrane proteins (ion channels) mediates this response, with the observed behavior resulting from a competition between voltage-gated potassium and sodium channels. The crown jewel of this subject is the Hodgkin–Huxley model of action potential propagation, which shows how cells can produce *propagating* pulses that serve as the key informational currency of a number of processes in multicellular organisms.

17.6 Problems

Key to the problem categories: • Model refinements and derivations • Estimates • Computational simulations • Model construction

• 17.1 Nernst potential

In the chapter, we derived the Nernst equation (Equation 17.3) using the Boltzmann distribution. Derive the same result, but this time by using the fact that in equilibrium the net flux of ions across the membrane is zero. Use Equation 13.56 (p. 530) to derive the contribution of the electrical force to the overall flux and show that by balancing this flux with the diffusive contribution you recover the Nernst equation.

• 17.2 A feeling for the numbers: charge transfer during depolarization

Equation 17.10 provided an estimate of the fractional charge associated with depolarization of a muscle cell. Use the same concept as in that estimate, but now produce the ratio given in Equation 17.10 as a function of the nerve radius and make a plot of the result.

• 17.3 The membrane as an electrical circuit

Show that the electrical circuit representing a patch of membrane given in Figure 17.11 can be substituted with an equivalent circuit with one effective conductance and one battery source. What are the effective conductance and voltage of these two elements in terms of the conductance and Nernst potential for Na^+ and K^+?

• 17.4 Relaxation to Donnan equilibrium

This problem explores what happens to the resting steady state of a membrane, after the ion pumps are suddenly turned off, say by addition of a neurotoxin. Namely, sodium ions are far from equilibrium in the resting state. The concentration of sodium inside the cell is 12 mM while outside the cell it is 145 mM; the membrane voltage is $V_{mem} = -90$ mV. This means that when pumps are turned off, sodium will rush into the cell.

(a) Using Ohm's law, find the current per unit area carried by sodium ions just after the pumps have been shut off. For the conductance of sodium channels per unit area, take the measured value of $0.13\,\Omega^{-1}\,m^{-2}$. Reexpress your answer as current per unit length along a giant axon, assuming a diameter of 1 mm.

(b) Find the charge per unit length of the axon contributed by all the sodium ions inside the axon. What would the corresponding quantity equal if the interior concentration of sodium matched the fixed exterior concentration?

(c) Given the current of sodium computed in (a), make a rough estimate of how long it will take for the concentration of sodium inside the axon to reach the value outside the cell? You may assume that the current is constant in time for this estimate.

(d) In the chapter, we described a nerve impulse as a membrane-depolarization event that lasts for about a millisecond. Compare with the time you just computed and comment on the size of the charge perturbation caused by a propagating action potential.

• 17.5 Single-channel recordings

Here we consider an electrical current through a single ion channel, as shown in Figure 17.25. We start by treating an ion channel as a cylindrical pore through the cell membrane. The channel is of length $L \approx 5$ nm and is filled with water. The concentration difference of Na^+ ions across the channel is $\Delta c \approx 100$ mM. This concentration difference produces a flux of ions through the channel given by $D\Delta c/L$, where D is the diffusion constant for ions in water. For small ions, like Na^+, the diffusion constant in water is of the order of $1\,\mu m^2/ms$.

(a) Assuming that the channel cross-section has an area $A \approx 1\,nm^2$, compute the number of sodium ions that pass through the channel every second due to diffusion.

(b) The figure shows the current recording from a single channel. Use the figure to estimate the number of ions passing through the channel per second, when it is open. How does your result compare with the result you derived in (a)?

(c) Using the figure of a single-channel current recording, estimate the probability that the channel is open. Sketch a graph of the current recording you expect to get from a channel that is open with probability 1/2.

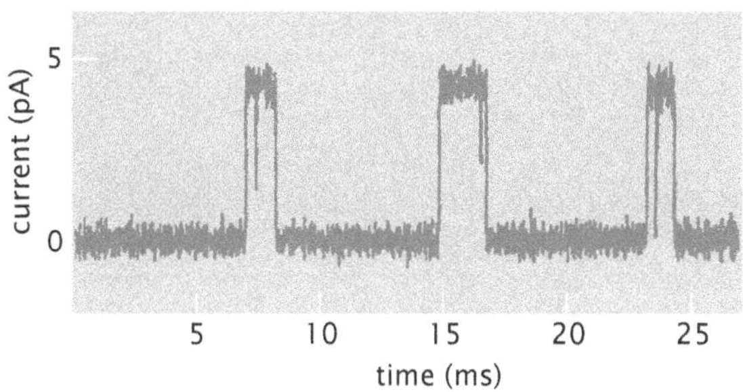

Figure 17.25: Current passing through a channel as a function of time. The channel switches between the closed and open states. When open, the channel permits the passage of ions, which is measured as a current.

17.6 Time-dependent cable equation

(a) Reproduce Figure 17.20 by carrying out a numerical solution of Equation 17.30 with Matlab or a similar program. Use the parameters given in the caption to Figure 17.20. For hints on how to solve this problem refer to the Computational Exploration in the chapter.

(b) Measure the speed of propagation of the front by collapsing the voltage profiles obtained for different times t after the initial voltage spike onto one master curve. The idea is to choose a speed v so that plotting the membrane voltage as a function of $x - vt$ makes the different voltage curves all fall on top of each other.

17.7 Propagation of spikes

Include sodium-channel inactivation in the time-dependent cable equation and reproduce Figure 17.24. For this calculation, you will need to numerically implement the kinetic equations that describe the time-dependent Na^+ conductance, Equation 17.46, and use the parameters given in the caption to Figure 17.24.

17.7 Further Reading

Kandel, ER (2006) In Search of Memory: The Emergence of a New Science of Mind, W. W. Norton. This outstanding book is a mix of autobiography and neuroscience and the early chapters give a wonderful account of the work of the heroes of neuroscience who figured out the anatomy and function of neurons.

Nelson, P (2004) Biological Physics: Energy, Information, Life, W. H. Freeman. Nelson's book is one of our favorite sources for the study of biological electricity. Nelson's discussion of membrane pumps inspired the estimate in the chapter.

Hille, B (2001) Ion Channels of Excitable Membranes, 3rd ed., Sinauer Associates. This is the "go to" book for learning about ion channels and their relation to the electrical properties of cells.

Katz, B (1966) Nerve, Muscle and Synapse, McGraw-Hill. Katz's book, though dated, still makes for enlightening reading.

Greenspan, RJ (2007) An Introduction to Nervous Systems, Cold Spring Harbor Laboratory Press. This fantastic book is written in the same tradition as Mark Ptashne's A Genetic Switch (Cold Spring Harbor, 2004) and is both interesting and accessible.

Benedek, GB, & Villars, FMH (2000) Physics With Illustrative Examples from Medicine and Biology: Electricity and Magnetism, 2nd ed., Springer-Verlag. This excellent book has a very useful discussion of the electrical properties of cells.

Fain, GL (1999) Molecular and Cellular Physiology of Neurons, Harvard University Press, and Fain, GL (2003) Sensory Transduction, Sinauer Associates. Both of Fain's books make fascinating reading on the subject of how cells communicate with each other and with the external world.

Levitan, IB, & Kaczmarek, LK (2002) The Neuron: Cell and Molecular Biology, 3rd ed., Oxford University Press. This book provides an extremely readable discussion of neurons and their electrical behavior.

Hodgkin, AL, & Huxley, AF (1952) A quantitative description of membrane current and its application to conduction and excitation in nerve, *J. Physiol.* **117**, 500. This remarkable paper laid out the entire quantitative basis for understanding the propagation of action potentials in neurons long before the molecular nature of ion channels was understood.

17.8 References

Alberts, B, Johnson, A, Lewis, J, et al. (2008) Molecular Biology of the Cell, 5th ed., Garland Science.

Kandel, ER, Schwartz, JH, & Jessell, TM, eds. (2000) Principles of Neural Science, 4th ed., McGraw-Hill.

Keynes, RD, Aidley, DJ, & Huary, CL-H (2001) Nerve and Muscle, 4th ed., Cambridge University Press.

Long, SB, Campbell, EB, & MacKinnon, R (2005) Crystal structure of a mammalian voltage-dependent shaker family K^+ channel, *Science* **309**, 897.

Reinagel, P, & Reid, RC (2002) Precise firing events are conserved across neurons, *J. Neurosci.* **22**, 6837.

Zhan, XJ, Cox, CL, Rinzel, J, & Sherman, SM (1999) Current clamp and modeling studies of low-threshold calcium spikes in cells of the cat's lateral geniculate nucleus, *J. Neurophysiol.* **81**, 2360.

Light and Life 18

Overview: In which we examine how light makes life on Earth possible

The sun has had a prominent role in mythology, religion, poetry, and science. This is not a surprise, since it is no exaggeration to say that the sun is the ultimate giver of life on planet Earth. In this chapter, we explore the way that energy from nuclear reactions in the sun is coupled to the synthesis of sugar molecules on our planet by living organisms. To analyze the process of photosynthesis, we first explore the quantum mechanics of confined electrons and show how simple models can be used to approximately compute the absorption properties of pigment molecules. Once electrons within these molecules are excited, processes of energy and electron transfer follow, with the eventual outcome being the storage of the energy of sunlight in the form of a transmembrane potential that can be used to power key biochemical reactions. We use a highly simplified quantum mechanical model to explore the dynamics and energetics of electron transfer. The energy stocked in these processes is then used in the fixation of carbon whereby inorganic carbon in the form of carbon dioxide is transformed into sugars. The importance of light to life is clearly not restricted to its role in photosynthesis. We use the study of vision as a second example of the interesting ways in which biological molecules absorb light and use such absorption as the basis of information transfer. Rather than just passively accepting the consequences of the visual appearance that organisms present to one another, some organisms go further and actively manipulate visual cues to control the content of communication, advertising their ripeness or desirability, hiding from the prying eyes of predators, and even in some circumstances generating light of their own.

> "We can scarcely avoid the inference that light consists in the transverse undulations of the same medium which is the cause of electric and magnetic phenomena."
>
> James Clerk Maxwell

18.1 Introduction

We have already described the various roles of energy in living systems, with a particular emphasis on chemical energy and on the storage of energy in chemical and electrical gradients across membranes. But our discussion so far has largely neglected the central role of electromagnetic energy for life on Earth, particularly that energy in the visible part of the spectrum. Those organisms living in the pitch-black bottom of the ocean or hidden beneath layers of rock or ice never see the sun and so visible light plays little or no role in their ecosystems. However, for the majority of surface-dwelling organisms, the sun and its light are literally the difference between life and death. Primary producers in all ecosystems are those that take energy from the nonliving world and capture it in forms that can be used for biological purposes, either for their own needs or to enrich the lives of those that eat them. The major form of primary production for surface-dwelling ecosystems is photosynthesis, where energy is captured from visible light and used to convert gaseous carbon dioxide into biologically accessible molecules such as sugars. In this chapter, we explore the process of photosynthesis in detail, in part as a way to illustrate the connections between quantum mechanics and the most basic necessities of life.

Though photosynthesis was probably the original form of life's exploitation of electromagnetic energy, given the opportunistic creativity of the living world, it is not surprising that organisms found other uses for this energy. As we will see, organisms also use electromagnetic energy to gain and deliver information both about their environments and themselves. Thus, many animals use and alter visible light cues to enhance the finer things in life, ranging from finding dinner to finding a desirable mate.

Originally, photosynthesizing organisms would bask in the sun and opportunistically use the photons that came their way. But as they developed motility and could swim up or down in the ocean to seek out higher rates of photon flux or preferred wavelengths such as shown in Figure 18.1, they essentially began to take advantage of the information content and not just the energy content of the light. This primitive form of information processing from the light available in the environment gave rise to progressively more sophisticated forms. The reflections of light bouncing off our fellow creatures give biological organisms rapid and sensitive information at a distance about events that may be of interest to them, including the appearance of prey or predators, as well as of potential mates.

The ability to respond to information in the environment provided an opportunity for organisms to manipulate the behavior of other organisms by changing the patterns of light they reflect. Pigmentation is used for many layers of communication in the living world. Plants use brightly colored flowers and fruits to attract animals useful for pollination and for spreading the resulting seeds. In the endless battle to avoid being eaten, pigmentation can also be used as camouflage to hide from predators. The ability to create elaborate patterns can also be used to make oneself *more* conspicuous when competing for the affections of the opposite sex. The power of visual cues in manipulating the behavior of others is particularly vividly illustrated by organisms that use chemical reactions to create their own light through the process of bioluminescence. Bioluminescent light can be used for all the same purposes as pigmentation, including hunting, camouflage, and sexual advertisement. However, since it is a

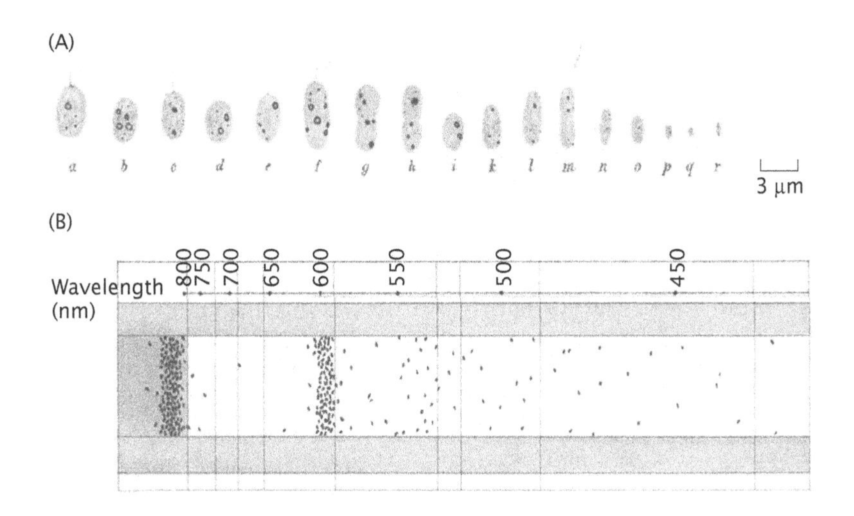

Figure 18.1: Phototaxis of bacteria. (A) In 1883, T. W. Engelmann described the light-dependent swimming behavior of bacteria from the Rhine river. This drawing shows some of the variety in shape and size of the motile bacteria he observed. (B) On exposing the bacteria to a spectrum created by focusing sunlight through a prism, Engelmann noted that the bacteria–seemed to prefer certain colors, with many clustering in the near infrared (800–900 nm wavelength) and others in a narrow band around 600 nm; these may be the wavelengths where photosynthesis is most efficient for these organisms. Engelmann also noted that the bacteria would switch their direction of swimming when they moved from a light region into a dark region, as if they were afraid of the dark, a behavior he termed "Schreckbewegung" (meaning "fright-movement"). (Adapted from T. W. Engelmann, *Arch. Gesam. Physiol. Mensch. Tiere* 30:95, 1883.)

great deal of trouble to make light, some organisms that have developed this ability find themselves captured and exploited by others. A famous example is the Hawaiian bobtail squid (*Euprymna scolopes*), which carries the bioluminescent bacteria *Vibrio fischeri* in specialized light organs that it can open and close at will.

18.2 Photosynthesis

The energy that enables living organisms to create order and complexity must ultimately be derived from the nonliving world. On Earth, there are at least several distinct sources of energy that have been harnessed by life. As a starting point, we can consider the sources of energy that human societies use to generate electricity. Geothermal energy and radioactivity are energy sources that are in a sense left over from the early turbulent days of Earth's formation. Tidal energy is driven by gravitational interactions between our planet and its celestial partners, especially the moon and the sun. All other sources of energy, including wind, waves and sunlight, are gifts to our planet from nuclear reactions in our nearby stellar benefactor. The output of these nuclear reactions arrives on Earth in the form of a constant barrage of photons of varying wavelengths as shown in Figure 18.2.

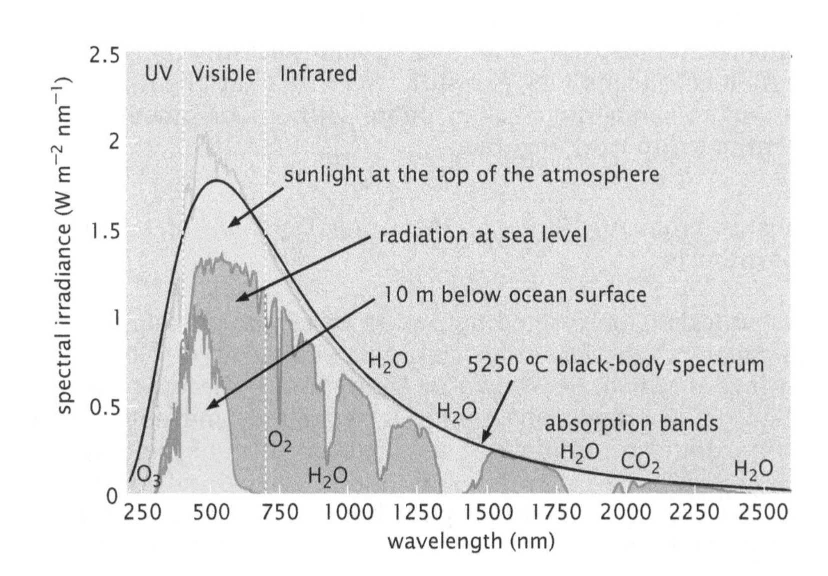

Figure 18.2: The solar spectrum. The spectral density of light as a function of wavelength at the top of the atmosphere (yellow) essentially reflects the spectrum of black-body radiation from the sun. This light is attenuated at the surface of the earth (red), largely due to absorption by H_2O, O_2, and CO_2 in the air. At 10 m below the surface of the ocean (blue), almost no infrared or red light is able to penetrate. (Adapted from the National Renewable Energy Laboratory.)

To get a sense of the amount of radiant energy arriving at the Earth's surface every second, we examine Figure 18.2 quantitatively.

ESTIMATE

Estimate: Solar Energy Fluxes Figure 18.2 provides the basis for a simple estimate of the incident power coming from sunlight that can potentially be used by living organisms. The units on the *y*-axis of the figure are W m^{-2} nm^{-1}. That is, this curve tells us how much energy arrives every second over a square meter area for each tiny bin of wavelengths. To find the total incident power, we need to sum up the power from each of these tiny bins, resulting in the end in units of W/m^2, or J m^{-2} s^{-1}.

To produce a crude estimate of the power density (W/m^2), we can replace the curve shown in Figure 18.2 with a rectangle starting on the left at 250 nm and ending on the right at 1250 nm. Choose the height of the rectangle to be 1 W m^{-2} nm^{-1}. If the reader performs this operation by tracing out such a rectangle with their finger on the page, it is clear that this is a reasonable approximation to the red shaded region corresponding to the light reaching the surface of the Earth. The area under our approximation to the solar spectrum in this case is 1000 W/m^2, a powerful rule of thumb that will serve us through the remainder of the chapter.

How can organisms take advantage of the massive energy flux beating down on them at about 1000 W/m^2? So-called primary producers use photon energy to convert carbon dioxide into sugars like glucose. For the purposes of our discussion as we try to understand in broad brushstroke form how photons, atmospheric carbon dioxide and water are used to build sugars, we will split photosynthesis into three fundamental processes. The first step is absorption of a photon by a biological molecule to generate an excited electronic state. The second step is harnessing the energy of the excited biomolecule to effect transport of energy to the reaction center and to effect charge transfer (usually electrons) across a membrane, resulting in an electrochemical gradient that stores the energy for future use in biosynthesis. The third step is the transfer of the separated electrons onto oxidized atmospheric carbon (i.e. carbon dioxide) to convert it into a reduced form where it can be incorporated into an organic carbohydrate molecule. These three steps are carried out by distinct macromolecular machines and involve somewhat different underlying physical mechanisms. As a result, we will treat them each separately, but remembering that in photosynthetic organisms they are all coordinated to work together.

Organisms From All Three of the Great Domains of Life Perform Photosynthesis

Photosynthesis is performed by a wide variety of organisms, including members of the bacterial, archaeal, and eukaryotic branches of life, a few of which are shown in Figure 18.3. The exact details of how light is absorbed, how charge is separated, and how carbon is fixed vary quite substantially. The particular form of photosynthesis that is of most interest to humans is that carried out by the green plants that provide food directly for humans and also for other animals that are in turn eaten by humans. To be fair, this particular form

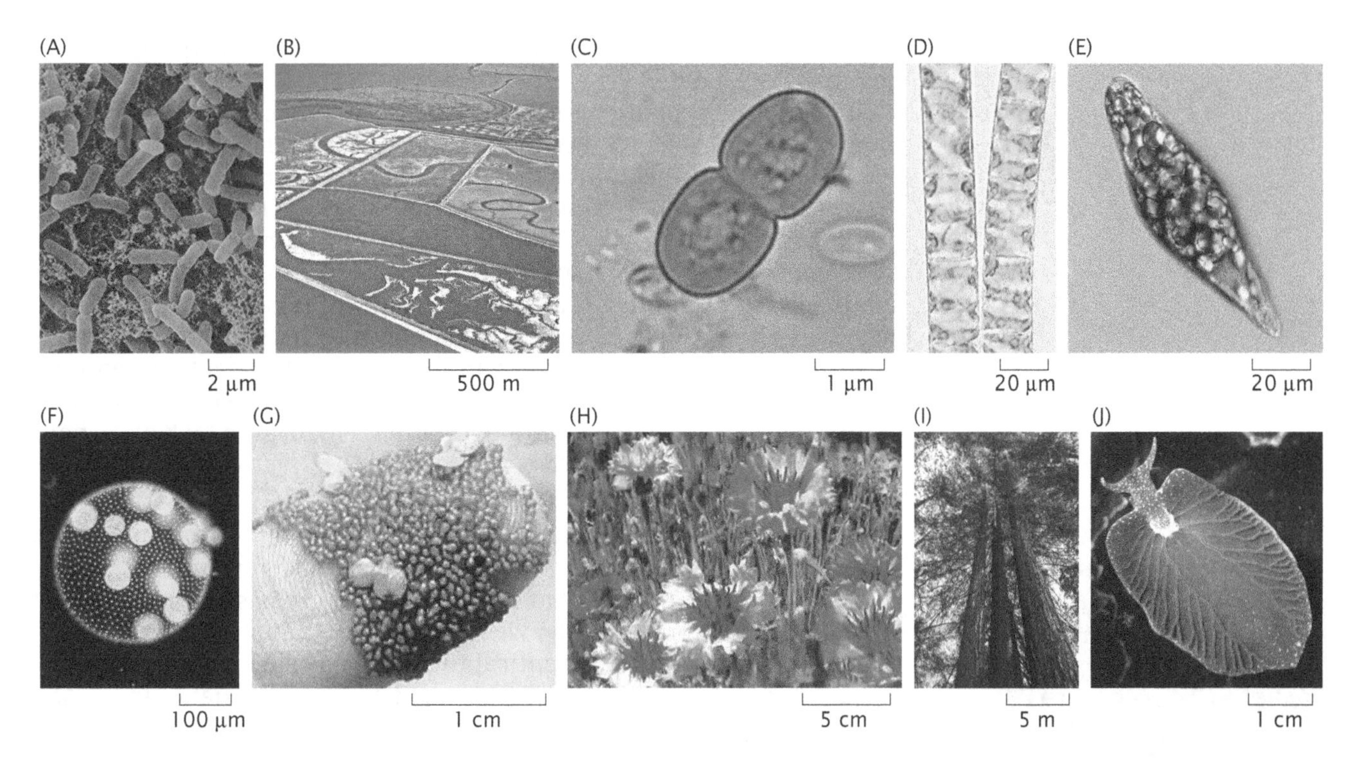

Figure 18.3: Gallery of photosynthetic organisms. (A) Green sulfur bacteria such as the *Chlorobium* shown here are among the most ancient photosynthesizers, carrying out a simple form of photosynthesis that uses chlorophyll to capture light energy but does not generate oxygen. (B) The purple archaeon *Halobacterium* captures light to generate energy in a way that is fundamentally different from the chlorophyll-containing organisms, using retinal isomerization in a manner similar to the way animals use retinal for vision. This salt-loving organism is found in high concentrations in evaporation ponds in San Francisco Bay where salt is made commercially. (C) The green chlorophyll-containing cyanobacterium *Synechococcus* performs complex, oxygen-generating photosynthesis, and is one of the most common microbes found in ocean waters. (D) *Spirogyra* is a simple alga, a eukaryote that benefits from the photosynthetic prowess of its chloroplasts, endosymbiotic organelles derived from captured cyanobacteria. (E) *Euglena* is a single-celled eukaryote that is both photosynthetic and motile. Besides using light for energy, it is also able to perceive the direction of light with its red eye spot and use this information to swim in the right direction. (F) *Volvox* is a colonial organism that is also photosynthetic and motile. The small green spheres within the colony are its offspring. (G) *Wolffia* is among the smallest of true plants. Each of the small green objects on this human fingertip is a different individual plant. (H) The cornflower is one of many medium-sized flowering plant species. (I) Redwood trees are among the largest photosynthetic organisms, converting carbon fixed using the energy of sunlight into trunks that may be nearly 100 meters tall. (J) *Elysia chlorotica* is a sea slug, an animal that has found a way to be photosynthetic. It preserves the chloroplasts from the algae that it eats and uses them for its own energy generation. Note: some of the scale bars for these photographs are estimates based on the known sizes of these organisms. (A, courtesy of Dennis Kunkel Microscopy, Inc.; B, courtesy of aroid (Flickr); C and E, courtesy of Dr Ralf Wagner; D, courtesy of Richard M. McCourt, Tree of Life; F, courtesy of Raymond E. Goldstein, University of Cambridge; G, courtesy of Ian Geldard; H, courtesy of Bro. Bob Resing; I, courtesy of Bernt Rostad; J, courtesy of Patrick Krug.)

of photosynthesis, called oxygenic photosynthesis, does not really belong to plants, but rather is an invention of cyanobacteria. All eukaryotic organisms that perform photosynthesis actually do so only with the help of specialized organelles derived from domesticated cyanobacteria, originally endosymbionts that have gradually evolved into organelles now known as chloroplasts.

Historically, much experimentation on the mechanism of photosynthesis has used plants or algae, so it is largely oxygenic photosynthesis that will serve as the focus of this chapter. It was known to man from the beginning of agriculture that plants need sunlight to grow. However, the elucidation of the mechanism depended upon a series of rigorous, quantitative experiments. As much as any other example in this book, the study of photosynthesis vividly illustrates the critical importance of addressing quantitative data with rigorously quantitative models.

Though green plants are familiar to all, much of the world's photosynthesis takes place in the oceans and is hosted by tiny cyanobacteria. We begin our quantitative dissection of oxygenic photosynthesis through a consideration of these micron-sized primary producers. In particular, we start by counting and measuring the critical components within cyanobacteria that perform each of the major subprocesses of photosynthesis. Light capture is performed by large arrays of chlorophyll in antenna complexes. Charge separation and electron transport is carried out in photosynthetic reaction centers, for cyanobacteria as for green plants. These reaction centers also generate molecular oxygen (O_2) as a byproduct. Finally, the task of carbon fixation is carried out in carboxysomes.

(A)

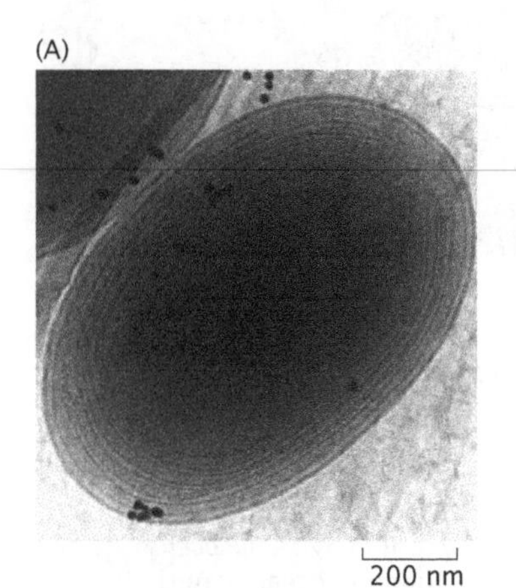

(B)

Figure 18.4: Structures of photosynthetic bacteria. (A) Cryo-electron microscopy image of a cyanobacterium (*Prochlorococcus*), revealing the series of membranes associated with the apparatus of photosynthesis. (B) Reconstruction of electron microscopy image highlighting the photosynthetic membranes in another species of cyanobacterium (*Synechocystis*). The plasma membrane is shown in red, the photosynthetic thylakoid membranes in green, and the carboxysomes in blue and purple. (A, adapted from C. S. Ting et al., *J. Bacteriol.* 189:4485, 2007; B, adapted from M. Liberton et al., *Protoplasma* 227:129, 2006.)

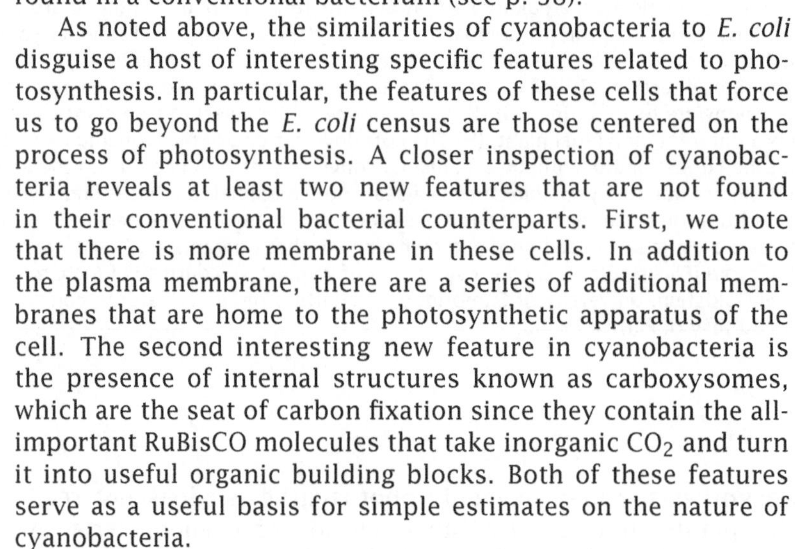

ESTIMATE

Estimate: Sizing Up Cyanobacteria In order to assess the molecular and organellar contents of a cyanobacterium, we refer to Figure 18.4. Note that in terms of the gross structure, if we don't look too carefully, we might be lulled into thinking that these cells are nearly identical to the *E. coli* cells that serve as our standard ruler and that we considered in such detail in Chapter 2. Indeed, in many respects, these cells are quite similar to their *E. coli* cousins. As seen in Figure 18.4, their size is comparable to our standard ruler, meaning that a cyanobacterium has a characteristic length scale of order one micron and a characteristic volume of roughly a femtoliter. This means that if we consider the number of proteins and water molecules in a cyanobacterium, we will obtain a similar inventory to that found in a conventional bacterium (see p. 38).

As noted above, the similarities of cyanobacteria to *E. coli* disguise a host of interesting specific features related to photosynthesis. In particular, the features of these cells that force us to go beyond the *E. coli* census are those centered on the process of photosynthesis. A closer inspection of cyanobacteria reveals at least two new features that are not found in their conventional bacterial counterparts. First, we note that there is more membrane in these cells. In addition to the plasma membrane, there are a series of additional membranes that are home to the photosynthetic apparatus of the cell. The second interesting new feature in cyanobacteria is the presence of internal structures known as carboxysomes, which are the seat of carbon fixation since they contain the all-important RuBisCO molecules that take inorganic CO_2 and turn it into useful organic building blocks. Both of these features serve as a useful basis for simple estimates on the nature of cyanobacteria.

To determine the membrane area and lipid content, we can imitate our earlier calculations on *E. coli*. Since the area of a bacterium of this size is roughly 5 μm^2, and the area per lipid is roughly 0.25 nm^2, the number of lipids in the outer membrane is roughly 10^7 (for the subtleties in the argument, see Section 2.1.2, p. 38). As seen in Figure 18.4, we can think of the membranes making up the photosynthetic apparatus as a series of concentric layers forming an onion-like structure, with each such layer contributing roughly 10^7 lipids. These pictures suggest a roughly tenfold increase in lipid area due to the various photosynthetic membranes relative to the amount of lipids found in a nonphotosynthetic bacterium such as *E. coli*.

As seen in Figure 18.5, another fascinating feature of these cells is the presence of their carboxysomes. As noted above, each such carboxysome is filled with complexes of RuBisCO, the enzyme that is able to take atmospheric CO_2 and begin the process of turning it into an organic substrate. As shown in the high-resolution reconstruction in Figure 18.6(A), each such carboxysome is filled with assemblies of RuBisCO. The inventory of these key molecules of carbon fixation can be taken further as shown in Figure 18.6(B), which gives a feel for the number of RuBisCOs per carboxysome, with the average number being roughly 250. As seen in Figure 18.5, there are of order ten carboxysomes per cyanobacterium, resulting in the insight that there are roughly 2500 RuBisCOs per cyanobacterium. In our estimates later in the chapter in which we consider the net primary productivity on Earth, we will crudely, but independently, estimate the number of RuBisCOs per cyanobacterium and find that it is in very nice accord with these direct measurements, supporting the claim that this protein is the most abundant on Earth.

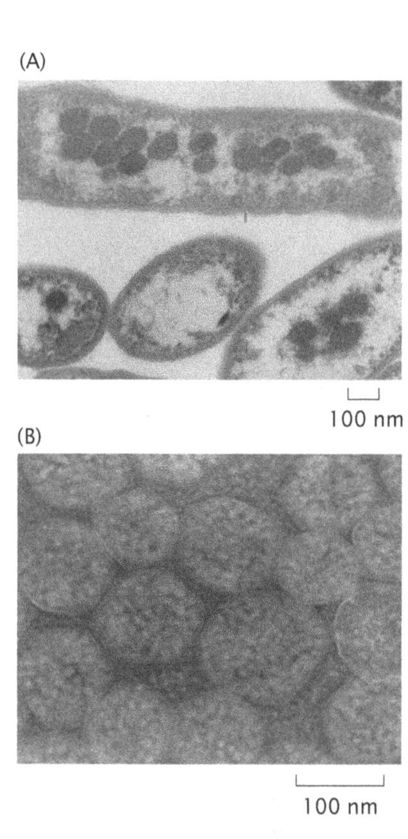

Figure 18.5: Carboxysomes in a photosynthetic bacterium. (A) Thin-section electron microscopy image of the sulfur-oxidizing bacterium *Halothiobacillus neapolitanus*, showing a dense collection of internal carboxysomes (red arrow). (B) Isolated carboxysomes from this same bacterium, negatively stained for electron microscopy. Individual RuBisCOs can be seen as light speckles within each carboxysome shell. (Adapted from G. C. Cannon et al., *Appl. Environ. Microbiol.* 67:5351, 2001.)

Our consideration of cyanobacteria gives us the chance to examine beautiful recent experimental work that follows the fate of individual carboxysomes during the cell division process. In keeping with our enthusiasm for the use of statistical arguments, studies of the statistics of partitioning of carboxysomes during cell division have been used as a tool to determine whether this partitioning is a passive or active process. As with our earlier discussion of partitioning of mRNA and proteins during cell division (see p. 44), the logic is that a binomial distribution of the relevant molecular players signals a passive partitioning mechanism, while more precise partitioning (like that seen for carboxysomes) indicates an active mechanism.

How many carboxysomes are found in each cell and how are they distributed during cell division? As seen in Figure 18.7, the mean number of carboxysomes is roughly four. More interestingly, as also shown in the figure (and explored more deeply in the problems at the end of the chapter), the segregation of the carboxysomes is not passive. This conclusion can be drawn from statistical arguments.

In Sections 2.1.2 (p. 38) and 8.2.1 (p. 313), we saw that the statistics of coin flips is described by the binomial distribution and can be used to help us understand both molecular partitioning during cell division and polymer conformations. In the example given in Figure 18.7, the partitioning of carboxysomes during segregation is viewed in terms of this same kind of binomial distribution. We calculate the consequences of a passive segregation mechanism in which the daughter cell that a given molecule arrives in is a random process. The argument can be stated simply. If the partitioning is purely random (like a coin flip), then the resulting distribution of N carboxysomes between two daughter cells should be binomial. Interestingly, as is seen in Figure 18.7(C), the distribution is sharper than that obtained from a binomial process. In this case, these observations led to experiments culminating in the identification of the active segregation mechanism responsible for distributing carboxysomes between the two daughter cells.

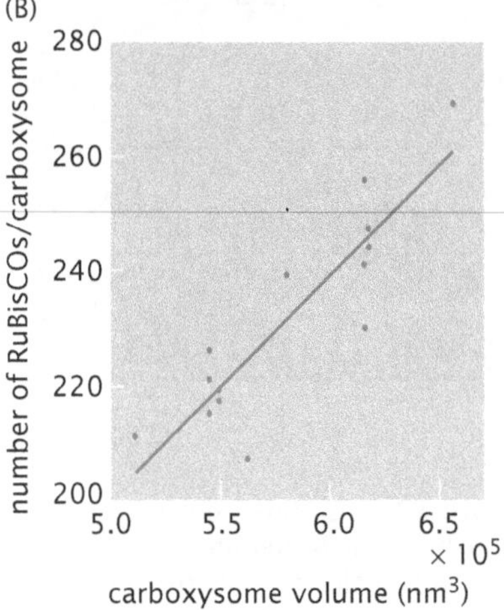

Figure 18.6: RuBisCO packing in the carboxysome. (A) Reconstruction from cryo-electron microscopy of the structure of an individual carboxysome, showing the RuBisCOs in the interior. (B) The number of RuBisCOs increases roughly linearly with the volume of the carboxysome. (Adapted from C. V. Iancu et al., *J. Mol. Biol.* 372:764, 2007.)

18.2.1 Quantum Mechanics for Biology

As our first foray into the physical basis of photosynthesis, we build up a minimal toolkit of ideas from quantum mechanics. The reason that it is necessary to go beyond the ideas from classical physics that have dominated the discussion until now is that photosynthesis involves several key processes such as the absorption of photons and the transfer of electrons that are intrinsically quantum mechanical events.

One of the puzzles that dominated early thinking on quantum theory was wave–particle duality, the observation that, depending upon the experiment in question, both light and matter exhibited properties that in our macroscopic experience would be called either "particle-like" or "wave-like." This is relevant to our story since the starting point of our discussion will be the way in which light interacts with pigment molecules. To understand this coupling, we appeal to the fact

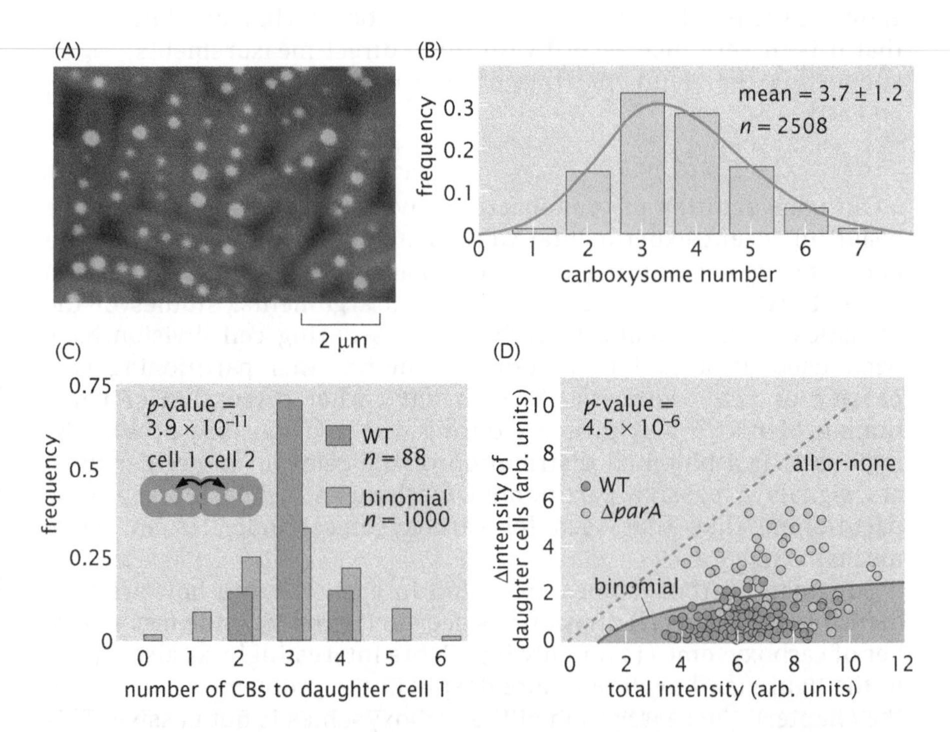

Figure 18.7: Carboxysome count and segregation. (A) Carboxysomes in this cyanobacterium (*Synechococcus*) are tagged in green using a fluorescent protein fused to one of the subunits of RuBisCO. The thylakoid membranes of the cells are intrinsically red-fluorescent because of their high chlorophyll content. The carboxysomes appear to be regularly spaced within each cell. (B) Counting the number of carboxysomes in a large population of these cyanobacteria gives a mean number of about 3.7 per cell. (C) Cells with exactly six carboxysomes were examined during cell division, and the number of carboxysomes (CBs) delivered to one daughter was recorded. The green bars show the observed segregation distribution and the beige bars show the expected binomial distribution for completely random segregation. The difference between the two is highly statistically significant, indicating that the normal segregation is not random. (D) Deletion of the gene encoding a cytoskeletal protein (*parA*) results in more random segregation of the carboxysomes. The solid line shows the expected average difference in carboxysome fluorescence for two daughter cells under random segregation (binomial distribution). Nearly all the wild-type (WT) cells (green circles) fall below this line, indicating better-than-random segregation. The mutant cells (beige circles) fall about equally on both sides of the line, suggesting that segregation has become more random. The dashed line shows the limiting case where one daughter cell receives all of the carboxysomes. (Adapted from D. F. Savage et al., *Science* 327:1258, 2010.)

that light can be thought of as a collection of photons, with the energy E of these photons and the wavelength λ of the light related through the simple but important relation

$$\lambda = \frac{hc}{E}, \tag{18.1}$$

where h is Planck's constant and c is the speed of light. The consequences of this simple relationship are shown in graphical form in Figure 18.8. Our attention is focused on the wavelengths around the visible part of the spectrum, since these are the most relevant to the photosynthetic processes of interest here, as is clear from Figure 18.2.

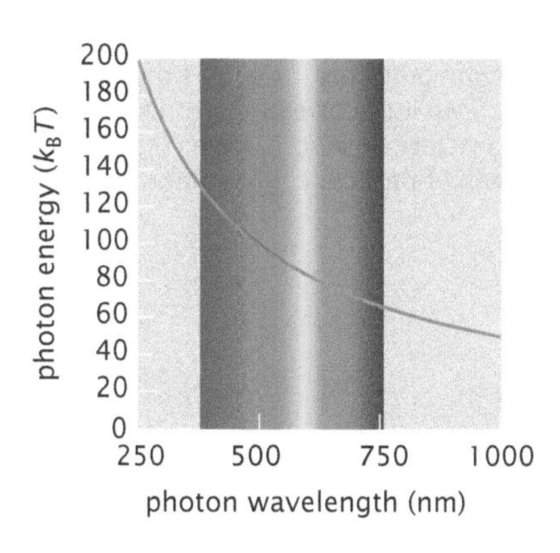

Figure 18.8: Relation between energy and wavelength of photons.

This same idea is not only useful for thinking about the energy associated with photons of different wavelength, but also provides the basis for a simple estimate of the energy of electrons that are confined to boxes with molecular dimensions. Treating molecules as electron boxes is offered in precisely the same spirit of simplification as the other highly simplified models discussed throughout the book and will help us understand the relation between molecular dimensions and absorption wavelength.

ESTIMATE

Estimate: Confinement Energies of Electrons In Figure 5.1 (p. 190), we showed the confluence of energy scales relevant to many biological processes. One of the lines shown in that figure corresponds to the energy of quantum mechanical confinement as a function of the length scale of the confining box. In particular, the idea is that the act of confining a quantum particle such as an electron to a box implies a certain energy cost that depends upon the size of that box. One way to estimate the energy scale of confinement is to exploit Heisenberg's uncertainty relation

$$\Delta p \Delta x \approx \hbar, \tag{18.2}$$

where we introduce $\hbar = h/2\pi$. The idea of Heisenberg's principle is that the uncertainties in the momentum and position are not independent. If the electron is confined to a region of size $\Delta x = a$, we can solve for the corresponding uncertainty in the momentum, resulting in the energy estimate

$$E = \frac{p^2}{2m} \approx \frac{\hbar^2}{2ma^2}. \tag{18.3}$$

The energies resulting from this simple estimate as a function of box size are shown in Figure 18.9. This simple result and the result shown in Figure 18.8 serve as the basis for developing intuition about the relation between molecular dimensions and the wavelengths of light that a given molecule will absorb. We return to this theme after a brief introduction to quantum mechanics.

Quantum Mechanical Kinematics Describes States of the System in Terms of Wave Functions

Throughout this book, so far, our description of the structures of the cell (at many different scales) has been entirely *classical*. That is, when thinking about the "state" of a given system, we have used quantities such as position and momentum to describe the disposition of the

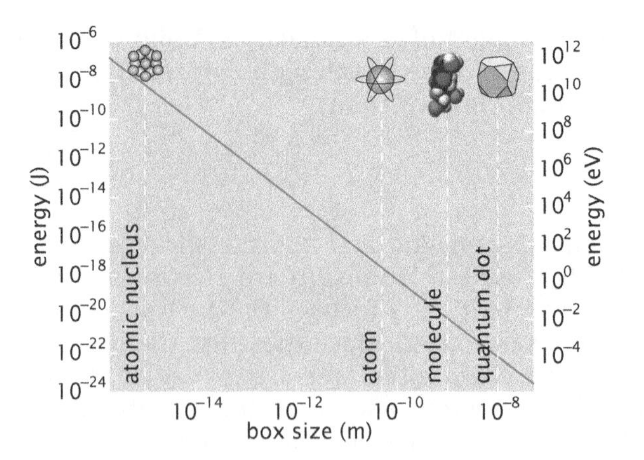

Figure 18.9: Energy of confinement as a function of box size. The dashed vertical lines correspond to the characteristic length scales for different kinds of particle confinement.

system of interest. Indeed, even when referring to complicated structures such as the entire *E. coli* genome, our treatment of the DNA polymer has resorted either to a deterministic structural description of the genome as a space curve $\mathbf{r}(s)$ or to a description as a random walk with specific nodal positions. Quantum mechanics changes all of that.

In those cases where we are forced to acknowledge the underlying graininess of matter and its quantum implications, the "state" of a system is described by a probability amplitude from which the probability of finding the particle at a specific position or with a specific momentum can be computed. This is in contrast to classical mechanics, where the state of a particle is given by specifying its position and momentum and where it is assumed that both of these can be determined with arbitrary precision. The governing equations of quantum mechanics recognize our inability to simultaneously specify the position and momentum of a particle. Precisely this insight is reflected in Heisenberg's uncertainty principle introduced earlier.

The need for a description of particle motion in terms of probability amplitudes is illustrated by the famous two-slit experiment. Imagine an electron (or photon, or proton, etc.) gun that fires electrons at an impenetrable wall with two slits, as shown in Figure 18.10. At some distance behind the wall is a detector that counts individual electrons that happen to fly into it over a specified detection time. By moving the electron detector along the x-axis, we can measure the number of electrons arriving at different locations. These measurements can then be used to construct a histogram, or probability distribution of electrons along the x-axis. The surprising outcome of such an experiment is that the measured probability $P_{12}(x)$ will show an interference pattern, just as one would observe if the experiment were done with a source of waves replacing the electron gun. In fact, this is nothing but Young's celebrated two-slit experiment that he used to confirm the wave nature of light and to measure its wavelength. What is unusual, and in fact impossible to reconcile with classical notions of particle motion, is that electrons, even though they are emitted by the electron gun and detected at the detector as individual particles, show the same interference pattern as a wave would, even when the electrons are arriving only one at a time. To describe (not explain!) this state of affairs, we associate with each of the two trajectories that the electron can take, passing through slit 1 or through slit 2, probability amplitudes $\psi_1(x)$ and $\psi_2(x)$. The probability of detecting the electron at position x if slit 2 (1) is closed is $P_1(x) = |\psi_1(x)|^2$ ($P_2(x) = |\psi_2(x)|^2$). Now, in the case when both slits

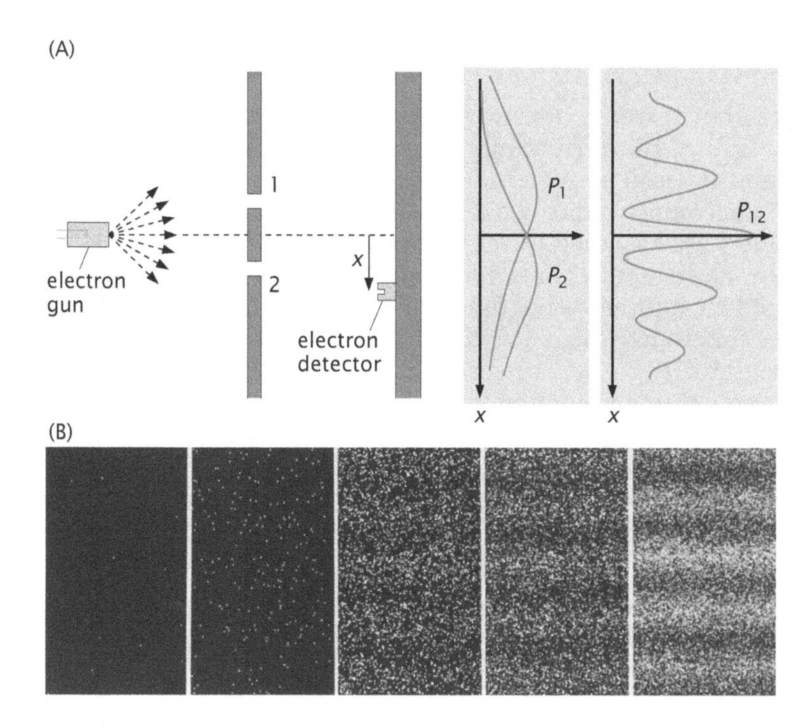

Figure 18.10: Two-slit experiment that demonstrates the wave-like nature of quantum particles. (A) Electrons are fired from an electron gun toward an impenetrable wall that has two slits through which electrons can pass. Electrons that make it through the slits are detected some distance away from the screen with an electron detector. Collected electrons along the x-axis show an interference pattern P_{12} characterized by successive peaks and valleys in the number of detected electrons as a function of x. When either of the two slits is closed, thus leaving either only slit 1 or only slit 2 for the electron to pass through, the positional distribution of detected electrons is given by P_1 and P_2. The essential weirdness of quantum mechanics is revealed by the fact that $P_1 + P_2 \neq P_{12}$. (B) Time course from an electron two-slit experiment performed by A. Tonomura and co-workers that shows the buildup of the interference pattern one electron at a time over about 30 minutes, starting with 10 electrons in the first image, and ending up with roughly 140,000 electrons in the last. (A, Adapted from R. P. Feynman et al., The Feynman Lectures on Physics. Addison-Wesley, 1965; B, from A. Tonomura et al., *Am. J. Phys.* 57:117, 1989.)

are open, the resulting probability amplitude is $\psi_{12}(x) = \psi_1(x) + \psi_2(x)$ and the associated probability is $P_{12}(x) = |\psi_1(x) + \psi_2(x)|^2$. Notably, $P_{12}(x) \neq P_1(x) + P_2(x)$ as would be expected for classical particles, for which each detected electron is assumed to have reached the detector following a definite trajectory that took it either through slit 1 or through slit 2. In fact, any experiment that would allow us to say with certainty which slit the detected electron went through (such as putting some sort of electron detector next to one of the slits) would end up destroying the observed interference pattern. A strange state of affairs! In fact, so strange that Einstein never accepted the fundamental role that probability plays in quantum mechanics, famously proclaiming "God does not play dice with the universe!" (Born & Einstein 2005).

The probability amplitude plays the same role in quantum mechanics that the amplitude of a classical wave, such as a water wave, a sound wave, or an electromagnetic wave traveling through a medium, does in classical mechanics. In the case of a classical wave, the square of the amplitude of the wave at a particular position in the medium tells us about the energy of the disturbance described by the wave at that position in space, while for a quantum wave it tells us about the probability of finding the quantum particle at that particular position. Unlike the amplitude of a classical wave, the probability amplitude itself has no simple physical meaning, and in fact it can be a complex number, which is why the probability is actually computed as the square of its modulus, $|\psi|^2 = \psi^*\psi$, where ψ^* is the complex conjugate of ψ. More precisely, $|\psi(x,t)|^2\,\mathrm{d}x$ is the probability that at time t the quantum particle is located between x and $x + \mathrm{d}x$. We can thus use the wave function to compute the uncertainty in the particle's position, $\Delta x = \sqrt{\langle x^2 \rangle - \langle x \rangle^2}$, where the averages are computed in the usual way that exploits the probability distribution function $|\psi(x)|^2$: $\langle \ldots \rangle = \int_{-\infty}^{\infty} (\ldots)|\psi(x)|^2\,\mathrm{d}x$.

Quantum Mechanical Observables Are Represented by Operators

Besides the position, the wave function can be used to determine the mean and the uncertainty of the momentum and energy of the quantum particle, or, for that matter, of any other dynamical variable. To accomplish this, we make use of "operators" that in quantum mechanics are assigned to each and every dynamical variable. Operators are mathematical "machines" that take a wave function as input and produce another wave function as output. In that sense, they are to wave functions as matrices are to vectors. For example, the momentum operator $\hat{P} = -i\hbar\partial/\partial x$ is essentially the derivative operation, and when applied to a wave function it returns its derivative with respect to position x multiplied by the factor $-i\hbar$. As a result of this definition, to compute the mean momentum of the quantum particle whose state is described by the wave function $\psi(x)$, we need to evaluate the integral

$$\langle\psi|\hat{P}|\psi\rangle = \int_{-\infty}^{\infty} \psi^*(x)[\hat{P}\psi(x)]\,dx\,. \tag{18.4}$$

Here, on the left-hand side of the equation, we have introduced the Dirac notation, which will considerably simplify bookkeeping in calculations later in the chapter. In fact, this notation evokes the inner product between two vectors, the row vector $\langle\psi|$ and the column vector $\hat{P}|\psi\rangle$. A more sophisticated mathematical exposition of the elements of quantum mechanics would go on to define an infinite-dimensional vector space called the Hilbert space, with wave functions as its elements, as the space of states of a quantum particle. Our focus though is not on the most general formulation but one that will allow us to develop in some quantitative detail simple models pertaining to quantum mechanical processes such as photon absorption and electron tunneling that are at the heart of photosynthesis.

A special role in quantum mechanics is played by the wave functions that are eigenfunctions of dynamical operators. These special states of the quantum particle are those with a definite value of the dynamical quantity in question, whether position, momentum, or energy. For example, the eigenfunctions of the momentum operator satisfy the equation

$$-i\hbar\frac{\partial\psi_p(x)}{\partial x} = p\psi_p(x)\,. \tag{18.5}$$

This equation demands that the derivative of the wave function be proportional to that wave function itself, a property held by exponential functions. We can see that the momentum eigenfunction is an exponential function of the form $\psi_p(x) = e^{ipx/\hbar}$, since such functions have precisely this property. Note that the modulus squared of this wave function is equal to 1, that is, $\psi_p(x)\psi_p^*(x) = 1$, which indicates that the probability of finding an electron anywhere on the x-axis is equally likely—in other words, its position is infinitely uncertain. This is in agreement with Heisenberg's uncertainty relation for a quantum particle that has zero uncertainty in its momentum.

Of all of the observable quantities represented by operators, we will focus most directly on the energy. The energy is represented by an operator known as the Hamiltonian. For a particle moving in a one-dimensional energy landscape characterized by a potential $U(x)$, we can write the Hamiltonian as

$$\hat{H} = -\frac{\hbar^2}{2m}\frac{\partial^2}{\partial x^2} + U(x). \tag{18.6}$$

We can think of the Hamiltonian as being built up from the kinetic and potential energies of the particle, with the first term giving the kinetic energy of the quantum particle and the second term giving the potential energy. If we recall that kinetic energy can be written in terms of momentum as $KE = p^2/2m$, we can obtain the kinetic-energy operator from the momentum operator as $\hat{P}^2/2m$. The average value of the kinetic energy in a quantum state ψ yields the same confinement energy we estimated earlier based on Heisenberg's uncertainty principle. The potential energy of the quantum particle due to some external field such as an electric potential is captured by $U(x)$. The real significance of the eigenfunctions of the Hamiltonian becomes apparent once we introduce the celebrated Schrödinger equation, which is the governing equation of quantum mechanics, in the same way that Newton's equation $F = ma$ is the governing equation of classical mechanics.

The Time Evolution of Quantum States Can Be Determined Using the Schrödinger Equation

The job of the governing equation of quantum mechanics is to predict the probability amplitude associated with a quantum particle at time t given that we know the amplitude at some earlier time. Such an equation was first written down by Schrödinger in the form of a rate equation (albeit one with complex coefficients) for the wave function, namely,

$$i\hbar \frac{\partial \psi(x, t)}{\partial t} = -\frac{\hbar^2}{2m} \frac{\partial^2 \psi(x, t)}{\partial x^2} + U(x)\psi(x, t). \tag{18.7}$$

Using the Hamiltonian operator defined in Equation 18.6, we can write the Schrödinger equation in a more compact form

$$i\hbar \frac{\partial \psi(x, t)}{\partial t} = \hat{H}\psi(x, t). \tag{18.8}$$

The eigenfunctions of the Hamiltonian satisfy the equation $\hat{H}\psi_E = E\psi_E$, which when substituted into the Schrödinger equation leads to

$$i\hbar \frac{\partial \psi_E(x, t)}{\partial t} = E\psi_E(x, t). \tag{18.9}$$

and

$$\psi_E(x, t) = \psi_E(x, 0)e^{-iEt/\hbar}. \tag{18.10}$$

In other words, the probability amplitude for a quantum particle in a state with definite energy evolves simply by a phase factor $e^{-iEt/\hbar}$, and therefore the probability of finding the particle at position x, $P(x) = |\psi_E(x, t)|^2 = |\psi_E(x, 0)|^2$, does not change in time. In this sense, the eigenstates of the Hamiltonian are stationary states of the Schrödinger equation. They are referred to as the energy levels of the quantum particle. The fact that energy levels of electrons confined to a finite region of space are quantized is what leads to discrete lines in the emission and absorption spectra of atoms and molecules. We now take up this issue in the context of a simple model of an electron confined to a one-dimensional box, the quintessential simplest quantum problem.

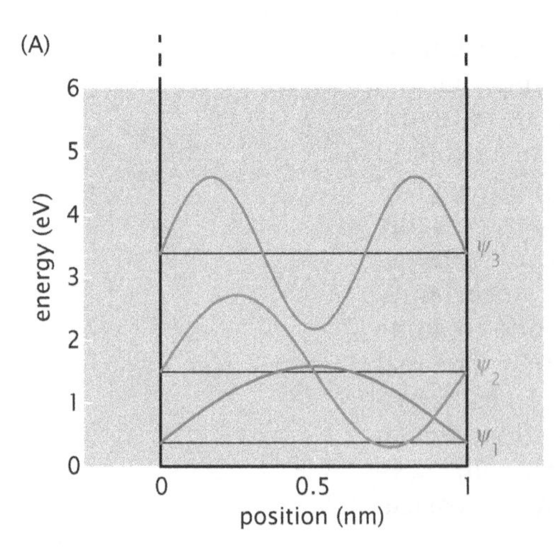

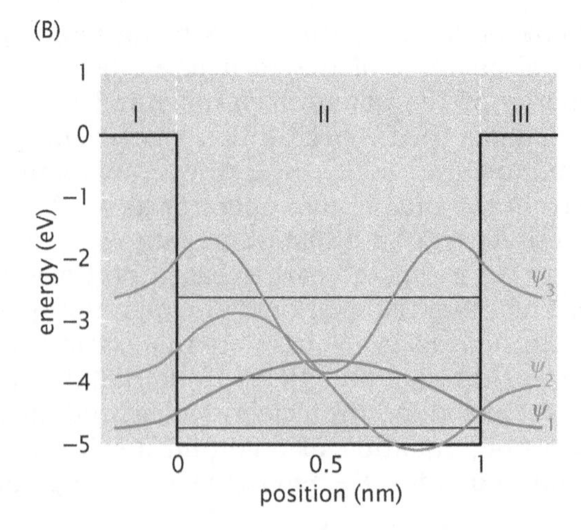

Figure 18.11: Particle-in-a-box models. (A) Potential well for a particle in a box with an infinite confining potential. The wave functions for the three lowest energy states are shown with their height in the box corresponding to their associated energies. (B) Finite-well particle in a box. The wave functions for the three lowest energy states are shown and, as with the wave functions in (A), the vertical position of the state corresponds to the energy of that state. Note that though the shapes of the wave functions are similar, the wave function for the finite box bleeds over into the classically forbidden region.

18.2.2 The Particle-in-a-Box Model

Now that we have the Schrödinger equation in hand, we can return to our estimates of the energy of electrons confined to boxes, but now treated using the full machinery of quantum mechanics. In particular, the estimate giving the energies of electrons confined to a box can be put on a firmer footing by explicitly computing the wave functions and energy states for an electron confined to a box like that shown in Figure 18.11(A). As will be shown in subsequent sections, by evaluating the energy levels for a particle in a box, we will be in a position to evaluate the wavelengths of light that are capable of exciting a molecule of interest, since these wavelengths have energies corresponding to the *differences* in energy between the energy states of the molecule.

The starting point for a quantum mechanical calculation (at least within the limited scope considered in this book) is the use of the steady-state Schrödinger equation, which for some potential-free region is given by

$$-\frac{\hbar^2}{2m}\frac{d^2\psi(x)}{dx^2} = E\psi(x). \tag{18.11}$$

Cursory inspection reveals that this is the same equation that arises in working out the dynamics of a spring (that is, the harmonic oscillator), and the solution can be written by inspection as

$$\psi(x) = A\cos kx + B\sin kx, \tag{18.12}$$

where we define $k = \sqrt{2mE/\hbar^2}$. Verification that this is the solution follows simply by differentiating $\psi(x)$ twice and plugging back into the equation.

One of the deepest results about both classical and quantum mechanical problems of this variety is the way in which constraints imply quantization. For example, when we pin a string at its ends, we force the vibrational frequencies to adopt only a limited set of values. Similarly, in the quantum mechanical setting, by insisting that the electron be confined to a one-dimensional box of length a, we have imposed the condition (see the edges of the solutions in Figure 18.11(A))

$$\psi(0) = \psi(a) = 0, \tag{18.13}$$

which will restrict the set of allowed energies to a special set of discrete values. The first of these conditions forces us to adopt $A = 0$. The second condition leads us to the realization that

$$ka = n\pi. \tag{18.14}$$

Invoking the definition $k = \sqrt{2mE/\hbar^2}$, we see that these solutions imply corresponding energy eigenvalues

$$E_n = \frac{\hbar^2}{2m}\left(\frac{n\pi}{a}\right)^2, \tag{18.15}$$

which, aside from numerical factors, is precisely the same energy we found from our simple estimate using the Heisenberg uncertainty principle. This result will serve us in our attempt to understand the connection between absorption wavelength and molecular dimensions.

Solutions for the Box of Finite Depth Do Not Vanish at the Box Edges

The particle-in-a-box model described above can be generalized to the more realistic case in which the bounding well is not infinitely high, resulting in some leakiness. This simple model will serve as the basis of our discussion of the electron transfer process during photosynthetic reactions. A schematic of the system is shown in Figure 18.11(B). In this case, we are once again interested in the energy eigenvalues. Intuitively, in this case it appears that the spectrum of bound states will be finite because once the energies get too large, the particle will no longer be bound. This should be contrasted with the infinite potential well considered in Figure 18.11(A), where there is an infinite spectrum of energy eigenvalues and their corresponding wave functions.

To solve the finite-well problem, we solve the Schrödinger equation in three distinct regions. For the two regions where the potential vanishes, the Schrödinger equation can be written in the same way as we did for the previous problem, namely,

$$-\frac{\hbar^2}{2m}\frac{d^2\psi(x)}{dx^2} = E\psi(x). \tag{18.16}$$

In the region where the potential is nonzero, the Schrödinger equation can be written as

$$-\frac{\hbar^2}{2m}\frac{d^2\psi(x)}{dx^2} - V_0\psi(x) = E\psi(x). \tag{18.17}$$

In all three regions, we can recognize the solutions as exponentials of the form

$$\psi_{\mathrm{I}}(x) = Ae^{\kappa x} \tag{18.18}$$

and

$$\psi_{\mathrm{II}}(x) = Be^{\kappa' x} + Ce^{-\kappa' x} \tag{18.19}$$

and

$$\psi_{\text{III}}(x) = De^{-\kappa x}. \tag{18.20}$$

The solutions in regions I and III are chosen to insure that the wave functions vanish appropriately as x approaches $\pm\infty$.

To complete the solution, we need to satisfy boundary conditions at the two interfaces at 0 and a. In particular, our aim is to match both the wave functions and their first derivatives at both boundaries, resulting in

$$A = B + C, \tag{18.21}$$

$$A\kappa = B\kappa' - C\kappa', \tag{18.22}$$

$$Be^{\kappa' a} + Ce^{-\kappa' a} = De^{-\kappa a}, \tag{18.23}$$

$$B\kappa' e^{\kappa' a} - C\kappa' e^{-\kappa' a} = -D\kappa e^{-\kappa a}, \tag{18.24}$$

where Equations 18.21 and 18.22 come from matching the wave functions and their derivatives at the interface between regions I and II and Equations 18.23 and 18.24 come from matching the wave functions and their derivatives at the interface between regions II and III. These four conditions can be rewritten in matrix form as

$$\begin{pmatrix} 1 & -1 & -1 & 0 \\ \kappa & -\kappa' & \kappa' & 0 \\ 0 & e^{\kappa' a} & e^{-\kappa' a} & -e^{-\kappa a} \\ 0 & \kappa' e^{\kappa' a} & -\kappa' e^{-\kappa' a} & \kappa e^{\kappa a} \end{pmatrix} \begin{pmatrix} A \\ B \\ C \\ D \end{pmatrix} = 0. \tag{18.25}$$

If we call the 4×4 matrix here $\mathbf{M}$, then the condition we must satisfy to determine the unknown coefficients A, B, C, and D is $\det \mathbf{M} = 0$. The determinant can be expanded, resulting in the following secular equation

$$e^{-(\kappa+\kappa')a}[(e^{2\kappa' a} - 1)\kappa^2 + 2\kappa\kappa'(e^{2\kappa' a} + 1) + (e^{2\kappa' a} - 1)\kappa'^2] = 0. \tag{18.26}$$

If we now define the quantity

$$\kappa' = i\sqrt{\frac{2m}{\hbar^2}(E + V_0)} = ik \tag{18.27}$$

and take the factor $e^{-\kappa' a}$ inside the square brackets in Equation 18.26, we are left with

$$e^{-\kappa a} 2i \sin ka \left(\kappa^2 + \frac{2\kappa k}{\tan ka} - k^2\right) = 0. \tag{18.28}$$

This can be simplified to the condition

$$\frac{2}{\tan ka} = \frac{k}{\kappa} - \frac{\kappa}{k}. \tag{18.29}$$

What this means is that we now need to find those values of the energy E (hidden within our parameters κ and κ') that solve this equation. If we define the parameters

$$\alpha = \frac{2mV_0 a^2}{\hbar^2} \tag{18.30}$$

and

$$x = -\frac{E}{V_0}, \tag{18.31}$$

we can simplify Equation 18.29 into the convenient dimensionless form

$$\frac{2}{\tan\sqrt{\alpha(1-x)}} = \sqrt{\frac{1-x}{x}} - \sqrt{\frac{x}{1-x}}. \tag{18.32}$$

For a given value of α, we are interested in finding those values of x for which the two sides of the equation are equal. Recalling the definition of x, it is clear that finding x is the same as determining the allowed energy eigenvalues. To make substantive progress with determining these values, we turn to numerical analysis as described in the following Computational Exploration.

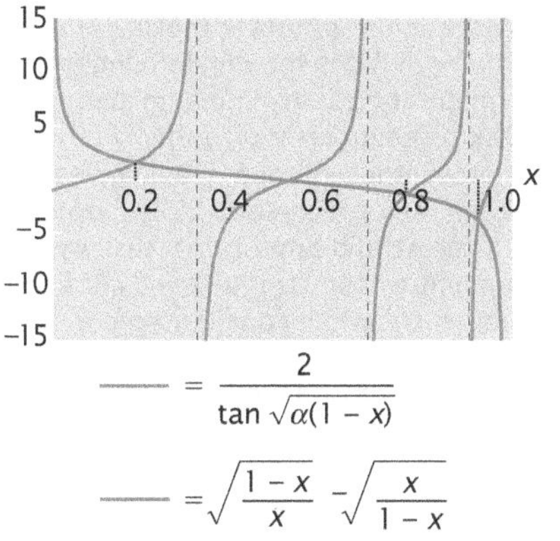

Figure 18.12: Energy eigenvalues. Graphical determination of the energy eigenvalues for a particle in a finite well. The solutions are found by plotting the two sides of Equation 18.32 and looking for those values of x at which the two curves intersect. The most deeply bound state is the solution to the *right* side of the graph.

COMPUTATIONAL EXPLORATION

Computational Exploration: Electrons in a Well of Finite Depth To find the eigenvalues for the Schrödinger equation for electrons in a potential well of finite depth, we have to resort to some numerical method. For this problem, we prefer graphical determination of the unknown values of x, though there are other "root-finding" schemes that are a powerful means to uncovering the relevant solutions.

The graphical strategy is shown in Figure 18.12. Note that this is precisely the same approach that we used earlier when solving numerically for Lagrange multipliers (p. 257). We are interested in determining the values of x for which the equation is satisfied *for a given choice of* α. Numerically, what this means is that we choose a value of α and then plot the two sides of Equation 18.32 as functions of x. The solutions correspond to those values of x for which the two curves intersect. To actually find those values of x, one unsophisticated but entirely useful approach is to use the feature in Mathematica or Matlab (or whatever software might be preferred by the reader) to plot the two functions over ever narrower regions.

Because the behavior of the tan function is periodic, there are multiple solutions, with each different solution corresponding to a different bound state. The reader is invited to make his or her own choice for the parameter α and to use this graphical strategy to find the energies and to compare them with those emerging from the infinite-well potential. The alternative scheme of numerical root-finding is explored in the problems at the end of the chapter.

18.2.3 Exciting Electrons With Light

The living world explodes with color, whether the shocking panoply of colors seen on the coral of the Great Barrier Reef or in our gardens every spring. These colorful displays are largely the result of the way key pigment molecules interact with light. In this section, our aim is to use the simplest model that has been developed to explain the interactions of these molecules with light using quantum mechanics.

In photosynthesis, the "boxes" that confine electrons are actually pigment molecules. These organic molecules are not proteins or nucleic acids or sugars or lipids, but highly specialized products of dedicated biosynthetic pathways. Figure 18.13 shows the structures of several of the most important pigment molecules associated with the photosynthesis process. The best-known of these photosynthetic pigments, chlorophyll, shares a similar porphyrin ring structure with the heme group found in hemoglobin (see Figure 4.6 on p. 146).

Figure 18.13: The molecules of light absorption in photosynthesis. (A) Chlorophyll *a* is the most abundant of pigment molecules in plants and, besides being critical to the photosynthesis process, also gives plants their green color. (B) β-Carotene is a carotenoid and is an accessory pigment whose structure we will find convenient when considering our "box model" of molecules. (C) Phycoerythrobilin is a red phycobilin, a light-absorbing molecule found in both cyanobacteria and red algae. (D) Absorption spectrum for biological pigments. Note that in the case of chlorophyll *a*, the *in vitro* absorption maxima are at wavelengths of roughly 430 nm (blue) and 660 nm (red) with almost no absorption in the green, thus conferring the familiar green color to plants. (Adapted from C. K. Mathews et al., Biochemistry, 3rd ed. Prentice Hall, 1999.)

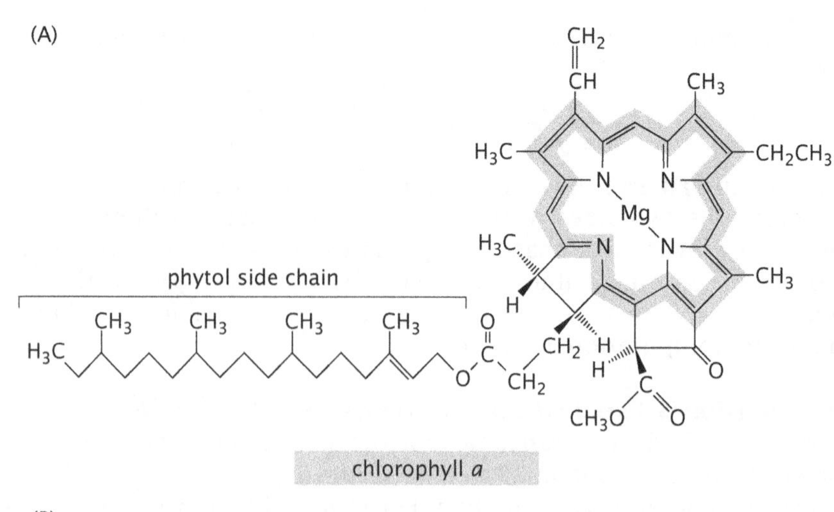

(B)

CH_3

H_3C CH_3 CH_3 CH_3 H_3C

CH_3 CH_3 CH_3

CH_3

β-Carotene

(C)

COO^- COO^-

CH_3 CH_2 CH_2 CH_2

H_3C CH H_3C CH_2 H_3C CH_3 H_3C CH

O N N N N O

H H H

phycoerythrobilin

(D)

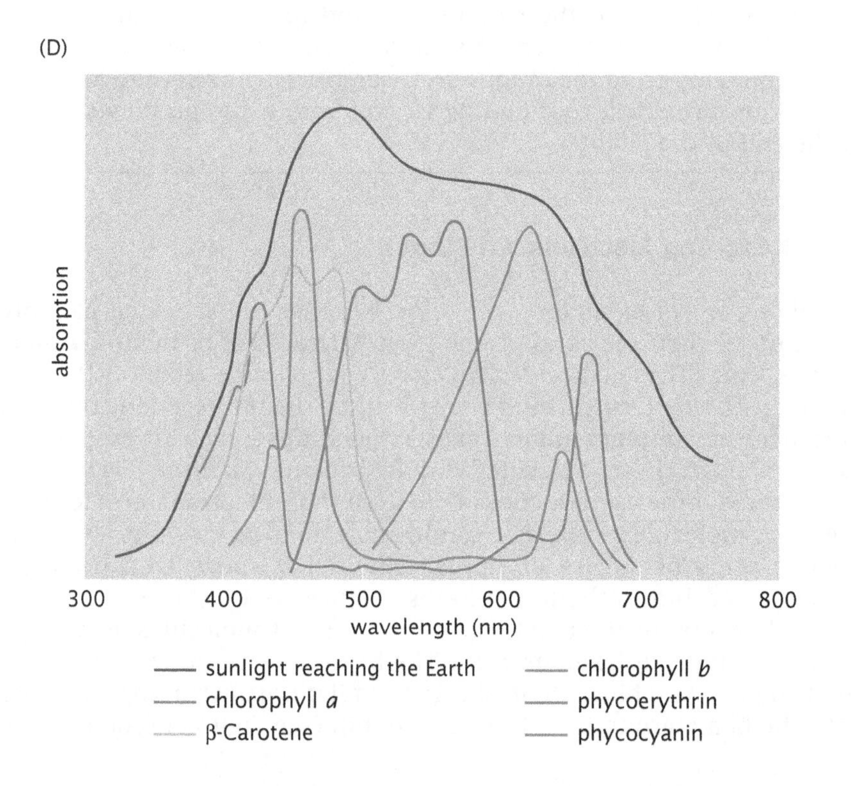

The most striking structural similarity among the different pigment molecules is that they all contain a relatively large system of conjugated double bonds (shaded in the figure). Within these conjugated systems, electrons associated with molecular wave functions known as pi orbitals are able to move freely. As we will see below, the size of these conjugated systems determines the energy level of the associated pi electrons, and is critical for determining the wavelengths of light that can be absorbed by each specific pigment. Figure 18.13 shows the differences among the various light absorption spectra for several photosynthetic pigments. We will now attempt to develop a simple quantitative model that can predict the absorption properties of pigment molecules based simply on the size of these conjugated double bond systems.

Absorption Wavelengths Depend Upon Molecular Size and Shape

The seeds of an intuitive understanding of the absorption properties of biological pigments can be sown by appealing to the particle-in-a-box problem worked out in Section 18.2.2 (p. 730). The key idea that is the basis of this section is that the absorption wavelength λ is related to the energy difference between the relevant molecular states through the equation

$$\frac{hc}{\lambda} = E_2 - E_1. \tag{18.33}$$

For example, we will often first be interested in the minimum-energy excitation to excite an electron in the lowest molecular energy state up to the next unoccupied energy state. This is sometimes referred to as the HOMO–LUMO gap, with the acronyms standing for "highest occupied molecular orbital" and "lowest unoccupied molecular orbital," respectively.

To compute these energy differences, we resort to the particle-in-a-box model already worked out in detail above, save that now we have to be more careful about the linear dimensions of the molecule and the corresponding number of electrons. Taking our cue from molecules such as β-carotene (see Figure 18.13), we begin by thinking of the pigment molecule of interest as a linear chain of N atoms and we assume that the mean spacing between atoms along the chain is given by a. This means then that the overall length of the chain is $(N-1)a$. From the standpoint of the quantization conditions worked out for the chain, this implies

$$k(N-1)a = n\pi. \tag{18.34}$$

We can then find the energy eigenvalues as $E = \hbar^2 k^2/2m$. To determine the absorption wavelength that corresponds to exciting an electron from the highest occupied energy state to the lowest unoccupied energy state, we use

$$\lambda = \frac{hc}{\Delta E}. \tag{18.35}$$

To make further progress, we now need to compute ΔE itself. How do we find the highest occupied energy level and its corresponding n? To do so, we need to count up the pi electrons in the molecule, which are those involved in the excitation process. Note that there is one such electron per atom (for a total of N electrons). Each energy state can accommodate two such electrons by the Pauli principle. The Pauli principle is both fundamental and deep, and here, instead of probing these depths, we opt for merely noting that the Pauli principle tells us that for systems with many electrons, we fill up the energy states one electron at a time from lowest energy to highest energy, with each state able to accommodate two electrons, one for each spin.

If N is an even number, the highest occupied state is given by

$$E_{N/2} = \frac{\hbar^2}{2m} \frac{(N/2)^2 \pi^2}{(N-1)^2 a^2}. \tag{18.36}$$

Using this strategy, we find that

$$\Delta E = E_{(N/2)+1} - E_{N/2} = \frac{\hbar^2 \pi^2}{2ma^2} \frac{N+1}{(N-1)^2}, \tag{18.37}$$

resulting in the wavelength

$$\lambda = \frac{8mc}{h} a^2 \frac{(N-1)^2}{N+1}. \tag{18.38}$$

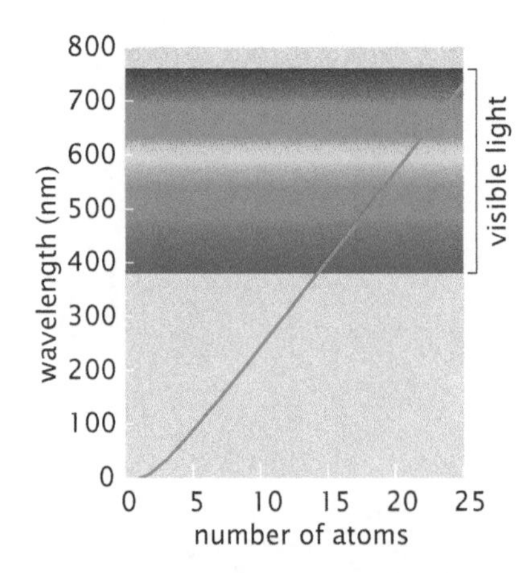

Figure 18.14: Wavelength of light absorption in a box model of a pigment molecule. This plot shows the results of treating a pigment molecule as a linear chain of atoms and computing the HOMO–LUMO gap and using that energy gap to compute a corresponding wavelength. The spacing between the atoms is taken as 0.1 nm.

The results of this analysis are shown in Figure 18.14, where we see that this toy model yields absorption of visible light for chains with a length of 15 to 25 atoms. This result agrees embarrassingly well with the actual size of the conjugated double-bond systems in chlorophyll (20 atoms) and β-carotene (22 atoms).

The model worked out in this section represents a vast simplification. The most important point to take away from the calculations done here is the way in which the landscape of confinement dictates the allowed energy levels. However, to really understand the detailed spectral properties of a given light-absorbing molecule requires a more thorough quantum mechanical treatment that explicitly accounts for the atomic positions as well as the underlying wave functions. In addition, we would have to go beyond the simple discrete states considered here to account for the broadening of the absorption as revealed in Figure 18.13(D).

ESTIMATE

Estimate: Number of Incident Photons Per Pigment Molecule One of the interesting points about the interplay between light absorption and electron transfer is the "stoichiometry" of the process. In particular, it is interesting to determine how many times each second a pigment molecule is subjected to excitation by an incident photon.

To begin, we make the simplifying assumption that all photons in the incident light have the same energy given by $E = hc/\lambda$, and taking $\lambda = 500$ nm results in an energy per photon of roughly 4×10^{-19} J. We can now use this number to find the approximate number of photons hitting a square meter every second by assuming that the full 1000 W/m^2 is being delivered. By dividing this incident power by the energy per photon, we can find the number of photons, namely,

$$\begin{aligned} &\text{number of photons per area per second} \\ &\quad \approx \frac{10^3 \text{ J m}^{-2} \text{ s}^{-1}}{4 \times 10^{-19} \text{ J/photon}} \\ &\quad \approx \frac{1}{4} \times 10^{22} \text{ photons m}^{-2} \text{ s}^{-1} \end{aligned} \tag{18.39}$$

Given this incident photon flux, we can now examine what this rough estimate tells us about the number of photons impinging on a pigment molecule every second. If we take the simple rule of thumb that each pigment molecule presents a cross-sectional area of roughly 1 nm^2, this means that there are roughly 2500 such photons making it to the pigment each second. However, the efficiency with which photons are absorbed means that only roughly 1 in 100 of these photons will actually be absorbed, leaving us with a modest several tens of photons exciting the pigment molecules each second.

Further details about this estimate are described in the treatment by Blankenship (2002); see the Further Reading at the end of the chapter.

18.2.4 Moving Electrons From Hither to Yon

Excited Electrons Can Suffer Multiple Fates

Thus far, we have discussed the physics of electron excitation in biological pigments as a result of the absorption of a photon; this is the critical process that allows living cells to capture the photon energy generated by nuclear reactions in the sun, which is roughly 150 million kilometers away from the surface of the Earth. In order to make biological use of the captured energy, the photosynthetic cell must convert the energy of the excited electron within the pigment molecule into some form that can be used for metabolic purposes. To this end, the cell must store the energy, essentially in the form of a charged biological battery, where there is a separation of charge such that an electrochemical gradient can be formed across a membrane. By using photon energy to transport charge, the captured energy of each photon can be stored additively rather than used immediately. Transmembrane electrochemical gradients of hydrogen ions can be used to generate ATP as we saw in Section 16.1.2 (p. 634). In the context of photosynthesis as practiced by cyanobacteria and by plants and algae containing chloroplasts, we will also examine how charge transport processes can result in the generation of molecules such as NADPH that carry high-energy electrons with the power to reduce oxidized compounds. The most important oxidized compound that must be reduced by cyanobacterial chloroplasts is carbon dioxide; their aspiration is to convert gaseous carbon dioxide into a sugar. The sugar will be used as a backbone for the biosynthesis of macromolecules by the photosynthetic organism itself and ultimately will provide biochemical building blocks for all organisms that make a living by grazing on the photosynthetic organisms and those that eat them and so on.

How can the energy of the excited electron in a pigment molecule be converted into a separation of charge across a membrane? Figure 18.15 shows the four general possible fates of electrons after the excitation process. The first and second of these possible fates, liberation of the energy in the form of heat or light, cannot directly lead to useful energy storage for biosynthesis (although liberation of energy in the form of light, better known as fluorescence, is extremely useful in many cases). However, the third and fourth possible fates involve the transfer of the electron excitation energy or even of the electron itself to another molecule; these are both critical in photosynthesis.

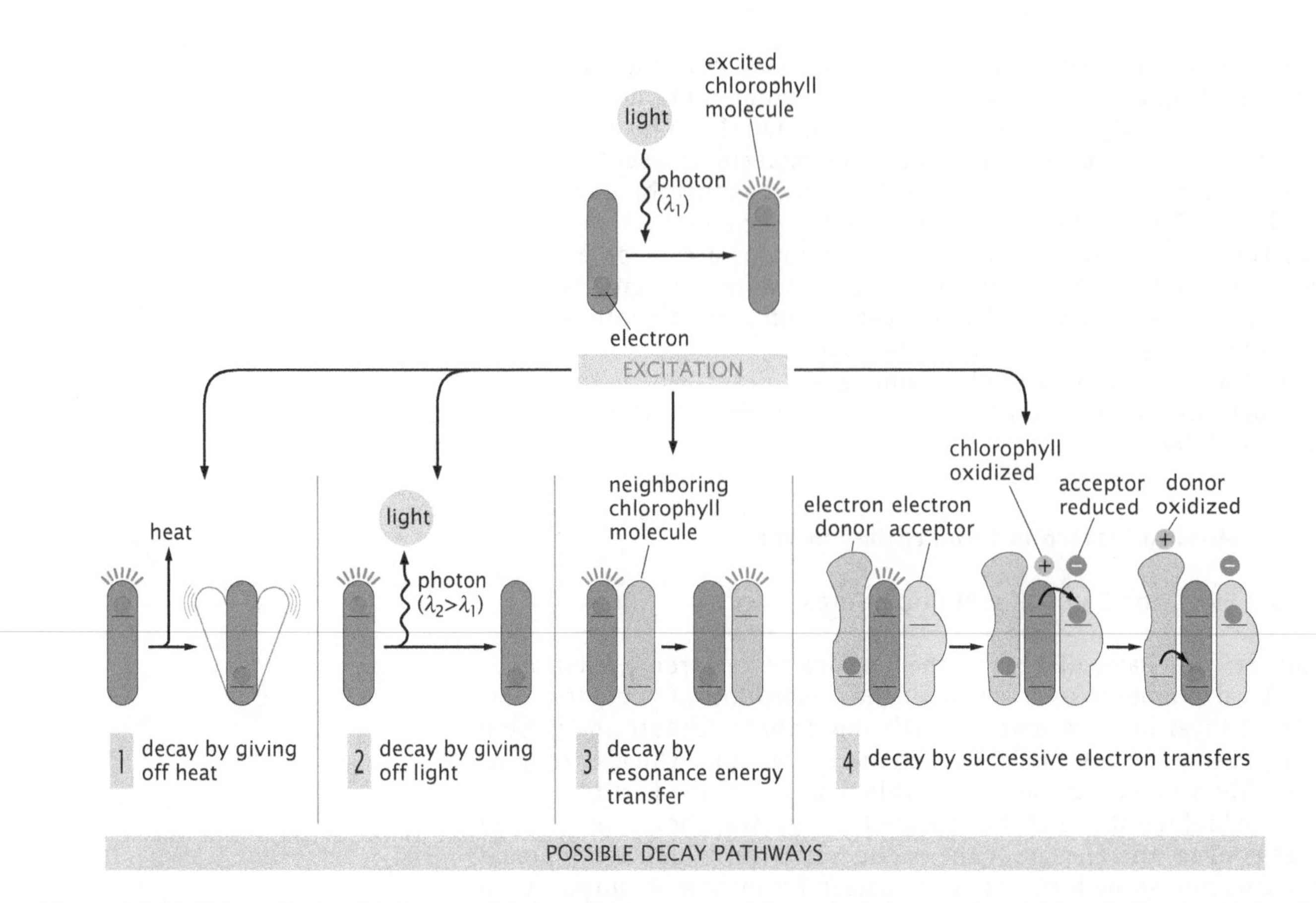

Figure 18.15: Schematic showing four possible fates for the energy of the excited electron after a chlorophyll molecule absorbs a photon of light. Pathway 3 is used during energy transfer from chlorophyll molecules in the antenna complex to the special pair at the reaction center. Pathway 4, used in the reaction center, generates a stable charge separation that can later be exploited for biosynthetic reactions.

First, we will consider the process of electron transfer from the excited donor molecule to a neighboring recipient, as this is the process that contributes most directly to the charge separation critical for photosynthetic biosynthesis. The process of electron transfer is energetically favorable as long as the second molecule has an available excited state at a lower energy level. If the first molecule (chlorophyll, for example) starts off in a neutral state, the product of the electron transfer reaction will be positively charged chlorophyll and a negatively charged acceptor. In the photosynthetic reaction center, the negatively charged acceptor transfers the electron to yet another acceptor that is physically further away from the donor fluorophore. In the meantime, the donor chlorophyll returns to its neutral state by claiming an electron from another donor. After several such transfer steps, the moving electron is effectively transferred from one side of the membrane to the other. Figure 18.16 shows the actual organization of the photosynthetic pigments within a simple reaction center used by purple bacteria. The proximity and relative orientation of all the pigments involved in electron transport determine the sequence and timing of the individual electron transfer reactions. After two cycles of photon absorption and electron transfer, an exchangeable quinone molecule carrying two electrons is released from the reaction center; at the same time, two electrons have been removed from the opposite side of the membrane. The charge imbalance is ultimately corrected by the transfer of hydrogen ions through

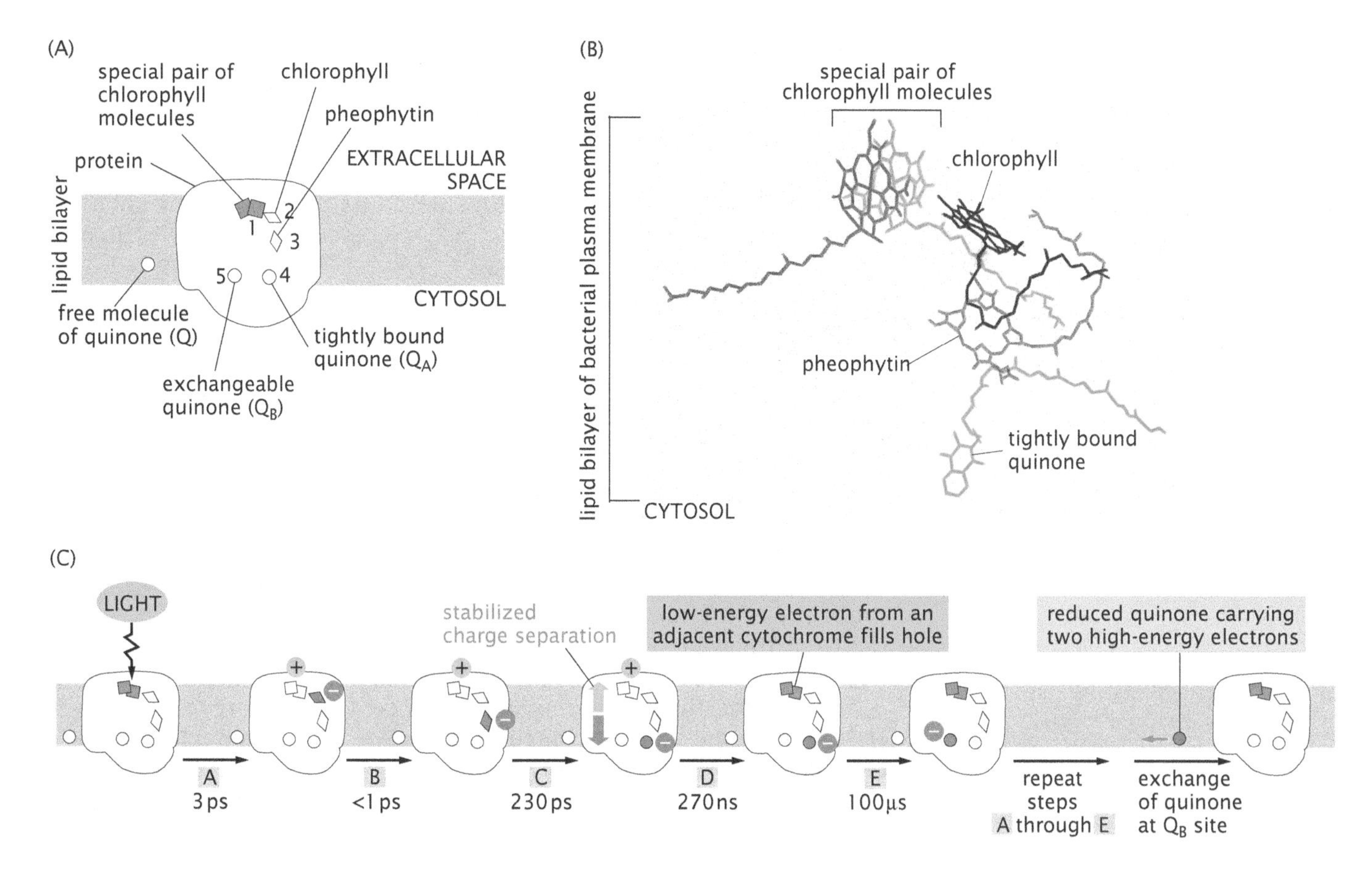

Figure 18.16: Pigment architecture and electron transfer in the photosynthetic reaction center. (A) Schematic of a bacterial photosynthetic reaction center, showing the approximate positions of pigments (1–5) embedded in the transmembrane protein (white). Electron transfer across the membrane is initiated by the excitation of pi electrons in the "special pair" of chlorophyll molecules shown at the top of the diagram. (B) Structural detail showing the relative positions of the pigments involved in initial charge separation (1–4 in part A). (C) Time scales and order of events leading to charge separation after excitation of the special pair chlorophylls. (Adapted from B. Alberts et al., Molecular Biology of the Cell, 5th ed. Garland Science, 2008.)

the transmembrane cytochrome–protein complex, resulting in a pH gradient that can be used to drive ATP synthesis.

Electron Transfer in Photosynthesis Proceeds by Tunneling

An important feature of electron transport through the photosynthetic reaction center is that the excited electron travels from one pigment to another in the series of pigments that serve in turn as the carriers of the transferred electron despite the fact that these pigments are not in direct physical contact. We will attempt to understand the electron transfer process by recourse to one of the most important ideas of quantum mechanics, namely, the idea of quantum mechanical tunneling. This idea made perhaps its most famous early appearance as a way of interpreting the radioactive decay of seemingly stable atomic nuclei. Tunneling is one of the counterintuitive results in which quantum mechanical effects are in direct conflict with the ideas that emerge from our everyday experience.

The essence of the tunneling concept is that the electron has a finite probability of appearing in what is known as the "classically forbidden region" as indicated schematically in Figure 18.11(B), where we see that there is a nonzero probability of finding the particle in a region where it is supposedly forbidden because of a lack of energy. If we imagine a particle moving on a potential-energy landscape, classical physics tells us that the highest potential energy will be attained when the entire kinetic-energy budget is converted into potential energy. In the quantum mechanical problem, these classical restrictions are no longer so severe and hence the particle can pass across energy barriers that cannot be breached when considered using classical mechanics. To get a feeling for how this all works, we resort to a simple estimate.

ESTIMATE

Estimate: The Tunneling Length Scale One of the most counterintuitive effects predicted by quantum mechanics is tunneling. In the world of macroscopic experience, tunneling is like driving a car into a wall and ending up on the other side without a crash. In the quantum setting, this effect can be seen in the case of a particle confined to a potential well (or a "box") like that shown in Figure 18.11(B), in which the particle still has a nonzero probability of being found outside the box. This is an impossibility according to classical physics since energy conservation demands that a particle that is capable of escaping from a potential well has an energy that is larger than the depth of the well. Violation of this condition would result in a negative kinetic energy upon leaving the well.

To explore this phenomenon, we once again resort to the Heisenberg uncertainty principle to make an estimate indicated schematically in Figure 18.17(A). Our goal is to determine the length scale ξ over which we expect to see an appreciable probability for the electron to be found in the classically forbidden region outside of the box. The reason this is interesting to the considerations of the present chapter is that neighboring protein molecules or domains within proteins partake in the electron transfer process and the extent to which such electron transfer occurs depends upon the "leakage" of the electron wave function from the donor site to the acceptor site. If the particle is found outside the box, the uncertainty in its position is set by the length ξ and the uncertainty in its momentum is determined by Heisenberg's principle and is given by $\Delta p \approx \hbar/\xi$. For the particle to escape from the box, it should have enough momentum to compensate for the difference between its energy E as measured from the bottom of the box and the depth of the potential well V_0, that is,

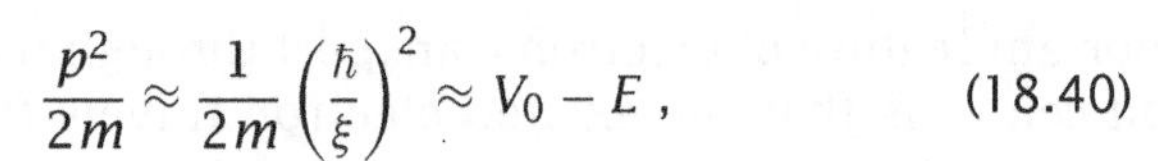

$$\frac{p^2}{2m} \approx \frac{1}{2m}\left(\frac{\hbar}{\xi}\right)^2 \approx V_0 - E\,, \tag{18.40}$$

where we have estimated the momentum p by its uncertainty Δp. From this equation, we arrive at the estimate

$$\xi \approx \frac{\hbar}{\sqrt{2m(V_0 - E)}}. \tag{18.41}$$

For an electron in the ground state, whose energy is close to zero, the tunneling length can be estimated as $\xi \approx 0.2\ \text{nm}/\sqrt{V_0}$, where V_0 is the depth of the well measured in electron volt units. This dependence of tunneling length on confinement energy is plotted in Figure 18.17(B).

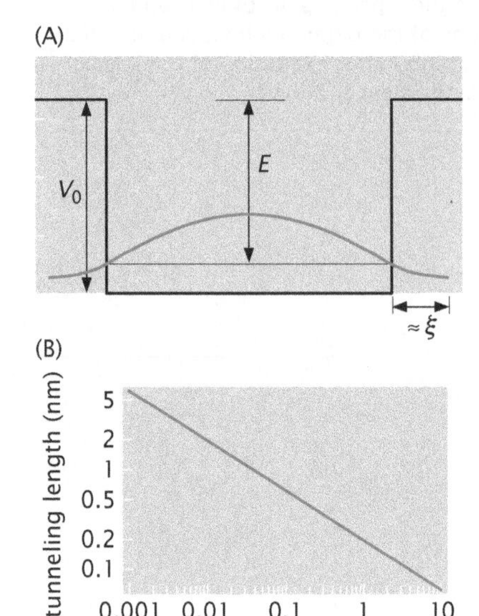

Figure 18.17: Tunneling length as a function of the well depth. (A) Schematic of the well showing the bleeding over of the wave function into the classically forbidden region. (B) The tunneling length is the distance from the well over which there is an appreciable probability of finding the electron, which is otherwise confined to the well.

Our aim is to get a feeling for how tunneling processes might operate in molecules. To that end, we work with a highly idealized one-dimensional box model of a molecule in which the molecular potential is represented as two adjacent square-well potentials like those introduced in our discussion of the particle in the box in Section 18.2.2. The reason this serves as a model molecule can be understood by appealing to simple diatomic molecules such as H_2 or more general heteronuclear molecules. In that case, electrons feel the total nuclear potential of two distinct hydrogen nuclei and the toy model of such a "molecule" shown in Figure 18.18 mimics such distinct nuclear potential wells. When considering tunneling between two different parts of a molecule, we treat the acceptor and the donor as having wells of different depths. In this case, tunneling refers to the idea that if an electron starts out confined to one of the wells, over time it has a finite chance of being localized in the second well.

Now that we have provided a verbal and graphical depiction of the electron transfer process and we have described the classic model of quantum mechanical tunneling, we seek a mathematical caricature of the process in molecules that captures some of the most essential features such as the distance dependence of the transfer rate. Our goal is to find the probability that if the electron starts in state A, it will end up in state B after a time t, a quantity we will label $P_{A\to B}(t)$. The details of our analysis are provided in the appendix at the end of the chapter (Section 18.5), while here we use broad brushstrokes to convey the key conceptual features of the calculation.

As shown in Figure 18.18(B), one of the useful approximations we make is that the total molecular wave function can be obtained by adding up the wave functions of the two distinct "atoms", that is, the solutions for the two isolated wells, with the relative contributions of ψ_A and ψ_B captured by as-yet unknown coefficients. In particular, we argue that

$$\psi_{tot} = a_A\psi_A + a_B\psi_B, \tag{18.42}$$

where a_A and a_B are the unknown coefficients that characterize the weighting of the two basis functions. The whole calculation then turns on minimizing the energy of the system as a way to find those unknown coefficients. For our purposes here, the key point is the extent to which the wave functions in the two wells overlap. Specifically, we are interested in how much of wave function ψ_A bleeds over into the well B and vice versa. This degree of overlap dictates the extent to which the tunneling between the two wells will occur and is captured in

$$H = -V\int_0^a dx\ \psi_A^*(x)\psi_B(x), \tag{18.43}$$

where V is the depth of the wells as shown in Figure 18.18(A). The tunneling rate itself is proportional to the square of this integral, with the form

$$\text{rate} \propto H^2, \tag{18.44}$$

and, as is shown in the appendix, using our box model of molecules given here, we find

$$\text{rate} = \frac{2\pi}{\hbar}H^2F^2\rho e^{-2\kappa d}, \tag{18.45}$$

where F is a distance-independent numerical factor that accounts for the details of the integral given above, the factor ρ will be discussed in

(A)

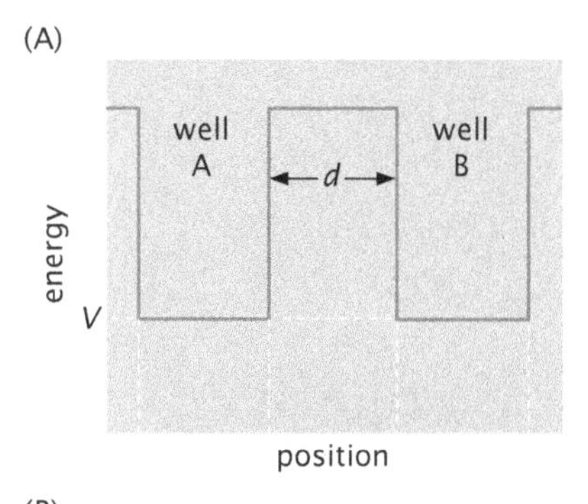

(B)

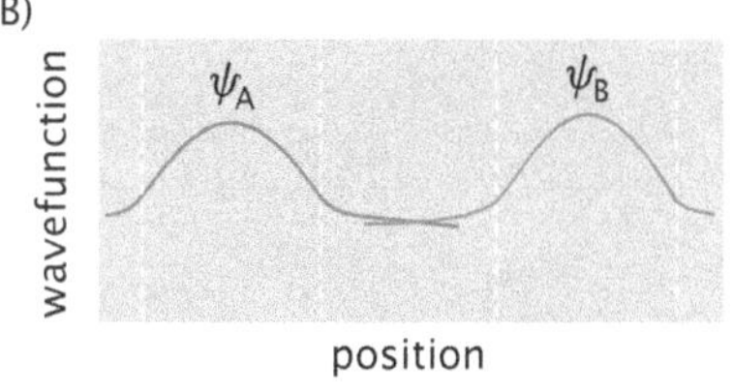

Figure 18.18: Box model of a molecule. (A) The "molecule" is characterized by two potential wells separated by a distance d. (B) The total molecular wave function is built up as a linear combination of the two "atomic" wave functions corresponding to the solutions for the individual potential wells.

more detail in the appendix and accounts for the density of accessible energy states, and the exponential dependence on the distance d, with a decay length κ, is an inheritance of the distance dependence of Equation 18.43. We now use this formula to estimate electron tunneling times in protein molecules.

EXPERIMENTS

Experiments Behind the Facts: Dynamics of Light and Electrons How do we know the order and timing of the electron transfer events as summarized in Figure 18.16? A beautiful set of experiments using time-resolved spectroscopy make it possible to use a laser pulse to excite the relevant molecules in the reaction center and then to measure the resulting fluorescence as a function of time.

Experiments on *Synechococcus* result in spectra like those shown in Figure 18.19. In these experiments, intact cells were excited with light with a wavelength of 580 nm, which resulted in the excitation of a phycobiliprotein. After this initial excitation event, fluorescent light is emitted from the various molecules in the order phycoerythrin (PE), phycocyanin (PC), allophycocyanin (APC), and chlorophyll *a* (Chl*a*), and these emissions can be used to trace the life history of the state of excitation of the various pigment molecules, with the idea being that as the shape of the spectrum changes over time, one can "read off" which molecule is currently giving off light.

Laser spectroscopy has also been used to measure the effect of distance between donor and acceptor on the tunneling time. To that end, a series of engineered proteins were constructed where an excited electron could be transferred from a copper ion in the native enzyme's active site to a ruthenium acceptor that could be artificially tethered at various locations within the protein structure, as illustrated in Figure 18.20(A). The concept of the experiment is to use two laser pulses, one to excite the system and a second one to query it as revealed by the change in charge state of the copper due to electron transfer.

The number of acceptors that have received an electron, N_A, is a function of time and satisfies the differential equation

$$\frac{dN_A}{dt} = k(N_0 - N_A), \tag{18.46}$$

where N_0 is the total number of proteins and k is the electron transfer rate for a given donor–acceptor distance, which we write as

$$k = k_0 e^{-2\kappa d}, \tag{18.47}$$

where $1/\kappa$ is the length scale over which the wave function decays in the classically forbidden region. We can solve for the number of acceptors that have received an electron up to time t as

$$N_A(t) = N_0(1 - e^{-kt}). \tag{18.48}$$

This kind of stereotyped response is exhibited in Figure 18.20(B). The probability $p(t)\,dt$, that a particular acceptor molecule receives an electron in the time interval $(t, t + dt)$ is the ratio of the number of molecules receiving an electron in this time, which is given by $N_A(t + dt) - N_A(t)$, and the total

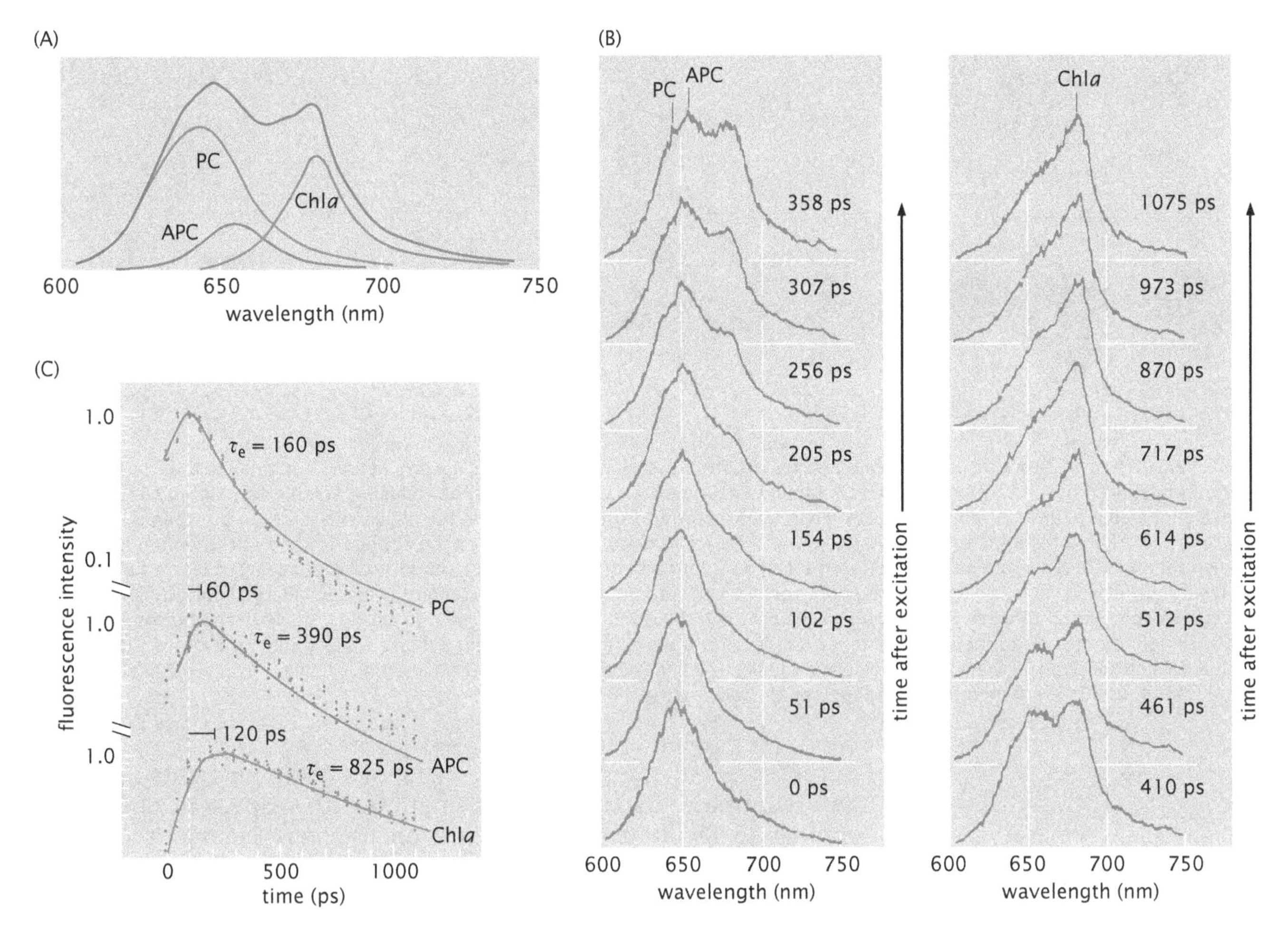

Figure 18.19: Time-resolved spectroscopy. (A) The cyanobacterium *Synechococcus* is an abundant photosynthetic organism in the ocean and has been widely used in laboratory studies of photosynthesis. Its overall fluorescence spectrum is shown as the red line. This spectrum is a combination of the fluorescence of the various photosynthetic pigments, the three most abundant being chlorophyll *a* (Chl*a*, green line), phycocyanin (PC, blue line), and allophycocyanin (APC, purple line). (B) Because the pigments fluoresce only when they have an excited electron, rapid time-resolved spectroscopy can be used to determine the order of electron transfer events in a culture of living cells. Here, the fluorescence spectrum was recorded about every 50 ps after initial delivery of a very brief (ps) laser pulse to the cell culture. At early time points, the shape of the spectrum corresponds to that of PC alone. Over about 1000 ps (1 ns), the fluorescence spectrum shifts through the APC peak to the Chl*a* peak. (C) Plotting peak fluorescence intensity at the characteristic wavelengths for each of the three pigments enables determination of the time delay in electron transfer to each pigment and the lifetime of each pigment's excited state. (Adapted from I. Yamazaki et al., *Photochem. Photobiol.* 39:233, 1984.)

number of molecules N_0. Using Equation 18.48 for $N_A(t)$ then leads directly to the expression $p(t)\,dt = ke^{-kt}\,dt$, the probability of a transfer event between times t and $t + dt$.

These experimental results also demonstrate that tunneling through protein over a given distance is substantially faster than tunneling through water or through a vacuum, as indicated in Figure 18.20(C). That is, the nature of the environment surrounding the two wells (for example, the dielectric constant of the medium) can influence the effective shape of the wells, and therefore the amount of overlap between the two wave functions associated with the presence of the electron in each of the two locations. Within the photosynthetic reaction center, the series of pigments that carry the excited electron are embedded in a large transmembrane protein complex (Figure 18.16). It seems that this arrangement serves two

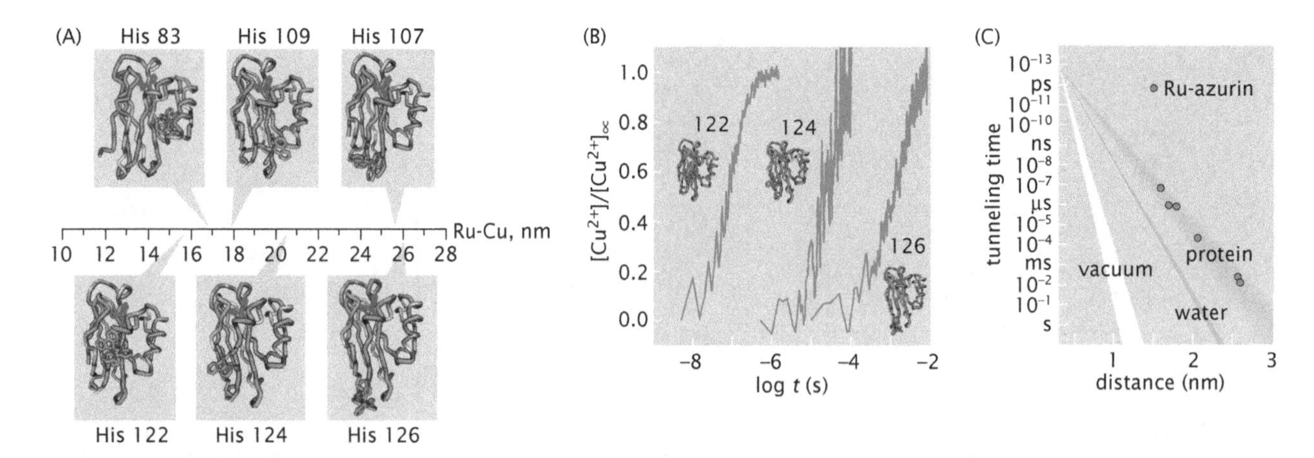

Figure 18.20: Azurin structure and tunneling times. (A) Structure of various genetically engineered proteins used to measure electron tunneling times. Azurin is a copper-containing enzyme that catalyzes a reaction reducing arsenic-containing compounds. In these engineered azurin variants, an electron can be transferred from the copper in the active site (blue) to a ruthenium acceptor attached to a histidine residue (orange). The distance in nanometers between the copper donor and ruthenium acceptor is shown for six different proteins. (B) Electron transfer as a function of time. The plot shows the amount of Cu^{2+} as a function of time resulting from the electron transfer reaction from Cu^{+} to ruthenium. The measurement is made spectroscopically and the three curves correspond to three different distances between the donor and acceptor within azurin, shown in the bottom row in (A). (C) Tunneling times as a function of distance for electron transfer through vacuum (white), water (blue), and protein (green); the blue dots correspond to the six proteins shown in (A). (A, C, adapted from H. B. Gray and J. R. Winkler, *Proc. Natl Acad. Sci. USA* 102:3534, 2005; B, adapted from R. Langen et al., *Science* 268:1733, 1995.)

purposes. First, the protein side chains hold the pigments at an appropriate distance and orientation with respect to one another to facilitate stepwise electron transfer. Second, the chemical environment within the protein may speed up the electron transfer process as compared with the baseline rate in water alone.

Estimate: Distance Dependence of Tunneling Times The ideas introduced above can be used to make a simple estimate of the time scales associated with electron tunneling. In particular, the most important feature of the toy model that we described is the realization that there is an exponential dependence of the tunneling rate on the distance between the acceptor and the donor.

ESTIMATE

Experimentally, clever measurements (described in more detail in the Experiments Behind the Facts section above) have been made in which the tunneling time is obtained as a function of distance between donor and acceptor as shown in Figure 18.20(C). To effect these measurements, different versions of the molecule were produced in which the distance between donor and acceptor was systematically varied. To estimate the parameter κ in Equation 18.45, we appeal to this data and note that effectively this unknown parameter is the slope of the rate-versus-distance curve on a semilog plot.

We can use the data of Figure 18.20(C) to estimate the parameter κ by using several data points as

$$\kappa \approx \frac{-1}{2(d_1 - d_2)} \ln \frac{\text{time}(d_2)}{\text{time}(d_1)}. \qquad (18.49)$$

Using the data points corresponding to $d_1 \approx 2.7\,\text{nm}$ and $d_2 \approx 1.6\,\text{nm}$ with mean transfer times of $10^{-2}\,\text{s}$ and $10^{-7}\,\text{s}$, respectively, we find that $\kappa \approx 5.25\,\text{nm}^{-1}$.

We calculate the parameter κ from "first principles" using our extremely simplified box model of a molecule in the appendix at the end of the chapter (Section 18.5).

Electron Transfer Between Donor and Acceptor Is Gated by Fluctuations of the Environment

So far, our discussion of the rate of electron tunneling has focused on the case where the energy levels in the two wells are essentially the same. As explained in the appendix (see Figure 18.53, p. 790), there is significant tunneling only when the energy levels of the initial and final states are nearly aligned. On the other hand, this raises the interesting question of how such tunneling works in real molecules, since, generically, the energy levels on donor and acceptor will be different. The key insight is that thermal fluctuations of the solvent lead to a change in the potential seen by the electron when it is on the donor molecule or on the acceptor, as shown in Figure 18.21. The time scales for these environmental fluctuations are typically much faster than the tunneling time. As a result, the net rate of electron transfer, k_{et}, can be computed as

$$k_{et} = kp_T, \quad (18.50)$$

where p_T is the probability of finding the donor–acceptor pair in the transition state characterized by degenerate electron levels (the middle state in Figure 18.21) and k is the tunneling rate from donor to

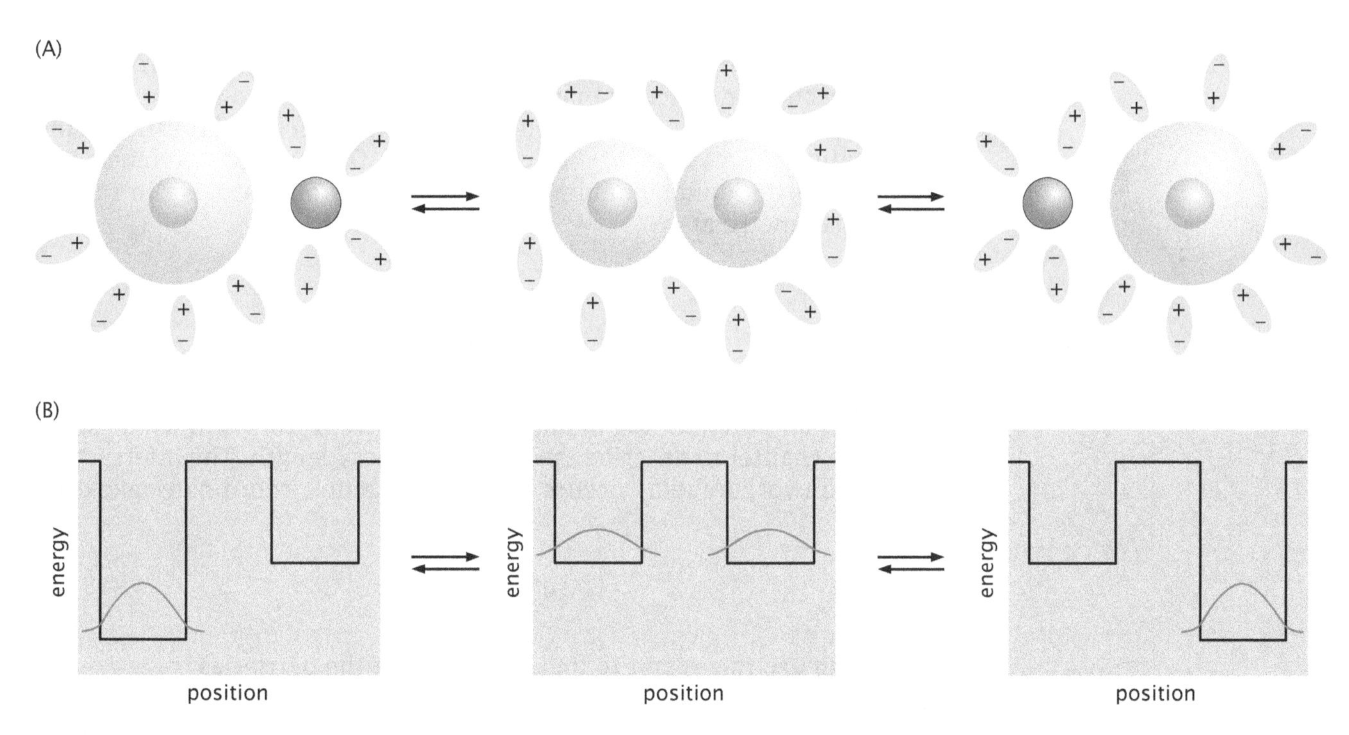

Figure 18.21: Schematic of the relation between the orientations of solvent molecules and the energy landscape of the electron. (A) In the left state, the electron cloud is located on the left atom, generating an electric dipole pointing to the right. As a result, the dipoles of the water molecules surrounding the system orient themselves in an energetically favorable way. A fluctuation in the water molecules can orient them such that the electron has the same interaction energy with the solvent regardless of which molecule the cloud is centered on. Finally, the solvent can relax to a set of configurations that stabilize the electron on the right molecule. (B) A simplified energy diagram of the reaction shown in (A) shows how both stable states (left and right) are connected through a transient fluctuation that makes both energy wells equally deep (center). (A, adapted from D. Chandler, in Classical and Quantum Dynamics in Condensed Phase Simulations, edited by B. J. Berne et al. World Scientific, 1998.)

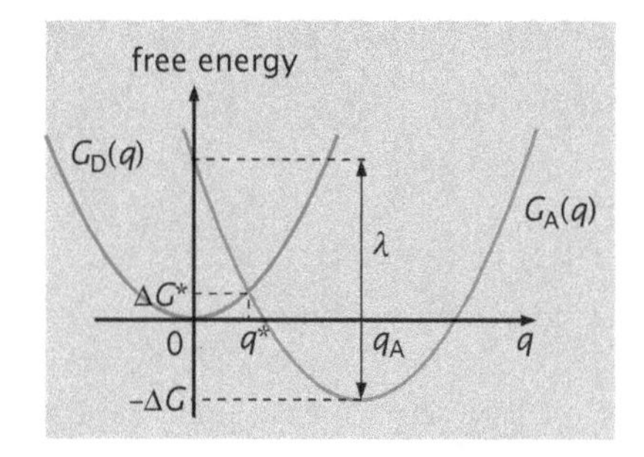

Figure 18.22: Model free energies of the donor–acceptor–solvent system for the situations when the electron is on the donor molecule $G_D(q)$ and when it is on the acceptor molecule $G_A(q)$. The reaction coordinate q represents the nuclear coordinates of the donor and the acceptor as well as the solvent degrees of freedom.

acceptor in the transition state, as was already described qualitatively in our discussion of tunneling and as is fleshed out in the appendix.

The quantity p_T can be estimated using the equilibrium statistical mechanics of a system with two states. In particular, to simplify matters, we assume that the donor–acceptor pair and the surrounding solvent can either be in the ground state of the donor or in the transition state. The free-energy difference between these two states is ΔG^*, as shown in Figure 18.22. Just as we found in the case of a simple ion channel (see Section 7.1.2, p. 286), the probability to be found in the transition state in this two-state model is

$$p_T = \frac{e^{-\beta \Delta G^*}}{1 + e^{-\beta \Delta G^*}}. \tag{18.51}$$

With this result in hand, we can now analyze both the temperature and driving-force dependence of the net electron transfer rate, which are inherited from the quantity p_T.

To develop further intuition about the probability p_T of a thermal fluctuation into the transition state, we now consider a simple version of Marcus theory. The idea is to introduce a simple model for the free energy of the donor–acceptor–solvent system as a function of a single reaction coordinate q, shown in Figure 18.22. To be concrete, this reaction coordinate captures the complexity of the solvent degrees of freedom as they jiggle around. The two parabolas $G_D(q)$ and $G_A(q)$ shown in the figure are an approximate representation of the free energy of the system when the electron that is being transferred is either on the donor molecule or on the acceptor molecule, that is, before or after the electron transfer reaction has occurred. The difference between the free-energy minima associated with $G_D(q)$ and $G_A(q)$ is the thermodynamic driving force for the electron transfer reaction, ΔG. The transition state is characterized by the reaction coordinate q^* at which the two parabolas intersect, and it satisfies the equation $G_D(q^*) = G_A(q^*)$. To compute the free-energy difference ΔG^* between the donor minimum and the transition state, we introduce the parametrization

$$\begin{aligned} G_D(q) &= \tfrac{1}{2}\kappa q^2, \\ G_A(q) &= \tfrac{1}{2}\kappa (q - q_A)^2 - \Delta G, \end{aligned} \tag{18.52}$$

where q_A is the position of the minimum of the acceptor free-energy parabola. Note that the parameter κ is different from that used earlier in the chapter to describe the tunneling decay length. The intersection of the two parabolas occurs when the reaction coordinate takes the value

$$q^* = \frac{q_A}{2} - \frac{\Delta G}{\kappa q_A}. \tag{18.53}$$

We can use this result to find the height of the barrier as

$$\Delta G^* = G_D(q^*) = \frac{(\Delta G - \lambda)^2}{4\lambda}, \tag{18.54}$$

where $\lambda = \kappa q_A^2/2$ is the reorganization energy introduced by Marcus. Physically, the reorganization energy is the free-energy difference between the acceptor state in which the reaction coordinate takes the value for which the donor state is at its minimum ($q = 0$) and the acceptor ground state ($q = q_A$).

Combining the result for the free energy of the transition state with the simple two-state model of the probability of reaching the transition state, we find

$$p_{\mathrm{T}} \approx \mathrm{e}^{-\beta(\Delta G-\lambda)^2/4\lambda}, \tag{18.55}$$

where we have made the additional assumption that the probability is much less than one, or, equivalently, that the free energy $\Delta G^* \gg k_{\mathrm{B}}T$. In light of this result, the net electron transfer rate is given by the expression

$$k_{\mathrm{et}} = \frac{2\pi}{\hbar}|H_{\mathrm{DA}}|^2 \rho \mathrm{e}^{-\beta(\Delta G-\lambda)^2/4\lambda}, \tag{18.56}$$

where H_{DA} corresponds to the tunneling matrix element that we introduced as H in Equation 18.43, though here with the full atomic-level realism demanded by the structure of the donor and acceptor and their environment. As mentioned earlier, the factor ρ will be discussed in detail in the appendix at the end of the chapter. This formula makes the surprising and counterintuitive prediction that the electron transfer rate can actually decrease with increasing driving force ΔG, as shown by the "inverted region" that can be seen in Figure 18.23. As seen in the figure, this bizarre theoretical insight has been confirmed experimentally.

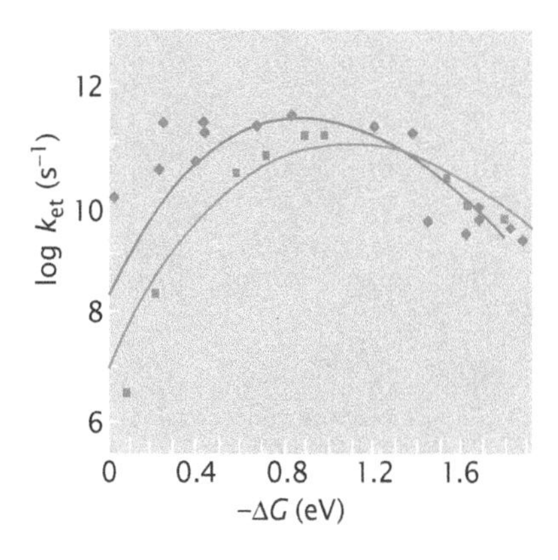

Figure 18.23: The "inverted region" and Marcus theory. In order to test the predictions of the Marcus theory, the free-energy change ΔG for electron transfer between a donor and acceptor in a synthetic molecular scaffold can be altered by changing the size and composition of the scaffold. This graph shows experimental measurements of the rate of electron transfer for two distinct series of donor–acceptor pairs, namely, iridium–pyridinium (blue squares) and porphyrin–quinone (red diamonds). For both molecular series, the rate of transfer initially increases with increasing driving force for lower values of ΔG, and then "inverts" for higher values of ΔG such that the rate actually decreases as the driving force increases. The lines show the fit to Equation 18.56. (Adapted from L. S. Fox et al., *Science* 247:1069, 1990 and M. R. Wasielewski et al., *J. Am. Chem. Soc.* 107:1080, 1985.)

Our description of the overall rate of electron transfer has been based upon two processes that can potentially have very different time scales: (i) the solvent rearrangement that gates the tunneling and (ii) electron tunneling itself. As illustrated in Figure 18.20 for azurin, the tunneling time is no shorter than microseconds.

Resonant Transfer Processes in the Antenna Complex Efficiently Deliver Energy to the Reaction Center

The favorable proximity, orientation, and chemical environment of the pigments in the photosynthetic reaction center thus work together to allow electron transport to be efficient and fast. As diagrammed in Figure 18.16, the rate-limiting step in this process takes about 100 μs for the reaction center of purple bacteria. However, our estimate in Section 18.2.3 predicted that the absorption of photons from sunlight by the chlorophyll molecules in the special pair at the top of the reaction center would happen only about ten times per second under most circumstances. Thus, if the only way that the special pair of chlorophyll molecules within the reaction center could achieve excitement were by direct collision with a photon, the photosynthetic apparatus would be sitting idle much of the time. In fact, photosynthetic organisms use a quantum process of resonance energy transfer, conceptually related to the process of electron tunneling, to dramatically increase the rate of energy delivery to the special pair.

An early clue to this mechanism was provided by a classic quantitative experiment, which aimed to measure the number of chlorophyll molecules used by green algae to generate a single oxygen molecule. The surprising result, shown in Figure 18.24, is about 300. If only one or two of these chlorophyll molecules are directly involved in the photosynthetic reaction center, what are the rest of them doing, and where are they located?

Figure 18.25 gives a sense of the architecture of the light-absorption apparatus in photosynthetic membranes. The most immediate qualitative observation about this architecture is that the vast majority

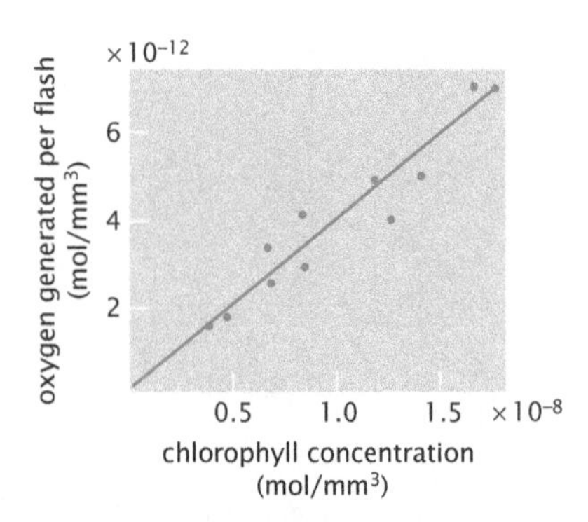

Figure 18.24: Measurement of the fraction of chlorophyll molecules directly involved in photosynthesis. In this classic experiment, a small culture of *Chlorella* algae was exposed to a single, saturating pulse of light for less than 10^{-4} s, and the amount of oxygen liberated was carefully measured; then the experiment was repeated for a variety of different cultures grown under different conditions so as to vary the amount of chlorophyll present in each culture. The slope of the measured line gives the average number of chlorophyll molecules required to generate one molecule of O_2 as 2480. Independent experiments under steady illumination also showed that the maximum conversion efficiency is about 8–10 photons per O_2 molecule. Dividing 2480 by 8 gives the approximate number of chlorophyll molecules associated with the production of a single O_2 molecule to be about 300. This is interpreted to mean that these 300 chlorophyll molecules are part of the antenna complex, which can absorb light but cannot directly participate in electron transfer. The total time required to complete one photochemical cycle was also measured in this experiment, and found to be about 0.02 s. (Adapted from R. Emerson and W. Arnold, *J. Gen. Physiol.* 16:191, 1932.)

of pigments are found outside of the reaction center, in beautifully arranged "antenna complexes" that surround the reaction center with its special pair. Each one of the several hundred pigment molecules in the antenna complex is capable of absorbing the energy of an impinging photon, vastly increasing the effective absorptive cross-sectional area of the reaction center. Although all photosynthetic reaction centers in organisms ranging from purple bacteria through green plants share roughly similar geometries in their arrangement of pigment molecules and supporting protein helices, and therefore probably share a single common evolutionary ancestor, the antenna complexes found throughout the photosynthetic world are extremely varied. Some antenna complexes contain multiple different pigments that preferentially absorb photons of different wavelengths, extending the reach of the reaction center in acceptable photon energy as well as in physical space.

Once one of the pigments in the antenna complex is excited, how can the energy travel to the special pair of chlorophylls in the reaction center? The antenna complex is effectively a solid-state device where pigment molecules are physically fixed in place by their supporting proteins, and many of the pigments are too far from the reaction center for electron tunneling to proceed at an acceptable rate. Instead, the energy is transferred from one pigment molecule to another through a process of energy transfer that does not involve the actual tunneling of an electron. In Figure 18.15, we indicated that resonance transfer was one of the four possible fates of an excited electron. In the antenna complex, a series of resonance transfer steps effectively allows the energy to hop from one pigment to another until it is finally delivered to the special pair, as shown in Figure 18.25(C). These energy transfer events are fast (on the order of picoseconds).

18.2.5 Bioenergetics of Photosynthesis

The biological purpose of the quantum mechanical acrobatics described above is simply to transfer electrons from relatively low-energy "donor" molecules to high-energy "acceptor" molecules that can then be used by the photosynthetic cell to drive metabolic processes, most critically the fixation of carbon dioxide to form sugar. The development of charge separation across a membrane through the process of electron transport also serves to generate a gradient of hydrogen ions that can be used in turn to synthesize ATP. In this section, we bring together what we have learned about entropy and electrostatic potentials to characterize this proton-motive force.

Electrons Are Transferred from Donors to Acceptors Within and Around the Cell Membrane

Different flavors of photosynthesis can be distinguished by their choices of initial electron donors and final electron acceptors. For example, in green sulfur bacteria, a single photosystem uses hydrogen sulfide (H_2S) as an electron donor and ultimately uses $NADP^+$ as an electron acceptor generating NADPH. As we will see in Section 18.2.6, NADPH plays a critical role in the fixation of carbon dioxide. In plants and cyanobacteria performing oxygenic photosynthesis, H_2S is not used. Instead, the electron donor for photosystem I is a reduced molecule called plastocyanin, which has been generated as the electron acceptor by a second photosystem. This second photosystem

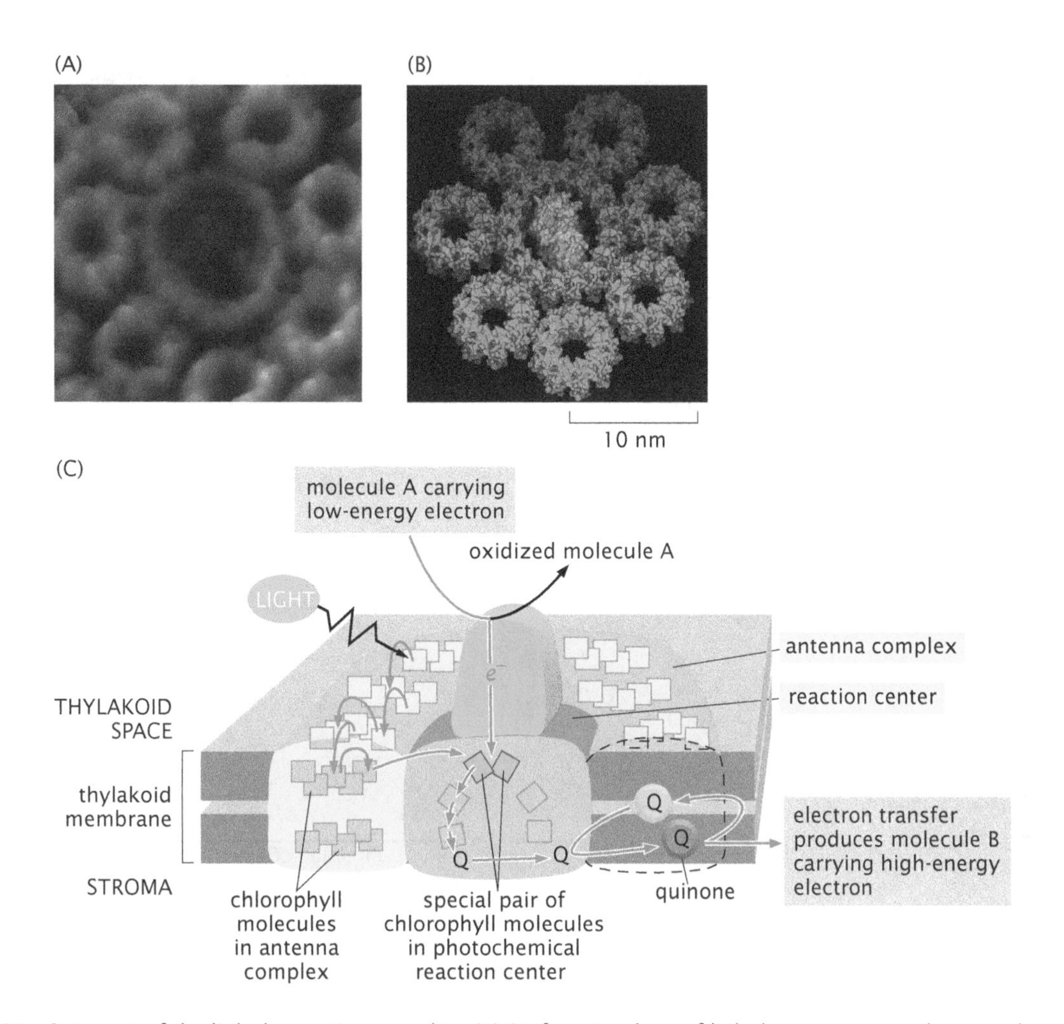

Figure 18.25: Structure of the light-harvesting complex. (A) Surface topology of light-harvesting complexes in the membrane of the photosynthetic bacterium *Rhodospirillum photometricum*, as determined using atomic-force microscopy. (B) Structural model of the assembly shown in (A), created by fitting high-resolution protein structures into the lower-resolution AFM structure. The photosynthetic reaction center (RC) is shown in orange, surrounded by a large green ring of antenna pigments (light-harvesting complex I; LHI) and seven smaller ring-shaped assemblies (light-harvesting complex II; LHII). The ratio of LHII to RC is variable and depends on the amount of light exposure during bacterial growth. (C) Role of the antenna complex in photosynthesis in plants. Although only the "special pair" of chlorophyll molecules in the reaction center can directly initiate electron transfer (blue), any of the chlorophyll molecules in the antenna complex can capture the energy of a photon and then transfer that energy by a series of resonance transfers (pink) to eventually excite one of the chorophylls in the special pair. This greatly increases photosynthetic efficiency. (A, adapted from S. Scheuring and J. N. Sturgis, *Science* 309:484, 2005; B, adapted from S. Scheuring et al., *J. Struct. Biol.* 159:268, 2007; C, from B. Alberts et al., Molecular Biology of the Cell, 5th ed. Garland Science, 2008.)

uses water as its ultimate electron donor and generates molecular oxygen as an oxidized by-product. Thus, ultimately, oxygenic photosynthesis transfers electrons from water to NADP, but goes through a bewildering array of intermediates and requires the energetic input of two separate photons captured by the two photosystems, as shown in Figure 18.26.

In biochemistry, reactions that involve the net transfer of electrons from one molecule to another are given the name oxidation–reduction reactions, or redox reactions for short. Because redox reactions involve the transfer of charged species, it is possible in a laboratory to set up a redox reaction that will generate a current between two beakers. Because of the electrical consequences of redox reactions, the free energy of reaction is traditionally measured in electron volt (eV) units rather than in kcal/mol or kJ/mol. This is only a minor

Figure 18.26: Changes in redox potential during photosynthesis. Oxygenic photosynthesis as practiced by plants and cyanobacteria requires the cooperation of two different photosystems, namely, photosystem II, which uses light energy to transfer electrons from water to a quinone (Q), and photosystem I, which receives electrons from the quinone (via intermediates) and ultimately transfers them to NADP+, generating NADPH, which can be used for carbon fixation. (Adapted from B. Alberts et al., Molecular Biology of the Cell, 5th ed. Garland Science, 2008.)

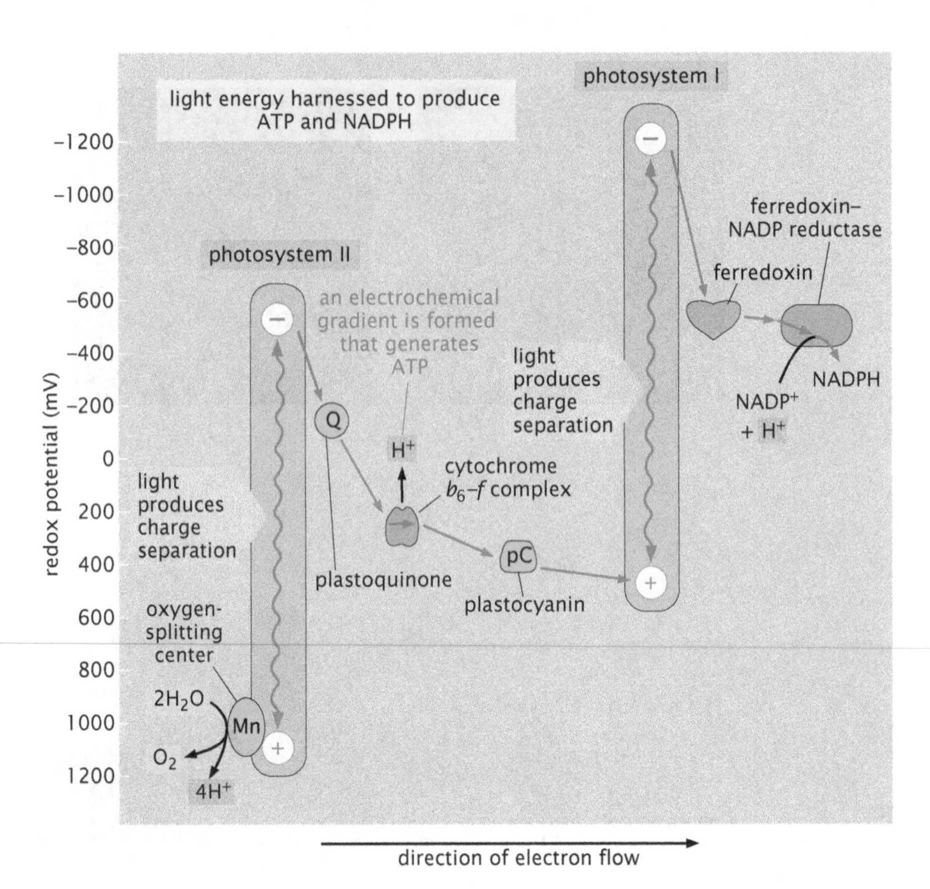

inconvenience, since eV, kJ/mol, and k_BT, units are all different choices of energy unit and can be readily interconverted.

Redox reactions occur because some atoms have greater affinity for electrons than others. Oxygen, for example, has an unusually strong affinity for its electrons and tends to steal them from all available sources. This is familiar to all of our readers through the annoying tendency of our favorite shiny metals to rust and serves as a geological reporter of the history of the atmosphere in the form of banded iron formations. Hydrogen has relatively low electron affinity and gives its electrons up easily. Carbon lies somewhere in between. In the organic biochemistry associated with life, a very large number of reactions concern increasing or decreasing the oxidation state of carbon. In the present-day atmosphere, because oxygen is abundant, oxidation of organic molecules occurs spontaneously, so the most fully oxidized state of carbon, that is gaseous carbon dioxide, is at the lowest energy state. To be used in any organic reaction, carbon dioxide must be reduced, that is, electrons must be delivered to the carbon atom. Fats are relatively more reduced than carbohydrates and are thus more potent energy sources.

Water, Water Everywhere, and Not an Electron to Drink

For a photosynthetic organism hungry to strip electrons from available donor molecules, water must be a desirable substrate. As compared with H_2S or other commonly used primary electron donors, water is outrageously abundant and present at very high concentration ($\approx$55.5 M). However, water is reluctant to give up its tightly bound electrons. The overall chemical reaction that must occur is

$$2H_2O \rightarrow O_2 + 4e^- + 4H^+. \tag{18.57}$$

The energy required is approximately $125k_BT$, which is substantially more than the energy carried by a typical visible-light photon. In the upper atmosphere, high-energy ultraviolet photons can trigger this reaction; this is one of the processes that contributes to the generation of our protective blanket of ozone. However, little ultraviolet light reaches the Earth's surface, where most photosynthesis occurs, as shown in Figure 18.2. Confined to working with photons of visible-light energies, early photosynthetic organisms were unable to strip the electrons from the abundant water in their environments. A particularly creative ancestor of the present-day cyanobacteria came up with an elegant solution to this problem. As summarized in Figure 18.27, the modern process of oxygenic photosynthesis proceeds by combining the energies of four visible-light photons.

Charge Separation across Membranes Results in a Proton-Motive Force

As seen in Figure 18.26, one of the consequences of all of the busy charge shuttling by the electron transport chain is the setting up of a transmembrane ion gradient. The total free energy stored in such gradients is based upon two distinct physical effects, namely, (1) the existence of a concentration gradient and (2) the setting up of an electric potential difference. To explore the free energy that is made available by setting up the transmembrane ion gradient, we imagine N protons on one side of the membrane, resulting in a free energy

$$G(N) = NqV - k_BT \ln \frac{\Omega!}{N!(\Omega - N)!}. \tag{18.58}$$

The first term accounts for the difference in electric potential as measured by the voltage V and the second term invokes a lattice model to capture the conformational degrees of freedom of the charges. Using the trusted Stirling approximation (see p. 221), we can rewrite this free energy as

$$G(N) = NqV - Nk_BT \ln \Omega + k_BTN \ln N - Nk_BT. \tag{18.59}$$

With this free energy in hand, we can now compute the free-energy difference across the membrane as

$$\Delta G = q\Delta V + k_BT \ln \frac{c_{\text{in}}}{c_{\text{out}}}, \tag{18.60}$$

where we have used the same reasoning exploited in Section 16.2.7 (p. 660). Given that the concentration of protons is related to the pH through $c = 10^{-\text{pH}}$, we can rewrite the free-energy difference as

$$\Delta G = q\Delta V - \Delta\text{pH}\; k_BT \ln 10, \tag{18.61}$$

also known as the proton-motive force. This free-energy difference can serve as a driving force for powering the synthesis of ATP. How much free energy this is actually worth is estimated in the problems at the end of the chapter.

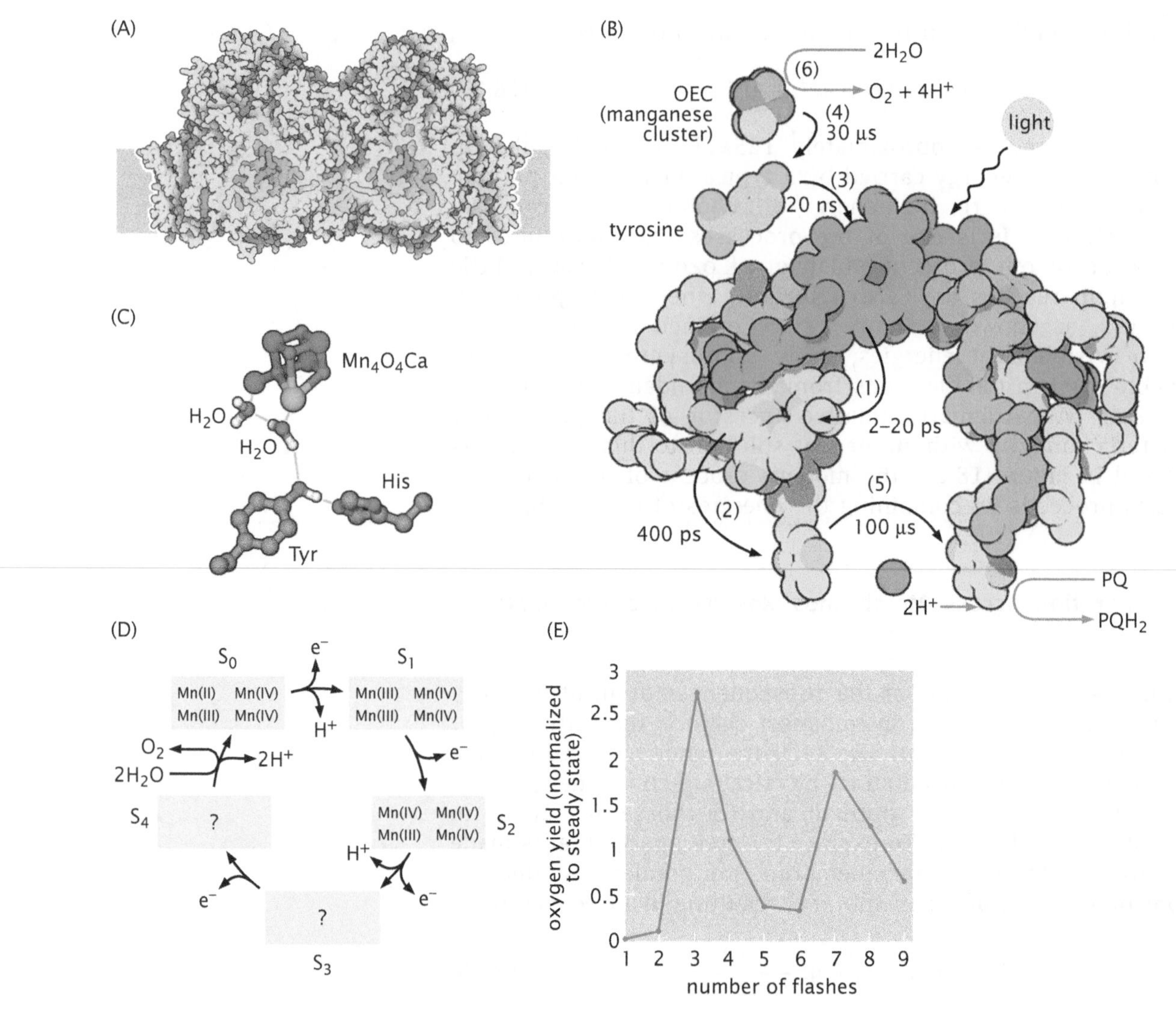

Figure 18.27: Making oxygen from water. (A) Overview of photosystem II, showing the position of the pigments (colors) within the transmembrane protein (gray). The manganese-containing cluster responsible for the conversion of H_2O into O_2 is shown in pink (PDB 1s5l). (B) Time scale and order of events in photolysis of water and electron transport. After excitation of the special pair of chlorophyll molecules and rapid transfer of the electron (1 and 2), the chlorophyll replenishes its electron hole by stealing one from the nearby tyrosine residue (3), which in turn takes an electron from the manganese cluster (4) (also known as the oxygen-evolving center, OEC). The electron from the chlorophyll is eventually transferred onto the exchangeable quinone (5), which also picks up an H^+ ion from the environment. After two cycles, the reduced quinone is released (PQH_2) and replaced by a fresh quinone (PQ). After four cycles, two reduced quinones have been generated, and the manganese cluster is completely depleted of exchangeable electrons (four in deficit). It then replenishes its electrons by stripping two water molecules and liberating O_2 (6). (C) Detail of the X-ray crystal structure of photosystem II showing the cubical structure of the manganese cluster together with the active tyrosine and two bound water molecules. (D) The four manganese atoms can exist in several different oxidation states (Mn(II), Mn(III), and Mn(IV)), allowing a total of five stable intermediates (S_0–S_4). Water is converted to oxygen only after the manganese cluster has given up four electrons. (E) Classic experiment showing that the energy of four photons is required to generate one molecule of oxygen. A culture of algae was exposed to a long, saturating pulse of light and then allowed to rest in the dark for five minutes. Next, the culture was given very brief (10^{-6} s) flashes of bright light, and the amount of oxygen liberated was measured after each flash. From the resting state (equivalent to S_1 in (D)), three flashes are required for oxygen to appear; thereafter, a spike of oxygen appeared after every four flashes. (A, B, courtesy of David Goodsell; B, C, D, adapted from A. W. Rutherford and A. Boussac, *Science* 303:1782, 2004; E, adapted from B. Kok et al., *Photochem. Photobiol.* 11:457, 1970.)

18.2.6 Making Sugar

Despite all of the quantum mechanical machinations that have peppered the chapter, we have yet to actually fix any carbon. Remember the goal of this process is to transfer electrons onto carbon dioxide to render the carbon into a more reduced state that is compatible with

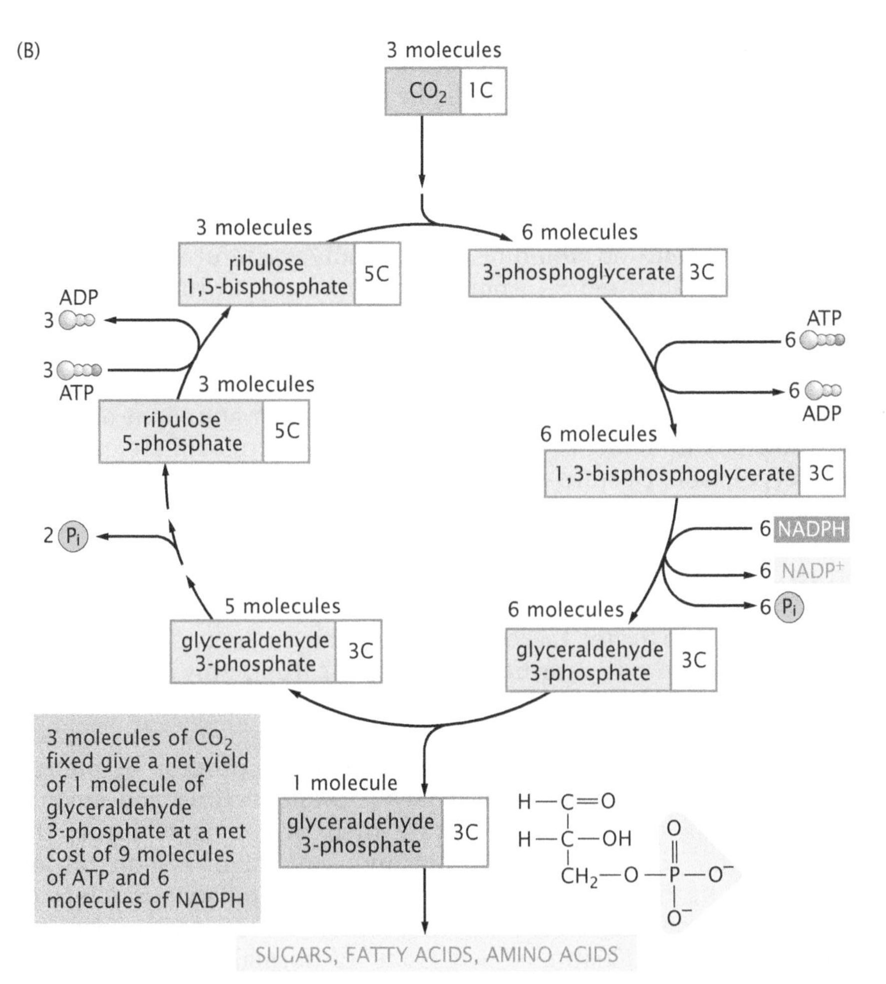

Figure 18.28: Biochemical reactions in carbon fixation. (A) The critical step in carbon fixation, catalyzed by RuBisCO, combines one molecule of carbon dioxide (CO_2) with a 5-carbon sugar (ribulose) to generate two 3-carbon sugars (glycerate). (B) In order to run as a cycle, this fixation step is combined with a large number of other steps to regenerate the ribulose. Three turns of the Calvin–Benson cycle are sufficient to generate a single 3-carbon sugar in a phosphorylated form (glyceraldehyde 3-phosphate) that can be used as a biosynthetic intermediate to generate other sugars or amino acids. The ATP and NADPH generated by the action of photosystem I and photosystem II are used within the cycle.

biosynthetic reactions. Fortunately, the final electron acceptor at the end of photosystem I was $NADP^+$ generating NADPH. As we mentioned in Section 5.1.2 (p. 190), NADPH is a commonly used carrier molecule whose biosynthetic function is to reduce other compounds delivering electrons where they are needed. The net process of delivering electrons from NADPH to CO_2 in order to generate carbohydrates involves an insanely complicated process that requires the action of over a dozen enzymes. This process is called the Calvin–Benson cycle and is shown in Figure 18.28. The net reaction of the first part of the Calvin–Benson cycle takes three 5-carbon sugars and attaches three CO_2 molecules. The 6-carbon intermediates are unstable and immediately split into 3-carbon units. These 3-carbon units are then reduced using six molecules of NADPH to fix the carbon. Of these six 3-carbon units, one is removed from the Calvin–Benson cycle and can be used directly for biosynthesis of carbohydrates or amino acids.

The remaining five 3-carbon units are then rearranged to regenerate the three 5-carbon units that started the cycle. The dizzying array of reactions includes intermediates with 3 carbons, 4 carbons, 5 carbons, and 7 carbons. Why are there so many intermediates in the cycle? It is rare for biochemical reactions to transfer single-carbon units between molecules. More commonly, carbon is transferred in groups of at least two or three. It is left as an exercise for the reader to figure out how three 5-carbon units can be most efficiently made out of five 3-carbon units without ever leaving a single carbon on its own. The series of transfer reactions is somewhat reminiscent of the famous riddle of how to transport a fox, a chicken, and a bag of corn across a river in a small boat when only one can be carried across at a time and the fox cannot be left alone with the chicken or the chicken alone with the corn. Several of the intermediate reactions are reminiscent of carrying the chicken backwards across the river. The simplest solution to this riddle is to allow the chicken to eat the corn and the fox to eat the chicken and then simply carry the happy fox in a single trip. However, this solution would be energetically wasteful and would fail to regenerate the three 5-carbon units necessary to run the Calvin–Benson cycle.

The rate-limiting step in the Calvin–Benson cycle is the reaction that attaches carbon dioxide onto the 5-carbon sugar. This reaction is catalyzed by an enzyme thought to be the most abundant on Earth and known as RuBisCO (short for ribulose 1,5-bisphosphate carboxylase oxygenase). Despite RuBisCO's abundance and importance, as an enzyme, it runs at a very slow rate and does not exhibit high specificity. The turnover rate of RuBisCO is roughly one per second. Furthermore, RuBisCO is indiscriminate in its substrate choice and can easily mistake O_2 for CO_2. When it accidentally performs its reaction with O_2 as a substrate, the original 5-carbon sugar is split into one 3-carbon unit and one 2-carbon unit. The 2-carbon unit is useless for biosynthesis.

The observant reader will note that O_2 has been generated as a waste product by the photoreactive center during the process of photon capture and electron transport. That is, cyanobacteria and eukaryotes containing chloroplasts generate their own poison during the process of photosynthesis. The poisonous effect of O_2 on carbon fixation can be thought of as analogous to the poisonous effect of carbon monoxide (CO) on human respiration. CO is a poison because hemoglobin cannot distinguish it from its normal substrate of O_2. Normally, this is not a problem, because CO is not abundant in the atmosphere, but can cause asphyxiation when a human is unfortunate enough to encounter a high concentration of CO in a closed space. Similarly, for organisms performing oxygenic photosynthesis, the production of O_2 was not a problem in the Earth's early atmosphere, when O_2 was extremely rare. The waste O_2 would simply diffuse away. However, as photosynthetic organisms became more abundant, the amount of atmospheric O_2 gradually rose, causing profound changes in the history of life's evolution on Earth. One consequence of atmospheric O_2 poisoning is illustrated in Figure 18.29. By now, O_2 has become such a fundamental component of the metabolic web of life on Earth, even though it began as a deadly poison, life as we know it would cease to exist if oxygenic photosynthesis were to be interrupted.

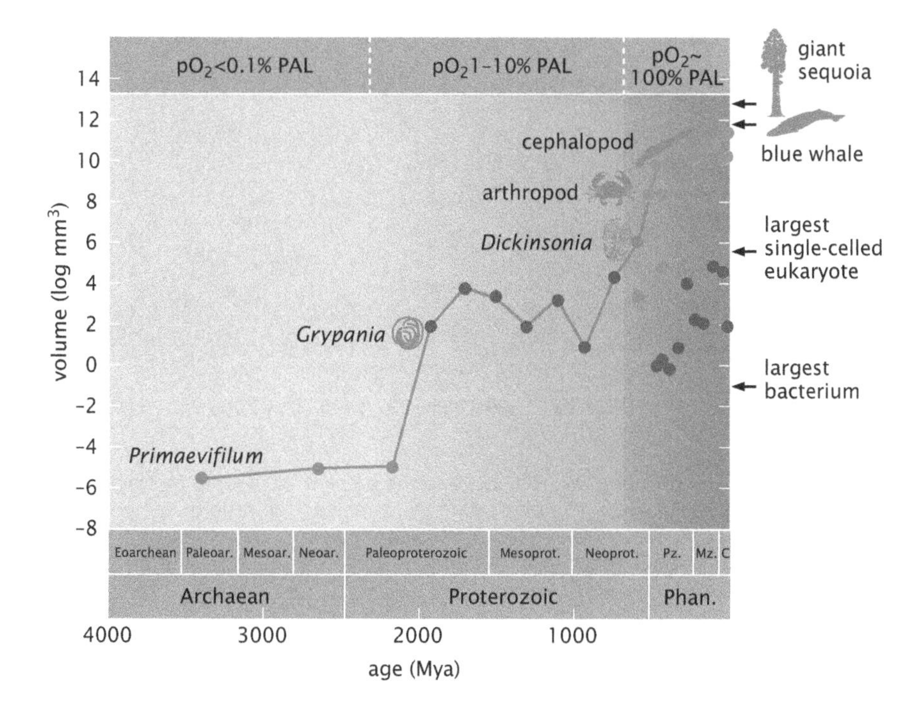

Figure 18.29: Atmospheric oxygen and organism size. This plot shows the sizes of the largest known fossil organisms throughout history. Putative prokaryotes are shown in orange, single-celled eukaryotes in brown, animals in blue, and plants in green. The sizes of the largest living organisms are indicated on the right. The maximum organismal volume has increased by about 16 orders of magnitude over 3.5 billion years, largely in two pronounced size jumps of about six orders of magnitude that correspond to increases in atmospheric oxygen levels. On the early Earth, the partial pressure of oxygen in the atmosphere was more than a thousand-fold lower than the present atmospheric level (PAL). Fossil organisms such as *Primaevifilum* were small and resembled present-day bacteria and archaea. In the early Proterozoic eon, the amount of atmospheric oxygen increased substantially, probably due to the invention of photosynthesis, and fossil size also suddenly increased. *Grypania* is an abundant fossil that resembles present-day filamentous algae. The earliest multicellular animal-like fossils, such as the Ediacaran organism *Dickinsonia*, appeared when atmospheric oxygen reached nearly current levels. (Adapted from J. L. Payne et al., *Proc. Natl Acad. Sci. USA* 106:24, 2009.)

As the figure indicates, the story of the chemical composition of the atmosphere is an old one. However, the atmosphere is continuing to change. In recent decades, the relative concentration of CO_2 has significantly increased while the relative concentration of O_2 has been going down. Over a similar time scale, humans have been burning increasingly large amounts of coal and petroleum for energy-generation, a process that notably consumes O_2 and liberates CO_2 as a waste product. It is sobering to reflect that our energy-generation habits may be as short-sighted as those of the cyanobacteria who felt that releasing O_2 as a waste was no cause for alarm.

Another striking feature of the measurements of atmospheric carbon dioxide, and oxygen shown in Figure 18.30, is their regular annual variation. Notably, the level of CO_2 is lowest during the months that correspond to summer in the northern hemisphere and highest during the months that correspond to winter. Keeling, who first measured this annual variation, realized that this pattern was likely to be a consequence of annual changes in photosynthesis by the forests of the northern hemisphere. This annual variation can be used as a starting point for estimates both of the total primary productivity and the number of RuBisCOs on planet Earth.

Estimate: Photosynthetic Productivity on Earth Biochemistry textbooks routinely assert the number of gigatons of carbon that are fixed each year in the process of photosynthesis. As a sanity check, can we estimate these numbers based on just a few foundational facts and assumptions? One idea is to appeal to the Keeling curve shown in Figure 18.30. Keeling's first discovery was not an increase in atmospheric CO_2 levels. Rather, his measurements were sufficiently good as to reveal the annual variation in CO_2 levels, and it is these variations that we will try to take advantage of as the basis of an order-of-magnitude (hopefully good to within a factor of ten) estimate of the primary productivity on Earth.

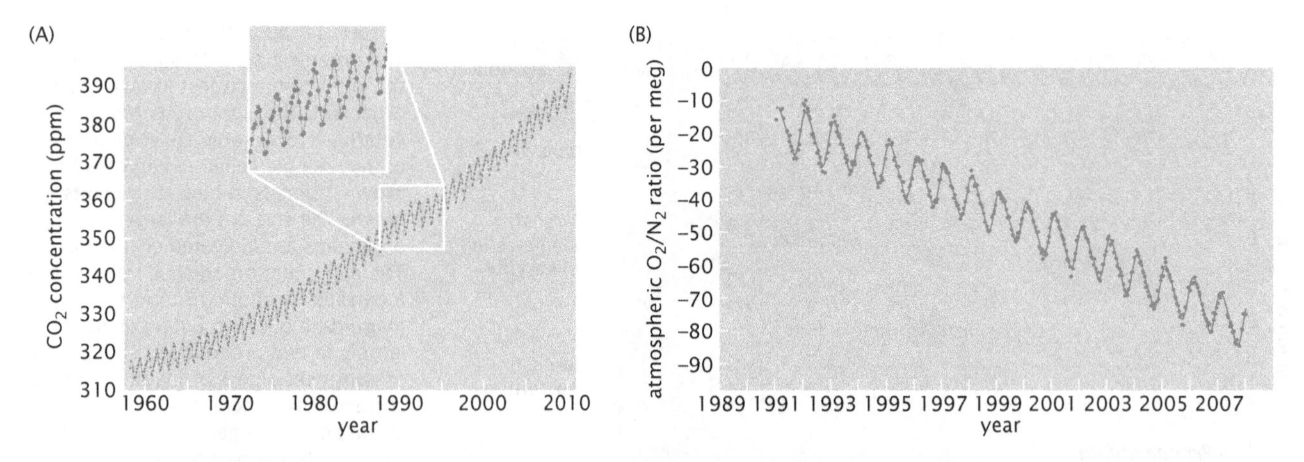

Figure 18.30: Recent changes in atmospheric levels of CO_2 and O_2. (A) CO_2 from 1960 to the present, measured by David Keeling at Mauna Loa, Hawaii. (B) O_2 levels in the atmosphere, measured by Ralph Keeling in Tasmania. The ratio of O_2 to N_2 is measured by comparison with a reference sample. The overall downward trend of −19 "per meg" per year corresponds to losing 19 O_2 molecules out of every 1 million O_2 molecules in the atmosphere each year. (Adapted from Ralph Keeling, Scripps Institution of Oceanography.)

We begin by working out the mass of the atmosphere itself by noting that the total weight of the atmosphere should be the same as atmospheric pressure times the area of the Earth's surface, resulting in

$$M_{\text{atmosphere}} = \frac{pA}{g} \approx \frac{10^5\ \text{N/m}^2 \times 4\pi(6 \times 10^6\ \text{m})^2}{10\ \text{m/s}^2} \approx 4 \times 10^{18}\ \text{kg}. \tag{18.62}$$

We can use this mass in turn to estimate the total number of molecules in the atmosphere as

$$\text{number of molecules} \approx \frac{M_{\text{atmosphere}}}{0.78 m_{\text{N}_2} + 0.2 m_{\text{O}_2}} \approx 10^{44}. \tag{18.63}$$

Using the Keeling curve, we see that the number of CO_2 molecules is roughly 400 parts per million, implying

$$\text{number of CO}_2\ \text{molecules} \approx \frac{400}{10^6} \times 10^{44} \approx 4 \times 10^{40}. \tag{18.64}$$

To determine the number of CO_2 molecules fixed each year in the process of photosynthesis, we make a crude approximation based on the annual variations in the Keeling curve, which reflect an asymmetry in the photosynthetic output of the northern and southern hemispheres. In particular, we note that the variability in atmospheric CO_2 is roughly ten parts per million and we make the guess that this variability is comparable to the scale of net carbon fixation itself, resulting in the estimate

$$\text{number of CO}_2\ \text{molecules fixed} \approx \frac{10}{10^6} \times 10^{44} \approx 10^{39}. \tag{18.65}$$

This number implies in turn that the mass of CO_2 fixed each year is roughly 10^{14} kg. In addition, we can make a rough estimate of the number of photons engaged in all of this photosynthesis. If there are roughly 10 photons used per carbon fixed, this implies that the number of photons engaged in this photosynthetic productivity is roughly 10^{40}.

We can say more using these numbers since they can also point us to a naive estimate of the total number of RuBisCO molecules churning out fixed carbon. The rate at which RuBisCOs are fixing carbon is given by

$$\text{rate of carbon fixation by RuBisCO} \approx \frac{10^{39}\ \text{molecules}}{3 \times 10^{7}\ \text{s}} \approx 10^{31}\ \text{s}^{-1}, \tag{18.66}$$

where we have used the fact that a year is roughly 3×10^{7} s. Again, we can make naive estimates about the number of RuBisCOs by using the rough figure that each such RuBisCO on average fixes one carbon per second, resulting in the estimate that the number of RuBisCO molecules is given by

$$\text{number of RuBisCO molecules} \approx 10^{31}, \tag{18.67}$$

a number that can be used to justify the claim that RuBisCO is the most abundant protein on earth.

18.2.7 Destroying Sugar

The interconnectedness of living organisms is a truism and nowhere is this more evident than in the consideration of the cycling of carbon in the reciprocal processes of photosynthesis and respiration. We have devoted much of our effort so far in this chapter to the consideration of the elegant and amazing question of how it is that nuclear reactions taking place 150 million kilometers away from the Earth can directly fuel the synthesis of sugars. Once those sugars are made, however, the journey of the carbons that make them up has only begun.

As we have seen, photosynthesis accomplishes two goals for the organisms that perform it. First, it captures energy from light in a way that can generate a transmembrane electrochemical potential that can be used as an energy source for other chemical reactions inside the cell, including synthesis of ATP. Second, it generates reducing power that can be used to fix carbon dioxide in order to make sugar and other carbon skeletons. Most nonphotosynthetic organisms on the planet consume the bodies of photosynthetic organisms directly or indirectly and essentially reverse this entire process. As we explored in the discussion of Section 2.1.2 (p. 38), a nonphotosynthetic organism such as *E. coli* uses some fraction of the organic molecules it eats as building blocks for its own macromolecules and uses the remainder to generate energy. The process of disassembling a carbohydrate or other carbon-containing molecule and combining it with oxygen to generate a transmembrane electrochemical gradient with CO_2 as a waste product is called *respiration*. As shown in Figure 18.31, the ongoing cycle of photosynthesis and respiration connects all life on earth.

The chemical reactions underlying respiration involve electron transfer from one carrier molecule to another in a manner very analogous to the processes we have already explored in our study of photosynthesis. In the form of respiration performed by animals, carbon-containing organic compounds such as sugar (or fat) are burned within the mitochondria in a stepwise process to liberate CO_2 as a waste product (which can then be taken up by plants and used for photosynthesis ... note that some people believe that talking to their houseplants improves growth—this may be true if the plants

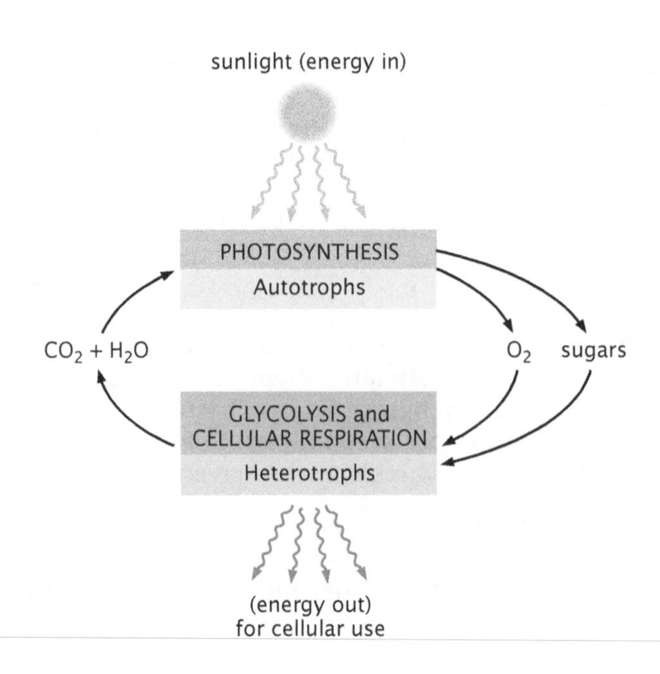

Figure 18.31: Interplay between photosynthesis and respiration. Autotrophs are organisms that can generate biological energy from nonbiological sources such as sunlight. Heterotrophs are organisms that absolutely rely on consuming biological products made by other organisms for their energy. Photosynthetic autotrophs (photoautotrophs) generate sugar and molecular oxygen, which are then used by them as well as by less self-sufficient chemoheterotrophs to generate ATP via glycolysis and respiration. The waste products of respiration, namely, CO_2 and H_2O, can be taken up again by photosynthetic autotrophs to be used in photosynthesis.

are taking up exhaled CO_2). In this process, electrons are transferred to NAD^+ (and to a related molecule, FAD) to generate the reduced high-energy electron-carrying forms of these molecules, NADH and $FADH_2$. Next, the electrons are ferried across the inner mitochondrial membrane by a series of electron carriers, and ultimately received by electron-hungry oxygen molecules on the outside. In accepting the electrons, O_2 also picks up hydrogen ions from its environment, and is converted into water through the reaction

$$O_2 + 4e^- + 4H^+ \rightarrow 2H_2O. \tag{18.68}$$

Note that this is exactly the reverse of the electron-stripping procedure performed by the manganese cluster during photosynthesis. In respiring mitochondria, the net effect of this electron transfer process is to create a pH gradient across the inner mitochondrial membrane. As we discussed in Section 16.2.7 (p. 660), this gradient of hydrogen ions is used to drive the rotation of ATP synthase, generating ATP for use by the animal.

Thus, all the work of photosynthesis, using light energy to convert CO_2 into sugar, and to strip electrons from water to generate O_2, is profligately undone by respiration.

18.2.8 Photosynthesis in Perspective

As a showpiece for physical biology, the process of photosynthesis covers many important topics. As shown in Figure 18.32, the events of photosynthesis span time scales from the femtoseconds required for an incoming photon to excite a pigment electron, through the picoseconds and nanoseconds of electron tunneling within the reaction center, through the microseconds required to strip water of its electrons to generate molecular oxygen, through the much longer time scales associated with carbon fixation, biosynthesis, and cell growth. Besides being a process of great scientific interest, as discussed eloquently in Morton's 2007 book *Eating the Sun* (see the Further Reading at the end of the chapter), photosynthesis is the heart and soul of life on Earth.

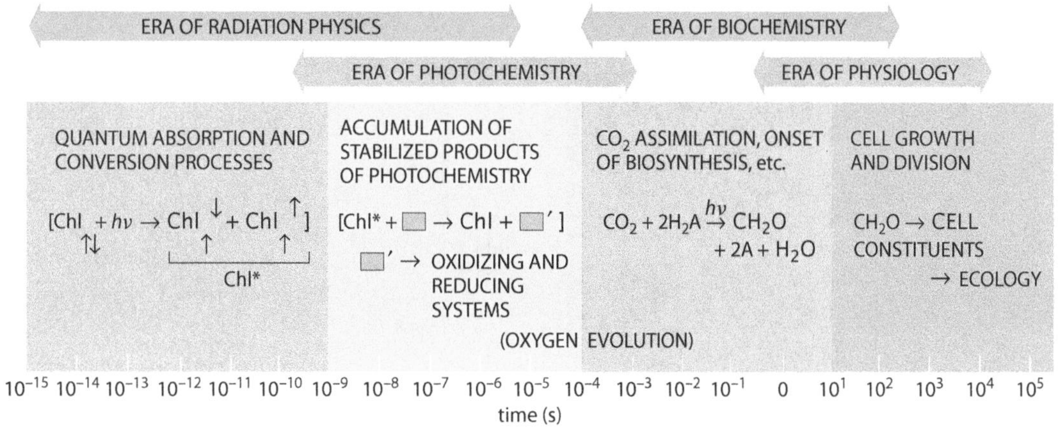

ERA OF RADIATION PHYSICS
ERA OF BIOCHEMISTRY
ERA OF PHOTOCHEMISTRY
ERA OF PHYSIOLOGY

QUANTUM ABSORPTION AND CONVERSION PROCESSES
ACCUMULATION OF STABILIZED PRODUCTS OF PHOTOCHEMISTRY
CO_2 ASSIMILATION, ONSET OF BIOSYNTHESIS, etc.
CELL GROWTH AND DIVISION

[Chl + $h\nu$ → Chl ↓ + Chl ↑] Chl*
[Chl* + □ → Chl + □′]
□′ → OXIDIZING AND REDUCING SYSTEMS
$CO_2 + 2H_2A \xrightarrow{h\nu} CH_2O + 2A + H_2O$
CH_2O → CELL CONSTITUENTS → ECOLOGY
(OXYGEN EVOLUTION)

10^{-15} 10^{-14} 10^{-13} 10^{-12} 10^{-11} 10^{-10} 10^{-9} 10^{-8} 10^{-7} 10^{-6} 10^{-5} 10^{-4} 10^{-3} 10^{-2} 10^{-1} 0 10^{1} 10^{2} 10^{3} 10^{4} 10^{5}
time (s)

Figure 18.32: Time scales associated with photosynthesis. The various events of photosynthesis, from electron excitation to cell growth, span nearly 20 orders of magnitude in time. (Adapted from M. Kamen, Primary Processes in Photosynthesis, Academic Press, 1963, as cited by Govindjee and H. Gest, *Photosynth. Res.* 73:1, 2002.)

18.3 The Vision Thing

In the first part of this chapter, we have been fixated on fixation, that is, how to use the energy available in photons for biochemical ends. However, photons also carry useful *information*. Though most photons come from the sun, once they arrive on Earth, they bounce off and are absorbed by objects in important and fascinating ways. Thus, if an organism was able to perceive the pattern of photons reflected from objects in its environment, it would be able to gain information about their location and nature that could be useful for sexual, aggressive, and defensive purposes. In animals, we call this ability vision. However, the ability to collect information from light and to use it to influence biological behavior long predates the evolution of animals and, indeed, is found in all the domains of the tree of life.

At the beginning of the chapter, we mentioned the early experiment by Engelmann that showed that some bacteria must have a primitive sort of vision, as they can tell the difference between light and dark and can detect specific colors of light (Figure 18.1). He trapped photosynthetic, motile bacteria in a thin glass chamber and illuminated them with sunlight that had been passed through a prism. Strikingly, the bacteria swam to form discrete bands within the rainbow spectrum, forming a living spectrometer. This result suggests that photosynthetic bacteria not only selectively absorb sunlight at specific wavelengths for use in electron transport, but also can somehow measure the wavelength of photons and use that information to influence their swimming behavior. A more direct demonstration of this capacity on the part of *Rhodospirillum centenum*, a swarming bacterium, is shown in Figure 18.33. Colonies of this organism glide toward a source of infrared light at their preferred wavelength for photosynthesis, but will glide away from visible light.

To make maximum use of light from the environment as a carrier of information, it is important for an organism to be able to detect not merely the intensity and wavelength of incoming photons, but also their location of origin. For multicellular organisms, animals in particular, this has been elevated into a fine art, with a multiplicity of designs of eyes capable of image formation whose technical specifications (sensitivity, dynamic range, and resolution) rival the cameras humans have succeeded in building. The underlying

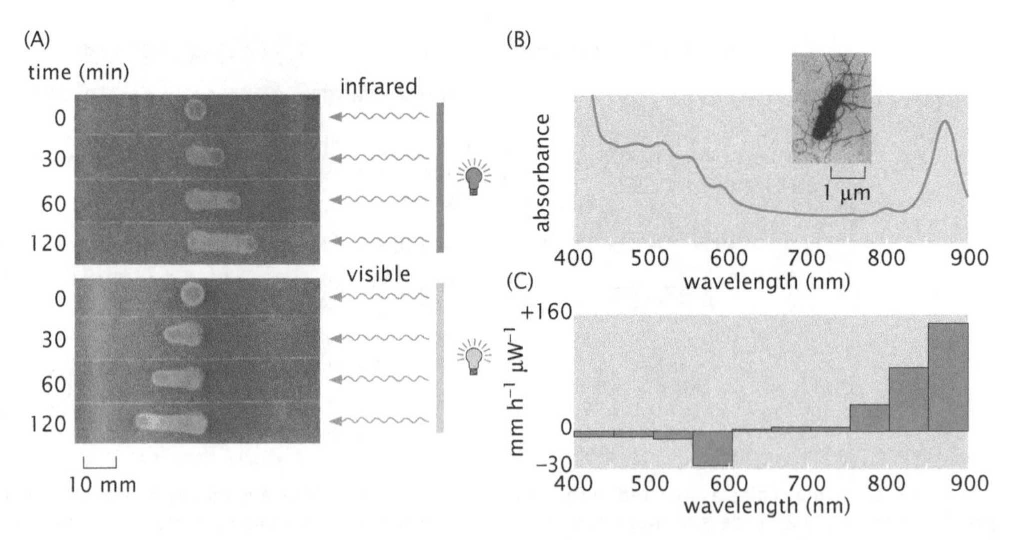

Figure 18.33: Wavelength-dependent movement of bacteria. (A) Swarming colonies of *Rhodospirillum centenum* move at a rapid rate toward a light source emitting infrared light (top), but move at a similar rate away from a visible-light source. (B) The absorption spectrum of this organism (top) shows two peaks: a broad range of visible light absorption below 600 nm and a narrower peak between 800 and 900 nm. (C) Over the same range of wavelengths, the action spectrum (bottom), a measure of how strongly a biological system responds to light, reveals a strong positive response only at the longer wavelengths, and a negative response that is strongest for wavelengths just under 600 nm. Inset: cells grown on agar develop numerous flagella all around their surfaces, which are used in swarming motility. (A, adapted from L. Ragatz et al., *Nature* 370:104, 1994; B, C, adapted from L. Ragatz et al., *Arch. Microbiol.* 163:1, 1995.)

mechanism of photon detection is conserved in all animal eyes and relies on the ability of absorbed photon energy to drive a conformational change in a small molecule called retinal, already introduced in section 15.2.2 (p. 581). In contrast to photosynthesis, where photon absorption, energy transfer, and electron transfer are all carried out by solid-state devices with no moving parts, photon processing in vision involves light-triggered protein conformational changes that have signaling consequences. In this section, we will explore the fundamental molecular and physical mechanisms in vision broadly construed.

18.3.1 Bacterial "Vision"

Much useful information on the fundamental mechanism of animal vision has been derived from studies of a salt-loving archaeon called *Halobacterium salinarum*, shown in its native habitat in Figure 18.3(B). This archaeon has a brilliant purple color caused by large arrays in its membrane of a protein called bacteriorhodopsin that carries a single retinal pigment. In *Halobacterium*, bacteriorhodopsin uses the energy of light to generate a transmembrane proton gradient, which can be used to synthesize ATP. In other words, the archaeon is using light to generate biochemical energy in a manner generally analogous to photosynthesis, although electron transfer is not involved. Indeed, the bacteriorhodopsin protein is not at all similar to the photosynthetic reaction center, and instead bears a striking structural and functional resemblance to the rhodopsins that underly animal vision.

The mechanism of light-driven proton transport is outlined in Figure 18.34. The key to this process is the ability of the small molecule retinal to undergo a light-dependent structural isomerization. We have already discussed this isomerization in Section 15.2.2 (p. 581). When the retinal is embedded in a protein, its light-dependent isomerization can drive a larger-scale conformational change in the protein, which can then have further biological consequences. As we will explore over the next few sections, the light-driven shape change of retinal has been harnessed for a variety of purposes.

In *Halobacterium*, a molecule of retinal is covalently attached in the middle of a seven-transmembrane-domain receptor. As shown in Figure 18.34(A), the retinal begins the photochemical cycle in the extended all-*trans* conformation. Absorption of a photon causes

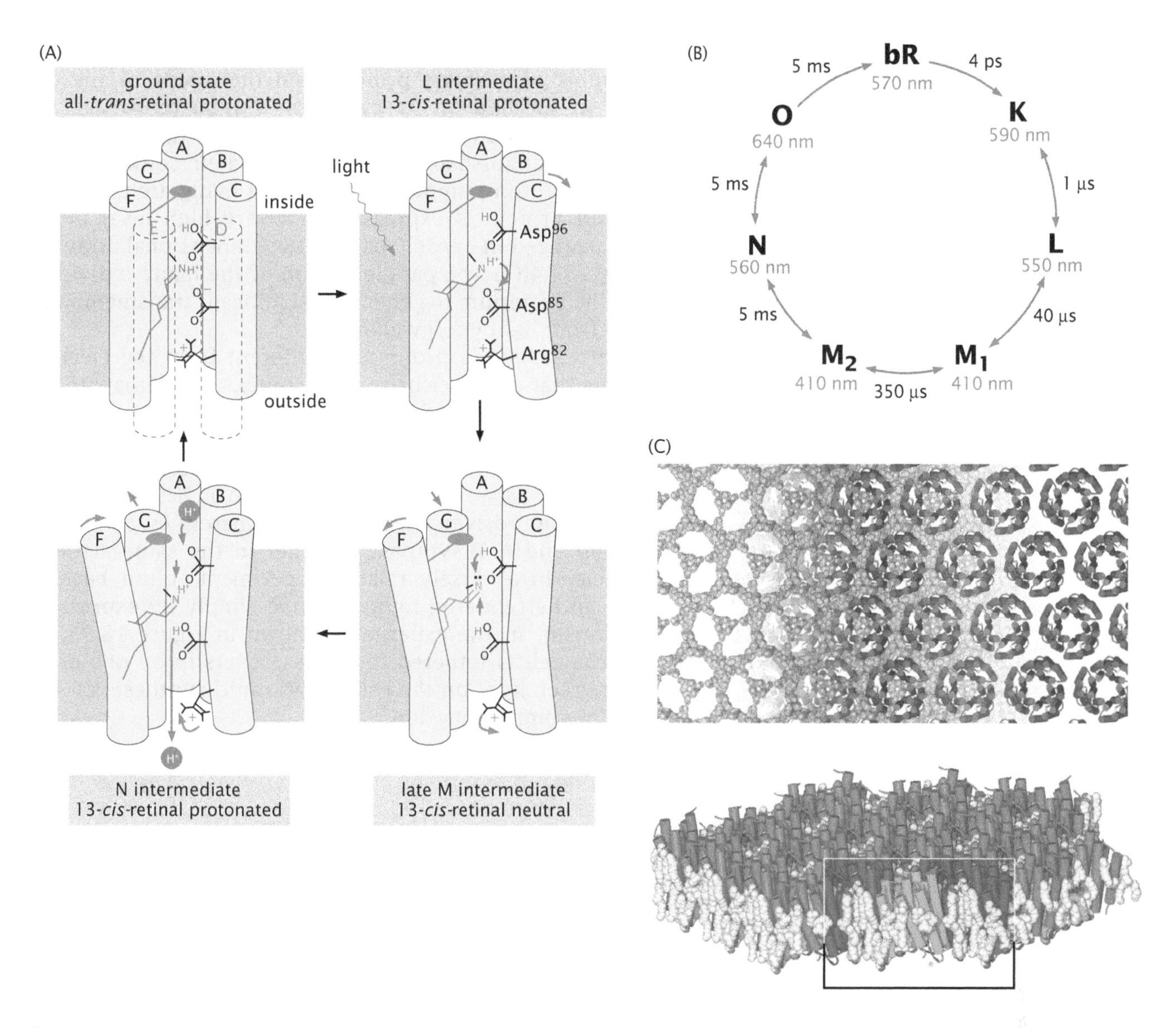

Figure 18.34: Proton transport by bacteriorhodopsin. (A) Each bacteriorhodopsin monomer has seven transmembrane helices, indicated with the letters A through G (for clarity, helices D and E are omitted in some diagrams). Before exposure to light, the helices are all fairly straight, and the covalently attached retinal molecule (shown in purple in this state) is in the all-*trans* configuration. An incoming photon drives the retinal into the 13-*cis* configuration. The protein relieves its internal strain with a kink in the C helix. Next, the F and G helices react with slight movements of their own. The bulky side chains attached to helix F (blue paddle) move to open the channel. This series of movements drives transfer of hydrogen ions along an axis made of two aspartate residues in helix C and the retinal itself. In order to return to the ground state, bacteriorhodopsin expels a hydrogen ion from Asp85, and picks up a hydrogen ion on Asp 96 from the solution on the opposite side of the membrane, resulting in the net transfer of one proton for each photon absorbed. (B) As the bacteriorhodopsin protein undergoes this series of conformational changes, the chemical environment of the retinal changes slightly, affecting its light absorption spectrum. This diagram shows the peak absorption wavelengths and lifetimes of each of the protein conformations that can be distinguished spectroscopically. The intermediates bR (ground state), L, M_2 (late M), and N are diagrammed in (A). (C) In the *Halobacterium* membrane, the bacteriorhodopsin monomers cluster together to form trimers, which are in turn packed into a nearly crystalline lattice. The upper diagram shows the relative density of the bacteriorhodopsin trimers (purple) and the membrane lipids (beige) as viewed looking down on the membrane. The lower digram shows an oblique view, with a single bacteriorhodopsin monomer highlighted in orange (boxed region). (A, adapted from W. Kühlbrandt, *Nature* 406:569, 2000; B, adapted from R. Neutze et al., *Biochim. Biophys. Acta* 1565:144, 2002; C, adapted from J. P. Cartailler and H. Luecke, *Annu. Rev. Biophys. Biomol. Struct.* 32:285, 2003.)

the retinal to switch to the 13-*cis* conformation. This causes strain within the protein's transmembrane helices. In relieving the strain, the protein undergoes a series of conformational changes that expose amino acid side-chain residues that can be protonated or deprotonated on either side of the membrane. The arrangement of a series of residues down the center of the seven-helix bundle facilitates the transport of a single proton from the inside of the cell to the outside. At the end of the cycle, the retinal returns to the lower-energy all-*trans* state. Figure 18.34(B) shows the number of spectroscopically distinguishable intermediate protein conformations in the light-driven proton pumping cycle. In the living archaeon, the bacteriorhodopsin protein is expressed at a very high level in the membrane and arranged in densely packed two-dimensional arrays as shown in Figure 18.34(C). The parallel action of the many individual pumps enables the bacterium to generate a significant transmembrane proton gradient from light energy alone.

Clearly, this mechanism is profoundly different from the electron transfer reactions that we described in the context of photosynthesis and appears to have evolved independently. But the net consequence is the generation of an electrochemical potential gradient across the membrane. *Halobacterium* is able to use the proton gradient to directly generate ATP. As we mentioned in Section 11.1.3, it is possible to reconstitute light-driven ATP production merely by combining bacteriorhodopsin and ATP synthase together in the same artificial membrane vesicle. In a closely related experiment, blind bacteria such as *E. coli* can be rendered light-sensitive simply by expression of bacteriorhodopsin. In the experiment shown in Figure 18.35, an ATP-starved *E. coli* cell is tethered to a glass coverslip by one of its flagella. Shining green light on the cell allows rapid synthesis of ATP, causing exuberant spinning motion.

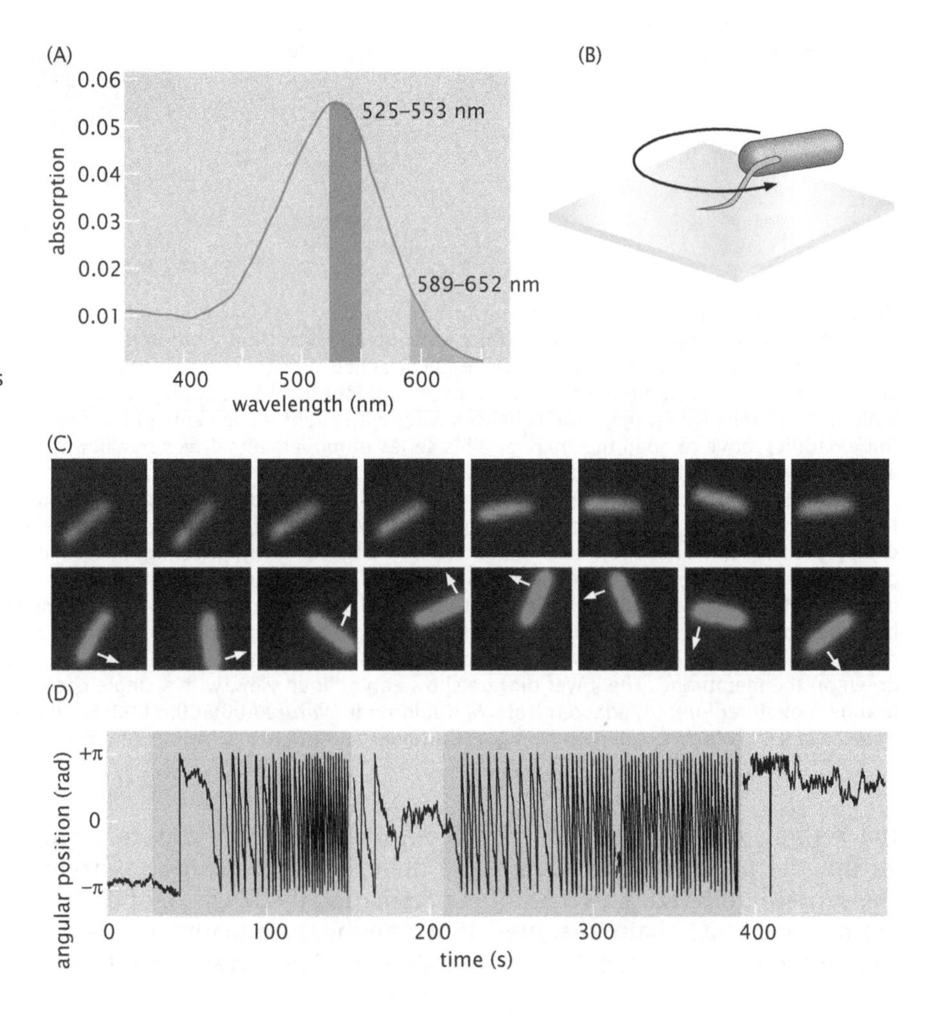

Figure 18.35: Light-powered *E. coli*. (A) A form of bacteriorhodopsin was expressed in a laboratory strain of *E. coli*. The red curve shows the absorption spectrum for this bacteriorhodopsin variant. Green light (525–553 nm wavelength) can maximally activate the hydrogen ion pumping activity, while red light (589–652 nm) has minimal effect. (B) The influence of the expressed bacteriorhodopsin on cell behavior was investigated by observing the rotation of cells tethered to glass coverslips by their flagella. The normal mechanisms used by *E. coli* to generate a transmembrane proton gradient were inhibited by poisoning the cells with sodium azide, leaving the bacteriorhodopsin as the only source. (C) In red light (top row), consecutive movie frames show that the bacterium barely moves. In green light (bottom row), the bacterium rapidly rotates counterclockwise, due to the spinning of its tethered flagellum. (D) A plot of angular position over time for a single bacterium shows rapid rotation only while green light is shining (green-shaded areas), and near paralysis in red light (gray-shaded areas), for repeated exposures. (Adapted from J. M. Walter et al., *Proc. Natl Acad. Sci. USA* 104:2408, 2007).

18.3.2 Microbial Phototaxis and Manipulating Cells with Light

Although the experiment described above involved a light-sensitive *E. coli* strain that was artificially created by scientists, it illustrates the important concept underlying the final section of this chapter, namely, that energy from photons can be used to transmit information to the cell about its environment. In the case of the light-sensitive *E. coli*, its swimming behavior changes depending upon the wavelength of light shone upon it. This is reminiscent of the phototactic behavior originally described by Engelmann.

Phototaxis is found in all branches of the tree of life and fundamentally requires that organisms be able to detect orientation, wavelength, or intensity of incoming light and alter their behavior in response. For example, the unicellular eukaryote *Euglena* (Figure 18.3E) has a specialized organelle appropriately called an "eyespot" that it uses to orient its swimming toward light. The mechanism of light detection for phototaxis is usually distinct from light absorption for photosynthesis, even in organisms such as *Euglena* that do both. The molecular basis of phototaxis in another eyespot-bearing unicellular eukaryote called *Chlamydomonas* was revealed when the light detectors were cloned and found to bear striking similarity to bacteriorhodopsin. In these particular proteins, called channelrhodopsins, the light-driven isomerization of retinal is not used to pump protons, but rather to open a cation channel that can allow sodium and potassium ions to freely cross the membrane. As we discussed in Section 17.3, opening a cation channel in a membrane will typically cause depolarization. As the eyespot is shaded from the back side, swimming in a helical path will result in different temporal light absorption patterns depending upon whether the motion is parallel or perpendicular to the light source.

More recently, the remarkable properties of channelrhodopsin have been exploited as a way to enable scientists to alter electrical signaling in neurons using light as a stimulus. As shown in Figure 18.36, channelrhodopsin is just one of a family of light-activated ion channels that various organisms use for phototaxis. A second particularly useful one is called halorhodopsin, which functions as a chloride channel. When coexpressed in a neuron, the combination of channelrhodopsin and halorhodopsin can be used to send detailed electrical messages. A pulse of blue light activating channelrhodopsin causes the neuron to fire while a pulse of yellow light activating halorhodopsin completely suppresses firing.

Using promoters that express transgenes only in specific populations of neurons, investigators have been able to create light-sensitive neural circuits in animals including *C. elegans*, zebrafish, and even mice. For example, Figure 18.37 shows the induction of a swimming escape response in a zebrafish triggered by directing a pulse of light to its brain. In addition to channelrhodopsin and halorhodopsin, there are many other exciting emerging technologies enabling nonperturbative light-driven manipulation of cell signaling and behavior in neurons and many other cell types.

18.3.3 Animal Vision

In multicellular organisms, the function of light detection is performed by specialized receptor cells that take the responsibility of absorbing photons and converting the information content they bear

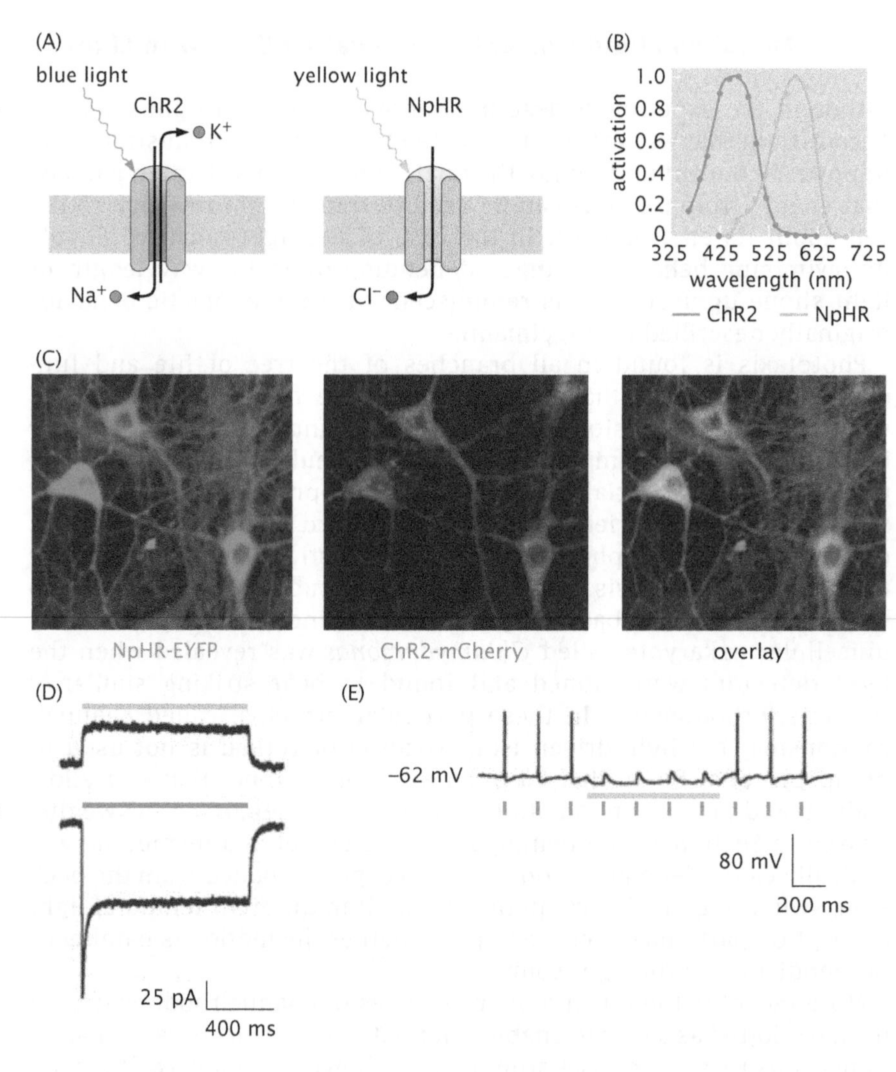

Figure 18.36: Optical control of neuronal activity. (A) Channelrhodopsin (left) is a light-activated cation channel, allowing transmembrane movements of potassium and sodium ions down their electrochemical gradient. Halorhodopsin (right) is a light-activated anion channel, primarily passing chloride ions. (B) The absorption spectra of channelrhodopsin and halorhodopsin are distinct. Channelrhodopsin is activated by blue light while halorhodopsin is activated by yellow light. (C) Both proteins can be coexpressed in cultured neurons. Here, the protein distribution is visualized using fluorescent tags. (D) These neurons can now be electrically stimulated by blue or yellow light. When held at a constant voltage, the exposure of the neurons to blue light (bottom) results in an inward current (shown here as downward) because the inward movement of positively charged sodium ions dominates. In contrast, the exposure of the neurons to yellow light (top) causes an inward flow of negatively charged chloride ions, and so results in a current of opposite sign. (E) Short pulses of blue light are capable of inducing action-potential firing in these cells. Each blue light pulse causes the cell to rapidly and dramatically depolarize from its resting potential of about −62 mV, and after each voltage spike the cell returns to its normal state. Exposure to yellow light completely blocks the action potentials triggered by the blue light pulses, as the influx of negatively charged chloride ions prevents depolarization. (Adapted from F. Zhang et al., *Nature* 446:633, 2007.)

into electrical signals that can be communicated to the nervous system. The simplest animal eyes consist of a single receptor cell. These eyes provide no spatial image, but can be used to detect the orientation, wavelength, and intensity of light much like the eye spot of *Euglena*. More ambitious animals have created fancier eyes that involve arrays of many photoreceptor cells, often in tandem with lenses, to generate true images. Animal vision presents a variety of interesting problems in physical biology. In this section, we will examine how the optical limits of light detection and the physical limit of the energies associated with photon-driven chemical processes play out in the eyes from a variety of creatures.

Image-forming eyes take an astonishing variety of forms, reflecting at least ten different engineering strategies within the animal kingdom. Two common types are shown in Figure 18.38. The familiar camera-style eye (or simple eye) is found among many vertebrates but also some invertebrates such as cephalopods (octopus and squid). In the very simplest cases, a two-dimensional array of photoreceptor cells is spread across the concave back of a cup-shaped invagination with an aperture facing the outside. Light through the aperture illuminates only a few of the photoreceptors, depending on the angle of the light source, enabling the animal to see position and shapes at low resolution. Incorporating a lens enables more efficient light capture, better resolution, and the opportunity to change focus. A second, profoundly different, but also common eye-building strategy is illustrated by the compound eyes found in many insects such as *Drosophila* and other arthropods. Here, the photoreceptors are arrayed on a convex surface, such that light sources from different angles can be detected directly. To improve the resolution in such eyes, it is common that the photoreceptors are collected at the bottom of roughly cylindrical light guides, or ommatidia, that are each capped with a small separate lens.

Several important specialized cell types are used to build eyes, in addition to the photoreceptor cells themselves. In most animals, the lens is formed by specialized epithelial cells that differentiate in a way that eliminates most of their internal membranous organelles (which would scatter light) and fills their cytoplasm with high densities of specialized proteins called crystallins. The critical feature of a protein that can enable it to perform this function appears to be its ability to be expressed at a very high concentration while remaining soluble (in contrast to the name, crystallin proteins do not form crystals). Some crystallins are related to heat-shock proteins, while others are related to common metabolic enzymes. The high density of protein in solution alters the refractive properties of the cytoplasm, bending the rays of incoming light to focus on the photoreceptor cells.

Figure 18.37: Control of zebrafish behavior using optically activated channels. (A) A young zebrafish is shown expressing fluorescently tagged channelrhodopsin in a subset of its neurons, including those responsible for a touch-sensitive escape response. (B) In these engineered animals, a pulse of blue light shined on the transparent head induces a rapid sideways flick of the tail identical to the natural escape response. (C) Several repeats of the same electrophysiological recording from one of the neurons expressing channelrhodopsin show that a brief (5 ms) pulse of blue light can reliably and reproducibly elicit an action potential in the cell within the living animal. (Adapted from A. Douglass et al., *Curr. Biol.* 18:1133, 2008.)

There Is a Simple Relationship between Eye Geometry and Resolution

Comparing the overall architecture of the two different types of eyes in Figure 18.38, it is clear that both represent workable solutions to the problem of how to resolve light sources at different locations in the animal's environment, but they have very different physical limitations. In Figure 20.30 (p. 942) we show a more detailed view a *Drosophila* eye, which we will address in the context of pattern formation in the fly.

One interesting example of the physical limits of vision comes from consideration of the lens size. In general, the angular resolution of a lens (that is, the smallest angle θ between two distant plane-wave light sources that can be resolved as distinct from one another) is an inverse function of the lens aperture diameter, specifically,

$$\sin\theta = 1.22\frac{\lambda}{D}, \tag{18.69}$$

where D is the lens aperture diameter and λ is the wavelength of the light of interest. This is the reason that powerful telescopes need to have huge lenses. For the human eye, with a lens aperture of a few millimeters, our angular resolution is about one arc-minute.

To see where this equation comes from, we quickly review diffraction through a circular aperture and ask the reader to flesh out the details in the problems at the end of the chapter. The key ideas are illustrated in Figure 18.39. In Figure 18.39(A), we show how we can keep track of the maxima and minima of an incident plane wave by appealing to the motion of the hand on a stopwatch. The position of the hand of the stopwatch will keep track of the *phase* of the wave at that particular point in space and time. We use the concept of these stopwatches to calculate the interference between multiple waves as described below. In particular, in Figure 18.39(B), we consider a plane wave originating from a very distant point source that hits a circular aperture of radius a. Each tiny patch of area on this circular aperture can be thought of as a source of outgoing spherical waves emanating radiation according to Huygens' principle. We sum up the amplitudes

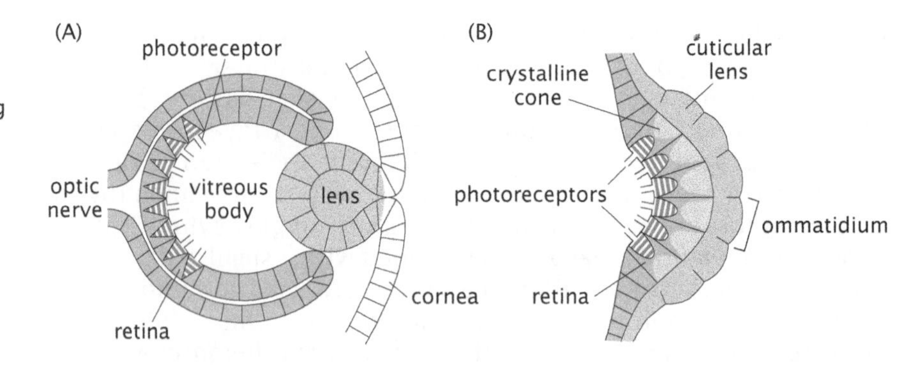

Figure 18.38: Different architectures for animal eyes. (A) Typical vertebrate eye, showing concave retina and single lens. (B) Typical arthropod eye, showing convex retina and multiple lenses. Photoreceptor cells are indicated with red stripes. (Adapted from M. F. Land and D.-E. Nilsson, Animal Eyes. Oxford University Press, 2012.)

from each of these little patches to evaluate the total wave amplitude on a plane at distance Z from the aperture. The key point of the analysis is that it demonstrates the way in which each of the little patches of area on the aperture sends out a spherical wave and how these waves interfere in a way that makes the resulting intensity in the far field have some structure. This interference can be thought of by imagining each patch of the aperture as sending out a ray that carries a stopwatch, with the regions of high intensity corresponding to those places in which the hands of the stopwatch are all more or less pointing in the same direction and those regions of low intensity corresponding to the stopwatches having hands pointing in different directions.

Using the strategy outlined above (the reader is encouraged to do this in detail in the problems at the end of the chapter), we find that

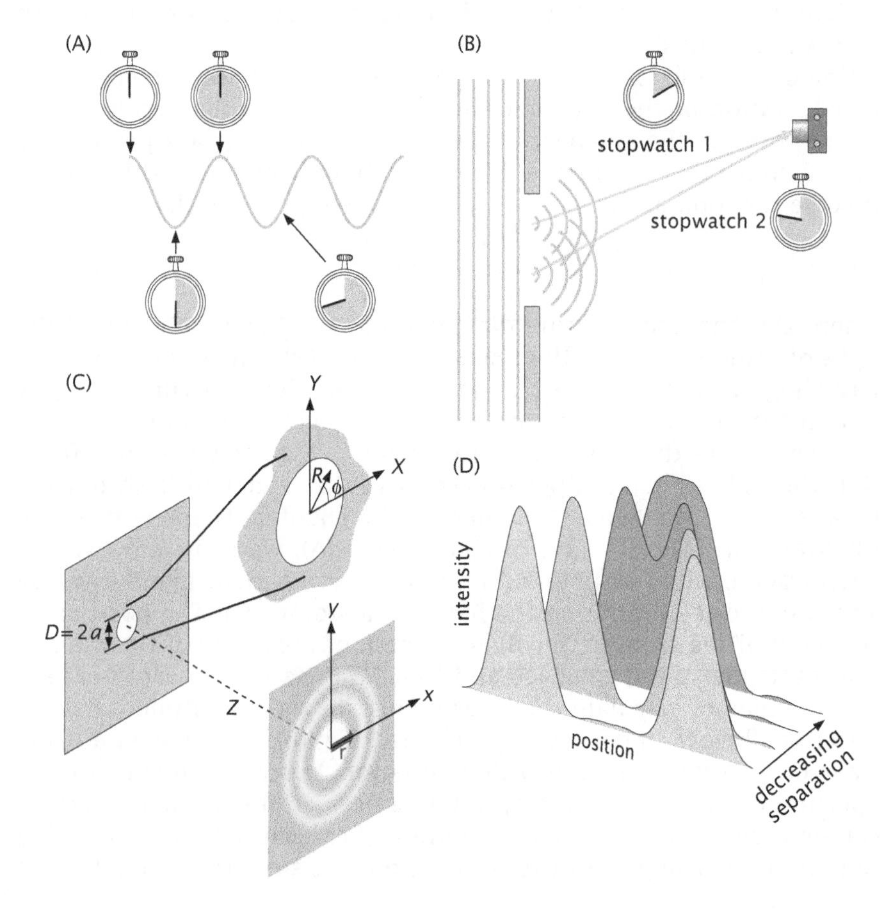

Figure 18.39: Optical resolution and circular apertures. (A) As a wave propagates, we can think of the repeated maxima and minima in the wavefront as reflecting the position of the hand of a stopwatch. This hand then keeps track of the *phase* of the wavefront. (B) Huygens' principle and the circular aperture. An incident plane wave passes through the circular aperture, which now serves as a source. We think of the aperture as an array of individual point sources. The waves emanating from each of these point sources will reach the detector with different positions of their corresponding stopwatches, leading to constructive or destructive interference. (C) Geometric parameters used to compute the light intensity on some plane of interest. The resulting Airy disc pattern is shown. (D) The image from two point sources lead to individual Airy discs. As these sources are brought closer, the Airy discs overlap, shifting the center of one past the minimum of the other and providing a criterion for the resolution limit.

the intensity of the radiation is described by the equation

$$I(r) = c\frac{a^2}{k^2 r^2} J_1^2\left(\frac{kar}{Z}\right), \tag{18.70}$$

where, as seen in Figure 18.39(C), r is the radial distance away from the center line of the point of interest, $k = 2\pi/\lambda$ is the wavevector of the incident radiation and tells us about its wavelength, a is the radius of the aperture, Z is the distance of the plane of interest from the aperture, and J_1 is the Bessel function of the first kind and first order. The intensity has azimuthal symmetry, but a damped oscillatory behavior as a function of distance from the center as shown in Figure 18.39(C). As a result of passing through the aperture, the intensity profile takes on a sophisticated form dictated by the shape and size of the aperture.

The diffraction effects set inherent limits on resolution as presented in Figure 18.39(D). Here, we consider two slightly separated point sources at a distance from the aperture that radiate plane waves propagating at a slight angle to one another. The plane waves are diffracted through our circular aperture. On the plane at position Z, the waves from each point source will lead to adjacent Airy discs. The question of resolution comes down to whether these Airy discs are sufficiently far apart so that we don't mistake them as a single bright spot. In order for each disc to be resolved individually, we will adopt the Rayleigh resolution criterion. This criterion states that the peak of the second source needs to be shifted relative to the peak of the first source by a sufficient amount that it falls outside of the first minimum of the other Airy disc. The factor 1.22 that appears in Equation 18.69 results from precisely this construction and in particular, arises because of the imposition of the condition

$$J_1\left(\frac{kar}{Z}\right) = 0, \tag{18.71}$$

resulting in

$$\frac{kar}{Z} = \lambda_{11}, \tag{18.72}$$

where λ_{ij} is the notation for the jth zero of the Bessel function of the ith order. For our case, $\lambda_{11} \approx 3.8317$. Note that we have the following geometric relationship, which can be read off Figure 18.39(C) as

$$\sin\theta = \frac{r}{Z} = \frac{\lambda_{11}}{ka} = \frac{3.8317\lambda}{2\pi a} = 1.22\frac{\lambda}{D}. \tag{18.73}$$

As mentioned above, if we consider incident radiation of wavelength 530 nm and an aperture size of 2 mm for the human eye, we find $\theta \approx 0.02^\circ \approx 1$ arc-minute. This means that we should be able to distinguish two adjacent points separated by as little as 0.5 mm from a distance of 1 m. Another interesting way to put this result in perspective is to address the resolution limit in the context of the separation of photoreceptors in the retina. Given the resolution limit, two resolvable point sources on the retina, which is located about 3 cm away from the pupil, would be separated by about 6 μm. As we will see in Figure 18.43, this distance is comparable to the separation between rods and cones in the retina. This could be interpreted as our eyes having just the right separation as to not to oversample what we perceive, since having them more tightly packed would not lead to any increase in resolution.

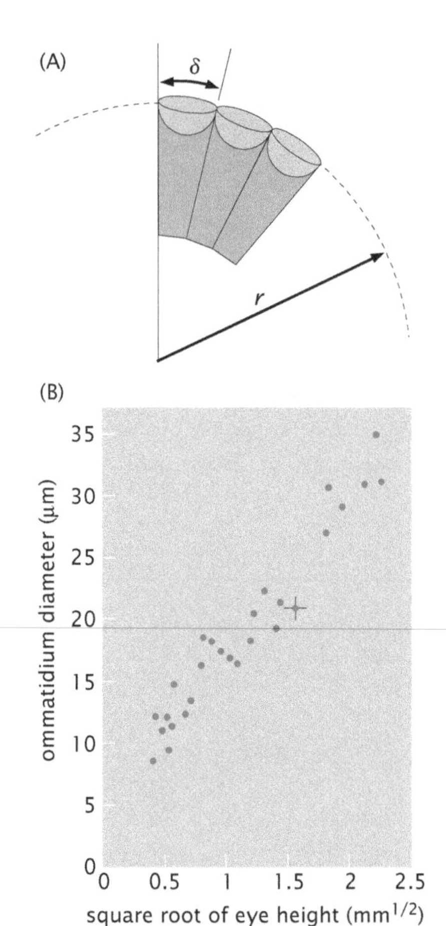

Figure 18.40: Scaling of size in bee eyes reflects the physical limit of resolution. (A) The schematic shows a compound eye of radius r and ommatidium size δ. (B) Measurements for ommatidium size and eye size for a variety of insects in the order *Hymenoptera*. Measurements for the bee are marked with a cross. (Adapted from H. B. Barlow, *J. Exp. Biol.* 29:667, 1952.)

The Resolution of Insect Eyes Is Governed by Both the Number of Ommatidia and Diffraction Effects

For compound eyes, where each ommatidium has its very own lens, the lenses are much smaller and therefore limit the resolution. The smallest insect ommatidia are a few microns in diameter. One way to increase the resolution of the lens is by increasing the size of the ommatidium. However, this would decrease the number of ommatidia that could be packed into an eye of fixed size, and therefore increase the angle between neighboring ommatidia. This would have the unwanted side effect of reducing the visual resolution experienced by the animal, as each ommatidium effectively acts as a single "pixel." The relevant geometry is shown in Figure 18.40(A). The competing demands of packing more ommatidia and having larger ommatidia to reduce diffraction effects suggest that there might be some optimal ommatidium diameter for an eye of a given size to maximize the final resolution. So, if it were really true that physical optics limited vision for insect eyes, then we should be able to predict the relationship between ommatidium size and eye size.

Feynman appears to have been inspired by studies like those of Barlow and others shown in Figure 18.40(B) in which the relationship between ommatidium size and eye size is examined, and in *The Feynman Lectures on Physics*, he gave a particularly simple treatment of the question of optimal ommatidium size as a function of the eye size. As shown in Figure 18.40(A), depending upon the relative dimensions of the individual ommatidia and the eye itself, from simple geometrical considerations the angle between adjacent ommatidia is given by

$$\Delta\theta_{\text{structure}} = \frac{\delta}{r}. \tag{18.74}$$

The characteristic sizes r of the eyes of interest here are between roughly 1 and 100 mm. Equation 18.74 tells us that to improve the angular resolution, we should decrease the size of ommatidia. But, at the same time, the diffraction effect canonized in Equation 18.69 implies a conflicting constraint on the resolution, namely,

$$\Delta\theta_{\text{diffraction}} \approx \frac{\lambda}{\delta}, \tag{18.75}$$

where we have neglected the prefactor.

As a result, as indicated in Figure 18.41, this means there is an optimal size, which we can find graphically by finding that ommatidium size δ_{opt} at which the curves for $\Delta\theta_{\text{structure}}$ and $\Delta\theta_{\text{diffraction}}$ intersect, corresponding to the condition

$$\frac{\delta}{r} = \frac{\lambda}{\delta}. \tag{18.76}$$

This result implies the simple scaling relation

$$\delta_{\text{opt}} = \sqrt{\lambda r}. \tag{18.77}$$

The scaling prediction of Equation 18.77 was tested for more than 20 diurnal hymenoptera with a range of different sizes, as shown in Figure 18.40(B). According to measurements made in 1928, the lenses over the ommatidia in the central part of the eye of the bee have an average diameter of 21.6 μm, corresponding to a resolving angle θ of about 1.6°. The angle between neighboring ommatidia in this region is about 0.97°. Similar measurements for 27 other insects of this same order are summarized in Figure 18.40(B). The observation that all the measurements fall nearly on a straight line (that is, that the ommatidium diameter goes as the square root of the eye size) indicates that the simple physical prediction works reasonably well for the insects examined, with eyes ranging in size from from only 1 mm to more than 60 mm.

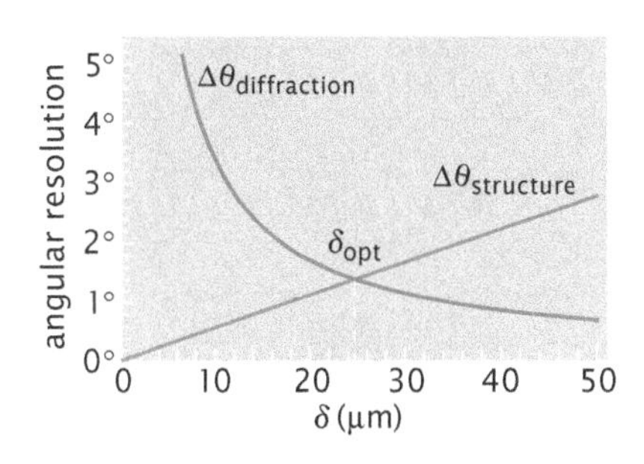

Figure 18.41: Angular resolution and the compound eye. The two curves show competing contributions to the angular resolution of the eye. $\Delta\theta_{structure}$ is governed by the number of ommatidia packed into the eye. $\Delta\theta_{diffraction}$ reflects the diffraction limit. The optimal ommatidium size is chosen to correspond to the case in which these two competing demands have the same $\Delta\theta$. Here, we have chosen $\lambda = 0.5$ μm and $r = 1$ mm.

The Light-Driven Conformational Change of Retinal Underlies Animal Vision

Despite all the anatomical variety of animal eyes and their different optical limitations, the fundamental light-driven chemical reaction that provides vision for all animals is conserved, and in fact is the same reaction used for light-driven proton pumping by bacteriorhodopsin, namely, the isomerization of retinal. For vision, the protein that acts as the receptor for photons is called opsin, and is well conserved at the level of amino acid sequence among all animal species. Similar to bacteriorhodopsin, the opsin associated with animal vision is a transmembrane protein with seven helices, which holds the retinal molecule covalently attached to a lysine residue. (There is very little amino acid sequence similarity between animal opsins and bacteriorhodopsin, so it is not possible to determine whether their rather similar overall structures are due to descent from a common ancestor or due to convergent evolution.) Unlike bacteriorhodopsin, which forms trimers in the archaeal cell membrane, animal opsins as shown in Figure 18.42 are found as monomers. Another intriguing structural difference is the initial configuration of the retinal molecule itself. Figure 18.42(C) shows how the close packing of the amino acid side chains surrounding the retinal constrain it in a bent configuration; here the lowest-energy state for the retinal is in the 11-*cis* form, not the all-*trans* form found in bacteriorhodopsin. As

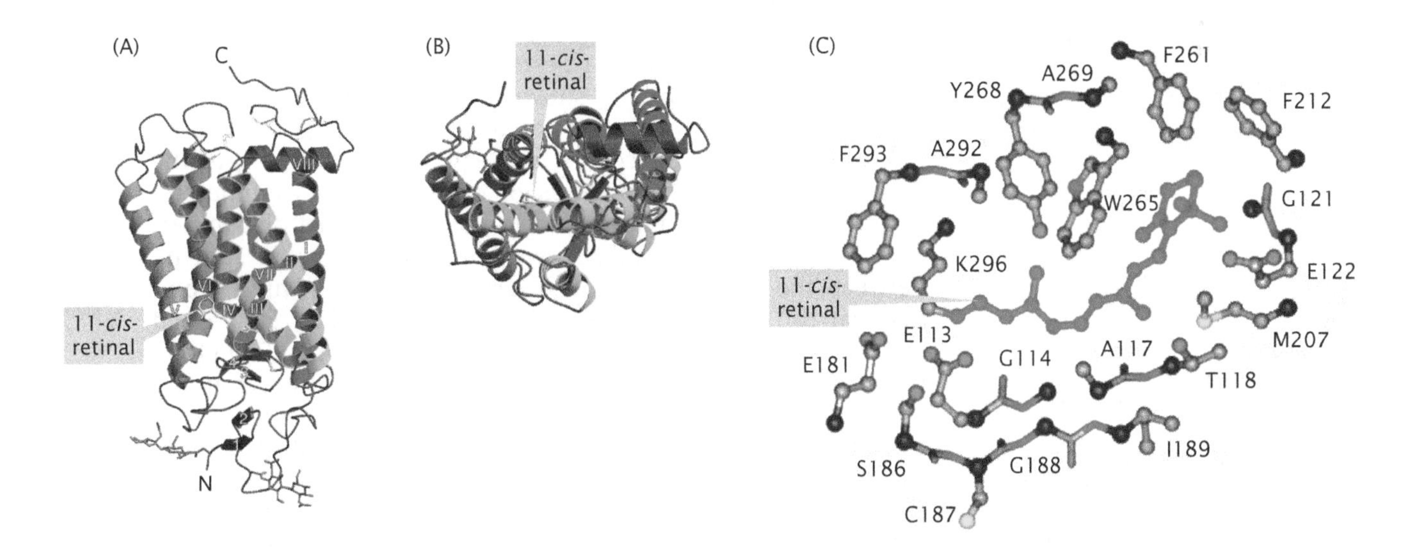

Figure 18.42: Structure of an animal opsin. (A) The crystal structure of rhodopsin (the form of opsin found in rod cells) from a cow reveals a transmembrane protein with seven helices in a bundle. (B) The view of rhodopsin from the cytoplasmic face shows the retinal molecule (yellow) buried in the middle of the seven helices. (C) The environment around the retinal is determined by the amino acid sequence of the protein. Tight packing of the side chains holds the retinal in the 11-*cis* configuration, which is its lowest-energy state (ground state) in this context. (Adapted from K. Palczewski et al., *Science* 289:739, 2000.)

we will discuss in more detail below, absorption of a photon by the 11-*cis*-retinal in an opsin causes it to flip into the all-*trans* form, leading to a protein conformational change that initiates a complex signal transduction cascade, eventually resulting in a change in the transmembrane potential for the cell.

In an animal eye, many copies of the opsin protein and its associated light-detecting retinal are expressed in specialized photoreceptor cells, whose ultimate function is to convert the informational signal received by absorbing photons into an electrical signal that can be communicated to neurons in the brain. For vertebrates that are able to distinguish different wavelengths of light and perceive them as different colors, there are typically at least two kinds of photoreceptor cells, called rods and cones. The rods and cones in a salamander eye are shown in Figure 18.43(A). Rods are very sensitive to low levels of light, and, as we will see below, they are capable of perceiving the presence of a single photon. Cones function well only at higher levels of light, but are usually specialized for detecting light of different colors.

How is it possible for a rod cell to detect a single photon? One important adaptation of these specialized cells appears to be a structural organization that maximizes the likelihood that a passing photon will be captured. In the context of photosynthesis, we have seen how the presence of a large antenna complex can increase the effective cross-sectional area for photon capture by a single photosystem reaction center. In contrast, there are no antennae in photoreceptor cells. Instead, the cell packs many copies of the opsin protein into a series of stacked membrane discs that are orthogonal to the direction of incident light. A small section of a rod cell is shown in Figure 18.43(B), illustrating the extremely dense and regular packing of the discs. Thus, an incoming photon that does not happen to collide with an opsin in the first disc has another chance to be absorbed when passing through the second disc and up to 2000 chances for the longest rod cells found in vertebrates.

(A)

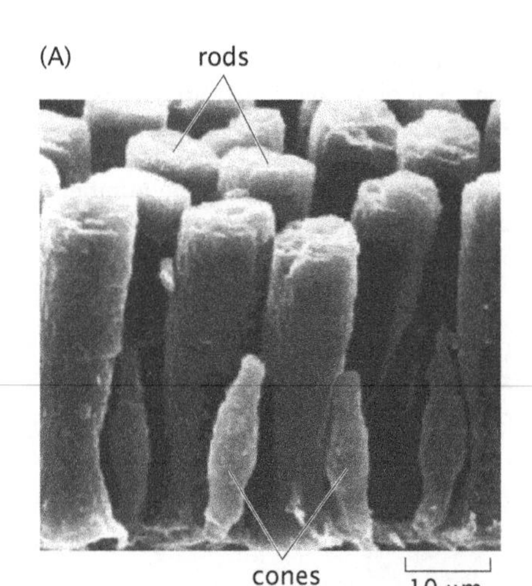

(B)

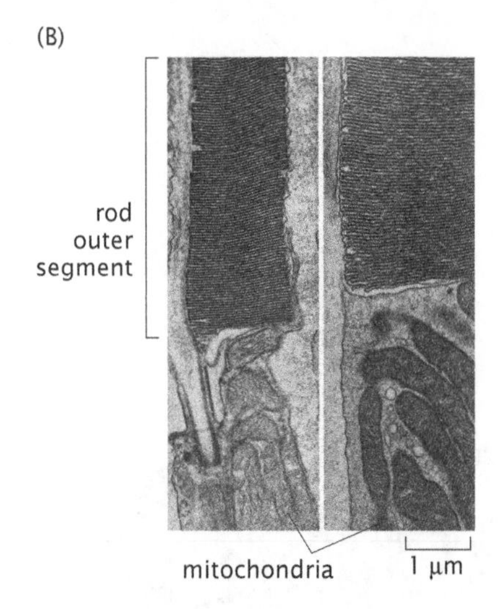

Figure 18.43: Animal photoreceptor cells. (A) This scanning electron micrograph shows the organization of rods and cones in the retina of a salamander. The relative sizes and numbers of rods and cones vary for different vertebrate animal species. (B) Electron micrographs of thin sections through a mammalian rod cell (left) and cone cell (right) reveal the close packing of the membrane discs within the outer segments that hold the rhodopsin. Below the outer segments, mitochondria pack the inner segments in order to provide the large amounts of ATP necessary for phototransduction. (A, courtesy of Scott Mittman and David R. Copenhagen; B, from D. W. Fawcett, The Cell, 2nd ed. W. B. Saunders, 1966.)

ESTIMATE

Estimate: Number of Rhodopsin Molecules Per Rod To get a sense of the possibility that a photon will encounter a rhodopsin molecule, in this estimate we further explore the geometry of the rod cells shown in Figure 18.43. As can be seen by using the mitochondrion as a ruler in Figure 18.43(B), the diameter of a rod cell is between 1 and 2 μm, while the length of its outer segment is of order 10 of its diameters. Again, by using the mitochondrion as a standard of comparison (see Figure 1.10, p. 18), we note that the mean spacing between the cristae in mitochondria is roughly 50 nm, implying that the spacing between the membrane discs in the rod cell is roughly half that value at 25 nm. Hence, for a 20 μm long rod cell outer segment we would estimate there are roughly 800 such membrane discs. In a given membrane disc, if we assume that the rhodopsin molecules are packed at a modest density with a mean spacing of roughly 10 nm (or 25,000 rhodopsins/μm^2), this still implies that there are over 100,000 rhodopsin molecules in every disc, leading to an estimate of in excess of 10^8 rhodopsin molecules in every rod cell.

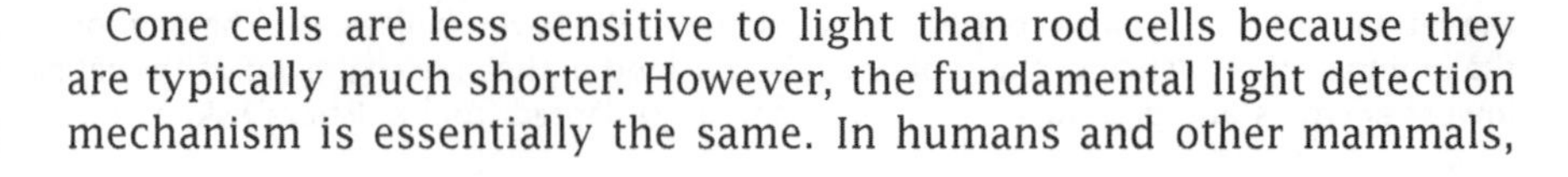

Cone cells are less sensitive to light than rod cells because they are typically much shorter. However, the fundamental light detection mechanism is essentially the same. In humans and other mammals,

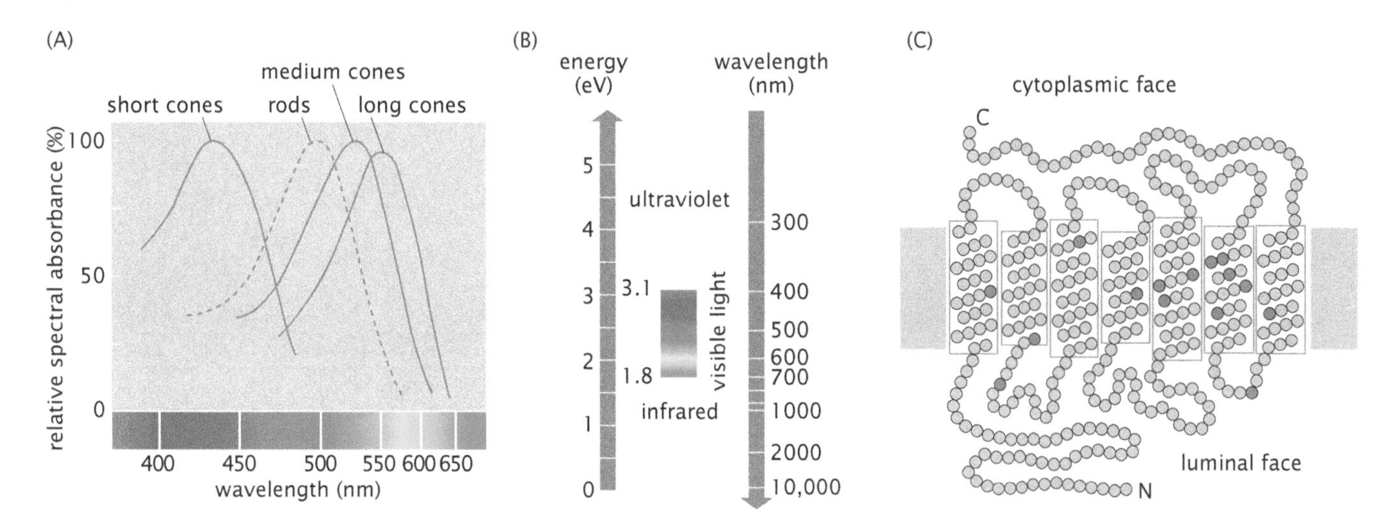

Figure 18.44: Color vision. (A) The absorption spectra for the four different types of human photoreceptor cells are shown; rods and the three color-specific cones (short or blue cones, medium or green cones, and long or red cones). (B) The range of visible light as perceived by humans runs from just under 400 nm to about 700 nm; this range is equivalent to a range of photon energies from 1.8 to 3.1 eV. Many animals have broader ranges of visual detection than humans. (C) The structure of human rhodopsin is cartooned here as a series of amino acid beads on a string, starting from the N terminus of the protein on the luminal face of the disc membrane and running through seven transmembrane helices to end at the C terminus on the cytoplasmic face. Here the amino acid sequences of the specialized rhodopsins found in medium versus long cones are compared; all of the sites where the amino acids differ are highlighted in red. This modest alteration in protein sequence is sufficient to significantly shift the absorption spectrum of the embedded retinal. (A, adapted from J. K. Bowmaker and H. J. Dartnall, *J. Physiol.* 298: 501, 1980. C, adapted from J. Nathans et al., *Science* 232:193, 1986.)

the sensitive rod photoreceptor cells carry a standard rhodopsin with maximal absorption in the green range of the visible spectrum, at about 500 nm. The cone cells are specialized for detecting different wavelengths of light, as shown in Figure 18.44. Mice have two kinds of cones, while humans have three; each kind of cone cell expresses a slightly distinct form of opsin, although every opsin uses the exactly identical 11-*cis*-retinal to detect light. How, then, can they perceive distinct colors? Recall that the configuration of the retinal within the opsin is constrained by the side chains of the amino acids surrounding it, as shown in Figure 18.42. In fact, the chemical environmental constraints imposed by the protein surrounding the retinal influence not only the configuration of its ground state but also its spectral properties. Interestingly, the number of amino acid changes necessary in rhodopsin to shift the absorption structure of the embedded retinal by an appreciable amount seems to be quite modest. Figure 18.44(C) shows the positions of the amino acids that differ between the two forms of opsin found in human cones specialized for detecting green light versus detecting red light. It thus seems very likely that variations in color vision could have arisen fairly easily throughout animal evolution, with the (presumably original) gene encoding rhodopsin as found in rod cells undergoing rounds of duplication and diversification to be optimized for detection of different colors. For humans, combinatorial information from the inputs of our various photoreceptors enables us to distinguish up to 10,000 distinct colors. Perhaps the world champion in color vision is the mantis shrimp, which inhabits very brightly colored coral reefs, with 12 different types of photoreceptor cells specialized for color vision (as well as others specialized for detecting polarized light). The mantis shrimp is thought to have the ability to detect up to 100,000 distinct colors.

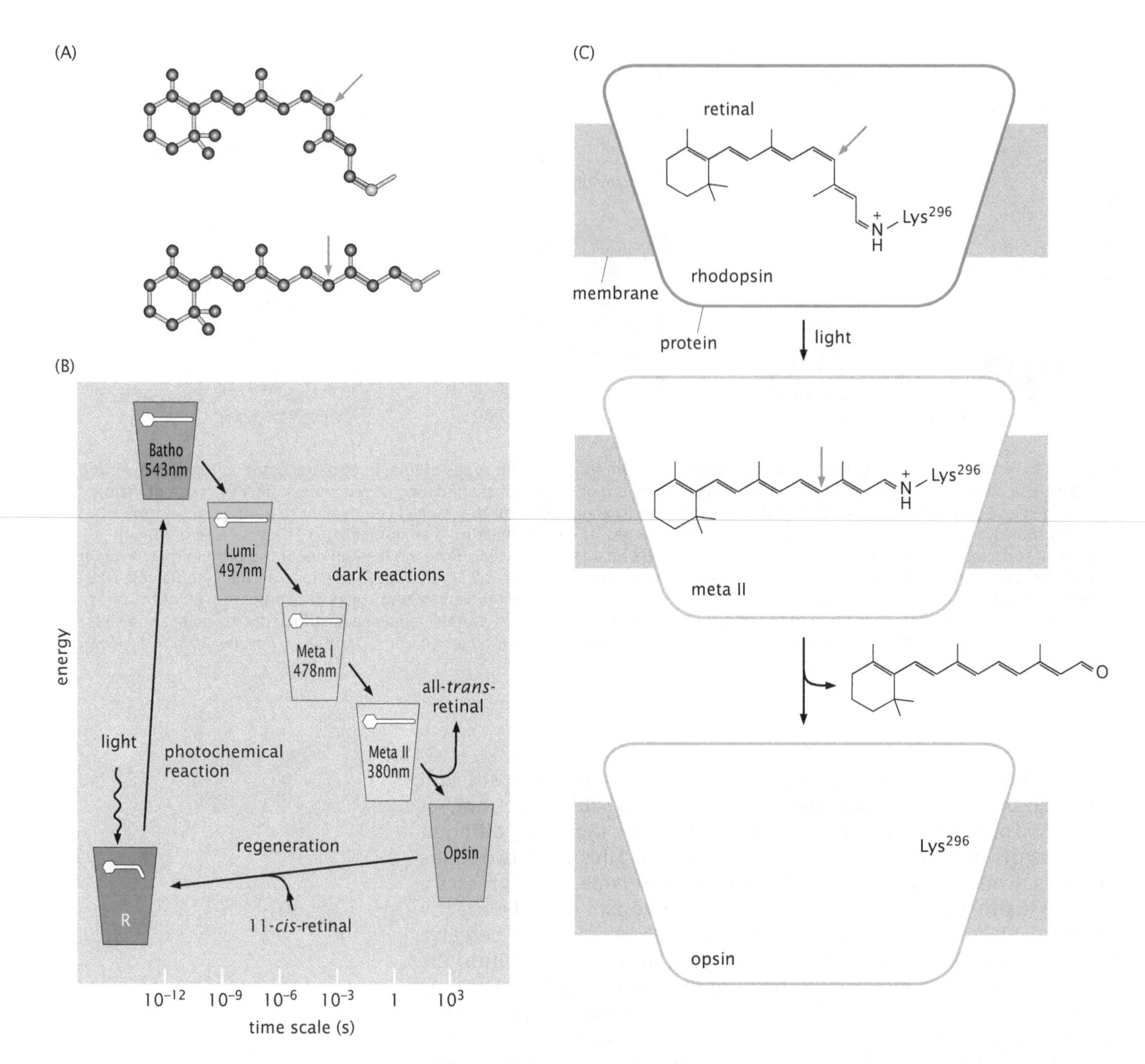

Figure 18.45: Retinal cycle in animal vision. (A) Within the context of the rhodopsin protein, the 11-*cis* form of retinal, shown at the top, is at a lower energy state than the all-*trans* form, shown below. The green arrow shows where the photoisomerization takes place. (B) Time scales of rhodopsin conformational changes after photon absorption. The relative energy levels of the conformational intermediates are indicated on the vertical axis. The maximum absorption wavelengths for each named intermediate are shown. (C) All-*trans* retinal cannot be re-isomerized into the 11-*cis* form within the context of the protein; instead, the retinal is cleaved from the lysine residue where it is covalently attached and must be replaced by a new 11-*cis*-retinal molecule. (B, C, adapted from K. Palczewski, *Annu. Rev. Biochem.* 75:743, 2006.)

The sequence of events that takes place after a retinal within an opsin protein absorbs the energy of a photon is illustrated in Figure 18.45 for the rhodopsin found in mammalian rod cells. Photon absorption converts the 11-*cis*-retinal buried in the rhodopsin into the all-*trans* form, a rapid event that takes only a few picoseconds. After this initial excitation, the protein undergoes a series of compensatory conformational changes analogous to those described for bacteriorhodopsin in Figure 18.34. Each of the intermediates can be distinguished spectroscopically, representing a gradual relaxation of the system. The long-lived metarhodopsin conformation is capable of activating the next step in the signal transduction cascade. Eventually (after about 300 s), the all-*trans*-retinal is cleaved from the rhodopsin protein and the protein is left without a pigment. Before it can detect another incoming photon, the retinal must be replaced by a fresh soluble molecule of 11-*cis*-retinal from the cytoplasm. In vertebrate eyes, the discarded all-*trans*-retinal can be transported out of the photoreceptor cell into a neighboring cell that harbors an isomerase enzyme, to convert it back into its 11-*cis* form, which can then be returned to the photoreceptor. Alternatively, fresh 11-*cis*-retinal can be derived from vitamin A (retinol), which is abundant in carrots and other vegetables that your mother tells you are good for your eyes.

Information from Photon Detection Is Amplified by a Signal Transduction Cascade in the Photoreceptor Cell

The central problem in animal vision is translating the conformational change of the retinal that has absorbed a photon into some kind of signal that can be perceived by the brain of the animal. As we have explored in Chapter 17, the informational transactions in animal nervous systems are mostly carried out by alterations in the transmembrane ion gradient, resulting in electrical signals. Vision is no exception. As we discussed above, bacteriorhodopsin uses the retinal structural change to pump hydrogen ions across the cell's membrane directly, resulting in a transmembrane ion gradient and voltage. In the visual system, however, the effect of light-induced retinal isomerization on transmembrane ion gradients is much more indirect. The reason for this is quantitative: in bacteriorhodopsin, the absorption of one photon results in the transmembrane transport of one hydrogen ion (i.e. one charge-equivalent). This amount of charge difference is orders of magnitude too low to be detected as a voltage change. For a rod cell, a change in membrane potential of 1 mV would require transport of some millions of ions. Furthermore, the rhodopsin in a rod cell is not located in the plasma membrane, where the relevant voltage change must take place, but rather in internal disc membranes. How can the isomerization of one tiny little retinal cause such a large change in cell behavior, on a different membrane compartment? Two issues are immediately apparent. First, there must be some mechanism to amplify and transmit the signal. Second, the amplification must require additional energy input.

The overall scheme used in vertebrate photoreceptors is shown in Figure 18.46. (Invertebrate photoreceptors work in an analogous way but some of the details differ.) The signal generated by the retinal conformational change is passed through several important intermediate signaling proteins before ultimately affecting the conductance of ion channels in the plasma membrane. As we have seen in Chapter 17, the resting membrane potential in most cells is established by ATP-consuming ion pumps that export sodium ions while importing potassium ions, coupled with potassium leak channels that allow some potassium ions back outside, resulting in a net transmembrane potential that is negative on the inside. In rod cells kept in the dark, a specific cation channel also remains open, allowing sodium and calcium ions to flow back into the cell, down their electrochemical gradients. Because of this extra leakiness, the resting potential of a rod cell membrane is only about −30 mV, much closer to neutral

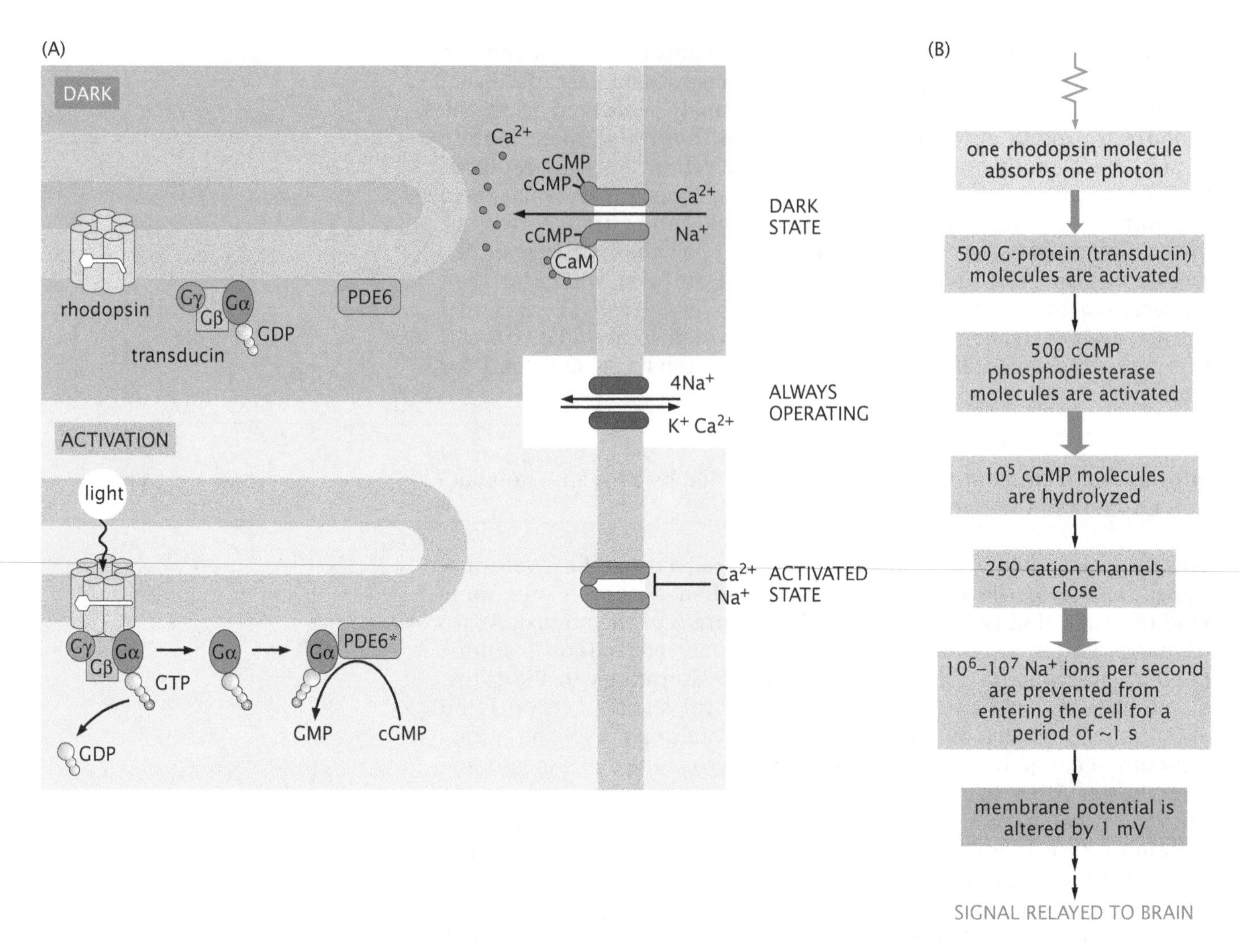

Figure 18.46: Signal transduction in the retina. Activation of rhodopsin results in the hydrolysis of cGMP, causing cation channels to close. (A) In the dark (top), the cGMP phosphodiesterase PDE6 is inactive, and cGMP is able to accumulate inside the rod cell. cGMP binds to a ligand-gated ion channel (dark green) that is permeable to both sodium and calcium ions. Calcium is transported back out again by an exchanger (shown in brown) that uses the energy from allowing sodium and potassium ions to run down their electrochemical gradients to force calcium ions to be transported against their gradient. When a photon activates a rhodopsin protein (bottom), this triggers GTP-for-GDP exchange on transducin, and the activated α subunit of transducin then activates PDE6, which cleaves cGMP. The ligand-gated channels close, and the transmembrane potential becomes more negative (hyperpolarized). (B) Signal amplification is achieved at several steps of the pathway, such that the energy of one photon eventually triggers a net charge change of about one million sodium ions (A, adapted from A. Stockman et al., *Journal of Vision* 8: 1, 2008. B, adapted from B. Alberts et al., Molecular Biology of the Cell, 5th ed. Garland Science, 2008.).

than the resting potential of a typical nerve or muscle cell. Critically, though, the channel allowing reentry of sodium and potassium is a ligand-gated channel, remaining open only when its ligand is present, an intracellular signaling molecule called cyclic guanosine monophosphate or cGMP (which is generated from GTP by a cyclase enzyme). When the concentration of cGMP drops, the leaking channels close, causing the membrane to slightly hyperpolarize (that is, the potential becomes more negative). The reader is invited to explore how the MWC model introduced throughout the book (see Section 7.2.4, p. 300 for an introduction to the MWC model) can be used to think about this gating in the problems at the end of the chapter.

So, how does rhodopsin activation alter the intracellular concentration of cGMP? The first of the series of signaling proteins that transmits information from the activated rhodopsin to the ion channels at the plasma membrane is called "transducin," one of a family of very important signaling proteins in animals that are generally known as "heterotrimeric G proteins" because they are typically composed of three distinct subunits, one of which binds GTP (or GDP). The G proteins are able to transduce signals initiated by the very large family of G-protein-coupled receptors, which include not only the opsins but also a huge number of other transmembrane proteins (almost always with seven transmembrane helices) that are involved in a wide variety of physiological functions. For example, the receptor that detects the hormone adrenaline, which floods our bodies when we are excited or frightened, is another G-protein-coupled receptor. It has been estimated that between 1% and 3% of the proteins encoded in animal genomes are members of this family. Their downstream effectors, the heterotrimeric G proteins, also come in a variety of forms. In most cases, the Gα subunit of the G protein is bound to GDP in the resting state, and also bound to the other two subunits Gβ and Gγ. The membrane-associated G-protein trimers are able to diffuse in the plane of the membrane. When a trimer comes into contact with an activated receptor (in this case, a rhodopsin protein with an all-*trans*-retinal), the Gα subunit releases the GDP in exchange for a GTP, and also unbinds from the Gβ and Gγ subunits. The free, GTP-associated Gα subunit can then interact with its downstream targets. For transducin activated by rhodopsin, the critical downstream target is the enzyme cGMP phosphodiesterase (PDE). This enzyme cleaves one of the bonds in cGMP, rendering it into the normal GMP form.

At this point, the cGMP concentration in the cytoplasm drops, allowing the ligand-gated ion channels to close, and consequently the electrical potential across the rod cell plasma membrane becomes slightly more negative. The rod cell couples directly to neurons at its base, and in the dark it continuously releases packets of neurotransmitters, as expected for a depolarized cell. The hyperpolarization brought about by cGMP destruction downstream of light activation causes the rate of neurotransmitter release to be slowed down. Thus, the downstream neurons, and eventually the brain, perceive light as a slight decrease in an otherwise constant neurochemical signal from the rod cell. Several energy-requiring amplification steps have occurred between the light-driven isomerization of the retinal and the alteration in neurochemical transmission, as summarized in Figure 18.46(B). First, the activated rhodopsin is able to trigger GTP exchange and subunit dissociation for some hundreds of transducin trimers during its lifetime. Indeed, recall that the activated metarhodopsin conformation is very long-lived ($\approx$300 s), so another complex set of energy-requiring enzymatic events is required to shut off its signaling capacity. These events include phosphorylation of the rhodopsin by a specific protein kinase, and binding of a protein called arrestin that competes with transducin for its binding site on rhodopsin. These processes are complete within a few 100 ms. Each activated transducin α subunit can activate only one phosphodiesterase, but that enzyme can cleave several hundred cGMP molecules before the switch is shut off again by hydrolysis of the GTP on the transducin α subunit, which returns the G protein and its effector enzyme to their resting states. The net loss of cGMP results in the closure of about 250 ligand-gated channels for about 1 s, which prevents the entry of up to 1 million sodium ions. Just as with an electric guitar amplifier, the amplification process requires significant energy input, which happens here at many levels. Every activated transducin goes through a cycle of GTP hydrolysis. Every cGMP molecule is made from a GTP. The resting transmembrane potential is initially set up by ATP-driven ion pumps. Thus, although the actual energy associated with a single photon is relatively small, around 2 or 3 eV, the downstream signal is much stronger and is actually large enough to be measured by an electrode.

This highly amplified system is therefore exquisitely sensitive to low light levels. However, highly amplified systems have an intrinsic drawback: the system is easily saturated, so that additional input (more light) cannot automatically result in more signal at the output. In fact, the vertebrate visual system is extremely good at adapting to levels of background light, despite the extreme built-in amplification, such that humans can perceive small changes in light intensity over a background range of about three orders of magnitude. The adaptation process relies on a scheme that measures the amount of calcium ions present in the cytoplasm. Because the cGMP-gated cation channels allow calcium as well as sodium to leak back into the cell, but the calcium is pumped out again through another transporter that always operates regardless of light level, the light-driven closing of the channels causes the calcium level to drop as well. This drop in calcium actually serves to counteract the closing of the cGMP-activated cation channels at several steps in the signal transduction cascade, so that even more light is required to generate a detectable electrical signal. Although the details of this adaptation mechanism are beyond the scope of this discussion, the quantitative issues raised by the necessity of maintaining sensitivity over a wide range of ambient backgrounds present particularly interesting challenges to the physical biology approach to understanding vision, and the history of both experimental and theoretical work in this area is filled with lovely examples.

The Vertebrate Visual System Is Capable of Detecting Single Photons

What is the evidence for the claim made above that vertebrates are capable of detecting one to several photons? A few particularly compelling quantitative experiments have established this proposition in vertebrate rod cells. The first that we will discuss involves simply measuring the hyperpolarization of isolated frog rod cells using electrophysiology. In complete darkness, occasional brief changes in the current can be observed, as shown in Figure 18.47(A). The spiking is suppressed in light. These spikes are strikingly uniform in size, but are much too big to be the result of the spontaneous opening or closing of a single ion channel. Instead, they are thought to be the cellular consequences of a spontaneous thermal isomerization of a single molecule of retinal within rhodopsin. We can use the frequency of these spontaneous events as a way to estimate the energy barrier associated with retinal isomerization. Recall from Section 16.2.3 (p. 647) that the rate of a reaction in which a molecule must overcome an energy barrier can usually be estimated using the Arrhenius relation $k = \Gamma e^{-\beta \Delta G_{\text{barrier}}}$. Because $\beta = 1/k_B T$, this equation predicts an exponential relationship between the rate of a reaction and the inverse of the temperature at which the reaction occurs. For the dark current observed in rod cells, the frequency of appearance of these spontaneous currents beautifully obeys this relationship, as shown in Figure 18.47(B). The slope of this line gives the size of the energy barrier, found here to be about 50 $k_B T$,

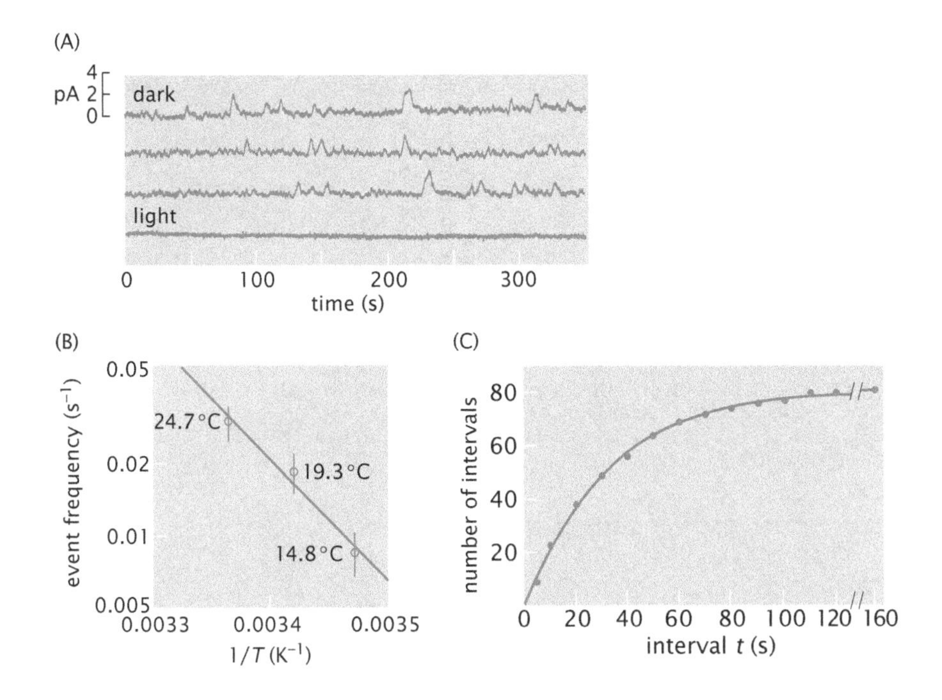

Figure 18.47: Thermally driven isomerization of retinal. (A) Electrophysiological measurements of the current across the membrane in an isolated rod cell from the retina of a frog, comparing the spontaneous events seen in the dark (top three traces) with the suppression of spontaneous events in the light (bottom trace). (B) Frequency of events as a function of the inverse of absolute temperature, plotted here on a semilog scale. The experimental temperatures in degrees Celsius are shown next to each corresponding data point. The Arrhenius relationship predicts a straight line, where the activation energy is the slope. (C) Cumulative distribution of intervals separating spontaneous events in the rod. The points show the measurements and the curve shows the fit to the expectation for a random process characterized by an exponential distribution of waiting times, with the average waiting time $\tau = 32$ s. (Adapted from D. A. Baylor et al., *J. Physiol.* 309:591, 1980).

consistent with the energy required for the light-driven isomerization of retinal (recall that the energy carried by a green photon is about 100 k_BT).

Another informative way of analyzing the frequency of these events is illustrated in Figure 18.47(C). Here, the cumulative distribution of intervals between events is plotted, which tells us how many of the total number of observed events were spaced by a time less than t. This cumulative distribution is fit to the equation $n = N(1 - e^{-t/\tau})$, where $N = 81$ is the total number of events for the experiment shown here. This functional form is precisely what is expected for a process characterized by the waiting time distribution $p(t) = e^{-t/\tau}$, as shown in the problems at the end of the chapter. Fitting the data gives the value of τ, the average time elapsed between events, which is of the order of 30 s for a frog rod cell. A simpler way to estimate this time scale is to examine the current traces in Figure 18.47(A), count the total number of events and the total elapsed time t_{tot}, and then compute $\tau \approx t_{tot}/N$. The large frog cells have about 2×10^9 copies of rhodopsin, so each individual rhodopsin undergoes a thermal or accidental isomerization event only about once every two thousand years.

Although the experiment shown above demonstrates that a single rod cell can have a measurable electrical response to the isomerization of a single retinal, it would not be useful for our brains to perceive a single isomerization event as important visual information. A human rod cell has about 10^8 rhodopsins, so (assuming that each retinal undergoes thermal isomerization about as frequently in the human as in the frog) each rod cell will have an accidental electrical event that could in principle be perceived as a single photon about once every 600 s (5 minutes). This does not seem too bad until we also consider that the human retina has about 90 million rod cells. If our brains noticed all of these events, then we would perceive complete darkness as a continuous sparkly field, with one rod cell going off inadvertently every few microseconds. In order to avoid this scenario, the human visual system uses several layers of information integration to ensure that information about light is only communicated to

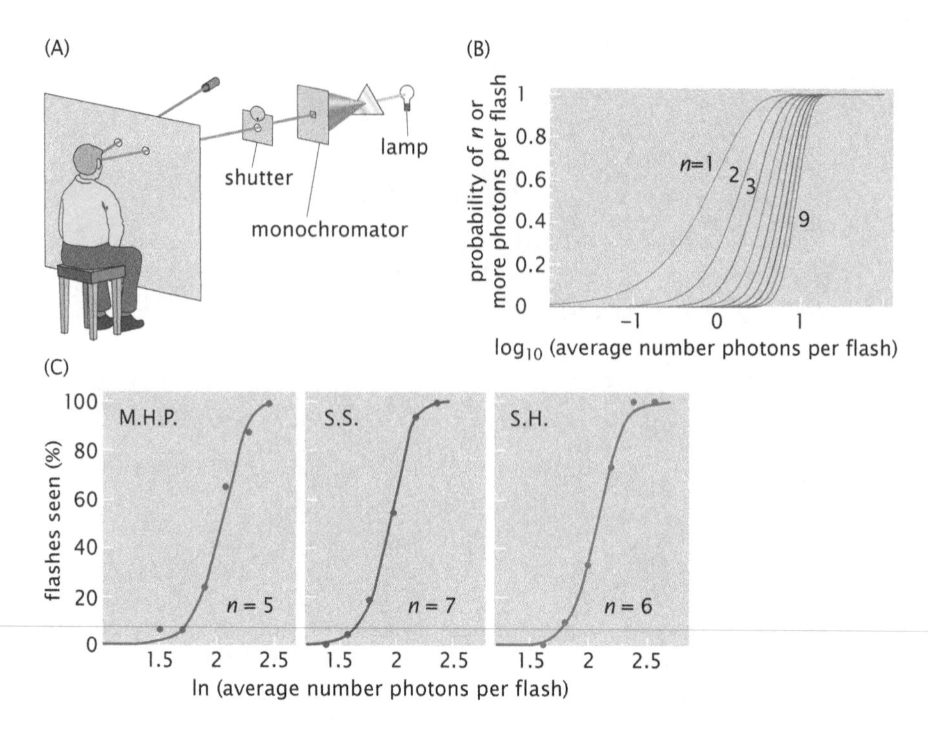

Figure 18.48: Limit of light detection in humans. (A) Optical system for measuring minimum energies necessary for vision. The dark-adapted observer sits behind an opaque screen with two small holes. Through the hole on the left, the observer focuses his attention on a small red light. On the right, light from a bright white lamp is passed through a monochromator to isolate a single wavelength, and then through a high-speed shutter that opens for only 1/1000 of a second. Forcing the observer to view the flash at a slight angle caused by looking at the red light directly ensures that the light will fall upon the most sensitive (rod-dense) part of the retina. The observer sits patiently for 300 test flashes of different intensities, and reports for each one whether or not he is able to see it. (B) Poisson probability distributions. For any average number of quanta ($h\nu$) per flash, the ordinates give the probabilities that the flash will deliver to the retina n or more quanta, depending on the value assumed for n. (C) Relation between the average energy content of a flash of light (in number of photons) and the frequency with which it is seen by three observers. Observer M. H. P. can detect a flash with 5 photons, S. S. needs 7, and S. H. splits the difference, requiring 6. The classic paper from which these results were reproduced had three authors with, curiously, these same initials. (Adapted from S. Hecht et al., *J. Gen. Physiol.* 25:819, 1942)

the brain if several physically nearby rods depolarize at nearly the same time, an event that should be characteristic of the perception of a true light stimulus. A clever experiment exploited the power of Poisson statistics (see The Math Behind the Models at the end of this section) to count the number of photons that can actually be perceived by a human being. The investigators, using themselves as subjects, sat for a long time in a dark room to render their eyes maximally sensitive. They then used a calibrated light source to generate flashes of photons in defined numbers and asked the observer whether or not the light had been flashed. The experimental setup is schematized in Figure 18.48(A).

As this experiment was repeated over and over again, the actual true number of photons generated by each flash would not be precisely identical each time. For example, a flash expected to produce 100 photons would sometimes produce 98 and sometimes 104. If we assume that there is a threshold of the number of photons that need to hit the retina to be perceived as light input, then we can use Poisson statistics to compute the likelihood that a pulse of a given strength will actually comprise at least as many photons as are necessary to cross that threshold. A family of curves illustrating this is shown in Figure 18.48(B). Note that as the number of photons in the assumed threshold increases from 1 to 10, the slope of the sigmoidal curve on the semilog plot becomes steeper and steeper. The experimental result for three different subjects is shown in Figure 18.48(C). By simply graphing the probability that a subject will be able to perceive a light flash as a function of the intensity of the light flash and then fitting the Poisson curves to that data, the threshold number of photons for each person could be estimated.

The actual minimal number of photons that is required for perception of a dim flash was thus measured to be around 6. However, the optical limitations of the human lens dictate that those 6 photons would be spread out over 100 μm^2 of retina, equivalent to the area covered by about 150 rod cells. Thus, it is highly unlikely that any two photons were actually detected within the same individual rod cell. As we have seen above, rod cells in the dark are subject to spontaneous small currents corresponding to thermal isomerization events, which are quantitatively indistinguishable from single-photon events. Thus, human perception of very dim light must require a coincidence detector that will pass the signal from the retina on to the brain if and only if several rods cells that are physically close to one another in the retina experience hyperpolarization at the same time, which would tend to indicate a true light signal rather than a thermal accident. Indeed, particular neurons in the retina called bipolar cells and horizontal cells seem to be capable of performing this integration function, receiving inputs from several nearby rod cells simultaneously. The neurons of the retina are capable of several other layers of rudimentary image processing as well, detecting edges and directed motion, so that visual signals will be passed on to the brain only if they are deemed by the retinal processing neurons to be truly worthy of notice.

The Math Behind the Models: The Poisson Distribution

MATH

Probabilistic reasoning is one of the centerpieces of modern science. Much of our understanding of how probability questions will play out is contained in a small set of hallowed probability distributions such as the binomial distribution, the Gaussian distribution, the exponential distribution, and the Poisson distribution. We have already seen in some of the problems of interest to us, such as the spatial extent of polymers, that the Gaussian distribution is a convenient continuous representation of the information contained in the binomial distribution. Interestingly, the binomial distribution can be used as the jumping-off point for deriving the Poisson distribution as well.

For concreteness, we imagine discretizing time into a series of steps each of length Δt and that, in every instant, one of two events can occur (such as for a molecule to decay or not or the arrival of a photon or not), with the probability of nothing happening (that is, no decay or no photon arrival) being overwhelmingly more likely. The probability that in these N time steps that we will have n photons is given once again by the binomial distribution

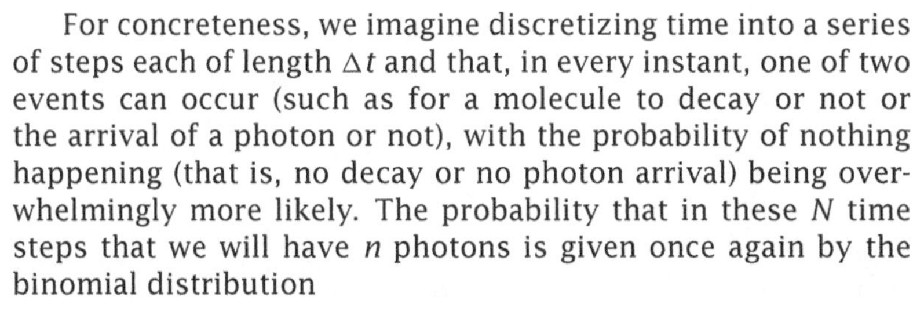

$$p(n, N) = \frac{N!}{(N-n)!n!} p^n q^{N-n}, \qquad (18.78)$$

but now with the proviso that $p \ll q$. If we exploit the fact that $N \gg n$, we can use our familiar approximation that

$$\frac{N!}{(N-n)!} \approx N^n, \tag{18.79}$$

resulting in

$$p(n,N) \approx \frac{N^n}{n!} p^n (1-p)^N, \tag{18.80}$$

where we have also exploited the fact that $p+q=1$.

For the binomial distribution, recall that the average value of n is given by $\langle n \rangle = Np$, which suggests that we further rearrange the expression given above as

$$p(n,N) \approx \frac{(Np)^n}{n!} \left(1 - \frac{Np}{N}\right)^N, \tag{18.81}$$

or

$$p(n,N) \approx \frac{\langle n \rangle^n}{n!} \left(1 - \frac{\langle n \rangle}{N}\right)^N. \tag{18.82}$$

Finally, by invoking the definition of the exponential function embodied in the equation

$$\mathrm{e}^{-x} = \lim_{N\to\infty} \left(1 - \frac{x}{N}\right)^N, \tag{18.83}$$

we see that our distribution adopts the Poisson form, namely,

$$p(n) = \frac{\langle n \rangle^n \mathrm{e}^{-\langle n \rangle}}{n!} \tag{18.84}$$

The binomial and Poisson distributions are compared in Figure 18.49 for different choices of p and N. Note that because $\langle n \rangle = Np$, once we have specified N and p, we have specified $\langle n \rangle$ itself. In Figure 18.49(A), we select conditions where the binomial and Poisson distributions are not equivalent, with $N=8$ and $p=1/2$. In Figure 18.49(B), we choose conditions in which the binomial and Poisson distributions are effectively equivalent, with $N=100$ and $p=1/25$.

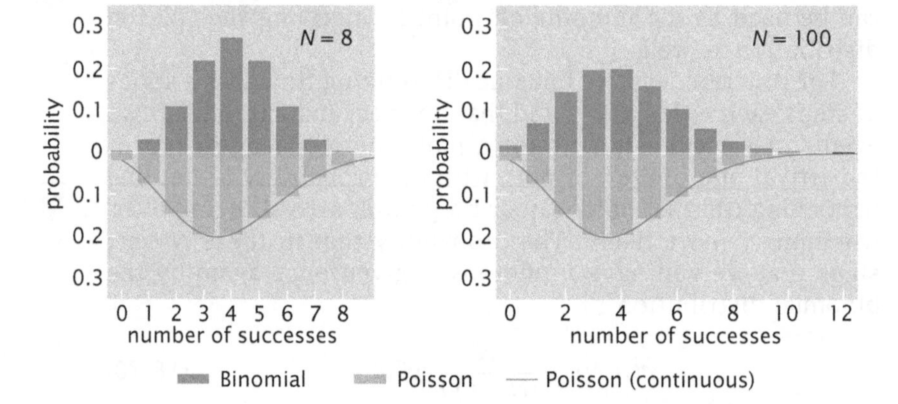

Figure 18.49: Comparison between the binomial and Poisson distributions. The two distributions are shown for the choices of parameters (A) $N=8$ and average $=4$ and (B) $N=100$, average $=4$.

18.3.4 Sex, Death, and Quantum Mechanics

Animals use the information about the world around them provided by their visual systems to find food and mates, and to avoid predators. It is not surprising, then, that living organisms have found ways to manipulate the light that bounces off them to influence the behavior of their fellow life forms. In this final section, we will touch briefly on a few of the many ways that multicellular organisms use visible light to engage in those favorite engines of evolutionary selection, fertilization and predation.

Starting with the rather more joyful process of fertilization, let us consider the birds and the bees. Monkeyflowers of the genus *Mimulus* are pollinated by small flying animals, and, like many other flowering plants, synthesize pigments in colors that serve to attract the attention of the pollinators (the pollinators are not acting altruistically; they usually get a meal of nectar or pollen in exchange for visiting the flowers). As illustrated in Figure 18.50, the pinkish *M. lewisii* monkeyflower (shown at the top left) is normally pollinated by bees, which visit it about 700 times more often than hummingbirds in the same environment. In contrast, the orangish *M. cardinalis* (shown at the bottom left) is visited about 1300 times more frequently by hummingbirds than by bees. A surprisingly definitive experiment has established that color alone has a significant influence on the flower preference of the pollinators. The production of yellow carotenoid pigments (similar in structure to the β-carotene illustrated in Figure 18.13) is determined in this flower genus by the expression of an enzyme encoded at a single locus, called *YUP*. Replacement of the *M. lewisii YUP* allele with that from *M. cardinalis* resulted in a somewhat more orangish flower, which became six times less attractive to bees, but 70 times more attractive to hummingbirds. Conversely, replacement of the *M. cardinalis YUP* allele with that from *M. lewisii* made it pinker, and, while not much changing its appeal to hummingbirds, made it 70 times more attractive to bees. This simple experiment illustrates that very small genetic changes can have a profound effect on the way that plants influence animal behavior, and possibly on their larger ecological niches.

Many other cases of pigment changes affecting fitness are known, including examples on the death side of the "sex and death" axis, where pigmentation enables camouflage and evasion of predators. In a particularly well-characterized case, the tiny (and apparently delicious) rock pocket mice of the American Southwest can be found in a range of colors from beige to nearly black. The darkness of the fur is largely determined by the amount of synthesis of the dark pigment melanin, which also is the origin for much of the variation in hair and skin color among humans. The light-colored mice are closely matched to the coloration of the sand in their desert environment, where they are subject to predation by carnivorous birds. About 1.7 million years ago, volcanic eruptions introduced large swaths of dark lava rock into the range of the rock pocket mice. In the present day, the black rocks are inhabited by dark mice and the sandy surroundings by light mice. Careful genetic mapping has shown that the dark color has arisen multiple times independently in the recent history of this species, and is apparently under strong selection to promote local adaptation such that mice closely match the color of their local environments. Although the exact rate of predation by owls for light and dark mice on different ground types has not yet been measured, this is clearly the most likely mechanism for such strong and rapid effects.

Figure 18.50: Allele substitution at a color locus alters pollinator behavior. (A) Wild-type *M. lewisii*. (B) *M. lewisii* with the yellow pigment-producing *YUP* allele from *M. cardinalis*. (C) Wild-type *M. cardinalis*. (D) *M. cardinalis* with the inactivated *YUP* allele from *M. lewisii*. The relative preferences of the two pollinators, bees and hummingbirds, for each flower are indicated below. A visit by a black bear to the observation area in California reduced the size of the data set. (Adapted from H. D. Bradshaw Jr and D. W. Schemske, *Nature* 426:176, 2003.)

The two preceding examples have involved cases where pigment generation by living creatures (flowers and mice) affects the behavior of other species (bees, hummingbirds, and owls). Of course, there are also many situations where the manipulation of visible light by living organisms is primarily targeted at conspecifics, members of the same species, for purposes of attracting mates and intimidating rivals. Often, synthesis of chemical pigments is manipulated for these purposes, but in some cases particularly vivid colors are generated by a completely different mechanism. Chemical pigments alter the appearance of visible light by selective absorption of certain wavelengths, but the appearance of color to observers can also be manipulated by selective scattering. The brilliant, iridescent color of the Blue Morpho butterfly (shown in Figure 18.51), for example, is not caused by the presence of a light-absorbing pigment, but rather by the structural properties of the material that makes up the scales covering the butterfly's wings. The scales of the butterfly are largely constructed of chitin, a long-chain polysaccharide made of repeating units of the sugar *N*-acetylglucosamine, embedded in a protein matrix. A cross-section through the scale, shown in Figure 18.51(C), reveals the regular, periodic arrangement of matrix material, generating alternating layers of material with different dielectric constants (air versus scale matrix), alternating at a length scale comparable to the wavelength of visible light. When made by humans, these kinds of structures are called "photonic crystals," and are prized for their ability to affect the propagation of light (or, more generally, electromagnetic waves) with wavelength selectivity. Humans have achieved expertise in fabricating photonic crystals only within the past 30 years, but animals have excelled at this for over 500 million years. Indeed, the trick of generating iridescent surfaces by regular arrangement of biological structures has apparently been developed many times independently, as photonic structures of myriad geometrical forms made of an equally wide variety of materials are found throughout all the major phyla of animals. Besides iridescent butterfly scales and bird feathers, the silvery reflections seen on the flanks of many fish arise from a similar physical principle, embodied in the assembly of crystals made from guanine.

Simple physical arguments shown schematically in Figures 18.51(D) and (E) explain how such structures work. As already described in Figure 18.39, constructive interference can be thought of as emerging when the phases of different light paths are roughly the same. One way to think of phase is by considering the total travel time of light in traversing a path from one point to another. Figure 18.51(D) shows the travel time for a number of different light paths from source to detector. Constructive interference occurs whenever two different

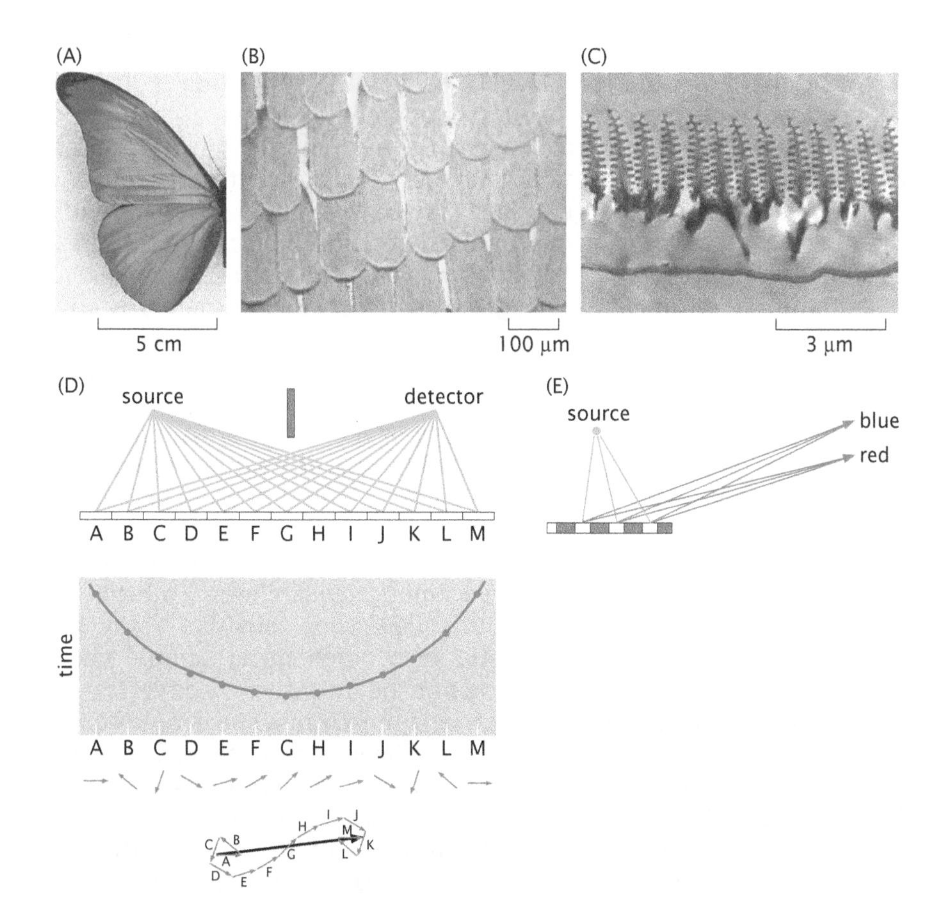

Figure 18.51: Coloration of the Blue Morpho butterfly. (A) The large butterfly *Morpho rhetenor* is brilliantly blue. (B) The blue color is generated by a layer of ground scales that contain no blue pigment molecules. (C) A cross-section through a ground scale, showing the regular periodicity of the matrix structure. (D) Light traveling from a source (for example, the sun) to an observer (the human eye) is reflected from a surface (the butterfly's scale). The arrows represent the phase of each reflected beam of light, as illustrated in Figure 18.39. Segments E through I correspond to the lowest travel time for the light and make the dominant contribution to the final arrow's length. (E) For a surface with a periodic alternation of reflective (white) and non-reflective (shaded) regions, constructive interference depends on the wavelength of light as well as on the angle. (A, C, adapted from P. Vukusic and J. R. Sambles, *Nature* 424:852, 2003; B, adapted from P. Vukusic et al., *Proc. R. Soc. Lond. B* 266:1403, 1999; D, E, adapted from R. P. Feynman, QED: The Strange Theory of Light and Matter. Princeton University Press, 1988.)

paths correspond to travel times that differ by an integer multiple of the time it takes for the phase "clock" to make one complete turn. As shown in Figure 18.51(E), grating structures are built by constructing structures such that only those paths that have the correct phase relations are permitted.

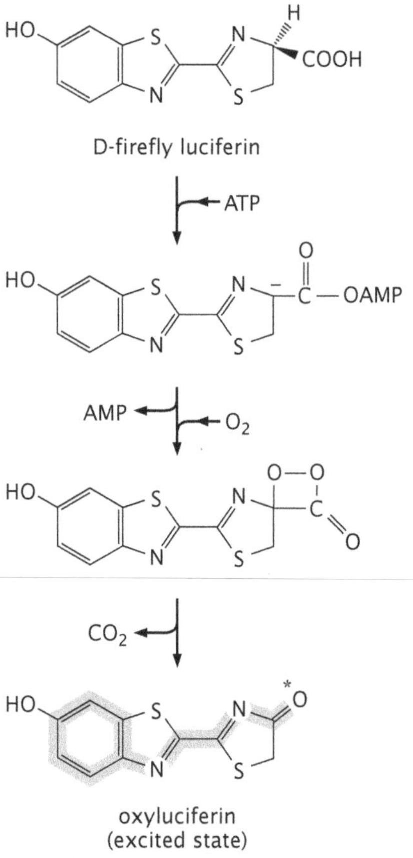

D-firefly luciferin

oxyluciferin
(excited state)

Figure 18.52: Light-production scheme for firefly luciferase. The conjugated double bond system in the excited state of the product oxyluciferin is highlighted at the bottom. The shaded regions in the molecule show the conjugated double-bond systems that are able to accommodate the excited pi electrons. Relaxation of the excited state of this molecule back to the ground state results in emission of a visible photon.

Let There Be Light: Chemical Reactions Can Be Used to Make Light

Using pigments and photonic crystals, organisms manipulate the photons arising from sunlight as a method for communicating or withholding information for animals with a visual system. In some cases, though, organisms actually generate their own light, typically also for purposes related to sex and death. Throughout this chapter, we have argued that the electromagnetic energy carried by photons can be converted into chemical energy, either resulting in electron separation as in the context of photosynthesis or leading to protein conformational changes eventually influencing transmembrane transport of ions as in vision. Conversely, organisms that produce bioluminescence demonstrate that this energy conversion can go both ways, such that biological chemical energy can produce photons. The most familiar terrestrial organisms that perform this feat are fireflies and glow worms, and many kinds of mushrooms also glow in the dark. Under the sea, bioluminescence is much more common, and is found among individuals in nearly all of the branches of the tree of life.

In order to produce bioluminescence, an organism must be able to use an energy-requiring chemical reaction to generate an excited state of an organic compound, in a configuration such that relaxation of the excited state back to a ground state will release a visible-light photon (that is, one carrying 2–3 eV of energy). The chemical reactions are catalyzed by enzymes called luciferases, which typically consume chemical energy in the form of ATP as well as converting molecular oxygen to carbon dioxide. The luciferase enzymes use a chemically complex substrate, luciferin, to create the light-generating excited state. Among light-producing organisms, both the luciferases and the luciferins are highly divergent, and may have evolved independently dozens of times. One scheme for light generation, used by fireflies, is shown in Figure 18.52. Adenylation of firefly luciferase, followed by oxidation with molecular oxygen, results in an extremely unstable high-energy intermediate compound with a four-membered cyclic peroxide ring. This molecule is quickly decarboxylated, releasing carbon dioxide and generating the excited product, oxyluciferin, which has effectively captured the chemical energy of the ATP and the molecular oxygen in its electronic excitation. The size of the conjugated double-bond system in oxyluciferin is similar to the size of the conjugated systems in the fluorescent pigments associated with photosynthesis (see Figure 18.13), so its relaxation generates a visible-light photon.

The luciferase reaction is beloved among researchers, both for its intrinsic elegance and for its utility. One very sensitive assay for measuring the concentration of ATP in a biochemical sample takes advantage of the fact that this reaction quantitatively converts ATP into visible-light photons which can be counted in a spectrophotometer. Important biological principles have been serendipitously uncovered due to the study of this mechanism. An interesting case involves the adorable little Hawaiian bobtail squid, *Euprymna scolopes*, which does not produce bioluminescence on its own but instead cultivates small monocultures of the bioluminescent bacterium *Vibrio fischeri* in specialized pockets on its mantle. The bacteria generate light only when they are present at a sufficiently high concentration; individual bacteria swimming free in seawater do not generate light, but the aggregated colony living on the squid do. In laboratory culture, *Vibrio fischeri* can also be induced to glow only by close proximity to a large number of neighbors. Fractionation experiments had suggested that a diffusible substance secreted by the bacteria was detected by the neighbors in a "quorum sensing" mechanism, and use of light emission by live bacteria as an assay enabled purification of the specific molecules that bacteria use to detect their own population density.

The squid family demonstrates other particularly interesting and creative uses of bioluminescence. The deep-sea cephalopod *Vampyroteuthis infernalis*, whose name means "vampire squid from hell," does not squirt out ink to confuse and obscure the vision of its predators the way that many other squid and octopus species tend to do. Instead, an alarmed *Vampyroteuthis* will eject a viscous fluid filled with glowing particles that presumably dazzles the predators long enough for the squid to swim away. Squirting out light must be much more effective than squirting out black ink in the deepest parts of the ocean where sunlight barely penetrates.

18.4 Summary and Conclusions

Light is the giver of life on Earth. This chapter focused on two key ways in which light interacts with living organisms. First, we examined how light powers living organisms through the process of photosynthesis. We go beyond previous chapters by expanding our theoretical toolkit to include quantum mechanics, which takes center stage in our study of light and its interaction with molecules such as chlorophyll. In particular, we used simple quantum mechanical models to explore the absorption of light and the transfer of energy and electrons during the process of photosynthesis. The second key thrust of the chapter explored the fascinating topic of how a variety of different organisms use light to perceive the world around them. The study of vision brings together many of the themes that have emerged throughout the book with concepts ranging from ligand-gated ion channels to the statistics of noisy processes.

18.5 Appendix: Simple Model of Electron Tunneling

In this appendix, we present the details of a toy model of electron tunneling in a "molecule" made up of the two boxes shown in Figure 18.18. To solve this problem, we resort to the same sort of variational calculations that we have used repeatedly throughout the book. Our philosophy with these calculations is almost always the same. There is some hard problem that we don't know how to (or want to) solve. Instead of facing the difficulties head on, we introduce a "trial solution" (effectively an intelligently designed guess) that depends upon some parameters and then we seek the choice of those parameters that minimizes the error in our guess. We have already solved the problem of the electron in a single finite potential well and will label the resulting wave functions $|1\rangle$ and $|2\rangle$ for the wells labeled 1 and 2, respectively. In Figure 18.18, we labeled our wells A and B, but here we favor the labels 1 and 2 since in this case the notational bookkeeping is much easier. Given these definitions, the total wave function can be written as

$$|\psi\rangle = a_1|1\rangle + a_2|2\rangle. \tag{18.85}$$

The essence of this approximation is that the wave function can be written as a linear combination of the "basis vectors" made up of the solutions to the problems of isolated wells in the two positions as shown in Figure 18.18(B). Our goal is to find the choice of a_1 and a_2 that minimizes the energy

$$E = \frac{\langle\psi|\hat{H}|\psi\rangle}{\langle\psi|\psi\rangle}. \tag{18.86}$$

Note that this section is in many ways the quantum mechanical inheritance of the ideas developed on two-level systems in Chapter 7.

To make further progress, we first need to examine the significance of the Hamiltonian in this case. In particular, we have

$$\hat{H} = \underbrace{-\frac{\hbar^2}{2m}\nabla^2}_{\text{kinetic energy}} - \underbrace{V_1(x-x_1) - V_2(x-x_2)}_{\text{potential energy in wells}} \tag{18.87}$$

where V_n is the potential centered on well n ($n = 1, 2$). We now note that we can write the quantities in Equation 18.86 as

$$\langle\psi|\hat{H}|\psi\rangle = a_1^* a_1 H_{11} + a_1^* a_2 H_{12} + a_2^* a_1 H_{21} + a_2^* a_2 H_{22}, \tag{18.88}$$

where we have introduced the notation $\langle i|\hat{H}|j\rangle = H_{ij}$, and

$$\langle\psi|\psi\rangle = a_1^* a_1 + a_2^* a_2, \tag{18.89}$$

where we have invoked the orthogonality approximation $\langle i|j\rangle = \delta_{ij}$. Given these definitions, we can now write the expression for the energy as

$$E(a_1, a_2) = \frac{a_1^* a_1 H_{11} + a_1^* a_2 H_{12} + a_2^* a_1 H_{21} + a_2^* a_2 H_{22}}{a_1^* a_1 + a_2^* a_2} \tag{18.90}$$

We are interested in minimizing this energy with respect to the unknown parameters a_1 and a_2. To do so, we invoke the conditions $\mathrm{d}E/\mathrm{d}a_1^* = 0$ and $\mathrm{d}E/\mathrm{d}a_2^* = 0$. To be explicit, let's consider one of these two derivatives. Using the quotient rule, we have

$$\frac{\mathrm{d}E(a_1, a_2)}{\mathrm{d}a_1^*} = \frac{H_{11}a_1 + H_{12}a_2}{a_1^* a_1 + a_2^* a_2} - \frac{a_1^* a_1 H_{11} + a_1^* a_2 H_{12} + a_2^* a_1 H_{21} + a_2^* a_2 H_{22}}{(a_1^* a_1 + a_2^* a_2)^2} a_1. \tag{18.91}$$

Note that, using Equation 18.90, we can rewrite this as

$$\frac{\mathrm{d}E(a_1, a_2)}{\mathrm{d}a_1^*} = \frac{H_{11}a_1 + H_{12}a_2}{a_1^* a_1 + a_2^* a_2} - E\frac{a_1}{a_1^* a_1 + a_2^* a_2} = 0. \tag{18.92}$$

The calculation for $\mathrm{d}E/\mathrm{d}a_2^* = 0$ is nearly identical (and all of the algebraic steps are identical). What we have learned is that the act of minimizing the expectation value of the energy with respect to the unknown coefficients a_1 and a_2 is equivalent to solving the two linear equations

$$H_{11}a_1 - Ea_1 + H_{12}a_2 = 0, \tag{18.93}$$

$$H_{21}a_1 + H_{22}a_2 - Ea_2 = 0. \tag{18.94}$$

These equations are conveniently summarized in matrix notation as

$$\begin{pmatrix} H_{11} - E & H_{12} \\ H_{21} & H_{22} - E \end{pmatrix} \begin{pmatrix} a_1 \\ a_2 \end{pmatrix} = 0 \tag{18.95}$$

The condition for the existence of a solution is that the determinant of the matrix in this equation in zero, that is, $\det(\mathbf{H} - \mathbf{I}E) = 0$, and results in

$$\det\begin{pmatrix} E_1 - E & H \\ H & E_2 - E \end{pmatrix} = 0\,, \tag{18.96}$$

where we have adopted the notation $H_{11} = E_1$, $H_{22} = E_2$, and $H_{12} = H_{21} = H$ (the significance of these matrix elements will be explored further when we look at the tunneling rate). Multiplying out the determinant results in a quadratic equation for the energies of the form

$$(E_1 - E)(E_2 - E) - H^2 = 0. \tag{18.97}$$

The resulting eigenvalues obtained by solving this quadratic equation are

$$E_{\pm} = \frac{E_1 + E_2}{2} \pm \frac{1}{2}\sqrt{(E_1 - E_2)^2 + 4H^2}. \tag{18.98}$$

With the energy eigenvalues in hand, we are almost ready to compute the time evolution of the system. However, to do so, we also need to work out the eigenfunctions and their corresponding time evolution. Our starting point is to plug the energy eigenvalues back into the original matrix equation and to solve for the coefficients a_1 and a_2. In particular, we have

$$\begin{pmatrix} E_1 - E_{\pm} & H \\ H & E_2 - E_{\pm} \end{pmatrix} \begin{pmatrix} a_1 \\ a_2 \end{pmatrix} = 0. \tag{18.99}$$

If we expand this and solve for a_1 and a_2 with the added proviso that the eigenvectors are normalized (that is, $a_1^2 + a_2^2 = 1$), we find

$$\begin{aligned} |+\rangle &= \frac{1}{\sqrt{H^2 + (-\varepsilon + \sqrt{\varepsilon^2 + H^2})^2}} \begin{pmatrix} H \\ -\varepsilon + \sqrt{\varepsilon^2 + H^2} \end{pmatrix} \\ &= \frac{1}{\sqrt{H^2 + (-\varepsilon + \sqrt{\varepsilon^2 + H^2})^2}} [H|1\rangle + (-\varepsilon + \sqrt{\varepsilon^2 + H^2}|2\rangle)] \end{aligned} \tag{18.100}$$

and

$$\begin{aligned} |-\rangle &= \frac{1}{\sqrt{H^2 + (-\varepsilon - \sqrt{\varepsilon^2 + H^2})^2}} \begin{pmatrix} H \\ -\varepsilon - \sqrt{\varepsilon^2 + H^2}, \end{pmatrix} \\ &= \frac{1}{\sqrt{H^2 + (-\varepsilon - \sqrt{\varepsilon^2 + H^2})^2}} [H|1\rangle + (-\varepsilon - \sqrt{\varepsilon^2 + H^2}|2\rangle)], \end{aligned} \tag{18.101}$$

where the + and − signs correspond to the plus and minus eigenvalues, respectively, and where we have introduced the quantity $\varepsilon = (E_1 - E_2)/2$.

Recall that we are interested in determining how the probability of finding the electron in the second well depends upon time. We denote this quantity as $P_{1\to 2}(t)$ and note that, using the rules of quantum mechanics, it can be written as

$$P_{1\to 2}(t) = |\langle 2|1, t\rangle|^2. \tag{18.102}$$

To obtain this quantity, we need to compute both $|2\rangle$ and $|1, t\rangle$ in terms of the eigenvectors of the Hamiltonian. By analogy to the way that we write a vector in the plane in terms of the basis vectors **i** and **j**, we can write our state vectors as linear combinations of the basis vectors as

$$|1\rangle = |+\rangle\langle +|1\rangle + |-\rangle\langle -|1\rangle. \tag{18.103}$$

As described in more detail in the Tricks Behind the Math at the end of this appendix, the way we obtain this probability is by computing the amplitude by taking what amounts to an analog of the dot product familiar from vector algebra. More to the point, we need to examine the time evolution of this wave function. To do so, we note that the time dependence of the state $|1\rangle$ inherits the time dependence of the eigenfunctions $|+\rangle$ and $|-\rangle$. This means that the time evolution of the 1 state is of the form

$$|1, t\rangle = e^{iE_+ t/\hbar}|+\rangle\langle +|1\rangle + e^{iE_- t/\hbar}|-\rangle\langle -|1\rangle \tag{18.104}$$

We recall that the basis vectors correspond to the wave functions centered on the two wells and are given by

$$|1\rangle = \begin{pmatrix} 1 \\ 0 \end{pmatrix} \tag{18.105}$$

and

$$|2\rangle = \begin{pmatrix} 0 \\ 1 \end{pmatrix}. \tag{18.106}$$

We can use these definitions to compute $\langle 2|1, t\rangle$, which is the key quantity needed to assess the transfer rate over time and is given by

$$\langle 2|1, t\rangle = \langle 2|(e^{iE_+ t/\hbar}|+\rangle\langle +|1\rangle + e^{iE_- t/\hbar}|-\rangle\langle -|1\rangle). \tag{18.107}$$

The coefficients in this expansion are given by

$$\langle +|1\rangle = \frac{H}{\sqrt{H^2 + (-\varepsilon + \sqrt{\varepsilon^2 + H^2})^2}}, \tag{18.108}$$

$$\langle -|1\rangle = \frac{H}{\sqrt{H^2 + (-\varepsilon - \sqrt{\varepsilon^2 + H^2})^2}}, \tag{18.109}$$

$$\langle 2|+\rangle = \frac{-\varepsilon + \sqrt{\varepsilon^2 + H^2}}{\sqrt{H^2 + (-\varepsilon + \sqrt{\varepsilon^2 + H^2})^2}}, \tag{18.110}$$

and

$$\langle 2|-\rangle = \frac{-\varepsilon - \sqrt{\varepsilon^2 + H^2}}{\sqrt{H^2 + (-\varepsilon - \sqrt{\varepsilon^2 + H^2})^2}}, \tag{18.111}$$

as can be seen by appealing to the eigenvectors themselves and by working through the resulting dot products. We can now rewrite the quantity $\langle 2|1,t\rangle$ as

$$\langle 2|1,t\rangle = c_{+}\mathrm{e}^{\mathrm{i}E_{+}t/\hbar} + c_{-}\mathrm{e}^{\mathrm{i}E_{-}t/\hbar}, \tag{18.112}$$

where we have defined the parameters

$$c_{+} = \frac{H(-\varepsilon + \sqrt{\varepsilon^2 + H^2})}{H^2 + (-\varepsilon + \sqrt{\varepsilon^2 + H^2})^2} \tag{18.113}$$

and

$$c_{-} = \frac{H(-\varepsilon - \sqrt{\varepsilon^2 + H^2})}{H^2 + (-\varepsilon - \sqrt{\varepsilon^2 + H^2})^2}, \tag{18.114}$$

for notational compactness. Computing the absolute square, we find

$$|\langle 2|1,t\rangle|^2 = c_{+}^2 + c_{-}^2 + 2c_{+}c_{-}\cos\frac{(E_{+} - E_{-})t}{\hbar}. \tag{18.115}$$

Using the definitions of c_{+} and c_{-}, a little algebra reveals that $c_{+} + c_{-} = 0$. This can now be used to rewrite $|\langle 2|1,t\rangle|^2$ as

$$|\langle 2|1,t\rangle|^2 = -2c_{+}c_{-}\left(1 - \cos\frac{2\sqrt{\varepsilon^2 + H^2}\,t}{\hbar}\right). \tag{18.116}$$

Trigonometry then tells us that this can be further simplified as

$$|\langle 2|1,t\rangle|^2 = 4c_{+}^2\sin^2\frac{\sqrt{\varepsilon^2 + H^2}\,t}{\hbar} \tag{18.117}$$

As written, our result tells us that the electron goes from state 1 to state 2 with a characteristic time scale of $h/\sqrt{\varepsilon^2 + H^2}$, but that, over time, it will return from whence it came, cycling back and forth between the two states. We cannot really call that electron transfer, since the word "transfer" conjures up images of permanence that are not present in this result. It is also important to recognize that this result depends upon the assumption that the energies of the two wells 1 and 2 are essentially identical; Figure 18.53 shows that the probability of this occurring becomes much less likely as the energies E_1 and E_2 become more different.

We can further simplify Equation 18.117 by introducing the variable $\Delta = \varepsilon/h$. If we make this simplification, then we have

$$|\langle 2|1,t\rangle|^2 = 4\frac{(-\Delta + \sqrt{1 + \Delta^2})^2}{[1 + (-\Delta + \sqrt{1 + \Delta^2})^2]^2}\sin^2\frac{\sqrt{1 + \Delta^2}Ht}{\hbar}. \tag{18.118}$$

If we now invoke the approximation that $\Delta \gg 1$,which states that the difference between the energies in each well is large compared with the coupling matrix element H, this simplifies to

$$|\langle 2|1,t\rangle|^2 \approx \frac{1}{\Delta^2}\sin^2\frac{\Delta Ht}{\hbar} = \frac{H^2}{\varepsilon^2}\sin^2\frac{\varepsilon t}{\hbar}, \tag{18.119}$$

where we have made use of the fact that $\sqrt{1 + \Delta^2} \approx \pm\Delta$ for $\Delta \gg 1$.

The irreversible transfer of the electron from 1 to 2 requires one further piece of physics, namely, the idea that there is a spectrum of possible states from and to which the electron can tunnel as a

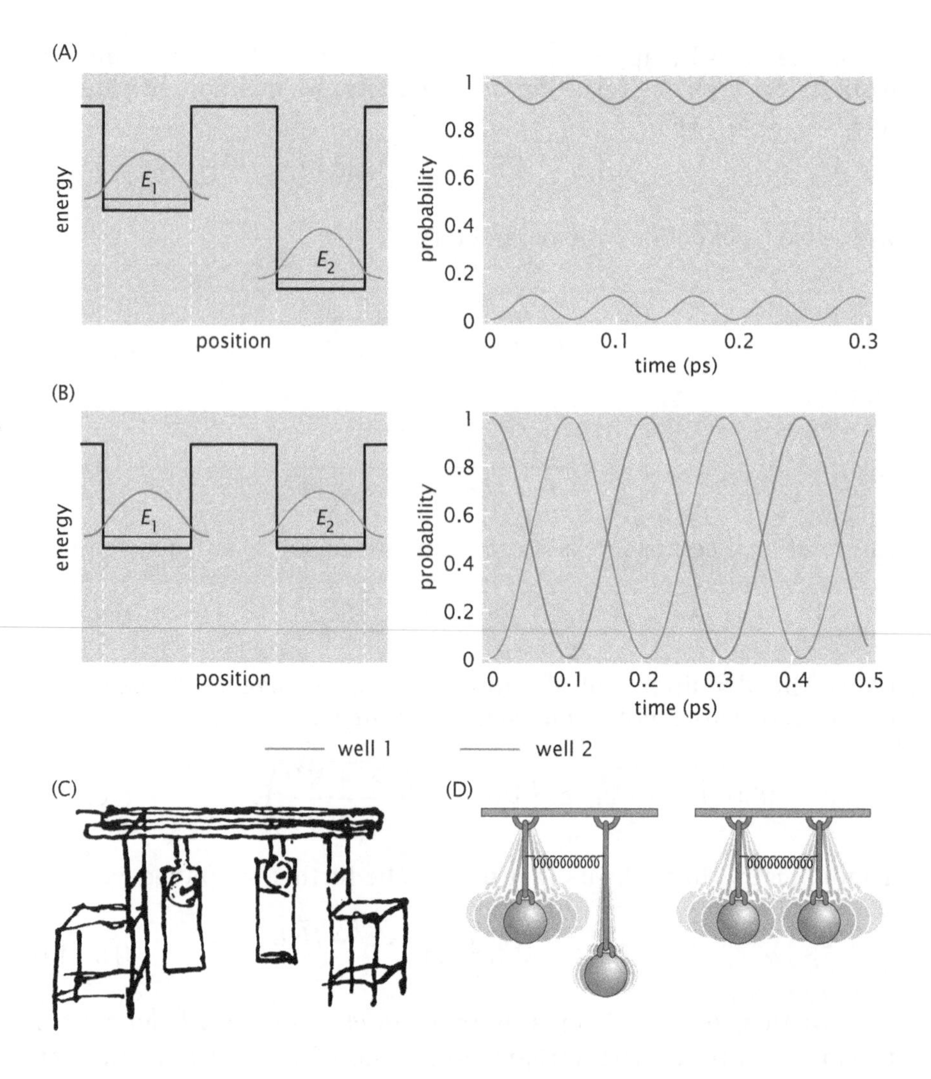

Figure 18.53: Tunneling probabilities for the two-well "molecule." (A) Case in which the energies in the two different wells are different. In this case, the probability of transferring the electron from well 1 to well 2 is low. (B) Case in which the energies in the two wells are matched. (C) Drawing from Christiaan Huygens showing the mechanical coupling of two pendula. (D) When the two pendula have similar frequencies, there is strong coupling. (C, adapted from Christiaan Huygens, Oeuvres complètes. Tome XVII. L'horloge à pendule 1656–1666 (ed. J. A. Vollgraff). Martinus Nijhoff, 1932.)

result of the vibrational broadening of the spectrum of allowed states. These states result from the fact that the molecules are vibrating and the quantum mechanical oscillations that describe these vibrations result in a series of equally spaced energies such as are shown in Figure 18.54. In the case in which we consider a spectrum of final states (for simplicity, ignoring the fact that the initial states are similarly broadened), the total transition probability is given by

$$P_{1\to \mathrm{f}}(t) = \sum_{n=0}^{N} |\langle 2_n|1, t\rangle|^2, \tag{18.120}$$

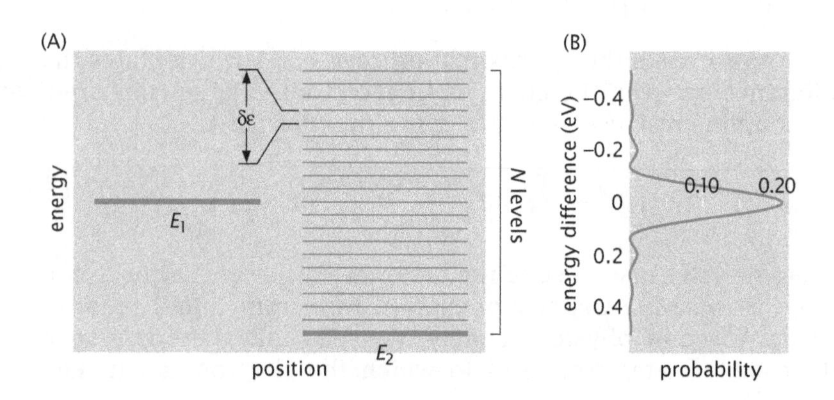

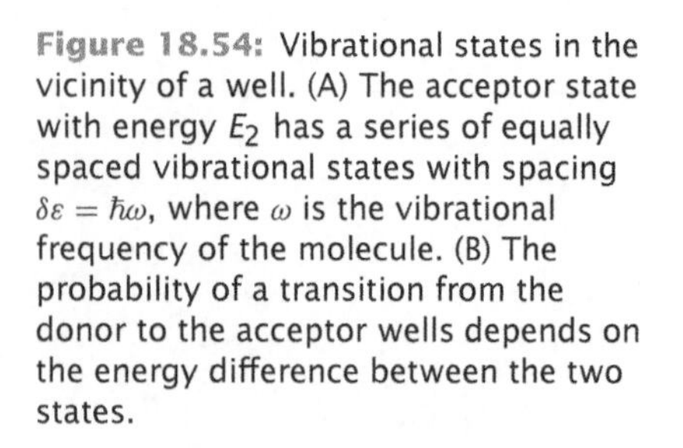

Figure 18.54: Vibrational states in the vicinity of a well. (A) The acceptor state with energy E_2 has a series of equally spaced vibrational states with spacing $\delta\varepsilon = \hbar\omega$, where ω is the vibrational frequency of the molecule. (B) The probability of a transition from the donor to the acceptor wells depends on the energy difference between the two states.

where we have now introduced the notation 2_n to describe the nth energy state in the neighborhood of the state 2 as indicated in Figure 18.54. If we assume an equally spaced ladder of such states (as one finds for quantum harmonic oscillators, for example), and using Equation 18.119, we can rewrite this expression as

$$P_{1\to f}(t) = \sum_{n=0}^{N} \frac{4H^2}{[E_1 - (E_2 + n\delta\varepsilon)]^2} \sin^2 \frac{[E_1 - (E_2 + n\delta\varepsilon)]t}{2\hbar}. \tag{18.121}$$

The sum in this expression can be converted into an integral by using the rule $\sum_{n=0}^{N} \approx (1/\delta\varepsilon) \int_0^{N\delta\varepsilon} \mathrm{d}E$, resulting in

$$P_{1\to f}(t) = \frac{4H^2}{\delta\varepsilon} \int_0^{N\delta\varepsilon} \frac{1}{[E_1 - (E_2 + E)]^2} \sin^2 \frac{[E_1 - (E_2 + E)]t}{2\hbar} \mathrm{d}E \tag{18.122}$$

This can be further simplified by making a change of variables $z = [E_1 - (E_2 + E)]t/\hbar$, resulting in the integral

$$P_{1\to f}(t) = \frac{4H^2}{\delta\varepsilon} \frac{t}{\hbar} \int_{2\varepsilon t/\hbar - N\delta\varepsilon t/\hbar}^{2\varepsilon t/\hbar} \frac{\sin^2(z/2)}{z^2} \mathrm{d}z. \tag{18.123}$$

From Figure 18.54, it is clear that the only significant amplitude comes when the energy of the donor and acceptor states are in close proximity. As a result, we can safely rewrite our integral as

$$P_{1\to f}(t) = \frac{4H^2}{\delta\varepsilon} \frac{t}{\hbar} \int_{-\infty}^{\infty} \frac{\sin^2(z/2)}{z^2} dz, \tag{18.124}$$

and then using the fact that the integral itself is $\pi/2$ results in

$$P_{1\to f}(t) = \frac{2\pi}{\hbar} |H|^2 \frac{1}{\delta\varepsilon} t. \tag{18.125}$$

For short enough time intervals, as we have seen a number of times throughout the book, we have that the probability of a transition in a short time is given by

$$\text{probability of transition} = \text{rate} \times \text{time}. \tag{18.126}$$

This means that for the electron tunneling problem considered here, the rate is given by

$$\text{rate} = \frac{2\pi}{\hbar} |H|^2 \rho(E), \tag{18.127}$$

where we use the more general notation $\rho(E) = \mathrm{d}N/\mathrm{d}E = 1/\delta\varepsilon$, known as the density of states.

In order to make contact with experiments on tunneling rates, we need to carry our calculations a little bit farther. In particular, how do the coupling matrix elements H_{fi} depend upon the distance between the two sites? To this end, we once again appeal to our toy model of a molecule and use the actual wave functions that we computed from this model to evaluate the overlap matrix elements explicitly. In particular, the relevant matrix element can be written as

$$H = \langle 1|\hat{H}|2\rangle = \int_{-\infty}^{\infty} \mathrm{d}x\, \psi_1^*(x) \left[-\frac{\hbar^2}{2m} \frac{\mathrm{d}^2}{\mathrm{d}x^2} + V_1(x) + V_2(x) \right] \psi_2(x). \tag{18.128}$$

This simplifies considerably, since $V_1(x)$ and $V_2(x)$ are both negative constants over a limited range of the x-axis and are zero elsewhere. Note that we have

$$\int_{-\infty}^{\infty} \mathrm{d}x\, \psi_1^*(x) \left[-\frac{\hbar^2}{2m} \frac{\mathrm{d}^2}{\mathrm{d}x^2} + V_2(x) \right] \psi_2(x) = E_2 \int_{-\infty}^{\infty} \mathrm{d}x\, \psi_1^*(x)\psi_2(x), \quad (18.129)$$

which we will ignore in the remainder of the discussion since it has no interesting dependence on the distance between the wells. This leads to the simplification

$$H = -V_1 \int_0^a \mathrm{d}x\, \psi_1^*(x)\psi_2(x). \quad (18.130)$$

By plugging in the functional forms for the wave functions deduced in Section 18.2.2, we find that this can be further simplified to

$$H = -V_1 \int_0^a \mathrm{d}x (C_1 \sin \kappa' x + D_1 \cos \kappa' x) A_2 \mathrm{e}^{-\kappa(d+a-x)}. \quad (18.131)$$

Recall that what we are most interested in evaluating is the distance dependence of the tunneling rate, which is inherited directly from these integrals, resulting in

$$\text{rate} = \frac{2\pi}{\hbar} V_1^2 F^2 \rho \mathrm{e}^{-2\kappa d}, \quad (18.132)$$

where F is a distance-independent numerical factor that results from evaluating the integrals that show up in the analysis of the coupling matrix element. The details of the calculations are left as an exercise for the reader at the end of the chapter.

In brief, this simple model for electron tunneling predicts that, within a given environment, the time required for an electron to tunnel from one location to another should simply increase exponentially with the physical distance between the two sites, a result already seen experimentally in Figure 18.20.

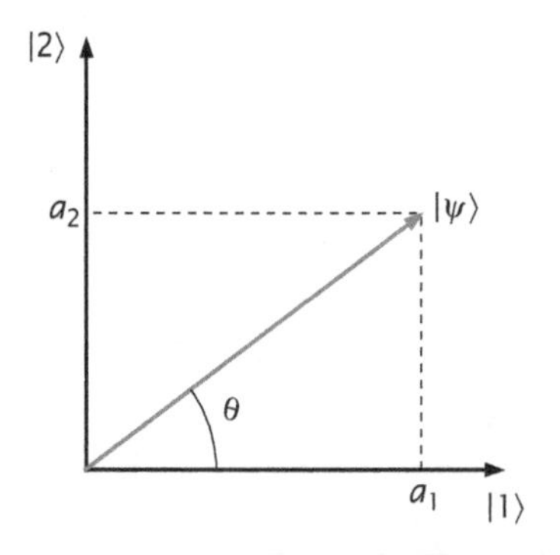

Figure 18.55: Inner product between quantum states. The quantum state vector $|\psi\rangle = a_1 |1\rangle + a_2 |2\rangle$ is a superposition of basis vectors $|1\rangle$ and $|2\rangle$, where the amplitudes a_1 and a_2 are obtained by taking the dot products, $a_1 = \langle 1|\psi\rangle = \cos\theta$ and $a_2 = \langle 2|\psi\rangle = \sin\theta$.

TRICKS

The Tricks Behind the Math: Dot Products to Find Amplitudes One of the most useful tricks when performing quantum mechanical calculations is to write a quantum state $|\psi\rangle$ as a sum over basis states $|i\rangle$, $i = 1, 2, \ldots$. For practical reasons, basis states are usually chosen as eigenstates of a particular operator, like the momentum operator or the Hamiltonian. To achieve this goal, we make use of the inner product between quantum states, which is equivalent to the more familiar dot product between vectors. In Figure 18.55, we demonstrate the mathematics of the inner product using this geometrical representation.

We begin by writing the state $|\psi\rangle$ as a linear combination of basis states $|1\rangle$ and $|2\rangle$ as

$$|\psi\rangle = a_1 |1\rangle + a_2 |2\rangle, \quad (18.133)$$

where for simplicity we have assumed that the space of quantum states is two-dimensional, as shown in Figure 18.55.

Furthermore, we assume that the basis states form an orthonormal set (which is the case in most applications), which means that they satisfy the relation $\langle i|j\rangle = \delta_{ij}$; the inner product is zero between two different basis states, and it is unity if we take the inner product of a basis state with itself. This is the property shared by the unit vectors $|1\rangle$ and $|2\rangle$ in Figure 18.55.

Our goal is to determine the amplitudes a_1 and a_2. Their physical interpretation is that $|a_1|^2$ and $|a_2|^2$ are the probabilities that the quantum particle in state $|\psi\rangle$ is either in states 1 or 2, respectively. This is like the infamous Schrödinger's cat that is equally dead or alive. In the example of the two-slit experiment described earlier, the two probabilities describe the likelihood that an electron that has reached the detector went through slit 1 or 2. To compute a_1, we take the inner product of the left- and right-hand sides of Equation 18.133 with the basis state $\langle 1|$ to obtain

$$\langle 1|\psi\rangle = a_1, \tag{18.134}$$

where we have made use of the orthonormality property of the basis states. Similarly, $a_2 = \langle 2|\psi\rangle$, and the quantum state $|\psi\rangle$ can be written as

$$|\psi\rangle = \langle 1|\psi\rangle\,|1\rangle + \langle 2|\psi\rangle\,|2\rangle\ . \tag{18.135}$$

This is a particularly useful trick when one is interested in computing the time evolution of a quantum state. Using the eigenstates of the Hamiltonian as the set of basis states simplifies this task greatly, as each of these basis states evolves simply as a phase, $e^{-iE_i t/\hbar}$, where E_i is the eigenvalue associated with eigenstate $|i\rangle$.

18.6 Problems

Key to the problem categories: • Model refinements and derivations • Estimates • Model construction

• 18.1 Energy of the hydrogen atom

The goal here is to make a crude order-of-magnitude estimate of the size of an atom by examining the "competition" between the electrostatic potential between the proton in the nucleus of the atom and the electron, on the one hand, and the kinetic energy cost of confinement, on the other. To do this, write a total energy of a toy model of a hydrogen atom as a function of the parameter a that is of the form

$$E_{\mathrm{tot}}(a) = E_{\mathrm{kinetic\ energy}}(a) + E_{\mathrm{electrostatic}}(a). \tag{18.136}$$

The kinetic-energy contribution can be estimated by using the Heisenberg uncertainty principle as was done in the chapter, and you should rederive it and explain how it scales with a. Then, write the interaction energy between the proton and the electron to obtain $E_{\mathrm{electrostatic}}(a)$. Next, minimize with respect to the unknown atomic size a and find an expression for the size of the atom as a function of key parameters such as $\hbar$, m, and e. How does your resulting expression compare with the Bohr radius?

• 18.2 Toy model of tunneling

In Figure 18.56, we show a simple model of electron tunneling through a finite barrier. Compute the transmission coefficient, the ratio between the incoming and outgoing amplitudes, for an incoming wave of energy E and a barrier of height $V > E$.

• 18.3 Approximate calculation of the energies for the finite potential well

For a deep but finite potential well, the wave functions corresponding to the lowest energy levels are well approximated by the simple, sinusoidal wave functions obtained in the case of an infinite well (see Figure 18.11). Therefore, a simple approximation for the energy spectrum of a particle in a finite well can be obtained by computing the expectation value of its Hamiltonian in the energy eigenstates for the infinite-well case. Here, we test the validity of this approximation by direct comparison with the exact energy spectrum computed in the chapter.

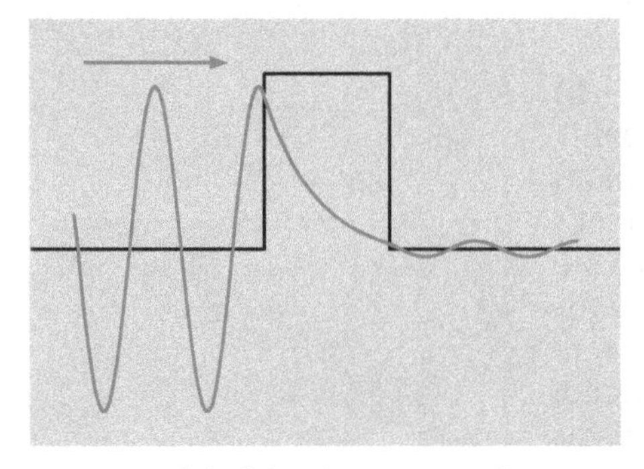

Figure 18.56: Toy model of the electron tunneling process.

(a) Compute the normalized energy eigenfunctions $\psi_n(x)$ for a particle in an infinite-well potential of width a by solving the Schrödinger equation, Equation 18.11, with the boundary condition given by Equation 18.13, and imposing the normalization condition $\int_0^a \psi_n(x)^* \psi_n(x)\,dx = 1$.

(b) Compute the approximate values of the dimensionless energies for a particle in a finite well, $\tilde{x}_n = -\tilde{E}_n/V_0$, where $\tilde{E}_n = \langle \psi_n|\hat{H}|\psi_n\rangle$, using the wave functions ψ_n from (a) and the Hamiltonian for the particle in a finite well described in the chapter. Express your result for $\tilde{x}_n$ in terms of the dimensionless parameter α defined in Equation 18.30.

(c) Following the strategy outlined in the Computational Exploration on p. 733, compute the exact values of dimensionless energies x_n for $n = 1, 2, 3, 4$ by solving Equation 18.32 for α that corresponds to a finite well with $V_0 = 10$ eV and $a = 1$ nm. Compare these values with those obtained in (b) and comment on the validity of the approximation.

• 18.4 Proton-motive force

(a) Use Equation 18.61 to make an estimate of the contribution of the H^+ concentration gradient to the proton-motive force in a cell. The typical pH difference is roughly 1 (about 7.7 inside and 7.0 outside).

(b) It takes about four protons to make one ATP from ADP. If the free energy needed is in the range 20–23 k_BT, how large must V be in order to generate an ATP assuming that the energy of the proton-motive force can be harvested with 100% efficiency? Recall that the definition of 1 eV ≈ 160 zJ $\approx 40\ k_BT$ is the energy gained by moving a charge of one electron through a potential difference of one volt. (Problem courtesy of Daniel Fisher.)

• 18.5 Electron tunneling

In this problem, the reader is asked to flesh out all of the details appearing in the appendix (Section 18.5). In particular, derive Equation 18.131 and compute what this implies about the tunneling rate.

• 18.6 Seeing the North star

Polaris has been known to generations of northern hemisphere navigators as a tool for finding latitude by simple geometrical measurements with a sextant. How much light actually reaches our eyes from a star like Polaris? Given that the luminosity of Polaris is roughly 2000 times that of the sun (1000 W/m^2 at the Earth) and that it is at a distance of 430 light years from Earth, work out the power output of Polaris, the number of photons crossing the pupil of your eye each second coming from this famous star and the mean spacing between these photons. What is the mean rate of arrival of photons to a single cone cell?
(Problem adapted from a beautiful discussion in R. W. Rodieck's 1998 book *The First Steps in Seeing* (see Further Reading), though we do not agree with all of his numbers.)

• 18.7 Eyes and the diffraction limit

(a) A point source emits an electric field that can be thought of as the real part of $\psi = (A/r)e^{i(\omega t - \mathbf{k}\cdot\mathbf{r})}$, where $\mathbf{r}$ is a vector pointing away from the source and $r = |\mathbf{r}|$. ω is the frequency and $\mathbf{k} = k\hat{\mathbf{r}}$ is the wave vector, which points in the radial direction. Notice that the amplitude A/r decreases as we move away from the source. Show that if we look at the field at a distance Z that is far away from the source, the incoming light can be thought of as a plane wave. This means that the phase is given by $\omega t - kZ$ everywhere on that plane.

(b) We now imagine an aperture as a series of tiny point sources, each serving to emit radiation as dictated by Huygens' principle. Write the field due to each point source on the plane located at a distance Z and position r on the x-axis as shown in Figure 18.39(C). Do this as a function of the position of the source within the aperture given by R and ϕ. Once again, use the approximation that the screen is very far away from the aperture in order to simplify your expressions.

(c) Add up the contribution of all point sources by integrating over the expression you obtained in (b). You will get a result in terms of Bessel functions. Calculate the intensity of the field at the screen by computing $I = \psi \times \psi^*$.

(d) For reasonable parameters, make a plot of the intensity profile on the screen using software such as Mathematica or Matlab. Find the first zero in intensity and relate it to the expression for the resolution limit given in Equation 18.69. What does this mean in terms of the human resolution limit? If you have two objects at about 1 m from you, what is the minimum separation between them such that you can still resolve them as separate?
(*Note*: More information on Bessel functions is provided in the hints on the book's website.)

• 18.8 Resolution of fossil insect eyes

Estimate the angular resolution of the insects whose fossil eyes are shown in Figure 18.57(A).

• 18.9 MWC model for signal transduction in the eye

Throughout the book, we have argued that the treatment of ligand-gated ion channels is a fascinating problem in statistical mechanics and that the Monod–Wyman–Changeux (MWC) model provides a great response to that challenge. Here, we explore this model in the context of cyclic-nucleotide-gated channels that are gated by cGMP.

(a) Write down an MWC model for the ligand-gated channel and calculate the probability of the channel being open as a function of the ligand concentration. Construct separate versions of this model by considering channels with 1, 2, 3, and 4 binding sites.

(A)

(B)

Figure 18.57: (A) Fossil from the compound eye of an arthropod from the Emu Bay Shale, Australia. (B) Compound eye from *Laphria rufifemorata* shown in an orientation hypothesized to be like that of the fossil. (Adapted from M. S. Y. Lee et al., *Nature* 474:631, 2011).

(b) For each version of the model, compare the resulting curve to the data shown in Figure 18.58 for the bovine retinal CNG channel. In order to do this, perform a "fit by eye." It might be useful to understand how each model behaves in the limits $L \to 0$, $L \to +\infty$, where L is the concentration of cGMP, and how the sharpness of the curve is affected by the choice of the various model parameters. What do you conclude about the number of binding sites for cGMP in this channel?

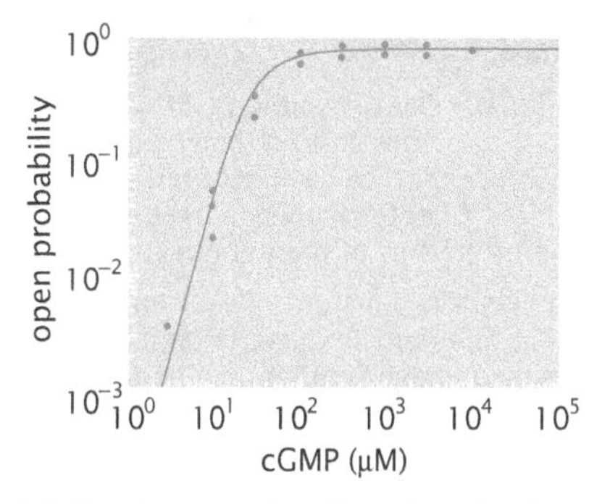

Figure 18.58: Open probability of the bovine retinal CNG channel. The probability of the channel being opened is measured as a function of the concentration of cGMP. The fit corresponds to an MWC model analogous to the model shown in Figure 7.26 for the case of four binding sites per channel. (Adapted from E. H. Goulding et al., *Nature* 372:369, 1994.)

• 18.10 Waiting time distributions

One informative way of analyzing the frequency of events is illustrated in Figure 18.47(C). Here, the cumulative probability of events is obtained by measuring how many events have occurred if we wait a time t. This cumulative probability is fit to the equation $n = N[1 - e^{-t/\tau}]$, where N is the total number of events. Show that his functional form is precisely what is expected for a process characterized by the waiting time distribution $p(t) = e^{-t/\tau}$.

• 18.11 Statistics of photon detection in the eye

Using the Poisson distribution, generate the curves in Figure 18.48(B).

18.7 Further Reading

Morton, O (2007) Eating the Sun: How Plants Power the Planet, Fourth Estate. An amazing description of the role of photosynthesis in life on Earth and the history of our understanding of the process.

Morison, MA, Estle, TL, & Lane, NF (1976) Quantum States of Atoms, Molecules, and Solids, Prentice-Hall. This excellent book champions simplified quantum mechanical models like those we have used in this chapter.

Nitzan, A (2006) Chemical Dynamics in Condensed Phases, Oxford University Press. The subject of energy and electron transfer is deep and subtle, and admits of a wide variety of variations on some of the basic themes covered in our chapter. Nitzan's book gives a much more thorough view of the many different ways that electron dynamics can be considered.

Rabinowitch, E, & Govindjee (1969) Photosynthesis, Wiley. Despite its age, this book is full of subtle insights and hence definitely still worth reading. It is available online at www.life.illinois.edu/govindjee/photosynBook.html.

Blankenship, RE (2002) Molecular Mechanisms of Photosynthesis, Blackwell Science. This interesting book is full of insights into photosynthesis and includes the estimate we did in the chapter for the number of photons striking a pigment molecule each second.

Falkowski, PG, & Raven, JA (2007) Aquatic Photosynthesis, 2nd ed., Princeton University Press. A beautiful book on photosynthesis in aquatic environments that makes us see the ocean in a different way.

Bayliss, NS (1948) A "metallic" model for the spectra of conjugated polyenes, *J. Chem. Phys.* **16**, 287 and Kuhn, H

(1949) A quantum mechanical theory of light absorption of organic dyes and similar compounds, *J. Chem. Phys.* **17**, 1198. In the absence of computational power, simple, intuitive models were important in trying to figure out how pigments absorb light. These clever papers give a deeper exploration of the kind of box models of pigment molecules considered in the chapter.

Lawlor, DW (2001) Photosynthesis, 3rd ed., BIOS Scientific Publishers. This book has many useful numbers relevant to photosynthesis. See Table 4.1 for a provocative and thorough "semi-quantitative analysis of the photosynthetic system in an 'average' C3 plant leaf."

Taiz, L, & Zeiger, E (2010) Plant Physiology, 5th ed., Sinauer Associates. This book provides an excellent pedagogical approach to many topics that we have covered in a cursory and superficial fashion. Chapter 7 gives a useful description of the light absorption properties of pigments.

Chandler, D (1998) Electron transfer in water and other polar environments, how it happens, in Classical and Quantum Dynamics in Condensed Phase Simulations, edited by BJ Berne, G Ciccotti, & DF Coker, World Scientific. This excellent article served as one of our main inspirations for this chapter and we have borrowed heavily from Chandler's approach for thinking about electron transfer and Marcus theory.

Hopfield, JJ (1974) Electron transfer between biological molecules by thermally activated tunneling, *Proc. Natl Acad. Sci. USA* **71**, 3640. A thoughtful treatment of electron transfer.

Rodieck, RW (1998) The First Steps in Seeing, Sinauer Associates. An outstanding introduction to the problem of vision with descriptions of the problem at scales ranging from molecules such as rhodopsin all the way to behavior. The notes on pp. 535–539 are an impressive compendium of numbers relevant to vision.

Feynman, RP, Leighton, RB, & Sands, M (1965) The Feynman Lectures on Physics, Addison-Wesley. Feynman's treatment of nearly everything is original and interesting. His discussion of compound eyes includes a description of the optimal size for ommatidia that is very clear.

Feynman, RP (1985) QED: The Strange Theory of Light and Matter, Princeton University Press. In this book, Feynman uses the idea of light waves as carrying stopwatches as a way of describing phase and does the most beautiful treatment of diffraction and gratings that we are aware of.

Dowling, JE (2012) The Retina: An Approachable Part of the Brain, The Belknap Press of Harvard University Press. A wonderful discussion of many features of vision and the retina.

Land, MF, & Nilsson, D-E (2012) Animal Eyes, 2nd ed., Oxford University Press. This books gives a wide-ranging discussion of eyes and how they work.

Nachman, MW, Hoekstra, HE, & D'Agostino, SL (2003) The genetic basis of adaptive melanism in pocket mice, *Proc. Natl Acad. Sci. USA* **100**, 5268. This paper describes the wonderful story of rock pocket mice and their coloration.

18.8 References

Alberts, B, Johnson, A, Lewis, J, et al. (2008) Molecular Biology of the Cell, 5th ed., Garland Science.

Barlow, HB (1952) The size of ommatidia in apposition eyes, *J. Exp. Biol.* **29**, 667.

Baylor, DA, Matthews, G, & Yau, K-W (1980) Two components of electrical dark noise in toad retinal rod outer segments, *J. Physiol.* **309**, 591.

Born, M, & Einstein, A (2005) The Born–Einstein Letters, 1916–1955: Friendship, Politics and Physics in Uncertain Times, Macmillan.

Bowmaker, JK, & Dartnall, HJ (1980) Visual pigments of rods and cones in a human retina, *J Physiol.* **298**, 501.

Bradshaw, HD Jr, & Schemske, DW (2003) Allele substitution at a flower colour locus produces a pollinator shift in monkeyflowers, *Nature* **426**, 176.

Cannon, GC, Bradburne, CE, Aldrich, HC, et al. (2001) Microcompartments in prokaryotes: carboxysomes and related polyhedra, *Appl. Environ. Microbiol.* **67**, 5351.

Cartailler, JP, & Luecke, H (2003) X-ray crystallographic analysis of lipid–protein interactions in the bacteriorhodopsin purple membrane, *Annu. Rev. Biophys. Biomol. Struct.* **32**, 285.

Douglass, AD, Kraves, S, Deisseroth, K, et al. (2008) Escape behavior elicited by single, channelrhodopsin-2-evoked spikes in zebrafish somatosensory neurons, *Curr. Biol.* **18**, 1133.

Emerson, R, & Arnold, W (1932) The photochemical reaction in photosynthesis, *J. Gen. Physiol.* **16**, 191.

Engelmann, TW (1883) Bacterium photometricum. Ein Beitrag zur vergleichenden Physiologie des Licht- und Farbensinnes, *Arch. Gesam. Physiol. Mensch. Tiere* **30**, 95.

Fawcett, DW (1966) The Cell, Its Organelles and Inclusions: An Atlas of Fine Structure, W. B. Saunders.

Fox, LS, Kozik, M, Winkler, JR, & Gray, HB (1990) Gaussian free-energy dependence of electron-transfer rates in iridium complexes, *Science* **247**, 1069.

Goulding, EH, Tibbs, GR, & Siegelbaum, SA (1994) Molecular mechanism of cyclic-nucleotide-gated channel activation, *Nature* **374**, 369.

Govindjee & Gest, H (2002) Celebrating the millennium: historical highlights of photosynthesis research, *Photosynth. Res.* **73**, 1.

Gray, HB, & Winkler, JR (2005) Long-range electron transfer, *Proc. Natl Acad. Sci. USA* **102**, 3534.

Hecht, S, Shlaer, S, & Pirenne, MH (1942) Energy, quanta, and vision, *J. Gen. Physiol.* **25**, 819.

Huygens, C (1932) Oeuvres complètes. Tome XVII. L'horloge à pendule 1656–1666, edited by JA Vollgraff, Martinus Nijhoff.

Iancu, CV, Ding, HJ, Morris, DM, et al. (2007) The structure of isolated *Synechococcus* strain WH8102 carboxysomes as revealed by electron cryotomography *J. Mol. Biol.* **372**, 764.

Kok, B, Forbush, B, & McGloin, M (1970) Cooperation of charges in photosynthetic O_2 evolution—I. A linear four step mechanism, *Photochem. Photobiol.* **11**, 457.

Kühlbrandt, W (2000) Bacteriorhodopsin—the movie, *Nature* **406**, 569.

Langen, R, Change, IJ, Germanas, JP, et al. (1995) Electron tunneling in proteins: coupling through a β strand, *Science* **268**, 1733.

Lee, MS, Jago, JB, & Garcia-Bellido, DC (2011) Modern optics in exceptionally preserved eyes of Early Cambrian arthropods from Australia, *Nature* **474**, 631.

Liberton, M, Howard Berg, R, Heuser, J, et al. (2006) Ultrastructure of the membrane systems in the unicellular cyanobacterium *Synechocystis* sp. strain PCC 6803, *Protoplasma* **227**, 129.

Mathews, CK, van Holde, KE, & Ahern, KG (1999) Biochemistry, 3rd ed., Prentice Hall.

Moser, CC, Keske, JM, Warncke, K, et al. (1992) Nature of biological electron transfer, *Nature* **355**, 796.

Nathans, J, Thomas, D, & Hogness, DS (1986) Molecular genetics of human color vision: the genes encoding blue, green, and red pigments, *Science* **232**, 193.

Neutze, R, Pebay-Peyroula, E, Edman, K, et al. (2002) Bacteriorhodopsin: a high-resolution structural view of vectorial proton transport, *Biochim. Biophys. Acta* **1565**, 144.

Palczewski, K (2006) G protein-coupled receptor rhodopsin, *Annu. Rev. Biochem.* **75**, 743.

Palczewski, K, Kumasaka, T, Hori, T, et al. (2000) Crystal structure of rhodopsin: a G protein-coupled receptor, *Science* **289**, 739.

Payne, JL, Boyer, AG, Brown, JH, et al. (2009) Two-phase increase in the maximum size of life over 3.5 billion years reflects biological innovation and environmental opportunity, *Proc. Natl Acad. Sci. USA* **106**, 24.

Ragatz, L, Jiang, ZY, Bauer, C, & Gest, H (1994) Phototactic purple bacteria, *Nature* **370**, 104.

Ragatz, L, Jiang, ZY, Bauer, C, & Gest, H (1995) Macroscopic phototactic behavior of the purple photosynthetic bacterium *Rhodospirillum centenum*, *Arch. Microbiol.* **163**, 1.

Rutherford, AW, & Boussac, A (2004) Water photolysis in biology, *Science* **303**, 1782.

Savage, DF, Afonso, B, Chen, AH, & Silver, PA (2010) Spatially ordered dynamics of the bacterial carbon fixation machinery, *Science* **327**, 1258.

Scheuring, S, Boudier, T, & Sturgis, JN (2007) From high-resolution AFM topographs to atomic models of supramolecular assemblies, *J. Struct. Biol.* **159**, 268.

Scheuring, S, & Sturgis, JN (2005) Chromatic adaptation of photosynthetic membranes, *Science* **309**, 484.

Sotckman, A, Smithson, HE, Webster, AR et al. (2008) The loss of PDE6 deactivating enzyme, RGS9, results in precocious light adaptation at low light levels, *Journal of Vision* **8**, 1.

Ting, CS, Hsieh, C, Sundararaman, S, et al. (2007) Cryo-electron tomography reveals the comparative three-dimensional architecture of *Prochlorococcus*, a globally important marine cyanobacterium, *J. Bacteriol.* **189**, 4485.

Tonomura, A, Endo, J, Matsuda, T, Kawasaki, T, & Ezawa, H (1989) Demonstration of single-electron buildup of an interference pattern, *Am. J. Phys.* **57**, 117.

Vukusic, P, Sambles, JR, Lawrence, CR, & Wootoon, RJ (1999) Quantified interference and diffraction in single *Morpho* butterfly scales, *Proc. R. Soc. Lond. B* **266**, 1403.

Vukusic, P, & Sambles, JR (2003) Photonic structures in biology, *Nature* **424**, 852.

Walter, JM, Greenfield, D, Bustamante, C, & Liphardt, J (2007) Light-powering *Escherichia coli* with proteorhodopsin, *Proc. Natl Acad. Sci. USA* **104**, 2408.

Wasielewski, MR, Niemczyk, MP, Svec, WA, & Pewitt, EB (1985) Dependence of rate constants for photoinduced charge separation and dark charge recombination on the free energy of reaction in restricted-distance porphyrin-quinone molecules, *J. Am. Chem. Soc.* **107**, 1080.

Yamazaki, I, Mimuro, M, Murao, T, et al. (1984) Excitation energy transfer in the light harvesting antenna system of the red alga *Porphyridium cruentum* and the blue–green alga *Anacystis nidulans*: analysis of time-resolved fluorescence spectra, *Photochem. Photobiol.* **39**, 233.

Zhang, Z, Wang, LP, & Brauner, M (2007) Multimodal fast optical interrogation of neural circuitry, *Nature* **446**, 633.

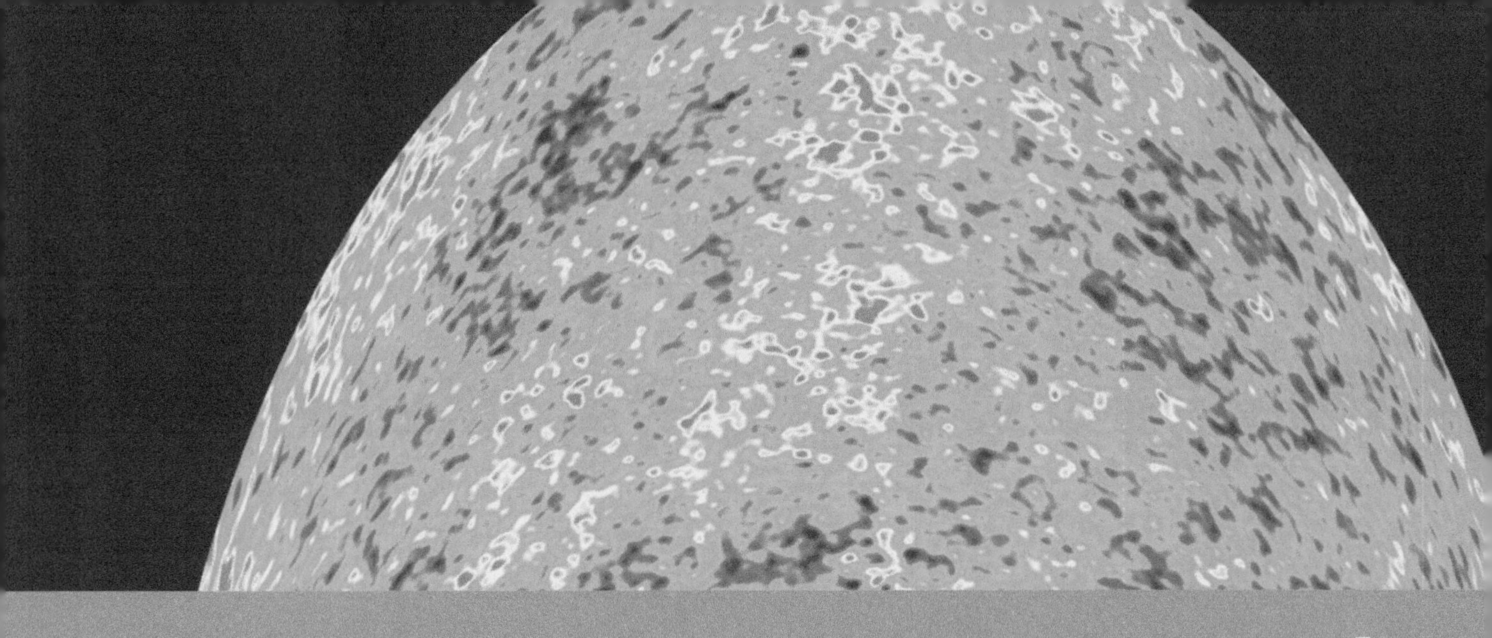

The Meaning of Life

PART 4

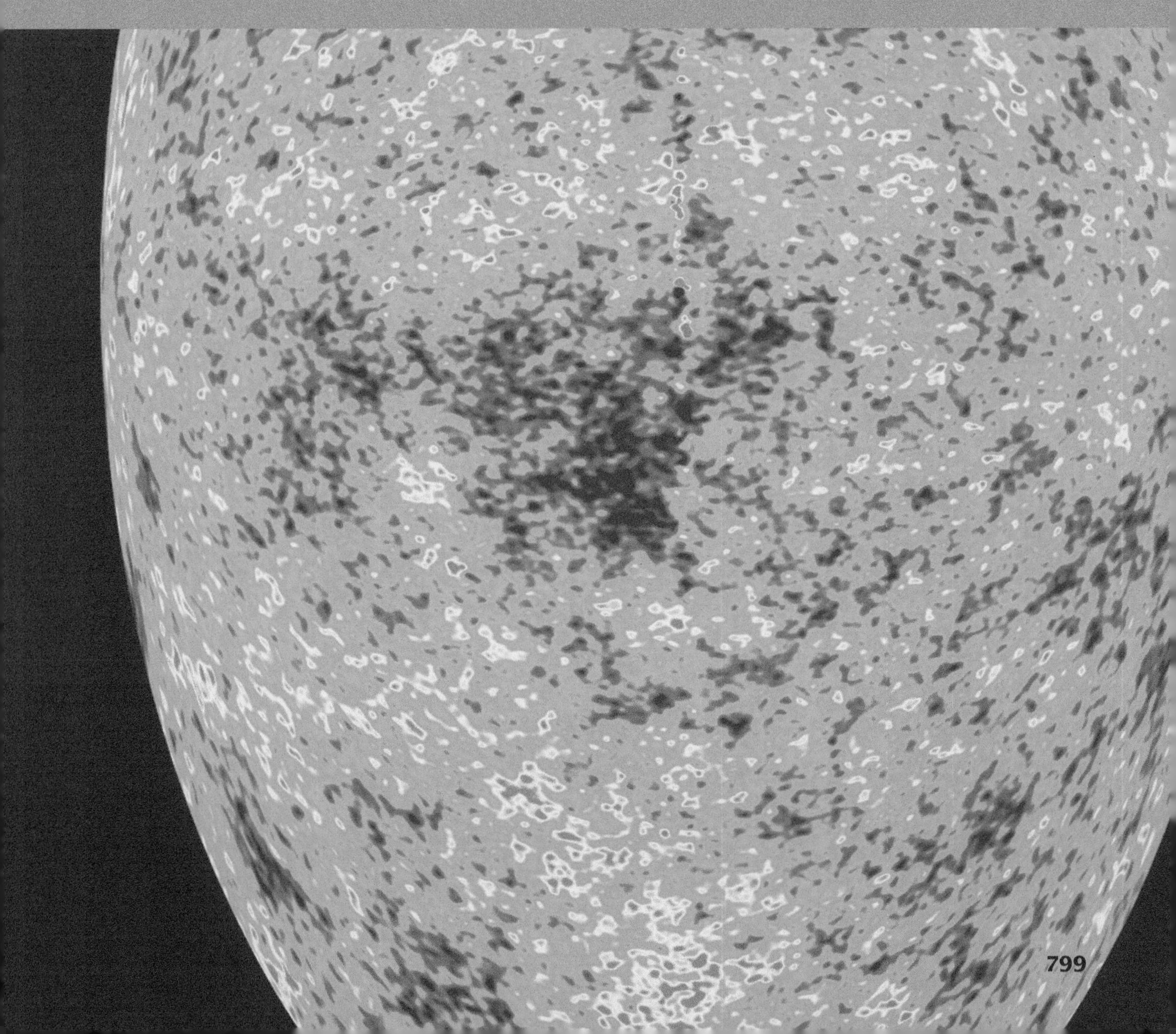

Organization of Biological Networks 19

Overview: In which statistical mechanics is used to study gene regulation

Specific genes are used only when and where they are needed. For example, we have made much of the classic example of the *lac* operon, which governs the enzymes responsible for lactose digestion. Similar control is exercised over genes in other bacteria, archaea, and eukaryotes. The tools worked out throughout the book leave us poised to consider important quantitative questions about gene regulation such as: how much is a given gene expressed, where in the cell (or the organism) is that gene expressed, and at what time during the cell cycle (or life history) of the organism? The key tools we will use to study these questions are statistical mechanics and rate equations. The statistical mechanical approach will use the probability of promoter occupancy as the key quantity of interest, whereas the rate equation approach will examine the concentrations of protein products over time. These same techniques will also be used to examine signaling with special emphasis on the "decisions" cells make about where to go.

"The human mind cannot go on forever accumulating facts which remain unconnected and without any mutual bearing and bound together by no law."

Alfred Russel Wallace

19.1 Chemical and Informational Organization in the Cell

Many Chemical Reactions in the Cell are Linked in Complex Networks

The reality of the chemical reactions that take place in the cell are a far cry from the relatively sterile and simple kinetic processes described in Chapter 15. In the discussion given there, we showed how to write the time evolution of the concentrations of a set of reactants and products. That theoretical machinery provides an appealing and useful picture for characterizing many of the beautiful *in vitro* experiments that have powered solution biochemistry. However, biochemistry in living cells has reactants and products linked in a complex set of lineages of biblical proportions where *A* begets *B*, which begets *C*, which in turn begets *D*, and so on, with the added nonanthropomorphic complication that *Z* might just beget *A* again. Indeed, the fact that *Z* can act back on *A* reflects the presence of feedback, which makes the dynamics even richer. Two of the most important classes of reaction that are central to the functioning of cells are those associated with gene regulation and signaling. Indeed, one of the features that most completely distinguishes the chemistry of a cell from that of solution biochemistry is the way in which the reactants are tuned by up- and down-regulation. Similarly, the reactions of the cell are also stimulated by external cues in the form of signaling cascades. In this chapter, we consider regulation and signaling by using a variety of tools developed throughout the book.

Genetic Networks Describe the Linkages Between Different Genes and Their Products

One of the most intriguing reasons why the chemistry of the cell cannot be viewed as a bag of reactants and products is the fact that this chemistry is under the strict control of the genetic machinery of the cell. In particular, if left to its own devices, some particular chemical pathway in the cell might just travel a path to eventual equilibrium. On the other hand, because of both external and internal cues, the machinery of the cell can receive orders via signaling pathways that lead, in turn, to the expression of some gene that results in a new reactant in the original chemical pathway that sends it off in some new direction.

The description of the informational pathways that dictate the cellular concentration profiles in both space and time of the various chemical reactants of interest is founded upon a higher level of abstraction. In particular, there are networks of genes that are linked together in sometimes horrifyingly complex arrays such as that shown in Figure 19.1. This network is an example of a particularly well-characterized genetic network that participates in the embryonic development of sea urchins. One important take-home message concerning this network is that it is a typical network and should leave the reader with a sense of the implied chemical complexity of these systems. In general, genetic networks like that shown in Figure 19.1 make no reference either to the passage of time or to the quantitative distributions of the molecules that mediate these networks. Rather, these networks are an abstraction that shows how genes (and their products) are linked to each other in both space and time. On the other hand, it is important to bear in mind that beneath the surface of these wiring diagrams are actual concentrations of the molecular players of these informational pathways.

Developmental Decisions Are Made by Regulating Genes

Often, genetic networks serve as the basis of the developmental decisions that send a cell or collections of cells down some developmental path. One of the intriguing features of multicellular organisms is that despite the overwhelming cellular diversity, generally, each cell carries the same genetic baggage. However, in general, cells only express a certain fraction of all the available genes. This differentiation is the

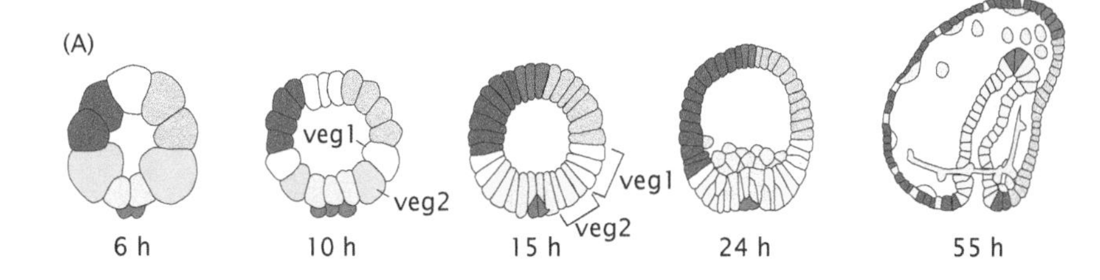

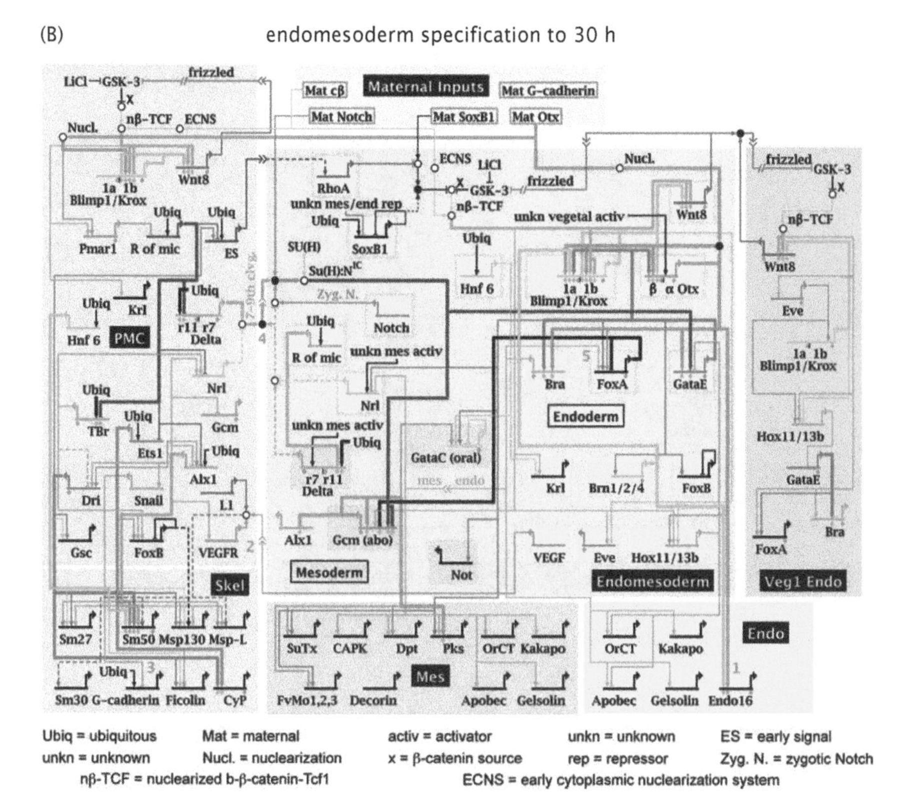

Figure 19.1: Genetic network associated with control of the developmental pathway of the sea urchin embryo. (A) Schematic of stages in the embryonic development of the sea urchin. (B) Genetic network associated with sea urchin development. (Adapted from S. Ben-Tabou de-Leon and E. H. Davidson, *Annu. Rev. Biophys. Biomol. Struct.* 36:191, 2007.)

basis of the development of embryos and the basis of the different structures found in multicellular organisms. The key point is that not all genes are being expressed all the time.

One of the most famous examples of a "developmental decision" is the lambda switch described in Chapter 4 and shown in Figure 4.10 (p. 152). After infecting an *E. coli* bacterium, lambda phage follows one of two developmental pathways. One pathway (the lytic pathway) results in the assembly of new phages and the lysis of the host cell. The second pathway, the lysogenic pathway, involves incorporation of the lambda genome into that of the host cell. Lysogeny can be reversed by damaging the cell with UV light, which triggers lytic replication.

Another compelling example of the role of developmental decisions is that of embryonic development in fruit flies. One of the most celebrated examples is that of the body plan along the long axis of the fly embryo, which is dictated by the distribution of certain proteins along the embryo. Figure 19.2 gives an example of the gradients in four key regulatory proteins that determine the anterior–posterior organization. These proteins determine the pattern of gene expression along the embryo, from which the Eve 2 stripe is the most well-understood example. These ideas were already introduced in Section 2.3.3 (p. 78).

Part of the hard-won wisdom of molecular biology is the recognition that there are many stages in the pathway between DNA and functional protein that can serve as regulatory points. Some of these

Figure 19.2: Regulatory proteins in the *Drosophila* embryo. The anterior–posterior (A–P) patterning of the fruit fly is dictated by genes that are controlled by spatially varying concentrations of transcription factors. (A) Schematic of the main transcription factors involved in the regulation of stripe 2 of expression of the *even-skipped* gene (eve). (B) Regulatory region of the stripe 2 of the *even-skipped* gene where the binding sites for each transcription factor have been identified. The binding site color on the DNA corresponds to the transcription factor color in (A). (C) Spatial profile of the morphogen gradients measured using immunofluorescence. The purple shaded region corresponds to the striped region shown in (D). (D) Resulting pattern of expression of the regulatory region shown in (B). (B, Adapted from S. Small et al., *EMBO J.* 11:4047, 1992.; C, adapted from E. Myasnikova et al., *Bioinformatics* 17:3, 2001; D, adapted from S. Small et al., *Dev. Biol.* 175:314, 1996.)

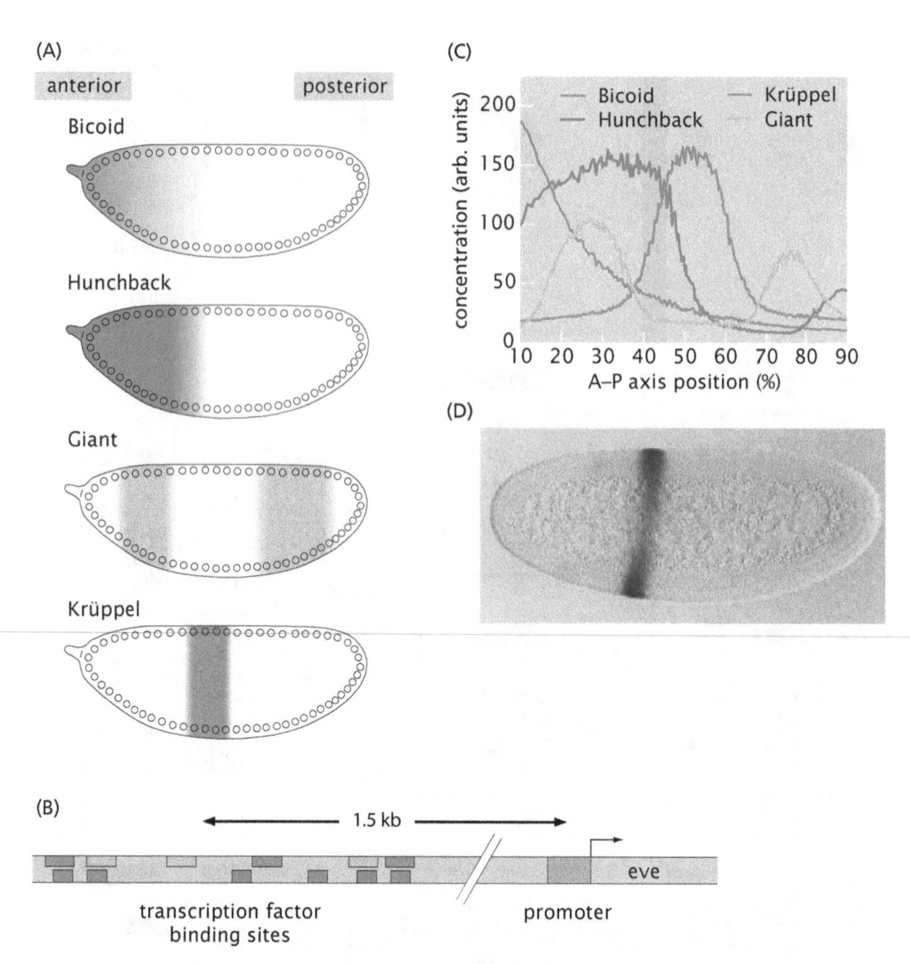

different regulatory mechanisms are shown in Figure 6.7 (p. 245). For the purposes of the present discussion, we will focus on one of the most common regulatory mechanisms, namely, transcriptional control, where the key decision that is made is whether or not to produce mRNA.

Gene Expression Is Measured Quantitatively in Terms of How Much, When, and Where

One of our main arguments is that gene expression is a subject that has become increasingly quantitative. In particular, it is now common to measure how much a given gene is expressed, when it is expressed, and where it is expressed. To carry out such measurements, there are a number of useful tools.

Experiments Behind the Facts: Measuring Gene Expression

Quantitative measurement of gene expression can be made at many stages between the decision to start transcription and the emergence of a functional protein product. As noted earlier, such measurements have provided a quantitative window on how much a given gene is expressed, where it is expressed spatially, and when.

One important way to characterize the activity of a gene is by virtue of its protein products. In particular, if the gene product has enzyme activity, that activity can be assayed as a reporter of the extent to which the gene has been expressed

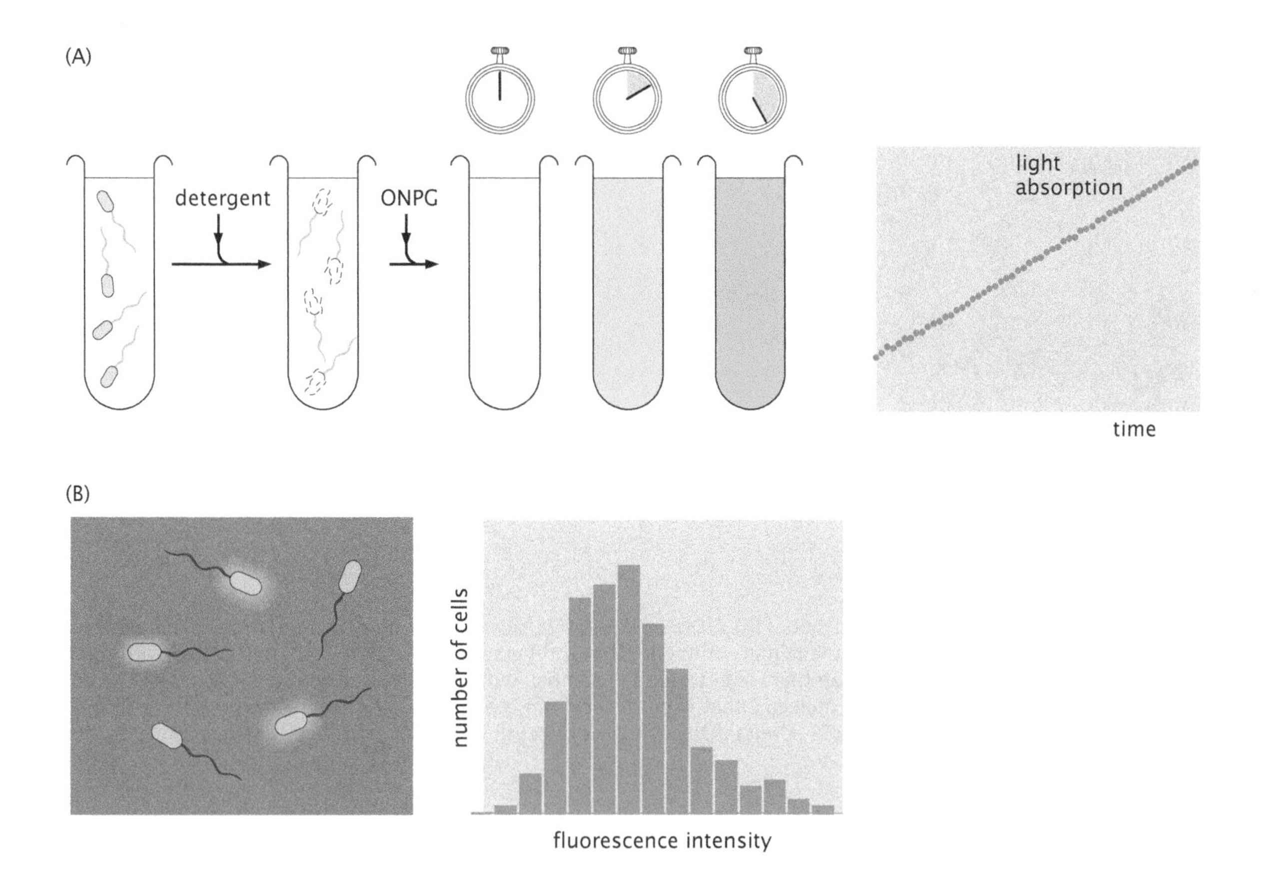

Figure 19.3: Measurement of gene expression. (A) Measurement of gene expression as a result of enzymatic activity. The promoter of interest drives the expression of an enzyme that can cleave a molecule that in the cleaved state is colored. The resulting rate of increase in light absorption is related to the amount of enzyme present in the cells. (B) The promoter of interest drives the expression of a fluorescent protein such as GFP. The amount of fluorescence per cell reports the extent of expression of the gene of interest.

as shown in Figure 19.3(A). Recall that β-galactosidase is the enzymatic product of the *lac* operon, as shown in Figure 4.13 (p. 155), and that the action of this enzyme is to clip lactose molecules. One of the impressive legacies of years of work on this system is a battery of substrates that respond differently to the enzymatic cleavage. One such substrate (ONPG) turns yellow upon cleavage, and measuring the rate at which a solution becomes yellow optically can provide a window on gene expression since it is proportional to the amount of enzyme (over some region of concentrations). By measuring the absorbance at the appropriate wavelengths, one obtains a picture of the amount of active enzyme. Such measurements are typically done on populations of cells. They also require lysing the cells, which means that only end-point assays can be performed with this technique. On the other hand, the sensitivity of this method is superb—to the point where the activity of less than one β-galactosidase molecule per cell can easily be measured. To carry out this kind of assay usually requires routine cloning in which sequences encoding the enzyme are inserted into the genome under the control of the transcription factors of interest.

From a molecular biology perspective, this same strategy of inserting a reporter into the gene of interest can be followed, but with the difference that the "reporter" molecule is a

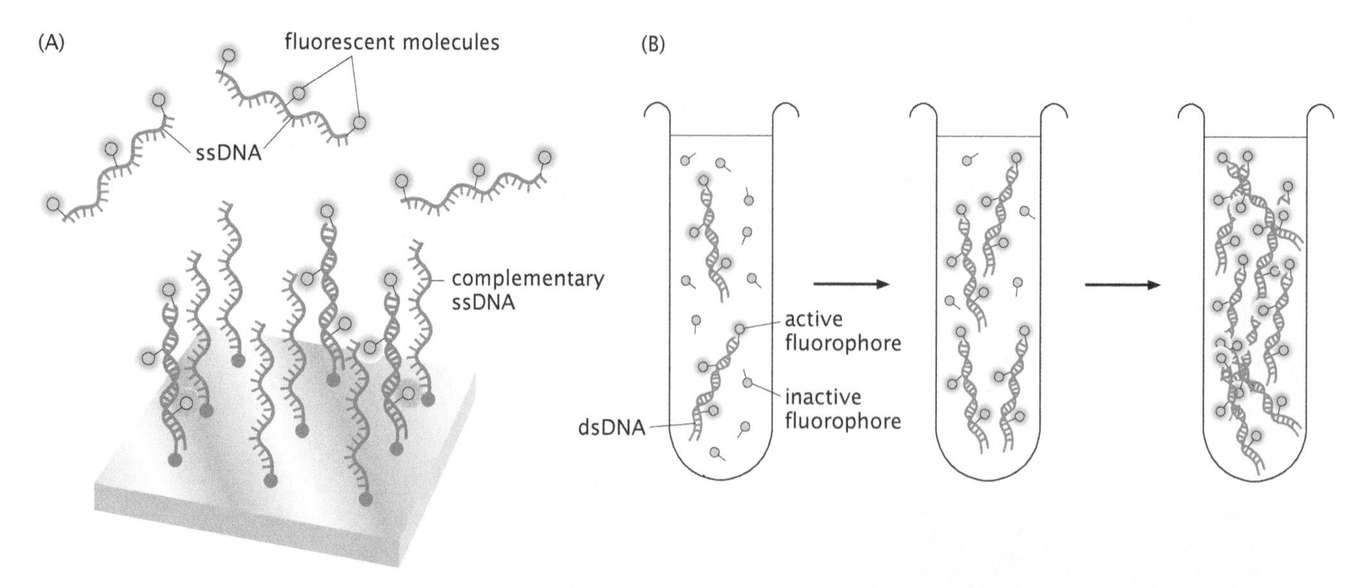

Figure 19.4: Measurement of mRNA concentration. (A) A DNA microarray uses a collection of different molecules on the surface of a slide, each of which has a sequence complementary to the mRNA (or reverse-transcribed ssDNA) associated with the gene of interest. By measuring how much hybridization there is between the sample and the molecules on the surface, one can count the mRNAs. (B) Quantitative PCR uses a template molecule that is produced from the mRNA using reverse transcription. The amount of template determines how many cycles of PCR it will take to reach a critical threshold of amplified DNA using fluorescence as a readout.

fluorescent molecule such as GFP rather than an enzyme. This case is shown in Figure 19.3(B). Relative fluorescence levels of reporters such as GFP are easy to characterize. As shown in Figure 3.3 (p. 93), GFP can be used to track the level of gene expression as a function of time in single living cells. This reporter has its disadvantages, as such fluorescent proteins are subject to photobleaching. Additionally, as we will see in the Computational Exploration on extracting levels of gene expression, the natural constituents of cells have an intrinsic fluorescence, which results in a cellular autofluorescence background that can potentially contaminate the readout from the GFP reporter.

A second scheme for characterizing the extent to which a given gene is expressed is by measuring how much mRNA from the gene of interest is present in the cell. One of the tools of choice for such measurements is the DNA microarray. DNA microarrays are built by labeling a surface with an array of different DNA molecules, each patch of which has small, single-stranded DNA (ssDNA) molecules with the same sequence, as shown in Figure 19.4. These sequences are chosen to be complementary to an entire battery of sequences corresponding to the genes of interest in the experiment. Cells are then broken up and their RNA (or DNA copies made from the RNA) is allowed to flow across the array and hybridize with the molecules on the surface. The various molecules extracted from the cell have been fluorescently labeled, so by looking at the fluorescence intensity at each point on the array, it is possible to read off how much RNA was present.

Another scheme for characterizing the amount of RNA is to use quantitative PCR. Once again, the cell is lysed and the mRNA molecules are turned into DNA using a reverse transcription reaction. Then these molecules are used as templates in a PCR, and it is seen how many cycles of PCR are needed before the quantity of DNA in the reaction exceeds some threshold. This cycle value is a direct reflection of the number of starting molecules, since starting with lots of template DNA will result in many more molecules at low cycle numbers than will starting with very little material. With quantitative PCR, one can detect mRNA copy numbers as low as 10.

Finally, with the advent of new sequencing technologies that make it possible to generate millions of sequence reads at a reasonable price, it has become commonplace to just sequence the complete mRNA content of cells. By doing so, one can simply count the number of mRNA molecules within the cell corresponding to the various genes of interest, resulting in genome-wide information in one experiment. As with the previous methods, this approach requires the conversion of all cellular mRNA into DNA in order to be sequenced.

As will be described in the remainder of this chapter, a useful surrogate for the actual question of the extent to which a given gene is expressed is to ask whether or not the promoter for the gene of interest is occupied. There are many *in vitro* and *in vivo* methods for finding out whether or not the promoter is bound to polymerase. *Chromatin immunoprecipitation* and *DNA footprinting* are two methods that are sensitive to promoter occupancy. For DNA footprinting, the idea is that the part of DNA where the transcriptional apparatus is bound will react differently when the system is exposed to agents such as restriction enzymes. The most common procedure is to try to digest the DNA using a restriction enzyme. It will not be able to access the DNA over which RNA polymerase is situated, leaving a "footprint" of a longer piece of DNA that can be easily detected. For chromatin immunoprecipitation, DNA is covalently crosslinked to bound proteins using reactive chemicals, and then the DNA is sheared into small fragments. Antibodies specific to polymerase are used to isolate the molecules of polymerase with their associated DNA fragments. Then, the chemical crosslinks are reversed, and the DNA fragments associated with polymerase are sequenced. This same technique can be modified to identify the specific DNA sequences that are associated with any other specific DNA-binding protein of interest, such as a repressor protein. These different methods can also be cleverly combined with the new sequencing technologies in order to perform such assays at the genome-wide scale, as we will see further below.

19.2 Genetic Networks: Doing the Right Thing at the Right Time

In "thermodynamic" models of gene expression, attention is focused on the probability that the promoter is occupied by RNA polymerase. In Section 6.1.2 (p. 244), we showed how the "bare" problem of polymerase molecules interacting with DNA could be solved using simple ideas from statistical mechanics. However, the shortcoming of that approach is that it ignores the existence of molecular gatekeepers that exercise strict control over the occupancy of promoters. We begin our dissection of gene expression with a consideration of these gatekeepers, which are known as transcription factors.

Promoter Occupancy Is Dictated by the Presence of Regulatory Proteins Called Transcription Factors

In Figure 6.8 (p. 246) we showed a cartoon of some gene of interest and the promoter and DNA upstream from it. As a first cut at the problem of promoter occupancy, we examined the probability of RNA polymerase binding as a competition between this promoter and nonspecific sites, both of which can be occupied by polymerase molecules. We now expand that discussion to account for the presence of a host of important accessory proteins that can either enhance (activate) or reduce (repress) the probability of promoter occupancy.

As before, we focus primarily on bacteria. What this means concretely is that we will treat RNA polymerase as a single molecule and ask the precise mathematical (but biologically oversimplified) question of whether or not the promoter is occupied by such an RNA polymerase molecule. In the eukaryotic case, this question is less easily posed, since the basal transcription apparatus consists of many parts, all of which need to be present simultaneously in order to start transcription.

19.2.1 The Molecular Implementation of Regulation: Promoters, Activators, and Repressors

Repressor Molecules Are the Proteins That Implement Negative Control

One of the key control mechanisms of genetic networks is negative regulation of transcription. What this means is that the decision to express the gene of interest is made very early on in the set of processes leading from DNA to protein, namely, at the point where RNA is synthesized. If there is little or no mRNA that codes for a given protein, then clearly the ribosomes are in no position to produce the corresponding protein. The molecular implementation of negative control is through protein molecules known as repressors, such as the Lac repressor introduced in Figures 4.13 (on p. 155) and 8.19 (on p. 334). In the case of bacteria, repressors can often be viewed as carrying out a blocking action in the sense that through DNA–protein interactions, they occupy the DNA in a region (called the operator) that overlaps the region where RNA polymerase binds (the promoter). The action of such repressor molecules is illustrated schematically in Figure 19.5. Note that the activity of repressors can, in turn, be regulated by small molecules, or inducers, that can bind and generate a conformational (or allosteric) change that alters the binding probability of the transcription factor for the DNA. Later in this chapter, we give a statistical mechanical interpretation of such cartoons.

It is important to recall that the point of cartoons like that in Figure 19.5 is to convey a conceptual picture and not a detailed molecular rendering of the explicit action of the various molecular participants. On the other hand, the fact that such cartoons can be constructed in the first place is often the result of having digested the significance of hard-won structural determinations from X-ray crystallography. Indeed, sometimes, not only the structures of the bare

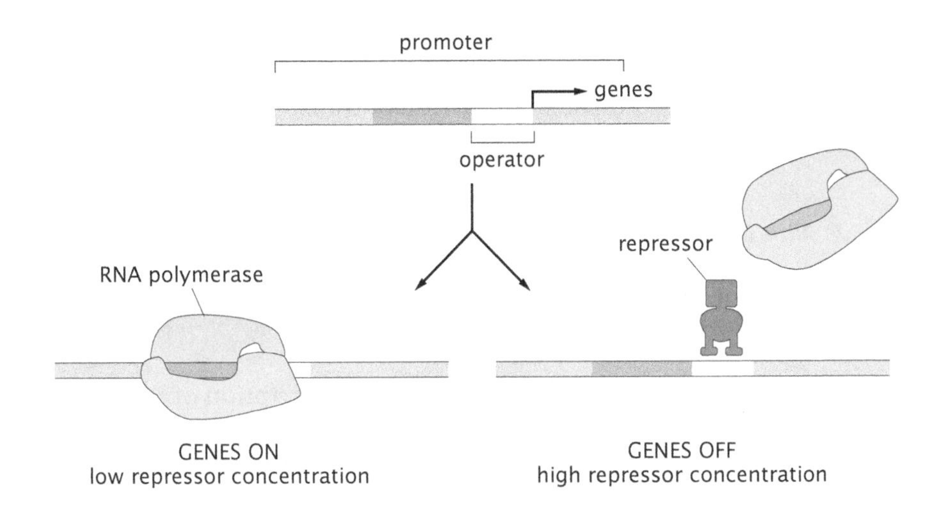

Figure 19.5: The process of repression. Cartoon representation showing the action of repressor molecules in forbidding RNA polymerase from binding to its promoter, or alternatively, if bound, from initiating transcription.

repressors are known, but even the structures of these repressors when complexed with DNA. In fact, there are a variety of structural implementations of repression, some famed examples of which are shown in Figure 19.6.

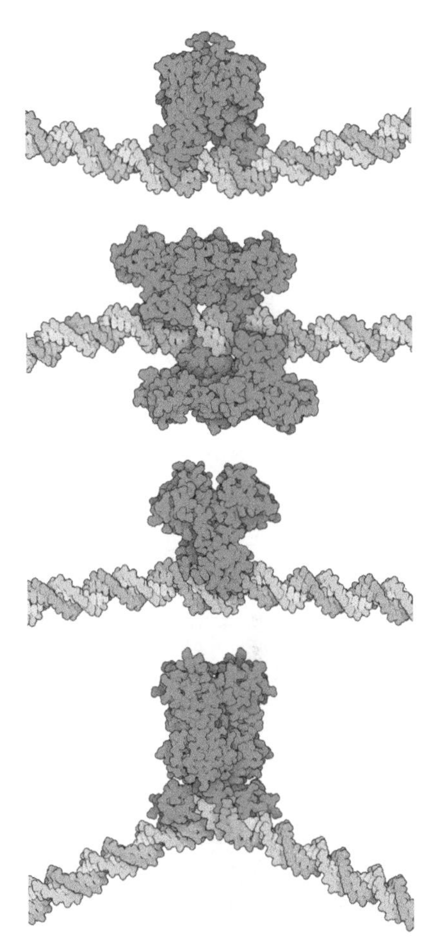

Figure 19.6: Examples of repressor molecules interacting with DNA. From top to bottom, the repressors are TetR (pdb 1QPI), IdeR (pdb 1U8R), FadR (pdb 1HW2), and PurR (pdb 1PNR). The point of the figure is to give an impression of the relative sizes of repressors and their target regions on DNA and to illustrate how these transcription factors deform the DNA double helix in the vicinity of their binding site. These drawings are renditions of actual structures from X-ray crystallography. (Courtesy of D. Goodsell.)

Activators Are the Proteins That Implement Positive Control

A second key mechanism for altering the extent to which a given gene is expressed is known as positive regulation of transcription, or, more provocatively, regulated recruitment. Here too, the idea is that the overall process of protein synthesis of a given gene product is regulated very early on where an accessory molecule enhances the probability of promoter occupancy by RNA polymerase. This mechanism is built around the idea of proteins other than RNA polymerase that bind to DNA and increase the probability that the RNA polymerase itself will bind the promoter. Just as repressors interfere with the ability of RNA polymerase to bind to its promoter, activators bind in the vicinity of the promoter and have adhesive interactions with RNA polymerase itself that enhance the likelihood of RNA polymerase binding. The key point is that the RNA polymerase molecule interacts not only with the DNA to which it is bound, but also through "glue-like" interactions with the activator molecule. A cartoon representation of the process of regulated recruitment (that is, activation) is shown in Figure 19.7.

As with the study of repressors, structural biology has permitted a range of atomic-level insights into the mechanisms of transcriptional activation. Figure 19.8 provides a gallery of some key activators, reveals their sizes relative to the DNA molecule, and illustrates the way in which they distort and occlude the DNA when bound.

Genes Can Be Regulated During Processes Other Than Transcription

Our discussion will focus primarily on transcriptional regulation. On the other hand, as shown in Figure 6.7 (p. 245), there are many points along the route connecting DNA to its protein products where gene expression can be controlled. Two of the most obvious and important ways in which the concentration of active protein is controlled are through the post-translational modifications phosphorylation and protein degradation. In addition, in recent years, a whole host of regulatory RNAs have been discovered that have greatly enriched the study of regulatory biology. For the moment, we focus on the way in which p_{bound} (the probability that the promoter is occupied by RNA polymerase) can be altered through the action of transcription factors such as repressors and activators.

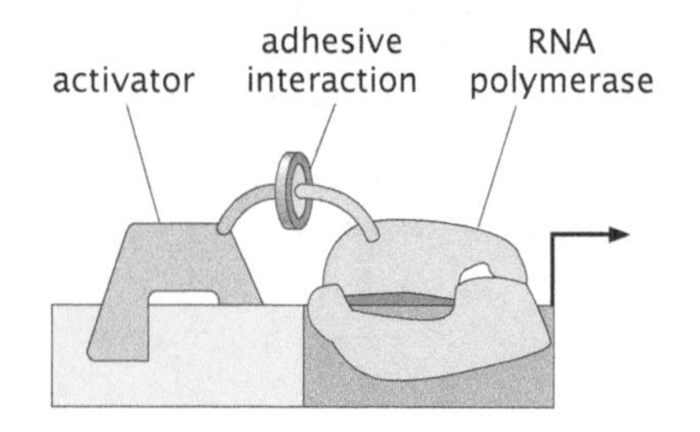

Figure 19.7: The process of activation. Schematic of the way in which activator molecules can recruit the transcription apparatus. Though both the activator and RNA polymerase have their own private interaction energies with the DNA, the enhancement in their occupancies is mediated by the adhesive interaction between them.

19.2.2 The Mathematics of Recruitment and Rejection

Recruitment of Proteins Reflects Cooperativity Between Different DNA-Binding Proteins

One of the key general ideas that pervade the description of transcriptional control (and beyond) is the idea of molecular recruitment. In the anthropomorphic terms suggested by the word "recruitment," the idea is that a given molecule that is bound on DNA summons some second molecule to the DNA, where it can then perform its task. For example, we think of RNA polymerase being summoned by some activator molecule such as a transcription factor (and vice versa) and exemplified by the CAP protein in the case of the *lac* operon. Though this colorful language is suggestive and conjures up a useful physical picture, from the perspective of the rules of statistical mechanics, this is nothing more than the well-worn idea of cooperativity cloaked in different verbal clothing.

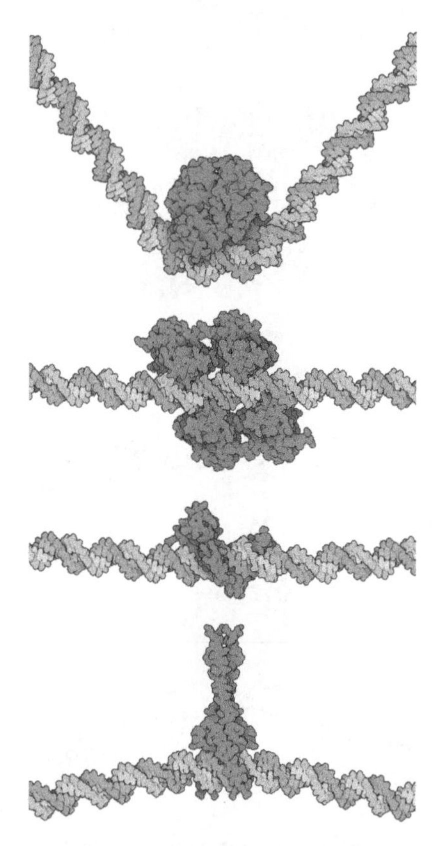

Figure 19.8: Structures of activator molecules. From top to bottom, the activators are CAP (pdb 1CGP), p53 tumor suppressor (pdb 3KMD), zinc finger DNA-binding domain (pdb 2GLI), and leucine zipper DNA-binding domain (pdb 1AN2). (Courtesy of D. Goodsell.)

Activators are proteins that regulate transcription by binding to a specific site on the DNA so as to recruit an RNA polymerase onto a nearby promoter site. It has been suggested that weak, nonspecific binding of the activator protein and the RNA polymerase can greatly enhance the probability of the polymerase binding to DNA, even for the very low concentrations of activator proteins typical of the cellular environment. To assess the feasibility of this strategy, we compute the probability of the polymerase being bound in the presence of an activator protein using a simple model that is depicted in cartoon form in Figure 19.9. The basic point of this cartoon is to show the different allowed states of polymerase and activator molecules and to use this enumeration of states to compute the probability that the promoter will be occupied. Indeed, this is the same "states-and-weights" mentality used throughout the book.

The first step in our analysis of this problem is to write the total partition function. Note that the partition function is obtained by summing over all of the eventualities associated with the activators and polymerase molecules being distributed on the DNA (both nonspecific sites and the promoter). As shown in Figure 19.9, there are four classes of outcomes, namely, both the activator site and promoter unoccupied, just the promoter occupied by polymerase, just the activator binding site occupied by activator, and, finally, both of the specific sites occupied. This is represented mathematically as

$$\begin{aligned} Z_{\text{tot}}(P, A; N_{\text{NS}}) = & \underbrace{Z(P, A; N_{\text{NS}})}_{\text{empty promoter}} + \underbrace{Z(P-1, A; N_{\text{NS}})\mathrm{e}^{-\beta\varepsilon_{\text{pd}}^{\text{S}}}}_{\text{RNAP}} \\ & + \underbrace{Z(P, A-1; N_{\text{NS}})\mathrm{e}^{-\beta\varepsilon_{\text{ad}}^{\text{S}}}}_{\text{activator}} \\ & + \underbrace{Z(P-1, A-1; N_{\text{NS}})\mathrm{e}^{-\beta(\varepsilon_{\text{ad}}^{\text{S}}+\varepsilon_{\text{pd}}^{\text{S}}+\varepsilon_{\text{ap}})}}_{\text{RNAP + activator}} . \end{aligned} \tag{19.1}$$

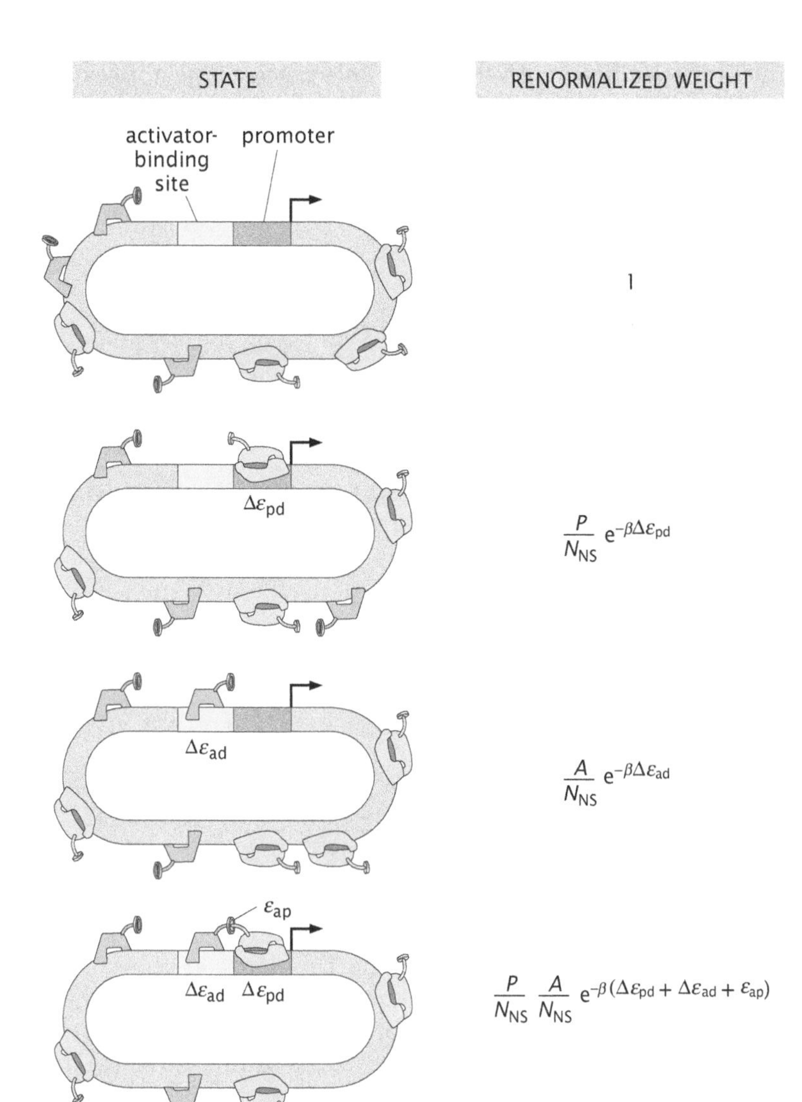

Figure 19.9: Schematic representation of the simple statistical mechanical model of recruitment. The states-and-weights diagram shows the different binding scenarios in the vicinity of the promoter of interest and the corresponding renormalized statistical weights obtained using statistical mechanics. We make the simplifying assumption that the nonspecific binding energy is *constant*. The large circular DNA is a cartoon representation of the bacterial genome.

Note that, notationally, the meaning of $Z(P, A; N_{NS})$ is that it is the partition function for P polymerase molecules and A activator molecules to be bound on the N_{NS} nonspecific sites and is given by

$$Z(P, A; N_{NS}) = \underbrace{\frac{N_{NS}!}{P!A!(N_{NS} - P - A)!}}_{\text{number of arrangements}} \times \underbrace{e^{-\beta P \varepsilon_{pd}^{NS}} e^{-\beta A \varepsilon_{ad}^{NS}}}_{\text{weight of each state}} . \tag{19.2}$$

We have also introduced the notation ε_{ap} to account for the "glue" interaction between the polymerase and activator. Like in Section 6.1.2 (p. 244) for the case of RNA polymerase, we introduce ε_{ad}^{S} and ε_{ad}^{NS} to characterize the binding energy of activator with its specific and non-specific DNA targets, respectively. Our expression involves a number of terms of the general form

$$\frac{N_{NS}!}{P!A!(N_{NS} - P - A)!} \times e^{-\beta P \varepsilon_{pd}^{NS}} e^{-\beta A \varepsilon_{ad}^{NS}} . \tag{19.3}$$

As we did earlier, we invoke a simplifying strategy that depends upon the fact that $N_{NS} \gg A + P$ and hence there will be almost zero chance of RNA polymerase and the activator finding each other on the same nonspecific site on the DNA. This permits the approximation $N_{NS}!/(N_{NS} - A - P)! \approx (N_{NS})^{A+P}$ introduced in Section 6.1.2 (see p. 244).

To compute the probability of promoter occupancy, we construct the ratio of all of those outcomes that are favorable (that is, polymerase bound to the promoter) to the total set of outcomes $(Z_{tot}(P, A; N_{NS}))$, namely,

$$p_{bound}(P, A; N_{NS}) = \frac{Z(P-1, A; N_{NS})e^{-\beta\varepsilon^{S}_{pd}} + Z(P-1, A-1; N_{NS})e^{-\beta(\varepsilon^{S}_{ad}+\varepsilon^{S}_{pd}+\varepsilon_{ap})}}{Z_{tot}(P, A; N_{NS})}. \quad (19.4)$$

We now propose to simplify this result by dividing both numerator and denominator by the numerator, resulting in

$$p_{bound}(P, A; N_{NS}) = \frac{1}{1 + [N_{NS}/PF_{reg}(A)]e^{\beta\Delta\varepsilon_{pd}}}, \quad (19.5)$$

where we introduce the regulation factor $F_{reg}(A)$, which is given by

$$F_{reg}(A) = \frac{1 + (A/N_{NS})e^{-\beta\Delta\varepsilon_{ad}}e^{-\beta\varepsilon_{ap}}}{1 + (A/N_{NS})e^{-\beta\Delta\varepsilon_{ad}}}, \quad (19.6)$$

and where we have defined $\Delta\varepsilon_{pd} = \varepsilon^{S}_{pd} - \varepsilon^{NS}_{pd}$ and $\Delta\varepsilon_{ad} = \varepsilon^{S}_{ad} - \varepsilon^{NS}_{ad}$. The details of the derivation are left to the problems at the end of the chapter. Note that in the limit that the adhesive interaction between polymerase and activator goes to zero, the regulation factor itself goes to unity. Further, note that for negative values of this adhesive interaction (that is, activator and polymerase like to be near each other), the regulation factor is greater than 1, which is translated into an effective increase in the number of polymerase molecules. The probability of RNA polymerase binding as a function of the number of activators is plotted in Figure 19.10.

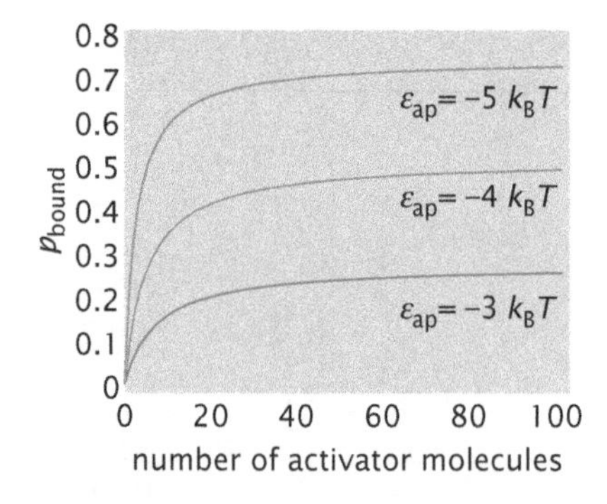

Figure 19.10: Illustration of the recruitment concept. This plot shows the probability of binding when the number of polymerase molecules is $P = 500$ and the binding parameters are $\Delta\varepsilon_{pd} = -5.3\ k_BT$ and $\Delta\varepsilon_{ad} = -13.12\ k_BT$. The three curves correspond to different choices of the adhesive interaction energy between polymerase and the activator.

The Regulation Factor Dictates How the Bare RNA Polymerase Binding Probability Is Altered by Transcription Factors

One of the intriguing claims that we will make is that a simple change in the effective number of RNA polymerase molecules $(P \rightarrow P_{eff})$ will suffice to capture the action of regulatory chaperones such as activators and repressors. This interpretation of the meaning of the regulation factor is shown in Figure 19.11. As a result of the presence of activators, it is as though the number of RNA polymerase molecules has been changed from P to $F_{reg}P$. For the case of activators, the regulation factor is greater than 1 and leads to an effective increase in the number of polymerase molecules. By way of contrast, we will show below that when repressors are present, they result in a regulation factor that is less than 1 and a concomitant decrease in the effective number of polymerase molecules.

In order for our calculations to really carry weight, we need to examine what they have to say about experiments. One of the primary measurables in *in vivo* experiments on regulation is the relative

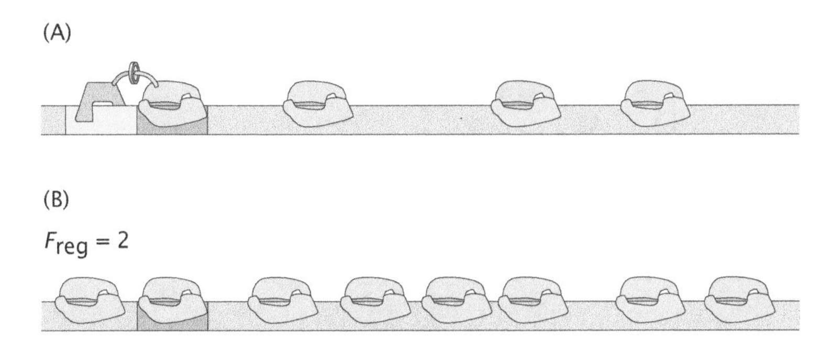

Figure 19.11: Regulation factor and the effective number of polymerase molecules. The presence of activators is equivalent to a problem with just polymerase molecules but a larger number of them. (A) The "bare" problem with activators and polymerase present. (B) The "effective" problem in which the presence of activators is treated as a change in the number of polymerase molecules.

expression for cases in which the transcription factor of interest is present or not. This qualitative notion is made quantitative by introducing the idea of the fold-change in activity, defined in the activation setting as

$$\text{fold-change} = \frac{p_{\text{bound}}(A \neq 0)}{p_{\text{bound}}(A = 0)} = \frac{1 + (N_{\text{NS}}/P)e^{\beta\Delta\varepsilon_{\text{pd}}}}{1 + [N_{\text{NS}}/PF_{\text{reg}}(A)]e^{\beta\Delta\varepsilon_{\text{pd}}}}. \quad (19.7)$$

What this expression reveals is how much more expression there is in the presence of activators relative to the "basal" state in which there is no activation.

As before, an inherent assumption in this analysis is the idea that the relative change in what is measured (for example, protein product, mRNA concentration, or promoter occupancy) is equal to the relative change in p_{bound}. Figure 19.12 illustrates the fold-change in gene expression for the problem of simple activation with a choice of parameters dictated by *in vitro* experiments for a value of $\Delta\varepsilon_{\text{ad}}$ in conjunction with an educated guess for ε_{ap} that results in typical fold-changes in activity reported *in vivo* of about 50. Note that a weak promoter satisfies the condition $(N_{\text{NS}}/P)e^{\beta\Delta\varepsilon_{\text{pd}}} \gg 1$, which implies that the fold-change in activity can be rewritten as

$$\text{fold-change} \approx F_{\text{reg}}(A). \quad (19.8)$$

Here we have also assumed that $(N_{\text{NS}}/PF_{\text{reg}})e^{\beta\Delta\varepsilon_{\text{pd}}} \gg 1$, which means that the promoter is not too strong even in the regulated case. The conclusion is that in the case of a weak promoter the actual details of the promoter, such as its binding energy, factor out of the problem.

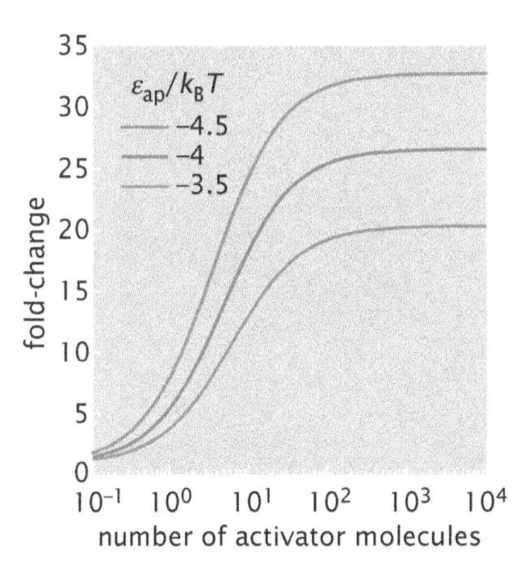

Figure 19.12: Fold-change due to activators. Fold-change in gene expression as a function of the number of activators for different activator–RNA polymerase interaction energies using $P = 500$, $\Delta\varepsilon_{\text{pd}} = -5.3\ k_BT$, and $\Delta\varepsilon_{\text{ad}} = -13.12\ k_BT$ based on *in vitro* measurements.

Activator Bypass Experiments Show That Activators Work by Recruitment

The simple picture of regulated recruitment introduced here is based in part upon a series of classic experiments known as activator bypass experiments. The key idea of such experiments is shown in Figure 19.13. These experiments involve a mix-and-match approach where the DNA-binding domain from one protein is fused with the activator domain of a second protein. A second version of this experiment is based upon direct tethering of the activator and the polymerase. After making the activator bypass constructs, it was found that the gene of interest was still activated. Our ambition here is to consider these experiments more quantitatively and to note that, if viewed from a mathematical perspective, these two classes of experiments lead to different quantitative outcomes that can be used to further test the full range of validity of the notion of regulated recruitment.

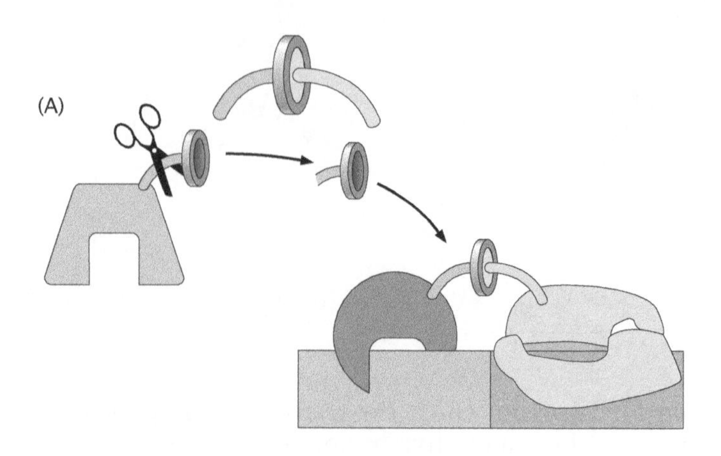

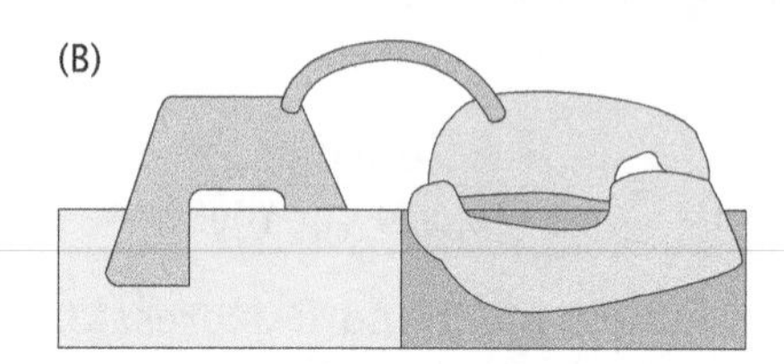

Figure 19.13: Schematic of activator bypass experiments. (A) Activator bypass type 1 in which activation is mediated by proteins with designer DNA-binding regions. (B) Activator bypass type 2 in which the activator is tethered directly to polymerase.

We have already worked out the regulation factor that is associated with activator bypass type 1 experiments. The only change relative to Equation 19.6 is that, by using different proteins, quantities such as $\Delta\varepsilon_{\mathrm{ad}}$ and $\varepsilon_{\mathrm{ap}}$ will have different numerical values, which means that the actual level of activation can be different in this experiment relative to its "wild-type" value. On the other hand, the entire functional form for the regulation factor is different in the case of activator bypass type 2. In this case, there are only two states we really need to consider, namely, polymerase with and without tethered activator bound at the promoter with weights $(P/N_{\mathrm{NS}})\mathrm{e}^{-\beta(\Delta\varepsilon_{\mathrm{pd}}+\Delta\varepsilon_{\mathrm{ad}})}$ and 1, respectively. This implies that the probability that polymerase will be bound is

$$p_{\mathrm{bound}}(P; N_{\mathrm{NS}}) = \frac{1}{1 + (N_{\mathrm{NS}}/P)\mathrm{e}^{\beta\Delta\varepsilon_{\mathrm{ad}}}\mathrm{e}^{\beta\Delta\varepsilon_{\mathrm{pd}}}}. \qquad (19.9)$$

This implies, in turn, that the regulation factor takes the particularly simple form

$$F_{\mathrm{reg}} = \mathrm{e}^{-\beta\Delta\varepsilon_{\mathrm{ad}}}, \qquad (19.10)$$

which amounts to the statement that the effective binding energy of polymerase is shifted and nothing more.

Repressor Molecules Reduce the Probability Polymerase Will Bind to the Promoter

The same logic that was introduced above to consider the case of pure activation (that is, recruitment) can be brought to bear on the problem of repression. Once again, we are faced with considering all of the ways of distributing the repressor and RNA polymerase molecules and it is convenient to introduce the partition function associated with the binding of these molecules to nonspecific sites as

$$Z(P,R:N_{\text{NS}}) = \frac{N_{\text{NS}}!}{P!R!(N_{\text{NS}}-P-R)!} \mathrm{e}^{-\beta P\varepsilon_{\text{pd}}^{\text{NS}}} \mathrm{e}^{-\beta R\varepsilon_{\text{rd}}^{\text{NS}}}, \tag{19.11}$$

which is formally identical to Equation 19.2, but where we have introduced the notation $\varepsilon_{\text{rd}}^{\text{NS}}$ to describe the nonspecific binding of repressor to DNA ($\varepsilon_{\text{rd}}^{\text{S}}$ will be reserved for the specific binding energy of repressor to its operator). In order to write the *total* partition function for all the allowed states, we now need to sum over the states in which the promoter is occupied either by a repressor molecule or by an RNA polymerase molecule. The set of allowed states in this simple model as well as their associated weights are shown in Figure 19.14. Note that in considering this particular model, we do not enter into structural fine points such as whether or not the RNA polymerase can be on its promoter at the same time as the repressor is bound to its operator—the model is intended to be the simplest treatment of the statistical mechanics of the competition between repressors and RNA polymerase.

The total partition function is given by

$$Z_{\text{tot}}(P,R;N_{\text{NS}}) = \underbrace{Z(P,R;N_{\text{NS}})}_{\text{empty promoter}} + \underbrace{Z(P-1,R;N_{\text{NS}})\mathrm{e}^{-\beta\varepsilon_{\text{pd}}^{\text{S}}}}_{\text{RNAP on promoter}} + \underbrace{Z(P,R-1;N_{\text{NS}})\mathrm{e}^{-\beta\varepsilon_{\text{rd}}^{\text{S}}}}_{\text{repressor on promoter}}. \tag{19.12}$$

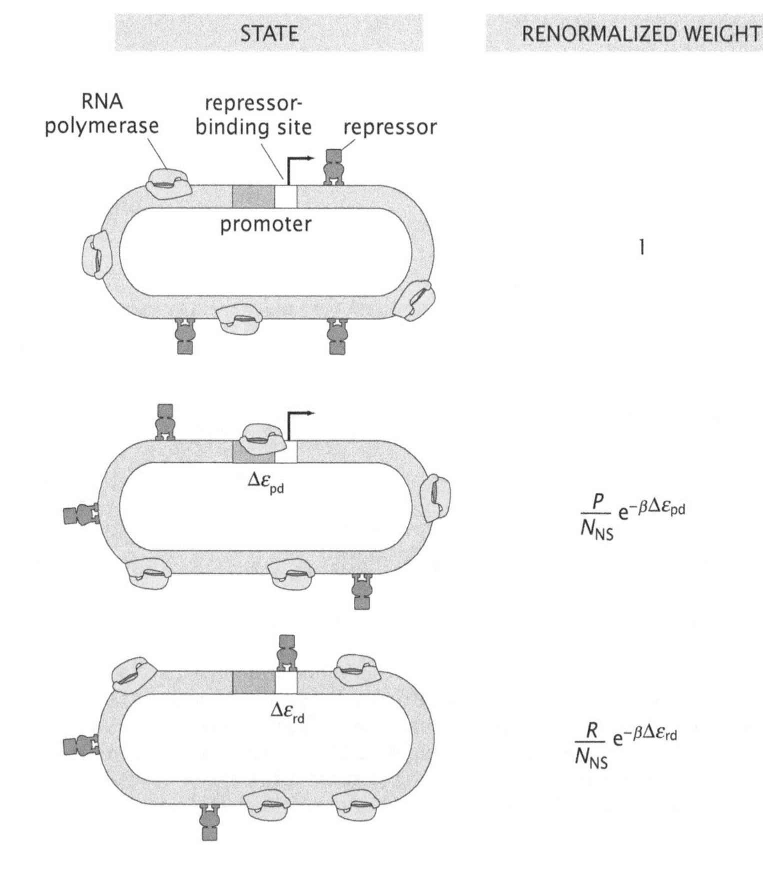

Figure 19.14: States and weights for the case of simple repression. The states of promoter occupancy are empty promoter, RNA polymerase on the promoter, and repressor on the promoter.

This result now provides us with the tools with which to evaluate the probability that the promoter will be occupied by RNA polymerase. This probability is given by the ratio of the favorable outcomes to all of the outcomes. In mathematical terms, that is

$$p_{\text{bound}}(P, R; N_{\text{NS}}) = \frac{Z(P-1, R; N_{\text{NS}})e^{-\beta\varepsilon_{\text{pd}}^{\text{S}}}}{Z(P, R; N_{\text{NS}}) + Z(P-1, R; N_{\text{NS}})e^{-\beta\varepsilon_{\text{pd}}^{\text{S}}} + Z(P, R-1; N_{\text{NS}})e^{-\beta\varepsilon_{\text{rd}}^{\text{S}}}}. \quad (19.13)$$

As argued above, this result can be rewritten in compact form using the regulation factor by dividing top and bottom by $Z(P-1, R; N_{\text{NS}})e^{-\beta\varepsilon_{\text{pd}}^{\text{S}}}$ and by invoking the approximation

$$\frac{N_{\text{NS}}!}{P!R!(N_{\text{NS}} - P - R)!} \approx \frac{N_{\text{NS}}^{\text{P}}}{P!}\frac{N_{\text{NS}}^{\text{R}}}{R!}, \quad (19.14)$$

which amounts to the physical statement that there are so few polymerase and repressor molecules in comparison with the number of available sites, N_{NS}, that each of these molecules can more or less

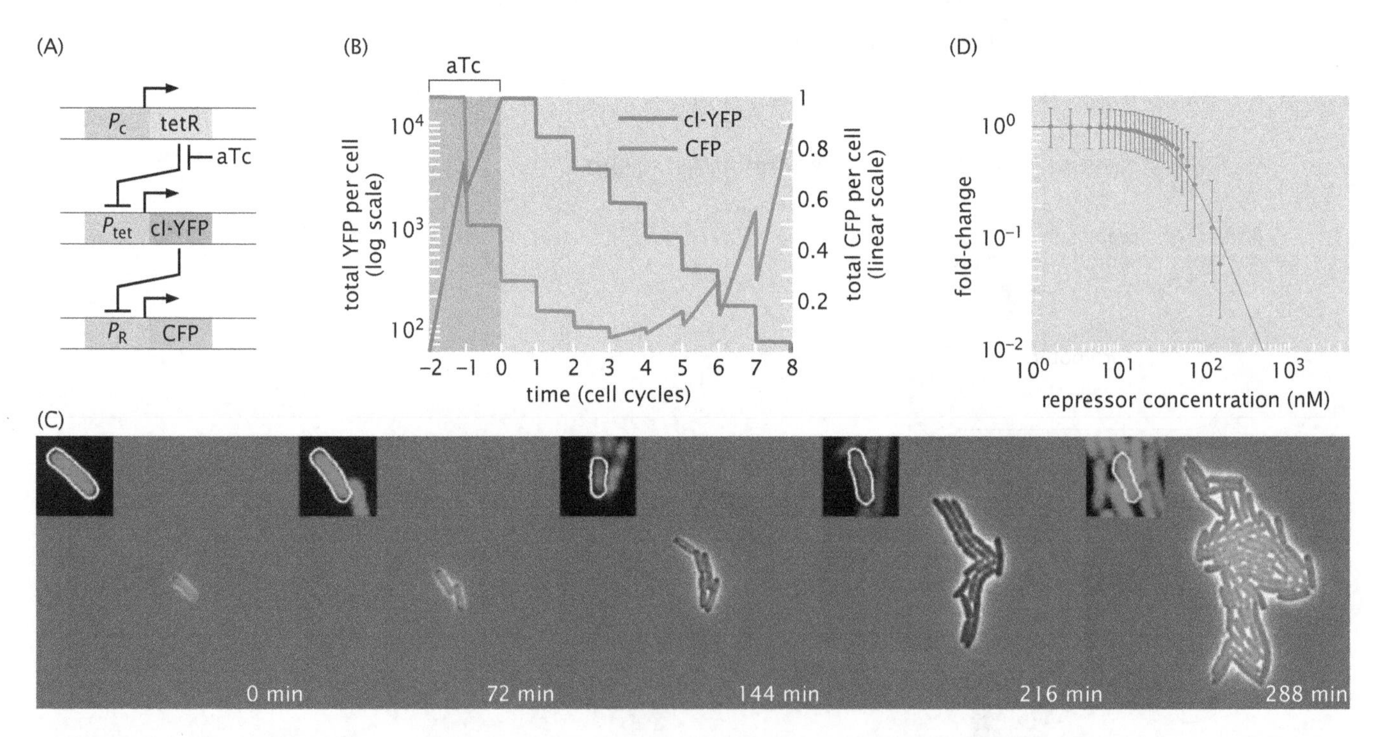

Figure 19.15: Dilution experiment and the measurement of fold-change in repression. (A) Diagram of the circuit. In the absence of the inducer aTc, the repressor TetR shuts down production of the transcription factor cI fused to YFP. This transcription factor, in turn, regulates the expression of the reporter CFP. (B) Schematic of the time course of an experiment. Adding aTc for a short period of time leads to the production of cI-YFP. Upon removal of aTc, no new cI-YFP is produced. As a result, in each new generation, there will be decreasing numbers of cI-YFP per cell, resulting in an ever-higher rate of expression of the downstream CFP gene. This dilution also permits the calibration of YFP fluorescence into absolute numbers of cI-YFP as discussed in the text. (C) Representative snapshots from the time course of an experiment. (D) Fold change (1/repression) as a function of cI repressor concentration measured using the dilution method. (Adapted from N. Rosenfeld et al. *Science* 307: 1962, 2005.)

fully explore those N_{NS} sites. The resulting probability is

$$p_{bound}(P, R; N_{NS}) = \frac{1}{1 + (N_{NS}/P)e^{\beta(\varepsilon_{pd}^{S} - \varepsilon_{pd}^{NS})}(1 + (R/N_{NS})e^{-\beta(\varepsilon_{rd}^{S} - \varepsilon_{rd}^{NS})})}. \tag{19.15}$$

This result can be couched in regulation factor language with the observation that the regulation factor itself is given by

$$F_{reg}(R) = \left(1 + \frac{R}{N_{NS}}e^{-\beta\Delta\varepsilon_{rd}}\right)^{-1}, \tag{19.16}$$

with $\Delta\varepsilon_{rd} = \varepsilon_{rd}^{S} - \varepsilon_{rd}^{NS}$. Note that the regulation factor in the case of repression satisfies the inequality $F_{reg} < 1$, which can be interpreted as a reduction in the effective number of RNA polymerase molecules. We explore this in more detail in Section 19.2.5 when discussing the particular case of the *lac* operon, though Figure 19.15 gives an example of an extremely elegant measurement of the effect of repression using the beautiful dilution method introduced in the Computational Exploration on p. 46.

COMPUTATIONAL EXPLORATION

Computational Exploration: Extracting Level of Gene Expression from Microscopy Images One way to determine the level of gene expression is to use microscopy images of cells expressing some fluorescent reporter. In this Computational Exploration, the reader is invited to use Matlab to extract the fluorescence intensities from a collection of cells and to use them to determine the fold-change in simple repression.

The logical progression associated with this analysis is introduced schematically in Figure 19.16. Note that we have images of the cells in two different channels. In particular, for each field of view, we have both a phase contrast image and a fluorescence image. Like with the example where we determined the cell cycle time of *E. coli* (p. 100), the first step is to find the cells in an automated fashion using some segmentation scheme. Additionally, we need to choose which one of the two images we want to do the segmentation with. Detecting cells using the fluorescence image is certainly appealing due to the absence of any other fluorescent objects. However, it is clear that for dimmer cells the segmentation might not work as well. As a result, we would risk biasing our segmentation based on the level of expression of the cells, the quantity we are actually interested in measuring! Instead, we choose to segment the phase contrast image, which should, in principle, not be subject to bias resulting from the level of fluorescence within each cell.

Following the procedure outlined in the example on the cell division time in *E. coli* (p. 100), once we have performed the thresholding, we will be left with a mask image with discrete regions that we identify as cells denoted by the different colors in Figure 19.16(C). These ideas are illustrated in the Matlab code associated with this exploration. Once the segmentation

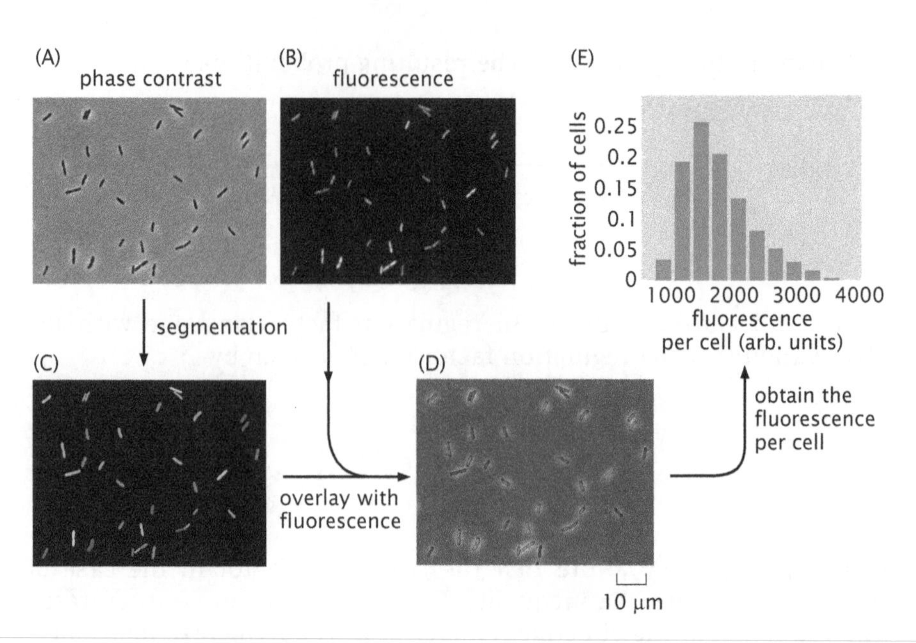

Figure 19.16: Schematic of the image segmentation algorithm to quantify levels of gene expression in bacteria. Two images of bacteria expressing a fluorescent protein are obtained, (A) one in phase contrast and (B) one in fluorescence. The phase contrast image is an imaging scheme that makes it possible to see the bacteria as dark objects. (C) These objects are automatically detected and segmented using computer software that assigns an identity to each segmented bacterium (represented by the different colors). (D) The mask generated by this procedure is applied to the fluorescence image in order to generate an overlay and integrate the fluorescence within the mask of each segmented cell. (E) By repeating this for multiple images and many cells, the distribution of fluorescence per cell can be computed.

process is complete, we can then obtain the fluorescence intensity in each of our cells. To do so, we use the segmented image from the previous step to find the individual cells and then, within each such cell, we ask for the fluorescence intensity of all of the pixels and sum them up. The result is a distribution of fluorescence per cell as shown in Figure 19.16(E). However, there is an extra subtlety that has to be taken into account when obtaining such fluorescence distributions. In particular, because of the intrinsic fluorescence of the cells themselves, there is a spurious contribution to the total fluorescence that we measure, F_{total}, which is given by

$$F_{\text{total}} = F_{\text{reporter}} + F_{\text{cell}}, \tag{19.17}$$

where F_{reporter} is the signal stemming from the fluorescent reporter while F_{cell} is the autofluorescence of the cell. As a result, we need to be able to subtract the cells' average autofluorescence if we want to report only on F_{reporter}. This can be easily done by following the steps outlined in Figure 19.16 and described above, but now for a strain of bacteria that lacks any fluorescent reporter. As a result, we will be able to measure the mean contribution of the cell autofluorescence to the total fluroescence, $\langle F_{\text{cell}} \rangle$, which can then be subtracted from the fluorescence values in the presence of the reporter.

With the fluorescence intensities in hand, we are now prepared to compute the fold-change itself so that we can examine the accord between the model of simple repression presented in Equation 19.16 and the data itself. The logic of this part of the analysis is presented in Figure 19.17. Here the idea is to use our mean fluorescence intensities, corrected for the fluorescence background, for both the regulated and unregulated promoters and then to construct the ratio of these means.

Examples of Matlab code that could be used to perform this Computational Exploration, as well as images of *E. coli* suitable for this analysis, can be found on the book's website.

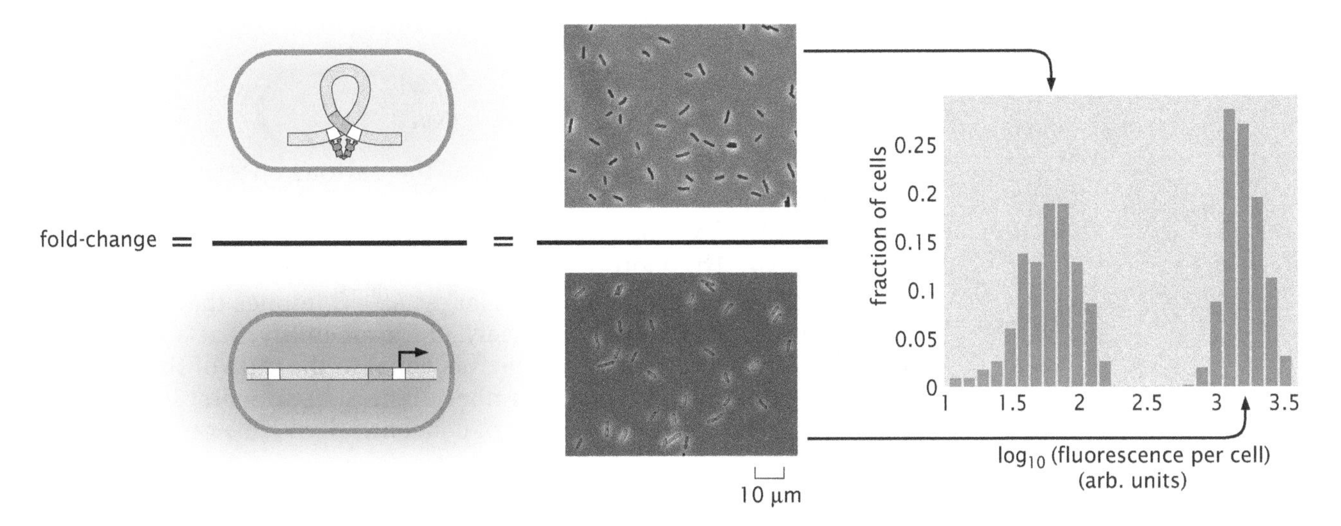

Figure 19.17: Converting image intensities to fold-change. The fold-change in gene expression is defined as the ratio of the levels of gene expression coming from a strain bearing the transcription factor of interest over a strain with a deletion of such transcription factor. For each one of these two strains, the procedure described in Figure 19.16 can be performed, leading to a distribution of fluorescence for each strain. Additionally, the cell autofluorescence is subtracted from each sample by analyzing a strain bearing no fluorescent protein. The means of each distribution can be divided in order to calculate the fold-change in gene expression.

19.2.3 Transcriptional Regulation by the Numbers: Binding Energies and Equilibrium Constants

We have heard it said that "physics isn't worth a damn unless you put in some numbers!" The abstract expressions obtained so far are much more interesting when viewed through the prism of particular measurements. Binding energies quantify the affinity of RNA polymerase or transcription factors for their DNA targets. In particular, RNA polymerase and transcription factors perform molecular recognition as a result of a rank ordering of their preferences for different sequences of nucleotides. Indeed, the sequence associated with a given promoter distinguishes it from some random sequence to which RNA polymerase would bind with a nonspecific binding energy ε_{pd}^{NS}. Specific binding energies can also be tuned. For example, even though there might be one very strong consensus promoter, that binding strength can be reduced by introducing mismatches in the sequence. A strong promoter, with a p_{bound} close to 1, will have a strong level of expression. On the other hand, by weakening a given promoter, cells can broaden their dynamic range by introducing a codependency on a battery of transcription factors that effectively tune the range of binding affinities and permit the regulation of promoter occupancy.

Equilibrium Constants Can Be Used To Determine Regulation Factors

In order to compute the regulation factors for the various regulatory scenarios under consideration in this chapter, we need to make estimates for the energy associated with binding protein X to the DNA, both specifically and nonspecifically; protein X can be a repressor or an activator. Binding energies are determined indirectly in experiments that measure the equilibrium constant for binding X to DNA (D). In particular, we consider the reaction

$$X + D \rightleftharpoons XD \tag{19.18}$$

with an equilibrium binding constant

$$K_X^{(\text{bind})} = \frac{[\text{XD}]}{[\text{X}][\text{D}]}. \tag{19.19}$$

Here, $[\cdots]$ denotes concentrations of the various species taking part in the reaction.

When a single X binds to DNA, there is an overall change in the free energy, Δf_{XD}. The more negative this quantity is, the more likely X will be bound to DNA. Similarly, a larger $K_X^{(\text{bind})}$ implies that the bound state is more likely. More precisely, the probability that a particular binding site on the DNA is occupied is equal to the ratio of the number of occupied sites to the total number of sites, as was first introduced in Section 6.4.1 (p. 270). In terms of concentrations, this can be written

$$p_{\text{bound}} = \frac{[\text{XD}]}{[\text{D}] + [\text{XD}]} = \frac{K_X^{(\text{bind})}[\text{X}]}{1 + K_X^{(\text{bind})}[\text{X}]}, \tag{19.20}$$

where the final expression follows from Equation 19.19. On the other hand, given that there are $[\text{X}]V_{\text{cell}}$ copies of protein X in the cell (V_{cell} is the volume of the cell), the probability of a DNA-binding site being occupied is

$$p_{\text{bound}} = \frac{[\text{X}]V_{\text{cell}}\mathrm{e}^{-\beta\Delta f_{\text{XD}}}}{1 + [\text{X}]V_{\text{cell}}\mathrm{e}^{-\beta\Delta f_{\text{XD}}}}. \tag{19.21}$$

Comparison of the two expressions for p_{bound} allows us to relate the microscopic and macroscopic views of binding through the relation

$$\frac{K_X^{(\text{bind})}}{V_{\text{cell}}} = \mathrm{e}^{-\beta\Delta f_{\text{XD}}}. \tag{19.22}$$

Using this relation, we can compute the binding free energies for RNA polymerase and the various transcription factors in *E. coli*, which provides an alternative description of the same underlying processes. Presently, we use these ideas to tackle the *lac* operon, which features both positive and negative regulation.

19.2.4 A Simple Statistical Mechanical Model of Positive and Negative Regulation

Real regulatory architectures in cells often involve both repression and activation simultaneously. In this case, we consider the five distinct outcomes shown in Figure 19.18 and captured through the total partition function

$$\begin{aligned}
Z_{\text{tot}}(P, A, R; N_{\text{NS}}) \\
&= \underbrace{Z(P, A, R; N_{\text{NS}})}_{\text{empty promoter}} + \underbrace{Z(P-1, A, R; N_{\text{NS}})\mathrm{e}^{-\beta\varepsilon_{\text{pd}}^{\text{S}}}}_{\text{RNAP}} \\
&+ \underbrace{Z(P, A-1, R; N_{\text{NS}})\mathrm{e}^{-\beta\varepsilon_{\text{ad}}^{\text{S}}}}_{\text{activator}} + \underbrace{Z(P-1, A-1, R; N_{\text{NS}})\mathrm{e}^{-\beta(\varepsilon_{\text{ad}}^{\text{S}}+\varepsilon_{\text{pd}}^{\text{S}}+\varepsilon_{\text{ap}})}}_{\text{RNAP + activator}} \\
&+ \underbrace{Z(P, A, R-1; N_{\text{NS}})\mathrm{e}^{-\beta\varepsilon_{\text{rd}}^{\text{S}}}}_{\text{repressor}} + \underbrace{Z(P, A-1, R-1; N_{\text{NS}})\mathrm{e}^{-\beta(\varepsilon_{\text{ad}}^{\text{S}}+\varepsilon_{\text{rd}}^{\text{S}})}}_{\text{activator + repressor}}.
\end{aligned} \tag{19.23}$$

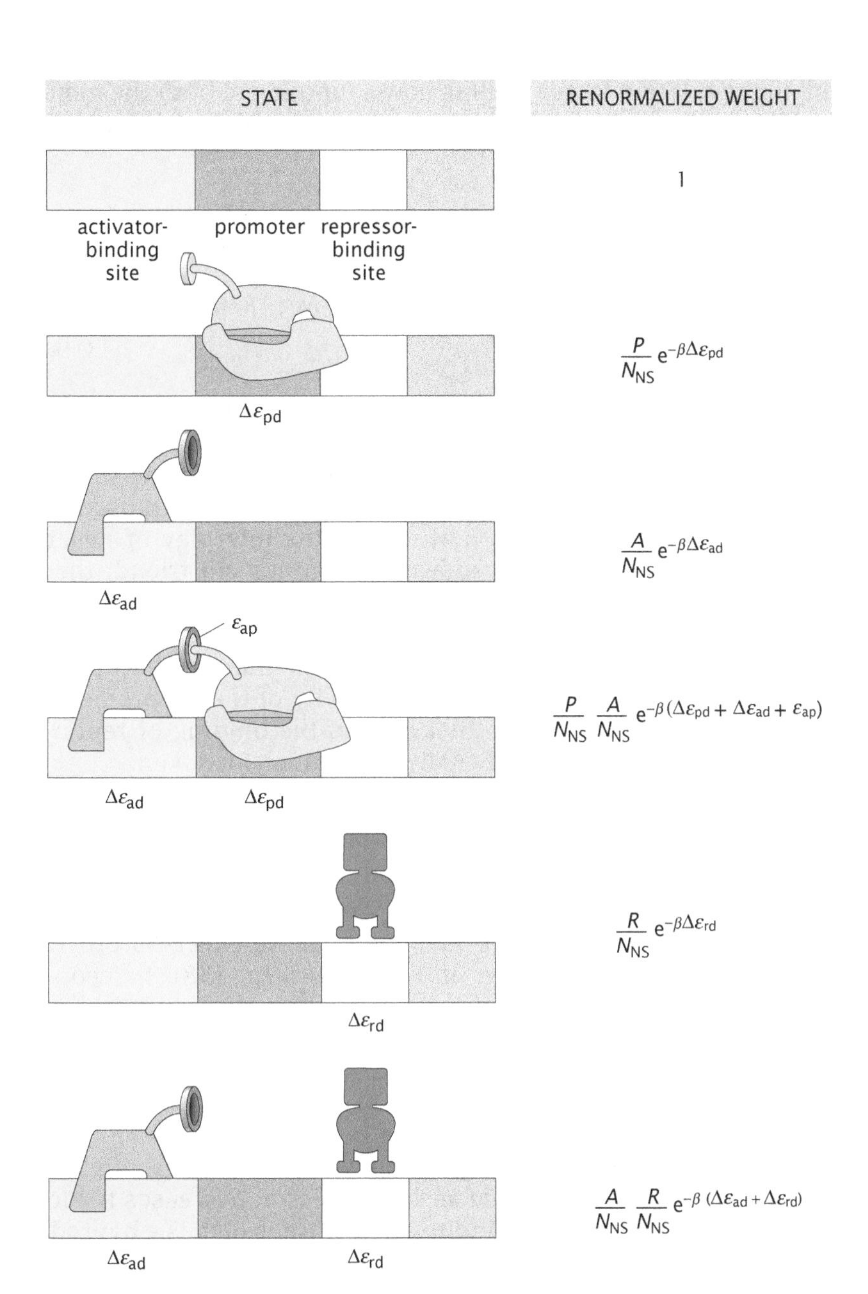

Figure 19.18: Schematic representation of the simple statistical mechanical model of recruitment and repression. States and weights for the case in which activation and simple repression act simultaneously.

Note that the cartoon shows a schematic representation of the different ways that the region in the vicinity of the promoter can be occupied and what the statistical weights are of each such state of occupancy. We can compute the probability of RNA polymerase binding by considering the ratio of favorable outcomes to the total partition function, resulting in

$$p_{\text{bound}}(P, A, R; N_{\text{NS}}) = \frac{Z(P-1, A, R; N_{\text{NS}})e^{-\beta\varepsilon_{\text{pd}}^{\text{S}}} + Z(P-1, A-1, R; N_{\text{NS}})e^{-\beta(\varepsilon_{\text{ad}}^{\text{S}}+\varepsilon_{\text{pd}}^{\text{S}}+\varepsilon_{\text{ap}})}}{Z_{\text{tot}}(P, A, R; N_{\text{NS}})}. \tag{19.24}$$

As before, perhaps the simplest way to interpret this result is with reference to the regulation factor, resulting in

$$p_{\text{bound}}(P, A, R; N_{\text{NS}}) = \frac{1}{1 + [N_{\text{NS}}/PF_{\text{reg}}(A, R)]e^{\beta(\varepsilon_{\text{pd}}^{\text{S}}-\varepsilon_{\text{pd}}^{\text{NS}})}}, \tag{19.25}$$

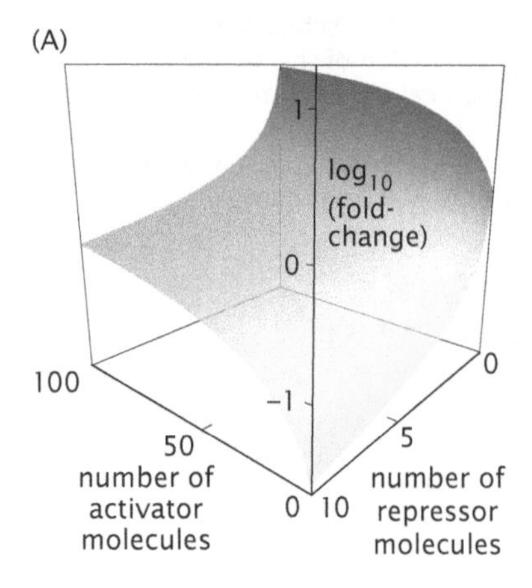

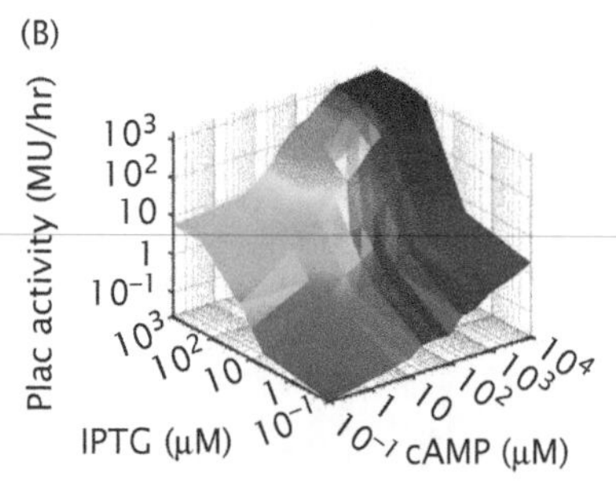

Figure 19.19: Combined regulation by repressor and activator. (A) The fold-change in gene expression as a function of the number of transcription factors shows their combinatorial action. The parameters used are $\Delta\varepsilon_{\text{ad}} = -10\ k_{\text{B}}T$, $\varepsilon_{\text{ap}} = -3.9\ k_{\text{B}}T$ and $\Delta\varepsilon_{\text{rd}} = -16.9\ k_{\text{B}}T$. (B) Activity of the *lac* operon measured in Miller units (MU) per hour as a function of the concentration of IPTG and cAMP, which regulate the binding of Lac repressor and CRP to the DNA, respectively. (B, adapted from T. Kuhlman et al., *Proc. Natl. Acad. Sci. USA* 104:6043, 2007.)

where the regulation factor itself is now a function of both the number of activators, A, and the number of repressors, R. In particular, the regulation factor is given by

$$F_{\text{reg}}(A,R) = \frac{1+(A/N_{\text{NS}})e^{-\beta(\Delta\varepsilon_{\text{ad}}+\varepsilon_{\text{ap}})}}{1+(A/N_{\text{NS}})e^{-\beta\Delta\varepsilon_{\text{ad}}}+(R/N_{\text{NS}})e^{-\beta\Delta\varepsilon_{\text{rd}}}+(A/N_{\text{NS}})(R/N_{\text{NS}})e^{-\beta(\Delta\varepsilon_{\text{ad}}+\Delta\varepsilon_{\text{pd}})}}. \tag{19.26}$$

The variation in fold-change in gene expression due to this regulatory architecture in the weak promoter approximation is shown in Figure 19.19(A). The objective of this figure is to illustrate the combinatorial control that can be reached when different transcription factors act in unison. Perhaps nowhere is this interplay of negative and positive regulation better known than in our old friend, the *lac* operon. In fact, Figure 19.19(B) reveals this interplay between activation and repression in the particular context of the *lac* operon. Here, instead of varying the intracellular number of transcription factors, the simpler approach of measuring the activity of the *lac* promoter as a function of the two inducers that control the binding of repressor and activator to DNA (IPTG and cAMP, respectively) is taken.

19.2.5 The *lac* Operon

Both repression and activation are key parts of the equipment of bacteria. Perhaps the most famous example of these effects is provided by the *lac* operon and is shown in Figure 4.15 (p. 158). Indeed, the *lac* operon has served as one of the central workhorses of the entire book, and the present section is the *denouement* of that discussion. In this case, the activator is the catabolite activator protein (CAP), also known as cyclic AMP receptor protein (CRP). In order to be able to recruit RNA polymerase, CAP has to be bound to cyclic AMP (cAMP), a molecule whose concentration goes up when the amount of glucose decreases. The repressor, known as Lac repressor, decreases the level of transcription unless it is bound to allolactose, which is a byproduct of lactose metabolism.

The *lac* Operon Has Features of Both Negative and Positive Regulation

Recall that the *lac* operon oversees the management of the enzymes that are responsible for lactose uptake and digestion. In particular, when *E. coli* cells find themselves simultaneously deprived of glucose and supplied with lactose, the genes of the *lac* operon are turned on so as to take metabolic advantage of the lactose. We have already described the way in which the Lac repressor forbids transcription of the genes associated with lactose digestion by binding on its operator. However, our earlier discussion was a bit too blithe, since we said nothing of what happens in the case where glucose and lactose are simultaneously available. If we were to adopt the picture of negative control described above, then our expectation would be that in this case there should be substantive transcription of the genes of the *lac* operon. However, there is a second element of positive control that completes the story. In particular, in the *absence* of glucose, the activator CAP binds to a site near the promoter (the RNA polymerase-binding

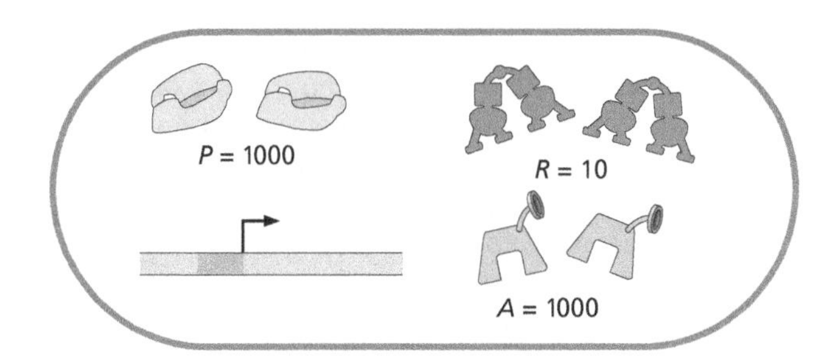

Figure 19.20: Census of the relevant molecular actors in the *lac* operon. The figure shows a rough estimate of the number of polymerase molecules, activators, and repressors associated with the *lac* operon.

site) as shown in Figure 4.15 (p. 158) and "recruits" RNA polymerase to the promoter. The census shown in Figure 19.20 gives a rough impression of the number of copies of some of the key molecules associated with the *lac* operon and illustrates the striking fact that some of the transcription factors exist with as few as 10 copies.

The geometry of the regulatory landscape for the *lac* operon is shown in Figure 19.21. Our discussion of Figure 4.15 (p. 158) was oversimplified in the sense that we ignored the presence of auxiliary binding sites for the Lac repressor that are revealed in Figure 19.21. In particular, there are two other binding sites for the Lac repressor. Specifically, there is a binding site known as O2 located 401 bp downstream from O1 and a second such site known as O3 situated 92 bp upstream. Part of our discussion will center on the subtle ways in which repression takes place in this system. Recall that the repressor itself is a tetramer with two "reading heads" that can each bind to a different operator, looping out the intervening DNA.

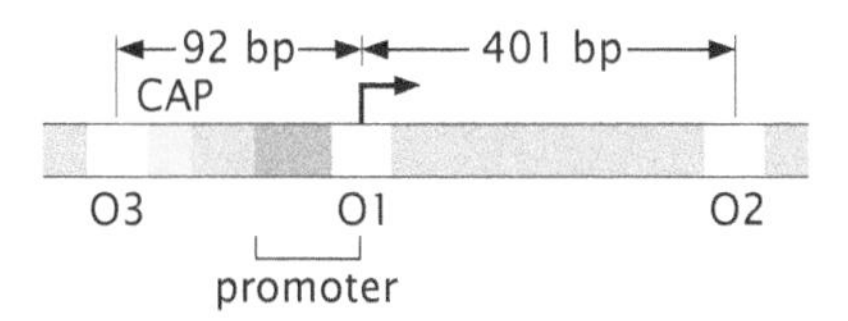

Figure 19.21: Position of the three *lac* operators and the CAP-binding site relative to the promoter. O1 is the main operator, while O2 and O3 are auxiliary binding sites for Lac repressor and are associated with DNA looping.

One of the most important roles for models like those described here is in providing a conceptual framework for thinking about both *in vivo* and *in vitro* data and in suggesting new experiments. A particularly compelling class of *in vivo* experiments using the *lac* operon measured the repression as a function of the strength and placement of the operator sites that are the targets of Lac repressor. In particular, *E. coli* cells were created that had only one operator for Lac repressor as well as mutants with different spacings between operators (a topic we return to below). The first set of experiments we consider are those in which only one operator was present for Lac repressor binding as shown in Figure 19.22. In these experiments, the repression was measured for cases in which the promoter was repressed by each of the operators O1, O2, and O3 individually. From the standpoint of the models considered here, all that is different from one experiment to the next is the binding energy of repressor for the DNA.

Recall that for a single repressor, the regulation factor is given by Equation 19.16. What is measured in the experiment is the ratio of the level of gene expression in the absence of repressor to that in the presence of repressor. For the purposes of our model, we replace

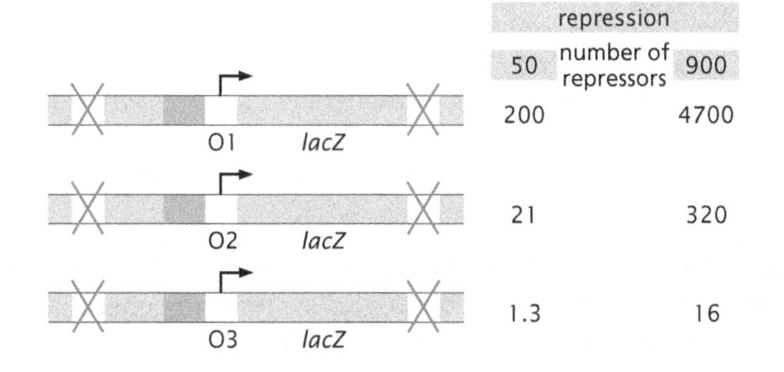

Figure 19.22: Repression in the *lac* operon. The DNA constructs used in these experiments deleted the auxiliary binding sites for repressor and tuned the strength of the main repressor-binding site. Repression, the inverse of the fold-change in gene expression, was measured in each construct for two different concentrations of Lac repressor. (Adapted from S. Oehler et al., *EMBO J.* 13:3348, 1994.)

this definition based upon a measure of protein content (that is, the product of the gene) with a definition based upon examining the probability that the promoter is occupied by RNA polymerase. The implicit assumption here is that the protein content is linearly related to the probability of promoter occupancy. More precisely, we define repression as the ratio of the probability of binding of RNA polymerase to the relevant promoter in the absence of repressor to the probability of such binding in the presence of repressor, namely

$$\text{repression} = \frac{p_{\text{bound}}(R=0)}{p_{\text{bound}}(R \neq 0)}. \tag{19.27}$$

Concretely, this result depends on the number of repressors, R, and their energy of binding to DNA. If we substitute for p_{bound} using Equation 19.15, we find that the repression can be written as

$$\text{repression}(R) = \frac{1 + (P/N_{\text{NS}})\text{e}^{-\beta\Delta\varepsilon_{\text{pd}}} + (R/N_{\text{NS}})\text{e}^{-\beta\Delta\varepsilon_{\text{rd}}}}{1 + (P/N_{\text{NS}})\text{e}^{-\beta\Delta\varepsilon_{\text{pd}}}}. \tag{19.28}$$

For the case of a weak promoter, this implies in turn that the repression level can be written as

$$\text{repression}(R) = [\text{fold-change}(R)]^{-1} \simeq [F_{\text{reg}}(R)]^{-1} = 1 + \frac{R}{N_{\text{NS}}}\text{e}^{-\beta\Delta\varepsilon_{\text{rd}}}. \tag{19.29}$$

One of the interesting opportunities afforded by this expression is the possibility of a direct confrontation with experimental data such as is shown in Figure 19.22.

In particular, the data of Figure 19.22 permit us to determine the only unknown in our expression for the repression, namely, the energy parameter $\Delta\varepsilon_{\text{rd}}$. Since the data reflect three different choices of binding strength, we find three different binding energies ($\Delta\varepsilon_{\text{rd}} = -16.9, -14.4$, and $-11.2\ k_{\text{B}}T$ for O1, O2, and O3, respectively). With these energies in hand, we can predict the outcome of repression measurements in which the number of repressors is tuned to other values as shown in Figure 19.23. Note that once the binding-energy difference has been estimated using one data point, it leads to a prediction for the behavior of the system for different numbers of repressor molecules in the cell and will serve as the basis for our analysis of the two-operator case as well.

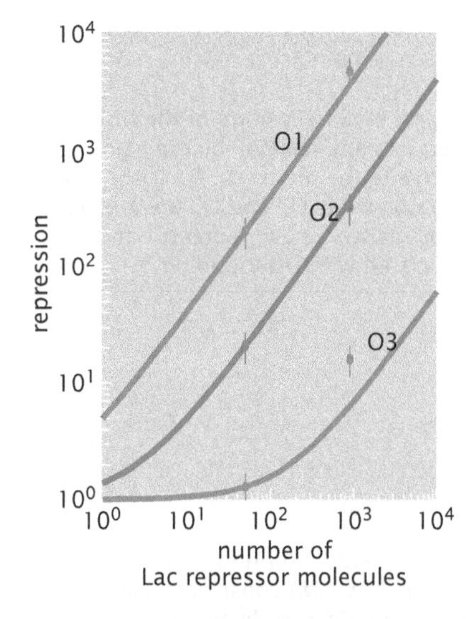

Figure 19.23: Repression model for the *lac* operon. Each curve shows how repression varies as a function of the number of repressor molecules in the cell for constructs with a single main binding site as shown in Figure 19.22. Different curves correspond to different main binding sites (operators) for the Lac repressor. (Data from S. Oehler et al., *EMBO J.* 13:3348, 1994.)

The Free Energy of DNA Looping Affects the Repression of the *lac* Operon

Our discussion of the *lac* operon from the statistical mechanical perspective has thus far ignored one of the more intriguing features of this system, namely, the presence of DNA looping. The behavior of the *lac* operon has been examined in great detail both *in vitro* and *in vivo*. One beautiful set of experiments that is particularly enlightening with reference to the class of models we have described thus far in the chapter examines the repression of the *lac* operon as a function of the spacing between the DNA-binding sites (the operators) for Lac repressor.

The data on repression as a function of interoperator spacing were introduced in Figure 1.11 (p. 19) as an example of the sophisticated quantitative data that exist on biological systems in general, and gene expression in particular. These beautiful experiments and others like

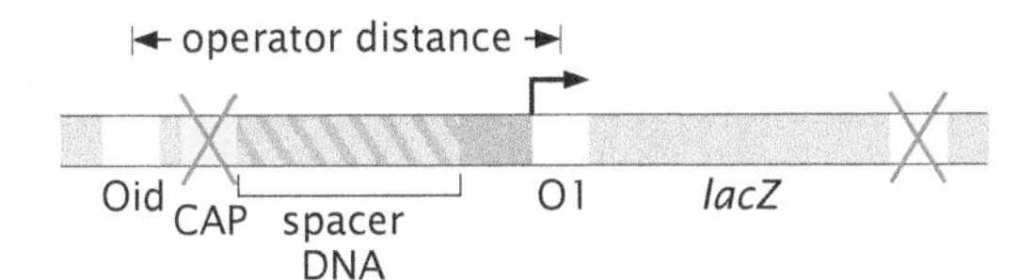

Figure 19.24: Construct used to measure repression in the presence of looping. The binding site for the activator CRP (shown as CAP in the diagram) was deleted, as was the third repressor-binding site. (Adapted from J. Müller et al., *J. Mol. Biol.* 257:21, 1996.)

them show a systematic trend in the promoter activity of the genes in question as a function of the distance between the binding sites for the repressor under consideration. One particularly telling feature of such data is the periodicity that results from the twist degrees of freedom and that reflects the need for particular faces of the DNA to be aligned in order to form a loop.

Figure 19.24 shows the DNA construct that was used to examine the *in vivo* consequences of DNA looping. In this construct, both the binding site for CRP and the operator O2 were deleted, while the promoter was replaced with a stronger promoter. The deletion of the CRP-binding site is intended to remove the question of activation from the problem. Note also that this construct permits the insertion of DNA sequences of arbitrary length between O1 and Oid, where Oid has replaced O3. Oid is a much stronger operator than O3, of approximately the same strength as O1. Finally, the deletion of O2 insures that looping will only occur between the two remaining operators.

In order to confront data like those shown in Figure 1.11 (p. 19), we need to expand our discussion of activators and repressors to include the effect of looping itself. In Figure 19.25, we show a minimal model of the states available to the system when RNA polymerase and Lac repressor are competing for the same region in the vicinity of the promoter. Note that this model permits different repressor molecules to occupy the two operators simultaneously, or a single molecule to occupy both sites and to loop the intervening DNA. We ignore the possibility of activator-binding since the activator-binding site was eliminated as shown in Figure 19.24. Note that this does not unequivocally rule out the possibility of nonspecific CAP binding, which might affect the results as well.

In order to proceed in quantitative terms, as usual, we need to write down the partition function that corresponds to assigning statistical weights to all of the allowed states depicted in Figure 19.25. Using exactly the same logic as in previous sections, the partition function can be written as

$$\begin{aligned} Z_{\text{tot}}(P,R;N_{\text{NS}}) = {} & \underbrace{Z(P,R;N_{\text{NS}})}_{\text{P}^{(0)},\ \text{O}^{(0)}_{\text{main}} \text{ and } \text{O}^{(0)}_{\text{aux}}} + \underbrace{Z(P-1,R;N_{\text{NS}})\text{e}^{-\beta\varepsilon^{\text{S}}_{\text{pd}}}}_{\text{P}^{(1)},\ \text{O}^{(0)}_{\text{main}} \text{ and } \text{O}^{(0)}_{\text{aux}}} \\ & + \underbrace{Z(P-1,R-1;N_{\text{NS}})\text{e}^{-\beta\varepsilon^{\text{S}}_{\text{pd}}}\text{e}^{-\beta\varepsilon^{\text{S}}_{\text{rda}}}}_{\text{P}^{(1)},\ \text{O}^{(0)}_{\text{main}} \text{ and } \text{O}^{(1)}_{\text{aux}}} \\ & + \underbrace{Z(P,R-1;N_{\text{NS}})\text{e}^{-\beta\varepsilon^{\text{S}}_{\text{rdm}}}}_{\text{P}^{(0)},\ \text{O}^{(1)}_{\text{main}} \text{ and } \text{O}^{(0)}_{\text{aux}}} + \underbrace{Z(P,R-1;N_{\text{NS}})\text{e}^{-\beta\varepsilon^{\text{S}}_{\text{rda}}}}_{\text{P}^{(0)},\ \text{O}^{(0)}_{\text{main}} \text{ and } \text{O}^{(1)}_{\text{aux}}} \\ & + \underbrace{Z(P,R-2;N_{\text{NS}})\text{e}^{-\beta\varepsilon^{\text{S}}_{\text{rdm}}}\text{e}^{-\beta\varepsilon^{\text{S}}_{\text{rda}}}}_{\text{P}^{(0)},\ \text{O}^{(1)}_{\text{main}} \text{ and } \text{O}^{(1)}_{\text{aux}}} \\ & + \underbrace{Z(P,R-1;N_{\text{NS}})\text{e}^{-\beta\varepsilon^{\text{S}}_{\text{rdm}}}\text{e}^{-\beta\varepsilon^{\text{S}}_{\text{rda}}}\text{e}^{-\beta F_{\text{loop}}}}_{\text{repressor/loop}}, \end{aligned} \tag{19.30}$$

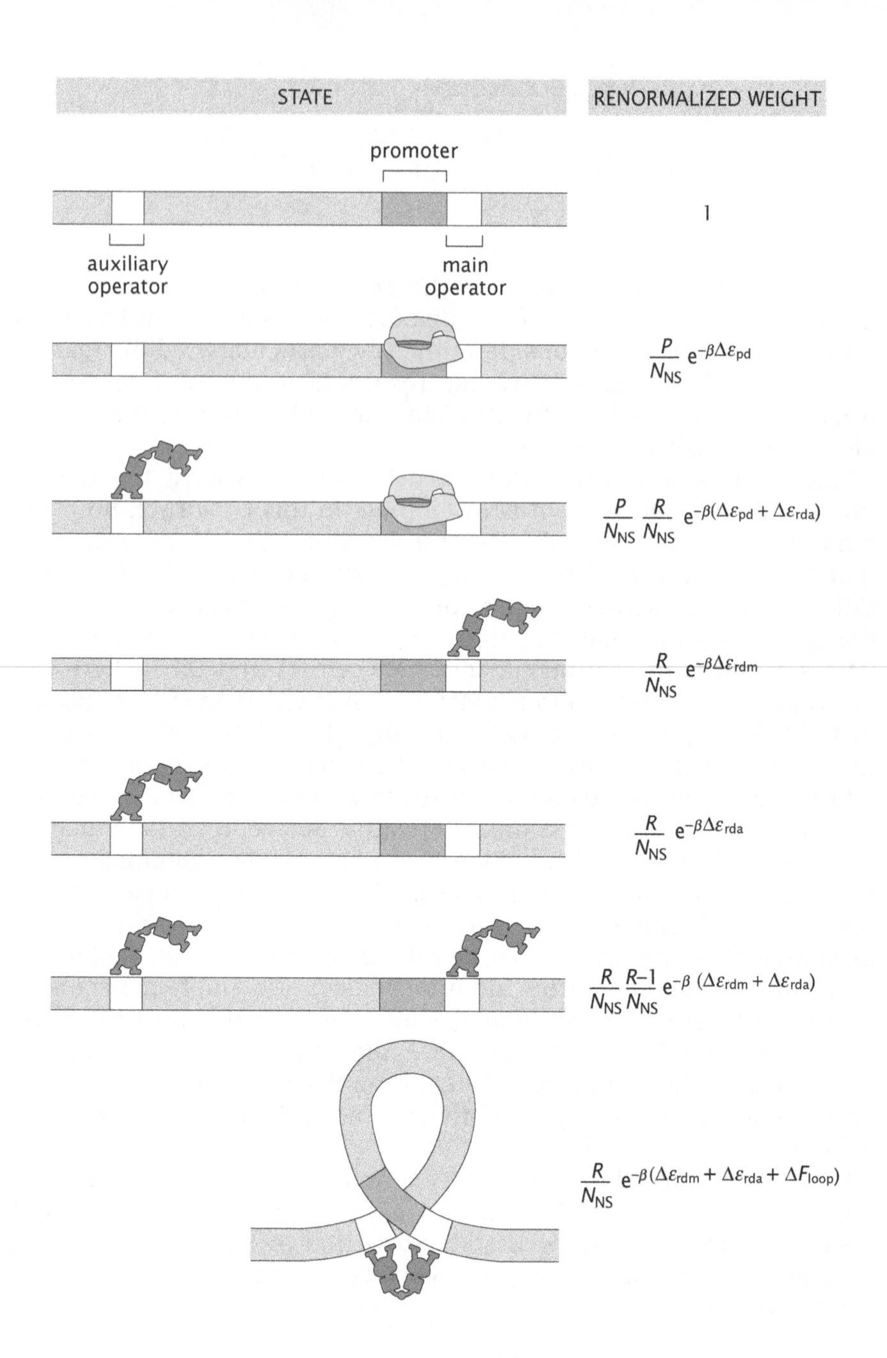

Figure 19.25: Looping states and weights in the *lac* operon. Each state corresponds to a different state of occupancy of the promoter and operators in the operon.

where ε_{rda} is the binding energy of the repressor for the auxiliary operator and ε_{rdm} is the binding energy of the repressor for the main operator. Our notation has clearly become more cumbersome and deserves explanation. First, we introduce $\text{P}^{(0)}$, $\text{O}^{(0)}_{\text{main}}$, and $\text{O}^{(0)}_{\text{aux}}$ to indicate that the occupancies of the promoter and main and auxiliary operators are zero, respectively. Next, the notation $\text{O}^{(1)}_{\text{main}}$ indicates that the main operator is occupied. The term with $\text{P}^{(0)}$, $\text{O}^{(1)}_{\text{main}}$, and $\text{O}^{(1)}_{\text{aux}}$ indicates the states for which there are distinct repressor molecules bound to the two operators and the final term accounts for the looped state.

One of the terms in the expression includes the looping free energy in the form

$$Z(P, R-1; N_{\text{NS}})\text{e}^{-\beta\varepsilon^{\text{S}}_{\text{rdm}}}\text{e}^{-\beta\varepsilon^{\text{S}}_{\text{rda}}}\text{e}^{-\beta F_{\text{loop}}}, \tag{19.31}$$

and the factor $\text{e}^{-\beta F_{\text{loop}}}$ deserves further comment. Recall that $Z(P, R-1; N_{\text{NS}})$ is itself already a sum over all of the possible ways of

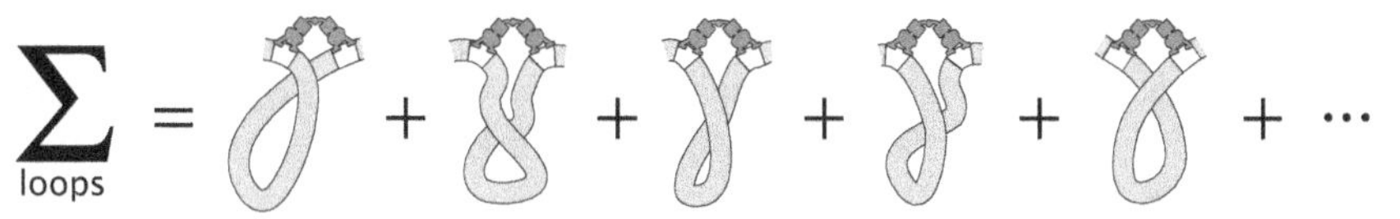

$$\sum_{\text{loops}} = \quad + \quad + \quad + \quad + \quad + \cdots$$

Figure 19.26: Summing over DNA loops. The sum $\sum_{\text{loops}}$ instructs us to sum over all conformations of the DNA loop as indicated schematically here.

distributing the P RNA polymerase molecules and the $R-1$ repressor molecules over the N_{NS} nonspecific binding sites on the DNA, with one of the repressors bound to both operators and looping the intervening DNA. However, for each and every one of these configurations, we have to sum over *all* of the possible geometries of the loop itself. That is, this contribution to the partition function is really of the form

$$Z_{\text{looped}}(P, R-1; N_{NS}) = \sum_{\text{loops}} Z(P, R-1; N_{NS}) e^{-\beta\varepsilon_{\text{rdm}}^{S}} e^{-\beta\varepsilon_{\text{rda}}^{S}} e^{-\beta\varepsilon_{\text{loop}}}, \tag{19.32}$$

where $\varepsilon_{\text{loop}}$ is the *energy* of a given loop configuration and $\sum_{\text{loops}}$ instructs us to sum over all of the possible loop configurations as schematized in Figure 19.26. Since most of the factors are independent of the looping geometry, we can rewrite this as

$$Z_{\text{looped}}(P, R-1; N_{NS}) = Z(P, R-1; N_{NS}) e^{-\beta\varepsilon_{\text{rdm}}^{S}} e^{-\beta\varepsilon_{\text{rda}}^{S}} \sum_{\text{loops}} e^{-\beta\varepsilon_{\text{loop}}}, \tag{19.33}$$

where we have pulled all terms out of the sum that do not depend upon the particular choice of looped state. One way to proceed at this point is to appeal to ideas about elasticity to determine $\varepsilon_{\text{loop}}$ and use the random walk as the basis for effecting the sum. On the other hand, a simpler scheme is to replace the sum by $e^{-\beta F_{\text{loop}}}$ and to treat F_{loop} as a phenomenological parameter as we have already done with the various binding energies.

With the partition function in hand, we can compute the probability of RNA polymerase binding by considering the ratio of favorable outcomes to the total partition function, resulting in

$$\begin{aligned} p_{\text{bound}}(P, R; N_{NS}) = {} & \frac{P}{N_{NS}} e^{-\beta\Delta\varepsilon_{\text{pd}}} \left(1 + \frac{R}{N_{NS}} e^{-\beta\Delta\varepsilon_{\text{rda}}}\right) \\ & \times \left[1 + \frac{P}{N_{NS}} e^{-\beta\Delta\varepsilon_{\text{pd}}} \left(1 + \frac{R}{N_{NS}} e^{-\beta\Delta\varepsilon_{\text{rda}}}\right)\right. \\ & + \frac{R}{N_{NS}} \left(e^{-\beta\Delta\varepsilon_{\text{rdm}}} + e^{-\beta\Delta\varepsilon_{\text{rda}}}\right) \\ & + \frac{R(R-1)}{(N_{NS})^2} e^{-\beta(\Delta\varepsilon_{\text{rdm}} + \Delta\varepsilon_{\text{rda}})} \\ & \left. + \frac{R}{N_{NS}} e^{-\beta(\Delta\varepsilon_{\text{rdm}} + \Delta\varepsilon_{\text{rda}} + \Delta F_{\text{loop}})}\right]^{-1}, \end{aligned} \tag{19.34}$$

where we have defined $\Delta F_{\text{loop}} = F_{\text{loop}} + \varepsilon_{\text{rd}}^{NS}$. From this expression, we can obtain the regulation factor

$$\begin{aligned} F_{\text{reg}}(R) = {} & \left(1 + \frac{R}{N_{NS}} e^{-\beta\Delta\varepsilon_{\text{rda}}}\right) \left[1 + \frac{R}{N_{NS}} \left(e^{-\beta\Delta\varepsilon_{\text{rdm}}} + e^{-\beta\Delta\varepsilon_{\text{rda}}}\right)\right. \\ & \left. + \frac{R(R-1)}{(N_{NS})^2} e^{-\beta(\Delta\varepsilon_{\text{rdm}} + \Delta\varepsilon_{\text{rda}})} + \frac{R}{N_{NS}} e^{-\beta(\Delta\varepsilon_{\text{rdm}} + \Delta\varepsilon_{\text{rda}} + \Delta F_{\text{loop}})}\right]^{-1}. \end{aligned} \tag{19.35}$$

To make contact with the results of Müller et al. (1996), we now need to write an expression for the repression as a function of the interoperator spacing. Recall that the repression is given by Equation 19.27 and takes the form

$$\begin{aligned} \text{repression}(N_{\text{bp}}) &= (F_{\text{reg}})^{-1} \\ &= \left[1 + \frac{R}{N_{\text{NS}}}(\mathrm{e}^{-\beta\Delta\varepsilon_{\text{rdm}}} + \mathrm{e}^{-\beta\Delta\varepsilon_{\text{rda}}}) \right. \\ &\quad + \frac{R(R-1)}{(N_{\text{NS}})^2}\mathrm{e}^{-\beta(\Delta\varepsilon_{\text{rdm}}+\Delta\varepsilon_{\text{rda}})} \\ &\quad \left. + \frac{R}{N_{\text{NS}}}\mathrm{e}^{-\beta(\Delta\varepsilon_{\text{rdm}}+\Delta\varepsilon_{\text{rda}}+\Delta F_{\text{loop}})}\right] \\ &\quad \times \left(1 + \frac{R}{N_{\text{NS}}}\mathrm{e}^{-\beta\Delta\varepsilon_{\text{rda}}}\right)^{-1}, \end{aligned} \tag{19.36}$$

where we have written repression(N_{bp}) as a function of the number of base pairs in the loop (N_{bp}) to signal that the looping free energy (and hence the repression) will depend upon the distance between the two operators. We have invoked the approximation that the promoter is weak (that is, $(N_{\text{NS}}/PF_{\text{reg}})\mathrm{e}^{\beta\Delta\varepsilon_{\text{pd}}} \gg 1$). In order to examine the significance of our results on looping, we consider the extent to which the model can be used to interpret existing data and to suggest new experiments. Notice that we already know all the parameters in the weights from the previous experiment, with the exception of ΔF_{loop}. We argue that for a given loop size, ΔF_{loop} is a parameter that should be indifferent to which combination of operators is used in these two-operator experiments and, as a result, once ΔF_{loop} is determined, the model is predictive. The results of a simple fit to the looping free energy are shown in Figure 19.27. To obtain these curves, any single data point is used to obtain the looping free energy itself and then the resulting curves are entirely predictive.

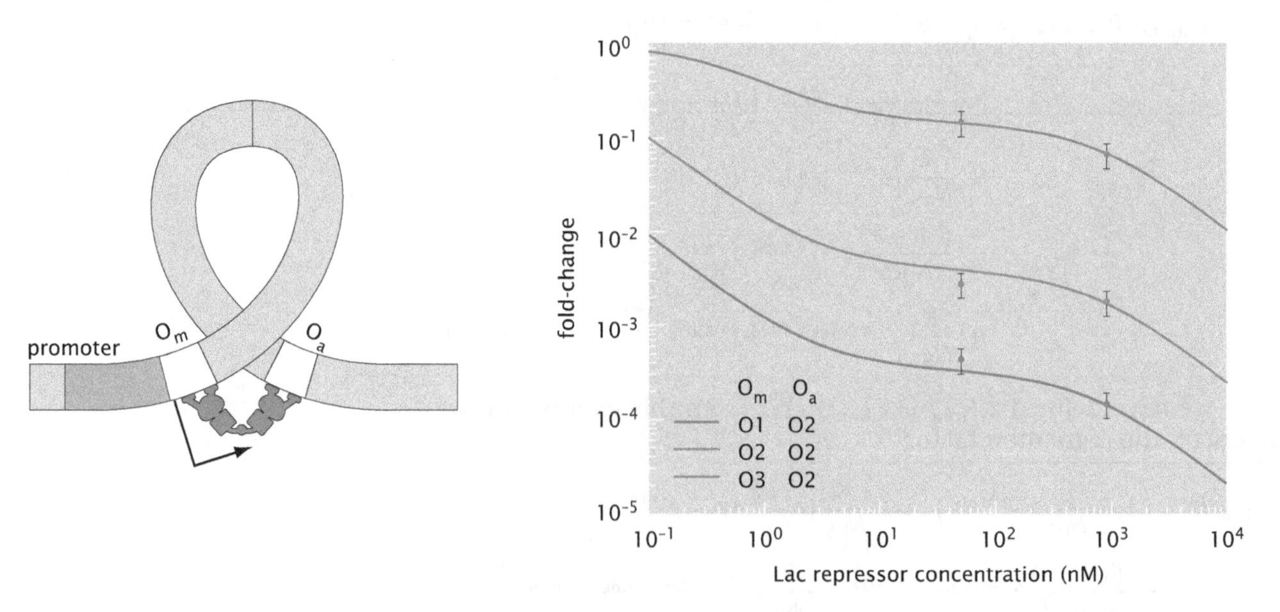

Figure 19.27: Repression and looping. A single fit to ΔF_{loop} giving 9.1 k_BT permits the investigation of multiple configurations of the different operators. (Data from S. Oehler et al., *EMBO J.* 13:3348, 1994.)

Inducers Tune the Level of Regulatory Response

Who regulates the regulators? So far, our story has been built around a set of transcription factors that are themselves above the law. But, in fact, we know that the action of these transcription factors is itself controlled by signaling processes and it is to this subject that we now turn, once again using the *lac* operon as the defining case study to set ideas. The famed Lac repressor is itself controlled by molecules known as inducers, with one of the most important examples being the synthetic inducer IPTG. IPTG binds to Lac repressor, thus reducing its affinity for DNA. In the natural context, the binding of an inducer (allolactose) to Lac repressor provides the feedback that eliminates repression and permits the synthesis of the enzyme that performs the chemical cleavage of lactose.

One of the ways in which the reduction in binding affinity of Lac repressor for DNA has been illustrated is through *in vitro* binding experiments like those schematized in Figure 19.28. This figure shows the *in vitro* occupancy of an operator by Lac repressor as a function of inducer (IPTG) concentration. However, when performing an analogous titration *in vivo* for the wild-type *lac* operon, a much sharper response for the level of gene expression as a function of the IPTG concentration is observed. The "sharpness" of the output signal (gene expression for the *in vivo* measurements or normalized binding probability for the *in vitro* measurements) can be quantified using the thermodynamically inspired Hill function introduced in Section 6.4.3 (p. 273) and given in this case by

$$\text{normalized gene expression} = \frac{\left(\frac{[IPTG]}{K_\text{d}}\right)^n}{1+\left(\frac{[IPTG]}{K_\text{d}}\right)^n}. \tag{19.37}$$

Here, $[IPTG]$ is the concentration of IPTG and K_d is an effective dissociation constant. This effective dissociation constant not only reflects the interaction of IPTG with Lac repressor, it also accounts for the change in affinity of Lac repressor to its operator DNA upon binding to the inducer, for the concentration of Lac repressors, and, in the *in vivo* scenario, also for any active pumping of IPTG into the cell. n is the Hill coefficient, a measure of the slope or sensitivity of the output signal with respect to the input concentration of IPTG.

As suggested above, modeling induction of the *lac* operon requires taking into account the passive and active transport by Lac permease (LacY) of IPTG into the cell as well as the binding and DNA looping of Lac repressor described in the previous sections. All these effects conspire to give an *in vivo* Hill coefficient that is significantly different than the *in vitro* counterpart. In Figure 19.28, it is shown how by creating different strains of bacteria bearing a deletion of Lac permease and lacking the auxiliary Lac repressor-binding sites, the *in vivo* and *in vitro* sensitivities can be reconciled.

19.2.6 Other Regulatory Architectures

The *lac* operon is one of the classic case studies in modern biology. Indeed, one of our arguments is that it is precisely the well-characterized biological examples that serve as fertile proving ground for physical biology approaches. However, there is much more to regulatory biology than the *lac* operon! One of the key questions that remains in light of successes with the thermodynamic approach is

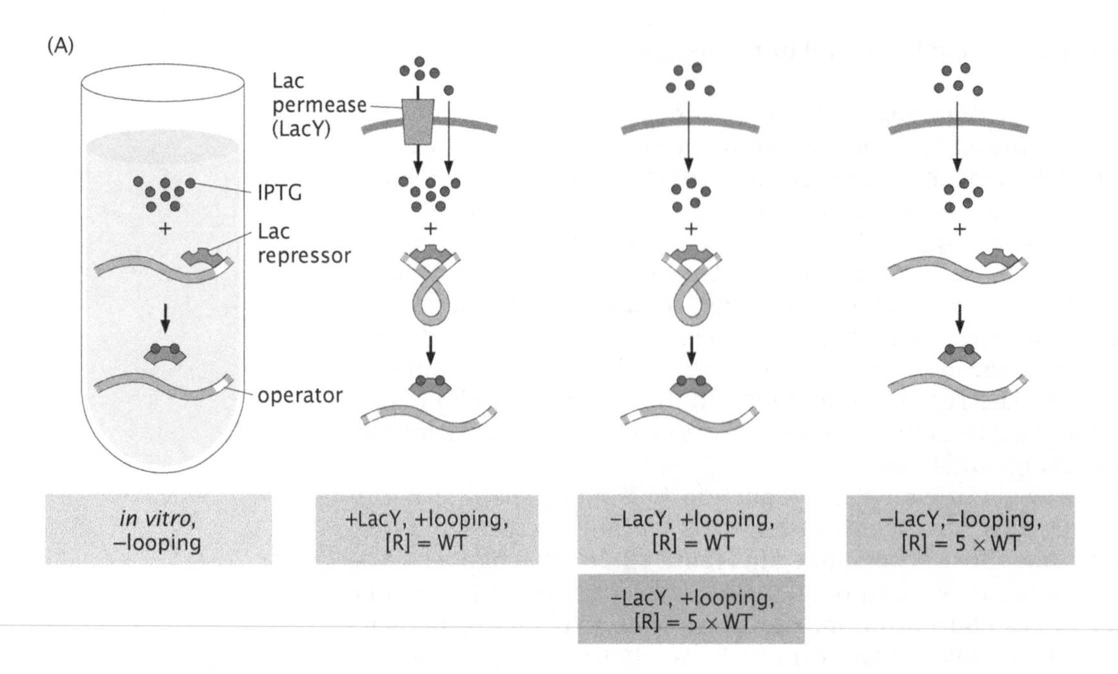

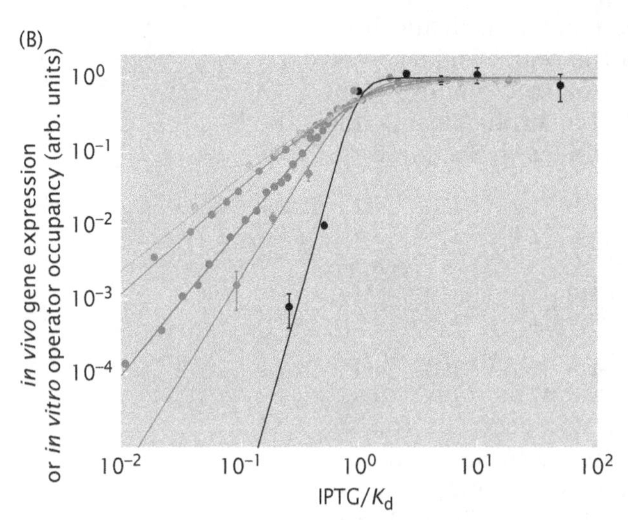

Figure 19.28: Sensitivity of induction in the *lac* operon. (A) The sensitivity can be characterized by measuring the *in vitro* occupancy of an operator as a function of inducer concentration for the simple construct containing only one Lac repressor-binding site. On the other hand, the *in vivo* level of gene expression and its resulting sensitivity can be characterized for the different *lac operon* mutants shown, where Lac permease (which pumps in inducer) or an auxiliary binding site for repressor on DNA were deleted or where the intracellular concentration of repressor, [R], is different from its wild-type (WT) value. (B) The resulting shape of the *in vitro* operator occupancy or *in vivo* level of gene expression are shown as a function of inducer concentration for the series of different mutants shown in (A), where the different curves correspond to the experimental conditions, with the gray box corresponding to the black line. As different elements of the system were deleted, the sensitivity of the induction neared that of the purified *in vitro* system. This sensitivity can be quantified by fitting to a thermodynamically inspired functional form such as the Hill function shown in Equation 19.37. Each curve has been normalized to its corresponding maximum in gene expression (*in vivo* data) or maximum in binding probability (*in vitro* data). (Data adapted from S. Oehler et al., *Nucleic Acids Res.* 34:606, 2006 and T. Kuhlman et al., *Proc. Natl. Acad. Sci. USA* 104:6043, 2007.)

the extent to which the same ideas can be used for other bacterial promoters, and, better yet, in the context of eukaryotic examples.

The Fold-Change for Different Regulatory Motifs Depends Upon Experimentally Accessible Control Parameters

So far, we have shown the cases of simple repression, simple activation, and DNA looping. In each of these cases, our experimental point of contact has been the fold-change, which tells us how expression changes in the presence of various regulatory motifs. Figure 19.29 summarizes how the thermodynamic models worked out so far in this chapter can be used to predict the input–output response of these regulatory architectures. Conceptually, what we are really after is descriptions of these networks in which we can identify the various regulatory "knobs" such as those shown in Figure 19.30, and tune these knobs both theoretically and in the context of quantitative experiments.

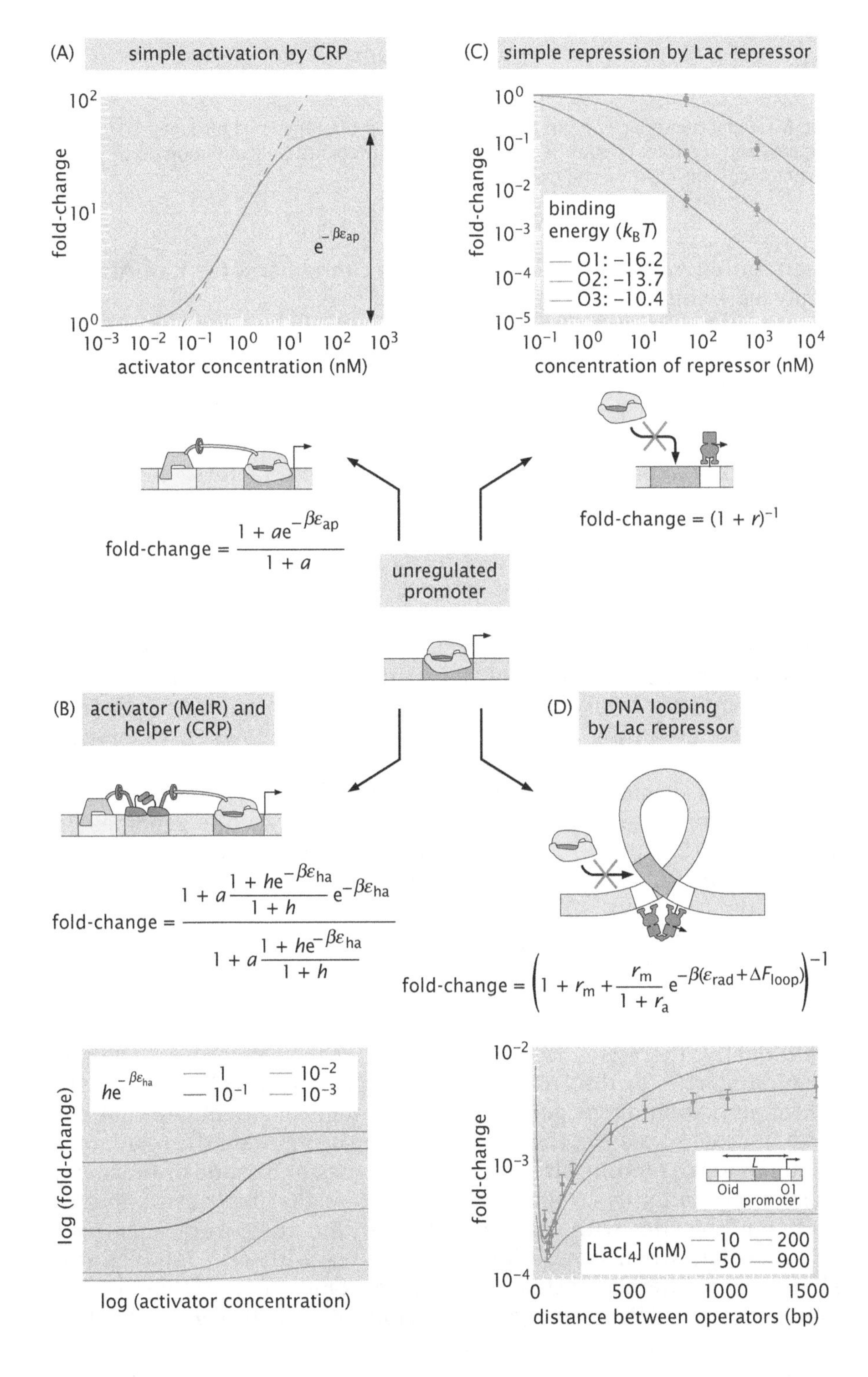

Figure 19.29: Thermodynamic models for diverse regulatory architectures. The thermodynamic models result in a predicted fold-change as a function of parameters such as operator binding strengths and transcription factor copy numbers. The resulting fold-change function serves as a governing equation dictating the regulatory output. The lowercase variables (a, h, r, etc.) correspond to $x = (X/N_{NS})e^{-\beta\Delta\varepsilon_{xd}}$, where X is the intracellular number of the particular transcription factor, $\Delta\varepsilon_{xd}$ its interaction energy with the DNA, and $N_{NS} = 5 \times 10^6$. (A) In simple activation, an activator recruits RNA polymerase to the promoter by interacting with it ($\Delta\varepsilon_{ad} = -13.8\ k_BT$, $\varepsilon_{ap} = -3.9\ k_BT$). (B) A helper molecule can recruit an activator to the promoter, which in turn can recruit RNA polymerase. (C) In simple repression, a repressor binds to a site overlapping the promoter, which results in the exclusion of RNA polymerase from that site. (D) Some repressors can also bind to multiple sites simultaneously, which results in an increase of the level of repression. The looping probability will depend on physical characteristics of the loop, such as its length ($\Delta\varepsilon_{rmd} = -16.7\ k_BT$, $\Delta\varepsilon_{rad} = -18.4\ k_BT$, and $\Delta F_{loop} = A/L + B \ln L - CL - E$, with $A = 140.6\ k_BT \times$ bp, $B = 2.52\ k_BT$, $C = 1.4 \times 10^{-3}\ k_BT$/bp and $E = 19.9\ k_BT$). We assume that one molecule per cell corresponds to 1 nM. (Adapted from L. Bintu et al., *Curr. Opin. Genet. Dev.* 15:125, 2005.)

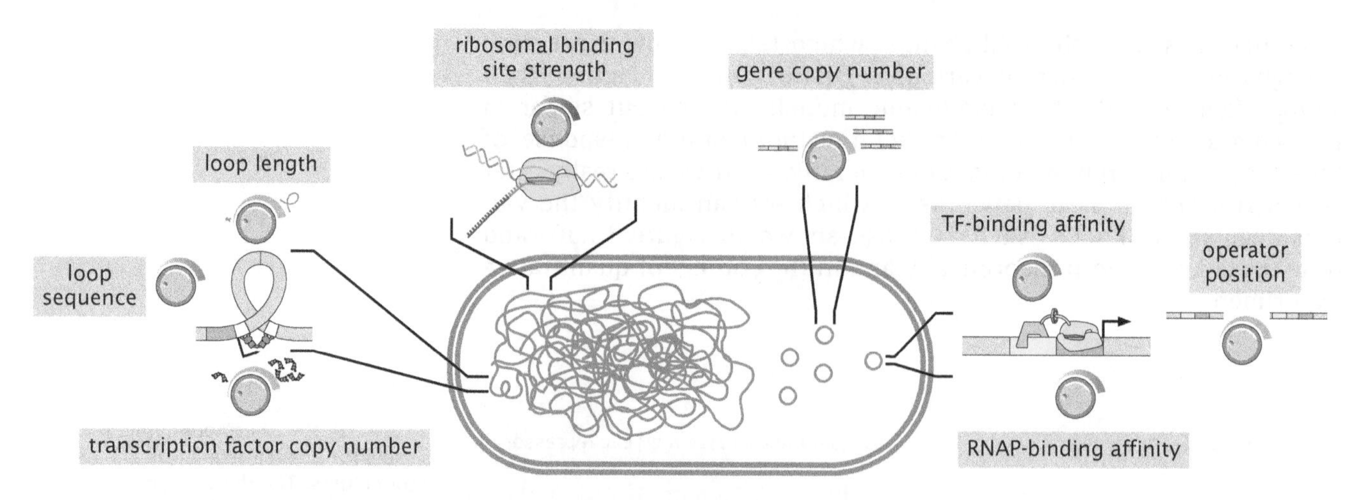

Figure 19.30: Dialing-in transcriptional output. Many parameters can be used to quantitatively tune the level of gene expression, including the number of copies of genes and the transcription factors that control them, as well as the binding strengths of a suite of different proteins that interact with the DNA during gene expression.

Quantitative Analysis of Gene Expression in Eukaryotes Can Also Be Analyzed Using Thermodynamic Models

Our simplified diagrams throughout the book have made it look as though the transcription apparatus is a single light-blue object (the polymerase) that binds to its promoter. However, even in bacteria, the transcription process is much more complicated since the basal transcription apparatus is a complex of multiple factors. In the case of eukaryotic organisms, a huge number of molecular species conspire together to drive transcription. As such, at first cut, one barely dares to use such streamlined "effective" models for which so much of the complexity is blatantly ignored. Nevertheless, the same thermodynamic description applied so far has been used to think about questions such as nucleosome occupancy and how it interferes with transcription and for a host of different eukaryotic regulatory architectures. In this brief discussion, we attempt to whet the reader's appetite by gossiping about several especially renowned eukaryotic examples.

As we will discuss thoroughly in Section 21.3.3 (p. 988), nucleosomes are found to be reliably positioned on eukaryotic genomes. In particular, they can be found in regulatory regions, with the resulting occlusion of transcription factor-binding sites as described in detail in Section 10.4.3 (p. 409). One of the systems where the interplay between nucleosomal occupancy and level of gene expression has been explored quantitatively is the PHO5 promoter in yeast. As shown in Figure 19.31(A), the PHO5 promoter is activated by PHO4, which binds to two sites, one within a nucleosome and the other not. Additionally, there is a TATA binding box at nucleosome −1. This is the binding site for the TATA box binding protein (TBP), which is critical for activation of the gene. Following a logic analogous to that put forth in order to dissect the *lac* operon, one can mutate the regulatory region to move binding sites into nucleosomes or outside of them and change their affinities. Figure 19.31(B) shows that the positioning of these binding sites with respect to those of the nucleosomes matters. In particular, when comparing architectures with sites of identical affinity, but different occlusion by nucleosomes, the features of the input–output function change appreciably. Moreover, Figure 19.31(C)

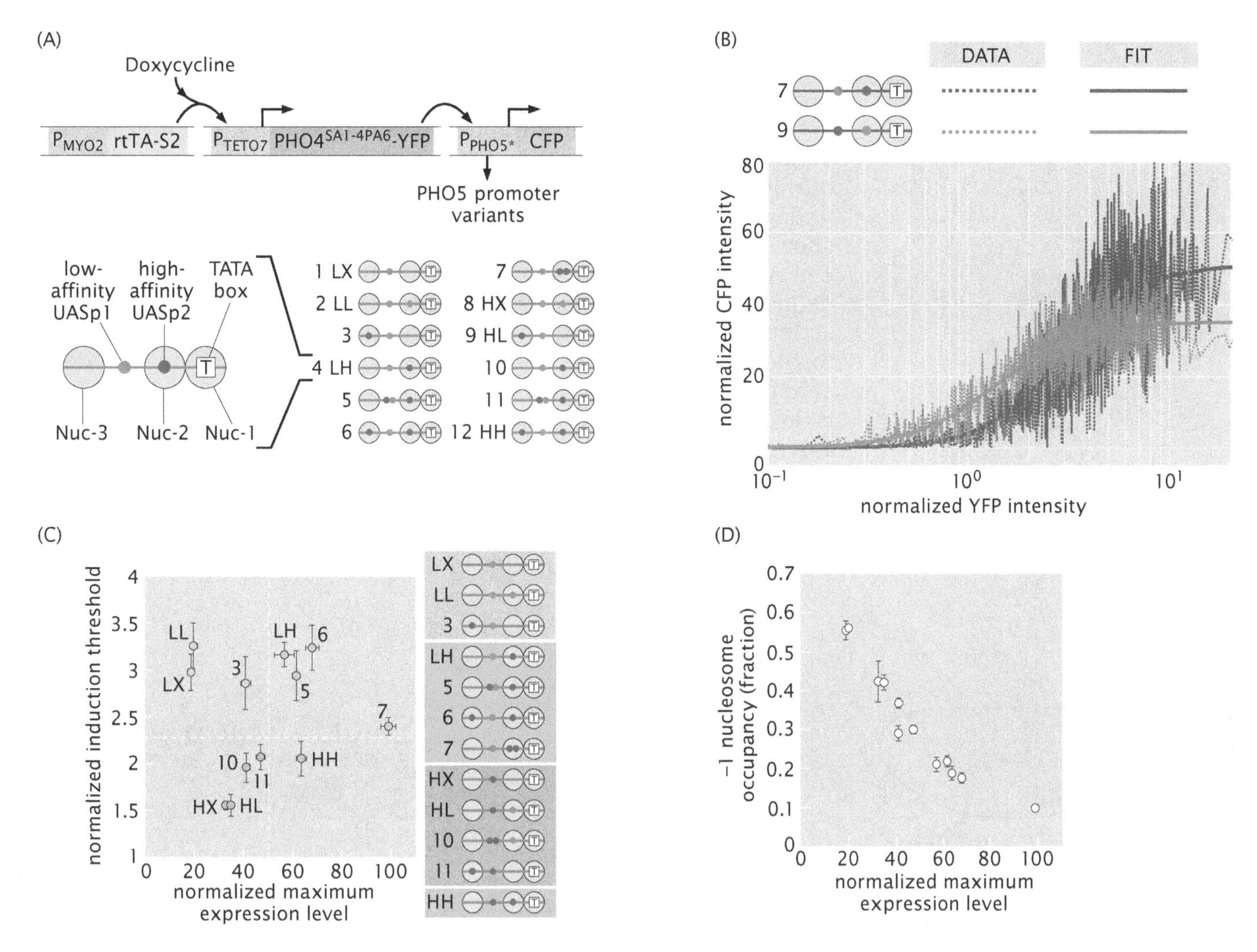

Figure 19.31: Role of nucleosomes in transcriptional regulation. (A) A PHO5 promoter expressing CFP is regulated by PHO4-YFP, with two target binding sites, UASp1 and UASp2. This promoter has a TATA box, which is bound by the TBP activator. The position of nucleosomes in this regulatory region has been mapped such that mutants can be generated where the various promoter features are occluded or not by nucleosomes. (B) Two architectures having the same binding sites but different occlusion geometries by nucleosomes show markedly different input–output functions. (C) Several variants for the PHO5 promoter result in input–output functions with drastically different induction thresholds and maximum levels of expression. (D) The maximum level of expression shows a strong correlation with the occupancy of nucleosome −1, which occludes the TATA box. (Adapted from Kim and O'Shea, *Nat. Struct. Mol. Biol.* 15:1192, 2008.)

shows how subtle changes in the regulatory region can lead to drastic changes in both the threshold and maximum level of expression of the input–output function.

Perhaps the role of nucleosome occupancy is most prominently revealed by Figure 19.31(D). In this case, the maximum level of gene expression shows a strong correlation with the occupancy of nucleosome −1 for each of the different regulatory architectures shown in Figure 19.31(A). Nucleosome −1 overlaps the TATA box, suggesting that one of the roles of PHO4 is to ultimately modulate the occupancy of −1 through the interaction with nucleosome remodeling complexes. This modulation determines, in turn, the absolute level of gene expression of the promoter. It is clear that thermodynamic models can be used to describe the many layers of regulation present in this type of problem. In particular, in Section 10.4.3 (p. 409), we have already shown how thermodynamic thinking can describe the probability of protein accessibility to a binding site that is buried inside a nucleosome. Further, in Section 21.3.3 (p. 988), we will show how statistical mechanics can lead to simple models that predict the probability landscape of nucleosome occupancy along the genome.

A further challenge in deciphering eukaryotic transcriptional regulation centers on how transcription varies in both space and time during development in multicellular organisms. Perhaps the most well understood such organism is the fruit fly *Drosophila melanogaster*. As shown in Figure 19.2, during the initial stages of development, the fly embryo expresses a battery of transcription factors in a cascade that defines sharper and sharper domains of expression. One of the transcriptional architectures that has been studied in most detail is related to the activation of the transcription factor Hunchback by the transcription factor Bicoid. As shown in Figures 19.2 and 19.32(A), Bicoid is expressed in an exponential profile along the anterior–posterior axis of the developing embryo. Activation by Bicoid is realized by binding to six sites of different strengths that lie upstream from the Hunchback promoter, as seen in Figure 19.32. The resulting pattern of Hunchback expression shown in Figure 19.32(C) presents a domain with a boundary at about 50% of the embryo length. The exquisite

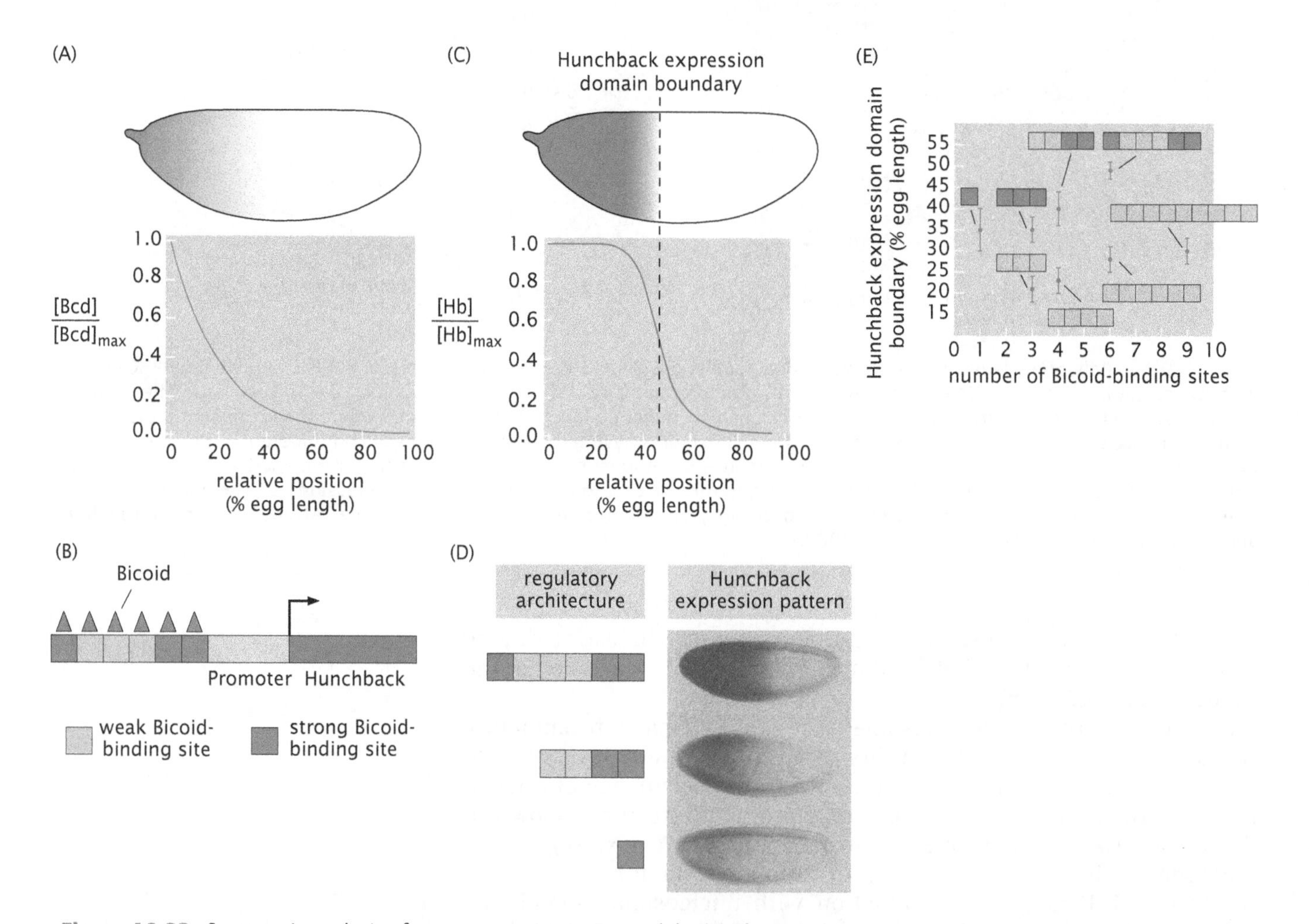

Figure 19.32: Systematic analysis of gene expression in *Drosophila*. (A) The Bicoid transcription factor is expressed in an exponential profile from the anterior to the posterior end of the fly embryo. (B) Bicoid acts as an activator of the Hunchback transcription factor by binding to six binding sites of different strengths located upstream from the Hunchback promoter. (C) The resulting pattern of Bicoid-dependent Hunchback expression domain presents a sharp boundary at about 50% of the embryo length. (D) By creating constructs with different numbers and affinities of binding sites, the boundary of the expression domain can be shifted systematically. (E) Hunchback domain boundary position for several regulatory architectures. (D, E, adapted from W. Driever et al., *Nature* 340:363, 1989.)

genetic control available in fruit flies permits us to query patterns of gene expression for regulatory architectures that have been mutated systematically in much the same fashion as the experiments we have described in the context of the *lac* operon and using the same perturbative philosophy by tuning knobs as shown in Figure 19.30. Figure 19.32(D) shows how different regulatory architectures driving a reporter gene can shift the position of the expression domain. In particular, fewer binding sites corresponds to a shift towards the anterior position, where Bicoid concentrations are higher. This kind of approach was followed systematically for several regulatory architectures, resulting in data like that shown in Figure 19.32(E).

In principle, we can use the same tools presented earlier in the chapter in order to understand the regulatory outcome of these experiments. Given a strength of the array of binding sites there will be a range of Bicoid concentrations over which there will be Hunchback activation. By reducing the number of binding sites or making them weaker, the size of this concentration range is reduced, resulting in a shift of the expression domain. The application of such thermodynamic ideas is spelled out in more detail in our discussion in the next chapter in Section 20.2.3. The work to be described there describes the state of the art in terms of experimental measurements of the dynamics of gene expression in embryonic development in flies and shows how simple thermodynamic models can be used as a basis for discussing these results, though these problems are hugely complex, and simple models like these are akin to highly distorted maps.

19.3 Regulatory Dynamics

19.3.1 The Dynamics of RNA Polymerase and the Promoter

Until now, our treatment of gene regulation has centered on the time-independent output of different regulatory motifs. On the other hand, as is clear from watching the development of any embryo, many of the most beautiful and important questions in regulation center on the orchestration of regulatory decisions over time. Another example that puts questions of the time dependence of gene expression front and center is the study of cells during the cell cycle. As was shown in Chapter 3, entire batteries of genes are expressed at different times during the cell cycle (see Figure 3.23 on p. 119 for a concrete example in the cell cycle of *C. crescentus*). Two of the key dynamical motifs that recur in organisms ranging from bacteria to humans are switches and oscillators. In the case of switches, depending upon some environmental cue, for example, a cell can change the regulatory state associated with particular genes from "off" to "on." Even richer behavior is exhibited by regulatory circuits that give rise to oscillations. So that we can see how switches and oscillators are constructed, we now take up the question of time-dependent gene expression.

The Concentrations of Both RNA and Protein Can Be Described Using Rate Equations

Our conceptual starting point for examining the dynamics of gene expression is the rate equation paradigm introduced in Chapter 15. In particular, we proceed by writing rate equations for the time evolution of the concentrations of various molecular participants in the regulatory problem of interest. The simplest scenario is to consider a dynamical description that refers only to the time development of the concentrations of the relevant proteins. On the other hand, sometimes it is convenient to characterize the time evolution of the mRNA transcripts as well. In either case, our strategy will be to consider some particular regulatory architecture in which different elements are linked and to write down a dynamical description of their concentrations.

One of the reasons we are forced to go beyond the thermodynamic models favored so far throughout the chapter is the advent of a new generation of experiments aimed at probing regulation. Recent advances on a number of different fronts have now made it possible to query the regulatory response of individual cells. Such experiments make it possible to watch the time evolution of both the mRNAs in individual cells and the proteins they code for.

Examples of the outcome of this recent generation of experiments for the mRNA content of cells are shown in Figure 19.33. One of the immediate insights coming from experiments like those leading to Figure 19.33(B) is that mRNA production is often "bursty." By watching individual cells over time, it becomes evident that there are periods of transcriptional silence occasionally punctuated by bursts of mRNA production. As shown in Figure 19.33(C), these experiments can be used to ask how noisy the gene expression process is, and the calculations in the remainder of the section will confront these questions

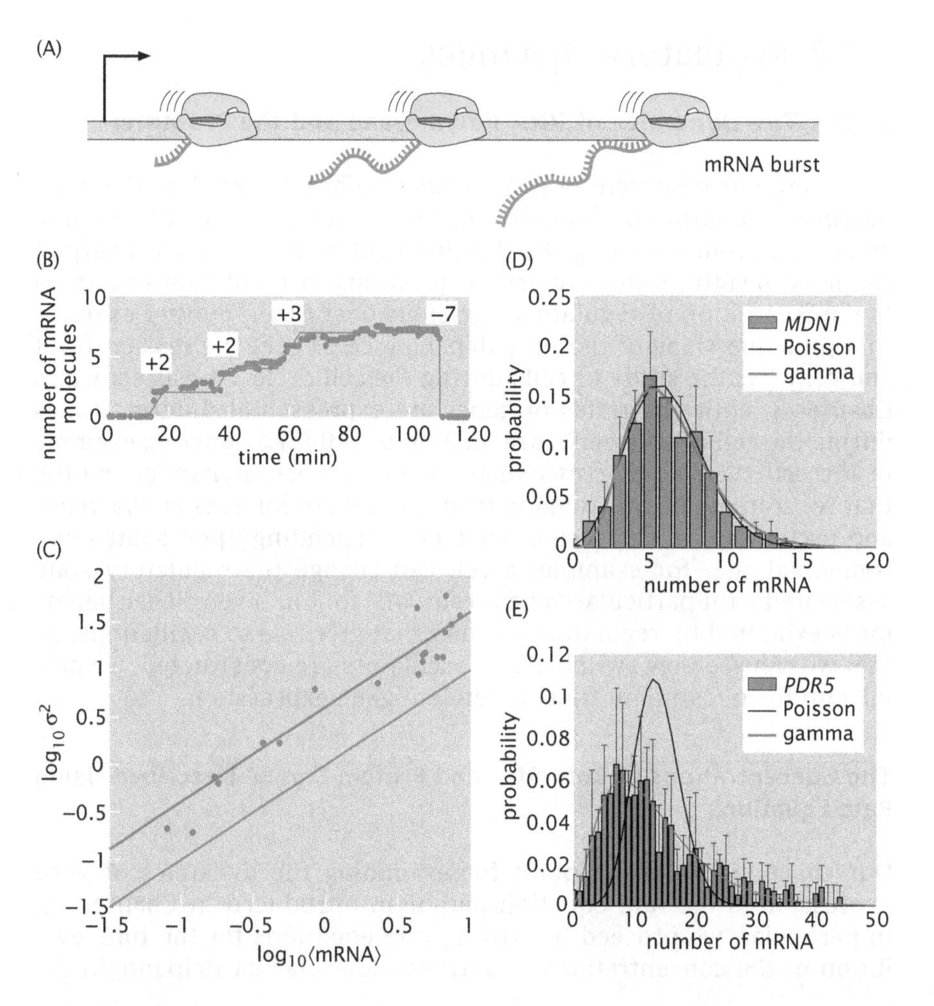

Figure 19.33: Time-dependent dynamics of transcriptional networks. (A) A burst of mRNA production. (B) Time history of the number of mRNA molecules in a given *E. coli* cell, revealing periods of no production punctuated by bursts of production with the size of the burst indicated by the numbers in the white boxes. (C) Noise in *E. coli* gene expression (measured by the variance) as a function of the mean level of gene expression. (D) Distribution of mRNA in yeast for the *MDN1* gene. (E) Distribution of mRNA in yeast for the *PDR5* gene. The distributions in (D) and (E) are fitted to various models considered in the section. (B, C, adapted from I. Golding et al., *Cell* 123:1025, 2005; D, E adapted from D. Zenklusen et al., *Nat. Struct. Mol. Biol.* 15:1263, 2008.)

head on. This noisiness is also revealed by evaluating the entire mRNA distribution over many cells as shown in Figures 19.33(D) and (E).

Before embarking on an analysis of the dynamics of particular regulatory architectures, we return to one of the most elementary questions that can be asked about regulatory dynamics. In particular, our use of statistical mechanics in the previous sections was predicated on the idea that the average amount of transcription from a gene of interest is proportional to the equilibrium occupancy of the promoter by RNA polymerase. This equilibrium assumption is justified when the binding of polymerase to the promoter occurs on a much faster time scale than the time it takes the polymerase to initiate transcription from the bound state. In that case, the polymerase is in rapid pre-equilibrium with the DNA and the amount of transcription is proportional to the fraction of time the polymerase is bound. This is very similar to the arguments put forward in Section 15.2.6 (p. 591), where we examined the conditions under which a chemical reaction can be treated as an equilibrium problem. Interestingly enough, this is not a necessary condition for the equilibrium assumption to hold, as we discuss in the estimate that follows.

Estimate: Dynamics of Transcription by the Numbers

ESTIMATE

The production of mRNA from a typical gene in *E. coli* occurs at a rate of about 10 per minute, while the average lifetime of an mRNA like that for the *lac* operon is a little more than 1 minute (for the distribution of mRNA lifetimes in *E. coli*, see Figure 3.14, p. 110). In steady state, the number of mRNAs created in the cell over any time interval must, on average, balance the number degraded. Since the mRNA molecules are created at 10 per minute, the same number of molecules must be degraded every minute, and we conclude that, on average, there will be 10 mRNA molecules per cell.

If we consider the process of transcription in some detail, it follows a number of biochemical steps. The key steps are RNA polymerase binding to the promoter regulating the gene of interest to form a closed complex with DNA; formation of an open complex in which the two strands of DNA are pulled apart allowing the RNA polymerase to read one of them; and promoter escape, when RNA polymerase begins transcribing the gene. These three steps can be represented by the reaction scheme

$$\mathrm{P}+\mathrm{D} \underset{k_-}{\overset{k_+}{\rightleftharpoons}} \mathrm{PD_C} \overset{k_{\text{open}}}{\longrightarrow} \mathrm{PD_O} \overset{k_{\text{escape}}}{\longrightarrow} \text{elongation}, \qquad (19.38)$$

where the escape into elongation leaves the promoter in the unbound state, ready to accept a new polymerase. Here, P is free RNA polymerase and D is unbound promoter. $\mathrm{PD_C}$ is the closed complex, while $\mathrm{PD_O}$ is the open complex whose formation is essentially irreversible.

For the binding and unbinding of polymerase to be in rapid pre-equilibrium, the condition $k_\pm \gg k_{\text{open}}$ needs to be satisfied. If RNA polymerase has time to bind and unbind from the promoter multiple times before open complex formation, then we can think of the first step as effectively an equilibrium step characterized by an equilibrium constant $K_\mathrm{P} = k_+/k_-$. Indeed, the binding of RNA polymerase to the *lac*UV5 promoter *in vitro* is so fast that the rates $k_\pm$ are not even measured in typical experiments. Instead, an equilibrium constant of $K_P \approx 200\ \mu M^{-1}$ is measured, while $k_{open} \approx 0.1\ s^{-1}$. This indicates that the rapid pre-equilibrium condition for polymerase binding to this promoter is met.

Once the RNA polymerase has initiated transcription, it elongates at a typical rate of about 50 nucleotides per second, which means that a typical gene of about 1000 nucleotides will be transcribed in about 20 seconds. Given that the average rate of production of mRNA is 10 per minute, this implies that there are about three RNA polymerases per gene at any given time. These numbers are typical for the production of messenger RNAs in *E. coli*, while ribosomal RNA, which is not translated and is one of the key components of ribosomes, is produced at rates of about 1 mRNA per second, almost an order of magnitude faster.

In the case of a regulated promoter such as *lac*UV5, we should also consider the rates at which transcription factors such as the Lac repressor come on and off the regulatory DNA. The diffusion-limited binding rate of Lac repressor to the operator DNA is about $0.003\ s^{-1}\ nM^{-1}$, which corresponds to an on rate of $0.03\ s^{-1}$, assuming 10 repressor molecules in the cell and using our rule of thumb that one molecule in *E. coli* corresponds to a concentration of 1 nM. The dissociation rate from operator DNA varies with the strength of the operator. For Lac repressor, this rate can range from $2\ s^{-1}$ for O3 all the way to $0.002\ s^{-1}$ for Oid. It is interesting to note that these rates are comparable and even slow when compared with the rate of transcription initiation, suggesting that the equilibrium assumption for this promoter might not be valid. However, similar reasoning for the chemical rate equations that are obtained by adding simple repression to the reaction scheme shown in Equation 19.38 leads to the conclusion that equilibrium reasoning is valid in this case as well.

19.3.2 Dynamics of mRNA Distributions

To write the dynamics for the simplest picture of mRNA production, we begin by elucidating the elementary processes that our promoter of interest can undergo in a time step Δt. The dynamics of such a promoter is represented in Figure 19.34, where we see that the

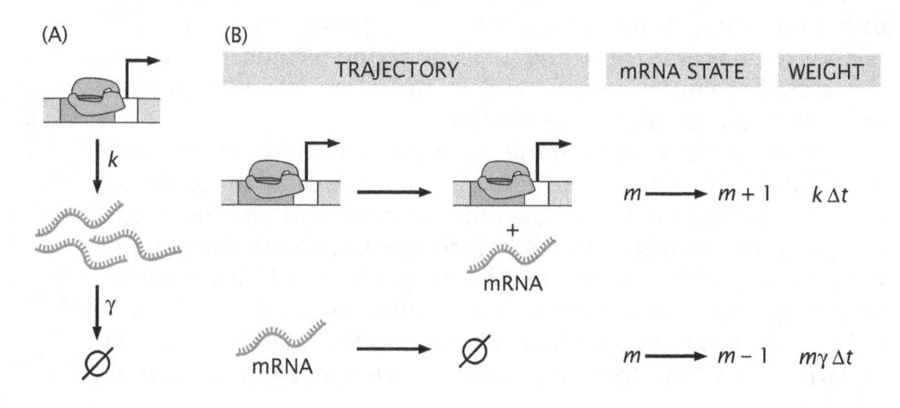

Figure 19.34: Trajectories and weights for the case of an unregulated promoter. (A) Schematic of the processes that can occur for the unregulated promoter, namely, mRNA production with rate k and mRNA decay with rate γ. (B) The individual processes that can occur in a time step of length Δt and their corresponding statistical weights.

mRNA count reflects a competition between the synthesis of new mRNAs with a rate k and their degradation with a rate constant γ. We are interested in determining the dynamics of the probability distribution $p(m,t)$ that tells us the probability of having m mRNA molecules at time t. Using the same "trajectories and weights" strategy adopted earlier and shown in Figure 19.34, we can write the time evolution as

$$\frac{\mathrm{d}p(m,t)}{\mathrm{d}t} = \underbrace{-kp(m,t)}_{m\to m+1} + \underbrace{kp(m-1,t)}_{m-1\to m} \underbrace{-\gamma mp(m,t)}_{m\to m-1} + \underbrace{\gamma(m+1)p(m+1,t)}_{m+1\to m}. \tag{19.39}$$

The first two terms correspond to the production of new mRNA molecules. Note that one of the terms occurs with a minus sign since the production of another mRNA molecule when we already have m of them leaves the system in a new state with $m+1$ molecules. The last two terms correspond to the situation in which we have m mRNA molecules decaying into $m-1$ molecules and $m+1$ molecules decaying to m molecules, respectively. Care must be taken in the $m=0$ case as the master equation for that state lacks the second term in Equation 19.39, given that the states with a negative number of mRNAs are unphysical. As we will see below, one way to implement this condition is by imposing the condition that $p(m<0,t)=0$.

With this dynamical equation in hand, there are a number of different strategies that can be adopted in order to learn what it implies. One idea is to establish dynamical equations for the moments of the distribution in anticipation of the more challenging situations that arise when we consider a regulated promoter. In those cases, analytic progress aimed at determining the full distribution $p(m,t)$ is very difficult, while the use of the moments such as $\langle m\rangle$ and $\langle m^2\rangle$ is both theoretically tractable and experimentally accessible. In general, we note that the jth moment is given by $\langle m^j\rangle = \sum_{m=0}^{\infty} m^j p(m,t)$. The low-order moments have an intuitive meaning, with the first moment providing the mean and the second moment some measure of the width of the distribution. A second strategy for characterizing our distribution that is also illuminating is to seek the steady-state properties of the distribution. We turn to both of these strategies in the pages that follow.

As our first exercise, we calculate the mean of the distribution $p(m,t)$ corresponding to the first moment of the distribution. To obtain the time evolution of the first moment, we multiply both sides of Equation 19.39 by m and sum over all possible values of m. This results in

$$\begin{aligned}\sum_{m=0}^{\infty} m\frac{\mathrm{d}p(m,t)}{\mathrm{d}t} = \sum_{m=0}^{\infty} m[&-kp(m,t)+kp(m-1,t)\\ &-\gamma mp(m,t)+\gamma(m+1)p(m+1,t)].\end{aligned} \tag{19.40}$$

Since the derivative is a linear operation, we can rewrite the left-hand side of this equation as

$$\sum_{m=0}^{\infty} m\frac{\mathrm{d}p(m,t)}{\mathrm{d}t} = \frac{\mathrm{d}}{\mathrm{d}t}\sum_{m=0}^{\infty} mp(m,t) = \frac{\mathrm{d}\langle m(t)\rangle}{\mathrm{d}t}, \tag{19.41}$$

which permits us to now rewrite Equation 19.40 as

$$\frac{\mathrm{d}\langle m(t)\rangle}{\mathrm{d}t} = -k\langle m(t)\rangle + k\sum_{m=0}^{\infty} mp(m-1,t) - \gamma\langle m^2(t)\rangle$$

$$+\gamma\sum_{m=0}^{\infty} m(m+1)p(m+1,t), \tag{19.42}$$

where we have invoked the definition of $\langle m^j\rangle$.

To make further progress, we need to come to terms with the two sums that are left in Equation 19.42. For the first of these sums, we introduce a new variable $m' = m-1$ such that the sum can be rewritten as

$$\sum_{m=0}^{\infty} mp(m-1,t) = \sum_{m'=-1}^{\infty} (m'+1)p(m',t). \tag{19.43}$$

As mentioned above, because it is unphysical to have a negative number of mRNA molecules, we impose $p(-1) = 0$, which allows us to start the summation over m' from 0 rather than from -1. As a result, we have

$$\sum_{m'=-1}^{\infty} (m'+1)p(m',t) = \sum_{m'=0}^{\infty} (m'+1)p(m',t)$$

$$= \sum_{m'=0}^{\infty} m'p(m',t) + \sum_{m'=0}^{\infty} p(m',t) = \langle m^1(t)\rangle + \langle m^0(t)\rangle. \tag{19.44}$$

Finally, we enforce the fact that the distribution is normalized to 1 at all time points, which means that $\langle m^0(t)\rangle = \sum_{m=0}^{\infty} p(m,t) = 1$. Using a similar strategy, we can now examine the second sum by introducing a variable $m' = m+1$, resulting in

$$\sum_{m=0}^{\infty} m(m+1)p(m+1,t) = \sum_{m'=1}^{\infty} (m'-1)m'p(m',t)$$

$$= \sum_{m'=0}^{\infty} (m'-1)m'p(m',t), \tag{19.45}$$

where we have used the fact that $(m'-1)m'p(m',t) = 0$ for $m' = 0$ in order to set the starting point of the sum to zero. As a result, we can rewrite the sum as

$$\sum_{m'=0}^{\infty} (m'-1)m'p(m',t) = \sum_{m'=0}^{\infty} m'^2 p(m',t) - m'p(m',t) = \langle m(t)^2\rangle - \langle m(t)\rangle. \tag{19.46}$$

We are now ready to put this all together and rewrite Equation 19.42 as

$$\frac{\mathrm{d}\langle m(t)\rangle}{\mathrm{d}t} = -k\langle m(t)\rangle + k\langle m(t)\rangle + k - \gamma\langle m^2(t)\rangle + \gamma\langle m(t)^2\rangle - \gamma\langle m(t)\rangle, \tag{19.47}$$

which results in

$$\frac{\mathrm{d}\langle m(t)\rangle}{\mathrm{d}t} = k - \gamma\langle m(t)\rangle. \tag{19.48}$$

This important result gives the rate of change of the mean number of mRNAs in a very simple form. One especially important outcome of this analysis is the insight that the steady-state mean level of mRNA expression is given by $\langle m \rangle = k/\gamma$.

Unregulated Promoters Can Be Described By a Poisson Distribution

One key question about these mRNA distributions is the functional form they adopt in steady state. In particular, what is the form of $p(m)$ after all the initial transients have decayed away? In the case of the simple promoter considered here, we make the educated guess that the distribution is of the Poisson form

$$p(m) = \frac{\lambda^m e^{-\lambda}}{m!}, \tag{19.49}$$

where λ is the mean of the distribution. To evaluate this mean, we resort to the definition

$$\langle m \rangle = \sum_{m=0}^{\infty} m \frac{\lambda^m e^{-\lambda}}{m!} = e^{-\lambda} \sum_{m=0}^{\infty} m \frac{\lambda^m}{m!}. \tag{19.50}$$

As we have done throughout the book, this can be evaluated by exploiting the trick of differentiating with respect to a parameter using

$$m\lambda^m = \lambda \frac{d\lambda^m}{d\lambda}. \tag{19.51}$$

In the context of Equation 19.50, this implies

$$\langle m \rangle = e^{-\lambda} \lambda \frac{d}{d\lambda} \sum_{m=0}^{\infty} \frac{\lambda^m}{m!} = e^{-\lambda} \lambda \frac{de^{\lambda}}{d\lambda} = \lambda. \tag{19.52}$$

We can now use these insights to directly substitute the trial solution into Equation 19.39 for the steady-state case, resulting in

$$0 = -k \frac{\lambda^m e^{-\lambda}}{m!} + k \frac{\lambda^{m-1} e^{-\lambda}}{(m-1)!} - \gamma m \frac{\lambda^m e^{-\lambda}}{m!} + \gamma (m+1) \frac{\lambda^{m+1} e^{-\lambda}}{(m+1)!}. \tag{19.53}$$

This can be simplified to the form

$$0 = \frac{1}{m!}(\lambda\gamma - k) + \frac{1}{(m-1)!}\left(\lambda^{-1} k - \gamma\right). \tag{19.54}$$

We see that if we now use the fact that $\lambda = k/\gamma$, the terms in both sets of parentheses are identically zero, confirming that the Poisson distribution is the appropriate steady-state distribution for this simplest dynamical model of transcription. Another interesting feature of this distribution is that its mean equals its variance.

As mentioned earlier, the study of noise in regulatory networks has become one of the central concerns of regulatory biology in recent years. One measure of this noise is provided by the so-called Fano factor, defined as

$$\text{Fano factor} = \frac{\langle m^2 \rangle - \langle m \rangle^2}{\langle m \rangle}. \tag{19.55}$$

Note that the Fano factor measures the relative size of the variance in mRNA number with respect to its mean. Poisson distributions like that considered above have the very special property that the Fano factor is equal to 1. One of the questions about the distributions like those shown in Figure 19.33 is whether they exhibit the Poisson form. To make further theoretical progress with this question in the context of this simplest of models of transcription, we need to compute the second moment.

We turn to the same trick as before by multiplying both sides of Equation 19.39 by m^2 and summing over all possible values of m. This results in

$$\sum_{m=0}^{\infty} m^2 \frac{\mathrm{d}p(m,t)}{\mathrm{d}t} = \sum_{m=0}^{\infty} m^2[-kp(m,t) + kp(m-1,t) - \gamma mp(m,t) + \gamma(m+1)p(m+1,t)] \quad (19.56)$$

which, using the same logic as before, can immediately be re-written as

$$\frac{\mathrm{d}\langle m^2(t)\rangle}{\mathrm{d}t} = -k\langle m^2(t)\rangle + k\sum_{m=0}^{\infty} m^2 p(m-1,t) - \gamma\langle m^3(t)\rangle + \gamma\sum_{m=0}^{\infty} m^2(m+1)p(m+1,t). \quad (19.57)$$

Once again, we have to treat the remaining summations in Equation 19.57. For the first of these sums, we make the change of variable $m' = m - 1$, resulting in

$$\sum_{m=0}^{\infty} m^2 p(m-1,t) = \sum_{m'=-1}^{\infty} (m'+1)^2 p(m',t) = \sum_{m'=0}^{\infty} (m'+1)^2 p(m',t), \quad (19.58)$$

where we have changed the starting m' of the sum. We now expand the binomial, resulting in

$$\sum_{m'=0}^{\infty} (m'+1)^2 p(m',t) = \sum_{m'=0}^{\infty} \left(m'^2 + 2m' + 1\right) p(m',t) = \langle m^2(t)\rangle + 2\langle m(t)\rangle + 1. \quad (19.59)$$

For the second sum, we make an analogous change of variable of the form $m' = m + 1$, resulting in

$$\sum_{m=0}^{\infty} m^2(m+1)p(m+1,t) = \sum_{m'=1}^{\infty} (m'-1)^2 m' p(m',t) = \sum_{m'=0}^{\infty} (m'-1)^2 m' p(m',t). \quad (19.60)$$

Once again, we have changed the lower limit of the sum since including the term $m' = 0$ does not change anything. We can now expand the binomial, resulting in

$$\sum_{m'=0}^{\infty} (m'-1)^2 m' p(m',t) = \sum_{m'=0}^{\infty} \left(m'^2 - 2m' + 1\right) m' p(m',t) = \langle m^3(t)\rangle - 2\langle m^2(t)\rangle + \langle m(t)\rangle. \quad (19.61)$$

We can now return to Equation 19.57 by substituting the outcome of our sums, resulting in

$$\frac{\mathrm{d}\langle m^2(t)\rangle}{\mathrm{d}t} = 2k\langle m(t)\rangle + k - 2\gamma\langle m^2(t)\rangle + \gamma\langle m(t)\rangle. \quad (19.62)$$

In steady state, this becomes

$$0 = (2k+\gamma)\langle m\rangle + k - 2\gamma\langle m^2\rangle. \quad (19.63)$$

If we now exploit the result of our calculation for the first moment, namely, $\langle m\rangle = k/\gamma$, this simplifies to

$$\langle m^2\rangle = \langle m\rangle^2 + \langle m\rangle. \quad (19.64)$$

As a result, we now see that the variance σ^2, the quantity that measures the spread of the data around its mean, is given by

$$\sigma^2 = \langle m^2\rangle - \langle m\rangle^2 = \langle m\rangle. \quad (19.65)$$

We see that in this case, our distribution has the special property that the variance is equal to the mean, implying that the Fano factor is equal to 1. This is a very important result because both the variance and the mean can be measured experimentally. As seen in Figure 19.33(C), the resulting variance as a function of the mean for the bacterial promoter measured in that experiment yields a Fano factor closer to 4 shown by the red line, with the blue signifying the Poisson prediction. This suggests that our simple model is wrong, or at least incomplete. In the next section, we try to find a way of resolving this discrepancy.

19.3.3 Dynamics of Regulated Promoters

How can we respond to data like that shown in Figure 19.33 and 19.35? As seen in Figure 19.35, the number of mRNA molecules in a cell as a function of time is not a smoothly increasing function. Rather, the production of mRNA is bursty. One of the insights emerging from our analysis of the thermodynamic models that could account for this burstiness is the fact that there are all sorts of regulatory interventions that can perturb the presumed steady production of mRNA envisaged in the model considered above.

This lesson is driven home simply by the case of simple repression, which we have considered throughout the chapter. The idea illustrated in Figure 19.36 is that the kind of time history shown in Figure 19.35 might result from the promoter switching back and forth between transcriptionally inactive (that is, repressor-bound) and active (that is, polymerase-bound) states.

The thermodynamic models showed us that promoters can exist in many different states: repressed, empty, occupied by RNA polymerase, activated, etc. The one-state model we explored in the previous section corresponds to the simplest situation in which RNA polymerase is always present at the promoter. When transcription starts and that polymerase molecule leaves the promoter, another RNA polymerase will take its place instantaneously. However, this clearly ignores the many regulatory interventions that are possible as a result of the vast array of transcription factors that are present in a cell.

Figure 19.35: Temporal history of mRNA production in *E. coli*. (A) Intensity of spots coming from labeled mRNA molecules as a function of time. Vertical white lines correspond to the cell division process. (B) Microscopy images of cells at various time points in the transcriptional history of the cells. Note that the mRNA molecules in this experiment tend to aggregate. As a result, when counting mRNA molecules, we must consider not only the number of puncta, but also their intensity. (Courtesy of Lok-hang So and Ido Golding, unpublished.)

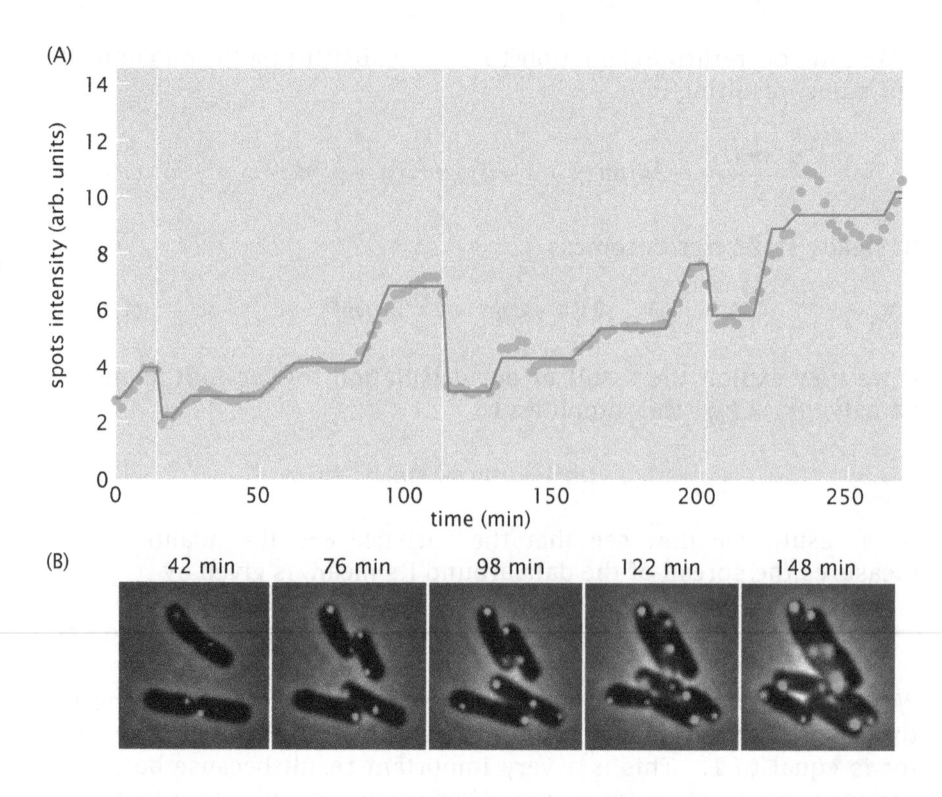

The Two-State Promoter Has a Fano Factor Greater Than One

To increase the realism of our treatment of the kinetics of our promoters, we now include the presence of these transcription factors, which we interpret as conferring different states of promoter activity between which the system can switch back and forth. As a first model, we consider simple repression in which the promoter can switch back

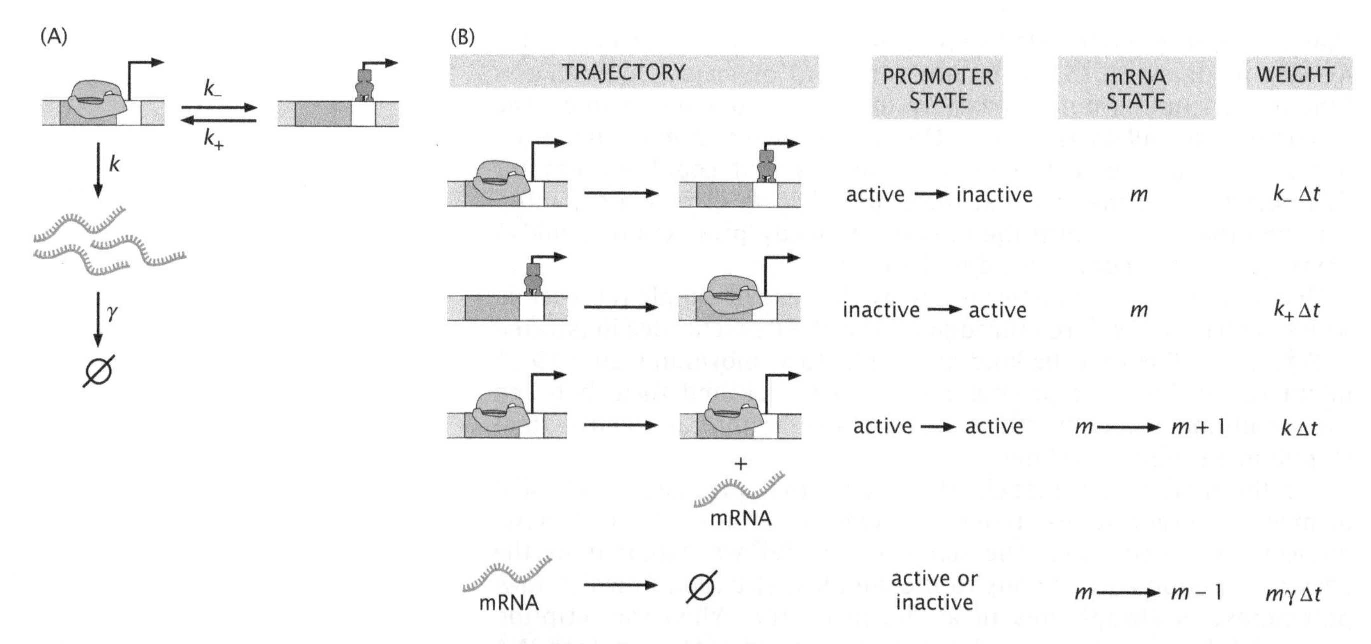

Figure 19.36: Trajectories and weights for the case of simple repression. (A) In an increment of time Δt, the system can suffer several different fates, including switching between the active and inactive states, degradation of an individual mRNA molecule, and production of a mRNA molecule while in the active state. (B) The individual trajectories available to the system in time Δt and their corresponding weights.

and forth between an inactive state (labeled "I") and an active state (labeled "A"). We can describe the kinetics of the promoter using the reactions

$$\mathrm{I} \underset{k_-}{\overset{k^+}{\rightleftharpoons}} \mathrm{A} \overset{k}{\rightharpoonup} \mathrm{mRNA} \overset{\gamma}{\rightharpoonup} \varnothing. \tag{19.66}$$

In the context of this new kinetic model, we now need to keep track of two variables, namely, the state of the promoter (that is, I or A) and the current number of mRNA molecules, m. To do so, we define two different probability distributions corresponding to the two promoter states. $p_{\mathrm{I}}(m,t)$ describes the probability of having the promoter in the inactive state with m mRNA molecules, whereas $p_{\mathrm{A}}(m,t)$ describes the probability of finding the promoter in the active state with m mRNA molecules.

As before, our goal is to write equations that describe the time evolution of these probabilities. Intuitively, we see that if we are thinking about how $p_{\mathrm{I}}(m,t)$ changes over time, there are only a few different processes that can transpire: (i) the promoter can switch from inactive to active, (ii) the promoter can switch from active to inactive, and (iii) an mRNA molecule can decay. This is expressed by the master equation

$$\frac{\mathrm{d}p_{\mathrm{I}}(m,t)}{\mathrm{d}t} = -\underbrace{k^+ p_{\mathrm{I}}(m,t)}_{\mathrm{I}\to\mathrm{A}} + \underbrace{k^- p_{\mathrm{A}}(m,t)}_{\mathrm{A}\to\mathrm{I}} - \underbrace{\gamma m p_{\mathrm{I}}(m,t)}_{m\to m-1} + \underbrace{\gamma(m+1)p_{\mathrm{I}}(m+1,t)}_{m+1\to m}. \tag{19.67}$$

The equation describing the time evolution of $p_{\mathrm{A}}(m,t)$ needs to account for the fact that mRNA molecules can be produced when the promoter is in this state. In this case, we have

$$\begin{aligned}\frac{\mathrm{d}p_{\mathrm{A}}(m,t)}{\mathrm{d}t} &= \underbrace{k^+ p_{\mathrm{I}}(m,t)}_{\mathrm{I}\to\mathrm{A}} - \underbrace{k^- p_{\mathrm{A}}(m,t)}_{\mathrm{A}\to\mathrm{I}} - \underbrace{k p_{\mathrm{A}}(m,t)}_{m\to m+1} \\ &\quad + \underbrace{k p_{\mathrm{A}}(m-1,t)}_{m-1\to m} - \underbrace{\gamma m p_{\mathrm{A}}(m,t)}_{m\to m-1} + \underbrace{\gamma(m+1)p_{\mathrm{A}}(m+1,t)}_{m+1\to m}.\end{aligned} \tag{19.68}$$

Our goal is to see what light this model sheds on the Fano factor. To answer this question, we need to evaluate the first two moments of the distribution, namely, $\langle m^1 \rangle$ and $\langle m^2 \rangle$. It is very convenient to define the partial moments $\langle m_{\mathrm{I}}^j \rangle = \sum_{m=0}^{\infty} m^j p_{\mathrm{I}}$ and $\langle m_{\mathrm{A}}^j \rangle = \sum_{m=0}^{\infty} m^j p_{\mathrm{A}}$ for the inactive and active states, respectively. These partial moments are a mathematical convenience that will allow us to calculate the actual moments of the distribution. In particular, by summing the partial moments of a given order, we have the full moments of the mRNA distribution, namely,

$$\langle m_{\mathrm{I}}^j \rangle + \langle m_{\mathrm{A}}^j \rangle = \langle m^j \rangle. \tag{19.69}$$

Solving for the moments with the equations discussed above requires long and cumbersome algebra. As a result, we now resort to a more general matrix scheme that generalizes to any promoter architecture. For this particular case, we begin by defining the vector

$$\mathbf{p}(m,t) = (p_{\mathrm{A}}(m,t), p_{\mathrm{I}}(m,t)). \tag{19.70}$$

In addition, we define three matrices $\mathbf{K}$, $\mathbf{R}$, and $\boldsymbol{\Gamma}$. The matrix $\mathbf{K}$ describes the transitions between the active and inactive states of the promoter and is given by

$$\mathbf{K} = \begin{pmatrix} -k^- & k^+ \\ k^- & -k^+ \end{pmatrix}. \tag{19.71}$$

The matrix **R** describes the production of mRNA,

$$\mathbf{R} = \begin{pmatrix} k & 0 \\ 0 & 0 \end{pmatrix}. \tag{19.72}$$

Finally, the matrix $\mathbf{\Gamma}$ describes the decay of mRNA and is given by

$$\mathbf{\Gamma} = \begin{pmatrix} \gamma & 0 \\ 0 & \gamma \end{pmatrix}. \tag{19.73}$$

Using this notation, the master equations for our two-state promoter can be written as

$$\begin{aligned} \frac{\mathrm{d}\mathbf{p}}{\mathrm{d}t} &= \mathbf{K}\mathbf{p}(m,t) - \mathbf{R}\mathbf{p}(m,t) + \mathbf{R}\mathbf{p}(m-1,t) \\ &\quad - m\mathbf{\Gamma}\mathbf{p}(m,t) + (m+1)\mathbf{\Gamma}\mathbf{p}(m+1,t). \end{aligned} \tag{19.74}$$

This can be simplified to the form

$$\frac{\mathrm{d}\mathbf{p}}{\mathrm{d}t} = (\mathbf{K} - \mathbf{R} - m\mathbf{\Gamma})\,\mathbf{p}(m,t) + \mathbf{R}\mathbf{p}(m-1,t) + (m+1)\mathbf{\Gamma}\mathbf{p}(m+1,t). \tag{19.75}$$

With these definitions in hand, we can now write matrix equations for the different partial moments. The details are left as an exercise and are discussed in the problems at the end of the chapter, but the general strategy is the same as before: multiply the governing equations by m^i and then sum over all m. Using this strategy, we find that the time evolution of the zeroth moment is given by

$$\frac{\mathrm{d}\langle \mathbf{m}^0 \rangle}{\mathrm{d}t} = \mathbf{K}\langle \mathbf{m}^0(m,t) \rangle. \tag{19.76}$$

For the first moment, similar manipulations result in

$$\frac{\mathbf{d}\langle \mathbf{m}^1 \rangle}{dt} = (\mathbf{K} - \mathbf{\Gamma})\langle \mathbf{m}^1(m,t) \rangle + \mathbf{R}\langle \mathbf{m}^0(m,t) \rangle. \tag{19.77}$$

The equation for the time evolution of the second moment then takes the form

$$\frac{\mathrm{d}\langle \mathbf{m}^2 \rangle}{\mathrm{d}t} = (\mathbf{K} - 2\mathbf{\Gamma})\langle \mathbf{m}^2(m,t) \rangle + (2\mathbf{R} + \mathbf{\Gamma})\,\langle \mathbf{m}^1(m,t) \rangle + \mathbf{R}\langle \mathbf{m}^0(m,t) \rangle. \tag{19.78}$$

We interest ourselves in the results for these various moments in steady state. This corresponds to setting the left-hand sides of our dynamical equations to zero and results in the collection of equations

$$1 = \mathbf{u} \cdot \langle \mathbf{m}^0 \rangle, \tag{19.79}$$

$$0 = \mathbf{K}\langle \mathbf{m}^0 \rangle, \tag{19.80}$$

$$0 = (\mathbf{K} - \mathbf{\Gamma})\,\langle \mathbf{m}^1 \rangle + \mathbf{R}\langle \mathbf{m}^0 \rangle \tag{19.81}$$

$$0 = (\mathbf{K} - 2\mathbf{\Gamma})\,\langle \mathbf{m}^2 \rangle + (2\mathbf{R} + \mathbf{\Gamma})\,\langle \mathbf{m}^1 \rangle + \mathbf{R}\langle \mathbf{m}^0 \rangle. \tag{19.82}$$

These equations involve the definition $\mathbf{u} = (1, 1)$ as a vector that sums over the components of $\langle \mathbf{m}^j \rangle$. In particular, we have used **u** in order to force the normalization of $\langle \mathbf{m}^0 \rangle$.

The idea of the analysis at this point is to bootstrap by successively determining higher-order moments in terms of those we have already determined. If we begin with Equations 19.79 and 19.80, we can determine the partial moments of zeroth order as

$$\langle m_A^0 \rangle = \frac{k^+}{k^- + k^+},$$
$$\langle m_I^0 \rangle = \frac{k^-}{k^- + k^+}. \qquad (19.83)$$

These can be rewritten in terms of the equilibrium constant for the inactive-to-active transition as

$$\langle m_A^0 \rangle = \frac{K}{1+K},$$
$$\langle m_I^0 \rangle = \frac{1}{1+K}, \qquad (19.84)$$

where we have defined $K = k^+/k^-$. In fact, what these two partial moments tell us is the probability of finding the system in either the inactive or active states.

With the zeroth moment in hand, we can now turn to Equation 19.81 to determine the first moment. We can rewrite this equation as

$$-(\mathbf{K} - \mathbf{\Gamma})^{-1} \mathbf{R} \langle \mathbf{m}^0 \rangle = \langle \mathbf{m}^1 \rangle. \qquad (19.85)$$

After some algebra, this leads in turn to

$$\langle m_A^1 \rangle = \frac{k}{\gamma} \frac{k^+ + \gamma}{k^+ + k^- + \gamma} \langle m_A^0 \rangle,$$
$$\langle m_I^1 \rangle = \frac{k}{\gamma} \frac{k^-}{k^+ + k^- + \gamma} \langle m_A^0 \rangle. \qquad (19.86)$$

This result tells us that the mean level of mRNA is given by

$$\langle m^1 \rangle = \langle m_A^1 \rangle + \langle m_I^1 \rangle = \frac{k}{\gamma} \langle m_A^0 \rangle. \qquad (19.87)$$

As expected, the mean level of expression is given by the probability of finding the promoter in the active state times a factor (k/γ) that tells us about the balance of the production and decay of mRNA.

To continue to the point where we can determine the Fano factor, we now need the second moment of the distribution. As a prelude, it is convenient to rewrite Equation 19.82 as

$$-(\mathbf{K} - 2\mathbf{\Gamma})^{-1} \left[(2\mathbf{R} + \mathbf{\Gamma}) \langle \mathbf{m}^1 \rangle + \mathbf{R} \langle \mathbf{m}^0 \rangle \right] = \langle \mathbf{m}^2 \rangle. \qquad (19.88)$$

From this equation, we obtain two rather complicated expressions for $\langle m_A^2 \rangle$ and $\langle m_I^2 \rangle$. However, when we add them together, which is equivalent to evaluating $\mathbf{u} \cdot \langle \mathbf{m}^2 \rangle$, we find

$$\langle m^2 \rangle = \frac{1}{2\gamma} \left(k \langle m_A^0 \rangle + 2k \langle m_A^1 \rangle + \gamma \langle m^1 \rangle \right). \qquad (19.89)$$

Finally, if we use the result from Equation 19.87, this can be rewritten as

$$\langle m^2 \rangle = \langle m^1 \rangle + \frac{k}{\gamma} \langle m_A^1 \rangle. \qquad (19.90)$$

Equation 19.90 can be further simplified by appealing to Equation 19.86. The result is given by

$$\langle m^2 \rangle = \langle m^1 \rangle \left(1 + \langle m^1 \rangle \frac{1}{\langle m_A^0 \rangle} \frac{k^+ + \gamma}{k^+ + k^- + \gamma}\right). \tag{19.91}$$

We are now in a position to compute the Fano factor itself, which is given by

$$\frac{\sigma^2}{\langle m^1 \rangle} = \frac{\langle m^2 \rangle - \langle m^1 \rangle^2}{\langle m^1 \rangle} = 1 + \langle m^1 \rangle \left(\frac{1}{\langle m_A^0 \rangle} \frac{k^+ + \gamma}{k^+ + k^- + \gamma} - 1\right). \tag{19.92}$$

This can be further simplified by using our result for $\langle m_A^0 \rangle$. Making the relevant substitution, we find

$$\frac{\sigma^2}{\langle m^1 \rangle} = 1 + \langle m^1 \rangle \frac{k^-}{k^+} \frac{\gamma}{k^+ + k^- + \gamma}. \tag{19.93}$$

This expression depends only upon the mean level of expression, the switching rates between the on and off states, and the rate of degradation of mRNA. Interestingly, the second term gives us the deviation from a pure Poissonian promoter. To develop intuition for this result, we note that $\gamma/(k^+ + k^- + \gamma)$ is always smaller than one.

To actually compare the results of this analysis with the data revealed in Figure 19.33(C), we have to determine the parameters that appear in our expression for the Fano factor. For the experiments shown in the figure, measurements have shown that these rates are given as follows. First, the mRNA degradation rate is $\gamma = 0.014\ \text{min}^{-1}$. This degradation rate corresponds to a lifetime of 70 minutes, which is clearly at odds with the average lifetime of mRNA molecules in *E. coli* as shown in Figure 3.14 (p. 110). The reason for this discrepancy is that the experiments performed in Figure 19.33(C) were done using an array of fluorescently tagged mRNA-binding proteins leading to the number of mRNA molecules per cell as a function of time as shown in Figure 19.35. However, the presence of these binding proteins, though useful to detect the production of mRNA molecules, makes the mRNA molecule very stable, such that the only "degradation" is due to dilution upon cell division. Hence, the lifetime of 70 minutes, corresponds to the length of the cell cycle in those experimental conditions. For the case in which the promoter was fully induced (saturating concentration of arabinose), the observed mean number of mRNAs is given by $\langle m^1 \rangle = 10$. In this case, the rates of switching between the on and off states were measured and are given by $k^+ = 0.03\ \text{min}^{-1}$ and $k^- = 0.2\ \text{min}^{-1}$. If we use these rates, we find that the Fano factor can be evaluated as

$$\begin{aligned}\frac{\sigma^2}{\langle m^1 \rangle} &= 1 + 10 \times \frac{0.2\ \text{min}^{-1}}{0.03\ \text{min}^{-1}} \frac{0.014\ \text{min}^{-1}}{0.03\ \text{min}^{-1} + 0.2\ \text{min}^{-1} + 0.014\ \text{min}^{-1}} \\ &\approx 1 + 3.6 = 4.6.\end{aligned} \tag{19.94}$$

This result is in reasonable accord with the observations reported in Figure 19.33(C). Of course, there are many effects that were not considered in our model. For example, one potential shortcoming of this model is related to the fact that we think about the effects of dilution due to cell division as an exponential process. In reality, this is a discrete process that occurs only at the small time interval corresponding to the separation of the mother cell into the two daughter cells. The reader is invited to explore the consequences of considering this refined decay mechanism by performing a numerical simulation using the Gillespie algorithm, which we describe in the Computational Exploration below.

Different Regulatory Architectures Have Different Fano Factors

Our use of the master equation approach advocated in this section has focused exclusively on unregulated promoters and the simple repression motif. However, just as with the thermodynamic models, different choices of regulatory architecture lead to different noise profiles (and Fano factors). Figure 19.37 shows the results of the same kind of analysis we have performed in this section to other regulatory architectures.

Computational Exploration: The Gillespie Algorithm and Stochastic Models of Gene Regulation Earlier in this section, we solved the simple case of an unregulated promoter using a master equation approach. By taking into account the different trajectories available to the system and their corresponding weights as shown in Figure 19.34, we were able to calculate both the evolution of the mean mRNA level and the higher moments of the mRNA distribution. Though such master equations are conceptually simple, they can quickly become intractable from an analytic perspective. An alternative approach is to directly integrate the equations numerically. Such a strategy was presented in the computational exploration in Section 3.1.3 (p. 99), where we showed how we can numerically integrate the logistic equation. As discussed there, when performing such numerical integrations, it is key to choose a time step Δt such that this time scale is shorter than all of the intrinsic time scales characterizing the system.

The stochastic nature of chemical reactions is typically revealed in cases where the number of molecular species is small such that random fluctuations are non-negligible. In cases like this, many integration steps Δt can go by without any reaction actually occurring. This has the annoying side effect that much of the computational time is spent effectively doing nothing "interesting." To put this in specific terms, we consider the simple reaction shown in Figure 19.38(A), where a species A can be converted to a species B and vice versa. For example, if we have only one molecule of species A and no molecules of species B, much time can go by without the reaction A $\rightarrow$ B occurring.

One of the strengths of the Gillespie algorithm for solving stochastic differential equations lies in its economy of computation. Instead of following the trajectory of a system by integrating with a Δt that is constant, it provides a strategy to adapt the Δt at each time step of the integration by randomly determining the time to the next possible reaction from a probability distribution, thus avoiding unwanted time steps in which no reaction occurs. The idea behind this algorithm is to construct a particular realization of the stochastic dynamics of the system. The concept is explored schematically for our simple reaction in Figure 19.38(A). To figure out what happens during each step in the simulation, we draw random numbers

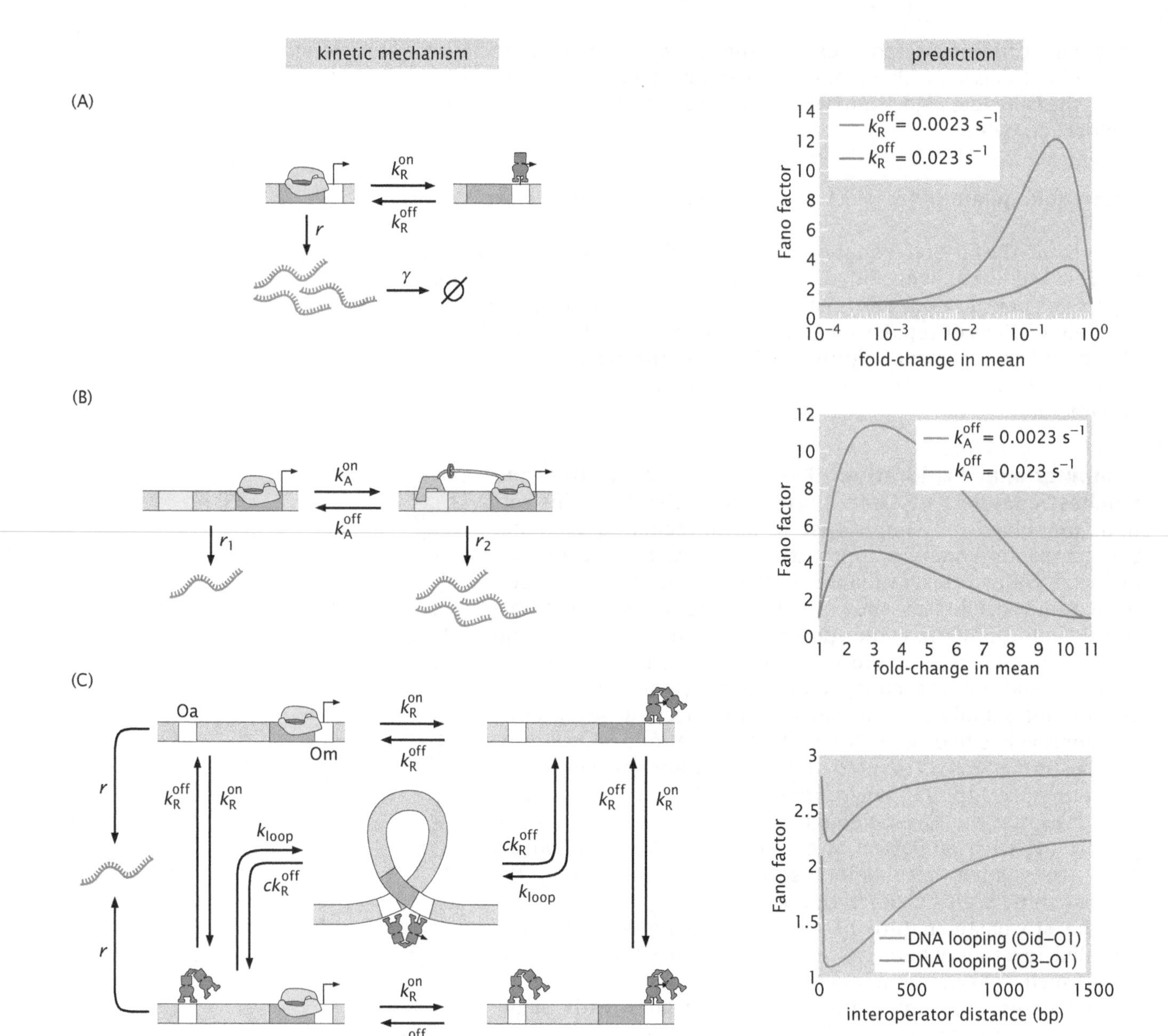

Figure 19.37: Stochastic models of transcriptional regulation for several regulatory architectures. The Fano factor is shown as a function of the fold-change in gene expression for (A) simple repression, (B) simple activation, and (C) repression by DNA looping. The kinetic models assumed for each architecture are shown on the left and their corresponding predictions are shown on the right. For all three figures, we use $r = 0.33$ mRNA s^{-1} and $\gamma = 0.011$ s^{-1}. The specific parameters for each figure are (B) $r_1 = r$ and $r_2/r_1 = 11$ and (C) $c = 1$, $k_{loop} = [J]k_R^0$, with $[J] = (\ln M)e^{-\beta\Delta F_{loop}}$ the same as given in Figure 19.29 and $k_R^0 = 2.7 \times 10^{-3}$ s^{-1} nM^{-1}, $k_R^{off}(\text{Oid}) = 1/(7\text{ min})$, $k_R^{off}(\text{O1}) = 1/(2.4\text{ min})$ and $k_R^{off}(\text{O3}) = 1/(0.47\text{ s})$. The fold-change in mean gene expression is obtained by varying k_R^{on} and k_A^{on} and is given by (A) $\left(1 + k_R^{on}/k_R^{off}\right)^{-1}$, (B) $\left[\left(k_A^{on}/k_A^{off}\right) r_1/r_2 + 1\right] / \left(k_A^{on}/k_A^{off} + 1\right)$, and (C) $\left[ck_R^{off}(k_R^{off} + k_R^{on})\right] / \left[k_{loop}k_R^{on} + c(k_R^{off} + k_R^{on})^2\right]$. The connection between fold-change in mean gene expression and Fano factor is explored in the problems at the end of the chapter. (Adapted from A. Sanchez et al., *PLoS Comput. Biol.* 7:e1001100, 2011.)

from specific distributions that are detailed below. One of the key insights of this algorithm is that we need to draw a random number twice per step in the reaction. First, we need to draw a random number in order to determine how much time Δt we need to wait until the next reaction takes place. The second drawing of a random number is analogous to a coin flip. We flip a coin (which is unfair) in order to determine which

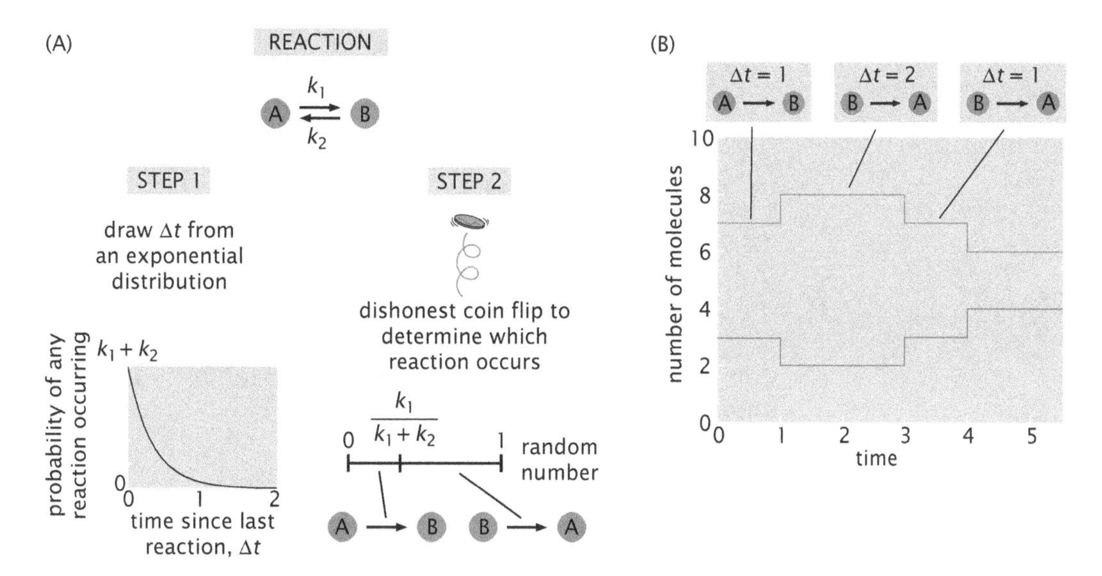

Figure 19.38: Concept of the Gillespie algorithm. (A) Schematic of the Gillespie algorithm for a simple chemical reaction. Here, two random decisions are made. The first step corresponds to drawing a random number from the exponential distribution given by Equation 19.101 in order to determine the time to the next reaction. The second step corresponds to a coin flip that determines which one of the reactions will occur. The bias of the flip is based on the magnitude of the rates corresponding to each possible reaction. (B) Example of a trajectory for the reaction shown in (A).

one of all the possible reactions actually occurred. An example of a realization of the approach for the reaction shown in Figure 19.38(A) is presented in Figure 19.38(B). Here we see how drawing a random number at each step leads to different values for the time interval until the next reaction, Δt, and for which reaction occurred at that time point (either species A to B or species B to A).

How can these ideas be applied to regulatory dynamics? In the unregulated promoter case shown in Figure 19.34(A), we have two possible reactions. First, an mRNA can be produced with a probability k per unit time. Second, an mRNA can decay with a probability γ per unit time and per mRNA molecule. Let's denote the mRNA production reaction as "1" and the mRNA decay reaction as "2" such that their generic rates per unit time will be k_i. This means that $k_1 = k$ and $k_2 = m(t)\gamma$, where $m(t)$ is the number of mRNA molecules at time t. For a time step Δt, we want to determine $P(i, \Delta t)\,\mathrm{d}t$, the probability of reaction i taking place in the time interval $(\Delta t, \Delta t + \mathrm{d}t)$. We construct this probability distribution by first noting that we need to impose that no reaction has already occurred between time points t and $t + \Delta t$. The probability of no reaction before Δt is written as $P_0(\Delta t)$, and hence the probability of the ith reaction occurring between Δt and $\Delta t + \mathrm{d}t$ is given by

$$P(i, \Delta t)\,\mathrm{d}t = P_0(\Delta t)k_i\,\mathrm{d}t, \tag{19.95}$$

where the term $k_i\,\mathrm{d}t$ corresponds to the probability of reaction i occurring in a time step $\mathrm{d}t$. But what is $P_0(\Delta t)$ (that is, what is the probability of no reaction occurring up until the time point Δt)? In order to calculate this, we write an expression for $P_0(\Delta t + \mathrm{d}t)$, the probability of no reaction occurring until the time point $\Delta t + \mathrm{d}t$. This is just the probability that no reactions took place in time Δt multiplied by the probability that no reactions take place in time $\mathrm{d}t$, namely,

$$P_0(\Delta t + \mathrm{d}t) = P_0(\Delta t)\left(1 - \sum_i k_i\,\mathrm{d}t\right), \qquad (19.96)$$

where the index i sums over all possible reactions. If we Taylor-expand $P_0(\Delta t + \mathrm{d}t)$ around Δt, we get

$$P_0(\Delta t + \mathrm{d}t) \approx P_0(\Delta t) + \frac{\mathrm{d}P_0(\Delta t)}{\mathrm{d}\Delta t}\,\mathrm{d}t. \qquad (19.97)$$

Comparing the terms in Equations 19.96 and 19.97 leads to the differential equation

$$\frac{\mathrm{d}P_0(\Delta t)}{\mathrm{d}\Delta t} = -P_0(\Delta t)\sum_i k_i. \qquad (19.98)$$

This equation can be solved to yield

$$P_0(\Delta t) = \mathrm{e}^{-\sum_i k_i \Delta t} = \mathrm{e}^{-k_0 \Delta t}, \qquad (19.99)$$

where we have used the initial condition $P_0(\Delta t = 0) = 1$, stating that at the beginning of our interval no reaction could have occurred yet. We have also defined $k_0 = \sum_i k_i$. As a result, we find

$$P(i, \Delta t)\,\mathrm{d}t = \mathrm{e}^{-k_0 \Delta t} k_i\,\mathrm{d}t. \qquad (19.100)$$

In order to make progress, we notice again that the probability distribution shown in Equation 19.100 can be thought of as the product of two probabilities. First, we determine what is the probability of any reaction occurring between time points Δt and $\Delta t + \mathrm{d}t$. In order to do this, we sum the distribution $P(i, \Delta t)\,\mathrm{d}t$ over all possible reactions i,

$$P(\Delta t)\,\mathrm{d}t = \sum_i P(i, \Delta t)\,\mathrm{d}t = \mathrm{e}^{-k_0 \Delta t} k_0\,\mathrm{d}t. \qquad (19.101)$$

In Figure 19.38(A), we show this distribution schematically as Step 1 of our algorithm. Here, we see that our algorithm picks the time to the next reaction in a random manner. In particular, the time Δt needs to be picked from the exponential distribution shown in Equation 19.101. To that end, we use a random number x that is uniformly distributed in the interval $(0, 1)$, which is the output of a random number generator in Matlab. From this number, we compute the time interval $\Delta t = (1/k_0)\ln(1/x)$. We can easily convince ourselves that Δt obtained this way is drawn from the exponential distribution shown in Equation 19.101. Namely, if $Q(x) = 1$ is the probability distribution for the variable x, then the probability distribution for Δt satisfies the equation

$$P(\Delta t)\,\mathrm{d}t = Q(x)\,\mathrm{d}x \qquad (19.102)$$

which simply states that the probability of finding x in the interval $(x, x + \mathrm{d}x)$ is the same as the probability of finding Δt in the interval that $(x, x + \mathrm{d}x)$ is mapped to by the function $\Delta t(x) = (1/k_0)\ln(1/x)$. (Note that this relationship is very general and very useful when switching from one random variable to another.) Substituting $Q(x) = 1$ and $x = e^{-k_0 \Delta t}$ into Equation 19.102, and noting that the probability distribution is necessarily positive, leads to $P(\Delta t) = k_0 e^{-k_0 \Delta t}$, as required.

Now that we know when the next reaction will occur, we can ask which one of all the possible reactions will take place. We calculate this probability by integrating Equation 19.100 over time,

$$P(i) = \int_0^{+\infty} P(i, \Delta t)\, dt = \frac{k_i}{k_0}. \tag{19.103}$$

An alternative way to obtain this result is by going back to the definition of k_i as a probability per unit time that reaction i will occur. Given a time interval Δt, the probability of reaction i taking place is then proportional to $k_i \Delta t$ such that

$$P(i) = \frac{k_i \Delta t}{\sum_j k_j \Delta t} = \frac{k_i}{k_0}. \tag{19.104}$$

We see that the probability of reaction i taking place is just given by the the relation between its rate, k_i, and the sum of the rates of all possible transitions in the system, k_0. This is shown schematically as Step 2 in Figure 19.38(A). The bias of our dishonest coin flip is then given by the magnitude of the rates corresponding to the different reactions. An example of how this bias is calculated from a random number picked in the interval [0,1] is shown in the figure.

We are now ready to implement this stochastic algorithm in order to solve for the dynamics of the unregulated promoter from Figure 19.34. For each iteration of the algorithm, we carry out the following steps:

1. Given the current number of mRNA molecules $m(t)$, calculate $k_1 = k$, which does not change as a function of mRNA number, and $k_2 = \gamma m(t)$, which does depend on time through $m(t)$.
2. Calculate a random number x uniformly distributed between 0 and 1 (this is the output of a random number generator). From it, compute the time interval to the next reaction, $\Delta t = (1/k_0) \ln(1/x)$, where $k_0 = k_1 + k_2$. Advance the clock by the time interval Δt.
3. Calculate a random number between 0 and 1. If the number is between 0 and k_1/k_0, then increase the mRNA number by one. If the number is between k_1/k_0 and 1, then decrease the mRNA number by one. Notice that we invoke a number in the interval $[0, 1]$ in an unbiased way. However, we split this interval into $[0, k_1/k_0]$ and $[k_1/k_0, 1]$. The result is that the coin flip is biased based on the values of the different rates.
4. Repeat.

In Figures 19.39(A) and (B), we show the results of the stochastic simulation for an initial condition in which no mRNA molecules are present in the system. In Figure 19.39(A), the time step to the next reaction, Δt, was drawn from the exponential distribution shown in Equation 19.101. In contrast, in Figure 19.39(B), the average time step stemming from that

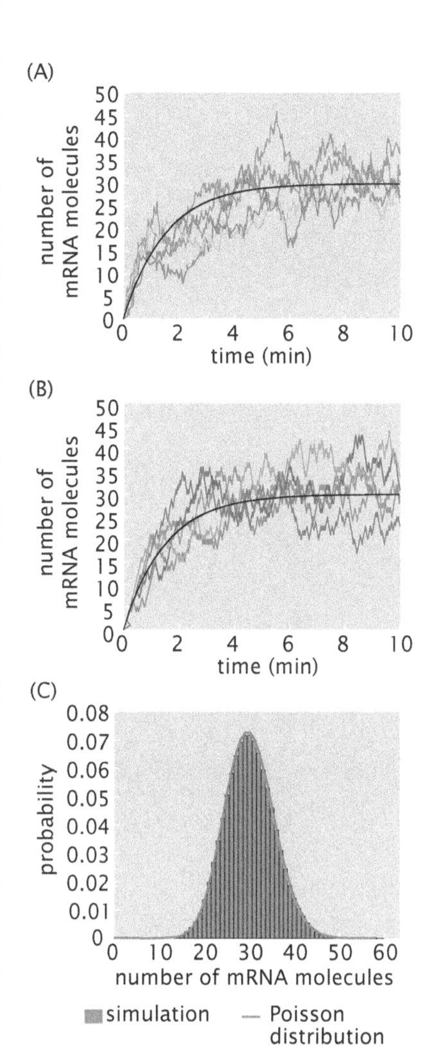

Figure 19.39: Numerical simulation of stochastic effects in the central dogma. (A) Various mRNA trajectories for the unregulated promoter shown in Figure 19.34 using mRNA = 0 as a starting condition. The deterministic solution for the mean number of mRNA molecules per unit time is included for comparison. For these simulations, the time step to the next reaction, Δt, was obtained from the distribution shown in Equation 19.101. (B) mRNA trajectories calculated using the average time step stemming from the exponential distribution shown in Equation 19.101. (C) Steady-state mRNA distribution obtained by the Gillespie algorithm and comparison with the exact solution given by a Poisson distribution. An mRNA transcription rate of 20 min^{-1} and a decay rate of 0.67 min^{-1} molecule^{-1} were used.

distribution was used to carry out the simulations. This is the suggested implementation of the Gillespie algorithm in this Computational Exploration. The reader is invited to explore the statistics of time-stepping for both approaches in the problems at the end of the chapter. Finally, notice how running the algorithm several times leads to different trajectories and how they all compare with the deterministic solution for the mean number of mRNAs given by

$$\langle m(t)\rangle = \left(1 - \mathrm{e}^{-\gamma t}\right)\frac{k}{\gamma}. \tag{19.105}$$

In the context of the discussion regarding Figure 19.34, we noticed that in steady state the probability distribution of mRNA molecules can be described by a Poisson distribution with mean k/γ. This can be verified using the Gillespie algorithm as shown in Figure 19.39(B). Here we have run a simulation at steady-state by setting the initial value of the simulation to the deterministic steady state value, $m(t=0) = k/\gamma$. From the resulting distribution, we determine the number of mRNA molecules as a function of time and generate the corresponding histogram. It is seen that the simulation is in excellent quantitative agreement with the Poisson distribution. This simple example gives a sense of how the simulations can be used as an alternative to the analytic treatment of these systems. The reader is invited to generate graphs like those shown in the figure for him- or herself.

19.3.4 Dynamics of Protein Translation

In the same way that the process of transcription is subject to variability, so are the other steps of the central dogma. For example, even though there might be a well-defined mean number of proteins arising from the translation of a single mRNA, the translation process is subject to variability. To see how this plays out, we start by considering a single mRNA in a cell of interest and compute the probability of n translation events taking place in the lifetime of this mRNA. Later, using this result, we will compute the steady-state protein distribution from a simple model of stochastic transcription and translation, thereby taking into account both key processes of the central dogma. In the simple kinetic model to be exploited here, over each interval of time Δt, two different processes can occur, namely, (i) the mRNA can be translated into a protein and (ii) the mRNA can decay. In Figure 19.40, we show these trajectories with their corresponding weights.

We do the bookkeeping on the mRNA state using the variable m, which keeps track of the number of mRNA molecules. This variable can adopt the values 1 or 0. Initially, we will have one mRNA, such

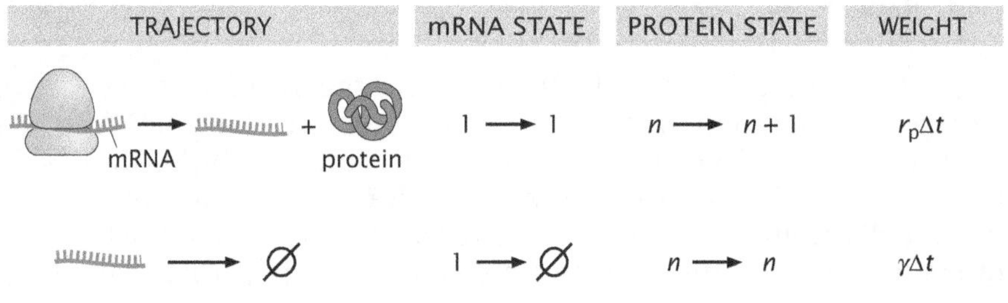

TRAJECTORY	mRNA STATE	PROTEIN STATE	WEIGHT
mRNA → mRNA + protein	$1 \rightarrow 1$	$n \rightarrow n+1$	$r_{\mathrm{p}}\Delta t$
mRNA → $\varnothing$	$1 \rightarrow \varnothing$	$n \rightarrow n$	$\gamma\Delta t$

Figure 19.40: Trajectories and weights for translation of a single mRNA molecule. During each interval of time Δt, the mRNA can either be transcribed or decay.

that $m = 1$. However, for long enough times, we know that $m = 0$, as the mRNA will eventually decay. The finite lifetime of mRNA molecules can be appreciated at the genome-wide level in both *E. coli* and yeast as shown in Figure 3.14 (p. 110). We also keep track of the number of proteins through the variable n. Our goal is to calculate the probability distribution $p(n, m; t)\,\mathrm{d}t$, namely the probability that during the time interval $(t, t + \mathrm{d}t)$, the number of mRNAs is m and the number of proteins is n. The master equation describing the time evolution of the state with one mRNA molecule is given by

$$\frac{\partial p(n, m = 1; t)}{\partial t} = -\gamma\, p(n, m = 1; t) + r_{\mathrm{p}}\left[p(n - 1, m = 1; t) - p(n, m = 1; t)\right]. \tag{19.106}$$

Like in the master equation for mRNA production from an unregulated promoter given in Equation 19.39, we need to be mindful of the $m = 0$ and $n = 0$ cases. For example, since $n < 0$ is unphysical, in the case where there are no proteins, $n = 0$, we need to drop the term in square brackets in Equation 19.106. This can be implemented by imposing $p(n < 0, m; t) = 0$. The state with zero mRNA molecules evolves according to the prescription

$$\frac{\partial p(n, m = 0; t)}{\partial t} = \gamma\, p(n, m = 1; t). \tag{19.107}$$

Solving these coupled differential equations is not straightforward. However, we are concerned with the function $P(n)$ defined as the probability that over the lifetime of the mRNA molecule, n proteins will have been produced. As we will see, the equation for this quantity is more tractable. The probability that the number of proteins synthesized during the lifetime of the mRNA is equal to n is given by $p(n, m = 0, t)$ at very long times, much longer than the decay time $1/\gamma$. In other words, $P(n) = p(n, m = 0, t)$ in the limit $t \to \infty$. We can compute this quantity from Equation 19.107 by integrating both sides from 0 to ∞ and noting that the number of proteins at $t = 0$ is zero (we start the clock when the mRNA is synthesized and no proteins have yet been produced). This results in

$$P(n) = \gamma \int_0^{+\infty} p(n, m = 1; t)\,\mathrm{d}t. \tag{19.108}$$

To make further progress with this, we must compute $p(n, m = 1; t)$ itself. To that end, we integrate both sides of Equation 19.106, resulting in

$$\int_0^{+\infty} \frac{\partial p(n, m = 1; t)}{\partial t}\,\mathrm{d}t = -\int_0^{+\infty} \gamma\, p(n, m = 1; t)\,\mathrm{d}t + \int_0^{+\infty} r_{\mathrm{p}}\left[p(n - 1, m = 1; t) - p(n, m = 1; t)\right]\mathrm{d}t. \tag{19.109}$$

Using the definition of $P(n)$ given above, we can write this as

$$p(n, m = 1; t \to +\infty) - p(n, 1; t = 0) = -P(n) + \frac{r_{\mathrm{p}}}{\gamma}\left[P(n - 1) - P(n)\right]. \tag{19.110}$$

This can be further simplified because we know that if we wait long enough, the mRNA will have decayed. As a result, we have $p(n, m = 1; t \to +\infty) = 0$. Further, we can use the initial condition that at time zero there is a single mRNA molecule and no corresponding protein resulting in the condition $p(n, m = 1; t = 0) = \delta_{n0}$. We are left with the equation

$$-\delta_{n0} = -P(n)\left(\frac{r_p}{\gamma} + 1\right) + \frac{r_p}{\gamma}P(n-1). \tag{19.111}$$

To solve this equation, we make the *ansatz* that $P(n) = A\lambda^n$. We start with the $n = 0$ case,

$$-1 = -P(0)\left(\frac{r_p}{\gamma} + 1\right), \tag{19.112}$$

which results in $A = \gamma/(r_p + \gamma)$. This can now be used in turn for the case when $n > 0$ for which we find $\lambda = r_p/(r_p + \gamma)$, resulting in the distribution

$$P(n) = \frac{\gamma}{r_p + \gamma}\left(\frac{r_p}{r_p + \gamma}\right)^n. \tag{19.113}$$

An interesting quantity to calculate is the mean number of proteins produced per mRNA. This is also called the "burst size," since the idea is that translation and mRNA decay occur on a time scale much faster than any protein decay. From the standpoint of proteins, the result of having one mRNA is a burst of protein production over a small amount of time. The mean can be shown to be

$$\langle n \rangle = \frac{r_p}{\gamma} = b, \tag{19.114}$$

where we have defined b as the burst size. Figure 19.41(B) shows the results from an experiment in which the number of β-galactosidase proteins produced per mRNA was measured in *E. coli.* The curve corresponds to a fit to the distribution calculated above with a mean burst size of five proteins per mRNA, revealing quite reasonable agreement with the distribution.

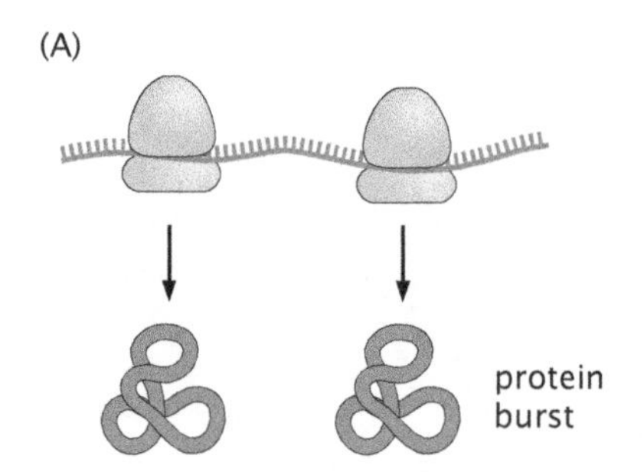

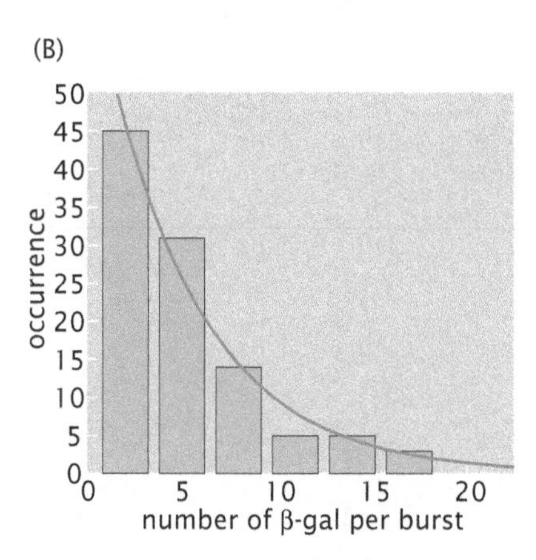

Figure 19.41: Time-dependent dynamics of protein production. (A) Bursts in translation where a single mRNA molecule can give rise to multiple proteins before it decays. (B) Distribution of burst sizes for the production of β-galactosidase in *E. coli* and fit to the probability distribution calculated in Equation 19.113. (B, adapted from Cai et al., *Nature* 440:358, 2006.)

In the calculation given above, we calculated the probability of obtaining n proteins as the product of translation of a single mRNA molecule. However, the total protein number within a cell is the result of the translation of multiple mRNA molecules. How do we determine the protein distribution in this more complicated case? The most straightforward approach is to write a master equation describing the evolution of both the number of proteins and the number of mRNA molecules within the cell. However, there are approximations that can be made that will simplify that task. For example, we can assume that the lifetime of an mRNA molecule is much shorter than the lifetime of the resulting proteins, as is clearly true for the case of *E. coli* as shown in Figures 3.14 and 3.15 (pp. 110 and 110). As a result, we will see no significant accumulation of mRNA over the life of a protein. Instead, every time an mRNA is transcribed, it will lead to a "burst" of protein production. The trajectories and weights corresponding to this model are shown in Figure 19.42. Note that we use our calculated $P(n)$ to account for the variability in protein production from a single mRNA.

Of course, one way to calculate the actual protein distribution is to solve the master equation stemming from the model shown in Figure 19.42. We define $P_{tot}(n, t)$ as the probability of having n proteins at time t. This equation is then

$$\frac{\partial P_{tot}(n,t)}{\partial t} = -n\gamma_p P_{tot}(n,t) + (n+1)\gamma_p P_{tot}(n+1,t) + \sum_{j=1}^{n} rP(j)P_{tot}(n-j,t) - \sum_{j=1}^{+\infty} rP(j)p(n,t). \tag{19.115}$$

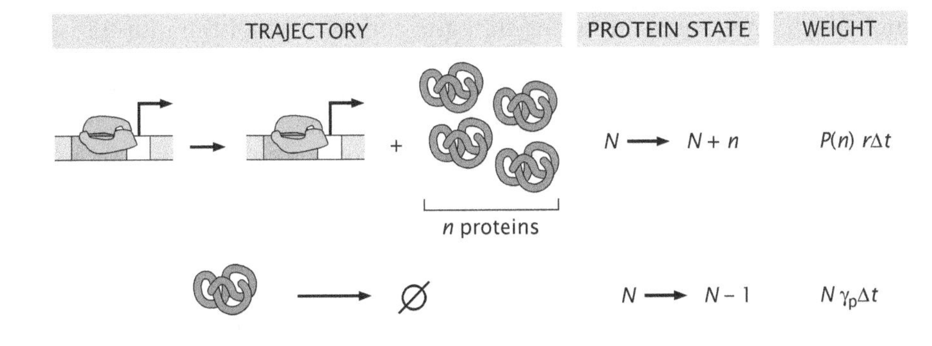

Figure 19.42: Trajectories and weights for a simple model of transcriptional and translational bursts. This model describes the production of mRNA through transcription and the subsequent protein production through the translation of mRNA in a burst before the mRNA decays. The size of the burst is given by $nP(n)$. The main assumption of this model is that the lifetime of an mRNA molecule is much shorter than the protein lifetime.

The last two terms in this equation are related to translation. In the first of these terms we account for all the ways we can go from having $n-j$ proteins to having n proteins from the translation of a single mRNA. This is the reason the probability $P(j)$ shows up in this term. The last term accounts for all the possible ways of leaving the state with n proteins due to the translation of more proteins.

We will solve this equation by making an educated guess for the distribution. We start by rewriting the distribution calculated in Equation 19.113 in terms of the burst size $b = r_p/\gamma$,

$$P(n) = \frac{b^n}{(1+b)^{n+1}}. \tag{19.116}$$

Now, let's say we have two mRNA molecules. We also assume that translation of one mRNA molecule occurs in a completely independent fashion from the second mRNA molecule. Under these conditions, the probability of producing N proteins from the translation of both such mRNA molecules is given by the product of the probability that the first molecule will produce n proteins and that the second will produce $N-n$ proteins, namely,

$$P_2(N) = \sum_{n=0}^{N} P(n)P(N-n). \tag{19.117}$$

Here we have simply used the fact that if one mRNA molecule produces n proteins, the other needs to produce $N-n$ if we want to have a total of N proteins translated. The operation carried out in Equation 19.117 is defined as the convolution of two functions. In particular, here we calculated the convolution of the function $P(N)$ with itself. In the Math Behind the Models at the end of this section, we describe the properties of convolutions in detail.

For three mRNA molecules, we can separate the problem into one and two mRNA molecules, namely,

$$P_3(N) = \sum_{n=0}^{N} P(n)P_2(N-n). \tag{19.118}$$

However, we already have an expression for $P_2(N)$ in terms of $P(N)$, which leads us to

$$P_3(N) = \sum_{n=0}^{N} P(n) \sum_{n'=0}^{N-n} P(n')P(n-n') = \sum_{n=0}^{N}\sum_{n'=0}^{N-n} P(n)P(n')P(N-n-n'). \tag{19.119}$$

What we have just shown is that if we want to calculate the probability distribution for the total number of proteins produced by m mRNA molecules, then we need to calculate the convolution of the individual probability distributions.

In order to make progress, we will assume that the number of proteins is large enough such that all the sums in the previous equations can be replaced by integrals. For example, for the case of two mRNA molecules, we have

$$P_2(N) = \int_0^N P(n)P(N-n)\,\mathrm{d}n. \tag{19.120}$$

This is again a convolution of $P(n)$ with itself just like in Equation 19.117, but now in integral form. Such integrals are much easier to solve in either Fourier or Laplace space. A careful description of Laplace transforms and how to use them in order to solve convolutions is presented in the Math Behind the Models below. After some algebra, we find the probability of having produced N proteins out of m mRNA molecules as

$$P_m(n) = \left(\frac{b}{1+b}\right)^n \left(\frac{1}{1+b}\right)^m \frac{n^{m-1}}{\Gamma(m)}, \tag{19.121}$$

where $\Gamma(m) = (m-1)!$ for m an integer. This distribution is called the negative binomial distribution (in the limit of large n and $b \gg 1$). Its continuous version is the more popular gamma distribution and can be obtained from Equation 19.121 by taking the limit of large n,

$$P_m(n) \to \frac{n^{m-1}\mathrm{e}^{-n/b}}{b^m\Gamma(m)}. \tag{19.122}$$

Does this negative binomial distribution solve the master equation shown in Equation 19.115? In order to do that, we need to plug our $P_m(n)$ into the master equation. We leave it as a problem to show that m, the mean number of mRNA molecules that a protein sees in its lifetime, is given by $m = r_\mathrm{p}/\gamma$. If a protein is very long-lived, then γ_p will be determined by the cell cycle, since decay will take place due to dilution by division. In that case, we can interpret r_p/γ as the number of mRNA molecules produced per cell cycle. Since each of these molecules leads to a burst of protein production, the inverse of this magnitude is often called the *burst frequency*. Together with the burst size $b = r_\mathrm{p}/\gamma$ the burst frequency fully determines the gamma distribution from Equation 19.122. One interesting thing is that, assuming that the proposed model in Figure 19.42 is correct, one can obtain dynamical information about the transcription and translation process from just looking at steady-state distributions. This concept is shown in Figure 19.43, where we present a strategy to perform such dynamical measurements leading to the values for the burst size and burst frequency. These values are to be compared with those obtained from fitting our gamma distribution to the experimental steady-state protein distribution. As we can see, there is reasonable qualitative agreement between the two techniques, suggesting that at least in an effective way it is valid to think of protein bursts in gene regulation.

The Math Behind the Models: Laplace Transforms and Convolutions Like its cousin the Fourier transform, which we have made use of repeatedly throughout the book, the Laplace transform is a very useful tool in physics for solving linear differential equations, such as those that appear in studies of mechanical and electrical phenomena. The basic premise here, like with other transforms, is to replace the sought function

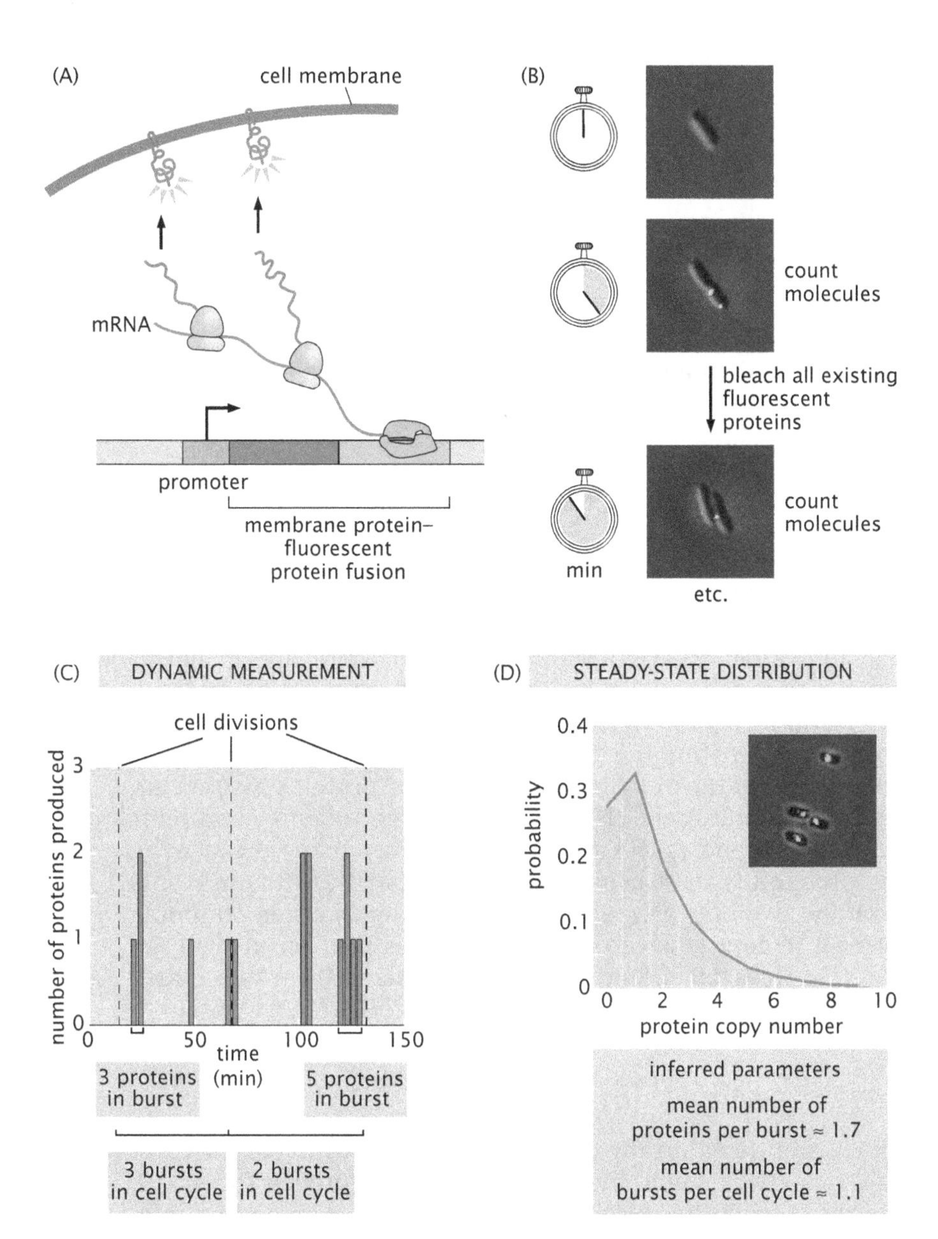

Figure 19.43: Protein bursting in *E. coli*. (A) A membrane protein can be fused to the fluorescent reporter YFP. The translated protein is localized to the membrane, where it can then be quantified. (B) If the level of expression is low, a snapshot of a cell can be taken in order to count proteins stuck to the membrane. The fluorophores can then be photobleached, making it possible to count the number of proteins produced in the next time step, which leads to a protein production rate. Using this method, the effective burst frequency and burst size can be measured. (C) An example of such a measurement of expression dynamics is shown. Here, each bar corresponds to the number of proteins observed on the membrane at that instant in time previous to photobleaching them in order to move to the next time point as shown in (B). Bursts of protein expression over several division cycles can be discerned as clusters of bars, which are associated with translation events off a single mRNA. From many such traces, the mean number of proteins per burst and the mean number of bursts per cell cycle can be calculated. (D) By fitting the steady-state distribution to the discrete gamma distribution from Equation 19.122, the burst frequency and burst size can also be estimated, yielding results comparable to the direct dynamical measurements. Inset: Representative field of view of the cells used to obtain the protein distribution. Note that the results from (C) and (D) correspond to different proteins and cannot be compared directly in a quantitative fashion. (B, C, adapted from J. Yu et al., *Science* 311:1600, 2006; D, adapted from Y. Taniguchi et al., *Science* 329:533, 2010.)

$f(t)$ with its transform $\tilde{f}(s)$, thereby turning the differential equation for $f(t)$ into a much simpler algebraic equation for its transform. The algebraic equation can then be solved for $\tilde{f}(s)$ which in turn can be used to obtain the original $f(t)$ by means of an inverse transform.

The Laplace transform of the function $f(t)$ is given by

$$\tilde{f}(s) = \int_0^{+\infty} f(t)\mathrm{e}^{-st}\,\mathrm{d}t. \tag{19.123}$$

The inverse transform is a more complicated matter, as it involves doing an integral of $f(s)$ in the complex plane, where s is assumed to be a complex number. In practice, one often uses tables of transforms and inverse transforms to solve the problem at hand.

The Laplace transform is particularly useful for solving convolution integrals, such as those that come up in the context of mRNA translation dynamics. The convolution integral of functions $f(t)$ and $g(t)$ is given by

$$C(t) = \int_0^t f(t')g(t-t')\,dt' \qquad (19.124)$$

and its Laplace transform is related to the Laplace transforms of $f(t)$ and $g(t)$ by the simple relation

$$\tilde{C}(s) = \tilde{f}(s)\tilde{g}(s). \qquad (19.125)$$

This can be demonstrated by taking the Laplace transform, as defined by Equation 19.123, of both sides of Equation 19.124, and then making a change of variables $u = t'$ and $v = t - t'$ in the two-dimensional integral that appears on the right-hand side. We leave the mathematical details as an exercise for the interested reader.

In the main text, we have derived the distribution of the number of proteins $P(n)$ obtained by repeated translation of a single mRNA molecule over its lifetime. In order to obtain the equivalent distribution $P_m(n)$ when there are m mRNA molecules in the cell, we make use of the convolution integral. For example, for the case $m = 2$, the sought distribution can be obtained by writing the probability of obtaining n proteins as the product of the probability of making n' proteins by translating the first mRNA and the probability of getting $n - n'$ proteins from the second mRNA molecule. The fact that we can write the probability as a product assumes that translation events from the two mRNAs are independent of each other. Summing over all n' between zero and n then takes into account all the possible ways of ending up with n proteins from two mRNA molecules. If we replace the sum with an integral, we obtain

$$P_2(n) = \int_0^n P(n')P(n-n')\,dn', \qquad (19.126)$$

which is nothing but the convolution of $P(n)$ with itself. We can now repeat the same divide-and-conquer approach for $m = 3$, and the distribution $P_3(n)$ will be given by the convolution integral of $P_2(n)$ and $P(n)$. Further iterations will then yield the sought distribution $P_m(n)$ as the convolution integral of $P_{m-1}(n)$ and $P(n)$.

The laborious procedure of calculating repeated convolution integrals is replaced by a simple multiplication using Laplace transforms. Namely, it follows from Equation 19.125 that $\tilde{P}_2(s) = \tilde{P}(s)\tilde{P}(s)$, and repeated use of Equation 19.125 then yields a simple formula for the Laplace transform of $P_m(n)$,

$$\tilde{P}_m(s) = \left[\tilde{P}(s)\right]^m. \qquad (19.127)$$

Therefore, if we want to compute the probability of getting n proteins out of m mRNA molecules, we simply need to calculate the inverse Laplace transform of $\tilde{P}_m(s)$ obtained from this equation.

So, let's start then by calculating the Laplace transform of $P(n)$,

$$\tilde{P}(s) = \int_0^{+\infty} \frac{b^n}{(1+b)^{n+1}} e^{-sn}\,dn. \qquad (19.128)$$

We can write this as

$$\tilde{P}(s) = \frac{1}{1+b}\int_0^{+\infty}\left(\frac{b}{1+b}\mathrm{e}^{-s}\right)^n \mathrm{d}n = \frac{1}{1+b}\left.\frac{\left(\frac{b}{1+b}\right)^n \mathrm{e}^{-sn}}{\ln\left(\frac{b}{1+b}\right) - s}\right|_0^{+\infty}, \tag{19.129}$$

which leads to

$$\tilde{P}(s) = -\left\{(1+b)\left[\ln\left(\frac{b}{1+b}\right) - s\right]\right\}^{-1}. \tag{19.130}$$

For m mRNA molecules, we need to calculate the inverse Laplace transform of $[\tilde{P}(s)]^m$. This inverse transform has an analytical form, but its calculation is cumbersome. We choose to just quote the result,

$$P_m(n) = \left(\frac{b}{1+b}\right)^n \left(\frac{1}{1+b}\right)^m \frac{n^{m-1}}{\Gamma(m)}, \tag{19.131}$$

where $\Gamma(m) = (m-1)!$ for m an integer. This distribution is called the negative binomial distribution (in the limit of large n and $b \gg 1$).

19.3.5 Genetic Switches: Natural and Synthetic

Switches are an important part of the genetic repertoire of all organisms. To explore the behavior of these switches more carefully, a synthetic version of such a switch was constructed in *E. coli* that had the convenient property that the gene product of the switch is a fluorescent reporter protein such that flipping of the switch can be read out by observing the fluorescent state of the cells. Data from this synthetic switch are shown in Figure 19.44.

The switch described above was constructed by using two repressor proteins whose transcription is mutually regulated as shown in Figure 19.45. This simple design allows us to see one of the most widespread regulatory features, namely, feedback. In particular, the protein that is the output from the first gene serves as a repressor

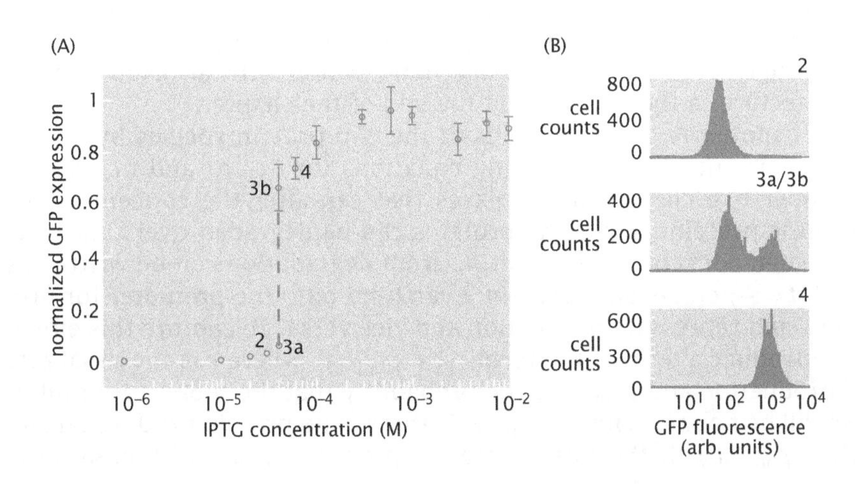

Figure 19.44: Data illustrating the flipping of the genetic switch in *E. coli* cells. (A) Average fluorescence of a population of *E. coli* cells harboring the genetic switch as a function of the concentration of an inducer molecule that flips the switch. In this case, IPTG (a lactose analog) is the inducer, which upon binding to Lac repressor produces an allosteric change that reduces its binding affinity. (B) Flow cytometry data showing the single-cell fluorescence distribution for different inducer concentrations. The labels correspond to points in the curve shown in (A). Bistability is revealed through the fact that there are two populations of cells at the same inducer concentration. (Adapted from T. S. Gardner et al., *Nature* 403:339, 2000.)

Figure 19.45: Regulatory architecture for a genetic switch. (A) There are two promoters that are under the transcriptional control of the gene product of the partner promoter. (B) States and weights for the two coupled genes making a genetic switch. For the case shown here, the Hill coefficient is $n = 2$ because the repressors bind as dimers. The more general case is considered in the text.

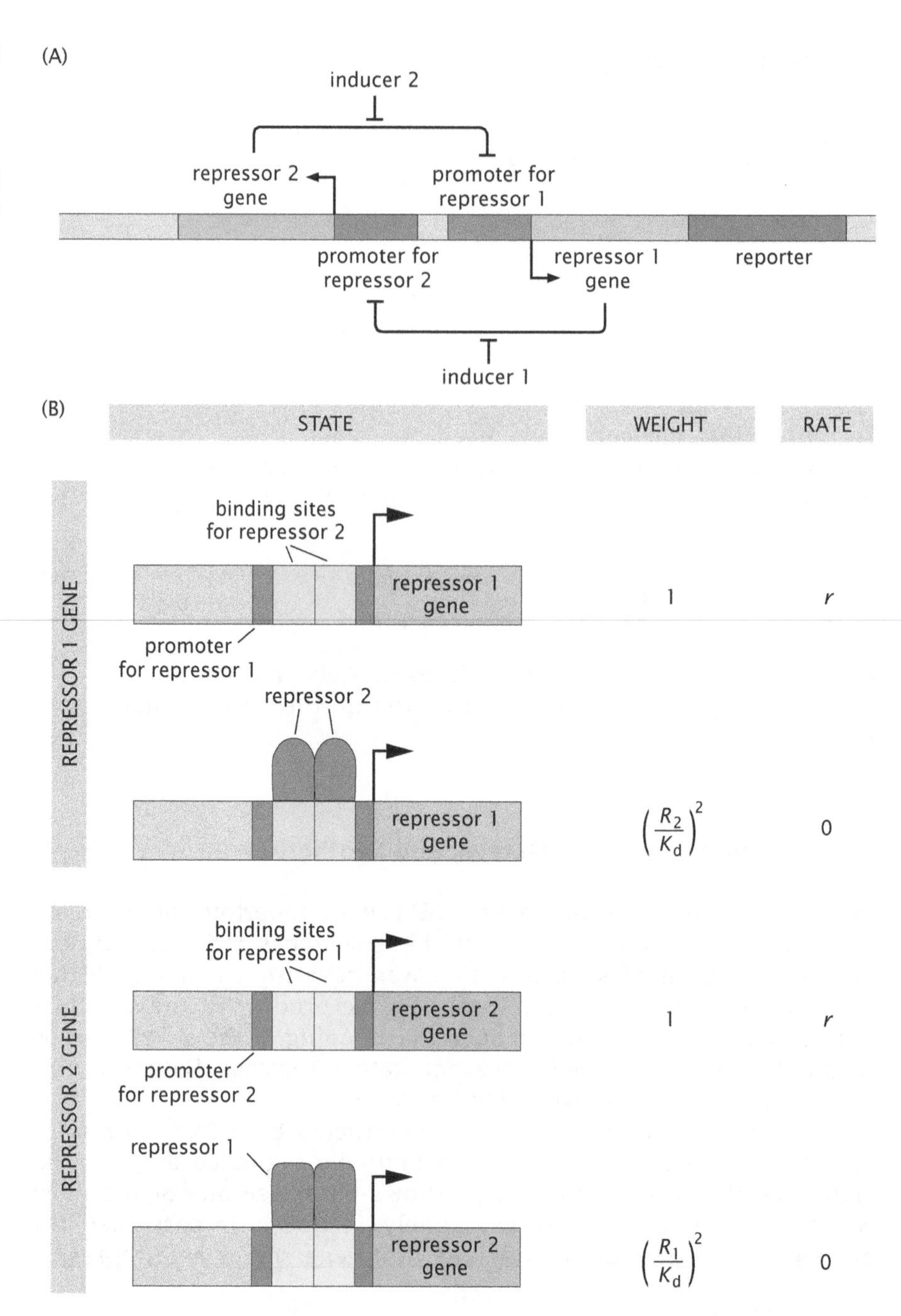

for the second gene. Conversely, the protein that is the output from the second gene serves as a repressor of the first gene. The reader is strongly urged to explore a genetic switch with an even simpler architecture in the problems at the end of the chapter.

We denote the concentrations of the two protein species by c_1 and c_2. We are interested in writing equations for dc_1/dt and dc_2/dt. We consider two classes of processes that can alter the concentrations of these proteins. First, the proteins can be degraded over time. The change in concentration resulting from degradation can be written as $dc_1/dt = -\gamma c_1$. Second, protein 2 can bind onto the promoter for protein 1 and repress its production and vice versa. To capture this effect, we introduce a term of the form $r(1 - p_{\text{bound}})$, where r is the basal rate of production and p_{bound} is the probability that the repressor of interest will be bound. When $p_{\text{bound}} = 1$, there is no protein production and when $p_{\text{bound}} = 0$, the rate of protein production takes its basal rate.

Recall from Chapter 6 that for binding described by a Hill function, we have

$$p_{\text{bound}}(c_1) = \frac{K_b c_1^n}{1 + K_b c_1^n}, \tag{19.132}$$

where K_b is the binding constant for the repressor. This implies in turn that the protein production rate for protein 2 is

$$r(1 - p_{\text{bound}}) = \frac{r}{1 + K_b c_1^n}. \tag{19.133}$$

A Hill function (see Section 6.4.3 on p. 273) rather than our statistical mechanical treatment has been used to model p_{bound} so that our treatment is consonant with the original literature. The reader will have the chance to explore the behavior of this circuit using p_{bound} as it has been considered throughout the book in the problems at the end of the chapter. Notice that our treatment of the binding constant here is slightly different than that favored in Section 6.4.3 and Figure 19.45, also for the purposes of consistency with the original literature.

Using the conceptual framework introduced above, the chemical rate equations for the genetic switch are

$$\begin{aligned} \frac{dc_1}{dt} &= -\gamma c_1 + \frac{r}{1 + K_b c_2^n}, \\ \frac{dc_2}{dt} &= -\gamma c_2 + \frac{r}{1 + K_b c_1^n}. \end{aligned} \tag{19.134}$$

The first terms on the right-hand sides of both equations correspond to protein degradation, and for simplicity we assume that the degradation rate (characterized by the parameter γ) of both proteins is the same. For proteins that are stable over time scales longer than the cell cycle (as is the case in the repressors used in this circuit), the dilution rate is determined by the cell doubling time and the subsequent dilution of the protein between the two daughter cells. Therefore, under these conditions, the effective protein degradation rate is the same and is set by the cell division time. The second terms on the right-hand sides of both equations characterize the rate of protein production. As introduced above, the *basal* rate of production is captured in the parameter r. However, this rate is reduced when the repressor is bound to the promoter of interest, as shown above. For simplicity, we assume that the basal production rates and the binding constants that characterize the affinity of the repressors for their binding site are the same for both genes. For a realistic circuit, these assumptions are not necessarily true, but will suffice here to describe the basic operation of the circuit. Another conceptual simplification implicit in these rate equations is the idea that the binding of the repressors is characterized by a Hill function with Hill coefficient n.

From a mathematical perspective, we wonder whether equations like Equations 19.134 yield switch-like solutions. Our assertion is that there are two regions in the space of parameters, one with a single stable solution corresponding to equal concentrations of the two species (decidedly not a switch) and another, more interesting regime, where we find two stable solutions distinguished by having one of the protein concentrations much larger than the other. For values of the parameters where the stable solutions are of this variety, the genetic network exhibits switch-like behavior.

In order to simplify the mathematical analysis of the circuit we resort to a dimensionless form for Equation 19.134. This is achieved by measuring c_1 and c_2 in units of $K_\mathrm{b}^{-1/n}$ and time in units of γ^{-1}. This reduces the circuit equations to

$$\begin{aligned}\frac{\mathrm{d}u}{\mathrm{d}t} &= -u + \frac{\alpha}{1+v^n},\\ \frac{\mathrm{d}v}{\mathrm{d}t} &= -v + \frac{\alpha}{1+u^n},\end{aligned} \qquad (19.135)$$

where the parameters $\alpha = rK_\mathrm{b}^{1/n}/\gamma$ and the Hill coefficient n are the only remaining dimensionless parameters. We have introduced the notation u for the dimensionless concentration of c_1 and v for the dimensionless concentration of c_2. At this point, our goal is to find the steady-state solutions of Equation 19.135 and analyze their stability for different values of α and the Hill coefficient.

To find the steady-state solutions to the rate equations, we set the time derivatives to zero. Since the equations are symmetric with respect to u and v, we immediately conclude that

$$u^* = v^* = \frac{\alpha}{1+v^{*n}} \qquad (19.136)$$

is always a solution. Clearly, this result does *not* exhibit the properties of a switch, since the concentrations of both proteins in this case are the same. Are there other solutions that exhibit switching behavior? The equations that determine the steady-state u^* and v^* are of the form $x = f(f(x))$, where $f(x) = \alpha/(1+x^n)$. To see this, solve the first equation for u and substitute that result into the second equation. Since the function f is monotonically decreasing (that is, larger values of x imply $f(x)$ is smaller) the composition $f \circ f$ will be a monotonically *increasing* function, like the function x itself. Therefore, there is the possibility that the two curves x and $f(f(x))$ intersect at more than one point, leading to multiple steady states. The detailed stability analysis is performed in the Math Behind the Models below.

To make these considerations explicit, we consider the case when the Hill coefficient n equals 2, which lends itself to analytic treatment. The steady-state equation for the repressor concentration u^* is

$$u^* = \frac{\alpha}{1+\left(\frac{\alpha}{1+u^{*2}}\right)^2}, \qquad (19.137)$$

and the same equation holds for v^*. A little bit of algebra transforms the above equation to a much simpler form given by a product of two polynomials

$$(u^{*2} - \alpha u^* + 1)(u^{*3} + u^* - \alpha) = 0. \qquad (19.138)$$

The steady-state solutions to the rate equations for the genetic switch, Equations 19.135, are therefore zeroes of the two polynomials appearing in the above equations.

The cubic polynomial has one real zero, which can be seen from Figure 19.46(A), where we plot the polynomial for different values of α. A mathematically rigorous way to show this is to note that

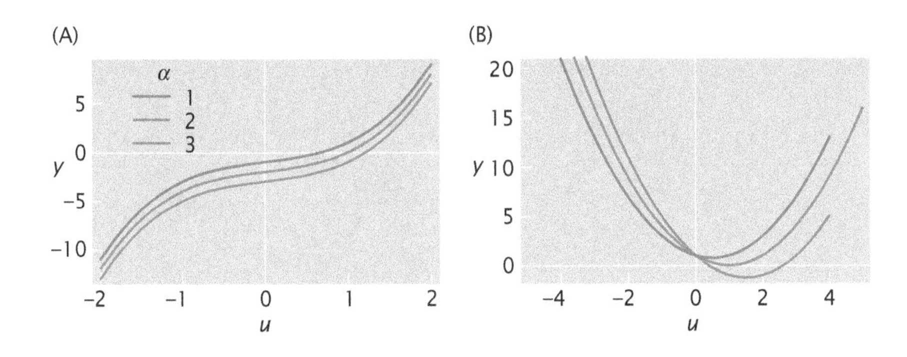

Figure 19.46: Steady-state solutions for protein concentrations in the genetic switch. (A) The function $y = u^3 + u - \alpha$ plotted for various values of α. The solution u^* corresponds to the point at which the curve crosses the u-axis. (B) The function $y = u^2 - \alpha u + 1$ plotted for various values of α. Depending upon the choice of α, there can be 0, 1, or 2 crossings of the u-axis.

the first derivative of this polynomial, $3u^{*2} + 1$, is always positive, which implies that the function is strictly increasing and can therefore intercept the u^*-axis at most once. The equilibrium state that corresponds to the zero of the cubic polynomial has equal concentrations of the two repressor species, since the equation $u^{*3} + u^* - \alpha = 0$ can be rewritten as $u^* = \alpha/(1 + u^{*2})$, and the right-hand side of this equation is v^*.

The quadratic polynomial in Equation 19.138 can have one, two, or no zeroes, depending on the value of α, as observed in Figure 19.46(B). For $\alpha < 2$, there are no zeroes; for $\alpha > 2$, the polynomial has two zeroes; while for $\alpha_c = 2$, the critical value of α, it has one zero at $u^* = 1$. For the two-solution case, the two steady-state values of u^* and v^* correspond to the two different ways of assigning the two roots to each of the dimensionless repressor concentrations. Namely, for a given u^*, the corresponding value of v^* can be calculated using $v^* = \alpha/(1 + u^{*2})$. For these values of u^* and v^*, the equality $u^* + v^* = \alpha$ is satisfied, assuming u^* is one of the zeroes of the quadratic polynomial in Equation 19.138.

In light of the general analysis done above, we see that for $\alpha < 2$ the genetic switch exhibits only one stable equilibrium state with $u^* = v^*$, while for $\alpha > 2$ it has two stable states and one unstable state. In the latter case, the unstable state is the one in which the concentrations of the two repressors are equal, while stable equilibrium states have either repressor u or repressor v in excess.

The dynamical behavior of a system of rate equations like those given in Equations 19.135 can be examined in a different way graphically using the idea of a phase portrait (the mathematics is explained in the Tricks behind the Math at the end of the section). The idea is that we can think of du/dt and dv/dt as the two components of a velocity vector and we can plot the velocity field at every point (u, v). The steady-state solutions will correspond to those points in the phase portrait where the vectors are zero. The solutions represented by those points are stable if for any small excursion away from that point, all the velocity vectors point towards the solution point. An example of this idea for several choices of α is shown in Figure 19.47. The phase portrait provides a convenient graphical representation of the dynamics of the genetic switch. Namely, for a given initial condition u_0, v_0, in order to see how the concentrations will evolve with time, all one has to do is follow the flow depicted by the arrows in the phase portrait. We therefore conclude that the stable steady states of the rate equations are associated with positions in the u–v plane where the phase flow converges from all directions, while diminishing in size, while unsteady states have at least one direction along which the flow is diverging.

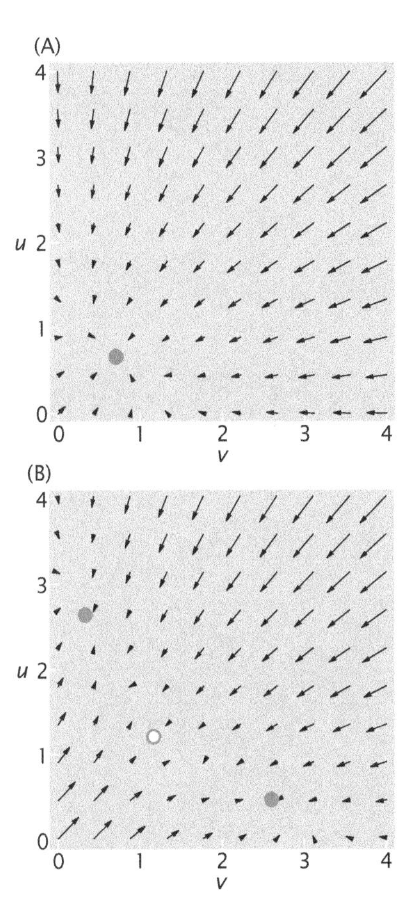

Figure 19.47: Graphical representation of the dynamics of the genetic switch. The phase portraits of the genetic switch for (A) $\alpha = 1$ and (B) $\alpha = 3$. Stable equilibria are represented by filled circles, while the unfilled circle corresponds to an unstable state.

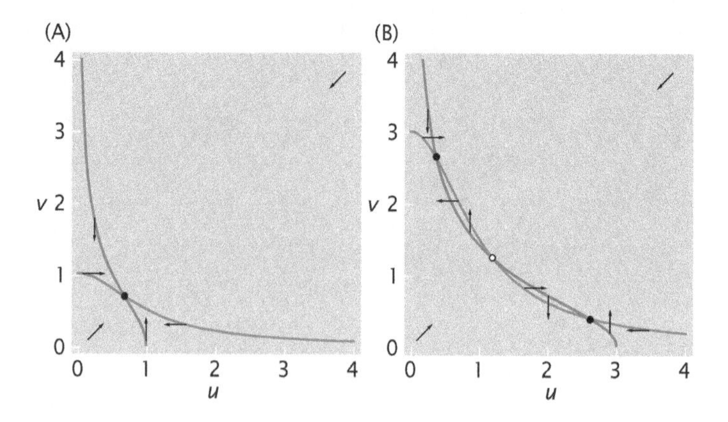

Figure 19.48: Graphical determination of the phase portraits for the genetic switch. Qualitative features of the phase portrait shown in Figure 19.47 can be constructed using the nullclines of the dynamical system described by Equations 19.135. The direction of the phase flow is first determined for large and small values of the dimensionless concentrations of the two repressors, and then the flow in the rest of phase space is determined by continuity. In particular, the direction of the u- or v- component of the flow can only change sign at the nullclines, which are the sets of points along which the rate of change of either u or v vanishes. The intersection of nullclines is a fixed point of the phase flow. (A) Nullclines and flow in phase space for the genetic switch with parameter $\alpha = 1$. The intersection of the two nullclines is a stable fixed point, indicating an absence of switch-like behavior. (B) Nullclines and phase flow in the case $\alpha = 3$. In this case, the two nullclines intersect at three positions, two of which are stable fixed points and one of which is unstable, as indicated by the phase flow. The two stable fixed points correspond to the states that the genetic switch can flip between.

In fact, the directions of all of the arrows shown in Figure 19.47 can be figured out by hand by using the nullclines as shown in Figure 19.48. The most important fact about each nullcline is that one of the two components of $(\mathrm{d}u/\mathrm{d}t, \mathrm{d}v/\mathrm{d}t)$ is zero on each of these nullclines by definition. For example, for the blue curve shown in Figure 19.48(A), we see that the $\mathrm{d}v/\mathrm{d}t = 0$, as evidenced by the horizontal arrows. The idea of a figure like this is that we can draw the arrows on the nullclines themselves and then, largely by exploiting the continuity of the vector fields, can figure out what the arrows are doing elsewhere.

The Tricks Behind the Math: Phase Portraits and Vector Fields As we have seen repeatedly in the book, there are many circumstances in which the dynamics of some system of interest involves *coupled* rate equations of the form

$$\frac{\mathrm{d}x}{\mathrm{d}t} = f(x, y), \qquad \frac{\mathrm{d}y}{\mathrm{d}t} = g(x, y), \tag{19.139}$$

where, in general, $f(x, y)$ and $g(x, y)$ are nonlinear functions. The idea of the phase portrait is to graphically depict the "flows" implied by the rate equations. In particular, we imagine a velocity vector field $\mathbf{v}(x, y) = (\mathrm{d}x/\mathrm{d}t, \mathrm{d}y/\mathrm{d}t)$, which depicts which way the system will "move" in the next time step. For a given initial condition (x_0, y_0), we can find the subsequent dynamics of the system by following the arrows.

One of the most classic examples of a dynamical system of the form described above is Lotka–Volterra population dynamics, in which a predator and a prey have their populations coupled. If we think of foxes (F) and hares (H), then the dynamics can be written as

$$\frac{dF}{dt} = FH - F, \quad (19.140)$$

$$\frac{dH}{dt} = -FH + H. \quad (19.141)$$

Effectively, what these equations say is that hares make more hares, and that the fox–hare interaction leads to an increase in foxes and a decrease in hares. An example of the nullclines for this system and the corresponding phase portrait are shown in Figures 19.49(A) and (B). In addition, this figure reveals some of the amusing observations on the dynamics of predator–prey systems.

From the standpoint of stability analysis, the most interesting points in a phase portrait are the fixed points. These are the points at which the vector field satisfies the condition $\mathbf{v}(x^*, y^*) = 0$. In other words, if we choose (x^*, y^*) as an initial

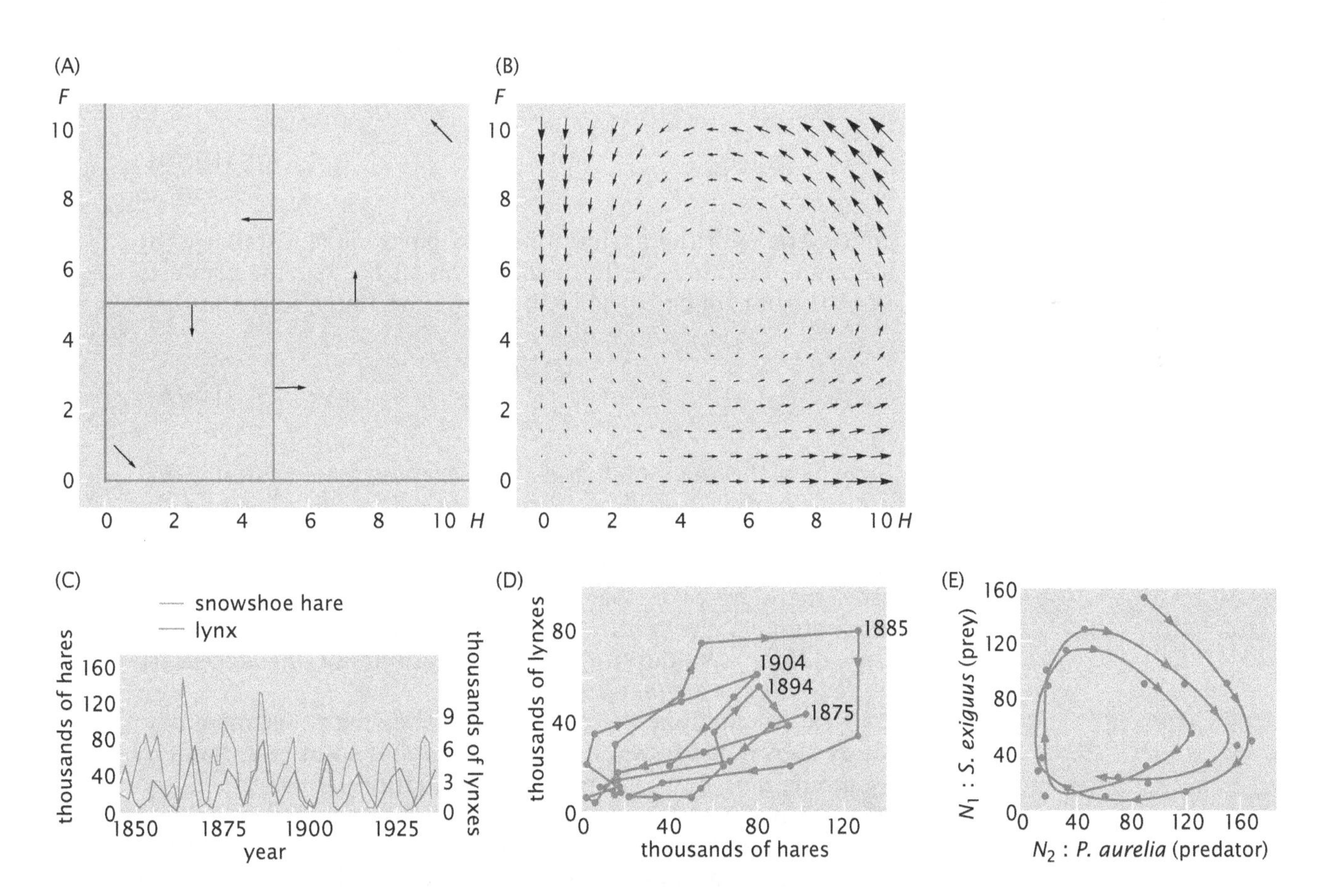

Figure 19.49: Lotka–Volterra model for predator–prey dynamics. (A) Nullclines are used to determine the directions of the flow. (B) Phase portrait for the predator–prey system of differential equations. (C) Population of lynx and hares as a function of time resulting from hunting records. (D) Alternative representation of the lynx and hare population over time, showing the oscillations. (E) Microbial example of population dynamics. (C, D, adapted from T. J. Case, An Illustrated Guide to Theoretical Ecology. Oxford University Press, 2000; E, adapted from G. F. Gause, The Struggle for Existence. Dover Publications, 2003).

condition, the system will stay put. Stability is determined by the directions of the arrows in the neighborhood of the fixed point. If the arrows *all* point back towards that fixed point, the point is said to be a stable fixed point. Otherwise, it is unstable. This type of graphical analysis is a powerful qualitative tool for examining the dynamics of nonlinear coupled equations.

MATH

The Math Behind the Models: Linear Stability Analysis for the Genetic Switch One of the key questions we can ask about the solutions found for the switch is the nature of their stability. Of course, one way to characterize the stability is to examine the phase portrait and to look at the directions of all of the little arrows around the various fixed points. We take an alternative approach here in which we search for steady-state solutions of the genetic switch by analyzing the case of α large and α small. First, assume $\alpha \gg 1$. We also assume that the solutions to the steady-state equations, namely,

$$u^* = \frac{\alpha}{1 + v^{*n}},$$
$$v^* = \frac{\alpha}{1 + u^{*n}}, \tag{19.142}$$

are such that $u^* \ll 1$. Then $1 + u^{*n} \approx 1$, and the steady-state values for the two concentrations that follow from Equation 19.142, to lowest order in $1/\alpha$, are

$$u^* = \alpha^{1-n},$$
$$v^* = \alpha, \tag{19.143}$$

consistent with the assumptions we have made. Similarly, by assuming that the solution to Equation 19.142 has the property $v^* \ll 1$, from which $1 + v^{*n} \approx 1$ follows, we find a new solution

$$u^* = \alpha,$$
$$v^* = \alpha^{1-n}, \tag{19.144}$$

for which the roles of u and v are exchanged. Assuming that both u^* and v^* are large leads to $u^* = v^* = \alpha^{1/1+n}$, while the assumption that both are small is inconsistent with Equation 19.142. We conclude that, in addition to the $u^* = v^*$ case, there are two other steady-state protein concentrations. Interestingly, the additional solutions are characterized by very different values for u^* and v^* providing the necessary ingredients for a genetic switch.

Next, we analyze the case $\alpha \ll 1$. Following the same analysis as above, we do not find any additional solutions. Namely, assuming $u^* \ll 1$, we compute from Equation 19.142 $v^* = \alpha$ and $u^* = \alpha$, since now $1 + \alpha^n \approx 1$. The same conclusions are reached assuming $v \ll 1$, while the assumptions $u^* \gg 1$ or $v^* \gg 1$ are not consistent with Equation 19.142. We conclude that there is a critical value of the parameter α, which will be of the order of 1 and dependent on the value of the Hill coefficient n, such that for values of α below the critical value the steady-state solution is unique, while for larger values of α there will be three steady states. Now, we examine the stability of these solutions, paying particular attention to the case when very different values for u and v are obtained in the steady state.

One of the most important requirements in carrying out an analysis like that given above is to assess the stability of the solutions to a given problem. What this means is that we perturb the system slightly from the steady state (that is, $u = u^* + \delta u$ and $v = v^* + \delta v$) and we ask if the perturbations grow or shrink in time. If the perturbations grow in time, the system is said to be unstable. If the perturbations shrink in time, the system is said to be stable. A favorite example for depicting this idea is to consider a particle on some potential-energy landscape. If the particle is at the bottom of a well (that is, the potential energy is locally of the form $\frac{1}{2}kx^2$), then a small disturbance of the particle from its equilibrium position will result in jiggling around the equilibrium point. Alternatively, if the particle is balanced at the point $x = 0$ on a potential-energy landscape of the form $-\frac{1}{2}kx^2$, then any slight disturbance to the particle will cause it to wander away from the equilibrium. The idea of our stability analysis in this case is the same—we ask whether a slight disturbance away from the steady-state concentration will lead to solutions that grow or decay in time.

To assess the stability of the steady state, we analyze the linear equations for the small deviations $(\delta u, \delta v)$ of the repressor concentrations away from their steady-state values. In particular, in Equation 19.135, we substitute $u = u^* + \delta u(t)$ and $v = v^* + \delta v(t)$ and then exploit the fact that $\delta u(t)$ and $\delta v(t)$ are small and Taylor-expand the nonlinear Hill functions in powers of δu and δv. The result of this analysis is

$$\frac{\mathrm{d}}{\mathrm{d}t}\begin{pmatrix}\delta u\\ \delta v\end{pmatrix} = \mathbf{A}\begin{pmatrix}\delta u\\ \delta v\end{pmatrix}. \qquad (19.145)$$

The matrix $\mathbf{A}$ given by

$$\mathbf{A} = \begin{pmatrix}-1 & f'(v^*)\\ f'(u^*) & -1\end{pmatrix} \qquad (19.146)$$

results from linearizing the rate equations, Equation 19.135, around the steady-state solution (u^*, v^*), and

$$f'(x) = -\frac{n\alpha x^{n-1}}{(1+x^n)^2}. \qquad (19.147)$$

At this point, the stability of this *linear* set of equations is queried by assuming solutions of the form $\delta u(t) = \delta u_0 \mathrm{e}^{\lambda t}$ and $\delta v(t) = \delta v_0 \mathrm{e}^{\lambda t}$. The essence of the analysis is to examine the sign of the parameter λ. If $\lambda < 0$, the perturbations decay in time, and if $\lambda > 0$, the perturbations grow in time. The behavior of λ is revealed by examining the eigenvalues of the matrix $\mathbf{A}$. The eigenvalues of $\mathbf{A}$ are both real and are given by

$$\lambda_{1,2} = -1 \pm \sqrt{f'(u^*)f'(v^*)}. \qquad (19.148)$$

For the steady-state solution to be stable, both λ_1 and λ_2 need to be negative. This will be the case if

$$f'(u^*)f'(v^*) < 1. \qquad (19.149)$$

Given this condition for the stability of the solutions, we can now revisit the different solutions found above and explicitly examine their stability. First we consider the single steady state, $u^* = v^* = \alpha$, that we found for $\alpha \ll 1$. In this case, using Equation 19.147, we find $f'(u^*)f'(v^*) = n^2\alpha^{2n} \ll 1$, and the stability condition, Equation 19.149, is satisfied. Next, we consider the three steady-state solutions found for $\alpha \gg 1$. For the solution $u^* = v^* = \alpha^{1/(1+n)}$, we find that $f'(u^*)f'(v^*) = n^2$. Since the Hill coefficient satisfies the condition $n > 1$, we conclude that this solution is unstable. A small perturbation will drive it to one of the other two solutions, which are stable. Namely, for $u^* = \alpha^{1-n}$ and $v^* = \alpha$, we see that $f'(u^*)f'(v^*) = n^2\alpha^{-n(n-1)} \ll 1$, and we conclude that the solution is stable. Since the third solution is obtained by u^* and v^* switching roles, it too will be stable.

The analysis above leads to the phase portrait shown in Figure 19.47 in terms of the parameter α. For α less than some critical value (which is of the order of 1), the rate equations at long times lead to a unique steady state in which the concentrations of the two repressor proteins are equal. On the other hand, for α larger than the critical value, at long times the system will settle into one of two stable states, with the concentration of one repressor dominating over the other. Which of the two steady states is reached depends on the initial conditions. In this regime the rate equations, Equation 19.134, describe a genetic switch.

19.3.6 Genetic Networks That Oscillate

In addition to switches, another dynamical element that is ubiquitous in cell dynamics is an oscillator where one or more chemical species in the cell vary in time in a periodic fashion. There are numerous ways that an oscillator can be built up from a collection of interacting genes and proteins, and here we examine a very simple example of a relaxational oscillator that makes use of two transcription factors, namely, a repressor and an activator. The repressor binds as a dimer and represses the production of the activator, while the activator increases its own production and that of the repressor, also binding as a dimer to the promoter DNA. The states and weights corresponding to this architecture are shown in Figure 19.50.

To write the chemical rate equation for the repressor and activator proteins, we assume that they are produced at a constant rate that depends on the particular state of the promoter, and that the probability of finding the promoter in one of its possible states is given by equilibrium considerations discussed earlier in this chapter. Using the states-and-weights diagrams in Figure 19.50 to compute the equilibrium probabilities for the different promoter states, we obtain the following rate equations for the evolution of the concentration of activator and repressor:

$$\frac{\mathrm{d}c_{\mathrm{A}}}{\mathrm{d}t} = -\gamma_{\mathrm{A}}c_{\mathrm{A}} + r_{0\mathrm{A}}\frac{1}{1+(c_{\mathrm{A}}/K_{\mathrm{d}})^2+(c_{\mathrm{R}}/K_{\mathrm{d}})^2} + r_{\mathrm{A}}\frac{(c_{\mathrm{A}}/K_{\mathrm{d}})^2}{1+(c_{\mathrm{A}}/K_{\mathrm{d}})^2+(c_{\mathrm{R}}/K_{\mathrm{d}})^2}, \tag{19.150}$$

$$\frac{\mathrm{d}c_{\mathrm{R}}}{\mathrm{d}t} = -\gamma_{\mathrm{R}}c_{\mathrm{R}} + r_{0\mathrm{R}}\frac{1}{1+(c_{\mathrm{A}}/K_{\mathrm{d}})^2} + r_{\mathrm{R}}\frac{(c_{\mathrm{A}}/K_{\mathrm{d}})^2}{1+(c_{\mathrm{A}}/K_{\mathrm{d}})^2}, \tag{19.151}$$

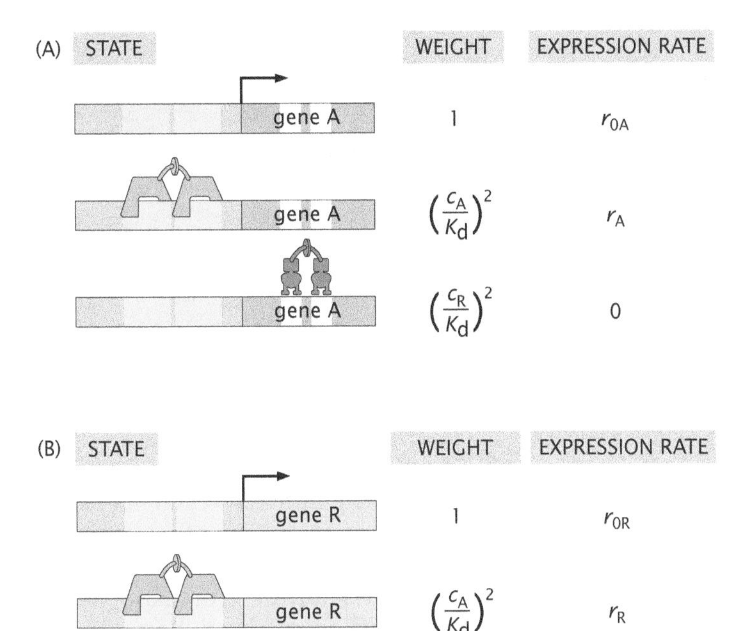

Figure 19.50: States and weights for the promoters that control the expression of activator and repressor proteins. (A) Regulation of the activator gene. The gene product has positive feedback and activates its own expression. (B) Regulation of the repressor gene. The presence of activator stimulates production of repressor.

where γ_R and γ_A are the degradation rates of the receptor and activator proteins, r_{0R} and r_{0A} are their basal rates of production, while r_R and r_A are the rates of production of these two protein species in the presence of activator bound to the promoter DNA. To simplify the analysis of the rate equations, we make the assumption that the same equilibrium dissociation constant K_d describes activator and repressor binding to promoter DNA.

As in the case of the genetic switch, we begin the analysis of the rate equations by writing them in dimensionless form. To that end, we use $1/\gamma_R$ as the unit of time and K_d as a unit of concentration. The rate equations for the dimensionless activator and receptor concentration are then given by

$$\begin{aligned}\frac{d\tilde{c}_A}{dt} &= -\tilde{\gamma}_A\tilde{c}_A + \frac{\tilde{r}_{0A} + \tilde{r}_A\tilde{c}_A^2}{1+\tilde{c}_A^2+\tilde{c}_R^2},\\ \frac{d\tilde{c}_R}{dt} &= -\tilde{c}_R + \frac{\tilde{r}_{0R} + \tilde{r}_R\tilde{c}_A^2}{1+\tilde{c}_A^2}.\end{aligned} \tag{19.152}$$

Oscillations can arise when there is a separation of time scales between the activator and repressor dynamics. To gain intuition about this, we plot the nullclines for the activator and repressor shown in Figure 19.51(A). The nullclines are the steady-state values of repressor and activator for fixed amount of activator and repressor, respectively. They are obtained by setting the time derivatives of the activator and repressor concentration in Equation 19.152 to zero and solving for the corresponding concentration of activator and repressor. This yields

$$\tilde{c}_R = \sqrt{-1 - \tilde{c}_A^2 + \frac{\tilde{r}_{0A} + \tilde{r}_A\tilde{c}_A^2}{\tilde{\gamma}_A\tilde{c}_A}} \tag{19.153}$$

and

$$\tilde{c}_R = \frac{\tilde{r}_{0R} + \tilde{r}_R\tilde{c}_A^2}{1+\tilde{c}_A^2} \tag{19.154}$$

for the two nullclines.

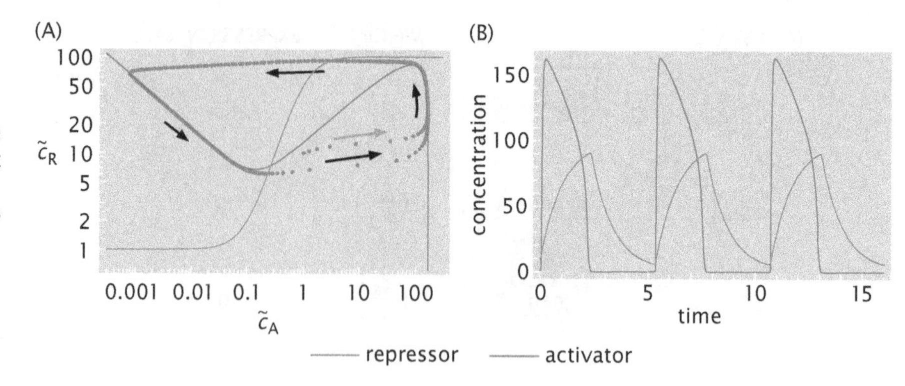

Figure 19.51: Dynamics of a genetic oscillator. (A) Nullclines for the two-component genetic oscillator for parameter values $\tilde{r}_{0R} = 1$, $\tilde{r}_R = 100$, $\tilde{r}_{0A} = 100$, $\tilde{r}_A = 5000$, and $\tilde{\gamma}_A = 30$. The light arrow indicates the initial transient and the dark arrows illustrate the limit cycle. (B) Solutions to the rate equations with initial conditions $\tilde{c}_A = 1$ and $\tilde{c}_R = 10$ using the same parameters as in (A).

If the repressor dynamics is much slower than the activator dynamics, then the activator dynamics will quickly reach its steady-state value for a given repressor concentration. In other words, at any instant in time, the amount of activator can be read off from its nullcline given the current concentration of repressor. Keeping this in mind, we can follow the progression of the dynamical system by starting initially with a small amount of repressor and activator, as shown in Figure 19.51(A). The activator concentration quickly reaches its steady state, which for a small amount of repressor is a large concentration of activator. Note that the green points in Figure 19.51(A) represent positions in phase space at equal time intervals, so a dense interval of points indicates slow phase flow, while a sparse one corresponds to fast flow. Once a large concentration of activator is obtained, this leads to a slow increase in the repressor concentration and the phase trajectory follows the right portion of the activator nullcline, as shown in Figure 19.51(A). When the repressor concentration rises above a critical value for which the steady-state activator concentration is small, the activator concentration quickly drops to this very small value, as indicated by the switch of the trajectory from the right to the left side of the activator nullcline. In response to this sudden drop in activator concentration, repressor concentration drops as well, but slowly, and the trajectory tracks the left side of the activator nullcline. Eventually, the repressor concentration drops below a critical value and the activator concentration jumps to a large, steady-state value (corresponding to the fast switch from the left to the right part of the activator nullcline) and the cycle repeats. Precisely this kind of progression, which is generally characteristic of relaxation oscillators, is shown in Figure 19.51(B) where we plot the concentrations of both activator and repressor over a few cycles.

19.4 Cellular Fast Response: Signaling

Gene regulatory networks are clearly of central importance to the functioning of organisms of all types. Of course, there are many aspects of biology where the dynamics of regulation is critical that do not involve gene transcription as an ultimate outcome. This is particularly obvious for biological behaviors that simply occur too quickly for transcription of new genes to have any useful impact. Rather, these *signaling* networks involve batteries of proteins and their partner ligands connected together such that their interactions affect the activity of some enzyme. For example, a membrane-spanning receptor might bind a ligand in the extracellular space. As a result of this binding event, there will be a concomitant structural change on the intracellular domain of this same protein, activating a protein kinase enzyme activity, which results in the phosphorylation of some other protein, rendering it active. The goal of the remainder of this chapter is to examine some examples of this kind of signaling and to construct simple models of their behavior.

19.4.1 Bacterial Chemotaxis

One fascinating and fairly well-understood example of signal transduction that we have mentioned briefly is the case of bacterial chemotaxis. Bacteria import small nutrients such as sugars and amino acids to use as building blocks, as we calculated in Chapter 3. A bacterial cell must take up a huge number (in excess of 10^9) of glucose molecules to go through a cycle of cell division. Obviously, this can be done more rapidly in areas of higher ambient glucose concentration. It therefore behooves the bacterium to actively seek out regions of its watery environment that contain the highest accessible concentration of glucose. An elegant and extraordinarily efficient system has evolved for this purpose. Several highlights of how we came to understand the workings of this system are sketched in the Experiments Behind the Facts below.

As we mentioned in Section 4.4.4 (p. 159), the motor used for swimming by the class of bacteria including *E. coli* and *Salmonella* is a rotary propellor that spins a long flagellum (each bacterial cell has several flagella that all work in synchrony). The only known control point the bacterium has for the rotor is to alter its direction of spin to be either clockwise or counterclockwise. Counterclockwise rotation of the flagella drives the bacterium forward in a nearly straight "run," while clockwise rotation causes the flagellar bundle to become disorganized and the bacterium "tumbles," randomly changing its direction. The chemotactic signal transduction machinery regulates this directional switching. If desirable nutrients are present at high concentrations, the bacterium tends to keep moving in a straight line, tumbling less frequently. If nutrient concentrations are low, the bacterium tends to tumble more frequently. *E. coli* is able to use the patterns of directional switching generated by this signal transduction network to swim up gradients of desirable nutrients. Some of the key elements of how this important and fascinating network works were indicated schematically in Figure 4.16 (p. 160).

How can a binary switch be used to detect the direction of a gradient? We can imagine at least two possibilities. First, the bacteria might be able to compare the signal coming from receptors located at the opposite poles of the cell, and switch in such a way as to swim toward the end with the higher signal, that is, sensing the gradient in space. Alternatively, the bacteria might be able to compare the signal being received at a given moment in time with the strength of the signal it received in the recent past, that is, sensing the gradient in time. As we will discuss below, the bacteria appear to use the time-based mechanism. The reader will have a chance to explore and compare these two possible schemes in the problems at the end of the chapter.

The cellular decision-making that attends chemotaxis is mediated by a signal transduction network that has been extremely well characterized. Our comments will center on the particular features of the *E. coli* chemotaxis network, which is an example of the two-component signaling systems introduced in Section 7.2.3 (p. 292). The key elements in this system are (i) membrane-spanning receptors that interact with the molecules in the environment (sugars, amino acids, etc.); (ii) CheW and CheA, proteins that bind to the intracellular domain of the receptor and change their activity depending on whether or not the receptor has a ligand bound (CheA is a protein kinase that can catalyze the attachment of phosphate groups to other target proteins, and CheW modulates CheA activity); (iii) a messenger molecule known as CheY that, when phosphorylated by CheA, can interact with the flagellar rotary motor to induce it to switch to clockwise (tumbling) rotation; (iv) CheZ, a phosphatase that can remove the phosphate from CheY; and (v) a pair of enzymes known as CheR and CheB that can respectively methylate and demethylate the receptors themselves, effectively tuning their affinity for their binding partners.

Experiments Behind the Facts: Measuring the Process of Chemotaxis Quantitative measurement of the behavior of bacteria engaged in chemotaxis has been performed in many elegant ways. Here we highlight a few of the key experiments that form the backdrop for our discussion. Video tracking microscopy was introduced to make it possible to perform single-cell analyses of bacteria engaged in their chemotactic response as shown in Figures 19.52(A) and (B). The idea of such experiments is easily stated, but the easy words mask what was an experimental *tour de force* when first introduced. Stated simply, the microscope stage is shifted constantly so that the cell of interest always stays in focus in the center of the field of view. A more recent version of the same experiment elects to hold the bacterium in an optical trap, with the runs and tumbles characterized by changes in the way the trapped bacterium jiggles about as shown in Figures 19.52(D) and (E).

Using tracking microscopy, it was possible to ask precise questions such as how fast are cells moving, how often do they tumble and what is their angular reorientation after a tumbling event? Figure 19.52 shows the outcome of such experiments. The advent of fluorescent proteins made it possible to observe cells engaged in these kinds of behaviors while simultaneously measuring the quantities (and even dynamics) of the molecules such as CheY-P that mediate the behavior. For example, in the experiment shown in Figure 19.53, the amount of CheY-P was monitored using fluorescence correlation spectroscopy (FCS) as introduced in Section 13.1.2 (p. 511). At the same time, as shown in the figure, by monitoring a fluorescent bead attached to one of the flagella on the immobilized bacterium, the direction of rotation could be observed, resulting in the ability to characterize the fraction of time (the so-called motor bias) the motor spends rotating in the opposite direction.

Recent FRET measurements have provided the kind of systematic, quantitative dissection of the chemotactic response that can really drive theoretical understanding forward. In the experiments shown in Figure 19.54(A), CheY-P and CheZ were each labeled with fluorescent molecules that serve as a FRET pair. As a result, the level of FRET serves as a direct readout of the amount of CheY-P as a function of the chemoattractant concentration because when there is lots of phosphorylated CheY, the interaction between CheY and CheZ is increased. As shown in Figure 19.54(B), the time history after stimulation

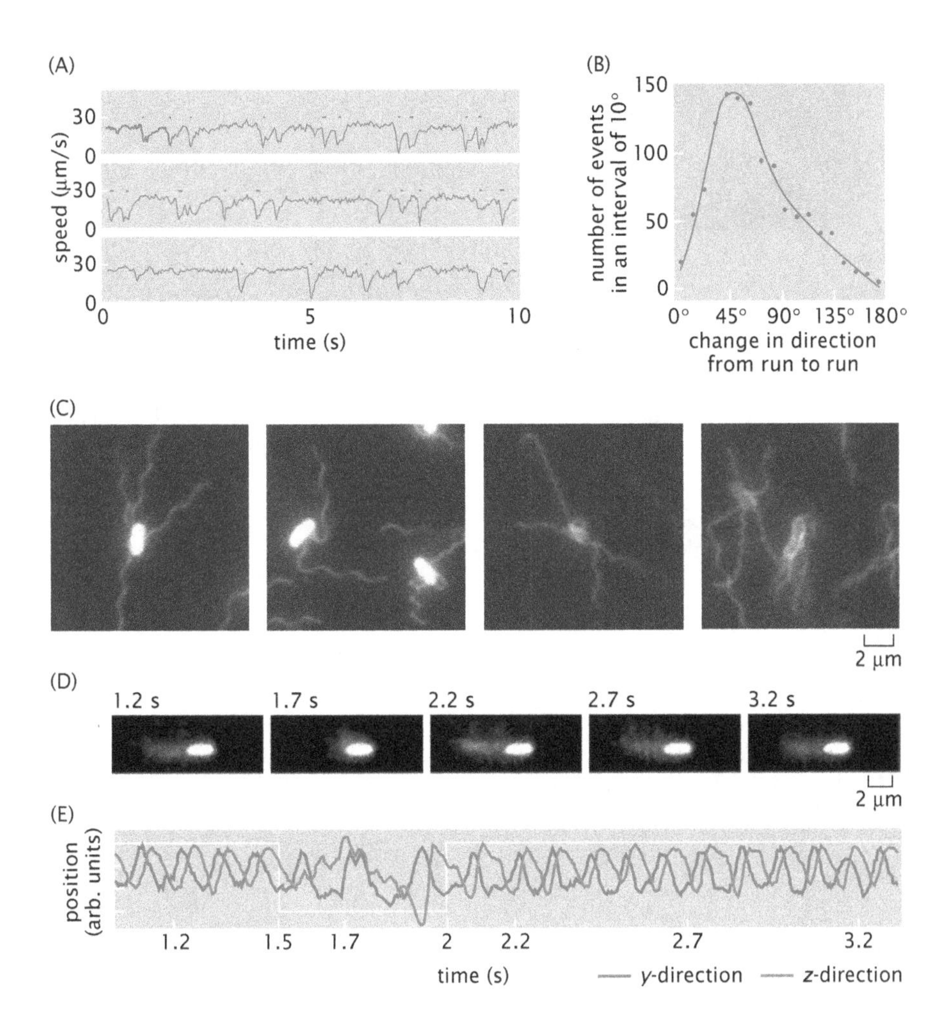

Figure 19.52: Chemotactic dynamics as observed using tracking microscopy and optical trapping. (A) Measurement of the speed of a bacterium as a function of time. The individual tumble events are shown by horizontal bars and are reflected by a marked reduction in the speed for a short interval. (B) Angular distribution of tumbles. (C) Images of tumbling bacteria illustrating the spreading apart of the flagella during the tumbling process. (D) Images of a bacterium held in an optical trap at various observation times. The fluorescently labeled flagella look different during the run and tumble events. (E) x- and y-positions of the bacterium as observed in the optical trap as a function of time. (A, B, adapted from H. C. Berg and D. A. Brown, *Nature* 239:500, 1972; C, adapted from L. Turner, W. S. Ryu, and H. C. Berg, *J. Bacteriol.* 182:2793, 2000; D, E, adapted from T. L. Min et al., *Nat. Methods*, 6:831, 2009.)

with a pulse of chemoattractant can be monitored directly with these experiments. Figure 19.54(C) shows how the activity of the chemoreceptor depends upon the concentration of chemoattractant for a number of different mutants that have their ability to adapt altered.

Examination of the tumbling frequency after exposure to a shift in concentration makes it possible to explore the question of adaptation. In particular, the time scale and precision of adaptation can be measured by watching cells after such a concentration jump and keeping track of their tumbling frequency. The results of such experiments are shown in Figure 19.55, where it is seen that the idea of "precise adaptation" is not a misnomer. It is also interesting to see how the adaptation time depends upon chemical details such as the concentration of CheR, while the precision itself does not.

Even for the relatively simple network that governs bacterial chemotaxis, it is hard to avoid getting lost in the alphabet soup of names, so we try to examine how the network works conceptually without focusing on the names of the molecules. In addition, we will take a hierarchical view, first explaining the overall functioning of the network and then taking up the fancy bells and whistles that make it work over such a wide range of concentrations, in the phenomenon

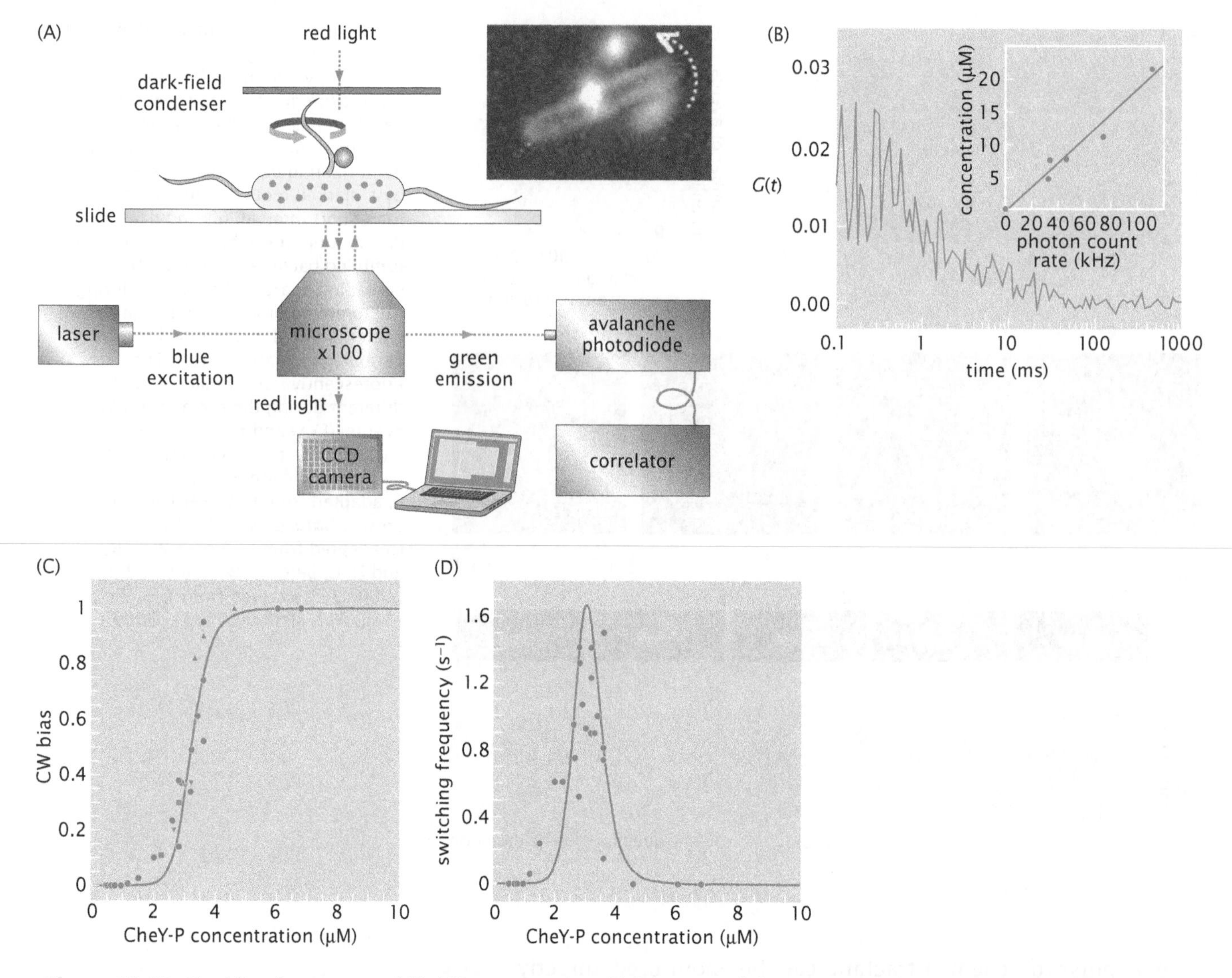

Figure 19.53: Tumbling frequency and CheY-P concentration. (A) Schematic of the experimental setup used to simultaneously quantify the amount of protein and the flagellar dynamics. (B) Measured correlation function as a function of time. This is related in turn to the concentration of protein (CheY-P-GFP). (C) Motor bias as a function of concentration of CheY-P. (D) Switching frequency and concentration of CheY-P. (Adapted from P. Cluzel, M. Surette, and S. Leibler, *Science*, 287:1652, 2000.)

known as adaptation. In simplest terms, the question of whether or not the cell will tumble (and hence change direction) comes down to the state of phosphorylation of the messenger molecule CheY. In order to be responsive to changes in the environment, the phosphorylation of CheY must be sensitive to whether or not there is a ligand bound to the receptor. In the presence of desirable attractant molecules, such as glucose or aspartate, the cell should repress tumbling, so we expect that the ligand-bound receptor will tend to be in the "off" form, where CheY is not phosphorylated, and the unbound receptor will tend to be in the "on" form, where CheY is phosphorylated. (Although *E. coli* is actually able to use the same chemotactic network to swim away from noxious chemicals, here we will only consider the happier problem of swimming toward delicious ones.) An idealization of these elements is shown in Figure 19.56(A), where we have combined the transmembrane receptor, CheW, and CheA into a single unit, and for the moment are ignoring the other components of the pathway.

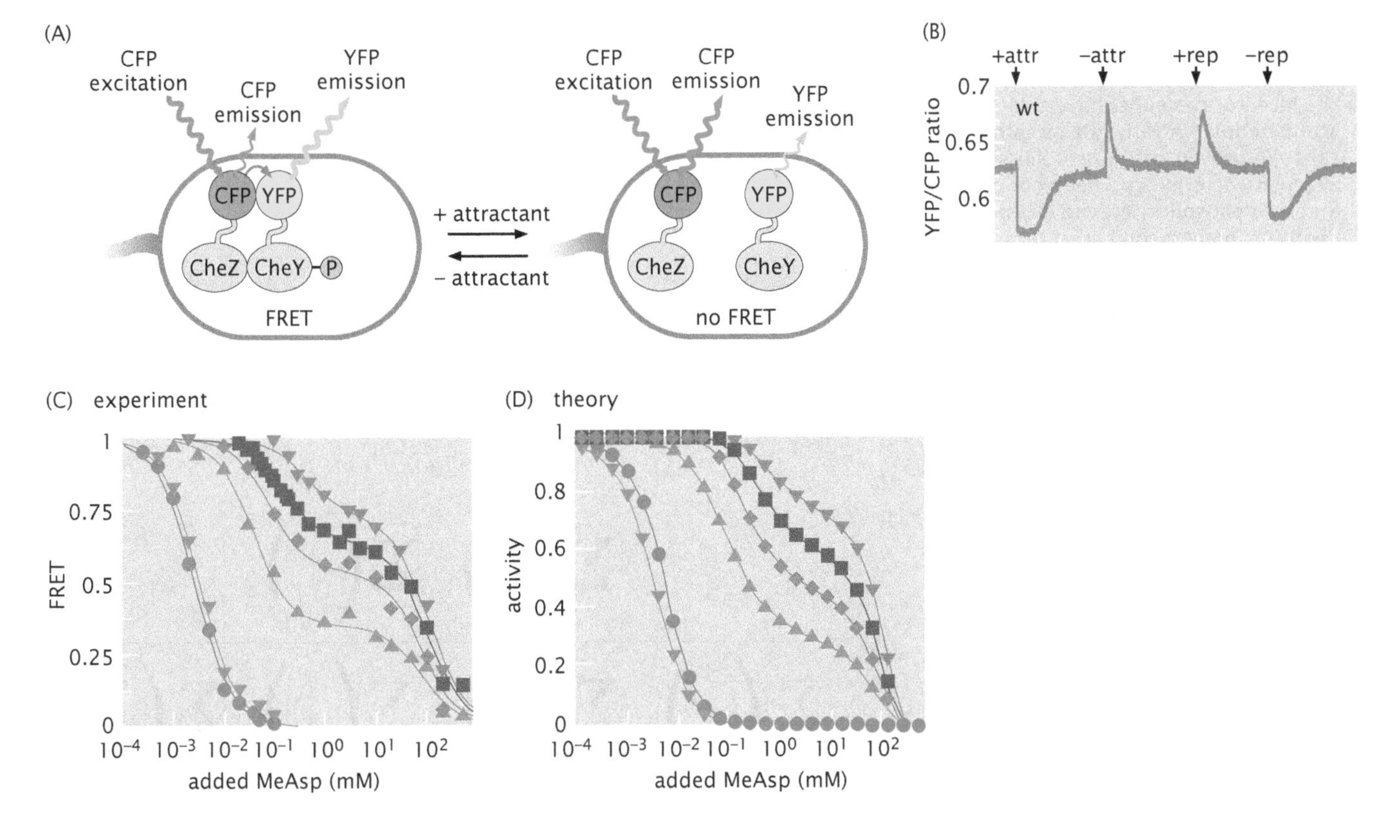

Figure 19.54: FRET measurements of chemotactic response. (A) Schematic of how the FRET measurements report on the chemotactic response of the system in the presence of chemoattractant. (B) FRET signal as a function of time. Addition of chemoattractant results in a decrease in the FRET signal corresponding to a reduction in the frequency with which the direction of rotation of the motor changes. After a certain adaptation time, the FRET signal (and the motor bias) go back to their unperturbed value. (C) Graph of concentration dependence of the "on" probability based on *in vivo* fluorescence resonance energy transfer (FRET) measurements. The different curves correspond to different bacterial strains. The wild-type response is shown as orange circles. The other symbols are for mutants that correspond to different states of receptor methylation, increasing from left to right. (D) The results of a calculation of the probability of the receptor being active as a function of the concentration of chemoattractant. The model reproduces many aspects of the living cell responses, including the complex behaviors of the methylation mutants. (A, B, Adapted from V. Sourjik and H. C. Berg, *Proc. Natl Acad. Sci. USA* 99:123, 2002; C, D Adapted from J. E. Keymer et al., *Proc. Natl Acad. Sci. USA* 103:1786, 2006.)

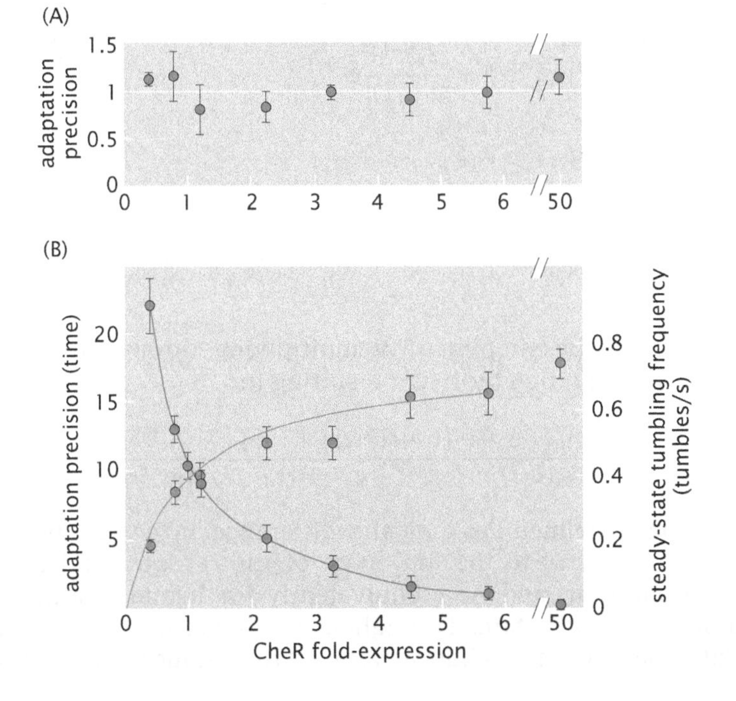

Figure 19.55: Tumbling frequency and adaptation. (A) Precision of the adaptation as measured by how precisely the rotational frequency returns to its original value. Rather than tuning the chemoattractant concentration in these experiments, the level of expression of CheR is controlled. (B) Tumbling frequency of the cells before stimulation is shown with blue data points and refers to the vertical axis on the right. Average adaptation time is shown by the red circles with reference to the vertical scale on the left. Both quantities are plotted as a function of CheR fold-expression. (Adapted from U. Alon, M. G. Surette, N. Barkai, and S. Leibler, *Nature*, 397:168, 1999.)

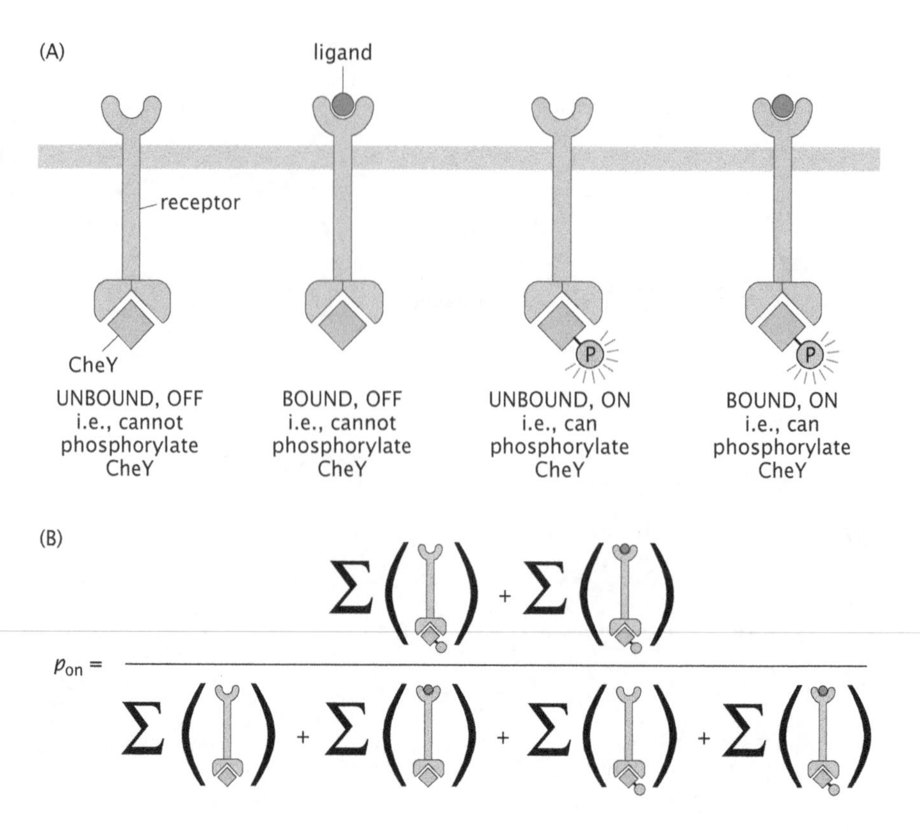

Figure 19.56: Probability that a receptor will be "on." (A) The receptor and its states of occupancy and activity. The receptor can either have a bound ligand or not. Similarly, the receptor can either be "on" or "off," where this state of activity determines whether or not it is able to phosphorylate the messenger CheY. (B) The probability that the receptor will be "on" is constructed as a ratio of the "on" states, appropriately weighted by their Boltzmann factors to the sum over the statistical weights of all states.

The MWC Model Can Be Used to Describe Bacterial Chemotaxis

We can treat this complex process approximately by appealing to our usual statistical mechanical formulation in which we imagine a rapid preequilibrium of the state of activity of the receptor. In particular, the quantity p_{on} measures the ability of the receptor to produce phosphorylated CheY, resulting in a change in the motor's direction of rotation. As we have done throughout the book, the statistical mechanics of this system can be examined by appealing to a states-and-weights diagram like that shown in Figure 19.57. The probability that the receptor will be active is obtained by constructing the ratio

$$p_{\text{on}} = \left[\frac{\Omega^L}{L!}\mathrm{e}^{-\beta L\varepsilon_{\text{sol}}}\mathrm{e}^{-\beta\varepsilon_{\text{on}}} + \frac{\Omega^{L-1}}{(L-1)!}\mathrm{e}^{-\beta(L-1)\varepsilon_{\text{sol}}}\mathrm{e}^{-\beta\varepsilon_{\text{on}}}\mathrm{e}^{-\beta\varepsilon_{\text{b}}^{\text{on}}}\right] \Big/ \left[\frac{\Omega^L}{L!}\mathrm{e}^{-\beta L\varepsilon_{\text{sol}}}\left(\mathrm{e}^{-\beta\varepsilon_{\text{off}}} + \mathrm{e}^{-\beta\varepsilon_{\text{on}}}\right) + \frac{\Omega^{L-1}}{(L-1)!}\mathrm{e}^{-\beta(L-1)\varepsilon_{\text{sol}}}\left(\mathrm{e}^{-\beta\varepsilon_{\text{off}}}\mathrm{e}^{-\beta\varepsilon_{\text{b}}^{\text{off}}} + \mathrm{e}^{-\beta\varepsilon_{\text{on}}}\mathrm{e}^{-\beta\varepsilon_{\text{b}}^{\text{on}}}\right)\right]. \tag{19.155}$$

This result can be simplified by multiplying through the top and bottom of the equation by $L!/\Omega^L$, resulting in

$$p_{\text{on}} = \frac{\mathrm{e}^{-\beta\varepsilon_{\text{on}}}[1 + (L/\Omega)\mathrm{e}^{-\beta\Delta\varepsilon_{\text{on}}}]}{\mathrm{e}^{-\beta\varepsilon_{\text{on}}}[1 + (L/\Omega)\mathrm{e}^{-\beta\Delta\varepsilon_{\text{on}}}] + \mathrm{e}^{-\beta\varepsilon_{\text{off}}}[1 + (L/\Omega)\mathrm{e}^{-\beta\Delta\varepsilon_{\text{off}}}]}. \tag{19.156}$$

Here we have defined $\Delta\varepsilon_{\text{on}}$ as the difference in energy between a single ligand bound to the "on" state of the receptor and the same ligand in solution, and $\Delta\varepsilon_{\text{off}}$ equivalently for ligand binding to the receptor in the "off" state. Throughout the book, we have repeatedly translated back and forth between the statistical mechanical language

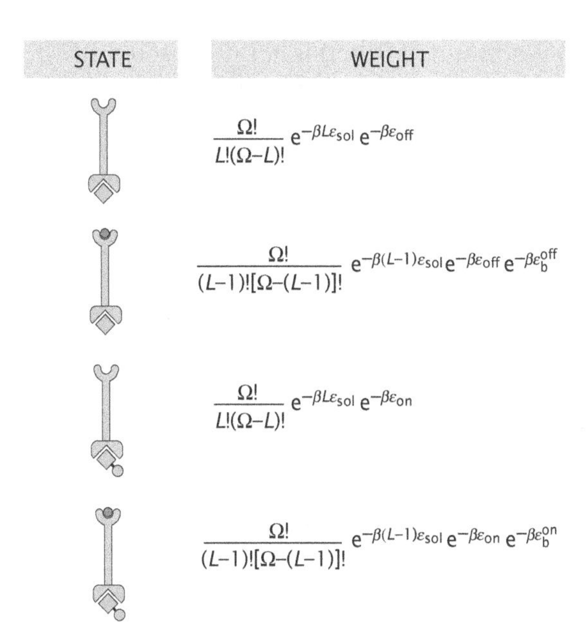

Figure 19.57: States and weights for a simple model of bacterial chemotaxis. The lower two states correspond to the case when the receptor is "on." The prefactors in front of the exponential terms correspond to the number of ways of rearranging the ligands in the lattice model of the solution.

used above and the thermodynamical language using equilibrium constants. By exploiting the relationship between energy differences and biochemical dissociation constants derived in Section 6.4.1 (p. 270), our expression for the probability that the receptor will be "on" can be rewritten using the dissociation constants as

$$p_{on} = \frac{1}{1 + e^{-\beta(\varepsilon_{off}-\varepsilon_{on})}\dfrac{1+[L]/K_d^{off}}{1+[L]/K_d^{on}}}. \tag{19.157}$$

This formula suggests that the probability of the "on" state depends on a few biologically important variables: the energy difference between the "on" and "off" states of the receptor in the absence of ligand, the affinities of the ligand for the "on" state and the "off" state of the receptor, and the amount of ligand itself. For attractive substances, binding of the ligand will tend to favor the "off" state (where CheY is not phosphorylated), that is, $K_d^{off} < K_d^{on}$.

Let us consider the implications of this result. In the absence of ligand (if $[L] = 0$), the equation simplifies to the familiar result for a two-state system such as an ion channel with the active and inactive states controlled by the relative values of ε_{off} and ε_{on}. Since in the absence of ligand the receptor is active for phosphorylation, we know that ε_{off} is larger than ε_{on}, thus favoring the "on" state. On the other hand, we expect that with increasing ligand concentration, the inactive state will predominate. This means within this model that $K_d^{off} < K_d^{on}$.

In order to modulate its response over a wide range of ligand concentrations and conditions, *E. coli* is actually able to move around in the parameter space of ε_{on} and ε_{off} by performing regulated covalent modifications of the receptor protein itself. This is the job of the methylase CheR and the demethylase CheB, which add and remove methyl groups on a series of glutamate residues present in the intracellular domain of the membrane-spanning receptor protein. The more highly methylated the receptor protein, the more likely it is to be in the "on" state. These modifications permit two impressive consequences. First, as mentioned above, *E. coli* can detect gradients of chemoattractants by comparing the strength of the signal it currently senses with the strength of the signal it detected in the recent past. Second, the bacterium is able to detect gradients in concentration over many orders of magnitude of absolute concentrations, a phenomenon known as adaptation. This corresponds to our own ability to whisper to someone else even in a crowded and noisy room, or our ability to see our surroundings either inside a darkened room or after stepping out into the bright sunshine. For the bacteria, both adaptation and time-sensing depend on the fact that the demethylase, CheB, is itself regulated by phosphorylation by CheA, and therefore depends on ligand binding to the receptor. If CheB is phosphorylated (that is, if the receptor is "on"), CheB will be more active as a demethylase, and will tend to convert the receptor into an "off" state, damping the response. Conversely, if CheB is dephosphorylated (that is, the receptor is "off"), more methyl groups will accumulate, tending to switch the receptor "on." This sequence of events takes some time, a few seconds. At the same time, ligand binding influences the activity state of the receptor. Therefore, receptor occupancy by ligand reflects current conditions, and the methylation state of the receptor reflects the conditions of a few seconds ago. The cell is able to swim up concentration gradients essentially by comparing these two signals.

Our calculations so far illustrate the key ideas, but they will not suffice to capture the full complexity of chemotactic behavior as revealed in Figure 19.54(C). In addition to the precise adaptation already discussed, the system exhibits a high degree of cooperativity. To account for cooperativity, our previous results can be amended to the form

$$p_{\mathrm{on}} = \frac{1}{1 + \mathrm{e}^{-n\beta(\varepsilon_{\mathrm{off}}-\varepsilon_{\mathrm{on}})} \dfrac{(1+[L]/K_{\mathrm{d}}^{\mathrm{off}})^n}{(1+[L]/K_{\mathrm{d}}^{\mathrm{on}})^n}}. \tag{19.158}$$

To see how this result emerges, Figure 19.58 resorts to our usual states-and-weights procedure in which we imagine a cluster of N receptors. The fate of the one is the fate of the many. Either all receptors are inactive or all are active. As in the usual MWC mentality, the relative energies of the inactive and active states are different and the K_{d} for the binding of ligands depends upon which of the two states the receptors are in. To see how the statistical weight of the active state arises, note that the number of *bound* ligands can be anything between 0 and N. The generic weight for the active state when it has n ligands bound is of the form

$$w_n = \mathrm{e}^{-\beta\varepsilon_{\mathrm{on}}} \frac{N!}{(N-n)!n!} \mathrm{e}^{-n\beta(\varepsilon_{\mathrm{b}}^{\mathrm{on}}-\mu)}. \tag{19.159}$$

However, we note that this is just the nth term in a binomial of order N (except for the prefactor $\mathrm{e}^{-\beta\varepsilon_{\mathrm{on}}}$), and hence, when we sum together all such terms, we find the overall statistical weight for the active state shown in the figure.

The inclusion of cooperativity sharpens the response of the system. Previously, we have considered cases of cooperativity such as oxygen binding to hemoglobin (Section 7.2.4, p. 298), where a single protein has multiple ligand-binding sites. In chemotaxis, the *E. coli* cell clusters essentially all its membrane-spanning receptors together in a single patch at one pole in a tight cluster as shown in Figure 13.23 (p. 537), such that binding of one ligand to one receptor can influence the conformational state of many other receptors, including distinct receptors that are able to detect different substances. A fully

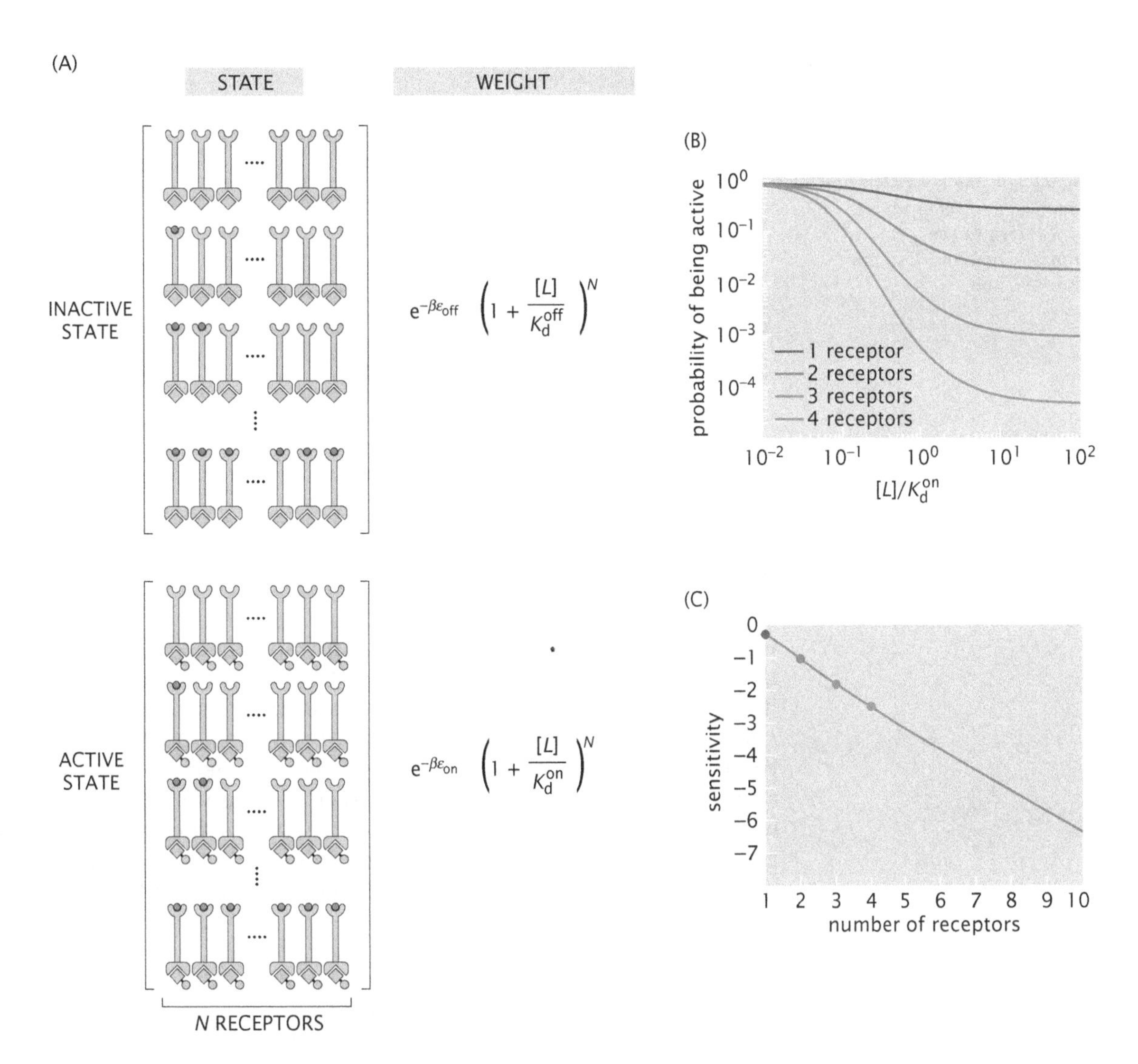

Figure 19.58: MWC model of bacterial chemotaxis. (A) States and weights for the MWC model in which there are N receptors in a cluster. (B) Probability that the receptors will be on as a function of the concentration of chemoattractant for different choices of the number of receptors in a cluster. (C) Sensitivity of the chemotactic response. For (B) and (C) a value of $K_d^{off}/K_d^{on} = 1/20$ was used.

detailed mathematical model that incorporates adaptation and cooperativity in mixed receptor clusters along with the basic two-state model derived above is able to reproduce many of the complex features of chemotactic receptor response, as illustrated in Figure 19.54(D).

Precise Adaptation Can Be Described by a Simple Balance Between Methylation and Demethylation

As already illustrated in Figure 19.55, bacteria that are exposed to a uniform change in concentration will temporarily respond as though they have been subjected to a gradient of chemoattractant. This response is characterized by a change in tumbling frequency. However, as seen in the figure, after some transient response time, they will faithfully return to their original tumbling frequency. An idea for how this takes place is illustrated schematically in Figure 19.59, where it is seen that the state of methylation of the receptor (here we examine a minimal model with only one site of methylation) is

Figure 19.59: Kinetic scheme for a toy model of precise adaptation. The concentration of unmethylated receptors is given by X, the concentration of active methylated receptors by X_A and the concentration of inactive methylated receptors is given by X_I. R refers to the concentration of CheR and B to the concentration of CheB.

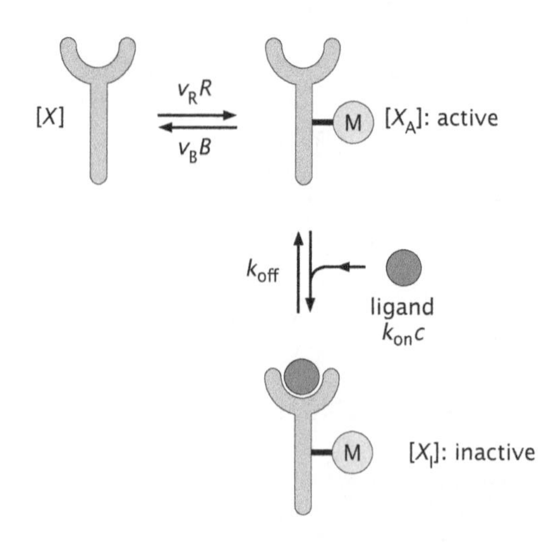

constantly tuned by the presence of CheR (methylation) and CheB (demethylation).

One of the key qualitative features of the model is captured by the topology of the various reactions and specifically by the fact that the demethylation can only take place from the active state. Hence, if the concentration of chemoattractant changes, it will temporarily change the number of active chemoreceptors, and this means that the balance between CheR and CheB will be perturbed, resulting in a net change in the number of methylated receptors depending upon whether the chemoattractant was decreased or increased. For example, if the amount of chemoattractant goes up, this will increase the number of inactive receptors. However, this will result in a concomitant increase in active receptors over time since CheR will win out over CheB in the coupled reactions they mediate.

The dynamics of this system of reactions is captured by three coupled dynamical equations. By inspection of Figure 19.59 we can read off these equations as follows. First, for the X concentration, we have

$$\frac{\mathrm{d}X}{\mathrm{d}t} = v_B B \frac{X_A}{K_A + X_A} - v_R R, \qquad (19.160)$$

where we assume that CheR is working at saturation (that is, there is an excess of X such that all of the CheR molecules are engaged in methlyation) and we have adopted the Michaelis–Menten form (see Section 15.2.7, p. 596) for the reaction of CheB. For the active state, we have

$$\frac{\mathrm{d}X_A}{\mathrm{d}t} = v_R R - v_B B \frac{X_A}{K_A + X_A} - k_{on} c X_A + k_{off} X_I, \qquad (19.161)$$

where we have included ligand binding and unbinding. Finally, for the inactive state, in this model, the only way to enter or exit the state is through ligand binding and unbinding, and this is captured through the dynamical equation

$$\frac{\mathrm{d}X_I}{\mathrm{d}t} = k_{on} c X_A - k_{off} X_I. \qquad (19.162)$$

As we have already seen, one of the most useful ways to examine the dynamics of low-dimensional dynamical systems like this is by appealing to the phase portrait. In this case, note that the equations for X_A and X_I make no reference to the concentration X. As a result, we can consider the dynamics of X_A and X_I independently of the dynamics

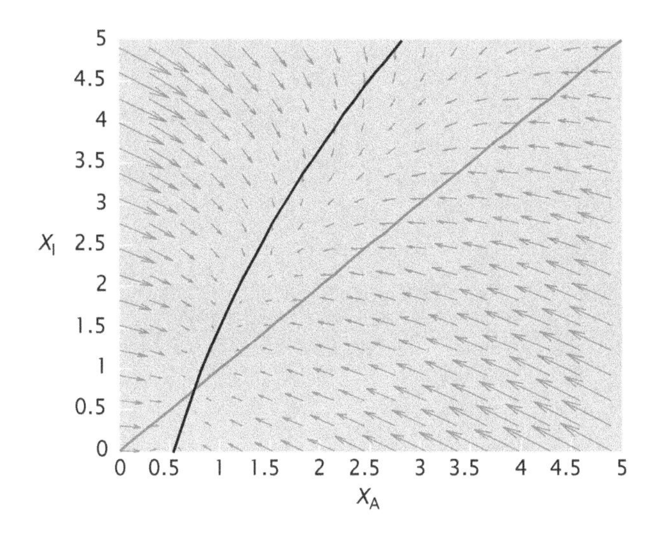

Figure 19.60: Phase portrait for simple model of precise adaptation. The straight line is the nullcline on which $dX_I/dt = 0$ and the other curve is the nullcline that shows the locus of points in the X_A–X_I plane where $dX_A/dt = 0$.

of X, which means we can resort to a two-dimensional phase portrait of the vector field $(dX_A/dt, dX_I/dt)$ as shown in Figure 19.60. The two nullclines, corresponding to $dX_I/dt = 0$ and $dX_A/dt = 0$, are shown on the phase portrait and their point of intersection is the fixed point.

The position of this fixed point can be solved for explicitly, resulting in

$$X_A^* = \frac{K_A v_R R}{v_B B - v_R R} \tag{19.163}$$

and

$$X_I^* = \frac{k_{on} c}{k_{off}} X_A^* = \frac{k_{on} c}{k_{off}} \frac{K_A v_R R}{v_B B - v_R R}. \tag{19.164}$$

In particular, by inspecting the functional form of X_A^*, we see that the concentration of the active form of the receptor does not depend upon the overall concentration c. As a result, when the concentration suffers an overall shift, the fixed point will shift up and down, but not to the left or right, illustrating that the fixed-point concentration of X_A^* is invariant, corresponding to the precise adaptation seen experimentally.

19.4.2 Biochemistry on a Leash

One of the most fundamental features of living organisms is movement. As noted in our discussion of chemotaxis, cells make "decisions" about where to go and these decisions in eukaryotes are implemented in the form of polymerization of actin filaments. Examples of actin polymerization organized in both space and time were shown in Figures 15.2 (p. 576) and 15.3 (p. 577). What chains of events link the detection of some external cue and the formation of new actin filaments in a motile cell? The advent of video microscopy in conjunction with a host of different classes of fluorescent markers has made the study of cell motility one of the most exciting areas of current research. As a particular case study that will allow us to flex several sets of muscles that we have developed throughout the book, we consider molecules that have the interesting feature that they include a tethered ligand and receptor pair that compete with free ligands. These tethering motifs are a common feature of signaling molecules.

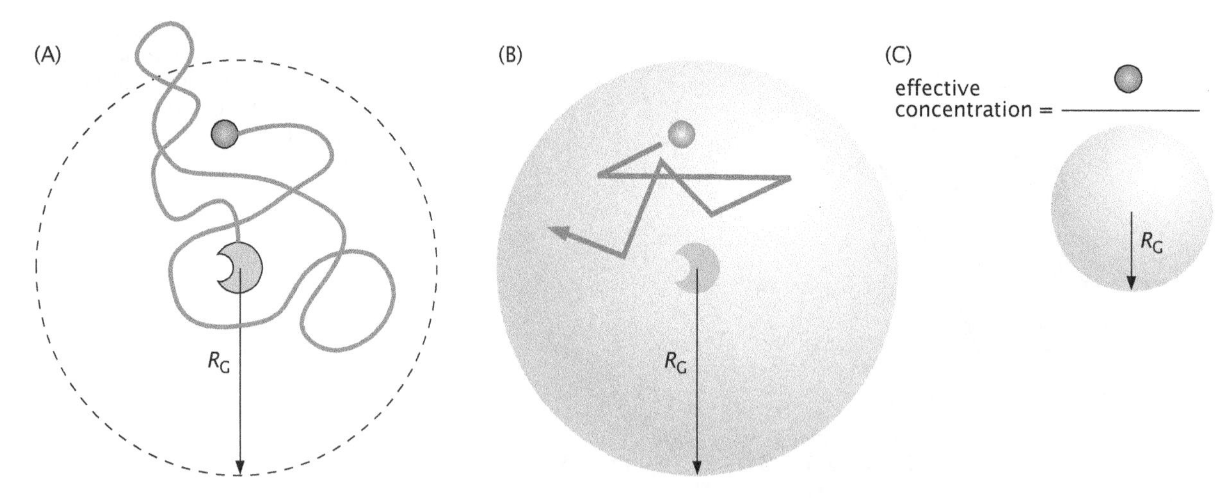

Figure 19.61: Tethering and effective concentration. (A) As a result of tethering, the ligand can only explore a limited region of space. (B) The concentration of the tethered ligand can be estimated by considering a sphere with a radius given by the radius of gyration of the tether. (C) To compute the effective concentration due to tethering, consider one ligand per volume given by a sphere with a radius equal to that of the radius of gyration.

Tethering Increases the Local Concentration of a Ligand

One simple way to see the significance of tethering is illustrated in Figure 19.61. The idea is that the tethered ligand is confined to a volume dictated by the length of the tether. In particular, if the tether has a length L resulting in a radius of gyration R_G, then the effective concentration of the tethered ligand can be estimated as

$$\text{effective concentration} = \frac{1}{\frac{4}{3}\pi R_G^3}. \tag{19.165}$$

To develop an intuitive sense of the significance of this tethering, this estimate can be used to roughly determine the concentration at which the free ligands compete with the tethered ligand. In particular, for the case in which a tethered ligand competes with free ligands for the attention of a tethered receptor, clearly at high enough concentrations, the free ligands will dominate the binding.

Signaling Networks Help Cells Decide When and Where to Grow Their Actin Filaments for Motility

The case of bacterial chemotaxis described above is but one of many examples where the motility of cells is dictated by the presence of environmental cues. In many cases, these environmental cues have the effect of inducing actin polymerization, which leads to changes in cell shape that are then coupled to motility. From the standpoint of cell signaling, a small signaling molecule can relay information to N-WASP, a protein that can interface with a complex of proteins called the Arp2/3 complex to create new actin filaments. The way in which this works is shown in Figure 19.62(A). In particular, the presence of two ligands, Cdc42 and PIP_2, activates N-WASP by binding to this protein in a way that then permits it to activate Arp2/3. The presence of Cdc42 and PIP_2 leads to the unbinding of GDB and B domains from the C domain and Arp2/3, and N-WASP begins to stimulate actin polymerization by recruiting (and perhaps appropriately orienting) actin monomers to the proximity of the Arp2/3. With the help of activated N-WASP, Arp2/3 promotes actin polymerization by providing heterogeneous nucleation sites. Here, our aim is to study this process quantitatively.

Synthetic Signaling Networks Permit a Dissection of Signaling Pathways

As with the analysis of genetic networks, one exciting way in which signaling pathways have been dissected is by rewiring such pathways to form various synthetic signaling networks. Figure 19.62(B) shows a synthetic activator of Arp2/3 in which a domain known as a PDZ domain is attached to the output domain that activates Arp2/3. On the other end of the construct is a peptide sequence that binds to PDZ. This synthetic protein mimics N-WASP and can be activated by soluble ligands that bind to the PDZ domain.

To analyze the function of this signaling process, we invoke statistical mechanics in the same spirit as we have earlier for considering gene regulation. The goal of our statistical mechanical model of the synthetic switch is to work out the probability that the molecule is in the active state. In particular, the active state corresponds to the case in which the tethered receptor is not bound to the tethered ligand. That is, the tethered ligand and receptor are separately flopping around freely. As usual, we resort to a states-and-weights diagram to work out the probability of the active state. As shown in Figure 19.63, there are three classes of states, each with their own corresponding statistical weights: (i) the switch is in the autoinhibitory state and the tethered ligand and receptor are bound to each other; (ii) the tethered ligand and receptor are both flopping around freely and the receptor has no bound free ligands; (iii) the tethered ligand and receptor are both flopping around freely, and the receptor has bound one of the free ligands. Our aim is to make falsifiable predictions for the signal dependence on, for example, the linker length and ligand concentration.

To develop an intuitive sense of how this situation plays out, the probability of finding the switch in the active state is represented

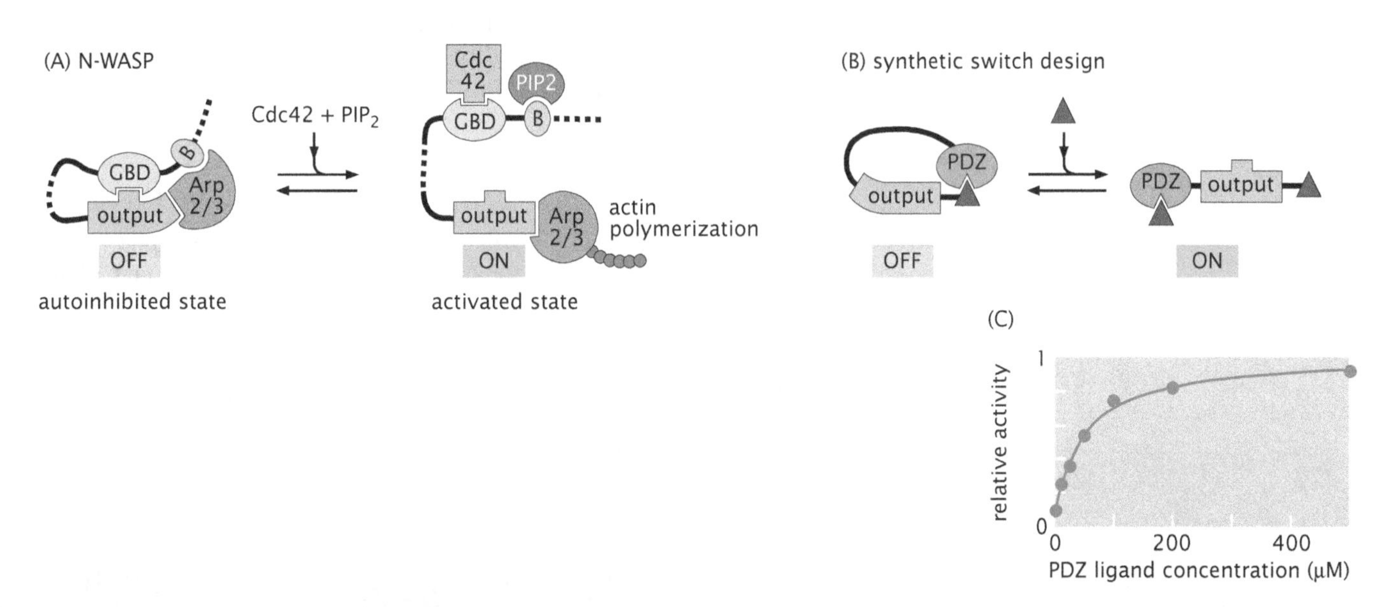

Figure 19.62: Schematic of the signaling process leading to actin polymerization. (A) Activation of Arp2/3 by ligands Cdc42 and PIP_2. (B) Synthetic switch constructed to activate Arp2/3 as a result of the presence of an alternative ligand. (C) Activity of the synthetic switch as a function of the signaling ligand. (Adapted from J. E. Dueber et al., *Science* 301:1904, 2003.)

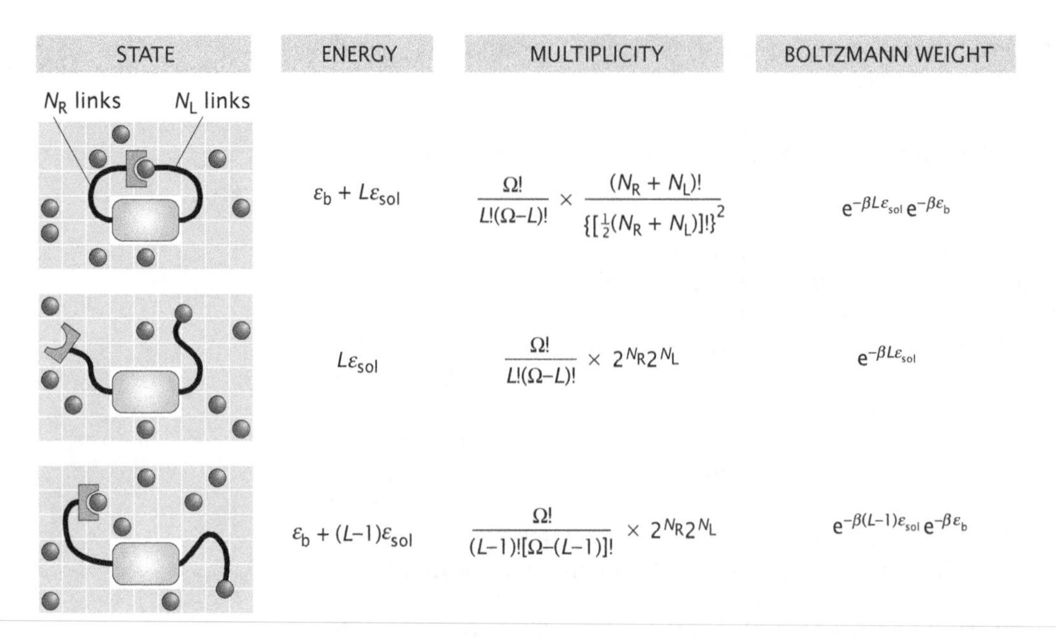

Figure 19.63: States and weights for the synthetic signaling problem.

schematically in Figure 19.64. The essence of the situation is that as the concentration of free ligand is increased, the probability that the receptor will be bound by one of the free ligands will increase until this outcome dominates the probability. From the standpoint of testing our understanding of such systems, one of the other design parameters that can be varied is the length of the flexible tethers. As will be shown explicitly when we demonstrate the contributions of the autoinhibitory state to the overall partition function, the length of the tether is a significant part of the overall free energy budget.

To make this calculation concrete, we resort here to simple one-dimensional ideas on the random walk introduced in Chapter 8 and show how the calculation generalizes to three dimensions, but leave the details for the reader as a problem at the end of the chapter. Our strategy will be to break the *total* partition function for this system down into three parts as reflected in Figure 19.63, where the sum can be written as

$$Z_{\text{tot}}(L, N_R, N_L) = \underbrace{Z_1(L, N_R, N_L)}_{\text{autoinhibitory state}} + \underbrace{Z_2(L, N_R, N_L)}_{\text{free tethers}} + \underbrace{Z_3(L, N_R, N_L)}_{\text{tether with ligand}} . \tag{19.166}$$

The parameter L is the number of ligands in the system, N_R is the number of Kuhn segments in the polymer tether that has the tethered receptor, and N_L is the number of Kuhn segments in the polymer tether that has the tethered ligand. Given these decompositions, we can then write the probability that the switch will be in the active state as

$$p_{\text{active}} = \frac{Z_2 + Z_3}{Z_1 + Z_2 + Z_3}. \tag{19.167}$$

The separate contributions to the total partition function can be worked out in much the way we have done in similar problems

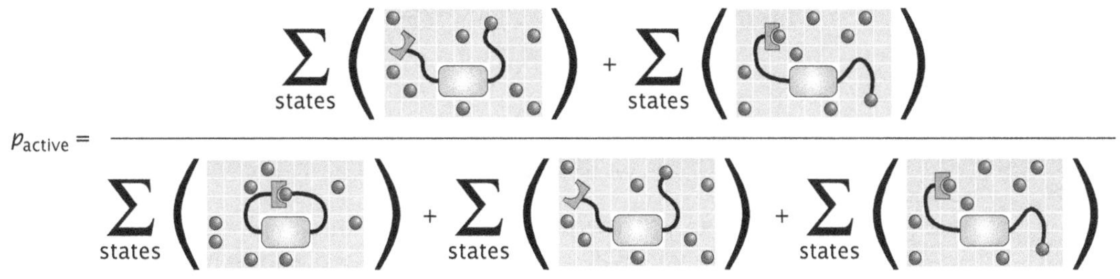

Figure 19.64: Probability of activation of Arp2/3. The numerator is the sum of the statistical weights of the active states.

throughout the book. The key point is that each class of state has a number of microscopically equivalent configurations and to find their contribution to the overall partition function, we need to multiply the Boltzmann weight for each class of state by its corresponding microscopic degeneracy (obtained by adding up all of the different ways of arranging the system). For example, the contribution from the states in which the tethers are flopping around freely and there is no free ligand bound is given by

$$Z_2 = \underbrace{\frac{N!}{L!(N-L)!}}_{\text{solution ligands}} \times \underbrace{2^{N_R} 2^{N_L}}_{\text{tether configs.}} \times \underbrace{e^{-\beta L \varepsilon_{sol}} e^{-\beta \varepsilon_{sol}}}_{\text{Boltzmann weight}} . \quad (19.168)$$

The treatment of the tether degrees of freedom is based on the simplest one-dimensional random walk in which we imagine that every segment in the tether can point either to the left or right and we do not worry about self-avoidance. It is straightforward to use a more robust model of the tethers, but we use this one for simplicity. What this means precisely is that each tether can be in one of 2^N different configurations, where N is the number of Kuhn segments in the tether of interest. We have also introduced the energy ε_{sol} for the energy of the ligands when they are free in solution and the parameter ε_{sol}^{lig} for the energy of the tethered ligand when it is in solution. The most interesting class of states are those that are associated with the autoinhibition of the switch and that involve the tethering ligand and receptor being linked. In this case, the contribution to the partition function is

$$Z_1 = \underbrace{\frac{N!}{L!(N-L)!}}_{\text{solution ligands}} \times \underbrace{\frac{(N_R + N_L)!}{\left\{\left[\frac{1}{2}(N_R + N_L)\right]!\right\}^2}}_{\text{tether closure}} \times \underbrace{e^{-\beta L \varepsilon_{sol}} e^{-\beta \varepsilon_b}}_{\text{Boltzmann weight}} , \quad (19.169)$$

where we have used the result from Section 8.2.4 (p. 333). The contribution from tether closure is the number of ways of making a closed loop out of a polymer of length $N_R + N_L$ Kuhn segments. The last contribution to the total partition function arises from those microstates in which one of the free ligands attaches to the tethered receptor. This means that the solution contribution to the partition function will only

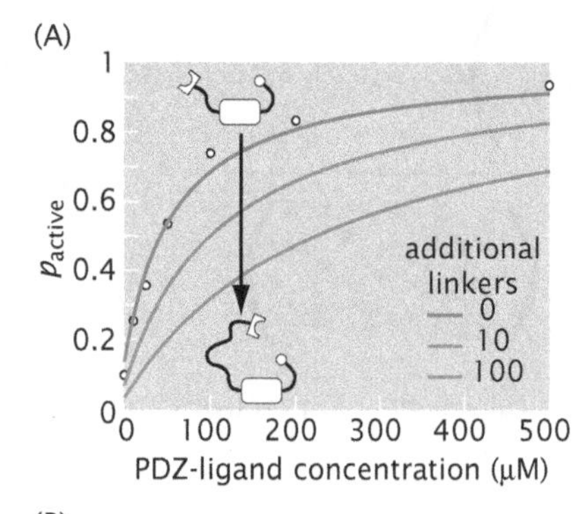

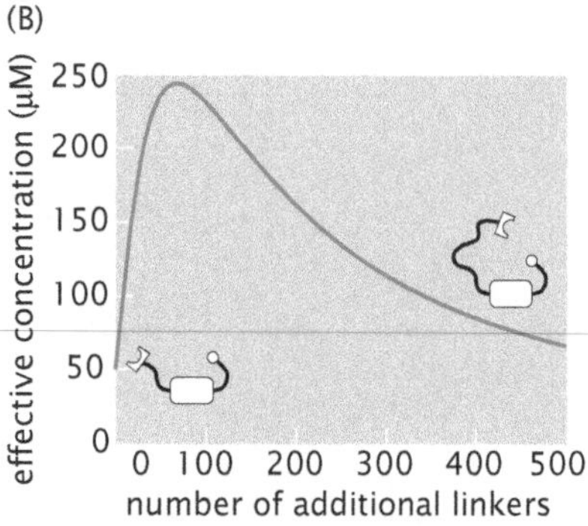

Figure 19.65: Prediction of dependence of activation on effective tail length. (A) p_{active} as a function of ligand concentration for different tether lengths. Experimental data are shown as small circles. (B) The effective concentration of tethered ligand as seen by the tethered PDZ domain as a function of tether length. (Data from J. E. Dueber et al., *Science* 301:1904, 2003.)

involve $L-1$ ligands. This term can be written as

$$Z_3 = \underbrace{\frac{N!}{(L-1)![N-(L-1)]!}}_{\text{solution ligands}} \times \underbrace{2^{N_R}2^{N_L}}_{\text{tether configs.}} \times \underbrace{e^{-\beta(L-1)\varepsilon_{\text{sol}}}e^{-\beta\varepsilon_b}}_{\text{Boltzmann weight}}. \qquad (19.170)$$

The actual formula for p_{active} can now be obtained by substituting the values for Z_1, Z_2, and Z_3 obtained above into Equation 19.167. The resulting expression is considerably simpler if we use an alternative form of this equation, namely,

$$p_{\text{active}} = \frac{1+(Z_3/Z_2)}{1+(Z_1/Z_2)+(Z_3/Z_2)}. \qquad (19.171)$$

This leads to an expression for p_{active} of the form

$$p_{\text{active}} = \frac{1+(c/c_0)e^{-\beta\Delta\varepsilon_1}}{1+p_{\text{loop}}e^{-\beta\Delta\varepsilon_2}+(c/c_0)e^{-\beta\Delta\varepsilon_1}}, \qquad (19.172)$$

where we have introduced $c = L/Nv$, $c_0 = 1/v$, and p_{loop}, which is the probability of forming a loop, and where $\Delta\varepsilon_1$ is the binding energy for a free ligand and $\Delta\varepsilon_2$ is the binding energy for the tethered ligand–receptor pair. For the one-dimensional model considered above, we have

$$p_{\text{loop}} = \frac{(N_R+N_L)!/\left\{\left[\frac{1}{2}(N_R+N_L)\right]!\right\}^2}{2^{N_R+N_L}}, \qquad (19.173)$$

which amounts to the ratio of the number of closed configurations for the polymer of length N_R+N_L to the *total* number of configurations. However, the one-dimensional model has outlived its usefulness and we can just as well use the result of a full three-dimensional analysis of p_{loop} using the Gaussian model of a polymer, for example. This calculation is left as an exercise for the reader.

The outcome of this kind of analysis is shown in Figure 19.65. There are several subtleties that were not accounted for in the calculation described above. First, as shown in the figure, the tethers do not emanate from the same point. This results in a fundamental difference in the behavior of p_{loop} as a function of tether length as shown in Figure 19.65(B). Second, in the figure, we used a three-dimensional Gaussian model for the tethers rather than the one-dimensional example worked out above.

19.5 Summary and Conclusions

Regulation and signaling are two of the most important ways in which cells orchestrate their behavior in space and time. The goal of this chapter has been to take stock of some of the key architectures of regulatory and signaling networks and to show how simple models using statistical mechanics and rate equations can be put forth to develop intuition and to make predictions about how these networks work. The so-called "thermodynamic models" of gene expression are predicated on the idea of using equilibrium statistical mechanics to examine the probability of promoter occupancy. A dynamical interpretation of these same questions uses rate equations to compute the concentration of both mRNA and their associated proteins.

19.6 Problems

Key to the problem categories: • Model refinements and derivations, • Estimates, • Data interpretation, • Computational simulations, • Model construction

• 19.1 Strong and weak promoters

In the chapter, we introduced repression as a quantitative measure of the reduction in the level of gene expression due to the action of a repressor molecule. For the simple model of repression introduced on p. 814, make a plot comparing repression in the case of a weak and a strong promoter. Show that, unlike the weak promoter case, in the case of the strong promoter, the repression depends upon the number of polymerase molecules in the cell.

• 19.2 Lac Repressor and the *lac* operon

A beautiful set of quantitative experiments on the *lac* operon were done by the Müller-Hill group in the 1990s, where repression of expression of the *lacZ* gene was measured in a population of different mutant *E. coli* cells. The mutant cells differed in the number, sequence, and position of the operator sites that bound the Lac repressor. In this problem, we explore how, using thermodynamic models of gene expression, these data can be used to obtain a number of quantities characterizing the Lac repressor–DNA interaction as well as DNA looping.

(a) Using the data from Oehler et al. (1994) shown in Figure 19.22 determine the *in vivo* binding energy of Lac repressor to each one of its operators and reproduce Figure 19.23.

(b) Use your results from (a), and the repression measured by Oehler et al. (1994) in cells with two operators present, which leads to DNA looping, in order to determine the looping energy and to reproduce Figure 19.27.

(c) As mentioned many times throughout the book, Müller et al. (1996) performed an experiment where the repression level was measured as a function of the distance between operators. The experiment and its results are shown in Figure 1.11 (p. 19). Based on their repression data and the thermodynamic models from the chapter, make a plot of the looping energy as a function of the interoperator distance. Show analytically that a maximum in repression corresponds to a minimum in looping energy. At what interoperator distance is the inferred looping free energy at a minimum? Is this consistent with the measured persistence length of DNA *in vitro*, which is 50 nm?

(d) Fit the looping energy obtained in (c) to the functional form $\Delta F_{\text{loop}} = a/N_{\text{bp}} + b\ln N_{\text{bp}} + cN_{\text{bp}} + e$. Use this looping energy to make predictions about the outcome of a hypothetical experiment similar to the one performed by Müller et al. (1996), but now using cells bearing 10, 200, and 900 Lac repressor molecules per cell.

Relevant data for this problem are provided on the book's website.

• 19.3 Sensitivity of the regulation factor

An important concept in gene regulation is the sensitivity, that is, how steep is the change in gene expression (for example, the steepness of the transition from the "off" to the "on" state in activation) in response to a change in the number of transcription factors. It can be quantified by obtaining the slope on a log–log plot of the level of gene expression versus the number of transcription factors at this transition. Using thermodynamic models of gene regulation, determine how the sensitivity depends on the relevant parameters for the following regulatory motifs in the case of a weak promoter:

(a) Simple activation.

(b) Simple repression.

(c) Two binding sites where the same species of repressor can bind. They can recruit each other and repress RNA polymerase independently. What happens when the interaction is turned off? For simplicity, assume that both binding sites have the same binding energy.

(d) Repression in the presence of DNA looping.

• 19.4 Plasmid copy number and gene expression

In this problem, we work out an expression for the repression for the case in which there are N plasmids, each harboring the same promoter subjected to repression by the simple repression motif.

(a) Write a partition function for P RNA polymerase molecules that can bind to the plasmids, resulting in expression of our gene of interest. Take into account the cell's nonspecific reservoir and assume that $P \gg N$. Calculate the mean number of plasmids occupied by RNA polymerase, $\langle N \rangle$. Could you just have predicted this result based on what you know about the $N = 1$ case?

(b) Work out an expression for the repression defined as

$$\text{repression} = \frac{\langle N \rangle (R = 0)}{\langle N \rangle (R \neq 0)}. \tag{19.174}$$

Make sure to take into account the distinct cases where $N < R$ and $N > R$, where R is the number of repressors, and assume that you are dealing with a weak promoter, namely $(P/N_{\text{NS}})e^{-\beta\Delta\varepsilon_{\text{pd}}} \ll 1$.

(c) Show that your result yields the same expression for simple repression in the case where $N = 1$ that we found in the chapter.

(d) Consider the case where there are two plasmids (that is, $N = 2$) and work out the repression as a function of the number of repressors and make a corresponding plot.

• 19.5 The transcriptional machinery in eukaryotes

In the thermodynamic models of gene regulation discussed in the chapter, the RNA polymerase is treated as a single molecular species. While this might be a reasonable assumption for transcription in prokaryotes, in eukaryotes tens of different molecules need to come together in order to form the transcriptional machinery. The objective of this problem is to develop intuition about the requirements for our simple model to apply in such a complex case by assuming that the transcriptional machinery is made out of two different subunits, X and Y, that come together at the promoter.

(a) Calculate the probability of finding the complex X + Y bound to the promoter in the case where unit X binds to DNA and unit Y binds to X. Can you reduce this to an effective one-molecule problem such as in the bacterial case?

(b) Calculate the fold-change in gene expression for simple repression using transcriptional machinery such as that proposed in (a). Explore the weak promoter assumption in order to reduce the expression to that corresponding to the bacterial case. Repeat this for the case where an activator can contact Y.

(c) Repeat (a) and (b) for a case where Y binds to a site on the DNA that is near the X-binding site, and there is an interaction energy between X and Y.

• 19.6 Induction of transcription factors

Even though experiments where the concentration of a transcription factor is varied are easier to interpret in terms of models, like those described in this chapter, the experiments that are the easiest to perform are those where the affinity of the transcription factor to its specific binding sites on the DNA is regulated by an inducer molecule. In the case of Lac repressor, allolactose or any of its analogs (IPTG, for example) can be used to reduce its specific binding energy to values similar to its nonspecific binding to DNA.

Assume a simple model of induction where one inducer molecule binds to the repressor, which then loses its ability to bind specifically to its operator site. Calculate repression in this case and plot it as a function of the number of inducer molecules in the cell.

• 19.7 Solving the unregulated promoter master equation

Solve the master equation for the unregulated promoter shown in Equation 19.39 in steady state by proposing a solution in terms of a generating function given by $f(s) = \sum_{m=0}^{+\infty} p(m)s^m$. In order to do this, you will have to multiply both sides of the equation by s^m and sum over all values of m in order to obtain a differential equation for $f(s)$.

• 19.8 Cell-to-cell variability as a function of fold-change

In the chapter, we derived the Fano factor for a promoter architecture regulated by a repressor that binds to a single site overlapping the promoter. In this case, the Fano factor depends on the mean absolute number of mRNA molecules per cell. An alternative way of looking at the Fano factor is as a function of the fold-change in gene expression, which, under the weak promoter approximation, is just the regulation factor. Reproduce the plot shown in Figure 19.37(A) by calculating the Fano factor as a function of the corresponding fold-change in the mean level of gene expression.

• 19.9 Separation of time scales and transcriptional regulation

For transcription to start, the RNA polymerase bound to the promoter needs to undergo a conformational change to the so-called open complex. The rate of open complex formation is often much smaller than the rates for the polymerase binding and falling off the promoter. Here, we investigate within a simple model how this state of affairs might justify the equilibrium assumption underlying thermodynamic models of gene regulation, namely that the equilibrium probability that the promoter is occupied by the RNA polymerase determines the level of gene expression.

(a) Write down the chemical kinetics equation for this situation. Consider three states: RNA polymerase bound nonspecifically on the DNA (N); RNA polymerase bound to the promoter in the closed complex (C); and RNA polymerase bound to the promoter in the open complex (O). To simplify matters, take both the rate for N → C and the rate for C → N to be k. Assume that the transition C → O is irreversible, with rate Γ.

(b) For $\Gamma = 0$, show that in the steady state there are equal numbers of RNA polymerases in the N and C states. What is the steady state in the case $\Gamma \neq 0$?

(c) For the case $\Gamma \neq 0$, show that for times $1/k \ll t \ll 1/\Gamma$, the numbers of RNA polymerases in the N and C states are equal, as would be expected in equilibrium.

• 19.10 Copy number and the Poisson promoter

The model of the Poisson promoter considered in the chapter assumed that the number of copies of the gene of interest was fixed at one. However, as a result of the replication of the chromosomal DNA, during some part of the cell cycle there will be two (or even more for rapidly dividing cells) copies of the gene of interest. In this problem, we imagine that during a fraction f of the cell cycle, there is one copy of our gene of interest and during the rest of the cell cycle there are two such copies.

(a) Write down the appropriate distribution $p(m)$ for m mRNA molecules as a function of the parameter f.

(b) Find $\langle m \rangle$.

(c) Find $\langle m^2 \rangle$ and use it to find the Fano factor.

(d) Plot the Fano factor as a function of f for different choices of the mean mRNA copy number for a single promoter. How "Poissonian" do you expect an unregulated promoter to be?

(Problem courtesy of Rob Brewster and Daniel Jones.)

• 19.11 Gillespie algorithm revisited

In the computational exploration, we showed how the mRNA evolves as a function of time.

(a) Plot the bias of the reaction-choice coin flip (that is, production or decay) as a function of time. Explain intuitively what is happening.

(b) Plot the time step as a function of time.

• 19.12 Mean protein burst size for a single mRNA

Using the probability distribution for a protein burst of size n from Equation 19.113 and the definition of the mean burst size as

$$\langle n \rangle = \sum_{n=0}^{\infty} nP(n), \qquad (19.175)$$

demonstrate that the mean burst size is given by the ratio of the protein translation rate to the rate of mRNA decay as in Equation 19.114:

$$\langle n \rangle = \frac{r_{\mathrm{p}}}{\gamma} = b. \tag{19.176}$$

• 19.13 A minimal genetic switch

In this problem, we consider a simpler switch than that considered in the chapter. For this switch, we consider an activator that activates its own production.

(a) Make a figure of the states, weights, and rates for an activator that activates itself by binding as a dimer and such that its binding to the DNA is characterized by a Hill function with Hill coefficient 2. Your states and weights should be analogous to those shown in Figure 19.45. Given these states and weights, write a rate equation for the time evolution of the activator. Include a term for a basal rate of production even in the absence of activator.

(b) Make a one-dimensional phase portrait by performing a graphical analysis of the differential equation based on the plot of dA/dt versus A. Use this phase portrait to characterize the existence of fixed points and their stability. Is it appropriate to refer to this as a switch?

• 19.14 Chemotaxis of *E. coli*

In chemotaxis experiments, a source of nutrient molecules can be introduced into the medium containing bacteria via a micropipette. The outward diffusion of the nutrient molecules creates a position-dependent concentration gradient, and the chemotactic response of the bacteria can be observed under a microscope.

(a) Estimate the nutrient gradient in steady state as a function of the distance from the micropipette r by assuming that it keeps the concentration fixed at c_0 for distances $r < r_0$. Make a plot of the concentration gradient as a function of r for typical values $c_0 = 1$ mM and $r_0 = 1\ \mu$m.

(b) Assuming that the bacterium makes two measurements of the concentration using one array of receptor proteins at one of its ends and another array at the other, estimate the maximum distance from the nutrient source for which the bacterium is still able to detect a gradient. Assume that the receptor array counts the number of molecules present in a cubic volume with side $a = 100$ nm. To solve this problem, you should recall that the counting error for N molecules is roughly $\sqrt{N}$, and in order to detect the difference in concentration between the two ends of the bacterium, the measurement error should be less than the difference itself.

(c) Now assume a different strategy, where one receptor is employed but the bacterium compares the concentration at two different positions along a run, separated by a distance of $10\ \mu$m. Compute the maximum distance from the nutrient source at which the bacterium will be able to detect the gradient in this case.

• 19.15 MWC model for heterogeneous receptor clusters

Develop an MWC model for the response of chemotactic receptor clusters where there are M molecules of one type of receptor and N molecules of the other type in a given cluster. The entire cluster is either active or inactive and the two different receptors are characterized by different affinities for the chemoattractant of interest. Specifically, derive an equation that is analogous to Equation 19.158 for the probability that the receptor cluster will be in the on state. To do so, construct a states-and-weights diagram like that shown in Figure 19.58.

• 19.16 N-WASP and biochemistry on a leash

In the last section of the chapter, we considered the action of N-WASP using a simple one-dimensional random walk model to treat the statistical mechanics of looping. Redo that analysis by using the Gaussian model of a polymer chain. First, assume that the loop has to close on itself and then account for the finite size of the protein domain. Compare your results with those obtained in the chapter.

19.7 Further Reading

Alon, U (2007) An Introduction to Systems Biology: Design Principles of Biological Circuits, Chapman & Hall/CRC. Alon's book gives a comprehensive and thoughtful discussion of regulation.

Bintu, L, Buchler, NE, Garcia, HG, et al. (2005) Transcriptional regulation by the numbers: applications, *Curr. Opin. Genet. Dev.* **15**, 125. Application of thermodynamic models to several different regulatory architectures.

Buchler, NE, Gerland, U, & Hwa, T (2003) On schemes of combinatorial transcription logic, *Proc. Natl Acad. Sci. USA* **100**, 5136. Excellent general discussion of thermodynamic models of gene regulation.

Walsh, CT (2006) Posttranslational Modification of Proteins: Expanding Nature's Inventory, Roberts and Company Publishers. This book is full of interesting insights into the phenomenology of post-translational modification. This is a reminder that there is more to regulation than transcriptional control.

Cherry, JL, & Adler, FR (2000) How to make a biological switch, *J. Theor. Biol.* **203**, 117. This article presents an interesting discussion of the issues that arise in designing biological switches.

Ptashne, M (2004) A Genetic Switch, 3rd ed., Cold Spring Harbor Laboratory Press. A beautiful book that focuses on ideas as opposed to facts and paints a picture of how gene regulation works.

Ptashne, M, & Gann, A (2002) Genes and Signals, Cold Spring Harbor Laboratory Press. This book provides an excellent overview of transcriptional regulation.

Michel, D (2010) How transcription factors can adjust the gene expression floodgates, *Prog. Biophys. Mol. Biol.* **102**, 16. This article gives a comprehensive discussion of the physics of gene expression.

Müller-Hill, B (1996) The Lac Operon: A Short History of a Genetic Paradigm, Walter de Gruyter. Müller-Hill's book is a fascinating and idiosyncratic account of the development of thinking on gene regulation in general and the *lac* operon in particular. The book is full of interesting touches such as Figure 3, which illustrates the ways in which synthetic analogs of lactose have played a role in the development of molecular biology.

Davidson, EH (2001) Genomic Regulatory Systems, Academic Press. Davidson's book is full of both interesting facts and provocative ideas. We particularly recommend it as a way to explore the complexity associated with eukaryotic gene regulation.

Ellner, SP, & Guckenheimer, J (2006) Dynamic Models in Biology, Princeton University Press. This book examines dynamical models and their relevance to biology and has a treatment of both the genetic switch and the repressilator.

Berg, HC, & Brown, DA (1972) Chemotaxis in *Escherichia coli* analysed by three-dimensional tracking, *Nature* **239**, 500. This paper uses a three-dimensional tracking technique to follow individual bacterial cells during chemotaxis and demonstrates how bacteria find their way by altering the timing of runs and tumbles.

Keymer, JE, Endres, RG, Skoge, M, Meir, Y, & Wingreen, NS (2006) Chemosensing in *Escherichia coli*: two regimes of two-state receptors, *Proc. Natl Acad. Sci. USA* **103**, 1786. This interesting paper is the basis of the model presented in Section 19.4.

19.8 References

Alon, U, Surette, MG, Barkai, N, & Leibler, S (1999) Robustness in bacterial chemotaxis, *Nature* **397**, 168.

Ben-Tabou de-Leon, S, & Davidson, EH (2007) Gene regulation: gene control network in development, *Annu. Rev. Biophys. Biomol. Struct.* **36**, 191.

Cai, L, Friedman, N, & Xie, XS (2006) Stochastic protein expression in individual cells at the single molecule level, *Nature* **440**, 358.

Case, TJ (2000) An Illustrated Guide to Theoretical Ecology, Oxford University Press.

Cluzel, P, Surette, M, & Leibler, S (2000) An ultrasensitive bacterial motor revealed by monitoring signalling proteins in single cells, *Science* **287**, 1652.

Driever, W, Thoma, G, & Nüsslein-Volhard, C (1989) Determination of spatial domains of zygotic gene expression in the *Drosophila* embryo by the affinity of binding sites for the bicoid morphogen, *Nature* **340**, 363.

Dueber, JE, Yeh, BJ, Chak, K, & Lim, WA (2003) Reprogramming control of an allosteric signaling switch through modular recombination, *Science* **301**, 1904.

Elowitz, MB, & Leibler, S (2000) A synthetic oscillatory network of transcriptional regulators, *Nature* **403**, 335.

Gardner, TS, Cantor, CR, & Collins, JJ (2000) Construction of a genetic toggle switch in *Escherichia coli*, *Nature* **403**, 339.

Gause, GF (1934) The Struggle for Existence, Williams & Wilkins (reprinted by Dover Publications, 2003).

Gillespie, DT (1977) Exact stochastic simulation of coupled chemical reactions, *J. Phys. Chem.* **81**, 2340.

Golding, I, Paulsson, J, Zawilski, SM, & Cox, EC (2005) Real-time kinetics of gene activity in individual bacteria, *Cell* **123**, 1025.

Kim, HD, & O'Shea, EK (2008) A quantitative model of transcription factor-activated gene expression, *Nat. Struc. Mol. Biol.* **15**, 1192.

Kuhlman, T, Zhang, Z, Saier, MHS, & Hwa, T (2007) Combinatorial transcriptional control of the lactose operon of *Escherichia coli*, *Proc. Natl Acad. Sci. USA* **104**, 6043.

Min, TL, Mears, PJ, Chubiz, LM, et al. (2009) High-resolution, long-term characterization of bacterial motility using optical tweezers, *Nat. Methods* **6**, 831.

Müller, J, Oehler, S, & Müller-Hill, B (1996) Repression of *lac* promoter as a function of distance, phase and quality of an auxiliary *lac* operator, *J. Mol. Biol.* **257**, 21.

Myasnikova, E, Samsonova, A, Kozlov, K, Samsonova, M, & Reinitz, J (2001) Registration of the expression patterns of *Drosophila* segmentation genes by two independent methods, *Bioinformatics* **17**, 3.

Oehler, S, Alberti, S, & Müller-Hill, B (2006) Induction of the *lac* promoter in the absence of DNA loops and the stoichiometry of induction, *Nuc. Acids Res.* **34**, 606.

Oehler, S, Amouyal, M, Kolkhof, P, von Wilcken-Bergmann, B, & Müller-Hill, B (1994) Quality and position of the three *lac* operators of *E. coli* define efficiency of repression, *EMBO J.* **13**, 3348.

Rosenfeld N, Young JW, Alon U, et al. (2005) Gene regulation at the single-cell level. *Science* **307**, 1962.

Small, S, Blair, A, & Levine, M (1992) Regulation of *even-skipped* stripe 2 in the *Drosophila* embryo, *EMBO J.* **11**, 4047.

Small, S, Blair, A, & Levine, M (1996) Regulation of two pair-rule stripes by a single enhancer in the *Drosophila* embryo, *Developmental Biology* **175**, 314.

Taniguchi, Y, Choi, PJ, Li, G-W, et al. (2010) Quantifying *E. coli* proteome and transcriptome with single-molecule sensitivity in single cells, *Science* **329**, 533.

Turner, L, Ryu, WS, & Berg, HC (2000) Real-time imaging of fluorescent flagellar filaments, *J. Bacteriol.* **182**, 2793.

Yu, J, Xiao, J, Ren, X et al. (2006) Probing gene expression in live cells, one protein molecule at a time, *Science* **311**, 1600.

Zenklusen, D, Larson, DR, & Singer, RH (2008) Single-RNA counting reveals alternative modes of gene expression in yeast, *Nat. Struc. Mol. Bio.* **15**, 1263.

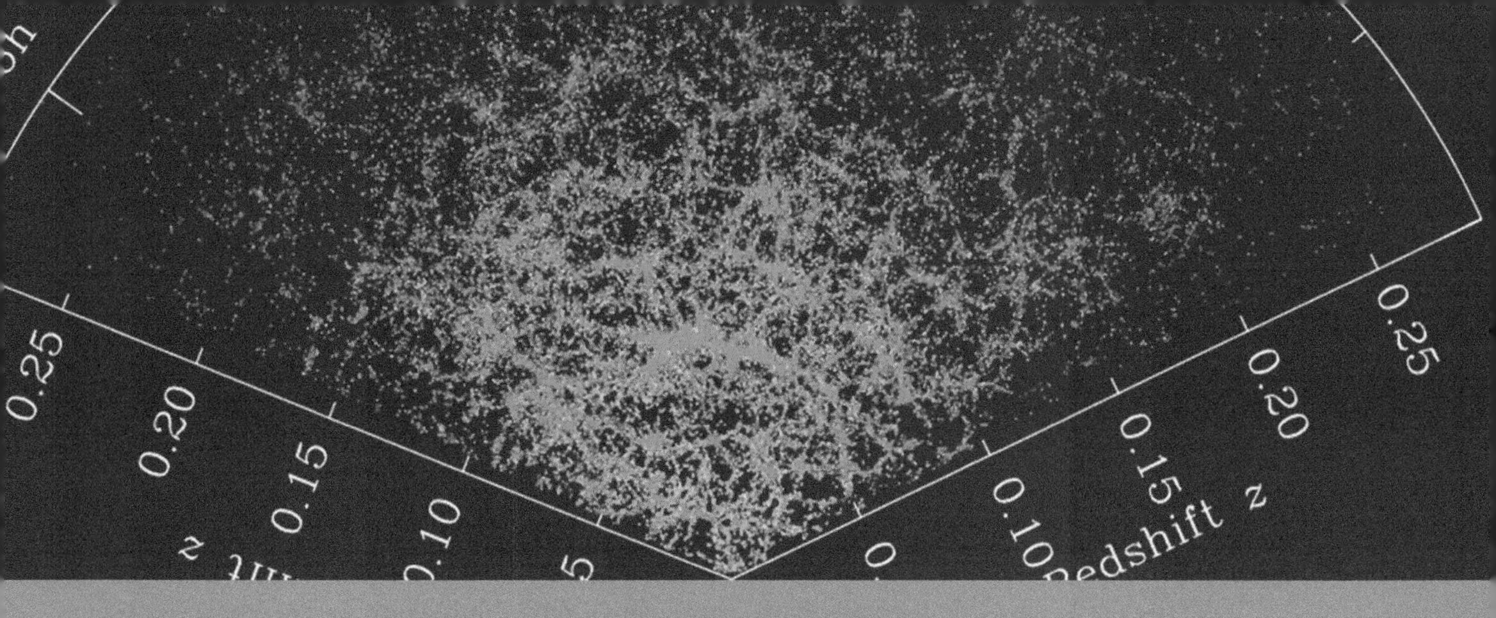

Biological Patterns: Order in Space and Time

20

Overview: In which we examine how living organisms make patterns in space and time

> "Tyger! Tyger! burning bright In the forests of the night, What immortal hand or eye Could frame thy fearful symmetry?"
>
> **William Blake**

Tigers have stripes and leopards have spots. Darwin's finches are renowned for the differences in their beak sizes and shapes. The developing fly embryo early after fertilization acquires "stripes" of its own in the form of precursors to its eventual segmented body plan. Even the formation of specific cells (heterocysts) responsible for nitrogen fixation in the long chains of cells in the microscopic photosynthesizer *Anabaena* reveal intriguing one-dimensional patterns. In each of these cases, specific molecular patterns formed in space and time give rise to the structures we use to identify and classify organisms. In this chapter, we take what we have learned about cellular decision-making to consider how cells lay down such patterns. We first examine a pageant of colorful and exotic patterns found in various organisms. With these phenomena sharply in focus, we then turn to several continuum models of pattern formation that explore how mathematical models that include diffusion, interactions, and reactions can give rise to various patterns and be used to suggest new experiments.

20.1 Introduction: Making Patterns

In *The Origin of Species,* Darwin devotes an entire chapter to the subject of instinct, describing how it works and the way in which evolution could have sculpted such remarkable behaviors. That chapter, in turn, is built around several key case studies that seem to have particularly fascinated the ever-observant Darwin, and one of these was the unmistakeable regular hexagonal patterns of the honeybee's hive. Darwin waxes eloquently and passionately on the subject, noting "He must be a dull man who can examine the exquisite structure of a comb, so beautifully adapted to its end, without enthusiastic admiration. We hear from mathematicians that bees have practically solved a recondite problem, and have made their cells of the proper shape to hold the greatest possible amount of honey, with the least possible consumption of precious wax in their construction. It has been remarked that a skilful workman with fitting tools and measures, would find it very difficult to make cells of wax of the true form, though this is effected by a crowd of bees working in a dark hive." A beehive has all of the hallmarks of the simplest of patterns: a series of identical repeating units with the property that once the position and orientation of one of those units is specified, there is no freedom left for the positions and orientations of all the rest, each bears a precise and simple mathematical relationship to all the others.

20.1.1 Patterns in Space and Time

There are many observable kinds of biological patterns at all spatial scales from molecules to cells to appendages to communities that follow some sort of geometrical rule. Many of the macromolecular assemblies already described throughout the book can be thought of as spontaneously self-organized patterns of interacting molecules, such as virus capsids and microtubules. In microtubules, each individual tubulin subunit makes specific contacts with only a handful of neighbors and yet the regularity of these rules of interaction enables the formation of a structure many orders of magnitude larger than the individual subunit. Filamentous structures like these can be used by cells as building elements for higher-order patterns that make functional structures such as the beating flagella shown in Figure 20.1(A) or the stereocilia on auditory hair cells that are organized with regularly graded lengths like the pipes of an organ as shown in Figure 20.1(B). Both of these are cases of regularly ordered patterns formed at the subcellular level, but the auditory hair cells also show a tissue-level pattern, where the stereocilia are all oriented in the same direction for all the cells in the field. Similarly, at still larger scales, collections of cells can exhibit similar geometric regularity, as strikingly demonstrated by the repetitive pattern of photoreceptors

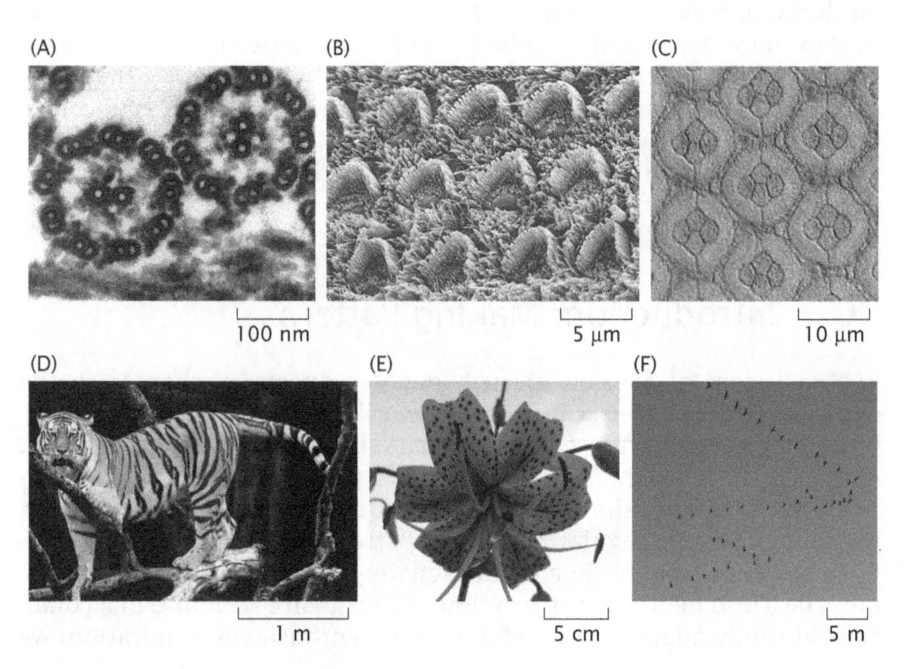

Figure 20.1: Biological pattern formation at many different length scales. (A) Microtubule array in 9/2 format in the cross-section of flagella from the unicellular alga *Chlamydomonas.* (B) Hair cells in the inner ear of a mouse. (C) Lens-making cells ordered hexagonally in a *Drosophila* eye. (D) Tiger showing striped coat pattern. (E) Tiger lily, exhibiting regular spots and geometrically arrayed petals and pistils. (F) Spatial order in a flock of geese. Note that some of the scale bars are estimates based upon the known sizes of the organisms. (A, courtesy of E. Smith, L. Howard and E. Dymek, Darmouth College; B, adapted from K. P. Steel and C. J. Kros, *Nature Genet.* 27:143, 2001; C, adapted from T. Hayashi and R. W. Carthew, *Nature* 431:647, 2004; D, courtesy of Mila Zinkova; E, courtesy of barkandbloom.com; F, courtesy of Eltjo Poort.)

and support cells in the fly retina as shown in Figure 20.1(C). Organisms ranging from tigers (Figure 20.1D) to tiger lilies (Figure 20.1E) also generate patterns of pigment and structure that range over many millions of cells. Even large groups of organisms such as flocking birds and schools of fish are able to exploit surprisingly simple locally operating rules to create large-scale dynamic collective behaviors, as shown in Figure 20.1(F).

As this short list of examples illustrates, pattern formation in biology is ubiquitous. Further, the notion of pattern transcends the geometrical ideas conjured up above, forcing us to move beyond the spatial patterns illustrated there to consider patterns in time. As shown in Chapter 3, biological responses often come in the form of patterns that are organized temporally. Whether we think about the regular pulses of a spike train in a neuron, or the much slower patterns of gene expression as cells progress through generation after generation of their cell cycle, or cyanobacteria undergoing their rhythmic patterns of gene expression that follow the daily cycles of night and day, temporal patterns are a key part of the repertoire of living cells.

20.1.2 Rules for Pattern-Making

The existence of pattern and order in the living world is striking to the most casual observer. Not only do living organisms generate regularly spaced structures and markings, like the stripes on a zebra or the petals on a flower, but the structures are also remarkably reproducible. Every maple leaf is similar in shape to every other maple leaf, but they are all dramatically different from the leaves of the giant sequoia. Every fruit fly looks very much like every other fruit fly, from the overall size to the number of wings to the patterns of bristles on their backs and legs. The extraordinary complexity and precision with which living organisms seem to be constructed has inspired comparisons between the workings of life and the workings of enormously complex human-built machines, such as airplanes or, famously, pocket watches. The eighteenth-century philosopher and theologian William Paley argued that the only way that he could imagine that the kind of order seen in a watch (or a human) could arise is by virtue of a maker that literally designed and assembled the complex structure. As we learn more about the actual mechanisms underlying biological pattern formation, the fantastic complexity of life becomes no less wondrous even as it is better understood. The process of laying down the body plan in a *Drosophila* embryo requires a precise series of many events involving communication of molecular spatial and timing information among cells and results in cell fate decision-making.

In some cases, cells or organisms take advantage of external information to control internal patterns, but in some of the most interesting cases, patterns are able to arise spontaneously. In the most general sense, *de novo* pattern formation can be thought of as a kind of symmetry breaking where spontaneous events within a uniform field of molecules or cells can cause the creation of large-scale order. In this chapter, we will narrow our scope to a particular kind of question in pattern formation: how do cells detect information about their spatial location relative to other cells in order to decide on a developmental fate? Most of the specific biological cases that we will consider involve multicellular animals and plants, although a few examples will be drawn from the bacterial domain of life as well. However, this

(A)

10 μm

(B)

10 μm

(C)

10 μm

(D)

10 μm

(E)

2 μm

selection is not at all comprehensive, and sadly we will neglect most of the fascinating cases where cells use molecular patterning information to build complex designs at the subcellular level. While we will touch on the mechanisms that vertebrate animals use to count out the correct number of vertebrae in their spines and mollusks use to build the fantastical patterns of seashells, we will not address the tiny cell-sized shells built from calcium carbonate such as those shown in Figure 20.2. The variety and precision of these exquisite structures can serve as an inspiration, though, for a drive to understand the mechanisms that we do address.

In this chapter, we will generally treat pattern formation as a problem in communication of spatial information. We will explore four distinct mechanisms that are used by various organisms for large-scale patterning: morphogen gradients (deriving spatial information from the concentration of a diffusible molecule that is made at a particular source location), Turing patterns (where diffusion coupled with certain kinds of chemical interaction enables large-scale patterns to emerge spontaneously), clock and wavefront mechanisms (where organisms use tricks of cell fate specification to turn a signal that oscillates regularly in time into a pattern of regular stripes in space), and finally lateral inhibition (where cells directly influence the fate choices of their nearest neighbors).

20.2 Morphogen Gradients

Some of the most compelling examples of biological pattern formation and those that have been studied in the most intensive molecular detail are found in animal development. All animals start out as a single undifferentiated cell, a fertilized egg, and over a period of days to months generate a much larger number of genetically identical cells that have assumed a wide variety of distinct fates. Direct observation of the process of animal development under a microscope reveals an astonishing series of cell movements, structural alterations, and tissue differentiation that follows a remarkably consistent sequence of events in all the embryos of a particular animal species. The question of how the genome is able to encode directions for this complex choreography of cell proliferation and differentiation in space and time is central to all of biology. Stated simply, the more general question is: how do all of the many cells that make up an organism, each of which harbors precisely the same genetic information (with some minor exceptions), turn out so differently?

20.2.1 The French Flag Model

In some tissue development processes, a group of cells that initially appear to be identical with respect to their contents and gene expression states will change over time to assume a series of different states that appear correlated with the cell's position within the

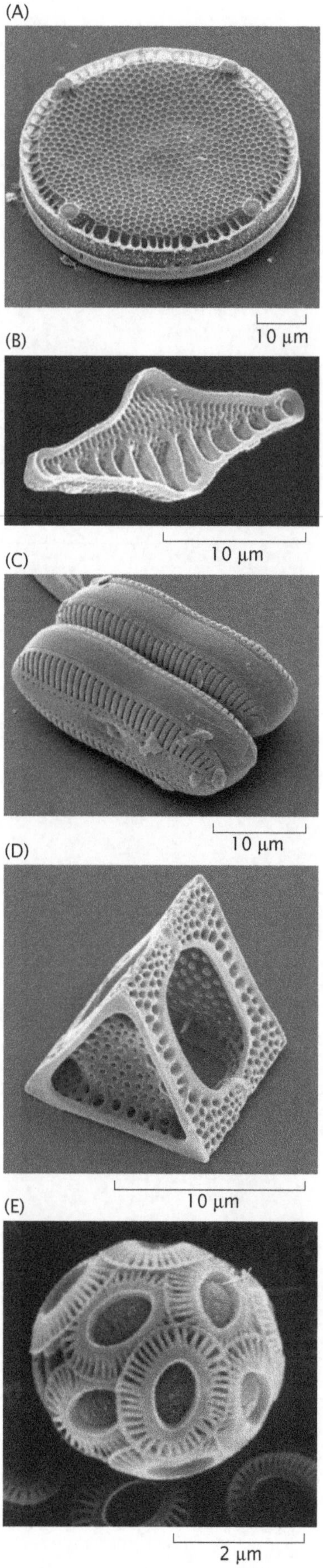

Figure 20.2: Cell-sized seashells of diatoms and coccolithophores. (A) *Eupodiscus radiatus*. (B) *Nitzschia* sp. (C) *Diploneis* sp. (D) *Triceratium* sp. (E) *Emiliana huxleyi*. (A, C, adapted from J. Bradbury, *PLoS Biol.* 2:e306, 2004; B, courtesy of Bowling Green State University Center for Algal Microscopy; D, courtesy of Steve Gschmeissner; E, courtesy of Jeremy Young.)

tissue. We can imagine at least two general mechanisms that might lead to this outcome. First, each cell could be preprogrammed for a certain fate depending upon its lineage within the embryo. Alternatively, each cell might harbor the potential to achieve any state but interprets some kind of positional information to make that choice. Transplantation experiments can in many cases distinguish between these possibilities—a preprogrammed cell will achieve the same fate even when it has been moved to a new location, but a cell performing information processing will achieve a new state depending on the site to which it is moved. Although both kinds of outcomes have been observed in particular cases, the second is probably more common and speaks to the remarkable plasticity of cell fates within an embryo.

How might the transplanted cell be able to determine its new position? A simple and powerful idea for how spatial information might be communicated across large fields of cells is illustrated in Figure 20.3. If there is a source of some diffusible signaling molecule that lies outside of the field of differentiating cells, the positions of the cells relative to that source are correlated with the local concentration of the signaling molecule. Many signaling molecules are known to have concentration-dependent effects on gene transcription, and, indeed, many of these concentration-dependent effects show threshold-like behavior. The general term used to describe a molecule that can influence the differentiation of a cell based on its concentration is "morphogen." The concept of the morphogen was strongly influential in the field of developmental biology long before there was compelling experimental identification of any specific molecules acting as morphogens in real organisms, but the idea that there must be such molecules to explain the kinds of processes seen in multicellular development was so persuasive that systematic effort by researchers working in many different systems has resulted in the identification of many real morphogen-like molecules. The key insight behind the morphogen concept is that spatial information can be communicated through the familiar mechanisms that cells can use to detect the *concentration* of a ligand in their surroundings, not just its presence or absence. In the "French flag model" illustrated in Figure 20.3, the cellular outcome of exposure to a signaling molecule is assumed to have two different thresholds. At a very low concentration, cells will retain some default state illustrated in red. An intermediate concentration induces a gene expression profile that converts cells to a distinct fate illustrated in white. At high concentration, additional genes will also be regulated, converting the cells to a third fate illustrated in blue. In this way, a single smoothly varying gradient of a single signaling factor can give rise to multiple different groups of differentiated cells with sharp boundaries between groups.

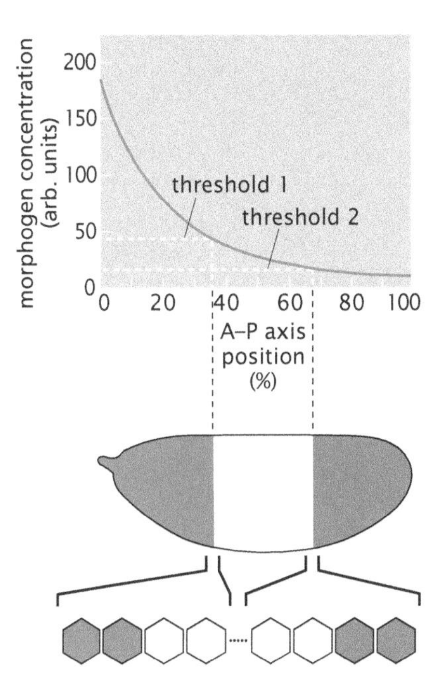

Figure 20.3: French flag model of positional information. A morphogen gradient is present along the long (anterior–posterior) axis of the embryo. When the concentration of the morphogen exceeds a first threshold, cells are in the "blue" state, while when they are below the first threshold but above a second threshold, they adopt the "white" state. (Adapted from L. Wolpert et al., Principles of Development, 3rd ed. Oxford University Press, 2007.)

Even in this oversimplified presentation, the power of the morphogen concept is apparent. However, in order for real morphogens to work in living systems, several challenges must be overcome. First, how can a concentration gradient be established that will be sufficiently uniform over time to influence cell fate decisions? Second, if the spatial dimensions of the gradient are set by fundamental physical processes (most often diffusion, as explained below), then how can the size of gradients be scaled for larger or smaller organisms? Third, given the general observation that the development of multicellular organisms is very similar from one individual to another within the same species, how can delicate systems such as concentration gradients of diffusible molecules remain robust to variations such as biological noise and environmental insults such as temperature changes? Finally, how do the cells that are receiving the morphogen signal measure its concentration with sufficient precision to ensure that they make the right developmental fate choices?

Each of these questions can be best addressed quantitatively. For the remainder of this section, we will explore cases where quantitative measurements in interesting model systems have yielded insights into the detailed mechanisms of morphogen gradient establishment and interpretation. As is often the case, some of the most critical insights arise not when we see how the data supports the simple quantitative models but rather when we see the precise points of disagreement between them.

20.2.2 How the Fly Got His Stripes

As highlighted throughout the book, a compelling example of the role of developmental decisions that might be profitably viewed from the perspective of ideas like that of the French flag model is that of embryonic development in fruit flies. One of the most familiar and conceptually important examples concerns the establishment of the anterior–posterior patterning of the fly body plan already introduced in Section 2.3.3 (p. 78). Positional information in this case is dictated by the distribution of certain proteins along the embryo. Figure 19.2 (p. 804) gives an example of the gradients in four key regulatory proteins that determine the anterior–posterior organization. These proteins determine the pattern of gene expression along the embryo, from which the expression of the second stripe of the *even-skipped* gene (*eve*) is one of the most well-understood examples.

An interesting question raised by the developmental processes in the early *Drosophila* embryo concerns how the spatial patterns of morphogens are generated. For example, the *eve* gene is expressed in seven stripes along the anterior–posterior axis of the embryo. The formation of the second stripe of *eve* expression is regulated by the minimal stripe element, a roughly 700 bp stretch of regulatory DNA, which converts the existing concentration profile of Bicoid, Hunchback, Giant, and Krüppel (see Figure 19.2, p. 804) into the expression of the *eve* gene. The patterns that serve as input to *eve* expression are themselves outputs of previous steps in the developmental program of the fruit fly. We are thus led to the questions: what is the initial condition, and how is it set up?

Bicoid Exhibits an Exponential Concentration Gradient Along the Anterior–Posterior Axis of Fly Embryos

The first morphogen whose concentration shows a pattern is Bicoid. The Bicoid protein is produced by maternal *bicoid* mRNA, which is found at the anterior end of the embryo. Once Bicoid is produced at the anterior end, it diffuses throughout the embryo. This protein is also degraded in time, and, as we show below, this combination of diffusion and degradation can give rise to an exponentially decaying profile of Bicoid concentration from the anterior to the posterior end. As seen in Figure 19.2 (p. 804), the concentration profiles of the other key molecular determinants are richer than the simple exponential profile exhibited by Bicoid.

In order to account simultaneously for the space and time dependence of transcription factors such as Bicoid, one useful approach is the use of so-called reaction–diffusion equations. These equations merge the thinking on diffusion developed in Chapter 13 with the ideas on rate equations introduced in Chapter 15. The general idea is that the various reactants and products can diffuse around and change their identity. Conceptually, the starting point of this analysis is that the concentrations of the various species of interest must be promoted to full functions of space *and* time. For example, in the case of Bicoid, we introduce the field $[\mathrm{Bcd}](\mathbf{r}, t)$. For simplicity, we imagine that the concentration profile is one-dimensional (that is, we consider a one-dimensional fly embryo) so that the concentration can be written as $[\mathrm{Bcd}](x, t)$. Further, the only "reaction" to which Bicoid is subjected is that it can decay over time. (As we will see in the next section, the idea of pattern formation by reacting and diffusing molecular species can be generalized to include more than a single diffusing species and more than one kind of reaction among them; these additional complexities will enable the formation of much more complex patterns.)

A Reaction–Diffusion Mechanism Can Give Rise to an Exponential Concentration Gradient

The rate equation for the concentration of Bicoid along the anterior–posterior axis is

$$\frac{\partial[\mathrm{Bcd}]}{\partial t} = D\frac{\partial^2[\mathrm{Bcd}]}{\partial x^2} - \frac{1}{\tau}[\mathrm{Bcd}]. \tag{20.1}$$

Here D is the diffusion constant of the protein, while τ is the mean lifetime. The second term on the right-hand side of the equation is a "sink" term and accounts for the fact that, even in the absence of diffusion, the concentration of Bicoid can change simply by virtue of decay processes characterized by the mean lifetime τ. Since the anterior region ($x = 0$) of the embryo acts as a source of Bicoid, we expect a steady state to develop, characterized by a concentration profile that does not change in time. In this case, the Bicoid concentration satisfies the equation

$$D\frac{\mathrm{d}^2[\mathrm{Bcd}]}{\mathrm{d}x^2} - \frac{1}{\tau}[\mathrm{Bcd}] = 0, \tag{20.2}$$

which has the solution

$$[\mathrm{Bcd}] = [\mathrm{Bcd}]_{\mathrm{max}}\mathrm{e}^{-x/\lambda} \tag{20.3}$$

where $[\mathrm{Bcd}]_{\mathrm{max}}$ is the concentration at the anterior end while $\lambda = \sqrt{D\tau}$ is the characteristic length over which the concentration decays by a factor of $1/\mathrm{e}$.

How does this predicted exponential profile correspond to what is seen experimentally, and how well does the predicted value of λ accord with its measured value? As seen in Figure 20.4, careful experiments have made it possible to measure the Bicoid gradients in a number of different species of fly. Figure 20.4(B) reveals the concentration profiles for Bicoid, which are well described by an exponential distribution like that derived above. Further, Figure 20.4(C) shows the measured values of the decay constant λ, which for *Drosophila melanogaster* has a characteristic value of roughly $\lambda \approx 100\ \mu\mathrm{m}$.

Though there remains uncertainty about the parameters D and τ, the diffusion constant for Bicoid appears to lie in the range 5–10 $\mu m^2/s$. Measurements of the degradation time yield approximately $\tau \approx 50$ minutes or $\tau \approx 3000$ s. As a result, we estimate

$$\lambda \approx \sqrt{5\ \mu m^2/s \times 3000\ s} \approx 120\ \mu m, \tag{20.4}$$

a number in reasonable accord with the measured value.

However, these experiments went even farther. As shown in Figure 20.4, similar measurements can be made for other fly species as well. Interestingly, embryos of fly species can differ in length by as much as a factor of five, and the fractional position at which the concentration falls off to $1/e$ of its value at the anterior end scales appropriately. From a biological point of view, this is necessary for proper scaling of the morphogen stripes and other patterns with embryo size (this in turn leads to the development of a proportional fly). Given the simple diffusion–degradation model that we have introduced, it is not clear at all how this comes about, since it would seem that the diffusion constant and the degradation rate should to a first approximation not be species-specific. Elaborations on this simplest model that attempt to come to terms with this scaling paradox are discussed in the Computational Exploration below. This highly simplified example is intended to whet the reader's appetite for the necessity of a full space–time description of the regulatory networks described in the previous chapter.

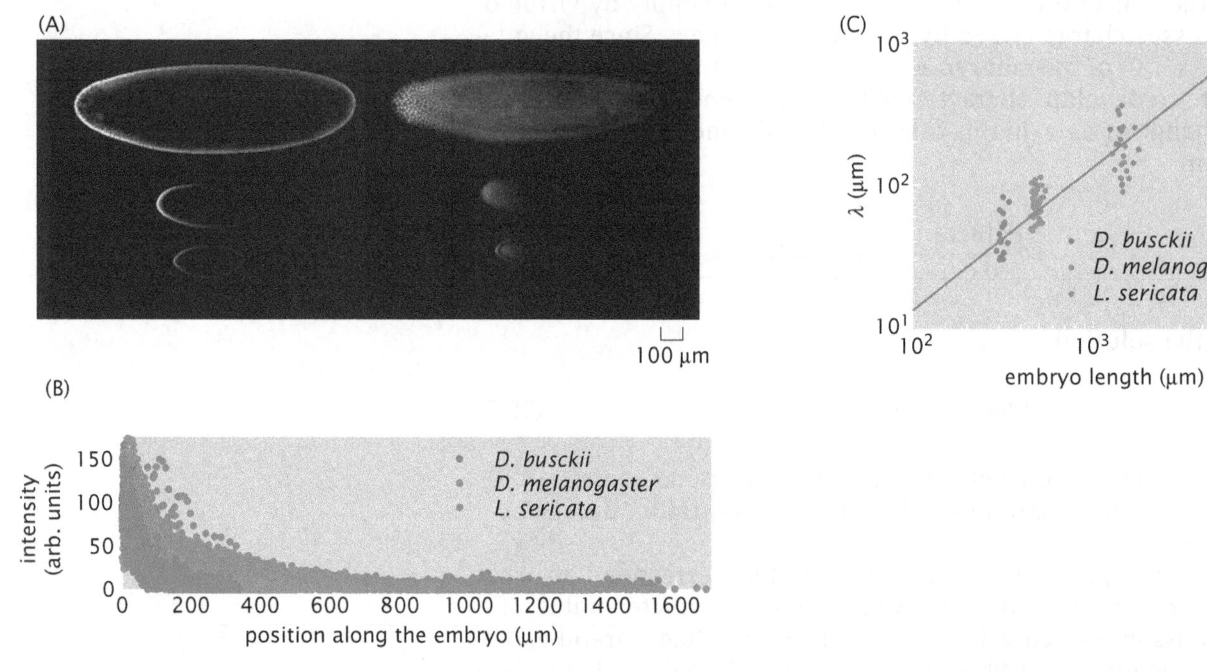

Figure 20.4: Scaling of the Bicoid gradient across different fly species. (A) Confocal images of Bicoid immunofluorescence staining for *L. sericata* (top), *D. melanogaster* (middle), and *D. busckii* (bottom). The same embryos were imaged through their middle (left) and from the top (right). (B) Intensity profiles obtained for several embryos of each species. (C) Fitting each Bicoid profile to an exponential yields a decay constant λ. This decay constant is plotted as a function of embryo length for all three species considered. The line corresponds to a linear fit with zero intercept. (Adapted from T. Gregor et al. *Proc. Natl Acad. Sci. USA* 102:18403, 2005.)

Computational Exploration: Scaling of Morphogen Gradients As we showed earlier in this section, in a reaction–diffusion model, the length scale associated with a morphogen gradient is set by the diffusion constant and the degradation rate of the morphogens themselves. However, in different fly species with significantly different embryo sizes, such as the three different species shown in Figure 20.4(A), the length scale of the Bicoid morphogen scales linearly with the embryo length. The measurement of this gradient is shown in Figure 20.4(B) for embryos from three fly species. Each of these gradients can be fitted to an exponential and its characteristic length plotted against the embryo length as shown in Figure 20.4(C). Despite the fact that both diffusion and degradation constants do not vary appreciably between the species, we see a linear scaling between the length scale of the gradient and that of the embryo. There must be something else setting the length scale of these exponential profiles. In this computational exploration, we consider one hypothesis for the role of nuclei in the establishment and scaling of the morphogen gradient.

We invoke a similar approach to that used in the computational exploration on the cable equation in Section 17.4.3 (p. 705). There, we solved a partial differential equation to compute the propagation of the electrical signal along an axon. In that case, we had to integrate over both temporal and spatial variables. In this computational exploration, we invoke the same tools in the sense that we divide space into small discrete boxes. We will consider the movement of morphogen molecules between boxes as well as degradation within each of them (and production at the source). However, unlike in the previous case, we now invoke the idea that different boxes have different properties. In particular, some of our boxes will correspond to nuclei and in these boxes the degradation

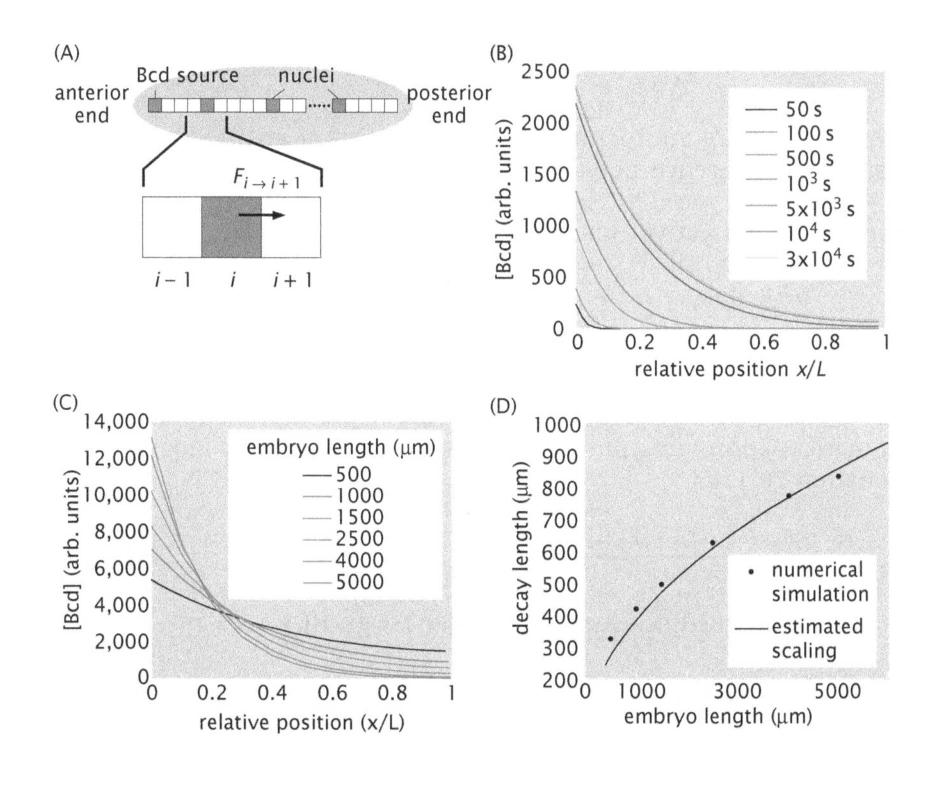

Figure 20.5: Numerical simulation of establishment of Bicoid gradient and scaling in a one-dimensional model of the embryo. (A) Discrete model for Bicoid diffusion along the embryo where we distinguish between cytoplasm and nuclei. (B) Bicoid profile for different times. This case corresponds to a model in which the properties of the embryo are uniform. (C) Steady-state Bicoid profiles for one-dimensional embryos of different lengths and containing 10 nuclei. (D) Decay lengths obtained from the simulation as a function of embryo length. The data is compared with the expected $\sqrt{L}$ scaling calculated for a one-dimensional embryo in the Computational Exploration.

properties are different than elsewhere. Since the different fly species have the same number of nuclei at their corresponding stages in development, the mean spacing between these nuclei is *different* in different species. The hypothesis put forth then is that the nuclei set the length scale through some property that is different than in the cytoplasm. In this particular case we explore a difference in degradation rate. Although the mechanism considered here hasn't been proved directly, we entertain this hypothesis as an example of how the nuclei can generate the observed scaling of the morphogen between species.

In order to gain intuition, we start by solving a problem with a source at the anterior end of the one-dimensional embryo and uniform diffusion and degradation throughout the embryo. The diffusion equation describing this situation was already given as Equation 20.1. In order to solve this equation numerically, we discretize the one-dimensional embryo into boxes. Given an embryo of length L, we consider $L/\Delta x$ boxes of size Δx. We begin by considering the diagram shown in Figure 20.5(A) in the case where all boxes are the same and we do not differentiate nuclei from the surrounding medium. What determines the number of Bcd molecules in box j at time point i? We can figure this out with reference to Equation 20.1. We discretize the different terms in the equation as follows. First, the time derivative can be approximated as

$$\frac{\partial[\mathrm{Bcd}](x,t)}{\partial t} = \frac{[\mathrm{Bcd}](x,t+\Delta t) - [\mathrm{Bcd}](x,t)}{\Delta t}, \tag{20.5}$$

where Δt is the time step used in our integration of the differential equation. In order to discretize the term involving the second derivative in x, we recall that the second derivative is gotten by taking the first derivative two times and can be written as

$$\frac{\partial^2[\mathrm{Bcd}](x,t)}{\partial x^2} = \frac{\partial}{\partial x}\left(\frac{[\mathrm{Bcd}](x+\Delta x/2) - [\mathrm{Bcd}](x-\Delta x/2)}{\Delta x}\right). \tag{20.6}$$

Note that we are approximating the derivative at position x by using the difference between the concentrations at positions $x+\Delta x/2$ and $x-\Delta x/2$. Using the same strategy a second time, the second derivative can be written as

$$\frac{\partial^2[\mathrm{Bcd}](x,t)}{\partial x^2} = \frac{1}{\Delta x^2}\left\{[\mathrm{Bcd}](x+\Delta x,t) - 2[\mathrm{Bcd}](x,t) + [\mathrm{Bcd}](x-\Delta x,t)\right\}. \tag{20.7}$$

As a result of Equations 20.5 and 20.7, we can write the numerical approximation to the diffusion equation for Bicoid, Equation 20.1, as

$$\begin{aligned}\frac{[\mathrm{Bcd}](x,t+\Delta t) - [\mathrm{Bcd}](x,t)}{\Delta t} &\\ = \frac{D}{\Delta x^2}\left\{[\mathrm{Bcd}](x+\Delta x,t) - 2[\mathrm{Bcd}](x,t) + [\mathrm{Bcd}](x-\Delta x,t)\right\} &\\ - \frac{1}{\tau}[\mathrm{Bcd}](x,t). &\end{aligned} \tag{20.8}$$

We solve this equation for the concentration at position x at time point $t + \Delta t$ by solving for $[\mathrm{Bcd}](x, t + \Delta t)$, using the Euler method, resulting in

$$[\mathrm{Bcd}](x, t + \Delta t) = [\mathrm{Bcd}](x, t) + \Delta t \left(\frac{D}{\Delta x^2} \{[\mathrm{Bcd}](x + \Delta x, t) - 2[\mathrm{Bcd}](x, t) + [\mathrm{Bcd}](x - \Delta x, t)\} - \frac{1}{\tau}[\mathrm{Bcd}](x, t) \right). \tag{20.9}$$

This tells us that in order to figure out the concentration at position x and time point $t + \Delta t$, we need to look at the concentration in the same box at the previous time step as well as the concentrations in the adjacent boxes. The difference in concentration between adjacent boxes is related to the flow of molecules between them through the diffusion constant as dictated by Fick's law.

In principle, given an initial condition that sets up $[\mathrm{Bcd}](x, t = 0)$, all we need to do in order to calculate the concentration profile at the next time step, $[\mathrm{Bcd}](x, t = \Delta t)$, is to use Equation 20.9. However, we also need to include information about the boundary conditions. For example, at $x = 0$, we know there is a source of Bicoid, resulting in the equation

$$[\mathrm{Bcd}](0, t + \Delta t) = [\mathrm{Bcd}](0, t) + \Delta t \left(\frac{D}{\Delta x^2} \{[\mathrm{Bcd}](\Delta x, t) - [\mathrm{Bcd}](0, t)\} - \frac{1}{\tau}[\mathrm{Bcd}](0, t) \right) + R\Delta t. \tag{20.10}$$

Because there is no box at $x = -\Delta x$, we have not included in this equation the flux coming from the left into our box at $x = 0$. Additionally, we have added a production term $R\Delta t$, where R is the rate of production of Bicoid protein.

The posterior end of the embryo is special as well because there can be no flux of Bicoid protein exiting the embryo at that end. As a result, we suppress the terms corresponding to a flux into the last box from the right, resulting in

$$[\mathrm{Bcd}](L, t + \Delta t) = [\mathrm{Bcd}](L, t) + \Delta t \left(\frac{D}{\Delta x^2} \{-[\mathrm{Bcd}](L, t) + [\mathrm{Bcd}](L - \Delta x, t)\} - \frac{1}{\tau}[\mathrm{Bcd}](L, t) \right). \tag{20.11}$$

We are now ready to carry out a numerical calculation of the evolution of the Bicoid profile. The Bicoid profile resulting from solving the equations as spelled out above is shown in Figure 20.5(B) for different time points. We see that by about 10^4 s ≈ 160 minutes ≈ 2.8 hours, the gradient seems to have reached a steady state. This time scale is slower than the 90 minutes that was actually measured for the establishment of the profile in the fly. Additionally, the decay length is roughly 120 μm, as expected from solving the problem analytically as shown in Equation 20.4.

Next, we generalize the description given above to account for the presence of nuclei and their possible impact on gradient establishment. A simple toy model makes the assumption that Bicoid only decays inside nuclei. In that case, the effective degradation rate $1/\tau_{\mathrm{eff}}$ is the product of the degradation rate when the Bicoid is in the nucleus, $1/\tau$, and the probability that the molecule finds itself in the nucleus, p_{nuc}. Assuming that the Bicoid molecules are uniformly distributed within a volume V, the probability of any given one being found in a nucleus can be expressed as $p_{\mathrm{nuc}} = N_{\mathrm{nuc}} v/V$, where N_{nuc} is the number of nuclei, while v is the volume of an individual nucleus. Putting all this together, we obtain a simple formula for the effective degradation rate, namely,

$$\frac{1}{\tau_{\mathrm{eff}}} = \frac{N_{\mathrm{nuc}} v}{V} \frac{1}{\tau}, \tag{20.12}$$

which leads to a formula for the decay length,

$$\lambda = \sqrt{D\tau_{\mathrm{eff}}} = \sqrt{\frac{D\tau V}{N_{\mathrm{nuc}} v}}. \tag{20.13}$$

One of the particularities of the embryos of different fly species shown in Figure 20.4 is that they all have the same number of nuclei, which are spread out in a volume $V = At$, where A is the surface area of the embryo and t is the thickness of the syncytial layer where the nuclei are located. Given that t is relatively constant between species, we only need to worry about how A changes between them. An interesting extra piece of information is that, though different species can have embryos that vary dramatically in size, the aspect ratio of the embryos is relatively constant. Given this observation, we can write the embryo surface area as

$$A = 2\pi RL = \pi\alpha L^2, \tag{20.14}$$

where R is the radius of the embryo, $\alpha = 2R/L$ is the aspect ratio of the embryo width to the embryo length, and we have assumed a cylindrical geometry for the embryo. As a result, we expect the decay length to be proportional to the embryo length. In particular, using Equation 20.14 to compute V and then substituting this result into the formula for the decay length given in Equation 20.13 leads to the result

$$\lambda = \sqrt{D\tau \frac{\pi\alpha t}{N_{\mathrm{nuc}} v}}\, L. \tag{20.15}$$

Hence, the simple idea that Bicoid degradation occurs within nuclei leads to a prediction for how the decay length should scale with the embryo size, which appears to be consistent with the experimental observation shown in Figure 20.4(C).

We can make a one-dimensional test of this kind of argument by performing a numerical solution of the reaction–diffusion equation with nonuniform degradation. To simplify matters, instead of considering the two-dimensional geometry of the cortical layer of the embryo, we once again simplify our analysis to a one-dimensional embryo. In this case, the same line of reasoning as above leads to an effective Bicoid degradation rate $\tau_{\mathrm{eff}}^{-1} = (N_{\mathrm{nuc}} l/L)\tau^{-1}$, where l is the size of an individual nucleus, and L is the size of the one-dimensional embryo, which in turn predicts a decay length that grows as

$\sqrt{L}$. As a result, in our simple simulation, if we keep the number of nuclei constant, we expect the decay length to scale with the square root of the embryo length.

To proceed, we assume that nuclei have a characteristic size $l = \Delta x$ such that a single box can be associated with a nucleus. In order to describe the evolution of the concentration in a box corresponding to a nucleus, we use Equation 20.9. However, for the remaining boxes corresponding to cytoplasm, we need to suppress the degradation term, resulting in

$$[\text{Bcd}](x, t+\Delta t) = [\text{Bcd}](x, t) + \Delta t\left(\frac{D}{\Delta x^2}\{[\text{Bcd}](x+\Delta x, t) - 2[\text{Bcd}](x, t) + [\text{Bcd}](x-\Delta x, t)\}\right). \quad (20.16)$$

In addition, we drop the decay terms corresponding to the boundary conditions, as we assume that there are no nuclei at those positions.

Figure 20.5(C) shows several Bicoid profiles calculated for embryos of different lengths, but all of them with 10 nuclei. For each one of these profiles, we calculate the corresponding decay lengths, which are shown in Figure 20.5(D). Note that, as argued above, the decay length scales approximately with the square root of the embryo length for this one-dimensional model. As a result, in its full three-dimensional complexity, the model presented here where protein decay is localized (or much faster) in the nuclei provides a plausible explanation for the observation that Bicoid gradients scale with embryo length as shown in Figure 20.4.

20.2.3 Precision and Scaling

The formation of spatial patterns of morphogen concentrations in the early *Drosophila* embryo follows a genetically determined program that is recursive in nature. Patterns of morphogens established in the previous division cycle of development are thought to lead to patterns of newly synthesized morphogens by acting together as transcription factors that regulate their synthesis. For example, Figure 19.2 (p. 804) shows how the concentration profiles of Bicoid, Hunchback, Krüppel, and Giant induce the expression of Eve in a narrow stripe. The position of the stripe is defined by the region of the embryo with large Hunchback and Bicoid concentrations and diminished Krüppel and Giant concentrations; the former pair of morphogens act as transcriptional activators while the latter pair repress the transcription of the *eve* gene.

This mechanism of generating spatial patterns of morphogens is different than the simple scheme proposed by Turing that will be taken up in the next section, and seems to be the one that is used in the development of multicellular organisms. A particularly fascinating example of this type of developmental program, and one that has been studied in considerable quantitative detail, is the expression of Hunchback following the establishment of the Bicoid gradient, shown in Figure 20.6.

Figure 20.6 shows the results of experiments where fluorescently labeled proteins were used to measure the concentration of Bicoid and Hunchback along the anterior–posterior axis of the embryo. Bicoid is the activator of Hunchback, and we can gain quantitative intuition

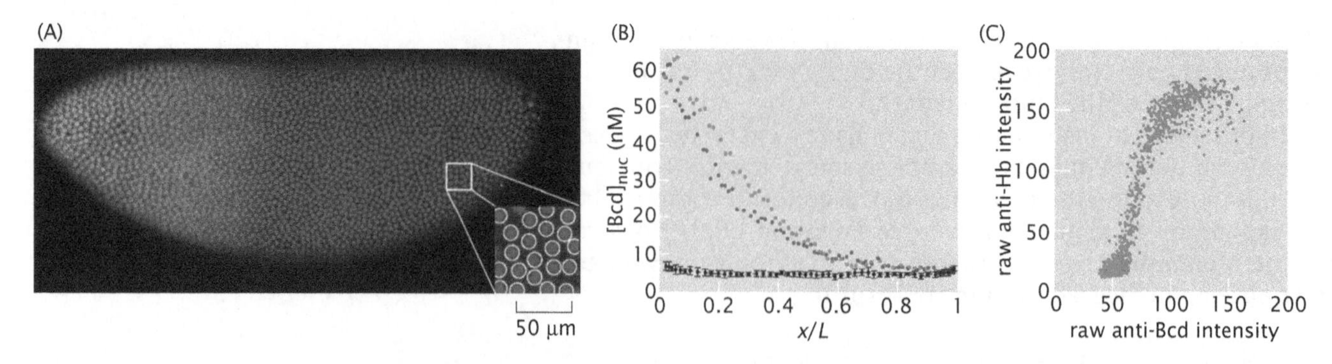

Figure 20.6: Spatial patterning of Bicoid and Hunchback. (A) Fluorescent image showing Bicoid (green) and Hunchback (red), which are both localized to the anterior half of the embryo. DNA is stained with a different color (blue), allowing for individual nuclei to be identified. (B) Quantification of the bicoid gradient as a function of relative position with respect to the egg length on the dorsal (red) and ventral (blue) sides of the embryo. The black points denote the background fluorescence intensity of an embryo that does not harbor fluorescent proteins. (C) Scatter plot of Hunchback versus Bicoid fluorescence intensity from 1299 identified nuclei in a single embryo shows a sharp onset of Hunchback with increasing Bicoid concentration. (Adapted from T. Gregor et al., *Cell* 130:153, 2007.)

about the establishment of the spatial pattern of Hunchback by using a simple model of *hunchback* expression, which is activated by the Bicoid protein. The idea of the calculation is to use thermodynamic models of gene expression to compute the amount of *hunchback* gene expressed at a specific position along the anterior–posterior axis of the embryo, given the amount of Bicoid at that same position. This is illustrated conceptually in Figure 20.7.

As noted above, Bicoid acts as a transcriptional activator of *hunchback*. It is thought to exercise this transcriptional control at the initial stages of embryonic development by binding to six distinct binding sites on the DNA region that regulate transcription of the *hunchback* gene. To account for this architecture of the control region on the DNA, we make use of a thermodynamic model that is equivalent in spirit to the MWC model already introduced several times in the book as the basis for thinking about hemoglobin (p. 300), ligand-gated ion channels (p. 305), and bacterial chemotaxis (p. 881). The idea of the model is that, as a result of the spatial compaction of the *hunchback* gene, it is either in the transcriptionally active "on" state, or in a compacted "off" state as depicted schematically in Figure 20.8. This spatial compaction could be due, for example, to the presence of nucleosomes that occlude the binding of transcription factors to DNA as discussed in Section 10.4.3 (p. 409). For Bicoid to act as a transcriptional activator, it must favor the "on" state, and we assume that in this state all six Bicoid-binding sites can be occupied, while in the "off" state the binding sites are not available. One possible mechanism by which this can be achieved is if the DNA that contains the binding sites is wound up in nucleosomes when the gene is in the "off" state, thus occluding the sites from Bicoid proteins, though more experiments are needed to really flesh out such a hypothesis. Our goal is simply to show how a model based upon this kind of hypothesis might work.

We use statistical mechanics to compute the probability of being in the on state (p_{on}), which we equate with the amount of Hunchback produced (that is, $[\mathrm{Hb}] = [\mathrm{Hb}]_{\mathrm{max}} p_{\mathrm{on}}$). We take the energy of the regulatory DNA when in the "off" state to be $\varepsilon_{\mathrm{off}}$, while its energy in the "on" state and in the absence of bound Bicoid is $\varepsilon_{\mathrm{on}}$; the difference between these two energies is the energy penalty for freeing up the DNA so that the Bicoid-binding sites are available for binding. If the *hunchback* gene is "on," the binding sites for Bicoid can be occupied

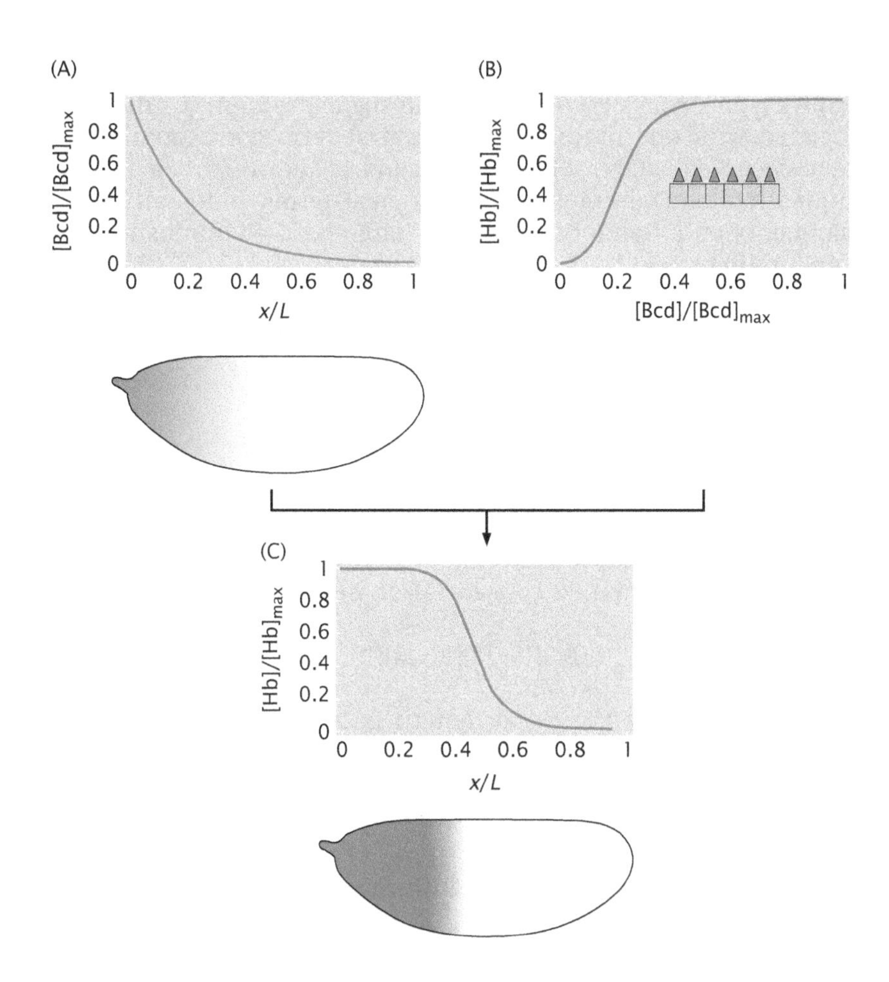

Figure 20.7: Establishment of the Hunchback profile. (A) Bicoid gradient along the anterior–posterior axis. (B) Amount of *hunchback* gene expressed as a function of the concentration of Bicoid. $[\mathrm{Hb}]/[\mathrm{Hb}]_{\max}$ is identified with p_{on}, which is given by Equation 20.17, where we have assumed $\varepsilon_{\mathrm{on}} - \varepsilon_{\mathrm{off}} = 5k_{\mathrm{B}}T$, and K_{d} is chosen to correspond to the concentration of Bicoid at $x/L = 0.5$, that is, $K_{\mathrm{d}} = 0.08[\mathrm{Bcd}]_{\max}$. (Experiments find that the Hunchback step occurs at roughly the midpoint between the anterior and the posterior end of the embryo.) The inset shows the six known Bicoid-binding sites responsible for Hunchback expression at this stage of development. (C) Concentration of Hunchback protein along the anterior–posterior axis is obtained by using the Bicoid gradient, Equation 20.18, as input into Equation 20.17.

by any number of Bicoid proteins, up to six. We assume that the dissociation constant K_{d} is the same for all six binding sites, and that there is no interaction between the Bicoid proteins when bound to the DNA. These ideas are summarized in the states-and-weights diagram shown in Figure 20.8. The total weight of the "on" state is computed by summing over all possible arrangements of the $0, 1, 2, \ldots, 6$ Bicoid proteins among the six sites. Since the binding to each site is independent, a simpler way to proceed is to compute the weight for one binding site, which is $1 + [\mathrm{Bcd}]/K_{\mathrm{d}}^{\mathrm{on}}$ and then raise it to a power equal

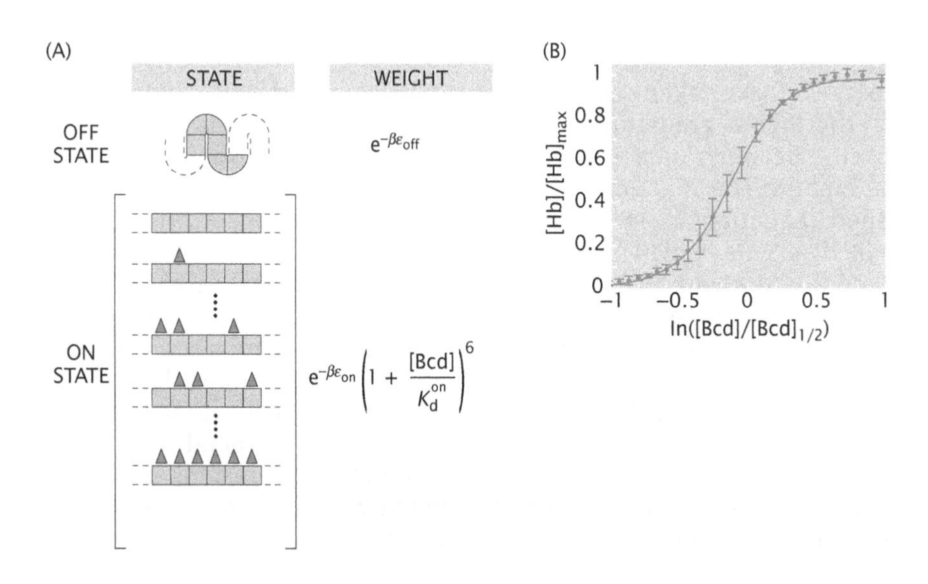

Figure 20.8: Bicoid activation of *hunchback*. (A) States and weights for the MWC model of regulation of *hunchback* expression by Bicoid. The DNA is imagined to have two states, one available for Bicoid binding and the other for which such binding is forbidden. There are six binding sites for Bicoid within the binding region. (B) Mean output level of Hunchback as a function of the input level of Bicoid averaged over nine embryos. (B, adapted from T. Gregor et al., *Cell* 130:153, 2007.)

to the number of binding sites, which in our case is six. We can understand why this works by noting that in the expression for the weight associated with one binding site, the first term corresponds to the "site-unoccupied" state, while the second is the weight of the "site-occupied" state. Then multiplying six such terms yields all possible assignments of "site-unoccupied" and "site-occupied" states to the six Bicoid-binding sites.

From the states-and-weights diagram shown in Figure 20.8, we can directly compute the probability of the *hunchback* gene being in the transcriptionally active state as

$$p_{\text{on}} = \frac{e^{-\beta\varepsilon_{\text{on}}}(1 + [\text{Bcd}]/K_{\text{d}}^{\text{on}})^6}{e^{-\beta\varepsilon_{\text{on}}}(1 + [\text{Bcd}]/K_{\text{d}}^{\text{on}})^6 + e^{-\beta\varepsilon_{\text{off}}}}. \tag{20.17}$$

We can account for the dependence of the Bicoid concentration on the position x along the anterior–posterior axis by exploiting the experimentally determined concentration profile, namely,

$$[\text{Bcd}] = [\text{Bcd}]_{\text{max}} e^{-x/\lambda}, \tag{20.18}$$

where we recall that the decay length is given by $\lambda \approx 100\,\mu\text{m}$ and, as can be seen from Figure 20.6(B), $[\text{Bcd}]_{\text{max}} \approx 60\,\text{nM}$. Despite the smoothly decaying Bicoid profile, the expression profile for Hunchback protein is much sharper, as seen in Figure 20.7.

Estimate: Bicoid Concentration Difference Between Neighboring Nuclei The production of Hunchback proteins in response to a given concentration of Bicoid proteins can be thought of as a "molecular detector" for counting Bicoid proteins within a single nucleus. We can then ask questions about the precision of this detector, in particular, if it is capable of discriminating between neighboring nuclei based on the difference in Bicoid concentration between them. To get a feeling for the numbers, we make a simple estimate using the concentration profile in Equation 20.18. The difference in Bicoid concentration between neighboring nuclei is given by $\Delta[\text{Bcd}] = [\text{Bcd}](x) - [\text{Bcd}](x + \Delta x)$. We can compute this explicitly as

$$\frac{\Delta[\text{Bcd}]}{[\text{Bcd}]} = 1 - e^{-\Delta x/\lambda} \approx \frac{\Delta x}{\lambda} \approx 0.1, \tag{20.19}$$

where we have taken into account the measured decay length for the Bicoid gradient, namely, $\lambda \approx 100\,\mu\text{m}$ and assumed a spacing between nuclei of roughly $10\,\mu\text{m}$. At the midpoint of the embryo ($x \approx 250\,\mu\text{m}$), where the Hunchback step is established, Equation 20.18 tells us that the Bicoid concentration is $[\text{Bcd}] \approx 5\,\text{nM}$, which, using a nuclear diameter of $6\,\mu\text{m}$, corresponds to about 500 molecules per nucleus. Here, we have used the rule of thumb that 1 nM corresponds to roughly one molecule per μm^3. As a result, the difference in number of Bicoid molecules between two neighboring nuclei is about 10% of 500, or 50 molecules. This leads to the conclusion that in this crude estimate, in order to discriminate between neighboring nuclei in the developing embryo, the molecular machinery responsible for making Hunchback must be able to count Bicoid molecules with enough precision to distinguish 500 from 450.

ESTIMATE

Experiments using fluorescently labeled Bicoid and Hunchback proteins, shown in Figure 20.6(C), reveal the precision with which Bicoid proteins are counted in the process of Hunchback expression. Namely, measurements of the nucleus-to-nucleus variability of Hunchback at the midpoint of the embryo, where the Hunchback concentration rises in a step-like fashion, were found to translate into a 10% uncertainty in the Bicoid concentration. This uncertainty in counting Bicoid proteins is on a par with the concentration difference of Bicoid proteins between neighboring nuclei at the midpoint of the embryo, indicating that the precision of Hunchback expression is close to the limit for being able to discriminate between them.

The establishment of the Hunchback profile in the developing fruit fly embryo with the observed amount of variability puts constraints on the *timing* of the processes responsible for the expression of the *hunchback* gene. The relevant time scale here is the developmental time, namely the mean doubling time for the nuclei in the early embryo, which is roughly 10 minutes. Therefore, we can ask a sharp question: is 10 minutes long enough for the Hunchback profile to be established with the measured precision?

To address the question of precision and timing posed above, we must first establish a framework within which we can make quantitative estimates about the precision with which a nucleus can count Bicoid molecules. To this end, we adopt the ideas first proposed by Berg and Purcell in the context of bacterial chemotaxis, where the question of interest was the detection of concentration gradients of chemoattractants by surface receptors on an *E.coli* cell. There the issue was how small a gradient a single cell can detect. Here, we consider a single Bicoid-binding site in the control element of the DNA that regulates *hunchback* expression as the detector. For this detector to successfully detect a 10% difference in Bicoid concentration between neighboring nuclei, its detection error must be less than 10%.

The mechanism by which the Bicoid detector detects is through binding. In particular, over a measurement time T (which we eventually wish to compare with the developmental time scale), the detector will switch back and forth between the state in which Bicoid is bound and that in which it is not bound, as illustrated in Figure 20.9. A trajectory, such as the one shown in Figure 20.9, provides a measurement of the probability that Bicoid is bound, given by the fraction of time that the detector is in the bound state,

$$p_{\text{bound}} = \frac{t_{\text{bound}}^{(1)} + t_{\text{bound}}^{(2)} + \cdots + t_{\text{bound}}^{(N)}}{T}. \tag{20.20}$$

The set of assumptions that led to Equation 20.17 assumed that activation by Bicoid occurs when any binding site of the control region or combination thereof is occupied. Here, we make a stronger simplifying assumption, which is that the regulatory function of Bicoid can be already realized when a single Bicoid molecule is bound to a single site.

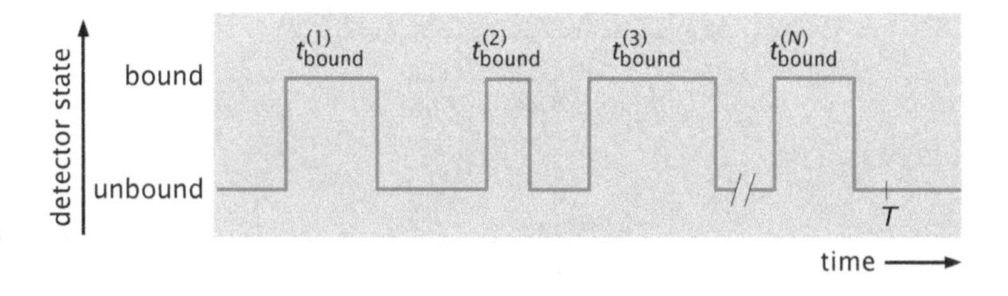

Figure 20.9: Trajectory of the Bicoid detector over a period of time T. The probability that a Bicoid is bound is estimated as the fraction of the time that the detector spends in the bound (on) state. The ith bound interval is characterized by the time $t_{\text{bound}}^{(i)}$.

We expect that as the measurement time T increases, the error in determining the probability that Bicoid is bound will decrease. Indeed, given that each period of time in the bound state $t^{(i)}_{\text{bound}}$ is an independent random variable with an identical distribution, the sum is also a random variable whose mean and variance are N times greater than the mean ($\bar{t}_{\text{bound}}$) and variance ($\Delta t^2_{\text{bound}}$) of each individual $t^{(i)}_{\text{bound}}$. This is a result of the celebrated law of large numbers, which we considered in Section 13.2.1 (p. 518). Therefore, the relative error made by the detector when measuring the probability of Bicoid bound is

$$\frac{\Delta p_{\text{bound}}}{\bar{p}_{\text{bound}}} = \frac{1}{\sqrt{N}} \times \frac{\Delta t_{\text{bound}}}{\bar{t}_{\text{bound}}}. \tag{20.21}$$

Assuming a simple dynamics of switching between the bound and unbound state of the detector that is completely analogous to the dynamics of the ion channel discussed in Section 15.2.5 (p. 589), we can compute the mean and the standard deviation of t_{bound} from the rate of switching from the bound to the unbound state (k_{off}). In particular, given that the distribution of bound times is exponential, as follows from assuming that transitions are described by a single rate constant, we have

$$P(t_{\text{bound}}) = k_{\text{off}} e^{-k_{\text{off}} t_{\text{bound}}}, \tag{20.22}$$

and we can compute the average values

$$\begin{aligned} \bar{t}_{\text{bound}} &= \int_0^\infty t_{\text{bound}} P(t_{\text{bound}})\, \mathrm{d}t_{\text{bound}} = \frac{1}{k_{\text{off}}}, \\ \overline{t^2}_{\text{bound}} &= \int_0^\infty t^2_{\text{bound}} P(t_{\text{bound}})\, \mathrm{d}t_{\text{bound}} = \frac{2}{k^2_{\text{off}}}, \end{aligned} \tag{20.23}$$

from which we can determine the variance as

$$\Delta t^2_{\text{bound}} = \overline{t^2}_{\text{bound}} - (\bar{t}_{\text{bound}})^2 = \frac{1}{k^2_{\text{off}}}. \tag{20.24}$$

Substituting the calculated mean and standard deviation of $\bar{t}_{\text{bound}}$ into Equation 20.21, we arrive at a simple result for the relative error made in estimating the probability that Bicoid is bound from a single trajectory taken by the detector,

$$\frac{\Delta p_{\text{bound}}}{\bar{p}_{\text{bound}}} = \frac{1}{\sqrt{N}}. \tag{20.25}$$

Here N is the number of observed transitions between the bound and the unbound state over time T, as illustrated schematically in Figure 20.9.

To estimate N, we take the mean residence times in the bound state ($1/k_{\text{off}}$) and the unbound state ($1/k_{\text{on}}$) to be roughly equal (this corresponds to a binding probability of Bicoid of roughly a half, as is the case at the midpoint of the embryo where the Hunchback step is observed), and hence

$$N \approx \frac{T}{\dfrac{1}{k_{\text{on}}} + \dfrac{1}{k_{\text{off}}}} \approx \frac{T}{\dfrac{2}{k_{\text{on}}}} = \frac{T k_{\text{on}}}{2}, \tag{20.26}$$

which gives

$$\frac{\Delta p_{\text{bound}}}{\bar{p}_{\text{bound}}} \approx \sqrt{\frac{2}{T k_{\text{on}}}}. \tag{20.27}$$

To numerically estimate the error that the detector makes in estimating the probability of Bicoid being bound, we take $T = 10$ minutes and we estimate the "on" rate by assuming it is diffusion-limited (see Section 13.3.1, p. 532). The diffusion-limited "on" rate is given by $k_{on} = 4\pi Da[\text{Bcd}]$, where $a = 3$ nm (10 bp of DNA) is roughly the linear size of the Bicoid-binding site (which we treat as a perfectly absorbing sphere of radius a), $D = 5\ \mu\text{m}^2/\text{s}$ is the measured diffusion constant of Bicoid in the embryo, and $[\text{Bcd}] = 8$ nM is the measured concentration of Bicoid at the position where the Hunchback step is established (roughly at the midpoint of the anterior–posterior axis). This leads to an estimate of the "on" rate $k_{on} = 0.9\ \text{s}^{-1}$, and we arrive at the final estimate for the relative error in the measurement of p_{bound},

$$\frac{\Delta p_{\text{bound}}}{\overline{p}_{\text{bound}}} \approx 0.06. \tag{20.28}$$

Note that this estimate is really a lower bound, as the rate of switching between the bound and the unbound state can easily be an order of magnitude smaller than the estimate based on the speed limit set by diffusion.

This error in p_{bound} propagates into an error in determining the concentration of Bicoid in the nucleus. The simple binding equation

$$p_{\text{bound}} = \frac{\dfrac{[\text{Bcd}]}{K_d}}{1 + \dfrac{[\text{Bcd}]}{K_d}} \tag{20.29}$$

relates the two. We can invert it to get the concentration of Bicoid as a function of p_{bound} as

$$\frac{[\text{Bcd}]}{K_d} = \frac{p_{\text{bound}}}{1 - p_{\text{bound}}}, \tag{20.30}$$

from which the relation between the two relative errors follows as

$$\frac{\Delta[\text{Bcd}]}{[\text{Bcd}]} = \frac{\Delta p_{\text{bound}}}{\overline{p}_{\text{bound}}} + \frac{\Delta p_{\text{bound}}}{1 - \overline{p}_{\text{bound}}}. \tag{20.31}$$

Using the result in Equation 20.28 and $\overline{p}_{\text{bound}} \approx 1/2$, we arrive at an estimate

$$\frac{\Delta[\text{Bcd}]}{[\text{Bcd}]} \approx 0.12, \tag{20.32}$$

which is similar to the observed difference in the Bicoid concentration between two neighboring nuclei (10%). Given that our estimate is at best a lower bound on the Bicoid counting error, and that it does not account for other sources of noise such as Hunchback transcription and translation, it raises the need for additional mechanisms responsible for establishing the Hunchback pattern with the observed precision. One idea that has been proposed, and for which some experimental evidence has been found in the form of correlations between Hunchback fluctuations between nuclei in close spatial proximity, is that the observed Hunchback concentration profile is established by spatially averaging over neighboring nuclei, which further reduces the variability and allows for shorter time averaging of Bicoid binding.

In considering the patterning of the early *Drosophila* embryo in the context of the morphogen gradient model, then, we have found both broad agreement with the predictions of the simple model and specific paradoxical conflicts between the quantitative data now available and the quantitative predictions of the model. As usual, the revelation of these paradoxes underlines the extreme importance of quantitative measurements. For example, in the case of Bicoid gradient formation, the shape of the gradient in *Drosophila* is roughly exponential, and the characteristic length scale is consistent with the molecular diffusion coefficient and degradation rate as measured in that system, but our apparent understanding of this phenomenon breaks down dramatically as we consider embryos that are larger or smaller, where the same physical processes appear to give rise to different length scales. Similarly, although it is clear that Hunchback expression depends on the ability of the responding cells to measure the local concentration of Bicoid in the gradient, the precision and speed with which the cells are able to perform this measurement defies the simple statistical mechanical approach. If we had only qualitative data in this system, or only a qualitative plausible model, we would think the problems understood. The more exacting quantitative approach usefully reveals the precise nature and dimensions of our ignorance, and can therefore effectively guide future research effort to resolve these paradoxes.

20.2.4 Morphogen Patterning with Growth in *Anabaena*

In the simplest cases of patterning by morphogen gradients, such as the Bicoid example described above, the diffusing morphogen influences the fate of cells or nuclei in a fixed spatial field. Much more commonly, though, the process of cell fate determination and differentiation takes place at the same time as growth, which introduces interesting complications into the problem of patterning.

Conceptually, one of the simplest cases of growth-coupled differentiation comes from the cyanobacterium *Anabaena*. This photosynthetic organism typically grows in a filamentous chain where cells remain physically connected to their siblings and cousins. In addition to being able to fix carbon, *Anabaena* is also capable of fixing nitrogen. However, the complex specializations required of a nitrogen-fixing cell are incompatible with further division. For an *Anabaena* filament, nitrogen fixation is carried out altruistically by occasional cells known as heterocysts that are found in fairly regular intervals along the cell chain as shown in Figures 20.10(A–C). The chemical products of nitrogen fixation can travel from the heterocysts to growing neighbors along the chain and, in return, the heterocysts are provided with fixed carbon from their vegetative partner cells. As cell proliferation proceeds, eventually some of the newborn cells will find themselves too far from the heterocysts to get enough fixed nitrogen. At this point, one of those cells must differentiate into a heterocyst for the good of the community.

This represents a simple distance measurement; for example, if heterocysts are spaced on average as one out of every 10 cells in the chain, then a new one should differentiate when the distance between two nearest neighbors has grown to 20. The spatial scale is set quantitatively by diffusion. The differentiation of heterocysts is controlled by a transcriptional activator, HetR, which activates its own transcription as well as that of genes involved in nitrogen fixation. In addition, HetR activates the transcription of inhibitory proteins that are processed to generate secreted regulatory peptides, which are able to diffuse away. These secreted peptides trigger the proteolytic degradation of HetR in nearby cells. Consequently, a gradient of HetR concentration can be observed, such that it is lowest in the

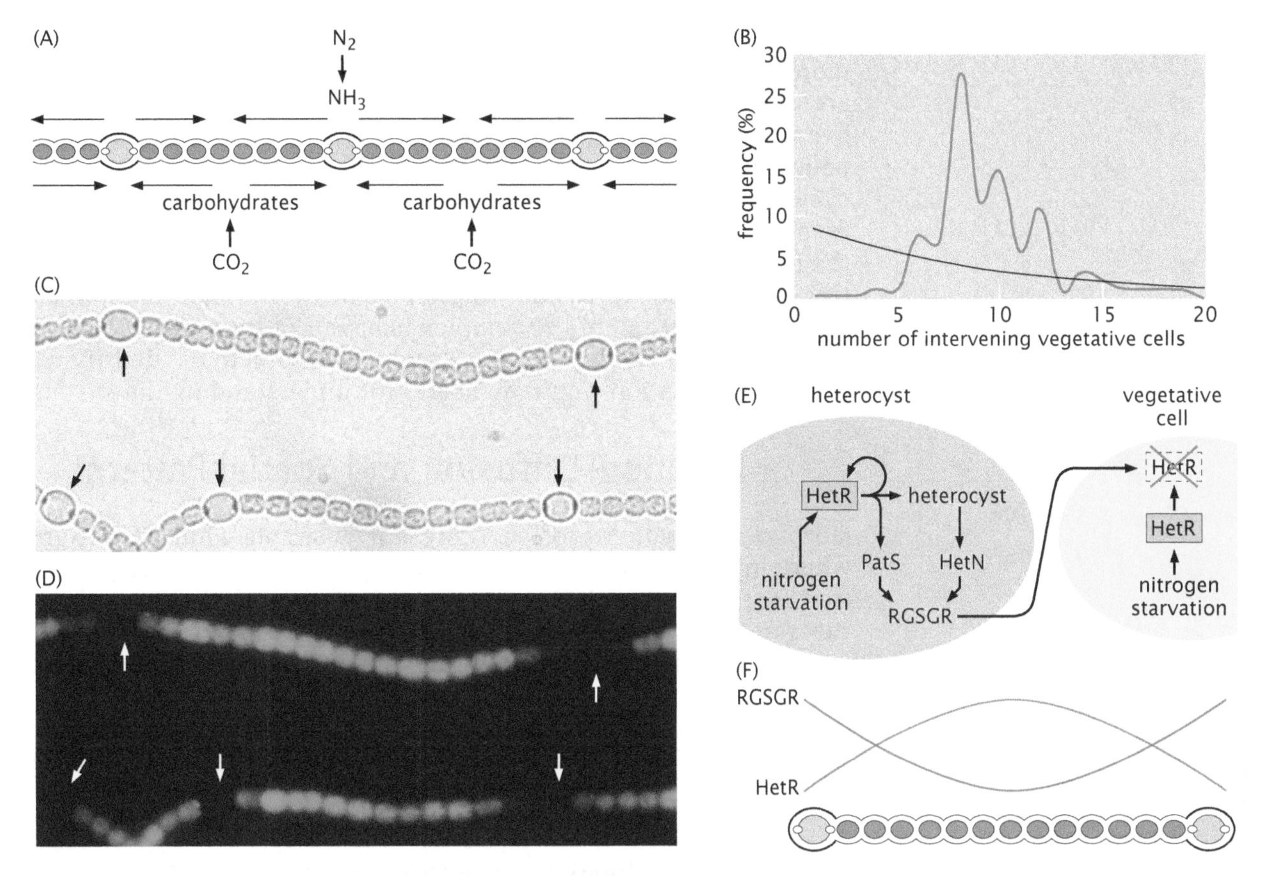

Figure 20.10: Structure and function of *Anabaena*. (A) Schematic showing the semi-regular distribution of the heterocysts that convert dinitrogen into ammonia. (B) Distribution of spacing between heterocysts. The black line shows the Poisson distribution with mean spacing 8, revealing the distribution for a case in which the spacing is determined by a random process. (C) Microscopy images showing both vegetative cells and heterocysts (labeled with arrows). (D) Same cells as shown in (C) but with GFP labeling of HetR showing the one-dimensional gradient. (E) HetR is expressed in *Anabaena* under conditions of nitrogen starvation. HetR is a transcription factor that stimulates its own transcription as well as that of genes necessary for differentiation into a heterocyst. Two of the targets of HetR are PatS and HetN. Both of these proteins can be proteolytically processed to release a pentapeptide fragment with the sequence RGSGR. The peptide is cell-permeable, so it can diffuse into neighboring cells. In the neighboring cells, RGSGR peptide promotes degradation of HetR, and also prevents it from upregulating its own expression. Therefore, in the vicinity of a differentiating heterocyst, there is a gradient of RGSGR peptide where the concentration is highest near the heterocyst and an inverse gradient of HetR in the non-heterocyst cells; this is shown in (F). (F) Schematic of the concentrations of key regulatory proteins. (B adapted from J. C. Meeks and J. Elhai, *Microbiol. Mol. Biol. Rev.* 66:94, 2002; C–F, adapted from D. D. Risser and S. M. Callahan, *Proc. Natl Acad. Sci. USA* 106:19884, 2009.)

cells immediately adjacent to the committed heterocyst, and highest in the cells furthest away. When the concentration of HetR exceeds some threshold, a new heterocyst will start to differentiate.

This simple model based on the idea that differentiated heterocysts secrete an inhibitor that prevents the differentiation of their neighbors is conceptually satisfying and does manage to reproduce some features observed in the biological system, such as the fairly regular spacing of the differentiated cells and the coupling of differentiation to chain growth. However, unsurprisingly, some features of the real biological system are not predicted. For example, if the secretion of the peptide inihibitors were the only factor suppressing differentiation of vegetative cells in a chain, then deletion of the genes encoding their precursors should result in a completely random spacing of heterocysts. In fact, such mutants still show nonrandom spacing, although the intervals are rather shorter than for wild-type cells. The spacing of the heterocysts is also affected by cellular metabolism, in ways that are not obviously directly connected to HetR expression. Furthermore, careful observation also suggests that not all cells in a chain are equally likely to become heterocysts, even when they have all been moved from nitrogen-rich to nitrogen-poor conditions simultaneously, perhaps because only cells at certain critical points in the cell cycle are able to initiate the differentiation program. Just as with the French flag model described above as a first attempt at explaining how the Bicoid gradient can be established in *Drosophila* embryos and control Hunchback expression, the failure of the simple diffusion-based model to predict all of the complex behavior of the real biological system serves as a useful focusing mechanism to help us identify where further information would be useful for understanding the system.

20.3 Reaction–Diffusion and Spatial Patterns

Morphogen gradients often represent a simple kind of patterning where information is external to the system that is being patterned, that is, the system that sets up the formation of the gradient is not the same as the cells that respond. Often, though, biological patterns can arise spontaneously and intrinsically, with the group of cells generating the spatial information being the same as the responders. How can well-organized, regular patterns arise spontaneously within an initially uniform field, when there is no external source of information?

20.3.1 Putting Chemistry and Diffusion Together: Turing Patterns

One of the first theoretical attempts at answering this question was provided in the 1950s by the famed British mathematician and computer scientist Alan Turing, who described a mechanism of pattern formation that relied only on chemical reactions between diffusing molecules. Turing also used the word "morphogen" to refer to the molecules in his model, because they are able to generate patterns and shapes. Turing's proposal has in the intervening years served as an inspiration for generations of models and experiments centering on pattern formation in biology. Interestingly, while the details of the pattern-forming systems in development, once worked out, have so far revealed mechanisms different from that imagined by Turing, this is a powerful example of a "wrong" model that has played an important role in guiding experiments. Even though the model is often dismissed as "wrong," its basic premise, that patterns of gene expression could be driven by diffusion and reactions between molecules (often the diffusing species are transcription factors that bind to DNA), has been proven correct over and over again.

To obtain intuition about the Turing mechanism of pattern formation, we make use of a toy model suggested by Turing himself in his seminal paper. The model has two cells, each of which can be thought of as its own chemical reactor with two molecular species, or "morphogens," X and Y. The chemistry performed by these morphogens is described by the rate equations

$$\begin{aligned} \frac{\mathrm{d}X}{\mathrm{d}t} &= 5X - 6Y + 1, \\ \frac{\mathrm{d}Y}{\mathrm{d}t} &= 6X - 7Y + 1, \end{aligned} \tag{20.33}$$

where all the integers appearing in the equations are reaction rates in units of inverse time. The important feature of these equations, which do not represent any particular reaction and are introduced simply to illustrate the Turing instability, is that the production of both morphogens is stimulated by X, while their decay is proportional to the number of Y molecules. In this sense, we refer to the morphogen X as the activator, while Y is the inhibitor.

As can be demonstrated instantly by substitution, the system of rate equations has a unique steady state, given by $X = Y = 1$. The stability of this state can be assessed in exactly the way we did in Chapter 19 for the genetic switch, by considering a small deviation away from the steady state and asking whether this perturbation grows or diminishes with time. Mathematically, this corresponds to introducing variables x and y that measure the deviation from the steady state, namely, $X = 1 + x$ and $Y = 1 + y$. The time evolution of x and y can be written in matrix form as

$$\frac{d}{dt}\begin{pmatrix} x \\ y \end{pmatrix} = \begin{pmatrix} 5 & -6 \\ 6 & -7 \end{pmatrix}\begin{pmatrix} x \\ y \end{pmatrix}. \qquad (20.34)$$

The question of stability is explored by imagining a time-dependent solution of the form $x(t) = x_0 e^{\lambda t}$ and $y(t) = y_0 e^{\lambda t}$ and examining whether these expressions grow or decay in time. More precisely, if we substitute this trial solution into Equation 20.34, we find that the parameter λ is determined as an eigenvalue of the matrix appearing on the right-hand side. In order for the initial perturbation away from the steady state to decay back to zero, the eigenvalues of the matrix have to both be negative. The eigenvalues are obtained from the quadratic equation

$$\det\begin{pmatrix} 5-\lambda & -6 \\ 6 & -7-\lambda \end{pmatrix} = \lambda^2 + 2\lambda + 1 = 0, \qquad (20.35)$$

from which we conclude that the two eigenvalues are degenerate and both equal to -1. This guarantees the stability of the steady state. In other words, if we start the system in any state, defined by the abundance of X and Y molecules, it will quickly decay into the steady state $X = Y = 1$.

A qualitatively new and seemingly counterintuitive situation develops when we go on to a system consisting of two cells, as shown in Figure 20.11(A), each containing some amounts of morphogens X and Y, subject to the same chemistry as that described by the rate equations given in Equation 20.33. The new twist to the story is that we allow the morphogens to freely diffuse between the two cells. Since diffusion tends to even out differences in concentration and the chemistry in each cell is such that $X = Y = 1$ is a stable steady state, one might expect that the uniform state in which both cells have this amount of morphogen would also be a stable steady state in the presence of diffusion. What Turing discovered was that this is not necessarily so! In particular, if the inhibitor, in this case morphogen Y, diffuses faster than the activator (morphogen X), the homogeneous steady state can become unstable toward a state with different concentrations of morphogens in the two cells. Intriguingly, diffusion can provide a mechanism by which a spatially nonuniform distribution of morphogens can develop. Turing suggested this as a mechanism by which nominally identical cells might suffer very different fates in the development process.

To explore how the Turing instability develops in mathematical detail in the toy model introduced above, we begin by writing down

(A)

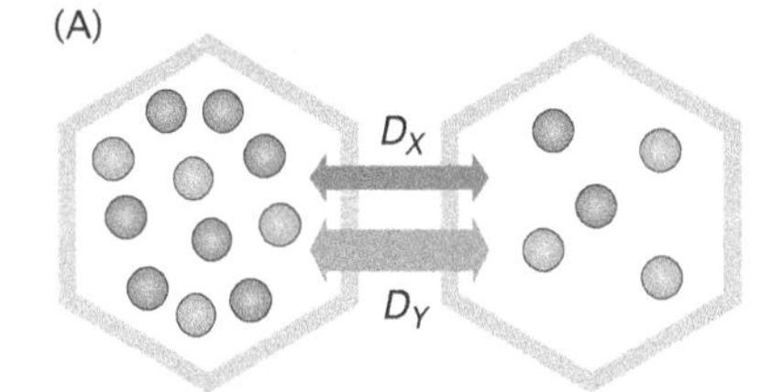

(B)

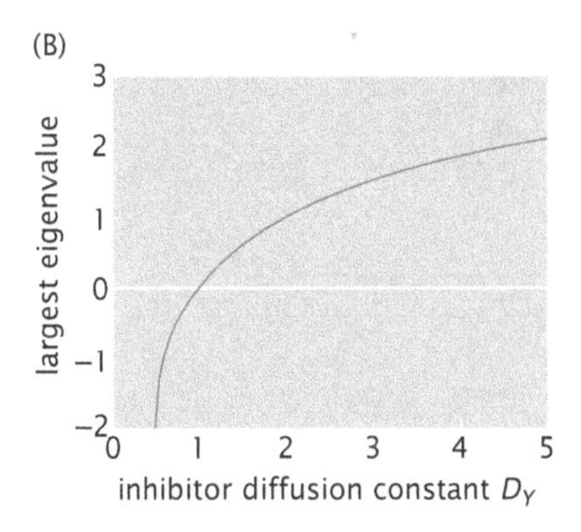

Figure 20.11: A two-cell model of the Turing instability. (A) Two cells, each containing two species of morphogen, are in diffusive contact. The morphogens in each cell participate in a chemical reaction that leads to a stable steady state in the absence of diffusion. (B) Increasing the diffusion constant D_Y of the morphogen that acts as an inhibitor eventually leads to an instability. This is signaled by the largest eigenvalue of the rate equation matrix becoming positive. For this example, $D_X = 0.5$.

the rate equations for the two cells in the presence of diffusion, namely,

$$\frac{dX_1}{dt} = 5X_1 - 6Y_1 + 1 + D_X(X_2 - X_1),$$
$$\frac{dY_1}{dt} = 6X_1 - 7Y_1 + 1 + D_Y(Y_2 - Y_1), \tag{20.36}$$

$$\frac{dX_2}{dt} = 5X_2 - 6Y_2 + 1 + D_X(X_1 - X_2),$$
$$\frac{dY_2}{dt} = 6X_2 - 7Y_2 + 1 + D_Y(Y_1 - Y_2), \tag{20.37}$$

where the state of cell 1 is given by (X_1, Y_1) while the state of cell 2 is specified by (X_2, Y_2). D_X and D_Y are the diffusion constants (with appropriate units for the discrete formulation) for the two morphogens. The chemical parts of the rate equations for both morphogens are the same as in Equations 20.33. However, the presence of diffusion results in an additional term that describes the diffusion current from one cell to the other; the discrete version of Fick's law tells us that this current is proportional to the difference in the morphogen content of the two cells and is in the direction of decreasing morphogen concentration. The previously identified steady state $X_1 = Y_1 = X_2 = Y_2 = 1$ remains so for the two-cell system, but, as we show below, its *stability* depends upon the values of the diffusion constants D_X and D_Y. To illustrate this, we compute the four eigenvalues of the rate equation matrix that governs the time evolution of the perturbation (x_1, y_1, x_2, y_2) from the steady state. Following precisely the same strategy as used for the case in the absence of diffusion, we find that the linearized rate equations are given by

$$\frac{d}{dt}\begin{pmatrix} x_1 \\ y_1 \\ x_2 \\ y_2 \end{pmatrix} = \begin{pmatrix} 5 - D_X & -6 & D_X & 0 \\ 6 & -7 - D_Y & 0 & D_Y \\ D_X & 0 & 5 - D_X & -6 \\ 0 & D_Y & 6 & -7 - D_Y \end{pmatrix}\begin{pmatrix} x_1 \\ y_1 \\ x_2 \\ y_2 \end{pmatrix}. \tag{20.38}$$

If we follow Turing in setting $D_X = 0.5$ and then vary the diffusion constant of the inhibitor between 0.5 and 5, we see that an instability arises. In Figure 20.11(B), we plot the largest eigenvalue (or, rather, real part) as a function of D_Y. A positive eigenvalue indicates an instability and it is clear from the figure that an instability emerges for $D_Y > 1$. For example, for $D_Y = 4.5$, the largest eigenvalue is 2 and the perturbation $(x_1, y_1, x_2, y_2) = (3\xi, \xi, -3\xi, -\xi)$ is unstable in the sense that ξ grows exponentially with time according to the formula $\xi(t) = \xi_0 e^{2t}$. This means that the amounts of morphogens X and Y in cell 1 grow over time, while in cell 2 they both diminish. If we think of the two morphogens as being transcription factors, we see how this instability can drive a different genetic program to be executed in the two cells. If we allow the dynamics to unfold according to the linearized dynamics worked out above, in the long-time limit we find the unphysical outcome that the amount of morphogen in cell 2 actually becomes negative. However, this unphysical situation is typically remedied by nonlinearities in the underlying rate equations. Recall that in this case the linear equations in Equation 20.38 arise by linearizing the rate equations in the vicinity of a steady state.

The Turing model with two cells described above showed that an instability leading to a different concentration of morphogens in the two cells can develop as a result of fast diffusion of the inhibitor.

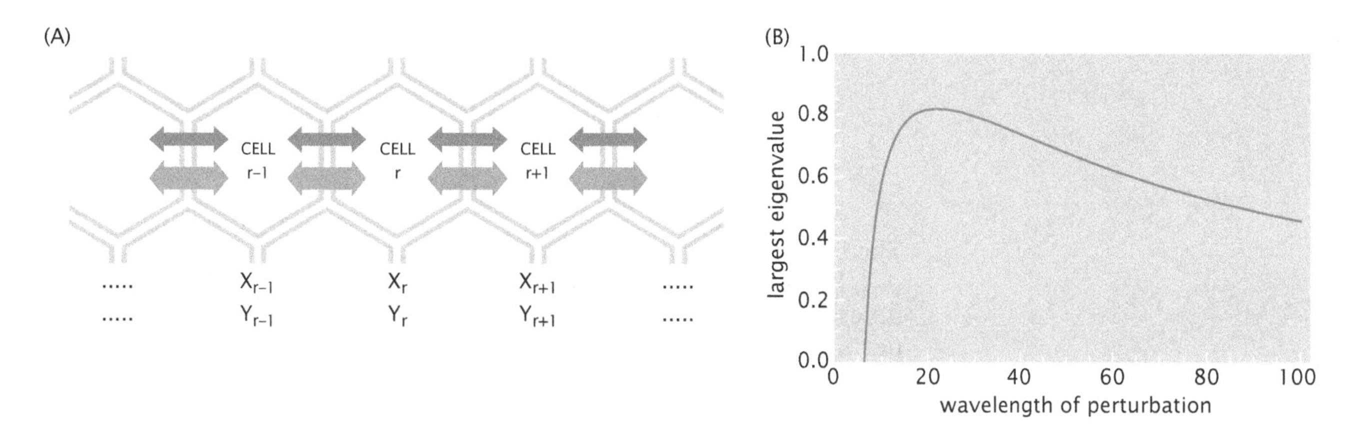

Figure 20.12: Turing waves. (A) One-dimensional lattice of cells each containing two reacting morphogen species, activators and inhibitors, which diffuse between cells. (B) The largest eigenvalue of the rate equation matrix depends on the wavelength of the perturbation from the homogeneous state. The perturbation with the largest eigenvalue is the dominant instability and leads to a periodic pattern of morphogen concentrations.

Similarly, and once again following Turing, we can study the emergence of spatially periodic patterns of morphogens by considering a one-dimensional lattice of cells, as shown in Figure 20.12. The morphogens engage in chemical reactions in each cell leading to a homogeneous (equal in all cells) steady state that is stable. As in the two-cell example, fast diffusion of the morphogen acting as an inhibitor can destabilize the steady state toward a spatially periodic pattern of morphogen concentrations.

We now generalize the highly simplified chemical reactions described earlier to allow for nonlinear interactions between the different species. In this case, the chemistry taking place in each of the cells is described by the rate equations

$$\begin{aligned} \frac{\mathrm{d}X_r}{\mathrm{d}t} &= f(X_r, Y_r), \\ \frac{\mathrm{d}Y_r}{\mathrm{d}t} &= g(X_r, Y_r) \end{aligned} \tag{20.39}$$

where X_r and Y_r are the numbers of activator and inhibitor morphogens in cell r. The functions f and g describe the reactions in each cell, and are in principle nonlinear functions of their arguments. Next, we assume that $(X_r, Y_r) = (h, k)$ is a stable steady state of the rate equations, just like $(1, 1)$ was a steady state for both cells in the two-cell example. We analyze the stability of the $(X_r, Y_r) = (h, k)$ state when diffusion between the cells is turned on. In this case, the rate equations become

$$\begin{aligned} \frac{\mathrm{d}X_r}{\mathrm{d}t} &= f(X_r, Y_r) + D_X(X_{r+1} + X_{r-1} - 2X_r), \\ \frac{\mathrm{d}Y_r}{\mathrm{d}t} &= g(X_r, Y_r) + D_Y(Y_{r+1} + Y_{r-1} - 2Y_r), \end{aligned} \tag{20.40}$$

where the diffusion term in each equation describes the sum of diffusive currents into cell r coming from the neighboring cells $r-1$ and $r+1$.

To assess the stability of the homogeneous steady state $(X_r, Y_r) = (h, k)$ we perform a linear stability analysis of Equations 20.40. We introduce the quantities (x_r, y_r) that describe the excursion of the morphogens away from the steady state, that is, $(X_r, Y_r) = (h + x_r, k + y_r)$, and assume the perturbation to be small. Expanding the reaction–diffusion equations, Equations 20.40, to linear order in x_r and y_r leads to a set of linear differential equations

$$\frac{dx_r}{dt} = A_1 x_r + B_1 y_r + D_X(x_{r+1} + x_{r-1} - 2x_r),$$
$$\frac{dy_r}{dt} = A_2 x_r + B_2 y_r + D_Y(y_{r+1} + y_{r-1} - 2y_r). \tag{20.41}$$

The coefficients A_1, B_1, A_2, and B_2 are partial derivatives of $f(X, Y)$ and $g(X, Y)$ with respect to X and Y evaluated at $(X, Y) = (h, k)$. In order to investigate the stability of the steady state with respect to a periodic perturbation, we consider a trial solution of Equations 20.41 in the form

$$x_r(t) = x(t)\,e^{i(2\pi r/\lambda)},$$
$$y_r(t) = y(t)\,e^{i(2\pi r/\lambda)}, \tag{20.42}$$

where λ is the wavelength of the perturbation in units of cell length. The physical perturbation is the real part of the expressions appearing on the right-hand sides of the above equations; the linearity of these equations makes this trick possible.

Substituting Equations 20.42 into Equations 20.41, we arrive at equations for the wave amplitudes $x(t)$ and $y(t)$ of the form

$$\frac{dx}{dt} = \left[A_1 + D_X\left(e^{i(2\pi/\lambda)} + e^{-i(2\pi/\lambda)} - 2\right)\right]x + B_1 y,$$
$$\frac{dy}{dt} = A_2 x + \left[B_2 + D_Y\left(e^{i(2\pi/\lambda)} + e^{-i(2\pi/\lambda)} - 2\right)\right]y. \tag{20.43}$$

Assuming that the wavelength $\lambda \gg 1$, we can use the Taylor expansion $e^x = 1 + x + x^2/2$ to simplify the above equations to

$$\frac{dx}{dt} = \left[A_1 - D_X\left(\frac{2\pi}{\lambda}\right)^2\right]x + B_1 y,$$
$$\frac{dy}{dt} = A_2 x + \left[B_2 - D_Y\left(\frac{2\pi}{\lambda}\right)^2\right]y. \tag{20.44}$$

As for the two-cell model, in order to determine the fate of the perturbation, we need to determine the eigenvalues of the rate matrix

$$\mathcal{R} = \begin{pmatrix} A_1 - D_X(2\pi/\lambda)^2 & B_1 \\ A_2 & B_2 - D_Y(2\pi/\lambda)^2 \end{pmatrix}, \tag{20.45}$$

which in this case will depend on the wavelength λ. The largest eigenvalue (or, rather, its real part) is plotted in Figure 20.12 as a function of λ for a particular choice of parameters, $A_1 = 1$, $B_2 = -1$, $A_2 B_1 = -1$, $D_X = 1$, and $D_Y = 100$, which were chosen so that morphogen Y is an inhibitor and it diffuses much faster than morphogen X. It is clear from the plot of the largest eigenvalue that there is a preferred wavelength $\lambda^* \approx 4$ for which the eigenvalue of $\mathcal{R}$ has a maximum, and is positive. This will be the dominant instability in the steady state. Namely, due to random fluctuations, the initial perturbation away from the steady state will be composed of periodic waves with all possible wavelengths, and the amplitude of the wave with wavelength λ^* will grow the fastest and dominate at long times. This is the mechanism by which a periodic spatial pattern of morphogens can arise spontaneously from a homogeneous steady state. An interesting prediction of this model is that the wavelength of the wave scales as the square root of the diffusion constant D_X (where $D_Y = 100D_X$). (Note that this is a simple consequence of dimensional analysis, since the diffusion constant has units of length^2/time, and the wavelength is a function of it and a bunch of rate constants that have dimensions of inverse time.)

The analytical approach that we have undertaken here is sufficient to predict the emergence of a standing wave of concentration in a one-dimensional lattice (shown in Figure 20.12). Expanding this concept to two dimensions becomes sufficiently complicated that most people (including Turing himself) prefer to approach it numerically. In two dimensions, the standing waves established by the Turing instability can be resolved as patterns of spots or stripes, depending on the exact parameters for transport and reaction. For various parameter choices, the spots can range from leopard-like to cow-like patterns, and the stripes from skunk-like to tiger-like. Figure 20.13 shows the dramatic variety of pigment patterns that can be generated by tweaking parameters in a simple reaction–diffusion simulation, paired with photographs of the coats of members of the cat family that are most closely matched by the simulated patterns. Mechanistically, however, it is unlikely that animal skin patterns actually arise from the mechanism that Turing envisioned. In general, the sizes of spots and stripes in animal skin are much too large to be consistent with molecular diffusion. Furthermore, as the developmental biology of animal patterning has been explored, it has become clear that the color patterns are due to migration and proliferation of dedicated pigment-carrying cells called melanophores that arise from specific cellular precursors and are not subject to a chemical "reaction" that can change their colors. Because these Turing-like patterns are most likely not generated by reaction–diffusion mechanisms, the Turing model is widely considered to be "wrong." But, as we will see below, the underlying concept is tremendously useful, and slight modifications of the idea do clearly operate in biological pattern formation. Furthermore, the elegant mathematics developed by Turing may well still apply to animal skin patterns, even if the underlying physical processes are distinct. For example, random cell migration may effectively take the place of diffusion.

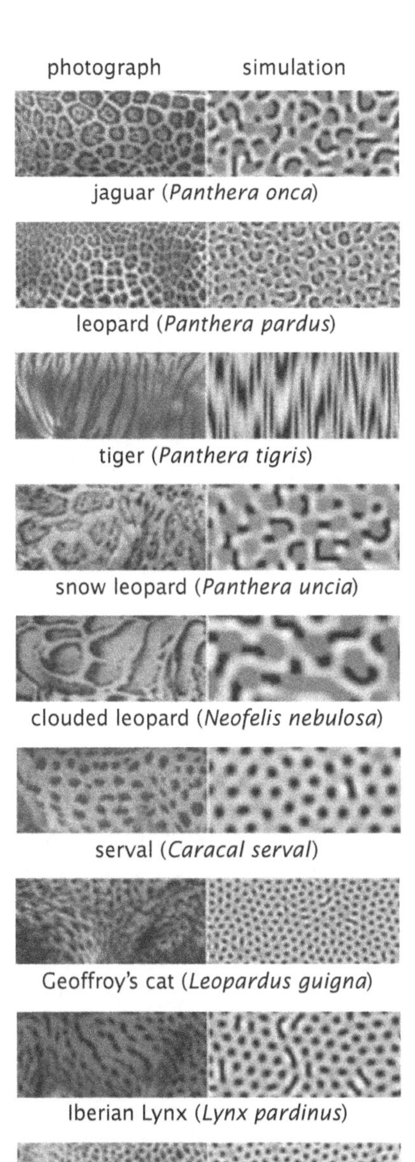

Figure 20.13: Coat patterns in a variety of felid species and their approximations by reaction–diffusion simulations. For all pairs, the simulated pattern is shown on the right and a size-matched photograph of animal skin is shown on the left. (Adapted from W. L. Allen et al., *Proc. R. Soc. Lond. B Biol. Sci.* 278:1373, 2011.)

The addition of a third morphogen to a Turing-like dynamical system allows the emergence of a new property: some solutions will generate traveling waves. The conditions necessary for an oscillatory solution with traveling waves are discussed in detail in Turing's original *tour-de-force* paper. In chemistry, an analogous system is the famous Belousov–Zhabotinsky reaction, which has delighted and amused generations of chemistry students. As we will explore in the next section, traveling waves of this kind have been found recently in biological systems.

In biology, the enduring reverberations of Turing's insight have included the appreciation that complex patterns can arise by self-organized processes from initially homogeneous conditions, that diffusion of molecular species can increase (rather than always decrease) local inhomogeneities, and that relatively small quantitative changes in parameters like reaction rate constants can yield impressive large-scale changes in the final resulting pattern. This last point is particularly important in conceptualizing the evolutionary basis of biological pattern change, as, for example, rate constants for particular molecular reactions can be easily altered by a few amino acid substitutions, and families of organisms including cats, fish, and seashell-making mollusks repeatedly demonstrate that evolution can fashion dramatic pattern differences among very closely related species. An open question in biology remains: how many apparently Turing-like patterns are actually generated by reaction–diffusion mechanisms? Throughout the remainder of this section, we will examine particular well-characterized cases where emergence of patterns from a homogeneous background depends on transport and reactions of known molecular species, and discuss how the actual details of these cases expand and reinforce the central ideas of Turing's legacy.

20.3.2 How Bacteria Lay Down a Coordinate System

Are there any real biological systems that show the features of Turing instabilities? The spots and stripes predicted by Turing look like the coat patterns of cheetahs and zebras, but the distance and time scales are much too large for any diffusion-based mechanism to be the real process establishing these lovely patterns. Recently, a bacterial system has been characterized that does show many Turing-like features at the appropriate molecular scale. The proteins involved are referred to as Min, and their biological role is to enable bacterial cells to identify the right place to build a division plane when they form two daughter cells. After chromosome replication is complete, rod-shaped bacterial cells such as *E. coli* typically divide at or very near their middles, so that the two daughters are of approximately equal size, and each inherits a complete genome. Indeed, careful measurements have shown that *E. coli* cells growing under normal conditions are capable of positioning their division plane nearly at the exact center of the cell with an accuracy of about 5%. In the absence of Min proteins, that precision is lost and the mutant cells regularly divide at inappropriate locations. If the division plane forms toward the end of the cell, one of the daughters may end up as a tiny "mini-cell" that is devoid of DNA. In fact, the name of the Min proteins is derived from the fact that they were originally identified from mutant cells that show this type of asymmetric division.

Three different Min proteins contribute to this process. MinC is an inhibitor of the assembly of FtsZ, the bacterial tubulin homolog that forms a ring at the center of the cell and recruits all of the other proteins necessary for construction of the septum between the two daughter cells (see Figure 3.7, p. 99). The goal of the Min system is to make sure that the local concentration of MinC remains lowest right at the exact center of the cell. MinC has no ability to localize itself, but it can hitch a ride on another protein, MinD. MinD has multiple interesting functions: it is able to bind and hydrolyze ATP, it can form dimers in solution and larger filaments when associated with membranes, and it can cycle between membrane-associated and soluble cytoplasmic forms. MinD appears to assemble on membranes cooperatively, such that new molecules are more likely to stably associate with membrane at the edge of an existing MinD-coated domain rather than on completely bare membrane. The third protein, MinE, is a critical regulator of MinD localization. MinE binding to ATP–MinD on the membrane leads to ATP hydrolysis and subsequent detachment of ADP–MinD from the membrane. In solution, ADP is released from MinD, which is now free to bind an ATP molecule and once again form dimers and then bind to the membrane. After MinE triggers MinD release from the membrane, MinE subsequently dissociates itself. Recent *in vitro* experiments using the Min proteins from *E. coli* have demonstrated

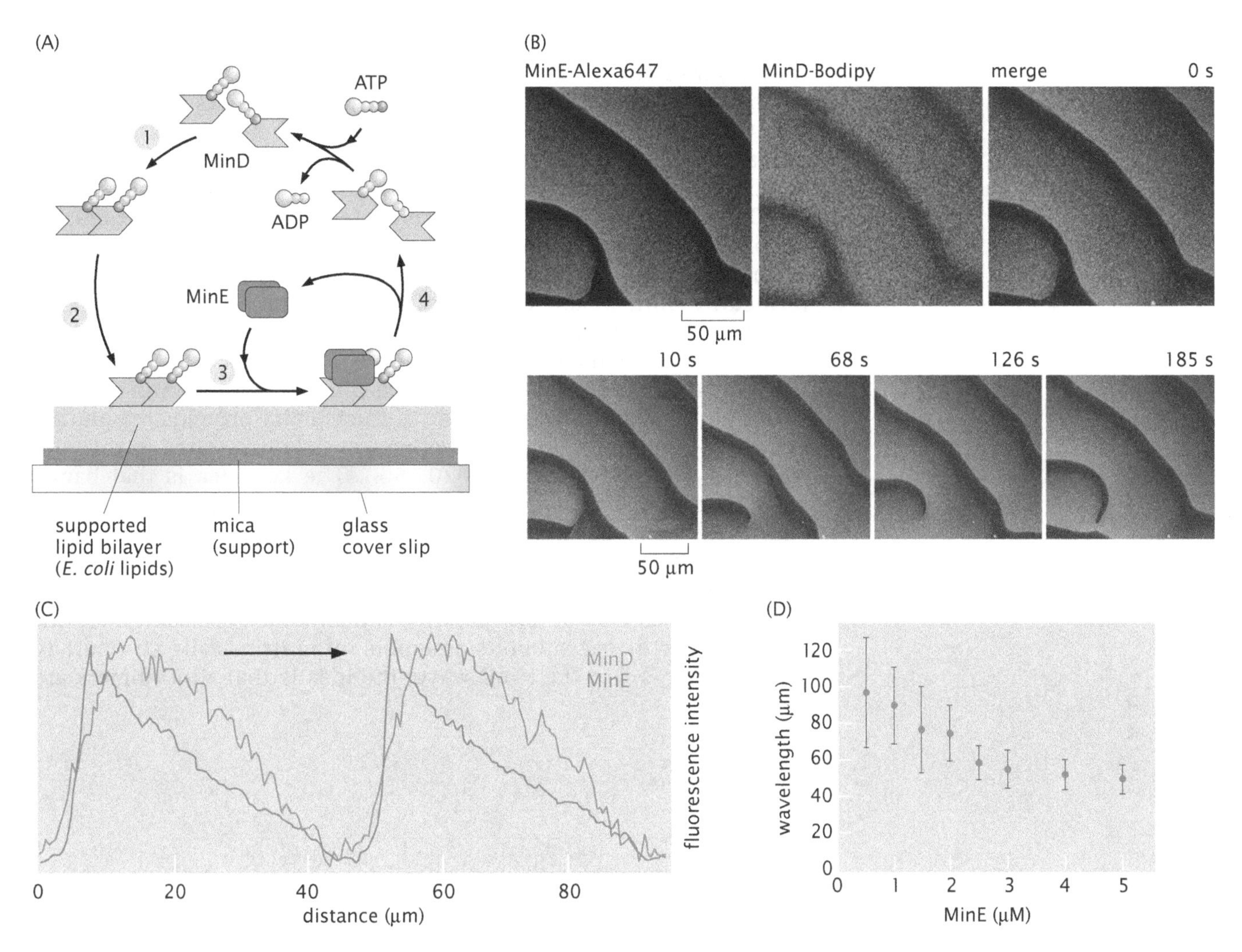

Figure 20.14: *In vitro* patterns in the Min system. (A) Reconstitution of MinD and MinE interaction with the membrane *in vitro*. The diagram illustrates key biochemical features: MinD can form dimers in solution and bind to the membrane, MinE binds MinD only on the membrane and causes it to dissociate and disassemble, followed by dissociation of MinE itself. (B) The spatial distribution of both proteins can be directly observed after tagging them with fluorescent molecules. The simple mixture of MinD, MinE, and ATP spontaneously forms dramatic wave-like patterns on lipid bilayers. As shown in the time series at the bottom, these waves migrate over time. (C) A cross-section through a wave showing fluorescence intensity suggests that accumulation of MinE delimits the rear of each MinD band. (D) The wavelength of the Min pattern decreases with increasing concentration of MinE. (A, B, adapted from M. Loose and P. Schwille, *J. Struct. Biol.* 168:143, 2009; C, D, adapted from M. Loose et al., *Science* 320:789, 2008.)

self-organized wave patterns of these proteins when they are bound to a lipid bilayer. Mixing together just the two proteins MinD and MinE, in the presence of ATP, will initiate the formation and lateral propagation of beautiful traveling waves on a supported membrane surface coated with *E. coli* lipids, as shown in Figure 20.14.

In many ways, this molecular system fulfills the important hallmarks of a Turing-like morphogen system. Using the terminology from the previous section, MinD is the activator X and MinE is the inhibitor Y. MinD recruits more MinD to the membrane (because of its cooperative assembly), and MinD also recruits its own inhibitor, MinE. MinE triggers the loss of MinD from the membrane, and, because MinE membrane association is destabilized after MinD has gone, MinE triggers its own dissociation as well.

The waves observed *in vitro* correspond to an areal coverage of the supported membrane by MinD and MinE proteins that is periodic in space and time. If MinD alone is present in the reaction, it forms a uniform layer on the membrane, and then the waves arise spontaneously when MinE is added. Importantly, direct measurements of diffusion of MinD and MinE along the surface of the membrane found that their lateral diffusion constants are much too slow to account for the rapid apparent movement of the waves, which propagate at rates around 0.3–0.8 μm/s. The diffusion of both proteins is about 50 times faster in solution than on the membrane, with a solution diffusion constant of about 60 $\mu m^2/s$. Thus, the wave of areal coverage is thought to be formed by diffusional exchange of Min proteins bound to the membrane and those in solution, and not by diffusion of Min proteins on the surface of the membrane.

A mechanism for how exchange of essentially immobile Min proteins bound to the membrane and freely diffusing Min proteins in solution can lead to the formation of traveling waves is illustrated in the context of a toy model in Figure 20.15(B). The key idea is that bands of MinD proteins evolve by adding additional proteins at the front end of the MinD band and, at the same time, MinD proteins detach from the back. This is precisely the mechanism that we described in Section 15.4.3 (p. 614) as being responsible for actin treadmilling, although the details of how polarity is maintained are somewhat different. For actin, the subunits assemble into intrinsically structurally polar filaments. For the MinD wave, the idea is that MinD dimers are

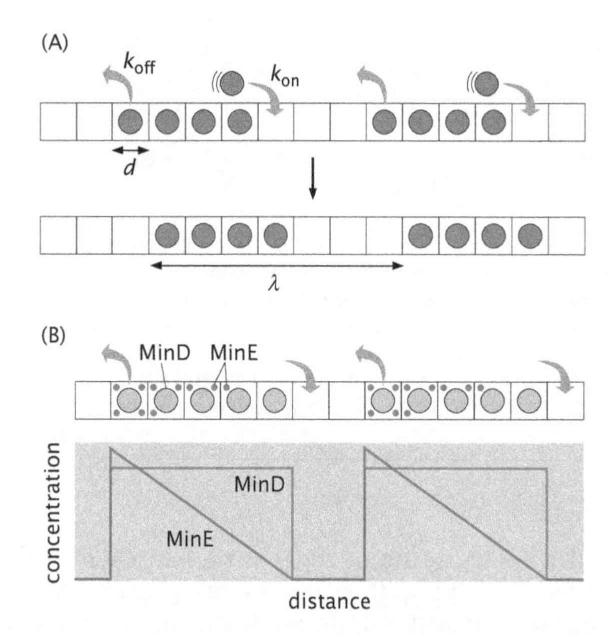

Figure 20.15: Toy model of Min waves. (A) Toy model of a traveling wave of particle density formed by exchange of particles bound to a one-dimensional lattice, with spacing d equal to the particle diameter, with those in solution. Particles attach at one end of the density band and detach at the other, with rates k_{on} and k_{off}, respectively. This is analogous to actin or microtubule treadmilling shown in Figures 15.32 and 15.33 (pp. 615 and 616). The two rates are required to be equal for the length of a band to remain constant in time. After a single attachment and detachment event, the band moves by a distance d that is related to the wave speed by $v = dk_{on}$. The wavelength of the pattern is λ. (B) Toy model of Min waves in one dimension. MinD proteins are added at the front end of the MinD band, where the MinE concentration is low. MinE binds only when MinD is present, and this leads to a concentration of MinE that increases toward the back end of the MinD band, as observed *in vitro* (see Figure 20.14). We assume that MinD dissociates when the MinE concentration has risen above a threshold value.

stabilized in their association with the membrane when they make physical contact with the edge of a MinD domain, but that this stabilization is reversed when the local bound concentration of MinE is high. Thus, new molecules of MinD stably associate only with the leading edge of the band, not with the rear. To an external observer, the bands of MinD with a wavelength λ will appear to travel at a speed v across the membrane (represented by the one-dimensional lattice in the toy model), even though each individual molecule of MinD is actually only coming on and off the membrane. At a given fixed point on the membrane, the local concentration of bound MinD will be periodic in time with a period T, where $T = \lambda/v$.

Like in the case of treadmilling actin filaments, the length of the MinD band will remain constant if and only if the "on" and "off" rates at the two ends are precisely balanced. Within the context of our toy model, this equality of rates can be achieved by having a MinD detachment rate that is stimulated by MinE, whose concentration increases from the front to the back end of the MinD band. According to the experimental measurement, the density of MinE appears to increase approximately linearly from the leading edge of the MinD band to the trailing edge (see Figure 20.14C). For our toy model, we assume that MinE proteins arriving diffusively from the solution phase are able to associate with immobilized (membrane-bound) MinD and then remain attached, and that the accumulation of MinE proteins on immobilized MinD proceeds at a constant rate. We further assume that when the local concentration of MinE passes some threshold (drawn in Figure 20.15 as four molecules per lattice unit), MinD is triggered to dissociate from the membrane and return to solution, followed by MinE. The linear increase of MinE from the front to the back of the MinD band is a simple consequence of the fact that the MinD proteins at the back end have been present on the one-dimensional lattice longer than those at the front end and thus will have accumulated more MinE proteins from solution, as shown in Figure 20.15(B). A simple prediction of this model is that the length of the MinD band and therefore the wavelength λ of the pattern will be reduced by having the MinE attachment rate increase, which can be achieved by increasing the concentration of MinE in solution. This prediction is in qualitative agreement with observations, as seen in Figure 20.14(D).

The proposed toy model can also be used to make a rough estimate of the speed v of the Min waves *in vitro*. As illustrated in Figure 20.15(A), the speed of the wave is $v = k_{\mathrm{on}}d$, which is the speed of length extension of the MinD band. The "on" rate for MinD proteins can be approximated as the diffusion-limited rate of capture $k_{\mathrm{on}} = Ddc$, where d is the lattice spacing (equal to the diameter of a MinD protein), D the diffusion constant of MinD in solution, and c its concentration in solution. Using the measured diffusion constant *in vitro* of $D = 60\ \mu\mathrm{m}^2/\mathrm{s}$ and the concentration of MinD employed in these experiments of $c = 1\ \mu\mathrm{M}$, and taking $d = 5$ nm, which is our rule of thumb for the size of a typical protein, we estimate $v \approx 0.7\ \mu\mathrm{m/s}$. This estimate is comparable to the range of velocities observed in experiments, $v = 0.3$–$0.8\ \mu\mathrm{m/s}$. It is notable that this traveling-wave speed is similar in magnitude to the diffusion-limited polymerization rates of actin filaments and microtubules.

So, *in vitro*, the Min system bears great resemblance to a Turing-like traveling wave. How is this dynamical property used by *E. coli* to regulate its division? Recall that the biological function of the Min proteins is to regulate the position of FtsZ ring assembly by ensuring that the concentration of the FtsZ inhibitor MinC (a passive passenger of MinD) is lowest right at the middle of the cell. In considering wave equations in the context of fluid mechanics or electromagnetism, it is well appreciated that the same governing dynamical equations that generate propagating waves in an extended medium will lead to standing waves when certain kinds of boundary conditions are imposed. Inspection of the dynamics of MinD localization inside of living bacterial cells (see Figure 20.16) suggests that its behavior is consistent with what might be expected when a traveling-wave system is confined to the small oblong box of the bacterial cell. As shown in Figure 20.16(A), MinD oscillates from one pole to another of a pre-divisional bacterial cell, with a period T of about 40 s, which is much shorter than the cell's division time (of the order of about 30 minutes). As the oscillating MinD carries MinC with it, the time-averaged concentration of the MinC protein will necessarily be lowest right at the cell's center, exactly where the FtsZ ring is supposed to assemble, and preventing FtsZ ring assembly near the cell poles, which would result in formation of minicells devoid of DNA. The nature of the MinD standing wave is illustrated more dramatically in long cells where division has been inhibited, as shown in Figure 20.16(B). For these filamentous cells, the standing-wave pattern shows a spatial periodicity that is about twice the normal length of an individual cell.

We can use the physical dimensions of the standing wave to estimate whether the Min traveling waves observed for the purified system *in vitro* really are a plausible physical underpinning for the center-finding mechanism in living cells. Mathematically, a standing wave can be described as two traveling waves, with the same amplitude, wavelength, and propagation speed, but moving in opposite directions. The wavelength λ of the *in vivo* standing wave is about 4–6 μm, which is twice the length of the pre-divisional cell shown in Figure 20.16(A), or the spacing between peaks shown in Figure 20.16(B). The period T is about 40 s. Setting $v = \lambda/T$, we can estimate v to be around 0.1 μm/s. This is the same order of magnitude as the wave propagation speed for the *in vitro* system, $v = 0.3$–0.8 μm/s, and the estimate becomes even closer if we recall that rates of diffusion for small molecules in cytoplasm are typically about fourfold slower than rates in water (see Figure 14.4, p. 549), which, according to our toy model, would decrease the propagation speed *in vivo* by a factor of four.

Overall the Min system in *E. coli* is a remarkable fulfillment of the biological promise of reaction–diffusion mechanisms as envisioned by Turing for establishing biological patterns. Certainly, there are a number of details in the system that he had not imagined (and we

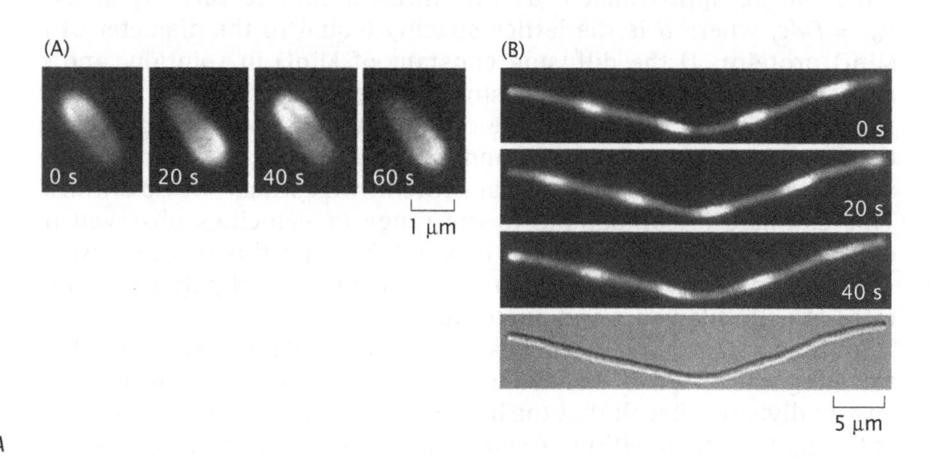

Figure 20.16: Min oscillations and how the bacterium finds its middle. (A) Fluorescently tagged MinD can be seen to oscillate rapidly from one pole of the cell to another with a period of only about 40 s, which is much less than the cell's division time. (B) In cells that are prevented from physically dividing, the Min oscillations continue, forming an oscillating stripe pattern throughout the entire length of the filamentous cell. The total size of the cell is shown in the transmitted-light micrograph in the bottom panel. (Adapted from D. M. Raskin and P. A. J. de Boer, *Proc. Natl Acad. Sci. USA* 96:4971, 1999.)

have not done full justice to the complexity and robustness of the real system in this brief treatment), and the addition of the boundary condition of the finite size of the bacterial cell also introduces new behavior. At present, molecular reaction–diffusion mechanisms with rather different boundary conditions are also thought to operate in other cellular systems. One common general concept that applies to many different kinds of systems is the idea of "local excitation and global inhibition." The application of this concept to the problem of neutrophil polarization is illustrated in Figure 20.17. For a neutrophil that needs to establish large-scale cell polarity in order to move toward an invading bacterium, the chemical signals emitted by the bacterium must be detected by receptors that are present on the cell's surface. Most signaling mechanisms associated with this kind of detection employ positive-feedback amplification, so that a small chemical signal from ligand–receptor binding on the surface of the neutrophil closest to the bacterium can be amplified in a way that will lead to large-scale polarization of the neutrophil's cytoskeleton. The problem with this is that positive feedback alone cannot lead to stable polarization. As the neutrophil begins to move closer to the bacterium, the local concentration of ligand will increase everywhere on the neutrophil's surface, including on the side facing away from

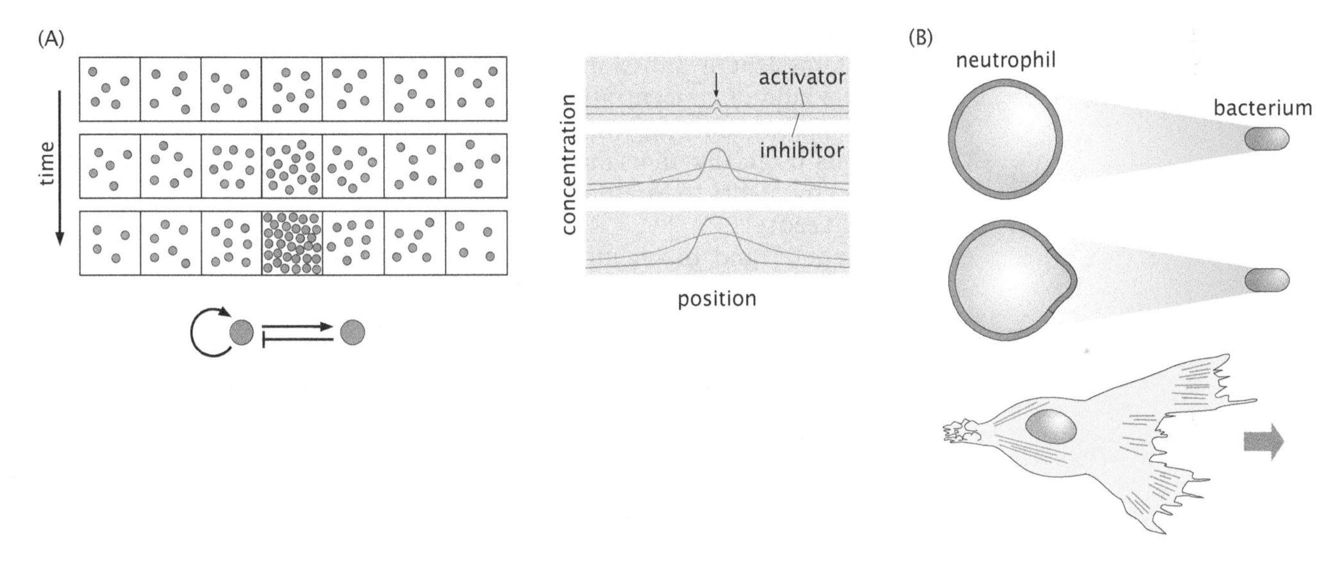

Figure 20.17: Local excitation and global inhibition in cell polarization. (A) Lattice model of local activation and global inhibition. Two interacting molecular species, shown here in red (activator) and blue (inhibitor) start off nearly uniformly distributed, with a small increase in activator leading to a sharply localized peak in activator concentration over time. The activator activates its own production (or, equivalently, the activity) and the production of inhibitor molecules, while the inhibitor represses the production of activator molecules. Both molecular species diffuse through the lattice, but the spread of inhibitor is much faster than that of activator. The graphs on the right show the time evolution of the position-dependent concentration of the two species, with an arrow indicating the initial local perturbation that transiently increases the concentration of activator. This small initial perturbation is amplified by the self-activation of the activator, which leads to a sharp increase in inhibitor concentration. The newly produced inhibitor molecules quickly diffuse away and repress activator production far from the position of the initial perturbation. The end result is that the activator dominates only closest to the signal, but the inhibitor dominates elsewhere. (B) A local excitation/global inhibition mechanism can contribute to large-scale cell polarization in response to external signals. Here, the bacterium is shedding peptide fragments that the neutrophil recognizes via a cell surface receptor. Although the concentration of the peptide is highest on the side of the neutrophil facing the bacterium, there is some peptide present all around the neutrophil. The receptor is postulated to initiate two kinds of intracellular signals, namely, a positive signal that promotes actin assembly and cell protrusion, and a negative signal that suppresses cell protrusion. As long as the positive signal acts locally while the negative signal acts globally (or, at least, over a longer distance than the positive signal), the positive signal (shown in red) can promote protrusion over the negative inhibitory signal only on the side of the neutrophil that is closest to the bacterium. The diagram at the bottom shows how directed cell migration can result from a positive signal that promotes branched actin filament network assembly (red) and a negative signal that acts to generate contractile myosin-actin bundles (blue). (A, adapted from H. Meinhardt, *J. Cell Sci.* 112:2867, 1999.)

the prey. One solution to this problem of runaway positive feedback is to have the system also include negative feedback. If the same chemical signal that triggers local signal amplification (in this case, protrusion of the cytoskeleton at the leading edge of the cell) can also induce an inhibitor that operates over longer distances, then the ratio of activator to inhibitor can be stably maintained at the cell front, as shown in Figure 20.17. This is fundamentally a Turing-like mechanism in that it depends critically on the interaction of activating and inhibiting molecular species where the mobility of the two species is different. As long as the inhibitor is able to act over a longer distance (either because it diffuses (or is transported) more quickly than the activator or because it is more stable than the activator), then a single spatially restricted signal that induces both activation and inhibition simultaneously can give rise to stable cell polarization. It seems likely that many slight variations on this theme are lurking around us.

20.3.3 Phyllotaxis: The Art of Flower Arrangement

The Turing model depends on diffusion as a concentration-dependent transport mechanism that, coupled with chemical interactions between different species, can generate patterns from an initially uniform field. However, the same general concept can apply even in systems where diffusion itself does not operate, as long as there is some kind of concentration-dependent transport. An important example of this concept, which fascinated Turing himself toward the end of his tragically shortened life, is phyllotaxis, the mechanism that is used by plants to establish regular spacing of branches, leaves, petals, and seeds.

Genetic and biochemical studies have established that a critical, general signaling molecule for plant cell fate determination is the hormone auxin. In the developing epidermis that gives rise to shoots, auxin may be synthesized in all cells, but it is then transported among the cells such that it ends up being concentrated in regularly spaced zones. The epidermal cells with the highest auxin concentrations are triggered to proliferate, forming a growing shoot known as a primordium. Depending on the type of plant and on other signals, the primordium may differentiate into a branch, a leaf, a petal, or some other plant structure. The beautifully regular geometries of plant structures, from sunflower heads to pine cones to tree limbs to rose petals, arise because of the regular rules of auxin accumulation and primordium growth that can give rise to even spacing during growth. Different plants have different divergence angles between their organs, but the most common angles are 180°, 90°, and 137.5°, the so-called "golden angle." Figure 20.18 shows how the emergence of primordia can lead to large-scale plant structures.

The molecular mechanism of pattern formation for primordia has been investigated genetically. One key player is the auxin transporter, PIN1. Critically, the amount of PIN1 at any cell boundary depends on the local concentration of auxin, as diagrammed in Figures 20.18(C–E), and the orientation of the transporter at the boundary is such that auxin is always transported up its net concentration gradient. In a sense, this form of concentration-dependent transport can be thought of as the opposite of diffusion, because in diffusive systems net movement of molecules ends up being down the concentration gradient. With the PIN1–auxin system, this type of polarized transport ends up

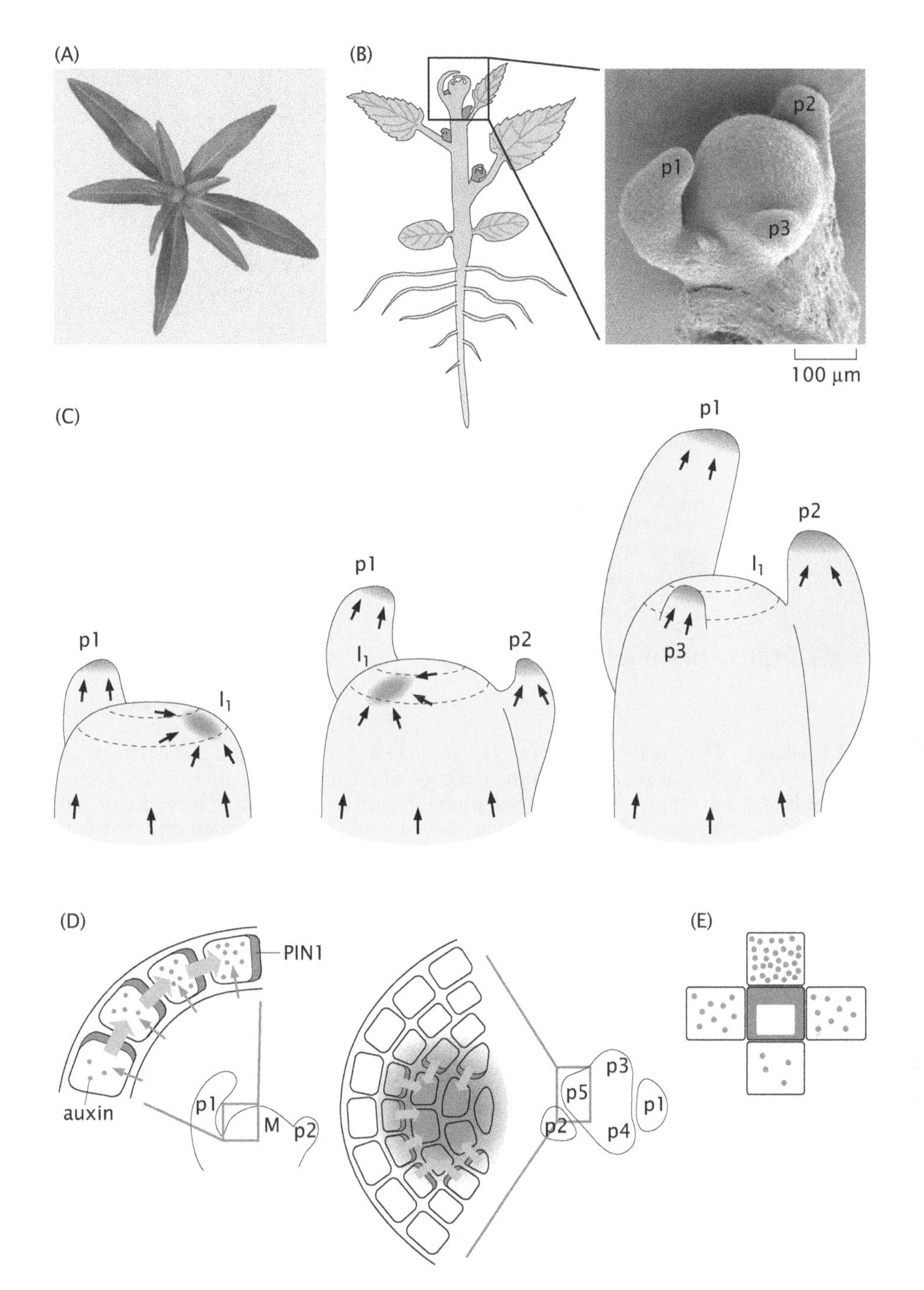

Figure 20.18: Phyllotaxis. (A) Viewed from above, the leaves of the snapdragon, *Antirrhinum majus*, emerge in a regular spiral pattern. Older leaves, near the bottom of the stem, are larger, and younger leaves, near the top of the stem, are smaller. (B) A diagram of a growing plant shows the location at the top of the shoot apical meristem (boxed) that gives rise to the primordia that will become the leaves. On the right, a scanning electron micrograph of a shoot apical meristem of the garden nasturtium, *Tropaeolum majus*, shows three primordia (p1, p2, and p3). The primordium p1 is the oldest of the three; the newborn p3 is just beginning to emerge. (C) Primordia emerge where auxin (green) is most concentrated. Auxin is synthesized by all the cells in the apical meristem, but is transported generally upward as the meristem grows (black arrows), and then concentrated (green spots). When the local auxin concentration has passed a threshold, the cells proliferate and the primordium begins to grow. After one primordium has sprouted out, a new primordium forms around the other side of the meristem (in this case, about 138° clockwise away). Successively emerging primordia are labeled as in (B). (D) At left, this cross-section of a small zone of the meristem shows a layer of epidermal cells. Auxin (green dots) is synthesized in all cells and transported into the epidermis (green arrows). Within the epidermis, auxin is transported upward (blue arrows) through the PIN1 transporter (red). At right, a top view of an emerging primordium shows how directional bias in the transport of auxin leads to the formation of a concentrated spot. (E) PIN1 localization. The cell in the middle has PIN1 (red) at all boundaries.

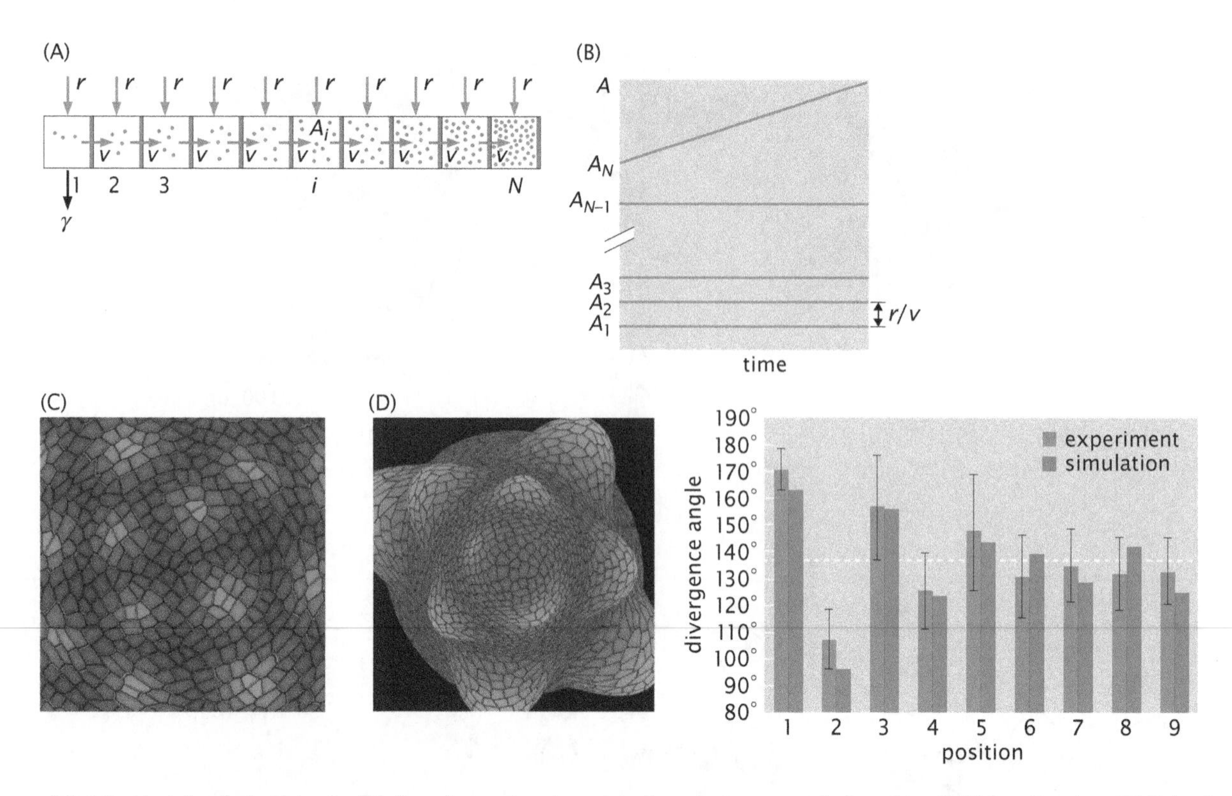

Figure 20.19: Models of phyllotaxis. (A) One-dimensional model of auxin transport induced by PIN1 localization. PIN1 (red) accumulates on one of the two cell walls that is oriented in the direction of the increase in auxin concentration, shown as green dots. Auxin is delivered to cells at a rate r and it is transported at rate v per molecule from one cell to the next. The primordium cell is characterized by a rate of auxin removal of γ per molecule. (B) The graph shows the predicted distribution of auxin molecules. The concentration in all the cells in the row is constant over time, with the exception of the last cell, N. Auxin concentration in cell N rises over time, until it crosses some threshold and cell N initiates a new primordium. (C) Result of a computational model of auxin transport in a two-dimensional array of cells. Concentrated patches of auxin appear in the cell layer at regularly spaced intervals. (D) Addition of cell growth to the model can lead to the observed geometrical patterns in plants. The image on the left shows a snapshot from a simulation that includes upward growth of the meristem and sprouting of the primordia. This geometrically realistic simulation can predict the divergence angles between successive primordia. The graph on the right shows that divergence angles computed from the model (red) compare favorably with those measured in *Arabidopsis* shown with standard error bars (blue). Starting from an unpatterned epidermis, the first primordia split off at angles near 180° and then near, 90°, but after a few oscillations they settle down to approximate the golden angle 137.5° (shown as a white, dashed line). (C, D, adapted from R. S. Smith et al., *Proc. Natl Acad. Sci. USA* 103:1301, 2006.)

acting as a positive-feedback loop, where regions of initially slightly higher than average auxin concentration further increase their auxin levels, and then differentiate to form a primordium.

The proposed mechanism of accumulation of auxin away from positions at which primordia are formed is made explicit in the toy model shown in Figure 20.19(A). Here we consider a one-dimensional array of cells that are receiving auxin at a rate r. One of the cells mimics the presence of a newly emergent primordium. Because the cells of the primordium rapidly proliferate and sprout outward, the primordium acts as a sink for the transported auxin. In our one-dimensional model, we portray this by having the auxin removed from the cell layer only at this one location, at a rate γ per particle. The walls of the cells are covered unevenly with transporter PIN1, leading to cell polarization. For our simple model, we assume that the transport rate of auxin through PIN1 is the same for all cells, but is polarized so that transport goes up the gradient. The net result is that there is a rate v per particle for transporting auxins from one cell to its nearest neighbors in the direction of increasing auxin gradient.

The rate equation describing the number of auxin molecules in cell i is given by

$$\frac{dA_i}{dt} = r + v\sigma_{i-1,i}A_{i-1} - v\sigma_{i,i+1}A_i, \tag{20.46}$$

and incorporates the two-state variable σ that keeps track of PIN1 polarization: $\sigma_{i,i+1}$ is +1 when $A_i > A_{i-1}$ and −1 in the opposite case, when the amount of auxin is smaller in cell i. Given the geometry of our cell array, namely that of a finite strip, the first and last cells in the strip have to be considered separately. We take cell 1 to be the one at which a primordium is formed initially. It therefore acts as a sink for auxins, which are removed at rate γ per molecule. For the case when $A_1 < A_2$, the polarization of the auxin transporter will lead to transport out of cell 1, resulting in

$$\frac{dA_1}{dt} = r - vA_1 - \gamma A_1, \tag{20.47}$$

while in the case $A_1 > A_2$, the transport will be into cell 1 from cell 2, and therefore

$$\frac{dA_1}{dt} = r + vA_2 - \gamma A_1. \tag{20.48}$$

For cell N, the time evolution of auxin number is

$$\begin{aligned} \frac{dA_N}{dt} &= r + vA_{N-1} \quad (A_N > A_{N-1}), \\ \frac{dA_N}{dt} &= r - vA_N \qquad (A_N < A_{N-1}), \end{aligned} \tag{20.49}$$

where, once again we have to distinguish between the two different possible polarizations of PIN1 at the cell membrane.

This set of equations will reach a steady state for all but the Nth cell, leading to a gradient of auxins that increases from cell 1 toward cell N. This is illustrated in Figure 20.19(A) and can be obtained from Equation 20.46 by setting the left-hand side to zero, and noting that in this case all the σ variables take the value +1, indicating a polarization of PIN1 toward the Nth cell. This leads to a recursive relation $A_i = A_{i-1} + r/v$ which results in auxins increasing in number linearly with cell position i. The Nth cell in this case will only have an auxin flux into the cell that is constant in time, leading to a linear increase of the auxin number with time. This accumulation of auxins in the Nth cell can drive the formation of a new primordium that now becomes a new auxin sink, which will drive the flux of auxins away from this primordium. The newly established flow of auxins will lead to an accumulation at a new cell, whose location depends on how long the initial primordium sink persists. If the original sink at position 1 has grown far enough away to no longer affect the epidermis, then the third auxin sink will appear at position 1 again and over time the primordia will emerge at alternate ends of the cell field. If, however, the primoridum at position 1 continues to act as a sink for a while, the third primordium will emerge roughly halfway between positions 1 and N.

In brief, the concentration-dependent directed transport mechanism is able to use auxin accumulation to trigger differentiation of auxin-rich domains into primordia that act as auxin sinks, redirecting the flow of auxin to points that are as far away from the currently emerging primordia as possible. If we imagine wrapping the row of cells shown in Figure 20.19(A) into a ring, which is growing upward to form a cylinder, then a short-lived sink lifetime would generate a system where the successive primordia emerge at 180° angles. As with the Turing model, generalizing this result to more realistic two- and three-dimensional cases is most easily done numerically rather than analytically. A computational implementation of this transport mechanism in a two-dimensional setting, leading to a pattern of cells that accumulate auxins in domains with roughly even spacing, is shown in Figure 20.19(C). Coupling the transport to tissue growth in the two-dimensional simulation can be made to produce the observed geometrical pattern of leaves, as shown in Figure 20.19(D).

Along with the pattern of primordium determination, the pattern of growth is important in determining the final distribution of the plant organs. So far, we have been considering the upward growth of a meristem, which, together with the auxin transport mechanism, can give rise to a spiral pattern of leaves along a vertical stem. If the plant tissue grows outward rather than upward, a flat spiral pattern can result, such as is seen most dramatically in a sunflower head, as shown in Figure 20.20(A).

The geometrical patterns observed in phyllotaxis, like the sunflower, can be generated in a simple physical system that uses magnetic fields and ferrofluid drops to mimic the emergence of primordia in the growing meristem. In the frame of the tip of the meristem, the primordia are formed periodically with period T, a distance R_0 away from the tip, and are advected away from the tip at a velocity v that corresponds to the speed of tissue growth, as illustrated in Figure 20.20(B). This simple mechanism of growth and advection can be implemented by introducing small drops of ferrofluid in the center of a dish containing a viscous fluid in the presence of a magnetic field perpendicular to the dish. The magnetic field magnetizes the drops, thus leading to their mutual repulsion, mimicking the auxin-driven growth of primordia that occurs in the interstitial regions formed by previously generated primordia (see Figure 20.18). A horizontal field gradient is used to produce a force on the drops that leads to a steady radial outward velocity v as a result of the frictional forces due to the viscous fluid. This set-up leads to the creation of steady-state patterns of drops that depend on the dimensionless constant $G = vT/R_0$, as shown in Figure 20.20(C). In particular, for $G \approx 0.15$, a spiral pattern of drops appears where successive drops are separated by the golden angle 137.5° as is typical of phyllotaxis. This value of G guarantees that a newly added drop is under the repulsive influence of three or more previously added drops. At higher values of v, the angle increases, reaching 180° when only two drops interact, placing them on opposite sides of the dish.

This clever experiment replaces the principles involved in phyllotaxis (polarized auxin transport, tissue growth, and primordium differentiation) with a completely different set of physical mechanisms (regular dripping and magnetic repulsion). But, by arranging the system appropriately, the same kinds of patterns can emerge, described by the same mathematics. Similarly, in many kinds of biological pattern formation, superficially similar patterns may be generated by distinct molecular mechanisms, but the superficial difference of the molecular behaviors may mask profound underlying unities of pattern formation concepts.

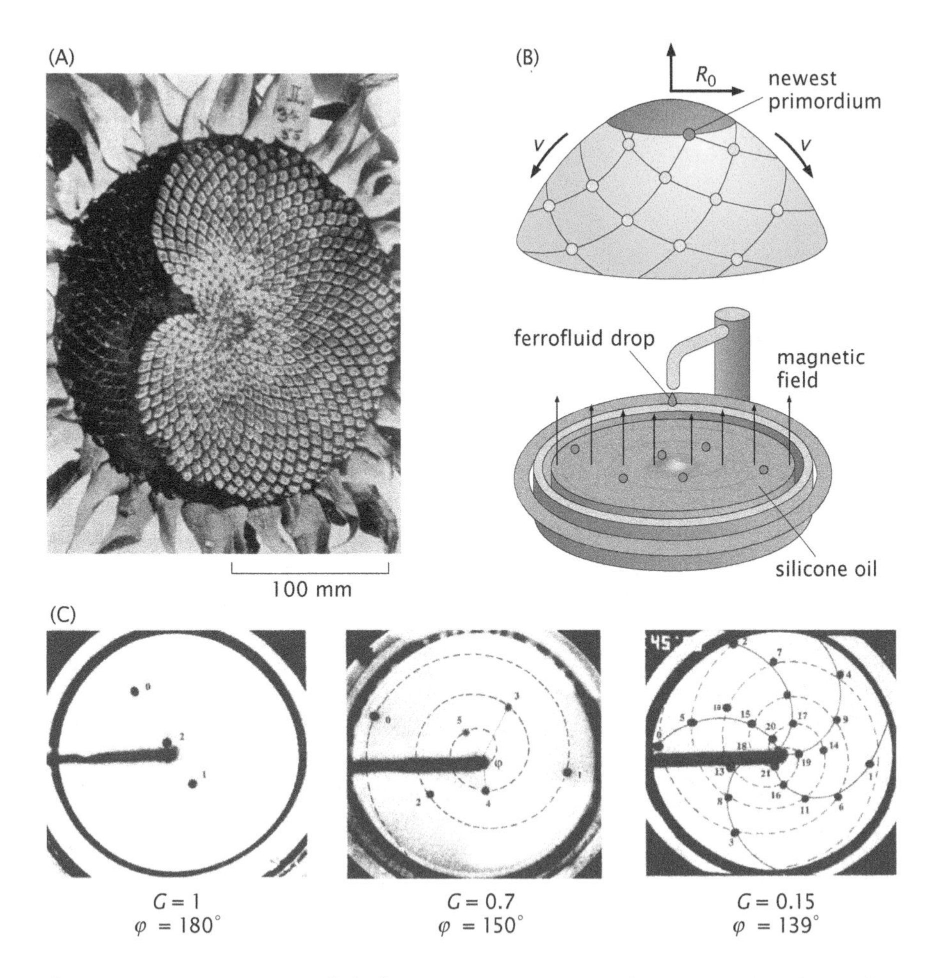

Figure 20.20: Emergence of phyllotactic patterns in two dimensions. (A) The seeds in the head of a sunflower form a spiral pattern, growing from the center outward. For this image, some of the seeds have been removed to emphasize the underlying structure. (B) Concept and implementation of an experimental system to imitate phyllotaxis. The diagram at the top shows the position of the newest primordium (red) and the outward growth of the sunflower head. At bottom, a flat disc covered with a layer of silicone oil is subjected to a magnetic field (arrows). The spigot at the top deposits drops of a colloidal suspension of metal particles, immiscible with the oil. These drops are deposited periodically; once on the surface, they repel each other. The net outward movement of the magnetic drops due to their mutual repulsion is analogous to the outward growth. (C) Experimental results from this system generate remarkably plant-like patterns. The nature of the pattern is controlled by the dimensionless combination $G = vT/R_0$, where v is the speed of advection, T the period of drop deposition, and R_0 the distance at which the drops are deposited; see (B). If the rate of drop deposition is relatively slow compared with their outward movement, the average angle between drops is close to 180°. Increasing the relative rate of dripping, or decreasing T, causes the system to converge to the golden angle, 137.5°. (A, adapted form A. H. Church, On the Relation of Phyllotaxis to Mechanical Laws. Williams & Norgate, 1904; B, C, adapted from S. Douady and Y. Couder, *Phys. Rev. Lett.* 68:2098, 1992.)

20.4 Turning Time into Space: Temporal Oscillations in Cell Fate Specification

Turing-type mechanisms can give evenly spaced spots and stripes, and also traveling waves, but this is not the only way to get regular striping of something. Recall the oscillatory circuit from Section 19.3.6 (p. 870); this puts cells in different states at different times. Is there some way to convert time into space?

20.4.1 Somitogenesis

The example of *Anabaena* showed cell differentiation coupled to one-dimensional growth. Our present example also involves an instance of one-dimensional growth, but in a much more complex system, namely, the development of the vertebrate body plan. Vertebrate animals are given this designation because of their regular patterns of bones that protect their spinal cord and provide structural support. Early in embryonic development, the regular size and spacing of these vertebrae are set by a mechanism that combines a temporal oscillator with a traveling wave of commitment to a particular developmental fate.

Rather than forming all at once, the vertebral segments form sequentially in a field of cells known as the pre-somitic mesoderm that is growing at the same time it is differentiating. The image in Figure 20.21(A) shows a chick embryo in the middle of this process. The segment precursors, called somites, can be seen on the left closer to the chick's head, regularly spaced on either side of the forming neural tube. The number and size of somites vary among vertebrate species: chicken embryos have 55, zebrafish 31, while snake embryos have as many as 315. This observation poses a challenge to any quantitative model of somitogenesis as it has to be able to reproduce this large variability in the observed somite patterns across species while being extremely regular and reproducible for a given species. This is similar to the scaling problem encountered in the early embryonic development of different species of flies discussed in Section 20.2.3. The solution that nature has come up with in this case seems to be quite different, though.

In 1976, Cooke and Zeeman described the clock-and-wavefront model of somitogenesis described in Figures 20.21(B,C). While the entire field of cells that make up the pre-somitic mesoderm is growing, all the cells are also thought to be undergoing a regular temporal oscillation of gene expression. The gene expression oscillations within a cell are driven by a delayed negative-feedback loop similar to the kind described in Section 19.3.6 (p. 870). Coupling of these cellular oscillators then leads to entrainment in which all the cells end up beating rhythmically with the same period. In addition to the oscillations, a traveling wave of gene expression is set up by morphogens in the pre-somitic mesoderm such that the front of this wave propagates from the anterior to the posterior end. As this traveling wave reaches the oscillating cells in the pre-somitic mesoderm, it arrests them and a differentiation program is initiated leading either to a rib precursor called a somite or to interstitial tissue. Which of these two different fates will be adopted is determined by the phase of the oscillation at the time of arrest. This is illustrated in Figure 20.21(C).

The clock-and-wavefront model makes a simple quantitative prediction that can be tested experimentally. In particular, if the ideas embodied in that model are correct, then the speed of the propagating front of commitment (V), the size of the somites (s), and the period of the clock (T) are related by the simple relation $V = s/T$. The key idea here is that the boundaries of a single somite are determined by a specific phase of oscillations in gene expression in the post-somitic mesoderm. During the time between two oscillation maxima or minima, which is the period T, the wavefront moves by a distance $s = VT$, setting this as the distance between neighboring somites. In one test of this prediction, measurements were performed for all three quantities for different vertebrate species and at different developmental

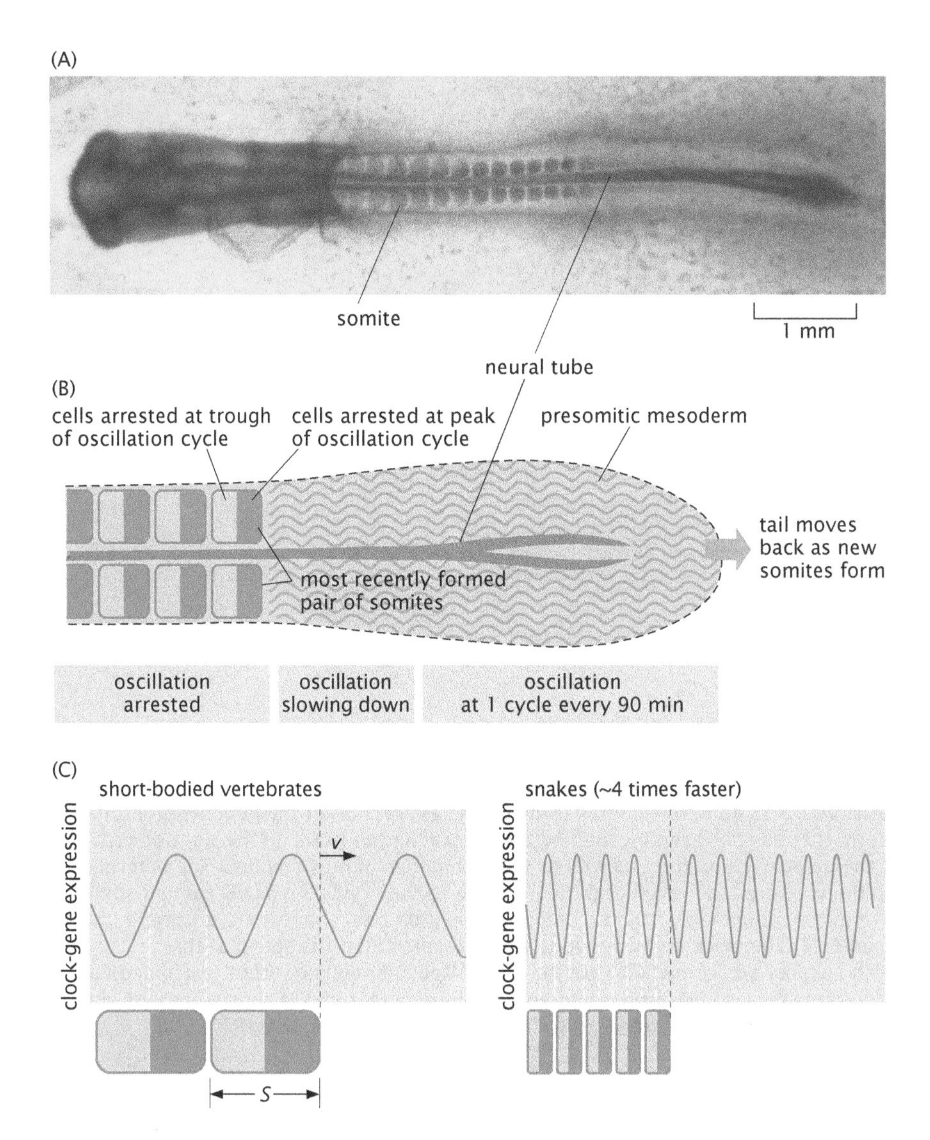

Figure 20.21: Somitogenesis. (A) A chick embryo at 40 hours post-fertilization is caught in the middle of forming its somites. (B) The temporal oscillation of gene expression in the presomitic mesoderm is converted into an alternating spatial pattern of gene expression in the formed somites. In the morphologically identifiable somites, the cells have arrested their gene oscillation, and their eventual fate depends on where in the cycle the individual cell has undergone arrest. As the wave of commitment moves posteriorly, subsequent pairs of somites are committed in turn. In this way, a temporal oscillation of gene expression traces out an alternating spatial pattern. (C) In snakes (right), the clock components seem to tick around four times faster (relative to growth rate) than in shorter-bodied animals (left), leading to the formation of fewer, larger somites in short-bodied animals and many, small somites in snakes. (A, from Y. J. Jiang et al., *Curr. Biol.* 8:R868, 1998; B, adapted from B. Alberts et al., Molecular Biology of the Cell, 5th ed. Garland Science, 2008; C, adapted from F. J. Vonk and M. K. Richardson, *Nature* 454:282, 2008.)

times. Figure 20.22(A) shows an example of the corn snake embryo imaged at the developmental stages with 165 and 202 somites. To measure the speed of the wavefront, the amount of gene expression for a gene (*MSGN1*) associated with oscillations in the pre-somitic mesoderm was imaged using *in situ* hybridization. It shows up as the blue smear in the posterior parts of the pre-somitic mesoderms in the two images in Figure 20.22(A). From these images, the speed of the wave front is determined by the relation $(W_b - W_a)/n$ where $W_{b,a}$ is the position of the wavefront at the two developmental times, as shown schematically in Figure 20.22(B), and n is the developmental time measured in the number of somites that have been formed between time a and time b ($n = 37$ in the example shown in the figure). The speed of somite formation, on the other hand, is D_S/n, where D_S is the length of the newly formed somite region between the two developmental stages. The prediction of the clock-and-wavefront model is that the ratio of these two speeds is 1 regardless of developmental time. This is indeed observed, as shown in Figure 20.22(B), where the ratio of the two speeds is plotted for different species and different developmental times. The remarkable collapse of all the data is a testament to the power and universality of the clock-and-wavefront mechanism

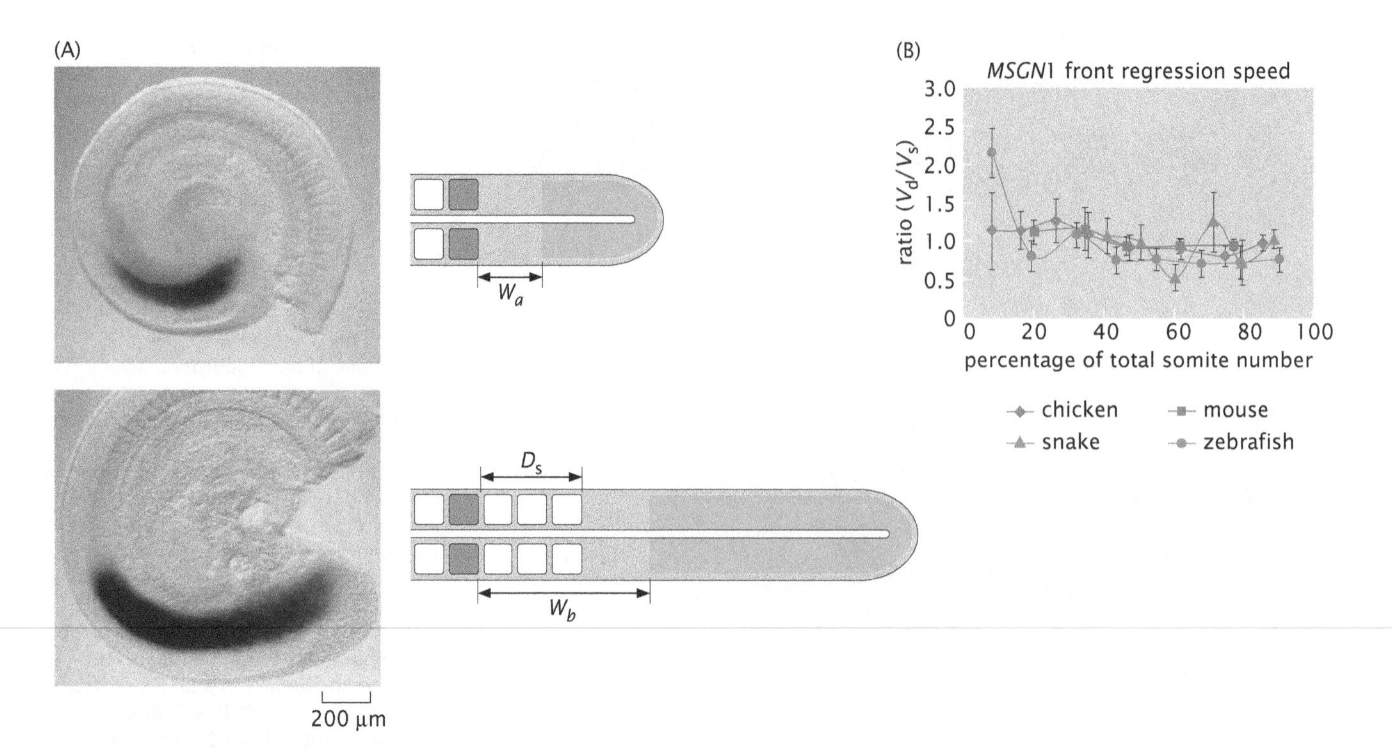

Figure 20.22: Experimental test of the clock-and-wavefront model of somitogenesis. (A) Images of the corn snake embryo at two different stages in the process of somitogenesis, with 165 and 202 somites formed. The expression of the gene *MSGN1* (blue) delineates the part of the post-somitic mesoderm that has not been invaded by the determination front. To the right of each image, the schematics show the extension of the determination front as well as the emergence of new somites. The determination front is represented by the boundary between the gray area to the left and the blue one to the right. The newly formed somite pair in the earlier stage of development, shown in the top panel, is colored green. Three additional somites are present in the later stage of development, shown in the bottom panel. (B) The clock-and-wavefront model predicts that the speed of the determination front (V_d) and the speed at which somites are formed (V_S) are the same. Measurements of these two speeds using images such as the one shown in (A) across different species and during different developmental times is consistent with this prediction of the model (that is, the ratio of the speeds is one). (Adapted from C. Gomez et al., *Nature* 454:335, 2008.)

as a conceptual framework for thinking about somitogenesis across different vertebrate species.

The clock-and-wavefront model leads to a somite size given by the speed of the front and the period of the oscillator. The number of somites, on the other hand, is determined by the clock period and the duration of the development period. To make a precise test of the clock-and-wavefront mechanism, a compelling experiment would be to vary the parameters of the model within one organism and to explore the consequences for somite formation. In a beautiful set of recent experiments, mutants were found in the clock period as shown in Figure 20.23(A), where the clock in the mutant zebrafish runs slower than in the wild-type animal. The clock-and-wavefront mechanism makes several interesting predictions about both somite size ($s = VT$) and number (number of somites = developmental time/T). However, for these predictions to hold, the developmental time for both the wild type and mutant must be the same as shown in Figure 20.23(B). Further, these predictions require that the determination front moves with the same speed in both the wild-type and mutant as shown in Figures 20.23(C, D, E).

Figure 20.24 shows the developmental consequences of the slower clock period for somitogenesis. Figures 20.24(A, B) illustrate how the somite length is changed in the mutant fish. Figures 20.24(C, D) go further by illustrating how the overall number of somites in the embryo

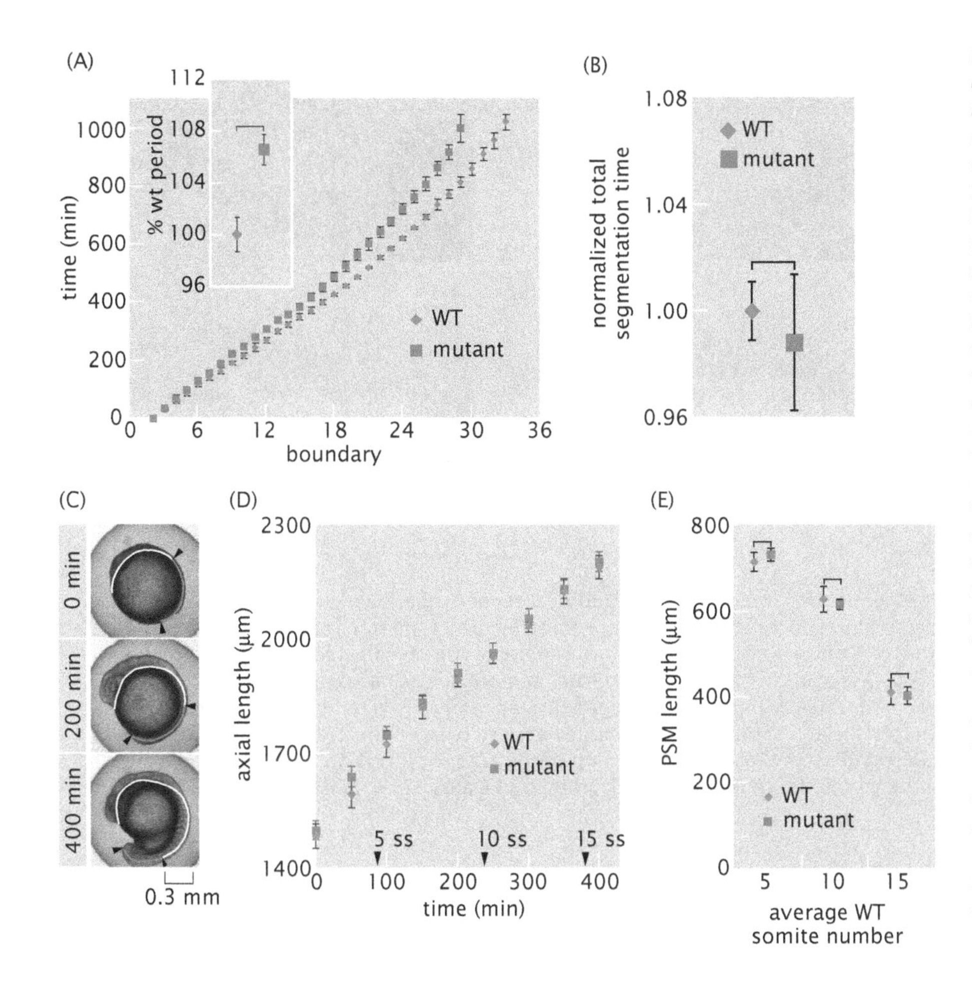

Figure 20.23: Experimental determination of the parameters appearing in the clock-and-wavefront model. (A) Measurement of the time of somite formation in a wild-type zebrafish and a mutant with a knock-out of the gene *hes6*. (inset) The period between formation of subsequent somites is increased in the mutant with respect to the wild-type. (B) Measurement of the total segmentation time in both the wild-type zebrafish and the *hes6* mutant showing that the duration is the same in both cases. (C) Images from time-lapse movies reveal total axial length based upon drawing a line (sum of white and red lines) from the anterior to the posterior end of the embryo. The presomitic mesoderm (PSM) length was measured along the same line (red line) by determining the distance between the most recently formed somite boundary (indicated by an arrow) and the posterior end of the mesoderm (also indicated by an arrow). (D) Comparison of axial lengths of the wild-type zebrafish and the *hes6* mutant. The arrows on the x-axis show the somite stage (ss) of the wild-type embryo. (E) Length of presomitic mesoderm for both the wild-type zebrafish and the *hes6* mutant. (Adapted from Schröter and Oates, *Curr. Biol.* 20:1254, 2010.)

and the number of vertebrae in the adult fish are reduced in the mutant in comparison with its wild-type counterpart.

The results of these experiments can be summarized by noting that the ratio of the somite lengths are given by

$$\frac{S_{\text{mutant}}}{S_{\text{wild-type}}} = \frac{VT_{\text{mutant}}}{VT_{\text{wild-type}}} = \frac{T_{\text{mutant}}}{T_{\text{wild-type}}} \tag{20.50}$$

and that the ratio of the number of somites is given by

$$\frac{n_{\text{mutant}}}{n_{\text{wild-type}}} = \frac{T_{\text{tot}}/T_{\text{mutant}}}{T_{\text{tot}}/T_{\text{wild-type}}} = \frac{T_{\text{wild-type}}}{T_{\text{mutant}}}. \tag{20.51}$$

Both of these predictions are consistent with the measurements put forth in Figure 20.24.

Even though the clock-and-wavefront model has proven to be quite successful in describing the process of somitogenesis, many interesting questions remain pertaining to the identities of the key molecular players and their interactions in setting up the oscillator and the wave.

20.4.2 Seashells Forming Patterns in Space and Time

Whether on the seashore or in our museums of natural history, the hardened remains of living organisms that synthesize beautiful exteriors are one of the most well-known examples of biological pattern formation. In the late seventeenth century, this fascination became an outright obsession for some, culminating in works such as Filippo

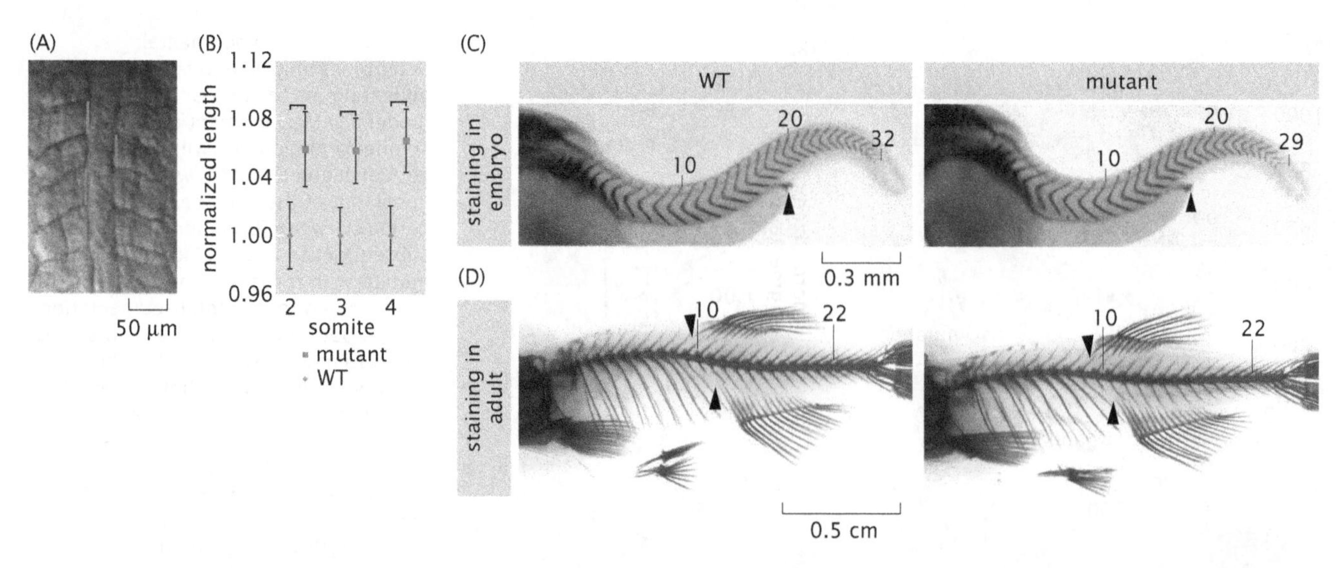

Figure 20.24: Testing the clock-and-wavefront model of somitogenesis through subtle perturbations in the segmentation network. (A) Measurement of the length of the somites. (B) Relative change in somite length in the mutant showing an increase with respect to wild-type. (C) Direct observation of embryos after completion of segmentation reveals a reduced total number of muscle segments (myotomes) in the mutant. The dissociation of clock period from anatomic regional identity is shown by comparing the relative position of the proctodeum (arrow) with respect to segment number. (D) This reduction in segment number is maintained through adulthood as shown by this skeletal staining presenting fewer vertebrae in the mutant. The differences between mutant and wild-type are further manifested by the relative positioning of the anterior insertion sites of the anal and dorsal fins (shown as arrows) with respect to the vertebrae. (Adapted from Schröter and Oates, *Curr. Biol.* 20:1254, 2010.)

Buonanni's 1681 work *Recreation for the Eyes and the Mind Through the Study of Shelled Animals.*

The previous two sections have shown how to use fundamental mechanisms of pattern formation to set up standing waves or traveling waves in space (Turing) and also oscillations in fate specification in time (somitogenesis). If we combine these concepts as applied to pigmentation with a growing field, we can make seashells. The mechanism behind such models is illustrated in Figure 20.25. At the growing edge of the seashell, the mollusk architect extends a small rim of its mantle over the shell lip. Sensory cells in the mantle are able to detect the pigmentation of the strip of shell laid down the day before, and communicate the previous day's pigment pattern to the secretory cells in order to dictate the desired pattern for the current day's secretion effort. The secretory cells then lay down another strip of shell, extending it outward and making it slightly larger. Day by day, the eventual shell pattern is thus laid down in rows, like the weft threads in a woven piece of cloth, or the courses of yarn in a knitted sweater. The sensory cells can also communicate with their neighbors laterally along the shell lip in order to establish longer-distance patterns. In the example shown in Figure 20.25(B), the sensory cells in the mantle have set up a lateral standing-wave pattern. A sensory cell that "tastes" a darkly pigmented region of the shell communicates to its nearby secretory cells the information that they should secrete pigmented shell; this is indicated as local excitation. At the same time, the pigment-stimulated sensory cell sends an inhibitory signal that is weaker than the excitatory signal but spreads over a broader field of secretory cells. Consequently, the net signal will be excitatory only in a very narrow band immediately next to the stimulated sensory cell, so pigment will be secreted in a thin and regular stripe. The shell grows to form a handsome and stylish pattern of regular horizontal pinstripes. As the shell grows larger over time, the spacing

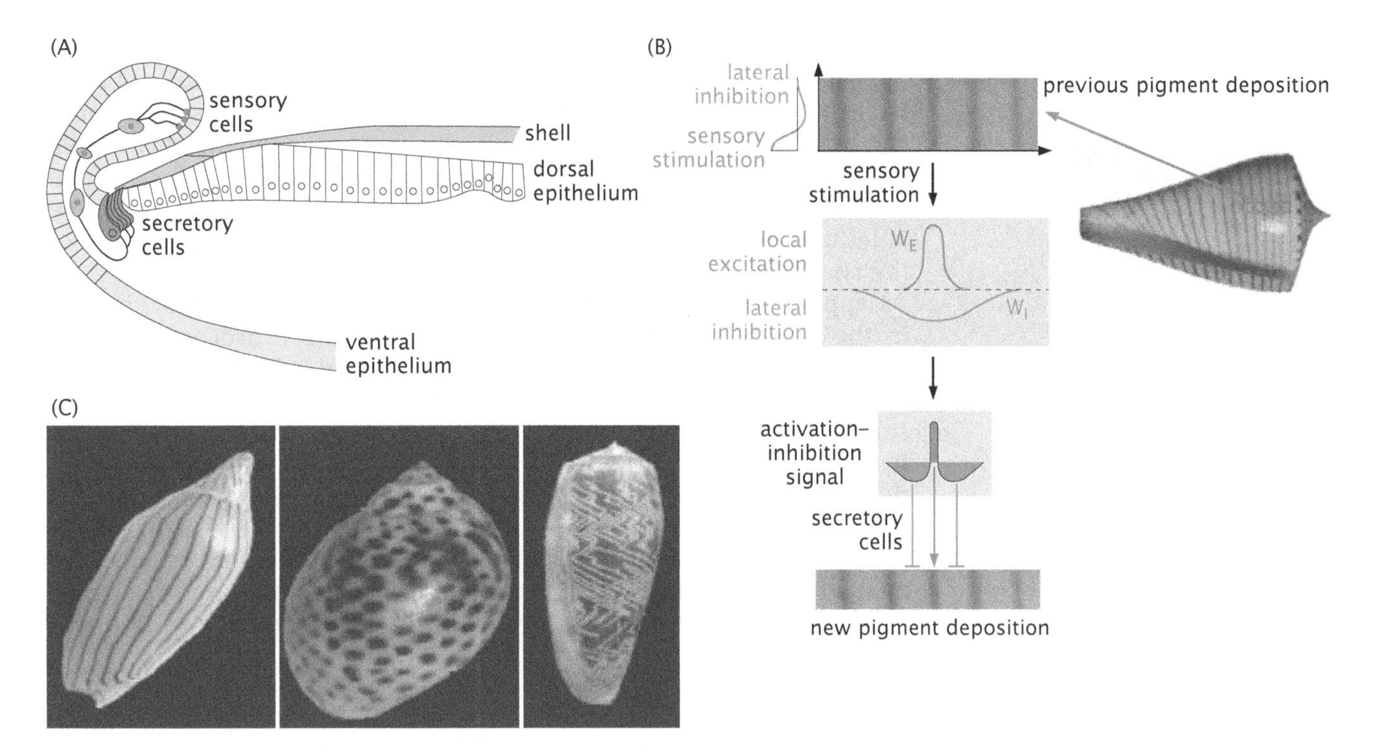

Figure 20.25: Model for seashell pattern formation. (A) A cross-section near the growing lip of a shell shows how the mantle of the mollusk can reach around the edge to "taste" the pre-existing pattern. The pigment pattern detected by the sensory cells is processed according to their pattern generation program, and communicated to the secretory cells, which lay down a new strip of appropriately pigmented shell. (B) Generation of horizontal pinstripes by propagation of a standing-wave pattern. The sensory cells detecting pigment generate both excitatory and inhibitory signals with respect to pigment generation by the secretory cells. The excitatory signals act over short distances and the inhibitory signals act over longer distances, maintaining narrow stripes. (C) Modifications of the processing rules followed by the sensory cells can create vertical stripes, spots, zigzags, or many, many other patterns. (Adapted from A. Boettiger et al., *Proc. Natl Acad. Sci. USA* 106:6837, 2009.)

between pinstripes is maintained (that is, the wavelength λ of the standing-wave pattern remains constant).

The ultimate pigment pattern on the shell depends on the intrinsic cellular program that the sensory cells use to communicate with one another and with their associated secretory cells. Figure 20.25(C) shows a few simple variations on the theme. If, instead of setting up a lateral standing wave, the sensory cells developed a temporal oscillator, they might all stimulate the secretory cells to generate pigmented shell on a particular day, followed by several days of unpigmented shell secretion. The net result would be vertical pinstripes rather than horizontal ones, as shown on the left. Combining standing waves with oscillators, or mixing in propagating waves, can result in patterns of spots or zigzags, as shown for two other shells in this figure.

One appealing aspect of this model for seashell pattern formation is that it is experimentally testable, resulting in measurements that are in reasonable accord with theory. An investigator can disrupt the ability of the mollusk's mantle to read the previous pigment pattern by judicious application of fine sandpaper to a small region of the cell's edge. For a given pattern generation rule, the animal should be able to eventually regenerate its normal pattern *de novo*, but the particular distortion of the pattern in the sanded region can be predicted computationally.

This description of pigment patterning rules can be modified to also apply to seashell texture and shape. Figure 20.26 shows two aspects of how small quantitative changes in the seashell growth program

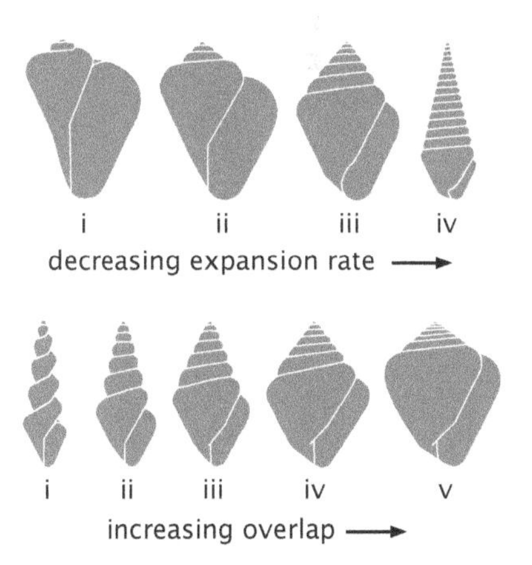

Figure 20.26: Geometrical patterns of shell growth. As mollusks lay down the progressive swirls of their spiral shells, the overall shape of the cell is set by a small number of geometrical parameters. These include the expansion rate and the angle of overlap. (Adapted from D. M. Raup, *Proc. Natl Acad. Sci. USA* 47:602, 1961.)

can lead to large-scale qualitative changes in cell shape. In the daily deposition of new shell, the mollusk can make the new strip of shell slightly longer than the old strip, in order to build itself a "more stately mansion." An impatient mollusk will expand quickly, while another might expand more slowly, resulting in very different final shapes. Similarly, slight changes in the angle of overlap between the older and newer parts of the shell can change the net aspect ratio of the shell.

Happily, the engine of evolution can easily generate modest changes in the quantitative parameters that govern these processes. Rapid evolution can often be seen most easily in island habitats, where introduction of a particular type of animal or plant into a new location can sometimes allow the organism and its descendants the opportunity to exploit a variety of underused ecological niches, resulting in adaptive radiation. Darwin's finches on the Galapagos Islands are a famous example of this idea. Also on the Galapagos, the land snails of the genus *Bulimulus* have undergone a spectacular radiation, resulting in more than 70 distinct endemic species sprinkled among the islands. The phylogenetic tree in Figure 20.27 shows the relative relatedness

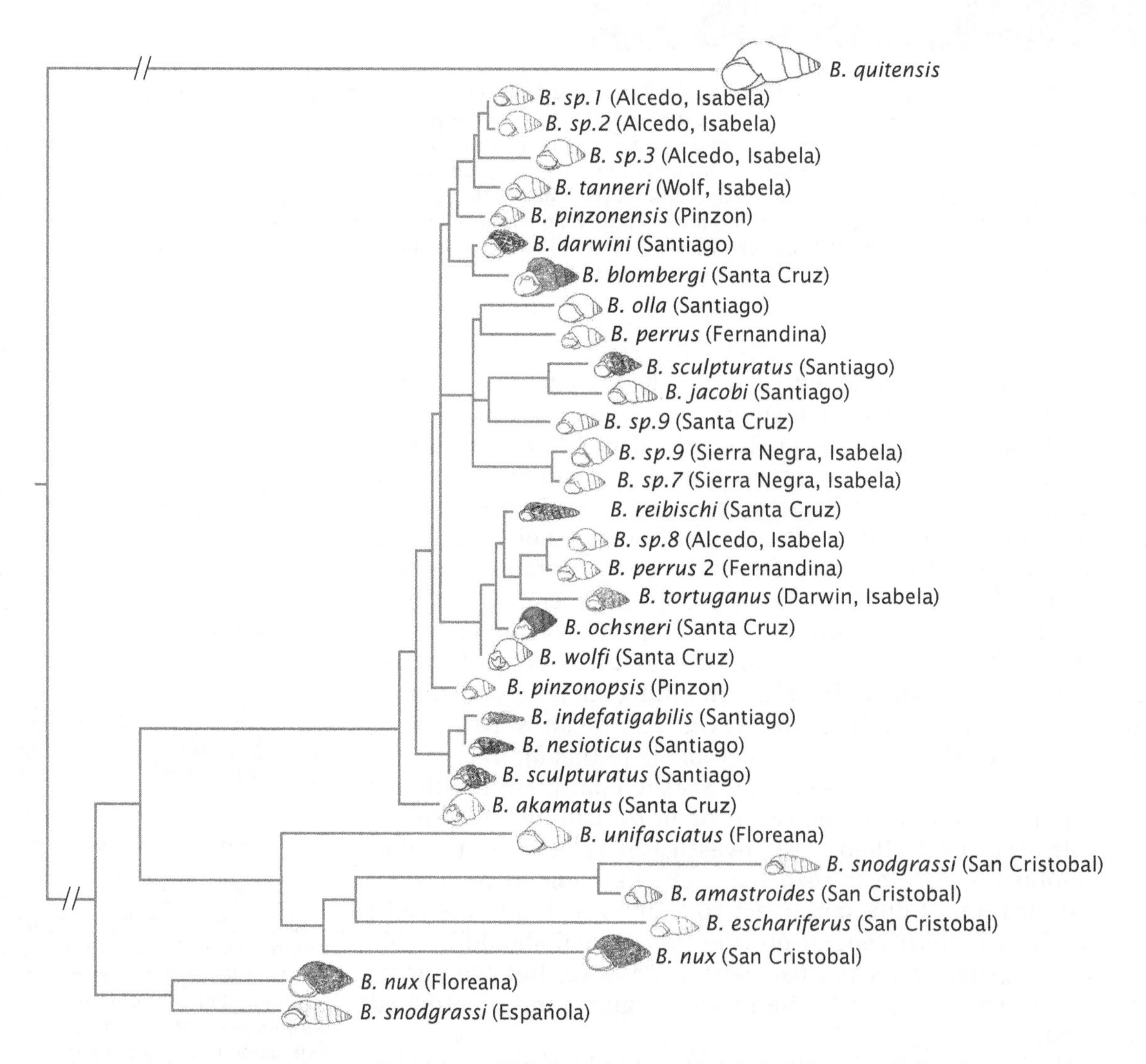

Figure 20.27: Phylogenetic tree of *Bulimulus* land snails on the Galapagos Islands. The lengths of the lines represent evolutionary distance as determined by DNA sequence similarity. Notably, species on older islands such as San Cristobal connect at deeper nodes. (Adapted from C. E. Parent et al., *Philos. Trans. R. Soc. Lond. B Biol. Sci.* 363:3347, 2008.)

of a subset of these species, together with a drawing of each shell and the name of the island where that particular species is found. It is immediately clear that very closely related species may show dramatic variations in size, shape, and pigmentation.

20.5 Pattern Formation as a Contact Sport

Among the most remarkable classes of patterns are those exemplified in Figure 20.1(C) where we see that in a field of cells, neighboring cells assume different fates resulting in generalized "checkerboard" patterns. How is it that one cell can assume one fate and adjacent cells another? Note that, in many cases, these decisions about fate reflect differences in patterns of gene expression from one cell to the next. That is, the emergence of the pattern is really a reflection of different programs of gene expression in different cells. In this section, we explore one molecular mechanism and an associated model for such local patterns.

20.5.1 The Notch–Delta Concept

One mechanism that has been considered for assembling complex spatial patterns is known as lateral inhibition. Lateral inhibition is a process in which one cell assumes a particular fate (which we will refer to as "primary") and then communicates that fate to neighboring cells, thus preventing them from doing the same and leaving them to a secondary fate. This process leads to a spatially inhomogeneous assembly of cells during development with the cells taking these different fates adopting patterns such as the checkerboard patterns shown schematically in Figure 20.28(A). This kind of direct cell–cell contact-mediated specification mechanism works particularly well for establishing "salt-and-pepper" fine-scale patterns, where individual cells within a larger tissue take on fates that are different from their near neighbors. Although we will focus in this section on the idea of lateral inhibition, similar molecular mechanisms can also operate to generate lateral activation, where one cell in a tissue will influence its near neighbors to adopt the same fate as it does. Lateral activation can help to make slightly noisy differentiation programs operate more smoothly, and coordinate the expression of near-simultaneous programs in neighborhoods of cells.

One of the best studied mechanisms of lateral inhibition is provided by the Notch–Delta system. In this case, the inhibition of the primary cell fate is accomplished by a molecule known as Notch. Notch is a membrane-bound receptor that is activated by the transmembrane ligand Delta supplied by a neighboring cell. As a result of this intercellular communication, there is a feedback mechanism that leads to amplification of small differences of Notch activity between neighboring cells, and therefore different cell fates for these cells. This feedback is established by decreased Notch activity, which results in stimulation of Delta activity in the same cell. A high level of Delta activity then stimulates Notch activity in neighboring cells, thus preventing them from assuming the primary fate, which is reserved for the cell with the low level of Notch activity. This mechanism of lateral inhibition is summarized by the cartoon in Figure 20.28(B).

We investigate the Notch–Delta mechanism in quantitative detail by resorting to a simple model that, like the schematic in Figure 20.28,

Figure 20.28: The mechanism of lateral inhibition. (A) In a patch of cells, lateral inhibition can lead to periodic patterns of cells that assume different cell fates. (B) Cells with a large Notch activity repress their primary cell fate. Notch activity is stimulated by Delta activity from a neighboring cell, while Notch activity within a cell down-regulates Delta activity in the same cell. One cell assumes its primary fate (shown in a darker shade of green) by having a reduced Notch activity, which stimulates a high Delta activity. The Delta in turn stimulates the Notch activity of the neighboring cell, which is inhibited from also assuming the primary fate. This leads to a situation where small differences in Notch activity are amplified and neighboring cells assume different fates. The color convention is such that colored molecules and lines indicate activity, whereas gray molecules and black lines signify inactive molecules.

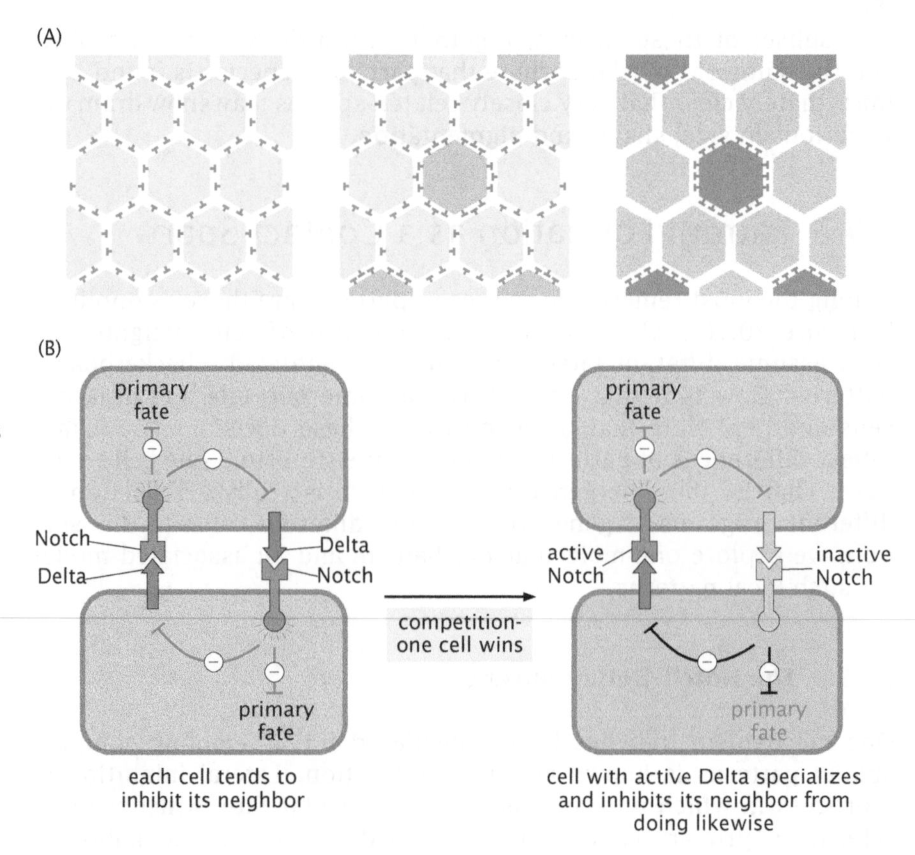

considers only two neighboring cells. Each cell is characterized by levels of Notch and Delta activity, which we denote by N_1, D_1 and N_2, D_2 for cells 1 and 2, respectively; we assume that all activities are normalized by their maximum value. The time evolution of the activities in the two cells can be written down as a system of rate equations, which follows from the trajectories-and-weights schematic shown in Figure 20.29 and builds upon precisely the same kind of arguments we have now used many times throughout the book.

There are four rate constants that describe this model of Notch and Delta and their interactions. The rate of Notch activation $F(D')$ is a monotonically *increasing* function of Delta activity in the *other* cell, denoted by D'. In a beautiful set of measurements, the characteristics of this function were measured directly by tethering Delta to a surface and using fluorescence to read out the Notch activity as shown in Figure 20.30. The rate of increase of Delta activity is given by the function $G(N)$, which is a *decreasing* function of Notch activity in the *same* cell. Finally, the rate constants γ_N and γ_D describe the rate of decay of Notch and Delta activities. The important qualitative conclusions that follow from this model are independent of the particular choice of functional form for the rates of Notch and Delta activation. In order to be concrete, a particular choice of these functions using the conventional Hill form is given in Figure 20.29.

The time evolutions of the Notch and Delta activities in cell 1 are

$$\begin{aligned}\frac{dN_1}{dt} &= F(D_2) - \gamma_N N_1, \\ \frac{dD_1}{dt} &= G(N_1) - \gamma_D D_1,\end{aligned} \qquad (20.52)$$

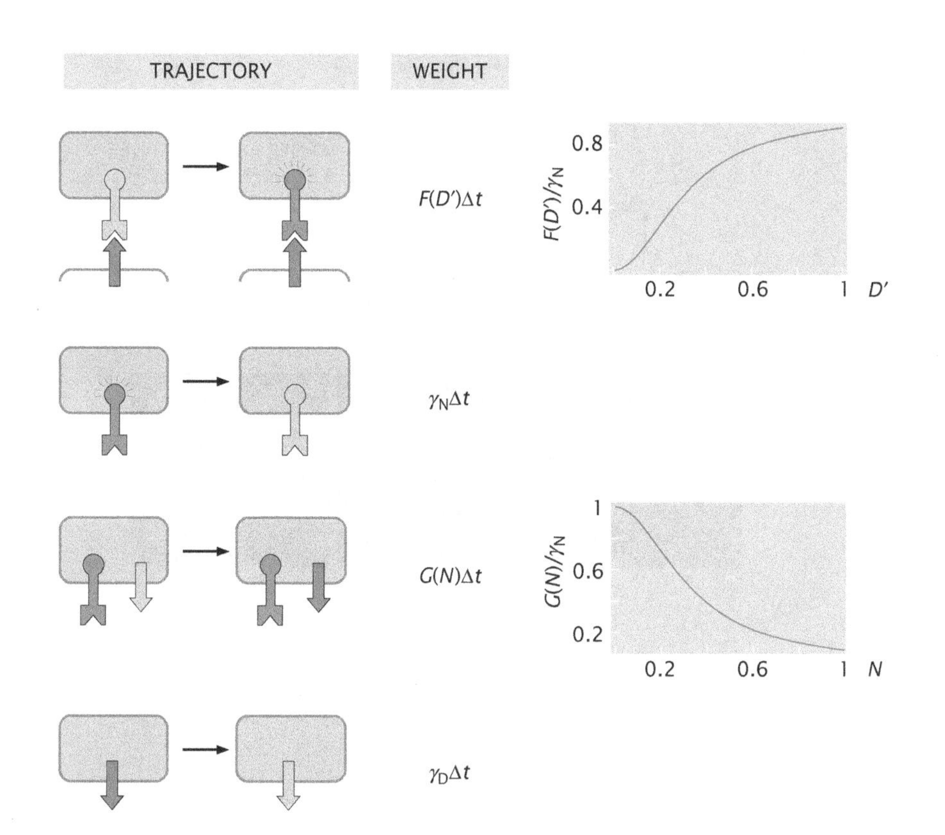

Figure 20.29: Trajectories and weights for a model of Notch–Delta lateral inhibition. Notch is stimulated by high Delta activity in the neighboring cell, indicated by a rate $F(D')$ that is a monotonically increasing function of the concentration D' of Delta in the neighboring cell. High Notch activity, on the other hand, leads to low Delta activity in the same cell, indicated by a rate $G(N)$ that is a monotonically decreasing function. Both activities have a constant rate of decay. The functional forms for the two rates that are plotted are conveniently described using Hill functions of the form

$F(x) = \gamma_N \frac{x^2}{0.1+x^2}$

and $G(x) = \gamma_D \frac{1}{1+10x^2}$.

and similarly we have

$$\frac{dN_2}{dt} = F(D_1) - \gamma_N N_2, \qquad \frac{dD_2}{dt} = G(N_2) - \gamma_D D_2 \tag{20.53}$$

for cell 2. We can render this system of equations dimensionless if we divide the left and right sides of each of the equations by the rate of Notch decay γ_N. The new set of equations is

$$\frac{dN_1}{d\tau} = f(D_2) - N_1, \qquad \frac{dD_1}{d\tau} = \nu[g(N_1) - D_1], \tag{20.54}$$

and,

$$\frac{dN_2}{d\tau} = f(D_1) - N_2, \qquad \frac{dD_2}{d\tau} = \nu[g(N_2) - D_2], \tag{20.55}$$

where $f = F/\gamma_N$ and $g = G/\gamma_D$ are dimensionless rates of Notch and Delta activity production, while $\tau = \gamma_N t$ is the dimensionless time and $\nu = \gamma_D/\gamma_N$ is the ratio of the two decay rates. The mathematical problem we are invited to solve is the long-time behavior of this system of rate equations given a set of initial conditions, in particular assuming an initially small difference between Notch activity in the two cells. More precisely, we are especially curious to know if there are stable solutions for this system corresponding to elevated Notch activity in one cell and not in the other.

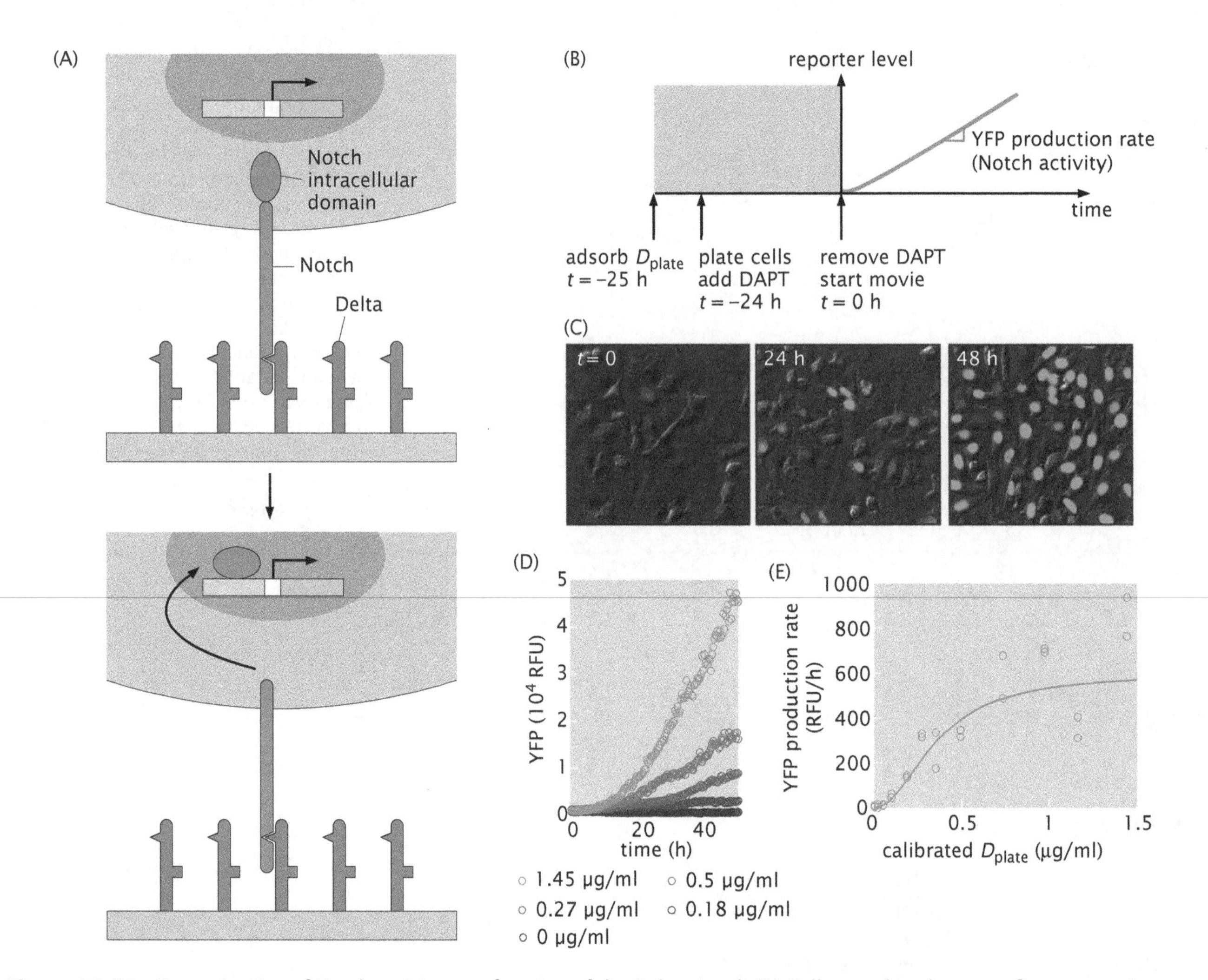

Figure 20.30: Determination of Notch activity as a function of the Delta signal. (A) Cells are placed on a surface presenting a uniform concentration D_{plate} of Delta. Upon binding to Delta, the Notch intracellular domain translocates to the nucleus and activates expression of the fluorescent protein gene (*YFP*). (B) Cells are initially placed on the surface in the presence of DAPT, a small molecule that inhibits Notch activation. The stopwatch responsible for timing the experiment begins (that is, $t = 0$) when the DAPT is washed out. (C) Time-lapse video of *YFP* activation by Notch–Delta interaction. (D) Time traces of YFP fluorescence are obtained for different concentrations of Delta on the surface. (E) YFP production rate, which is taken as a proxy for Notch activity, as a function of the concentration of Delta on the plate. (B–E, adapted from D. Sprinzak et al., *Nature* 466:86, 2010).

As written, note that we have four coupled nonlinear ordinary differential equations and we are interested in their long-time behavior as that will reveal insights into the fates assumed by the two cells. To simplify the problem of the long-time behavior, we analyze the situation in which the parameter $v \gg 1$; the reader is asked to consider the opposite limit, $v \ll 1$ in the problems at the end of the chapter. What this choice of the parameter v implies is that the values of D_1 and D_2 will "quickly" relax to their steady-state values and then the dynamics for N_1 and N_2 will unfold much more slowly. Effectively, there is a separation of time scales between the Notch and Delta degrees of freedom. In this case, the terms that multiply v in both equations are necessarily much smaller than 1 (of order $1/v$), which implies in turn that the terms in square brackets are nearly zero. This statement implies in turn that $g(N_1) \approx D_1$ and $g(N_2) \approx D_2$, which can now be used to eliminate reference to the variables D_1 and D_2 altogether. Substituting these approximations into the equations for the rate of change of

Notch activity in the two cells results in

$$\frac{dN_1}{d\tau} = f(g(N_2)) - N_1,$$
$$\frac{dN_2}{d\tau} = f(g(N_1)) - N_2. \tag{20.56}$$

By exploiting this separation of time scales, we are left with a 2 × 2 system of linear differential equations that we can analyze for its long-time behavior by constructing the phase portrait, which tells us the direction in which the vector $(dN_1/d\tau, dN_2/d\tau)$ is pointing for every choice of N_1 and N_2.

To set up the phase portrait, we first draw the nullclines given by the equations $f(g(N_1)) = N_2$ (shown in blue in Figure 20.31) and $f(g(N_2)) = N_1$ (shown in red in Figure 20.31). Those points at which the two curves intersect are fixed points, values of the Notch activities for which the rates of change of both Notch activities are zero. The vector representing the rate of change of (N_1, N_2) is vertical on the red line and horizontal on the blue line, while its direction (up or down, left or right) can be determined by appealing to the direction of the vector at the special points $(N_1, N_2) = (1, 1)$ and $(0, 0)$. In the first case both components of the vector $(dN_1/d\tau, dN_2/d\tau)$ are negative, while in the latter they are both positive. The fact that the sign of each component of the rate vector changes only at the nullclines means by continuity that we know the sign of the two components of the vector throughout the phase portrait. Indeed, this allows us to draw the rate-of-change vector on the nullclines, and, by continuity, everywhere else in the (N_1, N_2) plane, as shown in Figure 20.31. The phase portrait obtained in this way confirms our intuition: the steady state with Notch activity equal in both cells is unstable and small differences between the two cells are amplified, leading to new steady states in which the cell with an initially higher Notch activity ends up with a high Notch activity,

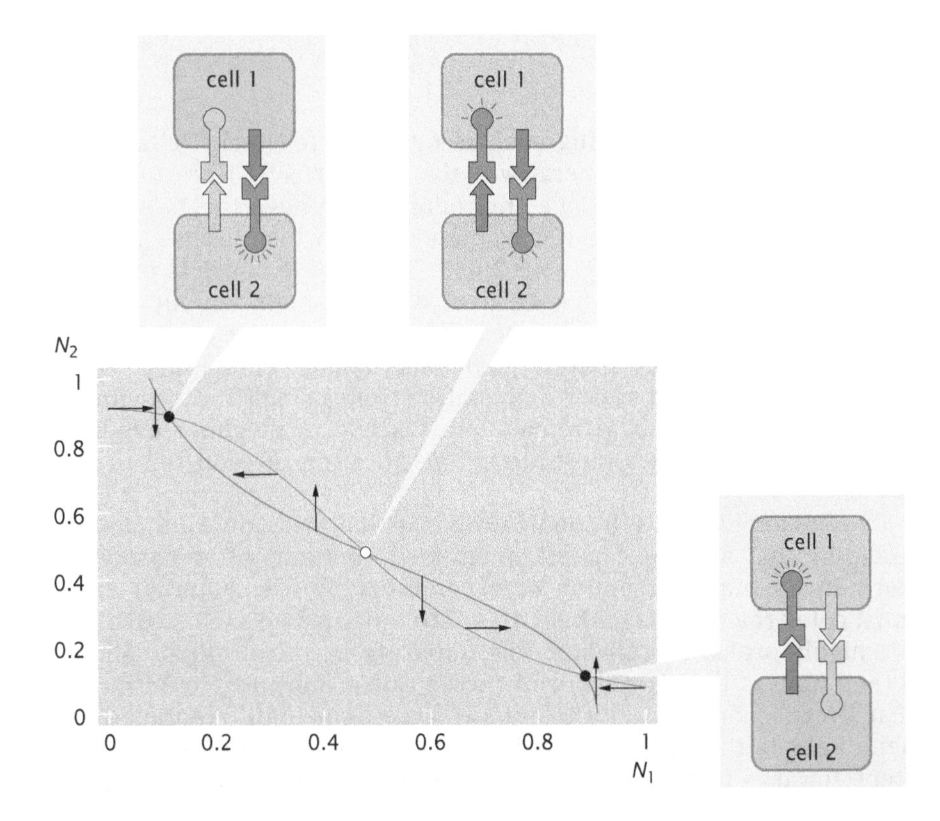

Figure 20.31: Nullclines and phase portrait for the Notch–Delta model. The red line is the nullcline along which the rate of change of Notch activity in cell 1 vanishes (that is, $dN_1/d\tau = 0$), while the blue line corresponds to $dN_2/d\tau = 0$. The black vectors show the direction of the vector $(dN_1/d\tau, dN_2/d\tau)$ that gives the rate of change of the Notch concentrations in different parts of the phase portrait. These vectors indicate two stable steady states in which the two cells exhibit large differences in Notch activity and an unstable steady state in which the Notch activity in the two cells is equal and moderate. The prediction of the phase portrait is that initial small differences in Notch activity will be amplified over time, leading to the cell with the initially slightly lower Notch activity assuming the primary cell fate, with the other cell remaining in the initial state corresponding to the secondary fate.

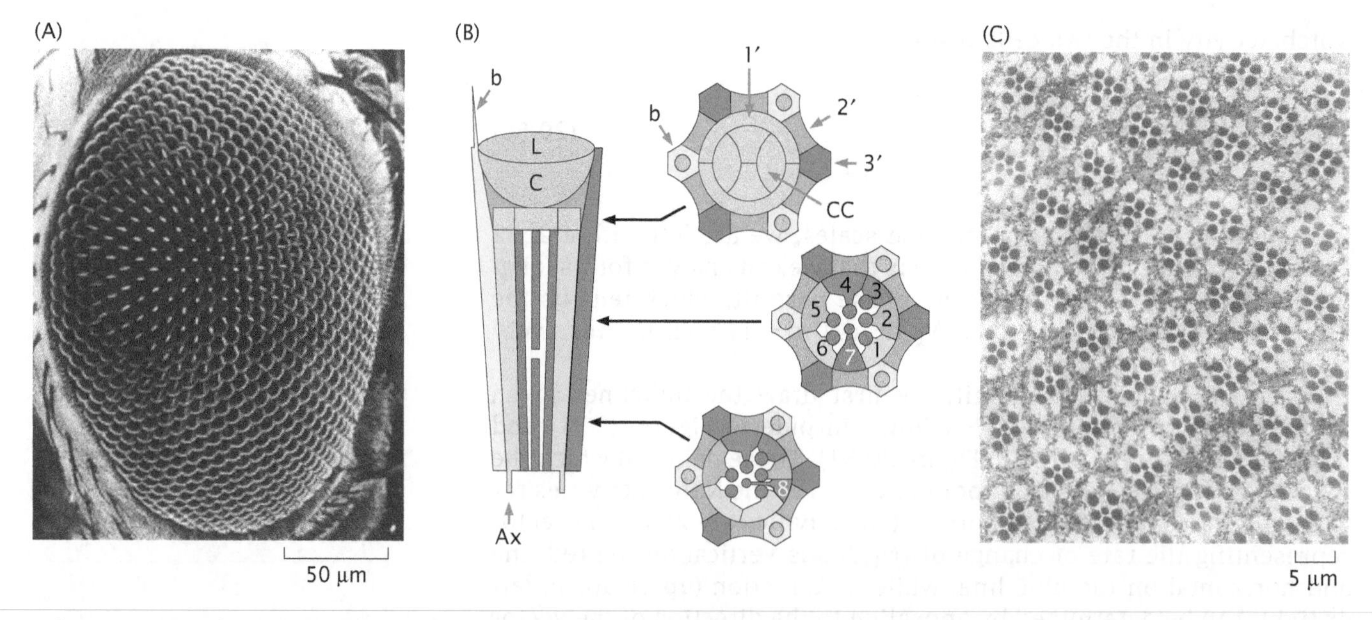

Figure 20.32: The *Drosophila* eye. (A) Scanning electron micrograph showing the regular hexagonal packing of the ommatidia. Tiny bristles protrude at three of the vertices of each hexagon. (B) Each ommatidium is made up of about 20 distinct cells, including eight photoreceptors (labeled 1 through 8), four lens-secreting cone cells (CC), three classes of pigment cells (1′ through 3′), and supporting cells, including those that form the bristles (b). The precise arrangement of the cells can be appreciated in three cross-sections at different heights through the sheet. The uppermost cross-section shows the diamond packing of the four lens-secreting cells (CC). The lens itself (L) and the fluid-filled pseudocone (C) are shown. The regular arrangement of the photoreceptors can be seen in cross-sections further down. An ommatidial axon is shown as Ax. (C) A thin-section electron micrograph at approximately the level of the second cross-section in (B). (A, C, adapted from D. F. Ready et al., *Dev. Biol.* 53:217, 1976; B, adapted from C. A. Brennan and K. Moses, *Cell. Mol. Life Sci.* 57:195, 2000.)

while the other cell assumes a low Notch activity. Thus, a small initial difference in Notch activity ultimately results in the two cells assuming different fates. This conclusion depends on the functional forms of the rates f and g, and we invite the reader to explore this in more detail in the problems at the end of the chapter.

20.5.2 *Drosophila* Eyes

Neighbor–neighbor signaling such as through the Notch–Delta mechanism is an important general way that similar cells from the same lineage and living in the same environment can nevertheless assume quite distinct fates. A dramatic example of this principle is seen in the *Drosophila* eye shown in Figure 20.32. The fly eye consists of about 800 units called ommatidia, each of which contains about 20 distinct cells. The ommatidia and the cells within them are arranged in a very precise repeating pattern that looks almost crystalline. All of these distinct cells arose during pupation from a single uniform undifferentiated epithelium. A remarkable cascade of neighbor–neighbor signaling events drives the ordered specification of each cell in this tissue.

The *Drosophila* larva, a food-eating machine, has only rudimentary eyespots that are not capable of image formation. After cocooning itself in its pupa, dreaming of becoming a fly, the pupating larva must convert a monolayer field of undifferentiated epithelial cells into the highly ordered adult eye. The initiating event in retinal cell differentiation is the movement of the so-called morphogenetic furrow that sweeps from posterior to anterior across the undifferentiated cell field. Cells in the furrow pause in their divisions and initiate the first steps towards fate commitment. Cell differentiation in the epithelial

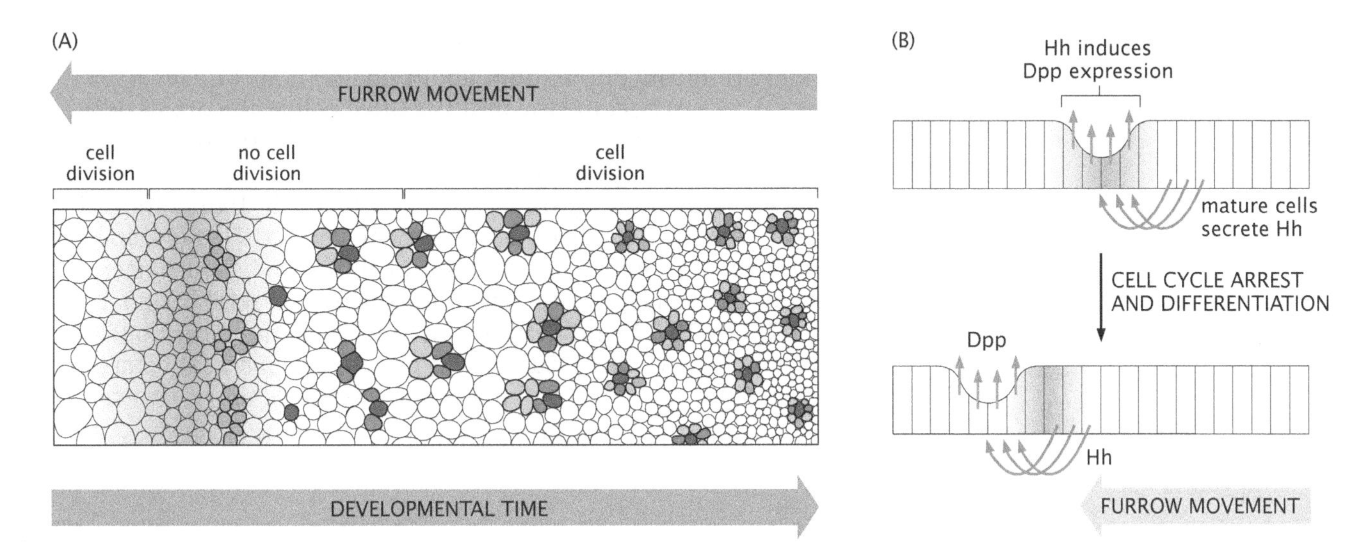

Figure 20.33: Progress of the morphogenetic furrow. (A) In this top view of the differentiating *Drosophila* retina, the morphogenetic furrow (gray band) has moved across the epithelial field from right to left. Cells begin to differentiate as the furrow passes, so cells on the left side are developmentally young or naive and cells become progressively more differentiated towards the right. The R8 cell (dark orange) is the first to differentiate. (B) As seen from the side, the progress of the morphogenetic furrow depends on the secretion of Hedgehog (Hh) by cells that have just begun to differentiate. This secreted signal induces neighbors to express decapentaplegic (Dpp), which arrests the cell cycle and also triggers a shortening of the cells that causes the appearance of the furrow. (Adapted from P. A. Lawrence, The Making of a Fly. Wiley-Blackwell, 1992.)

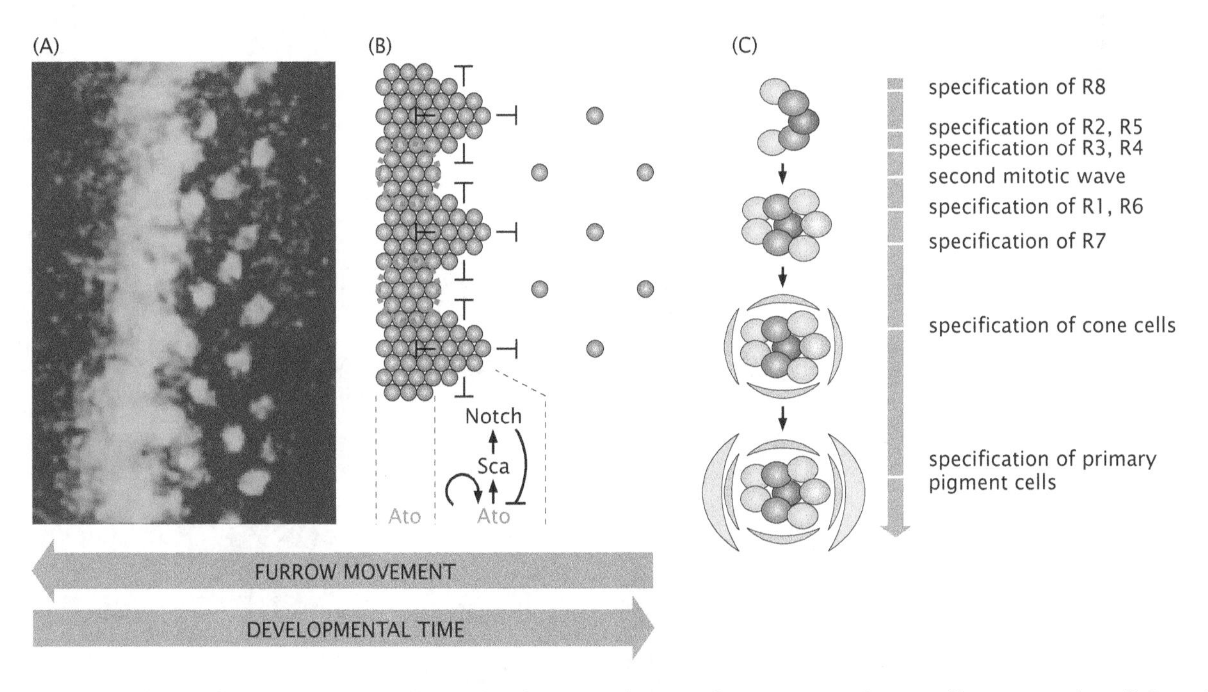

Figure 20.34: Cell specification in *Drosophila* eye development. (A) Atonal protein (green) is initially expressed in all the cells in the morphogenetic furrow. As these cells begin to differentiate, they signal to one another and shut down Atonal expression in their neighbors until only a few cells, evenly spaced, continue to express the protein. (B) A schematic diagram of Atonal expression and signaling. The clusters of cells shown in the red circles are furthest away from the inhibitory signals generated by the differentiating clusters, so these will retain Atonal expression longest and will give rise to the R8 cell in the next wave. This spatial signaling maintains the regular hexagonal spacing. Cells not expressing atonal are not shown. (C) The last cells continuing to express Atonal begin to differentiate into photoreceptor cells, type R8. These then initiate cell–cell signaling events that specify all of the remaining cells in the ommatidium. (A, adapted from M. E. Dokucu et al., *Development* 122:4139, 1996; B, adapted from B. J. Frankfort and G. Mardon, *Development* 129:1295, 2002; C, adapted from J.-Y. Roignant and J. E. Treisman, *Int. J. Dev. Biol.* 53:795, 2009.)

field occurs in an orderly fashion because the furrow progresses like a wave in a football stadium where the cells in each row await signals from their neighbors, as shown in Figure 20.33.

As the furrow passes, a vertical row of cells in the field begins to differentiate. Within this population, about one cell out of every 30 will win the local race to become the first specified cell in the ommatidium, the photoreceptor known as R8. The mechanism is analogous to that shown in Figure 20.28 (p. 940), except that in this context the relevant inhibitory signal is able to diffuse across a few cell diameters rather than simply signaling to the cells in physical contact. This is a specific instance of a more general mechanism of diffusion-based patterning that we have already explored in more detail in Section 20.3.1. As shown in Figure 20.34, the field of cells just past the morphogenetic furrow is now lightly sprinkled with R8 precursors in a neat hexagonal pattern. The R8 cells induce two immediate neighbors to assume

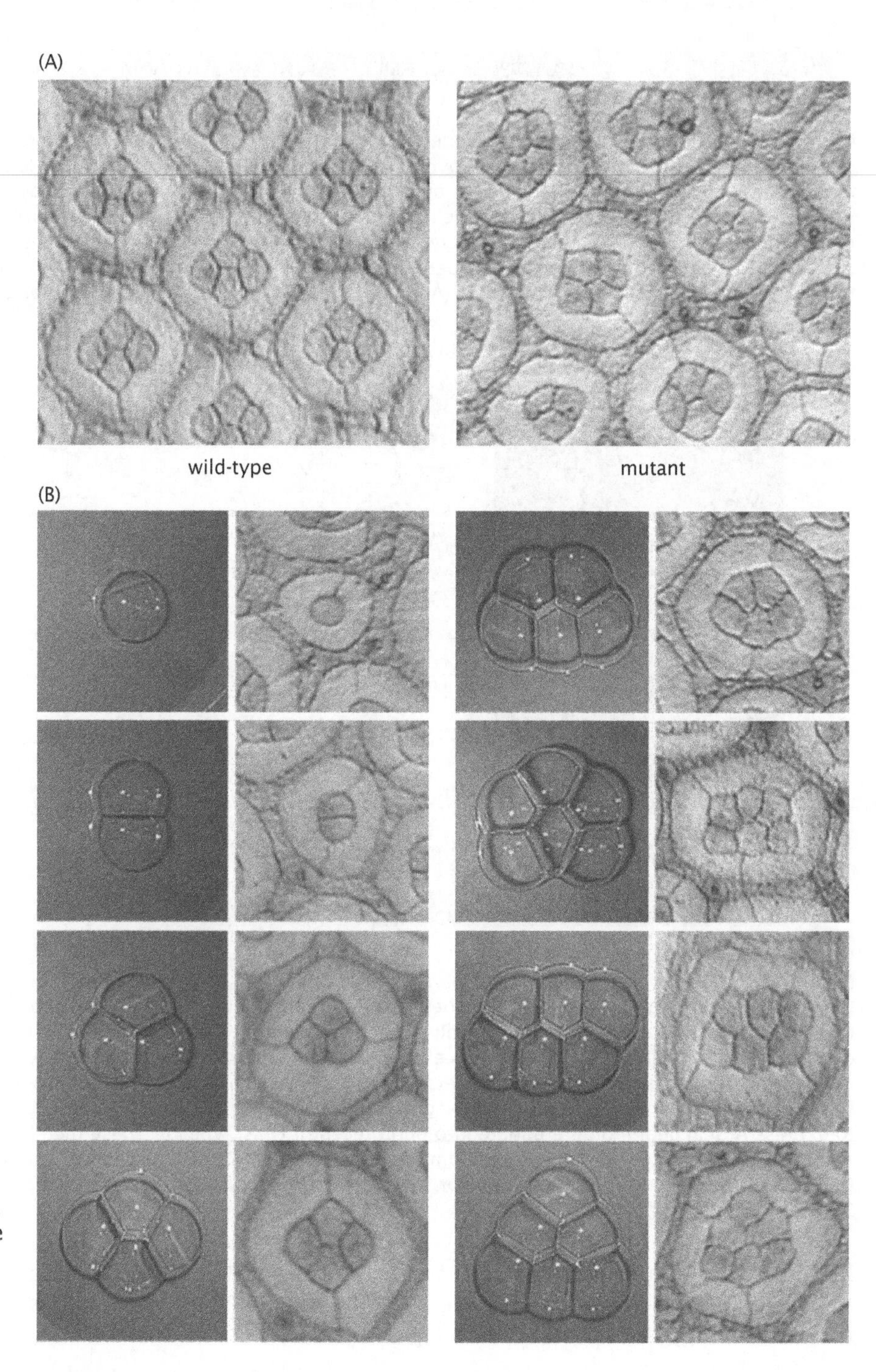

Figure 20.35: Cone cell patterning in the *Drosophila* retina. (A) Wild-type (left) and *Rough eye* (right) retinal cross-sections. (B) Individual ommatidia from the *Rough eye* mutant showing the range of observed packing patterns for one to six cone cells. (Adapted from T. Hayashi and R. W. Carthew, *Nature* 431:647, 2004.)

a different photoreceptor cell fate than R2 and R5. Stepwise, each differentiating cell triggers a neighbor to form the rest of the photoreceptors, the cone cells that will make the lens, and the two large primary pigment cells that form the boundaries of the ommatidium. Left over cells that do not receive a particular fate die by apoptosis.

The final establishment of the remarkably regular eye structure relies on the mechanics of cell–cell interactions as well as on their chemical signaling. This is particularly well illustrated in a series of experiments on the lens-making cone cells. Figure 20.35(A) shows a cross-section through a normal retina at the level of the top slice in Figure 20.32(B). Within each ommatidium, exactly four cells form a well-oriented diamond. To the right is a similar image from a particular mutant animal where cone cell differentiation is disrupted such that each ommatidium may have anywhere from one to six cone cells. Although the numbers of cells vary, they still form regular compact patterns. Figure 20.35(B) suggests that this packing may be due simply to the physical strength of cell–cell adhesion within these groups. For each pattern of cells shown in the right columns, a comparable conglomeration of soap bubbles is shown on the left.

The tremendous number of specific *Drosophila* mutants with slightly abnormal eyes has provided numerous such insights into the mechanisms of pattern formation in this tissue. For example, the *eyeless Russian* mutant shown in Figure 20.36 demonstrated that regular diamonds could form in place of regular hexagons. Rather than representing a completely different geometry for pattern formation, this mutant simply shows what happens when the pigment cells fail to enlarge and the shape of the ommatidium therefore defaults to the diamond packing of the cone cells.

Cell–cell signaling as exemplified by Notch–Delta is a powerful organizing principle. Although the simplest manifestation as we derived here enables cells to make only binary decisions between fate A and fate B, developing organisms don't get tired of using this same principle repeatedly to make very complex tissues with many cell types. Each local decision is still binary, A or not A, B or not B, etc., but the fact that these binary fate decisions are strung together in a cascade enables the specification of, in this case, 20 identifiably distinct cells packed into a highly regular geometry.

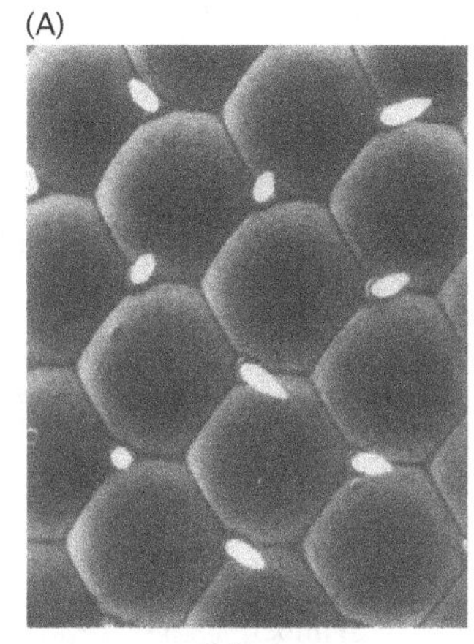

Figure 20.36: Packing patterns. As shown in scanning electron microscopy, wild-type ommatidia are hexagonal while *eyeless Russian* ommatidia are equally regular diamonds. (Adapted from D. F. Ready et al., *Dev. Biol.* 53:217, 1976.)

20.6 Summary and Conclusions

The formation of specific recognizable and reproducible patterns is one of the hallmark features of living organisms. The emergence of these patterns is a complex ballet of signaling, mechanical interactions, and gene expression, as well as many other important biological processes. In this chapter, we have given a brief sketch of some of the pioneering quantitative ideas that have been put forth as a conceptual framework to explain patterns. As the example of the fly's eye amply demonstrates, the patterning process is both subtle and complicated and the vast harvest of different mutants makes it clear that there is much molecular specificity that confers correct patterning. Nevertheless, many of the simple ideas introduced here can serve as a starting point for coming to terms with these beautiful features of living organisms.

20.7 Problems

Key to the problem categories: • Model refinements and derivations, • Estimates, • Model construction

• 20.1 Testing the French flag model

The French flag model shown diagrammatically in Figure 20.3 states that the spatial location of developmental decisions, such as the expression of a gene, are made by reading out different levels of a morphogen gradient. As a result, if we were to change the shape of the input morphogen, there should be a corresponding change in the downstream transcriptional decisions.

(a) Model the Bicoid gradient in *Drosophila melanogaster* as an exponential with decay constant λ and concentration at $x = 0$ of $[\text{Bcd}]_0$. Under these conditions, a certain gene is expressed at position x_0. What concentration of Bicoid is there at the point x_0? How would the position of the expression of the gene change if we were to change the overall Bicoid concentration by a factor α?

(b) Let's assume that a feature in *Drosophila* development such as the cephalic furrow, the boundary between the body of the fly and the structure that will give rise to its head, is determined solely by the Bicoid gradient. Given that for the wild-type case the furrow occurs at about 35% egg length, how would you expect the furrow to move if the dosage of the *bicoid* gene were changed by adding or subtracting copies of gene on the genome? This question was asked by Driever and Nüsslein-Volhard (1988) by creating mutant flies with different dosages of *Bicoid*, as shown in Figure 20.37. Compare their data for the displacement of the cephalic furrow with the model calculated in (a) by plotting them together. Does the model hold? Discuss the possible explanations.
(Relevant data for this problem can be found on the book's website.)

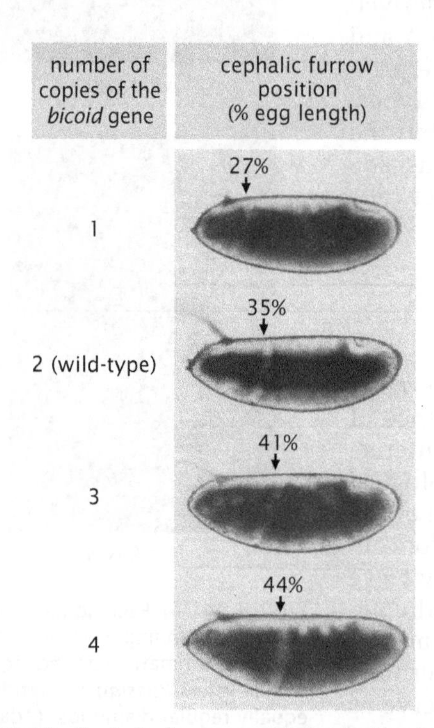

Figure 20.37: Cephalic furrow position as a function of Bicoid dosage. Mutant flies bearing different numbers of copies of the *bicoid* gene were created and the positioning of their corresponding cephalic furrows were measured. (Adapted from W. Driever and C. Nüsslein-Volhard, *Cell* 54:95, 1988.)

• 20.2 Transcription and translation in development

(a) The average length of a gene in *Drosophila melanogaster* is about 11 kb and the average elongation rate of a transcript is about 1.2 kb/min. How does the time to produce an average mRNA compare with the nuclear cycle times in the initial stages of fly development? For example, nuclear cycles 9 through 11 last no more than 9 minutes, while nuclear cycle 12 lasts about 12 minutes and cycle 13 lasts approximately 20 minutes. How do the genes that are actually expressed in this stage compare with the average gene in terms of their lengths? You can search for the size of these genes by going to http://flybase.org and searching for hb (*hunchback*), gt (*giant*), kr (*krüppel*), and kni (*knirps*).

(b) When *Drosophila* eggs are laid, they already contain mRNA for several "maternal factors." Bicoid is an example of such a factor. Its mRNA is localized at the anterior end of the embryo, serving as a source of Bicoid protein. It is essentially stable up until the end of nuclear cycle 14, when it gets actively degraded. In this problem, we want to estimate the number of mRNA molecules deposited in the embryo by its mother. To make this estimate, we appeal to measurements of the number of Bicoid proteins at nuclear cycle 14. Assume that all Bicoid is localized to the nuclei, which at cycle 14, approximately 120 minutes after the egg has been laid, have a radius of about 3.3 μm. Use the data shown in Figure 20.6(B) in order to estimate the total number of Bicoid molecules in the whole embryo at this point. Assuming that translation of *bicoid* mRNA is constant, estimate the number of mRNA molecules that led to your calculated number of Bicoid proteins. You might find it useful to estimate the number of ribosomes per kb on a transcript from Figure 3.13 (p. 108) and to use the translation rate discussed in Section 3.2.1 (p.107).

• 20.3 Phase portrait for a reaction–diffusion system

In Equation 20.33, we wrote the equations describing the dynamics of Turing's simplest reaction model. Construct a phase portrait that shows the dynamics of the system.

• 20.4 One-component Turing

In the chapter, we considered a one-dimensional lattice of cells, each containing two species of morphogens that undergo chemical reactions. Following Turing, we showed that diffusion of morphogens can destabilize a steady state described by a uniform concentration profile, leading to a spatially periodic pattern of morphogens. In this problem, we analyze the situation when there is only one morphogen species present.

(a) Rewrite the reaction–diffusion equation for the Turing system, Equation 20.40, for the case of a single morphogen whose concentration within a cell is Y_r. Then consider a small periodic perturbation of the uniform steady state

$Y_r = Y^*$ of the form $Y_r = Y^* + y(t)e^{i(2\pi r/\lambda)}$ and derive the dynamical equations for the amplitude $y(t)$.

(b) Assuming that there are N cells in the system and that they are arranged in a ring so that the $r = 1$ cell has the $r = N$ and $r = 2$ cells as its nearest neighbors, what are the allowed values of the wavelength λ for the periodic perturbation?

(c) Derive the conditions under which the uniform steady state is unstable to a small periodic perturbation. Argue that this one-component Turing system does not lead to spatially periodic pattern of morphogen concentration.

20.5 Wavelength of Turing patterns

Replace the discrete treatment of the Turing instability in a one-dimensional system discussed in the chapter with the corresponding continuum reaction–diffusion equations

$$\frac{\partial X}{\partial t} = D_X \frac{\partial^2 X}{\partial x^2} + f(X, Y), \tag{20.57}$$

$$\frac{\partial Y}{\partial t} = D_Y \frac{\partial^2 Y}{\partial x^2} + g(X, Y), \tag{20.58}$$

for the morphogen concentrations X and Y, which are both functions of position. Like for the case analyzed in the chapter, the functions $f(X, Y)$ and $g(X, Y)$ reflect the chemical interactions between the morphogens. Also, we introduce the notation

$$a_{11} = \left.\frac{\partial f}{\partial X}\right|_{(h,k)}, \quad a_{12} = \left.\frac{\partial f}{\partial Y}\right|_{(h,k)},$$

$$a_{21} = \left.\frac{\partial g}{\partial X}\right|_{(h,k)}, \quad a_{22} = \left.\frac{\partial g}{\partial Y}\right|_{(h,k)},$$

where h and k are steady-state values of X and Y. We are interested in the stability of the homogeneous steady state $(X, Y) = (h, k)$ with respect to a perturbation that is periodic in space, as an instability signals the spontaneous appearance of a spatially periodic pattern of morphogens.

(a) Linearize the reaction–diffusion equations in the vicinity of the fixed point $(X, Y) = (h, k)$. Consider the time evolution of a periodic perturbation of the form $(\delta X, \delta Y) = (\eta(t)e^{iqx}, \varepsilon(t)e^{iqx})$ in the vicinity of the fixed point (h, k), following the same strategy that we used in the chapter. In particular, derive expressions for the eigenvalues of the rate matrix $\mathbf{A}$, analogous to the one defined in Equation 20.45, in terms of its determinant ($\mathrm{Det}\,\mathbf{A}$) and trace ($\mathrm{Tr}\,\mathbf{A}$).

(b) Show that a necessary condition for linear stability of a fixed point in the absence of diffusion is that at least one of a_{11} and a_{22} is negative.

(c) Show that for a fixed point that is linearly stable in the absence of diffusion, a necessary condition for linear *instability* in the presence of diffusion is that exactly one of a_{11} and a_{22} is negative.

(d) Assuming that a_{22} is negative, show that at a Turing bifurcation (that is, at the onset of instability: if we were to continuously adjust parameters such as reaction rate constants or diffusion constants until the fixed point just becomes unstable), the characteristic length of the emerging pattern from the homogeneous steady state (fixed point) scales like

$$\lambda^{-2} \sim \lambda_X^{-2} - \lambda_Y^{-2},$$

where $\lambda_X \equiv \sqrt{D_X/a_{11}}$ and $\lambda_Y \equiv \sqrt{D_Y/(-a_{22})}$.

(e) Comment on the physical implications of these results. What is the wavelength of an oscillatory instability?
(Problem courtesy of Justin Bois.)

20.6 Notch–Delta model

In the chapter, we considered the dynamics of Notch and Delta in a two-cell system by taking the limit in which the decay rate of Delta is much greater than that of Notch. Here we examine the same problem in the opposite limit.

(a) Argue that in the limit $\nu = \gamma_D/\gamma_N \ll 1$, the Notch activity in the two cells quickly settles into a steady state. What are the resulting dynamical equations for the evolution of Delta that follow from Equations 20.54 and 20.55 in this limit?

(b) Using the functional forms for the Notch and Delta activation rates, $F(D')$ and $G(N)$, described in Figure 20.29, obtain a phase portrait for the dynamics of Delta in the two cells, analogous to the one shown in Figure 20.31. Based on your phase portrait, show that in the long-time limit, the system will settle into a steady state in which one cell assumes the primary fate while the other assumes the secondary fate.

20.8 Further Reading

Collier, JR, Monk, NAM, Maini, PK, & Lewis, JH (1996) Pattern formation by lateral inhibition with feedback: a mathematical model of Delta-Notch intercellular signalling, *J. Theor. Biol.* **183**, 429. This paper serves as the basis for our treatment of the Notch–Delta model.

Cross, M, & Greenside, H (2009) Pattern Formation and Dynamics in Nonequilibrium Systems, Cambridge University Press. This excellent book provides a deep physical and mathematical treatment of pattern formation that goes way beyond what we have included here.

Foe, VE, Odell, GM, & Edgar, BA (1993) Mitosis and morphogenesis in the *Drosophila* embryo: point and counterpoint, in The Development of *Drosophila melanogaster*, edited by M. Bate & A. Martinez-Arias, Cold Spring Harbor Press. This book chapter includes a beautiful table outlining the steps in *Drosophila* development, their duration, and their main features.

Gregor, T, Bialek, W, de Ruyter van Steveninck, RR, Tank, DW, & Wieschaus, EF (2005) Diffusion and scaling during early embryonic pattern formation, *Proc. Natl Acad. Sci. USA* **102**, 18403. This intriguing paper reports observations on the stripes along the anterior–posterior axis of different flies.

Koch, AJ, & Meinhardt, H (1994) Biological pattern formation: from basic mechanism to complex structures, *Rev. Mod. Phys.* **66**, 1481. A thorough review of thinking on pattern formation in biology.

Lewis, J (2008) From signals to patterns: space, time, and mathematics in developmental biology, *Science* **322**, 399. This article gives a brief but insightful description of issues surrounding pattern formation in biology.

Mirny, LA (2010) Nucleosome-mediated cooperativity between transcription factors. *Proc. Natl Acad. Sci. USA* **107**, 22534. This paper served as the template for the MWC model of genomic accessibility that we discussed in the context of Bicoid and *hunchback*.

Roellig, D, Morelli, LG, Ares, S, Jülicher, F, & Oates, AC (2011) SnapShot: the segmentation clock, *Cell* **145**, 800. This outstanding short paper explains models of somitogenesis and has excellent embedded videos that give a visual demonstration of the phenomenon and the model.

20.9 References

Alberts, B, Johnson, A, Lewis, J, et al. (2008) Molecular Biology of the Cell, 5th ed., Garland Science.

Allen, WL, Cuthill, IC, Scott-Samuel, NE, & Baddeley, R (2011) Why the leopard got its spots: relating pattern development to ecology in felids, *Proc. R. Soc. Lond. B Biol. Sci.* **278**, 1373.

Boettiger, A, Ermentrout, B, & Oster, G (2009) The neural origins of shell structure and pattern in aquatic mollusks, *Proc. Natl Acad. Sci. USA* **106**, 6837.

Bradbury, J (2004) Nature's nanotechnologists: unveiling the secrets of diatoms, *PLoS Biol.* **2**, e306.

Brennan, CA, & Moses, K (2000) Determination of *Drosophila* photoreceptors: timing is everything, *Cell. Mol. Life Sci.* **57**, 195.

Buonanni, F (1681) Ricreazione dell'occhio e della mente nell'osservazione della chiocciole, Rome.

Church, AH (1904) On the Relation of Phyllotaxis to Mechanical Laws, Williams & Norgate.

Dokucu, ME, Zipursky, SL, & Cagan, RL (1996) Atonal, Rough and the resolution of proneural clusters in the developing *Drosophila* retina, *Development* **122**, 4139.

Douady, S, & Couder, Y (1992) Phyllotaxis as a physical self-organized growth process, *Phys. Rev. Lett.* **68**, 2098.

Driever, W, & Nüsslein-Volhard, C (1988) The *bicoid* protein determines position in the *Drosophila* embryo in a concentration-dependent manner, *Cell* **54**, 95.

Frankfort, BJ, & Mardon, G (2002) R8 development in *Drosophila* eye: a paradigm for neural selection and differentiation, *Development* **129**, 1295.

Gleissberg, S, Groot, EP, Schmalz, M, et al. (2005) Developmental events leading to peltate leaf structure in *Tropaeolum majus* (Tropaeolaceae) are associated with expression domain changes of a YABBY gene, *Dev. Genes Evol.* **215**, 313.

Gomez, C, Ozbudak, EM, Wunderlich, K, et al. (2008) Control of segment number in vertebrate embryos, *Nature* **454**, 335.

Gregor, T, Bialek, W, de Ruyter van Steveninck, RR, Tank, DW, & Wieschaus, EF (2005) Diffusion and scaling during early embryonic pattern formation, *Proc. Natl Acad. Sci. USA* **102**, 18403.

Gregor, T, Tank, DW, Wieschaus, EF, & Bialek, W (2007) Probing the limits to positional information, *Cell* **130**, 153.

Hayashi, T, & Carthew, RW (2004) Surface mechanics mediate pattern formation in the developing retina, *Nature* **431**, 647.

Jiang, YJ, Smithers, L, & Lewis, J (1998) Vertebrate segmentation: the clock is linked to Notch signalling, *Curr. Biol.* **8**, R868.

Kuhlemeier, C (2007) Phyllotaxis, *Trends Plant Sci.* **12**, 143.

Lawrence, PA (1992) The Making of a Fly, Wiley-Blackwell.

Loose, M, Fischer-Friedrich, E, Ries, J, Kruse, K, & Schwille, P (2008) Spatial regulators for bacterial cell division self-organize into surface waves *in vitro*, *Science* **320**, 789.

Loose, M, & Schwille, P (2009) Biomimetic membrane systems to study cellular organization, *J. Struct. Biol.* **168**, 143.

Meeks, JC, & Elhai, J (2002) Regulation of cellular differentiation in filamentous cyanobacteria in free-living and plant-associated symbiotic growth states, *Microbiol. Mol. Biol. Rev.* **66**, 94.

Meinhardt, H, (1999) Orientation of chemotactic cells and growth cones: models and mechanisms, *J. Cell Sci.* **112**, 2867.

Parent, CE, Caccone, A, & Petren, K (2008) Colonization and diversification of Galpagos terrestrial fauna: a phylogenetic and biogeographical synthesis, *Philos. Trans. R. Soc. Lond. B Biol. Sci.* **363**, 3347.

Raskin, DM, & de Boer, PAJ (1999) Rapid pole-to-pole oscillation of a protein required for directing division to the middle of *Escherichia coli*, *Proc. Natl Acad. Sci. USA* **96**, 4971.

Raup, DM (1961), The geometry of coiling in gastropods, *Proc. Natl Acad. Sci. USA* **47**, 602.

Ready, DF, Hanson, TE, & Benzer, S (1976) Development of the *Drosophila* retina, a neurocrystalline lattice, *Dev. Biol.* **53**, 217.

Reinhardt, D (2005) Phyllotaxis—a new chapter in an old tale about beauty and magic numbers, *Curr. Opin. Plant Biol.* **8**, 487.

Risser, DD, & Callahan, SM (2009) Genetic and cytological evidence that heterocyst patterning is regulated by inhibitor gradients that promote activator decay, *Proc. Natl Acad. Sci. USA* **106**, 19884.

Roignant, JY, & Treisman, JE (2009) Pattern formation in the *Drosophila* eye disc, *Int. J. Dev. Biol.* **53**, 795.

Schröter, C & Oates, AC (2010) Segment number and axial identity in a segmentation clock and period mutant, *Curr. Biol.* **20**, 1254.

Smith, RS, Guyomarc'h, S, Mandel, T, et al. (2006) A plausible model of phyllotaxis, *Proc. Natl Acad. Sci. USA* **103**, 1301.

Sprinzak, D, Lakhanpal, A, LeBon, L, et al. (2010) *Cis*-interactions between Notch and Delta generate mutually exclusive signalling states, *Nature* **466**, 86.

Steel, KP, & Kros, CJ (2001) A genetic approach to understanding auditory function, *Nature Genet.* **27**, 143.

Tsiantis, M, & Hay, A (2003) Comparative plant development: the time of the leaf?, *Nat. Rev. Genet.* **4**, 169.

Vonk, FJ, & Richardson, MK (2008) Developmental biology: serpent clocks tick faster, *Nature* **454**, 282.

Wolpert, L, & Tickle, C (2011) Principles of Development, 4th ed., Oxford University Press.

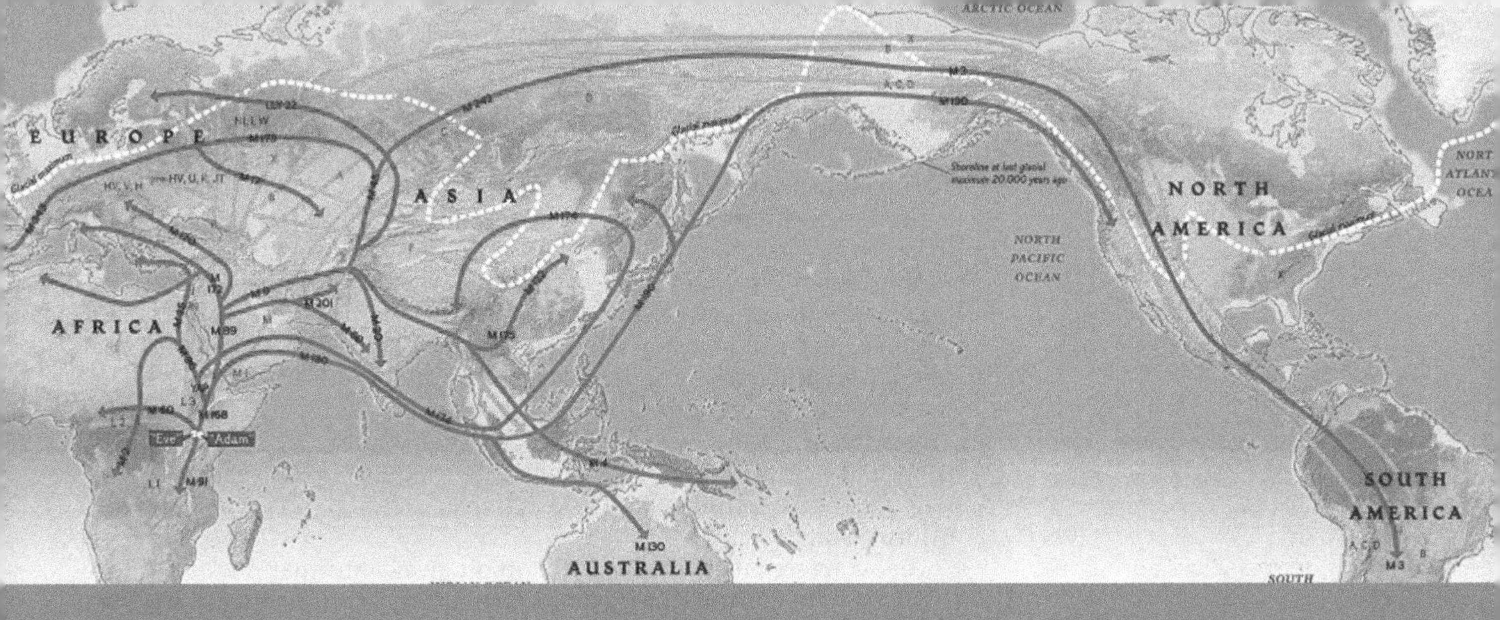

Sequences, Specificity, and Evolution

21

Overview: In which protein and DNA sequences are compared and the significance of sequence alignment is examined

> "So often in science and engineering, one finds that the difficulties anticipated from the armchair scarcely overlap those that confound one in the field or the laboratory."
>
> **Frederick Mosteller and David L. Wallace**

Evolution is the overarching conceptual framework that unifies biology. Historically, insights into the nature of evolution were largely based on macroscopic studies of form, experimental investigation into which pairs of unlike organisms are capable of producing offspring, and the examination of fossil remains of extinct species. However, modern biology is uniquely dependent upon the beautiful and subtle uses of genetics, which has provided a mechanistic underpinning for the study of evolution. One of the most important modern windows into biological evolution is the study of DNA and protein sequences. By comparing different sequences, it is possible to provide a quantitative measure of relatedness of both molecules and organisms. These studies have brought microscopic organisms into evolutionary focus in a way that studies of form could not. In 1958, Francis Crick foresaw this revolution, as evidenced by his remark: "Biologists should realize that before long we shall have a subject which might be called 'protein taxonomy,' the study of the amino acid sequences of the proteins of an organism and the comparison of them between species. It can be argued that these sequences are the most delicate expression possible of the phenotype of an organism and that vast amounts of evolutionary information may be hidden away within them." The aim of this chapter is to provide an introductory overview of sequences and ways of thinking about them using several simplified models, including the HP model already described on many occasions throughout the book (see Section 8.4.2 on p. 346). The ideas of sequence comparison are then used to consider several case studies such as antibiotic resistance and viral evolution. In addition, we examine how errors that are one of the substrates of evolution are made (and prevented) during the processes of the central dogma.

21.1 Biological Information

Throughout the book, we have examined biological organization from a number of different angles. However, one serious missing ingredient in our study has been an examination of how this organization is tied to the genomes of individual organisms. Any description of biological organization must ultimately appeal to the information content of a genome, and will be especially revealing in light of an estimate of the amount of information encoded in the genome of an organism. For example, the genome of *E. coli* is a sequence of letters (A, T, G, and C) 4.6×10^6 base pairs long, which could be printed in a book of about a thousand pages. The human genome is roughly 3×10^9 base pairs in length, and it would take a thousand *E. coli* books to hold all that information.

The convenient shorthand for biological information based on strings of letters often lulls us into thinking that biological information is similar in some way to information that we print in books or store on computer hard disks. However, there are important functional differences between the way that information is encoded and used in biological systems and the way that humans encode and use information for communication. One important aspect of biological information is revealed by this analogy. In human communication, the letter "A" contains the same information regardless of the font in which it is written or whether it is written in braille or converted into Morse code. Any form of the letter "A" can be translated into another form without any loss of information or change in its fundamental nature. This interchangeability is not generally true for biological information. When we use the letter "A" in reference to a DNA sequence, we are referring to the chemical structure of an adenine nucleotide, and no other chemical structure can be substituted for it without loss of information or alteration of the information that it encodes. Indeed, DNA damage is a precise example of this fact. A string of letters written in Roman type can be translated into Braille or Morse code and back into Roman letters without loss of information, but in biology a nucleotide sequence can be translated into an amino acid sequence, but not vice versa. As a result, we must be cautious in taking the information analogy too far. Nevertheless, many concepts such as signal-to-noise ratio, filtering, amplification, and feedback can be usefully applied to both biological information and human communication.

The DNA and protein sequences that are frequently referred to in the context of biological information storage and expression represent only a narrow subset of information and communication transmitted within and between living organisms. All forms of biological communication or information transfer involve interactions between molecules where their structure, conformation, or location can be modulated and effect some response, but this covers a very broad range of interactions. For example, the frequency with which a nerve cell fires its action potential and releases neurotransmitters to its targets conveys information to those downstream cells about its stimuli. Similarly, the amount of light received by a rod cell in the retina, the experience of pain resulting from the stimulation of a receptor in the skin, or the quality of a thought originating in the central nervous system can also be thought of as forms of information. The immune system famously processes information with both great breadth and specificity such that a memory T-cell can recognize an infection by a specific pathogen recurring decades after initial exposure and respond by stimulating the generation of perfectly targeted antibodies to kill that particular invader.

Information can also be transmitted vertically from a mother cell to its daughters not just in the form of its genome, but also in the form of heritable cell structures and an inherited differentiated state. Nearly all the cells in the human body contain essentially identical genomes, yet a dividing liver cell can normally only give rise to more liver cells, dividing skin cells to more skin cells, etc. This information on cell structure and state is mostly inherited epigenetically without altering the sequence of the genome itself. Similarly, structures may be perpetuated even as the material that makes them up is degraded and replaced. For example, we tend to think of the bones in our adult human bodies as being static structures—however, the matrix is continuously degraded by cells called osteoclasts and continually redeposited by cells called osteoblasts. Even with all of this busy turnover, the fundamental structural information of the size and shape of your bones is maintained with remarkable fidelity over many decades of adult life.

21.1.1 Why Sequences?

If biological information can be stored and transmitted in such a wide variety of forms, why do human investigators pay so much attention to nucleotide and amino acid sequences? Information encoded in the form of sequence does seem to hold a privileged place for cellular life forms on Earth. The strongest evidence for this is the universality of the genetic code (see Figure 1.4 on p. 10), which is nearly identical among all cells despite more than three billion years of evolutionary divergence. Similarly, the machinery of the central dogma that copies and processes sequence-based information has been largely conserved. Over the course of evolution, it seems that cells have entrusted much of their inheritance to faithful replication and segregation of linear sequences. Although it is certainly the case that daughter cells inherit other kinds of structure-based information from their mother (for example, the position of surface appendages and the organization of the endoplasmic reticulum), it is the DNA-based genome that is copied with greatest fidelity over the largest number of subsequent generations.

A critical feature facilitating the systematic analysis of biological sequence information is the observation that, to a first approximation, the ways that cells use and interpret sequence information are largely independent of context. When considered in the appropriate reading frame, UUU always codes for phenylalanine regardless of what comes before or after. Similarly, an RNA polymerase binding site is an RNA polymerase binding site regardless of where it is in the chromosome. A heptad repeat of certain amino acids is almost always able to make an α-helix in the context of a larger protein structure. Obviously, there are important exceptions to this general rule, but the context-independent interpretation of a particular piece of sequence information is usually an excellent starting point for understanding its significance. This relative context independence of sequence information stands in stark contrast to every other kind of information described above, all of which are strongly context-dependent. For example, the human retina converts the number of photons impinging on the eye into an electrical signal perceived as light by the brain, but the perceived intensity of light can vary by several orders of magnitude depending upon how much other light is present in the visual field and how much light impinged on the eye one second previously. The context independence and linear nature of biological sequences make it possible to devise quantitative metrics for degrees of similarity of this kind of information, which can be meaningfully applied to many different types of comparative studies. In this chapter, we will first develop the logic for measuring sequence similarities and differences, and then show several examples of how these kinds of measurements yield insight into biological function and evolutionary history.

EXPERIMENTS

Experiments Behind the Facts: Sequencing and Protecting DNA Over the past few decades, the technology for sequencing DNA has undergone revolutionary improvements in speed and cost-efficiency. Given the huge volume of DNA sequence data available today for a wide range of organisms across all the branches of the tree of life, it is surprising to recall that the very first complete genome sequence for any bacterium (*Haemophilus influenzae*, with a genome size of about 1.8 Mb) was reported only in 1995.

Before 1977, the only techniques available for DNA sequencing were extremely laborious and could be applied only to very short fragments of DNA. The first major technological breakthrough came in the form of the chain termination method, also known as dideoxy sequencing, which takes advantage of a particular property of DNA polymerase. The normal DNA elongation reaction catalyzed by DNA polymerase uses a free hydroxyl group in the sugar ring (located at what is conventionally called the 3′ position) at the end of the existing growing chain to form a covalent bond with the first phosphate group on the nucleoside triphosphate that is being incorporated. This reaction displaces the second and third phosphates as a single unit (called pyrophosphate). Next, the DNA polymerase shifts its position and is able to proceed using the 3′ hydroxyl group of the subunit it has just incorporated. However, DNA polymerase is equally well able to catalyze the elongation reaction using defective nucleoside triphosphates that lack the 3′ hydroxyl group and have a simple hydrogen at that position instead. (These are called "dideoxy" nucleotides because they are missing the hydroxyl at the 2′ position as well as at the 3′ position; the presence or absence of the 2′ hydroxyl distinguishes RNA (ribonucleic acid) from DNA (deoxyribonucleic acid).) After a dideoxynucleotide has been incorporated into the growing chain, DNA polymerase is unable to take another step, and stops the elongation reaction.

The way that this abnormal reaction can be used for DNA sequencing is illustrated in Figure 21.1 (left side). The most common current version of dideoxy sequencing uses fluorescently labeled defective nucleotides, such that A, C, G, and T each correspond to a different color. In the sequencing reaction, many identical copies of the same DNA sequence (for example, the products of a PCR reaction) are mixed with a primer to initiate polymerization, and then provided with DNA polymerase and a mixture of normal nucleotides along with a relatively small amount of the colorful dideoxynucleotides. When one of these dideoxynucleotides is incorporated into a growing chain, that chain stops at a certain length and is tagged with one of the four colors. Finally, the many products

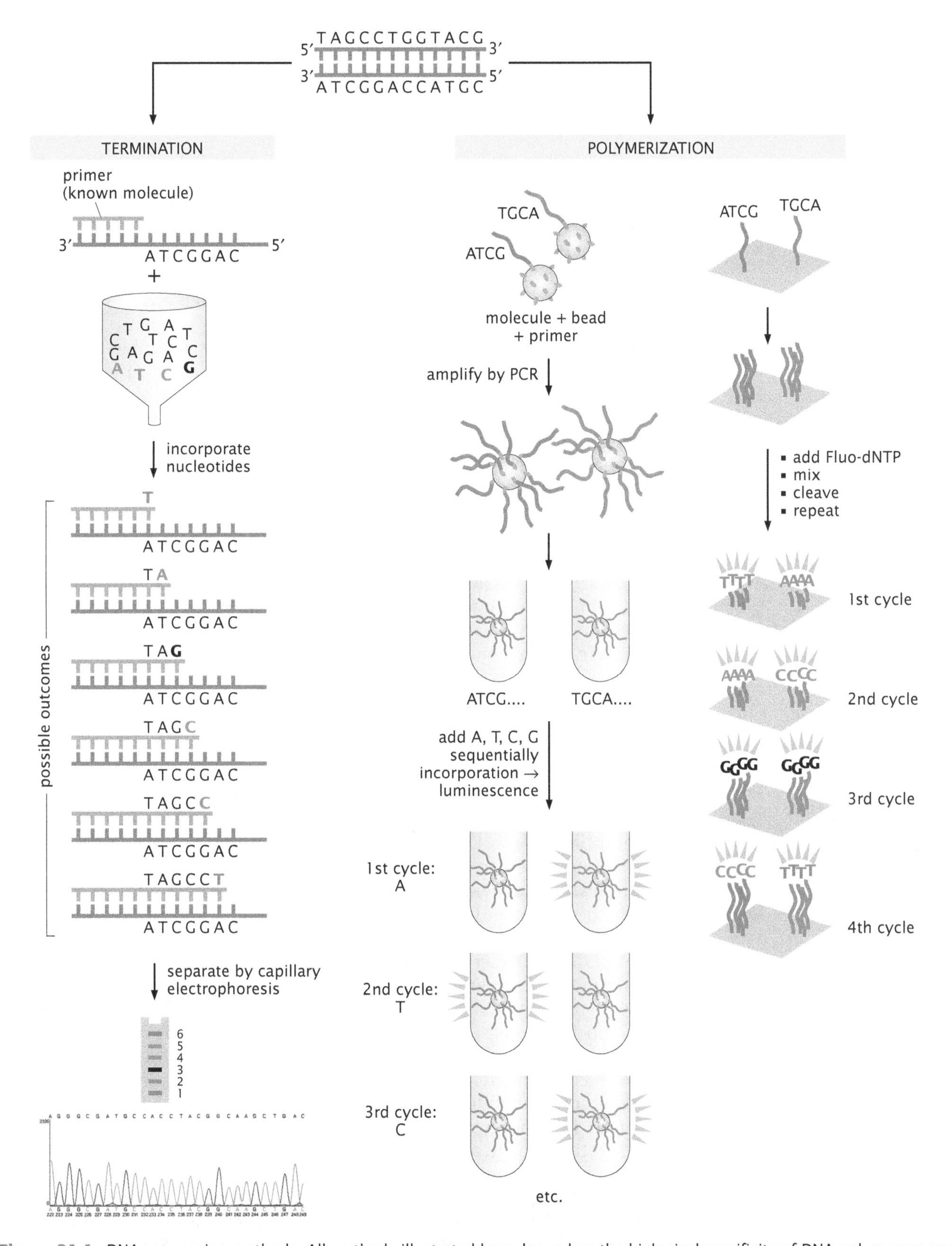

Figure 21.1: DNA sequencing methods. All methods illustrated here depend on the biological specificity of DNA polymerase and its ability to choose the correct base for incorporation into a growing DNA chain based on the complementary template strand. On the left, termination methods (also known as dideoxy methods) involve generating a large number of partially completed copies that are then separated based on their size. Polymerization methods, on the right, do not require a final separation step, but instead use various technologies to read out the identity of each base as it is incorporated (for simplicity, the second DNA strand is not shown in these diagrams).

of the elongation reaction are physically separated from one another based on their lengths (usually by gel or capillary electrophoresis), and the sequence is read from the shortest to the longest product according to the color associated with each fragment.

Dideoxy sequencing is conceptually and technically very simple, but is limited to sequencing just one DNA fragment at a time, using many identical copies of the same molecule to generate the populations of terminated chains that are separated by length. So-called "second-generation" sequencing methods, such as those shown on the right in Figure 21.1, rely on the ability of a machine to detect the identity of each base as it is incorporated, and avoid the need for length-dependent product separation after the elongation reaction is complete. An early version of second-generation sequencing, called "pyrosequencing," is shown in the middle. This method exploits the fact that the same pyrophosphate group is released by incorporation of any of the four nucleotides. As shown here, random fragments of DNA are attached to individual tiny beads and then PCR-amplified to make many identical copies of the same DNA sequence on each bead. The beads are then separated (for example, put into separate microwells) along with DNA polymerase and a mixture of other enzymes. A second group of enzymes catalyzes a series of reactions that can convert pyrophosphate into ATP and then into a photon of light by the action of luciferase. In order to determine the sequence of the DNA fragments on a single bead in a single well, a small amount of a single deoxyribonucleotide (for example, dGTP) is added to the well. If the next base in the complementary strand is a C, then the dGTP will be incorporated, and pyrophosphate will be released, triggering a tiny flash of light that can be detected. If the next base is A, G, or T, then the dGTP will not be incorporated, and no light will result. Then, the dGTP is enzymatically degraded, and another nucleotide is added. By this sequential tempting of the DNA polymerase with alternate nucleotides, the sequence can be determined. A simpler type of second-generation sequencing is shown on the right in Figure 21.1, which combines features of both earlier methods. Here, the DNA fragments are immobilized on a slide, and usually locally amplified as described above for fragments on beads. Then, a mixture of differently colored fluorescent versions of all four deoxynucleotides is added, in a form where the 3′ hydroxyl group is chemically protected, so that only one base can be added to each growing chain. The identity of the incorporated base is read out by detection of the fluorescent color that becomes immobilized at the site of the DNA molecules. The single incorporated bases are then chemically deprotected (in a reaction that also removes their fluorescent dyes) and another round of incorporation is allowed to proceed.

DNA sequencing is a rapidly evolving technology. The cost associated with sequencing one Mb of DNA has fallen from about $5000 in 2000 to about $0.20 in 2010. Even-newer sequencing methodologies that do not depend on the activity of DNA polymerase are currently being developed, and promise to reduce the cost and raise the speed of DNA sequencing even further. Our ability to interpret DNA sequence information is now limited more by the difficulties of handling computations on such large datasets than by the difficulty of DNA sequencing itself.

Techniques related to DNA sequencing can also be used to yield other kinds of biologically interesting information beyond the simple sequence itself. One important kind of information that has come up repeatedly is the locations where proteins (such as transcription factors) bind to DNA. A traditional method for locating the sites of specific DNA binding by protein factors, illustrated in Figure 21.2(A), is known as footprinting. In this technique, DNA that is radioactively labeled at one end is mixed with a sequence-specific binding protein, and the mixture is then digested with an enzyme that cleaves the DNA nonspecifically. For naked DNA, this results in a series of labeled digestion products of all possible lengths, which can be separated based on their length by gel electrophoresis, resulting in a ladder of bands where each step corresponds to a single base. When a specific site on the DNA is protected by a binding protein, the enzyme is unable to cleave there, and the result is a blank zone or "footprint" on the gel. As with dideoxy sequencing, the traditional method is limited to analysis of a single DNA sequence at a time.

As shown in Figure 21.2(B), this same concept can be readily adapted to higher-throughput DNA sequencing methods so as to give genome-scale information. For example, the distribution of nucleosomes across an entire eukaryotic genome can be analyzed by isolating the nucleosome-rich chromatin, digesting away the unprotected DNA, isolating single DNA-wrapped nucleosomes on a gel, and subjecting the associated DNA fragments to sequencing. Similarly, all of the binding sites for a given transcription factor can be quickly identified by crosslinking a mixed population of DNA-binding proteins to DNA *in vivo*, fragmenting the DNA by shearing or digestion, and then immunoprecipitating the particular transcription factor of interest, followed by sequencing of the associated DNA fragments. Furthermore, the concept of protection can also be applied to moving machineries such as RNA polymerase or ribosomes; in these cases, isolation and sequencing of the protected fragments of DNA or RNA respectively can give a "snapshot" showing the population distribution of the positions of all the copies of (for example) RNA polymerase in a given population of cells. Finally, long-range DNA–DNA interactions (usually mediated by protein binding partners) can be mapped by fragmenting protein-bound DNA, ligating any DNA fragments that find themselves together within the same complex, and sequencing the ligated fragments. This technique can reveal cases where portions of a genome that are distant from one another in actual linear sequence are found spatially nearby in a cell.

21.1.2 Genomes and Sequences by the Numbers

In Sections 8.2.2 (p. 321) and 10.4 (p. 398), we considered the physical size of genomes and the implications of this size for the strategies that organisms must use to pack and access their genomes. A related

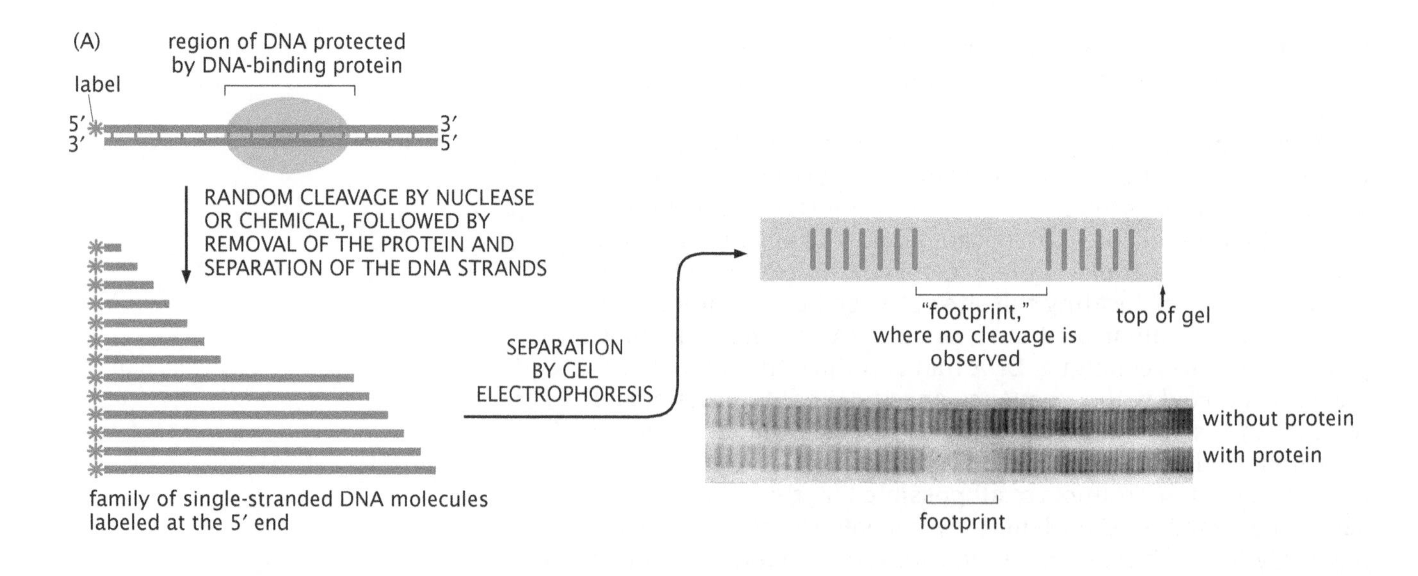

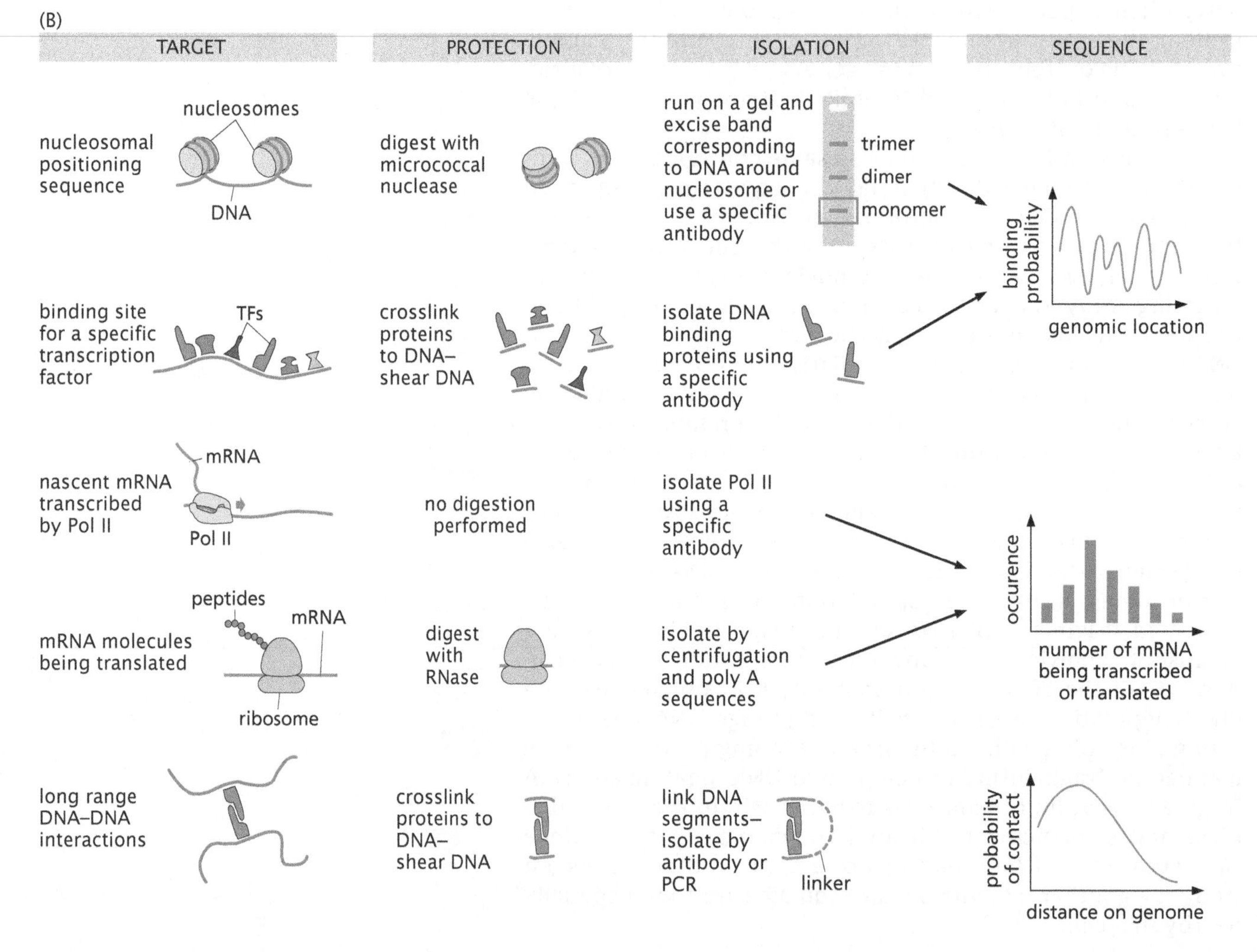

Figure 21.2: Experimental methods based on the concept of sequence protection. (A) Traditional *in vitro* footprinting can be used to find the binding site for a particular protein on a particular stretch of DNA. (B) An extended family of methods uses a similar series of target protection, isolation, and sequencing to find sequences of DNA or RNA associated with particular factors of interest, ranging from nucleosomes to transcription factors to ribosomes.

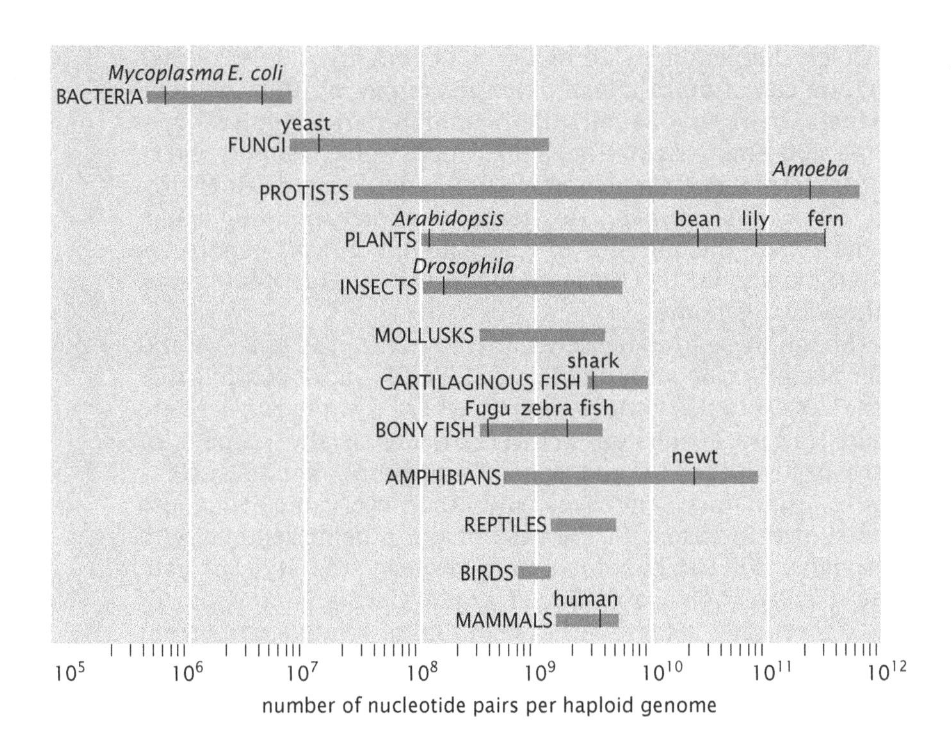

Figure 21.3: The size of genomes. This log-scale diagram indicates the tremendous range of genome sizes found even in closely related organisms. (Adapted from B. Alberts et al., Molecular Biology of the Cell, 5th ed. Garland Science, 2008.)

problem, perhaps of even greater interest, is to characterize the information size of a genome and to understand the significance of its information content. We have repeatedly emphasized the view that a feeling for the numbers is a central part of the development of intuition about a given problem. To that end, Figure 21.3 reports the genome sizes of some favorite organisms. What we see is that the genomes of different organisms vary in size by roughly six orders of magnitude.

It is tempting to think that larger genomes must contain more useful biological information; however, there is no evidence that this is the case. Organisms with the largest genomes are not clearly more complex than those with smaller genomes. Among animals, the largest genomes belong to newts and salamanders, which have genomes that are several orders of magnitude larger than the human genome, and even the salamander genome is dwarfed by those of some plants such as ferns. Even closely related species may have very different genome sizes. For example, the clawed frog *Xenopus laevis* has a genome (3.1×10^9 base pairs) approximately twice the size of its close cousin, *Xenopus tropicalis* (1.7×10^9 base pairs). The frogs are very similar in appearance and overall lifestyle except that *X. laevis* is somewhat larger. To better understand the significance of genome size, we turn now to an estimate of the gene content of genomes.

Estimate: Genome Size and the Number of Genes *E. coli* has served as our standard ruler throughout the book and as our reference organism for a host of different biological processes. We now use this organism as our first case study in the analysis of genomes. One challenge resulting from a knowledge of genome size is to attribute meaning to the entirety of the genomic DNA. A response to this challenge is revealing because it demonstrates both the advantages and pitfalls of simple estimates.

ESTIMATE

Given the genome size of *E. coli* at roughly 4.6×10^6 base pairs, we can attempt to estimate the number of different gene products. If we invoke our rule of thumb that a "typical" protein is 300 amino acids long, this implies that roughly every 1000 base pairs of the genomic DNA codes for a distinct protein. Given this estimate, this leads to a corresponding rough estimate that the number of genes in the *E. coli* genome is 4600. This should be contrasted with the roughly 4400 genes of the actual genome.

Though this estimate serves us well in the case of *E. coli* and other bacteria, the situation is considerably more complicated in the case of eukaryotic genomes, where a similar procedure would lead to a substantial overestimate of the number of genes as a result of the presence of noncoding regions such as introns, regulatory sequences, and other intergenic sequences of unknown function. In the case of the human genome with its roughly 3×10^9 base pair long genome, the style of estimate used in the case of *E. coli* would yield a gene count of 3×10^6, roughly a factor of 100 too large relative to current best estimates.

In fact, the shortcomings of our estimate concerning the gene content of eukaryotic genomes is instructive because it teaches us more about the sequence organization of eukaryotic genomes. One of the key lessons that emerges from this analysis is that eukaryotic genes are a complicated sequential arrangement of coding (exon) and noncoding (intron) regions (see Figure 4.28 on p. 182). A second sense in which we have to revise our picture of genome organization is to account for the large blocks of sequence surrounding coding regions that control when and where genes are expressed. These regulatory regions can be thousands of base pairs long. Overall, the different classes of noncoding regions can account for more than 90% of eukaryotic genomes.

Our brief introduction to biological information has emphasized the important role of universal sequence information in different organisms. To make these qualitative observations about sequence relationships concrete and quantitative, we need tools to compare different sequences, and we take up that topic in the next section.

21.2 Sequence Alignment and Homology

"All happy families are like one another, every unhappy family is unhappy in its own way." Nabokov answered Tolstoy's famous starting volley from *Anna Karenina* with his own, "All happy families are more or less dissimilar; all unhappy ones are more or less alike," as the opening line of *Ada*. To the practiced adult English speaker, the similarities and dissimilarities of these two sequences of words are at once evident, as are their quite distinct meanings. In contrast, a sequence with the same meaning as Tolstoy's but no sequence similarity might read "Parents and their smiling children paint a constant picture of fulfillment, while every teary-eyed domestic scene has its own unique history." This word play illustrates several serious points with respect to the pitfalls of sequence analysis.

When we examine naturally occurring sequences of nucleotides or amino acids, we are generally interested in their meaning, that is, in what they imply about the function of different macromolecules or in what they say about evolutionary history. Indeed, sequence similarity is often used as a surrogate for functional relationships among different proteins or for evolutionary relatedness. However, as the contrasting statements from Tolstoy and Nabokov illustrate, a similar sequence may encode a very different meaning, while a less eloquent but equally meaningful English sentence may convey the same sense as one of those without sharing any common words. Further, two sequences implying the same underlying "meaning" but without any obvious sequence similarity may signal a convergence of function molded by the demands of natural selection over evolutionary history; that is, they may not be related even though they have similar functions. In practice, it is often very difficult to tell whether two very different sequences that share similar functions arose by convergent evolution from distinct ancestors, or instead diverged from a common but distant ancestor. These puzzles can sometimes be resolved by seeking out and examining other related sequences or intermediate forms. The study of sequences in this fashion has turned biologists into molecular paleontologists, with different sequences serving as a new kind of living fossil that allows us all to look into the long and fascinating history of life on Earth. To see this, we examine an intriguing case study showing links between mammals such as mice and insects such as flies, and then turn to the concrete question of how to quantitatively compare different sequences.

Sequence Comparison Can Sometimes Reveal Deep Functional and Evolutionary Relationships Between Genes, Proteins, and Organisms

Because all living organisms on Earth are related to one another, we can often make reasonable guesses about the function of a particular protein by comparing it with a protein of similar sequence in a different organism where the function has been better explored. In some cases, the degree of sequence conservation between proteins from very different organisms can be quite surprising, even in cases where it is not clear that the function in one organism could still persist in the other. A famous example is the *Drosophila* gene *eyeless*, which, as its name suggests, was originally identified as the locus of a mutation that altered fly development in such a way that the adults never develop eyes. Figure 21.4(A) shows one of these unfortunate flies. The fly eye is a highly specific structure with a carefully programmed repeating pattern of photoreceptor cells. The *eyeless* gene appears to be a master regulator that is responsible for every aspect of development of this structure. Not only is its expression necessary for normal eye development, but using molecular tricks to force expression of the *eyeless* gene in other tissues as shown in Figure 21.4(B) is sufficient to convert them into ectopic eye-like structures (the word ectopic generally means "in the wrong place"). The flies cannot actually see out of these ectopic eyes because they

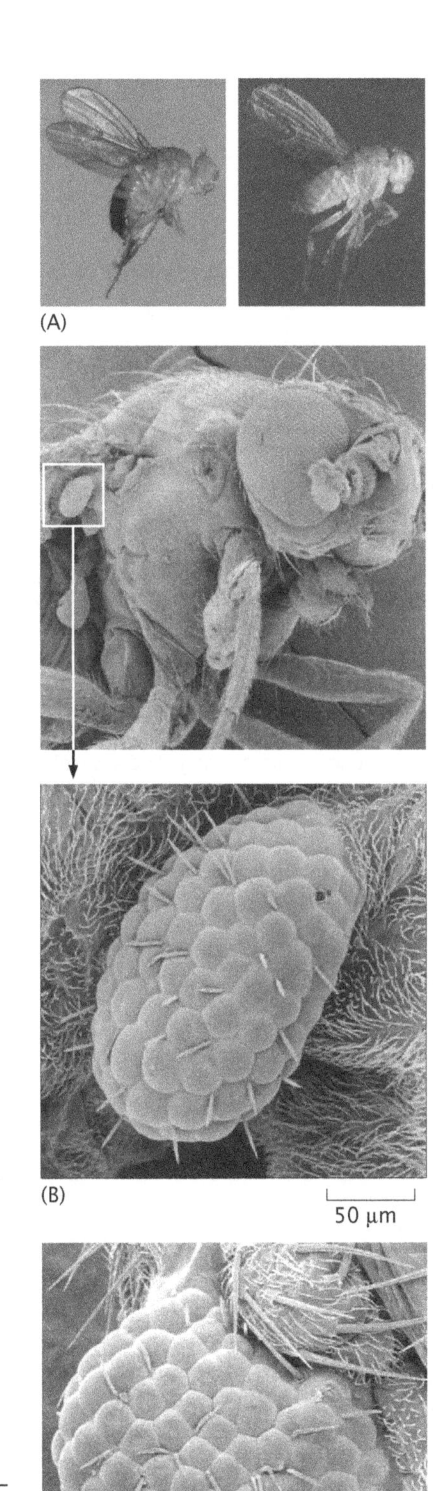

Figure 21.4: The master regulators of eye development. (A) Disruption of a single gene eliminates eye development in *Drosophila.* (B) Expression of *eyeless* in other tissues in *Drosophila* results in ectopic formation of eyes. In this case, an ectopic eye is formed just underneath the wing. (C) Expression of the mouse *eyeless* homolog *Pax6* in *Drosophila* also causes formation of ectopic eyes; this one formed on the fly's leg. (Adapted from G. Halder et al., *Science* 267:1788, 1995.)

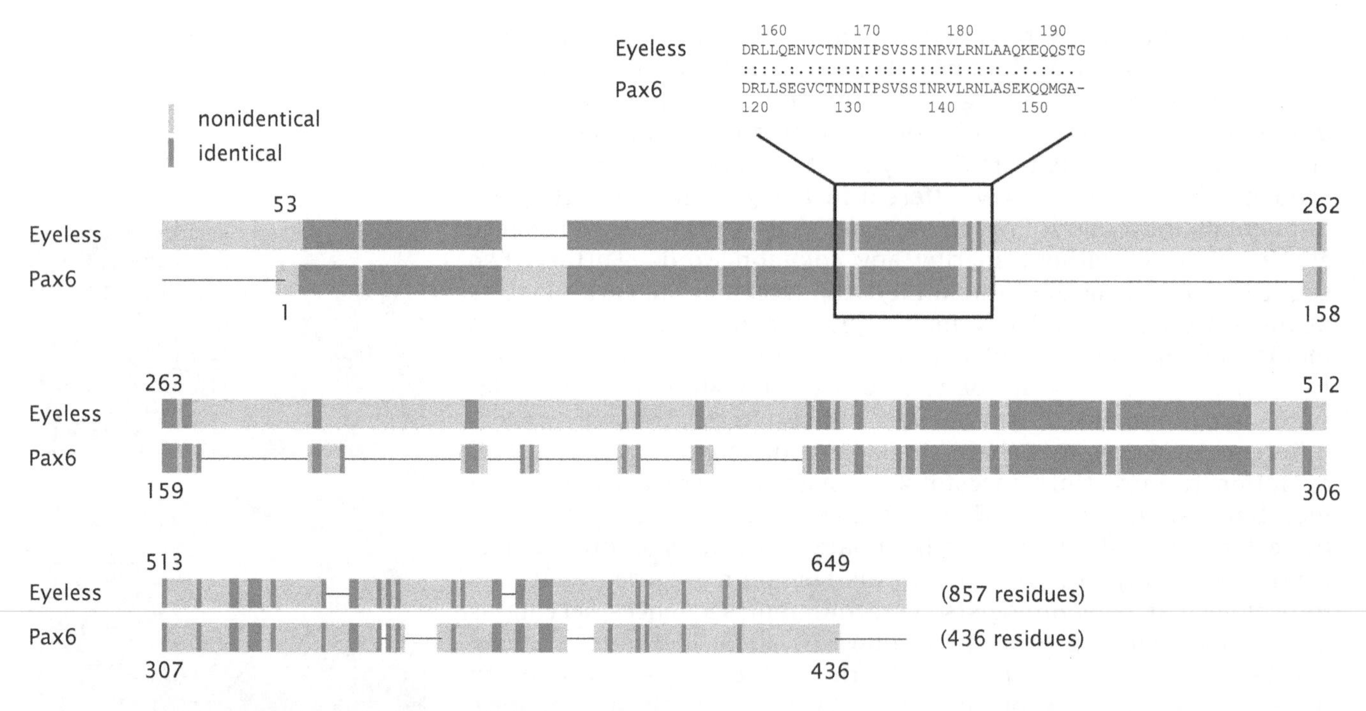

Figure 21.5: Comparison of amino acid sequences for Eyeless and Pax6. The sequences are compared using the Blosum62 scoring matrix and the SSEARCH alignment program from the FASTA suite. Rather than showing the detailed sequence comparison, this figure depicts the overall geography of the two sequences, with regions of exact sequence identity shown in red and nonidentical regions (or gaps) shown in light gray. The blown-up region at the top shows the actual amino acid sequence for the two proteins within one of the domains of strong sequence conservation. Double dots indicate that the amino acid is identical between the two sequences, and single dots indicate that the amino acids are chemically similar but nonidentical. The integers above and below the sequences identify particular residues as landmarks within the two sequences. Only the best matching subsequences are aligned. The accession numbers of the sequences used are 82069480 for Pax6 and 12643549 for Eyeless.

are not properly connected to the central nervous system, but they are otherwise well formed.

Mammals such as mice have eyes too, yet their eyes bear essentially no structural similarity to the eyes of the fly and the sequence of events in eye development is very different. Still, when the predicted protein sequences encoded by the mouse genome are lined up against the amino acid sequence for the protein encoded by *eyeless*, there is one mouse gene, called *Pax6*, whose sequence is strongly similar to the sequence of *eyeless*. Out of nearly 600 residues, more than 40% are identical.

To see how these genes are related at the level of proteins, Figure 21.5 shows a coarse-grained sequence comparison between the sequences for the Pax6 protein from mouse and the Eyeless protein from *Drosophila*. The alignment reveals long stretches of conserved amino acid sequence. The algorithms that underlie the construction of such sequence comparisons will be described later in the chapter. For the moment, we content ourselves with noting the overwhelming similarity of parts of these two genes from such distinct organisms.

Does this alignment reflect fundamental similarity or fundamental dissimilarity? Do *eyeless* and *Pax6* represent a happy family with similar functional meanings or an unhappy family with superficial resemblance but underlying discordance? The sequence similarity alone cannot tell us. For example, we might imagine that this highly conserved domain simply represents a binding site for some cellular structural element and the domain has been co-opted over evolutionary time to participate in the spatial localization of different proteins with different functions. Two experiments have resolved the question by demonstrating functional similarity of *eyeless* and *Pax6*. First, a transgenically altered mouse lacking normal expression of *Pax6* also fails to develop eyes, suggesting that *Pax6* may serve as a master regulator for eye development in the mouse. Second, and perhaps more amazing, as shown in Figure 21.4(C), overexpression of the mouse *Pax6* gene in a developing fly results in the ectopic formation of fly eye structures at different parts of the fly body. This remarkable result shows that the fly cells are able to interpret the presence of the mouse Pax6 protein as a signal to trigger the fly eye developmental program. Because the actual mouse eye developmental program is profoundly different from that operating in the fly, this result has been widely interpreted to mean that both flies and mice are descended from a common ancestor that used a protein similar to Eyeless/Pax6 to regulate the formation of some kind of light-sensitive tissue, and as the two separate lineages evolved and developed more elaborate eye structures the original regulatory signal was preserved.

Countless experiments of this nature demonstrating functional similarity stretching across evolutionary time and among very distantly related species have generally convinced us that a high level of sequence conservation tends to indicate similar protein function. However, the converse is not true. An inability to align two protein sequences does not prove that they have different functions. Indeed, it does not even prove that they have different three-dimensional structures.

In Section 15.3 (p. 599), we discussed the dynamics of polymerization of the cytoskeletal protein actin and mentioned its distantly related bacterial homolog ParM. The three-dimensional structures of these two proteins are nearly superimposable as shown in Figure 21.6. However, their primary amino acid sequences are very dissimilar, as will be shown later in the chapter (see Figure 21.12). Somehow, these very different amino acid sequences are able to communicate the same meaning, that is, nearly the same three-dimensional folded structure and very similar function (that is, to polymerize and form filaments). Can we determine whether ParM and actin are homologs that have undergone extreme divergence from a common ancestor or are two unrelated proteins that have converged on similar structures because of natural selection for similar functions? A macroscopic analogy might be the comparison of bird wings and bat wings. Although these two structures are roughly similar and serve the same purpose for the animals, it is clear from examination of bird and bat relatives and the fossil record that bird wings are derived from arm bones while bat wings are derived from hand bones, suggesting convergence rather than divergence. For the molecules under consideration, we have not yet found enough relatives and intermediate forms to answer the question definitively. Most researchers believe that ParM and actin are divergent homologs, but this proposition has not been proved.

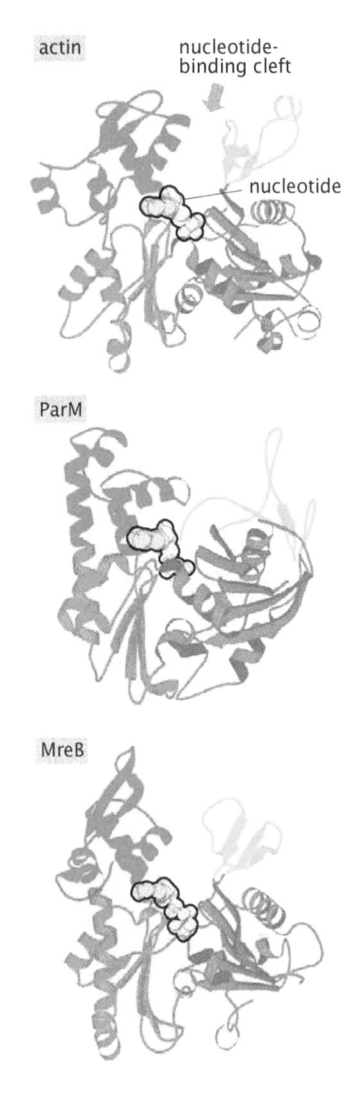

Figure 21.6: Comparison of actin-like proteins. The structures of actin and two of its bacterial homologs are shown as ribbon diagrams, with the bound nucleotide as a space-filling model. The four subdomains are each indicated by different colors. The overall shapes as well as the detailed patterns of connectivity between the α-helical and β-strand structural elements are very similar despite the fact that these proteins have essentially no recognizable sequence similarity. (Adapted from L. A. Amos et al., *Curr. Opin. Cell Biol.* 16:1, 2004.)

One of the critical outcomes of the genomic explosion of recent years has been the availability of huge amounts of sequence data for different organisms. As a result, the kinds of comparisons we have described are a mainstay of the bioinformatics agenda. As was shown above, it is sometimes possible to use sequence information to detect deep evolutionary similarities. To carry out sequence comparisons, we must first develop quantitative metrics for assessing sequence similarity and dissimilarity. To get a feeling for the types of algorithms that are used to carry out sequence alignments, we need to explore how different alignments are scored and then consider how to determine the alignment with the best score. We take up this challenge in the next section.

21.2.1 The HP Model as a Coarse-Grained Model for Bioinformatics

How do we construct an alignment of two sequences like that depicted in Figure 21.5? As is shown in the figure, only a part of the two sequences can really be overlapped in any meaningful sense. If we slid one of the two sequences relative to the other, it would reduce the number of aligned residues. Sequence alignments like those shown in Figure 21.5 result from comparisons of a huge number of different possible alternative alignments corresponding to sliding the sequences relative to each other and also to inserting gaps. The decision about which alignment of a nucleotide or amino acid sequence is "best" is determined by assigning scores to different alignments. The score reflects how many times in the alignment the two sequences have precisely the same or similar nucleotides or amino acids in register, how often the alignment requires us to shift one part of a sequence relative to another by introducing a "gap," and a cost for having mismatches (that is, different residues at the same position in the sequence).

The conceptual background for performing these alignments is an evolutionary hypothesis that asserts that the two sequences are the descendants of some ancestral common sequence. Over time, that ancestral sequence changes through a series of substitutions, deletions, and insertions. For the comparison of nucleic acid sequences, these mechanisms of sequence change are directly derived from our understanding of the kinds of biochemical errors that can be made by the DNA polymerase enzyme during replication, as well as other mechanisms generating mutations. Not all kinds of errors are equally likely; for example, it is more common that DNA polymerase makes the mistake of substituting a pyrimidine for a pyrimidine (that is, C for T, or T for C) and a purine for a purine (that is, A for G, or G for A) rather than replacing a pyrimidine with a purine (for example, A for C) or vice versa, because of the relative sizes of these different molecules. Certain kinds of chemical or ultraviolet light damage to DNA can also lead to mutation. Furthermore, not all changes are equally likely to be preserved through the action of natural selection. Individual residues or groups of residues whose chemical structure is critical for the particular biological function of a sequence (for example, serving as a recognition site for the binding of a transcription factor) are more likely to remain unchanged, because natural selection disfavors the replication of mutant organisms with an altered sequence. Nucleotide sequence changes can also occur at a much larger scale, through the wholesale insertion or deletion of long stretches of DNA by recombination or other processes. Nevertheless, the actual process of aligning nucleic acid sequences that have only four possible distinct residues at each particular location is conceptually straightforward.

Interpreting the alignment of amino acid sequences in proteins is somewhat more complex. Recall that each of the 20 amino acids is encoded by three nucleotides in a codon. An error by DNA polymerase that changes a particular nucleotide within a protein-coding sequence may result in a "synonymous" change to a different codon that still encodes the same amino acid or in a "nonsynonymous" change that alters the amino acid.

Inspection of the genetic code (see Figure 1.4 on p. 10) reveals that many nucleotide changes, particularly in the third position of a codon, are actually synonymous at the level of protein sequence. Some nonsynonymous changes are highly likely because only one nucleotide need be altered: for example, an aspartic acid residue (D) can be replaced with an arginine residue (R) by a single nucleotide change altering the codon GAC to AAC. However, some possible nonsynonymous amino acid changes demand multiple nucleotide changes, for example, a codon for phenylalanine (UUU or UUC) cannot be changed to encode lysine (AAA or AAG) with any fewer than three point mutations.

Insertions or deletions of single nucleotides, which are common kinds of errors made by DNA polymerase, can have drastic effects on protein sequence because the reading frame is shifted for the entire remainder of the sequence, often resulting in a nonfunctional protein. However, insertion or deletion of three nucleotides in a row will merely insert or delete a single amino acid in the protein sequence, which may not significantly affect the protein's function. Thus, even though insertion of a single nucleotide within a protein-coding sequence is a more common mutational event than insertion of three nucleotides in a row, we are much more likely to actually observe the triple-insertion event. Furthermore, because some amino acids share very similar chemical structures (for example, aspartic acid (D) and glutamic acid (E)), it is frequently the case that mutations resulting in substitutions of similar amino acids have less of an effect on protein structure or function than mutations resulting in substitutions of very different amino acids, which may be more strongly selected either negatively or positively during evolution. Some subtleties of protein structure also influence the likelihood that particular kinds of amino acid changes will have survived through natural selection; for example, insertion of multiple amino acids into an unstructured loop on the surface of the protein is less likely to significantly disrupt the folding and function of the protein than a similar insertion in a critical α-helix found within the protein's hydrophobic core. The complex interplay between mutation mechanisms and selection mechanisms renders the observable patterns of amino acid change during evolution extremely nonrandom.

In performing a comparison of two or more sequences of nucleotides or amino acids that were in fact derived by descent from a common ancestor, we would ideally wish to be able to reconstruct the actual series of mutational events that led to the observed changes. In reality, this is rarely possible, and instead we must make our best guesses as to the historical series of events by calculating an alignment score that includes some consideration of how likely and how important certain kinds of changes may be. The current alignment algorithms in daily use by biological researchers, including programs such as BLAST and ClustalW, are based on a sophisticated scoring scheme, to be discussed later in the chapter, that addresses these issues in part by using information gleaned from large databases of real biological sequences. Before we make an attempt to tackle the full complexity and subtlety of real sequence comparison, we will follow our usual approach of starting with a highly simplified and schematized toy model that will enable us to build intuition. To this end, we revisit the HP model already introduced in detail in Section 8.4.2 (p. 346), this time as the basis for comparing the sequences of different proteins. The essence of this model is to pare down the full 20-letter amino acid alphabet and to assign all amino acids to one of two categories based upon their physicochemical properties: H, for those amino acids that are hydrophobic, and P, for those that are polar (hydrophilic).

In Section 8.4.2, the HP model was introduced as a toy model for thinking about the problem of protein folding. Now we invoke this model as the basis of a simplified bioinformatic comparison of different proteins. Our rationale is that by using reduced alphabets in the bioinformatic setting, it is possible to make a trade-off between fidelity and ease of interpretation, precisely the same kind of trade-off that has been made repeatedly throughout the book nearly every time we have written down a model. In order to effect the HP model as an actual algorithm for bioinformatics, we need to recall how to assign amino acids to the H and P classes and we need to decide how to assign scores to different alignments. Figures 8.28 and 8.29 (p. 347) resolve the first issue and show the mapping between the full set of amino acids and their HP counterparts. For the purposes of sequence alignments, we are going to replace sequences of the full 20 amino acids by sequences of 1's and 0's, with an H residue given a 1 and a P residue given a 0. For example, if we have the amino acid sequence *SIRPRA* (a fragment of the *Pax6* and *eyeless* genes), in the HP mapping this will correspond to the sequence PHPHPH or 010101. This can be read off Figure 8.28 (see p. 347) directly. Within our HP model of bioinformatics, the goal will be to examine the alignments between different collections of H's and P's, which we will represent mathematically as 1's and 0's, in precisely the same spirit as we introduced state variables in Chapter 7.

21.2.2 Scoring Success

A Score Can Be Assigned to Different Alignments Between Sequences

In order to decide whether a given local alignment between two sequences is good or not (where "good" means statistically significant), we have to have some criterion for assigning a quantitative degree of plausibility (a score) that two different sequences are "related." Conceptually, the scoring scheme is meant to tell us whether, over evolutionary time scales, some ancestral sequence could, by a series of mutations, have resulted in the two sequences of interest that are being compared.

For example, let's compare two related proteins from different organisms that fulfill the same function. Given what we know about protein structure and evolution, if we find that a particular amino acid in a sequence is different from that at the corresponding position in a second sequence, we will be more surprised if the substituted amino acid has completely different physicochemical properties or is very different in size. By assigning scores to each of the possible substitutions, we have a quantitative measure of our degree of qualitative belief that two sequences have some common sequence ancestor. The choice of scoring matrices will be discussed in more detail below. For now, we examine the assignment of scores given a scoring matrix.

For simplicity, we start by considering two different sequences of equal length and assign a score to the degree of alignment between these two sequences. We adopt a notation in which σ_i refers to the ith element in the first sequence and σ_i' refers to the ith element in the second sequence. Given this scheme for identifying the elements in a given sequence, we can write the score in our HP model for the ith position in the alignment of the two sequences as

$$S_i(\sigma_i, \sigma_i') = \varepsilon_{\mathrm{HH}}\sigma_i\sigma_i' + \varepsilon_{\mathrm{PP}}(1-\sigma_i)(1-\sigma_i') + \varepsilon_{\mathrm{HP}}\sigma_i(1-\sigma_i') + \varepsilon_{\mathrm{HP}}(1-\sigma_i)\sigma_i'. \tag{21.1}$$

This equation says that when the two residues are both H's (that is, $\sigma_i = \sigma_i' = 1$), we assign a score to that position of $\varepsilon_{\mathrm{HH}}$; when they are both P's (that is, $\sigma_i = \sigma_i' = 0$), we assign a score $\varepsilon_{\mathrm{PP}}$; and when the two residues are *different*, we assign a contribution $\varepsilon_{\mathrm{HP}}$ to the score. In the HP scheme considered here, both $\varepsilon_{\mathrm{HH}}$ and $\varepsilon_{\mathrm{PP}}$ will be positive since we favor keeping the identity of a given residue the same. In contrast, $\varepsilon_{\mathrm{HP}}$ will be negative because in the HP model H's and P's have completely different physicochemical properties, and intuitively we expect that such substitutions are generally unlikely. We will comment further on how the actual numerical values in such a scoring matrix are determined below and the reader is invited to explore this further in the problems at the end of the chapter.

To illustrate how this scheme is used in practice, consider the two sequences 101010 and 101110. The total score that would be assigned to an alignment of these two sequences using Equation 21.1 is obtained by visiting each position in the sequence and computing the score at that position. The total score S_{tot} is obtained by adding up the contributions from each position in the sequence as

$$S_{\mathrm{tot}}(\{\sigma_i, \sigma_i'\}) = \sum_i^N S_i(\sigma_i, \sigma_i'). \tag{21.2}$$

We use the notation $\{\sigma_i, \sigma_i'\}$ to characterize the full set of letters that make up both sequences. For this case, if we do not shift one sequence relative to the other, we find $S_{\mathrm{tot}} = 3\varepsilon_{\mathrm{HH}} + 2\varepsilon_{\mathrm{PP}} + \varepsilon_{\mathrm{HP}}$ since there are three positions at which there are 1's at the same position, two positions at which there are 0's at the same position, and one position that has a *mismatch.* This kind of argument permits us to compare different sequences and to assign a number to the goodness of their alignment. Notably, this scheme makes important implicit use of the assertion that biological sequence information is largely context-independent; we assign the same energy penalty to a change of H to P regardless of what residues precede or follow it in the sequence.

Our treatment so far has ignored an important evolutionary feature, namely, the insertion and deletion of nucleotides or amino acids in a sequence. In practice, for performing real sequence alignments, we need to improve our analysis by allowing for gaps. This idea reflects the fact that a sequence might have been altered over evolutionary time by the insertion of a new nucleotide or a new amino acid residue. For example, if we compare the two sequences 110011 and 1101011, we find that if we insert a gap in the first sequence such that it is written as 110–011, then the sequences can be otherwise perfectly aligned. From the standpoint of our scoring scheme, this requires that we introduce a new parameter $\varepsilon_{\mathrm{gap}}$ that assigns a penalty to opening up a gap, giving a score $S_{\mathrm{tot}} = 4\varepsilon_{\mathrm{HH}} + 2\varepsilon_{\mathrm{PP}} + \varepsilon_{\mathrm{gap}}$. In most scoring schemes, the penalty for inserting a gap into a sequence is greater than the penalty for aligning a pair of mismatched residues, reflecting a mechanistic assumption that DNA polymerase is more likely to make errors that substitute one nucleotide for another than to insert or delete nucleotides. There may also be corrections that take gap size into account: for example, assigning a lower penalty for a single gap of six residues than for six independent single-residue gaps, reflecting the mechanistic assumption that some mutational events may result in large insertions or deletions of groups of nucleotides simultaneously.

Comparison of Full Amino Acid Sequences Requires a 20-by-20 Scoring Matrix

Our HP model of bioinformatics permits us to describe the mathematics of scoring simply, but is too coarse-grained to answer many questions of real biological interest. This HP perspective is primarily offered here as a way to consider the *meaning* of scoring without being encumbered by the full complexity of real sequences. For the comparison of real protein sequences, we require a scoring matrix that can assess the "cost" of changing any amino acid in the sequence to any other. There are a number of different scoring matrices; one of the most popular schemes, shown in Figure 21.7, is the so-called BLOSUM62 scoring matrix. The values of the scores assigned in this matrix are derived from analysis of real protein sequences, as described below. To assign a score to a given alignment, we proceed much the same as we did in Equation 21.2, by visiting each position in the two sequences of interest and assigning a score S_i to that position that is dictated by the matrix shown in Figure 21.7. If that particular position has a gap in one of the sequences, then we assign a gap penalty. When performing sequence alignments using BLAST (the reader is asked to do just that in the problems at the end of the chapter), there are

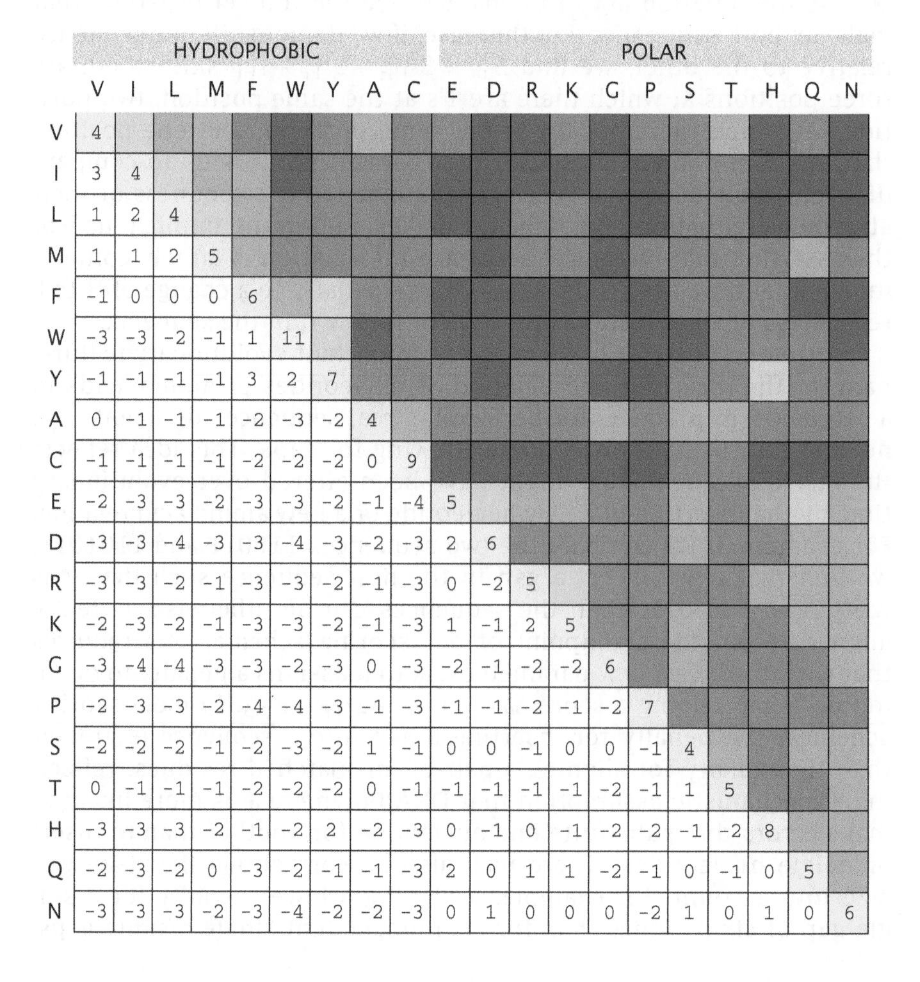

	V	I	L	M	F	W	Y	A	C	E	D	R	K	G	P	S	T	H	Q	N
V	4																			
I	3	4																		
L	1	2	4																	
M	1	1	2	5																
F	-1	0	0	0	6															
W	-3	-3	-2	-1	1	11														
Y	-1	-1	-1	-1	3	2	7													
A	0	-1	-1	-1	-2	-3	-2	4												
C	-1	-1	-1	-1	-2	-2	-2	0	9											
E	-2	-3	-3	-2	-3	-3	-2	-1	-4	5										
D	-3	-3	-4	-3	-3	-4	-3	-2	-3	2	6									
R	-3	-3	-2	-1	-3	-3	-2	-1	-3	0	-2	5								
K	-2	-3	-2	-1	-3	-3	-2	-1	-3	1	-1	2	5							
G	-3	-4	-4	-3	-3	-2	-3	0	-3	-2	-1	-2	-2	6						
P	-2	-3	-3	-2	-4	-4	-3	-1	-3	-1	-1	-2	-1	-2	7					
S	-2	-2	-2	-1	-2	-3	-2	1	-1	0	0	-1	0	0	-1	4				
T	0	-1	-1	-1	-2	-2	-2	0	-1	-1	-1	-1	-1	-2	-1	1	5			
H	-3	-3	-3	-2	-1	-2	2	-2	-3	0	-1	0	-1	-2	-2	-1	-2	8		
Q	-2	-3	-2	0	-3	-2	-1	-1	-3	2	0	1	1	-2	-1	0	-1	0	5	
N	-3	-3	-3	-2	-3	-4	-2	-2	-3	0	1	0	0	0	-2	1	0	1	0	6

Figure 21.7: A scoring matrix for the alignment of protein sequences. Each row and column correspond to a different amino acid, and a given entry in the matrix reveals the likelihood of replacing the amino acid in a given row with the amino acid in the corresponding column. The numbers along the main diagonal assign a positive score to maintaining a given residue in the sequence—the amino acid tryptophan (W) is the most favorable to maintain. The magnitudes of the substitution scores are graphically represented in the upper right half of the table by different colors; darker shades represent less favorable substitutions. The pattern reveals that it is generally more favorable to replace a polar amino acid (P) with another polar amino acid rather than with a hydrophobic amino acid (H). This particular choice of scoring matrix is known as BLOSUM62.

default choices for the scoring criteria that we have discussed here. Nevertheless, in performing sequence alignments, it is useful to keep the quantitative framework we have described here in the back of one's mind.

Different scoring schemes have different domains of applicability and can be thought of as providing stopwatches with different time units for examining evolutionary history. An interesting analogy is provided by the different tools used for measuring astronomical distances. For nearby objects, the parallax effect, familiar to us all from holding one finger in front of our eyes and then alternately closing one eye and then the other, takes advantage of the shift of an object relative to the background. For more distant objects, a class of stars that have periodic variations in their intensity known as Cepheid variables can be used as a "standard candle" for distance calibrations. For larger distances still, entire galaxies can be used as standard candles. The point is that there is not necessarily one universal distance standard for measuring astronomical distances, any more than there is a single sequence comparison scheme that sheds light on all evolutionary time scales.

The logic leading to a scoring scheme like that shown in Figure 21.7 is ultimately tied to ideas about the nature of evolution. The simplest idea might be that a given amino acid would be present at a given position in a sequence with a probability given by the frequency of occurrence of different amino acids. Those amino acids that occur more frequently have a higher chance of being found at the position in the sequence of interest. At the same time, as discussed above, we are aware that some kinds of mutational events are more likely to occur than others, and similarly that some substitutions of amino acids with similar chemical properties are more likely to survive natural selection than more radical changes. In brief, we want our scoring scheme to pick up correlations that are present in sequences as a result of the history imposed by evolution. To capture this idea, the score S_{ij} assigned to aligning residues i and j is based upon the so-called log-odds ratio, given as

$$S_{ij} = \log \frac{p_{ij}}{q_i q_j}, \tag{21.3}$$

where q_i is the probability of finding amino acid i based strictly on assessing amino acid usage frequencies and p_{ij} reports the frequency with which i is replaced by j in real sequences of related proteins. To construct a scoring matrix using these ideas, the key point is to generate the p_{ij}. In the case of the PAM scoring matrices, this was done by considering real protein sequences in closely related groups where only 1 in 100 of the amino acids had been altered. By searching through a large number of examples of nearly identical protein sequences, the frequencies of all the 210 different substitutions could be determined and used as the basis of constructing the log-odds ratio described above. The more familiar BLOSUM62 matrix results from examining alignments for sequences with much less sequence overlap, but is performed in a conceptually similar fashion. If we were in possession of a complete theoretical understanding of all of the mechanisms of both mutation and natural selection, we might be able to derive a scoring matrix from first principles that would be able to predict the likelihood of any particular kind of change. At present, our understanding is too limited for this, and we are more likely to be able to make useful sequence comparisons by relying on empirical analysis of actual protein sequences that have been found to exist in nature.

Returning to a more idealized world, the reader is invited to compute the log-odds ratios for the idealized HP model in the problems at the end of the chapter.

A useful heuristic analogy for thinking about these scoring schemes can be obtained by appealing to our study of cooperativity in Chapter 7. We think of the probabilities q_i as corresponding to Boltzmann weights, $e^{-\beta\varepsilon_i}$, where ε_i is the "energy" penalty for placing a particular amino acid i at a given position in the protein sequence. In the case when amino acid placements are random, we expect the energy penalty for placing amino acids i and j at the same position of two aligned sequences to be simply $\varepsilon_i + \varepsilon_j$. If the two sequences have a common evolutionary ancestor, we expect that there will be "cooperativity" associated with this placement characterized by an "interaction energy" ε_{ij}. In other words, the Boltzmann factor associated with having amino acids i and j at the same position in the two sequences is $e^{-\beta(\varepsilon_i+\varepsilon_j+\varepsilon_{ij})}$. In this view, the score S_{ij} is nothing but the interaction energy itself. Positive interaction energies correspond to negative scores and indicate that substitution of one of these amino acids by the other is less likely than suggested by chance. Similarly, negative interaction energies indicate a permissiveness that accepts swapping the identity of the two amino acids in question.

Even Random Sequences Have a Nonzero Score

Now that we know how to compute the score of a given alignment, we are prepared to examine the question of how good an actual score is. In particular, we note that we first have to assess how large a score we would get by just taking two random sequences. More importantly, if we compare some sequence of interest against all of the sequences found in some database, we have to have some way of saying whether or not the score we obtain might have arisen by pure chance. Obviously, we can only celebrate a high score once we have said that it is high relative to some meaningful standard.

How do we determine a standard score for this kind of comparison? To get a feel for that question, we consider two sequences, each with N members in the sequence and built up from our two-letter alphabet with each letter either a 1 or 0. We assume that the probability of finding a 0 (or a P) at a given position in the sequence is given by p, while the probability of a 1 (or an H) is given by $1 - p$. The score (assuming no gaps) that we assign to a given pair of sequences has already been given in Equation 21.1. Our goal now is to evaluate the average score that we should expect given that scoring function and assuming that the sequences of interest are strictly random. In particular, the average of interest is

$$\langle S_{\text{tot}}(\{\sigma_i, \sigma_i'\})\rangle = \left\langle \sum_{i=1}^{N} S_i(\sigma_i, \sigma_i') \right\rangle. \tag{21.4}$$

As usual, we use the notation $\langle S \rangle$ for the average of the quantity S. This expression may be simplified considerably by plugging in the score $S_i(\sigma_i, \sigma_i')$ as given by Equation 21.1, resulting in

$$\langle S_{\text{tot}}(\{\sigma_i, \sigma_i'\})\rangle = \sum_{i=1}^{N} \big[\varepsilon_{\text{HH}}\langle\sigma_i\rangle\langle\sigma_i'\rangle + \varepsilon_{\text{PP}}(1 - \langle\sigma_i\rangle)(1 - \langle\sigma_i'\rangle) + \varepsilon_{\text{HP}}\langle\sigma_i\rangle(1 - \langle\sigma_i'\rangle) + \varepsilon_{\text{HP}}(1 - \langle\sigma_i\rangle)\langle\sigma_i'\rangle\big]. \tag{21.5}$$

ALIGNMENT STATES		PROBABILITY	SCORE
σ_i	σ'_i		S_i
0	0	p^2	ε_{PP}
0	1	$p(1-p)$	ε_{HP}
1	0	$p(1-p)$	ε_{HP}
1	1	$(1-p)^2$	ε_{HH}

Figure 21.8: Scoring states and weights. This table shows the different outcomes at the *i*th position in a sequence comparison in the HP model, the probability of each such outcome, and the associated score.

The reason we were able to decouple the averages, namely, $\langle \sigma_i \sigma'_i \rangle = \langle \sigma_i \rangle \langle \sigma'_i \rangle$, is because we have assumed that the two sequences are random and independent and hence there is no sense in which σ_i depends upon σ'_i. We can effect these averages very simply through the recognition that

$$\langle \sigma_i \rangle = (1)(1-p) + (0)p, \tag{21.6}$$

where we have used the probability p that a given entry in the sequence is a 0. The probabilities of the different alignment states in this case are shown in Figure 21.8. As a result of these manipulations, we see that

$$\langle \sigma_i \rangle = 1 - p. \tag{21.7}$$

These simple calculations reveal the average score for a random sequence, which is

$$\left\langle S_{\text{tot}}(\{\sigma_i, \sigma'_i\}) \right\rangle = N[\varepsilon_{\text{HH}}(1-p)^2 + \varepsilon_{\text{PP}} p^2 + \varepsilon_{\text{HP}} 2p(1-p)], \tag{21.8}$$

as can be read directly from Figure 21.8. The reason such calculations are important is that they begin to help us think about the definition of the "noise" that we must fight in our attempt to find the "signal" in a given sequence alignment. That is, we need to have some way of knowing whether or not a given alignment is significant. To probe this question further, it is useful not only to have the average score expected for random sequences but also to know something more about the shape of the distribution of scores. In particular, we are going to pose our question of whether or not a given score is significant in terms of this distribution.

The Extreme Value Distribution Determines the Probability That a Given Alignment Score Would Be Found by Chance

The goal of our statistical analysis is to assess whether a given alignment signals something biologically meaningful such as functional similarity or common descent. To that end, we need a way of saying whether or not a particular score is exciting in its significance. To assess the statistical question that we face, we review the logic of a database search. We begin with our sequence of interest, and our goal is to search a database of sequences to find those that are related in some way. Operationally, one way to think of this is that we take our sequence of interest, visit each and every entry in the database, and find the best alignment (that is, highest scoring) that can be made with each such entry. As a result, to assess the significance of our presumed winners, we need to know something about the distribution of best scores. If getting a given best score is "easy," then our alignment should not be viewed as significant. On the other hand, if our best alignment score is hard to obtain, then we will be more tempted to attribute some significance to it.

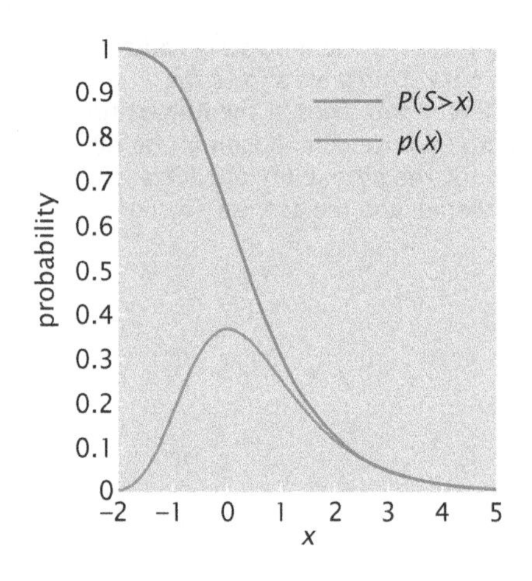

Figure 21.9: Extreme-value distribution for alignment scores. The figure shows both the probability density for finding a score x and the probability that a given score will be greater than x. The plot is made for the particular choice of distribution parameters $\lambda = 1$ and $\kappa = 1$.

What can we say about the distribution of scores that emerges from sequence alignments? Interestingly, just as we are accustomed to the emergence of the Gaussian distribution as a universal distribution associated with repeated measurements of a single random quantity, the distribution of highest scores (at least for the case of alignments without gaps) also follows a universal distribution, known as the "extreme-value distribution." This distribution applies not only to problems like the scoring problem considered here, but also to problems having to do with phenomena such as rainfall, maximum water levels, and even highest values in financial markets. To see how this works, imagine the following thought experiment. If we draw 100 samples from a Gaussian distribution and then keep only the highest number, and repeat this same process again and again, the distribution of highest numbers itself will follow its own very generic extreme-value distribution known as a Gumbel distribution. In particular, we can write the probability that the number will be greater than x as

$$P(x_{\text{max}} > x) = 1 - e^{-\kappa e^{-\lambda x}}. \tag{21.9}$$

This implies in turn that the probability density for finding a score between x and $x + dx$ is given by

$$p(x) = \kappa\lambda e^{-\lambda x - \kappa e^{-\lambda x}}. \tag{21.10}$$

The parameters κ and λ in the distribution determine its shape. Both of these expressions are plotted in Figure 21.9 for the case in which $\lambda = 1$ and $\kappa = 1$. The reason that this distribution is relevant to our scoring problem is that we can think of the alignment problem as an exercise in repeatedly determining the highest score. We start with our sequence of interest and a database that we wish to query. For each and every element of the database, we find the highest score. That is, for each sequence that we compare with our sequence of interest, we consider many different alignments and pick off that alignment with the highest score. The resulting distribution of highest scores (at least for the case of gapless alignments) can be shown to follow the extreme-value distribution as shown intuitively here.

The feature that makes the extreme-value distribution insidious for the purpose of determining the significance of scores obtained by sequence comparison is the long tail on the right-hand side of the curve. These tails can be described approximately by an exponential distribution, which decays more slowly than do the tails of a Gaussian (normal) distribution. The presence of these long tails means that there will be an appreciable number of high-scoring outliers that arise purely by chance. We need to understand the likelihood and abundance of these misleading outliers. We can use the extreme-value distribution to quantify how many times a search of our database would have yielded a score as good as that obtained in our alignment by evaluating

$$\text{Expect Value(score)} = N_{\text{DB}} \times \int_{\text{score}}^{\infty} p(x_{\text{max}})\, dx_{\text{max}}, \tag{21.11}$$

where N_{DB} is the number of entries in the database being searched and the integral is the probability that random chance would yield a score at least as high as our score of interest. The point of this Expect Value is to provide a quantitative measure of the likelihood that a given score would arise by chance. Note that the details of determining an Expect Value are more subtle than we have described here.

The easiest way to estimate the shape of the extreme-value distribution of alignment scores that is relevant for determining the Expect Value for a particular sequence alignment of interest is to generate a very large database of "random" sequences, attempt to align the sequence of interest to each of the random sequences in turn, and plot the resulting distribution of scores. The actual numerical Expect Value for any particular sequence alignment will of course depend on the size and nature of the random database that is being used as a basis for comparison, because the values of λ and κ are determined by fitting. If it were really completely true that biological sequence information is context-independent and that the only important pattern of amino acid distribution in protein sequences is their relative overall abundance, then a database of imaginary protein sequences built using a random number generator would be an adequate standard for significance comparisons. In reality though, there are subtle patterns of amino acid distributions within proteins that we only partially understand. Because the available databases of real protein sequences determined from genome sequencing projects are now extremely large (millions of sequences), it is in many ways just as easy and more relevant to generate a "random" database for these comparisons by simply permuting actual protein sequences. For example, one such database is generated by reversing the order of the amino acids in each individual protein within the real database. Using these kinds of permuted databases as the basis for comparison typically gives somewhat larger (that is, less significant) Expect Values than using a randomly generated database, and avoids certain kinds of artifacts. In performing actual sequence comparisons, it is worthwhile to keep these considerations in mind when deciding how to calculate Expect Values and how much confidence to place in the scores obtained.

False Positives Increase as the Threshold for Acceptable Expect Values (also Called E-Values) Is Made Less Stringent

Another way to probe the virtues and shortcomings of a given scheme for attributing meaning to sequence alignments is to use our tools to examine sequences for which the underlying relations are already known. In this case, we ask how well our scoring scheme does in distinguishing significant alignments from their spurious counterparts. To carry out such a test requires examining databases of sequences of known relation. For example, the DALI database classifies different proteins by alignment of their protein structure folds. Two other popular databases of this variety are SCOP and CATH. In choosing to use these structure alignment databases as test samples for our sequence alignment procedures, we are making the commonly accepted assumption that protein structure is a more accurate reporter of the evolutionary relatedness of proteins than is sequence similarity alone. It is important to recognize that, at present, any choice of a gold standard for "known" relationships between proteins depends on some kind of value judgment made by a human investigator; there is no completely objective way to determine whether two proteins are truly related or not, but some assertions of relatedness do merit great confidence because of the weight of many different kinds of experimental data on similarity of structure, function, and evolutionary lineage.

Given a scoring matrix such as BLOSUM62 or the HP matrix, we can use such a database to quantify the ability of that matrix to identify

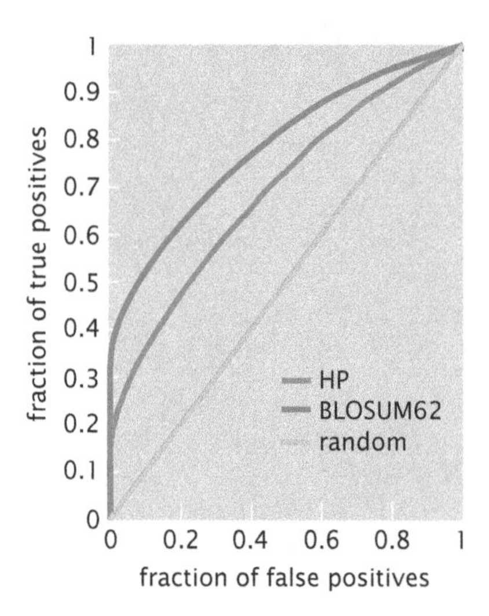

Figure 21.10: Ability of scoring matrices to identify related proteins. These ROC curves for the BLOSUM62 and HP scoring matrices show the ability of these matrices to identify true positives (proteins that share a fold) by assigning them a lower Expect Value than the false positives (proteins with unrelated structures). In the null model, true positives are chosen as often as false ones. A perfect model would be a line from (0, 1) to (1, 1); that is, all true positives are identified before false positives. Both HP and BLOSUM62 perform better than the null model, and the extra details in the full 20-by-20 BLOSUM62 matrix give it an advantage over the 2-by-2 HP matrix.

pairs of related proteins. One way to carry this out is first to align every protein in the database with every other protein using the scoring matrix and then rank the results in a list by their Expect Value. A perfect matrix would put all the related proteins (that is, true positives) at the top of this list, followed by all the unrelated ones (false positives). In practice, our scoring matrices will let in some false positives among the true positives and we would like to quantify to what degree a matrix allows false positives to creep in. The results in our list can be visualized by calculating the so-called Receiver Operating Characteristic (ROC) curve. This is constructed by choosing a cutoff Expect Value and then tallying the true and false positives that are lower than the cutoff. We start with a cutoff at the top of the list and move it down until all entries in the list have been included in the tally. The ROC curve is then a plot of true positives versus false positives, parameterized by our cutoff Expect Value. An example of such a curve is shown in Figure 21.10. Here we can see that the extra details included in the full 20-by-20 BLOSUM62 matrix have enabled it to perform better than an HP matrix. Nevertheless, the HP model holds some predictive value and performs better than a null model that chooses randomly between true and false positives and would produce a straight-line ROC curve from (0, 0) to (1, 1).

Ultimately, this section had two key objectives, namely, (i) to illustrate the statistical habit of mind that is implicit in any database search and comparison method, and (ii) to raise the flag of caution concerning the attribution of statistical significance to various sequence alignment methods without thinking hard about the assumptions that led to such attributions. In the following Computational Exploration, we give a sense of the work flow related to querying a protein database about a particular sequence. It is important to note though that throughout these sections we have skirted the important question of *how* these various manipulations are carried out in practice using real databases and real computers. These are extremely important topics and strike to the heart of how to construct search algorithms capable of dealing with the ever-growing databases of genomic information. The interested reader who wishes to learn more of the search techniques used in bioinformatics is invited to consult the Further Reading at the end of the chapter.

COMPUTATIONAL EXPLORATION

Computational Exploration: Performing Sequence Alignments Against a Database Throughout this chapter, we have given a detailed discussion of different strategies involved in comparing two protein or nucleotide sequences. In this computational exploration, we put all of that information together in order to illustrate what actually happens when a protein database is queried using some particular sequence.

We start with the query sequence corresponding to a protein fragment with a length of five residues as shown in Figure 21.11(A). This query sequence will be compared with every single protein in our database, some elements of which are also shown in Figure 21.11(A).

We start by calculating the score of all possible alignments between our query sequence and the first sequence in the database as shown in Figure 21.11(B). Each one of these alignments will result in an alignment score SCORE(i, j). Here, i corresponds to the protein our sequence is being aligned to (in the figure, we show $i = 1$), while j corresponds to each one of the possible alignments to protein i. Finally, for each

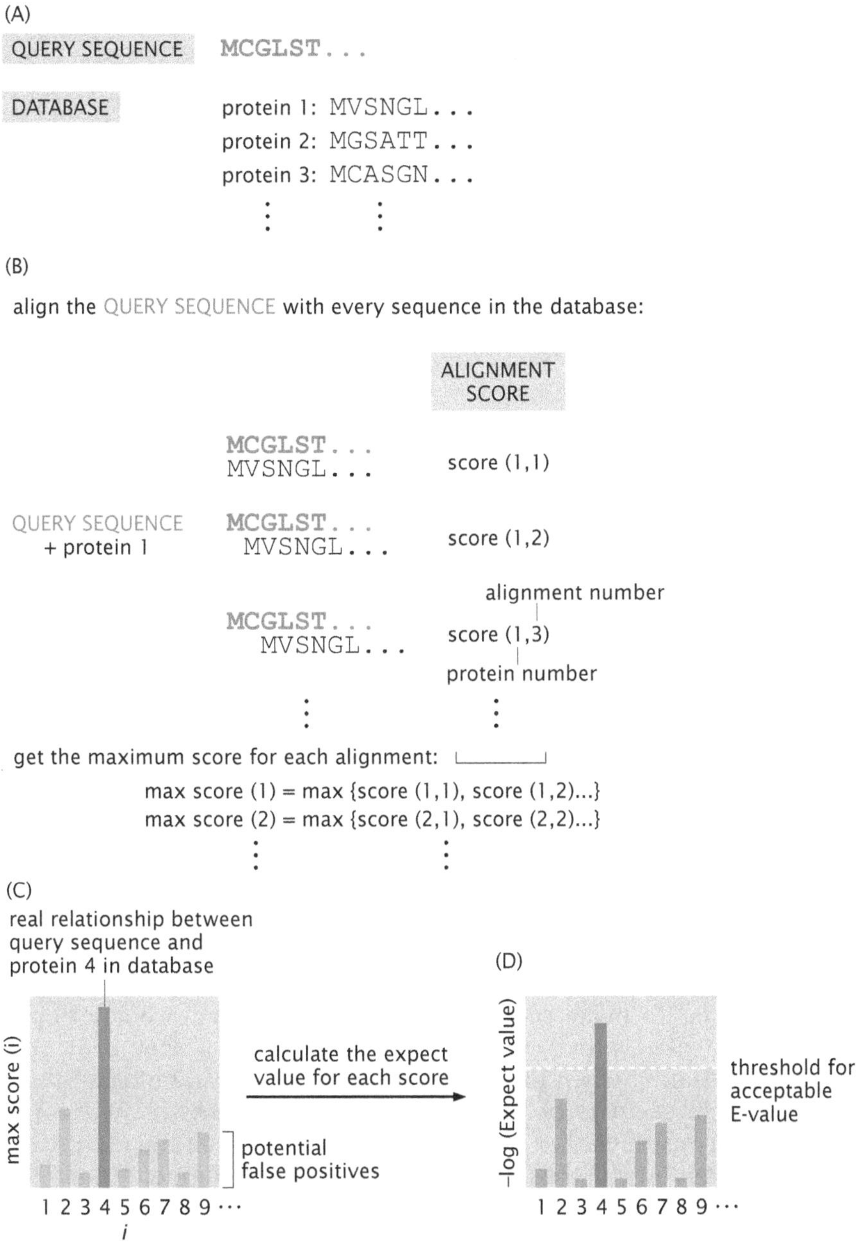

(A)

QUERY SEQUENCE MCGLST...

DATABASE
protein 1: MVSNGL...
protein 2: MGSATT...
protein 3: MCASGN...

(B)

align the QUERY SEQUENCE with every sequence in the database:

ALIGNMENT SCORE

QUERY SEQUENCE + protein 1

MCGLST...
MVSNGL...
score (1,1)

MCGLST...
MVSNGL...
score (1,2)

MCGLST...
MVSNGL...
score (1,3)

alignment number
protein number

get the maximum score for each alignment:

max score (1) = max {score (1,1), score (1,2)...}
max score (2) = max {score (2,1), score (2,2)...}

(C)

real relationship between query sequence and protein 4 in database

max score (i)

1 2 3 4 5 6 7 8 9 ···

i

potential false positives

calculate the expect value for each score

(D)

−log (Expect value)

threshold for acceptable E-value

1 2 3 4 5 6 7 8 9 ···

Figure 21.11: Querying a sequence against a protein database. (A) We start with a query sequence that is to be contrasted with every element of the database to be used. (B) For each element of the database, the score of each possible alignment against our query sequence is calculated. The maximum alignment score against each protein *i* in the database is stored in the variable MAXSCORE(*i*). For simplicity, we do not introduce gaps in the alignments. (C) The MAXSCORE value for each protein in the database is plotted, potentially revealing a real relationship between our sequence and protein number 4 in the database and also revealing a set of potential false-positive hits. (D) The Expect Value (E-Value) illustrates the probability of getting a particular score by random in the database. The lower the E-Value, the higher the chances that the hit is actually real as opposed to a false positive.

set of scores corresponding to protein *i*, we can find the best alignment obtained, resulting in the magnitude MAXSCORE(*i*).

In Figure 21.11(C), we plot all values of MAXSCORE corresponding to the alignment of our query sequence to the elements in our database. Here, we imagine the hypothetical scenario where our query sequence is actually related to protein $i = 4$ in the database. However, if we don't know this fact, what is a possible criterion to differentiate between a real hit in the database and the potential false positives? How do we determine if a score is actually relevant?

One way to determine the relevance of a particular score in the context of the database where the query is being made is the Expect Value or E-Value. In Figure 21.11(D), we plot −ln(E-Value) for each one of the MAXSCORE values shown in Figure 21.11(C). A smaller E-Value corresponds to a smaller chance of the particular score having occurred by random.

It is particularly amusing to perform such database searches against protein sequences based on the names of our favorite scientists. For example, we encourage the reader to perform a protein BLAST on the sequence "CHARLESDARWIN" against the protein database by going to http://blast.ncbi.nlm.nih.gov. Notice the different MAXSCORE values obtained for the proteins in the database and their corresponding E-values.

Structural and Functional Similarity Do Not Always Guarantee Sequence Similarity

In the remainder of the chapter, we will show how ideas on sequence alignment have been used to probe a variety of interesting biological questions. However, before embarking on this journey, we first explore one final cautionary case study that illustrates the limits of sequence analysis. An example that reveals the challenges of a sequence-based assessment of the relatedness of proteins from different species is given by comparing actin, ParM, and MreB. Recall from Section 15.3.2 (p. 600) that ParM is a bacterial analog of actin that participates in the segregation of plasmids. MreB is a distinct bacterial actin analog that is involved in cell shape determination. As shown in Figure 21.5, the structures of actin, ParM, and MreB are quite similar. To examine this structural similarity from a sequence perspective, Figure 21.12 takes the known *structure* alignment (that is, when the structures are aligned in three-dimensional space in the way that maximizes the spatial overlap between them) and compares the residues at the corresponding parts in the structure. It is clear that for the most part there is not a strong sequence similarity. Indeed, out of over 300 amino acids, only 10 are perfectly conserved in all three proteins. To make contact with our HP analysis, we color-code the residues that are hydrophobic and polar to see if they provide a better indication of when the structures will be similar. Indeed, in this case, there do seem to be small patches and patterns of hydrophobic and polar residues that are conserved, even though the identity of individual amino acids has changed. This example demonstrates the need for caution in using only sequence comparison as a window on functional relatedness; absence of evidence for sequence similarity is not evidence for structural or functional dissimilarity. Recognizing this problem, many researchers are now undertaking large-scale projects to determine the three-dimensional structure of as many proteins as possible, with the hope that protein structure databases may eventually approach sequence databases in scope, complexity, and richness.

21.3 The Power of Sequence Gazing

The ability to read and examine DNA sequences has revolutionized biology and already delivered on Crick's promise of ferreting out the "vast amounts of evolutionary information" that is "hidden away within them." Entire books are written around the study of sequence, and so our discussion can at best only be superficial and anecdotal. In the remainder of the chapter, we present a series of case studies, many of which harken back to themes already developed throughout the book, such as nucleosomes, the *lac* operon, hemoglobin, viruses,

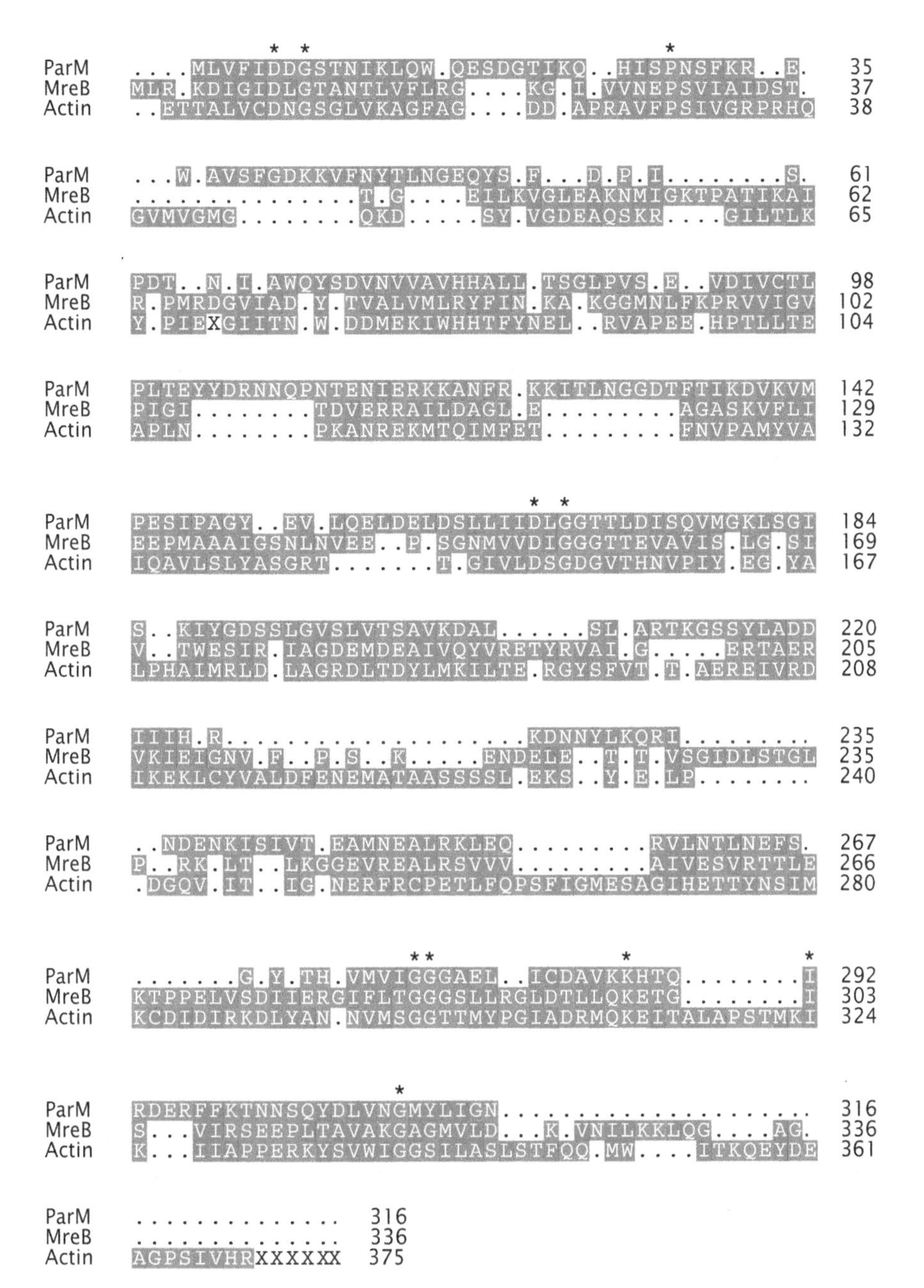

Figure 21.12: Comparison among sequences of bacterial and eukaryotic cytoskeletal proteins. The sequences are aligned by noting which parts of the proteins overlap structurally. The sequences of letters say which amino acids are found at particular positions within the structure. Amino acids marked with an asterisk are identical in all three sequences. The colors signify whether they are hydrophobic (red) or polar (blue). The dots in the alignment indicate regions in the sequences where amino acids have been inserted or deleted relative to the other sequences. The integers at the end of each row signify the position of that residue in the overall sequence. To access these sequences, the PDB codes are 1JCF for MreB, 1J6Z for actin, and 1MWM for ParM.

and the tree of life. The idea is to provide a sense of how the analysis of sequence can be used to gain insights into important biological questions.

21.3.1 Binding Probabilities and Sequence

We have already seen that by comparing sequences from different organisms we can learn much about the evolutionary history of life on Earth. But there are other ways that sequence gazing can tell us about how organisms work. For example, with a given genome sequence in hand, is there any way to determine how the genes within it are regulated? Until now, our statistical mechanical models of transcription factor binding (see Chapter 19) have treated the binding energy between the transcription factors and DNA as simple parameters without any real reference to the underlying sequence rules that give rise to them. But what happens when the sequence of letters making up such a binding site are slightly altered? Often, a change in one letter within some binding-site sequence does not eliminate binding, but rather alters the binding affinity of that site. As a result the definition of a binding site becomes much more flexible and our assessment of binding energies more subtle.

Position Weight Matrices Provide a Map Between Sequence and Binding Affinity

We begin with one of the most common classes of models in which the total binding energy can be thought of as being built up of a sum of contributions from each of the base pairs making up the binding site. This kind of model implies that the base pair at position one within the binding site contributes to the binding energy in a manner that is independent of the contribution of base pair two, for example.

To see how such models play out, consider an alphabet of only two bases or letters. In this toy sequence model, we denote purines (A and G) as R and pyrimidines (T and C) as Y. To be concrete, we consider a binding site with a length of five such bases. Now, let's say that we have a "consensus" site, which we define as the best possible binder. For the sake of argument, the consensus site is "YYYYY." Each deviation from this sequence will translate into a penalty to the total binding energy. Additionally, we will assume that the inner base pairs of the binding site contribute more to the binding energy than the outer ones. We represent the binding energies corresponding to the possible mutations of the binding site using a matrix referred to as a "position weight matrix" and given by

	1	2	3	4	5
Y	0	0	0	0	0
R	1	2	3	2	1

. (21.12)

Here, the columns represent the position along the binding site and the rows the different choices for bases at each position. The elements within the matrix give the energy change with respect to the consensus sequence in units of $k_B T$. Notice that we have assumed a symmetric site. Also, our statement about the importance of the center base pair with respect to the outer bases is reflected in the fact that the penalty for changing the center base pair is higher than that of all of the others.

With this matrix in hand, we can calculate the difference in binding energy between some sequence of interest and the consensus sequence. For example, for the binding site "YYRRY," we find

$$\varepsilon - \varepsilon_{\text{consensus}} = 3\ k_B T + 2\ k_B T = 5\ k_B T, \tag{21.13}$$

where we have simply exploited the fact that the third and fourth positions have been changed relative to their consensus values with an energy penalty given by Equation 21.12. More generally, when we are confronted with some new sequence, we can appeal to the position weight matrix to tell us its affinity relative to that of the consensus sequence. Using this approach, we can look at every possible 5-mer and determine its binding energy. More practically, this same information can be used to scan through some genomic sequence and to construct a binding-energy landscape as a function of position as shown in Figure 21.13(A). In this figure, we have located our consensus sequence in an otherwise random sequence. We move a

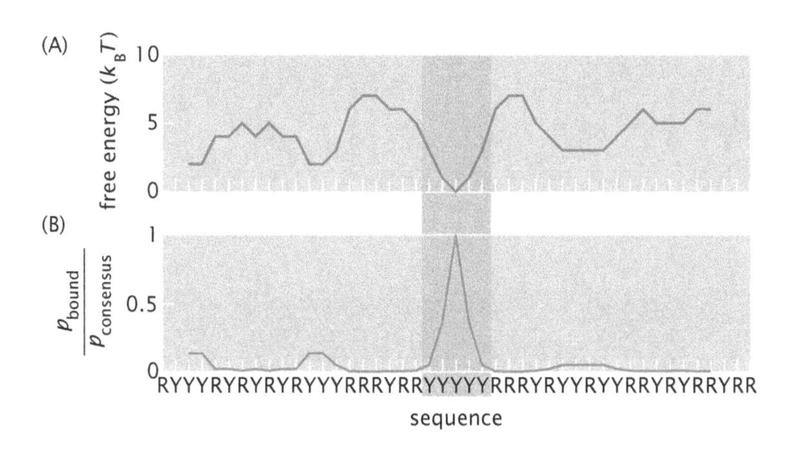

Figure 21.13: Toy model of a sequence landscape. A consensus site (YYYYY) has been surrounded by randomly generated bases. (A) The position weight matrix of Equation 21.12 has been used to compute the binding energy for each choice of central binding site along the sequence. (B) Plot of the relative probability of binding to each position with respect to the consensus site calculated using the energies presented in (A).

window of width 5 along this sequence and use the binding matrix shown in Equation 21.12 in order to calculate the binding energy. For each sequence, there is a corresponding binding energy and we see that for some positions the binding energy can be sufficiently small to have an appreciable binding affinity. A free-energy binding landscape such as that shown in Figure 21.13(A) reminds us of the concept of the nonspecific background used in our thermodynamic models of transcription-factor binding throughout the book, though now it is clear that the notion of a single nonspecific binding energy is a gross oversimplification.

An alternative way to look at the binding-energy landscape is by using it to compute the probability that a given site will be occupied relative to the probability of binding to the consensus site. This relative probability is precisely what the Boltzmann distribution tells us and is given by

$$\frac{p_{\text{bound}}}{p_{\text{consensus}}} = \mathrm{e}^{-\beta(\varepsilon-\varepsilon_{\text{consensus}})}. \tag{21.14}$$

In Figure 21.13(B), we show the resulting landscape of probabilities of binding as we slide the window along the sequence.

Frequencies of Nucleotides at Sites Within a Sequence Can Be Used to Construct Position Weight Matrices

Of course, the model captured in Equation 21.12 is totally made up. To use such models on real sequences, one has to have some scheme for assigning the energies making up the matrix. There are many sophisticated approaches to estimating such matrices for transcription factors based upon experimental binding data. To make these ideas concrete, we illustrate one example with reference to the CRP activator already discussed in Section 19.2.5. In Figure 21.14(A), we show a list of 23 binding sites for the activator CRP found in various regulatory sequences. The ability to make such a list in the first place results from the fact that this transcription factor acts on many different genes within the *E. coli* genome. Notice that some of the binding sites shown in the list correspond to those found in our beloved *lac* operon. CRP is a dimer that, to a first approximation, binds symmetrically to DNA. As a result the list of 23 binding sites can be extended by using their reverse complements to produce a list of 46 binding sites. For each position along the DNA-binding sequence, we can determine how often we find each base pair. One convenient way to look at the answer to this question graphically is through the use of a DNA logo

(A)

source	sequence
lac site 1	TAATGTGAGTTAGCTCACTCAT
lac L8	TAATGTAAGTTAGCTCACTCAT
gal	AAGTGTGACATGGAATAAATTA
ara site 1	AAGTGTGACGCCGTGCAAATAA
malT	AATTGTGACACAGTGCAAATTC
cat site 1	AAATGAGACGTTGATCGGCACG
*deo*P2 site 1	AATTGTGATGTGTATCGAAGTG
tnaA	GATTGTGATTCGATTCACATTT
ilvB	AAACGTGATCAACCCCTCAATT
malK	TTCTGTGAACTAAACCGAGGTC
malE	TTGTGTGATCTCTGTTACAGAA
*deo*P2 site 2	TTATTTGAACCAGATCGCATTA
lac site 2	AATTGTGAGCGGATAACAATTT
cat site 2	ACCTGTGACGGAAGATCACTTC
ara site 2	TGCCGTGATTATAGACACTTTT
pBR P4	CGGTGTGAAATACCGCACAGAT
ompA	AAGTGTGAACTCCGTCAGGCAT
crp	TAATGTGACGTCCTTTGCATAC
cya	AGGTGTTAAATTGATCACGTTT
*Col*E1 site 1	TTTTGTGAAAACGATCAAAAAA
*Col*E1 site 2	TTTTGTGGCATCGGGCGAGAAT
uxuAB	TGTTGTGATGTGGTTAACCCAA
tdc	ATTTGTGAGTGGTCGCACATAT

(B)

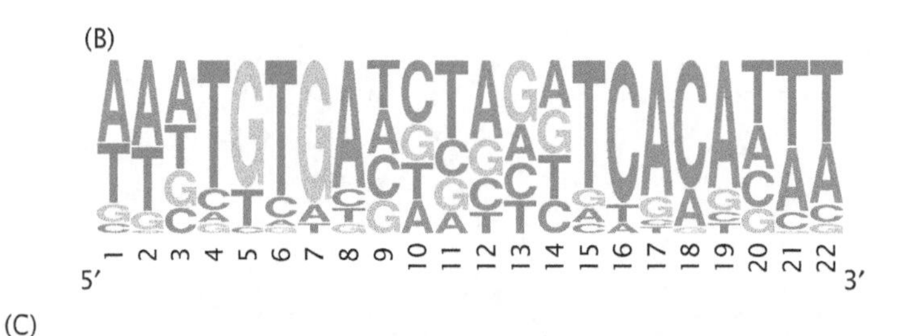

(C)

position	1	2	3	4	5	6	7	8	9	10	11	12	13	14	15	16	17	18	19	20	21	22
	binding energy per position (k_BT)																					
A	0	0	0	2.4	∞	2.9	2	0	0.1	0.6	1.3	0	0.5	0	2.1	2.5	0	1.2	0	0.3	0.3	0.3
C	2	3.1	0.9	1.7	2.8	1.8	∞	1.9	0.1	0	0.7	0.8	0.5	0.4	2.4	0	2.9	0	2.4	0.5	1.5	1.5
G	1.5	1.5	0.5	2.4	0	2.9	0	2.4	0.4	0.5	0.8	0.7	0	0.1	1.9	∞	1.8	2.8	1.7	0.9	3.1	2
T	0.3	0.3	0.3	0	1.2	0	2.5	2.1	0	0.5	0	1.3	0.6	0.1	0	2	2.9	∞	2.4	0	0	0

Figure 21.14: CRP-binding sites. (A) List of CRP-binding sites from various regulatory regions. (B) DNA logo obtained from the sequences shown in (A) and their reverse complements. The size of the bases at each position reflect how often that particular base was found in the list. (C) Position weight matrix calculated from the frequency of binding sites. The energies are given with respect to the consensus sequence. ∞ denotes that the particular base pair was not present at that position in any of the considered sequences. (A, adapted from O. G. Berg and P. H. von Hippel, *J. Mol. Biol.* 200:709, 1988)

such as that shown in Figure 21.14(B). For each position, the size of the base is proportional to its frequency in our list of binding sites.

In principle, each of the binding sites shown in Figure 21.14(A) has a different binding affinity for CRP. Yet, all of these binding sites are sufficiently strong that they are above a minimal threshold below which one would attribute "nonspecific binding." In fact, the standard argument is that evolution has selected for these sites to have a certain affinity that is useful for the function of the transcription factor. To proceed, assume that all sequences considered have the same affinity, ε. As a result, if we mutate the base pair at the first site such that the binding energy decreases by an amount $\Delta\varepsilon$, another mutation with an energy $-\Delta\varepsilon$ will have to be introduced somewhere else in the transcription factor footprint.

To determine the position weight matrix, we consider a pool of all possible sequences with a binding energy ε. This is analogous to considering all the possible microstates (each microstate being a realization of the sequence for the binding site) such that the sum from all of the bases adds up to the total energy ε. This is analogous to the situation put forth in Section 6.1.4, where we considered a system of n particles with energies such that the total energy always adds up to the same value. If we make an analogy between the particles introduced in Section 6.1.4 and each of the base pairs along the sequence, then the probability of finding a base b (which can be A, T, C, or G) at position i is given by the Boltzmann distribution, namely,

$$p(b, i) = \frac{e^{-\beta\varepsilon(b,i)}}{Z}. \tag{21.15}$$

Here, Z is the partition function and $\varepsilon(b, i)$ is the contribution of base b at position i to the total binding energy. Though β does not necessarily correspond to $1/k_BT$ as derived for the Boltzmann distribution, for convenience, we make that correspondence.

If Equation 21.15 truly describes the ensemble of possible sequences for the binding site, we can now relate the frequencies of appearance of bases at each position on the binding site to their corresponding contributions to the overall binding energy. For example, if we measure the energies with respect to the consensus sequence, we get

$$\frac{p(b, i)}{p(i)_{\text{consensus}}} = e^{-\beta[\varepsilon(b,i)-\varepsilon(i)_{\text{consensus}}]}, \tag{21.16}$$

where $p(i)_{\text{consensus}}$ is the probability of finding the consensus base at position i. We can now use the list of frequencies of each base pair for the CRP site shown in Figure 21.14(A) to estimate their contribution to the overall binding energy. The resulting energies can now be assembled into the CRP position weight matrix shown in Figure 21.14(C). Notice that the binding energies at each position are defined with respect to the binding energy for the consensus sequence, which by convention has an energy of zero. Also, since some bases did not show up in any of our sequences at a particular site, the model assigns them a binding energy of $+\infty$.

Can we now use this binding matrix to find new CRP sites in the *E. coli* genome? Do all binding sites that are found actually correspond to real binding sites? The reader is invited to confront the usefulness of this binding matrix in the following Computational Exploration.

COMPUTATIONAL EXPLORATION

Computational Exploration: Searching the *E. coli* Genome for Binding Sites Here, we use the position weight matrix for the transcription factor CRP shown in Figure 21.14(C) in order to scan the *E. coli* genome for binding sites. The goal of the analysis is to see how well we find binding sites on the genome by just using a position weight matrix. The resulting list of binding sites can be compared, in turn, with the tabulation of binding sites found in the RegulonDB database, which houses the majority of the current knowledge about transcriptional regulatory circuits in *E. coli*.

We start by applying this approach to the simple example of the *lac* operon. The sequence around the promoter is shown in Figure 4.13 (p. 155). Using the same approach that led to the binding landscape for our toy model in Figure 21.13, we scan through the regulatory sequence of the *lac* promoter using the position weight matrix shown in Figure 21.14(C). The result of this exercise is shown in Figure 21.15(A). Here, we can see how our position weight matrix can clearly detect the CRP-binding site. In principle, this should come as no surprise since we actually used the *lac* CRP site to calculate the position weight matrix in the first place (notice that it is one of the sites listed in Figure 21.14A)!

Now that we have seen that the position weight matrix can recover the known CRP-binding site within the *lac* operon, we pose the genome-wide question of what other CRP sites can be found in the entire *E. coli* genome. To accomplish this, we can run the position weight matrix along the *E. coli* genome and compare the result with the list of known CRP-binding sites from an online database such as RegulonDB. We start by downloading the complete genome of the *E. coli* strain MG1655. This can be done by going to the National Center for Biotechnology

Information (www.ncbi.nlm.nih.gov) and performing a search for MG1655 (you will have to navigate through the different genome options until you find the one we are looking for). Alternatively, it can also be accessed from the book's website. Next, we place our CRP position weight matrix at each base pair of the *E. coli* genome and ask for the resulting predicted binding energy.

On the other hand, we can obtain the actual documented CRP binding sites on the *E. coli* genome by going to RegulonDB (http://regulondb.ccg.unam.mx). Under "Downloads," we can obtain their data set "TF binding sites," which lists all known binding sites for transcription factors in this bacteria.

In Figure 21.15(B), we show a collection of binding sites with their corresponding probabilities of binding (measured with respect to the consensus sequence) as a function of the position along the *E. coli* genome. Notice that the scale of the *y*-axis is the same as in Figure 21.15(A), where we could clearly see the binding site located within the *lac* operon. When looking at a genome scale, we see that the vast majority of the positions along the genome give a small binding probability. One strategy for categorizing our results is to call all sites with a binding probability higher than 0.002 binding sites. However, this strategy has the shortcoming that it reveals a number of false positives, as seen by the predicted sites (blue points) that do not correspond to known sites (red circles), which were obtained from the RegulonDB database, in Figure 21.15(B).

Hence, though our position weight matrix succeeds in finding some of the known binding sites, there is this troubling list of spurious binding sites as well. There are many factors that led to this result. For example, because most of the binding sites in Figure 21.14(A) were determined using *in vitro* measurements, the position weight matrix doesn't take into account the potential interaction of CRP with other nearby transcription factors or the local conformation of DNA. Nevertheless, there is little doubt that position weight matrices are a good starting point to make an educated guess about the existence of binding sites in a DNA sequence of interest.

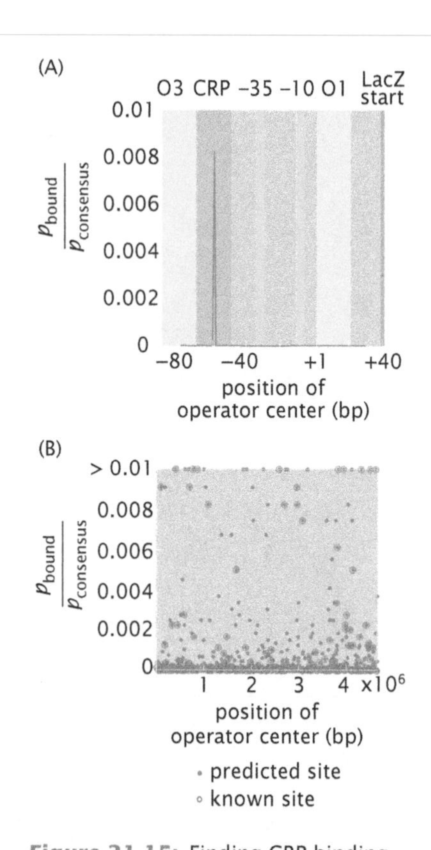

Figure 21.15: Finding CRP-binding sites in the *E. coli* genome. (A) The CRP position weight matrix is used to scan the regulatory region of the *lac* operon and to successfully identify the CRP-binding site within it. The colored regions correspond to the known features in the vicinity of the promoter such as the −10 and −35 boxes within the promoter, the binding sites for Lac repressor O1 and O3, and the *lacZ* transcription start site. (B) Comparison between predicted and known CRP sites across the whole *E. coli* genome.

Thus far, our analysis has focused on querying genomes to find binding sites. However, the same kind of calculational strategies can be used to ask other genome-wide questions, such as the distribution of nucleosome positions. In that case, the position weight matrix idea will be used to provide an energy scale with different nucleosomal locations. A more sophisticated treatment than that given here suppresses the assumption that all the binding sites listed in Figure 21.14(A) have the same affinity. In fact, it can be shown that this assumption can be relaxed with more or less the same results shown here. The reader is invited to explore this alternative derivation in the Further Reading at the end of the chapter. Additionally, a better way to estimate a position weight matrix invokes a combination of sequence information and binding affinity (as measured, for example, in terms of level of gene expression). In the following subsection, we present a clever method for harvesting this information from any regulatory region in bacteria in a high-throughput manner.

21.3.2 Using Sequence to Find Binding Sites

Knowledge of DNA sequence is really only a first step to understanding the hidden secrets of genomes. When we are given some freshly sequenced genome, one of the first things that we can try to come to terms with is the distribution of open reading frames (ORFs) by looking on both strands of the DNA and scanning it for start codons (that is, ATG) (see Figure 1.4 on p. 10) followed by long stretches of sequence that are not littered with any stop codons. Of course, this is harder than it sounds since, in the eukaryotic setting, these genes are themselves interrupted by introns that are removed from the eventual mRNA. The putative genes discovered by this method can then be "annotated" by using exactly the kind of sequence alignment tools described in the previous section to BLAST the gene against databases of sequences whose functions are already "known." Here too, there are many hidden assumptions, since there is the presumption that the known functions stay the same. For those cases in which the genes have been successfully identified, it is also of great interest to learn more about their promoter architectures.

As was made clear in Chapter 19, one of the features of the genome that is not transparent at all is its regulatory landscape. Where are the binding sites for the transcription factors that control when certain genes are expressed? What can be said about the distributions of regulatory RNAs? To make these points clearer, we return once again to one of our classic workhorses, the *lac* operon. So far, when we have talked about the *lac* operon (see Section 19.2.5, p. 822), the *Drosophila eve* minimal stripe element regulating the expression of stripe 2 (shown in Figure 19.2, p. 804), or any other regulatory element, we have simply stated the location and even the affinity of the different binding sites corresponding to the transcription factors of interest. But, how can we find the binding sites for the activator CRP shown in Figure 21.14(A) in the first place?

Our descriptions of promoter architecture thus far ignore the painstaking process involved in actually mapping those binding sites. When we are ignorant of where a transcription factor binds, or even of what transcription factors are involved in a given regulatory process, this calls for a new kind of sequence detective work. One key coarse-grained question simply asks about those stretches of DNA that exert a regulatory influence on the gene of interest. To discover these regulatory regions, one can make several reporter constructs where different chunks of the suspected regulatory real estate are located upstream from a promoter and then monitor the level of expression of that reporter. This process can be repeated iteratively to develop a picture of what proteins bind to the suspected regulatory region and, finally, to find out where they bind specifically. The binding of transcription factors to specific DNA sequences can then be confirmed using biochemical methods *in vitro*. It was precisely these kinds of experiments that led to the discovery of the minimal stripe element that is one of the determinants of anterior–posterior patterning in the fruit fly embryo.

As a proof of principle, an elegant alternative to this laborious procedure was recently applied to the *lac* operon. The approach merges clever experiments with precisely the kind of powerful sequence gazing described in this chapter. The experimental technique relies on two high-throughput techniques, namely, sorting and sequencing, the latter of which has only become available in the last few years. The particular case study mapped the regulatory region upstream from the promoter in the *lac* operon. As shown in Figure 21.16(A), this region includes the binding region for the activator CRP, discussed in the previous section in the context of position weight matrices. In the experiments used to carry out the regulatory detective work, this promoter controls the expression of GFP, allowing for the read-out of fluorescence in single cells. A library of different strains is constructed, each of which harbors a different version of the promoter that has roughly 10% of the base pairs mutagenized as shown diagrammatically in Figure 21.16(B). Because the regulatory architecture has been changed, each such strain will have a different level of expression as revealed by its fluorescence, and this fluorescence can be assayed one cell at a time by using fluorescence-activated cell sorting (FACS). In these machines, millions of cells flow past a laser and are queried and binned according to their level of expression. These populations can then be sequenced and the mutations corresponding to the level of fluorescence can be noted.

As a result of this experiment tens of thousands of reads of individual mutations with their corresponding expression levels are obtained. If a mutation is present in a base pair that is not part of any relevant binding sites, we expect the resulting level of expression to be relatively indifferent to this base-pair change. However, if the mutation is present within a binding site for CRP or RNA polymerase, by way of contrast, it will often lead to a substantial change in the resulting level of fluorescence. One way to express this observation in a quantitative way is through the use of mutual information. The

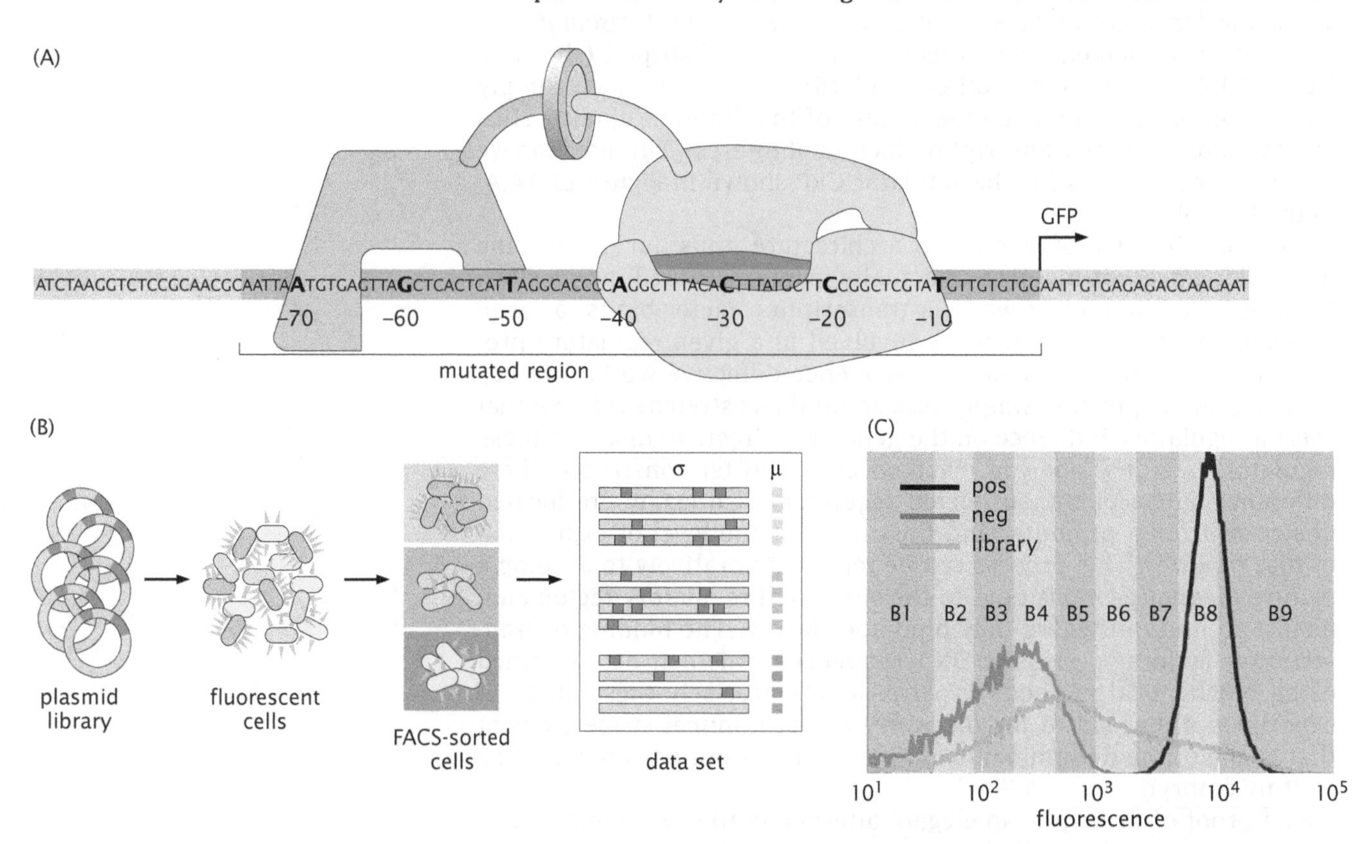

Figure 21.16: Finding binding sites by sequencing. (A) The regulatory region that was mutagenized in order to determine the regulatory importance of different base pairs. This construct uses GFP as a reporter of gene expression. (B) A library of mutagenized regulatory regions is inserted into *E. coli*. The cells are sorted in bins according to their level of fluorescence characterized by the parameter μ. (C) Histogram of single-cell intensities for the library, the nonmutagenized construct (pos) and a plasmid bearing no fluorescent reporter (neg). The different ranges used to sort the cells are indicated by the shaded areas in the plot. (Adapted from J. B. Kinney et al., *Proc. Natl Acad. Sci. USA* 107:9158, 2010.)

mutual information tells us how two different variables are linked. This goes beyond the "missing information" that we introduced in Chapter 6 by quantifying how much two variables depend upon one another, regardless of how that correspondence is realized. Like the missing information, the units of mutual information are in bits, and it tells us about how the missing information associated with one random variable x is decreased (on average) by knowing a different random variable y. In terms of missing information (or entropy), the mutual information of two random variables x and y is $I = \langle S_x - S_{x|y} \rangle_y$. Here S_x is the entropy associated with $p(x)$, the probability distribution of x, while $S_{x|y}$ is the entropy computed from the conditional probability $p(x|y)$ of x given y; $\langle \cdots \rangle_y$ refers to the average over the distribution $p(y)$. In particular, knowledge of y will allow us to win money by betting on the value of x when the mutual information is high. For the case of interest here, we ask how much information about the level of gene expression is encoded in each base pair. To answer this question, we consider two variables. First, we represent the particular base pair at position i along the coding region by b_i. Second, we represent the level of gene expression (that is, an index telling us which of the fluorescence bins our strain of interest is a member of) through the variable μ. The mutual information between the base pair at position i and the level of gene expression is given by

$$I(i) = \sum_{b_i, \mu} f(b_i, \mu) \log_2 \left[\frac{f(b_i, \mu)}{f(b_i) f(\mu)} \right], \tag{21.17}$$

where f denotes frequency. In the problems at the end of the chapter, we show that this formula for mutual information is equivalent to the one described earlier in terms of the average decrease in missing information. For example, $f(\mu)$ tells us how often we got the level of gene expression μ in our sample. $f(b_i)$ represents how often we got base pair b (which can be A, C, T, or G) at position i. Finally, $f(b_i, \mu)$ reveals in a quantitative way how often we found mutation b at position i associated with level of gene expression μ. Because we are using logarithms to base two, this means that the mutual information is measured in units of bits.

Though at first this definition of mutual information may seem cumbersome, in fact, it provides a very nice prism through which to view the data. To see this, we consider Figure 21.17, which shows the information footprint for two different experiments. Figure 21.17(A) shows the important base pairs for the case in which the activator CRP is not present. One of the immediate features of the information footprint that leaps out from the data is the presence of two special sequence motifs within the promoter. These motifs are known as the −10 and −35 boxes and are usually considered as the main determinants of the interaction strength between RNA polymerase and the promoter DNA. In the second experiment, shown in Figure 21.17(B), CRP was present. As a result, the mutual information reveals that the base pairs within the CRP-binding site encode information relevant to the level of gene expression. Naturally, we can see that the base pairs associated with the −35 and −10 boxes of the promoter, where RNA polymerase binds, remain extremely relevant.

To better understand the meaning and usefulness of expressing data in the language of mutual information, we turn our attention to a couple of instructive toy models. In this case, we will consider an alphabet composed of two letters, namely, H (for heads) and T (for tails). To start, we will think of a binding site composed of only two of these

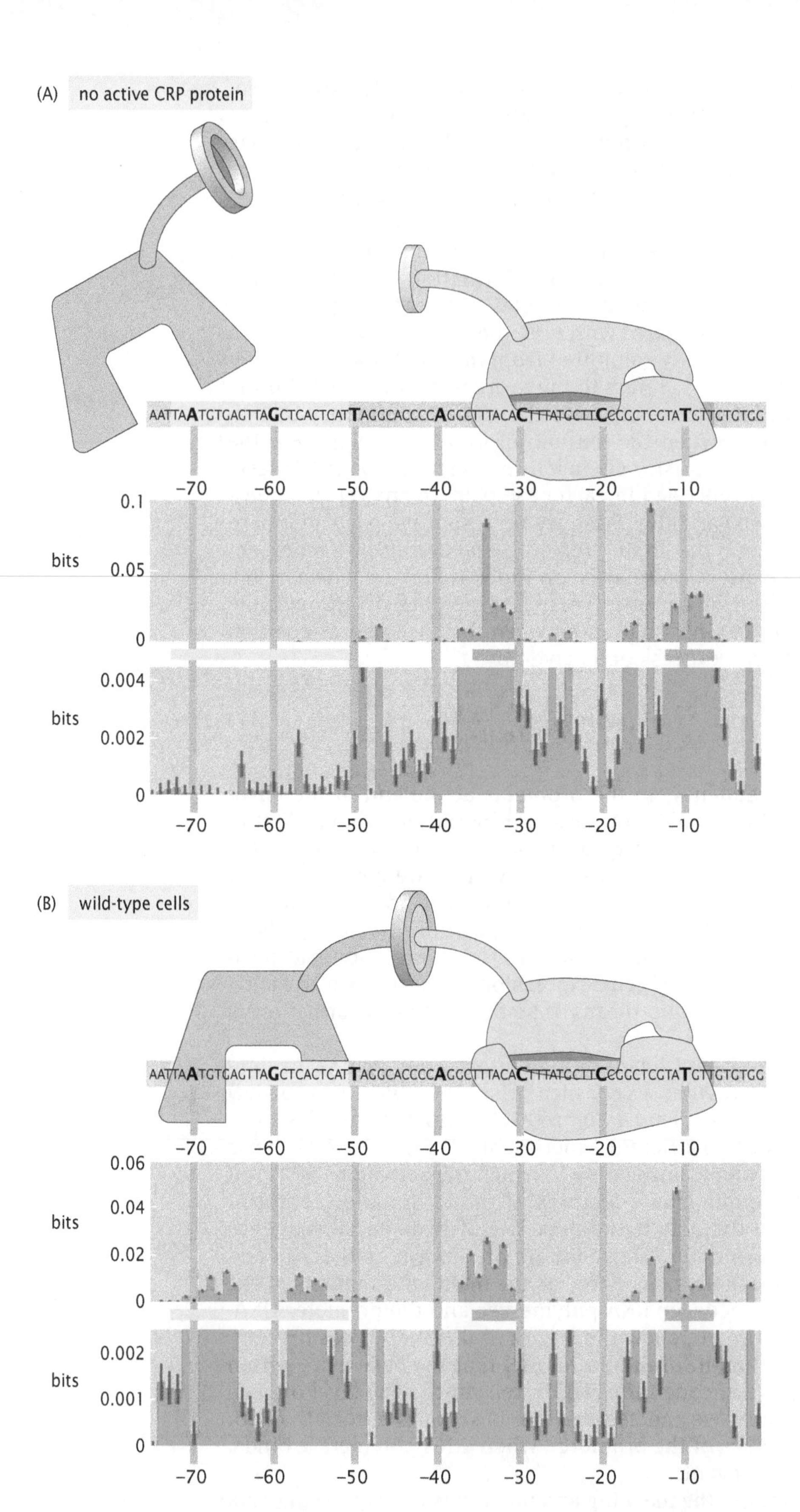

Figure 21.17: Information footprint for binding in the promoter region. The information footprint was calculated from the experiments schematized in Figure 21.16 using a formula conceptually similar to Equation 21.17. (A) For a strain where activator (CRP) is not active and cannot bind to DNA, the information footprint is dominated by base pairs in the region where RNA polymerase binds. (B) In the wild-type context, when CRP is present in the strain, the information footprint is dominated by the base pairs where RNA polymerase binds and those where CRP binds. (Adapted from J. B. Kinney et al., *Proc. Natl Acad. Sci. USA* 107:9158, 2010.)

letters. The sequences for the binding site in this toy model and their corresponding levels of gene expression are

Sequence	Level
HH	1
HT	1
TH	10
TT	10

(21.18)

We have chosen the values so that the first base pair determines the level of gene expression, since the choice of base pair at the second position changes nothing. Intuitively, this leads us to expect that all of the mutual information between sequence and gene expression is in the first base pair and no mutual information is to be found in the second one.

To see how this plays out, we invoke Equation 21.17 to calculate the mutual information. To carry out the calculations, we need three frequencies. Our model has two levels of gene expression, which we call high (Level = 10) and low (Level = 1). Each one shows up two times out of a total of four different realizations of the binding site. As a result, $f(\mu)$ is

f(High)	f(Low)
1/2	1/2

(21.19)

The frequency of each base at each position is

Position	f(H)	f(T)
1	1/2	1/2
2	1/2	1/2

(21.20)

Finally, we need to compute the frequency $f(b_i, \mu)$, which is given by

Position	f(H, High)	f(H, Low)	f(T, High)	f(T, Low)
1	0	1/2	1/2	0
2	1/4	1/4	1/4	1/4

(21.21)

We're now ready to compute the mutual information between each position on the binding site and the level of gene expression. For the first binding site, we get

$$I(1) = f(\mathrm{H}_1, \mathrm{High}) \log_2\left[\frac{f(\mathrm{H}_1, \mathrm{High})}{f(\mathrm{H}_1) f(\mathrm{High})}\right] + f(\mathrm{H}_1, \mathrm{Low}) \log_2\left[\frac{f(\mathrm{H}_1, \mathrm{Low})}{f(\mathrm{H}_1) f(\mathrm{Low})}\right]$$
$$+ f(\mathrm{T}_1, \mathrm{High}) \log_2\left[\frac{f(\mathrm{T}_1, \mathrm{High})}{f(\mathrm{T}_1) f(\mathrm{High})}\right] + f(\mathrm{T}_1, \mathrm{Low}) \log_2\left[\frac{f(\mathrm{T}_1, \mathrm{Low})}{f(\mathrm{T}_1) f(\mathrm{Low})}\right] \tag{21.22}$$

$$= 0 + \frac{1}{2} \log_2\left(\frac{1/2}{1/2 \times 1/2}\right) + \frac{1}{2} \log_2\left(\frac{1/2}{1/2 \times 1/2}\right) + 0, \tag{21.23}$$

where we have made use of the limit

$$\lim_{x \to 0} x \log x = 0. \tag{21.24}$$

This calculation tells us that the mutual information for the first site is given by

$$I(1) = \log_2 2 = 1. \tag{21.25}$$

For the second site, we find

$$I(2) = \log_2 \left(\frac{1/4}{1/2 \times 1/2} \right) = 0. \qquad (21.26)$$

As expected, we find that for the first site there is one bit of mutual information relating sequence and level of expression, while for the second site there is no mutual information. Intuitively, what this means is that by asking the question "What is the base in position 1?," we can at once learn what the level of expression is once the answer is given. On the other hand, a knowledge of the base at the second site brings us no closer to knowing the level of expression. Of course, our toy model describes a highly simplified situation, but it is exactly these kinds of ideas that go into constructing information footprints like those given in Figure 21.17. The reader is encouraged to further explore the mutual information in the problems at the end of the chapter.

Using these ideas, it is in principle possible to systematically search through genomes to characterize the information footprints for multiple genes by performing a similar experiment for each gene of interest. These information footprints, in turn, provide a window on the regulatory architecture for each of these genes. As noted throughout the book, another feature of the regulatory architectures found in eukaryotic genomes is the distribution of nucleosomes. We now take up the question of how sequence gazing is beginning to shed light on the rules governing nucleosome positioning in eukaryotic genomes.

21.3.3 Do Nucleosomes Care About Their Positions on Genomes?

The advent of sequencing methods has made it possible to ask genome-wide questions such as whether or not nucleosomes care about their positions on the genome or the complementary question of whether or not the genome itself cares about where nucleosomes are distributed. The reason that the answer to such questions matters is that the presence of nucleosomes has a direct influence on the expression of genes. For example, in the context of transcription, those parts of the genome that are occupied by nucleosomes are not accessible for binding by most transcription factors.

One extreme model for how the positioning of nucleosomes arises invokes ideas like those developed in Chapter 14 in which the DNA can be thought of as a one-dimensional "gas" of nucleosomes that cannot occupy the same genomic real estate simultaneously. At the opposite extreme is a picture in which nucleosomes adopt their positions based upon differences in the sequence-dependent free energy of formation such as we described in Chapter 10 (p. 407). Yet another mechanism for the positioning of nucleosomes is through the action of ATP-consuming enzymes that actually move nucleosomes along the DNA. As a prelude to examining these different ideas, we will examine what sequence has told us about the distributions of nucleosomes.

In Chapter 10 (p. 409), we introduced the fascinating and important topic of how individual nucleosomes are accessed, culminating in results like those shown in Figure 10.23 (p. 410). These experiments on single nucleosomes have been followed by experiments in which 17 nucleosomes are lined up on a fragment of DNA and then the site accessibility of the DNA on the middle nucleosome is measured, with results similar to those already found for the individual nucleosome in the sense that sites that are more deeply sequestered are harder to access. These studies on small collections of nucleosomes only confirm the need to better understand the nucleosomal distributions across genomes and the mechanisms of how genomic DNA is accessed for important transactions.

DNA Sequencing Reveals Patterns of Nucleosome Occupancy on Genomes

In recent years, a variety of studies have taken stock of where nucleosomes are found in the genomes of a host of different organisms. The idea of one approach was already introduced in Figure 21.2, where nucleosomal binding to a particular fragment of DNA protects that fragment from digestion by micrococcal nuclease, an enzyme that snips away base pairs from the unprotected ends of the linker DNA between nucleosomes. As shown in the figure, once the DNA that flanks the nucleosome is digested away, the protected fragment is then sequenced. More recently, a complementary technique has been introduced in which chemically modified histone H4 is prepared in such a way that it can perform a chemical cleavage of the DNA in the vicinity of the nucleosome dyad. The virtue of these complementary approaches is that in one case, the protected DNA yields a fragment bounded on the two ends by the extremities of the DNA associated with a given nucleosome, while in the chemical mapping case, the sequenced fragment is effectively from one nucleosome center to the next.

What emerge from these sequencing experiments are maps of nucleosome occupancy along the genomic DNA such as shown in Figure 21.18. In this case, the occupancy of a 20,000 base pair fragment of chromosome 14 in the yeast *Saccharomyces cerevisiae* is characterized by plotting

$$\text{occupancy}_i = \log \frac{N_i}{\langle N_{\text{genome}} \rangle}, \tag{21.27}$$

where N_i is the number of sequence reads in which the ith base pair shows up and $\langle N_{\text{genome}} \rangle$ is the genome-wide average number of reads per base pair. One of the most immediate features to emerge from such data is the clear depletion of nucleosomes in the vicinity of transcription start sites.

As is clear from the data of Figure 21.18, nucleosome maps provide a new window on how eukaryotic genomes are made available for the

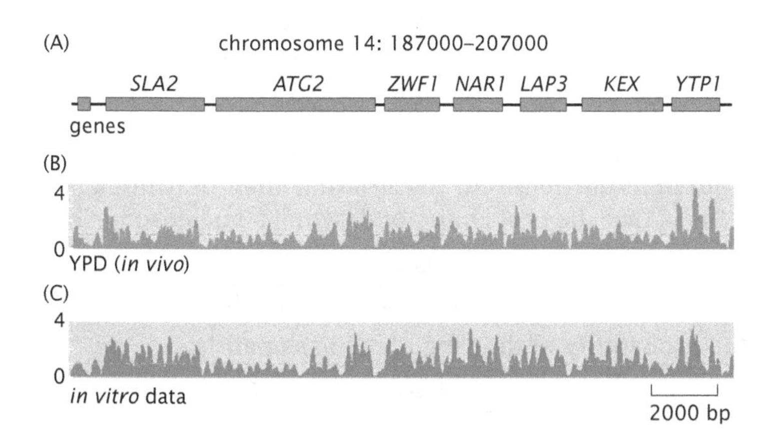

Figure 21.18: Map of nucleosome occupancy. (A) Schematic of a particular 20,000 bp region of chromosome 14 in yeast. (B) Nucleosome occupancy for an *in vivo* experiment where yeast cells were grown on YPD medium. (C) *In vitro* reconstitution of nucleosomes on the same region of yeast DNA shown in the schematic. (Adapted from N. Kaplan et al., *Nature* 458:362, 2009.)

many transactions such as replication, recombination, and transcription that require proteins to interact with specific genomic loci. For example, as shown in Figure 21.19, these maps provide an intriguing picture of the disposition of nucleosomes in the vicinity of promoters. In this case, all the promoters in yeast have been queried for their nucleosome occupancy, revealing a distinct and stereotyped signature of nucleosome depletion around transcription start sites. The same kind of data can be merged to provide a picture of the nucleosome-free regions as well as the oscillatory profile of nucleosome positions in the vicinity of these nucleosome-free regions.

In the spirit of much of the analysis carried out in this book, the kind of quantitative data described above calls out for a quantitative theoretical response. Indeed, there is an irony to the fact that data based strictly upon sequencing can be such a strong impetus for the kinds of physical models advocated throughout the book. A number of specific mechanisms, each intriguing in its own way, have been suggested as the basis for conferring the occupancy features seen in these genome maps. Figure 21.20 provides a schematic view of two extreme models for nucleosome occupancy that do not require the investment of energy. An alternative to these kinds of passive mechanisms is also shown in the figure and is based upon chromatin remodeling enzymes of various kinds that depend upon the consumption of ATP hydrolysis to rearrange nucleosomes.

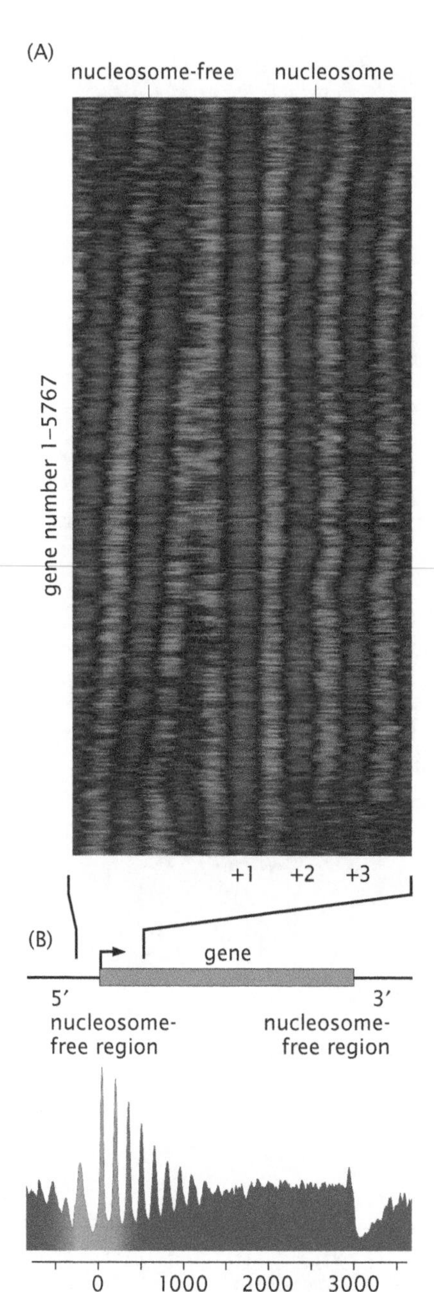

Figure 21.19: High-resolution picture of nucleosome occupancy near promoters in yeast. (A) Gene-by-gene depiction of nucleosome occupancy resulting from aligning the 5′ ends of 5767 open reading frames in yeast. Red characterizes those regions with a positive nucleosome signal and green those regions with a negative signal. (B) Average of the data from many different genes, with peaks and valleys corresponding to positioning relative to the transcription start site. (A, adapted from I. Whitehouse et al., *Nature* 450:1031, 2007; B, adapted from C. Jiang and F. Pugh, *Nat. Rev. Genet.* 10:161, 2009.)

A Simple Model Based Upon Self-Avoidance Leads to a Prediction for Nucleosome Positioning

To see how such models work, we explore the statistical positioning concept, which exploits the kinds of ideas already developed in Chapter 14 by thinking of nucleosomes as "crowders" with a finite size that are not permitted to overlap. One of the goals of such models is to explain the kinds of stereotyped features seen in Figure 21.19, such as the periodic array of nucleosomes in the vicinity of gene start sites. The conceptual idea of the statistical positioning model is that the DNA adjacent to the promoter of interest can be thought of as a two-species mixture of "free base pairs" and nucleosomes of width d (measured in base pairs). We denote the number of free base pairs by n_{bp} and the number of nucleosomes by n_{nuc}. Our goal is to compute the probability that a distance i away from the origin, the site on the genome will be covered by a nucleosome since this is what is revealed by the sequencing experiments described above.

Given an overall mean spacing $\langle L \rangle$ between nucleosomes on the genome, we can generate genomic configurations by drawing base pairs and nucleosomes from an urn like that shown in Figure 21.21 where the overall relative numbers of base pairs and nucleosomes within the urn are chosen so as to enforce the mean nucleosome density (or equivalently, mean linker length) on the genome. Note in particular that the number of base pairs $n_{\text{bp}} = \langle L \rangle n_{\text{nuc}}$. The way we generate configurations is by drawing from the urn and then replacing the just-picked item into the urn and then drawing again. What this means is that the probability of drawing a nucleosome is given by

$$p_{\text{nuc}} = \frac{n_{\text{nuc}}}{n_{\text{nuc}} + \langle L \rangle n_{\text{nuc}}} = \frac{1}{1 + \langle L \rangle}, \tag{21.28}$$

while the probability of drawing a free base pair is

$$p_{\text{bp}} = \frac{\langle L \rangle n_{\text{nuc}}}{n_{\text{nuc}} + \langle L \rangle n_{\text{nuc}}} = \frac{\langle L \rangle}{1 + \langle L \rangle}. \tag{21.29}$$

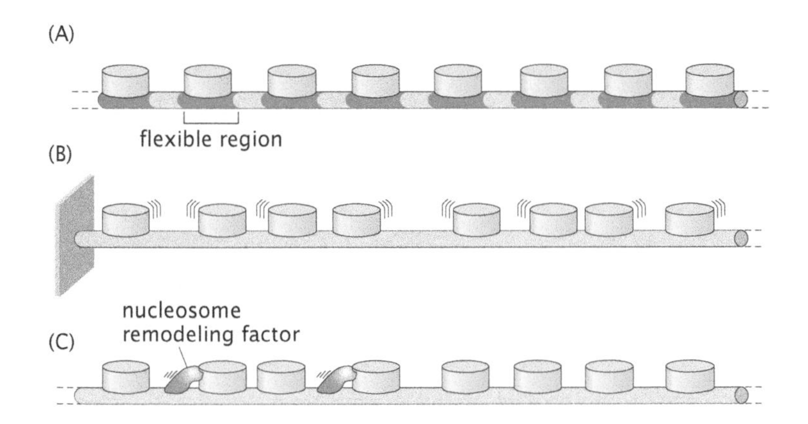

Figure 21.20: Mechanisms of nucleosome positioning within genomes. Different mechanisms for positioning of nucleosomes suggest different experiments. (A) Free-energy landscape model in which nucleosomes are positioned as a result of favorable interactions with specific sequence motifs that confer elastic flexibility, for example. (B) Nucleosome positioning as a gas. Excluded-volume interactions between adjacent nucleosomes induce correlations between positions of nucleosomes. (C) Active positioning by ATP-consuming enzymes.

A more streamlined way to view this is by thinking of our genomic region of length i as being in contact with a reservoir of base pairs and nucleosomes, a scenario we can describe by constructing our states and weights as shown in Figure 21.22 by using a chemical potential for the nucleosomes and base pairs as

$$\mu_{\text{nuc}} = k_B T \ln\left(\frac{1}{1+\langle L\rangle}\right) \tag{21.30}$$

and

$$\mu_{\text{bp}} = k_B T \ln\left(\frac{\langle L\rangle}{1+\langle L\rangle}\right). \tag{21.31}$$

Recall that we define the chemical potential of nucleosomes as

$$\mu_{\text{nuc}} = G(n_{\text{nuc}}, n_{\text{bp}}) - G(n_{\text{nuc}} - 1, n_{\text{bp}}) \tag{21.32}$$

The detailed derivation of these chemical potentials is given as an exercise for the reader in the problems at the end of the chapter, since the concept is identical to the earlier derivation in the context of gases and solutions.

With the chemical potentials in hand we turn to computing the probability $p(i)$ that the ith site of a genomic region of length N is occupied by a nucleosome. This probability is given by $p(i) = 1 - \bar{p}(i)$ where $\bar{p}(i)$ is the probability that the ith site is nucleosome free (i.e., it is a "base pair" in the sense of the urn model). This probability is given by

$$\bar{p}(i) = \frac{S(i-1)S(N-i)}{S(N)}, \tag{21.33}$$

where we have introduced the statistical weight

$$S(i) = \sum_{n_{\text{nuc}}=0}^{n_{\text{nuc}}^{\max}} \frac{[(i-n_{\text{nuc}}d)+n_{\text{nuc}}]!}{(i-n_{\text{nuc}}d)!\,n_{\text{nuc}}!} e^{\beta(i-n_{\text{nuc}}d)\mu_{\text{bp}}} e^{\beta n_{\text{nuc}}\mu_{\text{nuc}}}, \tag{21.34}$$

which captures the thermodynamic weight that takes into account all the possible ways of covering i sites on the genome with base pairs and nucleosomes. The numerator in the above expression is the weight of all the states in which the ith site is not covered by a nucleosome while the denominator is the weight of all the states regardless of whether the ith site is nucleosome free or not. The numerator is obtained by considering the ith site nucleosome free and then noting that this leaves the first $i-1$ sites before it and the $N-i$ sites after it free for occupying with nucleosomes or base pairs. Since the

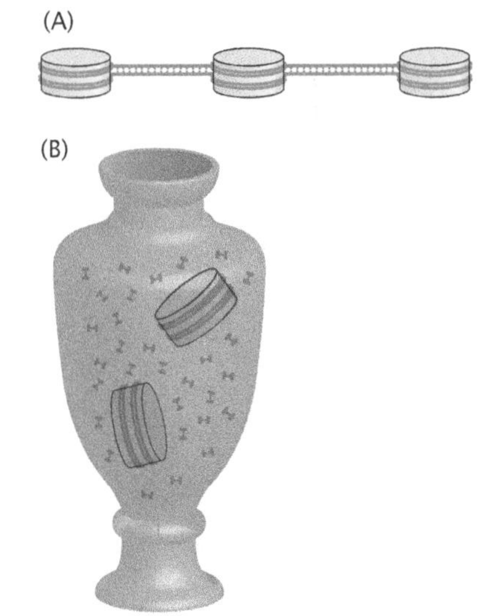

Figure 21.21: The nucleosome–DNA urn. (A) A particular allowed configuration and free base pairs on a region of length i. (B) An urn containing n_{nuc} nucleosomes and n_{bp} free base pairs is used to draw the elements of a one-dimensional array of nucleosomes and base pairs to generate nucleosomal sequences.

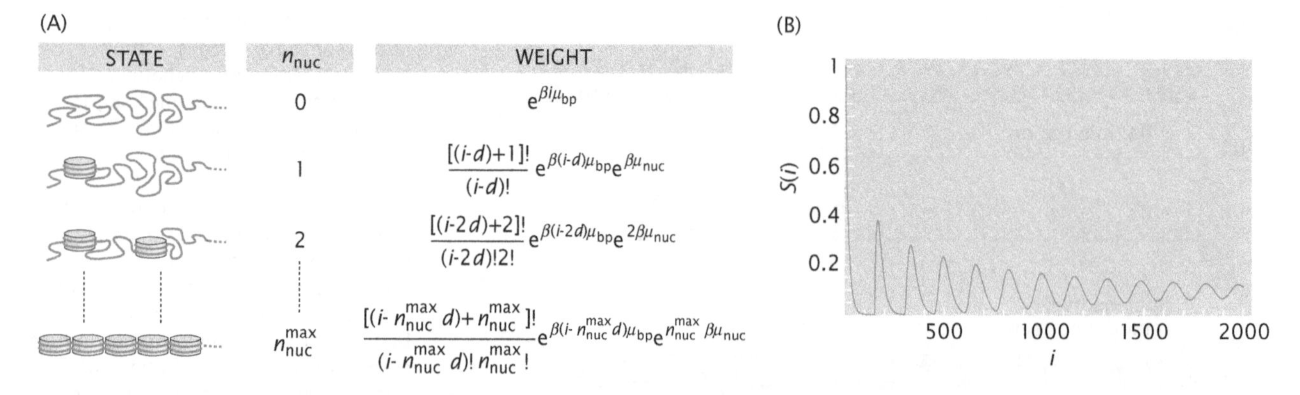

Figure 21.22: Nucleosome occupancy as a function of position. (A) States and weights for nucleosome–DNA configurations of length i. (B) Plot of the periodic nucleosome occupancy profile as a function of distance from the nucleosome-free region.

occupancy of these two stretches of genomic DNA with free base pairs and nucleosomes are independent, the total weight is the product of weights, $S(i-1)S(N-i)$. Note that for $N \gg i$, Equation 21.33 reduces to $\bar{p}(i) = S(i-1)$ independent of the length N of the genomic region under consideration. This means that by focusing on $S(i)$ itself, we can actually learn about the probability that a given base pair is occupied by a nucleosome.

The significance of these statistical quantities is elaborated in Figure 21.22(A) which shows the states and weights for the genomic region of length i of interest. The states and weights reflect the fact that we can have $0, 1, \ldots, n_{nuc}^{max}$ nucleosomes on that particular region, where n_{nuc}^{max} is given by the integer part of i/d.

In Figure 21.22(B), we show the outcome of plotting $S(i)$ as a function of i, the distance from the origin. Note that this results in a stereotyped oscillatory behavior, with the periodicity emerging from the fact that we can add nucleosomes in integer increments. Clearly, one of the most powerful predictions of this kind of model is that it implies a very specific change in the periodicity of occupancy if the density of nucleosomes (or equivalently, the mean linker length) is changed. This model is explored further in the problems at the end of the chapter.

A familiar and powerful approach to many problems in biology is the volley back and forth between the *in vivo* and *in vitro* characterization of the same problem. In the context of nucleosome positions, the power of the *in vitro* approach is that it makes it possible to isolate the different factors contributing to the observed cartographic trends seen in the maps. For example, by using purified histones and genomic DNA from yeast, it becomes possible to ask the question of how nucleosomes get distributed along genomic DNA in the absence of the many confounding (but fascinating) influences found in living cells. Figure 21.18(B) shows the results of applying this approach to genomic DNA from yeast.

The data provided in this section reveals yet another one of the powerful influences of modern sequence analysis. To complement this sequence data, we show the way a simple statistical positioning model (the Kornberg–Stryer model) can be constructed that makes polarizing predictions such as how the nucleosome positioning depends upon the average nucleosome density or linker length. As we have noted repeatedly throughout the book, such models, while probably a gross oversimplification, provide a conceptual starting point for developing intuition about the factors that give rise to observed regularities like those shown in Figure 21.19.

21.4 Sequences and Evolution

Why are metrics of sequence similarity so valuable in studying biology? A simple answer is that proteins with very similar sequences are likely to have very similar functions. But why should that be? Scientists generally agree that the great variety of living organisms that we see today are related to one another and are ultimately derived from a common ancestor through a series of divergences, specializations, and extinctions. Pairs of organisms that have diverged recently will generally tend to be more similar than pairs of organisms that have diverged in the distant past. Thus, dogs and wolves, whose lineages have diverged within the past 10,000 years, still retain many morphological and behavioral similarities and can even interbreed. Wolves and bears are less similar. Wolves and mice are even less similar, but are still more similar to each other than they are to redwood trees. These overt morphological similarities are consequences of similarities in gene product function, gene expression, and gene sequence. Species diverge from one another via mutational events (and other mechanisms such as recombination and lateral gene transfer), and through the subsequent operation of natural selection. Therefore, by comparing sequence similarities for a particular gene among many different organisms, we can reconstruct our best estimate of the evolutionary and mutational history that gave rise to their divergence.

For simplicity and clarity, our discussion will focus on evolutionary events that change the sequences of expressed proteins and RNA in diverging organisms. It is important to acknowledge that other kinds of evolutionary changes may be even more significant in generating organismal diversity. For example, changing the regulation of a gene encoding a protein that contributes to the specification of a tissue in a developing animal may result in a significant morphological change even if the protein structure is unchanged. The famous diversity of beak size and shape observed by Darwin in finch populations on the Galapagos Islands has been correlated with variations in the expression level of a particular protein, called BMP4 (bone morphogenetic protein) that is expressed at high levels in bird embryos destined to have large beaks and at lower levels in birds who develop smaller and more delicate beaks. Using molecular tricks to force the overexpression of BMP4 in a developing chick embryo results in a chicken with a remarkably large, finch-like beak. These striking changes do not need to involve any alteration in the amino acid sequence of BMP4, merely in its regulation. Ultimately, of course, these regulatory changes are also tied to mutational events in the DNA that alter the strength of promoter elements, etc., so many of the principles we illustrate here for the interpretation of sequence variation are generally applicable to other kinds of evolutionary questions.

In this section, we will use four case studies to examine useful applications of sequence analysis to understanding biological events. First, we will use hemoglobin and rhodopsin as windows onto evolutionary relatedness. Second, we will discuss the acquisition of antibiotic resistance in bacteria, an evolutionary phenomenon that is readily observed in the laboratory and is of increasing concern in the medical community. Third, we will explore the evolution of the human immunodeficiency virus (HIV) and see how sequence comparisons can help us understand how this deadly virus made the jump from primate hosts such as the chimpanzee to humans, and how it evolves both within individual humans and in the human population. Finally, we will tackle the ultimate evolutionary question of the relatedness of all cells on Earth. The goal of this series of vignettes will be to illustrate the scope of issues that have been touched in a significant way by the development of modern genome science. Some of these stories are expanded upon in the problems at the end of the chapter.

21.4.1 Evolution by the Numbers: Hemoglobin and Rhodopsin as Case Studies in Sequence Alignment

The previous subsections have highlighted *principles* relevant to sequence comparison primarily through the use of a simplified two-letter alphabet. With these principles in hand, we are now ready to tackle real biological sequence information. As a test case, we start with the α-chain of hemoglobin, a protein whose function we have considered in laborious detail in Sections 4.2 (p. 143) and 7.2.4 (p. 298). The function of hemoglobin as an oxygen-carrying protein is similar for almost all vertebrates. However, not all of the amino acid residues in the hemoglobin sequence are critical for its three-dimensional structure or its function. These residues are subject to alteration through mutation. Interestingly, even before the advent of modern sequence analysis, molecular comparisons between hemoglobins were used as the basis for deductions about molecular (and thereby organismal) evolution. In particular, by comparing the mobility of parts of hemoglobins from different species on two-dimensional gels, it was possible to estimate relatedness in species ranging from humans to sharks. Taking advantage of the availability of modern molecular sequences, we will revisit this analysis of hemoglobin, and then we will use comparison of rhodopsin sequences from a series of vertebrate animals in order to get a sense of what dinosaurs could see when they looked at the world.

Sequence Similarity Is Used as a Temporal Yardstick to Determine Evolutionary Distances

To examine the changes in hemoglobin over evolutionary time from the point of view of modern sequence analysis, we turn to sequence comparisons of the hemoglobins from eight different animals (humans, chimpanzees, gorillas, cattle, horses, donkeys, rabbits, and carp). The concept of the alignment is to take the known sequences of hemoglobin from these organisms and to do a "multiple alignment" of all of them based upon scoring ideas like those developed earlier in the chapter. The comparison of the primary amino acid sequence of hemoglobins from these different vertebrates is shown in Figure 21.23(A), and we see minor but significant variations. Interestingly, the hemoglobins of humans and chimpanzees have no differences. The comparison between humans and cows reveals several differences in sequence at particular sites.

Because these variations presumably arose by random mutational events over evolutionary time, we can also use these changes to reconstruct a rough tree where more similar hemoglobin sequences are grouped together, as shown in Figure 21.23(B). From this and many similar trees for other proteins, using a variety of assumptions, molecular phylogenists can infer the historical relatedness of modern species. In most cases, these kinds of molecular analyses are consonant with what was already known about relatedness on the basis of anatomical comparisons. One fascinating outcome of this

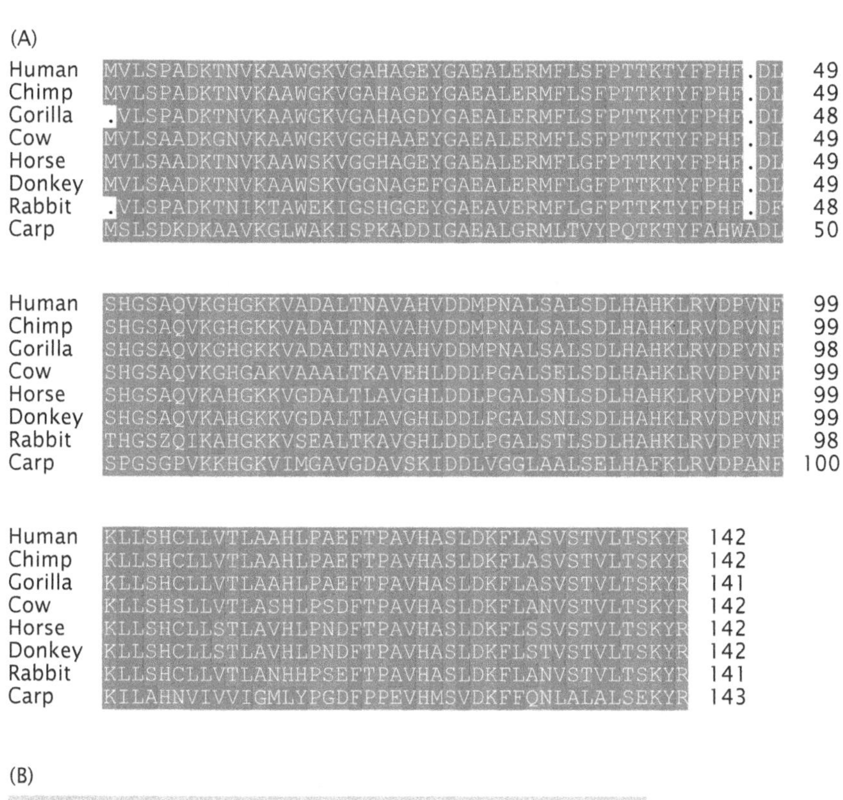

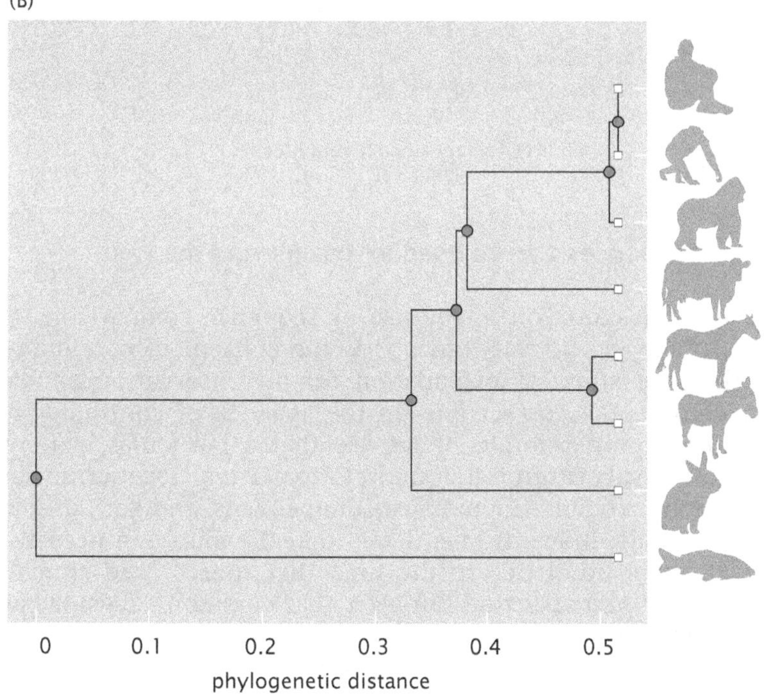

Figure 21.23: Comparison of hemoglobin sequences from different vertebrates. (A) From top to bottom, the sequences from the α-chain of hemoglobin are shown from humans (accession number 57013850, 142 amino acids long), chimpanzees (110835747, 142 aa), gorillas (122407, 141 aa), cattle (*Bos taurus*, 13634094, 142 aa), horses (*Equus caballus*, 122411, 142 aa), donkeys (*Equus asinus*, 62901528, 142 aa), rabbits (*Oryctolagus cuniculus*, 229379, 141 aa), and carp (*Cyprinus carpio*, 122392, 143 aa). The alignment was performed using ClustalW. Hydrophobic residues are indicated in red and polar residues in blue. (B) The tree diagram shows the degree of similarity among each of the sequences. Very similar sequences are connected by short branches; increasing branch length indicates decreasing sequence similarity.

kind of sequence detective work in conjunction with independent dating using the fossil record is the ability to use sequence similarity as a molecular clock. Figure 21.24 shows some classic studies on the molecular clock idea using hemoglobin and several other molecules (though it should be said that these results did not emerge by comparing the sequences of letters as we have done, but rather by measuring the propensity of related sequences to hybridize). The outcome of this and other studies is the recognition that sequence similarity can be used as the basis for deducing how long ago species diverged from their common ancestor.

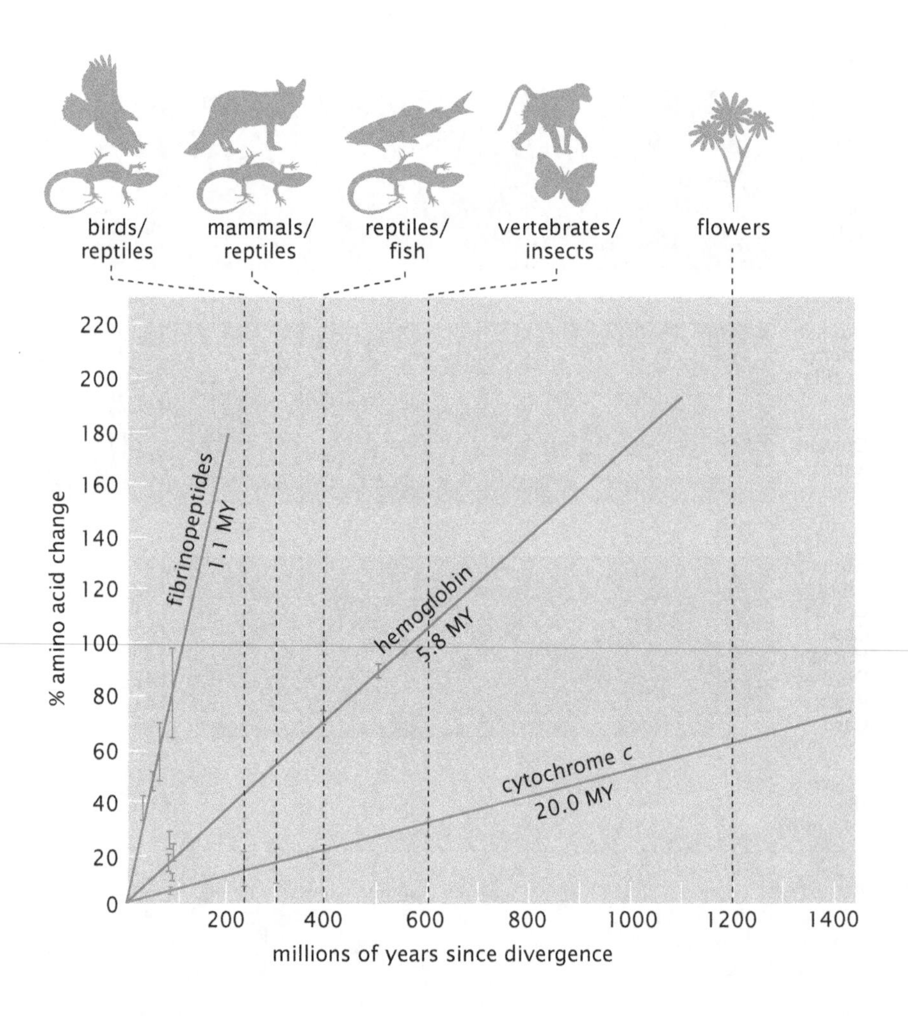

Figure 21.24: Using molecules as evolutionary clocks. This figure shows how using three different molecules (fibrinopeptides, hemoglobin, and cytochrome *c*) it is possible to characterize the divergence of species over evolutionary time. The vertical dashed lines indicate the approximate time when the indicated pairs of animals last shared a common ancestor. Animals and flowering plants shared a common ancestor 1200 million years ago. The times beneath the three lines represent the time needed for that particular protein to undergo a 1% change of sequence identity between two divergent groups. (Adapted from R. E. Dickerson, *Sci. Am.* 226:58, 1972.)

Modern-Day Sequences Can Be Used to Reconstruct the Past

The underlying assumption in the use of sequence comparisons to reconstruct phylogenetic relationships is the concept of parsimony, the idea that the simplest explanation for an observation should be assumed to be the correct one (in the absence of contradictory information). So, for example, if we see that a particular synonymous nucleotide substitution is found at a particular location in the hemoglobin gene sequence in humans, chimpanzees, and gorillas, but in no other animals, it seems most likely that the mutation occurred originally at some point before the time that these three animals last shared a common ancestor, but after that common ancestor had diverged from the ancestor that eventually gave rise to modern cows. Similarly, if a substitution is found only in humans and chimpanzees, but not in gorillas, it seems more likely that the mutation occurred after the common ancestor of humans and chimpanzees had diverged from the gorilla ancestor, rather than the (possible but less likely) scenario that the common ancestor of all three had the substitution, but the nucleotide changed back to the original state in the gorilla lineage only.

Thus, it is clear that comparing the protein amino acid and gene nucleotide sequences that we can easily obtain for modern-day organisms almost automatically leads us into speculation about the organisms of the past, and specifically about the sequences carried by common ancestors for modern organisms that have diverged from one another. Figure 21.25 shows an application of this speculation. As

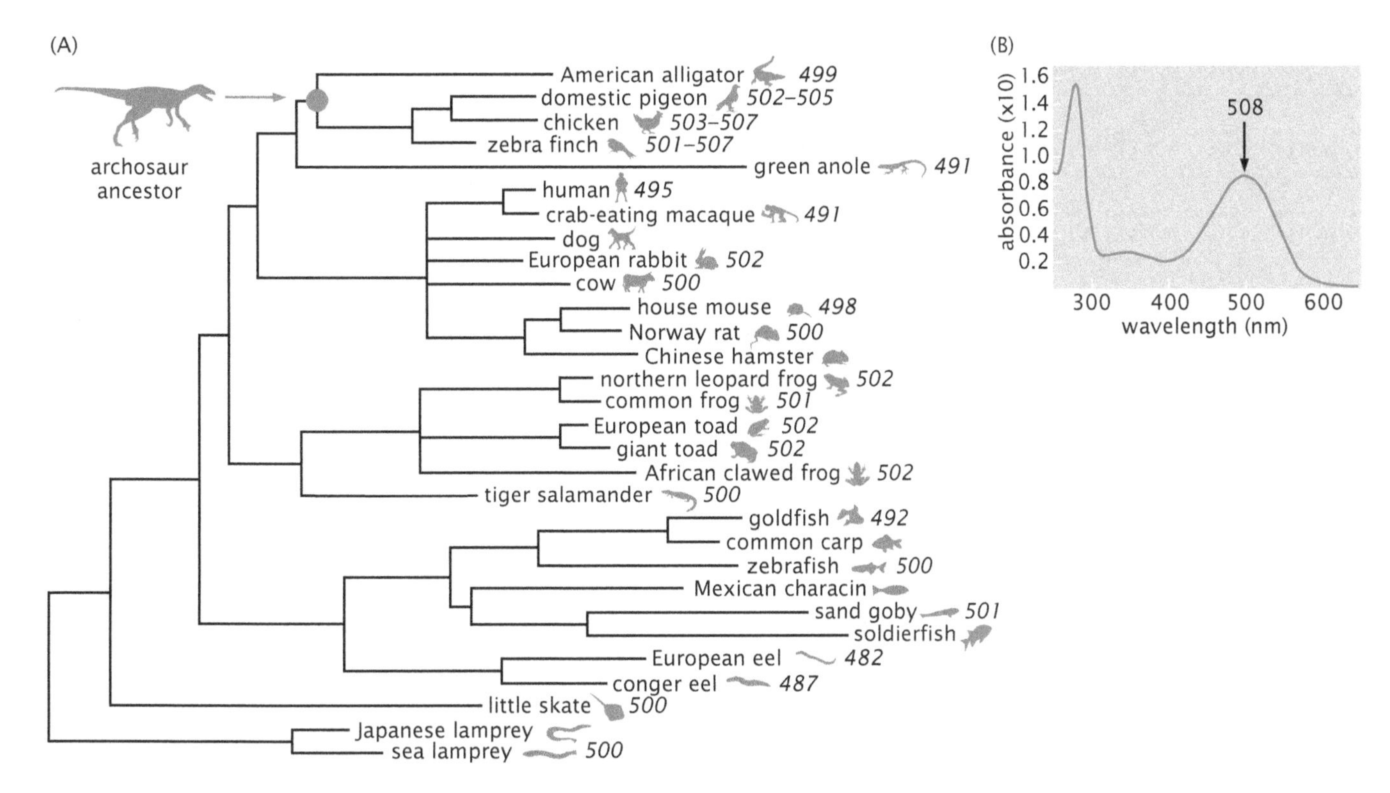

Figure 21.25: Reconstruction of dinosaur rhodopsin. (A) This phylogenetic tree shows the consensus evolutionary relationships for a series of vertebrate animals where the sequence for the rod photoreceptor rhodopsin has been determined. The wavelength of the maximum absorption peak in nm for each rhodopsin variant is indicated in italics. The sequence variation within the rhodopsins of the vertebrate clade was used to estimate the most likely sequence for the archosaur who was the last common ancestor of the present-day American alligator, domestic pigeon, chicken, and zebra finch. Various statistical methods for sequence reconstruction yielded very similar reconstructed sequences. (B) After synthesis of an artificial gene encoding the putative dinosaur rhodopsin sequence, the protein could be expressed and its absorption spectrum measured. The maximum of the visible light absorption peak for reconstructed dinosaur rhodopsin is 508 nm. (Adapted from B. S. W. Chang et al., *Mol. Biol. Evol.* 19:1483, 2002.)

shown in the phylogenetic tree, crocodilians such as the American alligator are more closely related to birds than either are to lizards such as the green anole, or to other vertebrates. The last common ancestor of the crocodilians and the birds was a dinosaur. Recall from Chapter 18 that vertebrate animals use the protein rhodopsin expressed in the rod photoreceptors of their retinas in order to see light. Curiously, the rod cells of different vertebrates are maximally sensitive to slightly different wavelengths of light; for example, European eels are maximally sensitive to shorter-wavelength blue–green light (about 482 nm wavelength), Norway rats to medium-wavelength blue–green light (about 500 nm), and chickens to longer-wavelength blue–green light (503–507 nm). Experimentally, it can be shown that these differences in wavelength sensitivity are attributable to slight differences in the amino acid sequences of the rhodopsin proteins in these animals. Each variant can be expressed and purified, and the absorption spectrum measured for isolated protein once it has been bound to retinal. The absorption spectra of the expressed and purified proteins closely match the absorption spectra of intact photoreceptor cells from these animals.

What wavelength of light could dinosaurs see best? The living descendants of dinosaurs, the crocodilians and birds, have maximum sensitivities ranging from 499 to 507 nm, and relatively closely related species such as the green anole have maximum sensitivities as low as 491 nm. Did dinosaurs see the world more like alligators, like birds, or like lizards? Because DNA degrades over time, we cannot directly sequence the DNA from a dinosaur fossil. However, using sequence analysis as outlined above, it is possible to make an educated guess as to what the sequence of rhodopsin most likely was for the dinosaur who was the common ancestor of the alligator and the chicken, but who was not an ancestor of the anole. The reconstructed dinosaur rhodopsin gene can then be artificially synthesized, the protein expressed, and its spectral sensitivity measured. The result of this is shown in Figure 21.25(B). The dinosaur appears to have been maximally sensitive to light at a wavelength of 508 nm. In other words, the dinosaur saw the world more like a modern bird than like either an alligator or a lizard.

21.4.2 Evolution and Drug Resistance

The germ theory of disease developed by Louis Pasteur, Robert Koch, and other great microbiologists of the nineteenth and early twentieth centuries provided the insight that serious human diseases can be caused and communicated by microscopic pathogens such as bacteria and viruses. Following this insight, it was clear that it should be possible to treat infectious diseases by selectively killing the pathogenic bacteria. In 1929, the British bacteriologist Alexander Fleming made a famous discovery, allegedly rooted in poor laboratory technique. As the story goes, he returned from a trip and glanced at bacterial plates left on the bench while he was away. On one of the plates, a spore of mold had germinated and given rise to a small fungal colony. Rather than casting away the contaminated plate in disgust as most modern bacteriologists would be inclined to do, Fleming noticed that the fungal colony had created a cleared zone around it on the plate where bacterial cells failed to grow or had been killed. This led Fleming to hypothesize that the mold was secreting some substance that could diffuse a short distance and kill bacteria. The purified form of this fungal product is called penicillin.

Penicillin first came into clinical use as an agent to treat bacterial infections in humans during World War II. Many soldiers wounded on the battlefield were treated with penicillin to prevent infections from bacteria entering open wounds. This medical advance resulted in a higher survival rate for wounded soldiers in World War II than in any previous war. Since then hundreds of other antibiotics have been identified and used to treat humans with bacterial infections. Together with vaccinations and public health measures such as the development of a clean water supply, widespread use of antibiotics has caused infectious diseases to drop from being the number one cause of death to the number six cause of death in the USA and other wealthy, industrialized countries. Worldwide, however, infectious and parasitic diseases still cause the plurality of human deaths because these medical advances are not available to all people.

Despite the extraordinary success of penicillin treatment in the 1940s and 1950s, it became clear almost immediately that penicillin could not be a universal magic bullet. Some people with serious bacterial infections simply failed to respond to treatment with the drug. Bacteria isolated from these patients also could not be killed with penicillin in the laboratory, even with very high concentrations that would easily wipe out others of the same bacterial species. These bacteria had somehow become resistant to the antibiotic.

In laboratory experiments, it was found that antibiotic resistance could arise spontaneously at a very low rate ($1{:}10^6$ to $1{:}10^8$) in a population of bacteria that had previously been susceptible to antibiotic treatment. We now understand that antibiotic resistance can be acquired by mutational events that change the sequence of the encoded proteins that are targets of the antibiotic and render them sufficiently altered in structure so that the drug no longer binds. As the bacterial population continues to grow, the very rare resistant individuals are able to replicate more rapidly and more successfully than the more abundant individuals that are susceptible to the drug. Over time, the entire population will be taken over by the antibiotic-resistant mutant. Under continuous selective pressure, further mutations can accumulate to render the bacteria even more resistant. Other kinds of genetic events also contribute to the acquisition of antibiotic resistance. Some gene products associated with antibiotic resistance are not merely altered drug targets, but specific enzymes that actively destroy or inactivate the drug. Bacteria also express energy-dependent transmembrane pumps that may export drugs from their cytoplasm, to prevent them from reaching the targets. Furthermore, bacteria are easily able to transfer DNA fragments to one another, even across species lines, on mobile elements such as plasmids and transposons. Plasmids carrying antibiotic resistance genes can spread rapidly in bacterial populations and confer resistance on previously susceptible individual cells even without any mutational events changing the cell's chromosomal DNA. Because the pressure of natural selection to favor resistant individuals is so strong, literally a matter of life and death, and because bacteria are able to divide rapidly, antibiotic resistance is quick to arise in the laboratory or in the body of a human patient being treated for a bacterial infection. This process is thought to be a small-scale and directly observable example of the same kinds of processes that contribute to large-scale diversification and change of organisms due to natural selection over evolutionary time.

One laboratory-based case study is shown in Figure 21.26. In this experiment, a wild-type strain of the bacterial pathogen *Mycobacterium tuberculosis* was grown in the presence of a fluoroquinolone antibiotic. Tuberculosis, the lung disease caused by this organism, is the most common life-threatening bacterial infection among humans and its clinical resistance to antibiotics is increasing worldwide. Fluoroquinolones are antibiotic drugs that target DNA gyrase, an enzyme that aids in separating the two DNA strands during bacterial DNA replication. Fluoroquinolones are frequently used as second-line antibiotics when primary treatment fails for patients with tuberculosis. After selection for *M. tuberculosis* able to grow in a low level of fluoroquinolones, followed by sequencing of the two genes encoding the two subunits of DNA gyrase, three independent point mutations were identified, as shown in Figure 21.26. These mutations were each able to confer a modest increase in resistance to their strains (2- to 10-fold). When these slightly resistant strains were challenged to grow on agar plates containing higher concentrations of the antibiotics, a series of secondary mutations resulted that conferred a more than 100-fold increase in the level of resistance, as also shown in Figure 21.26.

This kind of stepwise acquisition of increasing levels of resistance can also be observed in human patients with a bacterial infection undergoing antibiotic treatment. In one comprehensive study, a patient suffering from endocarditis (infection of the heart muscle) caused by *Staphylococcus aureus* was treated with a series of antibiotics. Throughout the course of the infection, physicians isolated *S. aureus* from the patient and these sequential isolates were

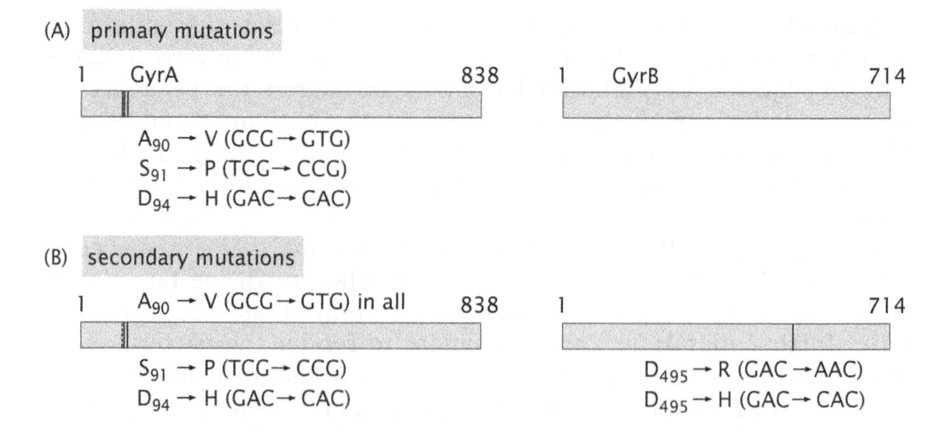

Figure 21.26: Stepwise increases in the level of antibiotic resistance in *M. tuberculosis.* (A) Selection with a low level of a fluoroquinolone antibiotic gave rise to three independent point mutations, all in the same small region of the gene encoding the A subunit of DNA gyrase. Each mutation is denoted with the one-letter code for the amino acid normally found at a particular position (for example, A_{90} means alanine at position 90) followed by the amino acid found at that position in the mutant strain. In parentheses are shown the nucleotides in the altered codon. (B) Subsequent selection of one of these mutant strains ($A_{90} \rightarrow V$) in progressively higher concentrations of antibiotic gave rise to strains carrying four different secondary point mutations, two at the same sites as found in the primary mutations and two at a site in the gene encoding the B subunit of DNA gyrase. Each of these mutant strains was significantly enhanced in its antibiotic resistance, over 100-fold compared with the parent strain. (Data adapted from T. Kocagöz et al., *Antimicrob. Agents Chemother.* 40:1768, 1996.)

subjected to whole-genome sequencing. Over a three-month period, a total of 35 point mutations accumulated, many of them in genes known to be associated with the acquisition of antibiotic resistance. Figure 21.27(A) shows a time course of the dates of antibiotic exposure and recoveries of individual bacterial isolates. Figure 21.27(B) shows the level of resistance to four antibiotics for each of these strains and the acquisition of point mutations in the chromosome.

These examples of the evolution of antibiotic resistance over short periods of time represent a microcosm of evolutionary theory. In observing the effects of mutational variation and natural selection in the context of antibiotic treatment over short stretches of time, we can begin to appreciate the long-term consequences of these processes acting on species. For antibiotics, we can draw a direct line between the known mechanisms of action of the drug, the sites of point mutations associated with resistance, and the phenotypic consequences for the survival of the bacterium.

21.4.3 Viruses and Evolution

Although the organization of viruses is profoundly different from bacteria, similar considerations govern their evolution. Viruses consist essentially of a nucleic acid information program (either DNA or RNA) wrapped in a protein shell and sometimes also a layer of membrane. Here, we concern ourselves with the change of the nucleic acid from one generation of the virus to the next. Many viral polymerases are extremely error-prone, and therefore sequence changes can accumulate rapidly. The speed of the viral evolutionary process is vividly illustrated by HIV, the causative agent of AIDS (acquired immune deficiency syndrome). The first antiviral drug approved for treatment of

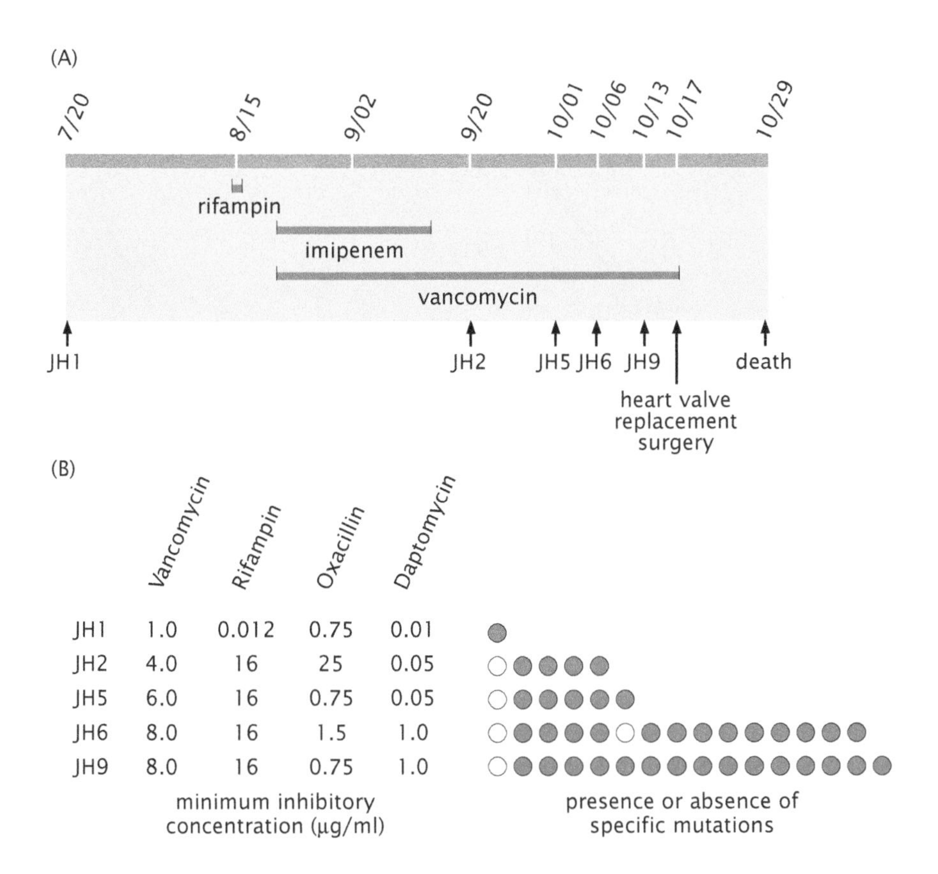

Figure 21.27: Stepwise acquisition of antibiotic resistance in a *S. aureus* infection. (A) A time line covering three months in the year 2000 indicates the time period when the patient was treated with various antibiotics, indicated as red bars, and various dates for several events denoted by black arrows. The different JH events denote bacterial isolates that were cultured from the site of infection. The dates of a surgical procedure and death are also indicated. (B) Summary of changes in antibiotic susceptibility for each of the five different bacterial isolates cultured on different dates, along with the sites of mutational events compared with the wild-type bacteria. Increasing values of the minimum inhibitory concentration indicate increasing drug resistance. A red dot indicates the appearance of a specific mutation in the genome of a given strain. A white dot indicates a reversion back to the wild-type sequence. Thus, strain JH1 has only a single point mutation relative to the wild type. JH2 has four new mutations, but the first mutation has reverted. The antibiotic susceptibility testing of the individual isolates shows that bacteria in the patient developed resistance to the three classes of antibiotics that had been used during treatment: rifampin, vancomycin, and beta-lactams (including imipenem and oxacillin). Interestingly the later isolates were also resistant to daptomycin, although this class of antibiotic had not been used to treat the patient. (Adapted from M. Mwangi et al., *Proc. Natl Acad. Sci. USA* 104:9451, 2007.)

HIV infection was AZT (zidovudine), a nucleoside analog that inhibits the reverse transcriptase enzyme required to copy the RNA genome of the virus into a DNA copy that can be inserted into the host genome. Viruses can become resistant to AZT by a single point mutation in the reverse transcriptase, so it was quickly found in the clinic that AZT treatment alone could briefly slow the progression of the disease, typically for only a few months, but could not cure the patient.

The Study of Sequence Makes It Possible To Trace the Evolutionary History of HIV

The rapid rate of mutation associated with HIV leads to sufficient variation in the virus infecting the human population today that it is often possible to tell which particular individual has transmitted

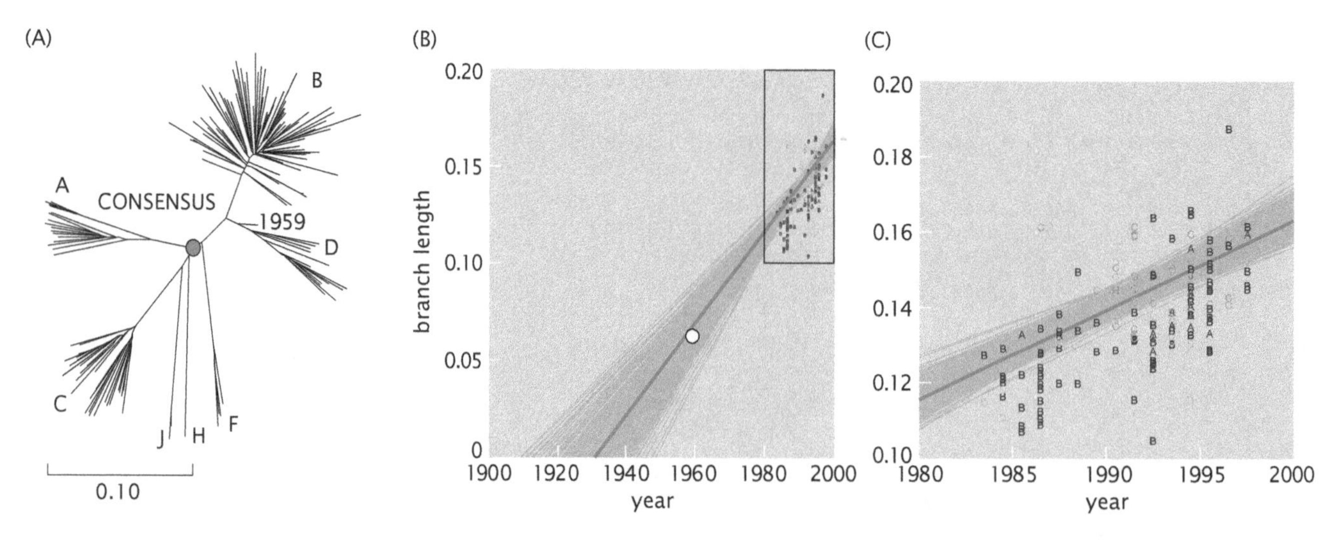

Figure 21.28: Evolutionary history of HIV. (A) The tree diagram illustrates the variation in protein sequence for the protein gp160, found on the surface of the viral envelope, for several hundred different viral isolates collected from patients between 1983 and 1997. The letters indicate different subgroups of the virus; for example, group B dominates the epidemic in the USA and Europe. (B) Plotting the evolutionary branch distance for each viral isolate relative to the consensus sequence as a function of the year of isolation shows that the protein sequences are becoming more divergent over time. Extrapolating this relationship backward in time yields the estimate that the single ancestral virus probably first caused human infection around 1930. The red shaded area indicates the confidence limits of this estimate. The single sequence indicated by the large circle was collected from a patient in Zaire in 1959. (C) The boxed region in (B) is expanded; note that strains within every subgroup exhibit increasing diversity over time. (Adapted from B. Korber et al., *Science* 288:1789, 2000.)

the infection to another person. This determination is made by comparing the sequences of the viruses. Comparing the sequences of many individual strains of HIV collected over time can similarly allow us to reconstruct an approximate view of its evolutionary history since entering the human population. As shown in Figure 21.28, the sequence of one particular protein encoded by viruses collected from different individuals between 1983 and 1997 has become progressively more divergent. Extrapolating this divergence backwards in time, we can estimate that all current strains of HIV1 were most likely descended from a single common virus ancestor that caused an early human infection sometime between 1920 and 1940. An interesting confirmation of this extrapolation comes from examination of the sequence of a virus isolate from a blood sample that was collected in Zaire in 1959, long before AIDS was recognized as a disease. This sequence is more similar to the consensus than any of the sequences from isolates collected later. Indeed, using the sequence divergence of this protein over time that resulted from fitting an extrapolation line to the viral sequences from modern patients results in the estimate that the sequence should have been isolated in 1957 if the linear model is correct. This remarkable agreement with the real time of collection increases our confidence that the simple linear model for diversification of protein sequence over time holds remarkably well for this HIV protein as well as for other proteins that change much more slowly over evolutionary time (see Figure 21.24). The reader is invited to carry out sequence studies on HIV proteins in the problems at the end of the chapter.

The Luria-Delbrück Experiment Reveals the Mathematics of Resistance

Together, viruses and their hosts are in a constant evolutionary struggle. In a classic experiment, Luria and Delbrück examined the

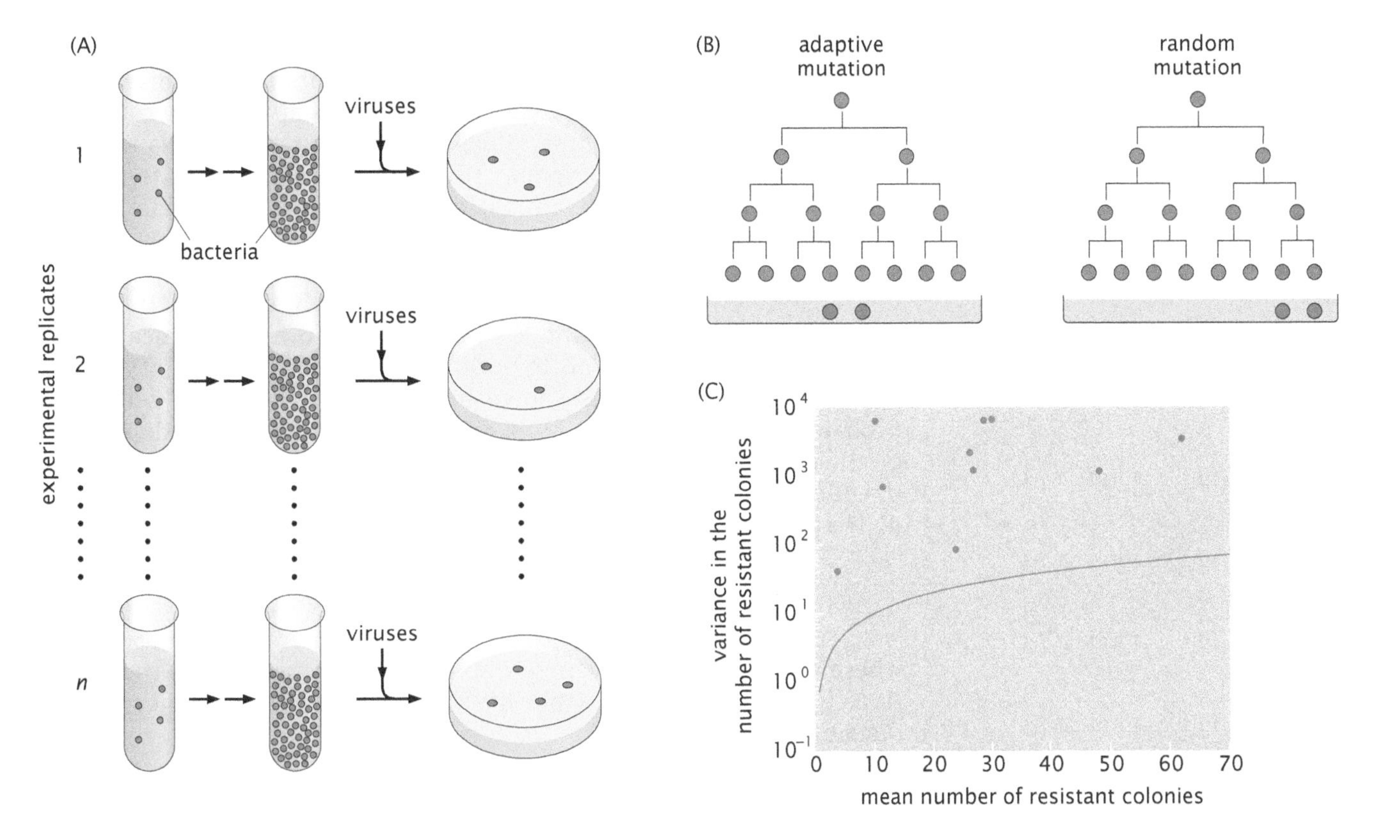

Figure 21.29: Schematic of the Luria–Delbrück experiment. (A) A starting culture with a small number (50–500) of bacteria is grown for about 15 generations. The bacteria are plated on plates containing excess quantities of bacteriophage. The existence of phage-resistant bacteria is reflected by the fact that some colonies grow on the plate. By performing this experiment repeatedly both the mean number of resistant colonies and the fluctuations around the mean can be calculated. (B) According to the adaptive mutation hypothesis, the resistant bacteria (red) do not appear until exposed to the virus. In contrast, according to the random mutation hypothesis, resistant bacteria can appear before the challenge with viruses and propagate their resistance to their offspring. (C) The variance in the number of resistant colonies is shown as a function of the mean number of resistant colonies for realizations of the experiment shown in (A). The observed fluctuations are much larger than those predicted by the adaptive mutation model (shown as a red curve) which predicts that the variance is equal to the mean.

dynamics of mutations in a population of bacteria subjected to viral infection. One of the key insights exemplified in our discussion both of antibiotic resistance and of the emergence of HIV is that, over time, there is always a fraction of the population (whether bacteria or viruses) that are resistant to strategies to destroy them. The goal of the Luria–Delbrück experiment was to determine whether the mutations giving rise to that resistance were simply present in the population or rather were induced by the infection process itself. We call the first hypothesis the "random mutation hypothesis" and the second the "adaptive mutation hypothesis."

In the experiment of Luria and Delbrück, the number of bacteria in a culture that are resistant to viral infection can be obtained by plating the culture on a plate containing viruses and counting the resulting number of colonies. At first, one might think that by counting the average number of colonies on a plate, one could only determine the total number of resistant bacteria, but not at what point those mutants appeared. However, by cleverly exploiting an understanding of fluctuations and armed with only the simple tool of plate counts, Luria and Delbrück were able to distinguish between these two hypotheses. The trick was to examine not only the *average* number of resistant bacteria found in a culture, but also the *fluctuations* in that average when the experiment is repeated multiple times.

The experimental strategy followed by Luria and Delbrück is schematized in Figure 21.29(A). The experiments start with a culture containing a small number of bacteria with typically 50–500 cells. The idea behind the small number of cells in the initial culture is to avoid introducing any bacteria that already have the resistance. After about 15 cell doublings or generations, the culture was spread on a plate containing excess amount of viruses. The resulting colonies correspond to single cells that were resistant to the virus. By repeating this experiment multiple times, one can readily calculate the mean number of resistant cells and the fluctuations around this mean.

In the adaptive mutation hypothesis, the cells develop resistance at the time of plating, as shown in Figure 21.29(B). Each cell has a probability a of becoming resistant and this process is independent of the destiny of the rest of the cells. If $a \ll 1$, the probability of m cells adapting and acquiring the mutation is given by the Poisson distribution introduced in Section 18.3.3 (p. 779), namely,

$$P(m) = \frac{\langle m \rangle^m e^{-\langle m \rangle}}{m!}, \tag{21.35}$$

where $\langle m \rangle = aN$ is the average number of cells presenting the mutation, with N the total number of starting cells. The variance of this distribution is given by

$$\langle (m - \langle m \rangle)^2 \rangle = \langle m \rangle. \tag{21.36}$$

The variance in the number of resistant colonies is then equal to the mean number of resistant colonies. In Figure 21.29(C), we show the data obtained by Luria and Delbrück overlaid with the calculation for the variance of the Poisson distribution from Equation 21.36. We can see from the figure that the fluctuations measured experimentally are much larger than those predicted by the adaptive mutation hypothesis.

Though we have just shown that the variance observed in the Luria–Delbrück experiment is inconsistent with the adaptive mutation model, we have not shown that it is consistent with the random mutation hypothesis. Intuitively, we can understand why the observed variance is higher than that of a Poisson distribution. At each point in time, the probability of a cell mutating into a resistant cell is given by the Poisson distribution. At that point in time, the variance in the number of new mutants is equal to its mean. However, now those new cells can give rise to progeny that will themselves be resistant. As a result, the fluctuations will be amplified.

In order to generate intuition about this we focus on the situation depicted in Figure 21.30. We think of a scenario where a new cell appears at time $t = 0$. In the random mutation case, mutations can occur at any cell division, whereas in the adaptive mutation scenario, this can only take place upon exposure to viruses at $t = 1$. The probability of mutation during a given cell division (random mutation) or upon virus exposure (adaptive) is denoted by a.

In the random mutation case, there are five possible trajectories. In order to calculate the mean number of resistant colonies, we multiply the number of resistant cells in each trajectory by its weight and divide by the sum of all weights. The mean number of resistant colonies is given by

$$\langle m \rangle_{\text{random}} = \frac{2a + 2a + 2a^2}{1 + 3a + a^2} = \frac{4a + 2a^2}{1 + 3a + a^2} \approx 4a, \tag{21.37}$$

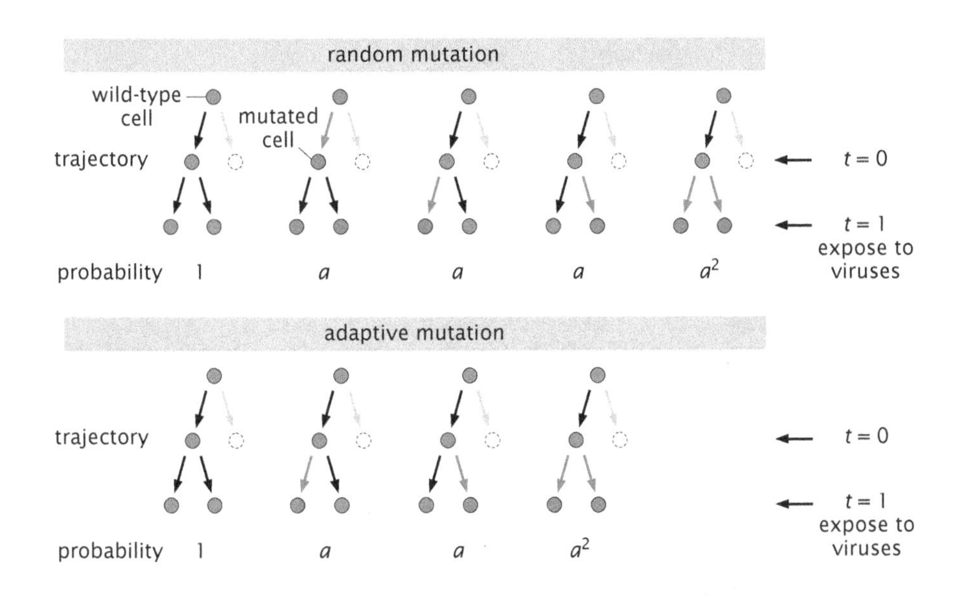

Figure 21.30: Toy model of the Luria–Delbrück experiment. Bacteria at $t = 0$ undergo a round of cell division. At $t = 1$, they are exposed to viruses. (A) Under the random mutation hypothesis, a mutation can occur in the cell division that led to the cell we are considering at $t = 0$ with a probability a. Additionally, either one or both of the daughter cells can acquire a mutation. (B) In the adaptive mutation hypothesis, cells can acquire a mutation upon exposure to viruses with a probability a.

where we have assumed that the probability of mutation is small, namely $a \ll 1$. The second moment is given by

$$\langle m^2 \rangle_{\text{random}} = \frac{4a + 2a + 4a^2}{1 + 3a + a^2} = \frac{6a + 4a^2}{1 + 3a + a^2} \approx 6a, \qquad (21.38)$$

which leads to the variance

$$\text{Var}_{\text{random}} = \langle m^2 \rangle_{\text{random}} - \left(\langle m \rangle_{\text{random}}\right)^2 \approx 6a. \qquad (21.39)$$

In the adaptive mutation case, there are four possible trajectories. The mean number of resistant colonies is given by

$$\langle m \rangle_{\text{adaptive}} = \frac{2a + 2a^2}{1 + 2a + a^2} = \frac{2a(1 + a)}{(1 + a)^2} = \frac{2a}{1 + a}. \qquad (21.40)$$

For the second moment, we get

$$\langle m^2 \rangle_{\text{adaptive}} = \frac{2a + 4a^2}{1 + 2a + a^2} = \frac{2a(1 + 2a)}{(1 + a)^2}, \qquad (21.41)$$

which leads to the variance

$$\text{Var}_{\text{adaptive}} = \langle m^2 \rangle_{\text{adaptive}} - \left(\langle m \rangle_{\text{adaptive}}\right)^2 = \frac{2a}{(1 + a)^2} = \langle m \rangle_{\text{adaptive}}. \qquad (21.42)$$

Not surprisingly, the variance is equal to the mean in the adaptive hypothesis.

We see then that under the random mutation hypothesis,

$$\frac{\text{Var}_{\text{random}}}{\langle m \rangle_{\text{random}}} \approx \frac{6}{4}, \qquad (21.43)$$

whereas for the adaptive mutation hypothesis, we have

$$\frac{\text{Var}_{\text{adaptive}}}{\langle m \rangle_{\text{adaptive}}} = 1. \qquad (21.44)$$

Already in one round of division we expect larger fluctuations under the random mutation scenario. Again, these fluctuations will become even larger as the size of the population increases. With this intuition

Figure 21.31: Trajectories and weights in the Luria–Delbrück experiment. The mutant population can increase either by mutation of a wild-type cell or by cell division of a mutant cell.

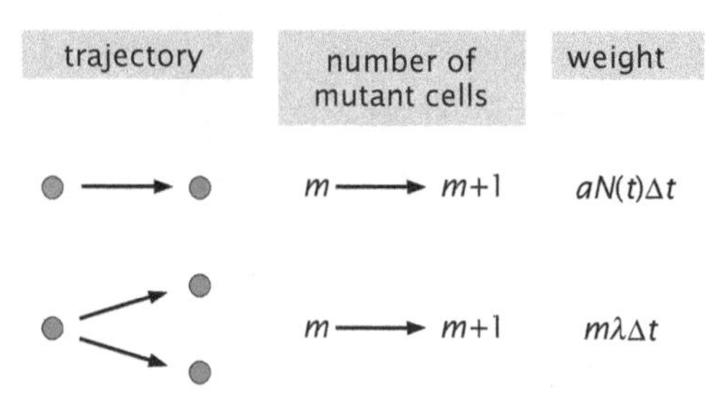

in hand, we now turn to calculating the variance as a function of the age of the culture.

In order to model the appearance and growth of the mutant population, we resort to a master equation that describes the probability of finding m mutant bacteria at time point t, $p(m,t)$. In doing so, we need to think of the possible trajectories of the system shown in Figure 21.31. In this figure, we can see that there are two ways the population of mutant cells can increase, namely, (i) by mutation of a wild-type cell and (ii) by cellular division of a mutant cell. The resulting master equation for $p(m,t)$ is

$$\frac{\partial p(m,t)}{\partial t} = \underbrace{\lambda(m-1)p(m-1,t) - \lambda m p(m,t)}_{\text{division of a mutant cell}} + \underbrace{aN(t)p(m-1) - aN(t)p(m)}_{\text{mutation of an already-existing cell}} . \tag{21.45}$$

Here, $N(t)$ is the total number of cells, where we assume that the number of mutants is small compared with the number of cells, namely $N(t) \gg m(t)$. a is the mutation rate and λ is the doubling rate. We also assume that the wild-type bacteria grow exponentially with the same rate as the mutant ones such that $N(t) = N_0 e^{\lambda t}$. Note that if we have $m-1$ mutants, we can gain one more mutant either by mutation of a wild-type cell or by division of a mutant cell.

We want to calculate the different moments of the distribution described by the master equation, Equation 21.45. In order to calculate the first moment, we multiply both sides of the equation by m and sum over all possible values of m, resulting in

$$\sum_{m=0}^{\infty} m \frac{\partial p(m,t)}{\partial t} = \frac{\partial \langle m(t) \rangle}{\partial t}$$
$$= \sum_{m=0}^{\infty} m[\lambda(m-1)p(m-1,t) - \lambda m p(m,t) + aN(t)p(m-1) - aN(t)p(m)] . \tag{21.46}$$

Here we have used the fact that $\sum_{m=0}^{\infty} m p(m,t) = \langle m(t) \rangle$. We need to calculate the different terms of the sum on the right-hand side of the equation. We begin with

$$\sum_{m=0}^{\infty} m(m-1)p(m-1,t) = \sum_{m=0}^{\infty} (m-1+1)(m-1)p(m-1,t)$$
$$= \sum_{m=0}^{\infty} \left[(m-1)^2 + (m-1)\right] p(m-1,t). \tag{21.47}$$

Since the probability of having a negative number of mutant cells is zero by definition, we can make the change of variables $m-1 \to m$ in the previous equation, resulting in

$$\sum_{m=0}^{\infty} m(m-1)p(m-1,t) = \langle m^2(t)\rangle + \langle m(t)\rangle. \tag{21.48}$$

The third term on the right-hand side of Equation 21.46 can be calculated from

$$\sum_{m=0}^{\infty} mp(m-1,t) = \sum_{m=0}^{\infty} (m-1+1)p(m-1,t)$$

$$= \sum_{m=0}^{\infty} [(m-1)+1]\,p(m-1,t) = \langle m(t)\rangle + 1, \tag{21.49}$$

where we have made use of the fact that the probability is normalized to 1, such that $\sum_{m=0}^{\infty} p(m,t) = 1$. When we put all of this together, we obtain a differential equation for the first moment,

$$\frac{\partial\langle m(t)\rangle}{\partial t} = \lambda\langle m(t)\rangle + aN(t). \tag{21.50}$$

It can be shown that the solution to this equation with an initial condition $p(m,0) = \delta_{m,0}$ is

$$\langle m(t)\rangle = atN(t). \tag{21.51}$$

The mutant population grows faster than the wild-type one. This is so because the mutants have two ways of increasing their numbers, namely, by mutation of a wild-type cell and by division of a mutant cell.

By following a similar strategy to that used for the first moment, we obtain a differential equation for the second moment of the distribution of mutant cells,

$$\frac{\partial\langle m^2(t)\rangle}{\partial t} = 2\lambda\langle m^2(t)\rangle + 2aN(t)\langle m(t)\rangle + \lambda\langle m(t)\rangle + aN(t). \tag{21.52}$$

This equation can be solved analytically given the initial condition $\langle m^2(t=0)\rangle = 0$, resulting in

$$\langle m^2(t)\rangle = \frac{ae^{\lambda t}N_0}{\lambda}\left(2e^{t\lambda} - t\lambda - 2\right) + \langle m(t)\rangle^2. \tag{21.53}$$

From here, we can calculate the variance for the random mutation hypothesis,

$$\mathrm{Var}_{\mathrm{random}}(t) = \langle m^2(t)\rangle - \langle m(t)\rangle^2 \approx \frac{2}{\lambda}e^{2t\lambda}aN_0 = \frac{2e^{t\lambda}}{\lambda t}\langle m(t)\rangle. \tag{21.54}$$

In this case, we have a dramatically different prediction for the variance than that offered by the adaptive mutation hypothesis, which calls for a variance of the form

$$\mathrm{Var}_{\mathrm{adaptation}}(t) = \langle m(t)\rangle. \tag{21.55}$$

This is to be contrasted with the variance for the random mutation hypothesis shown in Equation 21.54, which has a much stronger dependence on time.

After 15 generations (the typical waiting time used by Luria and Delbrück), in the random mutation hypothesis we expect a variance

$$\mathrm{Var}_{\mathrm{random}}(\lambda t = 15) \approx 4\times 10^5\langle m\rangle. \tag{21.56}$$

Though the variance is much greater than the mean, this simple model predicts that the variance will be many orders of magnitude more than the measured variance shown in Figure 21.29(C). The reason for this discrepancy is that there is a significant fraction of time during growth over which no mutants arise. Since we are dealing with a finite number of tubes in each experiment (depicted in Figure 21.29(A)), this outcome is very likely for the initial cell cycles. Of course, this would not be a limitation if the number of tubes used in the experiment were infinite. As a result, in the absence of mutants, there are no fluctuations to be amplified! In order to account for the finite size of the system and to confront the model with the actual data shown in Figure 21.29(C), the reader is referred to Luria and Delbrück's original paper and to Benedek and Villars (2000).

21.4.4 Phylogenetic Trees

One of the key ideas stemming in part from the molecular analyses of evolution is the formulation of a *universal* tree of life. The idea behind a bioinformatic investigation of the tree of life was to find a deeply conserved but slowly varying set of genes that are common to all life forms and to measure their sequence excursions over evolutionary time scales. To that end, the RNA structural content of the ribosome has served as a superb gold standard, because all living organisms have ribosomes with roughly similar structures, and it is widely accepted that the last common universal ancestor of all life on Earth must have used ribosomes that were comparable to the modern forms. By analyzing the ribosomal RNA genes from a broad variety of organisms, it is possible to classify them on the basis of their sequence similarity. Examples of these different sequences are shown in Figure 21.32.

One of the deeply interesting outcomes of the ribosomal RNA sequence hunt was the discovery of an entirely new branch on the tree of life. In retrospect, perhaps it is not so remarkable that as recently as the 1970s there were such gaping holes in our understanding of the diversity of life on Earth. The examination of the outward appearance (a strategy that has served us well in examining macroscopic forms) is remarkably ineffective at distinguishing microscopic organisms. In that sense, the addition of sequence analysis to the evolutionary biologist's repertoire is comparable in its significance to the telescope for the astronomer. As a result of sequence studies of microbes such as methanogens (methane-generating organisms that demonstrate the broad metabolic repertoire of microbes), an anomaly was noted. Many microscopic organisms, including the methanogens, that morphologically resemble bacteria (and had previously been classified as bacteria) were revealed by sequence analysis to be something else entirely. In particular, their ribosomal RNA sequences defy classification either as bacteria or eukaryotes and led to the assertion of a tree of life with a third branch corresponding to the Archaea. The resulting tree of life is shown in Figure 3.6 (p. 98). Many of the first archaea identified were "extremophiles," favoring oddball diets or living at very high or very low temperatures. Over time as more sequences have been collected from more diverse locations, it is clear that archaea are everywhere, not merely in extreme environments. For example, some archaea have been found living in human mouths, where they are associated with gum disease.

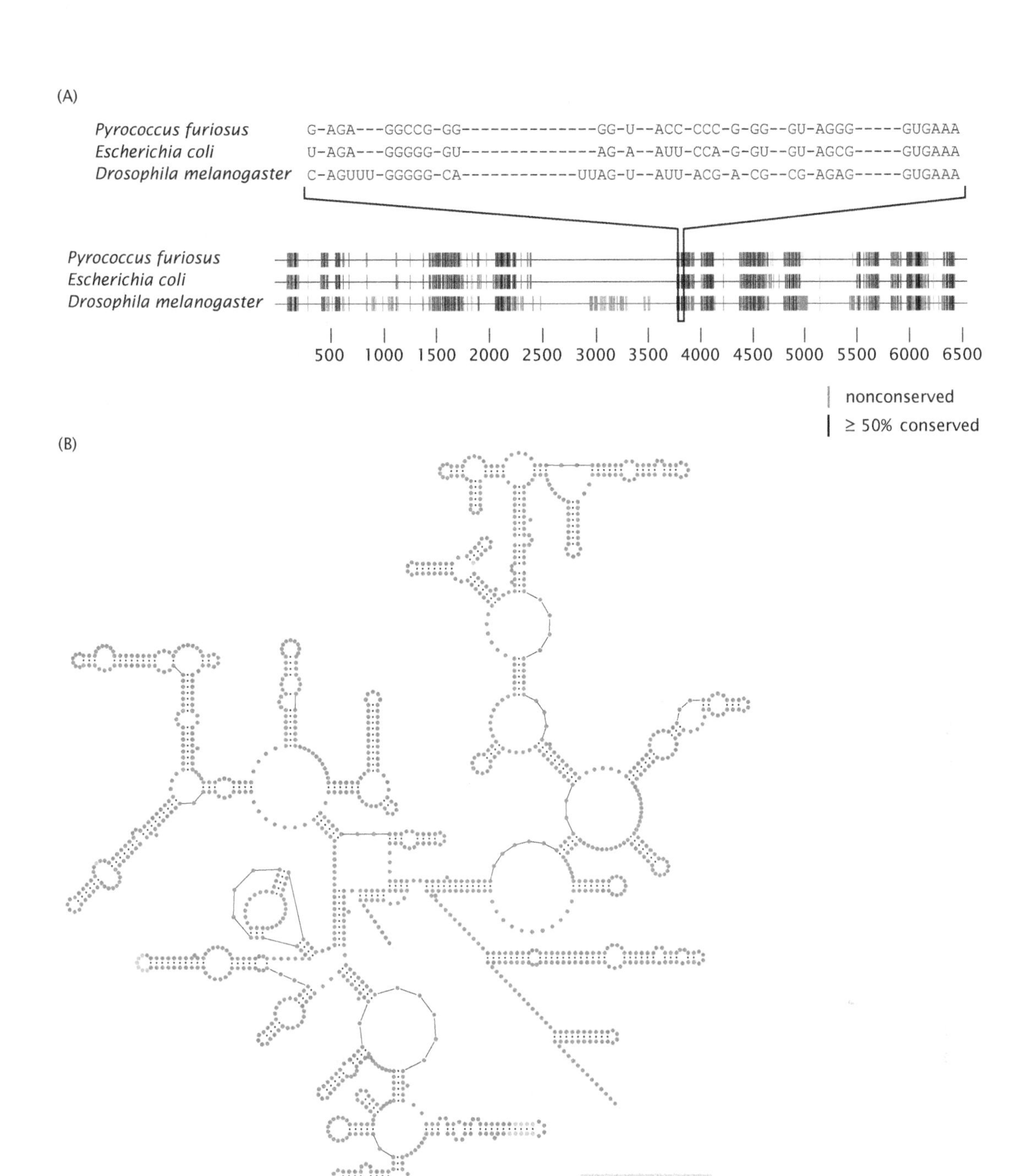

Figure 21.32: Sequences of ribosomal RNA. (A) Sequence alignments of ribosomal RNA from the small subunit of the ribosome. Each sequence is roughly 1500 nucleotides in length. At the top is shown a small segment at nucleotide resolution for three cells representing the three major domains of life: the archaeon *Pyrococcus furiosus*, the bacterium *E. coli*, and the eukaryote *D. melanogaster* (fruit fly). The dashes indicate the positions where nucleotides are found in some other species not shown here. Below, the overall structure of the ribosomal RNA is shown. Black corresponds to regions that are the same for more than 50% of the selected organisms. The alignment has over 6500 positions even though the average small subunit rRNA is only 1500 nucleotides long because the alignment must accommodate gaps corresponding to diversified loops that are present in only a few organisms. (B) Secondary structure of 16S RNA from *E. coli*, colored by variability rate. Bases shown as red dots are "hot spots" of nucleotide substitution, whereas those in blue are more conserved. (B, adapted from Y. Van de Peer et al., *Nucleic Acids Res.* 24:3381, 1996.)

As is standard, the length of the branches in this drawing of the tree of life is an approximate representation of the level of sequence similarity for the sequences at the end of each branch. Overall sequence distance between the ribosomal RNA for any two organisms can be estimated by measuring the total path length to go from one point to another on the tree. It is tempting to assert that these relatively shorter and longer sequence similarity distances are actually indicative of relatively shorter and longer divergence times in evolutionary history. However, there are at least two pitfalls associated with this assertion. First, it is not necessarily the case that identical sequences came from the same ancestor. For example, in a particular gene region, a mutational event might alter an A to a T and at some much later time another random event might change the T back to an A for a small number of offspring. This will make it appear as if the organisms carrying the second mutation are more closely related to the ancestral strain than those with a T although this is not in fact the case. Second, as we have discussed above, mutational events are never truly random. There are chemical and structural biases that make some nucleotide substitutions more likely than others. Furthermore, the influence of natural selection strongly determines which mutations are more or less likely to survive in real populations. In general, sequences that encode critical functions for an organism may appear to mutate slowly, while those regions that are less important or more specialized appear to mutate quickly.

Because of the central importance of ribosomes in life, these sequences have remained remarkably stable over the last several billion years, making it possible to include all organisms in this comparison, but a criterion based strictly on these ribosomal sequences may mislead. Construction of universal phylogenetic trees using other sequences such as ribosomal proteins, aminoacyl tRNA synthetases, or glycolytic enzymes yields trees with the same overall structure but significant differences in the details of branch length and exact branch pattern. Also, we must bear in mind that these trees at best trace evolutionary relationships for particular gene sequences, but not necessarily for the organisms in which they are found. Genes can be transferred horizontally from one organism to another, rendering the true phylogenetic tree of organismal life as an interconnected reticulum rather than a simple branched structure.

21.5 The Molecular Basis of Fidelity

Thus far in this chapter, we have concentrated on methods for analyzing and exploiting sequence variations in numerous different contexts. We have taken for granted the proposition that sequence variation exists and, indeed, have implicitly assumed that very similar sequences are likely to have arisen from a recent common ancestor. The idea that sequences diverge over evolutionary time by accumulation of mutations reflects the confluence of ideas from the central dogma of molecular biology and Darwin's theory of evolution. Given a particular protein coding sequence in the genome of an organism that is inherited by that organism's offspring, we can easily imagine how different lineages of the offspring will diverge from one another if occasional mistakes are made in copying the sequence, though there are more active processes such as recombination and gene transfer that are also in play.

A similar phenomenon is readily observed by scholars studying manuscripts from the era before the widespread use of the printing press. Manuscripts were usually housed in monasteries and laboriously copied by hand by generations of industrious monks who would pass their copies to other monastery libraries. No matter how conscientious the monk, an occasional mistake would appear in a manuscript, which would then be perpetuated by subsequent monks accurately copying the erroneous copy. Similarly, DNA is copied during each round of cell division and mistakes are perpetuated by further rounds of copying. These occasional mistakes are one of the raw materials on which the pitiless engine of natural selection operates to generate evolutionary diversity. Here we will consider in a quantitative way exactly when and how these copying mistakes arise. As we have emphasized previously, copying errors are by no means the only mechanism generating important DNA sequence alterations. We are focusing on this particular mechanism in part because there is a reasonably satisfying explanation for why certain kinds of errors occur at the observed rates that can be framed in the language of statistical mechanics developed throughout the book.

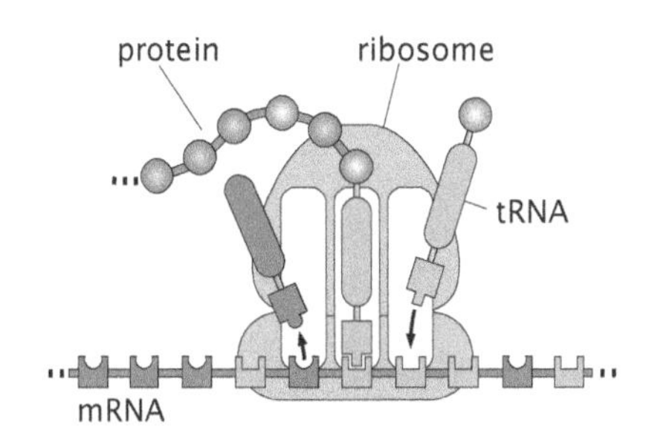

Figure 21.33: Translation in an HP world. In this hypothetical mechanism, mRNA has only two letters (codons), which can recognize two species of tRNA carrying either the H or P amino acids.

21.5.1 Keeping It Specific: Beating Thermodynamic Specificity

The Specificity of Biological Recognition Often Far Exceeds the Limit Dictated by Free-Energy Differences

One of the most remarkable features of many biological recognition events is their very high specificity. In particular, often the reactions involving either polymerization by monomers of different species (such as DNA replication and protein translation) or recognition (such as phosphorylation) take place with much higher fidelity than is expected on the basis of simple thermodynamic reasoning. For example, in the case of DNA replication, measured error rates are of the order of 10^{-9}, while in the case of translation, the error rates are of the order of 10^{-4}. In the following discussion, we illustrate why such rates exhibit higher fidelity than one would expect on the basis of thermodynamics.

As an exercise to illustrate the limits of thermodynamic specificity, consider the simplified model of translation shown in Figure 21.33. We consider the case in which there are only two codons in the nucleic acid alphabet (1 and 0) and two corresponding letters in the amino acid alphabet (H and P for "hydrophobic" and "polar," respectively). The tRNA charged with an H residue has the anticodon $\bar{1}$ while that charged with the P residue has the anticodon $\bar{0}$. We examine the probability that the incorrect amino acid would be incorporated as a result of the binding of a 1 codon on the mRNA with the $\bar{0}$ on the tRNA, for example. For simplicity, consider the case in which the energies associated with the 1–$\bar{1}$ and the 0–$\bar{0}$ bonds are equal and given by $\varepsilon_{\mathrm{corr}}$ ("corr" for correct) while the energies associated with the incorrect bonds 0–$\bar{1}$ and 1–$\bar{0}$ are given by $\varepsilon_{\mathrm{err}}$ ("err" for error) as shown in Figure 21.34.

Within this model, we can compute the probability of correct and incorrect incorporations by appealing to a lattice model like that shown in Figure 21.35. We consider a box with a single mRNA-binding site (that is, codon) for the H tRNA, L_{H} tRNAs charged with H as their amino acid cargo, and L_{P} tRNAs charged with P as their amino acid cargo. The goal of our statistical mechanical calculation is to compute

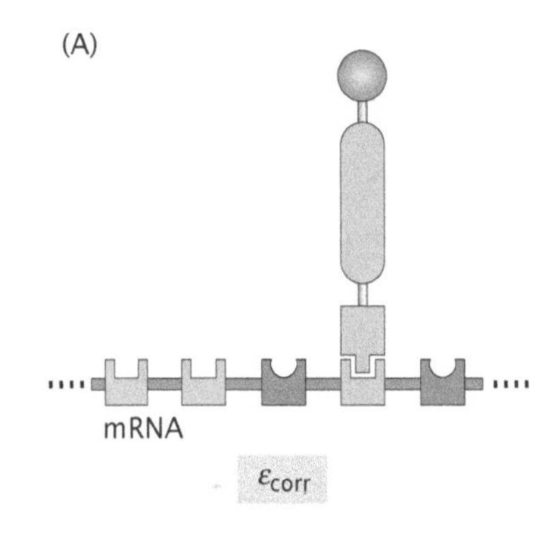

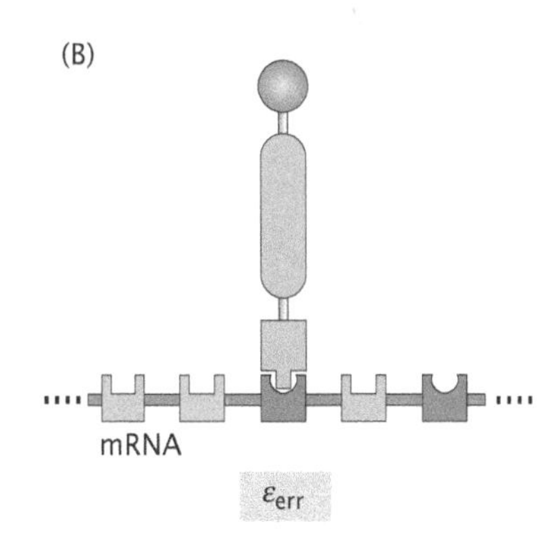

Figure 21.34: Energetics of tRNA and mRNA interaction. The figure shows a toy model of the recognition event between tRNA and mRNA in which correct and incorrect recognition are assigned different energies. (A) The correct tRNA anticodon binds to its codon partner on the mRNA with an energy $\varepsilon_{\mathrm{corr}}$. (B) The incorrect tRNA anticodon binds to its codon partner on the mRNA with an energy $\varepsilon_{\mathrm{err}}$.

Figure 21.35: States and weights for an HP model of translation. The mRNA can be unoccupied, occupied by the correct tRNA, or occupied by the incorrect tRNA.

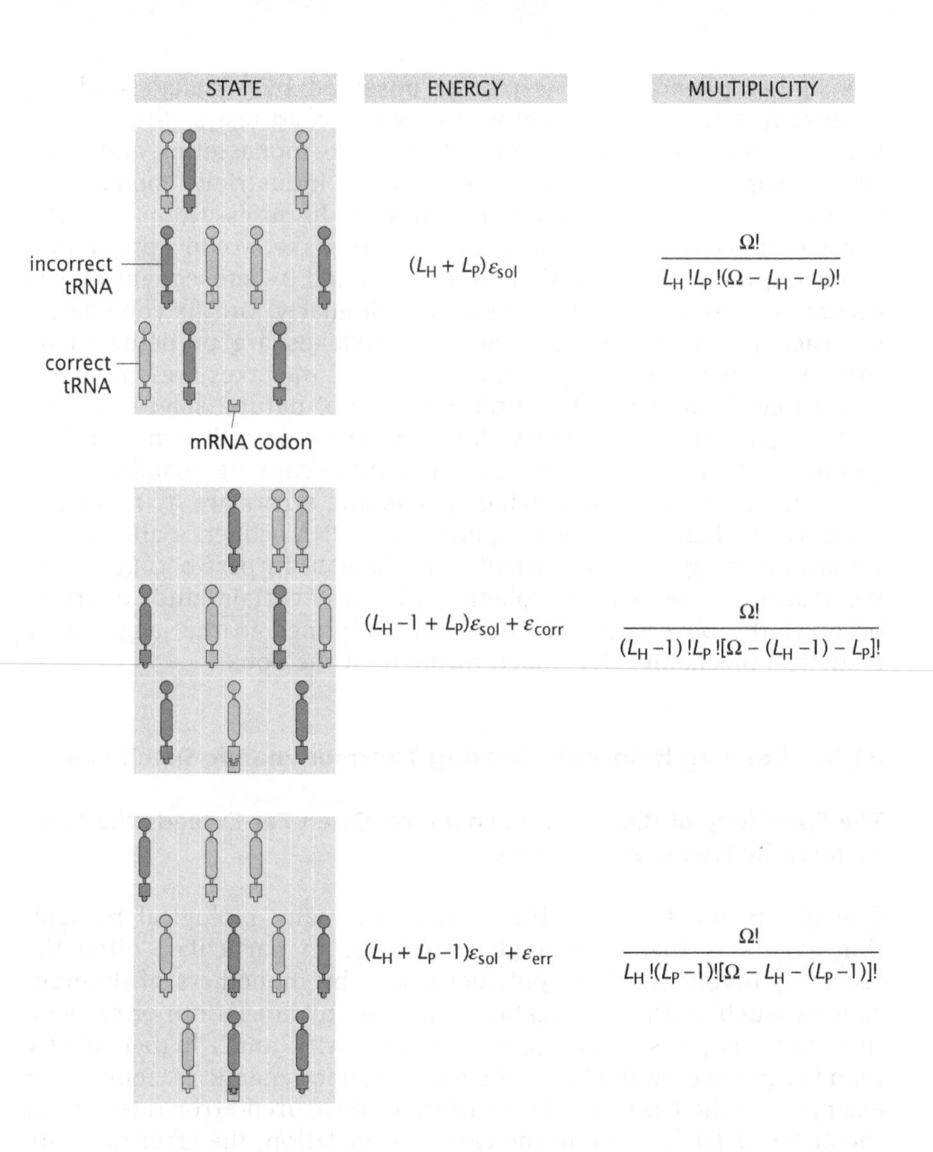

the relative probabilities of binding of the two different species of tRNA to the codon of interest. Once we have these probabilities in hand, the error rate in this simple model of translation will be given by the ratio p_{err}/p_{corr}.

As with all of our treatments of binding problems, Figure 21.35 instructs us to sum over all of the different arrangements of the system, each weighted according to its Boltzmann factor. The partition function is obtained by summing over the three classes of states where the mRNA is not bound to a tRNA, the mRNA is bound to the correct tRNA, and the mRNA is bound to the incorrect tRNA. Summing over these states results in

$$Z = \underbrace{\frac{\Omega!}{L_H!L_P!(\Omega - L_H - L_P)!} e^{-\beta(L_H+L_P)\varepsilon_{sol}}}_{\text{unoccupied}} + \underbrace{\frac{\Omega!}{(L_H - 1)!L_P![\Omega - (L_H - 1) - L_P]!} e^{-\beta(L_H-1+L_P)\varepsilon_{sol}} e^{-\beta\varepsilon_{corr}}}_{\text{correct}} + \underbrace{\frac{\Omega!}{L_H!(L_P - 1)![\Omega - L_H - (L_P - 1)]!} e^{-\beta(L_H+L_P-1)\varepsilon_{sol}} e^{-\beta\varepsilon_{err}}}_{\text{incorrect}} . \tag{21.57}$$

Once we have the partition function, we can compute the probabilities both of the correct tRNA being bound and of the incorrect tRNA being bound. In particular, the probability that the correct tRNA is bound is given by computing the ratio of the states in which the correct tRNA is bound to all of the states (that is, the partition function), resulting in

$$p_{\text{corr}} = \frac{(L_{\text{H}}/\Omega)\mathrm{e}^{-\beta(\varepsilon_{\text{corr}}-\varepsilon_{\text{sol}})}}{1+(L_{\text{H}}/\Omega)\mathrm{e}^{-\beta(\varepsilon_{\text{corr}}-\varepsilon_{\text{sol}})}+(L_{\text{P}}/\Omega)\mathrm{e}^{-\beta(\varepsilon_{\text{err}}-\varepsilon_{\text{sol}})}}. \tag{21.58}$$

Similarly, we can compute the probability of incorrect incorporation as

$$p_{\text{err}} = \frac{(L_{\text{P}}/\Omega)\mathrm{e}^{-\beta(\varepsilon_{\text{err}}-\varepsilon_{\text{sol}})}}{1+(L_{\text{H}}/\Omega)\mathrm{e}^{-\beta(\varepsilon_{\text{corr}}-\varepsilon_{\text{sol}})}+(L_{\text{P}}/\Omega)\mathrm{e}^{-\beta(\varepsilon_{\text{err}}-\varepsilon_{\text{sol}})}}. \tag{21.59}$$

Our interest now is in deducing the relation between the observed error rates and the energy difference for binding of the two different species. Using the results of Equations 21.58 and 21.59, we can compute the error rate as

$$\frac{p_{\text{err}}}{p_{\text{corr}}} = \frac{L_{\text{P}}}{L_{\text{H}}}\mathrm{e}^{-\beta(\varepsilon_{\text{err}}-\varepsilon_{\text{corr}})}. \tag{21.60}$$

For the case in which we assume that the concentrations of the two species of tRNA are equal, we find that the error rate is given by

$$\frac{p_{\text{err}}}{p_{\text{corr}}} = \mathrm{e}^{-\beta(\varepsilon_{\text{err}}-\varepsilon_{\text{corr}})}. \tag{21.61}$$

This result permits us to get a feeling for the error rates (or the required energy differences) dictated by strict thermodynamic specificity. The fundamental result is that the accuracy of translation must depend on the energy difference between binding the incorrect and the correct tRNA, that is, $\varepsilon_{\text{err}} - \varepsilon_{\text{corr}} = \Delta\varepsilon$. The error rate as a function of this energy difference is shown in Figure 21.36.

Because we know that codon recognition relies on the ability of complementary bases to form specific hydrogen bonds with one another, we can predict that the scale of this energy difference must be comparable to the energy associated with the formation of an individual hydrogen bond ($\Delta\varepsilon \approx 2k_{\text{B}}T$). Plugging this value for $\Delta\varepsilon$ into Equation 21.61 results in an estimate that the error rate for translation should be of the order of e^{-2}, or about 0.13. In other words, we expect an error to be made at about one out of every ten sites! Obviously the real ribosome is much more accurate than this, by about 1000-fold, with a true error rate of the order of 10^{-4}. One limitation of this estimate is that it assumes that the entirety of the favorable interaction between codon and anticodon is dictated by base pairing. In fact, there are also "stacking effects" that can raise the binding-energy difference, but these are still not enough to account for observed specificities. Something else must be going on.

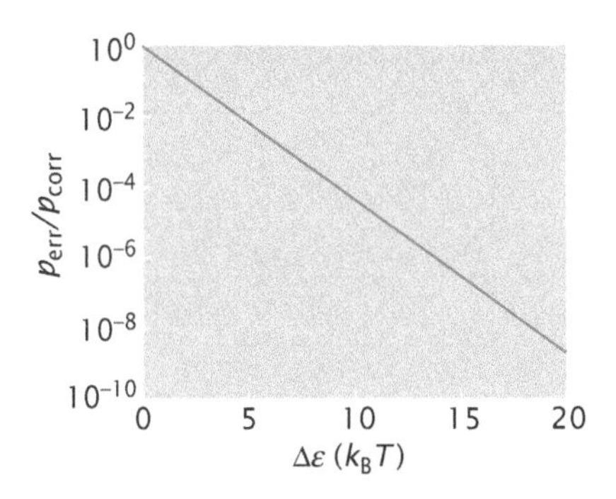

Figure 21.36: Plot of the error rate as a function of the difference in free energy of binding. When the tRNA binds to the mRNA, there is a favorable free energy of binding. If the anticodon on the tRNA is not the correct partner of the codon on the mRNA, there is a less favorable binding free energy. The plot shows how the error rate depends upon the free-energy difference between the binding of the correct and incorrect tRNA.

As an alternative point of view, we can take the known error rates and use them to deduce what energy difference they imply, using the formula $\varepsilon_{\text{err}} - \varepsilon_{\text{corr}} = -k_{\text{B}}T\ \ln(p_{\text{err}}/p_{\text{corr}})$. For example, if we use an error rate of 10^{-4}, Equation 21.61 implies that the energy difference between the incorrect and correct mRNA/tRNA binding is

$$\varepsilon_{\text{err}} - \varepsilon_{\text{corr}} = -k_{\text{B}}T\ \ln 10^{-4} = 4k_{\text{B}}T\ \ln 10. \tag{21.62}$$

If we use the approximate statement that $\ln 10 \approx 2.3$, we see that the energy difference of discrimination is large ($\approx 10\, k_{\text{B}}T$). When viewed from either perspective, the key outcome of this analysis is the clear recognition that thermodynamic specificity driven by hydrogen bond formation alone is not high enough to explain observed error rates.

A similar argument to that given above can be made using a kinetic description of HP translation. Indeed, the kinetic description of this model will serve as the jumping off point for our analysis of the mechanism of kinetic proofreading, the process that helps to close the gap between the thermodynamic specificity predicted above and the true specificity achieved by the ribosome. In this case, we consider reactions of the form

$$\mathrm{R} + \mathrm{T}_{\mathrm{corr}} \rightleftharpoons \mathrm{RT}_{\mathrm{corr}} \rightarrow \text{elongation} \tag{21.63}$$

and

$$\mathrm{R} + \mathrm{T}_{\mathrm{err}} \rightleftharpoons \mathrm{RT}_{\mathrm{err}} \rightarrow \text{elongation}. \tag{21.64}$$

We have introduced the notation R for the ribosome, $\mathrm{T}_{\mathrm{corr}}$ for the correct tRNA, and $\mathrm{T}_{\mathrm{err}}$ for the incorrect tRNA. These reactions can be characterized by rate equations like those introduced in Chapter 15. In particular, the rate at which ribosome–tRNA complexes are formed is given by

$$\frac{\mathrm{d}[\mathrm{RT}_{\mathrm{corr}}]}{\mathrm{d}t} = k_+^{\mathrm{c}}[\mathrm{R}][\mathrm{T}_{\mathrm{corr}}] - k_-^{\mathrm{c}}[\mathrm{RT}_{\mathrm{corr}}] \tag{21.65}$$

and

$$\frac{\mathrm{d}[\mathrm{RT}_{\mathrm{err}}]}{\mathrm{d}t} = k_+^{\mathrm{e}}[\mathrm{R}][\mathrm{T}_{\mathrm{err}}] - k_-^{\mathrm{e}}[\mathrm{RT}_{\mathrm{err}}], \tag{21.66}$$

where we have introduced the rate constants $k_\pm^{\mathrm{c}}$ and $k_\pm^{\mathrm{e}}$ for the formation of complexes with the correct tRNA (c) or an error (e) in the tRNA selection, respectively. To make contact with the treatment of the error rate given above, we consider the steady-state solution in which $\mathrm{d}[\mathrm{RT}]/\mathrm{d}t = 0$. Using the rate equations, this implies

$$\frac{[\mathrm{R}][\mathrm{T}]}{[\mathrm{RT}]} = \frac{k_-}{k_+} = K_{\mathrm{d}}. \tag{21.67}$$

Once again, we can compute the probability of an error in translation as

$$\frac{p_{\mathrm{err}}}{p_{\mathrm{corr}}} = \frac{[\mathrm{RT}_{\mathrm{err}}]}{[\mathrm{RT}_{\mathrm{err}}] + [\mathrm{RT}_{\mathrm{corr}}]} \Bigg/ \frac{[\mathrm{RT}_{\mathrm{corr}}]}{[\mathrm{RT}_{\mathrm{err}}] + [\mathrm{RT}_{\mathrm{corr}}]} = \frac{[\mathrm{RT}_{\mathrm{err}}]}{[\mathrm{RT}_{\mathrm{corr}}]} = \frac{k_+^{\mathrm{e}} k_-^{\mathrm{c}}}{k_+^{\mathrm{c}} k_-^{\mathrm{e}}} \approx \frac{k_-^{\mathrm{c}}}{k_-^{\mathrm{e}}}, \tag{21.68}$$

where in the last step we have invoked the approximate statement that the forward rates are the same for both the correct and incorrect tRNAs. One plausibility argument for this approximation is the idea that the "on" rates for the association of the charged tRNA and the ribosome are dictated by diffusion as described in Chapter 13 and that both species have the same diffusion constant. As before, the error rate is decided strictly on the basis of the energy difference between the two states and can be rewritten as

$$\frac{p_{\mathrm{err}}}{p_{\mathrm{corr}}} = \frac{K_{\mathrm{d}}^{\mathrm{corr}}}{K_{\mathrm{d}}^{\mathrm{err}}}. \tag{21.69}$$

Also as before, this result depends upon the assumption that the number of tRNAs of each species is the same. Because the effective dissociation constants for correct versus incorrect tRNA binding where the codon–anticodon interaction is altered by only a single hydrogen bond is approximately 10-fold, this kinetic analysis recovers the same conclusion that we had reached using analysis based on statistical mechanics, namely, that thermodynamic discrimination cannot account for observed very low error rates for translation.

High Specificity Costs Energy

The discussion given above provides thermodynamic specificity as a benchmark. Often, such specificity is biologically insufficient. One of the special tricks associated with many biological reactions is that they are coupled to energy-releasing hydrolysis reactions. This suite of ideas goes under the name of proofreading and refers to reactions in which substrate "verification" takes place more than once.

In the context of translation, indeed, there are various reaction intermediates including an intermediate hydrolysis reaction involving the elongation factor EF-Tu. The kinetic proofreading concept is that some energy in the form of hydrolysis can be paid in order to increase the specificity by altering the rate constants associated with the cognate and noncognate tRNAs. The particular details appropriate to translation are more complicated than the simplified reaction scheme considered here, but the basic idea of repetitive substrate verification has been corroborated experimentally and has stood the test of time.

A reaction scheme associated with translation including the hydrolysis step is given by

$$\begin{array}{l} \mathrm{R}+\mathrm{T}_{\mathrm{corr}} \underset{k_-^{(1)}}{\overset{k_+^{(1)}}{\rightleftharpoons}} \mathrm{RT}_{\mathrm{corr}} \underset{k_-^{(2)}}{\overset{k_+^{(2)}}{\rightleftharpoons}} \mathrm{RT}^*_{\mathrm{corr}} \rightarrow \text{elongation} \\ \qquad\qquad\qquad\qquad\quad \downarrow k_{\mathrm{d}} \\ \qquad\qquad\qquad\quad \mathrm{R}+\mathrm{T}_{\mathrm{corr}} \end{array} \tag{21.70}$$

and

$$\begin{array}{l} \mathrm{R}+\mathrm{T}_{\mathrm{err}} \underset{p_-^{(1)}}{\overset{p_+^{(1)}}{\rightleftharpoons}} \mathrm{RT}_{\mathrm{err}} \underset{p_-^{(2)}}{\overset{p_+^{(2)}}{\rightleftharpoons}} \mathrm{RT}^*_{\mathrm{err}} \rightarrow \text{elongation.} \\ \qquad\qquad\qquad\qquad\quad \downarrow p_{\mathrm{d}} \\ \qquad\qquad\qquad\quad \mathrm{R}+\mathrm{T}_{\mathrm{err}} \end{array} \tag{21.71}$$

The novel feature here is the existence of the intermediate state RT^*. The simple physical picture is that the hydrolysis interaction imposes a conformational change that alters the "off" rates in much the same way that the rates on microtubules are altered by hydrolysis. In particular, the noncognate tRNAs have a higher "off" rate than the cognate tRNAs, and this second stage complements the selectivity of the first recognition step during binding. To see how this plays out, we consider the kinetics of the reaction scheme for the correct tRNA,

$$\frac{d[\mathrm{R}]}{dt} = \frac{d[\mathrm{T}_{\mathrm{corr}}]}{dt} = -k_+^{(1)}[\mathrm{R}][\mathrm{T}_{\mathrm{corr}}] + k_-^{(1)}[\mathrm{RT}_{\mathrm{corr}}] + k_{\mathrm{d}}[\mathrm{RT}^*_{\mathrm{corr}}], \tag{21.72}$$

$$\frac{d[\mathrm{RT}_{\mathrm{corr}}]}{dt} = k_+^{(1)}[\mathrm{R}][\mathrm{T}_{\mathrm{corr}}] - k_-^{(1)}[\mathrm{RT}_{\mathrm{corr}}] - k_+^{(2)}[\mathrm{RT}_{\mathrm{corr}}] + k_-^{(2)}[\mathrm{RT}^*_{\mathrm{corr}}], \tag{21.73}$$

$$\frac{d[\mathrm{RT}^*_{\mathrm{corr}}]}{dt} = k_+^{(2)}[\mathrm{RT}_{\mathrm{corr}}] - k_-^{(2)}[\mathrm{RT}^*_{\mathrm{corr}}] - k_{\mathrm{d}}[\mathrm{RT}^*_{\mathrm{corr}}]. \tag{21.74}$$

A similar set of equations characterize the dynamics of the incorrect tRNA.

By examining the steady state, we can solve for $[\mathrm{RT}^*_{\mathrm{err}}]$ and $[\mathrm{RT}^*_{\mathrm{corr}}]$ and the error rate given by $f = [\mathrm{RT}^*_{\mathrm{err}}]/[\mathrm{RT}^*_{\mathrm{corr}}]$. To do so, we set the right-hand sides in all three rate equations equal to zero and solve for $[\mathrm{RT}^*]$, resulting in an error rate

$$f = \frac{k_-^{(1)} k_-^{(2)} + k_-^{(1)} k_\mathrm{d} + k_+^{(2)} k_\mathrm{d}}{p_-^{(1)} p_-^{(2)} + p_-^{(1)} p_\mathrm{d} + p_+^{(2)} p_\mathrm{d}}, \tag{21.75}$$

where the rate constants k are replaced with p for the reaction involving the incorrect tRNA. In the limit that the forward reactions are rapid compared with the back reactions, that is, the rates satisfy $k_-^{(2)} \ll k_\mathrm{d}$ and $k_+^{(2)} \ll k_-^{(1)}$, this expression for the error rate simplifies to

$$f = f_0 \frac{k_\mathrm{d}}{p_\mathrm{d}}, \tag{21.76}$$

where f_0 is the error rate in the absence of the proofreading step. This tells us that the error rate can be reduced beyond the thermodynamic specificity as a result of a second kinetic step. One of the experimental consequences of the proofreading hypothesis is that the rate of GTP hydrolysis per peptide bond formed should be higher for noncognate than cognate tRNAs.

Theoretical considerations like those presented here have served as a framework for trying to understand fidelity in translation (and other biological polymerization reactions). At the same time, a variety of experiments ranging from structural studies of ribosomes in complexes with attendant molecules to bulk *in vitro* biochemical investigations and single-molecule FRET techniques are still clarifying the *precise* mechanisms of translational fidelity. As a result, the two-stage proofreading model presented here should be seen as a demonstration of the proofreading principle as opposed to a concrete assertion of the mechanism of proofreading in translation. What appears to be an abiding conclusion of experimental work on translational fidelity is that noncognate codon–anticodon complexes are distinguished from their cognate counterparts at two steps, namely, (i) during an initial selection process prior to GTP hydrolysis and (ii) during proofreading after GTP hydrolysis, consistent with the conceptual underpinnings of the proofreading hypothesis introduced by Hopfield (1974) and Ninio (1975).

21.6 Summary and Conclusions

Biological sequences provide a window on the relatedness of nucleic acids and proteins. By performing sequence alignments, it is possible to infer relations between the functions of different proteins and, more importantly, to trace the evolutionary history of different organisms. This chapter has only scratched the surface of an enormous field. The goal of the chapter was to give a sense of the kinds of questions that must be faced in thinking about the data now being routinely obtained in biology and to sketch the ideas that allow these data to be examined. These ideas are relevant in problems ranging from the origins of antibiotic resistance to the phylogenetic organization of the tree of life.

21.7 Problems

Key to the problem categories: • Model refinements and derivations, • Estimates, • Data interpretation, • Model construction, • Database mining

• 21.1 Restriction enzymes and sequences

(a) Restriction enzymes are proteins that recognize specific sequences at which they cut the DNA. Two commonly used restriction enzymes are *Hind*III and *Eco*RI. Look up the recognition sequences that these enzymes each cut and make a sketch of the pattern of cutting they carry out. Consider the approximately 48,000 bp genome of lambda phage and make an estimate of the lengths of the fragments that you would get if the DNA is cut with both the *Hind*III and *Eco*RI restriction enzymes. There is a precise mathematical way to do this and it depends upon the length of the recognition sequence—a 5 cutter will have shorter fragments than an 8 cutter—explain that.

(b) Find the actual fragment lengths obtained in the lambda genome using these restriction enzymes by going to the New England Biolabs website (www.neb.com) and looking up the tables identifying the sites on the lambda genome that are cut by these different enzymes. How do these cutting patterns compare with your results from (a)?

(c) Plot the number of cuts in the lambda genome as a function of the length of the recognition sequence of several commercially available type II restriction enzymes. You can download the list of type II restriction enzymes from the book's website. Combine this plot with a curve showing your theoretical expectation.

• 21.2 Alignment tools and methods

HIV virions wrap themselves with a lipid bilayer membrane as they bud off from infected cells; in this viral membrane envelope are "spikes" composed of two different proteins (actually glycoproteins), gp41 and gp120. The gp denotes glycoprotein, and the number indicates their molecular weight in kilodaltons. gp120 and gp41 together form the trimeric envelope spike on the surface of HIV that functions in viral entry into a host cell. The primary receptor for gp120 is CD4, a protein found mainly on the white blood cells known as T-lymphocytes. gp120 avoids detection by the host immune system through a number of strategies, including rapid changes in sequence due to mutations.

In this exercise, you will grab a sequence for gp120 and "blast" it to find related proteins. Use the "Search database" on the LANL HIV Sequence Database site (www.hiv.lanl.gov/) to find a sequence for gp120; the complete gp120 molecule has about 500 amino acids, so a complete DNA sequence will have roughly 1500 base pairs.

With the BLAST website (www.ncbi.nlm.nih.gov/blast), open a new window and select "blastx" under the "Basic BLAST" heading. Copy and paste your approximately 1500 nucleotide sequence of gp120 into the top box under "Enter Query Sequence." For the database, select "Protein Data Bank proteins (pdb)." What we are doing is having BLAST translate the gp120 nucleotide sequence into an amino acid sequence and then compare it with amino acid sequences of proteins in the PDB. Note that there are some proteins that appear multiple times in the PDB because their structures have been analyzed and determined more than once or in different contexts. Finally, push the "BLAST" button and wait for your results to appear on a new page.

(a) How does BLAST determine the ranking of the results from your search?

(b) For your top search result, identify the percentage of sequence identity with your query sequence. Explain how this number is determined.

(c) You will notice that BLAST has used gaps in many of your alignments. What is the evolutionary significance of these gaps?

(d) For your top hit, how many alignments with this high a score or better would have been expected by chance?

(e) Looking at your ranked list of results and using only the "E-value," which hits do you expect to (possibly) have a genuine evolutionary relationship with your gp120 sequence and why?

• 21.3 Comparison of *Pax6* and *eyeless*

In this exercise, you will examine the sequences for both *Pax6* and *eyeless* and consider the differences and similarities between them. First, download the sequences for *Pax6* (82069480) and *eyeless* (12643549) from the NCBI Entrez Protein site using their accession numbers, given in parentheses.

Go to the BLAST homepage (www.ncbi.nlm.nih.gov/blast) and, choose "Align two sequences using BLAST (bl2seq)" under the "Specialized BLAST" heading. Instead of searching a large database as is typical with BLAST, we will only be aligning two sequences with one another. Paste your sequences for *Pax6* and *eyeless* into boxes for Sequence 1 and Sequence 2 and make sure that you choose blastp as the program. When all of this is done, push the Align button.

In your BLAST alignments there is a line between "Query" and "Sbjct" that helps guide the eye with the alignment; if "Query" and "Sbjct" agree identically, the matching letter is repeated in the middle; if they do not match exactly but the amino acids are compatible in some sense (a favorable mismatch), then a "+" is displayed on the middle line to indicate a positive score. Where there is no letter on the middle line indicates an unfavorable mismatch or a gap. The numbers at the beginning and end of the "Query" and "Sbjct" lines tell you the position in the sequence.

(a) Choose one of the alignments returned by BLAST and give a tally of the number of

(i) identical amino acids;
(ii) favorable mismatches;
(iii) unfavorable mismatches;
(iv) gaps.

(b) Give two examples of unfavorable mismatches and two examples of favorable mismatches in your chosen alignment. Based on what you know of the chemistry and structure of the amino acids, why might these amino acid pairs give rise to negative and positive scores, respectively?

(c) Choose one unfavorable mismatch pair and one favorable mismatch pair from your chosen alignment. What codons may give rise to each of these amino acids? What is the minimum number of mutations necessary in the DNA to produce this particular unfavorable mismatch? How many DNA mutations would be required to produce the particular favorable mismatch you chose?

• 21.4 Nucleosome urn and chemical potentials

(a) In our study of nucleosome positioning, we exploited a model in which the nucleosomes and base pairs were imagined to be drawn from an urn like that shown in Figure 21.21. Using the definition of chemical potential, work out the chemical potentials for both nucleosomes and base pairs.

(b) Using the results of part (a), compute the nucleosome occupancy as a function of position on the genome. Assume that base 0 is nucleosome free, and then compute the statistical weight $S(n)$ for covering bases 1 through n with any of the maximum allowed number of nucleosomes. Using $S(n)$ compute the probability $p(i)$ that the i-th base pair of a long stretch of N bases, labeled 1 through N, is covered by a nucleosome. Finally, plot this probability for average linker lengths of $\langle L \rangle = 15$, 30 and 50 bp and compare them by placing them all on the same graph.

• 21.5 Endogenous retroviruses and the human genome

Sequencing studies have led to the assertion that roughly 10% of the human genome is associated with endogenous retroviruses, that is, the fossil remnants of earlier viral infections by viruses like HIV that incorporate their genomes into the host genome. Assuming that a typical retrovirus has a genome length of roughly 10 kb, make an estimate of the number of retroviruses integrated within your own genome.

• 21.6 Sequence analysis of hemoglobin

In this problem, you will examine the changes in hemoglobin over evolutionary time from the point of view of modern sequence analysis with a comparison of the α-chain hemoglobins from eight different organisms (humans, chimpanzees, gorillas, cattle, horses, donkeys, rabbits, and carp). The concept of the problem is to take the known sequences of hemoglobin from all of these organisms, do a multiple alignment using ClustalW (www.ebi.ac.uk/clustalw) and then explore the differences and similarities between them.

Download the necessary hemoglobin sequences using NCBI Entrez Protein and the accession numbers given in the caption of Figure 21.23. Cut and paste the sequences above into a text file; before each sequence, make a line with a greater-than symbol and the name of the organism. For example,

```
> Human
mvlspadktn . . .
```

The ">" symbol is recognized by ClustalW as a comment that gives information about the sequence and is used to make your results more readable. Go to the ClustalW site and click the "Choose File" button. Navigate to your text file with the sequences, select it, and then run the alignment. After a few moments, the results will appear; the parts of greatest interest to us in this problem are the Alignment and the Cladogram, a graph that shows the degree of similarity among each of the sequences. Very similar sequences are connected by short branches; increasing branch length indicates decreasing sequence similarity.

(a) Examine the variability in the amino acids between the sequences and comment on the relatedness of the species in question. Do the results jibe with your intuition?

(b) Choose a position in the alignment that has at least three different amino acids in it among the eight different species. Make a list of all of the different amino acids used at this position and comment on their similarities or differences (that is, hydrophobic or polar, size, etc.).

(c) Now look at the Cladogram in your results that ClustalW generated based on the alignment. Does the tree correspond to your intuition about relatedness of species?

• 21.7 HP model

In this exercise, you should use a tool such as MATLAB to analyze the human, cow, and carp sequences shown in Figure 21.23. You will find it helpful to translate these sequences into H's and P's using Figure 21.7 (charged amino acids will be included in the P category). Download the three sequences from NCBI Entrez Protein using the accession numbers given in the caption of Figure 21.23.

(a) Using the human, cow, and carp sequences in the alignment shown in Figure 21.23, calculate the background frequency of H (hydrophobic) and P (polar) amino acids: that is, what fraction of letters are H and what fraction are P in those three sequences? You may ignore the gapped position in the alignment for all parts of this problem.

(b) Write down an equation for the expected probability of a given pair (H–H, P–P, H–P, or P–H) in terms of the background frequencies you found in (a). Use a computer to count the number of observed pairs of each kind and the total number of pairs. Using these results, calculate the expected and observed probabilities for the following pairs from the human, cow, and carp sequences in the alignment in Figure 21.23:

(i) H–H;
(ii) P–P;
(iii) H–P;
(iv) P–H.

Be careful to treat H–P and P–H equally, since the order of these pairs is not specified from the multiple alignment.

(c) Using the results of (b), give the HP log-odds scoring matrix in units of half-bits using Equation 21.3. A log-odds score is calculated by taking a logarithm of the ratio of the observed probability to the expected probability: $s_{AB} = \log[p_{AB}(\text{observed})/p_{AB}(\text{expected})]$. When a natural logarithm is used, the score is said to have units of "nats" of information; when a base-2 logarithm is used, the units are "bits." The popular BLOSUM62 scoring matrices are typically given in units of half-bits, $S_{AB} = 2\log_2[p_{AB}(\text{observed})/p_{AB}(\text{expected})]$.

(d) What is the expected score per letter in a (random) alignment using this matrix?

• 21.8 Mutations of bacteria in our gut

(a) The populations of the *E. coli* in the guts of a collection of humans can be large enough that multiple mutations can occur simultaneously in one bacterium. Suppose that a very particular combination of k point mutations is required for a pathogenic strain to emerge and that these must all arise in one cell division (as could be the case if the subsets of these mutations are deleterious). With the point mutation rate per base pair per cell division of μ, what is the probability m_k that this occurs in a single cell division? The simplest assumption is that the probabilities of the different mutations are independent.

(b) In a human large intestine, the density of bacteria is estimated to be about $10^{11.5}$ per milliliter, of which a fraction of about 10^{-4} are *E. coli*. Estimate how many *E. coli* per person this implies. In a population of N humans, with n *E. coli* in each of their guts, in T generations of the *E. coli*, estimate the total probability P_k that the particular combination of k mutations occurs at least once.

(c) With the population of Silicon Valley over one year, what are the chances this occurs for $k = 2$? For $k = 3$? Some crucial factors in your estimate are $\mu \approx 10^{-10} - 10^{-9}$ mutations per base pair per cell division and the generation time of *E. coli*: the standard lab result is that *E. coli* divide every 20 minutes. A low-end estimate for the division rate of *E. coli* in human guts is about once every few days. Why is this more realistic? Given these and other uncertainties, how big are the uncertainties in your estimates of P_2 and P_3?

(Problem courtesy of Daniel Fisher.)

• 21.9 The molecular clock

In eukaryotes, the majority of individual point mutations are thought to be "neutral" and have little or no effect on phenotype. Only a small fraction of the genome codes for proteins and critical DNA regulatory sequences. Even within coding regions, the redundancy of the genetic code is suffcient to render many mutations "synonymous" (that is, they do not change the amino acid, and hence the protein, encoded by the DNA). The slow accumulation of neutral mutations between two populations can be used as a "molecular clock" to estimate the length of time that has passed since the existence of their last common ancestor. In these estimates, it is common to make the simplifying approximations that (1) most mutations are neutral and (2) the rate of accumulation of neutral mutations is just the average point mutation rate per generation (that is, ignoring other kinds of mutations such as deletions, inversions, etc., as well as variations in and correlations among mutations).

(a) With a crude estimate of the point mutation rate of humans of 10^{-8} per base pair per generation, what fraction of the possible nucleotide differences would you expect there to be between chimpanzees and humans given that the fossil record and radiochemical dating indicate their lineages diverged about six million years ago? Compare your estimate with the observed result from sequencing of about 1.5%.

(b) Some parasitic organisms (lice are an example) have specialized and co-evolved with humans and chimps separately. A natural hypothesis is that the most recent common ancestor of the human and chimp parasites existed at the same time as that of the human and chimp themselves. How might you test this from DNA sequence data and other information? What are likely to be the largest causes of uncertainty in the estimates?

(Problem courtesy of Daniel Fisher.)

• 21.10 Fidelity of protein synthesis

The average mass of proteins in the cell is 30,000 Da, and the average mass of an amino acid is 120 Da. In eukaryotic cells, the translation rate for a single ribosome is roughly 40 amino acids per second.

(a) The largest known polypeptide chain made by any cell is *titin*. It is made by muscle cells and has an average weight of 3×10^6 Da. Estimate the translation time for titin and compare it with that of a typical protein.

(b) Protein synthesis is very accurate: for every 10,000 amino acids joined together, only one mistake is made. What is the probability of an error occurring for one amino acid addition? What fraction of average sized proteins are synthesized error-free?

• 21.11 Mutual information by another name

In the chapter, we introduced the concept of mutual information as the average decrease in the missing information associated with one variable when the value of another variable is known. In terms of probability distributions, this can be written mathematically as

$$I = \sum_y p(y) \left[- \sum_x p(x) \log_2 p(x) + \sum_x p(x|y) \log_2 p(x|y) \right], \tag{21.77}$$

where the expression in square brackets is the difference in missing information, $S_x - S_{x|y}$, associated with probability of x, $p(x)$, and with probability of x conditioned on y, $p(x|y)$. Using the relation between the conditional probability $p(x|y)$ and the joint probability $p(x, y)$,

$$p(x|y) = \frac{p(x, y)}{p(y)}, \tag{21.78}$$

show that the formula for mutual information given in Equation 21.77 can be used to derive the formula used in the chapter (Equation 21.17), namely

$$I = \sum_{x,y} p(x, y) \log_2 \left[\frac{p(x, y)}{p(x)p(y)} \right]. \tag{21.79}$$

• 21.12 Information content of the genetic code

The genetic code shown in Figure 1.4 specifies 20 amino acids and the STOP codon in terms of triplets of nucleotides. In order to develop intuition about the meaning of the mutual information, we would like to compute the mutual information between the first nucleotide of a codon and the amino acid that it specifies using the formula derived in Problem 21.11. This mutual information will tell us, in bits, the reduction in (on average) information needed to specify a particular amino acid, if the identity of the first nucleotide is known. One way to think about what the missing information represents is to imagine 64 balls, representing each codon, inside a bag. Each ball is associated with a particular amino acid. By drawing one of these balls randomly, how well can we guess what amino acid (or STOP codon) that ball will represent? The higher the missing information, the less predictable the outcome of this imaginary exercise will be.

(a) Assume a null model where each amino acid is encoded by the same number of codons (this can't be quite right, because 64 is not divisible by 21!). From this, compute the missing information associated with the genetic code in terms of bits.

(b) Using Figure 1.4, compute the probability for each of the 20 amino acids and the STOP codon, or in other words their frequency within the set of all possible (64) codons and the corresponding missing information. How does this missing information compare with that of the null model calculated in (a)?

(c) If the first nucleotide of the codon is A, there are still 7 possible outcomes corresponding to 7 different amino acids whose codon starts with A. What are the probabilities of each of these outcomes and what is the missing information associated with this conditional probability? How many bits of information have been gained by the knowledge that the first nucleotide of the codon is A?

(d) Repeat the above calculation for the other three choices of the first nucleotide. What is the average information gained by knowledge of the first nucleotide of the codon (or, equivalently, the mutual information between the first nucleotide and the amino acid)? What fraction of the missing information associated with the genetic code is this?

(Problem courtesy of Sharad Ramanathan.)

• 21.13 Open reading frames in *E. coli*

In this problem, we will search the *E. coli* genome for open reading frames. The actual genome sequence of *E. coli* is available on the book's website.

(a) Write a program that scans the DNA sequence and records the distance between start and stop codons in each of the three ORFs on the forward strand. You may skip the calculation for the reverse strand. You can find an example of this code implemented in Matlab on the book's website.

(b) Plot the distribution of ORF lengths L and compare it with that expected for random DNA calculated in Problem 4.7.

(c) Estimate a cut-off value L_{cut}, above which the ORFs are statistically significant, that is, the number of observed ORFs with $L > L_{cut}$ is much greater than expected by chance.

(Problem courtesy of Sharad Ramanathan.)

• 21.14 Protein mutation rates

Random mutations lead to amino acid substitutions in proteins that are described by the Poisson probability distribution $p_s(t)$. Namely, the probability that s substitutions at a given amino acid position in a protein occur over an evolutionary time t is

$$p_s(t) = \frac{e^{-\lambda t}(\lambda t)^s}{s!}, \qquad (21.80)$$

where λ is the rate of amino acid substitutions per site per unit time. For example, some proteins like *fibrinopeptides* evolve rapidly, and $\lambda_F = 9$ substitutions per site per 10^9 years. *Histones*, on the other hand, evolve slowly, with $\lambda_H = 0.01$ substitutions per site per 10^9 years.

(a) What is the probability that a fibrinopeptide has no mutations at a given site in 1 billion years? What is this probability for a histone?

(b) We want to compute the average number of mutations $\langle s\rangle$ over time t,

$$\langle s\rangle = \sum_{s=0}^{\infty} s p_s(t). \qquad (21.81)$$

First, using the fact that probabilities must sum to 1, compute the sum $\sigma = \sum_{s=0}^{\infty}(\lambda t)^s/s!$. Then, write an expression for $\langle s\rangle$, making use of the identity

$$\sum_{s=0}^{\infty} s\frac{(\lambda t)^s}{s!} = (\lambda t)\sum_{s=1}^{\infty}\frac{(\lambda t)^{s-1}}{(s-1)!} = \lambda t\sigma. \qquad (21.82)$$

(c) Using your answer in (b), determine the ratio of the expected number of mutations in a fibrinopeptide to that of a histone, $\langle s\rangle_F/\langle s\rangle_H$.

(Adapted from Problem 1.16 of K. Dill and S. Bromberg, Molecular Driving Forces, 2nd ed. Garland Science, 2011.)

• 21.15 The two-stage kinetic proofreading model

Derive the result of Equation 21.75.

21.8 Further Reading

Darwin, C (1859) On the Origin of Species, John Murray. Arguably the most important science book ever written. This remarkable synthetic opus is beautifully written, full of an amazing array of different ideas, observations, and experiments, and completely accessible to any interested reader. You must read this book.

Wallace, AR (2002) Infinite Tropics: An Alfred Russel Wallace Anthology, edited by A Berry, Verso. This beautiful tour of Wallace and his ideas and papers provides a fitting complement to Darwin's great work. Wallace's Sarawak and Ternate papers provide a compelling view of his ideas on both the fact and mechanism of evolution.

Carroll, SB (2006) The Making of the Fittest: DNA and the Ultimate Forensic Record of Evolution, W. W. Norton. This book describes with care and insight just how much can be learned from examining DNA sequences. Some of the most intriguing examples include fish that do not use hemoglobin and the evolution of color vision.

Lesk, AM (2008) Introduction to Bioinformatics, 3rd ed., Oxford University Press. This book is a very useful introduction to the topic and has some amusing examples such as comparing proteins from the woolly mammoth and modern day elephants.

Deonier, RC, Tavaré, S, & Waterman, MS (2005) Computational Genome Analysis: An Introduction, Springer. This wonderful book is full of interesting insights. Chapter 7 has an incisive discussion of scoring matrices.

Durbin, R, Eddy, S, Krogh, A, & Mitchison, G (1998) Biological Sequence Analysis: Probabilistic Models of Proteins and Nucleic Acids, Cambridge University Press. Though we have not read all of this book, several knowledgeable friends have independently recommended it highly.

Mount, DW (2004) Bioinformatics: Sequence and Genome Analysis, 2nd ed., Cold Spring Harbor Laboratory Press. Mount's tome is full of interesting and useful information on topics ranging from how databases are organized to how to think about sequence alignment.

Zvelebil, M, & Baum, JO (2008) Understanding Bioinformatics, Garland Science. This book does an excellent pedagogical job of explaining the issues and methods associated with bioinformatics.

Zuckerkandl, E, Jones, RT, & Pauling, L (1960) A comparison of animal hemoglobins by tryptic peptide pattern analysis, *Proc. Natl Acad. Sci. USA* **46**, 1349. This fascinating paper shows the logic of using molecules as documents of evolution.

Möbius, W, & Gerland, U (2010) Quantitative test of the barrier nucleosome model for statistical positioning of nucleosomes up- and downstream of transcription start sites, *PLoS Comp. Biol.*, **6**, e1000891. This beautiful article shows how the model by Kornberg and Stryer culminating in Figure 21.22 can be used to respond to real data on nucleosome positioning.

Neu, HC (1992) The crisis in antibiotic resistance, *Science* **257**, 1064. This article describes some of the different molecular mechanisms of antibiotic resistance.

Berg, OG, & von Hippel, PH (1987) Selection of DNA binding sites by regulatory proteins, *J. Mol. Biol.* **193**, 723. This paper describes the derivation of position weight matrices from binding site data.

Kinney, JB, Murugan, S, Callan, CG Jr, & Cox, EC, (2010) Using deep sequencing to characterize the biophysical mechanism of a transcriptional regulatory sequence *Proc. Natl Acad. Sci. USA* **107**, 9158. This paper provides an experimental method for determining the regulatory landscape of genomes, by examining the case study of the *lac* operon in *E. coli.*

Ogle, JM, & Ramakrishnan, V (2005) Structural insights into translational fidelity, *Annu. Rev. Biochem.* **74**, 129. Excellent discussion of translational accuracy and how insights into the structure of the ribosome shed light on this subject.

Rodnina, MV, & Wintermeyer, W (2001) Fidelity of aminoacyl-tRNA selection on the ribosome: kinetic and structural mechanisms, *Annu. Rev. Biochem.* **70**, 415. This paper describes experimental work that tests the proofreading hypothesis and provides the entire suite of relevant kinetic parameters.

Redondo, A (2006) Class notes prepared for the Latin American School on Bioinformatics, 5–9 June 2006, Mérida, Venezuela. Our treatment of proofreading essentially follows that described by Redondo.

Abzhanov, A, Protas, M, Grant, BR, Grant, PR, & Tabin, CJ (2004) Bmp4 and morphological variation of beaks in Darwin's finches, *Science* **305**, 1462. This paper demonstrates that variations in the timing and level of expression of a single developmental signaling protein, Bmp4, are strongly correlated with beak size and shape in developing finches from the Galapagos Islands. Moreover, forced expression of Bmp4 in a finch-like pattern in the beak of a developing domestic chicken converts its beak into a finch-like shape.

Sapp, J (2009) The New Foundations of Evolution: On the Tree of Life, Oxford University Press. Sapp provides a thorough and engaging story of the attempt to classify microscopic organisms and the way in which sequence analysis revolutionized our understanding of the evolution of life on Earth.

Dobzhansky, T (1973) Nothing in biology makes sense except in the light of evolution, *Am. Biol. Teacher* **35**, 125. A fascinating article that examines evolution from a variety of different angles.

Knoll, AH (2003) Life on a Young Planet: The First Three Billion Years of Life on Earth, Princeton University Press. This beautiful book gives an account of how the study of rocks and fossils has painted a picture of early life. Highly recommended to one and all.

Erwin, DH (2006) Extinction: How Life on Earth Nearly Ended 250 Million Years Ago, Princeton University Press. Another fascinating book that describes the role of extinctions in the history of life.

Pagni, M, & Jongeneel, CV (2001) Making sense of score statistics for sequence alignments, *Briefings Bioinf.* **2**, 51. Excellent, pedagogical discussion of extreme value statistics.

21.9 References

Alberts, B, Johnson, A, Lewis, J, et al. (2008) Molecular Biology of the Cell, 5th ed., Garland Science.

Amos, LA, van den Ent, F, & Löwe, J (2004) Structural/functional homology between the bacterial and eukaryotic cytoskeleton, *Curr. Opin. Cell. Biol.* **16**, 1.

Benedek, GB, & Villars, FMH (2000) Physics with Illustrative Examples from Medicine and Biology: Statistical Physics, 2nd ed., Springer. This book has an excellent treatment of the Luria–Delbrück experiment (as well as much else).

Berg, OG, & von Hippel, PH (1988) Selection of DNA Binding Sites by Regulatory Proteins. II. The Binding Specificity of Cyclic AMP Receptor Protein to Recognition Sites, *J. Mol. Bio.* **200**, 709.

Chang, BS, Jönsson, K, Kazmi, MA, et al. (2002) Recreating a functional ancestral archosaur visual pigment, *Mol. Biol. Evol.* **19**, 1483.

Crick, FHC (1958) On protein synthesis, *Symp. Soc. Exp. Biol.* **12**, 138.

Dickerson, RE (1972) The structure and history of an ancient protein, *Sci. Am.* **226**, 58.

Dill, K, & Bromberg, S (2011) Molecular Driving Forces, 2nd ed., Garland Science.

Emerman, M, & Malik, HS, (2010) Paleovirology—modern consequences of ancient viruses, *PLoS Biol.* **8**, e1000301.

Halder, G, Callaerts, P, & Gehring, WT (1995) Induction of ectopic eyes by targeted expression of the eyeless gene in *Drosophila, Science* **267**, 1788.

Hopfield, JJ (1974) Kinetic proofreading: a new mechanism for reducing errors in biosynthetic processes requiring high specificity, *Proc. Natl Acad. Sci. USA* **71**, 4135.

Jiang, C, & Pugh, BF (2009) Nucleosome positioning and gene regulation: advances through genomics, *Nat. Rev. Genet.* **10**, 161.

Kaplan, N, Moore, IK, Fondufe-Mittendorf, Y, et al. (2009) The DNA-encoded nucleosome organization of a eukaryotic genome, *Nature* **458**, 362.

Kocagöz, T, Hackbarth, CJ, Unsal, I, et al. (1996) Gyrase mutations in laboratory-selected, fluoroquinolone-resistant mutants of *Mycobacterium tuberculosis* H37Ra, *Antimicrob. Agents Chemother.* **40**, 1768.

Korber, B, Muldoon, M, Theiler, J, et al. (2002) Timing the ancestor of the HIV-1 pandemic strains, *Science* **288**, 1789.

Kornberg, R, & Stryer, L (1988) Statistical distributions of nucleosomes: nonrandom locations by a stochastic mechanism, *Nucleic Acids Res.* **16**, 6677.

Luria, SE, & Delbrück, M (1943) Mutations of bacteria from virus sensitivity to virus resistance, *Genetics* **28**, 491.

Mwangi, M, Wu, SW, Zhou, Y, et al. (2007) Tracking the *in vivo* evolution of multidrug resistance in *Staphylococcus aureus* by whole-genome sequencing, *Proc. Natl Acad. Sci. USA* **104**, 9451.

Ninio, J (1975) Kinetic amplification of enzyme discrimination. *Biochimie* **57**, 587.

Van de Peer, Y, Chapelle, S, & De Wachter, R (1996) A quantitative map of nucleotide substitution rates in bacterial rRNA, *Nucleic Acids Res.* **24**, 3381.

Whitehouse, I, Rando, OJ, Delrow, J, & Tsukiyama, T (2007) Chromatin remodeling at promoters suppresses antisense transcription, *Nature* **450**, 1031

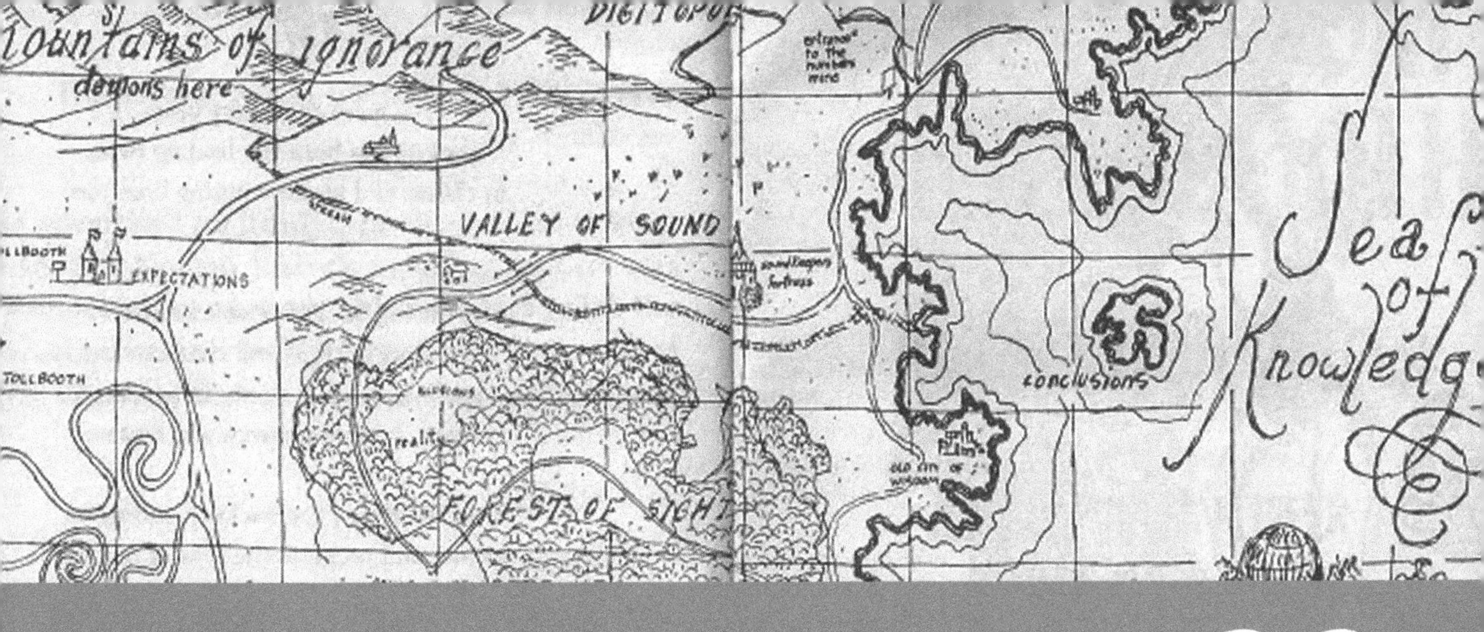

Whither Physical Biology? 22

Overview: In which we take stock of the role of quantitative analysis in the study of living matter

In this second edition of our long book, we have explored a number of different biological phenomena and discussed some productive avenues for how one can think about them quantitatively. The organizational thread that links our topics bears little resemblance to traditional biology texts, but rather is based upon the *physical* proximity of our topics. In this, our final chapter, we remind the reader of the philosophical foundations of the analysis we have presented. This is followed by a critical assessment of the many ways in which our discussion has fallen short. We close with a charge to the reader to go out and turn over a stone or look down a microscope and find a new biological problem that might be amenable to the kind of analysis we have described here.

> "'You must never feel badly about making mistakes,' explained Reason quietly, 'as long as you take the trouble to learn from them. For you often learn more by being wrong for the right reasons than you do by being right for the wrong reasons.'"
>
> **Norton Juster,** ***The Phantom Tollbooth***

22.1 Drawing the Map to Scale

This book has been an extended discussion and exploration of the virtues of a quantitative theoretical framework for thinking about the fascinating and diverse phenomena of the living world. The opening words of the book were borrowed from Darwin, one of the towering figures in the development of biological theory, who argued: "All observation should be for or against some view if it is to be of any service." In this book, we have specifically embraced the proposition that explicitly quantitative models, rather than qualitative models, are uniquely useful for the study of biology. We have attempted to give examples of the kinds of quantitative "views" that can breathe life into a wide range of biological observations and measurements. These fundamental quantitative models can serve as maps to help the curious navigate both familiar and unfamiliar areas of biology without getting lost.

For the science of biology, the relevant observations and measurements exclusively concern organisms that are currently living or have once lived on planet Earth. This narrow and perhaps introspective topical focus separates physical biology from many other domains of physics. Therefore, unlike other branches of theoretical physics that are content to explore how something *might* work, physical biology carries an explicit charge to focus on how living organisms *do* work. The models used in physical biology must therefore operate within a set of constraints including the long and idiosyncratic history of the evolution of our cellular tree of life, the universal use of water as a solvent, the limited chemical diversity of macromolecules available for construction, and the peculiar demands of natural selection in shaping the living world. At the same time, biological systems operating within these severe constraints offer scientists some of the most complex and enchanting behaviors exhibited by any kind of matter in the known universe.

Humans have been fascinated by the workings of the living world since the dawn of consciousness. Throughout much of the history of human investigation into biological phenomena, the kinds of observations that could be used to develop "views" in the Darwinian sense were limited by difficulty of observation. For Darwin himself, the central insight leading to the theory of evolution by natural selection arose by careful observation and consideration of living macroscopic organisms such as birds and flowering plants that could be readily observed by the naked eye of a nineteenth-century natural historian. Although this comprised a huge set of actual observations, it would be swamped by the total amount of biological information that would be available to Darwin today.

As experimental technologies have rapidly improved, biologists now find themselves frequently overwhelmed by the sheer magnitude of the number of experiments and measurements that they are able to perform on a daily basis. Finding informative patterns in a data-rich rather than data-poor environment becomes a very different kind of endeavor. Would the availability of so vastly much more data have eased Darwin's analytical process or hampered it? That is impossible to know. But for present-day biological scientists, the first decade of the new century has included a revolution in the nature of biological data. Thanks to technical advances such as high-throughput sequencing and automated microscopy and image analysis, a single investigator can now generate in a few days the volume of quantitative data that would have taken an entire research group months or years to generate even just a decade ago. As the amount of available biological data skyrockets, it becomes increasingly important to bear Darwin's statement about the value of observation in mind. It is ideas and concepts that turn a mountain of data into comprehensible, meaningful scientific results. In this book, we have adopted the philosophical perspective that very large and complicated sets of data can often reveal simple and unifying patterns when observed from the appropriate perspective. Choosing the most useful perspective for a given set of data requires intuition and luck, but also requires access to a toolbox of physical and mathematical fundamentals as we have illustrated in each chapter.

In the history of physics, one of our favorite examples of how a critical change in perspective can simplify and unify complicated observations involves the discovery of the law of universal gravitation. For many centuries, curious humans had tracked the apparent motions of the planets and stars in the night sky. Tycho Brahe realized as a teenager that more precise measurements were needed to develop a compelling mathematical model of these motions. As a result, he began the construction of ever larger tools such as his *quadrans maximus* (a giant protractor) and his cross staff for measuring the angular distances between heavenly bodies. Figure 22.1 shows the size of this instrument. These technical advances permitted the measurement of the positions of planets such as Mars with unprecedented accuracy. The tables of Brahe's data (available on the book's website) revealed an incomprehensible complexity of regular yet mysterious patterns as planets such as Mars appeared to move smoothly forward and then regress in their orbits only to turn around and move forward again. Kepler, considering this wealth of data, was able to reconcile these extremely complex observations with a simple physical picture in which the planets travel in elliptical orbits with the Sun at one of the foci of the ellipse. This simple change in perspective, or frame of reference, rendered precisely measured yet outrageously complicated quantitative measurements clearly understandable based on a one-line equation that could be written in the sand with a stick. With this as a starting point, Newton was able to figure out the inverse-square law of gravitation by comparing the rate at which objects fall near the Earth to the rate at which the Moon "falls" during its orbit around the Earth, using the kind of estimates we have attempted throughout the book to see that the accelerations suffered by the Moon and the object near the Earth's surface scale inversely as the square of their distance from the Earth's center. Together, Kepler's and Newton's ideas on gravity have had tremendous explanatory power for everything from simple torsional balances used to measure the gravitational constant all the way to the motion of galaxies. This broad explanatory power allowed the idea of gravitation to be elevated to the level of a scientific theory.

Figure 22.1: Tycho Brahe's *quadrans maximus*. This picture appeared in the 11-volume *Atlas Maior*, assembled in 1662 by the Dutch cartographer Joan Blaeu.

With this firm foundation in hand, deviations from the predictions of the theory of gravity became of great interest to scientists in the early twentieth century. In particular, Einstein wondered how Newton's theory would break down near massive objects, and also pondered anomalies in the motion of the planet Mercury. The outcome of Einstein's decade of musing on gravity resulted in the theory of general relativity.

Similarly, in the history of biology, the central theory of evolution has itself evolved by iterative cycles of simplification, expansion, and consequent modification. Although the theory of evolution is by now as well established as the theory of gravity, we by no means understand every detail of its workings and applications. Darwin, unaware of the work of the Austrian monk Mendel, posited that variable characteristics within a species must usually be passed unchanged from parents to offspring. Mendel's careful quantitative observation of inheritance patterns for observable characteristics in pea plants led to an extremely simple quantitative model with broad predictive power based upon the assumption that the units of inheritance, or genes, were immutable and independent of one another. Again, this radical change in perspective rendered complex inheritance patterns describable with a simple one-line equation. Later work by Sturtevant and Morgan, among others, proved that the simplification was, of course, an oversimplification; genes are not inherited independently of one another but are rather grouped together on chromosomes. The identification of DNA as the genetic material and the molecular biology revolution of the twentieth century led to later work demonstrating that even genes are not indivisible. These complexifications and extensions of the central theory of variation and inheritance relied on the initial insight of a simplifying perspective, so that deviations from the simplest expectation could themselves be profoundly informative, in the same way that Einstein could never have formulated the theory of general relativity without the initial simplifying insights of Kepler and Newton.

There is a long and productive tradition of biological model-building reaching from Mendel to the present day. In many cases, such as the analysis of action potentials shown in Chapter 17, this analysis has been quantitative from the outset, with scientific giants such as Helmholtz developing instrumentation permitting the experimental measurement of the velocity of action potentials. These measurements were followed by nearly a century of research into how nerve cells work, and, in particular, on the mechanism of action potential propagation. As a result of the quantitative nature of the data available on these systems, Hodgkin and Huxley were able to construct one of the great models in the history of biology. In many other cases, the biological model-building tradition has been to represent understanding in sophisticated diagrams and cartoons where important features and interactions are emphasized and unimportant features are deliberately ignored. These explanatory diagrams, commonly found as the last figure in experimental biology manuscripts, or as summary figures in reviews, typically incorporate the authors' accumulated knowledge about the system, comprising huge piles of positive and negative experimental results, analogies to other systems, and some degree of intuition. In many cases, the diagrams include implicit quantitative assumptions (for example, showing that the binding of a transcription factor to DNA represents a key regulated step in a cell's differentiation program), but the quantitative assumptions may often be shown qualitatively.

In a sense, one of our overarching goals in this book has been to show how implicit quantitative assumptions in biological model-building can be rendered explicit. Even a casual inspection of the current biological literature reveals that the results of biological experimentation are often presented in the form of graphs that reveal the kind of quantitative functional linkages among different variables that are the lifeblood of physics. We argue that, in those cases where the data exhibit quantitative relationships, verbal and cartoon descriptions are intrinsically insufficient. That is, if diligent and careful experimentalists are going to go to all the trouble to generate quantitative data, it is incumbent upon us that the "theory" that is either "for or against" these observations must answer those data on their own terms, that is, quantitatively. As noted by J. B. S. Haldane, "a mathematical theory may be regarded as a kind of scaffolding within which a reasonably secure theory expressible in words may be built up . . . without such a scaffolding verbal arguments are insecure." Throughout this book, we have constructed such scaffolding by using simple unifying physical perspectives that, in our estimation, have been able to link large amounts of quantitative, biologically relevant data to one another through concepts sufficiently simple to be written as single equations. This approach can convert a sketched map that shows only the relative positions of geographical features to a more useful map that is drawn truly to scale, giving an accurate quantitative representation of the distances between them. We hope it is clear that this approach is intended to serve as a leaping-off point in our maps of biological phenomena rather than as a destination.

The development of the kinds of models described above lends itself naturally to a style of interplay between theory and experiment that has been one of the central threads of the entire book. In particular, we have argued that this kind of quantitative volley back and forth between theory and experiment results in the formulation of questions that could not even be conceived in the absence of the underlying quantitative thread. For example, Figure 11.43 (p. 468) shows how the probability of opening for a mechanosensitive channel depends upon the lengths of lipid tails associated with the membrane where the channel has been reconstituted. Once such data exist, they suggest theories about the tension dependence of opening that make precise predictions about how this tension dependence *scales* with lipid tail length. These predictions in turn suggest new experiments. In this case, one prediction of this class of models is that multiple channels that are located near one another should exhibit cooperative gating. The nature of this cooperativity has to be worked out using mathematics; there is no verbal description that will permit us to deduce exactly how the cooperative opening of channels scales with membrane tension (for example, is the scaling linear or logarithmic?). If an experimental measurement proves the quantitative scaling prediction to be inaccurate, we will be forced to return to our modeling assumptions with that new insight and will be able to refine the model to better reflect biological reality, ideally proposing yet another round of definitive quantitative experiments.

22.2 Navigating When the Map Is Wrong

We do not believe that the proper goal of theory in biology is to provide an explanation for observations, instead, we subscribe to the view of Crick who noted: "The job of theorists is to make predictions." If particular predictions are confirmed by experiment, that provides the theorist with a warm feeling and the experimentalist can interpret a successful prediction as an indication that the theory is not completely on the wrong track, but in many cases it is the indisputable conflict between firm predictions and solid experimental results that leads to important new science. One of the impressions we have tried to fight in the book is that the only useful theory is a "right" theory. There may be a tendency when discussing theory in the biological setting to dismiss theoretical ideas too quickly as "wrong" if they do not "fit" all of the data. In our view, this is a way to miss many powerful opportunities to actually learn something. Indeed, if history is our guide, theories that lead to inconsistencies with the data are actually one of the most powerful tools for making new discoveries. Christopher Columbus' inaccurate idea of the map of the world led directly to the European discovery of the Americas. As a scientific example, one of the most important topics in statistical mechanics is the study of specific heats, that is, the amount of heat that is required to raise the temperature of a substance by 1 °C, for example. Interestingly, classical physics has a very concrete and universal prediction for the specific heat of crystalline solids that goes under the name of the law of Dulong and Petit, which says that the specific heat is $3R$, where R is the universal gas constant. Although this result is a good description of the high-temperature specific heat of many elemental solids, the low-temperature behavior is another matter altogether. Indeed, as the temperature drops, so too does the specific heat, in direct contradiction of the law of Dulong and Petit. This anomaly was one of the great challenges to the ideas of classical physics and resulted in one of the many key contributions of Einstein to the development of quantum theory. But even Einstein's model fell short of capturing the *scaling* of the low-temperature specific heat of solids, which resulted in another model from Debye that provided even deeper insights into the problem.

The point of this historical example is that simple models do not have to be consistent with every feature of the data to be powerful and useful, and the nature of the disagreement between the model and the data is often informative. The history of physics is full of important cases where useful models are understood even from their inception to be oversimplifications that are "wrong" in some respect, including the Bohr model of the atom, the Lorentz model of optical absorption, the electron gas picture of metallic solids, the Ising model of magnetism, and so forth. The "wrong" concept of the immutable, independently inherited gene put forth by Mendel is still the correct starting point for thinking about genetic inheritance in more realistic situations. The "wrong" concept of the MWC model of hemoglobin, envisioning exactly two interconverting structural states for this enormous and flexible molecular system that in reality can assume an uncountably large number of subtly distinct structures, nevertheless captures the fundamental essence of binding cooperativity in this system. Similar sentiments were put forth eloquently by G. E. P. Box (1979), who noted: "It should be remembered that just as the Declaration of Independence promises the pursuit of happiness rather than happiness itself, so the iterative scientific model-building process offers only the pursuit of the perfect model. For even when we feel we have carried the model-building process to a conclusion, some new initiative may make further improvement possible. Fortunately to be useful a model does not have to be perfect."

Because biological systems are so complex in terms of their molecular components and contingent evolutionary history, it is of course most likely to be the case that two different competing simple models for a single phenomenon are both partially correct, or may each be correct in some cases. The debate over whether communication between neurons is chemical or electrical was hotly debated for years. That is, is the communication between one neuron and another based on electrical signals or on some chemical message? Each hypothesis had its merits and suggested its own experimental strategies. Our current detailed and largely satisfying molecular-level understanding of the varieties of synaptic transmission is a direct consequence of the tension between these two classes of models, each of which can be correctly accused of being wrong in some details, but both of which also contain useful correct elements relevant to real biology.

22.3 Increasing the Map Resolution

The style of reasoning we have carried out throughout the book is based upon a hierarchical approach to model building. Often, our starting point has been what one might call order-of-magnitude biology, a low-resolution view of biological complexity that might be considered as a map of the world showing only the names of the continents. These models attempt to figure out the numerical magnitudes of the various parameters that are in play and to examine whether they make sense. Next, we turned to simple analytic models that allow us to derive scaling relationships between the various quantities of interest, emphasizing our philosophy of deriving analytical models that are sufficiently simple that they can be written with a stick in the sand. This increases the resolution of our conceptual maps so that individual countries can be identified. Although we firmly believe that this is the best approach for tackling an unfamiliar problem and for building intuition, it is nevertheless clear that there are many important instances in which this approach will fail. For every implementation of a simple analytic model, we have been forced to make explicit simplifications and approximations. Despite our awareness that these limit the quantitative applicability of the models to real-world systems, we have carefully chosen examples where the predictive power of these models is nonetheless impressive. But there are many categories of experiments where simple analytical models have so far been of limited utility, and a higher-resolution view is needed, to delineate individual cities and streets. In these cases, it is not necessary to abandon the physical and mechanistic approach that we have advocated, but only to increase its quantitative complexity. With the rapidly growing speed and computational power of even small and inexpensive computers, it has become realistic and even routine for scientists to carry out detailed computational simulations to help them interpret experimental data. The complexity of numerical simulations can range from straightforward expansions of the basic physical models that we have described here to extremely detailed attempts to use quantum mechanics to calculate electron distributions in protein structures.

The best computational models, ranging from molecular dynamics simulations of protein folding to ecological models of coral reef degradation, are all built by application of the fundamental physical models that we have described here, and in order to be valid must reflect the correct underlying physics. Used properly, the numerical power of computational approaches can be an important tool extending the reach of the physical biologist to problems where the complexity of the phenomenon is not compatible with a purely analytical approach. The computational explorations that we have included throughout the book have been intended to introduce some methods for extending analytical modeling to take advantage of the power of computation. We also hope that revealing the physical and mathematical bones underneath widely used computational tools such as BLAST will help users to always bear in mind the importance of understanding the assumptions behind the calculations.

One of the most useful aspects of computational approaches is in fact the impossibility of making vague statements and approximations. In a complex system with many interacting parts, for example in a dynamical simulation of flux through a metabolic pathway with numerous enzymes and intermediates, it is necessary to define precisely and quantitatively the exact nature of the interactions among all of the molecular players before the simulation can even be run. This exercise marvelously focuses the mind on which quantities are known or well measured and which require further experimentation. Of course, the hypothesized interactions that we assumed when setting up the simulation may simply not be correct; if this is the case, the model will tend to generate spurious results alerting us to the falsity of our assumptions and encouraging further thought. The necessity of making *some* assumption for representing incomplete knowledge is also found in cartography. A famous example is the Jagiellonian globe, a beautiful map of the world made in 1510 and currently housed in the university museum in Krakow, Poland, where

Nicolaus Copernicus studied as an undergraduate. The manufacturer wished to include AMERICA NOVITER REPERTA ("America newly discovered"), but did not know quite where to put it, and placed it in the Indian Ocean.

22.4 "Difficulties on Theory"

One of the most impressive chapters in Darwin's *Origin of Species* is Chapter VI, "Difficulties on Theory." In this chapter, Darwin took up the cause of stepping back to see where he might have slipped up in his thinking or where the data were thin and the conclusions tentative. Indeed, many of the so-called challenges levied against the theory of evolution by modern "critics" were actually put forth by Darwin himself, including his worries about the relative paucity of transitional forms in the fossil record as known at the time and about the origin of "organs of extreme perfection and complication" such as the eye. Most outsiders to science fail to realize that this kind of critical analysis is at the heart of scientific progress.

Often, in our scientific papers, we miss the opportunity to lay out the case against our own conclusions. But surely one of the best ways to make scientific progress is to openly acknowledge all of those places where confusion remains, where the data are tenuous, where the logic is unclear. To that end, we examine some of the many ways in which the development of the book has left us unsatisfied. The following annotated (but incomplete) list of our shortcomings is intended to give the reader a brief impression of areas where we see the greatest problems with our approach to physical biology.

The Problem with Spherical Cows

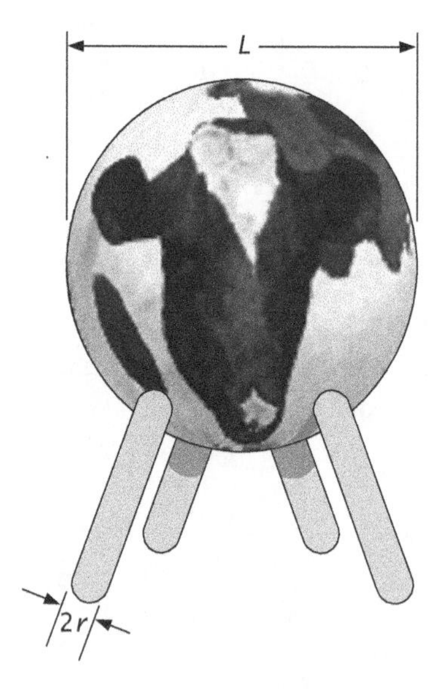

Figure 22.2: Spherical cow. The fine line between oversimplification and failure to omit irrelevant details is embodied in the parable of the spherical cow. A farmer hires a physicist to figure out how to increase milk production. The physicist works on the problem for six months and proudly returns to the farm to explain his solution. "First, we approximate the cow by a sphere...."

One of the difficulties with the ideas presented throughout the book is that it is often not clear how to navigate the narrows between throwing away important features when constructing a model and being overly enamored of the specific details of a given problem. The dangers of a model with too many details were highlighted by the Borges story in our first chapter. At the same time, it is clearly possible to go too far toward the opposite extreme. Indeed, this difficulty is the basis of the familiar joke that physicists are known for constructing models of spherical cows like that shown in Figure 22.2. A simple blanket statement concerning the vast majority of the models in this book is that they are caricatures of the real problem. In many cases, they strike us as providing object lessons in model construction, but still may be stripped too much to account properly for the biological complexity of a given phenomenon. Unfortunately, there is no single answer as to what level of detail is proper or sufficient for analysis of any biological problem, and indeed, different models at different levels of detail may be useful for the same biological system when different specific questions are being asked. We believe that a useful signpost is to explore models that are fertile in suggesting specific new experiments that can definitively distinguish between competing explanations. Here the inevitable tension between theory and experiment is likely to reach the level of conflict. Experimental scientists will not generally bother to perform experiments to test the predictions of a model when one of the underlying assumptions of the model is to ignore some aspect of the system that the experimentalist is convinced is important (for example, the cow's nonsphericity). However, model-builders will naturally resist the temptation to include every known detail. Actual scientific progress at this interface will require some concessions from both sides.

Modeler's Fantasy

In order to construct a mathematical model of a given problem, it is necessary in many cases to make guesses about the underlying mechanism that have not been suggested by prior experiment and may be unwarranted. This raises the suspicion that, at times, model-builders are just inventing stuff. For example, we have long been intrigued by the interaction between voltage-gated ion channels and the surrounding membrane. As part of that exercise, we examined limiting behaviors where α-helices move relative to the surrounding membrane in extreme ways. While this led to concrete predictions, it may be that none of our hypothesized mechanisms actually reflect reality. Hence, our treatment of the different models of voltage gating must be viewed as entirely provisional. In our view, the construction of such models is legitimized if they are used to make polarizing predictions that clearly distinguish different mechanisms, even if the proposed mechanisms have not yet survived experimental scrutiny. As with the spherical cows, any real progress will demand give-and-take between experimentalists and model-builders. The onus on the model-builders is to make their assumptions extremely explicit, and delineate exactly how critically their predictions depend on those particular assumptions. If an assumption is later shown by experiment to be incorrect, but is not very important to the outcome of the model, the model can probably be revised rather than thrown out. But, if an incorrect assumption is absolutely central to the model, we must submit to the reality as articulated by Darwin's friend and defender Thomas Henry Huxley, "Science is organized common sense where many a beautiful theory was killed by an ugly fact."

Is It Biologically Interesting?

Throughout the book, we have attempted to consider particular biological problems as seen through a quantitative prism. It is fair to wonder whether the effort is actually worth the trouble. Does understanding the forces associated with cytoskeletal polymerization offer any new insight into the biological workings of the cytoskeleton in cell motility, or does it merely confirm what we already know? Do we learn anything new about the biological uses of transcriptional regulation by exploiting statistical mechanics to make quantitative predictions about promoter activity based on RNA polymerase binding? One of the most important and astonishing biological breakthroughs of the past decade has been the realization that a previously invisible army of tiny RNA molecules is intimately involved in the regulation of gene expression. Not only did this set of discoveries involve essentially no mathematics or modeling at any stage, it also calls into question the entire basis of our statistical mechanical treatment for gene regulation. Put another way, we wonder whether the quantitative modeling approach we have explored in this book can ever become one of the foundational pillars of biological research on the brink of new discovery, standing alongside genetics and biochemistry as an exploratory method, or whether modeling will typically be relegated to the later stages of biological investigation, as a method for dotting the i's and crossing the t's for biological problems that are already fairly well understood, when the real intrepid explorers have moved on. In our view, the jury is still out on the extent to which the physical biology perspective will shed new light on important biological problems or will lead to new insights into physics. Ultimately, we believe that each individual has to follow his or her own curiosity and that, hopefully, now and then, this will lead to insights that are of wider interest.

Uses and Abuses of Statistical Mechanics

Statistical mechanics has been one of the main theoretical workhorses throughout the book. However, we have used statistical mechanics exclusively in its role as a theory about equilibrium states. We have repeatedly invoked simple *equilibrium* ideas with eyes wide open and attempted to map out the limitations of this perspective such as in the discussion of Section 15.2.6 (p. 591) where we explored methods for determining when it is legitimate to treat some subset of an overall chemical reaction network as being in equilibrium. Nevertheless, it is clear that great caution is needed so as not to get carried away with using equilibrium ideas just because we know how to do the associated mathematics and because the equilibrium assumption is convenient. Many of the problems described in the book from the equilibrium perspective should be revisited from the dynamic perspective.

Out-of-Equilibrium and Dynamic

As a follow-up to the previous objection, we are profoundly troubled by the fact that physicists are currently unable to provide a satisfactory general framework for thinking about nonequilibrium systems. There are important ideas for treating systems that are not "too far" from equilibrium, but no compelling and general formalism for treating nonequilibrium. Included in this shortcoming is the inability to really cope with the dynamics of living cells. Of course, this weakness is also a great opportunity and strikes us as one of the most interesting challenges at the interface between biology and physics. Because the living world offers many of the most compelling examples of observable nonequilibrium behaviors and also a plethora of tractable experimental systems for exploring the consequences of manipulating conditions, we hope that an enhanced fusion between physics and biology will help the physics community to develop the theoretical tools needed to overcome this gaping hole.

Uses and Abuses of Continuum Mechanics

The macromolecules of life are made of atoms. Nevertheless, for centuries, the tools of continuum mechanics have permitted us to ignore the graininess of matter when treating fluids and a myriad of different kinds of solids. In the twentieth century, these ideas were extended to new classes of matter such as liquid crystals, colloidal suspensions, and polymers. In Chapters 10–13, we ignored the graininess of matter and built our ideas on beams, membranes, fluids, and diffusion almost exclusively around a continuum picture. The limits of such continuum approaches are still being explored and the jury is still out as to how far we will actually be able to get in using such ideas to describe the behavior of cells and subcellular structures.

Too Many Parameters

Following the kind of dynamical treatment we used in Chapter 19 to characterize the behavior of genetic networks and signaling networks to its logical conclusion leads to writing down a vast array of coupled differential equations. For example, if one tries to write coupled rate equations to describe the time evolution of networks like that shown in Figure 19.1 (p. 803), they will involve literally tens to hundreds of concentrations and rate constants. In this case, we have to ask ourselves whether we can ever know all of the relevant parameters in such a description given the intrinsic uncertainties and systematic errors of our best measurement techniques, and the intrinsic variations found from one cell to another and even within a single cell over time. Even if we did know the correct values for all of these concentrations and rate constants to an arbitrary degree of precision, it is not clear that we would then feel justified in claiming that we truly understand the biological system in any meaningful way. An unknown scientist famously declared, "I am glad that the computer understands it, now can you explain it to me." It really remains unclear to us how to think properly about these kinds of problems, where our favored stick-in-the-sand analytical approach is woefully inadequate to produce any interesting insights, but the alternative fully realistic simulation results in an *in silico* system that is just as complicated as the real *in vivo* system, like Borges' "Map of the Empire whose size was that of the Empire."

Missing Facts

One of the difficulties with the mindset adopted in this book is that in order to make quantitative models of a given phenomenon, you have to have sufficient facts at hand to know what kind of mathematical treatment will be most appropriate. For example, our analysis of cooperative binding in hemoglobin was predicated on a long-standing series of experiments that demonstrated unequivocally the existence of cooperative binding. It is one of the most frustrating and exciting features of this current era in biology that now is the time when many of the key facts are being unearthed. Even for the well-characterized systems that we chose as examples in this book, it is likely that new experiments performed between the time we wrote our analysis and the time you read it will have led to new insights and perhaps invalidated the models. As illustrated in Figure 22.3, even the most carefully drawn map is limited by the information available to the cartographer. For any scientist, it remains a constant challenge to surf on the edge of rapidly moving knowledge without being drowned in unhelpful details. We are sure that we have missed many key facts.

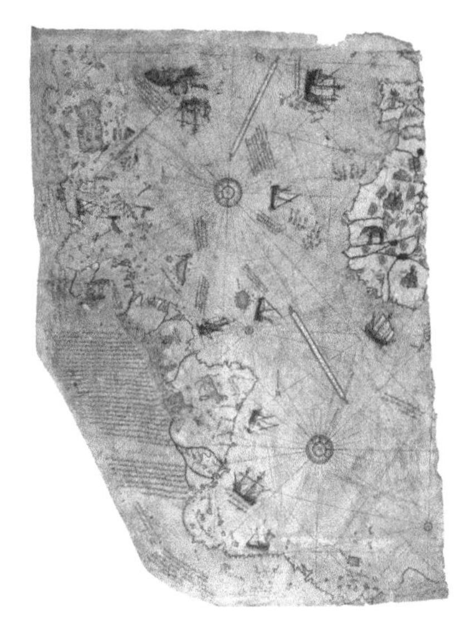

Figure 22.3: A map with missing information. This fragment of a world map was drawn in 1513 by the Ottoman admiral and cartographer Hadji Ahmed Muhiddin Piri, known as Piri Reis. The map shows the shape of the coastline of Brazil with remarkable accuracy, based on reports from European explorers. However, the interior of South America is uncharacterized and featureless beyond the mouths of a few rivers.

Too Much Stuff

Structurally, one of the most difficult and challenging aspects of trying to teach biology is succumbing to superficiality. As noted early on in the book, a fascinating tension in biology is to balance the quest for generality (for example, the idea that all living organisms are built around a single fundamental living unit—that is, cell theory—the universality of the genetic code, the theory of evolution, etc.) with the need to embrace each organism for its own unique properties and to welcome the diversity of life. Our book is much longer than we had originally planned, and this second edition is even longer than the first, and yet there remains the sense that much of what we have done has still been too superficial, that we have not dug sufficiently deeply into the most illuminating case studies. While trying to share our appreciation for the broad sweep of the living world, we have been guilty of taking a shallow view of it, and would perhaps have been wiser to narrow our focus and treat a smaller number of examples more rigorously.

Too Little Stuff

At the same time, one of the great charms and challenges of biology is its diversity. Whether at the level of the individual molecules that make cells tick, or the vast array of different classes of cells (Figure 2.17 shows some of the diversity presented by protists), or of multicellular organisms, there is nearly limitless opportunity to explore the particular, special features of any given problem. In this respect, the story told in our book is guilty of omissions at every turn. We have chosen a few individuals to stand in for great classes of diverse beings, such as using *E. coli* as our "standard" cell, and have essentially ignored huge branches of the tree of life, and huge disciplines within biology such as biochemistry. If we had infinite time, and you, our reader, infinite patience, we would have loved to provide to you a book of infinite length, in an attempt to do justice to all these wonders of the living world.

The Myth of "THE" Cell

Yet another unintended consequence of our effort to pursue generality in our biological descriptions is the myth of THE cell. Even for our "standard ruler," our old friend *E. coli*, there are a huge number of different strains, some of which are tamed and weakened laboratory strains and some of which can cause deadly infections. Among the laboratory strains, evolutionary drift has continued, so the exact nature of the *E. coli* used in one laboratory is different from that used in another laboratory down the hall. Within the colon of an individual human, the harbored *E. coli* may go through several hundred thousand generations, comparable to the number of generations that humans have gone through since we last shared a common ancestor with chimpanzees. Thus, each person's passenger *E. coli* is likely to be very different from every other person's. Even within one genetically defined strain of *E. coli* there are many different behaviors, depending upon external conditions and upon timing within the cell cycle. Experiments on "noise" further demonstrate the individual character of nominally identical cells even under identical conditions, and single-molecule experiments have unearthed the personalities of individual molecules as well. Further, our treatment of single-celled microorganisms has largely ignored the stunning discovery of an entirely new domain of life (the Archaea), which serves to further shatter the idea of a single bacterium like *E. coli* as a representative for all microbes. We still believe that the ultimate goal of science is to find ideas and concepts that unify our understanding of seemingly disparate phenomena. It is not clear how to reconcile that objective with the diversity and uniqueness of different organisms, cells, and even molecules.

Not Enough Thinking

We ignored the brain! Many have said that the twenty-first century will see the study of consciousness and the mind as the great frontier. Though we have touched on some of the particular molecular mechanisms and cell types that will form part of the substrate of that discussion with special emphasis on the retina and vision, effectively, we have completely ignored neuroscience and it is clear that there is much "physical biology" to be done in that setting as well.

Summing Up

Though any reader who has made it this far will surely find our book very long indeed, nevertheless, we had a large list of topics that were once part of the table of contents but were dropped either because we failed to understand the issues to our own satisfaction, or because we felt that the experimental data were insufficient, or for a myriad of other reasons. Some of the topics we especially regret not being able to treat in more detail (or at all) are the underlying driving forces that determine the shapes, fusion, and fission of organelles; the mechanisms whereby DNA-binding proteins are able to find their DNA targets, particularly in the very complex context of eukaryotic transcription; the biological uses of adhesion in forming multicellular organisms; the physics of interactions between organisms and their changing environments; and finally the underpinnings of evolution. Each of these areas seems to us to be a fertile ground that is ripe for analysis from the physical biology perspective, and indeed, in each of these cases, much deep and interesting progress has already been made.

The act of writing this book has convinced each of us that the study of living matter is one of the most exciting frontiers in human thought. Just as the makings of the large-scale universe are being revealed by ever more impressive telescopes, living matter is now being viewed in ways that were once as unimaginable as was going to the Moon. Despite the muscle-enhancing weight of this book, we feel that we have only scratched the surface of the rich and varied applications of physical reasoning to biological problems. Our overall goal has been to communicate a style of thinking about problems where we have done our best to illustrate the power of the style using examples chosen from biological systems that are well defined and usually well studied from a biological perspective. As science moves forward into the twenty-first century, it is our greatest hope that synthetic approaches for understanding the natural world from biological, physical, chemical, and mathematical perspectives simultaneously will enrich all of these fields and illuminate the world around us. We can only hope the reader has at least a fraction of the pleasure in answering that charge as we have had in attempting to describe the physical biology of the cell.

22.5 The Rhyme and Reason of It All

As we have emphasized in organizing this book around physical concepts rather than biological topics, we find it inspiring and comforting that a relatively small number of fundamental physical principles can be deployed to shed useful light on such a broad range of biological phenomena, covering time scales from ion channel opening to evolution, and size scales from individual proteins to the branching patterns of redwood trees. For the intrepid readers who have accompanied us all the way to the end of this long journey, we hope that you will find new ways to apply what you have learned with us, in contexts that neither we nor you may currently guess.

We leave you with the words of Norton Juster, with a more complete version of the quotation that we chose to open our final chapter. At the end of his classic book, *The Phantom Tollbooth*, the young hero Milo has completed a long and winding journey, having passed through such locations as the Foothills of Confusion, the Mountains of Ignorance, and the Doldrums, and even jumped to the Island of Conclusions, before finally reaching the Old City of Wisdom on the shores of the Sea of Knowledge, and witnessing the final battle where the massed armies of Wisdom defeat the demons of Ignorance.

"It *has* been a long trip," said Milo, climbing onto the couch where the princesses sat; "but we would have been here much sooner if I hadn't made so many mistakes. I'm afraid it's all my fault."

"You must never feel badly about making mistakes," explained Reason quietly, "as long as you take the trouble to learn from them. For you often learn more by being wrong for the right reasons than you do by being right for the wrong reasons."

"But there's so *much* to learn," he said, with a thoughtful frown.

"Yes, that's true," admitted Rhyme; "but it's not just learning things that's important. It's learning what to do with what you learn and learning why you learn things at all that matters."

"That's just what I mean," explained Milo, as Tock and the exhausted bug drifted quietly off to sleep. "Many of the things I'm supposed to know seem so useless that I can't see the purpose in learning them at all."

"You may not see it now," said the Princess of Pure Reason, looking knowingly at Milo's puzzled face, "but whatever we learn has a purpose and whatever we do affects everything and everyone else, if even in the tiniest way. Why, when a housefly flaps his wings, a breeze goes round the world; when a speck of dust falls to the ground, the entire planet weighs a little more; and when you stamp your foot, the earth moves slightly off its course. Whenever you laugh, gladness spreads like the ripples in a pond; and whenever you're sad, no one anywhere can be really happy. And it's much the same thing with knowledge, for whenever you learn something new, the whole world becomes that much richer."

"And remember, also," added the Princess of Sweet Rhyme, "that many places you would like to see are just off the map and many things you want to know are just out of sight or a little beyond your reach. But someday you'll reach them all, for what you learn today, for no reason at all, will help you discover all the wonderful secrets of tomorrow."

22.6 Further Reading

Ferguson, K (2002) Tycho & Kepler, Walker and Company. An interesting account of the development of precision measurement in early naked-eye astronomy and Kepler's response to the data.

Schutz, B (2003) Gravity from the Ground Up: An Introductory Guide to Gravity and General Relativity, Cambridge University Press. This book describes the long and fascinating history of the *refinement* of our understanding of gravity through successive approximation.

Longair, M (2003) Theoretical Concepts in Physics, 2nd ed., Cambridge University Press. Longair's beautiful book discusses many and important topics from a unique perspective. The chapter on Brahe and Kepler illustrates our point that the motion of the planets was indeed complex and was only seen to be "simple" when viewed from the right perspective. Similar simplification by change of perspective has been and will continue to be important in biology.

Valenstein, ES (2005) The War of the Soups and the Sparks: The Discovery of Neurotransmitters and the Dispute over How Nerves Communicate, Columbia University Press. This book describes the debate over synaptic transmission.

Darwin, C (1859) On the Origin of Species, John Murray. Darwin's chapter on "Difficulties on Theory" illustrates the spirit of searching for the truth by trying to find everything that might be wrong with his ideas.

Pais, A (1982) "Subtle is the Lord": The Science and the Life of Albert Einstein, Oxford University Press. Pais lays out the iterative search for understanding of the specific heats of solids and its role in the development of modern quantum theory.

Turing, AM (1952) The chemical basis of morphogenesis, *Philos. Trans. R. Soc. Lond.* **237**, 37. In this elegant paper, the famous mathematician and cryptographer turned his attention to pattern formation in biological systems. He argued that simple reaction–diffusion systems might be able to generate stationary waves in developing organisms that could account for the tentacle patterns on *Hydra* and the spotted black and white patterns on cows, among other things. Although the biological examples he chose to describe do not actually use this mechanism, this paper was hugely influential in the development of mathematical biology, and it stands as a beautiful example of the utility and power of "wrong" models.

Mogilner, A, Wollman, R, & Marshall, WF (2006) Quantitative modeling in cell biology: What is it good for?, *Dev. Cell* **11**, 279. This review article summarizes several cases where mathematical modeling at scales ranging from the application of simple algebraic equations to large-scale agent-based computer simulations have yielded important insights in cell biology.

22.7 References

Box, GEP (1979) Robustness in the strategy of scientific model building, in Robustness in Statistics, edited by RL Launer & GN Wilkinson, Academic Press.

Crick, F (1988) What Mad Pursuit: A Personal View of Scientific Discovery, Basic Books.

Haldane, JBS (1964) A defense of beanbag genetics, *Perspect. Biol. Med.* **7**, 343 (reprinted 2008 in *Int. J. Epidemiol.* **37**, 435).

Juster, N (1961) The Phantom Tollbooth, Epstein & Carroll.

Index

Page numbers with an "F" refer to a figure. Page numbers in italics to a problem.

B

C

Page numbers with an "F" refer to a figure. Page numbers in italics to a problem.

D

Page numbers with an "F" refer to a figure. Page numbers in italics to a problem.

E

Page numbers with an "F" refer to a figure. Page numbers in italics to a problem.

F

G

H

Page numbers with an "F" refer to a figure. Page numbers in italics to a problem.

I

J

K

L

M

Page numbers with an "F" refer to a figure. Page numbers in italics to a problem.

N

O

Page numbers with an "F" refer to a figure. Page numbers in italics to a problem.

P

Q

R

Page numbers with an "F" refer to a figure. Page numbers in italics to a problem.

S

T

Page numbers with an “F” refer to a figure. Page numbers in italics to a problem.

U

V

W

X

Y

Z